The Diversity of Fishes

The Diversity of Fishes

Biology, Evolution and Ecology

Third Edition

DOUGLAS E. FACEY
Emeritus Professor of Biology, Saint Michael's College,
Vermont, USA

BRIAN W. BOWEN
Research Professor, Hawaii Institute of Marine Biology,
University of Hawaii at Manoa, Honolulu, Hawaii, USA

BRUCE B. COLLETTE
Research Associate, NOAA, National Museum of Natural History,
Smithsonian Institution, Washington, DC, USA

GENE S. HELFMAN
Emeritus Professor of Ecology, Odum School of Ecology,
University of Georgia, Athens, Georgia, USA

WILEY

Edition History
Gene S. Helfman, Bruce B. Collette, Douglas E. Facey (1e, 1997); Gene S. Helfman, Bruce B. Collette, Douglas E. Facey, and Brian W. Bowen (2e, 2009)

Registered Offices
John Wiley & Sons, Inc., 111 River Street, Hoboken, NJ 07030, USA
John Wiley & Sons Ltd, The Atrium, Southern Gate, Chichester, West Sussex, PO19 8SQ, UK

Editorial Office
The Atrium, Southern Gate, Chichester, West Sussex, PO19 8SQ, UK

For details of our global editorial offices, customer services, and more information about Wiley products visit us at www.wiley.com.

Wiley also publishes its books in a variety of electronic formats and by print-on-demand. Some content that appears in standard print versions of this book may not be available in other formats.

Library of Congress Cataloging-in-Publication Data
Names: Facey, Douglas E., author. | Bowen, Brian W. (Brian William), 1957- author. | Collette, Bruce B., author. | Helfman, Gene S., author.
Title: The diversity of fishes : biology, evolution and ecology / Douglas E. Facey, Brian W. Bowen, Bruce B. Collette, Gene S. Helfman.
Description: Third edition. | Hoboken, NJ : Wiley, 2023. | Revised edition of: The diversity of fishes / Gene Helfman ... [et al.] 2nd ed. 2009. | Includes bibliographical references and index.
Identifiers: LCCN 2022030874 (print) | LCCN 2022030875 (ebook) | ISBN 9781119341918 (cloth) | ISBN 9781119341802 (adobe pdf) | ISBN 9781119341833 (epub)
Subjects: LCSH: Fishes. | Fishes—Variation. | Fishes—Adaptation.
Classification: LCC QL615 .F33 2023 (print) | LCC QL615 (ebook) | DDC 597–dc23/eng/20220810
LC record available at https://lccn.loc.gov/2022030874
LC ebook record available at https://lccn.loc.gov/2022030875

Cover Design: Wiley
Cover Images: Courtesy of Gene Helfman; Luiz A. Rocha, California Academy of Sciences; Jon Hyde & Kimberly Sultze, www.hydesultze.com; Andrew Nagy

Set in 9.5/12.5pt SourceSansPro by Straive, Pondicherry, India

SKY10076416_060324

To our parents, for their encouragement of our nascent interest in things biological;
To our spouses – Janice, RuthEllen, Sara, and Judy – for their patience and understanding during the production of this volume;
And to students, colleagues, and fish lovers for their efforts to preserve biodiversity for future generations.

Contents

Preface to the Third Edition

Thank you to all who helped make the first two editions of *The Diversity of Fishes* successful. We greatly appreciate the feedback and suggestions from those who have used them. And to those who have been asking us about an update – thanks for your patience. Here is what you can expect in the third edition.

- We have done our best to update information while recognizing that no textbook can represent the most current information in any large field. This book is intended as a starting point, not the final destination. Therefore, we encourage those using the text to consult more current sources for updated information.
- This edition includes more color photographs to better demonstrate the diversity and beauty of fishes that attract many to the field. We include photographs from public agencies, in part to acknowledge the valuable contributions to Ichthyology of those in natural resources and management organizations.
- Reorganization and consolidation of some topics has reduced the total number of chapters to 22.
- Each chapter begins with a Summary that provides a broad overview of the content of that chapter. This may be particularly useful for those using the text for a course and who do not intend to utilize some chapters in detail – students can read the summary of each of those chapters.
- Molecular genetics has transformed many aspects of ichthyology over the last few decades, and this is reflected throughout the text. Important concepts are introduced in Chapters 1 and 2, supported by an Appendix of terminology at the end of the book, and specific contributions of molecular genetics to the field of ichthyology are included in many chapters.
- Structure and function are addressed together rather than treated in separate chapters. For example, Chapter 3 addresses structure and function of the head (e.g. bones, muscles, breathing, jaw suspension, feeding, dentition), and Chapter 4 addresses structure and function of the trunk (e.g. bones, muscles, integument, scales, fins, locomotion).
- In chapters addressing the history of the fishes (Chapter 11) and systematics of the major groups of fishes (Chapters 12–15), we have largely adopted the approach of Nelson et al. (2016), which includes consideration of molecular phylogenetics. However, we have retained the approach of Nelson (2006) where conclusions from molecular data are quite different from those based on morphological evidence (see Chapter 15, for example). We note differences between the approaches and encourage readers to consult current sources for updated information and perspectives, as our collective understanding of the relationships among groups of fishes continues to evolve.
- Fishes as predators and prey are considered together in one chapter (Chapter 16). Many fishes are both predators and prey, and many physical and behavioral adaptations are rather similar and, in some cases, may have developed in response to one another.
- Our chapter on Zoogeography and Phylogeography (Chapter 19) uses global maps that more accurately represent the comparative sizes of oceans and landmasses than maps used in prior editions.
- Fish populations are covered in Chapter 20, including Population Ecology, Population Dynamics and Regulation, and Population Genetics.
- Fish interactions with other species within fish assemblages and broader communities, and their impacts on ecosystems are addressed in Chapter 21. This is a long chapter but avoids the redundancy of addressing similar types of interactions and impacts in multiple chapters.
- As in the past, we conclude with a chapter on Conservation (Chapter 22), but this now includes Conservation Genetics.

Preface to the Second Edition

The first edition of *The diversity of fishes* was successful beyond our wildest dreams. We have received constant and mostly positive feedback from readers, including much constructive criticism, all of which convinces us that the approach we have taken is satisfactory to ichthyological students, teachers, and researchers. Wiley-Blackwell has validated that impression: by their calculations, *The diversity of fishes* is the most widely adopted ichthyology textbook in the world.

However, ichthyology is an active science, and a great deal of growth has occurred since this book was first published in 1997. Updates and improvements are justified by active and exciting research in all relevant areas, including a wealth of new discoveries (e.g., a second coelacanth species, 33 more megamouth specimens, several new record tiniest fishes, and exciting fossil discoveries including some that push back the origin of fishes many million years and another involving a missing link between fishes and amphibians), application of new technologies (molecular genetics, transgenic fish), and increased emphasis on conservation issues (e.g., Helfman 2007). Websites on fishes were essentially nonexistent when the first edition was being produced; websites now dominate as an instant source of information. Many of the volumes we used as primary references have themselves been revised. Reflective of these changes, and of shortcomings in the first edition, is the addition of a new chapter and author. Genetics received insufficient coverage, a gross omission that has been corrected by Brian Bowen's contribution of a chapter devoted to that subject and by his suggested improvements to many other chapters. Brian's contributions were aided by extensive and constructive comments from Matthew Craig, Daryl Parkyn, Luiz Rocha, and Robert Toonen. He is especially grateful to John Avise, Robert Chapman, and John Musick for their guidance and mentorship during his professional career, and most of all to his wife, RuthEllen, for her forbearance and support.

Among the advances made in the decade following our initial publication, a great deal has been discovered about the phylogeny of major groups, especially among jawless fishes, sarcopterygians, early actinopterygians, and holocephalans. In almost all taxa, the fossil record has expanded, prompting reanalysis and sometimes culminating in conflicting interpretations of new findings. A basic textbook is not the appropriate place to attempt to summarize or critique the arguments, opinions, and interpretations. We have decided to accept one general compilation and synthesis. As in the 1997 edition, where we adopted with little adjustment the conclusions and terminology of Nelson (1996), we here follow Nelson (2006), who reviews the recent discoveries and clearly presents and assesses the many alternative hypotheses about most groups. Instructors who used our first edition will have to join us in learning and disseminating many changed names as well as rearrangements among taxa within and among phylogenies, especially Chapters 11–13. Science is continually self-correcting. We should applaud the advances and resist the temptation to comfortably retain familiar names and concepts that have been modified in light of improved knowledge.

Also, we have now adopted the accepted practice of capitalizing common names.

Acknowledgments

Thanks especially to the many students and professionals who corrected errors in the first edition (J. Andrew, A. Clarke, D. Hall, G.D. Johnson, H. Mattingly, P. Motta, L.R. Parenti, C. Reynolds, C. Scharpf, E. Schultz, M.L.J. Stiassny, and S. Vives proved particularly alert editors). Their suggestions alone led to many changes, to which we have added literally hundreds of new examples, facts, and updates. Wiley-Blackwell has provided a website for this second edition, www.wiley.com/go/helfman, through which we hope to again correct and update the information provided here. We encourage any and all to inform us wherever they encounter real or apparent errors of any kind in this text. Please write directly to us. Chief responsibilities fell on GSH for Chapters 1, 8–15, and 18–26 (genehelfman@gmail.com); on BBC for Chapters 2–4 and 16 (collettb@si.edu); on DEF for Chapters 5–7 (dfacey@smcvt.edu), and on BWB for Chapter 17 (bbowen@hawaii.edu). Once again and more than anything, we want to get it right.

Preface to the First Edition

Two types of people are likely to pick up this book, those with an interest in fishes and those with a fascination for fishes. This book is written by the latter, directed at the former, with the intent of turning interest into fascination.

Our two major themes are adaptation and diversity. These themes recur throughout the chapters. Wherever possible, we have attempted to understand the adaptive significance of an anatomical, physiological, ecological, or behavioral trait, pointing out how the trait affects an individual's probability of surviving and reproducing. Our focus on diversity has prompted us to provide numerous lists of species that display particular traits, emphasizing the parallel evolution that has occurred repeatedly in the history of fishes, as different lineages exposed to similar selection pressures have converged on similar adaptations.

The intended audience of this book is the senior undergraduate or graduate student taking an introductory course in ichthyology, although we also hope that the more seasoned professional will find it a useful review and reference for many topics. We have written this book assuming that the student has had an introductory course in comparative anatomy of the vertebrates, with at least background knowledge in the workings of evolution. To understand ichthyology, or any natural science, a person should have a solid foundation in evolutionary theory. This book is not the place to review much more than some basic ideas about how evolutionary processes operate and their application to fishes, and we strongly encourage all students to take a course in evolution. Although a good comparative anatomy or evolution course will have treated fish anatomy and systematics at some length, we go into considerable detail in our introductory chapters on the anatomy and systematics of fishes. The nomenclature introduced in these early chapters is critical to understanding much of the information presented later in the book; extra care spent reading those chapters will reduce confusion about terminology used in most other chapters.

More than 27,000 species of fishes are alive at present. Students at the introductory level are likely to be overwhelmed by the diversity of taxa and of unfamiliar names. To facilitate this introduction, we have been selectively inconsistent in our use of scientific versus common names. Some common names are likely to be familiar to most readers, such as salmons, minnows, tunas, and freshwater sunfishes; for these and many others, we have used the common family designation freely. For other, less familiar groups (e.g., Sundaland noodlefishes, trahiras, morwongs), we are as likely to use scientific as common names. Many fish families have no common English name and for these we use the Anglicized scientific designation (e.g., cichlids, galaxiids, labrisomids). In all cases, the first time a family is encountered in a chapter we give the scientific family name in parentheses after the common name. Both scientific and common designations for families are also listed in the index. As per an accepted convention, where lists of families occur, taxa are listed in phylogenetic order. We follow Nelson et al. (1994, now updated) on names of North American fishes and Robins et al. (1991, also now updated) on classification and names of families and of higher taxa. In the few instances where we disagree with these sources, we have tried to explain our rationale.

Any textbook is a compilation of facts. Every statement of fact results from the research efforts of usually several people, often over several years. Students often lose sight of the origins of this information, namely the effort that has gone into verifying an observation, repeating an experiment, or making the countless measurements necessary to establish the validity of a fact. An entire dissertation, representing 3–5 or more years of intensive work, may be distilled down to a single sentence in a textbook. It is our hope that as you read through the chapters in this book, you will not only appreciate the diversity of adaptation in fishes, but also consider the many ichthyologists who have put their fascination to practical use to obtain the facts and ideas we have compiled here. To acknowledge these efforts, and because it is just good scientific practice, we have gone to considerable lengths to cite the sources of our information in the text, which correspond to the entries in the lengthy bibliography at the end of the book. This will make it possible for the reader to go to a cited work and learn the details of a study that we can only treat superficially. Additionally, the end of each chapter contains a list of supplemental readings, including books or longer review articles that can provide an interested reader with a much greater understanding of the subjects covered in the chapter.

This book is not designed as a text for a course in fisheries science. It contains relatively little material directly relevant to such applied aspects of ichthyology as commercial or sport fisheries or aquaculture; several good text and reference books deal specifically with those topics (for starters, see the edited volumes by Lackey & Nielsen 1980, Nielsen & Johnson 1983, Schreck & Moyle 1990, and Kohler & Hubert 1993). We recognize however that many students in a college-level ichthyology class are training to become professionals in those or related disciplines. Our objectives here are to provide such readers with enough information on the general aspects of ichthyology to make informed, biologically sound judgments and decisions, and to gain a larger appreciation of the diversity of fishes beyond the relatively small number of species with which fisheries professionals often deal.

Adaptations versus adaptationists

Our emphasis throughout this text on evolved traits and the selection pressures responsible for them does not mean that we view every characteristic of a fish as an adaptation. It is important to realize that a living animal is the result of past evolutionary events, and that animals will be adapted to current environmental forces only if those forces are similar to what has happened to the individual's ancestors in the past. Such phylogenetic constraints arise from the long-term history of a species. Tunas are masters of the open sea as a result of a streamlined morphology, large locomotory muscle mass connected via efficient tendons to fused tail bones, and highly efficient respiratory and circulatory systems. But they rely on water flowing passively into their mouths and over their gills to breathe and have reduced the branchiostegal bones in the throat region that help pump water over their gills. Tunas are, therefore, constrained phylogenetically from using habitats or foraging modes that require them to stop and hover, because by ceasing swimming they would also cease breathing.

Animals are also imperfect because characteristics that have evolved in response to one set of selective pressures often create problems with respect to other pressures. Everything in life involves a trade-off, another recurring theme in this text. The elongate pectoral fins ("wings") of a flyingfish allow the animal to glide over the water's surface faster than it can swim through the much denser water medium. However, the added surface area of the enlarged fins creates drag when the fish is swimming. This drag increases costs in terms of a need for larger muscles to push the body through the water, requiring greater food intake, time spent feeding, etc. The final mix of traits evolved in a species represents a compromise involving often-conflicting demands placed on an organism. Because of phylogenetic constraints, trade-offs, and other factors, some fishes and some characteristics of fishes appear to be and are poorly adapted. Our emphasis in this book is on traits for which function has been adequately demonstrated or appears obvious. Skepticism about apparent adaptations can only lead to greater understanding of the complexities of the evolutionary process. We encourage and try to practice such skepticism.

Acknowledgments

This book results from effort expended and information acquired over most of our professional lives. Each of us has been tutored, coaxed, aided, and instructed by many fellow scientists. A few people have been particularly instrumental in facilitating our careers as ichthyologists and deserve special thanks: George Barlow, John Heiser, Bill McFarland, and Jack Randall for GSH; Ed Raney, Bob Gibbs, Ernie Lachner, and Dan Cohen for BBC; Gary Grossman and George LaBar for DEF. The help of many others is acknowledged and deeply appreciated, although they go unmentioned here.

Specific aid in the production of this book has come from an additional host of colleagues. Students in our ichthyology classes have written term papers that served as literature surveys for many of the topics treated here; they have also critiqued drafts of chapters. Many colleagues have answered questions, commented on chapters and chapter sections, loaned photographs, and sent us reprints, requested and volunteered. Singling out a few who have been particularly helpful, we thank C. Barbour, J. Beets, W. Bemis, T. Berra, J. Briggs, E. Brothers, S. Concelman, J. Crim, D. Evans, S. Hales, B. Hall, C. Jeffrey, D. Johnson, G. Lauder, C. Lowe, D. Mann, D. Martin, A. McCune, J. Meyer, J. Miller, J. Moore, L. Parenti, L. Privitera, T. Targett, B. Thompson, P. Wainwright, J. Webb, S. Weitzman, D. Winkelman, J. Willis, and G. Wippelhauser. Joe Nelson provided us logistic aid and an early draft of the classification incorporated into the 3rd edition of his indispensable *Fishes of the world*. Often animated and frequently heated discussions with ichthyological colleagues at annual meetings of the American Society of Ichthyologists and Herpetologists have been invaluable for separating fact from conventional wisdom. Gretchen Hummelman and Natasha Rajack labored long and hard over copyright permissions and many other details. Academic departmental administrators gave us encouragement and made funds and personnel available at several crucial junctures during production. At the University of Georgia we thank J. Willis (Zoology), R. Damian (Cell biology), and G. Barrett, R. Carroll, and R. Pulliam (Ecology) for their support. At St. Michael's College, we thank D. Bean (Biology). The personnel of Blackwell Science, especially Heather Garrison, Jane Humphreys, Debra Lance, Simon Rallison, Jennifer Rosenblum, and Gail Segal, exhibited patience and professionalism at all stages of production.

Finally, a note on the accuracy of the information contained in this text. As Nelson Hairston Sr. has so aptly pointed out, "Statements in textbooks develop a life independent of their validity." We have gone to considerable lengths to get our facts straight, or to admit where uncertainties lie. We accept full responsibility for the inevitable errors that do appear, and we welcome hearing about them. Please write directly to us with any corrections or comments. Chief responsibilities fell on GSH for Chapters 1, 8–15, and 17–25; on BBC for Chapters 2–4 and 16; and on DEF for Chapters 5–7.

Acknowledgments

Thank you to the many students and colleagues who provided constructive feedback on the first two editions. We hope that we were able to honor most of your suggestions in the preparation of this third edition.

We are profoundly grateful to colleagues who provided advice, reviews, encouragement, and logistic support during construction of the third edition, including A.M. Friedlander, M.A. Hixon, G. Orti, J.E. Randall, L.A. Rocha, R.J. Toonen. and the ToBo Lab at Hawai'i Institute of Marine Biology. Special thanks to R.C. Thomson for reviewing Chapter 2 (Phylogenetic Procedures), J. Webb for specific feedback on Chapter 6 (Nervous System and Sensory Organs), M. Wilson for clarifying some questions regarding Nelson et al. (2016), and to R. Hayden and the editing team at Wiley.

We also deeply appreciate the willingness of the following students and colleagues to share their artwork and photographs with us for use in the text: L. Allen, C.M. Ayers, C. Bauder, T. Berra, S.A. Bortone, E. Burress, R. Carlson, L. and C. Chapman, C. Clark, J. DeVivo, C. Cox Fernandes, B. and M. Freeman, J-F Healias, G. Hendsbee, Z. Hogan, M. Horn, K. Hortle, J. Hyde, T. Kelsey, R. Martel, A. Nagy, T.W. Pietsch, E.P. Pister, J. Randall, L.A. Rocha, R. Steene, K. Sultze, A. Summers, P. Vecsei, E. Widder, B. Young. We also thank the global community of photographers willing to share their work online, including employees of public agencies whose work is in the public domain.

Douglas Facey (dfacey@smcvt.edu) reorganized the topics and updated content throughout the book. Brian Bowen (bbowen@hawaii.edu) provided molecular genetics content and updates throughout the text, and also provided editorial input on multiple chapters. Bruce Collette (collettb@si.edu) assisted with updates of systematics in Chapters 11–15. Gene Helfman (genehelfman@gmail.com) took the lead role on the first two editions and provided editorial feedback on most chapters of this current edition.

About the Companion Website

This book is accompanied by a companion website.

www.wiley.com/go/facey/diversityfishes3

Resources include:

- Figures and tables from the book
- Links to supplementary, supporting content

PART I Introduction

The Lined Surgeonfish (*Acanthurus lineatus*) is a mainly herbivorous fish that eats mostly algae. It is found in the Indian Ocean to western Pacific Ocean, including Great Barrier Reef, Japan, Polynesia, and Hawaii. Photo courtesy of J. Hyde and K. Sultze, used with permission.

CHAPTER **1**

The Science of Ichthyology

Summary

Fishes account for more than half of all living vertebrates and exhibit remarkable evolution and diversity. There are over 35,000 living species of fishes (approaching 36,000 as we prepare this edition), of which over 100 are jawless (hagfishes, lampreys), approximately 1100 are cartilaginous (sharks, skates, rays), and the remaining are bony fishes.

A fish can be defined as an aquatic vertebrate with gills and with limbs in the shape of fins. There are, however, exceptions to such general rules, including hagfishes (nonvertebrate craniates) and some fishes that have lost their paired fins over evolutionary time (e.g. moray eels). Included in this definition is a tremendous diversity of sizes (from 8 mm gobies and minnows to the 12 + m Whale Shark), shapes, ecological functions, life history scenarios, anatomical specializations, and evolutionary histories.

Our current understanding of the relationships among the major extant lineages of fishes and other vertebrates shows that fishes include members of four different classes of vertebrates. The hagfishes (class Myxini) are craniates but not vertebrates. The major groups of living vertebrates include the lampreys (class Petromyzontida), sharks and other cartilaginous fishes (class Chondrichthyes), and the bony fishes and their descendants (class Osteichthyes). The Osteichthyes include the subclass Sarcopterygii (the lobe-finned fishes and their tetrapod descendants) and the subclass Actinopterygii (the ray-finned fishes).

CHAPTER CONTENTS

The Diversity of Fishes: Biology, Evolution and Ecology, Third Edition. Douglas E. Facey, Brian W. Bowen, Bruce B. Collette, and Gene S. Helfman.

Companion website: www.wiley.com/go/facey/diversityfishes3

Most (about 60%) living fishes are primarily marine, and the remainder live in freshwater; about 1% move between salt and freshwater as a normal part of their life cycle. The greatest diversity of fishes is found in the tropics, particularly the Indo-West Pacific region for marine fishes and tropical South America, Africa, and Southeast Asia for freshwater species.

Unusual adaptations among fishes include African lungfishes that can survive buried in dry mud for up to 4 years, Antarctic fishes that produce their own antifreeze compounds, deep-sea fishes that can swallow prey larger than themselves (some deep-sea fishes exist as small males that are entirely parasitic on larger females), species that live less than a year and other species that may live hundreds of years, fishes that change sex from female to male or vice versa, sharks that provide nutrition for developing young via a complex placenta, fishes that create an electric field around themselves and detect biologically significant disturbances of the field, light-emitting fishes, warm-blooded fishes, and at least one group, the coelacanths, that was thought to have gone extinct with the dinosaurs.

Historically important contributions to ichthyology were made by Linnaeus, Peter Artedi, Georges Cuvier, Achille Valenciennes, Albert Günther, David Starr Jordan, B. W. Evermann, C. Tate Regan, and Leo S. Berg, among many others.

Genetics and molecular techniques have become integral and essential components to understanding fish biology, evolution, and ecology. In many cases, similarities in DNA probably more accurately reflect evolutionary relationships among groups than do morphological similarities.

The literature on fishes is voluminous, including college-level textbooks, popular and technical books, and websites that contain information on particular geographic regions, taxonomic groups, or species sought by anglers or best suited for aquarium keeping or aquaculture. Scientific journals with a local, national, or international focus are produced in many countries. Another valuable source of knowledge is public aquaria. Observing fishes by snorkel or scuba diving will provide anyone interested in fishes with indispensable, first-hand knowledge and appreciation.

We encourage anyone interested in fishes to observe them closely and carefully, both in captivity and in their natural habitats. They are truly fascinating animals.

Introduction

We recognize the formal common names of fish species as proper names, and therefore they will be capitalized throughout this book. Hence, a green sunfish could be any sunfish (family Centrarchidae) that has a somewhat green color, whereas Green Sunfish refers only to *Lepomis cyanellus*. However, bluefin tuna is not capitalized because it could refer to any of the following three species: the Atlantic Bluefin Tuna (*Thunnus thynnus*), the Pacific Bluefin Tuna (*Thunnus orientalis*), and the Southern Bluefin Tuna (*Thunnus maccoyii*).

Fishes make up more than half of the over 60,000 species of living vertebrates. Along with this remarkable taxonomic diversity comes an equally impressive habitat diversity. Fishes are the cradle of vertebrate biodiversity, dating back more than 500 million years, giving rise to the amphibians, dinosaurs, modern reptiles, birds, and mammals, including you. Millions of years before humans emerged from Africa to spread across the planet, fish species had already attained global distributions in the planet's oceans, lakes, and rivers. Diminutive killifish flourish in the highest lakes of the Andes (Lake Titicaca at 3812 m elevation), and ghostly snailfish forage in utter darkness of the deepest ocean abyss (Mariana Trench at 8000 m depth), very near the physiological limit for life under pressure. In a good year, anchovies of the genus *Engraulis* will vastly outnumber humans, foraging for plankton in schools that can span 50 km. The largest fish, the Whale Shark (*Rhincodon typus*), can exceed 12 meters in length, whereas the minnow *Paedocypris progenetica* in the swamps of Sumatra measures less than 1 cm. From ice-covered polar oceans to oxygen-depleted swamps, from the deepest ocean to desert ponds that dry up for years, and through all the more benign environments in between, fishes have been the ecologically dominant vertebrates in aquatic habitats through much of the history of complex life. To colonize and thrive in such a variety of environments, fishes have evolved striking anatomical, physiological, behavioral, and ecological adaptations.

With an excellent fossil record, fishes are showcases of the evolutionary process, exemplifying the intimate relationship between form and function, between habitat and adaptation. These themes are the foundations for our journey through the diversity of fishes. However, a shadow darkens this narrative: some killifish species in Lake Titicaca are extinct due to introduced trout, anchovies are harvested in the millions of tons to supply fish oil, and the whale shark is no match for human impacts on the global environment. We hope that an appreciation for the diversity of fishes brings a mandate to protect this diversity, so that future generations of fishes can continue to generate remarkable adaptations, and future generations of students can appreciate this majesty.

What Is a Fish?

It may be unrealistic to define a "fish," given the diversity of adaptations that characterizes the multiple classes and thousands of species alive today, each with a unique evolutionary history going back over half a billion years. By recognizing this diversity, one can define a fish as an aquatic chordate with gills and often with paired limbs in the shape of fins. The term "fish" is not a formal taxonomic category but a convenient term for a variety of aquatic organisms as diverse as jawless hagfishes and lampreys; cartilaginous sharks and rays; primitive bony fishes such as lungfishes, sturgeons, and gars; and advanced ray-finned fishes.

Definitions are hazardous in science because exceptions may be viewed as falsifications of the definition (see Berra 2001). Exceptions to the preceding definitions do not negate them, but instead, reveal adaptations arising through powerful selection pressures. Hence the loss of scales and fins in eel-shaped fishes tell us that these structures are not beneficial to fishes with an elongate body. Similarly, although most fishes are ectothermic (the same temperature as the surrounding water), heterothermy (partial warm-blooded physiology) in tunas, lamnid sharks, and opahs indicates the metabolic benefits of elevated body temperature to continuously moving predators in open sea environments. Lungs or other air-breathing structures in lungfishes, gars, African catfishes, and gouramis demonstrate adaptations to environmental conditions where gills are insufficient for transferring adequate oxygen to the blood. Deviations from the definition of "fish" are not invalidations of the definition but are lessons about evolutionary innovations.

Vertebrate Classes

When we (the authors) were first learning our fishes, many textbooks listed five classes of vertebrates: Pisces (all fishes) and four classes of tetrapods – the amphibians, reptiles, birds, and mammals. But Nelson (1969) demonstrated that this five-class system was biased by a human perspective that overemphasized differences among tetrapods while neglecting many of the profound differences among the different evolutionary lineages of fishes. This approach minimized the large morphological and evolutionary gap between the jawless fishes (lampreys and hagfishes) and other groups of fishes and exaggerated the more recently evolutionary divergences between bony fishes and tetrapods. Thus "Pisces" was not a monophyletic group with a single evolutionary history but rather a term used for convenience to describe all non-tetrapod vertebrates, which included several groups that are not closely related to one another (see Chapter 11). More recently, however, modern phylogenetics with an emphasis on utilizing DNA-based molecular appraisals of evolutionary lineages has yielded a more accurate view of the relationships among the major groups of vertebrates (Fig. 1.1).

The great diversity of fishes includes several major branches, and there is some difference of opinion among experts regarding the major taxonomic groupings. Nelson et al. (2016) and our previous edition recognized five classes that included fishes. Nelson et al. (2016) recognize four classes, whereas Eschmeyer and Fong (2017) recognize eight. There is general agreement that the jawless hagfishes and lampreys each belong in their own class, with hagfishes not quite qualifying as vertebrates (hence the term "craniate"). Nelson et al. (2016) put all cartilaginous fishes with jaws in one class (Chondrichthyes), whereas Eschmeyer and Fong consider the Holocephalans (ratfishes, chimaeras) as a class apart from the sharks, skates, and rays (Euselachii). Nelson et al. 2016 consider all jawed, bony fishes (and their descendants, including tetrapods) as members of the class Osteichthyes, with the Sarcopterygii (lobe-fins) and Actinopterygii (ray-fins) as separate subclasses. (These each had separate class status in Nelson 2006.) Eschmeyer and Fong separate the lobe-finned fishes (and their descendants) into three classes and consider all ray-finned fishes as members of a single class. The differences are due to somewhat different interpretations of the level of differences among groups – but should not cause undue concern or confusion to

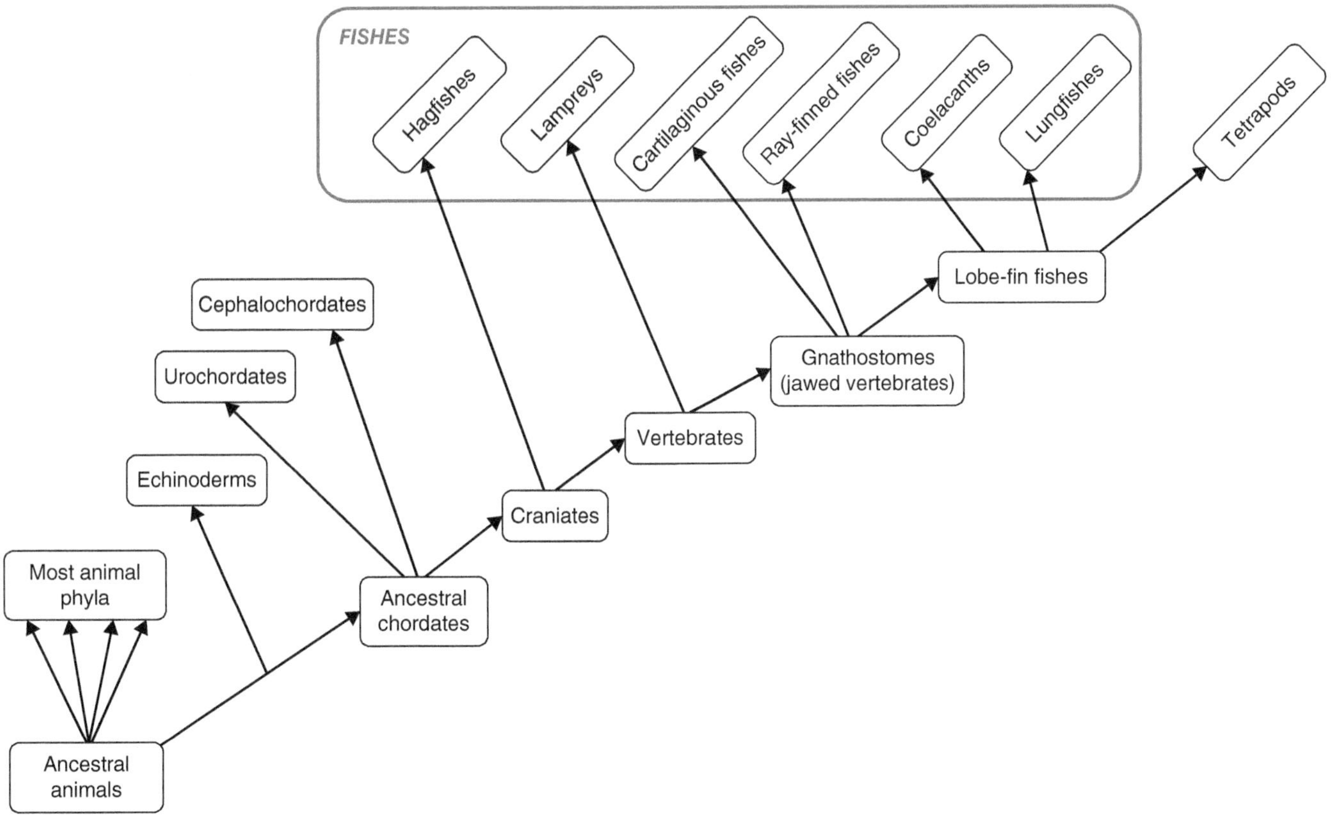

FIGURE 1.1 A cladogram showing the relationships of the major groups of chordates. Note that the fishes represent several major groups, and that the tetrapods are derived from a branch of the lobe-finned fishes.

TABLE 1.1 **The diversity of living fishes. Below is a brief listing of higher taxonomic categories of the Phylum Chordata, including living fishes, in phylogenetic order. This list is meant as an introduction to major groups of living fishes as they will be discussed in the initial two sections of this book. Many intermediate taxonomic levels, such as infraclasses, subdivisions, and series, are not presented here; they will be detailed when the actual groups are discussed in Part III. Only a few representatives of interesting or diverse groups are listed. Taxa mainly according to Nelson et al. 2016.**

- PHYLUM CHORDATA
  - Subphylum Cephalochordata – lancelets
  - Subphylum Craniata
    - Infraphylum Myxinomorphi
      - Class Myxini – hagfishes
    - Infraphylum Vertebrata
      - Superclass Petromyzontomorphi
        - Class Petromyzontida – lampreys
      - Superclass Gnathostomata – jawed vertebrates
        - Class Chondrichthyes – cartilaginous fishes
          - Subclass Holocephali – chimaeras
          - Subclass Euselachii – sharklike fishes
            - Infraclass Elasmobranchii –extant sharks and rays
        - Grade Teleostomi
          - Class Osteichthyes - bony fishes and tetrapod descendants
            - Subclass Sarcopterygii – lobe-finned fishes and relatives (includes Infraclass Tetrapoda), but only fishes listed here
              - Infraclass Actinistia (Coelacanthimorpha) – coelacanths
              - Infraclass Dipnomorpha – lungfishes
            - Subclass Actinopterygii – ray-finned fishes
              - Infraclass Cladistia – bichirs
              - Infraclass Chondrostei – paddlefishes, sturgeons
              - Neopterygii – unranked clade including Holosteans and Teleosts
                - Infraclass Holostei
                  - Division Ginglymodi – includes gars
                  - Division Halecomorphi – includes Bowfin
                - Infraclass Teleosteomorpha
                  - Division Teleostei
                    - Cohort Elopomorpha – tarpons, bonefishes, eels
                    - Cohort Osteoglossomorpha – bonytongues
                    - Cohort Otocephala
                      - Superorder Clupeomorpha – herrings
                      - Superorder Ostariophysi – minnows, suckers, characins, loaches, catfishes
                    - Cohort Euteleostei – further derived bony fishes
                      - Superorder Protacanthopterygii – salmons, smelts, pikes, stomiiforms (bristlemouths, marine hatchetfishes, dragonfishes
                      - Superorder Ateleopodomorpha - jellynose fishes
                      - Superorder Cyclosquamata - greeneyes, lizardfishes
                      - Superorder Scopelomorpha - lanternfishes
                      - Superorder Lampriomorpha - opahs, oarfishes
                      - Superorder Paracanthopterygii - troutperches, cods, toadfishes, anglerfishes
                      - Superorder Acanthopterygii - spiny rayed fishes: mullets, silversides, killifishes, squirrelfishes, sticklebacks, scorpionfishes, basses, perches, tunas, flatfishes, pufferfishes, and many others

students of ichthyology. Our understanding of fishes continues to evolve, as do the fishes themselves. In this text, we follow Nelson et al. (2016) and recognize four classes: Myxini, Petromyzontida, Chondrichthyes, and Osteichthyes (which includes the subclasses Sarcopterygii and Actinopterygii, Table 1.1).

Regardless of differences in higher taxonomic rankings, it is worth noting that in all interpretations of the phylogeny of major groups of vertebrates, the tetrapods (amphibians, reptiles, birds, and mammals, including humans) are a subgroup of the lobe-fins. This means that lungfishes and coelacanths are more closely related to you than they are to an Atlantic Bluefin Tuna or a Largemouth Bass. And all of the bony fishes are more closely related to you than they are to sharks, which are in a completely different class representing jawed vertebrates that have a skeleton of calcified cartilage instead of bone.

The Diversity of Fishes

There are well over 35,000 living species of fishes (Eschmeyer & Fong 2017), with the number approaching 36,000 as we complete this update of our text. (Among ichthyologists, the word "**fish**" is used to refer to a single individual or multiple individuals of a single species, whereas "**fishes**" is plural for more than one species, see Fig. 1.2). The great majority of these species are bony fishes with jaws, but there are also many jawless fishes (hagfishes and lampreys) and cartilaginous fishes with jaws (sharks, skates, rays, and chimaeras). Increased study of remote habitats, combined with modern genetic techniques, helps to continuously add new species to our lists – 7841 new species of fishes were described between 1998 and 2017 (Eschmeyer & Fong 2017). Consult the online *Catalog of Fishes* for updated information, including lists by major taxa.

One reason for the enormous diversity of fishes is the diversity of habitats in which fishes are successful. Water covers about 71% of the earth's surface, and in many places is very deep so the total volume of potential habitats is enormous. Fishes occupy nearly all aquatic habitats that have liquid water throughout the year, including thermal and alkaline springs, hypersaline lakes, sunless caves, anoxic swamps, temporary ponds, torrential rivers, wave-swept coasts, high-altitude and high-latitude environments, open expanses of large lakes

(A)

(B)

(C)

FIGURE 1.2 Fish versus fishes. By convention, "fish" refers to one or more individuals of a single species. "Fishes" is used when discussing more than one species, regardless of the number of individuals involved. (A) One fish, (B) two fish, (C) multiple fishes. (A) G. Helfman (Author); (B, C) D. Facey (Author).

and oceans, and deep ocean areas with no sunlight, low temperatures, and intense pressure. The altitudinal record is set by some nemacheiline river loaches that inhabit Tibetan hot springs at elevations of 5200 m; the record for unheated waters is Lake Titicaca in northern South America, where killifishes and pupfishes live at an altitude of 3812 m. The deepest living fishes are snailfishes and cusk-eels, which occur 8000 m down in the deep sea. When broken down by major habitats, a bit under 60% of fish species live in seawater, about 1% move between freshwater and the sea during their life cycles, and the rest live in freshwater. The highest diversities are found in the tropics. The Indo-West Pacific region known as the Coral Triangle (between Philippines, Indonesia, and New Guinea) has the highest marine diversity with an estimated 4000 species (Allen & Erdmann 2012), whereas South America, Africa, and Southeast Asia, in that order, contain the most species of freshwater fishes (Berra 2001; Lévêque et al. 2008).

Some fishes have adapted to life in water with little oxygen and can breathe air (see Chapter 3). There are even some species that make occasional excursions onto land (see Chapter 7). And we must not forget that over 530 million years of evolutionary adaptations have permitted some descendants of early lobe-finned fishes to become permanently terrestrial – but we call them tetrapods, reserving the term "fish" for those still mainly restricted to aquatic habitats.

Fishes show great variation in body length, ranging more than 1000-fold. The world's smallest fishes – and vertebrates – mature at around 7–8 mm and include the previously mentioned Indonesian minnow, *P. progenetica*, and two gobioids, *Trimmatom nanus* from the Indian Ocean and *Schindleria brevipinguis* from Australia's Great Barrier Reef. Parasitic males of a deep-sea anglerfish *Photocorynus spiniceps* mature at 6.2 mm, although females are 10 times that length. The world's longest cartilaginous fish is the Whale Shark *R. typus*, which can exceed 12 m, whereas the longest bony fish is the 8 m long (or longer) Oarfish *Regalecus glesne*. Body masses top out at 34,000 kg for whale sharks and 2300 kg for the Ocean Sunfish *Mola mola* (Fig. 1.3). Diversity in form includes relatively fishlike shapes such as minnows, trouts, perches, basses, and tunas, but also such unexpected shapes as boxlike trunkfishes, elongate eels

FIGURE 1.3 Some very large fishes. (A) The Ocean Sunfish (*Mola mola*) is estimated to be the heaviest of the bony fishes, with some estimated to weigh up to over 2000 kg . The Library of Congress / Public Domain. (B) The Oarfish (*Regalecus glesne*) is probably the longest bony fish in the world (can exceed 10 m) and is responsible for many reports of sea monsters . Photo by Wm. Leo Smith / Wikimedia Commons / Public Domain. (C) The Whale Shark (*Rhincodon typus*), a filter-feeding cartilaginous fish that can exceed 12 m and 21 tons, is the largest of the living fishes on the planet. Crystaldive / Wikimedia Commons / CC BY-SA 4.0.

and catfishes, globose lumpsuckers and frogfishes, rectangular ocean sunfishes, question-mark-shaped seahorses, and flattened and circular flatfishes and batfishes, in addition to the exceptionally bizarre fishes of the deep sea.

Superlative Fishes

A large part of ichthyology's fascination is the spectacular diversity of fishes. As a few examples:

- Coelacanths, an offshoot of the lineage that gave rise to the amphibians (and subsequently to all tetrapods), were thought to have died out with the dinosaurs at the end of the Cretaceous Period, about 65 million years ago. However, in 1938, fishers in South Africa trawled up a live Coelacanth. This fortuitous capture of a "living fossil" not only rekindled debates about the evolution of vertebrates but underscored the international and political nature of conservation efforts (see Chapter 13).
- Lungfishes can live in a state of dry "suspended animation" for up to 4 years, burying themselves in the mud and becoming dormant when their ponds dry up and reviving quickly when immersed in water (see Chapters 3, 13).
- Some Antarctic fishes don't freeze because their blood contains antifreeze proteins that prevent ice crystal growth. Some Antarctic fishes have no hemoglobin, and thus have clear blood (see Chapter 10).
- Deep-sea fishes include many forms that can swallow prey larger than themselves. Some deep-sea anglerfishes are characterized by females that are 10 times larger than males, the males existing as small parasites permanently fused to the side of the female, living off her bloodstream (see Chapter 10).
- Fishes grow throughout their lives, changing their ecological role several times. In some fishes, differences between larvae and adults are so pronounced that many larvae were originally thought to be entirely different species (see Chapter 9).
- Fishes have maximum life spans of as little as 10 weeks (African killifishes and Great Barrier Reef pygmy gobies) and as long as 400 years (Greenland shark; Neilsen et al. 2016). Several teleost fishes can live past 100 years (Randall & Delbeek 2009). Some short-lived species are annuals, surviving drought as eggs that hatch with the advent of rains. Longer-lived species may not begin reproducing until they are 20 or more years old (see Chapter 8).
- Gender in many fishes is flexible. Some species are simultaneously male and female, whereas others change from male to female or from female to male (see Chapter 8). Fishes, therefore, are the original genderfluid vertebrates and demonstrate that nature includes a broad range of gender identities.
- Fishes engage in parental care that ranges from simple nest guarding to mouth brooding to the production of external or internal body substances upon which young feed. Many sharks have a placental structure as complex as any found in mammals. Egg-laying fishes may construct nests by themselves, whereas some species deposit eggs in the siphon of living clams, on the undersides of leaves of terrestrial plants, or in the nests of other fishes (see Chapter 17).
- Many fishes can detect biologically meaningful, minute quantities of electricity, which they use to navigate and sense prey, competitors, or predators. Some groups can produce an electrical field and detect disturbances to the field, whereas others produce large high-voltage pulses to deter predators or stun prey (see Chapters 4, 16).
- Some fishes can produce light; this ability has evolved independently in different lineages and can be produced either by the fish itself or by symbiotic bacteria (see Chapter 10).
- Although the blood of most fishes is the same temperature as the water surrounding them, some pelagic fishes maintain body temperatures warmer than their surroundings and have circulatory systems that use countercurrent flow to conserve metabolic heat (see Chapter 7).
- Predatory tactics include attracting prey with lures made of modified body parts or by feigning death. Fishes include specialists that feed on ectoparasites, feces, blood, fins, scales, young, and eyes of other fishes (see Chapter 16).
- Fishes can erect spines or inflate themselves with water to deter predators. In turn, the ligamentous and levering arrangement of mouth bones allows some fishes to increase mouth volume by as much as 40-fold (see Chapters 3, 16).
- Some of the most dramatic demonstrations of evolution result from studies of fishes. Both natural and sexual selection have been experimentally manipulated in Guppy, swordtails, and sticklebacks, among others. These investigations show how competition, predation, and mate choice lead to adaptive alterations in body shape and armor, body color, vision, and feeding habits and locales (see Chapters 16, 20). Fishing has also proven to be a powerful evolutionary force, affecting the ages and sizes at which fish reproduce, body shape, and behavior (see Chapter 22).

Fishes have become increasingly important as model organisms for vertebrate research. Because of small size, ease of care, rapid growth and short generation times, and larval anatomical features, such species as Medaka, *Oryzias latipes*, and Zebrafish, *Danio rerio*, are used in studies of toxicology, pharmacology, neurobiology, developmental biology, cancer and other medical research, aging, genomics, and recombinant DNA methodology (e.g. Shima & Mitani 2004; Collins et al. 2010; Chang et al. 2013).

A Brief History of Ichthyology

Fishes would be just as diverse and successful without ichthyologists studying them, but what we know about their diversity is the product of the efforts of workers worldwide over several centuries. Science is a human endeavor and knowing something about historical ichthyologists and their contributions should help give a sense of the dynamics and continuity of this long-established science. Knowing the past can inform future directions.

Although natural historians, as well as many others, have studied fishes for millennia, the earliest treatment of fish behavior, reproduction, and classification in Western culture is Aristotle's *Historia Animalium*. Published about 2300 years ago, this wide-ranging text recognized cartilaginous and bony fishes and was hugely influential in zoology for two millennia. Modern science generally places the roots of ichthyology in the works of Carl Linne (Linnaeus), who produced the first organized system of classification. Zoologists have agreed to use the 10th edition of his *Systema Naturae* (1758) as the starting point for formal nomenclature. The genius of Linnaeus' system is what we refer to as **binomial nomenclature**, naming every organism with a two-part name based on **genus** (plural **genera**) and **species** (singular and plural, abbreviated **sp.** or **spp.**, respectively). Linnaeus did not care much for fishes, so his ichthyological classification, which put the diversity of fishes at less than 500 species, is based largely on the efforts of Peter Artedi (1705–1735), the acknowledged "father of ichthyology." Artedi fell into a canal in Amsterdam one night and drowned under suspicious circumstances (Pietsch 2010). Certainly not the last fish biologist to perish in the aquatic medium, he was buried in an unmarked pauper's grave, and it was left to his friend Linnaeus to publish the masterpiece *Ichthyologia* in 1738.

In the mid-1800s, the great French anatomist Georges Cuvier joined forces with Achille Valenciennes to produce the first complete list of the fishes of the world. During those times, French explorers were active throughout much of the world, and many of their expeditions included naturalists. Thus, the *Histoire naturelle de poissons* (1829–1849) includes descriptions of many species of fishes in its 24 volumes. This major reference is still of great importance to systematic ichthyologists, as are the specimens upon which it is based, many housed in the Museum National d'Histoire Naturelle in Paris.

A few years later, Albert Günther produced a multivolume *Catalogue of fishes in the British Museum* (1859–1870). Although initially designed to simply list all the specimens in the British collections, Günther included all the species of which he was aware, making this catalog the second attempt at listing the known fishes of the world.

The efforts of Linnaeus, Artedi, Cuvier and Valenciennes, and Günther all placed species in genera, and genera in taxonomic families, based on overall resemblance. A modern philosophical background to classification was first developed by Charles Darwin with his *On the Origin of Species* in 1859. His theory of evolution meant that species placed together in a genus were assumed to have had a common origin, a concept that underlies subsequent classifications of fishes and other organisms. In the same decade as Darwin's monumental publication, Woodward (1851) and Dana (1853) divided the waters of the world into biogeographic provinces based on species distributions, another fundamental contribution to understanding the origins of biodiversity (see Chapter 19).

A major force in American ichthyology was David Starr Jordan (1851–1931). Jordan moved from Cornell University to the University of Indiana and then served as founding president of Stanford University. He and his students and colleagues described the fishes collected during global expeditions in the late 1800s and early 1900s. In addition to a long list of papers, Jordan and his coworkers, including B. W. Evermann, produced several publications that form the basis of our present knowledge of North American fishes. This includes the four-volume *The Fishes of North and Middle America* (1896–1900), which described all the freshwater and marine fishes known from the Americas north of the Isthmus of Panama. Jordan (1923) published a list of all the genera of fishes that had ever been described, which served as the standard reference until it was updated and replaced by Eschmeyer et al. (1998).

Overlapping with Jordan was the distinguished British ichthyologist, C. Tate Regan (1878–1943), based at the British Museum of Natural History. Regan revised many groups, and his work formed the basis of most recent classifications. Unfortunately, this classification was never published in one place, and the best summary of it is in the individual sections on fishes in the 14th edition of the *Encyclopedia Britannica* (1929).

A Russian ichthyologist, Leo S. Berg, first integrated paleoichthylogy into the study of living fishes in his 1947 monograph *Classification of Fishes, both Recent and Fossil*, published originally in Russian and English. He was also the first ichthyologist to apply the **-iformes** uniform endings to orders of fishes, replacing the classic and often confusing group names.

In 1966, three young ichthyologists, P. Humphry Greenwood at the British Museum, Donn Eric Rosen at the American Museum of Natural History, and Stanley H. Weitzman at the US National Museum of Natural History, joined with an old-school ichthyologist, George S. Myers of Stanford University, to produce the first modern classification of the majority of present-day fishes, the Teleostei. This classification was updated in Greenwood's 3rd edition of J. R. Norman's classic *A History of Fishes* (Norman & Greenwood 1975), and is the framework, with modifications based on more recent findings, of the classification used by Joseph Nelson (*Fishes of the World*, now in a 5th edition) and followed in this book. Over the past 65 years, John E. Randall, based at the Bishop Museum in Honolulu, has formally described over 800 species of fishes, more than any ichthyologist including Linnaeus.

Details of the early history of ichthyology are available in D. S. Jordan's classic *A Guide to the Study of Fishes*, Vol. I (1905). For a more thorough treatment of the history of North American ichthyology, we recommend Myers (1964) and Hubbs (1964). An excellent historical synopsis of European and North

American ichthyologists can also be found in the introduction of Pietsch and Grobecker (1987); a compilation focusing on the contributions of women ichthyologists appears in Balon et al. (1994). Some more recent and important discoveries are reviewed in Lundberg et al. (2000).

Introduction to Fish Genetics

All fields of biology, from the conservation of rare species to human medicine, have benefitted tremendously in recent decades from advances in genetics, and ichthyology is no exception. Modern genetics has added significantly to our understanding of evolution, ecology, and conservation of fishes. These areas will be discussed throughout this text where appropriate, but for those readers who could benefit from some updating or reminding, we offer the following general background.

DNA is the blueprint of life, and evolution is the architect. Across the 530 million-year history of fishes, natural selection has continually modified the blueprint to accommodate challenges and new opportunities. The result today is over 35,000 species with unique genetic features, and within each species are thousands or millions of genetically unique individuals. Scientists have learned to read the DNA blueprint in recent decades and have begun to resolve the history written there. Genetic studies can resolve relationships from family pedigrees to the most ancient vertebrate lineages. Between these extremes, genetic surveys are useful for discovering new species and resolving biogeographic patterns (see Chapter 19) and management units in conservation (see Chapters 20, 22).

Genetics contains many specialties. For example, cytogenetics is the study of chromosomes, ecological genetics can reveal breeding behavior, population genetics is used to define management units (or stocks) for fisheries, evolutionary genetics demonstrates the basis of novel organismal traits, and molecular phylogenetics is the application of DNA data to resolve branches in the tree of life. Accordingly, sections on genetics will appear in many chapters of this text – but here, we introduce some of the basic principles and terminology.

The Jargon Barrier

To some students of fish biology, the topic of genetics may seem dry, jargon-laden, and lab-based. It is true that some fish geneticists wear lab coats instead of scuba gear and discuss base pairs instead of bait. However, genetics is also a diverse and exciting field that has much to offer ichthyologists, especially with respect to the focal point of this book, the diversity of fishes. Like many disciplines, the field of genetics has a vocabulary that can be challenging to those unfamiliar with the concepts and terminology. Therefore, Appendix 1 contains concepts and terms needed to understand the parts of the text that focus on genetics and molecular techniques.

Fish Genomics

Genomics is the study of the entire DNA sequence of an organism, which in fishes includes the small mitochondrial genome and the enormous nuclear genome that can contain over a billion **base pairs** (**bp**). The earliest explorations of fish genomics were **chromosome** counts and **karyotypes**, but genomics now refers primarily to the intensive efforts to record the entire nuclear genome of a species. In the late 1990s, the technology applied to the Human Genome Project was redirected toward fishes and other vertebrates, and a cottage industry of fish genome projects is emerging. The first fish genomes to be completely described come from Zebrafish (*D. rerio*), Medaka (*Oryzias* sp.), and two pufferfishes. More recently, **next generation sequencing (NGS)** has revolutionized the field, providing millions of bp from individual specimens. Whole genomes have been generated for many fishes (e.g. blacktail butterflyfish; DiBattista et al. 2016b), and the list extends to at least 20 bony fishes (including the coelacanth *Latimeria chalumnae*; Amemiya et al. 2013) and two cartilaginous fishes (including the Whale Shark *R. typus*; Read et al. 2017). To provide the maximum accessibility for studies of evolution, development, and genetic function, a genome must be annotated. **Genome annotation** is the process of identifying where each gene is located and what it does. In most cases, researchers don't need whole annotated genomes, and they only need to find variable sites to resolve relationships from close relatives (kinship) to deep evolutionary lineages (phylogenetics). These studies rely on thousands of **single nucleotide polymorphisms (SNPs)** scattered throughout the genome (Puritz et al. 2014).

Genome Size

Fish genomes include 21 to 100+ chromosomes in the nuclear genome (nDNA) with two copies of every gene in most (diploid) species. Some 58% of examined teleosts (334 out of 580 species) have 48 or 50 chromosomes (Naruse et al. 2004), and 48 is believed to be the ancestral state for ray-finned fishes. Despite this conservation of chromosome number, overall genome sizes can differ by more than two orders of magnitude, with the lungfish having the largest fish genome (81.6 pg of DNA in a set of chromosomes; 1 pg = 10^{-12} g), the bichir having the largest actinopterygian (ray-finned fish) genome (5.85 pg), and the tetraodontiforms (such as pufferfish) having the smallest genomes (0.35 pg, Table 1.2), compared to 3.4 pg in the human genome. The number of chromosomes and genome size can vary even within a single genus (see the *Oncorhynchus* examples in Table 1.2). Three trends in genome size are apparent:

1. There is a progressive reduction in DNA content from the earliest to the most derived bony fishes.
2. Apart from lungfishes, elasmobranchs have the largest fish genomes, with 3–34 pg of DNA (Stingo & Rocco 2001). In contrast, the holocephalans (the other major group

of cartilaginous fishes) have among the most compact genomes (1.2–1.9 pg; Venkatesh et al. 2005).

3. In general, freshwater fishes have larger genomes than marine fishes (Yi & Streelman 2005). This is attributed to smaller population sizes in freshwater fishes, which can reduce the power of natural selection to produce a compact genome.

It is notable that fishes with the most radically derived morphology (tetraodontiforms) have the smallest nuclear genome. The pufferfish (genus *Fugu*) provided an ideal candidate for the first fish genome study, having much the same set of genes observed in mammals but in a package eight times smaller than the human genome. The insights from the first round of fish genome studies are many, and highlights include:

- Genomes are dynamic with many rearrangements between species and sometimes within species. Segments of the genome that are similar between fishes and mammals are rarely longer than four consecutive genes.
- Previous estimates of the vertebrate genome ranged from 60,000 to 150,000 genes, but that number now appears to be 30,000–40,000 genes.
- Regions of the genome of unknown function are highly conserved (very similar) between fishes and mammals. High similarity between fishes and mammals indicates that these gene regions have important functions that are retained by strong natural selection.

Polyploidization and Evolution

Polyploidization is the wholesale duplication of the nuclear genome, and most authorities agree that such an event lies near the base of the ray-finned fish (Actinopterygian) evolutionary tree. Such events are rare but important in the evolution of fishes. Ohno (1970) proposed that gene duplication is essential for major evolutionary innovations in vertebrates, as opposed to the single nucleotide mutations that can distinguish populations and species. In this view, the duplicated genes are under relaxed selection pressure, because there are now four copies (instead of two) available to get the job done. When polyploidy occurs, the original function of the gene can be maintained, freeing the extra copies to develop new functions (neofunctionalization), or they can double the capacity of a crucial metabolic pathway. Over tens of millions of years, some of the duplicated genes will prove to be redundant and lose their function in a process known as diploidization (returning to the diploid state).

Dramatic support for Ohno's model of evolution came from the Zebrafish (*D. rerio*) genome, and the discovery of seven **HOX** genes, which are important regulators of morphological development during embryonic growth. In mammals there are four HOX genes on four chromosomes, whereas the Zebrafish lineage had eight HOX genes on eight chromosomes, followed by the loss of one copy (Amores et al. 1998). The pufferfish genomes provide additional evidence of gene duplication and neofunctionalization.

When did this early genome duplication occur? The duplication does not appear in sturgeon or gar but is shared by all surveyed teleosts (Hoegg et al. 2004). Hence this event must have occurred in the basal teleost lineage, on the order of 300–400 million years before present (mybp). This genome duplication may have provided a powerful toolkit for diversification of the teleosts, by providing twice as many genes as existed in ancestral fishes (Roest Crollius & Weissenbach 2005).

The ancient polyploidization in teleosts is not the only genome duplication in bony fishes. Additional whole-genome duplications have occurred in the ancestors of modern salmon 25–100 million years ago (mya) (Allendorf & Thorgaard 1984), in catostomids (suckers) about 50 mya (Uyeno & Smith 1972), and in carp about 12 mya (David et al. 2003). Partial duplications of the genome are probably more common, either as unequal exchanges during genetic recombination or the unequal sorting of chromosomes during meiosis. This phenomenon can occur when two species hybridize, as observed in poeciliiform fishes (guppies and mollies), occasionally giving rise to parthenogenic (unisexual) species (Vrijenhoek 1984; see Chapter 8).

Mitochondrial Genome

Mitochondria are the energy factories in living cells of all eukaryotes, and they have their own compact genome (**mtDNA**). In fishes and most other vertebrates, the mtDNA genome is about 16,500 bp, coding for 13 proteins, 22 tRNAs, 2 rRNAs, and has a distinct segment for the origin of gene replication known as the control region. In the first decades of molecular genetic studies, the mitochondrial genome was much more accessible than the nuclear genome because it has more copies per cell, a small size, and is constructed like a single chromosome. Hence the study of mitochondrial genomics has proceeded much more rapidly than nuclear genomics, with over 2900 complete fish mtDNA genomes resolved at this time (Miya et al. 2003; see MitoFish database at University of Tokyo: **http://mitofish.aori.u--tokyo.ac.jp/**). In contrast to the nuclear genome, the mtDNA has retained the same genes in the same locations, and genome size is very similar across the vertebrates (Table 1.2).

Transgenic Fishes

There has been considerable interest in genetically modified fishes to produce pharmaceutical products, novel aquarium pets, and faster-growing strains for human consumption. This was originally attempted by "shotgunning," injecting many copies of the desired gene into the nucleus of the eggs, a process with a very low success rate. Subsequent methods have grown more sophisticated, using viruses that can insert desired genes into a chromosome (Dunham 2004). With the advent of advanced CRISPR technology (Hsu et al. 2014), researchers

TABLE 1.2 **Chromosome number, nuclear genome size, and mitochondrial genome size in select fishes. C-values indicate the amount of DNA in a haploid complement (a single copy of the chromosomes), measured in picograms per cell. References for chromosome number and C-value are available from the Animal Genome Size database (http://www.genomesize.com). The nuclear genome sizes, in millions of base pairs (mb), are from Roest Crollius and Weissenbach (2005). References for mitochondrial genome sizes are given, and sizes are presented in thousands of base pairs (kb). In some cases, the chromosome number could not be obtained from the same species used to estimate genome size, so values are obtained from congeners (fish in the same genus) as follows: the chromosome number for *Anguilla japonica* is based on congeners *A. rostrata* and *A. anguilla*, for *Sardinops* it is based on *S. sajax*, for *Fugu* on *F. niphobles*, and for *Tetraodon* on *T. palembangensis*. The mtDNA genome size for dogfish is based on *Scyliorhinus canicula*.**

Species	Number of chromosomes	Nuclear genome size (C-value/mb)	Mitochondrial genome size (kb)
Sea lamprey *Petromyzon marinus*	168 (Vialli 1957)	2.44/n.a.	16.201 (Lee & Kocher 1995)
White Shark *Carcharodon carcharias*	82 (Schwartz & Maddock 1986)	6.45/n.a.	n.a.
Dogfish *Squalus acanthias*	60 (Pedersen 1971)	6.88/n.a.	16.696 (Delarbre et al. 1998)
Coelacanth *Latimeria* spp.	48 (Cimino & Bahr 1974)	3.61/n.a.	16.446 (Inoue et al. 2005)
Lungfish *Protopterus dolloi*	68 (Vervoort 1980)	81.6/n.a.	16.646 (Zardoya & Meyer 1996)
Bichir *Polypterus ornatipinnis*	36 (Bachmann 1972	5.85/n.a.w	16.624 (Noack et al. 1996)
Eel *Anguilla japonica*	38 (Hinegardner & Rosen 1972)	1.40/n.a.	16.685 (Inoue et al. 2001)
Sardine *Sardinops melanostictus*	48 (Ida et al. 1991)	1.35/n.a.	16.881 (Inoue et al. 2000)
Carp *Cyprinus carpio*	100 (Hinegardner & Rosen 1972)	1.70/n.a.	16.575 (Chang et al. 1994)
Zebrafish *Danio rerio*	48 (Hinegardner & Rosen 1972)	1.8/1700	16.596 (Broughton et al. 2001)
Stickleback *Gasterosteus* spp.	42 (Hinegardner & Rosen 1972)	0.70/675	
Chinook Salmon *Oncorhynchus tshawytscha*	56 (Ojima et al. 1963)	3.04/3100	16.644 (Wilhelm et al. 2003)
Rainbow Trout *Oncorhynchus mykiss*	60 (Rasch 1985)	2.60/2700	16.660 (Zardoya et al. 1995)
Medaka *Oryzias* sp.	48 (Uwa 1986)	0.95/800	
Pufferfish *Fugu rubripes*	44 (Ojima & Yamamoto 1990)	0.42/380	16.447 (Elmerot et al. 2002)
Green Pufferfish *Tetraodon nigroviridis*	42 (Hinegardner & Rosen 1972)	0.35/350	

can insert genes in the desired location with surgical precision, using nuclease (DNA cutting) enzymes and customized RNA segments. (CRISPR stands for "clusters of regularly interspaced short palindromic repeats.") The journal *Science* named CRISPR genome editing technology as the 2015 breakthrough of the year, because of the far-reaching implications for repairing defective genes, eliminating disease, and modifying the genomes of plants and animals. No doubt, these benefits will include revelations about fish biology and evolution.

The first genetically modified fish was announced in 1984, based on Rainbow Trout (*Oncorhynchus mykiss*) eggs injected with metallothionein (toxic metal resistance) gene (Maclean & Talwar 1984). Another early success was a transgenic Nile Tilapia (*Oreochromis niloticus*) modified with a gene for growth hormone. It attained three times the size of normal tilapia (Martínez et al. 1996). Rainbow Trout and Coho Salmon (*Oncorhynchus kisutch*) have also been modified with a salmon growth hormone gene that produced larger fish (Devlin et al. 2001). The first transgenic aquarium fish is the **glofish**®, a Zebrafish (*D. rerio*) with red, green, and orange fluorescent colors. Other species of glofish® have since been developed for the aquarium trade (see **http://www.glofish.com/**). This fluorescence gene may eventually be used to detect aquatic pollution by switching on a bright color in fishes that are exposed to environmental contaminants. There are now at least 50 transgenic fishes in research and aquaculture. In 2015, the US Food and Drug Administration approved the first transgenic fish for human consumption, the AquAdvantage salmon, an Atlantic salmon (*Salmo salar*) with a growth hormone gene from Chinook Salmon (*Oncorhynchus tshawytscha*) and a gene promoter from Ocean Pout (*Zoarces americanus*).

Many authorities regard transgenic fishes as a potential boon to aquaculture, allowing higher survival, faster growing fishes, and larger yields. Indeed, genetically modified fishes have the potential to alleviate hunger and promote human health in impoverished corners of the globe. However, other authorities warn of the hazards of transgenic fishes, especially escaping into the wild (Muir & Howard 1999, Bailey 2015). These hazards are similar to those of non-native fish introductions (see Chapter 22), with the additional threat that transgenic fishes could breed with native stocks. In a study of a transgenic Medaka (*O. latipes*), the modified males had greater mating success when introduced into a natural population, presumably because of their larger size, but their offspring had significantly lower survival (Muir & Howard 1999). In these circumstances, the accidental introduction of transgenic fishes might jeopardize future reproduction, possibly leading to extinction. Clearly, the benefits of this technology have to be balanced against significant risks.

Additional Sources of Information

This book is one view of ichthyology, with an emphasis on diversity and adaptation (please read the preface). It is by no means the final word nor the only perspective available. As undergraduates, we learned about fishes from other textbooks; all of these books are valuable. We have read or reread them during the production of this book to check on topics deserving coverage, and we frequently turn to them for alternative approaches and additional information. Among the most useful are Lagler et al. (1977), Bone et al. (1995), Hart & Reynolds (2002a, 2002b), Moyle & Cech (2004), and Barton (2006). For laboratory purposes, Cailliet et al. (1986) and Hastings et al. (2014) can be helpful. From a historical perspective, books by Jordan (1905, 1922), Nikolsky (1961), and Norman and Greenwood (1975) are informative and enjoyable.

Three references have proven indispensable during the production of this book, and their ready access is recommended to anyone desiring additional information and particularly for anyone contemplating a career in ichthyology or fisheries science. Most valuable is Nelson et al. *Fishes of the World* (5th edn, 2016). For North American workers, the current edition of Page et al. *Common and Scientific Names of Fishes from the United States, Canada, and Mexico* (7th edn, 2013) is especially useful. Finally, of a specialized but no less valuable nature, is Eschmeyer et al.'s *Catalog of Fishes* (**https://www.calacademy.org/scientists/projects/catalog-of-fishes**). The first two books, although primarily taxonomic lists, are organized in such a way that they provide information on currently accepted phylogenies, characters, and nomenclature; Nelson et al. (2016) is remarkably helpful with anatomical, ecological, evolutionary, and zoogeographic information on most families. Eschmeyer's volumes are invaluable when reading older or international literature because they give other names that have been used for a fish (**synonymies**) and indicate the family to which a genus belongs.

Of a less technical but useful nature are fish encyclopedias, such as Wheeler's (1975) *Fishes of the World*, also published as *The World Encyclopedia of Fishes* (1985), *McClane's New Standard Fishing Encyclopedia* (McClane 1974), or Paxton & Eschmeyer's (1998) *Encyclopedia of Fishes* (the latter is fact-filled and lavishly illustrated). Helfman & Collette (2011) address many common, and some not-so-common, questions about fishes. Species guides exist for most states and provinces in North America, most countries in Europe (including current and former British Commonwealth nations), and some tropical nations and regions. These are too numerous and too variable in quality for listing here; a good source for titles is Berra (2001, 2007). Two of our favorite geographic treatments of fishes are as much anthropological as they are ichthyological, namely Johannes' (1981) *Words of the Lagoon* and Goulding's (1980) *The Fishes and the Forest*. A stroll through the shelves of any decent public or academic library is potentially fascinating, with their collections of ichthyology texts dating back a century, geographic and taxonomic guides to fishes, specialty texts and edited volumes, and works in or translated from many languages. Among the better known, established journals that specialize in or often focus on fish research are *Ichthyology and Herpetology* (formerly *Copeia*), *Transactions of the American Fisheries Society*, *Environmental Biology of Fishes*, *North American Journal of Fisheries Management*, *US Fishery Bulletin*, *Canadian Journal of Fisheries and Aquatic Sciences*, *Canadian Journal of Zoology*,

Journal of Fish Biology, *Journal of Ichthyology* (the translation of the Russian journal *Voprosy Ikhtiologii*), *Australian and New Zealand Journals of Marine and Freshwater Research*, *Bulletin of Marine Science*, and *Japanese Journal of Ichthyology*.

The internet has developed into an indispensable source for technical information, spectacular photographs, and updated conservation information concerning fishes. Although websites come and go – and although web information often suffers from a lack of critical peer review – many sites have proven themselves to be both dependable and reliable. For general international taxonomic information, the Integrated Taxonomic Information System (ITIS, **www.itis.usda.gov/index.html**) and Global Biodiversity Information Facility (GBIF, **www.gbif.org**) are starting points. For user-friendliness and general information, FishBase (**www.FishBase.org**) is the unquestioned leader. Photographs and drawings are most easily accessed via Google Images (**http://images.google.com**). Wikipedia entries on individual species and groups are often written by experts in the field and corrected and updated frequently. For conservation status and background details, **www.redlist.org** is the accepted authority on international issues, and NatureServe (**www.natureserve.org**) is the most useful clearinghouse for North American taxa. Several museums maintain updated information on fishes; our favorites are the Australian Museum (**https://australianmuseum.net.au**), University of Michigan Museum of Zoology (**https://lsa.umich.edu/ummz**), Florida Museum of Natural History (**www.flmnh.ufl.edu/fish**, which is especially good for sharks), and the California Academy of Sciences (**www.calacademy.org/research/ichthyology**). For North American freshwater fishes, see the Texas Memorial Museum (**www.utexas.edu/tmm/tnhc/fish/na/naindex**) and the North American Native Fishes Association website (**http://nanfa.org/checklist.shtml**). The best sites provide links to many additional sites that offer more localized or specific information.

The internet also provides access to a seemingly endless and ever-changing array of videos that show what many of us find so fascinating about fishes – including their enormous diversity in form, function, and behavior. These videos range from segments of professionally produced programs to home-grown postings on YouTube ®. The nature of a printed text and the often changing nature of the web make it impractical to cite these website references in this book, as web postings often come and go.

Snorkeling and scuba diving are valuable methods for acquiring detailed information on fish biology. Time spent underwater observing fishes was formative and essential to each of us in developing our interest in and understanding of fishes. A full appreciation for the wonders of adaptation in fishes requires that they be viewed in their natural habitat, as they would be seen by their conspecifics, competitors, predators, and neighbors (it is fun to try to think like a fish). We strongly urge anyone seriously interested in fish biology to acquire basic diving skills, including the patience necessary to watch fishes going about their daily lives. **Public and commercial aquaria** are almost as valuable, particularly because they expose an interested person to a wide range of species, or to an intense selection of local fishes that are otherwise only seen dying in a bait bucket or at the end of a fishing line. Our complaint about such facilities is that, perhaps because of space constraints or an anticipated short attention span on the part of viewers, large aquaria seldom provide details about the fascinating lives of the animals they hold in captivity. Home aquaria are an additional source for inspiration and fascination, although we are ambivalent about their value because so many tropical fishes are killed or habitats destroyed in the process of providing animals for the commercial aquarium trade, particularly for marine tropical fishes. Advances in captive rearing are badly needed to alleviate the pressure on natural populations of marine ornamental fishes.

Supplementary Reading

Websites

Eschmeyer, WN and JD Fong. 2017. Species by family/subfamily in the Catalogue of Fishes. Catalogue of Fishes, California Academy of Sciences (**http://researcharchive.calacademy.org/research/ichthyology/catalog/SpeciesByFamily.asp**)

the Integrated Taxonomic Information System (ITIS, **www.itis.usda.gov/index.html**)

Global Biodiversity Information Facility (GBIF, **www.gbif.org**)

FishBase (**www.FishBase.org**)

www.redlist.org

NatureServe (**www.natureserve.org**)

Australian Museum (**https://australianmuseum.net.au**)

University of Michigan Museum of Zoology (**https://lsa.umich.edu/ummz**)

Florida Museum of Natural History (**www.flmnh.ufl.edu/fish**)

California Academy of Sciences (**www.calacademy.org/research/ichthyology**)

Texas Memorial Museum (**www.utexas.edu/tmm/tnhc/fish/na/naindex**)

the North American Native Fishes Association website (**http://nanfa.org/checklist.shtml**)

CHAPTER 2

Phylogenetic Procedures

Summary

Scientific terminology is critical to scientific communication and progress. There are over 35,000 species of fish, and in order for us to be able to discuss and understand them, we must know their names and how to identify them (taxonomy), and how to categorize them (systematics, or phylogenetics). Species are the fundamental unit of classification and can be defined as a single lineage of ancestor–descendant populations that maintains its identity from other such lineages. Species are usually reproductively isolated from other species.

Taxonomy deals with describing biodiversity (including naming undescribed species), arranging biodiversity into a system of classification, and devising guides to the identification. Systematics (or phylogenetics) is the study of relationships among species, genera, families, and higher taxa. Biologically, the most useful and informative approach to organizing species into categories is one based on the evolutionary history of the **species** and its relatives.

Cladistics, or phylogenetic systematics, is a system of morphological classification in which characters are divided into derived or advanced traits (apomorphies) and primitive or generalized traits (plesiomorphies). The goal is to find shared derived characters (synapomorphies) that define monophyletic groups or clades (groups containing an ancestor and all its descendant taxa).

CHAPTER CONTENTS

The Diversity of Fishes: Biology, Evolution and Ecology, Third Edition. Douglas E. Facey, Brian W. Bowen, Bruce B. Collette, and Gene S. Helfman.

Companion website: www.wiley.com/go/facey/diversityfishes3

Taxonomic characters can be meristic (countable), morphometric (measurable), morphological (including color), cytological, behavioral, electrophoretic, or molecular (nuclear or mitochondrial). Since about the 1990s, access to DNA data has revolutionized the field of phylogenetics, providing vast amounts of data for resolving evolutionary relationships and the age of those relationships.

Rules of nomenclature govern the use of taxonomic names. The International Code of Zoological Nomenclature promotes the stability of scientific names for animals. These rules deal with such matters as the definition of publication, authorship of new scientific names, and types of taxa.

Species and subspecies are based on type specimens housed in museum collections. Higher taxonomic categories are based on type taxa. Primary types include the holotype, the single specimen upon which the description of a new species is based. Secondary types include paratypes, which are additional specimens used in the description of a new species.

Importance of Classification

Scientific terminology is essential to scientific progress, although it is sometimes regarded as an impediment to broad communication. Things must be named and organized into categories for us to be able to talk about, compare, and understand them. This applies to cars, athletes, books, animals, plants, and students. For clarity, we must agree on the names of species (**taxonomy**) and the rules for creating and changing those names. The foundations of a taxonomically oriented discipline such as ichthyology include an organized, hierarchical system of names of fishes, which incorporate evolutionary hypotheses associated with those names. This structure provides a basis for identifying and discriminating among fish species and for understanding relationships among species and higher taxonomic categories. This enterprise is generally known as **phylogenetics** or **systematics**. In this chapter, we discuss the value, functions, and goals of phylogenetic procedures, different philosophies for classifying organisms, and how systematic procedures increase our understanding of fishes.

We need a system of classification because we cannot deal with all the members of a group (such as over 35,000 species of fishes) individually, so we must put them into some sort of classification. Different types of classifications are designed for different functions. For example, one can classify automobiles by function (sedan, SUV, pickup, etc.) or by manufacturer (Ford, General Motors, Toyota, etc.). Similarly, animals can be classified ecologically as grazers, detritivores, carnivores, and so forth. Good reasons exist for ecologists to classify organisms by these traits, but this is a special classification for special purposes. It does not reflect the evolutionary history and relationships among species. The most general classification is considered to be the **natural classification**, defined as the classification that best represents the evolutionary history of an organism and its relatives. A phylogenetic classification of taxonomic groups (**taxa**) holds extra information because the categories are predictive. For example, if one species of fish in a genus builds a nest, it is likely that other species in that genus also do so.

Species Concepts

Species are the fundamental unit of biological classification. The idea of recognizing species goes back millennia – many religious and cultural origin stories name distinctly different kinds of animals. Aristotle elaborated on the different kinds of animals in *Historia Animalium*. Ichthyologist C. Tate Regan (1926) defined a species as "A community, or a number of related communities whose distinctive morphological characters are, in the opinion of a competent systematist, sufficiently definite to entitle it, or them to a specific name." This practical, but somewhat circular definition of a species, now termed a **morphospecies**, is based on similarity but does not incorporate evolutionary relationships.

The late 1930s and early 1940s hosted the first major attempts to make classification compatible with evolution. Theodosius Dobzhansky (1935), a fruit fly geneticist, defined species as "a group of individuals fully fertile *inter se*, but barred from interbreeding with other similar groups by its physiological properties (producing either incompatibility of parents, or sterility of the hybrid, or both)." Julian Huxley (1940) integrated genetics with evolution in his book *The New Systematics*. Ernst Mayr (1942, p. 120) introduced the **biological species concept** in *Systematics and the Origin of Species*: "Species are groups of actually or potentially interbreeding populations which are reproductively isolated from other such groups." This was an important effort to move away from defining species strictly on the basis of morphological similarity. This definition has subsequently been modified to fit our improved understanding of evolution: an **evolutionary species** "is a single lineage of ancestor-descendant populations which maintains its identity from other such lineages and which has its own evolutionary tendencies and historical fate" (Wiley 1981, p. 25). In the age of genomics, Wu (2001) defined the species by "the fixation of all isolating genetic traits in the common genome of the entire population." An emerging (and still controversial) approach is to use genomic data and genetic divergence to identify species in a phylogeny (Flouri et al. 2018). The proliferation of species concepts remains unresolved. Ichthyologist Richard Mayden (1997) identified 22 species concepts, and Wilkins (2006) identified 26 species concepts ranging from agamospecies (for unisexual lineages) to taxonomic species (when a taxonomist says it is a species). Defining a species may be as difficult as defining a fish (Chapter 1), but the concepts of reproductive isolation and shared evolutionary history remain paramount.

Taxonomy Versus Systematics

These two words describe somewhat overlapping fields. **Taxonomy** deals with the theory and practice of describing biodiversity (especially naming species), arranging this diversity into a system of classification, and devising identification keys pragmatically based on morphology and DNA barcodes. It includes the rules of nomenclature that govern the use of taxonomic names. **Systematics** (or **phylogenetics**) is the study of relationships among species, genera, families, and higher taxa. Lundberg and McDade (1990) have presented a good summary of morphology-based systematics oriented toward fishes. Felsenstein (2004) provides a good overview of the statistical-based methods that dominate in molecular phylogenetics. Primary journals dealing with systematics of animals include *Systematic Biology* (formerly *Systematic Zoology*), published by the Society of Systematic Biologists, and *Molecular Phylogenetics and Evolution*, dedicated to Darwin's dream of having "fairly true genealogical trees of each great kingdom of Nature." For journals dealing with systematics of fishes, see Chapter 1, Additional Sources of Information.

Approaches to Classification

Prior to the middle 1900s, systematic classification was dominated by individual taxonomists who defined species based on their own criteria, usually focused on the characteristics of their taxonomic specialty. This led to inconsistencies in taxonomic identification, as the criteria for defining a beetle species (for example) could not be readily applied to a lizard species. "Because I said so" was an unsatisfying basis for classification, even if pronounced by foremost authorities, indicating a conspicuous lack of scientific rigor. This status quo changed in the second half of the 1900s with the emergence of three general philosophies of classification: cladistics, phenetics, and evolutionary systematics. All three approaches produce some sort of graphic illustration, usually a branching diagram or "tree," which depicts the different taxa, arranged in a manner that reflects hypothesized evolutionary relationships.

Cladistics, or **phylogenetic systematics** comprised a revolution in systematic methodology begun by a German entomologist, Willi Hennig, following publication of the 1966 English translation of an extensively revised version of his 1950 German monograph. His fundamental principle was to divide characters (usually morphological features) into two groups: those that are more recently evolved, derived, or advanced characters (**apomorphies**) and those that are more ancestral, primitive, or generalized (**plesiomorphies**). The goal in classification is to find shared derived characters (**synapomorphies**) that characterize groups containing all the descendants of an ancestral taxon. These monophyletic groups are called **clades**. Shared primitive characters (**symplesiomorphies**) are not useful for constructing phylogenetic classifications because primitive characters may be retained in a wide variety of distantly related taxa. Specialized characters that are present in only a single taxon (**autapomorphies**) are important in defining that species but are also not useful in constructing a phylogenetic tree.

In cladistics, taxa are arranged on a branching diagram called a **cladogram** (Fig. 2.1) depicting **monophyletic groups**, clusters of species that descended from a single ancestor. Cladistic classification tries to avoid **polyphyletic groups** – groups containing the descendants of different ancestors, and **paraphyletic groups**, groups that do not contain all the descendants of a single ancestor (Fig. 2.2).

An ideal example of how cladistics should work concerns the oceanic fish known as the Louvar (*Luvarus imperialis*). Most ichthyologists have classified the Louvar as a strange sort of scombroid fish (Scombroidei), the perciform suborder that contains the tunas, billfishes, and snake mackerels. However, a comprehensive morphological and osteological study (Tyler et al. 1989) showed clearly that the Louvar is actually an aberrant pelagic relative of the surgeonfishes (Acanthuroidei). This example is instructive because the study utilized 60 characters from adults and 30 more from juveniles (Fig. 2.1). Characters postulated to be reversals (return to original condition) or independent acquisitions (independently evolved) were minimal. With the cladistic approach, synapomorphies show that the relationships of the Louvar are with the acanthuroids, whereas noncladistic analysis overemphasized caudal skeletal characters, leading to placement among the scombroids.

Deciding whether a character is ancestral or derived is based largely on **outgroup** analysis, that is, finding out what characters are present in closely related **sister groups** outside the taxon (ingroup) under study. More than one outgroup should be used to protect against the problem of misinterpreting a derived trait in an outgroup as a shared ancestral trait. The **polarity** of a character or the inferred direction of its evolution (e.g., from soft-rayed to spiny-rayed fins) is determined using outgroup comparison. Problems arise when there are shared, independently derived similarities (**homoplasies**) such as parellelisms, convergences, or secondary losses. This problem is especially common in DNA data, where the four nucleotides (A,G,C,T) can independently switch in the same direction. For example, distantly related taxa may mutate from A to G at the same site on a chromosome, but these changes do not reflect shared evolutionary history.

Ideally, when constructing a classification, a taxon can be defined by a number of shared derived characters. However, conflicting evidence frequently exists. For morphological data, the principle used to sort out the conflicting evidence is **parsimony**: select the hypothesis that explains the data in the simplest or most economical manner, as discussed in the example of the Louvar. Phylogenetic programs based on parsimony algorithms include PAUP (phylogenetic analysis using parsimony; Swofford 2003). A thorough explanation of cladistic methodology is presented by Wiley (1981), and cogent, brief summaries can be found in Lundberg and McDade (1990) and Funk (1995).

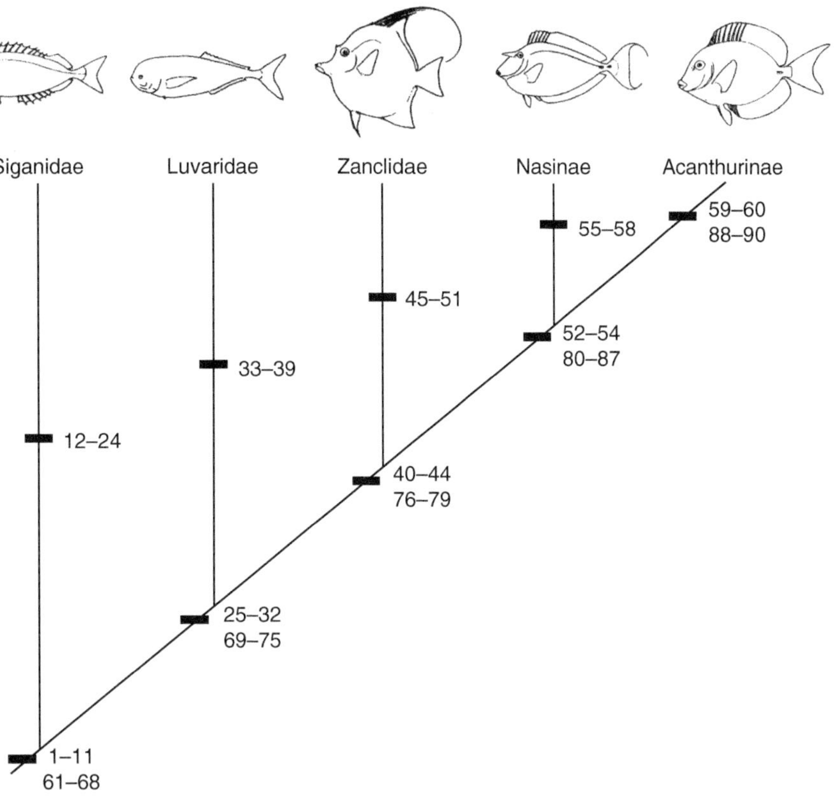

FIGURE 2.1 Cladogram of hypothesized relationships of the Louvar (*Luvarus*, Luvaridae) and other Acanthuroidei. Arabic numerals show synapomorphies: numbers 1 through 60 represent characters from adults, 61 through 90 characters from juveniles. Some sample synapomorphies include: 2, branchiostegal rays reduced to four or five; 6, premaxillae and maxillae (upper jawbones) bound together; 25, vertebrae reduced to nine precaudal plus 13 caudal; 32, single postcleithrum behind the pectoral girdle; 54, spine or plate on caudal peduncle; 59, teeth spatulate. Adapted from Tyler et al. (1989).

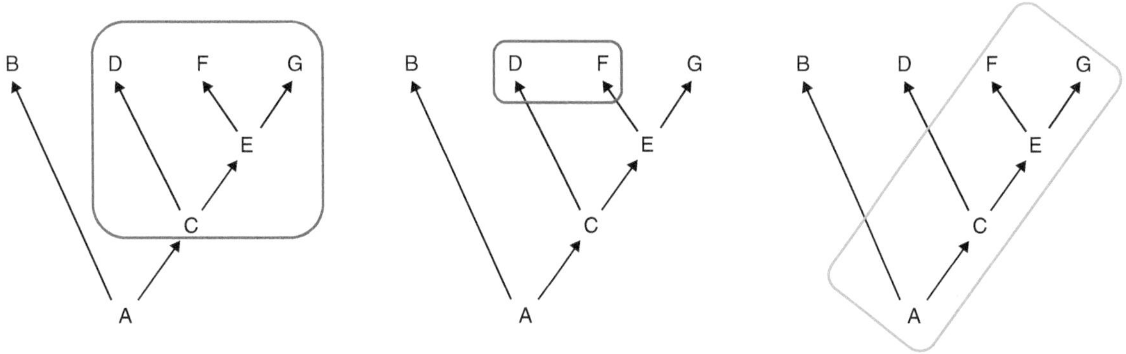

FIGURE 2.2 A hypothetical phylogeny is showing groups that are monophyletic, polyphyletic, and paraphyletic. The blue box shows a monophyletic group, which includes an ancestor and all of its descendants. This is the primary objective of cladistic classification. The red box shows a polyphyletic group, which includes descendants but not their common ancestor. The green box shows a paraphyletic group that includes an ancestor and only some of its descendants.

Cladistic techniques and good classifications based on these techniques have proved useful in analyzing the geographic distribution of plants and animals in a process called vicariance biogeography (see Chapter 19).

Phenetics, or **numerical taxonomy**, is a second approach to systematics which grew out of the availability of computers in the 1960s (Sokal & Sneath 1963). Phenetics starts with species or other taxa as **operational taxonomic units** (OTUs) and then clusters the OTUs on the basis of overall similarity, using an array of numerical techniques. Advocates of this school believe that the more characters used the better and more natural the classification should be (Sneath & Sokal 1973). Graphic representations in phenetics are known as **phenograms**, and relatedness is determined by comparing measured linear distances (or DNA sequence divergences) between OTUs.

Many systematists of the late 1900s were skeptical that using morphological characters, without distinguishing between ancestral and derived, would provide a robust classification. Hence cladistics was the dominant paradigm for phylogenetic studies for many years, and it seemed that numerical taxonomy might be consigned to the dust heap of scientific history. That all changed in the 1970s and 1980s with the rise of molecular phylogenetics. As noted earlier, in a data set comprised of just four character states (nucleotides A, C, G, and T), homoplasies will be very common, with distantly related taxa having the same character state by chance, not by co-ancestry. Although cladistic principles such as parsimony can be fruitfully applied to DNA data, numerical and statistical frameworks are preferable to analyze genetic data. The resulting genetic distances (expressed as percent DNA sequence divergence) are a powerful metric for determining the age of evolutionary partitions.

For DNA sequence data, which can contain thousands of nucleotide "characters," the dominant approach is **maximum likelihood**. Whereas the cladistic principle of parsimony requires a tree with the fewest steps, maximum likelihood finds the tree with the highest statistical likelihood, given what we know about how DNA mutates over time. A very useful modification of maximum likelihood is **Bayesian analyses**, which allows for statistical analyses using more complex and realistic models for how DNA mutates and evolves. Software to handle the enormous amount of data generated from molecular sequences within maximum likelihood and Bayesian approaches include RAxML (Randomized Accelerated Maximum Likelihood; Stamatakis 2014) and MrBayes (Ronquist & Huelsenbeck 2003).

Evolutionary systematics, as summarized by Mayr (1974), focuses on **anagenesis**, the amount of differentiation (or branch length) that has occurred since groups diverged on separate evolutionary pathways. For example, the lungfishes of Africa and South America diverged about 120 million years ago, and what began as species-level divergence is now recognized as two taxonomic families. The divergence along these two lineages must be taken into consideration along with **cladogenesis**, the process of branch or lineage splitting, to understand the evolutionary history of living organisms. Evolutionary relationships are expressed with a tree called a **phylogram**.

Most leading ichthyological theorists favor the cladistic approach for morphological data and the maximum likelihood (including Bayesian) approach for DNA sequence data.

Morphological Characters

Whichever system of classification is employed, characters are needed to differentiate taxa and assess their interrelationships. Characters, as Stanford ichthyologist George Myers once said, are like gold – they are where you find them. Characters are variations of a homologous structure and, to be useful, they must show some variation in the taxon under study. Useful definitions of a wide variety of characters were presented by Strauss and Bond (1990). Characters can be divided, somewhat arbitrarily, into different categories.

Meristic characters originally referred to characters that correspond to body segments (myomeres), such as numbers of vertebrae and fin rays. Meristic is used for almost any countable structure, including numbers of scales, gill rakers, and fin rays. These characters are useful because they are clearly definable, and usually other investigators will produce the same counts. Ideally, they are stable over a wide range of body sizes. Meristic characters can be analyzed statistically, so comparisons can be made between populations or species with a minimum of computational effort.

Morphometric characters refer to measurable structures such as fin lengths, head length, eye diameter, or ratios between such measurements. Some morphometric characters are harder to define exactly, and being continuous variables, they can be measured to different levels of precision and so are less easily repeated. Furthermore, there is the problem of **allometry**, whereby lengths of different body parts change with growth (see Chapter 9). Thus analysis of differences is more complex than with meristic characters. Size factors have to be compensated for through the use of such techniques as regression analysis, analysis of variance (ANOVA), and analysis of covariance (ANCOVA) so that comparisons can be made between actual differences in characters and not differences due to body size. Principal components analysis (PCA) also adjusts for size, as recommended by Humphries et al. (1981).

Widely used definitions of most meristic and morphometric characters were presented by Hubbs and Lagler (1964); some of these are illustrated in Fig. 2.3.

Anatomical characters include characters of the skeleton (osteology) and characters of the soft anatomy, such as the position of the viscera, divisions of muscles, and branches of blood vessels. Some investigators favor osteological characters because such characters are thought to vary less than other characters. Useful characters can include almost any fixed, describable differences among taxa. For example, color can include such characters as the presence of stripes, bars, spots, or specific coloration. Photophores are light-producing structures that vary in number and position among different taxa. Sexually dimorphic ("two forms") structures can be of functional value, including copulatory organs, like the gonopodium of a guppy (modified anal fin) or the claspers of chondrichthyans (modified pelvic fins). Cytological (including karyological), electrophoretic, serological, behavioral, and physiological characters are useful in some groups.

Molecular Characters

Nuclear DNA and mitochondrial DNA (mtDNA) have become increasingly useful at all levels of classification (Hillis & Moritz 1996; Page & Holmes 1998; Avise 2004). All organisms contain DNA, RNA, and proteins. Closely related organisms

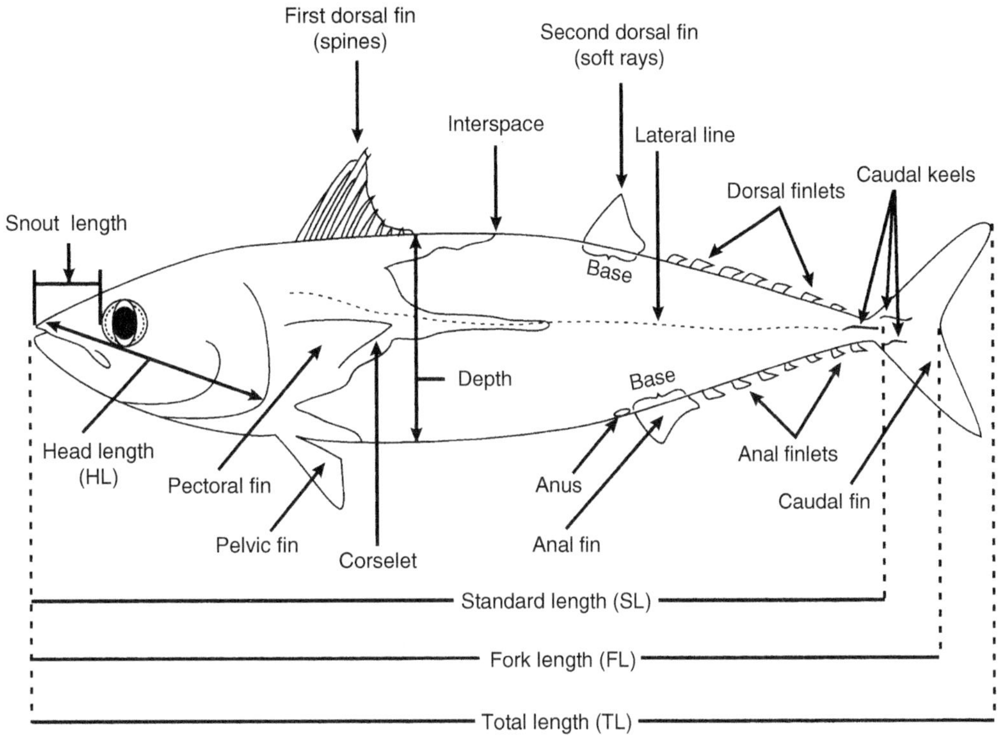

FIGURE 2.3 Some meristic and morphometric characters are shown on a hypothetical scombrid fish.

show a high degree of similarity in DNA sequences, and if the rate of mutation is known for a particular genetic sequence, a **molecular clock** can be applied to estimate the age of divergences. Molecular systematics uses such data to build trees showing evolutionary relationships. With the advent of next-generation sequencing, researchers can scan thousands of polymorphisms (SNPs as defined in Appendix 1) in the nuclear genome and whole mitochondrial genomes.

Just as morphology-based phylogenies can be plagued by homoplasies (similar structures that are not shared by co-ancestry), molecular characters have limitations for phylogenetic inference. Foremost among these is the distinction between "gene trees" and "species trees" (Maddison 1997). Different genomic regions may have different evolutionary histories, due to several important biological processes, including the sorting of genotypes during speciation, gene duplication, and hybridization. Alternate forms of a gene may diverge early in the speciation process, late in the speciation process, and even after the speciation process is complete via hybridization. For this reason, phylogenetic trees based on one or a few genetic loci should be viewed with caution. Many loci will provide a clearer view of the speciation process and evolutionary history (Fig. 2.4).

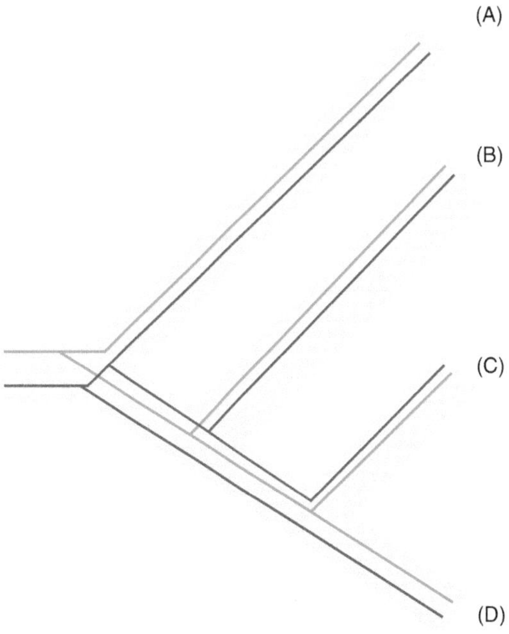

FIGURE 2.4 The shaded area represents a phylogenetic tree showing the evolutionary relationships of four taxa (A, B, C, D). Such a tree should be based on a preponderance of evidence from multiple sources, including multiple genes and loci. The red and green lines represent the evolutionary history of two different genes within the tree. The green lines follow the branching pattern of the tree and are therefore in agreement with most of the evidence used to create the tree, which suggests that B and C are more closely related to D than they are to A. However, the red lines suggest that B and C are more closely related to A than they are to D, and thus would lead to a different conclusion regarding the evolutionary history of the taxa represented in the tree.

The Role of Genetics in Systematics

Access to DNA sequence information has revolutionized the field of phylogenetics. Using models of DNA sequence evolution, researchers can sort out the relationships among fishes from the earliest lineages to the most recent speciation events.

Fishes arose approximately 530 million years ago (reviewed in Chapter 11), and three deep lineages survive today: the hagfishes (Myxiniformes), lampreys (Petromyzontiformes), and jawed vertebrates (Gnathostomes). Notably, that last category includes cartilaginous fishes (Chondrichthyes), extant bony fishes (Actinopterygii), and tetrapods (amphibians, reptiles, birds, mammals). The fossil record indicates that all these groups arose in the first hundred million years of fish history; however, the order in which they arose has been subject to extensive debate.

Takezaki et al. (2003) used over 27 kilobases of DNA sequence data from 35 nuclear genes to resolve the deepest lineages in the fish tree (Fig. 2.5). Despite fundamental morphological differences, the two jawless fishes (hagfishes and lampreys) appear to be each other's closest relatives. These data indicate that the cartilaginous fishes diverged next, followed by bony fishes/tetrapod radiation. Based on molecular studies, the coelacanths appear to diverge near the base of the bony fish/tetrapod bifurcation (Zardoya & Meyer 1996). These studies illustrate two points: First, molecular systematics is especially valuable in cases where the morphology is too divergent (or too similar) to make robust phylogenetic conclusions. Second, the lineage that gave rise to terrestrial vertebrates was the most recent of the major branches in fish history, demonstrating that you, the reader, are more closely related to lobe-finned fishes such as lungfishes and coelacanths, than those lobe-finned distant cousins of yours are related to other fishes such as sharks and tunas (see Chapter 11; see also Shubin 2008).

The most successful modern fishes are the teleosts. However, ray-finned fishes include four additional lineages,

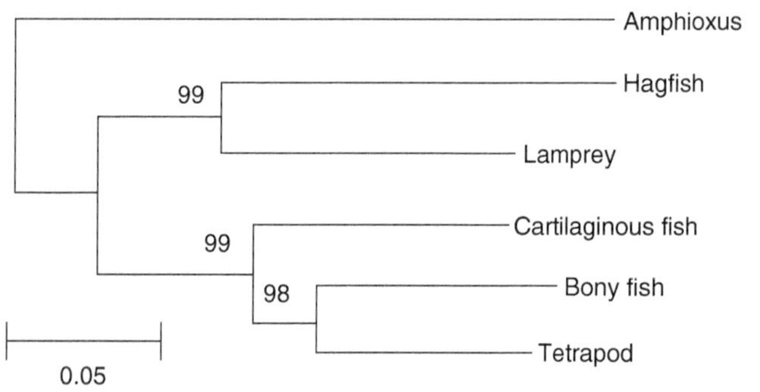

FIGURE 2.5 Phylogeny of the most ancient lineages of extant fishes. Previous studies had indicated that lampreys are more closely related to jawed fishes than hagfishes, based on shared primitive traits, including features of the nervous system, osmotic regulation, and a lens apparatus in the eye. In contrast, DNA sequence data indicate that the two jawless fish taxa are sister lineages, followed by the separation of the cartilaginous fishes from the lineage that gave rise to modern bony fishes (actinopterygians) and tetrapods. The outgroup (Amphioxus) has a notochord but not a true vertebral column, is united with vertebrates in the phylum Chordata, and is considered to be the closest extant relative of the vertebrates (see Chapter 13). This is a maximum likelihood tree based on 35 nuclear gene sequences; the scale bar indicates percent divergence in amino acid composition. Bootstrap support values are indicated above the primary branches. From Takezaki et al. (2003), used with permission.

known as the ancient actinopterygians. These include polypteriforms (bichirs and reedfish), acipenseriforms (sturgeons and paddlefish), lepisosteids (gars), and *Amia calva* (Bowfin). How they relate to teleosts, and each other, has been a matter of considerable debate, with systematists proposing almost every possible arrangement of relationships. However, most authorities have identified the polypteriforms as the oldest extant group of ray-finned fishes.

To address the evolutionary history of the ancient and modern ray-finned fishes, Inoue et al. (2003) analyzed entire mtDNA genomes from 12 of the ancient actinopterygians, 14 teleosts, and two elasmobranch outgroups (Fig. 2.6). This extensive DNA sequencing effort, approximately 16.5 kb per species, represents a growing trend in molecular phylogenetics fed by improvements in automated DNA sequencing technology, using entire genomes to reconstruct evolutionary relationships.

In keeping with earlier hypotheses, the polypteriforms appear to be the most ancient of the living ray-finned fishes. No doubt their persistence into the modern era is aided by unusual adaptations to arid conditions; for example, the bichir lives in semipermanent freshwater habitats in Africa, and their gas bladder functions as a primitive lung. They can obtain oxygen from air during periods of stagnation and drought and can move over land to another body of water if their lake or swamp dries up. Based on the mtDNA data, the sturgeons, paddlefish, gars, and Bowfin are a sister lineage to the teleosts.

The phylogeny of more derived teleosts (Percomorpha) has been investigated with 100 complete mtDNA sequences, and these data indicate many unexpected relationships, including a phylogenetic affinity between Lophiiformes (goosefish, long assumed to be an ancestral teleost) and Tetradontiformes (pufferfishes, long assumed to be among the most derived teleosts) (Miya et al. 2003). Clearly, these findings indicate a rich field for further investigation.

Molecular data can be used to test hypotheses of relationships among species that were formulated with morphological data. For example, the analyses of similar morphological data sets for the Scombroidei by Collette et al. (1984) and Johnson (1986) produced different cladograms resulting in very different classifications. Collette et al. (1984) postulated a sister-group relationship of the Wahoo (*Acanthocybium*) and Spanish mackerels (*Scomberomorus*) within the family Scombridae. In contrast, Johnson (1986) placed the Wahoo as sister to the billfishes within a greatly expanded Scombridae that includes billfishes as a tribe, instead of being in the separate families Xiphiidae and Istiophoridae. In part, the different authors reached alternate conclusions because they analyzed the data sets differently. Another part of the differences in classification centers on a large amount of homoplasy present. No matter which classification is employed, a large number of characters must be postulated to show reversal or independent acquisition. Molecular data, both nuclear and mtDNA (Orrell et al. 2006), supports the view that the Wahoo is a scombrid and strongly refutes a close relationship between billfishes and scombroids.

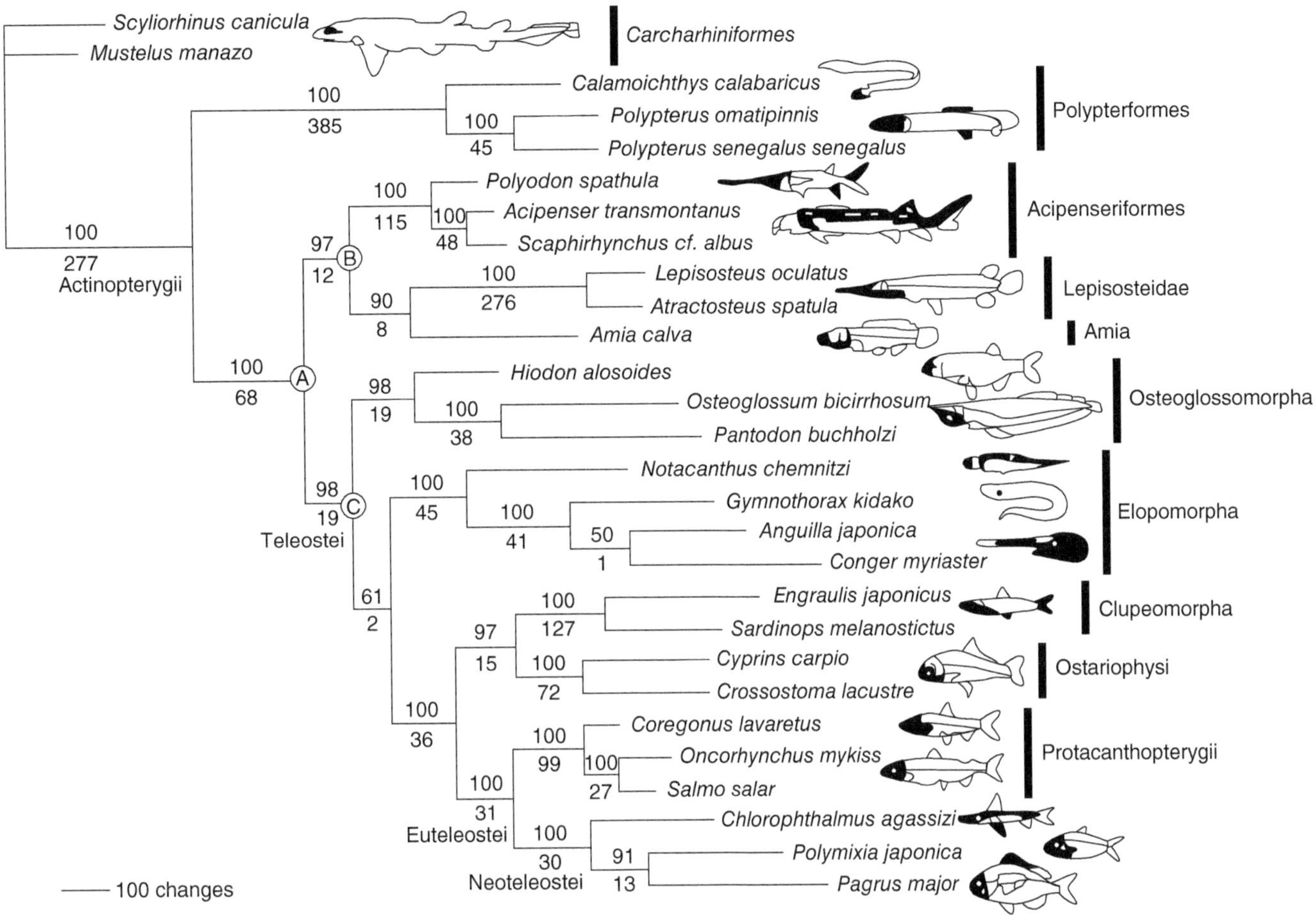

FIGURE 2.6 Phylogenetic relationships among actinopterygian fishes based on parsimony analysis of whole mtDNA genomes. The bar below indicates 100 mutational changes. Branch support is by bootstrap (above the branch) and Bremer decay (below the branch). Internal nodes A, B, and C denote the well-supported differentiation of teleosts from the other actinopterygian fishes. Tree topology indicates that the four lineages recognized as the ancient actinopterygian fish (polypteriforms, acipenseriforms, lepisosteids, and *Amia*) occupied the oldest positions in the phylogeny. These data do not support the proposal that the acipensiforms (sturgeons and paddlefish) are the sister group to the neopterygians (lepisosteids, *Amia*, and teleosts; Nelson 1969), but otherwise provide a good fit to previous phylogenetic hypotheses. From Inoue et al. (2003), used with permission.

Molecular phylogenies can be used to estimate the ages of lineages, from the earliest radiations to the most recent divergences within species, using a **molecular clock**. Based on the mutation rate of particular DNA sequences, researchers can deduce the timing of divergence (anagenesis) as well as the order of divergences (cladogenesis). As with morphological divergences, paleontological data and fossil calibrations are extremely valuable for establishing the rate of the molecular clock. However, molecular clocks can also be established with biogeographic events, such as the rise of the Isthmus of Panama about 2.8 million years ago (Lessios 2008). Many fish species were divided into West Atlantic and East Pacific cohorts by this event, most of which are now recognized as distinct species (see Chapter 19). Hughes et al. (2018) used genomic data and a molecular clock to establish that the ancestors of extant teleosts arose in the early Permian Period (about 300 million years ago) and that most living orders of ray-finned fishes such as Perciformes, Tetradontiformes, Scombriformes, and Cichliformes arose before the end of the Cretaceous Period (about 63 million years ago; Fig. 2.7).

Another use of molecular data is in what has been termed **barcoding**. This relies on differences between species in a relatively short segment of mtDNA, an approximately 655 base pair region of cytochrome oxidase subunit I gene (COI), which Hebert et al. (2003) have proposed as a global bioidentification system for animals. It has been likened to the barcodes that we see on items in grocery stores. For barcoding to be successful, within-species DNA sequences need to be more similar to each other than to sequences from different species. Successful barcoding will facilitate the identification of fishes, linking larvae with adults, forensic identification of fish fillets and other items in commerce, and identification of stomach contents. One potential problem is that using only a mitochondrial marker may fail to discriminate between species due to introgression of some maternally inherited characters, as has apparently happened between two species of western Atlantic Spanish mackerels, *Scomberomorus maculatus* and *S. regalis* (Banford et al. 1999; Paine et al. 2007).

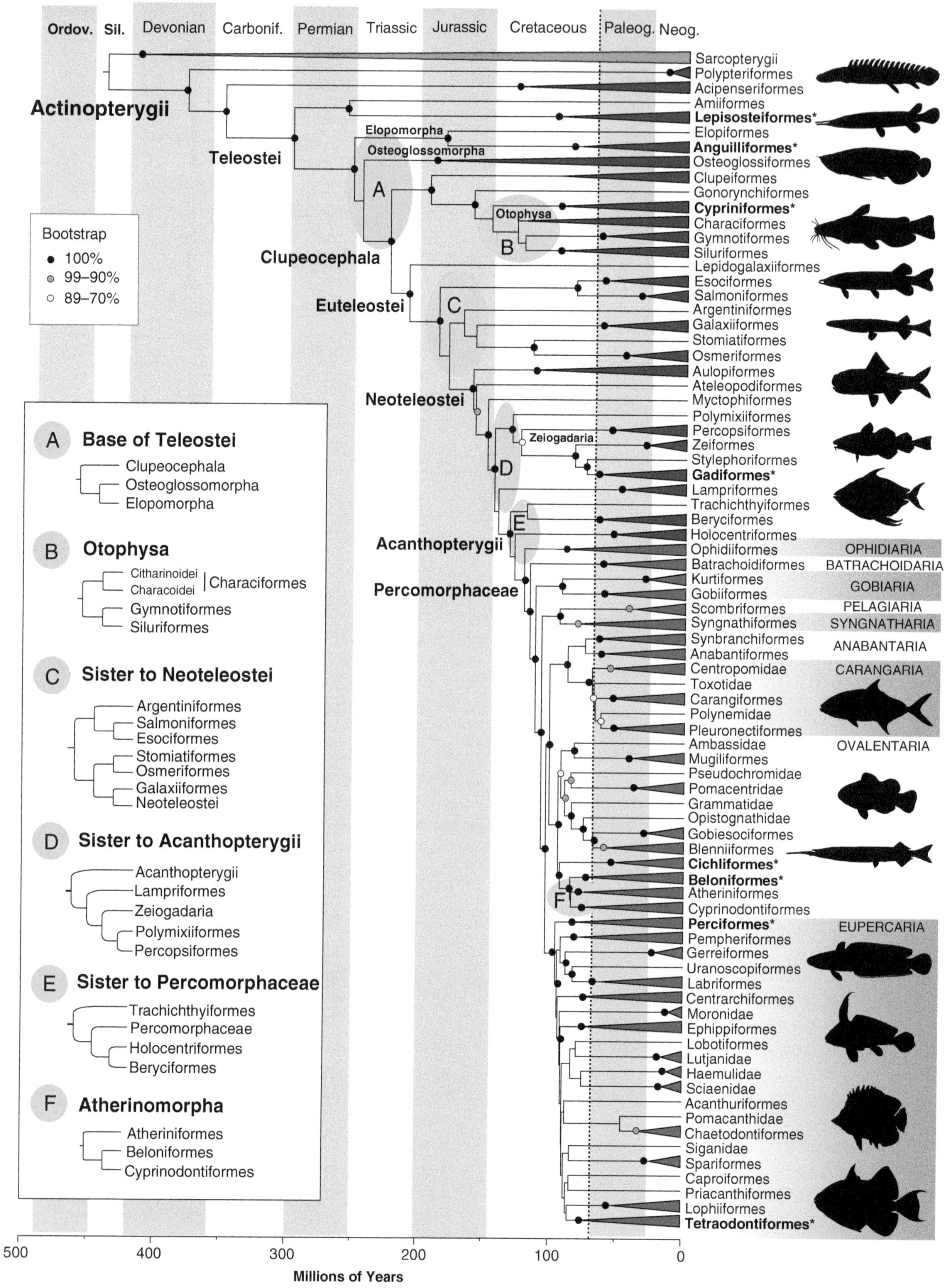

FIGURE 2.7 A maximum likelihood phylogeny for the ray-finned fishes (Actinoptyerigii), a group that includes more than 90% of living fishes. Three lobe-finned fishes (Sarcopterygii) are included as outgroups. This is a fossil time-calibrated tree that uses molecular clock estimates to evaluate the age of major evolutionary events (Hughes et al. 2018, see also Near et al. 2012b). The percomorph radiation is highlighted in red. Areas of major disagreement among gene trees are highlighted with yellow ovals A – F, with alternative topologies shown to the left. Dotted line indicates the Cretaceous – Paleogene extinction event. Taxa in bold with asterisks are characterized by complete genomes, and the entire data base includes 1105 nuclear loci from 303 species. From Hughes et al. (2018) with permission.

To test the utility of barcodes in fishes, Ward et al. (2005) analyzed 207 species of fishes, mostly Australian marine fishes. With no exceptions, all 207 sequenced species were discriminated. Similarly, except for one case of introgression, all 17 species of western Atlantic Scombridae were successfully discriminated with COI (Paine et al. 2007). Successes like these led to ambitious plans to sequence all species of fishes for the Fish Barcode of Life or FISH-BOL, fostered by the Consortium for the Barcode of Life and the Census of Marine Life. This is part of a grand scheme to produce a DNA global database for all species on our planet.

Units of Classification

Systematists use a large number of units to show relationships at different levels. Most of these units are not necessary except for the specialist in a particular group. For example, ray-finned fishes fall into the following units: kingdom Animalia; phylum Chordata (chordates); subphylum Craniata; infraphylum Vertebrata (vertebrates); superclass Gnathostomata (jawed vertebrates); grade Teleostomi; class Osteichthyes (bony fishes and their descendants); and subclass Actinopterygii (ray-finned fishes). Classification of three representative fishes is shown in Table 2.1.

Note the uniform endings for order (*-iformes*) and family (*-idae*) and that the group name is formed from a stem plus the ending. This means that if you learn that the Brown Trout is *Salmo trutta*, you can construct much of the rest of classification by adding the proper endings: Salmonidae is the family including the salmons, and Salmoniformes is the order containing the salmon-like fishes and their relatives. However, this is not always so easy – Sockeye Salmon (*Onchorhynchus nerka*) is also in the family Salmonidae and order Salmoniformes.

It is conventional to italicize the genus and species names of animals and plants to indicate their origin from Latin (or latinized Greek or other languages). Genus names are always capitalized, and species names are always in lowercase (unlike for some plant species names). The names of higher taxonomic units such as families and orders are never italicized but are always capitalized because they are proper nouns. Sometimes it is convenient to convert the name of a family or order into English (e.g., Clupeiidae into clupeid, Scombridae into scombrid), in which case the name is no longer capitalized. Common names of fishes have not typically been capitalized in the past, but this has changed, recognizing that the names are really proper nouns (Nelson et al. 2002). Capitalizing common names avoids the problem of understanding a phrase like "green sunfish." Does this mean a sunfish that is green, or does it refer to the Green Sunfish, *Lepomis cyanellus*?

International Code of Zoological Nomenclature

The **International Code of Zoological Nomenclature** is a system of rules designed to foster the stability of scientific names for animals. Rules deal with such topics as the definition of publication, authorship of new scientific names, and types of taxa. Much of the code is based on the **Principle of Priority**, which states that the first validly described name for a taxon is the name to be used. Most of the rules deal with groups at the family level and below. Interpretations of the code and exceptions to it are controlled by the International Commission of Zoological Nomenclature, members of which are distinguished systematists who specialize in different taxonomic groups.

Species and subspecies are based on type specimens, the specimens used by an author in describing new taxa. Type specimens should be placed in permanent archival collections, such as those discussed at the end of this chapter, where they can be examined by future researchers. **Primary types** include: (i) the **holotype**, the single specimen upon which the description of a new species is based; (ii) the

TABLE 2.1 Classification of Atlantic Herring, Brown Trout, and Chub Mackerel.

Taxonomic unit	Atlantic Herring	Brown Trout	Chub Mackerel
Division	Teleosteomorpha	Teleosteomorpha	Teleosteomorpha
Subdivision	Teleostei	Teleostei	Teleostei
Cohort	Otocephala	Euteleostei	Euteleostei
Superorder	Clupeomorpha	Protacanthopterygii	Acanthopterygii
Order	Clupeiformes	Salmoniformes	Scombriformes
Family	Clupeidae	Salmonidae	Scombridae
Genus	*Clupea*	*Salmo*	*Scomber*
species	*harengus*	*trutta*	*japonicus*
subspecies	*harengus*		

lectotype, a specimen subsequently selected to be the primary type from a number of **syntypes** (a series of specimens upon which the description of a new species was based before the code was changed to disallow this practice); and (iii) the **neotype**, a replacement primary type specimen that is permitted only when there is strong evidence that the original holotype specimen was lost or destroyed, and when a complex nomenclatorial problem exists that can only be solved by the selection of a neotype.

Secondary types include **paratypes**, additional specimens used in the description of a new species, and **paralectotypes**, the remainder of a series of syntypes when a lectotype has been selected from the syntypes. Among the many other kinds of types, mention should also be made of the **topotype**, a specimen taken from the same locality as the primary type and, therefore, useful in understanding variation of the population that included the holotype, and the **allotype**, a paratype of opposite sex to the holotype and useful in cases of sexual dimorphism (when males and females differ in some characteristic).

Taxa above the species level are based on type taxa. For example, the **type species** of a genus is not a specimen but a particular species. Similarly, a family is based on a particular genus.

PhyloCode

When Carl Linneaus developed the taxonomic classification system in the 1700s, Darwin's revolution was a century in the future, and there was no consideration of evolutionary relationships. As phylogenetics emerged as a robust science in the last 70 years, it became apparent that the higher taxonomic ranks of genus, family, order (and beyond) could be somewhat arbitrary and capricious. In the 1990s, a group of systematists proposed replacing the Linnean system with the PhyloCode based explicitly on evolutionary relationships (Cantino & de Queiroz 2004). The key concept is to replace nebulous taxonomic groupings (genus, family, order) with a phylogenetic classification based strictly on monophyletic lineages (clades). Initially, critics maintained that the Linnean system already organizes and conveys information about taxonomic categories and that replacing this system would not justify redefining millions of species and higher taxonomic levels (Harris 2005). In response, phylocode advocates have developed an International Code of Phylogenetic Nomenclature (de Queiroz & Cantino 2020) that adheres to species names as defined by the traditional International Code (see earlier). Hence the clade-based classification of phylocodes is meant to provide an alternative classification above the species rank, a classification that adheres to evolutionary groupings defined by monophyly. The PhyloCode is governed by the International Society for Phylogenetic Nomenclature. The PhyloCode is still controversial, and it remains to be seen how widely it will be followed.

Name Changes

Sometimes the scientific names of fishes change. This is not taken lightly by the ichthyological community, however, and it is intended to reduce or clarify confusion as more information is acquired. There are four primary reasons that systematists change names of organisms: (i) "**splitting**" what was considered to be a single species into two (or more); (ii) "**lumping**" fish that were considered distinct into one species; (iii) changes in classification (e.g., a species is hypothesized to belong in a different genus); and (iv) an earlier name is discovered and becomes the valid name by the Principle of Priority. Frequently, name changes involve more than one of these reasons, as shown in the following examples.

An example of "splitting" concerns the Spanish Mackerel of the western Atlantic (*Scomberomorus maculatus*), which was considered to extend from Cape Cod, Massachusetts, south to Brazil. However, populations referred to this species from Central and South America have 47–49 vertebrae, whereas *S. maculatus* from the Atlantic and Gulf of Mexico coasts of North America have 50–53 vertebrae. This difference, along with other morphometric and anatomical characters, formed the basis for recognizing the southern populations as a separate species, *S. brasiliensis* (Collette et al. 1978).

An example of "lumping" concerns tunas of the genus *Thunnus*. At one time, many researchers believed that the species of tunas occurring off their coasts must be different from species in other parts of the world. Throughout the years, 10 generic and 37 specific names were applied to the seven species of *Thunnus* recognized by Gibbs and Collette (1967). Fishery workers in Japan and Hawaii recorded information on their Yellowfin Tuna as *Neothunnus macropterus*, those in the western Atlantic as *Thunnus albacares*, and those in the eastern Atlantic as *Neothunnus albacora*. Large, long-finned individuals, the so-called Allison Tuna, were known as *Thunnus* or *Neothunnus allisoni*. Based on a lack of morphological differences among the nominal species, Gibbs and Collette (1967) postulated that the Yellowfin Tuna is a single worldwide species. Gene exchange among the Yellowfin Tuna populations was subsequently confirmed using molecular techniques (Scoles & Graves 1993), further justifying lumping the different nominal species. Following the Principle of Priority, the correct name is the **senior synonym**, the earliest species name for a Yellowfin Tuna, which is *albacares* Bonnaterre 1788. Other, later names are **junior synonyms**.

Genetic analysis plays an important role in lumping and splitting decisions, as exemplified by the bonefishes (genus *Albula*). These fishes inhabit sand flats in tropical and subtropical habitats, where they are widely sought by anglers because of their high-energy battles at the end of a fishing line. The bonefish was originally described by Linnaeus (1758), and subsequent taxonomic research contributed 23 species names for bonefishes around the world. However, as scientific communication improved in the 19th and 20th centuries, it became apparent that these regional "species" were very similar or indistinguishable and were therefore lumped until most bonefish were recognized as a single species (Whitehead 1986). This recognition of a single globally distributed bonefish began to unravel when Shaklee and Tamaru (1981) analyzed allozymes in Hawaiian bonefish and discovered two genetically distinct forms that occupy similar habitats and that only could be distinguished by careful examination of jaw structure. Subsequent comparisons with mtDNA cytochrome *b* revealed ancient genetic separations in the genus *Albula* ($d = 0.03–0.30$), indicating three species in the Caribbean and three in the East Pacific (Pfeiler et al. 2006; Bowen et al. 2007). Hidaka et al. (2008) discovered subtle morphological differences among Pacific bonefishes and split one widespread Pacific species into three regional species: *A. virgata* (a Hawaiian endemic), *A. argentea* (distributed from the central to West Pacific), and *A. oligolepis* (West Pacific to Africa). There are more than 10 bonefish species, although several have not been formally described (Fig. 2.8). The deepest genetic separation in the genus is between the two sympatric Pacific species *A. glossadonta* and *A. argentea*, with an mtDNA cytochrome *b* sequence divergence of $d = 0.26–0.30$. Based on a molecular clock calibrated for bonefish cytochrome *b* (1%/Ma), this corresponds to 26–30 million years. It is a remarkable finding that these two fishes, which are identical to the untrained eye and were considered a single species until recently, are five times older than the separation of gorillas and humans.

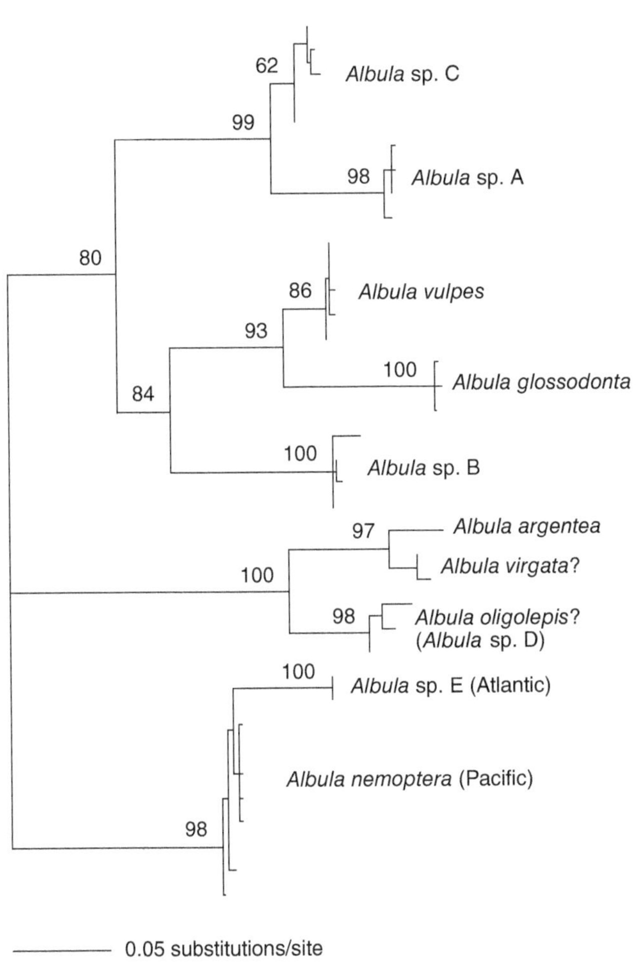

FIGURE 2.8 Phylogenetic relationships of bonefish species based on maximum likelihood analysis of mtDNA cytochrome *b*. Bonefish that occupy shallow sand flats were thought to be one species worldwide (*Albula vulpes*), with a second species occupying deeper water (*A. nemoptera*). However, allozymes and DNA studies demonstrate at least 10 evolutionary lineages in the genus *Albula*. Numbers above branches indicate bootstrap support. Some species have yet to be formally described (species A, B, C, and E), whereas others are only tentatively linked to a branch in the tree (*A. virgata*, *A. oligolepes*) pending DNA sequence analysis from voucher specimens. The scale bar indicates 5% sequence divergence. From Bowen et al. (2007), used with permission.

There are other reasons for changing names of fishes, as illustrated by tunas and the Rainbow Trout. Some researchers placed the bluefin tunas in the genus *Thunnus*, the Albacore in *Germo*, the Bigeye in *Parathunnus*, the Yellowfin Tuna in *Neothunnus*, and the Longtail in *Kishinoella*, almost a genus for each species. Gibbs and Collette (1967) showed that the differences are really among species rather than among genera, so all seven species should be grouped together in one genus. But which genus? Under the Principle of Priority, *Thunnus* South 1845 is the senior synonym, and the other, later names are junior. The name of the Rainbow Trout was changed from *Salmo gairdnerii* to *Oncorhynchus mykiss* in 1988 (Smith & Stearley 1989), affecting many fishery biologists and experimental biologists as well as ichthyologists. As with the tunas, this change involved lumping of species previously considered distinct and adhering to the Principle of Priority.

Collections

Important scientific specimens are generally stored in **collections** where they serve as **vouchers** to document identification in published scientific research. Collections are similar to libraries in many respects. Specimens are filed in an orderly and retrievable fashion. Curators care for their collections and conduct research on certain segments of them, much as librarians care for their collections. Historically most collections of fishes have been preserved in formalin and then transferred to alcohol for permanent storage. Now there is increasing attention to adding skeletons and cleared and stained specimens to collections to allow researchers to study osteology. Whole fish specimens are usually preserved in formalin and stored in alcohol. Unfortunately, the formalin preservation shatters DNA, so most specimens are not accessible with genetic assays. To address this lapse, many major fish collections, such as that at the University of Kansas, also house tissue collections, often in ethyl alcohol, some frozen at −2°C. Qualified investigators can borrow all of these materials from collections or libraries for their scholarly studies.

Collections may be housed in national museums, state or city museums, university museums, or private collections. The eight major fish collections in the United States (and their acronyms) include the National Museum of Natural History (USNM), Washington, DC; University of Michigan Museum of Zoology (UMMZ), Ann Arbor; California Academy of Sciences (CAS), San Francisco; American Museum of Natural History (AMNH), New York; Academy of Natural Sciences (ANSP), Philadelphia; Museum of Comparative Zoology (MCZ), Harvard University, Cambridge, Massachusetts; Field Museum of Natural History (FMNH), Chicago; and Natural History Museum of Los Angeles County (LACM). These eight collections contain more than

24.2 million fishes (Poss & Collette 1995). An additional 118 fish collections in the United States and Canada hold 63.7 million more specimens; at such locales, emphasis is often on regional rather than national or international fish faunas. These regional collections include the Florida State Museum at the University of Florida (UF), which has grown by the incorporation of fish collections from the University of Miami and Florida State University, and the University of Kansas (KU).

The most significant fish collections outside the United States are located in major cities of nations that played important roles in the exploration of the world in earlier times (Berra & Berra 1977; Pietsch & Anderson 1997) or have developed more recently. These include the Natural History Museum (formerly British Museum (Natural History); BMNH), London; Museum National d'Histoire Naturelle (MNHN), Paris; Naturhistorisches Museum (NHMV), Vienna; Royal Ontario Museum (ROM), Toronto; Rijksmuseum van Natuurlijke Historie (RMNH), Leiden; Zoological Museum, University of Copenhagen (ZMUC); and the Australian Museum (AMS), Sydney. Leviton et al. (1985) list most of the major fish collections of the world and their acronyms.

The use of museum specimens has been primarily by systematists in the past. This will continue to be an important role of collections in the future, but other uses are becoming increasingly important. Examples include surveys of parasites (Cressey & Collette 1970) and breeding tubercles (Wiley & Collette 1970); comparison of heavy metal levels in fish flesh today with material up to 100 years old (Gibbs et al. 1974); long-term changes in biodiversity at specific sites (Gunning & Suttkus 1991); and pre- and post-impoundment surveys that could show the effects of dam construction. Many major collections are now computerized (Poss & Collette 1995), and more and more data are becoming accessible as computerized databases, some linked together and available on the internet. An example is FISHNET (**http://www.fishnet2.net/index.html**), a distributed information system that links together fish specimen data from more than two dozen institutions worldwide.

Supplementary Reading

Avise J. 2004. *Molecular markers, natural history, and evolution*, 2nd edn. Sunderland, MA: Sinauer Associates.

de Carvalho MR, Bockman FA, Amorim DS, et al. 2007. Taxonomic impediment or impediment to taxonomy? A commentary on systematics and the cybertaxonomic-automation paradigm. *Evol Biol* 34:140–143.

Hebert PDN, Cywinska A, Ball SL, de Waard JR. 2003. Biological identifications through DNA barcodes. *Proc Roy Soc Lond B Biol Sci* 270:313–322.

Nelson JS, ed. 1999. The species concept in fish biology. *Rev Fish Biol Fisheries* 9:275–382.

Nelson JS, Starnes WC, Warren ML. 2002. A capital case for common names of species of fishes – a white crappie or a White Crappie? *Fisheries* 27(7):31–33.

Journals

Systematic Biology, Society of Systematic Biologists.

Molecular Phylogenetics and Evolution, published by Elsevier, Amsterdam, Netherlands

Websites

Catalog of Fishes, http://www.calacademy.org/scientists/projects/catalog-of-fishes for names, spellings, authorships, dates, and other matters.

FishBase, **http://fishbase.org/** for photos and information on fishes.

ITIS (Integrated Taxonomic Information System), **http://www.itis.gov/index.html** for authoritative taxonomic information on fishes (and other animals and plants).

PART II Form, Function, Ontogeny

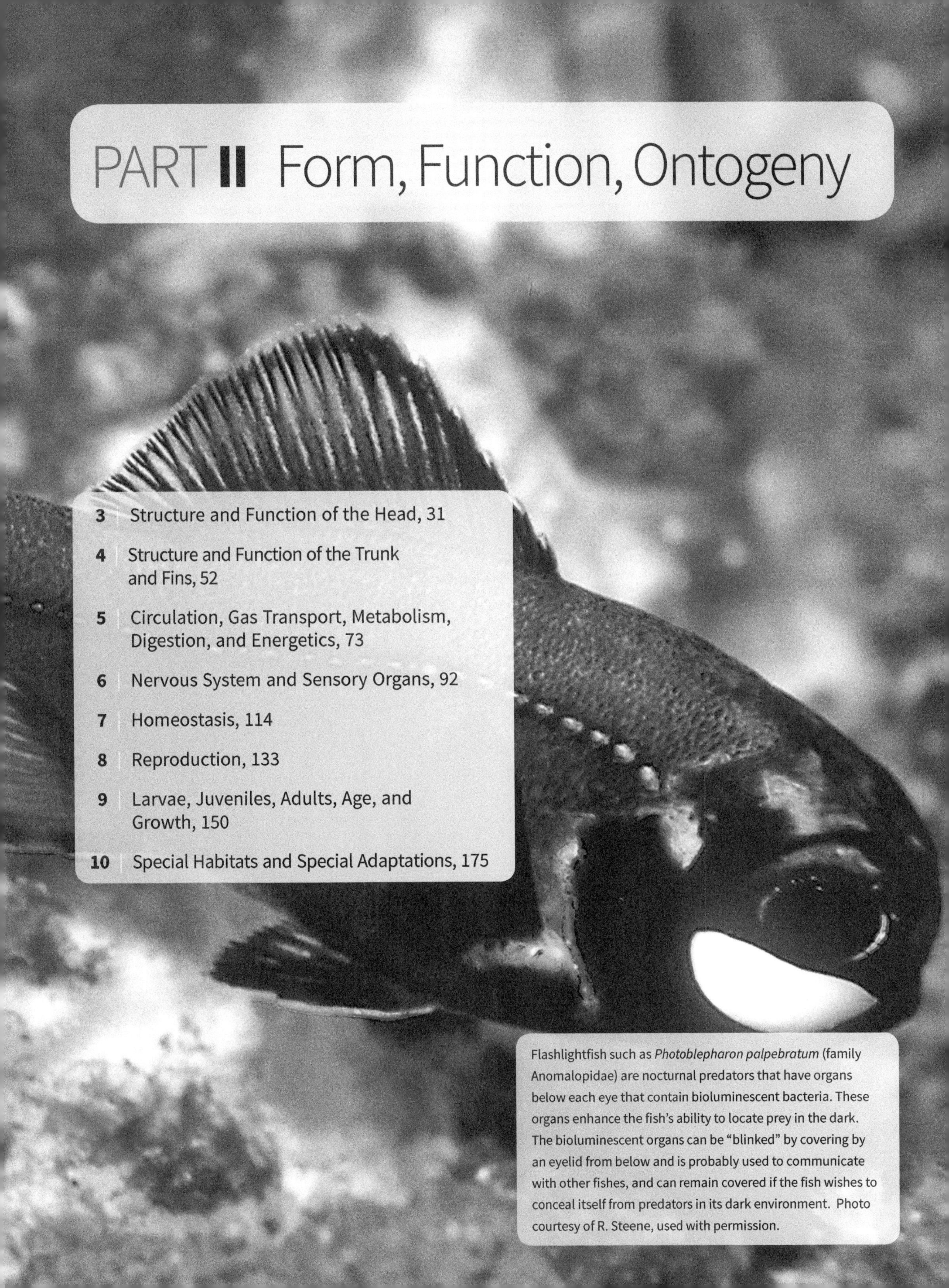

Flashlightfish such as *Photoblepharon palpebratum* (family Anomalopidae) are nocturnal predators that have organs below each eye that contain bioluminescent bacteria. These organs enhance the fish's ability to locate prey in the dark. The bioluminescent organs can be "blinked" by covering by an eyelid from below and is probably used to communicate with other fishes, and can remain covered if the fish wishes to conceal itself from predators in its dark environment. Photo courtesy of R. Steene, used with permission.

CHAPTER 3

Structure and Function of the Head

Summary

Osteology is the study of bones, which can be daunting in fishes because there are so many. In general, we see a reduction in the number of bones when comparing ancestral vertebrates to those that are more derived. Understanding the skeletal structure of fishes is critical to understanding how they function. The preparation of skeletons for study is a combination of art and science.

The ray-finned fishes (actinopterygians) have more skull bones than do the lobe-finned fishes (sarcopterygians), but in general, fish evolution has involved a reduction in bony elements of the skull. The skull encloses and protects the brain and is composed of the neurocranium and the branchiocranium. Actinopterygian skull bones can be divided into four regions: ethmoid, orbital, otic, and basicranial.

The branchiocranium consists of five series of endoskeletal arches (mandibular, palatine, hyoid, opercular, branchial) derived from gill arch supports. Fish gills provide a large surface area for gas exchange, and the countercurrent flow of blood and water across the lamellae maximizes the efficiency of gas exchange by diffusion.

Feeding in fishes involves adaptations of the jaw bones and muscles, teeth, pharyngeal arches, gill rakers, and digestive system, as well as modifications in body shape, sensory structures, and coloration. Food type can often be predicted from jaw and body shape and dentition

CHAPTER CONTENTS

The Diversity of Fishes: Biology, Evolution and Ecology, Third Edition. Douglas E. Facey, Brian W. Bowen, Bruce B. Collette, and Gene S. Helfman.

Companion website: www.wiley.com/go/facey/diversityfishes3

type, regardless of taxonomic position. Planktivorous fishes are usually streamlined, with compressed bodies, forked tails, and protrusible mouths that lack significant teeth. Lurking, fast-start piscivores are generally elongate, round in cross-section, with broad tails, posteriorly placed median fins, and long, tooth-studded jaws that grab prey. Alternatively, many piscivores that pursue prey for short distances are more robust, with fins distributed around the body outline and with large mouths for engulfing prey. There are, however, many specialists that depart from these general patterns.

An important feeding innovation among modern fishes, particularly in teleosts, was the development of protrusible jaws and the pipette mouth. Corresponding modifications to jaw bones, ligaments, and muscles allow a fish to rapidly project its upper jaw forward and increase the volume of the mouth cavity, both creating suction and increasing the speed with which a fish overtakes its prey.

In addition to anterior, marginal jaws, and dentition on the roof of the mouth and tongue, teleosts have gill arches modified into a second set of posterior, pharyngeal jaws. Pharyngeal jaws help move prey towards the throat and, in many fishes, serve to reposition prey for swallowing and for processing via crushing, piercing, and disarticulation. Pharyngeal teeth facilitate the eating of hard-bodied prey (mollusks, arthropods) and plant material.

Dentition type corresponds strongly with food type and is often repeated on the marginal jaws, vomer, palate, and pharyngeal pads. Piscivores and other predators on soft-bodied prey variously possess long, slender, sharp teeth, needlelike villiform teeth, flat-bladed triangular teeth, conical caniniform teeth, or rough cardiform teeth. Mollusk feeders have molariform teeth. Gill rakers also capture prey and may be numerous, long, and thin in plankton feeders, or widely spaced, stout, and covered with toothlike structures in predators on larger prey.

Mouth position also correlates with where a fish lives and feeds. Water column feeders typically have terminal mouths that open forwards, whereas surface feeders often have superior or supraterminal mouths that open upwards. Fishes that feed on benthic food types have subterminal or inferior mouths that open downward and that may generate suction forces that allow a fish to attach to hard substrates while feeding.

Osteology

The **osteology** (study of bones) of fishes is more complicated than that of other vertebrates because fish skeletons are made up of many more bones. The general evolutionary trend from primitive actinopterygians to more advanced teleosts and from aquatic sarcopterygians to tetrapods has been toward fusion and reduction in the number of bony elements (see Chapter 11, Trends During Teleostean Phylogeny). For example, a fossil chondrostean (actinopterygian) fish had more than 150 skull bones, whereas a primitive reptile (sarcopterygian) has 72, and humans (sarcopterygian) have 28 skull bones (Harder 1975).

Why do we need to know about the osteology of fishes? First of all, we cannot really understand such processes as feeding, respiration, and swimming without knowing which jaw bones, gill arch bones, and fin supports are involved. Much fish classification is based on osteology, making knowledge of the skeleton crucial for understanding relationships among fishes. Identification of bones is also important in paleontology, in identifying food of predatory fishes, and in zooarcheology for learning about human uses of fishes through time as revealed in kitchen middens, the waste pits of our ancestors.

If learning about fish bones is important, how does one go about studying them? Large fishes can be fleshed out and then either cleaned by repeated dipping in hot water or by putting the fleshed-out skeleton in a colony of flesh-eating (dermestid) beetles (Rojo 1991). Bemis et al. (2004) described a method requiring fairly complete dissection of the specimen followed by alcohol dehydration. Study of the osteology of small fishes and juveniles of large species was difficult until the development of techniques of clearing and staining. This technique, using the enzyme trypsin, makes the flesh transparent. Then the bones are stained with alizarin red S and the cartilages with Alcian blue. Dark pigments are removed with hydrogen peroxide bleach, and then the entire specimen is immersed in glycerin, which makes the skin and connective tissue invisible (Potthoff 1984; Taylor & van Dyke 1985; see Fig. 3.1).

The skeleton provides much of the framework and support for the remainder of the body, and the skin and scales form a transitional boundary that protects the organism from the surrounding environment. The general osteological description given here and many of the figures are based on members of a family of advanced perciform fishes, the tunas (Scombridae). Comparative notes on other actinopterygian fishes are added where needed. For a brief summary of the skeletal system, see Stiassny (2000), and for a dictionary of names of fish bones, see Rojo (1991). For a comprehensive treatment of fish anatomy, see Harder (1975); for brief updates on each of the organ systems, see the relevant chapters in Ostrander (2000).

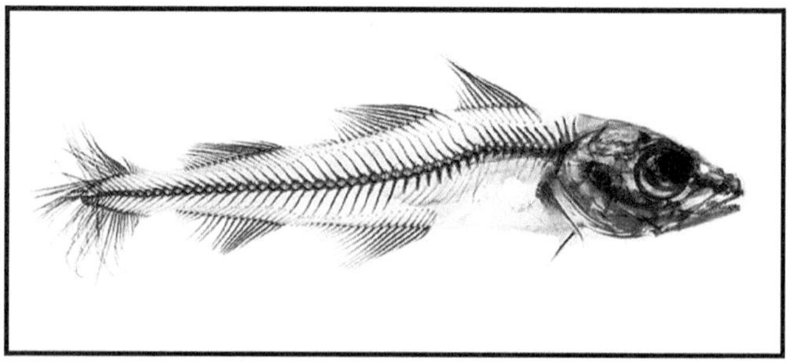

FIGURE 3.1 A cleared and stained specimen of Pacific Cod (*Gadus macrocephalus*). Courtesy of Adam Summers.

Cranial Skeleton

The **skull**, or cranium (Fig. 3.2), is the part of the axial endoskeleton that encloses and protects the brain and most of the sense organs. It is a complex structure derived from several embryological sources. The skull has two major components: the neurocranium and the branchiocranium. The **neurocranium** is composed of the chondrocranium and the dermatocranium. The **chondrocranium** derives from the embryonic cartilaginous braincase. Its bones ossify (harden) during ontogeny as cartilage is replaced by bone. The **dermatocranium** consists of dermal bones and is believed to have evolved from scales that became attached to the chondrocranium. Some bones, however, are of complex origin coming from both sources.

The **branchiocranium**, or visceral cranium, consists of a series of endoskeletal arches that formed as gill arch supports. The circumorbital, opercular, and branchiostegal bones overlie the branchiocranium, which abuts the neurocranium and pectoral girdle.

Skulls differ among the other basic groups of fishes. Hagfishes and lampreys ("agnathans") lack true biting jaws. Toothlike structures are present, but these are horny rasps, not true teeth (see Chapter 13). The round mouth with some internal cartilaginous support hence has led to agnathans also being called cyclostomes (literally, "round mouths"). It was once thought that lamprey jaws had been lost when lampreys evolved as parasites. However, the probable ancestors of the lampreys, the primitive cephalaspidomorphs (see Chapter 11), also lacked jaws, so lack of jaws is now thought to be a primitive character. The neurocranium of chondrichthyan sharks and rays is a single cartilaginous structure, whereas their jaws and branchial arches consist of a series of cartilages.

Neurocranium

The neurocranium consists of the chondrocranium and the dermatocranium. The chondrocranium of bony fishes is derived from cartilaginous capsules that formed around the sense organs. To clarify spatial relationships among a large number of bones in the skull, it helps to divide the skull into four regions associated with major centers of ossification. From anterior to posterior, these regions are the **ethmoid**, **orbital**, **otic**, and **basicranial**. For each region, the cartilage bones will be discussed first, followed by the dermal bones, which tend to roof over and often fuse with the underlying cartilage bones. Consult Harder (1975, pl. 1A-C) for a three-part plate of overlays that greatly helps visualize how the teleost skull bones fit together.

Ethmoid Region The ethmoid region remains variably cartilaginous even in adults of most modern bony fishes but there are also dermal elements fused to some of these bones. Two main sets of cartilage bones form the ethmoid region. Paired **lateral ethmoids** (or parethmoids) form the posterolateral wall of the ethmoid region and the anterior wall of the orbit (Figs. 3.3–3.5). The median chondral **ethmoid** (or **supraethmoid**) is the most anterodorsal skull bone. It often has a

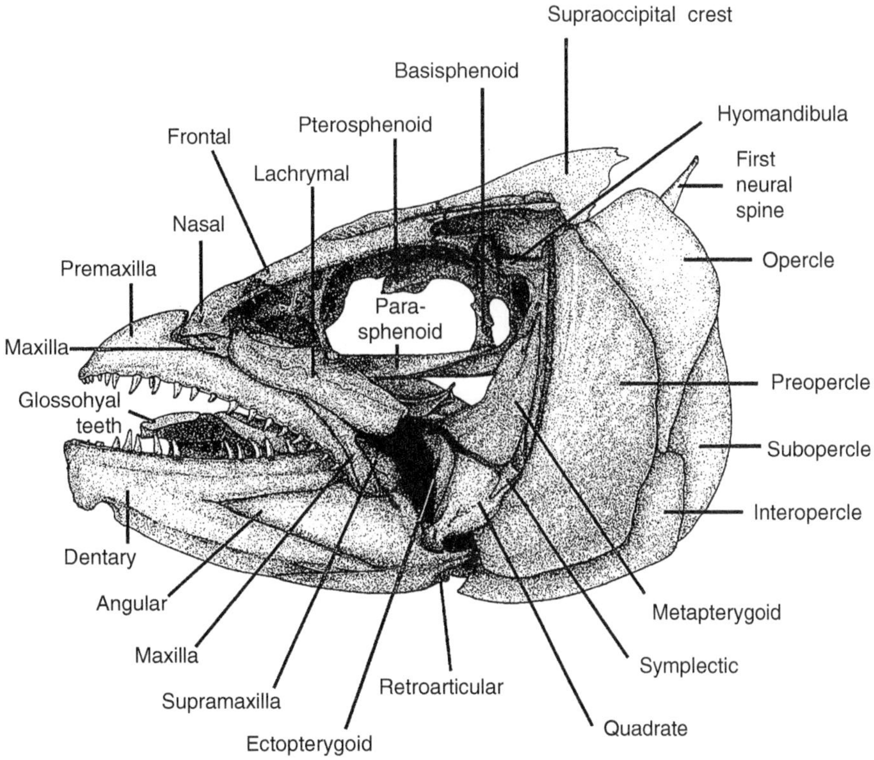

FIGURE 3.2 Lateral view of the skull of the Dogtooth Tuna (*Gymnosarda unicolor*). From Collette and Chao (1975).

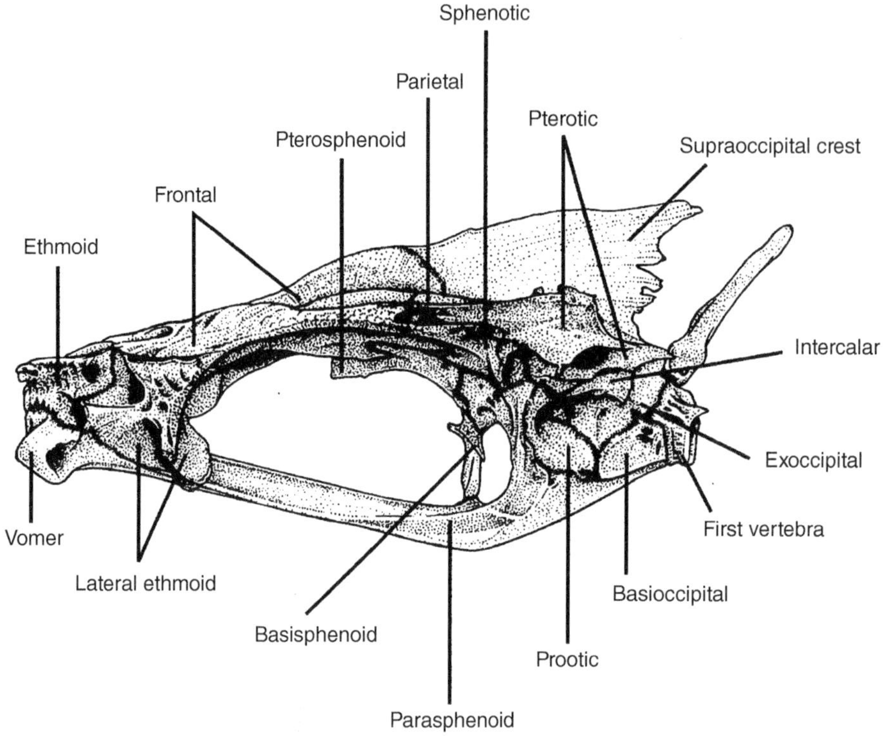

FIGURE 3.3 Lateral view of the neurocranium of the Dogtooth Tuna (*Gymnosarda unicolor*). From Collette and Chao (1975).

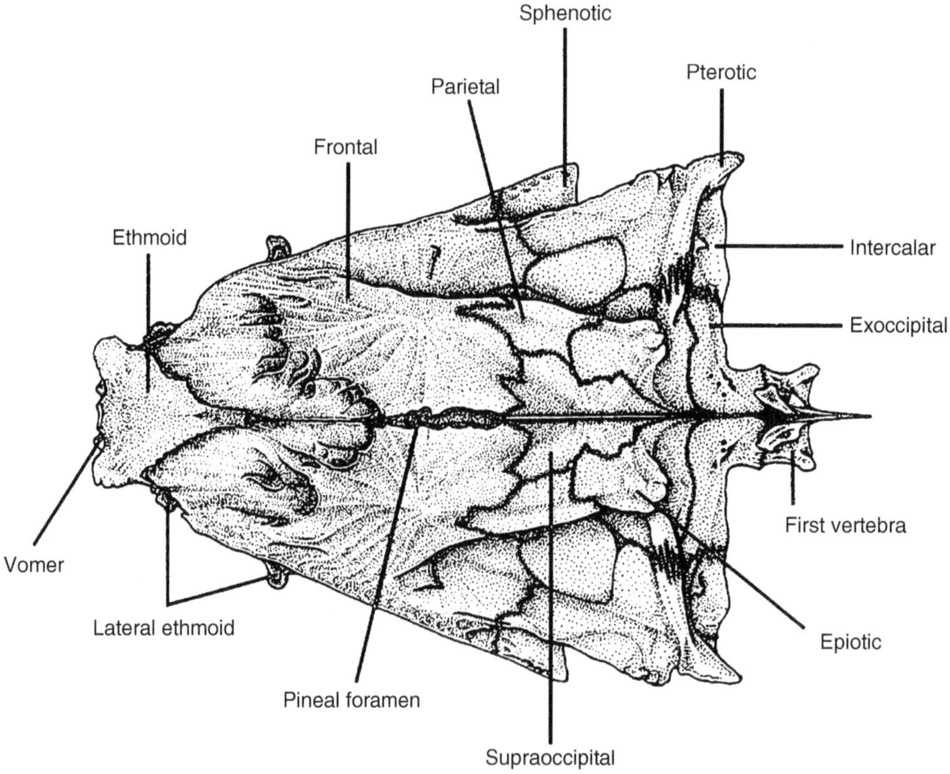

FIGURE 3.4 Dorsal view of the neurocranium of the Dogtooth Tuna (*Gymnosarda unicolor*). From Collette and Chao (1975).

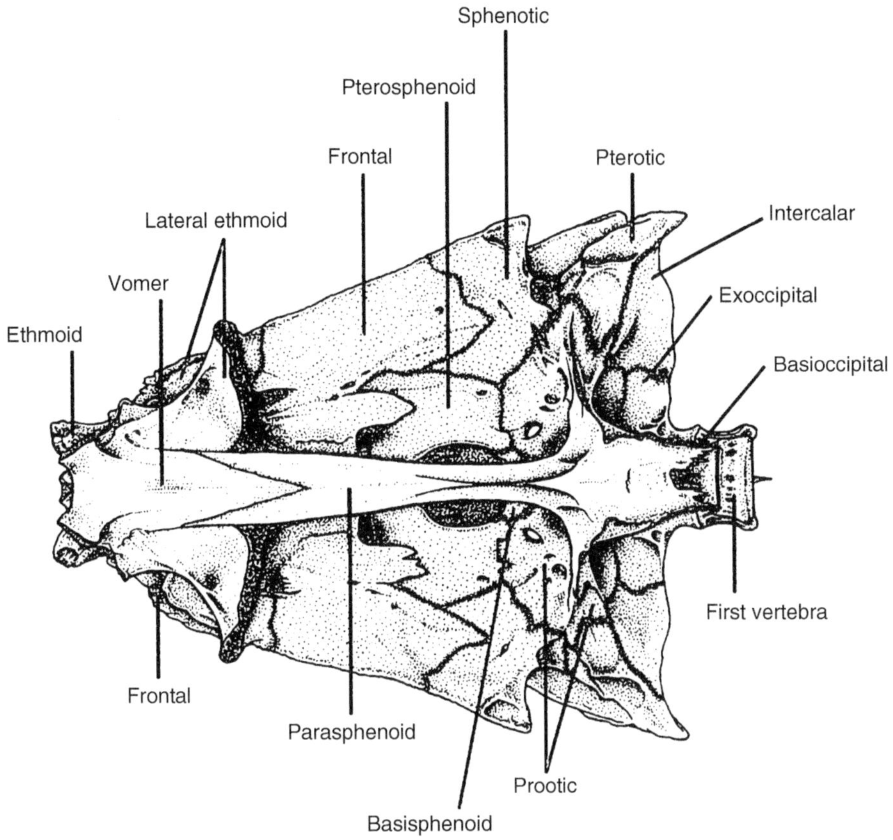

FIGURE 3.5 Ventral view of the neurocranium of the Dogtooth Tuna (*Gymnosarda unicolor*). From Collette and Chao (1975).

dermal element fused to it, in which case it is usually termed the mesethmoid. There are also two sets of dermal bones in this region. The median, often dentigerous (tooth-bearing), **vomer**, which may be absent in a few teleosts, lies ventral to the mesethmoid, whereas the paired dermal **nasals** are lateral to the ethmoid region, associated with the olfactory nasal capsule. The vomer is usually considered to be compound (chondral median ventral ethmoid + dermal vomer).

Orbital Region

The region that surrounds the orbit of the eye is composed of three sets of cartilage bones and two sets of dermal bones. Cartilage bone components include paired **pterosphenoids**, which meet along the ventral median line of the skull. The median **basisphenoid** extends from the pterosphenoids down to the parasphenoid and may divide the orbit into left and right halves. **Sclerotic** cartilages or bones protect and support the eyeball itself. Two sets of dermal bones are the paired **frontals**, which cover most of the dorsal surface of the cranium, and the **circumorbitals**. The circumorbitals form a ring around the eye in primitive bony fishes. However, this ring is reduced to a chain of small infraorbital bones under and behind the eye in advanced bony fishes. Advanced teleosts usually have **infraorbital 1** (IO_1); the **lachrymal**, or preorbital (IO_2), or jugal, or true suborbital (IO_3), which may bear a suborbital shelf that supports the eye; and the **dermosphenotic bones**, or postorbitals, which bear the infraorbital or suborbital lateral line canal (Fig. 3.6). Many primitive teleosts also have an antorbital and a supraorbital.

Otic Region

Five cartilage bones enclose each bilateral otic (ear) chamber inside the skull (see Figs. 3.3–3.5). Paired **sphenotics** form the most posterior dorsolateral part of the orbit roof. Paired **pterotics** form the posterior outer corners of the neurocranium and enclose the horizontal semicircular canal. Paired **prootics** form the floor of the neurocranium and enclose the utriculus of the inner ear. Paired **epiotics**, more recently called **epioccipitals**, lie posterior to the parietals and lateral to the supraoccipital and contain the posterior vertical semicircular canal. The median process of the posttemporal, by which the pectoral girdle is attached to the posterior region of the skull, attaches to the epiotics. The epiotics enclose part of the posterior semicircular canal. Paired **intercalars** (or opisthotics) fit between the pterotics and exoccipitals and articulate with the lateral process of the posttemporal. There is only one pair of entirely dermal bones in the otic region, the paired **parietals**, which roof part of the otic region and articulate with the frontals anteriorly, the supraoccipital medially, and the epiotics posteriorly.

Basicranial Region

Three sets of cartilage bones, one pair plus two median bones, form the cranial base. Paired lateral **exoccipitals** form the sides of the foramen magnum (Fig. 3.7), which is the passageway for the spinal cord. The

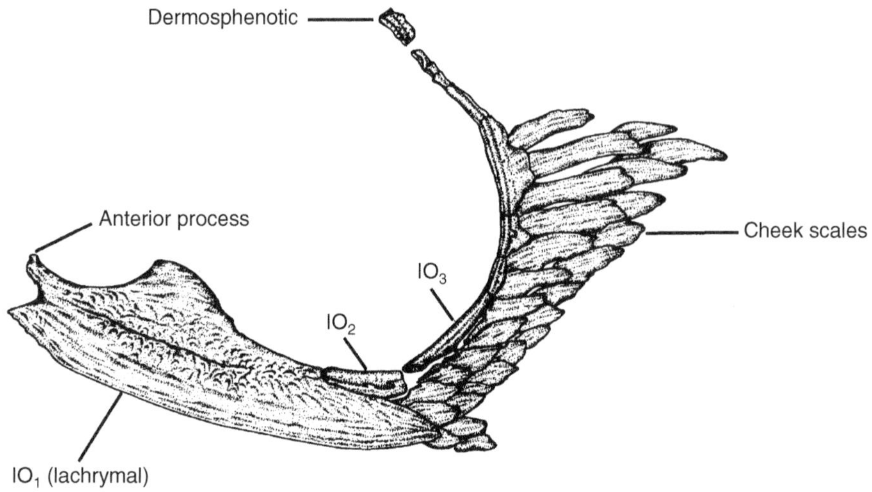

FIGURE 3.6 Left infraorbital bones in lateral view of the Spanish Mackerel (*Scomberomorus maculatus*). From Collette and Russo (1985b).

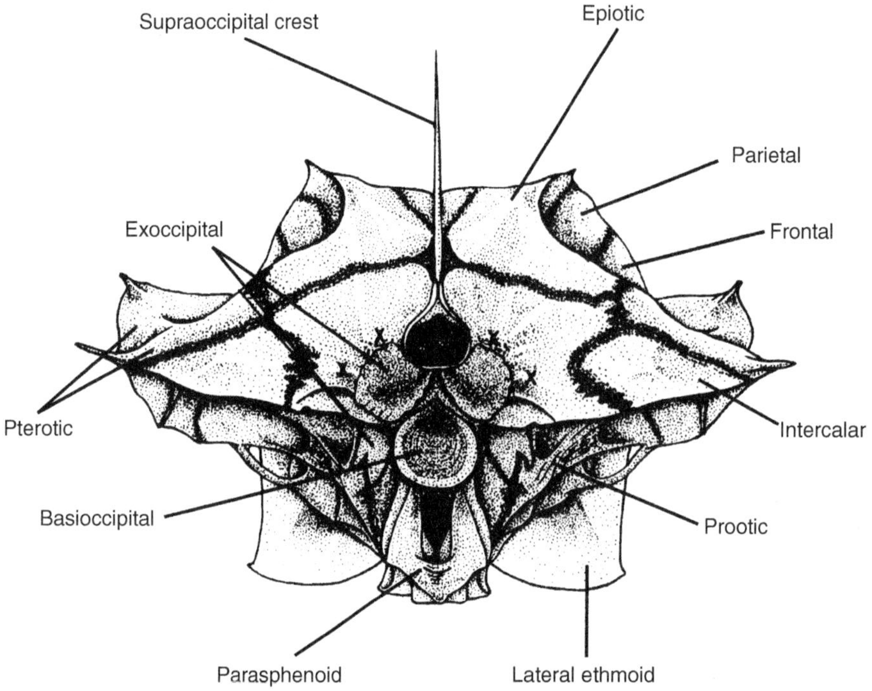

FIGURE 3.7 Rear view of the skull of a bonito (*Sarda chiliensis*). The crosses indicate points of attachment of epineural bones. From Collette and Chao (1975).

median **basioccipital** is the most posteroventral neurocranium bone and articulates (forms a joint) with the first vertebra. The dorsal median **supraoccipital** bone usually bears a posteriorly directed supraoccipital crest that varies among teleosts from a slight ridge to a prominent crest. The only dermal bone in the basicranial region is the median **parasphenoid**, a long cross-shaped bone that articulates with the vomer anteriorly and forms the posteroventral base of the skull.

Branchiocranium

The branchiocranium is divisible into five parts: the mandibular, palatine, hyoid, opercular, and branchial.

Mandibular Arch The **mandibular arch** forms the upper jaw and is known as the palatoquadrate cartilage in Chondrichthyes. It is composed entirely of dermal bones in bony fishes. The mandibular arch may have three sets of bones. The tooth-bearing **premaxillae** are the anteriormost elements. The **maxillae** are tooth-bearing in some soft-rayed fishes, but the maxilla is excluded from the gape (not a part of the jaw) in more advanced spiny-rayed fishes. The third bone that may be present is the **supramaxilla**. It is a small bone on the posterodorsal margin of the maxilla. Some teleosts, such as the herringlike fishes (Clupeoidei), have multiple supramaxillae.

The lower jaw consists of Meckel's cartilage in Chondrichthyes. In bony fishes, the dermal, tooth-forming

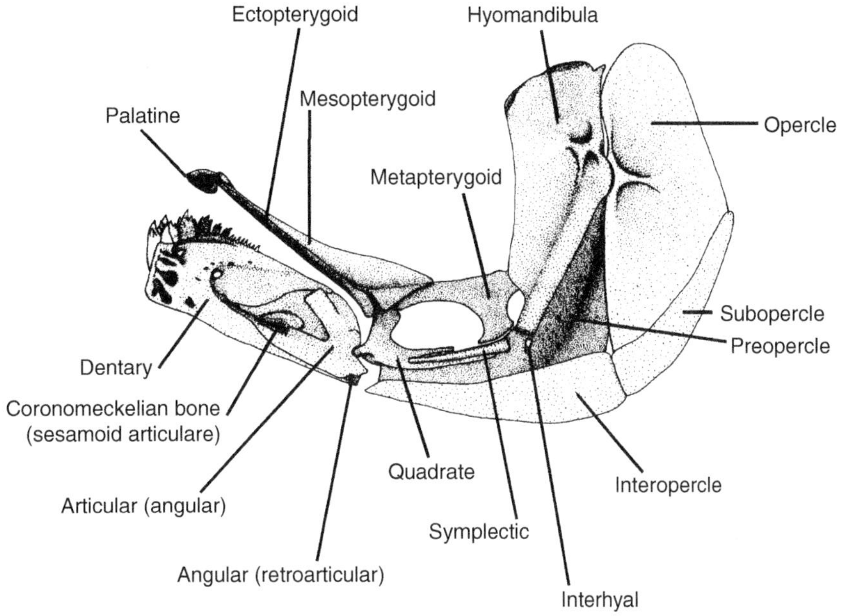

FIGURE 3.8 Lateral bones of the face and lower jaw suspension of generalized characin (*Brycon meeki*). From Weitzman (1962).

dentary bone (Fig. 3.8) covers Meckel's cartilage, which is reduced to a thin rod extending posteriorly along the inner face of the dentary to the angular. The **angular** (sometimes called articular) is a large, posterior dermal bone that fits into the V of the dentary. A ventroposterior dermal ossification forms the **retroarticular** (sometimes called angular), a small bone attached to the posteroventral corner of the angular. In most teleosts, the angular fuses with one or both articulars. The lower jaw forms a single functional unit in most bony fishes, but in the Kissing Gourami (*Helostoma temmincki*), African characins of the genus *Distochodus*, and some parrotfishes of the genus *Scarus*, there is a mobile joint between the dentary and the angular (Liem 1967; Vari 1979; Bellwood 1994).

Palatine and Hyoid Arches

The **palatine arch** consists of four pairs of bones in the roof of the mouth (see Fig. 3.8). The **palatines** are cartilage bones that are frequently tooth-forming. They have been called "plowshare" bones because of their characteristic shape. The dermal **ectopterygoids** are narrow bones, sometimes T-shaped, sometimes dentigerous. The dermal **entopterygoids** (or mesopterygoids) are thin bones that roof the mouth. The **metapterygoids** are cartilage bones, quadrangular-shaped and articulating with the quadrate and hyomandibula.

The **suspensorium** consists of a chain of bones that attach the lower jaw and opercular apparatus to the skull (see Fig. 3.8). The **hyomandibula** is an inverted L-shaped bone that connects the lower jaw and opercular bones to the neurocranium. The **symplectic** is a small bone that fits into the groove of the quadrate. The **quadrate** is a triangular bone with a groove for the symplectic; it has an articulating facet to which the lower jaw is attached (see following discussion of jaw suspensions).

The **hyoid complex** is a series of five pairs of bones (Fig. 3.9) that lie medial to the lower jaw and opercular bones and lateral to the branchiostegal rays that attach to them. The anteriormost bones are the dorsal and ventral **hypohyals** (or basihyals). They are followed by the anterior **ceratohyal**, a long flat bone that interdigitates with the posterior ceratohyal

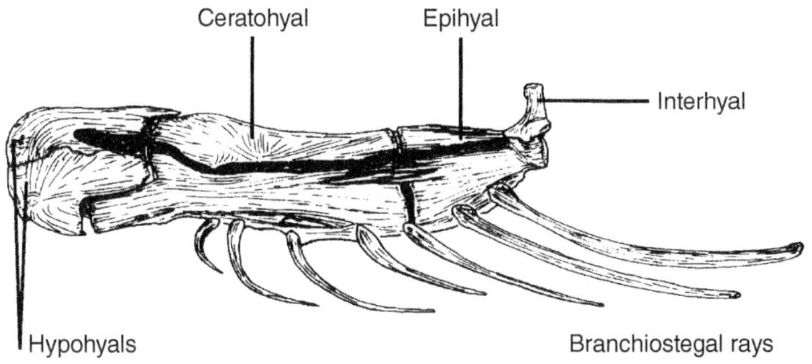

FIGURE 3.9 Left hyoid complex in lateral view of a Spanish mackerel (*Scomberomorus commerson*). From Collette and Russo (1985b).

posteriorly and to which some of the branchiostegal rays attach. The posterior ceratohyal (or epihyal) is a triangular bone to which some of the branchiostegal rays attach. The **interhyal** is a small, rod-shaped bone that attaches the hyoid complex to the suspensorium. The **glossohyal** is an unpaired flattened bone that lies over the anterior basibranchial and supports the tongue.

The dermal bones of the hyoid arch are the **branchiostegal rays**, elongate, flattened, riblike structures (Fig. 3.8) that attach to the ceratohyal and epihyal. They are important in respiration, particularly in bottom-dwelling species. Their number and arrangement are useful in tracing phylogenies (see McAllister 1968). The median **urohyal** is a flattened, elongated, unpaired bone that lies inside the rami (branches) of the lower jaw.

Jaw Suspension

Much interest and controversy have arisen over which of the gill arches of our jawless ancestors gave rise to the upper and lower jaws. Zoologists are not certain whether the jaw-forming arch was the first in the series, or whether it was posterior to a premandibular arch that has been lost (Walker & Liem 1994). Classically, four principal types of jaw attachment have been recognized (Fig. 3.10).

Amphistylic suspension is found in primitive sharks. The upper jaw is attached to the cranium by ligaments, and there is therefore little mobility of the upper jaw. The hyoid arch is attached to the chondrocranium and lower jaw and is involved in the suspension of both jaws.

Hyostylic suspension is found in most cartilaginous fishes (chondrichthyans) and all ray-finned fishes (actinopterygians), but Maisey (1980) found no dividing line between amphistyly and hyostyly in living sharks. Both jaws are suspended from the chondrocranium by way of ligamentous attachments to the hyomandibula, which is attached to the otic region of the neurocranium. This permits the upper jaw to be somewhat independent of the skull, allowing greater jaw mobility when feeding**.** **Methyostylic** suspension is a variety of hyostyly present in ray-finned fishes. Remnants of the second-gill arch (palatine and pterygoid bones) connect in the roof of the mouth. Dermal bones, the premaxilla and maxilla, form a new upper jaw. A new dermal anterior lower jaw element, the dentary, is connected with the angular, which is suspended from the otic capsule by hyoid derivatives.

Autostylic suspension is present in lungfishes and tetrapods. The processes of the palatoquadrate articulate with or fuse to the chondrocranium. The hyoid arch is no longer involved with jaw suspension. The hyomandibula becomes the columella of the inner ear in tetrapods.

Holostylic suspension is a variety of autostyly found only in the Holocephali (chimaeras and ratfishes). The palatoquadrate is fused to the chondrocranium and supports the lower jaw in the quadrate region. The name Holocephali means "whole head," in reference to the upper jaw being a part of the cranium.

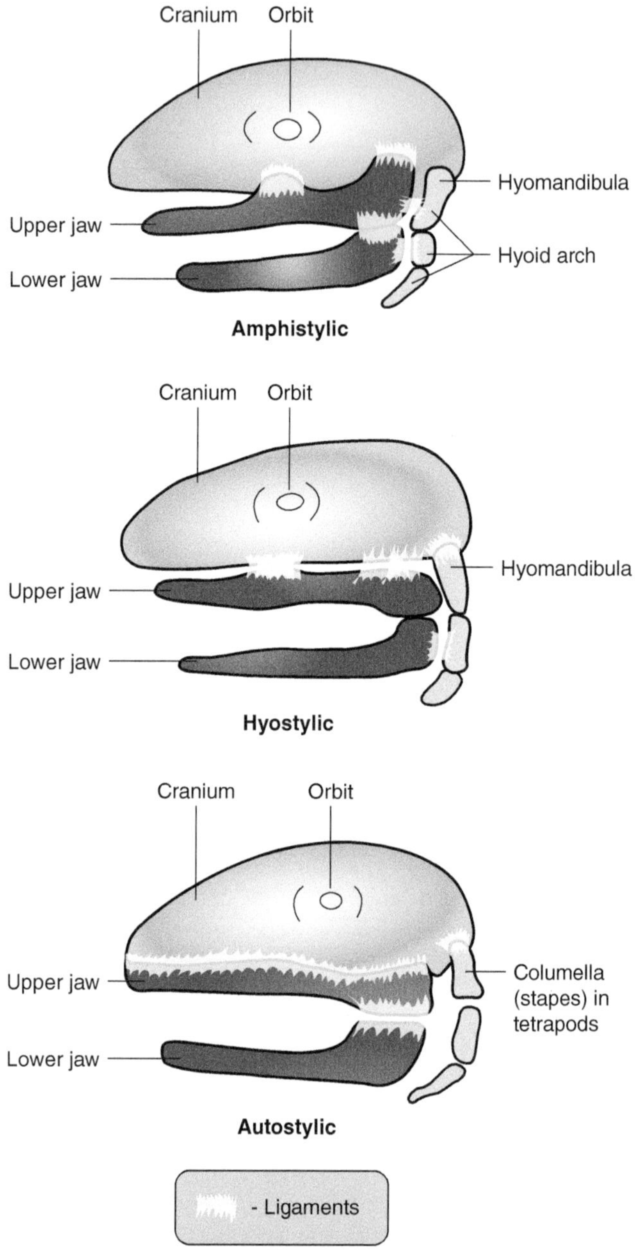

FIGURE 3.10 Major types of jaw suspension in fishes.

Note that the structural similarities between amphistylic and hyostylic jaws and between autostylic and holostylic jaws are convergent, as the groups with similar structural jaw suspensions are not closely related to one another phylogenetically.

Opercular and Branchial Series

The **opercular apparatus** consists of four pairs of wide, flat dermal bones that form the gill covers, protect the fragile underlying gill arches, and are involved in respiration and feeding. The **opercle** is usually more or less rectangular and is usually the largest and heaviest of the opercular bones (see Figs. 3.2, 3.8). It has an anterior articulation facet connecting with the hyomandibula.

The **subopercle** is the innermost and most posterior element. The preopercle is the anteriormost element. It overlies parts of the other three opercular bones. The **interopercle** is the most ventral bone.

The **branchial complex** consists of four pairs of gill arches, gill rakers, pharyngeal tooth patches, and supporting bones (Fig. 3.11). All elements of the gill arch are cartilage bones but may have toothed dermal elements incorporated. Three **basibranchials** form a chain from anterior to posterior. The first basibranchial is partially covered by the median **glossohyal**; the second and third serve as attachments for the hypobranchials and ceratobranchials. Three pairs of **hypobranchials** connect the basibranchials with the first three ceratobranchials; the fourth is cartilaginous. **Ceratobranchials** are the longest bones in the branchial arch and support most of the gill filaments and gill rakers. The fifth ceratobranchial is usually expanded, bears a tooth plate, and is sometimes called the **lower pharyngeal bone**. Four pairs of **epibranchials** attach basally to the ceratobranchials. They vary from being long and slender (like a short ceratobranchial) to short and stubby. Four pairs of **pharyngobranchials** attach to the epibranchials. The first is suspensory and attaches to the braincase. The other three may have dermal tooth patches attached to them and are then termed **upper pharyngeal bones**. For a detailed account of the gill arches and their use in tracing fish phylogeny, see Nelson (1969).

Muscles of the Head

Eye Muscles

Extrinsic eye muscles move the eye within its orbit. Eye muscles are evolutionarily conserved in that most vertebrates have the same three pairs of these striated muscles: inferior or ventral and superior or dorsal oblique; inferior or ventral and superior or dorsal rectus; and external or lateral and internal or medial rectus. Posteriorly, the eye muscles insert into dome-shaped cavities called myodomes in actinopterygian fishes. A suspensory ligament above the lens and a retractor lentis muscle below form the focusing muscle of the eye.

Some eye muscles have been converted into remarkable specialized structures in some fishes: an electric organ in the Electric Stargazer (*Astroscopus*, Uranoscopidae) and heater organs in two suborders of perciform fishes that include billfishes, swordfishes, tunas, and their relative (Xiphioidei and Scombroidei). The upper edges of the four uppermost eye muscles form an electric organ in the Electric Stargazer. Large stargazers can produce an electric discharge from these muscles strong enough to incapacitate a careless human handler. Their usual function is presumably to stun prey or deter predators. In billfishes (Istiophoridae and Xiphiidae), the superior rectus has been converted into a heat-producing muscle that keeps the eye warm during incursions into

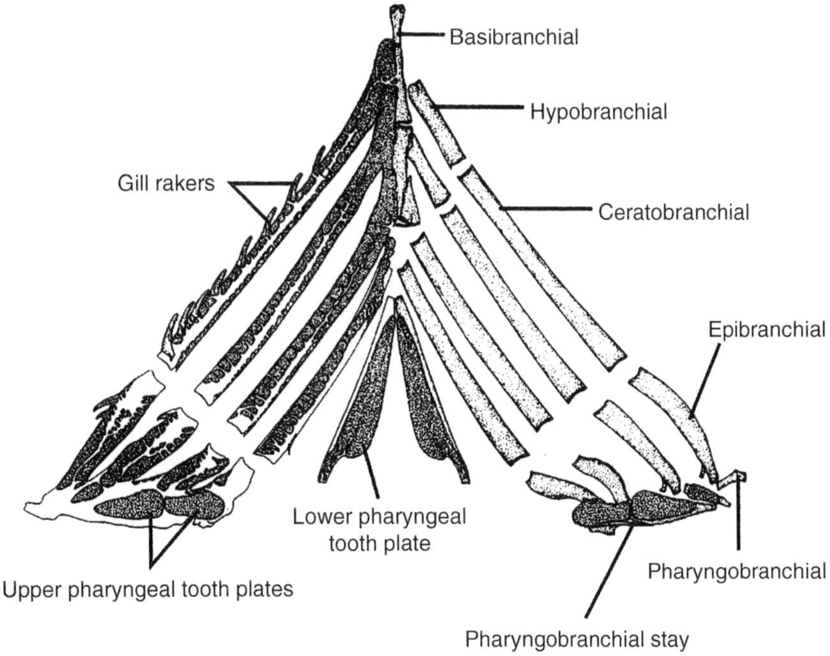

FIGURE 3.11 Branchial arch of a Spanish mackerel (*Scomberomorus semifasciatus*). Dorsal view of the gill arches with the dorsal region folded back to show their ventral aspect. The epidermis is removed from the right-hand side to reveal the underlying bones. From Collette and Russo (1985b).

deep, cold waters. In the Butterfly Mackerel *Gasterochisma* (Scombridae), the external rectus is the muscle that becomes the heater organ (Block 1991), showing independent evolution of this character (see Chapter 7, Endothermic Fishes).

Cheek Muscles

Seven principal muscles are involved in opening and closing the jaws, suspensorium, and operculum during feeding and breathing (Fig. 3.12). The major muscles are the **adductor mandibulae**, large muscles with several sections that insert on the inner surface of the upper and lower jaw and originate on the outer face of the suspensorium, the chain of bones that suspend the jaws from the neurocranium. The adductor mandibulae function to close the jaws. The **levator arcus palatini** occupies the postorbital portion of the cheek. The **dilator operculi**, the **adductor operculi**, and the **levator operculi** insert on the opercle. The **adductor arcus palatini** originates from the ventrolateral margin of the parasphenoid and underlies the orbit. The **adductor hyomandibulae** originates on the prootic and exoccipital and inserts on the hyomandibula. In addition, **pharyngeal** muscles, or **retractores arcuum branchialium**, run from the upper pharyngeal bones to the vertebral column and control the pharyngeal jaws.

Dorsal Gill-Arch Muscles

The dorsal gill-arch musculature, aspects of the associated gill-arch skeleton, the **transversus ventralis 4**, and the semicircular ligament were described for many species of fishes in over 200 families and over 300 genera of bony fishes in a massive, superbly illustrated study by Springer and Johnson (2004). They found that the transversus dorsalis was much more complex than previously recognized and was useful for defining various groups of fishes. A cladistic analysis of the dorsal gill-arch musculature and gill-arch skeletal characters (Springer & Orrell 2004) showed groups such as the Percopsiformes and the Ophidiiformes to be monophyletic, whereas other groups such as the Paracanthopterygii and the Labroidei are polyphyletic.

Breathing

Fish breathing is controlled by signals from the brain stem (medulla oblongata) to the muscles of the mouth and opercula (Taylor 2011). The **gills** of fishes are very efficient at extracting oxygen from the water because of the large surface area and thin epithelial membranes of the **secondary lamellae** (Fig. 3.13). Diffusion of gases across the gill membrane is enhanced by blood in the secondary lamellae flowing in the opposite direction to the water passing over the gills, thereby maximizing the diffusion gradient across the entire lamellar surface. This **countercurrent flow** ensures that as the blood picks up oxygen from the water it moves along the exchange surface to an area where the adjacent water has an even higher oxygen concentration, thus preventing loss of oxygen prematurely, before the blood reaches tissues in need of oxygenation.

Gills will function efficiently only if water is kept moving across them in the same direction, from anterior to posterior. This is accomplished in one of two ways. First, the great majority of fishes pump water across their gills by increasing and decreasing the volume of the **buccal** (mouth) **chamber** in front of the gills and the **opercular chamber** behind them. The expansion and contraction of these two chambers is timed so that the pressure in the buccal chamber is greater than the pressure in the opercular chamber, thereby ensuring that the water flows in the anterior to posterior direction throughout the breathing cycle (Fig. 3.14).

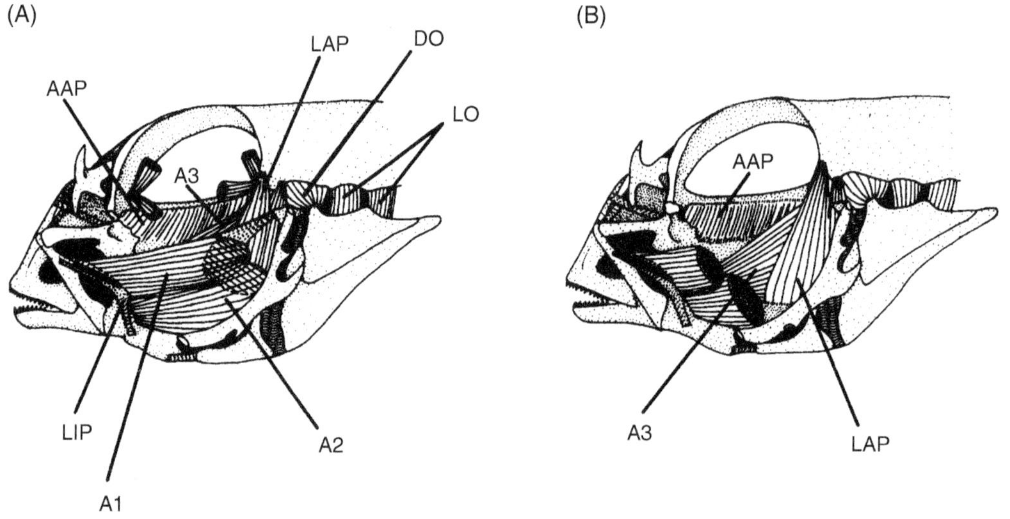

FIGURE 3.12 Cheek muscles of a sculpin, *Jordania zonope*. (A) Superficial musculature. (B) After removal of A1 and A2. A1, A2, and A3, adductor mandibulae; AAP, adductor arcus palatini; DO, dilator operculi; LAP, levator arcus palatini; LIP, ligamentum primordium; LO, levator operculi. Yabe (1985).

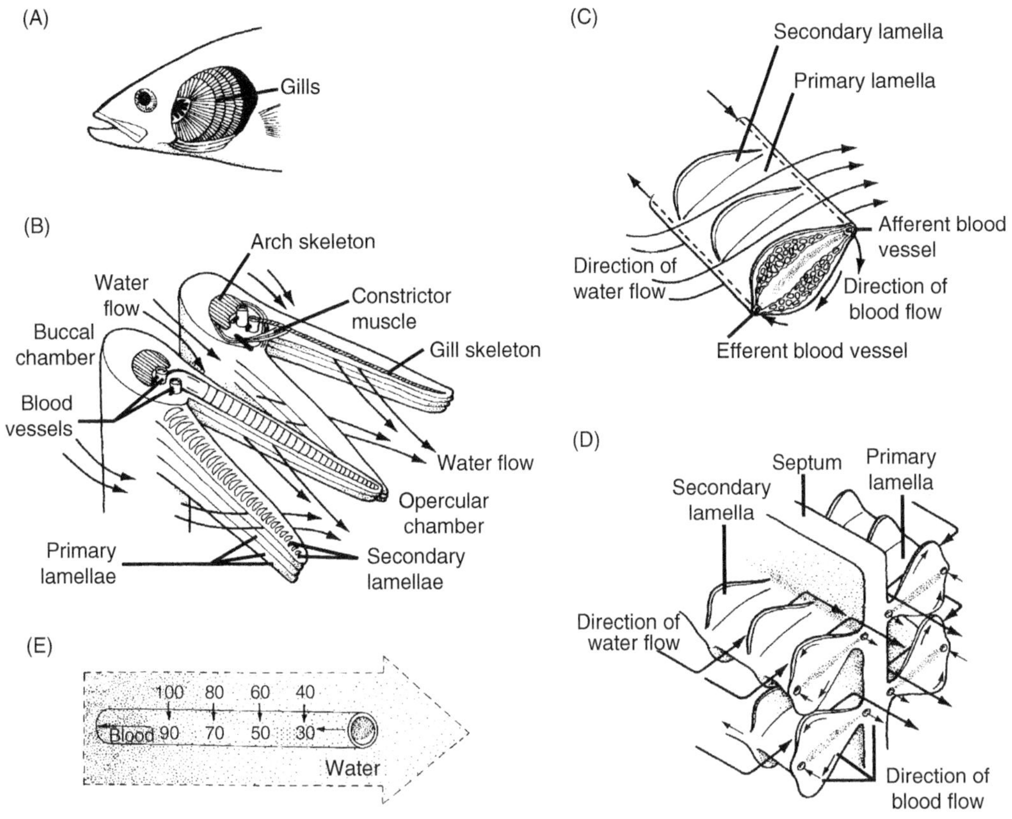

FIGURE 3.13 (A, B) The gill arches of a fish support the gill filaments (also called the primary lamellae) and form a curtain through which water passes as it moves from the buccal cavity to the opercular cavity. (C) As water flows across the gill filaments of a teleost, blood flows through the secondary lamellae in the opposite direction. (D) In elasmobranchs, even though septa create some structural differences in gill filaments, water flow across the secondary lamellae is still countercurrent to blood flow. (E) The countercurrent flow of water and blood at the exchange surface of the secondary lamellae ensures that the partial pressure of oxygen in the water always exceeds that of the blood, thereby maximizing the efficiency of oxygen diffusion into the blood.

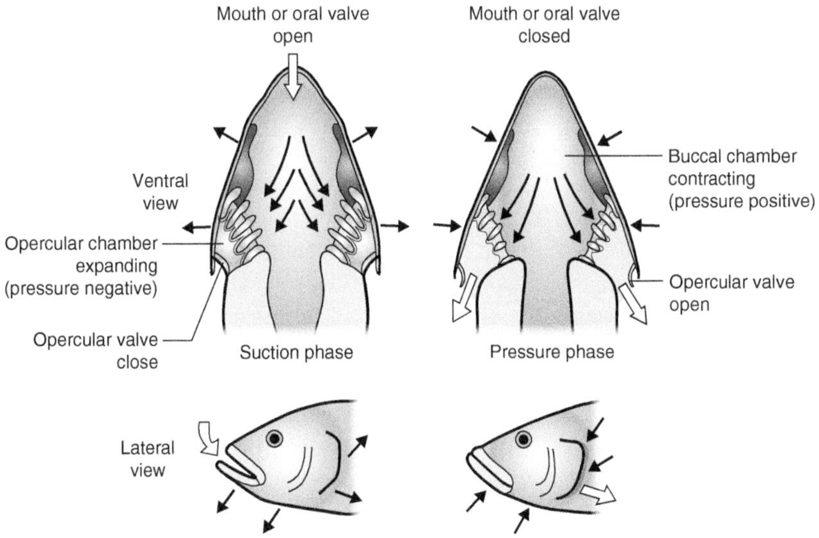

FIGURE 3.14 The timing of the expansion and contraction of the buccal (oral) and opercular cavities ensures that the pressure in the buccal chamber exceeds that of the opercular chamber throughout nearly all of the respiratory cycle. This creates a nearly steady flow of water from the buccal chamber to the opercular chamber, passing over the gill lamellae, which have blood flowing through them in the opposite direction. In the upper images, the fish is viewed from below. Adapted from Hildebrand (1988).

The second method of gill ventilation, called **ram ventilation**, consists simply of keeping the mouth slightly open while swimming. The forward movement of the fish keeps water flowing over the gills. This is an efficient way to ventilate the gills because the work of ventilation is accomplished by the swimming muscles, but it can only be used by strong swimmers moving at relatively high speeds. Some predatory pelagic fishes, such as tunas (Scombridae), rely exclusively on ram ventilation and must therefore swim constantly. At one time, it was thought that sharks also had to swim constantly in order to breathe. However, observations of so-called "sleeping" sharks on the ocean floor, including relatively sedentary species such as the Whitetip Reef Shark (*Triaenodon obesus*) and Nurse Shark (*Ginglymostoma cirratum*), indicate they too use a gill pumping mechanism similar to the one described earlier for teleosts. Many larger fishes use ram ventilation while swimming at moderate to high speeds but rely on pumping of the buccal and opercular chambers while still or moving slowly. As speed increases, they can switch from gill pumping to ram ventilation (Roberts 1975a).

The jawless fishes have a very different gill structure and rely on different means of ventilation. Hagfishes (Myxinidae) have a muscular, scroll-like flap known as a **velum** which moves water in through the single median nostril and over the gills (Fig. 3.15). When the hagfish's head is buried in food, water enters and leaves the gill area via the external opening behind the last gill pouch. Lampreys (Petromyzontidae) expand and contract the branchial area, causing water to flow in and out through the multiple gill openings. This method of ventilation is especially practical when the lamprey's mouth is attached to the substrate or a host organism.

Functional Morphology of Feeding

Feeding adaptations involve structures used in food acquisition and processing, such as jaw bones and muscles, teeth, gill rakers, and the digestive system. Less obvious, but also important, are morphological adaptations in eye placement and function, body shape, locomotory patterns, pigmentation, and lures. The functional morphology of feeding has intimate linkage to all aspects of fish evolution and biology.

For many fishes, a simple glance at jaw morphology, dentition type, and body shape allows accurate prediction of what a fish eats and how it catches prey. Small fishes with fairly streamlined and compressed bodies, forked tails, limited dentition, and protrusible mouths in all likelihood are planktivorous, feeding on small animals up in the water column. This generalization holds for fishes as diverse as osteoglossiform mooneyes, clupeomorph herrings, ostariophysine minnows, and representative acanthopterygian groupers (e.g., *Anthias*), snappers (*Caesio*), bonnetmouths (*Inermia*), damselfishes (*Chromis*), and wrasses (*Clepticus*). Large, elongate fishes with long jaws studded with sharp teeth and with broad tails adjoined by large dorsal and anal fins set far back on a round body are piscivores that ambush their prey from midwater with a sudden lunge (see Chapter 16). An alternative piscivorous morphology includes a more robust, deeper body, with fins distributed around the body's outline, and a large mouth with small teeth for ambushing and engulfing prey; this is the "bass" morphology of many acanthopterygian predators such as Striped Bass (Moronidae), groupers (Serranidae), black basses (Centrarchidae), and peacock basses (Cichlidae).

Our emphasis here will be on the functional morphology of structures directly responsible for engulfing and processing food. Moderate detail is provided, but we can only superficially discuss the diversity in structure, action, and interconnection among the 30 moving bony elements and more than 50 muscles that make up the head region of most fishes.

Jaw Protrusion: The Great Leap Forward

The evolution of closable jaws that could be used in feeding was undoubtedly the advance that drove early vertebrate evolution. One of the major advances made by, but not exclusive to, more advanced bony fishes is the ability to protrude the upper jaw during feeding. Jaw protrusion helps overtake a prey item, extending the mouth around the prey faster than the predator can move its entire body through the water. Attack velocity may thus be increased by up to 40%. Jaw protrusion also creates suction forces that can pull items from as far away as 25–50% of head length. As many as 15 different functions and advantages have been postulated for the protrusible jaw of teleosts. These advantages generally involve increased prey capture, but antipredator surveillance and escape may be enhanced (Lauder & Liem 1981; Motta 1984; Ferry-Graham & Lauder 2001).

Fossil evidence shows a strong increase in the estimated degree of jaw protrusion over time, indicating that jaw protrusibility was beneficial and that positive selection continued resulting in some species today with high degrees of protrusibility, perhaps up to 20% or more of body length. Significant jaw protrusion independently originated at least eight, and perhaps up to fifteen times, and the average estimated jaw protrusion among fishes increased from <1% SL to about 3% SL over the last 100 million years (Bellwood et al. 2015).

The elements involved in jaw protrusion include the bones of the jaw (premaxilla, maxilla, mandible), ligamentous connections of these bones to the skull and to each other (premaxilla to maxilla, ethmoid, and rostrum; maxilla to mandible, palatine, and suspensorium; mandible to suspensorium), and several muscles, notably the expaxials, levator operculi, hypaxials, adductor mandibulae, and levator arcus palatini (Fig. 3.16).

During jaw protrusion, the entire jaw moves forward and slightly up or down. Protrusion in a generalized percomorph (advanced teleost) occurs as the cranium is lifted by the epaxial muscles, and the lower jaw is depressed by muscles associated with the opercular and hyoid bone series. Movement of the mandible causes the maxillary to pivot forward, the suspensorium

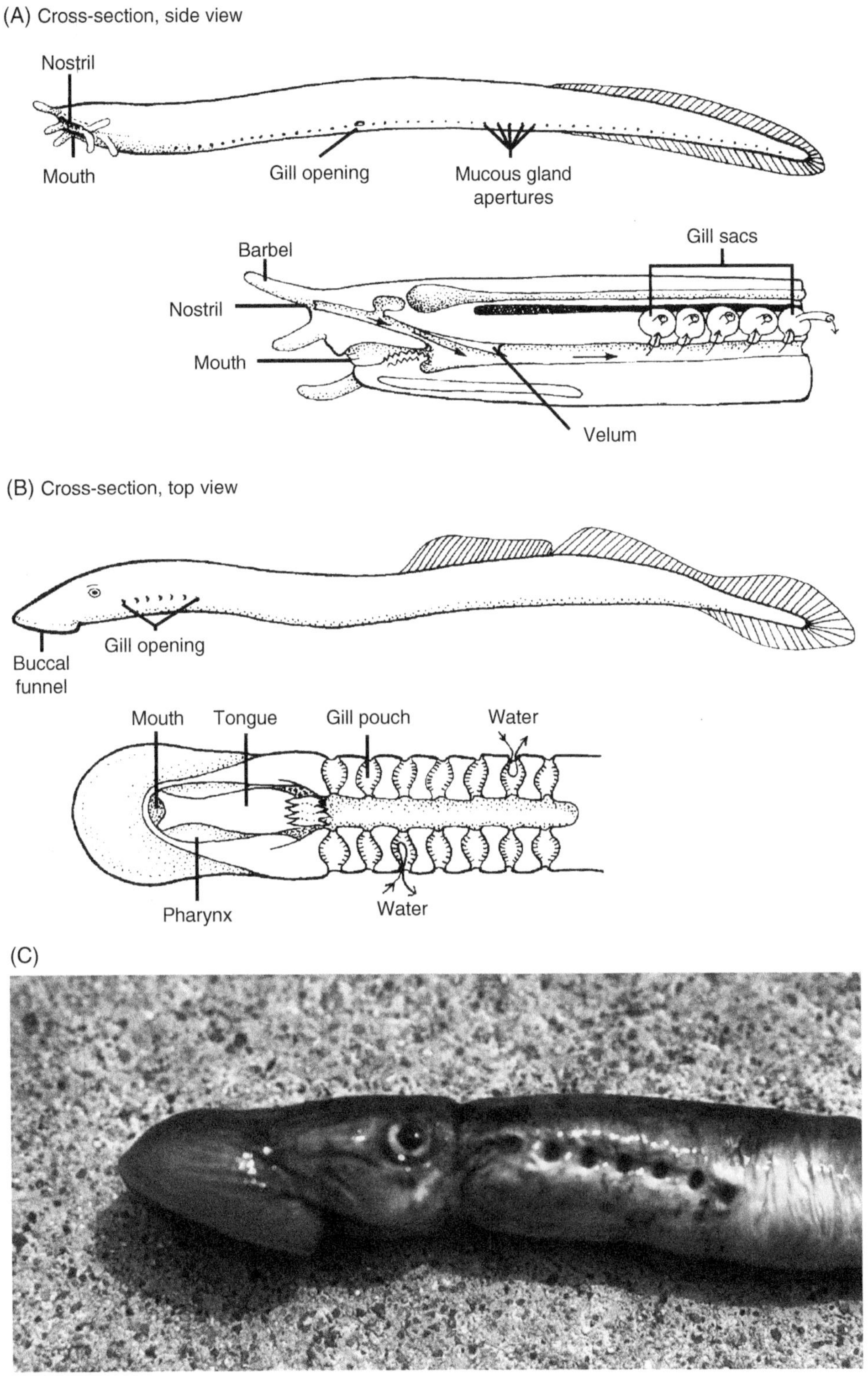

FIGURE 3.15 (A) Hagfishes have one or more external gill openings on each side. Movement of the scroll-like velum draws water in through the nostril and pushes it through the pharynx and branchial pouches. Excurrent branchial ducts then direct the water to the gill openings. (B, C) Lampreys have multiple external gill openings on each side. Expansion and contraction of the branchial pouches provide ventilation through each external opening. This permits continued breathing while the mouth is attached to the substrate or a host. Lee Emery / USFWS / Public Domain.

(the hinge joint that suspends the lower jaw from the cranium) contributing to maxillary rotation. The descending process of the premaxilla is connected to the lower edge of the maxilla, so the premaxilla is pushed forward, its ascending process sliding forward and down the rostrum. The jaw is closed through the actions of the adductor mandibulae muscle on the mandible, the levator arcus palatini on the suspensorium, and the geniohyoideus on the hyoid apparatus. Many variations on this

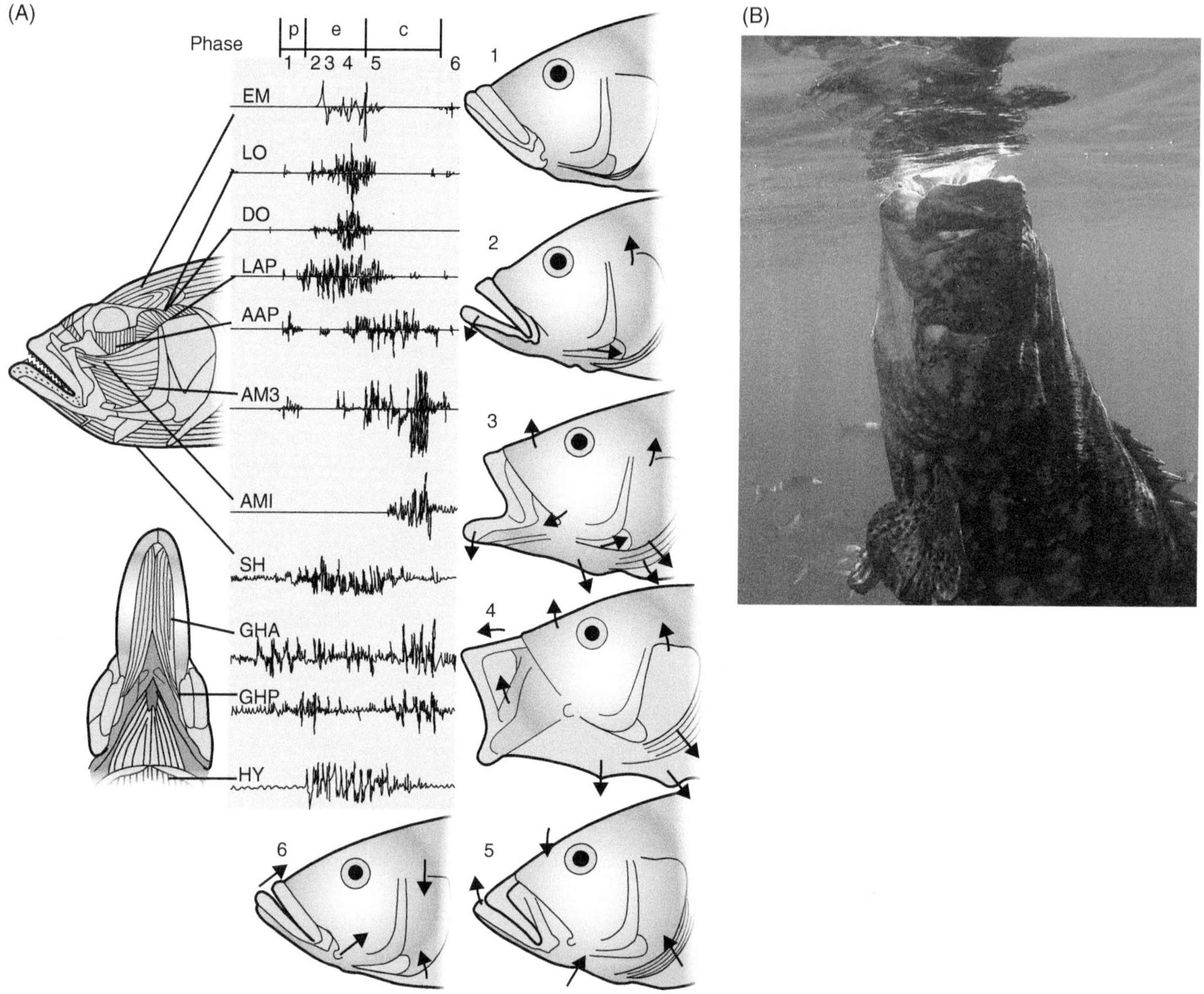

FIGURE 3.16 (A) Opening, protrusion, and closing of the jaw in most percoids, as demonstrated by Liem (1978) in the cichlid *Serranochromis robustus*. The major muscles involved are shown on the left and contraction of these muscles is shown by the electromyograms in the shaded box, which also indicates the major phases of preparation (p), expansion (e), and compression (c). A lateral view of the sequence of jaw protrusion and closing is represented along the right side and below, with stages 1–6 also shown at the top of the shaded box. Muscles shown include the adductor arcus palatini (AAP), adductor mandibulae (AM), dilatator opercula (DP), epaxial muscles (EP), geniohyoideus anterior (GA), geniohyoideus posterior (GP), hypaxial muscles (HP), levator arcus pelatini (LAP), levator operculi (LO), and sternohyoideus (SO). Jaw opening involves three major couplings of muscles, ligaments, and bones: epaxial muscles that lift the cranium; levator operculi muscles that move the opercular bones up and out and help depress the mandible; and hypaxial muscles that depress the mandible via actions of the hyoid apparatus. A, B, C are based on Liem 1978. (B) An Atlantic Goliath Grouper (*Epinephelus itajara*) displays the expansion phase of suction feeding to inhale a frozen baitfish at the surface. D. Facey (Author).

simplified description exist, differing among taxa in terms of twisting of jaw bones, points of attachment and pivot between structures, and actions of muscles and ligaments on particular elements (Motta 1984).

Jaw protrusion creates rapid water flow that carries edible particles, both small and large, into the fish's mouth. Suction velocity increases from 0 m/s to as much as 12 m/s in as little as 0.02–0.03 s (Osse & Muller 1980; Ferry-Graham et al. 2003). Fishes that feed on such different prey as phytoplankton, zooplankton, macroinvertebrates, and other fishes utilize suction to capture prey. Gill rakers, jaw teeth, and teeth on various non-marginal jaw bones (palate, vomer, tongue) prevent escape from the opercular chamber.

Suction pressures vary during a feeding event in advanced percomophs, increasing and decreasing four times. The four phases of suction feeding are preparation, expansion, compression, and recovery (Lauder 1983a; Liem 1978).

1. During **preparation**, as the fish approaches its prey, pressure in the buccal cavity increases as a result of inward squeezing of the suspensorium and lifting of the mouth floor.

2. The **expansion** phase is when maximal suction pressure develops; the mouth is opened to full gape via lower jaw depression, premaxillary protrusion, and expansion of suspensory, opercular, and mouth floor (hyoid) units. Expansion is the shortest phase during jaw activity, requiring only 5 ms in some anglerfishes. The negative pressures generated during expansion can reach −800 cmH_2O (0.7 atm) in the Bluegill (*Lepomis macrochirus*), approaching the physical limits imposed by fluid mechanics. Such rapidly achieved low pressure causes cavitation, which involves water vapor suddenly coming out of solution and forming small vapor-filled cavities (the bubbles produced behind an accelerating boat propeller result from cavitation) (Lauder 1983a). The popping noise made during feeding by Bluegill may result from the collapse of cavitation bubbles.
3. The **compression** phase occurs and pressure increases as the mouth is closed by reversing the movements of cranial bones, an activity that requires contraction of a different set of muscles (Fig. 3.16). The opercular and branchiostegal valves at the back of the head open up after the jaws close, which allows water to flow out of the buccal and opercular cavities while food is retained by the gill arches and rakers.
4. **Recovery** involves a return of bones, muscles, and water pressure to their pre-preparatory positions.

Modifications of this basic plan underscore some rather spectacular derivations that allow specialized feeding activities. In particular, cichlids show a diversity of foraging types unequaled in any other fish family (Goldschmidt 1996). In cichlids, jaw protrusion occurs as a result of movement of the suspensorium, independent of the maxilla. The consequence of this decoupling of suspensorium and maxilla is that the jaw can be protruded via four different pathways: lifting the neurocranium, abducting the suspensorium, lowering the mandible, or swinging the maxilla. Cichlids make use of different combinations of jaw elements and protrusion pathways to feed on different prey types or in different habitats (e.g., Waltzek & Wainwright 2003; Hulsey & De Leon 2005). The cichlid jaw is the closest that fishes have come to a prehensile feeding tool. It is likely that the derived trait of a decoupled suspensorium and the resulting trophic versatility have contributed greatly to their success (Liem 1978; Lauder 1981; Motta 1984; Liem & Wake 1985).

In most fishes, the suction pressure is produced via expansion of the buccal cavity. A generalized perciform such as the Yellow Perch increases its mouth cavity volume by a factor of six, creating considerable negative pressure. The apparent record for volume increase is held by a small (30 cm long), bizarre, elongate midwater fish, *Stylophorus chordatus*. During feeding, this fish throws its head back and thrusts its tubular mouth forward. Mouth volume increases almost 40-fold, creating pressures three times greater than in the generalized perch and engulfing copepods as water rushes in at a calculated velocity of 3.2 m/s (Pietsch 1978).

Another extreme of jaw protrusion occurs in the Sling-jaw Wrasse, *Epibulus insidiator* (Westneat & Wainwright 1989). Sling-jaws protrude their jaws up to 65% of their head length, which is twice the extension found in any other fish (Fig. 3.17). This extreme protrusion is accomplished via a major reworking of jaw elements. Several bones in the Sling-jaw's head have unique sizes and shapes, including the quadrate, interopercle, premaxilla, and mandible. Ligaments connecting these bones are unusually large, and a ligament found in no other fish links

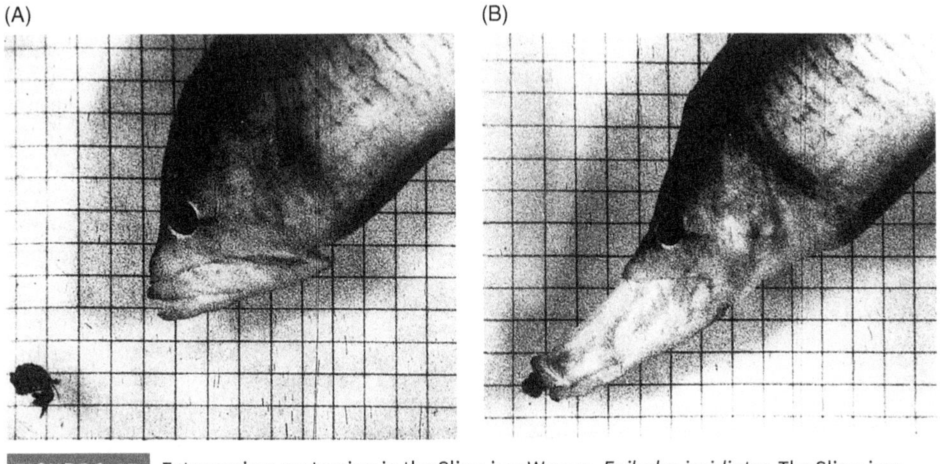

FIGURE 3.17 Extreme jaw protrusion in the Sling-jaw Wrasse, *Epibulus insidiator*. The Sling-jaw has novel bone shapes, extreme bone and ligament rotations, and ligament rotations, and has even invented a new ligament involved in jaw protrusion. (A) A 15 cm-long wrasse approaches its crustacean prey with its mouth in the retracted condition. Note that the posterior extension of the lower jaw, involving the articular and angular bones, extends as far back as the insertion of the pectoral fin. (B) During prey capture, the wrasse protrudes both its upper and lower jaws forward, extending them a distance equal to 65% of its head length. Jaw expansion creates suction forces that draw the prey into the mouth. Positions (A) and (B) are separated by about 0.03 s. Westneat and Wainwright 1989 / with permission of John Wiley & Sons.

the vomer to the interopercle. The modified bones undergo extreme and in some cases, unique rotations during jaw protrusion: the lower jaw moves forward during protrusion, a departure from the depression movement seen in all other fishes. The Sling-jaw shoots its mouth out at small fishes and crustaceans on coral reef surfaces, suctioning them into its mouth. It achieves a strike velocity of 2.3 m/s, but all of this speed is contributed by the jaw because the fish hovers almost still in the water while attacking prey. Extreme jaw protrusion in Sling-jaws involves the evolution of unique bones and ligaments, but the muscles of the jaw and skull have shapes, functions, and sequences of activity that differ little from generalized perciforms. Novel jaw function is therefore accomplished by drastic modification of some structures and the retention of primitive conditions in others. The Sling-jaw Wrasse exemplifies a widely made observation about the evolutionary process that every species represents a mosaic of ancestral and derived traits.

Suction feeding has evolved repeatedly and occurs in many non-teleosts as well as in primitive and specialized teleosts that cannot protrude their jaws. Elasmobranchs, including skates, rays, and such sharks as Nurse and Horn sharks, can generate suction forces as strong as −760 mmHg for feeding on buried mollusks or lobsters in reef crevices (Tanaka 1973; Motta & Wilga 1999, 2001). Lungfishes and Bowfin among nonteleosts, and anguillid eels, salmons, pickerels, and triggerfishes among teleosts do not protrude their jaws but use inertial suction for feeding. Suction in the non-protruding species is often accomplished by rapid depression of the floor of the mouth. Triggerfishes and other tetraodontiform fishes such as boxfishes can reverse this flow and forcefully expel water from their mouths (Frazer et al. 1991). Alternate blowing and sucking is used to manipulate food items in the mouth during repositioning for biting. Blowing is also used for uncovering invertebrate prey buried in sand or for manipulating well-defended prey items. A Red Sea triggerfish, *Balistes fuscus*, feeds on long-spined sea urchins. Triggerfishes swim up to an urchin sitting on the sand and blow a powerful jet of water at the urchin's base. The water stream lifts the urchin off the substrate and rolls it over, at which point the triggerfish bites through the unprotected oral disk, killing the urchin (Fricke 1973). Triggerfishes also use blowing to uncover buried prey such as sanddollars. Blowing involves compression of the mouth via actions of muscles associated with the opercular, mandibular, and hyoid bones (Frazer et al. 1991; Turingan & Wainwright 1993).

Pharyngeal Jaws

Depression of the mouth floor also creates water flow towards the throat, thereby helping push food items posteriorly. Here the prey encounters the second set of jaws, the pharyngeal apparatus characteristic of more advanced fishes (see Chapter 11). Pharyngeal jaws evolved from modified gill arches and their associated muscles and ligaments. The lower pharyngeal jaws are derived from the paired fifth ceratobranchial bones, whereas the upper jaws consist of dermal plates attached to the posterior epibranchial and pharyngobranchial bones. Dentition varies functionally among species that eat different food types but may also develop differently among individuals of a population as a function of the food types encountered by the growing fish. In the Cuatro Cienegas Cichlid of Mexico, *Cichlasoma minckleyi*, fish that feed on plants develop small pappiliform pharyngeal dentition, whereas those that feed on snails develop robust molariform dentition (Kornfield & Taylor 1983).

In their simplest action, pharyngeal jaws help rake prey into the esophagus. They may additionally reposition prey, immobilize it, or actually crush and disarticulate it. These actions involve at least five different sets of bones and muscles working in concert, including 10 muscle groups and bones of the skull, hyoid region, lower jaw, pharynx, operculum, and pectoral girdle. The main action is the synchronous occlusion (coming together) of the upper and lower pharyngeal jaws. In cichlids, prey is crushed between the anterior teeth of both pharyngeal jaws, pushed posteriorly by posterior movement of both jaws, and then bitten by the teeth of the posterior region of the jaws (Lauder 1983a, 1983b; Liem 1978).

Pharyngeal jaws influence feeding in another important manner. The constraint on prey size is determined by both mouth size and pharyngeal gape. If a prey item is too large to pass through the pharyngeal jaws, it is unavailable to the predator. Hence many predators can capture but not swallow a prey item because of pharyngeal gape limitation. In small-mouthed species, such as the Bluegill sunfish, oral and pharyngeal gape differ only by 20–30%. But in piscivores that use oral protrusion for prey capture, such as the Largemouth Bass, oral gape may far exceed pharyngeal gape, which means that usable prey size is considerably smaller than that which can be engulfed by the mouth. Posterior to the pharyngeal jaws is the throat, the width of which is determined by spacing between the cleithral bones of the pectoral girdles. Thus a predator can only eat prey that can pass through its oral jaws, pharyngeal jaws, and intercleithral space (Lawrence 1957; Wainwright & Richard 1995).

A crucial function of the pharyngeal apparatus in many species is therefore to crush prey to a size small enough to pass through the throat. Here prey morphology comes into play, because prey that is just small enough to fit between the pharyngeal pads may be too hard to crush and is thus unavailable to the predator. This interplay of structure, function, and the constraints created by the pharyngeal apparatus is shown nicely in Caribbean wrasses that feed on hard-bodied prey (Wainwright 1987, 1988a). The size of the muscles that move the pharyngeal jaws differs among three species, the Clown Wrasse (*Halichoeres maculipinna*), Slippery Dick (*H. bivittatus*), and Yellowhead Wrasse (*H. garnoti*). In all three species, muscle mass and pharyngeal gape increase with increasing body size (Fig. 3.18). At any size, Clown Wrasses have smaller pharyngeal musculature than the other two species. Small Slippery Dicks and Yellowhead Wrasses can crush and eat

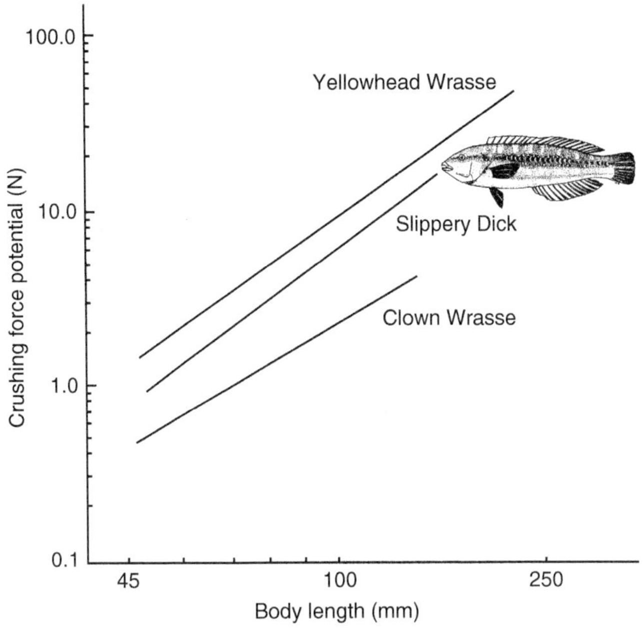

FIGURE 3.18 Crushing ability of the pharyngeal jaws in three related wrasses as a function of body size. Larger wrasses can crush larger snails because of their stronger pharyngeal jaws, and differences among species also influence preferred food types. Clown Wrasses have relatively weak jaws and feed on relatively soft-bodied prey, particularly when the fish are younger. Slippery Dicks and Yellowhead Wrasses have strong jaws and feed on shelled prey throughout their lives. After Wainwright (1988a); fish drawing from Gilligan (1989).

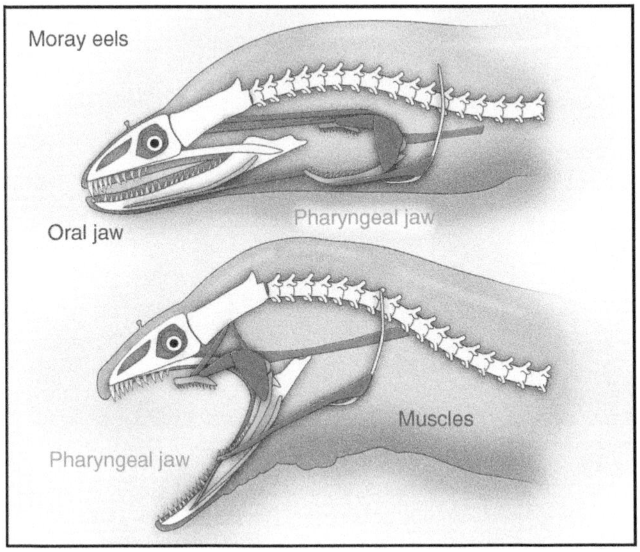

FIGURE 3.19 The protrusible pharyngeal jaws of moray eels (artwork by Zina Deretsky, US National Science Foundation, based on work by Rita Mehta – from NSF).

snails that are unavailable to larger Clown Wrasses. Small Clown Wrasses feed preferentially on relatively soft-bodied crabs and other invertebrates; they shift to snails only after attaining a body length of 11 cm, when they eat hard-bodied prey that are smaller than those taken by equal-sized fishes of the other two species. Slippery Dicks and Yellowhead Wrasses feed extensively on snails beginning at a relatively small fish body length of 7 cm. Pharyngeal crushing strength accounts for inter- and intraspecific differences in feeding habits in these fishes.

As is so often the case in evolution, an adaptation opens up selection pressures favoring additional innovations. Moray eels (Muraenidae) have weak suction pressures in the mouth cavity and feed in crevices and other tight places, which constrains jaw movements that would normally aid in swallowing. Morays have "solved" this dilemma by developing pharyngeal jaws that can project forward, reminiscent of the vicious subjects of the "Alien" movies. The upper (pharyngobranchial arch) and the lower (ceratobranchial arch) are equipped with sharp, highly recurved teeth like those in the oral jaws. In the resting position, the pharyngeal apparatus sits far back in the throat, behind the skull. But when prey are grasped by the oral jaws, a series of muscles push the pharyngeal jaws far forward into the oral cavity; these jaws then grasp the prey and retract back, pulling the prey toward the esophagus (Mehta & Wainwright 2007; Fig. 3.19).

Dentition

The prey a fish captures and eats are often predictable from the type of teeth the fish possesses. Even within families, species differ considerably in their dentition types as a function of food type and foraging mode as in butterflyfishes (Motta 1988), cichlids (Fryer & Iles 1972), and surgeonfishes (Jones 1968). Here we focus on general groups of foragers and how dentition corresponds to food type.

Piscivores and feeders on soft-bodied, mobile prey such as squid show five basic tooth patterns:

1. **Long, slender, sharp** teeth usually function to hold fish (mako, sandtiger, and angel sharks, moray eels, deep-sea viperfishes, lancetfishes, anglerfishes, goosefishes). In some groups (e.g., goosefishes, anglerfishes; also esocid pikes), elongate dentition is repeated on the palatine or vomerine bones. These medial teeth point backwards and may have ligamentous connections at their base, which allows them to be depressed as the prey is moved toward the throat but prevents escape back through the anterior jaws.
2. Numerous small, needlelike, **villiform** teeth occur in elongate, surface-dwelling predators such as gars and needlefishes, as well as in more benthic predators such as lizardfishes and lionfishes.
3. Flat-bladed, pointed, **triangular** dentition is usually used for slashing food items and is found in requiem sharks, piranhas, barracudas, and large Spanish mackerels. Piranhas have teeth that are remarkably convergent in shape with those of many sharks (Fig. 3.20). In sharks, the lateral margins of bladelike teeth are often serrated, which enhances their cutting function when the head is shaken,

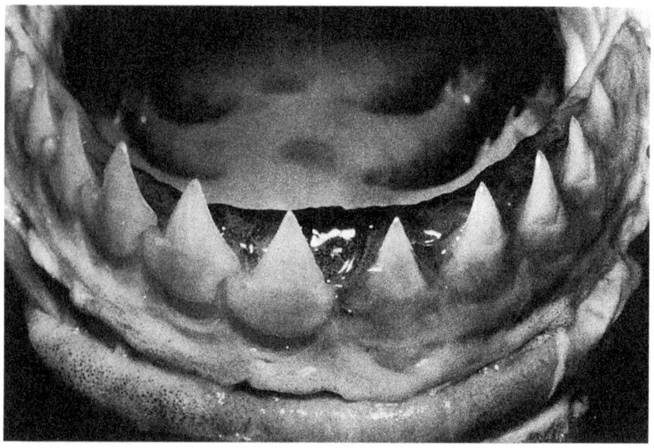

FIGURE 3.20 Convergence in dentition among predatory fishes. The triangular, razor-sharp teeth of a Red-bellied Piranha, *Pygocentrus nattereri*, are remarkably similar in shape and action to those of many sharks. Note the small lateral cusps at the base of the teeth, a feature also shared with many sharks. Piranhas also replace their teeth as do sharks, but piranhas alternately replace all teeth in the left or right half of a jaw, rather than replacing individual teeth or rows of teeth. The teeth in the left side of the jaw (= right side of photo) have recently erupted. Sazima and Machado 1990 / with permission of Springer Nature.

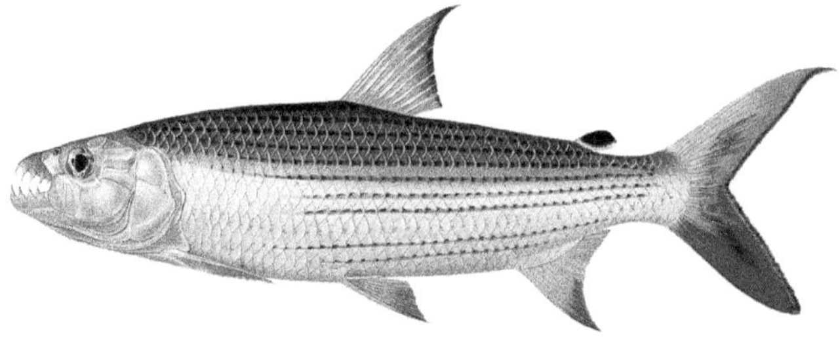

FIGURE 3.21 African Tigerfish, *Hydrocynus vittatus.*, are predatory alestids. Their dentition, is an extreme example of the triangular, flattened and fanglike teeth characteristic of many piscivorous fishes. Wikimedia Commons / CC BY 2.0.

or the jaws are opened and closed repeatedly. Sharks and piranhas, as well as other characins, have also converged on **replacement dentition**. Tooth replacement, regardless of dentition type, has evolved repeatedly and independently among bony fishes, occurring in brachiopterygian bichirs, amiiforms, lepisosteid gars, and most teleost superorders and orders, including osteoglossomorphs, elopomorphs, protacanthopterygians, ostariophysans, paracanthopterygians, and numerous acanthopterygians (Roberts 1967; Trapani 2001; Hilton & Bemis 2005).

4. Recurved, conical, **caniniform** teeth with sharp points characterize such piscivores as Bowfin, cod, snappers, and some seabasses. Sharp, conical dentition serves to grasp and hold. It reaches its extreme form in the almost triangular, fanglike, slightly flattened teeth of the African Tigerfish, *Hydrocynus* (Fig. 3.21).
5. Surprisingly, many highly predaceous piscivores have limited marginal **cardiform** dentition that has a rough sandpaper texture and consists of numerous, short, fine, pointed teeth (e.g., large seabasses, snook, Largemouth Bass, billfishes). The former species rely on large, protrusible mouths for engulfing prey fishes, whereas billfishes immobilize their prey by slashing or stabbing with the bill.

Often, a predator will have a mixture of dentition types, such as anterior canines followed by or intermixed with smaller, needlelike teeth (e.g., the Pike Characin *Hepsetus*), or long canines intermixed with smaller conical teeth (e.g., some wrasses). Ultimately, and regardless of location in the mouth and whether teeth are of one or several types, primary dentition type reflects food characteristics. Among 10 Australian ariid marine catfish species, piscivores have sharp, recurved palatine teeth, worm feeders have small, sharp, recurved palatine teeth, and molluskivores have globular, truncated palatine teeth (Blaber et al. 1994).

Fishes that feed on hard-bodied prey, such as mollusks, crabs, and sea urchins, often have teeth and jaw characteristics that represent a separation of the activities of capturing versus processing prey. Many such fishes have large incisors with flattened cutting surfaces (such as some Holocephali), or the teeth may be fused into beaks for scraping algae off corals, as in parrotfishes (Scaridae) and Pacific knifejaws (Oplegnathidae), or for biting crustaceans or echinoderms, as in blowfishes (Tetraodontiformes). The prey are then passed to flattened or rounded, **molariform** teeth for processing, located posteriorly in marginal or pharyngeal jaws. Convergence is apparent when comparing mollusk-eating fishes from different taxa, such as horn sharks and wolf-eels. Horn sharks (*Heterodontus*) have small conical teeth anteriorly, which grade posteriorly into broad, rounded pads for crushing and grinding. Wolf-eels have strong, conical canines anteriorly and rows of rounded molars posteriorly in each jaw (Fig. 3.22). Similar anterior–posterior differences occur in Freshwater Drum, Sheepshead, cichlids, and wrasses.

A suction versus chewing arrangement occurs in many fishes that feed on sand-dwelling mollusks. Catostomid suckers such as the River Redhorse, *Moxostoma carinatum*, are ostariophysans in which the molarlike teeth occur on the pharyngeal arches. Only the lower arch develops dentition, and these teeth usually occlude against horny pads in the roof of the mouth. In higher teleosts, the pharyngeal teeth are composed of both dorsal and ventral pharyngeal arch derivatives, such as in cichlids and the Redear Sunfish

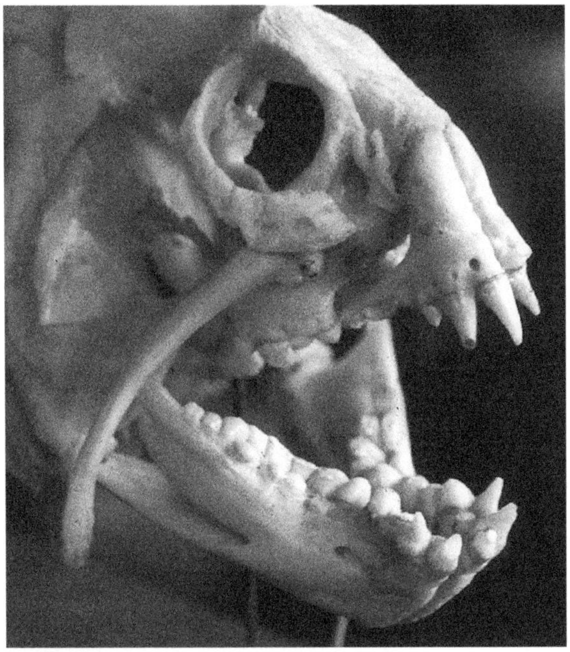

FIGURE 3.22 Fishes that feed on hard-bodied prey crush their prey with molariform teeth located far back in their mouths, but often have different tooth types in different parts of the jaw. In the Wolf-eel, *Anarrhichthys ocellatus*, caninelike anterior jaw teeth grasp prey and molariform teeth farther back in the marginal jaws crush the prey. G. Helfman (Author).

(*Lepomis microlophus*), locally referred to as the "shellcracker." Analogously, stingrays suction mollusks off the bottom and then crush them in pavementlike dentition. Fishes that remove attached invertebrate prey (such as sponges, ascidians, coelenterates, and chitons) from surfaces tend to have powerful oral jaws with incisorlike dentition (e.g., triggerfishes) or with teeth fused into a parrotlike beak (e.g., parrotfishes, pufferfishes). In parrotfishes, the beak bites off algae or pieces of coral that are then passed to the pharyngeal mill for grinding.

Another means of dealing with both soft- and hard-bodied prey has arisen in some sharks, such as the hemiscylliid bamboo sharks. These sharks have the classic sharp, spiky teeth expected of a feeder on soft-bodied prey such as fish and squid. However, when feeding on harder items such as crabs, ligaments at the base of each tooth allow it to hinge backward, overlapping the replacement tooth that sits immediately behind it in the jaw. The multiple rows of depressed teeth then form a functionally flat surface more appropriate for crushing hard prey. The teeth spring back up after a bite is taken (Ramsay & Wilga 2007; see also Summers 2006).

In addition to marginal, medial, and pharyngeal teeth, fishes may use **gill rakers** to assist in the capture or retention of prey. These are bony or cartilaginous projections that point inwards and forwards from the inner face of each gill arch. As with the various teeth, gill raker morphology corresponds quite closely to dietary habits. Piscivores and molluskivores, such as seabasses, black basses, and many sunfishes, tend to have short, widely-spaced gill rakers that prevent the escape of large prey out the gill openings. Fishes that eat zooplankton of large and intermediate size, such as the Bluegill and Black Crappie (*Pomoxis nigromaculatus*), have longer, thinner, and more numerous rakers. Feeders on small zooplankton, phytoplankton, and suspended matter have the longest, thinnest, and most numerous rakers; menhaden, *Brevoortia* spp., filter phytoplankton, detritus, and small zooplankters and have >150 rakers just on the lower limb of each gill arch. In North American whitefishes (Coregoninae), the Inconnu (*Stenodus leucichthys*) feeds on fishes and has 19–24 rakers; the Shortnose Cisco (*Coregonus reighardi*) feeds on smaller mysid shrimp, amphipods, and clams and has 30–40 rakers; whereas the Cisco (*C. artedii*) eats tiny zooplankters, midge larvae, and water mites and has 40–60 rakers (Scott & Crossman 1973). In most filter-feeding fishes, particles are captured by mechanical sieving, whereby large particles cannot pass through the narrow spaces between gill rakers. Electrostatic attraction, involving the capture of charged particles on mucous-covered surfaces, is also suspected (Liem 1978).

Mouth Position and Function

Mouth position, in terms of whether the mouth angles up, ahead, or down, also correlates with trophic ecology in many fishes (Fig. 3.23). The vast majority of fishes, regardless of trophic habits, have **terminal** mouths, which means that the body terminates in a mouth that opens forward. Deviations from terminal locations usually indicate habitat and feeding habit. Fishes that swim near the water's surface and feed on items at the surface often have mouths that open upwards, termed **superior** or **supraterminal** (e.g., African butterflyfishes, freshwater hatchet or flyingfishes, halfbeaks, topminnows, tarpon; see Fig. 3.24). Some predators that lie on the bottom and feed on prey that swim overhead also have superior mouths (e.g., stonefishes, weaverfishes, stargazers). Mouths that open downward, termed **subterminal** or **inferior**, characterize fishes that feed on algae or benthic organisms, including sturgeons, suckers, some minnows, suckermouth armored catfishes, Chinese algae eaters, some African cichlids, clingfishes, and loach gobies. Upside-down catfishes feed on the undersurfaces of leaves, but do so while swimming upside down and not surprisingly have inferior mouths. Fishes that do not have to visually fix on their prey (e.g., algal-scraping clingfishes, catfishes, loaches, cichlids), or that take somewhat random mouthfuls of sediments that are then sifted orally (e.g., suckers, mojarras), may gain an antipredator advantage by having an inferior mouth. A terminal mouth

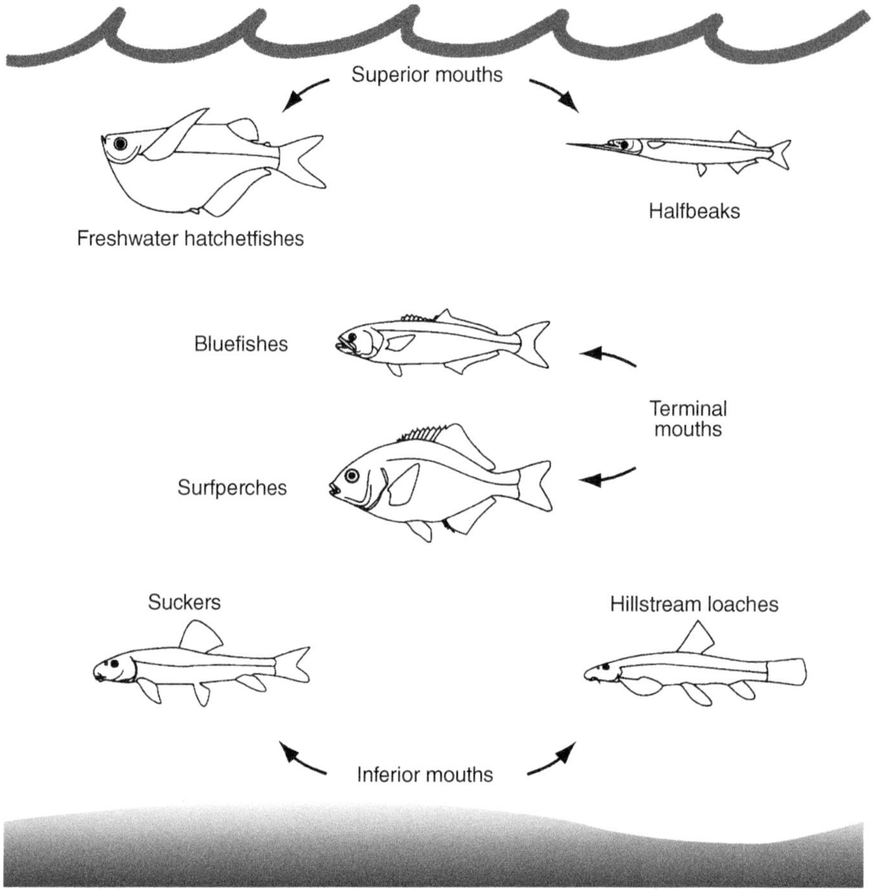

FIGURE 3.23 Correspondence among mouth position, feeding habits, and water column orientation in teleosts. Fishes with "superior" mouths frequently live near and feed at the surface, whereas fishes with "inferior" mouths often scrape algae or feed on substrate-associated or buried prey. Fishes with terminal mouths often feed in the water column on other fishes or zooplankton but are also likely to feed at the water's surface, from structures, and on the bottom. Adapted from Nelson (2006).

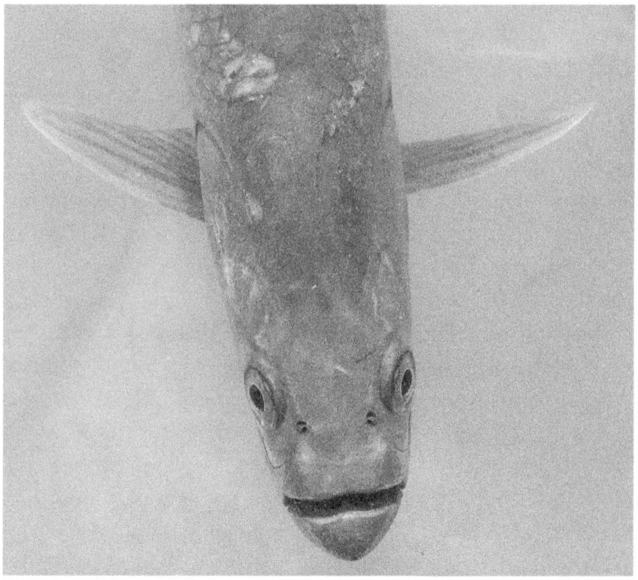

FIGURE 3.24 The supraterminal mouth of an Atlantic Tarpon (*Megalops atlanticus*) is apparent in this dorsal view. D. Facey (Author).

in such fishes would require that they angle head down each time they scraped or sampled the benthos, which would make them more conspicuous and less able to escape rapidly.

Specialized **suctorial** mouths characterize unrelated fishes that scrape algae from rocks, particularly if they also live in high-energy environments. This ecological grouping includes hillstream loaches, suckermouth armored catfishes such as the familiar Plecostomus of the aquarium trade, Southeast Asian algae eaters, and the loach gobies of Australia. The gyrinocheilid algae eaters live in swift streams where they rasp algae from rocks with their lips while remaining attached with their suctorial mouth. Gyrinocheilids have evolved an additional incurrent opening dorsal to the operculum that opens into the gill chamber. They breathe in through the dorsal opening and out through the operculum. Drawing water in through the mouth in the more normal manner would require the fish to detach from the substrate, at which moment it might risk being swept downstream. Mouths are not the only way for algae feeders to remain attached in wave-swept habitats. Gobiesocid clingfishes accomplish this via pelvic fins modified into a suction disk (Wheeler 1975; Nelson 2006).

Supplementary Reading

Alexander RMcN. 1983. *Animal mechanics*, 2nd edn. Oxford: Blackwell Science.

Duncker HR, Fleischer G, eds. 1986. *Functional morphology of vertebrates*. New York: Springer-Verlag.

Farrell AP, Stevens ED, Cech JJ, Richards JG. 2011. *Encyclopedia of fish physiology: from genome to environment*. Amsterdam, Boston: Elsevier/Academic Press.

Gerking SD. 1994. *Feeding ecology of fish*. San Diego: Academic Press.

Hildebrand M. 1982. *Analysis of vertebrate structure*, 2nd edn. New York: Wiley & Sons.

Hildebrand M, Bramble DM, Liem KF, Wake DB, eds. 1985. *Functional vertebrate morphology*. Cambridge, MA: Belknap Press.

Liem KF, Bemis WE, Walker WF, Grande L. 2001. *Functional anatomy of the vertebrates*, 3rd edn. Belmont, CA: Thomson/Brooks Cole.

Ostrander GK, ed. 2000. *The laboratory fish*. London: Academic Press.

Pough FH, Janis CM, Heiser JB. 2012. *Vertebrate life*, 9th edn. Upper Saddle River, NJ: Prentice Hall.

Schwenk K, ed. 2000. *Feeding: form, function and evolution in tetrapod vertebrates*. San Diego: Academic Press.

Vogel S. 1981. *Life in moving fluids*. Boston: Willard Grant Press.

Wainwright SA, Biggs ND, Currey JD, Gosline JM. 1976. *Mechanical design in organisms*. New York: J. Wiley & Sons.

Journal

Journal of Morphology.

Website

University of Massachusetts, Biology Department, **www.bio.umass.edu/biology/bemis/FAOV_PPTS/FAOV3.htm**.

CHAPTER 4

Structure and Function of the Trunk and Fins

Summary

The notochord of primitive chordates is replaced by the vertebral column in lampreys, Chondrichthyes, and bony fishes and tetrapods. Vertebrae form around the notochord where muscular myosepta intersect with dorsal, ventral, and horizontal septa.

Posterior vertebrae support the caudal fin in most fishes. In teleosts, hypurals (enlarged haemal spines) support the branched principal caudal fin rays. Three basic types of caudal fins are: (i) protocercal, the primitive undifferentiated caudal fin of adult lancelets, hagfishes, lampreys, and larvae of more advanced fishes; (ii) heterocercal, or unequal-lobed tail, in Chondrichthyes and primitive bony fishes; and (iii) homocercal, or equal-lobed tail, found in most advanced bony fishes.

Ribs (pleural ribs) attach to the vertebrae and protect the internal organs. Intermuscular bones are segmental, serially homologous ossifications in the myosepta of teleosts.

Hagfishes and lampreys lack pectoral and pelvic girdles. Sharks have a coracoscapular (pectoral) cartilage with no attachment to the vertebral column. In bony fishes, the pectoral girdle lacks a vertebral attachment but is connected to the back of the skull by the posttemporal bone.

The dorsal, anal, and adipose fins form the median or unpaired fins. Cartilaginous rods support the median fins of hagfishes and lampreys, whereas chondrichthyan fins are supported by ceratotrichia (fin rays

CHAPTER CONTENTS

The Diversity of Fishes: Biology, Evolution and Ecology, Third Edition. Douglas E. Facey, Brian W. Bowen, Bruce B. Collette, and Gene S. Helfman.

Companion website: www.wiley.com/go/facey/diversityfishes3

supported by stiff but flexible protein rods). In bony fishes, ceratotrichia are replaced during ontogeny by lepidotrichia, which are bony supporting elements derived from scales.

Primitive bony fishes have a single dorsal fin composed of soft rays. Advanced bony fishes usually have two dorsal fins: the anterior fin composed of spines and the posterior fin composed of soft rays; in some, these merge into a continuous fin with an anterior section supported by spines and a posterior section supported by rays.

Trunk musculature of most fishes is arranged in myomeres that create wave-like motion along the trunk while the fish is swimming. Trunk muscle fiber types include aerobic "red" muscles, anaerobic "white" muscles, and intermediate "pink" muscles. Muscles in some groups of fish have become modified as electric organs.

The skin and its derivatives, such as scales in fishes, provide external protection. The five basic types of scales are placoid, cosmoid, ganoid, cycloid, and ctenoid.

Locomotion in water presents very different physical challenges compared to those experienced by terrestrial animals. Density and drag are much greater in water, making locomotion energetically expensive and leading to the general hydrodynamic, streamlined shape of most fishes.

Swimming in fishes usually involves alternating contractions and relaxations of muscle blocks on either side of the body that result in the fish pushing back against the water and consequently moving forward. Many variations on this basic theme exist, and about 10 different modes of swimming have been identified that involve either undulatory waves or oscillatory back-and-forth movements of the body or fins. Body and fin shape correlate strongly with locomotory mode and habitat, the most extreme examples being the rapid swimming, highly pelagic mackerel sharks, tunas, and billfishes with streamlined bodies and lunate, high aspect ratio tails.

Sharks, being cartilaginous, cannot rely on muscles attached to a rigid bony skeleton for propulsion. They instead undulate via contractions of their body muscles, which are firmly attached to a relatively elastic skin that functions as an external tendon and provides propulsive force by rebounding. Some propulsive force comes from changing hydrostatic pressure inside the cylinder of the shark's body. The spacing of the two dorsal fins aids the tail in propulsion, and the tail works in concert with flattened ventral surfaces in the head region to counteract the weight of the body and to provide forward thrust.

Locomotory adaptations create trade-offs. Maneuverability is often achieved at a cost in fast starts and sustained speed and vice versa. Versatility is achieved by using different modes for different purposes (fin sculling for positioning, body contractions for fast starts and cruising), which causes most fishes to evolve generalist rather than specialized swimming traits. Highly specialized locomotion includes fishes that can "walk" across the bottom or on land, climb terrestrial vegetation, leap, and glide through the air.

Postcranial Skeleton

The **notochord** is primitively a supporting structure in chordates. It is a simple, longitudinal rod composed of a group of cells that, when viewed in cross-section, appear to be arranged as concentric circles. The most primitive chordate to possess a notochord is the "tadpole" larva of tunicates. The notochord provides support for an elongated body while swimming. Notochordal cells inside the notochord are few in number and contain large vacuoles. Turgor (hydrostatic pressure) of the notochordal cells provides rigidity. The notochord is found during embryonic development in all chordates, but intervertebral disks are all that remain of the notochord in most adults. However, it is present in adult lancelets, Chondrichthyes, Dipnoi, sturgeons (Acipenseridae), paddlefishes (Polyodontidae), and coelacanths. A 1 m long sturgeon may have a notochord nearly as long as its body and about 12 mm in diameter.

Vertebral Column

Vertebrae arise and form around the notochord where muscular myosepta intersect with dorsal, ventral, and horizontal septa. Vertebrae form from cartilaginous blocks called **arcualia**. Typically, there is one vertebra per body segment, the **monospondylous** condition. The basidorsal, interdorsal, basiventral, and interventral arcualia all fuse together to form a single vertebra. In the **diplospondylous** condition, the basidorsal fuses to the basiventral and the interdorsal fuses to the interventral, producing two vertebrae per body segment. Diplospondyly is present in the tail region of sharks and rays, in lungfishes, and in the caudal vertebrae of the Bowfin (*Amia*). Diplospondyly is thought to increase body flexibility.

Vertebrae are usually divided into **precaudal** (anterior vertebrae extending posteriorly to the end of the body cavity and bearing ribs) and **caudal vertebrae** (posterior vertebrae beginning with the first vertebra bearing an elongate **haemal spine** surrounding a closed haemal canal through which the caudal artery enters) (Fig. 4.1).

Vertebrae may have various bony elements projecting from them. Dorsally, there is an elongate **neural spine** housing a **neural arch** through which the spinal cord passes (Fig. 4.2A). Ventrally, there may be **parapophyses** that extend ventrolaterally and to which the ribs usually attach. The main artery of the body, the dorsal aorta, passes ventral to the precaudal vertebrae and enters the closed **haemal canal** (Fig. 4.2B) toward the end of the abdominal cavity, at which point it is referred to as the caudal artery. Other projections include **neural prezygapophyses** and **postzygapophyses** on the dorsolateral margins of the vertebrae and **haemal prezygapophyses** and **postzygapophyses** on the ventrolateral margins (Fig. 4.2D).

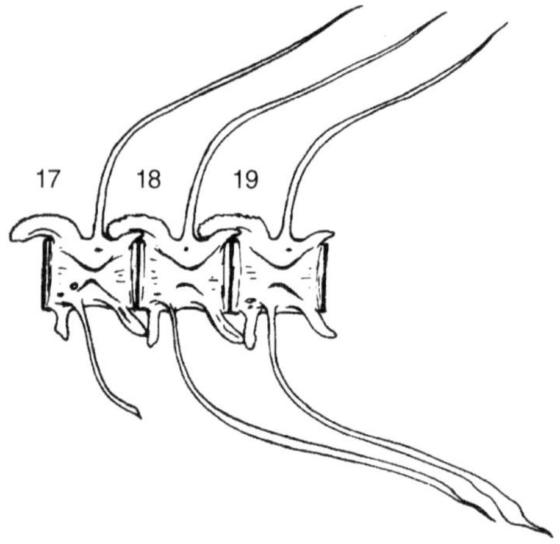

FIGURE 4.1 Junction of precaudal and caudal vertebrae in a left lateral view of the King Mackerel (*Scomberomorus cavalla*). The middle vertebra, with an elongate haemal spine, is the first caudal vertebra. Vertebrae numbered from the anterior. From Collette and Russo (1985b).

Ribs and Intermuscular Bones

Pleural ribs form in the peritoneal membrane and attach to the vertebrae, usually from the third vertebra to the last precaudal vertebra. They are distinct from **intermuscular bones** and serve to protect the viscera. Terminology used for these bones, and for ribs, was confused until Patterson and Johnson (1995) clarified the situation. Patterson and Johnson recognized three series of intermuscular bones: epineurals, epicentrals, and epipleurals. Primitively, ossified **epineurals** may be fused with the neural arches. Some are autogenous (unfused) and may develop an anteroventral branch as in characins (see Fig. 4.2D). Epineurals usually start on the first vertebra (sometimes on the back of the skull; see Fig. 3.6) and continue along the vertebrae well posterior to the ribs. **Epicentrals** lie in the horizontal septum and are primitively ligamentous. **Epipleurals** start medially and move anteriorly and posteriorly. They lie below the horizontal septum and are posteroventrally directed. Epicentrals and epipleurals have been lost in many advanced teleosts, leaving a series of short, straight epineurals lateral to the vertebral column and dorsal to the ribs. For these reasons, fillets taken from advanced teleosts such as perch and tuna contain fewer small bones than those from more primitive teleosts such as trout and herring.

Caudal Complex

The tail of a fish is a complex of vertebral centra, vertebral accessories, and fin rays that have been modified during evolution to propel the fish forward. The functional morphology of the fish tail and the history of its progressive change are discussed later in this chapter (Locomotory Types) and in Chapter 11 (Advanced Jawed Fishes I: Teleostomes). The teleostean caudal skeleton was largely neglected as a source of systematic characters until Monod (1968) surveyed the caudal skeleton of a broad range of teleosts and established a coherent and homogeneous terminology. Schultze and Arratia (1989) further showed the value of the caudal skeleton in the classification of teleosts. In primitive teleosts, a number of **hypurals** (enlarged haemal spines) support most of the branched **principal caudal fin rays** that form the caudal fin (Fig. 4.3). **Epurals** (modified neural spines) and the last haemal spine support the small spinelike **procurrent** caudal fin rays. In many advanced teleosts, the number of hypurals has been reduced to five. In some groups, such as atherinomorphs, sticklebacks, sculpins, the Louvar, tunas and mackerels, and flatfishes, the posterior vertebrae have been shortened, and some of the hypurals fuse to form a **hypural plate**. In scombrids, hypurals 3 and 4 are united into the upper part of the plate and hypurals 1 and 2 into the lower part (Fig. 4.4).

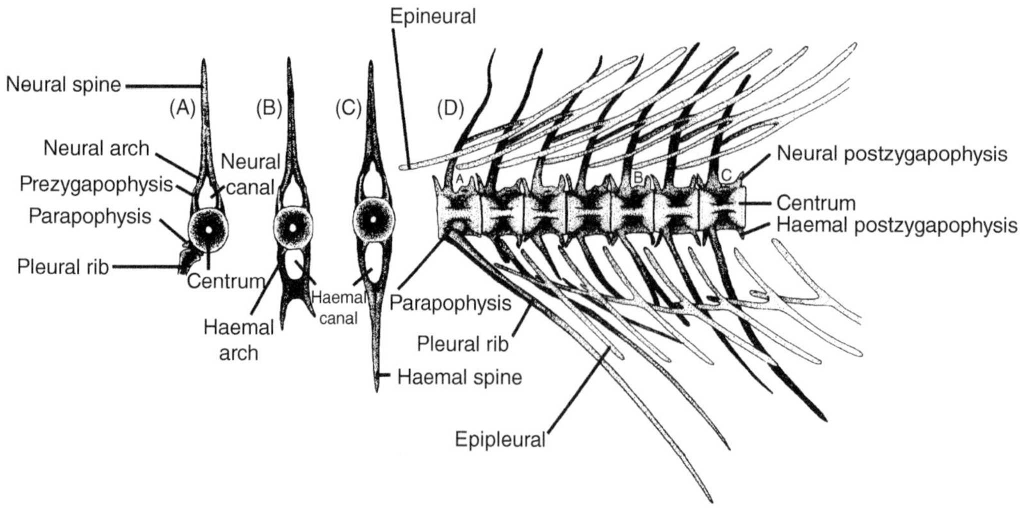

FIGURE 4.2 Representative precaudal and caudal vertebrae of generalized characin (*Brycon meeki*). (A) Anterior view of the 20th precaudal vertebra. (B) Anterior view of the 24th precaudal vertebra. (C) Anterior view of the second caudal vertebra. (D) Lateral view of the 20th precaudal through second caudal vertebrae. From Weitzman (1962).

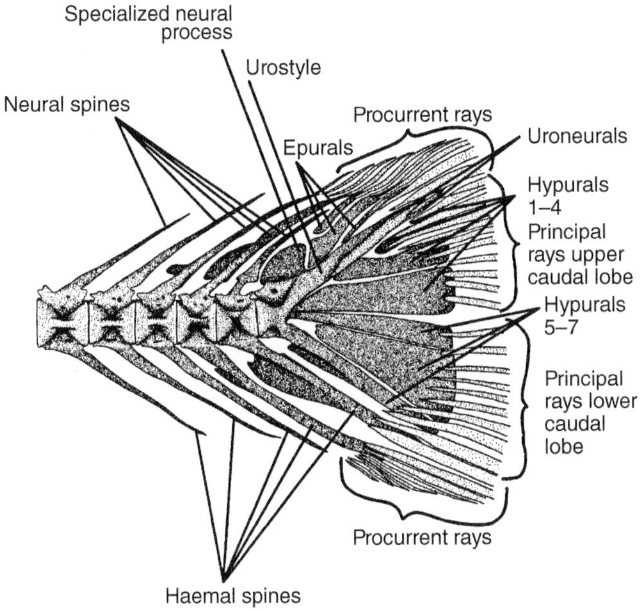

FIGURE 4.3 Posterior vertebrae and caudal complex of generalized characin (*Brycon meeki*). From Weitzman (1962).

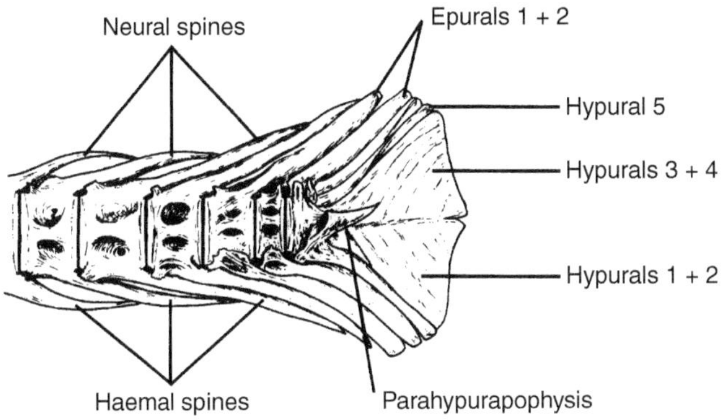

FIGURE 4.4 Caudal complex in left lateral view of a Spanish Mackerel (*Scomberomorus semifasciatus*). From Collette and Russo (1985b).

Caudal Fin Types

Caudal fins of fishes vary in both external shape and internal anatomy. The different types of caudal fins provide useful information about modes of swimming as well as about phylogeny. There are three basic types of fish tails, with an additional three types recognized for special groups of fishes.

The **protocercal** tail is the primitive undifferentiated caudal fin that extends around the posterior end in adult lancelets, agnathans, and larvae of more advanced fishes. In the **heterocercal**, or unequal-lobed, tail, the vertebral column extends out into the upper lobe of the tail. This type of tail is found in Chondrichthyes and primitive bony fishes such as sturgeons (Acipenseridae) and is still recognizable in gars (Lepisosteidae). *Amia*, the Bowfin, has what has been termed a hemihomocercal tail (Harder 1975), intermediate between heterocercal and homocercal, with external but not internal symmetry.

Most advanced bony fishes have a **homocercal**, or equal-lobed, tail (see Figs. 4.3, 4.4). In this type of tail, the caudal fin rays are arranged symmetrically and attached to a series of hypural bones posterior to the last vertebra that supports the caudal fin rays. These plates are ventral to the upward-directed urostyle, so this type of tail could be considered to be an abbreviated heterocercal tail.

The **leptocercal** (or diphycercal) tail resembles the protocercal in having the dorsal and anal rays joined with the caudal around the posterior part of the fish, but this is considered to have been secondarily derived, not primitive. This type of tail is found convergently in lungfishes (Dipnoi), coelacanths, rattails (Macrouridae), and many eel-like fishes.

The last vertebra of the **isocercal** tail, not the original urostyle, has been secondarily modified into a small flattened plate to which the caudal fin rays attach in the cods (Gadidae).

Ocean sunfishes (Molidae) have lost the posterior end of the vertebral column, including the hypural plate, i.e., they lack a true tail. A deep, abbreviated, caudalfin-like structure extends between the dorsal and anal fins and has been termed a **clavus** forming a **gephyrocercal** (or bridge) tail. By studying the ontogeny of the vertebral column and fins, Johnson and Britz (2005) have shown that the caudal fin is lost in molids and the clavus is formed by modified elements of the dorsal and anal fins. Because of this highly derived condition and other specialized osteological features, molids are considered to be the most advanced teleosts.

Appendicular Skeleton

Pectoral and pelvic girdles are primitively absent in the hagfishes and lampreys. Sharks have a coracoscapular cartilage that hangs more or less freely inside the body wall and has no attachment to the vertebral column. In rays, the pectoral girdle is attached to the fused anterior section of the vertebral column (synarchial condition) and also, by way of the propterygium of the pectoral girdle and antorbital cartilage, to the nasal capsules of the skull.

Pectoral Girdle

Unlike the condition in tetrapods, the **pectoral girdle** in bony fishes usually has no attachment to the vertebral column and instead attaches to the back of the skull via the posttemporal bone. Rather than dividing bones into cartilage and dermal, as done for the skull, it seems more practical to present the bones in sequence from the skull to the girdle bones themselves.

Three dermal bones are involved in the suspension of the pectoral girdle from the skull. The **posttemporal** usually has two anterior projections that attach to the epioccipital and intercalar bones on the back of the skull. The **extrascapular** (or supratemporal) is a thin tubular bone, sometimes two bones, that carry part of the lateral line canal onto the body.

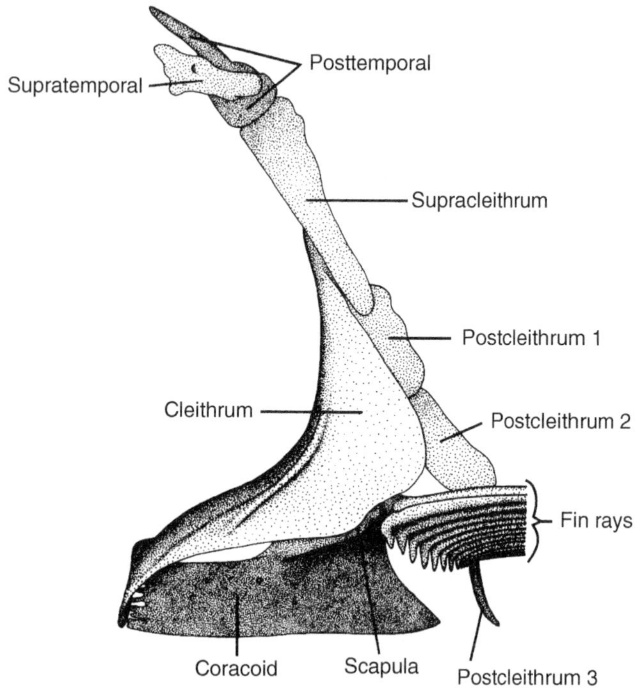

FIGURE 4.5 Left pectoral girdle of generalized characin (*Brycon meeki*). From Weitzman (1962).

They usually lie right under the skin dorsal to the posttemporal (Fig. 4.5). The **supracleithrum** is a heavy bone that lies between the posttemporal and the pectoral girdle.

The pectoral girdle is composed of two cartilage and one dermal bone in acanthopterygians. The dermal **cleithrum** is the largest, dorsalmost, and anteriormost element of the pectoral girdle. The **scapula** is a small bone, usually with a round scapular foramen, lying between the cleithrum and the radials. The **coracoid** is a long, thin bone that makes up the posterior part of the pectoral girdle and may support some of the pectoral fin radials. An additional element is found between the coracoid and cleithrum in many soft-rayed teleosts, the **mesocoracoid**. This bone is lost in spiny-rayed fishes as the pectoral fin moves up and assumes a vertical instead of oblique position.

The **actinosts** plus tiny **distal radials** are hourglass-shaped cartilage bones that connect to the pectoral fin rays. There are typically four in teleosts, attached to the coracoid and scapula.

Posterior and internal to the pectoral girdle are the dermal **postcleithra**. Soft-rayed teleosts typically have three: two are elongate and scalelike, and one is rodlike. Spiny-rayed teleosts typically have two: one scalelike, and the other more riblike.

Pelvic Girdle

The **pelvic girdle**, like the pectoral girdle, is usually not attached to the vertebral column in fishes as it is in tetrapods. In sharks, the pelvic girdle consists of the **ischiopubic cartilages** that float freely in the muscles of the posterior region of the body. In primitive bony fishes, there are paired pelvic bones, **basipterygia**, and radials to which the pelvic fin rays attach. In advanced bony fishes, both the pelvic bone itself and the radials are lost or fused so that the fin rays attach directly to the single remaining element, the basipterygium.

In soft-rayed teleosts, the pelvic fins are **abdominal** in position, ventrally located, and slightly anterior to the anal fin. The pelvic fins move forward to a thoracic position, directly below the pectoral fins, in spiny-rayed fishes. In some fishes (i.e., ophidiiform cusk-eels and gadiform cods), the pelvic fins lie anterior to the pectoral fins, a condition known as **jugular** pelvic fins. Jugular pelvic girdles may have attachments to the pectoral girdle.

Pelvic fin rays are frequently lost, and in some cases, such as eels (Anguilliformes), the South American needlefish *Belonion apodion*, and puffers, the pelvic girdle has also been lost.

Ligaments

Ligaments are nonelastic strands of fibrous connective tissue that serve to attach bones and/or cartilages to one another. Names of ligaments usually include their initial and terminal points. Some, however, are named after their shape or after persons. **Baudelot's ligament** is a strong white ligament that originates on the ventrolateral aspect of an anterior vertebra (usually the first) in lower teleosts or on the posterior part of the skull (usually the basioccipital) in advanced teleosts and inserts on the inner part of the cleithrum. Baudelot's ligament helps anchor the pectoral girdles to the sides of the fish.

Muscles

Fish muscle is structurally similar to that of other vertebrates, and fishes possess the same three kinds of muscles. **Smooth** muscle is nonskeletal, involuntary, and mostly associated with the gut but is also important in many organs and in the circulatory system. **Cardiac** muscle is nonskeletal but striated and is found only in the heart. **Skeletal** muscle is striated and comprises most of a fish's mass other than the skeleton.

Hagfishes and lampreys have a simple arrangement of striated skeletal muscles. These primitive fishes have no paired appendages to interrupt the body musculature. Skeletal muscle behind the head is uniformly segmental and is composed of shallow W-shaped **myomeres** (Fig. 4.6).

In jawed fishes, two major masses of skeletal muscle lie on each side of the fish, divided by the horizontal connective tissue septum. The **epaxial** muscles are the upper pair, and the **hypaxials** are the lower pair (see Fig. 4.6). A third, smaller, wedge-shaped mass of red muscle lies under the skin along the horizontal septum. This band of red muscle is poorly developed in most bony fishes although it is much more extensive, often more centrally located, and used for sustained swimming in fishes such as the tunas (see Chapter 7, Endothermic fishes).

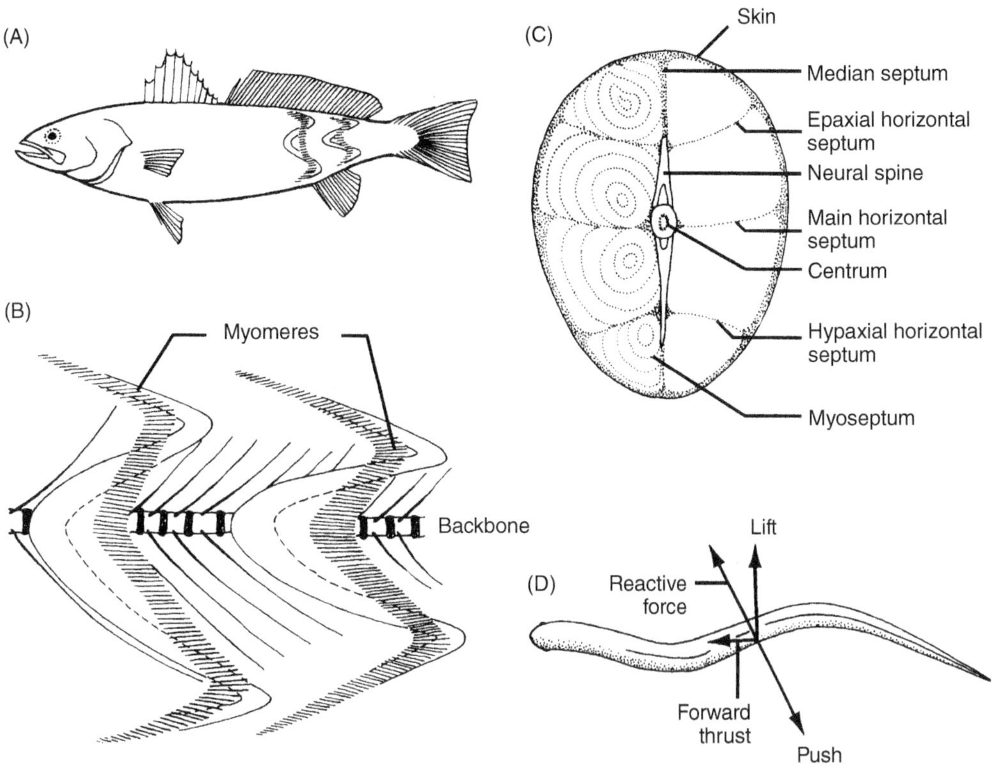

FIGURE 4.6 The anatomy of swimming in teleosts. (A) Lateral view of a Spotted Sea Trout, *Cynoscion nebulosus*, with the skin dissected away to show the location of two myomeres on the left side. (B) The same myomeres as they appear relative to the backbone in a sea trout. The hatched region is the part of the myomere located closest to the skin; the dashed line shows the interior portion of the myomere where it attaches to the vertebral column. The anterior and posterior surface of each myomere is covered by a myoseptum made of collagen fiber in a gel matrix, shown as a slightly thickened line. (C) Cross-section of a generalized teleost near the tail, showing the distribution of the various septa and their relationship to the backbone. Myosepta join to form median and horizontal septa. (D) How contractions produce swimming in a generalized fish (an eel is shown here). Progressive, tailward passage of a wave of contractions from the head to the tail push back on the water, generating forward thrust as one component of the reactive force. Sideways slippage (lift) is overcome by the inertia of the large surface area presented by the fish's head and body. Adapted from After Wainwright (1983) and Pough et al. (1989).

Faced with the conflicting demands of low-speed, economical cruising versus short bursts of maximum speed, fishes have solved the problem by dividing the locomotory system into two subsystems with different fiber types, white (glycolytic) and red (aerobic) (Bone 1978; Webb 1993). **White muscle** makes up the majority of the postcranial body of most fishes. It is used anaerobically in short-duration, burst swimming but fatigues quickly. White muscle is pale because its fibers lack myoglobin and because it has comparatively little vascularization and hence a limited oxygen supply. White muscle fibers have relatively few, small mitochondria, with energy resulting from anaerobic glycolysis. Muscle glycogen is depleted rapidly during contraction, producing large amounts of lactates that may require up to 12 h for full recovery after glycogen depletion (see Chapter 5, Gas Exchange and Transport). White muscle fibers are relatively large in diameter, up to 300 μm, and can therefore generate a great deal of power in a short period of time.

Red muscle usually forms a thin, lateral, superficial sheet under the skin between the epaxial and hypaxial muscle masses on each side of the fish. Red muscle is much better developed in muscles involved in sustained swimming, such as lateral red muscle in tunas and pectoral fin muscles in wrasses and parrotfishes. The red color is caused by abundant myoglobin and high blood volume (three times the number of capillaries of white muscle per unit weight). Mitochondria in red muscle are large and abundant and energy is supplied by the aerobic oxidation of fats. For these reasons, red muscle does not fatigue readily. During exercise, little change occurs in muscle glycogen or in the buildup of lactates; recovery after exercise using red muscle is rapid. In contrast with white muscle, red muscle has small-diameter fibers (18–75 μm). The strong taste of the prominent lateral red muscle in tunas (*chiai* in Japanese) leads to its being picked out from cooked tuna prior to canning for human consumption. It is, however, canned for cat food, which is why tuna cat food smells and tastes so strong.

Lamnid sharks and advanced tunas (tribe Thunnini) have more and deeper portions of red muscle than other fishes. A countercurrent heat exchanger system between the arterioles and venules of the cutaneous artery and vein ensures that the heat produced by muscular contraction remains in those tissues and is not carried off by the circulatory system to be lost at the gills (see Chapter 7, Endothermic Fishes). In at least the Atlantic Bluefin Tuna (*Thunnus thynnus*), this heat exchanger apparently helps regulate body temperature (Carey & Lawson 1973). Cross-sections of the body in representative scombrids show increasing development and internalization of the red muscles phylogenetically from mackerels to tunas (Sharp & Pirages 1978).

Red and white muscle fibers are extremes on a continuum that includes intermediate shades of pink muscle, which is intermediate in the other descriptive and metabolic qualities detailed earlier (Webb 1993). Like red muscle, pink muscle is used for sustained swimming and is recruited after red muscle but before white muscle. Another variation on muscle color and function occurs in the Antarctic notothenioid family Channichthyidae that lack hemoglobin, and typical red muscle, instead having yellow muscle in the heart and in the adductor and abductor muscles of the pectoral fin. The protein composition of yellow muscle is similar to that of normal red muscle in fishes with hemoglobin (Hamoir & Geradin-Otthiers 1980).

Muscles are arranged in pairs at the bases of the dorsal and anal fins: **protractors** erect the fins and **retractors** depress the fins. In addition, **lateral inclinators** function to bend the soft rays of the anal and second dorsal fins. For the paired fins, a single ventral **abductor** muscle pulls the fin ventrally and cranially. An opposing dorsal **adductor** muscle pulls the fin dorsally and caudally.

Only a few of the major muscles have been discussed here; see Stiassny (1999) and Winterbottom (1974) for a complete treatment.

Modified Muscles

Some fishes have specialized skeletal muscles that contract rapidly to create sound, and are therefore referred to as **sonic** muscles. Fish sonic muscles are the fastest muscles known among vertebrates (Parmentier et al. 2006). Based on their origins and insertions, there are two types of sonic muscles. Intrinsic sonic muscles completely attach to the wall of the gas bladder as in toadfishes (Batrachoididae, Fig. 4.7) and sea robins (Triglidae), whereas extrinsic sonic muscles have various origins and insertions, but generally, these paired muscles insert on the gas bladder or a neighboring structure. They are found in cusk-eels (Ophidiiformes), squirrelfishes (Holocentridae), and croakers (Sciaenidae).

Fishes in six different evolutionary lineages have developed **electric organs** from modified skeletal muscle cells that amplify the usual electrical production associated with muscle contractions. Caudal skeletal muscles, and sometimes lateral body muscles as well, are modified for electrogeneration in the rajid skates, mormyrid elephantfishes, gymnotiform knifefishes, and malapterurid electric catfishes. In torpedo rays (Torpedinidae, Narcinidae), hypobranchial muscles are involved, whereas an extrinsic eye muscle generates strong electrical discharges in the teleostean Electric Stargazer, *Astroscopus*.

FIGURE 4.7 Male Midshipman (*Porichthys*) attract females by rapid contraction of sonic muscles. Courtesy of C. Bauder.

The electric-generating cells of electric organs are referred to as **electrocytes**, and often are disklike modified muscle cells called **electroplaques**. When stimulated, ion flux across the cell membranes creates a small electric current, and because the cells are arranged in a column and discharge simultaneously, they produce an additive effect. A sizeable stack of cells can produce a considerable current – like many small batteries connected in series (Feng 1991). Although electrocytes of most electric fishes are modified muscle cells, South American electric fishes of the family Apteronotidae utilize modified neurons (Zupanc 2002).

The generation and detection of weak electric fields are particularly well developed in several groups of freshwater tropical fishes living in murky waters with poor visibility, such as the Gymnotiformes of South America and the Mormyridae of Africa. The electric organ discharge (EOD) of some species are brief pulses released at irregular intervals, whereas other species continuously produce oscillating, high-frequency waves of electricity (Zakon et al. 2002). The resulting electric field surrounds the fish, and any changes in the field are detected by the fish's tuberous organs (discussed in Chapter 6). Bending the body would distort the electric field, so these fishes typically rely on their extensive dorsal or anal fins for propulsion so that they can maintain a straight body posture.

The production of weak electric fields, as demonstrated by the gymnotids and mormyrids, requires considerable coordination by the central nervous system. In the South American gymnotid *Apteronotus*, the electric organs are controlled by pacemaker cells in the medulla, which are regulated by input from two clusters of neurons elsewhere in the brain (Zakon et al. 2002). The location and function of the pacemaker neurons of the African mormyrids are somewhat similar – a remarkable coincidence considering the two groups are believed to have evolved their EOD capabilities independently (Carlson 2002).

The African mochokid catfishes also are believed to produce and detect weak electric fields for object detection or communication (Hagedorn et al. 1990). The electric organ is located dorsally on these catfishes and has apparently evolved from one of the muscles associated with sound production, which occurs by stridulation of the pectoral spines.

Although most electric fishes generate only mild electric fields for communication and sensory purposes, others can generate currents strong enough to stun prey or ward off predators. The electric organs of an electric ray *Torpedo* (Torpedinidae) have about 45 columns of electrocytes (700 per column). The columns are oriented dorsoventrally, and the current is released dorsally because the dorsal surface of the organ and the overlying skin have lower resistance than the surrounding tissues. *Torpedo* can generate a discharge of 20–50 volts and several amps in seawater (Feng 1991) and stun prey 15 cm away (see Chapter 12). The Electric Eel *Electrophorus* (Electrophoridae, see Fig. 4.8), not a true eel but a close relative of the South American knifefishes, can generate pulses of 400 volts, or 1 amp (see Feng 1991), with its several electric organs, the largest of which consists of about 1000 electrocytes. These organs are embedded in the fish's lateral musculature. The two electric organs of the electric catfishes (Malapteruridae) are located on either side of the body, and each contains several million electrocytes. These organs generate a current of about 300 volts. Other fishes that emit strong electric currents include the stargazers (*Astroscopus*, Uranoscopidae), in which electroplaques are derived from ocular muscles.

Electric Eels are well known for their high-voltage outputs capable of deterring predators. This capability is not limited to fully immersed threats, however, as *Electrophorus* can thrust its body out of the water and deliver a powerful shock by making contact with the perceived threat (Catania 2017). *Electrophorus* also produces low-voltage pulses for sensing the surrounding environment, a trait found in other electric knifefishes (Gymnotidae). And *Electrophorus* also uses brief pulses of high-voltage outputs as a hunting strategy to activate motor neurons in prey from a distance, causing the prey to twitch and thus become detectable to the low-voltage sensory system, and also create temporary paralysis (Catania 2014). Small, immobilized prey can be quickly swallowed. Larger prey is grasped in the mouth and *Electrophorus* then curls its body, creating overlap in the electric field and more than doubling the voltage delivered to the prey, inducing tetany and fatiguing the prey's muscles, thus ensuring that the prey cannot struggle or escape while being manipulated and swallowed head-first (Catania 2015).

FIGURE 4.8 The Electric Eel, *Electrophorus electricus*, is closely related to the electrogenic gymnotid knifefishes of South America but are unique in their ability to produce strong electric pulses for defense and weak fields for electrolocation. Steven G. Johnson / Wikimedia Commons / CC BY-SA 3.0.

Integument

The **integument** is composed of the skin and skin derivatives and includes scales in fishes (and feathers and hair in birds and mammals). The integument forms an external protective structure parallel to the internal endoskeleton and serves as the boundary between the fish and the external environment. The structure of the skin in fishes is similar to that of other vertebrates, with two main layers: an outer epidermis and an inner dermis. See Elliott (2000) for a review of the integumentary system.

Epidermis

The epidermis is ectodermal in origin. In lampreys and higher vertebrates, the epidermis is stratified. The lowest layer is the **stratum germinativum**, composed of columnar cells. It is the generating layer that gives rise to new cells. In hagfishes, lampreys, and bony fishes, there is an outer thin film of noncellular dead cuticle (Whitear 1970). Males of some species display breeding tubercles, raised bumps often on the head or fins, and may also contain keratin, the fibrous protein that makes up much of the hair, feathers, and horns of mammals and birds (Wiley & Collette 1970).

The inner dermis contains blood vessels, nerves, sense organs, and connective tissue. It is derived from embryonic mesenchyme of mesodermal origin. It is composed of fibroelastic and nonelastic collagenous connective tissue with relatively few cells. Dermal layers include an upper, relatively thin layer of loose cells, the **stratum laxum** (or stratum spongiosum), and a lower, compact thick layer, the **stratum compactum** (Fig. 4.9). In adult fishes, the dermis is much thicker than the epidermis. The thickness of the integument depends on the thickness of the dermis. Scaleless species, such as catfishes of the genus *Ictalurus*, have relatively thick, leathery skin. The Ocean Sunfish (*Mola*) has the skin reinforced by a hard cartilage layer, 50–75 mm thick. Snailfishes (*Liparis*, Liparidae) have a transparent jellylike substance up to 25 mm thick in their dermis.

The chemical composition of fish skin is poorly studied, but some generalizations can be made. There is less water in fish skin than in fish muscle, a higher ash content, and similar amounts of protein. The main protein in skin is collagen, which is why fish skin has been used to manufacture glue. The chief minerals in fish skin are phosphorus, potassium, and calcium

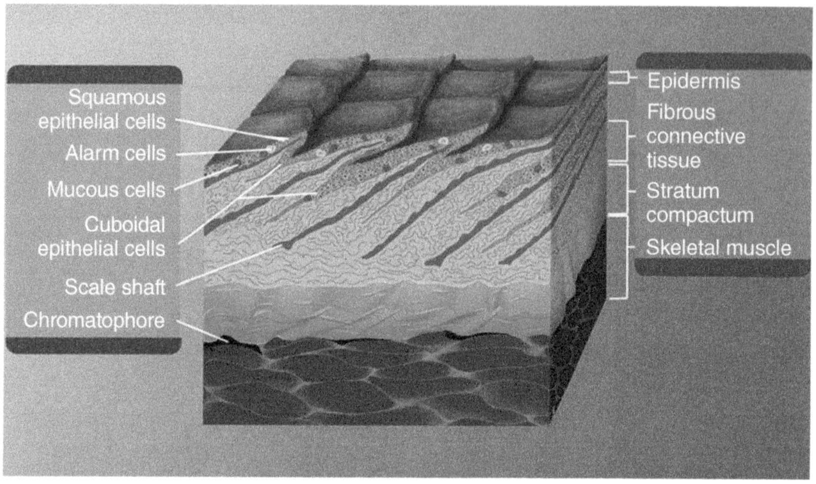

FIGURE 4.9 Structure of fish skin. Image courtesy of Stephanie Anne Hunczak.

(Van Oosten 1957). The ash composition of the skin of the Coho Salmon (*Oncorhynchus kisutch*) is:

P_2O_5	33%	CaO	14%
Cl	21%	Na_2O	9%
K_2O	17%	MgO	2%

Among the functions of the skin are mechanical protection and the production of mucus by epidermal mucous cells. **Mucin** is a glycoprotein, made up largely of albumin. Threads of mucin hold a large amount of water. It is possible to wring the water out of mucus, leaving threads of mucin. Among the first multicellular glands to evolve were the mucous glands of hagfishes (Myxinidae), called **thread cells** (Fernholm 1981). The oft-told story is that a hagfish + a bucket of water = a bucket of slime.

Other structures in the skin of fishes include epidermal venom glands associated with spines on fins (weeverfishes, Trachinidae; madtom catfishes, *Noturus*), opercles (venomous toadfishes, Thalassophryninae), and the tail (stingrays, Dasyatidae). **Photophores**, which produce bioluminescence, develop from the germinative layer of the epidermis. Color is due to **chromatophores**, which are modified dermal cells containing pigment. The skin also contains important receptors of physical and chemical stimuli.

Scales

Scales are the characteristic external covering of fishes. There are four basic types of scales.

1. **Placoid** scales are characteristic of the Chondrichthyes, although they are not as widespread in rays and chimaeras as they are in sharks. This type of scale has been called a "dermal denticle," but this is not accurate terminology because there are both epidermal and dermal portions, as in mammalian teeth. Each placoid scale consists of a flattened rectangular basal plate in the upper part of the dermis, from which a protruding spine projects posteriorly on the surface. The outer layer of the placoid scale is hard, enamel-like **vitrodentine**, derived from ectoderm. Vitrodentine is noncellular and has a very low organic content. The scale has a cup or cone of **dentine** with a pulp cavity richly supplied with blood capillaries, just as in mammalian teeth. Placoid scales do not increase in size with growth; instead, new scales are added between older scales. The teeth of elasmobranchs are evolutionary derivatives of placoid scales, and placoid scales are homologous with teeth in all vertebrates.
2. **Cosmoid** scales were present in fossil coelacanths and fossil lungfishes. The scales of recent lungfishes are highly modified by the loss of the dentine layer. Cosmoid scales are similar to placoid scales and probably arose from the fusion of placoid scales. Cosmoid scales are composed of two basal layers of bone: **isopedine**, which is the basal layer of dense lamellar bone, and cancellous (or spongy) bone, which is supplied with canals for blood vessels. Over the bone layers is a layer of cosmine, a noncellular dentinelike substance. Over the cosmoid layer is a thin superficial layer of vitrodentine. Growth is by addition of new lamellar bone underneath, not over the upper surface.
3. **Ganoid** scales were present in primitive fossil actinopterygian palaeoniscoids and are found in modern chondrosteans (sturgeons and paddlefishes), actinopterygian polypterids (bichirs), and holostean lepisosteids (gars). They are modified cosmoid scales, with the cosmine replaced by dentine and the surface vitrodentine replaced by **ganoine**, an inorganic bone salt secreted by the dermis. Ganoine is a calcified noncellular material without canals. Ganoid scales are usually rhomboidal in shape and have articulating peg and socket joints between them. The fossil palaeoniscoid scale is least modified in the bichirs, Polypteridae (three layers: ganoine, dentine, and isopedine). Ganoid scales are more modified in sturgeons

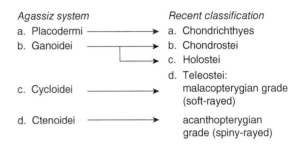

(Acipenseridae) and paddlefishes (Polyodontidae), in which lamellae of ganoine lie above a layer of isopedine. Sturgeon scales are modified into large plates, with most of the rest of the body naked.

Scales of gars (Lepisosteidae) are similar to Polypteridae in external appearance but are more similar to those of the Acipenseridae and Polyodontidae in structure. In the Bowfin (*Amia*) scales are greatly reduced in thickness to merely collagenous plates with bony particles, very similar to the cycloid scales of Teleostei.

4. **Leptoid** scales are completely dermal, bony, thin, and overlapping, which allows more flexibility in the fish's body. There are two main types: **cycloid** and **ctenoid**. There is no enamel-like layer except perhaps the **ctenii** (teeth on posterior border) and the most posterior and superficial ridges of the scale. These types of scales may have evolved from ganoid scales by loss of the ganoine and thinning of the bony dermal plate. Two major portions make up these scales: (i) a surface "bony" layer, which is an organic framework impregnated with salts, mainly calcium phosphate (as hydroxyapatite) and calcium carbonate; and (ii) a deeper fibrous layer, or **fibrillary plate**, composed largely of collagen.

Cycloid or ctenoid scales are present in the Teleostei, the vast majority of bony fishes. They have the advantage of overlapping like shingles on a roof, which gives great flexibility compared with cosmoid and ganoid scales. Small muscles pull unequally on the dermis, causing the anterior portion of the scale to become depressed in the dermis and covered over by the posterior margin of the preceding scale. Cycloid scales lack ctenii. Breeding tubercles and contact organs are present in many groups of fishes that lack ctenoid scales.

Including all scales with spines on their posterior margins under the term "ctenoid" is an oversimplification of the situation (Johnson 1984; Roberts 1993). Three different, general types of spined scales exist: (i) **crenate**, with simple marginal indentations and projections; (ii) **spinoid**, with spines continuous with the main body of the scales; and (iii) ctenoid, with ctenii formed as separate ossifications distinct from the main body of the scale (Roberts 1993). Crenate scales occur widely in the Elopomorpha and Clupeomorpha; spinoid scales occur widely in the Euteleostei; peripheral ctenoid scales (whole ctenii in one row) occur, probably independently, in the Ostariophysi, Paracanthopterygii, and Percomorpha; and transforming ctenoid scales (ctenii arising in two or three rows and transforming into truncated spines) are a synapomorphy of the Percomorpha.

As with fish skin, the chemical composition of scales is poorly known. About 41–84% is organic protein, mostly albuminoids such as collagen (24%) and ichthylepidin (76%). Up to 59% is bone, mostly $Ca_3(PO_4)_2$ and $CaCO_3$.

Phylogenetic Significance of Scale Types

Scales have been used as a taxonomic tool since the beginnings of systematic ichthyology (Roberts 1993). For example, Louis Agassiz divided fishes into four groups based on their scale type. More recent classifications are based on more characters but are similar to the system used by Agassiz.

Agassiz system		*Recent classification*
a. Placodermi	→	a. Chondrichthyes
b. Ganoidei	→	b. Chondrostei
	→	c. Holostei
		d. Teleostei:
c. Cycloidei	→	malacopterygian grade (soft-rayed)
d. Ctenoidei	→	acanthopterygian grade (spiny-rayed)

Whereas most groups of advanced acanthopterygian teleosts have ctenoid scales, some "ctenoid" groups may also have cycloid scales, and many species will have ctenoid scales on some parts of the body and cycloid scales on others. In the flatfishes, Pleuronectiformes, some species have ctenoid scales on the eyed side and cycloid scales on the blind side that is in contact with the bottom. Some flatfishes are sexually dimorphic, with males having ctenoid scales and females having cycloid scales.

Scale size varies greatly in fishes. Scales may be microscopic and embedded as in freshwater eels (Anguillidae), which led to their being classified as non-kosher because of the supposed absence of scales. Scales are small in mackerels (*Scomber*), "normal" in perches (*Perca*), large enough to be used for jewelry in Tarpon (*Megalops*), and huge (the size of the palm of a human hand) in the Indian Mahseer (*Tor tor*, a cyprinid gamefish reaching 43 kg in weight).

Development Pattern of Scales

In actinopterygian (ray-finned) fishes, scales usually develop first along the lateral line on the caudal peduncle, then in rows dorsal and ventral to the lateral line, and then spread anteriorly (see Chapter 9). The last regions to develop scales in ontogeny are the first to lose scales in phylogeny. Once the full complement of scales is attained in ontogeny, the number remains fixed. Therefore, the number of scales is a useful taxonomic character. Most scales remain in place for the life of the fish, which makes scales valuable in recording events in the life history of an individual fish, such as reduced growth that generally occurs during the winter or during the breeding season. Scales become deeply buried in the skin with age in the Swordfish, *Xiphias* (Govoni et al. 2004), leading some who wish to maintain a kosher diet to question if they can enjoy Swordfish, because kosher dietary laws require that a fish has both fins and scales.

Geographic variation can occur in the relative development of ctenoid scales in some species. For example, in the Swamp Darter (*Etheostoma fusiforme*) of the Atlantic Coastal Plain of the United States, the number of scales in the interorbital area increases from north to south (Collette 1962). In northern parts of the range, the few scales present are embedded and cycloid. Further south, there is an increase in number, and more scales have the posterior surface of the scale projecting through the epidermis, and these scales have more ctenii on them.

Modifications of Scales **Lateral line scales** have pores that allow vibrations in the surrounding water to penetrate into the canal beneath the scales and affect the neuromasts that are displaced by sound waves and send nerve impulses to the brain (see Chapter 6, Mechanoreception). Most fishes have **complete** lateral lines, that is, pored scales extend from behind the opercular region all the way to the base of the caudal fin. Some species, such as the Swamp Darter, have **incomplete** lateral lines, with the pores extending only part way to the caudal base. Other patterns include **disjunct** where there is an interruption between the upper and lower portions of the lateral line, as in most members of the large family Cichlidae, **multiple** with several lateral lines, and **absent**, where the lateral line is missing on the body (Webb 1989).

Some fishes have scales that are easily shed (referred to as **deciduous**). This is true of many species of herrings (Clupeidae) and anchovies (Engraulidae). It may be true of one species in a genus but not of another. For example, of two species of common Australian halfbeaks or garfishes (Hemiramphidae), scales remain in the River Garfish, *Hyporhamphus regularis*, but are easily lost in the Sea Garfish, *H. australis*.

Male darters of the genus *Percina* have **caducous scales**, a single row of enlarged scales along the ventral surface between the pelvic fins and the anus. Several structures in chondrichthyans may have arisen from fusions of modified placoid scales. These include the "spines" at the beginning of the first and second dorsal fins of the Spiny Dogfish, the prominent dorsal "spine" in some chimaeras (Holocephali), the caudal fin spine or barb in stingrays (Dasyatidae), and the teeth on the rostrum of sawfishes (*Pristis*).

As mentioned earlier, the structure of placoid scales in Chondrichthyes is the same as the structure of teeth in vertebrates, leading to the question: Which came first? Did some primitive chondrichthyan ancestor develop teeth that then spread over the body? Or did the ancestor first develop scales that then spread into the mouth and became modified into teeth? Apparently, the dermal armor of the earliest known vertebrates, the ostracoderms, broke up into smaller units, and some of these scales in the mouth evolved into teeth (Walker & Liem 1994).

In many teleosts, there is an external dermal skeleton in addition to the internal supporting skeleton. This is composed of segmented bony plates in pipefishes (Syngnathidae) and poachers (Agonidae) and bony shields similar to placoid scales with vitrodentine in several South American armored catfish families such as the Loricariidae. The body is enclosed in a bony cuirass (armor) in the shrimpfishes (Centriscidae) and is completely enclosed in a rigid bony box in the trunkfishes (Ostraciidae).

Many fishes have protective scutes or spines. The ventral row of scales is modified into **scutes** with sharp, posteriorly directed points in herrings, such as the river herrings (*Alosa*) and the threadfins (*Harengula*). Some jacks (Carangidae) have lateral scutes along the posterior part of the lateral line. Sticklebacks (Gasterosteidae) have bony lateral plates. These plates vary in number and size in *Gasterosteus aculeatus*, roughly correlated with the salinity of the habitat and the presence or absence of predators. Sharp erectable spines derived from scales are present in porcupine fishes (Diodontidae). Large bony "warts" characterize lumpfishes (*Cyclopterus*). Surgeonfishes (Acanthuridae) are so named because of the pair of sharp, anteriorly directed spines on the caudal peduncle.

Other modifications of scales are discussed elsewhere. Lepidotrichia, fin rays supporting the fins, probably originated from scales, and the superficial bones of the skull originated as scales and have become modified into dermal bones.

Scale Morphology in Taxonomy and Life History

For studying taxonomy and life history, various parts of the scale are distinguished. Cycloid and ctenoid scales can be divided into four fields (Fig. 4.10): anterior (which is frequently embedded under the preceding scale), posterior, dorsal, and ventral. The focus is the area where scale growth begins. The position and shape of the focus may vary, being oval, circular, rectangular, or triangular. Radially arranged straight lines called radii might extend across any of the fields. A **primary radius** extends from the focus to the margin of the scale. A **secondary radius** does not extend all the way out to the margin of the scale. Radii may be present in different fields: only anterior, as in pickerels (*Esox*); only posterior, as in shiners (*Notropis*); anterior and posterior, as in suckers (Catostomidae); or even in all four fields, as in barbs (*Barbus*). Ctenii may occur in a single marginal row or in two or more rows located on the posterior field.

Circuli are growth rings around the scale. Life history studies, particularly those dealing with age and growth, utilize such growth rings. This is especially useful in temperate waters where growth of body and scales slows significantly in fall and winter, causing the spacing between the circuli to decrease and thus leaving a band on the scales called an **annulus**. However, interpreting such marks as annuli requires caution because any retardation in growth may leave a mark. The stress of spawning, movement from fresh to salt water, parasitism, injury, pollution, and sharp and prolonged change in temperature may all leave marks on the scales similar to annuli. Scales grow in a direct relationship with body growth, making it possible to measure the distance between annuli and back calculate the age at different body sizes. Other hard structures also show growth changes (see Chapter 9, Age and Growth) and can be used for aging, such as fin spines, otoliths, and various bones such as opercles and vertebrae (DeVries & Frie 1996).

Scale morphology can also be useful in the identification of fragments such as scales found in archeological kitchen middens or in stomach contents. An example of the latter is Lagler's (1947) key to the scales of Great Lakes families. Scale morphology is also useful in classification, as shown by McCully's (1962) study of serranid fishes, Hughes's (1981) paper on flatheads, Johnson's (1984) review of percoids, Coburn and Gaglione's (1992) study of percids, and Roberts's (1993) analysis of spined scales in the Teleostei.

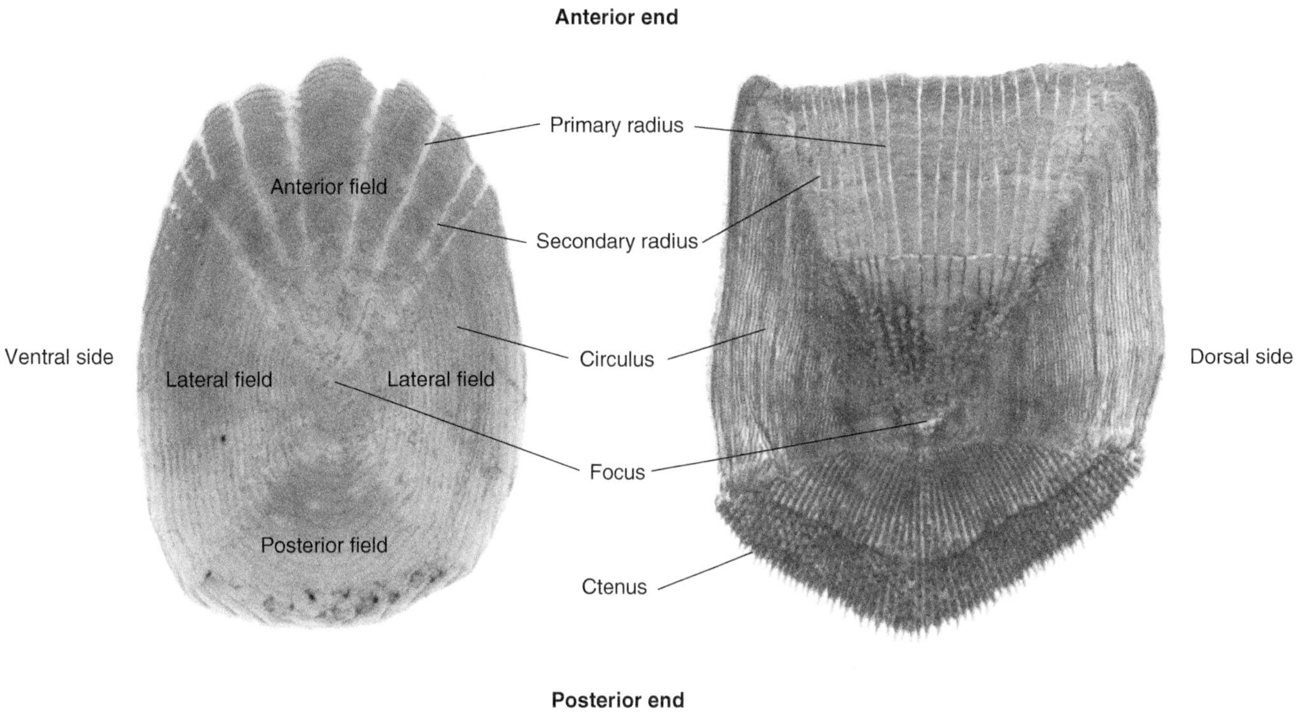

FIGURE 4.10 A cycloid scale (left) and a ctenoid scale (right). Note that the anterior field (upward in photograph) and lateral fields would be covered by other scales on the fish. The posterior field (exposed) is down in the photos. Courtesy of Z. Bräger.

Median Fins

The median or unpaired fins consist of the dorsal, anal, and adipose fins along with the dorsal and ventral profiles of the fish. In jawless fishes, cartilaginous rods support the median fins. In Chondrichthyes, the median fins are supported by **ceratotrichia**, fin rays composed of elastin and supported by dermal cells. Below the ceratotrichia are three layers of **radials** – rodlike cartilages that support the fin rays and extend inward toward the vertebral column. If a spine is present at the anterior end of a median fin in a chondrichthyan such as the Spiny Dogfish (*Squalus acanthias*), it is not a true spine such as is found in spiny-rayed fishes (Acanthopterygii) but is rather a fusion of radials.

In bony fishes, the ceratotrichia are replaced during ontogeny by **lepidotrichia**, bony supporting elements that are derived from scales. Ceratotrichia are present in lungfishes and larval actinopterygians. Primitive actinopterygians such as the Bowfin (*Amia*) still have three radials supporting each median fin ray, but these are reduced to two and then one in advanced teleosts. The remaining element is then known as an **interneural bone** if it is under the dorsal fins or **interhaemal bone** if it is above the anal fin.

Primitive soft-rayed teleosts have a single dorsal fin that is composed entirely of soft rays. Advanced teleosts usually have two dorsal fins, with the anterior one (**first dorsal fin**) composed of spines and the posterior one (**second dorsal fin**) composed largely of soft rays, although there may be one spine at the anterior margin of the fin. Some soft-rayed fishes such as the Common Carp, the Goldfish, and catfishes may have a single spine at the anterior end of the dorsal fin, but this is a bundle of fused rays, not a true spine.

True **spines** differ from soft **rays** in several characters. Spines are usually hard and pointed, unsegmented, unbranched, solid, and bilateral (with left and right halves). In contrast, rays are usually soft, not pointed, segmented, and often branched.

Some fast-swimming fishes, such as the mackerels and tunas, may have a series of **dorsal finlets**, small fins with one soft ray each, following the second dorsal fin.

Several groups of soft-rayed fishes have an additional fin posterior to the dorsal fin, the **adipose fin**, which varies greatly in size among different fishes. "Adipose" is a poor term for this fin because it is rarely fatty. The adipose fin usually lacks lepidotrichia and is supported only by ceratotrichia, although some catfishes have secondarily developed a spine, composed of fused rays, at its anterior margin. The function or functions of adipose fins remain something of a mystery, but their presence is useful in identifying members of five groups that usually have them: characins (Characiformes), catfishes (Siluriformes), trouts and salmons (Salmoniformes), lanternfishes and relatives (Myctophiformes), and trout-perches (Percopsidae).

The original function of the dorsal fin was as a stabilizer during swimming, but it has been modified in many different ways. It has been reduced or lost in rays (Batoidei) and South American knifefishes (Gymnotiformes). The dorsal and anal fins become **confluent**, joined with the caudal fin, around the posterior part of the body in many eels (Anguilliformes). The individual spines in the first dorsal fin have become shortened in fishes such as the Bluefish (*Pomatomus saltatrix*) and the Cobia (*Rachycentron canadum*). The first dorsal fin has been converted into a suction disk in the remoras (Echeneidae). The membranes between the spines have lost their attachment to each other in the bichirs (Polypteridae) and sticklebacks (Gasterosteidae). Venom glands have become associated with dorsal fin spines, and other spines, in fishes such as stonefishes (*Synanceia*), the weeverfishes (Trachinidae), and venomous toadfishes (Thalassophryninae). The spiny dorsal fin has been converted into a locking mechanism in the triggerfishes (Balistidae). It is depressible into a groove during fast swimming in the tunas (Scombridae). Perhaps the most extreme modification of a dorsal fin is the conversion of the first dorsal spine into an **ilicium**, or fishing rod, with an **esca**, or lure, at its tip in the anglerfishes (Lophiiformes).

The **anal fin** usually lies just posterior to the anus. In soft-rayed fishes, it is composed entirely of soft rays, as is the single dorsal fin of these fishes. In spiny-rayed fishes, the anal fin usually contains one or several anterior spines, followed by soft rays. Fast-swimming fishes that have dorsal finlets usually also have **anal finlets**, small individual fins following the anal fin.

The anal fin shows the least variation among fishes. It has been lost in the ribbonfishes (Trachipteridae). It is very long and serves as the primary locomotory fin in South American knifefishes (Gymnotiformes) and Afro-Asian featherfins (Notopteridae). The anterior part of the anal fin has been modified into a **gonopodium** for spermatophore transfer in male livebearers (Poeciliidae). It is also variously modified into what has been called an **andropodium** in males of *Zenarchopterus* and several related internally fertilizing genera of halfbeaks (Zenarchopteridae).

Locomotion: Movement and Shape

Body shape and locomotory behavior in fishes have evolved in response to the extreme density of water. Whereas locomotory adaptations in terrestrial and flying animals strongly reflect a need to overcome gravity, body and appendage shape in fishes reflects little influence of gravity because gas bladders or lipid-containing structures make most fishes neutrally buoyant (see Chapter 5, The Gas Bladder and Buoyancy Regulation). Fish locomotion is more constrained by the density of water and the drag exerted by it (Videler 1993).

Water is about 800 times more dense and 50 times more viscous than air. Locomotion through this dense, viscous medium is energetically expensive, a problem exacerbated by the 95% reduction in oxygen-carrying capacity of water as compared to air (see Chapter 5, Aquatic Breathing). The chief cause of added energetic cost is drag, which has two components, **viscous or frictional drag** involving friction between the fish's body and the surrounding water, and **inertial or pressure drag** caused by pressure differences that result from displacement of water as the fish moves through it. Viscous drag is not affected greatly by speed but more by the smoothness of a surface and by the amount of surface area, which is linked to body and fin shape; production of mucus reduces viscous drag. Inertial drag increases with speed and is therefore also intimately linked to body shape. Most fast-swimming fishes have a classic, streamlined shape that minimizes both inertial and viscous drag. A streamlined body is round in cross-section and has a maximum width equal to 25% of its length. The width : length ratio is 0.26 in some pelagic sharks, 0.24 in swordfish, and 0.28 in tunas. The thickest portion of a streamlined body occurs about two-fifths of the way back from the anterior end, another rule followed by large pelagic predators. Interestingly, these same streamlined fishes are also slightly negatively buoyant and hence sink if they cease swimming. They often have winglike pectoral fins that are extended laterally at a positive attack angle, thus generating lift (except sharks, see below). They minimize drag by retracting paired and median fins into depressions or even grooves in the body surface; a sailfish houses its greatly expanded dorsal fin "sail" in a groove on its dorsal surface during fast swimming (Hertel 1966; Hildebrand 1982; Pough et al. 2012).

Most fishes swim by contracting a series of muscles on one side of the body and relaxing muscles on the other. The muscle blocks, called **myomeres**, attach to collagenous septa, which in turn attach to the backbone and skin (Fig. 4.6). Depending on the swimming form involved (see below), contractions may progress from the head to the tail or occur on one side and then the other. The result of the contractions is that the fish's body segments push back on the water. Given Newton's Third Law of Motion concerning equal and opposite forces, this pushing back produces an opposite reactive force that thrusts the fish forward. Forward thrust results from combined forces pushing forward and laterally; the lateral component is canceled by a rigid head and by median fins and in some cases by a deep body that resists lateral displacement.

Locomotory Types

A general classification of swimming modes or types among fishes has been developed, building on the work of Breder (1926), Gray (1968), Lindsey (1978), and Webb (1984; Webb & Blake 1985). The chief characteristics of the different types are how much of and which parts of the body are involved in propulsion and whether the body or the fins undulate or oscillate. **Undulation** involves sinusoidal waves passing down the body or a fin or fins; **oscillation** involves a structure that moves back and forth. About one dozen general types are recognized: anguilliform, subcarangiform, carangiform, modified carangiform (thunniform), ostraciiform, tetraodontiform, balistiform, rajiform, amiiform, gymnotiform, and labriform; some of these are additionally subdivided. The names apply to the basic swimming mode of particular orders and families, although unrelated taxa may display the same mode, and many fish use different modes at different velocities.

The first four types involve sinusoidal undulations of the body and form a continuum regarding how much of the body undulates. **Anguilliform** swimming, seen in most eels, dogfishes, other elongate sharks, and many larvae occurs in fishes with very flexible bodies that are bent into at least one-half of a sine wave when photographed from the dorsal view (Table 4.1). All but the head contributes to the propulsive force (Muller et al. 2001). As a wave proceeds posteriorly it increases in amplitude. The speed (frequency) of the wave remains constant as it passes down the body and always exceeds the speed of forward movement of the fish because of drag and because of energy lost to reactive forces that are not directed forward. To swim faster, faster waves must be produced. Anguilliform swimmers are comparatively slow because of their relatively long bodies and involvement of anterior regions in propulsion; the same segments that push back on the water also push laterally and create drag because water pushes on these bent sections as the fish moves forward. Anguilliform swimming has its compensating advantages, including a greater ability to move through dense vegetation and sediments and to swim backwards (D'Aout & Aerts 1999). Anguilliform swimming in larval fishes, including such species as herrings that use carangiform swimming as adults, probably occurs because the skeleton of early larvae is unossified, and the fish is exceedingly flexible and anatomically constrained from employing other modes (see Chapter 9, Larval Behavior and Physiology).

To get around the self-braking inherent during anguilliform swimming, faster swimming fishes involve only posterior segments of the body in wave generation, using ligaments to transfer force from anterior body musculature to the caudal region. The progression of types from **subcarangiform** (trout, cod) through **carangiform** (Fig. 4.11, jacks, herrings) to **modified carangiform** or **thunniform** (mackerel sharks, billfishes, tunas) entails increasing involvement of the tail and decreasing involvement of the anterior body in swimming. One major advance in the carangiform and thunniform swimmers is the existence of a functional hinge that connects the tail to the caudal peduncle. This hinged coupling allows the fish to maintain its tail at an ideal attack angle of 10–20° through much of the power stroke. In anguilliform and subcarangiform swimmers, this angle changes constantly as the tail sweeps back and forth, producing less thrust at low angles and creating more drag at greater angles.

Thunniform swimmers also typically have a tail that originates from a narrow caudal peduncle that is often dorsoventrally depressed and may even have lateral keels that streamline it during side-to-side motion. This creates an overall more streamlined shape to the body and also reduces viscous drag and lateral resistance in a region of the body where they tend to be highest. The tail itself is stiff and sickle-shaped with a **high aspect ratio**, (tall, yet very narrow) which minimizes drag during rapid oscillation and is thus ideal for sustained swimming. The shape reduces viscous drag by reducing surface area and reduces inertial drag by having pointed tips that produce minimal vortices, which would increase drag. The efficiency of the system is increased by tendons that run around joints in the peduncle region and insert on the tail, the joints serving as pulleys that increase the pulling power of the muscle–tendon network. The thunniform mode of propulsion, involving a streamlined shape, narrow and keeled peduncle, and high aspect ratio tail, has evolved convergently in several fast-swimming, pelagic predators, including mackerel sharks, tunas, and billfishes, as well as porpoises and dolphins and the extinct reptilian ichthyosaurs (Chapter 10, The Open Sea). The fish and mammalian groups are also endothermic to some degree (Lighthill 1969; Lindsey 1978; Pough et al. 2012). Higher speed, sustained swimming in the mackerel sharks and tunas is also made possible by the large masses of red muscle along the fish's sides (explained earlier in this chapter). Location of the red muscle close to the fish's spine allows the body to remain fairly rigid and also permits the retention of heat generated by muscle contraction. Hence thunniform swimming and endothermy (see Chapter 7) are tightly linked.

Low aspect ratio tails are broad and flexible, such as those found in subcarangiform minnows, salmons, pikes, cods, and barracudas. This shape is better suited for rapid acceleration from a dead start and can also aid during hovering by passing undulations down their posterior edge. Intrinsic muscles associated with the tail in low aspect ratio species help control its shape. Rainbow Trout are able to increase the depth and hence produce a higher aspect ratio tail during high-speed swimming. **Fast start** predators, such as gars, pikes, and barracudas, hover in the water column and then dart rapidly at prey. These unrelated fishes have converged on a body shape that concentrates the propulsive elements in the posterior portion of the body: the dorsal and anal fins are large and placed far to the posterior, the caudal peduncle is deep, and the tail has a relatively low aspect ratio. Maximum thrust from a high-amplitude wave concentrated in the tail region allows for rapid acceleration from a standing start (see Fig. 4.12).

Ostraciiform swimming, as seen in boxfishes and torpedo rays, is extreme in that only the tail is moved back and forth while the body is held rigid; the side-to-side movement of the tail is more an oscillation than an undulation. In the weakly electric elephantfishes, body muscles pull on tendons that run back around bones in the caudal peduncle region and insert on the tail, causing the fish to swim with jerky tail beats. Such an arrangement is thunniform in anatomy but more ostraciiform in function. Weakly electric fishes, such as the elephantfishes and South American knifefishes mentioned below, often have devices for keeping their bodies straight while swimming. This relative inflexibility probably minimizes distortion of the electric field they create around themselves (see Chapter 6, Electroreception).

Ostraciid boxfishes carry this type of swimming to its extreme, having a rigid dermal covering that extends back to the peduncle area. Although a rigid, boxlike body propelled by a caudal fin seems an ungainly, even unlikely, means of getting around a coral reef, these active swimmers are elegantly constructed for dealing with water that flows past their bodies as they move or encounter currents. The flat surfaces and angular shelves and projections of the boxlike carapace generate vortices that counteract pitching and yawing that result from water flow, without active correction by fins, tail, or gas

TABLE 4.1 **Form, function, and locomotion in fishes. About 12 generalized types of swimming are recognized among fishes. The body part or fin providing propulsion is indicated by cross-hatching; the density of shading denotes relative contribution to propulsion. These locomotory patterns correlate strongly with body shape, habitat, feeding ecology, and social behavior. Convergence among unrelated fishes in terms of body morphology, swimming, and ecology demonstrates the evolutionary interplay of form and function.**

	Swimming type					
	Via trunk and tail			Via fins		
		Via tall				
	Anguilliform	Subcarangiform[a] Carangiform Thunniform	Ostraciiform	Tetraodontiform Balistiform Diodontiform[b]	Rajiform[b] Amiiform Gymnotiform	Labriform[c]
	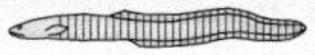				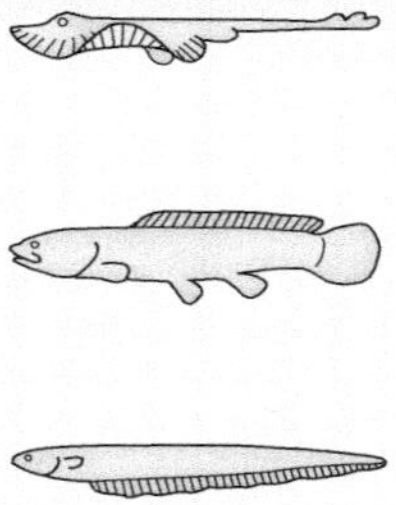	
Representative taxa	eels, some sharks, many larvae	salmon, jacks, mako shark, tuna	boxfish, mormyrs, torpedo ray	triggerfish, ocean sunfish, porcupinefish	rays, Bowfin, knifefishes	wrasses, surfperch
Propulsive force	Most of body	Posterior half of body	Caudal region	Median fin(s)	Pectorals, median fins	Pectoral fins
Propulsive form	Undulation	Undulation	Oscillation	Oscillation[d]	Undulation	Oscillation
Wavelength	0.5 to >1 wavelength	<1 (usually <0.5) wavelength			>>1 wavelength	
Maximum speed bl/s	Slow-moderate 2	Very fast–moderate 10–20	Slow?	Slow?	Slow to moderate 0.5	Slow 4
Body shape: lateral view cross-section	Elongate Round	Fusiform Round	Variable	Variable Often deep	Elongate Often flat	Variable
Caudal fin aspect ratio	Small Medium to low	Medium to large Low to high	Large Low	Small to medium Low	Variable Low	Large Low
Habitat	Benthic or suprabenthic	Pelagic, wc, schooling	Variable	wc	Suprabenthic	Structure associated

bl/s, body lengths per second attainable; wc, up in water column.

[a] In subcarangiform types (salmons, cods), the posterior half of the body is used, carangiform swimmers (jacks, herrings) use the posterior third, and thunniform or modified carangiform swimmers (tunas, mako sharks) use mostly the caudal peduncle and tail (see text).

[b] Rajiforms (skates, rays) swim with undulating pectoral fins, amiiforms (Bowfin) undulate the dorsal fin, and gymnotiform swimmers (South American knifefishes, featherfins) undulate the anal fin.

[c] Labriform swimmers use the pectorals for slow swimming but use the subcarangiform or carangiform mode for last swimming.

[d] Balistiform and diodontiform swimming are intermediate between oscillation and undulation; porcupinefishes also use their pectoral fins.

Adapted from Lindsey (1978), Beamish (1978), Webb and Blake (1985), and Pough et al. (2001).

FIGURE 4.11 The Greater Amberjack, *Seriola dumerili* (Carangidae), exhibits carangiform swimming. The back third to half of the body oscillates back and forth while the forward part of the streamlined body stays rigid, thereby minimizing turbulence. G. Helfman (Author).

FIGURE 4.12 Fast start predators, such as pike, gars, and the Great Barracuda (*Sphyraena barracuda*) shown here, typically possess long bodies, pointed heads, a broad low-aspect-ratio tail, and median fins placed relatively far back on the body. All of these traits contribute to their ability to dart rapidly at prey that they then grasp in tooth-studded jaws. G. Helfman (Author).

bladder. Hence, the "unfishlike" morphological features of the boxfish's body contribute to hydrodynamic stability (Bartol et al. 2005) (Fig. 4.13).

The last five swimming types employ median and paired fins rather than body–tail couplings. **Tetraodontiform** and **balistiform** swimmers (triggerfishes, ocean sunfishes) flap their dorsal and anal fins synchronously; their narrow-based, long, pointed fins function like wings and generate lift (forward thrust) continuously, not just during half of each oscillation. **Rajiform** swimmers hover and move slowly via multiple undulations that pass backwards or forwards along with the pectoral fins of skates and rays; in **amiiform** swimmers, undulations pass along the dorsal fin (Bowfin, African osteoglossomorph *Gymnarchus*, seahorses), whereas in **gymnotiform** swimming, undulations pass along the anal fin (South American and African knifefishes or featherfins). Rajiform and related swimming modes are slow but allow for precise hovering, maneuvering, and backing. The frequency with which waves pass along a fin can be very high, reaching 70 Hz in the dorsal fin of seahorses. **Labriform** swimmers (chimaeras, surfperches, wrasses, parrotfishes, surgeonfishes) row their pectoral fins, pushing back with the broad blade, then feathering it in the recovery phase. As some negative lift is generated during the recovery phase, these fish often give the impression of bouncing slightly as they move through the water. If rapid acceleration or sustained fast swimming is needed, labriform swimmers, as well as many other fin-based locomotors, shift to tail-based, carangiform locomotion.

Three final aspects of locomotory types deserve mention. First, the distinctiveness of the different locomotory types suggests that they are specializations, and specialization for one function usually produces compromises in other functions. Fishes that specialize in efficient slow swimming or precise maneuvering usually employ undulating or oscillating median fins. The long fin bases necessary for such propulsion (e.g., Bowfin, knifefishes, pipefishes, cutlassfishes) require a long body, which evolves at a cost in high-speed, steady swimming. Low-speed maneuverability can also be achieved with a highly compressed (laterally flattened), short body that facilitates pivoting, as found in many fishes that live in geometrically complex environments such as coral reefs or vegetation beds (e.g., freshwater sunfishes, angelfishes, butterflyfishes, cichlids, surfperches, rabbitfishes; see Drucker & Lauder 2001). These fishes typically have expanded median and paired fins that are distributed around the center of mass of the body and can act independently to achieve precise, transient thrusts, a useful ability when feeding on attached algae or on invertebrates that are hiding in cracks and crevices.

But a short, compressed body means reduced muscle mass and poor streamlining, whereas large fins increase drag. Again, such fishes achieve maneuverability but sacrifice rapid starts and sustained cruising. Relatively poor fast-start performance may be compensated for by deep bodies and stiff spines, which make these fishes difficult to swallow (see Chapter 16, Discouraging Capture and Handling); they also typically live close to shelter. At the other extreme, thunniform swimmers have streamlined bodies, large anterior muscle masses, and stiff pectoral and caudal fins that are extremely hydrodynamic foils. They trade-off exceptional cruising ability against an inability to maneuver at slow speeds. Although specialists among body types can be identified, optimal design for one trait – sustained cruising, rapid acceleration, or maneuverability – tends to reduce ability in the other traits. Because most fishes must cruise to get from place to place, must accelerate and maneuver to eat and avoid being eaten, many are locomotory generalists and lack highly specific, and therefore constraining, locomotory adaptations.

Second, this generalist strategy means that few fishes use only one swimming mode. Many fishes switch between modes depending on whether fast or slow swimming or hovering is needed. In addition, most fishes have median fins that can be erected or depressed, adding a dynamic quality to their locomotion. A Largemouth Bass can erect its first dorsal and anal fins to gain thrust during a fast-start attack, then depress these fins while chasing a prey fish to reduce drag, then erect them to aid in rapid maneuvering. Most groups, with the exception of the thunniform swimmers, are capable of hovering in midwater by sculling with their pectoral fins or by passing waves

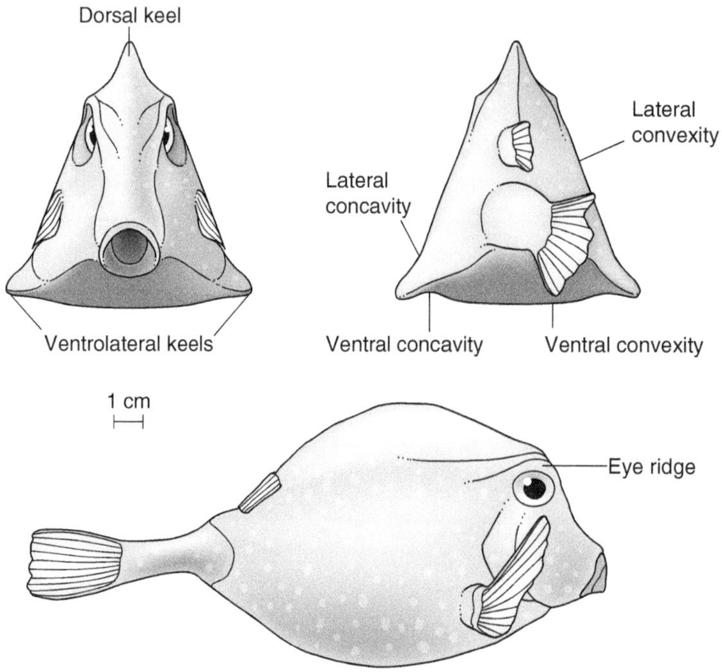

FIGURE 4.13 Anterior, posterior, and lateral views of a Smooth Trunkfish *Lactophrys triqueter*, showing its unusual body shape and protrusions, all which aid in hydrodynamics. After Bartol et al. (2003).

vertically along the caudal fin. When hovering, some forward thrust is generated by water exhaled from the opercles; this force is countered by pectoral sculling. The fin movement involved in hovering may be difficult to detect, both by human observers and potential prey, because fishes that use these techniques often possess transparent pectoral fins.

Third, not all fishes fit neatly into one of these categories, and many additional categories can and have been created to accommodate variations among taxa (see Lindsey 1978; Webb & Blake 1985; Videler 1993; Blake 2004; for more complete and alternative categorizations, and see next on sharks).

Swimming in Sharks: The Alternative Approach

Different fish lineages have evolved a variety of solutions to the challenges of locomotion in water. In the process, mutually exclusive specializations for cruising, rapid starts, or maneuverability have arisen (see above). The fossil record indicates that similar body morphologies and an apparent trend toward increasing concentration of activity in the tail region have appeared repeatedly during osteichthyan evolution (Webb 1982; see Chapter 11). These patterns and trends all capitalize on the substantial stresses that can be placed on a rigid, bony skeleton and the forces achievable by muscle masses attached directly or indirectly to bony structures. Elasmobranchs are, however, phylogenetically constrained by a relatively flexible and comparatively soft cartilaginous skeleton. Evolution of locomotion in chondrichthyans has, not surprisingly, taken a different albeit parallel path.

Most elasmobranchs swim via undulations, either of the body (sharks) or of the pectoral fins (skates and rays). Most sharks swim using anguilliform locomotion, although the amplitude of each wave in the caudal region is greater in swimming sharks than in eels. This exaggerated sweep of the posterior region probably capitalizes on the increased thrust available from the large heterocercal tail of a shark. Exceptions to anguilliform swimming include the pelagic, predatory mackerel sharks, which have converged in body form and swimming type with tunas, dolphins, and ichthyosaurs (discussed earlier). Skates and rays also undulate, passing undulations posteriorly along with the pectoral fins while the body is held relatively rigid. The exception in this group is the torpedo rays, which differ in that they have an expanded tail fin and swim via ostraciiform oscillations. In these strongly electrogenic rays, the pectoral region is unavailable for swimming because it is modified for generating electricity.

The mechanics of swimming in sharks are fascinating and somewhat controversial. Three topics have received the most attention, involving the functions of the median fins, skin, and tail during locomotion. Despite anguilliform movement, most sharks are active, cruising predators with relatively streamlined bodies. This would seem anomalous given the relatively low efficiency of the anguilliform mode and the apparent incompatibility of a fusiform body bent into long propulsive waves. However, sharks enhance the efficiency of their swimming mode in several ways.

Most sharks have two dorsal fins, the first usually larger than the second, separated by a considerable gap. The dorsal lobe of the pronounced heterocercal tail may be thought of as a third median fin in line with the dorsal fins, again separated from the second fin by a considerable gap. The distances between the three fins are apparently determined by the size of the fins, their shapes, and the waveform of swimming of the fish. Each fin tapers posteriorly, leaving behind it a wake as it moves through the water. This wake is displaced laterally by the sinusoidal waves passing down the fish, so the wake itself follows a sinusoidal path that moves posteriorly as the fish moves through the water. This wave is slightly out of phase with the fish's movements by a constant amount.

Calculations of the phase difference and wave nature of the wake suggest an ideal distance between fins that would maximize the thrust of the second dorsal fin and particularly of the tail. If timed correctly, the trailing fins can push against water coming toward them laterally from the leading fins. Such an interaction between flows would enhance the thrust produced by the trailing fin. Measurements of swimming motions and fin spacing in six species of sharks indicate just such an interaction (Webb & Keyes 1982; Webb 1984). Unlike bony fishes that use their median fins primarily for acceleration and braking but fold them while cruising to reduce drag, sharks use their median fins as additional, interacting thrusters (Fig. 4.14). Sharks are not alone in this interaction among fins. Studies of Bluegill (*Lepomis macrochirus*) indicate that the caudal fin also interacts with the vortices produced by the soft dorsal during steady swimming, thus providing additional thrust (Drucker & Lauder 2001).

The energy provided with each propulsive wave of muscular contraction is additionally aided by an interaction between the skin and the body musculature of a shark. The skin includes an inner sheath, the stratum compactum, made up of multiple layers of collagen fibers that are mechanically similar to tendons. The fibers form layers of alternately oriented sheets that run in helical paths around the shark's body, thus creating a cylinder reinforced with wound fibers, a structure that is exceptionally strong and incompressible but bendable (Motta 1977; Wainwright 1988b).

Inside the skin, hydrostatic pressure varies as a function of activity level. The faster the shark swims, the higher the internal hydrostatic pressure. Pressure during fast swimming is about 10 times what it is during slow swimming, ranging between 20 and 200 kPa (kilopascals: 1 Pa = 1 J/m^3 = 1 kg/m/s^2). Internal

FIGURE 4.14 Nurse Sharks, *Ginglymostoma cirratum*, are a good example of a shark in which both dorsal fins and the caudal fin provide thrust. The second dorsal fin of Nurse Sharks is unusually large, suggesting it contributes to cruising power and efficiency by pushing against the water shed by the first dorsal fin during its very anguilliform-type locomotion. G. Helfman (Author).

hydrostatic pressure develops from unknown sources, probably due to changes in the surface area of contracting muscles relative to skin area and to changes in blood pressure in blood sinuses that are surrounded by muscle. The shark's body is, therefore, a pressurized cylinder with an elastic covering.

During swimming, the higher the internal pressure, the stiffer the skin becomes, which increases the energy stored in the stretched skin. Body muscles attach via collagenous septa to the vertebral column and to the inside of the skin (for this reason, it is exceptionally difficult to remove the skin from the muscle of a shark). As the muscles on the right side of the body contract, muscles and skin on the left side are stretched. The stretched skin is very elastic, but stretched muscle is less so. As muscles on the right side relax, the energy stored in the skin on the left side is released, aiding muscles on the left side at a point when they can provide relatively little tension. Therefore, the skin may act in initiating the pull of the tail across the midline and increase the power output at the beginning of the propulsive stroke.

The faster the shark swims, the greater the elastic recoil from the stretched skin. Muscles attach to the relatively narrow vertebral column of calcified cartilage but also attach to the much larger surface area of stiff, elastic skin that encompasses the shark from head to tail and, in essence, forms a large, cylindrical, external tendon. The helically arranged fibers of the dermis extend onto the caudal peduncle and caudal fin, adding rigidity to both and perhaps storing elastic energy during each swimming stroke (Lingham-Soliar 2005). Muscles pulling on the skin provide propulsive energy that probably exceeds the thrust derived from muscles attached to the vertebral column (Wainwright et al. 1978; Wainwright 1983).

Most of the power in shark swimming comes from the heterocercal tail, which has a greatly expanded upper lobe (unlike the symmetrical tail found in most bony fishes). The expanded upper lobe of a shark's tail would seem to provide a lifting force to the posterior end of the body during horizontal locomotion. This lift should cause the posterior of the body to rise and drive the anterior end downward (Fig. 4.15). One long-held explanation is that the flat underside of the head and the broad stiff pectoral fins create lift at the anterior end to counteract the downward force. However, it seems inefficient for the tail and the pectoral fins to function against each other, the tail propelling and the pectoral fins continually breaking the shark's progress. Given the 400 million year success of elasmobranchs and the widespread occurrence of heterocercal tails in many previously speciose lineages of both bony and cartilaginous fishes, it is hard to imagine that heterocercal tails are inherently inefficient. This apparent dilemma has prompted an ongoing search for mechanisms that promote relatively straightforward propulsion.

The search has turned into something of a debate. The classic model, as described above, proposes that the tail pushes back and down, creating a reactive force that is countered by head shape and pectoral fins. An alternative explanation, based on interpretations of photographs and selective amputation of fin parts of tails held in a test apparatus (Simons 1970;

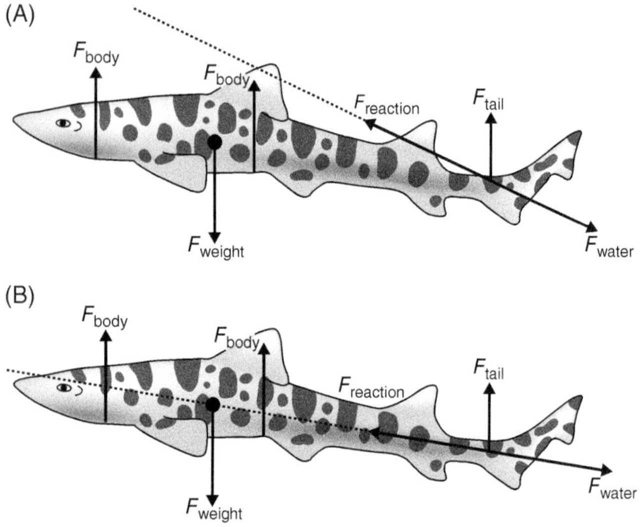

FIGURE 4.15 The two competing models that explain how horizontal locomotion is accomplished in sharks. (A) The modified classic model interprets the shape of the heterocercal tail as generating a downward and backwards thrust (F_{water}), lifting the tail up (F_{tail}); these produce a resultant force ($F_{reaction}$) that moves the body upwards and forwards. The flattened ventral profile of anterior body regions also provides lift (F_{body}). $F_{reaction}$ plus F_{body} counter the shark's tendency to sink because of its negative buoyancy (F_{weight}). The result is horizontal swimming. (B) In the alternative Thomson model, the upper and lower lobes of the tail provide counteracting forces that drive the fish directly ahead. The most recent research supports the modified classic model. However, the alternative model appears to explain locomotion dynamics in sturgeons, which also have heterocercal tails but which – unlike sharks – vary the flexibility and shape of dorsal and ventral tail lobes (Liao & Lauder 2000). After Wilga and Lauder (2002).

Thomson 1976, 1990), suggests that forward thrust is generated through the center of mass by differential movements of the upper and lower lobes of the tail (Fig. 4.15B). The classic model appears to be the most accurate description and is supported by video analysis (such as digital particle image velocimetry) and dye-tracer studies using free-swimming animals (Ferry & Lauder 1996). The classic model has been modified because shape and body angle, not pectoral lift, generate lift forces that are added to the lift exerted by the tail (Wilga & Lauder 2002). These forces are equal and opposite to the weight of the shark in the water. Braking by the pectoral fins is unnecessary.

The answer to how sharks climb, dive, and turn – rather than pivoting around their center of balance or rising in the water column – probably lies in their ability to continually adjust the relative angle of attack of their pectoral fins rather than altering thrust direction resulting from tail movement (Wilga & Lauder 2002). Maneuverability in bony fishes usually involves deep, compressed bodies and the use of median and pectoral fins; to accelerate, bony fishes increase the frequency of their tail beats. Sharks, with their streamlined bodies and relatively rigid fins, have taken a different evolutionary path to achieve maneuverability that may involve tail fin dynamics and paired fin adjustments. Sharks change speed by altering the tail beat frequency, but they also vary tail beat amplitude and the length of the propulsive wave passing down their body (Webb & Keyes 1982). Sharks have taken the relatively inefficient anguilliform swimming mode imposed by their flexible bodies, and combined elastic skin, rigid but carefully spaced median fins, and a heterocercal tail that produces a constant direction of thrust to achieve an efficient compromise between cruising, acceleration, and maneuverability. The actual mechanics of swimming in sharks and bony fishes is still a matter of debate and research, but our growing understanding underscores the intricacies and importance of locomotory adaptations in fishes.

Specialized Locomotion

Among the more interesting variations on locomotory type are fishes that have abandoned swimming for other means of getting around. A number of species walk along the bottom of the sea or leave the water and move about on land; these fishes have bodies that depart from a streamlined shape. Searobins (Triglidae) walk using modified pectoral rays that move independently of one another, in a manner that resembles the walking of an arthropod; the Flying Gurnard (Dactylopteridae) similarly tiptoes but uses modified pelvic rays instead. Frogfishes (Antennariidae) and batfishes (Ogcocephalidae) pull themselves along the bottom by moving their modified pectoral and pelvic fins; they also can "jump" up from the substrate by jet propulsion of water out their backward-facing, constricted opercles (Pietsch & Grobecker 1987). Australian handfishes (Brachionichthyidae), which get their common name because their pectoral fins are modified into an armlike appendage with an elbow and fingers, also use pectoral and pelvic fins to walk (Bruce et al. 1998).

Not all fishes are restricted to spending all of their time in water; some move about on land as well using a variety of different mechanisms. Climbing perches use paired fins and spiny gill covers to ratchet themselves along, whereas snakeheads row with their pectoral fins. So-called walking catfishes move across the land by lateral body flexion combined with pivoting on their stout, erect pectoral spines. Mudskippers (Fig. 4.16) swing their pectoral fins forward while supporting their body on the pelvic fins. They then push forward with their pectoral fins, somewhat similar to a person using crutches. Rapid leaps of 30–40 cm are accomplished by coordinated pushing of the tail and pectoral fins. Their unique pectoral fins are roughly convergent with the forelimbs of tetrapods, including an upper arm consisting of a rigid platelike region and a fanlike forearm and plantar surface (Gray 1968). Some species with anguilliform movement (moray and anguillid eels) are able to move across wet land employing their normal locomotion, which is similar to the terrestrial and aquatic movements of most snakes (Chave & Randall 1971; Lindsey 1978; Ellerby et al. 2001). Some fishes may also use the head or opercula for traction as the body wriggles (Pace & Gibb 2014).

Aerial locomotion grades from occasional jumping to gliding to actual flapping flight. Many fishes jump to catch airborne prey such as insects and birds (trout, Giant Tigerfish, Largemouth Bass); meter-long arawanas (Osteoglossidae)

FIGURE 4.16 Moving on land – while out of the water amphibious air-breathing fishes usually move about by either (i) side-to-side wriggling of the body (axial locomotion); (ii) wriggling the body and using the fins for traction (axial-appendage); or (iii) "crutching" by swinging the pectoral fins to drag the body such as this mudskipper. Some fishes may also use the head or opercula for traction as the body wriggles (Pace and Gibb 2014). Courtesy of J. Hyde and K. Sultze.

FIGURE 4.17 Sailfin Flyingfish, *Parexocoetus brachypterus*, showing the enlarged pectoral fins used for gliding and the elongate ventral caudal lobe with which the fish skulls rapidly along the ocean surface. These fish can remain in the air for over 45 s and 400 m, at estimated speeds exceeding 70 kph. Jordan and Evermann (1903) Warren Evermann / NOAA / Wikimedia Commons / Public Domain.

can leap more than a body length upward and pluck insects and larger prey, including bats, from overhanging vegetation. Other fishes take advantage of the greater speeds achievable in the air: needlefishes, mackerels, and tunas leave the water in a flat trajectory when chasing prey, and salmon leap clear of the water when moving through rapids or up waterfalls. Hooked fish jump and simultaneously thrash from side to side in an apparent attempt to throw the hook; such oscillation is less constrained by drag in the air than in water and therefore allows more rapid and forceful to-and-fro movement. Prey such as minnows, halfbeaks, silversides, mullets, and Bluefish jump when being chased.

Some fishes can remain airborne for more than just a few seconds, such as the flyingfishes (Exocetidae) and butterflyfishes (Pantodontidae), as well as hatchetfishes (Gasteropelicidae) that purportedly vibrate their pectoral wings to generate additional lift (Davenport 1994; see Chapter 16, Pursuit). The anatomy of the marine flyingfishes is highly modified for gliding (Fig. 4.17). The body is almost rectangular in cross-section, the flattened ventral side of the rectangle providing a flat surface that may aid during take-off. The ventral lobe of the caudal fin is 10–15% larger in surface area than the dorsal lobe and is the only part of the body in contact with the water as the fish becomes airborne. The pectoral fins are supported by enlarged pectoral girdles and musculature. The pectoral fins differ from normal teleost fins in the shape of and connections between the lepidotrichia, and the pectoral fin rays are thickened and stiffened, giving the leading, trailing, dorsal, and ventral surfaces more of a winglike than a finlike construction. In some flyingfishes, pelvic fins also contribute lift and are appropriately modified.

Some other atheriniform fishes, such as needlefishes and halfbeaks, also propel themselves above the water's surface by rapidly vibrating their tail, the lower lobe of which is the only part still in the water. Some halfbeaks have relatively large pectoral fins and engage in gliding flight. Gradations of pectoral fin length and lower caudal lobe strengthening and lengthening among atheriniforms provide a good example of apparent steps in the evolution of a specialized trait, in this case gliding (Lindsey 1978; Davenport 1994).

Supplementary Reading

Alexander RMcN. 1983. *Animal mechanics*, 2nd edn. Oxford: Blackwell Science.

Duncker HR, Fleischer G, eds. 1986. *Functional morphology of vertebrates.* New York: Springer-Verlag.

Farrell AP, Stevens ED, Cech JJ, Richards JG. 2011. *Encyclopedia of fish physiology: from genome to environment.* Amsterdam, Boston: Elsevier/Academic Press.

Gray J. 1968. *Animal locomotion.* London: Weidenfeld & Nicolson.

Hildebrand M. 1982. *Analysis of vertebrate structure*, 2nd edn. New York: Wiley & Sons.

Hildebrand M, Bramble DM, Liem KF, Wake DB, eds. 1985. *Functional vertebrate morphology.* Cambridge, MA: Belknap Press.

Hoar WS, Randall DJ, eds. 1978. *Locomotion. Fish physiology*, Vol. 7. New York: Academic Press.

Jordan DS, Evermann BW. 1903. The shore fishes of the Hawaiian Islands, with a General Account of the Fish Fauna. *Bull U. S. Fish Commission* XXIII: for 1903. Part I.

Liem KF, Bemis WE, Walker WF, Grande L. 2001. *Functional anatomy of the vertebrates*, 3rd edn. Belmont, CA: Thomson/Brooks Cole.

Ostrander GK, ed. 2000. *The laboratory fish.* London: Academic Press.

Pough FH, Janis CM, Heiser JB. 2012. *Vertebrate life*, 9th edn. Upper Saddle River, NJ: Prentice Hall.
Shadwick RE, Lauder GV. 2006. *Fish biomechanics. Fish physiology*, Vol. 23. New York: Academic Press.
Videler JJ. 1993. *Fish swimming*. London: Chapman & Hall.
Vogel S. 1981. *Life in moving fluids*. Boston: Willard Grant Press.
Wainwright SA, Biggs ND, Currey JD, Gosline JM. 1976. *Mechanical design in organisms*. New York: J. Wiley & Sons.
Webb PW, Weihs D. 1983. *Fish biomechanics*. New York: Praeger.

Journal

Journal of Morphology.

CHAPTER **5**

Circulation, Gas Transport, Metabolism, Digestion, and Energetics

Summary

The **cardiovascular system** is composed of the heart plus the network of arteries, veins, and capillaries that carry blood throughout the body. In fishes, deoxygenated blood from the body returns to the heart, which is ventral and posterior to the gills. The blood is then pumped through the ventral aorta to the gills, where it releases carbon dioxide and takes up oxygen. Oxygenated blood from the gills next travels to the body via the dorsal aorta, releases oxygen to the tissues, takes up carbon dioxide and returns to the heart via the post-cardinal vein. There are modifications of this basic plan in the many groups of air-breathing fishes.

The blood is mostly plasma but also contains white blood cells, which are part of the immune system, and red blood cells, which are important for most oxygen and carbon dioxide transport. As in other vertebrates, fishes also have a lymphatic system that returns extracellular fluids to the blood.

Water is denser and more viscous than air, making it energetically costly to move across respiratory surfaces. Water also contains considerably less oxygen than air, especially at elevated temperatures. Fish **respiratory systems** are primarily gills, although other organs may play an important role in gas exchange in some species. Fish gills are finely divided into filaments with numerous thin-walled lamellae to facilitate gas exchange. Blood moves through the lamellae in the opposite direction to the water flowing over the gills, maximizing the oxygen extraction from

CHAPTER CONTENTS

The Diversity of Fishes: Biology, Evolution and Ecology, Third Edition. Douglas E. Facey, Brian W. Bowen, Bruce B. Collette, and Gene S. Helfman.

Companion website: www.wiley.com/go/facey/diversityfishes3

the water by maintaining a strong diffusion gradient between oxygen concentration in the water and the blood. Gills also play an important role in ion and water exchange and excretion (see Chapter 7). Many groups of fishes have evolved mechanisms for absorbing oxygen from the air. Air breathing organs include skin, the mouth, structurally modified gills, different parts of the gut (esophagus, stomach, intestine), modified gas bladders, and in some cases, true lungs. These variations indicate that air-breathing has evolved independently many times in the fishes. The hemoglobin in red blood cells plays a critical role in the transport of oxygen, and the ability of this protein to bind oxygen is affected by temperature, pH, and levels of carbon dioxide. This allows the red blood cells to pick up oxygen at the gills and release it at the tissues where needed.

Fishes are primarily ectothermic, so their metabolic rates increase and decrease with water temperature. Metabolism is influenced by a variety of other factors, however, including the presence of food in the gut, activity, age, sex, and reproductive status. Therefore, metabolic studies of fishes acclimated to controlled laboratory conditions may not accurately represent the metabolic rates of fishes in nature.

Cardiovascular System

The cardiovascular system includes the heart and the system of arteries, veins, and capillaries that carry blood containing respiratory gases, wastes, excretory metabolites, minerals, and nutrients. Although the cardiovascular system serves all bodily functions, it is most closely associated with respiration, excretion, osmoregulation, and digestion. The basic pattern of blood flow in fishes is sometimes described as a single-pump, single-circuit system because blood travels only once through the heart per circuit around the body. (This is unlike the circulation of tetrapods in which blood travels from heart to lungs, then back to the heart for distribution to the rest of the body.) Deoxygenated blood from the body returns to the heart and is pumped to the gills, where it releases carbon dioxide and takes up oxygen. Oxygenated blood from the gills travels to the body, releases oxygen to the tissues, takes up carbon dioxide, and returns to the heart (Fig. 5.1). Blood returning from the gut includes nutrients from digestion for distribution throughout the body. Blood flowing to the kidneys allows wastes to be removed.

The Heart

Heart size as a proportion of body weight is lower in fishes than in other vertebrates. The heart is located posterior and ventral to the gills in all fishes, although it is located farther anterior in teleosts than in chondrichthyans. It lies in a membranous **pericardial cavity** that is lined with **parietal pericardium**. The basic fish heart consists of four sequential chambers. Venous blood enters (i) the **sinus venosus** (a thin-walled sac) from the **ducts of Cuvier** and the **hepatic veins**; it next flows into (ii) the atrium; then into (iii) the **ventricle**, a thick-walled pump; and finally, blood flows out of the heart into (iv) the **conus** or **bulbus arteriosus** (Farrell & Pieperhoff 2011). The conus arteriosus is a barrel-shaped elastic chamber with some cardiac muscle, present in Chondrichthyes and lungfishes (Dipnoi). Actinopterygian fishes have a bulbus arteriosus that is elastic but does not contain muscle tissue. Blood exiting the ventricle expands the conus or bulbus, dampening pressure oscillations, and the elastic recoil of the conus or bulbus provides a more consistent blood flow through the gills.

In lungfishes, the atrium is partly divided by a partition, partially separating oxygenated and deoxygenated blood, a step toward development of the divided heart of tetrapods. This division is most complete in the South American Lungfish (*Lepidosiren*), which is heavily dependent on breathing air, and least complete in the more ancestral Australian Lungfish (*Neoceratodus*), which uses air only as a supplement to gill respiration.

Heart valves help maintain pressure in the circulatory system and keep blood flowing in the right direction. **Sino-auricular valves** (usually composed of both endocardial and myocardial muscle) separate the sinus venosus and atrium. **Auriculoventricular** or atrioventricular **valves** vary in number depending on the group: Chondrichthyes and most bony fishes have two rows of valves, whereas primitive bony fishes have more; the Bowfin has four rows, the North American Paddlefish

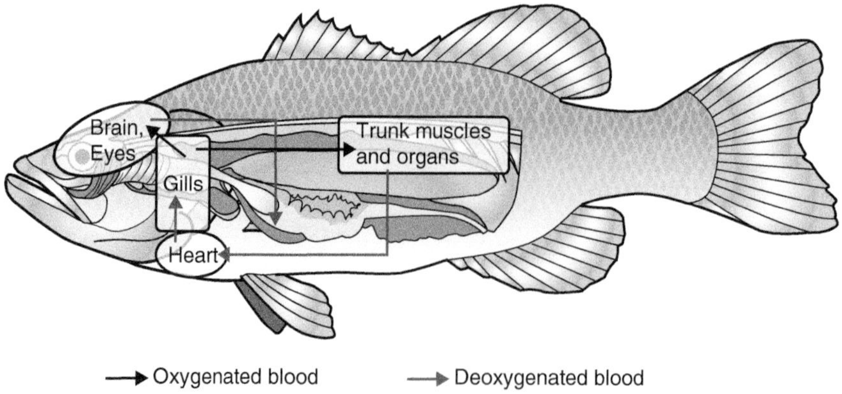

FIGURE 5.1 Basic circulatory pattern found in most fishes.

has five rows, and gars and bichirs (*Polypterus*) have six rows. These valves are absent in lungfishes, which instead have valves in the conus arteriosus. The number of **ventriculobulbar valves** is related to the length of the conus. There is usually one or, rarely, two in bony fishes, two to seven in Chondrichthyes, and up to 74 in eight rows in gars. Valves outside the heart region can occur in various parts of the circulatory system of fishes.

In many fishes, the heart does not have a separate blood supply, but instead gets its oxygen from the venous blood moving through the heart (Farrell 2011a). The elasmobranchs (sharks and relatives) and primitive bony fishes (such as sturgeons, gars, bowfin, lungfishes) have coronary circulation, but it is lacking in hagfishes, lampreys, and many teleosts. Those teleosts that have coronary circulation, however, are among the more active swimmers (such as salmon and tunas), or those that can thrive in low oxygen environments.

Major Blood Vessels

Afferent branchial arteries bring oxygen-deficient blood from the ventral aorta to the gill filaments for gas exchange. The number of arteries varies among different groups of fishes. Bony fishes and most Chondrichthyes have four (Fig. 5.2), but sharks and rays with more gills have more arteries. The Six-gilled Shark (*Hexanchus*) has five, and the Seven-gilled Shark, *Heptranchias* has six. Lungfishes have four to five branchial arteries, whereas hagfishes and lampreys have seven to 14, depending on the number of gill pouches.

Efferent branchial arteries bring oxygenated blood from the gills to the dorsal aorta. In addition, many fishes have a **pseudobranchs** (false gills), a small structure under the operculum composed of gill-like filaments that provide oxygenated blood directly to the eyes, which have high oxygen demand. Major veins such as the facial, orbital, postorbital, and cerebral join into paired **anterior cardinal veins**, which empty into the **common cardinal** (also called the duct of Cuvier) and then into the heart. The jugular vein collects blood from the lower head and also empties into the common cardinal in Actinopterygii.

The **dorsal aorta** is the largest and longest artery, and the main route of transport of oxygenated blood from the gills to the rest of the body (Fig. 5.3). It lies directly ventral to the vertebral column in the trunk region and gives off major vessels and segmental arteries. **Internal carotid arteries** run from the aorta to the brain. The **subclavian artery** goes to the pectoral girdle, the **coeliaco-mesenteric artery** supplies the viscera, and the **iliac** or **renal artery** supplies the kidneys. The dorsal aorta becomes known as the **caudal artery** upon entering the closed haemal canal of the caudal vertebrae. The major return route of blood from most of the body is the **postcardinal vein**. It is best developed on the right side and empties into the common cardinal or ducts of Cuvier, then into the sinus venosus and the heart.

In the advanced tunas (tribe Thunnini), an additional pair of large arteries, the **cutaneous arteries**, branch from the dorsal aorta and run laterally between the ribs. As these arteries approach the fish's skin, they divide into two vessels, each of which runs posteriorly, sending out arterioles to the underlying red muscle. Blood then flows toward the body core,

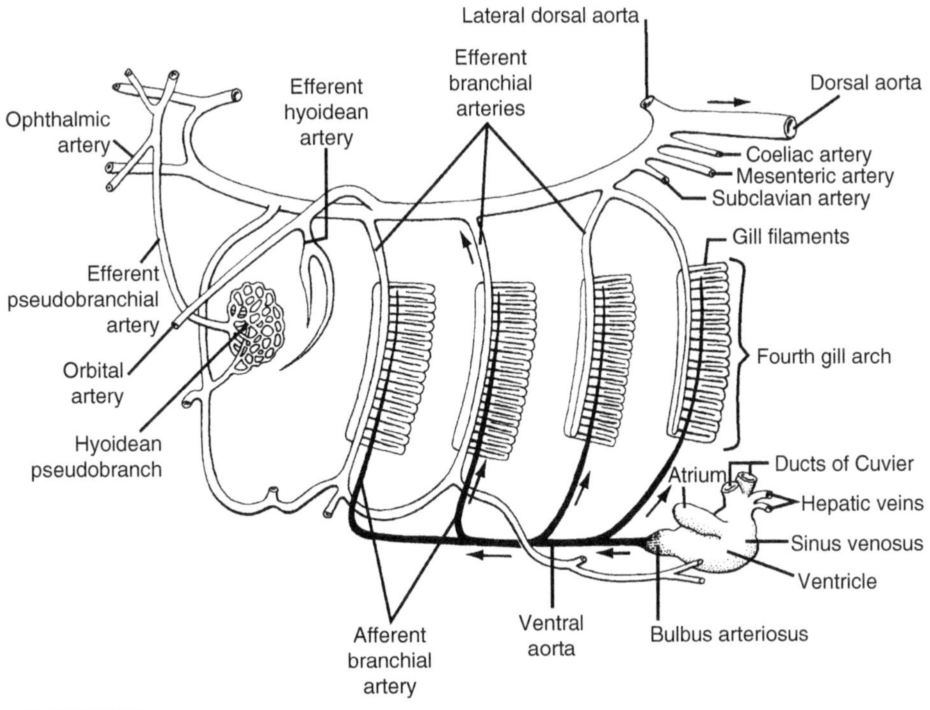

FIGURE 5.2 Gills and blood vessels of the head of a cod (*Gadus*). From Lagler et al. (1977).

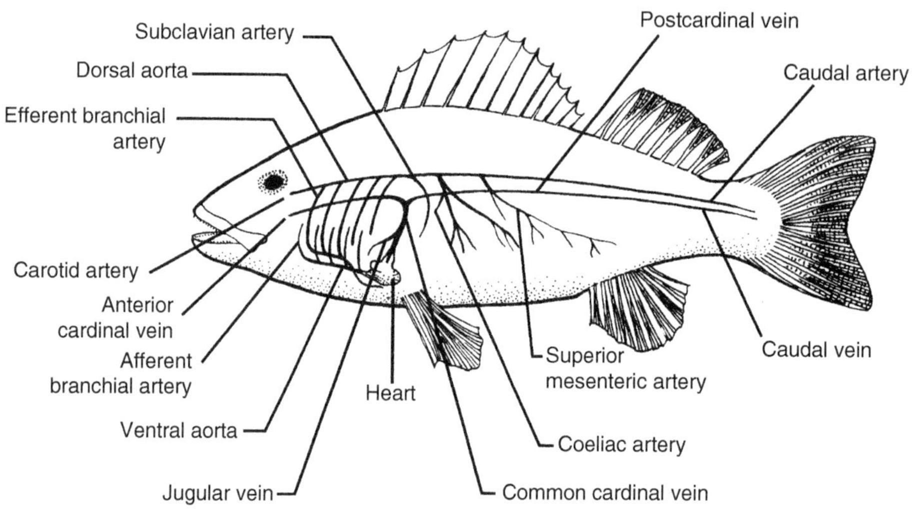

FIGURE 5.3 Main blood vessels of a bony fish. After Lagler et al. (1977).

passing alongside veins carrying warm deoxygenated blood, which warms the oxygenated blood coming into the muscles. This countercurrent heat exchanger helps keep the swimming muscles warm (see next section, and Chapter 7). Deoxygenated blood flows away from the core to the cutaneous vein and returns to the heart. Phylogenetically, the most advanced tunas show the greatest development of the subcutaneous circulatory system (Fig. 5.4).

Blood pressure is low in fishes; it is reduced as it passes through the gills after leaving the heart and before going to the rest of the body. Hagfishes have accessory hearts to assist with circulation, and some other fishes have modified veins that help propel blood when the surrounding muscles of the body contract (Farrell 2011b).

Countercurrent Networks

Many fishes have examples of countercurrent flow of blood that facilitate heat exchange or diffusion. Often called a rete mirabile ("wonderful net"). Retia (the plural term) are networks of small adjacent blood vessels that flow in opposite directions and are therefore also called countercurrent exchangers. They allow the transfer of heat, and also ions and gas molecules if the blood vessels are thin-walled capillaries. Retia are present in different parts of the circulatory system of a variety of fishes, indicating that they evolved multiple times for different purposes (see Stevens 2011).

As mentioned in the previous section, some large pelagic fishes have retia that conserve heat generated by swimming muscles and help maintain warm core body temperatures (see Chapter 7). This is found in some elasmobranchs and teleosts and is believed to have evolved independently in several groups about 60 million years ago. A rete associated with warming of the eyes and brain in billfishes evolved perhaps 15 million years ago.

Many teleosts have a choroid rete behind each eye that helps maintain high oxygen levels in blood to the retina. This rete is believed to have evolved once early in the teleosts (about 250 million years ago) and was then passed on to the more modern groups.

Blood

The volume of blood in teleosts is less than that of Chondrichthyes, and both have lower blood volumes than tetrapods. Hagfishes and lampreys have the greatest blood volume among fishes. Blood itself is mostly composed of plasma, red blood cells, and white blood cells (Farrell 2011c).

Plasma contains minerals, nutrients, waste products, enzymes, antibodies, and dissolved gases, but few detailed analyses of fish blood have been published. Solutes in the plasma lower its freezing point to about −0.5 °C in freshwater bony fishes, −1.0 °C in freshwater Chondrichthyes, −0.6 to −1.0 °C in marine bony fishes, and −2.2 °C in marine Chondrichthyes. The freezing point of seawater is about −2.1 °C. Antarctic notothenioids (see Chapter 10) have additional blood antifreeze glycoproteins that further reduce the freezing point of their blood, with some notothenioids showing freezing points as low as −3 °C.

Red blood cells (RBCs) account for nearly 99% of oxygen uptake at the gills, and transport the oxygen and release it at tissues. RBCs are nucleated, yellowish-red, oval cells in most fishes but are round in lampreys. Fishes have relatively fewer and larger RBCs than do mammals. Human RBCs, which lack nuclei, measure 7.9 μm across, whereas fish RBCs range from 7 μm in some wrasses to relatively giant 36 μm cells in the African lungfishes, *Protopterus*. RBCs are absent in some notothenioids (see Chapter 10) and in the leptocephalus larvae of eels.

White blood cells take up a very small proportion of the overall blood volume, but assist with blood clotting and immune system function. Four types of white blood cells perform different specific tasks: thrombocytes assist with blood clotting, lymphocyctes help the immune system recognize foreign material

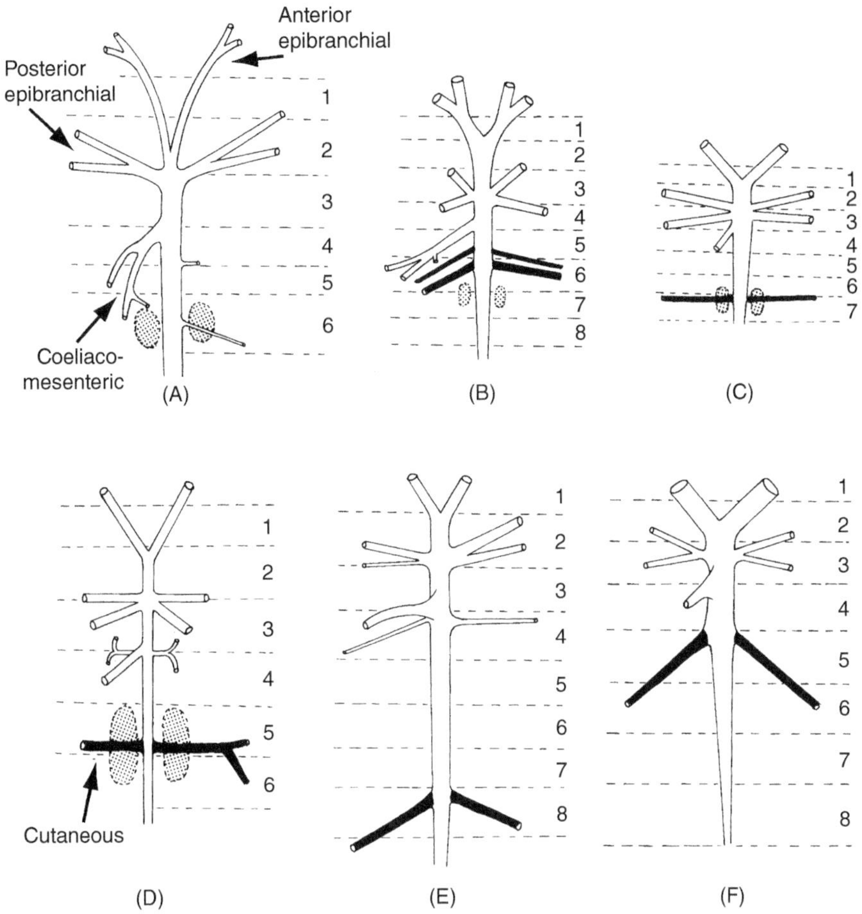

FIGURE 5.4 Anterior arterial system in ventral view in the Scombridae, showing the phylogenetic increase in the development of the subcutaneous circulatory system (darkened vessels). Numbers indicate vertebrae; stippled areas show where pharyngeal muscles originate. (A) Wahoo (*Acanthocybium*); (B) Frigate tuna (*Auxis*); (C) Little tuna (*Euthynnus*); (D) Skipjack (*Katsuwonus*); (E) Longtail Tuna (*Thunnus tonggol*); (F) Albacore (*Thunnus alalunga*). From Collette (1978/1979).

and elicit an immune response, monocytes and granulocytes engulf and digest foreign particles and cells (phagocytosis) (Farrell 2011c). Stress, such as exposure to some chemicals, can impair the function of white blood cells and thereby compromise the ability of fishes to fight infections and remain healthy.

Lymphatic System

Lymph is fluid similar to blood plasma, but that is no longer contained within the blood vessels. The lymphatic system is derived from the venous part of the blood vascular system. The lymph system may assist immune function and transport absorbed fats, but the primary role in fishes (and other vertebrates) seems to be the return of fluid that leaves the blood vessels (Olson & Farrell 2011). The bony fishes have lymph vessels and contractile "lymph hearts" in the body cavity and external tissues, often associated with the dorsal aorta or the main arteries branching from it. However, lymph vessels are not common in the gut or skeletal muscles of fishes. In tunas and some of their relatives (Scombridae) the lymph system has been repurposed as a hydraulic system for stiffening the anal and dorsal fins during quick turns (Pavlov et al. 2017). This is the only known example of the lymph system being used for biomechanical functions.

Gas Exchange and Transport

Fishes, like all eukaryotic life forms, require oxygen to support their metabolic needs. Although acquiring sufficient oxygen from water is challenging, fishes have evolved a range of morphological and physiological adaptations that increase the efficiency of oxygen uptake and delivery to help them succeed in a wide range of aquatic environments. Because fishes were the first vertebrates, these adaptations provided a physiological foundation upon which other adaptations eventually brought about the success of tetrapods and endothermy.

Fishes must extract oxygen from the water and distribute it to the cells of the body fast enough to meet the demands of metabolism. The oxygen maximizes the amount of adenosine triphosphate (ATP) that can be generated from glucose, the primary metabolic fuel of cellular metabolism, by permitting the aerobic completion of cellular respiration by oxidative phosphorylation.

If oxygen is not present, oxidative phosphorylation and the Krebs cycle cannot proceed, and the only energy available from the metabolism of glucose is from the small amount of ATP released during the initial glycolysis reaction. For glycolysis to continue producing some ATP, the pyruvate that also is produced is often converted to lactate and stored temporarily. If lactate levels get too high, however, glycolysis may be inhibited. Effect on swimming is temporary, however, as adult Pacific salmon (*Oncorhynchus*) exercised to exhaustion in a swim tunnel showed no decrease in swimming ability when tested a second time less than 1 hour after the initial test (Farrell et al. 2003). The less active Goldfish (Cyprinidae) can avoid lactate buildup altogether through an alternative biochemical pathway that converts excess pyruvate to alcohol which can then be excreted (Hochachka & Mommsen 1983; Hochachka & Somero 1984). This can be essential where Goldfish survive under ice with little or no oxygen through a long winter; Goldfish can continue producing ATP by glycolysis without suffering the problems associated with lactate buildup.

Fish breathing is controlled by signals from the brain stem (medulla oblongata) to the muscles of the mouth and opercula (Taylor 2011, see Chapter 3). Breathing rate increases in response to increasing levels of carbon dioxide and hydrogen ions (which will decrease pH) in the blood, and elevated levels of carbon dioxide in the water (Milsom 2011). Chemosensors in the blood vessels and on the gill arches monitor these and provide appropriate signals to the medulla to adjust the breathing rate.

Aquatic Breathing

Water contains much less oxygen than air – less than 1% by volume, as opposed to over 20% for air. Oxygen may be evenly distributed in flowing or turbulent water due to mixing, whereas still water may have more oxygen at the surface due to diffusion from the air. Some fishes take advantage of this by coming toward the surface to breathe when oxygen is limited. For example, Sailfin Molly (Poeciliidae) use aquatic surface respiration (ASR) as well as an increase in ventilation frequency to cope with hypoxic conditions (Timmerman & Chapman 2004). The use of ASR diminishes, however, after a period of acclimation to the low oxygen conditions. If fishes in an aquarium are congregating near the surface to breathe, the water in the tank may be oxygen depleted.

Gas solubility in liquids diminishes with increasing temperature. Warm water, therefore, contains less oxygen than cool water. Freshwater can hold about 25% more oxygen than seawater due to the diminished solubility of gases in water as the concentration of salts or other solutes increases. This **salting out effect** is true for all water solutions, including natural aquatic environments, blood plasma, cytoplasm, or a glass of carbonated beverage (just add some table salt and see what happens). The combined effects of temperature and salinity make oxygen availability especially low in warm, marine environments at the same time that metabolism is increased.

The relatively high density and viscosity of water means that more energy is required to simply move water across the respiratory surfaces relative to air breathing. A fish may use as much as 10% or more of the oxygen that it gets from the water simply keeping the breathing muscles going (Jones & Schwarzfeld 1974), whereas for air-breathing animals the relative cost is much lower, around 1–2%.

The **gills** of fishes are very efficient at extracting oxygen from the water because of the large surface area and thin epithelial membranes of the **secondary lamellae** (see Chapter 3, Fig. 3.13). Filaments from adjacent gill arches nearly touch one another, which maximizes gas extraction from the water. In addition, as fishes grow, the number of secondary lamellae on the filaments increases (Olson 2011a). As discussed in Chapter 3 (Breathing), diffusion of gases across the gill membrane is enhanced by blood in the secondary lamellae flowing in the opposite direction to the water passing over the gills, thereby maximizing the diffusion gradient across the entire lamellar surface. This, combined with unidirectional flow of water across the gills due to ventilation, helps to maximize oxygen uptake.

The total surface area of the gills is considerable; active fishes with higher metabolic demands generally have larger gill surface areas than less active fishes (Wegner 2011). For example, Skipjack Tuna are active pelagic predators and have about 13 cm^2 of gill area per gram of body weight (Roberts 1975b). Scup (Sparidae) are nearshore, active fish and have about 5 cm^2/g, whereas the sluggish, bottom-dwelling toadfish (Batrachoididae) has about 2 cm^2/g. Freshwater fishes tend to have somewhat smaller secondary lamellae and less gill surface area than salt-water fishes of similar size and level of activity, likely due to higher oxygen levels in freshwater (Wegner 2011).

Fishes control oxygen intake in several ways, including reduced or increased blood flow to gill filaments. Some fishes can modify the gill structure in response to oxygen level in the water (Nilsson 2011). For example, secondary lamellae of Crucian Carp (*Carassius carassius*, Fig. 5.5) appeared to be somewhat longer under warmer, lower oxygen conditions

FIGURE 5.5 Crucian Carp are reported to increase the effective length of secondary lamellae of their gills when exposed to low oxygen levels. Harka, Akos / USGS / Public Domain.

than when the fish were in colder, higher oxygen water (Sollid et al. 2003, 2005). The apparent cause was the regression of epithelial cells between the lamellae under low oxygen conditions, which increased the exposed surface area of the lamellae. Other species reported to make changes in exposed surface area of the lamellae include Mangrove Killifish (*Kryptolebias marmoratus*), European Eel (*Anguilla anguilla*), Largemouth Bass *(Micropterus salmoides)*, Rainbow Trout *(Oncorhynchus mykiss)*, and Brook Trout *(Salvelinus fontinalis)* (Nilsson 2011). The Amazonian *Arapaima gigas* also reduces exposed surface area of secondary lamellae as it develops into an air-breathing adult.

Many fishes will come to the surface to breathe when oxygen in the water is low. However, Tilapia (*Alcolapia graham*) in small, alkaline Lake Magadi in Kenya come to breathe at the surface in the afternoon, when oxygen levels in the water may be high due to photosynthesis. This behavior may be a response to irritation by hydroxide radicals, superoxide radicals, and hydrogen peroxide which can be produced in productive alkaline lakes with high sunlight (Johannsson et al. 2014). Surface-breathing exposes the fish to predation by birds, so the fish usually come to surface in groups (pods). In African air-breathing catfishes (Clariidae), time spent at the surface can be measured in fractions of a second. In addition, reducing blood flow or lamellar surface area under high oxygen conditions will reduce exposure to pathogens, toxins, and especially osmotic stress, as water and ion exchange are energetically expensive (see Chapter 6; Osmoregulation).

Although gills typically are identified as the respiratory organ of most fishes, any thin surface in contact with the respiratory medium is a potential site of gas exchange. Gas exchange across the skin (cutaneous respiration) can be important to some fishes, particularly in young fish whose gills have not yet developed fully. Newly hatched alevins of Chinook Salmon (Salmonidae) rely on cutaneous respiration for up to 84% of their oxygen (Rombough & Ure 1991). As the fish develop and their gills increase in size and efficiency, dependence on cutaneous respiration decreases to about 30% of total uptake in later stages. Adult eel (Anguillidae), plaice (Pleuronectidae), Reedfish (Polypteridae), and mudskippers (Gobiidae) gain 30% or more of their oxygen through their skin (Feder & Burggren 1985; Rombough & Ure 1991).

Air-Breathing Fishes

Air breathing evolved among fishes over 400 million years ago, and at least some members of extinct groups such as the placoderms and acanthodians may have been air breathers (Graham 1997a, 2011a). Early sarcopterygians gave rise to early tetrapods, which have since successfully colonized terrestrial habitats. But air breathing continued to develop independently in many groups of fishes. The evolution of air breathing was probably driven by two main factors: (i) persistent or occasional low oxygen levels in freshwater habitats; and (ii) emergence during low tides in marine and estuarine habitats (Graham & Lee 2004; Graham 2011a). In both marine and freshwater habitats, excursions onto land would provide access to additional resources. Lungs were present in many primitive fishes, and became more specialized and efficient among the sarcopterygians as they evolved, eventually giving rise to the modern tetrapods. As the actinopterygians evolved, the lung lost its respiratory function and became the gas bladder, which functions for buoyancy control (discussed later in this chapter) and enhances hearing in some fishes (see Chapter 6, Hearing). Subsequently, some of the more advanced fishes developed alternative mechanisms to once again take advantage of the oxygen available in air.

Close to 400 extant species of bony fishes in 140 genera spread over 50 families and 18 orders have some capacity to obtain oxygen from the air (Graham 2011a; Table 5.1). The broad diversity of known air-breathing fishes suggests that air breathing probably evolved independently at least 38 times, and quite possibly up to 70 (Graham 2011a). Most air-breathing fishes are found in freshwater, but some are found in brackish and marine habitats. Air-breathing fishes show great diversity in size, from as small as 3 cm up to more than 2 m, including Tarpon (Megalopidae) and *Arapaima gigas*, one of the largest freshwater fishes in the world. Notably, there are no known air-breathers among the cartilaginous fishes (Chondrichthyes).

Most air-breathing fishes remain in water all of the time (**aquatic air breathers**). Some only supplement gill respiration when necessary (**facultative air breathing**) whereas others must have access to air or to survive (**obligate air breathing**). Most air-breathing fishes live in tropical or sub-tropical habitats where high temperatures dramatically reduce dissolved oxygen levels in water. Many of these tropical air-breathing fishes live in freshwater habitats under thick forest canopies, in which high rates of decomposition decrease the amount of oxygen available, and shading inhibits aquatic photosynthesis, which would add some oxygen to the water. Some air-breathing fishes, such as the tarpons (*Megalops*), live in shallow marine and estuarine habitats, where the combination of high temperature and salinity can deplete oxygen. In addition, there are temperate air-breathing fishes such as Bowfin (*Amia*), gars (*Lepisosteus*), and mudminnow (*Umbra*), and one species that lives north of the Arctic Circle (Alaska Blackfish, *Dallia pectoralis*).

Despite the great diversity of air-breathing fishes, 39% of known species are found in just seven families (Graham 2011a) – the Callichthyidae and Clariidae (catfishes in the order Siluriformes), and the five families of anabantoids (gouramies in the order Perciformes). Among the anabantoids, changes in the jaws and branchial region that allow for air breathing also provide enhanced capabilities for sound reception and production, bubble-nest construction, and mouth brooding.

Some fishes also have the ability to survive, and even remain active, while out of the water due to their ability to breathe air (**amphibious air breathers**). These include some tropical freshwater species in habitats that may become dry seasonally, and marine intertidal species that leave the water to forage. Air breathing in these fishes is not solely a mechanism

TABLE 5.1 Diversity of fishes with air-breathing capabilities.

Order and family	No. genera/species	Habitat	Respiratory pattern
Cyprinodontiformes			
Aplocheilidae	1/5	F	AmV
Cyprinodontidae	1/4	F, M	AmV+AmS
Scorpaeniformes			
Cottidae	2/4	M	AmV
Perciformes			
Stichaeidae	4/5	M	AmS
Pholididae	3/5	M	AmS
Tripterygiidae	1/1	M	AmV
Labrisomidae	2/2	M	AmV
Blenniidae	7/32	M	AmV
Eleotridae	2/2	M	AF
Gobiidae	15/40	M, B	AF, AmV
Gobioididae	1/1	M, B	AF, AmS
Mastacembelidae	2/3	F, B	AmS
Anabantidae	3/24	F, B	AC, AmV+AmS
Belontiidae	12/44	F, B	AC, (AmV?)
Helostomatidae	1/1	F	AC
Osphronemidae	1/1	F	AC
Luciocephalidae	1/1	F	AC
Channidae	1/12	F	AC, AmS
Synbranchidae	3/14	F, B	AC+AF, AmV+AmS

Habitats: B, brackish; F, freshwater; M, marine. Respiratory patterns: AC, aquatic continuous; AF, aquatic facultative; AmS, amphibious stranded; AmV, amphibious volitional. Modified from Graham (1997a).

to survive low oxygen in the water; it also provides a means to take advantage of a habitat not available to other fishes.

Air-breathing organs of fishes today fall into three broad categories: (i) structures of the head and pharynx, such as modifications of the gills, mouth, pharynx, or opercles; (ii) organs that are derived from the gut, such as the lungs, gas bladder, stomach, or intestine; and (iii) skin, which can be very effective for gas exchange if it is well vascularized and moist (Graham 2011b).

Freshwater air-breathing fishes show a wide array of adaptations for aerial gas exchange. Gills are not well suited for aerial respiration because they collapse and stick together when not supported by water. There are, however, a few fishes that have modified gill structures that assist with aerial respiration, such as the modified treelike branches found above gill arches of the Walking Catfish (Clariidae) or the complex platelike outgrowths of the gill arches of anabantoids such as the Giant Gourami (Osphronemidae) and several other Asian perciforms (Fig. 5.6A, B). Other oral respiratory structures include highly vascularized surfaces of the mouth and pharynx, such as in the Longjaw Mudsucker (*Gillichthys mirabilis*; Martin & Bridges 1999).

The development of gill modifications suited for air breathing may be influenced by the availability of oxygen in the environment. Blue Gourami (*Trichopodus trichopterus*) juveniles raised in low oxygen conditions developed larger gill and labyrinth surface area than fish raised at higher oxygen levels. This shows that the development of the gills and the suprabranchial air breathing organ are affected by the environment (Blank & Burggren 2014).

Air-breathing organs also include modifications of the gut, such as the stomach, intestine, modified gas bladder, or lungs. Various modifications of the esophagus, stomach, or intestine probably evolved at least 10 independent times among the bony fishes, with five of those just among the catfish families Loricariidae and Scoloplacidae (see Nelson 2014). Gas bladders of most fishes are not well vascularized except in the regions

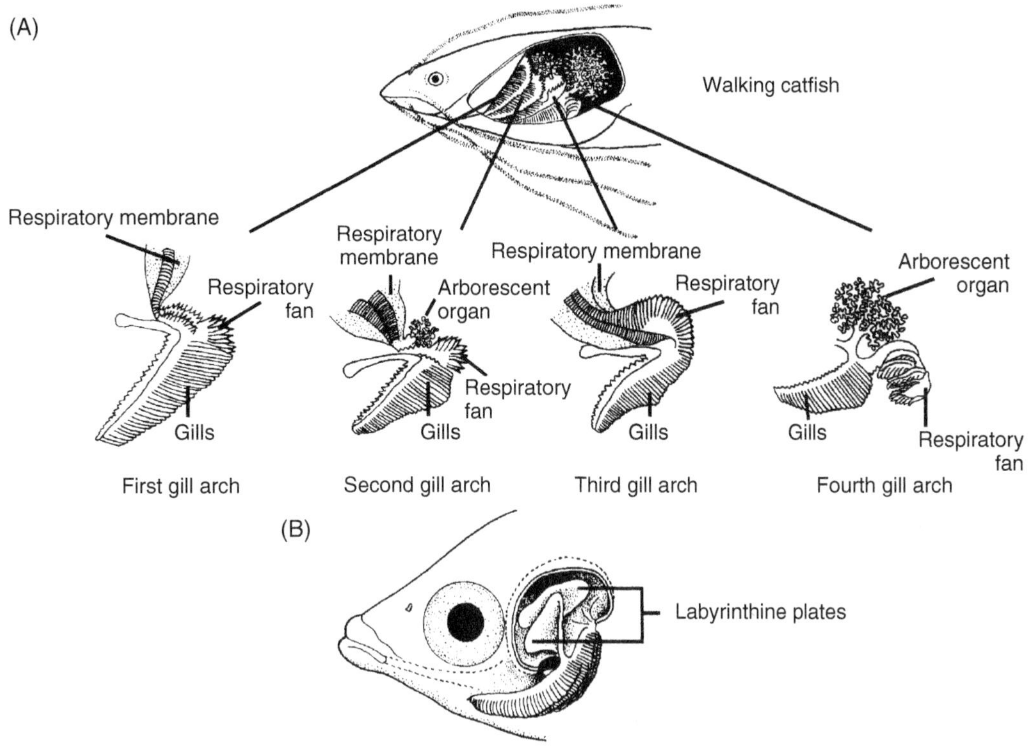

FIGURE 5.6 (A) Lateral views of the gill arches of the Walking Catfish (*Clarias batrachus*) showing the respiratory fans, respiratory membranes of the suprabranchial chamber, and treelike extensions (arborescent organs) that permit the fish to extract oxygen from the air when it is out of water. (B) A cut-away view of the branchial region of the Giant Gourami (*Osphronemus goramy*) shows a labyrinth of platelike extensions to accomplish the same goal. (A) based on Munshi (1976); (B) based on Peters (1978).

designed for gas deposition or removal (discussed later in this chapter), but several air breathers have highly vascularized and subdivided gas bladders designed for gas exchange. These include the very large South American *Arapaima* (Osteoglossidae), as well as the North American Bowfin (Amiidae) and gars (Lepisosteidae). Tarpons (*Megalops*) also use the gas bladder to augment respiration when oxygen levels in the water are low (Seymour et al. 2004). The gills of these aquatic air breathers are still important for getting rid of metabolic wastes, such as carbon dioxide and ammonia, and for regulating ionic and acid–base balance (see Chapter 7).

The six surviving lungfishes (Dipnoi) have true lungs. The Australian Lungfish (Neoceratodontidae) is a facultative air breather with a single lung, whereas the African and South American lungfishes (Protopteridae and Lepidosirenidae, respectively) are both obligate air breathers with bilobed lungs (discussed in Chapter 13). The gills of the South American Lungfish (*Lepidosiren paradoxa*) are of so little value for gas or ion exchange that the respiratory physiology of this species is more similar to that of an amphibian than to most other fishes (de Moraes et al. 2005).

The skin can be an effective organ for gas exchange in the air as long as it is well vascularized and kept moist. Some galaxiids, generally small freshwater fishes of the southern hemisphere, especially in Australia and New Zealand, rely on cutaneous respiration for both oxygen uptake and the excretion of carbon dioxide (Urbina et al. 2014). The small size and elongate body shape help by providing a large surface area to volume ratio, and the scaleless skin auguments gas diffusion. One species of galaxiid regulates cutaneous gas exchange by opening and closing pores in the skin on the abdomen, and also modifies its gills by thickening the epithelial layer for additional support when out of the water for several hours (Magellan et al. 2014).

A strong reliance on air breathing among some freshwater fishes aids survival in oxygen-poor habitats, but it also helps cope with drought. When rivers and ponds dry up, African lungfishes burrow into the sediment, dramatically slow their metabolism, and can remain in this torpid state for years. When the rains return and water levels rise, they leave their mud cocoons and become active (see Chapter 13). When the Walking Catfish (*Clarias*) is confronted with drought conditions, it "walks away" to find another pond, using a side-to-side lurching action supported by its stout pectoral spines.

Many intertidal fishes can breathe air, suggesting that this is part of the broad range of capabilities of fishes in this extreme habitat. Oxygen can become limited in tidepools due to increasing temperature and salinity and the ongoing respiration of animals and plants. Most marine air-breathing fishes evolved from relatively advanced fishes, so they do not have lungs and instead rely on modification of existing aquatic breathing structures such as gills and skin (Martin &

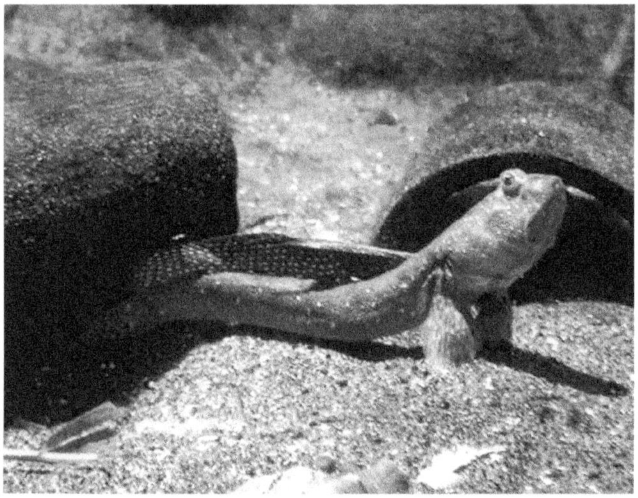

FIGURE 5.7 Mudskippers, such as *Boleophthalmus pectinirostris*, can exchange respiratory gases through their moist skin. Some will also carry a mouthful of air for gas exchange in the oral cavity or to help keep eggs in their burrow provided with oxygen. Wikimedia Commons / Public Domain.

Bridges 1999). Gills are modified with structural support to prevent collapsing in air, and the skin often is well vascularized, has few scales, and is kept moist. Mudskippers (Gobiidae, Fig. 5.7) absorb oxygen through their moist skin, but also supplement that by exchanging oxygen in the oral cavity. Some emergent intertidal species decrease their oxygen consumption rate by using anaerobic respiration, whereas other species simply reduce their activity until the next high tide.

Some of the mudskippers (Gobiidae), lay their eggs in incubation chambers out of the water to avoid low oxygen conditions that would be harmful to the developing embryos. Adults of some of these species augment the oxygen supply in the incubation chambers by bringing in mouthfuls of air (Graham 2011a, Toba & Ishimatsu 2014).

Although aquatic air breathers rely on gills and skin for the release of carbon dioxide, ion regulation, and nitrogen excretion (Shartau & Brauner 2014), the capability of some air-breathing fishes to tolerate extended periods of low oxygen availability, and in some cases aestivation, would require some biochemical means of either preventing or tolerating low blood pH (due to elevated carbon dioxide) and elevated levels of nitrogen wastes, which are usually eliminated by diffusion at the gills (Chew & Ip 2014). These alternative strategies for eliminating nitrogen wastes are discussed in Chapter 7 (Osmoregulation, Excretion, Ion and pH Balance).

Because air-breathing fishes can tolerate low oxygen levels, some have been used for high-density, low-maintenance aquaculture in warm climates. A review of the outcomes of some of these efforts, however, shows that although the fishes used can tolerate low oxygen in their natural environments, the high levels of metabolic wastes (e.g. carbon dioxide, ammonia, nitrites) and elevated levels of bacteria and other microbes in aquaculture ponds further depleted oxygen and may interfere with growth (Lefevre et al. 2014).

Gas Transport

Oxygen enters the blood at the respiratory surfaces and is transported via the circulatory system to tissues and released. Some oxygen is dissolved in the blood plasma. This is not enough, however, to support the metabolic needs of most fishes, except in some Antarctic icefishes (Channichthyidae). The red blood cells of most fishes and other vertebrates contain **hemoglobin**, a protein that increases the overall capacity of the blood to transport oxygen. Each hemoglobin molecule has four subunits, each of which can bind a single molecule of oxygen. The packaging of hemoglobin within red blood cells permits the binding and releasing of oxygen without affecting the plasma or other cells in the blood stream.

For hemoglobin to work well as an oxygen-transporting protein, it must bind oxygen at the respiratory surface and release it at the tissues elsewhere in the body. Like many proteins, hemoglobin is sensitive to physical and chemical conditions, such as temperature and pH. Hemoglobin has a higher affinity for oxygen (can bind more easily) at higher pH but has a lower affinity when pH decreases (Fig. 5.8A). This phenomenon, known as the **Bohr effect**, is caused by changes in the structure of the hemoglobin subunits that alter access to oxygen binding sites (see Berenbrink 2011). In some cases, the structure of hemoglobin can become altered so much that oxygen cannot bind to all potential binding sites, and the total capacity of the blood to carry oxygen is decreased (the **Root effect**, Fig. 5.8B; see Waser 2011). In the tissues, blood pH tends to be lowered by the presence of carbon dioxide, which combines with water to form carbonic acid (H_2CO_3). At the respiratory surfaces, however, carbon dioxide is released to the environment, thereby decreasing the level of carbonic acid in the blood and raising the pH. These phenomena become very important in understanding oxygen transport and also the function of the teleost gas bladder, discussed later.

Hemoglobin's affinity for oxygen decreases as temperature increases. This is one reason why cold-water fishes often cannot survive at higher temperatures. At these higher temperatures, the fish simply cannot pick up enough oxygen at its gills, and could therefore suffocate even though sufficient oxygen is present in the water. The blood of a coelacanth, for example, has highest affinity for oxygen at 15 °C, and the fish suffers from hypoxic stress at temperatures above 25 °C (see Fricke & Hissmann 2000). Many fishes have multiple types of hemoglobin (Verde et al. 2011), exhibiting different degrees of sensitivity to changes in temperature or pH, therefore ensuring oxygen transport even if environmental conditions change.

Hemoglobin varieties can have different affinities for oxygen. For example, the higher affinity of toadfish (Batrachoididae) hemoglobin makes it better adapted for low oxygen environments, whereas the lower affinity of mackerel (Scombridae) hemoglobin reflects an oxygen-rich environment and an active lifestyle (Hall & McCutcheon 1938). The higher oxygen affinity in Largemouth Bass (*Micropterus salmoides*, Centrarchidae) hemoglobin makes this species better adapted to warmer, lower oxygen environments, and less sensitive to hypoxia than

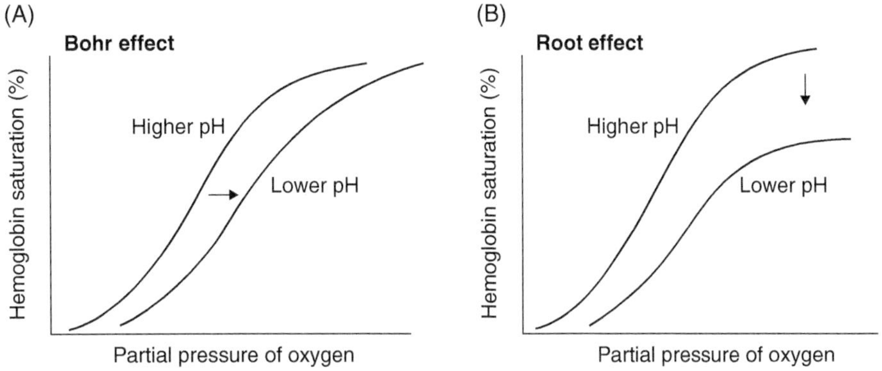

FIGURE 5.8 Oxygen dissociation curves. Vertical axes indicate the percent of total oxygen-binding sites that are occupied by oxygen. The horizontal axes indicate the concentration of oxygen dissolved in the solution surrounding the hemoglobin, typically blood plasma. A decrease in pH results in a shift of the curve to the right (the Bohr shift, A) and may also prevent full saturation of hemoglobin with oxygen (the Root effect, B).

its close relative, the Smallmouth Bass (*M. dolomieu*; Furimsky et al. 2003). Different fish hemoglobins also may show different temperature sensitivities. Some Antarctic fishes (Nototheniidae) possess hemoglobins that are effective at temperatures well below the effective temperature range of temperate fishes (Hochachka & Somero 1973). The hemoglobins of warm-bodied tunas and lamnid sharks are less sensitive to temperature changes than are hemoglobins of many other species (Nikinmaa 2011). This is adaptive because blood temperatures in these fishes may change by 10 °C or more as blood travels from the gills to the warm swimming muscles. If the hemoglobins were not thermally stable, arterial blood might unload its oxygen as it warmed in the countercurrent heat exchanger (and see Chapter 7, Endothermic Fishes), resulting in loss of oxygen to venous blood and depriving the highly active swimming muscles that need the oxygen most (Hochachka & Somero 1984).

In addition to transporting oxygen, the blood must pick up the carbon dioxide that is produced in cellular metabolism and transport it back to the gills or other respiratory surfaces for release. If excess carbon dioxide is not removed, blood and tissue pH will drop and interfere with normal metabolic processes. Because of this link between carbon dioxide levels and pH, the transport of carbon dioxide and oxygen are coupled (Brauner & Rummer 2011), as in other vertebrates.

Carbon dioxide can be carried in the blood in three forms. A relatively small amount is simply dissolved in the plasma. Some is bound to hemoglobin in red blood cells to form carbaminohemoglobin, but this occurs at a lower rate in fishes than in tetrapods (Nikinmaa 2011). The greatest proportion of carbon dioxide in the blood is carried as **bicarbonate ion** (HCO_3^-) resulting from the dissociation of carbonic acid.

At the active tissues, carbon dioxide diffuses into the blood. In the plasma, some carbon dioxide combines with water to form carbonic acid, which dissociates to bicarbonate and hydrogen ions. Most of the carbon dioxide, however, is drawn into the red blood cells, where this same reaction is taking place at a faster rate due to the presence of the enzyme carbonic anhydrase (Gilmour 2011). H^+ released from the dissociating carbonic acid inside the red blood cells binds to the hemoglobin and causes the release of oxygen, which then diffuses out of the red blood cells and into the tissues.

The dissociating carbonic acid also increases the concentration of bicarbonate (HCO_3^-) inside the red blood cell. Much of this HCO_3^- diffuses across the membrane of the red blood cell and into the plasma, keeping intracellular HCO_3^- levels from getting so high that they would inhibit further carbon dioxide uptake. The net result of all of these reactions is that the blood has taken up carbon dioxide, red blood cells have become slightly acidified, oxygen has been released from hemoglobin, the hemoglobin molecule itself has helped buffer against too much of a pH drop by taking up some carbon dioxide, and the bicarbonate level in the plasma has increased (see Brauner & Rummer 2011; Randall 2011).

When blood gets to the respiratory surface where carbon dioxide levels are low and oxygen levels are high, the reactions described in the previous paragraph occur in the opposite direction, resulting in the release of carbon dioxide, a slight increase in blood pH, and the binding of oxygen to hemoglobin within the red blood cells.

Metabolic Rate

Metabolism is the sum of all biochemical processes taking place within an organism. These reactions give off heat as a byproduct, therefore measuring the heat generated by an animal probably is the best way to measure its metabolism. This can be a difficult process, however, so frequently other indicators related to metabolism serve as indirect measures. In fishes, the rate of oxygen consumption is frequently used as an indicator of metabolic rate, but we must assume that no anaerobic metabolism takes place during the measurement period.

Metabolic rates can be influenced by a variety of factors, including age, sex, reproductive status, food in the gut, physiological stress, activity, season, and temperature. For this reason, it is useful to define metabolic terms. **Standard metabolic rate** is often defined as the metabolic rate of a fish while it is at rest and has no food in its gut (Nelson & Chabot 2011). However, many fishes under natural conditions feed regularly and therefore almost always have some food in the gut (Belokopytin 2004), so some amount of digestion is likely to be part of a fish's metabolism at all times. Fishes rarely remain still while metabolic rates are being measured, so the term **routine metabolic rate** is used to indicate that the rate was measured during routine activity levels. The resulting estimates of metabolic rate are therefore higher than what might be expected for a resting fish. Sometimes researchers will measure metabolism at several levels of activity and extrapolate back to zero activity to estimate standard metabolic rate.

Metabolic rate increases with activity until a fish is using oxygen as rapidly as its uptake and delivery system can supply it. This is its **maximum** (or active) **metabolic rate**. The difference between the standard metabolic rate and the maximum metabolic rate at any given temperature is known as the **metabolic scope**. The concept of metabolic scope can be important in trying to understand a fish's metabolic limits. Any factors that increase standard or routine metabolic rates, such as stress due to disease, handling, reproduction, or environmental conditions, narrow this scope and may limit other activities.

In general, fishes tend to have higher metabolic rates at higher temperatures, so as temperature increases a fish's need for oxygen also increases. Because the availability of oxygen in water decreases with increasing temperature, warm conditions stress most fishes. This stress probably was an important factor promoting the evolution of air-breathing in many tropical fishes.

Under laboratory conditions, fish acclimated to low temperatures consume less oxygen than fish of the same species acclimated to higher temperatures (see, for example, Beamish 1970; Brett 1971; Kruger & Brocksen 1978; DeSilva et al. 1986). The rates of many biochemical reactions increase with temperature, thereby increasing the need for oxygen to provide the energy for cellular metabolism. Under natural environmental conditions, however, the gradual **acclimatization** of a fish to seasonal changes involves many physiological processes, each of which can have an impact on overall metabolism. Therefore, the results of temperature **acclimation** studies during a single season may not accurately represent seasonal changes in metabolic rates (Moore & Wohlschlag 1971; Burns 1975; Evans 1984; Adams & Parsons 1998; Gamperl et al. 2002).

Temperature–metabolic rate generalizations based on studies of individual species acclimated to different temperatures should not be applied across species, especially those adapted to very different thermal environments. At low temperatures, for example, polar fishes have metabolic rates considerably higher than those of temperate species acclimated to the same low temperatures (Brett & Groves 1979). Metabolic rates of tropical fishes and those of temperate species acclimated to high temperatures differ only slightly.

Size also affects metabolism. Large fishes generally will have higher overall metabolic rates than small fishes, assuming other factors such as activity are constant. However, the metabolic rate per unit of mass, often called the mass-specific metabolic rate or **metabolic intensity**, is usually higher for smaller fishes. This relationship seems to hold true for other animal groups as well.

Swimming is metabolically costly to fishes. Because water is much denser than air, more energy is required to move through it. There is a trade-off, however, in that the density of water also provides buoyancy so that fishes do not have to resist gravity as they would in a less dense medium such as air. Under most conditions, fish swimming is energetically less expensive than bird flying (Bale et al. 2014). Oxygen consumption in fishes generally increases exponentially with swimming velocity, but oxygen consumption curves probably underestimate the true metabolic cost of swimming at high speeds because of the increased use of anaerobic metabolism at higher velocities.

The evolution of a torpedo-shaped, fusiform body undoubtedly is the result of energetic advantages. Fin shape and placement also are important considerations, as well as body flexion during the act of swimming. The fastest, most active swimmers are streamlined, with high, thin caudal fins that oscillate rapidly while the rest of the body remains fairly rigid. This eliminates the drag that would be created by throwing most of the body into curves while swimming forward. The relationship between body shape, fin placement, and swimming style are addressed in Chapter 4 (Locomotion: Movement and Shape).

Body shape and other morphological features also are important to the energetics of many benthic fishes. Bottom-dwelling stream fishes, for example, are able to hold their position in a high-flow environment without much energetic cost due to body shape and use of their fins. Mottled Sculpin (Cottidae) can use their pelvic fins to hold to the rocky substrate of swift mountain streams. They can even hold the position in a plexiglass swimming tunnel, apparently by using their large pectoral fins to create a downward force as the water flows over them (Facey & Grossman 1990). Their overall body shape of a large head and a narrow, tapering body may also help them remain on the bottom as water flows over them. These morphological adaptations give sculpins the ability to hold a position in moderate currents without a significant energetic cost. The bottom-foraging Longnose Dace (Cyprinidae) responds similarly at low to moderate velocities, showing no change in oxygen consumption. At higher velocities, however, it must resort to swimming to hold its position and its oxygen consumption increases dramatically (see Facey & Grossman 1990).

Energetics and Buoyancy

Swimming

Fishes have evolved a variety of mechanisms to minimize the cost of swimming. Variations in body shape, fin shape and location, and swimming style are addressed in Chapter 4. Fishes also can utilize vortices in their environment to reduce the cost of swimming (Liao et al. 2003). These vortices may be created by either water moving past an obstacle or by the movement of other fishes, such as those in a school (see Chapter 17). By carefully positioning themselves fishes can use the vortices to "slalom" ahead while reducing the activity of trunk muscles normally used in propulsion, thereby conserving energy.

The Gas Bladder and Buoyancy Regulation

The **gas bladder** is a gas-filled sac located between the alimentary canal and the kidneys (Jones 1957; Marshall 1960). Some refer to it as an "air" bladder, but this is not accurate because it is filled with oxygen, carbon dioxide, and nitrogen in different proportions than occur in the air. This structure is also often referred to as the "swim bladder," but it has nothing to do with generating propulsive forces for the act of swimming and instead saves energy by regulating buoyancy. For these reasons, we feel that the term gas bladder is more appropriate.

The original function of the gas bladder was probably as a lung, but in most fishes today, it functions mainly to control buoyancy. It also plays a more specialized role in respiration, sound production, and sound reception in some fishes. Some species in at least 79 of 425 families of extant teleosts have lost their gas bladders, at least as adults (McCune & Carlson 2004). Most of these fishes are either benthic or deep-sea species. Billfishes (Istiophoridae) and two genera of halfbeaks (all 10 species of *Hemiramphus* and one of two species of *Oxyporhamphus*) have a vesicular gas bladder composed of many discrete gas-filled vesicles (Tibbetts et al. 2007). Freshwater teleosts have somewhat larger gas bladders (about 7% of body volume) than do marine teleosts (about 5% of body volume, Alexander 2011), perhaps because salt water is denser and provides more buoyancy.

Embryologically, the gas bladder is a two-layered (tunica externa and tunica interna), specialized outgrowth of the roof of the foregut. The connection to the gut via the **pneumatic duct** is retained in some fishes (Pelster 2011). This **physostomous** condition (open gas bladder) is seen mainly in less derived teleosts, such as the Otocephala (including herrings and catfish) and Protacanthopterygii (including salmon). In **physoclistous** fishes, which include the more derived teleosts (Paracanthopterygii and Acanthopterygii), the gas bladder is initially open to the esophagus, but becomes sealed off once the gas bladder is initially filled during the larval stage.

The gas bladder is regulated by the sympathetic nervous system through a branch from the coeliaco-mesenteric ganglion and by branches of the left and right intestinal vagus (X) nerves; cutting the vagus prevents gas secretion into the gas bladder. Sensory nerve endings function as stretch receptors, responding to stretching or slackening of the gas bladder, thus providing information to the fish about the relative fullness of the gas bladder.

The secretory region of the gas bladder is located anteroventrally and contains the gas gland and a rete mirabile. The **gas gland** secretes lactic acid into the beginning of the capillary loop. This acidifies the blood, forcing hemoglobin to release oxygen, and also reducing the solubility of all dissolved gases, which helps add gas to the gas bladder. The rete associated with the gas gland helps build gas pressure to facilitate the diffusion of gases from the gas gland into the gas bladder. The resorptive region of the gas bladder, often called the oval, is usually located posterodorsally. It develops from the distal end of the degenerating pneumatic duct and consists of a thin, highly vascularized area. Circular muscles contract and close off the oval, preventing the release of gases. Longitudinal muscles contract and expose the oval, permitting gas escape to the blood. Most of the gas bladder is lined with a layer of cells containing crystals of guanine 3 μm thick, which decreases permeability by 40 times over an unlined membrane and thus limits gas escape except at the oval (see Pelster 2011).

For fishes that are not benthic, maintaining vertical position in the water column could be energetically expensive. This is not the case for most teleosts, however, because of their ability to regulate buoyancy by adjusting the volume of the gas bladder according to depth. Changes in depth are accompanied by changes in pressure, and the gas bladder expands and contracts accordingly. If a fish descends in the water column, the increase in pressure decreases the volume of the gas bladder, making the fish negatively buoyant and the fish begins to sink. Conversely, if a fish ascends in the water column, the gas bladder expands and the fish becomes positively buoyant. Both negative and positive buoyancy are energetically expensive, as the fish has to adjust by swimming. Therefore, to save energy fishes must be able to regulate the volume of the gas bladder by the release or addition of gases in order to maintain neutral buoyancy at a variety of depths. Czesny et al. (2005) showed that larval Yellow Perch (Percidae) that did not inflate their gas bladder fed less efficiently, used more energy, grew more slowly, were more susceptible to predation, and had higher overall mortality than those with properly inflated gas bladders.

Consider first the case of gas release. The gas bladder of a fish rising in the water column will continue to expand, so to remain neutrally buoyant the fish must release some of the gas. In physostomes, gas can be released directly via the pneumatic duct and esophagus. In some physostomes, however, such as eels (Anguillidae, Congridae), the pneumatic duct serves as a resorptive area for slow gas release via the blood but can release gas rapidly via the esophagus if necessary (Fig. 5.9A). In physoclists, gas must be released into the blood at the oval (Fig. 5.9B). The blood carries the excess gas to the gills, where it is released to the surrounding water. Fishes regulate the loss of gas by controlling the flow of blood to the oval and by using muscles to regulate the amount of gas entering the oval. Physoclistous fishes brought to the surface quickly, such as by angling or in commercial fishing gear, will have greatly expanded gas bladders and will float at the surface if released.

The addition of gas to the gas bladder is more complex. As a fish descends, the volume of the gas bladder decreases due to increasing pressure, and the fish must add gas to maintain neutral buoyancy. A physostome could theoretically swim to the surface, and take air into the gas bladder via the pneumatic duct. However, the change in pressure with depth would affect any air

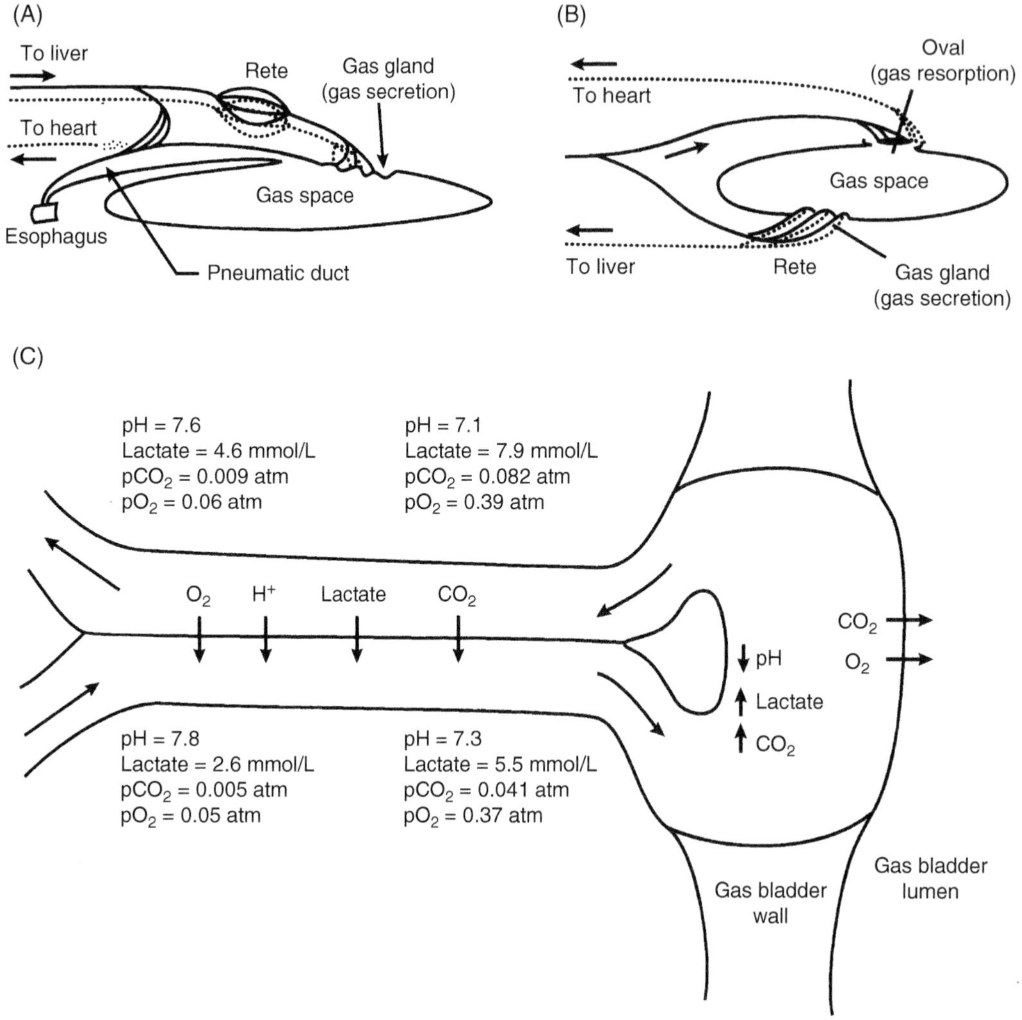

FIGURE 5.9 Schematic representation of the gas bladders of a physostome (A) and a physoclist (B). The pneumatic duct permits gas release via the esophagus in a physostome, whereas a physoclist must rely on a specialized area of the bladder wall for gas resorption. Both have gas glands with associated retia for gas addition. (C) Production of lactate and hydrogen ions by gas gland tissue triggers the hemoglobin's release of oxygen (the Bohr and Root effects) and a decrease in gas solubility (the salting-out effect). Countercurrent exchange of ions and dissolved gases in the rete creates very high gas pressures in the gas gland, thereby facilitating the diffusion of gases into the gas bladder. (A, B) after Denton (1961); data presented in (C) are for eels (*Anguilla*), from Koboyashi et al. (1989, 1990).

gulped at the surface, making this impractical, if not impossible. The addition of gas takes place by the diffusion of gases from the blood into the gas bladder at the gas gland. The process of inflating the gas bladder occurs by diffusion, so there must be a dramatic increase in the amount of gas in solution in the blood.

Three general physiological phenomena discussed earlier act together to bring this about (Fig. 5.9C). First is the effect of acidification on hemoglobin's ability to hold oxygen. The tissues of the gas gland produce lactic acid, which dissociates to lactate and hydrogen ions. The increase in hydrogen ion concentration decreases the blood pH, and the Bohr and Root effects cause unloading of oxygen from hemoglobin. This oxygen goes into solution in the blood plasma, increasing the amount of dissolved oxygen. The second phenomenon is the reduced solubility of gases as the concentration of lactate and hydrogen ions increases (the **salting-out effect**). This helps to drive the dissolved gases out of solution and into the gas bladder through the formation of small bubbles on the wall of the gas gland (Copeland 1969). The combined effect results in the diffusion of gas from the blood into the gas bladder. Elevated levels of plasma carbon dioxide also enhance the addition of this gas into the gas bladder (Pelster & Scheid 1992).

The third phenomenon that makes the gas gland so effective is the countercurrent exchange between the blood vessels leading to and from the gas gland through the rete mirabile. As blood leaves the gas gland and travels through the rete, lactate, hydrogen ions, and dissolved gases diffuse down their concentration gradients into the blood coming toward the gas gland. Hence, the countercurrent arrangement of the rete capillaries helps increase the levels of diffusible gases in the gas gland.

The reason that the blood can give up oxygen in the gas gland and return to the rete with a higher partial pressure of oxygen than it had when it entered the gas gland (as shown in Fig. 5.9C), is that partial pressure only indicates the amount of gas in solution; oxygen bound to hemoglobin is not in solution and therefore is not accounted for in the partial pressure. So although blood leaving the gas gland has less total oxygen than when it entered because some of the oxygen has diffused into the gas bladder, the partial pressure is higher because most of the oxygen that is present is in solution. Hemoglobin cannot bind much because the pH is low.

One other important factor is the timing of the release of oxygen by hemoglobin under acidic conditions (the **Root-off shift**) and the binding of oxygen by hemoglobin when pH increases (the **Root-on shift**). The Root-off shift occurs nearly instantaneously, whereas the Root-on shift takes several seconds. Therefore, hemoglobin in blood in the rete that is leaving the gas gland area does not increase its affinity or capacity for oxygen until it is already out of the rete.

Understanding how the rete mirabile functions to increase gas pressures in the gas gland helps explain why fishes with a long rete can create higher gas pressures than those with a shorter rete. Deep-sea fishes (see Chapter 10, The Deep Sea) must deposit gas under high-pressure conditions, and therefore have longer retia than do shallow-water fishes (Alexander 1993). Rattails (Macrouridae) and ophidioids living at abyssal depths of 4000 m and deeper have retial capillaries 25 mm in length or more; shallow water forms have retia only 1 mm long (Marshall 1971). The rete associated with the gas bladder of migratory eels (Anguillidae) lengthens as fish metamorphose from their shallow water, freshwater, or estuarine juvenile phase to their deep water, oceanic reproductive phase (Kleckner & Kruger 1981; Yamada et al. 2004).

Because the main function of the gas bladder is to maintain buoyancy at a given depth, several groups of teleosts find it more adaptive to have greatly reduced gas bladders, or none at all. Many benthic fishes, such as sculpins and flounders, either have gas bladders that are greatly reduced in size or lack gas bladders altogether. The absence of a gas float makes it that much easier to remain on the bottom. Fishes that change depth rapidly and frequently, such as some tunas, also lack gas bladders. Herring (Clupeidae) are marine physostomes that lack a gas gland. Their high body lipid content, however, also provides buoyancy, so decreasing gas bladder volume with depth is less of a problem (Brawn 1962).

Gas bladders are found only in bony fishes. Elasmobranchs must utilize other means to reduce their buoyancy (Alexander 2011). A cartilaginous skeleton helps because cartilage is much less dense than bone (the specific gravity of cartilage is 1.1, as opposed to 2.0 for bone), and the constant swimming of pelagic sharks helps prevent sinking by providing upward lift (see Chapter 4). Pelagic elasmobranchs also maintain high levels of low-density lipids in their large livers, which may make up 20–30% of their total body mass (Alexander 1993). The Basking Shark (Cetorhinidae) has a large liver that contains much squalene (specific gravity = 0.86), which is less dense than most other fish oils (specific gravities around 0.92); A Basking Shark that weighs 1000 kg on land weighs only 3.3 kg in water, or about 0.3% of its weight on land, due to the buoyancy provided by its large, oil-filled liver. Another low-density, oily compound, wax esters (specific gravity = 0.86), has been found in the livers of some benthopelagic sharks (Van Vleet et al. 1984). Livers of benthic sharks make up only about 5% of their body mass.

Some teleosts also reduce body density with lipids, wax esters, and other low-density compounds (see Alexander 2011). These include Oilfish (Gempylidae), coelacanths (*Latimeria*), mesopelagic lanternfishes (Myctophidae), and leptocephalus larvae of eels (Anguillidae). Other tactics to reduce body density include reduced ossification of bone and increased water content of tissues in some deep-sea fishes (see Chapter 10).

Energy Intake and Digestion

The diversity of feeding adaptations found among fishes is discussed in Chapter 3; the emphasis here is on post-ingestion processes. As in other vertebrates, the alimentary tract can be divided into regions. The most anterior part consists of the mouth, buccal cavity, and pharynx. Next is the foregut (esophagus and stomach), followed by the midgut or intestine, and then the hindgut. Voluntary striated muscle extends from the buccal cavity into the esophagus, involuntary smooth muscle from the posterior portion of the esophagus through the large intestine (see Olsson 2011 for a review of structure of the gut).

The **buccal cavity** (mouth) and pharynx are lined with stratified epithelium, mucus-secreting cells, and, frequently, taste buds, but lack the salivary glands present in mammals. This area is concerned with seizure, control, and selection of food. Most fishes lack a mechanism for chewing food in the mouth, so food items are swallowed whole or in large chunks and much of the physical breakdown takes place in the stomach. However, many fishes, such as minnows (Cyprinidae), suckers, croakers (Sciaenidae), cichlids (Cichlidae), wrasses (Labridae), and parrotfishes (Scaridae), have bony arches or toothed pads deep in the pharynx that are equipped with toothlike projections (pharyngeal teeth) that grind up food before it reaches the stomach (see Chapter 3, Pharyngeal Jaws).

The **esophagus** is a short, thick-walled tube lined with stratified ciliated epithelium, mucus-secreting goblet cells, and, often, taste buds. The anterior portion has striated muscles, the posterior part smooth muscles that produce peristaltic movement of food toward the stomach. The esophagus is very distensible, so choking is rare, but miscalculation of prey size or armament can lead to the death of the predator. In 2015 a White Shark that had apparently choked to death on a sea lion washed up on the shore in Australia.

The evolution of jaws presumably permitted the capture of larger prey, making a distensible **stomach** advantageous for temporarily storing food after it has been swallowed. Tough ridges along the internal wall of the stomach, along with contractions of the muscular wall, aid in the physical breakdown of foods. The stomach is lined with columnar epithelium with mucous-secreting cells and glandular cells that produce the proteolytic enzyme pepsin, as well as hydrochloric acid which makes the pepsin more effective by lowering pH. The combined physical and chemical activity of the stomach creates a soupy mixture that is released into the small intestine in small amounts (Krogdahl et al. 2011).

Although usually a fairly simple structure, evolutionary modifications of the fish stomach have led to some unusual functions. For example, pufferfishes and porcupinefishes (Tetraodontidae and Diodontidae) inflate themselves when threatened by filling the highly distensible stomach with water, thereby making them larger and more difficult to swallow (see Fig. 16.17); they will also inflate themselves with air when handled after capture. The stomach is modified into a grinding organ in sturgeons (Acipenseridae), gizzard shads (*Dorosoma*), and mullets (Mugilidae) and is used as an air-breathing organ in some of the South American armored catfishes (Loricariidae).

In hagfishes and lampreys, the absence of true jaws is correlated with the absence of a stomach. Some jawed fishes also lack true stomachs; a secondary condition with no simple ecological explanation (Kapoor et al. 1975). Fishes without true stomachs include chimaeras (Holocephali), lungfishes (Dipnoi), and several teleosts, including minnows (Cyprinidae) such as the European *Rutilus*, killifishes (Cyprinodontidae), wrasses (Labridae; see Chao 1973), parrotfishes (Scaridae), toadfishes (Batrachoididae), and the beloniforms (halfbeaks, needlefishes, flyingfishes). Characteristics of the stomachless condition are both cytological and biochemical. Cytologically, no gastric epithelium or glands are present. The stratified epithelium of the esophagus grades into the columnar epithelium of the intestine. Biochemically, no pepsin or hydrochloric acid is produced, making it impossible for these fishes to dissolve shells or bones, an interesting fact given that many wrasses and lungfishes eat molluscs, and parrotfishes munch on corals.

Digestion continues in the **small intestine**, aided by bile from the liver, which helps emulsify lipids, and by secretions from the pancreas that contain bicarbonate to neutralize the acid from the stomach and a variety of enzymes to complete the process of chemical digestion (Caruso & Sheridan 2011). The intestine of most fishes is lined with simple columnar epithelium and goblet cells, which secrete mucus. Usually, no multicellular glands are present. The chief exception to this is in the cods (Gadidae), which have small tubular glands in the intestinal wall. **Pyloric caecae**, fingerlike pouches that connect to the gut at the junction between the stomach and intestine, are often present. These may function in absorption or digestion, and they vary in number from only three in a scorpionfish (*Setarches*) to thousands that form a caecal mass in tunas. The number of pyloric caecae is useful in the classification of some groups, like the Salmonidae.

Although most fishes are carnivorous, herbivorous species can have a substantial effect on macrophyte or algal communities in both marine (Alcoverro & Mariani 2004) and freshwater (Nurminen et al. 2003) environments. Herbivorous fishes may depend in part on fermentation by symbiotic microorganisms in their guts to digest the plants they consume (Clements & Choat 1995, Fig. 5.10). Many primarily herbivorous tropical marine

FIGURE 5.10 Herbivorous fishes, such as Bermuda Chub (*Kyphosus sectatrix*) which feed on algae, may depend in part on fermentation by symbiotic microorganisms in their guts to digest the plants they consume. G. Helfman (Author).

fishes showed elevated levels of short-chained fatty acids (SCFAs) in the posterior gut segments. SCFAs are produced by microbial digestion of plant matter in the guts of terrestrial vertebrate herbivores. Most fishes examined also showed elevated SCFA levels in their blood, suggesting microbial fermentation of algae in the gut. Some planktivorous fishes studied also showed elevated SCFA levels. Some herbivorous fishes seem to rely on physical grinding or low stomach pH to break through plant cell walls (Lobel 1981). Herbivorous fishes with lower SCFA levels, suggesting limited microbial fermentation in the gut, possessed some mechanism for mechanically grinding ingested plant material (Clements & Choat 1995).

The small intestine is also the primary site of absorption of the products of digestion (Sundell 2011), and the length of the intestine varies and is generally correlated with feeding habits. Carnivores such as pickerels (*Esox*) and perches (*Perca*) have very short intestines, one-third to three-quarters of their body length. Elasmobranchs have a short, thick intestine with a large, spiraling fold of tissue (the **spiral valve**) to increase absorptive surface area (Fig. 5.11). In addition to sharks, some primitive bony fishes such as the coelacanths (*Latimeria*), lungfishes, gars, and Ladyfish (*Elops*) have a spiral valve intestine. Both hagfishes and lampreys have a straight intestine, but the surface area of the intestine is increased in the lampreys by the **typhlosole**, a fold in the intestinal walls. Teleosts generally have longer intestines, often with numerous pyloric caecae to increase the absorptive area (Buddington & Diamond 1987). The intestine is much longer in herbivores and detritus feeders to increase the opportunity to extract nutrients; long intestines are often coiled within the body (see Fange & Grove 1979; Lobel 1981). Some of these fishes, such as minnows, suckers, and topminnows (Cyprinodontidae), and several tropical marine fishes, including wrasses and parrotfishes, have reduced stomachs or lack them altogether (Fange & Grove 1979; Lobel 1981; Buddington & Diamond 1987). In the herbivorous North American

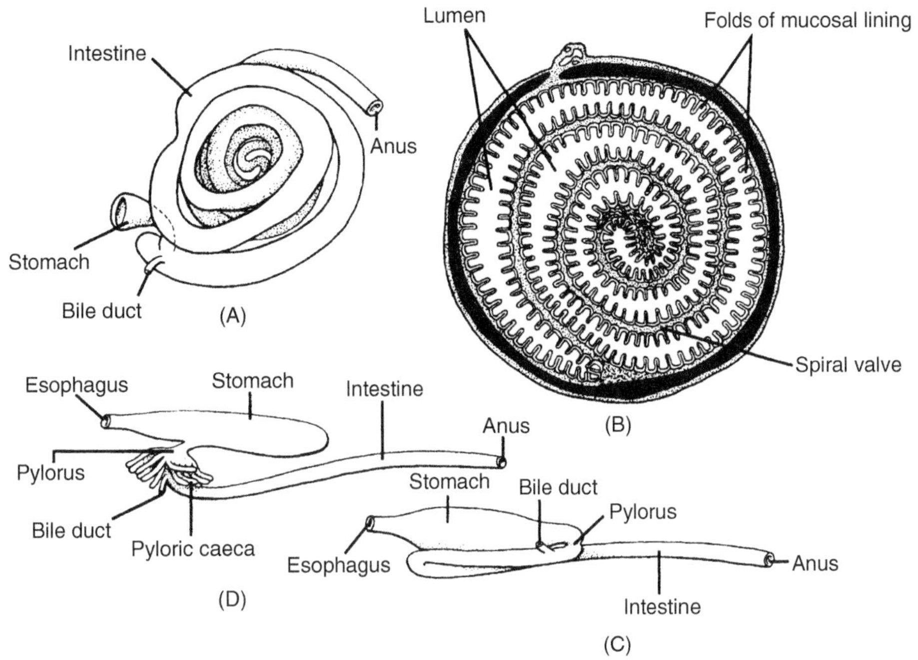

FIGURE 5.11 Variation in intestinal length and other features among carnivorous and herbivorous fishes. (A) An herbivorous catfish (Loricariidae). (B) Spiral valve in cross-section of intestine of a shark. (C) A carnivore, the Northern Pike (*Esox lucius*). (D) A carnivore, a perch (*Perca*). From Lagler et al. (1977).

stoneroller minnows (*Campostoma*), the intestine is very long and wrapped around the swim bladder. More surface area allows for more effective extraction of nutrients, especially for growth; transgenic Coho Salmon (Salmonidae) have more than two times the intestinal surface area than do their non-transgenic counterparts, which may help explain how these fish are so effective in extracting the nutrients needed to maintain their high rate of growth (Stevens & Devlin 2000).

As mentioned earlier, the liver and pancreas both participate in digestion. The **liver** develops as a ventral evagination of the intestine, as in other vertebrates. The anterior portion develops into the liver, and the posterior portion into the gall bladder and bile duct. The liver also stores fat in some fishes. Before humans produced vitamins A and D synthetically, cods and sharks were harvested for their liver oil, which is rich in these vitamins. The **gall bladder** is a thin-walled temporary storage organ for the bile. It empties into the beginning of the intestine by the contraction of smooth muscles. Bile is usually green due to bile pigments (biliverdin and bilirubin) resulting from the breakdown of blood cells and hemoglobin and also contains fat-emulsifying bile salts, which may assist in converting the acidity of the stomach to the neutral conditions in the intestine. The **pancreas** is both an endocrine organ and an exocrine organ that produces digestive enzymes (Caruso & Sheridan 2011). These enzymes include proteases such as trypsin, carbohydrases such as amylase and lipase, and, in some insect-feeding fishes, chitinase. The pancreas is a compact, often two-lobed structure in Chondrichthyes. It is a distinct organ in soft-rayed teleosts, but becomes incorporated into the liver as a hepatopancreas in most spiny-rayed teleosts (except for parrotfishes).

The **hindgut (large intestine)** and **rectum** is not as well defined externally in fishes as it is in tetrapods. Generally, the muscle layer near the rectum is thicker than in anterior regions, and the number of goblet cells in the large intestine increases in the rectal region. In Chondrichthyes, the hindgut is lined with stratified epithelium, contrasting with a single cell layer in the midgut. An iliocaecal valve between the small and large intestines is often found in teleosts, but this valve is absent in Chondrichthyes, Dipnoi, and *Polypterus*. Although there may be some nutrient absorption in the hindgut, this last major portion of the gut functions primarily in water absorption.

Once basic metabolic demands are met, excess nutrients can be accumulated. **Carbohydrates** are stored as **glycogen** either in the liver or in muscle tissue. **Lipids** and **proteins** also are stored, resulting in an increase in mass. Lipids tend to accumulate either in the liver, in muscles, or as distinct bodies of fat in the visceral cavity. Protein often goes into tissue growth. All of these potential energy sources are mobilized when needed, although carbohydrates are metabolized first. In prolonged periods of starvation, such as during the migration of salmonids, body lipids and proteins will also be used. Stored lipids yield considerably more energy per gram than stored carbohydrates or proteins; some sharks mobilize lipids from their livers during periods of starvation.

Bioenergetics Models

Bioenergetics models can aid in understanding energy intake and utilization. Models are complex, however, because the energetic costs and benefits of physiological activities are

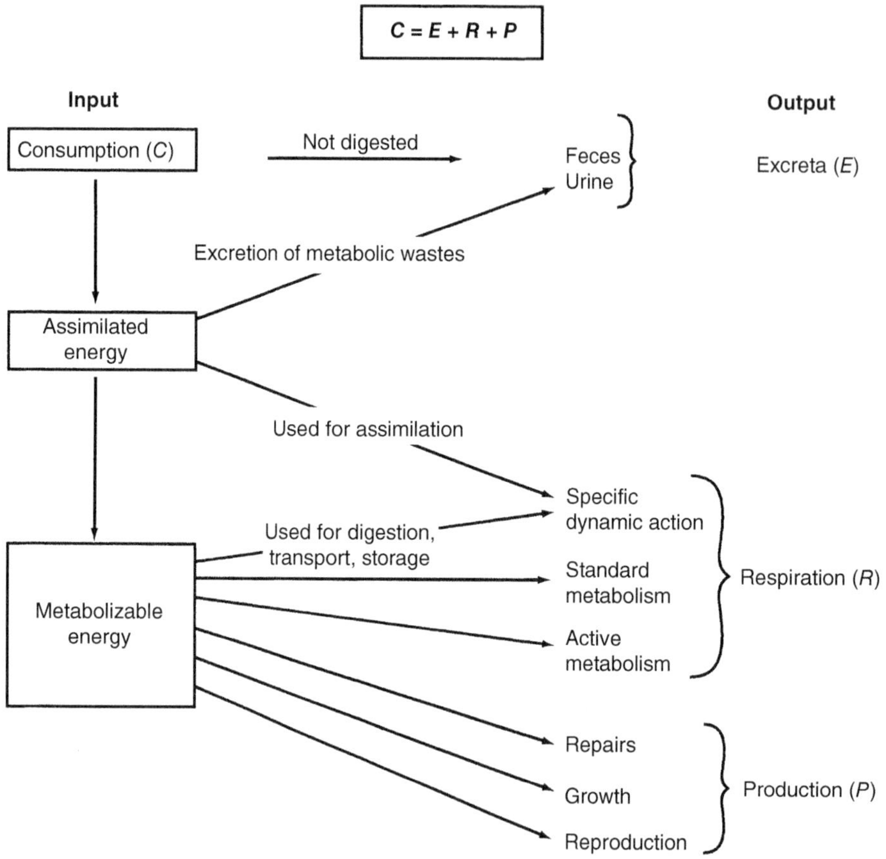

FIGURE 5.12 Partitioning of the energy consumed by a fish. Only energy not required to meet basic physiological needs (digestion, standard metabolism, repairs) or needed for activity is available for growth and gametes. Adapted from Videler (1993).

multifaceted, and must be accurately accounted for if the model is to provide a reasonable approximation of how energy is allocated. In addition, bioenergetic pathways differ on an individual basis. Therefore, bioenergetics models provide a broad conceptual framework rather than a precise prediction of what will happen in any particular organism. Bioenergetics models may, however, be useful in understanding how energy is allocated in populations and have been used to estimate the impacts of environmental alterations on rare species (Petersen & Paukert 2005). In addition, bioenergetics models of populations can be used to construct community bioenergetic models, thereby providing some understanding of energy flow through ecosystems and estimating how fish populations may be impacted by factors such as predators, invasive species, and climate change. But such models should be used cautiously. Bajer et al. (2003) applied two bioenergetic models to a controlled study of Yellow Perch, found deficiencies in both, and concluded that such models should be evaluated in lab and field studies before being applied.

Several methods can be used to determine the energetic content of food items, waste products, or components of fish growth such as tissue or gametes (see Wootton 1998). The energetic costs of different activities can be estimated either by direct calorimetry (measuring the heat produced by an organism) or some form of indirect calorimetry, such as measuring oxygen consumption (discussed earlier in this chapter). We can then construct a conceptual model (Fig. 5.12) to represent how energy may be partitioned. The energy equation is often represented as:

$$C = E + R + P,$$

where C is the energy **consumed**, E is the energy **excreted**, R is the energy used in **respiration**, and P is the energy remaining for **production**.

Some of the potential energy in food will never be digested and is therefore lost in the feces. The proportion that is digested is sometimes represented by the **absorption efficiency** (or "digestibility") and varies for different food types. Carnivorous fishes feeding on soft-bodied, highly digestible prey may have absorption efficiencies as high as 90% or more, whereas herbivores tend to have considerably lower absorption efficiencies (e.g., 40–65%; see Wootton 1998). In general, foods high in lipids and proteins have much higher absorption efficiencies than foods high in carbohydrates.

Of the energy that is absorbed during digestion, some are subsequently lost through the excretion of nitrogenous wastes. An additional 10–20%, depending on the amount and type of food consumed, is used in providing the energy needed for digestion (Jobling 1981). Larger meals and foods with higher protein content require more energy to digest and assimilate.

The remaining absorbed energy must be allocated among maintaining metabolism, swimming or other forms of activity, and the production of gametes or new somatic tissue (growth). Only energy remaining after other physiological maintenance needs have been met is available for growth or reproduction. Therefore, any factors that increase other metabolic demands can decrease growth or reproduction. Energy requirements for basic maintenance may increase with higher temperature, changes in salinity, or energy diverted to fighting infections, diseases, or parasites. In addition, exposure to contaminants that affect ion or water balance or that diminish the effectiveness of a fish's immune system can indirectly divert more energy away from growth and reproduction.

Supplementary Reading

Block B, Stevens E. 2001. *Tuna: physiology, ecology, and evolution. Fish physiology*, Vol. 19. New York: Academic Press.

Carrier, JC, Musick JA, Heithaus MR, eds. 2004. *Biology of sharks and their relatives*. Boca Raton, FL: CRC Press.

Diana, JS. 2003. *Biology and ecology of fishes*, 2nd edn. Travers City, MI: Cooper Publishing Group.

Eastman JT. 1993. *Antarctic fish biology: evolution in a unique environment*. San Diego: Academic Press.

Evans DH, Claiborne JB. 2006. *The physiology of fishes*, 3rd edn. Boca Raton, FL: CRC, Taylor & Francis.

Farrell AP, Steffensen JF. 2005. *Physiology of polar fishes. Fish physiology*, Vol. 22. New York: Academic Press.

Farrell AP, Stevens ED, Cech JJ, Richards JG. 2011. *Encyclopedia of fish physiology: from genome to environment*. Amsterdam, Boston: Elsevier/Academic Press.

Ishimatsu A. 2012. Evolution of the cardiorespiratory system in air-breathing fishes. *Aqua-BioScience Monographs* 5:1–28.

McKenzie DJ, Farrell AP, Brauner CJ. 2007. *Primitive fishes. Fish physiology*, Vol. 26. New York: Academic Press.

Milsom WK. 2012. New insights into gill chemoreception: receptor distribution and roles in water and air breathing fish. *Respir Physiol Neurobiol* 184:326–339.

Ostrander GK. 2000. *The laboratory fish*. London: Academic Press.

Perry SF, Tufts BL. 1998. *Fish respiration. Fish physiology*, Vol. 17. New York: Academic Press.

Val AL, De Almeida-Val VMF, Randall DJ. 2005. *The physiology of tropical fishes. Fish physiology*, Vol. 21. New York: Academic Press.

Journal

Journal of Morphology.

CHAPTER **6**

Nervous System and Sensory Organs

Summary

The nervous system and sensory organs of the earliest fishes provided the evolutionary basis for the nervous system and sensory organs of all other vertebrates. It is not surprising that these systems are especially effective in the aquatic environment, and have diversified among fishes over the last 450 million years or so. One derived group of early lobe-finned fishes became ever more specialized to various aspects of the terrestrial environment, resulting in the origin of tetrapods. Other groups of fishes became ever more adapted to a wide variety of aquatic habitats. The sensory environments in various underwater habitats are different from and much more diverse than what we experience in terrestrial habitats, so it is not surprising that the nervous system and sensory organs of modern fishes show enormous diversity in structure and function.

The nervous system of fishes, like that of other vertebrates, has two major components: the **central nervous system**, consisting of the brain and the spinal cord; and the **peripheral nervous system,** consisting of sensory neurons, sensory organs, motor neurons, and the autonomic nervous system that controls many involuntary physiological functions.

The brain of fishes is composed of three major parts – the forebrain, midbrain, and hindbrain. The forebrain consists of the telencephalon and the diencephalon. The **telencephalon** is the largest part of the fish brain.

CHAPTER CONTENTS

The Diversity of Fishes: Biology, Evolution and Ecology, Third Edition. Douglas E. Facey, Brian W. Bowen, Bruce B. Collette, and Gene S. Helfman.

Companion website: www.wiley.com/go/facey/diversityfishes3

It includes the processing area for the sense of smell and also includes regions important for learning and memory. The **diencephalon** includes the **pineal**, a light-sensitive structure that helps regulate daily and seasonal activity patterns. The pineal releases the hormone melatonin, which affects other hormonal functions through its effect on the pituitary. The diencephalon also includes the hypothalamus, which assists with regulation of many physiological functions through its neural and hormonal interaction with the pituitary.

The midbrain, or **mesencephalon,** includes the optic tectum (optic lobes) which receives input from the eyes. The hindbrain, or **metencephalon**, includes the cerebellum and is important in maintaining muscular tone and swimming equilibrium. The most posterior portion of the brain merges with the enlarged anterior part of the spinal cord and is called the **myelencephalon**, or brainstem (also called the **medulla oblongata**). It serves as the relay station for sensory input other than smell and sight, and contains nerve centers that control various functions including respiration and osmoregulation.

Fish brains are generally much smaller than the brains of birds or mammals of similar size, but sharks have considerably larger brains relative to their body size than bony fishes, and even larger than those of some birds and mammals. Among the teleosts, those with electroreceptive capabilities have particularly large brains.

The **peripheral nervous system** of fishes includes ten **cranial nerves**, which are defined like those in terrestrial vertebrates (I-X), as well as a series of lateral line nerves (anterior, middle, posterior). The **autonomic** nervous system regulates involuntary functions and is composed of sympathetic and parasympathetic ganglia and fibers.

Mauthner cells are large neurons that emerge from the brain, crossover near the hindbrain and run along the spinal cord. They are responsible for the rapid reaction of fishes to sudden stimuli, which often results in a quick bend of the body and sudden swimming away from the stimulus.

The sensory organs innervated by cranial sensory nerves are the eye, nose and taste buds, ear (including semi-circular canals), neuromasts of the mechanosensory lateral line system, and electroreceptors. These organs detect specific types of signals, such as light, sound, molecular shapes, orientation in the water column, electricity, or magnetic fields and convert them into nerve signals that are then carried by sensory neurons to the brain where the information is interpreted. Fish sensory organs show great diversity and specialization depending on the habitats and evolutionary history of the particular species, and many specific details and examples are provided later in the chapter.

The aquatic environment presents very different sensory challenges than the terrestrial environment. Water is a very good conductor of sound waves, but not long wavelengths of light (red, yellow, and orange). Fishes are surrounded by molecules in solution, so chemoreception (taste and smell) can take place anywhere on the body that has appropriate receptors. And water's conductive properties surround fishes with electric impulses and fields, making electroreception valuable for many species.

The **mechanoreception** senses detect vibrations in the water. These systems include balance and orientation, the lateral line system, and hearing. The basic mechanoreceptive sensory cell is the **hair cell**, so named because of the many microvilli on the apical surface. The position of the microvilli with respect to one another cause changes in nerve signals going from these cells to the brain. Hair cells line certain regions of the inner ear, including the fluid-filled **semicircular canals** and also small chambers that contain calcium deposits known as **otoliths** that rest against the hair cells. Movement or changes of a fish's orientation result in movement of the fluid in the semicircular canals and the position of the otoliths. Both of these are sensed by the hair cells and information regarding movement and vertical orientation is then relayed to the brain.

In the **lateral line system**, hair cells are typically grouped into clusters under a gelatinous covering, and the entire grouping is called a **neuromast** organ. Neuromast organs may be on the surface of fishes, but many are located in hollow canals beneath the scales of the trunk and some of the dermal bones of the head. Pores to the surface of the fish's skin allow changes in pressure of the water due to movement to cause distortion of the neuromast covering, stimulating the hair cells and causing changes in the nerve signals going to the brain. The lateral line system is particularly valuable to fishes in murky waters, caves or deep areas with little or no light.

Hearing in fishes is based on the detection of the vibrations of the otoliths by hair cells in the chambers of the inner ear. Hearing is enhanced in some fishes by anatomical linkage to the fish's gas bladder, which also vibrates in response to sound waves. Fishes get important cues from sounds in the environment, detecting suitable habitats and predators. Some also communicate by producing sounds by rubbing hard structures such as bones, teeth, or fin rays, and also vibrating muscles against the gas bladder.

Many fishes also have **electroreception** capabilities, which are based on sensory cells that evolved from the mechanoreceptive hair cells. Electroreception is believed to have been present in many early fishes and is found in some of the oldest living groups of fishes, including the Chondrichthyes (sharks and their relatives), coelacanths, lungfishes, sturgeons, and paddlefishes. The ability to detect electric fields has apparently evolved independently several times, and as a result it is found in distantly related groups. There are two general types of electroreceptors in fishes. **Ampullary receptors** are recessed in pits that contain an electroconductive gel. These sense low frequency electricity produced by other organisms and, therefore, can be quite useful in detecting nearby prey or potential predators. **Tuberous receptors** are embedded in the skin under a layer of loose epithelial cells and are sensitive to higher-frequency electrical signals produced by the electric organs of those fishes. These electric organs evolved from either modified muscles cells (in most groups) or nerve cells (in one family) and create an electric field used for communication and to sense physical disturbances in the surrounding environment. Many electric fishes evolved in tropical areas with poor water clarity, so producing and detecting electric fields have become essential for navigation and communication.

Magnetic reception has been demonstrated in some fishes, and is also likely an evolutionary derivative of mechanoreceptors. Fishes with sensitive ampullary electroreceptors, such as the sharks and their relatives, may have the ability to detect magnetic fields indirectly by sensing changes in surrounding electric fields created by the fish's movement through the Earth's magnetic field. This would provide valuable sensory information regarding location and orientation and be useful for navigation. Several species of bony fishes possess deposits of magnetite crystals embedded in their tissues; these crystals respond to magnetic fields and may provide valuable location and orientation information to assist with migration.

Vision is certainly an important sense for many fishes, but water can be quite variable as a visual environment. Aquatic habitats can range from totally clear to extremely murky depending on ions and chemical content, suspended sediments, and abundance of plankton. In addition, red light and UV radiation attenuate rapidly in water and therefore do not penetrate far below the surface. Hence, tropical fishes that appear colorful to us when artificially illuminated probably appear quite differently and even cryptic in their natural habitat; and some can see UV light which we humans cannot. Therefore, our visual perception of fishes in shallow tropical habitats is probably considerably different from what the fishes see. Blue light penetrates furthest into water and is therefore the only surface light available at greater depths. But many fishes live below the limits of even blue light penetration, in perpetual darkness of the deep sea. Some deep-sea fishes have **bioluminescent organs** that produce light, so vision can still play a role well below the depth of surface light penetration.

Fish eyes have the same basic structure as the eyes of other vertebrates, including humans. Light must pass through the cornea, pupil, lens, and central fluid-filled cavity before activating the sensory cells on the retina. These sensory cells may respond to somewhat specific wavelengths and intensities of light. Rods are activated by low light intensity, and are therefore helpful in poorly lit environments. Cones require brighter light, and different cones respond to different wavelengths, thereby providing color vision. Not surprisingly, fishes show a great degree of habitat specificity when it comes to photoreceptors. Many shallow water fishes have photoreceptors sensitive to red and UV light whereas deep-water fishes may have only receptors for blue light. There are some remarkable special cases, however, such as deep-sea dragonfishes that can produce and see red light, which plays a role in locating food and perhaps communication. It is important for us to have a basic understanding of both the physics of light in the aquatic environment and of what fish photoreceptors can detect in order to understand what fishes are actually seeing.

Chemoreception, which includes taste (gustation) and smell (olfaction), is based on the ability of molecules to bind to specific receptors based on their shapes. As terrestrial vertebrates, we typically associate taste with detection of molecules in solution in our mouths, whereas smell is the detection of molecules suspended in the air that come in contact with sensors in our nose. However, fishes are surrounded by molecules in solution that can come in contact with any part of the body – and, not surprisingly, many fishes have chemoreceptors in areas other than directly around the mouth and nostrils. Nevertheless, taste is still considered a contact sense utilizing taste buds primarily in or around the mouth to help a fish detect and sort potential food items. Taste buds may be especially concentrated on the barbels of some species, thereby providing a means for detecting potential food in the substrate. The sense of smell is due to chemosensory cells in the nostrils that receive water from around the fish. The sensitivity of fishes to chemicals varies greatly, with some detecting compounds at concentrations of less than 10^{-10} mol/L. This ability to detect incredibly low concentrations of certain chemicals helps fish sense potential predators, identify territories based on scent markings, find their way to suitable spawning areas, and locate mates.

There has been considerable debate over the years regarding whether or not fishes can feel **pain**, and we may never be able to resolve this question, depending on how we define the concept of pain itself. **Nociception** is the ability of an animal to detect a noxious stimulus – that is a stimulus that it apparently chooses to avoid. To some, this is an indication of irritation, but does not rise to level of pain, which involves a deeper and more subjective emotional response. Behavioral studies have shown that many fishes will change their behavior when exposed to certain stimuli, but we cannot evaluate the intensity of what the fish is experiencing, nor whether there is any cognitive recognition similar to what we consider pain.

Nervous System

The nervous system is composed of the central, peripheral, and autonomic nervous system. The **central nervous system** is further subdivided into the brain and the spinal cord (Healey 1957; Bernstein 1970; Northcutt & Davis 1983). The **peripheral system** is composed of the cranial and spinal nerves and the associated sense organs. The **autonomic** nervous system is composed of sympathetic and parasympathetic ganglia and fibers.

Central Nervous System

The central nervous system is composed of the brain, which is protected by the skull, and the spinal cord, which runs down the length of the fish, and is located within the neural canal of the series of vertebrae that form the spine.

The brain of fishes is composed of three major parts – the forebrain, midbrain, and hindbrain (Butler 2011). The forebrain consists of the telencephalon and the diencephalon. The **telencephalon** (which becomes the cerebrum of tetrapods) includes the olfactory bulbs (which receive inputs via the

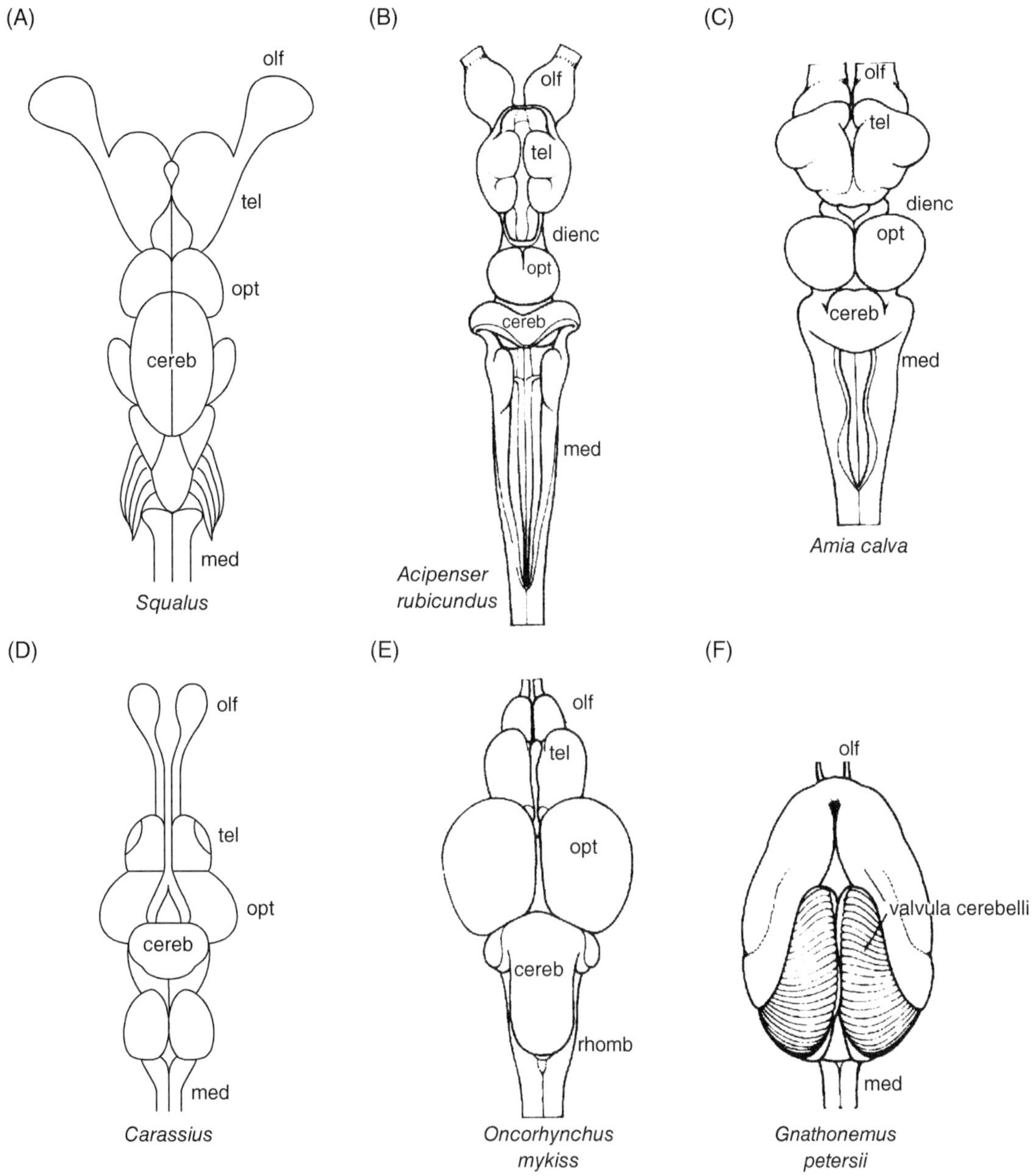

FIGURE 6.1 Dorsal views of brains of representative fishes: (A) shark (Squalidae); (B) sturgeon (Acipenseridae); (C) Bowfin (Amiidae); (D) carp (Cyprinidae); (E) trout (Salmonidae); and (F) elephantnose fish (Mormyridae). Major brain parts from anterior to posterior: olf, olfactory lobe; tel, telencephalon; dienc, diencephalon; opt, optic lobe; cereb, cerebellum; med, medulla oblongata (myelencephalon). In some species the diencephalon is obscured by large optic lobes. The enlarged valvula cerebelli of *Gnathonemus* (F) obscures other parts of the brain in this dorsal view. Based on Nieuwenhuys (1982, 2011).

olfactory nerve from the nose) and the regions important for learning and memory. The olfactory nerve is large in hagfishes and lampreys, huge in many sharks, and moderately large in many teleosts, which reflects the importance of olfaction in these fishes (Fig. 6.1).

Caudal to the telencephalon is the **diencephalon**, which includes the **pineal**, a light-sensitive structure that helps regulate daily and seasonal activity patterns. The diencephalon functions as a correlation center for incoming and outgoing messages regarding homeostasis and the endocrine system. The pineal is a well-vascularized light-sensitive structure on the dorsal surface of the diencephalon and it frequently lies beneath an unpigmented area of the cranium. The pineal plays a role in light detection, circadian and seasonal clock dynamics, as well as in color changes associated with background matching. It regulates daily and seasonal cycles by release of the hormone **melatonin**, which affects other hormonal regulation of physiological processes through its effect on the pituitary.

The midbrain, or **mesencephalon,** includes the optic tectum (optic lobes), which receives sensory input from the eyes (via the optic nerve, cranial nerve II) and other regions that process inputs from the ear and lateral line systems. The optic lobes are relatively large in species that are dependent on vision for prey detection, such as minnows and trout (Fig. 6.1D, E).

The hindbrain consists of the **metencephalon**, the cerebellum (important for muscle and swimming functions) and pons, and the medulla oblongata (myelencephalon or brainstem). The **cerebellum** is a large single median lobe that lies dorsal to the rest of the metencephalon. The enlarged cerebellum of the electrosensitive elephantnose fishes (Mormyridae) forms the valvula cerebelli (Fig. 6.1F), which extend over the dorsal surface of the telencephalon and plays a role in the interpretation of electrical input from electroreceptors.

The most posterior portion of the brain is the **medulla oblongata** (also called the brainstem or **myelencephalon**). Cranial nerves V through X emerge from this region. It also serves as the relay station for input from all of the sensory systems except smell (cranial nerve I) and vision (cranial nerve II). It contains nerve centers that control certain somatic and visceral functions, including respiration and osmoregulation.

Fish brains are on average about 7% the size of the brain of a bird or mammal of equal body size, but they vary in size relative to body size. Sharks have much larger brains relative to body size than teleosts and pelagic sharks have larger brains than pelagic teleosts (Lisney & Collin 2006). Elephantnose fishes (Mormyridae) have the largest brains among fishes (Nilsson 1996); this large brain is associated with electroreception, which will be discussed later in this chapter.

Despite their somewhat "primitive" reputation, the brains of sharks and their relatives are similar in size to those of birds and mammals of similar weight (Rodriguez-Moldes 2011) and more similar in structure to those of tetrapods than other fishes (Butler 2011). In addition, in fishes in general, the forebrain's pallium region is likely an area of more complex processing, including learning and memory (Braithwaite 2011). Therefore, the absence of the complex cerebral cortex found in mammals may not necessarily mean that fishes lack the capability for some of the complex brain processing associated with this region.

Peripheral Nervous System

The 10 **cranial nerves** in fishes are similar to those in other vertebrates. Cranial nerve I, the **olfactory nerve**, is a sensory nerve that runs from the olfactory bulb to the olfactory lobes. The **optic nerve** (cranial nerve II) runs from the retina to the optic lobes of the forebrain. As in other vertebrates, cranial nerves III (**oculomotor**), IV (**trochlear**), and VI (**abducens**) are somatic motor nerves that innervate the extraocular (striated) muscles of the eye: IV, the superior oblique; VI, the external rectus; and III, the remaining eye muscles. The **trigeminal**, V, and facial VII nerves are both mixed somatic sensory and motor nerve serving the interior and exterior sense organs and muscles of the head. Cranial nerve VIII, the **acoustic**, innervates the ear. The **glossopharyngeal**, IX, is a mixed nerve that supplies the gill region. It often fuses with cranial nerve X (the anterior ramus of the vagus nerve) as it exits the brain. The **vagus nerve** is a mixed nerve that innervates the viscera. In addition to these nerves, which are found in all vertebrates, fishes (and larval and aquatic adult amphibians) also have a series of lateral line nerves (anterior, middle, posterior; ALLn, MLLn, and Plln), which innervate the neuromast receptor organs on the head, trunk, and tail.

Mauthner Cells

Austrian medical student Ludwig Mauthner (1840–1894) described the unusually large neurons in the fish spinal cord that crosses over near the hindbrain (Pereda & Faber 2011). Subsequent study of these neurons, which now bear Mauthner's name, has revealed the important role that they play in the ability of fish to react very quickly and swim away from potential danger (Faber & Pereda 2011). An insulating myelin sheath combined with large diameter of the axon allow for rapid conduction of nerve impulses, resulting in an immediate reaction. Stimuli coming from one side of the fish elicit a reaction that involves a "C"-shaped bend of the fish's body followed quickly by extension of the tail, resulting in rapid movement away from the stimulus, and thus from potential danger (Fig. 6.2). Much of the input to the Mauthner cells seems to come from the auditory system, making many fishes especially responsive to sounds. Activation of the Mauthner cells, along with the rapid "C-start" response, has also been seen during feeding, suggesting that the response can be intentionally activated and may not be solely a reflexive response to danger (Canfield & Rose 1993).

Sensory Organs

Sensory organs detect specific types of signals in the environment and convert them into nerve signals that are carried by sensory neurons to the brain. The aquatic environment is a very different sensory environment than the terrestrial

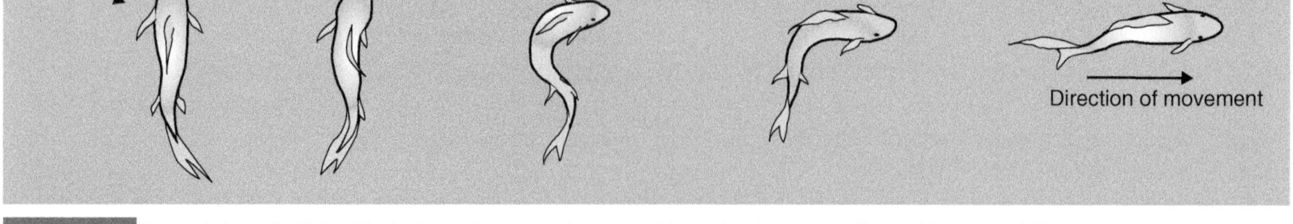

FIGURE 6.2 Dorsal view of a fish reflexively moving away from a sudden stimulus to rapidly avoid potential danger. The rapid response is facilitated by Mauther cells on the spinal cord that quickly communicate nerve signals to the trunk muscles.

environment, therefore fishes tend to emphasize different sensory systems than do terrestrial vertebrates. Fish sensory systems tend to focus on detecting vibrations (mechanoreception), electricity (electroreception), magnetic fields (magnetoreception) in some species, wavelengths of light (vision), and molecular shapes (chemoreception).

Mechanoreception

Mechanoreceptive sensory systems found in all fishes include: (1) the semicircular canals of the inner ear which are critical to maintaining position and equilibrium in 3-dimensional space, (2) the mechanosensory lateral line system, which responds to unidirectional water flows (e.g., currents) and oscillatory water flows (e.g., nearby vibrations, local disturbances in the water), and (3) hearing, which responds to acoustic cues (nearby vibrations and more distant sound pressure waves). All of these mechanoreceptive sensory systems are based on populations of sensory hair cells, which are directionally sensitive to water movements.

These hair cells are found within sensory epithelia located within the otolithic organs (sensory maculae) and semicircular canals (cristae) of the ear, and within a series of lateral line canals and on the skin of the head, trunk and tail (neuromasts). The hair cells (Fig. 6.3) include an array of cilia-like extensions on their apical surface. The short, and often graduated, stereocilia are microvilli (extensions of the cell membrane), whereas the much longer kinocilium is a true cilium composed of microtubules. Displacement of the shorter stereocilia toward the kinocilium increases the rate of nerve impulses sent to the brain by the nerve cells associated with a hair cell, whereas the rate of nerve impulses decreases if the stereocilia move in the opposite direction (Popper 2011).

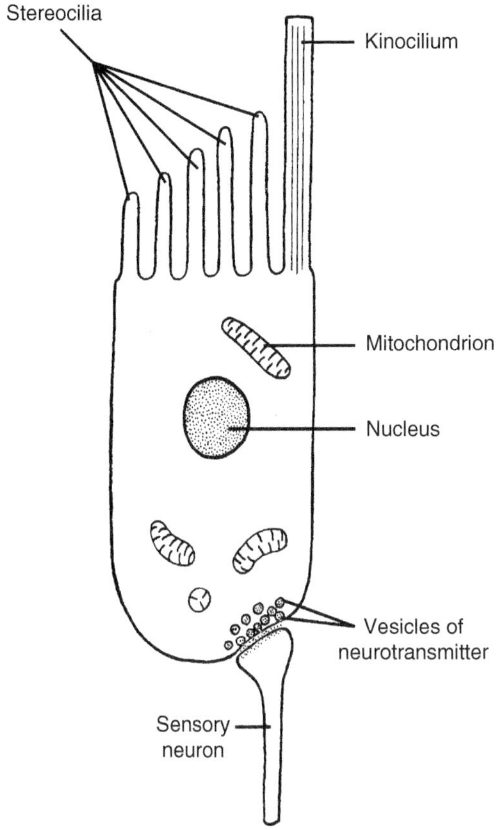

FIGURE 6.3 Mechanoreception involves sensory hair cells, which are found in the lateral line system of fishes and the inner ear of fishes and other vertebrates. The apical surface of a sensory hair cell usually has numerous stereocilia and a single, much longer kinocilium. Deflection of the stereocilia toward or away from the kinocilium causes an increase or decrease in the firing rate of the sensory neuron innervating the hair cell at its basal surface.

Equilibrium and Balance The hair cells in the semicircular canals detect movement of fluid and therefore play an important role in the ability of fishes to maintain their equilibrium and orient in 3-D space in the water column (Straka & Baker 2011). The semicircular canals are filled with a fluid (endolymph) and have a small population of sensory hair cells (the crista) in their terminal ampullae. Postural equilibrium and balance are maintained by **semicircular canals** (in the **pars superior**) and an otolithic organ known as the **utricle** (Fig. 6.4). Jawed fishes (and tetrapods) have three semicircular canals, lampreys (Petromyzontidae) have two, and hagfishes

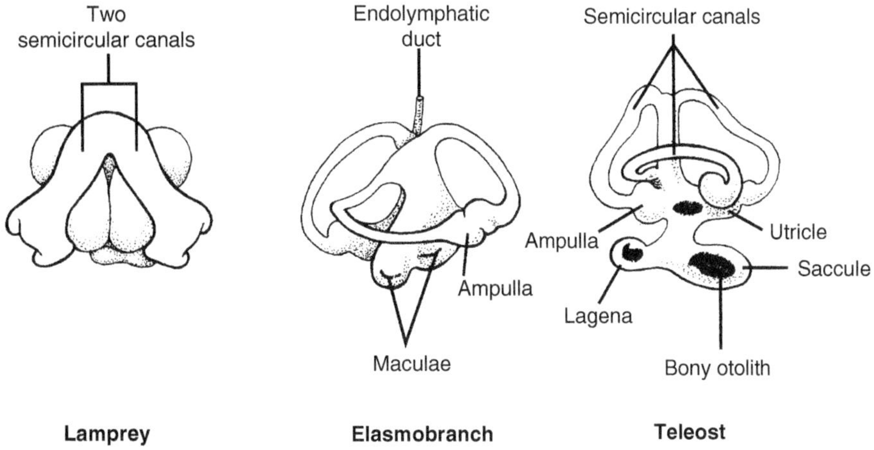

FIGURE 6.4 The inner ear of fishes. After Hildebrand (1988).

(Myxinidae) have one. Changes in acceleration or in orientation in 3-D space set the endolymph within the semicircular canals in motion causing displacement of the gelatinous cupula that covers the hair cells. This results in changes in the firing rate of the sensory neurons innervating the hair cells and signals the brain about changes in the fish's position in 3-D space. The utricle contains a calcium carbonate deposit, or **otolith** ("ear stone"), the lapillus, which rests on a bed of sensory hair cells. The downward pull of gravity on the lapillus triggers impulses from the sensory cells and provides the fish with information regarding its orientation in the water. The utricle works in coordination with the detection of light penetrating the water from above by the retina of the eye keeping the fish upright in the water (the **dorsal light reflex**). If the utricle and semicircular canals are removed on one side in Goldfish (*Carassius auratus*) they lose their ability to orient vertically, although this was often regained after several days (Ott & Platt 1988).

Lateral Line System The lateral line system is only functional in water and is found only in fishes and larval and permanently aquatic adult amphibians (see Webb 2013); the system is lost in amniotes (reptiles, birds, and mammals). The lateral line system is composed of sensory organs called **neuromasts**, which contain hair cells, like those of the inner ear. The ciliary bundles on the surface of each hair cell are embedded in a gelatinous cupula, (Fig. 6.5), which can be displaced by water movement, thereby moving the cilia of the hair cells and initiating a change in impulse sent to the brain via the sensory neurons of the lateral line nerves (Webb 2011). The cupula helps filter out background "noise" by preventing the hair cells from being affected by small vibrations – only vibrations strong enough to move the entire cupula will be detected by the hair cells within it. Neuromasts allow fishes to detect water flows or vibrations in the water that are generated by prey, predators, other fishes in a school, or environmental obstacles.

The lateral line system has two main subdivisions: superficial neuromasts, which are located on the skin; and canal neuromasts, which are located in pored, hollow canals beneath the scales of the trunk (Webb & Ramsay 2017) and within a subset of the dermal bones of the head (Fig. 6.5). The cranial and trunk canals are located in consistent locations around the head and along the side of a fish.

When fishes hatch, all neuromasts are found on the skin (superficial neuromasts) on the head and trunk. In most fishes, a subset of these neuromasts become enclosed in the lateral line canals, and the other neuromasts remain on the skin and may increase in number. Superficial neuromasts are more exposed, making them quite sensitive to water movement across the skin, which makes them particularly effective for low frequency water disturbances, especially in areas with little water velocity (Engelmann et al. 2000, 2002; Montgomery 2011; Braun et al. 2002). Superficial neuromasts are more abundant in fishes that are sedentary or slow swimmers and that inhabit quiet areas, such as Goldfish (*Carassius auratus*).

Canal neuromasts, in contrast, are shielded from constant stimulation by water moving across the skin and are better at detecting stimuli, whether the fish or the water around it is moving quickly. Therefore, they are more effective in detecting transient water flows of higher frequency (20–200 Hz; see

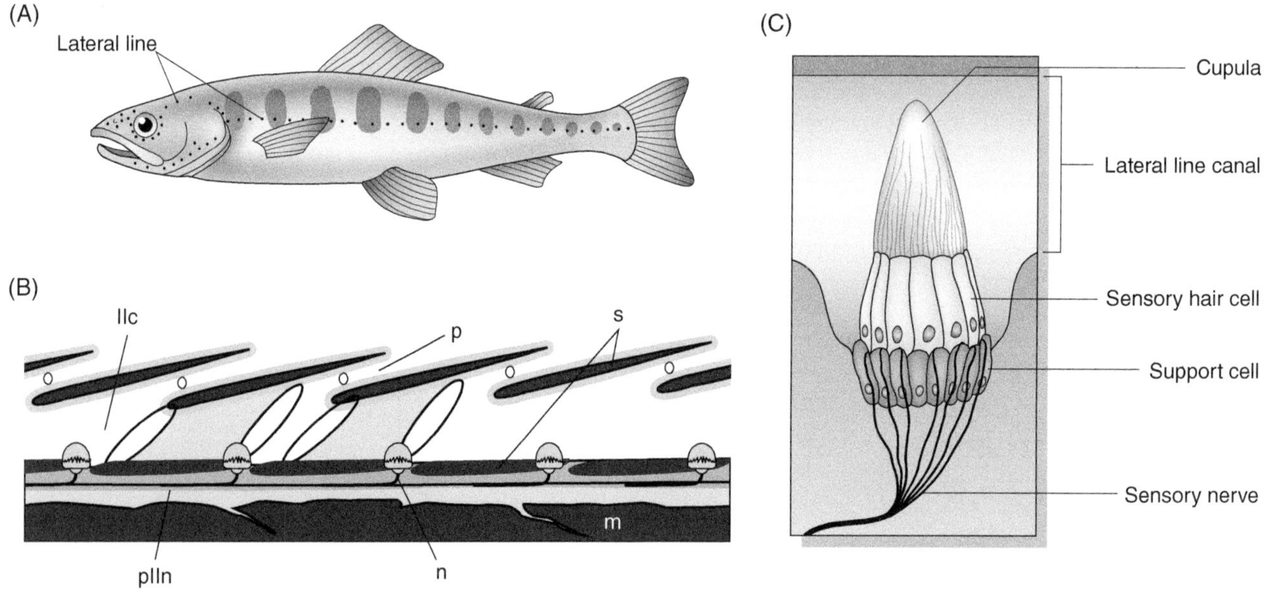

FIGURE 6.5 (A) Approximate location of lateral line in many fishes. (B) Cross-section of the lateral line on the trunk of a fish showing the distribution and innervation of neuromast receptors and the location of pores that connect the canal to the external environment. Abbreviations: llc, lateral line canal; m, trunk muscle; n, canal neuromast; p, pore; plln, posterior lateral line nerve; s, scale. Lateral line scales (black) are below the epidermis (gray) and embedded in the dermis (dark gray). Image adapted from Webb and Ramsay (2017) with permission. (C) Each neuromast is composed of several sensory hair cells, support cells, and innervating sensory neurons. The apical kinocilia and stereocilia project into a gelatinous cupula which overlays the entire neuromast.

Braun et al. 2002). The canals tend to be better developed in fishes that are fast swimmers or that live in fast or turbulent water. Rainbow Trout (*Oncorhyunchus mykiss*), for example, often inhabit running water and have well-developed neuromasts in narrow canals and very few superficial neuromasts (Engelmann et al. 2002). The canal neuromasts of the benthic Mottled Sculpin (*Cottus bairdi*), which often lives in fast-flowing streams, help the fish locate prey by filtering out background stimuli due to water currents (Kanter & Coombs 2003).

The lateral line system is sensitive to water flows and can therefore respond to movements of water caused by movements associated with feeding, gill ventilation, or swimming and the movements of the fins. To eliminate this interference, some fishes will stop swimming while focusing their attention on detecting prey or have the brain "tune down" the incoming signals from the lateral line while swimming (Montgomery 2011).

The structure of the lateral line system shows a good deal of anatomical variation among different species, which is interpreted as evolutionary adaptations for behaviors carried out in the particular habitat in which they live. For example, different fishes may have narrow, widened, branched, or reduced canals (Webb 1989). In species in which the canals are reduced, superficial neuromasts may be found in very high numbers, as in gobies (Family Gobiidae).

In many fishes there is one canal that runs along the midline of the trunk to the tail, but in other fishes the trunk canal may be incompletely formed, placed high or low on the body, or may be absent. When it is reduced or absent, superficial neuromasts may be present on the skin over the lateral line scales, or even on all scales of the body (e.g., gobies, cardinalfishes). Some of this variation is thought to be adaptive. For instance, the single canal on the trunk may be displaced dorsally in some benthic fishes (e.g., stargazers, Uranoscopidae) or ventrally in species that tend to live at the surface (e.g., flying fish, Exocoetidae).

Development of the lateral line system can be influenced by the flow conditions experienced by the juveniles (Mogdans 2019). For example, three-spine stickleback (*Gasterosteus aculeatus,* Gasterosteidae) from swift water habitats have more superficial neuromasts that conspecifics from calm water habitats (Wark & Peichel 2010). Wild juvenile Steelhead (*Oncorhynchus mykiss*) had more superficial neuromasts than conspecifics raised in hatcheries (Brown et al. 2013), and farm-raised Sea Bream (*Sparus aurata*) and Sea Bass (*Dicentrarchus labrax*) showed abnormal lateral line development as compared to wild conspecifics (Carillo et al. 2001; Sfakianakis et al. 2013). This developmental plasticity could compromise the fitness of captive-raised fish that are subsequently released into the wild. In addition, laboratory-raised Trinidadian guppies (*Poecilia reticulate*, Poeciliidae) raised in water with chemical cues from the predatory pike cichlid (*Crenicichla* sp.) developed more superficial neuromasts than conspecifics raised in tanks without the chemical cue (Fischer et al. 2013), indicating that external factors other than flow conditions may impact the development of mechanosensory capabilities.

The complex relationship between the structure of the lateral line system, fish behavior, and habitat is shown during the development of lateral line systems and vision in juveniles of three species of drums (family Sciaenidae) (Poling & Fuiman (1998). As in other fishes, the neuromasts that first appear are superficial and then some of them become enclosed in canals (canal neuromasts) as fish grow and as they become more active. However, the relative abundance of canal and superficial neuromasts differed among species and is correlated well with juvenile habitat and the relative role that vision might also play in behavior. Juvenile Spotted Seatrout (*Cynoscion nebulosus*), which inhabit shallow, murky inshore areas where mechanoreception would be more critical to predator and prey detection, have significantly more superficial neuromasts on their heads than do juveniles of Atlantic Croaker (*Micropogonis undulatus*), which settle offshore in clearer and deeper water and which have well-developed eyes. Juvenile Red Drum (*Sciaenops ocellatus*) are somewhat intermediate in both their habitat (bays and near-shore areas) and their sensory development.

There are many other examples of fishes in which mechanoreception helps compensate for a poor visual environment. Under experimental conditions, Lake Trout (*Salvelinus namaycush*) detected and followed the hydrodynamic trails left by prey fishes in total darkness. Their ability to capture prey was significantly inhibited when the lateral line system was rendered ineffective (Montgomery et al. 2002). Mottled Sculpin feed in low light conditions and rely on their lateral lines to detect and locate prey (Braun et al. 2002). Blind Cave Fish (*Astyanax mexicanus/jordani*) have many superficial neuromasts, as well as taste buds, on their heads to compensate for lack of vision (Schellart & Wubbels 1998). They rely more on their lateral line system than any other sense (Montgomery et al. 2001a). And at least 28 species of deep-sea fishes (order Stomiiformes) have more superficial neuromasts, as well as larger neuromasts with longer cupulae, than most fishes with better visual capabilities (Marranzino & Webb 2018).

The sensitivity of fishes to vibrations in the water can help them locate prey but also leaves them vulnerable to learned behavior of some predators. Several species of piscivorous birds, including herons and egrets (family Ardeidae) have been observed creating disturbances on the water's surface by tongue-flicking or bill-vibrating in order to attract fishes (Davis 2004).

Hearing The "soundscape" in aquatic habitats provides fishes with a great deal of important information (Fay 2011). Aquatic environments are quite varied in the levels of natural background sound, and hearing sensitivity of fishes tends to be well matched to their habitats. Fishes in noisy habitats, such as coastlines and swift rivers, tend to have higher sound thresholds and narrower ranges for sound detection than fishes in calmer, quieter habitats such as lakes and ponds (Schellart & Wubbels 1998). Coral reefs are known to be naturally noisy environments and detection of biologically relevant sounds (e.g., for mating) is often challenging (Tricas & Webb 2016).

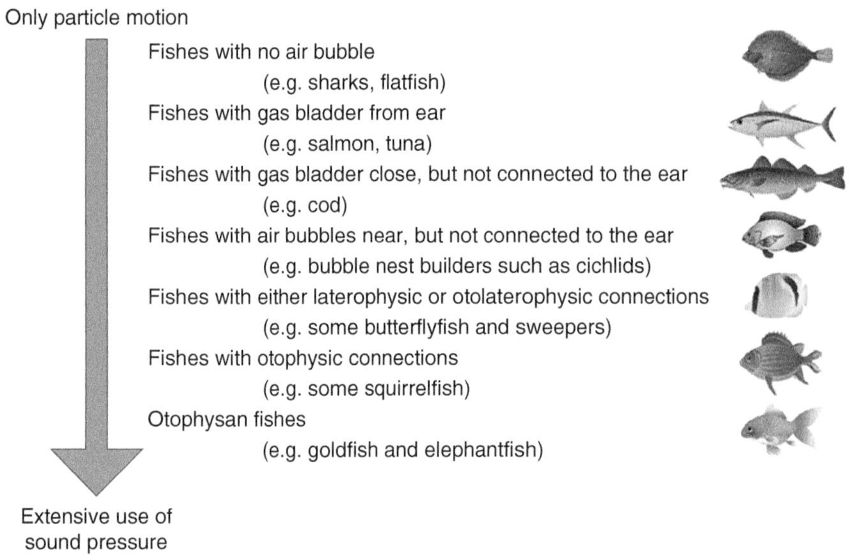

FIGURE 6.6 Fishes demonstrate a continuum of hearing capabilities. Those with the greatest hearing sensitivity have structural connections between internal gas spaces and the inner ear. The gas spaces resonate with sound waves passing through the fish and this can then be conducted to the sensory organs of the inner ear. Putland et al. (2019) / with permission of John Wiley & Sons.

Hearing in fishes is primarily the responsibility of the inner ear, including three otolithic organs – the **utricle** of the pars superior and the **saccule** and **lagena** of the **pars inferior** (Fig. 6.4). Each contains an otolith (the astericus, sagitta, lapillus, respectively) and has a patch of epithelial tissue composed of sensory hair cells (see Fig. 6.3) called the sensory macula. The bundle of cilia on the surface of all of the sensory hair cells projects into a gelatinous otolithic membrane, which provides a mechanical linkage between the hair cells and the otolith (see Putland et al. 2019).

The otoliths, composed of calcium carbonate, are characterized by complex shapes that vary among the three otolithic organs within a species, and each type of otolith varies in structure among species. Knowledge of species-specific otolith shapes can be valuable in estimating past abundance or use of fishes based on fossil and archaeological records. In addition, as a fish grows, layers of calcium carbonate are laid down daily, seasonally and annually in the otoliths resulting in an increase in otolith size and change in shape. Analysis of the number and thickness of growth rings in an otolith is routinely used to determine the age of a fish, which is essential in population studies.

Most fish tissue is transparent to sound pressure waves because its density is similar to that of the water. As sound waves pass through a fish, the otoliths move differently than the surrounding soft tissue (including the hair cells). Direct stimulation of the ear by the relative difference in movements by the sensory hair cells and the otoliths triggers impulses in the auditory nerve (Hastings 2011; Putland et al. 2019). In fishes with gas bladders, hearing is enhanced by the close proximity of the gas bladder to the ear. Sound pressure waves cause the gas in the **gas bladder** to oscillate (repeated expansion and compression) and the movement of the gas bladder wall causes fluid movements in the ear, thus enhancing hearing sensitivity).

Fishes display a wide range of hearing capabilities based on the extent of the connection between the gas bladder wall and the inner ear (see Popper & Fay 2011; Putland et al. 2019; Fig 6.6). Some fishes have anatomical specializations that even more effectively transmit oscillations of the volume of gas in the gas bladder to the inner ear (Yan 2003). For example, Squirrelfish (Holocentridae) and Clupeiformes (herrings, alewife, sardines, anchovies, etc.) have anterior extensions of the gas bladder that invade the ear (Akamatsu et al. 2003). African mormyrids (Mormyridae), also known for sensitive hearing, have extensions of the gas bladder that have separated from the main bladder leaving small bubbles of gas in intimate proximity to the ear that enhance sound detection (Yan & Curtsinger 2000). In gouramis (Anabantoidei) air-filled suprabranchial chamber in the roof of the mouth (used for air breathing) also enhances hearing sensitivity (Yan 1998). The butterflyfishes of the genus *Chaetodon* have anterior swim bladder extensions that come in close proximity of both the ear and the lateral line system in the head and enhance hearing (laterophysic connection; Tricas & Webb 2016).

Otophysan fishes, which includes over 60% of freshwater fish species, all have particularly acute hearing and pitch discrimination. This is due to the presence of an otophysic connection represented by the **Weberian apparatus**, a series of small bones derived from a few of the anterior-most vertebrae, which connect the anterior end of the gas bladder to the inner ear. These bones move in response to oscillations of the swim bladder wall (much like the middle ear bones of mammals) causing fluid movements in the ear. They confer the highest sensitivity and greatest frequency range of hearing known among fishes. Interference with the Weberian apparatus (elimination of one or more of the bones) or deflation of the gas bladder, results in decreased hearing capabilities (Yan et al. 2000).

Much less is known about hearing in elasmobranchs than in bony fishes even though the basic structure of the inner ear in chondrichthyans is similar to that in bony fishes (Casper 2011). One major difference is that instead of having otoliths composed of calcium carbonate, they have sand particles embedded in a gelatinous matrix in association with the sensory maculae. In addition, sharks have a connection from a depression on the back of the cartilaginous skull to the semicircular canals called the **fenestra ovalis** (Hueter et al. 2004). This connection is presumed to enhance hearing by directing sound to the macula neglecta, a small population of hair cells in addition to the three otolithic organs and the three semi-circular canals. Shark hearing is most sensitive at low frequencies, including those below 10 Hz, which are undetectable by humans. Sharks are most sensitive to pulsed sounds in this range, such as those emitted by the erratic swimming of an injured fish and can localize such sounds at distances of up to 250 m (Myrberg 1978; Myrberg & Nelson 1991; see Chapter 12, Subclass Euselachii).

Sources of Underwater Sound Sound can travel far in water, unlike light, so the soundscapes of aquatic habitats provide strong selection pressures for enhanced hearing in fishes. Auditory cues can identify potential mates, food sources, and predators. For example, dolphins use echolocation to locate prey, and several species of herrings (Clupeidae) apparently can detect these ultra-high frequency sounds (120 kHz) and may avoid predatory cetaceans trying to locate them (Plachta & Popper 2003; Kaatz 2011). Longspine Squirrelfish (*Holocentrus rufus*, Holocentridae) and Silver Perch (*Bairdiella chrysoura,* Sciaenidae) produce sounds associated with territorial displays or mating, and both stop these sounds when they hear the echolocation sounds produced by bottlenose dolphins (Luczkovich et al. 2011).

Another source of underwater sounds are the fishes themselves – many of which use auditory signals for inter- and intraspecific communication. Over 800 species of fishes in 109 families are able to produce sound in some way (Putland et al. 2019). Most fishes that produce sounds are Actinopterygians (ray-finned bony fishes); none are known among the hagfishes, lampreys, or Chondrichthyes (Ladich & Bass 2011). Many "vocal" fishes, such as male Plainfin Midshipman (*Porichthys notatus*, Batrachoididae), create humming or buzzing sounds by vibrating the gas bladder (Ladich & Yan 1998; Ladich & Bass 2011). This species is very abundant in the Sausalito district of San Francisco where, during the breeding season, they have been reported to keep awake humans who occupy houseboats along the shore.

Other fishes that use some variation of this basic concept include the cods (Gadidae), sea robins (Triglidae), drums and croakers (Sciaenidae), tigerperches (Terapontidae), pimelodid catfish (Pimelodidae), piranhas (Characidae), jaraqui (Prochilodontidae), thorny catfishes (Doradidae), upside-down catfishes (Mochokidae), and sea catfishes (Ariidae).

Another strategy for sound production is vibrating bony elements of the pectoral girdle or fins, as used by some catfishes (Siluriformes), croaking gouramis (Osphronemidae), and sculpins (Cottidae). Clownfish (Pomacentridae) create sounds with their mouth and teeth, whereas Seahorses (Sygnathidae) make clicking noises by rubbing together bones in the neck and back of the head.

Some schooling fishes (e.g., groupers, croakers, marine catfishes, and herrings) exhibit chorusing behavior, with multiple individuals responding to one another or creating sounds simultaneously (Kaatz 2011). Many species with well-known social systems use sounds as part of territorial or social dominance communication. Because so many fishes produce sounds associated with courtship and spawning, perhaps hydrophone recordings of these sounds could serve as a useful indicator of population densities.

Sounds are used in many species for attracting mates and during spawning; in some cases sound production is related to gender and maturity. The hermaphroditic Butter Hamlet (Serranidae) pairs up during mating. One of the pair emits sounds that initiate courtship and accompany the release of sperm while the other of the pair produces a different sound during the release of eggs. The two individuals then change both roles and sounds (Luczkovich et al. 2011; Kaatz 2011).

During the spawning season, mature male Weakfish/Spotted Seatrout (Sciaenidae) have well-developed drumming muscles not found in females or immature males. In another sciaenid, Atlantic Croaker, fish can produce sounds prior to maturity but the sounds become louder, of lower frequency, and at different pulse rates as the fish grow: the "voice" of these fish therefore changes as they mature. Female Plainfin Midshipman are more sensitive to the sounds created by dominant males during spawning season, indicating that hearing can also be affected by hormonal condition (Lu 2011).

Some fishes may use echolocation by producing sounds and then listening to their echos in order to help them orient in locations where vision is limited. Sea catfish (*Arius felis*, Ariidae) produce popping sounds that enable them to find their way through a maze without bumping into the walls, even when blinded. If the muscles used to produce these sounds are cut, they have difficulty navigating the maze (Luczkovich et al. 2011). Some deep-sea fishes (Macrouridae and Moridae) have sonic muscles, large otoliths, and modified ear structures that suggest they may use echolocation in their dark habitats.

Fishes also detect ambient sounds from their environment and alter their behavior accordingly. Biological sound, such as that produced by other fishes and invertebrates, appear to attract larval reef fishes to preferentially settle in areas with sounds that would indicate a suitable habitat (Putland et al. 2019). Ambient noise may also impact the evolution of hearing and sound production for communication. Males of two species of freshwater gobies (Gobiidae) that inhabit swift, rocky streams respond to and produce courtship sounds at frequencies around 100 Hz, within a "quiet window" of ambient noise (Lugli et al. 2003). The hearing of these fishes is most sensitive within this range, demonstrating that ambient sound is a selective force in the evolution of hearing.

The sensitivity of fishes to sound makes them potentially vulnerable to anthropogenic (human-generated) underwater noise (Hawkins 2011; Popper & Hawkins 2019). It is difficult, however, to assess the impact because controlled laboratory studies cannot fully replicate field conditions. High-intensity, low-frequency sound produced by air guns used in marine petroleum exploration can cause severe damage to the hair cells in the sensory epithelia (McCauley et al. 2003). Prolonged exposure of Goldfish (*Carrasius auratus,* an otophysan) to loud sound resulted in hearing loss (Smith et al. 2004). Lower intensity sounds can result in loss of hearing sensitivity in the Fathead Minnow (*Pimephales notatus,* an otophyan with Weberian Apparatus), although Bluegill (*Lepomis macrochirus*), which lacks an otophysic connection is less affected (Scholik & Yan 2002).

Electroreception

Electroreception is present in the oldest living groups of fishes – lampreys, Chondrichthyes (sharks and their relatives), coelacanths, lungfishes, bichirs and reedfishes, sturgeons, and paddlefishes. In addition, electroreception has evolved independently in several orders of teleost fishes (Jørgensen 2011). The receptor cells in the electrosensory organs that are responsible for detecting electric fields are closely related to the hair cells of the lateral line system (Kirschbaum & Denizo 2011). The electroreceptors are innervated by lateral line nerves, which project to an electroreceptive center in the hindbrain. Further processing of input from electroreceptors takes place in the same region of the midbrain responsible for processing input from the lateral line system (Fortune 2011).

There are two types of electroreception: passive and active. In passive electroreception fishes detect electric fields generated by other fishes or by physical and/or chemical processes in the environment. In active electroreception, electrogenic fishes that use electric organs to generate electric organ discharges (EODs) monitor the EODs of other fishes and changes in the electric field generated by their own EOD. **Ampullary organs** function in passive electroreception and **tuberous organs** function in active electroreception (Crampton 2019). About 16% of fish species (over 5600) exhibit passive electroreception, whereas active electroreception is found in about 1.5% (somewhat over 500 species).

Ampullary Organs Ampullary organs apparently evolved from a branch of the early lateral line in vertebrates and can detect DC and low frequency AC currents (Crampton 2019). They are located in deep layers of the skin (the dermis) and are connected to the skin surface by a canal filled with a conductive gel (Fig. 6.7A). They detect low-frequency electrical stimuli, including those that are a result of the physical environment and those of biological origin, such as muscle contractions and neuron activity generated by other fishes or invertebrate prey (Hofmann 2011).

Ampullary organs are found in Chondrichthyes (the Ampullae of Lorenzini), and in non-teleost (primitive) bony fishes – lungfishes (**Ceratodontiformes**), coelacanths (Coelacanthiformes), bichirs and reedfishes (Polypteriformes), and sturgeons and paddlefishes (Acipenseriformes). In addition, ampullary organs have evolved independently in three orders

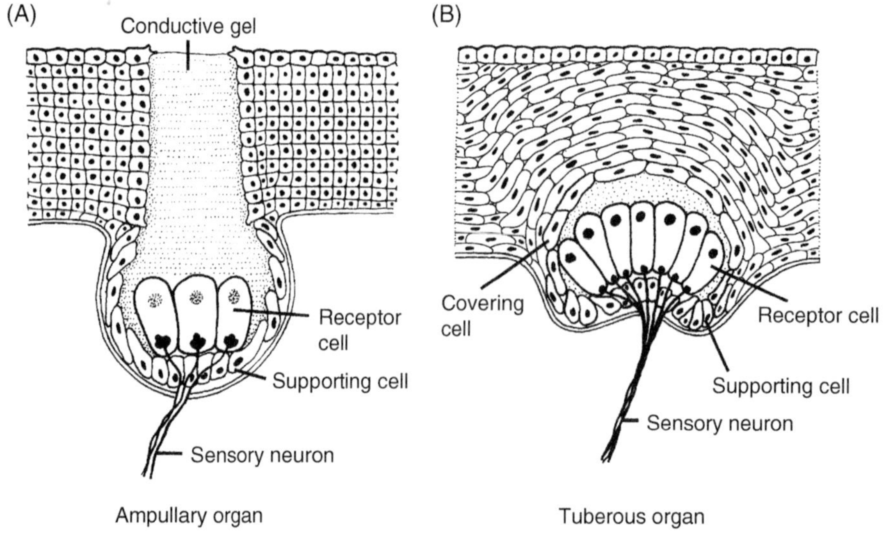

FIGURE 6.7 Schematic diagram of the structure of ampullary (A) and tuberous (B) electroreceptive organs. Both organs are surrounded by layers of flattened cells that join tightly to one another. This helps prevent current from bypassing the organs. Tight junctions between the receptor cells and supporting cells help focus incoming electric current through the base of the receptor cells, where they synapse with sensory neurons. Supporting cells in ampullary organs produce a highly conductive gel that fills the canal linking the sensory cells to the surrounding water. Adapted from Heiligenberg (1993), modified from drawing provided by H. A. Vischer.

of teleost fishes, Osteoglossiformes, Gymnotiformes, and Siluriformes (Jørgensen 2011).

For electroreceptive marine fishes (e.g., the Chondrichthyes), the ionic concentration of saltwater makes it a very good conductor of electricity, and the conductive gel in the canals allows the voltage current to easily reach the receptor cells across the less-conductive skin. Although fresh water does not conduct electricity as well as saltwater, ampullary receptors are also known among freshwater fishes, although they tend to have fewer receptor cells than marine species, and the receptor cells are closer to the surface of the skin, making the canals shorter. Freshwater fishes with ampullary receptors include the North American Paddlefish (*Polyodon spathula*), sturgeons (some of which are also marine or anadromous), the Mormyridae (in the Order Osteoglossiformes), Gymnotiformes, and some Siluriformes (Jørgensen 2011).

One of the main functions of ampullary receptors is prey detection. The ampullary receptors are often concentrated around the head, and in some species, such as the Eel-tailed Catfish (*Tandanus tandanus*, Plotosidae) they are especially abundant around the mouth and head (Whitehead et al. 2003; Collin & Whitehead 2004). The ability of the Paddlefish of the Mississippi River drainage of North America, to detect prey using electroreception is well studied. The elongated and flattened rostrum of this fish has the highest known density of ampullary receptors in freshwater fishes (Hoffmann 2011), and acts as an antenna that detects the relative densities of zooplankton (Fig. 6.8). This electrosensitive ability is particularly helpful because Paddlefish often inhabit areas with murky water and poor visibility.

Sharks have many ampullary organs concentrated in the head (Fig. 6.9), especially on the ventral side of the snout (Newton et al. 2019). The number of ampullary organs varies widely among species, from fewer than 150 in the Port Jackson Shark (*Heterodontus portusjacksoni*) to over 3000 in the Scalloped Hammerhead Shark (*Sphyrna lewini*). The distribution of ampullary organs along the broad head of the Scalloped Hammerhead Shark (*Sphyrna lewini*) may allow these fish to sample a wider area than sharks with narrower snouts (Kajiura & Holland 2002). The number of ampullary organs does not change over the life of a shark, but they get spread further apart with growth, so the extent of field sensed increases, but resolution may decrease.

The density and abundance of ampullary organs is related to habitat and feeding – benthic feeders tend to have more ampullae ventrally and around mouth; benthic fishes that ambush prey above them (Australian Angel Shark, Wobbegong Shark) have more ampullae dorsally; open water species have more even dorsal and ventral distribution; pelagic planktivores (Basking Shark, Megamouth Shark) have more dorsally above mouth and few ampullae overall (Newton et al. 2019).

The batoids (skates and rays) also have many ampullary organs around the mouth, but some species also have some along their expanded pectoral fins (Newton et al. 2019), which would further enhance their ability to detect benthic prey. Skates that feed on benthic prey have a higher density of ampullary receptor pores than skates that feed on more mobile prey, further supporting the role of electroreception in prey detection (Collin & Whitehead 2004). Skates that live in deeper water have more and larger ampullary organs than those that live in shallower areas, even within the same species, perhaps making the deeper fish more sensitive to bioelectricity and permitting them to detect prey from a greater distance (Raschi & Adams 1988). Sawfish (Pristidae) have a saw-like snout covered with ampullary pores that allow it to detect prey hidden beneath sand.

In a classic experiment, Kalmijn (1971) showed that sharks could detect and would attack living prey or electrodes emitting mild electrical signals but would ignore dead prey or live prey or electrodes that were covered by a barrier to electric fields. A similar experiment with Australian Lungfish (*Neoceratodus forsteri*) showed that they also use electroreception to detect prey (Watt et al. 1999).

Another use of passive electroreception is the detection of potential predators. Embryonic Clearnose Skates (*Raja eglanteria*) use their tails to move water through the egg case for respiration, but their muscular activity also generates electrical

FIGURE 6.8 The elongate, flattened rostrum of the Paddlefish (*Polyodon spathula*) has a high density of ampullary receptors, allowing them to detect the density of zooplankton, their primary food. Juvenile Paddlefish can detect individual zooplankters from as far away as 9 cm (Wilkens et al. 2002; Wilkens & Hofman 2007). USFWS/ Digital Library / Public Domain.

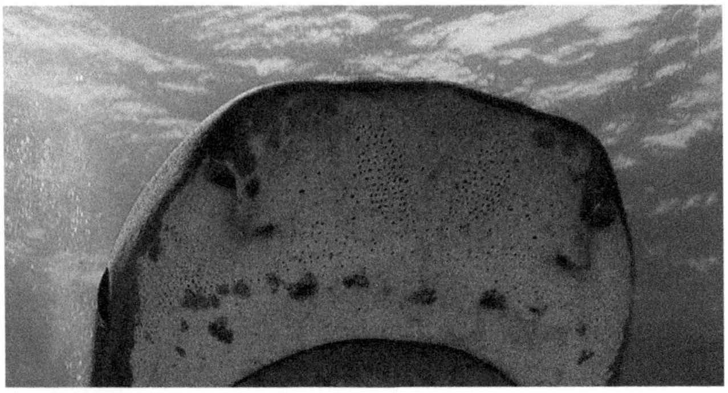

FIGURE 6.9 The small pores on the snout of this Tiger Shark (*Galeocerdo cuvier*) are openings of the ampullary electroreceptors. Albert Kok / Wikimedia Commons / CC BY-SA 3.0.

signals that could be detected by nearby predators. When the embryonic skates detect weak electrical stimuli from another source, such as those emitted by a nearby foraging shark, the tail movements stop (Sisneros et al. 1998). Similarly, when newly hatched small-spotted Catsharks (*Scyliorhinus canicula*) detect electrical stimuli that might represent a nearby predator, they temporarily cease gill ventilation movements to avoid detection (Peters & Evers 1985).

Ampullary receptors may also be important in social interactions, such as the recognition and location of conspecifics for mating. Some skates (Rajidae) have electric organs that produce weak output used during courtship, and male Round Stingrays (*Urolophus halleri*) can locate females buried in the sand based on the voltages produced by the female's respiratory muscles (Jørgensen 2011). Peters et al. (2002) proposed that the variability of the bioelectric field created by basic physiological processes of the Brown Bullhead (*Ameiurus nebulosus*, Ictaluridae) could provide a means of communication with conspecifics.

There has been considerable speculation regarding the role that electroreception may play in compass orientation among sharks. Ampullary receptors may permit some sharks to detect electric fields that are the result of movement of the fish, or water masses such as ocean currents, across the earth's magnetic field – thereby providing navigational cues for compass orientation (Kalmijn 2003; Newton et al. 2019; see section on Magnetoreception later in chapter).

The sensitivity of ampullary receptors may change during the life history of a fish, reflecting a change in functional role. For example, the ampullary receptors of the Atlantic Stingray (*Dasyatis sabina*) have response properties allowing detection of electric fields typical of large predators while a fish is young, but changes to being better able to detect prey and locate mates when the fish is older (Sisneros & Tricas 2002). A somewhat similar ontogenetic shift occurs in the Clearnose Skate (*Raja eglanteria*), a weakly electrogenic species that utilizes electrical communication for social and mating interactions (Sisneros et al. 1998).

The density of ampullary receptors may also change with age as the need for keen electroreceptive capability changes. Juvenile Scalloped Hammerhead Sharks feed in turbid water with poor visibility and have a very high density of ampullary pores on their heads. As the fish grows, the head broadens and the pores of the ampullary organs become more widely spaced as the fish also moves into more open water where visibility is better and detection of electric fields is presumably less important (Collin & Whitehead 2004). Similar trends are observed in the Bonnethead (*Sphyrna tiburo*) and Sandbar Shark (*Carcharhinus plumbeus*).

Tuberous Organs

Tuberous organs are also located in the dermis of the skin, but are covered with loosely packed cells with no canals or pores that open on the skin surface (Fig. 6.7B). They are able to detect higher frequency electric fields (50 Hz to > 2 kHz) produced by the fish's own electric organs (EOD discussed in Chapter 4), and are most sensitive to the frequencies produced by the fish's own EOD. Several different types of tuberous organs are found among species, but they fall into two main categories – those that encode timing (frequency) of the EODs, and those that encode the amplitude of the EOD (von der Emde & Engelmann 2011). They are only found in the teleosts that generate an EOD, such as the freshwater mormyrids, gymnarchids, and mochokid catfishes of Africa and the gymnotoids of South America. (Gymnotoids and mormyrids also have ampullary organs.)

Most of these fishes are considered weakly electric fishes, but others are strong electric fishes (e.g., the Electric Eel, *Electrophorus electricus*, and electric catfishes, *Malapterurus*) which can generate EODs strong enough to deter predators or to stun their prey. A small number of marine fishes generate electricity (e.g., stargazers, family Uranoscopidae; electric rays, order Torpediniformes), but the detection of EODs used for communication seems to be limited to freshwater fishes. This is likely because sea water is such a good conductor of electricity that generating and maintaining a stable electric field that can be interpreted by other fishes would be difficult. Electric organs evolved at least six times in fishes (Crampton 2019); weak organs appeared first, then stronger organs evolved in some groups.

The weakly electric fishes are primarily nocturnal and/or in turbid waters with low visibility. They use the generation of the EOD to locate hiding places to occupy during the day and to explore their environment at night (von der Emde 1998; Graff et al. 2004). Electroreception can also be used to locate prey and assist with navigation and orientation. A fish can detect objects moving into its electric field when those objects cause a change in the field, which results in a change in the rate of impulses received by the brain, such as when the fish encounters an object with different conductance than the surrounding water. This probably allows the fish to detect the size and distance of the object and may also permit discrimination between living and nonliving objects because their different electrical properties would create different distortions of the electric field. Some fishes have structures designed for especially acute electroreception, such as the elongated chin appendage of the elephantnose fishes (Mormyridae) of Africa (Fig 6.10). Because weak electrical signals attenuate rapidly in

FIGURE 6.10 The elephantnose fishes of Africa (*Gnathonemus*, Mormyridae) have an elongate snout with a high concentration of tuberous receptors. This "schnauzenorgan" moves rapidly while fish forages, scanning the environment for distortions in the fish's electric field that would indicate potential prey. The region of the brain receiving input from the schnauzenorgan is quite large, indicating its importance (von der Emde and Engelmann 2011). Courtesy of Todd Stailey.

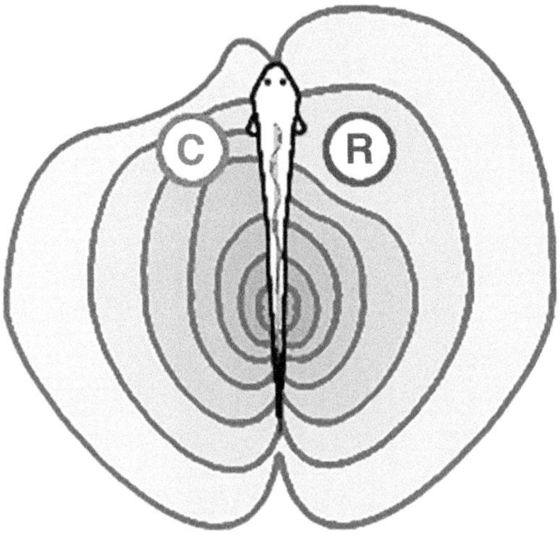

FIGURE 6.11 Electric organs, often made of modified muscles along the trunk, produce electric fields around some fishes, such as the elongate and laterally compressed South American knifefishes (order Gymnotiformes). The fish detects conductive (C) objects that concentrate the field and resistive (R) objects that spread the field, providing valuable information about their surroundings. Image by Huffers, CC_SA 3.0.

water, electrolocation is probably only effective within several centimeters (Crampton 2019).

The generation and detection of weak electric fields is particularly well developed in several groups of freshwater tropical fishes living in murky waters with poor visibility, such as the Gymnotiformes of South America and the Mormyridae of Africa. The EOD of some species are brief pulses released at irregular intervals, whereas other species continuously produce oscillating, high-frequency waves of electricity (Zakon et al. 2002). The resulting electric field surrounds the fish (see Fig. 6.11) and any changes in the field are detected by the fish's tuberous organs. Bending the body would distort the electric field, so these fishes typically pass waves down a single, long dorsal or anal fin for propulsion while maintaining a straight body posture (e.g., mormyrids, gymnotoids, *Gymnarchus niloticus*).

Electrical Communication Most weakly electric fishes produce EODs using electric organs derived from modified muscle cells. The gymnotoid ghost knifefishes (family Apteronotidae) are an exception – their electric organs are derived from modified nerve cells (see Chapter 4). EODs used for communication attenuate rapidly in water, so fishes must be within a few body lengths of one another (Crampton 2019) to communicate effectively. Two general types of EODs are generated by fishes: wave and pulse. Nearly all weakly electric African species and about one-third of weakly electric South American fishes use pulse signals, regardless of family.

EOD signals are quite species-specific (see Hopkins 2009), and electric fishes can distinguish EODs generated by individuals of their own species from those of others and usually respond more to their own species. They also recognize individuals within their species based on the signal, which demonstrates a combination of innate sensory processing as well as learning and memory. Electric communication is important in courtship of some species. EODs differ between males and females and vary with reproductive state and maturity. Sex hormones (androgens and estrogens) affect EOD by influencing the properties of the cells that produce the electric output (electrocytes in the electric organ) or the nerves that regulate them, which alters the characteristics of the EOD (Pappas & Dunlap 2011; Crampton 2019).

Within a species, the electric organs that produce EODs for communication typically produce a fixed type of signal (pulse or wave), but the brain controls the rate of firing. Electric field frequency shifts play an important role in dominance interactions in many electric fishes. South American knifefishes (Gymnotidae) have individual characteristic waveforms to their EODs. In *Gymnotus carapo*, rapid increases and decreases in frequency indicate threat, individuals with higher EOD frequencies are consistently dominant, and submissive individuals cease discharging (Black-Cleworth 1970). Among *Eigenmannia virescens* (Sternopygidae) dominant males typically use the lowest EOD frequencies. The Brown Ghost Knifefish (*Apteronotus leptorhynchus*) also demonstrates a variety of EODs that convey different meanings, including gender and social status (Zakon et al. 2002).

The elephantnose fishes (Mormyridae) of Africa use EODs for orientation, territorial interactions, species recognition, individual recognition, courtship, and to communicate social status (Carlson 2002; Terleph & Møller 2003). EODs are both species- and sex-specific among different life history stages, and interactions include cessation and frequency modulation of EODs ("bursts," "buzzes," and "rasps"), echoing, and duetting. In direct analogy to gymnotiform behavior, *Gymnarchus niloticus* is the only mormyriform to generate wave signals, and slightly lowers the frequency of its wave discharge as a submissive signal (C.D. Hopkins, pers. commun.).

The ability to produce electrical signals is believed to have originated in the African superfamily Mormyroidea and the South American order Gymnotiformes at about the time that these two continents separated from one another, approximately 120–100 million years ago (Crampton 2019). Both groups are nocturnal and thrive in tropical and subtropical waters with very low visibility.

Agonistic (aggressive) interactions include interference with a conspecific's electroreception. In *Gymnotus carapo*, dominant fish often shift their discharges to coincide with the short interval when a subordinate would be analyzing its own output, which could impair the subordinate's ability to electrolocate. Such interference is overcome in gymnotiforms such as *Eigenmannia* by a **jamming avoidance response** (JAR), in which fishes shift the frequency of their EOD when they get near one another, thereby preventing interference with one another's ability to electrolocate. Fish in a social group maintain a 10–15 Hz difference with their neighbors so that each individual has a "personal" discharge frequency (see

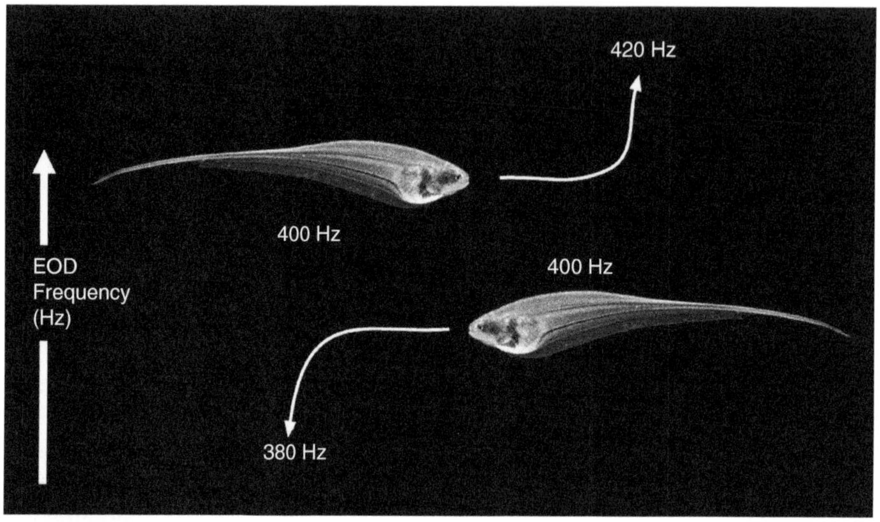

FIGURE 6.12 The jamming avoidance response (JAR) of two *Eigenmannia virescens*. Individuals typically each have separate frequencies, but in this example two electrically isolated fish produce electrical discharges of about 400 Hz. When nearby, or if tanks are connected electrically, the fish shift the frequencies of their EODs to avoid interfering with one another's signal. In this species during an aggressive interaction, the submissive individual shifts to a higher frequency whereas the dominant individual might remain at its normal frequency or shift to a lower frequency. C. Hopkins and M. Kawasaki / Wikimedia Commons / Licensed under CC BY 3.0.

Feng 1991). When several *Eigenmannia* were kept in separate tanks and all the tanks were connected by electrical wires, the fish shifted their frequencies to an average separation of 7 Hz (Fig. 6.12). The Brown Ghost Knifefish (*Apteronotus leptorhynchus*) also demonstrates a JAR (Zakon et al. 2002), as do quite a few species. In several other species, fish in a social group have nonoverlapping frequencies (Bullock et al. 1972; Hagedorn 1986).

Magnetoreception

At least some elasmobranchs and bony fishes respond to changes in magnetic fields (Walker 2011; Formicki et al. 2019), although the mechanism may differ between these two lineages of fishes. The ampullary organs of elasmobranchs may be able to detect the weak electric fields created as a fish moves through the earth's magnetic field. This ability would provide these fishes with an indirect way of sensing the earth's magnetic field and give them navigational information with respect to compass headings. Migrating sharks reportedly follow magnetic intensity gradients along topographic features in the sea floor as a means of orientation (Klimley 1993, 2015). In lab studies, Round Stingrays (*Urolophus halleri*) switched the location in which they searched for food when the electric field around them was artificially reversed, suggesting that geomagnetic cues might be used in daily activities (Kalmijn 1978). Wild caught Yellow Stingray (*Urobatis jamaicensis*) were conditioned in lab to respond to buried magnets and initiate foraging behavior, demonstrating their ability to detect these magnetic fields which could then be used for navigation (Newton and Kajiura 2017).

Among bony fishes, there is both behavioral and anatomical evidence for magnetoreception. Japanese Eel (*Anguilla japonica*) can be conditioned to respond to magnetic fields that are similar in magnitude to that of the earth (Nishi et al. 2004), larval Brown Trout (*Salmo trutta*) responded to magnetic fields (Formicki et al. 2004), and several studies have supported the suggestion that anadromous salmon use Earth's geomagnetic field for orientation as maturing adults return from open water feeding areas to their home river systems to spawn (see Ueda 2019). Other teleosts also have demonstrated magnetosensitivity (see Formicki et al. 2019) including yellowfin tuna (*Thunnus albacares*), European Perch (*Perca fluviatilis*), Common Roach (*Rutilis rutilus*), Rudd (*Scardinius erythmophtalmus*), Common Bleak (*Alburnus alburnus*), and Common Carp (*Cyprinus carpio*).

The ability to discriminate among different field strengths and inclinations and to orient to the directional polarity of the earth's magnetic field would aid in magnetic compass orientation and navigation. Sockeye Salmon (*Oncorynchus nerka*) returning from the Pacific Ocean to the Fraser River to spawn must go around Vancouver Island. Based on an examination of 56 years of fisheries data, Putman et al. (2013) showed that changes in the proportion of fish passing either north or south of the island was partly explained by fluctuations in the Earth's geomagnetic field.

Some speculate that the ability to detect geomagnetic fields may be associated with crystals of magnetite which have been extracted from the heads, sometimes in the nasal region, of some fishes, including Yellowfin Tuna, Chinook Salmon (*Oncorhynchus tshawytscha*), Chum Salmon (*O. keta*), and Rainbow Trout (*O. mykiss*) (Walker et al. 1984; Kirschvink et al. 1985; Ogura et al. 1992; Walker 2011).

Vision

The eyes of fishes are structurally similar to those of other vertebrates, including humans (Fig. 6.13). Light passes through a thin, transparent **cornea** and enters the eye through the **pupil**. The diameter of the pupil is fixed in teleosts and lampreys, but elasmobranchs have a muscular iris that controls its size and thereby regulates the amount of light entering the eye. The pupils of most deep-sea sharks are circular, whereas most other sharks have slit-like pupils; many skates and rays have crescent-shaped pupils, although there is considerable variability (Hueter et al. 2004).

Light next passes through the **lens**, which is denser and more spherical than the convex and flexible lens of terrestrial vertebrates (Hawryshyn 1998), although the lenses of some elasmobranchs are somewhat elliptical. The round lens helps focus light in a rather small eye, which is important because fish heads could not accommodate large eyes without having them protrude (Kröger 2011). Fishes focus on objects at different distances by moving the lens, thereby adjusting the distance between the lens and the retina. Light passes through the liquid-filled center of the eye to the retina, and then must pass through several layers of neurons before striking the photoreceptor cells (rods and cones) in the retina's deepest layer.

Fishes may have two types of sensory cells in the retina: rods and cones. These sensory cells respond to specific wavelengths and intensities of light. **Rods** are activated by low light intensity and are therefore helpful in poorly lit environments but provide low point-to-point resolution. Cones require brighter light, and different cones respond to different wavelengths, thereby providing color vision. Crepuscular species (those that are active at dawn and dusk) have high rod:cone ratios, and many nocturnal and deep-sea fishes have only rods. In addition, photomechanical movement of melanin in the retina exposes the rods in dim light and shields them in bright light (Hawryshyn 1998).

Cones are less sensitive than rods and therefore require brighter light for stimulation. Cone:rod ratios are highest in fishes which are active during the day (diurnal) and rely more on vision. There are several types of cones, each with a different photoreceptive glycoprotein (opsin) that responds to different wavelengths of light in the visible spectrum and in the UV range. A species may have only two or three types of cones, which is correlated with the quality of light in the fish's habitat (see Schweikert et al. 2019). **Porphyropsins** tend to be more common in fishes living in shallow areas or closer to the surface because they are sensitive to yellow-red light, which attenuates relatively quickly in water. **Rhodopsins** are more sensitive to blue-green light that penetrates further into the water and are therefore common in fishes inhabiting somewhat deeper areas. **Chrysopsins** are found in deep-sea fishes because they are most sensitive to deep blue light, which penetrates furthest into the water.

To understand what fishes are seeing in their aquatic environments, we must first understand the physics of light in water and the way in which photoreceptors detect particular light wavelengths. Water may be very clear in some habitats and extremely murky (turbid) in others depending on the

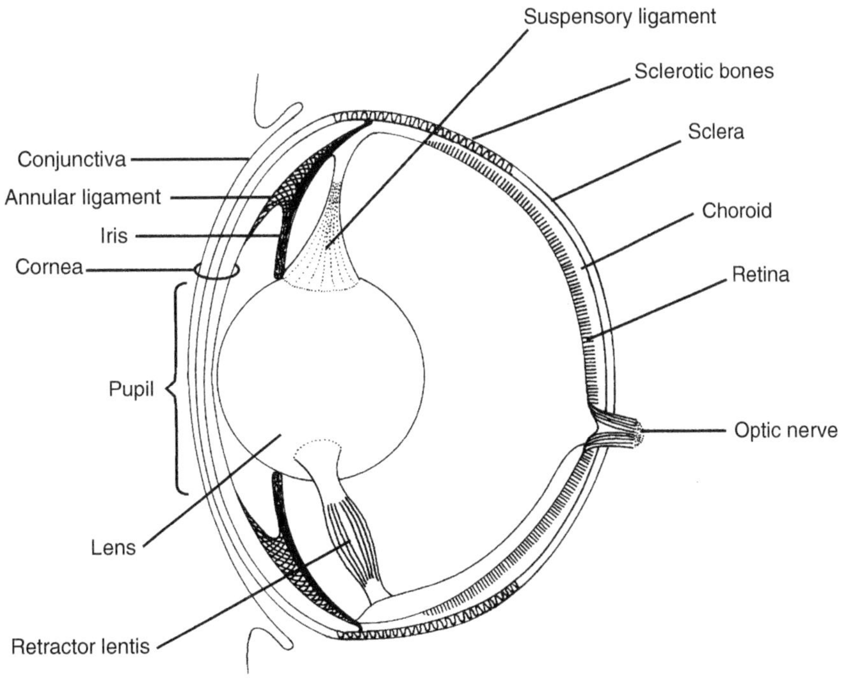

FIGURE 6.13 Cross-sectional view of the eye of a teleost. From Hildebrand (1988).

presence of suspended particles (sediments, detritus) and abundance of phytoplankton and zooplankton. The intensity of solar radiation decreases exponentially with depth, and the range of spectral light narrows as light penetrates the ocean. For instance, light in the red range (400–500 nm) and ultraviolet range (UV; >700 nm), at the two extremes of the spectrum, are absorbed by water molecules and do not penetrate far below the surface (e.g., <10 m). In contrast, green light is transmitted well in coastal waters, but blue light is the only light transmitted from the surface into the deep ocean (e.g., the mesopelagic). Because of these differences in light transmission in water, some fishes that appear colorful to us when artificially illuminated most likely appear quite different and even cryptic in their natural habitats (Marshall et al. 2019). For example, we often admire the color patterns of coral reef fishes, but these patterns are often only noticeable at close distances. From further away the patterns simply blend together and lose resolution. Therefore these patterns may be very useful for communication during close interactions, yet help the fish to blend into the background if seen from a few meters away. However, coral bleaching due to environmental stress in recent decades makes these fishes more conspicuous than they have been.

The damselfishes *Chromis viridis* and *C. atripectoralis* often appear blue-green in color, which may help them blend into the background color of the water. However, both also seem to change color as they move in position relative to the observer. This is due to the thin, multilayered structure of the scales, which causes different wavelengths of light to reflect differently depending on the angle. As a result, these fish appear blue when seen from below (against the blue sky), they appear green when viewed from the side, and appear yellow when viewed from above against the yellow-brown background of the corals (Marshall et al. 2019).

Not surprisingly, fishes show a great degree of adaptation in photoreceptors depending on what habitats they occupy (Schweikert et al. 2019). For instance, shallow-water fishes have receptors allowing them to see red light whereas deep-water fishes may have only receptors sensitive to the blue light that dominates in that habitat. Sensitivity to specific wavelengths of light can play an important role in feeding. For example, damselfishes (Pomacentridae) occupy shallow coral reef habitats, are herbivorous, and show high expression of genes responsible for retinal proteins sensitive to far-red light which is reflected by the algae upon which they feed (Stieb et al. 2017).

In some remarkable special cases (discussed later in this chapter), deep-sea dragonfishes (Stomiiformes) can see red light even though it does not penetrate to the depths at which they live, but they produce red light. Many deep-sea fishes have **bioluminescent organs (photophores, esca)** that produce light (often blue), allowing vision to be an important sense well below the depth to which surface light penetrates. But this light is only visible to other animals with appropriate photoreceptors and within the distance of the light's penetration.

Some fishes also have photoreceptors sensitive to UV light. UV-sensitive fishes, including some coral reef species, inhabit shallow water where UV light may be especially useful for close-range communication that would not be visible at greater distances. For example, Two-Bar Damselfish (*Dascyllus reticulatus*) have a UV-reflective patch on their dorsal fin that is used as an alarm signal for conspecifics (Losey 2003), and male Ambon Damselfish (*Pomacentrus amboinensis*) have UV-reflective patterns on their faces that are important in aggressive territorial interactions (Siebeck 2004). Among freshwater fishes, male Panuco Swordtail (*Xiphophorus nigrensis*) have UV-reflective markings that attract females (Cummings et al. 2003). In all three cases, these UV signals would not be detected by predators that are further away.

UV vision can be useful for prey capture as well. Juvenile salmonids can detect zooplankton more easily due to their ability to see UV, but this ability is lost as the fishes mature. As the fishes grow and move into deeper water, their diet changes and they become better adapted to see blue light (Bowmaker 2011).

Photoreceptors are not evenly distributed in the retina and typically occur in a mosaic pattern. In addition, photoreceptors do not occur in the same densities across the retina. The distribution of photoreceptors in the retina of the Striped Marlin (*Kajikia audax*) suggests that the region of the retina receiving light from above or in front of the fish provides greater acuity (due to a higher density of receptor cells) and color recognition, whereas the region of the retina receiving light from below is better suited for detecting dim, upwelling blue light (Fritsches et al. 2003). The pattern of photoreceptors of the European Smelt (*Osmerus eperlanus*) is similar (Reckel et al. 2003), and the retinal structure of the burrowing, deep-water Rufus Snake Eel (*Ophichthus rufus*) indicates three regions of high-receptor density (for higher visual acuity) that could help with locating food and burrows (Bozzano 2003).

Fishes may show **ontogenetic shifts** in the complement of different types of photoreceptors that correlate with changes in habitat at different life stages, as mentioned above for salmon. As Nile Tilapia (*Oreochromis niloticus*) grow and mature, changes in the retinal pigments cause the fish to become more sensitive to redder (long wavelength) colors, but lose the ability to detect UV (short wavelengths; Bowmaker 2011). Juvenile Lemon Sharks (*Negaprion brevirostris*) have porphyropsins, which are beneficial in shallow, turbid, and inshore habitats. The pigments change to rhodopsins as the sharks mature, which are more useful in the deeper, clearer open ocean (see Hueter et al. 2004). Similar porphyropsin–rhodopsin shifts occur in diadromous lampreys, salmon, and eels, which change pigments as part of the physiological changes needed to move from shallow freshwater habitats into the much deeper open sea (see Hawryshyn 1998; Bowmaker 2011). Yellowfin Tuna (*Thunnus albacares*) also may show changes in expression of visual pigments as they grow from planktivorous larvae to larger piscivores (Loew et al. 2002).

Not only can fishes detect a wide range of light wavelengths, including UV, but some fishes, such as anchovies (Engraulidae), cyprinids, salmonids (Salmonidae), halfbeaks (Hemirhamphidae), and cichlids (Cichlidae), can detect polarized light. This probably enhances the contrast of objects viewed underwater, permitting a better view of predators, prey, and potential mates, as well as providing directional information for migrating fishes. The ability to detect polarization may be most useful at dawn and dusk, when the polarization of light is highest (Hawryshyn 1992, 1998).

Lying between the photoreceptive retina and the tough, outer protective sclera of the eye is the **choroid**, which may contain a **tapetum lucidum**, a layer of reflective guanine crystals that enhances visual sensitivity under low light conditions by reflecting light not initially absorbed by the retina back into the eye. The tapetum causes the reflective appearance of the eyes of sharks and many nocturnal fishes. It is found in the Australian Lungfish (*Neoceratodus*), bichirs (*Polypterus*), most elasmobranchs, Holocephali, coelacanths, sturgeons, and some teleosts (Bone & Moore 2008). In some sharks that are active near the surface during the day, the tapetum is covered by dark pigment when light is abundant and uncovered when needed under low light conditions (Hueter et al. 2004).

The metabolically active tissues of the retina require a great deal of oxygen to function. In most fishes, a **choroid gland** helps maintain this through a countercurrent mechanism similar to that found in the gas gland associated with the gas bladder (discussed in Chapter 5). This U-shaped structure surrounds the optic nerve where it exits the eye and receives oxygenated blood from the pseudobranch, a gill-like structure on the inside surface of the operculum. Removal of the pseudobranch in trout (*Salmo*) results in decreased oxygen near the retina and a progressive loss of visual pigment (Ballintijn et al. 1977).

The outer layer of the eye (the **sclera**) is reinforced to protect the eye's delicate internal structures. The sclera of agnathans is fibrous and firm, chondrichthyans have cartilaginous plates in their sclera, and teleosts frequently possess sclerotic bones. These are well developed in the mackerels and tunas (Scombridae) and particularly in the billfishes (Istiophoridae and Xiphiidae), which have a bony stalk extending part way back along the path of the optic nerve to the brain.

Visual Adaptations for Special Habitats

Fishes that live at the water surface, or that occasionally find themselves totally out of the water, must be able to see through air. The eyes of mudskippers (Periopthalmidae), Atlantic Flying Fish (*Cypselurus heterurus*), Foureye Rock-skipper (*Dialommus macrocephalus*), and Galapagos Four-eyed Blenny (*Dialommus fuscus*) are modified to permit vision out of water (Brett 1957; Kröger 2011). The eyes of the surface-dwelling South American "four-eyed fish" *Anableps* (Anablepidae), permit simultaneous vision above and below the water (Fig. 6.14). Each eye has two pupils (one above and one below the surface of the water), an oblong lens, and a retina that is divided into dorsal and ventral sections. Light entering from above the water's surface enters the upper pupil, travels through the short axis of the oblong lens, and focuses on the ventral retina. Conversely, light from below the surface enters the lower pupil, travels the long axis of the lens, and is focused on the dorsal retina.

FIGURE 6.14 The divided pupil and oblong lens of the four-eyed fishes, *Anableps*, allows them to see above and below the surface, simultaneously. Harfus / Wikimedia Commons / CC BY-SA 3.0.

The deep sea is an optically challenging environment – the only available light is either dim blue light penetrating from above or point sources of bioluminescence generated by fishes and invertebrates living at these depths. Deep-sea fishes demonstrate a variety of adaptations that help to enhance vision in these vast areas with little light. In the mesopelagic zone (approximately 150–1000 m) light intensity decreases exponentially with increasing depth, so we see great variation in eye designs in fishes of this zone. Adaptations include changes in the size, shape, and orientation of the eyes, as well as changes in visual pigments, in order to maximize the capture and detection of the wavelengths of light reaching these depths (Douglas and Partridge 2011). The Brownsnout Spookfish (*Dolichopteryx longipe*) has eyes that collect light from both above and below (Fig. 6.15). Elongated tube-shaped eyes allow a longer focal length (cornea to lens to retina) without making the overall size of the eye too large for the head. A reflective tapetum maximizes the opportunity for light entering the pupil to reach the retina. Furthermore, the retinas of most deep-sea fishes have only rods, which are better for low light conditions and are most sensitive to blue light. Many of these same deep-sea fishes are bioluminescent, and use the light that they produce themselves to illuminate prey or communicate with conspecifics.

Dragonfishes (Stomiidae) live in deep ocean waters, produce their own blue-green bioluminescent light, and have photoreceptors in their retina that are sensitive to blue light. They also have receptors for far-red light, even though no far-red light from the surface will ever reach them. However, they produce far-red light in photophores below their eyes apparently

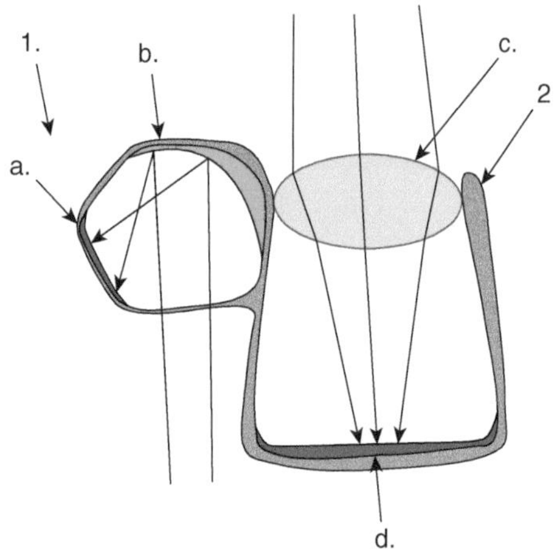

FIGURE 6.15 Each eye of the deep-sea Brownsnout Spookfish *(Dolichopteryx longipe)* has two components. The main chamber (2) gathers light from above and focuses it through a lens (c) onto the retina (d). Light from below and to the side enters through a different cornea, reflects off a layer of guanine crystals (b) and onto a separate layer of photosensory cells (a). Figure by Egmason / CC-SA 3.0.

FIGURE 6.16 A deep-sea dragonfish with photophores that emit reddish light can illuminate nearby prey and perhaps communicate with conspecific that have receptors that detect red light. But red wavelengths are invisible to most other fishes in the deep sea because their eyes lack the appropriate receptors. Courtesy of E. Widder / ORCA.

for local illumination of prey and communication with conspecifics (Kenaley et al. 2014, Fig. 6.16). But this light is not detectable by most other deep fishes because they lack receptors sensitive to that wavelength, so the dragonfishes apparently have their own, private wavelength for detecting prey and intraspecific communication. Interestingly, some lanternfishes (Myctophidae), which are often prey of the dragonfishes, also have some red photosensitivity, so they may be able to detect the presence of the predator (Douglas & Partridge 2011). This may be an example of a coevolutionary arms race between predators and prey.

Even faint deep blue light from the surface does not reach the bathypelagic zone (>1000 m) and with the exception of the bioluminescent lures of deep-sea anglerfishes, bioluminescence is rare. In this zone, small eyes seem to be a good solution because they are well suited for detecting rare point sources of light that are nearby, and therefore within range of bathypelagic fishes, which are weak swimmers. In addition, eyes are energetically quite expensive to maintain, and meals in the bathypelagic zone are few and far between – so small eyes are less of a drain on the fish's overall energy budget.

Deep-water adaptations of eyes are not only found in the oceans, although most deep habitats are marine. The species flock of sculpins and related fishes (suborder Cottoidei) of Lake Baikal, the world's deepest lake, provide an interesting example of parallel evolutionary adaptation of visual receptors. These fishes are believed to have evolved from a single shallow-water species over about 5 million years, and there are now over 20 species occupying a variety of habitats as deep as over 1500 m. The shallow-water species have photoreceptive pigments that can detect longer wavelengths of light (red/orange). Species occupying greater depths are more sensitive to shorter wavelengths (blue/green, then only blue) that penetrate deeper into the water column (Bowmaker & Hunt 2006).

Some fishes that live in perpetually dark habitats lack functional eyes as a result of degenerative evolution. The lack of eyes in hagfishes (Myxinidae) is likely a degenerative condition, as is the loss of functional eyes among some populations of cave-dwelling fishes (Hawryshyn 1998; see also Chapter 10). Although some cavefishes lack a cornea, lens, and iris, they still possess the genes that code for the proteins needed to detect light (Parry et al. 2003).

Chemoreception

Chemoreception, which includes the senses of taste (**olfaction**) and smell (**gustation**), is based on the ability of particular molecules to bind to specific receptors in the membranes of sensory cells. As terrestrial vertebrates, humans typically associate taste with detection of molecules in solution in our mouths, whereas smell is the detection of molecules suspended in the air that come in contact with sensory neurons in our nose. The aquatic environment is filled with a wide variety of chemical cues because so many chemicals dissolve in water. The sensitivity of fishes to chemicals varies greatly, with some detecting compounds at concentrations of less than 10^{-10} mol/L. Therefore, fishes can learn a great deal about their environment through chemoreception, which is often used for finding and identifying food, detecting and avoiding predators, locating suitable habitats, identifying territories based on scent markings, and communicating with conspecifics. The sense of smell helps fishes detect a broad range of chemical stimuli using chemical gradients emanating from a relatively distant source, whereas the sense of taste is primarily focused on recognition and acceptance of food (Hara 2011a; Kasumyan 2019).

Smell The olfactory organs of fishes are usually blind sacs with a flap of skin to separate incurrent and excurrent flow of water (Sorensen & Caprio 1998; Hara 2011b). In some fishes, such as the Bowfin (Amiidae) and eels (Anguillidae), the organs are tubular. Hagfishes and lampreys have only a single median olfactory sac nostril. In hagfishes, a nasohypophyseal duct connects with the pharynx so that hagfishes can smell water as it moves to the gills. In lampreys, however, the lone medial nostril leads to an olfactory chamber in a dead-end nasopharyngeal pouch on top of the head. In teleosts the olfactory chambers also are dead-end sacs that do not lead to the pharynx, except in a few cases such as stargazers (Uranoscopidae). Jawed fishes have a pair of olfactory organs, each of which has an incurrent and excurrent nostril (Fig. 6.17). The nares of elasmobranchs are located ventrally on the snout and also are not connected to the pharynx. Chimaeras (Holocephali) and lungfishes (Ceratodontiformes) have paired nares that connect to the oral cavity.

Each olfactory sac is lined with an **olfactory epithelium** composed of ciliated sensory receptor cells. The epithelium is typically folded, forming a "rosette" with increased surface area. Molecules of odorants dissolved in water bind to receptor proteins on receptor cell membranes and then nerve impulses are transmitted along their axons to the olfactory bulb in brain (Hara 2011b). The more extensive the lamellar folding, the greater the surface area available for sensory cells and this is thought to translate into sensitivity of the sense of smell. For example, freshwater eels (*Anguilla*), known for their extremely keen sense of smell, have far more folds in each rosette than do perch (*Perca*). However, behavioral and ecological correlates of the number of lamellae in the olfactory organ are not always clear. The movement of cilia (on non-sensory cells) may move water into and out of the olfactory organs, but other mechanisms may assist with nasal ventilation. Renewal of water within the olfactory organ ("sniffing") is essential for sampling olfactory cues in the environment. Many fishes have accessory olfactory ventilation sacs connected to the olfactory organs which compress periodically by repeated movements of the mouth causing water to be moved into and out of the olfactory organs (Kasumyan 2004).

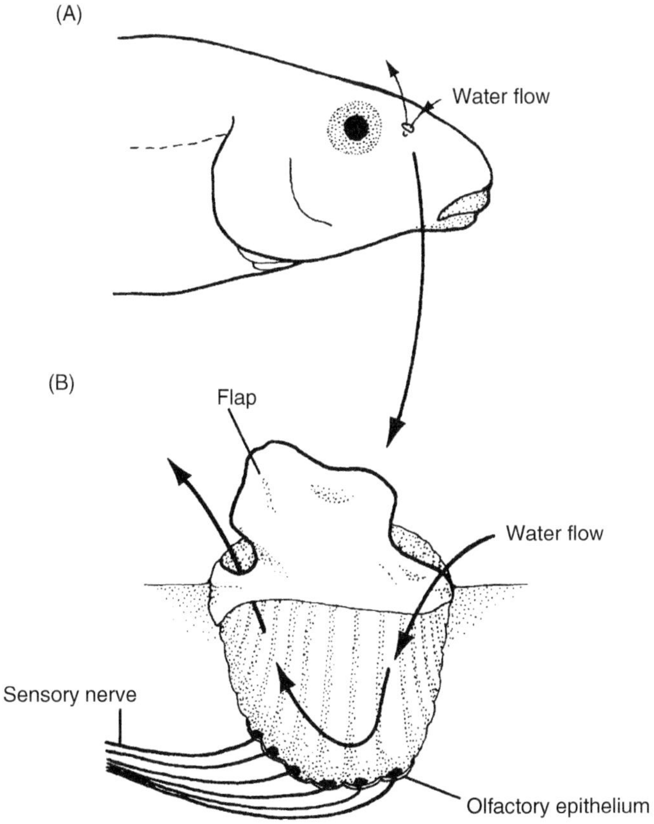

FIGURE 6.17 (A) External view of the nares of a fish. (B) The obvious flap of skin directs water across the sensory epithelium. Adapted from Lagler et al. (1977).

Fishes can detect four main categories of chemicals – amino acids, bile acids, sex steroids, prostaglandins – none of which are detectable by humans (Hara 2011c). Some fishes can detect these molecules in extremely low concentrations. Amino acids, particularly those of fairly simple structure and with certain attached groups, are detectable by many fishes at concentrations of around 10^{-10} mol/L (Hara 1993). Other compounds that are detectable by some fishes at very low concentrations include bile acids (10^{-9} mol/L), salmon gonadotropin-releasing hormone (10^{-15} mol/L), and some sex steroids (10^{-12} mol/L).

The ability to detect such small concentrations of certain chemicals makes olfaction valuable for habitat location for some fishes, including homing in salmon (Stewart et al. 2004; Ueda 2019). In one of the first studies to call attention to the important role of olfaction in salmon migration, Wisby and Hasler (1954) captured adult Coho Salmon (*Onchrhynchus kisutch*) migrating upstream and occluded the nares of about half the fish, blocking their sense of smell. The remaining fish were untreated, and all fish were displaced about a mile back downstream of the junction of the two river branches. The fish with unoccluded nares typically returned to the same stream in which they had been caught, whereas those with occluded nares did not. This suggested, and subsequent studies have confirmed, that migratory salmonids imprint on the scent of their home stream early in life, and then use their sense of smell to find their way back to spawn. Apparently, as the salmon mature, hormones that help trigger maturation of the gonads also affect development of sensory cells in the olfactory epithelium and make the fish more sensitive to certain chemicals associated with the home stream (Palstra et al. 2015; Ueda 2019).

Sea Lamprey (*Petromyzon marinus*) also are anadromous and rely on olfaction to identify a suitable spawning stream. A chemical signal released by juveniles (ammocoetes) provides a signal to adults that the stream apparently provides a suitable spawning and nursery habitat, and sexually mature males release another pheromone that attracts mature females (Wagner et al. 2006). This knowledge is being used in attempts to control parasitic Sea Lamprey by diverting them during their spawning migration (see Chapter 13, Petromyzontiforms).

Juvenile eels may use olfaction to locate suitable habitats as they migrate upstream after hatching at sea. Glass eels of both the Longfin Eel (*Anguilla dieffenbachii*) and Shortfin Eel

(*A. australis*) prefer water from their river of capture over well water, and the Shortfin Eels prefer water from lowland streams, where they tend to occur, over water from the mainstream of a river (McCleave & Jellyman 2002).

Olfactory cues also can be used to locate mates, and in some species males and females may exhibit different olfactory sensitivities. Male deep-sea ceratioid anglerfishes have much larger olfactory organs, olfactory nerves, and olfactory lobes than do females. In these fishes, females are much larger than the males and apparently release species-specific pheromones that attract the smaller, more mobile males. Males then attach themselves to the females and spend the rest of their lives as parasitic sperm factories (see Chapter 10, The Deep Sea). Gilthead Seabream (*Sparus aurata*) are sensitive to the excreted body fluids of sexually mature conspecifics (Hubbard et al. 2003), and male Brown Trout (*Salmo trutta*) and Lake Whitefish (*Coregonus clupeaformis*) both show courtship behavior when exposed to a prostaglandin released by females ready to spawn (Laberge & Hara 2003).

Olfaction may also be used to detect and avoid predators. Juvenile Lemon Sharks (*Negaprion brevirostris*) react to the odor of organic compounds produced by American crocodiles (*Crocodylus acutus*) that prey on small sharks where they co-occur (Hueter et al. 2004). Otophysan fishes respond to chemical alarm substance (CAS) released from club cells in the skin of injured conspecifics, or other prey species with which they occur (Hara 2011d; Maximino et al. 2019). Exposure to CAS in Zebrafish (*Danio rerio*) caused antipredator/evasive behavior (erratic, rapid bursts of swimming, and more time near bottom of tank); and after 5 min of exposure fish showed approximately an eightfold increase in average whole-body levels of the stress hormone cortisol (from 1.5 to 11.9 ng/g) above control fish (Barkhymer et al. 2019). Some non-Otophysan fishes also possess CAS that are apparently not released by club cells, suggesting that CAS may have evolved independently several times. There is also some evidence that the release of CAS may attract predators that are potential competitors for the initial predator, thereby distracting them from prey (see Maximino et al. 2019).

Chemical contaminants can interfere with olfaction and thereby disrupt important interspecific communication (Tierney 2011). For example, cadmium accumulates on the olfactory epithelium of Rainbow Trout (*Oncorhynchus mykiss*) and affects fish social behavior, including blocking their ability to detect alarm substance (Scott et al. 2003; Sloman et al. 2003).

Taste Taste is used primarily for food recognition. The chemosensory cells responsible for taste are mostly located in and around the mouth, including on the jaws, vomer, tongue, gills, lips, and barbels (Fig. 6.18). In addition, many fishes have external taste receptor cells on the skin of the head, trunk, and fins (Hara 2011e; Kasumyan 2019). Fishes have more taste receptor cells than other vertebrates, perhaps because these cells can be located anywhere on body. The taste region of brain can approach 20% of brain mass in some fishes, reflecting its high importance among fish sensory systems.

FIGURE 6.18 The chin barbels of goatfishes (Mullidae) have a high density of taste buds that are used to forage for food. Courtesy of Jon Hyde.

Taste receptor cells are often clustered into taste buds, which can contain 30–100 sensory cells, or they may occur individually on parts of some fishes. These solitary chemosensory cells can be numerous, with up to 4000 per mm^2 in some minnows (Cyprinidae), but they are not well studied.

When activated by the binding of a chemical, the receptor cells activate nerves that carry the signals to the gustatory centers of the brain (Kasumyan 2019). Fishes do not respond to the same types of chemicals that our taste buds recognize, but they do respond to low molecular weight compounds such as peptides, amino acids, and steroids (Hara 2011f). Many anthropogenic contaminants can harm or interfere with taste receptors, especially heavy metals and detergents (Kasumyan 2019). Biocides, low pH, hormones, and drugs can also interfere with taste, resulting in decreased feeding and subsequent loss of condition.

Do Fishes Feel Pain?

Sensory systems allow animals to respond to external stimuli and adjust their behavior in adaptive ways, including reactions to potentially threatening or injurious situations. As human predators on fishes, we have introduced such stimuli in the form of fishing. This brings up the biological and ethical issues of how fishes react to our actions, and, specifically, if fishes experience pain because of our catching them and whether we should adjust our behavior as a result. There has been considerable debate regarding whether or not fishes can feel **pain**, but our ability to resolve this question depends on how we define the concept of pain itself.

The argument has been made (Rose 2002, 2007) that fishes may not be capable of experiencing pain, at least in a similar way that we might experience it. This conclusion is based on a few main points: (1) Sensing pain requires a level of cognition and awareness that can only be achieved by animals that have a complex neocortex of the brain, which fishes do not have; (2) responses to noxious stimuli can occur without a level of awareness that could be perceived as pain and may be simply reflexive actions; and (3) the effects of analgesics on fish behavior should not be seen as evidence for an animal's ability to perceive pain because analgesics act at the subcortical level of the brain, and in the spinal cord. Those making this argument recognize that fishes can learn, but feel that the type of learning seen in fishes may occur without conscious awareness. They argue, therefore, that such learning should not be used as an argument that fishes possess higher level cognition. They conclude that fishes simply do not have the brain structure necessary to process and perceive pain. This position is based on a definition of pain that requires a conscious awareness of the stimuli, and also on the argument that this can only be achieved in a brain that has a neocortex, which fishes lack.

Nociception is the ability of an animal to detect a noxious stimulus – that is a stimulus that it apparently chooses to avoid. There is no disagreement regarding the fact that fishes can detect noxious stimuli, react to them, and have them affect their behavior (Sneddon 2011). To some, this is an indication of irritation, but does not rise to the level of pain, which evokes a deeper and more subjective emotional response.

Various studies have shown that fishes can learn to respond to noxious stimuli, these responses may be proportional to the strength of negative stimulus, and that responses to negative stimuli can be reduced with use of analgesic. For example, Rainbow Trout (*Oncorhynchus mykiss*) that had bee venom or acetic acid injected into their lip had significantly higher breathing rates and took much longer to resume feeding after treatment than did fish that were injected with saline, or fish that were just handled and had nothing injected (Sneddon 2003). The fish injected with acetic acid also rubbed their lips in the gravel and against the sides of the tank (not a reflexive response), and in a related study the effects of noxious stimuli were reduced when the fish were administered morphine. This has led some to conclude that bony fishes can feel pain (Sneddon 2004). Other studies have also shown that the responses to noxious stimuli often vary among species, may be different under different conditions, and that there is often considerable intraspecific variation in responses with some individuals exhibiting what human observers interpret as being more bold or aggressive than others (Sneddon 2011).

Studies of the fish brain indicate that the pallial region of the forebrain may provide the capacity for more advanced processing than humans have presumed in the past (Butler 2011), suggesting that a neocortex is not necessary to feel pain but that some other part of the brain could be involved. In addition, it would be anthropomorphic to presume that a fish's perception of pain is similar to ours, and we should make no presumptions about the emotional or psychological impacts of noxious stimuli on fishes, as we have no data to suggest that they have the capability for such cognition and their brain structure seems to make it unlikely that that they could.

Behavioral studies have shown that many fishes will change their behavior when exposed to certain stimuli, but we cannot evaluate the intensity of what a fish is experiencing, nor whether there is any cognitive recognition similar to what occurs when humans respond to pain. Iwama (2007) points out that we probably will never be able to know enough about the mental processing of stimuli to determine whether or not fishes may experience something similar to what we would describe as pain. He suggests, therefore, that the scientific community not be drawn into this debate, but instead focus on testable aspects of fish biology and physiology, while recognizing our "ethical responsibility to respect the life and well-being of all organisms." Other authors maintain that it is ethically irresponsible to not attempt to bring scientific understanding into this discussion.

Supplementary Reading

Carrier JC, Musick JA, Heithaus MR, eds. 2004. *Biology of sharks and their relatives.* Boca Raton, FL: CRC Press.

Eastman JT. 1993. *Antarctic fish biology: evolution in a unique environment.* San Diego: Academic Press.

Evans DH, ed. 1998. *The physiology of fishes*, 2nd edn. Boca Raton, FL: CRC Press.

Evans DH, Claiborne JB. 2006. *The physiology of fishes*, 3rd edn. Boca Raton, FL: CRC, Taylor & Francis.

Farrell AP, Stevens D, Cech JJ, Richards JG. 2011. *Encyclopedia of fish physiology: from genome to environment.* Boston, MA: Elsevier / Academic Press.

Hara TJ, Zielinski BS. 2007. *Sensory systems neuroscience. Fish physiology*, Vol. 25. New York: Academic Press.

Hueter RE, Mann DA, Maruska KP, Sisneros JA, Demski LS. 2004. Sensory biology of elasmobranches. *In* Carrier JC, Musick JA, Heithaus MR, eds. *Biology of sharks and their relatives*, pp. 325–368. Boca Raton, FL: CRC Press.

McKenzie DJ, Farrell AP, Brauner CJ. 2007. *Primitive fishes. Fish physiology*, Vol. 26. New York: Academic Press.

Ostrander GK. 2000. *The laboratory fish.* London: Academic Press.

Randall DJ, Farrell AP, eds. 1997. *Deep-sea fishes.* San Diego, CA: Academic Press.

Shadwick RE, Lauder GV. 2006. *Fish biomechanics. Fish physiology*, Vol. 23. New York: Academic Press.

Sloman KA, Wilson RW, Balshine S. 2006. *Behaviour and physiology of Fish. Fish physiology*, Vol. 24. New York: Academic Press.

Journal

Journal of Morphology.

CHAPTER 7

Homeostasis

Summary

In this chapter, we explore those processes that maintain physiological stability, often referred to as homeostasis. Specifically, we will investigate: (i) the roles of the endocrine system and the autonomic nervous system in controlling various physiological responses; (ii) the importance of body temperature and thermal relationships between fishes and their environments; (iii) the mechanisms involved in maintaining water, solute, and pH balance; (iv) how fish immune systems maintain health; and (v) how various forms of physiological stress can compromise fish health.

The endocrine system is responsible for most long-term regulation of physiological processes in fishes, including osmoregulation, growth, metabolism, color changes, development and metamorphosis, and stress responses. Many endocrine tissues are regulated by the pituitary, which is controlled by the hypothalamus in the brain. Some environmental contaminants can disrupt hormonally regulated physiological functions, such as sexual differentiation, because they mimic naturally occurring hormones. In some cases, these are a serious threat to fish populations as well as human health.

Involuntary physiological functions such as heart rate, blood pressure, blood flow to the gills and gas bladder, and the contraction of the smooth muscles of the gut, are controlled by the autonomic nervous system.

CHAPTER CONTENTS

The Diversity of Fishes: Biology, Evolution and Ecology, Third Edition. Douglas E. Facey, Brian W. Bowen, Bruce B. Collette, and Gene S. Helfman.

Companion website: www.wiley.com/go/facey/diversityfishes3

The body temperature of most fishes is similar to that of the surrounding water because of heat loss at the gills. Some large pelagic predators, such as tunas, billfishes, opahs, and lamnid sharks, can maintain elevated temperatures in some parts of the body by conserving internally generated heat (hence the terms regional endothermy and heterothermy). In most of these fishes, countercurrent heat exchange conserves the heat generated by the continuously active swimming muscles. A warmer body core allows faster and more efficient muscle contraction and digestion in cold water. Some of these fishes also use warm blood from metabolically active tissues to maintain warm eye and brain temperatures, which may result in better vision and faster neural processing. In billfishes and at least one species of tuna, one of the eye muscles has evolved into a specialized thermogenic organ. The specific morphology of the structures allowing for regional endothermy varies among some groups, providing several examples of convergent evolution.

Seasonal changes in water temperature can affect fish metabolism. However, fishes can compensate for some change by altering the concentration or form of certain enzymes to maintain essential biochemical processes in cold conditions. High water temperatures diminish the availability of oxygen in the water and can inhibit the function of important proteins such as hemoglobin and many enzymes. Hence, few fishes can survive warm water temperatures. The temperature of seawater in polar regions drops below the freezing point of the blood of most fishes. To avoid freezing, many polar fishes rely on supercooling or biological antifreeze compounds. The variations in structure among the five main types of antifreeze compounds provide yet another example of convergent evolution.

Ion and water exchange at the gills play a large role in maintaining proper water and ion balance in the blood. The large surface area of the highly permeable gill membrane allows for considerable exchange of water and ions between a fish's blood and the surrounding water. To maintain a stable internal osmotic condition, freshwater bony fishes produce dilute urine and take up ions through specialized transport cells in the gills. Saltwater bony fishes must drink seawater to replace water lost by diffusion at the gills, and they eliminate excess ions through their kidneys and specialized transport cells of the gill epithelium. Elasmobranchs have high levels of urea and TMAO (trimethylamine oxide) in their blood, and therefore gain water by diffusion across their gills. In all fishes, the exchange of water and ions across the gills and other organs such as the kidneys and gut is controlled by several hormones, including cortisol, prolactin, urotensins, and catecholamines (epinephrine and norepinephrine).

Most fishes eliminate nitrogenous wastes at their gills in the form of ammonia or ammonium, which readily diffuse into the surrounding water. Fishes also produce urea, which is excreted in the urine produced in their kidneys. Fish kidneys are paired longitudinal structures ventral to the vertebral column and are one of the primary organs involved in excretion and osmoregulation. They cannot, however, make urine more concentrated than the osmotic concentration of the blood plasma.

A fish's immune system acts to prevent the entry of pathogens or to destroy them if they do enter the body. The immune system starts with protective barriers including skin, scales, and mucus. Mucus is sticky and viscous to ensnare foreign material and contains chemicals that can destroy bacteria. If a pathogen gets past these barriers, it can be attacked and destroyed by immune system components such as phagocytic cells, cytotoxic cells, and complement proteins. The jawed fishes also show adaptive immune responses that involve recognition of an invader and the mounting of response specific to that invader. These responses can involve antibodies or cells that bring about the destruction of the potential pathogen. The adaptive response also involves the creation of memory cells that allow the immune system to respond more quickly if the fish is exposed to that same pathogen in the future; this is the basis for vaccinations to prevent disease.

The proper functioning of the immune system can be compromised by stress, such as that caused by handling or environmental factors including pollution. Stress from environmental factors also can result in a thicker mucous layer on a fish's gills, thereby inhibiting gas exchange, and cause a variety of other physiological impacts that can affect long-term energy balance and fish health. In addition, the responses to stress take energy away from other physiological functions, resulting in slower growth and diminished reproductive success.

There has been considerable advancement in recent years in identifying and understanding biomarkers, which are cellular and subcellular indicators of exposure to stress. These have proven useful in identifying stress in fishes before impacts result in declining populations and community-level disruptions. Monitoring some biomarkers has also been useful in evaluating remedial measures that diminished stress in fishes.

Coordination and Control of Regulation

The nervous and endocrine systems maintain communication among the various tissues in the body and regulate many physiological functions. The nervous system acts directly and rapidly due to neural circuitry. The endocrine system is well suited for long-term regulation of physiological processes, releasing chemicals that may affect nearby tissues and **hormones** that are transported by the blood. These signals may travel throughout the body but only affect those cells with the proper molecular receptors. The nervous and endocrine systems overlap considerably, particularly in the control of various endocrine tissues by the brain. As endocrinological research on fishes advances, it is often difficult to distinguish separate roles for these two regulatory systems.

The Endocrine System

Research continues to expand our knowledge of the endocrine systems of fishes, and it is not possible in this chapter to provide a complete synopsis of fish endocrine tissues, their hormones, and their effects. Instead, we will provide a brief summary of some of the hormones important to homeostasis but will not address the many other physiological functions of hormones in fishes.

Many endocrine functions are regulated by the pituitary, which is controlled by the hypothalamus of the brain. The pituitary has two main functional regions. The **posterior pituitary**, or **neurohypophysis**, is continuous with the hypothalamus and consists primarily of the axons and terminals of neurons that originate in the hypothalamus. The **anterior pituitary**, or **adenophypophysis**, lies in contact with the posterior pituitary, and in the actinopterygians these tissues fuse. Some fishes also have an intermediate lobe of the anterior pituitary, and elasmobranches have a ventral lobe below the anterior pituitary. There is considerable variation in the degree of association between the anterior and posterior pituitaries among the different major lineages of fishes. In general, the anterior and posterior pituitary are more closely associated with one another in fishes than in tetrapods, and therefore most fishes do not rely on blood vessels to carry chemical signals between these parts of the pituitary (McMillan 2011).

The posterior pituitary is a storage and release site of chemical messengers from the hypothalamus. Neuroendocrine cells (neurons that function as endocrine cells) begin in the hypothalamus and extend into the posterior pituitary where they release their chemicals, some of which are hormones that are released into blood vessels and trigger effects elsewhere in the body. Other chemicals released by the posterior pituitary may regulate the function of cells of the adjacent anterior pituitary and intermediate lobe and are referred to as releasing factors or releasing hormones. Some of these diffuse to the intended target cells in immediately adjacent sections of the pituitary, whereas others travel a short distance to target cells via blood vessels (McMillan 2011).

Many hormones produced by the anterior pituitary regulate the production of hormones by other tissues. In teleosts, for example, follicle-stimulating hormone (FSH) and luteinizing hormones (LH) help stimulate gonad maturation, thyroid-stimulating hormone (TSH) regulates hormone production by the thyroid, and adrenocorticotropic hormone (ACTH) stimulates the adrenal gland to produce the corticosteroids (cortisol, aldosterone). Some hormones from the pituitary have a more direct impact. Growth hormone (GH), promotes growth in many tissues of the body. Prolactin helps with salt-water balance by influencing the membranes of epithelial cells in the gills, gut, skin, kidney, and bladder. And melanocyte stimulating hormone (MSH) influences pigment granules in skin cells, thereby helping with color change (McMillan 2011).

Fishes are the only jawed vertebrates known to possess a caudal neurosecretory system. Located at the caudal end of the spinal cord, this region of neuroendocrine cells, the **urophysis**, is most highly developed in the ray-finned fishes and produces **urotensins** that help control smooth muscle contraction, osmoregulation, and the release of pituitary hormones (Takei & Loretz 2006). Urotensins play an important role in vasoconstriction, and therefore help regulate blood pressure and blood flow (Olson 2011b).

The thyroid tissue of most fishes is scattered as small clusters of cells in the connective tissue of the throat region. When stimulated by TSH from the anterior pituitary, these cells produce **thyroxin**, which plays a primary role in fish growth (Sheridan 2011), development, and metabolism. Thyroxin is important in development, including the sometimes extreme morphological and physiological changes associated with metamorphosis – such as the transformation of flounder larvae with an eye on each side of the head to flatfish with both eyes on one side of the head. It also initiates seaward migratory behavior of juvenile salmonids (Takei & Loretz 2006; see Chapter 9, Complex Transitions: Smoltification in Salmon, Metamorphosis in Flatfish). Thyroxin also seems to play a supplementary role in osmoregulation (McCormick 2011).

Maintaining proper calcium balance, including regulating calcium uptake at the gills, involves several hormones, including **stanniocalcin** from the corpuscles of Stannius embedded in the kidney, **calcitonin** produced by the ultimobranchial bodies in the back of the pharynx, and **prolactin** and **somatolactin** from the anterior pituitary (Takei & Loretz 2006).

The **interrenal tissues** of fishes are homologous with the adrenal glands of the tetrapods but are somewhat scattered in their location. The interrenal consists of two different types of cells, each of which produces different hormones. The **chromaffin cells** are located in the pronephros of agnathans, along the dorsal side of the kidney in elasmobranchs, and in the anterior, or head, kidney of teleosts. In response to stress or aggression, chromaffin cells produce and release the catecholamines **epinephrine** (adrenaline) and **norepinephrine** (noradrenaline) (Reid 2011). The catecholamines maintain or enhance the delivery of oxygen to body tissues by increasing gill ventilation rates and blood flow, heart rate, stroke volume, and blood pressure. They also increase oxygen transport capability by increasing the release of red blood cells from the spleen and increasing the attraction between hemoglobin and oxygen in the red blood cells. This increased blood flow to the gills may lead to increased ion exchange, which may explain why stressed fishes can experience significant osmoregulatory imbalances (discussed later in this chapter). The catecholamines are an important part of the "fight or flight" response associated with stress.

The second group of interrenal cells is steroid-producing cells located primarily in the pronephric, or head kidney, region. These manufacture and release two groups of corticosteroid hormones – the glucocorticoids, which are mainly involved with energy and glucose metabolism, and the mineralocorticoids, which play important roles in ion and water balance (Kiilerich & Prunet 2011). **Cortisol** is the primary corticosteroid, and in addition to its role in making more glucose available to sustain metabolism, it also plays an important role in maintaining electrolyte and water balance, growth, and development. It can affect multiple organs and influence the cardiovascular system, reproduction, appetite, immune suppression, and various aspects of metabolism. Cortisol levels increase with stress, which is why stress can have such a broad impact on physiological condition.

Many hormones in addition to the corticosteroids are involved in osmoregulation. For example, prolactin from the anterior pituitary, along with cortisol, is important in freshwater adaptation. Seawater adaptation involves cortisol, growth hormone from the anterior pituitary, vasopressin from the posterior pituitary, urotensins from the urophysis, atrial natriuretic peptide from the heart, and probably others (Takei & Loretz 2006).

Glucose metabolism is influenced by **insulin**, **glucagon**, and **somatostatin** from cells within the pancreas. Insulin enhances the transport of glucose out of the blood, promotes glucose uptake by liver and muscle cells, and stimulates the incorporation of amino acids into tissue proteins. Glucagon and related glucagon-like proteins seem to function in opposition to insulin, promoting the breakdown of glycogen and lipids in the liver and increasing blood glucose levels. Somatostatin also helps elevate blood glucose levels by promoting metabolism of glycogen and lipids, and by inhibiting the release of insulin (Takei & Loretz 2006).

Melatonin, produced by the pineal gland (near the top of the brain) and the retina of the eye, is secreted during the dark phase of daily light–dark cycles and helps regulate fish responses to daily and annual daylight cycles. This hormone influences many physiological processes and behaviors through its role in daily activity cycles (see Chapter 18, Circadian Rhythms). However, some fishes sustain these cycles even when the pineal is removed, so there must be other tissues and chemical signals involved (Reebs 2011).

Some hormones, when released into the environment, provide chemical signals to other fishes, and fishes can detect the difference between members of their own species and members of other species (Stacey & Sorensen 2011).

Endocrine Disrupting Compounds

As briefly summarized in the preceding section, the endocrine system regulates most physiological systems associated with maintaining homeostasis. Hormones also regulate sexual development and reproductive behavior, which affect the stability of fish populations and aquatic communities. This is why human-generated **endocrine disrupting compounds** (EDCs) can disrupt sexual development and fish population stability (see Sloman 2011b).

Endocrine disrupting compounds include a growing list of industrial chemicals, pharmaceuticals, and natural and synthetic hormones found in industrial effluent, agricultural and municipal runoff, and wastewater from municipal sewage treatment facilities. The list includes, but is not limited to, pesticides (e.g., aldrin, atrazine, chlordane, DDT, mirex, toxaphene), phthalates (found in cosmetics, plasticizers, adhesives, insecticides, printing inks, safety glass), and organohalogens (e.g., furans, polychlorinated biphenyls, dioxins). These chemicals make their way into surface waters, accumulate in fishes, and, because of their structural similarity to fish hormones, can interfere with hormonally controlled physiological processes, even in very small concentrations. For example, the ability of male Atlantic Salmon to respond to spawning cues released by females is compromised by exposure to some widely used pesticides (Sloman 2011b). Anti-inflammatory medicines such as salicylate (from aspirin) can interfere with cortisol synthesis, and industrial chemicals such as polycholorinated biphenyls (PCBs) can result in decreases in receptors for glucocorticoids such as cortisol (Wendelaar Bonga 2011).

The effects of EDCs can include altering levels of sex hormones, interfering with intracellular hormone receptors, altering secondary sex characteristics, altering gonad size and condition, creating intersex individuals (gonads containing both testicular and ovarian tissue), altering age or size of maturity, and affecting hatching success or incubation time. Some of the specific effects that have been noticed include the masculinization of female mosquitofish (Poeciliidae) exposed to effluent from pulp and paper mills (Bortone & Davis 1994; Fig. 7.1), noticeable changes in levels of androgens, estradiol, and vitellogenin in carp (Cyprinidae), and altered reproductive behavior in Goldfish (Cyprinidae) and guppies (Poeciliidae; see Greeley 2002). The synthetic estrogen used in birth control pills (17α-ethynylestradiol, EE2) is often in wastewater effluent. When added to an experimental lake in Ontario, it altered the sexual development of Fathead Minnow (Cyprinidae), including inducing the development of intersex males, and resulted in the collapse of the population within 2 years (Kidd et al. 2007). Exposure to wastewater treatment effluent with EE2 for just 21 days altered the sexual development of Fathead Minnow in laboratory studies (Filby et al. 2007).

Feminization of male Roach (Cyprinidae) by estrogenic compounds is widespread in rivers in the United Kingdom (Tyler & Jobling 2008). A study of Gudgeon (*Gobio gobio*) upstream and downstream of effluent from a pharmaceutical factory in France showed considerable evidence of endocrine disruption, including intersex fish and a male-biased sex ratio (Sanchez et al. 2011). In the United States, outmigrating juvenile Chinook Salmon from rivers in urban areas had higher levels of vitellogenin (VTG) than outmigrating juveniles from rural rivers, suggesting exposure to environmental estrogens in these urban rivers (Peck et al. 2011). An evaluation of fishes in nine widely spaced US river basins showed that about 3% of all fishes tested were intersex, and that this condition was particularly prevalent in male Largemouth Bass, *Micropterus salmoides* (18%), and Smallmouth Bass *M. dolomieui* (33%) (Hinck et al. 2009). The highest percentages of intersex bass were found in the southeastern United States. The intersex condition was most common in 1–3 years old Largemouth Bass but not found in Largemouth Bass over 5 years old. As many as 75% of male Smallmouth Bass (Fig. 7.1D) in branches of the Potomac River – the source of drinking water for Washington, DC – showed signs of feminization.

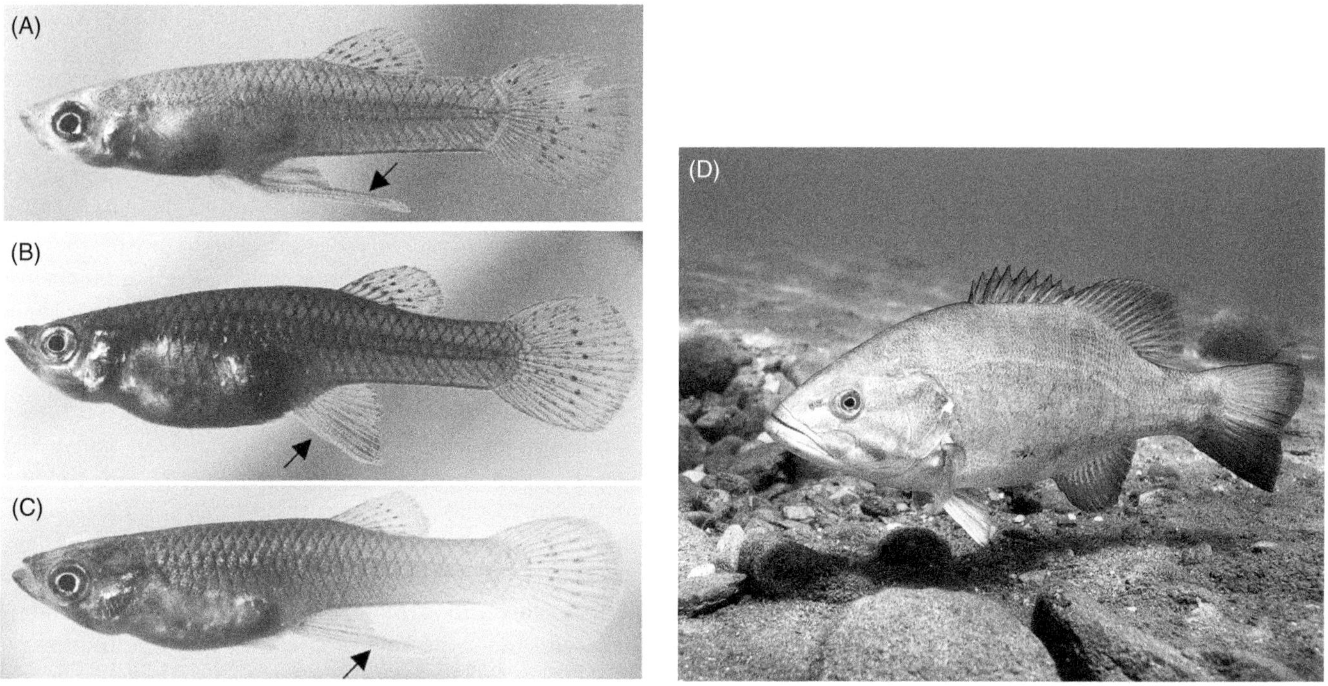

FIGURE 7.1 (A) The anal fin of a normal male *Gambusia* is elongated to form the gonopodium (arrow), an intromittent organ used to inseminate females. (B) In normal females, the anal fin is fan-shaped. (C) A masculinized female exposed to pulp mill effluent, in which the anal fin has developed into a gonopodium. (D) A survey of fishes in multiple rivers in the US showed feminization of up to 75% of male Smallmouth Bass in branches of the Potomac River. (A, B, C) – Courtesy of S.A. Bortone, used with permission; (D) – Smallmouth Bass photo by E. Engbretson / USFWS / Public Domain.

Any body function controlled by hormones can be altered by EDCs, and additional examples in fishes and other animals (including humans) appear regularly in the scientific literature. In 2009, The Endocrine Society released a report indicating that EDCs can contribute to human obesity, cancer, sexual disfunction, thyroid disfunction, and even brain development (Gore et al. 2015). Hence the control of environmental EDCs is a concern that extends far beyond fishes.

The Autonomic Nervous System

Involuntary physiological functions, such as control of internal organ function, are at least in part controlled by the autonomic nervous system or **ANS**. Neural signals from the central nervous system (brain and spinal cord) travel to the ganglia of the ANS that are located either along the spinal cord or at the target organs. Signals then travel from these ganglia to the target tissues. The ANS is primitive in agnathans, better developed in elasmobranchs, and well developed in the bony fishes (Nilsson 2011). The ANS often works together with the endocrine system to control involuntary physiological functions such as heart rate, blood pressure, blood flow through the gills, and many functions of the gastrointestinal system that are important to digestion and nutrition. The ANS also controls gas bladder volume, and therefore fish buoyancy, by regulating the absorption and secretion of gases and blood flow to various parts of the gas bladder (see Chapter 5, The Gas Bladder and Buoyancy Regulation). The dispersion and aggregation of pigment in melanophores are also partly controlled by the ANS, along with melanophore-stimulating hormone from the anterior pituitary.

Temperature Relationships

Most fishes are about the same temperature as the surrounding water. Therefore, fishes are not necessarily "cold-blooded." Most fishes are more appropriately termed **ectotherms**, because they lack mechanisms for heat production and retention. The outside, or ambient, temperature can change, but usually any change is slow due to the thermal stability and large volume of water. When blood flows through the gills, it becomes the same temperature as the surrounding water due to the thin gill membranes and then flows to the rest of the fish's body. There are, however, exceptions among some large fishes that produce and conserve heat in parts of their body, a condition referred to as either **heterothermy** or **regional endothermy** because some parts of the body are warm while the rest remains at ambient temperature. Fishes that can only live within a narrow range of temperatures are **stenothermal**, whereas those that can tolerate a broad range of temperatures are **eurythermal** (Schulte 2011a). Thermal tolerance limits of fishes can be compromised by stress, including stress due to chemical pollutants, parasites, and habitat degradation (Beitinger & Lutterschmidt 2011).

Coping with Temperature Fluctuation

Because the vast majority of fishes are ectotherms, their body temperature changes with that of the surrounding environment. Fishes that experience changing environmental temperatures, such as those characteristics of daily or seasonal changes, have several cellular and subcellular mechanisms for adapting to the new set of conditions. Many physiological adjustments are the result of switching on or off genes that are responsible for the manufacture of particular proteins. For example, acute heat stress initiates the synthesis of stress proteins, also known as **heat shock proteins** or **HSPs**, which maintain the structural integrity of proteins that otherwise would become denatured at higher temperatures, thereby allowing them to function biochemically (Currie 2011).

To compensate for the decreased rate of biochemical reactions at low temperatures, fishes may increase the concentration of intracellular enzymes by altering the rate of enzyme synthesis, degradation, or both. Increased cytochrome C concentration in Green Sunfish (Centrarchidae) moved from 25 to 5°C was due to a greater reduction in the degradation rate than the synthesis rate (Sidell 1977).

Some fishes produce **isozymes** to catalyze biochemical reactions more efficiently at different temperatures. Isozymes are alternative forms of enzymes that are regulated by switching on or off the different genes that control their production. Rainbow Trout (Salmonidae) acclimated to 2° versus 18°C exhibit different forms of acetylcholinesterase, an enzyme important to proper nerve function because it breaks down the neurotransmitter acetylcholine (Hochachka & Somero 1984). The ability of Longjaw Mudsuckers (Gobiidae) to tolerate wide temperature ranges is probably due to the regulation of isozymes of cytosolic malate dehydrogenase, an important enzyme in the Kreb's cycle (Lin & Somero 1995).

Polyploid species have multiple sets of chromosomes (see Chapter 1, Polyploidization and Evolution) that allow the development of novel functions in the "extra" set of genes, including isozymes that function at different temperatures. Hence polyploidy may allow the evolution of broad temperature tolerance. Among cyprinids, for example, Goldfish and Common Carp are both polyploid and can tolerate a wide range of temperatures, and the polyploid Barbel can acclimate to different temperatures better than the diploid Tinfoil Barb (O'Steen & Bennett 2003).

Laboratory acclimation studies in which a single variable, such as temperature is altered while other factors remain constant, can be helpful in understanding how fishes respond to change in that single variable. However, in their natural habitats, fishes usually adjust to simultaneous changes in several variables, such as temperature, photoperiod, and perhaps reproductive condition as seasons change. The absence of natural seasonal cues, such as changing photoperiod, may cause an artificially acclimated fish to respond differently than one that has been naturally acclimatized. For example, laboratory acclimated fishes typically have higher metabolic rates at higher temperatures (see Chapter 5, Metabolic Rate), yet seasonal reproductive cycles cause naturally acclimatized sunfish (Centrarchidae) to have higher metabolic rates in spring than in summer (Roberts 1964; Burns 1975). Other studies also have shown seasonal changes in metabolic rate that were independent of temperature in trout (Salmonidae; Dickson & Kramer 1971), two minnows (Cyprinidae; Facey & Grossman 1990), sunfish (Evans 1984), and sculpin (Cottidae; Facey & Grossman 1990).

FIGURE 7.2 Mummichog (*F. heteroclitus*) in the northern part of their range have a different form of the enzyme lactate dehydrogenase than conspecifics in the southern part of the range. NOAA / Public Domain.

Some fish species exhibit **allozymes**, alternative forms of the same enzyme that are controlled by different alleles of the same gene. Different populations of the species may produce different alleles depending on their geographic location. Livers of Mummichog (Cyprinodontidae, Fig. 7.2) along the east coast of the United States exhibit two allozymes of lactate dehydrogenase, an important enzyme in carbohydrate metabolism. In Maine, the frequency of the cold-functioning allele is nearly 100%, but that frequency declines progressively in populations further to the south (Place & Powers 1979). In Florida, the alternative warm-functioning allele has a frequency approaching 100%.

Acclimation to cold temperatures includes modifications at the cellular and tissue level as well. Fishes, as well as other organisms, can alter the ratio of saturated and unsaturated fatty acids in their cell membranes to maintain uniformity in membrane consistency (Moyes & Ballantyne 2011). The proportion of unsaturated fatty acids in the membrane increases as temperature drops, allowing the membranes to remain more fluid at colder temperatures. Some fishes also alter cholesterol levels in cell membranes to increase fluidity at lower temperatures. Fishes that live in very cold habitats, such as polar seas (see Chapter 10, Polar Regions), often show additional cellular-level metabolic adaptations. These include enzymes that function at low temperatures and more energy-producing mitochondria in their swimming muscles (Crockett & Londraville 2006).

Decreased muscle performance at low temperatures can be compensated by several mechanisms. Acclimation of Striped Bass (Moronidae) to low temperatures results in a substantial increase in the percent of red muscle cell volume occupied by mitochondria (Eggington & Sidell 1989), and an overall increase in the proportion of the trunk musculature occupied by red fibers (Jones & Sidell 1982); both of these adaptations would increase the aerobic capability of the fish. Muscle fibers of Goldfish (Cyprinidae) show an increased area of sarcoplasmic reticulum at lower temperatures (Penney & Goldspink 1980), which increases the calcium ions needed for muscle contraction.

At colder temperatures, fishes may utilize more muscle fibers to swim at a particular speed than they use at warmer temperatures (Sidell & Moerland 1989). Therefore, maximum sustainable swimming velocities are typically lower at low temperatures (Rome 1990).

Temperature changes may affect ion exchange at the gills in a few different ways (Crockett & Londraville 2006). Higher temperatures typically increase molecular activity, causing increases in diffusion rates. Changes in membrane fluidity due to changes in the saturation of fatty acids or concentration of cholesterol, as discussed earlier, can also affect membrane permeability – more fluid membranes tend to be more permeable. Many fishes show increases in activity of Na-K adenosine triphosphatase (ATPase) as temperature changes, suggesting increased metabolic activity to maintain osmotic balance.

Endothermic Fishes

Some large, active, pelagic marine fishes use internally generated heat to maintain warm temperatures in the swimming muscles, gut, brain, or eyes (Block 2011; Dickson 2011). This phenomenon is sometimes called **regional endothermy** because the heat is internally generated but only conserved in some regions of the body, whereas others use the term **heterothermy** because some regions of the body are warmer than others. Endothermy has evolved in only about 30 species of fishes (out of over 35,000). All of them are large predatory fishes in the open ocean and are in three main lineages. Most of the teleost endotherms are in the families Xiphiidae (Swordfish), Istiophoridae (marlins, sailfishes), and Scombridae (mackerels and tunas). One basal acanthomorph family, Lampridae (opahs) (Order Lampriformes), also has endothermy. The chondrichthyan endotherms are in the family Lamnidae (White, makos) and the closely related Alopidae (thresher sharks). These teleostean and chondrichthyan lineages have been separate for over 400 million years, yet some members of each evolved heat exchange systems independently about 60 million years ago (Stevens 2011). The physiological similarities of how endothermy is maintained are one of several examples of convergent evolution among the fishes.

The heat in endothermic fishes is generally the result of either swimming muscle activity or ocular muscles, which in some species have become modified into "heater organs." In all cases, the heat is retained by a modification of the circulatory system that forms a countercurrent exchange mechanism. The internal temperatures of endothermic fishes often are warmer than the surrounding water and remain fairly stable even as the fishes move from warm surface waters to colder deep-water (Block 2011). The higher muscle temperature apparently helps these fishes swim faster (Harding et al. 2021).

Laboratory studies of red (aerobic) and white (anaerobic) swimming muscles of endothermic fishes show that the red muscles contract more slowly and produce less power than do the white muscles (Syme 2011). However, the red muscles perform much better at warmer temperatures than they do when cold. It is important, therefore, to keep the red muscles warm in order for a fish to swim rapidly and not fatigue. Bluefin tunas (Fig. 7.3A) keep their muscle temperatures between 28 and 33°C while swimming through waters that range from 7 to 30°C (Carey & Lawson 1973). Yellowfin Tuna maintain muscle temperatures at about 3°C above ambient water, whereas Skipjack Tuna keep their muscles at about 4–7°C above ambient (Carey et al. 1971).

These warm-bodied fishes conserve heat from muscular activity through adaptations of their circulatory systems. In a typical ectothermic fish, blood returns from the body to the heart and then travels to the gills for gas exchange (see Chapter 5, Cardiovascular System). The large surface area and thin membranes of the gills permit heat to escape to the environment so that when the blood leaves the gills, it is the same temperature as the surrounding water. In a typical fish, this blood would then travel down the core of the fish via the dorsal aorta, keeping the core body temperature about the same as the surrounding water (Fig. 7.3B). In the large tunas, however, most of the cool blood leaving the gills is diverted to large peripheral vessels that run along the outside of the fish's body (Fig. 7.3C). As arterial blood flows toward the large swimming muscles near the core of the body, it passes through a network of small blood vessels where it runs countercurrent to warm blood leaving these muscles. This type of arrangement of blood vessels is referred to as a **rete mirabile** (see Chapter 5, Countercurrent Networks). The oxygenated blood is warmed as it passes through the rete and travels toward the swimming muscles. In this way the heat generated by the activity of the large swimming muscles is kept within the muscles themselves and is not transported via the blood to the gills where it would be lost to the surrounding water (Dickson 2011). Bigeye Tuna can regulate their body temperature by utilizing the heat exchange mechanism only in colder water when it is needed (Holland et al. 1992).

In most fishes, the red muscle tissue responsible for sustained swimming is located just below the skin, where it readily loses heat to the water. In the tunas, however, red muscle is located closer to the body core. This arrangement of the swimming muscles contributes to the very efficient swimming style, termed "thunniform," observed in the tunas and other endothermic fishes in which the high, thin tail oscillates rapidly while the body remains mostly rigid, thereby minimizing turbulence and reducing drag (see Chapter 4, Locomotory Types). The evolution of thunniform swimming and the accompanying displacement of the red swimming muscles toward the body core put an insulating layer of less-vascularized white muscle between the heat-generating red muscle and the surrounding water. This muscle arrangement may have been a prerequisite for the development of the circulatory adaptations necessary

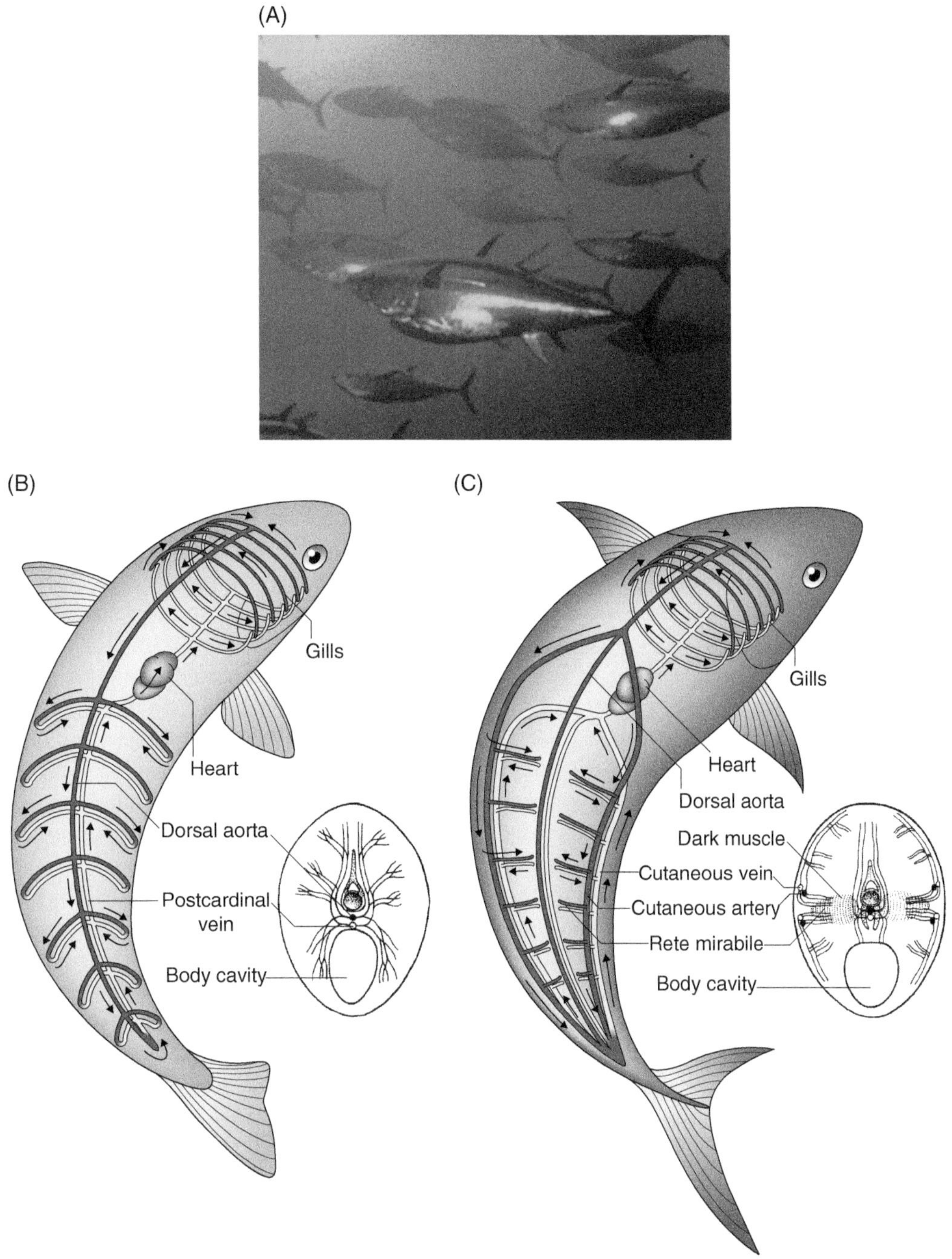

FIGURE 7.3 (A) Atlantic Bluefin Tuna (*Thunnus thynnus*) rely on warm, efficient red muscles for sustained swimming. D. Cedronefrom United Nations Food and Agriculture Organization / NOAA / Public Domain. (B) The circulatory system of a "typical" fish sends blood from the gills down the core of the fish, making it impossible to maintain an elevated core temperature in cold water. Arrows indicate the direction of blood flow. Carey FG. 1973. Fishes with warm bodies. Sci Am 228:36–44. (C) In the warm-bodied Atlantic Bluefin Tuna (*Thunnus thynnus*), most of the blood from the gills is shunted toward cutaneous vessels near the body surface and is carried through a heat exchanging rete en route to the active swimming muscles, which stay warm through this heat conservation mechanism. After Carey (1973).

to maintain elevated body temperatures (Block 2011). Swordfish (Xiphiidae) and endothermic sharks (Lamnidae, Alopidae) also have their red swimming muscles more centrally located and also possess heat exchangers, although there are some variations on the specific anatomical modifications involved (Stevens 2011). Bluefin tunas and lamnid sharks have cellular-level adaptations that allow the heart to beat particularly fast, which helps support elevated metabolism of the warm swimming muscles (Block 2011).

Smaller tunas also have retia (plural for rete) for heat exchange, but they tend to be located more centrally, below the vertebral column (Stevens 2011). Cool blood from the dorsal aorta is warmed as it passes through the rete and into the swimming muscles. This type of centrally located rete is found in smaller tunas that inhabit warmer oceans, whereas large tunas from colder regions have lateral retia, as shown in Fig. 7.2B. Large sharks of the family Lamnidae such as the White Shark, makos, and Porbeagle maintain elevated visceral and body core temperatures as much as 10–20°C above ambient via a heat exchanging rete located anterior to the liver (Carey et al. 1981). Small retia also have been observed in the viscera and red muscle of two species of thresher sharks (Alopiidae; Block & Finnerty 1994), which also maintain warm swimming muscles (Bernal & Sepulva 2005). Some sharks and tunas, then, take advantage of endothermy by conserving and recirculating heat that would have been lost to the environment, thereby avoiding the additional metabolic costs of specialized thermogenic tissues.

Another use of heterothermy in fishes is in warming parts of the central nervous system, especially the brain and eyes, which may enhance vision and neural processing in deeper, colder habitats (Block 2011). All endothermic fishes warm some part of their central nervous system, suggesting that this may have been a strong factor in the evolution of endothermy. This is accomplished by the generation of heat by special thermogenic tissues and by circulatory adaptations that use blood warmed in other parts of the body. The superior rectus eye muscle of Swordfish (Xiphiidae) and marlins (Istiophoridae), and the lateral rectus eye muscle of the Butterfly Mackerel (Scombridae), have lost the ability to contract and instead produce heat when stimulated by the nervous system. When stimulated these cells release calcium from the sarcoplasmic reticulum, which would trigger contraction in normal muscle cells. Instead, this calcium is rapidly transported back into the sarcoplasmic reticulum by ion pumps; the continuous release and pumping of calcium generate heat (Block 2011). This thermogenic organ (the only vertebrate thermogenic tissue known other than mammalian “brown adipose tissue”) seems to have developed for the particular purpose of generating heat for the brain and eyes. The billfish heater organs and associated retia are believed to have evolved about 15 million years ago (Stevens 2011). Eye muscles of other tunas and the lamnid sharks do not appear to be modified as heater organs (Block 2011), but they are highly active tissues, and retia near the eyes apparently help maintain elevated eye and brain temperatures.

In contrast to the regional endothermy described earlier, opahs (Lampridae) maintain whole-body endothermy based on a different set of swimming muscles. Opahs are oceanic predators that feed at depths of 50–400 m. Unlike the regionally endothermic fishes, these round-bodied fishes swim with their pectoral fins, like surgeonfishes and wrasses, using red muscle insulated by fatty tissue. This musculature produces heat that is retained by rete mirabile at the gills and distributed through the body to elevate temperature in the body core, heart, and cranium (Wegner et al. 2015). The North Atlantic Opah (*Lampris guttatus*) maintains a body temperature about 5°C above waters at 7.8–10.8°C.

The independent evolution of several different means to sustain systemic and cranial endothermy among large pelagic predatory fishes in three unrelated groups indicates that there are apparently strong evolutionary advantages to endothermy. Endothermic species have ecological ranges that include cold, and in some cases subpolar, ocean waters. The ability to swim faster, see better, react faster, and digest food rapidly evidently is a successful combination of characteristics for large, open-ocean predatory fishes.

Coping with Temperature Extremes

Extreme temperatures are dangerous to many living systems. High temperatures may cause structural degradation (denaturation) of proteins, including the enzymes that catalyze critical biochemical reactions. This can result in loss of biochemical function and potentially death. The physiological challenges of low temperature include compensating for reduced cellular metabolic rates and nervous function, and altered fluidity of cell membranes (Crockett & Londraville 2006). Probably the greatest potential danger at very low temperatures is intracellular formation of ice crystals, which can kill cells by puncturing cell membranes and organelles. Intracellular ice formation also causes osmotic stress because as water freezes the contained solutes remain dissolved in a decreasing volume of cytoplasm, causing osmotic concentration to increase.

Living in water protects most fishes from extreme environmental temperatures. Nevertheless, as water temperature increases, fishes experience decreased oxygen availability due to reduced gas solubility. Therefore, high temperatures can be especially stressful due to the combined effects of elevated oxygen demand due to increased metabolic rate and a temperature-induced Bohr effect that interferes with hemoglobin function, as discussed in Chapter 5 (Gas Transport). Not surprisingly, few fishes tolerate high temperatures (see Chapter 10, Deserts and Other Seasonally Arid Habitats).

Freshwater fishes are protected from dangerously cold temperatures because fresh water freezes at 0°C but is densest at 4°C. Therefore, ice forms on the surface of a lake or pond and forms an insulating layer between the liquid water and extremely cold air. Ions and other solutes depress the freezing point of the intracellular fluid of most fishes to around −0.7°C. Therefore freshwater fishes below the ice do not need special physiological mechanisms to cope with potentially freezing conditions, as long as they don’t touch the ice.

Marine fishes at high latitudes, however, are faced with different circumstances (see Chapter 10, Polar Regions). Seawater freezes at about −1.86°C, which is below the freezing point of the body fluids of most fishes. A marine fish could, therefore, be surrounded by water that is at a lower temperature than the fish’s freezing point – a potentially dangerous situation. Although some intertidal invertebrates and terrestrial vertebrates can survive freezing, fishes prevent ice formation through several different mechanisms.

One mechanism involves a physical property of crystal formation called **supercooling**, also called undercooling. Crystals will not grow unless a "seed" crystal exists to which other molecules can adhere. Under controlled laboratory conditions, Mummichog (*Fundulus heteroclitus*) were cooled to about −3°C, well below their normal freezing point, without ice formation (Scholander et al. 1957), but when touched with ice crystals the fish froze nearly instantaneously. Some marine fishes in polar environments live in a supercooled state (Fletcher et al. 2001; DeVries & Cheng 2005). The potential danger of contacting ice crystals is less of a problem for fishes that live in deep-water, where they are unlikely to encounter ice.

Many polar fishes come in direct contact with ice, however, and still do not freeze, indicating that they have physiological mechanisms to prevent internal ice formation (DeVries & Cheng 2005; see Chapter 10, Adaptations and Constraints of Antarctic Fishes). This protection generally involves the production of some type of **biological antifreeze**, triggered by genes that are activated seasonally (Fletcher et al. 2001). Antifreeze compounds, usually proteins or glycoproteins, can bring the freezing point of some Antarctic fishes, particularly the notothenioid icefishes, down well below the freezing point of seawater (Harding et al. 2003; Nath et al. 2013). These antifreeze compounds are produced in the liver and distributed throughout the body, and they also are produced in tissues likely to contact ice, such as the skin, gills, and gut. Species most likely to encounter ice have more copies of the antifreeze gene than fish that encounter less ice (Fletcher et al. 2001).

Five different types of biological antifreezes have been identified in fishes thus far, and all function by adhering to minute crystals as they form, thus preventing ice crystal growth (Harding et al. 2003). One type is antifreeze glycoproteins whereas the other four are different types of antifreeze proteins (types I-IV). Very similar antifreeze compounds may occur in unrelated species, demonstrating convergent evolution at the genetic and biochemical levels. For example, northern cods (superorder Paracanthopterygii, family Gadidae) and Antarctic nototheniids (superorder Acanthopterygii, family Nototheniidae) have very similar antifreeze glycoproteins, but the genes responsible for producing them do not appear to be the same.

In another example, herring (subdivision Clupeomorpha, family Clupeidae), smelt (subdivision Euteleostei, superorder Protacanthopterygii, family Osmeridae,) and sea ravens (subdivision Euteleostei, superorder Acanthopterygii, family Hemitripteridae) all have the same antifreeze protein coded by the same gene. The herrings and sea ravens have multiple copies of this gene, whereas the smelts have only one copy. Graham et al. (2012) evaluated the genome and evolutionary relatedness of these groups and concluded that the smelts may have acquired this trait by lateral gene transfer from the herrings. Perhaps during broadcast spawning fragments of DNA were taken up by smelt eggs or sperm, incorporated into the genome, and became fixed due to the evolutionary advantage of being able to produce a biological antifreeze. Lateral gene transfer is known among prokaryotes but not widely seen among eukaryotes – and especially not in more complex organisms such as vertebrates. However, the alternative explanations – (i) emergence of the same gene through mutation in three different groups, or (ii) inheritance of the gene from a distant shared ancestor with subsequent loss in the numerous other species in these lineages – seem even less likely.

The antifreeze shared by the herrings, smelts, and sea ravens is different, however than the antifreeze found in two sculpins, which are in a family closely related to the sea ravens. And each of these two sculpins (family Cottidae) have different antifreezes, indicating that antifreeze compounds have evolved independently and perhaps somewhat recently in the Cottidae (Fletcher et al. 2001).

The freezing point of body fluids also can be lowered by increasing the concentration of osmolytes (ions and other solutes) – the higher the concentration, the lower the freezing point. Notothenioids do this and achieve a slight (tenths of a degree) lowering of the freezing point. Other fishes rely strongly on increasing osmolytes to lower their freezing points in seawater.

Rainbow Smelt (Osmeridae) uses a combination of ice prevention tactics. They have an antifreeze in their blood to help prevent ice crystal growth. At very low temperatures, however, this antifreeze apparently is not enough protection, so they produce glycerol to increase the osmotic concentration of the blood and intracellular fluids, thereby further decreasing the freezing point (Raymond 1992). They begin increasing levels of glycerol and antifreeze protein in their blood in fall, when water temperatures decline to about 5°C (Lewis et al. 2004). As temperatures approach the freezing point of seawater, the glycerol concentration is so high that the smelt are nearly isosmotic to the ocean. This increase in glycerol concentration is more apparent in the colder winter months and may account for the reported sweeter flavor of these fish during that time of year.

Other fishes that live in areas that have warmer and colder seasons, such as Atlantic Cod (Gadidae), Shorthorn Sculpin (Cottidae), and Winter Flounder (Pleuronectidae), also exhibit increased levels of biological antifreezes during winter (Fletcher et al. 2001). Photoperiod seems to be the seasonal cue to increase or decrease antifreeze production. Because glycerol and protein or glycoprotein antifreezes are metabolically costly to produce, manufacturing them only when needed saves energy at other times of the year.

Antarctic fishes of the suborder Notothenioidei must maintain year-round protection from freezing because their environment rarely gets above −1.5°C, even in summer. In most fishes molecules as small as glycoprotein antifreezes would be lost in the urine. The fish would then need to produce more, at considerable energetic cost. The urine of notothenioids, however, does not contain these antifreezes because the kidneys of these fishes lack glomeruli, the small clusters of capillaries through which blood normally is filtered (DeVries & Cheng 2005; kidney function, including aglomerular kidneys, is discussed later in this chapter). Near et al. (2012a) estimated that antifreeze glycoproteins evolved in the nototheniids between 42 and 22 million years ago, which coincides with a period of geologic cooling. They also note that the adaptive radiation of the nototheniids into a species flock utilizing a wide variety of habitats in the Antarctic ecosystem took place much more recently, perhaps 11.6–5.3 million years ago, during a period of more extreme cooling around the South Pole.

Freeze protection strategies may not completely prevent ice formation within fishes. Small crystals of ice have been found in tissues that contact the surrounding water, such as the gills, skin, and gut. Ice also has been found in the spleen of some Antarctic fishes, perhaps carried there by macrophages that ingest small ice particles as part of the fish's immune response (DeVries & Cheng 2005).

Thermal Preference

For most fishes, ambient water temperature determines the thermal environment for critical biochemical and physiological processes, driving fishes to select environmental temperatures at which they can function most efficiently (Coutant 1987). Because different physiological processes may have different thermal optimums, the temperature selected by a fish often represents a compromise or "integrated optimum" (Kelsch & Neill 1990). Fishes select temperatures that maximize the amount of energy available for activity or metabolic scope (the difference between standard and maximum metabolic rates, as discussed in Chapter 5). Of course, habitat selection in the wild involves a compromise between temperature requirements and other important factors, such as dissolved oxygen levels, food availability, current velocity, substrate type, breeding opportunities, and avoidance of predators and competitors (see Coutant 1987). Temperature is, however, a very strong determinant of habitat choice by some fishes. Temperature registers implanted in trout reveal that when the water temperature of a New York stream exceeded 20°C, the fish selected cooler microhabitats within the river, such as tributary confluences and groundwater discharges. The body temperature of Brook Trout was up to 4°C below river temperature, whereas Rainbow Trout had body temperatures up to 2.3°C below river temperature (Baird & Krueger 2003).

Numerous laboratory investigations have shown that fishes select temperatures close to those to which they are acclimated (see Kelsch & Neill 1990). However, Chum Salmon (Salmonidae) and Blue Tilapia (Cichlidae) show very narrow and constant temperature preferences regardless of acclimation temperature. Species that evolved in areas with substantial seasonal changes in temperature, such as the Bluegill (Centrarchidae) of temperate North America, need the biochemical and physiological ability to shift temperature optima. Fishes that evolved in a more stable thermal regime, such as tilapia in the tropics, and salmonids in the cold temperate zone, probably lack the ability to make these kinds of adjustments, possibly putting them at peril during global climate change.

Temperature preferences can change as fishes grow, leading to different life stages of a given species utilizing different thermal niches. For example, juvenile Striped Bass (Moronidae) prefer temperatures around 25°C, whereas large adults will select cooler temperatures, around 20°C (Coutant 1985). This ontogenetic shift in temperature preference has important implications for the success of efforts to introduce this highly prized sport fish into various reservoirs and estuaries. A body of water that is ideal for the growth of young fish may be unsuitable for large adults, which may congregate in small areas of cooler water such as underground spring inputs or the deeper waters of stratified lakes and reservoirs (see Chapter 21, Temperature, Oxygen, and Water Flow). The thermal preference may be so strong that starving fish will not leave cooler deep waters to feed on abundant prey in warmer surface waters (Coutant 1985).

Strong thermal preferences can become a liability in the face of human alterations of aquatic environments. In summer, the deep, cooler hypolimnion of warm reservoirs can be attractive to large Striped Bass. As summer progresses, however, these deep waters can become oxygen depleted, leading to fish mortality (Coutant 1985). Thermal preferences also may cause fishes to congregate in areas with high levels of toxic pollutants, as has been reported for Striped Bass in the San Francisco Bay-Delta area. Uptake and bioaccumulation of some of these contaminants have been correlated with poor growth, high parasite loads, and decreased reproductive potential (Coutant 1985).

Understanding fish temperature preferences can be critical for evaluating the impacts of human activities. For example, power plant cooling systems discharge heated water into lakes and rivers, thereby altering their thermal structure. Fish may congregate near heated discharge areas in winter, but if the plant shuts down for a few days, fish acclimated to the warmer water can die. Similarly, hydroelectric dams often release deeper, cooler water from an upstream reservoir. Fishes that congregate in these cooler hypolimnetic waters may be drawn through the turbines and killed. The release of cooler water through a hydroelectric dam also can attract downstream fishes to the tailrace water during the warm summer months. The concentration of fish can create an attractive sport fishery, but it also can lead to overfishing and subsequent depletion of brood stock. In addition, some pumped-storage hydroelectric dams use large motors to run turbines in reverse to push water back to the upstream side of the dam when power is not needed. (When more electricity is needed, such as during periods of peak demand, this water is released again to generate electricity.) The attraction of fishes to the foot of the dam during periods of power generation can set the stage for high mortality if those fishes are drawn through the turbines (Helfman 2007).

The combination of cooler temperatures and high turbulence can cause water that is released from dams to become supersaturated with gases, especially nitrogen and oxygen. The blood of fishes living in these areas also can become supersaturated because of gas diffusion across the highly permeable gill membrane. When these fishes move to warmer, less turbulent areas, the gases come out of the solution and form bubbles in the blood. This **gas bubble disease** (similar to "the bends" in humans) can cause blocked and ruptured blood vessels, resulting in disorientation and death.

Because fish temperature preferences can vary with environmental factors such as season, time of day, food availability, health, and life stage, climate change will present physiological and ecological challenges to many fishes, especially those living near the margins of their thermal tolerance range (Crawshaw & Podrabsky 2011).

Osmoregulation, Excretion, Ion and pH Balance

Proper regulation of ion and water balance is critical to fishes, primarily involving the accumulation or excretion of the ions of sodium chloride (NaCl) salt. With the exception of the primitive hagfishes (Myxinidae), all vertebrates are **osmoregulators** – they regulate their internal osmotic environment within a fairly narrow range that is suitable for proper cellular function, even if the external osmotic environment fluctuates. The kidneys and gills are the primary organs for osmoregulation and salt excretion in fishes. Fishes that can tolerate only small changes in the solute concentration of their external environment are referred to as **stenohaline,** whereas those with the ability to osmoregulate over a wide range of environmental salinities are **euryhaline**. About 95% of teleosts are stenohaline, with the remaining 5% euryhaline (McCormick 2011).

The **kidneys** are paired elongated organs along the ventral side of the spine and above the gas bladder of most fishes. Left and right kidneys frequently join together to form soft, dark tissue under the vertebrae from the back of the skull to the end of the body cavity. Kidney tubules are involved with moving sperm in some fishes, so the kidneys and reproductive systems are sometimes discussed together as the urogenital system.

In most fishes, a **pronephros** is a transitional kidney that appears early in development and is replaced by a **mesonephros** as the fish grows. Adult hagfishes have an anterior pronephros and a posterior mesonephros, but it appears to be the mesonephros that is the functional kidney (Hickman & Trump 1969). Lampreys have pronephros until they reach about 12–15 mm when they develop mesonephros during metamorphosis.

The pronephros has **nephrostomes**, anterior funnels that empty into the body cavity by way of pronephric tubules. The **mesonephros** is more complex and consists of a number of **renal corpuscles**, each composed of a glomerulus surrounded by a Bowman's capsule. The **glomerulus** receives blood from an afferent arteriole from the dorsal aorta. The glomerulus acts as a filter and the filtrate, which includes water, salts, sugars, and nitrogenous wastes from the blood, is collected in the **Bowman's capsule**. The filtrate is then passed on to **mesonephric tubule** where water, sugars, and other solutes are selectively reabsorbed. In addition, waste products, excess ions, and other molecules that were not contained in the initial filtrate are added to the urine for elimination from the body. Marine and freshwater fishes differ considerably in kidney structure, reflecting the different problems faced by animals living in solutions of very different ion concentrations. Freshwater fishes must retain ions and rid themselves of excess water, and hence have larger kidneys with more and larger glomeruli, up to 10,000 per kidney and measuring 48–104 μm across (mean of several freshwater species = 71 μm). The glomeruli of marine fishes, which need to retain water and excrete excess ions, are smaller (27–94 μm across with mean of several species = 48 μm).

The basic process of **urine** formation in most fishes is similar to that of other vertebrates, but fishes cannot produce urine that is more concentrated than their blood. Urine contains water plus creatine, creatinine, urea, ammonia, and other nitrogenous waste products. Only 3–50% of the nitrogenous wastes are excreted through the urine, and much of this is ammonia. Most of the remaining nitrogen waste diffuses out as ammonia at the gills. In some fishes, urine drains from the tubules (nephrons) into collecting ducts and then into the bladder, where additional adjustments may be made to the ion and water balance of the urine prior to being excreted (Marshall & Grosell 2006). Freshwater fishes produce a large volume of dilute urine to eliminate excess water that diffuses through semipermeable membranes, such as the gills (McCormick 2011, see Fig. 7.4). Marine fishes lose water by diffusion at the gills and drink seawater to prevent dehydration; they excrete a low volume of concentrated urine.

Some fishes are **aglomerular**, lacking glomeruli in their kidneys. At least 30 species of aglomerular teleosts are known from seven different families of mostly marine fishes, such as Batrachoididae, Ogcocephalidae, Lophiidae, Antennariidae, Gobiesocidae, Syngnathidae, and Cottidae (Hickman & Trump 1969; Bone et al. 1995). Aglomerular kidneys can be advantageous for fishes the produce biological antifreezes because the antifreeze compounds will not be filtered out of the blood by the kidneys.

Gills are also an important osmoregulatory and excretory organ for fishes. Their large surface area, thin membranes, and highly specialized cell types make them well suited for this role. Nitrogen wastes are eliminated in the form of ammonia (NH_3) and ammonium (NH_4^+), both of which are soluble in the surrounding water. Diffusion of these wastes across the gills does, however, require immersion in water. Fishes that can survive out of water for extended periods convert ammonia to urea, which is less toxic and can be stored until the fish returns to the water. For example, African lungfishes (Protopteridae) produce ammonia when in the water but switch to urea production while estivating in a mud cocoon through long dry periods (Yancey 2001; discussed in Chapter 13). The amphibious mudskippers (*Boleopthalmus*, Gobiidae) increase mucus production by the skin and gills during terrestrial forays, and the mucus contains high levels of ammonia and urea (Evans et al. 1999). Ip et al. (2004a) found several different adaptations for protecting against ammonia toxicity among five air-breathing fishes, with most fish utilizing at least two mechanisms. These included reducing ammonia production by reducing amino acid catabolism, converting ammonia to less toxic compounds such as urea

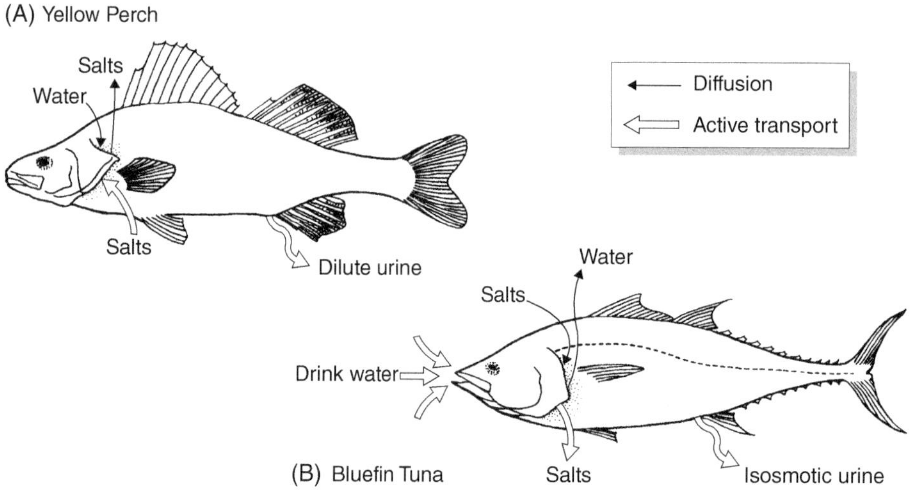

FIGURE 7.4 Maintaining osmotic balance in fresh versus seawater. (A) Freshwater bony fishes must produce a large volume of dilute urine to offset the passive uptake of water across their gills. They also must actively transport ions into the blood at the gills to compensate for the loss of these ions to the dilute freshwater environment. (B) Marine bony fishes passively lose water to their environment and gain salts by diffusion across their gills. They must, therefore, take in water through their food and by drinking seawater. Monovalent ions are actively transported out of the blood at the gills. Magnesium and sulfate ions, which are abundant in seawater, are excreted in the urine. Marine fishes conserve water by producing urine that is isosmotic to their blood.

or glutamine, excreting ammonia through the skin or digestive tract by increased volatilization, and increasing tolerance to ammonia at the cellular and subcellular level. Slender Lungfish (*Protopterus dolloi*) apparently convert ammonia to urea when exposed to air for 21–30 days (Wood et al. 2005). The Swamp Eel (*Monopterus alba*) converts ammonia to glutamine when exposed to air for 6 days, but suppressed ammonia production during aestivation in mud for 6 or 40 days (Chew et al. 2005). However, the African Sharptooth Catfish (*Clarias gariepinus*) survived 4 days of air exposure by tolerating very high levels ammonia in tissues (Ip et al. 2005). When exposed to elevated ammonia levels, the Giant Mudskipper (*Periopthalmodon schlosseri*) increased levels of cholesterol and saturated fatty acids in its skin, thereby decreasing skin permeability (Randall et al. 2004). In addition, exposure to the low oxygen or high sulfide on mudflats induces an enzyme system in some mudskippers to detoxify the sulfur (Ip et al. 2004b). There is still much more to learn about the physiological specializations of these amphibious fishes.

Osmoregulation in Different Types of Fishes

Hagfishes and Lampreys

Hagfishes (Myxinidae; see Chapter 13, Myxiniforms) are **osmoconformers**; their overall internal osmotic concentration is about the same as that of seawater. Because they live in fairly stable osmotic conditions in relatively deep water, they do not have to contend with internal osmotic instability. Although the overall internal osmotic concentration of hagfishes is the same as the ocean, there are differences in the concentrations of some individual ions. There is no difference, however, in the concentration of the two major ions, sodium and chloride, giving hagfishes the highest concentrations of these physiologically important ions among the craniates. Lampreys (Petromyzontidae), however, are osmoregulators and appear to utilize osmoregulatory strategies very similar to those of teleosts (Evans 1993).

Elasmobranchs

To prevent osmotic stress in seawater, marine elasmobranchs (see Chapter 12) retain high concentrations of **urea** in their blood. This, in addition to **trimethylamine oxide** (**TMAO**), which helps to stabilize proteins against the denaturing effect of urea, gives elasmobranch blood an osmotic concentration slightly higher than that of seawater. Elasmobranch gills are not readily permeable to urea, and this is probably enhanced by cells transporting urea back into the blood and thereby reducing the gradient between the cell and the surrounding water (Marshall & Grosell 2006). As a result of this urea retention, elasmobranchs are hyperosmotic to seawater and gain water by diffusion across their gills.

Elasmobranch gills have specialized transport cells, which may help with acid–base balance but apparently play no significant role in sodium or chloride balance. Instead, marine elasmobranchs rid themselves of excess sodium and chloride by active secretion via the **rectal gland**, which lies just anterior to the cloaca. Secretory tubules of the gland are lined with salt-secreting cells that are similar structurally and biochemically to the specialized transport cells of teleost gills. The rectal gland produces a solution that has about twice the NaCl concentration as the fish's extracellular fluids (Marshall &

Grosell 2006), and this solution drains into the lower intestine and is eliminated with other wastes.

Over 40 species of elasmobranchs, representing four families, are either euryhaline or exclusively freshwater. Those that are euryhaline tend to lose urea when they spend time in fresh water, and those that live exclusively in fresh water, such as the freshwater stingrays (Potamotrygonidae), do not produce much urea and rely on ammonia excretion to get rid of nitrogen wastes, as teleosts do (Marshall & Grosell 2006). The rectal glands of these fishes are also smaller and may become atrophied due to lack of use.

Marine elasmobranchs have glomerular kidneys, and their glomerular filtration rate is somewhat similar to those of freshwater fishes because the high urea content of the marine elasmobranchs causes them to gain water from their environment. The nephron is long, convoluted, and has specialized segments – the proximal segment, intermediate segment, distal segment, and collecting duct. Divalent ions, such as magnesium and sulfate, are actively transported from the blood into the proximal segment (as in marine teleosts), and the close proximity of the looping segments suggests that a countercurrent mechanism may be at work, perhaps to recover urea and TMAO (Marshall & Grosell 2006). A facilitated transporter for urea has been identified in the latter segments of the nephron.

Sarcopterygians

The coelacanths (Coelacanthidae) are marine sarcopterygians that also maintain elevated levels of urea and TMAO in their blood to offset the high ionic concentration of the external environment (see Yancey 2001), as do the marine elasmobranchs. The African and South American lungfishes (Dipnoi) are freshwater sarcopterygians that can survive long periods of drought by estivating in mud burrows. During this estivation period, they produce and retain high levels of urea, perhaps as a way of storing their nitrogen wastes in a form that is less toxic than ammonia and perhaps to help retain water. The phylogenetic distance between the sarcopterygians and the elasmobranchs, and the fact that both groups use urea as a nitrogenous waste and osmolyte, is an example of convergent evolution in the face of similar physiological challenges.

Freshwater Teleosts

Freshwater teleosts are hyperosmotic to their environment and therefore gain water and lose solutes by diffusion across the thin membranes of the gills and pharynx (Fig. 7.3A). Solutes also are lost in the urine. To maintain proper osmotic balance, freshwater fishes excrete a large volume of dilute urine and actively transport solutes back into their blood. Some of these solutes are recovered from urine as it is being formed in the kidney tubules. In addition, sodium and chloride ions are taken up from the surrounding water at the gills by specialized transport cells (referred to as "**mitochondria-rich cells**," "**ionoregulatory cells**," or "**chloride cells**" in some literature). Sodium and chloride uptake are apparently accomplished by different cells, and these processes may be linked with the secretion of hydrogen ions and bicarbonate ions, respectively (Marshall & Grosell 2006). The basolateral membranes of these specialized transport cells are inward extensions of the extracellular environment and come very close to the apical surface of the cell which is in contact with the surrounding water (Fig. 7.5A). This allows the transport mechanisms on the membranes to establish gradients of either hydrogen ions or bicarbonate ions high enough so that they diffuse from the cell to the surrounding water, and either sodium or chloride ions enter the cell as part of an ion exchange to maintain electrochemical balance (Fig. 7.5B, C).

The exchange of hydrogen ions for sodium ions, or bicarbonate ions for chloride ions, is likely enhanced by the enzyme carbonic anhydrase, which accelerates the conversion of water and carbon dioxide to hydrogen and bicarbonate ions. Tresguerres et al. (2006) propose that the exchange of chloride for bicarbonate may be achieved by the close linkage of carbonic anhydrase and ion exchangers on the apical and basolateral membranes, a mechanism which they call the "freshwater chloride-uptake metabolon." Sodium uptake is probably achieved by a protein that uses ATP to exchange incoming sodium ions for potassium ions (Marshall & Grosell 2006). Freshwater fishes also take up calcium at the gills by actively transporting calcium into the blood at the basolateral membrane, thereby decreasing the intracellular calcium concentration and encouraging the diffusion of calcium into the cell from the surrounding water (Perry et al. 2003).

Kidneys also play a role in osmoregulation and ion balance. In freshwater teleosts, glomerular filtrate passes into the proximal tubule where water and solutes, including sodium, chloride, and glucose, are recovered into the blood. Additional sodium and chloride may be recovered in the distal tubules, collecting ducts, and bladder before the urine is released from the body (Marshall & Grosell 2006).

Marine Teleosts

Marine teleosts face the opposite problem from that of freshwater teleosts. The high salt concentration of the ocean draws water out of the fish, and ions diffuse in across the permeable membranes (Fig. 7.4B). To counteract potential dehydration, marine teleosts drink seawater and actively excrete excess salts. The **mitochondria-rich cells** of the gills actively transport chloride ions from the fish's extracellular fluid into the cell along the extensive basolateral membrane. This increases the chloride concentration in the cell and results in chloride diffusing out of the cell at its apical surface and into the surrounding seawater. The increase of negatively charged chloride ions at the outside of the apical surface attracts positively charged sodium ions, which pass through the gill epithelium between the specialized transport cells and the adjacent accessory cells (Fig. 7.5D). Larger multivalent ions, especially magnesium and sulfate, which are abundant in seawater, are not readily absorbed in the gut and therefore are excreted (Marshall & Grosell 2006).

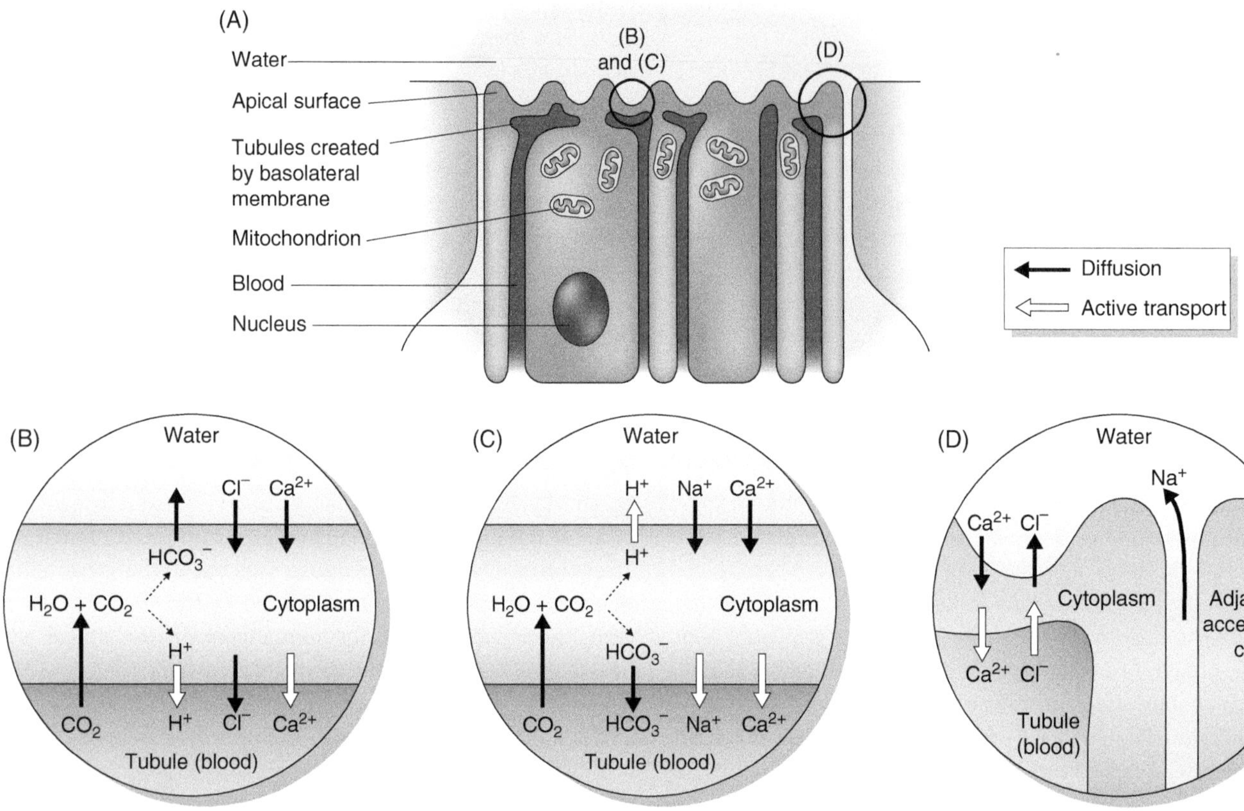

FIGURE 7.5 (A) In addition to the abundant mitochondria that provide the energy needed for high levels of active transport, mitochondria-rich (MR) cells have a highly infolded basolateral membrane that greatly increases surface area by creating a tubule system within the cell. This also brings the extracellular fluid within the tubules in close proximity to the apical surface of the cell, allowing the establishment of concentration gradients that assist with ion exchange. (B) The apical surface of some of the MR cells in the gills of freshwater fishes take up chloride ions from the surrounding water in exchange for the secretion of bicarbonate ions created by combining carbon dioxide with water. Proton pumps that transport hydrogen ions out of the cell and into the blood help to drive this process indirectly. (C) Other MR cells in freshwater fish gills take up sodium ions from the water by the excretion of hydrogen ions across the apical surface, removing sodium from the cell and transporting it into the blood at the basolateral surface. (D) Marine bony fishes actively transport chloride into the MR cells of the gills from the blood, creating a high intracellular concentration that results in the diffusion of the chloride across the apical surface and into the surrounding seawater. The outward flow of negatively charged chloride ions helps draw positively charged sodium ions out through the leaky membrane connecting the MR cell to an adjacent accessory cell. In B, C, and D, the active transport of calcium ions out of the cell into the blood helps to draw in more calcium from the surrounding water. After Marshall and Grosell (2006) and Tresguerres et al. (2006).

Most marine teleosts have glomerular kidneys, so urine forms initially by glomerular filtration. Some polar fishes, however, lack glomeruli and rely exclusively on active transport of solutes from the blood into the nephron to form urine. This means of urine formation prevents the loss of biological antifreezes (discussed earlier in this chapter). Whether glomerular or aglomerular, solutes such as sodium, chloride, magnesium, and sulfates, which are absorbed from the surrounding environment, are actively transported from the blood into the initial segment of the proximal tubule of the nephron to be eliminated in the urine. Marine teleosts also have a functionally distinct latter section of the proximal segment for the recovery of water, some sodium and chloride, and glucose. Urine passes from the proximal tubule through the collecting duct and into the bladder, where additional sodium and chloride can be recovered if needed. In addition, the bladder of marine teleosts is permeable to water, unlike that of freshwater teleosts, providing an additional opportunity for water recovery. This is why marine teleosts can produce urine that is isotonic with their blood, whereas freshwater teleosts can only produce dilute urine.

Diadromous Teleosts Teleosts that migrate between fresh and salt water, such as salmonids, must make appropriate adjustments in the specialized transport cells of the gill epithelium to physiologically adapt to the drastic change in osmoregulatory environment. For example, as Arctic Char migrate from the ocean into rivers, the membrane proteins of the gill epithelial cells responsible for sodium–potassium exchange increase in abundance, and so do the plasma sodium concentration and blood osmolarity (Bystriansky et al. 2007). This apparently is the result of increased activity of the genetic and molecular mechanisms responsible for creating these proteins and also provides evidence that this sodium–potassium exchange plays an important role in sodium uptake in freshwater fishes. Bystriansky et al. (2007) also note that there are apparently two forms of this sodium–potassium exchange protein, one that excretes excess sodium from saltwater fishes and another that assists with sodium uptake in freshwater fishes.

Control of Osmoregulation and Excretion

Like many homeostatic functions, osmoregulation is controlled by hormones. Some of these hormones act quickly to help fishes cope with rapid changes in the osmotic concentration of their environment by controlling the activity of existing cell membrane transporters or channels. Others act slowly and for longer time periods by restructuring osmoregulatory tissues and regulating the synthesis of proteins that create the channels and transporters (McCormick 2011). Cortisol is important for osmoregulation in both marine and freshwater fishes through the regulation of the proteins responsible for ion transport across the gills. Prolactin, Growth Hormone, and IGF1 also play important roles by regulating the growth and development of specialized transport cells responsible for ion transport.

Cortisol plays a key role in saltwater adaptation, by increasing the size and number of transport cells responsible for reducing levels of sodium and chloride in the blood, and modifying the lining of the intestine to increase water absorption (McCormick 2011). Blood levels of cortisol increase when euryhaline fishes are transferred to salt water. Cortisol also seems to help with sodium and chloride uptake in freshwater fishes. **Growth hormone** also increases the size and number of chloride-transporting cells and enhances the activity of the enzymes associated with sodium–potassium exchange. In addition, it enhances the expression of genes responsible for the protein involved in ion transport across epithelial cell membranes. Growth hormone also increases levels of insulin-like growth factor-I (IGF-I), which has been associated with increasing the ability of some fishes to tolerate increasing salinities. **Prolactin** appears to play a large role in adaptation to freshwater by decreasing the permeability of gill, kidney, bladder, and intestinal membranes to water and stimulating the uptake of sodium and chloride by the specialized transport cells of the gills. C-type natriuretic peptides also seem to help with sodium uptake and retention in low salinity environments.

pH Balance

Like all animals, fishes must maintain blood and tissue pH within certain limits because many enzymes that control critical biochemical processes are pH sensitive. Low or high pH can alter the configuration of these molecules, inhibiting their function. Fish blood is about pH 7.4, and levels below 7.0 can be lethal. In contrast, seawater is about 8.2, and typical freshwater varies from 6 to 8, although Magadi Tilapia (Cichlidae) can survive in alkaline lakes at pH 10. Hence fishes must maintain pH within a narrow range that is usually outside the ambient pH.

Blood pH is largely affected by metabolic byproducts such as carbon dioxide, which forms carbonic acid when in solution, and organic acids, such as lactic acid from anaerobic metabolism. Fishes cannot effectively lower blood pH by increasing ventilation to get rid of carbon dioxide, as terrestrial vertebrates can, in part because levels of carbon dioxide in water are higher than in air (Claiborne et al. 2002; Marshall & Grosell 2006). Fishes instead rely on epithelial transport of ions that affect pH, such as hydrogen ions and bicarbonate ions, and this is accomplished by specialized transport cells that typically are found in the gills but also may occur in the skin of some fishes (Marshall & Grosell 2006, McCormick 2011). Studies of Pacific Hagfish, *Eptatretus stoutii*, a basal craniate (considered an ancestral pre-vertebrate), suggest that the initial function of gills in early vertebrates may have been acid-base balance through ionoregulation, not respiratory gas exchange (Baker et al. 2015).

Carbon dioxide from cellular metabolism is mainly carried by the blood in the form of bicarbonate ions, and some is converted back to dissolved carbon dioxide at the gills, where it can easily diffuse into the surrounding water. Some of the specialized transport cells in the gill epithelium also seem to have the ability to exchange bicarbonate ions for chloride ions (see Fig. 7.5B). In addition, excess hydrogen ions are exchanged for sodium ions, also by some of the specialized transport cells (see Fig. 7.5C). Protein transporters designed for sodium/hydrogen ion exchange have been found on the gills of elasmobranchs, teleosts, and an agnathan (Claiborne et al. 2002). In addition, proteins that use ATP to transport hydrogen ions into the surrounding water may be important to pH regulation in freshwater teleosts and indirectly responsible for sodium uptake (the secretion of the hydrogen ions results in a charge imbalance, resulting in sodium entering the cells via sodium channels). These transport proteins occur in the gills of marine elasmobranchs and teleosts but do not seem to be as important as they are in freshwater fishes. Claiborne et al. (2002) review numerous studies in which the genetic and molecular mechanisms of the specialized transport cells control the activity and abundance of the protein transporters to allow fishes to maintain relatively stable blood pH. Perry et al. (2003) provide additional examples, particularly with respect to freshwater fishes, and also point out that kidneys play a role by regulating the amount of bicarbonate ion excreted in the urine.

The Immune System

The immune system plays an important role in homeostasis by maintaining animal health in both innate and adaptive ways (Rice & Arkoosh 2002). Innate mechanisms are found in all fishes and consist of immune factors that block invasion by potential pathogens. For example, the external layer of skin and scales is a physical barrier to infectious organisms. In addition, the sticky, viscous mucus secreted by fish epithelial cells probably helps to trap microorganisms, and the mucus can contain antibodies and chemicals that destroy or inhibit bacteria (Bernstein et al. 1997; Elliott 2011). The volume of mucus secreted may increase in stressful situations, indicating a response on the fish's part to shield itself from potentially harmful chemicals, microorganisms, or other agents. Phagocytic cells also move among the skin cells, engulfing potential pathogens and then slough off in the mucus (Elliott 2011).

Other parts of the innate response include inducible phagocytic cells that can attack and destroy potential pathogens, cytotoxic cells that destroy cells infected by viruses, and the complement system of proteins that attack the membrane of invading cells (Rice & Arkoosh 2002).

The lining of the gut provides another important barrier to pathogens (Jutfelt 2011). Mucus secretions and gut bacteria can inhibit the growth of potential pathogens. The epithelial cells are joined tightly, forming a physical barrier. Phagocytic cells in the gut lining can engulf and destroy invaders. However, a wide variety of factors may compromise the effectiveness of the gut as a barrier to potentially infectious agents. These include high water temperatures, poor water quality, low dissolved oxygen, vegetable lipids and proteins in the diet, stress, exposure to some viruses, some bacterial toxins, and inflammation.

Adaptive immune responses involve the detection of an invader and the creation of specialized response mechanisms to identify and destroy it. This response has not been seen in hagfishes and lampreys but is present in the gnathostomes (Bernstein et al. 1997). The organs primarily responsible for this response are the kidney, thymus, spleen, and gut.

The adaptive response includes both **cellular** and **humoral** components (Rice & Arkoosh 2002). The cellular component of the adaptive response includes **cytotoxic T cells** that can destroy native cells infected by viruses or that show signs of becoming cancerous. The humoral response involves the detection of specific invading compounds (antigens) and the production of antibodies that bind to them. The adaptive response is also assisted by phagocytic cells in the skin (Elliott 2011) and lymphocytes in the wall of the intestine (Jutfelt 2011). Cells lining the intestine constantly sample components of gut contents to build immunological resistance to potential pathogens.

Antibodies resulting from the adaptive response tag the antigenic particles for destruction by other components of the immune system, such as **macrophages** that engulf and digest the tagged antigens, or **complement proteins** that destroy tagged cells by puncturing their membranes. The Chondrichthyes were likely the first vertebrates to develop an adaptive immune response (Ballantyne & Robinson 2011), but the white blood cells that are made in bone marrow in other vertebrates are instead created in the Leydig's organ (near the esophagus) and the epigonal organ (usually near the gonads). The antibody structure of the Chondrichthyes is somewhat similar to that of the bony fishes and mammals. The structures of the genes responsible for antibodies are quite different, however, with those of the bony fishes somewhat intermediate between those of the Chondrichthyes and those of mammals (Bernstein et al. 1997).

In mounting an antibody response, the immune system also produces **memory cells** that remain in the bloodstream for extended periods (Rice & Arkoosh 2002). Memory cells help the animal's immune system react quickly if it encounters the same antigen in the future. Therefore, subsequent exposures prompt a rapid response, and the antigens are destroyed much more quickly than was the case during the initial exposure. Vaccinations, which have become important in fish culture, take advantage of memory cell development. By exposing fish to a less virulent form of a pathogen, the fish's immune system can defeat this initial infection and will retain memory cells to help it respond quickly and more effectively to subsequent exposures to a potentially more virulent form of the pathogen.

Stress

In a broad context, **stress** is a biological response that drives physiological systems outside their normal range. Fishes typically respond to short-term (acute) stress by mechanisms designed to maintain physiological function by compensating for the stress for a while, and when the stress passes the fish can return to its previous physiological state. If the stress is chronic (persistent), however, it may result in a readjustment of physiological set-points. This is sometimes referred to as **allostasis**, because rather than returning to its previous physiological state (homeostasis), the organism instead establishes a new baseline condition. This would include changes in gene expression that result in long-term alterations of proteins needed to maintain function under the new conditions (Iwama et al. 2006). Both acute and chronic stress can result in the diversion of energy resources away from growth, reproduction, and immune functions, thereby jeopardizing populations and communities as well as the health of individuals (Tort 2011).

Physiological responses to stress typically occur in three phases (Wendelaar Bonga 2011). The **primary response** is mainly the immediate release of epinephrine, followed by the release of **cortisol** in teleosts or **1α-hydroxycorticosterone** in elasmobranchs. Epinephrine release and the physiological responses that it initiates can occur in seconds but do not persist for long. The release of cortisol and the reaction to it, however, begin more slowly and are sustained for a longer period of time. Together, these hormones activate biochemical pathways that lead to a **secondary response**, which includes elevated levels of blood glucose to support increased metabolism. The secondary response also increases respiration rate, blood flow to the gills, and gill permeability. These increases help the fish to take in more oxygen to support elevated metabolism but also increase the diffusion of water and ions across the gill epithelium, thus creating more osmoregulatory stress and demanding more active transport, and therefore energy, for the fish to maintain osmotic balance.

Another part of the secondary response occurs at the cellular level – the induction of **stress proteins**, also called **heat shock proteins (HSPs),** because they were initially described as a response to elevated temperatures. These are a general cellular-level response to many types of stress, including temperature, various types of pollution, handling, hypoxia, and pathogens. There are three general categories of stress proteins based on their molecular weight. They help maintain the function of other proteins that are critical to cellular biochemical processes by protecting the shape of, helping repair, or helping control degradation of these other proteins. For example, the stress protein identified as HSP-90 apparently is important in protecting the function of the cellular receptor for cortisol, which would help the cell to respond to this important stress hormone (Iwama et al. 2006). Because stress proteins are a general response to many types of challenges, they may be used as indicators of a fish's exposure to unfavorable environmental conditions, including responses to temperature change, possibly a concern with climate change (Currie 2011).

If stress persists, the primary and secondary responses may lead to **tertiary responses** at the whole-animal or population level (Barton et al. 2002; Iwama et al. 2006; Moon 2011). Persistent elevated levels of the stress hormones, especially cortisol, can negatively affect fish growth, condition factor ($length^3$/mass), and behavior such as swimming stamina because the energy that would have been available for these functions have been diverted to dealing with stress (see Chapter 5, Bioenergetics Models). Stress can also divert energy away from reproduction, and stress hormones may affect the levels and effectiveness of sex hormones, further impacting reproduction (Tort 2011). Stress can delay sexual development in juvenile fishes and inhibit egg development, ovulation, spermatogenesis, and courtship behavior in adults (Wendelaar Bonga 2011).

Several factors can influence a fish's response to stress. These include gender, because the sex hormones themselves can affect the stress response, and the developmental stage of the individual, because juveniles and adults often respond differently. A fish's nutritional state or pre-existing stressors also can impact its response to subsequent stress (Barton et al. 2002). Responses to stress can also be seen at all levels of biological organization (Adams 2002; Hodson 2002). Short-term disturbances can lead to changes at the subcellular level as a fish tries to compensate physiologically, but these effects may not have implications at higher levels of organization, such as the overall health of the organism or the status of the population. Therefore, cellular and organismal responses to stress may be quite separate, and indicators of cellular-level responses may not necessarily indicate chronic stress at the organismal level (Wendelaar Bonga 2011).

Stress usually has negative impacts on fish immune systems. Acute stress may increase or decrease immune responses, but chronic stress typically leads to suppression (Tort 2011). Experimentally induced stress resembling capture significantly impacted the immune responses of Sablefish (*Anoplopoma fimbria*), so that those released as unwanted bycatch might have diminished capabilities to resist natural pathogens (Lupes et al. 2006). Chinook Salmon smolts exposed to elevated levels of ammonia for 96 h had lowered counts of lymphocytes, indicating increased susceptibility to disease (Ackerman et al. 2006). Environmental contaminants may also negatively affect fish immune systems by compromising the protective barriers of skin and mucus, affecting organs that filter pathogens from the blood, and interfering with intercellular signaling. For example, juvenile salmon from the polluted Puget Sound were more susceptible to pathogens because their immune responses were suppressed, and English Sole (*Parophrys vetulus)* may also be affected (see Rice & Arkoosh 2002).

Chronic stress can affect population and community structure through its impact on reproduction. A range of chemical contaminants has been identified as EDCs (discussed earlier in this chapter) because they interfere with some aspects of the hormonal signaling system that regulate the gonads and secondary sex characteristics (Greeley 2002). As more potential EDCs are identified in our surface waters, concern increases over the potential impacts on fish, other aquatic life, and adjacent terrestrial fauna (including us).

Indicators of Stress

Because chronic stress is not immediately lethal, it can be undetected until its effects influence fish populations and community structure. Interest in the early detection of stress in fishes has led to increased study of **biomarkers**, which are cellular and subcellular indicators of stress (Adams 2002; Amiard-Triquet & Amiard 2013). Stress can often be detected at the cellular level before it affects organismal or population health (Amiard-Triquet & Amiard 2013). Biomarkers, as well as biological indicators of stress at higher levels of biological organization, remain an active area of research and have proven useful in identifying environmental stress in fishes and in demonstrating reductions in physiological stress when environmental conditions have improved.

Environmental stressors can result in the alteration of DNA and interfere with the activity of some hormones (see Amiard-Triquet & Amiard 2013). Exposure to many chemicals can result in increased levels of liver enzymes responsible for their detoxification and metabolism and also the induction of stress proteins (discussed earlier). Therefore, levels of these biochemicals can be indicators of exposure to stress. Chronic stress can result in a variety of changes in cellular and tissue morphology in various organs, and biomarkers at this histopathological level may be good indicators because they show integrated, cumulative effects of physiological stress (Myers & Fournie 2002).

The liver is the primary organ of contaminant detoxification, so it is often the first to show damage due to environmental contaminants. Livers in fish from polluted areas are often larger and show higher levels of enzymes associated with detoxification than fish from less contaminated areas. The relative size of the liver (Hepatosomatic Index, HSI) of Rabbitfish (*Siganus oramin*) in Victoria Harbor, Hong Kong was larger in fish from more

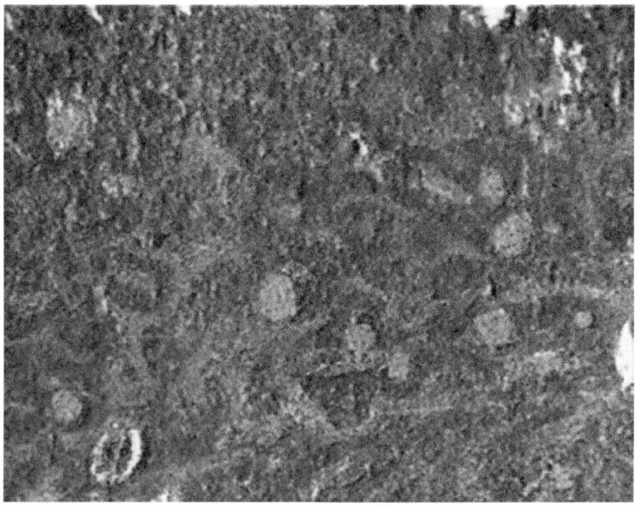

FIGURE 7.6 The yellowish clusters are aggregations of macrophages (a type of white blood cell) in the spleen of a Rock Bass (*A. rupestris*). Aggregations such as these in spleen and liver result from exposure to various stressors and can accumulate over the life of a fish. Their abundance and size can, therefore, provide some indication of stress to the fish. D. Facey (Author).

contaminated areas and seemed particularly sensitive to concentrations of heavy metals (Fang et al. 2009). Overall condition factor of the fish also was lower in areas of higher contamination. Myers et al. (2008) used biomarkers associated with liver function to show improvement in the health of English Sole (*P. vetulus*) in Eagle Harbor (Puget Sound, WA) after a PAH-contaminated site was capped with clean sediment. Liver biomarkers also were used to monitor condition and recovery of Rockfish (*Sebastes schlegeli*) and Marbled Flounder (*Pseudopleuronectes yokohamae*) after a 2007 oil spill along the west coast of South Korea (Jung et al. 2011).

The spleen also shows signs of environmental stress because of its important role in fish immune systems, as indicated by the presence of **macrophage aggregates**, also called melanomacrophage centers (Fig 7.6).

These are good biomarkers of multiple environmental stressors and also can be indicators of past exposure because they persist and accumulate with age. Several studies have supported the use of splenic macrophage aggregates as indicators of environmental stress (Wolke et al. 1985; Blazer et al. 1987; Macchi et al. 1992; Blazer et al. 1994). The number and size of splenic macrophage aggregates in 4 years old Rock Bass (*Ambloplites rupestris*) captured in Burlington Harbor (Lake Champlain, Vermont) in 1992 was higher than those in fish sampled in 1999, a few years after an upgrade of the local wastewater treatment facility (Facey et al. 2005).

A variety of other biomarkers have been used to evaluate physiological stress in fishes exposed to a range of contaminants. Juvenile European Seabass (*Dicentrarchus labrax*) exposed to the pesticide fenitrothion showed decreased levels of several enzyme biomarkers that could affect nerve function and negatively impact swimming (Almeida et al. 2010). Juvenile Delta Smelt (*Hypomesus transpacificus*), a federally threatened species, from the Sacramento-San Joaquin estuary exposed to elevated levels of copper, showed changes in the expression of genes that control the function of muscles, nerves, digestion, and the immune system (Connon et al. 2011).

Through these and other biomarkers and bioindicators, it is becoming possible to detect stress from a variety of agents, thereby permitting early remediation of potential impacts on fish physiology, health, growth, reproductive success, and community structure. Biomarkers have great promise to detect problems at the cellular level before they lead to irreversible damage at the organismal level.

Supplementary Reading

Block B, Stevens E. 2001. *Tuna: physiology, ecology, and evolution. Fish physiology*, Vol. 19. New York: Academic Press.

Carrier JC, Musick JA, Heithaus MR, eds. 2004. *Biology of sharks and their relatives*. Boca Raton, FL: CRC Press.

DiGiulio RT, Hinton DE. 2008. *The toxicology of fishes*. Boca Raton, FL: CRC Press.

Eastman JT. 1993. *Antarctic fish biology: evolution in a unique environment*. San Diego: Academic Press.

Evans DH. 1993. Osmotic and ionic balance. In: Evans DH, ed. *The physiology of fishes*, pp. 315–342. Boca Raton, FL: CRC Press.

Evans DH, Claiborne JB. 2006. *The physiology of fishes*, 3rd edn. Boca Raton, FL: CRC, Taylor & Francis.

Farrell AP, Steffensen JF. 2005. *Physiology of polar fishes. Fish physiology*, Vol. 22. New York: Academic Press.

Farrell AP, Stevens, ED, Cech JJ, Richards JG. 2011. *Encyclopedia of fish physiology: from genome to environment*. Amsterdam, Boston: Elsevier/Academic Press.

Iwama G, Nakanishi T. 1996. *The fish immune system: organism, pathogen, and environment. Fish physiology*, Vol. 15. New York: Academic Press.

McKenzie DJ, Farrell AP, Brauner CJ. 2007. *Primitive fishes. Fish physiology*, Vol. 26. New York: Academic Press.

Randall DJ, Farrell AP, eds. 1997. *Deep-sea fishes*. San Diego, CA: Academic Press.

Sloman KA, Wilson RW, Balshine S. 2006. *Behaviour and physiology of fish. Fish physiology*, Vol. 24. New York: Academic Press.

Val AL, De Almeida-Val VMF, Randall DJ. 2005. *The physiology of tropical fishes. Fish physiology*, Vol. 21. New York: Academic Press.

Wright P, Anderson P. 2001. *Nitrogen excretion. Fish physiology*, Vol. 20. New York: Academic Press.

CHAPTER **8**

Reproduction

This chapter is devoted entirely to reproduction, including sex determination, gender in fishes, gonads, gametogenesis, fertilization, and embryo development. The next chapter (Chapter 9) discusses early life history and fish growth. Courtship behavior among fishes is diverse and fascinating and is addressed in Chapter 17 as part of our treatment of fish behavior and social interactions.

CHAPTER CONTENTS

Summary

Reproductive development includes three very different processes: determination, differentiation, and maturation. Gender determination in most fishes is probably under genetic control and occurs at the time of fertilization. Differentiation occurs when recognizable ovaries or testes appear in an individual. Maturation is synonymous with achieving adulthood and occurs when a fish produces viable sperm or eggs. In some fishes, environmental conditions such as temperature and chemical pollutants can affect determination and maturation, and thereby influence reproduction.

The sexes in fishes are usually separate, although there are hermaphroditic fishes in some families. Gonads are usually paired and located in the dorsal part of the body cavity. Males have testes that produce sperm, and females have ovaries that produce eggs. Some hermaphroditic fishes possess

The Diversity of Fishes: Biology, Evolution and Ecology, Third Edition. Douglas E. Facey, Brian W. Bowen, Bruce B. Collette, and Gene S. Helfman.

Companion website: www.wiley.com/go/facey/diversityfishes3

mature testes and ovaries simultaneously and can change gender role during mating within minutes; others need more time for gonads to physiologically transform. In Chondrichthyes and primitive osteichthyans, eggs are shed into the body cavity. In gars and most teleosts, the lumen of the hollow ovary is continuous with the oviduct. The proportion of total body mass made up by gonads in mature individuals is typically far less among males (usually below 10%) than among females (up to 70%).

Gametogenesis describes the development of sperm and eggs through a series of definable stages. Fish sperm vary in size, shape, and number of accessory structures (flagella, acrosomes). Fish eggs develop as oocytes that contain yolk. Females are able to absorb unshed eggs and reuse the materials that go into egg production. The number of eggs a female produces (fecundity) generally increases with increasing body size within a species and varies among species from one or two in some sharks to many millions in large teleosts. Most marine fishes produce pelagic eggs that are fertilized externally and from which pelagic larvae hatch. Most freshwater fishes deposit eggs on vegetation or on the bottom, often in nests.

Reproductive effort is a measure of the energy allocated to reproduction by an individual and can be characterized in part by gonad weight/body weight, adjusted for the frequency with which an individual spawns and the phase of the spawning cycle. Females typically put more energy into their gonads than do males. However, total energy investment in reproduction is not just a matter of the size and number of gametes. Nest building, courtship, defending territories, and guarding eggs and juveniles are examples of other ways that fishes invest in the success of the next generation. Some species provide energetic provisioning for developing young in the form of nutritious secretions or by continuing to produce eggs that become food for young developing within the mother's reproductive tract.

Fishes in several different families undergo postmaturational sex reversal, changing from functional females to functional males (protogyny) or from male to female (protandry). The timing of sex reversal is affected by the social environment. Other factors include the number of members of each sex and their relative positions in dominance hierarchies. The evolutionary drive behind these shifts is likely the relative reproductive contributions of the sexes at different sizes. A few species are simultaneous hermaphrodites, functioning as males and females at the same time. Self-fertilization is rare, however, typically these fish pair up during spawning and fertilize each other's eggs. There are also some parthenogenetic fishes, especially among livebearers, that use sperm from donor males to activate cell division in eggs, but the male's genes are not incorporated into the genome of the offspring, or the paternal genes are lost in the next generation.

Anatomical differences between sexes are called sexual dimorphisms and include differences in body size, shape, color, dentition, or ornamentation. In most fishes males develop morphological changes during the breeding period, such as breeding tubercles or contact organs. Females are larger in most species, but not all. Sexually selected traits are those that result from intrasexual competition and from the breeding preferences of the opposite sex, hence the selective advantage of many dimorphic characteristics. A major cost of dimorphisms, however, is that conspicuous color or behavior is more easily noticed by predators.

Fertilization occurs external to the body of the female in most bony fishes and internally in most elasmobranchs and in about a dozen families of bony fishes. Physical and chemical barriers generally prevent multiple sperm from fertilizing one egg. If multiple sperm nuclei fuse with an egg nucleus, the resulting embryo usually will not complete development. Males of internally fertilizing species possess an intromittent organ, modified from fins or cloacal tissue, for injecting sperm into the female. Females of some internally fertilized species can store sperm for several months.

After fertilization the outer layer of the egg hardens somewhat, providing additional protection for the developing embryo. In most fishes cell division occurs in the small cap of cytoplasm on top of the yolk. As development proceeds, a gland on the head of the developing young secretes enzymes to break down the outer layer of the egg, allowing the young to emerge. Great variation exists among species regarding the degree of development at the time of hatching. Many species are identifiable at or just after the time of hatching, based on pigmentation and fin development. Most fishes hatch from eggs deposited into the environment (oviparity). However, some species incubate developing young within the body of an adult (viviparity). Among these, some provide nutrition through a placenta-like structure and give birth to live young (placental viviparity). Others incubate the fertilized eggs within the body of the parent (ovoviviparity). The incubating adult is usually the female, except among pipefishes and seahorses. There are also examples of intermediate conditions, with internally incubating embryos initially depending on the yolk of the egg but later gaining nourishment from secretions of the incubating adult.

During the early stages of development and growth, meristic (countable) traits may vary as a function of environmental conditions, especially temperature. Among most fishes, individuals that develop in colder water have greater numbers of scales, fin rays, and vertebrae. There are some species, however, that follow the opposite pattern (higher numbers at higher temperatures), whereas others show a pattern of highest or lowest counts at intermediate temperatures. It is important to keep this variability in mind if relying on meristic characteristics for species identification.

Determination, Differentiation, and Maturation

The development of a reproductively functional individual involves three processes (determination, differentiation, and maturation), each of which occurs at different life-history stages (Fig. 8.1). **Sex determination** is the process by which an individual becomes either male or female, usually during early ontogeny, and may be either genetically or environmentally

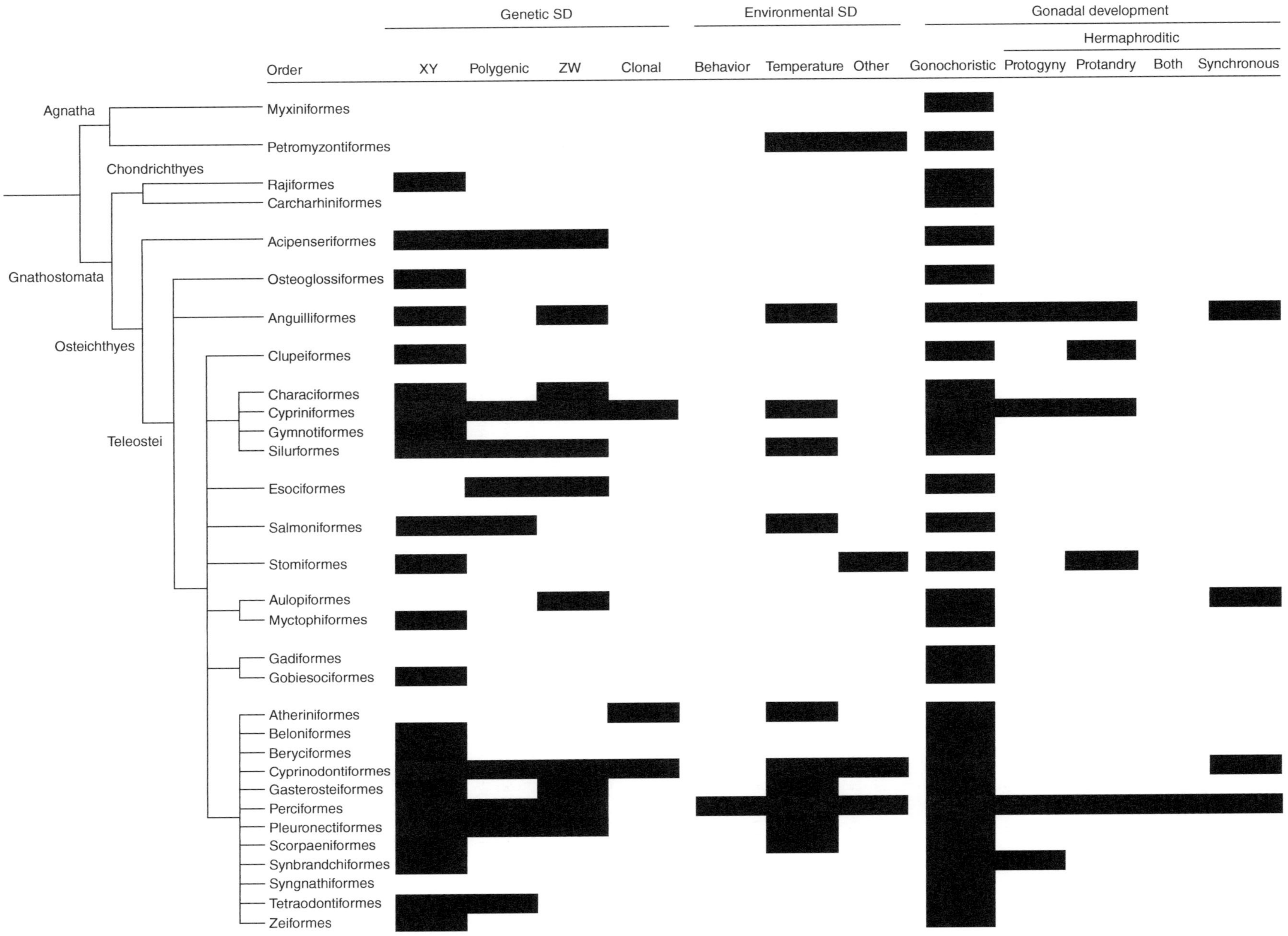

FIGURE 8.1 Patterns of sexual determination and maturation among fishes. Sex determination can be under genetic or environmental control, with sex-determining genes located on defined sex chromosomes (XY, ZW) or distributed among autosomal chromosomes. Among most sex-changing fishes, individuals mature as female first (protogyny) or male first (protandry). See text for details. From Devlin and Nagahama (2002), used with permission.

controlled. **Differentiation** involves the development of recognizable gonads (ovaries or testes) in an individual, although maturing gametes are not necessarily present. **Maturation** implies the actual production of viable gametes (spermatozoa or ova). An individual's gender may be determined at fertilization, the fish may differentiate as a juvenile, but it is not an adult until it matures.

Over 90% of fish species are **gonochoristic**, with sex defined by chromosomes (Piferrer 2011; Sloman 2011a). Where sex chromosomes are involved, the **heterogametic** gender can be either male (XY) or female (ZW) (Fig. 8.1). Most bony fishes and all elasmobranchs have an XX/XY chromosome system (Maddock & Schwartz 1996; Smith & Wootten 2016). These include the diverse families Cichlidae, Gobiidae, Percidae, Fundulidae, Balistidae, and Salmonidae. A minority of fishes have the ZZ/ZW system, in which females are heterogametic. This arrangement dominates in some widespread families, including Serranidae, Characidae, and Synodontidae. Both XY and ZW systems are found in the speciose families Cichlidae, Cyprinidae, and Poeciliidae, indicating that changes in sex determination systems can occur over relatively brief evolutionary timescales (Mank & Avise 2009).

Despite the prevalence of sex chromosomes and **genetic sex determination (GSD)**, the actual gender of individuals can be quite flexible, being influenced by genes on autosomal chromosomes as well as the sex chromosomes (Devlin & Nagahama 2002; Kikuchi & Hamaguchi 2013). Gender can be altered by hormonal responses to a variety of environmental cues, including temperature, season, and social status (**environmental sex determination, ESD**) (Godwin et al. 2003; Piferrer 2011). This lability has been exploited in aquaculture programs because it allows the production of monosex strains of economically valuable species where one sex grows faster or attains larger size than the other. A drawback of widespread ESD is that it makes fishes vulnerable to environmental degradation, including endocrine-disrupting chemicals and climate change (Strüssmann & Nakamura 2002).

Stressful environmental conditions and resulting increases in the stress hormone cortisol may affect sex determination. When juvenile Southern Flounder (*Paralichthys lethostigma*), a species known to exhibit temperature-dependent sex-determination, were raised in tanks with different background colors, fish raised in blue tanks had higher levels of cortisol and were more likely to develop as males than fish raised in black or gray tanks (Mankiewicz et al. 2013). In addition, fish fed a cortisol-enhanced diet were much less likely to develop as females than were control fish. In a somewhat similar study of Black Sea Bass (*Centropristis striata*), which are protogynous hermaphrodites, feeding juveniles a cortisol-enhanced diet resulted in partially masculinized female fish (Miller et al. 2019).

The timing at which gender is determined in fishes can be quite variable. Many fishes go through a **prematurational sex change**, initially differentiating as females, with some individuals later changing to males (Devlin & Nagahama 2002). Such ambivalence is not altogether surprising given that gonads in hagfishes, lampreys, and teleosts develop from a single structure, the epithelial cortex, which gives rise to ovaries in tetrapods. In sharks, ovaries develop from the cortex, whereas testes develop from the medulla. Sharks consequently show no sexual lability.

Temperature plays a strong role in sex determination in some fishes (Devlin & Nagahama 2002). Temperature extremes can affect sex determination in Sockeye Salmon, Ricefish or Medaka (*Oryzias latipes*), poeciliid livebearers, rivulines, and Siamese Fighting Fish (Francis 1992; Azuma et al. 2004). Experimental studies generally find **masculinization** of individuals or male-skewed sex ratios when the eggs or larvae of species of minnows, gobies, silversides, loaches, rockfishes, cichlids, and flounders are reared at higher temperatures, with the effect increasing as temperature rises. **Feminization** or female-biased sex ratios have resulted at higher temperatures in lampreys, salmon, livebearers, sticklebacks, and seabasses. The mechanisms underlying these effects appear to involve either altered enzyme activity or endocrine disruption (hormone synthesis or impaired steroid receptor function). Aromatase is an ovarian enzyme that converts testosterone to estradiol, a process vital to oocyte growth. In Nile Tilapia, *Oreochromis niloticus*, and Japanese Flounder, *Pleuronectes olivaceus*, elevated temperatures resulted in masculinization associated with reduced aromatase activity (Devlin & Nagahama 2002).

Other environmental factors that may affect sex determination include food availability and social interactions. Paradise Fish, *Macropodus opercularis*, begin as females and some later differentiate as males, but these changes occur prior to maturation. Final determination is based on social status: dominant individuals become male and subordinate individuals become female as a direct result of social interactions. Anguillid eels, despite having ZW heterogamety, produce more males in dense populations as an apparent response to crowding (Krueger & Oliveira 1999).

Northern populations of the Atlantic Silverside, *Menidia menidia* have a limited spawning season and exhibit genetic determination, but southern populations have a longer season and are more sexually labile. Southern larvae spawned in the spring at low temperatures tend to become females, whereas those spawned in summer at higher temperatures become male. Spring-spawned individuals have a longer growing period before the next spawning season than will later-spawned fish, and can therefore take advantage of the body size : egg number relationship and benefit more from larger body size as females than as males (Conover & Kynard 1981; Conover & Heins 1987).

Gonads and Gametogenesis

As in most other vertebrates, the sexes in fishes are usually separate (**dioecious**), with males having testes that produce sperm, and females having ovaries that produce eggs. Most fishes have paired gonads, although one member of the pair may be consistently larger than the other in some species, or only one gonad may be functional. Hagfishes and lampreys are unique in that only one ovary develops from the fusion of two primordia in lampreys and from the loss of one ovary in hagfishes.

Testes

The **testes** are usually paired, elongated structures in the dorsal part of the body cavity suspended by mesenteries known as **mesorchia** (Schultz & Nóbrega 2011a). They lie above the intestine and below the gas bladder when one is present. A variety of tubules and ducts, typically derived from the kidney, often conduct sperm to the outside through a genital pore.

Hagfishes and lampreys have a single testis. Sperm is shed into the peritoneal cavity and then passes through paired genital pores into a urogenital sinus and out through a urogenital papilla. Chondrichthyes exhibit internal fertilization; males using modified pelvic fins, termed **claspers**, to inseminate females. Sperm leave the testis through small coiled tubules, **vasa efferentia**, which are modified mesonephric (kidney) tubules. Sperm pass through **Leydig's gland**, which consists of small glandular tubules derived from the kidney. Secretions of Leydig's gland are involved in spermatophore production. The sperm then go through a **sperm duct**, which is a modified mesonephric duct, and into a **seminal vesicle**, a temporary storage organ that is also secretory.

Among Actinopterygii, the situation is similar, but no true seminal vesicles or sperm sacs are present. Marine catfishes (Ariidae), gobies (Gobiidae), and blennies (Blenniidae) have secondarily derived structures that have also been called seminal vesicles, but these are glandular developments from the sperm ducts and are not comparable to structures with the same names in tetrapods. These vesicles provide secretions that are important in sperm transfer or other breeding activities.

Lungfishes, sturgeons, and gars make varying use of kidney tubules and mesonephric (Wolffian) ducts. In the Bowfin (*Amia*), vasa efferentia bypass the kidney and go to a Wolffian duct. In *Polypterus* and the Teleostei, there is no connection between the kidney and gonads at maturity. The sperm duct is new and originates from the testes. Thus, the sperm duct of more primitive fishes such as the Chondrichthyes and Chondrostei is not homologous with that in the Teleostei.

Testes in immature males are typically reddish and take on a smooth texture and creamy-white coloration as the fish matures and spawning time approaches. The testes generally account for <5% of body weight, although this may be as high as 12% in some species. Follicles within the testes produce the developing spermatozoa (= **spermatogenesis**) through a series of meiotic and developmental transformations (see Schultz & Nóbrega 2011a, 2011b; Yaron & Levavi-Sivan 2011).

Fish sperm vary in size, shape, number of flagella (none to two), and presence or absence of acrosomes and other structures. Fish sperm are highly diagnostic of higher taxa and of some species (Jamieson 1991). Typical ejaculates during spawning contain millions of sperm. Sperm are released in seminal fluid in species with external fertilization, or in packets called **spermatophores** in internal fertilizers. It is commonly stated that males produce an excess of sperm, and consequently male reproductive success is limited more by access to females than by the ability to produce gametes (the opposite is considered limiting in females). However, under circumstances where males mate daily over a prolonged breeding season, **sperm depletion** can occur and mating may, in fact, be delayed until sperm stores are replenished (e.g., Nakatsaru & Kramer 1982; Jamieson 1991; see also Shapiro et al. 1994).

Ovaries

The **ovaries** are in the dorsal part of the body cavity suspended by mesenteries known as **mesovaria**, and typically ventral to the gas bladder if one is present. They are typically paired, but sometimes only one ovary is present in adults, as in some needlefishes (Belonidae). Ovary mass tends to increase with body size of individual females and can be as high as 70% of body weight.

Ovaries of hagfishes and lampreys have the same basic structure as do the male testes. There is a single ovary, and the eggs are shed into the body cavity and then pass through paired genital pores and out through a urogenital papilla.

In Chondrichthyes, the ovarian capsule is not continuous with the oviduct, so eggs are shed into the body cavity, the **gymnovarian** condition. The eggs enter the funnel of the oviduct, which is a **Müllerian duct**. The anterior part of the oviduct is specialized to form a **nidamental** or shell gland where fertilization takes place. The nidamental gland secretes a membrane around the fertilized egg. In **oviparous** (egg-laying) taxa, the membrane is composed of keratin. The nidamental gland may function as a seminal receptacle where sperm are nourished before fertilization. In **viviparous** (live-bearing) species, the posterior part of the oviduct is modified to form a **uterus**, which houses the developing embryos.

In osteichthyan fishes, the primitive gymnovarian condition is found in lungfishes, sturgeons, and the Bowfin. In gars and most teleosts, the lumen of the hollow ovary is continuous with the oviduct, termed the **cystovarian** condition. In trouts and salmons (Salmonoidei) and some other teleosts, the oviducts have been secondarily lost in whole or in part, so the eggs are shed into the peritoneal cavity and reach the outside through pores.

The development of eggs (**oogenesis**), is controlled by the endocrine system (see Yaron & Levavi-Sivan 2011), and progresses through various stages within the ovary. Oogonia develop from primordial sex cells in the germinal epithelium of the ovary wall. Proteinaceous yolk granules are deposited around primary oocytes during **vitellogenesis**, the precursors of yolk material being manufactured in the liver (see Reading & Sullivan 2011). Oil droplets are incorporated into the yolk. Ripe eggs pass from the ovary through the oviduct, which is a continuation of the ovarian tissue, to the outside via the cloaca. In elasmobranchs, no direct connection between ovary and oviduct exists, and hence eggs pass through the peritoneal cavity on their way to the oviduct. In several osteoglossomorph bonytongues, a loach (*Misgurnus*), anguillids, salmonids, and galaxiids, the oviduct is greatly reduced or absent, and eggs enter the body cavity prior to being shed (Blaxter 1969; Hempel 1979; Wootton 1990). Females that have spawned are termed **spent**; any residual eggs are resorbed by the ovary, and the proteins, fats, and minerals contained in them are reused by the female for maintenance, growth, or production of more eggs.

Fecundity, the number of eggs released by a female during a spawning cycle, varies from one to two in some sharks to tens of millions in the Tarpon, *Megalops atlanticus*, and European Ling, *Molva molva*, to 300 million in the giant Ocean Sunfish, *Mola mola*. Most larger temperate marine fishes produce tens of thousands to millions of eggs at a time. Fecundity generally increases with body size in an individual but decreases with increasing egg size and with increasing parental care. Mouthbrooders such as sea catfishes and some cichlids produce only about 100 eggs at a time, and livebearers such as the Four-eyed Fish, *Anableps*, contain about a dozen embryos. The relationship between egg number and body size is usually proportional to the mass of the female, reflecting the volume of a female's body that can carry the eggs (see Fig. 8.2). In addition to producing more eggs, larger females of many species produce bigger, better eggs that result in higher larval survival (e.g., in salmons, cod, haddock, Striped Bass, flounder) (Trippel 1995).

Exceptions show the premium placed on assuring the survival of young rather than just on producing large numbers of eggs. In mouth-brooding cichlids, fecundity increases in relation to the square of the length of the female because mouth size increases only linearly with increasing body length (Breder & Rosen 1966; Hempel 1979; Lowe-McConnell 1987). Because of resorption, fecundity estimates based on counts of ripe eggs may not necessarily indicate true **fertility**, which is the number of viable offspring produced. Fecundity estimates for livebearing fishes are further complicated by the consumption of eggs by developing embryos, a form of maternal provisioning (discussed later in this chapter).

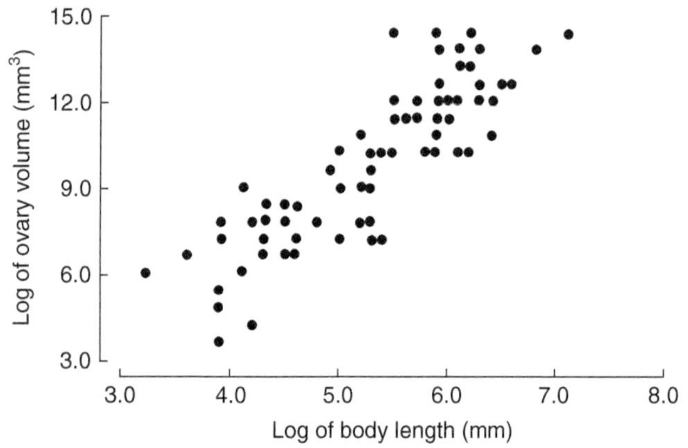

FIGURE 8.2 Bigger fish produce more eggs, both within and among species. Fishes grow throughout their lives (=indeterminate growth), therefore older fish are usually larger. Within a species, larger individuals generally produce more and often larger eggs. The generalization even holds for pregnant male seahorses, with larger individuals possessing larger pouches capable of holding more embryos (Woods 2005). Values are plotted for a variety of Canadian freshwater species. Redrawn after Wootton (1990).

There is a strong correlation between habitat and the characteristics of fertilized eggs. The females of most marine fish species, regardless of habitat or evolutionary relationships to other groups, release eggs into the water column, where they are fertilized and then float (Kendall et al. 1984). Eggs of pelagic spawners are generally spherical and have an oil globule for buoyancy. Egg diameters range from about 0.5 mm (*Vinciguerria*, Photichthyidae) to 5.5 mm (moray eels), with a modal size of about 1 mm. This remarkable convergence among distantly-related taxa indicates a common, adaptive set of solutions for eggs that disperse passively in a near-surface environment and that contain an embryo dependent on yolk for nutrition.

Most freshwater fishes and some coastal marine species diverge from this pattern and produce **demersal** eggs. Because they are deposited on the substrate, demersal eggs often have thick chorions and special coatings that may provide mechanical protection (ictalurid catfishes, North American minnows, killifishes, freshwater perch, blennies) (Boehlert 1984; Matarese & Sandknop 1984). Many demersal spawners construct nests and engage in some form of parental care (see Chapter 17). Demersal eggs tend to be relatively large, up to 7 or 8 mm in salmon, anarhichadid wolffishes, and zoarcid eelpouts. The largest teleostean eggs are produced by mouth-brooding marine catfishes and range from 14 to 26 mm, whereas the largest bony fish eggs are produced by the live-bearing Coelacanth, *Latimeria chalumnae*, with a diameter of about 9 cm. Shark eggs are generally larger than osteichthyan eggs.

Most fish eggs are spherical, but there are exceptions, such as the elongate eggs of some cusk-eels, anchovies, minnows, cichlids, parrotfishes, and gobies. Congrogadid eel blennies have strangely cross-shaped eggs, and eggs of some darters are deeply indented and appear almost heart-shaped. Although usually smooth, the outer vitelline membrane of the egg, termed the **chorion**, may be sculptured (lizardfishes, deep-sea hatchetfishes, mullets, some flounders), or have filaments, stalks, or spines (myctophiform lanternfishes and many atherinomorphs such as killifishes, flyingfishes, topsmelts, sauries, and halfbeaks). Filaments often help eggs attach to other eggs or to structures such as seaweeds, as in flyingfishes. The degree of segmentation and pigmentation of the yolk differs, with primitive teleosts such as many eels, herrings, and salmons having segmented yolks, whereas more advanced teleosts have homogeneous yolks. Pigmented yolks produce colorful eggs in gar, catfishes, and salmon. The occurrence, number, and location of oil globules in the yolk differ among species. Oil globules may serve as nutrition for embryos, as flotation mechanisms, and, when pigmented with melanin, may help protect sensitive developing structures from harmful ultraviolet radiation (F. D. Martin, pers. comm.).

A wide variety of fishes have been reported to produce toxic eggs that can result in serious symptoms if consumed (see **https://www.vapaguide.info/catalogue/NORAM-ANI-114**).

Energetic Investment in Reproduction

The effort and energy that different species and individuals expend on reproduction depends on the life history pattern a species has evolved (discussed in Chapter 20). **Reproductive effort** includes food intake and transfer of its energy to the gonads, as well as energy expenditure in somatic versus gonadal growth. In females, oocyte maturation involves mobilization of lipids and proteins from other parts of the body to the ovary. This maturation is accompanied by as much as a 10-fold increase in oxygen consumption by the ovary until the final stage of oogenesis, when ovarian mass increases via water accumulation.

A variety of indices are used to estimate reproductive effort. Some depict characteristics of an individual at a given point in time (**instantaneous measures**), others over the reproductive life of an individual (**cumulative measures**). One popular and simple instantaneous measure is the gonadosomatic index (**GSI**, also called the gonadal-somatic, or gonosomatic index). The GSI is the percentage of the body mass composed of gonad and is calculated as gonad weight/total body weight, or as gonad weight/(bodyweight – gonad weight, and sometimes minus gut-weight). Not surprisingly, GSI values generally reach their maxima just before spawning (Page 1983; Wootton 1990; Kamler 1991). A single maximum GSI value would indicate a defined spawning season, whereas several peak values would indicate a protracted spawning season. Values range greatly among species: cichlids <5%, darters 11–23%, pupfishes 2–14%, American plaice 5–20%, sticklebacks 20%, salmonids 20–30%, and European eels 47%.

The GSI of ripe females is a relatively accurate portrayal of energetic investment for fishes that spawn only once in a breeding season or lifetime, but it underestimates investment in fishes that spawn more than once because only a fraction of the eggs or oocytes a female will produce are present at any time (Heins & Rabito 1986). GSI calculations have been modified and improved to account for differences in body size among females and for females that do not shed all eggs at once (DeMartini & Fountain 1981; Erickson et al. 1985).

GSIs in males are generally much smaller than for females, reflecting less energy input. GSIs in male sticklebacks are 2% compared to 20% for females, and only 0.2% in male tilapia. Female Northern Pike, *Esox lucius*, allocate 6–18 times more energy to ovaries than males do to testes. Intraspecific variation in male GSI can reflect differences in reproductive tactics. In Bluegill, *Lepomis macrochirus*, some males guard nests and attract females with which they spawn; other males lurk at the periphery of the nests and sneak into the territory to join a spawning nest-guarder. GSI for territorial males is about 1%, but for sneakers is 4.5%. GSI in male Bluehead Wrasse, *Thalassoma bifasciatum*, varies between 3% and 5%, the larger value characterizing males that previously engaged in group spawning, where sperm competition among males is likely (Gross 1982; Shapiro et al. 1994).

Energetic investment in reproduction is not limited to gonads. The development of secondary sex characteristics in sexually dimorphic species, as discussed later in this chapter, requires some energetic investment. In addition, many fishes exhibit energy-intensive behaviors such as migration, nest construction, guarding territories, courtship, spawning, and defending eggs and juveniles, all intended to enhance successful reproduction (Katsiadaki & Sebire 2011). Courtship and parental care are addressed in Chapter 17.

Sex and Gender Roles in Fishes

Most fishes are **gonochoristic**, with sex determined at an early age and remaining fixed as male or female. However, many species display some form of hermaphroditism (see Fig. 8.1) and may change sex after initial maturation, referred to as **postmaturational sex change**. Although sex change is rare in most vertebrate groups, about 2% of teleost species are hermaphrodites (Smith & Wootten 2016). Sex reversal has evolved, apparently independently, in perhaps 34 families belonging to 10 orders, including moray eels (Anguilliformes), loaches (Cypriniformes), lightfishes (Stomiiformes), killifishes (Atheriniformes), swamp eels (Synbranchiformes), flatheads (Scorpaeniformes), boxfishes (Tetraodontiformes), and at least 24 perciform families (including snooks, seabasses, tilefishes, emperors, rovers, porgies, threadfins, angelfishes, bandfishes, damselfishes, wrasses, parrotfishes, and gobies) (Devlin & Nagahama 2002; DeMartini & Sikkel 2006).

Sex changers can be either: (i) **simultaneous hermaphrodites**, capable of releasing viable eggs or sperm during the same spawning; or (ii) **sequential hermaphrodites**, functioning as males during one life phase, and as females during another. **Protandrous** fishes develop first as males and then later change to females, whereas **protogynous** fishes mature first as females and then later become males. Variations on these patterns exist, such as protogynous populations with some males that develop directly from juveniles, or simultaneous hermaphrodites that function behaviorally as one sex at a time but can change roles quickly (Godwin 2011).

Changes in functional gender are controlled by the overall balance between female hormones (estrogens) and male hormones (androgens), both of which are produced by the gonads. Estrogens are converted to androgens by the enzyme cytochrome P450 aromatase (cyp19); therefore, cyp19 determines the balance between male and female hormones (Piferrer 2011). Different versions of cyp19 are found in the brain and the gonads, which helps explain why a fish's behavior at a specific time may not be consistent with its gamete production, as in some of the examples that follow.

Protogyny (female to male) is by far the most common form of hermaphroditism, accounting for 193 of the 235 species of sequential hermaphrodites surveyed by Devlin & Nagahama (2002) and exhibited in at least 17 tropical marine families, which is about one-fifth of reef families (DeMartini & Sikkel 2006). There are two general patterns of protogyny (Godwin 2011). **Monandric** fishes all mature initially as females and some later become males. This is common in gobies and groupers and is also seen in some damselfishes, wrasses, and parrotfishes. In contrast, **diandric** fishes are initially immature females and then later mature to either a reproductive male or female; either of these can subsequently become a brightly colored terminal phase male. This pattern is seen in some wrasses and parrotfishes (Fig. 8.3).

The Indo-Pacific Cleaner Wrasse, *Labroides dimidiatus*, forms harems of one large male and up to 10 females. Breeding access to the male is determined by a behavioral dominance hierarchy, the largest female dominating the next smallest, and so on. If the top (alpha) female is removed, the next largest female assumes her role, and the others move up a step. If the male is removed, the alpha female begins courting females within an hour and develops functional testes within 2 weeks (Robertson 1972; Kuwamura 1984).

Caribbean Bluehead Wrasse, *T. bifasciaftum*, usually begins life as predominantly yellow females or similarly colored males ("**initial phase**" coloration). Any of the initial phase fish can change into larger, "**terminal phase**" males, which develop a blue head, a black-and-white midbody saddle, and a green posterior region (Fig. 8.4). Large males set up territories over coral heads that females prefer as spawning sites. Some females are intercepted by and spawn with groups of up to 15 smaller males, but the largest, pair-spawning males have the highest fertilization success (Warner et al. 1975; Warner 1991). If the dominant terminal phase male is removed from the group, the largest of the initial phase fishes will almost immediately begin aggressive, dominating behavior toward the others, and will eventually become a sexually mature male. The fact that the dominating behavior precedes the transition to a sexually mature male indicates that the behavior is a response to social conditions and not the result of physiological changes that take longer (Godwin et al. 1996). Initial-phase fish that had their gonads removed still exhibited aggressive behavior, but this was suppressed by the introduction of estrogen-containing implants (Marsh-Hunkin et al. 2013).

Other well-studied protogynous species include the anthiine serranid, *Anthias squamipinnis*, a pair-spawning species that forms large aggregations in which females may greatly outnumber males. The precision of **social control of sex change** in this species is remarkable: if some males are removed from a large group, the same number of females change sex to replace them. Sex change to male in *Anthias* also occurs if the female : male ratio exceeds a threshold value (Shapiro 1979, 1987). The predominance of protogyny may reflect the fact that most teleosts, including gonochoristic species, differentiate first as immature females.

Protandry (male to female) has been reported in moray eels, loaches, lightfishes, platycephalids, snooks, porgies, threadfins, damselfishes, and crediid sandburrowers. The popular clown- or anemonefishes (*Amphiprion* spp., Pomacentridae; see Fig. 9.6) live in groups of two large and several small individuals in an anemone. Only the two largest fish in an anemone are sexually mature, the largest individual being female and the next largest being male. Although smaller fish may be as old as the spawning individuals, the behavioral dominance of the mature pair keeps these smaller males from maturing and growing, and a dominance hierarchy exists among the smaller males. If the female dies, the mature male changes sex to female and the next largest fish in the group takes over the

FIGURE 8.4 Terminal phase Bluehead Wrasse have the characteristic blue head and dark bands. Tibor Marcinek / Wikimedia Commons / Public Domain.

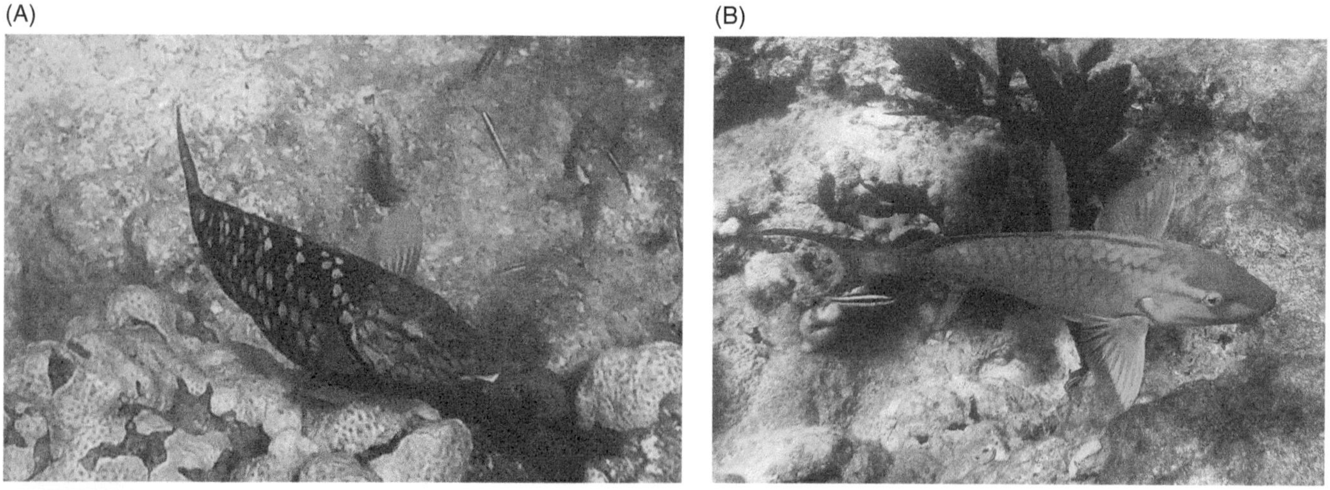

FIGURE 8.3 (A) Initial phase Stoplight Parrotfish, *Sparisoma viride*, have characteristic black, white, and red markings and can be either male or female. (B) Terminal phase males are bright green and dominate reproductive interactions with females. D. Facey (Author).

male role and grows rapidly (Godwin 2011). This inconvenient truth was judiciously sidestepped in the movie *Finding Nemo*; Nemo's father, Marlin, should have become female.

The Okinawa Rubble Goby (*Trimma okinawae*) can change sex multiple times. Adults have both ovarian and testicular tissue in different sections of the gonad. Social cues, primarily visual, are processed through the central nervous system and regulate genes that control levels of receptors for gonadotropic hormones in each part of the gonad. Gene activation can occur within about 8–12 h of visual cues, and the portion of the gonad being activated can produce the appropriate hormone within 1 day of gene activation (Kobayashi et al. 2009). Through this mechanism, the goby can change back and forth between being functionally male or functionally female depending on social interactions with conspecifics.

Simultaneous hermaphroditism (cosexuality, synchronous hermaphroditism) is least common, known from only four shallow water families (muraenids, rivulids, serranids, gobies) and most of the 16 families in the deep-sea order Aulopiformes (lizardfishes, Synodontidae, are the best-known exception) (Smith 1975; Warner 1978; St. Mary 2000; Devlin & Nagahama 2002). Two species of New World cyprinodontiform rivulids are capable of self-fertilization (*Kryptolebias hermaphroditus* of South America and *Kryptolebias marmoratus* of North and Central America). Self-fertilization in *Kryptolebias* is internal, producing clonal populations of homozygous hermaphroditic fish. Functional males can be produced depending on temperature and day length (Harrington 1971, 1975; Taylor 1992). Cyprinodontiform fishes are often the first fish colonists of small streams on islands and other seasonally adverse habitats (see Chapter 10). Self-fertilization assures that a single colonist can give rise to a new population, and maybe one means of assuring mates in low-density populations, a scenario that could also be applied to the deep-sea aulopiforms.

The other species of simultaneous hermaphrodites occur among the small hamlets (*Hypoplectrus*, *Serranus*). Each individual is physiologically capable of producing sperm and eggs at the same time, but behaviorally these fishes function as only one sex at a time while spawning. Caribbean hamlets (*Hypoplectrus*) engage in "egg trading" while they spawn in pairs over several hours (Godwin 2011). Members of a pair alternate roles, one fish releasing eggs while its partner releases sperm, and the two fish then switch roles. This behavior helps these simultaneous hermaphrodites prevent self-fertilization. The eastern Pacific *Serranus fasciatus* is a haremic, sex-changing, simultaneous hermaphrodite: one male guards and spawns with several hermaphrodites that act as females. If the male is removed, the largest hermaphrodite changes into a male (Fischer & Petersen 1987). Serranines have separate external openings for the release of eggs and sperm, which may prevent internal or accidental self-fertilization. Self-fertilization may occur in some serranines, but so far is observed only in aquaria (Thresher 1984).

In yet another variation, some fishes are **parthenogenetic,** with mature females producing female offspring without genetic contribution from a male. Livebearers in Mexico and Texas include all-female populations, but that require the sperm from gonochoristic males of other species to activate cell division in their eggs. Parthenogenesis in livebearers takes two forms: gynogenesis and hybridogenesis (Fig. 8.5). **Gynogenetic** females are usually triploid and produce eggs that are also 3N. These eggs are activated by sperm from other species, but no sperm genetic material is incorporated. **Hybridogenetic** females, in contrast, are diploid and produce haploid eggs that, during the reduction division of meiosis, keep the maternal genes and discard the paternal genes. Upon mating, these

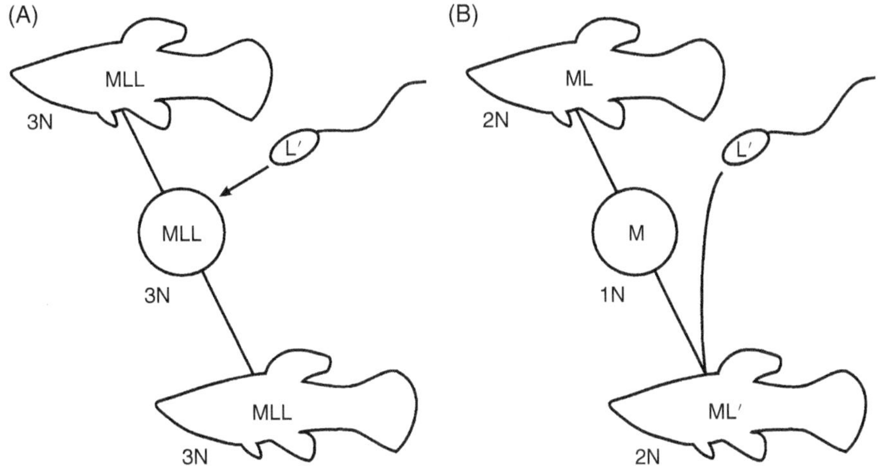

FIGURE 8.5 Parthenogenesis in Mexican livebearers. (A) In gynogenesis, a triploid female (designated MLL, shorthand for *Poeciliopsis monacha-lucida-lucida*) produces 3N eggs that are activated but not fertilized by sperm from a male *P. lucida* (L′). A daughter identical to the mother is produced. (B) In hybridogenesis, a diploid mother (ML, for *P. monacha-lucida*) produces haploid eggs (M) that contain only the maternal genome. Sperm from *P. lucida* (L′) combine to form a diploid daughter (ML′), but this male component will be discarded again during gamete production, and all future eggs will continue to have solely *monacha* genes. After Vrijenhoek (1984) and Allendorf and Ferguson (1990).

eggs unite with sperm from males of another species, forming a new, diploid hybrid daughter (no sons are produced). When the daughter mates, she again produces eggs that are haploid and "female." Hence the maternal lineage is conserved, and the male's genetic contribution is lost after one generation. These parthenogenetic populations are thought to have arisen originally as hybrids between *Poeciliopsis monacha* females and males of four congeners, *P. lucida*, *P. occidentalis*, *P. latidens*, and *P. viriosa*. The males of the four species are the usual sperm donors during mating. An additional species, the Amazon Molly, *Poecilia formosa*, is diploid and gynogenetic. Sperm from two other species (*P. mexicana* and *P. latipinna*) activate the eggs, but contribute no genetic material (Schultz 1971, 1977; Vrijenhoek 1984). Natural gynogenesis has also been reported for the cyprinid *Cyprinus auratus gibelio* (Price 1984). Among elasmobranchs, female Bamboo and Bonnethead Sharks in captivity have produced fertile eggs or given birth to live young without having mated (e.g., Mayell 2002, Feldheim et al. 2010). Fields et al. (2015) reported parthenogenesis among wild Smalltooth Sawfish (*Pristis pectinata*) in Florida, and speculate that it may be an adaptation to the critically low population density.

An immediate question that arises is how natural selection maintains males that waste gametes. Apparently, dominance hierarchies among "donor male" populations of livebearers exclude many males from mating with conspecific females. These are often the males that participate in the parasitized, heterospecific spawnings. **Satellite** or **peripheral males** that have very low reproductive output are characteristic of many vertebrate species (these are often the "sneakers" discussed in Chapter 17), providing an abundance of otherwise unused sperm (Moore 1984). Additionally, laboratory tests of mate preferences show that sexual females are more attracted to males courting gynogenetic females. Apparently a male can increase his chances of mating with a sexual female if he spends time courting asexual females, because sexual females copy the choices made by other females (Schlupp et al. 1994).

Certain generalities arise from surveys of sex change in shallow-water fishes, as do exceptions. Sex change is largely a tropical and subtropical, marine phenomenon (Policansky 1982c; Warner 1982). Cool temperate marine and freshwater sex changers are known (e.g., loaches, bristlemouths, swamp eels, wrasses, gobies) but are uncommon compared with tropical marine hermaphrodites. Patterns often follow familial lines, all members of a family being either protandrous or protogynous (although there are both protogynous and protandrous species among moray eels, seabasses, porgies, damselfishes, and gobies, and some serranids are clearly simultaneously hermaphroditic). However, population differences are becoming increasingly well known in sex-changing fishes. The Cleaner Wrasse, *L. dimidiatus*, is haremic under some conditions but forms pairs under others. Bluehead Wrasse are dominated by territorial-spawning males on small reefs with small populations, but by group-spawning males on large reefs with dense populations. Resource limitation, either food or habitat availability, and population size are frequent determinants of variation in mating systems. Clearly, sex change and mating systems respond to environmental variability (see Thresher 1984; Shapiro 1991; Warner 1991; Devlin & Nagahama 2002; Oldfield 2005).

The Evolution of Sex Change in Fishes

The topics of sex change and hermaphroditism have sparked a great deal of debate among biologists. Two questions dominate the discussion: (i) "why change sex?," and (ii) "when to change?" Answers lie primarily in the ecologies of individual species, greatly influenced by the relative reproductive success of males versus females at different sizes.

Sex change may be advantageous when, at a given size, the reproductive success of one gender becomes higher than the other gender (Ghiselin 1969; Warner 1975). There are, however, costs of changing, such as lost reproductive opportunities while undergoing the change and metabolic costs of altering gonads. This **size advantage model** assumes increased fecundity with increasing size; both are the norm among fishes. Reproductive success in females is generally limited by gamete production, whereas males are limited by the number of mates that they can acquire (Bateman's Principle). Males, including small males, generally produce a surplus of sperm, most of which never fertilize an egg. In contrast, egg production increases with growth in females, and most eggs are likely to be fertilized. These circumstances would favor protandry because small males can fertilize many eggs, but small females produce very few eggs. Such conditions would select for fish that began life as small but functional males and changed to female when they were large enough to produce more eggs than a small male could fertilize. For example, in pair-spawning, monogamous anemonefish, the lifetime egg production of the pair is maximized by having the larger female fish.

However, male fish often compete for females, and larger males tend to win. Hence one large, behaviorally dominant male can monopolize many females and fertilize their eggs, as in the Bluehead Wrasse and *Anthias* examples mentioned earlier. Under these circumstances, there is a great advantage to being the largest male, so selection might favor fish that are female first and then change to male when they gain the benefit of size. The age or size at which an individual should change sex is probably determined by an interaction between body size and social structure (Shapiro 1987). The size advantage hypothesis is complicated by exceptional situations, such as Bucktooth Parrotfish, *Sparisoma radians*, populations that contain large females and smaller, sex-changed males (see Munoz & Warner 2004).

Left unanswered is why more species of fishes and other vertebrates do not change sex. Ideas on this subject focus on the relative costs of changing sex in different taxa, the existence of dimorphic sex chromosomes (which are generally lacking in fishes), and differences in sex determination mechanisms

(Warner 1978; Devlin & Nagahama 2002). Add to these the realization that evolution is a predominantly conservative process. Reproductive systems are complex, given the behavioral, ecological, physiological, and anatomical components involved. Alterations to fine-tuned multifaceted systems are likely to reduce the stability that has evolved among the components. Hence the advantages of sex change would have to be very large to overcome fitness losses due to disruption of the coevolved gene complexes that code for the systems. Sex change is a viable alternative to gonochorism, but one that does not offer a sufficiently large advantage to overcome the costs of refitting the reproductive systems of most fishes.

FIGURE 8.6 Kype (hooked lower jaw) on a migrating male landlocked Atlantic Salmon, *Salmo salar.* D. Facey (Author).

Sexual Dimorphism

Some traits function primarily to attract mates or to aid in battles for access to reproductive opportunities. Such **sexually selected** traits confer a mating advantage to an individual and are a subset of natural selection. Sexually selected traits may serve no other purpose than mating and may even reduce fitness in other activities. Alternately, sexually selected traits may confer a fitness advantage, such as large size in males, which provides a physical defense from predators while simultaneously favored by females during mating (see section on Sexual Selection). Sexually selected traits are often referred to as **secondary** sexual characteristics. **Primary** characteristics are more directly involved in the mechanics of reproduction and include ovipositors, genitalia, and other copulatory structures such as claspers in elasmobranchs, gonopodia, or priapia in livebearers and brood pouches used in parental care. In some instances, a character may be both secondary and primary, serving both in mate attraction and in copulation or parental care. In sticklebacks, male sticklebacks are attracted to females with swollen bellies, but the swelling results from the female's ripe ovaries (Wootton 1976).

Secondary sexual characteristics have four general attributes: they are restricted to one sex (frequently the male), they do not appear until maturation, they often develop during a breeding season and then regress, and they generally do not enhance survival. **Sexual dimorphisms** are differences between sexes and are often secondary characteristics. These include morphological differences in body size, head shape, fin shape, dentition, and body ornamentation, or coloration (**dichromatism**). Some fishes exhibit electrical, chemical, and acoustic differences between the sexes. These are associated with differences in anatomy and physiology of these sensory modes (e.g., elephantfishes, salmons, minnows, gymnotid knifefishes, toadfishes, croakers, damselfishes, gobies). For example, male Plainfin Midshipman, *Porichthys notatus* (Batrachoididae), attract females by "humming," a sound produced by rapidly contracting and relaxing large muscles attached to the gas bladder walls. Males differ from females in having larger body size, different color, larger sonic muscles, and differing neural circuitry – involving larger cell bodies, dendrites, and axons in the brain – than are found in females (Bass 1996).

Many fish species are distinctly dimorphic, although the vast majority of fish species show no obvious sexual dimorphisms. Males are often the larger sex in salmon, sauries, wrasses, and clingfishes, whereas females are larger in mackerel sharks and Whale Shark, sturgeon, true eels (family Anguillidae), ceratioid anglers, sticklebacks, halfbeaks, silversides, livebearers, blennies, and billfishes. Male Dolphinfish (*Coryphaena hippurus*, Coryphaenidae) are larger and have a distinctly blunter head than females. Male anguillid eels develop larger eyes than do females during spawning migrations, and male salmon develop distinctly curved upper and lower jaws, called **kype**, that preclude feeding (see Fig. 8.6). Males in some species may have trailing filaments at the ends of their dorsal, anal, or caudal fins (characins, African rivulines, rainbowfishes, anthiine seabasses, cichlids, wrasses); enlarged median or paired fins (lampreys, bichirs, freshwater flyingfishes, minnows, characins, killifishes, livebearers, dragonets, gobies, climbing gouramis); elongate tail fins (livebearers); or have elongate, lobule-tipped gill covers or pelvic fins that are displayed in front of females during courtship (*Corynopoma*, Characidae; *Ophthalmochromis*, Cichlidae). In some species of elasmobranchs, males have longer or sharper teeth than females, which serve to grasp the female during courtship and copulation; females in turn may have thicker skin than males (discussed in Chapter 12).

Differences in body ornamentation include small bumps on the body, scales, and fins of mostly male fishes in 25 families of primarily soft-rayed teleosts. These bumps are called **nuptial** or **breeding tubercles** when of epidermal origin or **contact organs** when of dermal origin (Collette 1977) (Figs 8.7A, B). Scale differences also include scale type: male cyprinodontids have ctenoid scales, whereas females have cycloid scales (Berra 2001). Other dimorphic ornaments include small hooks on the anal fins and even tail of some male characins (a kind of contact organ), and rostral papillae in male blind cavefishes, photophore patterns in lanternfishes, and pigmented egg dummies on the anal fins of cichlids. Some male minnows, cichlids, wrasses, and parrotfishes develop bony or fatty humps on the front (nuchal region) of their head.

FIGURE 8.7 Breeding tubercles in fishes. Males of at least 25 different families develop keratinized bumps on their fins and body during the breeding season that may help maintain contact between spawning fish and stimulate females to spawn. (A) A breeding male Bluehead Chub, *Nocomis leptocephalus*, with prominent swollen head and breeding tubercles. (B) Internal structure of a tubercle from the snout of a gyrinocheilid algae eater. The tubercle consists of an outer cap of epidermal keratin, with concentrations of replacement keratin lying in a pit at the base of the tubercle that will replace the tubercle in the event of its loss. (A) Courtesy of E. Burress; (B) Wiley ML, Collette BB (1970).

Coloration differences are widespread and are usually expressed as more brightly colored males, either permanently (Bowfin, livebearers, killifishes, rainbowfishes, cichlids, wrasses, anabantids) or seasonally (minnows, sticklebacks, darters, sunfishes, cichlids). Male color change often involves the development of conspicuous bright or dark patches that break the rules of crypticity. Hence dark and light patches exist adjacent to one another, color transitions become sudden rather than gradual or reverse countershading develops (e.g., sticklebacks, sunfishes, temperate wrasses). Conspicuousness in the breeding season reinforces the premise that animals may risk an increased predatory threat for an improved chance to reproduce (Breder & Rosen 1966; Fryer & Iles 1972; Meisner 2005).

Some fishes exhibit internal fertilization, which requires that males possess an **intromittent organ** for introducing sperm into the female's reproductive tract. This can be both a primary and secondary sex characteristic because it is important for both sexual function and recognition. These structures have different names and are derived from different structures in different taxa. The pelvic fins of male elasmobranchs are modified into **claspers**. Among bony fishes, the anal fin forms the **gonopodium** (cyprinodotoids such as goodeids, anablepid four-eyed fishes, jenynsiids, and poeciliid livebearers); brotulids and surfperches have an enlarged **genital papilla**; and the phallostethoids have a **priapium** under the chin formed by the pelvic girdle, postcleithrum, and pectoral pterygial elements. The priapium is used for holding the female during copulation and fertilizing eggs just before they are released (Breder & Rosen 1966; Nelson 2006).

The importance of sexually selected, dimorphic traits as species-isolating mechanisms has become increasingly obvious in recent years as human alterations of the landscape interfere with mating patterns. In Lake Victoria, Africa, cichlids are separated taxonomically and apparently behaviorally by color variation in males. Females discriminate among closely related male suitors at least in part by color differences. This sexual selection based on color seems to have been driven both by color difference between males of two similar species (*Pundamilia pundamilia* and *P. nyererei*, Fig 8.8) and simultaneous evolution of the visual sensitivity in females (Seehausen et al. 2008). *P. pundamilia* males are more blue, and the females are more sensitive to light in the blue spectrum, whereas *P. nyererei* females are more sensitive to the red color of their conspecific males. The ability of females to detect the color difference is necessary to keep the species separate during mating (Maan et al. 2010; Selz et al. 2014). However, heavy runoff of nutrients from the surrounding deforested hillsides has increased turbidity and severely limited light transmission, especially at the long (red) and short (blue) ends of the visible light spectrum (Seehausen et al. 2008; Maan et al. 2010; Selz et al. 2014). This impairs mate selection, resulting in increased hybridization and jeopardizing the genetic integrity of many endemic species.

In Lake Malawi, female cichlid selection of males is also partially determined by visual input, but chemical communication also plays an important role (Plenderleith et al. 2005). Chemical signaling through altering patterns of urine release also plays an important role in signaling during agonistic interactions involving territory in the cooperatively breeding cichlid *Neolamprologus pulcher* (Bayani et al. 2017).

Other researchers have linked alterations and reductions in spawning behavior to increased turbidity. In clear water, female Sand Goby, *Pomatoschistus minutus*, show a strong preference for larger males, but this preference weakens as water clarity is reduced (Jarvenpaa & Lindstrom 2004). Other, related findings include a reduction in the intensity of red coloration in male three-spined sticklebacks in the Baltic Sea, linked to eutrophication-caused algal growth and reduced visibility (Candolin et al. 2007). Decreased spawning frequency, disrupted timing of spawning,

FIGURE 8.8 Turbidity-influenced light transmission is causing hybridization and loss of species in Lake Victoria. *P. pundamilia* (left) and *P. nyererei* (right) are closely related cichlids with overlapping ranges in Lake Victoria. Females select males based largely on their color patterns, and studies have shown differences in allele frequencies for visual pigments in the two species, with *P. pundamilia* having more alleles for detecting blue pigments, whereas *P. nyererei* have more alleles for pigments in the red spectrum (Seehausen et al. 2008). Pollution reduces water clarity in some areas, and also absorbs light in the blue spectrum. Behavioral and genetic study of five breeding populations near different islands showed that in areas with low visibility (about 0.5 m) both cichlids spawn in shallow areas, and female preference is compromised, resulting in hybridization. However, in areas with clearer water, female preference for color is much stronger and the two species remain separate because *P. pundamilia* spawns in shallow water whereas *P. nyererei* spawns at greater depths. Left photo of male *P. pundamilia*, by Oliver Selz / Wikimedia Commons / CC BY-SA 3.0; right photo of male *P. nyererei*, Kevin Bauman / Wikimedia Commons / CC BY 1.0.

and a 93% reduction in viable eggs produced by the Tricolor Shiner, *Cyprinella trichroistia*, are linked to disrupted visual cues associated with increased turbidity and resulting changes in light transmission (Burkhead & Jelks 2001).

Visual communication is not the only mode affected by anthropogenic impacts. Species recognition in the swordtail, *Xiphophorus birchmanni*, is chemically mediated, with females preferring conspecific males based on chemical cues. Fish tested in clean water maintain that species preference. Females tested in water subjected to sewage effluent and agricultural runoff did not discriminate between conspecific males and males of *X. malinche*, with which they can hybridize. The chemical most likely to interfere with communication was identified as humic acid, which is widespread and natural but elevated by human activities (Fisher et al. 2006). It seems reasonable to assume that other sensory modalities involved in courtship and species recognition, such as acoustic and electrical communication, could also be impaired by anthropogenic activities (e.g., Rabin & Greene 2002).

Sexual Selection

The major feature of **sexually selected traits** is that one sex bases its mating preferences on a character or set of characters in the other sex. The basis of mate choice in pair-mating fishes is complex, involving a combination of factors related to competition among members of the same sex and attraction by members of the opposite sex, but sexually dimorphic traits serve as immediate cues to a potential mate's ultimate reproductive success. In nest-guarding or otherwise territorial species, males typically compete for spawning sites and then females choose males based on male size and coloration, territory size and location, and quality of the substrate.

Females choose larger males in many nesting species in which males guard eggs and fry. Larger males are generally more effective nest guarders, and hence choosing a larger male is a means of insuring better protection from egg predators (Downhower et al. 1983; Fitzgerald & Wootton 1993). Bright coloration can serve both as a dominance signal and as a mate attractant; larger males are often the most colorful. Female damselfishes, wrasses, triplefins, blennies, and gobies select mates based in part on territory or spawning site characteristics. Given the positive correlations among the many traits that may indicate male quality (e.g., size, color, health, courtship intensity, dominance, territory size, quality of paternal care), it is not surprising that females often rely on a variety of characteristics rather than a single male trait in making their choices (Kodric-Brown 1990).

It is widely held that the sex that expends the greatest energy in gamete production will be in relatively limited supply, that the other sex will compete for the limiting sex, and that the limiting sex will be choosiest about mating partners. Brightly colored males and female mate choice are generally the rule in fishes, reflecting the greater cost of producing eggs than sperm. Pipefishes and seahorses (family Syngnathidae) reverse roles, with females depositing eggs into the brood sac of the males (Fig. 8.9). In pipefishes, females are often the more colorful sex,

FIGURE 8.9 A male seahorse (family Syngnathidae) with a brood pouch full of developing eggs. M. Al Momany / NOAA / Wikimedia Commons / Public Domain.

FIGURE 8.10 Male swordtail (*Xiphophorus helleri*, Poeciliidae) with an extended lower lobe of the caudal fin. Wojciech J. Płuciennik / Wikimedia Commons / CC BY-SA 4.0.

and males select mates based on female size and the intensity of courtship displays (Berglund et al. 1986a, 1986b, 2005). The female is also the more colorful sex in some cardinalfishes, damselfishes, and porcupinefishes on coral reefs. In the first two families, males mouth brood or defend the eggs.

Several well-studied instances of sexual dimorphism and sexual selection have given us insight into how such traits evolve and function. Male swordtails (*Xiphophorus* spp., Poeciliidae) develop a colorful, elongate, lower caudal fin extension, the sword, when they mature (Fig. 8.10). The sword continues to grow as the fish ages and may exceed the length of the rest of the body in older individuals. The sword serves no role in male–male interactions and may be a liability in predator avoidance. The only known function of the sword is during courtship, when the male displays the sword to females (Basolo 1990a). Females prefer to mate with males with longer swords, and the strength of the preference increases as the difference between sword lengths increases. Male swordtails also incur greater energy costs while swimming as a result of having an elongate caudal fin (Basolo & Alcaraz 2003).

Interestingly, swordlessness is the probable ancestral condition in the genus. Platyfish are primitive congeners of swordtails that do not develop swords. When plastic swords are surgically attached to the tails of male platyfish, female platyfish prefer males with artificial swords over normal, swordless males. Swords in male swordtails apparently evolved in response to a **preexisting preference** by females for elongate tails among males, which, when combined with natural variation in tail length, selected for males with progressively longer swords (Basolo 1990b).

As is often the case, strong selection for sexually dimorphic traits is achieved at a cost. The trade-off operating here is between mating success and conspicuousness to predators. For example, male Guppy attract females via a series of displays that are enhanced by brightly colored spots on the male's side. The largest, brightest, and most diverse spots are most attractive to females and predators alike. In Trinidad, predators are more common and diverse in lowland areas than in upland areas, and males in these predator-dense streams tend to have fewer, less intense, smaller, and less colorful spots. It makes little evolutionary sense for females in predator-dense areas to prefer to mate with males that were also very attractive to predators, if for no other reason than the female's sons would be relatively vulnerable. Not surprisingly, females from predator-dense areas generally do not show a preference for more colorful males (Houde & Endler 1990; J. Endler, pers. comm.); hence a compromise is struck between attracting mates and attracting predators (Endler 1991). Male Guppy with larger tails also exhibited poorer swimming performance than males with shorter tails (Karino et al. 2006). Related poeciliids show similar cost–benefit phenomena. Female *Gambusia* prefer males with larger gonopodia, but a larger gonopodium is grown at a cost in fast start swimming performance to escape predators. Male *Gambusia* in predator-free environments tend to have larger gonopodia (Langerhans et al. 2005).

Fertilization

Prior to fertilization, fishes participate in a wide range of mating and spawning behaviors. Because these are part of fish social interactions, we will address mating and spawning behavior in Chapter 17. Here we address the biology of fertilization, which occurs outside the body of the female in most fish species.

Fertilization occurs when a sperm enters the egg membrane via a funnel-shaped hole in the membrane called the **micropyle**, which is usually too narrow to allow the passage of more than one sperm (Murata 2003). Micropyle presence and size are diagnostic of different species. Eggs of most fish species have only one micropyle, but eggs of sturgeon and paddlefish have more. After sperm entry, the micropyle closes and the chorion hardens, preventing entry of more sperm (**polyspermy**). In most fishes, polyspermy is prevented by a variety of mechanisms. In Chondrichthyes more than one sperm may enter the egg, but only one sperm nucleus fuses with the egg nucleus. Fusion of a sperm nucleus with an egg nucleus forms a zygote, which will then begin developing into an embryo. If multiple sperm nuclei fuse with an egg nucleus, the resulting embryo is unlikely to survive.

In species with external fertilization, gametes remain viable for timespans ranging from less than a minute to as long as an hour, depending on temperature; longer viability generally occurs in colder water (Hubbs 1967; Petersen et al. 1992; Trippel & Morgan 1994). Studies of the proportion of eggs fertilized during natural spawning events indicate that at least 75% and often 90–95% of the eggs released are fertilized (Petersen et al. 1992, 2001; Marconato & Shapiro 1996). This number varies directly as a function of the volume of sperm released. In water-column spawners such as wrasses and parrotfishes, males can control sperm expenditure in response to female size and competition, releasing more sperm when spawning with larger females that release more eggs, or when other males are simultaneously attempting to spawn. In internally fertilizing species such as live-bearing mosquitofish, ejaculate volume increases in the face of competition (Evans et al. 2003). In benthic, territorial spawners, however, sperm expenditure does not appear to increase when females release more eggs (e.g., sticklebacks; Zbinden et al. 2001), or in the face of competition from other males (gobies and cyprinid bitterlings; Candolin & Reynolds 2002; Scaggiante et al. 2005).

Females of some internally fertilized species are able to store sperm in the ovary. Newborn male Dwarf Perch, *Micrometrus minimus* (Embiotocidae), are mature and inseminate newborn females. The females store this sperm for 6–9 months until they mature and ovulate, at which time fertilization occurs (Warner & Harlan 1982; Schultz 1993). Sperm storage is apparently common among sharks, with one captive female Brownbanded Bamboo Shark (*Chiloscyllium punctatum*) laying a fertile egg four years after mating (Bernal et al. 2014). Sperm storage is widespread in poeciliid livebearers, often involving more than one male partner (e.g., Evans & Magurran 2000; Luo et al. 2005). Some species store sperm and use it to fertilize multiple batches of eggs. Female Least Killifish, *Heterandria formosa*, store sperm from one copulation for as long as 10 months and use it to fertilize as many as nine different developing broods, several of which may be developing simultaneously – a phenomenon known as **superfetation**.

In a few species, activation of cell division is not synonymous with fertilization. Some poeciliid livebearers are **gynogenetic** in that females use sperm from males of other species to activate cell division, but no male genetic material is incorporated into the zygote (discussed earlier in this chapter). In some internally fertilized species, fertilization occurs but development may be arrested after a few cell divisions and then resumes when conditions are more favorable for hatching. In some annual fishes, such as the South American and African rivulines, eggs are fertilized and then buried; they spend the dry season in a resting state known as **diapause** (Lowe-McConnell 1987; see Chapter 10, Deserts and Other Seasonally Arid Habitats).

Internal fertilization is universal among sharks but is limited to about a dozen families of bony fishes; scientific names often emphasize this attribute. Most notable are the coelacanths; a silurid catfish; brotulids; livebearers; goodeids; three genera of halfbeaks (Zenarchopteridae); four-eyed fishes; the neostethids and phallostethids of Southeast Asia; scorpaenids in the genera *Sebastodes* and *Sebastes* (e.g., *Sebastes viviparus*); Baikal oilfishes; embiotocid surfperches; an eel pout, *Zoarces viviparus*; clinids; and two genera (*Xenomedea* and *Starksia*) of labrisomid blennies.

Embryo Development

After fertilization, the chorion of the egg stiffens (**water hardening**), which excludes additional sperm and also serves to protect the developing embryo. Embryogenesis in fishes proceeds as in most vertebrates. The embryo develops on top of the yolk; the yolk is concentrated at the vegetative (vegetal) pole and the fertilized egg is considered **telolecithal**. Hagfishes, elasmobranchs, and teleosts exhibit **meroblastic** cleavage – cell division occurs in the small-cap of cytoplasm that will develop into a fish, but not in the yolk. Lampreys have **holoblastic** cleavage, with the entire cytoplasm dividing. Bowfin, gar, and sturgeon exhibit an intermediate form, termed **semiholoblastic**. Cell division and differentiation continue in fairly predictable sequences, with many interspecific differences in the timing of appearance of different structures.

Organs and structures that are at least partially developed in many embryos prior to hatching include body somites, which are forerunners of muscle blocks; kidney ducts; neural tube; optic and auditory vesicles; eye lens placodes; head and body melanophores; a beating heart and functioning circulatory system, much of which is linked to circulatory vessels in the yolk; pectoral fins and median fin folds but not the median fins themselves; opercular covers, but not gill arches and filaments; otoliths; lateral line sense organs after which scales will form; and the notochord. By the time of hatching, the head and the tail have lifted off the yolk, the mouth and jaws are barely formed, fin rays may be present in the caudal fin, but little if any skeletal ossification has occurred. The nonfunctional gut is a simple, straight tube and the gas bladder is evident as a small evagination of the gut tube. Advanced embryos curl around on themselves, their bodies making more than a full circle. In fish that hatch at a relatively undeveloped stage, the eye is seldom pigmented, maintaining the transparency of the helpless, newly hatched, a free embryo with its very large and cumbersome yolk sac (Blaxter 1969; Hempel 1979; Matarese & Sandknop 1984; Lindsey 1988).

Just prior to emergence, **hatching gland cells** on the head of some fishes secrete proteolytic enzymes that break down the tough, outer chorionic membrane. Thrashing movements of the tail and body aid in the hatching process. Time spent in the egg prior to hatching varies greatly among species, from around 12 h in some coral reef fishes; to a few days in Striped Bass, centrarchid sunfishes, some cichlids, and surgeonfishes; to a week or two in smelts, darters, mackerels, and flatfishes; to several weeks or months in salmons and sharks. In all fishes studied, hatching time decreases with increasing temperature (Blaxter 1969; Kendall et al. 1984). "**Hatching**" is a convenient landmark during early development, but the exact developmental stage at which the embryo frees itself from the egg varies greatly among species and even varies within species depending on temperature and oxygen content of the water, among other factors (e.g., Balon 1975a).

Species identifications of unfertilized or recently fertilized eggs can be a difficult task, often requiring electron microscopy or molecular genetics. Embryos, especially those in advanced stages of development, are considerably easier to identify. Species-specific patterns of pigmentation often develop on the body and finfolds, particularly in marine species (e.g., codfishes, flatfishes). Other characters used for identification include the morphology of the head, gut, and tail, the number of myomeres, and the existence of elongate or otherwise precocially developed fin rays (e.g., flyingfishes, ribbonfishes, sandfishes, weeverfishes) (Matarese & Sandknop 1984).

Species can be assigned to reproductive categories based on where embryonic development occurs and whether developing embryos are dependent on maternal versus yolk provisioning. If the mother releases eggs that then rely on yolk for nutrition, the species is **oviparous** or egg-laying, as is the case in chimeras, some sharks and skates, and most bony fishes. If the young develop inside the mother and the mother provides nutrition via a placental connection, secretions, or additional eggs and embryos eaten by developing young, the species is **viviparous** or live-bearing. The young of viviparous fishes are generally born as juveniles, having bypassed a free larval stage (see Fig. 8.11). Viviparity characterizes about half of the 1200 species of Chondrichthyes and about 500 species of bony fishes (most of those with internal fertilization mentioned earlier). Nutritional provisioning among viviparous fishes can occur in several ways. Some sharks display oophagy, in which pregnant females continue to ovulate during gestation, and the additional eggs are eaten by the developing young (Sloman & Buckley 2011). Developing young of the Sand Tiger shark (*Carcharias taurus*) feed on other developing embryos, a condition known as adelphophagy (also referred to as embryophagy or filial cannibalism).

Some fishes are **ovoviviparous**, whereby young develop inside the mother but depend primarily on the yolk that was laid down during oogenesis (e.g., some sharks, scorpaenids, zoarcid eelpouts, and arguably, Guppy). However, many intermediate conditions exist involving both yolk and maternal secretions, blurring the distinctions (Breder & Rosen 1966; Hoar 1969; Kendall et al. 1984; Wourms 1988; Nelson 1994). Female cownose rays (*Rhinoptera bonasus*) have vascularized appendages (trophonemata) in the reproductive tract that release a secretion that nourishes developing embryos (Hamlett et al. 1985). Smooth Dogfish (*Mustelus canis*) embryos rely on yolk for about three months but are then supported by a placenta-like structure for the remaining 7–8 months of gestation (Sloman & Buckley 2011).

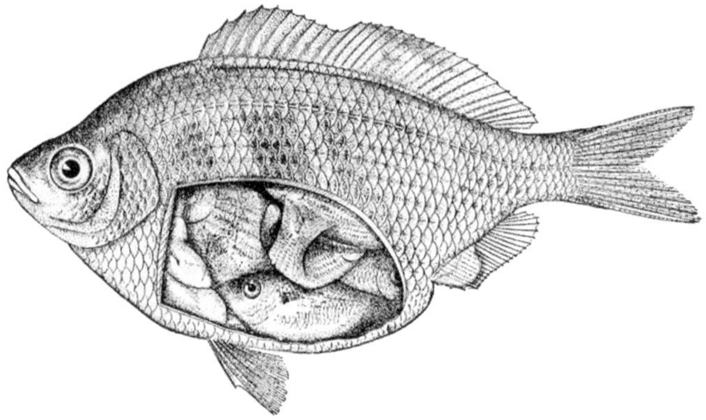

FIGURE 8.11 A pregnant female Shiner Perch, *Cymatogaster aggregata*, showing the well-developed embryos typical of the live-bearing family of surfperches (Embiotocidae) that are common to the Pacific coast of North America. Figure from *Bulletin of the United States Fish Commission* 1905, Public Domain.

Although live birth usually follows internal fertilization, the two are not synonymous. Many elasmobranchs (e.g., horn sharks, catsharks, skates, chimeras) have internal fertilization but lay eggs that develop for many months outside the female.

Meristic Variation

Embryonic development generally progresses according to instructions laid down in the genetic blueprint, but the timing and details of development are sensitive to environmental influences. Pollutants and chemical changes in the water cause larval abnormalities which can serve as indicators of environmental quality (see Longwell et al. 1992; Le Bihanic et al. 2014; Sfakianakis et al. 2015). Even natural variation in temperature, oxygen content, salinity, light intensity, photoperiod, or carbon dioxide can affect development. **Meristic** traits such as numbers of fin rays, vertebrae, lateral scale rows, myotomes, and gill rakers can vary due to environmental conditions.

The pattern of variation among meristic traits is not simple. The most commonly found relationship is for fin ray, vertebral, or scale numbers to increase with decreasing temperature. This inverse relationship exemplifies a general phenomenon, termed **Jordan's Rule**, which applies to latitudinal effects on meristic traits, although the actual determinant is water temperature (Lindsey 1988). However, an opposite pattern of increased meristic values with increased temperature has been observed for fin rays in Guppy and plaice. Some fishes also show patterns in which individuals raised at an intermediate temperature have more or fewer counts in some traits than fish raised at higher and lower temperatures. The actual quantitative difference between groups raised at different temperatures is often less than 3% difference in number of elements per Celsius degree of temperature.

A critical or sensitive period often occurs during embryonic or larval development when effects on meristic characters are strongest. Vertebral counts in Brown Trout are most sensitive to temperatures at around the time of gastrulation and again as the last vertebrae form. Vertebral number is most sensitive to temperature before hatching in herrings and killifishes, but this sensitivity occurs later in Paradise Fish and plaice. Vertebrae form before fin rays and have an earlier sensitive period (Blaxter 1969). Meristic characters may also be sensitive to events prior to fertilization, such as the temperatures at which parents are held and the age of the parents (Lindsey 1988). Causal mechanisms in these patterns are poorly understood. As an embryo develops, it differentiates via segment formation and grows via elongation. Environmental conditions may affect segment formation and elongation differently. Low temperatures may inhibit segment formation more than they inhibit elongation. Hence an embryo developing at low temperatures might be longer when differentiation occurred, causing more segments to be laid down and producing the pattern of more fin rays at lower temperatures (Barlow 1961; Blaxter 1969, 1984; Fahy 1982; Lindsey 1988; Houde 1989).

Supplementary Reading

Able KW, Fahay MP. 1998. *The first year in the life of estuarine fishes of the Middle Atlantic Bight.* New Brunswick, NJ: Rutgers University Press.

Avise JC, Jones AG, Walker D, et al. 2002. Genetic mating systems and reproductive natural history of fishes: lessons for ecology and evolution. *Ann Rev Genet* 36:19–45.

Blaxter JHS, ed. 1974. *The early life history of fish.* New York: Springer-Verlag.

Breder CM Jr, Rosen DE. 1966. *Modes of reproduction in fishes.* Neptune City, NJ: TFH Publishers.

DeWoody JA, Avise JC. 2001. Genetic perspectives on the natural history of fish mating systems. *J Heredity* 92:167–172.

Farrell AP, Stevens ED, Cech JJ, Richards JG. 2011. *Encyclopedia of fish physiology: from genome to environment.* Amsterdam, Boston: Elsevier/Academic Press.

Fuiman LA, Werner RG, eds. 2002. *Concepts in fisheries sciences: the unique contribution of early life stages.* Oxford: Blackwell Scientific Publishing.

Hoar WS, Randall DJ, eds. 1988. *The physiology of developing fish, Part B. Viviparity and posthatching juveniles. Fish physiology,* Vol. 11. San Diego: Academic Press.

Jamieson BGM. 1991. *Fish evolution and systematics: evidence from spermatozoa.* Cambridge, UK: Cambridge University Press.

Lasker R, ed. 1981. *Marine fish larvae.* Seattle: Washington Sea Grant Publications.

Leis JM, Carson-Ewart BM, eds. 2000a. *The larvae of Indo-Pacific coastal fishes. An identification guide to marine fish larvae.* Fauna Malesiana Handbooks. Leiden, the Netherlands: E. J. Brill.

Moser HG, ed. 1996. *The early stages of fishes in the California Current region.* CALCOFI Atlas No. 33. La Jolla, CA: California Cooperative Oceanic Fisheries Investigations. **www.calcofi.org**.

Moser HG, Richards WJ, Cohen DM, Fahay MP, Kendall AW Jr, Richardson SL, eds. 1984. *Ontogeny and systematics of fishes.* Special Publication No. 1. Lawrence, Kansas: American Society Ichthyologists and Herpetologists.

Journal

Any issue of the journal *Molecular Ecology.*

CHAPTER **9**

Larvae, Juveniles, Adults, Age, and Growth

Summary

Several classification systems have been developed to describe the life stages of various fish taxa, and terminology is not always consistent. Most fishes release eggs into the environment where they are fertilized and either remain in the water column (pelagic) or sink to the substrate (demersal), depending on the species. Some fishes have internal fertilization, and of these some incubate eggs and/or young internally. Many freshwater fishes have direct development into juveniles, but most marine fishes have a larval stage that can last for weeks (damselfishes), months (surgeonfishes), or years (*Anguilla* eels). When the larval yolk supply is exhausted, the young fish must forage on its own. Food availability during this transition period is critical to survival and may be a factor in determining many aspects of spawning timing and location. Most larvae die from starvation or predation during their first week of life. As a fish ages and continues to develop, physiological tolerance and sensitivity, ecological and behavioral competence, and survivorship all increase.

Marine fish larvae often bear little resemblance to the adults into which they grow. Larvae often possess large spines and trailing fins and appendages that provide protection from predators and help to maintain them in suitable habitat. Many fish larvae and juveniles were originally thought to be different species, genera, and even families before they were linked with the adults into which they develop.

CHAPTER CONTENTS

The Diversity of Fishes: Biology, Evolution and Ecology, Third Edition. Douglas E. Facey, Brian W. Bowen, Bruce B. Collette, and Gene S. Helfman.

Companion website: www.wiley.com/go/facey/diversityfishes3

Many marine fishes spawn offshore. Onshore movement of larvae may involve passive transport via wind-driven surface currents. As larvae grow, their swimming capability and sensory systems improve. Estuaries are valuable nursery habitats, and movement into an estuary can involve selective tidal stream transport, where a small fish moves up into the water column on flood tides and then hugs the bottom during ebb tides.

We often think of ontogeny as a series of specific phases, but the transitions that occur may be gradual and can involve complex changes in the anatomy and physiology of a fish, all of which require time and energy. In many fishes, juveniles are morphologically and physiologically very similar to adults. Some species, however, show complex and protracted metamorphoses, such as smoltification of salmon that move from freshwater to the ocean, and transformation of flatfishes from symmetrical, planktonic larvae to asymmetrical, bottom-dwelling juveniles.

Age at first reproduction and longevity vary greatly among fishes. Some male surfperches are sexually mature at birth, whereas sharks, sturgeons, and eels may not mature until they are older than 20 years. Longevity also ranges from less than 1 year in annual fishes to more than 100 years in sturgeons and rockfishes. Death among fishes often occurs due to disease or predation. However, some species, including lampreys and Pacific salmon, die after spawning once. Most fishes spawn multiple times, producing more and better eggs as they grow, although some species that spawn multiple times show gradual senescence that eventually leads to death. It is likely, however, that as these fish age and become weaker they will succumb to disease or predation.

The age of a fish can be estimated by counting growth rings on hard body parts such as scales, otoliths, vertebrae, and fin spines. The spacing of these growth rings shows periods of faster and slower growth, which often coincides with seasons, thus allowing the estimation of age. However, injury, stress, and changes in temperature, oxygen availability, or other environmental factors can lead to variation in growth and incorrect estimates of age. Changes in growth rates can also provide information about habitat shifts during ontogeny. Daily growth increments are often detectable on the otoliths of young fishes, allowing back-calculation to estimate spawning dates.

Size at a particular age varies greatly in fishes, even within a species. Growth curves that describe size/age relationships can be calculated using a number of equations, such as the von Bertalanffy growth equation. The condition of a fish, expressed as a ratio between weight and length, is often used as an indicator of fish health; heavier fish are interpreted to be healthier because they have accumulated more energy reserves. As a fish grows the ratio of the dimensions of its body change and can often be described by an allometric equation, revealing a nonlinear relationship.

Fishes continue to grow throughout their lives (indeterminate growth), although the rate of growth slows once they become mature and allocate energy toward reproduction. Life cycles of fishes are quite variable among species and are often complex. A fish must function physiologically, ecologically, and behaviorally at each stage in order to progress to the next stage. Transitional periods between stages can leave fishes vulnerable as physiological and morphological changes may not progress at the same rate, thereby compromising overall function. Habitat changes that may accompany changes in life stage can make these transitions even more challenging.

In some cases, juvenile traits may be retained in adults (paedomorphosis), and in other cases, juveniles may become sexually mature before acquiring otherwise adult characteristics (neoteny). Such changes likely explain the evolution of some of the smallest fish species as well as speciation in lampreys, elopomorphs, salmons, and deep-sea anglerfishes.

Early Life History

Several classification systems have been developed to describe the stages, states, phases, or intervals in the early life history of fishes, with each differing in terminology (Fig. 9.1). These schemes all attempt to subdivide about a dozen recognizable, general events during development into a coherent, descriptive progression. The simplest classification recognizes an **egg** (which after fertilization or activation contains a developing embryo), which hatches into a **larva**, which metamorphoses into a **juvenile**. Subdivisions of this basic sequence generally involve endpoint events, some of which occur quickly, others gradually (Fig. 9.1). Significant endpoint events include the closure of the blastopore and lifting of the tailbud of the developing embryo; absorption of the yolk sac, independent feeding, and flexion of the notochord of the larva; development of fin rays, scales, and pigmentation; and changes in body proportions of the juvenile (Fig. 9.2). These general descriptions overlie a more complicated sequence of events involving changes in the anatomy, physiology, behavior, and ecology of a developing fish (Fig. 9.3). Early life history stages have played an important role in fish systematics and classification (Cohen 1984; Moser et al. 1984).

Part of the confusion over developmental terminology arises from the great diversity of embryonic and larval types, developmental rates, and transitional stages or events that exist among the many lineages of fishes. Attempts at generalization are complicated by exceptions, nuances, and by whether research focuses on marine or freshwater species, pelagic or demersal young, live- or egg-bearers, and embryology or taxonomy. Some workers maintain that development is a continuous and gradual process and that designating exact stages is an arbitrary assignment. Others maintain that development is **saltatory**, with periods of gradual change punctuated by significant events or thresholds that allow for rapid change, such as the shift from dependence on yolk or maternal secretions to independent, exogenous feeding. This ongoing debate will not be resolved in the short space available here.

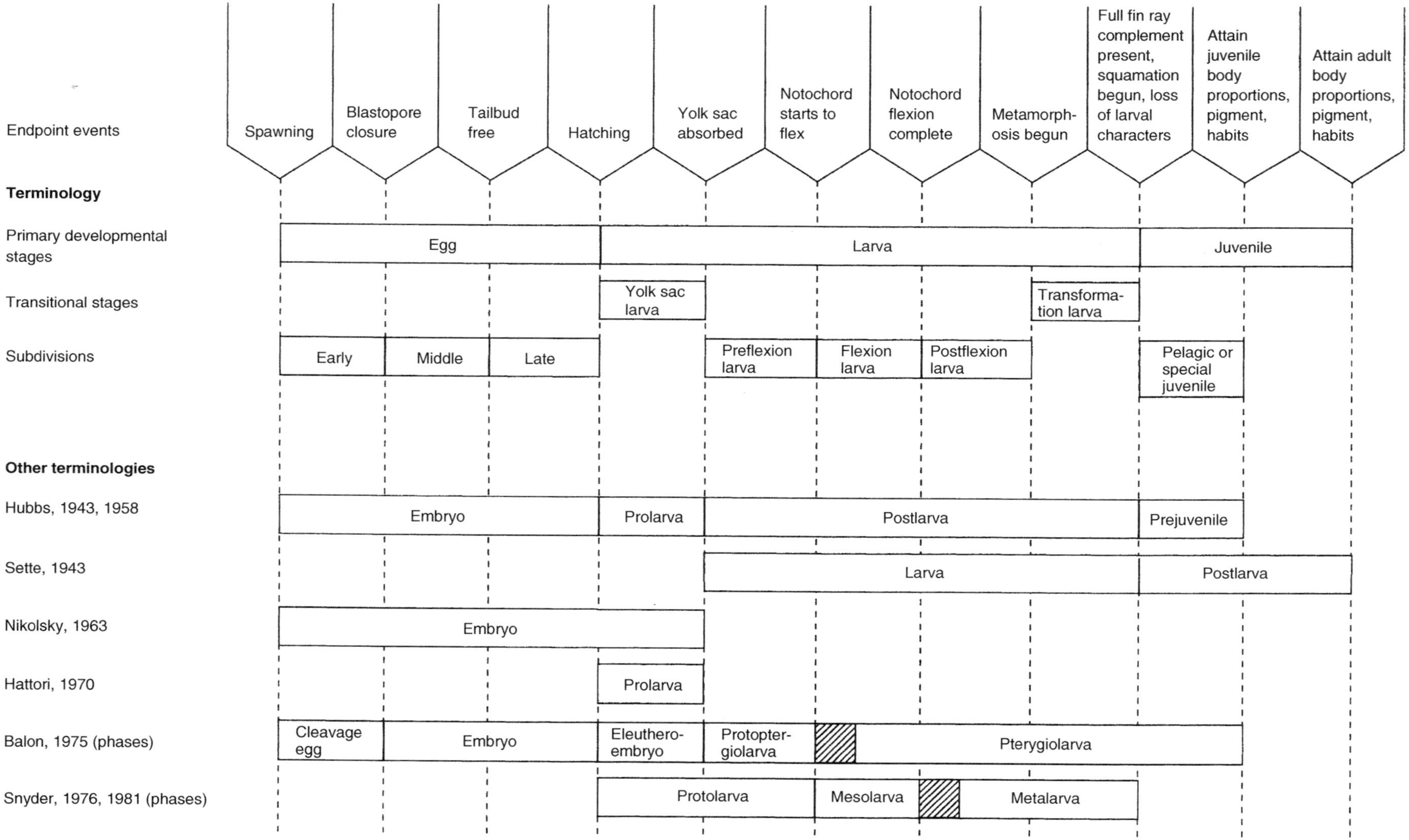

FIGURE 9.1 Events during, and terminology describing, the early life history of teleost fishes. The three basic stages – egg, larva, juvenile – can be further subdivided depending on definable events that occur during development and growth. The top half of the diagram summarizes one commonly used set of terminology, particularly for pelagic marine larvae. Alternative systems for describing these events are given in the lower half of the diagram (see Kendall et al. (1984) for reference details); the approach of Balon (1975b, second from bottom) may be more descriptive of freshwater taxa. From Kendall et al. (1984), used with permission.

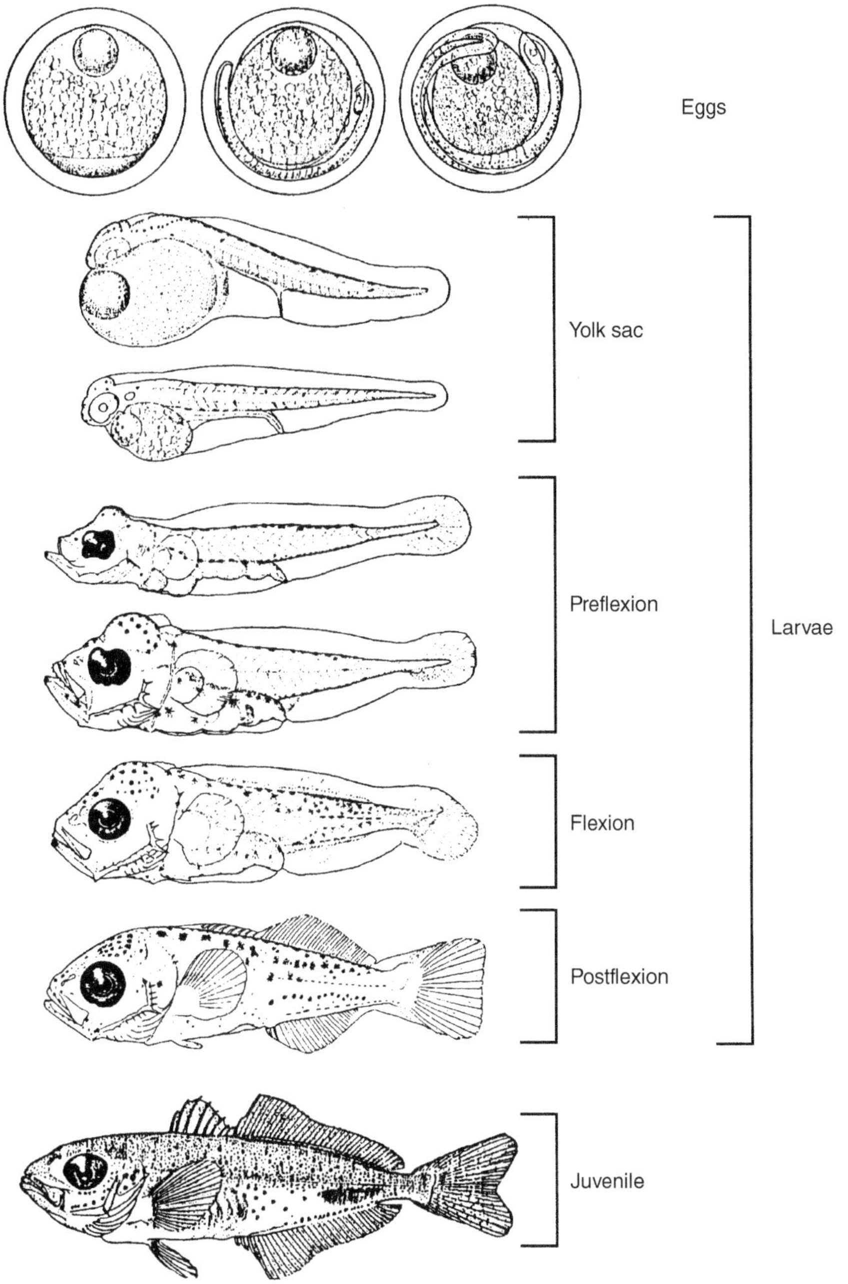

FIGURE 9.2 Stages during the early life history of the Horse Mackerel, *Trachurus symmetricus*. From Ahlstrom and Ball (1954).

Larvae

The presence of a larval stage sets fishes apart from the majority of other vertebrates and also underlies many of their more interesting adaptations. Many marine fishes emerge from an egg as a larva that bears little anatomical, physiological, behavioral, or ecological resemblance to the juvenile or adult into which the fish will eventually transform. The larval stage is probably the most thoroughly studied period of the early life history of fishes. Corresponding research challenges include the species identity of larvae, survival, habitat choice, and their role in determining the distribution and later abundance of many species, particularly those of commercial value.

Larval life generally begins as the fish hatches from the egg and switches from dependence on yolk reserves to external, planktonic food sources. The free-swimming young may still have a large yolk sac and be termed a **free embryo**, **eleutheroembryo**, or **yolk-sac larva** until the yolk is absorbed. **Fry** is a nonspecific term often used for advanced larvae or early juveniles; **swim-up stage** often refers to free embryos or larvae that were initially in a nest but have grown capable of swimming above the nest.

The larval stage continues until the development of the axial skeleton, fins, and organ systems is complete (Fig. 9.4). Median fin rays develop, first as short, fleshy interspinous rays; true fin rays and spines develop later between the interspinous rays. Scales develop, first along the lateral line near the caudal

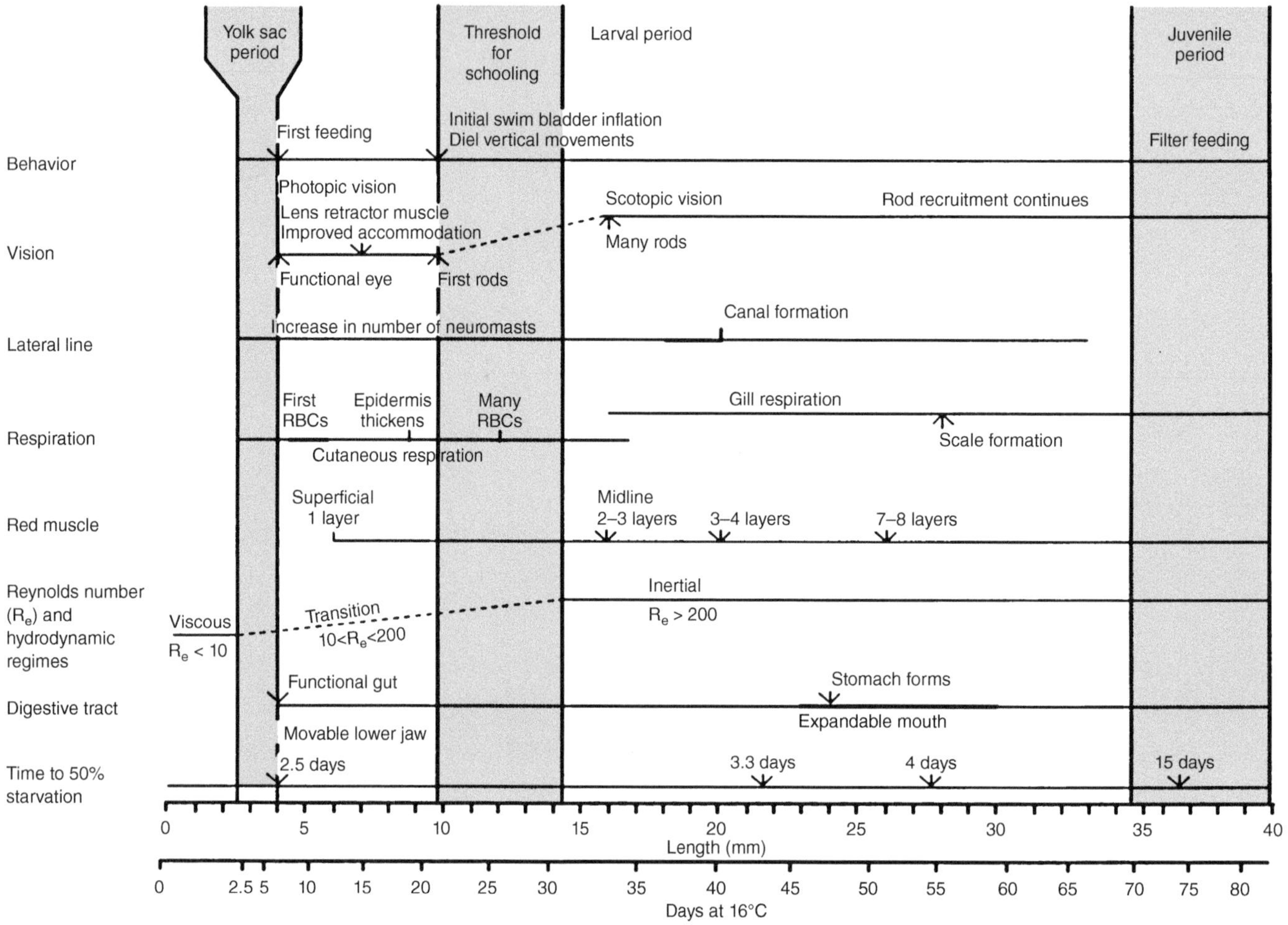

FIGURE 9.3 Behavioral, physiological, and anatomical events during the postembryonic early life history of a representative teleost, the Northern Anchovy, *Engraulis mordax*. The horizontal axis at the bottom represents days after hatching for larvae growing in 16°C water; events noted are those occurring after the larva begins exogenous feeding, when it is no longer solely dependent on yolk for energy. Photopic refers to daytime vision, scotopic refers to nighttime vision. Reynolds number is a measure of the difficulties that small larvae have with water viscosity (see Larval behavior and physiology). Time to 50% starvation refers to how long larvae can live without feeding, based on half of the larvae in an experiment dying after a given number of days without food. RBCs, red blood cells. From Hunter and Coyne (1982), used with permission.

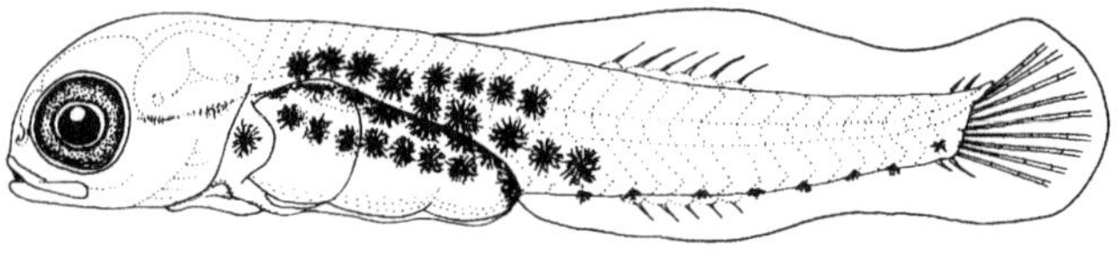

FIGURE 9.4 A recently hatched, marine teleost larva, as represented by a 6 mm clingfish, *Gobiesox rhessodon*. Note development of the mouth, eyes, median fin supports, and melanophores, and upward flexion of the notochord at the base of the tail. From Allen (1979).

peduncle, then in rows dorsal and ventral to the lateral line, and then anteriorly (Fig. 9.5). Once the full complement of scales is attained, the number remains fixed. The end of the notochord, termed the urostyle, flexes upwards and a triangular hypural plate develops just below it. Caudal rays grow posteriorly from the hypural plate elements. Characteristic larval pigmentation patterns develop, the eyes become pigmented, and the mouth and anus open and become functional. Until gill filaments develop, the larva relies on cutaneous respiration, largely involving oxygen absorption across the thin walls of the primordial finfolds. The circulatory system is at first relatively open, and corpuscle-free blood is pumped through sinuses around the yolk and between the fin membranes (Russell 1976).

Pelagic larvae are particularly common among marine fishes, but are also found among some freshwater species (e.g., whitefishes, temperate basses, percid pikeperches, and perches). The duration of the pelagic larval period varies widely, from 1 to 2 weeks in sardines and scorpaenid rockfishes, to about 1 month in many coral reef species, to several months or even years in anguillid eels. A lengthy larval existence aids in the dispersal of these young to appropriate habitats (discussed later in this chapter). Stream fishes, such as many minnows, darters, and sculpins, generally have demersal larvae with short larval periods. Young fish remain on the bottom among rocks or vegetation until they develop swimming and navigation abilities. Larvae of some rocky intertidal fishes (e.g., sculpins, pricklebacks, gunnels, clingfishes) may not disperse offshore but instead spend their entire larval existence

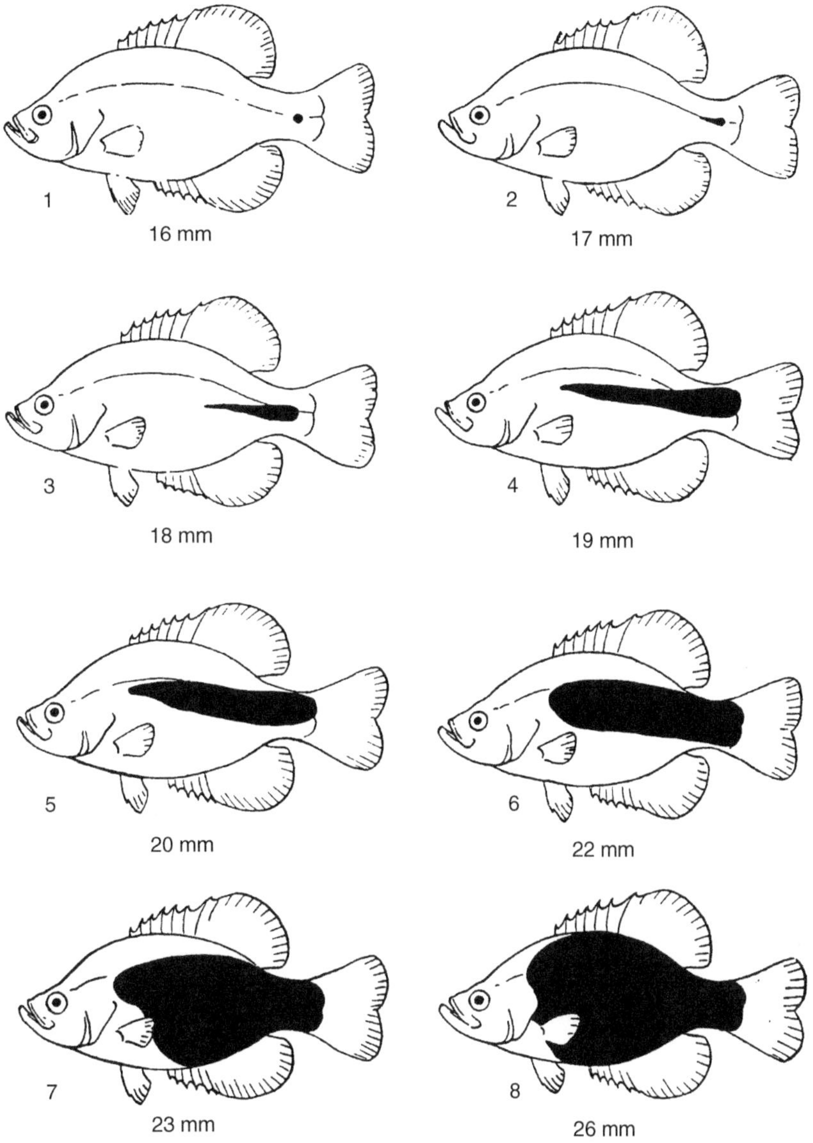

FIGURE 9.5 Development of scales in the Black Crappie, *Pomoxis nigromaculatus*, showing the general pattern of scales developing initially near the tail and spreading anteriorly. From Ward and Leonard (1954), used with permission.

within 5 m of the shoreline, which could guarantee their return to suitable habitat (Marliave 1986).

Larval periods are bypassed or very short in live-bearing fishes and in fishes that still possess considerable yolk reserves after hatching, such as salmonids. Arguably, no larval or juvenile phase exists in the viviparous Dwarf Perch, *Micrometrus minimus*; males are sexually mature when born (Schultz 1993). **Indirect development** involves a larval phase with distinct metamorphosis into the juvenile stage, whereas **direct development** occurs if a larval stage is very brief or not definable, i.e., the fish hatches into a miniature but immature adult, as in most freshwater and many coastal marine forms and in many cypriniforms, salmoniforms, and cottids (Youson 1988).

Larval Feeding and Survival

When yolk supplies are exhausted the fish becomes dependent on exogenous food sources, usually in the form of small planktonic organisms such as diatoms, larval copepods and mollusks, and adult euphausiids, amphipods, ciliates, tintinnids, appendicularians, and larvaceans. Even fishes that are herbivorous as juveniles and adults are usually carnivorous as larvae, as in rabbitfishes (Bryan & Madraisau 1977). This may reflect the difficulty of extracting energy and nutrients from plants.

The potential importance of food availability at the onset of exogenous feeding has greatly influenced our thinking about sources of larval mortality and the subsequent abundance of fish cohorts (year classes). Several influential hypotheses address the relationship between early larval biology, food availability, and adult population size in marine fishes. Hjort (1914) proposed the **Critical Period Hypothesis**, which stated that starvation at a critical period, perhaps the onset of exogenous feeding, was a strong determinant of later abundance. Blaxter and Hempel (1963) coined the phrase **point-of-no-return** to describe when larvae, as a result of starvation, are too weakened to take advantage of food even if it were available. Such irreversible starvation depends on larval condition and age: well-fed, young anchovy may last only 1–2 days before irreversible starvation sets in, whereas healthy, older flatfish larvae may be able to go 2–3 weeks without food. Cushing (1975) proposed the **Match–Mismatch Hypothesis**, which suggests that the timing of reproduction in many marine fishes has evolved to place larvae in locales where food will be available, i.e., that fish reproduction and oceanic production are synchronized. Since the cues of photoperiod and temperature that fish use to initiate spawning (see Chapter 18, Reproductive Seasonality) are not necessarily the same ones that determine plankton production, mismatches can result in high larval mortality (May 1974; Russell 1976; Hunter 1981; Blaxter 1984; Houde 1987); and this can be exacerbated by climate change.

Because of the relationship between larval survival and later population size, the actual causes and patterns of larval mortality are of considerable theoretical and practical interest. Literally, billions of larvae are produced annually by most populations of marine fishes. In most species, >99% of these larvae die from the combined effects of starvation and predation; the average fish probably lives less than a week (Miller 1988). Hence shifts in mortality rates can have major implications for later year class strength and for recruitment into older, catchable-size classes (Hobson et al. 2001).

The temperatures at which larvae develop represent a trade-off between growth rates, metabolic rates, larval duration, and mortality (Houde 1989). Across a 25°C temperature range characteristic of the difference between tropical and temperate conditions (5–30°C), growth rates are six times faster at the high end, but mortality rates can be four times greater as well. Larval duration at the lower end of the temperature range typically exceeds 100 days, but is 1 month or less at 25–30°C.

The metabolic requirements of small ectothermic animals such as fish larvae increase with temperature. **Gross growth efficiency** (weight increase/weight of food consumed) is constant despite temperature. At higher temperatures, a larva has a higher metabolic rate, and therefore must consume more food to achieve the growth rate of a larva at lower temperatures. This is additionally compromised because gross growth efficiency declines with increased ingestion, and **assimilation efficiency**, which is how much of the food is actually useful to the larva, declines with increased temperature. To maintain the same average growth rate, a tropical larva has to eat three times more food than a temperate larva of the same species. Mismatching larvae with food availability becomes more critical at higher temperatures.

Spawning patterns among species appear to represent adaptations to these temperature relationships. Tropical fishes (which experience minimum seasonality) typically spawn over an extended period, producing multiple batches of young, rather than releasing all their eggs in one large spawning session. This increases the probability that some larvae will experience the conditions necessary for successful growth, whereas a single spawning might lead to complete reproductive failure if food abundance is low, as it usually is in tropical seas. Temperate marine fishes that spawn in the summer, such as Atlantic Mackerel and White Hake, tend to spread their reproduction out over a longer period than do winter spawners such as herring and capelin at the same latitude (Houde 1989).

Pelagic larvae are common among coral reef species. Whereas many nearshore temperate species have short larval periods or retain their larvae near the adult habitat, dispersal via a pelagic stage is almost universal among coral reef fishes. Of the 100 or so families that commonly inhabit coral reefs, 97 have pelagic larvae. The exceptions are instructive in that it is easy to postulate historical constraints or adaptive disadvantages to dispersal. Marine plotosid catfishes are a freshwater-derivative family with highly venomous spines. The brightly colored young form dense, ball-shaped shoals and probably gain predator protection from this behavior, a lack of dispersal helping keep siblings together and facilitating the formation of monospecific shoals. Many of the viviparous brotulas live in fresh or brackish water caves near coral reefs, a habitat that could be difficult to relocate by a settling larva. One species of damselfish, the Spiny Chromis, *Acanthochromis polyacanthus*, lacks a dispersed larval stage and continues parental care after eggs hatch; interestingly, *Acanthochromis* larvae develop more slowly than most other damselfishes (Kavanagh & Alford 2003). Other non-dispersers include batrachoidid toadfishes, the monotypic convict blenny (Pholidichthyidae), and apparently reef species of the croaker family Sciaenidae.

The adaptiveness of a floating larva for the families with pelagic larvae probably results in part from selection for avoidance of abundant predators in benthic habitats. Larvae generally are not settling until they have developed avoidance capabilities. Dispersal may also reflect: (i) the possibility that successfully reproducing adults live in saturated habitats that offer few opportunities for settlement for their young; and (ii) the widespread and spotty distribution of coral reefs relative to immense oceanic expanses, necessitating the dispersal of offspring over as wide an area as possible. However, tagging, genetic, behavioral, and otolith chemistry methodologies have indicated that larvae are retained in nearshore gyres and currents and may return to parental regions. For example, marking and DNA genotyping studies of anemonefish on the Great Barrier Reef showed 15–60% of juveniles recruited back to their natal population, many settling <100 m from the anemone where they hatched (Jones et al. 1999b, 2005; see Fig. 9.6).

Pelagic existence is about the only thing these larvae have in common. Reproductive strategies include viviparity and oviparity, mouth-brooded eggs, nest builders, and demersal and floating eggs, some of the latter attached to seaweed. Larval periods range from 9 to >100 days, sizes at settlement from 8 to 200 mm. Some settle prior to metamorphosis to a juvenile stage, some after, and one family (Schindleriidae) is even mature at the time of settlement (Leis 1991).

FIGURE 9.6 Juvenile anemonefish (*Amphiprion*) disperse from the anemone where they hatch. Although they often settle nearby, they do not typically occupy the same anemone as their biological parents (Jones et al. 2005). This is genetically beneficial because if one of the adults is lost, one of the immature fish becomes mature to replace the absent adult. Because the younger fish is not likely to be an offspring of the other resident adult, it would not be mating with its own biological parent and risk weaker offspring due to inbreeding. A little something to think about for those of you familiar with the animated film, *Finding Nemo*. Courtesy of R. Martel.

Larval Behavior and Physiology

Larvae are not always passive and their behavioral capabilities diversify as they grow (Noakes & Godin 1988). Making behavioral observations on very small, transparent larvae is understandably difficult, but successes reveal the dynamics of larval development and the interdependence of morphology, physiology, and behavior. Larvae of reef fish swim, orienting to reefs and currents while moving vertically to remain in particular water masses (e.g., Paris & Cowen 2004). Atlantic salmon larvae hatch from eggs buried in gravel and the larvae remain in the gravel for almost another month but make behavioral responses to environmental cues (Fig. 9.7). The larvae can coordinate swimming movements involving both body and fins. A general geopositive and photonegative response, moving them down and away from light, keeps them buried. They also orient toward water currents. Emergence occurs as these responses reverse to photopositive and geonegative, prompting them to swim up and out of the gravel, but still into the current. They almost immediately react to food and interact with conspecifics, nipping at both, which helps obtain food and drive away potential competitors. They are also capable of swimming away if nipped.

The larvae of some reef fishes display surprisingly complex behavioral repertoires, especially as the time comes for them to drop out of the plankton and settle onto reefs. Surgeonfish larvae that have just settled from the plankton can school, can use topographic reef features to avoid strong currents, can avoid predators, and can swim actively against currents as strong as 15 cm/s (Sancho et al. 1997). The larva-juvenile transition of nearly half of 68 Caribbean reef species observed showed strong habitat specificity with respect to where they actually settled, and this specificity differed among species (Kaufman et al. 1992). Additionally, many species are able to delay or prolong settlement for several days or even weeks while seeking appropriate habitats. Settlement of larvae onto a coral reef as they transition to juveniles involves habitat choice, not a random, luck-of-the-draw process. Settlement habitat and selectivity vary among species even within taxonomic families (e.g., Vigliola & Harmelin-Vivien 2001; Leis & Carson-Ewart 2002).

Swimming capabilities in fish larvae depend on the relationship of body size to water viscosity and density. The viscosity of water provides a greater frictional obstacle for small organisms than for larger organisms. This is reflected in the **Reynolds number** (R_e = larval size × velocity × water density/ water viscosity) and means that small larvae function at low Reynolds numbers (<1) and must swim sporadically and energetically, approaching 50 tail beats per second in anchovy and mackerel larvae, followed by long periods of rest. These small larvae, however, will not sink rapidly when they stop swimming. Larger larvae are able to swim and glide via inertial forces because viscosity is less of a factor (Purcell 1977; Hunter 1981; see Body size, scaling, and allometry later in this chapter).

Feeding efficiency increases rapidly with development, as demonstrated in cichlids, herrings, carp, and seabass. Herring and anchovy larvae feed more efficiently as they grow, improving their success rate from 3–10% shortly after hatching to 60–90% in the later stages of larval life (Blaxter 1969; Hunter 1981; Kamler 1991).

In cichlids with parental care, larvae form cohesive shoals almost immediately upon hatching, promoting parental supervision. The larvae of some cichlids, like the beautiful discus (genus *Symphysodon*), feed on mucus produced by the parents. By 25–40 days after hatching, shoals disband as individuals begin to charge, ram, and otherwise display aggression towards their siblings. Territory defense follows shortly thereafter (Huntingford 1993). Larval feeding in Mottled Sculpin, *Cottus bairdi*, is strongly dependent on lateral line development. Sculpin as early as 6 days after hatching can detect vibrations produced by brine shrimp larvae in the dark, using free neuromasts that are located superficially on the skin of the head.

The ability to detect prey increases with age in most fishes, although some interesting specialized structures appear in larvae and then disappear during metamorphosis. In some weakly electric mormyrid elephantfishes of Africa, a special organ develops in larvae that produces electricity with a different discharge pattern than that shown in adults. This larval organ, which runs from the head to the caudal peduncle, degenerates after 2 months and is replaced by the adult caudal organ (Hopkins 1986).

The eyes of many adult fishes have a retina that contains both cones, which require bright light, and rods, which function well under low light (see Chapter 6, Vision). The eye in embryos and larvae generally contains only cones, limiting feeding and other activities to daytime (e.g., herrings, many salmons and cichlids, soles). Herring do not develop rods until metamorphosis, which also coincides with the beginning of shoal

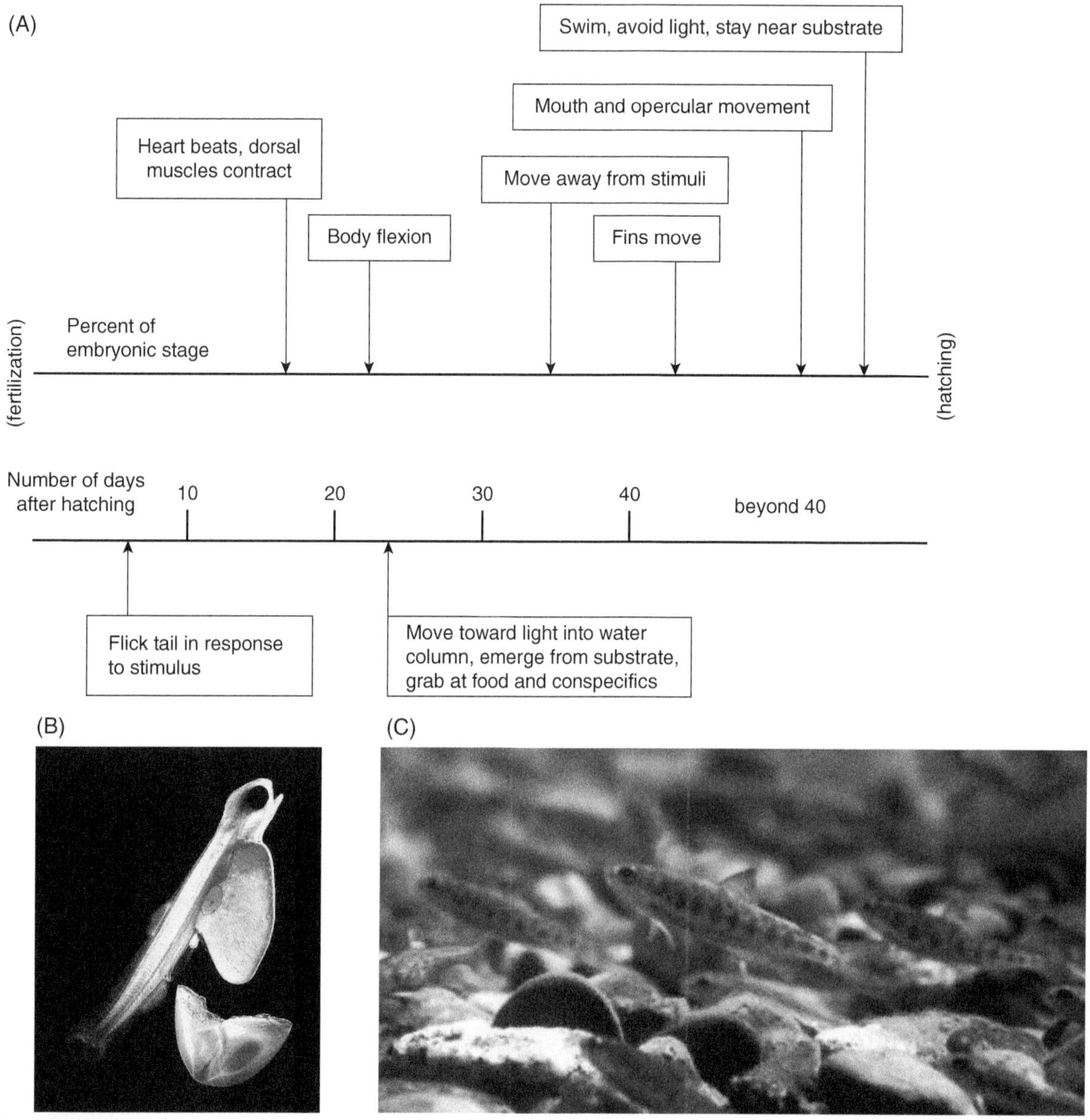

FIGURE 9.7 (A) Approximate developmental and behavioral milestones of Atlantic Salmon during the embryonic stage and several weeks after hatching. Note that embryos are capable of avoiding aversive stimuli via fin and body movements. Based on Abu-Gideiri (1966), Dill (1977b), and Huntingford (1986). (B) A salmon "alevin" or sac fry, the stage at which salmon hatch from the egg and eventually emerge from the gravel where the eggs were laid. Remains of the egg lie below the fish; the large golden structure attached to the belly of the fish is the yolk sac. Uwe Kils / Wikimedia Commons / Licensed under CC BY-SA 3.0. *(GNU Free Documentation License)* (C) Atlantic Salmon parr (note vertical parr marks) emerging from gravel to feed. E.P. Steenstra/ USFWS / Public Domain.

formation, as does the development of the lateral line, which is also involved in shoaling behavior. Many aspects of visual capability – in terms of sensitivity to light, size of objects detected, resolving power, dark adaptation, and range of wavelengths detected – improve with growth as a function of increased lens diameter, increased retina area, the addition of rods or sometimes of cones, the diversification and increased density of cones, and the addition of visual and screening pigments. Some fish larvae have unique visual structures not possessed by later stages. Stalked, elliptical eyes exist in larvae of 14 families of marine teleosts, including anguilliforms, notacanthiforms, salmoniforms, and myctophiforms (see Fig. 9.8A). Stalked eyes can increase 10-fold the volume of water viewable from any given spot, an obvious advantage when searching for food or avoiding predators. Stalked eyes are lost during metamorphosis to the juvenile stage (Blaxter 1975; Weihs & Moser 1981; Hairston et al. 1982; Fernald 1984; Noakes & Godin 1988).

The larval phase is characterized by the onset of function of most of the organ systems an individual will use for the rest of its life, except for reproduction. The marked vulnerability of larvae to starvation and predation, therefore, decreases as these organ systems become functional. Not surprisingly,

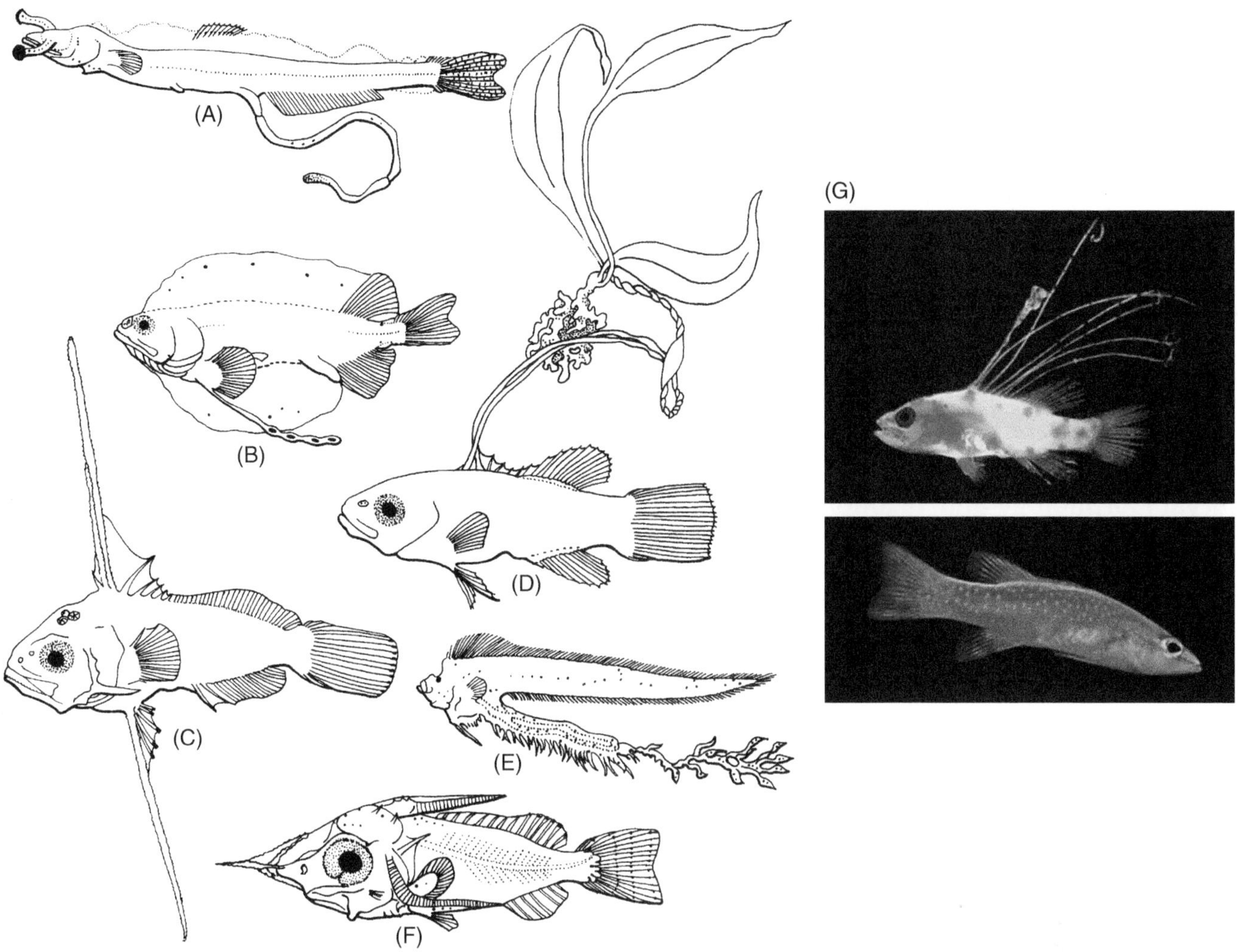

FIGURE 9.8 Larval diversity in marine fishes. Fish larvae often bear little external resemblance to the adults into which they grow. Spines probably make larvae harder to swallow, whereas trailing appendages could mimic siphonophore tentacles and therefore be avoided by predators, or may aid in flotation by slowing the sinking rate of the larva. (A) A 26 mm lantern fish larva, note eyes on stalks and trailing gut. (B) A 17 mm lanternfish larva (*Loweina*), note elongate pectoral ray, and dorsal and anal finfolds. (C) An 8 mm seabass larva, note serrated dorsal and pelvic spines. (D) An 11 mm seabass larva, note elongate dorsal spine. (E) A 64 mm "exterilium" larva of an unknown ophidiiform, note trailing gut. (F) An 8 mm squirrelfish larva, note spines on head and snout. (A) from Moser and Ahlstrom (1974); (B) from Moser and Ahlstrom (1970); (C) from Kendall (1979); (D) from Kendall et al. (1984); (E) from Moser (1981), used with permission of Washington Sea Grant Program; (F) from McKenney (1959); (G) The fish on the top is the planktonic larva of a deep-water serranid seabass, *Liopropoma olneyi*. Its DNA sequence did not match any known fish species and researchers were planning to describe it as a new species despite not knowing what the adult looked like. Then researchers working from a deep submersible off Curacao captured what was to become the holotype of *L. olneyi*, shown below. Its DNA sequence also matched no known serranids, but did match that of the mystery larva, connecting the two life-history stages. Modern genetic techniques quickly solved a puzzle that has plagued ichthyologists for centuries, namely linking together larvae and adults that are almost impossible to connect based on morphology alone. But this time the connection led to a species new to science (Baldwin & Johnson 2014). Cedric Guigand, University of Miami / Smithsonian Science News / Public Domain; Barry Brown / Smithsonian Science News / Public Domain.

anterior and posterior parts of larvae grow faster than middle sections, reflecting the larva's need to feed, respire, and swim (Fuiman 1983; Osse & van den Boogaart 1995). An obvious interrelationship exists between feeding ability and development of the jaws, digestive system, vision, and swimming musculature, and these linked systems develop in synchrony.

A less obvious pattern holds for the development of predator avoidance. In White Seabass, *Atractoscion nobilis* (Sciaenidae), larvae between 3 and 7 mm long (4–23 days old) show little difference in predator escape behavior. Only about 25% of larvae react to approaching predators, usually by a startle response involving **Mauthner cells**, large diameter motor neurons that allow a fish to rapidly swim away from a stimulus. Responses change markedly in slightly older fish. Between 7 and 10 mm (23–30 days), the visual and acoustic systems of the larvae improve markedly. Visual acuity and accommodation improve, the optic tectum of the brain where visual information is processed develops, the gas bladder inflates, and the number of free neuromasts on the head and body increases.

Both gas bladder and neuromasts are involved in detection of sound and water displacement, enhancing predator detection and individual orientation. As a result, White Seabass larvae become much more adept at avoiding ambushing and hovering predators. This behavior is augmented by a change to a demersal existence, which reduces the threat of predation by fast-swimming water column planktivores (anchovies, sardines, mackerel) to which they are still relatively vulnerable. Rapid development of neurosensory structures is therefore critical in the transition from a relatively passive target to an active larva that can avoid predators, over the course of less than 1 week (Margulies 1989). In herring (*Clupea harengus*) larvae, successful predator avoidance coincides with the appearance of lateral line neuromasts and the filling of the otic bulla with gas, both structures associated with hearing (Blaxter & Fuiman 1990).

Larval Morphology and Taxonomy

Whereas eggs tend to be generally similar across many taxa, the larvae that emerge are strikingly distinct and often rather bizarre when compared to adult morphology. The challenges to ichthyologists include identifying and linking larvae with their adult counterparts, and understanding the adaptive significance of the various structures that many larvae possess and then lose as they metamorphose into juveniles. One key is that larvae, although capable of locomotion, are at least initially relatively helpless and vulnerable. They are too slow to actively avoid most predators, other than those that also float with currents, such as various cnidarians with stinging cells. Many predators on pelagic larvae are small, gape-limited, and visually oriented. To counteract these predators, larvae rely on structures that make them spiny and increase their body dimensions, or they mimic potentially noxious planktonic animals such as siphonophores (Fig. 9.8). Structures such as extended fins, skin flaps, and gelatinous body coatings may also slow the sinking rate of larvae, keeping them in more nutrient-rich surface waters. Pigmentation patterns, which are often characteristic of larval stages and useful for identification, may screen harmful ultraviolet rays. This would apply particularly to the heavy melanistic pigmentation found on many species of surface-dwelling (**neustonic**) marine larvae (Moser 1981).

Knowledge of marine fish larvae has grown slowly and incrementally. In some cases many years of research were required to link larval and adult animals because the two stages are so different. The situation can be further complicated by intermediate stages between larvae and juveniles, sometimes called **prejuveniles**. For example, the amphioxides larva of branchiostomatid lancelets and the kasidoron larva of the deep-sea gibberfishes (Gibberichthyidae) were once considered separate families, the Amphioxididae and the Kasidoridae respectively. A number of small fishes that are now identified as larvae or prejuveniles of well-known taxa were once believed to be separate genera, such as *Ammocoetes* (lampreys), *Leptocephalus* (anguilliform eels), *Tilurus* and *Tiluropsis* (notacanthoid halosaurs), *Querimana* (mullets), *Vexillifer* (pearlfishes), *Rhynchichthys* (squirrelfishes), *Dikellorhynchus* (tilefishes), *Acronurus* (surgeonfishes), and *Ptax* (snake mackerels). The scutatus larva of the Big-eyed Frogfish, *Antennarius radiosus*, was initially described as its own genus and species, *Kanazawaichthys scutatus*. These and other distinctive larval stages are still given separate descriptive names, such as the exterilium ("external gut") stage of ophidioid cusk-eels (Fig. 9.8E), the "stalk-eyed" stylophthalmus larva of idiacanthid black dragonfishes, and the flagelloserranus larval stage of seabasses with elongate, ballooning second and third dorsal spines (Fig. 9.7D) (Richards 1976; Kendall et al. 1984; Pietsch & Grobecker 1987; Boschung & Shaw 1988; Eschmeyer 1990; Nelson 1994). Fortunately, the advent of DNA barcoding (Hebert et al. 2003) has greatly enhanced our ability to link larvae to the species that are classified based on adult morphology (see Chapter 20).

Larval Transport Mechanisms

Some inshore marine fishes in temperate and tropical environments spawn offshore but their larvae or juveniles use shallow habitats such as bays, mangroves, and other estuarine regions as nurseries (Beck et al. 2003). This characterizes many temperate species such as anguillid eels, croakers, porgies, Bluefish, scorpionfishes, and flatfishes. An important question therefore is how do such larvae, with their relatively limited swimming capabilities, move to shallow habitats? Active orientation, directed movement, utilization of favorable currents, and habitat choice are all implicated by the distribution and behavior of some species (Leis 1991; Kaufman et al. 1992; Cowen & Castro 1994; see earlier, Larval behavior and physiology).

Larval transport has three main components: movement towards shore, location of and movement into nursery areas, and retention in nursery areas (Norcross & Shaw 1984; Boehlert & Mundy 1988; Miller 1988) (Fig. 9.9). Most interpretations of distribution patterns and behavior propose a combination of passive and active mechanisms, with the degree of activity increasing with age. Although young larvae may rely largely on passive transport of the water mass in which they hatched, older larvae can actively seek particular water masses with which they move. This larval habitat choice results from a surprising ability and tendency to swim actively against all but the strongest oceanic currents.

For species that inhabit continental shelf waters, adults may spawn 100 km or farther offshore (e.g., anguillid eels, bonefishes, menhaden, scorpionfishes, croakers, bothid and pleuronectid flatfishes), and larvae face the challenge of reaching shallow coastal nursery grounds within 2 or 3 months. The spawning behavior of adults, especially their ability to place eggs in favorable locales, is an important starting point. Many species on both coasts of North America spawn in winter when wind-driven, onshore currents are common. Most marine eggs are buoyant and drift toward the surface, placing them in surface layers that are pushed shoreward by winds. Vertical movements by larvae into upper surface waters could also aid in retaining them in water masses that are moving shoreward (Norcross & Shaw 1984). Some larvae may be carried by major currents such as the Gulf Stream, and smaller water masses that spin off from the main current and move shoreward as "warm core rings" (Hare et al. 2002).

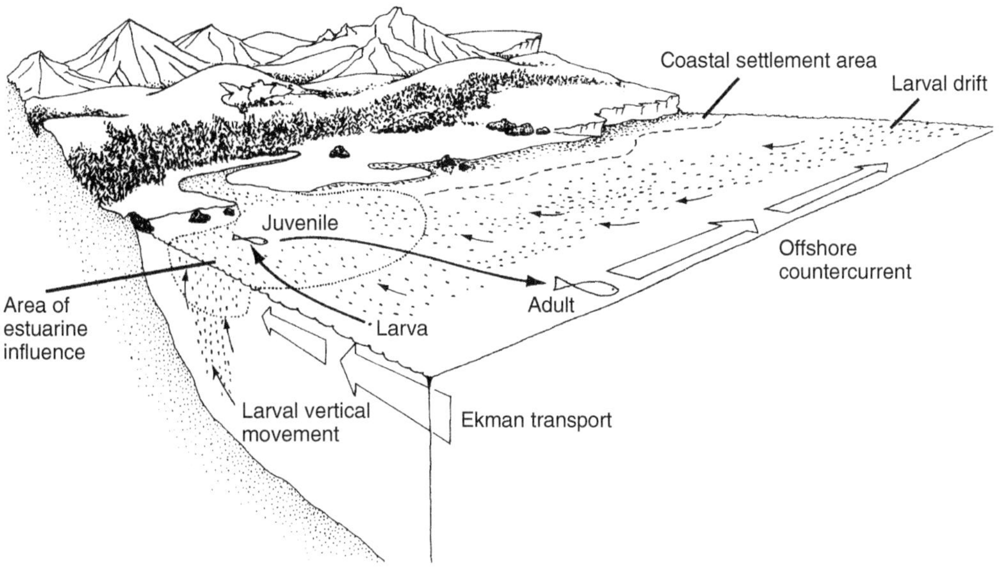

FIGURE 9.9 The general sequence of movement of marine larvae from offshore to inshore nursery grounds, as exemplified by events along the Oregon coast. Larvae are spawned offshore and carried onshore by shallow, wind-driven currents. They then move alongshore by drifting with nearshore currents until encountering stimuli from estuaries, which they enter probably via selective tidal stream transport. From Boehlert and Mundy (1988), used with permission.

Passive transportation alone may work for some tidal or estuarine species (Miller 1988), but cannot resolve the enigma of long-distance larval transport, and at some point active movement is required. Early-stage larva can swim less than 5 cm per second, but more advanced stages can move up to 47 cm per second (Leis et al. 2007). Among 11 common Great Barrier Reef families studied, larvae in the latter half of their larval phase could swim at speeds greater than the mean current speeds found around reef areas (Fisher 2005). This enhanced swimming ability is key to traversing long distance, navigating opposing currents, and successful recruitment to nursery areas. But the struggle doesn't end there. Once larvae find their way to a nursery ground, they frequently have to resist strong tidal, river, and wind currents that could flush them back out to sea.

Active behavioral mechanisms must also influence distribution because species that spawn in similar locales produce larvae that wind up in different places. For example, mullet, Bluefish, and dolphinfish spawn offshore, but larvae of the first two species use inshore nurseries, whereas dolphinfish larvae remain offshore.

Larvae need to feed, so egg placement must also be in areas that are productive. Species that develop offshore often spawn in productive regions that are relatively stable, such as gyres, upwellings, fronts, or other circulating patterns where food availability remains high (e.g., pollock, Dover Sole). Some evidence suggests that coral reef fishes spawn at times and places that tend to retain larvae in local circulation patterns, which would promote their return to parental or nearby locales. Larval retention has been at the heart of a debate over whether coral reefs are **self-recruiting** or are dependent on upstream larval sources, a consideration with direct application to the design and placement of marine reserves (Mora & Sale 2002; Sponaugle et al. 2002). There is now considerable evidence that at least for some locales and species dispersing larvae settle not far from where they hatched (Cowen et al. 2000; Hawkins et al. 2000; Taylor & Hellberg 2003; Buston & D'Aloia 2013). Given oceanographic features and currents and known larval behavior, an estimated 37% to 80% of snapper larvae self-recruit to Cuban waters (Paris et al. 2005). The pattern of currents and gyres around the Florida Keys would tend to aid retention of locally spawned larvae (Lee & Williams 1999). At Kimbe Island in Papua New Guinea, the likelihood of successful recruitment of Clown Anemonefish (*Amphiprion percula*) decreased dramatically as distance increased (Buston et al. 2011). For isolated island regions such as Hawaii, larval retention might be insurance against dispersal into vast and uninhabitable oceanic regions (Norcross & Shaw 1984; Lobel 1989). Hixon (2011) provides an overall historical review of coral reef ecology, including larval recruitment and retention.

After the journey towards shore, larvae accumulate along shorelines and at mouths of bays and estuaries, only to be found later inside these regions. Such pulsed, directional movement against the general net flow of water out of an estuary may involve **selective tidal stream transport** – small fish ride favorable currents and avoid unfavorable ones, usually by moving up into the water column on flood tides and down to the bottom during ebb tides (see Chapter 18, Tidal Patterns). This has been the suggested mechanism for movement into and retention within estuaries by young anguillid eels, herrings, shads, croakers, and plaice (Miller 1988).

Once in a nursery area, a fish must fight currents that would pull it offshore or to less desirable inshore habitats. This becomes less of a problem as a juvenile fish grows larger and stronger and can actively choose locales or currents, but remains a significant constraint for small larvae of species such as Atlantic Herring (*C. harengus*). Herring spawn in estuaries, and larvae first move upstream but later reside in downstream areas. This distribution is achieved by vertical movement with respect to different currents. In most estuaries, surface waters are less saline and therefore less dense and move downstream, and bottom waters are more saline, more dense, and move upstream with flood tides. In the St. Lawrence River estuary of Canada, young herring larvae remain near the bottom and are consequently carried upriver during flood tides, whereas older larvae tend to move up and down twice daily and hence hold the position in a relatively confined area (Fortier & Leggett 1983). Species that spawn in estuaries (e.g., various herrings, cods, flatfishes, wolffishes, sculpins, gobies) tend to have large, demersal eggs and brief larval stages, all characteristics that would minimize export out to sea (Hempel 1979; Norcross & Shaw 1984).

Larvae have to respond to environmental cues that informed them when they are approaching appropriate or inappropriate habitats. Cues such as odor, salinity, oxygen, turbidity, pH, geomagnetism, turbulence, light, food availability, temperature, and current speed and direction might be helpful. Coral reef species may orient toward sounds associated with reefs. Larval traps that incorporated underwater speakers that broadcast natural reef noises (reef-inhabiting snapping shrimp, fish vocalizations) attracted significantly more larvae and a greater diversity of larvae than silent traps, especially at night (Leis et al. 2003; Simpson et al. 2004; Tolimieri et al. 2004; Leis & Lockett 2005). Larvae are also attracted to odors associated with reefs (e.g., Atema et al. 2002). Awareness of predators may also play a role -juvenile grunts (*Haemulon spp.*) are more likely to recruit to reefs that lack predatory squirrelfish *(Holocentrus adscensionus)* (Beets 1997). Clearly, multiple cues facilitate attraction to and settlement in appropriate habitat (Kingsford et al. 2002; see Chapter 21). Responses to any such cues are likely influenced by tidal, circadian, or lunar rhythms (see Chapter 18).

Juveniles

Growth and change continue throughout the life of a fish, and the number of surviving individuals typically diminishes. Most commercially important fishes are exploited as adults, therefore growth and survivorship of juveniles and maturation to adulthood have been extensively studied.

Transitions and Transitional Stages

Hatching or birth and the onset of feeding represent two landmark events in the early life of a fish. Also of importance for many species is the change from larval to juvenile habitat, a transition that often involves settling from the water column and the assumption of a near-benthic existence. Traditionally, the larval phase is considered to end and the juvenile phase begin as larval characters are lost and the axial skeleton, organ systems, pigmentation, squamation, and fins become fully developed, at which time fish look essentially like a miniature adult. This transition can be brief and relatively simple, requiring minutes or hours in some damselfishes, or it can be very long and complicated, taking several weeks in salmons, squirrelfishes, gobies, and flatfishes (see next) (Kendall et al. 1984).

Some complex behavioral adaptations that are characteristic of major taxonomic groups do not appear until the juvenile phase. One example is the alarm reaction of the Ostariophysi, a group that includes about 70% of freshwater fishes (see Chapter 14, Cohort Otocephala, Superorder Ostariophysi; Chapter 16, Discouraging Capture and Handling). Minnows and other ostariophysan fishes have an alarm substance released by specialized skin cells in response to injury. Conspecifics that smell the substance exhibit an alarm reaction. The alarm reaction appears relatively late in development, after shoaling behavior develops and after fish can already produce alarm substance in their epidermal club cells. After 51 days post-hatching, minnows (*Phoxinus phoxinus*), react to alarm substances in the water the first time they encounter it, regardless of experience with predators (Magurran 1986a).

Although eggs and larvae are by far the most vulnerable life-history stages, the juvenile stage still requires successful food acquisition and predator avoidance. For example, juvenile Brook Trout, *Salvelinus fontinalis*, delay the smolt transformation characteristic of many other salmonids (discussed later in this chapter). Instead, they hatch in spring, take up residence in small, shallow streams, and acquire sufficient energy stores during their first summer to get through the winter period of low food availability. To acquire energy and to grow, they must establish and defend a feeding territory. The best territories are in relatively shallow water, which also exposes the fish to both aquatic and aerial predators. Predators can be avoided by remaining motionless, but motionless fish cannot chase prey or repel territorial invaders. Smaller Brook Trout take more risks and tend to feed more extensively and openly (Fig. 9.10), whereas larger fish are less willing to accept predation risks and are more willing to disrupt their feeding by taking evasive actions. The greater likelihood of winter starvation forces smaller juveniles to make the trade-off between predation

FIGURE 9.10 Juvenile Brook Trout take greater risks to acquire food much needed for growth, whereas larger fish tend to avoid risky exposure to predators even if it means less food at the moment. R. Hagerty/ USFWS / Public Domain.

risk and foraging differently from larger fish of the same age (Grant & Noakes 1987, 1988; see Chapter 16, Balancing Costs and Benefits).

Many fishes have transitional stages, making it difficult to pinpoint when fish change from one developmental form to another and also complicating the search for universally descriptive terminology about early life history. Transitional stages occur most dramatically between larval and juvenile, but also between juvenile and adult periods. The transitional phase between larva and juvenile in reef fishes has been variously referred to as post-larval, late-larval, new recruit, juvenile recruit, pelagic juvenile, transition juvenile, and settler. The transitional phase may be variable in length, even within a species because some young fish may not find an appropriate habitat and can delay transition into that stage.

Variability in larval period is evident in the Naked Goby, *Gobiosoma bosci*, which settles from the plankton and takes up a benthic, schooling existence for up to 20 days before transforming to solitary juveniles. Other gobies and a wrasse may have a 20–40 day period during which they can search for appropriate habitat as larvae without transforming into the more sedentary juvenile form. Flatfishes can delay transformation to the juvenile form if they do not encounter an appropriate juvenile habitat; they do this by alternating between settling on the bottom and swimming above it. Substrate preferences, which imply active search for appropriate habitat, have been observed in numerous larvae (e.g., Sale 1969; Kaufman et al. 1992; Sancho et al. 1997). Direct observations of settling coral reef species indicate that such flexibility may be relatively widespread, and that settlement and transition from larva to juvenile should not be viewed as an all-or-nothing decision. Once they are capable of settling, many larvae may have days or even weeks before they must transform into juveniles (Victor 1986; Breitburg 1989; Leis 1991; Kaufman et al. 1992).

Complex Transitions: Smoltification in Salmon, Metamorphosis in Flatfish

Metamorphosis brings major changes in the anatomy, physiology, and behavior of an animal. These alterations require restructuring embryonic and larval characteristics into adult structures that will function under very different environmental conditions. A brief example involves Sea Lamprey (see also Chapter 13, Petromyzontiformes). Larval lampreys, termed ammocoetes, are sedentary, blind, freshwater animals that reside in burrows in silty bottoms and filter suspended matter from the water. At metamorphosis to the juvenile stage, this animal is transformed into a parasite with a suctorial mouth, rasping tongue, salivary glands that secrete anticoagulants, functional eyes, tidal ventilation, and an ability to live in seawater (Youson 1988). Many fascinating examples of this transition are available, but two well-studied groups, salmons and flatfishes, will serve to exemplify the complex restructuring that accompanies the change from larval existence into later life-history stages.

Metamorphosis of Salmon: Smoltification

Widespread interest in salmonids has resulted in detailed knowledge and special terminology associated with different life-history stages. Typically, salmon and trout spawn in gravel pit nests in freshwater termed **redds**, the eggs hatch into **alevins** (yolk-sac larvae) that resorb the yolk and become **fry**. Fry develop species-typical patterns of vertical bars on their sides called **parr marks**, the fish now being called **parr** (Fig. 9.11). After a few months or years depending on species and population, the parr of anadromous species (Pacific and Atlantic salmons, Steelhead Trout) metamorphose into silvery **smolts** and move downstream to the sea. The processes associated with this change and the subsequent downstream migration of smolts are among the most intriguing and best studied biological aspects of the early life history of fishes.

Smoltification is a complex phenomenon involving reworkings of just about every characteristic of a young salmon. An interesting feature of the changes is that they are preparatory: they occur as the animal changes from a parr to a smolt in freshwater, anticipating the environmental conditions that the young fish will later encounter after it enters the ocean. Color changes from generally dark and barred to silvery, which is a better form of camouflage in the open sea (see Chapter 16, Invisible Fishes). The silvering results from an

(A)

(B)

FIGURE 9.11 (A) Atlantic Salmon parr in the Machias River, Maine. (B) An Atlantic Salmon transitioning from parr to smolt. EP Steenstra/ USFWS / Public Domain.

FIGURE 9.12 Some landlocked Atlantic Salmon (*Salmo salar*) stocked as smolts into tributaries of Lake Champlain return to the outlet of the hatchery in which they were raised, and not to the stream into which they were stocked. This trap was built on the hatchery outlet stream to facilitate the capture of these returning fish which are either used as broodstock for the hatchery or relocated. D. Facey (Author).

increase in the density of purine crystals, mostly guanine but also hypoxanthine, which are deposited beneath the scales and deep within the dermis. The fish also take on a slimmer, more streamlined shape that involves a reduction in condition as body lipids are metabolized. Despite a loss of lipids, smolts are more buoyant than non-migratory conspecifics due to increased gas volume in the gas bladder, which may reduce the energetic costs of migration. The complexity of hemoglobins in the blood increases, affecting oxygen affinity and the Bohr shift among other respiratory factors (see Chapter 5, Gas Transport).

These alterations prepare a migrating smolt for oceanic conditions that often include reduced availability of oxygen compared with the cold, turbulent waters of a stream or river. Many changes occur in gill function, including increased number of ionoregulatory cells and changes in ion permeability and enzymatic activity. These changes aid in the move from the **hypoosmotic** freshwater environment where ion loss is the major challenge to the **hyperosmotic** marine environment where water retention is the major challenge (McCormick & Saunders 1987; Hoar & Randall 1988; Wedemeyer et al. 1990; see Chapter 7, Osmoregulation, Excretion, Ion and pH Balance).

Behaviorally, Atlantic Salmon parr are highly territorial in shallow water, but as they become smolts they move into deeper water and form shoals, although a dominance hierarchy frequently exists in the shoals. Even this aggression decreases as fish start to move toward the sea. The movement is aided by a reversal in **rheotaxis**, the response to flowing currents that kept even embryos headed upstream. Positive rheotaxis (upstream orientation) disappears as fish in large shoals drift downstream with the currents (Hoar & Randall 1988; Noakes & Godin 1988; Huntingford 1993). During the parr-smolt transition young fish **imprint** on the odor of their home stream, enabling them to identify it when they return from the sea during the spawning migration. This is an important consideration if hatchery-raised fish are stocked into streams in an effort to create a naturally reproducing population. If hatchery conditions maximize growth and cause young fish to begin the transition prior to stocking, they may not return to the stocked stream when it is time to spawn (Fig. 9.12, also Chapter 18, Anadromy).

The transformations that occur during smoltification are driven by hormones (Fig. 9.13). Increases in corticosteroids, prolactin, and growth hormone respectively affect lipid metabolism, osmoregulation, and mineral balance. Cortisol and estradiol levels also increase. Thyroxine levels also increase naturally, and experimental injections of thyroid-stimulating hormone can induce many of the physiological and behavioral events of smoltification, such as purine deposition, gill enzyme activity, increased swimming activity, body growth, and lipid consumption. These changes indicate that thyroxine, interacting with photoperiod and endogenous rhythms, plays an important role in the process (Hoar & Randall 1988; Huntingford 1993).

Smoltification is not fixed in terms of age in a species or even a population. Atlantic Salmon may smolt at the ages of 1 to 7 years, depending on temperature and latitude. Onset of smoltification in siblings may vary by as much as a year. Well-fed individuals become smolts sooner, although genetic differences in feeding activity may cause some fish to cease feeding and consequently delay smoltification. Some evidence indicates a size threshold: Atlantic Salmon that do not attain a length of 10 cm by the fall of the first growing season are less likely to smolt the next year. Rate of growth and age interact

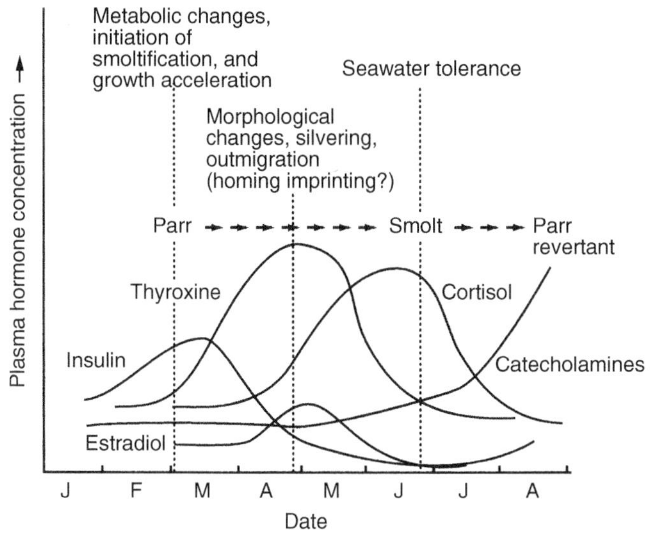

FIGURE 9.13 Sequence of events during smolting in the Coho Salmon, *Oncorhynchus kisutch*, and correspondence with changing levels of important hormones. Hormones implicated in the parr–smolt transformation tend to peak prior to, and may cause, the development of the various anatomical, physiological, and behavioral traits that characterize the smolt stage. One group, the catecholamines, remains low throughout the smolting process but climbs if the fish does not migrate to sea and instead reverts to the parr stage. Drawing by W. W. Dickhoff, in Hoar and Randall (1988), used with permission.

with this hypothesized threshold length. Faster growing fish are more likely to smolt, and older fish may smolt at a smaller size.

Timing is also important. The smolt stage itself lasts a few weeks; if a fish does not enter the sea, the process will reverse and the fish will return to the parr condition. Some individuals within a population, especially males, may bypass the smolt and migratory phases and remain behind in freshwater. When this happens, males may mature quickly at age 1 year and spawn with females that return the next season (see Chapter 17, Alternative Mating Systems and Tactics). In Atlantic Salmon, the proportion of such **precocious** males differs among populations, ranging from 5% to >50% of the males. The factors determining precocious maturation in male salmons have been debated, with evidence suggesting that food availability or genetic factors are determinant (Thorpe 1978; Hoar & Randall 1988). Finally, anadromous populations of Pacific salmon (*Oncorhynchus* spp.) and Atlantic Salmon can become **landlocked** by changes in river flow, land contours, or damming. These populations can adapt to freshwater life history, migrating from stream to lake instead of to the sea. Landlocked forms are very popular for stocking into lakes that lack salmon, and this has been done all over the world.

Metamorphosis of Asymmetrical Flatfish

Symmetry is an almost universal anatomical characteristic of animals. Most animals exhibit **bilateral symmetry** in their morphology, having roughly mirror-imaged structures to the right and left of the midline. Deviations from symmetry imply unexpected functions and adaptations. Biologists seek to understand the causation and function of asymmetry at the proximate level of genetic and environmental control of development and at the ultimate level of evolutionary adaptation.

Among the more startling examples of asymmetry is the "**handedness**" of flatfishes. The 14 families and over 770 species of the order Pleuronectiformes (e.g., flounders, halibuts, soles, plaice) are characterized by adults that lie on the bottom on one side of their body. Their flattened bodies are functionally analogous to many other benthic-living fishes such as angel sharks; skates; rays; banjo, suckermouth armored, and squarehead catfishes; ogcocephalid batfishes; platycephalid flatheads; and some scorpionfishes. The major difference is that all the other groups are flattened in a dorsal–ventral plane (**depressed**), whereas flatfishes are laterally flattened (**compressed**). Depressed fishes maintain their bilateral symmetry despite their extreme morphology. Most compressed fishes are deep-bodied, bilaterally symmetrical species that swim in the water column and use their flattened bodies to increase maneuverability or to increase their body depth against predators (e.g., serrasalmine characins, centrarchid sunfishes, many pompanos, monodactylid fingerfishes, butterflyfishes, ephippid batfishes and spadefishes, and surgeonfishes). Flatfishes are laterally compressed but lie on the bottom on either their right or left side. Among the most obvious accommodations to their unusual orientation can be seen in the structure and development of the head, especially the position of the eyes.

Flatfishes begin life as typical bilaterally symmetrical, pelagic larvae. In the Starry Flounder, *Platichthys stellatus*, larvae emerge from the egg when about 3 mm long and begin exogenous feeding. For the next month or two, they lead typical pelagic lives, until they reach a length of 7 mm. Then metamorphosis to a compressed shape begins (size at metamorphosis varies between 4 and 120 mm in different flatfishes). Most bones are incompletely ossified at this time, which makes the transformation easier. The skull grows asymmetrically causing one eye to move across the top of the head (Fig. 9.14). The result is a fish with both eyes on the side of the body that will orient upward when the fish lies on the substrate. In some species of bothids and paralichthyids, the eye moves through a slit that appears between the skull and the base of the dorsal fin. The dorsal fin remains in the midline or, in some species, grows forward until the first spine sits anterior to the eyes. The entire process happens quickly, over about a 5-day period in Starry Flounders, or in less than 1 day in some species.

Other asymmetries occur that reflect transformation to a compressed, benthic existence. The nasal organ on the blind side migrates to the dorsal midline, and the semicircular canals undergo a 90° displacement and the dorsal light reflex (see Chapter 6, Equilibrium and Balance) also changes appropriately for a fish lying on its side. At the time of metamorphosis or shortly thereafter, the fish takes up a benthic existence and loses its gas bladder. The side that is in contact with the substrate is usually unpigmented, may lack a lateral line, and has smaller pectoral and pelvic fins. In Windowpane, *Scophthalmus aquosus*, eye migration is coordinated with a number of other developmental events, all culminating at about the time the young fish takes up a benthic existence (Fig. 9.15).

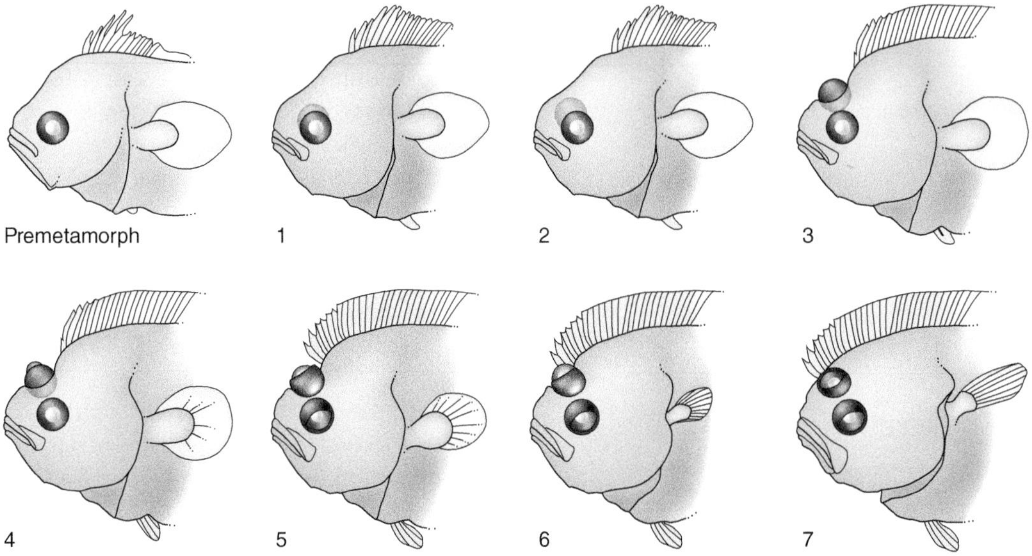

FIGURE 9.14 Progressive eye migration in a developing Summer Flounder, *Paralicthys dentatus*. When the flounder larva is about 10 mm long, the right eye begins to migrate over to the left side of the fish via a process that includes bone resorption and rotation of the fish's neurocranium. The entire process takes 3–4 weeks, during which time the larva grows 5–10 mm. The position of the right eye on the right side of the body is depicted in stages 1 through 3 (faint circle). Note other developmental changes, including development of eye structures, anterior migration of the dorsal fin, growth and elaboration of the pectoral and pelvic fins, and mouth growth. After Keefe and Able (1993), used with permission.

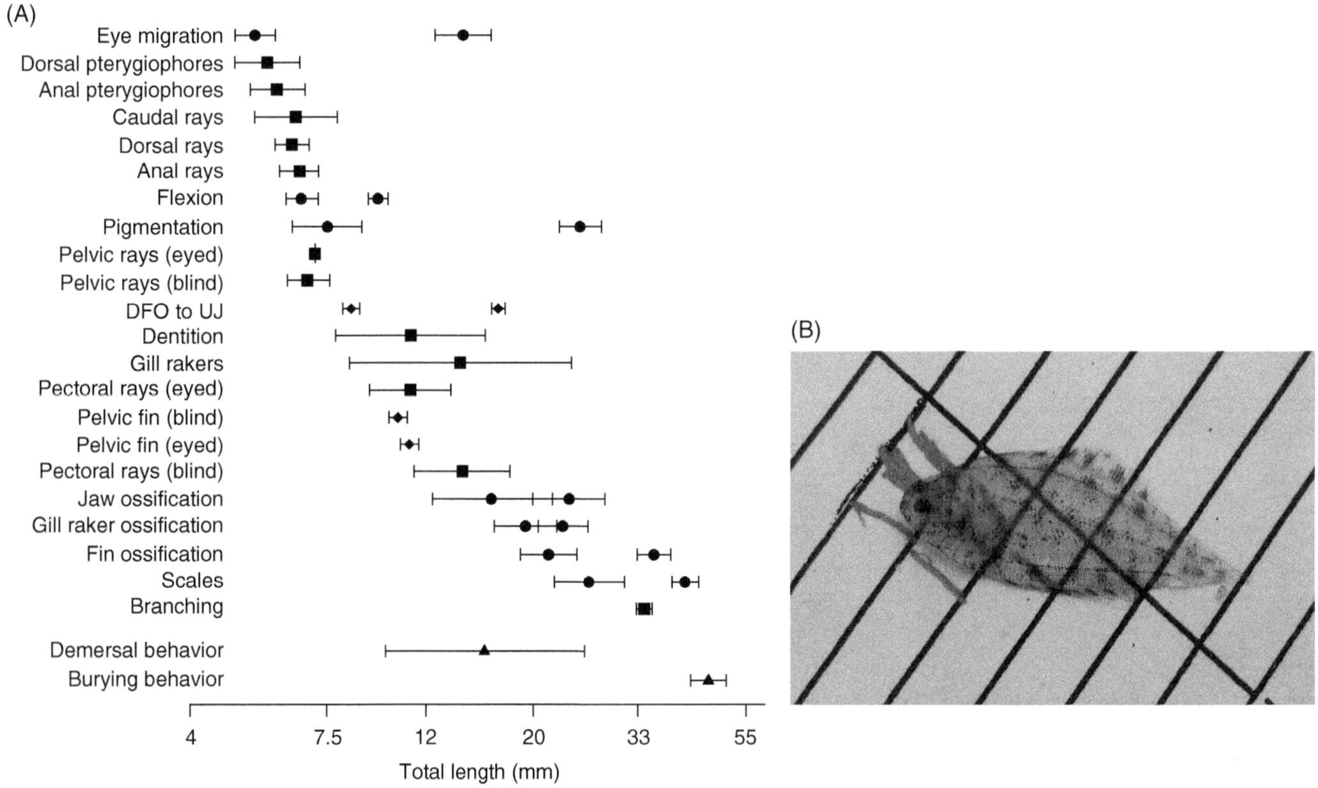

FIGURE 9.15 (A) Metamorphosis from a pelagic to a benthic life in flatfishes involves numerous traits and behaviors. A sequence of changes occurs along with eye migration, some at different times and at different rates. eyed, traits on the eyed side of fish; blind, traits on the blind side; DFO to UJ, distance from dorsal fin origin to anterior edge of upper jaw; Branching, branching of dorsal fin rays. Slightly modified from Neuman and Able (2002), used with permission. (B) A recently transformed flounder. The fish is about 1 cm long and still mostly transparent, having not yet developed from a transparent larva to a pigmented juvenile. The eye from the right side has not completely migrated to the left side. The fish is lying on its right side on lined paper, the lines showing through the transparent body. G. Helfman (Author).

As a rule, families are characterized by having both eyes on a particular side of the head. Hence left-eye flounders (Bothidae) lie on their right side and have both eyes on the left side, the right eye having migrated; this is termed the **sinistral condition**. There may be occasional developmental abnormalities, and individual members of right-eyed species may be left-eyed. In such individuals, viscera may also be twisted and color patterns abnormal. Regular variation in such handedness also occurs. Starry Flounders, although members of the right eye (Pleuronectidae) family and usually right-eyed or **dextral**, often include left-eyed individuals. In California, 50% of the individuals may be left-eyed, and in Japan 100% of these pleuronectids are left-eyed. That these nonconformist individuals are in fact abnormal is evident in the development of their optic nerves. In all vertebrates, normal development results in a crossing of the optic nerves leading from the eye to the brain, such that the right side of the brain receives information from the left eye and vice versa. In left-eyed Starry Flounders, the optic nerve crosses twice, literally twisting around itself, a condition that does not seem to have any adaptive advantage. Experimental crosses of individuals from different populations have established that the determination of handedness in flatfishes is under complex genetic control, but no evidence exists to suggest that one side is adaptively better than the other (Policansky 1982a, 1982b; Ahlstrom et al. 1984).

Adults

Maturation and Longevity

Not surprisingly, age at first reproduction and longevity vary greatly among fishes (Finch 1990), making it difficult to identify patterns or draw conclusions (for an excellent overview of various classifications of maturation stages, see Pusey et al. 2004, table 5). The adaptive significance of differences in **age at first reproduction** relates to trade-offs between committing energy to somatic growth versus reproduction, combined with expected mortality rates and the probability of living long enough to reproduce. Extremes in age at first reproduction include some embiotocid surfperches, the males of which are born producing functional sperm. Gobioid fishes in the genera *Schindleria* and *Paedogobius* mature in less than 2 months, *Schindleria* mature in as little as 3 weeks (Kon & Yoshino 2002a). Many small stream fishes mature in 1 year, being reproductively active the spawning season after they hatch (e.g., most darters), although maturation may take longer in populations at higher latitudes.

At the other extreme, sturgeons (which may live 80–150 years) and some sharks may take 10–20 years to mature. The slowest maturing shark is the Spiny Dogfish, *Squalus acanthias*, a species well known to students of comparative anatomy (Fig. 9.16) Spiny Dogfish do not mature until 20 or 30 years old and have one of the longest recorded life spans of a shark, upwards of 100 years. The record for naturally delayed reproduction among bony fishes is apparently held by American eels in Nova Scotia, which may not mature and undertake their spawning migration back to the Sargasso Sea until they are 40 years old.

FIGURE 9.16 Spiny dogfish (*Squalus acanthias*) is a commonly used dissection subject in comparative anatomy labs. These fish take about 20 years to reach maturity and have relatively few young, making population recovery very slow. D. Costa/ NOAA / Wikimedia Commons / Public Domain.

Longevity patterns are only slightly more definable. Larger fishes generally live longer than smaller fishes, e.g., White Shark (*Carcharodon carcharias*) longevity can exceed 70 years (Hamady et al. 2014), but there are many exceptions. The oldest teleosts known are scorpaenid rockfishes of the northeastern Pacific. Radioisotopic and otolith analyses indicate that Rougheye Rockfish (*Sebastes aleutianus*) live for 140 years, Silver-gray Rockfish (*S. borealis*) for 120 years, and Deepwater Rockfish (*S. alutus*) for 90 years (Finch 1990; Leaman 1991). The deep-sea Orange Roughy (*Hoplostethus atlanticus*) and Warty Oreo (*Allocyttus verrucosus*) can reach 125 years of age (Cailliet et al. 2001). Among common sport species, European Perch can live 25 years and Largemouth Bass can live 15–24 years (Das 1994; Boschung & Mayden 2004).

Numerous species live for a year or less, including the so-called annual fishes of South America and Africa (see Chapter 10, Deserts and Other Seasonally Arid Habitats). Several gobies have remarkably short generation times and life spans. The Australian coral reef goby *Eviota sigillata* spends 3 weeks as a planktonic larva, settles and matures within 1–2 weeks, and lives for no more than another 4 weeks (Depczynski & Bellwood 2005). The shortest known life span among freshwater fishes occurs in an African rivuline, the nothobranchiid *Nothobranchius furzeri*, with a life expectancy in the wild of a few months and a maximum life span in the laboratory of less than 12 weeks (Valdesalicil & Cellerino 2003). Other short-lived species include North American minnows in the genus *Pimephales* (Fathead, Bullhead, and Bluntnose Minnows), several galaxiid fishes from Tasmania and New Zealand, retropinnid southern smelts, Japanese Ayu, Sundaland noodlefishes (Sundasalangidae), a silverside, and a stickleback.

Death and Senescence

Death in fishes usually results from predation, accident, opportunistic pathogens, or accumulated somatic mutations that lead to a slow decline in health and increased susceptibility to environmental factors. However, some fishes age via the "programmed death" process of **senescence** that is more typical of mammals such as ourselves. Senescence refers to age-related changes that have an adverse effect on an organism and that increase the likelihood of its death (Finch 1990). Senescence includes the metabolic and anatomical breakdown that occurs in older adult animals following maturation and reproduction. Pacific salmon provide a dramatic example. Reproductively migrating fish in peak physical condition enter their natal river, mature, spawn, break down anatomically and physiologically, and die in a matter of weeks. Many of the anatomical and physiological changes that occur can be linked to the combined effects of overproduction of steroids and starvation. **Interrenal** cells, which are steroid-producing cells associated with the kidney, homologous with the adrenal cortex of mammals, secrete corticosteroids, producing blood levels of these substances five or more times higher than normal levels. This results in rapid degeneration of the heart, liver, kidney, spleen, thymus, and coronary arteries; the latter is strikingly similar to coronary artery disease in humans. The digestive tract including intestinal villi degenerates, fat reserves are depleted, and feeding ceases. Immune function also declines, in part due to elevated corticosteroids, increasing susceptibility to bacterial and fungal infections. In naturally spawning Pacific salmons, these side effects are irreversible (Fig. 9.17). Castrated males and females do not produce the elevated corticosteroids and do not spawn, but instead continue to grow to twice the length and live twice as long as intact fish. Precocious males, those that matured as parr and bypassed the smolt and marine phases, may survive spawning and breed again the next year (Finch 1990).

Equally spectacular senescence occurs in several other fish taxa. Reproduction in both parasitic and nonparasitic lampreys involves maturation accompanied by cessation of feeding and atrophy of most internal organs with the exception of the heart and gonads. Fats and muscle proteins are metabolized or transformed into gonadal products. Both males and females die shortly after spawning, probably from starvation. Anguillid eels live as juveniles for many years in rivers and lakes. They then undergo a reproductive metamorphosis that includes enlargement of eyes, changes in body coloration and fin proportions, gut degeneration, and cessation of feeding. After a reproductive migration to the sea that spans thousands of kilometers, the adults perish (see Chapter 18, Catadromy). Laboratory manipulation of hormone functions indicate that, as with salmons, rapid senescence results from elevated corticosteroids and starvation. During maturation, conger and snipe eels also experience gut atrophy and lose their teeth. The Ice Goby, *Leucosparion petersi*, which enters freshwater to spawn and then dies, develops enlarged adrenals, and undergoes splenic degeneration. More gradual senescence has been observed in many multiple-spawning species, such as herrings, haddocks, Guppy and other livebearers, annual killifishes, and Medaka. Anatomical and physiological indicators of gradual senescence include reduced or even negative length and weight change, reduced egg output, corneal clouding, disordered scales, malignant growths, spinal deformities, and impaired regenerative capability. Such senescent changes are more common in small, short-lived species (Lindsey 1988; Finch 1990; Kamler 1991).

FIGURE 9.17 Adults of the five species of Pacific salmon, such as this Sockeye Salmon (*Onchorhynchus nerka*), die after spawning, providing a valuable food source for scavenging birds and mammals. The decomposition of the adult salmon also represents an enormous input of nutrients to the stream and surrounding forest.
G. Helfman (Author).

Age and Growth

Age

Many of the phenomena described earlier include fairly precise statements of the age of the fish involved. How are such ages determined? Although size is generally correlated with age, most species show considerable variation in size at any particular age (see next section), making it difficult to estimate one from the other, especially in long-lived or slow-growing fishes. Researchers interested in determining a fish's age, therefore, look for structures that increase in size incrementally, in relation to some periodic environmental phenomenon. Many body parts meet this criterion, differing among fish species and among age groups. The most commonly used techniques involve counting naturally occurring growth lines on scales, otoliths (statoliths in lampreys), vertebrae, fin spines, eye lenses, teeth, or bones of the jaw, pectoral girdle, and opercular series. Representative growth patterns that are commonly used to age fishes include annual growth rings on scales and **daily growth increments** on otoliths (Brothers 1984).

Scales arise as bony plates in the dermis, and in most fishes begin to develop during the late larval stage or during metamorphosis to the juvenile stage. Bone-forming cells (**osteoblasts)** lay down layers of roughly concentric circles of bone, termed **circuli**, along the midbody, starting in the region of the developing lateral line and spread from posterior to anterior. Scales grow as more bone is added along their periphery, increasing in thickness but particularly in diameter. Diameter increase reflects body growth; circuli are closer together during periods of slow growth, such as winter at higher latitudes, and wider apart during rapid growth, such as during spring and summer (see Chapter 4), analogous to the growth rings of trees. This growth pattern creates alternating dark and light bands in the scale that correspond to periods of slow and fast growth, respectively, particularly when viewed with transmitted light (e.g., backlit). In a habitat with distinct growing and non-growing seasons, such as most temperate lakes, one thick and one thin band constitute a year's growth and is therefore referred to as an **annulus**. The number of annuli on a scale, therefore, gives a record of fish age in years.

However, many factors can interrupt annulus formation, thereby confounding age estimates. Growth typically slows down when fish enter spawning condition, reflecting the allocation of energy away from growth and into gamete production and reproductive behavior. Decreased growth occurs in many species that engage in parental care (see Chapter 17, Parental Care). Such **spawning checks** will appear as dense bands and can be mistaken for annuli, leading to overestimation of age. Bands resembling annuli, termed **false annuli**, can also result from multiple wet and dry seasons, as occur in many tropical locales, as well as from disease, parasites, recovery from injury, responses to pollutants, and forced periods of inactivity and nonfeeding. Feeding often slows down or ceases during summer periods of high water temperature and low oxygen.

Underestimation of age can also result if scales do not begin to develop until the fish is a few years old, as in anguillid eels, or if older fish reach a growth asymptote and hence grow little if at all. Based on scale ages, Pacific Sablefish, *Anoplopoma fimbria*, were generally thought to live 3–8 years and were managed as a fast-growing, short-lived, productive fishery. Subsequent studies, involving otolith sections examination in tagged fish, showed that the fish instead lived for 4–40 years, and some for as long as 70 years. However, older fish had essentially stopped growing, both in terms of increasing body and scale sizes, causing underestimation of age. These new analyses forced a major revision in management strategies. They point to a widespread realization, namely that different parts of a fish's body can grow at different rates (e.g., Casselman 1990). Validation of the annular nature of growth rings on scales and other structures is now generally recognized as essential (e.g., Hales & Belk 1992). Validation involves injection of dyes or radioisotopes that are incorporated into the ring formation in a scale or otolith. Fish are subsequently reexamined (or recaptured) to determine whether growth rings outside the marked ring accurately represent the time interval since injection (Beamish & McFarlane 1983, 1987; Stevenson & Campana 1992).

Scales lost due to physical injury will be replaced. The regenerated scales grow rapidly to approximately the same size as the scales that were lost, and will not display annual changes in growth shown by the original scales. The regenerated scales cannot, therefore, be used to estimate age prior to the injury.

The semicircular canals of the inner ear contain **otoliths**, which are calcareous structures of characteristic shapes and sizes depending on the species. Otoliths form earlier than scales, often appearing in the otic capsules of embryos prior to hatching (Brothers 1984). Otoliths grow via the accretion of layers of fibroprotein and calcium carbonate crystals. In many fishes, this deposition occurs on a daily basis, relatively independent of most environmental conditions. Hence a one-to-one correspondence of rings (**lamellae**) to days exists on the otoliths, allowing highly accurate estimates of fish age, particularly in larvae and juveniles (Pannella 1971; Brothers et al. 1976). The width of the daily increments can be a useful indicator of growth conditions and can offer valuable information about when significant events occur during the early life history of an individual, such as length of larval period or the transition to juvenile stage (Brothers & McFarland 1981). Of the three otoliths, the sagitta is usually the largest and the most useful for aging studies.

As fishes age, growth rates decrease. Hence growth rings are closer together and may become too close together to allow the resolution of daily increments. However, seasonal and annual records are still evident on these hard body parts. Changes in spacing of larger zones may indicate not only age but also when fish move among habitats that are more or less favorable for growth, as when eels migrate upriver or salmon smolts move from food-poor freshwater rivers to food-rich estuaries (Fig. 9.18). Because the chemistry of the accreted layers reflects the chemistry of the water in which a young fish develops, the otolith has been likened to an event recorder, allowing determination of when and which habitats growing fish occupy. Such information can be useful in determining the geographic origins of recruits to an area or into an exploited or depleted population, as well as periods of occupancy of different water masses and pathways of dispersal (Thorrold et al. 2002; Palumbi et al. 2003; Patterson et al. 2005).

In some cases, fish can be aged by taking advantage of transient environmental pollutants that are incorporated into scales or otoliths. For example, the atmospheric testing of thermonuclear bombs in the 1950s and 1960s produced a pulse of radiocarbon ($\Delta^{14}C$) that was quickly assimilated into the oceans and tissues of marine organisms. Over the ensuing decades, radiocarbon declined at a known rate. Therefore, assessing the concentration of radiocarbon at the center of an otolith can reveal the approximate year when that otolith was formed (Andrews et al. 2016). This "bomb radiocarbon" aging has been applied to a variety of elasmobranchs and teleosts, in some cases greatly extending longevity estimates. Based on vertebral rings, White Shark age estimates were no older than 23 years, but bomb radiocarbon shows that they may live longer than 70 years (Hamady et al. 2014).

Growth

Fishes have **indeterminate growth,** which continues throughout the life span of an individual although at a constantly decelerating rate. (This is in contrast to growth in most mammals and birds which is **determinate**, ceasing after an individual matures.) Therefore older fish are generally larger, all other things being equal. Growth in fishes is, however, quite variable and "**size at age**" differs enormously, whether we are

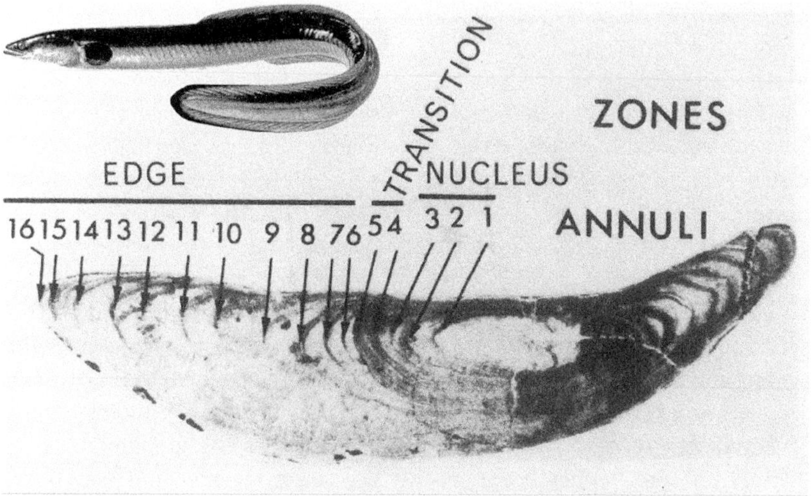

FIGURE 9.18 The correspondence between growth zones on an otolith and habitat use in an American eel, *A. rostrata*. This sagittal otolith indicates that the eel was 16 years old when captured. It spent 3 years at sea or in the estuary of the St. Lawrence River (fast-growth nucleus zone), migrated upriver over a 2-year period (slow-growth transition zone), and finally took up residence in the upper St. Lawrence–Lake Ontario area (fast-growth edge zone). Habitat use was confirmed by measuring strontium : calcium ratios in the different zones of the otolith, using a microprobe associated with an electron microscope. Different ratios arise when an animal inhabits oceanic versus freshwater. Adapted from Casselman (1983); American eel drawing from Bigelow and Schroeder (1953b).

comparing species, individuals within populations, or siblings from a single clutch.

Growth is regulated hormonally, primarily by Growth Hormone (GH) and Insulin-like Growth Factor (IGF), although other hormones (e.g., insulin, thyroid hormones, somatostain) are also involved (Sheridan 2011). The most important factors affecting growth are the quality and quantity of food. Fish appetite is controlled by the hypothalamus of the brain and is affected by over 20 different hormones (Volkoff 2011). A wide variety of biotic and abiotic environmental factors can also affect both appetite and growth, including temperature, food availability, nutrient availability, light regime, oxygen, salinity, pollutants, current speed, predator density, intraspecific social interactions, gender, reproductive status, and genetics (see Sheridan 2011; Volkoff 2011; Wootton 2011a). Nutrients digested and absorbed are available for basic maintenance, growth, or reproduction (Wootton 2011b; also see Chapter 5). Stress can reduce growth by diverting energy that might have otherwise been used to build more body tissue (Moon 2011).

The multiple factors affecting growth may offset one another. For example, fish often grow better at warmer temperature, as long as it is within the temperature tolerance of the species. However, higher temperatures also reduce levels of dissolved oxygen in the water, which can limit growth. The synergistic effects of these multiple factors can create large variations in the size of fishes of the same and different ages, and lead to so-called size-structured populations, age differences in ecological roles (ontogenetic niche), and cannibalism (Beverton & Holt 1959; Beverton 1987; see Chapter 20). Careful consideration and monitoring of these multiple factors are critical for successful fish culture operations (Jobling 2011)

When plotted against age, growth curves for fishes tend to flatten out at older ages, although the degree of flatness varies greatly among and within species (Fig. 9.19). This variation forms the basis of an equation commonly used to describe individual growth in most fishes, known as the **von Bertalanffy growth equation**, which in its simplest form can be written:

$$L_t = L_{max}\,(1 - e^{-gt}),$$

where L is length, t is a point in time, L_{max} is the maximum length attained by the species, e is the base of natural logarithms, and g is the all-important constant that describes the rate at which growth slows. The von Bertalanffy equation is based on bioenergetic considerations, viewing growth as a result of anabolic and catabolic processes by which a fish takes in oxygen and energy to build tissues, and uses up energy and tissue over its life. Many refinements of the equation have been made and alternatives proposed that take into account age- and weight-specific differences in growth, food consumption rates, temperature, and overall energy budgets (see Wootton 2011a).

The von Bertalanffy growth coefficient (K) can be useful in assessing fishery management approaches because slower growing fishes with lower K values tend to be more vulnerable

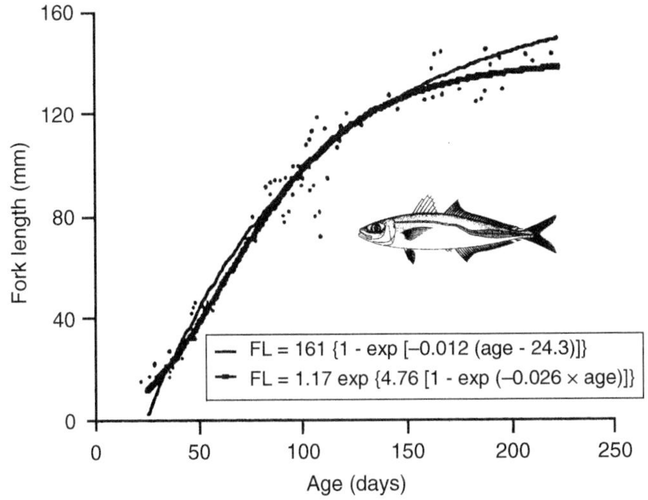

FIGURE 9.19 Growth curves and their statistical description. The plotted lines indicate growth over time for the Round Scad, *Decapterus punctatus*. The thin line and the upper equation are calculated from the von Bertalanffy equation; the thicker line and lower equation are based on a related calculation, the Gompertz equation. The von Bertalanffy equation predicts asymptotic growth; the Gompertz equation predicts a sigmoidal curve where growth increases and then decreases. The two lines are statistically similar, showing how growth slows with age and eventually approaches an asymptote. From Hales (1987), used with permission of the author; Round Scad drawing from Gilligan (1989).

to overfishing. Some typical values of *K* for a variety of fishes are as follows (data courtesy of J. Musick from various sources):

Anchovies, Engraulidae:	0.80–1.40
Tunas, Scombridae:	0.42
Menhaden, Clupeidae:	0.39
Flounder, Paralichthyidae:	0.32–0.40
Spanish mackerels, Scombridae:	0.17–0.35
Epinepheline grouper, Serranidae:	0.18
Porgy, Pagridae:	0.09
Swordfish, Xiphiidae:	0.09–0.19
Ground sharks, Carcharhinidae:	0.04–0.078

The relationship between the increasing mass and length of a fish involves a power function. Mass increases as a function of the cube of the length of the fish, reflecting that volume increases faster than surface area as overall size increases. Hence the equation for the relationship between mass and length is typically $M = aL^b$, where M is mass, L is length, and a and b are constants. The length exponent, b, is around 3.0 when the fish is growing isometrically, i.e., that its relative shape is remaining constant as it grows. Values greater or lesser than 3 indicate positive or negative allometric growth (see next section), and can serve as an indication of the relative health or condition of the fish.

The **condition factor, *K*,** (not the same as the von Bertalanffy growth coefficient, *K*) of a fish reflects the relationship between length and weight: $K = W/L^3$, where W is weight or mass and L is length. Population or cohort measures of *K* can indicate whether populations or subgroups are growing or feeding at expected rates. Changes in an individual's condition factor could indicate periods of good versus poor feeding success, disease, or imminent spawning. *K* is obviously a simplified indicator of general condition and lags far behind any actual events causing changes in relative condition. Other indices have also been proposed and debated (Ricker 1975, 1979; Anderson & Gutreuter 1984; Cone 1990; Wootton 1999). Richter et al. (2000) suggest a modified approach that also uses height in addition to length and weight. Froese (2006) provides a historical review, meta-analysis, and recommendations for condition factors and other applications of weight-length relationships in fishes.

Because the condition factor tells about an individual's history rather than its recent experience, measures that minimize the time lag between cause and effect have been developed, including biochemical analysis of protein uptake rate, energy content, or intermediary metabolism (RNA : DNA and ADP : ATP ratios), lipid content, and various chemical and biomarker indicators of stress (Busacker et al. 1990; Wedemeyer et al. 1990; Morgan & Iwama 1997; Schreck 2000).

Body size, Scaling, and Allometry

As emphasized repeatedly in this book, body size has an overriding influence on most aspects of fish biology. Some fish can grow from a larva a few millimeters long to an adult several meters long. An individual must perform all life functions at all sizes in order to reach the next stage; hence size-related phenomena are subject to constant selection pressures. Central to discussions of size are the concepts of **scale** and **allometry**, the latter topic forming the basis of a quantitative science of size (Gould 1966; Schmidt-Nielsen 1983; Calder 1984). **Scaling** refers to the structural and functional consequences of differences in size among organisms; allometry quantifies size differences among structures and organisms.

Changes in scale, whether over ontogenetic or evolutionary time, involve alterations in the dimensions, materials, and design of structures. A good example of scaling and its ramifications involves how an increase in body size affects the swimming speed and ability of large and small members of a species. The pelagic larvae of many marine fishes are small, elongate, and highly flexible, whereas adults take on a variety of shapes and swimming modes (see Chapter 4, Locomotion: Movement and Shape). The larvae of many herrings are almost eel-like and swim slowly, but adults have much deeper bodies and swim faster via the carangiform mode, in which the tail is the primary propulsive region. An increase in overall body mass, a **dimensional change**, requires the reworking of components. The internal skeleton changes from cartilage to bone, a **material change**. This corresponds to an increase in body musculature and a shift from anguilliform to carangiform swimming to take advantage of the stiffer nature of bone and the more efficient transfer of energy from contracting muscles to the propulsive tail.

This shift also corresponds to a **design change** from elongate with a rounded tail to a deeper, streamlined body with a forked tail, which is a more efficient morphology for a carangiform swimmer.

Allometry as a concept underscores a basic fact of growth and scaling, namely that the change in quantitative relationship between the sizes and functions of growing body parts is seldom linear. A doubling of the size of a fish will not necessarily lead to a doubling of its swimming speed. The relationship is more complex and depends on the measure of body size in question. For salmon, swimming speed increases approximately with the square root of the fish's length and with the 1/5th power of its mass (i.e., $length^{0.5}$, $mass^{0.2}$). Allometric relationships are described by equations of the nature $y = ax^b$ or $\log y = \log a + b \log x$. The exponent b describes the slope of the line that results when the relationship between the structures is plotted on log-log paper. For simple, linear proportionalities, $b = 1$, which is biologically rare. More often, b will take on positive or negative values for regression slopes greater or less than 1, respectively, indicating that a structure is increasing in size faster or slower than the increase in the trait to which it is being compared. The equations for swimming as a function of body size in Sockeye Salmon have exponents of 0.5 for body length and 0.17 for body mass (Schmidt-Nielsen 1983).

Numerous examples of allometric relationships in fishes can be given, emphasizing the far-reaching implications of size in fishes as well as convergence in selection pressures and solutions among disparate taxa. Focusing on locomotion and activity, the relative cost of swimming decreases with body size in most fishes, both within and among species. Such a relation indicates that it is more expensive for a small fish to move 1 g of body mass a given distance than it is for a larger fish to do the same (measured as oxygen consumed/g body mass/km, $b = -0.3$). Heart size in fishes increases with body size in an almost linear fashion, taking on values of about 0.2% of body mass and having a slightly positive exponent (heart mass = $0.002 \times$ body $mass^{1.03}$).

Not surprisingly, surface area of the gills relates to activity level. For example, active tunas have comparatively more gill surface than less active toadfishes. But within species and even among species, the surface area of the gills (m^2) increases allometrically and positively with body size (kg), with an exponent of 0.8–0.9. Locomotion and respiration relate to feeding activity, which is eventually translated into growth. Gut length increases allometrically with body length in many species, with an exponent of >1. Growth rate also scales with size, being faster in larger species, with an exponent of 0.61 (measured as change in mass/day relative to adult body mass) (Schmidt-Nielsen 1983; Calder 1984; Wootton 1999).

Questions about size, scaling, and allometry are often linked to the idea of **trade-offs**, another recurrent theme in this book. What constraints are imposed on an animal by changing its size, both ontogenetically and evolutionarily? What are the advantages and disadvantages of being very small as opposed to being very large? Large size may confer many advantages, but an individual must be small before it is large (except in some live-bearing species). During growth, an individual must incur the costs of small size early in ontogeny as well as the energetic and efficiency costs of reworking its size and shape during growth. Juveniles of a large species are often inferior competitors to adults of a small species. Rapid growth requires rapid feeding and high metabolic rate, which exposes a young fish to more predators and also often carries an increased risk of starvation. Size-related constraints also influence life-history attributes such as whether a species will produce many small or fewer large young, how extensive the parental care will be, and whether adults will mature quickly at a small size or slowly at a larger size.

Another relevant factor regarding size is that water is a dense medium, and overcoming drag has been a significant selective force in fish evolution. The shapes of fishes then become explainable in terms of **drag reduction** and which area of the body is used in propulsion. Both are intimately related to the mode of locomotion used. An important size-related attribute is the **Reynolds number**, which accounts for the size of an object, its speed, and the viscosity and density of the fluid through which it moves. Calculations of Reynolds numbers help explain swimming speed, body shape, and locomotory type. For very small fishes, including larvae, Reynolds number is small so the effects of drag overwhelm inertia. Larvae seldom glide because their mass relative to water viscosity prevents them from developing inertia as they swim. They must continue to expend effort to gain any forward progress. However, their challenges associated with overcoming inertia also mean that small fishes are less likely to sink. Large tunas, billfishes, and sharks have high Reynolds numbers and can use inertia to literally soar through the water, using their momentum to carry them forward.

The Ontogeny and Evolution of Growth

Much of the emphasis in this chapter has been on size relationships and the observation that indeterminate growth interacts intimately with many crucial aspects of fish biology. Growth processes – both general processes associated with length and mass increase but also in terms of changing body proportions – help explain many life history, behavioral, ecological, and physiological phenomena. We end this chapter by returning to the general question of how evolution has interacted with body growth processes to establish differences among life-history stages and species of fishes.

Ontogenetic Differences Within Species

Throughout the earlier discussions, we have emphasized anatomical and ecological differences among size classes of a species. Ontogenetic differences are detailed in several other chapters, such as the tendency for larger fish to occur deeper in a habitat, for populations to show age structure (see Chapter 20), for different size fish to interact with different predators and prey (see Chapter 16), for shoals to be sorted by size (see Chapter 17), and for different size fish to have different foraging capabilities (see Chapter 3). The major point here is that larvae, juveniles, and adults differ in habitat and ecology and must function both during definable stages as well as during transitional periods.

Adaptations appropriate to one stage may create **constraints** for other stages. Young fish may be constrained by structures that are adaptive later in life. For example, small juvenile Largemouth Bass are morphologically miniature adults. Instead of feeding on fishes, for which their morphology is well suited, they eat smaller zooplankton. This puts them in direct competition with juvenile and adult Bluegill, which are adapted to feed on zooplankton throughout their lives and hence have a competitive advantage over juvenile Largemouth Bass (Werner & Gilliam 1984). Conversely, later stages may retain characteristics of early ontogeny that may create impediments (see next section). Regardless, the differing selection pressures on larval, juvenile, and adult fish within a species help explain the general occurrence of individuals that appear and behave differently.

An additional conflict exists during ontogeny, brought about by the need for each stage to be immediately functional at a variety of tasks, including feeding, locomotion, and predator avoidance. All tasks are important, but the balance shifts as a fish ages. Hence predator avoidance may take precedence over feeding efficiency among younger, smaller fishes that are more vulnerable to predators. Such a trade-off has been shown in a range of species (e.g., salmonids, sculpins, cichlids) with respect to muscular and skeletal development and action. Juveniles exhibit relatively high levels of performance of locomotory and other defensive traits (e.g., fast-start escape responses) relative to their feeding and foraging abilities. The opposite applies to adults of the same species, in which feeding performance is maximized (Herrel & Gibb 2006).

The life history of a fish appears as a continuum of events from birth through maturation to death, with each phase preparing the fish for the next. However, sometimes adaptation to one phase inhibits progression into the next. For example, smolting and maturation in salmonids appear to be conflicting processes. Atlantic Salmon that smolt rapidly at 1 year of age may mature much later than fish that bypass the smolt stage and remain behind in freshwater. Administration of male hormones to young male Masu Salmon, *Oncorhynchus masou*, inhibits smoltification but causes maturation; castration of older fish causes them to undergo many of the transformations of smolting. The complexity, timing, and changes in habitat that occur during an animal's life cycle may function not only to prepare an individual for later phases but to also overcome the inhibitory or conflicting influences of previous phases (Thorpe 1978).

Evolution via Adjustments in Development: Heterochrony

Adjustments in developmental rates or timing may be a major pathway for speciation and even the novel morphological features that distinguish higher taxa (Cohen 1984; Mabee 1993). Such a process may explain several phenomena, such as why some adult fish have apparent larval or juvenile traits, why larval- or juvenile-appearing fishes are reproductively functional, or why closely related species may differ primarily in the duration of an early life-history stage, at the time when particular structures change or in the rate at which different structures grow.

Such alterations in the time of appearance and the rate of development of characters during ontogeny are referred to as **heterochrony**; they result from modification of regulatory genes and processes. Two general patterns of heterochrony are **paedomorphosis** ("child form"), which refers to the retention of juvenile traits in an adult, and **neoteny,** which refers to a juvenile form becoming sexually mature (Gould 1977; Youson 1988).

Within fish families, differing forms of heterochrony occurring at different stages in life history may have produced new species, as is suspected for gobioid fishes (Kon & Yoshino 2002b). Regardless, distinguishing among possibilities is not critical to appreciating heterochrony as a major evolutionary process (Fig. 9.20).

Heterochronic changes in transitions between developmental stages, such as the timing of metamorphosis from embryo to larva or from larva to juvenile, is one means by which new species evolve (Youson 1988). Variation in duration of larval life affects age or length at metamorphosis. Elopomorphs as a group are characterized by unique leptocephalus larvae that remain as larvae for long periods, up to 3 years in European eels (e.g., Miller & Tsukamoto 2004). In cladistic terms, this synapomorphy defines the group. Long larval life may be related to the apparently unique ability of leptocephali to absorb nutrients from the water across a very thin epithelium (Pfeiler 1986). Within the elopomorphs, further variations in developmental

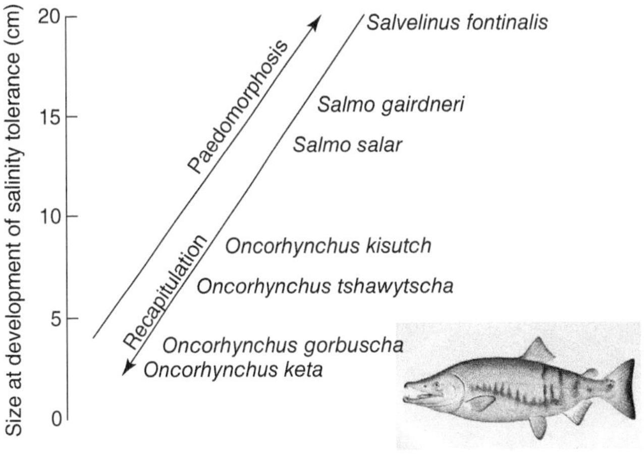

FIGURE 9.20 Salmonids differ in the minimum size at which they develop the necessary salinity tolerance to undergo the parr–smolt transformation. Observed differences among species could be explained by heterochronic shifts in the development of this trait. Such shifts might have changed the timing of various physiological processes involved in salinity tolerance. The ancestral condition is unknown, and so either acceleration or deceleration of timing could be responsible. Hence either paedomorphosis (increasing size at smoltification) or recapitulation (decreasing size at smoltification), or both, could have affected the evolution of this trait. From McCormick and Saunders (1987), used with permission. Color image of Chum Salmon, *Oncorhynchus keta*, from *The Fishes of Alaska*, Public Domain.

rate characterize distinctive species and may suggest processes that led to their separate evolution. For example, Tarpon, *Megalops atlanticus*, metamorphose at earlier ages and smaller lengths (2–3 months, 30 mm) than most other elopomorphs.

Speciation may result from or be maintained by heterochronic shifts in larval characteristics. Variation in larval period length in North Atlantic eels, *Anguilla rostrata* and *A. anguilla*, may have prompted the divergence of these closely-related species. Both species spawn at the same time and in the same Sargasso Sea locale (see Chapter 18, Catadromy) and larvae are transported by the Gulf Stream. American Eel (*A. rostrata*) leptocephali transform into juveniles and settle all along the Atlantic coast of North America, whereas European Eel (*A. anguilla*) larvae are carried past North America and on to Europe. Hence, the different larval durations, less than 1 year in American eels and 2–3 years in European eels, have a critical effect on their distribution and evolution.

The neotenic South American freshwater needlefish *Belonion* retains its juvenile "halfbeak" mouth structure because the upper jaw does not continue to grow as in other needlefishes.

The evolution of many lamprey species may have also occurred via heterochronic shifts. Many nonparasitic lamprey species are closely related to ancestral, parasitic forms. The major differences between ancestor and descendant species involve the length of larval versus adult life, with derived, nonparasitic forms typically having much longer larval periods, rapid metamorphosis, and a short, nonfeeding adult reproductive phase (see Fig. 13.6). A delay in the time of metamorphosis would result in just such a difference, essentially creating sexually mature larval forms that skip the parasitic feeding phase (Youson 1988; Finch 1990). For example, the parasitic Silver Lamprey (*Ichthyomyzon unicuspis*) and the nonparasitic Northern Brook Lamprey (*I. fossor*) are genetically extremely similar but have different life histories (Ren et al. 2014). The question of whether they are two species or different ecotypes of the same species that may have resulted from neoteny is a relevant management consideration when chemically treating streams to kill larvae of parasitic Sea Lamprey (*Petromyzon marinus*), which often co-occurs with these other lampreys. The Northern Brook Lamprey is rare and has protected status in some regions, which could be lost if it loses species status and is considered a variant of the more common Silver Lamprey. Should uncommon ecotypes be protected even if they are genetically very similar to more common ecotypes? Should we try to protect what appears to be speciation in progress?

Miniaturization among fishes may often evolve via paedomorphic processes (Weitzman & Vari 1988). Two of the smallest fish species known, a goby, *Trimmatom nanus*, and a cyprinid, *Danionella translucida*, reach sexual maturity at 10 mm length. They retain such larval features as incomplete scale formation, limited pigmentation, and partial ossification of the skeleton (Winterbottom & Emery 1981; Roberts 1986; Noakes & Godin 1988). One family, the subtropical and tropical Pacific Schindleriidae, has many neotenic characters, including a functioning pronephros (the early embryonic, segmented kidney of fishes that is drained by the archenephric duct rather than by the ureter), a transparent body, and large opercular gills. *Schindleria brevipinguis* matures at less than 8 mm, and *S. praematura* attains sexual maturity when only 1 cm long, even before it transforms completely from a planktonic "larva" (Leis 1991; Johnson & Brothers 1993; Watson & Walker 2004; see Chapter 15, Suborder Gobioidei). Adults of what may be the world's smallest vertebrate species, the 7.8 mm Southeast Asian cyprinid *Paedocypris progenetica*, possess a number of larval traits, including a long caudal peduncle with a skin fold along its lower edge, many unossified bones, a translucent body, and a neurocranium lacking several bones (Kottelat et al. 2006).

Characteristics of the deep-water ceratioid anglerfishes indicate that they evolved from shallow-water, benthic species via neoteny that involved an extended pelagic larval or juvenile phase. Many ceratioids have a gelatinous balloonlike skin as adults, which is also a larval trait. Mature males are distinctly larval in appearance, small, and are parasitic on females 10 times their size (Pietsch 2005). Larval-like males also occur in the deep-sea black dragonfishes and in the goby genus *Crystallogobius* (Moser 1981). Although large size confers an advantage in many species, the production of new species via heterochronic shifts to reduced body size shows that smallness is also advantageous under certain conditions.

Supplementary Reading

Diana JS. 2003. *Biology and ecology of fishes*, 2nd edn. Carmel, IN: Cooper Publishing Group.

Farrell AP, Stevens ED, Cech JJ, Richards JG. 2011. *Encyclopedia of fish physiology: from genome to environment*. Amsterdam, Boston: Elsevier/Academic Press.

Finch CE. 1990. *Longevity, senescence, and the genome*. Chicago: University of Chicago Press.

Gould SJ. 1977. *Ontogeny and phylogeny*. Cambridge, MA: Belknap Press.

Hoar WS, Randall DJ, eds. 1988. *The physiology of developing fish, Part B. Viviparity and posthatching juveniles. Fish physiology*, Vol. 11. San Diego: Academic Press.

Jobling M. 1995. *Environmental biology of fishes*. Fish and Fisheries Series No. 16. London: Chapman & Hall.

Miller MJ, Tsukamoto K. 2004. *An introduction to leptocephali: biology and identification*. Tokyo: Ocean Research Institute, University of Tokyo.

Moberg GP, Mench JA. 2000. *The biology of animal stress*. Wallingford, UK: CABI Publishing.

Pitcher TJ, Hart PJB. 1982. *Fisheries ecology*. London: Croom Helm.

Ricker WE. 1975. Computation and interpretation of biological statistics of fish populations. *Bull Fish Res Board Can* 191:1–382.

Weatherly AH, Gill HS. 1987. *The biology of fish growth*. London: Academic Press.

CHAPTER **10**

Special Habitats and Special Adaptations

Summary

Chapters 3 through 9 presented a foundation for understanding the basics of fish biology. But fishes inhabit a wide variety of aquatic habitats, including some that present considerable challenges and therefore require substantial adaptations. In this chapter, we will explore more closely some of the evolutionary adaptations that allow fishes to occupy almost all naturally occurring aquatic ecosystems that have any degree of permanence or at least predictability. Adaptations to extreme conditions help demonstrate why there are so many species of fishes, and also why fishes are such an incredibly fascinating group of animals to study.

The **Principle of Convergence** states that animals that evolve under similar selection pressures are likely to evolve similar adaptations. Strong selection pressures tend to produce strong similarities in unrelated animals. We will see multiple examples of this in several aquatic habitats that are special or extreme in some way, including deep sea, open waters, intertidal zones, areas of high turbulence, waters with extreme pH or salinity, polar regions, deserts, and caves.

It is estimated that between 10 and 15% of all fish species occupy deep ocean habitats – some in the open waters of the mesopelagic zone (between 200 and 1000 m), others in the bathypelagic (1000–4000 m), and still others even deeper. Although many of these fishes look somewhat similar, they have evolved independently from over 20 different orders

The Diversity of Fishes: Biology, Evolution and Ecology, Third Edition. Douglas E. Facey, Brian W. Bowen, Bruce B. Collette, and Gene S. Helfman.

Companion website: www.wiley.com/go/facey/diversityfishes3

of fishes. Convergence of deep-sea fishes has been largely in response to the combined selective forces of high pressure, low temperature, vast space, limited light, and little food available. Mesopelagic fishes are typically dark in color, use bioluminescence to attract prey or to hide their silhouette from predators, have large mouths, slender teeth, reduced skeletons and scales, low metabolism and enzyme activity, and gas bladders with long retia mirabilia to operate under high pressure. Their eyes are also more sensitive to shorter wavelengths of light, which penetrate to a greater depth. Many exhibit daily vertical migrations to feed at night in warmer waters closer to the surface where food is more abundant, and then return to greater depths at lower temperatures to conserve energy during the day. Bathypelagic fishes show stronger and more bizarre convergences, many of which are apparent adaptations to low energy availability. These include sex reversal, extreme skeletal and musculature reduction, eye loss, marked sexual dimorphism, behavioral energy conservation, and either very long gas-bladder retia for buoyancy regulation under very high pressure or loss of the gas bladder altogether and use of lipids for buoyancy.

Oceanic, pelagic fishes swim in the upper 100–200 m of water. This is the primary region for commercial fish production and is the habitat of herring-like fishes, sauries, carangoids, dolphinfishes, mackerels, tunas, and billfishes. Pelagic fishes are typically streamlined, silvery, and migratory, with a high proportion of red muscle for sustained swimming, and a deeply forked or lunate tail. Many of the larger marine pelagic species have fusiform bodies with a maximum girth about one-third of the way back from the head to minimize drag, have retractable fins, strongly lunate tails with a laterally keeled caudal peduncle, and some retain heat to keep their swimming muscles warm and more efficient. They respire efficiently and save energy by using ram-ventilation to move water across their gills. Life history differences between temperate and tropical species are influenced by seasonal and spatial food availability and lead to dramatic differences in year class fluctuations. Convergent evolution has led many freshwater pelagic species in large lakes to have some similar characteristics to those of pelagic marine species, such as body and tail shape and reflective scales, but no freshwater pelagic fishes are known to have the circulatory system adaptations needed to retain heat to keep their muscles warm.

Intertidal fishes inhabit a variety of coastal ecosystems such as estuaries, salt marshes, mudflats, and rocky shores. These are typically very productive habitats, but also physiologically challenging because temperature, dissolved oxygen, and salinity often change with tidal cycles. Fishes in these habitats must either move with the tides to stay in a relatively constant physiological environment or be able to make relatively rapid adjustments to the changing conditions. Some fishes inhabiting rocky intertidal areas remain in tidepools during low tide – but conditions can become warm, with decreasing dissolved oxygen, and also more saline due to evaporation. Other fishes may seek refuge in cool, damp spaces beneath rocks and algae, but they must continue to respire when not immersed in water. Fishes in estuaries and tidal marshes can be exposed to low oxygen and high levels of hydrogen sulfide due to decay of accumulated organic material. Fishes in coastal areas have the added challenge of being close to high levels of human activity and the resulting environmental impacts.

Fishes that inhabit high-energy zones such as the wave-swept intertidal zone or high-velocity rivers and stream beds have converged upon a small body with subterminal mouth, a body shape that is depressed dorso-ventrally, paired fins that are expanded, a reduced or absent gas bladder, and many have a ventral suction device adapted from fins. All these traits appear to facilitate the holding of position on the bottom despite strong water flow. The shape of the body helps water flow over the fish, and large pectoral fins may be spread to create downward pressure from the flowing water.

Oceans are relatively stable in salinity and pH, but some inland waters show considerable variation in these important environmental parameters that can limit the types of fishes that can live in those waters. For example, minnows (family Cyprinidae) are widespread in many North American freshwaters but are not often found in waters with a pH below about 4.5. Some Southeast Asia minnows, however, do fine at these somewhat low pH levels. Similarly, many fishes do not survive at pH above 8, but there are exceptions – some African fishes establish populations at pH 9 or above. Salinity is a strong determining factor in the distribution of most fishes. The physiological mechanisms of maintaining proper ion and water balance are rather specific and not easily changed for most fishes. Exceptions include some estuarine fishes that can make appropriate adjustments often and migratory diadromous species that undergo physiological metamorphic changes that allow them to move from freshwater to saltwater, or vice versa, at the appropriate stages in their life cycle. Human activities such as water withdrawal or diversion, or broader effects such as climate change influencing precipitation patterns, can have an impact on salinity.

The polar Arctic and Antarctic regions lie above 60° latitude. The Antarctic has been isolated much longer due to strong circumpolar currents and hence has more endemic, specialized fishes, half of which are in the icefish suborder Notothenioidei. These fishes are predominantly benthic, but some have evolved neutral buoyancy via reduced skeletal mineralization and increased lipid deposition. Antarctic fishes avoid freezing because their blood contains antifreeze compounds that prevent very small ice crystals from growing large enough to become biologically dangerous. One particular family of icefishes, the Channichthyidae, are unusually pale because they lack hemoglobin in their blood and myoglobin in their muscles. Absence of these oxygen-binding proteins is apparently due to a series of mutations, but the fishes survive because of their low metabolic needs and high oxygen solubility in such cold conditions. Other adaptations that have evolved since the loss of hemoglobin include highly vascularized fins, skin that lacks scales, and higher capillary densities in tissues, all of which help compensate for the low oxygen-carrying capacity of the blood. Arctic fishes are derived from different groups of fishes but have converged on some similar traits such as antifreeze compounds in the blood. The antifreeze compounds of Arctic and Antarctic fishes function the same, but they are different and have evolved independently at least several times. The long periods of daylight during Arctic summers affect behavior patterns of some fishes – those that are more nocturnal under shorter photoperiods tend to feed more actively when it is light. The prolonged twilight period expands foraging time for crepuscular foragers but also increases vulnerability to predation.

Desert freshwater fishes live on almost all continents in regions where water scarcity creates extreme conditions. Fishes that can survive dry periods often possess accessory respiratory structures for using atmospheric oxygen, such as skin or modified gills. Some fish populations survive drought periods via resting stages in their life cycle. Desert pupfishes complete their life cycle in seasonal pools and lay eggs that can remain in diapause until the following rainy season. The African lungfish burrows into the mud as water levels drop during the dry season and can estivate in a mucus-lined cocoon for several years. In addition to low oxygen, desert fishes often encounter extremes of salinity and alkalinity as water evaporates. The deserts of the southwestern US and western Mexico have a surprising diversity of endemic fishes, many of which are threatened.

Cave fishes live in lightless freshwater environments where food is scarce. Cave-adapted forms typically have low metabolic activity; reduced eyes, pigmentation, and scales; and increased chemosensory and lateral line development. Population densities and reproductive rates are low, often with a small percent of adults reproducing in a year. Those that do breed often have few, large eggs and exhibit parental care of eggs and juveniles; some are live-bearing. Their biology makes them especially vulnerable to habitat disturbances. Cave-dwelling fishes have converged on many of the traits evolved by deep-sea fishes, probably in response to food and light scarcity.

The Deep Sea

The oceanic depths exceed 11,000 m, but elasmobranchs are absent below 4000 m, and the deepest known teleosts are observed at 8400 m. These depths are subdivided into several different habitats, defined by light penetration and hydrostatic pressure (Bernal 2011). The upper approximately 200 m is termed the **epipelagic** or **euphotic zone**. This is the region where the photosynthetic activity of phytoplankton exceeds the respiration of the plants and animals living there (i.e., where production>respiration). The euphotic zone is the energy source for the deeper waters (Marshall 1971; Wheeler 1975; Nelson 1994; Castro & Huber 1997; Neighbors & Wilson 2006), and has its own distinctive subset of fishes which are discussed later in this chapter. The deeper areas, where light becomes limited or does not penetrate, consists of the **mesopelagic** (200–1000 m), **bathypelagic** (1000–4000 m), and **abyssal** (4000–6000 m) zones; deep-sea regions below 6000 m are referred to as **hadal** (Fig. 10.1). Bottom-associated species are considered **benthal**; those that swim

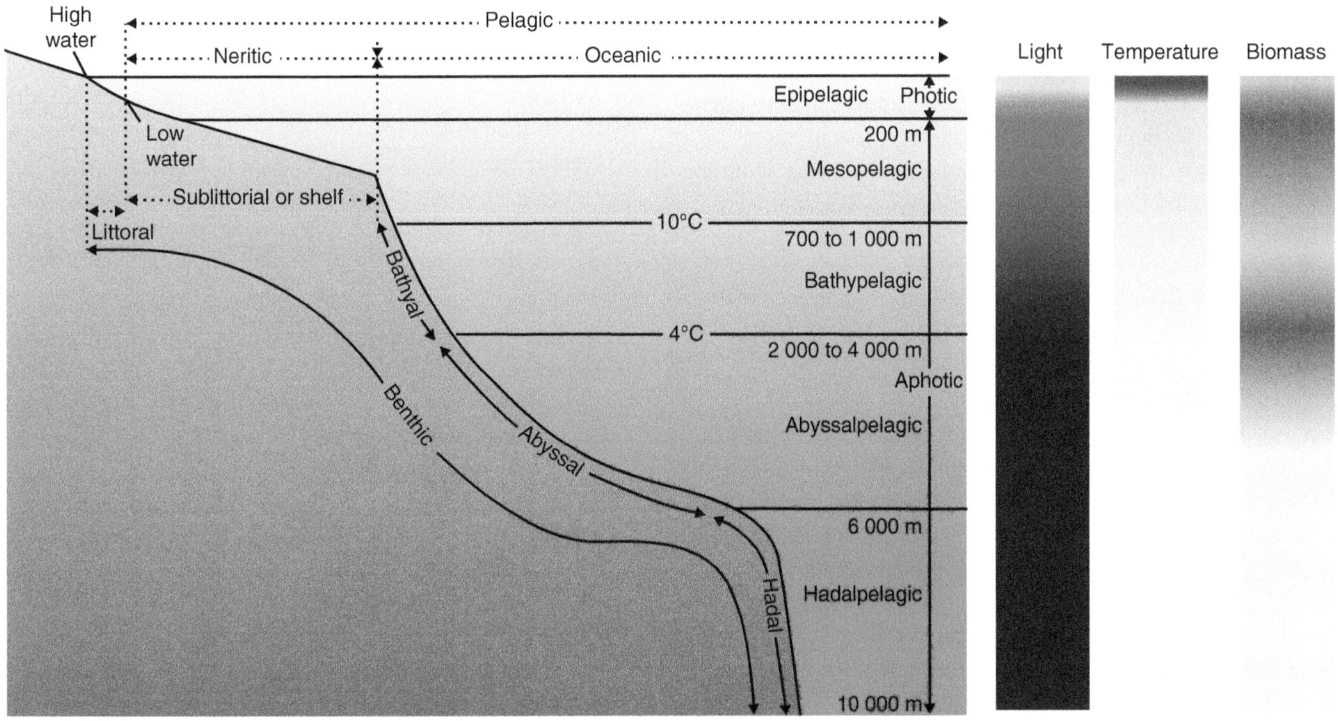

FIGURE 10.1 Regions and physical features of the deep-sea environment relative to depth. Available light and temperature decline with depth. Sinking plankton and organic matter are concentrated when they encounter colder, denser water. This attracts predators, resulting in zones of higher biomass in these thermal transition zones.

just above the bottom are **benthopelagic**, whereas those that live in contact with it along the upper continental slope at depths of less than 1000 m are **benthic**; benthal species in deeper regions are referred to as bathyal, abyssal, and hadal. The most diverse deep-sea fish assemblages occur between 40°N and 40°S latitudes, roughly between San Francisco and Melbourne, Australia in the Pacific Basin and between New York City and the Cape of Good Hope in the Atlantic Basin.

The deep-sea fishes of the mesopelagic and bathypelagic regions are readily recognized by just about anyone with a passing interest in fishes or marine biology. Deep-sea fishes often have light-emitting organs, termed **photophores**; large or long mouths studded with daggerlike teeth; chin barbels or dorsal fin rays modified as lures; reduced or eliminated bones; and eyes that are either greatly reduced or that are greatly enlarged and tubular to maximize the use of the little light available (Marshall 1954, 1971). Over 20 orders of fishes have species that have evolved for deep-sea existence, representing 10–15% of all fish species known (Farrell 2011d). More than 1000 species of fishes inhabit the open waters of the deep sea and another 1000 species are benthal, and there is good representation across orders of cartilaginous fishes and different superorders of bony fishes. Interestingly, the order Perciformes, which has shown great diversification in shallow and surface waters and contains more than one-third of all fish species, has few representatives in the deep sea (Farrell 2011d; also see Table 10.1). Deep sea fishes look alike because different ancestors invaded the deep sea from shallow regions and evolved similar anatomical and physiological solutions to the extreme environment. Understanding the convergent adaptations of deep-sea fishes requires that we first understand the physical environment of the deep sea and its influences on biota. Five characteristics of the deep sea appear to have been strong selective forces: (i) high pressure, (ii) low temperature, (iii) vast space, (iv) limited light, and (v) low energy (limited food) availability (Marshall 1971; Hochachka & Somero 1984; Farrell 2011d).

Factors Affecting Life in the Deep Sea

Pressure

Pressure increases constantly with depth at a rate of 1 atm/10 m of descent (1 atm = 1.03 kg/cm^2 or 14.7 lbs/in^2). Thus, pressure increases 20-fold, from 20 to 400 atm, between the top of the mesopelagic region at 200 m and the lower bathypelagic region at 4000 m. The deepest known fishes include the neobythitine cusk-eel, *Abyssobrotula galatheae*, collected at 8370 m in the Puerto Rico Trench, and a snailfish (family Liparidae) observed at 8143 m in the Mariana Trench (Nielsen 1977; Linley et al. 2016). At that depth, they experience pressures exceeding 800 atm, or *c.* 12,000 lbs/in^2.

Bathyal, abyssal, and hadal fishes have physiological adaptations to function under intense hydrostatic pressure, which affects the volume of water-containing compounds, the structure of proteins, and the rates of enzymatic reactions. Proteins in deep-sea fishes have numerous structural adaptations to function under high pressure (Somero 1992). The abundant osmolyte trimethylamine *N*-oxide (TMAO) is a chemical chaperone that stabilizes proteins under high water pressure. Concentrations of TMAO in teleost fishes increase with depth. However, the TMAO function has an upper limit at the point where concentrations would exceed isosmotic levels, switching the fish from hypoosmotic to hyperosmotic conditions. Salmon and other anadromous fishes can achieve this switch with a complex physiological reorganization, but deep-sea fishes cannot. The pressure limit for isosmotic TMAO corresponds to a depth of 8200 m, remarkably concordant with the deepest observations of cusk-eels and snailfish (Yancey et al. 2014). Hence the cusk-eel and snailfish mentioned above may be at the physiological limits of existence in the deep realm.

The tremendous pressures of the deep sea do not create as many problems for biological structures because fishes are made up primarily of water and dissolved minerals, which are relatively incompressible. The gas bladder is particularly affected because both volume relationships and gas solubility are sensitive to pressure. It is difficult to secrete gas into a gas-filled bladder under high pressure. Three gas bladder adaptations are known in deep-sea fishes to cope with the constraints of pressure:

1. The efficiency of gas secretion depends on the exchange surface of the capillaries of the **rete mirabile**, which increases the partial pressure of gases in the **gas gland**, the main gas-secreting organ (discussed in Chapter 5). Longer retia can build more gas pressure in the gas gland, therefore the length of the rete tends to increase with depth (Farrell 2011d). The retia of epipelagic fishes are usually less than 1 mm long, retia of upper mesopelagic fishes are 1–2 mm long, those of lower mesopelagic fishes are 3–7 mm long, and those of some bathypelagic fishes are 15–20 mm long.
2. Although mesopelagic fishes have large gas-filled bladders, most bathypelagic fishes have lost this structure. Flotation might therefore be a problem for these fishes, but their body musculature and skeletons are reduced as energy-saving mechanisms and they consequently approach neutral buoyancy. As long as a fish remains at relatively constant depths, it has minimal need for buoyancy control. However, many mesopelagic fishes undergo **diurnal vertical migrations.** These fishes have a greater need to adjust their buoyancy and have retained their gas bladders. Deep benthopelagic fishes are able to hover just above the bottom with minimal energy expenditure via a different mechanism – their gas bladders are filled with lipids that are relatively incompressible and less dense than seawater, thus providing flotation. Interestingly, the larvae of these fishes have gas-filled bladders, but these

TABLE 10.1 **Representative teleostean taxa from the three major deep-sea habitat types. Based on Marshall (1971, 1980); Wheeler (1975); Gage & Tyler (1991); Nelson (2006); Nelson et al. (2016). Also note that taxonomic groupings may change with additional evidence gathered over time.**

Mesopelagic (~750 spp.)
Cohort Elopomorpha
Notacanthiformes: Notacanthidae – spiny eels
Anguilliformes: Nemichthyidae – snipe eels; Synaphobranchidae – cutthroat eels
Cohort Otocephala
Superorder Alepocephali
Alepocephaliformes: Alepocephalidae – slickheads; Platytroctidae – tubeshoulders
Cohort Euteleostei
Superorder Protacanthopterygii
Argentiniformes: Microstomatidae – deepsea smelts; Opisthoproctidae – barreleyes
Stomiiformes: Gonostomatidae – bristlemouths; Sternoptychidae – hatchetfishes; Stomiidae – barbeled dragonfishes
Superorder Cyclosquamata
Aulopiformes: Evermannellidae – sabertooth fishes; Alepisauridae – lancetfishes; Paralepididae – barracudinas; Giganturidae – telescopefishes
Superorder Scopelomorpha
Myctophiformes: Neoscopelidae – blackchins; Myctophidae – lanternfishes
Superorder Paracanthopterygii
Stylephoriformes: Stylephoridae – tube-eyes
Superorder Acanthopterygii
Trachiniformes: Chiasmodontidae – swallowers
Scombriformes: Gempylidae – snake mackerels
Bathypelagic (~200 spp.)
Cohort Elopomorpha
Anguilliformes: Saccopharyngidae – swallower and gulpers; Eurypharyngidae – pelican eels; Nemichthyidae – snipe eels; Serrivomeridae – sawtooth eels
Cohort Otocephala
Superorder Alepocephali
Alepocephaliformes: Alepocephalidae – slickheads
Cohort Euteleostei
Superorder Protacanthopterygii
Stomiiformes: Gonostomatidae – bristlemouths
Superorder Paracanthopterygii
Gadiformes: Melanonidae – pelagic cods; Macrouridae – grenadiers and rattails
Superorder Acanthopterygii
Beryciformes: Melamphaidae – bigscale fishes; Stephanoberycidae – pricklefishes; Cetomimoidea – whalefishes
Trachichthyiformes: Anoplogastridae – fangtooths
Trachiniformes: Chiasmodontidae – swallowers
Ophidiiformes: Ophidiidae – cusk-eels; Bythitidae – viviparous brotulas
Lophiiformes: Ceratioidei – deep-sea anglerfishes, seadevils
Benthal[a] (~1000 benthopelagic and benthic spp.)
Cohort Elopomorpha
Notacanthiformes: Halosauridae – halosaurs; Notacanthidae – spiny eels
Anguilliformes: Synaphobranchidae – cutthroat eels
Cohort Euteleostei
Superorder Cyclosquamata
Aulopiformes: Synodontidae – lizardfishes; Chlorophthalmidae – greeneyes; Ipnopidae – spiderfishes and tripodfishes
Superorder Paracanthopterygii
Gadiformes: Macrouridae – grenadiers; Moridae – morid cods; Merlucciidae – merlucciid hakes
Superorder Acanthopterygii
Ophidiiformes: Ophidiidae – cusk-eels; Bythitidae – viviparous brotulas; Aphyonidae – aphyonids
Lophiiformes: Ogcocephalidae – batfishes
Scorpaeniformes: Liparidae – snailfishes
Perciformes: Zoarcidae – eel-pouts; Bathydraconidae – Antarctic dragonfishes; Caproidae – boarfishes

[a] Chimaeras and many squaloid sharks are benthopelagic. Most benthal fishes live above 1000 m, although some grenadiers and rattails live between 1000 and 4000 m, macruronid southern hakes live somewhat deeper, tripodfishlive to 6000 m, snailfishes to 8000 m, and neobythitine cusk-eels to 8000 m.

larvae, and the larvae of nearly all deep-sea fishes, are epipelagic, where the costs of gas secretion and buoyancy adjustment are much less. Benthopelagic squaloid sharks such as *Centroscymnus* and *Etmopterus* solve the buoyancy problem with livers that contain large quantities of the low-density lipid **squalene**. This organ can account for 25% of total body mass. Deep-water holocephalans also achieve neutral buoyancy via squalene and by reduced calcification of their cartilaginous skeletons (Bone et al. 1995).

3. Some deep-sea fishes belong to the less derived cohorts Elopomorpha and Otocephala, and the euteleostean superorders Protacanthopterygii and Paracanthopterygii – taxa which typically have a direct physostomous connection between the gas bladder and the gut. Deep-sea fishes are, however, "secondarily" physoclistous, having closed the pneumatic duct, thus preventing gas from escaping out of the mouth.

Temperature Water temperature at the surface changes daily and seasonally and is usually warmer than deeper waters. In the deep sea, however, temperature is a predictable function of depth. Water temperature declines with depth through the mesopelagic region across a permanent **thermocline** until the bathypelagic region, where temperature remains a relatively constant 2–5°C, depending on depth. The relatively rapid change in temperature with depth across the thermocline also affects water density, thereby creating a physical separation between warmer surface water and colder, denser deep water. This change in density concentrates organic matter sinking from the surface layers, creating a zone of high productivity and biodiversity, and often somewhat depleted oxygen due to decomposition. This layer also prevents a lot of organic matter from sinking further and providing energy to fishes below the thermocline.

Temperature is a strong predictor of distribution for different taxa of deep-sea fishes. Ceratioid anglerfishes and darkly colored species of the bristlemouths (*Cyclothone*) are restricted to the deeper region. Even within the mesopelagic zone, species sort out by temperature. Hatchetfishes, pale *Cyclothone*, and malacosteine loosejaws are restricted to the lower half at temperatures between 5 and 10°C, whereas lanternfishes and astronesthine and melanostomiatine stomiiforms occur in the upper half at 10–20°C. Latitudinal differences in temperature–depth relationships lead to distributional differences within species. Some species such as ceratioid anglers that are mesopelagic at high latitudes occur in bathypelagic waters at lower latitudes, a phenomenon known as **tropical submergence** that results from the warmer surface temperatures in the tropics.

Temperature is a minimal constraint on a fish that does not move vertically because the temperature remains constant at any given depth. However, vertically migrating mesopelagic species must swim through and function across a temperature range of as much as 20°C (see Fig. 10.1). Lanternfishes that migrate vertically have larger amounts of DNA per cell than do species that are non-migratory, perhaps allowing for multiple enzyme systems that function at the different temperatures encountered by the fishes (Ebeling et al. 1971).

Space The deep sea is an enormous area. Approximately 70% of the earth's surface is covered by ocean, and 90% of the surface of the ocean overlies water deeper than 1000 m; ocean deeper than 3000 m covers over 50% of Earth's surface. The bathypelagic region makes up 75% of the ocean and is therefore the most extensive habitat on earth. This large volume creates problems of finding food, conspecifics, and mates because bathypelagic fishes are never abundant. For example, female ceratioid anglerfishes are distributed at a density of about one per 800,000 m^3, which means a male anglerfish is searching for an object the size of a football in a space about the size of a large, totally darkened football stadium.

Deep-sea fishes show numerous adaptations that reflect the difficulties of finding potential mates that are widely distributed in a dark expanse. Unlike most shallow-water forms, many deep-sea fishes are **sexually dimorphic** in ways directly associated with mate localization. Mesopelagic fishes, such as lanternfishes and stomiiforms, have species-specific and sex-specific patterns of light organs, structures that first assure that individuals associate with the right species and then that the sexes can tell one another apart. Among benthopelagic taxa, such as macrourids, brotulids, and morids, males often have larger muscles attached to their gas bladders that are likely used to vibrate the bladder and produce sounds that can attract females from a considerable distance.

Some of the most bizarre sexual dimorphisms occur among bathypelagic species, where problems of mate localization are acute. The most speciose group of bathypelagic fishes is the ceratioid anglerfishes, of which there are 11 families and over 160 species (Bertelsen 1951; Pietsch 1976, 2005; Nelson 2006; Pietsch & Orr 2007). In several families, the males are dwarfed, reaching only 20–40 mm long, whereas females attain lengths 10 or more times that size, up to 1.2 m in one species. In five families, males attach temporarily to females, spawning occurs, and the males swim free (Pietsch 2005). In five other families, the males are entirely and permanently **parasitic** on the females, and males in these taxa may be as small as 6.2 mm, making them the smallest known sexually mature vertebrate (Fig. 10.2). Males attach most frequently to the ventral midline of the belly of the female, but may be attached on the sides, backs, head, and even the fishing lure of a female; as many as eight males have been found attached to a single female (including some species mismatches). In parasitic species, males attach by the mouth, his mouth tissue fuses

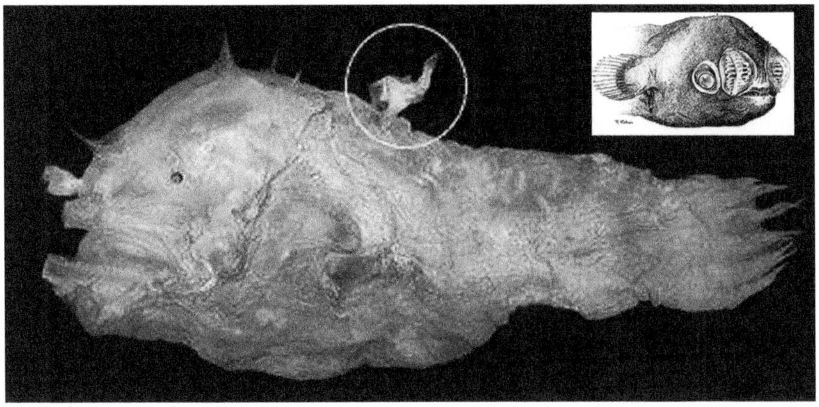

FIGURE 10.2 Size differences in male versus female anglerfishes. A 6.2 mm parasitic male *Photocorynus spiniceps* (Linophrynidae) (circled) attached to the dorsal area of a 46 mm female. Inset: a free-living, 18 mm male of *Linophryne arborifera* (Linophrynidae), showing the greatly enlarged eyes and olfactory lamellae apparently used in finding females. Courtesy of T. W. Pietsch.

with her skin, and he becomes parasitically dependent on her for nutrition. Many of his internal organs degenerate, with the exception of his testes, which can take up more than half of his body cavity. Females do not mature sexually until a male attaches to them (Pietsch 2005).

The premium placed on locating a mate is reflected throughout the anatomy and physiology of searching males. During this phase, males have highly lamellated olfactory organs and well-developed olfactory tracts, bulbs, and forebrains, whereas females have almost entirely degenerate olfactory systems. Males also have extensive red muscle fibers, the kind used for sustained swimming. Females have predominantly white muscle fibers, which usually function for short bursts of swimming. Males of some species possess enlarged, tubular eyes that are extremely sensitive to light, whereas females have small, relatively insensitive eyes. Males also have high lipid reserves in their livers, which they need because their jaw teeth become replaced by beaklike denticles that are useless for feeding but are apparently specialized for holding onto a female (the denticular jaws are derived embryologically from the same structures that in females develop into the fishing lure, discussed under Foraging Adaptations; Munk 2000). All this comparative evidence indicates that males are adapted for swimming over large expanses of ocean, searching for the luminescent glow and some olfactory cue emitted by females. Females in contrast are floating relatively passively, using their bioluminescent lures to attract prey at which they make sudden lunges, and releasing pheromones to attract males. Neither sex matures until the male attaches to the female, showing the coevolved nature of these traits.

The unrelated bathypelagic bristlemouths, which may be the most abundant vertebrates on earth, are quite similar, showing evolutionary convergence. Males are smaller than females, have a well-developed olfactory apparatus, extensive red muscle fibers, and larger livers and fat reserves. Although the males are not parasitic on the females, they are unusual in that they are **protandrous hermaphrodites**, maturing first as a male and later switching sex to female. Sex change theory predicts just such a switch because relative fitness favors being a male when small and a female when large (see Chapter 8, Sex and Gender Roles in Fishes). Cetomimid whalefishes – one of the few percomorph groups to occupy the bathypelagic region and second only to oneirodid anglerfishes in diversity there – have also converged on having dwarf males, although male whalefishes are not known to be parasitic on the larger females.

Light Below the euphotic zone, light is insufficient to support photosynthesis. Visible light to the human eye is extinguished by 200–800 m depth, even in the uniformly clear water of the mesopelagic and bathypelagic regions. Deep-sea fishes are 15–30 times more sensitive to light and can detect light down to between 700 and 1300 m, depending on water clarity. The mesopelagic region is often termed the **twilight zone**, whereas the bathypelagic region is continually dark. What little light that passes into the mesopelagic region has been differentially absorbed and scattered by water molecules and turbidity and is limited to relatively short, blue-green wavelengths around 470 nm.

Fishes of the permanently dark bathypelagic region and the low-light mesopelagic region have adaptations among both their eyes and photophores. Bathypelagic fishes, with the exception of male ceratioid anglerfishes, have greatly reduced eyes that probably function primarily for detecting nearby bioluminescence. Mesopelagic fishes have modifications to their eyes that generally increase their ability to capture what little ambient light is available, although different species appear to have emphasized capturing dim ambient spacelight versus brighter point sources from bioluminescence (Warrant & Locket 2004). Mesopelagic fishes have very large eyes, often measuring 50% of head length; most North American freshwater fishes have eye diameters that are only 10–20% of head length.

Mesopelagic fishes have only rods in their retinas, with visual pigments that are maximally sensitive at about 470 nm. This is a good match to the light environment at mesopelagic depths and also matches the light output from photophores, structures that are much more common among mesopelagic than bathypelagic fishes. In some fishes (dragonfishes), light is produced by a photophore or other intrinsic structure. In other fishes (ponyfish, flashlightfish) light is produced by bacteria symbionts. In all cases, it requires an oxidative reaction between the protein luciferin and an enzyme luciferase.

Bioluminescence has evolved independently at least 27 times in ray-finned fishes (Davis et al. 2016), in several superorders including over 45 families and about 190 genera of teleosts – as well as in dogfish sharks, squids, crustaceans, and other invertebrates.

Light organs are used in various ways, including prey attraction, prey illumination, identifying the species and sex, and hiding by obscuring a fish's silhouette (Mensinger 2011). Many mesopelagic bioluminescent fishes have ventral photophores that emit blue-green light (<515 nm) which would blend into background light that might penetrate from the surface. The light-producing organs may be a simple luminescent gland backed by black skin that emits on its own or contains bioluminescent bacteria. More complex circular photophores may be backed by the silvery reflective material with a lens through which light passes. In highly derived photophores, the lens may be pigmented and hence the light that is transmitted is of a different wavelength, as in the loosejaw dragonfishes which have a red filter over the subocular photophores and also have retinal reflectors and receptors sensitive to red wavelengths (e.g., Herring & Cope 2005). This unique combination of producing and detecting red light in a habitat where other species cannot detect this wavelength allows dragonfishes to communicate with each other without being detected by predators or prey. It could also serve to maximize illumination of mesopelagic crustaceans as prey (Lockett 1977; Denton et al. 1985; Sutton 2005). Photophores tend to flash on for 0.2–4 s, depending on the species. Different species of lanternfishes may have similar photophore patterns but different flash rates, suggesting a convergence in communication tactics between deep-sea fishes and fireflies (Mensinger & Case 1990).

Food

Limited light and huge volume mean that food is extremely scarce in most of the deep sea. Most marine food chains, except at **thermal vents**, originate in the euphotic zone, which makes up only 3% of the ocean. Food for bathypelagic fishes must therefore first pass through the filter of vertebrates, invertebrates, and bacteria in the mesopelagic zone; much of this food rains down weakly, unpredictably, and patchily in the form of carcasses, sinking sargassum weed, detritus, and feces. All deep-sea fishes are carnivorous, feeding either on zooplankton, larger invertebrates, or other fishes. Zooplankton biomass at the top of the bathypelagic is only about 1% of what it is at the surface, and densities of benthic invertebrates decrease with depth and distance from continental shores. High densities, diversities, and productivity of invertebrates at thermal vents on the deep-sea floor do not support a similar abundance or diversity of fishes. Only three species – bythitid brotula and two zoarcid eel-pouts – are endemic to and frequent vent areas (Grassle 1986; Cohen et al. 1990). A general scarcity of food in the deep sea puts a premium on both saving and obtaining energy.

Foraging Adaptations

Deep-sea fishes show a number of convergent foraging traits (Gartner et al. 1997). In general, zooplanktivores have small mouths and numerous, relatively fine gill rakers, whereas predators on larger animals have larger mouths and fewer, coarser gill rakers. **Daggerlike teeth** or some other form of long, sharp dentition is so characteristic of deep-sea forms that their family names often refer directly or indirectly to this trait, including such colorfully named groups as dragonfishes, daggertooths, bristlemouths, snaggletooths, viperfishes (Fig. 10.3), sabretooths, and fangtooths. Large, expandable mouths, hinged jaws, or distensible stomachs are also reflected in such names as gulpers, swallowers, and loosejaws. Saccopharyngoid gulper and swallower eels have enormous mouths that can expand to >10 times the volume of the animal's entire body, the largest mouth : body volume of any known vertebrate (Nielsen et al. 1989). Black dragonfishes, viperfishes, ceratioid anglerfishes, and sabertooth fishes can swallow prey larger than themselves (Fig. 10.3), as much as three times so in the case of the anglerfishes. Their swallowing abilities are increased because the pectoral girdle is disconnected from the skull, enlarging the intercleithral space of the throat (see Chapter 3, Pharyngeal Jaws). All of these anatomical specializations point to a strategy of taking advantage of any feeding opportunity that may come along, regardless of prey size.

A small number of shallow-water paracanthopterygian species, notably the goosefishes, frogfishes, batfishes, and anglerfishes, possess modified dorsal spines that are waved about to lure prey species within striking distance. Such lures reach their greatest and most diverse development among mesopelagic and bathypelagic fishes, where they occur on viperfishes, various dragonfishes, astronesthine snaggletooths, most ceratioid anglerfishes, and arguably as luminescent organs in the mouths of hatchetfishes, lanternfishes, and some anglerfishes and on the illuminated tail tip of the gulper eels. The typical anglerfish lure consists of an elongate dorsal spine, the **illicium**, tipped by an expanded structure called the **esca** (Fig. 10.4). Escae tend to have species-specific shapes, can regenerate if damaged, and are moved in a variety of motions that imitate the swimming of a small fish or shrimp (see Pietsch 1974).

Most mesopelagic fishes undertake evening migrations from the relatively unproductive mesopelagic region to the richer epipelagic zone to feed; they then return to the mesopelagic region at dawn. The migration to near the surface from as deep as 700 m, can take an hour or more, and may entail considerable energy expenditure. This movement is so characteristic of mesopelagic fishes, crustaceans, and mollusks that the community of organisms that migrates is referred to as the **deep scattering layer,** which is discernible by sonar signals that reflect off the gas bladders of the fishes. Hypotheses about the adaptiveness of the migration include: (i) a net energy gain from feeding in warm water and metabolizing in cold water; and (ii) exploiting surface currents that bring new food into the water column above the migrator. It is apparent that the migration serves a foraging purpose, given the 100-fold difference in plankton biomass between the two

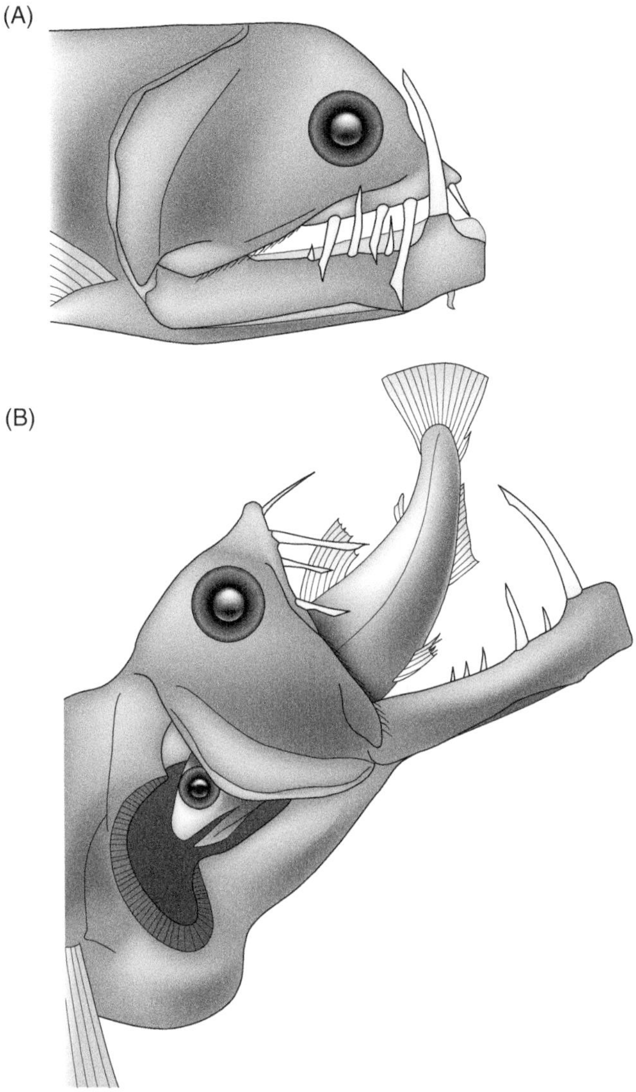

FIGURE 10.3 Extreme movements of the head and mouth during swallowing in the viperfish *Chauliodus sloani* (Stomiidae). (A) Mouth at rest, showing the premaxillary and mandibular teeth that sit outside the jaw when the mouth is closed. The maxillary and palatine teeth are small and slant backward. (B) Mouth opened maximally as prey is captured and impaled on the palatine teeth prior to swallowing. The anterior vertebrae and neurocranium are raised, the mandibuloquadrate joint at the back corner of the mouth is pushed forward, and the gill covers are pushed forward and separated from the gills and gill arches. The heart, ventral aorta, and branchial arteries are also displaced backward and downward. Such wide expansion of the mouth accommodates very large prey and is necessary for prey to pass between the large fangs. After Tchernavin (1953).

regions and also given that stomachs of migrators are empty in the evening before migration and full in the morning after migration. Not all fishes of the same genus, or even individuals of the same species, follow the same daily vertical migration pattern, however, as demonstrated in the dragonfishes (Stomiidae, Malacosteinae; Kenaley 2008). Vertical migration may be driven in part by hunger, with satiated individuals conserving energy by remaining at depth for a few days while digesting whereas their hungry conspecifics migrate to feed.

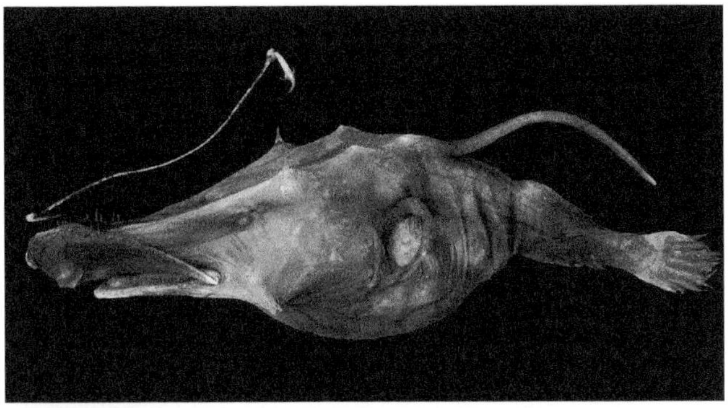

FIGURE 10.4 An adult female Wolftrap Angler, *Lasiognathus amphirhamphus* (Oneirodidae), about 15 cm long. The rodlike structure pointing tailward is the skin-covered caudal end of the dorsal spine that forms the illicium. The spine slides in a groove on the head, allowing the anglerfish to move it forward when fishing but to retract it otherwise. Courtesy of T. W. Pietsch.

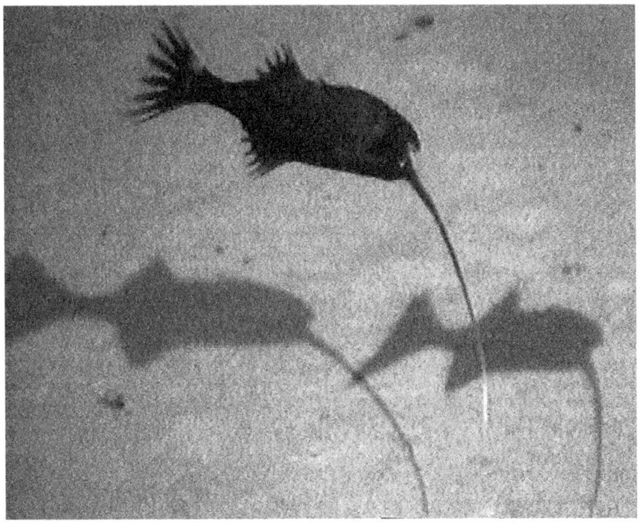

FIGURE 10.5 A 50 cm long whipnose anglerfishes presumably foraging just above the bottom at 5000 m depth. Its illicial lure is extended down toward the bottom (lower two profiles are shadows cast by photographic lights). Interestingly, in gigantactinids, the teeth of the lower jaw are elongated and curved, much like the upper jaw teeth of other anglerfishes, implying that upside-down foraging may be common in whipnose anglerfishes. Moore 2002 / with permission of American Society of Ichthyologists and Herpetologists.

The most extreme adaptations for opportunistic prey capture and energy conservation are found in the deepest regions of the bathypelagic zone. Fishes remain in place, perhaps because external cues of changing daylight are lacking or the energetic costs of migrating are too high. They instead attract prey with bioluminescent lures. Observations from submersibles suggest that bathypelagic forms adapt a "float-and-wait" foraging mode, hovering relatively motionless in the water column and making quick lunges at prey. This motionless hovering and luring even occurs when purportedly bathypelagic anglerfish forage near the bottom, as evidenced by fortuitous observations of a whipnose anglerfish, *Gigantactis*, swimming slowly upside-down just off the bottom, its illicium held stiffly in front in a slight downward-pointing arc (Moore 2002) (Fig. 10.5).

Energy Conservation Deep-sea fishes minimize their daily and long-term expenditure of calories in many ways. Biochemically, rates of enzymatic and metabolic activity and even levels of adenosine triphosphate (ATP) generating enzymes are lower in deep-sea fishes than in shallow-water relatives, which conserves energy used in locomotion, osmotic regulation, and protein synthesis (Somero et al. 1991). Energy savings are also accomplished via elimination or replacement of heavy components. Structurally, bathypelagic fishes are fragile compared with shallow-water, mesopelagic, and even deep-sea benthic fishes. Many of the heavy bony elements of shallow-water relatives have been eliminated. Pelvic fins are often missing or reduced to rudiments, bones of the head are reduced to thin strands, and many species are scaleless. Spines are rare among deep-sea fishes; even the few acanthopterygian groups that have managed to invade the deep sea, such as melamphaid bigscale fishes and chiasmodontid swallowers, have very feeble fin spines. Body musculature is also greatly reduced, by as much as 95% in the trunk and caudal regions compared with shallow-water forms.

Lacking trunk musculature, predator evasion becomes a problem. Most deep-sea fishes are colored in ways that should minimize their detection by potential predators. Mesopelagic fishes tend to be silvery or brown with ventral photophores that point downward. Silvery fishes effectively disappear in open water because they reflect the surrounding water. Ventral photophores may aid in breaking up the silhouette of the fish when viewed from below against the backdrop of weak downwelling light, as mentioned earlier. Bathypelagic fishes are generally dark brown or black, as would be expected where the background is black.

Additional energy savings are attained by replacing heavy structural components with less dense substances. Where glycerol lipids occur in shallow-water fishes, deep-sea forms have less dense wax esters. These structural changes save energy because metabolic costs of both construction and maintenance are reduced. In addition, elimination and replacement of heavy structures reduce the mass of the fish, making it closer to neutral buoyancy and eliminating costs associated with fighting gravity.

The substitution of wax esters for lipids in some fishes can have an unwelcome effect on an unwary consumer, however. For example, Escolar (*Lepidocybium flavobrunneum*, family Gempylidae) is widely distributed in tropical and temperate deep waters (200–885 m). It accumulates a high body content of wax esters from its diet. This provides some buoyancy and also gives the meat a smooth, buttery texture, and rich flavor. Escolar is sometimes sold as "butterfish" or "white tuna," and thus potentially confused with Albacore, *Thunnus alalunga*, which is sold as canned "white" tuna. But be warned – the wax esters in Escolar are not digestible, unlike triglycerides found in many other oily fishes. Consuming fish with wax esters can result in oily, orange diarrhea (keriorrhea – Greek for "flow of wax"), an undoubtedly most unpleasant experience.

Bathypelagic fishes as a group tend to have **free neuromasts** in their lateral lines, rather than having lateral line organs contained in protective canals, as in mesopelagic and benthic groups. Free neuromasts in shallow-water fishes, such as goosefishes, cavefishes, and many gobies, are usually associated with a very sedentary lifestyle, again suggesting a premium on energy-conserving tactics and an ability to detect minor water disturbances among bathypelagic species.

Convergence in the Deep Sea

The deep sea offers numerous striking examples of the Principle of Convergence. Benthopelagic fishes from at least 12 different families have evolved an eel-like body that tapers to a pointed tail, often involving the fusion of elongated dorsal and anal fins with the tail fin (Gage & Tyler 1991). Another aspect of convergence exemplified in the deep sea is that selection pressures can override phylogenetic patterns, producing closely related fishes that are biologically very different because they live in different habitats where they converge on the traits more appropriate to their habitat (Marshall 1971). *Gonostoma denudatum* and *Sigmops bathyphilus* (formerly *G. bathyphilum*) are Atlantic bristlemouths in the family Gonostomatidae. *G. denudatum* is a mesopelagic fish, whereas *S. bathyphilus*, as its name implies, is a bathypelagic species. *G. denudatum* is silvery in color and has prominent photophores, well-developed olfactory and optic organs and body musculature, a well-ossified skeleton, a large gas bladder, large gill surface per unit weight, large kidneys, and well-developed brain regions associated with these various structures. *S. bathyphilus*, in contrast, is black, has small photophores and small eyes, small olfactory organs (except in males), weak lateral muscles, a poorly ossified skeleton, no gas bladder, small gills and kidneys, and smaller brain regions. Similar comparisons can be drawn between other mesopelagic and bathypelagic gonostomatids, and between mesopelagic and bathypelagic fishes in general. Even bathypelagic forms derived from benthopelagic lineages, such as the macrourids and brotulids, have converged on bathypelagic traits (Marshall 1971).

The extreme demands of the deep-sea habitat have also led to convergence in non-teleostean lineages. The mesopelagic cookie-cutter sharks, *Isistius* spp., have a high squalene content in their livers that increases buoyancy. They also possess photophores and migrate vertically with the biota of the deep scattering layer. The diverse nature of bioluminescence, some fish producing their own light and others using symbiotic bacteria, is in itself a remarkable convergence. Deep-sea sharks and holocephalans also possess visual pigments that absorb light maximally at the wavelengths that penetrate to mesopelagic depths, as is also the case for another mesopelagic non-teleost, the Coelacanth, *Latimeria chalumnae*. Deep-sea crustaceans and mollusks have also evolved anatomical and physiological traits similar to those of fishes, including the emission of luminous ink (e.g., platytroctids, ceratioids, squids) (Marshall 1980; Hochachka & Somero 1984).

The adaptive morphologies of deep-sea fishes are especially remarkable because they may have evolved relatively recently. Anoxic conditions during the Cretaceous (~94 million years ago) made the waters below 1000 m uninhabitable for fishes. There have been at least six oceanic anoxic events during the 400 million year history of fishes, so the evolution of deep-sea fish fauna may be a balancing act between extinction (or extirpation) and recolonization (Priede & Froese 2013).

The Open Sea

The epipelagic region is technically the upper 200 m of the ocean off the continental shelves (see Fig. 10.1), but the terms epipelagic and pelagic are often used synonymously to describe fishes that swim in the upper 100–200 m of coastal and open sea areas (Bernal 2011). Pelagic fishes can be further divided into 12 subgroups based on the constancy of occurrence, relative depth, ontogenetic shifts, diel migrations, and use of structure; see Allen & Cross 2006). Common pelagic groups include many species of elasmobranchs (Mako, Whitetip, Silky, and Whale sharks), clupeoids (herrings, sardines, sprats, shads, pilchards, menhadens, anchovies), atherinomorphs (flying fishes, halfbeaks, needlefishes, sauries, silversides), opahs, oarfishes, Bluefish, carangids (scads, jacks, trevallies, pilotfishes), dolphinfishes, remoras, pomfrets, barracudas, scombroids (cutlassfishes, mackerels, tunas, swordfishes, billfishes), butterfishes, and tetraodontiforms (triggerfishes, molas). Diversity overall is estimated at around 325 species. The world's longest known bony fish is a pelagic species – the Oarfish, *Regalecus russellii*, which can exceed 7 m (Fig. 10.6)

The pelagic realm is unquestionably the most important and productive region of the sea as far as human consumption is concerned. Pelagic fishes have constituted nearly half of the 70–80 million tons of fish captured annually worldwide; coastal pelagics, particularly clupeoids, made up about one-third of the total, and offshore pelagics such as tunas and billfishes make up an additional 15% (Blaxter & Hunter 1982; Groombridge 1992; FAO 2004). The proportion of total global catch attributed to these groups has declined somewhat, however, as catches of some bottom-dwelling species have increased and fishing pressure in inland waters has expanded (FAO 2020).

Characteristic of the pelagic region are high solar insolation, variable productivity, large volume, and a lack of physical structure. The abundance and diversity of fishes in the open sea are made possible by the periodic high productivity that occurs due to convergence of major currents or **upwelling** of nutrient-rich cold water, promoting the bloom of algal plankton species and creating a **trophic cascade**, at least until the nutrients are used up. The greatest concentrations of fishes in the sea, and the largest fisheries, occur in such areas of upwelling; these areas may account for 70% of the world fisheries catch (Cushing 1975).

The anchovy fisheries of Chile and South Africa, and the sardine fisheries of California and Japan have been direct results of upwelling. These fisheries, with catches measured in millions of tons in peak years, are inherently unstable, booming and then collapsing with shifts in oceanographic conditions that reduced the magnitude of the upwelling (see Chapter 22, Commercial Exploitation).

Adaptations to the Open Sea

Many common threads run through the biology of pelagic fishes, suggesting convergent adaptation to pronounced and predictable selection pressures. In general, pelagic fishes are countershaded and silvery, round or slightly compressed, streamlined with forked or lunate tails, schooling, have efficient respiration and food conversion capabilities and a high percentage of red muscle and lipids, are migratory, and

FIGURE 10.6 United States Navy personnel hold a 23-foot (7.0 m), 300 lb (140 kg) Oarfish, *R. russellii*, found washed up on the shore near San Diego, California, in September 1996. Wm. Leo Smith / Wikimedia Commons / Public Domain.

account for all fish examples of **endothermy** (see Chapter 7). Differences in most of these characters correspond to how pelagic a species is; extreme examples are found amongst the migratory tunas, which have the fastest digestion rates, the highest metabolic rates, and the most extreme specializations for sustained levels of rapid locomotion of any fishes (Block 2011; Dickson 2011).

The optical properties of the scales of open-ocean fishes reflect, diffuse, and polarize light in a way that help make the fishes practically invisible in open water. Apparently, the arrangement of guanine platelets in the skin modifies polarization in an angle-dependent manner, which helps the fish reflect and polarize light and blend into their background (Brady et al. 2015).

Several superlatives apply to pelagic fishes and reflect adaptations to life in open water and an emphasis on continual swimming, often associated with long-distance migrations. Large sharks, salmons, tunas, and billfishes move thousands of kilometers annually, but even smaller coastal fishes can make annual migrations of 150 km (sprats) and even 2000 km (herring) (Cushing 1975). To sustain continual swimming, pelagic fishes have the highest proportion of red muscle among ecological groups of fishes. Within the mackerels and tunas, the amount of red muscle increases in the more advanced groups, which are also increasingly pelagic and inhabit colder water during their seasonal migrations. In more primitive mackerels, the red muscle is limited to a peripheral, lateral band of the body, whereas in advanced tunas the red muscle is more extensive, occurs deeper in the body musculature, and is kept warm by the countercurrent heat exchangers that are also more developed in advanced scombrids (see Chapter 7, Endothermic Fishes). Countercurrent exchangers have evolved convergently in tunas, opahs, and mackerel sharks – pelagic fishes that range into cold temperate and deep waters. This convergence suggests that heat conservation arose independently in these groups and allowed otherwise tropical fishes to expand their ranges into colder regions (see Block 2011; Wegner et al. 2015).

Body shapes and composition in pelagic fishes reflect the demands of continual swimming (Fig. 10.7). Unlike benthic fishes with depressed bodies and littoral zone fishes with deep, circular, compressed bodies, pelagic fishes tend to have **fusiform** shapes that minimize drag. This is accomplished with a rounder cross-section and by placing the maximum

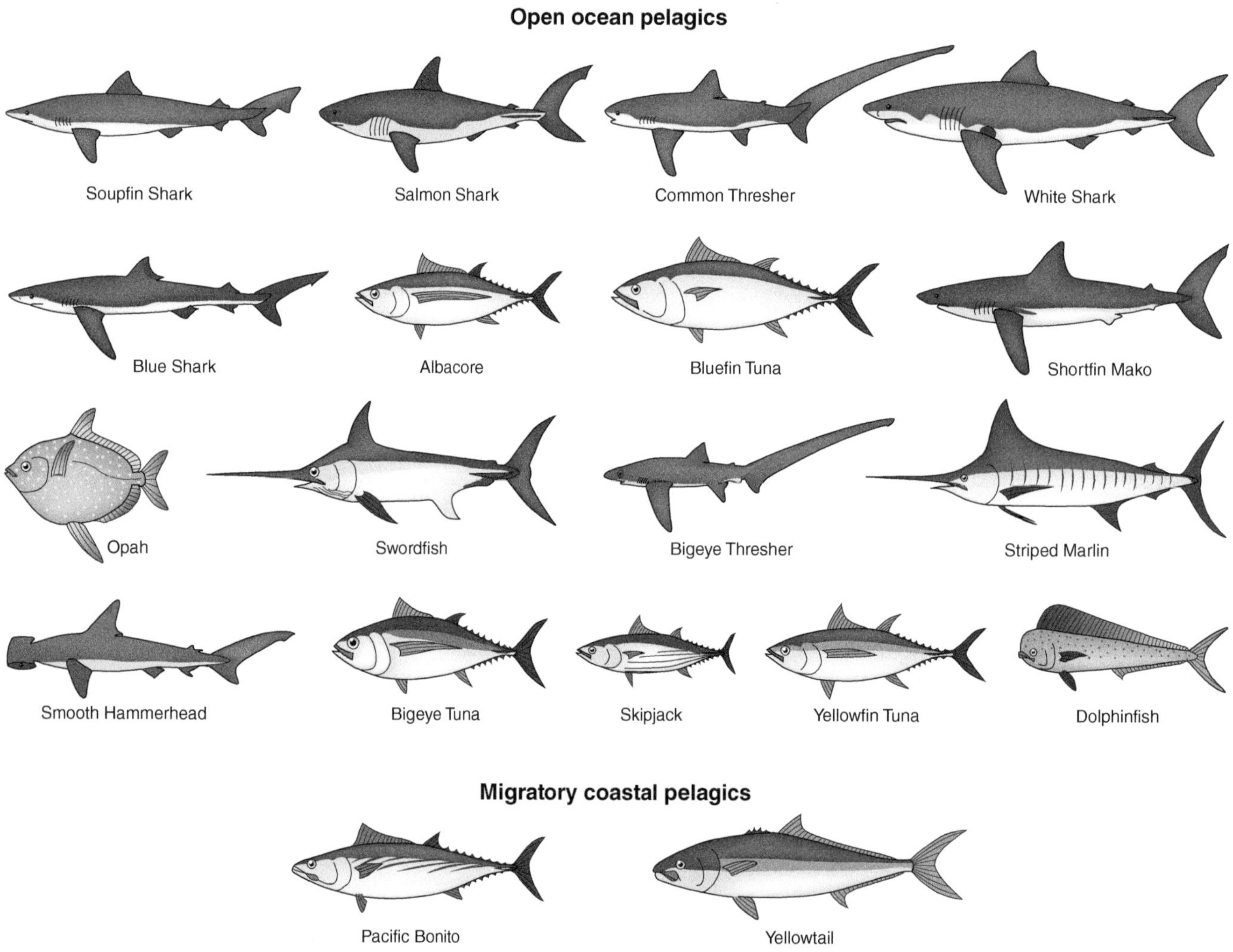

FIGURE 10.7 Open ocean and migratory coastal pelagic species of the California coast. Many of the open ocean species occur worldwide in temperate and especially tropical oceans. After Allen and Pondella (2006).

circumference of the body one-third of the way back from the head, an ideal **streamlined** shape also evolved convergently by pelagic sharks, whales, dolphins, and extinct ichthyosaurs (see Chapter 4, Locomotion: Movement and Shape). Streamlining is enhanced by having relatively small fins or having depressions or grooves on the body surface into which the fins can fit during swimming (e.g., tunas, billfishes). In high-speed fishes such as sauries, mackerels, and tunas, a series of small finlets occur both dorsally and ventrally anterior to the tail. These finlets may prevent vortices from developing in water moving from the median fins and body surfaces towards the tail, which would allow the tail to push against less turbulence. The extremely small second dorsal and anal fins of mackerel sharks, swordfishes, and billfishes could function analogously. The caudal fins of these fishes are typically tall, narrow, and stiff, thus providing a high aspect ratio suitable for rapid oscillation and maximum propulsion (Bernal 2011; and see Chapter 4).

Swordfish are built for speed and are reported to be among the fastest swimmers. Their overall body shape and fin structure are characteristic of most pelagic predators, and the sword may help reduce drag along the front of the fish. In addition, small oil-producing glands at the base of the sword, just in front of the eyes, secrete a methyl ester-based compound that may act as a lubricant to further reduce resistance (Videler et al. 2016).

Tunas have large scales around the anterior region of maximum girth (the corselet) that may reduce drag and thus create more favorable water flow conditions posteriorly, where actual propulsion occurs. In the region of the caudal peduncle and tail, sharks, jacks, tunas, Swordfish, and billfishes have lateral keels (Fig. 10.8) that reduce drag as the narrow peduncle is swept through the water. In the tunas, a single peduncular keel is supplemented by a pair of smaller caudal keels that angle towards each other posteriorly which may act as a nozzle that accelerates water moving across the tail, adding propulsive force (Collette 1978). Peduncular keels have evolved convergently in cetaceans, but the keels are oriented vertically – the same direction as the movement of the tail during swimming.

Many pelagic fishes swim continuously. In the Bluefish, jacks, tunas, Swordfish, and billfishes, this constant activity makes possible respiration by **ram ventilation** (see Chapter 3, Breathing). Instead of pumping water via a muscular buccal pump, pelagic fishes swim with their mouths slightly open while water flows across the gill surfaces. Ram ventilation requires that a fish swim continually at speeds of at least 65 cm/s, which is easily attained by any but the smallest tunas at their cruising speed of 1 body length/s. Ram ventilation conserves energy as the buccal pump mechanism used by other fishes may account for 15% of the total energy expended. Tunas and billfishes have lost the ability to pump water across their gills due to minimal branchiostegal development. They must therefore move continually to breathe. These fishes are also negatively buoyant and must move to keep from sinking (Roberts 1978).

The high levels of activity of pelagic fishes are fueled by an enhanced capacity for supplying oxygen to their muscles. For example, menhadens, Bluefish, and tunas have two to three times the hemoglobin concentration of typical inshore, sedentary fishes; hemoglobin concentration in tunas is more like that of an endothermic mammal than a typical fish. Tunas have large hearts that account for 2% of body mass and have concomitantly large blood volumes. The uptake of oxygen and release of carbon dioxide at the gills in herrings and mackerels are facilitated by exceedingly thin lamellar walls (5–7 μm thick) and numerous lamellae (>30/mm); comparable values for less active, inshore species are 10–25 μm and 15–25 lamellae/mm. The surface area of the gill lamellae relative to body weight is very high in mackerel sharks, menhadens, Bluefish, dolphinfishes, and tunas. The efficiency of the lamellae is enhanced by the fusion of adjacent lamellae and elaboration of the leading and trailing edges of the gill filaments. These modifications have occurred convergently in tunas, Swordfish, and billfishes but not in the less pelagic mackerels. Tunas remove more oxygen from the water as it passes over their gills than any other fish. This highly efficient oxygen uptake system is necessary to support their extremely high metabolic rates (Steen & Berg 1966; Collette 1978/1979; Blaxter & Hunter 1982). Tuna swimming muscles are also biochemically adapted to high levels of sustained activity with minimal fatigue and rapid recovery after bursts of activity (Bernal 2011).

Foraging

An open water existence limits the foraging options available to pelagic fishes. As a result, the fishes feed on phytoplankton, zooplankton, or each other. Many clupeoids swim through plankton concentrations with an open mouth, filtering the particles out of the water in a **pharyngeal basket** that has densely packed gill rakers (100–300/cm). Clupeoids also have an **epibranchial organ** that releases digestive enzymes while the food is still in the oral region. The digestive tract is long and has numerous pyloric caeca. Food passes very rapidly through this system, often taking less than an hour, but these fish can utilize a broad array of food types and are very efficient at converting food into energy and nutrients.

The foraging and migratory patterns of pelagic fishes such as tunas and billfishes become clearer when the nature of food availability in open tropical seas is considered. Estimates of zooplankton resources in the central Pacific indicate average densities on the order of 25 parts per *billion*. Large pelagic predators are feeding at even higher trophic levels, so their food is scarcer by one or two orders of magnitude. Since no animal is going to survive on food distributed evenly at such low densities, the success and rapid growth rates of many tunas attest to the extreme patchiness of food on the high seas. A nomadic lifestyle, driven by high metabolism and rapid swimming, makes sense when vast expanses must be covered in search of such patchily distributed resources (Kitchell et al. 1978).

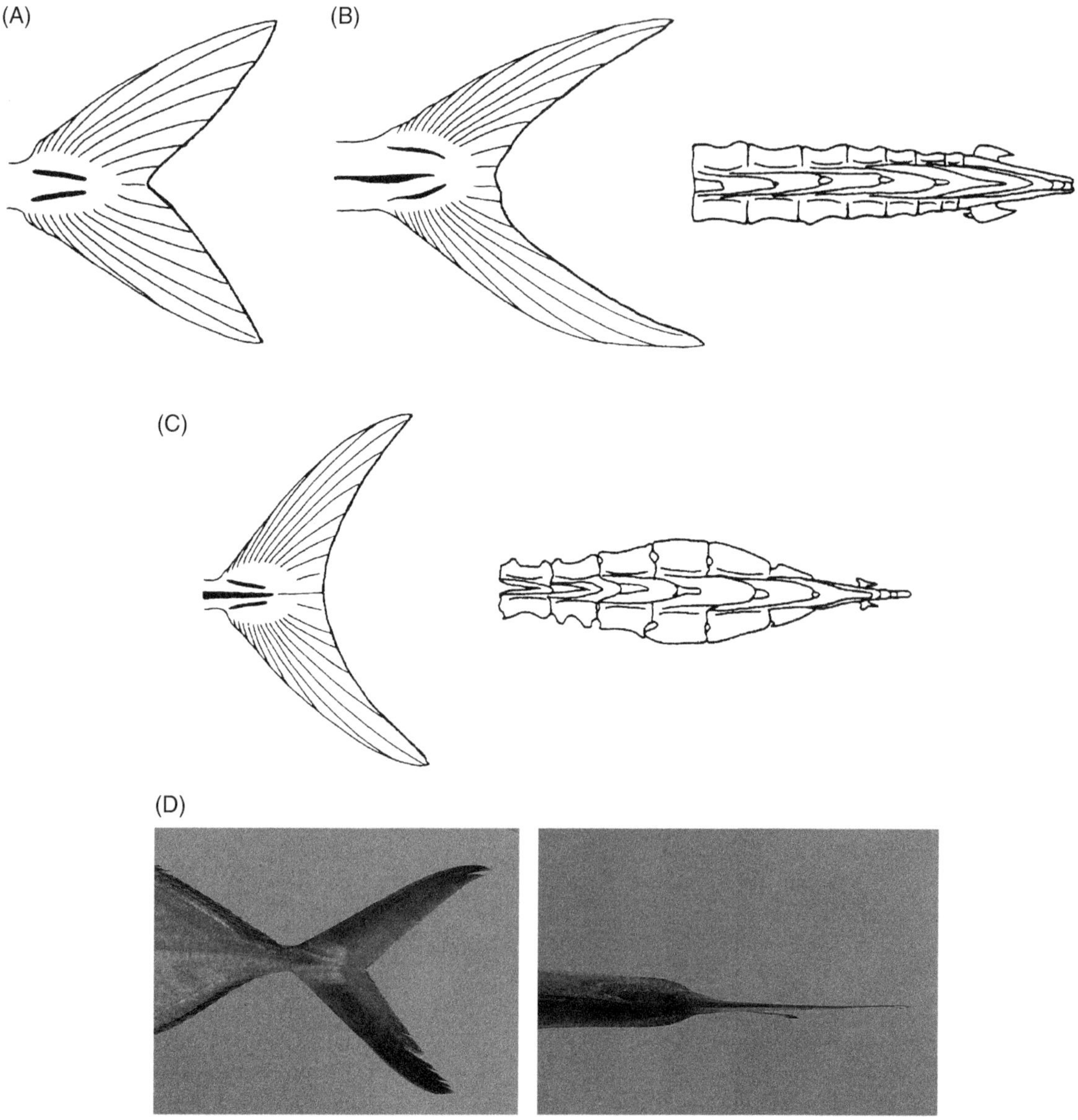

FIGURE 10.8 Keels and tails in open-water fishes. The evolution of mackerels and tunas has involved increasing degrees of pelagic activity. The more primitive mackerels and Spanish mackerels live inshore and swim more slowly and less continuously. More advanced high seas tunas swim continuously and faster and are more migratory. These ecological differences are reflected in tail shape and accessories, with more efficient, high aspect ratio tails and more elaborate keels characterizing the more pelagic tunas. (A) Mackerels have forked tails with one pair of fleshy caudal keels. (B) Spanish mackerels have a semilunate tail, caudal keels, and a median peduncular keel, but the peduncular keel is external only, lacking internal bony supports (right: dorsal view of peduncle skeleton). (C) Tunas have lunate tails and multiple keels, with lateral extensions of the peduncular vertebrae supporting the keels (shown on the right). Lunate tails and peduncular keels have also evolved in mackerel sharks, jacks, and billfishes. From Collette and Chao (1975) and Collette (1978), used with permission. (D) Lateral and dorsal views of peduncular keels of a small jack (Carangidae). D. Facey (Author).

Life History Patterns in Pelagic Fishes

Pelagic fishes occupy open ocean throughout all life-history stages. Two general patterns characterize the overall life histories of pelagic fishes, brought on by the relationship of parental versus larval food requirements, life span, spawning frequency, oceanic currents, and fish mobility. These patterns are referred to as cyclonic or anticyclonic.

Cyclonic patterns characterize higher-latitude species such as Atlantic Herring, in which the adults and larvae live in different parts of the ocean. Adults have a seasonal feeding area and tend to spawn once per year. Before they spawn, they migrate upcurrent to a region where food for larvae and juveniles will be particularly abundant. Larvae and juveniles drift with the currents to the adult feeding region. These fish invest considerable energy into each spawning episode, with the costs of both migration and egg production. Because of the spatial separation of adult and larval habitats, adults may not have reliable cues for predicting conditions at the spawning grounds, which leads to highly variable spawning success and large fluctuations in year-class strength (see Chapter 20, Population Dynamics and Regulation).

Anticyclonic patterns are more characteristic of low-latitude species such as tropical tunas and scads. The comparative aseasonality of tropical waters leads to less temporal fluctuation but extreme spatial variation in productivity. Adults move in a roughly **annual loop** through a major ocean basin, during which they spawn repeatedly (with the exception of the three species of bluefin tunas) rather than only in particular locales. Larvae and juveniles develop and feed along with adults, carried by the same current system in their relatively nomadic existence. The energy put into reproduction is spread out amongst several spawning episodes. Adults can use local environmental cues to determine the appropriateness of conditions for larvae, which is critical given the low productivity and patchiness of tropical open oceans. Hence anticyclonic species often show more stability in year-class strength. Tropical species tend to mature more quickly and live shorter lives.

Tunas evolved in the tropics but some species such as the Atlantic Bluefin Tuna (*Thunnus thynnus*) spend a large part of the year feeding in productive temperate locales (see Block & Stevens 2001). Bluefin show the phylogenetic constraint of their tropical history by returning to the tropical waters of the Gulf of Mexico or the Mediterranean Sea to spawn, forcing them into what is more of a cyclonic than an anticyclonic pattern (Rivas 1978). The same historical factors constrain anguillid eels such as the American, European, and Japanese species, which also return from temperate feeding locales to tropical breeding locales, but several years pass between the two life-history stages (see Chapter 18, Representative Life Histories of Migratory Fishes).

The high but periodic productivity of small planktonic animals in the open sea and the presence of major ocean currents have been contributing factors in the evolution of dispersive, planktonic larvae in most marine fishes, regardless of whether the adults are planktonic, pelagic, demersal, deep-sea, or inshore.

Flotsam

A special open ocean fauna occurs around what little structure is found in the open sea. Floating bits of seaweed (usually sargassum), jellyfishes, siphonophores, and driftwood almost always have fishes associated with them. Many **flotsam-associated** fishes such as filefishes and jacks are the juveniles of inshore or pelagic species; others such as sargassumfishes and driftfishes are found nowhere else, attesting to the reliability of encountering such objects. Flotsam also serves as an attractor for large predators, such as sharks, dolphinfishes, tunas, and billfishes (Gooding & Magnuson 1967); a floating log may have more than 400 tuna associated with it, often involving several species (Sharp 1978). Concentrations of flotsam are indicators of high productivity in the open sea because vertical circulation patterns (Langmuir cells) that accumulate flotsam also concentrate nutrients and zooplankton (Maser & Sedell 1994). The mechanisms by which pelagic fishes locate floating objects and their importance to fishes that do not feed around them remain a matter of conjecture. Objects made by humans have greatly increased the amount of flotsam, as evidenced by the vast Pacific Garbage Patch. This often-toxic trash can elevate fish mortality and promote dispersal of invasive species (see Chapter 22).

Evolution and Convergence

The greatest development of pelagic fish fauna is in the ocean. However, most major lakes have an open water fauna that consists partly of members typically associated with open waters as well as species whose ancestors were obviously inhabitants of nearshore regions. These **limnetic** fishes include osteoglossomorphs (Goldeye, Mooneye), clupeids (shads), characins, cyprinids (Golden Shiner, Rudd), salmonids (whitefishes, trouts, chars), smelts, silversides, moronid temperate basses, and cichlids. Many of these fishes live at the air–water interface and show specializations for this habitat, including upturned mouths, ventrally positioned lateral lines, and convergent fin placement and body proportions. These surface-dwelling traits occur in freshwater families, including characins, minnows, killifishes, and marine families including flying fishes (exocoetids and gasteropelicids), halfbeaks, and silversides (Marshall 1971). Regardless of ancestry, the same anatomical and behavioral themes that are seen in the ocean recur in freshwater limnetic species, including silvery color, compressed bodies, forked tails, schooling, high lipid content, and planktivorous feeding adaptations. Analogously, *Pleuragramma antarcticum*, a pelagic nototheniid in Antarctic waters, shows many traits characteristic of epipelagic fishes worldwide. Although derived from stocky, dark-colored, benthic ancestors, *Pleuragramma* has deciduous scales, a silvery body, forked tail, high lipid content for buoyancy, and is laterally compressed. The pelagic larvae of many benthic Antarctic fishes are also silvery, compressed, and have forked tails (Eastman 1993; see chapter section about Antarctic fishes). These examples of convergence indicate fairly uniform and continuous selection pressures characterize the open-water habitat.

With the exception of the clupeoids, most successful taxa of adult marine pelagic fishes are acanthopterygians. Missing among otherwise successful marine groups are elopiforms and paracanthopterygians, although both groups have proliferated in deep-sea mesopelagic and bathypelagic regions. These two groups may be phylogenetically constrained from inhabiting shallow open water regions, perhaps because of their tendency to be nocturnal. Other strongly nocturnal taxa are also missing from pelagic and limnetic habitats, including the otherwise successful catfishes, seabasses, croakers, grunts, and snappers. This is not to say that pelagic waters are devoid of life at night. The diel vertical migrations of many mesopelagic fishes bring them near the surface after sunset, where they can forage in the dark.

Intertidal Fishes

Intertidal areas include estuaries, salt marshes, mudflats, and rocky shores. These are very productive habitats, but also may be quite variable in temperature, oxygen, and salinity and therefore physiologically challenging for fishes (Schulte 2011b). In most coastal areas there are two high tides and two low tides per day; the magnitude of tidal fluctuation depends on the lunar cycle (greater during full and new moons) and geographic location. Irregular shorelines can create a channeling effect and result in tide changes of several meters. The Bay of Fundy, New Brunswick, can have tide changes greater than 10 m. Even in less extreme cases, these tidal changes shape the adaptations of fishes living in intertidal areas.

The general mechanism by which fishes deal with changes in salinity, dissolved oxygen, and temperature are discussed elsewhere (see Chapter 7, Homeostasis), but intertidal fishes must cope with frequent cyclical changes. Fishes can move with the tide, changing location but remaining in a relatively stable environment regarding salinity, dissolved oxygen, and temperature. However, fishes that remain in a specific physical location will experience changes in salinity, dissolved oxygen, and temperature of that location as tides rise and fall.

Coastal swamps, marshes, and mudflats typically have high deposits of organic material that decomposes and releases hydrogen sulfide (hence the notable "low tide" odor). Hydrogen sulfide interferes with hemoglobin's ability to carry oxygen and also alters mitochondrial biochemistry, restricting oxygen availability for fish. This can be a significant factor for fishes that live in burrows in these habitats (e.g., gobies, including mudskippers, and mangrove killifish). (The physiological challenges of high hydrogen sulfide levels are shared by some cave fishes, discussed later in this chapter.) It should not be surprising, therefore, that some intertidal fishes can breathe air, and will even emerge onto mudflats (e.g., mudskippers (Gobiidae) and rockskippers (Blenniidae)). These fishes have specialized organs to assist with aerial respiration, such as structural support in their gills, modifications of the buccal cavity or gut, or they may use cutaneous respiration while out of the water.

Rocky intertidal zones trap water in tidepools during low tide. These can provide a refuge for intertidal fishes, but also can experience considerable changes in physical conditions before they are inundated with the next rising tide. During the summer months, these tidepools can warm considerably over a few hours, triggering the production of heat shock proteins to stabilize biochemical reactions. The higher temperatures will also decrease dissolved oxygen levels, which can be further exacerbated by increasing salinity due to the loss of water by evaporation. Carbon dioxide, however, is more soluble in water than oxygen, and rising levels of CO_2 in the tide pool will lower pH. In winter months exposure to very cold air temperature may bring water temperatures in tide pools below the freezing point of the fishes in the pools. Crevasses in a rocky intertidal area also provide cool, damp refuges for fishes during low tide. Even if not fully submerged, some fishes can survive for several hours among the damp algae among the rocks. Many of these fishes have converged on an eel-like body form suited to wriggling in and out of these tight spaces (Fig. 10.9). Rocky coastlines can also be quite turbulent due to wave action, creating physical conditions similar to turbulence in high-velocity rivers and streams. Physical adaptations to these high-energy environments are discussed in the next section of this chapter.

FIGURE 10.9 The eel-like bodies of gunnels such as this Penpoint Gunnel, *Apodichthys flavidus*, as well as stichaeid pricklebacks, make it easier to find refuge among the moist, cool crevices of the rocky intertidal zone during low tide. Jeanne Luce / Wikimedia Commons / Public Domain.

Quite a bit of human activity is concentrated along shorelines and in coastal rivers and their estuaries. The multiple anthropogenic effects of human activity, such as changes in runoff patterns, reductions in river flow and freshwater influx, loss of wetlands, and chemical and thermal pollution add to the challenges for intertidal and other coastal fishes.

Strong Currents and Turbulent Water

Wave-swept rocky shores in the ocean and rapids in rivers and streams appear to be unlikely habitat for fishes because of the difficulties of remaining in place, let alone feeding and breeding under such conditions. Invertebrates successfully occupy such locales but, except for the groups that hide behind or under rocks, tend to be rather sessile or essentially glued in place (e.g., chitons, limpets, barnacles, and sea anemones in the ocean; caddis fly larvae, black fly larvae, and water penny beetle larvae in streams). With the exception of parasitic male anglerfishes, fishes have not evolved immobile forms and yet some species are exposed to the force of the waves and currents in these high-energy zones.

FIGURE 10.10 The depressed, terete body shape of sculpins (family Cottidae) helps these fishes cope with high-energy habitats such as freshwater streams and marine rocky intertidal areas. A Potomac Sculpin, *Cottus girardi*. Note the broad head, enlarged pectoral fins, and tapering body that help keep this fish on the bottom in a habitat that frequently includes fast-flowing water. R. Hagerty/ USFWS / Public Domain.

Fishes have converged on a general body shape, fin shape and distribution, and special devices for living in high-energy zones. Good examples in marine habitats are inhabitants of wave-swept, **intertidal surge regions**. Groups include various scorpaeniform cottid sculpins (Fig. 10.10), perciform blennioids (blenniid combtooth blennies, tripterygiid triplefins, clinid kelp blennies, labrisomid blennies), sicydiine gobies, and especially the aptly named clingfishes (Gobiesocidae). All tend to have bodies that are depressed dorsoventrally and somewhat tapered or teardrop-shaped (terete) when viewed from above, often with enlarged pectoral fins placed low on the body (e.g., Horn 1999; Boyle & Horn 2006). In some, the pelvic fins fuse and form a suction disk (e.g., clingfishes and some gobies). *Sicyases sanguineus*, a large (30 cm) Chilean gobiesocid, lives in and above the intertidal zone in locales exposed directly to waves (Paine & Palmer 1978; Cancino & Castilla 1988).

High-energy freshwater habitats have produced the most striking convergences, exemplified by fishes that live in **torrent zones** and share a body form clearly appropriate to maintaining position in strong, unidirectional currents (Fig. 10.11). The suite of anatomical traits on which these fishes have converged include:

- A dorsoventrally depressed, small (<15 cm) body, sometimes triangular or square in cross-section with a flattened ventral surface.
- Large, horizontally oriented pectoral fins are positioned low on the body; pelvic fins are also sometimes enlarged.
- A suction device, such as the mouth (e.g., suckermouth catfishes, algae eaters), or formed either by joined paired fins (hillstream loaches, gobies, clingfishes) or fins in combination with the ventral body surface (loach catfishes, loach gobies), sometimes with adhesive pads (sisorid catfishes).
- Subterminal or inferior mouths in just about all species.
- No gas bladder (psilorhynchids, amblycipitid loach catfishes).
- Modifications to respiratory behavior (rapid inhalations followed by a quiescent period of several minutes; Berra 2001) or respiratory structures, such as the incurrent opening at the top of the gill cover in gyrinocheilids, an analogous arrangement in astroblepid climbing catfishes, and a special fold of skin on which gill membranes rest in loach catfishes.

Many of these fishes live in mountain streams. Some are algae scrapers (gyrinocheilds, loricariid catfishes, parodontids, loach gobies), others are well known for ascending waterfalls (kneriids, astroblepid climbing catfishes, *Lentipes* gobies).

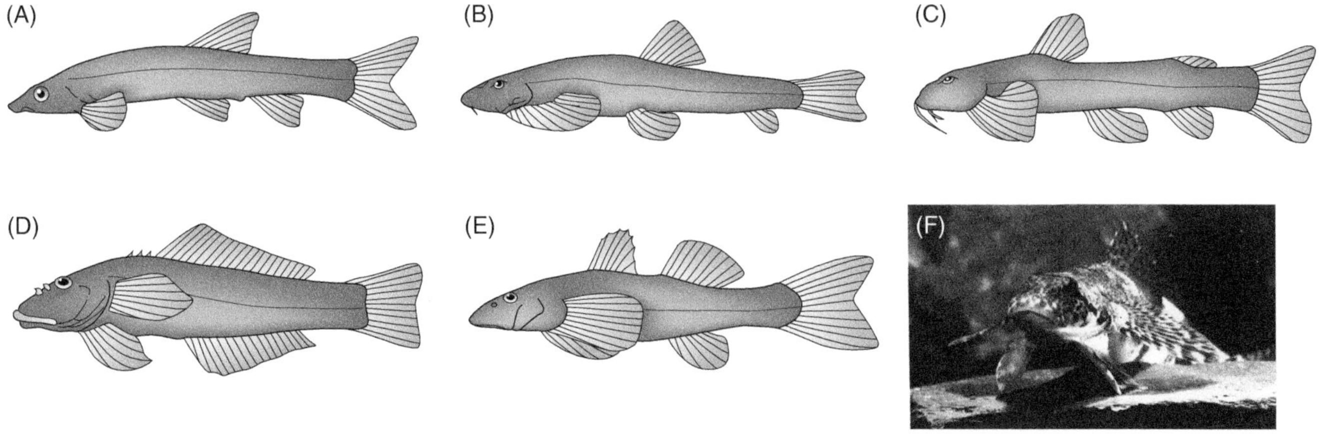

FIGURE 10.11 Convergence in body form among unrelated fishes that occupy swiftwater habitats in streams and rivers. (A) *Kneria*, an African kneriid (Gonorynchiformes). (B) *Gastromyzon*, an Asian balitorid hillstream loach (Cypriniformes). (C) *Amphilius*, an African amphiliid loach catfish (Siluriformes). (D) *Cheimarrichthys*, a New Zealand cheimarrichthyid Torrentfish (Perciformes, Trachinoidei). (E) *Rhyacichthys*, an Indo-Australian rhyacichthyid loach goby (Perciformes, Gobioidei). (F) Head-on photo of a Torrentfish, showing body profile and fin shape and placement characteristic of swiftwater fishes (*c.* 10 cm). (A–E) after Nelson (2006); (F) McDowall 2000 / with permission of Cawthron Institute.

TABLE 10.2 **A sampling of freshwater fishes that inhabit torrent and rapid zones of streams and rivers. Most if not all have converged on body shapes and proportions, fin arrangements and shapes, and other traits that reflect the need to hold the position on the bottom in swift-flowing water.**

Order	Family	Scientific name[a]	Common name
Gonorynchiformes	Kneriidae	*Kneria*	Knerias
Cypriniformes	Cyprinidae	*Rhinichthys cataractae*	Longnose Dace
Cypriniformes	Psilorhynchidae	*Psilorhynchus*	Mountain carps
Cypriniformes	Gyrinocheilidae	*Gyrinocheilus*	Algae eaters
Cypriniformes	Balitoridae	*Balitora, Gastromyzon*	Hillstream loaches
Characiformes	Parodontidae	*Parodon*	Parodontids
Siluriformes	Amphiliidae	*Amphilius*	Loach catfishes
Siluriformes	Nematogenyidae	*Nematogenys inermis*	Mountain Catfish
Siluriformes	Astroblepidae	*Astroblepus*	Climbing catfishes
Siluriformes	Loricariidae	*Otocinclus, Farlowella*	Suckermouth armored catfishes
Siluriformes	Amblycipitidae	*Amblyceps*	Torrent catfishes
Siluriformes	Sisoridae	*Sisor rheophilus*	Sisorid catfishes
Perciformes	Cheimarrhichthyidae	*Cheimarrichthys fosteri*	New Zealand Torrentfish
Perciformes	Gobiesocidae	*Gobiesox fluviatilis*	Mountain Clingfish
Perciformes	Rhyacichthyidae	*Rhyacichthys*	Loach goby
Perciformes	Gobiidae	*Lentipes concolor*	O'opu Alamo'o

[a] The specific name is given for representative or monotypic species, and the generic name is given when several species exist.

Many have scientific or common names that reflect specialized morphologies or suggest habitat preferences. The list includes species from perhaps 16 families and at least five different orders of teleosts (Table 10.2); other taxa in North American streams that use similar habitats and show some of the modifications include catostomid hognose suckers (*Hypentelium*), scorpaeniform sculpins (*Cottus*), and several percid darters (*Etheostoma, Percina*).

The relationship between form and function in many of these is fairly obvious. The depressed-flattened shape of the body as well as the large, horizontally oriented paired fins help flowing water push the fish down against the substrate. An adhesive or suction device similarly prevents being dislodged. Subterminal mouths allow for algae scraping or benthic feeding, whereas opening a terminal mouth creates drag. Many have reduced or lost the gas bladder, a broadly convergent trend among benthic fishes in general but clear advantage in swift-flowing water.

Undoubtedly, some anatomical characteristics reflect phylogeny as much as adaptation to habitat, although phylogeny can preadapt organisms to particular habitats as well as constrain them from occupying others (addressed later in this section). Preadaptation may help explain the abundance of catfish families in Table 10.2; catfishes as a group are freshwater benthic dwellers and somewhat depressed in body shape. Existing adaptations probably facilitated the invasion of high-energy, freshwater habitats by marine gobiesocid clingfishes, given their flattened, teardrop-shaped bodies, benthic habits, absent gas bladder, and pelvic fins fused into a suction disk. Most of the seven clingfishes that inhabit freshwater (among 167 species in the family) live in high-velocity stream zones at moderate to high elevations of Central and northern South America, regions where few of the other fishes listed in Table 10.2 occur. These clingfishes likely encountered an available, relatively unoccupied **adaptive zone** for which they already possessed appropriate traits (e.g., Guzman et al. 2001). Similar circumstances may help explain the successful invasion of insular freshwaters by the other perciforms in Table 10.2, such as the cheimarrhichthyidae Torrentfish (New Zealand), rhyacichthyid loach gobies (New Guinea, New Caledonia), and the gobiid O'opu Alamo'o (Hawaii).

Extreme pH and Salinity

The acidity or alkalinity of a water body strongly determines the existence and types of fishes that occur there. Seawater is naturally buffered against shifts in hydrogen ion content (pH) and hence pH is seldom a concern for marine fishes; seawater usually has a pH of about 8.0–8.3. Freshwater, in contrast, is readily affected by substances that alter pH. Freshwater fishes normally live in water with a pH range between 6 and 8, a pH of 7 being neutral. Acidic conditions (pH < 7) often result from the decay of organic matter that is not filtered through soil or further broken down. The black or tea-stained coloration of many swamps, and the black water rivers of the southeastern US, and of major tributaries of the Amazon such as the Rio Negro and of many African rivers, are examples of naturally occurring low pH water (pH 3.8–4.9). Such "soft" waters are also low in dissolved substances and inorganic ions, but high in organic acids such as humic and fulvic acids (Lowe-McConnell 1987). Some fishes have evolved under conditions of low pH and do best in slightly acidic waters (e.g., many tetras), whereas other groups are intolerant of acidic waters. Minnows, which are widespread throughout North America, are often missing from river systems where the pH falls below 4.5 (Laerm & Freeman 1986), although cyprinids in Southeast Asian inhabit waters with low pH. Acidic precipitation, a lowering of pH that results from industrial pollution, causes reproductive failure in many fishes and has eliminated fishes from the poorly buffered lakes of the Adirondack Mountains in New York and in many lakes throughout Scandinavia (Baker & Schofield 1985; Helfman 2007).

High pH is caused by an abundance of hydroxyl (OH) groups, producing **alkaline** conditions. High alkalinity occurs naturally in waters that run through limestone rocks, or where extensive evaporation occurs. Some fish have adapted to alkaline conditions that are lethal to most other animals. A small (<8 cm) African cichlid, *Alcolapia grahami*, is the only fish that can live in Lake Magadi in Kenya under conditions of extreme alkalinity, salinity, and temperature. Water flows into the lake from hot springs at a pH of 10.5, a salinity of 40 parts per thousand (ppt), and a temperature of 45°C. The water has a high load of sodium bicarbonate, sodium chloride, sodium sulfate, and sodium fluoride and has a conductivity of 160,000 µmho/cm (most African lakes have a pH of 7–9 and a conductivity of 100–1000 µmho/cm). The fish occupy pools and graze on algae at temperatures up to 41°C. Their upper-temperature limit creates a distinctive browse line between 41° and 45°, where algae are safe from fish grazing (Coe 1966; Fryer & Iles 1972).

Salinity determines the distribution of many if not most fish families. Biogeographic categories of freshwater fishes focus on whether taxa can tolerate salinities greater than a few ppt. In one approach (see Briggs 1995; Berra 2001), freshwater fishes are classified as **primary** (those that cannot cross saltwater boundaries, such as minnows, characins, most catfishes, pike), **secondary** (those that can cross at least short saltwater regions, e.g., cyprinodontoids, cichlids), and **peripheral** (those derived from marine families or that spend part of their lives in the ocean, e.g., salmons, sculpins). Please refer to Chapter 19 for further discussion on this.

Physiological limitations prevent fishes from moving between regions of high and low salt concentration. Freshwater fishes must conserve salts and eliminate water, whereas saltwater fishes must conserve water and eliminate salts (see Chapter 7, Osmoregulation, Excretion, Ion and pH Balance). Extremes of and rapid changes in ionic concentration can cause **osmotic stress**. Hypersalinity occurs in many areas, either as a result of heated water flowing through easily soluble rocks, or due to daily or seasonal evaporation and concentration of salts as water courses dry up during low tides or droughts.

Diadromous fishes that move between freshwater and marine habitats as part of their reproductive cycles must undergo physiological changes as part of their metamorphosis. These include anadromous salmonids (Salmonidae), which spawn in freshwater but spend much of their lives feeding and growing in the oceans (Cooke et al. 2011), and catadromous eels (Anguillidae) which have the opposite life cycle (Righton & Metcalfe 2011). The specific changes associated with these life cycles are addressed elsewhere in this text.

Some of the most widely distributed families in freshwater turn out to be those that show a high tolerance to both rapid fluctuations and extreme conditions of salinity. Many cyprinodontoid killifishes and pupfishes can tolerate ranges of salinity from 0 to 35 ppt (seawater) and appear to tolerate rapid shifts in salinity from high to low concentration, such as those brought on by rainstorms. Some, such as the Mediterranean *Aphanius* and several North American *Cyprinodon*, live in water two to three times saltier than seawater. These capabilities have preadapted them for life in isolated habitats such as desert springs and pools (Roberts 1975b). Similar abilities characterize cichlids and gobies, two of the world's largest families of fishes. *Oreochromis amphimelas*, a cichlid, inhabits Lake Manyara in Africa, where the sodium content is twice that of seawater and is increased ionically by abundant potassium salts (Fryer & Iles 1972). Certain large inland water bodies are too saline to support even the most osmotically tolerant species, including the Dead Sea of the Middle East and the Great Salt Lake in Utah, where salinities exceed 200 ppt. The salinity of the Salton Sea of California, exceeding 44 ppt, has steadily increased due to evaporation without water replacement, making it uninhabitable by a progression of fishes introduced for angling (Barnum et al. 2017). Water withdrawal due to human activities can cause salinization of a lake and threaten the fishes there, as has occurred in the Aral Sea on the border of Kazakhstan and Uzbekistan.

Polar Regions

The far north (**Arctic**) and south (**Antarctic**) polar regions are roughly the areas above 60° latitude. They have much in common, primarily related to cold water temperatures and short growing seasons, but they differ geologically and environmentally and support very different biotas, including fishes. The Arctic is a frozen oceanic region surrounded almost entirely by land, whereas the Antarctic is a frozen continent surrounded by ocean (Fig. 10.12). Frigid conditions began about 3 million years ago in the Arctic, whereas the Antarctic fauna have been evolving under current conditions for over 25 million years (Briggs

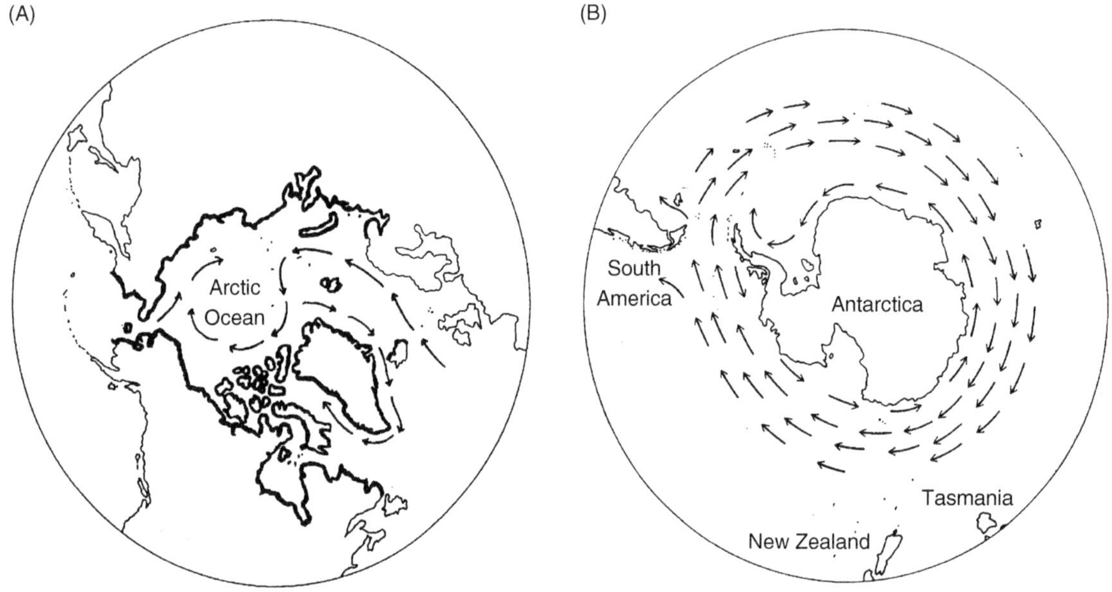

FIGURE 10.12 North and south polar regions. General oceanic circulation patterns are shown by arrows. (A) The Arctic Ocean centers on the North Pole; the southern limits of the region are indicated by the dark continental borders. (B) The Southern Ocean surrounds Antarctica. Some of the islands on the periphery of the south polar region are indicated. From Briggs (1995).

& Bowen 2012). Freshwater fishes are lacking from the Antarctic because most water bodies have permanent ice cover and many freeze to the bottom during the winter. High Arctic lakes and rivers have a limited fish fauna and most of these are cold temperate species at the northern edge of their range (Scott & Crossman 1973). Freshwater fishes at high latitudes adjust behaviorally to the strong effects that seasonality has on light levels, day length, and growing season. Polar oceans are in a liquid state below the first few meters and have more fishes, but the superabundance of ice at the surface, plus scouring by ice or ice anchored to shallow bottoms limit the distribution and behavior of polar fishes, which have developed remarkable adaptations to avoid freezing to death, as discussed in Chapter 7 (Coping with Temperature Extremes). The ocean habitats near the north and south poles support a wide variety of species of different lineages. In the following sections, we address several ways in which polar fishes show convergence in structure and function, especially prevention from freezing.

Antarctic Fishes

Antarctica is surrounded by at least 900 km of the deep Southern Ocean. Strong **circumpolar currents** and distinct temperature differences occur between the polar and subpolar regions, delimited by a region known as the **Antarctic Convergence** at 50–60°S. This region creates a distance, depth, and thermal barrier to interchange between the cold-adapted species of the Antarctic region and temperate-adapted species to the north. Antarctic fishes have also had sufficient time to adapt and speciate. The Drake Passage between South America and Antarctica opened about 30–34 million years ago. As the distance increased, circumpolar currents became established, further isolating Antarctica and making it extremely cold. The current frigid conditions have persisted for about 10–14 million years (Cheng & Detrich 2012). Spatial and temporal seclusion and climatic extremes have resulted in a diverse fish fauna dominated by endemic notothenioid thornfishes, cod icefishes, channichthyid crocodile icefishes, plunderfishes, and dragonfishes, as well as several non-notothenioid groups (Farrell & Steffensen 2005; Fukuchi et al. 2006).

Notothenioids as a group are benthic fishes and half of all species live in less than 1000 m of water (Fig. 10.13). They lack gas bladders, are dark in coloration, and are round or depressed in cross-section with a square or rounded tail. Many seek cover inside sponges, either as a refuge from predatory mammals or as a spawning substrate. Eggs placed inside hard sponges such as hexactinellid glass sponges are probably protected from most predators (Dayton et al. 1974; Konecki & Targett 1989). Larvae are pelagic and show adaptations specific to shallow, open water existence, including silvery coloration, relatively compressed bodies, and forked tails (as discussed earlier in this chapter). Notothenioids have also radiated into most nonbenthic niches and consequently show substantial variation in body form and behavioral tactics, starting from a shared body plan.

A few species, including the abundant Cod Icefish, *P. antarcticum*, are pelagic zooplanktivores. So-called **cryopelagic fishes** live in open water just below the ice. The food chain for these fishes starts with ice algae, which is eaten by amphipods and euphausiids, which are in turn eaten by the fishes. Cryopelagic fishes have a uniform light coloration that may help them blend in with the icy background against which they would be viewed. They also possess better chemical defenses against freezing and have greater buoyancy than benthic relatives.

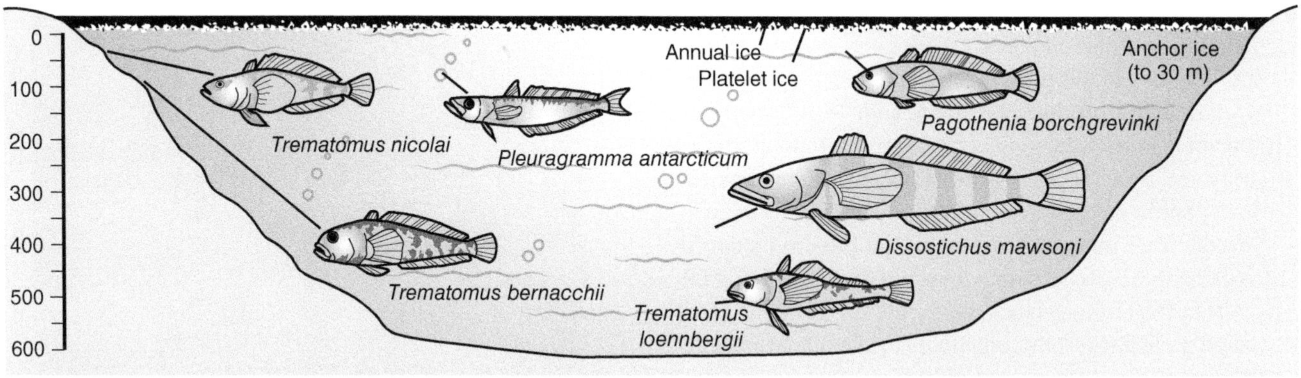

FIGURE 10.13 Body form and habitat types of common Antarctic nototheniid fishes. The lines show the preferred depths and habitats. From Eastman and DeVries (1986). Reproduced with permission.

Notothenioids produce a small number of relatively large, 2–5 mm eggs during a short, 1–2-month spawning season. The unhatched larvae have developmental periods of 2–6 months, followed by a long, slow-growing pelagic stage that lasts a few months to 1 year. Many benthic species exhibit parental and biparental guarding (Daniels 1979; Kellermann & North 1994).

Notothenioids are opportunistic feeders, taking a wide range of prey types, with many pelagic and mesopelagic juveniles and adults feeding on the ubiquitous krill, *Euphausia superba*, which is also the major prey of whales, penguins, and other seabirds. Although the annual temperature variation in Antarctica is seldom more than 4°C (−2 to +2°C), and in some locales as little as 0.1°C, fish feeding rates decline in winter. They are, however, still relatively high during winter (e.g., 65% of summertime intake in *Harpagifer antarcticus*; Targett et al. 1987), compared with temperate locales where many fishes cease feeding in winter.

Mesopelagic fishes are particularly abundant throughout the water column of the Southern Ocean. Lanternfishes are the most diverse group of mesopelagic fishes at lower latitudes but are epipelagic in the Antarctic. The myctophid lanternfish *Electrona antarctica* is the most common fish above 200 m. It feeds heavily during the day, in contrast with the typical mesopelagic pattern of nocturnal foraging of lanternfishes at lower latitudes. Mesopelagic species are also an important component of the community living near the ice edges or "oceanic marginal ice zone." Large numbers of lanternfishes are eaten in the open sea and at the edge of the pack ice by seabirds, whales, and seals. Commercial midwater trawl fisheries for mesopelagic species have harvested thousands of tons of the lanternfish *Electrona carlsbergi* from the South Georgia Island region. As with their more northerly, low-latitude relatives, deep-living mesopelagic fishes in the Antarctic show lower enzyme activity and slower metabolic rates than shallow-water forms, which is an adaptation to low food availability at depth (Kellermann & North 1994).

The plunderfishes (Harpagiferidae) are advanced perciform fishes, yet are remarkably similar in morphology and behavior to the more primitive scorpaeniform sculpins of northern temperate waters. Plunderfishes and sculpins have converged to fill similar niches in their respective communities (Wyanski & Targett 1981). Their similarities represent adaptations to a predominantly benthic existence, and include a relatively depressed, elongate, tapering body; large, spiny head with large eyes and a large, terminal mouth; long dorsal and anal fins; large pectoral fins; rounded caudal fin; and a dorsally located lateral line. Both groups show ecological and behavioral similarities as well, using a sit-and-wait feeding mode on relatively large, mobile benthic invertebrates.

Adaptations and Constraints of Antarctic Fishes

Notothenioids are best known for two adaptations to the cold, often energy-limited waters of the area, where water temperatures average −1.87°C and total darkness prevails for 4 months each year. First, their blood contains remarkably effective **antifreeze compounds** that limit ice crystal formation and make it possible for them to live in water that is colder than the freezing point of most fish blood, including their own. Second, some have evolved **neutral buoyancy** without a gas bladder, which has permitted these species to move off the crowded bottom where most notothenioids live and into the water column.

No known species of fish can actually tolerate having its tissue freeze. The major threat to fishes in the Antarctic is ice, which floats at the surface in the form of bergs, sheets, and platelet ice, but also attaches to the bottom in water less than 30 m deep in a form called anchor ice. The greatest danger comes from ice crystals penetrating or propagating across the body and seeding the formation of ice inside the fish, which would rupture cells. Amazingly, many Antarctic fishes live in water that is colder than their blood's freezing point. Fishes from lower latitudes typically freeze when placed in water colder than −0.8°C, whereas Antarctic fishes can live in water as cold as −2.19°C. They accomplish this because their blood contains the ions normally found in fish blood plus glycopeptide antifreeze compounds (also discussed in Chapter 7, Coping with Temperature Extremes). The glycopeptides function by keeping the ice from propagating across the fish's skin. A notothenioid can be cooled as low as −6°C without freezing, as long as free ice is not in the water. In most cold-adapted fishes, antifreeze compounds are produced in the liver, but in the notothenioids this occurs in the pancreas and anterior portion of the stomach (Cheng & Detrich 2012), suggesting an independent evolutionary origin. Molecular evidence shows that species living in the coldest conditions exhibit gene duplication, allowing the production of more antifreeze. Conversely, Notothenioids in New Zealand, where water temperatures do not get below zero, are derived from those in Antarctica but have fewer antifreeze genes.

Several other adaptations accompany the production of antifreeze compounds. Notothenioids are unusual among teleosts in that their kidneys lack glomeruli, the structures that filter small molecules from body fluids to the urine for excretion. Glomeruli would remove the antifreeze glycopeptides, which would be energetically expensive to continually replace. A fairly strong correlation exists between antifreeze effectiveness and the frequency with which a species encounters free ice. For example, the shallow water bathydraconid dragonfishes frequently come in contact with ice and have the highest levels of antifreeze compounds. Within the cod icefish genus *Trematomus*, shallow-water species that live in the coldest water and rest in ice holes or on anchor ice have freezing points of −1.98 to −2.07°C, whereas deeper living species that seldom encounter ice crystals freeze at −1.83 to −1.92°C. Even within species, shallow-water populations have significantly more freezing resistance than deeper-water populations (DeVries 1970).

Neutral buoyancy has developed in at least two water column dwelling members of the family Nototheniidae, the Antarctic silverfish, *P. antarcticum*, and its large predator, the Antarctic Toothfish, *Dissostichus mawsoni*. Whereas most Antarctic fishes are 15–30 cm long, toothfish reach lengths of 1.6 m and weights of over 70 kg. Neutral buoyancy allows these fishes to occupy the underutilized water column, thus taking them away from threatening anchor ice crystals and into a region of seasonally abundant food sources such as fish larvae and krill. Both species evolved from benthic ancestors and lack a gas bladder, which presents a buoyancy challenge to a fish living in open water. Neutral buoyancy in these two nototheniids is achieved via several mechanisms. Toothfish have cartilaginous skulls, caudal skeletons, and pectoral girdles, which reduces their mass because cartilage is less dense than bone. The skeleton itself is less mineralized than in benthic relatives, by a factor of six in the toothfish and 12 in *Pleuragramma*. Bone is also reduced in the vertebral column, which is essentially hollow except for the notochord. Additional buoyancy is achieved by lipid deposits around the body, including a blubber layer under the skin, and fat cells or sacs located between muscle fibers or muscle bundles (Eastman & DeVries 1986; Eastman 1993). Weightlessness via analogous routes of weight reduction and replacement is also seen convergently in bathypelagic fishes, another water column dwelling group where evolution has placed a strong premium on energy-saving tactics, and in the unrelated Chondrichthyes.

The icefishes (family Channichthyidae, Fig. 10.14) are nototheniids with the unique trait of having no hemoglobin in their blood and no myoglobin in their muscles, giving them a very pale appearance. Hence, these fishes are sometimes referred to as "white blooded" or "bloodless." These proteins are critically important in every other known group of vertebrates for oxygen transport (hemoglobin in red blood cells) and oxygen storage (myoglobin in muscles). Molecular studies support the hypothesis that a series of mutations disrupted the gene responsible for producing hemoglobin in ancestral icefish (Near et al. 2006; Cheng & Detrich 2012). Whereas other vertebrates would not survive such a mutation, apparently these fish were able to survive without hemoglobin due to the highly oxygenated, cold water. However, living with limited oxygen delivery to body tissues was apparently still a challenge, and the channichthyids evolved a number of other characteristics to compensate, including loss of scales and increased vascularization of skin and fins to increase gas exchange, an increase in heart size and cardiac output, increased blood volume, and relatively low metabolic rates (reduced protein synthesis, reduced activity, slow growth), and denser capillary beds in body tissues (Hemmingsen 1991; Cheng & Detrich 2012). Six of 16 species in this family also lack myoglobin in their heart muscles, which is compensated by increased volume of mitochondria. These compensatory adaptations may have evolved in response to elevated levels of nitric oxide (NO) in body tissues due to the loss of hemoglobin and myoglobin, which would break down the nitric oxide to nitrate (Cheng & Detrich 2012).

Some non-channichthyid nototheniids have increased blood volumes and reduced hemoglobin concentrations, perhaps reflecting an intermediate stage in the response to respiratory conditions in the Antarctic that have led to the hemoglobin-free condition of the channichthyid icefishes (Wells et al. 1980).

FIGURE 10.14 The Antarctic icefishes (Channichthyidae) are the only known vertebrates that lack hemoglobin, the protein in red blood cells needed to carry oxygen. The inability to produce hemoglobin due to mutation of the gene needed to make it is an example of regressive evolution – similar to the loss of eyes and vision in cavefishes addressed later in this chapter. Photo of Chaenocephalusaceratus by Ambiederman / Wikimedia Commons / CC BY-SA 4.0.

Arctic Fishes

The Arctic has fewer endemic fishes than the Antarctic. This is not surprising as the Arctic is less geographically isolated and has not been in its current cold condition as long. The oceanic environment between **boreal** (subarctic) and Arctic areas is fairly continuous. On the Pacific side, the Bering Sea flows into the Arctic Ocean and has done so since the Bering Strait opened up 3.5 million years ago. Similarly, on the Atlantic side, the Arctic Ocean is directly connected to the Greenland Sea. Hence, Arctic fishes are either species that evolved there since the current climate developed or are cold-tolerant Pacific or Atlantic species that dispersed from source areas rather than being endemic to the Arctic itself. The Arctic has undergone repeated warming and cooling until about 3 million years ago when the present cold conditions stabilized, leaving less time for evolutionary adaptation and speciation than in the Antarctic (Briggs & Bowen 2012).

Arctic fishes have evolved convergently with Antarctic fishes in the production of antifreeze compounds (Farrell & Steffensen 2005). Glycoprotein antifreeze occurs in Arctic and Greenland Cod, whereas Warty Sculpin, Canadian Eel-pout, and Alaska Plaice possess peptide antifreezes (Clarke 1983). Arctic Cod are frequently observed resting in contact with ice and taking refuge inside holes in ice, so their potential for encountering seed crystals is very high. In some of these fishes, kidney glomeruli are convergently reduced to help retain antifreeze compounds (Eastman 1993). Several boreal cods, sculpins, eel-pouts, and flatfishes whose ranges extend into Arctic water also have antifreeze compounds in their blood.

Water temperatures show greater annual and latitudinal variation in the Arctic than in the Antarctic, which means that fishes are likely to encounter extreme winter cold but also relatively high summer temperatures. Winter temperatures commonly drop to −1.8°C as in the Antarctic, but water can reach 7 or 8°C during the summer. Few Antarctic fishes can tolerate water temperatures above 7 or 8°C regardless of acclimation temperature, whereas Arctic species have upper temperature tolearances of 10–20°C depending on species and acclimation temperature (DeVries 1977). Several north polar species produce less antifreeze during the summer, particularly among boreal fishes that may encounter temperatures well above freezing. Winter Flounder, *Pseudopleuronectes americanus*, has a blood volume of 3% antifreeze in winter and 0% in summer. Reduced antifreeze production during warmer months probably saves energy and may also increase the blood's capacity to carry oxygen or nutrients.

The Effects of High Latitude on Activity Cycles and Predator–Prey Interactions

Most fishes feed either during daylight or darkness, with a small number primarily active during dawn and dusk. These cycles of activity are based on a strong internal clock and can be maintained for some time under laboratory conditions of constant light or darkness. However, in nature, the activity cycles are calibrated by the rising and setting of the sun.

The situation at high latitudes presents a very different light regime and corresponding selective pressures. Above the Arctic Circle, light levels never reach "nighttime" during midsummer, and growing seasons are short and intense. Winter brings a time of continual darkness and low food availability. Summer and winter therefore present extreme and opposite light conditions. Do fishes maintain strict diurnality or nocturnality under such variable and extreme conditions, or do they adjust their activity patterns to the changing seasons?

Laboratory studies with species whose natural ranges extend beyond the Arctic Circle have produced some striking and seemingly adaptive departures from the standard picture developed at lower latitudes. At intermediate latitudes below the Arctic Circle, Burbot (*Lota lota*) are nocturnal throughout the year. However, at higher latitudes, the fish are continually active during the summer, whereas during the winter they shift to diurnal behavior. During spring and fall they are primarily nocturnal. Similar activity cycles have been observed in other nocturnal or crepuscular species, including sculpins and Brown Trout, and can be induced experimentally in Brown Bullheads.

Interpreting these patterns is not immediately easy. It seems that the change to arrhythmic, continual behavior in summer is a means of taking advantage of high, continuous productivity. Limiting activity to the short nighttime period each day during summer would severely restrict an animal's intake. Nocturnality during spring and fall may represent a return to the normal, evolved response of the species as day length and twilight length closely approximate the more usual and widespread conditions at lower latitudes. The switch to diurnality during winter in an animal well adapted to function in the dark remains puzzling. Regardless, changes in the length of, and light intensity during, twilight provide the apparent cues that lead to the phase shifts observed in these fishes (Muller 1978a, 1978b).

The influence of **twilight length** at high latitudes is also shown in the predator–prey relations of marine fishes. Dawn and dusk at low latitudes are the times when fish switch between feeding and resting and are often times of maximal predator activity. If twilight is a dangerous time for prey fishes at low latitudes where twilight lasts for a relatively short time, we might expect the prolonged twilight that occurs at higher latitudes to be even more dangerous.

Conducting extensive underwater observations at high latitudes can be uncomfortable and few such studies have been attempted. In the one instance where the question of twilight interactions was addressed, observers found that extended twilight meant extended periods of predation. Hobson (1986) watched sculpins, greenlings, and flatfishes preying on Pacific Sand Lances, *Ammodytes hexapterus*, in Alaska. Sand lances school and feed on zooplankton during the day and bury in the sand at night. Schools of sand lances are relatively immune to these benthic predators during daylight, and the predators do not occur at night in the limited resting areas that the sand lances use. However, during twilight, the predators aggregate in the resting area under the schools as they break up. The predators are particularly effective at capturing sand lances that have just entered the sand or that re-emerge shortly after burying because of apparent dissatisfaction with their initial choice of resting site. The twilight transition from schooling to resting appears to be the most dangerous time for the sand lances.

Twilight conditions at the date and latitude of observation (May, 57°N) lasted about 2 h; about twice as long as at tropical latitudes where similar observations have been made with different predators and prey. The period of intense predation in Alaska is also about twice as long as that observed at tropical locales. The longer days of spring and summer at high latitudes mean that diurnal fishes experience a much longer foraging period, but this benefit comes with a cost of increased predation during the lengthened twilight periods.

Deserts and Other Seasonally Arid Habitats

Deserts are difficult to define because they differ in altitude, temperature range, amount of rainfall, and seasonality of water availability, among other traits. Many treatments define a desert as an area that receives less than 30 cm of rainfall annually. A more general definition is that a desert is an area where "biological potentialities are severely limited by lack of water" (Goodall 1976), a definition that stresses the common thread of water scarcity as the significant selection factor and can therefore apply to areas with seasonal droughts, such as swamplands that dry up periodically. Such habitats are inhospitable for fishes. However, algae and invertebrates capitalize on the periodic water in arid regions. It is not surprising then to find a small number of fishes capable of surviving under conditions of periodic **dewatering** in desert regions around the world, presenting dramatic examples of adaptation and convergent evolution.

For fishes, the disappearance of water is preceded by a continuum of increasingly stressful conditions. As water evaporates, temperatures generally rise, dissolved substances such as salts become more concentrated, oxygen tension drops, carbon dioxide increases, and competition and predation intensify. Desert fishes must therefore be tolerant of widely varying and extreme salinity, alkalinity, temperature, and depleted oxygen. They may also have to be able to outcompete other fishes and avoid predators despite physiological stress. Desert stream fishes also must withstand periodic flash flooding. Desert-adapted fishes, not counting species that migrate to more permanent habitats when waters recede, often show three general adaptations: (i) an annual life history involving egg deposition in mud during the wet season, an egg resting period (**diapause**) during the dry season, death of the adults, and egg hatching when water returns the next year; (ii) **accessory respiratory structures** for using atmospheric oxygen (lungs, gill and mouth chambers, cutaneous respiration); and (iii) in perennial species, **estivation**, where adults pass the dry season in some sort of resting state.

Deserts occur on all major continents and many of these deserts contain fishes. Africa has many habitats that dry up seasonally and that contain fishes with desert adaptations. Among the most successful groups in Africa are cyprinodontiform killifishes and rivulines, which are popular aquarium species. Many of these are **annual** fishes (e.g., *Fundulosoma*, *Nothobranchius*, *Aphyosemion* spp.), living for 8 months in mud holes, swamps, and puddles (Fig. 10.15). They mature after only 4–8 weeks, spawning daily and burying eggs as much as 15 cm deep in muddy bottoms, a remarkable feat for fishes that seldom exceed 5 cm long. The adults die and the eggs spend the dry season in a state of arrested development until the next rains come. Some eggs can remain in such a state of diapause for up to 5.5 years. An annual life history effectively maintains a permanent population in a temporary habitat (Wourms 1972; Simpson 1979).

African and Asian clariid or walking catfishes are capable of leaving drying water bodies and moving across up to 200 m of moist grass in search of water. They will also bury themselves as deep as 3 m in sandy sediments as water levels drop. They can survive by employing aerial respiration via treelike **suprabranchial organs** over the second and fourth-gill arches (see Chapter 5, Air-breathing Fishes), although they cannot survive if the sand dries up (Bruton 1979).

The African lungfishes (Protopteridae) are true estivators. During a drought, they burrow into mud, secrete a cocoon, and enter a torpid condition in dry mud until the next rains, an event for which they can wait up to 4 years (discussed in Chapter 13). Many other fishes in African swamps are adapted to the deoxygenation that accompanies seasonal dry periods, using a variety of air-breathing mechanisms (see Table 5.1). Mochokid catfishes, killifishes, and *Hepsetus odoe*, the Pike Characin, are surface dwellers, taking advantage of higher oxygen tensions near the air-water interface. Lungfishes and bichirs use lungs, clariid catfishes have gill chamber organs, anabantids have labyrinth organs, snakeheads have pharyngeal diverticula, and featherfin knifefishes and phractolaemids have alveolar gas bladders.

In South America, drought resistance has evolved in parallel to the African examples. Many fishes of the Amazon region have evolved means of using atmospheric oxygen when drought or vegetative decay lower oxygen levels (Kramer et al. 1978). Surface swimmers, such as arawanas and some characids (pacus, *Brycon*), have vascularized lips. Modifications of the alimentary tract to absorb oxygen are common, including the mouth region of Electric Eels and swamp eels, air-filled stomachs in loricariid catfishes, a vascularized hindgut in callichthyid armored catfishes, and a vascularized gas bladder in lungfish, Arapaima, and erythrinid trahiras. As with the walking catfishes, South American species are reported to abandon drying pools and cross small stretches of wet vegetation or mud in an apparent search for new and wetter habitats (e.g., erythrinids such as *Hoplias* and *Hoplerythrinus*, callichthyid catfishes, some rivulines) (Lowe-McConnell 1987).

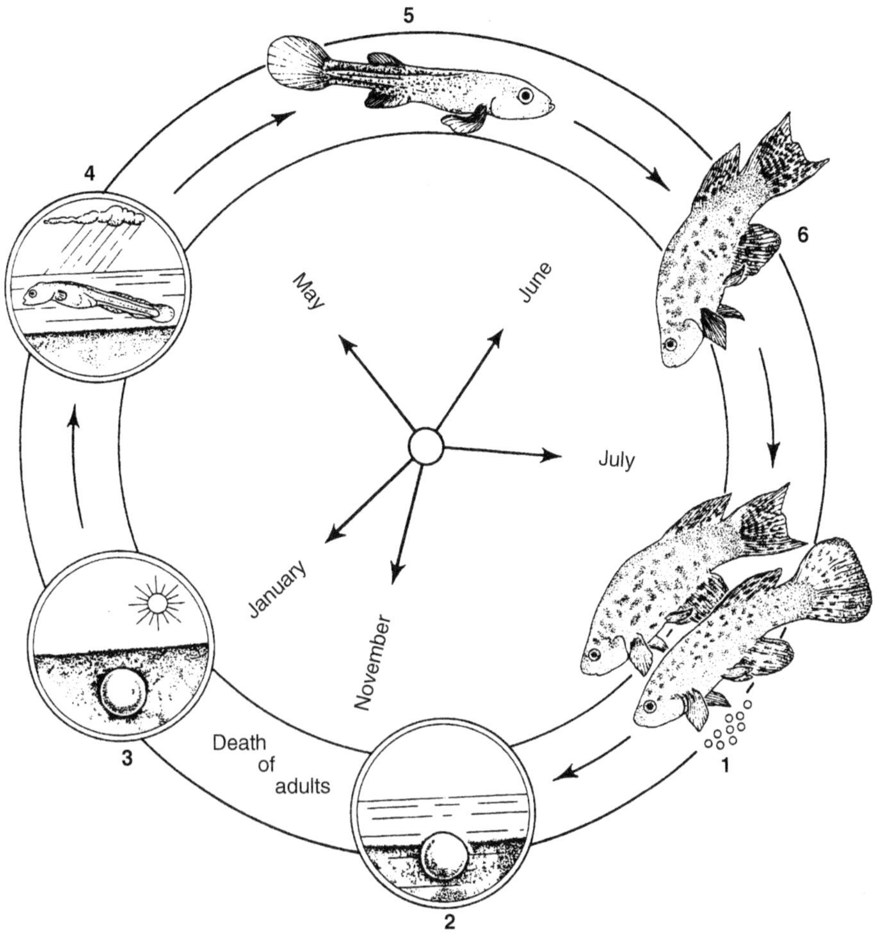

FIGURE 10.15 Life cycle of annual cyprinodontoids, as shown by the Venezuelan *Austrofundulus myersi* (Aplocheilidae): 1, spawning occurs over a protracted period; 2, shelled eggs are deposited in the mud; 3, as the water dries up, adults die but eggs remain viable in an arrested developmental stage; 4, with the return of the rains, eggs hatch; 5, larvae and juveniles grow rapidly; 6, maturation occurs after only a month or two, followed by spawning. From Wourms (1972), used with permission.

Conventional desert areas also exist in South America. The Chaco region of northwestern Paraguay receives less than 30 cm of water annually, with a normal 3-month winter drought period that can last as long as a year (Smith 1981b). During the annual drought, aquatic habitats become isolated and dry up. During the rainy season, these habitats are often repopulated by fishes from overflowing portions of the Paraguay River. Adaptations to drought include estivation in mud by juvenile and adult lungfish (*Lepidosiren*), accessory respiratory structures for using atmospheric oxygen (lungfishes; catfishes, *Hoplosternum*, *Pterygoplicthys*; characiforms, *Hoplias*), and annual life histories and diapausing eggs among cyprinodontiforms, which are also successful throughout much of tropical South America (e.g., *Cynolebias*, *Rivulus*, *Austrofundulus*). Localized extirpation occurs annually in species that invade the Chaco region during the wet season but that lack the abilities to overcome drought conditions.

Australia is largely a desert continent. Its freshwater fish fauna is dominated by **marine derivatives**, such as eels (Anguillidae), catfishes (Plotosidae), rainbowfishes (Melanotaeniidae), barramundi (Latidae), temperate basses and pygmy perches (Percichthyidae), grunters (Terapontidae), gobies (Gobiidae), and sleepers (Eleotridae). Several Australian fishes show distinct adaptations to periodic drought. The endemic Salamanderfish *Lepidogalaxias salamandroides* occurs commonly in southwestern Australian habitats that dry up during the annual summer drought. As waters recede, the fish burrows into bottom sediments and surrounds itself with a thick mucus coat (Berra & Allen 1989; Pusey 1990). Salamanderfish conserve water by absorbing it from the surrounding soil until soil moisture content approaches zero. They avoid the build-up of toxic nitrogenous wastes by metabolizing lipids rather than proteins; the endpoint of lipid metabolism is carbon dioxide, not nitrogen compounds (Pusey 1989).

Several species in the related family Galaxiidae in Australia and New Zealand occur in similar temporary habitats and may also estivate during dry periods. Some gobies and hardyhead silversides that live in desert springs in central Australia are exceptionally tolerant to high temperatures, high salinities, and low dissolved oxygen. The Desert Goby, *Chlamydogobius eremius*, typifies desert-adapted species in its ability to survive an extreme range of conditions, including salt concentrations from zero to 60 ppt (seawater is 35 ppt), temperatures between 5 and 40°C, and oxygen concentrations below 1 ppm. To avoid lethal conditions in thermal springs or high summer temperatures, it seeks cooler vertical or lateral portions of springs, buries itself in cooler silt, and even emerges from the water to capitalize on evaporative cooling and aerial respiration (Glover 1982).

North American Deserts

Some of the best-studied desert fishes occur in the southwestern United States. The Basin and Range Province of North America contains four different deserts, the Great Basin, Mojave, Sonoran, and Chihuahuan deserts (Naiman & Soltz 1981). The province, which includes such seemingly inhospitable areas for fishes as Death Valley, Ash Meadows, Salt Creek, and Devil's Hole, constitutes almost 10% of the total land area of North America. Although desert conditions have existed periodically in the region for approximately 70 million years, the southwestern deserts as they exist today are relatively young, no more than 12,000 years having passed since the last wetter, "pluvial" period when the area contained abundant, interconnected standing and running water. Despite their relative youth and small size, the southwestern deserts contain 182 native species, 149 of which are endemic to the basin and many of which are endemic to single locales (the area includes both US and Mexican endemics). Endemicity in the fishes of the desert southwest is the highest of any place in North America.

Two major types of desert habitat are occupied by fishes: (i) isolated pools and basins supplied by underground springs that have fairly regular flow; and (ii) intermittent marsh and arroyo habitats along flowing watercourses that originate in wetter areas such as mountainous highlands and that flow into arid regions. The native fishes that occur there belong to five principal families and segregate according to fish size, habitat size, and environmental extremes. Small livebearers (Poeciliidae) and even smaller desert pupfishes (Cyprinodontidae) live in the most extreme or isolated habitats such as intermittent streams and spring basins; these fishes include 20 desert-adapted species and subspecies in the genus *Cyprinodon*. Many of these endemic pupfishes are derived from a single group that dispersed among the connected waters of the Death Valley system over 3.5 million years ago, became isolated as the region dried up, and are found today in only a single isolated pool or stream system (Hillyard 2011).

Small streams contain small minnows (Cyprinidae) that are <6 cm long; larger streams and small rivers support medium-sized suckers (Catostomidae) and trout (Salmonidae). The largest fishes, such as large suckers and the Colorado Pikeminnow (*Ptychocheilus lucius*, up to 2 m), live in major drainages. Body size is intimately tied to habitat size (Smith 1981b). The smallest pupfish, the endangered 2 cm **Devils Hole Pupfish**, *Cyprinodon diabolis*, lives on an 18 m^2 shelf in a spring basin in the smallest habitat of any known vertebrate (Fig. 10.16). In contrast, the Colorado Pikeminnow is the largest minnow in North America and lives in the area's largest habitat, the Colorado River.

FIGURE 10.16 Devils Hole, Nevada, natural home of the Devils Hole Pupfish, the first fish listed under the US Endangered Species Act. Visible are water-level monitoring equipment and a platform for people to walk on while doing fish counts. J. Barkstedt / US National Park Service / Public Domain.

The fishes in marshes and small streams experience the harshest conditions and show the strongest adaptations to desert existence. Desert pupfishes show extraordinary tolerance to environmental extremes. Some can live in water with dissolved oxygen below 2.5 ppm, although there is variation among species. Most fishes show stress at <5 ppm, depending on water temperature. Although these are freshwater fishes, some desert pupfishes can tolerate salinities higher than seawater; the pupfish of Cottonball Marsh in Death Valley tolerate not only very high temperatures but also salinity of 45 ppt (Hillyard 2011). Pupfishes persist in water temperatures that vary from freezing in winter to above 40°C in summer, the highest recorded for a habitat containing live fishes. The temperature ranges for successful reproduction, however, are a bit lower than those for adult survival (Hillyard 2011). The desert pupfishes are the only fish in many of these extreme habitats, and many have lost their lateral lines and pelvic fins, which may be energy-saving responses in isolated habitats that lack predators.

As an example of a behavioral adjustment to drought conditions, adult Longfin Dace, *Agosia chrysogaster*, of the Sonoran Desert move into moist algae during hot days and emerge during cooler nights to forage in only a few millimeters of water (Minckley & Barber 1971).

Although most emphasis is given to periods of low water in deserts, a major influence on stream- and river-dwelling fishes are periodic flash floods, when waters can change from nearly stagnant to raging torrents in a matter of seconds (Naiman 1981). Colorado River endemics (see Fig. 22.5), such as the Humpback Chub (*Gila cypha*), Bonytail Chub (*G. elegans*), and Razorback Sucker (*Xyrauchen texanus*), have evolved physical features that provide hydrodynamic stability during periods of high or turbulent flow. These include flattened heads, keeled napes, cylindrical bodies, small scales, elongate, narrow caudal peduncles, and anterior humps. The large humps could also be a convergent response to gape-limited predation by the endemic Colorado Pikeminnow (e.g., Portz & Tyus 2004).

High-flow adaptations are not restricted to large fishes. The Threatened 5 cm Gila Topminnow, *Poeciliopsis occidentalis*, has been extirpated through much of its range due to predation by introduced Mosquitofish, *Gambusia affinis*. However, topminnows are able to coexist with Mosquitofish in streams that experience periodic flash floods because the topminnows show instinctive behavioral adaptations to high discharge, including rapid movement to shoreline areas as waters rise, and proper orientation to strong currents. Mosquitofish, which evolved in southeastern regions that lack flash floods, behave inappropriately and are flushed out of rivers when floods occur (Meffe 1984).

Although the desert pupfishes and other fishes survive and reproduce in the extreme conditions of the desert southwest, these fishes do not exhibit several other traits common to many desert forms, such as estivation, air-breathing, or diapausing eggs. Adaptations of the desert pupfishes are most likely extensions of capabilities possessed by ancestral lineages rather than being newly evolved. Cyprinodontids are small fishes that frequently inhabit estuaries where temperature, salinity, and oxygen availability vary widely. Adaptation to such estuarine conditions would constitute **preadaptation** for desert conditions. Working against the evolution of desert-specific adaptations are the comparative youth of the region, as well as the periodic connection of desert watercourses and pools with each other and with estuarine and riverine areas that serve as sources of new immigrants. Selection for desert adaptations would be relaxed during wetter periods, and dilution of such adaptations would also occur due to gene flow from ancestral areas. In addition, some of the different species and subspecies may hybridize, further underscoring the vulnerability of these fishes.

Caves

Among the more extreme aquatic environments imaginable are underground water systems where no light penetrates and where food availability depends on infrequent replenishment from surface regions. However, **cave living** also has advantages, including a scarcity of competitors and predators and a constant, relatively moderate climate. Fishes have evolved independently in caves around the world and, not surprisingly, similar adaptations to cave life have evolved repeatedly in different phylogenetic lineages. The darkness, low productivity, and in some cases high atmospheric pressure of cave environments have also led to some surprisingly strong convergences between cave fishes and deep-sea fishes.

Caves usually develop in limestone formations because of the solubility of carbonaceous rock, although caves exist in other rock types such as lava tubes on volcanic slopes. Caves include places where water dives underground and resurfaces after a short distance, or where springs upwell near the surface and are illuminated by dim but daily fluctuating daylight (technically a **cavern**). The classic cave environment is a continually dark, subterranean system where fluctuations in temperature, oxygen, and energy are minimal and where little interchange occurs with other areas. The biota of caves are especially interesting because a continuum of habitats exists between the surface, caverns, and deep caves. We can often identify closely related and even ancestral organisms from which cave populations and species evolved. This allows the comparison of cave and surface forms to illuminate the processes that have produced cave adaptations.

Over 135 species in 19 families and 10 orders of teleosts have colonized caves. These unusual fishes – termed variously **hypogean**, **troglobitic**, **phreatic**, and **stygobitic** – occur in scattered locales at tropical and warm temperate latitudes on all continents except Antarctica and Europe (Proudlove 1997a, 2006; Weber et al. 1998). With the exception of some bythitid cusk-eels and gobies, the families are restricted to freshwater. Most cave fishes are ostariophysans (characins, loaches, minnows, and eight catfish families), which is not surprising given the overwhelming success of this superorder in freshwater habitats. This is also a group with connections between the gas bladder and inner ear to enhance hearing, which could be quite useful in a dark cave environment. The remaining four families are either paracanthopterygian (ambloypsid cavefishes) or acanthopterygian (poeciliid livebearers, synbranchid swamp eels, and cottid sculpins). Only one family, the amblyopsid cavefishes, consists primarily (four of six species) of cave-dwelling forms. Many are known from only one or a few locations, although sampling difficulties prevent accurate population estimations. But isolation seems to be commonplace: at least 48 species are known from only one location.

Adaptations to Cave Living

Typical cave-adapted fishes are characterized by **troglomorphism,** a suite of regressive and constructive adaptations for cave living that include a lack of pigmentation, reduced scales, a reduction or loss of light receptors (involving eyes and the pineal gland) (Fig. 10.17) and optic region of the brain, loss of schooling behavior, the dorsal light reaction, and circadian

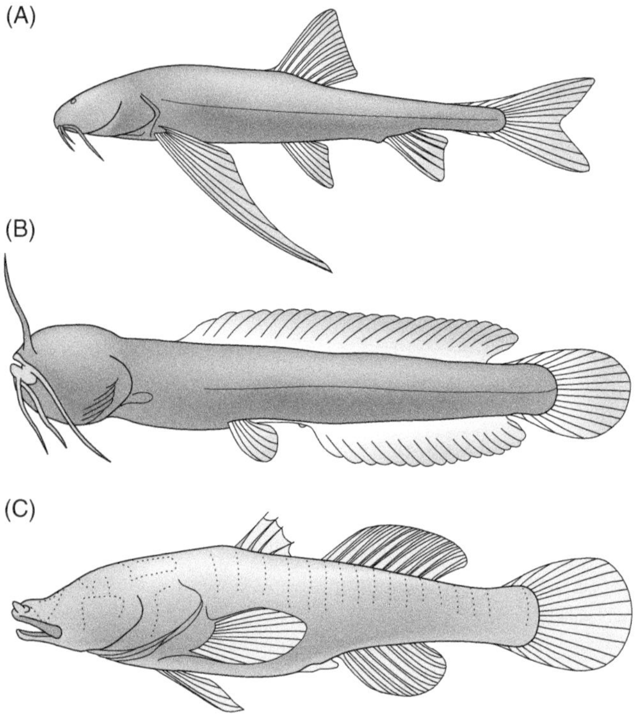

FIGURE 10.17 Cave fishes from three different orders, showing convergent loss of eyes, among other oddities. (A) A balitorid river loach, *Triplophysa xiangxiensis* (Cypriniformes), from China. (B) A clariid catfish, *Horaglanis krishnai* (Siluriformes), from India. (C) An eleotrid sleeper, *Typhleotris madagascariensis* (Perciformes), from Madagascar. After Weber et al. (1998).

rhythms (Wilkens 1988; see Chapters 6, 18). Enhanced traits include greatly expanded lateral line and external chemosensory receptors coupled with increases in brain areas associated with mechanoreception and chemoreception (Yamamoto & Jeffery 2011). These enhanced non-visual traits are shared by many cave-adapted animal taxa. For example, taste buds in surface-dwelling *Astyanax fasciatus* are generally restricted to the mouth region, whereas in cave-adapted populations of the same species they cover the lower jaw and ventral areas of the head. The cave-adapted *A. fasciatus* are about four times more effective in finding food on the bottom of a darkened aquarium than are the surface forms due to their enhanced chemoreceptive ability.

Adaptations to unpredictable or irregularly occurring food supplies also exist. When fed as much as they can consume, cave *Astyanax* build up about four times more fat reserves than surface forms (37% of body mass vs. 9%). Parallel comparisons can be made within the cavefish family Amblyopsidae; those living in caves swim more efficiently, have lower metabolic rates, and find prey quicker and at greater distances in the dark than closely-related surface forms. The water in some cave habitats has high levels of hydrogen sulfide, and fishes in these habitats may have elevated metabolic rates due to the need to detoxify sulfide and may compensate with reduced body size (Passow et al. 2015).

Cave forms are also better at avoiding obstacles and at memorizing the locations of objects than are the surface fish. Cave catfishes (blindcats; Ictaluridae) show parallel changes with respect to eye loss, absence of pigmentation, pineal reduction, enlarged lateral line pores and canals, and brain modifications. Many analogous adaptations have been observed in other cave-adapted taxa, including beetles, amphipods, crickets, crayfishes, shrimps, and salamanders (Poulson 1963; Poulson & White 1969; Culver 1982; Langecker & Longley 1993; Parzefall 1993).

Adjustments to cave existence also occur in the life history traits of cave-dwelling fishes. Cavefishes (Amblyopsidae) produce fewer but larger eggs with greater yolk supplies, delayed hatching, later age at maturation and longer life spans (Bechler 1983). Reproductive rates of cave populations are surprisingly low. Only about 10% of the mature fish in a population of cavefishes may breed in any one year, each female producing 40–60 large eggs. These eggs are incubated and the resulting fry protected in the mother's gill cavity for 4–5 months, long after the young are free-swimming. This may be the longest period of parental care for an externally fertilized fish species. Many of these characteristics are what one would expect in a habitat where adult mortality and interspecific competition are low, environmental conditions stable, and food scarce (Culver 1982; see Chapter 20).

The degree of anatomical and behavioral change in a cave population is often correlated with the age of cave colonization. Eye loss, characteristic of cave-adapted forms, shows some responsiveness to light availability. When young *A. fasciatus* from caves of different presumed ages are raised in the presence of light, individuals from old cave populations do not develop eyes, surface populations develop eyes, and young cave populations vary in eye size (Parzefall 1993). Not surprisingly, visual displays are generally lacking during courtship of cave species, even in taxa such as livebearers and characins where they occur commonly in surface forms.

Food sources in caves are rather limited. No photosynthesis can occur in a sunless cave, therefore food can only arrive if brought in by other animals or carried in by percolation through the rock or by water currents, such as during occasional floods. Common food types differ among families, but bat and cricket guano, bacteria, algae, small invertebrates (isopods, amphipods, copepods), and conspecifics are the common food types of most groups (Parzefall 1993). In Mexican caves containing the livebearer *Poecilia mexicana*, bat guano is supplemented by bacteria associated with sulfur springs in the cave, an interesting analog to deep-sea vent communities. Cave fishes respond to chemical or mechanical cues given off by the food; a clay ball dropped into the water will induce active swimming and searching by cave fishes within 1 m.

Cave fishes usually live at low densities, particularly those in isolated deep caves; most populations involve hundreds or at most thousands of individuals. Population density is strongly correlated with food availability, which again correlates with the degree of isolation. Typical population densities of such fishes as the amblyopsid cavefishes are low, ranging from 0.005 to 0.15 fish/m^2. The Blind Cavefish, *A. fasciatus*, can reach densities of 15/m^2 and *P. mexicana* can reach densities

FIGURE 10.18 *C. thamicola*, a waterfall-climbing, cave-dwelling fish from Thailand. Known only from two locales and designated Vulnerable by IUCN because of small populations and limited distribution, this remarkable 30 mm fish shows classic specializations for both cave and swift water. ChulabushKhatancharoen / Wikimedia Commons / CC BY 2.5.

of 200/m^2 where sulfur springs occur, and near-surface caves that contain bats as an energy source host even higher densities of cave-dwelling fishes.

At the pinnacle of this discussion of evolution in special habitats is a specialized freshwater fish that truly exemplifies the Principle of Convergence. The waterfall climbing cave fish *Cryptotora thamicola* from Thailand is a member of the hillstream loach family (Balitoridae) along with several torrent-dwelling fishes. Its morphology is similar to that of other torrent dwellers: greatly enlarged pectoral and pelvic fins with adhesive pads, and a short, blunt, sloping forehead. It occurs in fast-flowing, cascading water where it has been observed to climb waterfalls (Kottelat 1988; Trajano et al. 2002; Proudlove 2006), except the waterfalls are in caves. In addition to the adaptations to rushing turbulent water, *Cryptotora* is a classic cave dweller: naked, eyeless, and colorless (Fig. 10.18). Strong selection pressures produce predictable adaptations, and adaptation to one selective regime does not preclude simultaneous adaptation to other selective regimes.

Preadaptation, Evolution, and Convergence in Cave Fishes

Adaptation to the cave environment often involves two contrasting trends in the development of structures. Organs that may have been useful to surface ancestors but are of limited use in the cave, such as eyes and pigment, are gradually lost, a process known as **regressive evolution** (Jeffery 2009). They are compensated for by **hypertrophied** ("overdeveloped") structures, such as enlarged lateral line and chemosensory receptors, and their neural correlates. The mechanisms and agents of selection leading to regressive evolution – namely the relative importance of selection, pleiotropy, energy economy, population size, time since isolation, and gene flow – remain a matter of active debate (Culver 1982; Gross 2012).

Some groups possess preadaptations that may have made the transition to cave life quicker. Surface-dwelling Mexican characins show reduced eye development when raised in the dark, and blinded surface fish are as effective at avoiding obstacles as are cave-adapted fish. The multiple recognized forms of the blind Mexican cavefish, *Astyanax mexicanus* were likely established by multiple independent invasions from at least two ancestral surface-dwelling populations (Gross 2012). At least 10 cave families commonly contain nocturnal species; nocturnality and its attendant emphasis on non-visual sensory modes would be an important preadapation for cave living. Some cave-dwelling characins develop taste buds outside the mouth. This pattern also exists in surface-dwelling ictalurid catfishes; in fact, taste buds are more numerous on the barbels and general body surface than in the mouth of ictalurids, which could ease the transition to a cave environment. An elongate body and other eel-like features occur in nearly one-third of cave forms, such as the synbranchid swamp eels, cusk-eels, clariid catfishes, loaches, trichomycterid catfishes, and arguably the amblyopsid cavefishes themselves. Seven acanthopterygian species (i.e., non-anguilliforms) are eel-like. Anguilliform swimming may be advantageous in the narrow confines of caves (see Chapter 4, Locomotory Types). Evolution of eel-like bodies has occurred in several dozen non-anguilliform fishes, another case of convergent evolution.

Several authors have noted the similarities in traits between cave fishes and bathypelagic deep-sea forms, known as the **deep-sea syndrome**. Similar adaptations in the two habitat types include losses of pigmentation, scales, and light receptors, expanded lateral line and chemosensory receptors, and attendant modifications in the brain. In the blind catfishes, which live deeper than most other cave fishes (400–500 m), additional convergences occur in reduced body size, gas bladder regression, large lipid deposits, and reduction of musculature and skeleton. These changes can be viewed as adaptations to overcome challenges associated with energy conservation in an environment with limited food availability (Langecker & Longley 1993). These parallels underscore once again the descriptive power of the Principle of Convergence: if selection pressures and processes are strong and analogous, convergence can occur not just among species within a habitat but also between habitats with similar challenges.

Supplementary Reading

Block B, Stevens E. 2001. *Tuna: physiology, ecology, and evolution. Fish physiology*, Vol. 19. New York: Academic Press.

Briggs JC. 1995. *Global biogeography*. Amsterdam: Elsevier.

Castro P, Huber ME. 1997. *Marine biology*, 2nd edn. Dubuque, IA: Wm. C. Brown/Time-Mirror.

Culver DC. 1982. *Cave life: evolution and ecology*. Cambridge, MA: Harvard University Press.

Eastman JT. 1993. *Antarctic fish biology: evolution in a unique environment*. San Diego: Academic Press.

Farrell AP, Stevens ED, Cech JJ, Richards JG. 2011. *Encyclopedia of fish physiology: from genome to environment*. Amsterdam, Boston: Elsevier/Academic Press.

Farrell AP, Steffensen JF. 2005. *Physiology of polar fishes. Fish physiology*, Vol. 22. New York: Academic Press.

Fukuchi M, Marchant HJ, Nagase B. 2006. *Antarctic fishes*. Baltimore, MD: The Johns Hopkins University Press.

Gage JD, Tyler PA. 1991. *Deep-sea biology: a natural history of organisms at the deep-sea floor*. Cambridge, UK: Cambridge University Press.

Marshall NB. 1971. *Explorations in the life of fishes*. Cambridge, MA: Harvard University Press.

Marshall NB. 1980. *Deep sea biology: developments and perspectives*. New York: Garland STPM Press.

Naiman RJ, Soltz DL, eds. 1981. *Fishes in North American deserts*. New York: Wiley & Sons.

Priede IG, Froese R. 2013. Colonization of the deep sea by fishes. *J Fish Biol* 83:1528–1550.

Proudlove GS. 2006. *Subterranean fishes of the world. An account of the subterranean (hypogean) fishes described up to 2003 with a bibliography 1541–2004*. Moulis, France: International Society for Subterranean Biology.

Randall DJ, Farrell AP, eds. 1997. *Deep-sea fishes*. San Diego, CA: Academic Press.

Sharp GD, Dizon AE, eds. 1978. *The physiological ecology of tunas*. New York: Academic Press.

Websites

Convention on the Conservation of Antarctic Marine Living Resources, **www.ccamlr.org**.

Desert Fishes Council, **www.desertfishes.org**.

International Society for Subterranean Biology, **www.area.fi.cnr.it/sibios**.

Pelagic Fish Research Group, **http://pelagicfish.ucdavis.edu**.

The Congo Project (torrent fishes), **http://research.amnh.org/ichthyology/congo**.

PART III Taxonomy, Phylogeny, and Evolution

Stingrays and skates are cartilaginous fishes (Class Chondrichthyes) of the family Rajidae. They are related to the sharks, but are much more flattened, with the spiracles on the top for water intake while breathing and gill slits on the underside where exhaled water exits. Pictured here is a Bluespotted Ribbontail Ray, *Taeniura lymma*, in the Saudi Arabian Red Sea. Photo courtesy of L. Rocha, used with permission.

CHAPTER 11

A History of Fishes

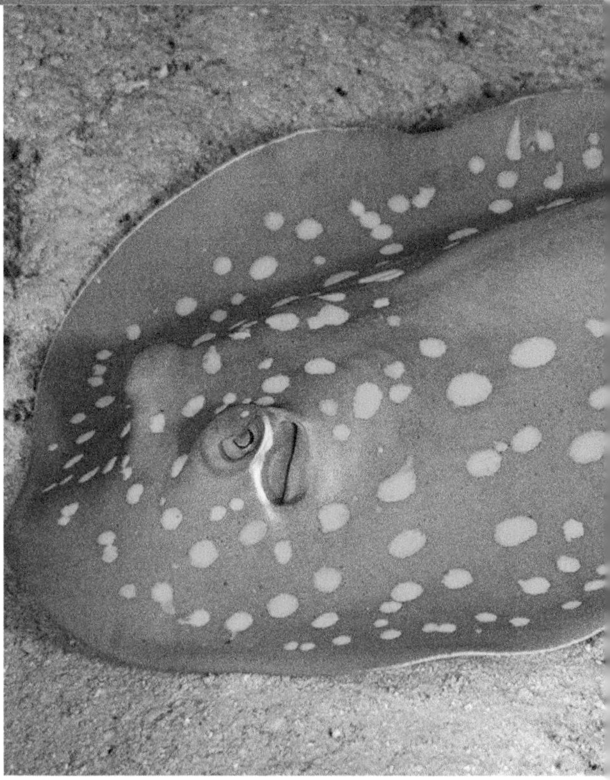

Summary

Fishes emerged from the Cambrian biodiversity radiation over 500 million years before present (mybp). Some fossil groups can be linked with extant taxa, some extant taxa lack obvious fossil antecedents, and numerous groups arose, prospered, and became extinct. The first fishes detected as fossils occurred during the Early Cambrian Period (~541 – 485 mybp) and persisted into the Devonian Period (~419 – 359 mybp). They lacked jaws but possessed bony armor and had a muscular feeding pump. Five major extinct groups of diverse jawless fish-like chordates are recognized: conodonts, pteraspidomorphs, anaspids, thelodonts, and osteostracomorphs. Conodonts (subphylum Conodontophorida) were well known from toothlike structures that fossilized abundantly during Precambrian and later times but could not be linked to any particular body form. Four centimeter long body outlines containing the conodont tooth apparatus were finally discovered in Scotland and Wisconsin in the 1980s. The latter four groups are vertebrates and frequently referred to as "ostracoderms" in reference to a bony shield that covered the head and thorax. Ostracoderms lived in both salt- and freshwater habitats.

The development of jaws was a critical step in the evolution of fishes. However, the ancestry of jawed fishes is unclear because no intermediate fossils between jawed and jawless forms have been found. Placoderms were early jawed fishes that arose in the Silurian Period (~444 – 419 mybp), disappeared by the Early Carboniferous (~359 – 299 mybp), and left no

CHAPTER CONTENTS

The Diversity of Fishes: Biology, Evolution and Ecology, Third Edition. Douglas E. Facey, Brian W. Bowen, Bruce B. Collette, and Gene S. Helfman.

Companion website: www.wiley.com/go/facey/diversityfishes3

apparent modern descendants. They had bony plate-like skin, and many had a hinge at the back top of the head that allowed for greater opening of the mouth. Placoderm teeth consisted of dermal bony plates attached to jaw cartilage and could not be repaired or replaced. Many were predators and quite large (up to 6 m).

The first advanced jawed fishes were the acanthodians. Called "spiny sharks" by some due to their superficial resemblance to modern sharks, but they are not related. Acanthodians were water column swimmers. Many of their traits suggest they share a common ancestry with modern bony fishes, and they are often placed with the osteichthyians in the grade Teleostomi.

Two subclasses form the class Osteichthyes, the Sarcopterygii and the Actinopterygii. These subclasses arose during the Silurian and Devonian and gave rise to modern bony fishes. Sarcopterygians diversified into several infraclasses, including the Actinistia (coelacanths), Onychodontida, Dipnomorpha (several orders that include lungfishes), Rhizodontida, Osteolepida, and Elpistostegalia, and Tetrapoda (stem tetrapods, amphibians, reptiles, birds, and mammals). Elpistostegalians are the most likely ancestors of tetrapods, in that they share skull and neck characteristics and fin patterns with stem tetrapods. Actinopterygians diversified into several extinct orders, plus the Infraclass Cladistia (bichirs), Infraclass Chondrostei (palaeoniscoids, sturgeons, and paddlefishes), and the Neopterygii, which includes the Infraclass Holostei (gars and Bowfin), and Infraclass Teleosteomorpha (teleosts).

Actinopterygians arose during the Silurian. An early, successful group was the palaeoniscoids, which had a triangular dorsal fin, heterocercal tail, paired ray-supported fins with narrow bases, and ganoid scales. Important structural changes occurred in the jaw apparatus that strengthened the bite, increased the gape, and created suction forces. Mobility also improved with lighter scales, vertebral ossification, and an increasingly symmetrical tail.

The Neopterygii are the modern ray-finned fishes, and the most diverse group of vertebrates. They first appeared in Late Permian Period (~299–252 mybp) and radiated extensively during the Mesozoic Era (~252 – 66 mybp). Two extant pre-teleostean groups are the lepisosteiforms (gars) and amiiforms (Bowfin). Teleostean evolution largely repeats and extends trends that originated with the ancestral palaeoniscoids, particularly with respect to advances in jaw and fin structure and function. Convergence in body form and presumably ecological function is striking across palaeoniscoid, neopterygian, and teleostean lineages.

The earliest teleosts were the pholidophoriforms. Four distinct lineages arose from these ancestors: the tarpon and true eel elopomorphs, the bony tongue osteoglossomorphs, the herringlike and minnowlike otocephalans, and the euteleosts, which contain most modern bony fishes. Five major trends characterize teleostean evolution: reduction of bony elements, shifts in position and function of the dorsal fin, placement and function of paired fins, caudal fin and gas bladder modifications, and improvements to the feeding apparatus.

Chondrichthyans (cartilaginous fishes) include several extinct orders, three extinct superorders (Cladoselachimorpha, Ctenacanthimorpha, Xenacanthimorpha), and two subclasses that include extinct and extant groups – the Euselachii (sharklike fishes) and the Holocephali (chimaeras). Sharklike elasmobranchs first appeared in the Late Ordovician Period (~485 – 444 mybp), underwent tremendous diversification, and are represented today by a comparatively depauperate group of specialized modern sharks and rays. Earlier successful radiations included the cladoselachimorphs and xenacanthimorphs, the latter a largely freshwater group. Modern sharks arose during the Mesozoic, showing improvements in jaws, dentition, vertebrae, and fins that paralleled locomotory and feeding changes in bony fishes.

Holocephalans may date back to the Devonian. They share with euselachians a calcified skeleton and pelvic fin claspers but differ by having non-protrusible jaws in which the upper jaw is fused to the braincase, and by a single opercular opening. Holocephalans, whose exact relationships remain a matter of debate, were tremendously successful and diverse through the Mesozoic but are represented today by a small subset of chimaeras.

The First Vertebrates

Fishes were the first vertebrates. Understanding the evolutionary history of fishes is therefore important not only for what it tells us about fish groups, but for what it tells us about evolution of the vertebrates and ultimately our own species. Innovations during fish evolution that were passed on to modern vertebrates (including humans) include dermal and endochondral bone and their derivatives (vertebral centra, bony endoskeletons, brain cases, teeth), jaws, brains, appendages, and the internal organ systems that characterize all vertebrate groups today. During 500 million years of evolution, fishes colonized and dominated the seas and fresh waters and eventually emerged, at least for short periods, onto land. Major clades prospered and vanished or were replaced by newer groups with presumably more efficient adaptations.

Extant ("living") fishes therefore represent the most recent manifestations of adaptations and lineages that have their roots in the early Paleozoic Era (~541 – 251 mybp). The more than 35,000 species of extant fishes constitute only a fraction of the diversity of fishes that has existed historically, as should be evident from the long lists of extinct forms given here (which in turn represent a select fraction of the diversity of former taxa). Many of the extinct forms are exotic in their appearance, whereas others are remarkably similar to living forms, at least in external morphology. A major challenge to ichthyology involves unravelling the evolutionary pathways of both modern and past fish taxa in the process of determining relationships among groups. Which of the many fossil groups represent ancestral types? Which were independent lineages that died out without leaving modern forms? What are the links between and among groups of the past and present? What do fossilized traits tell us about ancient environments? Where do similarities represent inheritance, convergence, or coincidence among extinct and living groups? And how have past adaptations influenced and perhaps constrained present morphologies and behaviors?

The focus of this chapter is on fishes that lived during the Paleozoic and Mesozoic eras, and on modifications that occurred during the evolution of different, major extinct groups, leading to the dominant bony and cartilaginous fishes of today. We deal first with jawless fishes, then with ancestors of modern bony fishes because these occur earlier in the fossil record, and finally with the cartilaginous sharks, skates, rays, and chimaeras. This presentation focuses on extinct rather than extant fishes, recognizing that the distinction is artificial, that many lineages arose hundreds of millions of years ago and still have modern, living representatives, and that direct ancestors of some extant forms arose before other groups that have since gone extinct (see last section of this chapter, Continuity in fish evolution). We follow the basic organization of Nelson et al. 2016, including corrections and updates at **https://sites.google.com/site/fotw5th/**, because of its synthetic and broad approach, recognizing that their conclusions are one of many alternative interpretations of the literature.

Extinct Jawless Fishes

Phylum Chordata[a]
- Subphylum Conodontophorida
  - †Class Conodonta
- Subphylum Craniata
  - Infraphylum Vertebrata
    - †Superclass Pteraspidomorphi
      - Class Pteraspidomorpha (Diplorhina)
        - Subclass Astraspida
          - Order Astraspidiformes
        - Subclass Arandaspida
          - Order Arandaspidiformes
        - Subclass Heterostraci
          - Orders Cardipeltida, Corvaspidida, Lepidaspidida, Tesseraspida, Traquairaspidiformes, Tolypelepidida, Cyathaspidiformes, Pteraspidiformes
    - †Superclass Anaspidomorphi
      - Class Anaspida
    - †Superclass Thelodontiomorpha
      - Class Thelodonti
        - Orders Archipelepidiformes, Furcacaudiformes, Thelodontiformes,
    - †Superclass Osteostracomorphi
      - Class Cephalaspidomorphi (Monorhina)
        - Orders Cephalaspidiformes, Galeaspidiformes, Pituriaspidiformes

[a] Classification based on Nelson et al. (2016), including corrections and updates at **https://sites.google.com/site/fotw5th/**
† Extinct group. All subgroups within an extinct major taxon are also extinct.

The very first fishlike vertebrates undoubtedly evolved from invertebrates, perhaps a cephalochordate. However, the first "fishes" left no fossil record and their form and relationships remain a mystery. By the time fishlike fossils appear in Early Cambrian deposits, roughly 530 mybp (Fig. 11.1), complex tissue types had evolved, including filamentous gills, V-shaped myomeres, and a distinct dorsal fin. *Myllokunmingia fengjiaoa*, perhaps the oldest recognized fishlike vertebrate, was found in the Chengjiang geological formation of Yunnan Province in southwestern China. *Myllokunmingia* was 3–4 cm long and is thought to be allied with (is a sister group to) ancestors of modern lampreys, although agreement is far from universal (Xian-guang et al. 2002; Shu et al. 2003).

If modern cephalochordates such as the lancelets (*Branchiostoma*) are considered fishlike – if not exactly fishes – then the ancestry of fishes can be traced farther back to the cephalochordate-like yunnanozoans (*Haikouella* and *Yunnanozoon*) from the Early Cambrian, or to the much-heralded *Pikaia* with its dorsal nerve cord and notochord, from the Middle Cambrian Burgess Shale of British Columbia (see Chapter 13, Amphioxiformes).

Clearly recognizable fish specimens, such as the arandaspid pteraspidomorph *Sacabambaspis janvieri* from Bolivia, appear later, dating to 470 mybp (Gagnier et al. 1986; Gagnier 1989). This and related **jawless**, finless forms inhabited shallow seas or estuarine habitats in tropical and subtropical regions of the Gondwanan and Laurasian

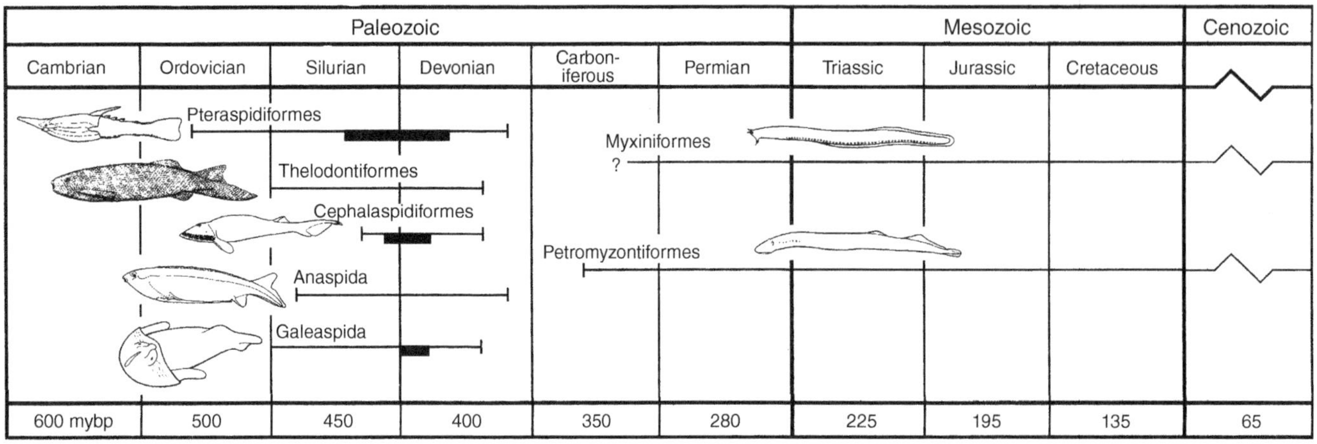

FIGURE 11.1 Periods of occurrence of major jawless fish taxa based on the fossil record. Thickened portions of lines indicate periods of increased generic diversity within a group. Time periods are not drawn to scale (e.g. the Cretaceous lasted almost 50 million years longer than the Silurian, but both are given equal space). Early Cambrian fossils that were arguably fishlike are not included (see text). Fossils are lacking for myxiniforms and petromyzontiforms during the Mesozoic. Data largely from Carroll (1988), Pough et al. (1989), Nelson (2006), and references therein.

supercontinents. Their innovations include **true bone** (probably evolved independently in several ancestral groups) and a **muscular feeding pump**. Bone existed only as an external covering and would have provided protection from predators, as well as serving as a metabolic reserve for calcium and phosphate and an insulator of electrosensory organs (Northcutt & Gans 1983; Carroll 1988). A muscular feeding pump would have been more efficient for moving food-bearing water through a filtration mechanism than was the ciliary feeding mechanism of protochordates. Another major advance over the cephalochordates that preceded them was that, although lacking jaws, the early fossilized fishes were craniates. They had a head region containing a brain with specialized sensory capsules and cranial nerves, all contained in a protective skeletal braincase (Maisey 1996).

Subphylum †Conodontophorida – The Conodonts

Between Late Proterozoic and Late Triassic times (600 to 200 mybp), a group of animals known as **conodonts** ("cone-shaped teeth") arose, proliferated, and died in seas worldwide. The fossil remains, referred to as conodont "elements," consist of toothlike structures generally about 1 mm long and made from calcium phosphate (Fig. 11.2). Known since the mid-1800s, their abundance allowed them to serve as stratigraphic landmarks in determining the age of fossil beds. It was not until the 1980s that fossilized soft body parts were discovered, allowing speculation on true relationships (Briggs et al. 1983; Smith et al. 1987). Before these discoveries, the elements were identified variously as copulatory structures of nematodes, as radulae of snails, and as jaws of annelid worms, among other things. More conservative authors generally placed the animals in a separate, extinct phylum, the Conodonta, with uncertain relationships (Clark 1987).

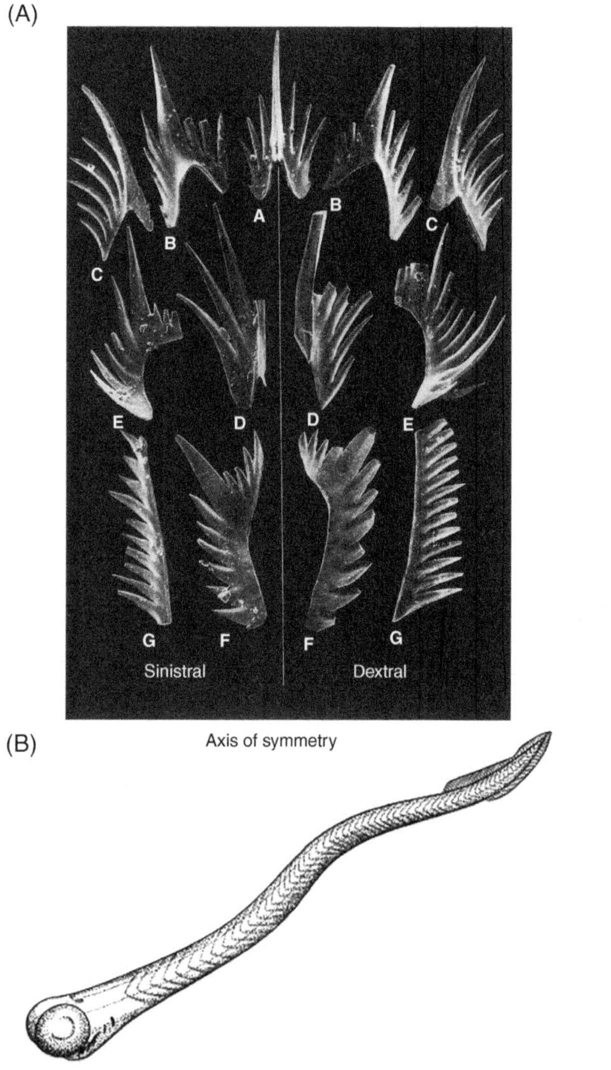

FIGURE 11.2 Conodonts. (A) Conodont apparatus. The various elements (A–G) occur on the right (dextral) and left (sinistral) sides of the head region of the conodont animal and function as the feeding apparatus. (A) From Clark (1987), used with permission of John Wiley & Sons. (B) A conodont, as reconstructed by Aldridge et al. (1993).

The recent discoveries and subsequent reanalyses indicate that the earliest "protoconodonts" of the Paleozoic and Early Cambrian may likely have been invertebrates aligned to chaetognaths (Donoghue et al. 2000). Later euconodonts ("true conodonts") that arose in the Late Cambrian are true chordates, with V-shaped muscle blocks, a bilobate head and cartilaginous head skeleton, eyes contained in otic capsules, extrinsic eye muscles, a compressed body, axial lines suggestive of a notochord, and unequal tail fins supported by raylike elements (Donoghue et al. 1998) (Fig. 11.2B). The total body length ranged between 4 and 40 cm. The conodont elements were contained in the head region and apparently functioned as teeth. Eyeballs and extrinsic eye muscles, chevron-shaped muscle blocks, and apparent bone cells in a dermal skeleton strongly indicate that not only were conodonts chordates, but they may even be classified as vertebrates (Gabbott et al. 1995; Janvier 1995; Purnell 1995).

What are the affinities then of this ancient, highly successful, tooth-bearing, primitive chordate/vertebrate? Following the discovery of the actual animal, the popular interpretation was that the conodont elements were dentition homologous to the rasping, book-closing action of modern hagfishes, placing conodonts near the base of the hagfish lineage (Helfman et al. 1997; see Chapter 13). However, more complete, cladistic analysis incorporating multiple structures and taxa indicates that conodonts constitute a separate, extinct superclass and class that arose after the earliest myxine hagfishes and petromyzontomorph lampreys, i.e. they are more derived than hagfishes and lampreys and may even be basal to the jawed fishes that arose later (Donoghue et al. 1998, 2000).

Subphylum Craniata, Infraphylum Vertebrata

Vertebrate craniata possess, among other features, a **dermal skeleton** and **neural crest**, the latter describing regions of the developing nerve cord that are precursors to gill arches, pigment cells, connective tissue, and bone. Within the vertebrates are six superclasses of fishes, four of which are extinct.

The first fishes, conodonts aside, were historically termed **ostracoderms** ("shell-skinned"), in reference to a bony shield that covered the head and thorax. But ostracoderm is now considered an artificial designation (not monophyletic) that includes perhaps four distinct superclasses of jawless craniate fishes, the **Pteraspidomorphi**, **Anaspidomorphi**, **Thelodontomorphi**, and **Osteostracomorphi**. Relationships between ostracoderm groups and modern jawless fishes such as hagfishes and lampreys remain speculative, with revised interpretations appearing as new fossil discoveries are made.

†Pteraspidomorphi Pteraspidomorphi (or Diplorhina = "two nares") derive their alternate name from impressions on the inside of the head plates indicating two separate olfactory bulbs in the brain. Pteraspidomorphs were jawless filter feeders in both marine and freshwater environments; they occurred from the Early Silurian until the end of the Devonian. Three subclasses of pteraspidomorphs are recognized, the Astraspida, Arandaspida, and Heterostraci. Primitive forms, such as the Ordovician *Astraspis*, *Arandaspis*, and arandaspid *Sacabambaspis*, had symmetrical tails, full body armor, and multiple branchial openings (Fig. 11.3).

Heterostracans ("those with a different shell") had dermal armor that extended from the head almost to the tail, necessitating swimming by lashing the tail back and forth, much like a tadpole. The tail in most forms was **hypocercal**, in that the notochord extended into the enlarged lower lobe of the tail. Their body form, armor, and tail morphology suggest that heterostracans plowed the bottom, pumping sediments into the ventral mouth, and filtering digestible material through the pharyngeal pouches. The armor is generally **sutured** and shows **growth rings**, indicating incremental growth. Early pteraspidiforms were small (*c.* 15 cm), but some heterostracans reached 1.5 m. Two orders, seven families, and more than 50 genera are recognized (see Denison 1970; Carroll 1988).

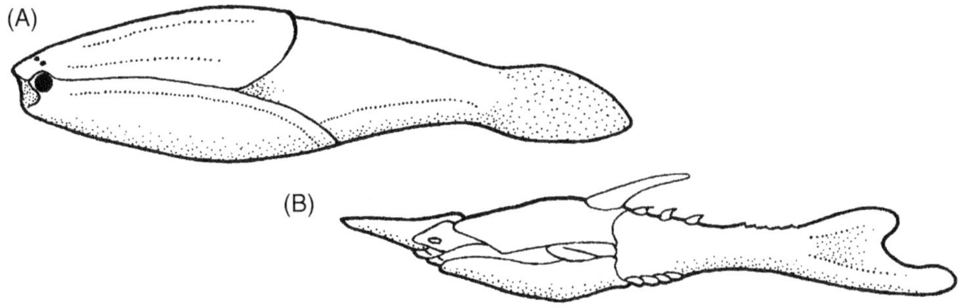

FIGURE 11.3 The earliest known fishes were jawless pteraspidomorphs with armored head shields. Pteraspidomorphs included such small, primitive forms as (A) *Arandaspis* (subclass Arandaspida) from Australia, as well as more advanced forms such as (B) *Pteraspis* (subclass Heterostraci) from Devonian Europe. (A) after Rich and van Tets (1985); (B) after Moy-Thomas and Miles (1971).

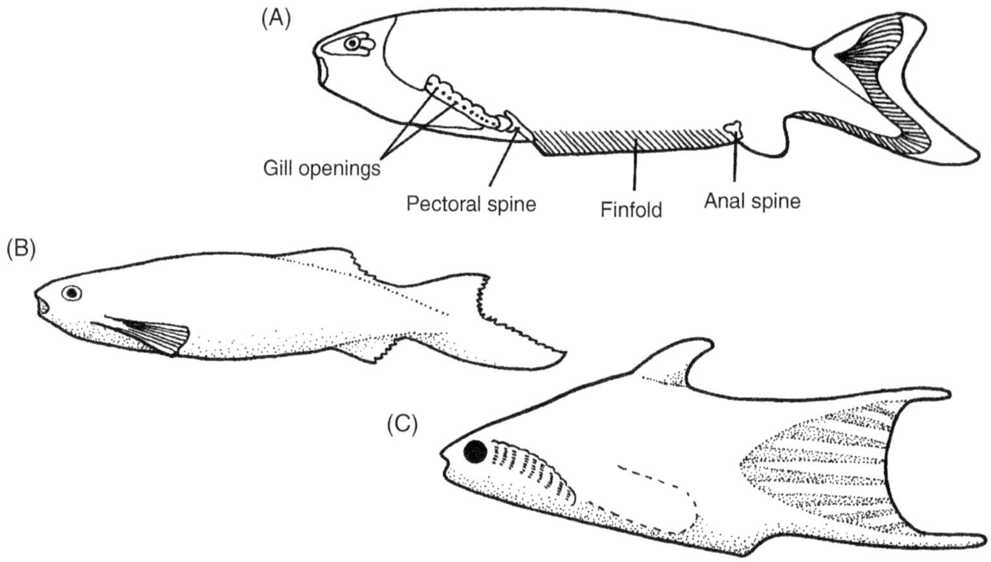

FIGURE 11.4 Other jawless fishes are placed in the superclasses Anaspidomorphi and Thelodontomorphi. (A) Anaspids, such as *Pharyngolepis*, were convergent in body form with the thelodonts, but probably led a benthic existence. (B) Thelodonts were more streamlined, such as *Phlebolepis* with its hypocercal tail. (C) The furcacaudiform forktail thelodonts may be among the first fishes to occupy the water column. (A, B) After Moy-Thomas and Miles (1971); (C) Modified from Wilson and Caldwell (1993).

Later heterostracans, such as the pteraspidiform *Pteraspis* from the Early Devonian, had hypocercal tails, fused dorsal and ventral head plates, and single branchial openings (Fig. 11.3B). The Devonian also produced highly derived forms, such as the sawfish-like *Doryaspis* and the tube-snouted, blind *Eglonaspis*. Trends in the development of pteraspidiform lineages include the reduction of armor through fusion of plates, narrowing of the head shield, and development of lateral, presumably stabilizing, projections (**cornua**). These changes all suggest strong selection for increased mobility and maneuverability. While these anatomical changes were taking place, pteraspidiforms invaded freshwater habitats (Carroll 1988).

†Anaspidomorphi The more fusiform, compressed anaspidiforms, such as *Pharyngolepis* (Fig. 11.4A), occurred from the Late Silurian through the Late Devonian. They were seldom larger than 15 cm and had pronounced hypocercal tails and terminal mouths. Anaspids originated in nearshore marine habitats and gradually entered fresh water. The anaspid body was covered largely with overlapping, tuberculate scales. One advance was the development of flexible, lateral, finlike projections that had muscles and an internal skeleton, thus giving these small fishes considerable maneuverability for (suprabenthic) life in the water column.

†Thelodontomorphi Thelodonts ("nipple teeth," also known as coelolepids or "hollow scales") were diminutive (10–20 cm), fusiform, jawless fishes that were covered with denticles rather than bony plates (Fig. 11.4B). They were abundant and widespread, their denticles/scales serving as stratigraphic indicators in paleontological studies. Most were depressed (flattened), with a horizontal mouth, asymmetrical hypocercal tails, and a detectable lateral line that ran the length of the body. Many thelodonts had dorsal and anal "fins." Their mode of life was probably similar to pteraspidiforms, namely skimming and filtering small organisms from bottom sediments while swimming, although genera with a fusiform body shape and terminal mouths suggest they may have been water column swimmers. A suprabenthic existence is almost certain for the furcacaudiform (literally "forktail") thelodonts of northwestern Canada (Fig. 11.4C). These were shaped like minnows or pupfishes and had compressed bodies, symmetrically forked tails, tubular mouths, and a stomach (Wilson & Caldwell 1993).

Three orders and perhaps 14 families are recognized, with representatives from the Late Ordovician to the Late Devonian. Early thelodonts appear in marine deposits but later groups invaded fresh water. Nelson (2006) summarized the diversity of viewpoints that exist about thelodont position and relationships.

†Osteostracomorphi The superclass Osteostracomorphi contains one class and three orders of jawless fishes. The highly diverse class Cephalaspidomorphi (or Monorhina = "single nostril") first appears in the Late Silurian, approximately 100 million years after the appearance of the pteraspidiforms. They too flourished until the end of the Devonian. They had two semicircular canals and evidence of true bone cells.

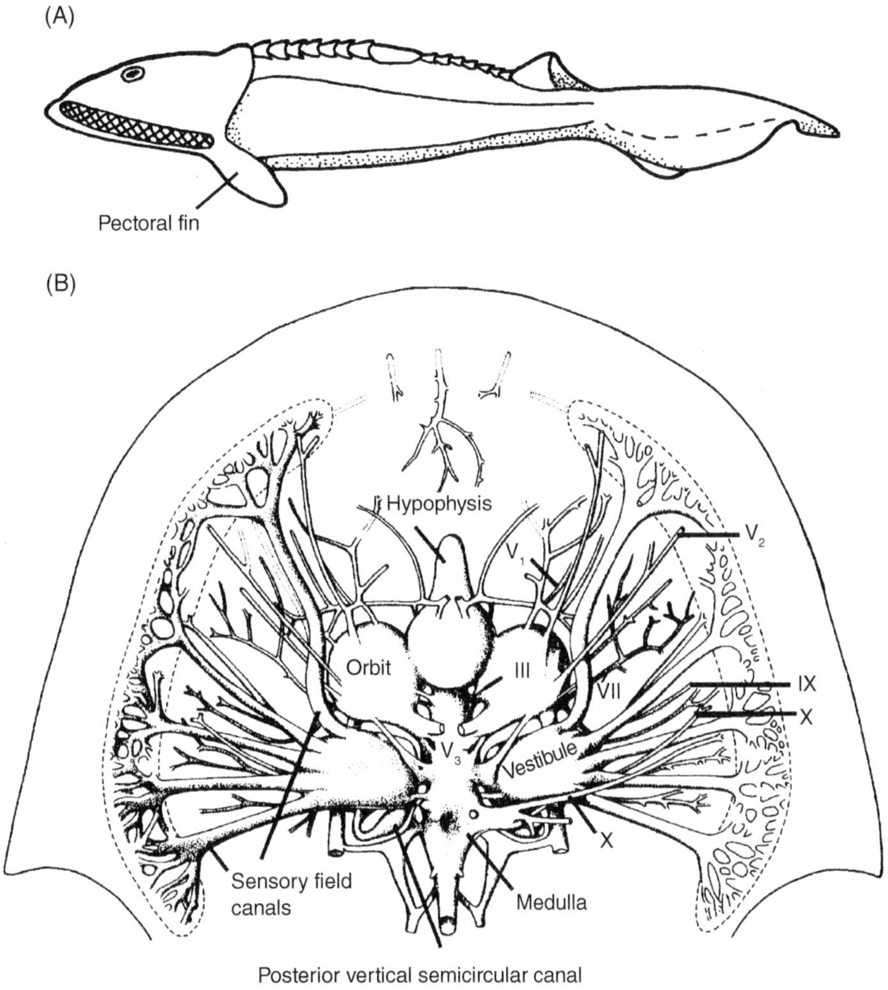

FIGURE 11.5 Cephalaspidomorphs (superclass Osteostracomorphi) were diverse jawless forms that appeared during the Silurian and lasted into the Devonian. The largest order was the cephalaspidiforms, including (A) *Hemicyclaspis*. (B) Thin sections of headshields clearly show brain differentiation and cranial nerves (roman numerals), organized similarly to modern lampreys. (A) After Moy-Thomas and Miles (1971); (B) from Stensiö (1963), used with permission.

The best known cephalaspidomorphs are a predominantly freshwater group, the **Cephalaspidiformes** (Fig. 11.5). These were abundant and diverse fishes; nearly 100 species have been described just in the genus *Cephalaspis* (see Jarvik 1980). Rather than acellular bone, cephalaspidiform armor was cellular. Another cephalaspidiform innovation, also evolved in jawed vertebrates, is ossification of the endoskeleton. Paired lateral appendages in cephalaspidiforms are thought by some to be homologous to gnathostome pectoral fins (Nelson 2006). Unlike the armor of pteraspidiforms, cephalaspidiform head shields are sutureless and lack any apparent growth rings. All fossils of many species are the same size, indicating a naked (nonfossilizing), growing larva that metamorphosed into an armored adult of fixed size. The head shield included one medial and two lateral regions (sensory fields) of small plates sitting in depressions and connected to the inner ear by large canals, for which either an acousticolateralis, electrogenerative, or electroreceptive function has been suggested (Moy-Thomas & Miles 1971; Carroll 1988; Pough et al. 1989). The tail was heterocercal, which may have made skimming along the bottom easier by counteracting the upward lift that the lateral appendages and flattened underside of the head would have generated.

The internal anatomy of the cephalaspidiform head shield is remarkably well known. Swedish paleontologist Erik Stensiö and colleagues painstakingly sectioned rocks containing cephalaspidiforms and worked out the anatomical details of the braincase and cranial nerves (Fig. 11.5B). These efforts allowed identification of such structures as the olfactory lobes, diencephalons, and myelencephalon, the relationship of the hypophyseal sac to the olfactory opening, the relative sizes of the right and left ganglia, the alternation of cranial nerves, the separation of dorsal and ventral nerve roots, the location of blood vessels, the existence of two vertical semicircular canals, and other details (Moy-Thomas & Miles 1971).

The other two cephalaspidomorph orders are the **galeaspidiforms**, with 10 families found only in China, and the **pituriaspidiforms**, with two species found only in Australia. In galeaspidiforms, the median nasohypophyseal opening is large and anterior to the eyes. Paired gill compartments are numerous, up to 45 in number, which is the extreme among vertebrates. Bone was acellular rather than cellular (Halstead et al. 1979; Janvier 1984; Pan 1984).

Based on the detailed anatomical studies of Stensiö, subsequent workers interpreted many cephalaspidiform head structures to be homologous with modern lampreys, concluding that an ancestral–descendant relationship existed. However, more recent analyses indicate that the osteostracomorphs are the closest jawless relatives to jawed vertebrates or gnathostomes, constituting a sister group (i.e. osteostracomorphs and gnathostomes are more closely related to each other than they are to any other clade, including lampreys).

Later Evolution of Primitive Agnathous Fishes

Although much has been written about possible descendants of the early agnathans, additional discussion of the interrelationships of these primitive groups is beyond the scope of this book, mainly because authorities disagree as to where the relationships lie. Different authors consider different characters as ancestral, derived, or convergent, and consequently arrive at different conclusions about relationships between and among jawless and jawed forms. One interpretation gaining acceptance, and the one presented here, is summarized in Figure 11.6. For an historical overview of this controversy, the reader is referred to Jarvik (1980), Carroll (1988), Forey and Janvier (1993), Long (1995), Janvier (1996, 2001), Maisey (1996), Forey (1998), Donoghue et al. (2000), Clack (2002), Pough et al. (2005), Nelson (2006), and Nelson et al. (2016).

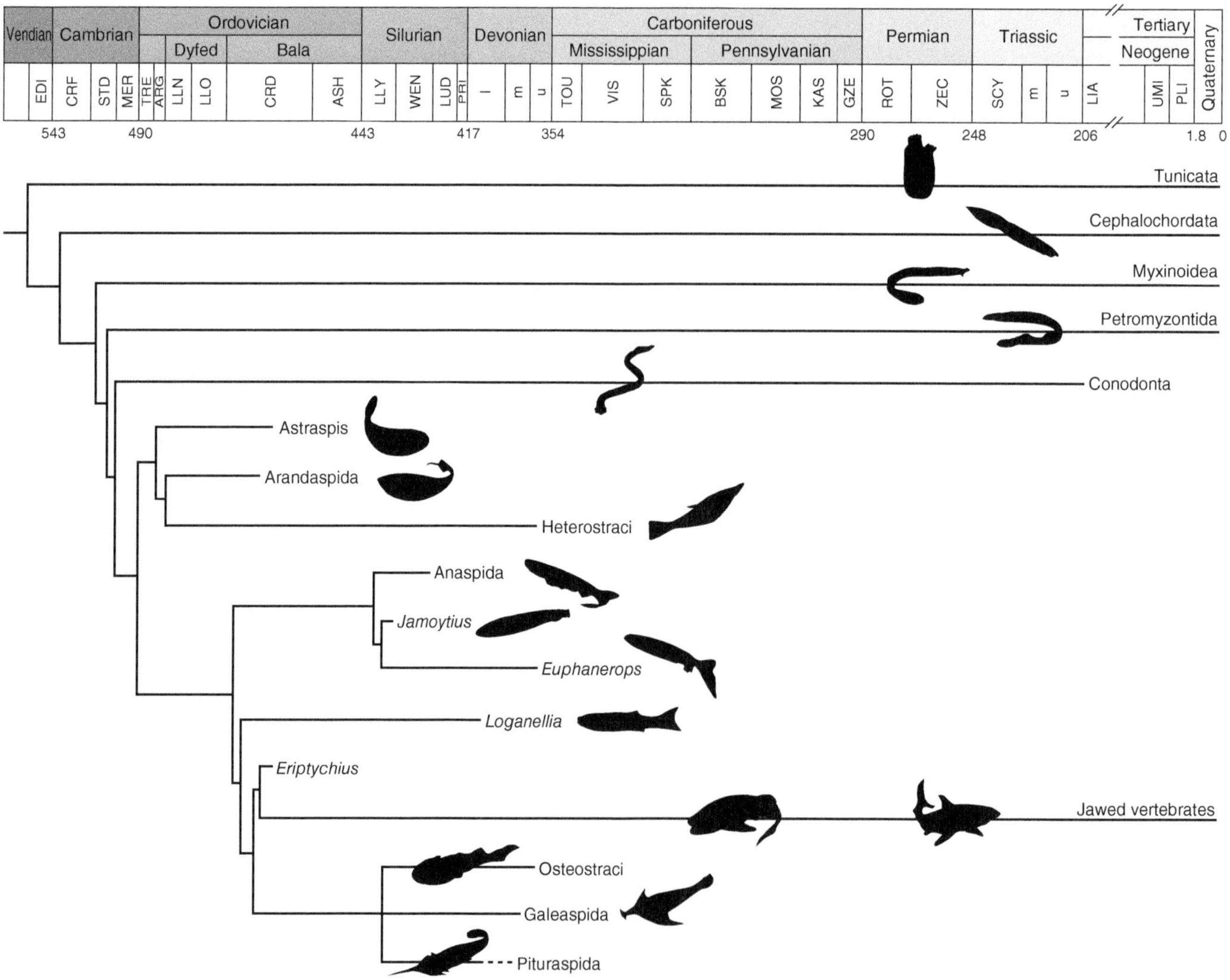

FIGURE 11.6 One view of relationships among early agnathous fishes, modern jawless forms, and jawed vertebrates. Notable here is the stem or sister position of cephalochordates relative to all craniates, of conodonts relative to all jawless vertebrates except lampreys, and of osteostracomorphs (Osteostraci, Galeaspida, Pituriaspida) relative to jawed fishes. Major geological time periods are given at the top of the figure, with abbreviated subdivisions immediately below. The time scale is millions of years before present. Most groups depicted are discussed in the text. From Donoghue et al. (2000), used with permission.

Early Gnathostomes: the First Jawed Fishes

Phylum Chordata
Subphylum Craniata
Superclass Gnathostomata
†Class Placodermi
Orders Pseudopetalichthyiformes, Acanthothoraciformes, Rhenaniformes, Antiarchiformes, Petalichthyiformes, Ptyctodontiformes, Arthrodiriformes

† Extinct group.

The superclass **Gnathostomata** is characterized by several innovations lacking in jawless forms. **Jaws** are present, derived from gill arches. **Paired limbs** with skeletal support are usually present, as is **endochondral bone**, **three semicircular canals**, and **dentine-based** teeth. However, no clearly intermediate fossils between jawed and jawless forms have been found. The origins of jaws and the other structures that characterized the early gnathostomes are buried in the fossil record, belonging to some group yet to be discovered. Homologies between the gill arches of osteostracomorphs and the jaws of later groups are unclear, and the early fossils of jawed fishes already possessed jaws, teeth, scales, and spines. To further complicate our understanding of chronology and phylogeny is the age of different fossils versus the widely held view that placoderms preceded acanthodians and may have been ancestral to chondrichthyans. However, chondrichthyan scales and denticles have been found in Late Ordovician deposits, acanthodians show up later in the Early Silurian, and the earliest placoderms do not appear "until" the Middle Silurian (Nelson 2006).

There is, however, an abundance of fossils that gives us a clear picture of the diversity of forms that the innovation of jaws must have permitted, with groups proliferating and with many early groups giving rise to extant taxa (Fig. 11.7). The evolutionary importance of true jaws cannot be overemphasized: "perhaps the greatest of all advances in vertebrate history was the development of jaws and the consequent revolution in the mode of life of early fishes" (Romer 1962, p. 216).

This revolution included a diversification of the food types that early fishes could eat. Large animal prey could be captured and dismembered, and hard-bodied prey could be crushed. Agnathous fishes were probably limited to planktivory, detritivory, parasitism, and microcarnivory. Stomachs for storage of food evolved, probably because of jaws that could bite off pieces of food. With the advent of jaws, both carnivory and herbivory on a grand scale became possible, as reflected in the size of the fishes that soon evolved. Jaws also allowed for active defense against predators, leading to de-emphasis on armor, which in turn meant greater mobility and flexibility. This increase in agility was greatly enhanced by the development of paired, internally supported pectoral and pelvic appendages, "the most outstanding shared derived character of the gnathostomes besides the jaw" (Pough et al. 1989, p. 235).

†Class Placodermi

Placoderms ("plate-skinned") had tremendous success and diversity. Their name refers to the peculiar bony, often ornamented, plates that covered the anterior 30–50% of the body. Most placoderms had depressed, even flattened bodies, suggesting benthic existence. They may have preyed upon, and eventually replaced, pteraspidiform and cephalaspidiform fishes. As in ostracoderms, placoderms occurred first in marine habitats but later moved into fresh water. As in both ostracoderms and acanthodians (see next section), many placoderm groups show an evolutionary trend toward reduced external armor, leading to a mobile existence in the water column. Placoderms had ossified haemal and neural arches along the unconstricted notochord and three semicircular canals. Placoderms arose in the Late Silurian, flourished worldwide in the Devonian, and disappeared by the Early Carboniferous. Their disappearance often correlates with the proliferation of chondrichthyans at the end of the Devonian, and ecological replacement is suspected.

Seven orders, 25–30 families, and perhaps 200 genera of placoderms are recognized (Fig. 11.8). **Pseudopetalichthyiformes** and **Acanthothoraciformes** both date from the Early Devonian are therefore the oldest known jawed vertebrates. **Arthrodiriformes** (arthrodires, "jointed neck") is the largest order, containing about 170 genera. They possessed a unique hinge at the back top of the head between the braincase and the cervical vertebrae, termed the **craniovertebral joint**. This joint allowed opening of the mouth by both dropping the lower jaw and raising the skull roof, thus increasing gape size. As the group evolved, this joint became larger and more elaborate, and dentition diversified into slashing, stabbing, and crushing structures. Arthrodires were among the largest of the placoderms. *Dunkleosteus* (Fig. 11.8E) was perhaps 6 m long, with a head more than 1 m high; some fossils indicate that *Dunkleosteus* may have reached twice that size (Young 2003). Their large size and impressive dentition implicate the arthrodires as major predators of Devonian seas. Arthrodires (e.g. *Groenlandaspis*) have also been found with fossilized silver and red pigment cells distributed in a pattern indicative of countershaded coloration. Red pigment cells suggest that color vision had already evolved in fishes more than 350 mybp (Parker 2005).

None of the other placoderm orders attained the success of the arthrodires. **Rhenaniforms** were extremely dorsoventrally depressed and bore a striking resemblance to modern skates, rays, and angel sharks (e.g. *Gemuendina*, Fig. 11.8D), although their lateral fins were too heavily armored to be undulated or flapped in the manner

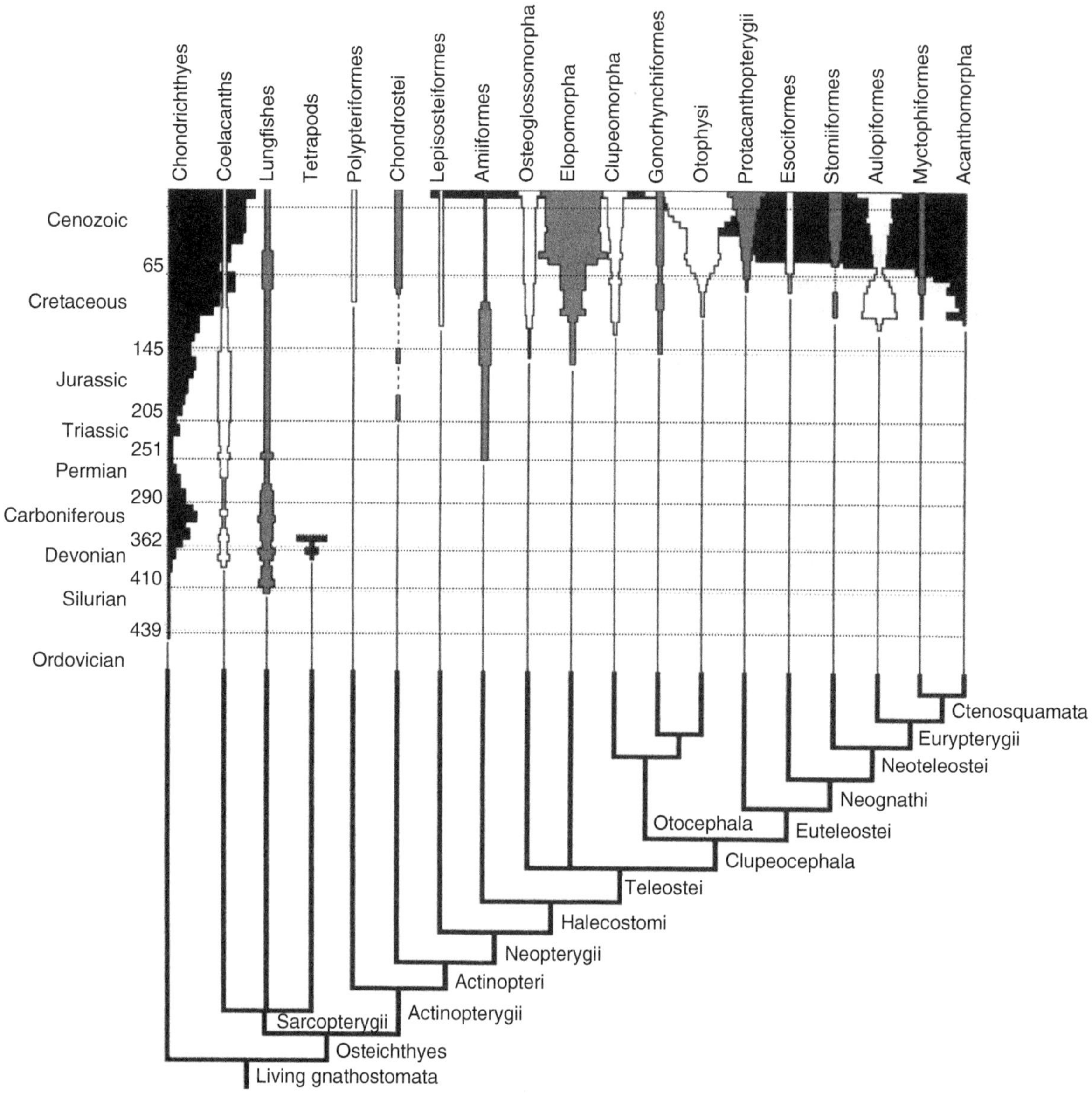

FIGURE 11.7 Periods of occurrence of major jawed (gnathostome) fish taxa based on the fossil record. Column width represents familial diversity within a group (only half of chondrichthyan and acanthomorph diversity is shown). The time scale is millions of years before present. A glance at the figure reveals why the Devonian is commonly referred to as the Age of Fishes: during the Middle Devonian, most major groups discussed in this chapter, including jawless forms shown in Figure 11.1, were represented. Slightly modified from Stiassny et al. (2004).

characteristic of modern skates and rays. The antiarchiforms (**antiarchs**, e.g. *Bothriolepis*, Fig. 11.8C) were predominantly freshwater, heavily armored, benthic fishes with a spiral valve intestine and jointed, arthropod-like pectoral appendages that had internal muscularization. **Ptyctodontiforms** greatly resembled modern holocephalans in body form (see Fig. 11.8B) and are the first fishes known to possess apparent male intromittent organs in the form of claspers associated with the pelvic fins, an indication of internal fertilization.

Although the craniovertebral joint of many placoderms afforded increased jaw mobility compared to forms with a fixed upper jaw, placoderms lacked **replacement** dentition. Placoderm "teeth" consisted of dermal bony plates made up of a unique dentine-like material attached to jaw cartilage. This bone was often differentiated into sharp edges and points, producing "fearsome blade-like jawbones, which wore away during growth like self-sharpening scissors, to leave a hardened core forming massive stabbing blades" (Young 2003, p. 988). However, these blades were subject to breakage and wear, with no apparent repair or replacement mechanism. Placoderm jaw morphology also prohibited them from developing suction forces when feeding. The innovations of serially replaced teeth and of jaws that could create suction characterize the fish taxa that evidently replaced the placoderms and acanthodians.

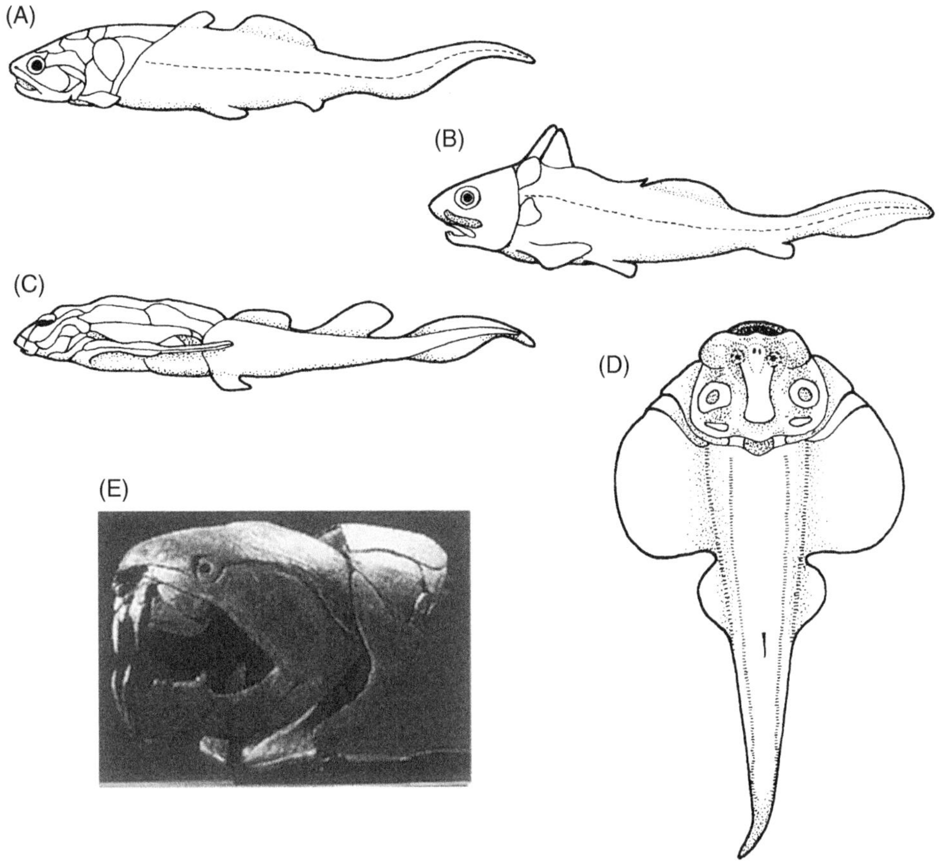

FIGURE 11.8 Placoderms. (A) The coccosteomorph *Coccosteus*, (B) the ptyctodontid *Rhamphodopsis*, (C) the antiarch *Bothriolepis*, (D) the rhenanid *Gemuendina*, and (E) *Dunkleosteus*, a giant arthrodire placoderm from the Devonian. In *Dunkleosteus*, the meter high head was followed by a proportionately large body, but actual lengths are unknown because fossilized remains of the posterior skeleton are lacking. (A–D) After Jarvik (1980) and Stensiö 1963; (E) Courtesy of Chip Clark.

Advanced Jawed Fishes I: Teleostomes

Phylum Chordata
- Subphylum Craniata
  - Superclass Gnathostomata
    - Grade Teleostomi
      - †Class Acanthodii
          - Orders Climatiiformes, Diplacanthiformes, Ischnacanthiformes, Acanthodiformes
      - Class Osteichthyes
        - Subclass Sarcopterygii
          - Infraclass Actinistia (Coelacanthimorpha)
            - Order Coelacanthiformes (coelacanths)
          - Infraclass Dipnomorpha
            - †Superorder Porolepimorpha
              - Order Porolepiformes
            - Superorder Dipnoi
              - †Orders Diabolepidiformes, Dipnorhynchiformes, Dipteriformes, Ctenodontiformes
              - Order Ceratodontiformes (includes living lungfishes)
          - †Infraclass Rhizodontida
          - †Infraclass Osteolepidida
          - †Infraclass Elpistostegalia
          - Infraclass Tetrapoda (the tetrapods)[a]
        - Subclass Actinopterygii (the ray-finned fishes) – *these are addressed in next section of this chapter*

†Extinct group.

[a] Nelson (2006), among others, recognized the tetrapods (including the approximately 30,500 living species) as an infraclass of the fleshy finned sarcopterygians, declaring them a "divergent sideline within the fishes that ascends onto land and into the air and secondarily returns to water" (p. 87). Nelson et al. (2016) concur. It will be interesting to see how students of these higher vertebrate groups respond to this depiction.

The first bony fishes are represented by fragments and microfossils from the Late Ordovician. From these ancestors two distinct groups arose: the class **Acanthodii** and the class **Osteichthyes** - together they make up the grade **Teleostomi**. Osteichthyes has two subclasses, the **sarcopterygians** and **actinopterygians**. Osteichthyes literally means "bony fishes," but this class includes the tetrapods (an infraclass of the sarcopterygians) as they are descendants of the bony fishes.

Teleostomes are grouped together because they share cranial, scale, and fin similarities, but especially because both acanthodians and actinopterygians possess **three otoliths** (sarcopterygian lungfishes have two otoliths and coelacanths have only one). Acanthodians diversified in the Silurian and Devonian and lasted through the Permian. The **Actinopterygii** ("ray-fins") are known first from scales in Late Silurian deposits, whereas **Sarcopterygii** ("fleshy or lobe fins") appear in the Early Devonian.

The actinopterygians and sarcopterygians share numerous characteristics, including the bone series in the opercular and pectoral girdles, the pattern of their lateral line canals, fins supported by dermal bony rays, a heterocercal tail with an epichordal (upper) lobe, replaceable dentition, and a gas bladder that developed as an outpocket of the esophagus. Sarcopterygians diversified into extinct and modern coelacanths, lungfishes, extinct rhizodontimorphs, osteolepidimorphs, and elpistostegalians, and also gave rise to tetrapods. Actinopterygians underwent tremendous multiple radiations, producing the cladistian bichirs, the chondrosteans (many fossil groups plus modern sturgeons and paddlefishes), and neopterygians, including gars and related fossil groups, Bowfin and related fossil groups, and ancient and modern teleosts.

Although we tend to view the more derived fishes as improvements over the primitive taxa, in part because the former are represented today, placoderms and acanthodians existed literally side by side with the "more advanced" forms for more than 100 million years. At some point, for climatic or biological reasons that are unclear, the innovations of the more derived gnathostomes, or the evolutionary constraints placed upon the more primitive groups, led to a replacement of one group by the other. The result was an incredible series of species radiations in four or five very different lineages, derived forms of which are still alive today.

†Class Acanthodii

The oldest fossils of relatively advanced, jawed fishes belong to **Acanthodians**, or "spiny sharks," from Late Ordovician deposits. Their Latin name refers to the stout median and paired spines evident in most fossils; their similarity to sharks is largely superficial and few current authors feel they are related to modern chondrichthyans (but see Jarvik 1980). Acanthodians were found mainly around Laurasia, the ancient "super-continent" made up of modern North America, Europe, and Asia (except India). They were generally small (20 cm to 2.5 m); occurred in both salt and fresh water, had cartilaginous skeletons; a body covered with small, non-overlapping scales; large heads; and large eyes. Their streamlined, round bodies, reduced armor when compared to ostracoderms, subterminal mouths often studded with teeth (including teeth inside the mouth and on the gill rakers), and fin placement suggests that they were water column, not benthic, feeders. Given the success of ostracoderms in benthic habitats, it is not surprising that the next fish group to evolve would occupy the relatively unexploited water column.

Four orders, 12 families, and at least 60 genera of acanthodians have been described, many from isolated spines and teeth (Nelson et al. 2016). All four orders show interesting parallels in evolution. Early acanthodians had multiple gill covers, broad unembedded spines anterior to all fins except the caudal, as well as additional spine pairs between the pectoral and pelvic fins (Fig. 11.9A). More advanced species had single gill covers and lost the ancillary paired spines, the remaining spines being

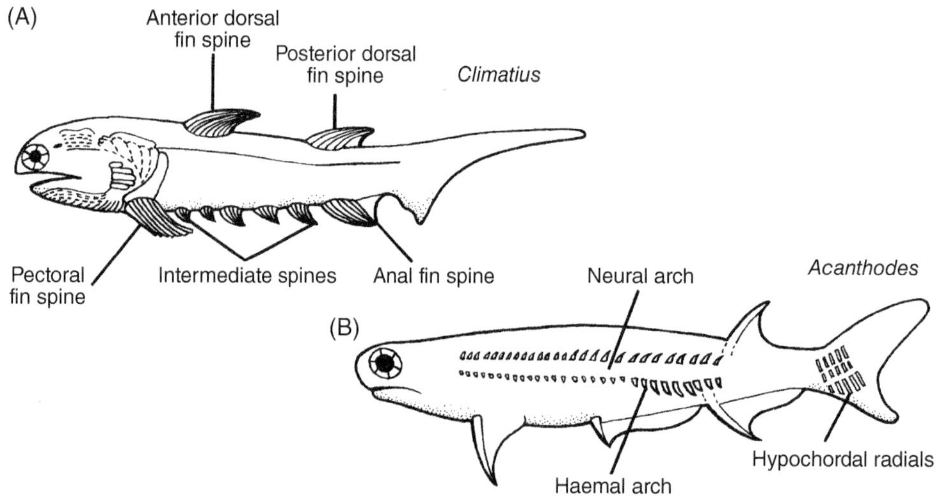

FIGURE 11.9 Acanthodians. (A) *Climatius*, a primitive acanthodian with multiple gill covers and multiple, unembedded spines. (B) The more advanced *Acanthodes*, with fewer, thinner, more deeply embedded spines, a single gill cover, and a more symmetrical caudal fin. After Moy-Thomas and Miles (1971).

thinner and embedded in the body musculature (e.g. *Acanthodes*, Fig. 11.9B). Acanthodiforms were toothless and had long gill rakers, indicating a planktivorous habit. Because acanthodians possessed a third (horizontal) semicircular canal and neural haemal arches associated with the unconstricted vertebral column, and other shared, derived traits (otoliths, lateral line canals, ossified operculum, branchiostegals, cranial and jaw series, including the new interhyal bone), they are included within the Teleostomi (Lauder & Liem 1983; Maisey 1986). Acanthodians survived until the Early Permian, outlasting the major ostracoderm groups by 100 million years.

Class Osteichthyes

Subclass Sarcopterygii This subclass includes seven infraclasses: the **Actinistia** (coelacanths), four extinct infraclasses (**Onychodontida**, **Rhizondontida**, **Osteolepidida**, **Elpistostegalia**), the **Dipnomorpha** (a variety of extinct fishes with stout bodies and paddlelike paired fins and the extant specialized, modern lungfishes), and the **Tetrapoda**, which emerged onto land to become amphibians, reptiles, birds, and mammals.

Ancestral Sarcopterygians Ancestral sarcopterygians are among the most actively studied fossil groups of fishes, in no small part because of their place in tetrapod evolution. Discoveries have often prompted reanalysis of relationships among fossil and extant groups. Agreement is far from universal: Forey (1998) summarized major hypotheses, presenting 13 different phylogenies proposed by different authorities in the last 20 years of the twentieth century. Forey's concluding analysis is presented in Figure 11.10 and is largely followed here. The debate revolves largely around the relative positions of lungfishes, coelacanths, and the

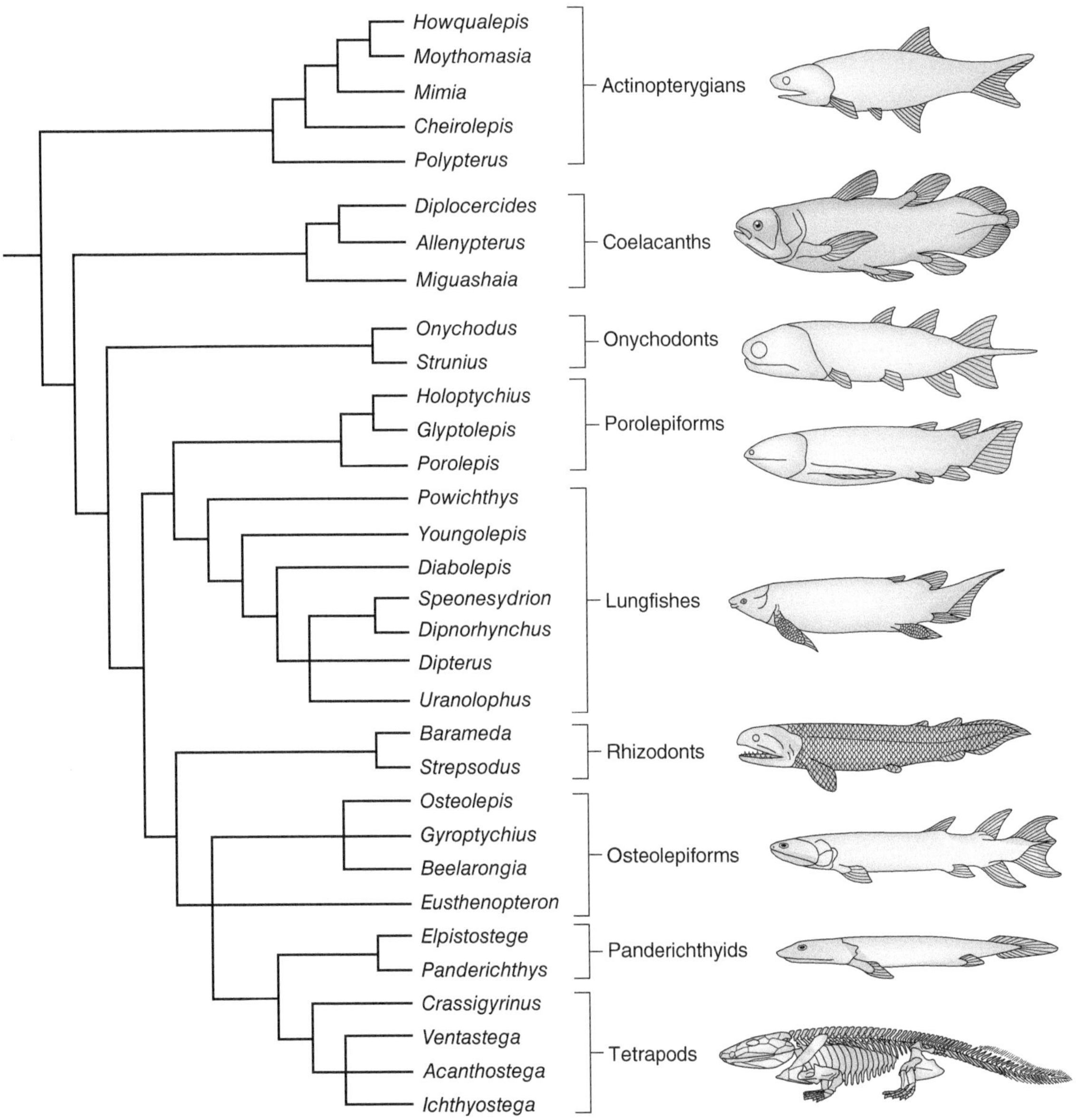

FIGURE 11.10 One view of relationships among bony fishes (class Osteichthyes), showing actinopterygians as the sister group to the various extant and extinct sarcopterygian taxa. After Forey (1998).

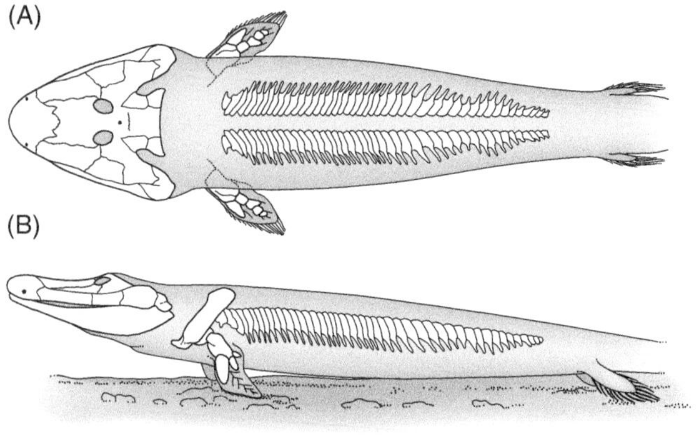

FIGURE 11.11 Dorsal (A) and lateral (B) views of the reconstructed elpistostegalian fish, *Tiktaalik roseae*. Features evident are the lack of opercular bones, the tetrapod-like arrangement of elements in the pectoral fins/limbs, and the stout ribs forming a rib cage that may have protected the lungs. From Daeschler et al. (2006), used with permission.

osteolepiform–porolepiform–panderichthyid lineages relative to tetrapods. Discovery of *Tiktaalik roseae*, a Devonian fish intermediate between the elpistostegalian/panderichthyid group and early tetrapods has influenced the debate.

The search for the putative missing link between piscine sarcopterygians and early tetrapods (the latter no longer classified as amphibians) was greatly clarified with publication in 2006 of the description of *Tiktaalik roseae* and a discussion of its place in the vertebrate lineage, as discussed earlier. Few fossil discoveries, aside from those involving hominid ancestors, have received as much media attention.

Fossils of *Tiktaalik roseae* (Fig. 11.11) were discovered in Nunavut Territory of Arctic Canada and announced to the world in early April 2006 with considerable fanfare (Daeschler et al. 2006; Shubin et al. 2006). This fossil bridged an anatomical gap of approximately 20-million-years between *Panderichthys* (Middle Devonian, 385 mybp) and *Ichthyostega* (Late Devonian, 365 mybp) (Ahlberg & Clack 2006). *Tiktaalik* grew to almost 3 m and is identifiable as an elpistostegalian (= panderichthyid) because of a number of sarcopterygian, fishlike traits. It possessed the dorsally placed eyes, gill arches, scales, pectoral and pelvic fin rays, lower jaw, and palate of those advanced sarcopterygians. But it also possessed the shortened skull roof, otic skeleton, mobile neck, and most significantly, the functional wrist joint of the later stem tetrapods such as *Acanthostega* and *Ichthyostega*.

Several other skeletal features are intermediate or more tetrapod-like (e.g. loss of opercular and subopercular bones; reduced fin rays; elongate, crocodile-like snout; stout, interlocking ribs suggestive of a lung cage; widened spiracle and broadened skull also suggestive of lung function). Equally important, *Tiktaalik* fossils come from Late Devonian strata 382–383 million years old, precisely between the fish and tetrapod groups. The fossils occurred in freshwater alluvial deposits typical of meandering stream systems. Other animals found in the same deposits included an antiarch placoderm, lungfish, porolepiforms, and osteolepidid and tristichopterid sarcopterygians.

Reconstructions of *Tiktaalik* indicate a heavy-bodied organism without dorsal fins but with teeth, neck, wrist, and digits. These were ". . . large, flattish, predatory fishes with crocodile-like heads and strong limb-like pectoral fins that enabled them to haul themselves out of the water" (Ahlberg & Clack 2006, p. 748). The pectoral skeleton is especially striking in that it is clearly transitional between a fish fin and a tetrapod limb in terms of both structure and function (Shubin et al. 2006) (Fig. 11.12). Although still sporting fin rays, *Tiktaalik*'s distal fin structure includes transverse joints and digitlike elements (e.g. a primordial wrist and digits, ". . . transversely aligned and capable of flexion and extension" (Shubin et al. 2006, p. 768)). This structure would be capable of supporting the fish on its "fingertips," presumably to hold itself up above the water surface and perhaps support itself to some extent on land, actions unlikely among earlier sarcopterygians given their internal limb skeleton. The cladogram of relationships places *Tiktaalik* firmly between the lobefin fishes and stem tetrapods of the Late Devonian (Fig. 11.13).

Some considered *Tiktaalik* as representing a sarcopterygian sister group of the stem tetrapods, whereas others suggest that fossil trackways indicate that tetrapods may have emerged prior to *Tiktaalik* (see Lucas 2015). It does appear, however, that *Tiktaalik* demonstrates transitional anatomical features between lobe-finned fishes and tetrapods (hence the term "fishapod" by co-discoverer Neil Shubin), on par with *Archaeopteryx* in linking birds with ancestral reptiles.

Infraclass Actinistia Actinistians are all in the order Coelacanthiformes. Fossil actinistians, also referred to as coelacanthimorphs due to their coelacanth-like body form, appeared in the Middle Devonian and are not known after the Late Cretaceous. They occurred worldwide in both marine and fresh water. The fossil record of the group is extensive: at least 83 valid species in 24 genera and perhaps nine families are recognized. Diversity was maximal during the Early Triassic, when 16 described species existed (Forey 1998; Fig. 11.14). All but one family and two species are extinct.

Coelacanths are in many respects more specialized than other sarcopterygians, possessing a unique spiny rather than a lobate first dorsal fin; a three-lobed caudal fin with a middle fleshy, fringed lobe (the term "coelacanth" describes the hollow nature of the fin rays that support the tail); a rostral organ involving a rostral cavity with several openings on the snout associated with electroreception; and lacking internal choanae, cosmine in the scales, branchiostegals, and a maxilla. Most evolution in the group occurred during the Devonian (some early genera had a heterocercal rather than a diphycercal tail), and later species are surprisingly unchanged in body shape and jaw morphology from the early representatives, although trends of change (reduction in some bones, increases in others) have occurred (Cloutier 1991). Rapid evolution and morphological variation occurred in early coelacanths, including an eel-like species from the Middle Devonian (Friedman & Coates 2006). Prior to the discovery of the living Coelacanth in 1938 (see Chapter 13), coelacanthimorphs were of interest primarily to paleontologists as a specialized, extinct

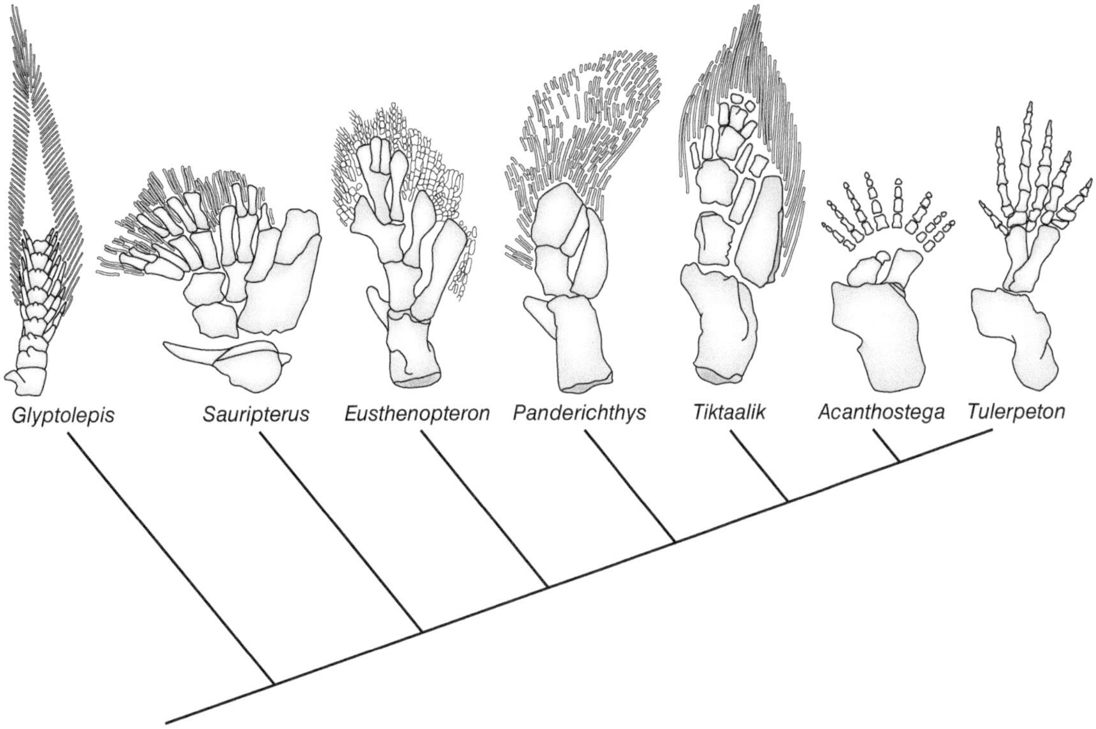

FIGURE 11.12 A cladogram of relationships among sarcopterygians and tetrapods, evidenced by changes in the pectoral fin and limb. *Tiktaalik* retains the central axis of enlarged endochrondral bones of more primitive sarcopterygians, but has fewer lepidotrichia (fin rays) and more radial elements than ancestral fishes. *Tiktaalik* is more advanced in its proliferation of transverse joints across the distal region of the fin, allowing for propping up and moving the body. *Glyptolepis* was a porolepiform dipnomorph related to lungfishes; its archipterygial fin is representative of the basal condition. From Shubin et al. (2006), used with permission.

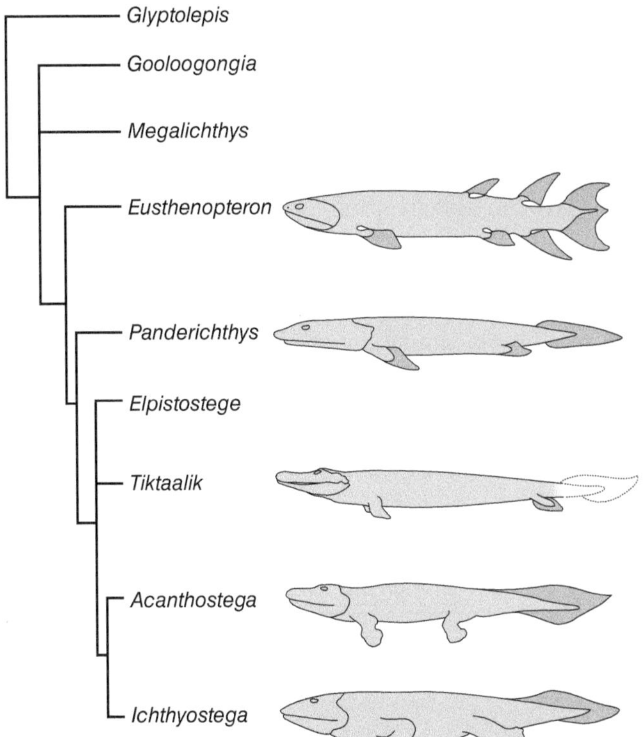

FIGURE 11.13 Cladogram (strict consensus tree) of relationships among sarcopterygians and tetrapods, showing *Tiktaalik*'s intermediate position as a sister group (with *Elpistostege*) to the early tetrapods *Acanthostega* and *Ichthyostega*. The cladogram was calculated from an analysis of 114 characters and nine taxa. After Daeschler et al. (2006).

group notable for its conservativism and its relationship to the reputed ancestors of tetrapods.

Infraclass Dipnomorpha Dipnomorphs are made up of fishes in the extinct superorder **Porolepimorpha** (one order, two families from the Devonian), plus the Superorder **Dipnoi,** which includes several extinct orders of lungfishes and the order **Ceratodontiformes** (three fossil and two extant families). Lungfishes as a group have been generally referred to as **Dipnoi** or dipnoans ("double-breathing"). They arose in the Devonian in marine environments, expanded into freshwater habitats, and died out by the end of the Triassic. Primitive lungfishes were characterized by two dorsal fins; fleshy, scale-covered, paired, leaflike **archipterygial** fins with a bony central axis and with fin rays coming off the central axis; a lack of teeth on the marginal jaw bones, but with tooth plates inside the mouth, and with the premaxilla, maxilla, and dentary missing; a solid braincase; and a pore-filled, cosmine coating on the dermal bones that covered the skull and scales and that may have been associated with electroreception (Fig. 11.15). Later species occupied fresh water, and trends in lungfish evolution include loss of the first dorsal fin, fusion of the median fins (second dorsal, caudal, anal) to form a symmetrical tail (earlier forms had heterocercal tails), elaboration of the tooth plates and development of replaceable dentition, replacement of ossified centra with cartilage, fusion of skull bones, and concomitant loss of the cosmine covering.

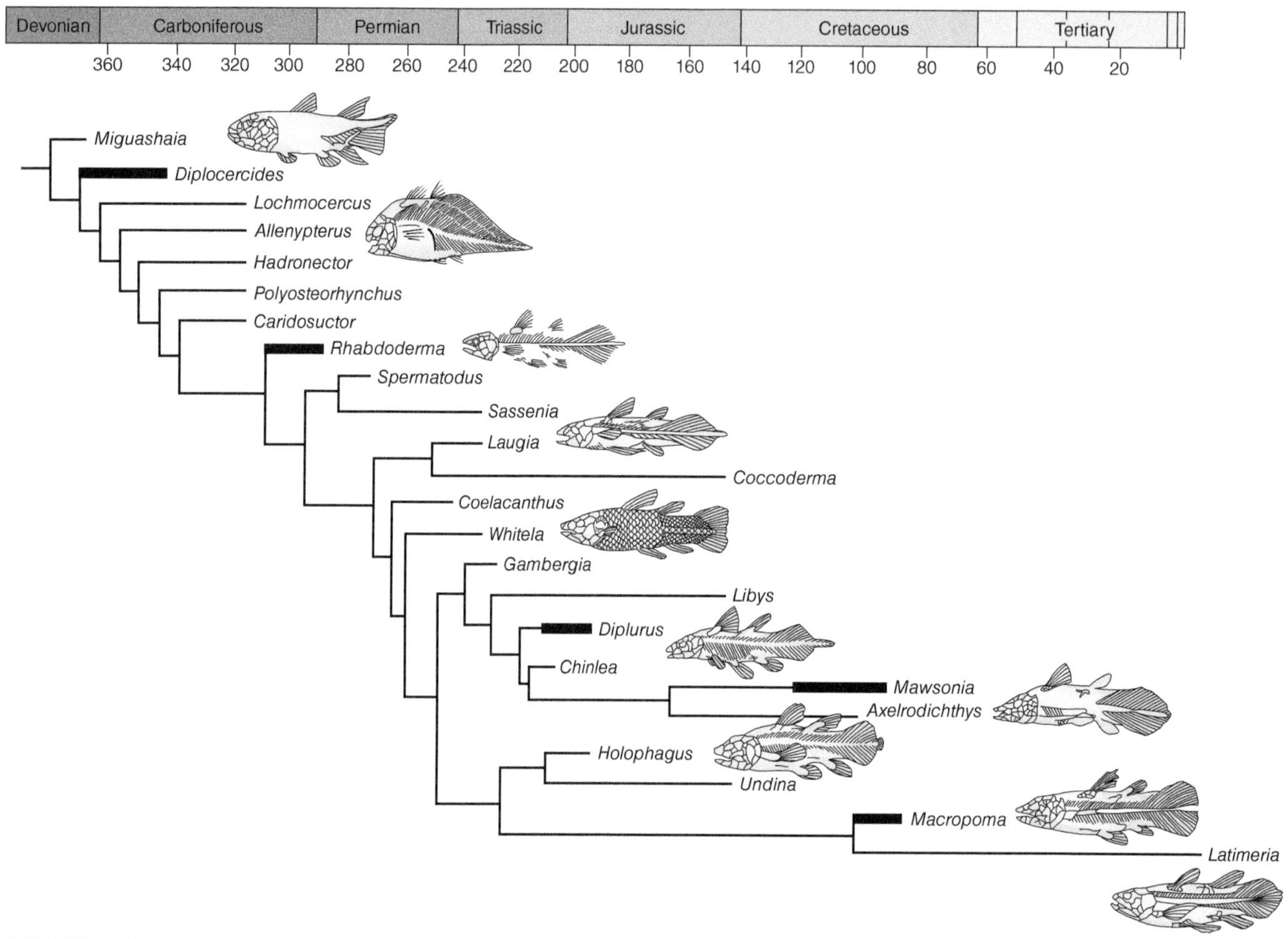

FIGURE 11.14 Phylogenetic relationships and fossil occurrence among the 24 coelacanth genera. Thick vertical bars show time ranges of occurrence for long-lived genera. The time scale is million years before present. Coelacanths are among the best-studied fossil groups, stimulated in part by the discovery of a living species after an 80-million-year hiatus in the fossil record. After Forey (1998).

Ceratodontimorphs appear first in the Early Triassic and are represented today by the freshwater order **Ceratodontiformes**, containing three families and six species of lungfishes in Australia (Ceratodontidae, one species), South America (Lepidosirenidae, one species), and Africa (Protopteridae, four species) (see Chapter 13). Modern lungfishes take the anatomical trends to the extreme, having eel-like, largely cartilaginous bodies, lacking any cosmine bony layers, and possessing diphycercal tails. The modern Australian lungfish resembles the heavier bodied dipnoans of the Paleozoic and Mesozoic. Although limited to fresh waters on three continents today, fossil ceratodontids occupied North and South America, Africa, and Madagascar, many recovered from marine deposits.

Lungfishes underwent extensive diversification during the Devonian, evolving more than 60 genera and 100 species, 80% of which occurred during the Late Devonian (Marshall 1987). Numbers diminished substantially during the Carboniferous. Many lungfish species are known only from fossilized toothplates and other structures found in fossilized lungfish burrows. These finds indicate that air breathing and **estivation** (entering torpor and burrowing in mud during drought) evolved as early as the Devonian, a fortuitous (for paleontologists) instance of fish waiting for rains that never came (Moy-Thomas & Miles 1971). Some ceratodontids were quite large; a North American Jurassic species, *Ceratodus robustus*, was 4 m long and may have weighed as much as 650 kg (Robbins 1991). The modern genus *Neoceratodus* occurs as early as the Late Cretaceous in Australia. The lepidosirenid lungfishes of Africa and South America represent a family that goes back to the Late Carboniferous, but members of the two extant genera do not appear until the Eocene and Miocene, on the same continents where they occur today (Carroll 1988).

Much controversy has swirled around the ancestry of lungfishes, as well as a possible dipnoan ancestry for terrestrial vertebrates (see reviews in Carroll 1988; Pough et al. 1989). Some of this speculation originated with the early misidentification of lungfishes as amphibians (see Chapter 13). Other arguments have focused on shared aspects of the lungs, limb-like fins, and internal nostrils (e.g. Rosen et al. 1981). However, workers in this area have increasingly reached the conclusion that the ancestry of tetrapods is more closely linked to another group of sarcopterygians, the infraclass Elpistostegalia.

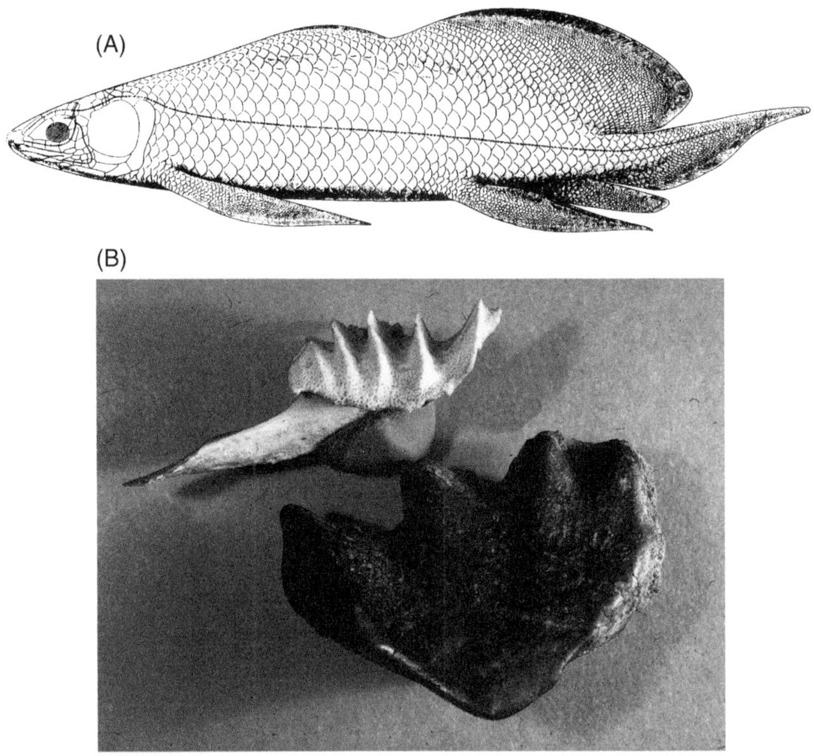

FIGURE 11.15 Extinct and extant lungfishes. (A) *Scaumenacia*, a Late Devonian lungfish from eastern Canada; (B) Toothplates from a fossil lungfish, *Ceratodus*, from the Late Triassic (*c.* 5 cm wide) and from the extant Australian lungfish, *Neoceratodus* (upper structure). The Australian lungfish is considered to be more similar to ancestral forms than are the living African and South American species. The *Neoceratodus* toothplate is mounted on a piece of modeling clay. (A) From Jarvik (1980), used with permission; (B) G. Helfman (Author).

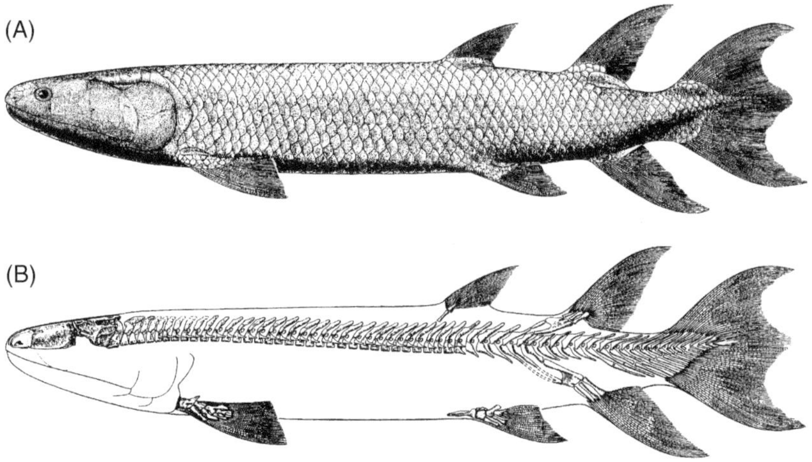

FIGURE 11.16 *Eusthenopteron foordi*, a well-known osteolepidiform and member of a lineage considered close to the direct ancestor to tetrapods. (A) The full restoration, and (B) the neurocranium, endoskeleton, and fin supports. Note the large mouth, large symmetrical tail, and posteriorly placed median fins, all characteristics of active predators. From Jarvik (1980), used with permission.

Tetrapodomorphs: Tetrapod Ancestors Appearing in the Early Devonian with the dipnoans are three infraclasses referred to by some collectively as tetrapodomorphs because of their tetrapod-like body forms. First to appear were the rhizodonts (infraclass **Rhizodontida**, order Rhizodontiformes, family Rhizodontidae), with at least eight genera. Next were the osteolepidiforms (infraclass **Osteolepida**, order Osteolepidiformes; osteolepidiforms in Fig. 11.10), including five

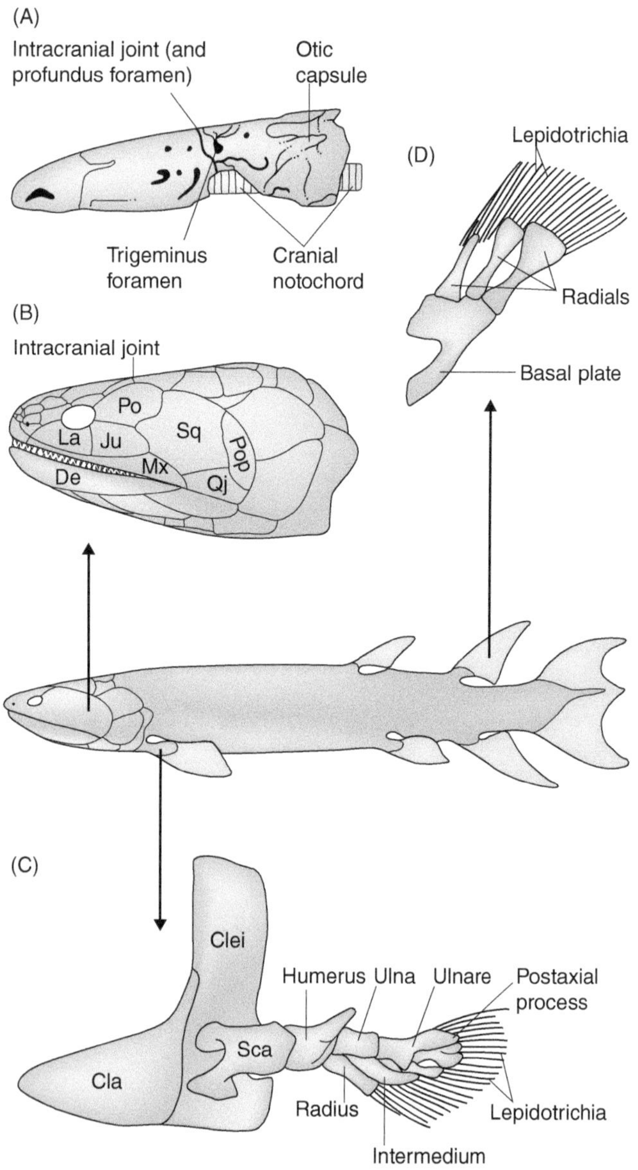

FIGURE 11.17 Presumed key traits that characterized the sarcopterygian ancestors of tetrapods, as evidenced by *Eusthenopteron*. Among the traits are (A) an intracranial joint in the skull roof associated with the profundus nerve foramen; (B) the arrangement of the dermal skull bones; (C) axial elements of the pectoral fin skeleton (e.g. humerus, ulna, ulnare); and (D) support skeleton of the second dorsal fin. After Ahlberg and Johanson (1998).

families. Finally, the infraclass **Elpistostegalia** appeared, with its important (to us as tetrapods) genera *Elpistostege* and *Panderichthys* (and *Tiktaalik*, discussed earlier) in the family Elpistostegalidae.

Tetrapodomorphs as a group were large, predatory fishes characterized by sarcopterygian traits such as two dorsal fins, cosmine covering of the bones and scales, kinetic (jointed) skulls, lobed fins, and replacement teeth on the jaw margins. They remained common throughout the latter half of the Paleozoic, and most forms disappeared by the end of the Permian. Some were large (up to 4 m long), cylindrically shaped predators that occurred primarily in shallow, freshwater habitats (Fig. 11.16). Evolutionary trends include reduction in dermal bone thickness, a change from diamond- to round-shaped scales, and an increasingly symmetrical tail. The latter trait is often considered indicative of a hydrostatic function for the gas bladder (Moy-Thomas & Miles 1971).

Of the three groups, we know most about the osteolepidiforms and especially the tristichopterid *Eusthenopteron foordi* because of exceedingly well-preserved material painstakingly prepared by Stensiö and associates (Fig. 11.16). One specimen alone required 6 years of serial grinding and many more years of analysis to characterize just the anatomy of the skull of this fish. Jarvik (1980) commented that we probably know more about the skeletal anatomy of *Eusthenopteron* than we do about most extant fishes. This knowledge is fundamental to our understanding of the anatomical transitions that occurred as sarcopterygians changed from purely aquatic forms capable of breathing atmospheric oxygen to semiterrestrial forms capable of movement on land and no longer dependent on gills (Fig. 11.17).

Although osteolepidiforms possessed many homologies with later tetrapods, these fishes were unlikely to have been transitional forms to living on land, even temporarily. It is the elpistostegalians that are generally considered the most likely sister group of modern tetrapods. Focus has been placed on many apparent homologies, including eye position, skull roof bones, paired fins, dentition, and vertebral accessories (Pough et al. 1989, 2005; Forey 1998). Elpistostegalians and tetrapods both have eyes set close together on the top of the skull facing upwards, with eyebrowlike ridges. The median series of skull roof bones – frontals, parietals, and nasals – may be homologous, although not all workers agree on terminology. The paired fins of the osteolepidiforms and elpistostegalians are very similar to those of the stem tetrapods of the Late Devonian, such as *Ichthyostega* (Fig. 11.18). This fin type contains bones homologous to the proximal elements of tetrapod fore- and hindlimbs (humerus, radius, ulna; femur, tibia, fibula), unlike the axially arranged, leaflike archipterygial fins of the dipnoans. The tetrapodomorph fin could provide improved body support for benthic locomotion, perhaps including movement across land. The dentition of both osteolepidiforms and early tetrapods was very similar, consisting of conical teeth with numerous infoldings of the dentine, termed labyrinthodont dentition, although this may have been a convergent trait among large carnivorous vertebrates (see Pough et al. 2005). Both groups also had ossified neural spines that grew dorsally from ring-shaped, ossified, vertebral centra.

Infraclass Tetrapoda The tetrapods are one of several infraclasses of the Class Osteichthyes. This branch of the Osteichthyes has, to varying degrees, made a successful transition to a terrestrial environment and are therefore not considered fishes. But they are descended from the bony fishes and therefore more closely related to the sarcopterygiians and actinopterygians than those groups are to the other major lineages of animals that we label as fishes.

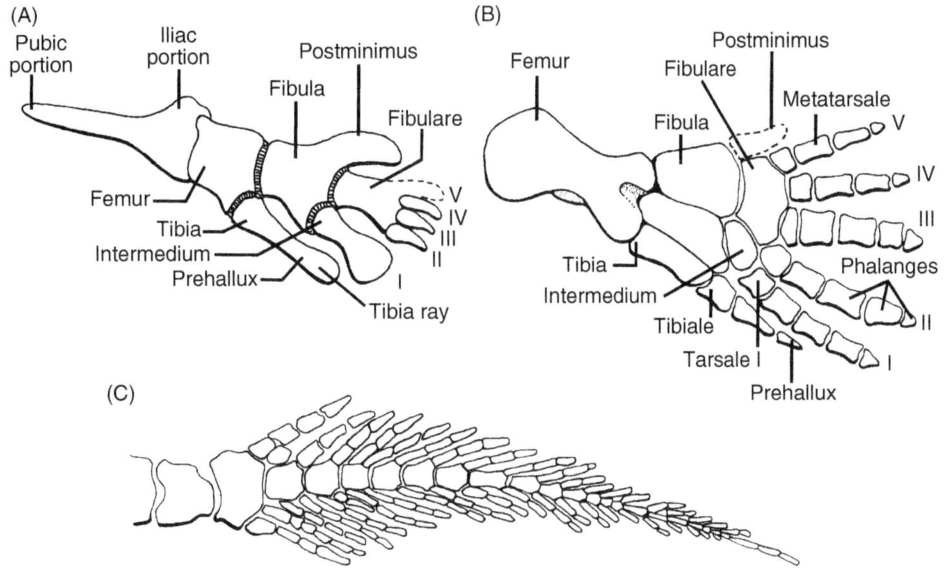

FIGURE 11.18 Comparative pelvic appendages of: (A) *Eusthenopteron*, a Devonian osteolepidiform fish; (B) *Ichthyostega*, a Devonian stem tetrapod; and (C) *Neoceratodus*, a modern lungfish. Note the apparent homologous bone series of the osteolepidiform and tetrapod limb, as compared with the less similar central axis and radials of the "archipterygial" lungfish fin. (A, B) From Jarvik (1980), used with permission; (C) Semon (1898) / Biodiversity Heritage Library / Public Domain.

Subclass Actinopterygii

Superclass Gnathostomata
- Grade Teleostomi
  - Class Osteichthyes
    - Subclass Actinopterygii
      - †orders of extinct stem-group actinopterygians without higher taxonomic ranking - Cheirolepidiformes, Palaeonisciformes, Dorypteriformes, Platysomiformes, Tarsiiformes, Guildayichthyiformes, Phanerorhynchiformes, Saurichthyiformes, Redfieldiiformes, Ptycholepidiformes, Pholidopleuriformes, Perleidiformes, Luganoiformes
      - Infraclass Cladistia
        - Order Polypteriformes (bichirs, see Chapter 13)
      - Infraclass Chondrostei
        - †Order Chondrosteiformes
        - Order Acipenseriformes (sturgeons, paddlefishes, see Chapter 13)[a]
    - Neopterygii *(an unranked clade)*
      - †Order Pycnodontiformes
      - Infraclass Holostei
        - Division Ginglymodi –
          - †Orders Dapediiformes, Semionotiformes, Macrosemiiformes
          - Order Lepisosteiformes (one extinct family, plus the extant gars; see Chapter 13)[b]
        - Division Halecomorphi
          - †Orders Parasemionotiformes, Ionoscopiformes
          - Order Amiiformes (extinct families, plus extant Bowfin; see Chapter 13)[b]
      - Infraclass Teleosteomorpha
        - †Division Aspidorhynchei - orders Aspidorhynchiformes, Pachycormiformes
        - DivisionTeleostei
          - †Orders Pholidophoriformes, Dorsetichthyiformes, Leptolepidiformes, Crossognathiformes, Ichthyodectiformes, Tselfatiiformes, Araripichthyiformes
          - Supercohort Teleocephala (see Chapters 14, 15)
            - Cohort Elopomorpha - 4 extant orders
            - Cohort Osteoglossomorpha – 1 extinct, 2 extant orders
            - Cohort Otocephala - 3 superorders; 1 extinct, 7 extant orders
            - Cohort Euteleostei – 8 superorders; over 40 extant orders

†Extinct group
[a] Acipenseriforms appear in the fossil record before some extinct chondrosteans.
[b] Fossil Bowfin and gar relatives appear in the fossil record before some extinct neopterygians.

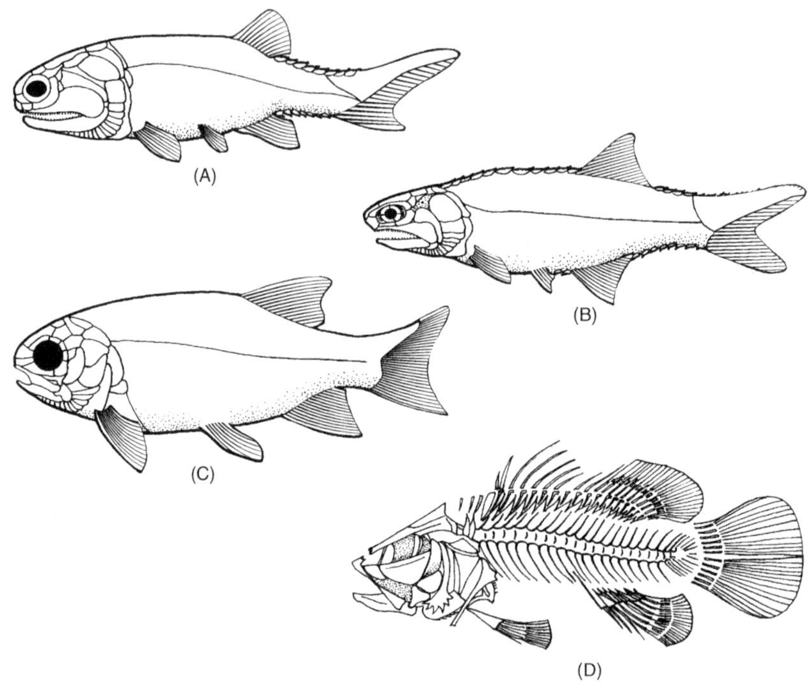

FIGURE 11.19 Actinopterygian fishes at different grades of development. (A) *Moythomasia* and (B) *Mimia*, two primitive palaeoniscoid fishes from the Late Devonian, with thick rhomboidal scales extending onto the fins, broadly triangular dorsal and anal fins, fulcral (ridge) scales along the back, a long mouth, and an asymmetrical heterocercal tail. (C) *Parasemionotus*, a pre-teleostean neopterygian from the Triassic, showing more flexible fins, shorter mouth, and abbreviate heterocercal tail. (D) *Eolates*, an advanced euteleost from the Early Eocene, with characteristic teleostean diversified dorsal and anal fins, shortened vertebral column, premaxillary dominated upper jaw, and homocercal tail. (A) After Jessen (1966); (B) after Gardiner (1984); (C) after Lehman (1966); (D) after Sorbini (1975).

The primitive fish groups discussed so far are interesting for their antiquity and diversity and for the effort required by paleontologists to slowly unearth and interpret features of their design. Yet these fishes bear little resemblance to most modern groups and are at most only distantly related to the familiar fishes of today. Speculation about the natural history, behavior, and ecology of extinct forms is based on scant information, much of it difficult to interpret. It is consequently challenging to imagine these animals as the living creatures that they were. These difficulties do not apply however to the ancestors of the Actinopterygii, the most successful of today's fishes. Although just as ancient as most of the other groups, primitive ray-finned fishes are similar in size and shape to many extant fishes, and many of their fossils are very well preserved. We can therefore equate many fossil and extant actinopterygians in terms of descendancy, form, and possibly function.

Early, Stem-Group Actinopterygians There are at least 13 extinct orders of early actinopterygians. In prior treatments (see Nelson 2006, Helfman et al. 2009) these were grouped with the chondrosteans, but more recently they have been separated from that infraclass (see Nelson et al. 2016).

The origins of the Actinopterygii are obscure. Scale fragments appear in Late Silurian marine deposits, which may mean that the group is older than the sarcopterygians and as old as placoderms and elasmobranchs. Among the bony, jawed fishes, only the acanthodians are older, supporting speculation of an acanthodian ancestry for modern bony fishes. However, complete fossil actinopterygians do not appear until the Mid to Late Devonian, when the group had expanded into a variety of marine, estuarine, and freshwater habitats. These early fishes, collectively known as **palaeoniscoids,** were relatively small (5–25 cm) and were distinguished from sarcopterygians by the presence of a single triangular dorsal fin, a forked heterocercal tail with no upper lobe above the unconstricted notochord, paired fins with narrow rather than fleshy bases, dermal bones lacking a cosmine layer, scales joined by a peg-and-socket arrangement and covered with ganoine ("ganoid" scales), relatively large eyes, and a blunt head (Fig. 11.19A, B). The term "ray-fin" refers to the parallel endoskeletal fin rays that were derived from scales. These rays supported the median and paired fins, which were moved by adjacent body musculature. In contrast, the fins of the Sarcopterygii had a thick, bony central axis and muscles contained in the fin itself (see Fig. 11.18C).

The most primitive group, the Devonian cheirolepidiforms, includes a species with the distinction of possessing the largest number of pelvic fin rays known among fishes, living or otherwise. *Cheirolepis canadensis* had 124 such rays versus six or fewer in living teleosts. The most diverse order, the Palaeonisciformes, contained four suborders and 17 families of well-represented fishes that showed tremendous morphological diversity (Figs. 11.19 and 11.20). The other orders of early chondrosteans are often lumped together as "palaeoniscoids" despite taxonomic differences, and palaeoniscoids are then treated as ancestral to later neopterygians and therefore teleosts.

Among the other orders are the Carboniferous tarrasiiforms, which were remarkably convergent with many modern eel-like forms, possessing an elongate body, dorsal and anal fins continuous with the caudal fin (the latter being diphycercal in this group), and pelvic fins and scales reduced or absent (Fig. 11.20). Saurichthyiforms converged on a needlefish body shape and are thought to have been similarly predatory on small fishes, and phanerorhynchiforms bore a superficial resemblance to modern sturgeons.

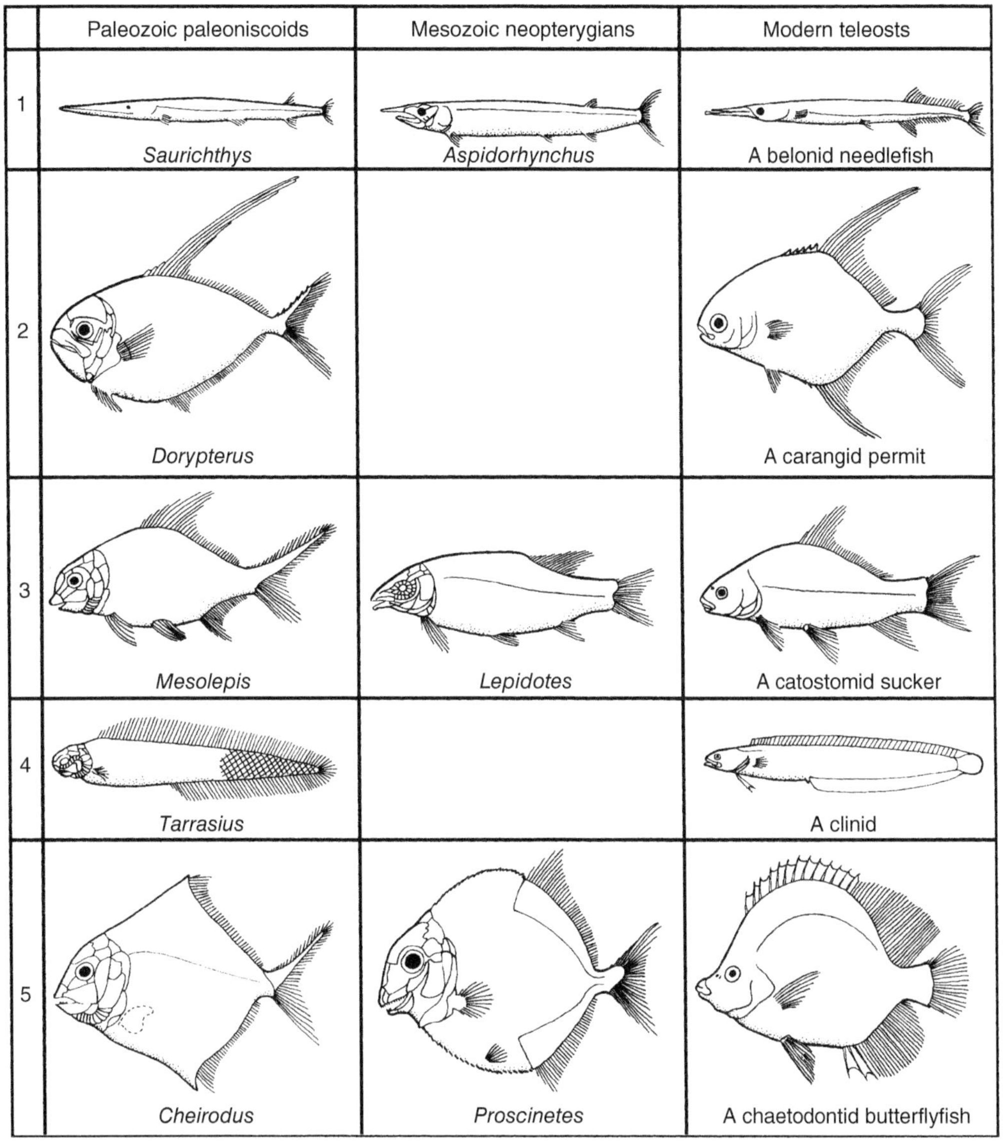

FIGURE 11.20 Morphological (and ecological) convergence in fish evolution. Palaeoniscoids were ancestral to early neopterygians, which were ancestral to modern teleosts. Certain body designs or plans have apparently been repeatedly favored in actinopterygians, leading to convergent designs among unrelated lineages. These striking convergences in body shape and presumably function are depicted for representative palaeoniscoids, early neopterygians, and teleosts. 1, Elongate piscivores with long, tooth-studded jaws and dorsal and anal fins placed posteriorly for rapid starts; 2, compressed-bodied, predatory, shallow water fishes with deeply forked tails and trailing fins; 3, broad-finned bottom feeders with subterminal mouths; 4, eel-like benthic forms; and 5, compressed, circular forms with large fins for maneuverability in shallow water habitats with abundant structure. Gliding fishes such as the Triassic chondrostean *Thoracopterus* (Fig. 11.21) can also be equated with modern teleostean flyingfishes. Pough et al. (1989).

A considerably derived order from the Late Triassic, the perleidiforms, included *Thoracopterus*, a genus with expanded paired fins thought capable of biplane gliding, as occurs in modern exocoetid flyingfishes (Tintori & Sassi 1992). *Thoracopterus* possessed the enlarged pectoral and pelvic fins, reinforced rays in the paired fins, asymmetrical caudal fin, expanded caudal neural spines for muscle insertion, posterior position of dorsal and anal fins, and head shape of modern gliding forms (Fig. 11.21), which leave the water to escape predators (addressed in Chapter 16). Such convergence, remarkable in itself, would have required substantial reduction in the heavy armoring characteristic of the early chondrosteans.

Palaeoniscoid Trends Palaeoniscoids flourished throughout the latter Paleozoic. Meanwhile, ostracoderms, acanthodians, and placoderms disappeared and sarcopterygians diminished in abundance. This correlation suggests ecological interaction among groups, and possible replacement of

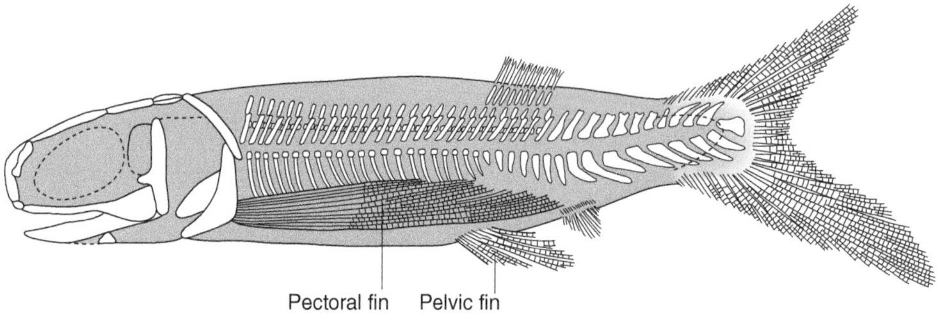

FIGURE 11.21 *Thoracopterus magnificus*, a 6 cm-long perleidiform from the Triassic adapted to gliding. Most notable are the expanded pectoral and pelvic fins and asymmetrical caudal fin with its larger lower lobe. These and other traits are strongly convergent with features that allow modern exocoetid flyingfishes to engage in biplane gliding. From Tintori and Sassi (1992), used with permission.

primitive jawed and jawless fishes with more advanced actinopterygian and chondrichthyan lineages. What innovations did the ray-finned fishes possess that might have given them ecological superiority? The available evidence strongly suggests that, once again, changes in jaw and fin structure leading to diversified feeding habits and increased mobility were critical to actinopterygian success and dominance.

Changes in the mechanics of jaw opening and closing during actinopterygian phylogeny have been the subject of intensive study (e.g. Lauder & Liem 1981; Lauder 1982; see also Carroll 1988, Pough et al. 2005 for reviews). The highly ossified braincase of the early actinopterygians makes it possible to determine the origins, insertions, and approximate sizes of the different muscle masses involved in jaw function, from which we can estimate the forces in operation.

During actinopterygian evolution, culminating in advanced teleosts, changes in the angles and connections between the skull case, dermal bones, muscles, and ligaments of the head and jaws have been most influential. In particular, the hyomandibula has been reoriented from oblique to vertical, the posterior end of the maxilla has been freed from the cheek bones, and the jaw musculature has increased in size and complexity. These changes increased the speed and strength of the bite. They also allowed for enlarging of the mouth both vertically and laterally. Hence, when the mouth was opened, its volume increased and it assumed a more tubular shape. In this modified condition, when the mouth is opened, water and prey are sucked in; when the mouth is closed, instead of water being pushed back out through the jaws, flow continues posteriorly through the gill slits, thereby trapping prey inside rather than pushing it back out of the mouth. Transport of water over the gills during breathing may also have been facilitated by these modifications.

Apparent improvements in other skeletal components are no less important. Palaeoniscoid *scales* changed from heavy, interlocking, diamond-shaped units to thinner, lighter, circular, cycloid structures. This reduction was accomplished by elimination of the dentine, vascular, and ganoine layers. Because palaeoniscoid fins consisted of jointed scales, reduction in scale thickness meant increased flexibility in fins; fins became mobile structures composed of **dermal fin rays** that could be erected or lowered and also moved laterally. Associated with scale reduction was increased ossification of the **vertebral column**, leading to recognizable centra with dorsal and ventral accessory structures (neural and haemal arches). These accessory structures are closely related to modifications in the **caudal region**, where a major trend has been toward an increasingly symmetrical, homocercal tail (see Fig. 11.19). Caudal fin rays became supported by a series of ventral accessory, hypural bones.

All these changes during palaeoniscoid phylogeny imply increased reliance on locomotion, integral for both escape and prey capture. Heavy ganoid scales offer passive protection against predators but do not function until a predator has already captured a prey individual, a risky event best avoided by potential prey. Lighter scales mean a lighter, more flexible body, capable of more rapid swimming and quicker turns. Greater reliance on a gas bladder for attaining neutral buoyancy has also been suggested, which also frees fins to provide propulsion and maneuverability. Weight reduction of fins allows them to better serve as propellants, or as brakes and flaps for swimming, stopping, and turning (and gliding?). The correlation between dermal armor reduction and increased vertebral ossification may indicate a shift from reliance on an external elastic/hydrostatic skeleton to an internal, muscular/tendinous system. Increased speed and mobility, combined with the already mentioned improvements in mouth structure, would mean that more advanced actinopterygians would not only be better at avoiding predators but also at capturing prey. These trends in palaeoniscoid evolution reoccur later in more advanced actinopterygian lineages (see Fig. 11.20).

Infraclass Cladistia After considerable discussion and controversy in the past (see Nelson 1994; Helfman et al. 1997), the **polypteriforms** (bichirs and reedfish; see Chapter 13) were recognized as a separate infraclass and the sister group of the other three subclasses of actinopterygians, the Chondrostei, Holostei, and Teleosteomorpha. Also referred to as **brachiopterygians**, fossil cladistians are known only as far back as the Middle Cretaceous of Africa and Late Cretaceous of South America. This represents a dramatic gap in the fossil record for a group considered more primitive than other actinopterygians, which are known from the Devonian (cheirolepidiform and palaeonisciform chondrosteans) and the Triassic (semionotiform neopterygians).

Infraclass Chondrostei As mentioned earlier, many of the extinct groups of early actinopterygiians were at one time considered chondrosteans, but more recent consideration (Nelson et al. 2016) has separated them from this group. The infraclass **Chondrostei** includes one extinct order – the Chondrosteiformes, with one family (Chondrosteidae). The other order of chondrosteans, the Acipenseriformes, includes one extinct family (Peipiaosteidae), and the extant Acipenseridae (sturgeons) and Polydontidae (paddlefishes). These modern chondrosteans have fossil representation in the Jurassic and Early Cretaceous, respectively. These groups are addressed in detail in Chapter 13.

Neopterygii (Unranked Clade) "In their great numbers and degree of anatomical diversity, the modern ray-finned fishes may be considered the most successful of all vertebrates" (Carroll 1988, p. 136). Just as improvements in feeding and locomotion may have created competitively superior, primitive actinopterygians, continued evolution of these same traits probably led to the replacement of early actinopterygians by more advanced forms. These descendants, termed **Neopterygii** ("new fins"), first appear in the fossil record during the Late Permian. They underwent an initial radiation in the Triassic and Jurassic and then expanded more extensively in the Late Cretaceous. Many of the orders of modern teleostean fishes, the dominant group of bony fishes alive today, are represented in this late Mesozoic radiation. In fact, of the more than 40 recognized living orders of teleosts, half have fossil records that date back to the Cretaceous, with only about seven orders arising more recently than the Eocene (younger than 50 million years old).

Pre-teleostean neopterygians include eight orders, six of which are extinct. Jurassic **semionotiforms** were quite diverse, radiating into species flocks in eastern North America; some analyses place this order on a direct line to modern gars (Lepisosteiformes). **Pycnodontiforms** were another diverse group of shallow water marine forms. **Aspidorhynchiforms** converged on a needlefish-like body form, as did the saurichthyiform palaeoniscoids before them (see Fig. 11.20). The **Pachycormiformes**, with one family and 11 genera of Jurassic to Late Cretaceous fishes, are considered by some to be a sister group to early teleosts. A giant pachycormid, *Leedsichthys problematicus*, has been discovered in Middle to Late Jurassic marine deposits in what is now England, Western Europe, and Chile (Fig. 11.22). Reconstructions suggest a total length in excess of 15 m, making it the largest bony fish, and perhaps the largest fish, to ever exist (Martill 1988; Liston 2004; see also **www.big-dead-fish.com**). Anatomical features indicate that – like the modern Whale, Basking, and Megamouth sharks – *Leedsichthys* was planktivorous, another example of convergence of form and function across taxa and time (e.g. Fig. 11.20).

The extant neopterygians are in the infraclasses **Holostei** and **Teleosteomorpha**.

Infraclass Holostei The two extant pre-teleostean, neopterygian groups, the gars (Division Ginglymodi, Order Lepisosteiformes) and Bowfin (Division Halecomorphi, Order Amiiformes), are intermediate between palaeoniscoids and teleosts in a number of structures: gars retain the ganoid-like scales of primitive neopterygians, Bowfin have a primitive gular plate under the head, and both groups have identifiably heterocercal tail elements. In most other respects, they are quite specialized, as would be expected for fishes that have existed as recognizable taxa since the Mesozoic. They differ sufficiently in derived traits to generally justify their placement in separate orders, although some analyses indicate that similarities among gars, Bowfin, and their fossil relatives justify their placement together in a separate group, the **Holostei** (e.g. Olsen & McCune 1991), which is now ranked as an infraclass (Nelson et al. 2016). Neither gars nor Bowfin are considered to be on a direct line to the teleosts.

FIGURE 11.22 *Leedsichthys problematicus*, perhaps the world's largest fish ever. This 15 m+ zooplanktivorous pachycormid is known from fragments and several partial skeletons discovered in clay deposits from the Middle–Late Jurassic. Photo from Dmitry Bogdanov / Wikimedia / Public Domain.

Infraclass Teleosteomorpha According to Nelson et al. (2016), this infraclass includes the extinct and pre-teleost genus *Prohalecites,* the extinct pre-teleost Division Aspidorhynchei (orders Aspinorhynchiformes and Pachchormiformes, both mentioned above), and the Division **Teleostei.**

Teleosts ("perfect bone") far outnumber all other living fish groups, accounting for more than 29,500 species – more species than in all other vertebrate classes combined. Because Chapters 14 and 15 are devoted to characterizing different teleostean groups, they will only be briefly described here. For the present discussion, it is important to realize that teleostean evolution largely repeats and extends trends that originated with the ancestral palaeoniscoids and were continued in early neopterygians. Refinements in the structure and function of mouths and fins appear to explain much of the success of the group. Evidence of these trends is preserved both in the fossil record and in the ancestral traits retained by recognizably primitive teleostean taxa.

Teleosts, despite their incredible diversity, form a definable group with a recognizable ancestry. On cladistic grounds, at least 27 anatomical synapomorphies support teleosts as a monophyletic group (see Nelson 2006, Nelson et al. 2016). Chief among these are ural neural arches elongated to form the **uroneurals** of the tail support, unpaired **basibranchial toothplates**, a distinctive **urohyal**, and the prevalence of a **mobile premaxilla**.

Teleosts arose in the Middle or Late Triassic (215 mybp), followed by major diversification into modern groups in the Cretaceous. Teleostean evolution apparently involved four major radiations, three that each gave rise to distinct, primitive subdivisions, and a fourth that produced the major advanced groups alive today (between three and six other radiations died out during the Mesozoic). Multiple radiations imply that modern teleosts as a group could be regarded as polyphyletic, more a developmental grade than a single clade. Yet shared traits among the modern groups imply a monophyletic clade (Lauder & Liem 1983) (Fig. 11.23).

Separate ancestors are postulated for the different radiations, but all may have been derived from the **pholidophoriforms**, an early mainstem teleost group, now extinct. Five families of pholidophoriforms are recognized. The extinct early teleostean order **Ichthyodectiformes** included four families of highly predatory fishes, including the 4 m *Xiphactinus*

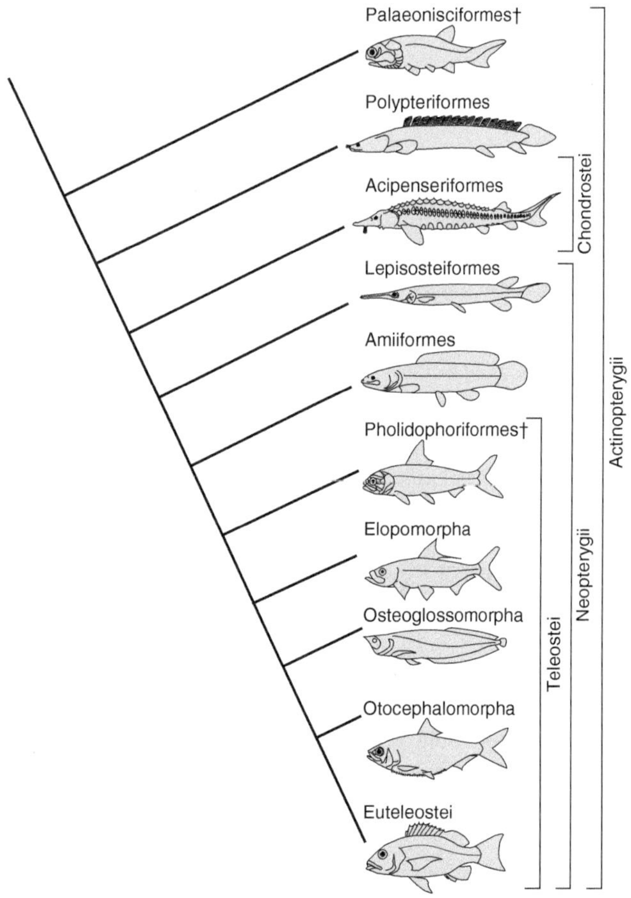

FIGURE 11.23 Phylogenetic relationships among actinopterygian fishes. Pholidophoriforms are one of several possible groups ancestral to modern teleosts. Daggers indicate extinct groups. Based largely on Nelson (2006), Nelson et al. (2016), and papers cited in those publications.

previously considered as possible osteoglossomorphs (Helfman 2007) but are now ranked as pre-teleocephalans. Two other poorly understood teleostean Mesozoic orders are the **leptolepidiforms** and **tselfatiiforms**.

The first three major radiations of modern teleosts produced the **elopomorphs** (tarpons and true eels), **osteoglossomorphs** (bony tongues), and **otocephalans** or **ostarioclupeomorphs** (herrings and minnow relatives). These groups stand separately as cohorts of the Teleostei, apart from the larger, more advanced, fourth radiation, the cohort **Euteleostei**. Elopomorph eels and tarpons are contained in four orders, all extant. Osteoglossomorphs include two living orders and one extinct Jurassic and Cretaceous order, the Lycopteriformes. The cohort Otocephala is divided into three superorders. The **Clupeomorpha** (herrings and anchovies) contain one living order, the **Clupeiformes**, and one extinct order, the Cretaceous to Eocene Ellimmichthyiformes. The superorder **Alepocephali** consists of one extant order (Alepocephaliformes). The other otocephalan superorder **Ostariophysi** contains five orders, all living. Euteleosts include the advanced, living, bony fishes, divided into at least eight superorders, over 40 orders, a few hundred families, and more than 17,000 species (see Chapters 14, 15 and Fig. 14.1). Most groups are well represented in early Cenozoic deposits, such as the famous Eocene sites in Green River, Wyoming, and Monte Bolca, northern Italy (see Frickhinger 1995; Long 1995; Maisey 1996).

Trends During Teleostean Phylogeny Although numerous derived traits characterize teleostean groups, trends in five areas can be readily linked to functional improvements that contributed to teleostean success. These trends include reduction in bony elements, repositioning and elaboration of the dorsal fin, change in placement and function of paired fins, structural modifications to and interaction between the caudal fin and gas bladder, and jaw improvements.

Reduction of Bony Elements Teleosts show a general reduction in bony elements as compared to pre-teleostean groups (see Nelson 1994, 2006). This reduction occurred through fusion or actual loss of bones. For example, more derived teleosts have the following features.

1. There are fewer, more ossified vertebrae (in general 60–80 in many elopomorphs and clupeomorphs, 30–40 in ostariophysans, 30–70 in protacanthopterygians, 20–35 in paracanthopterygians, and 20–30 in most percomorphs). A shorter, more ossified axial skeleton would allow for attachment of stronger trunk musculature, thus enhancing locomotion.
2. There are fewer vertebral accessories, such as intermuscular bones and ribs, and the replacement of numerous small intermuscular bones with fewer, thicker zygopophyses (compare the "boniness" of fillets from a herring or trout with that from a tuna or flatfish).

3. There are fewer bones in the skull (e.g. the orbitosphenoid is missing in perciforms; there are 10–20 branchiostegals in elopomorphs, osteoglossomorphs, and clupeomorphs, 5–20 in ostariophysans and protacanthopterygians, and 4–8 in paracanthopterygians and acanthopterygians).
4. There is a reorganization and reduction in the number of bones of the tail, including fusion of the supporting bones (epurals, hypurals, centra) and a reduction of the number of fin rays in the tail (most less derived teleosts have 18 or 19 principal fin rays, never more than 17 in perciforms; see also below).
5. There is a reduction of the number of biting bones in the upper jaw from two to one. The maxilla becomes *excluded from the gape* in paracanthopterygians and acanthopterygians. In more primitive groups, it is a tooth-bearing bone, whereas in the two spiny superorders, it pivots with the elongate premaxilla to create a tubular mouth (see 'pipette mouth', discussed later under Feeding Apparatus Modifications).
6. There is a reduction in the number of fin rays in paired fins (six or more soft pelvic rays in most less derived teleosts, six or fewer in most paracanthopterygians, and one spine with five or fewer rays in most acanthopterygians).
7. There is a reduction in the amount of bone in the scales (compare the heavy cycloid scale of a tarpon, Megalopidae, or Arapaima, Osteoglossidae, with the thin ctenoid scales of most paracanthopterygians and acanthopterygians). A trend toward reduction in armor is familiar by now, as it also occurred during the evolution of several groups (Table 11.1). One possible interpretation is that mechanical protection against predators was of paramount importance when several of these taxa arose, but a premium on mobility soon developed because lighter, quicker fishes yielded improvements in both predator avoidance and feeding success.

TABLE 11.1 **Repeated trends in fish evolution. Although fishes represent diverse and heterogeneous assemblages assigned to at least five different classes, certain repeated trends have characterized the evolution of these groups or of major, successful taxa within them. The following list summarizes traits or characteristics common to the evolution of several groups.**

1. *Origin in oceans, radiation into fresh water*: thelodonts, pteraspidiforms, cephalaspidiforms, anaspids, placoderms, dipnoans, actinopterygians, teleosts, elasmobranchs
2. *Feeding and locomotion improvements*:
   - **A.** Diversification of dentition: acanthodians, placoderms, dipnoans, palaeoniscoids, teleosts, elasmobranchs
   - **B.** Improved inertial suction feeding: elasmobranchs, chondrosteans, neopterygians, teleosts
   - **C.** Increased caudal symmetry: dipnoans, osteolepidimorphs, coelacanthimorphs, palaeoniscoids, teleosts (reversed in pteraspidiforms and elasmobranchs)
   - **D.** Decreased external armor: pteraspidiforms, acanthodians, placoderms, dipnoans, osteolepidiforms, palaeoniscoids, teleosts
3. *Bases of spines become embedded in body musculature*: acanthodians, elasmobranchs
4. *Fusion of skull bones*: pteraspidiforms, acanthodians, placoderms, dipnoans, teleosts
5. *Bone preceded cartilage as skeletal support*: cephalaspidiforms (if ancestral to lampreys), dipnoans, acipenseriforms
6. *Electroreceptive ability*: pteraspidiforms, cephalaspidiforms, acanthodians, placoderms, dipnoans, actinistians, cladistians, chondrosteans, teleosts, elasmobranchs (for extinct groups, based largely on morphology of pits and canals in head and body; reinvented in modern teleosts) (see Pough et al. 1989; Chapters 6, 12, 13)

Shifts in Position and Use of the Dorsal Fin The dorsal fin in less derived teleosts is a simple, spineless, fixed, single, midbody keel that prevents rolling and serves as a pivot point for fishes that typically swim in open water situations (e.g. mooneye, tarpon, bonefish, herrings, minnows, trouts) (Fig. 11.24). In more derived teleosts, the trend is for the dorsal fin to become elongate and diversified. This is usually manifested as two fins, the anterior portion spinous and the posterior portion soft-rayed. Diversification of a fin into an anterior, hardened spinous portion and posterior, flexible portion maintains the protective function of the fin without sacrificing its role in maneuverability. Stability is still provided when the fin is erect, but many other functions can be served. The erected spiny dorsal provides protection from predators by increasing the body dimensions of the fish; folding the spinous dorsal against the body enhances streamlining. Rapid raising and lowering of the dorsal serves as a social signal in many fishes (similar diversification and actions in the anal fin serve the same purposes). The soft dorsal, through its flexibility, can function as a rudder when slightly curved and as a brake when greatly curved. It can also provide mobility if sinusoidal waves are passed down its length (various knifefishes) or if it is flapped in conjunction with the anal fin (triggerfishes, pufferfishes, ocean sunfish).

Truly bizarre modifications of the dorsal fin are seen in many acanthopterygians. In anglerfishes (order Lophiiformes), the first spiny ray is modified into an elongate, ornamented lure to attract prey. In some groups, filaments and fleshy growths increase the resemblance to seaweed or other structures (e.g. Sargassumfish, family Antennariidae, order Lophiiformes). Scorpionfishes (order Scorpaeniformes) use the spiny dorsal as a venom delivery system for protection against predators. Long, trailing filaments of probable social function (mate attraction, school maintenance) characterize many acanthopterygian

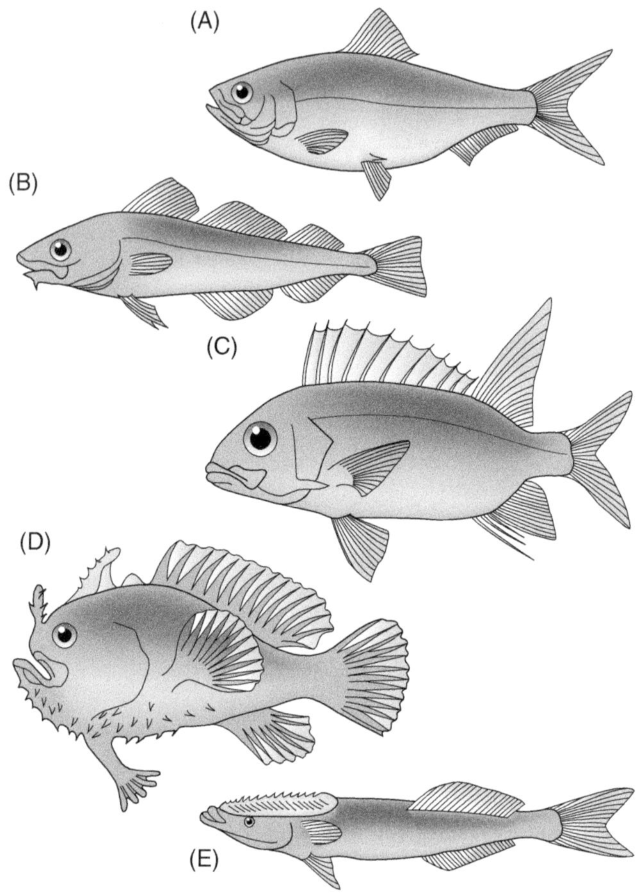

FIGURE 11.24 Diversification of the dorsal fin in modern teleosts. (A) Primitively, the dorsal fin is a single, spineless, subtriangular structure that serves as an antiroll device and pivot point during swimming, such as in the herrings (Clupeidae). However, this simple fin has been greatly modified in more advanced groups and can serve in locomotion, predator protection, and a variety of other functions. (B) In cods (Gadidae), three dorsal fins exist. (C) More commonly, a spiny anterior and soft-rayed posterior separation occurs, as in the squirrelfishes (Holocentridae). (D) In frogfishes (Antennariidae), modified dorsal spines serve as lures and as camouflage. (E) The sucking disk of the sharksucker (Echeneidae) is derived embryologically from the spiny dorsal fin. Modified from Nelson (2006).

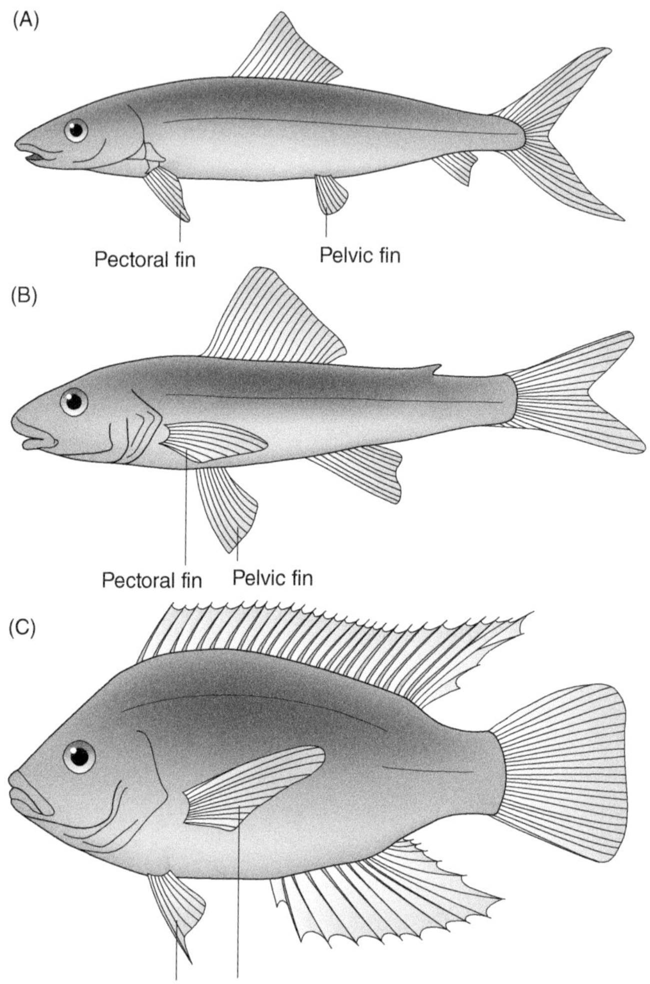

FIGURE 11.25 The phylogeny of paired fin locations in teleosts. The locations and functions of the pectoral and pelvic girdles have changed during evolution of the Teleostei. Pectoral fins move from a ventral to a lateral position and the pectoral fin base changes its orientation from horizontal to vertical. Pelvic fins move from abdominal to thoracic and even jugular locations. Extant representatives of phases in this observed trend are represented by (A) an elopomorph (bonefish, Albulidae), (B) a primitive paracanthopterygian (Troutperch, Percopsidae), and (C) a generalized acanthopterygian (cichlid, Cichlidae). This trend is by no means absolute: many specialized, relatively primitive teleosts have laterally placed pectorals (e.g. catfishes) and advanced teleosts may have pelvics in abdominal positions (e.g. atherinomorphs), but overall the trends describe a progressive change during teleostean phylogeny. Modified from Nelson (2006).

fishes (e.g. carangids, angelfishes, cichlids). The suction disk of sharksuckers (remoras, family Echeneidae, order Carangiformes) is derived from the first dorsal fin.

Placement and Function of Paired Fins In basal teleosts, pectoral fins are oriented horizontally and located in the **thoracic** position, below the edge of the gill cover; pelvic fins occur at midbody in an **abdominal** location (Fig. 11.25). In this configuration, both fins act primarily as planes that help stabilize movement up and down (pitch) or from side to side (roll), as well as providing some braking force. During teleostean phylogeny, pectoral fins move up onto the sides of the body and their base assumes a vertical orientation; pelvic fins move into **thoracic** or even **jugular** (throat) position. These relocations have several apparent functions (see Webb 1982). Pectoral fins on the side can be sculled for fine movement and positioning, such as slow swimming, and hovering and backing in midwater. As these fins are often transparent, their use in locomotion might be less obvious to a potential prey animal than would be lateral undulations of the body. Placement of the pelvic fins forward helps in braking and reduces pitching; their location under the spinous dorsal, in combination with spinous armament, increases the effective body depth of a fish at the point at which it is most likely to be attacked by a predator (Webb & Skadsen 1980).

Caudal Fin and Gas Bladder Modifications Actinopterygian evolution is characterized by a progressive increase in symmetry of the caudal fin (tail) (see Fig. 11.19). Caudal fins became externally and functionally homocercal fairly early in the group's history. Fossil impressions of the tails of late Paleozoic palaeoniscoids show that the upper and lower lobes were equal, in contrast to the heterocercal and abbreviate heterocercal tails of earlier palaeoniscoids. Symmetry becomes more pronounced in the teleosts, reflecting the internal modifications that followed. These internal changes include notochord and body shortening and the reworking of large bones and sets of bones that support the caudal fin rays. In particular, teleosts developed a series of **hypural bones** from several haemal arches. Some of these bones fused to form a ventral **hypural plate**, continuing a trend evident in palaeonisciforms. Fusion and reduction of number of vertebrae, reduction of intermuscular bones, and increased tail symmetry all correlate with a greater role of the caudal region in locomotion. The trend is toward an increased dependence on high power caudal swimming, culminating in steadily swimming fishes with lunate tails, such as jacks (Carangidae), tunas (Scombridae), and billfishes (Istiophoridae) (Webb 1982). Less derived teleosts used sequential contraction of trunk musculature throughout the body, producing a wave of contraction from head to tail (see Chapter 4). By focusing muscle contraction on the tail and its supporting structures, more derived teleosts could swim faster and more efficiently (Carroll 1988; Lauder 2000). Hydrodynamic attributes and implications of heterocercal and homocercal tails are discussed in Chapter 4 (Locomotion: Movement and Shape).

In apparent conjunction with tail and paired fin modifications, an additional teleostean trend is added control over **gas bladder function**. It can be debated whether gas bladders arose initially as breathing or buoyancy control structures (discussed in Chapter 5), but the latter function has taken precedence in teleosts. Living pre-teleosteans and less derived teleosteans have a **physostomous** gas bladder, in which a pneumatic duct connects the gas bladder with the gut and ultimately the mouth. The gas bladder is filled with gas by gulping air; gas is expelled largely via the same route. Fine adjustments are difficult in this system, and the fish is somewhat dependent on access to the surface. More derived teleosts (paracanthopterygians, acanthopterygians) are **physoclistous**, having lost the pneumatic duct and the link to atmospheric air. They instead rely more on internally generated and absorbed gases to fill and empty the gas bladder and are capable of finer control of buoyancy (physostomous fishes have gas secretion capabilities, but are usually not as refined as in physoclists).

Although gas bladders do not normally fossilize, the co-development of a gas-filled, internally controlled gas bladder, a homocercal tail, and paired, multifunctional, flexible fins is taken as a strong indication that they evolved as a suite of characters. A gas bladder can make an otherwise dense fish neutrally buoyant, which means that small, precise adjustments in body orientation and movement will not be counteracted by continuous sinking. Thus, a fish can remain at a fixed point in the water column and turn on its body axis without moving forward. This trend toward a combination of a functionally homocercal tail and some sort of internal hydrostatic organ was previously evident in the three major sarcopterygian groups. Lungfishes, osteolepidimorphs/elpistostegalians, and coelacanths evolved symmetrical, diphycercal tails. The first two groups had functional gas bladders (lungs), and early coelacanths possessed a gas bladder (Lund & Lund 1985), although it is small and fat-filled in the living coelacanths. Interestingly, the feeding mode of the living coelacanths involves hovering relatively still in open water by a combination of paired fin movements and buoyancy control (Fricke et al. 1987).

Feeding Apparatus Modifications Two major changes have characterized the anatomy of foraging in teleosts.

1. Teleosts continue a trend seen in neopterygians with respect to increasing suction capabilities. Teleosts developed a protrusible **pipette mouth**, capable of generating powerful, directed, negative pressures. The pipette mouth results from enlargement of the muscles and modifications to the bones in the jaw apparatus, most notably the maxilla and hyomandibula, but also involves connections with the mandibular, opercular, and pectoral bone series. Early in teleostean evolution, the rear portion of the maxilla was freed from its connection with other cheek bones, allowing it to swing forward and allowing the reoriented hyomandibula to move outward, thus increasing mouth volume. Skin folds developed along the lateral margins of the jaw bones, creating a hole-free tubular apparatus that prevents lateral escape by small prey. In the more derived groups (Paracanthopterygii and particularly the Acanthopterygii), the premaxilla develops an **ascending process** which is basically a vertical extension at its anterior tip that slides along the front of the skull, thus allowing the premaxilla to shoot forward at prey as the mouth is opened (see Chapter 3, Functional Morphology of Feeding).

   The end product of action in this complex of bones, muscles, ligaments, and pivot points is very rapid expansion of the orobranchial chamber. In anglerfishes, mouth volume can increase 13-fold over the course of 7 ms (7/1000 of a second) (Pietsch & Grobecker 1987). Maximal expansion of the gape is one direction these changes take, particularly in predators on other fishes. Many derived teleosts that feed on zooplankton and other small prey have reduced gape width to increase suction power. And some groups have maximized the speed of mouth extension without suction. For example, zooplanktivorous *Chromis* damselfish (Pomacentridae) can fully protrude their jaws in as little as 6 ms to capture evasive copepods (Coughlin & Strickler 1990). During jaw protrusion in the Pikehead,

Luciocephalus pulcher (Luciocephalidae), head length is increased by one-third at a rate of 51 cm/s, thus increasing the speed of attack by almost 40%, with no appreciable suction force generated; velocity increases of up to 89% have been recorded in Largemouth Bass (Nyberg 1971; Lauder & Liem 1981). Observations of mouth extension without suction have fueled a debate as to whether the pipette mouth developed primarily to generate suction power for inhaling prey or to rapidly extend the anterior portion of the body to overtake prey (Lauder & Liem 1981; Lauder 1982). Regardless, the primary selection pressure driving these modifications in the jaw was undoubtedly prey capture.

2. Once prey are captured, they are passed back into the mouth to be manipulated. For soft-bodied prey, including most small fishes, manipulation primarily involves positioning the prey to facilitate head-first swallowing, thus avoiding the teleostean fin spines that might cause choking or blockage if prey are swallowed tail first. Later digestion of fish prey requires little more than chemical breakdown in the gut. However, many potential prey have extremely effective physical defenses that are relatively impervious to gut chemistry. The hard, calcareous shells of mollusks, the chitinous exoskeletons of crustaceans, and the cell walls of plants all require mechanical rupturing before digestive enzymes can have much effect. A protrusible mouth is effective for initial capture of prey, but apparently protrusibility evolved at a sacrifice in up-and-down chewing motion. Therefore mechanical rupturing must occur elsewhere. In teleosts, it has been the dentition and musculature of the **pharyngeal** "**jaws**" that have diversified to serve this chewing, crushing, and grinding function (Lauder 1982), which is addressed in some detail in Chapter 3.

Pharyngeal **pads** lie posterior to the marginal jaws, just anterior to the esophagus, and are derived from dermal tooth plates in the pharynx. During teleostean phylogeny, the function of these pads has elaborated from holding prey prior to swallowing to manipulation and preparation that facilitates digestion. The pads have become armed with a variety of dentition types and have fused to dorsal and ventral elements of the gill arches. The branchial musculature has been reworked and a new muscular connection from the anterior vertebral column has been made, to bring upper and lower plates together in complex, powerful movements. Hence in acanthopterygian groups with this **pharyngognathous** condition, we find the mollusk-feeding croakers and drums (Sciaenidae) with molariform dentition, parrotfishes (Scaridae) with pharyngeal jaws capable of grinding up coral rock to expose the algae contained therein, and the highly successful cichlids with a variety of pharyngeal tooth and jaw arrangements that allow their food to be "crushed, triturated, macerated, compacted or in other ways prepared" (Liem & Greenwood 1981, p. 93). The development and diversification of pharyngeal jaws and dentition has undoubtedly broadened the diet of teleosts to include hard-bodied prey and, more importantly, plant material; herbivory is essentially unknown in non-teleostean fishes. This diversification probably extended teleost foraging capabilities far beyond what was possible with the early actinopterygian dependence on more anterior jaw elements. It is more than coincidence that several of the most successful modern teleostean families (cyprinids, cichlids, labrids) have both highly protrusible front jaws and diversified pharyngeal jaws.

Although our emphasis above has been on identifying five general areas that changed during teleostean phylogeny, it is important to remember that these traits changed in concert, that anatomical trends during teleostean phylogeny represent a suite of adaptations. Modification of one trait probably enhanced the effectiveness of other traits. The greatest manifestation of the trends is evident in the acanthopterygian fishes, with their ctenoid scales, diversified yet spiny fins, symmetrical tails, fine maneuverability via pectoral fin sculling, physoclistous gas bladders, greatly expandable mouth volumes, and effective pharyngeal teeth. The result "has been increased swimming speed combined with maneuverability . . . without significant loss of defensive structures" (Gosline 1971, p. 152). In other words, more derived teleosts represent quick, spiny fish with a highly efficient feeding apparatus that can catch and eat small, hard prey items.

Also note that these trends generally describe different taxonomic groups but in no way preclude the possibility of a less derived group acquiring specializations characteristic of a more derived taxon. For example, true eels (a relatively less derived teleostean order), as well as other eel-like fishes, regardless of taxonomic position, have expanded dorsal and anal fins, greatly reduced or absent scales, and often lack pelvic and even pectoral fins. Elaborate median fins are not found solely in more derived superorders. Many osteoglossomorphs are highly derived, specialized fishes that use their own electrical output to locate objects and locomote via an elongate dorsal (gymnarchids) or anal (gymnotids) fin (see Chapter 14). Many adult deep-sea fishes (stomiiform dragonfishes, ateleopodiform jellynose fishes, aulopiform lizardfishes, myctophiform lanternfishes) are physoclistous like more derived teleosts, probably to prevent gas loss via the gut and because they never go to the surface to gulp air. Elaborate pharyngeal dentition, a hallmark of the Acanthopterygii, is used widely in the relatively less derived minnows and suckers (Ostariophysi). A protrusible mouth – brought on by an ascending, sliding premaxillary process – characterizes acanthopterygians and the closely related paracanthopterygians, but was evolved independently and differently in a less derived group, the ostariophysan cypriniforms, as well as in elasmobranchs and sturgeons. Environmental conditions determine the selection pressures operating on a lineage; groups that evolve more effective adaptations will be favored.

Advanced Jawed Fishes II: Chondrichthyes

Superclass Gnathostomata
  Grade Chondrichthyomorphi
    Class Chondrichthyes
      †Superorder Cladoselachimorpha
        Orders Cladoselachiformes, Symmoriiformes
      †Superorder Ctenacanthimorpha
        Orders Ctenacanthiformes, Squatinactiformes
      †Superorder Xenacanthimorpha
        Orders Bransonelliformes, Xenacanthiformes
    Subclass Euselachii
      †Infraclass Hybodonta
        †Order Hybodontiformes
      Infraclass Elasmobranchii (modern sharks and rays)
        Division Selachii (sharks - see Chapter 12)
          Superorder Galeomorphi
            Orders †Synechodontiformes, Heterodontiformes, Orectolobiformes, Lamniformes, Carchariniformes
          Superorder Squalomorphi
            Orders Hexanchiformes, Squaliformes, †Protospinaciformes, Echinorhiniformes, Squatiniformes, Pristiophoriformes
        Division Batomorpha (rays - see Chapter 12)
            Orders Torpediniformes, Rajiformes, Pristiformes, Myliobatiformes
    Subclass Holocephali
            †Orders Iniopterygiformes, Orodontiformes, Eugeneodontiformes, Petalodontiformes, Debeeriiformes, Helodontiformes
          Superorder Holocephalimorpha
            †Orders Psammodontiformes, Copodontiformes, Squalorajiformes, Chondrenchelyiformes, Menaspiformes, Cochliodontiformes
            Order Chimaeriformes (chimaeras)

†Extinct group.

The lineages of bony fishes can be traced with fair certainty back to the Silurian. Their success is evidenced by the diversity of forms found throughout the late Paleozoic and Mesozoic, and because of the overwhelming dominance of teleosts today. However, another group of fishes also arose during the early Paleozoic that radiated in an altogether different direction during the Mesozoic, and is well represented today. These are the **Chondrichthyes** ("cartilaginous fishes"), a group that rapidly specialized as marine predators. By the Carboniferous, sharks made up as much as 60% of the species of fishes in some shallow tropical habitats (Lund 1990).

Although traditionally thought of as "primitive" because of their cartilaginous skeleton, it turns out that many of the characters of modern Chondrichthyes are secondarily derived and represent specializations for a very different, parallel mode of life in water. In addition to several extinct orders and 3 extinct superorders (Cladoselachimorpha, Ctenacanthimorpha, Xenacanthimorpha), two major subclasses of chondrichthyans have representatives still living today – the Holocephali and the Euselachii. The two groups are united by several synapomorphies, chief among which are a prismatic type of **calcification of endoskeletal cartilage** and the presence of **pelvic claspers** in males (Grogan & Lund 2004).

The common ancestor of the two groups remains to be discovered, and many "sharklike" fossils do not fit well into known groups, or are the subject of debate. Our knowledge of early chondrichthyan phylogeny is constrained by the availability of fossil skeletal material; by its nature, cartilage does not fossilize readily and hence our ideas concerning many basal groups rest on incomplete specimens. Accordingly, interrelationships among the Chondrichthyes are, once again, the subject of considerable discussion. Fortunately, the last few decades have seen an upsurge in discoveries, clarifying if not solving many earlier points of contention but leaving others unresolved (see Nelson 2006 for a review).

The Early Chondrichthyes

Definitive sharklike fossils first appear in the Early Devonian, including teeth (418 mybp) and an intact shark fossil (409 mybp); scales or dermal denticles are known from the Late Ordovician (455 mybp). The chondrichthyians have undergone several major radiations, with much controversy surrounding interrelationships. At least eight orders with origins in the Paleozoic arose and disappeared by the Triassic (Compagno 1990b). Identification of the various lineages is based largely on tooth, scale, and spine morphology, and the fossil evidence indicates that, as with bony fishes, foraging and locomotor improvements characterize successive groups.

There are several extinct orders in addition to three extinct superorders (**Cladoselachimorpha**, **Ctenacanthimorpha**, and **Xenacanthimorpha**) of early chondrichthyians. These three superorders are no longer included within the subclass Euselachii (Nelson et al. 2016). Cladoselachimorphs included two orders and three families. One family, the Cladoselachidae (Fig. 11.26A) had five gill slits and a terminal mouth. Their dentition, referred to as **cladodont**, consisted of multicuspid teeth in which the central cusp was usually larger. The teeth were made of enamel-covered dentine and were homologous with scales. These fishes were often large (2 m), pelagic, marine predators with an unconstricted notochord protected by calcified cartilaginous neural arches, and with small precaudal, lateral keels analogous to those found in modern pelagic sharks (Moy-Thomas & Miles 1971). The dorsal fins were often preceded by a spine that may have been supportive or protective in function. Caudal morphology was functionally symmetrical, although the notochord extended into the dorsal lobe of the fin (Fig. 11.26A). Cladoselachids were recognizably sharklike in appearance.

The xenacanthomorphs were common in tropical waters from the Early Devonian into the Triassic. Recognized families include xenacanthids, bransonellids, and diplodoselachids. Xenacanths had a tooth type different from the cladoselachids termed **pleuracanth**, in which the two lateral cusps were large, and the median cusp was smaller. Xenacanthids invaded fresh water and developed an eel-like morphology (Fig. 11.26B). Some xenacanthid sharks had pectoral fins reminiscent of the archipterygium of the dipnoans and may have been bottom dwellers. Xenacanthids also were unusual in possessing two distinct anal fins. The ctenacanthiforms of the Middle Devonian to the Late Triassic had two dorsal fins, each with a prominent spine, an anal fin set far back on the body, and a slightly overhanging snout along with a terminal mouth (Fig. 11.27A).

Another order of extinct chondrichthyians, the **Hybodontiformes**, are now included within the infraclass **Hybodonta** of the subclass **Euselachii** and are seen as a sister group to the infraclass **Elasmobranchii** (Nelson et al. 2016). The hybodonts date back to the Triassic and Jurassic. Unlike modern euselachians, hybodonts retained the terminal mouth of the ctenacanth sharks, rather than the subterminal mouth evolved by elasmobranchs in the Jurassic (Fig. 11.27B). Hybodont teeth represented an innovation over more primitive sharks in that hybodonts had **multicuspid** teeth that were often differentiated into anterior grasping and posterior crushing types, functionally analogous to the marginal and pharyngeal teeth of modern teleosts (and such modern forms as the heterodontid bullhead sharks). Hybodontoid paired fins were flexible and mobile, probably giving them a maneuverability that was not possible with the stiffer appendages of the earlier sharks. Caudal fins became increasingly heterocercal, a reverse of the trend seen in bony fishes. Paralleling a trend seen during acanthodian phylogeny, the spines that precede the dorsal fins became more deeply embedded in the body musculature.

Subclass Euselachii

As mentioned above, the Euselachii include the extinct infraclass **Hybodonta**, and their sister group the infraclass **Elasmobranchii**. Although hybodonts were notably diverse during the Triassic and Jurassic, occupying perhaps as many adaptive zones as modern sharks, neither they nor any of the earlier shark groups survived beyond the Mesozoic. They were

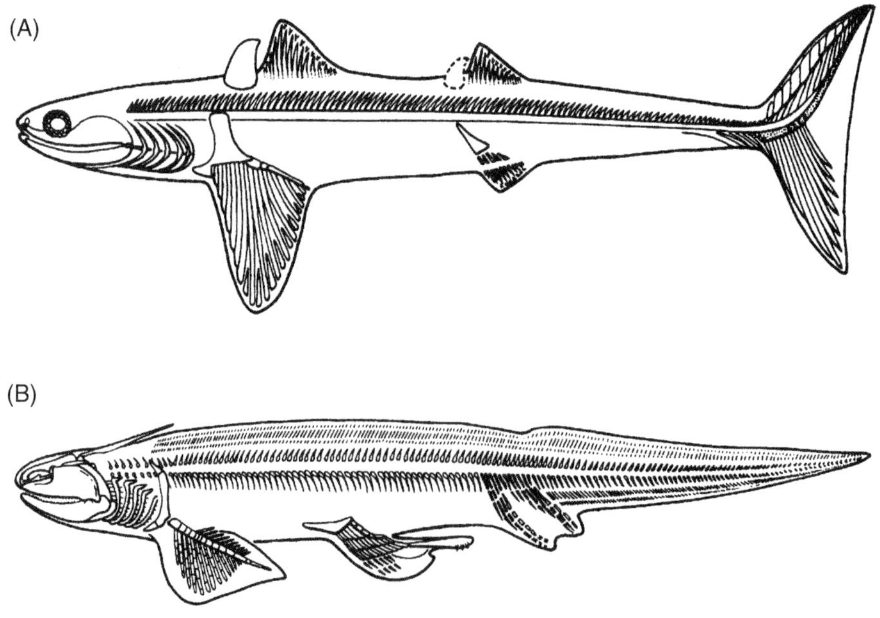

FIGURE 11.26 Diversity in the body form of Paleozoic sharks from the two extinct superorders. (A) *Cladoselache*, a cladoselachid (Cladoselachimorpha); (B) *Xenacanthus*, a freshwater xenacanthid (Xenacanthiimorpha). (A) From Schaeffer (1967); (B) from Schaeffer and Williams (1977), used with permission.

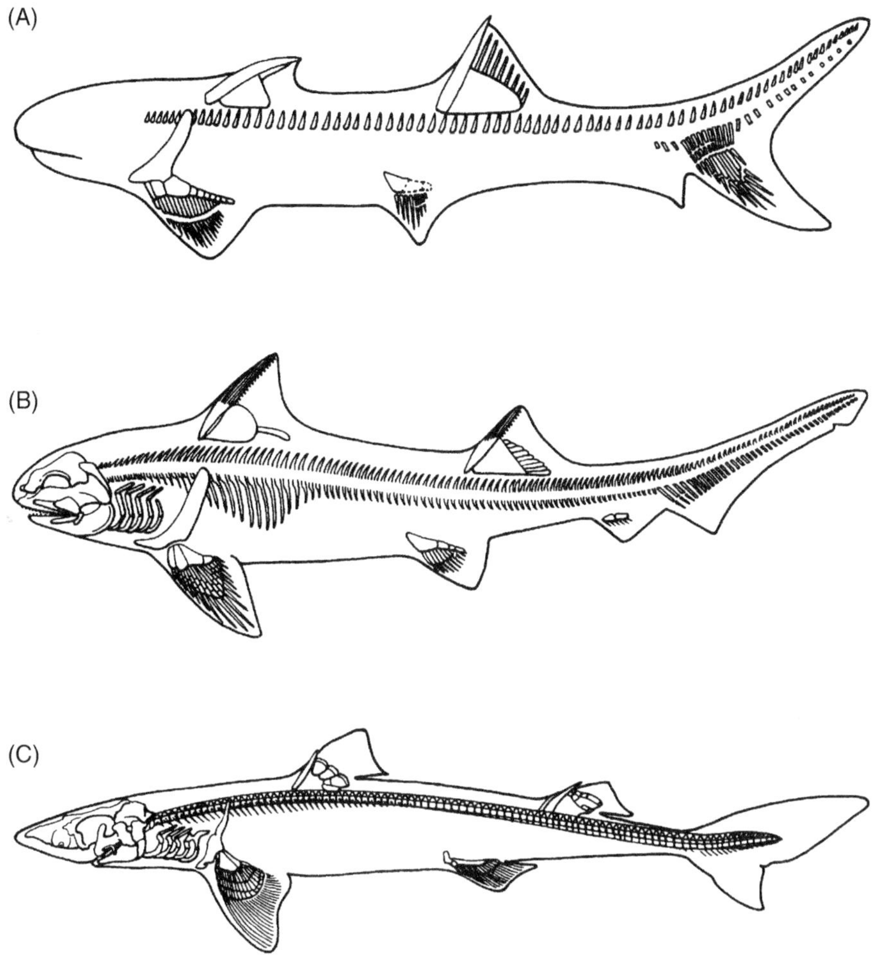

FIGURE 11.27 Examples of similar structure among some ancient and modern chondrichthyians. (A) *Ctenacanthus*, a Late Devonian ctenacanthid (superorder Ctenacanthimorpha, order Ctenacanthiformes); (B) *Hybodus*, a hybodontid, representative of the subclass Euselachii, Infraclass Hybodonta, order Hybodontiformes, the most diverse elasmobranch group in the Triassic and Jurassic; and (C) *Squalus*, a modern squaliform shark in the subclass Euselachii, infraclass Elasmobranchii, division Selachii, superorder Squalomorphi. From Schaeffer and Williams (1977), used with permission; with updated classification from Nelson et al. (2016).

replaced by more derived sharks in marine habitats and by neopterygians in freshwater regions. The more derived euselachiians first appear in the Early Triassic, contemporaneous with the hybodonts. By the Early Jurassic, recognizably modern sharks are found (Fig. 11.27C). One major distinction between modern and earlier sharks is the characteristic overhanging snout of the modern sharks, producing a ventral rather than terminal mouth. The overhanging snout results from an enlarged rostral area that encases a larger olfactory system. Modifications in jaw suspension, jaw–pectoral girdle linkage, and jaw-opening muscles create a protrusible upper jaw and the generation of suction forces, paralleling the trend in teleosts. **Calcified vertebral centra** largely replaced the unconstricted notochord of earlier groups, and fin supports changed from multiple basal cartilages with cartilage radiating out to the fin margins to smaller, fused basal supports (usually three), and flexible rays termed **ceratotrichia** composed of keratin-like protein supporting the web of the fin. This combination of vertebral and fin modifications should have provided for faster swimming and greater maneuverability.

Regardless of the radiation in question, several elasmobranch innovations probably gave them a selective advantage over the other early gnathostomes. In contrast to the placoderms and most acanthodians, sharks quickly evolved a **tooth replacement** mechanism. Teeth grew in whorls or spiral bands (Fig. 11.28), with the functional, exposed tooth backed up by several replacement teeth embedded in the jaw cartilage. As embedded teeth grew, they moved along the whorl until they erupted at the jaw periphery, only to be later replaced by younger teeth. Dentition replacement patterns differ among lineages of modern sharks (see Chapter 12), but in all likelihood teeth were regularly shed and replaced spontaneously in primitive groups, as happens in modern elasmobranchs. This

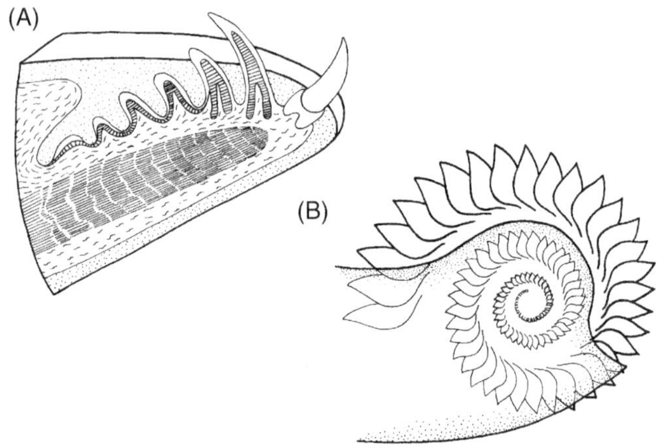

FIGURE 11.28 Tooth replacement in chondrichthyans. (A) Cross-section through the jaw of a modern shark, showing a functional tooth backed by rows of developing replacement teeth. Variations on this mechanism are found in many fossil groups. (B) Symphysial (middle) portion of the lower jaw of the late Paleozoic edestoid *Helicoprion*, thought to be a holocephalan, showing its spiral replacement tooth whorl. After Carroll (1988) and Pough et al. (1989).

arrangement took on some relatively bizarre forms, as in the Permian edestoid holocephalan, *Helicoprion* (Fig. 11.28B).

Other characteristics of modern sharks had origins in late Paleozoic groups. Sexually dimorphic males had pelvic fins modified as intromittent organs for sperm deposition, indicating that **internal fertilization** evolved early in chondrichthyans (see Fig. 11.26B). A strong dependence on electroreception, highly acute olfactory capabilities (associated with the longer snout that houses the nasal capsules), increased buoyancy through oil accumulation in the liver (paralleling gas bladder evolution in bony fishes), and large brain and body size have all contributed to the position of modern elasmobranchs among the top predators in marine habitats (see Chapter 12).

Subclass Holocephali

A calcified cartilaginous skeleton and internal fertilization, among other traits, link the **Holocephali** ("whole heads") with the elasmobranchs (see Chapter 12). Holocephalans, represented today by chimaeras and ratfishes, arguably date back to at least the Late Devonian (Fig. 11.29). Holocephalans and elasmobranchs are considered to form a monophyletic unit with the shared, derived traits of prismatic endoskeletal calcification and pelvic claspers. The stem group is debated, but a Middle Devonian braincase from Bolivia of an animal named *Pucapampella* has been proposed as ancestral.

Regardless, holocephalans differ in many respects from elasmobranchs. Most notable is the position and structure of the gill chamber, which is located further forward than in sharks and has a **single opercular opening** covering four gill openings. Holocephalans have non-protrusible jaws because the palatoquadrate (upper jaw) is fused to the braincase (a version of **autostylic suspension**, named **holostylic** because it is found only in the Holocephalans, as discussed in Chapter 3; in modern elasmobranchs the upper jaw gains mobility via a posterior hyomandibula and an anterior ligamentous connection to the chondrocranium (**hyostylic suspension**). Most fossil and all modern chimaeras and ratfishes have **tooth plates** on the jaw margins that continue to grow during ontogeny; iniopterygians and eugeneodontiforms had replacement dentition. Tail form in holocephalans is variable but often of a diphycercal nature, hence the "ratfish" designation for extant chimaeras.

Chimaeras are a truly ancient group of fishes, the living members of which represent a very small subset of a previously diverse clade. (The group gets its name from the chimaera in Greek mythology, an imaginary monster constructed of parts from other animals.) Research has identified (with contention) 13 orders of holocephalans, 12 of which are extinct. Even two of the three suborders of Chimaeriformes, the order containing all modern forms, are extinct. None of the extinct forms existed into the Cenozoic, and all three modern holocephalan families have fossil records dating back to the Jurassic and Cretaceous. Early holocephalans showed a tremendous diversity of form, including orodontids that reached 4 m in length and debeeriids that did not exceed 10 cm in length. Some petalodontiforms such as *Janassa* were skate-like in morphology and others such as *Belantsea* were globose and almost pufferfish-like in shape. *Chondrenchelys* was eel-like (Fig. 11.29C).

The past few decades have seen an impressive increase in discoveries of fossil holocephalans, largely through the untiring efforts of Eileen Grogan and Richard Lund (see **www.sju.edu/research/bear_gulch**). In our 1997 edition, we anticipated "fossil discoveries [that would help] develop a more meaningful synthesis" of relationships among holocephalans. This synthesis is now underway, but interpretations of the new findings have proliferated (e.g. Grogan & Lund (2004) refer to two subclasses, the Euchondrocephali, recognized here as subclass Holocephali, and the Holocephali, recognized here as the superorder Holocephalimorpha). We have chosen to follow the more traditional terminology and organization laid out in Nelson (2006) and Nelson et al. (2016) until workers in this dynamic area approach a consensus.

A History of Fishes: From Bones to Genomes

As should be obvious, the gaps in our knowledge about ancient fishes and their relationships to one another and to modern groups are large and plentiful. Such gaps are the initiation points for future research. For starters, four topics arise from these unanswered questions and deserve further exploration.

(A)

(B)

(C)

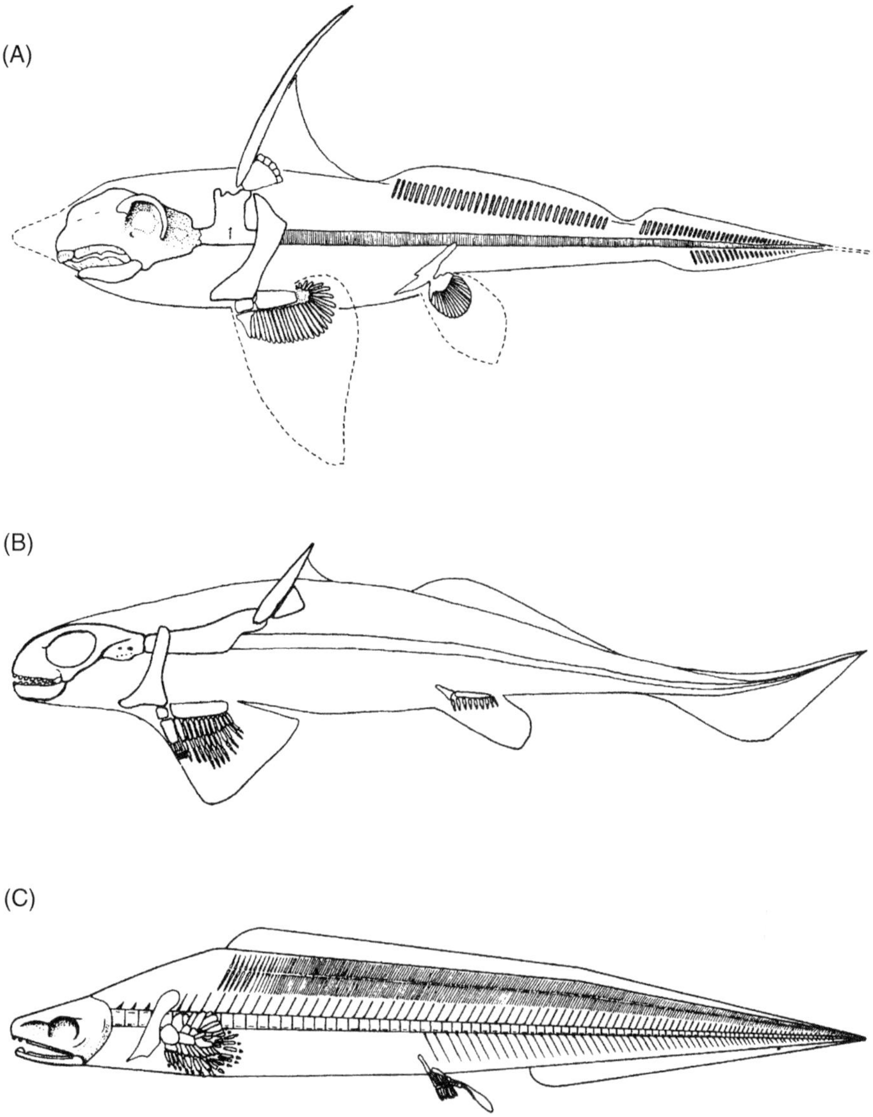

FIGURE 11.29 Extinct holocephalans. (A) *Ischyodus*, a Jurassic callorhinchid in the same family as modern plownose chimaeras; (B) *Helodus*, a Late Devonian helodontiform; and (C) *Chondrenchelys*, a Early Carboniferous chondrenchelyiform. Note the convergence in body form between *Chondrenchelys* and the actinopterygian *Tarrasius* and the clinid in Figure 11.20. From Patterson (1965), used with permission.

The Diversity of Fossil Fishes

We speak of the success of different ancient groups and compare among them and between modern teleosts and extinct forms. The preceding discussion is dependent on the fossil record, because fossils rarely provide DNA for molecular phylogenetic analyses. But how accurately does the fossil record represent the diversity of fossil fishes? How many fishes would we estimate are alive today if we were forced to rely on fossils? As of 1988, approximately 333, or about 8% of modern teleostean genera, are represented by Recent fossils (Carroll 1988; Nelson 1994). Significantly, the number of fossils available for study decreases with time because geological processes tend to destroy fossilized material. Therefore, the fossil record of Recent fishes is at best an optimistic underestimate of the accuracy with which earlier groups are represented.

Fossilization is a chance process, compounded by the relatively small surface of the earth available for paleontological discovery. Inadequacy of sampling is obvious when we realize that most fossils are recovered from only the top few meters of rock, and much of the surface land of the Mesozoic and early Cenozoic has been subducted by tectonic processes (see Chapter 19). Our limited sample size is aggravated by inaccessibility of major areas of the earth's surface; recall that 70% of our planet is under water, where very little paleontological exploration occurs. Significantly, about 2400 species, or 10%, of living fishes occur in water deeper than 200 m

(Cohen 1970), yet few of the recognized pre-teleostean groups are postulated as having occupied the deep sea. Deep-sea fishes, regardless of taxon, are highly convergent in body form and structure (see Chapter 10); such adaptations should be obvious in fossils and such fishes should be assignable to the deep-sea habitat. However, the fossil record for living deep-sea groups is understandably limited. For example, stomiiforms are among the most abundant of the deep-sea orders, with >50 recognized genera, but only five of these, or about 10%, have a fossil record (Nelson 1994; Carroll 1988). Fossil rarity may also reflect convergence on the trait of reduced ossification, reducing further the likelihood of fossilization. The deep sea is one of the most stable aquatic habitats on earth, and it seems unlikely that living in the deep sea is a teleostean innovation. Pre-teleostean diversity in deep-sea habitats is obviously underestimated.

Compound these problems further with the realization that many fossil species are described based on a single, often fragmentary, specimen. How many of these fragments remain undiscovered and, more importantly, how many rare species never fossilized? In our search for antecedents of modern groups, how does this selective preservation of forms affect our interpretations of lines of descent, particularly if there exists at best a one-in-10 chance that an ancestor will fossilize? Our optimistic hope is that the fossil record is somehow a proportional and representative subsample of history, that we accept that we have grossly underestimated the diversity of primitive fishes, and that many more future researchers will take up the challenges of paleontology.

The Role of Molecular Phylogenetics

Looking to the future, molecular phylogenetics offers an exciting new companion to the fossil-based history of fishes. Until late in the 20th century, the evolutionary saga of fishes was resolved by a few dozen palaeontologists spending many thousands of painstaking hours recovering and interpreting the fossil record. Rather than digging in the ground for clues, genomic technology allows us to dig into genomes to detect evolution events that could never be recovered from fossils. For example, sequencing of the fugu (pufferfish) genome reveal that the common ancestor of teleosts had a whole genome duplication about 350 mybp (Christoffels et al. 2004). This duplication is believed to be the foundation for the teleost diversification of novel forms and functions as documented in the fossil record. In other cases, molecular phylogenetics can complement previous morphological analyses to resolve remaining uncertainties about the history of fishes. Examining the genome of the coelacanth (*Latimeria chalumnae*) indicates that the lungfish, rather than the coelacanth, is the closest living relative of tetrapods (Amemiya et al. 2013). A third advantage of molecular phylogenetics is to address the missing data conundrum identified above. Near et al. (2012b) analyzed 232 species from all extant lineages of Actinopterygiian fishes to conclude that elopomorphs (eels and tarpons) are the sister group to all other teleosts. Hughes et al. (2018) analyzed genomic data from 66 of the 72 extant orders of ray-finned fishes to demonstrate that most major lineages of living fishes were already established in the Mesozoic (> 65 mybp). Molecular phylogenetics offers a great toolbox to investigate the hypotheses constructed by our fossil-bearing predecessors. The corresponding limitation is that molecular phylogenetics depends almost entirely on extant species. Fossils as a rule do not yield intact DNA, and so the history of extinct lineages is not recoverable. The beautiful synergy between fossils and genomes is that each informs the other: to calibrate molecular clocks, and estimate the age of diversifications, Near et al. (2012b) used 36 fossil age estimates, and Betancur et al. (2013) used 60 fossil calibrations.

The Tangled Web of Early Vertebrate Relationships: Primitive Does Not Necessarily Denote Ancestral

It is intellectually frustrating to have major living taxa, e.g. modern agnathans, jawed fishes, and gnathostomes in general, for which we can find no clear ancestral lineages. Such phylogenetic problems beg for solution. As a result, considerable effort has been extended attempting to link modern agnathan groups with Paleozoic forebears, and for that matter, modern gnathostomes with ancestors among the diversity of fishes that proliferated during the Devonian.

Plausible alternative explanations exist. First, the ancestors of modern groups may have died out without leaving fossil remains, at least none that we have found so far. Second, Paleozoic lineages were extinguished, period. Hence, similarities between ancient and extant groups result from convergence and perhaps some retention of primitive characteristics derived from a common, distant (unfossilized) ancestor. The latter scenario is perfectly reasonable given the rather advanced condition of bones in the agnathous fishes and of jaws and other supporting bony elements in the early gnathostomes when they first appear in the fossil record. Groups ancestral to these early lineages must have existed for millions of years but lacked the necessary mineralized structures to fossilize. Extinction is a universal characteristic of species; it has been estimated that the average "life span" of a species is around 10 million years (Raup 1988). The mass extinctions that have occurred during the history of life (e.g. the Burgess Shale fauna in the Precambrian, and the Permo-Triassic and Cretaceous–Tertiary extinction events) have been particularly disastrous for shallow marine faunas, wiping out 50–100% of the species in existence at the time (Raup 1988). It seems reasonable to assume that fish lineages were as susceptible to mass extinctions as were contemporaneous invertebrate groups; declines in diversity of actinistians, amiiforms, hybodonts, holocephalans, and perhaps neopterygians at the end of the Cretaceous may attest to the vulnerability of fish groups to mass extinction.

Continuity in Fish Evolution

This chapter focuses on the antecedents of modern fishes. Implied in the organizational approach we have taken here is that fossil fishes should be dealt with separately from living forms. This separation is, however, arbitrary and superficial. It is more of a stylistic convenience for organizing a textbook than a statement of philosophy. Students of fish evolution should quickly recognize that modern fishes are extensions of fossil groups, which is one reason that molecular studies of current fishes can help us understand some aspects of the past. As was pointed out earlier, the majority of modern fish families already existed in the Mesozoic if not earlier. Although some primitive groups that are unrepresented today (e.g. "ostracoderms," placoderms, acanthodians, osteolepidiforms, palaeoniscoids) probably deserve separate treatment from modern forms, it makes just as much sense to treat truly ancestral forms, such as primitive dipnoans, actinistians, neopterygians, and chondrichthyans, together with their modern derivatives. To paraphrase paleontologist A. R. McCune, why should mode of preservation – in rocks or in alcohol – be the primary determinant of how we deal with a taxonomic group? If modern, "primitive" species (e.g. the living coelacanths) were to become extinct through human neglect, would they immediately have to be placed only in a discussion of extinct fishes? It is our hope that students of ichthyology will recognize the continuity that exists between less derived ("primitive") and more derived ("advanced") groups and not view them as separate entities but rather as a continuum of organic change within lineages.

The following chapters on chondrichthyans and living representatives of primitive taxa focus on species that have strong, direct ties to the extinct (we think) groups discussed above. Where one lineage grades into another is in reality an undefined segment in a line drawn in geological time.

Supplementary Reading

Ahlberg PE, Johanson Z. 1998. Osteolepiforms and the ancestry of the tetrapods. *Nature* 395:792–794.

Carroll RL. 1988. *Vertebrate paleontology and evolution*. New York: W. H. Freeman.

Clack JA. 2002. *Gaining ground: the origin and evolution of tetrapods*. Bloomington, IN: Indiana University Press.

Donoghue PCJ, Forey PL, Aldridge RJ. 2000. Conodont affinity and chordate phylogeny. *Biol Rev* 75:191–251.

Forey PL. 1998. *History of the coelacanth fishes*. London: Chapman & Hall.

Frickhinger KA. 1995. *Fossil atlas – fishes*. Malle, Germany: Hans A. Baensch.

Gosline WA. 1971. *Functional morphology and classification of teleostean fishes*. Honolulu: University Press of Hawaii.

Janvier P. 1996. *Early vertebrates*. Oxford, UK: Oxford University Press.

Jarvik E. 1980. *Basic structure and evolution of vertebrates*, Vol 1. London: Academic Press.

Jørgensen JM, Lomholt JP, Weber RE, Malte H, eds. 1998. *The biology of hagfishes*. London: Chapman & Hall.

Lauder GV. 2000. Function of the caudal fin during locomotion in fishes: kinematics, flow visualization, and evolutionary patterns. *Am Zool* 40:101–122.

Lauder GV, Liem KF. 1983. The evolution and interrelationships of the actinopterygian fishes. *Bull Mus Comp Zool* 150:95–197.

Liem KF, Lauder GV, eds. 1982. Evolutionary morphology of the actinopterygian fishes. *Am Zool* 22:239–345.

Long JA. 1995. *The rise of fishes*. Baltimore, MD: Johns Hopkins Press.

Maisey JG. 1996. *Discovering fossil fishes*. New York: Henry Holt & Co.

Matsen B, Troll R. 1995. *Planet ocean: dancing to the fossil record*. Berkeley, CA: Ten Speed Press.

Moy-Thomas JA, Miles RS. 1971. *Palaeozoic fishes*, 2nd edn. London: Chapman & Hall.

Norman JR, Greenwood PH. 1975. *A history of fishes*, 3rd edn. New York: Halstead Press.

Pough FH, Janis CM, Heiser JB. 2005. *Vertebrate life*, 7th edn. Upper Saddle River, NJ: Pearson Prentice Hall.

Journal

Ichthyolith Issues.

Websites

www.ageoffishes.org.au.
www.devoniantimes.org.
www.palaeos.com.

CHAPTER **12**

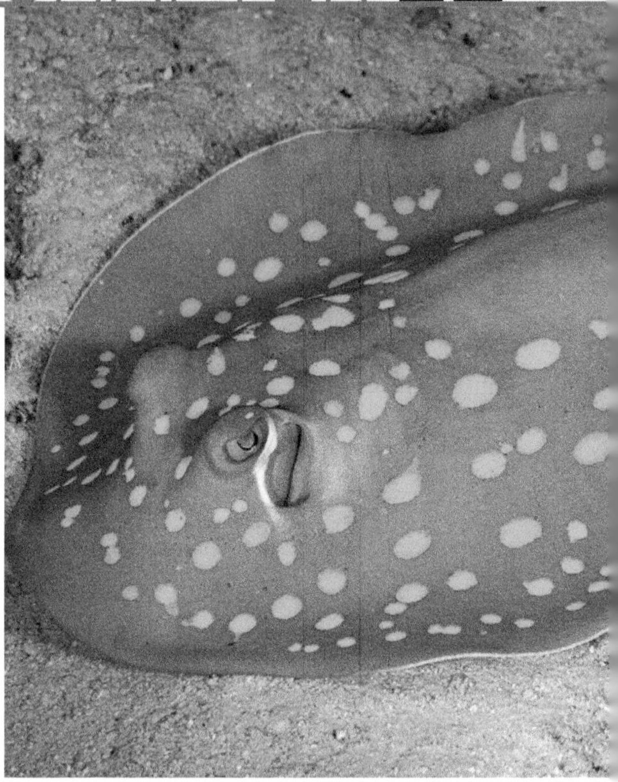

Chondrichthyes: Sharks, Skates, Rays, and Chimaeras

Summary

The Chondrichthyes contain two living subclasses, the Euselachii (sharks and rays) and the Holocephali (chimaeras, ratfishes). The Euselachii are represented by one living infraclass, the Elasmobranchii, which contains nine orders and over 550 species of sharklike fishes in two superorders, and four orders and over 660 species of skates and rays. Carchariniform requiem sharks and squaliform dogfish sharks are the most diverse shark orders, and rajiform skates and myliobatiform stingrays are the most diverse batoid orders.

Elasmobranch vertebrae consist of calcified cartilage. Teeth are replaced throughout life and are not fused to the jaws. The mouth is usually subterminal and the upper jaw is not fused to the braincase, allowing protrusion during feeding. Fin rays consist of ceratotrichia composed of a keratin-like protein. Most elasmobranchs have five external gill slits. In general, elasmobranchs are marine, mobile predators that grow slowly and have slow metabolism, rely strongly on nonvisual senses, and produce small numbers of young.

Sharks are relatively large, many exceeding 1 m in length. The largest sharks are the Whale Shark (>12 m), Basking Shark (9 m), and hammerhead, thresher, sleeper, and Tiger sharks (5–6 m). The largest verified White Shark was 6 m long and weighed 3300 kg. The extinct Megatooth Shark was 16 m long and may have weighed approximately 48,000 kg.

CHAPTER CONTENTS

The Diversity of Fishes: Biology, Evolution and Ecology, Third Edition. Douglas E. Facey, Brian W. Bowen, Bruce B. Collette, and Gene S. Helfman.

Companion website: www.wiley.com/go/facey/diversityfishes3

There are some small species, however, such as deep-water Lantern Sharks (<20 cm long).

Most sharks live in shallow, marine habitats, but a few occupy depths beyond 3000 m. Two families of rays inhabit fresh water, and sawfishes and Bull Sharks frequently move into rivers, the latter occurring as far as 4000 km from the ocean. Elasmobranchs that inhabit fresh water osmoregulate to counteract influx of water and loss of salts.

Sharks are active predators with relatively large home ranges. Many coastal and oceanic species undertake migrations of 1000–16,000 km. Locomotion is very efficient. Placoid scales are shaped to minimize drag, and dorsal fins work in conjunction with the heterocercal tail and pectoral fins to maximize propulsion. The large, lipid-filled liver provides buoyancy. Sharks have slower metabolism and require less food for a given body weight than do bony fishes and therefore do not need to feed as frequently. Slow metabolism leads to slow growth and old age in many species.

Except for the largest sharks and manta rays, which are plankton feeders, most sharks use their protrusible jaws and sharp, often serrated teeth to dismember prey. Teeth are replaced every few days; a shark may produce 30,000 teeth during its life. Feeding specializations include suction feeding and molariform teeth for crushing mollusks (many rays), elongate tails or snouts for striking and incapacitating prey (thresher sharks, sawfishes, sawsharks), and muscles modified for electricity production to stun prey (torpedo rays).

Sharks have good vision, particularly at night. They are exceptionally sensitive to chemical stimuli, can localize sound, and can detect weak electrical and geomagnetic cues, which they use to localize prey and may assist with navigation in the open ocean. Sharks have relatively large brains compared to bony fishes.

Elasmobranchs overall grow slowly and mature at a relatively old age compared to bony fishes. Fertilization is internal; some species lay eggs, whereas others gestate young internally. Gestation is long, and the young are small replicas of adults. No parental care is given after birth, but female investment during gestation is very high, particularly in those species with complex placental structures. Relatively few young are produced at a time. Sharks are very susceptible to overfishing due to their slow growth, slow maturation, and low fecundity. Shark fisheries typically boom then quickly bust and do not recover. Shark populations have been in decline for decades and a general moratorium on the capture of many species is needed.

Holocephalans (chimaeras, ratfishes, rabbitfishes) include six genera and about 48 species of cartilaginous fishes. They differ from sharks by having the upper jaw fused to the braincase, a single gill cover, separate anal and urogenital openings, and an erectable dorsal spine. They are entirely marine, inhabiting shallow to moderate depths. As in the sharks, holocephalans were much more diverse during the Paleozoic and Mesozoic than they are today.

Class Chondrichthyes, Subclass Euselachii

The subclass Euselachii includes the extinct infraclass Hybodonta and the extant infraclass Elasmobranchii. Although thought of as "primitive" by many, modern sharks, skates, and rays are highly derived, specialized fishes that differ dramatically from the abundant, diverse elasmobranchs that dominated marine and even freshwater habitats through much of the Mesozoic (see Chapter 11). Many traits that characterize elasmobranchs – such as a cartilaginous skeleton, placoid scales, internal fertilization, replacement dentition, and multiple gill slits – appeared early in the 400+ million-year history of the group. However, modern sharks, skates, and rays exhibit tremendous variation in these and other characteristics, and have developed additional anatomical, life history, and behavioral adaptations that set them apart from bony fishes and make them surprisingly vulnerable to human exploitation. Only in recent years has the uniqueness and vulnerability of elasmobranchs received recognition and adequate attention.

Definition of the Group

Modern elasmobranchs are generally large (>1 m) predatory fishes with a calcified but seldom ossified skeleton, including distinctive **calcified vertebral centra**. They differ from bony fishes in that the skull lacks sutures and their teeth are not fused to the jaws but are instead embedded in the connective tissue of the jaws. **Teeth**, which have the same embryonic origin as and may be derived from **placoid scales** (discussed in Chapter 4) are replaced serially; such replacement is less common in osteichthyans. The biting edge of the upper jaw is formed by the **palatoquadrate cartilage**, rather than by the maxillary or premaxillary bones. The palatoquadrate is free from the braincase, creating a protrusible upper jaw during feeding. The mouth is subterminal (= ventral). Nasal openings are ventral and incompletely divided by a flap into incurrent and excurrent portions; bony fishes generally have completely separated, dorsally positioned nasal openings. Fin rays in elasmobranchs are unsegmented **ceratotrichia** composed of a keratin-like protein.

Typical sharks, skates, and rays usually have five, and sometimes six or seven, external gill slits on each side. The first gill slit of elasmobranchs is often modified as a **spiracle**, supported by the hyoid arch and first functional gill arch. Elasmobranchs lack lungs and gas bladders, but possess large livers with low-density oil that provides buoyancy. Their intestines have a "spiral valve" that increases absorptive surface area. Internal fertilization is universal to the group; males possess pelvic fin-derived intromittent organs (**myxopterygia** or **claspers**) and females either lay eggs or nourish embryos internally for several months before giving birth. Chloride ions and

Class Chondrichthyes[a]
- Subclass Euselachii (sharklike fishes)
  - Infraclass Elasmobranchii (Neoselachii) (all extant sharks and rays)
    - Division Selachii (sharks)
      - Superorder Galeomorphi
        - Order Heterodontiformes (9 species, marine): Heterodontidae (bullhead and horn sharks)
        - Order Orectolobiformes (44 species, marine): Rhincodontidae (Whale Shark), Parascyllidae (8 species of collared carpet sharks), Brachaeluridae (2 species of blind sharks), Orectolobidae (12 species of wobbegongs), Hemiscylliidae (17 species of bamboo sharks), Stegostomatidae (2 species of zebra sharks), Ginglymostomatidae (2 species of nurse sharks)
        - Order Lamniformes (15 species, mostly marine): Odontaspididae (3 species of sand tiger sharks), Mitsukurinidae (Goblin Shark), Pseudocarchariidae (Crocodile Shark), Lamnidae (5 species of mackerel sharks), Megachasmidae (Megamouth Shark), Cetorhinidae (Basking Shark), Alopiidae (3 species of thresher sharks)
        - Order Carcharhiniformes (374 species, mostly marine): Pentanchidae (90 species of deep-water sharks), (about 150 species of cat sharks), Proscylliidae (7 species of finback sharks), (4 species of false cat sharks), Leptochariidae (Barbeled houndshark), Triakidae (46 species of houndsharks), Hemigaleidae (8 species of weasel sharks), Carcharhinidae (at least 58 species of requiem sharks), Sphyrnidae (10 species of hammerhead sharks)
      - Superorder Squalomorphi
        - Series Hexanchida
          - Order Hexanchiformes (6 species, marine): Hexanchidae (4 species of cow sharks), Chlamydoselachidae (2 species of frill sharks)
        - Series Squalida
          - Order Squaliformes (123 species, marine): Dalatiidae (9 species of kitefin sharks), Etmopteridae (47 species of lantern sharks), Somniosidae (17 species of sleeper sharks), Oxynotidae (5 species of rough sharks), Centrophoridae (16 species of gulper sharks), Squalidae (29 species of dogfish sharks)
        - Series Squatinida
          - Order Echinorhiniformes (2 species, marine): Echinorhinidae (2 species of bramble sharks)
          - Order Squatiniformes (22 species, marine): Squatinidae (22 species of angel sharks)
          - Order Pristiophoriformes (8 species, marine): Pristiophoridae (8 species of saw sharks)
    - Division Batomorphi (Batoidea) (skates and rays)
      - Order Torpediniformes (65 species, marine): Narcinidae (including Narkidae, 42 species of numbfishes), Torpedinidae (including Hypnidae, 23 species of torpedos or electric rays)
      - Order Rhinopristiformes (67 species, marine and freshwater): Trygonorrhinidae (8 species of fiddler or banjo rays), Rhinobatidae (35 species of guitarfishes), Rhinidae (10 species of bowmouth guitarfishes), Glaucostegidae (9 species of giant guitarfishes), Pristidae (5 species of sawfishes)
      - Order Rajiformes (299 species, marine): Rajidae (158 species of skates), Arhynchobatidae (108 species of softnose skates), Gurgesiellidae (18 species of leg or pygmy skates), Anacanthobatidae (14 species of smooth skates)
      - Order Myliobatiformes (238 species, marine and freshwater): Zanobatidae (2 species of panrays), Hexatrygonidae (Sixgill Stingray), Dasyatidae (97 species of stingrays), Potamotrygonidae (38 species of freshwater stingrays), Urotrygonidae (18 species of American round stingrays), Gymnuridae (14 species of butterfly rays), Plesiobatidae (Deepwater Stingray), Urolophidae (29 species of round stingrays), Aetobatidae (5 species of pelagic eagle rays), Myliobatidae (40 species of eagle rays, including Aetobatidae, Rhinopteridae, and Mobulidae)

[a] Higher classification groups based on Nelson et al. (2016), but see corrections and updates at **https://sites.google.com/site/fotw5th/**. Orders, families, and numbers of species from Fricke et al. (2020) but refer to **https://researcharchive.calacademy.org/research/ichthyology/catalog/SpeciesByFamily.asp** for updates.

metabolic waste products in the form of urea and trimethylamine oxide (TMAO, an ammonia derivative) are concentrated in the blood and serve in osmotic regulation. A single **cloaca** serves as an anal and urogenital opening.

Historical Patterns

Most orders of living chondrichthyans appeared by the Upper Jurassic, and all orders appeared by the end of the Cretaceous. Some extant genera have been found in Upper Cretaceous deposits, with little change in some species since the Miocene (Compagno 1990a). Most extant groups have evolutionary histories much younger than actinopterygian and neopterygian fishes (see Chapter 11, Advanced Jawed Fishes II: Chondrichthyes). Cladoselachians died out by the Permo-Triassic transition and xenacanths died out during the Triassic. Ancestral groups such as ctenacanths and hybodonts disappeared during the Mesozoic.

Although some living sharks have morphologies like ancestral Paleozoic and Mesozoic species, these similarities reflect convergent evolution. The modern groups are very different with respect to features of the cranium, vertebral column, fin skeletons, tooth structure, and squamation. The greatest departure from a generalized body form exists in the highly successful batoids, the most derived of which are the large-brained, filter-feeding manta and devil rays (Mobulidae). Among the sharklike elasmobranchs, the most derived species include the hammerhead sharks (Sphyrnidae), angel sharks (Squatinidae), and saw sharks (Pristiophoridae) (Compagno 2001; Compagno et al. 2005a, 2005b). Shark systematics is an active field and relationships among most groups are well established, although some groups remain unresolved and await further study (de Carvalho 1996; McEachran et al. 1996) (Fig. 12.1).

Modern Elasmobranch Diversity

Over 1200 species of elasmobranchs exist today, including over 550 described sharklike species and over 650 skates and rays (Fricke et al. 2020, but check **https://researcharchive.calacademy.org/research/ichthyology/catalog/SpeciesByFamily.asp** for more recent updates) (Fig. 12.2). Sharks (division **Selachii**) can generally be distinguished from rays (division **Batomorphi / Batoidea**) by the following features. Sharks have (i) gill openings on the sides of the body; (ii) the anterior edge of the pectoral fin not attached to the side of the head; (iii) the anal fin present in galeomorphs but absent in squalomorphs (except for the five species of hexanchiforms); and (iv) small

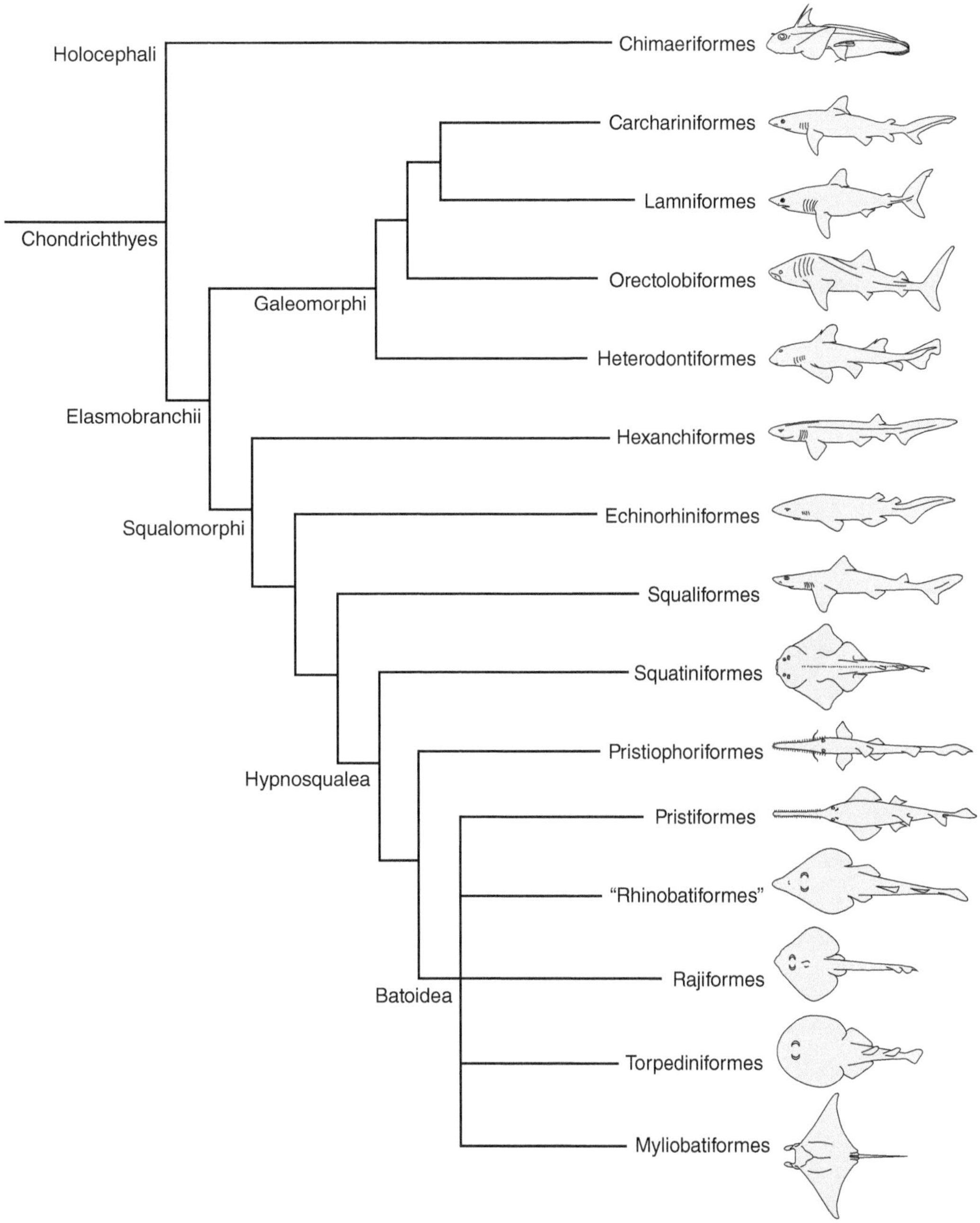

FIGURE 12.1 Phylogenetic relationships among living chondrichthyans. Relationships among the batoid rays remain a matter of debate, including discussion of whether the rhinobatiform guitarfishes are in fact monophyletic. From Stiassny et al. (2004), used with permission.

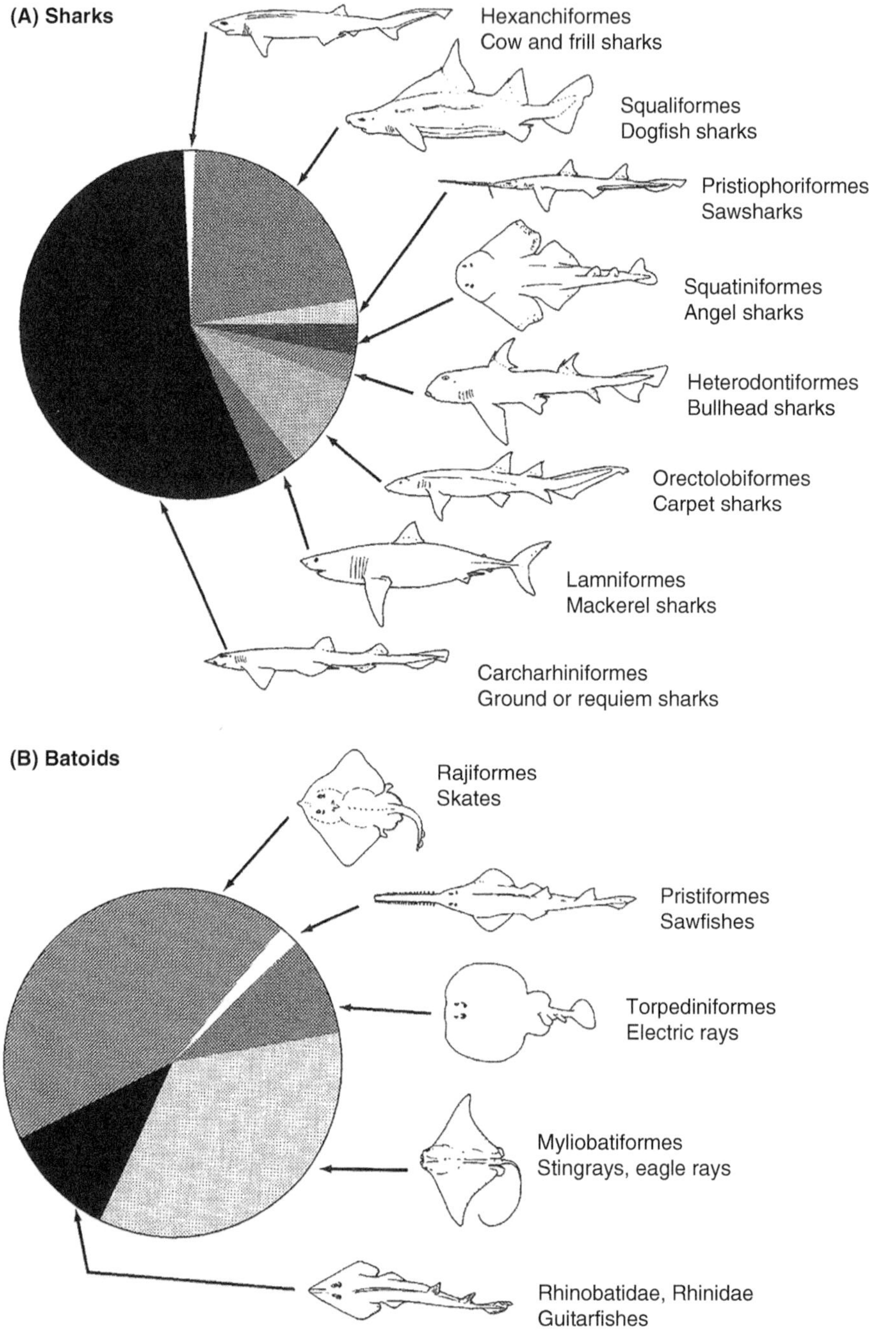

FIGURE 12.2 Taxonomic distribution and representative orders of over 1250 species of modern sharks, skates, and rays. (A) Sharklike fishes in nine orders constitute 40% of modern euselachian species, with the carcharhiniform (ground or requiem) sharks outnumbering all other orders combined. The echinorhiniform bramble sharks, with two species, are not shown. (B) Raylike batoids make up 60% of the Euselachii, dominated by skates and stingrays; the four recognized orders are shown. Guitarfishes (Rhinobatidae, Rhinidae) are diverse members of the Rajiformes. Adapted from Compagno (1990b), used with permission.

lateral spiracles compared with large dorsal spiracles in rays. Rays in contrast have (i) ventral gill openings; (ii) the anterior edge of the enlarged pectoral fin attached to the side of the head; (iii) the anal fin absent; and (iv) the intake of water for breathing chiefly through enlarged dorsal spiracles (except in water column species).

Among the sharks, the **requiem** or ground sharks (Carcharhiniformes) make up more than half the species and are particularly diverse in tropical and subtropical, nearshore habitats. Offshore, pelagic sharks include **lamniform** species such as mako, White, thresher, and Basking sharks, whereas the **squaliform** dogfishes are particularly successful in the North Atlantic,

North Pacific, and deep-sea regions. The **batoids** are concentrated in four orders, the torpediniform torpedo rays, the rhinopristiforms (sawfishes, guitarfishes, banjo rays), the rajiform skates, and the myliobatiform stingrays (Fricke et al. 2020). Skates are most diverse and abundant in deep water and at high latitudes, whereas stingrays are most diverse in tropical, inshore waters. Most skates have one or two dorsal fins and long, slender claspers that are depressed at their distal end, whereas stingrays have a serrated tail spine (the "sting"), lack dorsal fins, and have short, stout claspers that are cylindrical or only moderately depressed.

Amidst this diversity, certain general patterns emerge that emphasize the unique traits and fascinating adaptations of elasmobranchs. These trends include (i) large size; (ii) a marine habitat; (iii) mobility; (iv) slow metabolism and slow growth; (v) predatory feeding habits; (vi) reliance on nonvisual senses; (vii) low fecundity and precocial (independent) young; and (viii) vulnerability to exploitation (see Compagno 1990b; Gruber 1991).

Body Size

When compared with bony fishes, sharks as a group have always been relatively large. Modern sharks range from the 15 g, 17 cm Dwarf Lantern Shark, *Etmopterus perryi* (Etmopteridae), and several sharks in the 22–25 cm range (e.g. dalatiid pygmy sharks, *Squaliolus laticaudus* and *Squaliolus aliae*; proscyliid Pygmy Ribbontail Catshark, *Eridacnis radcliffei*) to the 4000 kg, 10 m Basking Shark, *Cetorhinus maximus*, and the 12,000+ kg, 14 m long Whale Shark, *Rhincodon typus* (Rhincodontidae), the largest fish in the world. At least 90% of living sharks exceed 30 cm in body length, 50% reach an average length of about 1 m, and 20% exceed 2 m (Springer & Gold 1989). Maximum sizes of sharks, particularly the maximum size reached by the superpredatory White Shark (*Carcharodon carcharias*, Lamnidae), is a subject plagued by misinformation and exaggeration; part of that exaggeration is the frequent addition of the word "great" preceding the proper species name, which is simply White Shark.

Large size is intimately linked with the feeding and reproductive ecology of sharks. As predators on other fishes, including other elasmobranchs, large size confers an advantage in terms of greater swimming speed during pursuit or long-distance cruising, and allows for larger mouth size and larger jaw muscle attachment. Such traits make sharks effective predators on smaller fishes and also decrease their own vulnerability to predators, either via rapid escape or active defense. It is suggested that sharks larger than 1 m long are relatively immune to shark predation, and it is not surprising that birth sizes of many sharks are close to the 1 m critical length (e.g. Sand Tiger, Odontaspididae; White and Longfin Mako, Lamnidae; Dusky, Carcharhinidae).

Sharks that give birth to smaller young often have relatively large litters or short intervals between reproduction (e.g. Atlantic Sharpnose, Carcharhinidae; Scalloped Hammerhead and Bonnethead, Sphyrnidae). Predation also affects nursery ground location and interacts with growth rate. Sharks that give birth in offshore or beachfront areas that are frequented by large sharks tend to have relatively rapid growth rates of 30–60 cm during the first year (e.g. Thresher, Alopiidae; Shortfin Mako; Blue, Tiger, Spinner, and Sharpnose, Carcharhinidae; Bonnethead). Sharks that release their young in relatively predator-free inshore nursery areas such as bays, sounds, estuaries, or shallow reef flats tend to grow only 15 cm in the first year (e.g. Bull, Sandbar, and Lemon, Carcharhinidae; Scalloped Hammerhead) (Branstetter 1990, 1991).

The Mismeasure of Man Eaters

Maximum sizes of shark species are a matter of much speculation and imagination. Researchers tend to be conservative and therefore accept only documented measurements with an accurate measuring tape and weighing scale, preferably accompanied by the preserved specimen, or at least by a photograph with a ruler for scale. However, very large animals are difficult to preserve and harder to store, and photographs can be doctored or just misleading because of problems with parallax. Hence, verified maximum sizes and reported maxima ("bigger than the boat") vary considerably.

For example, the longest recorded Whale Shark is 12 m, but the species is known to grow much larger, perhaps as large as 18 m. Basking Sharks (Cetorhinidae) have been reliably measured at 9.76 m, but lengths of 12–15 m have been reported. Other accepted (vs. reputed) lengths for large predatory shark species include: Shortfin Mako (3.3 vs. 4.0 m), Great Hammerhead (5.5 vs. 6.1 m), Thresher Shark (5.7 vs. 7.6 m), Greenland Shark (6.4 vs. 7 m), and Tiger Shark (5.9 vs. 7.4 and 9.1 m) (Springer & Gold 1989; Herdendorf & Berra 1995).

Nowhere is the potential for sensationalism greater than in the case of the White Shark, *Carcharodon carcharias* (and note that the name of this species does not include "Great"). "Verified" lengths reported for this species often include an Australian record of 36.5 ft (11.1 m). Some authors have taken the liberty of rounding off that measurement to 40 ft (12.3 m). Re-examination of the teeth and jaws from the reputed 36.5 ft specimen suggest that it was only 16.5 ft (5 m) long and that the reported length resulted from a typographical error. The largest reliably measured White Shark was a 19.5 ft (5.944 m) long female caught off Ledge Point, Western Australia in 1984 (Randall 1973, 1987; Mollet et al. 1996). This length stands in contrast to a photograph published in *The Guinness book of animal facts and feats* (Wood 1982), of a purported 29.5 ft (9.1 m) Azores shark, but the photograph suggests a much smaller animal and no verification of the measurements has been possible.

Extrapolations from jaw dimensions of known sharks indicate that bite marks on dead whales could come from sharks larger than 6 m and several specimens in the 7 m range (all female) have been reported, but no such giants have been authenticated (Randall 1973, 1987; Ellis & McCosker 1991; Mollet et al. 1996). The heaviest White Shark reliably weighed had a mass of 3324 kg (Springer & Gold 1989). White Sharks are born at a length of around 100 cm and a mass of 13 kg (Ellis & McCosker 1991).

If the extant White Shark can attain a length of 6 m and weigh in excess of 3000 kg, then how large was the biggest member of the genus, the widespread "Megatooth" Shark, *Carcharodon megalodon*, that lived during the Mid-Miocene through Late Pliocene, 16 to 1.6 million years ago? Teeth from this giant are common at many fossil-bearing locales in Europe, Africa, Australia, India, Japan, and North and South America (Bruner 1997). Enamel heights (the vertical distance from the base of the enamel portion to its tip, Fig. 12.3) in excess of 100 mm are not unusual (the largest White Shark teeth are about 60 mm high); the largest *C. megalodon* tooth found had an enamel height of 168 mm (Compagno et al. 1993; see also Applegate & Espinosa 1996; Gottfried et al. 1996) (some researchers place the Megatooth Shark in the genus *Carcharocles*).

Paleontologists, and others, have assembled these teeth into reconstructed jaws of this shark and then extrapolated to total body length based on jaw dimensions. These reconstructions have been notoriously inaccurate. The most famous was produced by the American Museum of Natural History in 1909 (Fig. 12.3A). The jaws of this reconstruction were oversized because (i) the preparators created a wider-than-accurate jaw by using all anterior (midline front) teeth of equal size across the jaws, whereas most sharks, including *C. carcharias*, have smaller lateral and posterior teeth at the sides; and (ii) the cartilaginous jaw of a shark is generally no broader than the enamel height of the biggest tooth. In the American Museum reconstruction, cartilage breadth was four times enamel height, creating a larger jaw. The two errors produced a jaw about 30% larger than it should have been, which created a larger shark.

Length estimates extrapolated from that jaw, influenced by tooth size: body length ratios of the mis-measured 36.5 ft Australian specimen, have ranged between 60 and 100 ft (18.5 to 31 m), which has been rounded to 120 ft (37 m) in some popular books. It was a Megatooth Shark that terrorized the New England town of Amity in Peter Benchley's (1974) novel *Jaws*. Given that snout length is 6% of total length in White Sharks, and assuming the ill-fated swimmer on the cover of the paperback version of the novel is 1.7 m tall, the Amity shark was a conservative 21 m long. The mechanical shark (named "Bruce") used in the movie version of *Jaws* depicted a White Shark about 7.3 m long (Stevens 1987).

Recent reconstructions of *C. megalodon* (Fig. 12.3B) have used more quantitative methods in estimating size, such as the statistical relationships of tooth enamel height and jaw dimensions to body length and mass from known White Shark specimens (Fig. 12.3C). Extrapolating from *C. carcharias* to *C. megalodon*, a Megatooth Shark with a tooth enamel height of 168 mm would be about 16 m long and weigh approximately

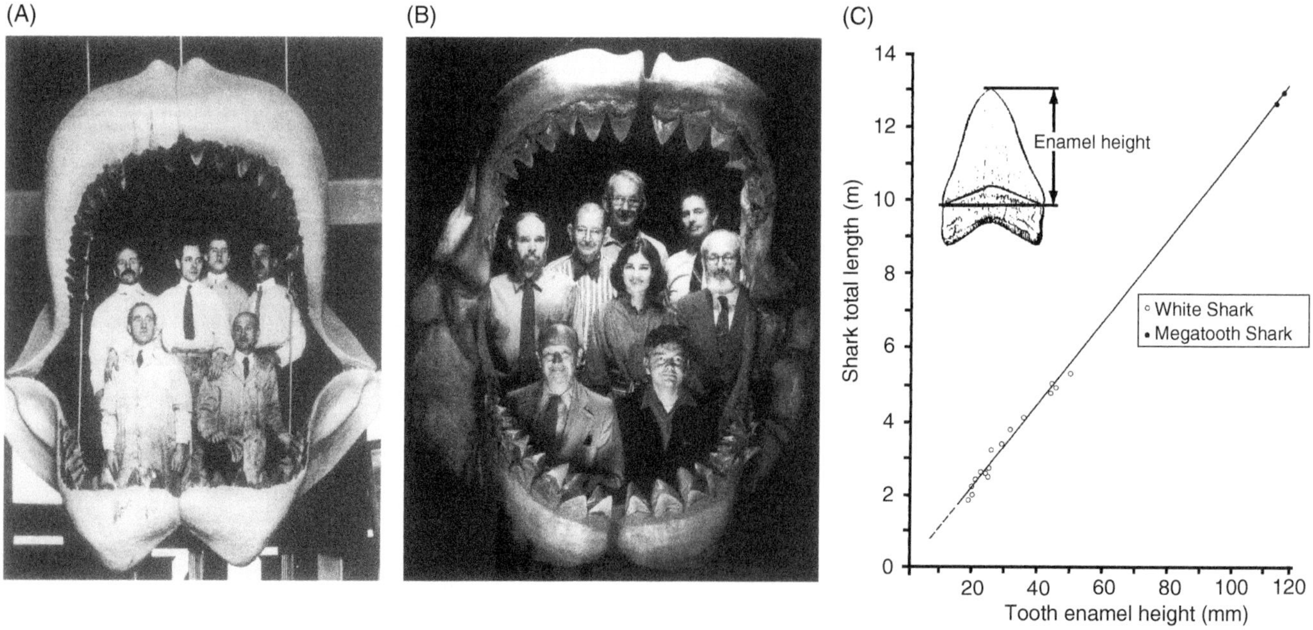

FIGURE 12.3 Reconstructing the jaws and estimating the size of the extinct Megatooth Shark, *Carcharodon megalodon*. (A) Jaw reconstruction as inaccurately prepared in 1909. The jaws are about one-third too large because equal-sized, anterior teeth were used throughout the jaws, and the cartilage is about four times broader than in living sharks. (B) Reconstruction by the Smithsonian Institution, suggesting a body length of about 13 m. (C) Calculating the lengths of White and Megatooth sharks. Total body length is directly related to maximum tooth size (enamel height) in White Sharks (*Carcharodon carcharias*); hence body length can be estimated for sharks from which only teeth are available. This gives a maximal size of 6 m for the White Shark. Assuming a similar relationship existed for the extinct Megatooth Shark, placement of two of the larger known teeth along the same regression line (closed circles) suggests a body length of about 13 m; the largest tooth found indicates lengths up to 16 m. The approximate equation for calculating total length from tooth height is: Total length (m) = 0.096 (enamel height, mm) – (0.22). Data from Randall (1973), Compagno et al. (1993), and Gottfried et al. (1996). (A) American Association for the Advancement of Science, © 1971, used with permission; (B) Courtesy of Chip Clark.

48,000 kg (Compagno et al. 1993; Gottfried et al. 1996). If body proportions were similar to those of extant White Sharks, the jaws would be >1 m wide, the dorsal fin would be 1.4 m high, and the tail would be 1.75 m tall.

The Megatooth Shark, although probably the largest shark to ever live, occurred with other relatively gigantic Miocene/Pliocene predators, including the Speartooth Mako, *Isurus hastalis* (estimated at 6 m long), a hemigaleid, *Hemipristis serra*? (5 m), as well as the extant White Shark (Compagno 1990b).

Habitats Most elasmobranchs are marine organisms of relatively shallow temperate and particularly tropical waters, although all oceans except the Antarctic have one or more species. Most inhabit continental and insular shelves and slopes: 50% of all species occur in <200 m of water and 85% occur in <2000 m. Only about 5% dwell in the open ocean, including a few myliobatiforms such as manta rays; another 5% occupy fresh water. Truly cold water, high-latitude sharks (not counting those that live in the perpetually cold, deep sea) are limited to the relatively large hexanchids (Sixgill Shark), Basking Shark and sleeper sharks (Cetorhinidae, Somniosidae), smaller squaliform sharks, and some catsharks (Scyliorhinidae).

Although bony fishes inhabit ocean depths to 8000 m, sharks do not occur as deep. Deep-sea sharks, those frequently captured in aphotic (lightless) waters below 1000 m, include several squaliform species (Portuguese, kitefin, lantern, sleeper, and rough sharks), carcharhiniform cat and false catsharks, and the sixgill. Record depths include sightings of the Portuguese Shark, *Centroscymnus coelolepis*, at 3690 m (Clark & Kristof 1990); an unidentified spiny dogfish at 4050 m; and a disputed depth of 9938 m for the squalid *Euprotomicrus bispinatus* (see Herdendorf & Berra 1995). But these are exceptional occurrences because Chondrichthyes in general seldom occur in abyssal regions deeper than 3000 m, probably because of low food supplies and the high metabolic costs of maintaining a large, oil-rich liver used to maintain buoyancy (Priede et al. 2006). Hence, sharks are restricted from 70% of the world's ocean volume and have no natural deep-water refuges from human exploitation.

Sharks have also failed to occupy other habitats characterized by environmental extremes of temperature, oxygen, turbulence, drought, and salinity, unlike bony fishes. Among the few exceptions are *Pentanchus* cat sharks that inhabit deep-water basins that are characterized by relatively low oxygen levels and relatively high temperatures and salinities. These species have elongated gill regions and expanded gill filaments (Compagno 1984).

Although the ocean is home to most elasmobranchs, at least 170 species in 34 families can enter estuarine and freshwater habitats, and 47 species are restricted to such regions (Martin 2005). Two families of rays (Potamotrygonidae, Dasyatidae) are truly freshwater, seldom if ever entering marine conditions (Compagno 1990b). The potamotrygonid river stingrays include about 38 species restricted to fresh waters of Atlantic drainages of South America. Eight species of dasyatid whiptail stingrays in the genera *Dasyatis* and *Himantura* are restricted to rivers in Africa, Southeast Asia, New Guinea, and Australia, and another six dasyatid species are euryhaline, spending most of their lives in fresh water (Martin 2005). Included among the obligate freshwater forms is the Giant Freshwater Stingray or whipray, *Himantura chaophraya* of tropical Australia, New Guinea, Borneo, and Thailand (Fig. 12.4). *Himantura chaophraya* may be the world's largest stingray and perhaps the world's largest freshwater fish, reaching a length of 5 m, a width of 2.4 m, and a mass of 600 kg (Roberts & Monkolprasit 1990; Compagno & Cook 1995a).

FIGURE 12.4 The Giant Freshwater Whipray *Himantura chaophraya*. This endangered species occurs in rivers of Southeast Asia, New Guinea, and tropical Australia. Courtesy of Z. Hogan.

Three species of carcharhinid river sharks in the genus *Glyphis* are obligate freshwater inhabitants of rivers in India, Borneo, and tropical Australia. A few carcharhinid sharks enter fresh water periodically, the most notable example being the Bull Shark, *Carcharhinus leucas*. Bull Sharks have been captured as far as 4200 km up the Amazon River, and more than 1200 km up the Mississippi River at Alton, Illinois (Thorson 1972; Moss 1984). Bull Sharks regularly traverse the 175 km long Rio San Juan between the Caribbean Sea and Lake Nicaragua in Central America and occur in other rivers and lakes of Mexico and Central and South America. Bull Sharks are the most likely perpetrators of attacks on humans in rivers worldwide (e.g. Coad & Papahn 1988). Finally, six species of pristid sawfishes (Pristidae) are strictly euryhaline. These include the Largetooth Sawfish, *Pristis perotteti*, which also moves regularly between lakes and the ocean in Central and South America and has established genetically distinct, reproducing populations in some lakes (Montoya & Thorson 1982; Thorson 1982) (Fig. 12.5).

Sharks and sawfishes that move freely between marine and freshwater habitats can adjust the osmotic concentration of their blood appropriately. Whereas in salt water, the major problem is salt accumulation and water loss, in fresh water the problems reverse (discussed in Chapter 7). A Bull Shark in fresh water reduces the salt concentration of its blood by about 20% and urea concentration by about 50%. This cuts osmotic pressure of the blood in half to about 650 mOsm. To accommodate additional water flowing in from the environment due to osmotic pressure, urine production increases 20-fold. Salt concentration of the urine is reduced by the same factor, thus conserving salts. The rectal gland, which functions in salt water to secrete excess NaCl back into the environment, shuts down. All these changes are quickly reversed upon return to the ocean.

FIGURE 12.5 Sawfishes are among the most imperilled marine and estuarine fishes in the world. Although little directed fishing occurs for sawfishes, they are frequently entangled in nets of all types. Such bycatch remains the major threat to the US federally listed Smalltooth Sawfish *Pristis pectinata* and its congener, the Largetooth Sawfish *Pristis perotteti*. Both were once common from the Gulf of Mexico up the east coast from Florida to Cape Hatteras (Simpfendorfer 2000). Sawfishes in the USA now occur only in peninsular Florida, primarily in the Everglades region. Shown here are results from a fishing tournament, *c.* 1920, in Key West, Florida. Some of the fish were said to have weighed 765 kg. Courtesy of Matthew McDavitt.

The truly freshwater potamotrygonid stingrays are restricted to low salinity conditions. Unlike all other elasmobranchs, they have lost the ability to concentrate urea (although the enzymes for urea production still occur in their livers), and they lack functioning rectal glands. These stingrays die in water with more than 40% of the salt concentration of sea water (about 15 ppt). More typical marine elasmobranchs, including those that spawn in estuaries, can survive salinities as low as 50% of sea water if acclimated slowly. Unlike Bull Sharks, these typical elasmobranchs achieve an osmotic balance by reducing only the urea concentration of their blood, without reduction in salt concentration (Thorson et al. 1967, 1973; Moss 1984; Thorson 1991).

Because humans and their destructive activities are concentrated along rivers and estuaries, freshwater and estuarine elasmobranchs – as top predators with typical and vulnerable elasmobranch life histories – are even more threatened than is generally the case for elasmobranchs (Compagno & Cook 1995b; Martin 2005; see Shark conservation, later in this chapter). The Giant Freshwater Stingray was once listed as Vulnerable by the International Union for the Conservation of Nature (IUCN 2004; **www.redlist.org**) due to directed and bycatch fishing, habitat destruction, and range fragmentation from dams. Compagno and Cook (1995a) suggested that its status be elevated to Critically Endangered, but more recently it was listed as Data Deficient (IUCN 2020). Of the 47 obligate euryhaline and freshwater elasmobranch species known in 2004, 18 had been assigned to high-risk categories (Critically Endangered, Endangered, or Vulnerable) by the IUCN (2004); another five were Data Deficient, indicating insufficient information to determine their status (Martin 2005).

Movement and Home Ranges

Water as a medium for locomotion exacts a high energetic price on any organism. The long evolutionary history of sharks is characterized by the development of anatomical and physiological traits that appear to favor movement at the lowest possible energetic cost. Many of the features of the integument, fins, buoyancy devices, and swimming behavior of sharks, as well as short- and long-term movement patterns, reflect possible adaptations to these energetic constraints.

Most sharks have heterocercal tails, with asymmetry in both the internal support and external appearance. The typical heterocercal tail is associated with an active lifestyle above the bottom, as in most requiem sharks and hammerheads. Diversity in tail fin shape is considerable, however (Bone 1988). Symmetrical tails preceded by lateral keels characterize high-speed, pelagic sharks such as the Mako, White, and Porbeagle

(Lamnidae); large body size and symmetrical tails characterized the presumably pelagic, predatory edestoid holocephalans of the Carboniferous (see Chapter 11). Lateral keels are also found convergently on such pelagic teleost predators as tunas (Scombridae) and billfishes (Istiophoridae).

Extreme heterocercality in contrast is usually found in relatively inactive, benthic sharks such as the wobbegongs and nurse sharks (Orectolobidae) and cat sharks that essentially lack a lower tail lobe. A specialized, extreme heterocercal tail occurs in active swimmers such as the thresher sharks, in which the dorsal tail lobe, which may constitute 50% of the body length, is purportedly used for herding and stunning prey. Many rays lack a tail fin (e.g. stingrays, eagle rays, manta rays) and swim by flapping or undulating their scaleless pectoral fins. In angel sharks the lower tail lobe is enlarged.

The placoid scales of sharks apparently serve a streamlining function by reducing drag (Fig. 12.6). Unlike the relatively flat scales of many bony fishes, each placoid denticle of a shark has a pedestal and an expanded top, which often has ridges running parallel to the body of the shark. It has been postulated that this particular shape, which is mimicked in the winged keels of high-performance sailboats, helps reduce swimming-induced drag by reducing turbulence along the shark's body. The reduced turbulence would reduce the energetic cost of swimming and may make sharks harder to detect by potential prey. Hydrophones detect much less noise from swimming sharks than from swimming bony fishes, suggesting that turbulence reduction also enhances stealth in sharks. Not surprisingly, the scales of benthic and slow-swimming sharks, as well as many rays, lack apparent streamlining features and are instead enlarged for protection or are absent (e.g. Bramble Shark, Echinorhinidae; Thorny Skate, Rajidae) (Moss 1984). The sting or barb on the dorsal surface of the tail of a stingray is a modified, elongate placoid scale with serrate edges and a venom gland at its base.

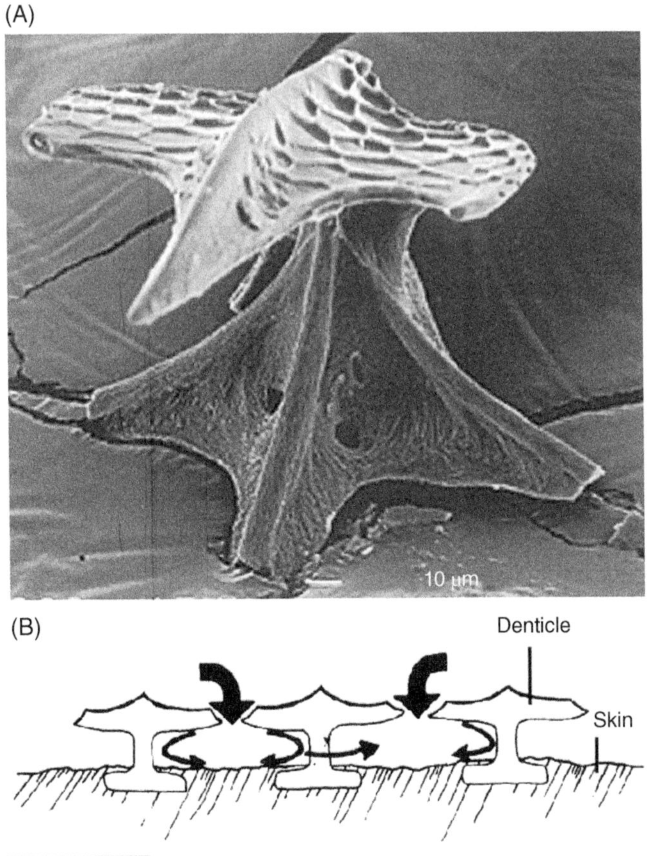

FIGURE 12.6 The role of scales in drag reduction in sharks. (A) Scanning electron micrograph of a single denticle from a scyliorhinid cat shark, showing the pedestal and winged keel arrangement thought to absorb turbulence, which reduces drag. (B) Cross-sectional representation of placoid scales, showing reduction of turbulence along the body. Strength of water flow corresponds to thickness of the black arrows. (A) Konstantinou et al. 2000 / with permission of Springer Nature; (B) from Moss (1984), used with permission.

Maintenance of a constant depth is potentially expensive if a fish must constantly swim to overcome gravity. Bony fishes control their buoyancy by adjusting volume of a gas bladder (discussed in Chapter 5). Sharks have arrived at a completely different solution to the challenge of buoyancy control: they both reduce their body weight and fill their body with low-density substances. Weight reduction comes largely from a skeleton made of cartilage, which has 55% of the specific gravity of bone (1.1 vs. 2.0). Buoyancy can also be enhanced by the oils contained in the large liver, which may constitute up to 30% of the weight of the fish. Deep-sea squaloid sharks and some pelagic species such as Whale, Basking, and White Sharks have large livers that are as much as 90% oil. These sharks are almost neutrally buoyant (e.g. a Basking Shark that weighs 1000 kg in air weighs only 3.3 kg in water, or about 0.3% of its air weight). Typical bony fishes with their gas bladders deflated have in-water weights about 5% of air weights, i.e. a 100 kg fish weighs 5 kg in water. The oil in a shark's liver is primarily **squalene**, which has a specific gravity of 0.86 (the specific gravity of sea water is 1.026) (Baldridge 1970, 1972; Moss 1984; Bone 1988).

An oil-based buoyancy system may have advantages over a gas-filled system. To maintain neutral buoyancy, bony fishes must adjust volume of the gas bladder or they will sink rapidly or float uncontrollably. The physics of gas secretion and absorption limit the rate at which bony fishes can change depth. Oil, unlike gas, is incompressible and provides constant buoyancy regardless of depth and pressure. Sharks can therefore move up and down repeatedly over tens or hundreds of meters daily without having to make adjustments in their buoyancy control mechanism. Telemetry studies have shown vertical movements to depths of more than 400 m in hammerheads, 590 m in Bigeye Thresher Sharks (*Alopias superciliosus*), 750 m in Basking Sharks, and to at least 980 m in Whale Sharks and White Sharks (e.g. Sims et al. 2003; Weng & Block 2003; Bonfil et al. 2005; Wilson et al. 2006b; see also Carey and Scharold (1990) on Blue Sharks and Carey and Clark (1995) on Sixgill Sharks).

Such rapid and repeated vertical migration, with its attendant exposure to tremendous pressure and temperature changes, is uncharacteristic of most bony fishes. An additional energetic advantage of an oil-filled liver is that it serves as an energy reserve; many sharks metabolize their liver oils when starved. An apparent downside of this adaptation is that producing lipids for liver storage is energetically costly.

Activity spaces, or home ranges, of most sharks are poorly known. Advances in electronic telemetry have permitted the fitting of ultrasonic transmitters to sharks and tracking them through several days and even weeks of activity (Nelson 1990). Juvenile Lemon Sharks less than 1 m long in Bimini initially patrol shallow beach areas of about 0.7 km^2. This area increases with age: a 1.8 m long shark may have a range of 18 km^2, and a 2.3 m long shark's range encompasses more than 90 km^2. Habitats change with age also, as the shark moves from the shallow nursery area to open sand flats and finally to the reef and beyond (Sundstrom et al. 2001). Daily homing movements, involving daytime aggregation at seamounts and nocturnal foraging in open water several kilometers away, have been demonstrated in Scalloped Hammerheads, *Sphyrna lewini* (Klimley & Nelson 1981).

On a larger scale, many sharks make extensive movements and migrations on the order of hundreds and even thousands of kilometers. Based on tag-and-recapture data, sharks can be classified as local, coastal, or oceanic. Local sharks, including Bull, Nurse, and Bonnethead sharks, spend most of their adult lives in a relatively confined, nearshore area of a few hundred square kilometers. Coastal species, such as Sandbar, Blacktip, and Dusky sharks, stay near continental shelves but move 1600 km or more. Bigeye Threshers move over 2700 km from waters off New York to the eastern Gulf of Mexico. Sandbar Sharks move between the northern Atlantic region of the USA and the Yucatan Peninsula of Mexico, a distance of about 5600 km. Even comparatively small, coastal sharks can cover large distances during their lifetimes. A *c.* 1.5 m long School Shark, *Galeorhinus galeus*, tagged off New Zealand's South Island was caught 4940 km away off South Australia, 3.5 years later (Hurst et al. 1999). A Spiny Dogfish, *Squalus acanthias*, tagged off Washington state moved 8704 km during the 10-year period between release and recapture (see Kohler & Turner 2001 for an extensive review).

Oceanic species may cross entire ocean basins, sometimes repeatedly. Distances travelled by oceanic species in the North Atlantic include 7800 km (Blue Shark), 6700 km (Tiger Shark), 4500 km (Mako), and 2800 km (Oceanic Whitetip); these are almost certainly underestimates of actual distance because they record only start and end points. Telemetered movements using satellite tracking are even more impressive. A Whale Shark tagged in the Sea of Cortez, Mexico travelled 13,000 km in 37 months to the western North Pacific Ocean (Eckert & Stewart 2001). One Whale Shark tagged in the Sulu Sea of the Philippines travelled to the South China Sea off Vietnam 4567 km distant in 2.5 months; another tagged off the north-western coast of Malaysia moved 8025 km over a 4-month period (Eckert et al. 2002).

Most spectacularly, some sharks recross major oceans. Blue Sharks make return trips between North America and Europe, a distance that exceeds 16,000 km (Casey & Kohler 1991). White Sharks are now known to migrate between central California and Hawaii, a minimum distance of 3800 km (Boustany et al. 2002). A 4 m female White Shark (nicknamed Nicole after Australian actress Nicole Kidman, an advocate of shark conservation) was followed via satellite telemetry across the Indian Ocean and back between South Africa and Australia. The round trip took 9 months and covered a minimal distance of 22,000 km (Bonfil et al. 2005) (Fig. 12.7). These and other large-scale movements across political boundaries have substantial conservation implications because management regulations and enforcement vary greatly among nations.

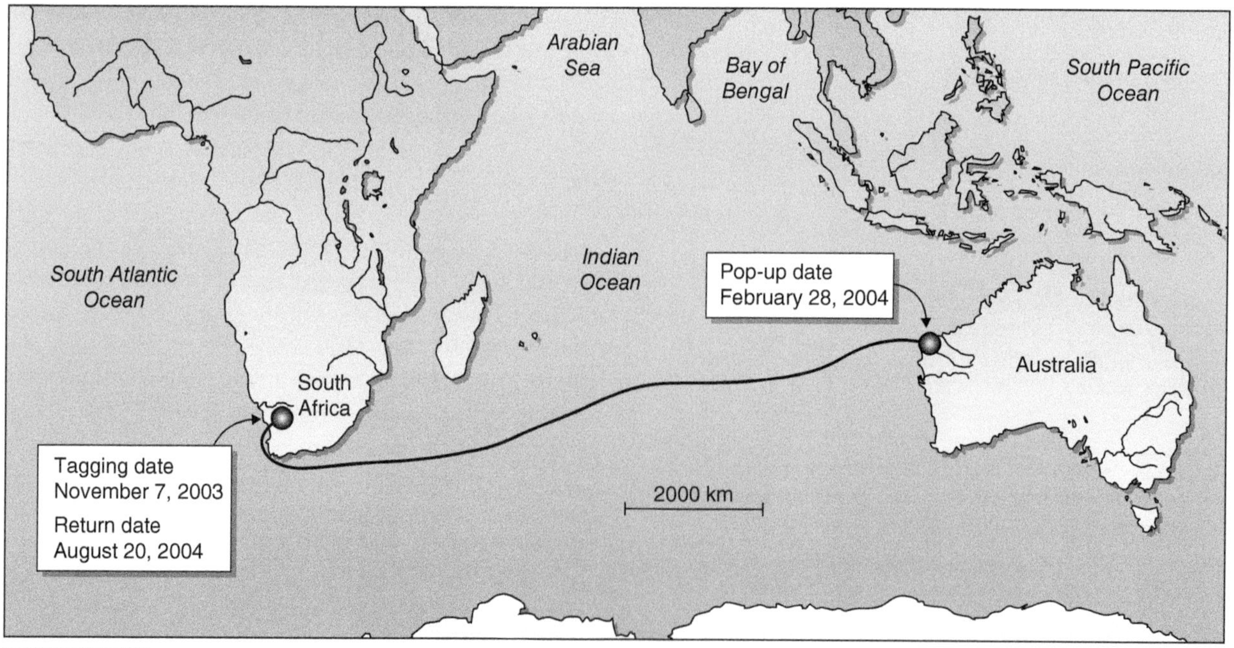

FIGURE 12.7 The track of Nicole, a 4-m White Shark that had been seen over a 6-year period in South Africa. Nicole was then followed via satellite telemetry from South Africa to Australia between November 2003 and February 2004, and was then seen again off South Africa in August 2004. Her minimal roundtrip distance was 22,000 km. Modified from Bonfil et al. (2005).

Metabolism and Growth Rate: Life in the Slow Lane Many aspects of the biology of sharks point to a strong emphasis on **efficient energy use** when compared with bony fishes. In addition to the anatomical features such as fin and scale morphology mentioned previously, physiological attributes of sharks indicate a premium placed on energy conservation. Resting metabolic rates of a 2 kg Spiny Dogfish average 32 mg O_2/kg body weight/h, about one-third of the average resting rate for comparably sized teleosts. In dogfish, the active metabolic rate is only triple that of the resting rate, whereas teleost active rates often go up 10-fold (Brett & Blackburn 1978). Extrapolated prey intake rates indicate that a 2 kg dogfish would need only 8 g fish prey per day for maintenance, whereas a similar-sized salmon would require four times that amount.

Comparable data for larger sharks are unavailable, for obvious logistic reasons. Estimated oxygen consumption suggests that sharks consume about half the oxygen of equivalent-sized bony fishes (Moss 1984). For example, calculations of energy consumption have been made for a 4.5 m White Shark that was harpooned with a temperature-recording ultrasonic transmitter (Carey et al. 1982). Based on the rate at which the temperature of the shark's muscle mass changed as it passed through a thermocline (region of rapid temperature change), its oxygen consumption rate was calculated to be about 60 mg O_2 or 0.2 kcal/kg body weight/h (a 60 kg human consumes about 1.6 kcal/kg/h).

Low metabolic requirements may translate into reduced energetic needs compared to bony fishes. White Sharks feed commonly on dead whales. Based on the caloric value of 30 kg of whale blubber found in a 940 kg White Shark's stomach and on the above calculations of metabolic rate, it was estimated that White Sharks can maintain themselves by feeding on whales only once every 6 weeks (Carey et al. 1982).

Food consumption rates vary greatly among shark species but are apparently most closely related to degree of activity (Duncan 2006). Relatively sedentary sharks, such as the Nurse Shark, consume 0.2–0.3% of their body weight per day and digest an average meal over at least 6 days. Moderately active sharks, such as Sandbar and Blue sharks, consume 0.2–0.6% of body weight per day, digesting a meal in only 3–4 days. Very active sharks such as mako sharks, which are "warm-bodied" (see Chapter 7, Temperature Relationships), eat 3% of their body weight per day, digesting their meals in 1.5–2.0 days. Translated into annual consumption rates, a mako shark eats about 2 kg/day, or about 10–15 times its body weight per year. These estimates are about half the annual consumption of an individual teleost, emphasizing the relative energy efficiency of sharks.

Another possible energy-saving mechanism of some sharks is heat conservation. Lamnid sharks and to a lesser extent thresher sharks can conserve some of the heat generated during muscle contraction and thereby maintain their muscle and stomach temperatures at about 7–10°C above ambient water conditions (Carey et al. 1982; McCosker 1987). Heat generated during muscle contraction in most sharks is dissipated because cold, oxygenated blood coming from the gills moves into the deeper parts of the body. In lamnid sharks, a **countercurrent exchange** arrangement helps warm arterial blood flowing from the gills (see Chapter 7). Higher body temperatures may permit maintenance of a higher metabolic level and generation of more muscle power, thus facilitating the capture of fast swimming prey (including endothermic marine mammals) and may increase the rate at which food is digested (Carey et al. 1982; Bone 1988). The warmer body temperature does not, however, extend the ability of these large predators to invade cool waters at high latitudes as some have speculated (see Harding et al. (2021)).

Low metabolic demands may be linked to relatively **slow growth** and **long life spans**. After an initial rapid growth phase of 15–60 cm increase per year, growth in most sharks slows considerably. Growth rates of juveniles and adults of 12 species of medium- to large-sized sharks averaged only 5 cm/year (Thorson & Lacy 1982; Branstetter & McEachran 1986). Longevity estimates vary considerably among and within species, but chondrichthyans on average live longer than bony fishes (Cailliet & Goldman 2004). Among batoids, sawfishes live 30–44 years, stingrays 3–28 years, and skates 9–24 years (except in Europe's largest skate, the Critically Endangered Common Skate, *Dipturus batis*, that can live to 50 years). Sharks are similarly long-lived, again with much variation. Angel sharks live to 35 years, carpet sharks 19–35 years, dog sharks 12–70 years (including the longest lived species, the deep-water *Centrophorus squamosus*), mackerel sharks 10–25 years, and ground sharks 4–32 years (most chimaeras live 5–10 years, with a maximum longevity of 29 years in *Chimaera monstrosa*). Age estimation in sharks is frustrated by a lack of retained, calcified structures; growth rings in vertebrae are the most used indicator of age when more direct measurements are unavailable (Cailliet 1990; Cailliet & Goldman 2004).

Feeding Habits Sharks are apex predators throughout the world, stationed at the top of the food webs in which they occur. All elasmobranchs are carnivorous, taking live prey or scavenging on recently dead animals. There is no evidence of herbivory or detritivory among sharks, and the only departures from feeding on relatively large prey are the huge, filter-feeding, zooplanktivorous Basking, Megamouth, and Whale sharks and manta rays. Bony fishes constitute 70–80% of the diet of most shark species. Those feeding on other prey types include carcharhinid Tiger, Bull, and Galápagos sharks (other sharks); hammerheads (stingrays); White, Tiger, Sleeper, and Cookie Cutter sharks (marine mammals); Tiger Sharks (seabirds); and Leopard, Nurse, and Green Dogfish sharks and most skates and rays (invertebrates). Somniosid sleeper sharks in Antarctica feed on the ocean's largest invertebrates, the giant and colossal squids (Cherel & Duhamel 2004).

The characteristic ventral mouth of most sharks (exceptions include the Megamouth, Frill, Whale, and angel sharks and the manta rays) is apparently linked to dependence on bite strength rather than suction during feeding. Nurse sharks,

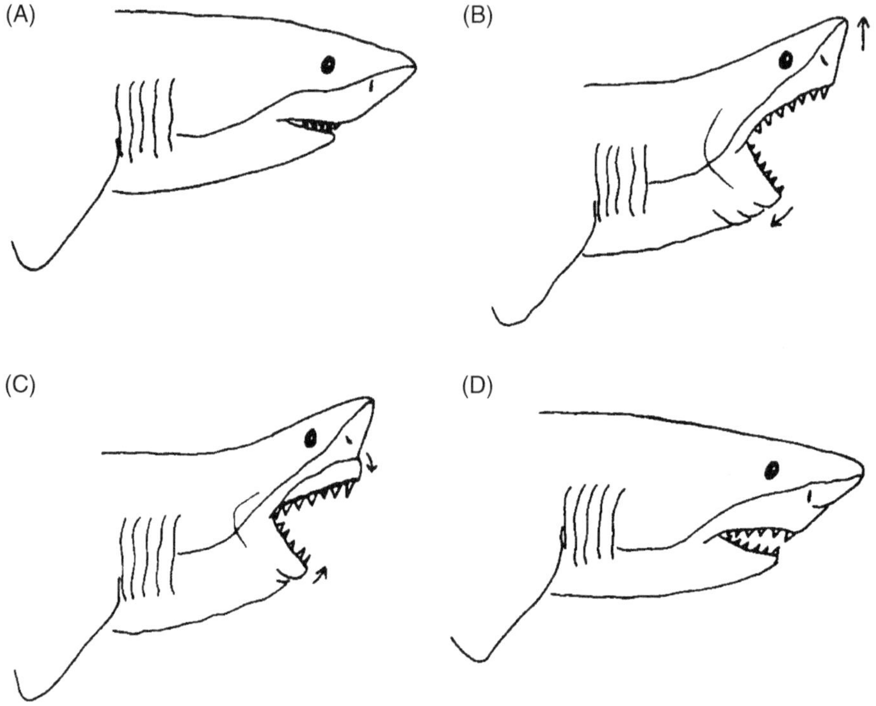

FIGURE 12.8 Head and jaw movements associated with feeding in the White Shark, *Carcharodon carcharias*. (A) Normal resting position; (B) snout is lifted and lower jaw depressed, achieving maximal gape; (C) lower jaw is lifted forward and upward and the palatoquadrate (teeth-bearing upper jaw) is rotated forward and downward, thereby closing the jaws (= the bite); (D) snout is dropped back down and palatoquadrate retracted to resting position. The bite (component C) occurs quickly, requiring on average 0.8 s. From Tricas and McCosker (1984), used with permission.

as well as some heterodontiform sharks and many rajiforms, utilize suction extensively while feeding, but negative pressures are produced by expansion of the enlarged orobranchial cavity rather than enlargement of the mouth via jaw protrusion. Underslung jaws in sharks provide larger regions for muscle attachment than in fishes whose jaw bones extend all the way to the tips of their snouts. The shark configuration allows for both protrusion of the palatoquadrate (upper jaw) and the generation of the powerful biting forces required to cut through the skin, bone, and muscle of their prey (Fig. 12.8). Hence sharks are not limited to prey that can be swallowed whole, whereas the vast majority of bony fishes are "**gape-limited**" and risk suffocation if they attack prey larger than they can swallow whole.

Separation of the palatoquadrate from the cranium serves as more than a diagnostic character for identifying elasmobranchs. This separation also allows upper jaw teeth to be retracted during nonfeeding periods, which aids in streamlining of the head profile. **Protrusion** during the bite increases bite efficiency, jaw closure speed, and prey manipulation (Wilga et al. 2001). Hence, prey can be efficiently cut by the closing action of the jaws and serrated teeth, rather than merely grasped.

In many carcharhiniform and lamniform sharks, lower jaw teeth are spikelike, whereas upper jaw teeth are flat bladed with serrated edges. In a typical jaw closing sequence (Fig. 12.8), the lower jaw is lifted first, impaling the prey, and then the upper jaw is brought down, once or repeatedly. Once prey are grasped, many sharks also shake their head and upper body, or even rotate the entire body about its long axis. During this movement, the upper jaw teeth slice through the prey, eventually removing a chunk of flesh (Moss 1977, 1984; Tricas & McCosker 1984). Jaw protrusion is extreme in some rajiforms due to the loss of ligamental connections between the palatoquadrate and cranium. These fishes dexterously pick up food objects from the bottom with their jaws through a combination of suction and grasping (Moss 1977, 1984).

One departure from this norm occurs in some squaloid sharks, in which the lower teeth are more bladelike than the uppers and are arranged in a flat band across the lower jaw, perpendicular to the axis of the body. This group includes two species of highly specialized, 40 cm long, bioluminescent cookie cutter sharks, *Isistius brasiliensis and Isistius plutodus* (Dalatiidae), the only ectoparasitic elasmobranchs (Fig. 12.9). Craterlike wounds about 4 cm across are frequently found on tunas, dolphins, whales, and even Megamouth Sharks, all animals that spend part of their time at mesopelagic depths (200–1000 m). These wounds match precisely the mouth size of a cookie cutter shark and it has been deduced that the small sharks lurk among small mesopelagic organisms and prey on the larger predators that move into this region to feed (Jones 1971; Shirai & Nakaya 1992). The pattern and nature

(A)

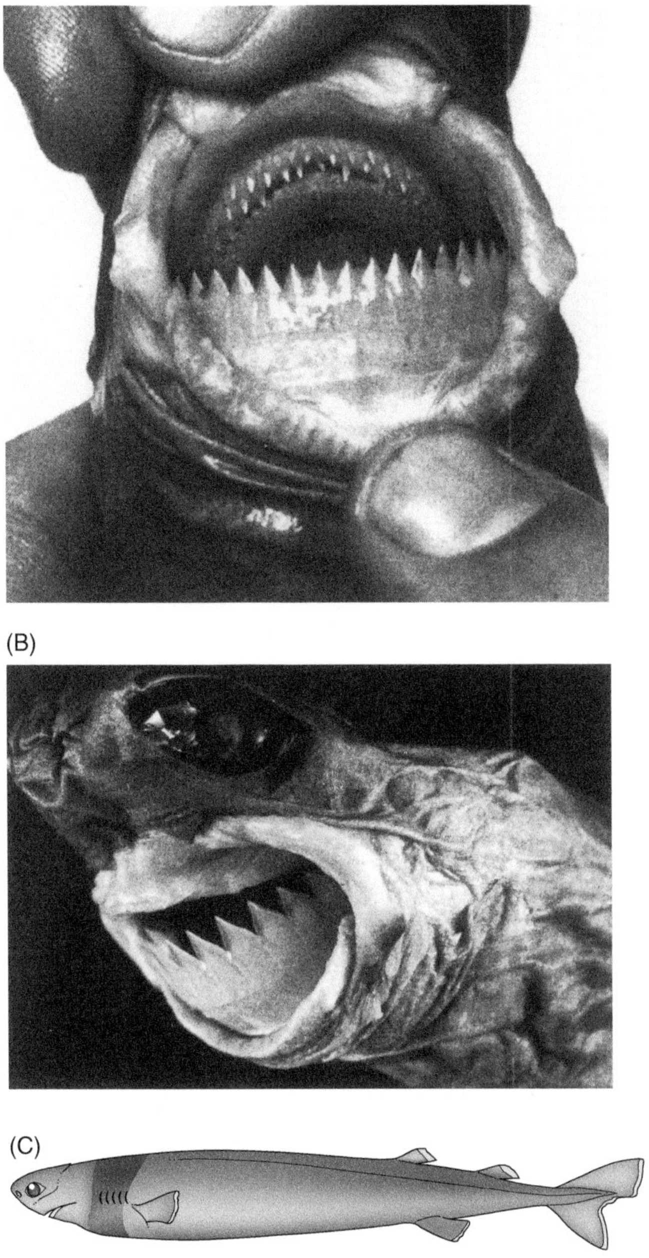

(B)

(C)

FIGURE 12.9 Cookie cutter sharks. (A) *Isistius brasiliensis*, the Cookie Cutter Shark, is a small (about 40 cm) tropical species that lives at mid-ocean depths and parasitizes tunas, other fishes, Megamouth Sharks, and marine mammals, gouging circular plugs of flesh out of their sides with its specialized dentition. (B) The congeneric Largetooth Cookie Cutter Shark, *I. plutodus*, has the largest teeth for its body size of any known shark. Its teeth are twice as large relative to body size as a White Shark's teeth. (C) Drawing of the Cookie Cutter Shark, *I. brasiliensis*. (A) Springer and Gold 1989 / with permission of Smithsonian Institution Press; (B) Compagno 1981 / with permission of Kluwer Academic Publishers; (C) after P. Vecsei.

of the green-emitting photophores (light organs) on *Isistius* suggest that they are essentially invisible because its light output closely matches the spectrum of background light at the depths where it swims (Widder 1998). The symmetrical craterlike wounds may be formed when the shark attaches to the side of its victim and then spins about its long axis, removing a flat cone of tissue (E. Clark, pers. comm.). Similar but much larger wounds on the sides of seals and beluga whales may be the result of attacks by large Greenland Sharks (Somniosidae), which have a dentition pattern like that of *Isistius* (A. Fisk, pers. comm.).

Dentition morphology, an important taxonomic characteristic for identifying species, correlates strongly with food type in most sharks (Fig. 12.10). Predators on fishes and squids, such as mako, Sand Tiger, and Angel sharks, have long, thin, piercing teeth for grasping whole prey, which they often swallow whole. Most requiem sharks, which are also piscivores, have such piercing teeth in the lower jaw and bladed teeth with finely serrated edges for cutting prey in the upper jaw (discussed earlier). Sharks and rays that feed on hard-shelled prey such as mollusks and large crustaceans have specialized, broad dentition for crushing (Fig. 12.11). Bamboo Sharks can feed on hard-bodied prey with sharp teeth by bending the teeth backwards such that the anterior surfaces crush rather than pierce the prey.

Whereas most sharks engage only the anterior, peripheral row or rows of teeth while feeding, in mollusk feeders most of the posterior rows participate in the crushing action. Filter-feeding sharks typically have greatly reduced teeth that may function minimally during feeding; they instead use their gill rakers to trap small prey. In bony fishes, teeth are attached directly to the bone of the jaws, whereas in sharks the teeth are embedded in the connective tissue. Sharks have teeth only on their jaw margins, not attached to other bones and structures in the mouth, as in bony fishes. Shark teeth are basically enlarged scales, derived embryologically from the same tissues as the dermal denticles.

Dentition **replacement** is characteristic of all sharks, although detailed information on patterns of replacement is limited to a few species. Some sharks, such as White Sharks and hammerheads, replace teeth individually as they are worn out or lost. In contrast, Spiny Dogfish, Greenland, and Cookie Cutter sharks apparently replace entire rows. Replacement occurs regardless of use; nonfunctional teeth in interior rows continue to grow and move forward, eventually displacing or replacing functional teeth. Teeth also grow as the shark grows and hence teeth in the most internal, nonfunctional rows are larger than the functional teeth about to be replaced. Turnover rates have been calculated for a few species in captivity. Lemon Sharks replace a functional tooth about every 8–10 days, Sand Tigers every 2 days, Leopard Sharks every 9–12 days, Nurse Sharks every 9–28 days, and Horn Sharks every 28 days (Motta 2004). Given these numbers, it has been estimated that a shark may produce on the order of 30,000 teeth during its lifetime (Moss 1967; Springer & Gold 1989; Overstrom 1991).

Some sharks use structures other than jaw teeth to capture prey. Thresher sharks possess a long, scythe-like upper caudal lobe of their tail. Threshers herd fish into tight schools by circling and splashing with the tail, then stunning prey with quick whips of the tail. Evidence that thresher sharks use their tails in prey capture comes from observations of many threshers

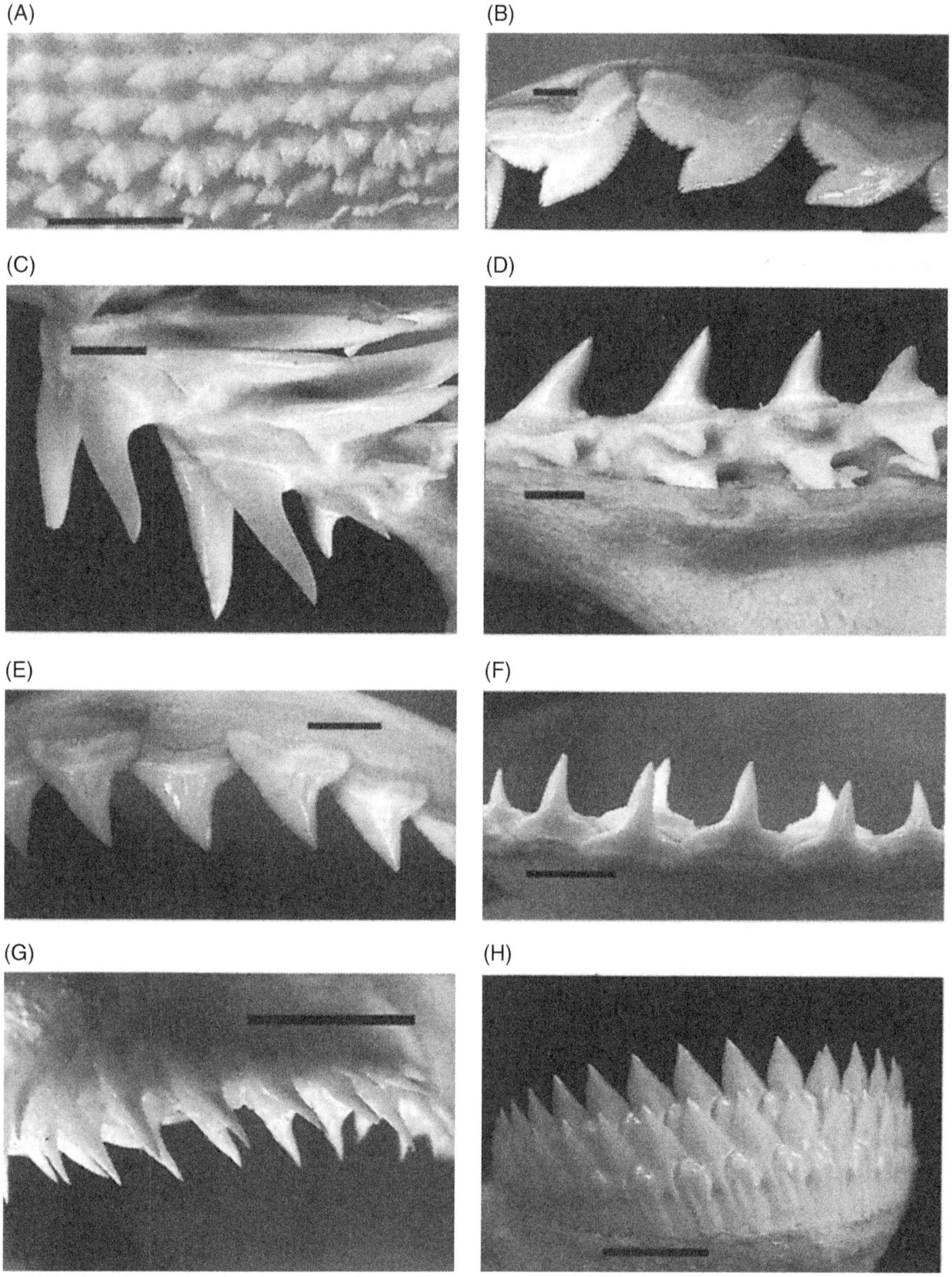

FIGURE 12.10 Representative tooth types of modern sharks: (A) Nurse Shark; (B) Tiger Shark; (C) Shortfin Mako, upper jaw; (D) Shortfin Mako, lower jaw; (E) Sandbar Shark, upper jaw; (F) Sandbar Shark, lower jaw; (G) Kitefin Shark, upper jaw; and (H) Kitefin Shark, lower jaw. All except the Nurse Shark feed largely on fish and squid; Nurse Sharks eat a variety of reef invertebrates such as lobsters. Black bars are 1 cm. Motta 2004 / with permission of Taylor & Francis.

caught by the tail on baited hooks (Compagno 1984). The lateral projections of the rostral cartilage of hammerhead sharks have long intrigued anatomists. Observations by divers indicate that hammerheads, which tend to specialize on stingrays, can use their broad head to pin their stingray prey to the bottom while taking bites from the margins of the stingray's enlarged pectoral fins that are critical to its swimming (Strong et al. 1990; see Fig 16.14).

In a fine example of convergent evolution, two unrelated families of elasmobranchs in different orders have evolved a long, flattened extension of the snout armed with lateral teeth for slashing and disabling prey (see Fig 12.5). **Sawfishes** are very large members of the ray family Pristidae (order Rajiformes), whereas **sawsharks** are much smaller members of the shark family Pristiophoridae (order Pristiophoriformes).

(A)

(B)

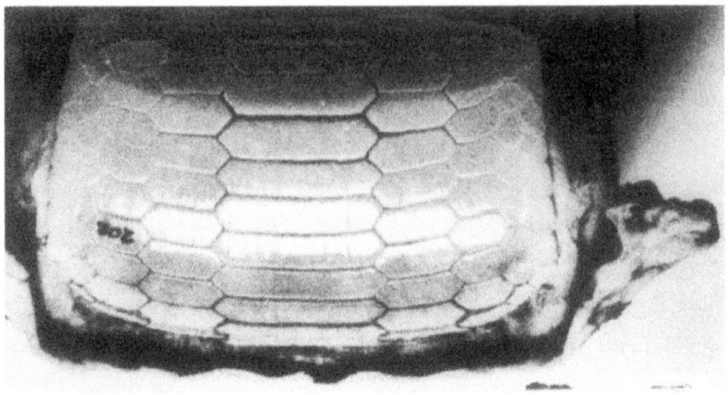

FIGURE 12.11 Pavement or molariform, crushing teeth characterize sharks and rays that feed on hard-bodied prey. (A) Lower jaw of the Horn Shark, *Heterodontus francisci* (Heterodontidae); the anterior teeth grasp and the posterior teeth crush prey. (B) Lower jaw dentition of a Cownose Ray, *Rhinoptera bonasus*, a predator on clams. (A) Motta 2004 / with permission of Taylor & Francis; (B) Case 1973 / with permission of Pioneer Litho Co.

FIGURE 12.12 The Megamouth Shark, *Megachasma pelagios*. This 4–5 m long zooplanktivore was first captured in 1976, northeast of Hawaii (Taylor et al. 1983). It became entangled at a depth of 160 m in a parachutelike sea anchor of a naval research vessel over much deeper water. As of July 2008, only 40 individuals were known, mostly from tropical and subtropical regions. Source: Marko Steffensen / Alamy Stock Photo.

A spectacular non-oral prey capture device is evident in the strong electric discharges of torpedo rays (Torpedinidae). These batoids possess modified hyoid and branchial musculature capable of emitting electric discharges of up to 50 volts and 50 amps, producing an output approaching 1 kW. Although electric discharges can occur when the torpedo is disturbed and hence may serve as a predator deterrent, the more interesting function appears to be in prey capture. Torpedos will lie on the bottom and ambush prey during the daytime, but at night they swim or drift slowly in the water column. Upon encountering a potential prey fish, they envelope it with their pectoral fins and emit pulsed electric outputs that are modified in terms of rate and duration in response to prey reaction. The stunned, immobilized prey fish is then pushed toward the torpedo's mouth via water currents produced by undulations of the pectoral disk (Bray & Hixon 1978; Fox et al. 1985; Lowe et al. 1994). Torpedos differ from most other batoids in that they possess a well-developed caudal fin that is used during locomotion. Most batoids use their pectoral fins for locomotion, but this anatomical region has been usurped for electric organ function in torpediniforms.

Filter feeding occurs in four different groups of sharks and rays and in each has taken a slightly different evolutionary route. The Megamouth Shark, *Megachasma pelagios* (Megachasmidae), is the most spectacular shark species discovered in the twentieth century (Fig. 12.12). Megamouth Sharks feed on euphausiid krill, jellyfish, and small schooling fishes. Megamouth cartilage is poorly calcified and body musculature is relatively soft, suggesting a sluggish filter feeder with a terminal mouth. Prey are ingested via suction and then captured on dense gill raker papillae (Compagno 1990c; Yano et al. 1997).

Manta rays, which can be over 6 m wide, filter feed via **ram feeding**, which involves swimming with the mouth open. Small crustaceans and fishes are additionally guided into the pharynx with the **cephalic lobes** on either side of the mouth that are anterior subdivisions of the pectoral fins; prey are then caught on pharyngeal filter ridges. Whale Sharks use ram feeding while swimming and a **suction/gulp-and-drain** mechanism when stationary to feed on plankton or small fish, which are filtered through cartilaginous rods (Compagno 1984, 1990c). Basking Sharks swim at a constant speed of about 1 m/s with their mouths open, using passive ram filtration that employs gill raker denticles and mucous to capture very small zooplankters. Basking Sharks were thought to shed raker denticles, cease feeding, and enter a torpid state when zooplankton abundances declined in winter. However, behavioral observations, plankton density measurements, tracking with archival tags, and energetics calculations indicate that most retain their denticles, can feed well into the winter, and migrate to regions of high plankton productivity during colder months and continue feeding (Sims 1999; Sims et al. 2003).

Although sharks swim actively throughout the diel cycle and will capitalize on opportunities to feed at any time, catch statistics, ultrasonic telemetry, and direct observation indicate that most species are crepuscular or nocturnal foragers. Catches of sharks on baited hooks are greater by night than by day. Activity cycle data from telemetered Lemon Sharks indicate a doubling of swimming speed during twilight as compared to daytime or night-time speeds (Sundstrom et al. 2001). Increased nocturnal swimming has also been found in Gray Reef, Blue, Scalloped Hammerhead, Horn, angel, and Swell sharks, as well as torpedo rays (Myrberg & Nelson 1991). In contrast, the White Shark is primarily a diurnal feeder, at least where its major prey is diurnally active marine mammals (Klimley et al. 1992; Klimley 1993).

Sensory Physiology

As most sharks are primarily nocturnal foragers, it is not surprising that non-visual senses are particularly well developed. **Olfactory sensitivity** has long been recognized as extreme in sharks. Fish extracts can be detected by Lemon Sharks at levels as low as 1 part per 25 million parts sea water, and in Blacktip and Gray Reef sharks at 1 part per 10 billion, equivalent to about one drop dispersed in an Olympic-sized swimming pool (50 m × 25 m × 2 m deep). This sensitivity, which is greatest to proteins, amino acids, and amines, is achieved at several levels of operation. Water flows into the shark's nasal sacs located under the front of the snout, perfusing the underlying, large olfactory organs. Receptor cells in the organs receive stimuli, which are then transmitted via the olfactory nerve to the olfactory bulb and lobes of the forebrain for integration. Olfactory lobes/bulbs in the shark brain are greatly enlarged, indicating their importance.

Using this sensitivity, sharks are exceptionally adept at following odor trails in the water, even without currents to guide them upstream to an odor source (Montgomery 1988). Odor localization may be accomplished by comparing the intensity of stimulation in the nasal sacs. This scenario suggests another possible function of the expanded rostral cartilage of hammerhead sharks. Lateral displacement of the nostrils on the margins of the hammer might provide improved stereo-olfaction, making odor localization easier. Olfaction also plays a role in breeding, as evidence exists that sexual behavior is mediated by chemical sexual attractants (pheromones) produced by females (Hueter & Gilbert 1991).

Sharks reportedly have good vision up to about 15 m, depending on water clarity. The shark retina is dominated by rods, as would be expected of fishes that are chiefly active at night and twilight. Nocturnal sensitivity is enhanced by an additional layer of reflective guanine platelets behind the retina, called the **tapetum lucidum**. The tapetum reflects light back through the eye (hence the "eye shine" of sharks and other nocturnal animals), which allows light entering the eye to strike the retinal sensory cells twice, increasing the likelihood of detection. Platelets are angled such that reflected light passes through the same receptors as incoming light, thus preserving acuity.

Sharks also typically have low densities of cone cells in their retinae; these cells generally increase daytime acuity and are associated with **hue discrimination** (color vision). Color vision has been demonstrated in Lemon Sharks. Shark eyes differ from those of most other fishes in that they possess a **pupillary response** to changing light levels, which means they can regulate the amount of light entering their eyes (similar to the eyes of mammals). In many batoids, an additional eyelid-like structure called the **operculum pupillare** expands under bright light conditions to shade the eye and perhaps increase responsiveness to movement.

Acoustic sensitivity and **sound localization** capabilities are also greatly developed in sharks. Sharks hear sounds below 600 Hz (the fundamental of E below high C on a piano), including infrasonic sound below 10 Hz. Hearing is centered in three chambers within the inner ear that contain the **maculae** (sing. = macula). The maculae contain specialized nerve cells that are linked to small granules of calcium carbonate that vibrate in response to sound stimuli. The chambers are connected to the environment via two small pores, the endolymphatic ducts, on the top of the shark's head. A fourth macula, the **macula neglecta**, is located below an opening in the skull. The three major maculae are apparently involved in detecting low-frequency pulsed sounds, whereas the macula neglecta may provide information on the direction of the sound (Hueter & Gilbert 1991; Myrberg & Nelson 1991).

In addition to sensing vibratory cues in the surrounding medium (sound), sharks are also sensitive to minor variations in water displacement. This "**distant touch**" **sensitivity** is accomplished via mechanoreceptors distributed both along canals such as the lateral line and as independent "pit organs" scattered about the body. Distant touch sensitivity also has a directional component, suggesting that as a shark moves through the water, it can detect the presence, location, direction of movement, and relative speed of moving objects that displace water, or of stationary objects that reflect water moving off the body of the shark.

Electroreception provides sharks with an additional channel for sensory input, and is anatomically and developmentally related to mechanoreception (Collin & Whitehead 2004) (see Chapter 6). Input for electroreception begins at numerous small pores on the shark's head, snout, and mouth. The pores lead to conductive, gel-filled canals that terminate in ampullary receptor cells termed **ampullae of Lorenzini**. The receptor cells, which anatomically resemble hair cells of the lateral line, respond to weak electric fields, sending afferent fibers via the lateral line nerve to regions in both the mesencephalon and telencephalon of the brain.

Elasmobranchs as a group have evolved and fine-tuned electrical sensitivity to an extraordinary level. Many biological activities generate weak electrical fields, including muscular contraction, such as heart function and breathing, nerve conduction, and the voltage created by ionic differences between protoplasm and water. A resting flatfish (Pleuronectiformes) creates a low-frequency direct current bioelectric field with a strength of more than 0.01 μV/cm (1/100th of a microvolt) measured 25 cm away. Predatory sharks use these weak electrical cues to locate prey. The electrical sensitivity of large sharks is truly amazing. Human sensitivity is on the order of 0.1 volt. Sharks have demonstrated **detection thresholds** of 1×10^{-9} V/cm, or 1 billionth of a volt, approximately 10 times more sensitive than the 0.01 μV output from prey.

Elasmobranchs are sufficiently electrosensitive that they can probably detect the earth's magnetic field and currents induced by their swimming through that field. Sharks could therefore determine their compass headings during transoceanic migrations, and sharks reportedly follow magnetic intensity gradients along topographic features in the sea floor as a means of orientation during migration (see Klimley 1993, 2015). Stingrays in the lab can learn to orient in uniform direct current fields weaker than the earth's field. In learning trials, stingrays also reverse the location where they search for food when the electric field around them is artificially reversed, suggesting that geomagnetic cues can be used in normal daily activities (Kalmijn 1978). **Magnetite** in the inner ear has been implicated as a component of geomagnetic orientation (Vilches-Troya et al. 1984), but electroreception likely also plays a role.

It has been noted that the ampullary electroreceptors are geometrically centered around the mouth of many elasmobranchs. This positioning could allow a shark or ray to home precisely on a potential food source solely by electroreception, effectively aligning the food in its receptor field and then engulfing the prey. In this way, sharks can detect prey buried in the sand or sit motionless in the dark and snap up prey that swim nearby (Kalmijn 1982; Tricas 1982).

Extreme sensitivity to environmental stimuli is of no use to an animal unless the information can be collected, processed, and acted upon. Such **integration** is the role of the central nervous system, particularly the brain. The ratio of brain to body weight in sharks is greater than for most bony fishes, except perhaps the electrogenic elephantfishes (Mormyridae). The ratio of brain to body weight in sharks is more comparable to that of some birds and marsupials (Fig. 12.13A). Within feeding types and habitat zones, many sharks have larger brains than ecologically comparable bony fishes such as pelagic, predatory billfishes and tunas, which have relatively large brains compared to other teleosts (Lisney & Collin 2006) (Fig. 12.13B). Requiem and mackerel sharks have large forebrains and complex cerebellums; eagle rays and other stingrays have the most complex brains among elasmobranchs (Northcutt 1977).

Reproduction and Development

A few generalizations can be made about shark reproduction (Carrier et al. 2004; Musick & Ellis 2005). Ages at maturation vary widely among shark species but are typically older than for most teleosts. Most sharks mature in 6–l8 years, although much greater ages are not uncommon. For example, Lemon Sharks (*Negaprion brevirostris*) in south Florida mature at an average of 24 years, and Spiny Dogfish (*Squalus acanthias*) in British Columbia mature at an average age of 35 years (Saunders & McFarlane 1993). Longer lived species tend to mature at greater ages: sawfishes at 10 years old, Angel Sharks and Common Skates at 11 years old, Nurse and Whale Sharks at 16–25 years, and some dog sharks at 30 and 45 years (Cailliet & Goldman 2004).

Sharks have a long gestation period and produce relatively few, large, well-developed young that are small replicas of the adults. No parental care is given after egg laying or birth, except for a bullhead shark (Heterodontidae) in which the female picks up her recently laid eggs in her mouth and wedges them in crevices in rocks or plants.

Fertilization in all sharks is internal and has been throughout the known evolutionary history of the class. Male sharks possess intromittent organs in the form of modified pelvic fins termed myxopterygia or **claspers** (the term "claspers" apparently arose from Aristotle's misconception that the structures were used to hold the female, rather than to inseminate her). Females of some species can store sperm in the shell gland for years.

Postfertilization development and embryonic nutrition vary among taxonomic groups (Wourms et al. 1988; Compagno 1990b; Pratt & Castro 1991; Wourms & Demski 1993a; Musick & Ellis 2005). The major contrasts involve whether young develop internally or externally and whether and how the mother provides nutrition for the growing embryo (Table 12.1). The ancestral condition in chondrichthyans, and the most common condition among extant species, is **viviparity**, or live-bearing (Musick & Ellis 2005). In about half of live-bearing species, the developing embryo is retained in the uterus and nourished with yolk provided by a yolk sac that is attached directly to the digestive system of the embryo (termed variously **yolk sac**, **lecithotrophic**, or **aplacental viviparity**, and also **ovoviviparity**).

Some viviparous sharks have evolved an additional form of nourishment. After about 3 months, yolk reserves are exhausted and the young then feed directly on eggs ovulated by the female (**oophagy**). Oophagous sharks include Threshers, Whites, Makos, and Sand Tigers. Sand Tigers have carried this one step further. The first embryo to consume its yolk then consumes its siblings (**embryophagy**) before assuming an oophagous existence (Gilmore et al. 1983; Gilmore 1991). At birth, Sand Tiger litters are composed of two large (1 m long) young, one in each uterus.

The most complex developmental pattern, **placental viviparity**, characterizes advanced members of the most diverse modern family of sharks, the Carcharhinidae, as well as the closely related hammerheads (Sphyrnidae). After the yolk is absorbed, the spent yolk sac attaches to the uterine wall to form a **yolk-sac placenta**. In a construction strongly analogous to the mammalian condition (although involving different embryonic tissues), the stalk of the yolk sac, which is attached to the embryo between the pectoral fins, forms an umbilical cord that transports nutrients and oxygen to the embryo and carries metabolic wastes to the mother. In some sharks, such as the Sharpnose and hammerheads, the umbilical cord diversifies further and develops **appendicula** or outgrowths that serve as additional sites for exchange of materials, including

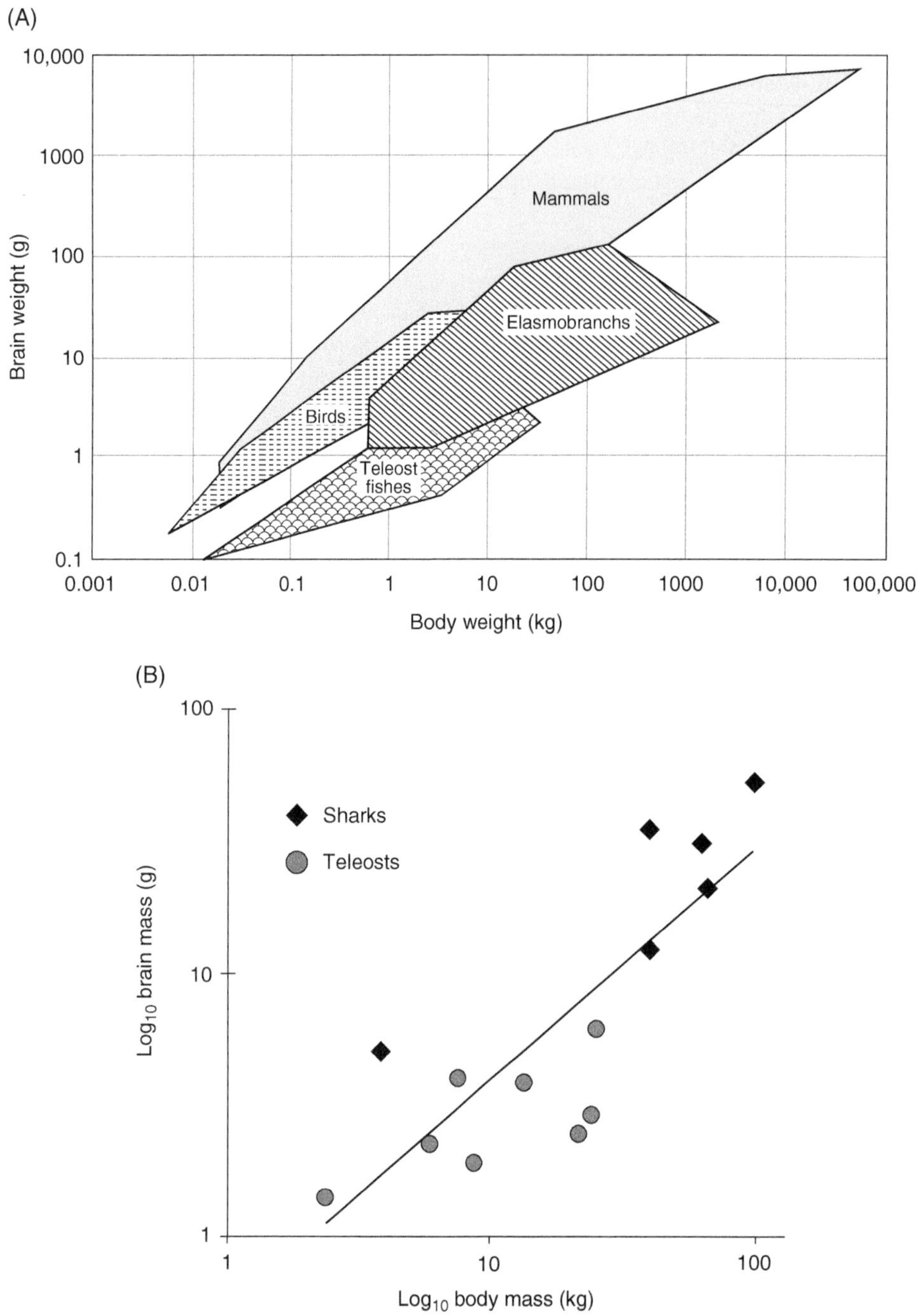

FIGURE 12.13 Brain size in sharks. (A) Sharks have relatively large brains for their body size, overlapping in this respect with birds and mammals as much as with bony fishes. (B) Among pelagic, predatory fishes, many sharks have relatively large brains for their body mass. (A) From Springer and Gold (1989), based on Northcutt (1977) and Moss (1984), used with permission; (B) after Lisney and Collin (2006).

uptake of nutrients in histotroph or "uterine milk" secreted by special cells in the wall of the uterus (Fig. 12.14). Uterine milk may be absorbed through the skin and mouth and also via modified gill filaments that exit from the spiracle and gill slits of the developing embryo. In the myliobatoid stingrays and manta rays, nourishment is solely via ingestion of uterine milk without a placental connection (**uterine viviparity**).

Clutch sizes in viviparous sharks range from two (Sand Tigers, threshers, Longfin Mako) to as many as 70–135 (Tiger Sharks, Blue Sharks), with an average of around 8–10 (Branstetter 1990). Whale Sharks, once thought to be oviparous, have been found with as many as 300 embryos inside, making that species the most fecund shark known. A gravid female may contain smaller embryos still in their egg cases, whereas larger individuals are free-living. Whale Shark pups are born at around 60 cm long (Joung et al. 1996).

Gestation periods in sharks average 9–12 months but may be as short as 3 or 4 months (Bonnethead) or as long as

TABLE 12.1 A summary of embryonic development and nutrition in chondrichthyans.

I. All nutrition from yolk sac	
	A. Yolk sac viviparity (= lecithotrophic viviparity, ovoviparity): all living orders except heterodontiforms, lamniforms, and rajiforms
	B. Yolk sac oviparity (= lecithotrophic oviparity): all living holocephalans, all heterodontiforms, and all Rajidae
II. Some nutrition from mother (= matrotrophy)	
	A. Nutrition from uterine secretions (= histotrophy): many squaliforms and carchariniforms, and all myliobatiforms
	B. Nutrition from eating unfertilized eggs (= oophagy): all lamniforms and some carchariniforms (includes embryophagy of *Carcharias taurus*) and pseudotriakids

From Nelson (2006), after Musick and Ellis (2005).

2 years (Spiny Dogfish) to perhaps 3.5 years (Basking Shark, Frilled Shark *Chlamydoselachus anguineus*). Production of young apparently exacts a large energy cost on females, which appear emaciated at the end of the gestation period. Females of many species reproduce only in alternate years, suggesting it takes at least a year for the female to recover from her last clutch. Such long **reproductive intervals** slow the potential rate at which shark populations can grow and thus recover from exploitation.

Certain accommodations must be made in live-bearing sharks to facilitate the passage of relatively large young through the birth canal. The expanded lateral lobes of the cranium of hammerhead sharks are soft and pliable at birth and then stiffen shortly after. Spines on embryonic Spiny Dogfish are covered with pads of tissue until after the young are born. The toothy rostrum of developing sawfish is at first soft and contained in a rubbery envelope which protects both the rostrum and the mother during gestation and birth.

The alternative condition is egg-laying (**oviparity**), with the embryo deriving all nutrition from its large yolk reserves. About 40% of living elasmobranchs are oviparous, including bullhead sharks (Heterodontiformes), many nurse sharks (Orectolobiformes), as well as all skates (Rajiformes). Clutch sizes among oviparous sharks and batoids both average about 60 eggs per year (Musick & Ellis 2005). Shark and skate egg cases, often called **mermaid purses**, are large (2–4 cm) and are protected by a tough, keratinoid shell secreted by a maternal nidamental or shell gland. The egg case, which contains a single embryo, is attached to seaweeds or other structures and the embryo develops for a relatively long period (several weeks to 15 months) and emerges at a relatively large size.

The protective nature of these egg cases is attested to by instances of their being transported in air for several days and still hatching out healthy young (Lagler et al. 1977). However, they are not free from predation and have been found in the stomachs of other sharks, flatfishes, and elephant seals, and are actively preyed upon by boring gastropod mollusks such as whelks (Cox & Koob 1993). A small number of sharks retain the encased eggs in the oviduct before laying them (termed **retained oviparity**). Clutch size in oviparous sharks is difficult to assess because females may lay two eggs at a time repeatedly over several months (Lineaweaver & Backus 1970).

Shark courtship and copulation commonly involve the male grasping the female with his teeth, while the claspers are inserted (Pratt Jr & Carrier 2001). This pattern has been observed in a half dozen sharks and an equal number of batoids and is inferred from bite marks in many others. Bite scars are common on the pectoral fins and flanks of adult females but

FIGURE 12.14 Placental viviparity in advanced sharks. A new-born Atlantic Sharpnose Shark (Carcharhinidae) showing the umbilical cord with outgrowths ("appendicula") for nutrient uptake. The cord terminates in a placenta that attaches to the uterine wall of the mother. Hamlett 1991 / with permission of American Littoral Society.

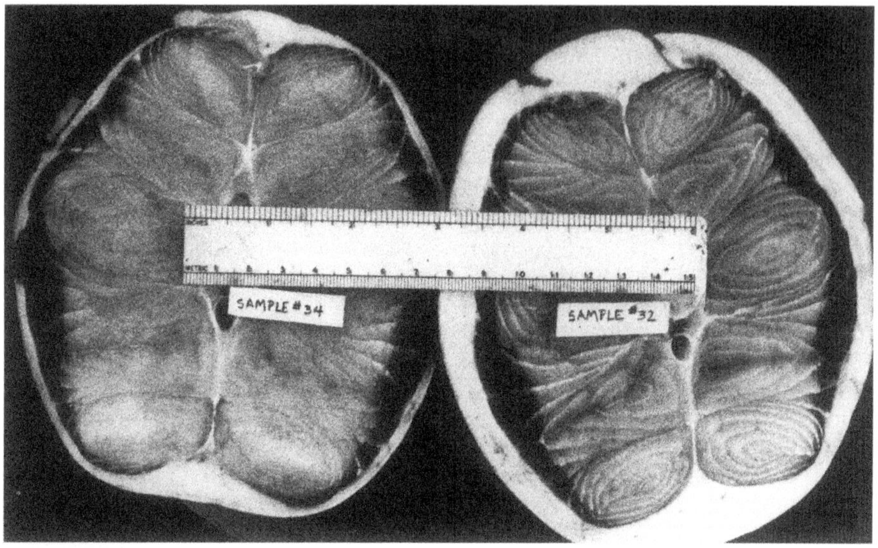

FIGURE 12.15 Sexual dimorphism in the skin thickness of sharks. On the left is a cross-section through a male Blue Shark, on the right a female. Female sharks often have thicker skin than males, probably because during mating males typically bite and hold females. Pratt and Carrier 2001 / with permission of Springer Nature.

appear to do little actual damage because the skin of females is typically much thicker than that of males. In Blue Sharks, the female's skin is three times as thick as the male's and, more importantly, thicker than the male's teeth are long (Fig. 12.15). In some ray species, males develop more pointed dentition upon maturation and during the mating season. Males may also use their electrosensory system to detect responsive females, as has been shown in the Round Stingray *Urolophus halleri* (Tricas et al. 1995).

Shark Conservation Shark populations globally have decreased dramatically since the 1970s. Steepest declines occurred in the late 1980s, when international fishing accelerated, driven in part by demand for shark fins in some Asian countries (Rose 1996, 1998; Fong & Anderson 2002). In the Gulf of Mexico and northwest Atlantic, both already heavily overfished, shark populations continued to decline between 1985 and 2000 (Baum et al. 2003; Baum & Myers 2004): Silky Sharks declined 90%, hammerheads were reduced 89%, thresher sharks 80%, White Sharks 79% (>6000 captured!), Oceanic Whitetips 70–99%, Tiger Sharks 65%, the coastal species complex 61% (range 49–83%), and Blue Sharks 60%. Only mako sharks suffered losses of less than 50%. Although disagreement exists on the exact magnitude of the declines (e.g. Baum et al. 2005; Burgess et al. 2005), it is generally accepted that these numbers are likely representative of trends in other oceans, indicating that, ". . . overfishing is threatening large coastal and oceanic sharks . . . [and] several sharks may also now be at risk of large-scale extirpation" (Baum et al. 2003, p. 390). The decline in populations of these apex predators can have broad ecological impacts on marine food webs through trophic cascades.

Many of the life history features of sharks explain this **vulnerability to exploitation**. As apex predators, sharks seldom occur in the large numbers that characterize more resilient fish species (e.g. herrings and mackerels feed much closer to the base of the marine food chain). Importantly, sharks replace themselves slowly. Slow growth and maturation rates, long gestation periods, long intervals between reproductive bouts, and relatively small clutch sizes suggest that shark species have evolved in circumstances where juvenile mortality is typically low compared to the much more prolific bony fishes on which sharks feed. Under natural circumstances, sharks have few predators aside from other sharks, and most of these can be avoided by attaining large size. The unfortunate outcome of this life history strategy is that shark populations are not capable of overcoming the high mortality rates imposed by commercial exploitation.

Shark Fisheries Sharks are exploited for meat, fins for soup, skin, teeth and jaws (White Shark jaws have sold for as much as US$10,000; Heneman & Glazer 1996), and organs for medicine (fallaciously). Harvesting of shark fins has driven much of the market demand for shark products. Fins are dried and processed to make shark-fin soup, which has had a long, cultural history in China as a luxury product to be served at special occasions such as weddings (Kuang 1999). Although shark meat is sold and consumed, shark fins have been one of the world's most valuable fishery products, worth $55/kg wet, and when processed and sold in Asia, as much as $1650/kg. A bowl of shark fin soup has sold for as much as $150 (Rose 1996; Vannuccini 1999). Hong Kong imported 7800 metric tons of shark fins in 1996, estimated to be worth $250 million. In that year, the US shark fishery for sharks other than dogfish similarly weighed 7000 metric tons but was worth only $10 million (Branstetter 1999; Kuang 1999).

The dramatically higher market value of shark fins over shark meat explains the practice of **finning**, in which fins are cut off a captured live shark which is then tossed overboard, and unable to swim, it sinks and slowly dies. Processing and storing shark meat takes time and considerable space. Fins, however, can be removed quickly, take up minimal space, and have a much higher market value. The fins are sold and resold through dealers, processed, and eventually used to make shark fin soup. Ironically, shark fin soup often contains very little actual shark product. The processed fins function primarily as a thickener, with most of the soup flavor coming from chicken or other meat stock. The fin parts may be filtered out of the soup and the solid fin parts discarded.

Since about 2005 demand for shark fins has been in decline, due largely to attention to the issue of shark finning by several international organizations. The anti-finning movement gained traction in 2006 when NBA basketball star Yao Ming was enlisted as a spokesperson. Others celebrities, business people, and students followed, leading to efforts to ban the dish from official Chinese government banquets and tarnish the image of this previously highly sought-after luxury item. Market demand and value for shark fins subsequently declined.

Besides fins and meat, commercial shark products have included powdered cartilage, promoted as a cancer cure. Because some believed that sharks are immune to cancer, it was thought that shark cartilage could prevent cancers in humans. But sharks do get cancer, including carcinomas of cartilage (Borucinska et al. 2004; Ostrander et al. 2004; G. K. Ostrander, pers. comm.). That inconvenient truth, along with US Food and Drug Administration and Federal Trade Commission injunctions against companies making anticancer claims for shark cartilage, have not deterred unscrupulous individuals from capitalizing on the demand for cancer cures (Barrett 2000). Shark cartilage powder has sold for as much as $145/g (Vannuccini 1999), and a single plant in Costa Rica alone processed 235,000 sharks/month to make cartilage pills (Camhi 1996).

The presumed link between shark cartilage and cancers is that cancers grow because of increased blood supply, and processed cartilage can inhibit blood vessel proliferation (e.g. Berbari et al. 1999; Cho & Kim 2002). However, clinical trials have shown no significant pharmacological effect of cartilage pills on cancerous cells themselves, leading to "unsatisfactory patient outcome[s]" (Gonzalez et al. 2001). Shark cartilage is also believed by some to help with arthritis, psoriasis, and some other inflammatory ailments. These claims are also unproven, but sharks were still processed for their cartilage, and capsules were available online in 2021.

Some drugs derived from shark liver, stomach, and gall bladder may be effective in treating lung and ovarian cancers. The active substance is the aminosterol **squalamine,** which is produced synthetically by reputable drug manufacturers. There is no need to kill sharks to obtain squalamine (Bhargava et al. 2001; Zhang et al. 2003a). But the fact that sharks are a real and potential source for the research and discovery of useful drugs, combined with their ecological importance in aquatic ecosystems, justifies conservation efforts.

The history of commercial shark fishing is a history of collapsed fisheries. Examples include fisheries for thresher sharks in California, School Sharks in Australia, Spiny Dogfish in the North Atlantic, Porbeagles off Newfoundland, Basking Sharks off Ireland, Bull Sharks and sawfish in Lake Nicaragua, and Soupfin Sharks off the US Pacific coast. Two examples typify the boom-and-bust pattern that characterizes commercial exploitation of shark populations. The Porbeagle (*Lamna nasus*) fishery of the western North Atlantic has been well documented. The fishery was wiped out in only 7 years. In 1961, uncontrolled exploitation began and about 3,500,000 pounds (1,575,000 kg) were caught. Catch peaked only 3 years later at 16 million pounds, then crashed. By 1968, only a few hundred sharks could be found. Twenty years later, populations had not returned to pre-exploitation levels (Campana et al. 2002).

Before 1937, about 600,000 pounds or 6000 individual Soupfin Sharks, *Galeorhinus zyopterus*, were landed annually in California. With the development of a market for shark liver oil, this fishery expanded rapidly to 9,000,000 pounds (90,000 fish) in 1939, fell to 5,000,000 pounds in 1941, and by 1944 was back to 600,000 pounds, despite continued intensive effort. Importantly, catch rates fell from 60 sharks per set in 1939 to one shark per set in 1944, indicating a significant decline in the population. Thirty years later, populations had still not returned to pre-1939 levels (Moss 1984; Anderson 1990; Manire & Gruber 1991).

Most shark populations cannot withstand a fishing mortality even as low as 5% removal of the existing population each year (Pratt & Castro 1991). Between 1986 and 1990, commercial shark landings doubled every year in the USA. All indicators suggest that North American shark populations are in decline and that management plans, including a moratorium on the capture of some species, are desperately needed. Globally, annual harvests of sharks, skates, rays, and chimaeras increased from 1950 to 2000, but declined from 2000 to 2017 (Okes & Sant 2019).

The repeated scenario of large initial catches, rapid decline, and slow if any recovery highlight the need for careful management of all exploited shark stocks. Interestingly, sharks exhibit a remarkably close relationship between **stock size and recruitment**. Fisheries managers can predict future recruitment into populations based on existing reproductive stocks. This degree of predictability characterizes few other commercial species. The unfortunate fact is that shark fishing has been one of the least regulated commercial fishing activities. Management plans have been proposed but not implemented in many countries. The USA did not put a management plan into effect until 1993, and those regulations proved insufficient and had to be tightened further in 1997 (Poffenberger 1999). But even with implementation, shark conservation is complicated by shark biology. Local management is not adequate to protect shark populations because so many species undergo long distance movements through international waters. International efforts at conservation, which are historically difficult to negotiate, are crucial.

Conservation Efforts Despite the overall decline and mis- (or non-existent) management of shark populations, there are some successful efforts to conserve sharks, such as the anti-finning campaign discussed earlier. The January 2019 update of the IUCN's Shark Specialist Group's evaluation of 1060 species of sharks, skates, and rays, classified 199 as threatened: 23 critically endangered, 53 endangered, and 124 vulnerable; and 440 were listed as data deficient (**https://www.iucnssg.org/uploads/5/4/1/2/54120303/rl_assessment_results_190124.pdf**). Such attention often leads to protective legislation at national and international levels. For example, in 2020, the Convention on International Trade in Endangered Species of Wild Fauna and Flora (CITES) listed over 15 sharks and rays in Appendix II, regulating and restricting capture and trade in these species; the sawfishes (family Pristidae) were listed in Appendix I, indicating that they are threatened with extinction (see **https://cites.org/eng/app/appendices.php**).

Ironically, in many countries, including those where fishing provides an important livelihood for a large population segment, live sharks are worth considerably more than dead sharks because of the contribution of ecotourism-related **shark viewing** to the local economy. Some tourists will spend considerable amounts of money to travel to places where they can watch sharks and will also utilize local hotels, restaurants, and other services while visiting. Whale Shark watching is popular in the Maldives, Western Australia, Philippines, and Mexico, and in many locales, divers spend large sums to watch reef sharks, manta rays, and stingrays. A shark can be captured once but can be viewed many times. Individual sharks in the Bahamas may be worth US$750,000 alive but only $40–50 dead (S. Gruber, pers. comm., in Daves & Nammack 1998). An economic analysis for the Republic of Maldives indicated an impressive 100- to 1000-fold difference in value (Anderson & Waheed 2001). Numerous Caribbean island tourist destinations promote opportunities for visitors to feed stingrays while snorkelling as fun for the whole family. Live elasmobranchs can provide sustainable commercial activity, whereas harvesting for commercial use does not.

Class Chondrichthyes, Subclass Holocephali

Class Chondrichthyes
- Subclass Holocephali[a]
  - Superorder Holocephalimorpha (six extinct orders and modern chimaeras)
    - Order Chimaeriformes (modern chimaeras) (56 species, marine): Callorhinchidae (3 species of plownose chimaeras), Chimaeridae (45 species of shortnose chimaeras or ratfishes), Rhinochimaeridae (8 species of longnose chimaeras)

[a] We follow Nelson et al. (2016) here for higher taxonomy. Order, families, and numbers of species from Fricke et al. (2021) but refer to **https://researcharchive.calacademy.org/research/ichthyology/catalog/SpeciesByFamily.asp** for updates.

Chimaeras

Although knowledge of holocephalan taxonomy has increased markedly in recent years, surprisingly little remains known about their general biology and natural history (for reviews see Didier 2005). Most of the characters that define the elasmobranchs also describe the Holocephali, suggesting a common albeit unknown ancestor. Chimaeras, also known as **ratfishes** or **rabbitfishes**, share with sharks a cartilaginous skeleton and male intromittent organs, a sutureless skull, ceratotrichial fins, and spiral valve intestine. Holocephalans similarly lack lungs and gas bladders, also using an oil-filled liver for buoyancy. Development is again direct, without a larval stage (Bigelow & Schroeder 1953a; Compagno 1990b).

In contrast to sharks in which the upper jaw has a ligamentous connection to the cranium (the **amphistylic** condition), chimaeras have upper jaws that are immovably attached to the braincase (the **holostylic** condition). The name Holocephali ("whole head") refers to this fusion of palatoquadrate and neurocranium in modern chimaeras. Teeth differ from those of sharks, being continually growing, crushing or cutting plates instead of replaceable dentition. Chimaeras have three pairs of hypermineralized toothplates, two vomerine and palatine toothplate pairs in the upper jaw and a large pair of mandibular toothplates on the bottom (hence the name rabbitfish). The anterior plates are blade-like, whereas the posterior plates are flattened for crushing hard-bodied foods.

As in bony fishes, a single-gill flap covers four internal gill openings, rather than the five or more external gill slits of sharks. Chimaeras lack distinct vertebral centra and a spiracular gill opening (although embryonic chimaeras have a spiracle). Instead of a cloaca, they possess separate anal and urogenital openings. Males have sharklike pelvic claspers that are extensions of the pelvic fins, through which sperm is transferred. Anterior to the pelvic girdle are prepelvic **tenaculae** contained in prepelvic pouches. Tenaculae consist of a single row of stout spines and are used to anchor the female during copulation. Males of some species also have an additional frontal tenaculum on the head that is used to grasp the posterior edge of the female's pectoral fin during copulation (Fig. 12.16). Chimaeras are oviparous, laying a few, 10 cm long eggs with shells composed of a keratin-like protein like those of sharks (e.g. Moura et al. 2004); development may take 5–10 months prior to hatching.

Holocephalans lack scales except for small dermal denticles along the midline of the back and on the claspers of the males. The first dorsal fin has a poison-laden spine and is erectable, not fixed. The body generally tapers posteriorly to a pointed tail, hence the alternative common name ratfish. Chimaeras swim via a combination of body undulations and pectoral fin flapping.

Three families, six genera and 56 species of chimaeras are recognized (Fricke et al. 2021, **https://researcharchive.calacademy.org/research/ichthyology/catalog/SpeciesByFamily.**

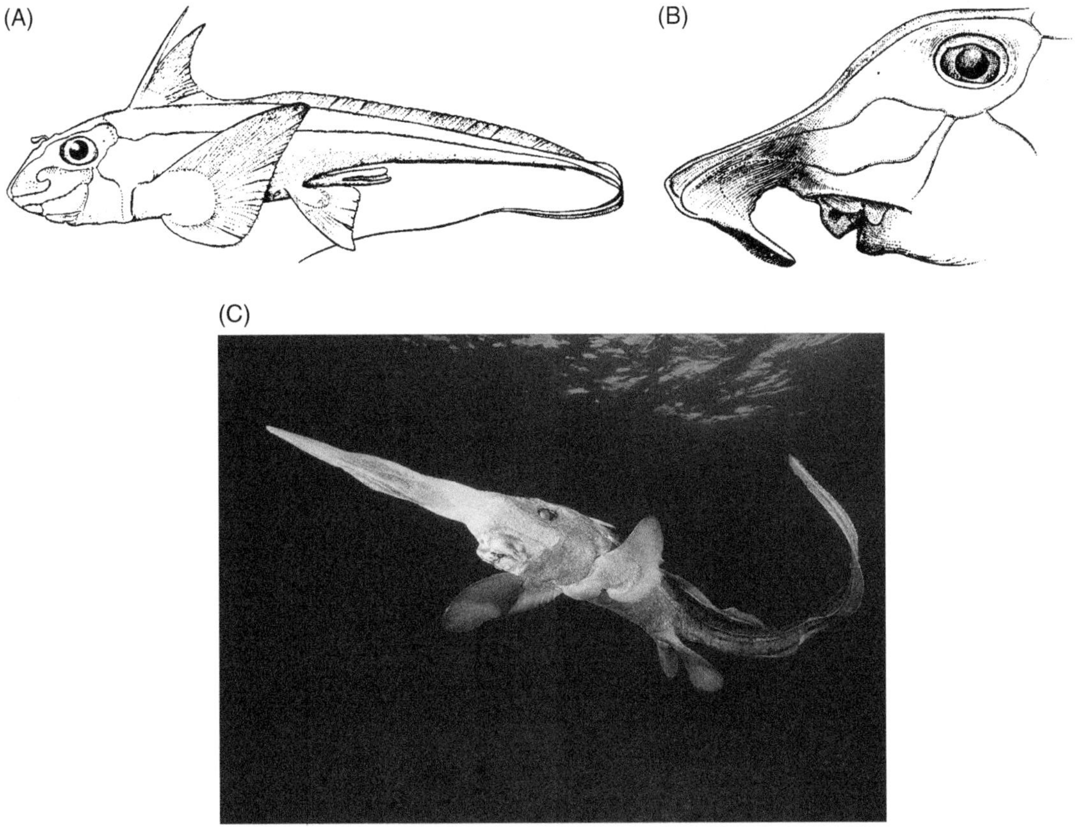

FIGURE 12.16 Modern holocephalans. (A) *Chimaera cubana*, a 50-cm-long Caribbean chimaerid; note the pelvic claspers and also the frontal clasper or tenaculum on the forehead of the male. (B) Head of a callorhinchid chimaera, *Callorhinchus milii* (the Elephant Fish), showing the unique hoe-shaped proboscis. (C) The Pacific Spookfish, *Rhinochimaera pacifica*, can grow to 1.5 m; the electroreceptive fleshy snout helps locate food. (A, B) From Bigelow and Schroeder (1953a), used with permission; (C) Kelvin Aitken/VWPics/Alamy Stock Photo.

asp), with perhaps more species remaining to be named. Adult size ranges from 60 to 200 cm, with females often larger than males. Chimaeras are cool water, marine fishes. Although geographically widespread, low-latitude species occur in deeper water. As a group they are mostly found below 80 m to as much as 2600 m and are usually captured close to the bottom. In contrast to many bony fishes, younger chimaeras often occupy deeper water than adults, the latter partaking in seasonal, inshore migrations (Didier 2005). Chimaeroids feed mostly on hard-bodied benthic invertebrates, which they crush with their toothplates.

Extant holocephalan species represent a small fraction of a previously successful and diverse group (Lund 1990; Grogan & Lund 2004; see Chapter 11). Although only three families in a single order are alive today, 6 fossil orders, represented by perhaps 9 families, lived during Paleozoic and Mesozoic times. Several groups had more sharklike dentition, or dentition unlike any living or other extinct chondrichthyans. The modern families have fossil records dating back to the Jurassic and Cretaceous (Carroll 1988; Nelson et al. 2016).

Supplementary Reading

Budker P. 1971. *The life of sharks.* New York: Columbia University Press.

Carrier, JC, Musick JA, Heithaus MR, eds. 2004. *Biology of sharks and their relatives.* Boca Raton, FL: CRC Press.

Castro JI. 1983. *The sharks of North American waters.* College Station: Texas A&M University Press.

Compagno LJV. 1984. *Sharks of the world. An annotated and illustrated catalogue of shark species known to date.* FAO Fisheries Synopsis No. 125, Vol. 4. Rome: Food and Agricultural Organization.

Compagno LJV. 2001. *Sharks of the world. An annotated and illustrated catalogue of shark species known to date. Vol. 2. Bullhead, mackerel and carpet sharks (Heterodontiformes, Lamniformes and*

Orectolobiformes). FAO Species Catalogue for Fishery Purposes, No. 1, Vol. 2. Rome: Food and Agricultural Organization.

Compagno LJV, Dando M, Fowler S. 2005a. *A field guide to sharks of the world*. London: Collins.

Compagno LJV, Dando M, Fowler S. 2005b. *Sharks of the world*. Princeton, NJ: Princeton University Press.

Ellis R. 1976. *The book of sharks*. New York: Grosset & Dunlop.

Ellis R, McCosker JE. 1991. *Great white shark*. New York: Harper Collins.

Gruber SH, ed. 1991. *Discovering sharks*. American Littoral Society Special Publication No. 14. Highlands, NJ: American Littoral Society.

Hamlett WC, ed. 1999. *Sharks, skates, and rays: the biology of elasmobranch fishes*. Baltimore, MD: Johns Hopkins Press.

Hodgson ES, Mathewson RF, eds. 1978. *Sensory biology of sharks, skates, and rays*. Arlington, VA: Office of Naval Research.

Klimley AP, Pyle P, Anderson SD. 1996. Tail slap and breach: agonistic displays among white sharks? In: Klimley AP, Ainley DG, eds. *Great white sharks: the biology of Carcharodon carcharias*, pp. 241–260. San Diego: Academic Press.

Lineaweaver TH, Backus RH. 1970. *The natural history of sharks*. Philadelphia: J. B. Lippincott.

Moss SA. 1984. *Sharks. A guide for the amateur naturalist*. Englewood Cliffs, NJ: Prentice-Hall.

Pratt HL, Jr., Gruber SH, Taniuchi T, eds. 1990. *Elasmobranchs as living resources: advances in the biology, ecology, systematics, and the status of fisheries*. NOAA Technical Report No. 90. Washington, DC: National Oceanic and Atmospheric Administration.

Shuttleworth TJ, ed. 1988. *Physiology of elasmobranch fishes*. Berlin: Springer-Verlag.

Springer VG, Gold JP. 1989. *Sharks in question*. Washington, DC: Smithsonian Institution Press.

Stevens JD, ed. 1987. *Sharks*. New York: Facts on File Publishers.

Wourms JP, Demski LS. 1993b The reproduction and development of sharks, skates, rays and ratfishes: introduction, history, overview, and future prospects. *Env Biol Fish* 38:7–21.

Websites

www.elasmo.com.

www.elasmo-research.org.

www.eulasmo.org.

www.flmnh.ufl.edu/fish/Organizations/aes/aes.htm.

www.sharktrust.org.

CHAPTER 13

Living Representatives of Primitive Fishes

Summary

A small number of anatomically primitive and unusual fish species occur on all the earth's major continents, often in tropical or subtropical, swampy habitats. These fishes represent the last remaining representatives of groups that dominated aquatic environments during the Paleozoic and Mesozoic periods. The ancestry of several species can be easily traced to otherwise extinct groups, whereas other species are obviously highly derived, specialized fishes whose close affinities can only be surmised from anatomical similarities. These "living fossils" include some of the most spectacular and controversial ichthyological discoveries of the past two centuries and support the idea that every species is a mixture of ancestral and derived characteristics.

Teleosts are the most diverse group of vertebrates alive today, but a few highly derived species represent the successful primitive groups of pre-fishes and fishes of the past. These are the lancelets, hagfishes, lampreys, coelacanths, lungfishes, sturgeons, paddlefishes, bichirs, gars, and Bowfin. The challenges of understanding the evolutionary relationships among these groups, and thus classifying them, reminds us that nature represents a continuum of successful possibilities and often is not easily divided into clear categories.

CHAPTER CONTENTS

The Diversity of Fishes: Biology, Evolution and Ecology, Third Edition. Douglas E. Facey, Brian W. Bowen, Bruce B. Collette, and Gene S. Helfman.

Companion website: www.wiley.com/go/facey/diversityfishes3

Cephalochordate lancelets are fish-like yet lack most chordate structures, and are thus considered invertebrates. They are filter-feeding bottom dwellers. Hagfishes and lampreys are jawless fishes that appear somewhat similar, likely due to convergent evolution, but differences in mouth position, tooth and tongue morphology, embryology, pineal complex, and gill structure suggest separate ancestries. Hagfishes are entirely marine, high-latitude predators and scavengers that lack larvae, produce copious slime, and can tie themselves into knots. Commercial "eelskin" comes from hagfishes. Lampreys are primarily freshwater, temperate, often parasitic fishes with complex life cycles. Numerous nonparasitic species have evolved from parasitic ancestors.

Coelacanths had been thought to have gone extinct about 80 mybp, until a live one was captured in 1938 off South Africa. Today, a small, endangered population of perhaps 200–600 exists in the Comoros Islands, and a second species with unknown population size has been located in Indonesia, South Africa, Madagascar, and along eastern Africa. The living coelacanths are very much like their Paleozoic ancestors, with lobe fins, diphycercal tail, hollow spines, a specialized notochord, and a jointed skull.

Living lungfishes are a small subset of a widely distributed, diverse Paleozoic and Mesozoic subclass. The Australian lungfish is most like earlier species; the South American and African species are highly derived in many respects. Lungfishes lack jaw teeth but have unusual toothplates on the mouth roof and floor. African lungfishes can estivate in dried mud for up to 4 years.

The most primitive actinopterygian fishes are the highly derived, relict, chondrostean sturgeons and paddlefishes. They share many traits (cartilaginous skeleton, heterocercal tail, few scales, numerous fin supports, unique jaw suspension), but differ in most respects. Sturgeons are large, freshwater and anadromous, long-lived fishes of North America, Europe, and Asia that are highly prized for their eggs (caviar) and have been heavily overfished. Two species of paddlefishes, in two different genera, occur in large rivers of North America and China. Paddlefishes have a long rostrum, and the North American Paddlefish likely uses it to detect weak electric fields of its zooplankton prey.

The bichirs and Reedfish of Africa have been variously classified with the lungfishes, lobefins, and rayfins because they have larvae with gills, lobe-like fins, ganoid scales, and a modified heterocercal tail. But they have uniquely constructed median, caudal, and paired fins and an unusual chromosomal arrangement, and are therefore placed in their own infraclass, the Cladistia or Brachiopterygia.

Two living orders represent close ancestors of teleosts. The lepisosteiform gars are predaceous fishes that occur in North and Central America, where they occupy backwaters and swamps. They breathe atmospheric oxygen via a highly vascularized gas bladder. Unusual traits include interlocking ganoid scales, opisthocoelous vertebral centra (convex anteriorly, concave posteriorly), an abbreviate heterocercal tail, and poisonous eggs. Closer to the teleosts is the monotypic Bowfin (Amiiformes), which is restricted to eastern North America. They can also breathe atmospheric air and are predaceous. Bowfin have cycloid scales, biconcave vertebra, a large gular plate, an elongate dorsal fin, and the males guard the young for an extended period.

Lancelets and Jawless Fishes (Hagfishes and Lampreys) – Living Groups

Phylum Chordata[a]

- Subphylum Urochordata – The tunicates are invertebrates that have a swimming larval phase with chordate characteristics that are lost during metamorphosis to a sedentary adult.
- Subphylum Cephalochordata
    - Order Amphioxiformes (lancelets) (30 species, marine, tropical and temperate): Branchiostomatidae, Epigonichthyidae
- Subphylum Craniata
    - Infraphylum Myxinomorphi
        - Class Myxini (living hagfishes) (81 species, temperate marine)
            - Order Myxiniformes: Myxinidae (hagfishes) (87 species)
    - Infraphylum Vertebrata
        - Superclass Petromyzontomorphi
            - Class Petromyzontida
                - Order Petromyzontiformes (living lampreys) (48 species, temperate fresh water and anadromous): Petromyzontidae (42 species of northern lampreys), Geotriidae (3 species of pouched lampreys), Mordaciidae (3 species of southern topeyed lampreys)

[a] Higher classification groups based on Nelson et al. (2016), but see corrections and updates at **https://sites.google.com/site/fotw5th/**. Orders, families, and numbers of species from Fricke et al. (2021) but refer to **https://researcharchive.calacademy.org/research/ichthyology/catalog/SpeciesByFamily.asp** for updates

Amphioxiforms

Lancelets are not fishes because they lack many diagnostic characters of the group. However, cephalochordates – which are sometimes referred to as invertebrate chordates along with urochordates and hemichordates – seldom receive treatment in invertebrate textbooks. Their evolutionary and anatomical affinities are much closer to the vertebrates (see Northcutt & Gans 1983; Gans et al. 1996), and lancelets are quite fish-like and studied primarily by ichthyologists, thus their inclusion in an ichthyology textbook.

Lancelets are small (up to 8 cm), slender organisms that as adults occupy sandy, usually shallow bottoms in all major oceans (Fig. 13.1). They commonly bury themselves in the sediments with just the anterior portion of the body protruding from the bottom. Lancelets filter diatoms and other small food items from the water via cilia that transport water through the mucus-laden mouth and pharynx and out through the **atriopore**. Food-trapping mucus is produced by the **endostyle**, a pharyngeal organ that also functions in iodine uptake and may therefore be homologous to the thyroid of vertebrates (Nelson 2006).

Spawning in most species occurs in early summer. Spawning adults swim up into the water column, the females following the males. Larvae metamorphose after 2–5 months. Larvae are free swimming, ciliated, planktonic animals that have the mouth and anus on the left side of the body. The mouth eventually moves to the middle, but the anus remains on the left side. Larvae can be very abundant at times, reaching densities of $1000/m^3$ in regions of upwelling. The larvae settle on sandy or sandy-shell bottoms, mature in 2 or 3 years, and live as adults for 1–4 years depending on the species. Both larvae and some immature adults undergo diel vertical migrations, moving to surface waters at night (Bigelow et al. 1948; Boschung & Shaw 1988).

Lancelets are intriguing in their lack of many typical chordate structures. They differ from fishes by lacking most parts of a head (e.g. there is no cranium, brain, complex eyes, external nostrils, or ears); hence they are sometimes called **acraniates**. Lancelets also lack vertebrae, scales, genital ducts, a heart, red blood cells, hemoglobin, and specialized respiratory structures (gills), and have only one cell layer in the epidermis. Lancelets have up to 25 pairs of gonads as compared to one in lampreys and hagfish and two in most other fishes. The number of internal gill clefts increases throughout life in lancelets, whereas in fishes the number is fixed at birth. The notochord, a definitive chordate structure, extends beyond the anterior end of the dorsal nerve tube (e.g. beyond the "brain"), to the anterior end of the body.

Lancelets are economically valuable in two respects. *Branchiostoma lanceolatum*, a European and Mediterranean species, is the "amphioxus" commonly used in biology laboratories as a study animal. Besides its utility as an example of primitive chordate features, it is morphologically convergent with the ammocoete larva of the lampreys (discussed later in this chapter), thus providing comparative material and popular questions for laboratory practicals. Lancelets are also targeted by a seasonal fishery in southern China, where they are dredged from the bottom with scoops at a rate of about 30,000 kg, or l billion lancelets, annually.

The fossil record for cephalochordates is limited, but recent findings suggest ancestry among the cephalochordate-like yunnanozoans (*Haikouella* and *Yunnanozoon*) of the Lower Cambrian, or perhaps in *Pikaia* from the Middle Cambrian Burgess Shale of British Columbia, which had a dorsal nerve cord and notochord (discussed in Chapter 11). A recognizable cephalochordate fossil has been found in Early Permian deposits in South Africa (Oelofsen & Loock 1981). The evolutionary importance of the lancelets lies in their presumed place as ancestors of advanced chordates, as revealed in their embryology.

Two lancelet families are recognized, each containing one genus (Branchiostomatidae, *Branchiostoma*; Epigonichthyidae, *Epigonichthyes*). The larva of a *Branchiostoma* species that remains planktonic after metamorphosis was at one time thought to be a third family (Amphioxidae, Boschung & Shaw 1988). Traditionally, species have been designated based on myomere and fin chamber counts, position of atriopore and anus, gonadal characteristics, and notochord and caudal fin shape. Statistical analysis has revealed considerable meristic variation in such taxonomic features (Poss & Boschung 1996). Additionally, genetic analysis using 12S rRNA gene comparisons and mitochondrial DNA sequences indicates the existence of previously undetected species and lack of monophyly in recognized genera (Nohara et al. 2005; Xu et al. 2005).

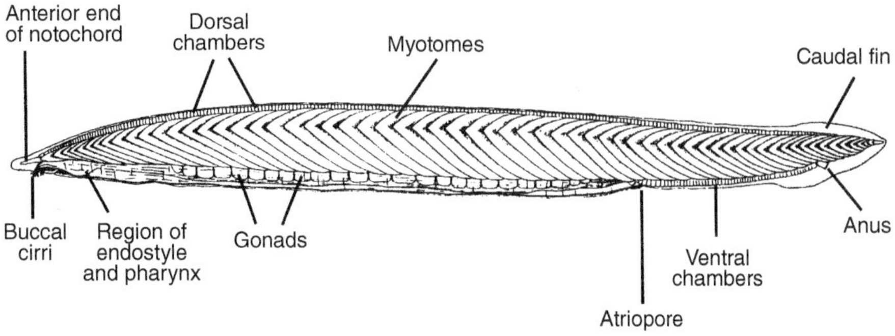

FIGURE 13.1 An adult lancelet, Branchiostoma longirostrum; actual length, 43 mm. Adapted from Boschung (1983), used with permission.

Additional study will likely reveal that cephalochordates are richer in biological information and species than previously recognized.

Lancelets – Not Quite Fishes, but Close There was considerable debate in the late 1800s about the place of lancelets in vertebrate evolution. Hundreds of papers about lancelet anatomy and embryology were published, largely motivated by a debate among three influential German biologists, Ernst Haeckel, Karl Sempel, and Anton Dorn. Sempel and Dorn proposed that vertebrates evolved from some wormlike ancestor, whereas Haeckel saw the ancestors of the lancelets as our most likely ancestors. Haeckel based his conclusions on similarities (presumed homologies) between lancelets and fishes, including the perforated pharynx, segmented axial muscles, a notochord, a dorsal hollow nerve cord, and a postanal tail (Stokes & Holland 1998).

Haeckel wrote that we should regard lancelets, "with special veneration" because of their place in our ancestry; Sempel, Dorn, and religious authorities countered that reverence for lancelets caused the "dignity of humanity to be trodden underfoot." Haeckel's homologies and views have proven correct, brought about by a resurgence in interest in lancelet embryology and application of new analytical technologies (Stokes & Holland 1998). The lancelet's mucus-secreting feeding structure, the endostyle, is now considered a homolog of the vertebrate thyroid gland. Lancelet nephridia are simple versions of the vertebrate kidney, and a photosensitive, pigmented structure in the anterior portion of the lancelet nerve cord may be homologous to vertebrate eyes. Significantly, several developmental genes that code for particular somites in the lancelet embryo – specifically *Hox-1* and *Hox-3* – code for the same body sections in the vertebrates, including neural structures in the head (i.e. our brain). Finally, a cladistic analysis based on ribosomal DNA sequences places lancelets as the sister group to vertebrates, with a common ancestor in *Pikaia* or a related form.

Hagfishes and Lampreys: Evolutionary Relationships

Ancestor–descendant relationships of jawless fishes are fraught with controversy. Just about every conceivable permutation on relationship among hagfishes, lampreys, and jawed fishes has been proposed at some time, including hagfishes, lampreys, or gnathostomes as ancestral to the other groups (see Hardisty 1982). The possible ancestors of jawless fishes are well represented in the Silurian and Devonian (see Chapter 11). But the extinct groups are very different from one another, as are extant jawless fishes from presumed ancestral groups and from each other.

Although traditionally treated as related orders of **cyclostomes** ("round mouths") – a hypothesis supported by some molecular studies (see Nelson et al. 2016) – similarities in the body morphology of modern hagfishes and lampreys are thought to reflect convergent evolution. It is probably wisest to deal with them individually and independently and appreciate them for the unique yet primitive organisms that they are. Similarly, the term "**agnatha**" ("without jaws"), previously given superclass status, is now recognized as being paraphyletic; the term is still used as an informal adjective for jawless fishes. Hagfishes are considered a more primitive, separate, non-vertebrate group within the subphylum **Craniata,** but in their own infraphylum, the **Myxinomorphi**, constituting the sister group of vertebrates and the basal craniate taxon (Nelson et al. 2016). Lampreys are placed in the infraphylum **Vertebrata** as one of six superclasses of extinct and extant jawless and jawed fishes except the hagfishes. Vertebrates possess essential traits in common, especially dermal skeletal elements (secondarily lost in lampreys) and neural crest tissue (the embryonic nerve cord tissue that develops into gill arches, connective tissue, and bone), among others (discussed in Chapter 11).

Lampreys and hagfishes share a host of anatomical, physiological, and biochemical traits but have an even greater number of differences. Although both groups are scaleless, lampreys lack the mucus-producing capability of hagfishes. Lampreys have one or two dorsal fins supported by radial muscles and cartilage, whereas hagfishes have a single continuous caudal fin. Lampreys have a terminal mouth, hagfishes a subterminal mouth. Lampreys have a larval stage; hagfishes have direct development. In adult lampreys, the external opening of the nasohypophysis is dorsal and the tract ends internally in a blind sac above the branchial region; in hagfishes, the external opening is terminal, and the internal opening is into the pharynx. Lampreys have two semicircular canals, hagfishes only one. Lampreys have a pineal organ and functional eyes, hagfishes possess neither.

A major similarity between the two groups involves their immune responses. All jawed vertebrates, including fishes, have immune systems that involve immunoglobulin-type antigen receptors that produce pathogen-specific antibodies in response to infectious agents such as microbes. Lampreys and hagfishes also produce pathogen-specific defensive substances, but instead of antibody proteins, jawless fishes produce different kinds of proteins called **variable lymphocyte receptors**. Hence, "two strikingly different modes of antigen recognition . . . have evolved in the jawless and jawed vertebrates" (Alder et al. 2005, p. 1970).

Among the differences between the groups, lampreys possess lateral line neuromasts that are touch sensitive; these are lacking in hagfishes. All lampreys have seven gill openings, hagfishes vary between 1 and 16. Although the tongue possesses keratinous, replaceable teeth in both groups, it is anatomically and functionally different. Hagfishes use the tongue for biting and tearing, whereas lampreys use it for rasping and suction. These and other differences in embryological, skeletal, neuromuscular, respiratory, cardiovascular, endocrine, osmoregulatory, chromosomal, and reproductive features all point out the disparate nature of the two groups (Hardisty 1982; Fernholm 1998; Nelson 2006).

Myxiniforms

Hagfishes, otherwise known as slime eels or slime hags, derive their alternative names from the copious mucus they produce via 70–200 ventrolateral pairs of slime glands (Fig. 13.2). Mucus production is the combined result of these holocrine slime glands, as well as merocrine exudates from the epidermis itself (Spitzer & Koch 1998). The slime glands contain both mucous cells and thread cells, the latter being a unique trait in hagfishes that may strengthen the slime. Each slime gland is surrounded by connective tissue and striated muscle fibers that help exude the slime upon stimulation. The mucus itself consists of a protein plus a carbohydrate that binds to water and expands to form a loose jelly. A 50 cm hagfish can fill an 8 L bucket with slime in a matter of minutes (Fig. 13.3).

Not surprisingly, slime production may serve multiple functions, some speculative. Authors have suggested that (i) hagfishes produce slime when attacking dying fish, perhaps hastening suffocation of the prey by clogging its gills; (ii) mucus could protect the hagfish from digestive enzymes when feeding inside the body of a prey animal; (iii) mucus is repulsive to other scavengers such as sharks or invertebrates and thus serves to overcome competition; and (iv) slime stabilizes burrow walls in the muddy bottoms in which hagfishes live (Bigelow & Schroeder 1948a; Brodal & Fange 1963; Hardisty 1979; Smith 1985a). Martini (1998), among the few to observe hagfishes in both the field and lab, found no evidence that burrow walls were stabilized with mucus or any other substance.

Hagfishes typically produce slime in response to being disturbed or handled. Mucus undoubtedly serves some antipredator function, perhaps by making the fish too slippery to handle or by clogging the gills of a potential predator and threatening it with suffocation. Slime has its drawbacks, however. A hagfish covered in its own slime will suffocate after a few minutes.

FIGURE 13.3 A single hagfish can produce prodigious quantities of slime when disturbed. Courtesy of J. Meyer.

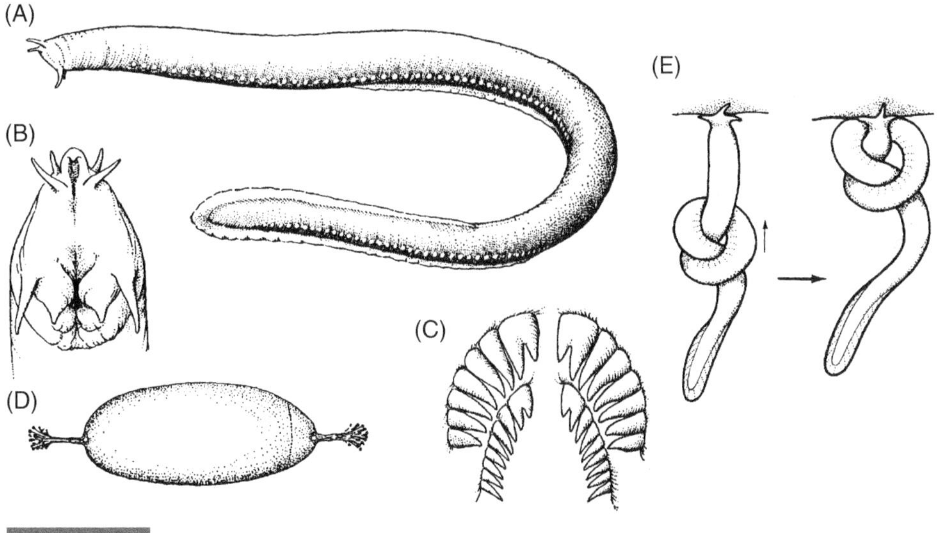

FIGURE 13.2 Hagfishes. (A) Adult Atlantic Hagfish, *Myxine glutinosa*, 38 cm long. Porthole-like structures along the side are mucous glands. (B) Ventral view of the head region of an Atlantic Hagfish. The upper orifice is the nasal opening, and the lower orifice is the mouth. (C) The lingual (tongue) teeth of a hagfish. (D) Hagfish egg, approximately 40 mm long. (E) A hagfish pressing a knot against the side of its prey to gain leverage when tearing off flesh. (A–D) From Bigelow and Schroeder (1948a), used with permission; (E) Based on Jensen (1966).

Hagfishes rid themselves of slime by tying an overhand knot in their tail and then sliding the knot forward along the body, pushing the mucus ahead until the knot and mucus reach the anterior end and the fish can back away from the slime mass. A hagfish is also capable of backflushing its gills and nostril with water to rid them of slime (Conniff 1991).

Hagfishes are highly specialized animals and should not be considered as simply "primitive" fishes. They possess **four rudimentary hearts**: a primary, three-chambered branchial or systemic heart posterior to the gills, and three auxiliary, single-chambered hearts located just behind the mouth (the paired cardinal heart), at midbody (the portal heart), and at the end of the tail (the paired caudal heart). These multiple pumping stations beat at different rates; the branchial and portal hearts contract via intrinsic muscles, whereas the cardinal and caudal hearts are squeezed by surrounding, extrinsic skeletal muscle. The auxiliary hearts are necessary to re-establish blood flow in venous vessels after blood leaves several sinuses where blood flow slows. The sinuses take the place of capillary beds, giving the hagfish a partially open circulatory system, more an invertebrate than a vertebrate trait. Contraction of body wall musculature during activity also aids in pushing blood from the sinuses into adjoining vessels. Hagfish blood is unique among craniates in being **isosmotic** with sea water, making it about three times saltier than the blood of bony fishes and lampreys. Hagfish kidneys are much simpler than those of other fishes, including lampreys, which may explain why hagfishes are restricted to a narrow range of salinities (Jensen 1966; Hardisty 1979, 1982).

Other hagfish peculiarities characterize the respiratory, digestive, immune, and sensory systems. Oxygen uptake in hagfishes occurs both at the gills and at capillary beds in the skin. Unlike most fishes, hagfishes inspire water through their nostril and then pump it via the mouth to the gill sacs. Cutaneous respiration comes into play when a hagfish has its nostril and gills buried deep in the carcass of a prey fish. Cutaneous respiration is undoubtedly facilitated by the oxygen-rich nature of the cold waters that hagfishes normally frequent, although the mud in which they bury is often anoxic. As an apparent adaptation to anoxic conditions, mud-burrowing species are exceedingly hypoxia tolerant, able to exist in anoxic conditions for hours or longer (Malte & Lomholt 1998). Low oxygen consumption and very low basal metabolic rates appear to characterize hagfishes. Smith (1985a) calculated that Black Hagfish, *Eptatretus deani*, could maintain itself for 1 year after only 1.5 h of feeding on a high-energy source such as a carcass.

Hagfishes lack a true stomach, having instead an intestine that begins at the pharynx and ends at the anus, with an anterior muscular subdivision that prevents water inflow. Hagfishes do not bleed when their skin is cut, nor do such wounds become infected. Their immune system produces complement-like factors instead of immunoglobulins; hagfishes lack a defined thymus, spleen, or bone marrow, which are the usual sites of antibody production in vertebrates. Hagfishes also lack complete eyes but have photosensitive receptors in their head (which contain retinal structures but no lens and are probably incapable of image formation); they also lack a cloacal region. Fossil material from the Late Carboniferous indicates that Paleozoic hagfishes possessed more developed eyes than recent forms (Bardack 1991). Apparently, visual sensory input has been lost over time in the deep, dark habitats that hagfishes occupy. Food is found largely through olfaction and touch, the six barbels around the mouth serving both functions.

Hagfishes are nocturnal predators on a wide variety of small, benthic invertebrates, but are better known for their scavenging behavior (Shelton 1978; Smith 1985a; Martini 1998). Hagfishes enter a dead or dying fish or other animal via some orifice or by digging through the skin and then consuming their prey from the inside, leaving only the skin and bones. The knot tying action that hagfishes use to deslime their bodies is also employed during feeding. A hagfish grasps a prey item by everting, retracting, and closing its toothplates. It will then pass a knot forward along its body and then press the knot against the prey as a means of levering off a piece of flesh (see Fig. 13.2E). Such knot-feeding is also seen in moray eels. Food is removed via a repeated evert–grasp–retract–release cycle of the toothplates (Martini 1998).

Reproduction in hagfishes remains something of an enigma. Both sexes contain only a single gonad, rather than the paired gonads found in most jawed fishes. In immature animals, this gonad is differentiated anteriorly as ovarian tissue, and posteriorly as testicular tissue. Upon maturation, one cell type prevails, and no evidence of functional hermaphroditism has been found, but spawning has never been observed. Fertilization is thought to be external because males possess no intromittent organ and females do not contain fertilized eggs. Females produce eggs in batches, depositing about 20–30, 1.5–4.0 cm long, heavily yolked, sausage-shaped eggs covered by a keratin-like shell (see Fig. 13.2D). These comparatively large eggs attach to each other and to the ocean floor. Incubation takes about 2 months, development is direct with no larval stage, and the young emerge as 45 mm long replicas of the adults. Levels of hormones associated with reproduction suggest that the Atlantic Hagfish (*Myxine glutinosa*) may have a seasonal reproductive cycle (Powell et al. 2005), but other hagfishes show no obvious seasonality in spawning. However, actual spawning times, frequencies, places and behaviors, embryological details, ages at maturity, and reproductive life spans are unknown for most species.

Hagfish species occur almost worldwide in temperate and cold temperate ocean waters above 30° latitude in both hemispheres, although hagfishes are uncommon in polar seas (Hardisty 1979). Few hagfish species occur shallower than 30 m, being limited by both the low salinities and high temperatures found at shallower depths; 34 ppt and 20°C appear to be the minimum salinity and maximum temperatures tolerated (Krejsa et al. 1990a, 1990b). The few tropical species occur in deep water, hagfishes having been captured as deep as 2700 m and photographed at 5000 m. Until recently, hagfishes had little commercial value and were largely viewed as nuisance species that scavenged on more valuable fishes.

Hagfishes are preyed upon by dolphins, porpoises, seals, sea lions, and octopus, sometimes accounting for 25–50% of the diet of individual predators (Martini 1998). Human consumption seems to be localized to Asia, where broiled hagfish, called "Anago-yaki" in Japan, is a marketable commodity (e.g. Honma 1998).

Hagfish taxonomy is based on the arrangement of efferent gill ducts (one vs. more excurrent openings), number of slime pores, finfolds, and tentacle and dentition patterns (Fernholm 1998). Some authors recognize two families, the Myxinidae with a single external gill aperture and the Eptatretidae with multiple external gill openings; Fernholm et al. (2013) recognize three subfamilies within the Myxinidae. Maximum lengths range between 25 and 100 cm, except for a recently described giant hagfish from New Zealand that attains a length of at least 127 cm and a mass of 6.2 kg (Mincarone & Stewart 2006). Ongoing analyses indicate several undescribed species in areas where only a single species was thought to occur, although disputes over the validity of some new species exist (e.g. Wisner & McMillan 1990; Nelson 2006).

The only fossil hagfish known is a small, 7 cm long specimen, found in Pennsylvanian (300 million years before present (mybp)) deposits in Illinois (Bardack 1991). This species, *Myxinikela siroka* – notable for its functional eyes, anteriorly placed gill pouches, and apparent lack of slime pores – is otherwise very similar to extant forms. Its discovery underscores the conservative nature of the hagfish lineage, a clade that may trace its ancestry into Early Paleozoic times via the conodonts (Krejsa et al. 1990a, 1990b).

Commercial Value of Hagfishes Few consumers probably know that hagfish are the source of "eelskin" wallets, purses, and briefcases. Leather workers in South Korea developed a method for tanning hagfish skin in the late 1970s (Conniff 1991). The product, often marketed as "conger eel" is a soft, supple-yet-strong, thin leather of considerable economic value. A substantial fishery, valued at $100 million annually, developed off Korea, Japan, and surrounding waters. More than 1000 boats and dozens of leather processing plants were involved in the mid 1980s.

Hagfishes are caught primarily at night with baited bamboo or plastic traps at depths of 30–500 m, where the main species captured are *Paramyxine atami*, *Myxine garmani*, and *Eptatretus burgeri* (Gorbman et al. 1990). Effort peaked in 1986, when daily catches averaged 5000 kg per boat. However, this success soon fell to less than 1000 kg/boat/day, as hagfish populations experienced overfishing pressure from the working fleet. Boats also routinely lost 200 traps per month, creating a tremendous competing "ghost fleet" that was still catching and killing hagfish. Given the low fecundity and apparent infrequent reproduction of hagfishes, and a total lack of knowledge of population size or replacement rate, an unregulated effort was bound to lead to a collapsed fishery (overfishing was apparently responsible for a fishery collapse in Japan during World War II and again in the early 1990s; Honma 1998).

In the late 1980s, Korean leather companies sought other sources of hagfish leather to feed a growing worldwide demand. Fisheries opened up along the west coast of North America and the Atlantic coast of Canada, where commercial catches were previously nonexistent. In California alone, 1989 landings of Pacific hagfish, *Eptatretus stouti*, exceeded 2,000,000 kg and involved boats from 19 different ports (Nakamura 1991). Both eastern Pacific and western Atlantic fisheries have experienced obvious signs of overfishing, including population declines and decreased catch-per-unit-effort (Martini 1998). Biological information has lagged behind the economic efforts, and regulatory legislation has been slow to develop. Moreover, eastern Pacific hagfishes produce a thinner skin of lesser quality and durability than the western species, which has affected the desirability of "eelskin" products (Gorbman et al. 1990).

What are the possible ecological consequences of overexploitation of hagfish populations? Hagfishes are not viewed as particularly charismatic by most people and it is unlikely that the environmental movement in any country will adopt the hagfish as a symbol of the need for preservation efforts and of loss of biodiversity. However, hagfishes can be exceedingly abundant in some areas. Densities for both Pacific and Atlantic hagfish, *E. stouti* and *M. glutinosa*, have been estimated at as high as 400,000–500,000/km^2 (*c.* 0.5 hagfish/m^2) (Nakamura 1991; Martini 1998). At such densities, their impacts as predators and scavengers, as bioturbators of sediment, and as recyclers of nutrients could make them a critical ecosystem component in soft-bottom benthic regions, the most abundant habitat type in the world's oceans. In addition, hagfishes and Sea Lampreys are an important part of the diet of dolphins and several pinnipeds (seals and sea lions), animals of definite concern to the informed public. Substantial reductions in hagfish populations would have unpredictable ecological consequences, an unhappy situation with a familiar ring.

Petromyzontiforms

Whereas hagfishes have a reputation as scavengers, many lampreys are parasitic on other vertebrates. Lampreys superficially resemble hagfishes in general body form (Fig. 13.4). As with hagfishes, lampreys lack constricted vertebrae, the body being supported by a notochord. They also lack paired fins, jaws, a sympathetic nervous system, and a spleen. They too are scaleless, have a single nostril, and have keratin-like teeth on the tongue. However, adult lampreys possess functional eyes, dorsal fins, an additional semicircular canal, a cerebellum, separate dorsal and ventral roots of the spinal nerves (an innovation among vertebrates), and a spiral-like rather than a straight intestine (Hardisty 1982).

Among the most striking differences between the two jawless fish groups is mode of reproduction. Whereas hagfishes presumably spawn repeatedly during their lives and produce a few large eggs each time, lampreys produce many small eggs and the

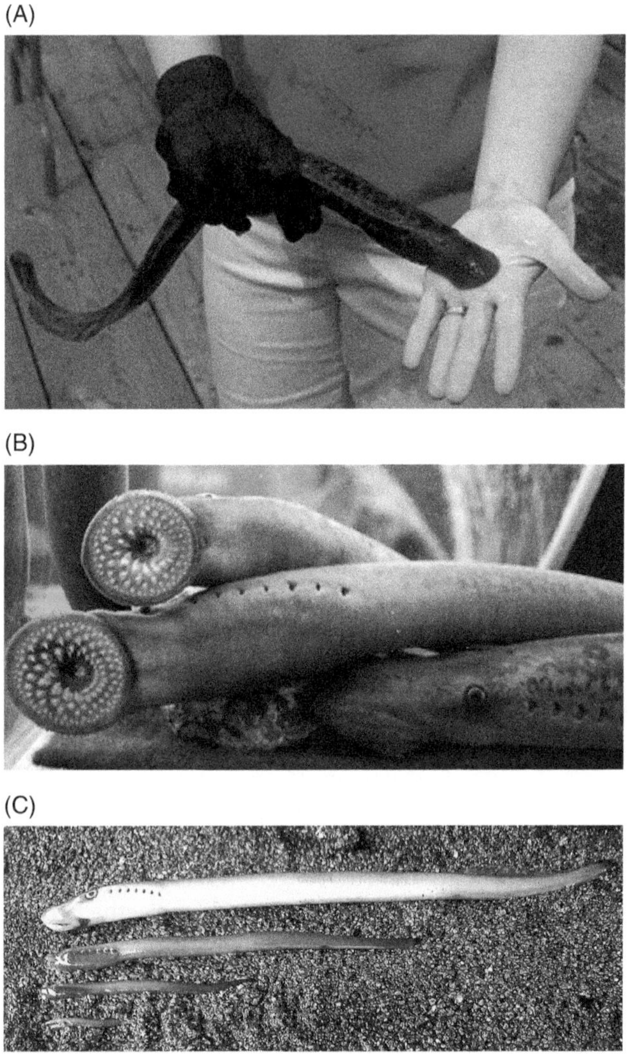

FIGURE 13.4 Lampreys. (A) An adult parasitic Sea Lamprey (*Petromyzon marinus*) captured in Lake Erie. Marisa Lubeck / USGS / Public Domain (B) Sea Lampreys in a tank show their oral disk for holding onto hosts and the teeth on their tongue (at center of oral disk) used to rasp a hole in the skin of the victim. Openings on the side behind the eyes are for the gills. Joanne Gilkeson / USFWS / Public Domain (C) Four age classes of Sea Lamprey from the Ausable River, NY – three ammocoetes and a transformer metamorphosing to the parasitic adult stage. Specimens were collected in fall and represent ages 15 months, 27 months, 39 months, and 51 months. Courtesy of Brad Young.

adults die after spawning. Fecundity varies from about 1000 eggs in nonparasitic species to a few hundred thousand in the larger parasitic species. Hatching in lampreys occurs after 12–14 days and young emerge as a 6 mm larval **ammocoete** (Fig. 13.4C).

In all lampreys, the free-living, blind, toothless ammocoetes, which were not definitively linked to adult lampreys until the mid-1800s, typically burrow into the bed of a silty stream or river. Ammocoetes are burrowed with their heads protruding from the bottom, filtering microscopic organisms from the water column by capturing them on mucus produced in the pharynx. This mode of feeding, possession of an **endostyle** that later develops into a thyroid gland, and the structure of the pharynx are strongly reminiscent of adult lancelets, supporting hypotheses of relationship between the groups (although lancelets move water through the pharynx via ciliary action, whereas ammocoetes pump water via pharyngeal musculature). Under favorable conditions, ammocoetes can achieve high densities, on the order of $30/m^2$ (Beamish & Youson 1987). Ammocoetes may live and feed for up to 7 years, achieving a maximum size of about 10–15 cm.

Transformation to the adult stage takes place in summer and fall in most species. In nonparasitic species, called **brook** or **dwarf lampreys**, the adult is seldom larger than the ammocoete. After an adult life span of about 6 months, during which no feeding occurs, the adult spawns and dies. In parasitic species, free-living ammocoetes turn into parasitic adults that may live for 1–3 years before spawning and dying.

Parasitic species are relatively large, up to 120 cm. Parasitic adults attach to the sides of host fishes using the toothed oral disk, rasp a hole in the skin, and live off the blood or body fluids (*Ichthyomyzon*, *Petromyzon*, *Mordacia*) or flesh (*Lampetra*, *Geotria*) of their host (Potter & Gill 2003) (Fig. 13.4B). Blood loss by the host can be substantial, amounting to 30% of the weight of the lamprey per day. Parasitic lampreys can contribute significantly to the mortality of host species. The River Lamprey, *Lampetra ayresi*, may kill 18×10^6 kg of herring and 10% of the salmon off coastal British Columbia annually (Beamish & Youson 1987).

These natural levels of mortality may have relatively little effect on host population success under normal conditions, but where lampreys have been accidentally introduced, the effects can be catastrophic. The Sea Lamprey, *Petromyzon marinus*, invaded the upper Laurentian Great Lakes of North America via canals constructed to connect the lakes with river systems that reach the Atlantic Ocean. Sea Lampreys have contributed to the decline or extirpation of several fish species, such as Lake Trout, whitefishes, and Blue Pike (Fuller et al. 1999; Daniels 2001). Extensive lamprey control strategies, involving chemical poisons (which also kill some nontarget aquatic species) and various methods for trapping adults in spawning tributaries, have apparently helped reduce lamprey populations (Smith 1971; Hanson & Swink 1989; Youson 2003). Ironically, Sea Lampreys are considered imperilled in several European countries because of overfishing, migration blockage from dams, and sedimentation of spawning habitat. In France, among other nations, the species is "highly esteemed for the table" (Keith & Allardi 1996, p. 38).

Anatomical differences between lampreys and hagfishes are strongly reflected in their different foraging tactics. Unlike hagfishes, lampreys have **circumferential teeth** on the oral disk that aid in grasping live prey (see Fig. 13.4B). The olfactory and respiratory pathways are necessarily separated because lampreys feed and breathe while attached to the exterior of their prey. The nasohypophyseal opening carries water to a blind olfactory organ dorsal to the gill pouches. Attachment involves sealing the contact region between the prey and oral disk via secretion of mucus and the reduction of pressure in the buccal cavity. A vacuum is created as muscles in the mouth and pharyngeal region expel water out of the gill openings. A velar flap then seals the branchial chamber off from the buccal and pharyngeal regions, thus maintaining low pressures in the buccal cavity, allowing water to be pumped in and out of the gills for breathing purposes while also keeping food out of the branchial chamber. Negative pressure in the mouth helps maintain the hold on the prey and promotes the flow of body fluids from the prey to the lamprey once the rasping tongue has gone to work. Uptake of blood and fluids is further aided by anticoagulants in the lamprey's saliva (Hardisty 1979).

Blood circulation in the adult lamprey differs from that of the hagfish in several respects. Although lamprey circulation is also "open" in that it is characterized by sinuses connecting arterial and venous systems, these sinuses are not as prevalent as in the hagfish. The primary lamprey sinuses are located in the branchial region and are associated with blood-gas exchange. Lampreys lack the multiple hearts of hagfishes, having instead a large, single, vagally innervated heart in a pericardial cavity, located in the typical fish position posterior to the gills. Blood pressure is five times higher in lampreys than in hagfishes.

The reproductive biology of lampreys is well known compared to that of hagfishes. Lampreys also undergo a period of sexual intermediacy when both testicular and ovarian tissues can be found in the developing single gonad, but this period is confined to the ammocoete. During sexual maturation, lampreys undergo radical behavioral, anatomical, and physiological changes that parallel those found in salmon and anguillid eels, other families that die after reproduction. These changes cause or at least foretell an inability to live beyond the spawning period. Feeding ceases, the gut atrophies, osmoregulatory function shifts, dentition deteriorates, the body shrinks, the eyes and liver degenerate, hematopoesis (blood production) decreases, and lipid and glycogen stores are reduced. Secondary sex characters, such as thickened fins and genital papillae, are formed.

Upon maturation, adults undertake a spawning migration again reminiscent of the migrations of salmon and eels. Distances moved may range from a few kilometers in nonparasitic or landlocked species to more than 1000 km in species that move from the ocean to fresh water. Spawning locales are typically the upper regions of streams where bottom types are dominated by gravel and cobbles. These locales are often used in successive years by new generations of lampreys, although little evidence suggests that lampreys return to spawn in the stream of their birth, as salmon do. Adults on spawning migrations are attracted to suitable areas by detecting a bile acid, petromyzonol sulfate, produced by larvae (Fine et al. 2004). Males then produce a pheromone that is highly attractive to females (Johnson et al. 2006).

Spawning begins after a male constructs a nest pit by attaching to a rock with the mouth and then carrying the rock downstream or by holding onto large cobbles and thrashing the downstream area. The pit thus created is ringed by large and small cobbles. A female will also take part in nest construction, but site selection appears to be initiated by the male. Nonparasitic species may engage in group spawning in a single nest, whereas larger, parasitic species may engage in pair spawning and male defense of the nest site. When the nest pit is finished, the female attaches to one of the large upstream cobbles, the male attaches to the anterior portion of the female, coils his body around hers, and the two thrash while the male squeezes the eggs out of the female while he releases sperm. Spawning occurs repeatedly over 2–9 days and both sexes die within a few days after spawning. Eggs remain in the nest pit for about 2 weeks, and larvae remain in the nest for an additional week. The ammocoetes then drift downstream to areas of slow current and silty or muddy bottoms, where they burrow and begin feeding (Hardisty & Potter 1971; Hardisty 1979).

Lampreys, like hagfishes, are cool water species that seldom occur at latitudes below 30° or in water temperatures above 20°C in either hemisphere. Only two low-latitude lamprey species are known and both occur at high elevations in Mexico, forming an interesting mirror image to the submergence of tropical species among the entirely marine hagfishes. Nonparasitic forms are entirely confined to fresh water, whereas parasitic forms may occupy fresh water or may be anadromous. **Anadromous** species hatch in fresh water where they live as larvae, move into coastal marine habitats as metamorphosed adults, and then return to fresh water to spawn (Hardisty 1979, 1982; Nelson 1994).

Lamprey taxonomy is based largely on mouth, tentacle, and dentition characteristics (Gill et al. 2003). Lampreys are taxonomically unique in that they have the largest diploid chromosome number of any vertebrate, between 140 and 170 in many northern hemisphere species. The Petromyzontidae of North America and Europe form a monophyletic clade that separated early from the Geotriidae and Mordaciidae of South America, Australia, and New Zealand (Potter & Gill 2003). The petromyzontids include two subfamilies (Petromyzontinae, Lampetrinae) with eight genera and 42 species (Nelson et al. 2016). Two species of an extinct family, the Mayomyzontidae, have been found in 300–320-million-year-old Carboniferous deposits of North America. The more primitive *Hardistiella montanensis* had a hypocercal tail and lacked an oral sucker, whereas *Mayomyzon pieckoensis* was relatively small and lacked teeth everywhere except the tongue (Janvier et al. 2004). *Mayomyzon* is similar to modern petromyzontids despite its antiquity (Nelson 1994) (Fig. 13.5).

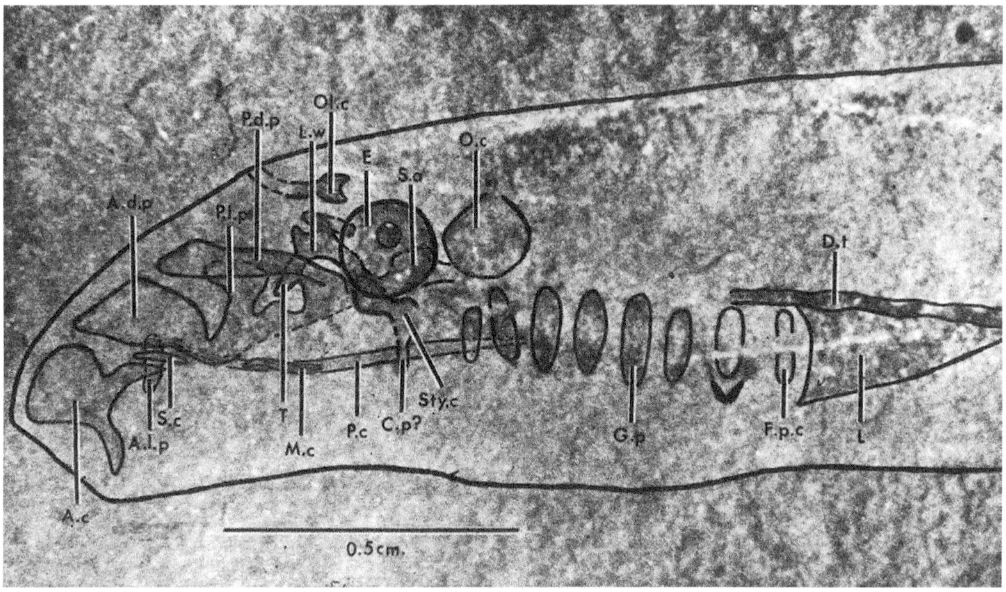

FIGURE 13.5 Reconstruction of the Carboniferous lamprey, *Mayomyzon pieckoensis*, from Illinois. The fossil, seen in lateral view, bears a striking resemblance to modern petromyzontid lampreys. Several recognizable relevant anatomical features are outlined in black: Ac, annular cartilage; Dt, digestive tract; E, eye; Gp, gill pouch; L, liver; Lw, lateral wall of braincase; Oc, otic capsule; Olc, olfactory capsule; Pc, piston cartilage. From Bardack and Zangerl (1971), used with permission.

Paired Species in Lampreys

Lamprey evolution provides a rare glimpse of ongoing speciation processes: ancestral and derived species exist contemporaneously and often in the same river system (Salewski 2003). At least 18 nonparasitic species, or about half of all lampreys, can be matched to ancestral parasitic forms. The phenomenon has been repeated by 11 species of northern hemisphere, petromyzontid, ancestral species and one southern hemisphere, mordaciid lamprey (*Mordacia mordax* gave rise to *M. praecox*). In some instances, a parasitic river lamprey ancestor has evidently given rise to two or three nonparasitic brook lamprey species. In each pair, the ammocoetes are almost indistinguishable, except that nonparasitic ammocoetes may grow larger. Although adults of the nonparasitic species are smaller, both species often have the same number of myomeres. Dentition in the adults of the parasitic species is relatively constant in number and shape and in being functionally hooked and sharp. In the nonparasitic, nonfeeding adult species of a pair, dentition is variable and blunt.

Derivation of nonparasitic forms apparently occurred via extension of the larval period and a shortcutting of the metamorphosis process, both intimately linked to the thyroid gland and its hormones (Youson & Sower 2001) (Fig. 13.6). Parasitic species may spend 4 years as larvae and then take 2 years to feed and mature. A corresponding nonparasitic form has a larval period of 6 years, followed by a relatively short, 6 month or less, maturational period. Hybridization between members of a pair is exceedingly rare; this reproductive isolation is maintained largely by size differences. Nonparasitic males, being smaller, cannot coil around and squeeze parasitic females while their vents are in proximity to one another, actions that are necessary for the extrusion of eggs and proper fertilization. Adoption of a shortened, nonparasitic mode of adult existence may expand the range where a species can occur: small brooks may have abundant food resources for larvae but an insufficient supply of potential host fishes for a feeding adult.

Although the parasitic and nonparasitic species of a pair are reproductively isolated and ecologically distinct, they may be genetically very similar because the separation is relatively recent in time and we are seeing evolution in progress. This can result in challenging management decisions if one species of the pair is protected due to its rarity and the other is more common and therefore not protected. If there is little or no genetic distinction, should they be managed as two species, or two ecomorphs of one species?

The phenomenon of repeated, parallel evolution of such **paired** or "**satellite**" species is unique among vertebrates (Hardisty & Potter 1971; Vladykov & Kott 1979; Beamish & Neville 1992), and may result from sympatric speciation, where new species arise due to behavioral and reproductive isolation, but without geographic separation, from an ancestral species (Salewski 2003).

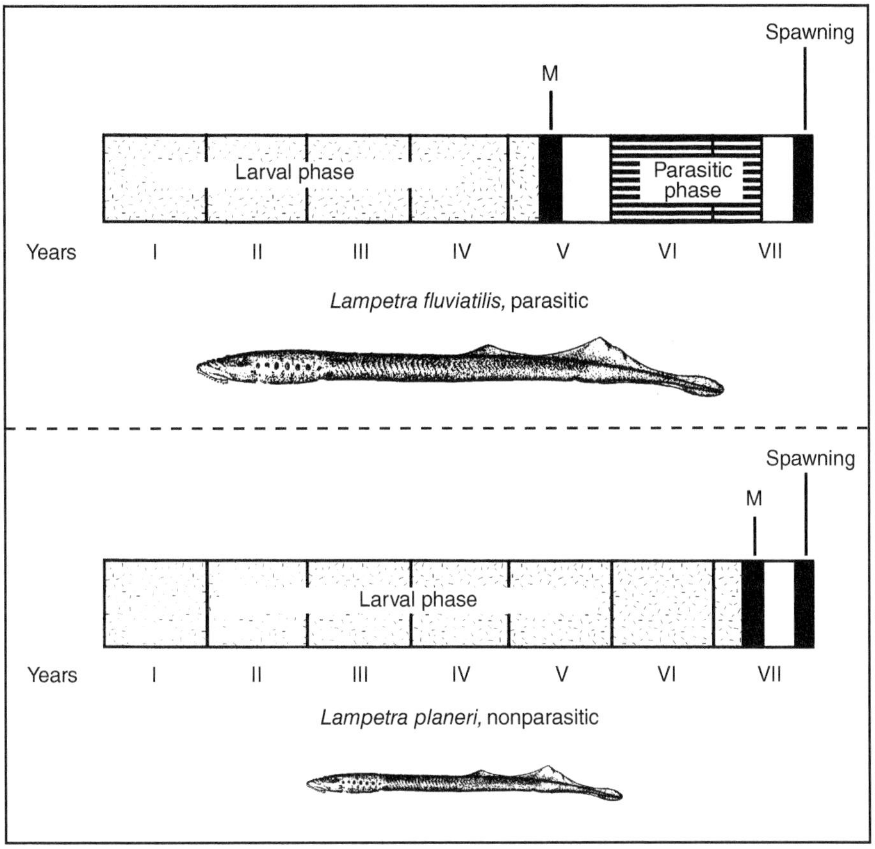

FIGURE 13.6 Comparative life histories of a species pair of European lampreys. (A) The parasitic ancestor, the River Lamprey *Lampetra fluviatilis*. (B) The nonparasitic derived species, the Brook Lamprey *L. planeri*. The evolution of nonparasitic from parasitic forms involves a lengthening of the larval phase and a shortening of the maturational period. The onset of metamorphosis is denoted by M; unshaded areas represent nonfeeding periods. Adapted from Hardisty (1979).

Primitive Bony Fishes – Living Groups

Subphylum Craniata[a]
- Superclass Gnathostomata
  - Grade Teleostomi
    - Class Osteichthyes
      - Subclass Sarcopterygii
        - Infraclass Actinistia (Coelacanthida)
          - Order Coelacanthiformes: Latimeriidae (coelacanths, two extant species, marine)
        - Infraclass Dipnomorpha
          - Superorder Dipnoi
            - Orders Ceratodontiformes: Neoceratodontidae (Australian Lungfish, fresh water), Lepidosirenidae (South American lungfish, fresh water), Protopteridae (African lungfishes, four species, fresh water)
        - Infraclass Tetrapoda (the tetrapods, therefore not addressed in this text)
      - Subclass Actinopterygii (ray-finned fishes)
        - Infraclass Cladistia
          - Order Polypteriformes (Brachiopterygii): Polypteridae (bichirs and Reedfish, 14 species, fresh water)
        - Infraclass Chondrostei
          - Orders Acipenseriformes: Acipenseridae (sturgeons, 27 species, coastal and fresh water), Polyodontidae (paddlefishes, two species, fresh water)
        - Neopterygii – unranked clade, includes Holostei and Teleosteomorpha (Ch 14, 15)
        - Infraclass Holostei
          - Orders Lepisosteiformes: Lepisosteidae (gars, 7 species, fresh and brackish water), Amiiformes: Amiidae (Bowfin, fresh water)

[a] Higher classification groups based on Nelson et al. (2016), but see corrections and updates at **https://sites.google.com/site/fotw5th/**. Orders, families, and numbers of species from Fricke et al. (2021) but refer to **https://researcharchive.calacademy.org/research/ichthyology/catalog/SpeciesByFamily.asp** for updates.

Subclass Sarcopterygii, Infraclass Actinistia (Coelacanthida): The Living Coelacanths

> *There can be few episodes in the history of ichthyology to rival the excitement following the announcement in the* East London Dispatch *of 20 February 1939, declaring that a coelacanth had been captured off... South Africa.*
>
> Forey 1998, p. 1

If you open an ichthyology text published prior to the 1940s (e.g. Günther 1880; Jordan 1905; Norman 1931), you will find passing mention of a relatively obscure group of extinct fishes that represented a side branch of the lineage that presumably gave rise to the tetrapods. These were the coelacanths, conservative sarcopterygian fishes that had gone unchanged in many respects since the Devonian (Jarvik 1980; Forey 1998). Their fossil record stretched nearly 300 million years, from the Middle Devonian, 360 mybp, to near the end of the Cretaceous, 80 mybp, when they and then the dinosaurs disappeared.

Imagine the world's surprise when, just before Christmas 1938, a living coelacanth was trawled from a depth of 70 m off the east coast of South Africa (Smith 1939, 1956; Weinberg 2000). ***Latimeria chalumnae***, described by J. L. B. Smith, retains many of the characteristics that had defined the coelacanths since the establishment of their lineage: a thin bony layer encasing the vertebral spines and fin rays (the name *coel-acanth*, meaning "hollow spines," refers to the hollow nature of the fin rays that support the tail); an unconstricted and unossified notochord, modified as a strong-walled elastic tube; fleshy, lobed pectoral, pelvic, anal, and second dorsal fins (= Sarcopterygii); a symmetrical, three-lobed, diphycercal tail with an epicaudal fringe portion extending beyond the midline; relatively large, thick, bony scales; a double gular plate under the lower jaw; a dorsal intracranial articulation (a joint in the braincase that functions to increase gape size); and numerous other osteological features (Fig. 13.7).

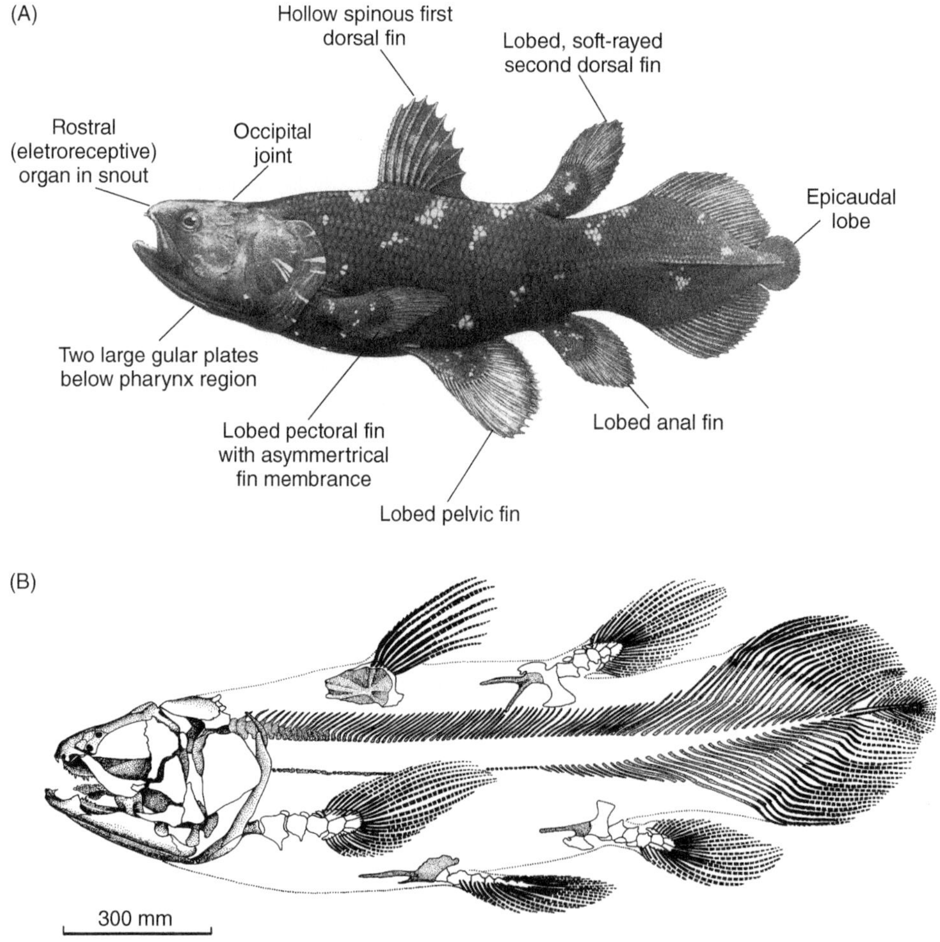

FIGURE 13.7 The living African Coelacanth, *Latimeria chalumnae*, an extant member of a group thought to be extinct for about 80 million years. Discovered in 1938, the two known species occur in small populations restricted to volcanic slopes off East Africa and northern Indonesia. Both are recognized internationally as endangered. (A) External anatomy; some traits that distinguish coelacanths from other living fishes are noted. (B) Skeletal anatomy; note skull joint, arrangement of bones of the fins, and the unconstricted notochord. (A) Drawing by S. Landry, from Musick et al. (1991); (B) from Forey (1998), used with permission.

The first and subsequent specimens also confirmed speculation about other aspects of coelacanth biology, including reproductive mode. One paleontological finding of a Jurassic species showed skeletal impressions of small coelacanths inside a larger one, suggesting that coelacanths were viviparous (Watson 1927). A later fossil indicated eggs inside a coelacanth, suggesting oviparity and implicating cannibalism in the case of Watson's specimen. More recently, dissections of a gravid female *Latimeria* have revealed 5–26 well-developed young with yolk sacs or yolk sac scars (Smith et al. 1975; Bruton et al. 1992). *Latimeria* is a lecithotrophic live-bearer: young develop in the oviducts from the largest eggs of any known bony fish (9 cm diameter, >300 g) and gain all their nutrition from the large, attached yolk sac (Fricke & Frahm 1992). Watson's original interpretation was correct.

Coelacanths are not ancestral to the tetrapods but represent an offshoot lineage within the sarcopterygians (see Chapter 11). The elpistostegalian tetrapodomorphs are the most likely ancestral group. Elpistostegalians apparently had well-developed lungs, as befits a tetrapod ancestor. In contrast, *Latimeria* has a fat-filled gas bladder that is no more than a vestigial outpocket of the gut. It is obviously used for hydrostatic control and is not a functional "lung," not surprising for a fish that lives between 100 and 250 m depth and seldom if ever ventures near the surface. The blood vessel that drains the gas bladder returns blood to the sinus venosus at the back of the heart, as in other fishes. In tetrapods, this vein carries oxygenated blood to the left side of the heart and then to the rest of the body. The coelacanth heart itself is characteristically fishlike in that it has no divisions into left and right sides. The gut has a spiral valve, also typical of primitive fishes and not found in tetrapods; the spiral valve in *Latimeria* has parallel spiral cones rather than a scroll valve as found in ancestral gnathostomes. *Latimeria* lacks internal choanae (nostrils with an excurrent opening into the roof of the mouth); tetrapods possess internal choanae.

Recent behavioral findings have further clarified our understanding of *Latimeria*'s ecology. Smith (1956) called the coelacanth "Old Four Legs," in reference to the leglike appearance of the paired fins. This led to speculation that *Latimeria* literally walked along the bottom on its pectoral and pelvic fins. Motion pictures taken from small submarines indicate that *Latimeria* almost never touches the bottom (Fricke et al. 1987, 1991b). It instead drifts in the water column with the currents, sculling with its paired fins in an alternating diagonal pattern: when the left pectoral and right pelvic fins are moved anteriorly, the right pectoral and left pelvic fins move posteriorly. This is the pattern of locomotion shown by tetrapods, and interestingly, also by the lungfish *Protopterus* when moving across the bottom with its paired fins (Greenwood 1987).

Latimeria is highly electrosensitive – as are some other primitive fishes – detecting weak electric currents via a unique series of pits and tubes in the snout called the **rostral organ**. This structure bears similarities to the enlarged ampullae of Lorenzini of sharks (Bemis & Hetherington 1982; Balon et al. 1988). During underwater observations, weak electric currents were induced in a rod placed near drifting *Latimeria*, and the fish responded by orienting in a vertical, head-down manner. As is characteristic of many nocturnally active fishes, the living coelacanth forms daytime resting aggregations, with as many as 17 fish occurring together in a single small cave. The fish have large, overlapping home ranges and return to the same caves repeatedly (Fricke et al. 1991b). These observations suggest that the electrical sense of *Latimeria* could serve not only for prey detection but also for nocturnal navigation while moving through the complex lava slopes that these fishes inhabit (Bemis & Hetherington 1982).

Coelacanths have an extensive, well-studied fossil record, dating back to the Middle Devonian. As many as 121 different species have been described, of which 83 are probably valid, constituting 24 genera and perhaps nine families (Cloutier & Forey 1991; Forey 1998). Diversity was maximal during the Early Triassic, when 16 described species existed in both marine and fresh water.

The Living Coelacanths, At Least for Now When Marjorie Courtenay-Latimer went down to the docks of East London, South Africa, to wish the crew of the trawler *Nerine* a happy Christmas, she could not have had a notion of how this friendly gesture would completely change her life and the course of twentieth century natural science. Captain Goosen had saved several fishes from his recent catch that he thought she might want for the East London Museum's collections. Included in the pile was a curious, 1.5 m long fish that was "... pale mauvy blue with iridescent silver markings. ... Was it a lungfish gone balmy?" (Courtenay-Latimer 1979, p. 7).

Ms. Courtenay-Latimer sent a rough drawing and description of the fish (Fig. 13.8) to Dr. J. L. B. Smith, a South African chemist turned ichthyologist. The Christmas mail and summer rains delayed communication between Courtenay-Latimer and Smith, and it was almost 2 weeks before a telegram arrived from Smith desperately urging Courtenay-Latimer to preserve as much of the fish as possible. Smith suspected the fish was a coelacanth, but it seemed so implausible. Unfortunately, the size of the fish, the summer heat, and bad luck conspired against them and only the skin was preserved and mounted by a taxidermist. On February 16, 1939, Smith finally managed to drive to East London and view the mount and confirm that the fish was without doubt, "scale by scale, bone by bone, fin by fin... a true Coelacanth" (Smith 1956, p. 41). Smith named the fish *Latimeria chalumnae* in honor of Ms. Courtenay-Latimer and the Chalumna River off which the fish was captured. The hunt for a second, more complete specimen began immediately.

Despite a sizable promised reward, intensive collecting efforts along much of the eastern coastline of Africa, and deep-sea trawling around the world, a second specimen was not obtained for 14 years. The second coelacanth was slightly different in that it lacked a first dorsal fin and a caudal fringe, probably having lost them to a shark. Smith erected a new genus, *Malania*, in honor of the then Prime Minister of South Africa, D. F. Malan, who loaned Smith a plane to fly to the capture locale and snatch the fish away from French authorities. As Malan was also the architect of the racial separation doctrine of apartheid in South Africa, Smith's "patronymic" was viewed as a distasteful

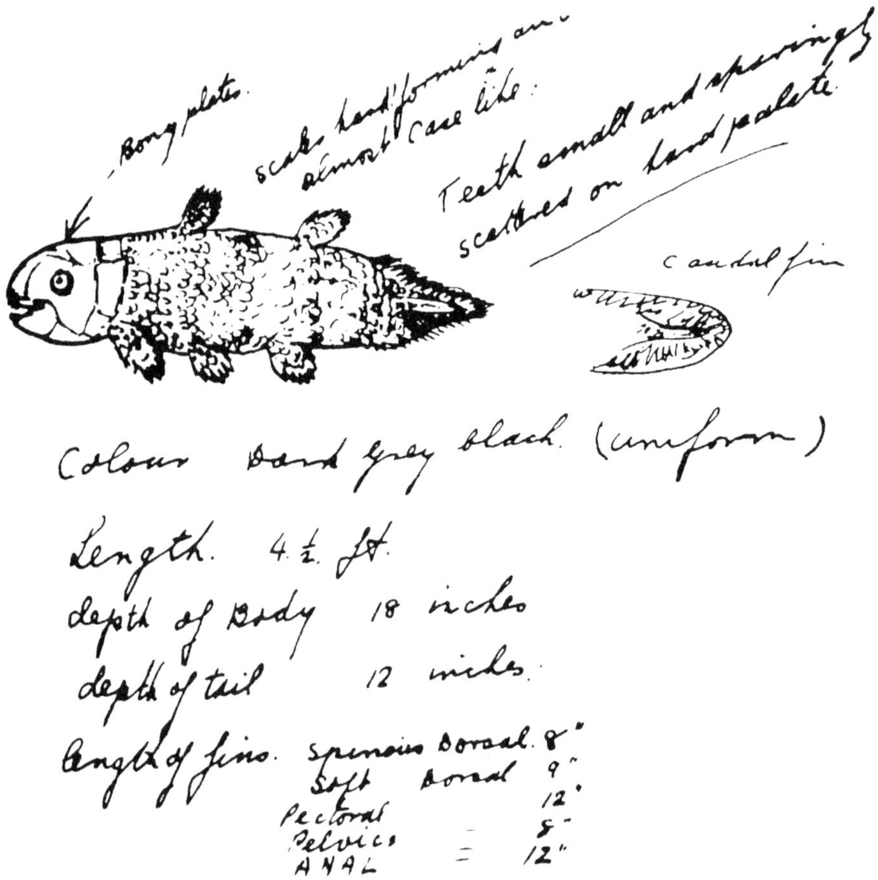
Bony plates
Teeth small and sparingly scattered on hard palate
caudal fin
Colour dark grey black (uniform)
Length 4 ½ ft
depth of body 18 inches
depth of tail 12 inches
length of fins
spinous Dorsal 8"
soft Dorsal 9"
Pectoral 12"
Pelvics 5"
ANAL 12"

FIGURE 13.8 Marjorie Courtenay-Latimer's drawing and description of the first coelacanth, as sent to J. L. B. Smith. Key features pointed out by Courtenay-Latimer included bony plates on the head and the extra median lobe in the caudal fin. From Smith (1956), used with permission.

political expediency by many outsiders. Later analysis and additional specimens confirmed that they were the same species, thus *Malania* was abolished in favor of *Latimeria* (in accordance with the principle of priority, see Chapter 2).

The second and all but a half dozen of the known 175 specimens of *L. chalumnae* have been caught off the coast of the Comoros Islands (now the Union of the Comoros), a small island group in the Indian Ocean that lies between the island of Madagascar and Mozambique in East Africa. The fish have been captured by hook-and-line off the western coasts of two islands, Grand Comoro and Anjouan. The fish are usually captured as bycatch of the fishery for Oilfish (*Ruvettus pretiosus*, Gempylidae). The coelacanth has the native name "**Gombessa**" and is not a desirable food fish (the often-cited use of the scales for roughening bicycle tire tubes is erroneous; Stobbs 1988). The fish are limited to areas of relatively recent, steep lava flows that are perforated with small caves. By day the fish rest in caves at depths between 180 and 250 m (Fricke et al. 1991b). In the evening, they move into deeper water (200–500 m) to feed on small fishes, which they capture via a suction–inhalation mechanism, much like a Giant Sea Bass (Fricke & Hissmann 1994). The relatively restricted depth range may relate to temperature preferences of 18–23°C and reflect the oxygen saturation properties of coelacanth blood, which functions poorly in warmer, less oxygen-rich surface waters (Hughes & Itazawa 1972).

Specimens range in size from 42 to 183 cm in length and weigh from 1 to 95 kg, the largest individuals being female (Bruton & Coutouvidis 1991). Age estimates indicate that coelacanths live from 20 to as much as 40–50 years (Bruton & Armstrong 1991). Females do not mature until 15 years old, and gestation may require 3 years, the longest of any known vertebrate (Froese & Palomares 2000). Intensive efforts have yet to reveal other populations around the Comoros Islands, although individual animals have been caught in trawls and gillnets off Mozambique, southern Madagascar, Kenya, and the Tanzanian coast (De Vos & Oyugi 2002; **www.dinofish.com**). An alarming 29 fish – including six in one night – were captured off Tanzania between 2003 and 2006 (Tony Ribbink, pers. comm.). In 2000, a second East African population was discovered by divers off the KwaZulu-Natal, South Africa coast (Venter et al. 2000).

Excitement surrounding the discovery, naming, and further pursuit of *Latimeria chalumnae* continued, and in September 1997, Mark and Arnaz Erdmann spotted a coelacanth in a fish market in Sulawesi, northern Indonesia, fully 10,000 km east of the Comoros locale. Targeted fishing produced another specimen in July 1998 at depths and habitat types similar to those in the Comoros (Erdmann et al. 1999), and additional fish were found to the west and southwest. While Erdmann and colleagues were engaged in a detailed anatomical and biochemical analysis, tissue samples from the 1998 specimen were literally hijacked and used in describing the Indonesian fish as a new species, *L. manadoensis* (Pouyaud et al. 1999b). Subsequent, thorough studies – by Erdmann and colleagues – confirmed the uniqueness of *L. manadoensis* (Holder et al. 1999). More recent comparisons of the mitochondrial genome of the two species indicate the lineages may have separated as long as 30–40 million years ago (Inoue et al. 2005). [Unfortunately, the Principle of Priority in the Zoological Code (see Chapter 2) does not disqualify names on account of piracy, so the Pouyaud et al. description stands as first published – see Weinberg (2000) for an account of these shenanigans and much more.]

The world took notice of *Latimeria* in a big way, perhaps too big. The hype and publicity surrounding the Comoran coelacanths have posed a serious threat to their continued existence. The total Comoran population is estimated at 200–600 individuals and was thought to be declining (Fricke et al. 1991a; Fricke & Hissmann 1994; Hissmann et al. 1998). Small clutch size and late maturation indicate a slow reproduction rate, which means individuals are replaced slowly in a population. Between 1952 and 1992, at least 173 individuals were captured, most as research and display material for museums (Bruton & Coutouvidis 1991). Unfortunately, a black market for coelacanths also developed because of the animal's freak appeal (Stobbs 1988; Bruton & Stobbs 1991). Celebrity transformed a bycatch fishery into a directed fishery; a single coelacanth was worth US$150, or about 3–5 years' income to a fisher. The fish eventually sold for $500–2000 on the open market. This directed fishery was eliminated when the Comoran government outlawed the capture of coelacanths, but incidental captures undoubtedly still occur (estimated in the 1990s at the rate of 5–10 fish per year, which could represent as much as 5% of the adult population captured annually; H. Fricke, pers. comm.).

All these circumstances – slow growth and maturation, small clutch size, limited habitat and geographic range, limited recruitment, small and perhaps decreasing population size, intense exploitation – indicate that coelacanths are particularly vulnerable and threatened by extinction. International conservation efforts were initiated: the coelacanth was listed as Critically Endangered by the International Union for the Conservation of Nature (IUCN) and placed in Appendix I of the Convention on International Trade in Endangered Species of Wild Fauna and Flora (CITES), thereby outlawing commercial trade by signatory nations. A Coelacanth Conservation Council was formed to coordinate and promote research on and conservation of coelacanths; this organization evolved into the African Coelacanth Ecosystem Programme. Efforts have also focused on providing alternative fishing methods and species for Comoran fishers (see Coelacanth Rescue Mission, **www.dinofish.com**) and to discourage ongoing, well-financed efforts at capturing live specimens for display in public aquaria.

The Coelacanth Conservation Council proposed that the coelacanth be adopted as the international symbol of aquatic conservation, equivalent to the panda's status for terrestrial conservation, because ". . . Coelacanths occupy a unique place in the consciousness of man: they represent a level of tenacity and immortality which man will never achieve during his short stay on earth" (Balon et al. 1988, p. 274) (Fig. 13.9).

(A)

(B)

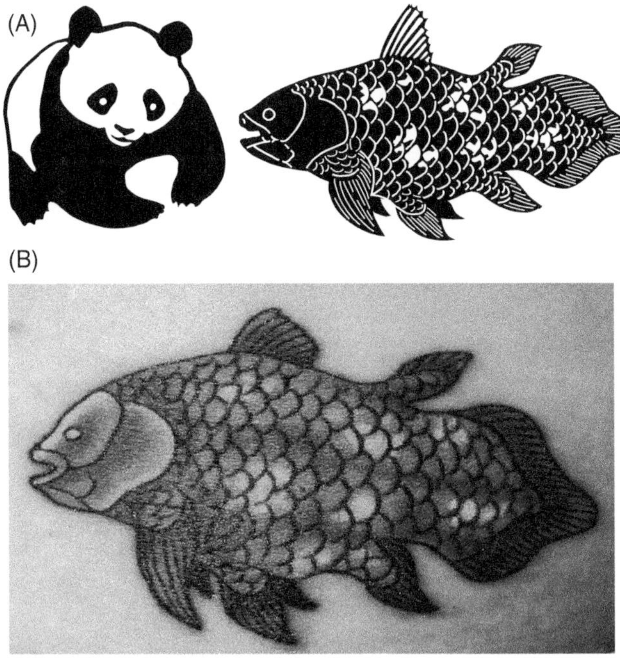

FIGURE 13.9 Coelacanths are as cuddly as pandas. (A) The Coelacanth Conservation Council's (CCC) image of a coelacanth, proposed to serve as the World Wildlife Fund's symbol for marine conservation, the panda representing terrestrial conservation. (B) An ichthyology student was moved by the plight of the coelacanth and had the CCC image tattooed on her hip. G. Helfman (Author).

Subclass Sarcopterygii, Infraclass Dipnomorpha, Superorder Dipnoi, Order Ceratodontiformes: The Lungfishes

Lungfishes, commonly referred to as "dipnoans" because of their two methods of breathing, are well represented in the fossil record on all major continents, including Antarctica. They arose early in the Devonian and were widespread and diverse until the Late Triassic. Today, they are represented by three genera that date back to the Cretaceous, with six remaining species in South America, Africa, and Australia. All extant lungfishes occupy freshwater habitats, although most of the 60 described fossil genera were marine.

Lungfishes possess a mosaic of ancestral and derived traits that initially clouded their taxonomic position (Conant 1987). The South American species, *Lepidosiren paradoxa*, reveals in its specific name some of the confusion its mixture of traits must have caused. It was first described in 1836 and thought to be a reptile because of the structure of its lung and the placement of the nostrils near the lip. An African species, *Protopterus annectens*, was discovered the next year and proclaimed to be an amphibian based on its heart structure. Both species were very different from fossil lungfishes and relationships between extant and extinct forms were not obvious. After about 30 years of debate, the systematic position of lungfishes among the Sarcopterygii was generally accepted, with recognition that lungfishes had "singularly embarrassed taxonomists" (Duvernoy 1846, in Conant 1987).

Tetrapods are now viewed by many (at least among ichthyologists) as another infraclass within the Sarcopterygii (see Nelson et al. 2016). Because lungfishes are a sister group to the tetrapods, lungfishes are phylogenetically closer to tetrapods (including us) than to other bony fishes (see Chapter 11, also Figure 1.1).

A general and distinctive characteristic of lungfishes is the existence and location of massive **toothplates**. Teeth are not attached to the jaw margins as in most other living fishes, but instead occur only on interior bones (Bemis 1987). These toothplates are often quite large and apparently function in crushing aquatic insects, crustaceans, and particularly mollusks; the toothplates are better developed in the Australian than in the South American and African species. It is the toothplates that most commonly fossilize and which form the basis of much of our understanding of evolution in the group.

The living African and South American lungfishes are placed in the families **Protopteridae** and **Lepidosirenidae,** respectively (Fig. 13.10A–C). The four African species of the genus *Protopterus* are widely distributed through Central and

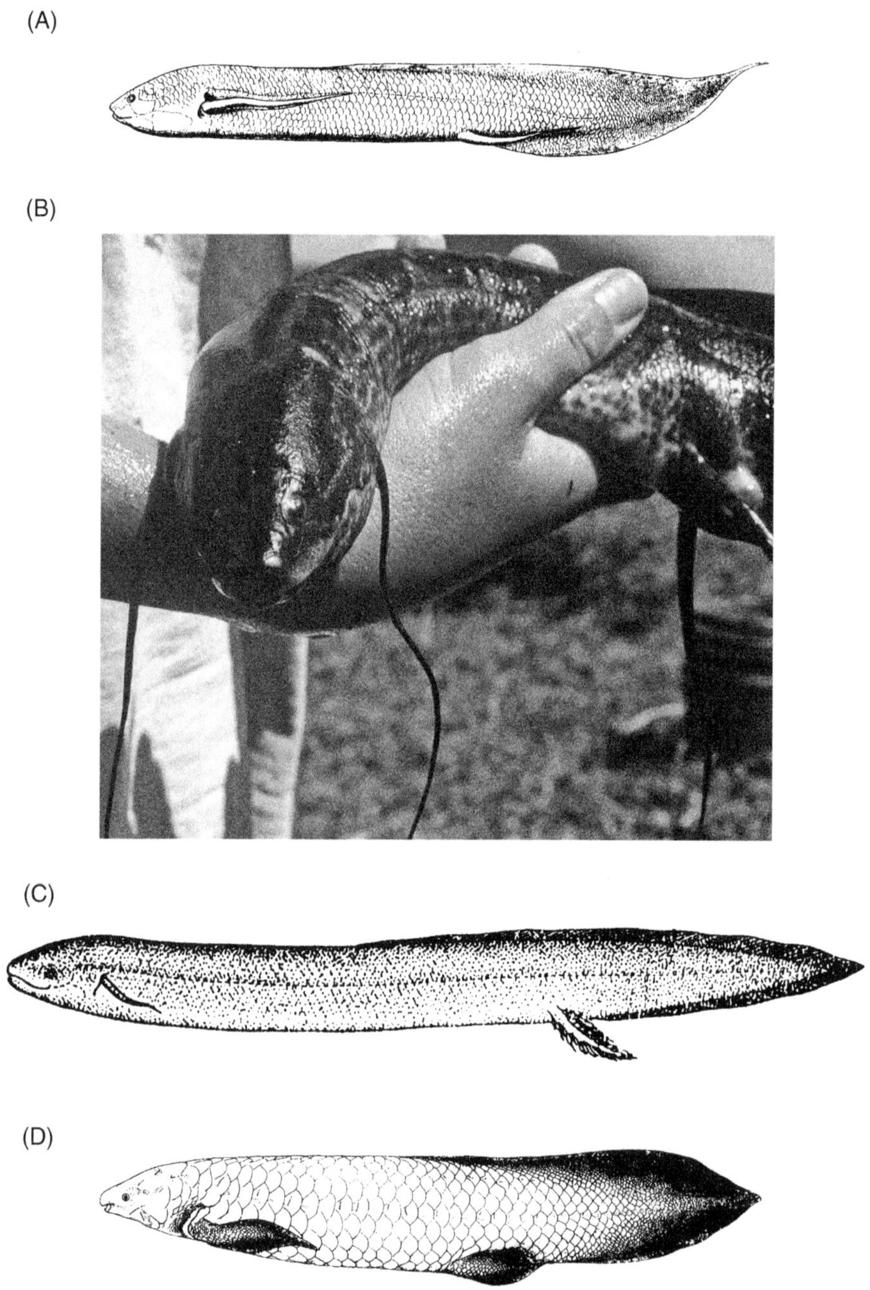

FIGURE 13.10 Modern lungfishes. (A) An African lungfish, *Protopterus annectens*, one of four species in the genus. (B) A live *Protopterus*; note the filamentous pectoral and pelvic fins. (C) The South American Lungfish, *Lepidosiren paradoxa*, showing the vascularized pelvic fins that develop on males during the breeding season. (D) The Australian lungfish, *Neoceratodus forsteri*. (A, D) From Jarvik 1980; (B) Courtesy of L. and C. Chapman; (C) from Norman (1931), used with permission.

South Africa, occurring in both lentic (still) and lotic (flowing) habitats of major river systems, including a variety of swamp habitats (Greenwood 1987). Maximum sizes range from 44 cm (*Protopterus amphibius*) to 180 cm (*P. aethiopicus*). Young fishes are active by night, adults by day, and food includes a variety of hard-bodied invertebrate taxa, with mollusks predominating (Bemis 1987). Protopterids are obligate air breathers throughout their post-juvenile life, obtaining 90% of their oxygen uptake via the pulmonary route.

African lungfishes are best known for their ability to survive desiccation of their habitats during the African dry season. Such **estivation** behavior, as described for *P. annectens*, involves construction of a subterranean mud cocoon (Greenwood 1987). As water levels fall, the lungfish constructs a vertical burrow by biting mouthfuls of mud from the bottom, digging as deep as 25 cm into the mud. As the swamp dries, the lungfish ceases taking breaths from the water surface, coils up in the burrow with its head pointing upward, and fills the chamber with secreted mucus. This mucus dries, forming a closely fitting cocoon, and the fish becomes dormant (Fig. 13.11). This dormant period normally lasts 7 or 8 months, but can be extended experimentally for as much as 4 years in *P. aethiopicus*. During estivation, lungfish rely entirely on air breathing, the heart rate drops, they retain high concentrations of urea and other metabolites in the body tissues, metabolize body proteins, and lose weight. With the return of rains, the lungfish emerges from the burrow and resumes activity, which includes cannibalizing smaller lungfish that have also just emerged from their burrows.

African lungfishes build burrow-shaped nests, often tunnelling into the swamp bottom or bank. Eggs and young are guarded by one parent, presumably the male. The male has no specialized structures to aid in oxygenating the water in the nest as reported for *Lepidosiren* (see below), although male *P. annectens* have been observed "tail lashing" near the nest, which may serve the same purpose. Young African lungfishes have external gills (Fig. 13.12), one of the traits that caused many nineteenth century biologists to consider them amphibians.

The South American Lungfish, *L. paradoxa* (Fig. 13.10C), is considered to be the most recently derived member of the family (Greenwood 1987). Surprisingly, little is known of its natural history compared to the African and Australian species. It occurs in swampy regions of the Amazon and Parana river basins (Thomson 1969a) and grows to about 1 m in length. As with *Protopterus*, adults have reduced gills and are obligate air breathers. Estivation in burrows occurs but is poorly documented. *Lepidosiren* is best known for its reproductive behavior, although conjecture exceeds information. Eggs are deposited in a burrow nest as in *Protopterus* and guarded by the male. During the breeding season, after egg deposition, the male's pelvic fins develop vascularized filaments, in apparent response to increased testosterone levels (Cunningham & Reid 1932; Urist 1973). These sexually dimorphic structures are purportedly used to supplement the respiratory needs of the young in the burrow, although actual behaviors and measurements during breeding have yet to be detailed. Young

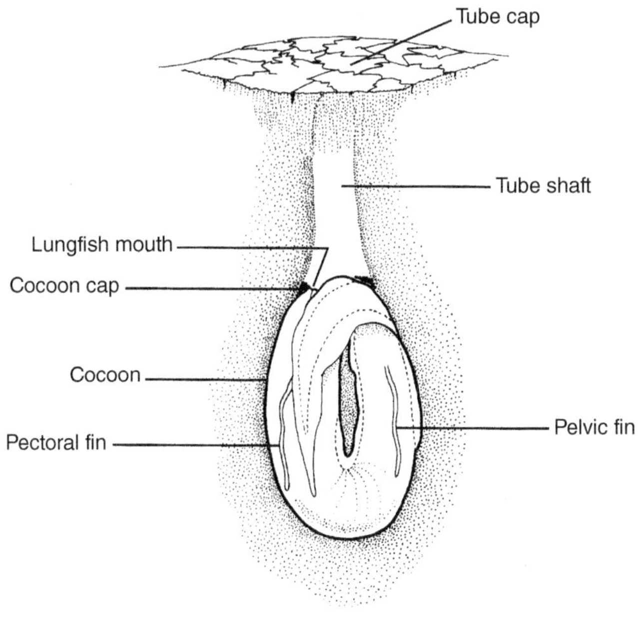

FIGURE 13.11 An African lungfish estivating in its mud and mucus cocoon, viewed from the ventral surface of the fish. Redrawn from Greenwood (1987).

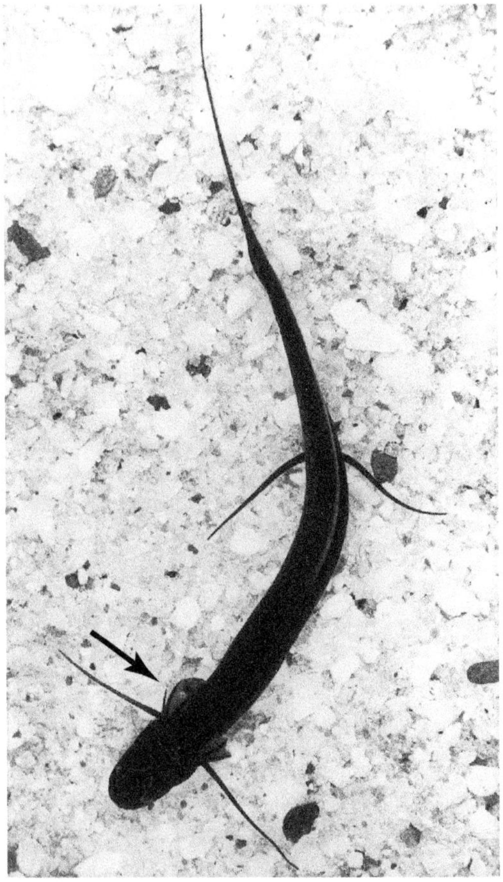

FIGURE 13.12 A young African lungfish. The arrow indicates the external gills that misleadingly caused lungfishes to be classified as amphibians. Herald 1961 / with permission of Chanticleer Press, Inc.

lepidosirenid lungfishes are not obligate air breathers, a trait that may reduce their exposure to a variety of predators while they are small and exceedingly vulnerable.

The Australian species, *Neoceratodus forsteri* (Fig. 13.10D) was the last lungfish to be described scientifically, in 1870. It has a very limited native distribution, restricted primarily to the Burnett, Fitzroy, and Mary river systems of Northeastern Australia, with transplanted populations in the Brisbane River and several small reservoirs (Kemp 1987; Pusey et al. 2004). Among living lungfishes, *Neoceratodus* is closest to the ancestral forms in many anatomical respects, including a large (up to 150 cm long), relatively stout body (to 20 kg); large cycloid scales covering the entire body; flipperlike "archipterygial" fins; pectoral fins inserted low on the body; a broad diphycercal tail; and a single lung. Fossilized toothplates undistinguishable from those belonging to *N. forsteri* have been found in Early Cretaceous deposits of New South Wales, indicating the species is at least 140 million years old. This makes *Neoceratodus* not only the oldest living lungfish but perhaps "the world's oldest living vertebrate species" (Pusey et al. 2004, p. 59).

Neoceratodus feeds in the late afternoon and maintains activity during the night, capturing benthic crustaceans, mollusks, and small fishes that it crushes with its distinctive toothplates. It is able to locate live animals by detecting the electric field emitted by the prey, adding it to the list of primitive fishes that are highly electrosensitive (Watt et al. 1999). Unlike the South American and African species, *Neoceratodus* is a **facultative air breather** that relies on gill respiration under normal circumstances. Its lung may serve more as a hydrostatic than a respiratory organ (Thomson 1969a). Uptake of oxygen through the skin occurs, at least in juveniles. No special adaptations to avoid desiccation have been observed, and the fish must be kept moist and covered by wet vegetation or mud to survive out of water.

Sexes show only slight dimorphic coloration during the breeding season and are otherwise indistinguishable. In the native riverine habitat, spawning is rather unspecialized, involving deposition of eggs on aquatic plants in clean, flowing water at any time of the day or night. Fish spawn in pairs, when females deposit 50–100 eggs per spawning; no parental guarding occurs. Development of the young is direct and gradual, with no obvious larval stages or distinct metamorphosis. Young are born without external gills. Maturation does not occur until fish are 15–20 years old, and specimens in captivity in public aquaria have lived at least 65–70 years.

As a result of spawning and nursery habitat destruction brought about by impoundment construction, pollution, and perhaps interactions with introduced species, *Neoceratodus* populations have declined in some areas. The lungfish was granted Vulnerable status in 2003 under Australia's federal Environment Protection and Biodiversity Conservation Act. Fish have been transplanted into several Queensland rivers and reservoirs to aid the species' recovery.

A Seventh Lungfish? No discussion of extant lungfishes would be complete without at least brief mention of *Ompax spatuloides*, initially described as another lungfish in Australia (Fig. 13.13). *Ompax* was described based on a 45 cm specimen served to the director of the Brisbane Museum during a trip to northern Queensland in 1872, 2 years after the scientific discovery of the first Australian lungfish. The fish was reputed to occur with *Neoceratodus* in a single water hole in the Burnett River. It had a body covered with large ganoid scales, small pectorals, and an elongate, depressed snout, "very much the form of the beak of the *Platypus*" (Castelnau 1879, p. 164). The Director had a sketch made of the fish, but ate it nonetheless. The sketch and notes were sent to a prominent regional ichthyologist, Count F. de Castelnau, who described the species and speculated that it was most closely related to the gars of North America. *Ompax* appeared in Australian faunal lists as a ceratodontid for 50 years, even though a second specimen was never found. Finally, in 1930, an anonymous report appeared

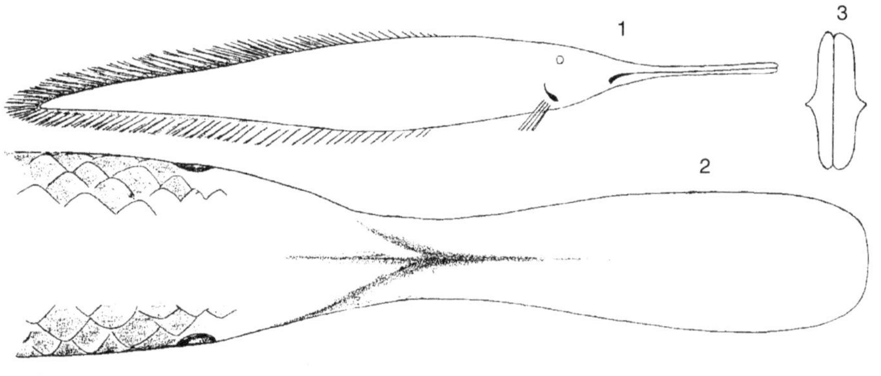

FIGURE 13.13 A seventh lungfish, *Ompax spatuloides*? This is the illustration that appeared in the original species description by Castelnau (1879). It shows (1) lateral view, (2) dorsal view of the head, and (3) presumably a cross-section of the bill, but unlabelled in the original illustration. Castelnau (1879) / Proceedings of the Linnean Society of New South Wales / Public Domain. An anonyomous report in 1930 revealed that this had been fabricated from the nose of a platypus, the head of a lungfish, the body of a mullet, and the tail of an eel.

in a Sydney newspaper recounting how the specimen was in fact a ruse and had been fabricated from the nose of a platypus, the head of a lungfish, the body of a mullet, and the tail of an eel (Herald 1961).

Subclass Actinopterygii, Infraclass Cladistia, Order Polypteriformes: Bichirs and Reedfish

Taxonomic relationships among and within most relict groups, both in terms of affinities with other living fishes and identification of ancestral lineages, are reasonably well understood. Lungfishes, coelacanths, chondrosteans, gars, and Bowfin all have well-defined, relatively extensive fossil records with which modern species can be associated. In addition, derived traits are either unique to a group or shared with other groups in ways that confirm evolutionary hypotheses of relationship (Table 13.1). Although healthy debate on the details of relationship among these fishes exists, most researchers agree on the general patterns of interrelatedness.

The **cladistians** stand out as an exception to this pattern of consensus. Over the years, workers have cited anatomical similarities to justify placing them variously with lungfishes, closer to the stemline sarcopterygians, or squarely among the Actinopterygii as another chondrostean (Patterson 1982). Other taxonomists emphasized unique characteristics and placed them in their own subclass, the Brachiopterygii. But fossil discoveries in Middle–Upper Cretaceous deposits of southeastern Morocco established definitive polypterid lineages at least as far back as 91–95 mybp (Dutheil 1999). Admittedly, this is still relatively recent for what is thought to be the basal actinopterygian group (i.e. chondrosteans, thought to have arisen later, have a fossil record that goes back to the Devonian; see Chapter 11).

Modern cladistians are represented by two genera confined to west and central tropical Africa, including the Congo and Nile river basins (Fig. 13.14). Fifteen species, referred to as **bichirs** (pronounced bih-shéars), belong to the genus *Polypterus*; the remaining species is the Reedfish or Ropefish,

TABLE 13.1 Characteristics of extant relict fishes. Presence (+) or absence (−) of a trait, or its condition, is indicated in the body of the table. Shared characteristics among unrelated forms are strong evidence of convergent evolution, since these groups have long histories of demonstrated, separate evolution.

Trait	Lungfishes		Coelacanths	Chondrosteans		Polypterids	Gars	Bowfin
	Australian	S. Am./Af.		Sturgeons	Paddlefishes			
Scales	Cycloid	Cycloid	Cycloid[a]	Scutes[b]	–[c]	Ganoid	Ganoid	Cycloid[d]
Gular plates	–	–	2	–	–	2	–	1
Spiracle	–	–	–	+	+?	+	–	–
Larva ext gills	–	+	–	–	–	+	–	–
Lungs[e]	Sing vent	Dbl vent	Fatfill gb	Dorsal gb	Dorsal gb	Dbl vent	Vasc gb	Vasc gb
Spiral valve	+	+	+	+	+	+	+ (remnant)	+ (remnant)
Centra	–	–	–	–	–	+	+[f]	+
Tail	Diphy	Diphy	Diphy	Hetero	Hetero	Hetero[g]	Abb hetero	Abb hetero
Lobed fins	+	–	+[h]	–	–	+	–	–
Electroreceptors	+	+	+	+	+	+	–	–
Chromosome 2N	54	38/34	48[i]	112	120	36	68	46

abb, abbreviate; Af., Africa; dbl, double; diphy, diphycercal; ext, external; hetero, heterocercal; gb, gas bladder; sing, single; S. Am., South America; Vasc, vascularized; vent, ventral.

[a] Coelacanths are sometimes said to have cosmoid scales, however no extant fishes have scales containing cosmine (Jarvik 1980).

[b] Sturgeons have five longitudinal rows of bony scutes, plus “dermal ossifications” scattered around the body (Vladykov & Greeley 1963, p. 25). These scutes contain ganoin and could be considered ganoid.

[c] Paddlefishes are mostly naked, with four types of scales (fulcral, rhomboid, round-based, and denticular) scattered on the head, trunk, and tail; the histology of these scales is unclear. Trunk scales are more abundant on *Psephurus* than on *Polyodon* (Grande & Bemis 1991).

[d] Bowfin “cycloid” scales are convergent not homologous with those of teleosts (Grande & Bemis 1998).

[e] Outpocketings of the esophagus are gas bladders, but are often called lungs when their primary function is breathing atmospheric air.

[f] Gar centra are opisthocoelous (concave on rear face, convex on front).

[g] Lower lobe of brachiopterygian tail created by rays coming off the ventral surface of the notochord.

[h] Coelacanth fin bases are lobed except for first dorsal.

[i] Coelacanth chromosomes are more like those of ancient frogs than of other sarcopterygians such as lungfishes (Bogart et al. 1994).

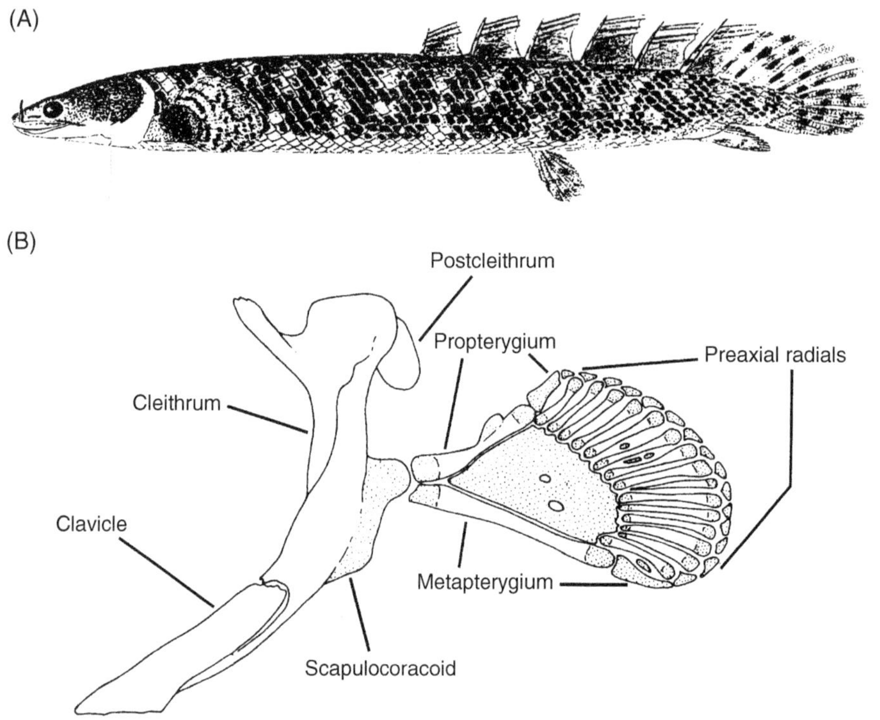

FIGURE 13.14 (A) A 29 cm long bichir, *Polypterus palmas polli*, from the Ivory Coast. Note the lobe-like pectoral fin base and the horizontal flaglike fin rays that extend from the distal portion of each dorsal fin spine. (B) The "peculiar and overelaborated" pectoral fin of a bichir, showing the wishbone-like basal structure (propterygium and metapterygium) that supports the radials and fin rays. (A) From Hanssens et al. (1995), used with permission; (B) from Rosen et al. (1981), courtesy of the Department of Library Services, American Museum of Natural History.

Erpetoichthyes (formerly *Calamoichthyes*) *calabaricus*. Bichirs and Reedfish grow to 90 cm, although most bichir species are shorter. All are predatory and inhabit shallow, vegetated, and swampy portions of lakes and rivers.

In poorly oxygenated water, bichirs are obligate air breathers and up to 93% of their air breaths come from spiracular breathing (Graham et al. 2014). They will drown if denied access to the surface. Bichirs are unique in that they inhale through their mouths and exhale spent air from the highly vascularized and invaginated lungs through their dorsally placed spiracles; the spiracles serve no apparent aquatic respiratory function (Abdel Magid 1966, 1967). Polypterids are additionally unique in that they inhale by **recoil aspiration** (Brainerd et al. 1989), using the elastic energy stored in their integumentary scale jacket during exhalation to power inhalation of atmospheric air. The existence of similar bony scale rows in some Paleozoic amphibians suggests that the evolution of air breathing and perhaps eventual terrestriality may be linked to recoil aspiration that originated in fishes.

Controversy over taxonomic position arises because brachiopterygians exhibit superficial anatomical traits that have been used to justify their inclusion in almost every one of the major taxa discussed in this chapter (see Table 13.1). Cladistians possess lobe-like fins (a sarcopterygian trait), ganoid scales (a palaeoniscoid or lepisosteiform trait), two gular plates (as does the coelacanth), spiracles (in common with sturgeons), feathery external gills when young and double ventral lungs (in common with lepidosirenid lungfishes), a modified heterocercal tail (as in gars and Bowfin), and a spiral valve intestine (shared by all major groups).

However, the internal structure of many of these seemingly shared primitive characteristics is very different from those of other taxa, indicating convergence on the traits and not homology. The external gills are only analogous to, not homologous with, the gills of young lungfishes. The tail is heterocercal in structure but symmetrical in external appearance; the medial and lower portions are created by rays coming off the ventral surface of the notochord, unlike any other fishes. Confusion often arises whenever we attempt to compare among living fishes, each well adapted to environmental conditions of the recent past. Many anatomical traits, including those most critical to systematic analyses, are retained from ancestors, whereas other traits represent recent derivations that have evolved in response to conditions greatly changed from the ancestral selection pressures. Hence we have the existence of a mosaic of primitive and derived traits in every living species, homology and analogy intertwined, with difficulty in knowing the proportions of the two. Each attempt at linking a trait in cladistians with a counterpart trait in another group becomes a possible apples-and-oranges comparison.

Cladistian Autapomorphic and Synapomorphic Traits The bichirs have a remarkable number of autapomorphic (unique, derived) traits. Their median and paired fins are unlike those of any other major taxon. Bichirs are also referred to as "flagfins" because the 5–18 dorsal finlets each consist of a vertical spine to which are attached horizontal rays, giving them a "flag and pole" appearance. In all other ray-finned fishes, the dorsal fin rays emerge as vertical bony elements from the body of the fish. The pectoral fin is lobe-shaped but constructed differently from the lobe fins of lungfishes and crossopterygians, or for that matter, any other fish, living or extinct. The supporting structures of the pectoral fin are shaped like a wishbone with a flat plate (Fig. 13.14B). A. S. Romer, a leader of modern vertebrate paleontology, referred to the polypterid pectoral fin as a "peculiar and over-elaborated development" (Romer 1962, p. 198).

Other apparent autapomorphic traits include relatively few and small chromosomes (Denton & Howell 1973); the structure, arrangement, replacement, and differentiation of teeth (Wacker et al. 2001); and possession of only four rather than five gill arches, the fifth having been lost (Britz & Johnson 2003). And recoil aspiration breathing is performed by no other known extant group.

The placement of cladistians at the base of bony fish phylogeny is justified by a number of derived traits shared with the rest of the Actinopterygii. These characters include egg structure, nuclear DNA-coded genes, *Hox-A* gene sequences, mitochondrial DNA and amino acid sequences, and cranial skeleton morphology and function (Bartsch 1997; Bartsch & Britz 1997; Venkatesh et al. 2001; Chiu et al. 2004; Kikugawa et al. 2004). These and other synapomorphies make the Cladistia "the sister group of all other actinopterygians" (Nelson 2006, p. 88), rather than a sarcopterygian or a chondrostean, settling the issue.

Subclass Actinopterygii, Infraclass Chondrostei, Order Acipenseriformes: Sturgeons and Paddlefishes

Although considered primitive actinopterygians, the extant acipenseriform sturgeons and paddlefishes are highly derived, relict species that bear little resemblance to ancestral chondrosteans. The two families probably diverged from each other during the Jurassic, but they still share a number of characteristics such as a cartilaginous skeleton, heterocercal tail, reduced squamation, more fin rays than supporting skeletal elements, unique jaw suspension, and a spiral valve intestine. Although largely cartilaginous, their skeletons are secondarily so: their probable Early Mesozoic ancestors (palaeoniscoids) were bony.

Acipenseridae There are over 25 species of sturgeons, all in the family Acipenseridae and restricted to the northern hemisphere (Binkowski & Doroshov 1985; Williams & Clemmer 1991; Bemis et al. 1997; Vecsei 2001; Van Winkle et al. 2002). Four genera are recognized, *Acipenser*, *Huso*, *Scaphirhynchus*, and *Pseudoscaphirhynchus*. All species spawn in fresh water, although some species move seasonally between marine and fresh water and some are technically anadromous. Species restricted to fresh water include the North American Lake Sturgeon (*Acipenser fulvescens*) and three river sturgeons (*Scaphirhynchus* spp.), the latter occurring only in larger rivers such as the Mississippi and Missouri. Anadromous species, those spending part of their lives at sea but returning to fresh water to spawn, include the Atlantic Sturgeon, *Acipenser oxyrhynchus*, the White Sturgeon, *A. transmontanus* (the largest North American freshwater fish, attaining a length of 3.6 m and a weight of 800 kg), and the Beluga of eastern Europe and Asia, *Huso huso* (the largest and economically most valuable freshwater fish in the world, attaining a length of 8.6 m and a weight of 1300 kg, and not to be confused with the toothed whale of the same common name). As with other anadromous species (discussed in Chapter 18), landlocked populations of sturgeons can develop.

Anatomically, sturgeons can be identified by the four barbels in front of the ventrally located mouth, five rows of bony scutes (large bony shields) on a body otherwise covered with minute ossifications, a heterocercal tail, elongate snout, a single dorsal fin situated near the tail, no branchiostegal rays, and a largely cartilaginous endoskeleton, including an unconstricted notochord (Fig. 13.15). Although generally slow-swimming feeders on benthic invertebrates, the protrusible mouth can be extended very rapidly, allowing larger individuals to feed on fishes (Carroll & Wainwright 2003). Vision plays at best a minimal role in prey detection, with touch, chemoreception, and probably electrolocation via rostral ampullary organs being more important (Buddington & Christofferson 1985; Gibbs & Northcutt 2004).

The life history traits of sturgeon make them unique and susceptible to overexploitation by humans. They are exceptionally long-lived: Beluga have been aged at 118 years, and White Sturgeon at 70–80 years (Nikol'skii 1961; Scott & Crossman 1973; Casteel 1976). As is often the case with long-lived vertebrates, sexual maturity is attained slowly. In the Atlantic Sturgeon, both sexes mature after 5–30 years, the older ages characterizing individuals at higher latitudes. After maturation, females may only spawn every 3–5 years (Smith 1985b), and even longer intervals may characterize other sturgeon species. Fecundity is relatively high: ovaries may account for 25% of the body mass of a female, making a large female exceedingly valuable. A female Beluga captured in 1924 from the Tikhaya Sosna River of Russia weighed 1227 kg and yielded 245 kg of caviar (**www.guinnessworldrecords.com**). High-grade caviar can sell for more than US$150/oz or $5000/kg, making that fish potentially worth in excess of $1 million.

Sturgeon are also commercially valuable as a smoked product, and the gas bladder was processed into isinglass and used for gelatin, clarifying agents, and as a commercial art glue. Natural predators beyond the juvenile stage are rare; parasitic lampreys are one of the few organisms capable of attacking

(A)

(B)

FIGURE 13.15 Sturgeons. (A) An Atlantic Sturgeon, *Acipenser oxyrhynchus*. Note the rows of bony scutes on the body, distinct heterocercal tail, and elongate snout with barbels preceding the ventral mouth. (B) A Lake Sturgeon is released in the Niagara River by USFWS Fisheries worker. (A) Artwork by D. Raver/ USFWS / Public Domain. (B) B. Billings / USFWS / Public Domain.

an adult sturgeon (Scott & Crossman 1973). Hence natural mortality rates of adults were historically low, creating vulnerability when such species are subjected to the high mortality rates associated with commercial exploitation.

It is therefore not surprising that sturgeons worldwide have declined due to overexploitation, dam building, habitat destruction, and pollution. Large Atlantic Sturgeon were at one time sufficiently abundant in North American coastal rivers that navigation by canoes and small boats was sometimes hazardous, particularly given the fish's habit of leaping 1–2 m out of the water. Commercial landings exceeded 3 million kg annually in 1890, but 100 years later, landings were reduced by 99% (Smith 1985b).

Lake Sturgeon have been extirpated from a large part of their native range (ironically, Lake Sturgeon disappeared from the Sturgeon Falls area of the Menominee River, Wisconsin around 1969; Thuemler 1985). The Shortnose Sturgeon *Acipenser brevirostrum* of North America and the Atlantic Sturgeon *Acipenser sturio* are both listed in the CITES 2022 Appendix I, and all other *Acipenser spp.* are listed in Appendix II (**https://cites.org/eng/app/appendices.php**). Internationally in 2000, nine sturgeon stocks or subspecies were considered Critically Endangered, 25 Endangered, and 13 Vulnerable (**www.redlist.org**). Although sturgeon fishing is highly regulated nationally and internationally, high economic values have promoted rampant poaching and black markets, at the same time that fishery management and enforcement programs have collapsed (Vecsei 2005; Helfman 2007).

Part of the vulnerability of sturgeons results from an interaction between habitat degradation and the reproductive biology of these large, slow maturing fishes. Spawning is hampered by siltation and contamination of clean gravel and rock areas, and by dam construction that blocks migrations and limits access to spawning habitat. The spawning period in several species may be very short, on the order of 3–5 days, and if environmental conditions are inappropriate, spawning may be abandoned for that year (Buckley & Kynard 1985; Williot et al. 2002). Recruitment of new fish into the population is further prevented by overharvest of mature individuals and also of fish before they reach reproductive age (sometimes as a result of incidental bycatch of juveniles in gillnets set for anadromous shad or salmon). Given late maturation and the infrequency of spawning, stocks driven to low numbers have a difficult time recovering, requiring extreme management solutions and justifying captive propagation of many species (Binkowski & Doroshov 1985; Billard & Lecointre 2001; Pikitch et al. 2005).

Acipenseroid fishes are generally regarded as highly modified descendants of palaeoniscoids that lived during the Permian and Triassic. Recognizable acipenseriforms have been found in Permian deposits in China (Lu et al. 2005), and early, recognizable sturgeon fossils date to the Upper Cretaceous of Montana (Wilimovsky 1956; Choudhury & Dick 1998). A related, extinct family, the Chondrosteidae, is known from fossils from the Lower Jurassic to Lower Cretaceous periods.

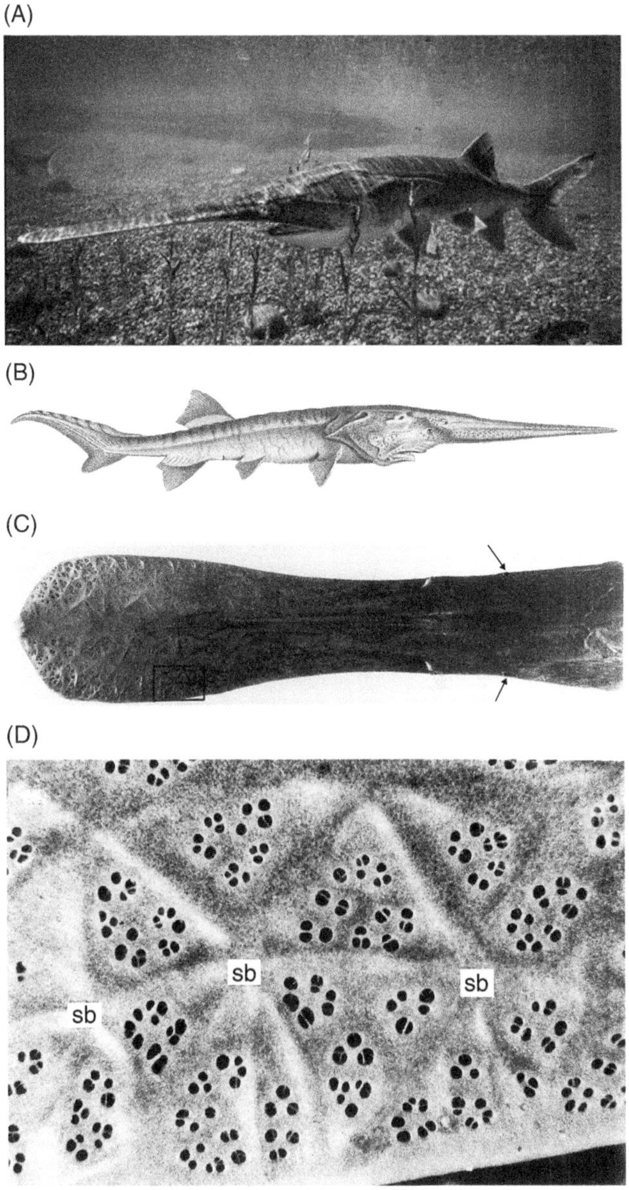

FIGURE 13.16 Paddlefishes. (A) The North American Paddlefish, *Polyodon spathula*. (B) The Chinese Paddlefish, *Psephurus gladius*, a poorly known and now probably extinct chondrostean that had been restricted to the Yangtze River system of China. (C) The rostral paddle of the North American Paddlefish in dorsal view; arrows indicate position of the eyes. (D) Area at lower left of (C) enlarged, showing the stellate bones (sb) that support the paddle, and the ampullary organs, which are the dark circular holes in the paddle that reportedly serve as electroreceptors. (A) U.S. Department of the Interior / Public Domain; (B) Wikimedia / Public Domain; (C, D) Grande and Bemis 1991 / with permission of Taylor & Francis.

Polyodontidae

Paddlefishes also date back at least to the Early Cretaceous (Grande et al. 2002), but only two species remain, the Paddlefish of North America, *Polyodon spathula*, and the Chinese Paddlefish, *Psephurus gladius* (Fig. 13.16A, B). They have larvae similar to those of sturgeons and retain the heterocercal tail, unconstricted notochord, largely cartilaginous endoskeleton (with ossified head bones), spiracle, spiral valve intestine, and two small barbels. They differ from the acipenserids in most other respects. Paddlefishes have no bony scutes and the body is essentially naked except for patches of minute scales. In addition, Paddlefishes are not benthic swimmers but instead move through the open waters of large, free-flowing rivers, feeding on zooplankton or fishes.

The North American Paddlefish, or spoonbill cat, prefers rivers with abundant zooplankton. Adult Paddlefish typically swim through the water both day and night with the nonprotrusible mouth open, straining zooplankton and aquatic insect larvae indiscriminately through the numerous, fine gill rakers. Food size is limited by gill raker spacing, as small zooplankters escape the mechanical sieve of the Paddlefish's mouth (Rosen & Hales 1981). This picture of the Paddlefish as a passive filterer is confused by the occasional benthic and water column fishes, such as darters and shad, found in its stomach (Carlander 1969). Small juveniles, in which neither gill rakers nor the paddle are well developed, pick individual zooplankters out of the water column.

The **rostral paddle**, which accounts for one-third of the body length in adults, has abundant ampullary receptors on the surface of the paddle which detect biologically generated electricity (Fig. 13.16C, D). Paddlefish, especially juveniles, use the ampullary receptors to detect weak electric fields created by zooplankton such as water fleas (*Daphnia*) from distances of up to 9 cm (Wilkens et al. 2002). The paddlefish rostrum is therefore equivalent to "an electrical antenna, enabling the fish to accurately detect and capture its planktonic food in turbid river environments where vision is severely limited" (Wilkens et al. 1997, p. 1723).

North American Paddlefish may live for 30 years and attain 2.2 m length and 83 kg mass, although fish of this size are now exceedingly rare. Diminishing populations are evidenced by changes in the species' range. Although currently restricted to the Mississippi River drainage system, populations of Paddlefish historically occurred in the Laurentian Great Lakes and have been extirpated from at least four states (Gengerke 1986). Causes of population decline are similar to those affecting sturgeon. Paddlefish are long-lived but do not mature until they are 7–9 (males) or 10–12 (females) years old, and then spawn only at 2–5-year intervals. Loss of spawning habitat, which is fast-flowing water with clean, gravel bottoms, is a major problem. Appropriate spawning areas are degraded by damming, which decreases water flow and leads to siltation. Paddlefish are sought commercially and recreationally for their flesh and eggs; overfishing has been frequently implicated in population declines (Russell 1986). Reservoirs created by dams are productive feeding habitats for adults but do not provide appropriate spawning areas. Although not federally protected in the USA, all US states along the Missouri River have prohibited commercial fishing for it (Graham 1997b). The species has been listed in Appendix II of CITES, thus providing a mechanism to curtail overfishing and illegal trade, especially of Paddlefish caviar (Jennings & Zigler 2000).

The Chinese Paddlefish, *Psephurus gladius* (see Fig. 13.16B), had been exceedingly rare and is now probably extinct. It is the more primitive of the two species and differs primarily in head and jaw morphology and body size. The paddle is narrow and more pointed, not broad and rounded. *Psephurus* also has fewer but thicker gill rakers that resemble those of sturgeons, a protrusible mouth, and grows larger (over 3 m and 500 kg, erroneously reported to 7 m). It inhabited the Yangtze River system of central China and fed primarily on small, water column and benthic fishes (Nichols 1943; Nikol'skii 1961; Liu & Zeng 1988). Historically it also occurred in the Yellow River. Relatively little is known about its biology, including spawning habits, locales, or habitat (Liu & Zeng 1988; Grande & Bemis 1991; Birstein & Bemis 1995; Wei et al. 1997).

Psephurus was highly prized for its caviar but was considered the most endangered fish in China because of overfishing, habitat destruction, and dam construction that blocked adults from reaching spawning grounds. It was probably anadromous, adults moving upriver to spawn and juveniles moving down to the East China Sea to grow. Gezhouba Dam on the Yangtze, completed in 1981, essentially cut the Paddlefish's habitat in half and blocked spawning migrations. The species has had full protection in China since 1983 but no recruitment to the population was thought to be occurring, and fewer than 10 adult Paddlefish had been found annually below the dam from 1988 to 1994 (Wei et al. 1997). The massive Three Gorges Dam, completed in 2006 (with other aspects of the project completed in 2009), was expected to drive the species into extinction (Fu et al. 2003). The last live specimen was seen in 2003, and the species likely went extinct between 2005 and 2010 (Zhang et al. 2019).

The fossil record for polyodontids is limited to four known species and some fragments, the most primitive from the Lower Cretaceous of China and the others from the Upper Cretaceous, Paleocene, and Eocene of North America (Grande & Bemis 1991; Grande et al. 2002). The jaws and gill arches of the oldest species, *Protopsephurus liui*, resemble those of *Psephurus*, indicating that piscivory is the ancestral condition and that planktivory as observed in *Polyodon* is a derived trait (Grande et al. 2002).

Subclass Actinopterygii, Infraclass Holostei: The Gars and Bowfin

The gars and Bowfin are descendants of the palaeoniscoids that dominated fresh and marine waters for 200 million years from the Mid Devonian into the Mesozoic Era. Gars and Bowfin are considered members of the infraclass **Holostei**, which also includes a variety of extinct fishes. Some analyses concluded that holosteans are paraphyletic, more a grade of development than a true clade, but Grande (2010) once again advocated recognition of the Holostei. The two modern groups differ in many important respects and their relationships to the palaeoniscoids, and position in the lineage leading to modern teleosts, are a matter of discussion. Both groups are considered neopterygian because of shared jaw, tail, and dermal armor characteristics. Most workers consider gars to be more primitive and place them in their own division (**Ginglymodi**), but some view *Amia* (division **Halecomorphi**), as the more primitive group (Normark et al. 1991; Olsen & McCune 1991; Grande & Bemis 1998).

Division Ginglymodi, Order Lepisosteiformes

All seven species of living gars are in the family **Lepisosteidae**, four in the genus *Lepisosteus* and three in *Atractosteus* (Fig. 13.17). These elongate, predatory fishes are restricted to North and Central America and Cuba; five species occur east of the Rocky Mountains in North America, the remaining species occur in Central America. Gars typically inhabit backwater areas of lakes and rivers, such as oxbows and bayous. Oxygen tension in such habitats is often low and gars must breathe atmospheric air at these times, using their compartmentalized, highly vascularized gas bladder as a lung (Smatresk & Cameron 1982; Smith & Kramer 1986).

Gars have entirely ossified skeletons. Their primitiveness is evident in their hinged, diamond-shaped, interlocking ganoid-like scales and abbreviate heterocercal caudal fin. Ganoin is an enamel-like material on the upper surface of the scale and characterized the squamation of the Paleozoic and Early Mesozoic palaeoniscoids, which are thought to be ancestral to (or a sister group of) modern teleostean groups. The gars have retained this primitive trait. The same logic applies to the caudal skeleton. Abbreviate heterocercal tails characterized the later holosteans but have given way to the homocercal tail of the teleosts. A constricted and ossified notochord may be a derived innovation in lepisosteids rather than an indication of ancestral status to teleosts, since lepisosteid vertebral centra are essentially unique among living fishes. Gar centra are **opisthocoelous**, being concave on their posterior surface and convex on the anterior surface, allowing for a "ball-and-socket" articulation (most fishes have **amphicoelous** vertebrae in which both surfaces are concave; only one blenny species, some tailed amphibians, and a few birds have opisthocoelous vertebrae) (Suttkus 1963; Wiley 1976). The division name Ginglymodi refers to the hinged articulation between the vertebrae.

The Alligator Gar, *Atractosteus spatula*, is the largest member of the family and one of the largest freshwater fishes

(A)

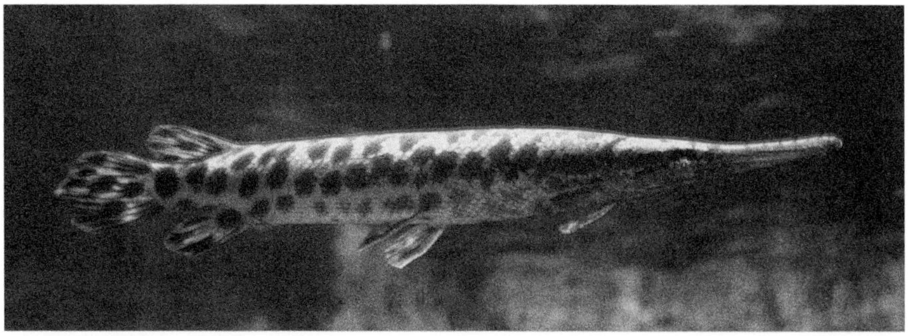

(B)

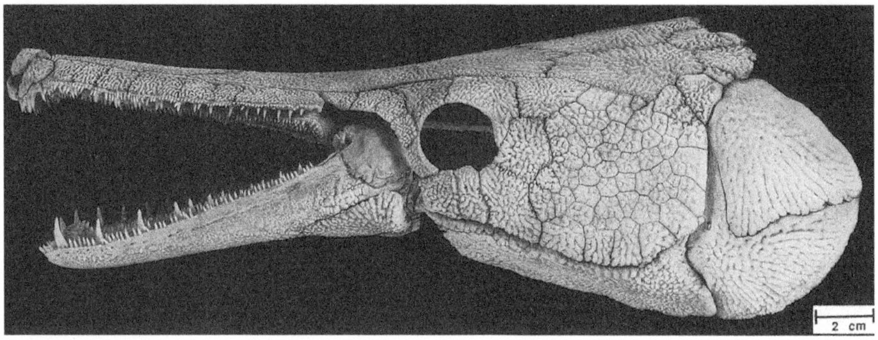

FIGURE 13.17 Gars. (A) A Spotted Gar, *Lepisosteus oculatus*, showing the distinctive elongate, tooth-studded snout and posteriorly placed dorsal and anal fins characteristic of this family of North and Central American predators. (B) Head of the large Alligator Gar, *Atractosteus spatula*. Note the numerous bones in the head and cheek and the myriad needle-like teeth. (A) B. Gratwicke / Wikimedia Commons/ CC BY 2.5; (B) Grande and Bemis 1998 / with permission of Taylor & Francis.

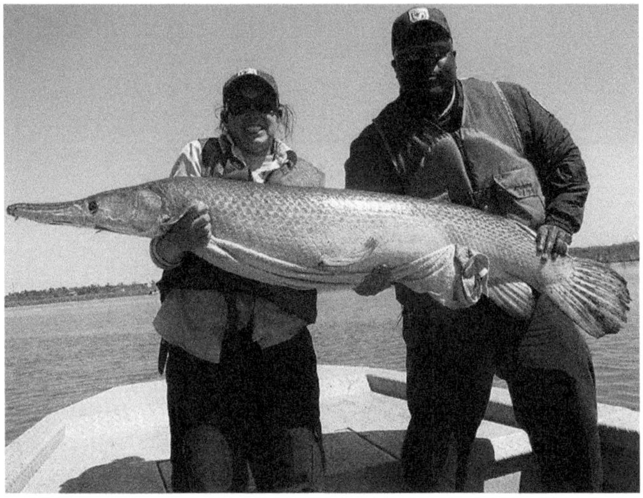

FIGURE 13.18 An Alligator Gar caught in Louisiana by the Baton Rouge Fish and Wildlife Conservation Office. USFWS / Public Domain.

in North America (Fig. 13.18). It attains a length of 3 m and a weight of 140 kg (Suttkus 1963). Although most gars are considered water column predators on other fishes and often hover just below the surface, Alligator Gars also feed extensively on bottom-dwelling fishes and invertebrates, and scavenge on benthic food (Seidensticker 1987). Alligator Gars, as well as other species, frequently enter estuarine regions (Suttkus 1963).

Comparatively little is known about the life history and general biology of gars. This is unfortunate because they are ecologically important in many fish assemblages, often becoming quite abundant in rivers and backwaters. Gars are additionally interesting in that they are the only freshwater fishes in North America with toxic eggs. The eggs are distinctly green in color and can cause sickness and even death when eaten by chickens and mice. However, the possible ecological function of this toxicity, and whether it actually affects fish or invertebrates that might feed on the eggs, remains undetermined (Netsch & Witt 1962).

Seven fossil species of gars are recognized, dating back to the Lower Cretaceous of North America, Europe, Africa, and India, and indicating a widespread Pangean distribution (Wiley 1976; Stiassny et al. 2004).

Division Halecomorphi, Order Amiiformes: The Bowfin The Bowfin, *Amia calva*, is generally considered more derived than the gars (Fig. 13.19). *Amia* and its extinct relatives in the order Amiiformes and two related orders make up the division **Halecomorphi**. *Amia* retains the abbreviate heterocercal tail and rudimentary spiral valve intestine of more primitive groups, but has teleost-like amphicoelous vertebrae as

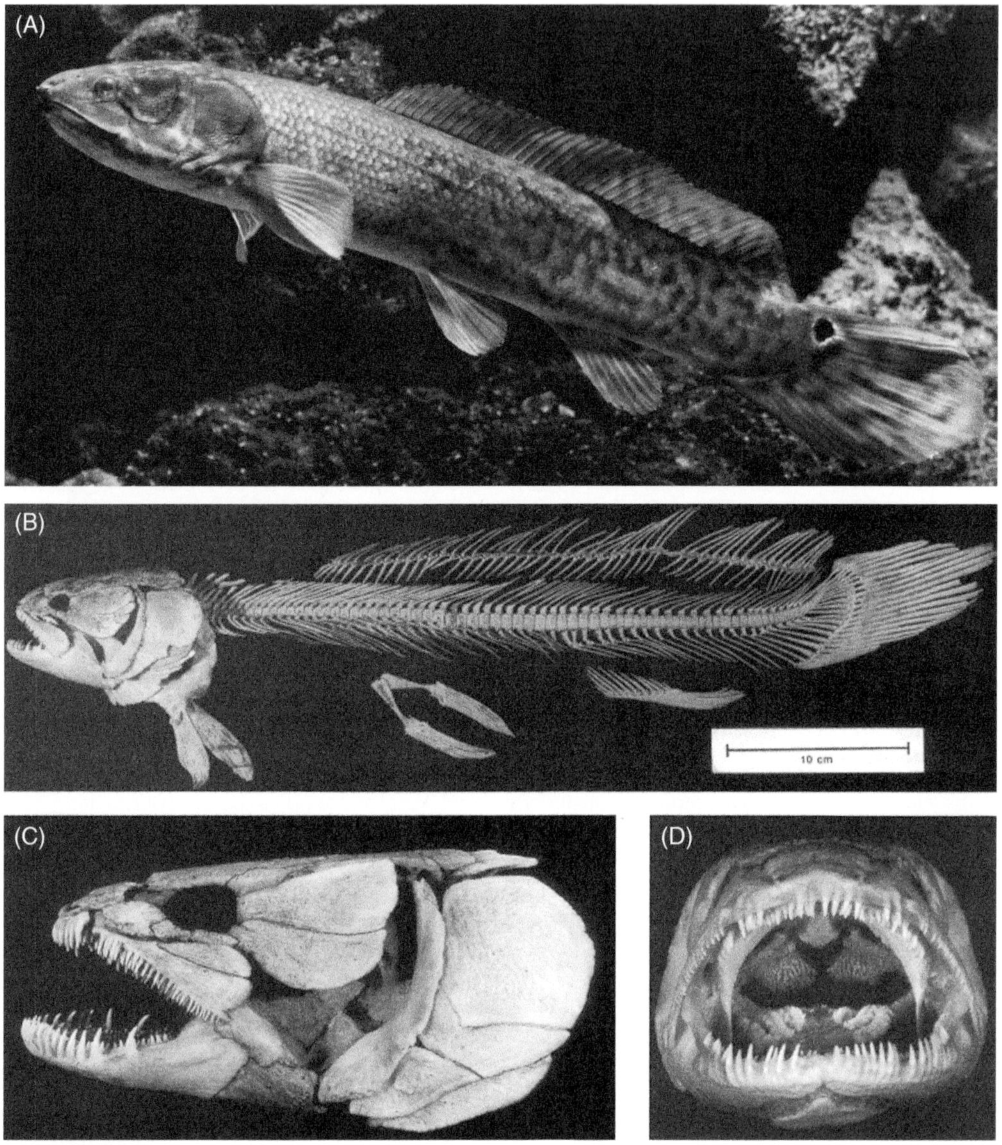

FIGURE 13.19 (A) The Bowfin, *Amia calva*, a member of a monotypic order endemic to North America. (B) Entire skeleton; note the elongate dorsal fin used in slow forward and backward locomotion and the upturned caudal vertebrae forming the abbreviate heterocercal fin. (C) Skull showing the multiplicity of bones that are later lost or fused in teleosts. (D) Anterior view looking into the mouth; the abundant, large teeth are evident, as is the single gular plate on the underside of the head. Grande and Bemis 1998 / with permission of Taylor & Francis.

well as cycloid scales, a scale type in which the ganoid and dentine layers have been lost, leaving only a reduced bony layer. The Bowfin's cycloid scales resemble those in teleosts but are probably convergent and not homologous with teleostean cycloid scales (Jarvik 1980; Stiassny et al. 2004). The Bowfin's head is exceptionally bony, invested in massive dermal bones that are greatly reduced in teleosts.

The Bowfin is distinct among all living fishes in possessing a single, median **gular plate** on the underside of the head (Fig. 13.19D). It is the only non-teleostean fish to swim via undulations of its long dorsal fin, which allows it to move slowly both forward and backward with stealth. Rapid swimming is accomplished by more conventional body and tail movements (Scott & Crossman 1973; Becker 1983).

The Bowfin is widely distributed throughout much of the eastern half of North America from southern Quebec and Ontario to eastern Texas. It is most common in vegetated lakes and backwater areas of large rivers, occupying deeper waters by day and moving into shallows at night to feed. It has abundant, sharp, conical, slightly curved teeth on both the jaws and palate (the internal structure of the teeth is unique among vertebrates); strong jaw musculature; large size (to 1 m and 9 kg); and opportunistic, predatory habits. Bowfin feed on invertebrates, fishes, frogs, turtles, snakes, and small mammals, which they engulf via suction, whereas gar impale food on their small, sharp teeth (Lauder 1980).

Bowfin males build nests in shallow water by clearing a circular depression on the bottom about 0.5 m across. The males also engage in parental care, guarding the young vigorously until they are relatively large (10 cm). The male has a distinct black spot at the base of its caudal fin; such nonseasonal sexual dimorphism does not occur in other living primitive bony fishes, although males and females differ in many teleosts.

Amia is incapable of surviving in warm, deoxygenated water without access to atmospheric oxygen. As in gars, Bowfin gulp air and pass it to a highly vascularized gas bladder. Some controversy has developed over whether Bowfin are capable of lungfish-like estivation in drying conditions. Anecdotal evidence suggests that Bowfin can bury in mud and survive for periods of weeks (e.g. Green 1966), whereas experimental laboratory findings suggest that Bowfin are physiologically incapable of surviving more than 3–5 days of air exposure (McKenzie & Randall 1990).

Amiiform fishes have been distinct since the Early Jurassic, amiids appeared in the Late Jurassic, and the genus *Amia* dates back at least to the Early Eocene (Grande & Bemis 1998). The fossil record reveals 11 genera and 27 other amiid species, including three other species in the genus *Amia*; representatives occurred in North and South America, Europe, Asia, and Africa (Grande & Bemis 1998). Many were marine fishes and almost all were piscivorous, as evidenced by fish remains in their stomachs. One Eocene giant, *Maliamia gigas*, from West Africa may have attained a length of 3.5 m (Patterson & Longbottom 1989).

Conclusions

Trends in the characteristics of the living members of ancient groups, and comparisons with the recently successful teleosts, raise a number of intriguing questions. As anatomically and taxonomically diverse as these relict fishes are, certain convergent similarities in morphology, behavior, and ecology suggest interesting evolutionary patterns that may have characterized the evolution of major fish groups (see Table 13.1). The success of extant lungfishes, gars, the Bowfin, and the enigmatic bichirs in swampy, seasonally evaporating, tropical or semitropical environments underscores the question of evolutionary succession among major fish lineages.

Are we seeing the relegation of remnant, competitively inferior groups to marginal habitats, or do we instead observe continued superiority of ancient groups in the habitats where they originally evolved and in which they had an evolutionary head start? What explains retention or independent evolution of the spiral valve in most of these primitive groups and the Chondrichthyes, but its replacement in higher bony fishes with a linear intestine? Why is electroreception retained in primitive groups (except *Amia* and the gars) but lost in most modern higher taxa, except for a few which have independently re-evolved and elaborated the electrical sense (see Chapter 6)? Are the trends that characterize fish evolution in general (see Table 11.1) – reduction in bony armor, development of the pipette mouth and pharyngeal dentition, elaboration of the dorsal fin, relocation of pelvic and pectoral girdles, an increasingly symmetrical caudal fin – necessarily improvements on the primitive design? If so, how have the relict species managed to hold on in the face of competition and predation by the more derived teleosts? And finally, why have these few species, among the thousands of ancestral species and their derivatives, survived so long while their relatives succumbed to the ultimate fate of all organisms?

Supplementary Reading

Balon EK, Bruton MN, Fricke H. 1988. A fiftieth anniversary reflection on the living coelacanth, *Latimeria chalumnae*: some new interpretations of its natural history and conservation status. *Env Biol Fish* 23:241–280.

Bemis WE, Burggren WW, Kemp NE, eds. 1987. *The biology and evolution of lungfishes*. New York: Alan R. Liss.

Bigelow HB, Farfante IP, Schroeder WC. 1948. Lancelets. In: *Fishes of the western North Atlantic*, Vol. 1, Part 1, pp. 1–28. New Haven, CT: Sears Foundation Marine Research, Yale University.

Binkowski FP, Doroshov SI, eds. 1985. *North American sturgeons: biology and aquaculture potential. Developments in environmental biology of fishes*. Dordrecht: Dr. W. Junk.

Birstein V, Bemis W. 1995. Will the Chinese paddlefish survive? *Sturgeon Q* 3(2):12.

Brodal A, Fange R. 1963. *The biology of Myxine*. Oslo: Scandinavian University Books.

CITES (Convention on International Trade in Endangered Species). 2001. *CITES identification guide – sturgeons and paddlefish*. Ottawa: Minister of Supply and Services, Canada. **www.cws-scf.ec.gc.ca**.

Dillard JG, Graham LK, Russell TR, eds. 1986. *The paddlefish: status, management and propagation*. American Fisheries Society Special Publication No. 7. Columbia, MO: American Fisheries Society North Central Division.

Gans C, Kemp N, Poss S, eds. 1996. The lancelets (Cephalochordata): a new look at some old beasts. The results of a workshop. *Israel J Zool* 42:1–446.

Hardisty MW. 1979. *Biology of the cyclostomes*. London: Chapman & Hall.

Hardisty MW, Potter IC. 1971–1982. *The biology of lampreys*, Vols 1–4b. New York: Academic Press.

Jarvik E. 1980. *Basic structure and evolution of vertebrates*, Vol 1. London: Academic Press.

Jørgensen JM, Lomholt JP, Weber RE, Malte H, eds. 1998. *The biology of hagfishes*. London: Chapman & Hall.

McCosker JE, Lagios MD, eds. 1979. The biology and physiology of the living coelacanth. *Occ Pap Calif Acad Sci* 134:1–175.

Musick JA, Bruton MN, Balon EK, eds. 1991. The biology of *Latimeria chalumnae* and evolution of coelacanths. *Env Biol Fish* 32:1–435.

Smith BR, ed. 1980. Proceedings of the 1979 Sea Lamprey International Symposium (SLIS). *Can J Fish Aquat Sci* 37:1585–2214.

Smith JLB. 1956. *The search beneath the sea.* New York: Henry Holt & Co.

Thomson KS. 1969b. The biology of the lobe-finned fishes. *Biol Rev* 44:91–154.

Thomson KS. 1991. *Living fossil. The story of the coelacanth.* New York: Norton.

Weinberg S. 2000. *A fish caught in time: the search for the coelacanth.* New York: Harper Collins.

Websites

IUCN Sturgeon Specialist Group, **www.iucn.org/info_and_news/press/sturgeon.html**.

World Sturgeon Conservation Society, **www.wscs.info**.

CHAPTER **14**

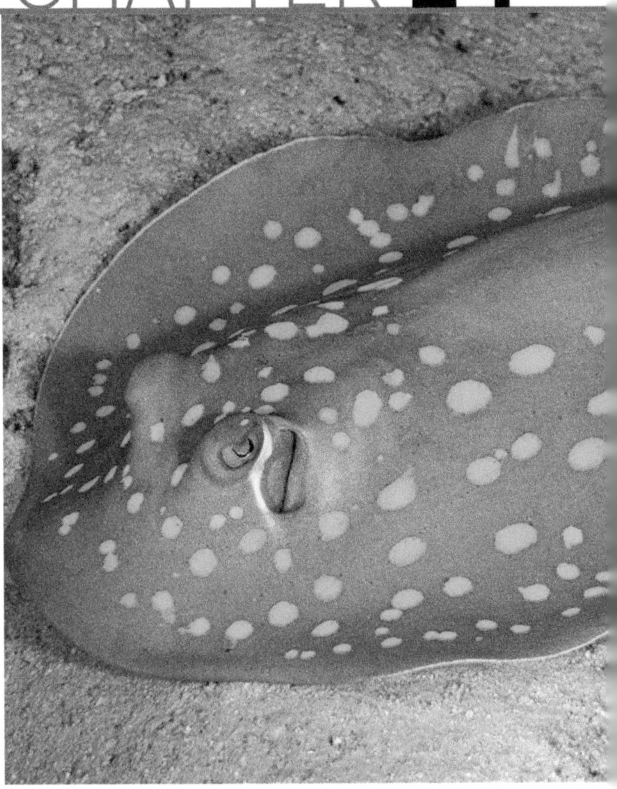

Teleosts I: Elopomorpha Through Paracanthopterygii

CHAPTER CONTENTS

Summary

About 30,000 species of living fishes are members of the division Teleostei. The name **teleost** means roughly "perfect bone," referring to their evolutionary position as the most advanced of the living, bony fishes. Bone mass in teleosts is reduced from the pre-teleostean condition, but internal cross-struts in the bone give it exceptional strength without great mass. Teleosts account for 96% of all living fishes, including most major fishery species. They inhabit the widest range of habitat types and show the greatest variation in body plans and foraging and reproductive habits of any fishes. By comparison, extant groups introduced in Chapters 12 and 13 are carnivorous and occur in a limited number of habitats. Chondrichthyans are 99% marine, whereas lungfishes, gars, Bowfin, and sturgeons are largely big river or swamp dwellers.

Teleosts occur in nearly every imaginable freshwater and marine habitat, from ocean trenches to high mountain lakes and streams, from polar oceans at −2°C to alkaline hot springs at 41°C, from torrential rivers and wave tossed coastlines to stagnant pools. There are flying, walking, and immobile teleosts, and annual teleosts that emerge from resting eggs when it rains and then breed and die. Some teleosts brood their eggs and young in their mouths, others lay eggs inside mussels, and some jump out of the water to lay eggs on the undersides of terrestrial plant leaves and periodically splash them to keep them moist. Trophically, teleosts feed on other fishes, carrion,

The Diversity of Fishes: Biology, Evolution and Ecology, Third Edition. Douglas E. Facey, Brian W. Bowen, Bruce B. Collette, and Gene S. Helfman.

Companion website: www.wiley.com/go/facey/diversityfishes3

invertebrates, mammals, scales, eyes, eggs, and zooplankton; teleosts include the only fishes known to feed on vegetation in various forms, including phytoplankton, cyanobacteria, algae, detritus, and vascular plants and their seeds. The only truly endoparasitic vertebrates are teleosts. Some teleosts produce either light or electricity. Teleosts are the most diverse and diversified taxon of all the vertebrates, having radiated into more niches and adaptive zones than all the other vertebrate groups combined.

This text will not attempt to provide detailed information on even a subset of the approximately 30,000 living teleostean species. Our objectives in Chapters 14 and 15 are to provide a feeling for (i) what characterizes teleosts and separate them from the less derived fishes discussed in earlier chapters; (ii) what characteristics separate different taxa within the teleosts and represent evolutionary advances within the division; (iii) which groups have been successful in what regions and habitat types; and (iv) what are some of the more interesting species and adaptations in this exceptionally successful group. Our focus is on living fishes, but it should be recalled that teleosts have existed since the Mesozoic and that the taxonomy of many of these groups is strongly influenced by characteristics of relatives known only from fossils.

Teleosts arose in the Early Mesozoic and radiated as modern elopomorphs, osteoglossomorphs, otocephalans, and euteleosts. Over 40 orders and 450 families are recognized, defined largely by skull and tail modifications that improved feeding and locomotion.

The cohort Elopomorpha is characterized by a ribbon-shaped leptocephalus larva and include tenpounders, tarpons, bonefishes, spiny eels, and true eels. Elopomorphs are predominantly marine fishes that occur from very shallow to very great depths. The anguilliform true eels include 15 families of elongate fishes, one of which is catadromous, spawning at sea but growing in fresh water.

The cohort Osteoglossomorpha is chiefly tropical, freshwater fishes in which teeth on the tongue bite against the mouth roof. Some species are highly prized in the aquarium trade. The diverse African elephantfishes produce and detect weak electric fields.

The cohort Otocephala includes the superorders Clupeomorpha and Ostariophysi. Clupeomorphs are generally small, schooling fishes of pelagic marine and occasionally freshwater habitats. They are characterized by a connection between the gas bladder and inner ear, which enhances hearing, and by bony scutes on the belly. The Clupeomorpha include herrings and anchovies, which are exceedingly important fisheries species. The species-rich Ostariophysi contains predominantly freshwater fishes such as catfishes, milkfishes, minnows, suckers, characins, loaches, and South American knifefishes. Ostariophysans have enhanced hearing due to a chain of small bones (the Weberian apparatus) that connects the gas bladder to the inner ear (absent in the more primitive Anotophysi). They produce and respond to alarm substances (*Schreckstoff*). Cypriniforms possess pharyngeal jaws and dentition used in manipulating and crushing prey and vegetation. Characins are highly successful South American and African fishes such as piranhas and tetras. Siluriform catfishes include 39 families of primarily benthic, nocturnal fishes with barbels, spines, and adipose fins. Gymnotiform knifefishes have converged with the osteoglossomorph elephantfishes in the production and detection of weak electric fields.

The fourth cohort, the Euteleostei, contains three-quarters of the families and two-thirds of the species of teleosts. Protacanthopterygians are loosely related marine, freshwater, and diadromous fishes that include the deep-sea argentiniforms, salmons, and smelts. Whitefishes, graylings, salmons, and trouts constitute the salmoniforms, characterized by an adipose fin behind the dorsal fin, a small axillary process at the base of each pelvic fin, and the configuration of the last three vertebrae. Many salmonids undergo extensive migrations between fresh and salt water during their lives. Osmeriforms are generally small, silvery, elongate, water column dwelling fishes such as freshwater smelts, deep-sea barreleyes and slickheads, and southern hemisphere Salamanderfish and galaxiids. The esociform pikes and pickerels are now also included within the Protacanthopterygii. These are mainly northern hemisphere freshwater predators, although the mudminnows (Umbridae) are now also included in the Esociformes. The deep-sea dragonfishes (order Stomiiformes), which are characterized by large mouths, long teeth, and photophores, are now also considered protacanthopterygians.

Euteleosts more derived than protacanthopterygians are collectively placed in the unranked taxonomic category Neoteleostei. The neoteleosts share similarities in the articulation between the skull and first cervical vertebra, two muscles that move the pharyngeal jaws, and jaw teeth that can be depressed posteriorly. Three superorders of neoteleosts are primarily deep-sea or pelagic fishes. Ateleopodomorph jellynose fishes swim just above the bottom and have cartilaginous skeletons. Cyclosquamates are primarily deep-sea forms, such as the bizarre giganturid telescopefishes and tripodfishes, but include the shallow water lizardfishes. Scopelomorphs are primarily lanternfishes, which have species-specific photophore patterns.

Neoteleosts more derived than scopelomorphs possess true fin spines and are termed acanthomorphs. Other acanthomorph advances include strengthened vertebral accessories and tail structures that improved swimming, pharyngeal tooth diversification, and improved jaw protrusion. Lampriomorphs (tube-eyes, oarfish) and polymixiomorphs (beardfishes) are less derived acanthomorphs; oarfish may exceed 8 m in length and are the world's longest teleost.

The superorder Paracanthopterygii includes beardfishes, freshwater cavefishes, and trout-perches, but is mostly marine, benthic, nocturnal fishes, including zeiforms, stylephorids, and commercially important cods. Several groups formerly included in the Paracanthopterygii, such as the deep-sea ophidiiforms, acoustically active toadfishes, anatomically specialized batfishes and handfishes, and the diverse bathypelagic lophiiform anglerfishes have been moved into the Acanthopterygii (see Chapter 15).

Teleostean Phylogeny

Teleosts arose in the Middle Mesozoic (probably Late Triassic, *c.* 200 million years ago), from a neopterygian ancestor, possibly a pachycormiform (discussed in Chapter 11).

The earliest teleosts were probably pholidophoroids or leptolepoids, groups that may have been ancestral to more than one of the main lineages of teleosts, including the elopomorphs and osteoglossomorphs. Monophyly of the Teleostei is supported by both morphological (de Pinna 1996) and molecular evidence (Near et al. 2012b, Betancur-R et al. 2013). The important point to remember, reiterating the phylogenetic account given in Chapter 11, is that modern teleosts arose during four major radiations that produced the cohorts **Elopomorpha**, **Osteoglossomorpha**, **Otocephala**, and **Euteleostei**, the latter being by far the largest (Fig. 14.1).

There is remarkable diversity among the teleosts. They include over 60 orders, 470 families, and 4600 genera – although the specific numbers of taxa continue to change as we learn more. Despite their amazing diversity, teleosts share several characters that indicate **common ancestry**, particularly in the euteleosts. The primary shared derived (synapomorphic) characters that unite the teleosts involve numerous bones of the tail and skull. Importantly, the ural neural arches of the tail are elongated into **uroneural** bones. This means that in the tail base region, the neural arches that sit dorsal to the vertebral column fuse into elongate bones termed uroneurals. These new bones serve as basal supports for the rays that form the upper lobe of the tail fin and thus help stiffen it; their number and shape change during teleostean phylogeny. In the skull, among other characters, teleosts have a **mobile premaxillary** bone rather than having the premaxilla fused to the braincase. A mobile premaxilla is essential for upper jaw protrusion and allows a fish to shoot its mouth forward during prey capture, creating suction pressures and overtaking prey.

In sum, major changes that define the teleosts contributed to the advances in locomotion and feeding that apparently led to their success, as detailed in Chapter 11. Most of the characteristics described here are discussed in more detail in Wheeler (1975), Berra (1981, 2001), Carroll (1988), Nelson (1994, 2006), and Nelson et al. (2016). The overall classification follows, with many of the characteristics described, and the numbers of species provided for different orders. The major taxonomic groups are based on Nelson et al. (2016) – but see corrections and updates at **https://sites.google.com/site/fotw5th/**). Orders, families, and numbers of species are from Fricke et al. (2020) – but refer to **https://researcharchive.calacademy.org/research/ichthyology/catalog/SpeciesByFamily.asp** for updates.

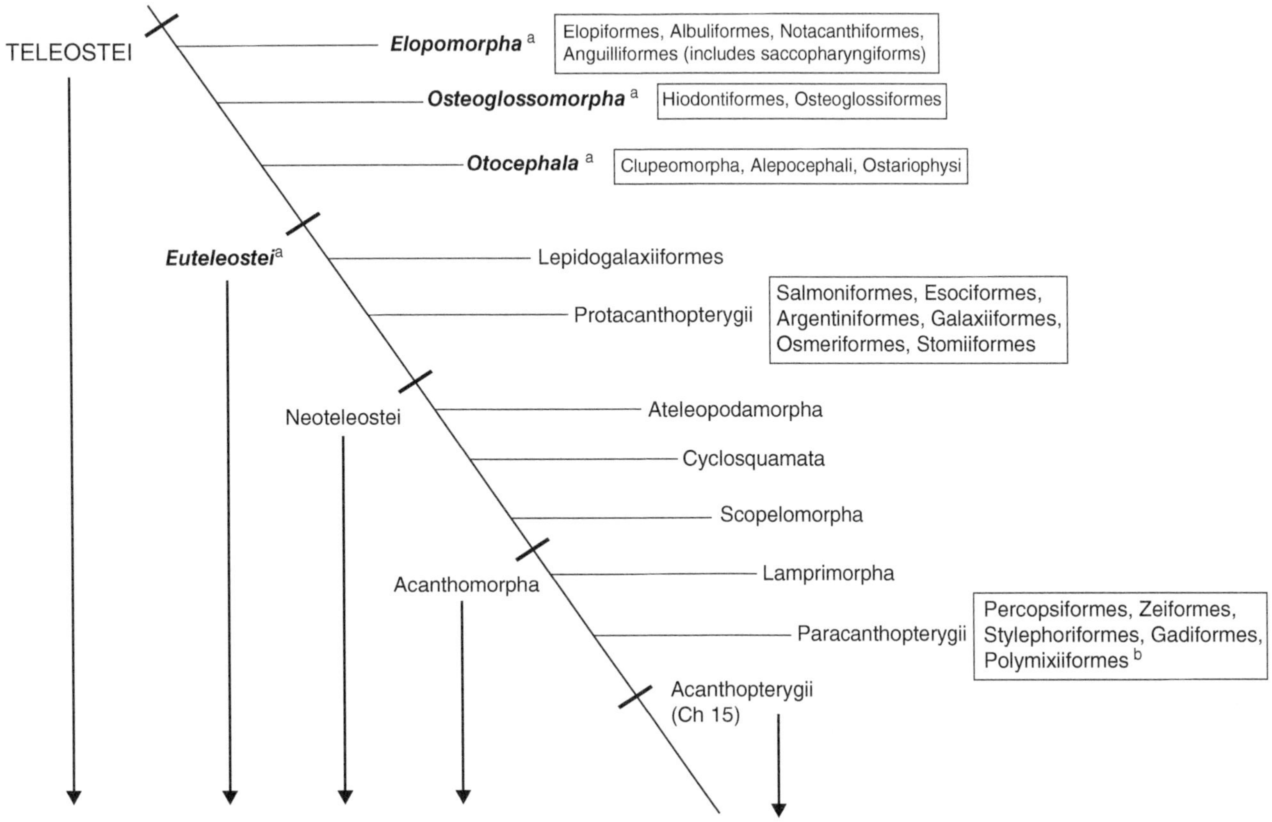

[a] We follow Nelson et al. 2016 and others in labeling these groups as cohorts of the division Teleostei (of the Infraclass Teleosteomorpha). In our prior edition (Helfman et al. 2009), these cohorts were labeled as subdivisions.
[b] Betancur-R et al. 2017 places Polymixiiformes after Paracanthopterygii but before Acanthopterygii.

FIGURE 14.1 Increased genetic information in recent decades has resulted in reconsideration of the phylogenetic relationships among many groups of living teleosts. This figure is largely based on Nelson et al. 2016, with some notations of differences with Betancur-R et al. 2017. Readers are encouraged to consult other, more current, assessments. Cohorts[a] are in bold italics. Neoteleostei has no formal rank (below cohort and above superorder) and includes the more derived superorders among the Euteleostei. Acanthomorpha also has no formal rank (below cohort and above superorder) and refers to the teleosts with true spines in the dorsal, anal, and pelvic fins. The remaining terms are superorders and orders, with orders ending in "-iformes".

A Survey of Living Teleostean Fishes

Class Osteichthyes – a review of major groups
Subclass Sarcopterygii – *addressed in Chapter 13*
Subclass Actinopterygii
Infraclass Cladistia – *addressed in Chapter 13*
Infraclass Chondrostei – *addressed in Chapter 13*
Neopterygii
Infraclass Holostei – *addressed in Chapter 13*
Infraclass Teleosteomorpha – *addressed in this Chapter (14) and Chapter 15*
Division Teleostei
Cohorts Elopomorpha, Osteoglossomorpha, Otocephala, Eutelostei

Cohort Elopomorpha

Cohort Elopomorpha[a]
Order Elopiformes: Elopidae (7 species of tenpounders and Ladyfish), Megalopidae (2 species of tarpons)
Order Albuliformes: Albulidae (13 species of bonefishes)
Order Notacanthiformes: Halosauridae (16 species of halosaurs), Notacanthidae (11 species of spiny eels)
Order Anguilliformes (966 species): Protanguillidae (Cave Eel), Synaphobranchidae (about 49 species of cutthroat eels), Heterenchelyidae (8 species of mud eels), Myrocongridae (4 species of myroconger eels), Muraenidae (about 215 species of moray eels), Chlopsidae (25 species of false moray eels), Colocongridae (8 species of shorttail eels), Derichthyidae (3 species of longneck eels), Ophichthidae (about 345 species of snake and worm eels), Muraenesocidae (9 species of pike congers), Nettastomidae (about 46 species of duckbill eels), Congridae (about 168 species of conger eels), Moringuidae 15 (species of spaghetti eels), Nemichthyidae (9 species of snipe eels), Serrivomeridae (10 species of sawtooth eels), Anguillidae (17 species of freshwater eels), Cyematidae (Bobtail Eel), Monognathidae (15 species of onejaw gulpers), Neocyematidae (Bobtail Snipe Eel), Eurypharyngidae (Gulper Eel), Saccopharyngidae (10 species of swallowers).

[a] Higher classification groups based on Nelson et al. (2016), but see corrections and updates at **https://sites.google.com/site/fotw5th/**. Orders, families, and numbers of species from Fricke et al. (2020) but refer to **https://researcharchive.calacademy.org/research/ichthyology/catalog/SpeciesByFamily.asp** for updates.

For many years, morphological evidence had led many to consider the Osteoglossomorpha as the most primitive living teleosts. But more recent molecular and morphological research indicates the **Elopomorpha** (Fig. 14.2) as the most primitive group, and hence the sister group of the other three cohorts (Near et al. 2012b, Betancur-R et al. 2013). A distinct pelagic larval form, termed a **leptocephalus** ("pointed head"), unites this speciose marine group. Leptocephali are typically lanceolate or ribbon-shaped (Fig. 14.2B) and many of them shrink during metamorphosis to the juvenile form. For many years, the link between larval and adult species was not made and hence the two life history stages were placed in very different taxa. The leptocephali of elopiform tarpons and bonefishes have a forked tail, whereas eel larvae have a pointed tail. Leptocephali are exceedingly long-lived, remaining as larvae for as long as 2–3 years in some anguillid species. During this time, they are dispersed by currents over large oceanic expanses, perhaps gaining nutrition from dissolved organic matter that they absorb through their skin or feeding on gelatinous zooplankton (e.g. Mochioka & Iwamizu 1996). They are thin and fragile in appearance, this effect heightened by a lack of red blood cells, which makes them translucent.

Elopomorphs are also distinguished by the development of thin, riblike epipleural intermuscular bones that extend from the vertebral column into the surrounding trunk musculature. These are the small bones in the meat of primitive teleosts (conger eels, herrings, carps, trouts) that make them difficult to filet and eat. Long epipleural and epineural ribs become less common in higher teleosts such as paracanthopterygians and acanthopterygians, which rely more on stouter, more firmly attached zygapophyses. These differences make both cleaning and eating easier.

The **elopiform** ladyfishes and tarpons retain a primitive characteristic, namely a **gular** bone or splint on the underside of the throat; this structure is well developed in the more primitive Bowfin, coelacanths, and bichirs but is lost in all other teleosts (except perhaps for an anabantoid, the Pikehead). Other justifications for considering elopiforms as primitive teleosts include (i) a large number of branchiostegal rays in the throat (23–35 vs. 5–7 in many higher teleosts); (ii) inclusion of the maxilla in the gape, giving them two biting bones in the upper jaw rather than one; and (iii) heavy, bony scales that contain ganoin, a bone layer otherwise only found in gars and bichirs. The Atlantic Tarpon, *Megalops atlanticus*, is a legendary gamefish that reaches a length of 2.5 m and a mass of 150 kg. A large (65 kg) female may contain more than 12 million eggs, making tarpon one of the most fecund fishes.

Albuliform bonefishes are also popular gamefishes that occupy sandy flats in shallow tropical waters; recent molecular studies suggest about 23 species exist where historically only one was recognized (discussed in Chapter 2, Fig. 2.8). The notacanthoids (halosaurs and spiny eels) are an offshoot suborder between the albuliforms and the anguilliforms; they develop from leptocephalus larvae but otherwise stand in marked anatomical and ecological contrast to other members of the albuliforms. These deep-sea, benthic eels occur down to 5000 m, making them among the deepest living fishes known.

The 16 families of **anguilliforms** are "true" eels, i.e. those with a leptocephalus larva, as distinguished from the approximately 45 other families of "eel-like" fishes that have converged on an elongate body and other anatomical and behavioral traits. An eel-like body facilitates forward and

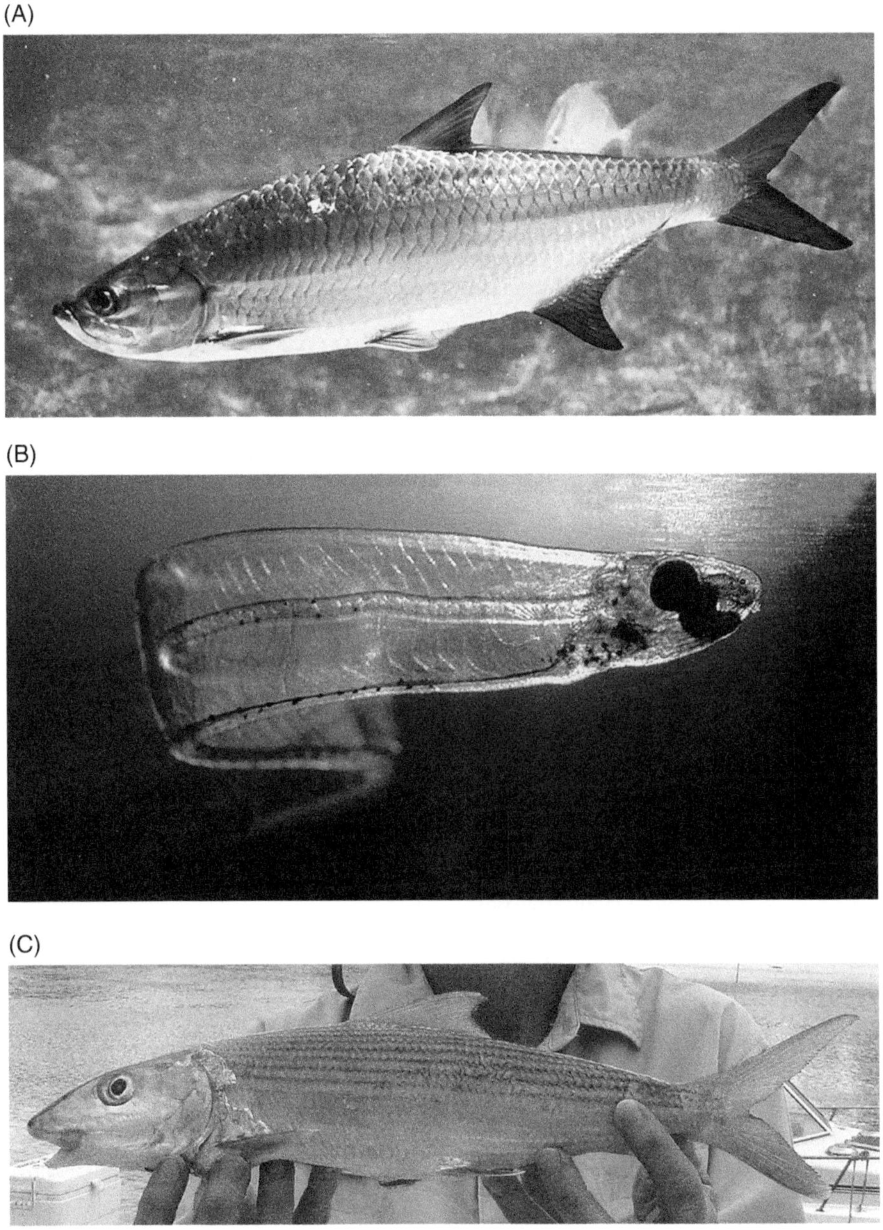

FIGURE 14.2 Elopomorphs. (A) A Tarpon, *Megalops atlanticus*. Citron / Wikimedia Commons / CC BY-SA 3.0. (B) A leptocephalus larva of a conger eel. U Kils / Wikimedia Commons / CC BY-SA 3.0. (C) An Atlantic Bonefish, Albula vulpes. G. Helfman (Author).

backward movement into and out of tight places and soft bottoms. Some anguilliforms are open water, pelagic forms, despite the relatively slow locomotion imposed by an anguilliform swimming mode. Anguilliforms are distinguished by loss of the pelvic girdle and by a modified upper jaw that is formed by fusion of the premaxilla, vomer, and ethmoid bones. The 17 species of anguillid eels are catadromous, spawning at sea but spending most of their lives feeding and growing in fresh water.

Muraenid moray eels and their relatives (about 215 species) are largely marine, tropical and warm temperate species best known from coral reefs. To many human observers, they appear dangerous, perhaps even sinister, in part because their sedentary habits require them to hold their fang-studded jaws open while actively pumping water over their gills and out a constricted opercular opening. Although capable of inflicting serious wounds, morays are more dangerous as agents of ciguatera food poisoning, a toxin that originates in a dinoflagellate alga and is magnified in piscivores that eat prey contaminated with the toxin. The congroid eels (nearly 500 species) include fossorial (burrowing) forms such as garden eels, worm eels, and snake eels, the latter burrowing into sediments backward with a hardened, pointed tail. Benthic conger eels are similar ecologically to morays. Other deep-sea

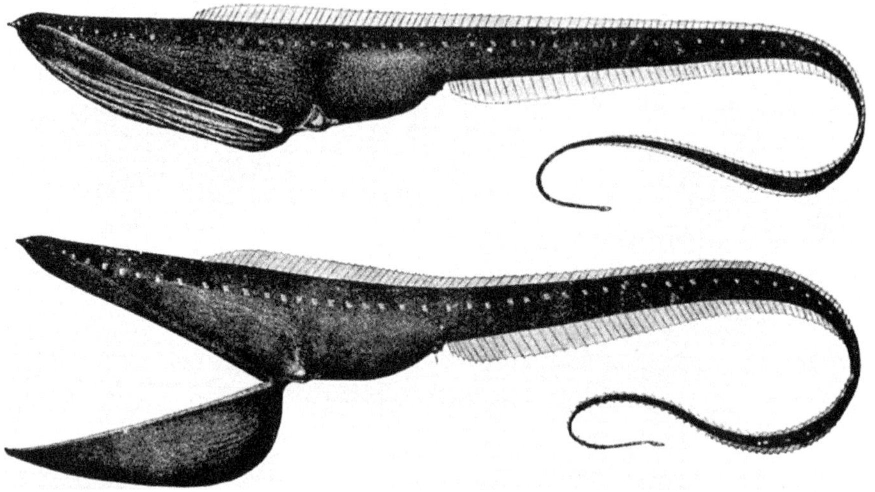

FIGURE 14.3 A Gulper or Pelican Eel, *Eurypharynx pelecanoides*. Ironically, this highly specialized, 40 cm long bathypelagic fish feeds on surprisingly small prey, which they capture by opening their huge, dark mouths that probably generate little suction pressure. The related swallower eels can feed on prey larger than themselves. Gulper Eels are unique among teleosts because they have five gill arches and six visceral clefts. Wikimedia Commons / Public Domain.

mesopelagic and bathypelagic congroids include longneck eels, snipe eels, and sawtooth eels. One family of congroids, the synaphobranchid cutthroat eels, contains a facultative parasitic species, the Snubnose Parasitic Eel, *Simenchelys parasiticus*. Although often a scavenger, *Simenchelys* sometimes burrows into the flesh of bottom-living fishes such as halibut. Two 20 cm long individuals were found lodged in the heart of a longline captured, 500 kg Shortfin Mako Shark, where they had been feeding on blood. Histological features of the inhabited heart suggested that these eels had possibly been living in the shark's heart prior to its capture, pointing to a truly parasitic relationship (Caira et al. 1997).

The saccopharyngiforms, once considered in their own order, are now a suborder of the Anguilliformes (Nelson et al. 2016, Betancur-R et al. 2017). They include the truly bizarre deep-sea gulper and swallower eels and their relatives (Fig. 14.3). These species are distinguished not only by elaborate, extreme specializations of the head and tail, including an extremely long jaw, but also for a lack of features normally found in teleosts. Among the missing structures are the symplectic and opercular bones, branchiostegals, maxilla and premaxilla, vomer and parasphenoid, scales, pelvic or pectoral fins, ribs, pyloric caeca, and gas bladder. Some early authors argued that saccopharyngoids were not really bony fishes. Nelson (1994, p. 115) considered the saccopharyngoids "perhaps the most anatomically modified of all vertebrate species." Another saccopharyngoid family, the monognathids, contains 15 species with rostral fangs and apparent venom glands, a unique feature among fishes (Bertelsen & Nielsen 1987).

Cohort Osteoglossomorpha

Cohort Osteoglossomorpha[a]
- Order Hiodontiformes: Hiodontidae (2 species of mooneyes)
- Order Osteoglossiformes (247 species): Pantanodontidae (Freshwater Butterfly Fish), Osteoglossidae (16 species of bonytongues), Notopteridae (10 species of featherfin knifefishes), Mormyridae (about 224 species of African elephantfishes), Gymnarchidae (African Aba)

[a] Higher classification groups based on Nelson et al. (2016), but see corrections and updates at **https://sites.google.com/site/fotw5th/**. Orders, families, and numbers of species from Fricke et al. (2020) but refer to **https://researcharchive.calacademy.org/research/ichthyology/catalog/SpeciesByFamily.asp** for updates.

The **osteoglossomorph** bonytongues (Fig. 14.4) and their relatives occur in fresh water on all major continents except Europe, although only Africa has more than a few species (see Chapter 19, Archaic Freshwater Fish Distributions). Anatomically, the group gets its common name from well-developed **teeth on the tongue** that occlude (bite) against similarly toothed bones (parasphenoid, mesopterygoid, ectopterygoid) in the roof of the mouth. Although chiefly a tropical group, two species (the Mooneye, *Hiodon tergisus*, and Goldeye, *H. alosoides*) occur in major river systems of northern North America. The Arapaima or Pirarucu of South America (*Arapaima gigas*) is one of the world's largest freshwater fishes, reaching a length of 2.5 m. Arapaima have been stocked in lakes and reservoirs in Southeast Asia, where they are actively sought as sport fish.

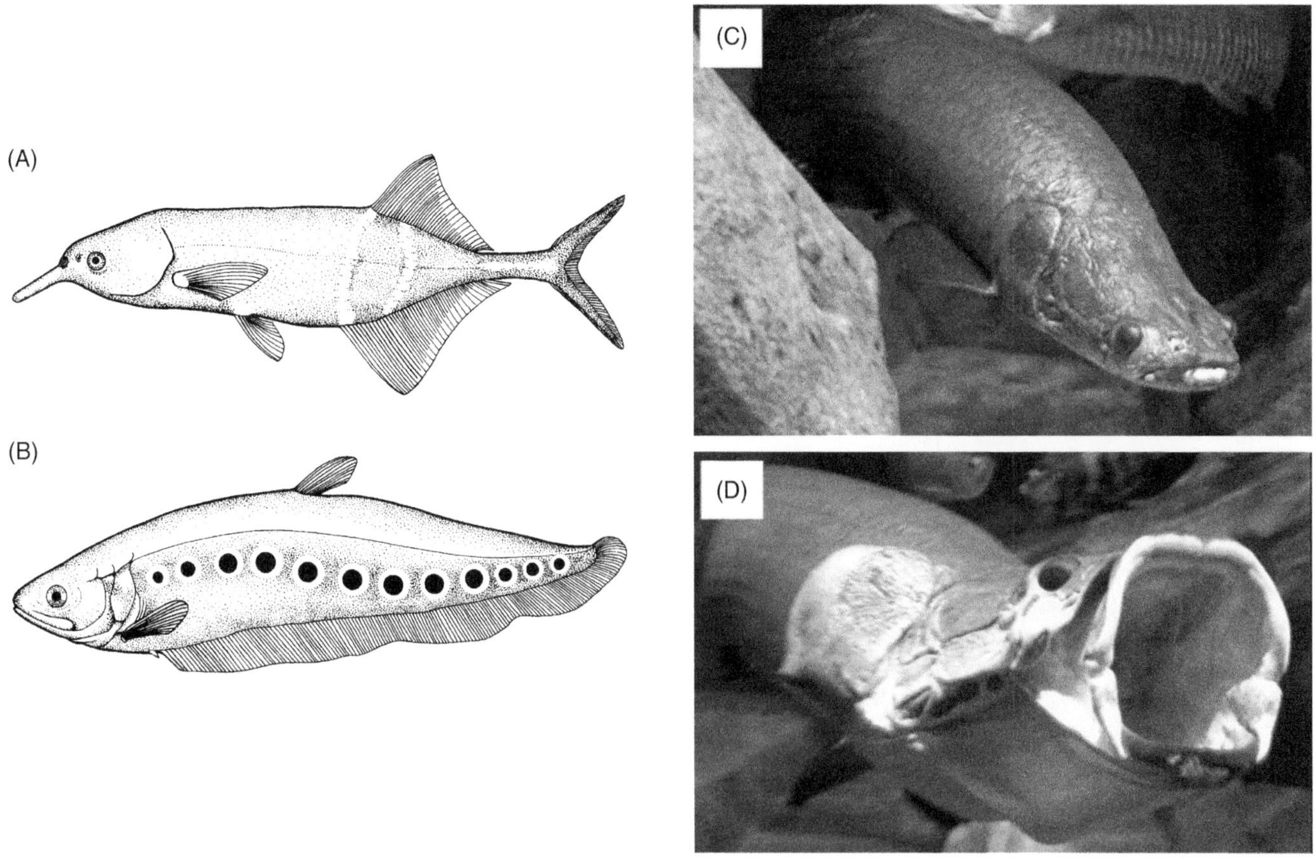

FIGURE 14.4 Osteoglossomorphs. (A) A mormyrid elephantfish, *Gnathonemus petersii*, from Africa. (B) A notopterid featherfin or knifefish, *Chitala chitala*, from Asia. (C, D) The South American Arapaima or Pirarucu, *Arapaima gigas*, a large predator. (A, B) After Paxton and Eschmeyer (1994), with permission; (C, D) G. Helfman (Author).

The South American Arawana (*Osteoglossum bicirrhosum*), the African Butterflyfish, *Pantodon buchholzi*, the notopterid featherfins or Old World knifefishes, and the mormyrid elephantfishes are popular aquarium species. The Asian Arawana or Golden Dragonfish, *Scleropages formosus* (Fig. 14.5), has been depleted in the wild due to overcollecting and is now protected in Appendix I of the Convention on International Trade in Endangered Species of Wild Fauna and Flora (CITES). Mormyridae, the most speciose family in the cohort with >200 species, and the related *Gymnarchus niloticus*, possess a highly evolved electrical sense that involves both the production and detection of weak electric fields, an appropriate sense for fishes that are nocturnally active and typically occur in turbid waters. The electrical sense is used to localize objects and is also important during social interactions (see Chapter 6, Electroreception; Chapter 17, Electrical Communication); analysis of electric organ discharges suggests that many cryptic species exist that are only separable on the basis of the wave patterns of their electric discharges (e.g. Arnegard & Hopkins 2003). Mormyrids have the largest cerebellum of any fish and a brain size: body weight ratio comparable to that of humans; the **mormyrocerebellum** is the neural center for coordinating electrical input. Mormyrids have a large learning capacity and are reported to engage in play behavior, a rarity among fishes, although not as unusual as might be expected (Burghardt 2005). Mormyrids are also important food fishes in Africa, with some attaining a length of 1.5 m.

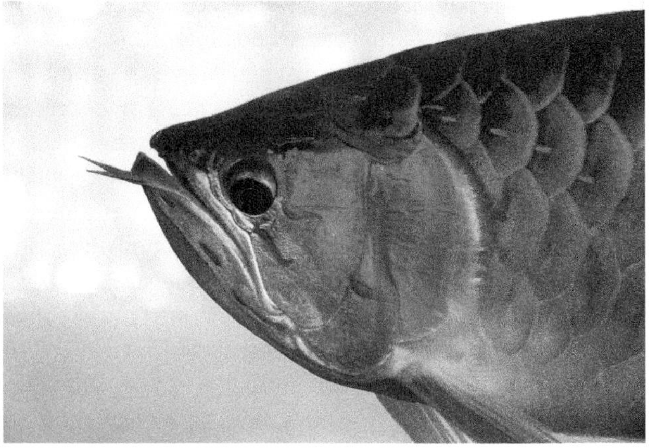

FIGURE 14.5 The Asian Arawana or Golden Dragonfish, *Scleropages formosus*. Overcollecting for the aquarium trade pushed this species to the brink of extinction. Marcel Burkhard / Wikimedia Commons / CC BY-SA 3.0.

Cohort Otocephala (Ostarioclupeomorpha)

The past decade has seen considerable reanalysis of relationships among teleosts more derived than the elopomorphs and bonytongues. It is now widely agreed that herrings, slickheads, and minnow-like fishes, earlier classified separately, belong in the same cohort, the **Otocephala** (or its tongue-twisting but descriptive synonym Ostarioclupeomorpha) (Johnson & Patterson 1996; Arratia 1997, Betancur-R. et al. 2013).

Superorder Clupeomorpha

Superorder Clupeomorpha[a]
Order Clupeiformes (416 species): Denticipitidae (Denticle Herring), Spratelloididae (9 species of small round herrings), Clupeidae (about 197 species of herrings). Dussumieridae (19 species of round herrings), Engraulidae (159 species of anchovies), Chirocentridae (2 species of wolf herrings), Pristigasteridae (38 species of longfin herrings).

[a] Higher classification groups based on Nelson et al. (2016), but see corrections and updates at **https://sites.google.com/site/fotw5th/**. Orders, families, and numbers of species from Fricke et al. (2020) but refer to **https://researcharchive.calacademy.org/research/ichthyology/catalog/SpeciesByFamily.asp** for updates.

Among the most abundant and commercially important of the world's fishes are the herringlike **clupeiforms**; large fisheries exist (or existed) for California Sardines, Peruvian Anchoveta, Atlantic and Gulf Menhaden, Atlantic Herring, and South African Sardine and Anchovy (Hutchings 2000a, 2000b; Hilborn 2005). Almost all are open water, pelagic, schooling forms, 80% of which are marine. Clupeomorphs are distinguished by a gas bladder that extends anteriorly up into the braincase and contacts the utriculus of the inner ear and in some species extends posteriorly to the anus; the gas bladder also has extensions to the lateral line canals. This **otophysic** ("ear-to-gas-bladder") condition apparently increases the hearing ability of these fishes by increasing their sensitivity to low-frequency (1–1000 Hz) sounds. Low-frequency sounds of 3–20 Hz are typically those produced by tail beats of other fishes, such as neighbors in a school and attacking predators (Blaxter & Hunter 1982). Clupeomorphs also typically possess a series of sharp, bony scutes along their ventral edge and some also have scutes anterior to the dorsal fin. These scutes may make these fishes harder for predators to capture and swallow, although direct proof is lacking. Of phylogenetic significance, the superorder Clupeomorpha (modern clupeiforms and extinct, related orders) possess evolutionary advances over elopomorphs in terms of a modified joint at the posterior angle of the jaw (angular fused to articular rather than to retroarticular) and caudal skeleton reduction (reduced first ural centrum and reduction to six in number of hypural bones). These derived traits foreshadow the continued changes in jaw and tail structures that occurred during the evolution of higher teleostean groups.

The engraulid anchovies are relatively elongate zooplanktivorous clupeoids with large mouths made possible by an elongate maxillary that extends considerably behind the eye. Anchovies range in size from a minute Brazilian species (*Amazonsprattus*, 2 cm) to a piscivorous riverine New Guinea anchovy (*Thryssa scratchleyi*, 37 cm). The largest clupeids are the chirocentrid wolf herrings, *Chirocentris dorab* and *C. nudus*, with an Indo-Pacific to South Africa distribution. Wolf herrings are herrings gone mad. They reach a length of 1 m (the next largest clupeoid is a 60 cm Indian clupeid) and have fanglike jaw teeth plus smaller teeth on the tongue and palate, which they use to capture other fishes.

The largest family in the superorder is the Clupeidae, which includes about over 190 species of herrings, round herrings, shads, alewives, sprats, sardines, pilchards, and menhadens. Clupeids can be marine, freshwater, or anadromous. Landlocked forms are common - anadromous shads, alewives, and herrings have become established in lakes and reservoirs and in rivers trapped behind dams. Whereas herrings, sardines, and menhaden are important commercially, some of the larger shads are popular sportfish (e.g. the American Shad, *Alosa sapidissima*; McPhee 2002). Probably the best-known fish fossil in the world is †*Knightia*, a freshwater Eocene herring from the Green River shale formations of Wyoming (Fig. 14.6).

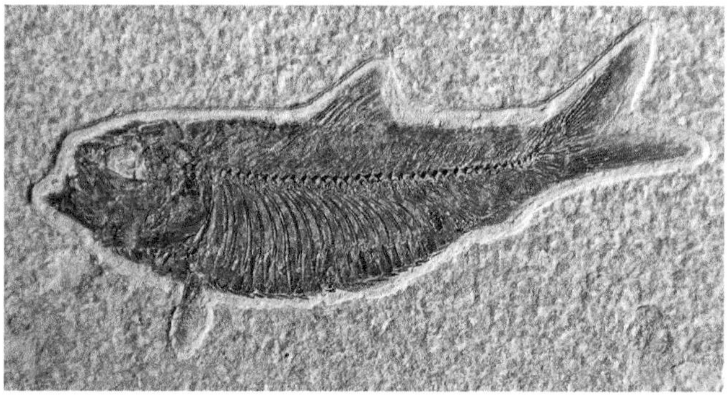

FIGURE 14.6 *Knightia alta*, an Eocene herring from the Green River formation of Wyoming (actual length 12 cm). Excellent fossils of *Knightia*, such as this one in which the characteristic abdominal scutes of clupeids are clearly visible, are abundant and are sold as curios. G. Helfman (Author).

Superorder Alepocephali

Superorder Alepocephali[a]
Order Alepocephaliformes (3 families, about 143 species): Platytroctidae (40 species of tubeshoulders), Bathylaconidae (3 species of bathylaconids), Alepocephalidae (at least 100 species of slickheads)

[a] Higher classification groups based on Nelson et al. (2016), but see corrections and updates at **https://sites.google.com/site/fotw5th/**. Orders, families, and numbers of species from Fricke et al. (2020) but refer to **https://researcharchive.calacademy.org/research/ichthyology/catalog/SpeciesByFamily.asp** for updates.

Earlier morphological studies such as Lauder and Liem (1983) united alepocephaloids with the Argentiniformes based on possession of a crumenal (epibranchial) organ – extra cartilage and gill rakers found in some filter-feeding fishes. However, numerous molecular studies such as Betancur-R. et al. (2013) place them within the Otocephala. The position of alepocephaloids within the Otocephala is uncertain, whether closer to the Clupeomorpha or Ostariophysi (Nelson et al. 2016). Alepocephaloids are dark-colored fishes that lack both an adipose fin and a swimbladder. They are deep-sea marine fishes found in tropical and temperate waters.

Superorder Ostariophysi

Superorder Ostariophysi[a]

- Series Anotophysi
  - Order Gonorhynchiformes (37 species): Chanidae (Milkfish), Gonorhynchidae (5 species of beaked sandfishes), Kneriidae (about 30 species of shellears), Phractolaemidae (African Snake Mudhead)
- Series Otophysi
  - Order Cypriniformes (23 families, about 4,449 species): Gyrinocheilidae (3 species of algae eaters), Catostomidae (82 species of suckers), Botiidae (59 species of pointface loaches), Vaillantellidae (3 species of longfin loaches), Cobitidae (about 221 species of spined loaches), Barbuccidae (2 species of scooter loaches), Gastromyzontidae (134 species of hillstream loaches), Serpenticobitidae (3 species of snake loaches), Balitoridae (102 species of river loaches), Ellopostomatidae (2 species of squarehead loaches), Nemacheilidae (about 741 species of brook loaches), Paedocyprididae (3 species of tiny carps), Psilorhynchidae (28 species of mountain carps), Cyprinidae (about 1,723 species of barbs and carps), Sundadanionidae (9 species of tiny danios), Danionidae (351 species of danionids), Leptobarbidae (5 species of cigar barbs), Tincidae (Tench), Acheilognathidae (76 species of bitterlings), Gobionidae (218 species of gobionids), Tanichthyidae (4 species of mountain minnows), Leuciscidae (679 species of minnows)
  - Order Characiformes (24 families, about 2,247 species): Citharinidae (8 species of citharinids), Distichodontidae (about 105 species of distichodontids), Crenuchidae (97 species of South American darters), Alestiidae (about 117 species of African tetras), Hepsetidae (6 species of African pikes), Tarumaniidae (Muckfish), Erythrinidae (19 species of trahiras), Parodontidae (about 32 species of darter tetras), Cynodontidae (8 species of sabertoothed characids), Serrasalmidae (98 species of piranhas and pacus), Hemiodontidae (about 32 species of hemiodontids), Anostomidae (at least 148 species of toothed headstanders), Chilodontidae (8 species of headstanders), Curimatidae (about 115 species of toothless characiforms), Prochilodontidae (21 species of flannel-mouths), Lebiasinidae (75 species of pencil fishes), Ctenoluciidae (7 species of pike-characins), Chalceidae (5 species of tucan fishes), Triportheidae (21 species of hatchet characins), Gasteropelecidae (9 species of freshwater hatchetfishes), Bryconidae (50 species of bryconids), Iguanonectidae (34 species of iguanonectids), Acestrorhynchidae (26 species of pike characins), Characidae (at least 1,193 species of characins)
  - Order Siluriformes (39 families, about 4,116 species): Nematogenyidae (Mountain Catfish), Trichomycteridae (about 320 species of parasitic catfishes), Callichthyidae (about 221 species of armored catfishes), Loricariidae (about 996 species of suckermouth catfishes), Scoloplacidae (6 species of spiny dwarf catfishes), Astropblepidae (82 species of climbing catfishes), Diplomystidae (7 species of velvet catfishes), Cetopsidae (43 species of whalelike catfishes), Chacidae (4 species of squarehead catfishes), Plotosidae (about 42 species of eeltail catfishes), Ailiidae (25 species of Asian schilbeids), Horabagridae (10 species of imperial catfishes), Bagridae (about 220 species of bagrid catfishes), Akysidae (57 species of stream catfishes), Amblycipitidae (42 species of torrent catfishes), Sisoridae (at least 283 species of sisorid catfishes), Pangasiidae (29 species of shark catfishes), Siluridae (about 103 species of sheatfishes), Kryptoglanidae (Indian Cave Catfish), Aspredinidae (46 species of banjo catfishes), Auchenipteridae (124 species of driftwood catfishes), Doradidae (96 species of thorny catfishes), Heptapteridae (222 species of heptapterid catfishes), Phreatobiidae (3 species of cistern catfishes), Pimelodidae (at least 114 species of long-whiskered catfishes), Pseudopimelodidae (49 species of bumblebee catfishes), Clariidae (about 122 species of airbreathing catfishes), Heteropneustidae (4 species of airsac catfishes), Ariidae (about 153 species of sea catfishes), Anchariidae (6 species of anchariid catfishes), Austroglanididae (3 species of rock catlets), Cranoglanididae (3 species of armorhead catfishes), Ictaluridae (51 species of North American freshwater catfishes), Lacantuniidae (Chiapas Catfish), Amphiliidae (102 species of loach catfishes), Malapteruridae (19 species of electric catfishes), Mochokidae (223 species of upside-down catfishes), Claroteidae (about 85 species of claroteid catfishes), Schilbeidae (35 species of schilbeid catfishes)
  - Order Gymnotiformes (256 species): Apteronotidae (about 96 species of ghost knifefishes), Sternopygidae (about 51 species of glass knifefishes), Gymnotidae[b] (45 species of naked-back knifefishes), Hypopomidae (36 species of bluntnose knifefishes), Rhamphichthyidae (28 species of painted knifefishes)

[a] Higher classification groups based on Nelson et al. (2016), but see corrections and updates at **https://sites.google.com/site/fotw5th/**. Orders, families, and numbers of species from Fricke et al. (2020) but refer to **https://researcharchive.calacademy.org/research/ichthyology/catalog/SpeciesByFamily.asp** for updates.

[b] The Electric Eel or electric knifefish, *Electrophorus electricus*, now considered a gymnotid, was previously placed in its own family, the Electrophoridae.

Freshwater habitats worldwide are dominated in terms of numbers of both species and individuals by ostariophysans, which account for about 68% of all freshwater species. Ostariophysans include such disparate taxa as milkfish, minnows, carps, barbs, suckers, loaches, piranhas, tetras, catfishes, and electric eels, but two unique traits characterize most members of this massive taxon. With the exception of the gonorhynchiforms, ostariophysans possess a unique series of bones that connect the gas bladder with the inner ear, an **otophysic** condition. This **Weberian apparatus**, named after the German anatomist who first described it, involves a set of bones derived from the four or five anterior (cervical) vertebrae and their neural arches, ribs, ligaments, and muscles. The superorder gets its name from this complex structure (*ostar* = small bone, *physa* = a bladder; "otophysic" basically means "ear" and "bladder"); ostariophysans with the apparatus are referred to as the **Otophysi**. When sound waves contact the fish, the gas gas bladder vibrates, and this vibration is passed anteriorly to the inner ear by the Weberian apparatus thus enhancing hearing (see Chapter 6). Unrelated taxa have convergently evolved connections between the gas bladder and the inner ear, either by an otophysic extension of the gas bladder anteriorly (elephantfishes, clupeoids, cods, Roosterfish, porgies, some cichlids); by a bony connection involving the pectoral girdle or skull (squirrelfishes, triggerfishes); or, in chaetodontid butterflyfishes, by connections between anterior extensions of the gas bladder and the lateral line canal system (Webb et al. 2006). Many of these families are known sound producers, and it is assumed they derive auditory advantages via their specialized structures. Gonorhynchiforms, the **Anotophysi**, possess a primitive homolog of the Weberian apparatus consisting of three modified anterior vertebrae associated with cephalic rib bones.

The second shared derived trait that helps define the Ostariophysi is the **alarm response**, which involves the production of an **alarm substance** (***Schreckstoff***), and a behavioral **alarm reaction** to the presence of the substance in the water (***Schreckreaktion***). The alarm substance is given off when specialized dermal club cells are ruptured, as when a predator bites down on a prey fish. Nearby individuals, most likely schoolmates, sense the chemical in the water and take a variety of coordinated escape actions, depending on the species. Possession of the alarm response was a factor contributing to the inclusion of the gonorhynchiforms within the Ostariophysi.

Some ostariophysans lack one or both parts of the response and reaction for apparently adaptive reasons. Piranhas lack the alarm reaction, which makes sense as many of their prey are also ostariophysans and it would be counterproductive for a predator to flee each time it bit into prey. Some nocturnal, non-schooling, or heavily armored ostariophysans lack both parts of the alarm response, including Blind Cave Characins, electric knifefishes, and banjo and suckermouth armored catfishes. An interesting seasonal loss of the production end of the response occurs in several North American minnows. Nest building and courtship in these fishes often involves rubbing by males against the bottom and between males and females, during which time the skin and its breeding tubercles may be broken. It would be counterproductive to the male if he produced a substance that frightened females away during these activities. Males resume the production of alarm substance in the fall, after the breeding season. As with the Weberian apparatus, convergent evolution of alarm substances and responses have evolved in other teleostean groups, including sculpins, darters, and gobies (Smith 1992).

Ostariophysans encompass two series, the Anotophysi with one order and the Otophysi with four orders. These taxa are too diverse to allow much detail here, and many aspects of their biology are treated in other chapters of this book. The most primitive order is the **Gonorhynchiformes**, which includes the Milkfish, *Chanos chanos* (series Anotophysi, family Chanidae), and three other relatively small tropical families. Milkfish are an important food fish in the Indo-Pacific region and are often cultured in brackish fishponds, where juveniles are raised to edible size on an algae diet. Milkfish grow to almost 2 m and 25 kg and is a popular sportfish in some areas.

The series Otophysi contains the bulk of freshwater fish species globally. The **Cypriniformes** constitutes the largest order and probably contains the most familiar species of the superorder. The Cyprinidae, the largest family of freshwater fishes and the second largest family (after the gobies) of all fishes, contains about 1,723 of the > 4,449 cypriniform species. Among the better-known cyprinids are the minnows, shiners, carps, barbs, barbels, gudgeons, chubs, dace, pikeminnows, Tench, Rudd, bitterlings, and bream; also such popular aquarium fishes as the Southeast Asian "sharks" (Redtail Black Shark, Bala Shark), Goldfish, Koi (domesticated common carp), Zebra Danios, and rasboras. The Zebrafish or Zebra Danio, *Danio rerio*, has become a standard laboratory animal in developmental genetics, toxicology, and medical research (Westerfield 2000; Gong & Korzh 2004; see also Zebrafish Information Network, **http://zfin.org**). Zoogeographically, cyprinids are most diverse in Southeast Asia, followed by Africa, North America (where there are 300 species according to Berra (2001)), and Europe. Cyprinids are absent from Australia and South America, their ecological roles filled largely by osmeriforms and atheriniforms in the former and by characins in the latter (see Chapter 19).

In the cyprinids, we see the first real development of pharyngeal dentition, a second set of jaws in the throat region that are derived from modified, tooth-bearing pharyngeal arches. Specifically, the fifth ceratobranchial (=pharyngeal) bone occludes against an enlarged posterior process of the basioccipital bone to form the pharyngeal bite. Cyprinids are also the first teleosts to develop a highly protrusible upper jaw and to eliminate the maxillary bones from the biting bones and gape of the mouth, both trends that are increasingly developed in more advanced teleostean taxa. Exclusion of the maxilla from the gape is a characteristic of all fishes more derived than the salmoniforms, although the bite of salmoniforms and their relatives involves the maxilla. The exclusion versus inclusion of the maxilla in cyprinids versus salmoniforms has led to some controversy over which group is more derived. The bulk of

FIGURE 14.7 The giant and imperiled Asian Carp, *Catlocarpio siamensis*, native to the Mekong River basin is among the world's largest minnows. Courtesy of Jean-Francois Healias.

the evidence favors salmoniforms as the more derived clade ("minnows before trout"; Smith 1988); maxilla inclusion in salmoniforms may be a secondarily evolved trait.

Some cyprinids have chromosomes in the polyploid condition, an unusual occurrence among fishes. The normal diploid 2N condition of most cyprinids is 48 or 50, although tetraploid (2N = 100), hexaploid, and even octaploid species occur, as is the case for the goldfish (Buth et al. 1991). Polyploidy is linked with large size in minnows; the world's largest species are the Southeast Asian *Catlocarpio siamensis*, a tetraploid (Fig. 14.7), and the Indian Mahseer, *Tor putitora*, both of which reach 2.5–3 m in length. The largest minnow in North America is the piscivorous Colorado Pikeminnow, *Ptychocheilus lucius*. Exceptional size in cyprinids is also often accompanied by predatory habits, as implied by the scientific names of such large (>2 m) species as *Elopichthys bambusa* and *Barbus esocinus*. Most cyprinids, however, are quite small (<5 cm), and the smallest freshwater fish, and perhaps vertebrate, is an Indonesian cyprinid, *Paedocypris progenetica*, that matures at 7.8 mm.

The superfamily Cobitoidea includes the disparate families of algae eaters, suckers, loaches, and river loaches. Gyrinocheilid algae eaters have modifications to the mouth and gill apparatus that allow them to scrape algae from rocks in areas of strong current. The mouth is modified into a sucking organ that helps them cling to rocks while scraping off algae. The fish breathes by inhaling water dorsally and exhaling it ventrally through small apertures in the gill opening. Suckers (Catostomidae) include about 72 species of relatively large (50–100 cm), chiefly North American fishes (Figs 14.8, 14.9). One species, the Chinese Highfin "Shark," *Myxocyprinus asiaticus*, occurs in eastern China, and another species, the Siberian Longnose Sucker, *Catostomus catostomus rostratus*, occurs in Alaska and northeastern Siberia (Scharpf 2006). Species include the buffaloes, Quillback, carpsuckers,

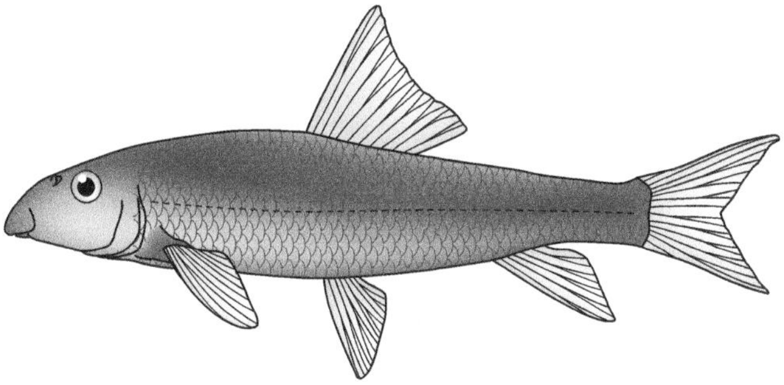

FIGURE 14.8 The extinct Harelip Sucker, *Moxostoma lacerum*. Once abundant in 13 eastern US states, this may have been the first American fish driven to extinction, around 1900 (see Jenkins & Burkhead 1993). Actual coloration is not known because no live fish were ever drawn or photographed; only one adult was preserved. Trautman (1981).

Blue Sucker, redhorses, jumprocks, and the extinct Harelip (Fig. 14.8) and Snake River suckers. Most suckers are benthic feeders in flowing water with inferior mouths and plicate or papillate lips. Exceptional are some lake suckers in the western USA (e.g. *Chasmistes*) that are midwater planktivores with more terminal mouths. Suckers can be quite confusing taxonomically because they frequently hybridize. Taxonomic confusion is further confounded by the scientific and common name combinations in this family, such as the Quillback, *Carpiodes cyprinus*, and the River Carpsucker, *Carpiodes carpio*, which can be mixed up with the Common Carp, *Cyprinus carpio*, which is a cyprinid.

Loaches (Cobitidae) constitute 221 species of predominantly Eurasian fishes that have their highest diversity in Southeast Asia. Included are such popular aquarium fishes as the Kuhli, Clown, and Skunk loaches, the weatherfishes, and the Golden Dojo. Weatherfishes (*Misgurnus*) obtained their name

FIGURE 14.9 A juvenile Robust Redhorse, *Moxostoma robustum*. Growing to large size (80 cm, 8 kg), this rare catostomid endemic to Atlantic slope rivers of the southeastern USA went unrecorded for over 120 years. Rediscovered in 1991, a cooperative effort among government, corporate, and nongovernmental organizations succeeded in captive propagation, release into the wild, and establishment and reproduction by propagated fish (see Helfman 2007; www.robustredhorse.com/h/reportpubs.html). G. Helfman (Author).

because they become restless when atmospheric pressure drops preceding a storm. Their sensitivity to barometric fluctuations may somehow relate to their air-breathing abilities, which involve gulping air and passing a bubble to the intestine where gaseous exchange occurs. Balitorid river or hillstream loaches are a highly diverse family (102 species), many of which are specialized for life in fast-flowing mountain streams of India and Southeast Asia. Their paired fins tend to be oriented horizontally, are enlarged, and have adhesive pads on their ventral surfaces. In addition, their bodies are depressed dorsoventrally, and their mouths are ventral, all anatomical adaptations to life in swift or turbulent water (see Chapter10). One large family of balitorids, the Nemacheilinae, includes several cave-dwelling species (see Chapter 10).

The **characiforms** are another large order (about 2,247 species) of primarily tropical otophysans characterized (usually) by an adipose fin, well-armed mouths and replacement dentition (e.g. piranhas; see Fig. 14.10), and peripheral ctenoid as opposed to the cycloid scales found in less derived teleosts. This is a remarkably diverse order anatomically and ecologically, including predators, zooplanktivores, scale eaters, detritivores, and herbivores; the latter category includes fishes that feed on seeds, leaves, and fruits. Characiforms may be surface, water column, or benthic dwellers, although most species are found in midwater, many in shoals. Body sizes range from very small (13 mm adult tetras) to quite large (e.g. 1.5 m long tigerfishes) and body shapes range from long, slender, almost darter-like benthic fishes (e.g. South American darters, *Characidium*) to deep-bodied, compressed piranhas and hatchetfishes (*Gasteropelecus*). Numerous popular, colorful aquarium fishes belong to this order, including *Distichodus, Prochilodus*, headstanders, spraying characins, freshwater hatchetfishes, blind characins, pencilfishes, tetras (*Cheirodon, Hemigrammus, Micralestes, Paracheirodon*), and silver dollars, as do important food fishes (*Prochilodus, Colossoma, Brycon*). Currently, 24 families are recognized, although past classifications have recognized as few as one. The great majority of species (*c.* 1300) are South American, about 200 are African, a small number live in Central America, and one species, the Mexican Tetra, *Astyanax mexicanus*, extends naturally into southwestern Texas. Another 10 species, including piranhas, have been introduced into the USA. Because of its large size and tropical nature, the order has undergone considerable taxonomic revision, much of which is still in progress.

FIGURE 14.10 Piranhas, *Serrasalmus* spp., are representative of the speciose tropical order of characiform fishes. G. Helfman (Author).

The most primitive characiforms are the 113 species of African citharinoids (*Distichodus, Citharinus*), attesting to an African origin for the order and the connection between Africa and South America prior to the breakup of Gondwanaland in the Mesozoic. African characiforms also include the advanced alestiids (about 117 species). Two of the largest characiforms are predaceous African species. The Pike Characin, *Hepsetus odoe* (Hepsetidae), is an impressive predator that reaches

65 cm in length and has fanglike teeth. It is remarkably convergent with the alestiid tigerfishes, *Hydrocynus* spp. Not far behind in the dentition and size categories are the cynodontid wolf characins or payara of South America (*Hydrolycus*: 1.2 m, 18 kg), a much-sought sportfish.

Apart from various popular aquarium species, certainly the best known or at least most notorious characids are the piranhas (Fig. 14.10). This family, the Serrasalmidae, contains about 98 species, some of which are predatory (*Serrasalmus*), others which are scale-eating opportunists and specialists (*Pygocentrus*, *Catoprion*), and some that are largely herbivorous, such as the pacus and silver dollars (*Colossoma*, *Metynnis*) (Sazima 1983; Nico & Taphorn 1988). Despite their reputation and potential for doing damage, many purported attacks on humans by piranhas actually result from post-mortem scavenging on drowning victims (Sazima & Guimaraes 1987). In recent years, however, this picture has changed as a result of a proliferation of dams in southeastern Brazil that created ideal piranha spawning habitat. More than 85 attacks on bathers by nest-guarding piranhas have been reported, many resulting in serious injury (Haddad & Sazima 2003). The large, herbivorous serrasalmine species are important food fishes in the Amazon basin and are also important dispersers of seeds during the wet season, particularly because they use their massive dentition to husk seeds, which may aid germination (Goulding 1980). The Characidae contains 13 subfamilies with at least 1193 described species.

The diversity of catfishes (**Siluriformes**) amazes most everyone (Burgess 1989; Arratia et al. 2003) (Fig. 14.11). Approximately 39 families and at least 4116 species of catfishes are recognized, and it is not surprising that catfish systematics remains active, controversial, and unsettled (see the

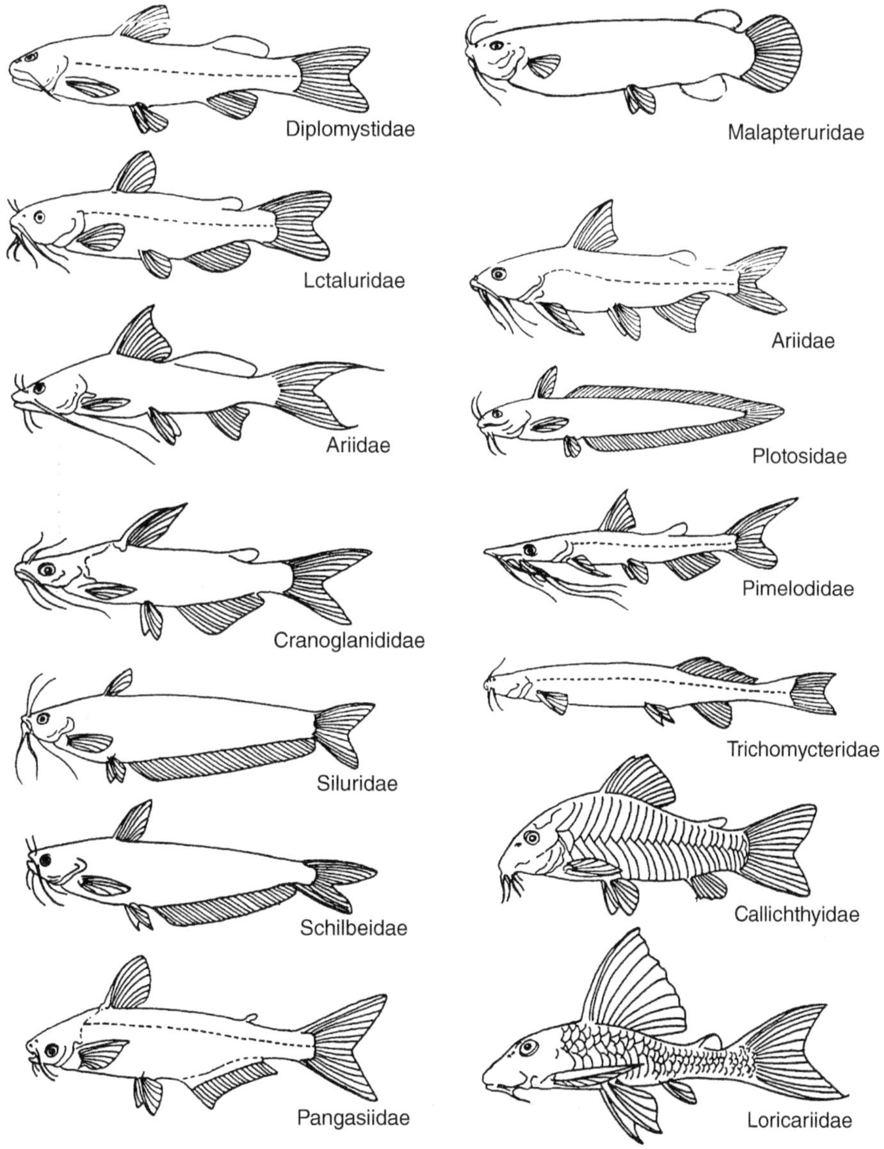

FIGURE 14.11 Selected catfishes, showing some of the array of body types and shared characteristics among the 39 families. Drawings by John Quinn, in Burgess (1989), used with permission of TFH Publications.

All Catfish Species Inventory, **http://silurus.acnatsci.org** for regular updates and photos). Commonalities among the families include: fusion, reduction, or loss of a number of skull bones found in less derived teleosts, including the maxilla; teeth on the roof bones of the mouth (vomer, pterygoid, palatine); an adipose fin, sometimes with rays or a spine; an unsegmented, spine-like ray at the front of both the dorsal and pectoral fins that in some families is covered by a toxin produced by epidermal glandular cells; the dorsal spine is often preceded by a shorter spine that helps lock the larger spine in the erect position; a lack of scales, often combined with the presence of bony plates or tubercles; small eyes (and nocturnal, benthic foraging habits); and one to four pairs of barbels associated with both the upper and lower jaws that serve both chemosensory and tactile functions.

Catfishes are known from all continents, including Antarctica during the Oligocene. They reach their greatest diversity in South America, where the largest families occur (loricariid suckermouth armored catfishes with about 996 species, trichomycterid pencil catfishes with 320 species, and callichthyid armored catfishes with about 221 species); the bagrids of Africa/Asia and the African mochokid upside-down catfishes are close behind with over 220 species. The most primitive catfishes are the South American diplomystids, which have a well-developed, toothed maxillary bone, a trait they share with the extinct Eocene hypsidorids. As with the vast majority of ostariophysans, most catfishes are confined to fresh water, including many cave species (e.g. Proudlove 2006). However, two families of catfishes are primarily marine, the widespread sea catfishes (Ariidae) and the highly venomous Indo-Pacific eeltail catfishes (Plotosidae). The plotosids have diversified in several respects. Juvenile lined catfishes, *Plotosus lineatus*, are tightly schooling, diurnally active fishes in contrast to the solitary, nocturnal behavior of most other families. Both families contain members that occur in fresh water, such as the plotosid tandan catfishes of Australia (*Tandanus*), which probably represents secondary evolution of the use of freshwater habitats (i.e., freshwater ancestor gave rise to a marine species which then reinvaded fresh water). Marine plotosids have highly venomous spines and relatively bold coloration, suggesting an aposematic warning function, which does not appear to deter their chief predators, the highly venomous seasnakes (Voris et al. 1978).

Most catfishes are naked, lacking true scales. But in some families, different parts of the body are covered with bony plates (armorhead catfishes, loach catfishes, sea catfishes, thorny catfishes, callichthyid armored catfishes, suckermouth armored catfishes), or thorn-like projections, tubercles or "odontodes" (sisorids, thorny catfishes, aspredinid banjo catfishes, spiny dwarf catfishes). The popular "plecostomus" catfishes of the aquarium trade are loricariid suckermouth armored catfishes; most are in the genus *Hypostomus* because, regrettably, *Plecostomus* is no longer a valid scientific name. Relatives in the genus *Pterygoplichthys* have become a nuisance introduction in Mexico and Florida where their populations have exploded.

Some of the world's largest freshwater fishes are catfishes, including the predatory European Wels (*Siluris glanis*, Siluridae) at 5 m and 330 kg; the herbivorous, Critically Endangered, Mekong Giant Catfish (*Pangasianodon gigas*, Pangasiidae) at 3 m and 300 kg; and the 2.8 m, 150 kg long-whiskered pimelodid Piraiba (*Brachyplatystoma filamentosum*) of South America (Fig. 14.12). Piraiba are importantly economically and ecologically in the Amazon basin and are legendary for their annual migrations (Barthem & Goulding 1997). The largest catfishes in North America are the Flathead and Blue catfishes, *Pylodictis olivaris* and *Ictalurus furcatus*, which reach about 1.5 m and 50–68 kg. Very small catfishes such as spiny dwarf catfishes (Scoloplacidae) and whalelike catfishes (Cetopsidae) are only 20–25 mm long as adults. The pencil or parasitic

(A) (B)

FIGURE 14.12 Large catfishes. (A) The Mekong Giant Catfish, one of the world's largest catfishes. This specimen was caught from a stocked population in Bung Sam Lan Lake, Thailand. (B) Two *c.* 20 kg Flathead Catfishes, North America's second largest catfish. (A) Courtesy of Jean-Francois Healias; (B) G. Helfman (Author).

catfishes (Trichomycteridae) of South America include species that eat mucus and scales from other fishes or that pierce the skin or gill cavities of other fishes and feed on blood. At least one parasitic species, the Candiru (*Vandellia cirrhosa*), also called the Toothpick Fish or Vampire Fish, has been reported to enter the vaginal canal or urethra (urinary tract) of bathers and become lodged in place due to its opercular spines – but many of these reports have been called into question (Burgess 1989; Spotte 2002).

Most catfishes are benthic, but some silurid sheatfishes, schilbeids, ageneiosid bottlenose catfishes, and *Hypophthalmus* lookdown catfishes normally swim above the bottom (the relatively bizarre lookdowns are probably filter feeders). Occupation of the water column has produced some striking convergences in species that hover in open space. For example, the Glass Catfish, *Kryptopterus bicirrhus*, of Southeast Asia is a 10 cm long, transparent silurid with one pair of long barbels that protrude outward from its head, a long anal fin, no adipose fin, a forked tail, and only a single small ray in its dorsal fin (hence the name *kryptopterus* = "hidden fin"). It tends to hover tail down in the water column, often in shoals. The African Glass Catfish, *Parailia pellucida*, is morphologically and behaviorally similar.

Several unique modifications occur in the different families, far too many to detail here (see Burgess 1989 and several chapters in Arratia et al. 2003). Among these traits are accessory air-breathing structures and terrestrial locomotion in airbreathing catfishes, airsac catfishes, and callichthyid armored catfishes; generation of electric impulses in the African electric and upside-down catfishes; climbing ability in climbing catfishes and jet propulsion in banjo catfishes; use of lures in angler catfishes; and mouth-brooding of large eggs in sea catfishes.

The most derived ostariophysans are the **gymnotiforms**, which show internal anatomical similarities to the siluriforms and probably share a common ancestor. However, gymnotiforms are distinct from catfishes and all other ostariophysans, as well as from all other teleosts except osteoglossiform mormyrids and gymnarchids, in that they produce and receive weak electric impulses. Gymnotiforms are known collectively as South American knifefishes because of their strong resemblance to the African knifefishes (Notopteridae); the latter have electrogenic relatives among the mormyrids and gymnarchids but do not produce electricity themselves.

Gymnotiforms are restricted to Central and South America and consist of 256 species in five families. Anatomically, they are characterized by an elongate, compressed body, an extremely long anal fin that reaches from the pectoral fin to the end of the body, no dorsal or caudal fin, small eyes (and nocturnal foraging), and electrogenic tissue combined with modified lateral line organs for detecting weak electric fields. Gymnotiforms range in size from 9 cm long hypopomid bluntnose knifefishes to the 2.2 m long Electric Eel. The electrogenic tissue is derived from modified muscle cells in four families, and, curiously, from nerve cells in the apteronotid ghost knifefishes

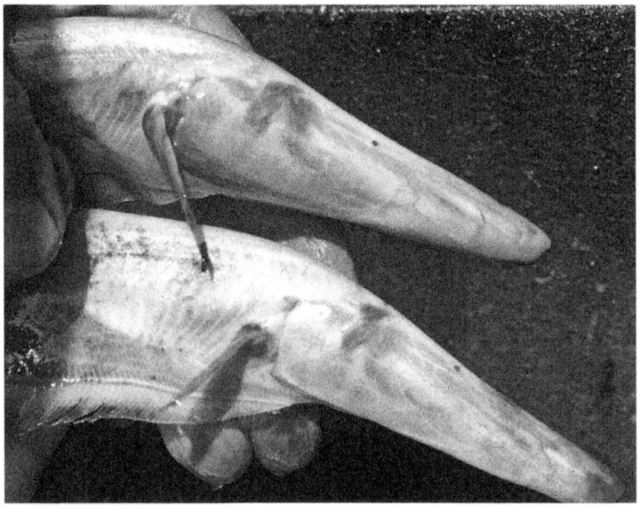

FIGURE 14.13 Two individual *Orthosternarchus tamandua*, an apteronotid knifefish from the Amazon basin. The small black dot on the head is the greatly reduced eye. These predators occur at depths of 6–10 m where they feed on insect larvae (Fernandes et al. 2004). Courtesy of C. Cox Fernandes.

FIGURE 14.14 Gymnotiform electric knifefishes – an Electric Eel (*Electrophorus electricus*). S. Johnson / Wikimedia Commons / CC BY-SA 3.0.

(Alves-Gomes 2001). Apteronotids also depart from the rest of the order by having a distinct tail fin and include some truly bizarre species (Fig. 14.13). Electrical output in gymnotiforms is continual at high frequencies, as compared to the pulsed, low-frequency output of mormyrids. The electrical output is usually very weak, on the order of fractions of a volt, except in the notorious electric eels. The well-known *Electrophorus electricus* (Fig. 14.14), puts out a weak field for electrolocation purposes and strong pulses over 600 volts for stunning prey or deterring predators (including people). De Santana et al. (2019) reported two new species of electric eels, one of which can generate electric pulses over 850 volts.

Cohort Euteleostei

Cohort Euteleostei[a]

Order Lepidogalaxiiformes. Lepidogalaxiidae (Salamanderfish)

Superorder Protacanthopterygii

Order Salmoniformes. Salmonidae (228 species of trouts, salmons, and whitefishes).

Order Esociformes. Esocidae (7 species of pikes and pickerels), Umbridae (7 species of mudminnows)

Order Argentiniformes (202 species): Argentinidae (about 28 species of argentines or herring smelts), Microstomatidae (about 28 species of pencilsmelts), Bathylagidae (25 species of deepsea smelts), Opisthoproctidae (21 species of barreleyes)

Order Galaxiiformes. Galaxiidae (66 species of galaxiids)

Order Osmeriformes. Osmeridae (15 species of smelts), Plecoglossidae (Ayu), Salangidae (18 species of icefishes), Retropinnidae (5 species of New Zealand smelts), Microstomatidae (pencilsmelts), Platytroctidae (tubeshoulders), Bathylaconidae (bathylaconids), Alepocephalidae (slickheads)

Order Stomiiformes (444 species): Gonostomatidae (33 species of bristlemouths), Sternoptychidae (about 78 species of marine hatchetfishes), Phosichthyidae (24 species of lightfishes), Stomiidae (about 309 species of barbeled dragonfishes)

[a] Higher classification groups based on Nelson et al. (2016), but see corrections and updates at **https://sites.google.com/site/fotw5th/**. Orders, families, and numbers of species from Fricke et al. (2020) but refer to **https://researcharchive.calacademy.org/research/ichthyology/catalog/SpeciesByFamily.asp** for updates.

All the teleosts more derived than the ostariophysans are generally placed together in the cohort **Euteleostei**, the "true" teleosts. This designation underscores the conclusion that teleostean phylogeny involved four major radiations, the first three producing less derived and separate groups (elopomorphs, osteoglossomorphs, otocephalans), and the fourth containing a vast array of more derived, euteleostean fishes (346 families, 2935 genera, and over 17,400 species).

The monophyly of the euteleosts as well as the organization of the cohort remain an area of debate because unique characters common to all or most members are lacking (or are confused by exceptions), and evidence from different approaches (e.g. anatomical, developmental, and molecular) supports differing conclusions. Johnson and Patterson (1996) and Wiley and Johnson (2010) proposed a scheme, followed here, that emphasizes three shared traits:

1. Similarities in the pattern of embryonic development of the supraneural bones among euteleosts, which are the small T-shaped or rod-like bony or cartilaginous elements that lie within the musculature between the cranium and the dorsal fin.
2. The presence of an outgrowth on the stegural bone, a structure associated with the neural arches of the vertebral centra of the tail base.
3. The presence of caudal median cartilages, cartilages that lie between the hypural bones of the caudal base.

Strong support for Euteleostei is also seen in molecular-sequence studies such as Betancur-R. et al. (2013) and others listed in Nelson et al. (2016). Johnson and Patterson's analysis eliminates reliance on even more problematic structures for defining the euteleosts. Older summaries emphasized the shared existence of structures that had probably evolved convergently among taxa or that were shared with groups clearly primitive to euteleosts, or that were missing from important clades. One such trait was the presence of epidermal **breeding tubercles** and dermal **contact organs**, small bumps on the fins and bodies that develop during the breeding season in 25 families (Collette 1977). Another, highly enigmatic trait was the presence of an **adipose fin**. This small, usually rayless fin sits posterior to the first dorsal, often just anterior to the caudal peduncle. It can be a small vertical flap or bump, or it can be a long, substantial structure confluent with the caudal fin, as in many catfishes. It occurs in characiforms, catfishes, smelts, the deep-sea stomiiforms, aulopiforms, and scopelomorphs, the salmoniforms (salmons, trouts, whitefishes), and trout-perches. It is generally absent in more derived fishes, particularly those with a spiny first dorsal fin. Its functions are a mystery.

Protacanthopterygians as a group have undergone repeated revision, taxa being removed and added to the superorder as new information is obtained and old data are reinterpreted. The name is used here because it provides an organizational category for a series of apparently related fishes (Fig. 14.15). Based primarily on molecular evidence, such as Betancur-R. et al. (2013), Nelson et al. (2016) considered the Lepidogalaxiiformes, containing only the strange Salamanderfish *Lepidogalaxias salamandroides*, to be the primitive sister group of the Protacanthopterygii. Previously, it was included in, or as sister to the Galaxiidae. We follow Nelson et al. (2016) in including esociform pikes and mudminnows in the Protacanthopterygii, rather than aligning them with more derived teleostean groups.

A truly unique fish is the Salamanderfish *Lepidogalaxias salamandroides* (Fig. 14.16). This benthic, elongate fish inhabits seasonally dry ponds of southwestern Australia, where it buries in the mud and lives in a torpid state after ponds dry up, re-emerging with the next rains. Salamanderfish

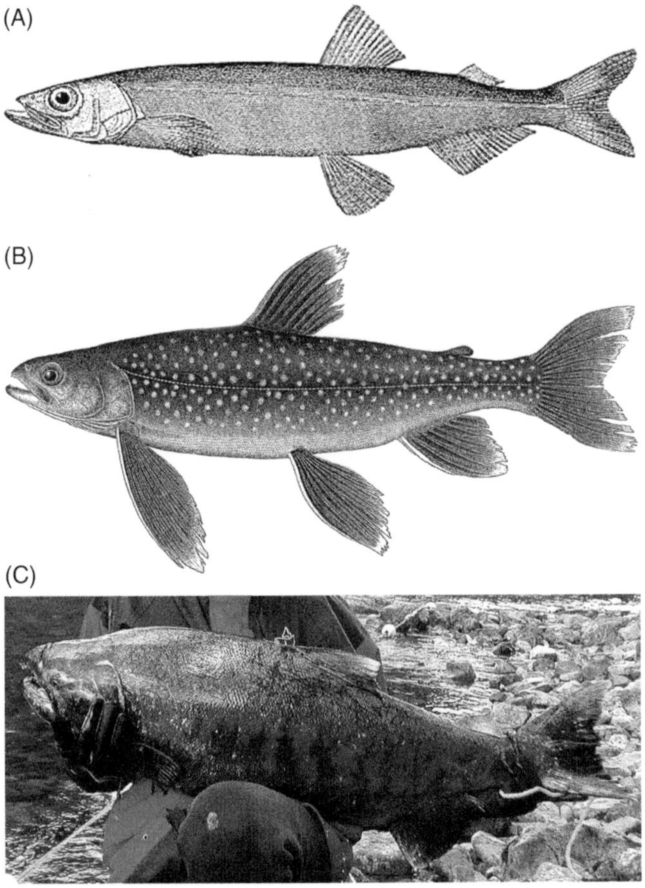

FIGURE 14.15 Protacanthopterygians. (A) An osmerid, the Capelin, *Mallotus villosus*. (B) The Longfin Svetovidov's Char, *Salvethymus svetovidovi*, of Lake El'gygytgyn, Siberia. This long-lived, small char (30 years, 30 cm) is threatened by pollution and fishing. (C) A large Chinook Salmon, *Onchorhynchus tshawytscha*, from British Columbia. (A) From Evermann and Goldsborough (1907). *The fishes of Alaska*. Bull. U.S. Bur. Fish, Public Domain; (B) drawing by Paul Vecsei, used with permission; (C) courtesy of R. Carlson.

lack eye muscles but instead have a flexible neck joint that allows them to bend their neck at a right angle to the side, a very unusual ability in fishes. Neck bending is made possible by large gaps between the back of the skull and the first cervical vertebrae and between the first and second cervical vertebrae. Salamanderfish also lack a normal gas bladder but instead have a gas-containing structure made up of simpler mesentery-like tissue and collagen fibers (Berra & Allen 1989; Berra et al. 1989). This seems to be a case of convergent evolution that resulted in an alternative type of physostomous gas bladder.

Superorder Protacanthopterygii

There is only one family within the order **Salmoniformes,** the **Salmonidae** (although the osmerid smelts have sometimes been included in the order). Salmonids are an important group commercially, ecologically, and aesthetically, and a fascinating group evolutionarily. They are the focus worldwide of fisheries classes and of popular and technical books. Different taxonomic treatments recognize as many as three different families (coregonid whitefishes, thymallid graylings, salmonid salmons and relatives), but we follow Nelson et al. (2016) and treat them as subfamilies of a larger Salmonidae. Their general anatomical similarities include an adipose fin, no spiny fin rays, a triangular flap at the base of the pelvic fin (**pelvic axillary process**), gill membranes free from the ventral side of the head, maxilla included in the gape, a physostomous gas bladder, and vertical barring (parr marks) on the sides of most young. Internally the last three vertebrae angle up toward the tail, and a myodome, or area of the skull where the extrinsic eye muscles insert, is present.

The **coregonine** whitefishes and ciscoes consist of three genera with as many as 88 species of relatively large-scaled salmonids that lack teeth on the maxillary bone. They are zooplanktivorous fishes in high-latitude lakes of North America and Eurasia that show a great deal of within-species variation and specialization, forming **species complexes** that differ from lake to lake (Bernatchez et al. 2010). Kottelat (1997) documented at least five apparent extinctions and several more extirpations of unique coregonines in Europe. During the mid-twentieth century, two coregonines native to the Laurentian Great Lakes, the Deep Water Cisco, *Coregonus johannae*, and the Blackfin Cisco, *C. nigripinnis*, disappeared as the result of overexploitation, pollution, siltation, competition with introduced species, and perhaps predation by introduced Sea Lamprey. Whitefishes exemplify the vulnerability of taxa that develop localized specializations (the Wild Salmon Center and Ecotrust maintains a website at **www.wildsalmoncenter.org** that includes accounts of endangered salmonids based on information from the International Union for the Conservation of Nature (IUCN)'s Salmon Specialist Group). The **thymalline** graylings, *Thymallus*, are a small group of about five species of northern hemisphere riverine salmonids that are easily identified by an elongate, flowing dorsal fin.

The subfamily **Salmoninae** contains seven Eurasian and North American genera (*Brachymystax, Acantholingua, Salmothymus, Hucho, Salvelinus, Salmo, Oncorhynchus*) that differ from other salmonids by having small dorsal fins, small scales, and teeth on the maxillary bone (Fig. 14.15B, C). Most species of economic importance are in the latter three genera, although the Siberian Taimen, *Hucho taimen*, is the world's largest salmonid at 2 m and 70 kg (a commercially caught individual weighed 114 kg). North American Salmoninae are currently divided into three genera and approximately 20 species, the names and relationships of which have been the subject of considerable debate (Fig. 14.17). The chars (or charrs) include the Lake, Brook, and Bull trouts, Arctic Char, and Dolly Varden, are all in the genus *Salvelinus*. The northernmost-living freshwater fish in the world is the Arctic Char, *Salvelinus alpinus*, of Lake Hazen, on Ellesmere Island in Canada (80°N); many char are anadromous, moving into the sea to feed and grow and then back into fresh water to spawn. Arctic Char in isolated lakes frequently differentiate into distinct ecomorphological types,

FIGURE 14.16 Neck flexibility in the Australian salamanderfish, *Lepidogalaxias salamandroides*. This unusual benthic fish is able to bend its neck sideways and downward due to a unique arrangement of spaces between the skull and the cervical vertebrae. A lack of ribs throughout the vertebral column probably aids in neck bending and also allows the fish to make sinuous movements. (A) A 35 mm long salamanderfish in the bent-neck position. (B) A cleared and stained, 49 mm long salamanderfish showing intervertebral gaps and lack of ribs. (C) A comparison specimen of the related *Galaxiella munda* (Galaxiidae, 46 mm), showing the tightly coupled vertebrae and more elongate ribs. Note also the well-developed pelvic girdle of *Lepidogalaxias*, which is used as a prop during resting. Berra and Allen 1989 / with permission of Taylor & Francis.

forming functionally different ecological populations within taxonomic species.

The remaining salmonines are the Atlantic basin salmon and trout (*Salmo salar*, the Atlantic Salmon; *S. trutta*, the European Brown Trout, with numerous subspecies and races; and three other European species), and the 11 species of Pacific basin trouts and salmons in the genus *Oncorhynchus* (Behnke 2002). Two *Oncorhynchus* species, *O. masou* and *O. rhodurus*, are endemic to Japan. Pacific trouts and salmons include narrowly distributed, landlocked forms such as Golden (Fig. 14.18) and Gila trouts, and species that are spectacularly anadromous, such as the Chinook, Chum, Coho, Pink, and Sockeye salmons (*O. tshawytscha*, *O. keta*, *O. kisutch*, *O. gorbuscha*, and *O. nerka*, respectively). Latinized names of Pacific salmon defy the usually predictable logic of binomial nomenclature. The mystery is partially solved when one learns that these spellings represent Russian names transliterated into English and later converted into Latin over a 60-year period by a series of German and British ichthyologists (Moyle 2002). Some Pacific salmon undergo oceanic migrations of thousands of kilometers before returning to their birth river to spawn and die (see Chapter 18; Representative Life Histories of Migratory

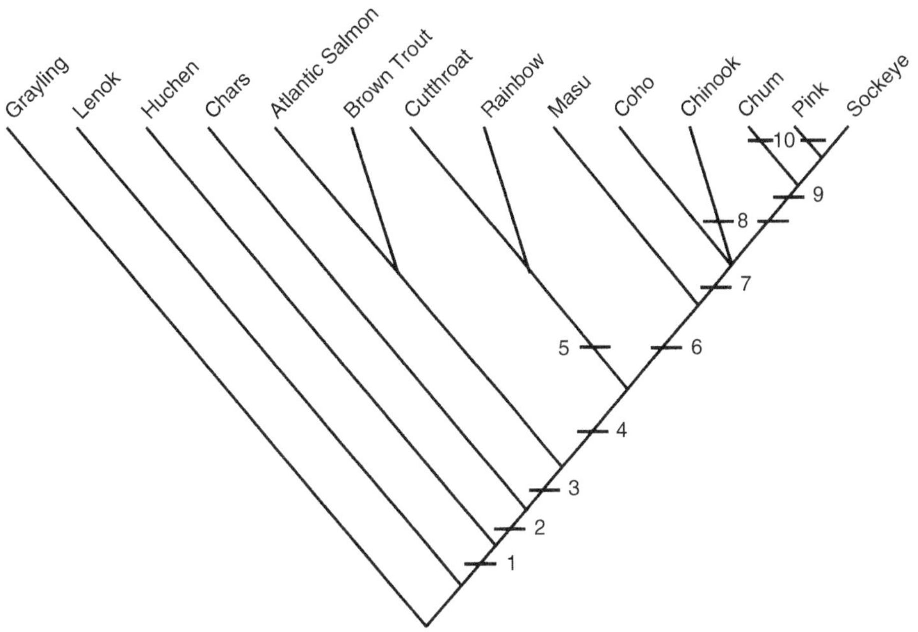

FIGURE 14.17 Phylogeny of the salmonids. A cladogram of most living salmonids based on life history traits shows the evolution of the various species. The same cladogram is constructed if anatomical and biochemical traits are used. The first four lineages represent separate genera (*Thymallus*, *Brachymystax*, *Hucho*, *Salvelinus*), Atlantic Salmon and Brown Trout are in the genus *Salmo*, and the remaining eight species (those above trait 4) are all in the genus *Oncorhynchus*. The more primitive coregonine whitefishes would come off to the far left of the cladogram and are not shown. The following life history and reproductive characteristics are the shared derived characters that were used to construct the cladogram. The listed characters correspond to the numbered branch points in the cladogram. Groups to the right and above the number possess the trait, those to the left do not. From Smith and Stearley (1989), used with permission. 1, egg diameter greater than 4.5 mm; females dig redds (nests); large males have hooked jaws (kype) during breeding season; 2, fall spawners; 3, commonly undergo long oceanic migrations; 4, most spawners undergo irreversible hormonal changes; 5, spring spawners; 6, anadromous forms die after spawning; 7, non-migratory individuals tend not to reproduce; 8, most smolt in first year, some go to sea as even younger fry; 9, juveniles are strong schoolers, parr are slender; 10, the freshwater phase is reduced; young migrate soon after emerging from gravel.

FIGURE 14.18 A Golden Trout, *Oncorhynchus aguabonita*, the California state fish. Greatly depleted by introduced species and habitat destruction, Golden Trout are now the focus of restoration programs that include eradication of the European Brown Trout that were actively stocked in past decades. Courtesy of E. P. Pister.

Fishes). The actual number of species, subspecies, and distinct races of Pacific salmons is a matter of considerable and important debate because of the wholesale destruction of stocks in various rivers of the Pacific Northwest region of the USA (Lichatowich 1999; Williams 2006). Many of these stocks are reproductively isolated and genetically distinct and therefore are viewed as unique evolutionary taxa, termed evolutionary significant units (ESUs) for conservation purposes. This is an extremely important consideration when considering stocking programs to try to rebuild populations.

The order **Esociformes** consists of two temperate families of freshwater fishes in which the maxillary bone is included in the gape but is toothless, and in which the median fins are located relatively far back on the body. Esocids include the Grass, Redfin, and Chain pickerels, the Northern and Amur pikes, and the superpredatory 1.4 m Muskellunge, *Esox masquinongy* (Fig. 14.19). The Northern Pike, *E. lucius*, has the most widespread natural east–west distribution of any completely freshwater fish, occurring across the northern portions

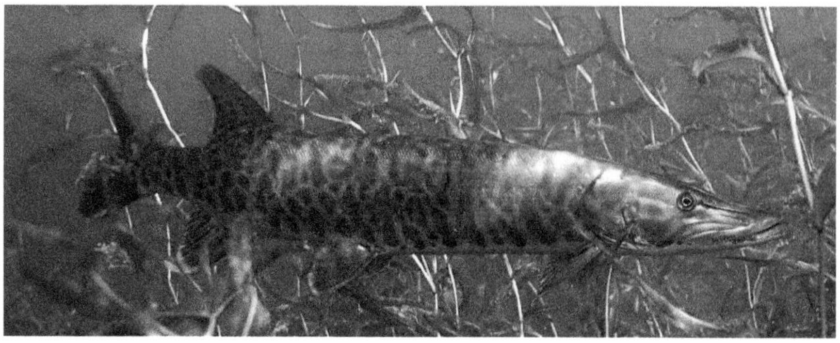

FIGURE 14.19 The esocid Muskellunge, *Esox masquinongy*, one of the largest predatory freshwater fishes in North America. E. Engbretson / USFWS / Public Domain.

of North America, Europe, and Asia (= circumpolar distribution). The family goes back into the Cretaceous, and one Paleocene species is very similar to the Northern Pike despite a 62-million-year separation. Anatomical and behavioral convergences between freshwater pikes and saltwater barracudas, the latter a more derived acanthopterygian, are remarkable. The mudminnows and blackfishes (Umbridae) are also scattered across northern North America, Europe, and Asia. The Central Mudminnow, *Umbra limi*, has a remarkable ability to survive in high-latitude lakes that become ice-covered and oxygen-poor through much of the winter.

Argentiniforms are four families of primarily deep-sea inhabitants. The barreleyes, Opisthoproctidae, are bizarre looking deep-sea fishes with elongate tubular eyes that point upward. The platytroctids are also deep-sea fishes that exude a blue-green luminous fluid from a papilla located under the shoulder girdle, perhaps analogous to squid ink; many species also possess photophores (light organs), a convergent trait among deep-sea species.

The **Galaxiformes** occur in Australia and New Zealand but also in southern South America and Africa. Galaxiids constitute important commercial whitebait fisheries in New Zealand. They have complex life cycles, exhibiting all major types of diadromy, including anadromy, catadromy, and amphidromy. Some are also semelparous, spawning only one time before dying.

Among the **Osmeriformes** is the family **Osmeridae**, known as smelts, which includes 15 species of marine, freshwater, and diadromous (migrating between fresh water and the sea) species that inhabit shallow waters. Osmerids are generally small, silvery, elongate fishes that occupy the water column (Fig. 14.15A). They have a single, soft-rayed dorsal fin and sometimes an adipose fin. The maxilla is included in the gape, and most jaw bones possess teeth. Although their pelvic fins are abdominal, as is the case for most of the preceding teleostean groups, some osmerids have their pectorals located higher on the body than is typical in the less derived groups. Osmeriforms include commercially important species such as Capelins (*Mallotus*), Eulachons (*Thaleichthys*), Asian Ayu (*Plecoglossus*), and Rainbow Smelt (*Osmerus*), many of which are superficially similar to the more advanced silversides (Atherinidae). Another osmeroid is a second family that dominates cold freshwater environments of the southern hemisphere: the Retropinnidae. These are small (10–35 cm), marine and anadromous (spawn in fresh water, grow in the sea) fishes of New Zealand and Australia known as New Zealand smelts. They sometimes establish landlocked populations in lakes. Retropinnids and galaxiids as a group have suffered numerous extirpations and extinctions as a result of habitat destruction and the stocking of non-native trouts. One retropinnid, *Prototroctes oxyrhinchus*, is known to be extinct.

The order **Stomiiformes** was previously placed in the Neoteleostei but, based largely on genetic studies such as Betancur-R et al. 2013, are now considered to be better placed as sister to osmeriforms in the Protacanthopterygii. All are deep-sea fishes of the mesopelagic and bathypelagic regions (open water, between 200 and about 4000 m depths) and are often characterized by long teeth uniquely attached to jaw bones, large mouths, histologically unique photophores (light organs) that include a duct, and a peculiar ventral adipose fin ahead of the anal fin in some. Gonostomatid bristlemouths (*Cyclothone*) and phosichthyid lightfishes (*Vinciguerria*) may be the most abundant and widely distributed vertebrates on earth. The sternoptychid marine hatchetfishes possess several structural specializations that emphasize the vertical plane and body compression. The mouth opens vertically, the photophores point down, a preopercular spine points upward, the pelvic bones are oriented vertically, the lateral compression of the body is heightened by an abdominal keel-like structure, and a bladelike structure preceding the dorsal fin is made up of dorsal pterygiophores that project through the back of the fish. Pterygiophores normally serve as the basal support for median fins, not as elements of a fin itself. Idiacanthine black dragonfishes have a larval form with eyes at the ends of elongated stalks. Stomiiforms show many traits common to other, unrelated deep-sea fishes that are generally viewed as convergent adaptations to the light- and food-limited conditions of the deep sea (see Chapter 10).

Neoteleostei

All of the fishes after the stomiiforms are considered to be **neoteleosts**, a category without formal rank that lies somewhere between a cohort and a superorder (see Fig. 14.1). Six neoteleostean superorders are recognized, including four relatively specialized deep-sea and pelagic superorders and two very diverse, highly derived superorders, the paracanthopterygians and acanthopterygians. Neoteleosts as a taxon are considered to be monophyletic on the basis of four skull and jaw characters that are lacking in other teleosts:

1. The manner in which the vertebral column connects to the back of the skull changes. In the neoteleosts, the first vertebra articulates with three bones of the skull (basioccipital and the two exoccipitals), whereas in less derived teleosts, it articulates with only the unpaired basioccipital.
2. A muscle, the retractor dorsalis, connects the vertebral column with dorsal elements of the upper pharyngeal jaws and pulls those jaws posteriorly.
3. A shift occurs in the insertion position of another muscle, one of the internal levators that originates on the base of the skull and lifts the pharyngeal jaws.
4. A unique hinged manner of attaching teeth to the jaws develops, allowing the tooth to be depressed toward the back of the mouth.

In addition, a trend toward more anteriorly located pelvic fins and more laterally located pectoral fins is evident during neoteleostean phylogeny, and acellular bone (skeletal material lacking bone cells) occurs in most neoteleosts, whereas less derived groups have bone cells (Smith 1988).

Superorder Ateleopodomorpha

Superorder Ateleopodomorpha[a]
Order Ateleopodiformes: Ateleopodidae (11 species of jellynose fishes)

[a] Higher classification groups based on Nelson et al. (2016), but see corrections and updates at **https://sites.google.com/site/fotw5th/**. Orders, families, and numbers of species from Fricke et al. (2020) but refer to **https://researcharchive.calacademy.org/research/ichthyology/catalog/SpeciesByFamily.asp** for updates.

The **ateleopodiform** jellynose fishes are an unusual group of bulbous-headed, elongate species that swim just above the bottom in deep water. Their skeleton is largely cartilaginous. Their large and pointed head, exaggerated anal fin, and relatively pointed tail are all traits they share with other deep, benthopelagic fishes such as chimaeras, spiny eels, halosaurs, eucla cods, rattails, and grenadiers (see Chapter 10).

Superorder Cyclosquamata

Superorder Cyclosquamata[a]
Order Aulopiformes (285 species): Aulopidae (15 species of flagfins), Chloropthalmidae (21 species of greeneyes), Paraulopidae (14 species of cucumber fishes), Ipnopidae (33 species of deep-sea tripod fishes), Bathysauroididae (Largescale Deep-sea Lizardfish), Scopelarchidae (18 species of pearleyes), Notosudidae (17 species of waryfishes), Giganturidae (2 species of telescopefishes), Synodontidae (83 species of lizardfishes), Bathysauridae (2 species of deep-sea lizardfishes), Paralepididae (about 62 species of barracudinas), Anotopteridae (3 species of daggertooths), Evermanellidae (7 species of sabertooth fishes), Omosudidae (Omosudic), Alepisauridae (2 species of lancetfishes), Pseudotrichonotidae (4 species of sand-diving lizardfishes)

[a] Higher classification groups based on Nelson et al. (2016), but see corrections and updates at **https://sites.google.com/site/fotw5th/**. Orders, families, and numbers of species from Fricke et al. (2020) but refer to **https://researcharchive.calacademy.org/research/ichthyology/catalog/SpeciesByFamily.asp** for updates.

Fishes more advanced than ateleopodomorphs are sometimes referred to as **eurypterygians** ("wide or broad fins"). The least derived superorder in this group is the Cyclosquamata ("cycloid scales"), which contains one large order of almost entirely deep-sea fishes. **Aulopiforms** have a surprisingly extensive fossil record (approximately 6 families and perhaps 20 genera) for a group that today is largely open water and deep v. The extant deep-sea groups include the truly bizarre giganturid telescope fishes (Fig. 14.20A) which undergo a spectacular metamorphosis during which the premaxillary, palatine, orbitosphenoid, parietal, symplectic, posttemporal, supratemporal, hyoid, and cleithral bones in the head region are lost, as are several gill arches, the gill rakers, and the gas bladder. The larvae possess an adipose fin, pelvic fin, and branchiostegal rays that are also lost during metamorphosis. The adults possess large tubular eyes, a huge mouth, flexible teeth, an expandable stomach, pectoral fins located exceptionally high on the body above the eyes, loose skin, and a peculiar tail with the ventral lobe extending far beyond the dorsal lobe.

Chlorophthalmid greeneyes are the first fishes encountered in this survey that are hermaphroditic, as are members of other families in this order. All of the less derived groups so far have been distinctly gonochoristic, which means that an individual is only one sex throughout its life. Hermaphroditism of some form is surprisingly common in higher teleosts. This dichotomy between less derived and more derived teleosts suggests that sexual lability – either in terms of initial sexual determination or an ability to change sex later in life – represents a derived trait that became differentially retained or independently evolved after the neoteleostean or eurypterygian level of evolution was reached.

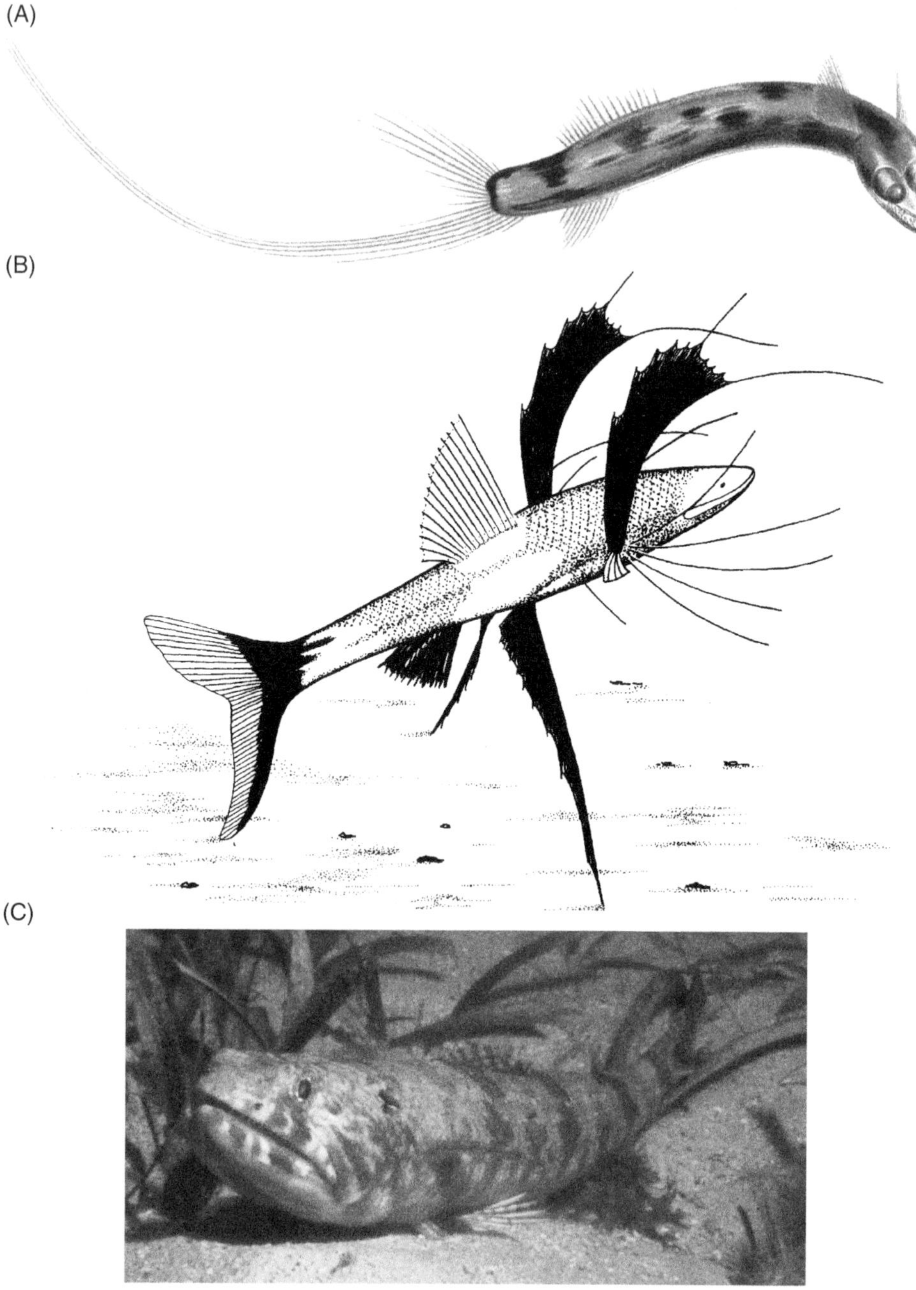

FIGURE 14.20 Aulopiforms, from deep and shallow water. (A) Giganturid telescopefishes are mesopelagic, water column dwellers. (B) Ipnopid spiderfishes are deep-sea benthic dwellers. (C) Synodontid lizardfishes are benthic, shallow water, sand bottom, lurking predators around coral reefs and in some subtropical areas. (A) Wikimedia Commons / Public Domain; (B) after Heezen and Hollister (1971); (C) G. Helfman (Author).

Included among the ipnopids are the often-illustrated spiderfishes or tripodfishes, which have greatly elongated pectoral, pelvic, and caudal rays; they use the pelvics and caudal to form a tripod for resting on the deep ocean floor (Fig. 14.20B). The best-known shallow representatives of this order are the synodontid lizardfishes (Fig. 14.20C), which are common benthic inhabitants of coral reefs worldwide (and whose family name should not be confused with the synodid upsidedown catfishes). Lizardfishes are closely related to the secondarily pelagic Bombay ducks (*Harpadon*) that support an important fishery in the Indian Ocean. The order includes the alepisaurid lancetfishes, which are large (up to 2 m) mesopelagic predators on other fishes. Lancetfishes are distinguished by their large, sail-like dorsal fin that extends from the head almost to the caudal peduncle (function unknown). These fishes, which look very much like scombroid snake mackerels, have proven a great boon to deep-sea taxonomists as several species of mesopelagic fishes have been described from the stomachs of alepisaurids.

Superorder Scopelomorpha

Superorder Scopelomorpha[a]
Order Myctophiformes: Neoscopelidae (6 species of blackchins), Myctophidae (at least 252 species of lanternfishes)

[a] Higher classification groups based on Nelson et al. (2016), but see corrections and updates at **https://sites.google.com/site/fotw5th/**. Orders, families, and numbers of species from Fricke et al. (2020) but refer to **https://researcharchive.calacademy.org/research/ichthyology/catalog/SpeciesByFamily.asp** for updates.

All fishes more derived than the Cyclosquamata have lost the fifth pharyngeal toothplate and the muscle that lifts it (Johnson 1992). These groups have predominantly ctenoid scales, and are often termed the Ctenosquamata. The superorder **Scopelomorpha** is composed of one order of abundant deep-sea fishes, the **myctophiforms** (Fig. 14.21). Myctophiforms have lost the fifth pharyngeal toothplate but most still have cycloid scales, justifying their more ancestral status among ctenosquamates. This status is also shown in their retention of an adipose fin, but they are derived in that the maxilla is excluded from the gape. The myctophid lanternfishes, with at least 252 species, are an important group of mesopelagic deep-sea fishes in terms of diversity, distribution, and numbers of individuals. They occur in all seas, from the Arctic to the Antarctic, and are the prey of numerous other fishes, as well as of marine mammals. They make up a large fraction of the deep scattering layer – a diverse assemblage of fishes and invertebrates that live at mesopelagic depths (below 200 m) during the day and migrate toward the surface at dusk. Myctophid taxonomy is often based on otolith structure and species-specific patterns of photophores, characters that are even preserved in some fossils.

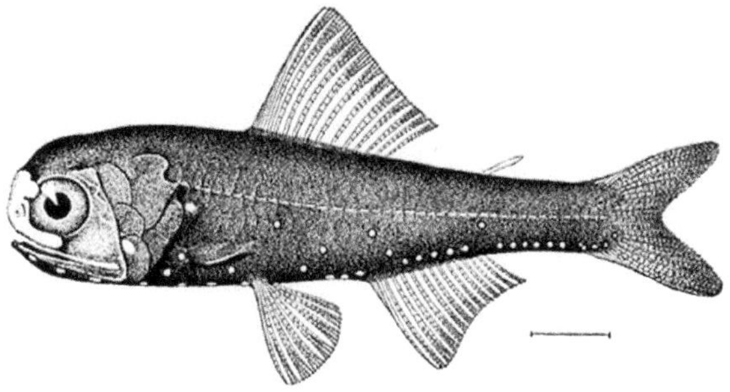

FIGURE 14.21 A Spothead Lanternfish, *Diaphus metopoclampus*. The round structures along the ventral half of the body are light-emitting photophores. Wikimedia Commons / Public Domain.

Acanthomorpha: The Spiny Teleosts

Neoteleosts more derived than the scopelomorphs possess true fin spines and are termed acanthomorphs. The appearance of true fin spines, rather than hardened segmented rays, marks a major evolutionary step in the evolution of bony fishes. True spines occur in the dorsal, anal, and pelvic fins of more derived teleosts. Spines develop when the two halves of the primitively paired and jointed dermal fin rays fuse into a single, unsegment structure. Several other characteristics, mostly associated with improved locomotion and feeding, also characterize teleosts beyond the ctenosquamate level of development (see Chapter 11). Locomotion was improved by the strengthening of vertebral accessories (zygapophyses), providing body stiffening and better attachment for muscles. These changes allowed a shift from slow, sinusoidal motion of the entire body to rapid oscillation of the tail region, driven by tendons attached to the tail base. The tail itself also underwent considerable modification. Pharyngeal teeth diversified and the maxilla shifted from a tooth-bearing bone to a structure that helps pivot the premaxilla, making mouth protrusion and suction more effective. These modifications probably made possible the explosive radiation of spiny-rayed Acanthomorpha during the Early Cenozoic. The acanthomorphs include the superorders Lamprimorpha, Paracanthopterygii (which now includes the polymyxiiforms), and Acanthopterygii (this last group addressed in Chapter 15).

Superorder Lamprimorpha

Superorder Lamprimorpha[a]
Order Lampriformes (30 species): Lampridae (6 species of opahs), Veliferidae (2 species of velifers), Lophotidae (5 species of crestfishes), Radiicephalidae (2 species of tapertails), Trachipteridae (about 11 species of ribbonfishes), Regalecidae (3 species of oarfishes)

[a] Higher classification groups based on Nelson et al. (2016), but see corrections and updates at **https://sites.google.com/site/fotw5th/**. Orders, families, and numbers of species from Fricke et al. (2020) but refer to **https://researcharchive.calacademy.org/research/ichthyology/catalog/SpeciesByFamily.asp** for updates.

As is often the case, the least derived members of a major taxon retain certain ancestral traits but possess others indicative of more derived status. **Lampriforms** lack true spines, but the maxilla helps move the premaxilla and bares no teeth. The connection between the two upper jaw bones and the way they slide to protrude the mouth are unique, so much so that the group used to be referred to as the Allotriognathi, or "strange jaws." The seven families of lampriforms are almost all open water oceanic fishes with unusual body and fin proportions. Opahs (family Lampridae) are relatively large (up to 1.8 m, 70 kg), oval-shaped, colorful pelagic predators on squids and other fishes (Fig. 14.22A). They have evolved circulatory mechanisms to maintain warm swimming muscles, which improves muscle performance (discussed in Chapter 7). The remaining five families contain mostly large and rare, elongate, pelagic fishes with long dorsal fins. Crestfishes and radiicephalids have ink sacs that they discharge through their cloaca. In trachipterid ribbonfishes, the caudal fin is made up only of the upper lobe, which sits inclined at

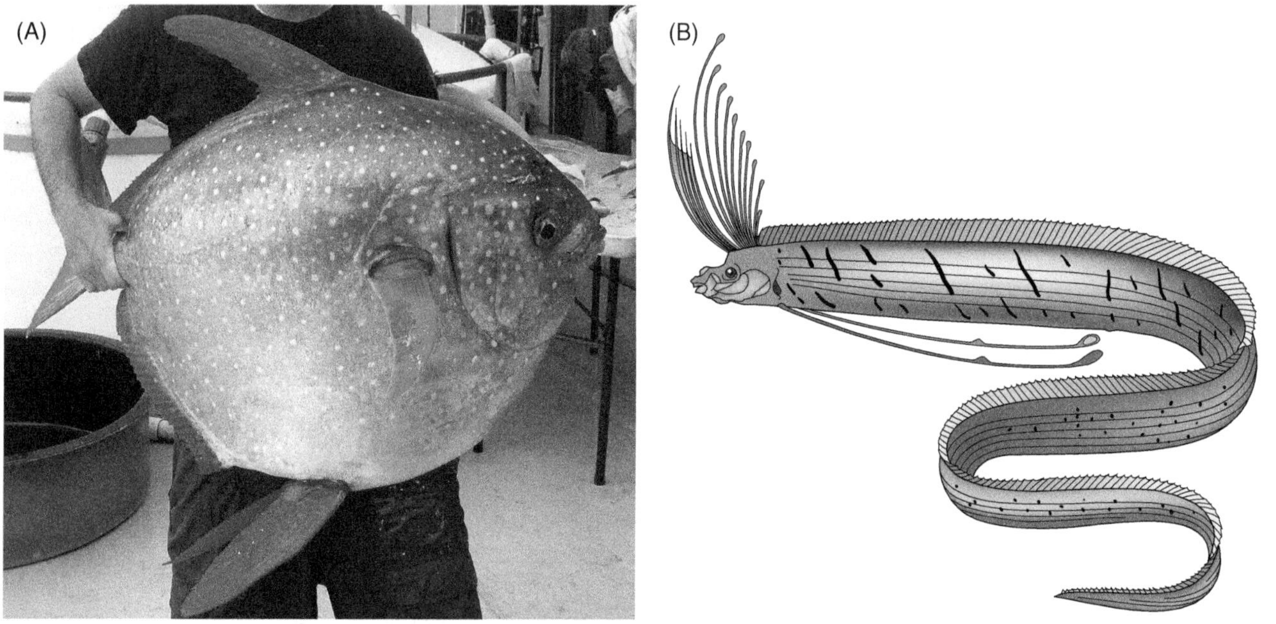

FIGURE 14.22 Lampriformes. (A) An opah or moonfish, *Lampris regius*. This large pelagic predator has red fins, a silvery-white body, with a bluish back. NOAA / Wikimedia Commons / Public Domain. (B) The oarfish, *Regalecus glesne*, the world's longest teleost, reaching lengths of 7 m to as much as 11 m. Whole specimens are rare, but see Fig. 10.6. Courtesy of T. Roberts.

a right angle upward. The enigmatic oarfish, *Regalecus*, reaches a confirmed length of over 7 m and perhaps as much as 11 m, making it the longest extant teleost; uncertainty arises because intact specimens are seldom found (Fig. 14.22B). Despite their size, they are apparently planktivores that hover head up in the water column, swimming via rapid undulations of the long dorsal fin. Their name comes from bladelike expansions at the end of the pelvic fins. Specimens are uncommon, many obtained after storms when the bodies of these strange-looking fishes are tossed up on beaches. The oarfish, with its bluish-silvery body, scarlet dorsal crest of elongate fin rays, and deep red fins, is likely responsible for many sea serpent sightings, particularly those referring to monsters "having the head of a horse with a flaming red mane" (Norman & Fraser 1949, p. 113).

Superorder Paracanthopterygii

The 21 families currently recognized as paracanthopterygians represent a major and contentious side branch during the evolution of more derived acanthomorphs. They are weakly defined by a number of characters, chiefly involving the caudal skeleton and holes in the skull through which cranial nerves pass (Nelson 1994), but ". . . there is no firm basis to believe [the superorder] is monophyletic" (Nelson 2006, p. 232). Since the original description of the Paracanthopterygii by Greenwood et al. (1966), various authors have moved many of the original groups into the Percomorphi (Nelson et al. 2016). Their ecological success is largely as benthic, marine fishes that are nocturnally active or live in permanently darkened waters, such as the bathypelagic region of the deep sea or in caves. Only 23 relatively less derived paracanthopterygian species live in fresh water. Many paracanthopterygians produce sounds with sonic muscles on their gas bladders.

Superorder Paracanthopterygii[a]

- Order Polymixiiformes (10 species): Polymixiidae (beardfishes)
- Order Percopsiformes (12 North American species): Percopsidae (2 species of Trout-perches), Aphredoderidae (Pirate Perch), Amblyopsidae (9 species of cavefishes)
- Order Zeiformes (34 species): Cyttidae (3 species of lookdown dories), Oreosomatidae (10 species of oreos), Parazenidae (4 species of smooth dories), Zeniontidae (7 species of armoreye dories), Grammicolepididae (3 species of tinselfishes), Zeidae (7 species of dories)
- Order Stylephoriformes: Stylephoridae (*Stylephorus chordatus*, Tube-eye).
- Order Gadiformes (610 species): Muraenolepididae (8 species of eel cods), Bregmacerotidae (at least 14 species of codlets), Euclichthyidae (Eucla Cod), Macrouridae (about 405 species of grenadiers or rattails), Moridae (about 106 species of morid cods), Melanonidae (2 species of pelagic cods), Gadidae (about 25 species of cods), Lotidae (22 species of hakes), Phycidae (11 species of phycid hakes), Merlucciidae (18 species of hakes)

[a] Higher classification groups based on Nelson et al. (2016), but see corrections and updates at **https://sites.google.com/site/fotw5th/**. Orders, families, and numbers of species from Fricke et al. (2020) but refer to **https://researcharchive.calacademy.org/research/ichthyology/catalog/SpeciesByFamily.asp** for updates.

The taxonomic status of the enigmatic family Polymixiidae has been the subject of considerable debate (see Johnson & Patterson 1993a, 1993b; Nelson 1994, 2006). Beardfishes possess derived characters such as four to six true spines in the dorsal fin and four spines in the anal fin, and their pelvic fins are located forward on the body. Yet they retain two sets of intermuscular bones, the epineurals and epipleurals, a characteristic of less derived taxa, and have a unique arrangement of jaw muscle ligaments. This combination of ancestral, derived, and unique characteristics complicates the resolution of their relationships. At different times, they have been classified as a sister taxon to the remaining acanthomorphs, with the less derived paracanthopterygian percopsiforms (trout-perches and relatives), or with less derived acanthopterygian beryciforms (squirrelfishes and relatives). We follow Nelson et al. (2016) and include them within the Paracanthopterygii instead of between lamprimorphs and paracanthopterygians as Nelson (2006) did. Beardfishes have large eyes, chin barbels, are about 30 cm long, and live at moderate depths (180–640 m).

Near the base of the paracanthoptgerygians are the **percopsiforms**, small (<20 cm), freshwater fishes, eight of nine of which live in eastern North America. The Trout-perch, *Percopsis omiscomaycus* (Fig 14.23A), and its Columbia River congener, *P. transmontana*, are the most derived fishes with an adipose fin. This somewhat ancestral feature is interesting given that modern percopsiforms possess several traits suggesting a reversal of the evolutionary trends of derived teleosts, including fossil percopsiforms. Modern species have fewer fin spines, more vertebrae, and a more posteriorly located pelvic girdle than occurred in fossil forms. The aphredoderid Pirate Perch, *Aphredoderus sayanus*, is a swamp dweller with the distinction of having its anus relocate from just anterior of the anal fin in juveniles to the throat region of adults. Speculation abounded on the functional reasons for this anatomical anomaly. Recent observations reveal that Pirate Perch are nocturnally active fish that hide and spawn in dense vegetation and root mats (Fletcher et al. 2004; Tiemann 2004). Females push head first into such root mats to deposit eggs, and males enter the same spot shortly thereafter to deposit sperm; hence the anterior position of the anus allows deposition of eggs and sperm deep into a protected, otherwise inaccessible area (Fig. 14.24). By day, fish emerge from vegetation refuges only far enough to expose the head and anus, at which time they defecate. The jugular position of the anus would permit defecation without altering water quality in the refuge and would also minimize exposure to predators. As is so often the case, seemingly specialized traits often confer multiple functional benefits.

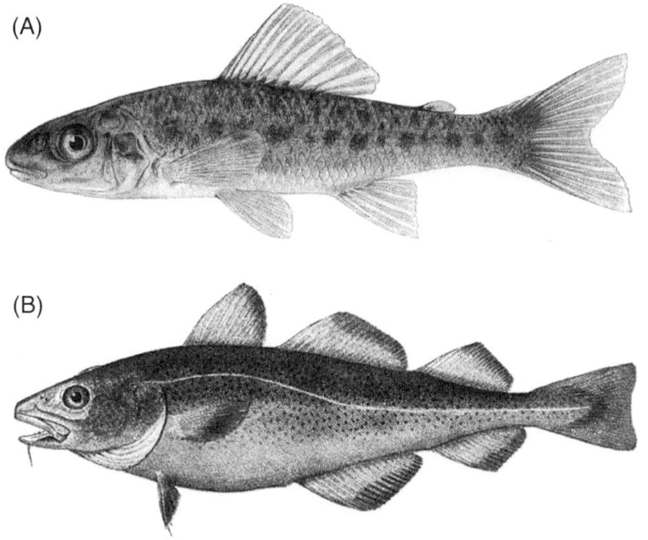

FIGURE 14.23 Paracanthopterygiians. (A) A percopsid, the Trout-perch, *Percopsis omiscomaycus*; (B) A gadiform, the Atlantic Cod, *Gadus morhua*. (A) E. Edmondson and H. Chrisp / Wikimedia Commons / Public Domain; (B) image from NOAA photo library.

FIGURE 14.24 Spawning behavior of the Pirate Perch, *Aphredoderus sayanus*, a fish with a jugular-positioned anus. Pirate Perch spawn in dense root mats, the female and then the male pushing into narrow canals to deposit eggs and sperm. Inset: a female in the process of laying eggs. From Fletcher et al. (2004), used with permission.

The amblyopsid cavefishes include nine species of highly modified, often blind and scaleless forms that show numerous adaptations for cave life (see Chapter 10, Caves). Their isolated, easily disturbed habitats and bizarre appearance has made them vulnerable to both collecting and habitat disruption, and several species are threatened and hence protected nationally and internationally.

Zeiforms are a confusing assortment of primitive marine paracanthopterygians that have highly protrusible mouths and a unique caudal skeleton. There is strong molecular and some morphological support for placing the Zeiformes within the Paracanthopterygii (Nelson et al. 2016). Included in this order are such commercial species as the European John Dory, *Zeus faber*.

Among the **stylephoriforms**, the Tube-eye, *Stylephorus chordatus* is formerly nested within lampriforms but recent molecular studies indicate that it is closely related to gadiforms (Betancur-R. et al. 2013, Nelson et al. 2016). It swims in a vertical position, head up in depths of 300–800 m. The 30 cm long Tube-eye is capable of an almost 40-fold enlargement of its mouth volume during feeding, which may be a record among vertebrates.

The **gadiforms** include some of the most important commercial fishes in the world such as the cods, haddocks, hakes, pollocks, and whitings (Fig. 14.23B). Gadiforms lack true spines but have various configurations of fin rays. The long dorsal fin is relatively diversified compared with most primitive groups; it is often divided into two or three parts, an anterior ray that is sometimes spinous (grenadiers) or elongate and even filamentous (morid cods, codlets, eel cods); true cods (family Gadidae, subfamily Gadinae) have three dorsal fins and two anal fins. Pelvic fins of the gadiforms are thoracic or jugular in position and are sometimes modified into filaments with a possible sensory function (e.g. Eucla Cod, codlets, phycid hakes). Many have chin barbels (grenadiers, morid cods, eel cods, phycid hakes, cods), a convergent trait in benthic or near-benthic fishes. Gadiforms are northern marine fishes with the solitary exception of the Burbot, *Lota lota*, which is a freshwater gadid of Holarctic (high latitude, northern hemisphere) distribution. The commercially important Atlantic Cod, *Gadus morhua*, is the largest species in the order. Specimens once reached lengths of 1.8 m and weighed over 90 kg, although fish over 10 kg are now rare due to extensive overfishing. Previously "inexhaustible" cod fisheries have crashed throughout much of the North Atlantic, affecting entire ocean ecosystems and the fishery economies dependent on them (Mowat 1996; Kurlansky 1997). In 1989 more than 6 million tons of North Pacific Walleye Pollock, *Theragra chalcogramma*, were harvested, the largest food fishery in the world at the time.

Superorder Acanthopterygii

This most derived and extremely diverse superorder of teleosts is addressed in Chapter 15.

Supplementary Reading

Berra TM. 2001. *Freshwater fish distribution*. San Diego: Academic Press.

Betancur-R R, Wiley EO, Arratia G, et al. 2017. Phylogenetic classification of bony fishes. *BMC Evol Biol* 17:162.

Carroll RL. 1988. *Vertebrate paleontology and evolution*. New York: W. H. Freeman.

Johnson GD, Patterson C. 1996. Relationships of lower euteleostean fishes. In: Stiassny MLJ, Parenti LR, Johnson GD, eds. *Interrelationships of fishes*, pp. 251–332. San Diego: Academic Press.

Nelson JS. 2006. *Fishes of the world*, 4th edn. Hoboken, NJ: Wiley & Sons.

Nelson JS, Grande TC, Wilson MVH. 2016. *Fishes of the world*, 5th edn. Hoboken, NJ: Wiley. 707 pp.

Paxton JR, Eschmeyer WN, eds. 1998. *Encyclopedia of fishes*, 2nd edn. San Diego, CA: Academic Press.

Websites

All Catfish Species Inventory, **http://silurus.acnatsci.org**.

Wild Salmon Center and Ecotrust, **www.stateofthesalmon.org**.

www.fishbase.org.

Zebrafish Information Network, **http://zfin.org**.

Fricke R, Eschmeyer W, Fong JD. most recent update. *Genera/Species by family/subfamily in Eschmeyer's catalog of fishes*. **https://researcharchive.calacademy.org/research/ichthyology/catalog/SpeciesByFamily.asp**.

CHAPTER 15

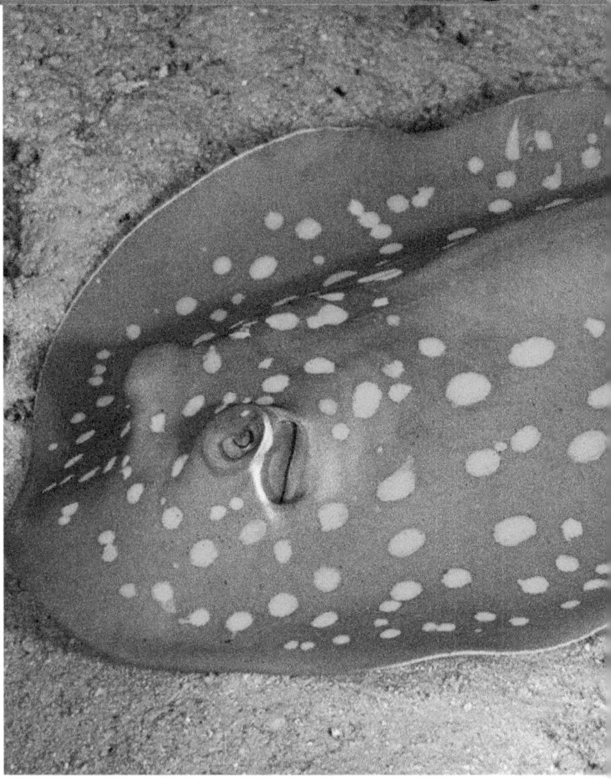

Teleosts II: Spiny-Rayed Fishes

Summary

Most fishes (about 17,000 species in over 250 families) belong to the superorder Acanthopterygii and have highly protrusible jaws, a complex pharyngeal apparatus, two dorsal fins, and spines in the first dorsal, anal, and pelvic fins. The phylogeny of the acanthopterygiians has been a topic of research and discussion for many years, with different authors reaching somewhat different conclusions on the organization of the group. We recognize three major subgroups, which we rank as series: the Berycimorphaceae, Holocentrimorphaceae, and Percomorphaceae. We further divide the Percomorphaceae into five subseries – the Ophidiida, Batrachoidida, Mugilomorpha, Atherinomorpha, and Percomorpha.

The berycimorphs include two orders that are either nocturnal or occupy deep-sea habitats. The holocentrimorphs are nocturnal fishes of relatively shallow, tropical marine habitats; they are included among the berycimorphs by some authors.

The Percomorphaceae include thousands of species occupying a broad range of habitats. The Ophidiida and Batrachoidida had both been considered paracanthopterygians, but molecular evidence has led to their inclusion among the acanthopterygii. Ophidiids typically occupy benthic, dark habitats – many are marine, with some found at great depths, but the group include some cave-dwelling freshwater species as well. Batrachoids are well-camouflaged primarily benthic marine fishes.

CHAPTER CONTENTS

The Diversity of Fishes: Biology, Evolution and Ecology, Third Edition. Douglas E. Facey, Brian W. Bowen, Bruce B. Collette, and Gene S. Helfman.

Companion website: www.wiley.com/go/facey/diversityfishes3

Mugilomorph mullets are marine and freshwater fishes with unconnected pectoral and pelvic girdles. Atherinomorphs (silversides, needlefishes, flyingfishes, halfbeaks, killifishes, and livebearers) are shallow water, marine, or freshwater fishes that live near the surface and have a unique jaw protrusion mechanism.

The remaining groups of acanthopterygiians are percomorphs. Gasterosteiforms are small fishes with dermal armor plates, small mouths, and unorthodox propulsion. The Sygnathiforms include the pipefishes and seahorses, in which males carry developing embryos in a ventral pouch. Synbranchiforms are eel-like inhabitants of swampy freshwater habitats; the swamp and rice eels of Africa, Asia, and Central and South America can breathe air. Scorpaeniforms are mostly marine, benthic fishes, except for freshwater cottid sculpins; many have head spines and venomous fin spines (e.g. turkeyfishes, stonefishes, scorpionfishes).

There has been considerable research and discussion regarding the grouping of the Perciformes, and some authors have divided the group into multiple orders. We retain the group as a single order, but with 18 suborders, that includes most marine and freshwater fishes of littoral (near-shore) zones. Perciformes have abdominal pelvic fins, lateral pectoral fins, and fewer than 18 caudal rays. The most perch-like perciforms are in the suborder Percoidei. These include snooks, temperate basses, and the diverse seabasses. Centrarchid sunfishes and percids (darters, perches, pikeperches) are important freshwater percoids of North America. Other percoids include predatory carangoids, such as carangid jacks and pompanos, Cobia, dolphinfishes, and the shark-sucking remoras. Heavy-bodied, tropical, benthic, predatory families (snappers, groupers, sparoids), as well as fishes with barbels or feelers (croakers, threadfins, goatfishes) are placed here. Other percoid families include archerfishes that shoot insects out of overhanging vegetation, the colorful coral reef butterflyfishes and angelfishes, and many other families. Many families among the suborder Percoidei are placed in separate orders based on molecular phylogenies.

The suborder Labroidei has several speciose families. Cichlids are primarily freshwater fishes that have undergone explosive speciation, forming species flocks in Central African lakes. Other labroid families are the primarily tropical damselfishes, wrasses, and parrotfishes, and the temperate surfperches and odacids. Zoarcoids are tidepool and deep-sea, benthic fishes (eel pouts, gunnels, wolf-eels). Notothenioid icefishes dominate the Antarctic and have many adaptations to very cold water. Blennioids (clinids, blennies) and gobioids (sleepers, gobies) are large suborders of generally small, benthic, marine species.

The suborder Acanthuroidei includes the spadefishes, rabbitfishes, and surgeonfishes, the latter two being important families of coral reef herbivores. Xiphioids and scombroids (billfishes, mackerels, tunas) are the fastest and largest predatory bony fishes in the sea. Tunas and billfishes have independently evolved endothermy and heat conservation. Stromateoids are also largely pelagic marine fishes that associate with floating objects. The most derived perciform suborders include the Anabantoidei (gouramies) and Channoidei (snakeheads), which are African and Asian freshwater fishes, many with specialized gill structures for absorbing atmospheric oxygen.

The three remaining orders of percomorphs are considered the most derived. Pleuronectiform flatfishes begin life as pelagic, symmetrical larvae but metamorphose into adults that lie on the bottom on one side and displace many body organs, most noticeably the eyes. The Lophiiforms are a marine group that had previously been placed as the most derived order among the paracanthopterygiians. The lophiiforms include the goosefishes, frogfishes, and deep-sea anglerfishes; many of these use a modified dorsal spine as a lure to attract prey, which are then inhaled into their gaping mouth. Tetraodontiforms are primarily tropical reef dwellers, including triggerfishes, boxfishes, puffers, and porcupinefishes. They often have beaklike jaws, many fused skull and axial bones, and spiny, leathery skin. The most derived tetraodontiforms are the giant Ocean Sunfishes, which have highly cartilagenous skeletons.

Although potentially frustrating to students and nonsystematists, relationships among acanthopterygians are a matter of active research, particularly with molecular phylogenies that lack morphological support. Many groups are only provisionally placed in the acanthopterygian phylogeny, and much remains to be learned about this, the most successful and speciose group of modern vertebrates.

Superorder Acanthopterygii: Introduction

Most modern bony fishes belong to a single highly derived superorder, the **Acanthopterygii**. The group is so diverse and its members are so important from all standpoints that a full chapter is needed to discuss them, although no one chapter or one book can do them justice. Several genera and families, such as the sticklebacks, livebearers, darters, black basses, perches, butterflyfishes, cichlids, damselfishes, tunas, and billfishes are the subjects of one or several books themselves, and so the treatment in this chapter is understandably cursory. The phylogeny and taxonomy presented here, as well as many of the aspects of biology of different groups, are taken largely from Nelson (2006), Nelson et al. (2016), and Betancur-R et al. (2017), references that should be consulted for additional details. And as Nelson points out repeatedly, taxonomic work on most groups is ongoing and is often unsettled.

Given the remarkable diversity of the more derived spiny-rayed fishes – approximately 17,000 species in over 250 families – it is a tribute to their successful suite of adaptations that they are generally recognized as a coherent group. Although controversy about relationships and taxonomic position among the various orders and families abounds, certain generalities can be made about the group as a whole and the characteristics that define it. Two primary innovations are shared by most lineages of acanthopterygians:

1. Upper jaw mobility and protrusibility are maximal in this group. This is achieved by the development of a dorsal extension of the anterior tip of the premaxilla, termed the **ascending process**. This process slides along the rostral cartilage on the snout of the fish, shooting the upper jaw forward and downward. Protrusion is aided by a cam-like connection between the maxilla and premaxilla, the maxilla rotating and helping push the premaxilla forward (Lauder & Liem 1983; discussed in Chapter 11).
2. Pharyngeal dentition and action reach their highest level of development. Improved function is aided by a redistribution of the attachments of muscles and bones in the pharyngeal apparatus. The retractor dorsalis muscle (see preceding chapter) now inserts on the third pharyngobranchial arch, and the upper pharyngeal jaws are supported principally by the second and third epibranchial bones.

Acanthopterygians also typically have: ctenoid scales (with numerous exceptions); a physoclistous gas bladder; maxilla excluded from the gape; two distinct dorsal fins, the first of which is spiny and the second of which is soft-rayed; at least one spine in the anal fin; pelvic fins located anteriorly, containing one leading spine and five or fewer soft rays; pectoral fins placed laterally on the body; and an externally symmetrical tail fin supported by fused basal elements. Several other trends in feeding, locomotion, and predator protection characterize the higher spiny-rayed fishes and show progressive change during acanthopterygian phylogeny. Most of these were discussed in Chapter 11 and will only be summarized here as particularly good examples or striking exceptions are encountered among the taxa. An important point to be remembered is that these are the most derived and diverse of today's fishes, dominating the shallow, productive habitats of the marine and freshwater environments.

Acanthopterygian Phylogenies: Can't We All Just get Along?

A student or a non-systematist fish researcher, upon encountering the bewildering diversity, unfamiliar names, and labyrinthine relationships of more derived teleosts, can understandably desire a straightforward presentation of the facts and a single classification to learn and use. Ah, were it that simple.

Systematics draws on multiple sources of information to form the building blocks of a classification; traditional morphology (augmented by chemical and radiological techniques) and molecular genetics are but two such sources. Correctness often hinges as much on force of argument as on strength of data. Different authors using different techniques can draw different conclusions, as can different authors using the same data; two authors in the same paper looking at the same data may even disagree (e.g. Springer & Johnson 2004, and your illustrious authors).

Over the years, there has been considerable debate regarding the classification of the acanthopterygians. There is agreement on the major composition of the superorder Acanthopterygii, but differences on the placement of some groups, as well as the ranking and terminology used for some of the taxa. Figure 15.1 shows three different interpretations, in the form of cladograms, about relationships among the acanthopterygians. Figure 15.1A is the arrangement used in this text in which we incorporate some aspects of the cladograms based more on molecular data (Fig. 15.1B, C) when that information corroborates interpretations based on morphology (see Nelson 2006). Figure 15.1B (based on Nelson et al. 2016) and Figure 15.1C (based on Betancur-R et al. 2017) rely much more heavily on molecular characters. Over time, and with additional information and discussion, perhaps a modified or composite view will emerge. In updating his *Fishes of the World* from third to fourth edition, Nelson (2006) lamented that, given "so much conflicting information . . .. no comprehensive synthesis seemed possible" (Nelson 2006, p. 261). He chose to remain with a classification scheme more similar to his previous edition (Nelson 1994). The subsequent fifth edition of *Fishes of the World* (Nelson et al. 2016) presents some different phylogenetic relationships among acanthopterygian taxa, in part because stronger emphasis is placed on interpretations of genetic information. Betancur-R et al. (2017) also rely strongly on genetics and present an organization rather similar to that of Nelson et al. (2016).

The classification of the Acanthopterygii has been based primarily on morphological characters as summarized by Nelson (2006) and Wiley & Johnson (2010), but recent classifications such as Nelson et al. (2016) and Betancur-R et al. (2017) rely almost entirely on molecular and genomic data and frequently lack morphological support. Chapters 11–13 of this text rely on paleontological and morphological evidence. Molecular and morphological evidence were reasonably consistent up through the Paracanthopterygii in Chapter 14. However, the new molecular classifications among the acanthopterygians bear little resemblance to classical morphological phylogeny creating a problem of which classification to follow in this chapter. We follow the organization of Nelson et al. (2016) and Betancur-R et al. (2017) for some families in the first few series presented, but then return to the classification of Nelson (2006) for the remaining taxa. We hope that future research and discussion will clarify the contradictions between molecular and morphologically based classifications and bring some resolution of the differences between the older classifications and the various new molecular classifications. As Nelson stated in his fourth edition of *Fishes of the World* – "*Students of ichthyology should study these works as examples of how researchers can arrive at different conclusions*" (Nelson 2006, p. 225).

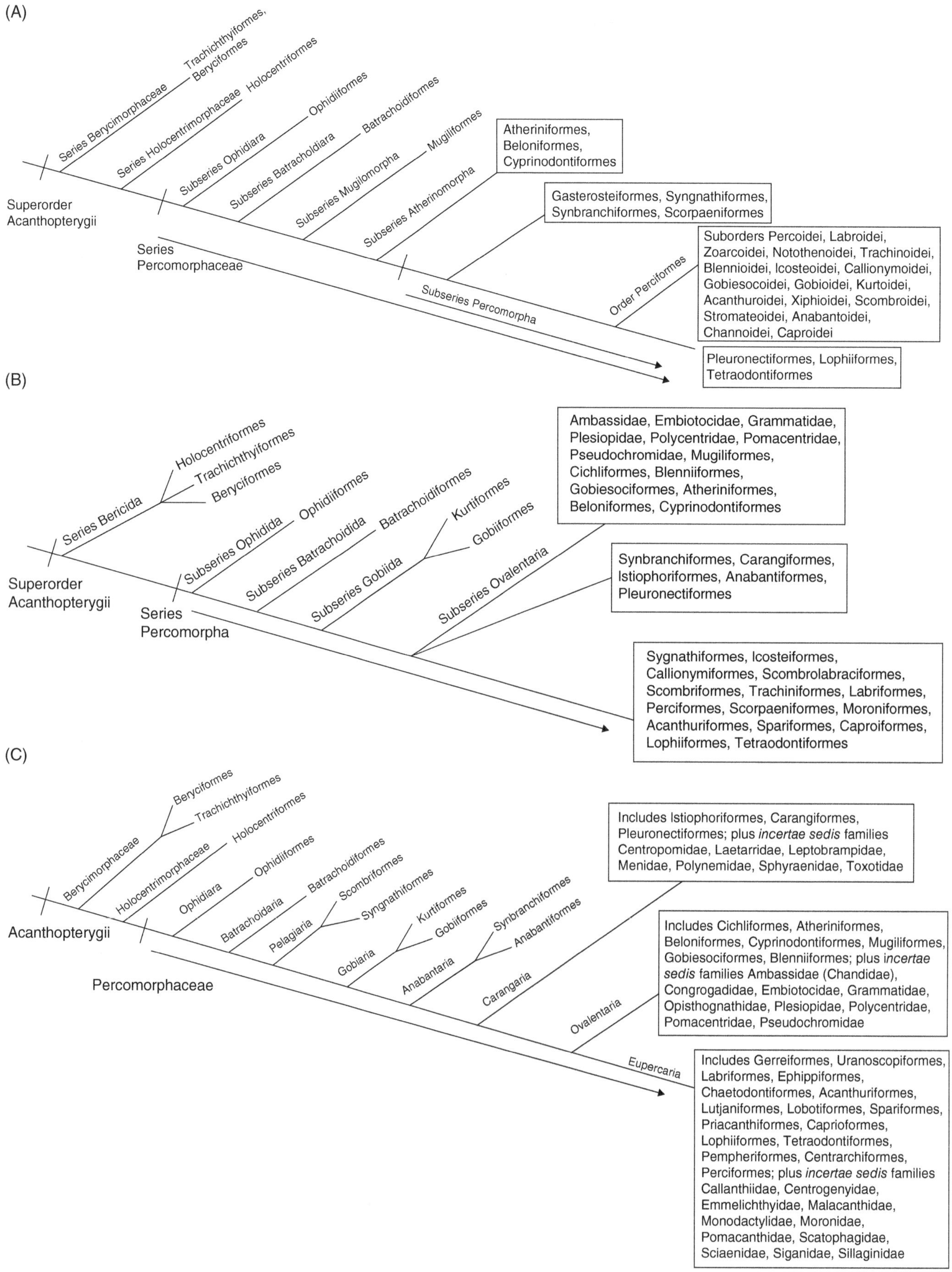

FIGURE 15.1 Three different interpretations of the relationships among the acanthopterygians. The top cladogram (A) represents the presentation in this text. (B) (based on Nelson et al. 2016) and (C) (based on Betancur-R et al. 2017) rely more strongly on molecular data. The Latin term *incertae sedis* (used in C) means that the broader phylogenetic relationship of the designated groups is unknown based on the methods used.

Series Berycimorphaceae (Berycida)

Order Trachichthyiformes: Anomalopidae (9 species of flashlight fishes and lanterneye fishes), Anoplogastridae (2 species of fangtooths), Diretmidae (4 species of spinyfins), Monocentridae (3 species of pinecone fishes), Trachichthyidae (48 species of roughies, slimeheads)

Order Beryciformes: Berycidae (10 species of alfonsinos), Melamphaidae (72 species of bigscale fishes), Barbourisiidae (Red Whalefish), Cetomimidae (including Mirapinnidae and Megalomycteridae (about 26 species of whalefishes, tapetails, and bignoses), Rondeletiidae (2 species of redmouth whalefishes), Stephanoberycidae (4 species of pricklefishes), Gibberichthyidae (2 species of gibberfishes), Hispidoberycidae (hispidoberycids)

Higher classification groups based on Nelson et al. (2016), but see corrections and updates at **https://sites.google.com/site/fotw5th/**. Orders, families, and numbers of species from Fricke et al. (2020) but refer to **https://researcharchive.calacademy.org/research/ichthyology/catalog/SpeciesByFamily.asp** for updates.

At the base of the acanthomorphs are two orders of either deep-sea or nocturnal fishes, the beryciform pricklefishes and relatives, and the trachichthyiform roughies and the flashlight fishes (Fig. 15.2). These large-headed, round fishes have many percomorph characteristics, except the tail fin has a primitively large number of rays (18 or 19) as compared to the 17 caudal rays that typify most advanced percomorphs. **Trachichthyiforms** often have the large eyes typical of nocturnal fishes and possess strong spines on the head or gill covers. They include such relatively shallow water luminescent forms as pinecone fishes and flashlight fishes, and the commercially important Orange Roughy, *Hoplostethus atlanticus*. The **beryciforms** (gibberfishes, pricklefishes, cetomimoid whalefishes) are largely deep-sea forms characterized by luminescent organs, weak or absent fin spines, and reduced scale coverage. The cetomimid flabby whalefishes are second only to the oneirodid anglerfishes (order Lophiiformes, later in this chapter) in species diversity among bathypelagic forms and may exceed anglerfishes in abundance in deeper waters. Interestingly, cetomimids have converged with anglerfishes in having dwarf males, although whalefishes are not known to be parasitic on the larger females. Beryciforms are well represented in the fossil record, dating back to the Late Cretaceous.

(A)

(B)

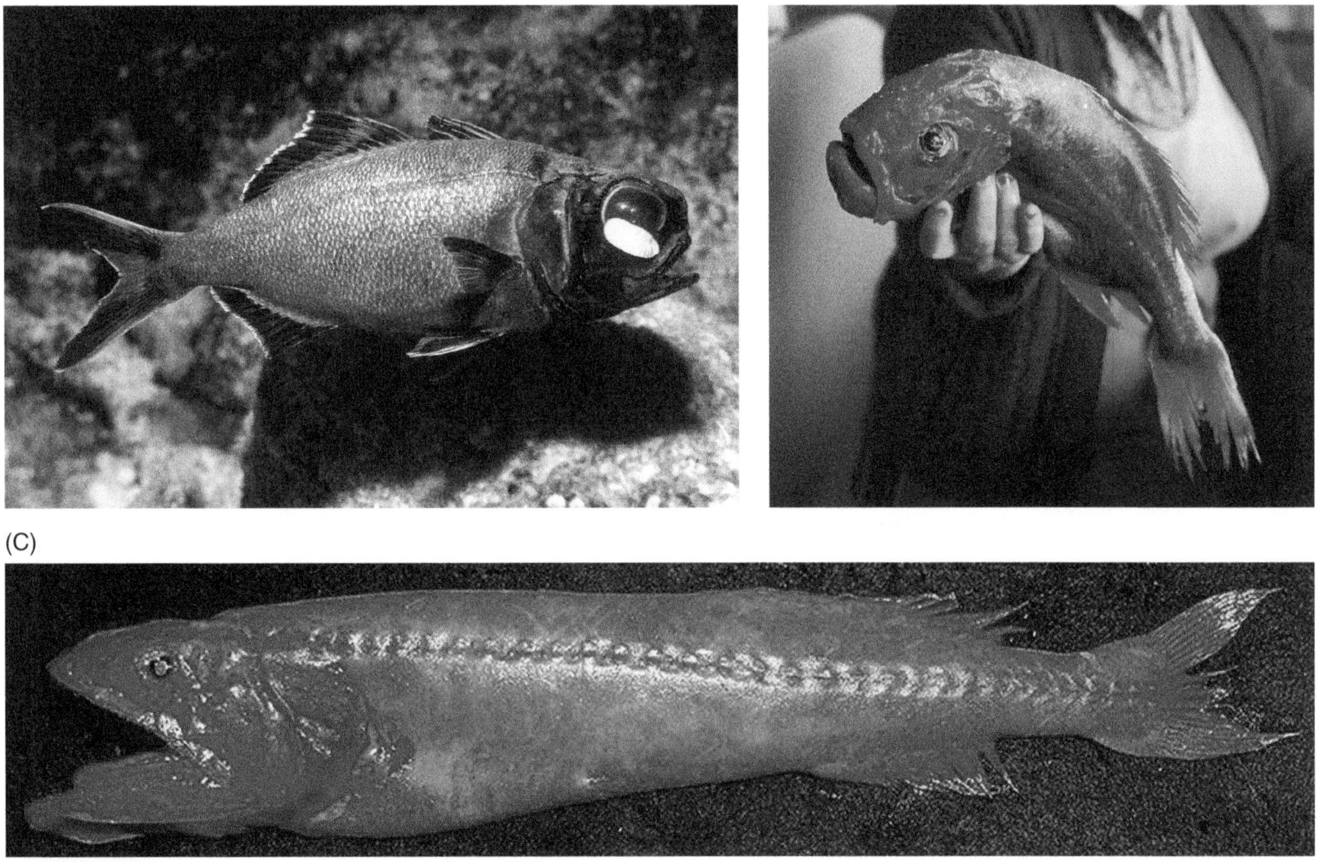

(C)

FIGURE 15.2 Berycipmorphs. Red light does not penetrate to the depths at which Orange Roughy (B) and whalefishes (C) live. Therefore, these fishes will appear black in their dark natural habitat. (A) - A flashlight fish (*Anomalops katoptron*) showing the bioluminescent organ below the eye. Courtesy of R Steene (B) - An Orange Roughy, *Hoplostethus atlanticus*, a trachichthyid beryciform. G. Helfman (Author) (C) - A whalefish. NOAA / Public Domain.

Series Holocentrimorphaceae

Order Holocentriformes: family Holocentridae (90 species of squirrelfishes)

Orders, families, and numbers of species from Fricke et al. (2020) but refer to **https://researcharchive.calacademy.org/research/ichthyology/catalog/SpeciesByFamily.asp** for updates.

Nelson et al. (2016) present the holocentriforms as an order within the series Berycimorpha (Berycida), but Betancur-R et al. (2017) have them separated from the berycids and as a sister group to the series Percomorpha. We concur with Betancur-R et al. and have assigned them series status to set them apart from the Berycimorphs and Percomorphs. The group contains only the Order **Holocentriformes** and the family Holocentridae (squirrelfishes, Fig. 15.3). Squirrelfishes are a tropical marine group with representatives in the Atlantic, Pacific, and Indian oceans. They are typically found at depths under 100 m and some species are close to shore and often hide in crevasses and other shaded areas among coral reefs during daylight hours.

Series Percomorphaceae

Subseries Ophidiida
Subseries Batrachoidida
Subseries Mugilomorpha
Subseries Atherinomorpha
Subseries Percomorpha

The series Percomorphaceae is a huge group of more than 17,000 species. It is the same group as the Percomorpha of previous authors and does have some morphological support (Johnson & Patterson 1996; Wiley & Johnson 2010). Based on molecular characters, Betancur-R et al. (2017), divide this series into nine subgroups above the level of order. However, as addressed earlier, we are reluctant to use some of these groups because they are not supported by morphological data; we look forward to additional research to illuminate these relationships. Therefore, we will adhere more closely to the groupings in Nelson (2006) and follow the outline of five subseries and will address the orders in each of those groups.

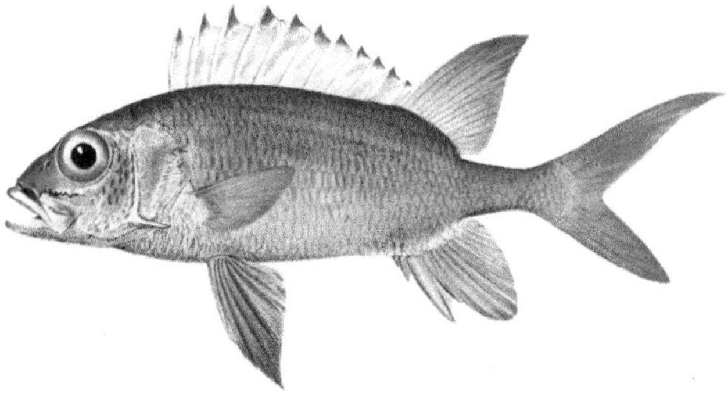

FIGURE 15.3 Formerly included among the beryciphorms, squirrelfishes (family Holocentridae) are now considered in their own order, the Holocentriformes. They typically occupy habitats with low-light conditions, or remain in shaded areas when in shallow water such as coral reefs. NOAA / Public Domain.

Subseries Ophidiida

Order Ophidiiformes: Ophidiidae (278 species of cusk-eels), Bythitidae (127 species of viviparous brotulas), Dinematichthyidae (117 species), Carapidae (36 species)

(Orders, families, and numbers of species from Fricke et al. (2021) but refer to **https://researcharchive.calacademy.org/research/ichthyology/catalog/SpeciesByFamily.asp** *for updates.)*

Ophidiiforms were previously placed in the Paracanthopterygii but have been moved into the Percomorphaceae based primarily on molecular evidence (Nelson et al. 2016). They have the pectoral fins high up on the body with a vertical orientation. The pelvic fins, when present, are located anteriorly under the head in what is termed the mental or jugular position. Pelvic fin loss in this group probably relates to their eel-like bodies; many eel-like fishes, regardless of taxonomic position, have reduced or absent pelvic fins and girdles. Ophidiiforms inhabit what must be viewed as marginal or at least exceptional habitats for fishes. Pearlfishes are **inquilines** (tenants), living inside the body cavities of starfishes, sea cucumbers, bivalves, and sea squirts; some may be parasitic, feeding on the internal organs of their hosts. The common name originated with the discovery of the oyster's revenge, where an individual Pacific Pearlfish, *Encheliophis dubius*, became entombed in a black-lip oyster (*Pinctada mazatlanica*) (Fig. 15.4A). Pearlfishes are nearly unique among fishes in that they have two distinct larval stages, a "vexillifer" pelagic stage followed by a "tenuis" demersal stage during which they search for a host. Ophidiid and bythitid cusk-eels and brotulas (Fig. 15.4B) include blind species in freshwater caves of the Caribbean basin and Galápagos Islands, and infaunal coral reef species that hide deep within crevices. The depth record for a fish is held by a cusk-eel, *Abyssobrotula galatheae*, taken 8370 m down in the Puerto Rico Trench. Bythitid brotulas and parabrotulid false brotulas are live-bearers, a rare derivation among percomorphs.

Subseries Batrachoidida

Order Batrachoidiformes: Batrachoididae (84 species of toadfishes, midshipmen)

(Orders, families, and numbers of species from Fricke et al. (2021) but refer to **https://researcharchive.calacademy.org/research/ichthyology/catalog/SpeciesByFamily.asp** *for updates.)*

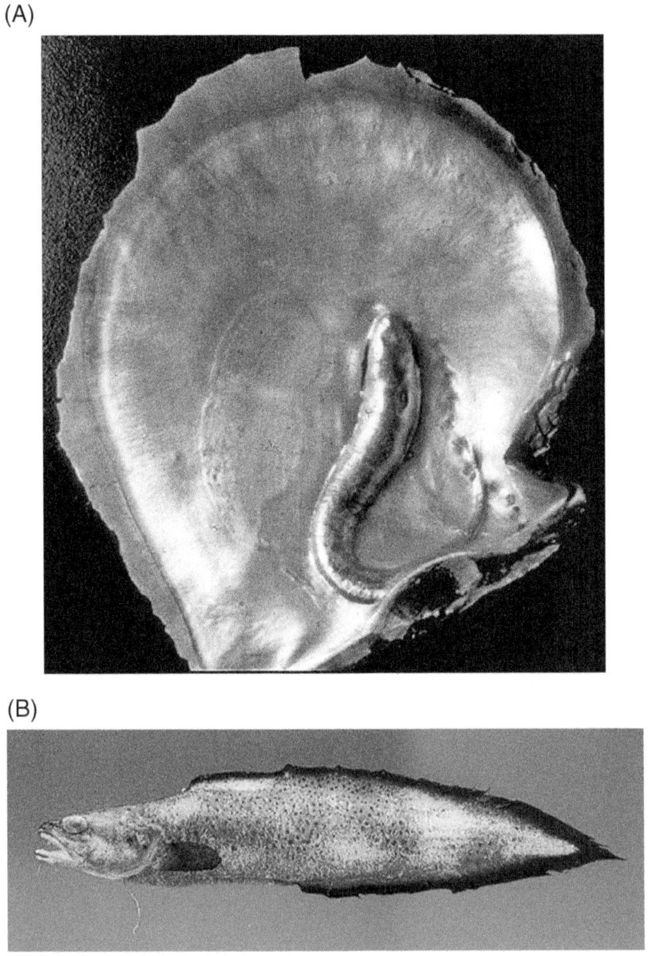

FIGURE 15.4 (A) A Pacific Pearlfish, *Encheliophis dubius*, entombed in a black lip pearl oyster. Courtesy of the Museum of Comparative Zoology and Harvard University. (B) An ophidiiform, the Bearded Brotula, *Brotula barbata*. NOAA / Public Domain.

Batrachoidiforms were also previously placed in the Paracanthopterygii, but have been moved to the Percomorphaceae based on molecular evidence (Nelson et al. 2016). They are well-camouflaged, benthic marine fishes with eyes placed high on the head, flattened heads and large mouths, relatively elongate dorsal and anal fins (Fig. 15.5), multiple lateral lines, and only three pairs of gills (rather than the usual five pairs). Their dorsal fins have two or three stout spines. Midshipmen (*Porichthys*) have 600–800 lateral photophores; unusual among shallow water fishes. Fishes in this order are often quite vocal, producing a variety of sounds with their gas bladders. The muscles that vibrate the gas bladder of toadfishes are the fastest contracting muscles known among vertebrates (Rome et al. 1999). Male midshipmen have been the focus of complaints by houseboat dwellers in San Francisco Bay during the (midshipmen's) breeding season, when the males produce a sustained, low frequency "hum" (Ibara et al. 1983). In the venomous toadfishes of the advanced subfamily Thalassophryninae, dorsal and opercular spines are part of a complex system that injects a powerful venom. Toadfishes are unusual zoogeographically because they are a shallow, warm water family that is most diverse in the Americas, whereas most tropical marine families have their greatest diversity in the Indo-Australian region (Collette & Russo 1981; see Chapter 19, Marine Zoogeographic Regions). Three South American species are restricted to fresh water. In morphology, ecology, and venom production, toadfishes are convergent with the scorpaeniform stonefishes and perciform weeverfishes.

For the remainder of this chapter, there is reduced morphological support for some of the groups recognized by molecular systematists such as Pelagiara and Ovalentaria, so we decided to return to the morphologically based classification utilized by Nelson (2006) and our second edition with notes referring to the current molecular rearrangement of fish groups. We feel that it is premature to utilize groups such as the "Pelagiara," which is postulated to contain 17 families that have never been placed together: "No morphological diagnosis exists for pelagiarians, representing a case of significant incongruence between morphological and molecular data" (Betancur-R et al. 2017, p. 22).

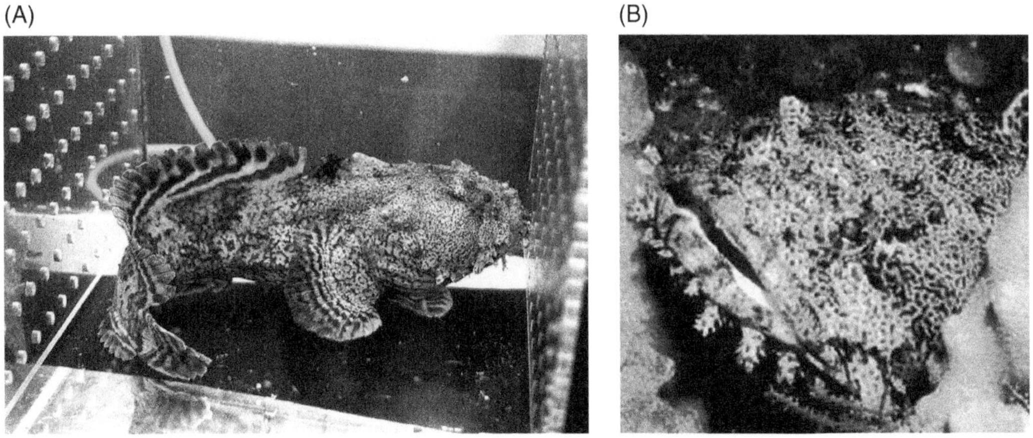

FIGURE 15.5 Batrachoidiforms – The Oyster Toadfish (*Opsanus tao*) is easily visible in a tank (A) NASA / Wikimedia Commons / Public Domain, and we can see how well it would blend into its habitat, especially when hiding in the substrate. (B) NOAA / Public Domain.

Subseries Mugilomorpha

Order Mugiliformes: Mugilidae (80 species; mullets, grey mullets)

(Orders, families, and numbers of species from Fricke et al. (2021) but refer to **https://researcharchive.calacademy.org/research/ichthyology/catalog/SpeciesByFamily.asp** *for updates.)*

The **mugilid** mullets (not to be confused with the mullid goatfishes) are a family of nearshore, catadromous fishes of considerable economic importance and of some taxonomic controversy. Their distinctly separated spiny and soft dorsal fins and spines in the pelvic and anal fins are used to justify their inclusion with the other acanthopterygians (Fig. 15.6). They are considered somewhat less derived because some have cycloid scales or scales intermediate between cycloid and ctenoid, and the pelvic girdle lacks any direct ligamentous or bony connection to the cleithral region of the pectoral girdle. In most more derived groups, the two girdles are connected. Mugilids are considered to be part of the molecular-based group Ovalentaria by Nelson et al. (2016). Many mullets are detritivorous, feeding on the organic silt that covers the bottom and digesting the minute plants and animals in such ooze with a gizzard-like stomach. Mullets frequently leap from the water for inexplicable reasons; one study showed that the frequency of such jumping increases when dissolved oxygen levels are low (Hoese 1985).

FIGURE 15.6 **(A)** A striped mullet, *Mugil cephalus*. From Jordan (1905). (B) A school of mullet swimming in the Wacissa River, Florida. R Haggerty / USFWS / Public Domain.

Subseries Atherinomorpha

Order Atheriniformes: Atherinidae (79 species of Old World silversides), Bedotiidae (16 species of Malagasy rainbowfishes), Melanotaeniidae (112 species of rainbow fishes), Pseudomugilidae (19 species of blue eyes), Telmatherinidae (18 species of sailfin silversides), Notocheiridae (Surf Silversides), Isonidae (5 species of surf sardines), Dentathrinidae (Tusked Silverside), Phallostethidae (23 species of priapiumfishes), Atherinopsidae (108 species of New World silversides)

Order Beloniformes: Adrianichthyidae (39 species of adrianichthyids), Scomberesocidae (5 species of sauries), Belonidae (44 species of needlefishes), Hemiramphidae (about 61 species of halfbeaks), Zenarchopteridae (63 species of viviparous halfbeaks), Exocoetidae (about 74 species of flyingfishes)

Order Cyprinodontiformes: Aplocheilidae (16 species of Old World rivulines), Nothobranchiidae (307 species of African rivulines), Rivulidae (468 species of New World rivulines), Pantanodontidae (2 species of African freshwater spine killifishes), Fundulidae (42 species of topminnows and killifishes), Cyprinodontidae (103 species of pupfishes), Profundulidae (12 species of Middle American killifishes), Goodeidae (47 species of goodeids and splitfins), Fluviphylacidae (7 species of American lampeyes), Poeciliidae (275 species of livebearers), Anablepidae (19 species of four-eyed fishes), Aphaniidae (42 species of oriental killifishes), Valenciidae (3 species of Valencia toothcarps), Protcatopodidae (81 species of African lampeyes)

(Orders, families, and numbers of species from Fricke et al. (2021) but refer to **https://researcharchive.calacademy.org/research/ichthyology/catalog/SpeciesByFamily.asp** *for updates.)*

The most successful fishes at the surface layer of the ocean and of many freshwater habitats are in the three orders of the Atherinomorpha: Atheriniformes, Beloniformes, and Cyprinodontiformes. Such well-known surface dwellers as silversides, needlefishes, sauries, flyingfishes, halfbeaks, killifishes, topminnows, and livebearers all belong to this group (Fig. 15.7). Anatomically, the atherinomorphs are set aside from the rest of the acanthopterygians in part because they have a unique way of protruding the jaw. The premaxilla does not articulate directly with the maxilla. Protrusion instead occurs by an intervening linkage between premaxilla and maxilla via the rostral cartilage. Atherinomorphs typically have terminal or superior mouths, as would be expected of surface-feeding fishes. Internal fertilization and live-bearing of young have evolved repeatedly within the group; many of the egg-laying families have chorionic filaments that protrude from the demersal, adhesive eggs and help them attach to plants and other structures. These filaments were the basis for naming the Ovalentaria, a molecular clade of 40 families (including the Atherinomorpha) by Wainwright et al. (2012). Atherinomorphs are sexually unique in that the only known unisexual (all female) fishes are members of this group, namely populations of a silverside and some poeciliid live-bearers.

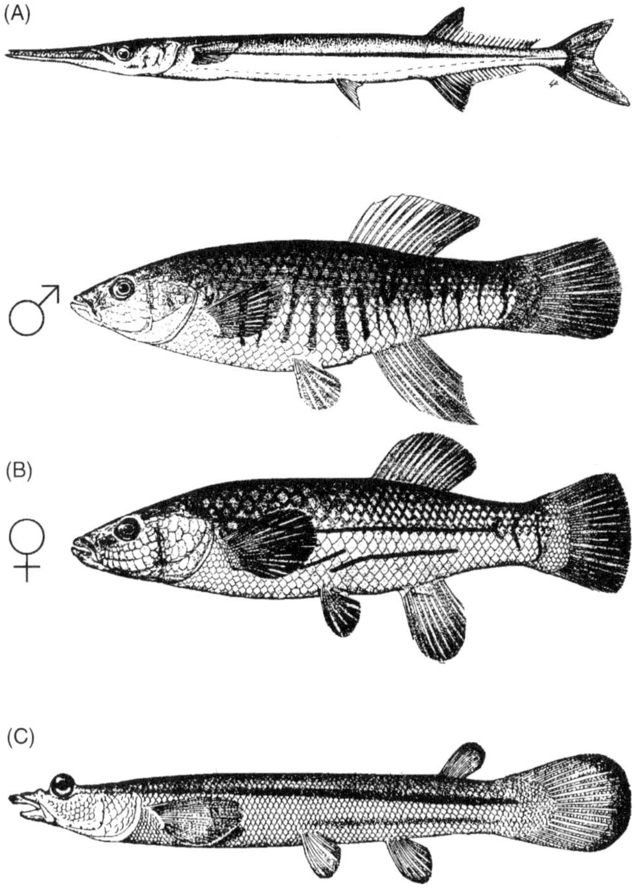

FIGURE 15.7 Atherinomorphs. (A) A belonid needlefish, *Tylosurus crocodilus*. (B) Fundulid Striped Killifish, *Fundulus majalis*, showing the male above, and female below. (C) The four-eyed fish, *Anableps*. (A) From Collette (1995), used with permission; (B) drawing by H. L. Todd, in Collette and Klein-MacPhee (2002); (C) from Jordan (1905).

Within the order **Atheriniformes** are six families of generally small, silvery fishes. The **atherinopsid** New World silversides are widespread freshwater and marine fishes that normally occur in schools in shallow water. Atherinopsids include the grunions (*Leuresthes* spp.) of southern and Baja California, which ride waves up beaches to spawn in wet sand every 2 weeks during the summer (see Chapter 18, Semilunar and Lunar Patterns). Sexual determination is under environmental control in some species, best studied in the Atlantic Silverside, *Menidia*. A radiation of 18 atherinopsid species, including apparent piscivorous species, has developed in lakes of the Mexican plateau. The **melanotaeniid** rainbowfishes of Australia and New Guinea are strongly sexually dimorphic freshwater fishes. Males have brighter colors and longer fins than females, traits that make them popular aquarium species. Such pronounced sexual dimorphism is rare in more primitive groups outside of the breeding season, except in some deep-sea fishes. The small (<4 cm) **phallostethids** are peculiar Southeast Asian atheriniforms in which the pelvic girdle and other structures of the males are modified into a complex clasping and intromittent organ for holding onto females and fertilizing their eggs internally. Females lack a pelvic girdle; they are also unusual in that they lay fertilized eggs rather than having young develop internally, which is the more normal course for fishes with internal fertilization.

Beloniforms are predominantly silvery, marine fishes active at and sometimes above the surface of the water. The **adrianichthyids** include the medakas or ricefishes, *Oryzias*, that are used extensively in genetic, embryological, and physiological investigations. Some beloniforms have several anatomical features that show precursors and intermediate conditions during the evolution of rather specialized traits. The lower lobe of the caudal fin in less derived beloniforms has more principle rays than the upper lobe. A rounded or square tail in less derived groups has changed into a forked tail fin with a slightly elongate lower lobe in the **belonid** needlefishes, species which can leave the water in short, arcing leaps (Fig. 15.7A). The lower lobe is very pronounced in the exocoetid flyingfishes, which use it as a sculling organ to accelerate during take-offs and to extend their gliding flights that can last hundreds of meters (Davenport 1994; see Chapter 16, Prey, and Fig. 16.13). A tendency for elongation of the lower or both jaws occurs in all belonoid groups, expressed as a garlike prey capture structure in piscivorous needlefishes and as unequal jaw lengths in sauries and particularly in halfbeaks. Different families show different developmental rates for the two jaws before the adult condition is reached, suggesting that evolution within the group has involved alterations in developmental rate of the jaws (Boughton et al. 1991; see Chapter 9, Evolution via Adjustments in Development: Heterochrony). For example, the lower jaw of some juvenile needlefishes is at first longer than the upper jaw, which later catches up. In some flying fishes, the lower jaw is at first elongate but later in life both jaws are essentially equal in length and neither projects forward. Halfbeaks, despite their predatory appearance, use their elongate lower jaw to feed on floating pieces of seagrasses; halfbeak herbivory is especially notable given that hemiramphids lack a true stomach, grinding up plant material in a pharyngeal mill. Some freshwater species take insects at the water surface.

The **cyprinodontiforms** are a major group of freshwater fishes, many of which show a high tolerance for saline and even hypersaline conditions. They are largely surface swimmers, preying on insects that fall into the water, which they detect using lateral line pores on the upper surface of the head. Life history traits in different cyprinodontiform families take on extreme conditions. Some of the South American and African **aplocheilid** and **nothobranchiid** rivulines are annual fishes that live in temporary habitats, spawn during the rainy season and die, their genes preserved in eggs that lie in a resting state in bottom muds until the next rains; the eggs are sufficiently drought resistant to survive for over 1 year (see Chapter 10, Deserts and Other Seasonally Arid Environments). *Kryptolebias* (formerly *Rivulus*) *marmoratus* of southern Florida and the West Indies is the only fish species known to be self-fertilizing. Sexual dimorphism reaches extremes in the elongate, brightly colored median fins of male **rivulines** (e.g. lyretails, panchax), **poeciliids** (sailfin mollies, guppies, swordtails), and pupfishes. Some **goodeids** have a placenta-like connection between the mother and the internally developing young (see Chapter 17, Parental Care and Fig. 17.6).

The **cyprinodontid** pupfishes are environmentally tolerant fishes that can live in water of highly variable salinity and temperature, characteristics that have allowed them to invade fluctuating environments such as saltmarshes, springs, and desert ponds (see Chapter 10). The isolated nature of many such habitats fuels rapid speciation but also makes the inhabitants extremely vulnerable to environmental disturbance; many pupfishes and their relatives have been extinguished or currently have Critically Endangered status according to the International Union for the Conservation of Nature (IUCN 2004; see also **www.desertfishes.org**). Many species of *Orestias* have evolved in Andean lakes, including Lake Titicaca, which at 4570 m above sea level is the highest natural body of water populated by fishes. **Anablepid** four-eyed fishes, a phylogenetically intermediate family within the cyprinodontiforms, is extraordinary in its eye structure (Fig. 15.7C). Four-eyed fishes are surface dwellers that swim with their protruding eyes half out of water. The pupil of the eye itself is physically divided into dorsal and ventral halves, the upper half capable of forming focused images of objects in air and the lower half simultaneously forming images of objects underwater (an intertidal labrisomid, the Galápagos four-eyed blenny, *Dialommus fuscus*, has converged on a similar eye structure). Some of the live-bearing poeciliids originated through hybridization and today do not include functional males; females instead mate with males of other species to activate embryogenesis, male genetic material being excluded from future generations (Meffe & Snelson 1989; Houde 1997; Uribe & Grier 2005; see Chapter 8, Sex and Gender Roles in Fishes). Poeciliids are an important ecological component of freshwater habitats on islands of the tropical western Atlantic and Caribbean, as well as coastal, tropical streams.

Subseries Percomorpha: Overview

The most derived euteleostean clade is the **Percomorpha**, a diverse and varied taxon that contains more than 13,000 species of largely marine families, although several successful freshwater groups also belong in this lineage. Percomorphs have in common an anteriorly placed pelvic girdle that is connected to the pectoral girdle directly or by a ligament; the pelvic fin also typically has an anterior spine and five soft rays, larger numbers of rays occurring in primitive percomorph taxa. This subseries, and similarly the order Perciformes, have been divided into multiple groups by some authors based on molecular characters (Fig. 15.1, and see Nelson et al. 2016, Betancur-R et al. 2017). We largely retain the groups of Nelson (2006) but will point out some of the differences as we address the groups.

We address the Percomorpha as three main subgroups: (i) the basal, or less derived percomorphs, (ii) the perch-like fishes (order Perciformes) – a group which some have divided into multiple orders (e.g. Nelson et al. 2016), and (iii) the more derived orders. The less derived groups are those showing generally fewer modifications from an ancestral percomorph condition; some refer to these groups as the "primitive" percomorphs. Similarly, the term "advanced" is often used in describing the more derived groups – those that show the greatest degree of evolutionary modification from the original stem groups.

Subseries Percomorpha: Basal (Less Derived) Orders

Order Gasterosteiformes: Hypoptychidae (1 species - Sand Eel), Aulorhynchidae (2 species of tubesnouts), Gasterosteidae (20 species of sticklebacks), Indostomidae (3 species of armored sticklebacks)

Order Syngnathiformes: Pegasidae (7 species of seamoths), Dactylopteridae (7 species), Solenostomidae (6 species of ghost pipefishes), Syngnathidae (322 species of pipefishes and seahorses, placed in separate subfamilies), Aulostomidae (3 species of trumpetfishes), Fistulariidae (4 species of cornetfishes), Macroramphosidae (9 species of snipefishes), Centriscidae (4 species of shrimpfishes)

Order Synbranchiformes: Synbranchidae (27 species of swamp eels), Chadhuriidae (11 species of earthworm eels), Mastacembelidae (93 species of freshwater spiny eels)

Order Scorpaeniformes: Scorpaenidae (233 species of scorpionfishes and rockfishes), Aploactinidae (50 species of velvetfishes), Pataecidae (3 species of Australian prowfishes), Gnathanacanthidae (1 species - Red Velvetfish), Congiopodidae (16 species of pigfishes), Triglidae (132 species of sea robins and gurnards), Peristediidae (45 species of armored searobins); Bembridae (12 species of deepwater flatheads), Platycephalidae (86 species of flatheads), Hoplichthyidae (17 species of ghost flatheads), Anoplopomatidae (2 species of sablefishes), Hexagrammidae (10 species of greenlings), Normanichthyidae (1 species - Barehead Scorpionfish), Rhamphocottidae (1 species - Grunt Sculpin), Ereuniidae (3 species of deepwater sculpins), Cottidae (294 species of sculpins, Baikal oilfishes, and deepwater Baikal sculpins), Trichodontidae (2 species of sandfishes), Hemitripteridae (8 species of searavens), Agonidae (49 species of poachers), Psychrolutidae (37 species of fathead sculpins), Bathylutichthyidae (2 species of Antarctic sculpins), Cyclopteridae (32 species of lumpfishes), Liparidae (444 species of snailfishes)

(Orders, families, and numbers of species from Fricke et al. (2021) but refer to **https://researcharchive.calacademy.org/research/ichthyology/catalog/SpeciesByFamily.asp** *for updates.)*

Gasterosteiforms are generally small marine and freshwater fishes with dermal armor plates, small mouths, and unorthodox propulsion (Fig. 15.8). Sticklebacks are among the world's most intensively studied fishes behaviorally, physiologically, ecologically, and evolutionarily (Wootton 1984; Bell & Foster 1994; Roberts Kingman et al. 2021). Separate populations of sticklebacks often diverge in anatomical traits and may constitute distinct genomes. The extent of predator avoidance, spines, and dermal plates often vary in relation to the threat from predators experienced by a population, making them showcases of the evolutionary process, as is the repeated,

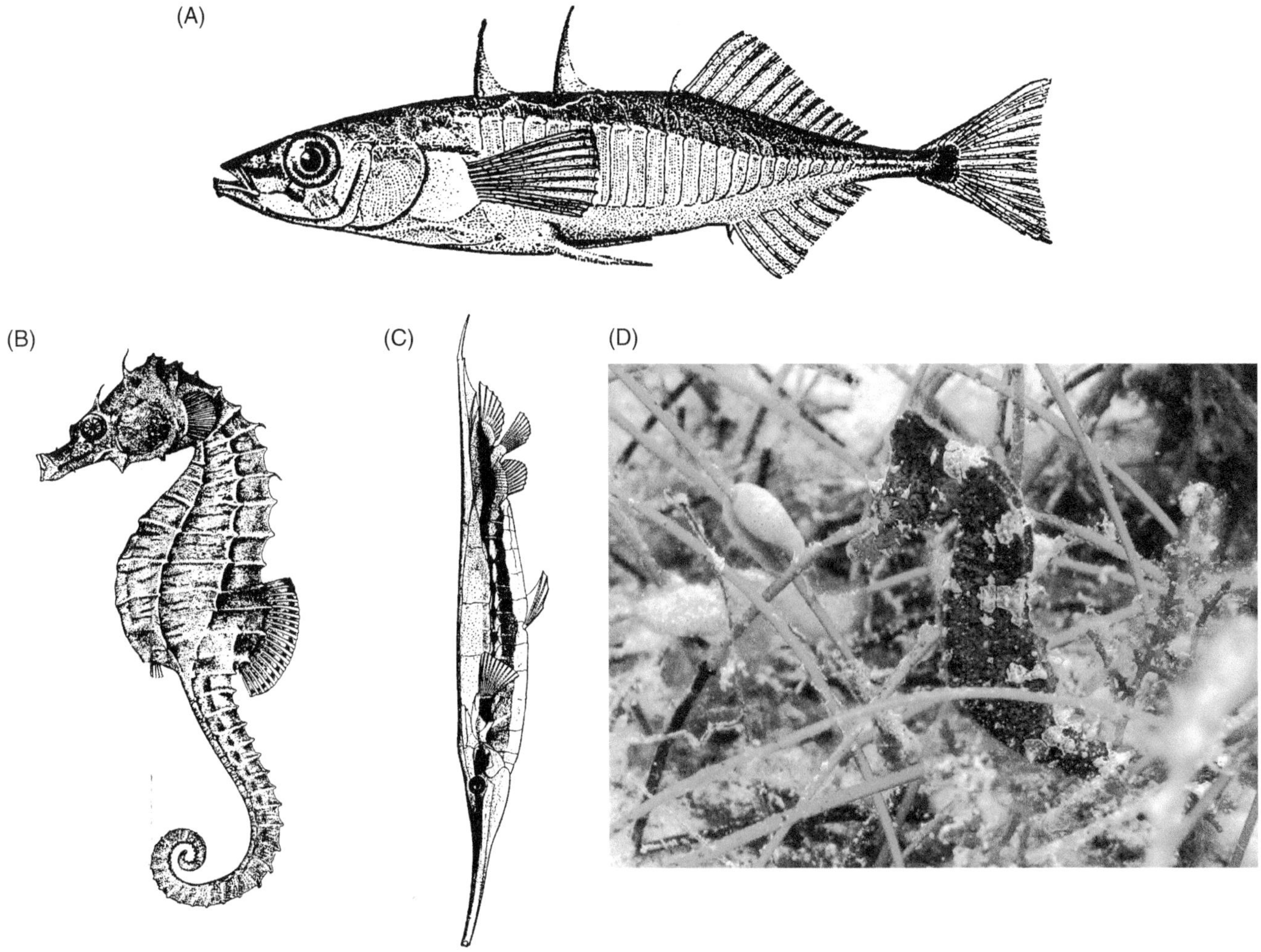

FIGURE 15.8 Gasterosteiforms and Syngnathiforms. (A) A Three-spined Stickleback, *Gasterosteus aculeatus*. (B) A sea horse, *Hippocampus erectus*. (C) A 13 cm centriscid shrimpfish, *Aeoliscus strigatus*. Centriscids are extraordinary gasterosteiforms that often hover head-down among sea urchin spines, where they are particularly well camouflaged. The first dorsal spine forms the posterior end of the body while the second dorsal, caudal, and anal fins are directed downward. (D) The shape and color of seahorses and their relatives often help them blend into their surroundings. Also see Figures 16.3 and 16.4A for other examples of camouflage among sygnathiforms. A, B, C from Jordan (1905). (D) NOAA / Public Domain.

independent appearance of species pairs in multiple lakes (e.g. Rundle et al. 2003; Boughman et al. 2005).

The **Syngnathiforms** include several unusually shaped fishes encased in bony rings, including indostomiid armored sticklebacks, pegasid seamoths and syngnathid pipefishes, sea dragons, and seahorses (Fig 15.8). In the primitive solenostomid ghost pipefishes, the female carries developing eggs in a brood pouch formed by pelvic fins fused to the ventral body surface. In more derived groups, sexual role reversal is the norm, and syngnathid pipefishes and seahorses are the only vertebrates in which the male literally becomes pregnant. An evolutionary gradient of degrees of male parental care exists within the family and correspond to the recognized phylogeny of the group. Pipefish taxonomy is based in part on whether eggs are embedded or attached to the male's ventrum, whether the pouch is sealed or open, and whether plates or membranes protect the eggs. In less derived species, eggs are attached externally to the male's ventral surface where they develop and hatch. In more derived species, the eggs are deposited within a pouch, fertilized, and the embryos develop within the pouch, where they obtain protection, oxygenation, osmoregulation, and nutrition from the male. Such role reversal in reproductive behavior includes females that actively court and compete for males (e.g. Rosenqvist 1990). Correlated with a poorly developed olfactory apparatus, the fewest functional genes involved with olfaction, 15, were found in the Broad-nose Pipefish, *Syngnathus typhle* (Policarpo et al. 2021). Locomotion is accomplished by rapid undulation of the small dorsal fin; the caudal fin is lacking in seahorses, which have transformed the caudal peduncle region into a prehensile structure for holding onto structures. Because of their attractiveness and small size, seahorses are highly sought for the aquarium trade, but their feeding and water quality requirements are such that few survive for long in glass boxes. A greater threat is unregulated and unsustainable harvest for traditional Asian medicinal preparations (Vincent et al. 2011). Because they mate for life and reproduce slowly, seahorses are particularly vulnerable to overharvest. Their genus, *Hippocampus*, is considered at risk by the IUCN (International Union for Conservation of Nature) and is listed in Appendix II of the

Convention on International Trade in Endangered Species (CITES; **www.cites.org**) (2021, **https://cites.org/eng/app/appendices.php**; see also **http://seahorse.fisheries.ubc.ca**).

Also exceptional within the sygnathiforms are the **aulostomoids** – the trumpetfishes and cornetfishes, which are elongate, large (up to 1 m), lurking and stalking piscivores with very expandable mouths. The intriguing centriscid shrimpfishes of the Indo-Pacific are small, extremely compressed fishes with the shape and proportions of an edible peapod encased in thin bone (see Fig. 15.8C). Due to an almost right-angle flexure in the vertebral column, their second dorsal, caudal, and anal fins all point ventrally, and they tend to swim with their dorsal edge leading while oriented head down. The fish typically hover head down among the spines of long-spined sea urchins where they are protected and difficult to see due to their thinness and a long black lateral stripe. The **dactylopterids** (flying gurnards) had been considered in the order Scorpaeniformes but molecular evidence places them in the Sygnathiformes, and in the same suborder as the aulostomids. These fishes have huge pectoral fins which they expand as they walk along the bottom on their elongate pelvic fins; it is unlikely that adult "flying" gurnards ever leap out of the water or for that matter ever swim far above the bottom.

The **synbranchiforms** are a small order of primarily freshwater, eel-like fishes (Fig. 15.9). The synbranchid swamp and rice eels are air-breathing fishes found in Africa, Asia, and Central and South America. They have many unusual derivations, including loss of pectoral, pelvic, dorsal, anal, and, in some, caudal fins. Most are protogynous (female first) hermaphrodites. Synbranchids also have a unique upper jaw arrangement in which the palatoquadrate attaches at two points to the skull, termed **amphistylic suspension** and not known in any other teleosts. The highest number of OR genes associated with olfaction, 429, were found in a synbranchid, the Zig-zag Eel, *Mastecembelus armatus* associated with a well-developed olfactory rosette (Policarpo et al. 2021). Swamp and rice eels released by irresponsible aquarium keepers constitute a growing threat as an invasive species in Florida, Georgia, and Hawaii (see **www.invasivespeciesinfo.gov/aquatics/swampeel.shtml**).

The **scorpaeniforms** are a large order of predominantly marine fishes (Fig. 15.10). Their exact position within the percomorphs is a matter of considerable debate. Most have spines projecting from different bones on the head, including a posteriorly directed spine derived from a bone below the eye, giving them the name "mail-cheeked fishes." Many scorpaeniforms lack scales as part of a general suite of adaptations to benthic living. The scorpaenid scorpionfishes and rockfishes are a diverse group of benthic marine fishes with large mouths and venomous spines in their dorsal, anal, and pelvic fins. The sebastine rockfishes (a subfamily of over 130 species) are a diverse, important commercial group of often live-bearing, long-lived (up to >140 years), overexploited species of the temperate North Pacific (e.g. Boehlert & Yamada 1991; Love et al. 2002). Other subfamilies within this group include the colorful and venomous lion or turkeyfishes (e.g. *Pterois*), as well as the camouflaged and highly venomous stonefishes (*Synanceia*) – the latter purported to possess the deadliest of fish venoms in their spines. The hexagrammid greenlings are littoral zone and kelp associated fishes endemic to the North Pacific. The family includes the highly edible, predatory, and significantly overfished Lingcod, *Ophiodon elongatus*.

FIGURE 15.9 Synbranchiforms such as the Asian Swamp Eel (*Monopterus albus*) are native to Asia, Africa, and Central and South America, but have become introduced to Hawaii and some southern states of the United States. USGS / Wikimedia Commons / Public Domain.

(A)

(B)

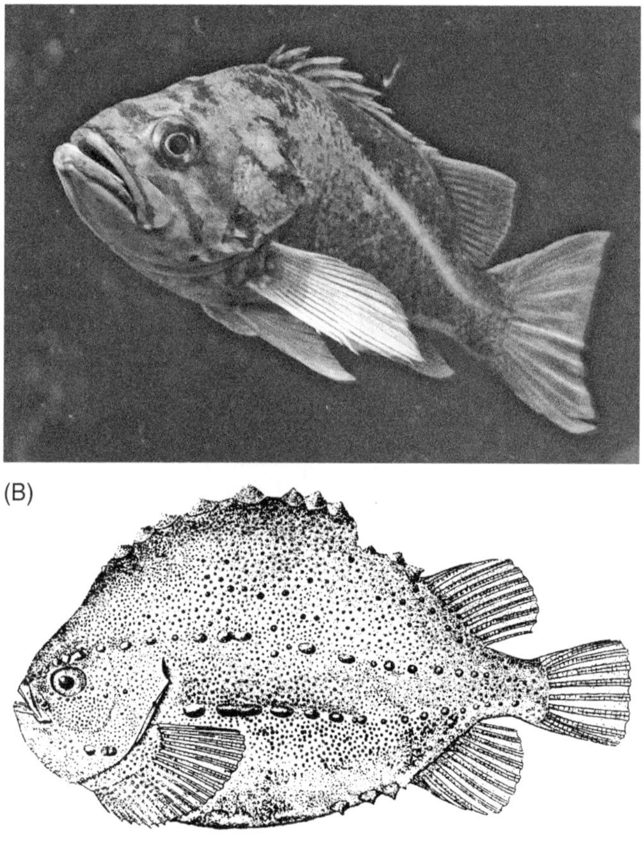

FIGURE 15.10 Scorpaeniforms. (A) A Canary Rockfish, *Sebastes pinniger*, one of the numerous sebastine rockfishes of the North Pacific. (B) The Lumpfish, *Cyclopterus lumpus*, of the North Atlantic. (A) G. Helfman (Author); (B) drawing by H. L. Todd, in Collette & Klein-MacPhee (2002).

The only freshwater scorpaeniforms are in the suborder Cottoidei, which includes the cottid sculpins of North American headwater streams and tidepools, as well as a species flock of comephorid oilfishes and other cottoid species in Lake Baikal in Asia. Many cottoids lack scales but have prickly skin. The Cabezon, *Scorpaenichthys marmoratus*, of the Pacific coast of North America is apparently unique outside the tetraodontiform teleosts and holostean gars (Chapter 13) in having toxic eggs. The Trichodontidae (2 species of sandfishes), formerly considered trachinforms, are now included within the scorpaeniforms (Nelson et al. 2016). The most derived scorpaeniforms are the cyclopteroid lumpfishes and snailfishes. The globose lumpfishes have bony tubercles arranged in rows around their body and a sucking disk made from modified pelvic fins, an unusual trait for fishes that do not frequent high-energy zones. The Lumpfish of the North Atlantic, *Cyclopterus lumpus*, is highly prized for its caviar and has been seriously depleted in parts of its range. They are also valued as cleaner fishes to help decrease parasitic copepods in salmon farming operations (see Chapter 22, Is Aquaculture an Answer?). Liparid snailfishes, which also have pelvic suction disks, occur broadly geographically and ecologically. They are found in most oceans from the Arctic to the Antarctic and can inhabit tidepools or benthic regions deeper than 7000 m.

Subseries Percomorpha, Order Perciformes: The Perch-Like Fishes

In the phylogenetic structure of the more derived teleosts that we present, the largest order in the Percomorpha, and for that matter of vertebrates, is the **Perciformes**, containing over 100 families and well over 10,000 species, more than a third of all fishes.

The classification of the perciforms is a subject of considerable discussion, which is not surprising in such a diverse taxon. Nelson et al. (2016) placed many families previously included in the order into other orders within the Percomorpha, retaining just 62 families (with about 365 genera and 2,248 species) in the Perciformes (see Fig. 15.1B). As mentioned at the beginning of this large section on the series Percomorphaceae, although these distinctions are supported by molecular data, many are not supported by morphological characters. We decided, therefore, to follow Nelson (2006) and keep all of these groups within the Perciformes at this time. We encourage readers to consult Nelson et al. 2016 (including the updates at **https://sites.google.com/site/fotw5th/**), Betancur-R et al. 2017, and more recent research for phylogenetic re-evaluations of the Percomorphs.

Our discussion will focus on selected families within 18 perciform suborders. The success of perciforms is greatest in, but by no means limited to, coral reef habitats, where six of the eight largest families abound (gobies, wrasses, seabasses, blennies, damselfishes, cardinalfishes). Two other large families, the cichlids and the croakers, reach their maximum diversity in tropical lakes and nearshore temperate marine habitats, respectively. The fossil record for perciforms dates back to the Early Cenozoic, and recognizable members of most suborders had evolved by the Eocene, indicating very rapid evolution and diversification over a period of about 20 million years (Carroll 1988).

Suborder Percoidei

Suborder Percoidei: Centropomidae (13 species of snooks), Ambassidae (about 54 species of Asiatic glassfishes), Perciliidae (2 species of southern basses), Latidae (14 species of giant perches), Moronidae (6 species of temperate basses), Percichthyidae (23 species of temperate perches), Acropomatidae (13 species of lanternbellies), Symphysanodontidae (13 species of slopefishes), Polyprionidae (4 species of wreckfishes), Serranidae (about 598 species of seabasses), Centrogenyidae (1 species - False Scorpionfish), Ostracoberycidae (3 species of ostracoberycids), Callanthiidae (18 species of splendid perches), Pseudochromidae (155 species of dottybacks), Grammatidae (18 species of basslets), Plesiopidae (about 51 species of roundheads), Opistognathidae (89 species of jawfishes), Dinopercidae (2 species of cavebasses), Banjosidae (3 species of banjofishes), Centrarchidae (45 species of sunfishes, black basses, and pygmy sunfishes), Percidae (242 species of darters and perches), Priacanthidae (21 species of bigeyes), Apogonidae (385 species of cardinalfishes), Epigonidae (47 species of deepwater cardinalfishes), Sillaginidae (about 38 species of sillagos), Malacanthidae (about 16 species of tilefishes), Lactariidae (1 species - False Trevally), Dinolestidae (1 species - Long-finned Pike), Scombropidae (3 species of gnomefishes), Pomatomidae (1 species - Bluefish), Nematistiidae (1 species - Roosterfish), Coryphaenidae (2 species of dolphinfishes), Rachycentridae (1 species - Cobia), Echeneidae (8 species of remoras), Carangidae (151 species of jacks), Menidae (1 species - Moonfish), Leiognathidae (53 species of ponyfishes), Bramidae (20 species of pomfrets), Caristiidae (18 species of manefishes), Emmelichthyidae (18 species of rovers), Lutjanidae (113 species of snappers), Caesionidae (23 species of fusiliers), Lobotidae (15 species of tripletails), Gerreidae (about 54 species of mojarras), Haemulidae (136 species of grunts), Nemipteridae (75 species of threadfin breams), Lethrinidae (43 species of emperors), Sparidae (164 species of porgies), Polynemidae (42 species of threadfins), Sciaenidae (297 species of croakers and drums), Mullidae (about 101 species of goatfishes), Pempheridae (85 species of sweepers), Glaucosomatidae (4 species of pearl perches), Leptobramidae (2 species of beachsalmons), Bathyclupeidae (10 species of bathyclupeids), Monodactylidae (6 species of moonfishes), Toxotidae (10 species of archerfishes), Arripidae (4 species of Australasian salmons), Dichistiidae (2 species of galjoen fishes), Kyphosidae (14 species of sea chubs), Drepaneidae (3 species of sicklefishes), Chaetodontidae (about 134 species of butterflyfishes), Pomacanthidae (90 species of angelfishes), Enoplosidae (1 species - Oldwife), Pentacerotidae (13 species of armorheads), Nandidae (7 species of Asian leaffishes), Polycentridae (5 species of South American leaffishes), Terapontidae (61 species of grunters), Kuhliidae (12 species of flagtails), Oplegnathidae (7 species of knifejaws), Cirrhitidae (35 species of hawkfishes), Chironemidae (6 species of kelpfishes), Aplodactylidae (5 species of marblefishes), Cheilodactylidae (2 species of morwongs), Latridae (32 species of trumpeters), Cepolidae (45 species of bandfishes), Sphyraenidae (27 species of barracudas)

(Families and numbers of species from Fricke et al. (2021) but refer to **https://researcharchive.calacademy.org/research/ichthyology/catalog/SpeciesByFamily.asp** *for updates.)*

The largest perciform suborder is the Percoidei, and it shows considerable diversity (Fig. 15.11). Percoids, in contrast with less derived teleosts such as ostariophysans and protacanthopterygians (and continuing trends in acanthopterygians and percomorphs), are characterized by: (i) the presence of spines in the dorsal, anal, and pelvic fins; (ii) two dorsal fins (never an adipose fin); (iii) ctenoid scales; (iv) pelvic fins in the abdominal position; (v) laterally placed and vertically oriented pectoral fins; (vi) maxilla excluded from the gape; (vii) physoclistous gas bladder; (viii) absence of orbitosphenoid, mesocoracoid, epipleural, and epicentral bones; (ix) acellular bone; and (x) never more than 17 principal caudal fin rays.

The basal families of percoids are what would generally be considered bass-like fishes. **Centropomids** are primarily large, piscivorous fishes of lakes, estuaries, and nearshore regions, including the snooks (*Centropomus*) of tropical America. Related **latids** include the Barramundi of Australia and Nile Perch of Africa (*Lates* spp.). Nile Perch and their relatives have been widely introduced in African lakes and have caused the extinction of hundreds of native species including endemic cichlids (see Chapter 22, Species Introductions). The **moronid** temperate basses include lake-dwelling and anadromous predators in North America such as the White Bass and Striped Bass (*Morone* spp.). The closely related **polyprionid** wreckfishes include such commercially important species as the Atlantic Wreckfish (*Polyprion americanus*) and the Giant Sea Bass of California (*Stereolepis gigas*), the latter reaching lengths of 2 m and weighing up to 250 kg. The seabass family

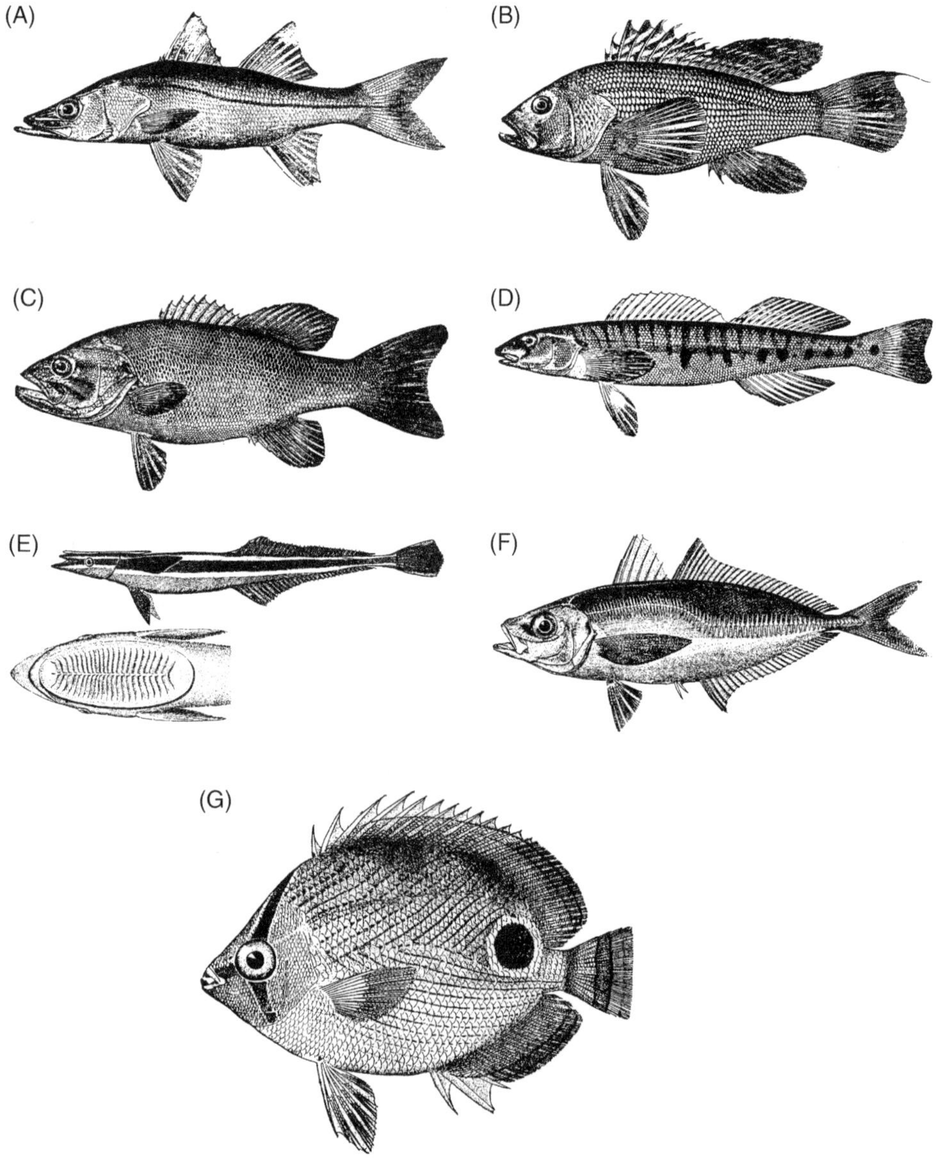

FIGURE 15.11 Representative percoid fishes. (A) A centropomid Snook, *Centropomus undecimalis*. (B) A serranid Black Sea Bass, *Centropristis striata*. (C) A centrarchid Smallmouth Bass, *Micropterus dolomieu*. (D) A percid darter, the Log Perch, *Percina caprodes*. (E) An echeneid Sharksucker, *Echeneis naucrates*, with a top view of the first dorsal fin that forms a suction disk. (F) A carangid Rough Scad, *Trachurus lathami*. (G) A chaetodontid Foureye Butterflyfish, *Chaetodon striatus*. (A, C, D, G) from Jordan 1905; (B, E, F) drawing by H. L. Todd, in Collette & Klein-MacPhee (2002).

Serranidae contains nearly 600 species and is one of the largest fish families (as many as 15 subfamilies have been variously recognized). It contains a tremendous diversity of sizes and shapes of fishes that have three spines on the opercle but may differ in many other characters. Serranids vary in size from 3 cm long planktivorous anthiines to the 3 m long, 400 kg Goliath Grouper, *Epinephelus itajara*, which eats lobsters and small turtles as well as fishes. Three subfamilies are recognized, the first two (Serraninae and Anthiinae) consisting of mostly small forms such as hamlets, sand perches, and the colorful *Anthias*. The subfamily Epinephelinae is defined by long, stout or filamentous dorsal and/or pelvic fin spines in the larvae. It contains most commercially important species such as groupers, hinds, coneys, gag, and scamp but also includes diminutive and striking basslets and the chemically protected soapfishes (e.g. *Rypticus*, *Grammistes*), which exude a soap-like toxin from their skin when disturbed. Many serranids are hermaphroditic, usually starting as female and then later becoming male (protogyny), although some hamlets and members of the genus *Serranus* function simultaneously as either sex. Many groupers reproduce in large spawning aggregates which are easily located by fishers, leading to massive overfishing (e.g. Nassau Grouper, *Epinephelus striatus*), prompting the need for highly regulated fisheries (Beets & Friedlander 1999). To confuse the taxonomic issue, most species known as basslets belong to the related family **Grammatidae**, which also includes the neon-colored Royal Gramma that lives under ledges in the Caribbean and is popular among divers and aquarists.

Two important percomorph families in North American fresh waters are the centrarchids and the percids. The **Centrarchidae** contain 45 species, including numerous sunfishes, crappies, and rock basses (*Lepomis, Pomoxis, Ambloplites*), the seven pygmy sunfishes (*Elassoma*), and the seven black basses in the genus *Micropterus* (e.g. Largemouth Bass, Smallmouth Bass, plus undescribed species in the southeastern USA). Centrarchids are the dominant carnivores in most lakes in the USA and southern Canada and are also known for their nesting behavior, the males digging and defending circular nests on the bottom through much of the summer. Centrarchids are native to the region east of the Rocky Mountains with the exception of the Sacramento Perch, *Archoplites interruptus*, a California endemic sunfish with a shrinking native range (Moyle 2002) but widely introduced elsewhere in the western United States. Centrarchids have been introduced so widely and successfully that their importation is now outlawed or severely regulated in several countries (Cowx 1997; Helfman 2007). The pygmy sunfishes (*Elassoma*) are interesting because of their miniaturization (maximum length 45 mm, but several species are smaller than 20 mm); the seven species are primarily dwellers of swampy habitats in southeastern USA (e.g. Everglades Pygmy Sunfish, Okefenokee Pygmy Sunfish). Many males take on iridescent blue coloration during the breeding season. The taxonomic position of this enigmatic group is a matter of debate, with different authors placing them outside the perciform order and more closely aligned with less derived mugilomorphs, atherinomorphs, gasterosteiforms, or synbranchiforms. Other authors, such as Nelson et al. (2016) return them to the sunfish family Centrarchidae as a separate subfamily.

The family **Percidae** is one of the most successful non-ostariophysan freshwater families in the world. There are at least 240 species of percids, most of which occur in North America. The dominant lake forms are larger species such as the Yellow Perch (*Perca flavescens*) and Walleye (*Sander vitreus*) (Colby 1977; Craig 1987). Yellow Perch have their counterpart in the Eurasian Perch, *P. fluviatilis*; three species of *Sander* pikeperches also occur in Europe (Collette & Banarescu 1977). North American streams contain more than 184 species of darters, mostly in the genera *Percina* and *Etheostoma* (Fig. 15.12). The greatest diversity is in the southeastern USA; the state of Tennessee alone houses 90 darter species. These small, benthic fishes feed primarily on aquatic insect larvae and other invertebrates in fast flowing, clean water where males defend nests and court females. During the breeding season the males take on color patterns that rival the brightest poster colors of tropical fishes (e.g. Page 1983; Etnier & Starnes 1993; see **www.cnr.vt.edu/efish/families/percidae.html**). Darters are disproportionately imperiled by siltation and other forms of pollution and habitat modification, problems that are exacerbated by their small sizes, small geographic ranges, headwater habitats, and, especially, benthic breeding and feeding habits. Somewhere between one-half and two-thirds of all darter species are considered to be at risk of extinction, and the Maryland Darter, *Etheostoma sellare*, is thought to be extinct (Warren et al. 2000; Jelks et al. 2008).

(A)

(B)

FIGURE 15.12 (A) The tangerine darter, *Percina aurantiaca*, the second largest of the nearly 200 species of darters endemic to North America (*c*. 18 cm). Courtesy of J. DeVivo. (B) A Greenbreast Darter, *Etheostoma jordani*, in Holly Creek in northwest Georgia. Courtesy of A. Nagy.

The **apogonid** cardinalfishes are a speciose (385 species) family of small (<10 cm), nocturnal coral reef fishes. Their large eyes, large mouths, distinctly separated dorsal fins (the second with a single spine), deep bodies, and relatively pointed heads distinguish them from most other reef fishes that hover motionlessly just above or in structure. Some cardinalfishes mouthbrood their eggs, the male or female being responsible in different species. Many cardinalfishes live in close association with invertebrates (e.g. sea urchins, branching corals, anemones) and use them as refuges. A few Indo-Pacific species enter estuaries, and several New Guinea species are restricted to fresh water. The dramatically colored and narrowly distributed Banggai Cardinalfish, *Pterapogon kauderni*, was once thought to be extinct in the wild – a very rare occurrence among marine fishes (Fig. 15.13). It has apparently recovered because of captive breeding and perhaps inadvertent releases from aquarium holding facilities (Vagelli & Erdmann 2002). The **malacanthid** tilefishes are a marine group that inhabits burrows. Sand tilefishes (e.g. *Malacanthus*) are tropical species that live over shallow, sandy areas and dig complex burrows which they reinforce with shell and coral fragments, pieces of which are piled in a mound at the burrow's entrance. Their engineering activities create hard bottom patches that are used by small fishes and invertebrates that would otherwise not colonize sandy regions (see Chapter 21, Fishes as Producers and Transportersof Sand, Coral, and Rocks). The larger temperate latiline tilefishes (e.g. *Caulolatilus*) are commercially sought species that inhabit large burrows in deeper soft bottom regions, although it is unknown whether they construct the holes themselves.

FIGURE 15.13 Banggai Cardinalfish, endemic to the Banggai Islands of Indonesia, were depleted due to collecting for the aquarium trade. However, captive animals in a holding facility escaped and multiplied and have even increased the originally limited natural range. See Helfman (2007) for details. G. Helfman (Author).

The next dozen or so percomorph families are generally active, marine, water column dwellers with relatively compressed, often silvery bodies. The larger species are piscivores and the smaller ones are zooplanktivores. **Lactariid** false trevallies and **dinolestid** long-finned pikes have converged on the body shapes and habits of predatory jacks and barracudas, respectively. The cosmopolitan **pomatomid** Bluefish, *Pomatomus saltatrix*, which occurs in most major ocean basins except the eastern Pacific, has a well-deserved reputation for voraciousness (see Hersey 1988). This aggregating predator will enter a school of prey fish and slash and dismember far more individuals than are actually eaten; attacks on humans unfortunate or foolish enough to be in the water during such feeding frenzies are well documented. The family **Nematistiidae** is sometimes combined with the next four families (dolphinfishes, cobia, remoras, and jacks) to form a clade known as carangoids or the order Carangiformes of Nelson et al. (2016). The colorful Roosterfish, *Nematistius pectoralis*, is a monotypic piscivore of warm, eastern Pacific, inshore areas. It looks like an amberjack with a cockscomb of seven elongate dorsal spines. It is actively sought as a gamefish and attains lengths of 1.5 m and can weigh 50 kg. The **coryphaenid** dolphinfishes or mahimahis include two species of open water, surface-oriented predators that are often found in association with floating structure or seen chasing flyingfishes. Male Dolphinfish (bulls) have a square head profile involving expansion of the bony portion of the supraoccipital region (forehead) (Fig. 15.14), whereas in females the forehead slopes more gradually; such obvious skeletal sexual dimorphism is rare in acanthopterygians. The golden coloration of corphyphaenids has earned them the Spanish name *dorado* and their color-changing habits when brought on board a boat are legendary; Yann Martel in *Life of Pi* likens subduing a mahi-mahi with a club as "beating a rainbow to death" (Martel 2003, p. 185).

The next two families are somewhat similar in body form (Fig. 15.15), although the monotypic Cobia, *Rachycentron canadum* (**Rachycentridae**), is larger than the **echeneid** remoras. Cobia are a popular sportfish, and anglers frequently locate them by fishing near manta rays, but the nature of the association is unexplored. Cobia are also raised in offshore pens as a food fish. The **echeneid** remoras or sharksuckers are a highly specialized group of eight species in which the first dorsal fin has been modified into a sucking organ for attachment to sharks, billfishes, whales, turtles, and an occasional diver. The adhesive suction disk is a lamellar structure supported and controlled by a complex series of muscles and skeletal elements that function to erect and depress the laminae and create suction pressures. A force more than 17 newtons was required to dislodge a remora from shark skin; earlier studies indicated that a 67 mm long common sharksucker could support a pail of water that weighed 11 kg (see Fulcher & Motta 2006). Some remoras such as the large (1 m) *Echeneis naucrates* are frequently seen free-swimming, whereas smaller species are almost always attached to hosts, but detach to feed on scraps from the predatory host's meals and have also been observed cleaning the host's mouth, teeth, and gills.

(A) (B)

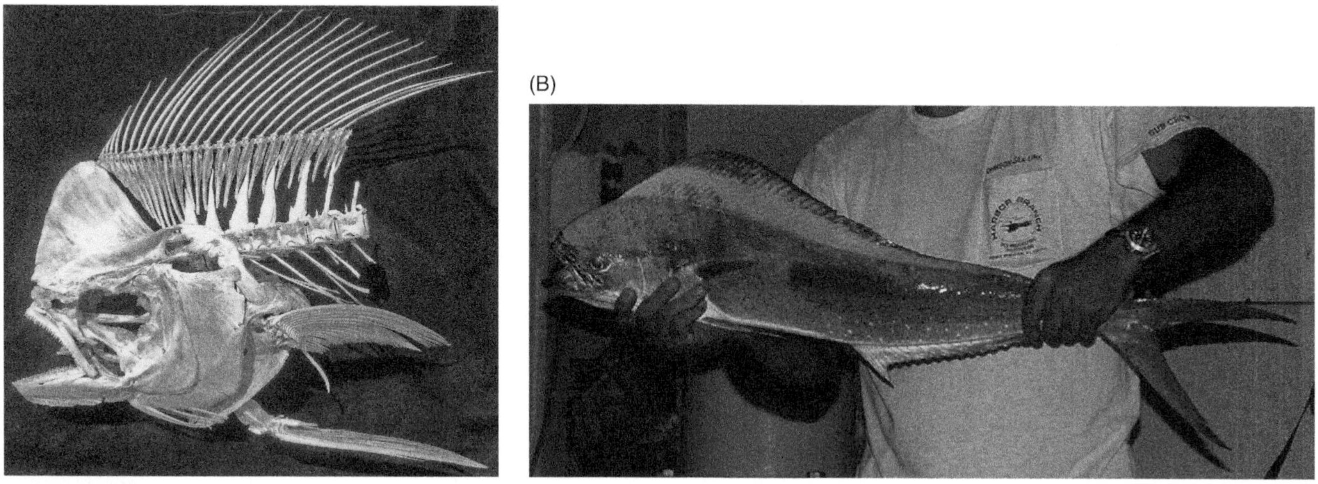

FIGURE 15.14 (A) Dolphinfish, dorado, or mahimahi, *Coryphaena hippurus*, exhibit extreme sexual dimorphism. The male shown here has a greatly enlarged supraoccipital bone in its forehead, which gives it the characteristic square-headed appearance. Females have a much narrower and streamlined profile. Skeletal preparation Grant Stoecklin. (B) A Dolphinfish shortly after capture shows its brilliant iridescence. NOAA / Public Domain.

The **carangid** jacks and pompanos are a large family (151 species) of tropical nearshore and pelagic predators and zooplanktivores that range in size from the small scads (*Decapterus*, *Selar*, *Trachurus*) to the large amberjacks and pompanos (*Seriola*, *Caranx*). Carangids tend to be slightly to very laterally compressed, the extreme occurring in the Lookdown, *Selene vomer*, which nearly disappears when it faces an observer head-on. Carangids of all sizes are often found in shoals and evidence of cooperative hunting exists in a few large species (e.g. *Caranx melampygus*; see Chapter 16, Attack and Capture). Carangids engage in a form of highly efficient and powerful locomotion, termed **carangiform swimming**, involving side-to-side movement of primarily the tail; thrust is transferred from the body musculature to the tail via tendons that cross the caudal peduncle region (Chapter 4, Locomotion: Movement and Shape; see Fig. 4.11).

A subsequent group of families (lutjanids through nemipterids) consists of generally heavy-bodied, tropical fishes that swim near the bottom and feed on large invertebrates and fishes (with some notable exceptions). The **lutjanids** are a large family (113 species) of ecologically diverse but generally carnivorous marine fishes inhabiting shallow to moderate depths in tropical and warm temperate seas and estuaries (e.g. Polovina & Ralston 1987). The typical snapper is a fairly large (up to 1 m), heavy-bodied, suprabenthic, nocturnal or crepuscular predator with large canine teeth, such as the Gray, Red, Mangrove, or Mutton snappers (*Lutjanus* spp., *Pristopomoides* spp.) (Fig. 15.16A). However, many snappers live in the water column and are more streamlined, including the Vermilion and Yellowtail snappers (*Rhomboplites*, *Ocyurus*). A closely related family, the **caesionid** fusiliers, are small, streamlined, and brilliantly colored planktivores with forked tails and protrusible mouths that school near drop-offs and reef edges on coral reefs of the Indo-West Pacific. **Lobotid** tripletails get their name from the unusual arrangement of soft dorsal, soft anal, and caudal fins of nearly equal size and rounded shape, the dorsal and anal fins placed posteriorly on the body overlapping the tail fin. Tripletails are heavy-bodied, bass-like fishes of estuarine and fresh waters worldwide in temperate and tropical waters. Adults can reach a meter in length and are uncommon; juveniles are more frequently encountered floating leaflike on their sides in mangrove regions. The silvery mojarras (**Gerreidae**) are common inhabitants of sandy or silty regions near coral reefs and other shallow, warm water habitats worldwide; some species enter fresh water. Their body shape and feeding habits are somewhat incongruous. They have a forked tail and the mouth protrudes slightly downward, which might suggest zooplanktivory, but they are typically observed foraging head-down with their extremely protractile mouths extended into the bottom sediments. When they emerge, they typically expel clouds of sediment out their gill openings, having retained benthic invertebrates with their gill rakers.

Grunts (**Haemulidae**) are moderate-sized coral reef fishes that are unusual in that they are more diverse in the New World rather than the Old World tropics. Most grunts form shoals as juveniles and some such as the Porkfish, *Anisotremus*, which is common in the Florida Keys, continue to shoal as adults. They are typically nocturnal feeders on benthic or grassbed associated invertebrates, undertaking distinctive migrations between shaded daytime resting areas (see Fig. 15.16B) and nighttime feeding regions. The **inermiid** bonnetmouths are haemulid derivatives adapted for zooplanktivorous feeding. They have the slender bodies, forked tails, and protractile mouths typical of many zooplanktivorous fishes. Porgies (**Sparidae**) are gruntlike in appearance but are more diversified in their feeding than the haemulids. The western Atlantic Sheepshead, *Archosargus probatocephalus*, has massive pharyngeal dentition used for crushing hard-bodied prey such as mollusks. Several sparids, such as the Pinfish, *Lagodon rhomboides*, feed extensively on plants, making this one of the

(A)
(B)
(C)

FIGURE 15.15 (A) This female Cobia, *Rachycentron canadum* (Rachycentridae), will be used as brood stock for aquaculture. NOAA / Public Domain (B) Remoras (sharksuckers, Echenieidae) resemble Cobia, but are in a different family. NOAA / Public Domain (C) The dorsal fin of remoras has become modified and serves as a sucking disk that allows them to adhere to large fishes and sea turtles. Tibor Marcinek / Wikimedia Commons / Public Domain.

few percoid families to include strongly herbivorous species (herbivory becomes more common in more derived groups). Together with the **centracanthids**, **lethrinids**, and **nemipterids**, the porgies form a superfamily of related families known as sparoids; the sparids are the only family in the group in the western Atlantic region.

The next three families contain fishes that are frequently seen swimming just above and probing into the bottom with modified appendages. **Polynemid** threadfins are tropical marine fishes with highly specialized pectoral fins that are divided into two parts. The upper webbed portion is located laterally and shaped like a normal pectoral fin, whereas the ventral portion consists of three to seven long, unconnected rays that extend down from the throat region and are used to feel for prey on the bottom. The mouth is subterminal, as befits a bottom feeder. **Sciaenids** are a widespread tropical and temperate family and are particularly diverse in the southeastern USA. Some sciaenid croakers and drums also have a subterminal mouth. Many species have one or several small chin barbels. This large family (297 species) includes such important commercial and sportfishes as the Red Drum (spot tail bass), Black Drum, croakers, weakfish, sea trouts, kingfishes, White Seabass, corbinas, and the endangered Mexican Totoaba and Chinese Bahaba (the last two species can exceed 100 kg in mass). The common names for the family come from sound production habits that involve the vibration of muscles attached to the gas bladder. As is frequently the case with sound-producing fishes, sciaenid otoliths are exceptionally large. The role of acoustic stimuli in the biology of sciaenids is also reflected in their very extensive lateral line, which extends posteriorly onto the tail and anteriorly as numerous pits and canals on the head. Although a predominantly marine family, freshwater species are common in South America, and one species, the Freshwater Drum, *Aplodinotus grunniens*, may have the largest natural latitudinal range of any freshwater fish, occurring throughout the Mississippi River and adjacent drainages of North America and into Central America, from southern Saskatchewan and Quebec to Guatemala. The **mullid** goatfishes are the third bottom-oriented family. This tropical family of medium-sized, nearshore marine predators has two highly prehensile chin barbels that the fishes use to probe bottom sediments for prey (see Fig. 6.18). Their foraging activities frequently flush invertebrates from the sand, and it is not unusual to see wrasses and carangids following goatfishes and capturing escapees.

Monodactylid fingerfishes and **toxotid** archerfishes are brackish water families of chiefly Indo-Pacific distribution. The monodactylids, or monos, are popular aquarium fishes. Their silvery-white, laterally compressed bodies are exaggerated by extremely tall dorsal and anal fins, making some species twice as deep as they are long. Adult *Monodactylus* lack pelvic fins, although juveniles possess them. They are convergent in shape and ontogenetic pelvic loss with the very compressed carangid Germanfish, *Parastromateus niger*. The archerfishes are a well-known and unique group of small, surface-dwelling estuarine and freshwater fishes that feed actively on terrestrial prey. Insects are shot out of overhanging vegetation with bullets of water produced by compressing the gill covers and shooting water drops along a groove created by the tongue and palate. This behavior is all the more fascinating because the fish corrects for the curving trajectory of its propelled droplets and target movement as well as for light refraction at the water's surface, its eyes being submerged during hunting (Dill 1977a; Schuster et al. 2006). **Kyphosid** sea chubs (also called rudderfishes) are an herbivorous family of 14 reef species that swim actively in shoals relatively high above the reef compared to most other herbivores. Kyphosids are unique among fishes in that at least two western Australian species contain symbiotic bacteria in their guts that break down algae via fermentation

(A) (B)

FIGURE 15.16 (A) A Mutton Snapper (*Lutjanus analis*, Lutjanidae) resting behind a gorgonian coral in St. Croix. Snappers are twilight and nocturnal predators on reef fishes throughout the tropics. G. Helfman (Author). (B) Grunts (Haemulidae) such as these French Grunt (*Haemulon flavolineatum*), also take advantage of shade during the day and feed at twilight or after dark. Courtesy of J. Hyde and K. Sultze.

(A) (B)

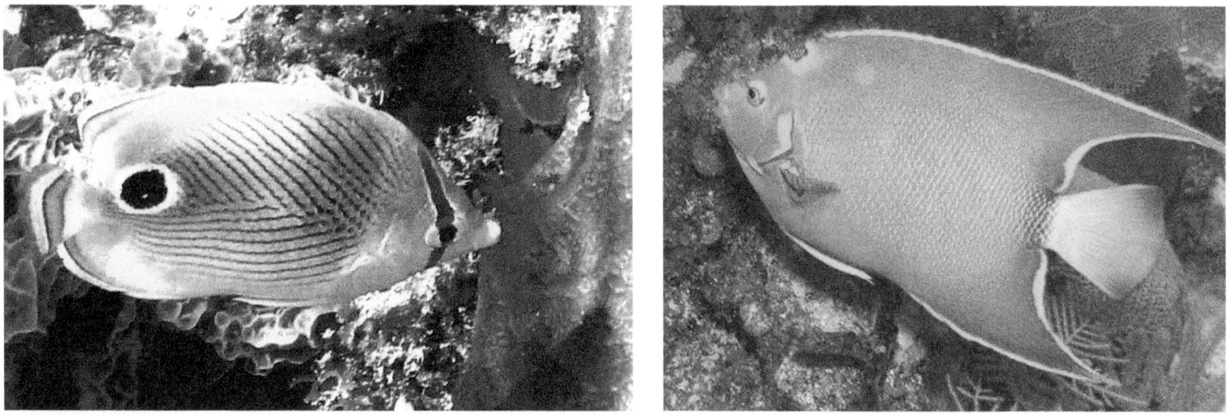

FIGURE 15.17 Butterflyfishes (A, Chaetodontidae,) and angelfishes B, (Pomacanthidae) are two well-recognized families of coral reef fishes. (A) Courtesy of J. Hyde and K. Sultze; (B) J. Bedford / NOAA Photo Library / Public Domain.

(Rimmer & Wiebe 1987). Although a predominantly tropical family, two temperate derivative species, the Opaleye, *Girella nigricans*, and the Halfmoon, *Medialuna californica*, extend into cooler California waters.

Some authors feel that the kyphosids, monodactylids, and toxotids form, with the next five or so families, an unranked group known as the Squamipinnes, a name that refers to the rows of scales that cover the base of the dorsal and anal fins. The best-known families in this group are the butterflyfishes and angelfishes (Fig 15.17). The **chaetodontid** butterflyfishes include 134 tropical shallow water species. Their center of diversity is in the Indo-Pacific region where about 100 species occur. The tropical Atlantic contains 13 species and the eastern Pacific has only four species. In many people's minds, butterflyfishes are synonymous with coral reefs (Burgess 1978; Motta 1989), and their presence and abundance is proposed as a measure of the health of coral reef habitats (Ohman et al. 1998). Butterflyfishes are colorful and swim conspicuously about the reef during the daytime, often in pairs or small shoals, residing for long periods on the same reefs and with the same partners. Trophically, they fall into several categories of microconsumers, feeding either on coral polyps, small invertebrates hidden in crevices in the reef, tube worms, or zooplankton. Anatomically they are deep-bodied, highly compressed forms, their body shape being exaggerated by stout dorsal, pelvic, and anal spines and a slightly to greatly elongated snout region. Closely related to and often mistaken for butterflyfishes are the similarly or larger sized angelfishes (**Pomacanthidae**). A major distinguishing feature between the two is the existence of a stout, posteriorly projecting spine at the angle of the preopercular bone and the absence of a pelvic axillary process in angelfishes. Many angelfishes undergo dramatic ontogenetic color changes, several species having similar but striking patterns as juveniles that change to species specific and still-striking adult patterns. Larger species such as the Caribbean French and Gray angelfishes frequently form pairs. Trophically, angelfishes differ from butterflyfishes in consuming sessile, benthic invertebrates such as sponges, tunicates, and anthozoans. Some species are known to follow sea turtles and feed on their feces, which may explain their disconcerting habit of hovering near latrines associated with undersea habitats such as Tektite and Hydrolab, before such submarine structures had internal plumbing. Again, about three-quarters of angelfish species occur in the Indo-Pacific. Angelfishes and butterflyfishes are among the few marine species in which hybrids are frequently discovered (e.g. Pyle & Randall 1994; Hobbs et al. 2009).

Most of the remaining families in the suborder Percoidei are relatively small. Among the more speciose groups are the **nandid** and closely related **polycentrid** leaffishes of South American, African, and southern Asian fresh waters. They are best known for their striking morphological resemblance to floating leaves, a deception they enhance behaviorally by drifting slowly through the water toward unsuspecting prey, which they engulf with a remarkably expandable mouth. **Terapontid** grunters are a marine and freshwater family containing 61 species that have a unique means of producing sounds. Paired muscles run from the back of the skull to the dorsal surface of the gas bladder; in other sound-producing fishes that utilize muscles to vibrate the gas bladder, such as gadoids, triglids, and sciaenids, the muscles are derived from trunk musculature and originate in the body wall. **Kuhliid** flagtails are predominantly marine and estuarine inhabitants but some species have evolved to fill ecological niches in freshwater habitats on oceanic islands.

Five subsequent families are placed in the superfamily Cirrhitoidea, united by elongate and unbranched lower five to eight rays in their pectoral fins. The colorful **cirrhitid** hawkfishes are small to medium reef predators that are best known for sitting absolutely still on tops of corals in seemingly conspicuous locales, waiting for potential prey fishes to either not notice them or to habituate to their presence (a Hawaiian hawkfish is the only predator known to have eaten a Cleaner Wrasse). Hawkfishes look like a cross between a small seabass and a scorpionfish, but they are readily identified by the presence of filamentous tufts or cirri at the top of each spine in the first dorsal fin and by the elongate pectoral rays characteristic of the superfamily. One fairly deep-water reef species, the Longsnout Hawkfish, *Oxycirrhites typus*, occurs almost exclusively in black coral trees. Its deep-water habits may more closely reflect depletion of its preferred perch, which has been removed from accessible shallow locations due to the jewelry trade. The next to last family of percoids are the **cepolid** bandfishes, which vary in their morphology from elongate, eel-like forms to fairy bassletlike, deep-water forms.

The barracudas (Sphyraenidae, Fig. 15.18) appear to be percoids, although some morphological data suggest that they may be less derived, basal members of the Scombroidei, the suborder that contains the tunas (Orrell et al. 2006).

FIGURE 15.18 A sphyraenid, the Great Barracuda, *Sphyraena barracuda*. G. Helfman (Author).

Nelson et al. (2016) place the barracudas along with the swordfishes (Xiiphidae) and billfishes (Istiophoridae) in the order Istiophoriphormes, whereas we keep the sphyraenids within the suborder Percoidei and have the two billfish families in the suborder Xiphioidei, which is closely related to the Scombroidei (both addressed later in this chapter). Twenty-seven species of barracuda inhabit tropical and subtropical regions of the Atlantic, Pacific, and Indian oceans. Most barracudas are schooling predators, an important exception being the usually solitary Great Barracuda, *Sphyraena barracuda* (Paterson 1998). Great Barracuda approach 2 m in length (a topic of considerable controversy), have fanglike, flattened teeth (see Fig. 16.16) capable of slicing cleanly through most prey, and have the unnerving habit of following divers around the reef, perhaps motivated by either curiosity or territoriality.

Suborder Labroidei

Suborder Labroidei: Cichlidae (1,729 species of cichlids), Embiotocidae (23 species of surfperches), Pomacentridae (423 species of damselfishes), Labridae (564 species of wrasses), Odacidae (12 species of cales), Scaridae (100 species of parrotfishes)

(Families and numbers of species from Fricke et al. (2021) but refer to **https://researcharchive.calacademy.org/research/ichthyology/catalog/SpeciesByFamily.asp** *for updates.)*

Although some of the remaining perciform suborders contain very speciose families (e.g. blennies, gobies), by far the numerically most successful suborder is the Labroidei. Labroids are predominantly tropical, marine fishes (e.g. damselfishes, wrasses, parrotfishes), with a few species in the first two families inhabiting warm temperate waters. Two additional families, the surfperches and odacids, are temperate and marine. The most successful family in the suborder is the tropical freshwater cichlids, although no single biological generality applies to all members of this fantastically speciose and varied family of fishes. These six families are united primarily on the basis of pharyngeal jaw morphology, involving features of both the upper and lower jaws by Nelson (2006) and Wiley & Johnson (2010). However, Nelson et al. (2016) restrict the group to three families: Labridae, Scaridae, and Odacidae, and move the Embiotocidae, Cichlidae, and Pomacentridae to the Ovalentaria as recommended by Wainwright et al. (2012).

Cichlids (initial "ci" sounded as in "pop*si*cle") are viewed as the basal group of the suborder. Among the abundant cichlid species are many aquarium fishes that have achieved popularity because of their small size, colorfulness, and willingness to behave and breed within the confines of an aquarium. Familiar South American species include freshwater angelfishes (*Pterophyllum*), discus (*Symphysodon*), oscars (*Astronotus*), convict cichlids (*Archocentrus*), Peacock Bass (*Cichla*), and gravel-eaters (*Geophagus*). One species, the Rio Grande Cichlid, *Cichlasoma cyanoguttatum*, gets as far north as Texas and has invaded Lake Pontchartrain in adjacent Louisiana. Numerous cichlids, particularly African tilapias, have been deliberately or accidentally introduced in Florida, California, and Hawaii, chiefly for aquaculture purposes. Although diverse in the Western Hemisphere (*c.* 400 species), the great majority of cichlids occur in Africa, where they have radiated explosively into numerous species flocks (discussed in Chapter 19). In the absence of many other teleost groups, some of these cichlids evolved into forms similar to other families (Fig. 15.19), thus occupying ecological niches that evidently were available. Old World cichlids include fishes in the genera *Haplochromis*, *Lamprologus*, *Oreochromis*, *Pseudotropheus*, *Sarotherodon*, and *Tilapia* (and many others). Most cichlids build nests and many African forms brood the eggs and young in the mouth of either the male or more usually the female. A few species occur in Israel and Iran and also in India and Sri Lanka, where they commonly inhabit estuaries (e.g. *Etroplus*). Cichlids are convergent in behavior, morphology, and ecology with centrarchid sunfishes. The two families can be distinguished by two nostrils on each side of the head and a continuous lateral line in sunfishes, whereas cichlids have only one nostril on each side and an interrupted lateral line (Fryer & Iles 1972; Keenleyside 1991; Lévêque 1997; Barlow 2000; Kullander 2003; and many others; see **www.cichlidpress.com** for spectacular photos and additional references).

The **embiotocid** surfperches look very much like some of the larger, deep-bodied cichlids. However, they are an entirely (with one exception) marine family of small to medium, inshore fishes that occur most commonly around kelp beds, rocky reefs, surf zones, and tidepools. Twenty of the 23 species occur along the Pacific coast of North America, the other three living in Korea and Japan. They are the only labroids that are live-bearers, the female giving birth to fully developed, large young (see Fig. 8.11). In a few species, males are even born reproductively mature. Trophically, some are specialized as zooplanktivores, whereas most pick invertebrates from the bottom or off plants. The **pomacentrid** damselfishes are generally smaller, more colorful, tropical marine equivalents of the surfperches. Many damselfishes are herbivorous (e.g. *Stegastes* spp.), whereas others are zooplanktivorous and show the usual adaptations associated with life in the water column above structure, namely a fusiform body, forked tail, and highly protrusible mouth (e.g. *Chromis* spp.). Herbivorous damselfishes are typically territorial, guarding a small patch of reef substrate in which they feed, hide, and in the case of males, court females and guard developing eggs. No post-hatching care of young is shown except in one Indo-Pacific species, *Acanthochromis polyacanthus,* which also lacks a larval stage. Some species are intimately associated with invertebrates, such as the anemonefishes (*Amphiprion*, *Prems*) (see Fig. 17.22). This is a tropical family containing 423 species that reaches its highest diversity in the Indo-Pacific region, with a few species in each ocean basin occurring in warm temperate waters (e.g. the Sergeant Major (*Abudefduf saxatilis*) in the Caribbean, and in California, the Blacksmith, *Chromis punctipinnis*, and the state marine fish, the Garibaldi, *Hypsypops rubicundus*) (Emery & Thresher 1980; Allen 1991; see also papers in Allen et al. 2006) (Fig. 15.20).

The next three families are closely related. The largest family is the wrasses, **Labridae**, a remarkably diverse and widespread marine taxon of at least 564 species that occurs in all tropical seas. Many temperate and even cool temperate members occur in the Pacific and Atlantic oceans, such as the eastern Pacific California Sheepshead (Fig. 15.21A) and Senorita (*Semicossyphus*, *Oxyjulis*), the western Atlantic Tautog and Cunner (*Tautoga*, *Tautogolabrus*), and several eastern Atlantic wrasses (*Labrus* spp.). Wrasses range in size from 5 cm (many species) to the Giant Humphead (or Maori or Napoleon) Wrasse of the Indo-Pacific, *Cheilinus undulatus* (Fig. 15.21B), which can be over 2 m long, weigh 200 kg, and has the unlikely diet of cowries and crown-of-thorns starfish. It is depleted almost everywhere it is found and many Pacific island nations now outlaw its export and even capture (Sadovy et al. 2019). Pharyngeal jaws are especially diversified among labrids, several species able to handle well-protected prey such as crabs, mollusks, and echinoderms. Labrids typically bounce along in the water column using labriform locomotion (See Chapter 4, Locomotion), a paddling motion of their pectoral fins, stopping momentarily above the bottom to capture prey with protractile jaws and stout teeth, picking zooplankters out of the water column, or removing external parasites from other fishes. The razorfishes (*Hemipteronotus*, *Xyrichtys*) are very compressed and escape disturbances by diving rapidly into bottom sediments. Wrasses as a group are strongly diurnal and enter sandy bottoms or reef crevices at night to sleep. Many wrasses change sex, most starting off life as females and later changing to very differently colored and shaped males (see Chapter 8, Sex and Gender Roles in Fishes).

The **odacids** are a small family of 12 species limited to the temperate waters of New Zealand and southern Australia. They are intermediate in appearance between wrasses and parrotfishes, having the elongate wrasse body form but the non-protractile jaws and fused teeth of parrotfishes. The **scarid** parrotfishes include 100 species of tropical marine fishes best known for their fused teeth that form a parrotlike beak (Fig. 15.21C, D). The beak is used for biting off algal fronds or pieces of dead coral or scraping the surface of live coral, which is then passed to massive pharyngeal mills for grinding and extracting algal cells from the coral matrix. As with wrasses, parrotfishes generally change color and sex, from initial phase females to terminal phase males. Parrotfishes are generally larger than wrasses, with some species such

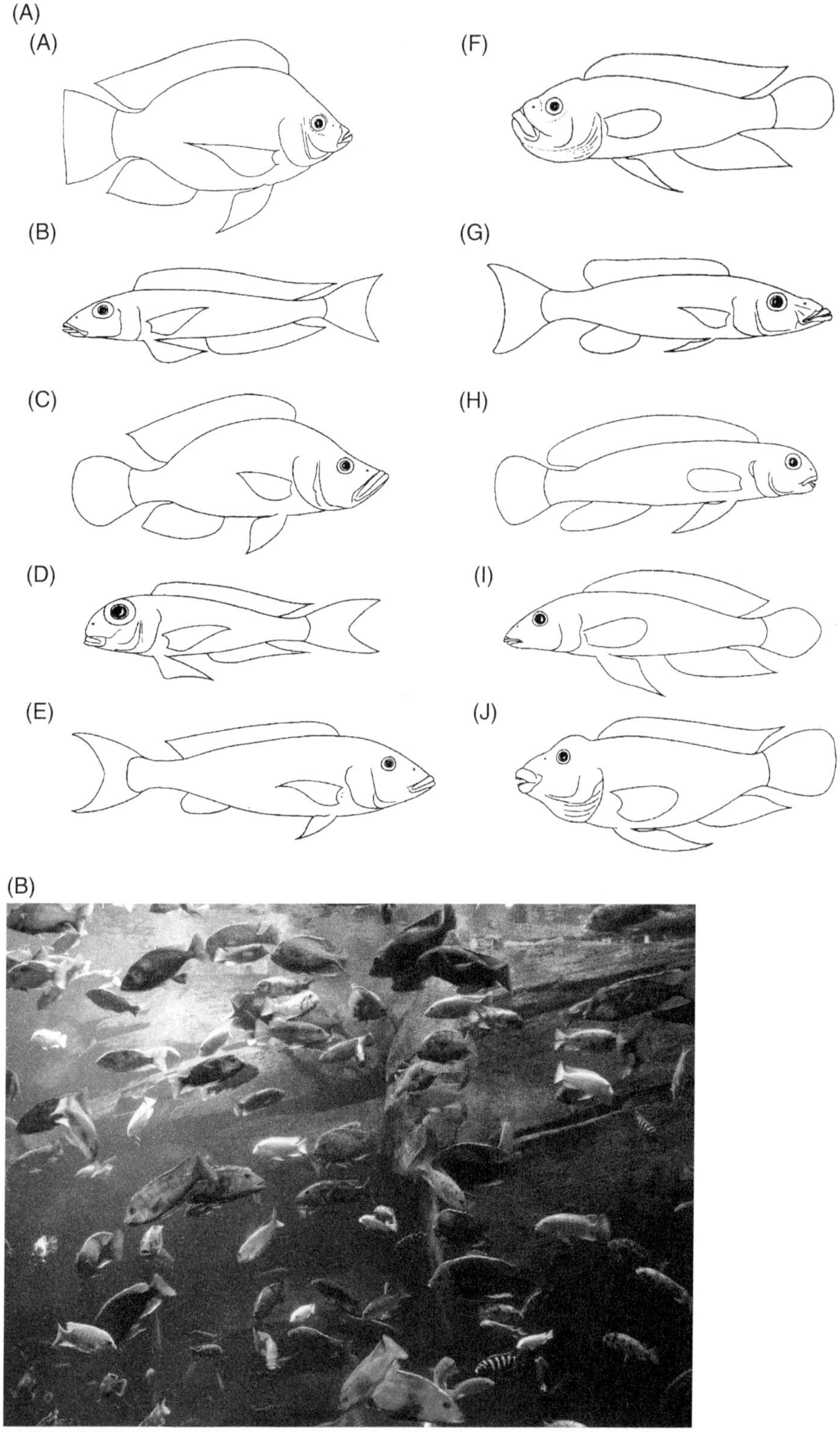

FIGURE 15.19 (A) Diversity in body shape among African cichlids. These fishes belong to several different genera of cichlids, yet are roughly similar in body form to several other teleostean families. Cichlid genera and suggested convergences are: (A) Tilapia versus a centrarchid sunfish; (B) *Xenotilapia* versus a malacanthid tilefish; (C) *Serranochromis* versus a serranid seabass; (D) *Xenotilapia* versus a gobiid goby; (E) *Boulengerochromis* versus a lutjanid red snapper; (F) *Telmatochromis* versus a batrachoidid toadfish; (G) *Rhamphochromis* versus a centropomid snook; (H) *Telmatochromis* versus an opistognathid jawfish; (I) *Julidochromis* versus a labrid wrasse; and (J) *Spathodus* versus a scarid parrotfish. From Fryer & Iles (1972), used with permission. (B) This aquarium at the Georgia Aquarium is somewhat representative of a scene over a rocky area of Lake Malawi where there has been an explosive speciation among cichlids. To view some of the spectacular body shape and color variation among African cichlids, go to **www.cichlidworld.com/photo.html**. G. Helfman (Author).

FIGURE 15.20 Pomacentrid damselfishes. (A) the Sergeant Major, *Abudefduf saxatilis*, found in tropical and subtropical waters of the Atlantic Ocean, with similar congeners in the Indo-Pacific, and (B) the Garibaldi, *Hypsypops rubicundus*, a temperate pomacentrid damselfish common in kelp beds of southern California; it really is that orange. (A) Courtesy of J. Hyde and K. Sultze; (B) G. Helfman (Author).

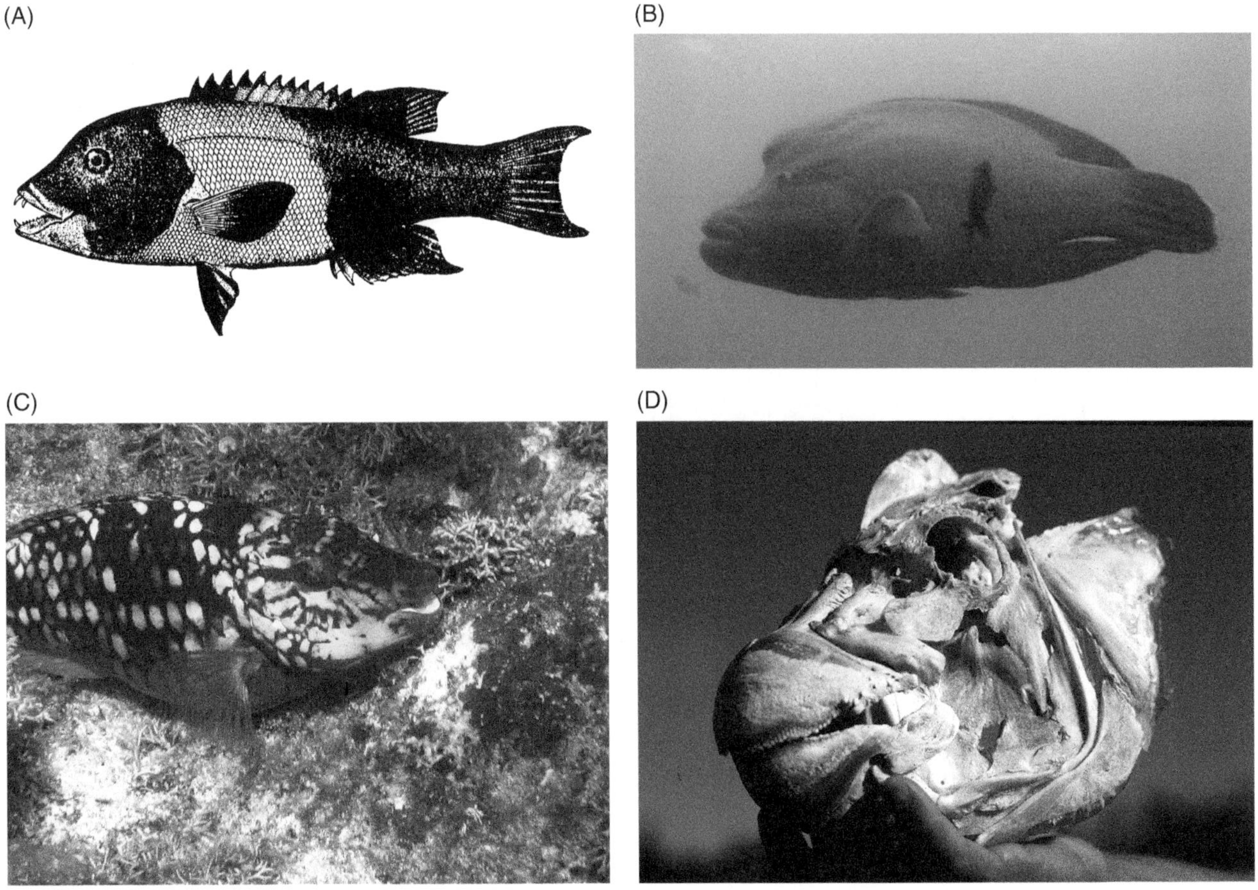

FIGURE 15.21 Representative labroids. (A) A California Sheephead (*Semicossyphus pulcher,* Labridae). (B) A Napoleon or Humphead Wrasse (*Cheilinus undulatus*, Labridae), the largest member of the speciose wrasse family. (C) An initial phase Stoplight Parrotfish (*Sparisoma viride*, Scaridae), (D) skull of a Rainbow Parrotfish, *Scarus guacamaia*, showing the fused, parrotlike beak. (A) From Jordan (1905); (B) NOAA / Public Domain; (C) Courtesy of J. Hyde and K. Sultze; (D) G. Helfman (Author).

as the blue and rainbow parrotfishes of the Atlantic (*Scarus* spp.) and the bumphead parrotfish of the Pacific (*Bulbometopon*) attaining a meter in length. These spectacular, large species are becoming rare due to overfishing and are seldom seen outside of protected areas (Dulvy & Polunin 2004).

Suborder Zoarcoidei

Suborder Zoarcoidei: Bathymasteridae (7 species of ronquils), Eulophiidae (4 species of eulophiids), Zoarcidae (4 subfamilies with 309 species of eel pouts), Stichaeidae (about 39 species of pricklebacks), Cryptacanthodidae (4 species of wrymouths), Pholidae (15 species of gunnels), Anarhichadidae (5 species of wolffishes), Ptilichthyidae (1 species of Quillfish), Zaproridae (1 species of Prowfish), Scytalinidae (1 species of Graveldiver)

(Families and numbers of species from Fricke et al. (2021) but refer to **https://researcharchive.calacademy.org/research/ichthyology/catalog/SpeciesByFamily.asp** *for updates.)*

The zoarcoids as a group are generally elongate fishes of the North Pacific that occupy benthic habitats ranging from tidepools to abyssal depths. The **zoarcid** eel-pouts are eel-like fishes with round heads, a single nostril, long dorsal and anal fins, and pointed tails. Some eel pouts give birth to live young after eggs develop internally, a condition referred to as ovoviviparity (see Chapter 8, Embryo Development). Zoarcids inhabit soft bottoms at moderate to great depths (20–3000 m). Most species occur in the North Pacific, some in the North Atlantic, in deep tropical regions of both oceans, and about 10% of the species occur in the southern oceans near Antarctica. In contrast, the **stichaeid** pricklebacks and **pholid** gunnels are most common in intertidal and shallow, nearshore habitats, primarily in North Pacific waters. Members of both families are primarily microcarnivores, although two pricklebacks, *Cebidicthys violaceus* and *Xiphister mucosus*, are herbivorous year-round, an uncommon trait at high latitudes for any fishes (Horn 1989). The **anarhicadid** wolffishes or wolf-eels are the anatomical and ecological equivalents of moray eels at high latitudes (Fig. 15.22). Attaining lengths of 2.5 m and weighing 45 kg, these large benthic predators of the North Pacific and Atlantic often live under rocks and have large anterior conical canines and massive lateral and palatine molars for catching and crushing crustaceans, clams, sea urchins, and fishes.

Suborder Notothenioidei

Suborder Notothenioidei: Bovichtidae (12 species of temperate icefishes), Pseudaphritidae (1 species of catadromous icefish), Eliginopsidae (1 species of Patagonian Blenny), Nototheniidae (about 59 species of cod icefishes), Harpagiferidae (12 species of spiny plunderfishes), Artedidraconidae (36 species of barbeled plunderfishes), Bathydraconidae (17 species of Antarctic dragonfishes), Channichthyidae (21 species of crocodile icefishes)

(Families and numbers of species from Fricke et al. (2021) but refer to **https://researcharchive.calacademy.org/research/ichthyology/catalog/SpeciesByFamily.asp** *for updates.)*

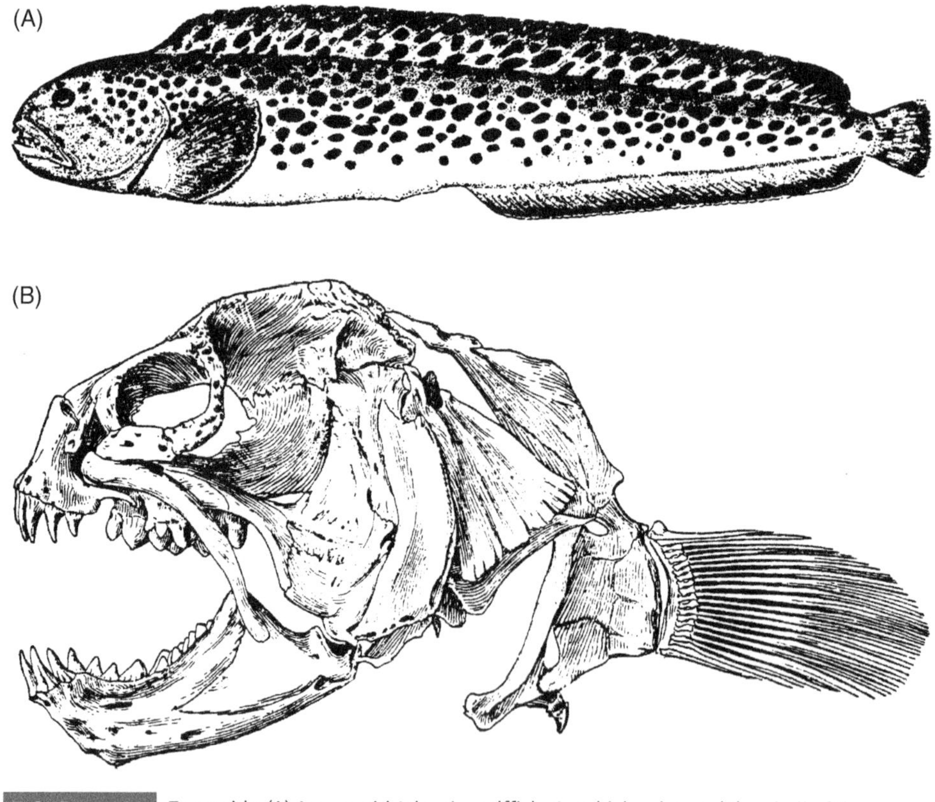

FIGURE 15.22 Zoarcoids. (A) A zoarcoid Atlantic wolffish, *Anarhichas lupus*. (B) A skull of the related Pacific Wolf-eel, *Anarrhichthys ocellatus*, showing the massive, diversified dentition of these predators (see also Fig. 3.22). From Jordan (1905).

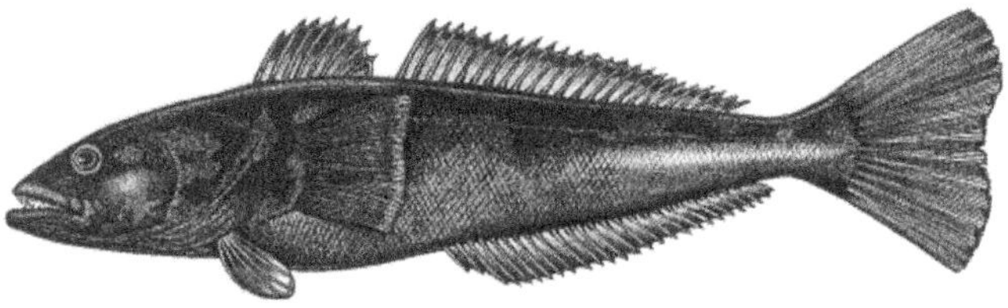

FIGURE 15.23 The Patagonian Toothfish, *Dissostichus eleginoides*, a large, long-lived nototheniid of southern oceans subjected to considerable illegal, unreported, and unregulated (IUU) fishing. Take a pass on eating the Patagonian Toothfish. Drawn by Bruce Mahalski, from Lack & Sant (2001), used with permission.

The notothenioids are commonly referred to as the icefishes (Fukuchi et al. 2006). The suborder is restricted primarily to high latitudes of the southern hemisphere, with greatest diversity in benthic habitats of Antarctica. These cold-water fishes show numerous physiological and behavioral adaptations to prevent their tissues from freezing, including the production of a variety of glycoprotein antifreezes (see Chapter 7, Coping with Temperature Extremes). Many fascinating characteristics of this group are detailed in Chapter 10 (see section on Antarctic Fishes) and will not be repeated here. The **bovichtids** of Australia, New Zealand, and southern South America are considered the stem group for the rest of the suborder. **Nototheniid** cod icefishes are predominantly benthic, with some secondarily pelagic species that achieve neutral buoyancy by depositing lipids in their muscles and by reduction of skeletal material, two adaptations to water column existence that occur convergently in other families derived from benthic ancestors (e.g. cottoid Baikal oilfishes, and many deep-sea forms). The best known nototheniid is the Patagonian Toothfish (or Chilean seabass), *Dissostichus eleginoides*, a large (to 2.4 m, 130 kg), long-lived (to 50 years), slow reproducing, and grossly overfished benthopelagic predator (Fig. 15.23). The **channichthyid** crocodile icefishes (discussed in Chapter 10) are well studied because they lack red blood cells, hemoglobin, and myoglobin, making their blood and flesh colorless. These traits are an adaptation to the high amount of dissolved oxygen in cold Antarctic waters.

Suborder Trachinoidei

Suborder Trachinoidei: Chiasmodontidae (36 species of swallowers), Champsodontidae (13 species of gapers), Pinguipedidae (100 species of sand perches), Cheimarrichthyidae (1 species - New Zealand Torrentfish), Trichonotidae (10 species of sanddivers), Creediidae (18 species of sand burrowers), Percophidae (1 species of duckbill), Leptoscopidae (5 species of southern sandfishes), Ammodytidae (35 species of sand lances), Trachinidae (9 species of weeverfishes), Uranoscopidae (60 species of stargazers), Pholidichthyidae (2 species of convict blennies)

(Families and numbers of species from Fricke et al. (2021) but refer to **https://researcharchive.calacademy.org/research/ichthyology/catalog/SpeciesByFamily.asp** *for updates.)*

Trachinoids are mostly benthic, questionably related marine fishes that are retained as an order by Nelson et al. (2016) except for the Trichodontidae, which has been moved to the order Scorpaeniformes (suborder Cottoidei). Several species of trachinoids sit buried in the sand throughout the day or seek refuge in the sand when not feeding. **Chiasmodontid** swallowers depart from the suborder norm in being one of the few acanthopterygians to occupy mesopelagic and bathypelagic depths. They show convergent traits with other deep-sea fishes, including a large mouth, long teeth, slender jawbone elements, distensible mouth and stomach, black coloration, and photophores. The **Cheimarrichthyidae** consist of a single New Zealand species, *Cheimarrichthys fosteri*. It is known as the Torrent fish, reflecting its daytime habitat in turbulent streams. Its body form, inferior mouth, large horizontally placed pelvics, and broad flattened head converge with other swift water fishes such as Longnose Dace, balitorine hillstream loaches, African kneriids, amphiliid loach catfishes, clingfishes, and rhyacichthyid loach gobies (see Chapter 10, Strong Currents and Turbulent Waters). **Trichonotid** sanddivers share a peculiarity with some elasmobranch rays by having protuberant eyes and a dorsal eyelid of sorts made up of an iris flap with strands that extend over the lens. Both groups rest on or bury in the sand in shallow water with only their eyes visible.

Ammodytid sand lances are small, elongate, shoaling fishes that feed on zooplankton in the water column by day and spend nighttime buried in the sand. The **trachinid** weeverfishes are well-known eastern Atlantic and Mediterranean benthic fishes with highly venomous opercular and dorsal spines. The **uranoscopid** stargazers are another venomous family, with two grooved spines and an accompanying gland sitting just behind the gill cover and above the pectoral fins. Stargazers also lie on the bottom or bury in the sand, with their dorsally located eyes exposed. Their incurrent nostrils are directly connected to the mouth, which may allow them to breathe while buried. A fleshy filament extends upward from the floor of the mouth and is used to lure prey. Stargazers include some of the only marine teleosts that are electrogenic, strong pulses of electricity (up to 50 volts) being produced by highly modified extrinsic eye muscles. They discharge when captured and may also use electricity to stun prey. Stargazers are convergent in body form and habits with paracanthopterygian toadfishes and blennioid dactyloscopid sand stargazers.

Suborder Blennioidei

Suborder Blennioidei: Tripterygiidae (183 species of triplefin blennies), Dactyloscopidae (48 species of sand stargazers), Blenniidae (405 species of combtooth blennies), Clinidae (87 species of kelp blennies), Labrisomidae (130 species of labrisomid blennies), Chaenopsidae (93 species of tube blennies)

(Families and numbers of species from Fricke et al. (2021) but refer to **https://researcharchive.calacademy.org/research/ichthyology/catalog/SpeciesByFamily.asp** *for updates.)*

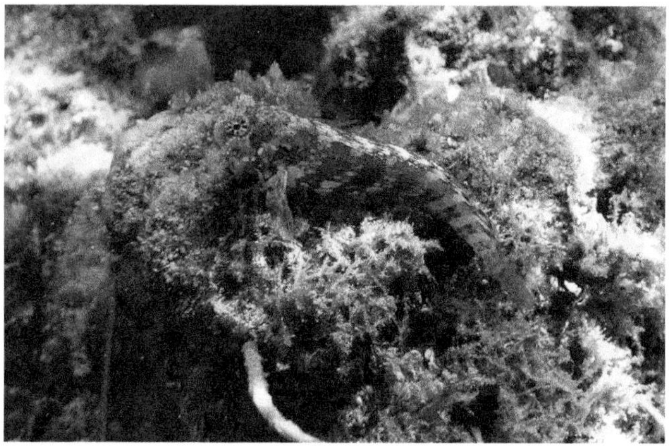

FIGURE 15.24 There are over 400 species in the family Blenniidae; the fish in the photo is probably a Pearl Blenny (*Entomacrodus nigricans*). Courtesy of J. Hyde and K. Sultze.

Blennioids comprise six families of small, benthic, marine fishes of tropical and subtropical regions. Nelson et al. (2016) upgraded this group to the order Blenniformes, but we keep them as a suborder (as in Nelson 2006). Blennioids generally possess long dorsal and anal fins and fleshy flaps termed cirri on some part of the head. Triplefin blennies (**Tripterygiidae**) derive their name from having a soft dorsal fin plus a spiny dorsal fin divided into two parts (hakes and cods are the only other fishes with three distinctive dorsal fins). **Dactyloscopid** sand stargazers look like miniaturized (to 15 cm) uranoscopid stargazers with their oblique mouths and stalked eyes. They similarly have a specialized breathing mechanism probably related to their burying habits. In most fishes, water is brought into the mouth and out the gills by the combined actions of buccal and opercular pumps (discussed in Chapter 3). Sand stargazers move water via a branchiostegal rather than an opercular pump. Fingerlike projections inside the mouth may keep sand out of the gills. Some dactyloscopid males care for eggs by carrying them under the axil of each pectoral fin.

The combtooth blennies of the family **Blenniidae** are very diverse, accounting for over 400 species of mostly small, benthic fishes in tropical and subtropical waters worldwide (Fig 15.24). The comblike teeth are used to crop algae in many species. Many of the more derived nemophine sabre-toothed blennies (*Aspidontus, Meiacanthus*) swim freely in the water column and mimic other fishes. The best-known example is the cleanerfish mimic, *Aspidontus taeniatus*. This sabre-toothed blenny strongly mimics the blue and black coloration and bobbing solicitation dance of labrid cleanerfishes, particularly the Cleaner Wrasse, *Labroides dimidiatus*. When allowed to approach a cleaner-seeking fish, rather than cleaning the host, *Aspidontus* bites off a piece of fin (juvenile *L. dimidiatus* are also mimicked by juvenile fang blennies, *Plagiotremus rhinorhynchos*). Other sabre-toothed blennies attack passing fish and remove scales or pieces of fin. They are also known to attack human divers, generally biting once the diver has passed overhead, which is always a surprising, sometimes painful, and a distinctly unsettling experience. Some of the sabre-toothed species are referred to as poison-fanged blennies because their hollow lower canines can inject a toxin. Poison fang blennies (*Meiacanthus*) may be mimicked by similarly colored blennies (e.g. *Ecsenius, Plagiotremus, Runula*) and thus gain protection from predators (Losey 1972; Springer & Smith-Vaniz 1972; Moland & Jones 2004).

Clinids, the kelp blennies, are shallow water, benthic forms associated closely with structure in both temperate southern and northern hemispheres. Most are small, but the predatory Giant Kelpfish, *Heterostichus rostratus*, reaches 60 cm. **Chaenopsid** pike blennies and tube blennies are generally small, tropical, New World fishes that are most often found living in or with corals. Many tube blennies are essentially infaunal. Shortly after settling from the plankton, a tube blenny will take up residence in an available polychaete worm tube and is likely to remain there for the rest of its life. The intriguingly named Sarcastic Fringehead (*Neoclinus blanchardi*) is the largest of the chaenopsids (at 30 cm) and is known to defend its territory from conspecifics by jousting with its huge, colorful mouth (videos are available on the web).

Suborders Icosteoidei, Callionymoidei, and Gobiesocoidei

These three suborders are recognized as orders by Nelson et al. (2016). The monotypic suborder Icosteoidei and family **Icosteidae** contains the very peculiar North Pacific Ragfish, *Icosteus aenigmaticus*. Elliptical in shape and highly compressed, spineless, scaleless, without pelvic fins as an adult, and with a largely uncalcified cartilaginous skeleton, these 2 m long pelagic predators look like a free-swimming flatfish with a limp body. They are a reported favorite prey of sperm whales.

Two families make up the suborder Callionymoidei, the **callionymid** dragonets (201 species) and the poorly studied, deeper water **draconettid** slope dragonets (15 species). Dragonets are a chiefly marine family of small, shallow water fishes in the Indo-West Pacific. Some species are pale white and live over sand, whereas others associated with hard bottoms are quite colorful. The callionymids include a popular aquarium species, the green and orange Mandarin Fish or Splendid Dragonet, *Synchiropus splendidus*.

The suborder Gobiesocoidei contains 180 species, all in the clingfish family **Gobiesocidae**. Gobiesocids are a shallow water to amphibious family of small marine fishes often found in high-energy wave zones. The pelvic fins are modified into a sucking disk, the body is depressed, the head is rounded and flattened, and the skin is smooth and scaleless (Fig. 15.25A). They have a unique pectoral girdle and vertebral rib arrangement. A relative giant in the family, the 30 cm Chilean *Sicyases sanguineus* feeds on snails, barnacles, chitons, and other high intertidal prey, as well as many kinds of algae, often preferring the wave-splashed supratidal region to more regularly inundated depths (or heights) (Paine & Palmer 1978). Its flesh is reputed to have aphrodisiac qualities.

(A)

(B)

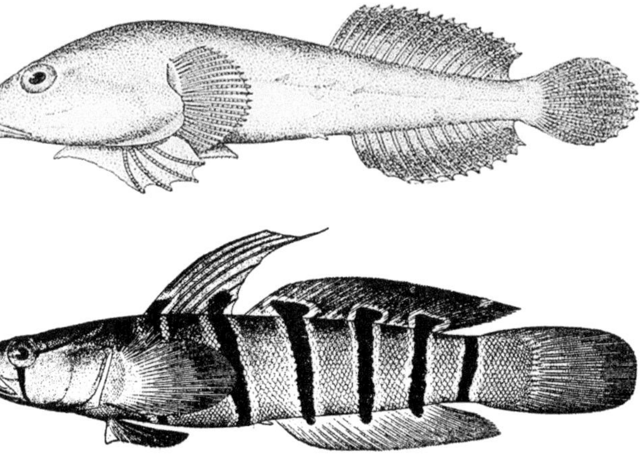

FIGURE 15.25 Clingfishes (A, family Gobiesocidae) and many gobies (B, family Gobiidae) have the pelvic fins joined to form a suction disk to hold position on substrates. (A) from Wikimedia/Library of Congress/Public Domain; (B) from Jordan (1905).

Suborder Gobioidei

Suborder Gobioidei: Rhyacichthyidae (3 species), Odontobutidae (23 species), Eleotridae (189 species including the subfamilies Eleotrinae, Milyeringinae, and Butinae), Thalasseleotrididae (3 species), Gobiidae (1953 species including subfamilies Oxudercinae, Sicydiinae, Gobionellinae, Amblypopinae, Gobiinae), Kraemeriidae (9 species), Microdesmidae (94 species including subfamilies Microdesminae, Ptereleotrinae), Xenisthmidae (16 species), and Schindleriidae (8 species)

(Families and numbers of species from Fricke et al. (2021) but refer to **https://researcharchive.calacademy.org/research/ichthyology/catalog/SpeciesByFamily.asp** *for updates.)*

Based largely on molecular evidence, Nelson et al. (2016) and Betancur-R et al. (2017) placed the gobioids apart from the subseries Percomorpha and in their own subseries, the Gobiida containing two orders, Kurtiformes and Gobiiformes. We retain the gobioids in the Percomorpha, in the suborder Gobioidei of the order Perciformes. The phylogeny of the gobioids will continue to be an ongoing field of research, demonstrated in part by Fricke et al. (2021) re-classifying some of the families from Nelson et al. (2016) to subfamilies.

Gobioids are usually small, benthic or sand-burrowing fishes, mostly marine but about 10% inhabit fresh water. Gobioids, as is common among benthic fishes, lack a gas bladder. Different families show differing degrees of fusion of the pelvic fin. Two families, the sleepers and gobies, account for 95% of the species in the suborder, the latter family being by far the largest. In the stream-dwelling **rhyacichthyid** loach gobies of Indo-Australia, the flattened anterior third of the body in combination with the pelvic fins form a sucking disk for holding position in fast-flowing water. The **eleotrid** sleepers are small to medium (to 60 cm), widely distributed estuarine and stream fishes of tropical and subtropical regions. They are often the major predators of stream systems on oceanic islands such as Hawaii and New Zealand and have a complex life history that includes a marine planktonic larva, reflecting a probable marine ancestry for the family. Pelvic fins are usually separated.

The **gobiid** gobies (Fig. 15.25B) constitute the largest family of marine fishes in the world, with more than 1950 species (the distinction of largest fish family overall is contested among gobies, cyprinids, and cichlids, all with around 2000 species). Gobies usually have their pelvic fins united, some species using them as a suction disk for clinging to hard substrates. Many species live on mud or sand or in association with invertebrates such as sponges, sea urchins, hard and soft corals, and shrimps (see Chapter 17, Interspecific Relations: Symbioses). The Neon Goby, *Gobiosoma oceanops*, is an important cleanerfish in the Caribbean and is convergent in coloration with labrid cleanerfishes of the tropical Pacific. Other derivative species include the essentially amphibious **oxudercid** mudskippers (*Periophthalmus, Boleophthalmus*); these are so successful out of water that they are sometimes referred to as land gobies (see Fig. 4.16). Gobies are generally small (< 10 cm) and some are among the world's smallest fishes. Diminutive species include an Indian Ocean species, *Trimmatom nanus*, which matures at 8–10 mm, and several species in the genera *Eviota*, *Mistichthys*, and *Pandaka* that mature at about 10 mm. The largest goby is a western Atlantic/Caribbean form, the Violet Goby, *Gobioides broussonetti*, a purplish eel-like fish about 50 cm long with elongate dorsal and anal fins. Gobies are predominantly marine, but some of the freshwater species are important members of island stream assemblages, including species capable of ascending waterfalls and occupying headwater regions far from the ocean (e.g. Hawaiian *Lentipes*). Goby systematists include among their ranks the former Emperor of Japan (e.g. Akihito 1986).

The remaining gobiids are mostly small, eel-like, tropical marine fishes that live on or in sand. Elongate **kraemeriid** sand gobies often rest in the sand with just their head exposed, frequently in wave-tossed areas. The **microdesmid** wormfishes are similar to kraemeriids, burrowing in sand and mud. The related, spectacularly colored **ptereleotrins** are also known as hover gobies, dartfishes, or firefishes, names which describe their coloration and their habit of hovering above the bottom and diving rapidly into a burrow when disturbed. The **schindleriid** infantfishes are an enigmatic family of three species of small (2 cm) pelagic fishes that are neotenic or paedomorphic, which means that they are essentially adults that retain larval traits or larvae that have developed functional gonads. Retained larval characteristics in schindleriids include a larval-type kidney (pronephros), lack of pigmentation, and unossified skeleton. The extremely small size of many gobiids may represent convergent neotenic pathways (Johnson & Brothers 1993). Many fish families contain species that have evolved through heterochronic alterations in developmental sequences (discussed in Chapter 9); an example analogous to schindleriids involves nettastomatid duckbill eels, in which the leptocephalus larvae possess developed ovaries (Castle 1978).

Suborder Kurtoidei

Suborder Kurtoidei: Kurtidae (2 species of nurseryfishes)

(Families and numbers of species from Fricke et al. (2021) but refer to **https://researcharchive.calacademy.org/research/ichthyology/catalog/SpeciesByFamily.asp** *for updates.)*

Nelson et al. (2016) and Betancur-R et al. (2017) place the nurseryfishes (Kurtidae), along with the cardinalfishes (Apogonidae), in the order Kurtiformes under the subseries Gobiida. We retain the kurtids in the subseries Percomorpha, in the suborder Kurtoidei of the order Perciformes, and we retain the apogonids in the suborder Percoidei of the Perciformes.

The **kurtids** of the Indo-Malay and northern Australia regions are interesting because of the peculiar way males care for the eggs. Males have a hooklike growth of the supraoccipital crest on the top of their heads to which the eggs are attached and where they are carried until hatching (see Fig. 17.7). The means of attachment and time at which fertilization occurs are apparently unknown (Berra & Humphrey 2002).

Suborder Acanthuroidei

Suborder Acanthuroidei: Ephippidae (15 species of spadefishes), Scatophagidae (3 species of scats), Siganidae (32 species of rabbitfishes), Luvaridae (1 species - Louvar), Zanclidae (1 species - Moorish Idol), Acanthuridae (85 species of surgeonfishes)

(Families and numbers of species from Fricke et al. (2021) but refer to **https://researcharchive.calacademy.org/research/ichthyology/catalog/SpeciesByFamily.asp** *for updates.)*

Nelson et al. (2016), based on molecular evidence, has reclassified some of these groups. They retain the scatophagids and siganids in the order Perciformes, but in the superfamily Siganoidea and not in the suborder Acanthuroidei although they indicate a probable relationship to this suborder. They place the ephippds in the order Moroniformes, along with the Moronidae (temperate basses) and the Drepaneidae (sicklefishes). The Luvaridae, Zanclidae, and Acanthuridae are placed in the suborder Acanthuroidei of the order Acanthuriformes, which also includes the suborder Sciaenoidei with the families Sciaenidae (drums) and Emmelichthyidae (rovers). We feel that this reorganization is premature and in need of further research, and will therefore retain the groups within the suborder Acanthuroidei.

The Acanthuroidei contains, with one exception, a group of medium-sized, compressed fishes with small mouths that usually form shoals over coral reefs or in nearby habitats. The **ephippid** spadefishes (e.g. *Chaetodipterus*, western Atlantic) and batfishes of the genus *Platax* are conspicuous inhabitants of drop-offs and passes around reefs, although young spadefishes also frequent sandy beaches along the Atlantic coast of the USA. (The reef-dwelling ephippid batfishes, genus *Platax*, should not be confused with the benthic ogcocephalid batfishes, discussed later as part of the order Ophiiformes.) Juveniles of the ephippids look and act remarkably like floating leaves, and juvenile batfishes are popular in the aquarium trade. Their growth to large size however makes them less desirable pets, prompting aquarists to release Indo-Pacific batfishes into south Florida waters (Semmens et al. 2004). The **scatophagid** scats bear a slight resemblance to serrasalmine piranhas, but as their name implies, their feeding habits tend more toward feces and detritus than to live prey. They are inhabitants of estuaries and the lower portions of rivers in the Indo-Pacific, where they are reputed to aggregate near sewage outfalls. Rabbitfishes (**Siganidae**, Fig. 15.26A) are reef, grassbed, and estuarine herbivores with a unique pelvic formula of I, 3, I, reflecting the hard spine at either edge of the fin. Rabbitfishes are in fact very spiny fishes, their first dorsal spine projecting forward rather than upward, and many of the spines possessing a painful toxin (the forward projecting spine frequently impales the uninitiated fisher). Although most rabbitfishes are counter-shaded species that shoal in seagrass and mangrove areas, reef-dwelling species such as the Foxface

(A) (B)

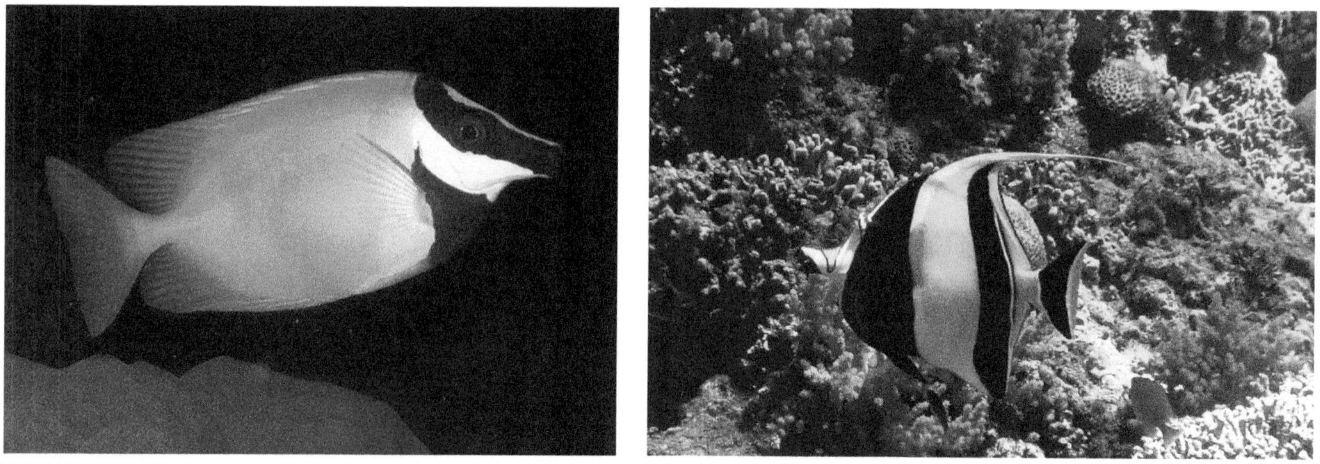

FIGURE 15.26 A – A Foxface Rabbitfish (*Siganus vulpinus*, Siganidae). Dr. Dwayne Meadows / NOAA / Public Domain; B- A Moorish Idol (*Zanclus canescens*, Zanclidae). Courtesy of J. Hyde and K. Sultze, used with permission.

Rabbitfish (*Siganus vulpinus*), converge in coloration and habitat with butterflyfishes and even form apparently monogamous pairs, as happens in several butterflyfishes. The monotypic Louvar, *Luvarus imperialis* (**Luvaridae**), is a pelagic derivative of the suborder (see Fig. 2.1). It is a large (to 1.8 m, 140 kg), non-shoaling fish with extremely high fecundity, a large female containing nearly 50 million eggs. Louvars converge with the pelagic scombroids (see below) in having a lunate tail and a lateral keel on the caudal peduncle, and the posteriorly set dorsal and anal fins resemble the finlets of the tunas and mackerels (Tyler et al. 1989). The head shape looks more like a dolphinfish, another pelagic species. Louvars feed on jellyfishes, salps, and ctenophores.

The Moorish Idol, *Zanclus canescens* (**Zanclidae**, Fig 15.26B) is another monotypic species related to the surgeonfishes. It is a strikingly shaped and colored Indo-Pacific and eastern Pacific reef fish that is remarkably convergent with butterflyfishes in body form, coloration, and behavior, including elongate dorsal spines and projectile horns above the eyes, as in the butterflyfish genus *Heniochus*. The 80 species of **acanthurid** surgeonfishes, unicornfishes, and tangs are most easily distinguished by the knife-blade present on the caudal peduncle. This blade is a modified scale and can exist as fixed, laterally projecting plates in *Prionurus* and the unicornfish genus *Naso*, or as single, forward-projecting knives that are exposed as the fish flexes its body. The blade is often covered with a toxic slime, the strength of the toxin apparently directly related to the length of the blade. The peduncular blade makes surgeonfishes among the few fishes that should not be grasped by the tail. Unicornfishes derive their name from a long bony protuberance on the head of some species that serves an unknown function. Surgeonfishes are often beautifully colored fishes, the color changing with age. As a group, they are herbivorous (except for planktivorous unicornfishes); species differ in dentition, jaw mechanics, and the body angles at which they remove algae from the reef.

Suborder Xiphioidei

Suborder Xiphioidei: Xiphiidae (1 species - Swordfish), Istiophoridae (9 species of billfishes)

(Families and numbers of species from Fricke et al. (2021) but refer to **https://researcharchive.calacademy.org/research/ichthyology/catalog/SpeciesByFamily.asp** *for updates.)*

Some authors such as Nelson et al. (2016) put these two families in the order Istiophoriformes, and also include the barracudas, family Sphyraenidae in that order.

The temperate and warm-temperate **xiphiid** Swordfish, *Xiphias gladius*, and the more tropical **istiophorid** Sailfish, spearfishes, and marlins (*Istiophorus*, *Istiompax*, *Tetrapturus*, *Kajikia*, and *Makaira*) are among the fastest and largest predators in the sea (Fig. 15.27A). Some are large and fast enough to be chief predators on relatively large tunas. The bill in both groups consists of an expanded premaxillary bone that is depressed and smooth in the swordfish and more rounded and pricklier in marlins and their relatives. Other differences include absence of pelvic fins, a single caudal keel, and relatively stiff, sharklike pectoral, dorsal, and anal fins in the swordfish. The istiophorids in contrast have long pelvic filaments, flexible pectorals, double keels, and a long, depressible spiny dorsal fin that reaches its extreme expression in the sailfishes. The bill can serve as a spear, cutlass, or club in capturing prey. Billfishes have independently evolved cranial endothermy (discussed in Chapters 7 and 10). Swordfish attain sizes of 530 kg, whereas both Blue and Black marlin grow to 900 kg.

Suborder Scombroidei

Suborder Scombroidei: Scombrolabracidae (1 species - Longfin Escolar), Gempylidae (26 species of snake mackerels), Trichiuridae (47 species of cutlassfishes), Scombridae (51 species of mackerels, Spanish mackerels, tunas)

(Families and numbers of species from Fricke et al. (2021) but refer to **https://researcharchive.calacademy.org/research/ichthyology/catalog/SpeciesByFamily.asp** *for updates.)*

Based on molecular characters, Nelson et al. (2016) place the Scombrolabracidae in its own order (Scombrolabraciformes). They also remove the Gempylidae, Trichiuridae, and Scombridae from the order Perciformes and place them in the suborder Scombroidei of the order Scombriformes. They are joined by several other families that Nelson et al. place in the suborder Stromateoidei (Amarsipidae, Centrolophidae, Nomeidae, Ariommatidae, Tetragonuridae, Stromateidae). Scombroids, stromateoids, and several other families have been placed in the Series Pelagiara based solely on molecular characters (Betancur-R et al. 2017; Arcila et al. 2021) although not by Nelson et al. (2016) or Pastana et al. (2021).

Scombroids (Orrell et al. 2006) include some of the largest and most economically valuable predators in the sea, namely the tunas. The suborder is characterized by a non-protractile mouth in more derived groups, a secondary modification given the general trend in teleosts toward increasing protrusibility. Several families have independently evolved some form of endothermy and heat conservation (discussed in Chapters 7 and 10).

The monotypic *Scombrolabrax heterolepis* (**Scombrolabracidae**) is a peculiar 30 cm long, deep-water oceanic fish with a protractile jaw and a unique gas bladder arrangement that includes numerous bubblelike projections that fit into depressions of expanded vertebral accessories. It is sometimes placed in its own suborder (or order, as in Nelson et al. 2016). The **gempylid** snake mackerels include 26 species of pelagic and deep-water predators characterized by an elongate body, large mouth with long teeth, a long spiny dorsal fin, and a series of dorsal and ventral finlets just ahead of the tail. The family includes the cosmopolitan Oilfish, *Ruvettus pretiosus*, a large (1.8 m, 45 kg) predator of moderate depths. It is sometimes referred to as the castor-oil fish because of the

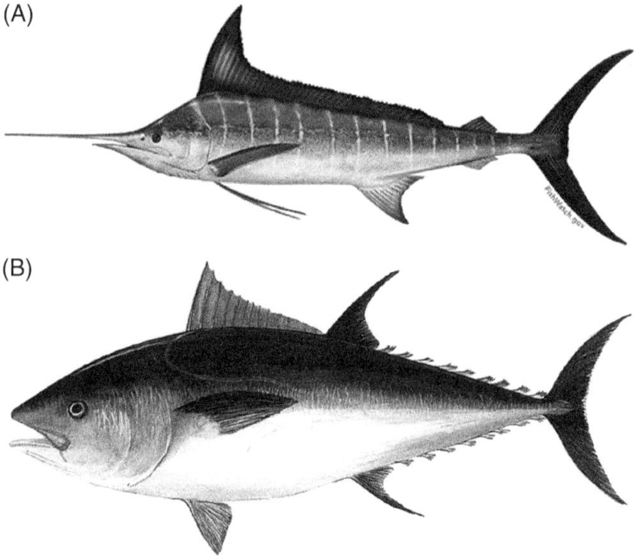

FIGURE 15.27 (A) Striped Marlin (*Kajikia audax*) are among the larger members of the family Istiophoridae. NOAA / Public Domain. (B) Atlantic Bluefin Tuna (*Thunnus thynnus*) are the largest members of the family Scombridae. NOAA / Public Domain.

purgative quality of its meat. An active fishery for Oilfish in the Comoro Islands off eastern Africa captures endangered Coelacanths as bycatch. The **trichiurid** cutlassfishes look like very compressed, silvery snake mackerels that have lost their pelvic fins, most of their anal fin, finlets, and most of the tail. The fanglike teeth belie a diet of large zooplankton, at least in smaller individuals. Trichiurids can be remarkably abundant and constitute major fisheries. The Largehead Hairtail, *Trichiurus lepturus*, had consistently ranked in the top 10 species in global fishery landings, averaging close to 1.5 million metric tons annually (FAO 2004) and this has continued to as recently as 2009; although there are signs of overfishing in some areas (FAO 2020).

The **scombrid** mackerels and tunas are highly adapted for a mobile, open sea existence in terms of anatomy physiology, and behavior (Fig. 15.27B; see Chapter 10, The Open Sea). Less derived members of the group, such as the mackerels, Spanish mackerels, and relatives (*Scomber*, *Rastrelliger*, and *Scomberomorus*) tend to live closer to shore, whereas more derived members are highly pelagic and nomadic (Collette & Graves 2019). Although chiefly a tropical and subtropical family, several species move into cold waters for feeding. Sizes range from relatively small, 50 cm mackerels (*Scomber*, *Auxis*) to the giant Bluefin Tuna, *Thunnus thynnus*, at >3 m and 650 kg. Most are schooling fishes of tremendous commercial importance (e.g. Sharp & Dizon 1978). Several tuna species are overfished, especially the three bluefin species (Atlantic Bluefin, *T. thynnus*; Pacific Bluefin, *T. orientalis*, and Southern Bluefin, *T. maccoyii*), which have been depleted throughout their ranges because of their extreme economic value (Collette 1999; Maggio 2000; Safina 2001b). In 2019, the first Pacific Bluefin Tuna of the year at the Tsukiji Central Fish Market in Tokyo sold for the equivalent of 3 million US dollars.

Suborder Stromateoidei

Suborder Stromateoidei: Amarsipidae (1 species - Amarsipa), Centrolophidae (30 species of medusafishes), Nomeidae (17 species of driftfishes), Ariommatidae (8 species of ariommatids), Tetragonuridae (3 species of squaretails), Stromateidae (19 species of butterfishes)

(Families and numbers of species from Fricke et al. (2021) but refer to **https://researcharchive.calacademy.org/research/ichthyology/catalog/SpeciesByFamily.asp** *for updates.)*

Stromateoids are generally tropical and warm temperate fishes of the open sea that often associate as juveniles with floating or swimming objects, particularly with siphonophores and jellyfishes. They are sometimes included with the Scombroidei as a second suborder of an order Scombriformes (see Nelson et al. 2016) or as part of the Pelagiara (see Betancur-R et al. 2017), a large molecularly defined group of marine fishes. In contradiction to all large-scale molecular phylogenies of percomorphs, Pastana et al. (2021) conclusively recover a monophyletic Stromateiformes of six families supported by eight unambiguous synapomorphies, which are highly unlikely to have originated independently. According to Pastana et al. (2021), the Centrolophidae is not monophyletic with several of its members arranged as successive sister-groups to a family containing the other stromateoid families. Amarsipidae is the sister group to the other four families of stromateiforms.

A characteristic of the stromateoids is a thick-walled sac in the pharyngeal region that contains "teeth" made from hardened papillae. **Centrolophid** medusafishes bear a superficial resemblance to the icosteid Ragfish. **Nomeid** driftfishes, as the name implies, hover around and under floating logs, siphonophores, jellyfishes, and seaweed as juveniles and occur in deeper water as adults. Juveniles of the Man-of-war Fish, *Nomeus gronovii*, live with impunity among and even feed on the stinging tentacles of the Portuguese man-of-war. **Ariommatids** superficially resemble carangid scads (i.e. *Decapterus*). **Tetragonurid** squaretails are elongate fishes encircled by ridged scales and with a long caudal peduncle that has a single keel on either side formed from scale ridges. They feed on pelagic cnidarians and ctenophores, which they bite with specialized knifelike teeth. **Stromateid** butterfishes and harvestfishes are round or elliptical in profile with a forked tail and are similar in shape to some carangids; they are sometimes referred to as pompanos, a name more correctly applied to several carangids. As with many open sea groups, butterfishes lack pelvic fins as adults (for unknown reasons).

Suborder Anabantoidei

Suborder Anabantoidei: Anabantidae (34 species of climbing gouramies), Helostomatidae (1 species - Kissing Gourami), Osphronemidae (132 species of gouramies)

(Families and numbers of species from Fricke et al. (2021) but refer to **https://researcharchive.calacademy.org/research/ichthyology/catalog/SpeciesByFamily.asp** *for updates.)*

Nelson et al. (2016) place the Anabantoids (suborder Anabantoidei) and the suborder Channoidei in the order Anabantiformes, instead of keeping them in Perciformes as we have. Anabantoids are also called labyrinth fishes because of a complexly folded, auxiliary breathing structure derived from the epibranchial of the first gill arch located above the gills in the gill chamber (see Chapter 5, Air-Breathing Fishes). Functionally, the suprabranchial or labyrinth organ is the primary breathing structure for many species, and anabatoid fish in well-aerated aquaria will die if not allowed to gulp air at the surface. In most anabantoids, the male exhales a nest of mucous-covered bubbles among which eggs are laid and which he guards. **Anabantid** climbing gouramis or climbing perches are African and Asian freshwater fishes that derive their name from their ability to move across wet ground (and supposedly even up wet tree trunks), jerking along by thrusts from the tail while the pectoral fins and gill covers act as props. The Kissing Gourami, *Helostoma temminckii*, is the sole member of the family **Helostomatidae**. The peculiar kissing behavior of this species is derived from its feeding habits that involve scraping algae from surfaces using keratinous teeth on distinctive lips. The function of kissing, in which two individuals repeatedly press their open mouths against each other, is poorly understood but may be an aggressive behavior.

The family **Osphronemidae** is divided into four subfamilies with over 130 species. The osphronemine Giant Gourami, *Osphronemus goramy*, reaches 80 cm in length and is a popular food fish that is cultured throughout Southeast Asia. Its air-breathing abilities help keep it alive in fish markets. The Macropodusinae includes the Siamese fighting fishes and paradisefishes. Bettas (Siamese Fighting Fish, *Betta*) are used extensively in behavioral and genetic studies. Males are exceedingly pugnacious toward each other. They are bred and fought like fighting cocks, making them one of the few fishes cultured for reasons other than food, appearance, or research. Fights to the death in the confines of an aquarium do not reflect real-life situations where a subordinate fish can flee from a dominant individual. Luciocephalines in the genus *Colisa* shoot water droplets at terrestrial insects, in a manner analogous to that of the toxotid archerfishes (Dill 1977a). The luciocephaline Pikehead, *Luciocephalus pulcher*, is an elongate stalking predator on small fishes with a body form characteristic of other such piscivores (elongate jaws, slender body, dorsal and anal fins set far back on body, rounded tail. As befits a derived percomorph, Pikeheads have the most protrusible mouth of any teleost. When feeding, the mouth is shot forward rapidly, surrounding the prey. Pikeheads have an interesting bone in the gular region of their throats that is analogous to the gular plate(s) of the primitive coelacanth, Bowfin, bichirs, and some elopomorphs; whether this reinvented gular bone functions in oral incubation of eggs or in mouth protrusion is unclear (Liem 1967).

Suborder Channoidei

Suborder Channoidei: Channidae (56 species of snakeheads)

(Families and numbers of species from Fricke et al. (2021) but refer to **https://researcharchive.calacademy.org/research/ichthyology/catalog/SpeciesByFamily.asp** *for updates.)*

Snakeheads (**Channidae**) are highly predatory freshwater fishes of tropical Asia and Africa. They have a suprabranchial breathing organ reminiscent of that of the anabantoids and are primarily swamp dwellers. The robust, elongate bodies, long dorsal and anal fins, and ringed eyespot on the caudal peduncle (Fig. 15.28) give some channids a superficial resemblance to a Bowfin (Amiidae – see Chapter 13). Some snakeheads reach over 1.8 m in length and 20 kg in mass and are prized food fishes. Small species and young individuals of larger species are sold in the pet trade, and live individuals are sold as food fish. Because they grow large and are predatory, or because people want to establish desirable food species outside their native range despite the ecological consequences, snakeheads have been released into the wild in numerous Asian and North American locales and are reproducing at least in Florida, Hawaii, and Maryland (Fuller et al. 1999; Courtenay & Williams 2002). Courtenay and Williams (2002) analyzed the potential negative impact snakeheads could have in regions where they are or were likely to become established and concluded the potential for significant ecological harm was high. Snakehead importation and transfer across state lines is now illegal in the USA.

Suborder Caproidei

Suborder Caproidei: Caproidae – 18 species of boarfishes *(according to Nelson et al. 2016)*

Our treatment of the order Perciformes ends appropriately with an enigmatic group whose exact placement remains something of a puzzle. In many earlier treatments, the **caproid** boarfishes were thought to be pre-perciforms, most closely allied with the zeiform dories, which they resemble. Tyler et al. (2003) supports that conclusion, whereas others consider them a more derived perciform, a decision Nelson (2006) follows but with uncertainty. Nelson et al. (2016) places this group in its own order, the Caproiformes, with a single family, the Caproidae. These are medium-sized (to 30 cm), reddish, deep-bodied, rhomboid, schooling fishes at moderate depths (50–600 m).

Subseries Percomorpha: The More Derived Percomorph Orders

The final three orders of the series Percomorpha are highly derived groups that exhibit some rather dramatic morphological adaptations suitable for rather specific habitats.

(A)

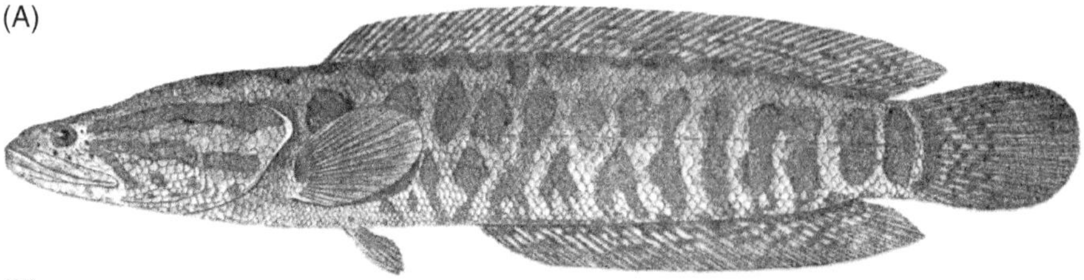

(B)

FIGURE 15.28 Snakeheads. (A) The Northern Snakehead, *Channa argus*. This Southeast Asian native was established in a pond in Maryland in 2002. (B) The Giant Snakehead, *Channa micropeltes*, a large (up to 1 m and 20 kg) freshwater predator of Southeast Asian still waters. (A) Courtenay and Williams (2002) / USGS / Public Domain (2002); (B) Courtesy of Jean-Francois Healias.

Order Pleuronectiformes

Order Pleuronectiformes: Psettodidae (3 species of spiny turbots), Citharidae (6 species of largescale flounders), Scophthalmidae (9 species of turbots), Bothidae (168 species of lefteye flounders), Paralichthyidae (64 species of sand flounders), Pleuronectidae (63 species of righteye flounders), Achiridae (35 species of American soles), Paralichthodidae (1 species - Measles Flounder), Rhombosoleidae (20 species of South Pacific flounders), Achiropsettidae (4 species of southern flounders), Oncopteridae (1 species - Remo Flounder), Samaridae (30 species of crested flounders), Poecilopsettidae (21 species of bigeye flounders), Soleidae (180 species of soles), Cynoglossidae (167 species of tonguefishes)

(Families and numbers of species from Fricke et al. (2021) but refer to **https://researcharchive.calacademy.org/research/ichthyology/catalog/SpeciesByFamily.asp** *for updates.)*

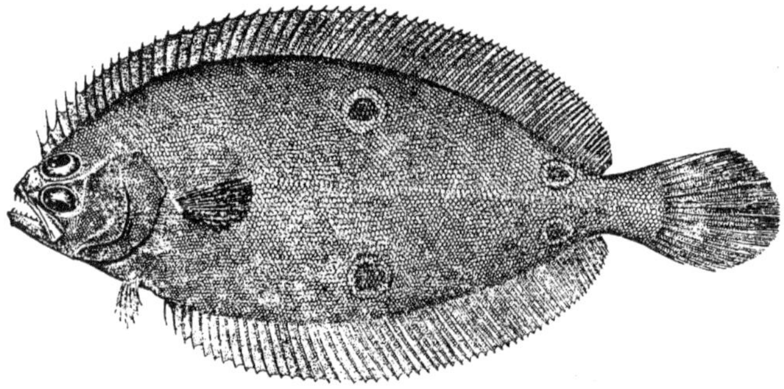

FIGURE 15.29 A pleuronectiform, the bothid Fourspot Flounder, *Paralichthys oblongus*. Drawn by H. L. Todd in Collette & Klein-MacPhee (2002). Also see Fig 16.4C, D for examples of flatfishes blending into their surroundings.

Flatfishes are distinctive, compressed acanthopterygians that share certain features, most noticeably a marked asymmetry that includes having both eyes on the same side of the head in juveniles and adults (Fig. 15.29). Flatfishes begin life as bilaterally symmetrical, pelagic fishes but during the larval period or shortly thereafter, asymmetrical growth of the skull causes one eye to migrate to the other side and the fish settles to the bottom, lying on its blind, more flattened side (see Chapter 9, Metamorphosis of asymmetrical flatfishes). Eye movement is made more complicated by the position of the anterior portion of the dorsal fin, which often originates above or ahead of the eyes. Teeth, scales, paired fins, and pigmentation also typically differ between sides. Families are generally either right-eyed or left-eyed defined by which eye stays put. Flatfishes are benthic, carnivorous, marine fishes (with perhaps 10 freshwater derivative species) that lack a gas bladder; most live at shallow to moderate depths in Arctic, temperate, and tropical locales. Some species are able to change the coloration of the eyed side to match the shading pattern of the background upon which they rest (see Fig. 16.4C, D). Flatfishes constitute important fisheries for forms known as dab, flounders, halibuts, plaice, sole, tonguefishes, turbots, and whiffs. Most species are in five large families. Ancestral, intermediate, and closely related forms to flatfishes are not known and the fossil record is limited.

The primitive **psettotid** spiny turbots show the least movement of the eye during metamorphosis. Many authors, such as Nelson et al. (2016), place this family in its own suborder, the Psettodoidei. The **Paralichthyidae** include the Summer Flounder (*Paralichthys dentatus*) and California Halibut (*P. californicus*), the latter species reaching a size of 1.5 m and 30 kg. The **pleuronectid** righteye flounders include the larger halibuts such as the Atlantic and Pacific halibuts. Females of the Pacific Halibut, *Hippoglossus stenolepis*, may live 40 years and reach barndoor proportions of 3 m long and 200 kg mass, but may not mature until they are 16 years old. The fishery for Pacific Halibut, unlike that of the depleted Atlantic species, is one of the better-managed fisheries in the world. Several other commercially important flatfish species are pleuronectids, including the Arrowtooth Flounder, Petrale Sole, Rex Sole, Winter Flounder, Yellowtail Flounder, and English Sole. Among the **bothid** lefteye flounders is the Peacock Flounder (*Bothus lunatus*) of the Caribbean. An **achirid** American sole, the Hogchoker, *Trinectes maculatus*, commonly invades rivers of Florida and the Atlantic coast of the USA. True soles are in the family **Soleidae**, which are usually right-eyed. Among the soleids is the Red Sea Moses Sole, *Pardachirus marmoratus*, which exudes a toxin, pardaxin, reported to be a natural shark repellant. The Common Sole, *Solea solea*, of European waters is reported to raise its black-tipped pectoral fin when disturbed in an action that mimics the raising of the dorsal fin of venomous trachinid weeverfishes. The **cynoglossid** tonguefishes are the most elongate of the flatfishes and also show considerable variation in habitat, including shallow water and burrowing forms (e.g. Blackcheek Tonguefish, *Symphurus plagiusa*), several species that occur as deep as 1900 m, as well as purely freshwater forms (three species from Indonesia). One species, *Symphurus thermophilus*, occupies toxic sulfur ponds at deep hydrothermal vents.

Order Lophiiformes

Order Lophiiformes: Lophiidae (30 species of goosefishes), Antennariidae (52 species of frogfishes) Tetrabrachiidae (2 species of tetrabrachiid frogfishes), Lophichthyidae (1 species - Lophichthyid Frogfish), Brachionichthyidae (14 species of handfishes), Chaunacidae (29 species of coffinfishes or sea toads), Ogcocephalidae (90 species of batfishes), Caulophrynidae (5 species of fanfins), Neoceratiidae (1 species - Toothed Seadevil), Melanocetidae (6 species of black seadevils), Himantolophidae (22 species of footballfishes), Diceratiidae (7 species of double anglers), Oneirodidae (71 species of dreamers), Thaumatichthyidae (9 species of wolftrap anglers), Centrophrynidae (1 species - Deepsea Anglerfish), Ceratiidae (4 species of seadevils), Gigantactinidae (24 species of whipnose anglers), Linophrynidae (27 species of leftvents).

(Families and numbers of species from Fricke et al. (2021) but refer to **https://researcharchive.calacademy.org/research/ichthyology/catalog/SpeciesByFamily.asp** *for updates.)*

The **lophiiforms were** previously considered to be the most derived order within the Paracanthopterygii but have been moved to the Percomorpha based primarily on molecular evidence, which also suggest that they are a sister group to the Tetraodontiformes (the next group addressed). Lophiiforms are a diverse and often bizarre-looking group of marine fishes that include benthic, shallow water dwellers but that have evolved many highly modified, open-water, deep-sea forms. Many if not most of them use a modified first dorsal spine as a lure for catching smaller fish. The basal group is the **lophiid** goosefishes, known commercially as monkfish. Goosefishes occur on both sides of the Atlantic and also in the Pacific and Indian oceans. The western North Atlantic Goosefish, *Lophius americanus*, can exceed 1 m in length and 40 kg in mass, and has a huge mouth with long, recurved teeth that point back into the mouth. Goosefishes prey on other fishes and on diving seabirds. **Antennariid** frogfishes also rest on the bottom and are well-camouflaged, globose fishes that can walk across the bottom on their pectoral and pelvic fins (Fig 15.30A). (An old name for the lophiiforms, **Pediculati**, refers to the elbowlike bend in the pectoral and the footlike appearance of the pelvic fins). The esca or lure of frogfishes can be quite ornate, mimicking a small fish, shrimp, or worm. When not waved in front of potential prey, the esca sits in a protective depression between the second and third dorsal spines. If an esca is bitten off, it apparently can regenerate back to its species-specific form (Pietsch & Grobecker 1987).

Geographically restricted fishes make up the family of **brachionichthyid** handfishes (Fig. 15.30B). These small, colorful (red, orange, or pink with dark spots) benthic fishes occur only in southeastern Australia, with five of the eight known species restricted to Tasmania. Unlike most marine fishes, handfishes lack a pelagic larval stage, fully formed juveniles emerging from eggs that are guarded by the female for 7–8 weeks. Because of their size, restricted range, attractive coloration, threats from introduced predators, and low dispersal of both young and adults, most species are considered highly endangered. The Spotted Handfish, *Brachionichthys hirsutus*, has Critically Endangered status with IUCN.

The **ogcocephalid** batfishes are among the least fish-like fishes (other candidates include seahorses, shrimpfishes, boxfishes, and ocean sunfishes). The flattened, rounded head accounts for more than half the length of the body; it tapers quickly behind the expanded pectoral fins, giving the fish the appearance of a rounded axe with a short handle (Fig. 15.30C). Batfishes alternate walking on their pectorals with swimming via jet propulsion of water expelled from their round, backward facing opercular openings. As modified as the batfishes are, they are rivalled in strange appearance by the 11 families of deep-sea anglerfishes, suborder **Ceratioidei** (Fig. 15.31). The ceratioids are the most speciose fishes of the vast bathypelagic region, comprising 160 species. Among their other derived traits, 23 species in five families have very small males that fuse to and become parasitic on the larger females, the difference in length between male and female being as much as 60-fold (Pietsch 2005; Pietsch & Orr 2007; see Chapter 10, The Deep Sea). Male parasitism appears to have evolved independently perhaps seven times in the suborder.

FIGURE 15.30 Lophiiformes. (A) Striated Frogfish (*Antennarius striatus*, family Antennariidae), note the elaborate esca or lure at the end of the modified first dorsal spine or illicium (also see fig 16.1); NOAA / Public Domain. (B) Spotted Handfish (*Brachionichthys hirsutus*, family Brachionichthyidae); Rick Stuart-Smith/ Wikimedia Commons / CC BY 3.0. (C) A batfish (family Ogcocephalidae); NOAA / Public Domain.

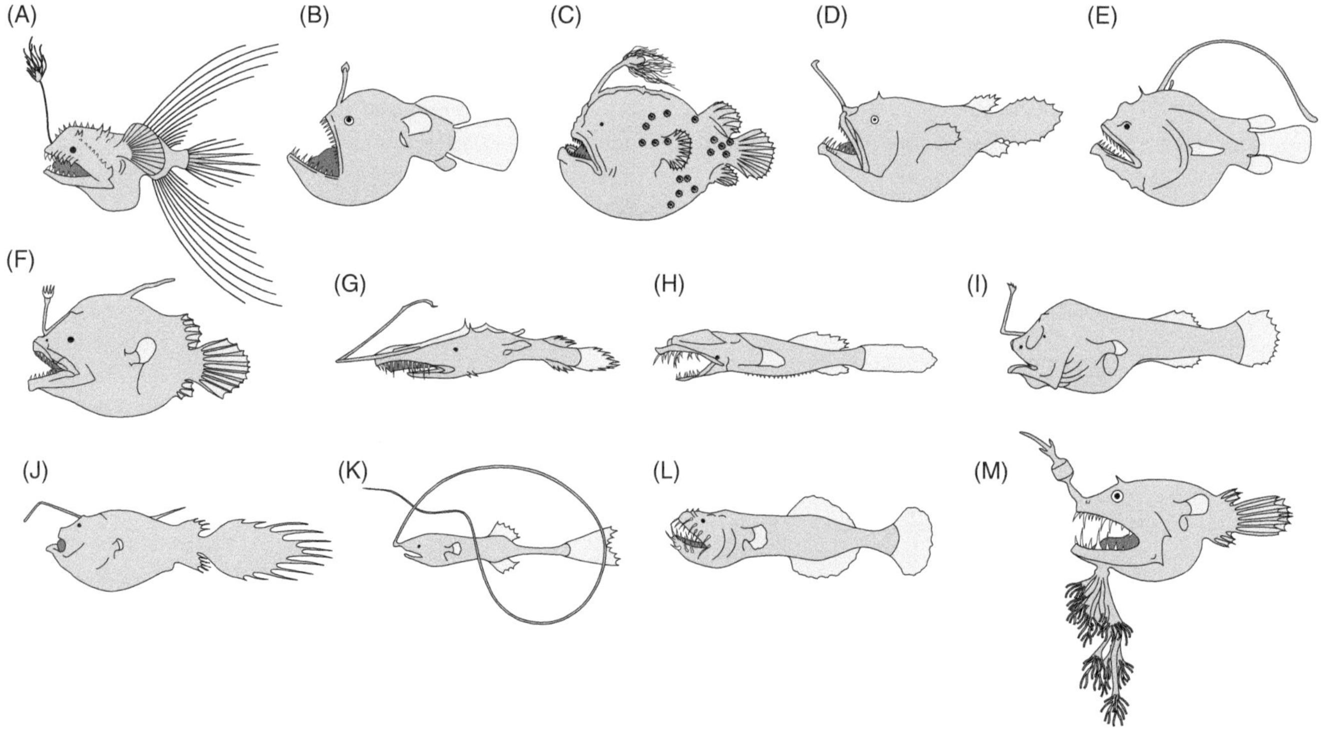

FIGURE 15.31 Ceratioid anglerfishes. Shown are females from the 11 ceratioid families: (A) Caulophrynidae; (B) Melanocetidae; (C) Himantolophidae; (D, E) Diceratiidae; (F) Oneirodidae; (G, H) Thaumatichthyidae; (I) Centrophrynidae; (J) Ceratiidae; (K) Gigantactinidae; (L) Neoceratiidae; (M) Linophrynidae. See Pietsch and Orr (2007) for color photos. After Pietsch (2005), used with permission.

Order Tetraodontiformes

Order Tetraodontiformes: Triacanthodidae (24 species of spikefishes), Triacanthidae (7 species of triplespines), Balistidae (41 species of triggerfishes), Monacanthidae (108 species of filefishes), Ostraciidae (24 species of boxfishes, trunkfishes, and cowfishes), Aracanidae (13 species of deepwater boxfishes), Triodontidae (1 species - Threetooth Puffer), Tetraodontidae (193 species of puffers), Diodontidae (19 species of porcupinefishes), Molidae (5 species of ocean sunfishes)

(Families and numbers of species from Fricke et al. (2021) but refer to **https://researcharchive.calacademy.org/research/ichthyology/catalog/SpeciesByFamily.asp** *for updates.)*

The most highly derived fishes are in the order Tetraodontiformes (Fig. 15.32). The name refers to the common pattern of four teeth in the outer jaws of pufferfishes (family **Tetraodontidae**). Fishes in the order are characterized by a high degree of fusion or loss of bones in the head and body. In the head region, such bones as the parietals, nasal, infraorbital, and posttemporal are commonly absent, both hyomandibular and palatine bones may be firmly attached to the skull, and the maxilla is fused to the premaxilla. The pelvic fins and lower vertebral ribs are often missing, and the vertebral number is reduced from the common acanthopterygian condition of 26 to as little as 16. The skin has a thick, leathery feel and is covered by scales that are modified into spines, bony plates, or ossicles, some of rather spectacular proportions. Many tetraodontiforms can also rotate their eyes independently. The fossil record for the group goes back at least to the Early Eocene and quite likely the Late Cretaceous, once again pointing out that modern bony fishes have a long evolutionary history. Some tetraodontiforms (triggerfishes, puffers) are predators on sessile benthic invertebrates, others are water column swimmers above the reef that feed on zooplankton (Black Durgon Triggerfish), and some are large, offshore planktivorous species (Gray Triggerfish) or jellyfish feeders (Ocean Sunfish). All but 20 species of pufferfishes are marine.

The suborder Balistoidei contains the triggerfishes (**Balistidae**) and the filefishes (**Monacanthidae**). In many species, the first dorsal spine is particularly long and stout and can be locked in the erect position via an interaction with the second spine (see Fig. 15.32B). The base of the smaller second spine protrudes forward and fits into a groove on the posterior edge of the first spine, locking the first spine into position. Depressing the second spine releases the lock, hence the name triggerfish. Sound production in this group is common, produced by grinding of the teeth or vibration of the gas bladder via the pectoral spine; the legendary Hawaiian name for the triggerfishes *Rhinecanthus aculeatus* and *Rhinecanthus rectangulus* is *humuhumu nukunuku apua'a*, which means "the fish that sews with a needle and grunts like a pig." In boxfishes (**Ostraciidae** and **Aracanidae**), the entire body except the fins and caudal peduncle are encased in a bony box, which is triangular or rectangular in cross-section. Stout spines sometimes protrude anteriorly just above the eyes and posteriorly just ahead of the anal fin. Swimming is accomplished via undulations of median fins (see Chapter 4, Locomotion : Movement and Shape).

The suborder Tetraodontoidei contains the puffers and ocean sunfishes. Instead of individual teeth, all members of this group have beak-like dentitions. Pufferfishes can deter predators by inflating their body by filling the stomach with water (see Chapter 16, Fig. 16.17). Three families of pufferfishes are generally recognized based on the number of toothlike structures in each jaw: a single three-toothed species (**Triodontidae**), the smooth and sharpnose pufferfishes with four teeth (**Tetraodontidae**), and the spiny pufferfishes, burrfishes, and porcupinefishes with two fused teeth (**Diodontidae**). Tetraodontids have prickly skin. Because of bone loss and fusion, tetraodontids produce tasty, boneless fillets, but many species concentrate a powerful neurotoxin, tetrodotoxin, in their viscera or other organs such as the skin, which can cause death in humans. Specially licensed and supervised *Fugu* restaurants in Japan serve the meat of puffers of the genus *Takifugu*, which contains small amounts of tetrodotoxin and provides a narcotic high. Despite such chemical protection, pufferfishes are commonly eaten by sea snakes. Diodontids have spines of varying length that are erected when the fish inflates, creating a large, round, and essentially inedible pincushion.

The most extreme tetraodontiforms are the five species of temperate and tropical molas (**Molidae**). The body is oval in lateral view with tall, thin dorsal and anal fins that propel the fish (Fig. 15.32C). They lack a true tail but instead have a clavus, formed by dorsal and anal fin rays (Johnson & Britz 2005). Early accounts describe ocean sunfishes as sluggish, however, new data from tagging studies (e.g. Nakamura et al. 2015) show that molids are well-adapted and efficient swimmers that make deep foraging dives to target gelatinous prey. In between deep dives to cold environments, they rewarm at the surface by basking. The Southern Ocean Sunfish, *Mola alexandrini*, gets to be nearly 3 m long, and weighs as much as 2300 kg (Thys et al. 2021). Fecundities of 300 million eggs have been reported, an apparent record among fishes. Ocean sunfishes have the fewest number of olfactory receptors genes, and similarly few olfactory lamellae, suggesting they have a very poorly developed sense of olfaction (Policarpo et al. 2021).

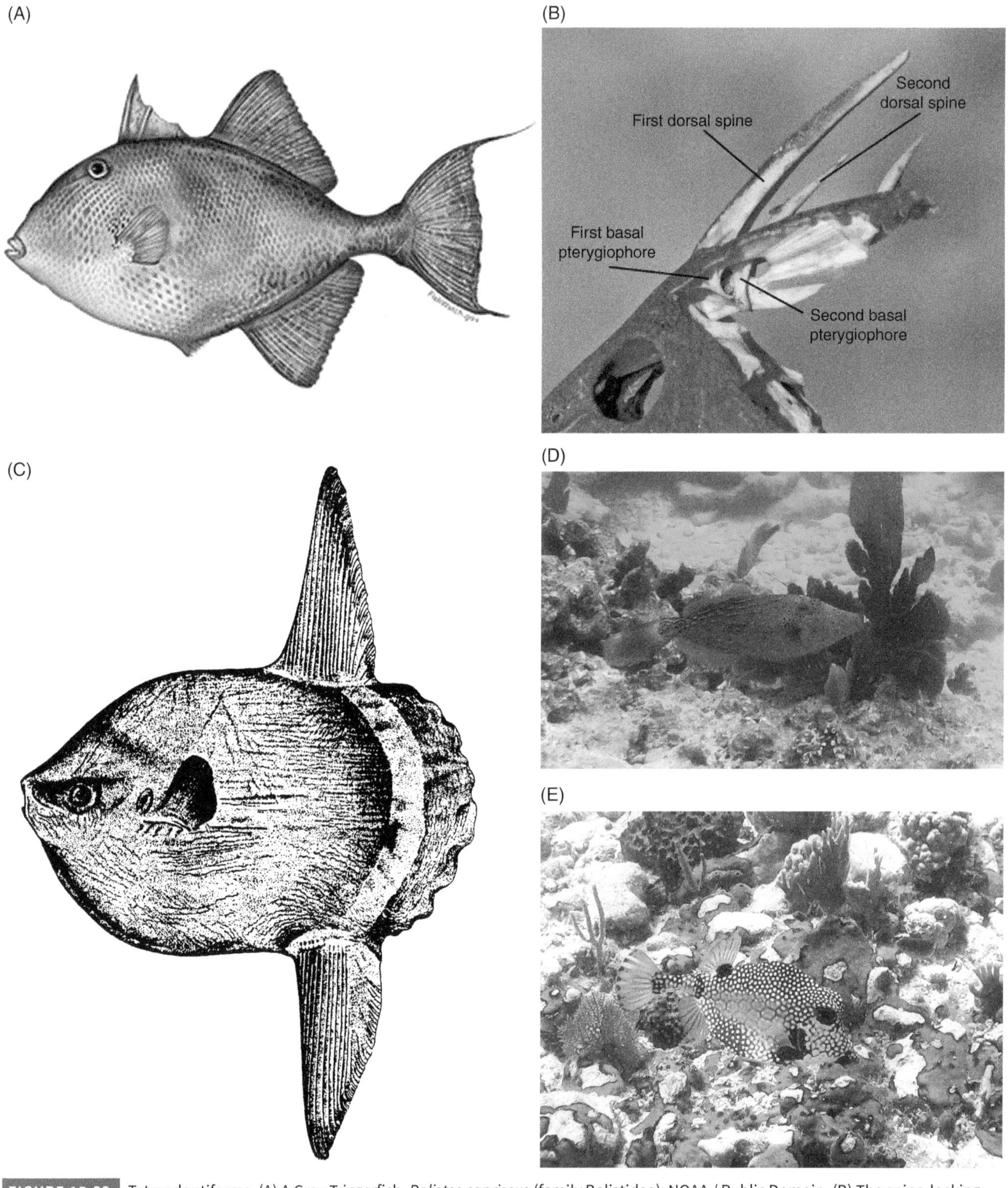

FIGURE 15.32 Tetraodontiforms. (A) A Gray Triggerfish, *Balistes capriscus* (family Balistidae); NOAA / Public Domain. (B) The spine-locking mechanism of triggerfishes, showing how the second dorsal spine fits into and helps lock the first spine in the erect position. Pushing posteriorly on the second spine causes the second basal pterygiophore to push against the first basal pterygiophore, releasing the first spine; G. Helfman (Author). (C) The Ocean Sunfish, *Mola mola*, a member of the family Molidae, considered the evolutionarily most derived of all teleosts; drawn by H. L. Todd in Collette & Klein-MacPhee (2002). The tetraodontiforms also include the filefishes (D, family Monacanthidae) and boxfishes (E, family Ostraciidae); D. Facey (Author).

Supplementary Reading

Arcila D, Hughes LC, Meléndez-Vazquez F, Baldwin CC, White WT, Carpenter KE, Williams JT, Santos MD, Pogonoski JJ, Miya M, Ortí, and Betancur-R R. 2021. Testing the utility of alternative metrics of branch support to address the ancient evolutionary radiation of tunas, stromateoids, and allies (Teleostei: Pelagiaria). Systematic Biology DOI:**https://doi.org/10.1093/sysbio/syab018**.

Betancur-R R, Wiley EO, Aratia G, Acero A, Bailly N, Miya M, Lecointre G, and Ortí G. 2017. Phylogenetic classification of bony fishes. BMC Evolutionary Biology 17:162.

Carroll RL. 1988. *Vertebrate paleontology and evolution*. New York: W. H. Freeman.

Nelson JS. 2006. *Fishes of the world*, 4th edn. Hoboken, NJ: Wiley & Sons.

Nelson, JS, Grande TC, and Wilson MVH. 2016. Fishes of the world. Fifth edition. Wiley & Sons, 707 pp.

Paxton JR, Eschmeyer WN, eds. 1998. *Encyclopedia of fishes*, 2nd edn. San Diego, CA: Academic Press.

Stiassny MLJ, Parenti LR, Johnson GD, eds. 1996. *Interrelationships of fishes*. San Diego: Academic Press.

Websites

www.cichlidpress.com.

www.desertfishes.org.

www.dragonsearch.asn.au.

www.FishBase.org.

http://gobiidae.com.

www.oceansunfish.org.

http://seahorse.fisheries.ubc.ca.

Fricke R, Eschmeyer W, Fong JD. 2021. *Genera/Species by family/subfamily in Eschmeyer's catalog of fishes*. San Francisco, CA: California Academy of Sciences. **https://researcharchive.calacademy.org/research/ichthyology/catalog/SpeciesByFamily.asp**. Electronic version accessed November 2021.

PART IV Fish Behavior

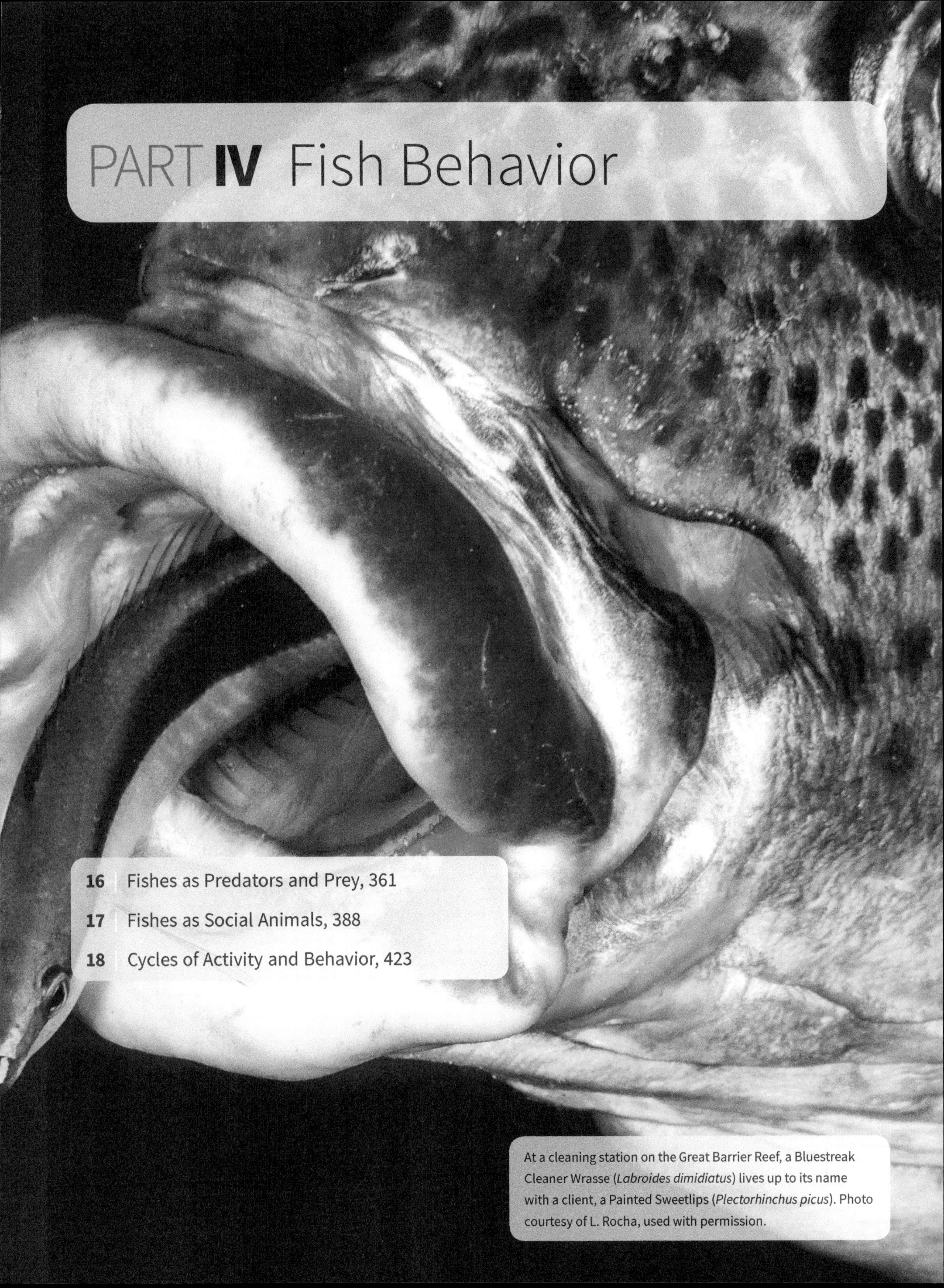

At a cleaning station on the Great Barrier Reef, a Bluestreak Cleaner Wrasse (*Labroides dimidiatus*) lives up to its name with a client, a Painted Sweetlips (*Plectorhinchus picus*). Photo courtesy of L. Rocha, used with permission.

CHAPTER **16**

Fishes as Predators and Prey

Summary

This chapter explores the behavior and ecology of feeding in fishes. Most fishes are both predators and prey, hence factors affecting a fish's ability to acquire food and avoid being eaten have a strong evolutionary influence on morphology, behavior, and ecology. Successful predators usually search, pursue, attack, capture, and handle prey using different structures and behaviors at different stages of the predation cycle. Not surprisingly, prey often have evolved countermeasures to limit the effectiveness of various predation strategies and behaviors.

Search may be active or passive, and detection can depend on multiple senses. In active search, a fish moves through the water, whereas in passive search, a sedentary, camouflaged predator waits for prey to come within striking distance. Some predators have accessory structures that attract prey to the predator. Fishes that forage in groups often find food faster than solitary fish, and being part of a group can help decrease the likelihood that an individual will be attacked by a predator. Some predators cooperate in conspecific groups to facilitate more efficient hunting and acquisition of prey. A study in the Red Sea documented cases of grouper soliciting cooperative hunting with moray eels to the benefit of both predators.

To escape predators, prey must avoid detection, evade pursuit, prevent or deflect attack and capture, and discourage handling if caught.

CHAPTER CONTENTS

The Diversity of Fishes: Biology, Evolution and Ecology, Third Edition. Douglas E. Facey, Brian W. Bowen, Bruce B. Collette, and Gene S. Helfman.

Companion website: www.wiley.com/go/facey/diversityfishes3

Different behaviors and structures achieve these different functions. Detection is avoided by various types of camouflage that make a prey unrecognizable, appear inedible, blend into its background, or just disappear. Tactics employed include protective resemblance, mimicry, disruptive coloration, countershading, mirror sides, and transparency. Shoaling may increase the ability of prey to detect and assess approaching predators. Many fishes also use shade to decrease the likelihood of being seen while also improving their likelihood of seeing an approaching predator or prey.

Pursuit places a predator close enough to attack prey. Open-water predators rely on sustained speed to overtake and capture prey, and thus have the appropriate morphological and physiological features to reduce drag and sustain speed without tiring. Some open-water predators work collectively to keep prey in tight groups near the surface, where they are more vulnerable to attack. Ambush predators are designed for remaining still either in the water column or near the bottom and generating sudden bursts of speed to capture prey that come near. If there is any pursuit at all, it would be brief.

To discourage pursuit, a prey fish can demonstrate or feign unpalatability, outdistance or outmaneuver the predator, or seek shelter in the bottom, in vegetation, or in the water column. Some fishes with toxic skin or spines advertise these structures via color or behavior (surgeonfishes, lionfishes, weeverfishes, and pufferfishes). Many fishes dive into holes in coral or rocks or into the sand when pursued. Open-water species must outdistance a predator; flyingfishes leap from the water and glide at twice their swimming speed.

Attack and capture are often synonymous in fishes. Predators on grouped prey generally try to separate an individual from the group before attacking. Most fishes use their rapidly expandable oral cavity to suck in prey. In more derived fishes, this is assisted by protrusible jaws, but even more ancestral groups with non-protrusible jaws can generate considerable suction by expanding the oral chamber laterally and downward. Some predators incapacitate prey prior to capture. In a few species the incapacitating attack may include electrical discharges, whereas others use morphological adaptations of the head or tail that can be used as weapons. Some fishes blow streams of water to overturn or uncover prey, and a few species squirt water to knock terrestrial insects into the water.

An attack can be prevented by quick evasive action, or by active or passive defense. Rapid, fast start moves are almost universal among fishes from early ontogeny on. Schooling fishes undergo a series of maneuvers when attacked; aggregating functions particularly well during attack because multiple targets presented by the prey confuse the predator.

Handling can improve the ability of the predator to swallow and digest prey, and includes positioning prey so it can be swallowed head-first, or removal of spines and shells via chewing. Some fishes exhibit tool use by using features of the environment to improve access to food, such as slamming hard-bodied invertebrates against a rock to crack open their shells. Final digestion is mostly chemical in action, although mechanical grinding occurs in those fishes with a gizzardlike structure.

Prey use a variety of tactics to counter capture and handling by predators. Easily shed scales may allow prey to slip out of the mouth of a predator. Slime and toxic or distasteful skin secretions also discourage capture (moray eels, toadfishes, soapfishes, soleid soles, and trunkfishes). Specialized skin cells in ostariophysan and a few other fishes secrete an alarm substance that warns schoolmates of an attack. Alarm substances and alarm calls may also attract secondary predators, which could frighten the first predator into releasing a victim. Capture and swallowing can be discouraged by prey that increase their effective body size by erecting spines. Puffers and porcupinefishes discourage predators by inflating themselves with water, and some have spines that become erect when the fish is inflated.

Many predatory fishes supplement their diets by scavenging, and some specialists rely primarily on recently dead animals or detritus for food. Handling in scavengers requires the separation of edible from inedible, either in the mouth or stomach. With few exceptions, fishes are gape-limited and do not attack food larger than they can swallow. Large prey must be dismembered, either by chopping or crushing with the jaw and pharyngeal teeth or, in the case of many eel-like fishes, tearing off pieces by gripping food in the jaws and twisting or spinning the body.

Some fishes have endosymbiotic bacteria in the gut that help digest plants, although most fishes lack this capability. Herbivory is more common in tropical than temperate habitats. Herbivores must be able to identify whether plants are edible, and overcome mechanical and chemical defenses via chewing or chemical digestion. Plant material takes longer to digest, hence herbivores tend to have longer guts. Territoriality on coral reefs is common among herbivorous fishes, some of which will defend their "gardens," and is often overcome by shoaling behavior in competitors.

Foraging places fishes at risk of becoming prey themselves. Thus many fishes must trade-off their own foraging against the risk of predation. Experimental studies have shown that fishes often give up foraging opportunities when threatened by predators, but will risk greater threats when exceptionally hungry or when the rewards are high enough. Prey are sensitive to the degree of threat presented by a predator and is able to take evasive action proportional to the degree of threat.

Introduction

Predation has an overriding influence on the morphology, behavior, and ecology of fishes. The selection forces in operation are obvious and strong: fishes with a relative feeding advantage may grow faster, but fishes that get eaten are eliminated from the gene pool. This chapter explores the behavior and ecology of feeding in fishes, with emphasis on the evolutionary interplay between predatory and escape tactics, the so-called "predator–prey arms race." For organizational purposes, adaptations are classified by where they appear to fit in the predation cycle, which is the sequence of events involving **searching/detecting**, **pursuing**, **attacking**, **capturing**, and **handling** prey (Curio 1976). Often the distinction between phases is blurred; for example, pursuit and attack may occur simultaneously, as can attack and capture. Structures employed in feeding and their functions are detailed in Chapter 3. An oft-cited adaptation in foraging contexts is the formation of groups, termed **shoaling** when swimming is unorganized, but referred to as **schooling** when individuals are polarized, swimming parallel and in the same direction (Pitcher & Parrish 1993). Groups function to both increase feeding success and deter predators, and function also changes at different phases of the predation cycle.

The critical tasks facing prey are to avoid detection, evade pursuit, prevent or deflect attack and capture, discourage handling, and ultimately escape from a predator. Just as predators have evolved different adaptations at different phases in the predation cycle, so have prey developed **antipredator tactics** that correspond to cycle phases. Many of these defenses are structural, involving modified body parts or adaptive use of coloration. Other defenses are behavioral, and many defenses combine actions with structures (e.g., Godin 1997a). Defenses generally function to **break the predation cycle**, the earlier the better: an attribute such as camouflage that makes it difficult for the predator to find the prey carries less risk of injury than an attribute such as toxic skin that deters the predator during handling. As an integration of topics covered in most of this chapter, we end with a brief discussion of the trade-offs fishes face when their own feeding activities expose them to the threat of predation.

Search

Predators

Predators can search for prey actively or passively. **Active search** implies locomotion while the predator scans the environment with any of the sensory modes discussed in Chapter 6. Water column searchers, such as herrings, anchovies, minnows, tunas, and billfishes rely heavily on vision, as do nocturnal plankton feeders. Olfaction, gustation, and hearing are also important for some water column searchers, particularly sharks. Low-frequency sounds of 20–300 Hz are especially attractive to sharks, whereas amino acids elicit feeding responses in many predatory fishes. Smell, taste, touch, or electrolocalization (passive or active) are employed extensively by benthic and nocturnal foragers such as eels, catfishes, gymnotid knifefishes, sea robins (triglids), goatfishes (mullids), and threadfins (polynemids), with polyodontid paddlefishes apparently using electrical cues to identify large concentrations of plankton. Chemoreception and touch are used by other groups that possess barbels, such as sturgeons, minnows, cods, and croakers. Some fishes search by speculation, rather than waiting until prey are detected. Goatfishes move along the bottom probing into sediments with their muscular barbels that are equipped with abundant taste receptors; some goatfishes flush prey by inserting their mobile barbels into refuge holes where prey have sought shelter (Hobson 1974). Boxfishes (Ostraciidae) and triggerfishes (Balistidae) expel jets of water from their mouths to blast sand away from buried prey. Logperch (Percidae) roll stones with their snouts in search of hidden insect larvae. These speculating foragers frequently have attendant species that follow them and snap up prey disturbed by the forager's activity.

The energy expended in active search can be saved by camouflaged predators that lie in wait on the bottom or in other structure. Such camouflage is often termed protective resemblance when hiding from predators, or aggressive resemblance when lying in wait (the latter usage is inaccurate behaviorally since "aggression" should be reserved for combat situations between animals, not for predatory activities). Benthic, camouflaged predators lie on rocks or soft bottoms or can be slightly (or greatly) buried by sediment. Their skin is colored to resemble algae-covered rocks, tunicates, sponges, and other bottom types. Wartlike and other fleshy outgrowths of skin and fins are common (Fig. 16.1). These fish rush explosively from the bottom to capture prey or open their typically large mouths rapidly and inhale prey, which is made easier by protrusible jaws (see Chapter 3, Jaw

FIGURE 16.1 "Aggressive" mimicry in the Striated Frogfish *Antennarius striatus*. The white, worm-like lure, or esca, sits at the end of the elongate first dorsal spine, termed the illicium. The lure is waved by movements of the illicium, thereby attracting potential prey fishes toward the mouth of the well-camouflaged frogfish. Jens Petersen / Wikimedia Commons / CC BY-SA 3.0.

Protrusion: The Great Leap Forward). Many scorpionfishes (Scorpaenidae), flatheads (Platycephalidae), seabasses (Serranidae), and hawkfishes (Cirrhitidae) rest exposed on the bottom, whereas lizardfishes (Synodontidae), stonefishes (synanceine scorpaenids), stargazers (Uranoscopidae), and flatfishes (Pleuronectiformes) lie with only their eyes exposed above the sediment. For such lie-in-wait predators, vision is the primary sense mode by which prey is detected, except for the elasmobranchs which may also use electrical cues. Many benthic, immobile ambushers appear surprisingly conspicuous, at least to a human observer. They may rely on prey habituating to their presence and thus growing careless.

Some water column predators, including countershaded or silvery-sided fishes such as gars (Lepisosteidae), pikes (Esocidae), and barracuda (Sphyraenidae), also lie in wait, floating motionless and darting at prey that fail to recognize them (Fig. 16.2).

FIGURE 16.2 Variations on a theme: convergence in morphology among fast start predators. Lurking predators that swim in the water column tend to be elongate with long mouths, sharp teeth, and fins set far back on the body. Examples from six different orders and eight families are shown, including one extinct form. (A) Lepisosteiformes, *Lepisosteus* (Lepisosteidae), gar. (B) Characiformes, *Ctenolucius* (Ctenoluciidae), Gar Characin. (C) Esociformes, *Esox* (Esocidae), pike artwork by Timothy Knepp, USFWS. (D) Beloniformes, *Ablennes* (Belonidae), needlefish. (E) Cyprinodontiformes, *Belonesox* (Poeciliidae), Pike Killifish. (F) Perciformes, *Sphyraena* (Sphyraenidae), barracuda, **https://commons.wikimedia.org/wiki/File:Barracuda_laban.jpg**. (G) Perciformes, *Luciocephalus* (Osphronemidae), Pikehead. (H) †Osteolepiformes, *Eusthenopteron* (Eusthenopteridae), a Devonian tetrapodomorph. (A) Brett Billings / USFWS / Public Domain; (B) Rufus46 / Wikimedia Commons / CC BY-SA 3.0; (C) Timothy Knepp / USFWS / Public Domain; (D) Hamid Badar Osmany / Wikimedia Commons / CC BY-SA 3.0; (E) L. G. Nico / USGS / Public Domain; (F) Laban712 / Wikimedia Commons / Public Domain; (G) Pikehead photo by BEDO (Thailand) / Wikimedia Commons / CC BY-SA 4.0; (H) Dr. Günter Bechly / Wikimedia Commons / CC BY-SA 4.0.

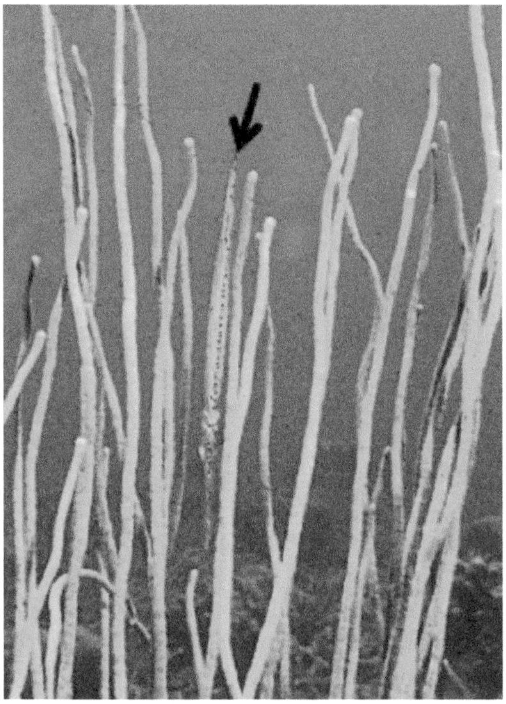

FIGURE 16.3 Some fishes have evolved morphological and behavioral traits that make them easy to mistake for other objects in their habitat. Here an Atlantic Trumpetfish (*Aulostomus maculatus*) (arrow) adopts a pale body coloration while hovering head-down among gorgonian corals. The trumpetfish even sways in unison with the corals as surges pass. G. Helfman (Author).

This group also includes substrate- and leaf-mimicking species such as trumpetfishes (Aulostomidae, Fig. 16.3) and leaffishes (Nandidae). Many predators shift among search patterns. Trumpetfish lie in wait among gorgonian corals to ambush roving prey, hide behind swimming herbivores such as parrotfishes, or swim actively in the water column and attack relatively stationary schools of zooplanktivores. By day, torpedo rays erupt from the sand at prey that have wandered over them, whereas at night they swim actively above the bottom in search of swimming or resting prey. Prey behavior and density often determine which search mode will be employed. For example, young lumpfish (Cyclopteridae) cling to rocks with their modified pelvic fins and make short excursions to feed on nearby zooplankton when prey densities are high. At low prey densities, the larvae swim through the water column searching for and feeding on plankton, thereby incurring the greater costs of active search but avoiding starvation (Brown 1986; Helfman 1990).

Zooplanktivorous fishes exhibit a variety of search tactics and detection capabilities. The electrosensitive rostrum of the Paddlefish (*Polyodon*) detects high densities of zooplankton and thereby provides information for opening the mouth to filter feed (see Chapter 6). Most fishes, however, lack this capability and swim through the water column visually scanning an area ahead of them that is shaped approximately like a hemisphere, the widest part being closest to the fish. The volume of this search space, the distance from objects at which fish react, and the size object that a fish is capable of detecting change with fish size, water clarity, illumination level, and velocity. Large juveniles can detect smaller objects than can small juveniles, and most fishes react farther away in clearer water or after light levels exceed some threshold value (Hairston et al. 1982). Zooplanktivores that feed in currents employ searching tactics that vary as a function of current speed. Fish remain in place and wait for food objects to approach them; upon detection, fish then swim toward prey at low current velocities (10–14 cm/s) but fall back with the current at higher speeds (McFarland & Levin 2002).

Reaction distance is heavily dependent on prey size, to the extent that most zooplanktivores will react to and pursue the largest appearing prey in their visual field. This means that a small zooplankter near a fish may be taken preferentially to a larger plankter farther away because the smaller prey appears larger (the **apparent size hypothesis**). However, prey immobility and location also affect selection, smaller prey being preferred if they are mobile or are more directly in front of the forager (O'Brien et al. 1985; O'Brien 1987). The speeds at which fish search vary as a function of fish size and food concentration and appear to approach the optimal in terms of maximizing intake relative to energy expense (Ware 1978; Hart 1993).

Cooperative Feeding

Various forms of group hunting are exhibited among some fishes. Apparent **cooperative feeding**, involving some form of coordinated herding or driving of prey by circling or advancing predators, has been observed in several shark species, including the Blacktip Reef, Lemon, and Oceanic Whitetip sharks (Carcharhinidae), sand tiger sharks (Odontaspididae), and thresher sharks (Alopiidae); the latter using their long caudal lobe for both herding and stunning prey (Motta & Wilga 2001; Motta 2004). Detailed observations of apparent cooperative feeding have been made on Sevengill Sharks, *Notorynchus cepedianus*, surrounding and then attacking seals off South Africa (Ebert 1991). These sharks form a loose circle around the prey, and the circle gradually tightens until one and then the group of sharks attacks the seal.

Cooperative feeding has been reported among other species, including piranhas, jacks, and Yellowtail, and some large pelagic predators such as billfishes and tunas hunt cooperatively in groups, herding prey species such as herring into "bait balls" near the surface and pursuing and then attacking the school of prey very effectively (Bigelow & Schroeder 1948b; Hiatt & Brock 1948; Voss 1956; Potts 1980; Partridge 1982; Sazima & Machado 1990; Steele & Anderson 2006). If the prey can be kept in a tight formation available for attack, much less time and energy is needed to search. The benefits of predators forming schools or shoals may be countered by **intragroup competition** for food, with competition increasing as the group size increases.

Grouped fishes may search more successfully than individuals. Foragers in groups may locate food sooner, ingest food faster, have more time available for foraging, and grow faster than solitary foragers. For example, in minnows (*Phoxinus phoxinus*, Cyprinidae), Goldfish (*Carassius auratus*, Cyprinidae), and Stone Loaches (*Noemacheilus barbatulus*, Cobitidae), shoal members spend less time before finding food than do solitary individuals, and the benefit increases with increasing shoal size (Pitcher & Parrish 1993). Accelerated rates arise because a fish in a shoal can search for food while simultaneously watching for signs of successful feeding in shoal mates, thus increasing the area over which it effectively searches. Also, the time each individual spends scanning for predators may decrease, leaving more time for feeding.

The weakly electric Cornish Jack (*Mormyrops anguilloides*) of sub-Saharan Africa hunts nocturnally and usually in groups. The fish use their electric organ discharges to locate prey and also maintain contact with one another. Foraging groups are often made up of the same individuals from one day to the next. This group hunting yields considerably better results for all fish than hunting alone (Arnegard & Carlson 2005).

In a particularly fascinating example of cooperative foraging, Bshary et al. (2006) documented Roving Coralgrouper (*Plectroporus pessuliferus*) recruiting the cooperation of Giant Moray (*Gymnothorax javanicus*) to improve foraging success in the Red Sea. The grouper hunts during the day, mainly near the bottom and close to coral reefs, but its prey often seek refuge among the crevices of the corals where the grouper is unable to follow. The moray lives among the crevices, but is typically nocturnal and rests among the crevices during the day. Grouper seeking assistance in hunting approach a resting moray, assume a head-down posture and shake their head from side-to-side. The moray leaves its resting cavity, follows the grouper, and the two hunt together. Prey that seek refuge in the crevices are pursued by the moray, which usually gets an easy meal but sometimes flushes the prey out of the corals where it is eaten by the grouper. The moray typically gets the prey more often than the grouper, but grouper were far more successful in this cooperative hunting arrangement than if hunting alone. By cooperating with the grouper, the moray received some relatively easy meals during the day, when it otherwise would be resting. In a variation of the recruiting behavior, sometimes a grouper that was hunting alone assumed the head-down, head-shaking behavior over a cavity into which a prey fish had sought refuge. This attracted the attention of a nearby moray, or sometimes another predator, which would explore the crevice and either get a meal, or flush the prey out where it was vulnerable to the grouper. In these examples, cooperative hunting was always initiated by the grouper and was more likely to occur if the grouper had not eaten within 2 h, suggesting that the incentive to recruit help may be driven, in part, by hunger.

Prey

Prey fishes can improve their chances of survival by either **avoiding detection** by the predator or **detecting the predator first**. In the former case, some form of camouflage is used; in the latter case, the all-important element of surprise is eliminated. Most fishes are both predators and prey, and must therefore simultaneously search for food and try to avoid being detected by predators.

Fishes that avoid detection by predators will not have to flee. Avoiding detection by visually hunting predators can be accomplished by either being unrecognizable, or becoming invisible with respect to the background. In both categories, deception is accomplished through either reduction of **photocontrast** with the background or **disruption of the outline** of the fish. A common form of the first tactic is called **protective resemblance,** in which a fish matches its background so well that the fish is no longer distinguishable. Resemblance is achieved through constant or variable coloration and epidermal body growths that match surrounding objects. As most predatory animals are highly sensitive to movement, protective resemblance is usually enhanced by immobility.

Many examples among fishes of remarkable resemblances to background structures can be given. Seahorses (Syngnathidae) and their relatives such as the Leafy Seadragon (Antennariidae), demonstrate some of the most spectacular examples of protective resemblance and readily "disappear" among seaweed (Fig. 16.4A). Clingfishes (Gobiesocidae), shrimpfishes (Centriscidae), and cardinalfishes (Apogonidae) have long black stripes and hover among the spines of sea urchins. Yellow-spotted Gobies (Gobiidae) with greenish bodies match both the green stalks and yellow polyps of the antipatherian sea whips on which they rest. Agonid sea poachers have rugose bodies covered in brown, orange, black, white, and red that match the sponge- and algae-covered bottom on which they are found. Green pipefishes (Syngnathidae) and wrasses (Labridae) live among green-stemmed sea grasses. Predatory fishes also employ protective resemblance and relative immobility as they lie in wait on the bottom for prey (or hide from their own predators). These predators include lizardfishes, goosefishes, stonefishes, scorpionfishes, toadfishes, sculpins (Fig. 16.4B), flatheads, and stargazers. Flatfishes are masters of camouflage, changing color and pattern to resemble a variety of bottom types (16.4C, D; Fujimoto et al. 1991).

An alternative to disappearing by blending into the background is to be obvious but to appear as an inedible object. This is achieved by mimicking distasteful or otherwise inedible organisms or objects. Juvenile sweetlips (Haemulidae) and batfish (Ephippidae) have the coloration and unfishlike swimming behavior of flatworms and nudibranchs, and juvenile burrfishes (Tetraodontidae) mimic opisthobranch mollusks. The invertebrates that are mimicked possess skin toxins, are brightly colored, behave conspicuously, and are avoided or rejected by most predatory fishes. Several small fishes mimic small floating sticks, blades of grass, or dead leaves, including juvenile needlefishes, halfbeaks, ephippid batfishes, lobotid tripletails, and adult nandid leaffishes (Randall & Randall 1960; Randall & Emery 1971; Moland et al. 2005; Randall 2005). Predation pressure is often strongest on young fishes, and many of these examples involve juvenile fishes that "grow out" of a mimetic stage and are less cryptic as adults.

Some special cases involve mimicry of dangerous fishes by otherwise harmless species. The plesiopid *Calloplesiops*

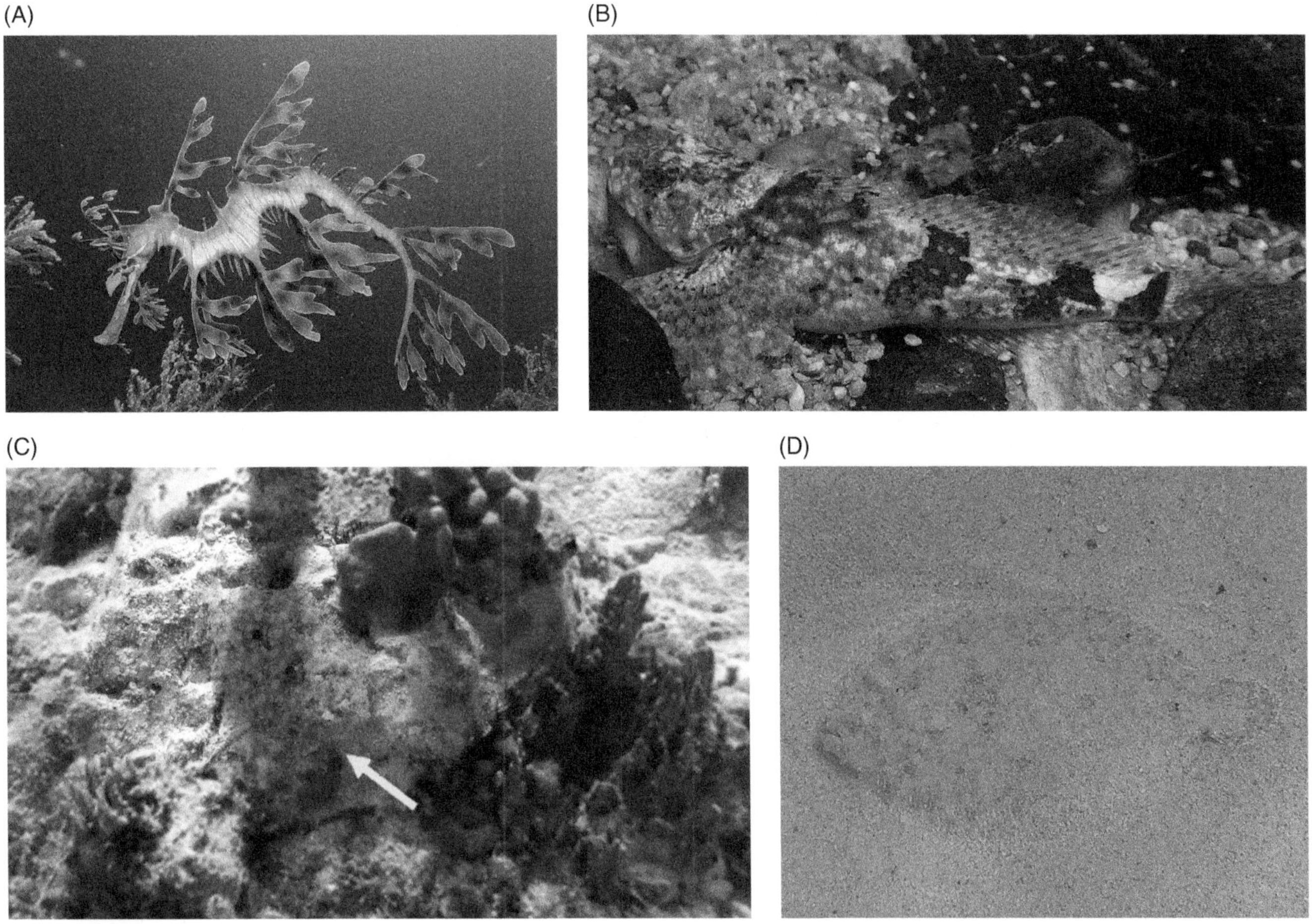

FIGURE 16.4 Some fishes avoid detection by predators by blending in with objects in their surrounding habitat. (A) A Leafy Sea Dragon (*Phycodurus eques*) hovering amongst vegetation in southern Australia. (B) A Banded Sculpin (*Cottus carolinae*) blends into the substrate in the Canasauga River. (C, D) Flatfishes can adjust their coloration to match background, and will also cover themselves with substrate to further blend in to the habitat. The yellow arrow in B points to the head. (A) James Rosindell / Wikimedia Commons / CC SA-4.0; (B) Courtesy of A. Nagy; (C and D) Photos by D. Facey (Author).

altivelis has a dark body with small white spots and a white-ringed ocellus or eye-spot at the posterior base of its dorsal fin (Fig. 16.5). When frightened, it swims into a crevice but leaves the posterior portion of its body in the open, expanding its dorsal, caudal, and anal fins. In this posture, it appears remarkably similar to the protruding head, eye, and mouth of a predatory Moray eel, *Gymnothorax meleagris* (Muraenidae), and may thus intimidate potential predators (McCosker 1977).

Disruptive coloration is another means of visual deception. Vertebrates recognize organisms by their outlines and by the gradual shading differences that exist among regions and features within an outline. A disruptively colored fish has areas of contrasting color, usually black and white, on the body that break up the outline of the fish, making it unrecognizable. This is one explanation for the bold coloration of some reef fishes such as humbug damselfishes, Rock Beauty Angelfish, and some croakers, as well as the vertical barring and dorsal spots of many shallow water species (sculpins, sunfishes, darters, cichlid angelfishes, jacks, barracudas, and tunas) that will be viewed by the flickering light of vegetated areas or created by wavelets passing overhead (McFarland & Loew 1983) (Fig. 16.6A). Even strikingly colored reef fishes, along with their less colorful, shallow water counterparts on reefs and in temperate lakes and kelp beds, assume a dark and light, blotchy coloration when resting at night. This disruptive pattern presumably breaks up their outlines and makes them more difficult to discern at low light levels. Disruptive coloration is also a reasonable explanation for the common occurrence of split-head coloration in many lurking predators (Fig. 16.6B).

Invisible Fishes

Protective resemblance and disruptive coloration are camouflage tactics seen in many organisms in a variety of habitats (Cott 1957; Edmunds 1974; Lythgoe 1979). Coloration that makes an animal literally disappear from view, rather than blend in with its background, exploits unique features of the distribution of light underwater. In air, all portions of the visible spectrum, from deep blue to deep red (*c.* 400–700 nm) are well represented. Hence objects of all possible colors can be found in

FIGURE 16.5 The Comet (*Calloplesiops altivelis*) is a tropical reef fish with a spotted pattern and eyespot near the tail. A, Ewen Roberts / Wikimedia Commons / CC BY 2.0. When a Comet is threatened, it hides its head in the surrounding reef and leaves its tail exposed. B, Vincent C. Chen / Wikimedia Commons / CC BY-SA 4.0. The tail, along with the eyespot, resembles the head of the White-mouthed Moray (*Gymnothorax meleagris*, C, David Burdick, NOAA / Wikimedia Commons / Public Domain.

most habitats. In addition, brightness varies substantially and irregularly as a function of sun and viewing angle. The brightest part of the sky may be the horizon during early morning and late afternoon. The ground, vegetation, or objects below an observer or viewed laterally may be as bright, or brighter, than objects overhead.

In water, however, light has a much more predictable distribution, particularly in open water situations where the bottom is not visible. Sunlight is refracted at the water's surface, being bent downward even at relatively low sun angles. Hence the brightest light is consistently directly overhead, or **downwelling**. Water molecules act as a powerful filter, both **absorbing** and **scattering** light; light attenuation is even stronger if dissolved or suspended particles that cause turbidity are present. Since all light underwater, with the minor exception of bioluminescence, originates as sunlight, objects viewed from above or horizontally will reflect light that has passed through the water filter. **Upwelling** light consists of light photons that have passed down and then back up again through the water column and is the weakest component; upwelling light is typically only about 1% as strong as downwelling light. Horizontal **space light** is intermediate in strength, but again consists of light that has first passed vertically down and then horizontally through the water; space light is on average about 5% as strong as downwelling light.

The attenuation of light with depth is also very **symmetrical around the vertical**, which means that a diver measuring light at 45° from the vertical will see the same quantity and quality of light whether the meter is pointed at a 45° angle north, south, or in any other direction. Similarly, light measured 90° off vertical (i.e., horizontally) will be identical ahead of and behind a viewer. These two physical characteristics of light in water – **uniform reduction with depth** and **uniform attenuation around the vertical** – are primary influences on fish coloration, particularly in the context of camouflage tactics that render the fish invisible.

Invisibility can be accomplished via three mechanisms: countershading, silvery sides, and transparency. **Countershaded** fishes grade from dark on top to light on the bottom. The actual color or shade of the fish is less important than: (i) the strength of the light that the fish reflects, which differs at different angles from the horizontal; (ii) the background light against which the fish will be compared, and (iii) the viewing angle of the observer. Countershading is easiest to understand when viewing a fish from above. A fish with a dark dorsum absorbs bright downwelling light, thus presenting a dark target

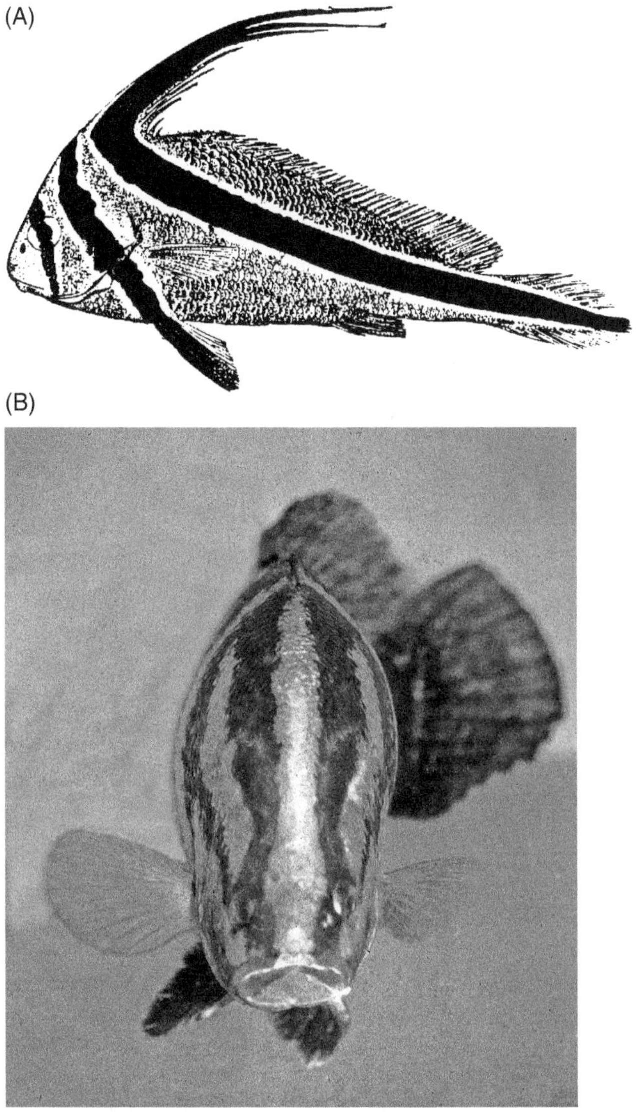

FIGURE 16.6 Examples and functions of disruptive coloration in fishes. (A) The Jacknife Fish, *Eques lanceolatus*, may use boldly contrasting, dark and light regions to emphasize those parts of its outline that are not fishlike in appearance, thus momentarily confusing a potential predator. (B) Many lurking predators possess a dark or light interorbital stripe that could disrupt their head outline when viewed head-on, making recognition by prey momentarily difficult. The Japanese Snook (*Lates japonicus*) has a distinctive, light-colored stripe along the midline of the head. (A) From Cott (1957), used with permission; (B) Courtesy of J. DeVivo.

against the dark background of dim upwelling light. However, most fishes are viewed by their predators or prey from the side, and here the intricacies of countershading work best. When viewed from slightly above the horizontal, the darker dorsolateral surface of the fish absorbs relatively bright downwelling light, creating a dark target that is seen against the darkened background of slightly upwelling light. Similarly, if viewed from slightly below the horizontal, a light-colored ventrolateral surface reflects weak upwelling light, creating a relatively bright target seen against the lighter background of slightly downwelling light. A countershaded fish disappears into the background because the gradation of its color is opposite to the distribution of light in water, which creates a target that is identical to the background. The fish reflects light that is roughly equivalent to the background against which it is seen at all viewing angles, dark against dark, light against light, and intermediate against intermediate. The same effect is reached in the mesopelagic region by fishes with dark backs and ventral bioluminescence (discussed in Chapter 10).

Reverse countershading occurs in fishes, but these exceptions prove the rule that countershading is camouflage. Many male fishes, such as sticklebacks, sunfishes, cichlids, and wrasses, take on bright dorsal or dark ventral colors during the breeding season, a time when conspicuousness helps them attract females and repel territorial intruders. The best proof-by-exception comes from the reversely countershaded mochokid upside-down catfishes which feed on the undersides of leaves and even swim in open water in an upside-down orientation. A predacious Lake Malawi cichlid, *Tyrannochromis macrostoma*, exhibits reverse countershading and often attacks prey while upside down (Stauffer et al. 1999). Colorful reef fishes often superimpose their bright coloration over a countershaded body and vary the dominant color pattern depending on whether they are engaged in social interactions or avoiding predators.

It is not immediately obvious that silvery sides make a fish invisible. **Mirror-sided** fishes include some of the world's most abundant and commercially important species, including herrings, anchovies, minnows, salmons, smelts, silversides, mackerels, and tunas. These and other mirror-sided fishes are predominantly open water, pelagic species that take advantage of the unique light conditions that prevail underwater. To understand how mirror sides work, one must imagine a piece of plate glass suspended in midwater (Fig. 16.7A). The glass is invisible because the background light passes right through it; an observer sees the water column background, not the glass. Because light attenuates uniformly with depth and is distributed symmetrically around the vertical, a flat mirror suspended underwater achieves the same effect as clear glass. The mirror reflects light of an intensity and color that is identical to the light that would be passing through a piece of glass suspended at the same locale, i.e., as if the mirror were not there. Light coming from a 45° angle above the horizontal and reflecting off the mirror and into the eyes of an observer located 45° below horizontal is identical to light that would pass through the mirror if it were clear glass. An observer comparing the light reflected off the fish with the background light sees no difference; the mirror, or fish, consequently disappears into the background.

Crucial to the function of mirror sides is that the fish maintains a vertical orientation at all times, since any deviation from verticality will reflect light that is either brighter or darker than the background (Johnsen & Sosik 2003). Anyone who has watched a school of forage fishes has witnessed periodic bright flashes as individuals deviate from vertical swimming. Mirror-sided fishes maximize verticality by being laterally compressed. The guanine and hypoxanthine crystals that actually reflect the light are embedded in the scales and skin and are stacked together in platelets. The reflecting crystals are separated by a

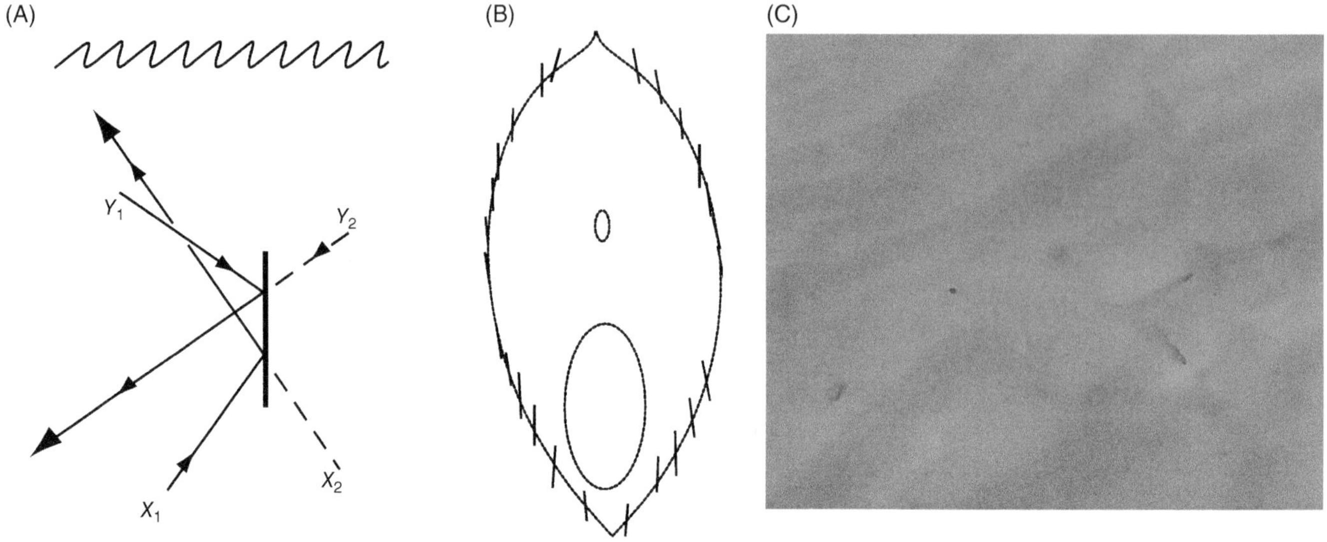

FIGURE 16.7 The functional morphology of mirror sides in fishes. (A) A clear plate of glass suspended in water is invisible because background light passes directly through it (extensions of dashed lines X_2 and Y_2 to the observer's eyes); an observer sees no difference between the glass plate and its background. A mirror suspended in water also disappears because light reflected off the mirror (solid lines X_1 and Y_1) is identical to the background light that would pass through the object if it were clear (dashed lines). (B) Cross-section through the body of a Bleak (*Alburnus alburnus*) to show orientation of the reflective platelets in silvery fishes. The platelets are embedded in the skin and scales and are oriented vertically, even along the curved surfaces of the fish. After Denton and Nicol (1965). (C) Reflective sides, faint wavy vertical bars, and countershading help this jack (Carangidae) blend into the sandy background. D. Facey (Author).

space equal to about one-quarter of the wavelength of the light usually reflected, the theoretical optimum spacing for achieving reflectivity. Whereas the scales and skin conform to the curvature of the body, the **reflecting platelets** in the scales have a vertical orientation, even in regions where the body is curved (Denton & Nicol 1962, 1965; Denton & Land 1971) (Fig. 16.7B).

The third means of achieving invisibility is via relative **transparency**. This is a characteristic of fishes that live in very clear water immediately below the surface where the effects of sun angle are strongest and symmetrical distribution around the vertical is weakest. Halfbeaks and needlefishes fall into this category, along with some specialized freshwater forms such as the X-ray Tetra, *Pristella maxillaris*, and Glass Bloodtail, *Prionobrama filligera* (Characidae); African and Asian Glass Catfish (Schilbeidae, Siluridae); bagrid catfishes in the genera *Chandramara* and *Pelteobagrus*; a gymnotid knifefish, *Eigenmannia*; and the Glass Fish, *Chanda* (Ambassidae).

Larvae and young juveniles of many fishes are pelagic and transparent, their pigmentation developing along with their later habitat preferences (see Chapter 9). The body musculature and to some extent the bones of such fishes are translucent and these fishes are consequently difficult to see. However, certain structures, most notably the eye, brain, and gonads, apparently cannot function in a transparent state; the gut also often contains prey that is opaque or quickly turns opaque when digested. These pigmented organs often have a silvery coating. The function of this silver film is poorly understood. It might reflect light as do the silver platelets in scales as described above, or the coating could shield delicate structures from harmful ultraviolet radiation that penetrates into clear, shallow water.

Early Detection

Predator–prey interactions often occur over a period of tens of milliseconds. With such rapid reaction times, predators generally require an **element of surprise** to be successful. The element of surprise can be eliminated if prey detects the predator before the predator detects the prey, or at least if the predator is seen before it gets within striking distance. Early detection can be achieved through the collective vigilance of a shoal: fish in shoals detect an approaching predator more quickly than do solitary fish (Magurran et al. 1985; Milinski 1993).

In open water, the direction and angle of the sun can have a significant impact on vision and visibility, especially near the surface. This may be the reason that the majority of attacking approaches by White Sharks (*Carcharodon carcharias*) at the surface come with the sun behind the shark, although these sharks approached potential prey from any direction when investigating (Huveneers et al. 2015). Attacking from the direction of the sun would likely make the prey easier for the shark to see and also make it difficult for prey looking toward the sun to see the approaching predator. On sunny days, the direction of the attacks shifted from morning to afternoon based on the position of the sun, but under cloudy conditions attacks came from any direction. The pattern was not seen in all sharks observed, however, so there does appear to be some interindividual variation.

Another example of a predator becoming "invisible" is demonstrated by needlefishes (Beloniformes), which typically lurk and hunt at the water's surface. In addition to attacking small fish prey from the side, they also leap from the water briefly and attack from above. The attack is typically at a low angle which may make the attacking predator more difficult to see by the prey because the low angle of attack would cause light to reflect off the surface rather than penetrate into the water (Day et al. 2016).

Many shoals, as well as solitary fish, gain a **relative visual advantage** over approaching predators by hovering under structures such as floating logs or vegetation, undercut banks or coral ledges, or overhanging trees or artificial structures such as docks and bridges. Under appropriate conditions, a shaded fish can see an approaching fish in sunlight as much as twice as far away as the sunlit fish can see the shaded fish (Fig. 16.8). This phenomenon can be easily experienced by a diver approaching a ledge or dock; objects in the shadow of the ledge are difficult to discern until the observer swims into the shade of the object. The implications for predators lurking or prey hiding in shade are obvious; a predator approaching a shaded prey fish will lose all of the advantages of surprise because the predator will be spotted long before it can see the prey. Prey careless enough to pass near shaded structure are likely to be captured by predators lurking undetected within the darkened region.

The visual advantages and disadvantages around shade arise as a result of physical and physiological phenomena associated with the way in which the vertebrate eye responds to stimuli of different strengths. Sensitivity of the eye to light is determined by the brightest features of the environment, so dimly lit objects are difficult to see against a bright background. An observer in a darkened room is hidden from the view of people in sunlight outside the room, but the sunlit people are easily seen by the shaded observer. In water, this effect is heightened by turbidity. As light enters water, it reflects off particles such as phytoplankton and silt. When looking horizontally underwater on a sunny day, particles are particularly obvious as bright blotches. Particles closest to the observer's eyes are the brightest because the light reflected off them travels the shortest distance. Turbidity therefore creates a bright region adjacent to the observer, a **veil of brightness** that intervenes between the eye and more distant targets (Lythgoe 1979). The eye quickly adapts to this bright region, and objects farther away in the darker background of space light become more difficult to detect (Fig. 16.9).

The eye's adaptation to bright background and veiling brightness combine to explain the relative visual advantage of a

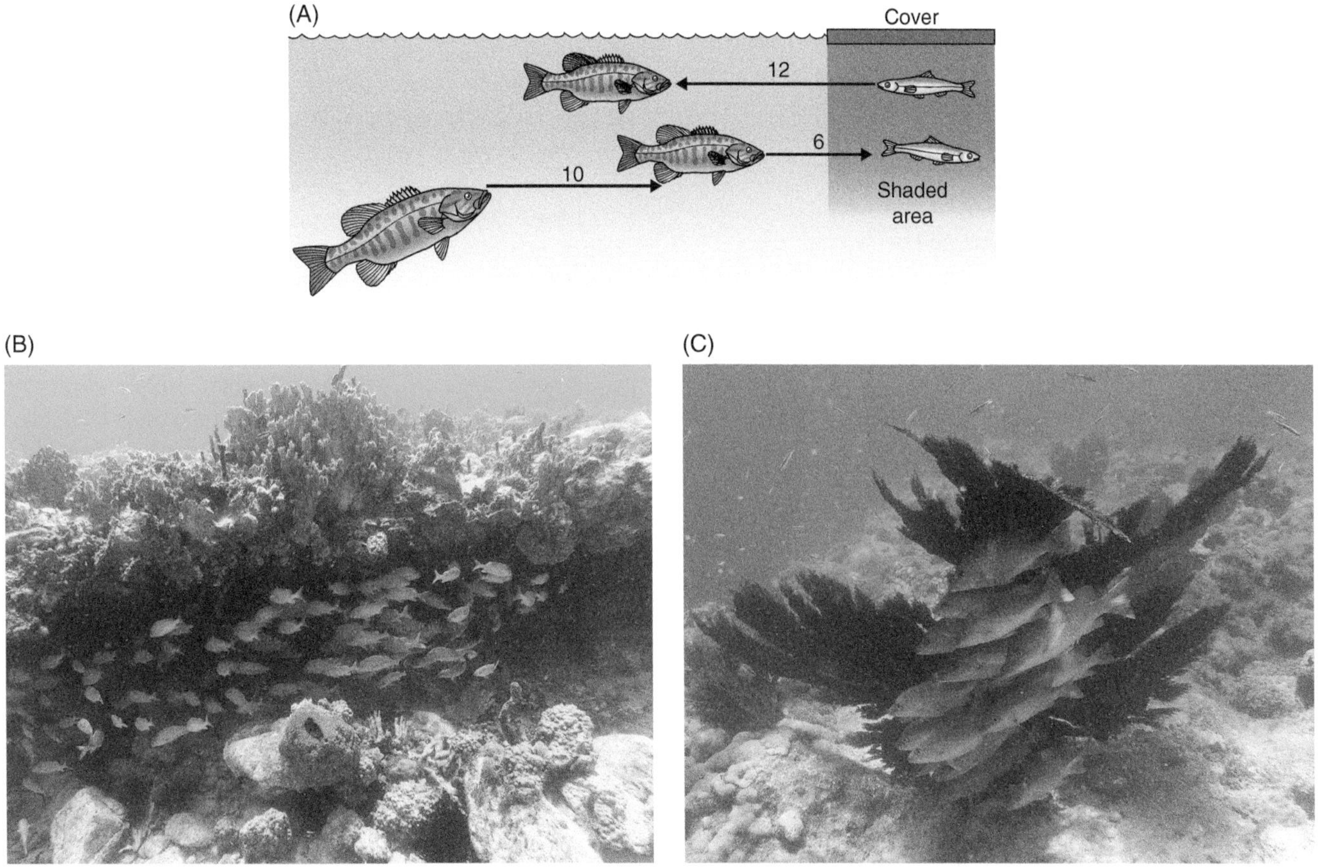

FIGURE 16.8 (A) The advantage to fishes of hovering in shade. On a sunny day in a lake, a shaded observer has a relative visual advantage over sunlit observers. When horizontal visibility is 10 m, a shaded observer can detect sunlit objects 12 m away, which is approximately 1.2 times better visibility than experienced by a sunlit observer viewing a sunlit target (= a 20% advantage over ambient conditions). More significantly, the sunlit observer cannot see an object in the shade until it is 6 m away, which gives the shaded observer a 100% advantage over the sunlit observer. The relative visual advantage decreases on cloudy days, as does the attractiveness of overhead objects. (A) After Helfman (1979b, 1981a); (B, C) Coral reef fishes congregate in shaded portions of the reef on a sunny day; Photos (both B and C) by D. Facey (Author).

FIGURE 16.9 Countershading, silvery sides, and turbidity from phytoplankton and suspended particles from wave action make Atlantic Tarpon (*Megalops atlanticus*) difficult to see in the shallow waters of the Dry Tortugas. D. Facey (Author).

fish in shade. Hovering in shade is a tactic commonly employed by resting fishes (Fig. 16.8B, C). Many nocturnally active fishes form **daytime resting schools** in shaded regions by day, including various herrings, silversides, squirrelfishes (Holocentridae), glasseyes (Priacanthidae), snappers, and copper sweepers (Pempheridae). Diurnally active fishes also hover in shade when resting, including various suckers (Catostomidae), centrarchid sunfishes, jacks (Carangidae), and goatfishes (Mullidae). The relative advantage accrues to predators as well as prey and it is not unusual for solitary, lurking predators to hover in shaded areas and strike at prey that pass by (e.g., trout, pickerel, snook (Centropomidae), Largemouth Bass, and barracuda).

Shoaling and Search

The antipredation benefits of group formation apply to all phases of the predation cycle, including search. Fish in a shoal have a lower probability of being found by a predator than the same fish distributed solitarily (Brock & Riffenberg 1960). Shoals are undoubtedly more conspicuous than solitary fish, so provide no camouflage value. However, depending on distance and visibility a shoal may be mistaken for a large fish and therefore be avoided by an approaching predator (Pitcher & Parrish 1993). Shoal formation is probably common in prey fishes because of the necessity to move and find food, particularly among herbivorous and planktivorous fishes. Highly evolved protective resemblance is not an option for such fishes; hence group formation is an alternative.

Upon detection of a predator, fish in shoals typically shift to **polarized, schooling tactics**. Behaviors are emphasized that preserve the integrity of the threatened group (Pitcher & Parrish 1993). Subgroups stream toward the main group (but move as coordinated units, not as individuals), interindividual distances decrease, and movements become synchronized among school members. Heterospecific shoals (those containing more than one species) sort out by species, conspecifics associating with individuals of their own species and size. If few conspecifics exist, members of the minority species may seek shelter rather than wind up as the odd members of a school (e.g., parrotfish, Scaridae; Wolf 1985).

In some situations, members of the prey group may move away from the shoal, approach the predator, and then return to the shoal. These **predator inspection visits** have been witnessed in Mosquitofish and Guppy (Poeciliidae), sticklebacks (Gasterosteidae), Bluegill (Centrarchidae), and gobies (Gobiidae). The behavior may: (i) allow prey fish to assess the identity, motivational state, or other traits of the predator; or (ii) inform the predator that it has lost the element of surprise and that an attack is unlikely to be successful (Magurran 1986b).

Prey can also discourage a searching predator by behaving aggressively. Several prey species actually attack potential predators and drive them from the area. This behavior, best known from bird studies and commonly called **mobbing**, has been documented for individuals or groups of squirrelfishes, snappers, grunts, goatfishes, butterflyfishes, damselfishes, wrasses, and surgeonfishes interacting with predatory moray and snake eels, lizardfish, trumpetfish, scorpionfish, stonefish, flatheads, barracuda, and flatfish, and for Bluegill, Longear Sunfish, and Largemouth Bass interacting with turtles and water snakes. Mobbing fish may contact the head or tail of the predator, or may display in front of the predator by swimming in place and erecting dorsal spines and rolling the body. Mobbing reduces the predation rate in an area because mobbed predators take longer to return to an area than do predators that are ignored (Motta 1983; Ishihara 1987; Hein 1996). Predators may leave an area because the physical attacks of the mobbing fish are injurious or because the actions of the mobbers notify other prey individuals to the presence of the predator, which lowers the predator's potential success in the area, analogous to the alarm calls of birds and small mammals (Helfman 1989).

Either inspection or mobbing might explain why some prey converge on or follow predators immediately after a successful attack on the group. This action has been observed in Yellow Perch attacked by pike, in snappers attacked by jacks, in Bluegill attacked by pickerel, in territorial damselfish attacked by several predators, and in planktivorous damselfish attacked by trumpetfish (Nursall 1973; Potts 1980; Dominey 1983; Ishihara 1987; G. S. Helfman, pers. obs.).

The focus of this discussion has been on avoiding detection by visual predators. However, many nocturnal predators and those that live in turbid habitats rely heavily on acoustic, bioelectrical, and chemical cues to find prey. Pacific Herring, *Clupea pallasii*, respond to sounds such as those emitted by echolocating dolphins by ceasing to feed, dropping in the water column, and schooling actively; fish already in schools drop in the water column and increase their swimming speed (Wilson & Dill 2002). Another clupeid, the American shad, *Alosa sapidissima*, first moves away from an echolocation sound and then swims erratically if the sound strengthens (Popper et al. 2004).

Prey species may change their behavior based on prior exposure to sensory cues from predators. Juvenile damselfish,

Pomacentrus wardi, that had been trained in the lab to recognize sight and odor of predators were less likely to take risks, and therefore more likely to survive, when released to a reef (Lönnstedt et al. 2012). Hunger played a role, however, with well-fed damselfish taking less risk than hungrier individuals.

Pursuit

Predators

Pursuit places a predator close enough to attack prey, and two dramatically different categories of pursuing predators have evolved – each of which has evolved independently in many different lineages. One strategy focuses on maximizing speed while overtaking fleeing prey; the other requires minimal aerobic output but a proliferation of deceptive tactics followed by a brief burst of activity resulting in capture.

Cursorial, chasing predators are capable of high-speed sustained chases of rapidly swimming prey. Morphologically, these are the most streamlined fishes, having bodies that are round in cross-section and taper to a thin, laterally keeled caudal peduncle, with the greatest body depth one-third of the way back from the head. The tail is narrow with a high aspect ratio (height : depth, see Chapter 4, Locomotion: Movement and Shape), and the median and paired fins typically fit into grooves or depressions during high-speed swimming. These fishes include the apex pelagic predators (lamnid sharks, tunas, and billfishes). Nearer to shore, where prey can escape into structure, streamlining is sacrificed to allow for rapid braking and improved maneuverability. Bodies are more oval in cross-section, fins are larger, and tails broader. Examples include salmons (Salmonidae), snook (Centropomidae), Striped Bass (Moronidae), black basses (Centrarchidae), and large-bodied cichlids (e.g., Peacock Bass).

In **lurking** or lie-in-wait predators that swim above the bottom, pursuit is synonymous with attack. These fishes have converged on a general body morphology that permits fast starts at a sacrifice in sustained speed and maneuverability. These fishes have elongate, flexible bodies; long snouts with many sharp teeth; broad, symmetrical dorsal and anal fins placed far back on the body opposite one another; and relatively large caudal fins with low aspect ratios. The group includes gars, pikes and pickerels, needlefishes, barracudas, and specialized fishes in such diverse families as the characins and cichlids. The low body profile of these fishes serves additionally in evoking a slower response by prey than do predators with deeper body profiles (see Dominici & Blake 1997). These fishes typically hover high in the water column or lurk motionless on the edges of vegetation beds, relying on their camouflage to gain access to prey. Additional piscivores that have converged on this morphology include the Australian endemic Long-finned Pike, *Dinolestes lewini* (Perciformes, Dinolestidae), and some of the world's largest minnows (Cyprinidae), such as the Colorado Pikeminnow, *Ptychocheilus lucius*, of North America, and the Kanyu or Yellowcheek, *Elopichthys bambusa*, of Asia – both of which can exceed 2 m and 40 kg.

A third general category of predator does not involve pursuit, but instead depends on luring prey within range of being engulfed. These fishes are typically benthic, and luring can involve all or part of the predator's body (e.g., Randall 2005). In goosefishes, anglerfishes, and frogfishes (Lophiiformes; discussed in Chapter 10), the first dorsal spine is elongated and its end is highly modified into a species-typical **esca** or lure that resembles a small fish, shrimp, or worm which is wriggled in a lifelike manner to attract potential prey (Fig. 16.1). The body of the predator is camouflaged to resemble the bottom or, in the case of deep-sea anglers with bioluminescent lures, the dark surrounding waters. Small fishes approach the lure and are quickly inhaled by the large mouth, and their escape is often prevented by long, backward facing teeth (Pietsch & Grobecker 1978, 1987).

Luring has evolved independently in other groups of fishes. A scorpionfish (Scorpaenidae) also uses a modified dorsal spine for a lure; hatchetfishes (Sternoptychidae), lanternfishes (Myctophidae), some anglerfishes (Ceratioidei), and stargazers (Uranoscopidae) have lures in their mouths; barbeled plunderfishes (Artedidraconidae) use chin barbels, chacid catfishes use maxillary barbels, and snake eels (Ophichthidae) have a lingual (tongue) lure; and in gulper eels (Eurypharyngidae) the tail tip is illuminated (Randall & Kuiter 1989).

Nimbochromis livingstonii, a large predatory cichlid from Lake Malawi, Africa, capitalizes on the tendency of many small cichlids to scavenge on recently dead fishes. The predator lies on its side on the bottom and assumes a blotchy coloration typical of dead fish. When scavengers come to investigate and even pick at its body, the predator erupts from the bottom and engulfs them. This is the only known example of **death feigning** (thanatosis) in fishes (McKaye 1981a).

Camouflage helps make some predators invisible to prey. Stonefish (Synanceiidae) lie buried on the bottom and strike at prey fishes in a narrow zone directly above their mouths. If a potential prey fish swims between the mouth and the dorsal fin, the stonefish will raise its dorsal fin, herding the fish back into the strike zone (Grobecker 1983).

Approach to a prey fish is also facilitated by camouflage. Although most predators and their prey are countershaded or silvery, such coloration disguises a fish only when it is seen from the side. This is not the view that a prey fish has of an approaching predator. A convergent coloration trait shared by slow-stalking predators is the **split-head color pattern** (Barlow 1967). A dark or light line that contrasts with general body coloration runs from the tip of the snout along the midline between the eyes to the top of head or dorsal fin (see Fig. 16.6). This coloration is evident in pickerel (Esocidae), some soapfishes and seabasses (Serranidae), the tigerperch *Datnioides* (Lobotidae), the Leaffish, *Polycentrus schomburgkii* (Polycentridae), and some hawkfishes (Cirrhitidae), piscivores which are otherwise protectively colored and which approach their prey slowly and head-on. The split head pattern seems to be a form of **disruptive coloration**, dividing the head into halves and disrupting its outline. Prey may consequently fail to recognize the pattern as the head of a predator. Because predator–prey interactions occur on a timescale of tens of milliseconds, a moment's delay in recognition may be all that a predator requires to attack successfully. Prey fishes are often frightened by general, head-on, facial characteristics of predators (e.g., Dill 1974; Karplus et al. 1982), hence disguising the face would eliminate critical cues used in predator recognition (Fig. 16.10).

Prey

Once a predator finds and recognizes prey, pursuit may begin. Antipursuit tactics involve discouraging the predator due to real or feigned unpalatability, shelter seeking, outdistancing or outmaneuvering the predator, or disappearing into the background.

Some fishes possess stout, sharp, sometimes poison-laden spines that can be used in defense against predators. Others possess toxic chemicals in their skin and internal organs. Noxious prey typically advertise their unpalatability with coloration and movement that make them and their defenses quite evident. Such **aposematic** ("warning") coloration or behavior is typical of animals that are dangerous or inedible (e.g., many bees, wasps, caterpillars, butterflies, bufonid toads, porcupines, and skunks). Deliberate, slow movements aid learning of the warning signal without eliciting attack from predators that are otherwise conditioned to pursue rapidly fleeing prey. By advertising their inedibility, prey short-circuit the predation cycle at an early phase, saving the energetic costs of flight and the

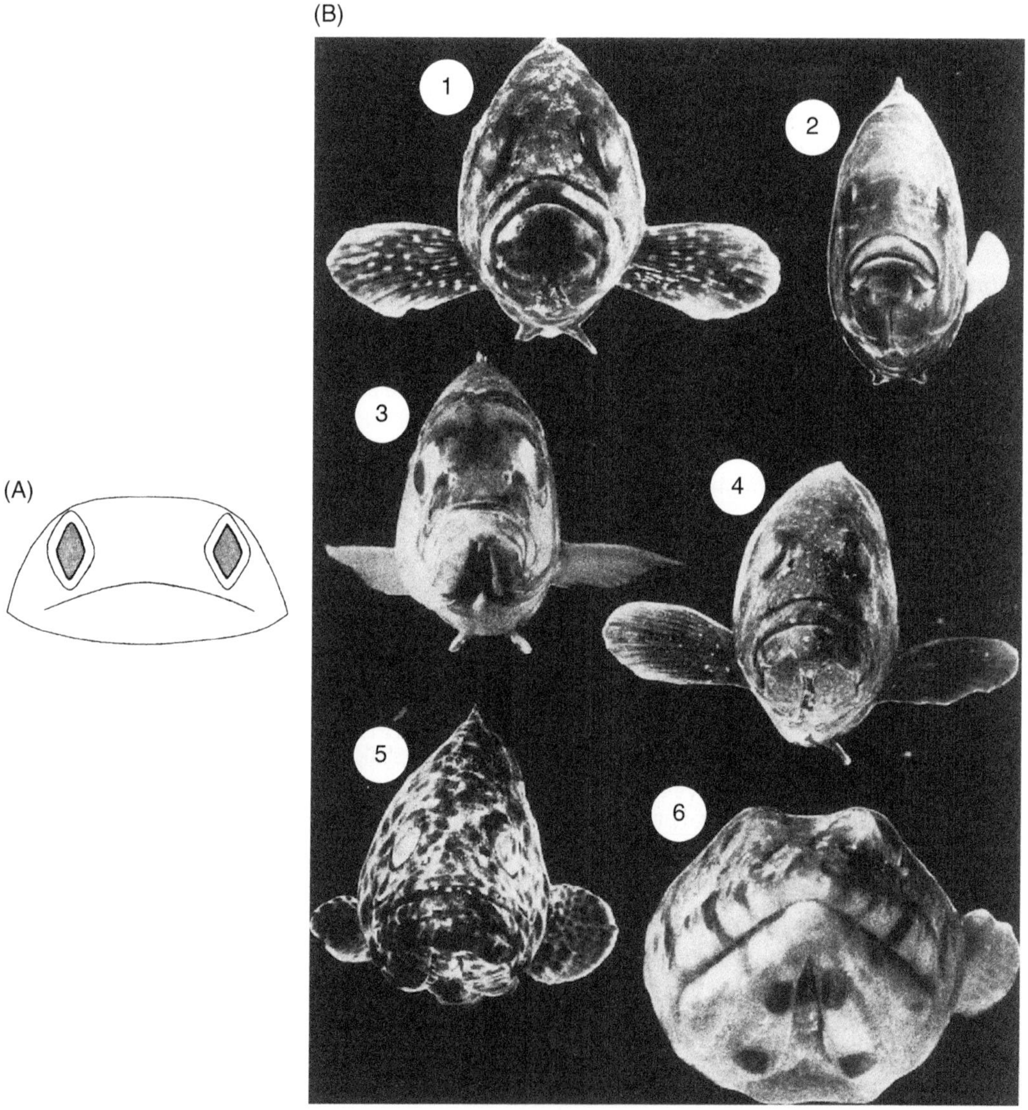

FIGURE 16.10 Is this the face of death? (A) The general features of a predatory face that elicit fright responses in prey fishes are a broad head; a wide, downturned mouth; and ringed, broadly elliptical eyes, as shown in drawing on the left. (B) Head-on views of piscivorous reef fishes: 1, *Epinephelus summana*, a seabass; 2, *Cheilinus trilobates*, a wrasse; 3, *Lutjanus kasmira*, a snapper; 4, *Cephalopholis argus*, a seabass; 5, *Epinephelus fario*, a seabass; 6, *Synodus variegatus*, a lizardfish. (A) from Karplus et al. (1982); (B) Karplus and Algom (1981) / with permission of John Wiley & Sons.

possible injury costs of being handled; predators in turn save time and energy and avoid potential injury or possible death.

Many fishes with obvious defenses have bold or bright coloration that could provide a warning to predators. The scalpel-like enlarged scales in the caudal peduncle of surgeonfishes (Acanthuridae) may be surrounded by a bright yellow or orange patch. Lionfish (Pteroidae) have contrasting red, black, and white fins and a stately posture that accentuates their poisonous fin spines (Fig. 16.11). Weeverfishes (Trachinidae) erect their dark-colored, highly venomous dorsal fins when disturbed. Many pufferfishes (Tetraodontidae, and other tetraodontiforms), including the famous *fugu* puffers served in exclusive Japanese restaurants, contain powerful **tetrodotoxins** in their skin, liver, and gonads. These fishes have contrasting rather than countershaded body markings and move about in exposed locations during the day. Whereas most catfishes are inactive during the day, some plotosid marine catfishes with exceedingly powerful spine venoms and contrasting dark and light coloration shoal conspicuously during the day.

Evolution of enlarged lower canine teeth of fangblennies (tribe Nemophini) may have initially provided a benefit for micropredation, such as feeding on the scales and fins of larger fishes, but subsequent evolution of grooves in the teeth accompanied by venom glands also provide a defense mechanism (Casewell et al. 2017). This venom does not appear to inflict discomfort or pain, but instead interacts with opioid receptors and produces strong hypotensive effects including dizziness, as well as mild neurotoxic effects, perhaps creating a sufficiently disorienting and unpleasant experience that the affected predator would learn to not select fangblennies as prey in the future. Whatever the mechanism, it is apparently effective – supported in part by the evolution of Batesian mimicry by several nonvenomous fishes.

Many zooplanktivores take shelter in bottom structure when pursued by predators. Hence, anthiine serranids, fusiliers (Lutjanidae), butterflyfishes (Chaetodontidae), damselfishes (Pomacentridae), wrasses (Labridae), and surgeonfishes (Acanthuridae) dive toward the refuge of coral when disturbed. Morphology and behavior correlate strongly with vulnerability among these fishes. Small fishes feed closer to the bottom to compensate for their slower swimming speeds. Species that forage farther from the bottom tend to have more fusiform bodies and more deeply forked tails, both characteristics of faster swimming fishes (Davis & Birdsong 1973; Hobson 1991) (Fig. 16.12). Many fishes are permanently associated with holes, cracks, or tubes in the bottom to which they retreat when threatened (e.g., garden eels, Congridae; jawfishes, Opisthognathidae; tilefishes, Malacanthidae; tubeblennies, Chaenopsidae; hover gobies, Gobiidae). A few, such as the razorfishes (*Hemipteronotus*, Labridae), dive into sand with incredible speed. Fishes associated with macrophyte beds in lakes and on reefs typically use the vegetation as shelter when pursued (e.g., centrarchid sunfishes, cichlids, rabbitfishes, and kelpfishes).

Evading a pursuing predator in open water requires superior speed or maneuverability. The odds here favor the predator, since predators are typically larger than their prey and larger fishes can usually swim faster. Some empirical

FIGURE 16.11 Lionfishes, native to Indo-Pacific regions, are among the few fishes that are thought to be colored to inform other species of their toxic characteristics, in this case their poisonous dorsal spines. Alexander Vasenin / Wikimedia Commons / CC BY-SA 3.0.

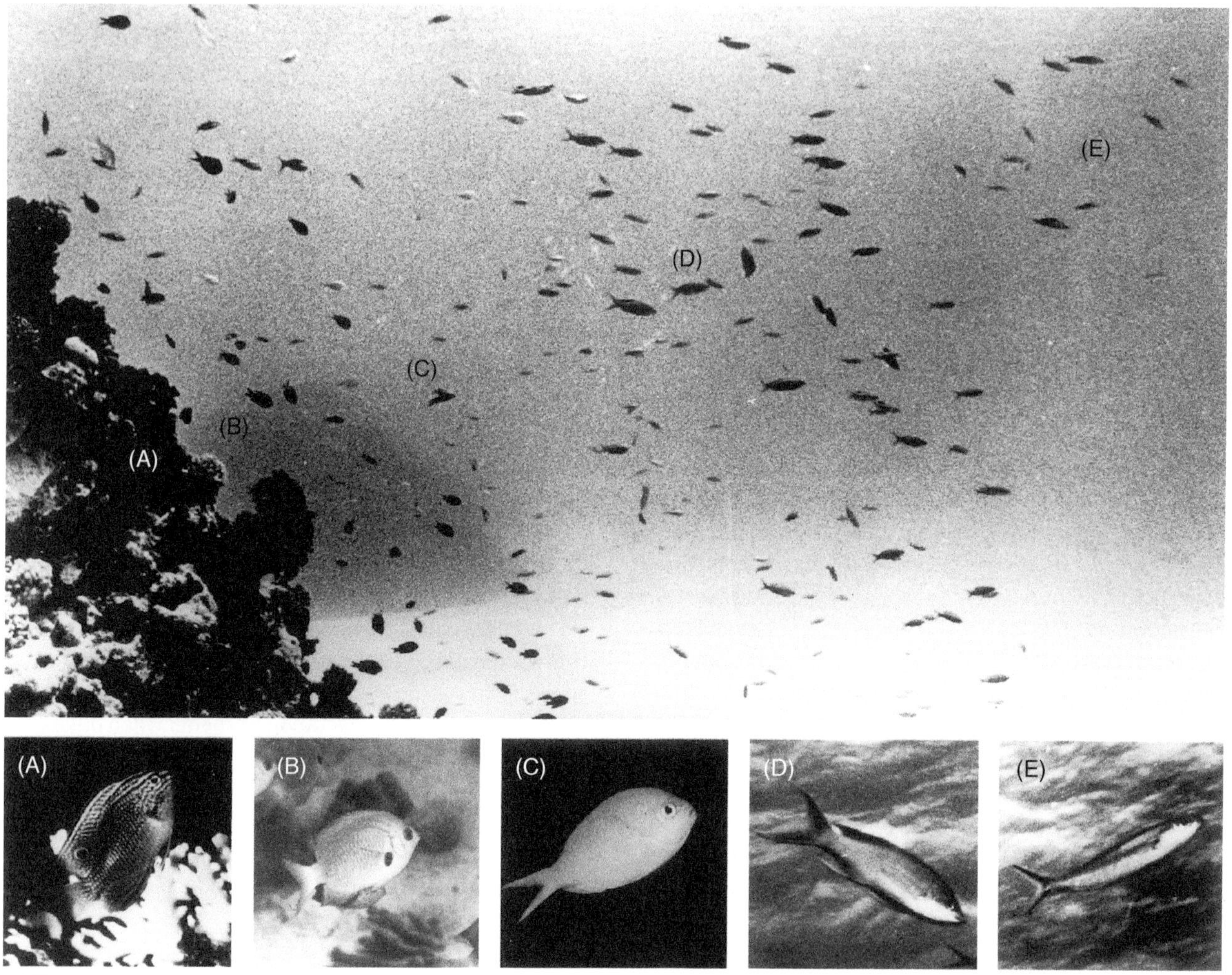

FIGURE 16.12 Predator avoidance has shaped body morphology in zooplanktivorous reef fishes. Fishes that feed close to protective structure tend to be more deep-bodied with square or rounded tails, those that forage higher in the water column are more streamlined with forked tails. All these fishes dive for the coral when threatened by predators. Streamlining may also facilitate holding the fish's position in the stronger currents higher above the reef. Lettering on the photograph of the reef indicates the zones where the five different fishes typically feed: (A–C) damselfishes (pomacentrids), (D) *Anthias* (a serranid), (E) *Pterocaesio* (a lutjanid). From Hobson (1991), used with permission.

studies show that prey accelerate faster than predators, perhaps because escape often involves relatively quicker C-start maneuvers (discussed in Chapter 6, Mauthner Cells) whereas a predatory attack utilizes S-start activities (Dominici & Blake 1997). Small size also enhances maneuverability and permits tactics unavailable to large predators.

To outdistance a predator, small pelagic prey can take advantage of the drag reduction that can be gained by becoming airborne, a tactic used most effectively by the exocoetid flyingfishes (Davenport 1994; Fig. 16.13). Flyingfishes commonly double their speed after emerging into the air, accelerating from about 36 km/h in water to as much as 72 km/h while airborne. They typically take off into the wind and travel for 30 s and as far as 400 m in a series of up to 12 flights. Multiple flights are interspersed with periods of rapid taxiing when only the beating, elongate lower lobe of the tail fin contacts the water surface. Fish may reach an altitude of 8 m. Refraction at the water surface makes them undetectable by predators except when a calm sea and bright sun create a visible shadow. Hence the flyingfish's reentry point would be largely unpredictable, particularly if it followed a curving flight path. Flyingfishes glide rather than fly, gliding efficiency depending largely on wing surface area divided by body mass. Flyingfishes come in two-winged varieties with enlarged, curved pectoral fins and four-winged varieties that also have enlarged pelvic fins. These fishes are small, never exceeding 50 cm in length. A 100 g flying fish has a total pectoral fin surface area of about 200 cm^2 (Howell 2014).

Other "flying" fishes include the African freshwater Butterflyfish, *Pantodon* (Osteoglossidae), and the South American freshwater hatchetfishes (Gasteropelecidae), the latter species generating flying forces by vibrating its pectoral fins via pectoral muscles that may account for 25% of its body weight. The greatly enlarged pectoral fins of adult dactylopterid flying gurnards are expanded during cruising over the bottom; these fish have never been observed airborne. Many fishes leap into the air when escaping predators, including minnows

FIGURE 16.13 Flying fish become airborne to escape predators. NOAA / Public Domain.

(Cyprinidae), halfbeaks (Hemiramphidae), needlefishes (Belonidae), sauries (Scomberesocidae), cyprinodontids, atherinids, Bluefish (Pomatomidae), mullets (Mugillidae), and tunas and mackerels (Scombridae). And some predators (dolphinfishes, mackerels, tunas, and billfishes) leap into the air in horizontal pursuit of prey. Flyingfishes do have one advantage over such airborne predators: a flyingfish's cornea is flattened, which gives it the ability to focus both in water and air. Other fishes have a curved cornea, which only allows focusing in water. A remarkable convergence on the "flyingfish morphology" occurred among at least one chondrostean in the Triassic; the perleidiform, *Thoracopterus*, possessed expanded pectoral and pelvic fins, a longer lower caudal lobe, and other traits suggesting that it too glided out of the water (see Chapter 11).

Attack and Capture

An actual attack takes place as the predator launches itself at the prey and engulfs the prey in its mouth. The actual strike of pikelike predators is short and fast, involving a prestrike S-shaped bending of the body and maximal forward propulsion driven by the combined surface area of the median and caudal fins; pike acceleration in this phase has been measured at more than 150 ms^{-2} (or over 500 km/h, Webb 1986; Dominici & Blake 1997). Prey are impaled on the sharp jaw teeth, manipulated into a head-first position, and swallowed; in barracuda, large prey can be cut into smaller pieces for swallowing. In most predators, the attack is focused on the center of mass of the prey's body, because the escape response involves a pivoting on the center of mass and hence the center moves least relative to other prey body parts.

The evolutionary advance of fishes is synonymous with the development of jaws (discussed in Chapters 3, 11), which function to surround, impale, or inhale prey and then pass the prey posteriorly for processing (discussed later in this chapter). Active prey are brought into the mouth by overtaking, extending the mouth, and suction, often in combination. Fast start predators overtake or intercept their prey and impale them on sharp teeth. Overtaking may involve rapid swimming via body musculature, but many fishes also rapidly project the mouth out to create suction and surround the prey, as in Largemouth Bass (Nyberg 1971). The advantages of protrusible jaws and a pipette mouth are addressed in more detail in Chapter 3.

Predatory fishes occasionally take terrestrial prey, particularly if the prey land on the surface of the water. However, African Tigerfish (*Hydrocynus vittatus*) can feed on Barn Swallows (*Hirundo rustica*) in flight as the birds fly low over the water. Apparently the Tigerfish make the needed adjustments to account for different refraction of light in air and water as they follow their prey, and then leap just far enough out of the water to grab the birds (O'Brien et al. 2014).

Many benthic, lie-in-wait predators have particularly large mouths that can be opened and closed rapidly. Frogfishes, *Antennarius* (Antennariidae) can expand their oral cavity up to 14-fold, engulfing prey in less than 6 ms. Stonefish, *Synanceia*, engulf prey in 15 ms. Speed is not solely characteristic of large fish feeding on large prey. Zooplanktivorous damselfishes (*Chromis*, Pomacentridae) project their mouths to capture individual zooplankton in 6–10 ms; jaw protrusion followed by suction is used during capture (Coughlin & Strickler 1990). Slower movements suffice for nonevasive planktonic prey. Whale Sharks, Basking Sharks, manta rays, herrings, anchovies, and mackerels swim through plankton concentrations with their mouths open, passively filtering prey out of the water with their fine gill rakers. North American Paddlefish use electrosensitive cells in the rostrum to detect the density of zooplankton, which may signal when it is a good time to filter feed (Wilkens et al. 1997, see Chapter 13). Some cyprinids and cichlids filter concentrated plankton by remaining in one place and pumping water in and out of their mouth, again using the gill rakers as a sieve (Drenner et al. 1987; Ehlinger 1989).

Specialized Strategies for Subduing Prey

A few specialized predators **immobilize** prey before engulfing them. Electricity-generating predators such as the electric ray *Torpedo*, and electric eels (*Electrophorus*), stun individual prey with a powerful discharge (Bray & Hixon 1978; Lowe et al. 1994). In torpedo rays, the predator encompasses the prey with its pectoral fins and discharges its electric organ. The immobilized prey is then grasped in the mouth.

Other fishes utilize anatomical extensions to immobilize prey. Thresher sharks (Alopiidae) use the long upper lobe of their heterocercal tail in a whip-like manner to stun prey fishes. This is achieved either by direct contact or by concussion of the whip-like tail in the water immediately adjacent to the prey (Aalbers et al. 2010). Sawfishes and sawsharks reportedly swing their broad, tooth-studded bills among prey fishes, impaling some in the process (Wueringer et al. 2012). Hammerhead sharks (Sphyrnidae) get their name from the expanded cephalic lobes of the head. This broadened head may serve several purposes, including enhancement of vision, olfaction, and electroreception, but it is also used to subdue prey. Hammerheads often feed on stingrays, and the Great Hammerhead, *Sphyrna mokarran*, appears to be something of a stingray specialist. Hammerheads have been seen attacking stingrays from above, using the broad head to pin the stingray down, then pivot and take bites from the expanded fins that the stingray needs to swim (Fig. 16.14). The stingray is now unable to escape, and the hammerhead can finish its meal. A shark with a more typical narrow snout would not be able to have the same efficient result in attacking a large stingray (Strong et al. 1990).

Billfishes belong to two closely related families, and members of both have bills that develop as forward growths of the upper (premaxillary) jaw bone. In the Xiphiidae (Swordfish), the sword is flattened and smooth, whereas in the Istiophoridae (marlins, spearfishes, and Sailfish), the bill is rough and round in cross-section. Observations of these fish hunting support the conclusion that these bills can be used to stun prey with lateral slashes, or spear prey with a straight-on strike. Sportfishers frequently observe marlin knocking prey such as small tuna into the air, or find their lures with characteristic scratch marks indicative of strong sideways blows of the rough spear. Tuna and dolphinfish perforated with spear holes have been found in marlin stomachs and observers have watched marlin spear hooked tuna prior to swallowing them. One diver temporarily became a screen between a marlin (probably a Black Marlin, *Makaira indica)* and the Amberjack (*Seriola lalandi,* Carangidae) it was pursuing. The Amberjack had been injured by divers who were spearfishing. It pulled free of the spear, and was then pursued by the marlin. After hiding for a moment behind the diver, the amberjack tried to escape, only to be overcome and impaled by the marlin, which then shook its prey free and swallowed it (van der Elst & Roxburgh 1981). The Swordfish, *Xiphias gladius*, evidently uses its smooth, flat bill primarily to decapitate cephalopod prey and slash them into swallowable pieces. Slashing also occurs as a swordfish enters a shoal of prey; maimed fish are picked up on subsequent passes.

The billfishes can also use their bills defensively. An unlucky researcher discovered this in April 2003, when he entered the water off Maui to videotape false killer whales attacking a 3 m^{+} marlin. The marlin speared the diver through the right shoulder, causing considerable tissue damage (Honolulu Advertiser, 17 April 2003). Swordfish also may use their bills defensively or during territorial encounters: broken Swordfish (and other billfish) bills have been found embedded in boat hulls and other objects. In 1967 the deep submersible *Alvin* was attacked and skewered by a 60 kg Swordfish at a depth of 600 m; the fish was still stuck when the sub was brought to the surface (Wisner 1958; Ellis 1985; photo available on the web).

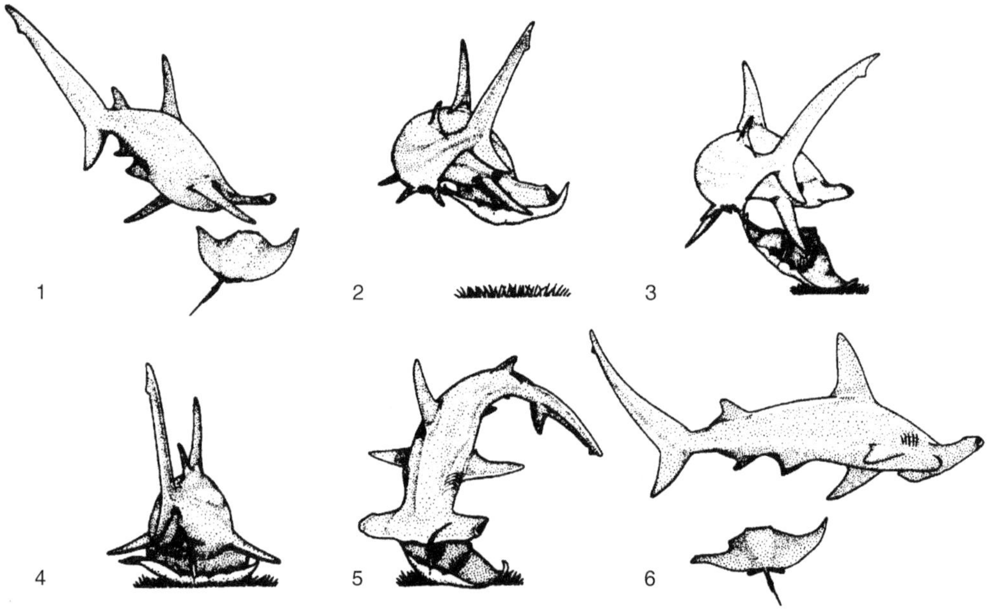

FIGURE 16.14 A 3 m long hammerhead shark captured and bit a piece off of a l m wide Southern Stingray by pinning the ray against the bottom with its broad head while the shark pivoted around and bit off the front margin of its pectoral fin. The following sequence is shown: 1, shark chases the ray; 2, shark strikes downward across the back of the ray with the flat underside of its head; 3, ray bounces off the bottom from the force of the blow while the shark brakes with its pectoral fins; 4, shark delivers a second downward blow across the back of the ray; 5, shark pivots while holding the ray against the bottom and takes a bite from the front of the left pectoral fin; 6, injured ray attempts to swim off followed by the shark. From Strong et al. (1990); used with permission.

Tool Use by Fishes These previous examples of predatory fishes using unusual anatomical features to subdue prey are somewhat akin to tool use, except in those cases the "tool" is part of the fish itself. There are fishes, however, which use aspects of their environment to help them acquire food, which falls more closely into the tool use category of behavior.

Several species of wrasses (Labridae) carry bivalve molluscs or crustaceans to rocks in their habitat and repeatedly smash the prey against the rock to break open its shell (Bernardi 2012; Brown 2012; Dunn 2016). Although the fish does not pick up and manipulate the rock, as expected in some definitions of "tool," the fish are clearly using the rock to accomplish a task that would otherwise not be possible.

Water itself can be employed as a tool by some fishes to alter their environment with the goal of obtaining food. Archerfish have long been recognized for their ability to blow streams of water into the air to knock down insect prey. The tongue and a groove in the top of the mouth form a tube that is used to propel water. Trials have demonstrated that archerfishes can learn to hit targets 30–50 cm above the water, moving as fast as 20 cm/s, with up to 50% accuracy. This task requires correcting not only for refraction at the water's surface and the arc and decelerating velocity of the fired droplet but also for the change in the three-dimensional location of the target (Dill 1977a; Schuster et al. 2006; Brown 2012; Newport et al. 2014). What makes this behavior even more astounding is that individuals apparently can learn to adjust their shots by observing the efforts of shoalmates. Several species of gouramis (*Trichogaster* spp, Osphronemidae) also use stream of water to obtain food (Brown 2012), a behavior that was labeled "prey spitting" by Vierker (1975).

Streams of water can also be an effective tool under water, as shown by the triggerfishes (Balistidae) which blow water to overturn sea urchins so the less spiny underside can be attacked. Several species of sharks and rays have also been reported to blow water to uncover prey (Hueter et al. 2004), and Kuba et al. (2010) reported that fresh water stingrays (*Potamotrygon castexi*, Potamotrygonidae), were able to figure out that they could utilize jets of water to remove food items from a plastic tube.

Mudskippers also utilize water as a tool for feeding. They carry water in their opercular and buccal cavities while moving about on land and this water forms a "hydrodynamic tongue" while feeding to facilitate the swallowing of food, with movements of the hyoid similar to that of some terrestrial feeding vertebrates such as newts (Michel et al. 2015; Heiss et al. 2018). In some instances, the water from the mudskipper's buccal cavity protrudes a bit from the mouth, comes in contact with the prey, and is then sucked back into the mouth. The ability to capture and swallow prey on land provides mudskippers with a greater degree of terrestrial feeding capability than amphibious fishes which must return to the water to swallow. If the buccal water volume in a mudskipper is reduced, they are less successful in swallowing, or they must grab prey in the jaws and return to the water to swallow. Mudskippers make the needed adjustment in feeding by keeping the gill covers closed while feeding on land, whereas they open during aquatic feeding (Michel et al. 2016).

Prey Responses to Predator Attack

When attacking prey in shoals, the major obstacle to successful prey capture is the presence of other shoal members; a direct relationship exists between likelihood of escape and numbers of individuals in the shoal. Hence many predators engage in tactics that separate prey individuals from the group. For example, when Pike (*Esox lucius*, Esocidae) attack minnow (*Phoxinus phoxinus*) schools, the predator's strike often leaves individual prey separated from the school. The predator then preferentially chases these individuals, which may account for 89% of the predator's success (Magurran & Pitcher 1987). Similar **group-separating tactics** are employed by other predators, such as Blue Jack, *Caranx melampygus* (Carangidae), when attacking mixed species schools of snappers, and piranha, *Serrasalmus spilopleura* (Characidae), attacking cichlid shoals (Potts 1980; Sazima & Machado 1990). Stragglers are up to 50 times more likely to be attacked than fish within a group, and success rates for attacks on stragglers can be four times higher than for attacks on the main shoal (Parrish 1989a, 1989b).

Aggressive mimicry also occurs in a group context, the victims either being fish and invertebrates exterior to the shoal or even unsuspecting members of the shoal itself. In both cases, shoal membership by the mimic allows the predator to get close enough to attack. Juveniles of the Indo-Pacific grouper, *Anyperodon leucogrammicus* (Serranidae), swim with and resemble adult females of four similar species of wrasses (Labridae). Small fishes have no reason to fear the wrasses, which feed on a variety of small benthic invertebrates. However, the grouper is a piscivore and has been observed snaring small damselfishes while swimming among the shoaling wrasses. Similarly, hamlets of the West Indian genus *Hypoplectrus* (Serranidae) resemble damselfishes and angelfishes, the models here being zooplanktivores or herbivores. The hamlets are carnivores and their presumed mimicry could allow them to sneak up on invertebrate prey that do not distinguish them from the otherwise harmless, and usually more numerous, model species (Moland et al. 2005; Randall 2005). Some scale-eating fishes attack from within shoals, their primary prey being shoalmates that they resemble. These predators include a characid, *Probolodus heterostomus*, which feeds on schooling characids in the genus *Astyanax*, and a cichlid, *Corematodus shiranus*, which resembles and schools with its prey, a tilapiine cichlid, *Oreochromis squamipinnis* (Sazima 1977; Thresher 1978).

Prey choice is also affected by relative numbers of different kinds of individuals in the prey shoal. Oddity in appearance or behavior often stimulates attack. In mixed schools, the minority species is often attacked disproportionately, such as when Gafftopsail Pompano (Carangidae) feed on anchovetas (Engraulidae) that are schooling with more numerous Flatiron Herring (Clupeidae) (Hobson 1968). In experimental trials, Largemouth Bass are much more successful predators when one or two blue-dyed minnows are added to small minnow shoals; the odd individuals are taken preferentially (Landeau & Terborgh 1986). Predation on odd individuals within a school may result from them standing out against the background of the more common type or, as suggested by Landeau and Terborgh (1986), because oddity alleviates the confusion effect created by a mass of similar-appearing prey animals. Whether predatory fishes are more likely to attack injured individuals within a group remains unclear. Certainly sharks and barracudas are attracted to injured or erratically swimming prey, and the literature on mammalian predators (e.g., lions, wolves, and hyenas) suggests that fishes would also feed preferentially on injured individuals. However, the only quantitative study of the subject indicates, if anything, an avoidance of injured prey (Major 1979). Regardless, predation on odd individuals indicates strong selection for uniformity of appearance and behavior within a shoaling species.

Preventing and Deflecting Attacks

When actually attacked, a prey fish can make a quick evasive move, employ an active defense, rely on passive structural defenses to deflect the predator, or use a combination of actions. Rapid, **fast start** escape movements that lead to maximal acceleration away from the attacking predator are almost universal among fishes, developing early in the ontogeny of larvae and continuing to function into adulthood (Webb 1986). Fast start escape movements occur in response to visual, acoustic, tactile, electrical, and water displacement stimuli. The reaction to water displacement is significant because many predators of small fishes use suction feeding, which means that a larva would experience the water around it suddenly moving toward a predator. Anatomical and behavioral features of this escape response indicate that it operates near the physical and temporal limits of nerve conduction and muscle contraction, emphasizing the importance of predation during early life as well as later in most fishes.

Responses of Aggregated Prey

Shoals under attack perform a series of identifiable maneuvers, elements of which have been noted for groups of minnows and shiners (Cyprinidae), Yellow Perch (Percidae), snappers (Lutjanidae), and sand lances (Ammodytidae), among others (Pitcher & Parrish 1993). The tactic employed is dependent on the type of predator and the intensity of its attack (Fig. 16.15). Shoals generally avoid a slowly approaching predator by maintaining about a 5–15 prey body length space between the predator and the group. Often predators will swim slowly through a prey school, at which time the school separates ahead of the predator and then closes back together behind it, basically creating a prey-free vacuole around the predator. The function of the predator's slow maneuver is unclear; possibly the predator is testing for injured prey or hoping to catch an inattentive individual off guard. In more concerted attacks, prey move rapidly out from the point of attack, scattering in different directions and fleeing the scene or seeking refuge in nearby structure.

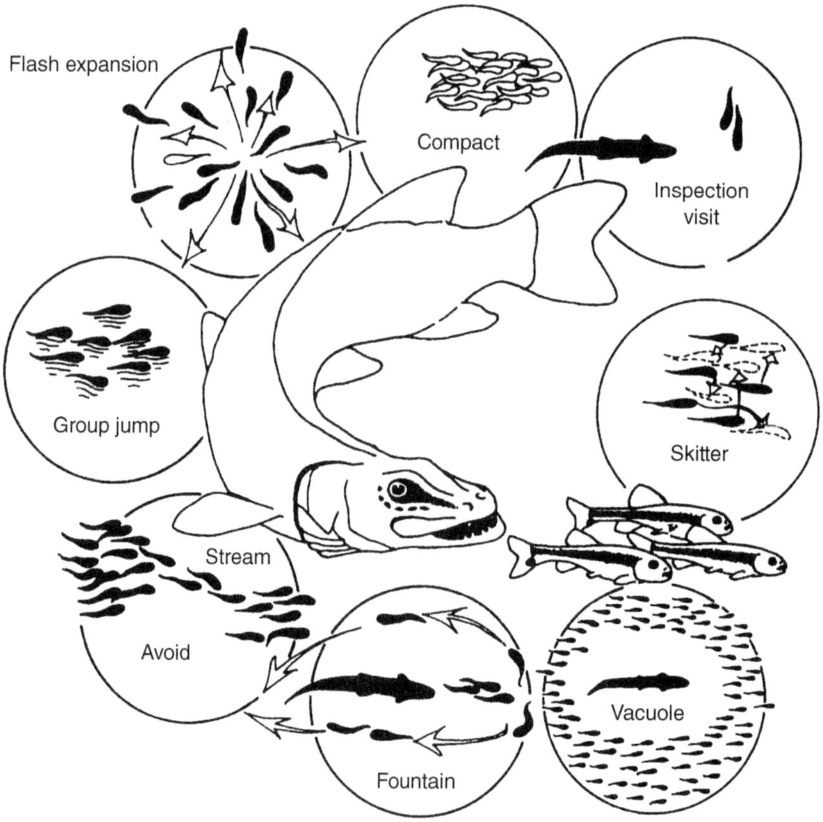

FIGURE 16.15 Examples of the responses of minnows under attack. Responses increase in intensity as the predator's actions become more threatening. The hierarchy of responses begins at the top ("compact") and proceeds clockwise to "flash expansion." From Magurran and Pitcher (1987), used with permission.

Group membership may reduce the statistical likelihood of an individual being attacked by a predator, producing a **dilution** or **attack abatement effect**. If a predator only consumes one or a few prey once it encounters a group, then the likelihood of any one individual being eaten decreases as group size increases. An additional benefit of grouping includes the passage of information about a predator to individuals unexposed to the predator. This is termed the **Trafalgar effect**, after the Battle of Trafalgar, when Admiral Nelson sent information through his fleet using flag codes, informing ships on the far side of the fleet about the enemy's actions (Pitcher & Parrish 1993).

The most widely demonstrated cause of a decrease in predator success is the **confusion effect**. In studies with Largemouth Bass feeding on minnows, Pike feeding on perch, sticklebacks feeding on water fleas, and jacks feeding on anchovies (Godin 1986; Landeau & Terborgh 1986), the predators caught fewer prey from large than from small schools. Success declined because the predator switches targets as it moves through the school, apparently confused by the number of multiple, edible objects moving across its field of vision. Specific prey behaviors seemingly function to increase confusion. In **skittering**, as displayed by minnows, an individual accelerates rapidly, rises in the water column, and then quickly rejoins the group. **Protean** behavior, seen in anchovies and silversides (Engraulidae, Atherinopsidae, Atherinidae), involves quick, uncoordinated up-and-down movements by several adjacent individuals just prior to resumption of polarized schooling. **Roll-and-flash**, often seen in herring (Clupeidae) schools, occurs when an individual rotates on its long body axis and reflects bright sunlight; it then returns to a normal upright position. The eye is quickly drawn to the point of the flash, but the fish seemingly disappears when upright orientation is resumed (a similar distractive function has been proposed for ink-squirting in octopods and squid under attack). The physiological basis of the confusion effect is poorly understood, although it may relate to an "information overload" problem whereby the predator's capacity for processing information is exhausted by the sheer number of objects in the visual field (Milinski 1990). Both invertebrate and vertebrate predators are subject to the confusion effect, as anyone who has ever attempted to net individual fish from a school in a large aquarium can attest. Not surprisingly, prey fish usually join the larger of two schools when given a choice, and the speed at which this decision is made increases when predators are present (Hager & Helfman 1991).

Handling

Handling includes any post-capture manipulation required to subdue prey and make it ingestible and digestible. Fishes that feed on hard-bodied prey or on prey with primary external defenses such as spines, shells, bony scales, or toxic skins must usually spend time and energy in handling, as do scavengers that feed on prey too large to be swallowed whole.

Prey that are small enough to be swallowed and possess no exceptional defenses (e.g., bony armor, poisonous spines, or toxic skin secretions) are usually manipulated into a head-first orientation for swallowing by jaw and head movements of the predator (Reimchen 1991). Such manipulation is less important with nonspiny fishes such as clupeiforms, most ostariophysans, and salmoniforms, but even soft-rayed fishes can become lodged in the throat if they are large and swallowed tail-first. Head-first swallowing facilitates depression of the dorsal, anal, and pelvic fins, all of which can anchor themselves in the predator's mouth or throat. Additionally, head-first swallowing reduces the likelihood of escape from the mouth since few non-eel-like fishes can swim backwards effectively. Many predators will attack prey long enough that the tail or more of the prey fish protrudes from its mouth. Such a large meal obviously represents an energetic bonanza, but is obtained at a potential cost if it hampers the predator's ability to escape its predators. On occasion, one finds a dead predator floating with a large, dead prey fish protruding from its mouth, testimony to the importance of gape limitation and the evolution of an ability to accurately estimate the size of potential prey.

Teeth assist with preparation of prey for swallowing, and dentition type is a reliable indicator of both prey type and foraging tactics (see Chapter 3, Dentition). Piscivores either hold their prey with large, sharp-pointed teeth or numerous needlelike teeth, or they chop prey up with flat, bladelike teeth. These teeth may be in the marginal jaws or on the palate and tongue. Insect feeders generally have moderately stout, conical, recurved teeth, again marginally or as part of the pharyngeal apparatus (see Chapters 3, 14). Most fishes that feed on molluscs or echinoderms crush the shell in the mouth using molarlike teeth. Fishes with parrotlike beaks (parrotfishes, puffers) feed on tough sponges, algae, or coral, with supplemental crushing in a pharyngeal mill in parrotfishes. Gill rakers also characterize different foraging types, functioning either for holding prey in the throat or as mechanical barriers to escape. Numerous, long, thin, gill rakers filter out small plankton as water passes through the mouth and out the opercular openings; gill raker spacing is usually directly related to prey size. Fish eaters have harder, stouter, more widely spaced rakers that prevent escape through the gill opening (e.g., seabasses, Largemouth Bass).

Most fishes are **gape-limited**, so their diets are constrained to include only items that can be swallowed whole. Swallowing entails movement of prey down the throat, which cannot expand to a width greater than the space between the cleithral bones. Feeders on hard-bodied prey can be limited by the relatively small gape of their pharyngeal jaws (Wainwright 1988a). There are exceptions, however, and some fishes have teeth and jaw structures that allow them to take bites out of prey or cut prey items into pieces. Sharks as a group are well known for the capability. A few species of bony fishes also have this capability, notably piranhas (Characidae), African Tigerfish (Alestiidae), Bluefish (Pomatomidae), and barracuda (Sphyraenidae, Fig. 16.16) – all fishes with specialized cutting or chopping dentition and powerful jaw muscles. Some advanced

FIGURE 16.16 Sharp, blade-like teeth of the Great Barracuda allow it to cut prey into pieces small enough to swallow. G. Helfman (Author).

coral reef species use powerful jaws and teeth to tear pieces out of sponges, such as various pufferfishes (Tetraodontidae, Diodontidae). Others can take small pieces of fins or flesh from prey, including some characins and cichlids, sabre-toothed blennies (Blenniidae), and at least one species of Forcipiger butterflyfish (Chaetodontidae).

Probably the most common tactic for overcoming gape limitation is **nibbling**, as seen in many small-mouthed shallow water marine and freshwater fishes such as centrarchid sunfishes, cichlids, damselfishes, wrasses, and surgeonfishes. But the great majority of fishes, those that utilize suction regularly during feeding, lack both the dentition and the jaw strength to nibble effectively. However, eel-like fishes and other elongate, aquatic vertebrates can grasp food in the mouth and spin rapidly around their long body axis, thus tearing chunks from the larger mass of a prey item. The American Eel, *Anguilla rostrata*, and other members of the family Anguillidae lack both the dentition and jaw musculature necessary for chopping or nibbling food into smaller pieces. If they are unable to swallow a large food item, they give it a few shakes and tugs. If the item still does not yield, the eel will hold onto the food and rotate rapidly, up to 14 rotations per second. This action twists the food and shears off a smaller piece. In some cases the eels will wedge the food item against the bottom or in a crevice and start spinning again until a small enough piece is removed. Such **rotational feeding** has been documented for more than 20 fish species, including other anguillids, moray eels (Muraenidae), snake eels (Ophichthidae), conger eels (Congridae), clariid catfishes, rocklings (Phycidae), a rattail (Macrouridae), rice eels (Synbranchidae), a Sablefish (Anoplopomatidae), greenlings (Hexagrammidae), sculpins (Cottidae), pricklebacks (Stichaeidae), gunnels (Pholidae), cod icefishes (Nototheniidae), and dabs (Pleuronectidae). Moray eels can add another option to their food-handling. Although capable of shaking and spinning, morays also tie themselves in knots. A moray will grab a live prey fish, tie an overhand knot in its tail, quickly run the knot up to its head and lever the knot against the prey. The combined force of the strong jaws, sharp teeth, and pressure from the knot can decapitate a prey fish, disabling it and also making it small enough to swallow. Hagfishes also use knots when tearing chunks from dead fish, although the process is considerably slower. Both rotational feeding and knotting are closely linked to an elongate body form. Having evolved an eel-like form, species apparently reap the additional benefit of being able to spin about their long body axis, an action not available to more conventionally shaped fishes. Without sacrificing the use of suction forces for feeding on common, small prey, rotational feeders can also overcome gape limitation and avail themselves of the occasional jackpot that a recently dead or dying fish represents (Helfman & Clark 1986; Miller 1987, 1989; Helfman 1990; Measey & Herrel 2006; De Schepper et al. 2007).

Some armored and otherwise defended prey require special handling tactics. Sea urchins are abundant and their internal organs are edible, but the defensive spines must first be removed. Special methods for removing the spines from urchins include plucking individual spines off to expose the outer test (triggerfishes, Balistidae), blowing water jets to roll the urchin over, exposing the relatively spineless ventral surface (triggerfishes), or picking the urchin up by a spine and bashing it open on a rock (wrasses, Labridae). Wrasses also smash crabs against rocks to remove a leg or claw, which they then crush in their pharyngeal jaws (Wainwright 1988a). The use of environmental features such as rocks to access food is considered by some to be tool use by fishes, as mentioned earlier in this chapter (see Bernardi 2012; Brown 2012).

Some predators apparently wash distasteful substances off the surface of prey by manipulating them in the mouth. For example, a Largemouth Bass fed whirligig beetles that secrete noxious chemicals or meal worms dipped in distasteful chemicals will repeatedly slosh the worm in its mouth and spit it out several times before finally swallowing it. Undipped worms are simply swallowed (T. Eisner, pers. comm.).

Final handling occurs in the stomach and intestines (see Chapter 5, Energy Intake and Digestion). Chemical breakdown via acids and enzymes is the rule, supplemented by mechanical grinding in the gizzards of Gizzard Shad (Clupeidae) and mullets (Mugilidae), and the gizzardlike stomach of Milkfish (Chanidae), some characoids (Prochilodontidae, Curimatidae), butterfishes (Stromateidae), and surgeonfishes (Acanthuridae). Deep-sea fishes such as black swallowers (Chiasmodontidae) have highly distensible stomachs that expand to accommodate prey considerably longer than the body of the predator (see Chapter 10, The Deep Sea).

Discouraging Capture and Handling

In fishes, capture involves taking prey into the mouth. Defenses against capture exploit the gape limitation that constrains most predators to feeding on prey small enough to be swallowed whole. Hence many anti-capture adaptations involve permanent or temporary increases in prey body size and elaboration of body armor that make it difficult to: (i) bring prey into the mouth; (ii) close the mouth once the prey are there; or (iii) swallow captured prey. Many fishes have dorsal and anal fins that appear out of proportion to their bodies, or bodies with exaggerated depth, such as citharinids, silver dollars (Characidae), veliferids, snipefishes (Macrorhamphosidae), crappies (Centrarchidae), fanfishes (Bramidae), manefish (Caristiidae), butterflyfishes (Chaetodontidae), tangs (Acanthuridae), Moorish Idols (Zanclidae), and spikefishes (Triacanthodidae). The exaggerated body humps on endemic Colorado River suckers and minnows have been interpreted as an evolved defense against gape-limited Colorado Pikeminnow (Portz & Tyus 2004). Greatly elongate dorsal, pelvic, and anal fins in many larval fishes (e.g., ribbonfishes, seabasses) may also reduce predation.

Puffers and porcupinefishes (Diodontidae) have multiple adaptations of muscles, skin, scales, stomach, and peritoneal cavity that allow them to inflate with water (or air if out of the water) when attacked or handled (Fig. 16.17; Brainerd 1992; Wainwright et al. 1995). This makes them difficult, if not impossible, to be swallowed by many predators. Spines of the porcupinefish pose an additional hazard to a predator. Inflation is made possible by specialized pumping muscles in the mouth and throat, a stomach that has lost its digestive function and expands 50 to 100-fold, and skin that is incredibly stretchable due to many small folds. Spines, if present, are modified scales and strongly anchored in the tight skin when inflated.

The evolutionary development of spines that defines the Acanthopterygii accomplishes a similar defense. Predators focus their attacks on the center of mass of the body, which is often where the prey's body is deepest; this depth is increased by erectable spines. A temporary increase in depth can be achieved by erecting the dorsal, pelvic, and anal fins with their stiff armament. A Bluegill (Centrarchidae) increases its body depth by about 40% by erecting its fins, making it a larger and hence less desirable food item for many predators. Erecting fins as a predator approaches may be a way of discouraging the predator before it attacks.

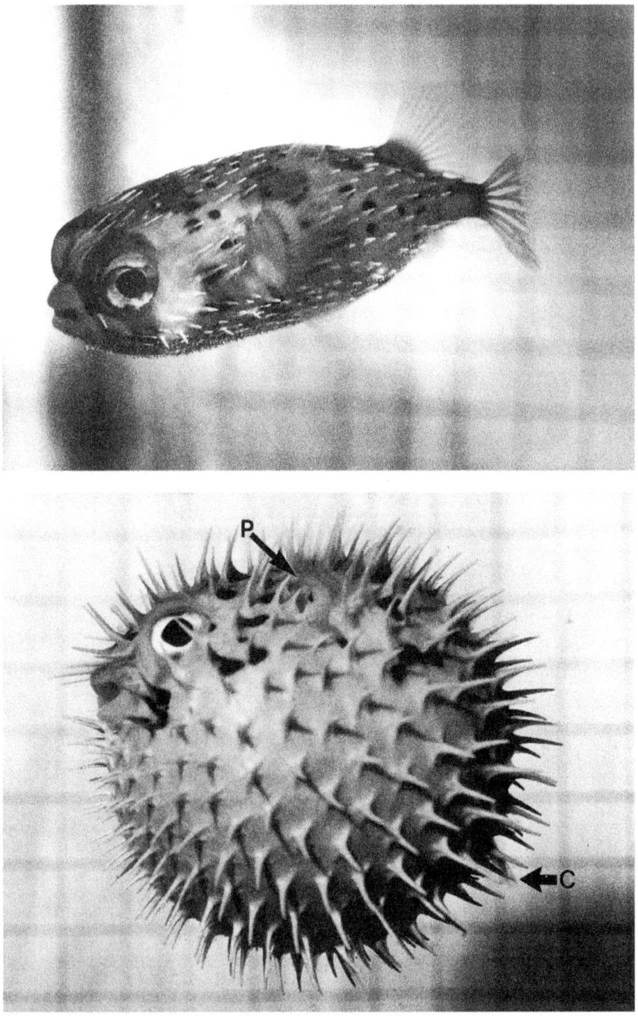

FIGURE 16.17 When members of the family Diodontidae, such as pufferfish, balloonfish, and porcupinefish, are threatened, they inflate themselves by pumping water into their highly expandable stomach. This dramatically increases their volume, making them difficult to swallow. Some have spines embedded in a highly derived, stretchable skin, presenting additional challenges for a would-be predator. Extremities that might offer a predator a grasping point, such as the caudal (C), pectoral (P), and other fins, sit largely within the protective framework of the spines when the fish is inflated. Brainerd (1992) / with permission of John Wiley & Sons.

The effectiveness of erected spines in preventing passage of prey towards the predator's throat can be enhanced by additional structures. Sticklebacks lodge themselves in the mouths of predators such as pike by locking their dorsal and pelvic spines, forcing the predator to break the spines before swallowing can occur. In leiognathid ponyfishes, the dorsal and anal fin spines, positioned opposite one another where the body depth is greatest, have a locking mechanism (Nelson 2006). Triggerfishes link the first two dorsal spines to prevent depression of the dorsal fin. Triggerfishes can wedge themselves into a crevice or a predator's mouth and lock the spines; the second spine (the "trigger") has to be pushed posteriorly to depress the dorsal fin (see Chapter 15, OrderTetraodontiformes and Fig. 15.32B). Indo-Pacific rabbitfishes (Siganidae) possess several unusual spine adaptations. The first dorsal spine points forward instead of up, which could inhibit head-first swallowing by predators, and each pelvic fin has hard spines at the leading and trailing edge of the fin. These are difficult fish to handle without getting punctured and the spines are covered with a toxic slime that causes painful wounds, at least in humans.

Most defensive morphological traits appear to be evolved responses to the threat of predation, such as pelvic spine length and degree of armor plating of sticklebacks in populations that vary in predation pressure (e.g., Vamosi 2005). However, other adaptations involve more immediate phenotypic changes induced in individuals by predators. Crucian Carp (*Carassius carassius*), Eurasian Perch (*Perca fluviatilis*), and Roach (*Rutilus rutilus*) have been shown experimentally to react to the presence of predators by changing body proportions or fin placement and shape. Carp and perch increase body depth during growth, whereas roach move the dorsal fin posteriorly and the pelvic fins anteriorly and widen the anal fin. Although these responses are complicated by intervening variables of food availability and population density, the evidence indicates that the changes are induced by chemicals released by predators. Increased body depth would make prey harder to swallow, and the fin changes in roach are thought to improve swimming ability during escape (Holopainen et al. 1997; Eklöv & Jonsson 2007).

Dermal and epidermal defenses also play an important role in resisting capture and in complicating handling. Many fishes exude mucus upon capture. This slime may make the fish slippery and harder to hold (hagfishes, anguillid eels), but in many the slime or other skin secretions contain distasteful substances that cause rejection by the predator (some moray eels, Muraenidae; marine catfishes, Ariidae; toadfishes, Batrachoididae; clingfishes, Gobiesocidae; soapfishes, Serranidae; gobies, Gobiidae; trunkfishes, Ostraciidae) (Hori et al. 1979; Smith 1992; Shephard 1994). Some coral reef gobies secrete skin toxins that cause loss of equilibrium and even death in predators (Schubert et al. 2003). The toxins are water soluble, thus maximizing their detectability to nearby potential enemies. Toxicity varies across species, the most toxic gobies being the most active and brightest colored, again corresponding with predation risk. Goby skin toxins may also affect the attachment behavior of external parasites (Munday et al. 2003). In certain flatfishes, toxic steroid aminoglycosides secreted from glands at the base of the dorsal and anal fins have a repellent effect on predators such as sharks (Primor et al. 1978; Tachibana et al. 1984).

External defenses in some species include a thickened or hardened dermis (e.g., the ganoid scales of lepisosteid gars and polypterid bichirs, the carapace made up of scale plates in ostraciid boxfishes, and the toughened skin of balistid leatherjackets). Populations of Three-spine Sticklebacks, *Gasterosteus aculeatus* (Gasterosteidae), that co-occur with predators have more lateral bony scutes and longer dorsal spines than do comparatively predator-free populations (Reimchen 1983; FitzGerald & Wootton 1993). Many shoaling species have easily dislodged, deciduous scales which may allow them to slip away from predators, analogous to the easily shed wing scales of moths and butterflies.

A special case of a handling-induced antipredator response in shoaling fishes and a few other species is the production of and reaction to alarm chemicals. **Alarm reactions** are best known in ostariophysans, where they were first discovered (see Chapter 14, Cohort Otocephala, and Superorder Ostariophysi). Substances and reactions also occur in some salmonids, livebearers, sculpins, darters, Yellow Perch, cichlids, and gobies, and are suspected in galaxiids, killifishes, and silversides (Smith 1986, 1992; Chivers & Smith 1998; Brown 2003). The **alarm substance** is released when the skin of a fish is broken, such as during a predatory attack.

Reactions depend on the species and the situation. Shoaling fishes often react by schooling tightly and moving away from the area where the alarm substance is released. Some solitary cyprinids sink to the bottom, whereas benthic species (gudgeon, Cyprinidae; loach, Cobitidae; suckers, and Catostomidae) freeze in place, utilizing their cryptic coloration to avoid detection. When alarm substances are in the water, overhead predators cause shiners (Cyprinidae) to hide in vegetation, whereas fish predators elicit a strong schooling response. The alarm reaction spreads as additional individuals detect the alarm substance or as they react visually to schoolmates. Many fishes show an alarm reaction to water in which predators have been kept, indicating again the probable importance of chemical interactions among fishes (see Chapter 17, Chemical Communication); some minnows even show alarm reactions when exposed to the feces of a predator that has fed on conspecifics (Brown et al. 1995). Juvenile convict cichlids, *Archocentrus nigrofasciatus*, show increasing levels of alarm reaction to increased concentrations of alarm substance (Brown et al. 2006).

Some fishes use nonchemical channels to transmit alarm signals. Visual signals induced by predators include increased fin flicking rates in schooling characins and in parental cichlids guarding young, head bobbing by gobies, and inspection visits and mobbing as discussed above. Many fishes emit **distress sounds** when held, prodded, or speared (e.g., catfishes, grunts, drums, and triggerfishes). At least three families of fishes (cods, squirrelfishes, and groupers) produce distinctive sounds when confronted with predators. Squirrelfishes produce a staccato sound that causes conspecifics to take refuge or inspect the predator (Myrberg 1981; Smith 1992).

The adaptive significance of responding to an alarm signal is obvious: it is advantageous to know that a predator is active in an area and to take appropriate action. Coordinated flight behavior within a school lessens a predator's chance of additional success. Further exploration of prey responses have shown additional benefits, including facilitated learning and recognition of predators and dangerous habitats, induced morphological changes in prey, and adaptive shifts in life history characteristics (e.g., Chivers & Smith 1998).

Scavengers, Detritivores, and Herbivores

Many fishes **scavenge** on dead and dying animals. A few species obtain most of their nutrition through scavenging (e.g., hagfishes) or detritivory (e.g., some minnows and suckers, curimatids, prochilodontids, mullets, and some Old World cichlids), whereas others supplement predation and omnivory with scavenging (e.g., catfishes, anguillid eels). Importantly, most predators will not pass up freshly dead prey (otherwise bait would not work in hook-and-line fisheries) and most scavengers and herbivores will take advantage of easily captured live prey. In essence, although dietary specializations certainly exist, fishes are highly opportunistic and will eat available prey of the appropriate size. At Johnston Atoll in the tropical Pacific, discarded doughnuts are eaten readily at the surface by such carnivores as snake eels, butterflyfishes, and flounders, and by such herbivores as damselfishes, parrotfishes, and surgeonfishes (D. A. Mann, pers. comm); the effects of such scavenging on fish girth or cardiac health has not, to our knowledge, been evaluated.

For scavenging animals, the predation cycle is usually shortened to search, wait, manipulate, and handle, whereas for detritivores and herbivores the waiting is eliminated. One task confronting **detritivores** is that of separating edible, fine particulate organic matter from any refractory, inedible sediments ingested. Ridges in the mouth and a maze of passageways associated with the gill rakers and epibranchial organs accomplish this in characoids. A **winnowing** process occurs in the orobranchial chambers as fishes pick up a mouthful of bottom material, sift it in the mouth, and expel inedible sediments back out the mouth or out the gill openings. Detritivores have some of the longest or most complexly folded intestines of any fishes, attesting to the resistance of detritus to enzymatic digestion (Bowen 1983; and see Chapter 5, Energy Intake and Digestion).

Herbivory is less common in fishes than among mammals and birds. Non-teleostean fishes are exclusively carnivorous, with the possible exception of limited herbivory in the Australian Lungfish, *Neoceratodus forsteri*. In teleosts, we find the evolution of **pharyngeal mills** and **gizzards** – mechanisms for rupturing cell walls and digesting plant matter. The most diverse freshwater fish taxa include substantial numbers of herbivorous species (characoids, minnows, catfishes, and cichlids), and herbivores on coral reefs are among the most abundant fishes there (e.g., halfbeaks, parrotfishes, blennies, surgeonfishes, and rabbitfishes). Digestion of vegetation is more efficient at higher temperatures (see Chapter 5, Energy Intake and Digestion), thus herbivory among fishes is more common in tropical waters (see Chapter 21, The Effects of Fishes on Vegetation).

Herbivory requires accurate search and efficient handling. Herbivores, particularly those that browse on upright macroalgae and do not graze on finer algal turfs, appear to use visual cues for selecting edible versus inedible species. Herbivory is consequently a primarily daytime activity. Targeted search is necessary because vegetation can defend itself by being tough or by producing chemicals, often in the form of halogenated terpenoids. Herbivorous fishes show strong preferences among algal types, feeding preferentially on species that lack structural and chemical defenses, while avoiding limestone-encrusted species or algae that contain deterrent chemicals. Some of these chemicals can slow growth or cause death in fishes (Horn 1989; Hay 1991).

Specializations for handling vegetation relate to the difficulty with which cell walls are disrupted, cellulose is digested, or defensive structures and chemicals are overcome. Herbivorous fishes typically have long guts, high ingestion rates, and rapid gut transit times. Large quantities of vegetation are passed through the gut and relatively little nutrition is assimilated from each ingested fraction. Cell walls are broken down in pharyngeal mills or lyzed in highly acidic (pH as low as 1.5) stomachs, although conclusive evidence of enzymes capable of digesting cellulose (i.e., cellulase) is lacking. Most fishes lack endosymbiotic bacteria and other microbes that aid in the digestion of vegetation, but exceptions include surgeonfishes, which contain bacteria, flagellates, and peculiar protist-like organisms, and sea chubs (Kyphosidae), which possess unique digestive tract morphology and hindgut microflora that aid in digestive fermentation (Fishelson et al. 1985; Rimmer & Wiebe 1987). Interestingly, some sea chubs feed heavily on brown algae that are avoided by most other herbivores (Horn 1989; Kramer & Bryant 1995).

Herbivory on coral reefs is intimately linked to both shoaling and territoriality. Most herbivores either defend exclusive territories (e.g., damselfishes, adult parrotfishes, blennies, and surgeonfishes) or roam about the reef in monospecific or heterospecific shoals (sea chubs, parrotfishes, surgeonfishes, and rabbitfishes). Territorial defense is very successful against solitary foragers but less so against grouped foragers. Individuals in large groups sustain fewer territorial attacks and have higher feeding rates than solitary foragers or members of small groups. Hence territoriality by some fishes promotes aggregation behavior in others (Robertson et al. 1976; Foster 1985).

Balancing Costs and Benefits

Foraging vs. Predatory Threat

It is important to realize that most predators are also prey, therefore foraging decisions often must be made in the context of danger to the feeder. These conflicting demands create a **foraging–predation risk trade-off**, and there is evidence suggesting that foraging fishes take into account risk to their own survival when choosing among food types, locales, and methods. Sticklebacks feed more slowly and gobies eat less in the presence of predators, even when the predators are behind a transparent partition (Magnhagen 1988; Milinski 1993). Juvenile Coho Salmon are less willing to travel long distances to intercept floating prey when presented with photographs of large predatory trout (Dill & Fraser 1984). Juvenile Black Surfperch (Embiotocidae) shift from feeding in low-growing, exposed algae when predatory kelp bass (Serranidae) are absent to feeding in tall, bushy algae when predators are present (Holbrook & Schmitt 1988a). Herbivorous minnows and loricariid catfishes abandon shallow areas of high algal productivity for deeper, less productive areas to avoid both predatory birds and fishes (Power 1987). Although juvenile Bluegill grow fastest when feeding in open water on zooplankton, in the presence of Largemouth Bass smaller, more vulnerable Bluegill move into vegetated areas where they have lower mortality rates from predation but also grow more slowly because of lower food intake. Bluegill too large to be swallowed by bass remain in the open water areas (Werner et al. 1983). Similar trade-offs and attendant costs have been shown in such fishes as Ocean Pout and Guppy (e.g., Botham et al. 2006; Killen & Brown 2006).

The balance between foraging and predation risk is shown by studies that vary the degree of threat and the strength of the reward. Minnows and surfperch will risk feeding in patches where predators are present if food densities are high; otherwise they avoid patches with predators (Gilliam & Fraser 1987; Holbrook & Schmitt 1988b). Additional factors can affect the decision process. Hunger or parasite loads cause sticklebacks to resume feeding sooner after exposure to predators and also cause fishes to feed closer to potential predators (Godin & Sproul 1988; Milinski 1993). When offered food in the presence of a cichlid predator, female Guppy will accept more risk for more reward whereas males avoid the predator regardless of food availability, implying that reproductive output is more dependent on food intake in females than in males (Abrahams & Dill 1989).

Finally, prey fish do not feed or avoid predators in an all-or-nothing fashion. They appear to weigh the potential threat from a predator and to take action appropriate to the degree of threat. This **threat sensitivity** is evident in individuals and groups. Territorial damselfishes respond more strongly to larger than to smaller predators and to predators in a strike pose than to searching predators. The strength of the flight behavior elicited also increases as the predator draws closer (Helfman 1989). Minnows in shoals also employ a graded series of escape responses that increase in strength and effectiveness as a pike escalates its attack (see Fig. 16.15), and sticklebacks spend progressively less time foraging as predatory trout increase in number (Fraser & Huntingford 1986; Magurran & Pitcher 1987). Threat sensitivity – to chemical and visual cues and sometimes involving learning by observing avoidance behavior in conspecifics – has also been shown in rainbowfish, sticklebacks, sculpins, and cichlids (Bishop 1992; Brown & Warburton 1997; Chivers et al. 2001; Brown et al. 2006). Threat sensitivity makes good Darwinian sense. A prey individual that is capable of assessing just how threatening a predator is and that is able to devote an appropriate amount of time and energy avoiding the predator will have more time for other fitness-influencing activities (feeding, breeding, defending a territory, or young) than will an individual that flees or hides any time a predator arrives in the area.

Optimal Foraging

Natural selection should favor animals that forage efficiently, selecting foods and feeding activities that maximize the ratio of benefits to costs (see Hart 1993). Benefits include calories and nutrients ingested, whereas costs involve energy used up, time lost to other activities, or exposure to predators or parasites.

Fishes may perform **optimally** when choosing food types, feeding locales and times, and foraging modes. When Bluegill were presented with water fleas (*Daphnia*) of different sizes and at differing densities, the largest prey were eaten when prey were abundant, but all prey were eaten when prey were scarce (Werner & Hall 1974). At intermediate prey densities, the largest zooplankters are consumed first, then the intermediate prey, and finally the smallest. This classic study showed that under controlled conditions, fish could be most selective when presented with an overabundance of high-quality food and progressively less selective as food becomes less abundant or lower in quality. The apparent size hypothesis, discussed earlier in this chapter, can have some impact on foraging; however, and may account for some apparent deviation from optimal foraging. Small items appear large if they are close, and fish may therefore at times select small items even when larger items are available but farther away.

Fishes have also shown an ability to assess the relative profitability of different food patches and to switch among patches as resources are depleted. When South American cichlids, *Aequidens curviceps*, were presented with two food patches of different profitability, they aggregated in the more profitable patch in direct proportion to the difference in food availability. Fish moved between patches periodically, feeding most where food was most abundant, then switching as food was depleted. Similar results have been obtained in studies of minnows, Guppy, and sticklebacks (Godin & Keenleyside 1984; Abrahams 1989).

Natural selection should also produce foragers that choose a method of food handling that gives them the **greatest relative return** for their effort. American Eel employ three modes for handling food. Small pieces of food (<85% of jaw width) can be suctioned into the mouth and swallowed. Larger pieces require dismembering. Large but soft pieces are grasped and shaken until a piece is removed, whereas large, firm foods are grasped while the eel spins. In terms of net energy return and growth rate, suction is the most profitable and spinning the least profitable feeding mode, with shaking falling somewhere between. When offered food types in a two-way choice situation, eels consistently met the expectations of the cost–benefit approach, by preferring food that required the least additional handling (Helfman & Winkelman 1991; Helfman 1994).

Supplementary Reading

Cott HB. 1957. *Adaptive coloration in animals*. London: Methuen.
Curio E. 1976. *The ethology of predation*. Berlin: Springer-Verlag.
Edmunds M. 1974. *Defence in animals*. New York: Longman.
Feder ME, Lauder GV, eds. 1986. *Predator–prey relationships: perspectives and approaches from the study of lower vertebrates*. Chicago: University of Chicago Press.

Gerking SD. 1994. *Feeding ecology of fish*. San Diego: Academic Press.

Godin JJ. 1997a. Evading predators. In: Godin JJ, ed. *Behavioural ecology of teleost fishes*, pp. 191–236. Oxford: Oxford University Press.

Godin JJ, ed. 1997b. *Behavioural ecology of teleost fishes*. Oxford: Oxford University Press.

Howell S. N. G. 2014. *The amazing world of flyingfish*. Princeton Univ. Press.

Ivlev, VS. 1961. *Experimental ecology of the feeding of fishes* (Scott D, transl.). New Haven, CT: Yale University Press.

Keenleyside MHA, ed. 1991. *Cichlid fishes: behaviour, ecology and evolution*. London: Chapman & Hall.

Kerfoot WC, Sih A, eds. 1987. *Predation: direct and indirect impacts on aquatic communities*. Hanover, NH: University Press of New England.

Krebs JR, Davies NB, eds. 1991. *Behavioural ecology: an evolutionary approach*, 3rd edn. London: Blackwell Science.

Lythgoe JN. 1979. *The ecology of vision*. Oxford: Clarendon Press.

Matthews WJ. 1998. *Patterns in freshwater fish ecology*. New York: Chapman & Hall.

Noakes DLG, Lindquist DG, Helfman GS, Ward JA eds. 1983. *Predators and prey in fishes*. Developments in Environmental Biology of Fish No. 2. The Hague: Dr. W. Junk.

Pitcher TJ, ed. 1993. *The behaviour of teleost fishes*, 2nd edn. London: Chapman & Hall.

Reebs S. 2001. *Fish behavior in the aquarium and in the wild*. Ithaca, NY: Cornell University Press.

Sale PF, ed. 2002. *Coral reef fishes: dynamics and diversity in a complex ecosystem*. San Diego, CA: Academic Press.

Steele MA, Anderson TW. 2006. Predation. In: Allen LG, Pondella DJ, Horn MH, eds. *The ecology of marine fishes: California and adjacent waters*, pp. 428–448. Berkeley, CA: University of California.

Wootton RJ. 1999. *Ecology of teleost fishes*, 2nd edn. Fish and Fisheries Series No. 24. New York: Springer-Verlag.

CHAPTER 17

Fishes as Social Animals

Summary

Reproductive success is the ultimate determinant of adaptations. Factors that characterize the breeding systems of fishes include frequency of mating, number of partners, and gender role of individuals. Not surprisingly, fishes show considerable inter- and intra-taxonomic variation in all factors. Most fishes spawn repeatedly, but some spawn only once during their lives. Most fishes have multiple breeding partners, although there are exceptions.

Many fishes spawn in nests that may be simple depressions in the bottom or more elaborate structures made of rocks or vegetation. These are typically constructed by the male, occasionally glued together with body secretions. Most fishes show some choice in selecting mating partners; truly random mating is rare. The spawning act itself often involves communication by elaborate body and fin movements, color change, and chemical and sound production.

Many fishes show no parental care once eggs are released and fertilized. Among those exhibiting parental care, the male most commonly guards the eggs until they hatch. Internal gestation by females occurs in most elasmobranchs and in a few bony fishes. Mouth brooding is practiced by many cichlids, some catfishes, cardinalfishes, and a few other families. Eggs may also be carried in male brood pouches (seahorses) or attached to the head (nurseryfishes). Some cichlids provide food for young in the form of epidermal secretions. Parental care increases survival of the young but

CHAPTER CONTENTS

The Diversity of Fishes: Biology, Evolution and Ecology, Third Edition. Douglas E. Facey, Brian W. Bowen, Bruce B. Collette, and Gene S. Helfman.

Companion website: www.wiley.com/go/facey/diversityfishes3

limits the future spawning activities of the parents and exposes them to increased predation. In some cichlids, young from previous broods assist in guarding and maintaining a nest, and receive some protection within the home territory in return.

Not all members of a population use the same reproductive tactics. In species in which dominant males have preferential access to females (e.g., salmons, sunfishes, wrasses, and gobies), smaller males often lurk on the edge of a spawning area and dash in rapidly ("sneaking") while a territorial male is spawning. In some sunfishes, "satellite" subordinate males without their own spawning territory may resemble females to gain access to nesting areas, then release sperm while the dominant male is spawning with a female. We do not know whether such alternative tactics are genetically fixed in individuals or represent behavioral adjustment to social conditions.

Molecular genetics has provided a powerful tool in better understanding mating systems, and shows that these systems can be quite variable within a species in terms of maternity and paternity of eggs and larvae in a brood. For example, the juveniles in a nest protected by a dominant male sunfish likely come from several different mothers, and may also represent the young of some subordinate "sneaker" or "satellite" fathers.

Social aggregations, except where fish incidentally converge on a resource, require communication to be maintained; territorial defense similarly requires communication. Fishes use multiple senses to communicate with one another. Static and dynamic visual displays, using colors and movement of fins or gill covers or turning on and off photophores, are common. The multiple functions of bright coloration in reef fishes have been long debated. They may have evolved in part because clear water makes predators detectable at a distance and the reef provides many places to hide from predators, thus eliminating a major cost of being colorful.

Fishes use sound when grasped by a predator, when spawning, defending territories, and in maintaining shoals. Sounds are produced by vibrating swim bladders, by rubbing bones or teeth together, and by the movement of the fish through the water. Sounds intended to communicate with conspecifics, however, may also be detected by potential predators.

Chemical production and detection functions during food finding, predator avoidance, mating, migration, parental care, territoriality, individual recognition, and aggregation. Pheromones are chemicals produced for intraspecific communication. Shoaling species are attracted to water that has contained conspecifics, and aggression can be reduced by production of specific chemicals in catfishes. The skin of many cyprinids releases chemicals if injured, subsequently providing a warning to others of potential danger. Tactile communication is limited primarily to mating, parent–offspring activities, and during extreme fights. Electrical communication is used extensively in families that have evolved the ability to produce and detect weak electric fields (South American knifefishes, elephantfishes); electrical output is often species, sex, and size specific.

Agonistic interactions involve aggression and submission, usually between conspecifics interacting in dominance hierarchies or during territorial encounters. Territoriality is common in fishes, which defend feeding, breeding, resting, and predator refuge territories.

Activity in fishes is often limited to a fairly defined area, termed a home range. Home ranges may be as small as a few square meters or as large as many square kilometers; larger species and individuals generally move over larger ranges. Individuals have an internal map of their range and a highly developed ability and strong tendency to return to their home range when experimentally displaced.

Many fishes aggregate in loosely organized shoals or tightly organized schools. Aggregations seem to mainly decrease the success of predators, but also can assist with reproduction, finding food, and can also save energy. Most aggregations form and break up repeatedly, but some have long-term stability that may even exceed the life span of individual members and are thus traditional.

Symbiotic relationships between species include parasitism, mutualism, and commensalism. Three fish families are known internal parasites of other fishes or of invertebrates (cutthroat eels, candiru catfishes, and pearlfishes). Mutualistic relationships include the many species that pick external parasites off other fishes, clownfish–anemone associations, and shrimp–goby pairs. Some fishes may also be in commensal relationships with invertebrates or other fishes, with only one member of the relationship receiving a benefit. Closer examination, however, has revealed at least some of these relationships to be more mutualistic. For example, remoras (Echeneidae, also called sharksuckers) are often seen as "hitchhikers" attached to large elasmobranchs – but they can also serve as cleaners, removing food particles and parasites, activities that could be beneficial to the host.

Reproduction

Evolutionary success is determined by an individual's ability to place genes into future generations, relative to the success of conspecifics. To transmit genes, most fishes must mate with conspecifics (exceptions are a few parthenogenetic species). Many fish species are relatively solitary as adults, but must overcome individualistic habits and seek out potential mating partners during breeding seasons. They must find suitable spawning habitats and substrates, perhaps modify these into nests, synchronize activities of males and females, and avoid hybridization with similar species that may be spawning in the same areas. Aggregations of reproductively active individuals may compete for spawning sites and partners, eliciting territorial and mate choice behavior. Courting and spawning may make participants more vulnerable to predators. After spawning, many species engage in varying degrees of parental care. All these activities and characteristics constitute the diverse mating systems of fishes. Our emphasis here will be on the diversity of mating systems, with focus on patterns and adaptations where they can be identified.

Reproductive Patterns Among Fishes

Important components of a **breeding system** include frequency of mating, number of partners, and gender role of average individuals (Table 17.1). Fishes show greater diversity in these traits than do other vertebrates. Most fishes remain one gender throughout adult life, but some fish species change sex; others are parthenogenetic, producing young from unfertilized eggs (see Chapter 8). Some fishes retain a single mate, perhaps for life, others mate promiscuously, and a few are haremic (a dominant mature individual of one sex mates with multiple individuals of the opposite sex within a group guarded by that dominant individual). In addition, different individuals of the same species may use different breeding systems.

TABLE 17.1 **A summary of components of breeding systems in fishes, with representative taxa. Accurate categorization is often hampered by the difficulty of following individual fish over extended periods in the wild. Although families are listed for some components, exceptions are common within a family.**

I. *Number of breeding opportunities*
A. Semelparous (spawn once and die): lampreys, river eels, some South American knifefishes, Pacific salmons, Capelin
B. Iteroparous (multiple spawnings):
1. A single, extended spawning season: most annuals (rivulines)
2. Multiple spawning seasons: most species (elasmobranchs, lungfishes, perciforms)
II. *Mating system*
A. Promiscuous (both sexes with multiple partners during breeding season): herrings, livebearers, sticklebacks, greenlings, epinepheline seabasses, damselfishes, wrasses, surgeonfishes
B. Polygamous:
1. Polygyny (male has multiple partners each breeding season): sculpins, sunfishes, darters, most cichlids; or polygyny (haremic): serranine seabasses, angelfishes, hawkfishes, humbug damselfishes, wrasses, parrotfishes, surgeonfishes, trunkfishes, triggerfishes
2. Polyandry (female has multiple partners each breeding season): anemonefishes (in some circumstances)
C. Monogamous (mating partners remain together for extended period or the same pair reforms to spawn repeatedly): bullheads, some pipefishes and seahorses, *Serranus*, hamlets, jawfishes, damselfishes, tilefishes, butterflyfishes, hawkfishes, cichlids, blennies
III. *Gender system*
A. Gonochoristic (sex fixed at maturation): most species (e.g., elasmobranchs, lungfishes, sturgeons, bichirs, bonytongues, clupeiforms, cypriniforms, salmoniforms, beryciforms, scombroids)
B. Hermaphroditic (sex may change after maturation):
1. Simultaneous (both sexes in one individual): *Kryptolebias*, hamlets, *Serranus*
2. Sequential (individual is first one sex and then changes to the other):
a. Protandrous (male first, change to female): anemonefishes, some moray eels, *Lates calcarifer* (Centropomidae)
b. Protogynous (female first, change to male): *Anthias*, humbug damselfishes, angelfishes, wrasses, parrotfishes, gobies
C. Parthenogenetic (egg development occurs without fertilization):
1. Gynogenetic: *Poeciliopsis*, *Poecilia formosa* (no male contribution, only egg activation)
2. Hybridogenetic: *Poeciliopsis* (male contribution discarded each generation)
IV. *Secondary sexual characteristics (traits not associated with fertilization or parental care)*
A. Monomorphic (no distinguishable external difference between sexes): most species (clupeiforms, carp, most catfishes, frogfishes, mullets, snappers, butterflyfishes)
B. Sexually dimorphic:
1. Permanently dimorphic (sexes usually distinguishable in mature individuals): *Poecilia*, anthiine seabasses, dolphinfishes, *Cichlasoma*, some angelfishes, wrasses, parrotfishes, chaenopsid blennies, dragonets, Siamese Fighting Fishes

TABLE 17.1 **(Continued)**

2. Seasonally dimorphic (including color change only during spawning act): many cypriniforms, Pacific salmons, sticklebacks, lionfishes, epinepheline seabasses, some cardinalfishes (female), darters, some angelfishes, damselfishes, wrasses, blennies, surgeonfishes, porcupinefishes (female)
3. Polymorphic (either sex has more than one form): precocial and adult male salmon; primary and secondary males in wrasses and parrotfishes
V. *Spawning site preparation (see Table 17.2)*
A. No preparation: most species of broadcast spawners (e.g., herring)
B. Site prepared and defended: sticklebacks, damselfishes, sunfishes, cichlids, blennies, gobies
VI. *Place of fertilization*
A. External: most species (lampreys, lungfishes, Bowfin, tarpons, eels, herrings, minnows, characins, salmons, pickerels, codfishes, anglerfishes, sunfishes, marlins, flatfishes, pufferfishes, porcupinefishes)
B. Internal: elasmobranchs, coelacanths, livebearers, freshwater halfbeaks, scorpionfishes, surfperches, eel-pouts, clinids
C. Buccal (in the mouth): some cichlids
VII. *Parental care (see Table 17.2)*
A. No parental care: most species
B. Male parental care: sea catfishes, sticklebacks, pipefishes, greenlings
C. Female parental care:
1. Oviparity with post-spawning care: *Oreochromis*
2. Ovoviviparity without post-spawning care: rockfishes (*Sebastes*)
3. Viviparity without post-spawning care: elasmobranchs, *Poecilia*, surfperches
D. Biparental care: bullheads, discus, *Cichlasoma*, anemonefishes
E. Juvenile helpers: some African cichlids (*Lamprologus, Neolamprologus, Julidochromis*)

Modified from Wootton (1990).

Lifetime Reproductive Opportunities

Most fishes are **iteroparous**, spawning more than once during their lives. Some well-known species, however, are **semelparous**, dying after spawning once. Most salmon of the genus *Oncorhynchus* (Pink, Chum, Chinook, Coho, and Sockeye salmon) hatch in fresh water, migrate to the sea for a period of 1–4 years, and then return to their natal (birth) stream where they spawn and die. Although the life cycle of females appears to be relatively fixed across a species, male life cycles may vary within a population (addressed later in this chapter, Alternative mating systems and tactics).

Other semelparous fishes include lampreys, anguillid (freshwater) eels, and the osmeriform southern smelts (Retropinnidae) and galaxiids of Australia and New Zealand (McDowall 1987). American Shad (*Alosa sapidissima*) are semelparous in southern locations (30–33°N), largely iteroparous at northern latitudes (41–47°N), and variably iteroparous at intermediate latitudes (Leggett & Carscadden 1978). With the exception of such annual fishes as aplocheiloid rivulines, semelparous fishes are diadromous or include at least one major migratory phase in their life cycle. Anguillid eels show sex-based differences in tactics within an overall semelparous strategy. Males often mature rapidly (*c.* 3–6 years) and at a uniformly small size (30–45 cm) regardless of locale, whereas females are consistently longer (35–100 cm) and may mature quickly (4–13 years) at low latitudes or slowly (6–43 years) at high latitudes. Slow maturing females grow larger and produce more eggs than smaller, faster maturing females. In American eels, males have a relatively restricted geographic distribution, occurring primarily in estuaries of the southeastern United States, whereas females are found throughout the North American range of the species and in all habitats. As far as is known, all members of an anguillid species migrate to the same oceanic region to spawn and die (Sargasso Sea in the western Atlantic for American and European eels, the Philippine Sea for Japanese eels) (Helfman et al. 1987; Jessop 1987; see Chapter 18, Representative Life Histories of Migratory Fishes).

Mating Systems

Mating systems are defined by the number of mating partners an individual has during a breeding season, and there are three common categories (Table 17.1). **Promiscuous** breeders show little or no obvious mate choice and both males and females spawn with multiple partners, either at one time or over a short period. Such spawning has been documented for the Baltic Herring (Clupeidae), Guppy (Poeciliidae), Nassau Grouper (Serranidae), humbug damselfish colonies (Pomacentridae), cichlids, and the Creole Wrasse (Labridae) (Thresher 1984; Barlow 1991; Turner 1993).

Polygamy, where only one sex has multiple partners, takes multiple forms. **Polyandry**, where one female mates with several males (and presumably not vice versa), is relatively uncommon but is seen in an anemonefish (Pomacentridae) (Moyer & Sawyers 1973). Polyandry might also be descriptive of female ceratioid anglerfishes which have more than one male attached (see Chapter 10, The Deep Sea). In **polygyny,** which is more common, males are the polygamous sex. Territorial males that care for eggs and young are frequently visited by several females, as in sculpins, sunfishes, darters, damselfishes, and cichlids. Polygyny can also develop into **harem formation**, where a male has exclusive breeding rights to a number of females that he may guard. Harems have been observed in numerous cichlids and in several coral reef families (e.g., tilefishes, anthiine serranids, damselfishes, wrasses, parrotfishes, surgeonfishes, and triggerfishes).

Lekking, in which males congregate to display to females, is common among polygynous birds and mammals in which only the female provides parental care (Emlen & Oring 1977). Among fishes, some African cichlids come close to forming true leks. Large numbers (*c.* 50,000) of male *Cyrtocara eucinostomus* congregate along a shallow 4 km long shelf in Lake Malawi and build sand nests and display to passing females each morning. Females spawn and then mouth-brood eggs elsewhere. The male aggregations break up each afternoon, when fish feed (McKaye 1983, 1991). Some fishes form "leklike" aggregations of males (e.g., Arctic Char, Atlantic Cod, damselfishes, wrasses, parrotfishes, and surgeonfishes), but the display ground is also an appropriate place for launching or caring for eggs, which stretches the definition of lekking (Loiselle & Barlow 1978; Moyer & Yogo 1982; Figenschou et al. 2004; Windle & Rose 2007). In a unique variation on leklike behavior, female triggerfish (*Odonus niger*, Balistidae) form a communal display ground for 1 day before spawning, after which they all mate with a single, nearby male (Fricke 1980).

Monogamous fishes mate with the same individual repeatedly and exclusively, and some may remain together during non-mating times. Strongly pairing species include North American freshwater catfishes, many butterflyfishes and angelfishes, most substrate guarding and some mouth-brooding cichlids, and anemonefishes; in the butterflyfishes, pairs may remain together for several years and probably mate for life (Reese 1975). Monogamous coral reef fishes commonly spawn with the same partner on a daily basis over an extended period without ensuing care of young, whereas freshwater species such as cichlids spawn over a limited time and then both parents typically care for the young. Monogamy has evolved independently in many groups, often in conjunction with territoriality and paternal care (Whiteman & Côté 2004). Monogamy is also known in freshwater bonytongues, bagrid and airsac catfishes, and snakeheads, and among at least 18 marine families, including pipefishes and seahorses, hermaphroditic hamlets, jawfishes, cardinalfishes, tilefishes, hawkfishes, damselfishes, wrasses, blennies, gobies, wormfishes, surgeonfishes, triggerfishes, filefishes, and pufferfishes (Barlow 1984, 1986; Thresher 1984; Turner 1993, Whiteman & Côté 2004).

Courtship and Spawning

As discussed in Chapter 8, sex in most fishes is determined early in life and does not change. Some species, however, change from male to female (protandry) or from female to male (protogyny). In many fish species, males and females look alike (at least to us), but others demonstrate sexual dimorphism (also discussed in Chapter 8). Sexual selection may be based on both primary and secondary sex characteristics, and can result in these characteristics or behaviors becoming more prominent in subsequent generations.

Spawning Site Selection and Preparation

Many fishes spawn in nests, which may be no more elaborate than a simple depression in the substrate excavated by either or both sexes using their fins or body (e.g., lampreys, Bowfin, trahiras, and sunfishes). In lampreys and salmon, the eggs are covered with additional sand or gravel and then abandoned, whereas in other species the male remains to guard the exposed eggs. Many species spawn in a rock crevice or space on or under a shell or rock or in a hollowed log that has been excavated or picked clear of growth and debris (e.g., ictalurid catfishes, sculpins, poachers, darters, cichlids, damselfishes, clingfishes, sleepers, and gobies). Some damselfishes "sandblast" a surface clean by spitting sand against it and then fanning it with their fins. The European Wrasse, *Crenilabrus melops*, lines a rock crevice with different types of algae. Softer algae at the back of the crevice serve as a substrate for spawning, whereas tougher, stickier coralline algae are packed into the outer portion of the crevice to protect the eggs.

Burrows are excavated and later guarded by lepidosirenid lungfishes, ictalurid catfishes, jawfishes, tilefishes, and gobies. Males of some stream minnows construct nest mounds by piling as many as 14,500 small stones, 6–12 mm in diameter, that are carried in the mouth to the nest site from as far as 4 m away (e.g., *Nocomis*, *Semotilus*, and *Exoglossum*). Eggs deposited by the female fall into the interstices between the stones and are covered by additional stones; the nest is guarded and kept free of silt by the male (Breder & Rosen 1966; Keenleyside 1979; Potts 1984; Thresher 1984; Wallin 1989, 1992; Johnston 1994). In African cichlids, males build nests in sand, some of elaborate design (Fig. 17.1). These structures are primarily display, courtship, and spawning stations from which the female picks the eggs up in her mouth almost as quickly as they are fertilized (Fryer & Iles 1972; McKaye 1991).

Nests can also be constructed of intrinsically produced materials, sometimes combined with extrinsically gathered objects. Gouramis and Siamese Fighting Fishes (Anabantoidei) and Pike Characins produce bubble or froth nests, consisting of mucus-covered bubbles that stick together in a mass. Some anabantoids add plant fragments, detritus, sand, and even fecal

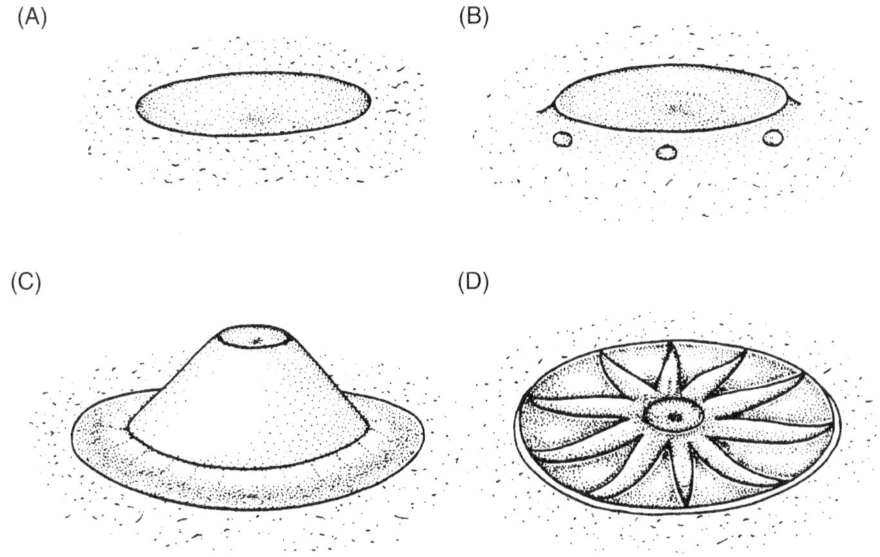

FIGURE 17.1 Spawning nests or bowers of African cichlids. Male cichlids construct sand structures that vary from simple pits to complex structures, where females deposit eggs just prior to picking them up in their mouths for brooding: (A) *Oreochromis andersonii* and *O. niloticus*; (B) *O. variabilis*; (C) *O. macrochir* from Lake Bangweulu; (D) *O. macrochir* from Lake Mweru. In the latter, radiating spokes are created by the male plowing through sand with his open mouth from the focal point to the edge. Structures vary from 15 to 150 cm across. From Fryer and Iles (1972), used with permission.

particles to the bubble mass. Male sticklebacks pass a mucoidal, threadlike substance manufactured in the kidneys out through the cloaca, using it to bind together pieces of leaves, grasses, and algal filaments that create a cup or tunnel nest in which a female deposits eggs. All of these structures are tended by the male following spawning (Wootton 1976; Keenleyside 1979).

Courtship Patterns

Courtship is the series of behavioral actions performed by one or both members of a mating pair just prior to spawning; it has several functions that maximize spawning efficiency. Courtship aids in species recognition, pair bonding, orientation to the spawning site, and synchronization of gamete release. Courtship is often necessary to overcome territorial aggression by the male, who might otherwise drive the female away from the site (in many species, males already have eggs in their territories and must guard against predation by conspecifics of both sexes). Courtship may be relatively simple (as in herring), or may involve a large number or a complex progression of displays and signals by one or both members (e.g., *Corynopoma*, Characidae; Guppy; sticklebacks).

During courtship, individuals frequently change color from their normal, counter-shaded patterns to bolder, contrasting color patterns. In many species (e.g., minnows, silversides, and cichlids) existing body coloration intensifies or the head becomes dark relative to the remainder of the body. Sound production during courtship, usually by the male, occurs in many fish families (sturgeons, minnows, characids, codfishes, toadfishes, sunfishes, grunts, sciaenids, darters, damselfishes, cichlids, blennies, and gobies), often in accompaniment with visual displays involving exaggerated or rapid swimming patterns, erection of fins, and jumping out of the water (Fine et al. 1977; Myrberg 1981, 2002; Lugli et al. 1997; Johnston & Johnson 2000; Lobel 2001; Johnston & Phillips 2003). Atlantic cod (*Gadus morhua*) and Pollack (*Pollachius pollachius*) utilize variations in sound duration and intensity as acoustic signaling as part of spawning (Wilson et al. 2014), and Burbot (*Lota lota*), a freshwater gadoid, also produce sounds associated with the spawning season (Cott et al. 2014, Fig. 17.2). The Longsnout Seahorse (*Hippocampus reidi*) produces sound during courtship (Oliveira et al. 2014), and Lake Sturgeon (*Acipenser fulvescens*) create a variety of sounds during the spawning season (Bocast et al. 2014). Female cichlids are attracted to sound-producing male conspecifics, but sound alone did not attract females unless a male fish was present (Estramil et al. 2014).

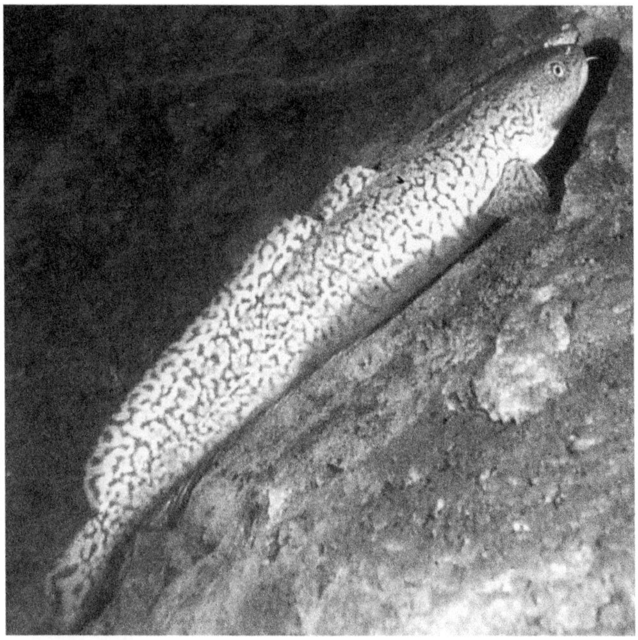

FIGURE 17.2 Some of the cods (Gadidae) are known to vocalize during courtship, utilizing rapidly contracting drumming muscles associated with the gas bladder. Cott et al. (2014) captured pre-spawning Burbot (*Lota lota*) and placed them with a recording device in enclosures under the ice in Great Slave Lake (Northwest Territories, Canada). When spawning season arrived, a variety of sounds, from slow knocks to rapid buzzes, were recorded – similar to the sounds of some of their marine cousins. Burbot photo by Achim R. Schloeffel / Wikimedia Commons / CC BY-SA 3.0.

Courtship and spawning sounds accounted for about one-third of the 85 different sound patterns produced by 45 species of Hawaiian coral reef fishes (Tricas and Boyle 2014).

Chemical stimulants are also involved. Male Goldfish, Zebra Danios (Cyprinidae), and gobies begin courtship activities when exposed to water that held a gravid female, and gravid female gobies are attracted to male-produced androgynous substances (Hara 1982; Stacey & Sorensen 1991). Species and sex recognition during courtship in cichlids occurs more quickly when individuals receive both visual and chemical cues from potential mates (Barlow 1992). Clearly, communication through a variety of sensory modes is important in courtship rituals, and is discussed further later in this chapter (Aggregation, aggression, and cooperation; Communication).

Some appreciation of the evolutionary premium placed on successful courtship can be gained by realizing that the gas bladder muscles that produce the boatwhistle mating call of the male Oyster Toadfish, *Opsanus tau* (Batrachoididae), contract at a rate of 200 Hz, making them among the fastest contracting vertebrate muscles known; the shaker muscles of a rattlesnake's tail contract at only half that rate (Rome et al. 1996).

FIGURE 17.3 Indigo Hamlets (*Hypoplectrus indigo*) are simultaneous hermaphrodites that spawn multiple times in a few minutes during sunset. For each spawning event they rise in the water column, embrace each other, and one fish releases some eggs and its partner fertilizes them. They then change roles and repeat the process several times. Courtesy of L. Rocha.

Spawning

The act of spawning may take place in the water column, above the bottom, in contact with plants and rocks, and in some special cases, out of water. Among fishes exhibiting external fertilization, behaviors associated with spawning often involve rapid swimming, quivering, vibrating, fin spreading, and enfolding of the female with the male's fins or body. The breeding tubercles and contact organs common in many fishes may help maintain contact between members of a pair and may also stimulate the female. Internally fertilizing species also engage in elaborate courtship sequences. Male Guppy perform a variety of actions involving following, luring, biting, and sigmoid swimming that display their fins and body coloration until a female allows them to approach and copulate. The sequence and types of displays by the male serve as a species-isolating mechanism in that females reject males of the wrong species after viewing their courtship displays (Keenleyside 1979).

Species-specific sounds may also be produced during the spawning act itself. For example, in the simultaneously hermaphroditic hamlets (Fig. 17.3), the fish that will release its sperm emits a courtship call and the fish that will release its eggs emits a spawning call. As individuals switch roles during a prolonged spawning bout, they also switch the sounds they produce (Lobel 1992).

Although the great majority of fishes spawn as part of large groups, **pairing** of individual males and females within these groups is common. Short-term pair formation probably assures efficient gamete release and fertilization; haphazard release of gametes could result in a large proportion of eggs going unfertilized because sperm become inviable and are rapidly diluted in open water, and eggs become unfertilizable within minutes after release (Hubbs 1967; Petersen et al. 1992). Codfishes spawn in large aggregations, but males establish small territories, actively court individual females using visual, tactile, and acoustic signals, and the pair moves synchronously to the surface where gametes are released while the genital openings of both fish are in close contact (Brawn 1961). Aggressive defense of females and pair spawning also occurs in schooling tunas (Magnuson & Prescott 1966). Pair spawning characterizes most epinepheline seabasses, which also form large breeding aggregations.

Group spawning, involving more than two fish, usually involves one female accompanied by several males (Fig. 17.4). This is the pattern in group-spawning minnows, suckers, salmon, smelt, wrasses, and surgeonfishes. In groups of Bluehead Wrasse, males release sperm in direct proportion to the number of eggs released by the female and the number of competing males in the group (Shapiro et al. 1994). In fishes with alternative mating systems, such as wrasses and parrotfishes on coral reefs, some individuals spawn as pairs whereas others spawn in multiple male groups. Truly random spawning associations, as described for promiscuous species, occur most frequently in water column spawners or in such benthic spawners as herring (Keenleyside 1979). One of the more fascinating cases of group spawning – because of its visibility – is in grunions. These small fishes ride waves of high spring tides up onto beaches of southern and Baja California. Females dig a hole tail-first into the sand and release eggs as multiple males surround her and release sperm. (The eggs hatch at the next spring tide, when waves once again reach the eggs – see Chapter 18, Semilunar and Lunar Patterns.)

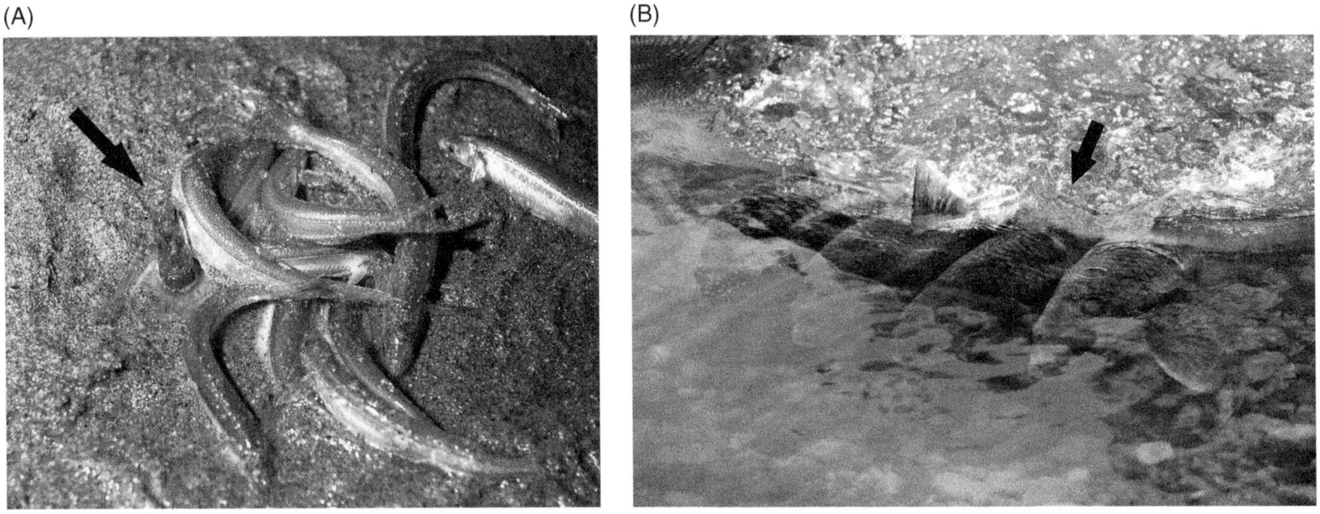

FIGURE 17.4 Spawning frequently involves multiple males and a single female. (A) California grunion (*Leuresthes tenuis*, Atherinopsidae) spawn on beaches at the top of the tide zone, a single female (arrowed) assuming a head-up position in the sand while males encircle her and release milt. (B) Robust Redhorse (*Moxostoma robustum*, Catostomidae) typically spawn over gravel and cobbles in groups of three, a single female (arrowed) flanked on either side by a male. In this group, the female has one male on her left and two on her right. (A) Courtesy of M. Horn; (B) Courtesy of B. Freeman.

Water column spawners on coral reefs often rush rapidly upward and release their gametes at the top of the rush, sometimes near the surface. Speeds approach 40 km/h in the Striped Parrotfish, *Scarus croicensis* (Colin 1978). This pattern has been observed in more than 50 species in over 18 families, but its purpose is not clear. Movement up in the water column places the eggs out of reach of many benthic or near-benthic egg predators and into currents that promote dispersal. However, by moving away from the reef, spawning adults face the conflicting threat of exposure to predation (Sancho et al. 2000a, 2000b). Not surprisingly, fishes that can defend themselves (e.g., larger seabasses, trunkfishes, and porcupinefishes) spawn higher in the water or more slowly than smaller, more vulnerable species (Thresher 1984). For many species, the spawning rush may serve as a final synchronizing event in the courtship sequence and may also help evade the sneakers that abound close to the reef (e.g., Sancho 1998, and see Alternative mating systems and strategies later in this chapter).

Substrate-spawning fishes are less likely to form large groups than water column spawners; they also release fewer eggs at each spawning. Males typically set up territories over appropriate spawning substrate, chase away intruding males, and court passing females. Females enter the territory and deposit one or a few adhesive eggs while the male folds his body or fins around her and presses her against the substrate. Sperm release occurs almost immediately, again in part because sneaker males are always nearby. In many cichlids, females deposit eggs on the substrate, then pick up the eggs in their mouth, and then facilitate fertilization by taking sperm into the mouth (Fig. 17.5).

An interesting variation on oral fertilization occurs in callichthyid catfishes of the genus *Corydoras*, the popular armored catfishes of the aquarium trade. In these catfishes, the female places her mouth over the genital opening of the male and drinks his sperm. She then passes the sperm rapidly through her digestive system, extrudes eggs which are held between her pelvic fins, and releases the male's sperm to fertilize the eggs, which are then deposited on the substrate. Sperm drinking could be one way of a female maximizing control over which male fertilizes her eggs. Sperm viability in the female's gut may be facilitated by the specialized nature of the callichthyid intestine, which is modified for air breathing (see Chapter 5, Air-Breathing Fishes). Callichthyids pass air bubbles rapidly from their mouths to their intestines, perhaps preadapting them for passing sperm through quickly and unharmed (Kohda et al. 1995).

Fishes that spawn on the bottom generally use available structure to protect their eggs. Many eggs are adhesive and stick to plants, rocks, woody debris, shells, or other hard substrates (e.g., herring, silversides, and cichlids). Some eggs have tendrils or projections that wrap around plants and debris (e.g., skates, halfbeaks, and flyingfishes). Eggs of the Port Jackson Shark (Heterodontidae) have an auger-like whorl around their exterior. Females lay these eggs in cracks and water motion apparently serves to screw the egg deeper into the substrate. As discussed earlier, California Grunion (*Leuresthes tenuis*, Atherinopsidae) spawn on sandy beaches after high tides on dark nights following a full or new moon. Capelin (*Mallotus villosus*, Osmeridae) also ride waves up beaches and deposit their eggs in the sand, although subtidal spawning is more common.

A few fish species use live invertebrates as a spawning site. Marine snailfishes, *Careproctus* spp. (Liparidae), lay their eggs inside the gill chambers of various crabs. Species of bitterling, *Rhodeus* (Cyprinidae), use freshwater mussels as a spawning site. The male first defends and displays over a particular mussel. The female deposits eggs into the gill chamber of the bivalve using her long ovipositor, after which the male releases sperm over the incurrent siphon of the mussel. Eggs develop inside the mussel and emerge as free-swimming young (Breder & Rosen 1966).

The act of spawning brings together fish that may normally be solitary, territorial, or extremely sensitive to predators, often at locales they seldom frequent. Spawning aggregations

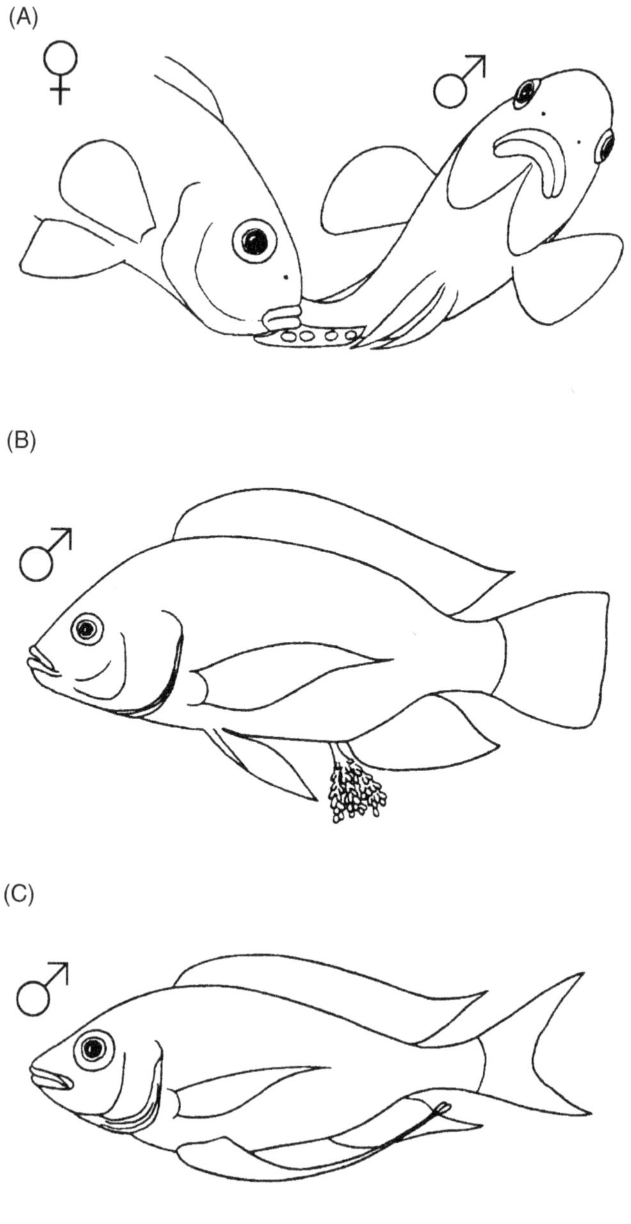

FIGURE 17.5 Fertilization occurs in the mouth of some female African cichlids. (A) Males of many African cichlids have round spots, termed egg dummies, on their anal fins. During spawning, the female repeatedly deposits a few eggs on the spawning site and then immediately takes them up in her mouth. The male spreads his anal fin against the bottom and the female mouths the egg dummies as the male ejaculates. (B) In some species, females instead mouth "genital tassels," which are elongate, orange lobules that grow from the genital region. (C) Other males have greatly elongate pelvic fins with enlarged, conspicuous tips that reach to the cloaca. All such structures and behaviors may facilitate fertilization, assure paternity, and minimize predation on newly laid eggs. From Fryer and Iles (1972), used with permission.

are especially common among coral reef species, engaged in by at least 164 species in 26 families, the best known being seabasses, snappers, wrasses, parrotfishes, and surgeonfishes (Claydon 2005). As a possible mechanism to help overcome behaviors that would be counterproductive at this critical moment in a fish's life, many fishes exhibit **spawning stupor**. When in a spawning aggregation or mode, species that are normally difficult to approach or are very active instead move slowly, in an almost trancelike state. They take little or no evasive action when approached by predators or divers. Spawning stupor has been observed in minnows, suckers, mullets, silversides, seabasses, pompanos, and snappers (Johannes 1981; Helfman 1986; Johannes et al. 1999). Most observations are anecdotal, leaving open an excellent opportunity for quantified or manipulative investigations that have grown in importance because aggregating spawners are especially vulnerable to overexploitation (Fewings & Squire 1999, Sadovy & Cheung 2003).

Parental Care

Extent and Diversity of Care

Parental care is surprisingly common, widespread, and diverse in fishes (Breder & Rosen 1966; Blumer 1979, 1982; Baylis 1981; Potts 1984; Keenleyside 1991; Sargent & Gross 1994; DeMartini & Sikkel 2006; Balshine and Sloman 2011). Although most species scatter or abandon eggs upon or after fertilization, approximately 90 of the *c.* 460 families of bony fishes include species that engage in some form of defense or manipulation of eggs and young (Table 17.2; Mank et al. 2005 put the number closer to 30%, or 150 families). Parental care is more practical, and perhaps necessary, for demersal or adhesive eggs that are likely to be found by predators searching along the bottom or among plants or other structure.

A wide range of behaviors and levels of energy investment are all considered forms of parental care. These include: construction and maintenance of a nest; burying eggs once deposited in the nest; chasing potential egg and fry predators from the nest; fanning or splashing eggs or young with the mouth, fins, or body to provide oxygen and to flush away sediments and metabolic wastes; removal of dead or diseased eggs; cleaning eggs by taking them into the mouth and then returning them to the nest; carrying eggs or young in the mouth or gill chambers, in a ventral brood pouch, or externally on the head, back, or belly; coiling around the egg mass to prevent desiccation; retrieving eggs or young that wander from the nest or aggregation; accompanying foraging young and providing refuge or defense when predators approach; secreting specialized mucus that inhibits pathogen growth or that free-swimming young eat; and provisioning young or aiding them in the capture of food.

Internal gestation of young is a special type of parental care shown by many Chondrichthyes and a limited number of bony fishes. Placental connections, common in Chondrichthyes, occur in only one osteichthyan family, the Mexican goodeids (Blumer 1979; Wourms 1988) (Fig. 17.6), although transfer of nutrients directly from mother to developing young (**matrotrophy**) has been demonstrated in poeciliid livebearers and sebastine scorpaenids (Marsh-Matthews et al. 2001; DeMarais & Oldis 2005; DeMartini & Sikkel 2006).

Although much guarding involves simple behavior derived from everyday activities, such as fanning eggs with the pectoral

TABLE 17.2 A classification of reproductive guilds in teleost fishes, based largely on spawning site and parental care patterns. Specific examples of many groups are given in Table 20.1.

I. *Nonguarding species*
A. Open substrate spawners:
1. Pelagic spawners
2. Benthic spawners:
a. Spawners on coarse bottoms (rocks, gravel): (i) pelagic free embryo and larvae; (ii) benthic free embryo and larvae
b. Spawners on plants: (i) nonobligatory; (ii) obligatory
c. Spawners on sandy bottoms
B. Brood hiders:
1. Benthic spawners
2. Cave spawners
3. Spawners on/in invertebrates
4. Beach spawners
5. Annual fishes
II. *Guarders*
A. Substrate choosers:
1. Rock spawners
2. Plant spawners
3. Terrestrial spawners
4. Pelagic spawners
B. Nest spawners:
1. Rock and gravel nesters
2. Sand nesters
3. Plant material nesters:
a. Gluemakers
b. Non-gluemakers
4. Bubble nesters
5. Hole nesters
6. Miscellaneous materials nesters
7. Anemone nesters
III. *Bearers*
A. External bearers:
1. Transfer brooders
2. Forehead brooders
3. Mouth brooders
4. Gill chamber brooders
5. Skin brooders
6. Pouch brooders
B. Internal bearers:
1. Ovi-ovoviviparous
2. Ovoviviparous
3. Viviparous

Adapted from Moyle and Cech (2004) and Wootton (1990, 1999), based on Balon (1975a, 1981).

fins to maintain adequate oxygen levels (Green & McCormick 2005), some parental activities represent surprisingly unique specializations. For example, in many seahorses and pipefishes, females deposit their eggs at the entry to brood pouches on the male's belly, a structure that varies in complexity within the family. The male fertilizes, retains, and protects the eggs and young inside the pouch, helping them osmoregulate and providing them with oxygen and perhaps nutrition until they reach a relatively advanced stage of development (Wilson et al. 2003a). Birth involves contractions and contortions by the male that expel the young from the pouch. In the kurtoid nurseryfishes, an advanced perciform suborder of the southwestern Pacific, males develop a unique, downward-bent, hook on their foreheads to which the eggs are attached and carried until hatching (Fig. 17.7). The hook develops as a modification of the supraoccipital crest of the skull and is covered by highly vascularized, folded skin. Just how the eggs get there is a matter of conjecture (Berra & Humphrey 2002).

One of the most unusual parental patterns is shown by the Spraying Characin (*Copella*, Lebiasinidae), which deposits its eggs out of water. The male and female line up under a leaf and leap together as much as 10 cm into the air, turning upside down and adhering to the leaf's underside momentarily. In this manner, a dozen or so fertilized eggs are stuck repeatedly to the leaf. Over the next 2–3 days, the male moistens the egg mass by splashing it at 1 min intervals with flips of his tail (Fig. 17.8). Newly hatched young fall into the water (Krekorian & Dunham 1972).

Preventing desiccation of eggs exposed to air also explains unusual parental care in intertidal species. Several small, elongate intertidal fishes such as pricklebacks, gunnels, and wolf-eels coil their bodies around the egg mass as the tide goes out, thus trapping a small pool of water around the eggs (Blumer 1982). Other fishes that spawn in the intertidal, such as temperate wrasses and sculpins, cover the eggs with algae, thus controlling desiccation during low tides (Potts 1984). Rockhopper Blenny, *Andamia tetradactyla*, spawn in supralittoral nests in crevices that are above the high water mark for about 12 h daily, the male remaining with the eggs throughout the day (Shimizu et al. 2006). Depositing eggs above the intertidal – covered with algae, deposited among rocks, or buried in sand – has many surprising advantages, once desiccation is prevented. Higher incubation temperatures, higher oxygen concentrations, and reduced predation are among the considered benefits (DeMartini 1999).

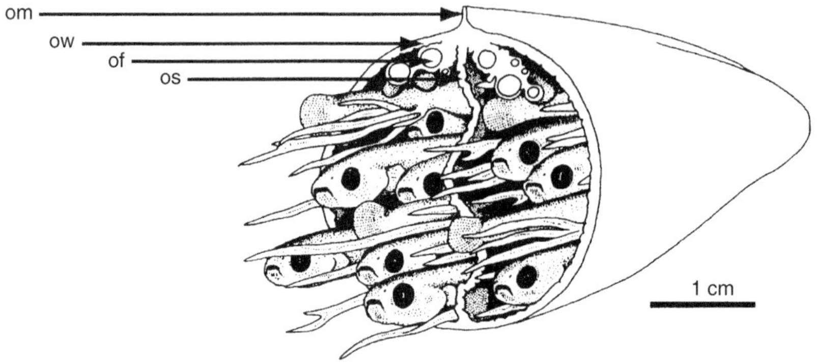

FIGURE 17.6 Well-developed (near-term) embryos in the ovary of a Mexican goodeid, the Butterfly Splitfin, *Ameca splendens*; 13 embryos are visible. The anterior third of the ovary is not shown. Fingerlike extensions projecting forward from the ovary are trophotaenia, which are epithelial structures that grow from the embryos' anal regions and serve to take up nutrients provided by the mother. Trophotaenia have evolved convergently in goodeids, ophidioids, and embiotocid surfperches. of, oocytes; om, ovarian mesentery; os, ovarian septum; ow, ovarian wall. Regrettably, *A. splendens* is considered to be extinct in the wild, although it is commonly kept and bred in aquaria (IUCN 2006). Drawing by J. Lombardi, from Wourms et al. (1988), used with permission.

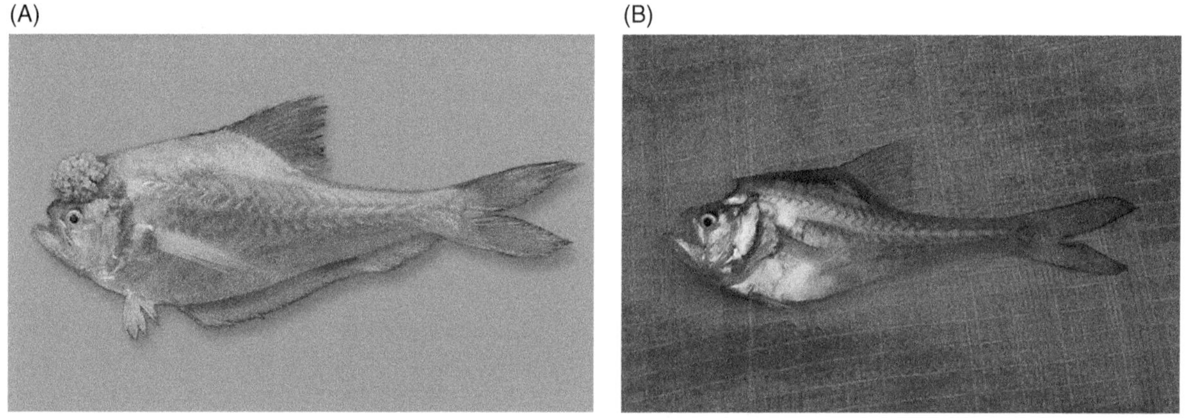

FIGURE 17.7 (A) A preserved male Nurseryfish, *Kurtus gulliveri*, with eggs attached to his occipital crest. This specimen, estimated to be c. 210 mm SL, was collected in Papua New Guinea in 1984 using rotenone and photographed by Kent Hortle (photo used by permission of publisher, Springer, and Kent Hortle). (B) The bony spur on the occipital crest of the males keeps the egg mass in place (*photo by Tim Berra, used with permission*). Males with eggs have been captured by subsequent sampling efforts using nets, but the eggs became dislodged upon capture. Berra and Wedd (2017) / with permission of Springer Nature.

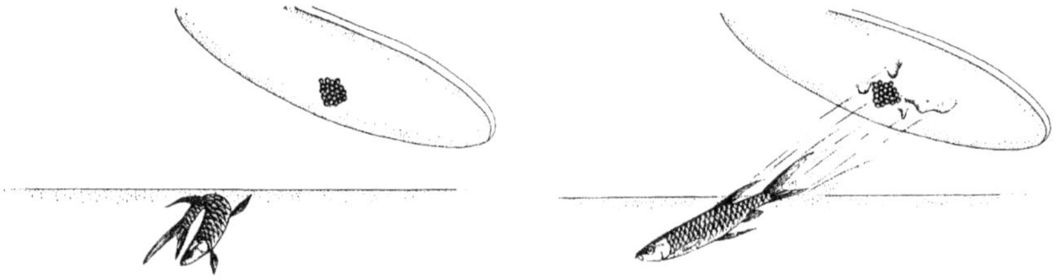

FIGURE 17.8 Parental care in the Spraying Characin, *Copella* sp. Eggs in this species are deposited on the undersides of overhanging vegetation, out of the water. The male guards the eggs, splashing them periodically with his tail to keep them moist. Krekorian and Dunham (1972), used with permission.

A few fishes provide food for their young via epidermal secretions. Such **trophic provisioning** has been observed in bagrid catfishes and in several cichlids (see Balshine and Sloman 2011, Sloman and Buckley 2011), including discus (*Symphysodon*), Midas Cichlid, as well as members of the genera *Aequidens*, *Etroplus*, and *Oreochromis* (Fig. 17.9A). This form of parental care is suspected in numerous other cichlids, and in a bonytongue and a damselfish. Thickened scales and increased mucus production have been identified in the adults of provisioning species. The importance of provisioning relative to

(A)

(B)

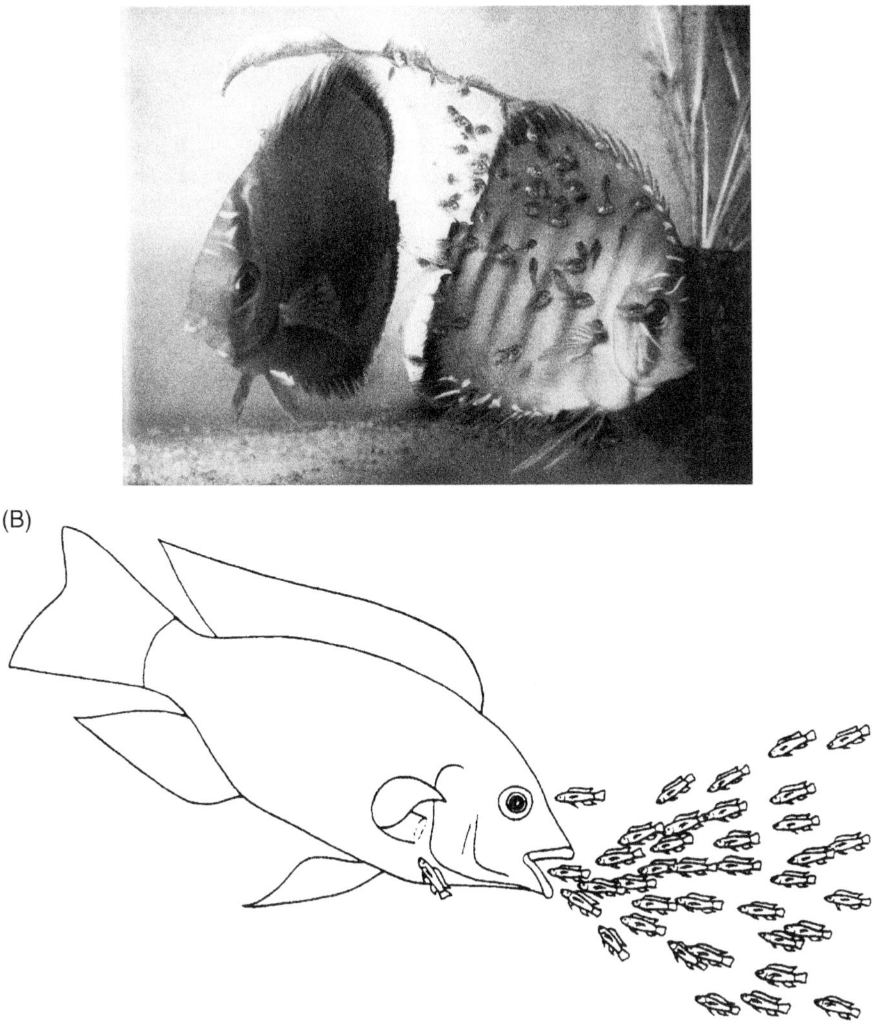

FIGURE 17.9 Parental care in cichlid fishes. Two of the more striking forms of parental care exhibited by members of the diverse cichlid family are shown. (A) Provisioning young. Few fishes with external fertilization actually provide nutrition for their young. Many cichlids are suspected of provisioning, but the behavior is best known in the discus, *Symphysodon*. A pair of discus is shown, with the young feeding on the mucus secretions of the female. (B) Mouth brooding. A female opens her mouth after signaling danger to a shoal of young. Mouth brooding of eggs is fairly widespread in fishes, but brooding of free-swimming young is relatively rare. (A) Herald (1961) / with permission of Penguin Random House LLC; (B) From Fryer and Iles (1972), used with permission.

other food sources of the young is unclear (Noakes 1979). Provisioning in the form of mucus secretions may also provide some benefit to immunity; immunoglobulin has been found in mucus of some species (Sloman and Buckley 2011).

Many sharks, and perhaps the living coelacanths, produce trophic eggs that are eaten by developing embryos prior to hatching (Wourms 1981; Heemstra & Greenwood 1992; Sloman and Buckley 2011). One bagrid catfish in Lake Malawi, Africa, feeds trophic eggs to free-living juveniles. *Bagrus meridionalis* young position themselves under the vent of the guarding female and apparently ingest eggs as they are extruded by the mother; 40% of the young in a nest may have such eggs in their stomachs. Circumstantial evidence indicates that the parental male also helps feed the young by uncovering invertebrates present in the nest or even by spitting out invertebrates captured elsewhere into the nest (LoVullo et al. 1992).

Epidermal secretions serve additional, care-giving functions. Male Fringed Darters, *Etheostoma crossopterum*, coat eggs with mucus secreted from their heads. This mucus has both antibacterial and antifungal activity, and egg clutches with a guarding male present had lower mortality and significantly reduced rates of fungal and bacterial infection (Knouft et al. 2003). Male Redlip Blennies, *Ophioblennius atlanticus*, and Peacock Blennies, *Salaria pavo*, produce a mucus enriched with antimicrobial substances from their specialized, sexually dimorphic, anal glands (Giacomello et al. 2006). When tending eggs, the males frequently rub their anal region over the nest surface, which could serve to transfer mucus to the eggs. Antimicrobial activity is turning up increasingly as different taxa are explored, and a new class of antibiotic peptides called **piscidins** has been isolated in fish mucus (Noga & Silphaduang 2003). It is a short adaptive jump from producing mucus to using such mucus to aid the survival of developing young, and it would not be surprising to find such parental specializations in other species.

The bottom of the ocean, lake, or a stream is a relatively hazardous environment for defenseless eggs, and many species carry the eggs rather than leave them deposited on the substrate. Mouth or oral brooding is the most common form of egg-carrying, having been documented in at least six families (sea catfishes, lumpfishes, cardinalfishes, cichlids, jawfishes, and gouramis); gill chamber brooding occurs in North American blind cavefishes. Eggs are picked up, usually by the male, shortly after fertilization. In the case of some cichlids, eggs are fertilized in the female's mouth, where they are retained. In cichlids, oral brooding extends well beyond hatching. Free-swimming young forage as part of a shoal near the female. When predators approach, the female signals the young by backing slowly with the head down. The young swim toward her head and she sucks them into her mouth (Fig. 17.9B). Some predatory cichlids will ram the head region of females that are carrying young, forcing them to spit a few out, which are then eaten by the predator (McKaye & Kocher 1983). Forms of external carrying include attachment of eggs to the male's lower lip (suckermouth armored catfishes) or head (nurseryfishes), or the belly of either parent (bagrid and banjo catfishes) (Breder & Rosen 1966; Balon 1975a; Blumer 1982; Berra & Humphrey 2002).

The Gender of Care-Givers

The most common care-giver in fishes is the male. Males alone or in combination with females (biparental care) account for approximately 80% of 77 families in which the sex of the care-giver is known; males alone care for young in almost 40 families (Blumer 1979, 1982; Mank et al. 2005). The predominance of male parental care in fishes contrasts markedly with its occurrence in other vertebrates, where care by females (amphibians, mammals) or both parents (birds) is more common (post-hatching care is uncommon in reptiles). Male guarding may be explainable as an evolutionary result of external fertilization and a male's way of assuring he alone fertilizes a batch of eggs (**paternity assurance**) (Ah-King et al. 2005). To accomplish this, a male should: (i) provide a suitable locale where females will lay eggs to be fertilized; and (ii) guard the eggs so that no other male can fertilize them.

Paternity assurance was the likely driving force behind the evolution of brood pouches in pipefishes and seahorses. The female deposits eggs in the male's abdominal pouch, where only his sperm are likely to reach the eggs. In most species, extending care beyond the fertilization stage greatly increases the probability of successful hatching and dispersal, thereby increasing the likelihood that offspring will live to reproduce. Egg and larval predators are common in all environments, as are fungal infections. A guarding male can chase off fishes and invertebrates that might eat the eggs, and can remove diseased or dead eggs, thus slowing the spread of fungi and other infectious pathogens.

Males may care for young longer because females are more likely to spawn with males that already have eggs or young in the nest (e.g., Fathead Minnow, Three-spined Stickleback, Painted Greenling, River Bullhead, Tessellated Darter, and Browncheek Blenny). An unexpected outcome of a female preference for males with eggs is **nest and egg usurpation** (also known as **allopaternal** or **alloparental** care). Male sticklebacks will raid other males' nests, steal eggs, and deposit these eggs in their own nest. Male Fathead Minnow evict males from existing nests and then guard the acquired eggs. In **brood piracy**, a large male may usurp the nest of another male, spawn, and then abandon the nest to be guarded by the original territory holder (Van den Berghe 1988; Magnhagen 1992).

A related phenomenon is interspecific **brood parasitism** or **egg dumping**, where one species spawns in a nest constructed by and guarded by another species. Several species, including gars and minnows, spawn in nests guarded by male sunfishes, and Golden Shiner are known to spawn in nests of Bowfin and Largemouth Bass (Katula & Page 1998). Small minnows also spawn over the mound nests built by larger minnows, such as Bluehead Chub; the eggs are guarded by the large male chub. The chub may benefit by a dilution effect whereby predators are likely to eat the more numerous minnow eggs, whereas the minnow eggs receive the protection of a nest guarded by a large male. A dilution effect probably explains why bagrid catfish tolerate and guard cichlid young in their nests. Mistaken identity is unlikely – the guarding catfish parents selectively chase cichlid young to the periphery of the nest, exposing the cichlids to higher predation rates and decreasing mortality in the catfish young. The young cichlids benefit from the protection of two large catfish plus the mother cichlid that remains nearby (McKaye 1981b; Unger & Sargent 1988; McKaye et al. 1992).

Other examples of brood parasitism include the mochokid Cuckoo Catfish, *Synodontis multipunctatus*, of Lake Tanganyika, which parasitizes broods of several mouth-brooding cichlids by laying its eggs on the substrate as the female cichlid is picking up her own fertilized eggs. The young catfishes eventually eat the cichlid larvae (Sato 1986; Barlow 2000) (Fig. 17.10). Clariid catfishes in Lake Tanganyika are known to dump eggs in nests of auchenoglanidid catfishes (Ochi et al. 2001), and mixed species brooding has been observed in Lake Baikal sculpins (Munehara et al. 2002).

Brood parasitism carries obvious potential costs, as evidenced by the Cuckoo Catfish example, so it is not surprising that some species have tactics that apparently counteract such parasites. A Japanese fish, the taxonomically uncertain Aucha Perch, *Siniperca* (or *Coreoperca*) *kawamebari* (Sinipercidae), is parasitized by a native minnow with a shorter spawning season; egg dumping leads to higher predation rates on perch eggs. Female Aucha Perch, which normally prefer to spawn in nests with more eggs, avoid perch nests with high numbers of eggs during the minnow's spawning season (Baba & Karino 1998).

If male care evolved to ensure that no other males fertilized the eggs, then males would not be expected to provide care in 21 teleostean families with internal fertilization (Mank et al. 2005). This is almost universally true and even applies to species that are exceptional relative to the familial norm. For example, most sculpins have external fertilization and male parental care. In *Clinocottus analis* and *Oligocottus* spp., fertilization is internal and male parental care is absent (Perrone & Zaret 1979). A cardinalfish, *Apogon imberbis*, exhibits internal fertilization and male care; spawning occurs repeatedly over an extended 5-day period, the male chases off other males, and

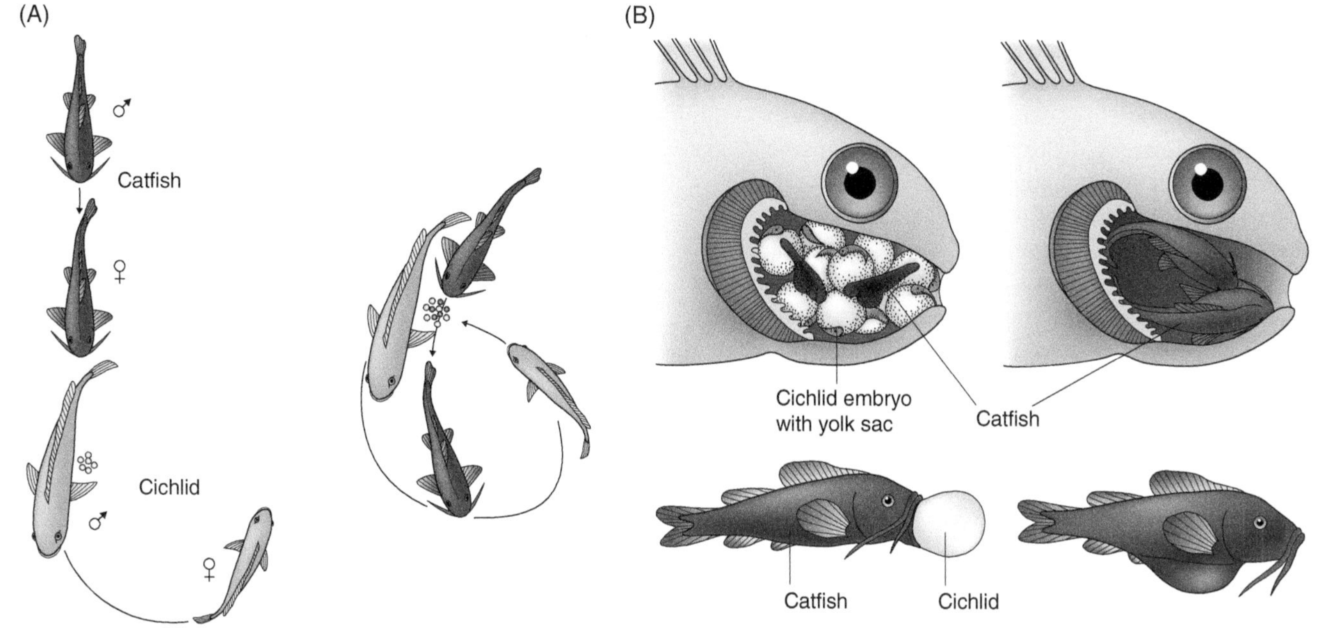

FIGURE 17.10 (A) Cuckoo Catfish pairs (darker fish) follow close behind spawning, mouth-brooding cichlids, laying their eggs amongst the fertilized cichlid eggs. The female cichlid picks up the catfish eggs along with her own. (B) In the mouth of the mother cichlid, catfish young hatch earlier and develop faster than cichlids, eating first the yolk sacs and eventually entire cichlid larvae. After Sato (1986) and Barlow (2000).

he also picks up eggs in his mouth immediately after they are deposited (Blumer 1979).

A final category of care deserving attention is the phenomenon of **cooperative breeding**, or helpers at the nest. Nonparental care-givers, usually young from a previous breeding episode, remain with the parents and feed and defend new young or defend and maintain the territory. Such helpers occur in over 150 bird and 25 mammal species and in at least 19 species of Lake Tanganyika cichlids (Taborsky 1984; Heg & Bachar 2006). Two of the best-studied Tanganyika species are *Neolamprologus pulcher* and *Lamprologus brichardi* (Stiver et al. 2006). In *L. brichardi*, helpers remain for about a year through two to three subsequent breeding cycles. They clean and fan eggs, larvae, and fry, remove sand and snails from the breeding hole, and defend the parental territory. Helpers suffer slower growth rates than nonhelping individuals, but receive protection from predators due to territorial shelters and the protective activities of larger family members. Females with helpers produce more fry.

Helping generally imposes a cost because helpers do not reproduce directly while remaining with their parents. However, helpers may promote their fitness (contribution of genes to future generations) more by raising siblings than by attempting to breed on their own. Helping may, therefore, be an example where kin selection explains an apparently altruistic activity. It is somewhat remarkable that cooperative breeding in fishes has as yet only been observed in related cichlids in Lake Tanganyika (helpers among anemonefishes are suspected but to date been found lacking, e.g., Buston 2004). Cooperative breeding among cichlids is additional evidence of the tremendous ecological and evolutionary plasticity of that family (Kawanabe et al. 1997).

The Costs of Care

Caring for offspring by parents requires trade-offs. A guarding parent often has reduced opportunity to feed, which may reduce later gamete production (e.g., in salmon, ricefishes, and livebearers; Blumer 1979). Male Smallmouth Bass, *Micropterus dolomieu*, fan their nests 24 h a day when they have eggs and young there and rarely leave the nest to feed; mortality among males during their first breeding season can exceed 90% in some lakes (Wiegmann & Baylis 1995). Gillooly and Baylis (1999) measured change in the whole-body composition of male Smallmouth Bass in the field across an 8-day parental care period. Nest-guarding males lost an average of 3% of lean mass, a potentially significant amount given that breeding occurs shortly after fish come out of the winter starvation period. Other fishes known to suffer energy loss, decreased growth, delayed reproduction, compromised immune function, or higher mortality due to parental care include other centrarchids, Three-spined Stickleback, cichlids, cardinalfishes, and gobies (Chellappa & Huntingford 1989; Lindstrom 2001; Okuda 2001).

Some costs can be overcome in part if the male eats some of the eggs, a phenomenon known as **filial cannibalism** (Fitzgerald 1992; Manica 2002; DeMartini & Sikkel 2006). Such cannibalism is known in at least 17 teleost families, with starvation avoidance by the adult the most likely cause (Manica 2002). Feeding while guarding young is relatively rare in mouth-brooding species. Another cost incurred when nest guarding is lost opportunity to spawn, which compromises future reproductive output. The decision of when to abandon current progeny will therefore be influenced by how much a parent's guarding activities can reduce mortality in the current clutch versus what opportunities for breeding exist in the near future (Perrone & Zaret 1979; Sargent & Gross 1994). Short breeding seasons, scarce additional mates, and short lifetimes would favor parental care of existing offspring over searching for additional spawning opportunities. Females can exploit this dilemma by preferring males that are already guarding eggs (discussed earlier in this chapter).

Caring for young also carries predation risks. Brood defense may reduce predation on the young but simultaneously increases the parent's exposure to predators. Guarding parental sticklebacks, Pumpkinseed sunfish, and gobies take more risks as their offspring grow, indicating that the value of the brood can increase relative to parental survival during the parental care phase (Colgan & Gross 1977; Pressley 1981; Magnhagen & Vestergaard 1991). Finally, an inverse relationship often exists between degree of care and number of eggs produced. Pelagic egg scatterers produce hundreds of thousands or millions of tiny eggs that they abandon, whereas species that participate in extensive parental care characteristically produce relatively small clutches of dozens to a few hundred larger eggs. High-quality care may only be possible when small numbers of young are produced. The ultimate evolutionary product, however, is how many offspring make it into the next breeding generation. The existence of **alternative tactics** within and among species attests to the fact that no single reproductive system is universally optimal.

Alternative Mating Systems and Tactics

The literature on social and reproductive behavior in fishes has increasingly focused on the variety of tactics that fishes use, both among and within species. That interspecific differences should arise is not surprising given the different ecologies and evolutionary histories of different lineages. Intraspecific variation is more puzzling because we tend to think in terms of species characteristics and "species-typical" behavior.

In addition to initial and terminal males in sex-changing wrasses, small alternative males also exist among gonochoristic fishes such as minnows, salmons, midshipmen, sticklebacks, livebearers, topminnows, sunfishes, cichlids, wrasses, blennies, and gobies (Taborsky 1994) (Fig. 17.11). In the Bluegill, *Lepomis macrochirus*, larger, older **parental** males (17 cm long, 8.5 years old) construct nests, court females, and then guard the eggs that they fertilize. Two forms of younger males take advantage of the territorial behavior of the dominant parental males. **Satellite** males are intermediate in size and age (9 cm, 4 years old); they mimic female coloration and behavior and hence gain access to a nest, interposing themselves between the parental male and the female during spawning, thereby fertilizing some eggs as they are released. Smaller, **sneaker** males (7 cm, 3 years old) lurk in nearby vegetation and dart through nests during spawnings, releasing sperm as they swim near the mating couple. Thus, three options arise from two discrete alternative life histories: parental males that delay maturation, grow large, and begin spawning when they are older than 7 years old versus males that mature as small 2-year-old, acting first as sneakers and then later (when they achieve the size of reproductive females) as satellite males.

Pacific salmon of the genus *Oncorhynchus* demonstrate an analogous pattern (Fig. 17.11B). Male Coho Salmon, *O. kisutch*, occur as two types in spawning streams. Large (52 cm long, 2.5 years old), colorful **hooknose** males court the females that have dug linear nests (redds) in the gravel bottom. Breeding success is directly related to male size and proximity to females; larger fish fight successfully to be closest and thereby spawn most. A second group of males, called **jacks**, are smaller and younger (34 cm long, 1.5 years old). These males hide in nearby stream debris and dash onto the redd as the hooknoses are spawning with females. Intermediate-size fish are relatively rare, perhaps because they are too small to fight successfully and too large to hide successfully and because the extra year feeding at sea results in a significant increase in size. Again, sneaking provides a spawning opportunity for small males that otherwise could not compete with large males for access to females (Gross 1984).

In both examples, it is still unknown whether the alternative tactics are brought about by genetic or environmental influences or a combination of the two. Are sneaker and satellite males genetically programmed to behave as such, do they develop in response to immediate environmental conditions – including the density of larger, parental or hooknose fish – or do genes and environment combine to determine proportions of males with different mating habits? A combination of influences is indicated by the salmon data. Jacks develop from fish that grow faster when young. Clearcutting around streams typically raises stream temperatures and increases the amount of debris, promoting faster growth and thereby producing more jacks and more habitat favorable to jacks. Hatcheries favor and produce faster growing fish, and intense fishing pressure also targets the larger fish. Ironically, human activities appear to be selecting for smaller fish, which are less desirable from a fisheries management viewpoint (Gross 1991).

Alternative mating tactics representing variations on the above patterns exist in other fishes (e.g., Fallfish minnows, cichlids, Peacock Wrasses, blennies, gobies; Ross 1983; Van den Berghe 1988; Barlow 1991; Magnhagen 1992; Henson & Warner 1997; Taborsky 2001; Neat et al. 2003). In many instances, the existence of one pattern creates the conditions that favor the development of the other: sneakers depend on territorial males to provide them with breeding opportunities. However, the advantages of sneaking decrease as the density of sneakers increases due to competition. Alternatively, no single strategy may confer a consistently greater selective advantage over others, so each is favored at different times and is consequently maintained at least at low frequencies in a population. Finally, differing reproductive modes may represent nothing more than alternative, equally adaptive responses to similar environmental forces (Fischer & Petersen 1987).

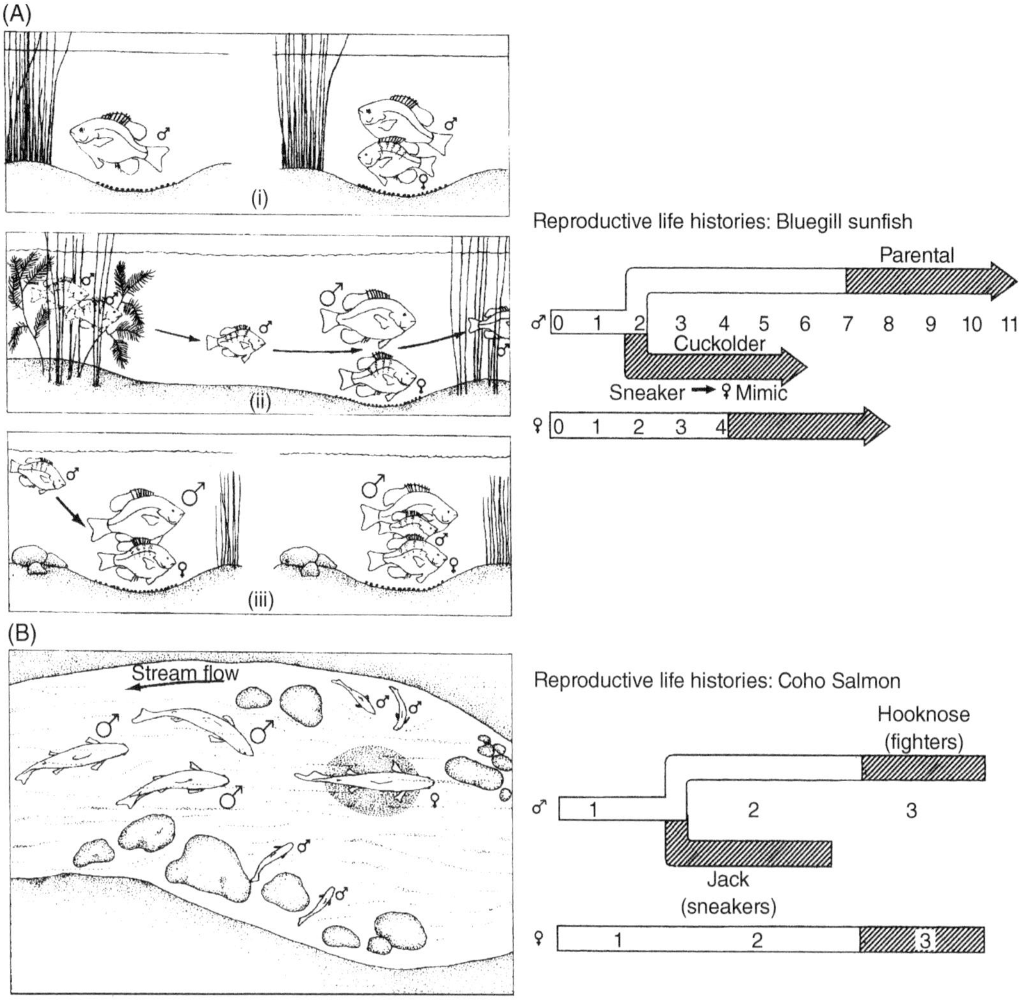

FIGURE 17.11 Variations in life history and behavior result in alternative mating tactics in male Bluegill sunfish and Coho Salmon. Both species are characterized by large, territorial males that court females and fight other males versus smaller males that interject themselves during spawnings by larger males. (A) Bluegill males occur as: (i) large males that dig nests, spawn, and guard eggs; (ii) small sneaker males that hide in vegetation and dart past spawning pairs, quickly depositing sperm; and (iii) intermediate satellite males that mimic female coloration and behavior and thus gain access to spawning pairs. The life history alternatives for male Bluegill are to mature at a young age and small size and adopt a sneaker and then later a female mimic role, or to mature later at a larger size and adopt a courting and parental role. All females mature at an intermediate size and age. (B) Coho Salmon males occur as large, hooknosed males that fight for access to females in midstream, or as smaller jacks that hide near structure or in shallow water and sneak copulations. Life history alternatives for the salmon are to mature at a young age and small size and adopt a sneaker role, or to mature later at a larger size and adopt a fighting role. All females mature at a relatively large size and old age. Modified from Gross (1984, 1991).

Genetic Resolution of Mating Systems

Molecular genetics, and particularly microsatellite markers, have launched a renaissance in the field of reproductive biology. Previous conclusions about breeding systems that have accrued over many decades, often requiring labor-intensive observations, can now be efficiently tested with individual-specific genetic markers. Questions about **monogamy** (couples mating only with each other), multiple paternity and maternity in egg clutches, egg thievery, and brood parasitism can be resolved. Microsatellites also allow genetic reconstructions of family pedigrees with a high degree of certainty. These genetic tools have highlighted the distinction between social mating systems, as defined by behavior, and genetic mating systems, as defined by relationships in a DNA-based pedigree. For example, genetic analysis of broods of Convict Cichlid *(Amatitlania siquia)*, a species that appears to be socially monogamous, show that juveniles being cared for by two parents as a single brood may in fact be from several different parents (Lee-Jenkins et al. 2015, Fig. 17.12).

Multiple paternity or maternity in a clutch of eggs can be readily detected based on the number of alleles observed at microsatellite loci. The methodology is straightforward in

FIGURE 17.12 Convict Cichlids raise young as socially monogamous pairs of parents. A detailed analysis of juveniles, however, revealed that broods being cared for by a pair of adults did not all have the same parents – 79% of broods evaluated had some "adopted" young (Lee-Jenkins et al. 2015). S. Olkowicz / Wikimedia Commons / CC BY 2.5.

diploid organisms: survey individuals in a brood (eggs or offspring) with microsatellite markers. At each **autosomal** (not sex-linked) locus, the maximum number of alleles in the offspring of a monogamous brood is four (two from the mother, two from the father). If five or six alleles are detected, at least three parents (usually including two fathers) are contributing, if seven or eight alleles are detected then at least four parents are contributing. Usually these assays are conducted with three or more microsatellite loci to attain reliable estimates of the number of parents.

To accurately reconstruct family relationships and corresponding breeding systems, it is preferable to genetically survey all candidate parents for a brood of offspring. Although this may be achievable in large mammals (including humans), it is seldom practical with fishes, and is impossible in pelagic-spawning marine fishes. In these cases, statistical methods can be employed to estimate parental assignments (Bernatchez & Duchesne 2000).

Marine Fish with Pelagic Larvae

The level of multiple paternity/maternity in marine fishes with **pelagic** (oceanic) larval dispersal is unknown. In fishes that spawn in aggregations (including many pelagic fishes), monogamy could be uncommon. Fishes that breed as stable pairs (including many coral reef fishes), however, could have a high degree of monogamy. For these cases, researchers have long wondered whether siblings could stay together during the pelagic larval phase, and recruit to the subadult habitat as a group of related individuals (Shapiro 1983). This runs counter to the long-held view that marine fish larvae are highly dispersive. The first genetic test of kinship in young-of-year reef fishes, based on three allozyme loci, found no evidence of related individuals in the Red Sea serranid *Anthias squamipinnis* (Avise & Shapiro 1986), and the same conclusion was reached for the clownfish *Amphiprion percula*, based on seven microsatellite loci (Buston et al. 2007).

However, several lines of evidence indicate that fish larvae have advanced swimming and navigational skills (Leis & Carson-Ewart 2000b) and some can recruit back to their region of origin (Jones et al. 2005). These observations resurrect the possibility that kin groups (siblings) might remain together in the pelagic phase and settle out on the same reef habitat. Planes et al. (2002) used allozymes to survey juveniles of the Unicornfish (*Naso unicornis*) recruiting to Pacific reefs and observed high relatedness within these groups. Pujolar et al. (2006) found high relatedness within some cohorts of the catadromous eel (*Anguilla anguilla*) recruiting to European streams. This is remarkable given that European eel larvae may spend more than a year in the pelagic zone prior to transforming into juveniles. Selkoe et al. (2006) conducted microsatellite surveys to assess recruits of the Kelp Bass (*Paralabrax clathratus*) on the West coast of North America and found siblings and half-siblings (sharing one parent) in seven out of 40 samples. Hence, evidence of kinship among recruits of marine fishes has been increasing. However, these studies indicate that the phenomenon is not consistently observed in all groups of recruits, even within a single species and region.

Nesting Fishes

Among the known egg-laying (**oviparous**) fishes, nest guarding (usually by males) is most common in freshwater species but also occurs in marine and anadromous fishes (Balshine & Sloman 2011, Katsiadaki & Sebire 2011). Here, the genetic surveys support previous suggestions that monogamy is frequently subverted by **sneaker males** (those that do not maintain a nest but deposit sperm into other nests), nest takeovers, and egg thievery. Furthermore, genetic studies have revealed the success rate of these alternative breeding strategies. In a review of the genetic literature on the mating systems of fishes, DeWoody and Avise (2001) reported that when males guard the nest, on average they retain about 70–95% of paternal contributions. The remainder can be either from males that maintain nearby nests, or sneaker males. In this review of 10 species and 177 nests, every species exhibited some level of multiple paternity within nests.

In addition to multiple paternity in male-guarded nests, egg contributions from multiple females are common. In those genetic surveys, microsatellite surveys were augmented with maternally inherited mtDNA, whereby the number of haplotypes indicates a minimum number of mothers. Based on a summary of 10 species, DeWoody and Avise (2001) reported a range of one to 10 mothers per nest, with an average of 3.1 females/nest. In the same survey, the authors reported that eggs from a single female are routinely found in multiple nests.

Egg thievery is a puzzling phenomenon wherein nesting males steal clumps of fertilized eggs from other nests – eggs that have no genetic contribution from their new guardian. In a survey of 24 nests of Fifteen-spine Stickleback (*Spinachia spinachia*), four had eggs that were probably stolen, as indicated by no maternal or paternal affiliation with nest mates (Jones et al. 1998). Why would a male deliberately guard and hatch eggs that are not his own? The most accepted explanation is that stolen eggs "prime" the nest for subsequent egg laying. Neophyte males may pose as successful breeders and guardians, thereby increasing their attractiveness to discriminating females.

Live-Bearing (Viviparous) Fishes

Internal fertilization guarantees that the caretaker is the biological mother. However, rates of multiple paternity are variable across the (primarily freshwater) fishes that bear live young. Chesser et al. (1984) used allozymes to survey broods of the Mosquitofish (*Gambusia affiinis*), concluding that 56% of females contained embryos from multiple males. However, a reexamination of this species with microsatellites revealed that the multiple paternity rate is near 100% (Zane et al. 1999). The available evidence indicates that multiple paternity is common and widespread in the live-bearing fishes, as originally predicted by Chesser et al. (1984).

All elasmobranchs have internal fertilization (excluding cases of parthenogenesis, Fig. 17.13) and most give birth to live young, although a minority, including skates (Rajiformes), horn sharks (Heterodontiformes), and Chimaeras (Chimaeriformes), lay egg sacks. Regardless of the oviparous or viviparous pathway, internal fertilization again guarantees that the female is the biological mother, and also seems to promote multiple paternity. Daly-Engel et al. (2006) used microsatellite data to detect multiple paternity in two out of three surveyed members of genus *Carcharhinus* (requiem sharks), indicating that the phenomenon may be widespread in elasmobranchs. Microsatellite surveys demonstrated that about 40% of Sandbar Shark (*Carcharhinus plumbeus*) litters in Hawaii have multiple fathers (Daly-Engel et al. 2007), compared to 86% of Lemon Shark (*Negaprion brevirostris*) litters in the Bahamas (Feldheim et al. 2004), and about 19% of Bonnethead (*Sphyrna tiburo*) litters in the Gulf of Mexico (Chapman et al. 2004). Microsatelite analysis of a female Bluntnose Sixgill Shark (*Hexanchus griseus*, Hexanchidae) and her 71 near-term pups revealed that they had been fathered by nine different males (Larson et al. 2011). Growing evidence indicates that multiple paternity is common but highly variable in elasmobranchs (Fitzpatrick et al. 2012).

FIGURE 17.13 Parthenogenesis has only rarely been observed among vertebrates in the wild. Fields et al. (2015) report the discovery of this form of asexual reproduction among Smalltooth Sawfish (*Pristis pectinata*). They speculate that this may have been prompted by the rarity of the species and the subsequent lack of available males. To quote the fictional Ian Malcolm (played by actor Jeff Golblum in "Jurassic Park") – "Life. . . finds a way". D. R. Robertson / Wikimedia Commons / Public Domain.

Mouth-Brooding Fishes

Fishes in the family Cichlidae have independently evolved mouth brooding in several genera. Fry are retained, primarily in the mother's mouth, after hatching (Goodwin et al. 1998, Balshine and Sloman 2011). The few genetic surveys conducted to date demonstrate both multiple paternity and, surprisingly, multiple maternity in female mouth-brooders. In the Blue Cichlid (*Pseudotropheus zebra*), microsatellite markers demonstrate multiple paternity in six of seven broods, and the female brooding the eggs was the mother in all cases (Parker & Kornfield 1996). In the Lake Tanganyika mouth-brooder *Tropheus moorii*, however, 18 of 19 broods examined with microsatellites had a single father (Egger et al. 2006). In the Lake Malawi mouth-brooder *Protomelas spilopterus*, microsatellite analyses reveal that four of six mouth-broods in females contained unrelated young at frequencies of 6% to 65% (Kellogg et al. 1998). In other words, females are brooding young from other females of their species. While this may be a simple mix-up between adjacent females, or maladaptive behavior, hypothesized benefits include attraction of mates, increased survivorship of siblings by dilution effect, and kin selection (aiding close relatives).

It is not clear that the mating behavior of mouth-brooders should be different from other egg-laying (oviparous) fishes. However, mouth brooding apparently reduces genetic mixing among populations more than other reproductive behaviors by eliminating the larval stage and reducing juvenile dispersal. Both the mouth-brooding Banggai Cardinalfish (*Pterapogon kauderni*, one of the few marine mouth-brooders) and the mouth-brooding tilapia (*Sarotherodon melanotheron*) show strong genetic separations among populations (Pouyaud et al. 1999a; Hoffman et al. 2005).

Pouch Brooding and Sex Role Reversal

The remarkable natural history of the family Syngnathidae (pipefishes and seahorses) has elicited much attention because of the "pregnant" (pouch-brooding) males. Just as internal fertilization guarantees that the viviparous female is the mother of offspring, the pouch brooding by male syngnathids assures that males brood only their own offspring. However, microsatellite studies indicate that the rate of monogamy varies from 10% to 100%, as males may carry eggs from a single female or from as many as six females (Jones & Avise 2001). These same studies indicate that females may contribute eggs to more than one male pouch (polyandry).

In most fish species, females make the greater investment in reproduction, and males must compete for the limiting resource, specifically access to egg-laying females. Sexual selection theory maintains that the gender competing for the limited resource will have more pronounced secondary sexual characteristics (such as bright coloration), will be under stronger sexual selection, and will show a tendency toward multiple mating. The sex role reversal of the syngnathids offers a rare mirror image of typical sex roles, and an opportunity to test sexual selection theories (Vincent et al. 1992). In most (but not all) syngnathids, the males' pouches, rather than the females' eggs, are the limiting resource. Hence females compete for space in these pouches and, consistent with theory, display characteristics that are usually associated with males. For example, when sexual dimorphism is apparent in syngnathids, it is usually the females that display the conspicuous ornamentation (Dawson 1985). Secondly, in the few cases where sexual selection (for reproductive success) has been measured in syngnathids, it is higher in females than males (Jones et al. 2001). Additionally, although there is considerable variation in syngnathid mating systems, microsatellite surveys show a range from monogamy to **polyandry** (multiple males mating with a single female), rather than the predominant **polygyny** (multiple females mating with a single male) observed in nesting fishes (Avise et al. 2002). The research to date generally confirms sexual selection theories that were originally formulated in the realm of male sexual selection and polygyny.

Fish Aggregation, Aggression, and Cooperation

Fishes associate during non-mating periods and may help or hinder one another. Social interactions, involving aggression or cooperation, occur between individuals of the same species (**intraspecific** interactions) as well as between different species (**interspecific** interactions). Nonreproductive social patterns in fishes involve solitary or territorial individuals, pairs, loose aggregations, and relatively permanent schools or colonies that may change daily, seasonally, and ontogenetically. Fishes keep apart or together through communication that may involve several sensory modes. In the rest of this chapter we review examples of non-mating social interactions in fishes, particularly the influence that communication has on patterns of aggregation, spacing, aggression, and cooperation, to show the diversity of evolved solutions to problems of survival and, ultimately, reproduction.

Communication

Communication involves the transfer of information between individuals during which at least the signal sender derives some adaptive benefit (Myrberg 1981). To send information, the signal sent must **contrast** with the background. Although this is often most obvious to many of us visually, such as when bright objects are most easily seen against dark backgrounds, or vice-versa, the contrast principle applies to all sensory modes. Background noise, be it visual, acoustic, chemical, tactile, or electric, will mask a signal. Information is transmitted when the signal exceeds the noise. Conversely, an animal becomes cryptic if it blends in with the background. The message sent usually results in repulsion or attraction or may inform the signal receiver about the physiological state or behavioral motivation of the sender. Frequently, signals from several modes are combined to enhance the message and reduce ambiguity.

Visual Communication

Vision plays a critical role in fish communication in most environments (see Chapter 6). Coloration is dependent on **hue** (wavelength mixtures), **saturation** (wavelength purity), and **brightness** (light intensity) (Hailman 1977; Levine et al. 1980). Coloration is incorporated into scales, skin, fins, and eyes as the product of pigments, achromatic elements, or structural colors. Pigmented cells (chromatophores) in the dermis contain carotenoids and other compounds that reflect yellow, orange, and red. Achromatics are black and white. Black coloration results from the movement of melanin granules within melanophores; dispersed melanin darkens a fish, whereas melanin concentrated in the melanophores makes the fish appear lighter in color. White coloration comes from light reflected by guanine crystals in leucophores and iridophores. Greens, blues, and violets are generally structural colors produced by light refracted and reflected by layers of skin and scales; the color depends on the thickness of the layers relative to the wavelength of the light (Lythgoe 1979; Levine et al. 1980).

The diversity of color in fishes is essentially unlimited, ranging from uniformly dark black or red in many deep-sea forms, to silvery in pelagic and water column fishes, to countershaded in nearshore fishes of most littoral communities, to the strikingly contrasted colors of tropical freshwater and marine fishes. Visibility (and invisibility) depends on a combination of **fish color**, the **transmission qualities** of water in specific habitats, **background** characteristics, and the **visual physiology of the eye**, especially the retina (Losey et al. 2003; Marshall et al. 2003a, 2003b).

Melanin and guanine reflect light across the entire visible spectrum and are therefore potentially useful in almost all habitats. Black and white are among the most commonly used colors in fishes (e.g., minnows, characins, catfishes, sunfishes, damselfishes, butterflyfishes, grunts, drums, cichlids, gobies, and triggerfishes). In the clear waters of a coral reef or tropical lake, yellow and its complement indigo blue are most visible; these are the colors commonly found on butterflyfishes, angelfishes, grunts, damselfishes, parrotfishes, and wrasses on reefs and on characins, minnows, Guppy, rainbowfishes, and cichlids in tropical waters. Nearshore temperate habitats, particularly in fresh water, tend to be stained with organic compounds that give them a yellowish tinge. Red and its complement blue-green are more visible under these conditions and it is not surprising that the breeding colors of minnows, salmonids, sticklebacks, darters, and sunfishes often incorporate these (Lythgoe 1979).

Colors on a fish's body may be used in **static** or **dynamic displays**. Static coloration generally provides information about the species, sex, reproductive condition, or age of a fish. Species identification is achieved through a combination of body form and color; ichthyologists as well as fishes use this combination for species' identification (Thresher 1976). In the myctophid lanternfishes, the number and pattern of photophores (light organs) is species specific and probably aids in schooling and in sexual identification. The taxonomic skills of many fishes are quite good; the Beau Gregory Damselfish can apparently distinguish among approximately 50 different species of reef fishes that intrude on its territory (Ebersole 1977).

Sexual dimorphism in coloration and body morphology is common in fishes, occurring as a permanent distinction in many tropical species or more seasonally in temperate fishes; generally males are the more distinctive sex. Ontogenetically distinctive coloration may aid in the identification of potential schoolmates, augmenting the tendency of fishes to aggregate with members of equal size. Juveniles and adults have different color patterns in at least 18 coral reef families (Thresher 1984). As French grunts (Haemulidae) settle from the plankton and take up residence on a coral reef, they develop at least four distinctive color phases associated with changes in habitat and behavior (McFarland 1980).

Dynamic displays involve either rapid exposure of colored, previously hidden structures or changes in color, and can include movements of the body, fins, operculae, and mouth. Often fin erection or gill-cover flaring exposes patches of color that contrast sharply with surrounding structures. Grunts open their mouths in head-to-head encounters to expose a bright red mouth lining. Many fishes flare their gill covers during aggressive, head-on encounters; gills and gill margins often contrast with the rest of the body (salmonids, centrarchid sunfishes, cichlids, labrids, and Siamese fighting fish). Fin erection and coloration play a dominant role in visual displays, probably because the movement associated with their erection is particularly eye catching. As a result, differential coloration of fins (inclusion of spots and stripes) is common. Fin flicking serves in calling young to parents, as a schooling signal, and during **agonistic** interactions which include both aggressive and submissive activities.

A special case of dynamic display involves the flashing of bacterially produced light by ponyfishes (Leiiognathidae), in which males "shine" their light inward toward a reflective coating of the gas bladder (Fig. 17.14). The light then passes outward through transparent skin and a moveable, muscular shutter in the body wall. Males in schools sometimes coordinate light flashing in spectacular, synchronized displays (Woodland et al. 2002; Sasaki et al. 2003).

Changeable colors can advertise alterations in the behavioral state of a fish, or conceal a fish from aggressors or predators. During agonistic, predator–prey, and breeding interactions, individuals will blanch or darken and develop bars or spots on a moment-to-moment basis (e.g., minnows, dolphinfishes, rudderfishes, cichlids, damselfishes, surgeonfishes, tunas, and flatfishes). One can often predict the winner of a territorial encounter by observing differences in body shading. Color may change with time of day – even the most colorful fishes by day turn relatively dull or blotchy at night. For example, neon and cardinal tetras (*Paracheirodon*, Characidae), which are brilliant blue-green and red by day, assume an inconspicuous pinkish tinge as they rest on the bottom at night (Lythgoe & Shand 1983). Such changes suggest that many visually mediated agonistic interactions cease with nightfall, but that many piscivorous fishes are still capable of locating prey at night using visual cues (Helfman 1993).

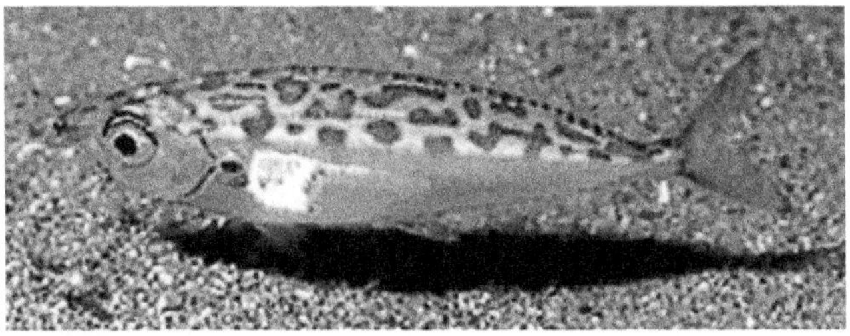

FIGURE 17.14 A nighttime photograph of a male ponyfish, *Leiognathus elongatus*, emitting light from its specialized circumesophageal light organ. The light display is the bright rectangular area just posterior of the pectoral fin. Light emission involves a complex series of structures and behaviors including bacterial light production, internal reflection, and transmission through a muscular shutter in a transparent section of the body wall. Four different light displays have been described in which duration and intensity of light emission are varied. Sasaki et al. (2003) / with permission of Springer Nature.

Short-term color change is primarily under the immediate control of the nervous system, whereas longer term ontogenetic and seasonal changes are more likely controlled by hormones. Seasonal color change is most often associated with the onset of breeding activity, when territorial males develop bright, contrasting coloration. In the spring, North American minnows and darters assume color patterns that rival fishes of tropical reef or river assemblage (Fig. 17.15). Females in many of these species undergo less dramatic seasonal changes. Interesting ontogenetic changes occur in migratory salmonids and anguillids. Many juvenile salmonids live in streams and combine countershading with vertically oblong, dark "parr" marks that may be disruptive in function. These fish migrate to the open ocean as smolts and develop a silvery coloration that is more effective camouflage in open water, pelagic situations. Upon returning to their natal (birth) stream, many species assume a bright, boldly contrasting breeding coloration that is the antithesis of camouflage. Anguillid eels change from transparent pelagic larvae to countershaded stream- and lake-dwelling juveniles, to bronze or silvery oceanic, reproducing adults.

The spectrum of human-visible light falls between 400 and 700 nm, perceived as violet to red colors. UV-A radiation lies between 320 and 400 nm and is invisible to humans. However, many fishes, from elasmobranchs to higher teleosts, have retinal pigments with maximal absorption characteristics well in the UV range (Losey et al. 1999a, 2003).

FIGURE 17.15 Some temperate freshwater fishes rival tropical fishes for bright colors – especially during the spawning season. (A) Two male Speckled Darter (*Etheostoma stigmaeum*) spar over spawning territory; (B) Spawning Rainbow Shiner (*Notropis chrosomus*). Courtesy of A. Nagy.

Hence, **ultraviolet** (**UV**) **reflectance** and **detection** is part of the visual world of fishes (Siebeck et al. 2006). Zamzow et al. (2008) showed that cones in the eyes of eight of 38 Hawaiian marine teleosts tested responded to UV light, and 10 other species could detect violet light, which has a lower wavelength than the blue light detected by human eyes.

Although UV light is scattered rapidly in water, biologically useful amounts of UV light penetrate clear aquatic environments to at least 100 m depth. UV light can be especially useful for detecting zooplankton against an open water background (Jordan et al. 2004). In a social context, the rapid scattering of UV light means that skin pigments that reflect UV, which have been found on the fins, head, and bodies in at least 21 families of reef fishes, will be visible only over short distances. This creates an ideal condition for social signaling at short range while minimizing recognition by other species such as predators (Cummings et al. 2003; Losey 2003; Zamzow et al. 2008). UV reflection and detection has increasingly proved to play a role in fish social behavior, including mate choice in Guppy and Three-spined Stickleback (Smith et al. 2002; Rick et al. 2006), shoaling decisions in sticklebacks (Modarressie et al. 2006), and territorial encounters in damselfishes (Siebeck 2004). UV detection may be important in the ability of fishes to detect **polarized light**, providing additional opportunities for target discrimination in foraging and social signaling as well as affecting orientation ability (Mussi et al. 2005). The visual world of fishes is very different from ours, and our attempts at interpreting visual and other signals require capitalizing on developing technologies and keeping an open mind.

Visual agonistic displays often involve highly **stereotyped** movements. Combat may involve lateral displays, where two fish swim in place with fins spread, oriented either parallel or antiparallel (head to tail) (Fig. 17.16). As an interaction

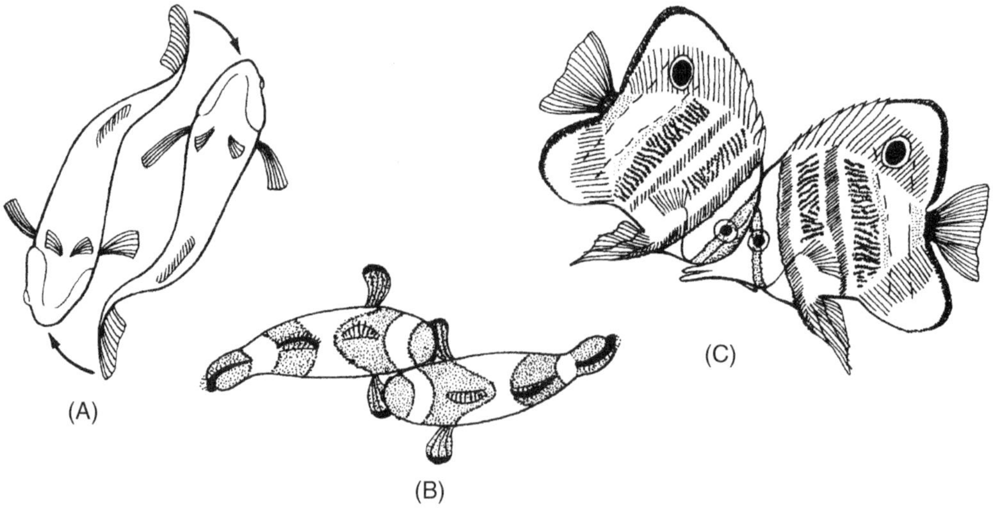

FIGURE 17.16 Lateral and frontal displays in fishes. During agonistic interactions, fish may line up parallel, antiparallel, or head-to-head and remain stationary, spread fins or operculae, change colors, and swim in place or circle one another. (A) Typical swimming-in-place lateral display when water currents (arrows) are directed at the head of the opponent, as happens in many cichlids. (B) Lateral display in the clownfish, *Amphiprion*, during which individuals strike each other with their pectoral fins. (C) Head-to-head pushing in the butterflyfish, *Chelmon rostratus*. (A) Chiszar (1978)/With permission of John Wiley & Sons; (B, C) Eibl-Eibesfeldt (1970)/with permission of Holt McDougal.

escalates, fish may begin body beating, a vigorous swimming-in-place that pushes water at an opponent and that may indicate the relative strengths of the combatants. Hence tactile and acoustic, near-field information may be added to the visual display. Antiparallel fish may strike one another with the pectoral fins (as in the anemonefish, *Amphiprion*) or may "carousel," swimming in tight circles around one another. Carouseling can lead to biting of caudal fins or chasing. Color changes frequently accompany lateral displays, and "color fights" occur in some species as different color phases indicate different levels of aggression (e.g., the nandid, *Badis badis*; Barlow 1963). Frontal displays, sometimes with fish facing each other head on and even grabbing each other's mouth, are also common (e.g., in grunts, Corkwing Wrasse, Kissing Gouramis).

Ritualized combat can decide the outcome of an interaction without actual physical fighting. It is in the best interests of both opponents to settle a dispute without incurring injury. The potential for such injury obviously varies among species, but can be considerable, as has been discovered by scuba divers who ignored the distinctive, ritualized, head-swinging displays of Gray Reef Sharks that had been approached too closely (Johnson & Nelson 1973) (Fig. 17.17). White Sharks also engage in apparently ritualized, agonistic displays toward other White Sharks, including parallel swimming and slapping the tail on the surface in the direction of another White Shark while feeding (Klimley et al. 1996).

A particularly good example of the multiple functions of visual displays involves the Eyelight Fish, *Photoblepharon palpebratus* (Anomalopidae; Morin et al. 1975). This 6 cm long, nocturnally active fish lives in the shallow waters of the Red Sea. It possesses a semicircular luminous organ just below each eye that contains continuously emitting bioluminescent bacteria. A muscular lid can expose or conceal the light organ. The Eyelight Fish is unique in that it forms shoals at night and uses its light for feeding, predator avoidance, and in behavioral interactions. The light organ is exposed to attract zooplankton prey and then to illuminate prey. If approached by a predator, the Eyelight Fish swims with the light organ exposed and then conceals it and changes direction, thus moving to a location contrary to its former direction of movement. In a social context, shoals form at night when small groups swim close enough to see each other's lights. Male–female pairs hold territories over the reef. If an intruding *Photoblepharon* approaches, the female swims up to it with her light organ concealed and then exposes it literally in the face of the intruder, causing it to depart (Morin et al. 1975).

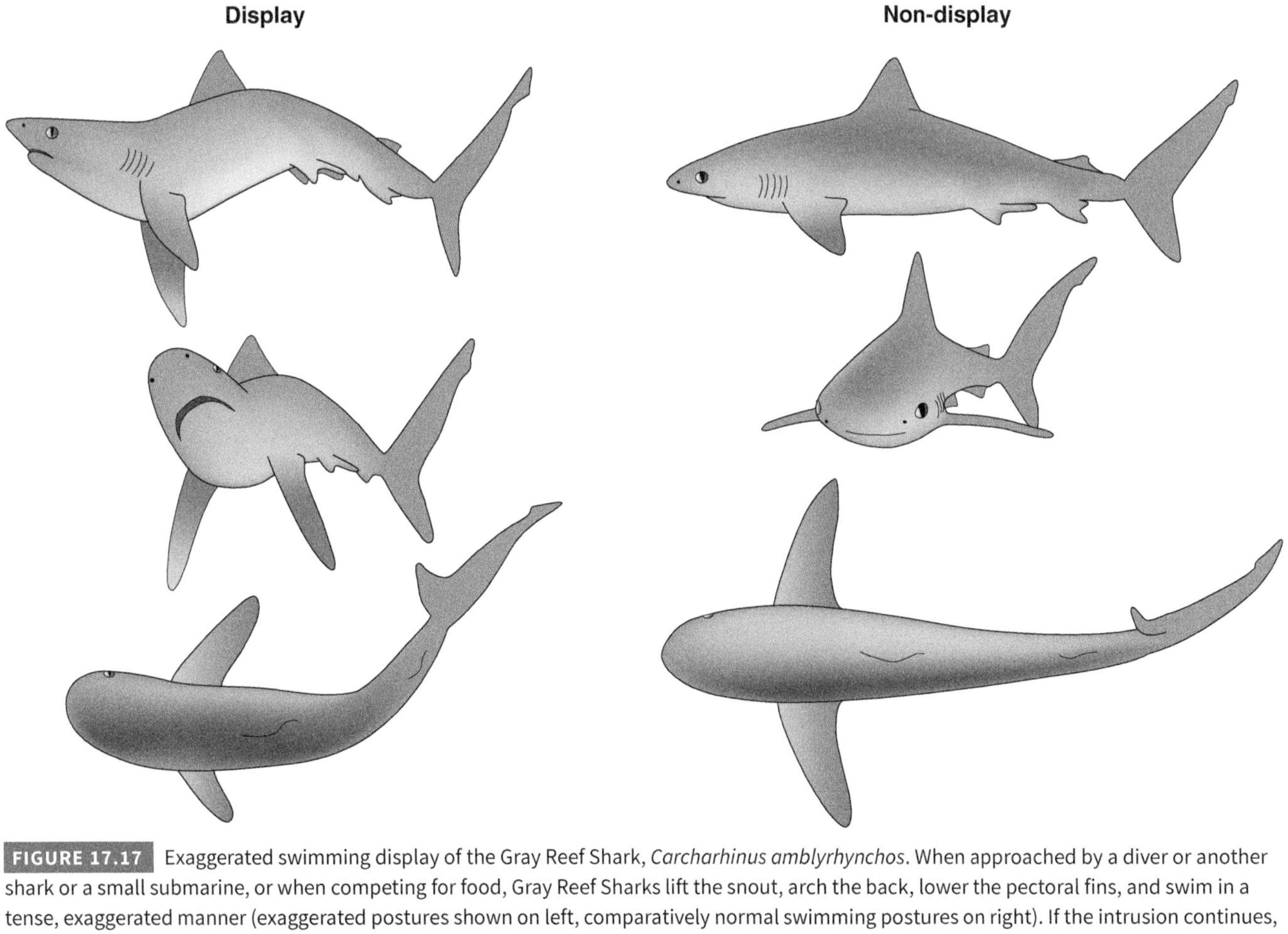

FIGURE 17.17 Exaggerated swimming display of the Gray Reef Shark, *Carcharhinus amblyrhynchos*. When approached by a diver or another shark or a small submarine, or when competing for food, Gray Reef Sharks lift the snout, arch the back, lower the pectoral fins, and swim in a tense, exaggerated manner (exaggerated postures shown on left, comparatively normal swimming postures on right). If the intrusion continues, the displaying shark may attack the intruder. Similar displays, without attacks, have been observed in Galàpagos, Silky, Lemon, and Bonnethead sharks, and a Bull Shark performed exaggerated S-turn swimming before attacking a small boat that had been following it (S. Gruber, pers. comm.). After Johnson and Nelson (1973), used with permission.

Coloration in Coral Reef Fishes No single topic has dominated the literature on coloration in fishes more than the question of why many coral reef species are so brightly and boldly ornamented. Most hypotheses have focused on coloration serving an **informational** or **anti-informational** function. Anti-informational, camouflage explanations were favored by earlier workers who felt that bright coloration helped fish avoid detection against the brightly colored background of corals, sponges, tunicates, and algae on the reef (Longley 1917). However, background matching (aside from countershading) more often characterizes benthic animals that are relatively immobile. Active reef fishes would continually change the background against which they were viewed, often making them a contrasting, conspicuous target.

Bright coloration may also serve to provide information, particularly among reef fishes, many of which are highly social and interact intra- and interspecifically. The clear waters of the reef offer an opportunity for visual signals to evolve. Reef fishes have been referred to as **poster colored** to emphasize their conspicuousness and the possible advertising function of their color patterns (Lorenz 1962; see also Breder 1949). The species-specific nature of color patterns also argues for their role in helping individuals tell species apart during mating, aggregating, or territorial encounters (Harrington 1993). Color patterns often differ among individuals, allowing **individual recognition** of territorial neighbors or of partners as a pair moves across the reef (Reese 1981; Wilson et al. 2006c). The placement of yellow, red, and black patches posteriorly may help the trailing member of a pair maintain visual contact in the complex reef environment (Kelley & Hourigan 1983). Some angelfishes and surgeonfishes are colorful and territorial as juveniles, but both color and aggressiveness fade later in life. Such an ontogenetic correlation between agonism and poster coloration is additional support for an informational function of reef fish coloration (Thresher 1984). Also, the eye of many reef fishes is often exaggerated and highlighted, sitting at the convergence of radiating lines or outlined in bright, contrasting colors (e.g., some seabasses, angelfishes, damselfishes, wrasses, jawfishes, clinids, surgeonfishes; also some centrarchid basses and cichlids).

Undisturbed reef systems contain an abundance of large, visually hunting predators (lizardfishes, trumpetfishes, cornetfishes, scorpionfishes, flatheads, groupers, hawkfishes, jacks, snappers, emperors, barracuda, flatfishes; Hobson 1994). How can small reef fishes afford to be conspicuous? The answer may lie partly in the reef structure itself. Few other habitats contain the variety and number of hiding places of a healthy coral reef. With adequate refuge sites available, the coloration of fishes that live close to the reef is less constrained by predators than in related, non-reef species. Such correlations hold well for families like damselfishes, wrasses, parrotfishes, rabbitfishes, dragonets, and gobies in which water column-, sand-, or grassbed-dwelling species are often countershaded or drab whereas near-reef species are more boldly colored.

Water clarity can also work against predators, whose activities are conspicuous to potential prey at considerable distances. Diurnally active prey fishes typically have eyes containing dense arrays of small cones that are ideal motion detectors during bright illumination (see Chapter 6, Vision). Any predatory movements will be detected at great distances, allowing prey to take flight or hide long before an attack occurs. Not surprisingly, nocturnally active fishes are not typically poster colored but instead possess relatively uniform coloration. Diurnal fishes seek shelter at night and assume subdued colors that include blotchy, presumably disruptive, camouflage hues. Bright illumination, clear water, and abundant refuges have apparently served to liberate coloration in diurnal reef fishes from its usual anti-informational, cryptic function to an informational, communicative function (Thresher 1977). An analogous pattern holds for African cichlids, where small species that live over and take refuge in complex substrates (rocks, snail shells) tend to be much more boldly colored than larger relatives that live over sand or in the water column (Barlow 1991).

It seems likely that reef fishes are both cryptically and conspicuously colored (Marshall et al. 2003b). Many early studies attempted to explain reef fish coloration from dead specimens or from photographs of live individuals taken with unnaturally powerful lights striking fish at unnatural angles. Behavioral observations may provide the best answers. Agonistic encounters occur between neighboring fishes, whereas predatory encounters occur over larger distances. The bright and contrasting colors that many reef species use in their signals and displays may not be visible over the distances at which predator–prey interactions occur because red, orange, and yellow wavelengths are attenuated much more quickly in clear water than are blues and greens (Marshall 2000). With the brighter colors less visible, such patterns as countershading may conceal a potential prey individual from a searching predator.

In addition, reef fish color is not static. Butterflyfishes are among the most colorful of the reef fishes. They and many other boldly colored reef fishes are also countershaded. During aggressive intraspecific encounters in the Raccoon Butterflyfish, *Chaetodon lunula*, the countershading fades and the yellow coloration intensifies. Intensification of species-specific coloration occurs in many other reef fishes as well as in temperate marine and freshwater species during mating and agonistic encounters (Thresher 1984). A reef fish can mask its poster colors to hide from a predator or to appease a competitor, or it can intensify coloration to intimidate the competition. We must conclude, therefore, that reef fish coloration is dynamic and multifunctional (Hamilton & Peterman 1971; Ehrlich et al. 1977; Marshall et al. 2003a, 2003b).

Acoustic Communication

Sound production occurs in well over 50 families of cartilaginous and bony fishes (Myrberg 1981, 2002; Hawkins & Myrberg 1983; Hawkins 1993; Ladich & Fine 2006, Kasumyan 2008). Sound production most commonly involves: (i) prey responses to being startled or handled by predators; (ii) mate attraction, arousal, approach, or coordination sounds; (iii) agonistic interactions with competitors for mates and resources; and (iv) attraction of shoal mates.

Startle and **release calls** occur in a wide variety of fishes and are elicited when a fish is grabbed, poked, or even surprised. A sudden grunt, croak, or drumbeat might distract a predator, perhaps causing it to release its grip on the prey or hesitate in its attack long enough for the prey to escape. A release call could also attract additional predators, including predators on the individual holding the signaler. A small predator with prey in its mouth could be handicapped in its own efforts to evade a larger predator and might abandon its meal rather than risk becoming one (Mathis et al. 1995). Release sounds could also function as alarm calls that notify conspecifics of a predator's presence and activity (see Chapter 16, Discouraging Capture and Handling).

Sound is an integral part of the **courtship and spawning** behavior of many fishes, as discussed earlier in this chapter. Some sounds produced by male damselfishes (Pomacentridae) and European croakers (Sciaenidae) drive off intruding males. Territorial males also produce vocalizations to bring females closer during courtship (e.g., toadfishes, centrarchid sunfishes, and gobies). Signaling rate frequently increases as a female draws nearer, or during the spawning act itself (cods, serranids), suggesting that acoustic communication synchronizes activities between members of a pair. In at least one species of an African mouth-breeding cichlid, male vocalizations stimulate gonadal activity in females, paralleling a widely observed phenomenon in seasonally breeding birds (Myrberg 1981; Lobel 1992). Marine and freshwater cods (Gadidae), some seahorses (Syngnathidae), sturgeon (Acipenseridae), cichlids (Cichlidae), minnows (Cyprinidae), and darters (Percidae) are all reported to use sound communication during courtship and spawning (Holt and Johnston 2014).

During **agonistic encounters** associated with territorial behavior, sounds are usually produced by an aggressive or dominant animal and the submissive animal typically retreats. Sound production during agonistic interactions occurs in many teleosts, including sea catfishes (Ariidae), loaches (Cobitidae), squirrelfishes (Holocentridae), butterflyfishes (Chaetodontidae), damselfishes (Pomacentridae), gouramis (Osphronemidae), and triggerfishes (Balistidae). Unique structures in butterflyfishes probably aid in detection of agonistic vocalizations – the laterophysic connection is formed from extensions of the anterior gas bladder that connect with the lateral line and project toward the inner ear (Webb 1998; Webb et al. 2006). In anemonefishes and other pomacentrids a unique sound-producing "sonic ligament" connects the hyoid bar (ceratohyal) and the inner part of the mandible and helps the fish close its mouth rapidly, bringing its teeth together and producing popping sounds (Parmentier et al. 2007). Submissive animals also produce sounds that may reduce aggression in an opponent, as recorded from anemonefishes (*Amphiprion*, Pomacentridae) (Myrberg 1981; Hawkins 1993). The catalog of sound-producing fishes and interesting acoustic adaptations continues to grow as more studies are conducted (Parmentier and Fine 2016).

Sound production also functions during **shoal formation and maintenance**. Most group maintenance sounds are produced by vibrating the gas bladder or stridulating of teeth, bones, and fin spines (see Rice & Lobel 2004; Amorim 2006). However, other mechanisms exist. Pacific and Atlantic Herring, *Clupea pallasii* and *C. harengus*, create trains of pulsed sounds by releasing gas from a duct near the anus. These sounds have been named **fast repetitive ticks** (FRTs), and may last up to 7 seconds. FRTs are emitted more often at night and FRT frequency increases as school size increases, suggesting that they serve to maintain contact between schoolmates (Wilson et al. 2003b). (Yes, some herring communicate through fish FRTs.)

Other group maintenance sounds result from water displacement by fins and bodies during swimming and are detected via the lateral line of neighboring fish. Such water displacement informs aggregating fishes of their location relative to schoolmates, serving as a minor repulsive force that combines with visual input to maintain distance between individuals. Pollack (Gadidae) that are experimentally blinded swim slightly further from schoolmates than when intact. Fish that are not blind but that have had acoustic information eliminated through severing the lateral line nerve swim closer than normal to schoolmates (Pitcher et al. 1976).

Eavesdropping by predators may be a significant cost of sound production. Many predatory fishes (sharks, groupers, snappers, black basses, jacks, barracuda, tunas) are attracted to the incidental, low-frequency sounds produced by feeding or injured fishes. Bottlenose dolphins (*Tursiops truncatus*) include a high proportion of sound-producing fishes (e.g., croakers, grunts, and toadfishes) in their diet (Barros & Myrberg 1987). And some believe that predators such as dolphins can hear the FRTs produced by herring, a prey species. Interception of signals, whether by predators, competitors, or potential prey, is always a potential cost affecting the evolution and use of communication signals by a species.

Chemical Communication

Exchanges involving chemicals primarily involve the release and reception of **pheromones**, which are chemicals secreted by a fish, detected by conspecifics, and produce a particular behavioral or developmental response in the receiving individual (Hara 1982, 1993; Liley 1982). Chemicals are sensed by both **gustation** (taste) and **olfaction** (smell) in fishes. Sensory receptors are often located not only in the mouth and nostrils, but also on the barbels or even the body surface in many fishes (see Chapter 6, Chemoreception), or on filamentous, muscularized pelvic fins (e.g., gouramis, Osphronemidae).

Chemicals play an important role in food finding and predator avoidance (see Chapter 16), mating, migration (Chapter 18), parental care, species and individual recognition, aggregation, and aggression in fishes. Cichlids rely on chemical communication for a variety of social interactions, including courtship, spawning, recognition of mates and offspring, social hierarchies, and warning conspecifics through the release of alarm substances (Keller-Costa et al. 2015). In salmonids, skin mucus contains species-specific amino acids that are used for individual and sexual recognition. Species recognition in other species is also mediated by chemicals in skin mucus (Hara 1993). Bullhead catfishes (Ictaluridae) and Eurasian Minnow (*Phoxinus phoxinus*) can recognize individual conspecifics based on odor. Several schooling species (herring, minnows, plotosid catfish, and young salmonids) show an attraction response to water that has contained conspecifics (Pfeiffer 1982). Chemical cues help monogamous Pebbled Butterflyfish (*Chaetodon multicinctus*) distinguish mates from other conspecifics within their feeding territories, but visual signals are important as well (Boyle and Tricas 2014).

Chemically mediated agonistic interactions include scent marking of territories or shelters, which takes advantage of the persistence of chemical signals relative to other sensory modes (Hara 1993). Members of sexual pairs of Blind Goby, *Typhlogobius californiensis*, defend a burrow against individuals of the same sex. Recognition of burrow mates and of intruders is based on chemical cues. Yellow Bullhead catfishes develop dominant–subordinate relationships that are mediated by chemical secretions. Bullheads also produce an aggression-inhibiting pheromone when living in groups. Fighting by aggressive individuals even decreases when they are exposed to water in which a communal group was living. In Siamese Fighting Fish, *Betta splendens* (Osphronemidae), males display more actively in front of mirrors when placed in water that had contained another male (Todd et al. 1967; Hara 1993).

Tactile Communication

Tactile information is transmitted at close range, when fish are in contact. Accurate information about the relative strength of combatants can be exchanged during pushing matches or mouth fighting. Many fights escalate into and end in biting. Anemonefish strike each other with their pectoral fins during antiparallel lateral displays (see Fig. 17.16). Fish frequently touch each other with tactile sensors, such as barbels in catfishes, loaches, and goatfishes, but also with long, filamentous pelvic fins (e.g., gouramis); searobins (Triglidae) have touch receptors in their separated anterior pectoral fin rays. Nuptial tubercles, epidermal bumps on the body and fins of many fishes, are used to stimulate potential mates and maintain contact between a pair during breeding (see Chapter 8). Courtship and copulation in many sharks involves the male biting the female, and he often holds onto the female's fins or body for prolonged periods (Pratt & Carrier 2001).

Touching between fishes is fairly uncommon, except during extreme fights, mating, and parent–offspring interactions. In parental species, young frequently contact the parent, usually using the mouth to feed on parental tissue or mucus. Such behavior also serves to maintain cohesion between parent and young, to promote parental behavior (perhaps by stimulating the production of parental behavior-inducing hormones), and to communicate the behavioral state of the young, such as hunger. Parent-touching behavior occurs in bonytongues (Osteoglossidae), catfishes (Bagridae), damselfishes, and more than 20 species of cichlids (Noakes 1979).

Electrical Communication

Fishes are unique in that some species both produce and receive electrical information based on very weak electrical output (see Chapter 6, Electrical Communication). The **electric organ discharge** (EOD) is species and often sex specific in South American gymnotiform knifefishes and African mormyriform elephantfishes. A fish can modify amplitude, frequency, pulse length, or interpulse length of its discharge, or alter parts of its EOD such as the fundamental frequency or peak power frequency. Fish can thus exchange information about species, sex, size, maturational and motivational state, location, distance, and individual identification. Electric discharges are used commonly during agonistic interactions (Bullock et al. 1972; Westby 1979; Hagedorn 1986; Hopkins 1986).

Much research has been conducted on the social context and function of EODs during courtship and territorial encounters in both groups (Møller 2006). Most but not all species have sexually dimorphic EODs. In apteronotid knifefishes, the male emits at a higher frequency in some species but in others it is the female that has a higher frequency discharge (Zhou & Smith 2006). Isolation of male hypopomid knifefish, *Brachyhypopomus pinnicaudatus*, led to a gradual decrease in the sexually dimorphic component of the duration and amplitude of its waveform. The differences were restored when a second fish was introduced to the test animal, suggesting that maintaining sexual differences in EOD comes at some cost, perhaps explaining why sexual dimorphism is not universal (Franchina et al. 2001). If two knifefish are emitting at the same frequency, the overlap can cause interference (jamming). A **jamming avoidance response** (see Chapter 6) is well known in gymnotids, whereby fish avoid jamming by shifting their EOD frequency away from that of nearby conspecifics. The Brown Ghost Knifefish, *Apteronotus leptorhynchus*, actively jams the output of others during competitive interactions (Tallarovic & Zakon 2005). Both male and female Brown Ghosts presented with actual or simulated (via electrical playback) intruders with a higher EOD frequency than their own raise their EOD frequencies to within potential jamming range.

In mormyriform fishes, shifts in EOD duration and phase amplitudes occur during agonistic encounters in juvenile as well as adult fishes, regardless of gender. EODs are used during interactions in combination with other display modes, utilizing multisensory communication systems that enhance signal transmission and reception (Schuster 2006). Interacting fish will head butt one another and also swim parallel and in place, which could push water and sound waves at the other fish as well as providing visual and tactile cues (Terleph & Møller 2003; Terleph 2004).

Agonistic Interactions

Aggressive interactions can result from competition or potential competition for valuable resources. Defendable resources include food and feeding areas, refuge and resting sites, mates and mating grounds, eggs, and young. Defense can produce dominance hierarchies in aggregating fishes or territoriality in more solitary species. In addition, a hierarchy can exist among neighboring territory holders, and dominant–subordinate relationships often exist when solitary fish meet.

Behavioral Hierarchies

Dominance hierarchies ("peck orders") can be linear or despotic. In **linear** hierarchies, an alpha animal dominates all others, a beta animal is subordinate to the alpha but dominates lower ranked individuals, and so on down to the last, or omega, individual. Such a hierarchy exists in harems of the sex-changing Cleaner Wrasse, *Labroides dimidiatus* (see Chapter 8). A single male dominates up to six females, which in turn have their own linear hierarchy. Linear hierarchies also exist in salmonids, several livebearers, and centrarchid sunfishes (Gorlick 1976). In **despotic** situations, a single individual is dominant over all other individuals, while subordinate animals have approximately equal ranks. In captive anguillid eels, a single large individual can monopolize 95% of a 300 L aquarium, relegating 25 other individuals to the remaining area where they mass together in continual contact. Despotic hierarchies have also been observed in Coho Salmon, *Oncorhynchus kisutch*, and bullhead catfishes, *Ameiurus* spp. (Paszkowski & Olla 1985).

Dominance can be determined by size, sex, age, prior residency, and previous experience. In general, large fish dominate over smaller, older over younger, and residents over intruders. In many species, males usually dominate females, whereas in others, such as the Guppy, females dominate males. Previous experience, in terms of recent wins and losses, often determines the outcome of future interactions; victorious fish tend to be aggressive and defeated fish submissive. Dominant fish typically occupy the most favorable microhabitats, relegating subordinates to suboptimal sites with respect to cover availability, current velocity, or prey densities. As a consequence, dominant individuals will have higher feeding rates, which ultimately lead to faster growth, better condition, and higher fitness (e.g., salmonids; Bachman 1984; Gotceitas & Godin 1992). Dominance hierarchies have also been observed in requiem and hammerhead sharks, minnows, ictalurid catfishes, amblyopsid cavefishes, cods, ricefishes (Oryziidae), topminnows, livebearers, centrarchid sunfishes, cichlids, labrids, blennies, and boxfishes (Ostraciidae).

Territoriality

Territoriality implies a defended space around some resource (Grant 1997). A territory may encompass several resources, as in male pomacentrid damselfishes in which the territory provides food (algae), a spawning site and the eggs spawned there, and refuge holes from predators where the territory holder also rests at night. Territories are often subunits of the larger **home range** occupied by an individual (see next section).

Territoriality is widespread in fishes, occurring in such diverse groups as anguillid eels, cyprinids, ictalurid catfishes, gymnotid knifefishes, salmonids (affecting stocking programs and the effects of introduced species), frogfishes, sticklebacks, pupfishes, rockfishes, sculpins, sunfishes and black basses, butterflyfishes, cichlids, damselfishes, wrasses, barracuda, blennies, gobies, surgeonfishes, and anabantids. In elasmobranchs, the threat responses of Gray Reef Sharks (see Fig. 17.17) may represent defense of personal space.

Territorial defense often involves displays such as fin and gill spreading, lateral displays and exaggerated swimming in place, vocalizations, chasing, and finally biting. Prolonged exchanges of displays frequently occur at territorial boundaries. When territories are being established or contested (as opposed to temporary trespassing), territory holders usually defeat intruders, previous winners defeat previous losers, and large fish defeat small fish. Territories near one another can create "territorial mosaics" of several contiguous territories (e.g., salmonids, pomacentrids, mudskippers, blennies; Keenleyside 1979).

Larger territories typically cost more to defend, in terms of energy and time expended, exposure to predators, and resource loss to competitors while defending distant portions of a territory. Food production affects territory size because a territorial animal must often meet its daily energy requirements from the resources available within its territory. As would be expected, increased food density leads to a decrease in territory size (e.g., in Rainbow Trout; rockfishes, Scorpaenidae; surfperch, Embiotocidae; several damselfishes; Hixon 1980a, 1980b). Interestingly, in Beau Gregory Damselfish, males decrease territory size with increasing food but females respond by increasing territory size. Larger females can produce more eggs and hence increased energy intake apparently overcomes the costs of defending a larger territory (Ebersole 1980).

Territoriality is often flexible. Territorial boundaries and intensity of defense can vary as a function of the relative impacts of different intruders. Herbivorous damselfishes defend a larger space against large competitors such as parrotfishes and surgeonfishes than against small damselfishes; damselfishes also tolerate large competitors for shorter times inside the territory, attacking them more aggressively. The strongest attacks are directed at potential egg predators, which have the greatest relative impact on the reproductive success of the damselfish. Juvenile Coho Salmon also defend a larger territory against larger conspecific intruders. Butterflyfishes chase species with which they overlap in diet but tolerate the presence of non-competitors (Myrberg & Thresher 1974; Reese 1975; Ebersole 1977; Dill 1978).

Territoriality may also vary over time at several levels. Juvenile grunts (Haemulidae) form daytime resting shoals over coral heads. Individuals stake out small territories of about 0.04 m^2 within the shoal; the territories often contain refuge sites from predators. These territories are defended vigorously with open mouth displays, chases, and biting. In the evening the shoal becomes a polarized school that moves from the reef to adjacent grassbeds to feed, and territoriality is not expressed during this migratory period. Once in the grassbeds, the shoals break up and the fish occur as widely spaced, foraging individuals, implying space-enforcing behaviors (McFarland & Hillis 1982). Territoriality changes with age in many species. Young Atlantic Salmon are territorial in streams. As they grow and their food requirements shift to larger prey, they move into deeper water and join foraging groups that have dominance hierarchies rather than territories (Wankowski & Thorpe 1979). In some species, agonistic interactions occur during the breeding season, with fish aggregating peaceably at other times (e.g., codfishes).

Home Ranges

Territories are usually a spatial subset of the larger area that a fish uses in its daily activities. These larger areas are **home ranges** or activity spaces and are common in fishes, which move over the same parts of the habitat at fairly predictable intervals, often daily but also at other timescales (Lowe & Bray 2006). Home range is dependent on fish size and species. Larger species and individuals generally move over larger ranges, although range size may decrease with growth in an individual (e.g., Bocaccio Rockfish, *Sebastes paucispinis*, Starr et al. 2002; Greasy Grouper, *Epinephelus tauvina*, Kaunda-Arara & Rose 2004). Home range may be very restricted, as in the few square meters around a coral head (e.g., gobies, damselfishes) or contained in a tide pool (e.g., pricklebacks) (Sale 1971; Horn & Gibson 1988; Kroon et al. 2000). Some benthic stream fishes may have ranges of 50–100 m^2 (Hill & Grossman 1987) whereas others may range over hundreds of meters (Albanese et al. 2004). Intermediate ranges of a few hundred square meters characterize many lake, riverine, kelpbed, and reef species, although many large coral reef fishes are relatively sedentary, utilizing concentrated reef resources. Home ranges of many large species such as groupers and snappers may not exceed 0.1 km^2 (Pittman & McAlpine 2003). Pelagic predators such as tunas, salmons, large sharks, and billfishes cross entire oceans seasonally or repeatedly. But even these oceanic wanderers show evidence of periodic residence in certain areas on a seasonal basis (see Chapter 18, Annual and Supra-Annual Patterns: Migrations).

Although a fish may spend 90% of its time each day within its home range, it is common to encounter individuals many meters or even kilometers away from their usual activity space. Such movements characterize fishes in most habitats (Fausch et al. 2002). These periodic excursions imply a well-developed **homing ability** in many species. Numerous studies, involving experimental displacements of tagged individuals, have repeatedly shown a strong tendency to return to home sites in many fishes. The Tidepool Sculpin, *Oligocottus maculosus*, can be displaced as far as 100 m from its home tide pool and will find its way back, using either visual or chemical cues. Older fish can still remember the way home after 6 months in captivity. Younger fish have shorter memory spans and require both visual and chemical cues to find home successfully, whereas older individuals do not require both types of information (Horn & Gibson 1988). In some species, adults find their way around the home range by identifying landmarks, creating a **cognitive map** of the locale (Reese 1989). In general, older fish have a stronger homing tendency and often occupy smaller home ranges than younger individuals which is not surprising because juveniles are the colonists that most often invade recently vacated or newly created habitat (Gibson 1993).

The use of a home range is affected by several components of a fish's biology. Normal ranges are often deserted during the breeding season. This may involve no more than a female damselfish having to leave her territory to lay eggs in the adjacent territory of a male, but can also involve long-distance movements of 100 km or more to traditional group-spawning areas, such as occurs in many seabasses on coral reefs. Colorado Pikeminnows, *Ptychocheilus lucius* (Cyprinidae), make annual round-trip movements of as much as 400 km between traditional spawning and normal home range areas. The home range also interacts with shoaling behavior in some species and can differ among individuals within a species. Yellow Perch, *Perca flavescens*, form loose shoals (see next section) of many individuals that forage in the shallow regions of North American lakes. Home range size is directly correlated with the amount of time individuals spend in shoals. Individuals with strong shoaling tendencies also have larger home ranges. As a shoal enters the residence area of an individual, the resident fish joins the shoal until the shoal moves to the boundary of the home range. Home ranges and fidelity to particular sites have probably arisen because intimate knowledge of an area increases an individual's ability to relocate productive feeding areas or effective refuge and resting sites, reducing the amount of energy expended and risk incurred while searching for such locales (Helfman 1984; Tyus 1985; Shapiro et al. 1993).

Knowledge of home range size has important implications for fish conservation, especially with regard to the creation of reserves and protected areas. Reserves must be large enough to encompass the home ranges of both sedentary and mobile species; without specific knowledge of daily, seasonal, and ontogenetic movements, a reserve might fail to encompass the range of habitats or the actual areal expanse needed to protect most species and most life history stages (Kramer & Chapman 1999; Cooke et al. 2005; Sale et al. 2005). The likelihood of spillover of individuals into adjacent areas, an anticipated benefit of reserve creation, also depends on movement and will vary in relation to reserve design and species behavior. Relationships between home range size and reserve design have been examined in tropical and temperate locales involving taxa as diverse as seabasses, sparids, goatfishes, wrasses, and surgeonfishes, to name a few (Meyer et al. 2000; Egli & Babcock 2004; Meyer & Holland 2005; Popple & Hunte 2005; Topping et al. 2005).

Aggregations

Shoals and Schools

The most obvious form of social behavior in fishes is the formation of groups, either unorganized **shoals** or organized, polarized **schools** (Fig. 17.18). By convention, some social attraction among individuals is required for a group to be considered a shoal or a school, whereas fish that are mutually attracted to food or other resources are an **aggregation** (Freeman & Grossman 1992). Shoals involve social attraction, coordination, and numbers. Two fish do not constitute a shoal because one fish often leads and the other follows. However, when three or more fish co-occur, each fish reacts to the movements of all adjacent fish. A shoal, therefore, is a group of three or more fish in which each member constantly adjusts its speed and direction in response to other group members; if the behavior is highly synchronized and fish swim parallel, the group is a school (Partridge 1982).

As many as half of all fish species may form aggregations at some time in their life. The chief functions seem to be to reduce the success of predators, increase foraging success, synchronize breeding behavior, and increase hydrodynamic efficiency, although aggregations can serve several purposes concurrently (see Parrish & Hamner 1997). Some species shoal throughout their lives (e.g., many herrings, anchovies, minnows, and silversides), others only as juveniles (e.g., Bowfin, plotosid catfishes, surgeonfishes, and pufferfishes). Some species aggregate when young, disband as juveniles or adults, and reaggregate to spawn, either in groups or as pairs (many salmonids, seabasses). Foraging aggregations may turn into breeding aggregations as fishes migrate to traditional spawning locations and are joined by members of other aggregations (e.g., Yellow Perch, grunts, rabbitfishes). Normally solitary adults may congregate during the winter and such aggregations probably remain together through a spring spawning season (e.g., carp).

Schooling tendencies may change with predation intensity. Guppy in predator-dense habitats school throughout their lives, but where predators are rare only juveniles school. Eurasian Minnow that co-occur with predators inherit a stronger schooling tendency than do minnows without predators (Magurran 1990). The prevalence of shoaling behavior particularly among juveniles and small species, both of which are more susceptible to predation, supports the antipredator function of most aggregations (Shaw 1970, 1978; Magurran 1990; Pitcher & Parrish 1993; see Chapter 16).

Regardless of function, most fish shoals are relatively unstable. Few fishes, at least in inshore locales, maintain

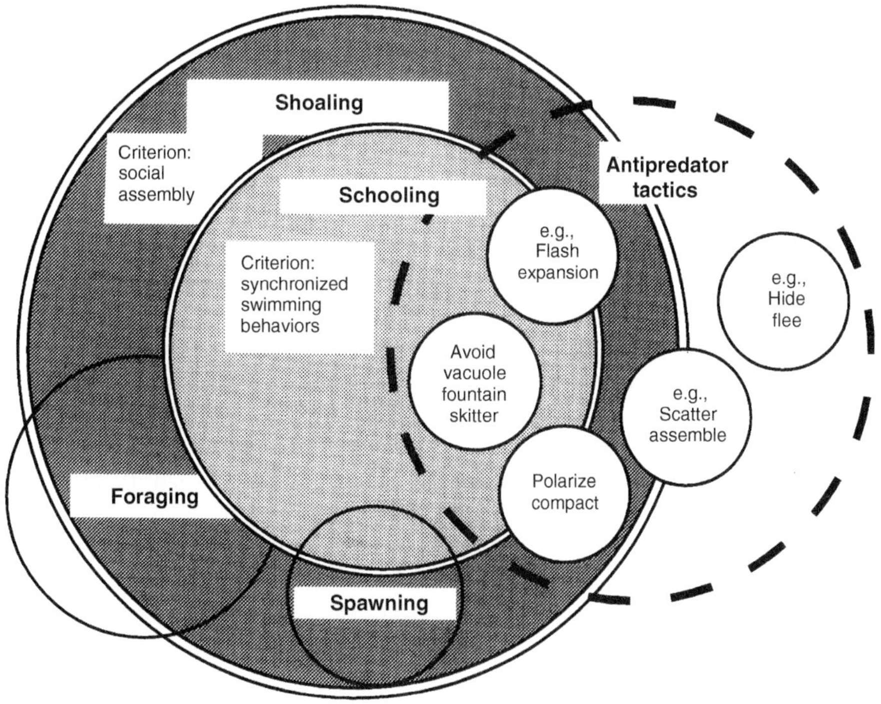

FIGURE 17.18 Types and activities of fish aggregations. Shoals contain fish attracted to one another but whose activities are only loosely coordinated. In schools, behavior is synchronized: fish often swim parallel, in the same direction, with fairly uniform spacing (the "polarized school" of many authors). Foraging and spawning groups generally form shoals, whereas predator avoidance often results in highly synchronized schooling activities. In this figure, five common antipredator actions and their relationship to grouping behavior are shown in the smallest circles. After Pitcher (1983), used by permission of the publisher Academic Press Ltd, London.

their groups through an entire 24 h period. Many shoals form each morning, disband at night, and reform the following morning but with different individuals. In fact, available evidence indicates little shoal fidelity in most fishes: dace, minnows, killifish, Yellow Perch, surgeonfishes, parrotfishes, and Bluegill sunfish join and leave foraging shoals frequently (Freeman & Grossman 1992; Hoare et al. 2000). Climbing Perch, *Anabas testudineus* (Anabantidae), even show a preference for shoaling with unfamiliar individuals, especially if unfamiliar fish are part of a larger group (Binoy & Thomas 2004), and Eurasian Minnow, will prefer shoals of unfamiliar fish when the unfamiliar fish are in a significantly larger shoal (Barber & Wright 2001). However, advances in molecular genetic techniques are revealing a degree of permanence in the form of **stable kin groups** in some wild fish schools (e.g., Brook Char, *Salvelinus fontinalis*; Fraser et al. 2005; Eurasian Perch, *Perca fluviatilis*, Behrmann-Godel et al. 2006).

Fidelity may also be strong in nocturnal fishes that reaggregate each dawn and form daytime resting schools at fixed refuge locales (e.g., squirrelfishes, grunts, copper sweepers, some bullhead catfishes; Hobson 1973; Helfman 1993). Cornish Jack (*Mormyrops anguilloides*) show group fidelity in forming nocturnal foraging groups, using their electric impulses to locate prey and also to communicate with and recognize one another (Arnegard and Carlson 2005). In some nocturnal species, relatively complex social structure and interactions develop that rival the societies of birds and mammals (see later section on Social transmission of cultural traditions). Migratory schools (e.g., large tunas, Bluefish) may also show strong group fidelity, but definitive information is lacking.

Fishes space themselves fairly regularly within fish schools (Partridge et al. 1983; Abrahams & Colgan 1985). In a theoretical ideal spatial distribution, school members should be 0.3–0.4 body lengths apart, five body lengths behind and centered between preceding fish, with neighbors beating their tails in antiphase (opposite directions). In this arrangement fish could gain a 65% energy saving from the wakes and vortices generated by fish around them (Weihs 1975). Although few, if any, groups achieve the proposed ideal structure (Pitcher & Parrish 1993; Parrish & Turchin 1997), considerable energy savings may accrue in a school because the fish are doing more than simply "drafting" in a region of reduced flow created by the fish ahead of them (Liao et al. 2003). Some fishes contract only anterior body muscles, allowing a passive wave of undulation to propagate posteriorly along the body, much as waves pass down a flag in the wind. A fish thus progressively slaloms between the vortices shed by the fish swimming ahead of it, matching its swimming movements in phase with oncoming drag vortices, rather than expending energy pushing off the vortices (similar energy savings are experienced by fish undulating behind a rock or other object in flowing water).

Earlier authors had proposed that a hydrodynamic advantage may also develop through drag reduction when one fish swims through the mucus produced by fish ahead of it in a school (Breder 1976). However, insufficient mucous is produced, even in large schools, to affect drag significantly (Parrish & Kroen 1988). Predation also disrupts spacing because individuals under attack should attempt to place schoolmates between themselves and a predator (the **selfish herd** phenomenon; Hamilton 1971). Again, however, experimental tests call into question whether central locations are in fact safer (Parrish 1989b). Many factors (size, sensory input to the lateral line, visibility of neighbors, swimming speed, species composition, vulnerability to and behavior of predators, and social status) must contribute to the exact (and variable) structure of schools.

Colonial fishes form essentially stationary aggregations. Colonies may exist for breeding, as when male sunfishes and cichlids and female triggerfishes aggregate and construct nests or set up display sites. Some damselfishes set up contiguous territories in suitable habitat patches on a coral reef. Three-spot and Bicolor damselfishes inhabit areas of a few square meters, even though adjacent, similar reef areas contain no such fishes (Schmale 1981). Garden eels (Congridae) occupy small burrows a few centimeters apart on sandy regions among coral reefs (Fig. 17.19). Jawfishes (Opisthognathidae), another burrowing coral reef form, tend to form colonies of two to nine individuals on rubble-strewn sandy bottoms (Colin 1973).

Optimal Group Size

Predatory success decreases as prey group size increases (Neill & Cullen 1974; Landeau & Terborgh 1986). Larger prey groups also experience greater competitive success, foraging efficiency, and hydrodynamic efficiency. Consequently, selection should favor prey fish that join and maintain large shoals. But benefits of larger groups also may be offset by competition for food and mates, by interference among individuals avoiding predators (e.g., confusion among group members due to collisions, indecisiveness, or obstructed views), and increased conspicuousness of large versus small groups (Pitcher & Wyche 1983; Abrahams & Colgan 1985; Parrish 1988). Given these costs and benefits, can an optimal group size be determined? Do fishes tend to form or join optimally sized shoals?

Optimal group size is complicated because antipredator functions probably favor larger optima than do feeding groups. When Fathead Minnow were allowed to choose between shoals of different sizes, they consistently chose the larger of the two shoals, particularly when one shoal was relatively small (less than eight fish) and a predator was present. Zebra Danio, *Danio rerio*, females also consistently chose a larger shoal, although male Danios showed no such preference. Banded Killifish, *Fundulus diaphanous*, chose larger shoals when under predatory threat but smaller shoals when feeding cues were present. If dominance hierarchies exist, dominant individuals, with their preferential access to resources, may have a larger optimum

FIGURE 17.19 Garden eels (Congridae) live in colonies of several hundred individuals on sand bottoms near coral reefs. Individuals feed on zooplankton during the day, often extending just the anterior portion of their bodies out of burrows, the sides of which are cemented with mucus produced by the fish's skin. Withdrawal of one individual into its burrow stimulates withdrawal of all other members of the colony. G. Helfman (Author).

than would subordinate members. Subordinate animals must decide between sustaining the costs of a large group versus being alone. Fishes alter their decisions with respect to shoal size in response to changes in the social and ecological context of shoaling (Hager & Helfman 1991; Hoare et al. 2004; Ruhl & McRobert 2005).

Social Transmission of Cultural Traditions in Fishes

To biologists, **culture** involves the behavioral transmission of information, and fishes in aggregations have the opportunity to observe and learn from one another. **Traditions** are social behaviors maintained across generations, either by inheritance or by learning, as when young individuals are taught by or observe and copy the actions of older individuals. Such social learning has been well documented in numerous non-fish species (Bonner 1980).

Fishes also exhibit a number of behavioral traditions, and the number of examples of **social learning** increases as studies expand (Brown & Laland 2003). The same breeding locales are frequently used year after year in both marine and freshwater fishes. Although part of this continued use relates to site-specific, appropriate conditions for spawning, dispersing, or caring for larvae, seemingly adequate, nearby sites are ignored while the traditional site continues to be used. Traditional breeding locales have been found in numerous fish species (e.g., herring, groupers, snappers, surgeonfishes, rabbitfishes, parrotfishes, wrasses, mullets; Loiselle & Barlow 1978; Johannes 1981; Thresher 1984; Turner 1993). The return of salmon to their natal stream to spawn is not included because information is obtained through individual imprinting and memory and may have a hereditary component (McIsaac & Quinn 1988).

The process by which traditions are established and maintained has been investigated with respect to breeding sites in wrasses and twilight migration routes in grunts. Female Bluehead Wrasses, *Thalassoma bifasciatum*, mate with solitary large males that hold territories. The locations of these territorial mating sites may remain stable for more than 12 years, encompassing four wrasse generations. Adjacent, seemingly appropriate sites are not used. Female choice of sites, rather than choice of males, determines where males establish territories. To test whether traditions were maintained by a genetic response or through social transmission, Warner (1988, 1990) removed entire populations from reefs and replaced them with naive individuals. He found that breeding sites chosen by transplanted groups were a random sample of the available locales and that newly used territories eventually became traditional breeding sites. Hence former traditional sites were likely maintained by social convention. Interestingly, if an additional removal/replacement manipulation was performed on the same reefs, the second group of transplants tended to prefer the same sites as the first transplanted group. This suggests that the fish assessed site quality and had definite preferences, which did not necessarily include the original, traditionally used spawning locales. Hence tradition is powerful enough that a breeding locale may continue to be used even though that locale may not be the best available in the habitat.

Juvenile grunts (*Haemulon*), as well as many other nocturnal reef fishes, undergo a remarkably predictable migration at dawn and dusk each day that probably thwarts the success of twilight predators (McFarland et al. 1979). Grunts feed on invertebrates at night in the grassbeds adjoining patch and fringing reefs. By day, they form resting schools over coral heads. The locations of such schools and the routes taken by the school resident over a coral head represent traditional activities. Over more than a 3-year period, a school will take the same approximate route to and from the grassbed, even though no individual grunt in the school is more than 2 years old. How do younger fish know the correct route between resting site and grassbed?

To test for the relative influences of genetic and social transmission, Helfman and Schultz (1984) transplanted individuals between schools of juvenile French Grunt. After mapping established migratory routes, members of distant groups were added to resident schools. Transplanted fish (identified by small injected paint marks) were allowed to follow residents for four twilights. Then all residents were removed and the migration of the transplanted fish was observed. The route was similar to the one taken previously by the residents. To test for a possibly innate response (i.e., given the terrain, any grunt at this locale would take the same route regardless of experience), new transplants were given no opportunity to observe migrating fish. These control fish migrated in a variety of directions. Hence, the social traditions of resting site locale and twilight migration route in grunts are established via **cultural transmission**. Learning through cultural transmission of information has been shown in several other contexts in several other fish species (Brown & Laland 2003; Kelley & Magurran 2003; Griffin 2004).

Interspecific Relations: Symbioses

Symbiosis is the living together of two unrelated organisms. In **parasitism** one member of a pair benefits and the other suffers a reduction in fitness, in **commensalism** one benefits and the other is neither harmed nor helped, and in **mutualism** both members of a pair benefit. Mutualistic relationships are particularly interesting because they suggest a relatively long period of co-evolution between the species. Fishes form mutualistic relationships with other fishes and with a variety of invertebrate species.

Parasitism

The abundance and diversity of external and internal parasites of fishes is tremendous and beyond the scope of this discussion (Dogiel et al. 1961; Sinderman 1990; Gabda 1991; Bush et al. 2001; Benz & Bullard 2004). Three families of fishes – synaphobranchid eels, trichomycterid catfishes, and carapid pearlfishes – include members that are internal parasites on fishes and other animals (carapids are discussed under mutualism and commensalism). The Snubnose Parasitic Eel, *Simenchelys parasiticus* (Synaphobranchidae), burrows into the flesh of bottom-living fishes such as halibut and has even been found in the heart of a Mako Shark (see Chapter 14, Cohort Elopomorpha). The trichomycterids feed on tissue and blood in a host's gill cavity. (The trichomycterid Candiru (*Vandellia cirrhosa*) of the Amazon River Basin has been reported to enter the urogenital orifice of some human bathers.)

Many fishes, including marine catfishes, characins, tigerperches (Theraponidae), carangids, sea chubs (Kyphosidae), sparids, cichlids, blennies, and spikefishes (Triacanthodidae) fall into a category somewhere between parasitism and predation by removing scales or taking pieces out of fins of other fishes (Losey 1978; Noakes 1979; Zander et al. 1999). Populations of scale-eating cichlids from Lake Tanganyika (*Perissodus*) contain even numbers of individuals whose mouths twist left or right, facilitating scale removal from the right or left sides of their prey, respectively (Hori 1993). Cookie cutter sharks, *Isistius* spp. (Squalidae), remove plugs of flesh from a variety of fishes and marine mammals (and rarely a human swimmer), and lampreys rasp through the skin of numerous fishes and feed on tissues and body fluids (see Chapters 12, 13). At least two species, the Cutlips Minnow, *Exoglossum maxillingua*, of North America, and the Eyebiter Cichlid, *Dimidiochromis compressiceps*, of Lake Malawi, reportedly remove and eat eyes from other fishes.

Access to food sources for many of these **partial consumers** depends on deceit. For example, small piranhas resemble and school with other characins and then bite the tails off their schoolmates; several juvenile carangids (e.g., *Scomberoides*, *Oligoplites*) resemble their silverside and anchovy schoolmates, whose scales they remove; sabretooth blennies mimic cleanerfishes (see next section); and cookie cutter sharks lure their victims using bioluminescence to resemble potential prey that live in the mesopelagic region (Losey 1978; Sazima & Machado 1990; see Chapter 10, The Deep Sea).

Mutualism and Commensalism

Some of the best studied examples of **mutualism** involve fishes that pick external parasites from other fishes. Cleaning behavior exists in almost all aquatic environments and involves dozens of shrimp and more than 111 fish species in 29 families (Sulak 1975; DeMartini & Coyer 1981; Lucas & Benkert 1983; Tassell et al. 1994; Côté 2001; Zander & Sotje 2002). As many as 18 different species of nearshore, primarily kelpbed species in California are known to clean other fishes (McCosker 2006). Mutualistic, co-evolved relationships are most obvious on coral reefs (Limbaugh 1961; Feder 1966; Losey 1987; Losey et al. 1999b). Juveniles of a number of wrasse, butterflyfish, damselfish, and angelfish species clean other fishes, but cleaning specialists

FIGURE 17.20 A Yellowstripe Goatfish, *Mulloidichthys flavolineatus* at Kona, Hawaii is being cleaned by two Hawaiian Cleaner Wrasses, *Labroides phthirophagus*. B. Inaglory / Wikimedia Commons / CC BY-SA 3.0.

occur among the cleaner wrasses (*Labroides*) of the Pacific (Fig. 17.20) and the neon gobies (*Gobiosoma*) of the Caribbean. Cleaners are usually territorial, occupying well-defined and often prominent coral heads or other locales referred to as **cleaning stations**. Recognizable, stereotypical behavior is important in communication between host and cleaner. Host fishes of numerous species approach these stations and pose, frequently assuming head-up or head-down positions while hovering in the water column, blanching in color, spreading their fins, and opening their mouths. Cleaners approach in a bouncing or tail wagging manner, frequently contacting a host during this dance, and then pick over the host's body surface, often entering the mouth or gill covers of herbivores and piscivores alike (Grutter 2004).

Parasites, particularly copepods, are removed, as are mucus and pieces of tissue around wounds; parasite loads are rapidly reduced following cleaning bouts (Grutter 1999; Sikkel et al. 2004; Cheney & Côté 2005). Hosts without parasites or wounds will solicit cleaning, thus tactile stimulation alone by the cleaner must attract some fishes (Losey 1979; Bshary & Wurth 2001), and hosts that are parasite-free may allow cleaners to pick them over and feed on mucus as a means of maintaining a relationship that is more valuable at other times. A cleaning bout is terminated when the cleaner leaves or the host fish shudders or snaps its mouth closed and open.

Cleaning relationships have commercial applications – Lumpfish (*Cyclopterus lumpus*) and temperate wrasses (Labridae) such as European Corkwing Wrasse (*Symphodus melops*) and Ballan Wrasse (*Labrus bergylta*) reduce external parasite infestations on Atlantic Salmon (*Salmo salar*) kept at extremely high densities in aquaculture pens (Sayer et al. 1996, Brooker et al. 2018).

Many cleaners have converged on coloration patterns involving bold stripes, and a dark median lateral stripe appears to be important for host fish recognition of the cleaner guild (Stummer et al. 2004; Arnal et al. 2006). Both cleanerfishes and shrimp appear to be largely immune to predation and consequently service such predators as moray eels, seabasses, snappers, and barracuda (Côté 2001). In the Indo-Pacific, at least one wrasse species, *Diproctacanthus xanthurus*, cleans damselfishes that do not leave their territories and hence cannot take advantage of cleaning stations (Randall & Helfman 1972).

Two species of sabre-toothed blennies, *Aspidontus taeniatus* and *Plagiotremus rhinorhynchus*, mimic the coloration and behavior of *Labroides* spp. and apparently gain access to posing hosts from which they take pieces of fins and body tissue. The deception, referred to by some as aggressive mimicry, is most successful with young hosts. Further study of this phenomenon by monitoring of diets of blennies that strongly resemble cleaner wrasses revealed that dependence on fin-biting as a food source varied by location. Blennies observed on the Great Barrier Reef, in the Red Sea, Japan, and Indonesia rarely bit fins of other fishes, and those that did were mainly smaller blennies and fin pieces did not make up a large part of their diets (Cheney et al. 2014; Fujisawa et al. 2018). In these areas blennies did not mimic the behavior of cleaner fishes, and the occasional fishes that were victims of fin-biting did not typically solicit cleaning by their posture. In French Polynesia, however, the Sabre-tooth Blenny relied more on fin-biting as a food source and often did attack fishes that appeared to be soliciting cleaning by their posture.

The importance of cleanerfishes in reef fish dynamics may differ at different locales. The experimental removal of cleaners from a Caribbean reef led to a decrease in host fish density and an increase in parasitic infections, whereas a similar removal in Hawaii had no apparent effect (Losey 1978; Gorlick et al. 1987). A more extensive, multi-reef, 6-month removal experiment on the Great Barrier Reef of Australia also found no detectable effect on total fish abundance or on fish species diversity (Grutter 1997). However, a survey and removal/addition investigation of *Labroides dimidiatus* in the Red Sea found little short-term impact after removal but a significant decline in fish diversity after 4–20 months; immigration or experimental addition of cleaners also resulted in a significant increase in fish diversity within the first few weeks (Bshary 2003). Additionally, an 18-month study that excluded *L. dimidiatus* cleaners from small Australian reefs showed that exclusion reefs had half the species diversity and one-fourth the abundance of fishes of reference reefs (Grutter et al. 2003). The strongest impact in both long-term studies was on large, mobile species – the same species that are likely to affect other reef organisms via predation and grazing. Hence cleanerfishes, although small and relatively rare, may act as **keystone** species in some fish communities (Grutter et al. 2003).

Some interspecific associations involve exploitation of one species' feeding habits to the benefit of another species, and are therefore probably examples of **commensalism**. Fishes that dig in the substrate, such as stingrays, goatfishes (Mullidae), suckers (Catostomidae), and Yellow Perch are commonly followed by other fishes that feed on invertebrates disturbed by the digger's activities. Such following and scrounging has also been observed among European wrasses, the producer species being much larger than the scrounger (Zander & Nieder 1997). The purportedly commensal relationship between shark-sucking remoras (Echeneidae) and large hosts such as sharks and rays is probably of this nature, involving feeding by the remora on leftovers following a host's meal. However, actual interactions between host and hitchiker are seldom observed. Some remoras may clean parasites off their hosts, which would be mutualistic, whereas others may create a hydrodynamic burden, particularly when attached to relatively small hosts, creating a parasitic situation. Certain postural changes seen in sharks suggest they are trying to get a remora to move to a less sensitive part of the shark's body (Ritter 2002), although other observations suggest that sharks and rays that accelerate or jump clear of the water may be attempting to dislodge remoras (Brunnschweiler 2006).

Symbioses with Invertebrates

Symbiotic interactions with non-fish species generally involve the use of invertebrates as spawning substrates, as predator refuges, to avoid extreme climatic environmental conditions (e.g., protection from desiccation or wave action), and as shoal mates. As discussed earlier in this chapter, female European Bitterling (*Rhodeus amarus)* lay eggs in the mantle cavity of a freshwater mussel while males guard territories over the mussel.

The use of an invertebrate as a structural refuge against predators is common. Shrimpfish (Centriscidae), clingfishes (Gobieosocidae), cardinalfishes (Apogonidae), and juvenile grunts hover among the spines of long-spined sea urchins or rest beneath the urchins. Such fishes are usually clear or white with black stripes, allowing them to hide among the sea urchin's spines. Some fishes seek shelter inside living invertebrates, a habit called **inquilinism** or **endoecism**. The Caribbean conchfish, *Astrapogon stellatus* (Apogonidae), lives by day in the mantle cavity of a queen conch, a large gastropod. Individual conchfish forage at night on small crustaceans and enter the siphon canal of a live conch an hour before sunrise. Other cardinalfishes live with crown-of-thorns starfish, sea anemones, and sea urchins. Members of the elongate pearlfish family, Carapidae, similarly live by day inside mollusks (see Fig. 15.4A) and various echinoderms such as sea cucumbers and pincushion starfish; as many as 15 individual pearlfish have been found in a single sea cucumber host. Pearlfishes, the more primitive species of which are free-living, are also nocturnal foragers on small invertebrates but cross the line from commensalism to parasitism by consuming the viscera of their host (Thresher 1980, 1984; Parmentier & Vandewalle 2005). Dependence on finding a host can be costly to settling larvae, as indicated by the large number of pearlfish tenuis larvae that have been found in the stomachs of adult pearlfishes (Tyler et al. 1992). Other inquiline species include liparid snailfishes (e.g., *Liparis inquilinus*) and Red Hake, *Urophycis chuss* (Phycidae), which inhabit sea scallops and can be found together in a single scallop (Luczkovich et al. 1991).

Many gobies live among sponges and corals, sea whips, and brain corals or share burrows created by worms or various shrimplike crustaceans (e.g., Arrow Goby, *Clevelandia ios*; Blind Goby, *Typhlogobius californiensis*; McCosker 2006). Truly mutualistic partnerships occur in tropical gobies that co-habit with alpheid shrimp, the goby serving as a sentry while the prawn digs and maintains the burrow (Fig. 17.21). Communication between partners is primarily tactile: the essentially blind shrimp maintains antennal contact with the goby's tail and senses tail flicks executed by the goby when predators approach (Preston 1978; Karplus 1979).

Symbioses between fishes and cnidarians are common, including fish that live on soft and hard corals (various gobies), or among the tentacles of jellyfish and Portuguese man-of-war (e.g., the Man-of-war Fish, *Nomeus gronovii*, Nomeidae). Although many fishes associate with sea anemones (Randall & Fautin 2002), including eastern Pacific Painted Greenlings *Oxylebias pictus* (Elliott 1992), the most highly evolved relationships are between pomacentrid anemonefishes and large sea anemones in the tropical Pacific and Indian oceans (Fig. 17.22). Approximately 30 fish species in the genera *Amphiprion* and *Premnas* and 10 species of anemones are involved. Details differ among species of fishes and anemones, but basically any other fish that touches the anemone's tentacles is likely to be stung by nematocysts (stinging cells), paralyzed, and consumed, whereas anemonefish frequently contact the tentacles and are not stung. Although the exact mechanism that protects anemonefishes from the nematocysts remains unclear, the mucus coating on the fish is believed to play an important role (Elliott & Mariscal 1997).

An anemonefish's intimacy with its host is considerable. Embryonic anemonefish imprint on the smell of host

FIGURE 17.21 Some gobies live in mutualistic relationships with burrowing shrimp. The shrimp maintain the burrow and the gobies defend the area. Courtesy of J. Randall.

FIGURE 17.22 Anemonefish (subfamily Amphiprioninae) move among the tentacles of an anemone. Stinging cells in the anemone's tentacles would paralyze other fishes but are not discharged when contacted by a resident anemonefish. Courtesy of J. Randall.

anemone species prior to dispersing as larvae, which influences their choice of settlement sites after the planktonic period (Arvedlund et al. 1999). Most individuals seldom move more than a few meters from their host, and adults remain with a single host for life. The relationship is considered mutualistic because the fish gains protection from predators on both itself and its eggs (which are laid on coral rock under the anemone) and also consumes other anemone symbionts and even the anemone itself. In turn, anemonefish chase away predatory butterflyfishes that eat anemones. The fish may also remove feces and debris from the anemone's upper surface, may drop food onto the anemone, may consume anemone parasites, and the fish's excreted waste products may stimulate the growth of symbiotic algae (zooxanthellae) within the anemone (Mariscal 1970; Allen 1975; Fautin 1991). Due to the combined benefits, anemones that harbor anemonefishes have faster growth rates, higher asexual reproduction (fission) rates, and lower mortality rates than anemones that lack the symbionts (Holbrook & Schmitt 2005).

For fishes that use invertebrates as protection, such as gobies in invertebrate burrows and clownfishes in anemones, refuges may be in short supply and territorial defense of the structure is fairly common (Grossman 1980). This is the probable explanation for the complex social system, territorial defense, and sex reversal of anemonefishes (see Chapter 8), populations of which appear to be limited by the number of available anemones.

Interspecific Shoaling

Many fish aggregations contain members of more than one species, forming **heterospecific shoals**. When several abundant, morphologically similar species school together, they tend to segregate by species, either associating with conspecifics more closely or even creating horizontal layers that are relatively monospecific. Hence each fish gains the added benefit of being in a large school but avoids the risk of being the odd individual among another species (Allan 1986; Parrish 1988, 1989a). Individuals of different size, shape, color, or behavior are likely to be preferentially attacked by predators.

Uniformity among members of a school provides advantages in terms of hydrodynamics and predator avoidance. However, fishes that aggregate for foraging reasons are not as constrained by the need to be similar. For example, foraging schools of parrotfishes and surgeonfishes in the Caribbean frequently include trumpetfishes, hamlets, butterflyfishes, goatfishes, and wrasses. Surgeonfishes and parrotfishes feed on algae and thus benefit from the large numbers that overwhelm territorial herbivorous damselfishes. Carnivorous species may consume invertebrates flushed by the activities of the herbivores or may also capitalize on territorial swamping and feed on invertebrates that live in algal mats of the territory or on the eggs of the damselfish. Larger predators, such as trumpetfish, may use the school or its members as moving blinds that conceal the predator and allow it to feed on the damselfish itself (trumpetfish will change color to match that of large or abundant school members). The presumed costs that small carnivores might suffer due to increased conspicuousness in a heterospecific shoal are apparently outweighed by gaining access to otherwise defended resources. The trade-off is underscored by the evasive maneuvers that minority fish take when a mixed species shoal is threatened. Rather than flee with the school, odd fish abandon the school and seek nearby shelter (Robertson et al. 1976; Aronson 1983; Wolf 1985).

Finally, fishes also form shoals with non-fish species. Many tunas school with or below various dolphin species in the tropical Pacific. The dolphin may then serve as an indicator for commercial fishing operations seeking tuna, which surround the schools with large purse seine nets thereby injuring, or even drowning, some dolphin as tuna are harvested. On a less grand scale, postlarval French Grunt school with dense clouds of mysid shrimps shortly after the grunts settle from the plankton and onto coral reefs. Both species are similar in size (8–13 mm) and appearance, but the mysids greatly outnumber the grunts. Grunts benefit from the antipredation function of the schools, affording them a degree of protection probably related to the number of mysids in a school. As the grunts grow, they school more on the periphery of the mysid aggregation and feed on the mysids. What began as a commensal or mutualistic relationship turns into a predator–prey interaction (McFarland & Kotchian 1982).

Supplementary Reading

Almada VC, Oliveira RF, Gonçalves EJ, eds. 1999. *Behaviour and conservation of littoral fishes.* Lisboa, Portugal: Instituto Superior de Psicologia Aplicada.

Barlow GW. 2000. *The cichlid fishes: nature's grand experiment in evolution.* Cambridge, MA: Perseus Publishing.

Breder CM, Jr., Rosen DE. 1966. *Modes of reproduction in fishes.* Neptune City, NJ: TFH Publications.

Fryer G, Iles TD. 1972. *The cichlid fishes of the Great Lakes of Africa.* Edinburgh: Oliver & Boyd.

Godin JJ, ed. 1997. *Behavioural ecology of teleost fishes.* Oxford: Oxford University Press.

Houde AE. 1997. *Sex, color, and mate choice in guppies.* Princeton, NJ: Princeton University Press.

Huntingford FA, Torricelli P, eds. 1993. *Behavioural ecology of fishes.* Ettore Majorana International Life Sciences Series, Vol. 11. New York: Harwood Academic Publishers.

Keenleyside MHA. 1979. *Diversity and adaptation in fish behaviour.* Berlin: Springer-Verlag.

Keenleyside MHA, ed. 1991. *Cichlid fishes: behaviour, ecology and evolution.* London: Chapman & Hall.

Ladich F, Collin SP, Møller P, Kapoor BG, eds. 2006. *Communication in fishes,* Vols 1 and 2. Enfield, NH:Science Publishers.

Matthews WJ. 1998. *Patterns in freshwater fish ecology.* New York: Chapman & Hall.

Pitcher TJ, ed. 1993. *The behaviour of teleost fishes,* 2nd edn. London: Chapman & Hall.

Potts GW, Wootton RJ, eds. 1984. *Fish reproduction: strategies and tactics.* London: Academic Press.

Reebs S. 2001. *Fish behavior in the aquarium and in the wild.* Ithaca, NY: Cornell University Press.

Reese ES, Lighter FJ. 1978. *Contrasts in behavior.* New York: Wiley-Interscience.

Reinboth R. 1975. *Intersexuality in the animal kingdom.* Berlin: Springer-Verlag.

Thresher RE. 1984. *Reproduction in reef fishes.* Neptune City, NJ: TFH Publications.

Uribe MCA, Grier HJ, eds. 2005. *Viviparous fishes.* Homestead, FL: New Life Publications.

Wootton RJ. 1999. *Ecology of teleost fishes,* 2nd edn. Fish and Fisheries Series No. 24. New York: Springer-Verlag.

CHAPTER 18

Cycles of Activity and Behavior

Summary

Biological systems are cyclical in nature. Most fishes are diurnal, being active by day and resting at night. Fewer species are active at night (nocturnal), and some predators are crepuscular, being active primarily during dusk and dawn twilight periods. For many fishes, the external cues for activity and inactivity appear to be sunset and sunrise. Distinctive activity cycles are most pronounced in tropical environments, and less so at higher latitudes.

Circadian rhythms have a period of approximately 24 h, driven by internal (endogenous) clocks. It appears that many cells and tissues within fishes have intrinsic molecular clocks, and that overall organismal activity patterns are coordinated by the release of the hormone melatonin from the pineal gland in response to light and dark cycles.

Other endogenous cycles include circatidal patterns in intertidal fishes that correspond to high and low tides. Intertidal fishes either move in and out with tides (juveniles using the shallows as a nursery area) or remain in the intertidal at low tide and seek shelter to avoid heat, desiccation, and oxygen stress (resident families such as gunnels, sculpins, blennies, and gobies). Nursery species gain access to tidal areas by moving up into the water column with flooding currents and by hugging the bottom on ebbing currents.

CHAPTER CONTENTS

The Diversity of Fishes: Biology, Evolution and Ecology, Third Edition. Douglas E. Facey, Brian W. Bowen, Bruce B. Collette, and Gene S. Helfman.

Companion website: www.wiley.com/go/facey/diversityfishes3

Some fishes spawn on a biweekly (semilunar) or monthly (lunar) cycle. For example, grunion spawn every 2 weeks during very high tides in the summer, laying their eggs in the sand high on the beach just as the tide turns. Their eggs then hatch 2 weeks later when the next very high tides cover them. Many coral reef species spawn biweekly or monthly when tides and therefore currents are maximal, which may help move the eggs away from abundant, reef-dwelling zooplanktivores. Nonreproductive migrations in many fishes (eels, salmons) are also tied to lunar cycles.

Fishes at nontropical latitudes spawn more seasonally, usually during spring, probably to allow larvae and juveniles to take advantage of the food available during spring and summer blooms of plankton. Cycles are set in motion by changing day length. Winter activities of many temperate species are not well understood; aggregating in deeper water and burying in the substrate have been documented. Many freshwater tropical fishes spawn seasonally in response to changing rainfall regimes, often migrating into flooded forests and swamps to reproduce.

The life cycle of many fishes includes migration over long distances, either for reproduction or to take advantage of seasonal changes in food availability. Movement between fresh and salt water is called diadromy; anadromous species spawn in fresh water but grow in the ocean (lampreys, sturgeons, and salmons), catadromous species spawn at sea and grow in fresh water (eels, mullets, and temperate basses), and amphidromous species move between habitats more than once (galaxiids, southern graylings, and sleepers). Anadromy is more common at temperate, northern locales, whereas catadromy occurs more at southern locales and at low latitudes. These patterns place early life history stages in habitats most favorable for growth.

Fishes navigate across distances by orienting to cues of light, geomagnetism, currents, odors, and temperature. Sun compasses are used by many species, as is polarized light. Elasmobranchs, eels, salmons, and tunas are sensitive to the earth's magnetic field. Some fishes have extremely sensitive chemoreception capability and can use their sense of smell in migration. Pacific and Atlantic salmons imprint on the odor of the stream in which they are born and then return to that stream after years at sea using olfactory cues once within the home river system. Catadromous eels begin life in tropical seas, ride ocean currents to continental areas as larvae, grow in fresh water, and then migrate thousands of kilometers back to their open ocean spawning region using orientation cues that remain a mystery.

Introduction to Cycles of Behavior

Biological systems are cyclical, and physiological and behavioral cycles exist at numerous temporal scales. Hearts beat and some nerves discharge spontaneously on a regular rhythm. Hormone production, respiration, locomotor activity, and photomechanical movements within the eye show distinct cycles. Many of these processes are driven by a neural pacemaker or are linked to external cues. Hence the rising and setting of the sun, the phases of the moon, and the annual orbit of the earth around the sun create periodic physical stimuli such as illumination, climatic, and tidal cycles that in turn determine the onset, timing, and periodicity of many activities in fishes. It is not surprising that such cycles are common; evolutionary adaptation is facilitated by constant or at least predictable selection pressures, and the external cues for such cycles as day length, tides, and seasonal climate have been distinct and relatively predictable (Schwassmann 1980), at least prior to human-induced climate disruptions. In this chapter, we review some pronounced biological cycles in fishes that are driven by internal (endogenous) clocks, by external (exogenous) cues, and by a combination of factors. We will focus on daily, semilunar or biweekly, monthly or lunar, seasonal, and annual patterns of activity, particularly those that involve foraging, migration, and reproduction.

Diel Patterns

The 24 h, **diel**, or daily periodicity of the earth's rotation creates a predictable pattern of light and darkness that has a profound effect on the biology of almost all animals and plants. Organisms are cued externally by sunrise and sunset; by day, night, or twilight length; or their activities are determined by an internal clock with a roughly 24 h period that may be reset by some external light cue.

Light-Induced Activity Patterns

Activity patterns in fishes generally represent a direct response to changing light levels, but are also affected by the activity patterns of their predators and prey (McFarland et al. 1999). In most environments, fishes are **diurnal** and tend to feed primarily during the day or are **nocturnal** and feed by night, whereas some feed primarily during **crepuscular** periods of twilight and fewer still show no periodicity (Table 18.1). On average, about one-half to two-thirds of the species in an area will be diurnal, one-quarter to one-third are nocturnal, and about 10% are crepuscular (Helfman 1993; Lowe & Bray 2006) (Fig. 18.1).

These distinctions are sharpest at tropical latitudes, where families can often be characterized as diurnal, nocturnal, or crepuscular (Hobson 1991; Helfman 1993). On **coral reefs**, herbivorous fishes are almost exclusively diurnal; perhaps they need to visually identify edible and inedible algae. Parrotfishes, surgeonfishes, rabbitfishes, and sea chubs roam the reef, often in large shoals. They feed on algae and seagrass, some of which is defended by territorial damselfishes, parrotfishes, and blennies. Fishes that feed primarily on encrusting sponges, tunicates, corals, and hydrozoans are also largely diurnal; this group includes angelfishes, butterflyfishes, pufferfishes, and triggerfishes. Many wrasses, butterflyfishes, goatfishes, mojarras, and

TABLE 18.1 **Diel activity patterns – defined as when fishes feed – of better known groups and families of teleostean fishes. For many families, activity patterns are only known for a few species. Some large families appear under more than one heading because of intrafamilial variability.**

All or most species diurnal
Acanthuridae (surgeonfishes), Ammodytidae (sandeels), Anthiinae (anthiine seabasses), Atherinopsidae (surfsmelts), Chaetodontidae (butterflyfishes), Characoidei (characins), Cichlidae (cichlids), Cirrhitidae (hawkfishes), Cyprinodontidae (killifishes), Embiotocidae (surfperches, except Walleye and Rubberlip), Esocidae (pikes), Gasterosteidae (sticklebacks), Gobiidae (gobies), Kyphosidae (sea chubs), Labridae (wrasses), Mugillidae (mullets), Mullidae (goatfishes), Percidae (perches, darters, except pikeperches), Pomacanthidae (angelfishes), Pomacentridae (damselfishes), Scaridae (parrotfishes), Siganidae (rabbitfishes), Synodontidae (lizardfishes)
All or most species nocturnal
Anguilliformes (most true eels, some morays, and congers are diurnal), Anomalopidae (flashlight fishes), Apogonidae (cardinalfishes), Batrachoididae (toadfishes), Clupeidae (herrings), Diodontidae (porcupinefishes), Grammistidae (soapfishes), Gymnotoidei (South American knifefishes), Haemulidae (grunts), Holocentridae (squirrelfishes), Kuhliidae (aholeholes), Lutjanidae (snappers), Mormyridae (elephantfishes), Ophidiidae (cusk-eels), Pempheridae (sweepers), Priacanthidae (glasseye snappers), Sciaenidae (drums), Siluriformes (catfishes)
Both diurnal and nocturnal species
Carangidae (jacks), Catostomidae (suckers), Centrarchidae (sunfishes), Congridae (conger eels), Cyprinidae (minnows), Gadoidei (cods), Leiognathidae (ponyfishes), Mullidae (goatfishes), Pleuronectiformes (flatfishes), Salmonidae (salmon, trout), Scorpaenidae (rockfishes: diurnal juveniles, nocturnal adults), Serranidae (groupers), Sphyraenidae (barracudas)
Several crepuscular species[a]
Carangidae (jacks), Elopidae (tarpons), Fistulariidae (cornetfishes), Gadoidei (cods), Lutjanidae (snappers), Serranidae (groupers)
Several species without distinct activity periods
Aulostomidae (trumpetfishes), Muraenidae (moray eels), Pleuronectiformes (flatfishes), Scombridae (mackerels, tunas), Scorpaenidae (scorpionfishes, rockfishes), Serrandiae (groupers)

[a] Also active at other times.
Adapted from G. Helfman (1993) and Lowe and Bray (2006).

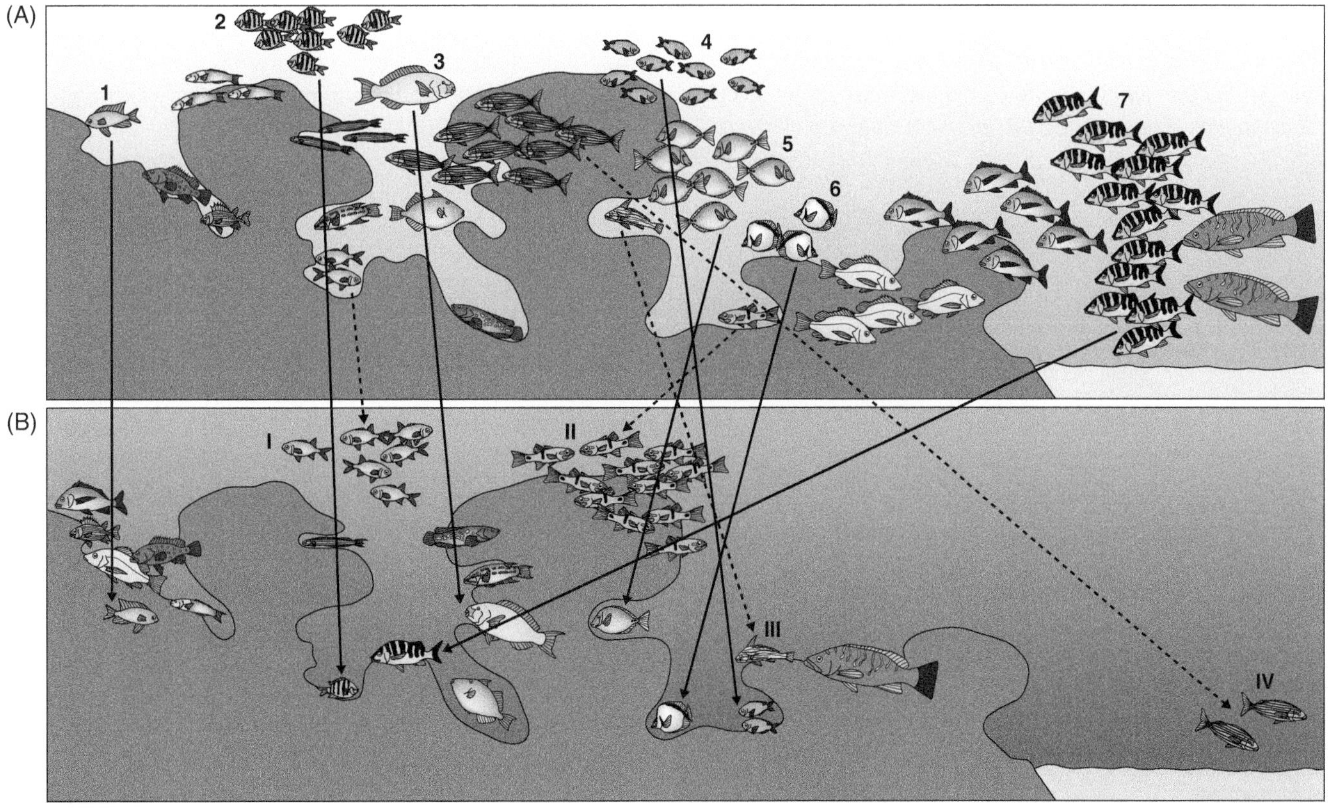

FIGURE 18.1 Day (A) versus night (B) distribution of families in a rocky reef area, Baja California, Mexico. Solid lines show habitat differences of diurnal fishes, dashed lines of nocturnal groups. Diurnal species: 1, benthic damselfishes; 2, Sergeant Major damselfishes; 3, parrotfishes; 4, zooplanktivorous damselfishes; 5, surgeonfishes; 6, butterflyfishes; 7, Graybar Grunts. Nocturnal species: I, squirrelfishes; II, cardinalfishes; III, drums; IV, grunts. After Lowe and Bray (2006), based on Hobson (1968).

small seabasses eat mobile or buried invertebrates (e.g., small crustaceans and polychaete worms) and are also diurnal.

Zooplanktivores are particularly abundant and conspicuous during the day, including anthiine seabasses, damselfishes, wrasses, fusiliers, and butterflyfishes. Zooplanktivores also form large shoals, but these aggregations often remain over a particular section of the reef and wait for currents to bring planktonic prey to them; congrid garden eels similarly "aggregate" their burrows and feed on passing zooplankton. A small group of piscivores, including lizardfishes, trumpetfishes, cornetfishes, scorpionfishes, jacks, hawkfishes, barracuda, and flatfishes, are active primarily during daylight. Cleaning fishes (wrasses, gobies, juvenile angelfishes, and butterflyfishes) that pick external parasites from other fishes are active by day (see Chapter 17, Interspecific Relations: Symbioses). However, some of the largest shoals of fishes encountered by day in tropical waters are nonfeeding, **resting shoals** of nocturnal foragers. Sometimes numbering in the millions, aggregations of zooplanktivorous silversides, anchovies, and herrings frequently hover near structure or in sandy embayments (Parrish 1992). They may be subjected to attacks by roving seabasses, jacks, and tunas. Better protected are the daytime resting shoals of invertebrate-feeding squirrelfishes, copper sweepers, cardinalfishes, and grunts, which occur over coral or in caves (Fig. 18.2).

The prey resources available by day on a coral reef change dramatically at night (Hobson 1991, 2006). Many invertebrates bury themselves in the sand or in small holes in the coral, whereas others come out of hiding and move about the reef and into the water column. The maximum size of zooplankton increases substantially from relatively small (<1 mm long) diurnal forms to larger (>2 mm long) nocturnal animals; these larger invertebrates make up most of the diet of nocturnal zooplanktivores.

FIGURE 18.2 A daytime resting school of juvenile French and White Grunts hover amidst *Diadema* sea urchins. At dusk, grunts move away from coral areas and feed in sand and grassbed regions, returning the next morning to the same daytime locales. G. Helfman (Author).

Almost all nocturnally active fishes are carnivorous, feeding on mobile invertebrates that they locate and engulf with relatively large eyes and large mouths. Grunts, snappers, porgies, and emperors are generally found close to the bottom, whereas zooplanktivorous anchovies, herrings, silversides, squirrelfishes, cardinalfishes, glasseye snappers, and copper sweepers forage higher in the water column. Nocturnal zooplanktivores rely heavily on vision to find their prey and feed successfully even at relatively low light levels (Holzman & Genin 2003).

Predation pressure on fishes at night compared to day apparently varies as a function of habitat, with near-bottom predation greater but water column predation lessened (e.g., Danilowicz & Sale 1999; but see Sancho et al. 2000a). The piscivores that roam the reef after nightfall include various eels, sharks, squirrelfishes, snappers, groupers, and jacks (Young & Winn 2003). An apparent reduction in predation in the water column is reflected in the morphology of nocturnal planktivores, which tend to be less streamlined than their diurnal counterparts, and also in a general lack of shoaling behavior at night by either nocturnal or diurnal species.

Successful feeding by piscivores occurs primarily during the transitional periods of **evening and morning twilight**, when diurnal and nocturnal groups essentially replace one another ecologically (see next section). Crepuscular predators include tarpon, cornetfishes, groupers, snappers, and jacks, but can also include smaller species such as lizardfishes, squirrelfishes, grunts, and pinguipedid sandperches (Holbrook & Schmitt 2002). Although their activities are concentrated at this time, most of these fishes, and predatory fishes in general, are highly opportunistic and will take prey any time of the day or night.

Fishes in other habitats and latitudes vary in the predictability of their daily activity cycles. Tantalizingly little is known about tropical freshwater assemblages (Lowe-McConnell 1987). Characins, cyprinids, and cichlids are predominantly diurnal, catfishes nocturnal, but beyond such generalizations our knowledge is relatively limited. Poeciliids as a group are usually considered diurnal, but work in Trinidad suggests greater flexibility than that usually portrayed. Guppy in streams that contain predators such as the erythrinid Trahira or Wolffish, *Hoplias malabaricus*, feed during the day and cease activity and move to stream edges at dusk. But where *Hoplias* does not occur, Guppy feed actively at night, show better growth rates, and engage in more courtship (Fraser et al. 2004). Guppy are among the best studied fishes behaviorally, including both field and laboratory manipulations. Perhaps when such manipulations or greater studies are performed on other groups that we view as invariant in their diel activities, we will discover greater plasticity (Reebs 2002).

At higher latitudes, activity cycles are similar in some respects but notably different in others. In *temperate lakes*, familial distinctions are weaker than on coral reefs. Diurnal zooplanktivores are abundant (minnows, sunfishes, and perches) and have nocturnal counterparts among the herrings, minnows, whitefishes, and sunfishes (Helfman 1981b, 1993). Herbivores are relatively rare. Diurnal invertebrate feeders include minnows, suckers, mudminnows, topminnows, sunfishes, and perches. Their nocturnal counterparts include eels, catfishes, trout, sculpins, sunfishes, and drums. Piscivores, which again have an activity peak at twilight, are represented during the day by pickerels, Northern Pike, and black basses and at night by Bowfin, salmonids, Burbot, temperate basses, sunfishes, and pikeperches. Nocturnal fishes rest by day amidst vegetation or other structure or form daytime resting shoals. Diurnal fishes often sink to the bottom and rest in relatively exposed locations at night.

Nearshore California kelp beds and rocky reefs also contain diurnal, nocturnal, and crepuscular species, but again familial distinctions are blurred (Ebeling & Hixon 1991; Lowe & Bray 2006). By day, shoaling zooplanktivores are abundant (silversides, seabasses, surfperches, damselfishes, wrasses, and scorpionfishes), as are diurnal invertebrate feeders (seabasses, surfperches, clinids, and gobies); these groups have nocturnal equivalents among the scorpionfishes, grunts, croakers, and surfperches that feed on relatively large prey both near and above the bottom. Herbivores are relatively rare, although they are more abundant in shallower, intertidal areas (blennies, pricklebacks, and gobies; Horn 1989). Piscivores (seabasses, scorpionfishes, and greenlings) are active primarily during twilight or nighttime; many scorpaenid rockfishes are diurnal as juveniles but nocturnal as adults (Lowe & Bray 2006). Nocturnal fishes form daytime resting aggregations and diurnal fishes rest at night either in holes or in exposed locales.

The above discussion focuses on active animals. However, inactivity accounts for half the diel cycle in most fishes, which often prompts the question of whether or not fish sleep. **Sleep** occurs when a fish assumes a typical resting posture for a prolonged period, uses some form of shelter, and is relatively insensitive to disturbance (Reebs 1992, 2001, 2002). By this definition, many species sleep. Some parrotfishes and wrasses secrete a mucus envelope around themselves at night while sleeping. The mucus cocoon may thwart roving nocturnal predators such as moray eels by concealing odors or serve as an early warning, given that the parrotfish dashes out of the mucus envelope when it is touched (Videler et al. 1999). It may also have antibacterial or antiparasitic properties (Shephard 1994). The barrier, however, fails as a deterrent to attacks by snails, such as the dwarf triton *Colubraria*, which pierces the cocoon and the fish with its proboscis and sucks blood from the sleeping parrotfish (Bouchet & Perrine 1996).

The adaptive significance of sleep among fishes remains a matter of debate. One likely function is immobilization during a period when an animal is relatively inefficient at both foraging and predator avoidance. Hence energy is conserved and predators avoided, assuming some refuge is found before a quiescent state is assumed (Reebs 1992).

Twilight Changeover

A striking sequence of composition of the fishes around a coral reef takes place at dawn and dusk. This sequence has been observed at several Pacific and Caribbean locales and probably occurs in most coral reef assemblages. Similarities among sites, despite dissimilar species, suggest that common selection pressures are operating, leading to a convergence in the behavior of the fishes. The twilight changeover period involves approximately four different phases of activity (Hobson 1968, 1972, 1991; Collette & Talbot 1972; Helfman 1993):

1. *Diurnal fishes move to the reef.* Beginning about an hour before sunset, zooplanktivorous fishes (e.g., anthiine serranids, butterflyfishes, and damselfishes) descend from the water column to the reef, and large herbivores (e.g., parrotfishes, surgeonfishes) migrate from daytime feeding locales to nighttime resting locales along predictable paths.
2. *Diurnal fishes seek cover.* From just before sunset until about 20 min after sunset, diurnal fishes seek shelter in the reef. Small individuals enter holes and cracks in the reef, whereas larger fish nestle under overhangs and in depressions. A species sequence exists; wrasses are among the first to seek shelter, followed by zooplanktivorous damselfishes, butterflyfishes, larger damselfishes, and parrotfishes. The time at which a species seeks cover is constant, with only a few minutes variation from one evening to the next. Many diurnal species continue to seek cover during the next phase.
3. *The quiet period: evacuation of the water column.* Beginning about 10–15 min after sunset, the level of activity and number of fishes above the reef drops precipitously. For the next 15–20 min, activity by small fishes in the water column comes to a standstill. Hobson (1972) termed this phase the quiet period, when neither diurnal nor nocturnal fishes are moving about. All activity does not cease, however. Predatory fishes, such as groupers, jacks, and snappers, are active at this time, generally swimming close to the bottom and striking up at prey fishes that remain in the water column. This predatory tactic undoubtedly capitalizes on the difficulty prey have seeing dark-colored predators below them against the background of the darkened reef, whereas the predators are striking up at targets that are silhouetted against the lighter evening sky.
4. *Emergence and migration of nocturnal fishes.* The end of the quiet period is marked by the movement of nocturnal fishes up into the water column and along the reef face. Bigeyes, cardinalfishes, and croakers appear over the reef about half an hour after sunset and begin feeding on invertebrates. Grunts, squirrelfishes, and copper sweepers migrate along predictable paths from daytime resting locales to nighttime feeding areas. The water column above and around the reef is now occupied by active fishes.

At dawn, the evening sequence is repeated in reverse. Nocturnal fishes migrate back to resting locales and seek shelter, often in the exact same spot they occupied the previous day (Marnane 2000). A morning quiet period occurs when predators are most active, and then diurnal fishes reoccupy the water column and migrate to their daytime feeding locales. The predictability of times and locales, cued primarily by specific light levels, is striking.

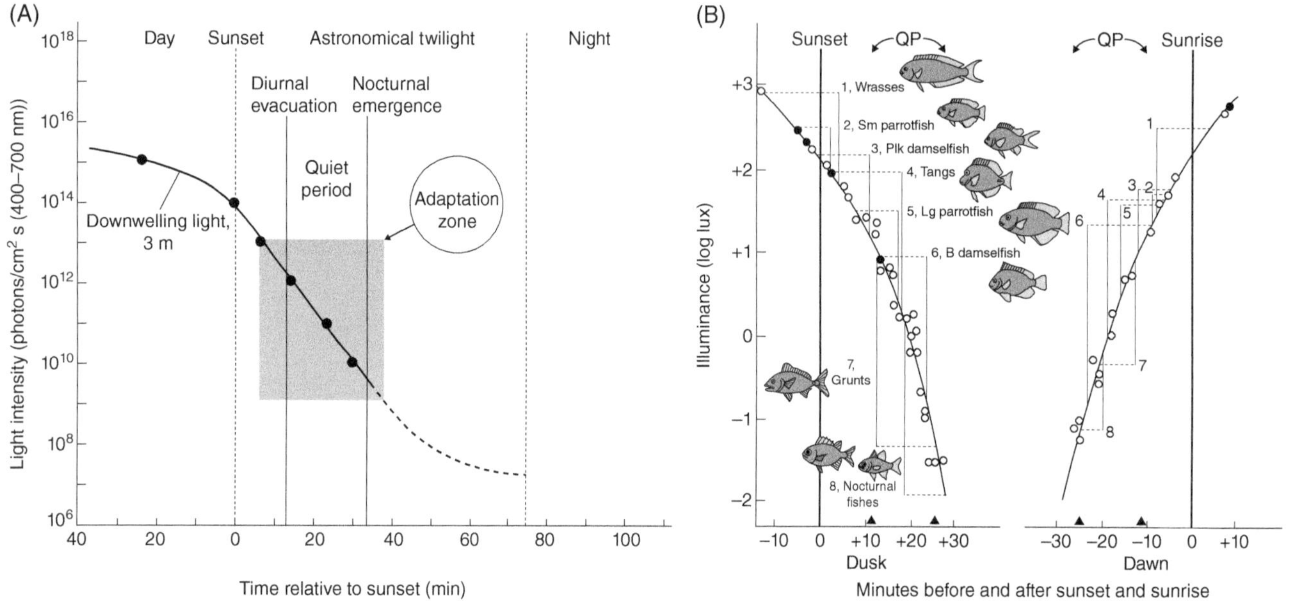

FIGURE 18.3 Light availability, dark adaptation, species changeover, and predator–prey interactions at evening twilight on a coral reef. (A) Available light (curved line) decreases maximally during the period from 13 to 33 min after sunset. This is the time (stippled area) when (i) diurnal eyes are dark-adapting, (ii) predators are maximally active and successful, and (iii) diurnal and nocturnal prey species abandon the water column, creating the quiet period. Approximate lux values for light units are 10^{14} photons = 10,000 lux, 10^{8} photons = 0.0001 lx. (B) Temporal sequence of cessation of activity of reef families at dusk, and onset at dawn. Families are coded by number. B, benthic; Lg, large; Plk, planktivorous; QP, quiet period; Sm, small. (A) after Munz and McFarland (1973), used with permission; (B) after McFarland et al. (1999), used with permission.

Crepuscular predators are the apparent key to understanding the predictable, convergent nature of events during twilight on coral reefs. Predatory threat results from a combination of environmental, physiological, and behavioral factors unique to crepuscular periods (Fig. 18.3). During twilight, light declines from daytime levels of about 10,000 lux to nighttime levels of about 0.0001 lux. Eye characteristics of 265 species of teleost reef fishes, representing 43 families, show that nocturnal fishes had traits to improve vision in low light conditions, including large eyes and pupils, but had reduced depth of focus. In contrast, diurnal species had much greater diversity of optical capabilities (Schmitz & Wainwright 2011). The light-adapted, cone-dominated eyes of many diurnal species cannot dark-adapt quickly enough and thus become ineffective at capturing light in the changing, dimmer conditions of dusk. At the same time, conditions are still too bright for the sensitive, rod-dominated eyes of nocturnal fishes that are highly effective at capturing light.

Twilight is therefore a period of intermediate conditions, when cones still function, but not with great efficiency. Many reef predators have fewer but larger cones than are found in diurnal eyes and more but smaller cones than are found in nocturnal eyes. This intermediate eye provides less visual acuity than a diurnal eye during the day and is less effective at light capture than a nocturnal eye at night (Munz & McFarland 1973). However, the intermediate eye is relatively better during the changing conditions of twilight, when neither diurnal nor nocturnal eyes function well. The light-capturing photopigments in the retinae of reef fishes also indicate an influence of twilight conditions. Diurnal, nocturnal, and crepuscular fishes have rod pigments that are most sensitive to light in the blue-green portion of the spectrum (about 490 nm), which matches prevailing wavelengths during twilight better than it matches the dominant greener nighttime light at 580 nm. Both diurnal and nocturnal fishes appear to sacrifice nocturnal vision in favor of being able to capture light during the dangerous crepuscular periods.

These anatomical and physiological differences, combined with the predatory tactic of striking up at backlit prey in the water column, helps explain why the post-sunset minutes are so dangerous for potential prey species. The quiet period of inactivity by small diurnal and nocturnal fishes appears to be a direct result of the threat of being eaten by predators at that time, rather than physical limitations involving visibility of their own prey (Rickel & Genin 2005).

At temperate latitudes, twilight changeover patterns are more variable compared to coral reef species. Activity patterns of temperate lake and kelp bed fishes are less precise in that: (i) many species feed both diurnally and nocturnally; (ii) reputedly diurnal and nocturnal species overlap in activity times; (iii) species within a family vary in major periods of activity; and (iv) individuals within a species vary in twilight changeover activities. Such variation could result from longer twilight lengths at higher latitudes, where dark adaptation of a fish's eye keeps pace with the rate of light change and hence diurnal species can maintain activity well into twilight and nocturnal species can commence activity before twilight ends. A latitudinal gradient in twilight length could interact with reduced predation pressure, reduced species diversity, or greater climatic instability to produce the relatively "unstructured" temporal patterns at higher latitudes (Helfman 1993; Lowe & Bray 2006).

Vertical Migrations

An entirely different daily rhythm of migration that appears largely dependent on light levels is the vertical migration undertaken by numerous fish species in both marine and freshwater habitats (see Chapter 10, The Deep Sea). In most fishes, this movement involves an upward migration at dusk to feed and a downward migration at dawn. For example, Alewife (*Alosa pseudoharengus*, Clupeidae) migrate upward in lakes in the evening at a rate that parallels the migration of their prey, a mysid shrimp (Fig. 18.4).

Zooplankton often migrate to the surface at night to feed and take advantage of reduced visual acuity in zooplanktivorous fishes. Predator avoidance could also explain vertical movements in many larval and juvenile fishes (e.g., Sockeye Salmon, Walleye Pollock); by remaining in dark, deep waters by day, vertical migrators can avoid diurnal predators. Vertical migration could also increase a fish's encounter rate with plankton if surface currents replenish the food supply in surface waters. In addition, fishes may gain an energetic advantage by moving into warm surface waters to feed actively and then returning to cooler, deeper waters to metabolize and grow (McLaren 1974; Janssen & Brandt 1980; McKeown 1984; Nielson & Perry 1990). Fishes exhibiting diel vertical migration may benefit from multiple factors, but there has been growing evidence in support of the predator avoidance hypothesis (see Stockwell et al. 2010).

Not all diel activity cycles relate only to feeding and predator avoidance. The timing of spawning is quite predictable for many species and even families. Diurnal spawners include many minnows, sunfishes, darters, cichlids, and wrasses. Twilight spawning characterizes some damselfishes (dawn) and butterflyfishes, wrasses, parrotfishes, and bothid flounders (dusk). Nocturnal spawning, not surprisingly, is difficult to observe but is known in the Yellow Perch, a strongly diurnal feeder (see below).

FIGURE 18.4 Zooplantivorous fishes often follow the vertical migrations of their prey, such as the Opossum Shrimp, *Mysis relicta*, which feeds many fishes in deep, coldwater lakes. Photo of *Mysis relicta* by Per Harald Olsen / Wikimedia Commons / CC BY-SA 3.0.

Circadian Rhythms

A **circadian** rhythm is a pattern of activity governed by an internal clock with a period of roughly 24 h (see Reebs 2011). The onset of activity may be shifted each day by some external stimulus (*zeitgeber,* German for "time giver") such as sunrise. Such an environmental reset mechanism is important because day length changes during different seasons, especially farther from the equator. Tides and feeding events can also serve as *zeitgebers.* Activity rhythms in many teleosts can become established (entrained) if a meal is provided at a fixed time each day. Fish then develop an activity rhythm that anticipates the time of feeding, even in the absence of food and in constant light (Spieler 1992). In the absence of a *zeitgeber,* such as during experimental conditions of constant light or darkness, a free-running rhythm is often maintained at slightly more or less than 24 h.

Free-running rhythms, involving either diurnal activity and nocturnal inactivity or the converse, have been demonstrated in a number of fishes, including hagfishes, swell sharks, anguillid eels, minnows including Goldfish, salmonids, suckers, South American knifefishes, burbots (Gadidae), killifishes, moronid temperate basses, and wrasses (Boujard & Leatherland 1992; Reebs 1992, 2002; Gerkema et al. 2000). Many fishes that show such patterns also exhibit considerable inter- and intraindividual variation in the rhythms (Reebs 2002).

Normally distinct activity cycles can be disrupted by experimental additions of predators or by the removal of resting structure. Distinct cycles also often break down during the breeding season and when fish migrate. Many strongly diurnal reef fish species spawn late into evening twilight (Sancho et al. 2000b), and normally diurnal minnows, Yellow Perch, and gobies spawn at night. Daily activity rhythms may be lost in species that demonstrate parental care, in part because eggs and larvae must be guarded and fanned throughout the diel cycle, not just when the parents are normally active. Studies of several species, including catfishes, sticklebacks, centrarchid sunfishes, cichlids, and damselfishes indicate that parental care is also provided during the time period when adults would normally be inactive (Reebs 1992, 2001).

Circadian rhythms control many other aspects of fish behavior, morphology, and physiology. Most fish tissues and organs have inherent molecular clocks that operate on approximately 24 h cycles, and will retain this even when removed from the fish because it is driven by approximately 24 h cycling of *clock* gene expression (see Reebs 2011; Idda et al. 2016). However, these separate intrinsic clocks seem to be coordinated by levels of the hormone **melatonin**, which is released from the **pineal organ** on the dorsal surface of the brain in response to periods of light and dark detected by the pineal and the eyes. This can then influence coloration, locomotor activity, and social behavior, as well as longer-term effects such as seasonal control of reproduction, sexual maturation, development, and growth. Melatonin levels are typically high at night and lower during the day. Secretion of hormones, such as prolactin, estradiol, progesterone, cortisol, testosterone, thyroxine, and triiodothyronine also follow endogenous (internally generated) circadian, semilunar, or lunar periodicities that are in turn affected by day length, temperature, and other hormone concentrations. Changing the light or temperature regime, or injecting a fish with hormones or hormone precursors, will cause changes in swimming activity and rest, temperature and salinity selection, reproduction, fat deposition, weight gain, and other aspects of growth. Hence the light-dark cycle can affect the timing of neural and hormonal cycles, which then entrain cellular rhythms in tissues, all governing the activity and behavior of the fish.

There does not appear to be a specific location of a particular master clock in fishes, unlike mammals in which a cluster of nerve cells (the suprachiasmatic nucleus) in the hypothalamus of the brain serves as a master clock.

Tidal Patterns

Tidal cycles are caused by the gravitational pull of the moon, and to a lesser extent the sun, on the oceanic water mass. Most coastlines experience a **semidiurnal** tidal regime that involves two high tides and two low tides each day, highs and lows being separated by about 6.2 h. Relatively strong, **spring** and relatively weak, **neap** tides occur at biweekly intervals. Daily tidal ranges can vary from a few centimeters to several meters depending on locale. During low tide, animals inhabiting the intertidal zone typically either remain in a tide pool, hide beneath large algae masses, move out to deeper water, or are physiologically able to withstand the period of exposure to air and the accompanying temperature change. Movements in general have to be synchronized with the fall and rise of tides. A drop in hydrostatic water pressure from a smaller overlying water column could serve as an external cue of a falling tide, and flooding of a pool could indicate an incoming tide. However, most intertidal animals, including fishes, appear to anticipate tidal changes via an internal clock that is reset by external cues, as discussed above.

Shallow intertidal areas – mud flats, saltmarshes, seagrass beds, mangroves, reef flats, or the rocky intertidal – are among the most productive regions in the sea. Marine algae, large and small, grow rapidly in warm, shallow water, resulting in an abundant food base. Shallow depths mean that large aquatic predators are relatively scarce. These conditions create a relative bonanza for fish species that can adapt to the physiological conditions within the intertidal region. Tidal regions are by definition fluctuating environments, but most fluctuations occur with predictable periodicity. Hence animals can capitalize on the fluctuations, or at least adapt to predictable environmental constraints. Falling tides in particular create numerous problems for fishes, including desiccation, rapid changes in and exposure to extreme temperatures, pH, and salinity, and exposure to terrestrial and aerial predators. Fishes generally follow one of two courses in dealing with low tide conditions: they either abandon shallow water with falling tides, or they remain in the intertidal area and seek shelter in cracks, algae, under rocks, or in pools (Gibson 1992, 1993; Horn & Martin 2006).

The former, termed **visiting** species, migrate in and out with the tides. This is particularly common with the many species that use intertidal areas as nursery grounds or refuges for juvenile fishes. Saltmarsh creeks along the Atlantic coast of North America serve as such nurseries for numerous species of worm eels, herrings, croakers, porgies, mullets, and flatfishes, as well as housing adults of dozens of other species. Approximately 80% of the commercial landings from the US Atlantic and Gulf of Mexico fisheries consist of species that spawn offshore and use saltmarshes for nurseries (Shenker & Dean 1979; Weinstein 1979; Miller et al. 1985). Juvenile fishes are carried in and out of the intertidal zones and salt-marsh creeks with flooding and ebbing tides. Analogous tidal nursery situations exist in many parts of the world, such as mangroves in most tropical areas and the tidal swamps of western and northern Australia, which were used by juveniles of 24 families of fishes (Davis 1988). Adults of coral reef species also show on-off reef movements that correspond with tides. Low tides on hot days can create hot, anoxic conditions over large sections of reef and sandy tidal flats. Such areas are commonly avoided during low tides but reoccupied by fishes that move back onto the reef from the deeper reef face or from channels as cooler, oxygenated water floods the region during incoming tides.

Intertidal **resident** species remain in the intertidal zone at low tide and hide in areas insulated from complete desiccation, or make periodic visits into water or spray zones. Residents show the greatest degree of adaptation to the intertidal environment. Most are relatively small (<20 cm), which allows them to hide in holes and cracks or under piles of vegetation (pricklebacks, gunnels, sculpins, clingfishes, blennioids, and gobies), and also presents less surface area to turbulence. Bodies are either thin and elongate (gunnels, pricklebacks, and clinids) or depressed (sculpins, clingfishes) (Fig. 18.5).

FIGURE 18.5 Thin, elongate fishes, such as this Monkeyface Prickleback, *Cebidichthys violaceus*, can occupy damp, cool crevasses among the rocks during low tide in a rocky intertidal region. NOAA / MBARI / Wikimedia Commons / Public Domain.

Elongate bodies are effective at wriggling into tiny places. The intertidal zone is frequently exposed to breaking waves, particularly during high tides, and depressed body morphologies of many species are convergent with those of fishes that live in other high-energy environments such as river rapids (see Chapter 10, Strong Currents and Turbulent Water). Suction cups, formed by fused pelvic fins, occur convergently in clingfishes, snailfishes, and gobies. Also convergent in fishes dwelling in high-energy and bottom areas is negative buoyancy, achieved through a missing or greatly reduced gas bladder. Some intertidal residents have evolved extreme tolerance to water loss: clingfishes can live up to 4 days out of water if humidity exceeds 90% and can sustain as much as 60% loss of total water content. This tolerance exceeds that of many amphibians. Gobiid mudskippers of the tropical Indo-Pacific spend 80–90% of their time out of water, being submerged and inactive only during high tides (Horn & Gibson 1988; Gibson 1992).

Tidal activity cycles in many fishes have an apparent endogenous basis. Several species, including Shanny (*Lipophrys pholis*, Blenniidae) and Rock Goby (*Gobius paganellus*, Gobiidae) held in the laboratory under constant conditions still show an activity pattern that corresponds to a semidiurnal tidal regime (see Reebs 2011). They rest at the local time of low tides and swim actively at the expected times of high tides (see Fig. 18.6). Similar patterns, also known from anguillid eels, Tomcod (Gadidae), clingfishes, killifishes, sculpins, mudskippers, and flatfishes, appear to reflect an internal **circatidal** clock with a period of about 12 h, but one cannot completely rule out the possibility that the fishes are sensing fluctuations in the gravitational pull of the moon and sun. The

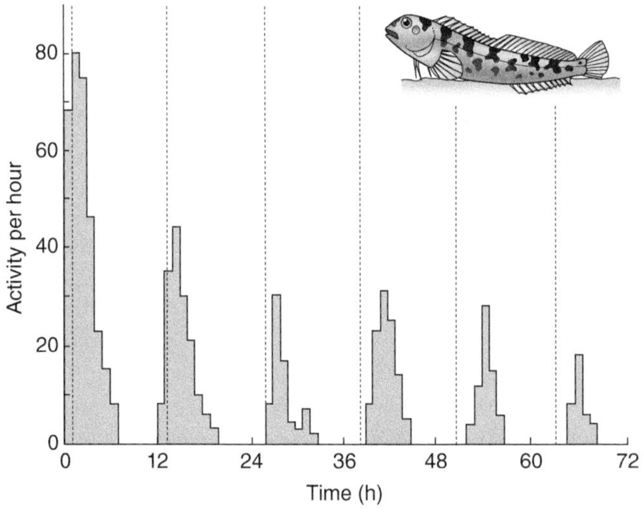

FIGURE 18.6 Endogenous, circatidal activity rhythm of Shanny, *Lipophrys pholis*, held in a laboratory under continuous light. Activity is indicated by the darkened histograms, times of high tide where the fish was captured are denoted by the vertical dashed lines. Shanny are normally active at high tide. In the absence of tidal stimuli, the activity cycle "free runs" with a period of 12.4 h, displacing it slightly from predicted high tides with each cycle. In the field, the fish's clock would be recalibrated (reset) by the hydrostatic pressure of high tide, which would keep the fish's activity synchronized with actual tides. Adapted from Gibson (1993), after Northcott et al. (1990).

clock may be reset by changing water depth, which in the field would correlate with fluctuating hydrostatic pressure caused by alterations in the weight of the water column above the fish (Gibson 1982, 1992).

Activity cycles in many intertidal species are constrained by the cyclical nature of oxygen availability. Photosynthesis during the day oxygenates the water, but both vegetation and animals consume oxygen at night. Hence many intertidal fishes have an ability to breathe air (Horn & Martin 2006) and many reduce their activity, and their oxygen consumption, at night (Horn & Gibson 1988). Tides may override normal activity patterns for species that occur in a variety of habitats. American Eels are strongly nocturnal in nontidal habitats, and in tidal regions travel with tidal currents by day and swim against them while foraging at night. Both eels and killifish capitalize on tidal flooding to gain access to the food resources of the saltmarsh surface, eels by night and killifish by day (Weisberg et al. 1981; Helfman et al. 1983). Conversely, intertidal fishes that live in regions with very minimal tidal ranges, as in the Mediterranean, synchronize their activity patterns with day–night cycles instead of with tides (Gibson 1993).

Utilization of inshore areas by larval and juvenile fishes creates a particular logistic problem related to tidal cycles. Water flow is favorable for entry into such areas during only half of the tidal cycle; for much of the time, small fishes must fight outflowing currents of several knots, impeding or reversing any progress they may have made. Fishes overcome this problem by engaging in **selective tidal stream transport** or modulated drift (Wippelhauser & McCleave 1987; Forward & Tankersley 2001). Inwardly migrating fishes move up into the water column on incoming tides, but move down close to the bottom on outgoing tides. They are consequently carried inshore with incoming currents, but minimize slipping back offshore by taking advantage of reduced ebb currents as water is slowed by bottom topography and friction. Selective tidal stream transport has been observed in post-larval American eels, spot (Sciaenidae), and flounders. Adult anguillid eels, cod, and flatfishes on spawning migrations use similar transport mechanisms to move along the shore or in open water. Selective tidal stream transport could be adaptive as a directional aid and also reduces the energy and time required to reach a particular locale; the response might be driven by an endogenous circatidal clock, as discussed in the previous section (Miller 1988; Gibson 1992; Forward & Tankersley 2001; Metcalfe et al. 2006).

Semilunar and Lunar Patterns

One postulated function of cyclical behavior is that it may help individuals **synchronize** their behavior with that of conspecifics. Nowhere is such synchronization more necessary than in reproduction. Not only must both sexes aggregate at the same locale to release gametes, but preparatory events of gametogenesis (gamete production) and secondary sex character development must also occur with similar timetables that converge on the same small time window. Predictable external cues that occur over biologically appropriate time periods are prime candidates as drivers of such cycles. The monthly orbit of the moon is a particularly common *zeitgeber* – it is predictable, and also directly links to lunar phase, nocturnal illumination, and tidal and current strength.

An example of this interrelationship involves the grunion *Leuresthes tenuis* (Atherinopsidae), that spawn literally on the beaches of southern and Baja California. Grunion spawn after high tides on the three or four dark nights following a full or new moon during the spring and summer, the time when highest tides occur at night (Walker 1952; Martin & Swiderski 2001; Martin et al. 2004). These lunar periods correspond to spring tides, when water is pushed to its maximum height up the beaches. The same wave cover will not occur for at least 2 weeks. Females ride waves up the beach, dig with their tails into the wet sand, and deposit eggs that are fertilized by sperm released by males around the female. The eggs normally develop in the moist sand for about 10 days, but they will not hatch until the next spring tides when waves once again cover the higher portions of the beach and the agitation of the breaking waves stimulates hatching (Griem & Martin 2000). If waves do not reach the eggs, they can delay hatching for an additional 2 weeks (gonad maturation is, however, on an 18-day, not 14-day, cycle). Spring tides promote synchronization of male and female behavior, allow a sufficient 2-week interval for embryogenesis, and facilitate spawning under the cover of darkness, which may lessen predation on adults. The eggs are deposited where further wave action will not expose them for 2 weeks, again potentially reducing predation. A second grunion species, *L. sardina*, engages in similar behavior but spawns both day and night. In its Sea of Cortez locale, spawning coincides with the period of highest tides (Clark 1925; Gibson 1978).

Grunion reproduction is an example of **semilunar** periodicity; such cycles usually involve a 14.7-day interval. Synchronization with a period of maximum spring tides is common in fishes with semilunar, lunar, or longer cycles. Two groups with such rhythms are intertidal spawners and coral reef spawners. Intertidal spawners include the grunions as well as a southern hemisphere Common Galaxias (Inanga), *Galaxias attenuatus*, Surf Smelt (*Hypomesus*), Capelin (*Mallotus*), a stickleback (Gasterosteus), silversides (*Menidia*), killifishes (*Fundulus*), four-eyed fish (*Anableps*), and a Japanese pufferfish (*Fugu*) (Martin & Swiderski 2001). Intertidal spawners deposit eggs during spring high tides on sand or pebble beaches or amidst algal and root mats, leaf axils, and bivalve shells. Spawning often occurs on the days or nights following full or new moons. The eggs are exposed to air sometime during the next 2 weeks before the next spring tides, at which time they hatch. Smelt and killifish eggs require aerial incubation and will die if continually immersed, apparently because of relatively low oxygen concentrations in water. A reduction of aquatic predation on the eggs is often postulated as the primary selection pressure favoring intertidal spawning, although predation on puffer eggs is minimal due to their toxicity (Gibson 1978; Taylor 1984, 1990; Leatherland et al. 1992).

Many coral reef fishes exhibit lunar or semilunar synchronization in their spawning activities (Johannes 1978; Gladstone & Westoby 1988; Robertson 1991; Domeier & Colin 1997; Takemura et al. 2004), although some species show no correlation with lunar period (Sponaugle & Pinkard 2004). Most larger coral reef species that spawn in the water column do so at twilight or at night during new or full moon or both, often on high or ebbing tides. Larger species move to deep water, often to form spawning aggregations at predictable locales. Smaller species rush momentarily into the water column above the reef. Both groups have pelagic eggs and larvae, and larvae return to reefs primarily during spring tides. Johannes (1978) postulated that such behavior serves to move eggs and larvae out of the range of the abundant benthic and demersal predators on the reef, basically exporting the eggs out of the adult habitat but not necessarily out to sea.

Other explanations for the periodicity and timing of coral reef fish spawning either focus on larval or adult biology. Larval hypotheses, in addition to those proposed by Johannes, include maximized dispersal of larvae to distant habitats, swamping or saturation of predators (such as Whale Sharks; Heyman et al. 2001), synchronization of larval production with production of invertebrate larvae on which they feed, reduction of competition among different larval cohorts, nocturnal spawning to minimize ultraviolet damage to floating eggs, and optimization of the timing of larval settlement in appropriate reef habitats. Adult biology hypotheses focus on synchronization of activities among spawners, optimization of the conditions under which adults spawn, and improvement of conditions for egg guarding in species that show parental care.

No single hypothesis explains reproductive timing in all species (Barlow 1981; Taylor 1984; Robertson et al. 1990; Gibson 1992; Sancho et al. 2000b), but larval biology hypotheses may pertain more to water column spawners and adult biology hypotheses may better explain selective forces acting on benthic spawners that guard their eggs. Proof by exception comes from studies of species with unpalatable eggs, larvae, and adults. The Sharpnose Puffer, *Canthigaster valentini*, is protected by noxious chemicals at all life history stages (Gladstone & Westoby 1988). It spawns near the bottom during the day throughout the year, shows no definitive cycling, has an unhurried courtship display, exhibits no parental care, and embryos hatch on incoming tides. Apparent liberation from predation has also removed the selection pressures that induce periodicity and other more usual spawning behaviors in reef fishes. Few direct tests of most hypotheses have been attempted and much exciting work remains to be done on this topic.

A variety of adaptive scenarios can be postulated for lunar cycles synchronized with spring tides in nearshore marine and estuarine fishes. More puzzling are lunar spawning cycles in some freshwater and high seas fishes. For example, two Lake Tanganyika cichlids spawn during full moons, which might minimize diurnal predation on eggs while enabling adults to monitor the activities of their own, nocturnal, catfish predators. An apparent semilunar spawning cycle in an offshore gadoid, *Enchelyopus cimbrius*, is even more puzzling, aside from synchronization of reproductive behavior among adults (Gibson 1978; Taylor 1984; Leatherland et al. 1992).

Lunar cycles have been found in other aspects of the biology of fishes. White Sucker show an endogenous lunar rhythm of temperature preference, selecting relatively high temperatures during the new moon and lower temperatures during the full moon. Guppy show a change in the spectral sensitivity of their retinae that has a lunar periodicity. Semilunar cycles have been found in feeding rate, body mass, body length, scale growth, otolith deposition, RNA/DNA content, and in concentrations of various plasma constituents in fishes. These phenomena may all be interrelated; feeding rate could affect all the other growth and condition parameters. The downstream migration of young salmonids is also cued by lunar events. Newly emerged Coho Salmon fry (*Oncorhynchus kisutch*) move downstream from spawning redds at night during the new moon. Cover of darkness may reduce predation, whereas synchronization could aid in forming shoals and also swamp any predators that were present. Older smolts move downstream by day during the new moon, which delivers the migrants at the river mouth during spring low tides, which in turn aids movement out of the river and into the sea. Models explaining smoltification and migration in several salmonids (see Chapter 9) suggest that the full moon initiates a process of morphological, behavioral, and physiological changes (perhaps mediated by thyroid hormones) that prime the animal for eventual downstream migration during the new moon. The new moon is also the primary time of upstream migration by elvers and downstream migration of maturing adults of anguillid eels. Again, the underlying mechanisms and clocks driving these periodicities remain a mystery (Leatherland et al. 1992).

Seasonal Patterns

Activity and Distribution

Most fishes are ectotherms, and are therefore affected by seasonal fluctuations in temperature. At temperate and polar latitudes, food availability, vegetative cover, turbulence, oxygen availability, and water clarity all vary greatly among seasons. Ice cover on high-latitude lakes may lead to oxygen depletion and winterkill conditions; thermal stratification at lower latitudes creates analogous summerkill conditions. Hence fishes in these habitats characteristically move into and out of shallow, nearshore zones with the progression of the seasons.

For example, a typical pattern in a lake in the northeastern United States finds the shallow regions devoid of vegetation and fishes in early spring after ice melt and until the surface water warms above about 10°C. With warming water, minnows, catfishes, pickerel, sunfishes, black basses, killifishes, and Yellow Perch move into nearshore regions. At water temperatures of around 15°C, sunfishes and many others spawn and vegetation growth is apparent. In late spring and early summer, as temperatures exceed 20°C, vegetation is well established and fishes are distributed throughout the littoral zone, including deeper portions such as drop-offs down to the thermocline. In late summer and early fall, as temperatures fall below about 15°C and plants begin to die back, fishes first move from the deeper littoral zones to the shallower regions. As temperatures fall below 10°C and vegetation becomes sparse, fishes abandon nearshore regions presumably for deeper water. If periods of warm weather occur during the fall, fish will reoccupy and then abandon the shallows as the water warms and recools (Hall & Werner 1977; Keast & Harker 1977; Emery 1978; Helfman 1979a).

We know much about fish distribution and activity during spring, summer, and fall. However, winter biology remains poorly understood, although information is growing. In North American temperate and arctic lakes, many fishes feed actively despite thick ice cover. These include smelt, numerous salmonids, esocids (Northern Pike, Chain Pickerel), percids (Yellow Perch, Walleye, and Sauger), and centrarchid sunfishes. Where many lake fishes go in winter remains a mystery. Deeper water is a logical choice because vegetation, which provides shelter for both fishes and their prey during warm months, disappears from the shallows in winter. Also, winter storms make shallow regions unstable before ice cover develops, and cause ice grinding when ice breaks up. Small and large temperate North American lakes experience a net decrease in both diversity and abundance of fishes in shallow waters during the winter. Centrarchids as a group move into deeper water. Large Common Carp (*Cyprinus carpio*) and Bigmouth Buffalo (*Ictiobus cyprinellus*) in Lake Mendota, Wisconsin move to two traditional winter aggregation areas in relatively deep (5–7 m) water. Gill netting in those areas during one winter caught 43,000 kg of Common Carp and 3000 kg of Bigmouth Buffalo whereas nets set at other locales caught nothing. Northern Pike show a tendency to occupy deeper water and to swim farther offshore under ice.

Some species remain in the shallows or move into them from deeper water. Minnows remain in the littoral zone of lakes and occupy piles of twigs, small cracks in rocks and logs, or even bury themselves as much as 0.5 m down in gravelly bottoms. Salmon and trout, whitefishes, Burbot, and sculpin occupy deeper water by summer but move into shallower water to feed under the ice. In the Laurentian Great Lakes, fishes abandon the shallows in fall and early winter, but some (whitefishes, herrings, salmonids, Troutperch, sculpins, and suckers) return after the ice cover develops and filamentous algae appear. Again, early winter storms make the shallows down to about 10 m depth a turbulent and unstable habitat. At very high, Arctic latitudes, there are little day/night or summer/winter differences in the relatively depauperate fish faunas of lakes (Diana et al. 1977; Johnsen & Hasler 1977; Emery 1978; Helfman 1979a).

The extreme conditions that develop during winter in ponds and small lakes lead to different behavior patterns. As ice and snow cover develop, deoxygenation occurs, beginning at the lake bottom and moving up through the water column.

At very low oxygen tension levels (e.g., <0.5 mg/L), lake species that are most resistant to winterkill mortality (e.g., mudminnows, Umbridae; Northern Pike; Yellow Perch) engage in behaviors that enhance their survival. They move up in the water column and take up positions immediately under the ice where it is thinnest and where oxygen concentrations are greatest, with their noses in contact with the ice. They seek out gas bubbles and inhale water from around the bubbles. Mudminnows will even engulf air bubbles that have been squeezed out of ice as it freezes or are exhaled by aquatic mammals such as beavers and muskrats. Sunfishes, such as Bluegill, swim throughout the water column and frequently encounter deoxygenated water, and may therefore be among the first to die under winterkill conditions (Petrosky & Magnuson 1973; Klinger et al. 1982).

Among stream fishes, salmonids are as usual the best studied with regard to winter behavior. Several species (especially juvenile Sockeye and Atlantic salmon and Rainbow, Brown, Cutthroat, and Bull trout) remain active but switch from constant activity or diurnal foraging and nocturnal refuging to nocturnal foraging and resting by day (Thurow 1997; Valdimarsson & Metcalfe 1998; Bremset 2000; Jakober et al. 2000; Steinhart & Wurtsbaugh 2003). Daytime shelter use often entails settling under boulders or in the spaces between boulders and cobbles, thus saving energy by resting in areas of low current flow, protecting fish from physical damage from ice moving in the water column, and probably also concealing them from predators. Suitable refuge habitats – either cobbles or slow-flowing water – may be limiting, and inter- and intraspecific competition for appropriate refuge sites is reflected in territorial combat between fishes when they move into shelters at dawn (Harwood et al. 2001, 2002). Overwinter survival of Rainbow and Cutthroat trout and Chinook and Coho salmon is higher in stream sections that contain cobbles or large woody debris (Solazzi et al. 2000). Minnows also switch from diurnal to nocturnal activity as temperatures drop, occupying cobble substrates during diurnal resting periods (Cunjak 1996; Greenwood & Metcalfe 1998). These observations underscore the importance of maintaining high habitat diversity – especially clean, complex bottoms with cobble or woody debris but also of pools and backwater areas – as a means of improving year round survival of a variety of stream fishes (Cunjak 1996; Jakober et al. 1998; Brown et al. 2001).

Temperate marine fishes also exhibit cycles of small-scale, seasonal movements that relate to temperature and climatic changes (longer migrations are discussed below). Many fishes abandon shallower waters when large algae die back in winter. In central Californian kelp beds, juvenile fishes inhabit understory kelp during spring and summer, using it for shelter and eating the invertebrates that live there. Juveniles disappear each fall and winter as the understory dies back or is reduced by periodic storms. Adults of resident species, particularly surfperches (Embiotocidae) and predatory Kelp Bass (*Paralabrax*, Serranidae) tend to remain in the area year round but undergo changes in diet and foraging locale as the resource base shifts. Southern Californian bays and estuaries undergo a marked cycle of species richness and individual abundance, both of which peak in summer and are lowest in winter. The fauna contains resident (topsmelts, surfperches, gobies, and flatfishes) and seasonal (anchovies, mullets) species. Seasonal movements in and out of the bays are strongly linked to changes in temperature, salinity, and the productivity of macroalgae. In Puget Sound, Washington, which is relatively protected from winter storms, rockfishes (*Sebastes*, Scorpaenidae) school in midwater and move down a few meters to slightly deeper water in the winter. Benthic species remain in kelp bed and reef areas year round (Ebeling & Laur 1985; Horn & Allen 1985; Ebeling & Hixon 1991; Stephens et al. 2006).

On the Atlantic coast of North America, Summer Flounder (*Paralichthys dentatus*, Bothidae) spend warmer months nearshore along the coastline and in bays and migrate offshore in the fall to deeper (30–200 m) water to spawn. In contrast, Winter Flounder (*Pleuronectes americanus*, Pleuronectidae) migrate to deeper water in the summer and then return to bays as the water cools; they also spawn in winter. Other species undergo seasonal movements that differ by individual age. Adult Tautog (*Tautoga onitis*, Labridae) move offshore in the fall as water temperatures drop below about 10°C, whereas young Tautog and Cunner (*Tautogolabrus adspersus*) move from grass and algal beds that are dying back to other shallow habitats that provide greater shelter before these fishes enter a winter torpid state. A pattern in many temperate marine environments is a dependence on algae as a refuge or as an indirect or direct food source. As colder months approach and algal beds cease productivity and lose their "above-ground" parts, many species abandon these regions for deeper waters or waters that will provide cover during months of low food production (Bigelow & Schroeder 1953b; Olla et al. 1979; Rogers & Van Den Avyle 1982).

Reproductive Seasonality

The most notably seasonal activity in fishes is reproduction. Successful reproduction requires careful synchrony in physiology, anatomy, and behavior of both sexes. Spawning occurs when both sexes have completed gametogenesis, gamete maturation, secondary sex character development, and spawning readiness and arrive at the proper spawning locale at the same time. A series of environmental cues are likely to trigger each stage of a reproductive cycle. Seasonally dependable cues, particularly ones that may ensure survival of larvae (plankton blooms, sea temperature changes, and alterations in currents) are the most likely cues to be used and are usually associated with seasonal, cyclical climatic events such as monsoonal rains, oceanic surface, and upwelling currents (e.g., El Niños), and temperature cycles. Although environmental cues influence timing, flatfishes, and seabass held under constant laboratory conditions still show predictable seasonality in their gonadal cycles, indicating an endogenous basis to reproductive cycles (Bye 1990).

Seasonal cycles occur in most families, but "seasons" are defined differently in temperate versus tropical habitats. Most species in temperate latitudes spawn in spring or summer, a few in fall and winter. Conditions favoring larval growth and survival appear to be primary determinants of the phasing of reproduction. In temperate locales, spring and summer are times of maximal food productivity, and are also periods when protective vegetation is maximally available. Although many fishes in tropical and even subtropical regions breed year round (e.g., livebearers, numerous reef species), even these species show periods of peak reproductive activity that occur at relatively predictable times of the year.

Temperate freshwater fishes undergo reproductive cycles that are influenced strongly by **changing photoperiod** (day length) and **temperature**. Because gametogenesis is a complicated and lengthy process (see Chapter 8), environmental conditions at the time of initiation of gametogenesis will be different from those in effect when spawning occurs. Hence different cues are used at different phases in the cycle. In salmonids, spawning time is heritable, occurring at the same time each year over a period of 2–6 weeks in a particular genetic strain (Scott 1990). However, timing may differ among stocks in geographically nearby rivers or even in different streams flowing into the same river, reflecting locally adapted genotypes (e.g., NRC 1996b; Stewart et al. 2002) (Fig. 18.7). The rhythm is circa-annual, endogenous, and entrained by environmental cues, primarily photoperiod, but can be modified by temperature. Salmonids are generally divided into fall (September to December) and winter (January to March) spawners. Most species are fall spawners, including Brown Trout, Brook Trout, Lake Trout, and Atlantic Salmon; among Pacific *Oncorhynchus* species, spawning occurs in late summer through early winter, with much latitudinal variation (Groot & Margolis 1991; Augerot 2005; Quinn 2005). Rainbow Trout are generally late winter spawners. In both groups, the reproductive cycle is initiated during the previous springtime in response to increasing day length.

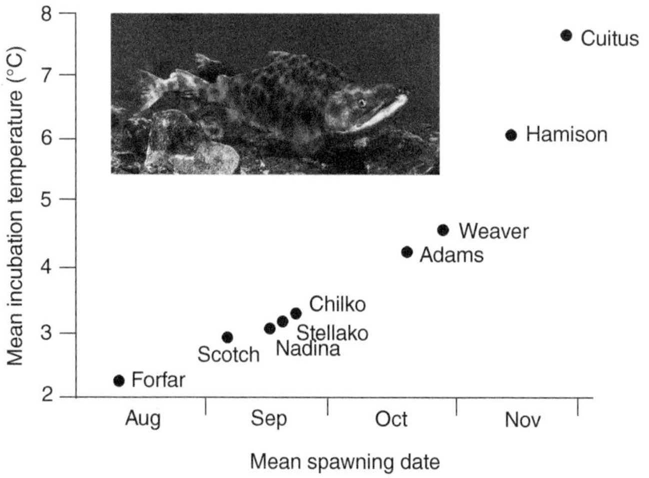

FIGURE 18.7 Genetically based, local adaptation in Fraser River Sockeye Salmon. Among stocks (named dots), fish spawn on different dates, and eggs are incubated at different mean temperatures. These differences lead to emergence dates favorable to juvenile feeding. Different spawning dates also help to coordinate migrations among smolts originating at different distances from the sea. After Brannon (1987); inset M. Love / USGS / Public Domain.

Temperate cyprinids, such as the Golden Shiner (*Notemigonus crysoleucas*), Goldfish (*Carassius auratus*), Humpback Chub (*Gila cypha*), and Lake Chub (*Couesius plumbeus*), all spawn in late spring and early summer. Gametogenesis begins in the fall in response to decreasing temperature and shortening day length, advances slowly during the winter, and then accelerates and is completed in spring in response to increasing day length and rising temperature. Sticklebacks have a similar cycle, as do most spring/early summer spawning fishes in temperate locales. Common Carp show a variant cycle, involving gonad development in late summer, quiescence in winter, and then final maturation of oocytes and spawning in the spring. The European Tench, *Tinca tinca*, is unusual in that it spawns in the fall. Late fall, winter, and early spring in temperate lakes are too unproductive for the small larvae of most species and consequently spawning does not normally occur during those seasons. The exceptionally large size of eggs and physiological tolerance of cold temperatures in salmonids may explain their fall–winter spawning and success at high latitudes (Baggerman 1990; Hontela & Stacey 1990).

Temperate fishes use **photoperiod** as a proximate environmental indicator of current and future climate. Typically, long days (e.g., >13 h light) and warm temperature cause gonadal **recrudescence** (resumption of gametogenic activity), whereas short days inhibit recrudescence, regardless of temperature. Available evidence suggests that many temperate fish species have an endogenous, circa-annual clock that drives reproductive activities, and that this clock is affected by another, circadian clock of photosensitivity. A critical piece of information is day length; days shorter than some minimum cause both initiation and cessation of reproductive behavior. The circadian clock can tell a fish if day length is increasing, but the fish must be most sensitive to light at a time of the day when daylight would indicate increasing day length (Fig. 18.8). Daylight during the first 8–10 h after sunrise could occur at just about any time of year, but daylight 10–12 h or more after sunrise will not occur during winter. Hence fish have a clock that tells them how many hours have passed since sunrise, and they tend to be insensitive to light during the first 10 h or so after sunrise. Light encountered after that period, during the "photoinducible phase" of photosensitivity, has a strong influence on gonad development. The existence of a photoinducible phase and a photosensitive circadian rhythm was discovered by exposing fish to 2 h pulses of light at different times of the day. Sticklebacks exposed to light 14–16 h after sunrise showed greater rates of sexual maturation than fish experiencing light at other times of a light–dark cycle. The position and length of the photoinducible phase change with season and temperature (Baggerman 1990; Taylor 1990).

Seasonality among freshwater fishes at tropical latitudes (between 30°N and 30°S latitude) is defined more by rainfall than by temperature (Goulding 1980; Lowe-McConnell 1987;

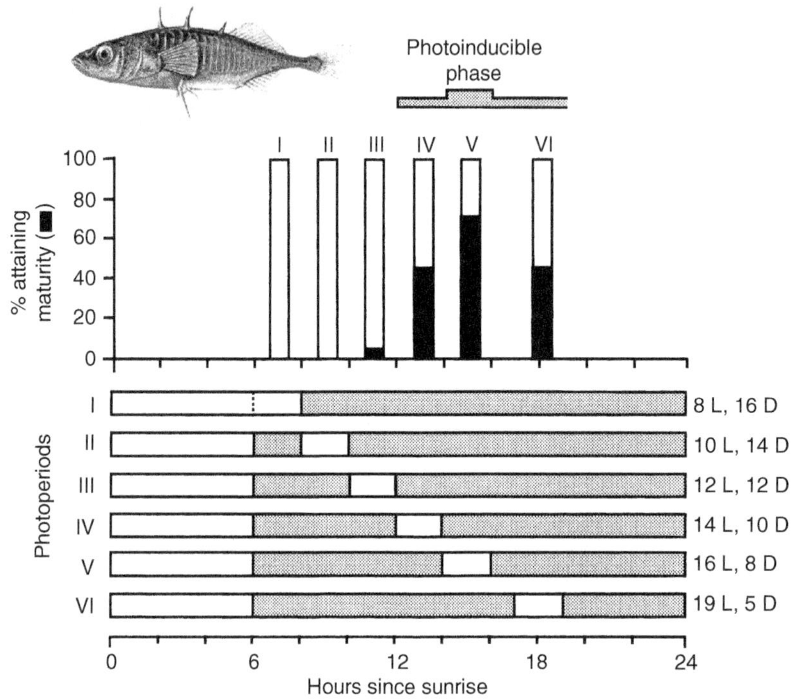

FIGURE 18.8 Sticklebacks have a daily rhythm of photosensitivity that is maximal about 14–16 h after sunrise. During this maximally sensitive period, exposure to light induces sexual maturation. The photoinducible phase helps the fish determine if day length is increasing, as would happen during spring and summer. Experimental manipulations of daily light/dark cycles can pinpoint the existence and position of the photoinducible phase by providing a 2 h pulse of light at different times after sunrise. The open bars at the bottom indicate when lights were on, the darkened portions when lights were off (0 = sunrise). Light 14–16 h after sunrise would be naturally experienced when days consisted of 16 h of light (L) and 8 h of dark (D), as would happen during summer. Maximum levels of sexual maturity (bar V) are found in this light regime. From Baggerman (1985), used with permission; stickleback drawing from www.seagrant.wisc.edu.

Munro 1990a; Winemiller & Jepsen 1998) (Fig. 18.9). Regions between 15° north and south of the equator generally have two rainy seasons per year, whereas higher tropical latitudes have one rainy and one dry season. The floodplains, lakes, and seasonal swamps created by a rising river are common spawning and nursery grounds in many locales, and many riverine fishes have reproductive cycles that coincide with seasonal inundation of gallery forests and swamps, perhaps cued by rainfall or rising water levels (Lim et al. 1999; Agostinho et al. 2004; de Lima & Araujo-Lima 2004). Newly inundated areas are advantageous spawning locales because: (i) accumulated nutrients are released, which creates plankton blooms and food for progeny; and (ii) predation is minimized by abundant vegetation for refuging and because large flooded expanses minimize contact with predators. In contrast, receding water and reduced habitat space during the dry season means that predators and competitors are concentrated, which also leads to deoxygenated water. Dry season and aseasonal spawners often provide extensive parental care, including provisioning of young, and/or possess secondary breathing structures (e.g., lungfishes, bagrid catfishes, cichlids, and anabantoids).

Adults of many tropical freshwater species, including osteoglossid arapaima, mormyrids, large characins, cyprinids, many catfishes, and gymnotid knifefishes, migrate up tributaries and onto floodplains to spawn (Fig. 18.9); many such migrations cover much more than 100 km (Munro 1990a; Lucas & Baras 2001; Welcomme 2003). Lake-dwelling species and populations in these families move into tributary streams, whereas lacustrine herrings, silversides, and percomorphs often spawn within the lake itself. Seasonality in other riverine species involves migrations upriver to headwater regions in anticipation of seasonal rains (e.g., large characins, catfishes). Many small characins, killifishes, livebearers, and cichlids reproduce year round, although peaks in recruitment often correspond with high water.

Worldwide, predatory species often spawn earlier than their prey, thus assuring a food source for young predators. For example, the South American characin, *Hoplias malabaricus*, is predatory throughout life and breeds earlier than most other species. In contrast, juvenile piranhas are omnivorous and adults breed along with other nonpredatory species, although predatory species that used to be restricted to

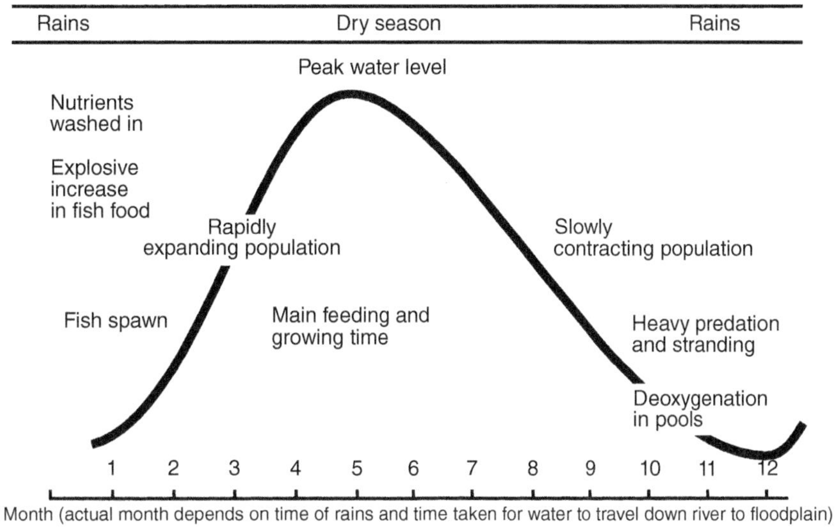

Fish movements	Up river	Disperse on floodplain	Back to river	Confined to pools
Fish biomass	Production of young	Very rapid growth	Heavy losses	Much reduced
Fishing	Of upstream migrants (destructive)	Diffcult: fish dispersed, much cover	Intensive as fish move back to river	In pools (dry season refuges)

FIGURE 18.9 The seasonal progression of events for many fishes in large tropical rivers. The dark curved line indicates relative water levels. Seasonal flooding of highly productive gallery forests and swamps opens these areas up to lateral migration, feeding, and spawning by fishes. Many regions show two rainy seasons and both result in lateral migrations. Fat stores increase as fishes capitalize on the food abundance in flooded regions. From Lowe-McConnell (1987), used with permission.

seasonal spawning in inundated floodplains now regularly find favorable spawning habitat behind dams (e.g., Haddad & Sazima 2003; see Chapter 22, Dam Building). Regardless, spawning migrations upriver or onto flooded areas must be preceded by gonadal recrudescence that anticipates seasonal rainfall by several months. The cues that stimulate gonadal growth and gametogenesis are poorly understood, but may include photoperiod and temperature changes (particularly at higher latitudes in the tropics), social interactions, food availability and energy stores, as well as endogenously controlled rhythms, perhaps entrained by previous spawning itself (Munro 1990a).

Temperate marine teleosts have restricted spawning periods that vary by species, locale, and genetic stock (Bye 1990). At any particular locale, however, a stock is likely to have a fairly short and predictable time period during which most spawning occurs. Typically, pelagic species spawn over a 4-month period, with a shorter period of maximal activity. For example, Atlantic Cod in the North Sea off the northeast coast of England spawn between January and May, with 70% of eggs produced during 6 weeks of that period. Peak spawning of European Plaice (*Pleuronectes platessa*) in the Southern Bight of the North Sea occurred within 1 week of January 19 over the 39-year period between 1911 and 1950.

The locale of spawning is also fairly predictable and defined by oceanic phenomena such as thermoclines or frontal areas, which are transition regions between differing water masses. American and European eels migrate to spawn in a region of the Sargasso Sea where a persistent frontal zone, defined by marked horizontal differences in temperature, salinity, and water density, exists every spring (McCleave et al. 1987). Inshore, temperate marine species are strongly seasonal. Coastal California species spawn mostly in the spring and summer (e.g., Grunion, surfperches, halibut, Sheephead, Blacksmith, croakers, seabasses, and pricklebacks), some spawn in winter or early spring (rockfishes, Starry Flounder, Lingcod), and a few spawn in the fall (greenlings, Cabezon). Predominantly spring and summer spawning also typifies Atlantic temperate fishes (halibut, killifishes, most flatfishes, seabasses, porgies, many croakers, and wrasses), with a few winter spawners (some flounders, sculpins, and some croakers) (Ferraro 1980; Holt et al. 1985).

Coral reefs experience less extreme seasonal variation than temperate habitats and are less subject to the vagaries of rainfall than tropical freshwater systems. As a consequence, many coral reef fishes spawn through most or all of the year, particularly at low latitudes (e.g., many damselfishes, wrasses, parrotfishes, grunts, and surgeonfishes). Nevertheless, seasonal reproduction is also common among many families, including groupers, snappers, damselfishes, rabbitfishes, gobies, and pufferfishes. Seasonal spawning peaks have been found in most tropical locales, including sites in the Indian Ocean, South China Sea, and tropical Atlantic and Pacific oceans. "Springtime" peaks are most common, followed next by two periods of major spawning activity in the spring and fall. The most common environmental correlate of these peaks is that they occur when major currents around islands are weakest. Spawning during slack current periods would minimize the long-distance dispersal of larvae away from home reefs (Johannes 1978; Sale 1978; Robertson 1991).

Recruitment of larvae to the reef is also cyclical and seasonal. Even species that breed throughout the year show seasonal peaks in the arrival of larvae and episodic pulses of larval arrival. French Grunt, *Haemulon flavolineatum*, in the Caribbean breed year round. Larvae arrive in semilunar pulses over an 8-month period, with greatest recruitment in May, June, and November. Damselfishes in the southern Great Barrier Reef breed during a 5-month, summertime period and larvae are recruited in pulses, with one or a few major pulses accounting for most arrivals. Larvae arrive on a lunar cycle, the major feature being that little recruitment occurs around the time of the full moon. Most settlement of fish larvae on coral reefs occurs at night, suggesting a strong influence of visual predators. Avoidance of full moon periods would have a similar function. Evidently, spawning and recruitment do not necessarily follow the same timetables. Larvae can be produced, but conditions for settling after the larval period of a month or two may not be favorable. In fact, abundance of larvae offshore and settlement of larvae on the reef do not necessarily coincide. For example, Nassau Grouper, *Epinephelus striatus*, larvae can be found along the Bahamas Bank of the western Atlantic over a 2- to 3-month period, but actual larval settlement occurs almost entirely over a 4- to 6-night period when storm-driven currents push water and larvae into shore (McFarland et al. 1985; Doherty & Williams 1988; Doherty 1991; Shenker et al. 1993).

Seasonal Reproduction: Proximate and Ultimate Factors

Munro (1990b) has proposed a classification of the proximate cues that determine the occurrence of different portions of the reproductive cycle. He recognizes four factors that control the development and synchrony of breeding cycles.

1. **Predictive cues** are general periodic environmental events that a fish can use to predict that the spawning season is approaching. Changing day length and temperature are predictive cues that are likely to trigger the onset of gametogenesis and secondary sex character development. Gametogenesis may have an endogenous circa-annual rhythm that is entrained by some predictive environmental cue.
2. **Synchronizing cues** signal the arrival of spawning conditions. Typically, the presence of a suitably appearing and behaving mate, perhaps releasing pheromones, may serve as such a cue, causing final gamete maturation and release. The pheromone may even be produced by another species, as in the case of minnows that are nest associates of other species and spawn only in the presence of the host species (Rakes et al. 1999). The presence of vegetation or other spawning substrates plays a role in some species. Synchrony is important to ensure contact between the sperm and eggs, and to prevent hybridization. In many species, gametes decline in fertility rapidly after ovulation and spermiation. Hence, a small temporal window of spawning receptivity and opportunity exists.
3. **Terminating cues** signal the end of the spawning period. Because breeding conditions remain optimal for a short period, including the above-mentioned changes in gamete viability, breeding seasons are typically short. Gonad regression occurs after breeding in response to environmental cues (i.e., changes in predictive cues), exhaustion of gametes, or the departure or changes in behavior of conspecifics. Nest guarding species may respond to the presence of eggs in a nest, causing hormonal changes that inhibit spawning and encourage egg care and aggression.
4. The first three categories of cues can all be modified by secondary factors such as water quality, lunar cycle, adult nutrition, predator presence, and social interactions. These **modifying factors** are the causes of intraspecific variation in breeding at different latitudes or in different habitats.

Evolutionarily, why is seasonal breeding so prevalent in fishes? Gamete production, particularly in females, is energetically expensive. Gametes are usually released in batches; time and energy are required to replenish gametic products, even in males (Nakatsuru & Kramer 1982; Shapiro et al. 1994). Courtship and spawning, and parental care where it occurs, require time and energy and expose participants to predators. Few fishes can therefore afford to reproduce year round. Hence a decision in evolutionary terms must be made about the optimal time to reproduce, optimality being defined in terms of the relative costs and benefits of current versus future reproduction (see Chapter 20, Life Histories and Reproductive Ecology). The conditions for egg dispersal, larval survival and growth, and larval recruitment vary through the year and are dependent on seasonally driven climatic variation. In most species, spawning appears to be synchronized with periods most favorable for the survival of young. In temperate marine fishes with pelagic larvae, food availability is one critical determinant. Spawning coincides with seasonal blooms of zooplankton, thus maximizing the chances that larvae will encounter prey during the **critical period** shortly after they use up the energy stores of their yolk supply (the Match–Mismatch Hypothesis of Cushing (1973); see Chapter 9, Larval Feeding and Survival). Individuals that spawn at times when the probability of egg, larval, and their own survival are higher will be more successful than individuals that spawn at less suitable times (Munro 1990a).

Annual and Supra-Annual Patterns: Migrations

Many fishes engage in periodic long-distance movements and there is vast literature on various aspects of migratory behavior (Harden-Jones 1968; Leggett 1977; Baker 1978; Northcote 1978; McCleave et al. 1984; McKeown 1984; Dodson 1997; Lucas & Baras 2001; Secor 2015). Our focus will be on species that undergo fairly large-scale migratory cycles with an annual or greater period, either in the ocean or between the ocean and fresh water, with lesser treatment of the so-called potamodromous fishes that undergo reproductive migrations within fresh water (see Lucas & Baras 2001; Welcomme 2003).

Migrations take several general forms. Reproductive migrations take animals from a habitat that is optimal for adult survival to one that is better for larval or juvenile survival. Fishes that spawn several times in their lives (the **iteroparous** condition) may undergo this migration more than once (e.g., Atlantic Sturgeon, American Shad, Atlantic Salmon, and the world's largest salmon, the Taimen Salmon of Siberia, *Hucho taimen*, which may weigh 70 kg). **Semelparous** fishes, those that spawn once and die, undergo the migration only once (e.g., sea lampreys, anguillid eels, Pacific salmons, and some galaxiids).

Inherent in reproductively migrating species is the complementary migration that juveniles take to juvenile and adult feeding areas. In some species, nonspawning juveniles and adults also migrate between feeding and spawning areas along with reproductively active individuals (e.g., sturgeon). Reproductive migrations may involve movement between lakes and tributary streams or between different parts of a river system, as occurs in large tropical characins and catfishes. Adults of the prochilodontid Coporo, *Prochilodus mariae*, in the Orinoco region migrate from Andean piedmont tributary rivers to wet-season spawning and feeding habitats in lowland floodplains, returning to tributaries as river levels fall. All such species are decimated by dam construction that blocks these extensive migrations (Barbarino-Duque et al. 1998; Lucas & Baras 2001). Other reproductive migrations involve fishes that move between the sea and fresh water (diadromy, see below), or may entail movements within ocean basins in a roughly circular or back-and-forth pattern (Bluefish, tunas). Additional species engage in transoceanic, seasonal migrations that do not appear linked directly to reproduction, but instead probably place adult fish in optimal locales to intercept seasonally available food sources (pelagic sharks, billfishes) or may move individuals away from climatically unfavorable areas to regions that are less harsh (e.g., Summer and Winter flounder).

Diadromy

Many species of migratory fishes move predictably between fresh and salt water at relatively fixed times in their lives. These **diadromous** fishes include about 160 species, or a little less than 1% of all fish species, but many of them are very important commercially and their complex life histories are fascinating (Table 18.2). Diadromy takes three different forms, anadromy, catadromy, and amphidromy (Fig. 18.10). **Anadromous** fishes such as lampreys, sturgeons, shads, Pacific salmons, smelts, and Striped Bass spend most of their lives in the ocean and then migrate to fresh water to spawn. Many anadromous species develop **landlocked** populations that never migrate to the sea but instead spawn in inlet streams to large lakes. **Catadromous** fishes such as anguillid eels, mullets, temperate basses, and some sculpins spend most of their lives in fresh water and then return to the ocean to spawn. **Amphidromous** fishes (Ayu,

TABLE 18.2 Families of known diadromous fishes. Modified from McDowall (1987).

Anadromous	Catadromous	Amphidromous
Petromyzontidae, lampreys	Anguillidae, true eels	Clupeidae, herrings
Geotriidae, southern lampreys	Galaxiidae, galaxiids	Plecoglossidae, Ayu
Mordaciidae, southern lampreys	Scorpaenidae, scorpionfishes	Prototroctidae, southern graylings
Acipenseridae, sturgeons	Moronidae, temperate basses	Galaxiidae, galaxiids
Clupeidae, herrings	Centropomidae, snooks	Syngnathidae, pipefishes
Ariidae, sea catfishes	Kuhliidae, aholeholes	Cottidae, sculpins
Salmonidae, salmons	Mugilidae, mullets	Mugiloididae, sandperches
Osmeridae, smelts	Bovichthyidae, bovichthyids	Eleotridae, sleepers
Retropinnidae, New Zealand smelts	Pleuronectidae, righteye flounders	Gobiidae, gobies
Galaxiidae, galaxiids		
Gadidae, cods		
Gasterosteidae, sticklebacks		
Cottidae, sculpins		
Moronidae, temperate basses		
Gobiidae, gobies		
Soleidae, soles		

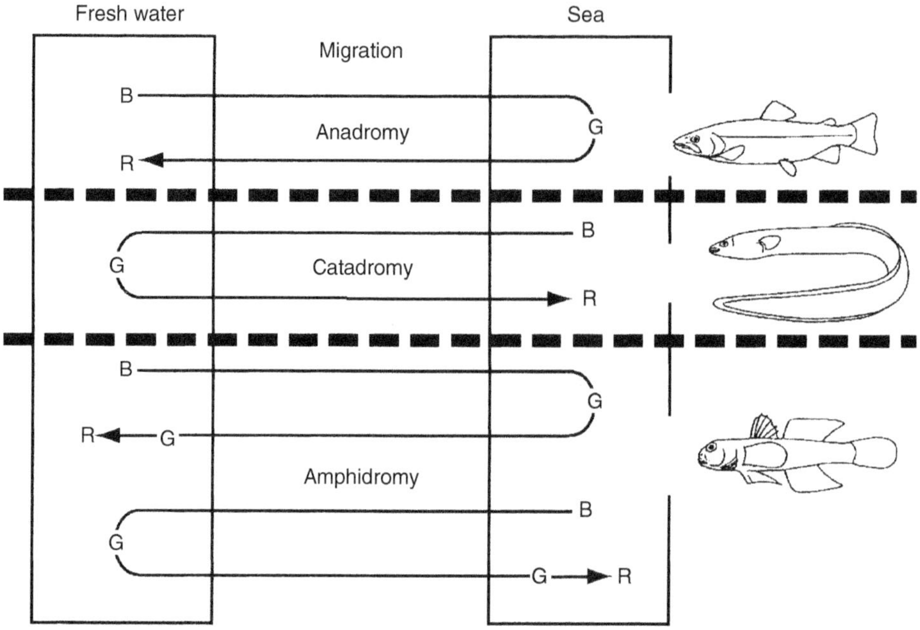

FIGURE 18.10 Diadromy takes three general forms: anadromy, catadromy, and amphidromy. In anadromy, adults spawn in fresh water, juveniles move to salt water for several years of feeding and growth, and then migrate back to fresh water to spawn. In catadromy, adults spawn at sea, juveniles migrate to fresh water for several years to feed, and return to the sea to spawn. In amphidromy, spawning can occur in either fresh or salt water (but usually fresh), larvae migrate to the other habitat for an initial feeding and growth period, then migrate to the original habitat as juveniles where they remain for additional feeding and growth prior to spawning. B, birth; G, growth; R, reproduction. Modified from Gross (1987).

galaxiids, southern graylings, sandperches, sleepers, and gobies; Fig. 18.11) move between marine and fresh water at certain phases of their lives, but the final migration occurs long before maturation and spawning occur. The chief distinction between amphidromy and anadromy is that the migration into fresh water usually occurs in the adult stage in anadromy, but in the juvenile stage in amphidromy (McDowall 2007a). About half of all diadromous fishes are anadromous, the other half equally divided between catadromous and amphidromous forms (McDowall 1987, 1988, 1999).

The geographic distribution of the different forms of diadromy is interesting because it provides insight into the evolution of the behavior (Fig. 18.12). Anadromy is largely a northern hemisphere, high-latitude phenomenon, catadromy is more common at low latitudes and in the southern hemisphere, and amphidromy has a bimodal distribution at middle latitudes in both hemispheres, with greater representation in the southern hemisphere (McDowall 1987). Interpretations of why different forms of diadromy prevail at different latitudes are confounded by phylogenetic histories, but one appealing analysis views diadromous migrations as complex adaptations that function to place both larvae and adults in environments where food is most abundant (Gross 1987; Gross et al. 1988; see Dodson 1997 for a critique). For a long-distance migration to

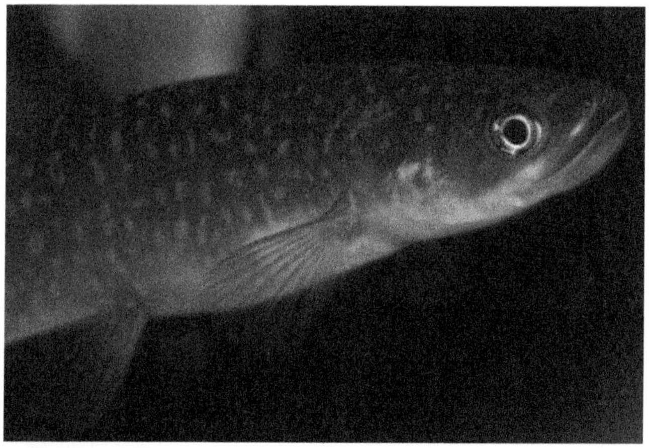

FIGURE 18.11 Many familiar diadromous fishes are either anadromous (such as salmon) or catadromous (such as anguillid eels). Some galaxiids (Galaxiidae), however, many of which are found in southern Australia and New Zealand, are amphidromous. They spawn in freshwater, move into salt water when very young, and then migrate back into rivers while still juveniles. They complete their amphidromous life cycle by spawning in these rivers. Galaxiids, such as this Giant Kokopu, *Galaxia argenteus*, of New Zealand, are threatened with extinction due to predation and competition from introduced non-native trout. Blueether / Wikimedia Commons / CC BY-SA 3.0.

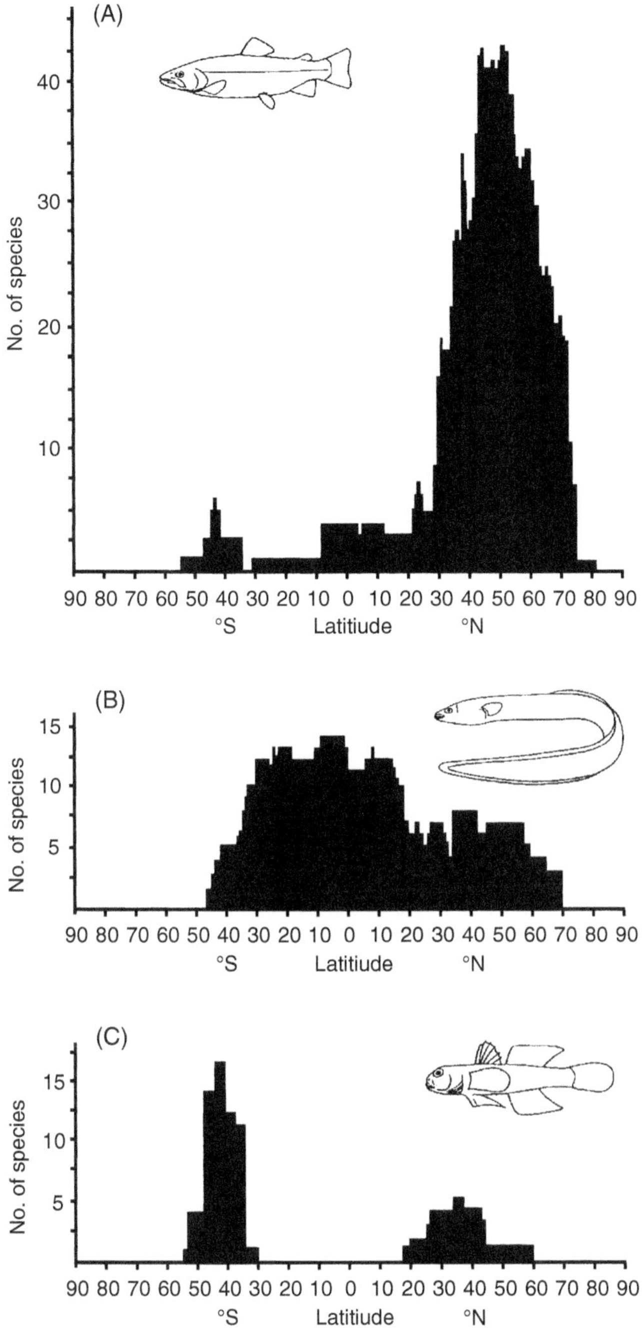

FIGURE 18.12 The latitudinal distribution and frequency of different forms of diadromy among major fish groups. The number of species employing each tactic is plotted as a function of latitude, showing: (A) anadromy to be largely a northern, temperate and polar phenomenon; (B) catadromy to be more tropical and subtropical in distribution; and (C) amphidromy to be more bipolar and temperate. Modified from McDowall (1987).

evolve, the gains in fitness from moving must exceed the fitness an individual would have achieved had it remained in its original habitat. Gains have to be sufficiently large to also overcome losses – including osmotic costs, energy and time lost, and predation risk – incurred while migrating. If an individual migrates during its life, it should ideally (i) spawn in a place of low predator density to minimize egg mortality but where (ii) larvae can drift passively to locations of higher productivity appropriate to their growth needs, and (iii) place juveniles and young adults in areas where they can maximize their feeding, thus allowing them to build up energy stores necessary for (iv) the long migration back to the optimal (low productivity) spawning locale.

Available evidence indicates very different growth and reproduction rates in the different habitats. Juvenile Pacific salmon may increase their daily growth rate by 50% during their first week in the ocean. Sockeye Salmon, *Oncorhynchus nerka*, are referred to as Kokanee when landlocked and seldom attain 25% of the body size of anadromous individuals. Comparisons between diadromous and non-diadromous stocks of the same salmonid species (e.g., Cutthroat Trout, Rainbow Trout, Atlantic Salmon, and Brown Trout) indicate that diadromous stocks produce on average three times as many eggs as non-diadromous conspecifics, probably because diadromous fish grow to larger size as a result of increased feeding in the ocean. Mortality rates are difficult to estimate, but the three-fold reproductive advantage of diadromy would more than make up for greater mortality at sea before the benefits of diadromy were negated.

Combining such growth and reproduction differences with information about relative productivity of fresh waters and oceans at different latitudes suggests that the temperate prevalence of anadromy and more tropical occurrence of catadromy are evolutionary logical. In temperate regions, oceans tend to be more productive than fresh waters, whereas fresh waters are more productive than the ocean in tropical regions. Hence the primary feeding habitat for a migratory species should be the ocean at high latitudes versus rivers and lakes at lower latitudes. The survival of young may be enhanced if spawning occurs where productivity, and also presumably predation, are least: streams and rivers at high latitudes, the ocean at lower latitudes. As long as the costs of movement between spawning and growth habitats are not excessive, the existence of anadromy in colder climates and catadromy in warmer climates appears adaptive.

The relatively high frequency of amphidromy among fishes on oceanic islands suggests strong evolutionary benefits to this life style. Many of the fishes inhabiting freshwaters of islands such as Guam, Papua New Guinea, the Galapagos, Fiji, Samoa, New Zealand, the Philippines, and Indonesia are amphidromous (McDowall 2004). The ability for juveniles to move from freshwater to the oceans early in life provides not only opportunities for growth in productive marine waters, but also is likely responsible for the ability of these fishes to spread among islands. How else would freshwater fishes come to occupy oceanic islands, (other than introduction by humans)? All native stream fishes of the Hawaiian archipelago are amphidromous gobioids (McDowall 2007b) and the lack of genetic differences suggests that there must be somewhat regular gene flow among the islands. The ability to move among islands during a juvenile dispersal period would also account for recovery of populations after significant natural disturbances such as geologic or volcanic activity, or flash-floods in the steep island terrain.

Mechanisms of Migration

Fishes may move thousands of kilometers through the open and seemingly landmark-free ocean. A great deal of research has focused on the means by which fish undertake long-distance migrations, specifically how they **orient** toward and **locate** their ultimate destinations. Research has identified numerous possible cues used in orientation, including sun and polarized light, geomagnetic and geoelectric fields, currents, olfaction, and temperature discontinuities and isolines (Leggett 1977; McCleave et al. 1984; McKeown 1984).

Birds use a **sun compass** and **internal clock** to orient. An animal must be able to sense the time of day, the altitude, azimuth (angle with the horizontal), and compass direction of the sun at a given time and date, correcting for the 15°/h movement of the sun across the sky. Experimental evidence suggests that some fishes use such a mechanism. Swordfish (*Xiphias gladius*) can maintain a constant compass heading in the open sea for several days. Displaced parrotfish return relatively directly to their home locations on sunny days. When the sun is obscured, when fitted with eyecaps, or when held in darkness such that their internal clocks have been shifted 6 h, displaced fish are disoriented or move in a direction appropriate for a 6 h clock shift. Juvenile Sockeye Salmon have a sun compass which they complement with a magnetic compass at night or during overcast conditions. **Polarized light** can also provide directional cues, and Sockeye Salmon are able to detect and discriminate between vertically and horizontally polarized light, which could aid them particularly during dawn and dusk migrations toward the sea, when light is maximally polarized. Other salmonids, minnows, halfbeaks (Hemiramphidae), damselfishes, and cichlids can also sense polarized light, which often involves detection of ultraviolet radiation undiscernible to the human eye (Quinn & Brannon 1982; McKeown 1984; Hawryshyn 1992; Mussi et al. 2005).

A magnetic compass implies a sensitivity to the earth's magnetic fields. Such a sensitivity has been demonstrated in elasmobranchs, anguillid eels, salmonids, and tunas (Collin & Whitehead 2004; see Chapter 6, Magnetoreception; Chapter 12, Sensory Physiology). Sharks are theoretically capable of navigating using geomagnetic cues, since they can detect fields 10 to 100 times weaker than the earth's magnetic field, as well as fields created by ocean currents moving through the earth's magnetic field, or fields induced by their own movement. An induced field would change as the animal's compass heading changed, being strongest when moving east or west and weakest when heading north or south, thus giving it directional information. A magnetic compass could be useful in transoceanic migrations undertaken by large pelagic sharks (e.g., Blue, White, and Tiger sharks, see Chapter 12).

Orientation abilities are also needed for homing, as happens when Scalloped Hammerhead Sharks, *Sphyrna lewini*, return daily to small seamounts in the Sea of Cortez after foraging offshore at night. Scalloped Hammerheads may use a combination of directional cues, including visual landmarks, auditory cues produced by fishes and invertebrates, electrical cues induced by site-specific currents, and geomagnetic fields at seamounts. The use of multiple cues and redundant systems are a general feature of migratory animals. Redundant information increases the accuracy of the information, and backup systems provide information when conditions interfere with or negate the use of other cues (Kalmijn 1982; Klimley et al. 1988; Klimley 1995; Meyer et al. 2005).

Water currents transport fish eggs, larvae, and adults, but may also provide orientational information. Where currents border on other water masses, differences in water density, turbulence, turbidity, temperature, salinity, chemical composition, oxygen content, and color could all act as landmarks to a migrating fish. And once inside a current and out of sight of or contact with the bottom or other stationary objects, it is difficult to imagine that a fish could sense the water's movement, unless the fish could detect induced magnetic fields as discussed earlier. In shallow waters, many fishes show a positive or negative **rheotactic** response that causes them to move up- or downstream, respectively. The strength and direction of response may change with season and ontogeny. Selective tidal stream transport (discussed earlier in this chapter) is such a response, whereby a fish moving upriver in an estuary swims actively against an ebbing tide and drifts passively with a flooding tide. Olfactory cues are often carried on currents. Homing of salmon to chemicals in the streams in which they were spawned (discussed in Chapter 6, and later in this chapter) probably applies to many stream and intertidal fishes (e.g., minnows, sculpins, and blennies), although the age at which a fish learns the chemical fingerprint of a water body will vary. Some fishes are extremely sensitive to familiar chemicals even at low concentrations ($1 : 1 \times 10^{-10}$, and perhaps 10^{-19} depending on species), so just a few molecules of a substance are sufficient for detection (Hara 1993).

Seasonal movement is induced or directed by **temperature** changes in several migratory species. American Shad, *Alosa sapidissima*, move north along the Atlantic seaboard in the spring, staying in their preferred water temperatures of 13–18°C. Individuals may winter as far south as Florida and spawn in Nova Scotia, 3000 km away. Some oceanic species follow specific isotherms during seasonal migrations. Albacore Tuna, *Thunnus alalunga*, move north during the summer along the Pacific coast of North America, staying within a fairly narrow 14.4–16.1°C temperature zone; east–west movements are contained within a temperature range of 14 and 20°C. Onshore arrival of water masses of the preferred temperature serve as predictors of the arrival of the fish. Many other tuna species also migrate to stay within fairly narrow temperature ranges.

Many pelagic fisheries, which rely on oceanic migrations to bring fish into regions on a seasonal basis, are highly dependent on water masses of the correct temperatures moving into specific areas. Cod and Capelin (*Mallotus villosus*) in the Barents Sea of northern Europe are available to Finnish fisheries in cold years when fish migrate farther west to warmer waters. In warm years, fish restrict their movements to the eastern side of the basin and are then exploited in the Murmansk area. The response to temperature may be a direct, behavioral one involving thermal preference, or an indirect response related to food abundance. Often, plankton blooms are associated with changing water temperatures and hence fish may be tracking food availability that responds to temperature. Herring in the Norwegian and Greenland seas migrate in response to the inflow of warm Atlantic water, which in turn stimulates plankton growth and food availability (Leggett 1977; McKeown 1984; Dadswell et al. 1987).

Representative Life Histories of Migratory Fishes

Among vertebrates, fishes stand out in terms of the complexity of their life histories, and migratory fishes have among the most complex life histories. Here, we highlight details of a few of the better known and more interesting species.

Anadromy Some of the most spectacular examples of highly evolved, complex migrations involve fishes that spawn in fresh water but spend most of their lives at sea. Included among anadromous fishes are lampreys, sturgeons, shads and herrings, salmons and trouts, and Striped Bass (see Table 18.2). The classic case involves Pacific salmons; Chinook Salmon, *Oncorhynchus tshawytscha*, (also called King Salmon) can serve as an example. Chinook Salmon spawn in streams of the Pacific northwest coast of North America during the summer and fall, depending on locale (Quinn et al. 2002). Eggs are buried in gravel nests and hatch into alevins (yolk-sac larvae), which emerge and make their way downstream, eventually transforming into silvery smolts after a few months to 2 years, depending on when they were spawned. Smolts move out into the ocean, grow into juveniles and adults, and move in a series of counterclockwise ellipses through the Northeast Pacific that may carry them as far north and west as the Aleutian Islands of Alaska or as far south as northern California, covering distances of several thousand kilometers. (Sockeye Salmon, *O. nerka*, migrate even farther from land and in larger circles in the open sea, and may cover tens of thousands of kilometers.)

After 1–8 years, maturing Chinook adults return to the nearshore area. A coastal migration eventually carries them to the mouth of the river from which they migrated as smolts. They enter this river and work their way up, bypassing hundreds of potentially usable streams and swimming past seemingly insurmountable barriers such as rapids, waterfalls, and predators (Fig. 18.13). In this final stage, they cease feeding, change to a reddish color, and the males develop the characteristic hooknosed appearance known as a **kype**. They ultimately find the natal stream in which they were hatched, and even the exact place where they were incubated; here they spawn and die (Netboy 1980; Healey & Groot 1987; Brown 1990; Groot & Margolis 1991; Augerot 2005; Quinn et al. 2006).

At each juncture in this complicated journey, fish make directional decisions (Keefer et al. 2006). Numerous mechanisms, which vary depending on life history stage and habitat, have been proposed to provide directional information for a migrating fry, smolt, juvenile, and adult (Fig. 18.14). Movement by young fish from spawning sites in natal rivers to the ocean involves a combination of responses to light (including a sun compass and discrimination of polarized light), geomagnetic cues, and water currents. The fish must also **imprint** on the chemical fingerprint or **home stream olfactory bouquet** of its

FIGURE 18.13 Pacific salmon begin their lives in streams, move downstream into the ocean (or large lakes in landlocked populations) and then find their way back home to the stream of their origin to spawn. This final spawning migration involves moving past obstacles such as waterfalls and predators. A male brown bear eating salmon at Katmai National Park, Alaska. P. Hamel, U. S. National Park Service / Wikimedia Commons / Public Domain.

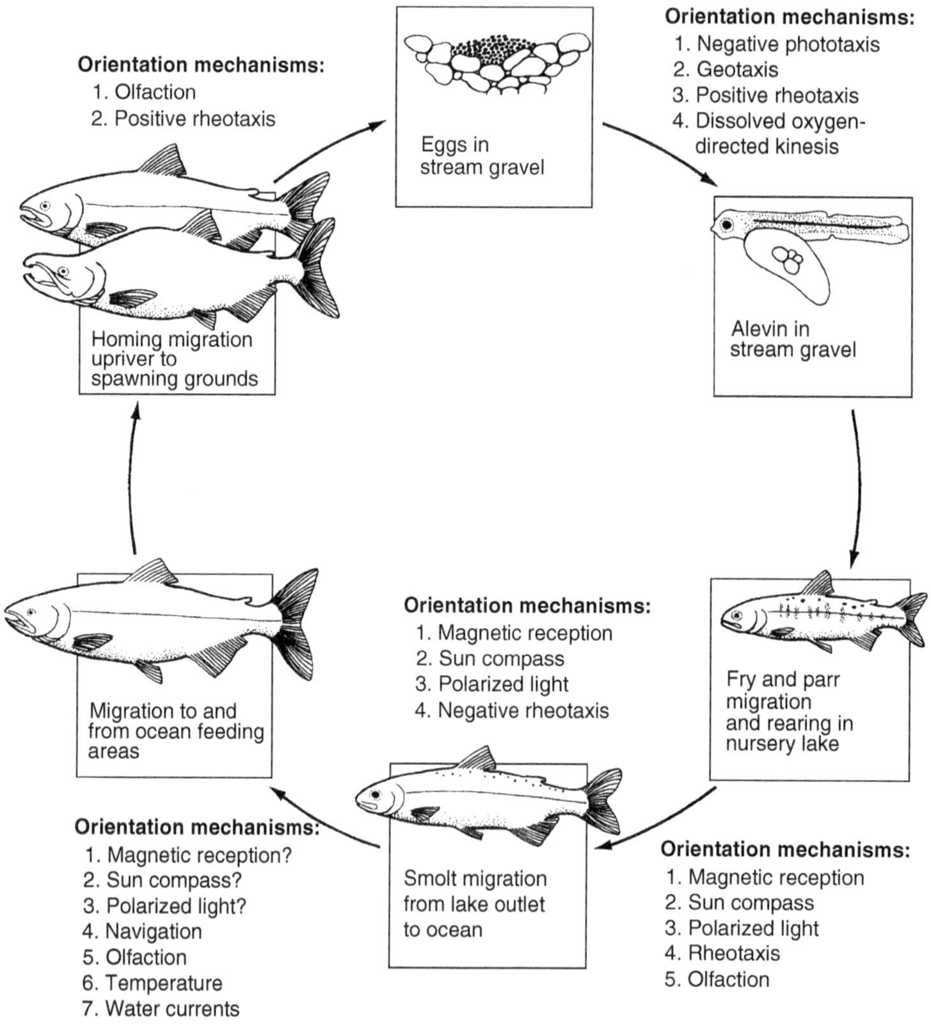

FIGURE 18.14 Characteristic life history of a Pacific salmon, as seen in the Sockeye Salmon, *Oncorhynchus nerka*. At different stages, different orientation mechanisms are likely to come into play to help the developing, growing, or maturing fish find its way to, through, and away from the sea and back to its natal river to spawn. Responses to light, gravity, and current are initially important for recent hatchlings. Later, sun compass and magnetic detection, backed up by other cues, aid a fish moving downstream and into the ocean. Finally, the detection of home stream chemicals on which the juvenile imprinted lead the maturing adult back to its spawning grounds. From Quinn and Dittman (1990) used with permission.

home stream and river, or even of multiple, sequential habitats (Dittman & Quinn 1996; Carruth et al. 2002).

Open ocean migration and eventual home stream selection offer very different problems that probably require different orientation systems. Quinn (1982) proposed a combined **map–compass–calendar** system to explain movements on the high seas. The map would involve learned or genetic knowledge of the distribution of the earth's magnetic field (which has also been mapped by oceanographers and is predictable). Compass directions, provided by celestial and magnetic cues, can be used to maintain directional headings. The calendar would require an assessment of day length or change in day length, with input from an endogenous circa-annual clock. Integration of all this information would tell a fish where it was, where it was going, and how long it would take to get there, forming the basis of a navigational system. The case for geomagnetic navigation is supported by evaluation of over 50 years of sockeye salmon catch data around Vancouver Island which revealed shifts in primary navigation routes that corresponded with changes in the strength of the Earth's magnetic field (Putman et al. 2013).

Once maturing adult salmon arrive in the coastline region of their home river, they shift to an olfactory guided response to natural chemicals contained in different rivers. Fish move upriver and pass tributaries that did not have the appropriate bouquet. Upon encountering the correct chemical cues, fish move upstream in that system until they arrive at the appropriate spawning site. Recognition of the appropriate odor may be due to hormones associated with maturation that also influence sensory cells in the olfactory epithelium and make the fish more sensitive to chemicals associated with the home stream (discussed in Chapter 6, Palstra et al. 2015; Ueda 2019).

Juvenile Coho Salmon, *O. kisutch*, transplanted into Lake Michigan were imprinted on synthetic chemicals in hatchery water and released. Eighteen months later, most chemically imprinted salmon that entered streams chose streams containing the same chemicals (Hasler & Scholz 1983; Quinn & Dittman 1990). Olfactory homing during migration is so strong that Chinook Salmon migrating up the Columbia River moved preferentially through fish ladders along the shore that carried the strongest scent of their home stream (Keefer et al. 2006).

Although as many as 98.6% of Chinook Salmon may home correctly to the Cowlitz River in Washington, the same species may show 10–13% straying rates in California rivers. Tagging studies have shown that as many as 47% of fish may wind up in a non-natal stream. The pattern of straying is, however, adaptive. High fidelity (low straying) rates characterize species and populations that spawn in large, stable rivers, whereas straying is more common in fish that come primarily from small, unstable rivers with variable flow characteristics, where juvenile survival is also more variable. Straying can then be viewed as an alternative life history trait that functions as a bet-hedging tactic to ensure survival of some offspring in situations where the natal river may become uninhabitable (Quinn 1984; see Chapter 17 for other examples of alternative reproductive tactics in salmonids).

Home stream return, perhaps involving olfactory guidance, also occurs in Striped Bass, *Morone saxatilis*, which form distinct stocks along the US Atlantic coast that are associated with major river systems. Fish migrate north in the spring and south in the fall along the Atlantic coast, but return each spring to spawn in their natal rivers. American Shad, Alewife, and Blueback Herring also home to their natal rivers to spawn (Boreman & Lewis 1987; Loesch 1987; Quinn & Leggett 1987).

Catadromy All 15 species of the eel family Anguillidae are thought to spawn in the sea but grow up in fresh water. The best known species are the European, Japanese, and American eels, all of which undergo larval and adult migrations of truly epic proportions. The American Eel, *Anguilla rostrata*, can serve as an example.

American Eel spawn in the Sargasso Sea, an unproductive region of the western Atlantic northeast of Hispaniola and the Bahamas. The exact locale of spawning remained a mystery until the 1920s, when Danish biologist Johannes Schmidt analyzed 25 years of oceanic plankton tows and determined that the smallest eel larvae of both American and European species were captured in this area. (These larvae are known as **leptocephali** and were once thought to be a different species of fish altogether.) Schmidt's results have been subsequently confirmed by captures of even smaller larvae (<7 mm long) of both species at the same time from the same locale. American and European eels spawn in overlapping areas during early spring; eggs hatch and leptocephalus larvae drift northward with major ocean currents. American Eel leptocephali metamorphose after about 1 year, and ride ocean currents westward to the North American coast. European larvae apparently do not metamorphose until they are 2–3 years old, hence they drift past the coast of North America. Interestingly, hybrids between the European and American species stop halfway, in Iceland. Mysteriously, leptocephali are not thought to feed, or at least they have nonfunctional guts during most of their larval phase. Leptocephali are next attracted by the mixture of organic materials dissolved in outflowing fresh waters and migrate upriver, moving by selective tidal stream transport (discussed earlier in this chapter) and transforming into transparent **glass eels** about 50 mm in length. As they move upriver, they become pigmented and are called **elvers**.

Elvers grow into juvenile **yellow eels** that take up residence in estuaries and fresh water for periods that range from 3 to 40 years, the time depending on sex and latitude. Males are more abundant in southerly latitudes and in estuaries. They generally do not grow larger than 44 cm and usually mature after 3–10 years. Females are likely to be found throughout a river system, from the estuary all the way up to the headwaters. Female American Eel probably have the widest geographic and environmental range of any nonintroduced freshwater fish anywhere in the world. Their habitats include rapidly flowing, clear, headwater streams, large lakes and rivers, underground cave springs (Fig. 18.15), lowland rivers and swamps, down to estuarine saltmarshes. American Eel are found from Iceland to Venezuela, including most Caribbean islands and Bermuda, and range up the Mississippi River to its headwaters and as far west as the Yucatan Peninsula.

FIGURE 18.15 Yellow phase American eels in a cave spring in northern Florida. These animals have a strong circadian rhythm, resting in total darkness during the day and emerging from the cave in the evening to feed in the nearby shallows of a cypress swamp. G. Helfman (Author).

Maturation may take from 4 to 13 years at southerly locales and as much as 43 years in Nova Scotia (Jessop 1987). In general, more northerly populations and those farther from the Sargasso Sea contain older, larger, and usually female animals. As the eels mature, they turn a silvery-bronze color, the pectoral fins become pointed, the eyes enlarge (particularly in males), and fat stores are accumulated. These nonfeeding, **silver eels** then migrate back to the Sargasso Sea to spawn; migrations beginning earlier for animals farther from the spawning grounds, which apparently synchronizes the time of arrival at the spawning grounds.

Silver eels travel as much as 5000 km to spawn and then apparently die. There has been considerable conjecture regarding the spawning area (Tesch 1977; McCleave et al. 1987; Helfman et al. 1987; Avise et al. 1990). Data from migrating American Eel fitted with pop-up satellite archival tags (PSATs) and released on the Scotia shelf suggest that the maturing adult eels travel northeasterly in relatively shallow water along the edge of the continental shelf for a considerable distance before making a rather direct southward approach, crossing the Gulf Stream, to the spawning grounds in the Sargasso Sea (Béguer-Pon et al. 2015; Castonguay et al. 2018).

Oceanodromy Oceanodromous fishes migrate within ocean basins, usually in a circuit, traveling with major ocean currents. The migration serves to place different life history stages in seasonally appropriate locales. The range may therefore include an area for spawning from which eggs and larvae float to a nursery area, winter and summer feeding areas for juveniles and adults, and also migratory zones through which a stock moves. Juveniles may move between seasonal feeding areas for several years before maturing and migrating to spawning grounds. The great tunas (*Thunnus*, Scombridae), particularly those living in temperate waters, are representative. Subspecies or stocks have been suggested for different ocean basins in the past, but movement across ocean basins and probable mixing of stocks tends to eliminate genetic differences (e.g., for the highly migratory Albacore Tuna, *Thunnus alalunga*; Graves & Dizon 1987).

There are three recognized species of bluefin tuna: the Northern Bluefin Tuna, *Thunnus thynnus*, in the Atlantic and Mediterranean; Pacific Bluefin, *T. orientalis*; and Southern Bluefin, *T. maccoyii*, off Australia, New Zealand, and South Africa. Northern Bluefin tagged off Florida have been recaptured in Norway, involving a minimum migration distance of 10,000 km. In addition, fish of different sizes have different migratory patterns, and adults of different sizes may spawn at different times and places. In the western North Atlantic, the largest Northern Bluefin (120–900 kg) have a migratory cycle that begins on summer feeding grounds (May to September) over the continental shelf from Cape Hatteras to Nova Scotia. This is followed by fall and winter movements offshore and south to wintering grounds that include the Bahamas, Greater Antilles, and into the Caribbean and Gulf of Mexico. In the spring, the giants move northward in oceanic waters and onto the continental shelf in late spring, and then back to the summer feeding grounds. Spawning occurs in southern waters (Gulf of Mexico, Straits of Florida) in May and June, and in the Mediterranean and Black seas during the warm summer months. Mixing may involve as many as 30% of fish crossing the Atlantic from west to east (mixing from east to west is less well known) (McClane 1974; Richards 1976; Rivas 1978; Lutcavage et al. 1999; Block et al. 2001; Block & Stevens 2001).

Differentiation into **stocks**, some that mix and some that do not, appears common among oceanodromous fishes. Atlantic Herring (*Clupea harengus*) are subdivided into several spawning groups or stocks (six alone in the Northeast Atlantic) that may be further subdivided into isolated stocks in various estuaries and inlets. Migrations carry different stocks to overlapping feeding areas but spawning occurs at separate times and places. Atlantic Cod (*Gadus morhua*, Gadidae) and Plaice (*Pleuronectes platessa*, Pleuronectidae) are also differentiated into several migratory stocks with distinct spawning grounds. Bluefish (*Pomatomus saltatrix*, Pomatomidae) occur worldwide in warmer oceans except for the eastern Pacific. Schools apparently migrate onshore and offshore with the seasons, perhaps following baitfish. Along the US Atlantic coast, this migration involves an inshore migration in spring and summer and a return to offshore locales in fall and winter, which corresponds with movements of a primary prey species, Menhaden (*Brevoortia tyrannus*). Bluefish movements occur progressively later as one travels north, and the pattern is complicated by a degree of north–south migration (McKeown 1984; Hersey 1988). Larger predators, such as Blue Marlin (*Makaira nigricans*, Istiophoridae), also are oceanodromous, move seasonally, and form local but wide-ranging stocks (see also Chapter 12 on pelagic sharks).

Supplementary Reading

Ali MA, ed. 1980. *Environmental physiology of fishes*. New York: Plenum Press.

Ali MA, ed. 1992. *Rhythms in fishes*. New York: Plenum Press.

American Fisheries Society. 1984. Rhythmicity in fishes. *Trans Am Fish Soc* 113:411–552.

Augerot X. 2005. *Atlas of Pacific salmon. The first map-based status assessment of salmon in the North Pacific*. Berkeley, CA: University of California Press.

Baker RR. 1978. *The evolutionary ecology of animal migration*. New York: Holmes & Meier.

Block B, Stevens E. 2001. *Tuna: physiology, ecology, and evolution. Fish physiology*, Vol. 19. New York: Academic Press.

Brown B. 1995. *Mountain in the clouds: a search for the wild salmon*, reprint edn. Seattle, WA: University of Washington Press.

Dadswell MJ, Klauda RJ, Moffitt CM, Saunders RL, Rulifson RA, Cooper JE, eds. 1987. Common strategies of anadromous and catadromous fishes. *Am Fish Soc Symp* 1.

Gauthreaux SA, Jr., ed. 1980. *Animal migration, orientation and navigation*. New York: Academic Press.

Goulding M. 1980. *The fishes and the forest. Explorations in Amazonian natural history*. Berkeley, CA: University of California Press.

Groot C, Margolis L, eds. 1991. *Pacific salmon life histories*. Vancouver, BC: University of British Columbia Press.

Harden-Jones FR. 1968. *Fish migration*. London: Edward Arnold Ltd.

Hersey J. 1988. *Blues*. New York: Random House.

Lowe-McConnell RH. 1987. *Ecological studies in tropical fish communities*. London: Cambridge University Press.

Lucas MC, Baras E. 2001. *Migration of freshwater fishes*. Oxford: Blackwell Publishing.

McCleave JD, Arnold GP, Dodson JJ, Neill WH, eds. 1984. *Mechanisms of migration in fishes*. New York: Plenum Press.

McDowall RM. 1988. *Diadromy in fishes*. London: Croom Helm.

McKeown BA. 1984. *Fish migration*. London: Croom Helm.

Munro AD, Scott AP, Lam TJ, eds. 1990. *Reproductive seasonality in teleosts: environmental influences*. Boca Raton, FL: CRC Press.

Quinn TP. 2005. *The behavior and ecology of Pacific salmon and trout*. Seattle, WA: University of Washington Press.

Reebs S. 2001. *Fish behavior in the aquarium and in the wild*. Ithaca, NY: Cornell University Press.

Secor DH. 2015. *Migration ecology of fishes*. Johns Hopkins University Press.

Thorpe JE, ed. 1978. *Rhythmic activity of fishes*. London: Academic Press.

PART V Fish Distribution, Ecology, and Conservation

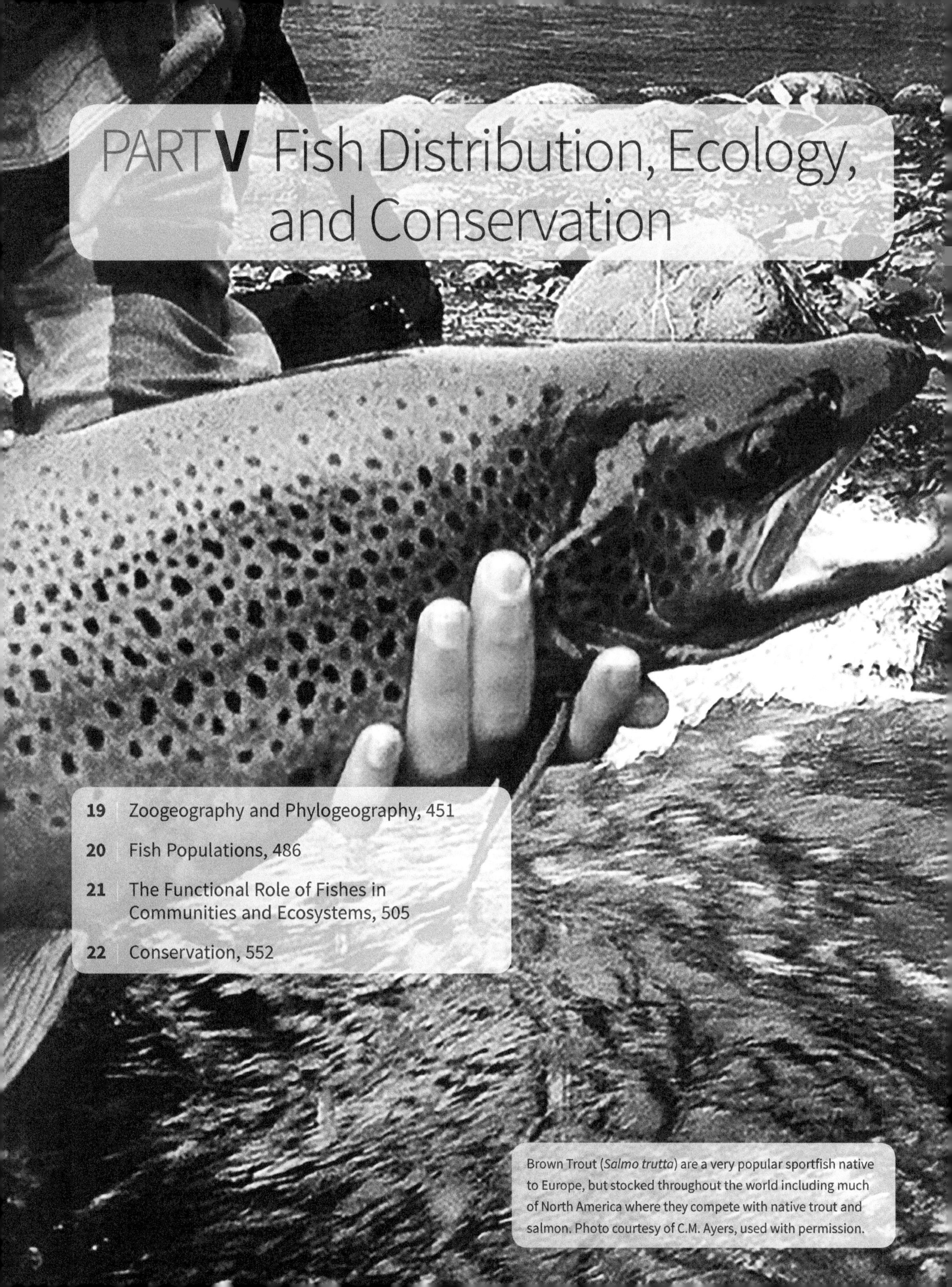

Brown Trout (*Salmo trutta*) are a very popular sportfish native to Europe, but stocked throughout the world including much of North America where they compete with native trout and salmon. Photo courtesy of C.M. Ayers, used with permission.

CHAPTER **19**

Zoogeography and Phylogeography

Summary

Biogeography is the field that documents the geographic distribution of species past and present. Zoogeography is the subspecialty that applies to fishes and other animals. Phylogeography is the geographic distribution of genetic lineages. There are two primary models that explain the distribution of organisms and genetic lineages. In the vicariance model, species distributions are shaped by geographic isolation, and evolutionary histories can be reconstructed according to geologic events. In contrast, the dispersal model is based on fishes dispersing from regions of origin and becoming established in other locations. Vicariance is most effective for explaining the evolutionary history of freshwater fishes, whereas dispersal is a nearly-ubiquitous feature of marine fishes. Both are important for our understanding of the distribution and evolution of fishes.

Phylogeography bridges the fields of population genetics and phylogenetics. Very often the biogeographic separations defined by species distributions are supported by surveys of genetic diversity within widespread species. Phylogeographic surveys can reveal rare dispersal events that are difficult to detect in field studies, but very important for illuminating the diversity of fishes.

Understanding the present distributions of fishes requires knowing how the arrangement of the continents has changed. All present continents were part of a single landmass, Pangaea, from over 430 million years

CHAPTER CONTENTS

The Diversity of Fishes: Biology, Evolution and Ecology, Third Edition. Douglas E. Facey, Brian W. Bowen, Bruce B. Collette, and Gene S. Helfman.

Companion website: www.wiley.com/go/facey/diversityfishes3

ago until about 180 million years ago. Pangaea then split into a northern portion, Laurasia, which later became Eurasia and North America, and a southern portion, Gondwana, which was subsequently divided into South America, Africa, Australia, and Antarctica, about 90 million years ago.

Marine fishes account for 50% of the 35,000+ known species of fishes, 49% are freshwater species, and 1% move between the two habitats. Marine fishes have four main ecological domains: (i) epipelagic, surface-dwelling species (about 1% of all fishes); (ii) deep pelagic species (4%); (iii) deep benthic species (5%); and (iv) inshore or continental shelf species (40%). Inshore marine fishes occur in four major biogeographic regions, in the order of decreasing biodiversity: Indian-West Pacific, western Atlantic, eastern Pacific, and eastern Atlantic.

Fresh water constitutes less than 0.01 % of the planet's water, yet hosts almost half of the world's fish species. Globally, there are six major freshwater regions: (i) Nearctic region (North America except tropical Mexico), with at least 14 families and about 950 species; (ii) Neotropical region (South and Middle America), with 43 families and more than 7000 species; (iii) Palearctic region (Europe and Asia, north of the Himalayas), with 27 families, with many minnows and loaches; (iv) African (or Ethiopian) region, with 95 families and 3360 species, including many primitive species; (v) Oriental region (India, southern China, Southeast Asia, the Philippines, and the East Indies to Wallace's Line), with 28 families, including 12 families of catfishes and four cypriniform families, and about 3000 species; and (vi) Australian region (Australia, New Guinea, and New Zealand), with only two primary freshwater species, both ancient relicts of a much wider archaic distribution pattern.

Six groups of primitive primary freshwater fishes have ancient distributions best explained by vicariance and continental drift: lungfishes, bichirs, paddlefishes, gars, Bowfin, and bonytongues.

Most primary freshwater fishes belong to the Ostariophysi, which includes the Cypriniformes, Characiformes, Siluriformes, and Gymnotiformes. Cypriniform carps, minnows, loaches, and suckers are found primarily on the northern continents. Characiform characins comprise 10–16 families, with the greatest diversity in South America (200 genera, more than 1000 species) but with 23 genera and 150 species in Africa. Siluriform catfishes, which include 34 families and about 2000 species, occur on all continents. Gymnotiform electric fishes comprise six families restricted to South America.

Continental Drift, Tectonic Plates, and Fish Distributions

Continents and ocean basins have changed dramatically in location and size during the earth's history. Continents are blocks of largely granitic and sedimentary rocks. Continents "drift" because they literally float on top of the earth's denser basaltic crust. New crust develops at midoceanic ridges as basalt upwells from the earth's mantle. This basalt flows outward, causing spreading of the basaltic seafloor plates and widening of ocean basins. The moving plates carry the overlying continents with them (**plate tectonics**; Fig. 19.1). Crust finally is subducted (dives under) at oceanic trenches and plate

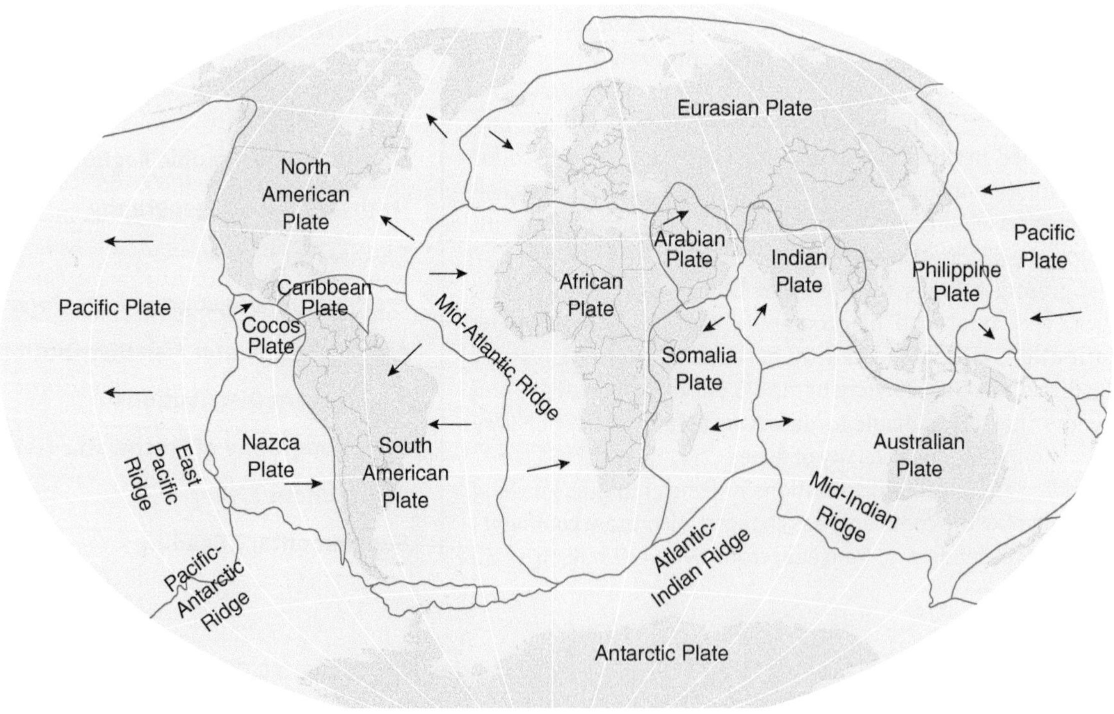

FIGURE 19.1 The earth's important tectonic features. Dashed lines denote margins of major lithospheric plates; arrows indicate direction of plate movements. Adapted from Springer (1982).

margins. The notion of **continental drift** was first formally proposed by Alfred Wegener (1966, but was first published in German in 1915) to explain the fit that the west coast of Africa makes with the east coast of South America. Wegener's concept was ridiculed by scientists for many years but gained acceptance as geophysical and paleontological evidence accumulated (Romano et al. 2016).

The present continents were at one time all part of a single landmass, **Pangaea**, which had coalesced by the Silurian (430 million years ago). About 180 million years ago, during the Mesozoic, Pangaea split into a northern portion, **Laurasia** (Eurasia and North America), and a southern portion, **Gondwana**. Gondwana later split into South America, Africa, Australia, and Antarctica about 90 million years ago. Present widespread distributions of several fish taxa, such as lungfishes, osteoglossomorphs, and ostariophysans (discussed later in this chapter), may well have been formed when the southern continents were still connected, before the breakup of Gondwana.

Vicariance and Dispersal

Zoogeography is the science of understanding the distributions of different groups of animals. Fishes live almost everywhere water occurs, but even a casual glance at species lists from different localities demonstrates that few places have the same kinds of fishes. The challenge then becomes one of discerning patterns in the present distribution of different species, genera, families, and higher taxa, and then trying to understand how these patterns are related to the evolution of the different groups. Basically, we are asking how and why fish faunas differ and how different fishes got where they are today.

Understanding fish evolution through zoogeography has been based primarily on two models: **vicariance** and **dispersal.** According to the **vicariance** model, species distributions are shaped by geographic isolation, rather than by active dispersal. In this framework, the geographic distributions of organisms can be interpreted through the sequence of breakups and collisions of land masses over time. Although this emphasis on vicariance revitalized the field of biogeography, it was also dominated by some who denied the primary alternative model, **dispersal** (and colonization), as a viable explanation of species distributions. Some who promoted the vicariance view regarded dispersal as trivial or unprovable (de Queiroz 2005). In particular, the advocates for vicariance biogeography claimed that plate tectonics and other geological processes provided a testable set of expectations because they could be linked to geological events, whereas rare dispersal events did not readily fit into a hypothesis testing format. However, the emergence of molecular techniques in evaluating lineages has provided some of the physical evidence for dispersal that had been lacking (Bowen & Grant 1997).

Vicariance is an excellent model to explain distributions of many freshwater fishes, while dispersal models are a better fit for many marine fishes, given their large ranges and high potential for dispersal as both larvae and (in the case of tunas, billfishes, and pelagic sharks) swimming adults. However, there are exceptions to this trend, especially among marine fishes that lack a pelagic larval stage, such as the Spiny Damselfish (*Acanthochromis polyacanthus*; Bay et al. 2006), Banggai Cardinalfish (*Pterapogon kauderni*; Vagelli et al. 2009), and seahorses (genus *Hippocampus*; Lourie et al. 2005). Vicariance models work well when population structure is shaped primarily by geographic barriers rather than life history or ecology. The importance of both the vicariance and dispersal models in understanding fish distributions and the evolutionary relationships among populations will be a recurring theme in this chapter.

Examples of Vicariance

The process by which phylogenetics and zoogeography complement each other and inform us about a group's phylogeny can be demonstrated with the *Scomberomorus regalis* species group of Spanish mackerels. This group is defined as monophyletic based on the unique presence of nasal denticles, toothlike structures within the nasal cavity (Collette & Russo 1985a, 1985b). Pertinent vicariant barriers include the widening gap between East and West Atlantic (>80 mybp) and the rise of the Isthmus of Panama between West Atlantic and East Pacific (~3.5 mybp). There are six Atlantic and East Pacific species relevant to this discussion: *tritor*, *maculatus*, *concolor*, *sierra*, *brasiliensis*, and *regalis* (Fig. 19.2).

Firstly, a comparison of the distribution of the six species (Fig. 19.3) with that of the phylogeny (Fig. 19.2) indicates that the eastern Atlantic species (*tritor*) is the plesiomorphic sister species of the rest of the species group. Secondly, the western Atlantic *maculatus* and then the two eastern Pacific species, *concolor* and *sierra*, suggest speciation following elevation of the Isthmus of Panama. The two most derived species, *brasiliensis* and *regalis*, are found in the western Atlantic, with *regalis* occupying coral reefs, an unusual habitat for Spanish mackerels. This case illustrates that the oldest partitions in the phylogeny correspond to the oldest barrier, followed by a more recent separation of West Atlantic and East Pacific lineages.

These morphology-based patterns among Spanish mackerels can then be compared with patterns of other species in Table 19.1 to look for concordance in the distribution of evolutionary lineages. Are the single eastern Atlantic species of the halfbeak *Hyporhamphus*, the needlefish *Strongylura*, and the toadfish *Batrachoides* the plesiomorphic sister species of the western Atlantic and eastern Pacific species in these genera? Did these patterns arise from the widening of the Atlantic Ocean as the plate containing the Americas and the plate containing Eurasia and Africa (Fig. 19.1) moved farther apart? Molecular genetic studies are available for some of the groups and support this pattern for Spanish mackerels (Banford et al. 1999) but not for the needlefishes (Banford et al. 2004).

The trumpetfishes (genus *Aulostomus*) provide another example of support for the vicariance explanation of species

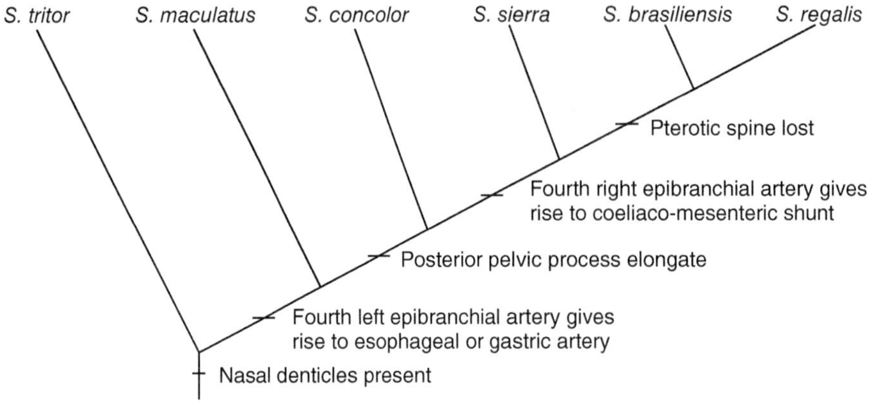

FIGURE 19.2 Cladogram of the *Scomberomorus regalis* group of Spanish mackerels. From Collette and Russo (1985a). With permission of John Wiley & Sons.

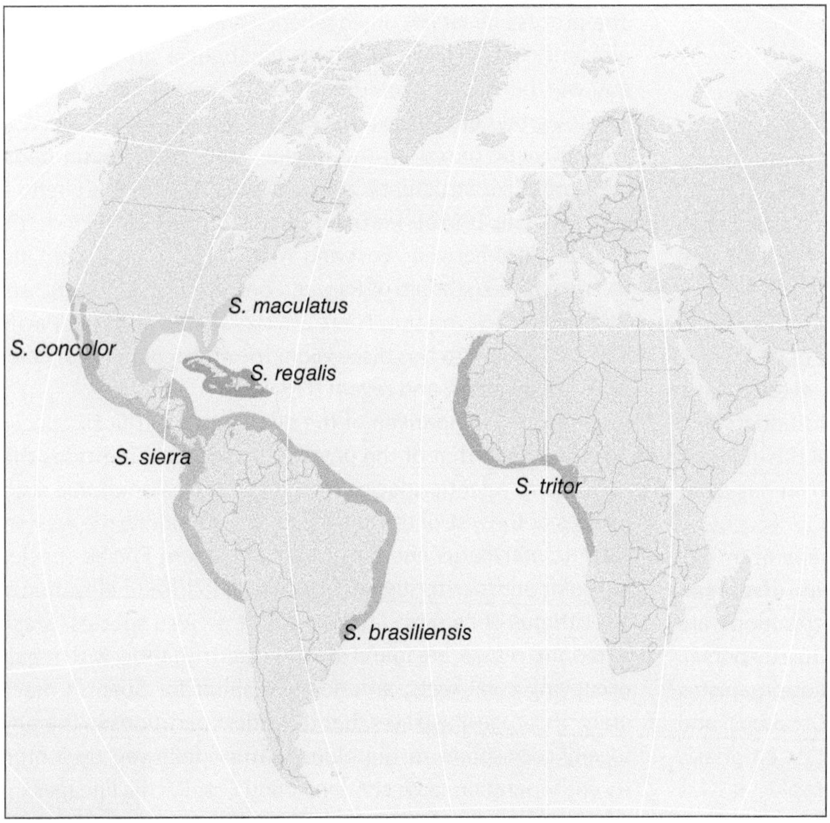

FIGURE 19.3 Ranges of the regalis group of Spanish mackerels (*Scomberomorus*). NOAA / Public Domain.

TABLE 19.1 Numbers of species in selected genera of inshore fishes from the western Atlantic, eastern Pacific, and Gulf of Guinea (numbers in parentheses indicate freshwater species of marine origin).

Family	Genus	Western Atlantic	Eastern Pacific	Eastern Atlantic
Scombridae	*Scomberomorus*	4	2	1
Belonidae	*Strongylura*	3 (+1)	2 (+1)	1
Hemiramphidae	*Hyporhamphus*	3 (+2)	4	1
Batrachoididae	*Batrachoides*	3 (+1)	3	1
Totals		13 (+4)	11 (+1)	4

distribution. This group inhabits tropical reefs and rocky habitats around the world. They have a pelagic larval and juvenile stage that can exceed 100 days, hence they have extensive potential for dispersal.

Wheeler (1955) recognized three species: West Atlantic *Aulostomus maculatus*, East Atlantic *A. strigosus*, and Indian-Pacific *A. chinensis*. He noted that the West Atlantic and Indian-Pacific species were well-differentiated morphologically, but that the East Atlantic species was very similar to the Indian-Pacific species. In this biogeographic enigma, the closest relative to the East Atlantic species is not in the West Atlantic, as proximity would indicate, but in the Indian (and Pacific) Ocean.

What are the barriers that could produce this curious outcome? Tropical regions of the West Atlantic and Pacific oceans have been separated by the Isthmus of Panama for about 3.5 million years, whereas the tropical Atlantic and Indian oceans have been isolated for 2.5 million years by the frigid Benguela upwelling system in southern Africa (Shannon 1985). Prior to that time, roughly coinciding with the onset of modern glacial cycles 2.6–2.8 mybp (Dwyer et al. 1995), a warm water corridor connected the Indian and South Atlantic oceans.

The history of these two oceanic barriers, in combination with the morphological data, invoke a biogeographic scenario in which West Atlantic and Indian-Pacific trumpetfishes were initially isolated by the Panama barrier (3.5 mybp), followed by isolation of Indian-Pacific and East Atlantic populations by the Benguela Current (2.5 mybp). Bowen et al. (2001) tested predictions of this biogeographic model, both the timing and the order of events, with mtDNA cytochrome *b* and a conventional molecular clock (2%/MY). In a molecular phylogeny (Fig. 19.4), the deepest partition is between the West Atlantic and Indian-Pacific species (d = 6.3–8.2%), yielding a timeframe of about 3–4 million years. The divergence between Indian-Pacific and East Atlantic species (d = 4.4–5.4%) indicates a timeframe of about 2–2.5 mybp. In this case, the mtDNA data confirm a vicariant model based on morphology and the timing of barriers between ocean basins. It is also notable that the 3.5-million-year separation of *A. maculatus* and *A. chinensis* is accompanied by diagnostic morphological differences, whereas the 2.5-million-year separation of *A. chinensis* and *A. strigosus* is not.

Examples of Dispersal

Historically it was difficult to make a strong case for the dispersal model due to a lack of supporting evidence until molecular genetic techniques emerged in the mid-1980s. The field of **phylogeography** is concerned with the geographic distribution of genetic lineages, usually at the level of deep population structure, species, and genera. This molecular perspective was prompted by the advent of mtDNA technology for wildlife studies in the 1980s, culminating in the seminal publication *Intraspecific phylogeography: the mitochondrial DNA bridge between population genetics and systematics* (Avise et al. 1987). As the title implies, this field is at the junction of population genetics and systematics (phylogenetics), with additional foundations in biogeography. The key innovation with mtDNA sequence data, later extended to nDNA sequence data (Karl & Avise 1993), is that the differences between alleles or haplotypes are known. Previously, allozyme studies could compare the frequency of alleles, but researchers did not know whether the alleles were different by 2 mutations, or 20 mutations. Therefore, allozyme studies could not determine whether those alleles arose 1 million years ago or 10 million years ago. The age of these alleles can reveal important information: When two different alleles are at 100% frequency in separate populations, the age of the populations can be estimated with a **molecular clock** that uses predicted mutation rates to estimate time since divergence. Often these estimates of the age of populations (and species) are linked to known biogeographic events (such as the Panama barrier, discussed elsewhere in this chapter).

Phylogeographic methods, and their immediate precursors in population genetics, have provided evidence to support the dispersal model in some cases. As McDowall (1978) noted, dispersal events are very difficult to document directly (especially in fishes), nonetheless these events yield clear genetic signals. This was dramatically demonstrated by Rosenblatt and Waples (1986), who used allozymes to test the prediction of an ancient vicariant separation between marine fishes of the East Pacific and central Pacific, separated by over 5000 km of open ocean. The Pacific Ocean sits atop a geological plate that is over 100 million years old, and corresponding genetic divergences should be very deep. Instead, the allozymes revealed much more recent connections, on the order of thousands of years rather than millions of years. Lessios and Robertson (2006) revisited the issue 20 years later with mtDNA data, and found that 19 of 20 species either shared haplotypes across the barrier, or had haplotyes that were a few mutations apart. The exception was the pipefish *Doryrhamphus excisus*, a member of the Syngnathidae, known to have low dispersal ability from population genetic assessments.

Sardines (*Sardinops* spp.) occupy upwelling zones in the cold temperate corners of the Pacific and Indian oceans: South Africa (*S. ocellatus*), southern Australia (*S. neopilchardus*), Japan (*S. sagax melanostictus*), Chile (*S. s. sagax*), and California (*S. s. caeruleus*). On an east–west axis across the Pacific Ocean, there are vast expanses of ocean that are inhospitable to sardines, but potentially breached by drifting pelagic larvae. On a north–south axis, tropical waters above 27°C are lethal to sardines and prohibit movement across the equator except during the coldest (glacial) conditions. The dispersal model predicts recent separations or ongoing gene flow across the Indian and Pacific oceans (Parrish et al. 1989). The vicariance model mandates ancient separations based on the breakup of the continent Pacifica, on the order of tens of millions of years, between East and West Pacific (Nelson 1985).

In the **parsimony network** which illustrates the relationships among sardine haplotypes (Fig. 19.5), South African and Australian "species" share haplotype C and Californian and Chilean "subspecies" share haplotype M, indicating shallow population structure rather than ancient species. A molecular clock for the mtDNA control region (15–20%/MY) indicates that

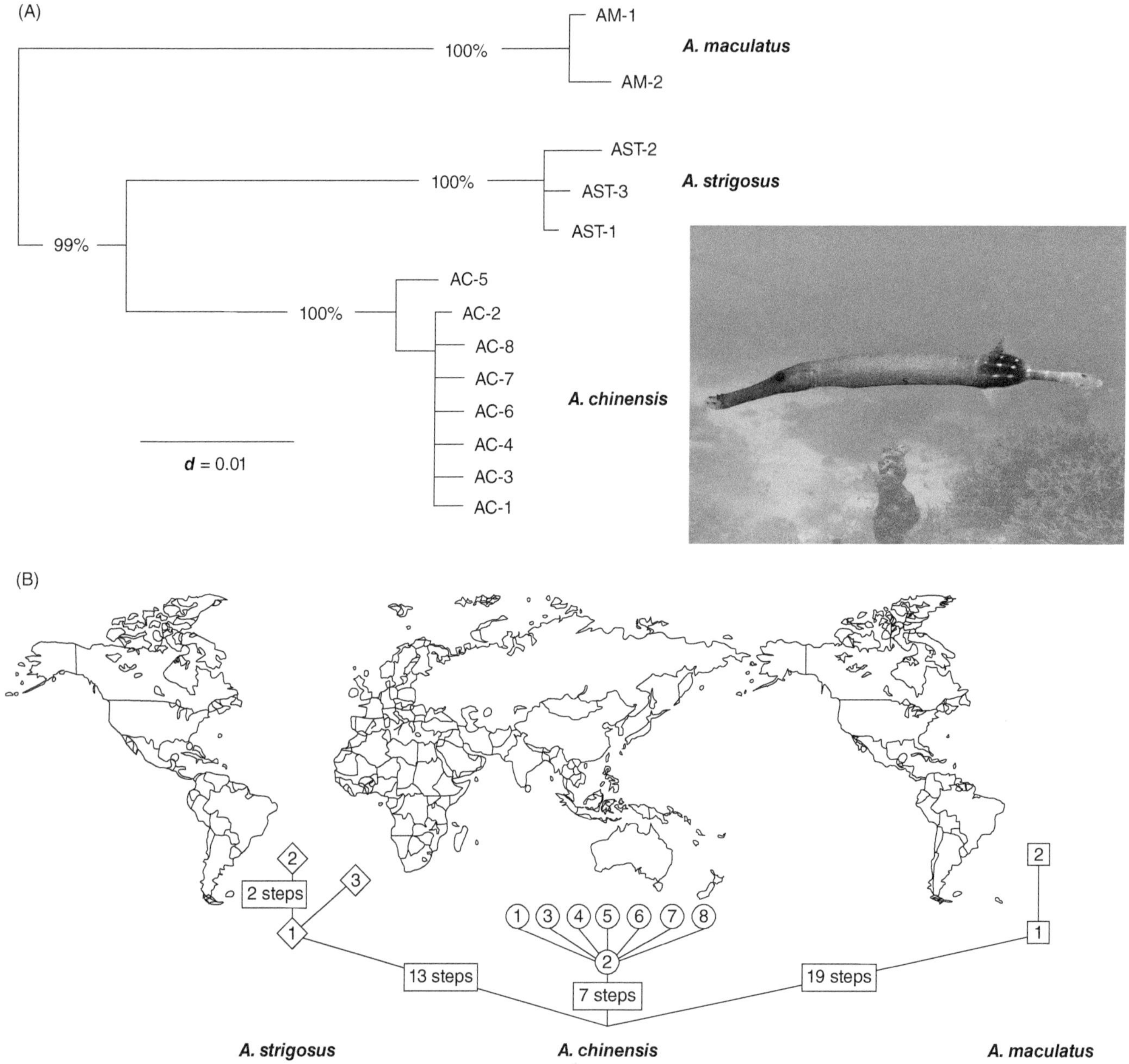

FIGURE 19.4 A neighbor joining tree (A) and parsimony network (B) based on mtDNA cytochrome *b*, showing the relationships among trumpetfish species across the range of genus *Aulostomus*. The scale indicates 1% sequence divergence. The data indicate a 3–4-million-year-old separation between West Atlantic (*A. maculatus*; haplotypes AM-1 and AM-2) and Indian-Pacific (*A. chinensis*; haplotypes AC-1 to AC-8) species, followed by approximately a 2.5-million-year separation between the Indian-Pacific and the East Atlantic species (*A. strigosus*; haplotypes AST-1 to AST-3). These separations correspond to major vicariant events separating the tropical fauna of each ocean basin (see text). From Bowen et al. (2001), used with permission. Photo of *Aulostomus chinensis* taken by A.A. Rahman/Wikimedia Commons/CC BY-SA4.0.

all these sardines share a common ancestor at approximately 300,000–500,000 years ago (Bowen & Grant 1997). Based on the same comparisons with an allozyme molecular clock, the common ancestor is aged at 200,000 years ago (Grant & Leslie 1996). With either of these timeframes, the vicariance model of ancient separations is refuted. Sardines have crossed both the Pacific and the equator in recent evolutionary history. It is notable that the connection across the equator is in the East Pacific, which has a steep continental shelf and deep, cold water even in the tropics. In contrast, the tropical zone between Japan and Australia is generally shallow, warm, and apparently impenetrable to cold-adapted sardines.

Fish Distributions in Major Zoogeographic Regions

Historically, the study of fish distribution in the major zoogeographic regions has been divided into marine and freshwater components. Marine fishes comprise about 50% of the over 35,000 species of fishes, whereas freshwater fishes make up about 49%; the remaining fishes move between marine and freshwater habitats (Fricke et al. 2022). Since the previous edition of this text, nearly 4000 new species of fishes have been described.

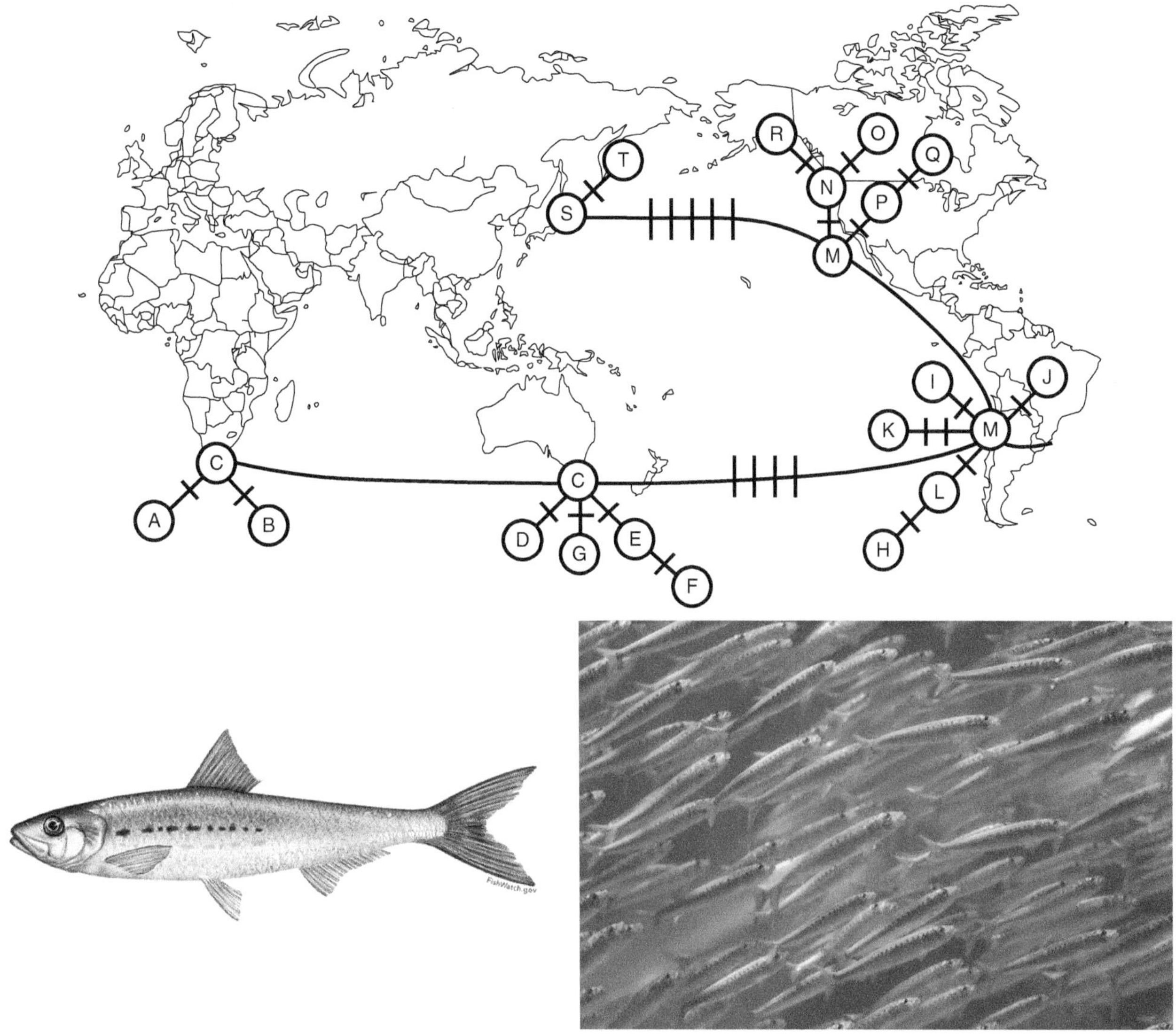

FIGURE 19.5 The sardines support the dispersal model of fish speciation and distribution. A parsimony network of mtDNA control region sequences, illustrating relationships among sardines (genus *Sardinops*) in five temperate upwelling zones of the Indian and Pacific oceans (hashmarks indicate multiple mutations along a branch). The 20 haplotypes are labeled A to T. Haplotype C occurs at both South Africa and Australia, and haplotype M occurs at both Chile and Mexico, indicating shallow population structure between these regions, and recent colonization around the rim of the Indian-Pacific Basin. The five regional forms were previously regarded as separate species, a taxonomy that is not supported by the mtDNA analysis. From Bowen and Grant (1997), used with permission. Fish images are from NOAA / Public Domain.

Most of the newly described species of freshwater fishes are from Southeast Asia, southeastern Europe, western Africa near the equator, and the Amazon and Parana basins of South America. Many newly described marine species are in coastal Australia, the southern China Sea, northwestern Indian Ocean, and southwestern Caribbean. Many new fish species await discovery in the deep reefs, below the depths accessible with scuba gear.

Marine Fishes

Although we refer to our planet as Earth (or *Terra* in Latin, and some science fiction), it is really Planet Ocean. Not only is 71% of the planet's surface covered with water, but because water supports life from the surface down to the oceans' greatest depths (about 11,000 m), the total oceanic living volume is 300 times greater than the terrestrial. Anyone first observing our planet from outer space would be struck by this and surely would name the planet for its blue water cover, unique in the solar system.

Biological diversity is especially rich in tropical areas, in part because of high productivity, habitat complexity (such as coral reefs), and the enormous area encompassed by habitats near the equator. And the closer we look, the more we find. As mentioned earlier, many of the newly described marine fishes are in tropical or subtropical waters.

Four main ecological divisions are recognized among the marine fishes:

1. **Epipelagic** fishes, which dwell from the surface down to 200 m, make up 1.3% of the total number of marine fish species.

2. **Deep pelagic** fishes include about 4% of the total. These water column dwelling fishes can be further subdivided into **mesopelagic** fishes, which live between 200 and 1000 m, and deeper dwelling **bathypelagic** fishes.
3. **Deep benthic** fishes comprise about 5% of the total.
4. **Littoral or continental shelf** species are shallow-dwelling fishes that inhabit the shore and shelf above 200 m. They are the largest group, constituting 40% of the total number of marine fish species.

Notably, at least 284 fishes have achieved global distributions, mostly in the pelagic zones (Gaither et al. 2016).

Epipelagic Fishes

Many epipelagic species are worldwide in distribution. However, many inshore epipelagic species have more restricted distributions. One member of a family may be confined to one side of an ocean and be represented by another, allopatric species (a closely related species not occurring in the same area), living on the other side of the ocean.

As examples, consider the distribution patterns of some tunas (Scombridae), halfbeaks (Hemiramphidae), and needlefishes (Belonidae). Most species of tunas of the genus *Thunnus* are widespread offshore. Several species, including Albacore (*T. alalunga*), Yellowfin (*T. albacares*), and Bigeye (*T. obesus*) have global distributions, indicating that genetic interchange occurs among populations in the Atlantic, Indian, and Pacific oceans. Little tunas of the genus *Euthynnus* have a different distribution pattern, more closely associated with the shore. One species occurs in the Atlantic (*E. alletteratus*), one in the Indian-West Pacific (*E. affinis*), and one in the eastern Pacific (*E. lineatus*). Among Spanish mackerels, *Scomberomorus*, distributions of species are even more shore associated, with allopatric species in the Atlantic, Indian-West Pacific, and eastern Pacific (Collette & Russo 1985b).

Among halfbeaks (Hemiramphidae), species of the genus *Hemiramphus* are more widespread than species of the more inshore genus *Hyporhamphus*. For example, two species of *Hemiramphus* are found on both sides of the Atlantic, and one species (*He. far*) is widespread throughout the Indian-West Pacific (and has even invaded the eastern Mediterranean Sea through the Suez Canal). In contrast, all species of *Hyporhamphus* in the western Atlantic differ from those in the eastern Atlantic, Indian-West Pacific, and eastern Pacific.

Similarly, needlefishes (Belonidae) of the genera *Ablennes* and *Tylosurus* are much more widespread than species of the genus *Strongylura* (Cressey & Collette 1970). *Ablennes* is worldwide, and two species of *Tylosurus* (*T. acus* and *T. crocodilus*) are nearly worldwide, with different subspecies recognized in parts of their ranges. Species of *Strongylura* are more numerous and have more restricted distributions, like species of *Hyporhamphus*.

Distributions of epipelagic inshore fishes may be limited by temperature, either directly or indirectly. For example, for needlefishes (genus *Strongylura*; Fig. 19.6), a clear relationship exists between temperature and the northernmost and southernmost distribution records.

Deep-Water Fishes

Many species of deep-water fishes are also widespread. In looking at their distributions, we cannot rely on surface maps, because ocean basins may have underwater **sills**, ridges that act as barriers to the distribution of deep-water fishes. Sills act as barriers because they physically inhibit the movement of fishes and they also restrict the mixing of waters. For example, the Mediterranean Sea is continuous with the Atlantic Ocean at the surface via the 12.9 km wide Straits of Gibraltar. However, at 1200 m, the Mediterranean Sea is 14°C, whereas the adjacent Atlantic Ocean is 2.5°C at the same depth and these depths are interrupted by a 286 m deep sill. Another sill at 350 m separates the western from the eastern Mediterranean at the Strait of Sicily (Patarnello et al. 2007). Similarly, the Red Sea is 23°C at 125 m, whereas the Indian Ocean is 2.5°C at the same depth, and the two areas are separated by a shallow sill. Deep-water fishes adapted to the cool temperatures of the Atlantic and Indian oceans may not be able to penetrate the Mediterranean and Red seas because they cannot tolerate the warm temperatures at the sill that separates the ocean from the adjacent sea.

Littoral Fishes

Temperature is also a major limiting factor for the distribution of shallow water fishes. The greatest diversity of marine fish species is in tropical waters. Part of this great biodiversity is associated with the coral reefs that provide habitat for the fishes and their prey. Reef-building corals are restricted to depths above 150 m in clear waters generally warmer than 18°C; corals reach their maximum extent at 23–25°C. Common groups of coral reef fishes include moray eels (Muraenidae); squirrelfishes (Holocentridae); several families of percoids such as seabasses (Serranidae), grunts (Pomadasyidae and Haemulidae), snappers (Lutjanidae), cardinalfishes (Apogonidae), and butterflyfishes (Chaetodontidae); and some more advanced families such as damselfishes (Pomacentridae), wrasses (Labridae), parrotfishes (Scaridae), gobies (Gobioidei), blennies (Blennioidei), surgeonfishes (Acanthuridae), triggerfishes (Balistidae), and boxfishes (Ostraciidae).

Marine Zoogeographic Regions

Global biogeographic patterns in the marine environment were outlined by Ekman (1953, but first published in German in 1935) and Briggs (1974). Briggs' patterns focused on a system of coastal and shelf provinces defined by a minimum of

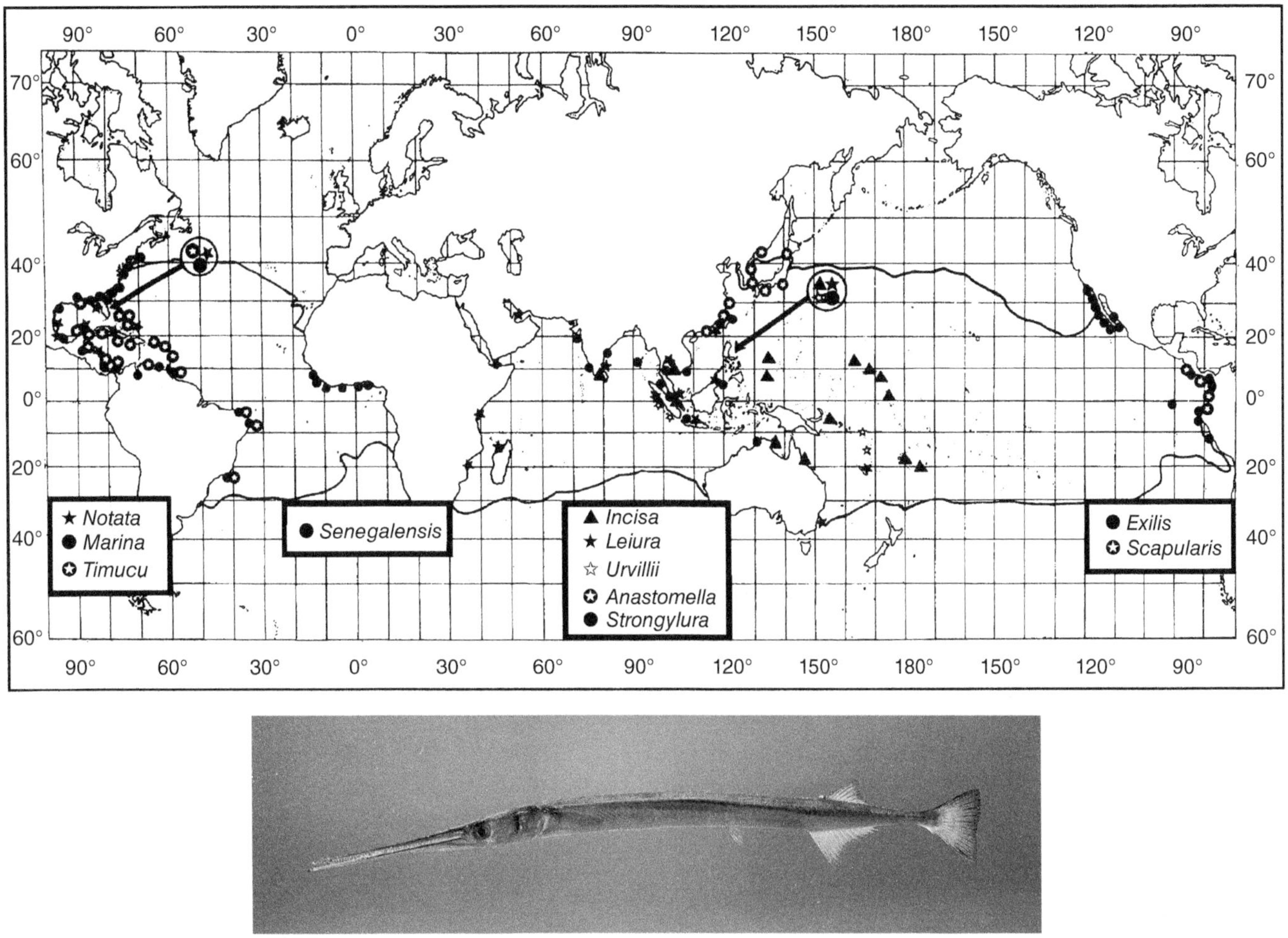

FIGURE 19.6 Distribution of marine populations of 11 species of needlefishes of the genus *Strongylura* in relation to the 23.9° isothere. Cressey and Collette (1970). Photo is Atlantic Needlefish (*Strongylura marina*) from NOAA / Public Domain. SEFSC Pascagoula Laboratory / Wikimedia Commons / CC BY 2.0.

10% endemism (Fig. 19.7). A system of 64 large marine ecosystems (LMEs) was developed by Sherman et al. (2005). LMEs are relatively large areas of 200,000 km^2 or greater characterized by distinct bathymetry, hydrography, productivity, and trophically dependent populations. About 90% of the world's annual yield of marine fisheries is produced within the boundaries of the 64 LMEs. Spalding et al. (2007) provide a hierarchical nested global system for coastal and shelf areas, Marine Ecosystems of the World (MEOW), which includes 12 realms containing 62 provinces and 232 ecoregions. Further details of 37 pelagic province contained within four broader realms (Southern Coldwater, Northern Coldwater, Atlantic Warmwater, and Indian-Pacific Warmwater) are described in Spalding et al. (2012).

The simplest division of the distributions of inshore marine fishes is into four major marine regions, in the order of decreasing biodiversity: (i) Indian-West Pacific, (ii) western Atlantic, (iii) eastern Pacific, and (iv) eastern Atlantic. These regions are separated from each other either by continents or by large expanses of open ocean, and each has been subdivided into different units by different authors.

Indian-West Pacific Region

The Indian-West Pacific region – from South Africa and the Red Sea east through Indonesia and Australia to Hawaii and the South Pacific Islands, all the way to Easter Island – contains about one-third of the species of shallow marine fishes, about 4000 species, compared to no more than 1200 in any other region. Multiple datasets show global maxima of marine biodiversity to be in the Coral Triangle also known as the Indo-Australian Archipelago (Fig. 19.7). Analysis of distribution data for 2983 marine species of fishes, other vertebrates, and invertebrates reveals a pattern of richness on a finer scale and identifies a peak of marine biodiversity in the central Philippine Islands and a secondary peak between peninsular Malaya and Sumatra (Carpenter & Springer 2005). Biodiversity is also high in many other marine taxa in this region (Briggs 1974). There are about 1000 species of hermatypic (reef-building) corals in this region, which is 10 times the number of species present in the western Atlantic (Allen 2008). Among other groups, the Indian-West Pacific contains about 1000 species of bivalve mollusks (including all the giant clams, Tridacnidae), twice that

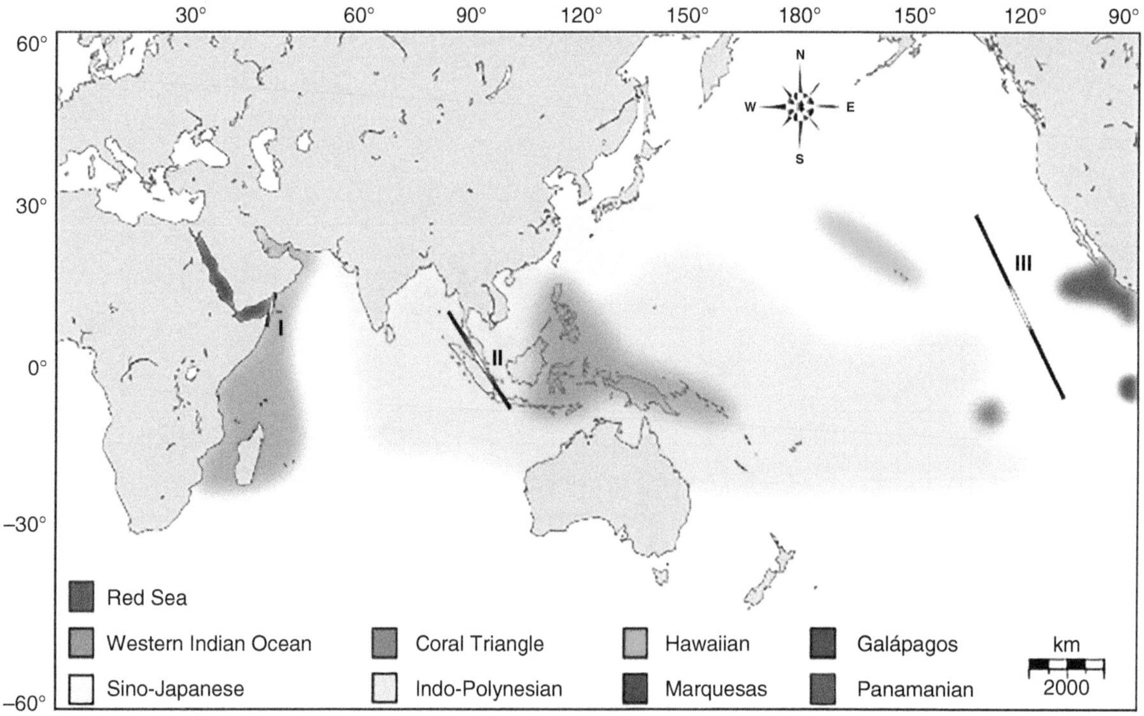

FIGURE 19.7 Biogeographic provinces of the shallow tropical Indo-Pacific as defined by 10% endemism. Primary barriers are indicated as I) Red Sea Barrier, which can close off to isolate the Red Sea from the Indian Ocean, II) Indo-Pacific Barrier, which forms a near-complete land bridge between Southeast Asia and Australia during low sea levels associated with glacial maxima, and III) East Pacific Barrier, a large stretch of open ocean with no shallow water habitats. Light blue indicates the vast Indo-Polynesian Province that spans almost half the planet. Connectivity across this range is believed to be maintained by the network of thousands of islands and atolls, leaving no gap in shallow water habitat greater than 900 km. Embedded in the light blue is the Coral Triangle in dark blue, the pinnacle of marine biodiversity. From Bowen et al. (2016) with permission.

in the western Atlantic; and 49 of the 50 species of sea snakes (Hydrophiidae), compared to one species in the eastern Pacific. Some families of fishes such as the whitings (Sillaginidae) and rabbitfishes (Siganidae) are endemic to the Indian-West Pacific.

The number of fish species (and other tropical marine species) declines to both the east and west of the Coral Triangle in a steep biodiversity gradient. For example, over 3000 fish species inhabit the Coral Triangle, but that number declines to about 630 in Hawaii and to about 140 at Easter Island, the far eastern edge of the vast Indo-Polynesian biogeographic province (Randall 2007; Randall & Cea 2011; Allen & Erdmann 2012). The reasons for this gradient are still debated, but marine phylogeographic studies (addressed later in this chapter) have revealed much of the puzzle.

The Pacific Plate (Fig. 19.1) is a major biogeographic feature of the Indian-West Pacific. The Pacific Plate is the largest of the earth's lithospheric plates and occupies most of the area that has been referred to as the Pacific Basin (Springer 1982). The number of taxa decreases sharply as one proceeds eastward across the western margin of the plate. In addition, there is a high degree of endemism on the plate.

An instructive example of the kinds of distributions one finds associated with specific regions and plates occurs with Spanish mackerels of the genus *Scomberomorus* (Collette & Russo 1985b). Of the 18 species in the genus, 10 occur in the Indian-West Pacific, but they are noticeably absent from the Pacific Plate (Springer 1982, fig. 40). One species, *Scomberomorus commerson*, is widespread throughout much of the Indian-West Pacific. This distributional pattern cannot tell us much because widespread species are not as informative about the causes of the distribution patterns as those with more restricted distributions.

In contrast, the ranges of three species, *S. guttatus*, *S. koreanus*, and *S. lineolatus* (Fig. 19.8), stop at the continental margin, at what is known as Wallace's Line (discussed later, under Oriental region). Australia and southern New Guinea, east of Wallace's Line, have a Spanish mackerel fauna that consists of four different species: *S. multiradiatus*, *S. semifasciatus*, *S. queenslandicus*, and *S. munroi*. These four species do not extend into the East Indies or even to the north coast of New Guinea, although they easily could swim that far. This distribution pattern is obviously not simply a result of present ecological factors but instead must be historical, related to the earlier evolution and dispersal of the genus. The present island of New Guinea resulted from the collision of two plates that may have contained two different fish faunas.

Western Atlantic Region

The western Atlantic region includes the temperate shores of North America, the Gulf of Mexico, the tropical shores of the Caribbean Sea, and the tropical and temperate shores of South

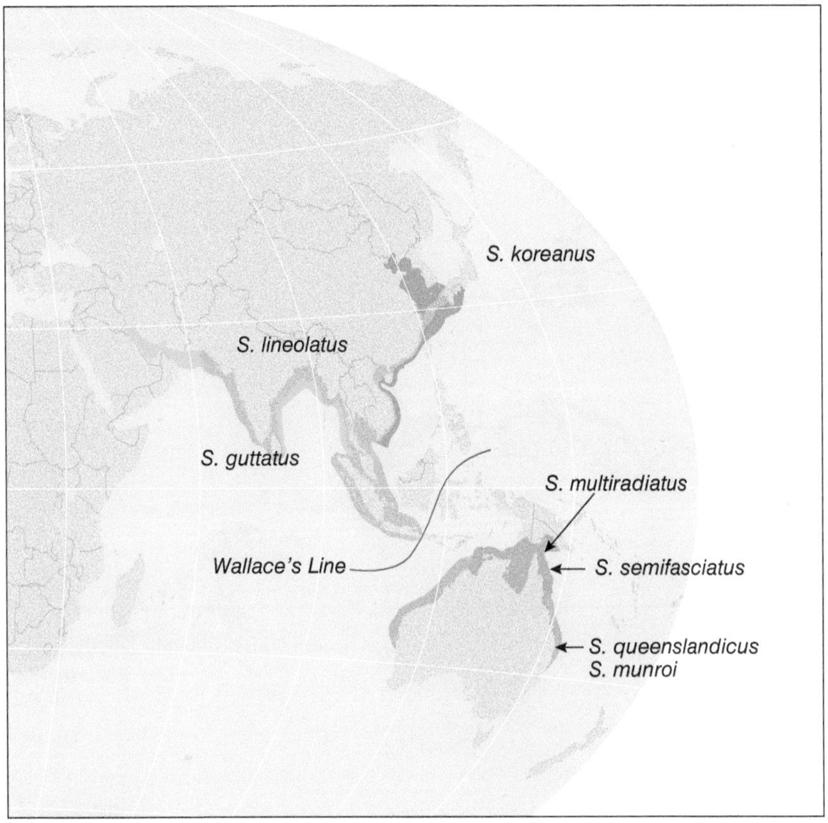

FIGURE 19.8 The ranges of seven Indian-West Pacific species of Spanish mackerels (*Scomberomorus*), three continental and four Australian, with reference to Wallace's Line delimiting the continental margin. Adapted from Collette and Russo (1985b).

America. The fish fauna of the western Atlantic region comprises about 1000 species, partially divided into northern and southern parts by the freshwater outflow of the Amazon River. The biodiversity hotspot for this region is the Caribbean with about 750 species, whereas the Brazilian biogeographic province has about half that number, with many species shared between the two provinces (Floeter et al. 2008). Bottom trawling below the freshwater outflow of the Amazon River in 1975 closed part of the supposed "gap" in fish distributions (Collette & Rützler 1977). At 14 benthic stations off the mouth of the Amazon, under the superficial freshwater layer, a typical reef fish fauna of 45 species was found, but these "coral reef" species were associated with 35 species of sponges in water too turbid for coral growth (Fig. 19.9).

Sponges provide the necessary structural habitat for "coral reef" species, which allows genetic continuity between the two supposedly separated populations. Rocha et al. (2002) showed the variable effects of the Amazon outflow on three species of the surgeonfish genus *Acanthurus*. The Amazon outflow is a strong barrier to dispersal of *A. bahianus*, a modest barrier for *A. coeruleus*, and has no discernable effect on *A. chirurgus*, which has been collected on deep soft bottoms with sponge habitats under the Amazon outflow. Both *A. bahianus* and *A. coeruleus* live in shallower waters and are not as tolerant of silt as *A. chirurgus*.

Although many groups show their maximum diversity in the Indian-West Pacific, a few show maximum diversity in the Americas. Two-thirds of the species of toadfishes, Batrachoididae, occur in New World waters (Collette & Russo 1981). The most generalized subfamily, the Batrachoidinae, is worldwide. However, the two most specialized subfamilies, the luminous midshipmen (Porichthyinae) and the venomous toadfishes (Thalassophryninae), are restricted to the western Atlantic and eastern Pacific (plus a few freshwater species derived from Atlantic or Pacific marine species).

Eastern Pacific Region

The eastern Pacific region was isolated from the West Atlantic 3–5 mybp by the rise of the Isthmus of Panama, and many sister species pairs show the evolutionary legacy of this event, such as the Spanish mackerels, *Scomberomorus sierra* of the eastern Pacific, and *S. brasiliensis* found in the Caribbean Sea. Some sister species have clearly differentiated from a morphological perspective, such as Spanish mackerels and toadfishes of the genus *Batrachoides*. Others, such as the halfbeaks of the genus *Hyporhamphus* are less well differentiated morphologically. This region contains fewer species of fishes than are present in the western Atlantic. Some widespread Atlantic taxa, such as the Bluefish (*Pomatomus saltatrix*) and Cobia (*Rachycentron canadum*), are absent.

On the other side, the **eastern Pacific barrier** extends across 5000 km of open water between the central Pacific islands and the American mainland, acts as a distance barrier

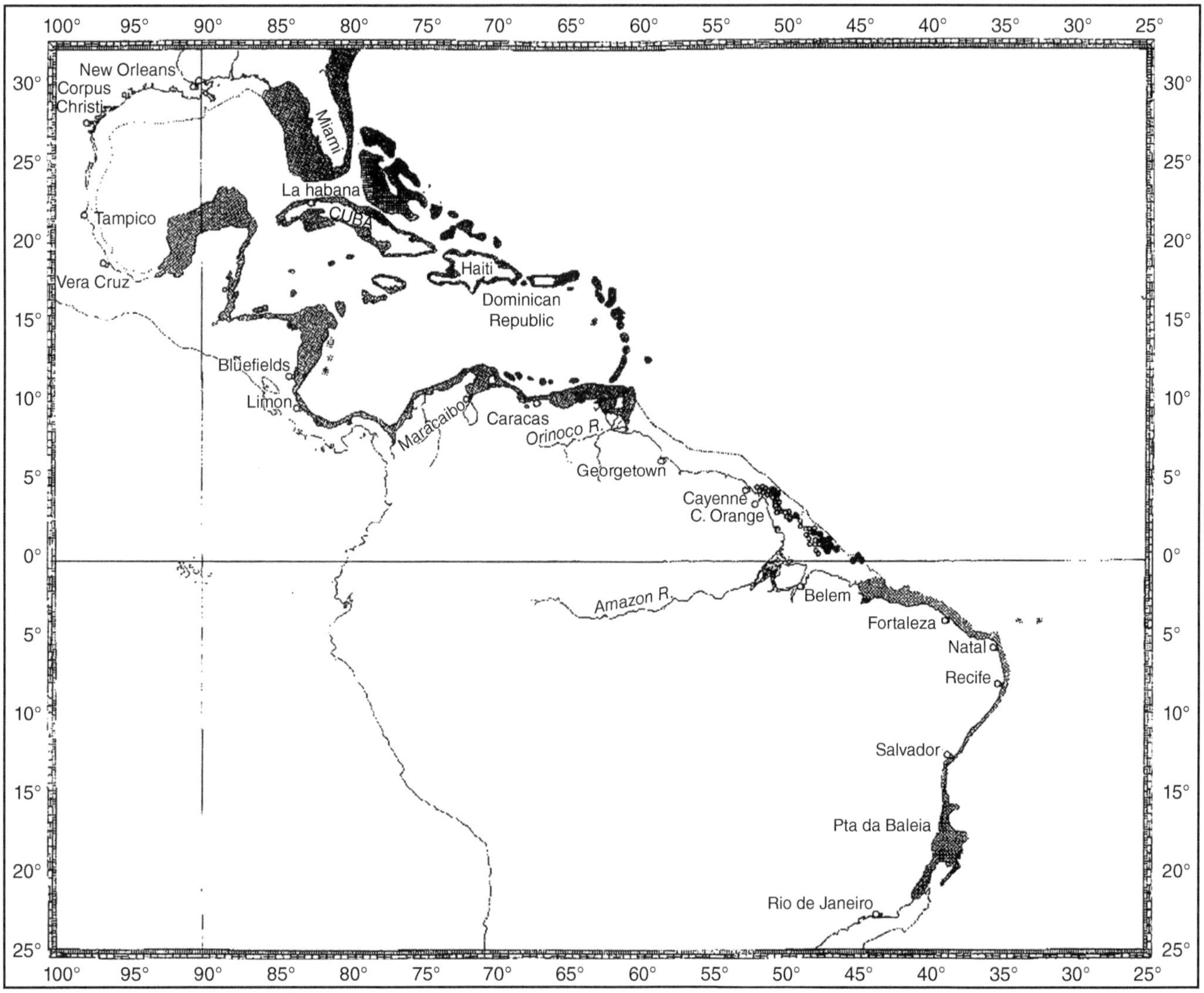

FIGURE 19.9 Distribution of coral reef fishes (shaded areas) in the tropical western Atlantic. Black dots indicate 14 stations where coral reef fishes were caught in association with sponges. Collette and Rützler (1977).

limiting the dispersal of shore species (Briggs 1974). However, the eastern Pacific barrier is not impermeable, as genetic studies indicate at least occasional dispersal from Hawaii to the eastern Pacific (Lessios & Robertson 2006). Several Indian-West Pacific shore fishes cross the eastern Pacific barrier and are found at offshore islands such as the Revilligedos, off the coast of Mexico, and Clipperton and Cocos, off the coast of Costa Rica. Distributions of a species of mackerel (*Scomber australasicus*) and a needlefish (*Tylosurus acus melanotus*) extend from the western Pacific through the Hawaiian Islands to these islands, but these species are replaced by related forms (*S. japonicus* and *T. pacificus*, respectively) along the eastern Pacific coast of Middle America. These exceptions to the completeness of the eastern Pacific barrier may be related to habitat differences between the offshore islands and the mainland.

The Panama Canal connects the eastern Pacific with the western Atlantic. However, unlike the Suez Canal, the Panama Canal is not at sea level. It contains a freshwater Lake Gatun in its middle, and fresh water is used to raise the water level in a series of locks to lift ships up to the lake and then down to the ocean on the other side. This freshwater barrier prevents marine species from moving between the two oceans, with the exception of a few species that tolerate a wide range of salinities (McCosker & Dawson 1975). A proposed sea-level canal would allow mixing of the two different faunas and might have grave effects on the fishes and marine invertebrates on both sides of the isthmus. Diseases, parasites, and aggressive Indian-West Pacific species that pose little current danger in the eastern Pacific, such as the crown-of-thorns starfish and a sea snake, might damage coral reefs and the fish fauna of the western Atlantic.

Eastern Atlantic Region

In the eastern Atlantic Ocean, tropical shore fishes are restricted to the Gulf of Guinea, a relatively small area that extends from Dakar, Senegal, to Angola, and includes offshore islands such as the Cape Verde Islands, Annobón, and Fernando Po. Coral cover is sparse in the tropical part of the eastern Atlantic, partly due to the large amount of freshwater runoff and accompanying sediment that flows out of such rivers as the Congo, Niger, and Volta. The eastern Atlantic is depauperate in many fish and invertebrate groups and contains only about 500 species of shore fishes. A few families, such as the porgies (Sparidae), have radiated in the eastern Atlantic.

Comparisons among genera of shore fishes of the western Atlantic, eastern Pacific, and eastern Atlantic demonstrate the relative depauperate nature of many eastern Atlantic groups. For example, four genera that contain two to four species each in the western Atlantic and eastern Pacific have only a single species in the eastern Atlantic (Table 19.1). Such comparison of patterns of diversity among different families can tell us much about not only the zoogeography of a group but also its probable phylogenetic history, particularly if we apply modern approaches to both zoogeography and phylogeny (e.g., the Spanish mackerels, *Scomberomorus*, discussed earlier).

Mediterranean Sea

The Mediterranean Sea is a somewhat depauperate extension of the eastern Atlantic Ocean, with about 720 species of fishes. Drying out during the Messinian Salinity Crisis millions of years ago eliminated most fishes from the Mediterranean, and cooler temperatures in the Straits of Gibraltar prevented warm-water fishes in the Gulf of Guinea from moving into the warm waters of the eastern Mediterranean when the Atlantic and Mediterranean were reconnected at the end of the Messinian Salinity Crisis about 5.3 million years ago (Patarnello et al. 2007). In 1869, a sea-level route, the Suez Canal, was opened, connecting the warm but depauperate eastern Mediterranean with the Red Sea, the latter being part of the rich Indian-West Pacific region. For some time after construction, faunal transfers between the Red Sea and the Mediterranean were inhibited by the saline waters of the Bitter Lakes in the middle of the canal. In the first edition of *A history of fishes* in 1931, Norman reported that 16 species of Red Sea marine fishes had moved through the Suez Canal and established themselves in the eastern Mediterranean Sea. In the ensuing decades, the number has increased to at least 100 **Lessepsian migrants** (named after Ferdinand Lesseps who was in charge of constructing the canal), fishes having moved in one direction, from the Red Sea into the eastern Mediterranean. As an example of how successful a migrant can be, the Brushtooth Lizardfish (*Saurida undosquamis*) was first taken in the Mediterranean in 1952. By 1955, 266 tons of this lizardfish were landed by local trawlers, constituting close to 20% of the trawler catch in Israeli waters (Golani 1993).

Why have these movements been virtually unidirectional? First, the diversity of inshore fishes is greater in the Red Sea, part of the Indian-West Pacific fish fauna, than in the Mediterranean, which means fewer ecological opportunities for new immigrants. Second, there appears to be "empty niches" in the eastern Mediterranean, associated with water temperatures, with temperatures again being warm enough for warm-water fishes. In addition, many of the species that penetrated the canal are widespread species, adapted to a wide variety of living conditions. Consider the distributions of three of the invading species, a halfbeak and two mackerels. *Hemiramphus far* is the most widespread member of its genus, known from South Africa across the Indian Ocean north to Okinawa, south to Australia, and east to Tonga and Fiji. It successfully moved through the canal and established populations that have now spread west and north. The two mackerels, the Narrow-barred Spanish Mackerel (*Scomberomorus commerson*) and the Indian Mackerel (*Rastrelliger kanagurta*), are the most widespread members of their genera, occurring from South Africa north to the Red Sea, east to China and Japan, and south to Australia and Fiji (Collette & Nauen 1983, maps on pp. 49 and 63). Such generally successful colonist species could have been predicted as the most likely taxa to take advantage of the opportunities in the eastern Mediterranean, because they are adapted to a wide range of ecological conditions.

Red Sea

The Red Sea is shallow (137 m) with minimal freshwater input and a narrow connection to the Indian Ocean. Like the Mediterranean, it has undergone dessication and salinity crises, but on a younger timescale of 100,000 years. The Red Sea has at least 1078 documented fish species (Golani & Bogorodsky 2010), with 14% endemism overall, but much higher (50%) in butterflyfishes (Chaetodontidae). Sea level during glacial maxima dropped as much as 140 m, sufficient to cut off the connection to the Indian Ocean at the Strait of Bab al Mandab. Some authorities believe the hypersaline conditions predominated during these glacial maxima, extirpating most marine life. However, the presence of many endemic species indicates that the Red Sea fauna persisted across glacial cycles, perhaps in the adjacent Gulf of Aden (reviewed in DiBattista et al. 2016).

Arctic and Antarctic Fishes

Marine shore and continental shelf species down to 200 m from Arctic and Antarctic waters account for about 5.6% of the total fish fauna. The two polar regions contain at least 538 species of fishes, 289 in the Arctic and 252 in Antarctica (Møller et al. 2005). Most of these groups are found in only one polar region, indicating that these assemblages, and the adaptations needed for life in extreme cold (see Chapter 10, Polar Regions), evolved independently of one another. Only 12 of 214 polar fish genera and 10 of 72 polar fish families were reported in both areas.

The Arctic region north of 60° in the Pacific (approximately Nunivak Island, Alaska) to Newfoundland and northern Norway in the Atlantic has 20–25% endemism (Briggs 1974) and contains at least 242 species in 45 families with about 20% endemism. Six families in Cottoidei with 72 species and five families in Zoarcoidei with 55 species account for about half the species (Mecklenburg et al. 2011). Additional members of the fish fauna include gadiforms, salmonids, pleuronectiforms, and chondrichthyans. Arctic species include skates, herrings, greenlings, poachers, snailfishes, pricklebacks, wolffishes, and gunnels. Most of these groups originated in the Pacific after the Bering Strait (between Alaska and Russia) opened 3–3.5 mybp.

Antarctica and the surrounding Southern Ocean contain at least 322 species of fishes in 50 families (Eastman 2005). The immediate Antarctic region has 174 species in 13 families, with a much higher endemism (88%) than the Arctic. Of the fishes in the immediate Antarctic region, six families in the suborder Notothenioidei (icefishes) account for 55% of the species and more than 90% of the individuals. Primitive notothenioids, such as the Bovichtidae, occur in southern hemisphere habitats of Australia, New Zealand, and South America. Some families occur in both Antarctica and the surrounding continents (Nototheniidae and Channichthyidae), some occur in Antarctica and nearby oceanic islands such as Las Malvinas (the Falklands) (Harpagiferidae), and one family (Bathydraconidae) is restricted to Antarctica. Of the notothenioids, 97% are Antarctic endemics; even 70% of the non-notothenioids are endemic. Six other families that contribute multiple species to the region are, in order of species diversity (Eastman 2005): snailfishes (70 species), eel pouts (24), skates (eight), and eel cods, deep-sea cods, and southern flounders (four species each).

In some cool-water species, such as the chub mackerels, *Scomber japonicus,* and *S. colias*, distributions are interrupted by warm low-latitude regions. Such species are considered to have **antitropical distributions**, in that they are present in temperate waters on either side of the equator (Hubbs 1952). Other cold-water species show **tropical submergence**, that is, they continue their ranges into tropical regions by moving into deeper waters that have temperatures similar to the cold waters of Arctic and Antarctic regions.

Marine Fish Phylogeography

Marine fishes have few barriers to dispersal, and can show low population genetic separations across vast regions of the planet. This has implications for speciation that will be explored later in the text (see Chapter 20, Population Genetics). Here, we review a few of the major biogeographic barriers, and how genetic studies have illuminated the nature and history of these barriers.

Transarctic Interchange Between the North Pacific and North Atlantic

Approximately 3.5 mybp the Bering Strait opened and allowed a cold temperate waterway between the North Pacific and North Atlantic basins, as indicated by paleontology and geology (Vermeij 1991). This opening persisted for more than half a million years and colonization proceeded in both directions, but with most movement of fishes from the highly diverse Northeast Pacific to the relatively depauperate Atlantic (Briggs 1970). The directionality of exchange is usually inferred from the fossil record, which shows that hundreds of species moved from the Northwest Pacific into the Atlantic. The initial interchange has been followed by perhaps three more openings of lesser duration and impact, including one event about 2 million years ago, and one following the last glacial period (<12,000 years ago) (Nikula et al. 2007).

Several widespread and abundant Atlantic fishes are the product of this Pacific invasion. The Atlantic Herring (*Clupea harengus*) is distinguished from the Pacific Herring (*C. pallasi*) in allozyme surveys, indicating a separation time of 3.6–6.6million years (Grant 1986).

One important invasion may have occurred in the other direction: molecular clock estimates for allozymes and mtDNA indicate that Atlantic Salmon (genus *Salmo*) invaded the Pacific during this interval, giving rise to the Pacific salmon genus *Oncorhynchus* (Kitano et al. 1997). For the anadromous and freshwater smelts (*Osmerus* spp.), mtDNA data indicate an exchange approximately 2–2.5 mybp, possibly linked to the second opening of the transarctic waterway (Taylor & Dodson 1994).

Regardless of the direction of exchange, it is clear that the transarctic interchange had a tremendous effect on biodiversity, especially in the North Atlantic. Genetic assays with allozymes and mtDNA complement the paleontological investigations, to provide a relatively complete picture of this great natural invasion.

The Panama Barrier

The Isthmus of Panama eliminated contact between the Atlantic and Pacific about 3.5 mybp. This is an impassable barrier except for marine species that can tolerate the freshwater conditions of the Panama Canal (built by the United States and opened in 1914), such as the Atlantic Tarpon (*Megalops atlanticus*; McCosker & Dawson 1975). A number of studies has compared sister species across this barrier, first with allozymes and later with mtDNA, and these comparisons have been especially useful for calibrating molecular clocks against a reliable geological event (Lessios 2008). For example, the genetic differentiation between Atlantic and Pacific trumpetfishes (genus *Aulostomus*) fits a timeframe for the establishment of the Panama barrier (discussed earlier in this chapter).

The Indian–Atlantic Barrier

Southern Africa is another barrier to tropical species, where warm waters of the Indian Ocean (Agulhas Current) collide with cold upwelling in the South Atlantic (Benguela Current). However, the barrier is not absolute. Contemporary dispersal may be possible in warm-core gyres from the Indian Ocean that occasionally cross the frigid Benguela Current and transport tropical species into the Atlantic, as indicated by the occasional arrival of Indian Ocean biota at the island of St. Helena in the South Atlantic (Edwards 1990).

In addition to the trumpetfish example, several studies have demonstrated colonization events from the Indian to Atlantic oceans. In the Blue Marlin (*Makaira nigricans*),

Buonaccorsi et al. (2001) report two divergent mtDNA lineages. One lineage is restricted to the Atlantic, the other occurs in both the Atlantic and Indian-Pacific. Apparently Blue Marlin populations in the two ocean basins were isolated for an extended period, followed by a rare colonization of the Indian–Pacific lineage into the Atlantic. Molecular clock estimates put the initial Atlantic–Indian divergence at about 600,000 years ago.

The goby genus *Gnatholepis* contains a single Atlantic member, the Goldspot Goby, *G. thompsoni*, which is indistinguishable from *G. scapulostigma* in the Indian Ocean (Randall & Greenfield 2001). Hence comparative morphology indicated a recent colonization event. Rocha et al. (2005b) determined that this Atlantic species was indeed closely related to *G. scapulostigma* (d = 0.0054 with mtDNA cytochrome *b*), indicating a colonization into the Atlantic approximately 150,000 years ago (Fig. 19.10).

Like the transarctic exchange described above, the route around southern Africa is an important biogeographic pathway for marine fishes, enhancing biodiversity in the Atlantic Ocean. Studies of the Indian–Atlantic barrier have focused on colonization into the Atlantic, consistent with prevailing currents. However, colonization in the other direction is possible for species that can overcome the water flow, as indicated by two recent examples from sharks. In the Scalloped Hammerhead Shark (*Sphyrna lewini*) and Blacktip Shark (*Carcharhinus limbatus*), haplotypes in the Indian Ocean are recently derived from the Atlantic populations (Duncan et al. 2006; Keeney & Heist 2006). Unlike most marine fishes, wherein dispersal is accomplished primarily by larvae, most sharks are active swimmers from birth. Swimming is the only form of dispersal available, and may eventually prove to be a fundamental difference between the phylogeography of bony fishes and elasmobranchs.

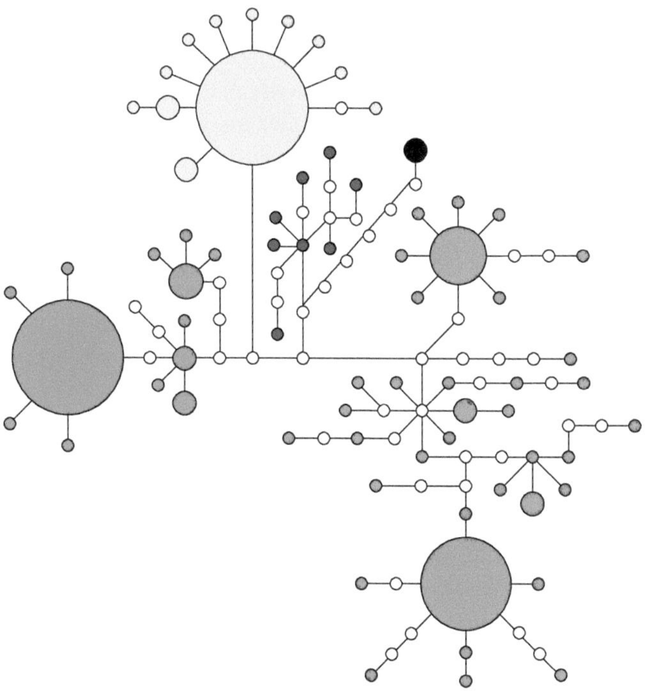

FIGURE 19.10 Parsimony network for mtDNA cytochrome *b* showing that the Goldspot Goby, *Gnatholepis thompsoni*, of the Atlantic Ocean is the product of a recent colonization from the Indian Ocean. Populations in the western, central, and East Atlantic are indicated by blue, green, and yellow coloration. The sister species (*Gnatholepis scapulostigma*) is indicated in red (South Africa) and black (Pacific). Breaks in the branches (small circles) indicate mutation events, and unbroken branches indicate a single mutation regardless of length. The size of the circle indicates the frequency of each haplotype. From Rocha et al. (2005b), used with permission.

The Coral Triangle Biodiversity Hotspot

As mentioned earlier, the Coral Triangle between Indonesia, New Guinea, and the Philippines is the biodiversity pinnacle for marine fishes and other tropical marine organisms. Several theories exist to explain this marvel of fish diversity, including a *center of origin* theory that posits the Coral Triangle as a species producer and exporter. Alternately this region could be a *center of accumulation*, between Pacific and Indian Oceans. Phylogeographic studies have uncovered evidence of both. Phylogenetic trees of three speciose families of fishes (Labridae, Pomacentridae, and Chaetodontidae) indicate that 60% of the global biodiversity in these families originated in the Coral Triangle (Cowman & Bellwood 2013). This strongly supports the center of origin theory. Other studies have shown that centers of high endemism on the periphery of the Indo-Pacific, including Hawaii and the Red Sea, can export diversity toward the center (Eble et al. 2011; Coleman et al. 2016). This strongly supports the center of accumulation model. Bowen et al. (2013) proposed that both models can be correct. Species arising in the Coral Triangle evolve into new species in peripheral habitats, and colonize back toward the center in a process dubbed *biodiversity feedback* (Fig. 19.11).

In considering evidence from fossils, phylogenies, and phylogeography, David Bellwood and colleagues proposed that the Coral Triangle biodiversity is the product of 30 million years of geological stability. Initially the region accumulated species from other regions, incubated new species through the Miocene (23–5 mybp), and became a net exporter of species in the last 5 million years (Cowman & Bellwood 2013 and references therein).

Phylogeographic studies continue to unravel the evolutionary history of fishes. They are most powerful when evidence from multiple species is assembled to resolve general trends (e.g., Lessios & Robertson 2006), and when the evidence from DNA is combined with biogeographic information, morphology, earth history, and ecology.

Freshwater Fishes

Freshwater fishes make a much larger contribution to biodiversity than might be expected based on area alone. About 49% of the world's fish species live in fresh waters (Fricke et al. 2022)., which is quite remarkable when we realize that fresh waters comprise only 0.0093% of the water on the planet (Horn 1972). Therefore, nearly half of all fish species live in less than 0.01%

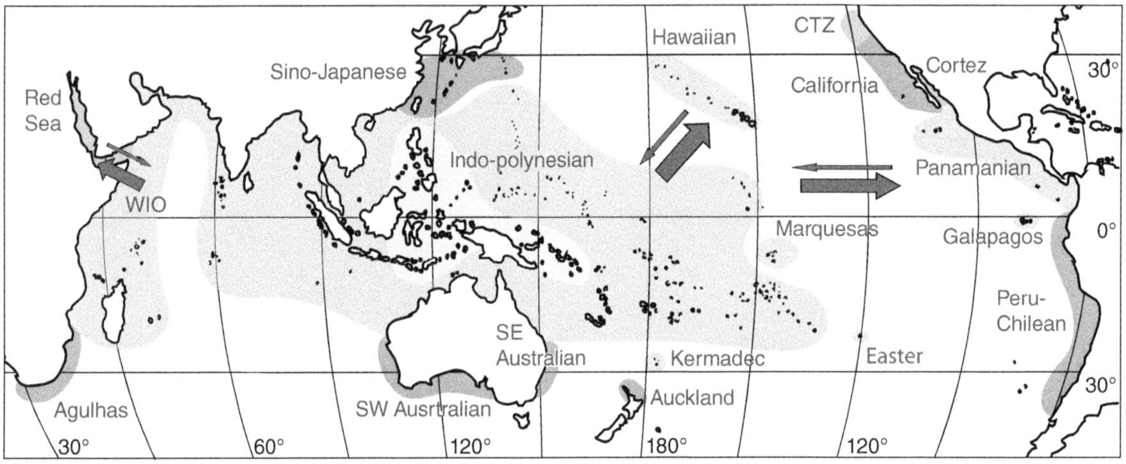

FIGURE 19.11 Map of the Indo-Pacific showing warm-temperate biogeographic provinces (in grey) and tropical provinces (in gold). Arrows at the North Pacific and Red Sea indicate a primary flow of species out to peripheral biogeographic provinces (center of origin model), but also a secondary flow from the peripheral provinces in towards the biodiversity hotspot of the Coral Triangle (center of accumulation model). The Coral Triangle biodiversity hotspot is indicated by lighter gold. The California Transition Zone (CTZ) within the Oregon and California provinces is indicated by light grey. Abbreviation: WIO, Western Indian Ocean. From Briggs and Bowen (2013), with permission.

of the world's water supply. Another way of looking at this is to calculate the mean volume of water per species of marine versus freshwater fishes. The average marine species has 113,000 km^3 available to it, whereas the average freshwater species has only 15 km^3.

The probable causes of this 7500-fold disparity in biodiversity in the two major habitat types are undoubtedly very complex, involving ecological as well as historical (phylogenetic and geological) factors. Two likely influences are productivity and isolation. Shallow waters receive significant sunlight, allowing photosynthesis, which forms the base of food webs. Most fresh waters are shallow and relatively productive, whereas most water in the world's oceans lies well below the euphotic zone where primary productivity occurs. Shallow marine waters are productive and support a diverse fauna, as is evident in the coastal zones that support 45% of all fish species.

Also influencing diversity is the potential for isolation, a historical factor that differs greatly between marine and freshwater habitats. Marine habitats are broadly continuous; significant faunal breaks occur primarily where continental landmasses, large rivers, or sills occur, and where major oceanic currents act as geographic boundaries. Fresh waters, in contrast, are frequently and readily broken up into isolated water bodies. A variety of geologic events, as well as some biological activity such as dam building by beavers, can lead to a body of water losing its connections with other bodies, which in turn isolates the fishes in that body from gene flow with other areas. Genetic isolation is a driving force of evolution, leading to such dramatic events as explosive speciation and formation of species flocks, such as cichlids in Lake Malawi, Lake Victoria, and Lake Tanganyika of Africa. Isolating events are, therefore, much more common in fresh waters, and it is therefore not surprising that so many species of freshwater fishes have arisen in such little space.

Species Flocks

When a volcano erupts, an earthquake causes uplifting, a landslide blocks a river or divides a lake, or drought and flood cycles accompany longer term climatic changes, freshwater habitats can become **isolated**, separating small numbers of fishes from their conspecifics, predators, and competitors. The reproductive future of such an isolated individual or individuals is usually rather bleak, as potential mates and resources may also be in short supply in the newly created habitat. But such events occasionally lead to an evolutionary bonanza, as evidenced by several so-called species flocks or species swarms. A **species flock** is a group of closely related species that share a common ancestral species and that are endemic to an isolated region such as a single lake or island (Greenwood 1984). Flocks evolve when a newly created habitat with essentially open niches is colonized and the colonizing species experiences a relaxation of the many selection pressures that normally kept its population in check. Descendants of the founders disperse, differentiate, and fill the open niches. The end product is **rapid speciation** – a literal explosion of speciation events – and the production of several descendant species in the area.

Some of the most spectacular assemblages of fishes and other animals worldwide represent the end products of explosive speciation. Non-fish examples include such showcases of evolution as the fruit flies, land snails, and honeycreepers of the Hawaiian Islands, Darwin's finches in the Galápagos Islands, and the amphipods of Lake Baikal in Asia. Fish examples are numerous, instructive, and involve many different teleostean as well as some non-teleostean taxa. In addition, the fossil record shows us that the process of explosive speciation has operated dramatically in the distant as well as recent past.

The best-known fish flocks occur among the cichlid fishes of the three Great Lakes of Africa: Malawi, Tanganyika, and

Victoria. The first two lakes are long, deep, rift valley lakes, situated along a split in the earth's crust that is nearly 3000 km long. Lake Victoria, the largest lake in Africa, is more round and shallow, having been created as rivers were blocked by slow uplifting of the basin. These lakes vary in size between 28,000 and 69,000 km^2, putting them within the size range of the Laurentian Great Lakes of North America (19,000–82,000 km^2). Although of similar sizes, the species diversities of the lakes are totally dissimilar. The five Laurentian Great Lakes together contain about 235 fish species, whereas their African counterparts contain more than five times that number. The African Great Lakes each contain more fish species than any other lakes in the world (Fryer & Iles 1972; Greenwood 1981, 1991; Ribbink 1991; Goldschmidt 1996; Kornfield & Smith 2000; Snoeks 2000; Lamboj 2004).

The exact number of species in each African lake is difficult to determine because the region is remote from academic institutions that specialize in fish taxonomy, which means that many species remain to be collected or described, although concerted efforts continue. In addition, anatomical differences among some species are subtle (identification of species requires cytogenetic and biochemical analysis), and there is good evidence that environmental degradation and introduced species have recently wiped out many species (discussed in Chapter 22). Approximations for the three lakes indicate the following distributions (L. R. Kaufman, pers. comm.):

	Lake Malawi	Lake Tanganyika	Lake Victoria
Cichlids	700–1000	180–250	400–500
Endemic cichlids	695–995	178–248	395–495
Non-cichlids	53	113	65

Conservatively, there are perhaps 75 genera and 1300–1750 species of cichlids in the three lakes (estimates vary widely), as well as smaller flocks in smaller lakes and the rivers of Central Africa (e.g., Greenwood 1991; Stiassny et al. 1992), and flocks of non-cichlids such as catfishes in Lake Malawi, and of mastacembelid eels, claroteid and mochokid catfishes, and latid perch in Lake Tanganyika (Agnese & Teugels 2001; Lévêque et al. 2008).

The relevant points here are that most of the fishes in each lake are cichlids, most or all of the cichlids are endemic, and in each lake it is likely that most or all of the endemic cichlids share a common ancestor (Meyer et al. 1990; Meyer 1993), although some assemblages undoubtedly resulted from multiple invasions involving multiple ancestors (e.g., Kaufman 2003). The morphological and ecological divergence from such an ancestor is astounding, given that Tanganyika and Malawi are only 2–10million years old. The largest and smallest cichlids in Lake Tanganyika are *Boulengerochromis microlepis*, a predator, which attains a length of 80 cm, whereas small planktivores of the genus *Lamprologus* may be only 4 cm long as adults. The difference in mass between adults of the two genera is about 8000-fold. Morphological diversity includes African cichlids that artificially resemble many different families of teleosts and occupy habitats and niches that parallel those of the other teleosts (see Fig. 15.19). More spectacular still, sedimentation, radiocarbon dating, and mitochondrial DNA data all indicate that Lake Victoria may have been completely dry as recently as 12,500 years ago, which means that 300 endemic species there evolved in a very short period (Johnson et al. 1996). Age analysis of lakes, as with most aspects of African cichlid biology, is an active field with ever-changing conclusions (e.g., Seehausen 2002).

Trophically, African cichlids do it all. Trophic groups include species that specialize in eating phytoplankton, sponges, sediments, periphyton, leaves, mollusks, benthic arthropods, zooplankton, fish scales and fins, fish eyes, eggs and embryos, and other fishes. Major anatomical adaptations associated with different trophic habitats are found in the lips, marginal dentition, gill rakers, and particularly the pharyngeal jaws of the different trophic groups. Two flocks-within-flocks occur in Lake Malawi, where each subflock has differentiated into a particular feeding type. Approximately 27 (but perhaps as many as 200) species of closely related *mbuna* cichlids live over rocky areas and feed on the algae and associated microfauna of the algae, a food type known as *aufwuchs*. An additional flock of approximately 17 *utaka* cichlids live together and feed on zooplankton (Fryer & Iles 1972; Ribbink 1991; Lévêque 1997).

Whereas the African cichlids form the largest species flocks among fishes, other examples are often as dramatic (Lévêque et al. 2008). The following is a partial list of well-known flocks and some of their interesting characteristics:

1. The oldest extant species flock of fishes occurs in Lake Baikal, Russia, the oldest and deepest lake in the world. Here, sculpin-like cottoid fishes have differentiated into perhaps three families and approximately 33 species in 12 genera, including four recently described species (N. Bogutskaya, pers. comm.). Highly derived members of this group include two species of live-bearing, pelagic, comephorid Baikal oilfishes, a marked difference from the ancestral benthic, egg-laying sculpins (see also Berra 2001). Lake Baikal has also produced a flock of amphipods and has an endemic, freshwater species of monk seal. Habitat degradation in the lake has unfortunately pushed several Baikal endemics onto endangered species lists.
2. As many as three genera and 18 species of cyprinids form a flock in Lake Lanao of the Philippines, which sits above an uplifted waterfall 18 m high. This dramatic and controversial flock includes fishes with "**supralimital jaw specializations,**" indicating that derived species have jaw characteristics outside the normal variation found within the rest of the family. The flock is also unique in that the presumed ancestor, *Puntius binotatus*, still occurs in lowland streams below the waterfall. The validity of the flock and its traits are obscured by the destruction of holotypes during World War II and subsequent, multiple introductions of game and forage species that have displaced the native fishes; only three of the original cyprinids still occur in the lake (Kornfield & Carpenter 1984).

3. Eighteen species of atherinopsid silversides occur in a few lakes of the Mesa Central of Mexico. Diversification in this group includes a wide range in adult sizes and feeding types, from relatively typical, small (6 cm) zooplanktivores to piscivorous giants 30 cm long with specific names like *lucius* and *sphyraena* ("pike" and "barracuda") (Barbour 1973; Echelle & Echelle 1984).
4. A complex of flocks occurs among killifishes in Lake Titicaca and the surrounding lakes of the Peruvian and Bolivian Andes. Most of these species belong to one widespread genus, *Orestias*. Because several lineages are involved, the killifish assemblage is actually made up of several flocks rather than a single flock, the largest being about 15 species. The species have diversified into deep water, midwater, planktivorous, piscivorous, miniaturized, and broad-headed forms, a departure from the surface-dwelling, insect-feeding killifish norm (Parenti 1984).
5. Eight species of coregonid ciscoes evolved from a common ancestor in the Laurentian Great Lakes of North America. It is also likely that smaller flocks of coregonids have arisen in lakes of the western USA, Canada, and northern Europe, totaling perhaps 45 species, with repeated convergent modifications in eye size, body size and shape, gill raker number and morphology, snout shape, migration patterns, and spawning behavior (Smith & Todd 1984; Kottelat 1997).
6. Approximately, 15 species of cyprinid fishes in the genus *Labeobarbus* co-occur in Lake Tana of Ethiopia. They are in part unusual in that eight species are piscivores, which is an unorthodox feeding pattern for minnows (Nagelkerke et al. 1994; Palstra et al. 2004).
7. Historical continuity is evident in flocks of semionotid fishes that occupied the rift valley lakes of what is now the northeastern coastline of North America between North Carolina and Nova Scotia (McCune et al. 1984; McCune 1990). Semionotids were very successful Jurassic neopterygians that may have been ancestral to modern gars (Lepisosteidae) (see Chapter 11). Semionotid flocks of up to 17 species formed and were extinguished repeatedly as lakes filled and evaporated on a 21,000-year cycle over a period of 33 million years. Within the genus *Semionotus*, body shape varied substantially, including elongate pike-like forms, rounded sunfish-like forms, and intermediate shapes. The setting and speciation patterns directly parallel flock formation in the modern African rift valley lakes. Other "fossil flocks" include a radiation of eight sculpin species during the Pliocene in Lake Idaho in the western USA (Smith & Todd 1984).

The diversity of species that exist, their ecological relationships and innovations, and the repeatability of the process are remarkable examples of speciation and adaptation, even occurring in taxa that we do not normally think of as highly variable, speciose, or particularly rapidly evolving. How can such speciation occur, particularly in lakes that may only be a few thousand years old? A small lake near the edge of Lake Victoria may provide a clue. Lake Nabugabo sits 3 km away and 15 m above Lake Victoria, draining into the larger lake through a swamp. Six species of *Haplochromis* cichlids occur in Nabugabo, five of which are endemic and have close relatives in Victoria. Charcoal dates from former strandlines indicate that Nabugabo is only 4000 years old and that Victoria has repeatedly risen and overflowed into Nabugabo and other surrounding lakes, providing colonists for the smaller lakes. As waters receded, these colonists were isolated from competitors and predators and would have been able to occupy the niches of the newly created small lakes, speciating in as little as 4000 years. As the larger lake rose again, the new species could now swim into the larger, ancestral lake, increasing its diversity if they were unable to interbreed with their former conspecifics. As this scenario was repeated in numerous small, satellite lakes around Victoria, the generation of many species that would eventually occupy the larger lake is imaginable. It is through such **allopatric** processes of isolation and differentiation that species flocks in most lakes are likely to have developed.

For Lake Victoria in particular, where rock-dwelling species are abundant, the "satellite lake hypothesis" is not entirely satisfactory because few of the satellite lakes have rocky habitats. Other processes, involving within-lake (**sympatric**) development of species with minimal dispersal, have been proposed (e.g., Greenwood 1991; Galis & Metz 1998). Sexual selection of females for different colored males, reinforced by morphological and behavioral plasticity that leads to feeding specializations, is likely to have contributed (see Chapter 8, Sexual Dimorphism, Fig. 8.8). Regardless of the mechanisms resulting in this diversity, African cichlids represent "the most explosive speciation and adaptive radiation in vertebrate evolution yet described" (Galis & Metz 1998, p. 2).

Freshwater Fishes Versus Fishes in Fresh Waters

Up to this point in our discussion, we have used the term *freshwater fishes* a little carelessly. Historically, there was also much confusion surrounding the term until Myers (1938) distinguished between **primary** freshwater fishes, which are very strictly confined to fresh water, and **secondary** freshwater fishes, which are generally restricted to fresh water but may occasionally enter salt water. Most families of primary freshwater fishes have had a long evolutionary history of physiological inability to survive in the sea. The term **peripheral** is used for a number of genera and species of marine families that have taken up more or less permanent residence in fresh water or that spend part of their life cycle in fresh water and another part in marine habitats (diadromous fishes). There are about 85 families of primary freshwater fishes, 11 of secondary, and more than 30 of peripheral freshwater fishes.

The origins of freshwater fishes, and the importance of distinguishing among the different types, become particularly clear when the composition of fishes in the fresh waters of islands is considered. Continental islands that have been connected to the adjacent mainland, such as Trinidad, have the same kinds of fishes as are present on the adjacent mainland of South America. Oceanic islands such as Bermuda, the West Indies, and Hawaii, which have never been connected with continents, have no native primary freshwater fishes. All the native fishes in their fresh waters are secondary or peripheral species, or perhaps derived from secondary or peripheral ancestors. Oceanic islands are among the cases in which dispersal models provide better explanations for distributions of freshwater fishes than do vicariance models, as discussed earlier in this chapter.

Freshwater Zoogeographic Regions

One way to understand the distribution of freshwater fishes is to recognize six regions or realms, proposed by Alfred Russel Wallace in 1876: (i) Nearctic (North America except tropical Mexico); (ii) Neotropical (Middle and South America, including tropical Mexico); (iii) Palearctic (Europe and Asia north of the Himalayan Mountains); (iv) African; (v) Oriental (Indian subcontinent, Southeast Asia, the Philippines, and most of Indonesia); and (vi) Australian (Australia, New Guinea, and New Zealand). He et al. (2020) evaluated freshwater fishes of China and adjacent areas and concluded that their data suggest that the delineation between the Palearctic and Oriental biogeographic regions should be "the north of the Amur River basin, along the Khentii-Yablonovy-Stanovoy Range-Dzhugdzhur Mountains," rather than further south (He et al. 2020, p. 397); this would include most of China in the Oriental region.

Maps of the distributions of freshwater fish families are presented by Berra (2007), and Lévêque et al. (2008) discussed the freshwater fishes of these regions. Ongoing studies continue to increase the number of identified species, including many endemics, especially in the tropical regions. We suggest, therefore, consulting recent sources for updated information.

Nearctic Region

The Nearctic region consists of North America, south to the Mexican plateau. The North American freshwater fish fauna is the best known and has been mapped by Lee et al. (1980) and discussed thoroughly by Hocutt and Wiley (1986) and Mayden (1993). There are at least 14 families of primary freshwater fishes and a total of about 950 species of fishes in the region. The most speciose families include three families of ostariophysans – the Cyprinidae, Catostomidae, and Ictaluridae (the only North American family of Recent catfishes) – plus two percoid families – the Percidae (especially the darters) and the Centrarchidae. The ranges of five Nearctic families – the Cyprinidae, Catostomidae, Ictaluridae, Percidae, and Centrarchidae – extend south well into Middle America. Included in the Nearctic region are the species-rich Southern Appalachian river drainages, with about 350 species (e.g., Lydeard & Mayden 1995).

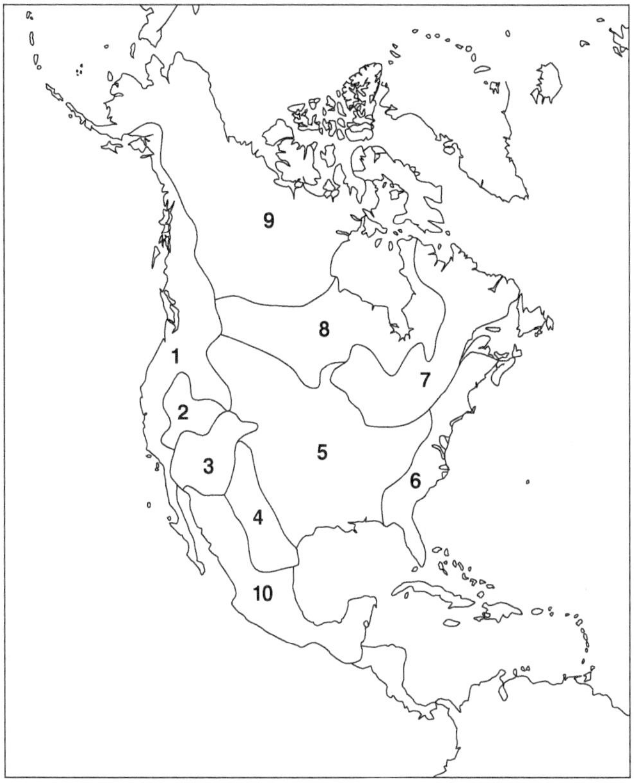

FIGURE 19.12 Main ichthyological provinces in North and Middle America (from Lévêque et al. 2008): (1) Pacific Coastal, (2) Great Basin, (3) Colorado, (4) Rio Grande, (5) Mississippi, (6) Atlantic Coastal, (7) Great Lakes–St. Lawrence, (8) Hudson Bay, (9) Arctic, and (10) Mexican Transition.

This region can be divided into 10 provinces (Fig. 19.12) based mainly on distribution data for freshwater fishes, mussels, and crayfish (Abell et al. 2000, as modified by Lévêque et al. 2008). The Mississippi Province has the most species (over 375, with about 130 endemic), followed by the Mexican Transitional (about 200 endemics) and the Rio Grande (over 150 species, about 80 endemic).

Neotropical Region

The Neotropical region consists of South America and Middle America, into which some North American species have moved. This region has the largest freshwater fish fauna in the world, with 43 families of primary freshwater fishes, and an estimate of over 8000 species, at least 3000 of which inhabit the Amazon River basin (Albert & Reis 2011; Reis et al. 2016) which supports the highest freshwater fish diversity in the world. There are no minnows or suckers in South America, but their ecological equivalents may be in the eight families of characins with over 1200 species. Other ostariophysans include 13 families of catfishes with about 1300 species, and six families of Gymnotiformes.

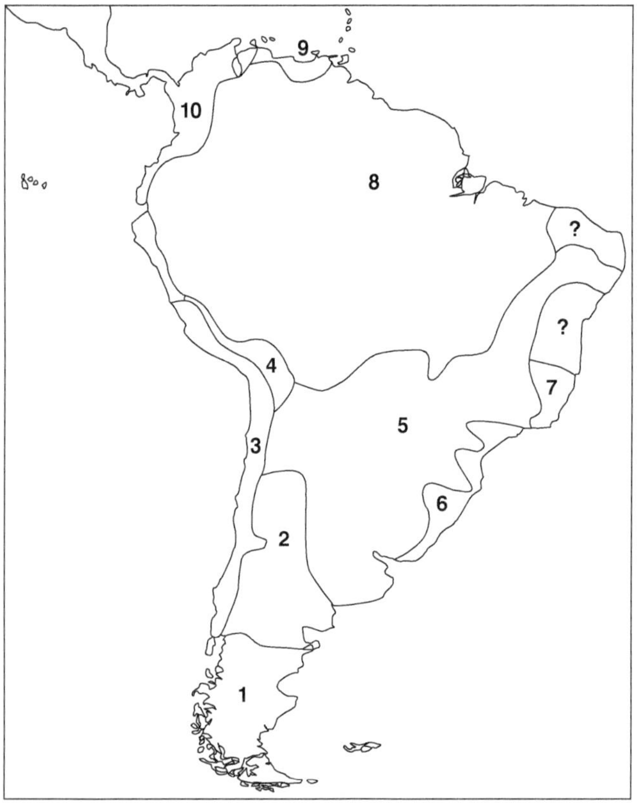

FIGURE 19.13 South American ichthyological provinces (from Lévêque et al. 2008): (1) South Patagonian, (2) North Patagonian, (3) Trans-Andean (South), (4) Lake Titicaca, (5) Paranean, (6) South-East Brazilian, (7) East Brazilian, (8) Guianean-Amazonian, (9) North Venezuelan, and (10) Trans-Andean.

Cichlids (about 150 species) are the most speciose perciform group. Representatives of many marine families have also become established in freshwaters of South America: herrings (Clupeidae); toadfishes (Batrachoididae); needlefishes (Belonidae, three endemic genera and seven species); croakers (Sciaenidae); soles (Achiridae); and freshwater stingrays of the family Potamotrygoninae, which are endemic to South America (freshwater stingrays of the family Dasyatidae are found in Africa, Asia, and Australia). Lévêque et al. (2008) recognized 10 ichthyofaunal provinces in this region (Fig. 19.13). The Guianean-Amazonian has by far the greatest diversity (about 3000 species, 2000 of which are endemic), followed by the Paranean (about 850 species, 515 endemic) and the Trans-Andean (about 425 species, 325 endemic).

African Region

The African region contains about 95 families of primary and secondary fishes (Lévêque et al. 2008), with over 3300 species as of 2020. Below the Palearctic-influenced region of Northern Africa there are 45 families of freshwater fishes, 15 of which are endemic. Africa contains fewer species of freshwater fishes than South America, but a larger number of archaic fishes. The fish fauna is dominated by ostariophysans, with more than 300 species of minnows (Cyprinidae), 190 characins (Characidae), and more than 360 catfishes from six families. Two secondary freshwater fish groups, the Cyprinidontiformes and Cichlidae have extensive radiations in Africa. Lévêque & Paugy (2006) suggested 12 ichthyofaunal provinces for Africa (Fig. 19.14) but Lévêque et al. (2008) showed 10, with two small areas (Eburneo-Ghanean, Ethiopian) being incorporated into the adjacent larger Nilo-Sudan region. The Maghreb Province north of the Sahara region has rather few species, with many similar to Europe. The greatest freshwater fish diversity is found in and around the Great Lakes in the East Coast/Orientale province, with an explosive cichlid radiation and 95% endemism.

Palearctic Region

The Palearctic region includes Africa north of the tropics, and Eurasia including Russia and Central Asia. The section in western Europe (Fig. 19.15A) contains only about 27 families of primary freshwater fishes with many minnows and loaches but only about 10 species of catfishes from four families. Europe was historically thought to be relatively depauperate in freshwater fishes, but reanalysis suggests about 550 species, with considerable endemism in the southern regions (Kottelat & Freyhof 2007). The Ponto-Caspian and Central Peri-Mediterranean provinces have the highest species counts, with about 100 each, whereas the other provinces each have about 40 to 65 species. The Central Peri-Mediterranean and Iberian provinces have the highest rates of endemism, with each over about 60%. The Eastern Peri-Mediterranean has about 30% endemism, whereas the remaining provinces are each under 10%

In an evaluation of freshwater fishes in China and adjacent regions of Asia, He et al. (2020) identified four major areas (Fig. 19.15B): Palearctic, High Central Asia, East Asia, and South Asia. The study of this broad area reported over 2060 species, representing 20 orders, 57 families, and 411 genera; most of these species (over 1650) were in China, which we will include in the Oriental region.

Oriental Region

We will consider most of China in the Oriental region, which also includes India, Southeast Asia, the Philippines, and the East Indies out to Borneo and Bali. According to He et al. (2020) China has over 1650 species of freshwater fishes (360 genera, 55 families, 20 orders); over 60 genera and 1030 species were endemic to China. About 94% of the species in China were in three orders: Cypriniformes (over 1200 species), Siluriformes (almost 200 species), and Perciformes (about 130 species). About 20% of the species recorded in China had been reported since 2000, with most of these in the biodiversity hotspot of Southwest China.

Alfred Russel Wallace (1860, 1876) proposed a boundary between the Oriental and Australian terrestrial faunas that Thomas Huxley named **Wallace's Line** (Fig. 19.16A). Some authors extend the line even farther to the east (Weber's Line)

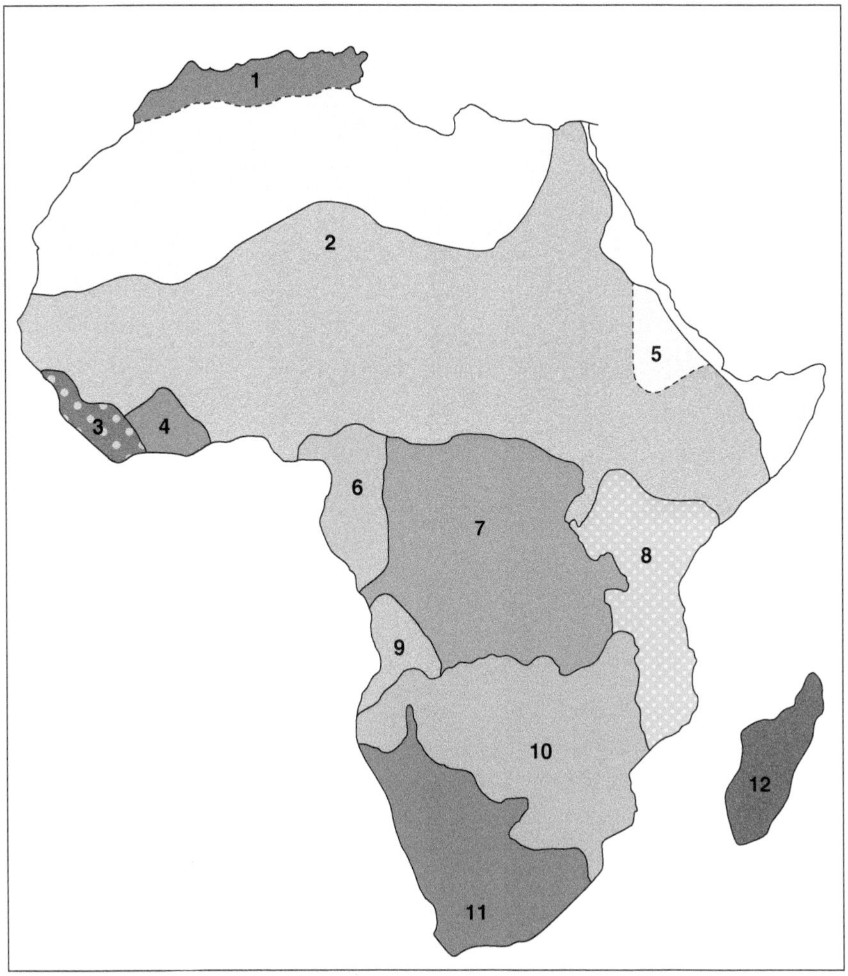

FIGURE 19.14 Main ichthyological provinces in Africa (based on Lévêque & Paugy 2006; Lévêque et al. 2008): (1) Maghreb, (2) Nilo-Sudan, (3) Upper Guinea, (4) Eburneo-Ghanean, (5) Ethiopian, (6) Lower Guinea, (7) Congo, (8) East Coast, (9) Quanza, (10) Zambezi, (11) Southern, and (12) Madagascar. Modified from Lévêque et al. 2006.

to also include the Celebes (now Sulawesi) and some other Indonesian Islands in the Oriental region. The region contains 28 families of primary freshwater fishes with 12 families of catfishes and four families of cypriniform ostariophysans: minnows (Cyprinidae), loaches (Cobitidae), algae eaters (Gyrinocheilidae), and river loaches (Balitoridae), which are endemic to the region. Non-ostariophysan families include snakeheads (Channidae), spiny eels (Mastacembelidae), labyrinth fishes (Anabantoidei), and a few cichlids. Only two species of primary freshwater fishes occur east of Wallace's Line. All other fishes in fresh waters east of the line have been derived from marine groups.

Australian Region

The Australian region has only two species of primary freshwater fishes, both ancient relicts of a much wider archaic distribution pattern (discussed in next section). Two families of secondary and 16 families of peripheral freshwater fishes of marine origin include freshwater species of catfishes (two marine families, Ariidae and Plotosidae), silversides (Atherinidae), rainbowfishes (Melanotaeniidae), halfbeaks (Zenarchopteridae), needlefishes (Belonidae), Teraponidae, Centropomidae, Percichthyidae, and gobies (Gobiidae). Australia has 10 ichthyological provinces (Fig. 19.16B, see Lévêque et al. 2008). The Northern province has the largest number of species at 75, of which 38 are endemic. The Northern province has 34 species (25 endemic) in common with New Guinea, indicating the connection between these land masses when sea levels were lower (shaded region in Fig. 19.16A). The very small Southern Tasmanian province has only eight freshwater species, all endemic to this area and with very restricted ranges.

Archaic Freshwater Fish Distributions

The term **archaic** is used to refer to the distribution of six groups of primitive primary freshwater fishes that date back long enough that their present distribution may be based on a different

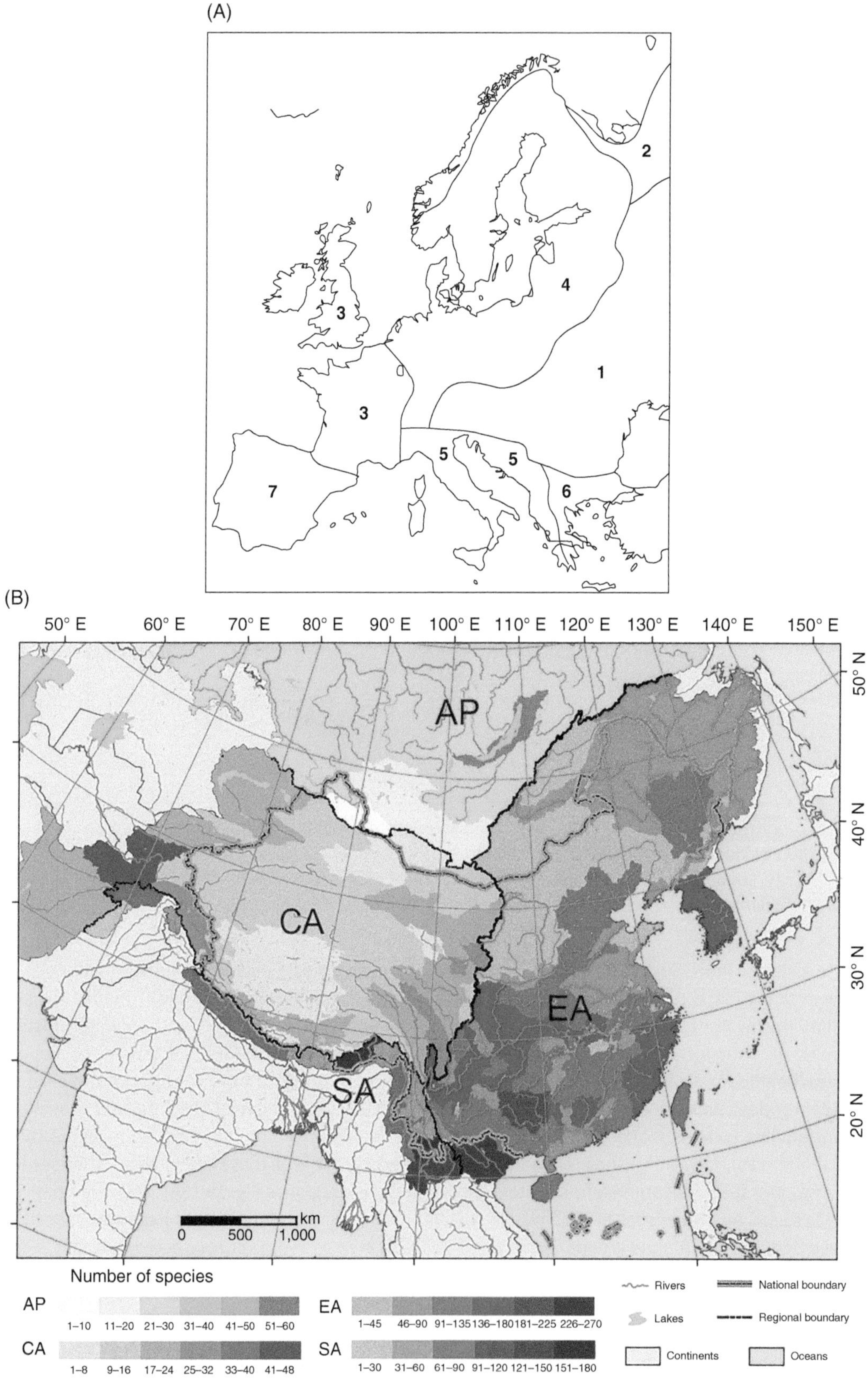

FIGURE 19.15 (A) Main ichthyological provinces in western Europe (from Lévêque et al. 2008): (1) Ponto-Caspian; (2) Northern Europe; (3) Western Europe; (4) Central Europe; (5) Central Peri-Mediterranean; (6) Eastern Peri-Mediterranean; and (7) Iberian Peninsula. (B) Four main zoogeographic regions for freshwater fishes in China and adjacent regions (from He et al. 2020): Palearctic (AP), High Central Asia (CA), East Asia EA), and South Asia (SA).

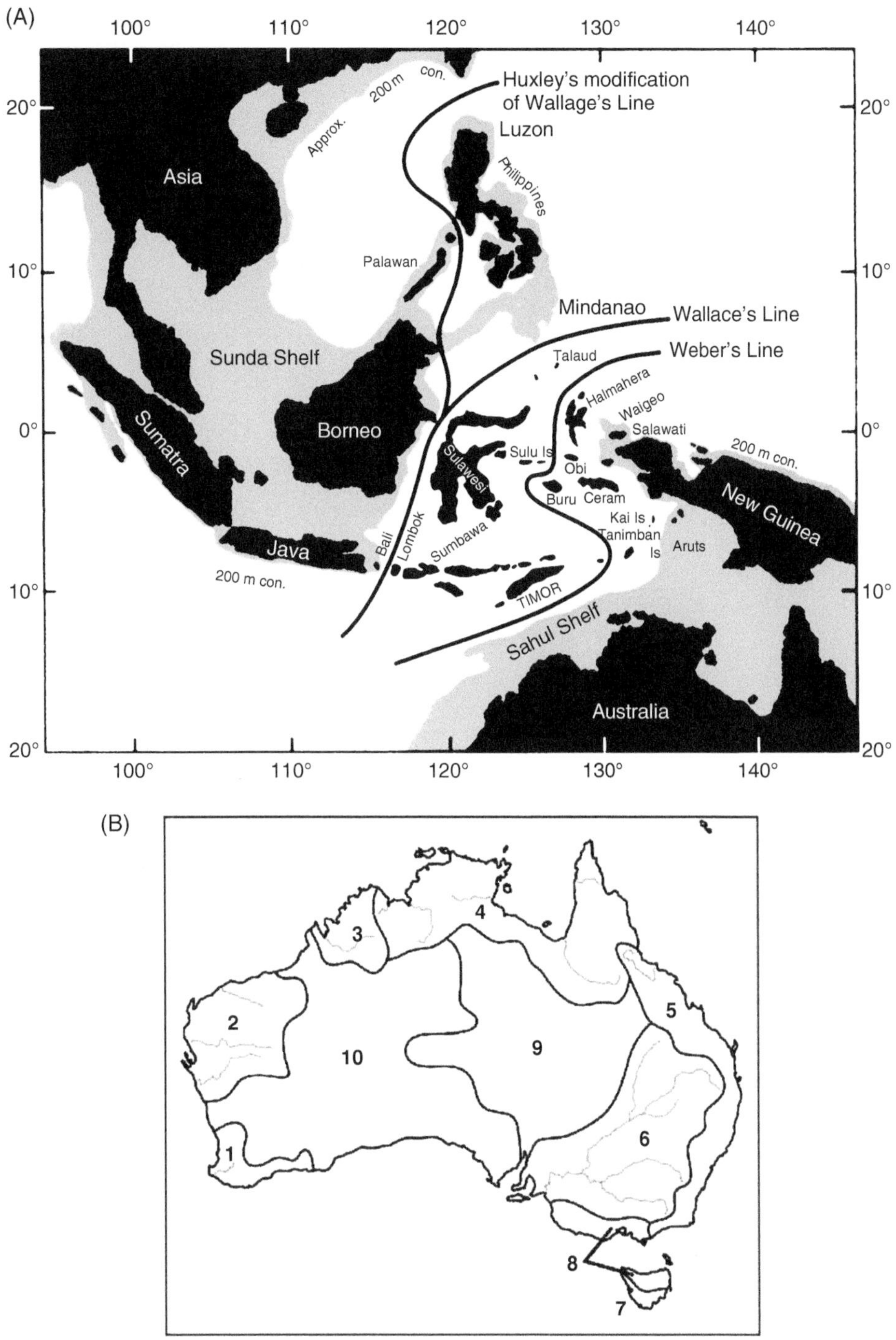

FIGURE 19.16 (A) Wallace's and Weber's lines separating the Oriental and Australian regions. The shaded areas show how the major landmasses would be connected if the sea retreated to the 200 m line. Adapted from Berra (2007). (B) Ten ichthyological provinces have been identified in Australia (see Lévêque et al. 2008): (1) Southwestern, (2) Pilbara, (3) Kimberley, (4) Northern, (5) Eastern, (6) Bass, (7) Southern Tasmanian, (8) Murray-Darling, (9) Central Australia, and (10) Paleo.

arrangement of the continents. These groups include lungfishes (Dipnoi), Polypteriformes, Polyodontidae, Lepisosteidae, Amiidae, and Osteoglossomorpha.

There are three living genera of lungfishes: the South American *Lepidosiren*, African *Protopterus*, and Australian *Neoceratodus*. Placement of *Lepidosiren* and *Protopterus* together in a single family, Lepidosirenidae, instead of in separate families, emphasizes their close relationships (Lundberg 1993). *Neoceratodus* is the most divergent lungfish, morphologically and physiologically, and has a relict distribution, restricted to portions of the Burnett and Mary rivers in southeastern Queensland (see Chapter 13).

Other archaic groups include the bichirs, Polypteridae, which consist of two living African genera – *Polypterus* with 11 species and the monotypic *Erpetoichthys* (previously *Calamoichthys*) – and a fossil genus (†*Dajetella*) from the Late Cretaceous and Paleocene of Bolivia (Lundberg 1993; Froese & Pauly 2011). There are two species of paddlefishes, family Polyodontidae: one (*Polyodon spathula*) from the Mississippi River of North America and the other (*Psephurus gladius*) from the Yangtze River of China. The seven species of gars, Lepisosteidae, are usually considered as secondary freshwater fishes and comprise two genera, *Lepisosteus* and *Atractosteus* in North America, Central America, and Cuba, plus fossils known from India and Europe. Only one Recent species of Bowfin, Amiidae, survives today (*Amia calva* of the United States), but fossils of this archaic group have been found on all the continents except Australia (Grande & Bemis 1998) and show that the present-day distribution of these groups is a relict of their original, much wider distribution.

Another archaic group is the Osteoglossomorpha, the most primitive suborder of the Teleostei (Fig. 19.17). It includes six families. There are four genera in the Osteoglossidae, two in each of two subfamilies (or families; Lundberg 1993): Heterotinae, *Heterotis niloticus* in the Nilo-Sudan Province of Africa and *Arapaima gigas* from the Amazonian lowlands and Guianas

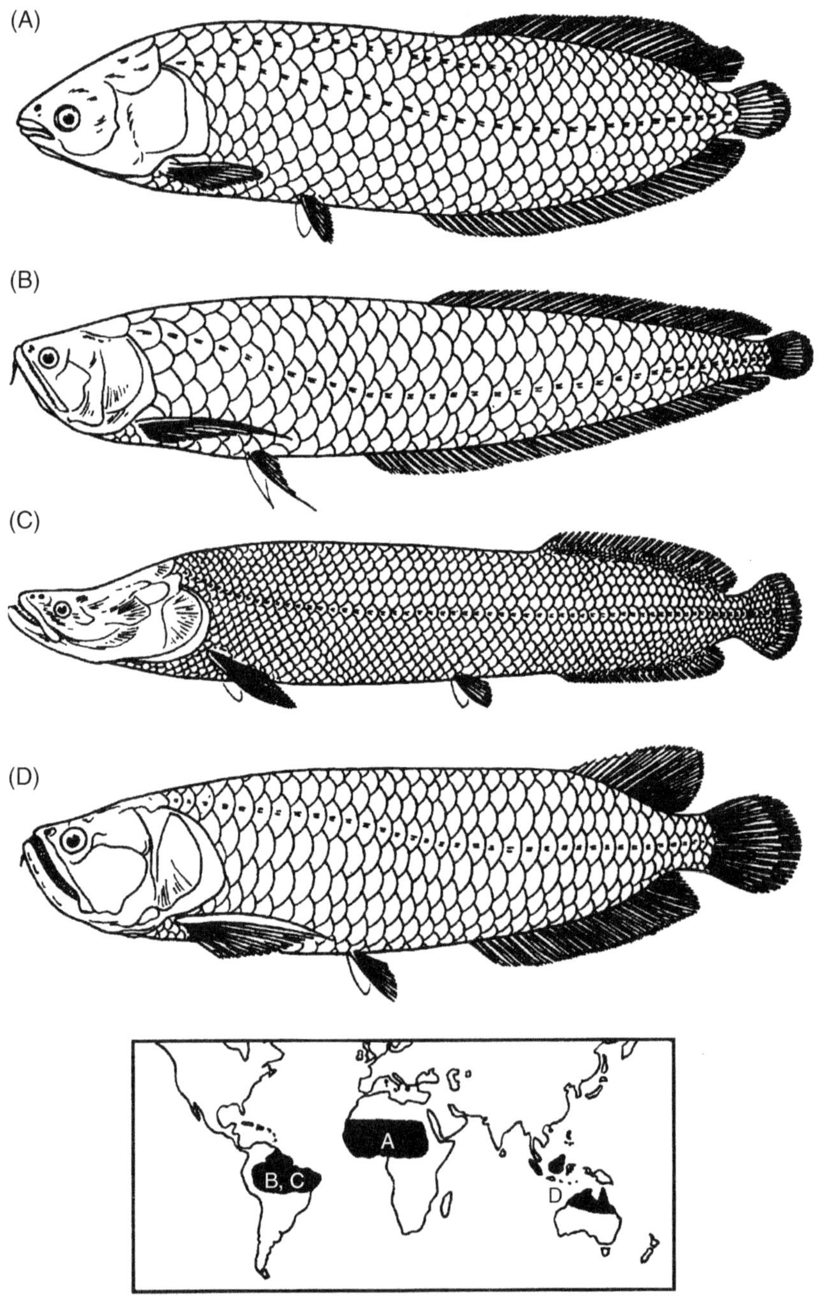

FIGURE 19.17 Osteoglossidae and their distribution (inset below). (A) *Heterotis*. (B) *Osteoglossum*. (C) *Arapaima*. (D) *Scleropages*. From Norman and Greenwood (1975) used with permission.

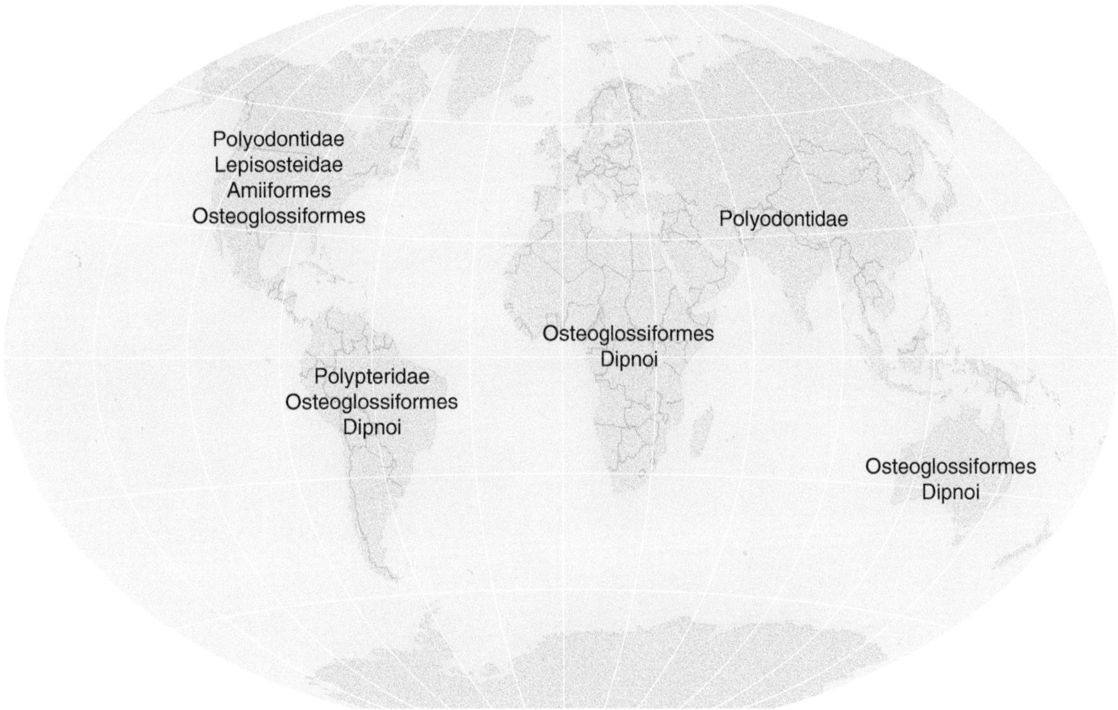

FIGURE 19.18 Summary of archaic freshwater fish distributions.

of South America; and Osteoglossinae, two species of *Osteoglossum* from South America and three species of *Scleropages* from Queensland, New Guinea, and Southeast Asia. The lungfish *Neoceratodus forsteri* and *Scleropages* are the only native primary freshwater fishes found in Australia.

Other osteoglossomorphs include the African freshwater butterflyfish *Pantodon* (Pantodontidae), sister group of the Osteoglossidae; the North American Hiodontidae, the Goldeye *Hiodon alosoides* and the Mooneye *H. tergisus*; the African knifefishes (Notopteridae), and the African Mormyriformes, Mormyridae (elephantfishes), the largest family of osteoglossomorphs with about 200 species; and the monotypic Gymnarchidae (*Gymnarchus niloticus*). Archaic fish distributions are summarized in Fig. 19.18.

More Recent Distributions

Five groups of (mostly primary) freshwater fishes have more recent distributions than the six archaic groups. These include the pickerels and relatives, the darters and perches, the sunfishes, the cichlids, and the Ostariophysi.

The order Esociformes contains the family Esocidae with the pikes and pickerels (*Esox*) including 7 species from North America and Eurasia, and two mudminnow (*Umbra*) species from eastern and western United States – and one relict species, *Umbra krameri*, from the Danube River in Europe (Fig. 19.19). The mudminnows, *Umbra*, had been considered a separate family, Umbridae, within the Esociformes but are now included in the Esocidae (see Page et al. 2013). The Northern Pike, *Esox lucius*, ranges across northern North America, Europe, and Asia, giving it the broadest natural distribution of any freshwater fish in the northern hemisphere.

Three families of acanthopterygian fishes are of major importance in fresh waters. The Percidae, the perches and darters, includes over 200 species, a few of which are Palearctic (European) and more than 160 of which are Nearctic (American), including the darters, subfamily Etheostomatinae, with over 200 species endemic to North America (Fig. 19.20). The black basses and sunfishes (Centrarchidae) include 34 species, 33 from eastern North America and one relict species, *Archoplites interruptus*, from California (Fig. 19.21). The Cichlidae ecologically replaces the Centrarchidae and Percidae in Central and South America, Africa, Madagascar, and southern India (Fig. 19.21). This large family, with more than 2000 species, is usually considered a secondary freshwater family because some species show salinity tolerance. Global distributions of the esocids, percids, centrarchids, and cichlids are summarized in Figure 19.22.

The largest group of freshwater fishes is the series Otophysi of the superorder Ostariophysi, which includes four orders: Cypriniformes, Characiformes, Siluriformes, and Gymnotiformes.

The Cypriniformes includes three large and two small families found primarily in the northern continents. The Cyprinidae, the carps and minnows, is one of the largest families of freshwater fishes, with about 2000 species. It is found in North America, Africa, Europe, and Asia (Fig. 19.23). The highest diversity of cyprinids is found in Asia. The 78 species of suckers, Catostomidae, are confined to North America, except for a relict genus in China, *Myxocyprinus*, and a recent reinvasion

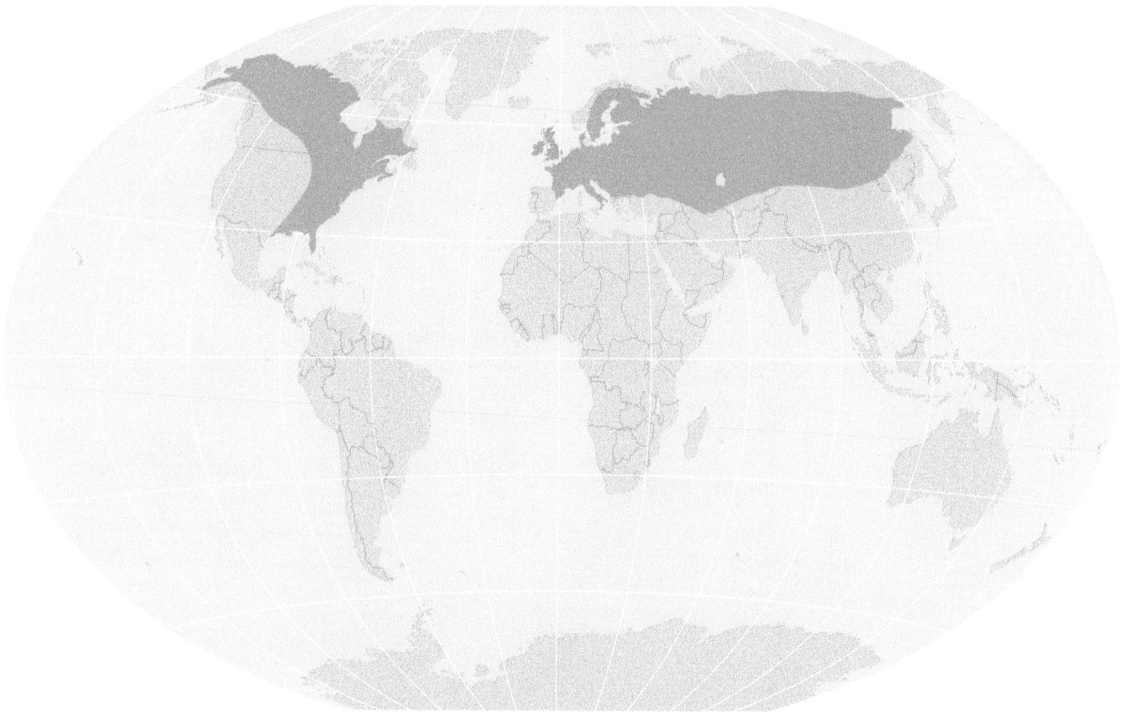

FIGURE 19.19 Distribution of the pickerels, Esocidae. Adapted from Lagler et al. (1977).

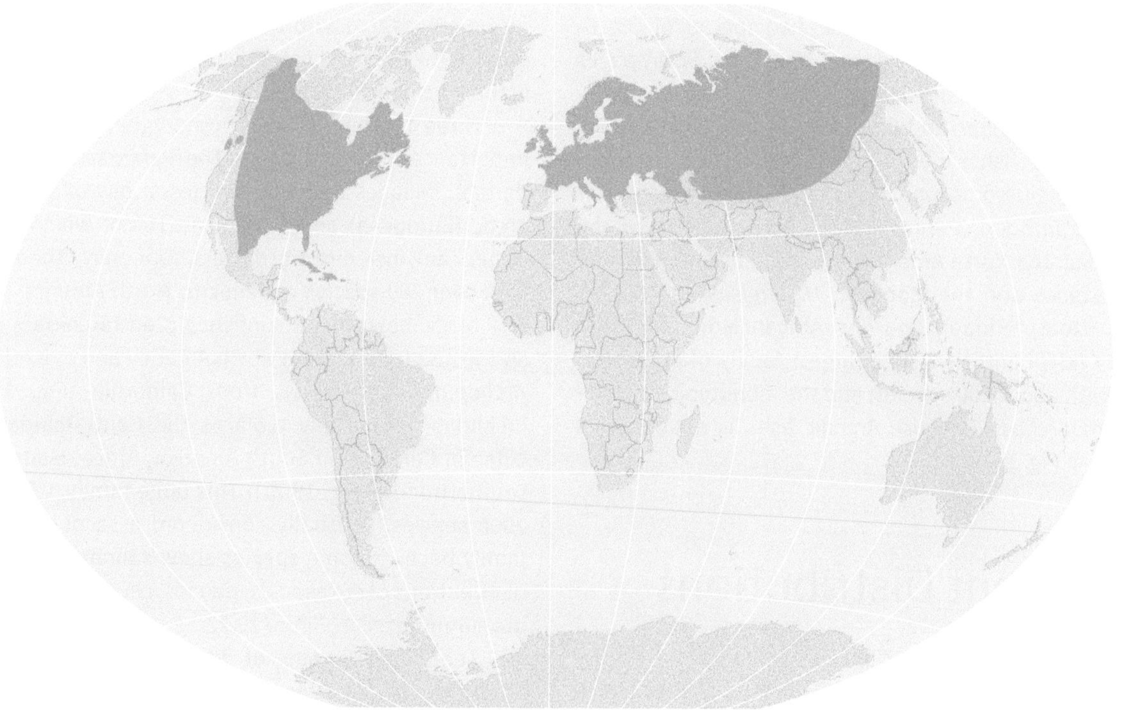

FIGURE 19.20 Distribution of the darters and perches, Percidae. Adapted from Norman and Greenwood (1975).

of Siberia from North America by *Catostomus catostomus* (Fig. 19.24). The loaches, Cobitidae, include about 260 species found in Eurasia. The two smaller families, the Gyrinocheilidae and the Balitoridae, occur in Southeast Asia.

The Characiformes, the characins, comprises 18 families of tetras and relatives, including the famous pirhanas and the diminutive tetras population in the aquarium trade. The greatest diversity of the order is in South America, with 200 genera and over 1000 species. Characins are also widespread in Africa, with over 200 species, currently placed in four families.

The Gymnotiformes comprises five families and more than 250 species, with over 100 species of electric fishes restricted to the Neotropic region (South America).

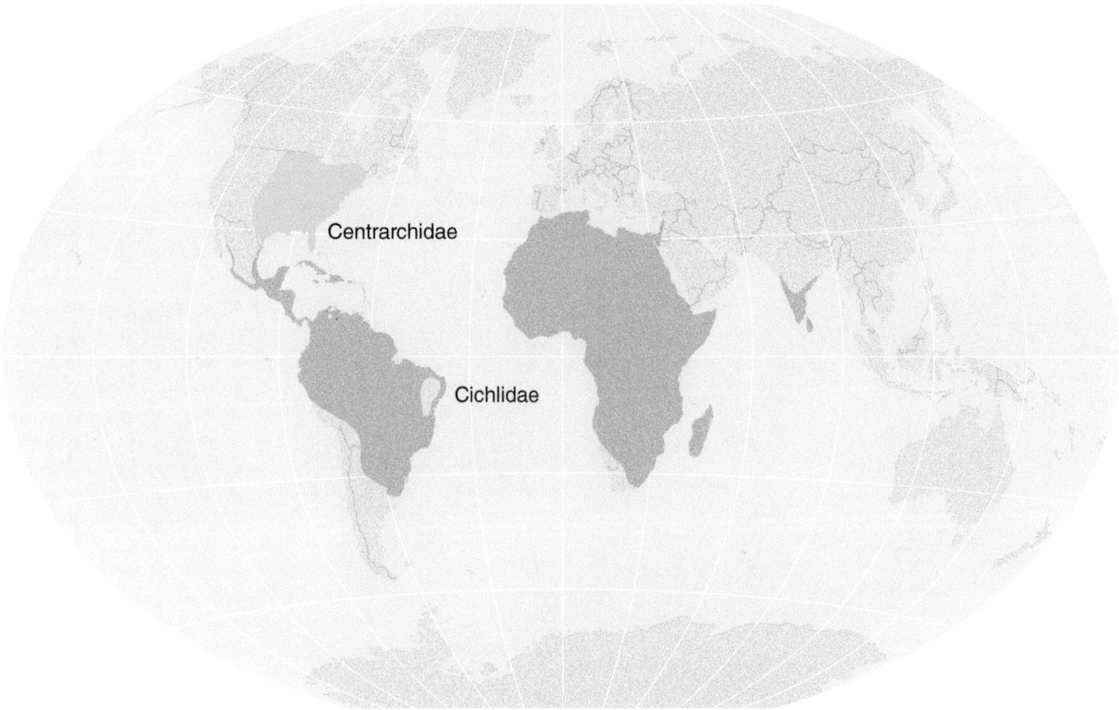

FIGURE 19.21 Distribution of the sunfishes (Centrarchidae) and cichlids (Cichlidae). Adapted from Lagler et al. (1977).

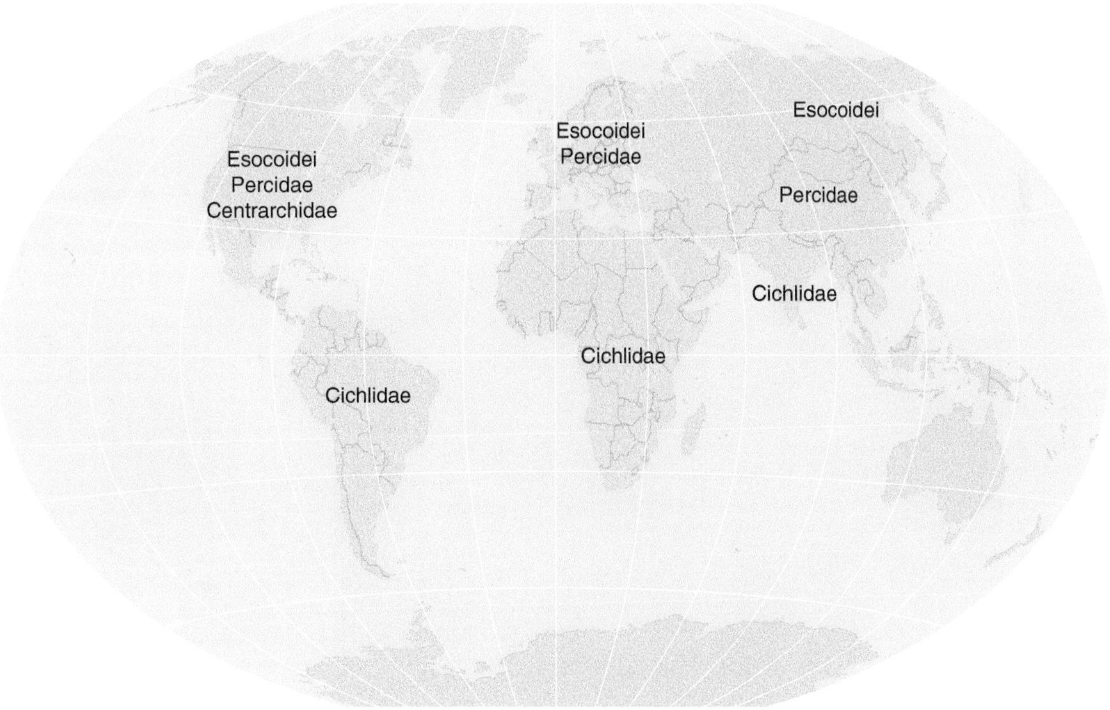

FIGURE 19.22 Summary of recent distributions of primary freshwater fishes of the Centrarchidae, Esocoidei, and Percidae, and the Cichlidae which is considered a secondary freshwater family.

The catfishes, order Siluriformes, include 36 families of diverse fishes and over 3000 species. There is only one freshwater family in North America, the Ictaluridae, including *Ictalurus*, *Ameiurus*, and madtoms of the genus *Noturus*. There are at least 14 endemic families and more than 1200 species in South America (Lundberg 1993). Some of the families are the suckermouth catfishes, Loricariidae and Astroblepidae, the popular aquarium fishes in the Callichthyidae, and the parasitic catfishes in the Trichomycteridae. Africa has six freshwater siluriform families with about 400 species. Europe has only the Siluridae with two species of *Siluris*, including the huge *S. glanis*, reaching 5 m in length and 300 kg. There are several

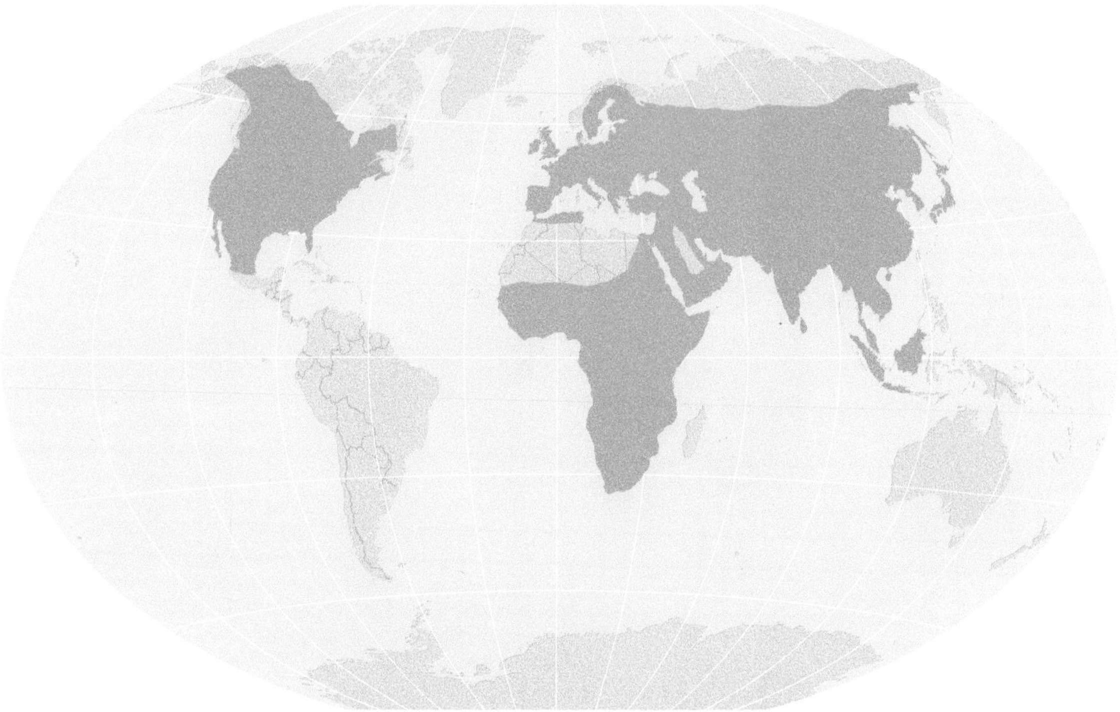

FIGURE 19.23 Distribution of the minnows and carps, Cyprinidae. Adapted from Lagler et al. (1977).

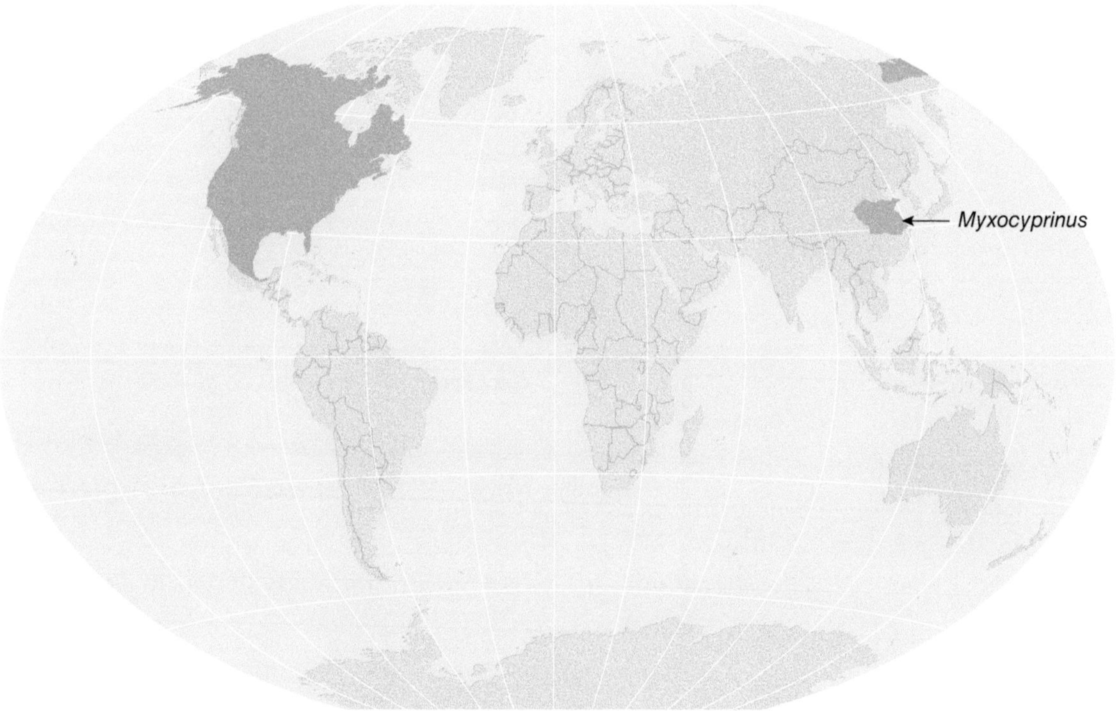

FIGURE 19.24 Distribution of the suckers, Catostomidae. Adapted from Lagler et al. (1977).

families in Asia. The catfishes also contain two marine families, Ariidae and Plotosidae, making them exceptions to the primary freshwater fish nature of the Siluriformes. To further complicate the issue, the Plotosidae has secondarily invaded fresh waters of Australia and New Guinea. The dominance of otophysans among primary freshwater fishes is summarized in Fig. 19.25.

Similarities Between South American and African Freshwater Fishes

The primary freshwater fishes of South America and Africa are remarkably similar. The Dipnoi and Osteoglossomorpha among archaic fishes link South America, Africa, and Australia. The more recent distributions of the Characiformes and

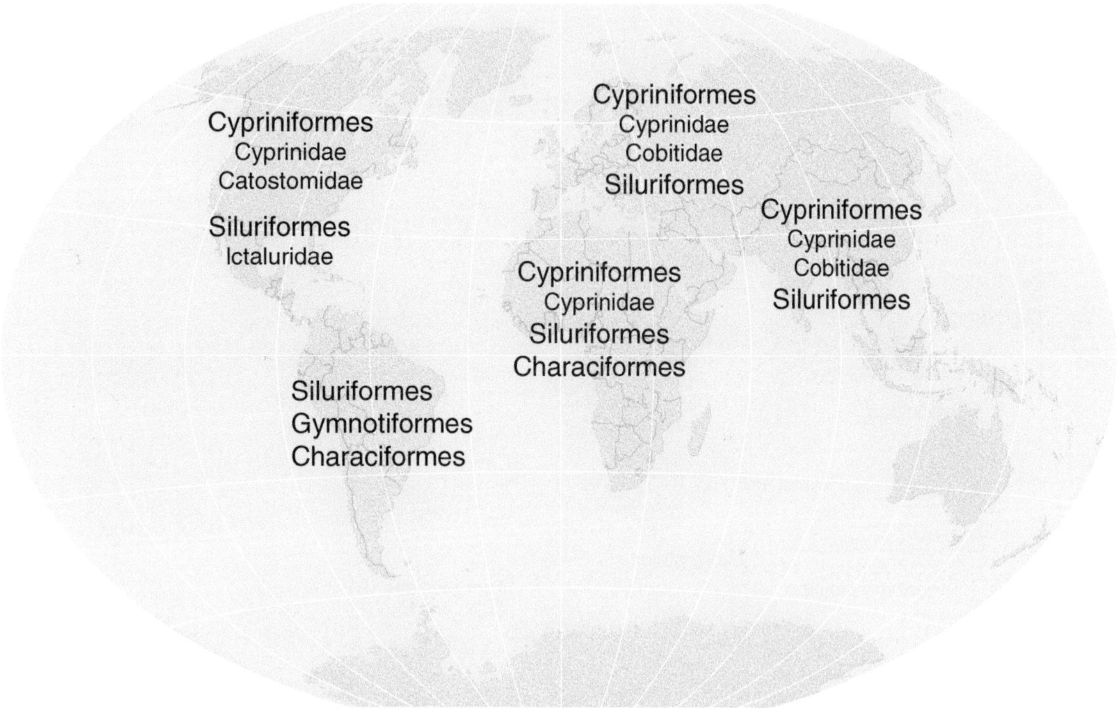

FIGURE 19.25 Summary of otophysan distribution.

Cichlidae also link South America and Africa. A question arising from these parallels is whether this similarity is due to dispersal or vicariance. Some researchers believe that the Ostariophysi originated in Southeast Asia and dispersed to South America from Africa across a direct land bridge. However, land bridges work particularly poorly for freshwater fishes because rivers seldom run lengthwise along such bridges. Other researchers favor dispersal of Ostariophysi through North America, but this is not supported by recent distributions or by fossil evidence. Among 13 putative African–South American clades, only three clades – lepidosirenid lungfishes, polypterid bichirs, and doradoid catfishes – clearly fit the simple continental drift–vicariance model, with a common ancestor inhabiting fresh waters of the African–South American landmass before the opening of the Atlantic Ocean (Lundberg 1993). Distributions of other groups, such as cichlids and characins, are more difficult to explain this way because they have evolved more recently, so perhaps some dispersal has taken place in addition to vicariant events.

An explanation for the distribution of a freshwater galaxiid in South America and Australia is a good example of the argument over dispersal versus vicariance. *Galaxias maculatus* is a small diadromous fish with a highly disjunct distribution in streams in eastern and Western Australia, New Zealand, South America, and some oceanic islands (Berra et al. 1996). Rosen (1978) considered galaxiid fishes to be part of a pan-austral Gondwanan biota that was fragmented by the movement of the southern continents during the Mesozoic, and therefore concluded that the present distribution resulted from vicariant events. *Galaxias maculatus* is the only galaxiid that breeds in brackish not fresh water, and its transparent whitebait larvae grow in the ocean before returning to fresh water. McDowall (1978), therefore, found it simpler to accept a relatively recent oceanic dispersal by a fish with a marine juvenile stage. Recent allozyme electrophoresis shows that *G. maculatus* is a single species with surprisingly little genetic variation across its entire range (Berra et al. 1996). These data tend to support a dispersal hypothesis in this case, but mitochondrial DNA analysis is needed to further test this hypothesis.

Middle American Freshwater Fishes

The freshwater fish faunas of North and South America are very different. Minnows, suckers, ictalurid catfishes, darters, and sunfishes predominate in the north, whereas cichlids, characins, gymnotoids, and a wide array of different catfish families are common in the south. What happens in the region between North and South America, in Middle America?

Middle America acts partly as a **filter barrier** slowing the movement of North American fishes south and South American fishes north, thus allowing invasion by marine groups. A few representatives of North American families such as Ictaluridae, Catostomidae, and *Lepisosteus* extend south to Costa Rica (Fig. 19.26). A few representatives of South American families such as Cichlidae (*Cichlasoma*) and Characidae (*Astyanax*) extend north to the Rio Grande. Of the approximately 456 species of freshwater fishes in Central America, over 75% comprise secondary freshwater fishes

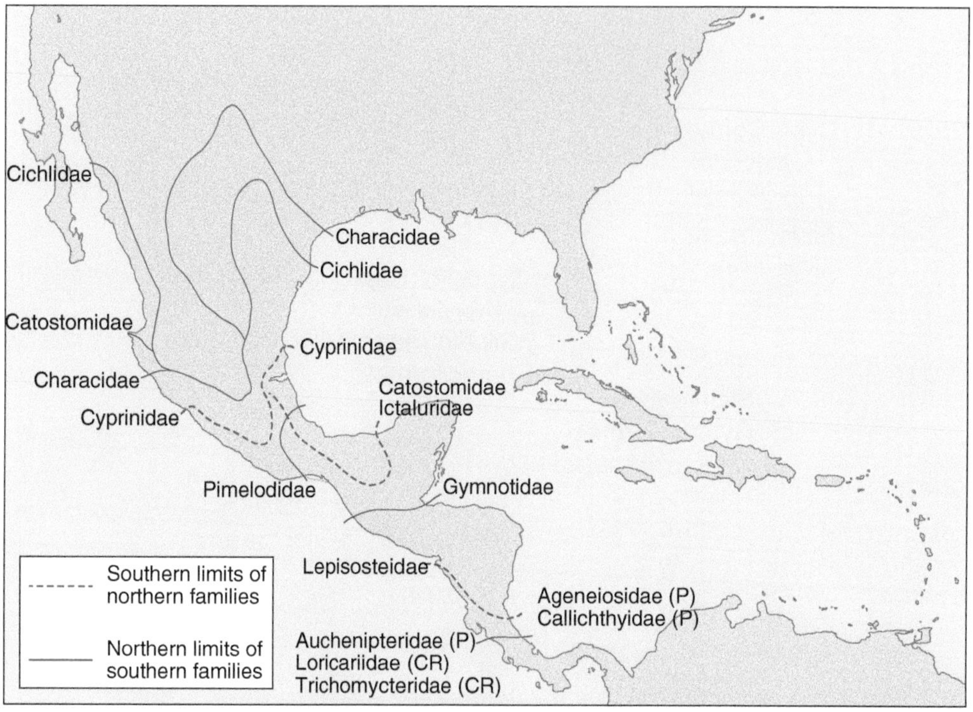

FIGURE 19.26 Distributional limits of certain primary and secondary freshwater fishes in Central America. A single characin (*Astyanax mexicanus*) and a single cichlid (*Cichlasoma cyanoguttatum*) reach north to the Rio Grande. CR, Costa Rica; P, Panama. Adapted from Miller (1966, fig. 1).

such as Cyprinodontidae, Poeciliidae, Cichlidae, and marine invaders, and only 104 species are primary freshwater fishes (Miller 1966).

The general pattern of contributions from both major regions and of many secondary freshwater fishes also holds for more localized faunas in the region. For example, the **Usumacinta Province**, comprising the Grijalva and Usumacinta rivers of Guatemala and Mexico, contains a mix of North American and South American fishes totaling over 200 species (Miller 1966). Two secondary groups, the cyprinodontoids and the cichlids, comprise about 90 species. There are also a large number of marine derivatives (18 species in nine families). Among these are a few endemics, including species from three marine families, Belonidae (*Strongylura hubbsi*), Hemiramphidae (*Hyporhamphus mexicanus*), and Batrachoididae (*Batrachoides goldmani*). Freshwater species of these three families occur elsewhere, but what factors have led to this remarkable parallel derivation of endemic freshwater species? Part of the explanation is due to the depauperate nature of the primary freshwater fishes of the region, but this is true of many other rivers south through Panama. More information is needed on the nature of the water in the Usumacinta Province. Is it high in ions, as is true of southern Florida fresh waters that also contain a number of marine species (but not endemics derived from marine species)? Have there been historical factors or vicariant events involved? Here is an interesting problem involving phylogeny, biogeography, and physiology that awaits solution.

Phylogeography of Freshwater Fishes

Fishes cannot get out of water and travel over dry land to the next drainage, with rare exceptions such as "walking catfish" in the genus *Clarias*. For most freshwater fishes, opportunities for dispersal are few, the genetic differences between drainages are high, and vicariance models generally work well to explain evolutionary patterns. Geographic and oceanographic barriers can explain the majority of **sister species** (species that are each other's closest relative) relationships, although differences in freshwater characteristics (such as the Andes-derived white water, and the lowland-derived black water of the Amazon) may be a factor as well. Looking beyond these geographic and oceanographic barriers, four primary factors shape the phylogeography of freshwater fishes:

1. Changes in drainage routes. Stream captures are the most widely studied phenomenon, where erosion, earthquakes, or other geographic changes divert a stream from one drainage to another drainage. The high diversity of freshwater fishes in central North America may be due in part to streams in the Appalachian Mountain Range that switched from flowing toward the Atlantic coast to flowing toward the Mississippi River (Hocutt & Wiley 1986). Flooding, which in essence causes temporary stream capture, may also transfer fishes between drainages.

2. Glaciation. During these "ice ages" the temperate fishes (distributed between 25° and 65° latitude) were massively displaced by the cooling and advancing glaciers. At the end of each glacial epoch, enormous **proglacial lakes** formed at the retreating edge of ice sheets, some larger than the contemporary Great Lakes of North America. These large and shifting water masses provided extensive opportunities for dispersal (Hocutt & Wiley 1986; Behnke 1992).
3. Coastal opportunities for dispersal. Some freshwater fishes are tolerant of high salinity conditions and can survive for extended periods (days or weeks) in coastal waters. For example, the freshwater cichlids are members of the suborder Labroidei that includes surfperches, damselfishes, wrasses, and parrotfishes, all marine groups (Streelman & Karl 1997; see Chapter 15). Hence it is no surprise that some cichlid species can tolerate salt water. The other coastal opportunity for dispersal occurs during periods of heavy rainfall. Chesapeake Bay is a 300 km long estuary that usually contains ocean water at one end and fresh water at the other. However, in the aftermath of hurricane events, fresh water extends out of the mouth of the bay, as do the freshwater fishes that are usually confined to individual rivers.
4. Plate tectonics, wherein the movements of continents can separate or join populations of freshwater fishes.

Here, we briefly examine three of these phenomena, whereas the fourth (coastal dispersal) is covered in the following section (Anadromous fishes).

Reconstructing Stream Captures

The South Island of New Zealand has two primary drainage systems, the Clutha in the north and Southland in the south. These two historically isolated drainages retain distinct faunas: a phylogeographic survey of galaxiid mudfishes reveals a number of **cryptic species** among taxa that were previously believed to span both drainages (Waters et al. 2001). Furthermore, these mtDNA studies have demonstrated two stream captures in this glacially influenced region. The older stream capture involved the Nevis River, which changed course from a southern to a northern drainage system, introducing a lineage of the mudfish *Galaxias gollumoides* that is characteristic of the Southland drainage (Waters et al. 2001). Molecular clock estimates indicate an ancient colonization, on the order of 300,000–500,000 years ago. In contrast, geological studies indicate that the Von River changed course (also from south to north) during the most recent glacial interval, about 12,000 years ago. The corresponding mtDNA survey indicates the presence of another mudfish derived from the Southland Drainage (Burridge et al. 2007). These colonizations of the northern drainage are significant, as they comprise two of the nine native freshwater fishes.

Northwest South America has a complex geological history that includes the uplift of the central Andes during the Miocene (5–23 mybp). This tectonic event isolated the Pacific drainages from the central Amazonian basin, a shift that should be detectable in the phylogeography or regional fishes. Rincon-Sandoval et al. (2019) examined five Ostariophysi species with complete mtDNA genomes and over 1000 nuclear loci to document river shifts and headstream captures that would be difficult to demonstrate with geological data.

Glacial Eradication and Recovery

The most recent glaciation affected North America severely, with a greater ice sheet than the Asian and European glaciations combined. This ice sheet reached as far south as the 44° latitude (where Toronto, Ontario and Bangor, Maine are today) about 23,000 years ago, followed by deglaciation 15,000 to 8000 years ago. The current distributions of several salmonids, in particular Lake Trout (*Salvelinus namaycush*) and Lake Whitefish (*Coregonus clupeaformis*), were almost completely covered by this ice sheet. These species, now broadly distributed from Alaska to the Atlantic drainages, must have persisted in refugia (perhaps at the fringe of their current range) for thousands of years. Hence patterns of genetic diversity can help to identify glacial refugial and recolonization pathways.

Bernatchez and Dodson (1991) used mtDNA to resolve four major lineages among populations of Lake Whitefish. These correspond to refugia in northern Eurasia, Beringia (Siberia-Alaska), the Mississippi valley, and perhaps two Atlantic locations (Fig. 19.27). The Mississippi haplotypes occupy the majority of the current range of Lake Whitefish, observed from New York to the Yukon. This is consistent with other fish distributions in indicating that the Mississippi fauna had the greatest opportunities for dispersal through proglacial lakes.

In a review of phylogeographic studies for 42 North American freshwater fishes, Bernatchez and Wilson (1998) observed a significant decline in mtDNA nucleotide diversity with increasing latitude, a clear indication of southern refugia during the last glacial period. Regression analysis indicates a steep (five-fold) drop in diversity from 25° to 46°N latitude, then consistently low diversity from 46° to 65°N (Fig. 19.28). It is remarkable that the analysis indicated the 46°N boundary for reduced genetic diversity, closely paralleling the southern limit of North American glaciers at 44°N. Almost universally, North American freshwater fishes are genetically depauperate in the deglaciated areas above the 44° to 46°N boundary. A similar pattern is apparent in Europe. Of the five mtDNA lineages observed in European Brown Trout (*Salmo trutta*), only one has colonized previously glaciated areas (Bernatchez 1995).

Plate Tectonics: The Mystery of the Asian Arowana (*Scleropages formosus*)

The distribution of arowanas (Osteoglossomorpha) has been a longstanding biogeographic mystery, as these primary freshwater fishes occur on four continents that are isolated

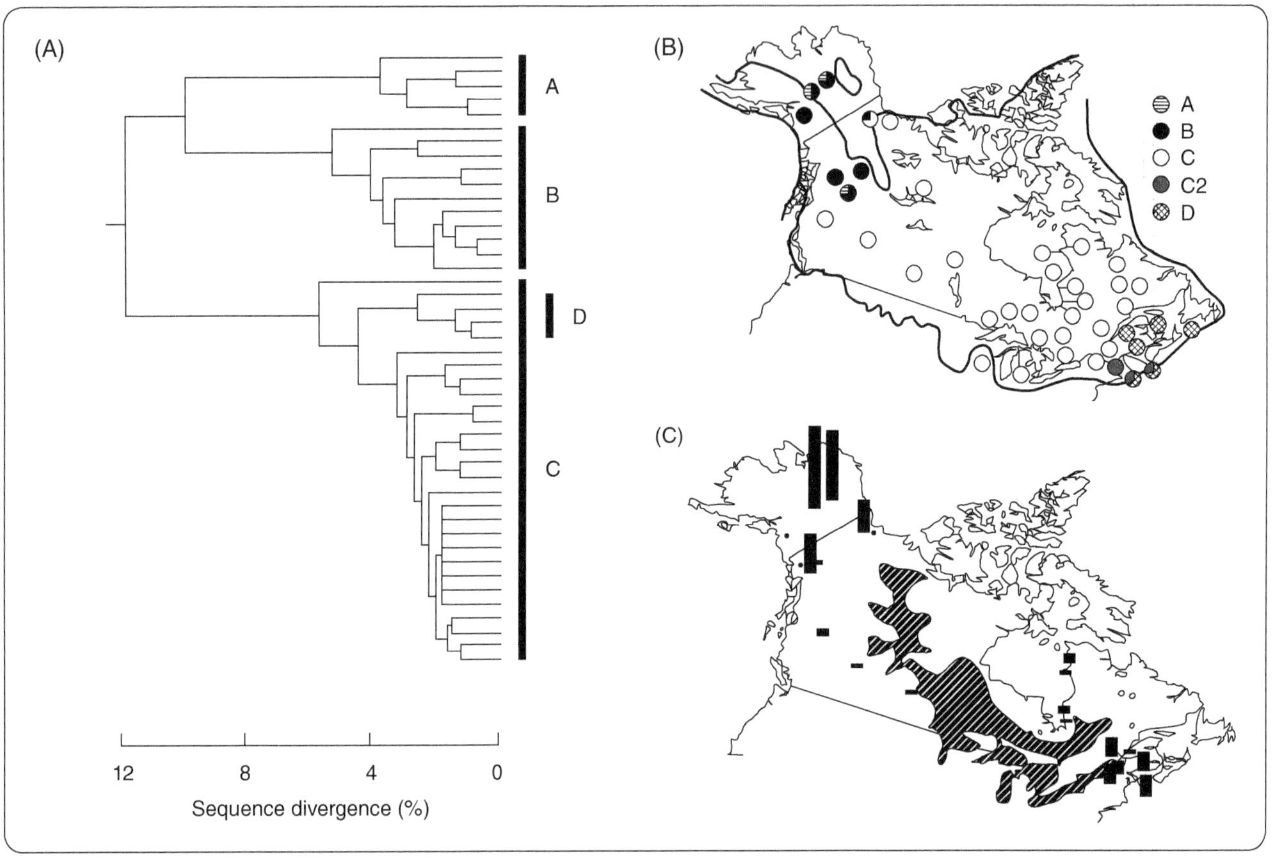

FIGURE 19.27 Phylogeographic data for the Lake Whitefish (*Coregonus clupeaformis*) based on mtDNA sequence data from 41 populations across the species range. (A) The phylogeny of Whitefish lineages corresponding to four glacial refugia. The scale bar indicates sequence divergence. (B) Distribution of the four lineages (A, B, C and C2, D) following postglacial dispersal. (C) Nucleotide diversity of sampled areas, in relation to the area formerly inundated by major glacial lakes (shaded area); the height of bars indicates the level of nucleotide diversity. From Bernatchez and Wilson (1998), used with permission.

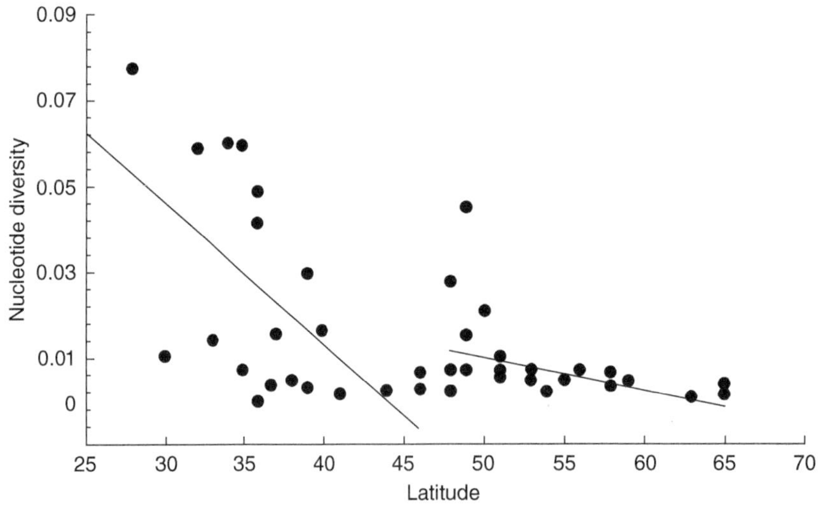

FIGURE 19.28 The relationship between nucleotide diversity and latitude for North American freshwater fishes, showing a general trend of reduced genetic diversity in areas that were under glacial ice. Two lines with two different slopes fit the data, with the break between the lines corresponding closely to the 46°N latitude, near the southern limit of the most recent glaciation. Each dot represents a different species. From Bernatchez and Wilson (1998), used with permission.

by formidable marine barriers (see Fig. 19.17). Their distribution in South America, Africa, and Australia can be explained by the breakup of the southern supercontinent Gondwana (including South America, Africa, Antarctica, Madagascar, and India) about 150 mybp. However, the distribution in Southeast Asia is hard to explain without a marine dispersal event from Australia. Despite the strict freshwater requirements of arowanas, this was the favored explanation until recently. The Australia to Asia dispersal hypothesis was given further support by taxonomic studies based on morphology, which united in one genus the Asian *Scleropages formosa* with the Australian *S. jardinii* and *S. leichardtii*.

Based on a molecular clock for two mtDNA genes calibrated with several bony fishes, Kumazawa and Nishida (2000) estimate that the Asian and Australian arowanas actually diverged about 140 mybp. This timeframe coincides with the separation of India from the southern supercontinent, and subsequent transport into the northern hemisphere. India connected with Asia by about 40 mybp, and may have allowed the colonization of Asia at that time. This possibility is supported by the presence of fossil *Scleropages* in Sumatra, dating to the Eocene (35–57 mybp). Hence the biogeographic mystery of the arowanas unraveled when molecular clock data showed that the *Scleropages* species in Asia and Australia diverged in the Early Cretaceous, much farther back than is typical for members of the same genus.

Anadromous Fishes

Anadromous fishes typically show strong site fidelity when they return from the ocean to natal streams to spawn, but with an error rate that is high enough to allow colonization of adjacent rivers. Anadromous salmon are the subject of much scientific interest because their life history (especially the site fidelity of spawning adults) is a compelling focus for ecological and genetic studies, and because of the wildlife management and conservation issues associated with salmon fisheries (see Chapter 22). Life history and population genetic studies provide a scientific foundation for resolving stocks, populations (in the genetic sense), and evolutionary significant units (ESUs, see Chapter 20, Population Genetics).

The seven species of anadromous salmon and trout (genus *Oncorhynchus*) on the Pacific coast of North America are the most widely studied in the world, with extensive allozyme, mtDNA, and microsatellite inventories. Here there is a management mandate to define "distinct population segments" under the US Endangered Species Act, using ecology, life history, and genetics (Waples et al. 2001). Note that since genetics is not the sole criterion, differences in behavior (especially the timing of migrations to and from the ocean) are incorporated as well.

Currently the salmon and trout in US waters are divided into 12 ecologically distinct regions (Fig. 19.29), mostly corresponding to major tributaries and drainages. Within these 12 regions are a total of 58 designated ESUs (Table 19.2). Chinook

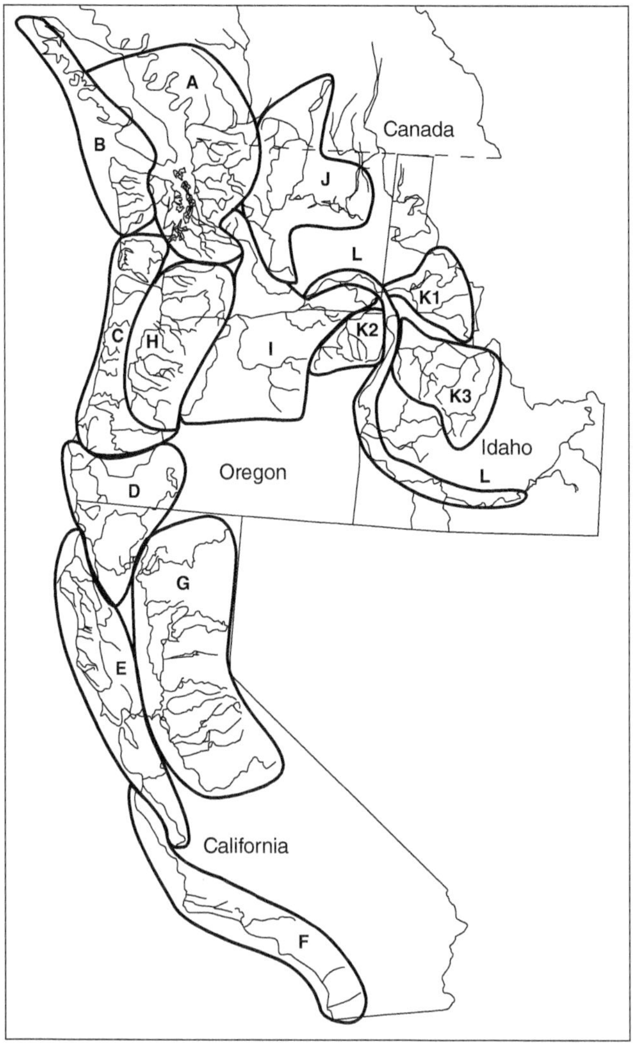

FIGURE 19.29 Drainage basins of western North America that support at least one spawning population of Pacific salmon (see Table 19.2). Letter codes correspond to the following ecosystems: A, Georgia Basin; B, temperate rainforest; C, north coast; D, Klamath Mountains; E, northern California; F, southern California; G, Central Valley; H, Willamette/Lower Columbia River; I, mid-Columbia River; J, upper Columbia River; K, Snake River tributaries; L, mainstem Snake River. From Waples et al. (2001), used with permission.

Salmon (*O. tshawytscha*) and Steelhead/Rainbow Trout (*O. mykiss*) have the widest range and the most subdivisions, with 17 and 15 ESUs, respectively. Chinook, Steelhead, Sockeye (*O. nerka*), and Coastal Cutthrout (*O. clarki clarki*) also have considerable diversity in life history, particularly in the timing of smolting and spawning migrations, and all but Chinook have non-anadromous, "landlocked" populations. Some anadromous Sockeye populations use lakes for juvenile development, whereas others use a riverine habitat or migrate quickly to the sea. In contrast, Chum (*O. keta*), Pink (*O. gorbuscha*), and Coho Salmon (*O. kisutch*) are relatively inflexible in life history traits, including a very brief freshwater stage in the first two species (Groot & Margolis 1991; Quinn 2005).

Three species have anadromous and non-anadromous (remaining in fresh water) forms that inhabit the same drainages:

TABLE 19.2 **The number of management units and evolutionary significant units (ESUs) within each species of salmon on the West coast of the United States. Note that management units can be defined with ecology and life history, as well as shallow population genetic structure. The higher level designation of ESU requires deep population structure or other evidence of evolutionary divergence. From Waples et al. (2001).**

Species	Management units				ESUs
	Genetics	Ecology	Life history	Total	
Pink	2	2	1	5	2
Chum	2	4	1	7	4
Sockeye	9	4	6	19	7
Coho	2	6	1	9	7
Chinook	10	11	7	28	17
Steelhead	7	11	7	25	15
Cutthroat	3	6	2	11	6

Sockeye/Kokanee, Steelhead/Rainbow Trout, sea-run/freshwater Cutthroat Trout. The genetic surveys demonstrate that in each case, the two forms within each drainage are closely related populations, relative to populations with the same behavior in other drainages (Foote et al. 1989; Utter et al. 1989). These life history variants, including the option of remaining in fresh water, arose independently in each of the drainages.

The exception to this pattern is the Chinook Salmon in the Columbia and Snake Rivers, where deep genetic partition exists between stream-maturing and ocean-maturing forms (Myers et al. 1998). Spring-spawning salmon of the stream-maturing type co-occur with fall-spawning salmon of the ocean type, but they do not interbreed, another example of complex population structure (discussed further in Chapter 20).

Pink Salmon are hard wired to a 2-year breeding cycle, so that even-year spawners never encounter odd-year spawners. The result is two distinct ESUs in a single species that inhabit the same feeding areas, mate in the same location at the same time of year, but never interbreed (Churikov & Gharrett 2002).

Sockeye Salmon and Cutthroat Trout have strong population structure, and in both cases this structure is linked to higher dependence on freshwater habitat. The Sockeye is anadromous but requires a lake habitat for juvenile development, whereas the Cutthroat is non-anadromous throughout much of its inland range. In the populations that have access to the ocean, many individuals stay in fresh water, and sea-running individuals do not migrate far from their river of origin (Johnson et al. 1999). Hence Cutthroat may be described as the least anadromous member of the genus *Oncorhynchus* in western North America. Freshwater fishes typically have more population structure than anadromous fishes, which have more structure than marine fishes (discussed further in Chapter 20). It makes sense, therefore, that the least anadromous fish would have the highest genetic structure (Waples et al. 2001).

Finally, the conservation genetics of these Pacific salmon are a matter of much concern. Perhaps 30% of the ESUs (corresponding to 27% of the genetic diversity) that existed prior to European contact are extinct, and another third are listed as endangered or threatened (Gustafson et al. 2007). Much has been lost, but enough remains to rebuild these stocks under prudent management regimes. However, widespread aquaculture of salmon has the potential to genetically compromise native populations. In addition, interbreeding with fish that escape from captivity has the potential to introduce maladaptive genetic traits into local populations (Utter 2004).

Supplementary Reading

Zoogeography

Albert J, Reis RE. 2011. *Historical biogeography of neotropical freshwater fishes*. Berkeley: University of California Press.

Briggs JC. 1974. *Marine zoogeography*. New York: McGraw-Hill.

Briggs JC, Bowen BW. 2012. A realignment of marine biogeographic provinces with particular reference to fish distributions. *J Biogeogr* 39:12–30.

Humphries CJ, Parenti LR. 1999. *Cladistic biogeography*, 2nd edn. New York: Oxford University Press.

Lomolino MV, Riddle BR, Whittaker RJ. 2017. *Biogeography*, 5th edn. Sunderland, MA: Oxford University Press.

Nelson G, Platnick N. 1981. *Systematics and biogeography. Cladistics and vicariance*. New York: Columbia University Press.

Phylogeography

Avise JC. 2000. *Phylogeography, the history and formation of species.* Cambridge, MA: Harvard University Press.

Avise JC. 2004. *Molecular markers, natural history, and evolution,* 2nd edn. Sunderland, MA: Sinauer Associates.

Bermingham E, Martin AP. 1998. Comparative mtDNA phylogeography of neotropical freshwater fishes: testing shared history to infer the evolutionary landscape of lower Central America. *Mol Ecol* 7:499–517.

Bowen, BW, Gaither MR, DiBattista JD, et al. 2016. Comparative phylogeography of the ocean planet. *Proc Nat Acad Sci USA* 113:7962–7969.

Floeter SR, Rocha LA, Robertson DR, et al. 2008. Atlantic reef fish biogeography and evolution. *J Biogeogr* 35: 22–47.

Rocha LA, Craig MT, Bowen BW. 2007. Phylogeography and the conservation genetics of coral reef fishes. *Coral Reefs* 26:501–512.

Journals

Diversity and Distributions. A journal of conservation biogeography. Blackwell Publishing.

Global Ecology and Biogeography. A journal of macroecology. Blackwell Publishing.

Journal of Biogeography. Blackwell Publishing.

Website

http://research.amnh.org/ichthyology/congo/.

CHAPTER 20

Fish Populations

Summary

In this chapter, we address fish populations – individuals of the same species that share the same resources and gene pool. The success of populations over time depends on the development of life history strategies that are successful for the local or regional environmental conditions.

Life history strategies result from differences in the allocation of energy and resources to the often conflicting demands of maintenance, growth, migration (if applicable), and reproduction. Large size is advantageous in fishes because larger fishes produce more eggs and escape more predators. Reproduction at an early age and small size incurs a substantial cost in future reproduction; delayed reproduction means more eggs or sperm produced, but at the risk of dying before spawning. Theory accurately predicts the effects of mortality on reproductive age, size, interval, and allotment; individuals in populations with high adult mortality reproduce earlier and have higher fecundity and shorter reproductive intervals. As a result, commercial and sport harvesting of large fish can create evolutionary pressure toward smaller fish in subsequent generations, referred to as fisheries-induced evolution (FIE). Behavior of some individuals within a population makes them more susceptible to different harvesting methods, and studies show that behavior is at least in part heritable. Commercial or sport fishing can, therefore, selectively remove individuals based on their

The Diversity of Fishes: Biology, Evolution and Ecology, Third Edition. Douglas E. Facey, Brian W. Bowen, Bruce B. Collette, and Gene S. Helfman.

Companion website: www.wiley.com/go/facey/diversityfishes3

behavior, raising the concern that strong fishing pressure may affect the range of behaviors expressed in future generations.

Populations increase and decrease as a result of age-specific reproduction and survivorship rates. Because of dispersing larvae, migration into populations (recruitment and colonization) has a strong influence on year-class strength. Different age fishes differ substantially in size and feeding habits, making cannibalism a frequent cause of mortality. Production is a measure of how much biomass a population produces yearly and is important in determining sustainable exploitation rates for commercial fishes. Most fish populations produce <10 g/m²/year, most of which occurs in younger age classes.

Population genetic structure, the level of isolation between populations of the same species, is highest in freshwater fishes and marine fishes that lack a pelagic larval stage. Among fishes with a pelagic stage, there is no simple relationship between the length of the pelagic stage and the extent of dispersal. Ecosystem specialists, with highly restricted habitat or feeding, tend to have higher population structure than generalists. Comparisons of maternally inherited mitochondrial DNA (mtDNA), and biparentally inherited nuclear DNA (nDNA) can reveal differences in dispersal between males and females, a common outcome in migratory marine fishes, especially sharks.

Many populations are relatively isolated from other populations of the same species, which allows for genetic differentiation. Genetically distinct populations can exist in neighboring lakes (e.g. whitefishes, sticklebacks), and Pacific salmons often occur as genetically isolated stocks in adjacent rivers. Genetic differences without apparent geographic separation occur within populations of Arctic Char. In contrast, hybridization between species results when species-specific spawning habitat is unavailable or degraded, or when disproportionate numbers of one species exist. Hybridization is more common among freshwater fishes.

Population Ecology

Life Histories and Reproductive Ecology

Ecological adaptations are traits of an individual that ensure its survival and reproduction in response to selection pressures from the biotic and abiotic environment. It is important to emphasize that natural selection operates at the level of the individual, favoring individuals of one genotype while selecting against individuals with less favorable genotypes. We can then ask if survival and reproduction are enhanced by how an individual selects an appropriate habitat in which to live (discussed in the next chapter), how it budgets its time and energy among the activities and conflicting demands presented on a daily basis, and how it eventually partitions energy into growth versus reproduction.

A **life history** can be viewed as how individuals divide up their time and resources among the often-conflicting demands associated with maintenance, growth, reproduction, mortality, and migration. Life history characteristics or **traits** are measurable aspects of an individual's life history and include age- and size-specific birth rates (and associated characteristics such as clutch size, egg size, offspring provisioning, and clutch frequency), and the probabilities of death and migration (Congdon et al. 1982; Dunham et al. 1989). These traits vary among species, among populations within a species, and among individuals and sexes within a population in ways that make evolutionary sense, indicating adaptive responses to selection. The challenge to biologists is to identify trends in life history traits, identify the likely selection pressures causing variation, and interpret the adaptiveness of the variation (Potts & Wootton 1984; Stearns 1992; Winemiller & Rose 1992; Matthews 1998). Many life history traits are correlated, which means that they are inherited together and change in direct relationship with one another, making it somewhat difficult to isolate the exact interaction between environment and adaptation.

Analyses of life history traits focus on females, in part because female reproductive effort produces eggs, each of which has a much greater likelihood of becoming a new individual than is the case for the millions of sperm produced by a male. Over a dozen life history characteristics or traits have direct links with reproduction and have been identified and quantified in many fishes. Background detail on reproductive biology and anatomy is presented in Chapter 8. A thorough treatment of life history characteristics can be found in Matthews (1998).

1. *Age and size at maturation.* A complex but fascinating trade-off exists between **early versus late maturation**; the trade-off depends on the probability of successful reproduction versus the risk of death. A female that delays maturation until she is larger and older will produce more eggs at each spawning but runs the risk of dying before she ever reproduces (Fig. 20.1). A fish that spawns at an earlier

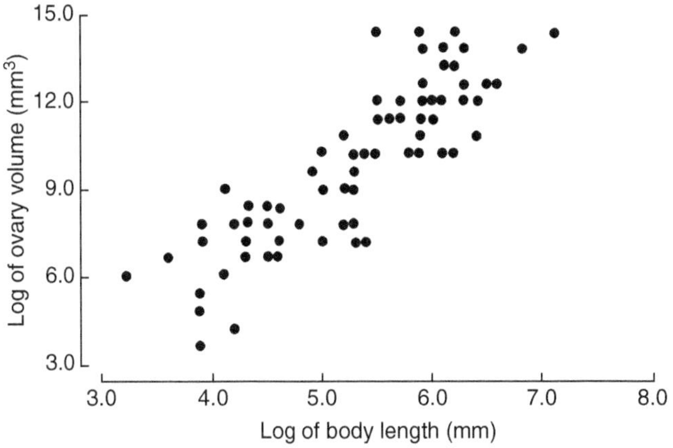

FIGURE 20.1 Bigger fish produce more eggs, both within and among species. Since fish grow throughout their lives (= indeterminate growth), older fish are usually larger. Within a species, larger individuals generally produce more eggs and often larger eggs. The generalization even holds for pregnant male seahorses, with larger individuals possessing larger pouches capable of holding more embryos (Woods 2005). Values are plotted for a variety of Canadian freshwater species. Redrawn after Wootton (1990).

FIGURE 20.2 Strong selection against large adults in a fish population, such as pressure from predators or human harvesting, can cause future generations to mature at younger ages and smaller sizes, as has been demonstrated in the well-studied Guppy (male above, females below). Per Harald Olsen / Wikimedia Commons / CC BY 3.0.

age stands a greater chance of getting some genes into the next generation at least once. However, younger fish are smaller and hence produce fewer and often smaller eggs, which lessens the chance that any will make it past egg and larval predators and starvation. Also, by allocating energy to reproduction, the earlier spawning fish has slower somatic (body) growth and is then more subject to predation because of smaller body size (Werner et al. 1983). Additionally, reproduction uses up much energy, potentially placing a smaller fish with lesser energy stores in a weakened condition, which reduces the chances of future reproduction. Theoretically, females in populations where adult survival is poor should reproduce at an earlier age than in populations where survival is better. And, as predicted, female Guppy (*Poecilia reticulata*) in downstream locales in Trinidad, where predators are abundant, mature earlier than upstream populations with fewer predators (Reznick & Endler 1982, Fig. 20.2). Similarly, individuals in commercially exploited fish populations, particularly those where adults are targeted by the fishery, often reproduce at younger ages than fish in unexploited populations (e.g., O'Brien et al. 1993; Heath & Speirs 2012; also addressed later in this chapter).

2. *Body size.* Even very large predators eat relatively small prey, but only large predators can eat large prey. Therefore, larger fishes are susceptible to predation by fewer predators while at the same time larger fish can catch and swallow a broader range of prey types. Larger fishes are also able to store more energy, swim faster and farther, and better overcome harsh abiotic conditions such as strong currents (Karr et al. 1992). Size determines territorial interactions as well as male mating success in many fishes. Also, larger fish emerge from winter with greater energy stores and in better condition than do smaller individuals (Cargnelli & Gross 1997), and larger fish produce eggs with higher hatching success and higher larval survival (Trippel 1995). This premium on large size comes, however, at a cost because energy allocated to somatic growth is unavailable for immediate reproduction.

3. *Longevity.* The longer an individual lives, the more reproductive opportunities it should have, discounted by how long it waits until first reproduction (point 1 discussed earlier) and how long an interval exists between reproductive periods (point 10 discussed later).

4. *Clutch size.* The number of eggs a female produces at each spawning varies as a function of body condition and size, age, egg size, and spawning frequency. Egg number is referred to as **fecundity** and commonly refers to the number of eggs or young produced per year, but it can be subdivided into batch, breeding season, or lifetime fecundity. Combining clutch size with egg size (point 5 discussed later) gives a measure of **reproductive allotment**, which is the percentage of a female's weight devoted to eggs or embryos. Theoretically, females in populations where adult survival is poor should devote more energy to reproduction than in populations where survival is better. Reproductive allotment in female Guppy in predator-dense populations is 30% greater than in females subject to less predation (Reznick & Endler 1982). Commercially exploited species (e.g. pike, halibut) show increased fecundities as compared to unexploited populations, a change in a life history trait which could compensate for high levels of predation or exploitation (Policansky 1993b).

5. *Egg size and size at birth.* Both the mean and range of egg size vary within species and within individuals (Matthews 1998; Wootton 1999). Eggs spawned early in a season for multiple spawners tend to be larger. In the Least Killifish, *Heterandria formosa* (Poeciliidae), the more broods a female produces, the smaller the young from each brood will be. In the Orangethroat Darter, *Etheostoma spectabile* (Percidae), offspring hatched from larger eggs are bigger and less likely to starve (Marsh 1986). Greater investment in each egg, largely in terms of amount of yolk, increases the chances of survival for that offspring. A larger larva is better able to avoid predators and feed independently (Richards & Lindeman 1987). The volume of the ovarian space in a female determines fecundity, producing an inverse relationship between egg size and number of eggs. Fish that produce larger eggs have lower batch fecundities.

6. *Time until hatching and exogenous feeding.* For egg-laying species, which is most fishes, eggs are deposited on substrates or in the water column and are essentially defenseless, either immobile or floating. After hatching, yolk-sac fry are inefficient swimmers. The longer a larva spends growing inside the egg or absorbing yolk resources, the larger it will become before having to obtain its own food. A larva trades off the increased vulnerability it experiences while being passive against the advantages it will have in finding food and avoiding predators once it achieves independence from the egg shell and yolk sac.

7. *Larval growth rate and interval length.* Rapid growth provides a larva with the same advantages as large egg size or yolk supply, namely achieving a larger size earlier. However, rapid growth requires more energy and higher metabolism, which in turn demands more efficient or

faster feeding and an increased likelihood of starvation. A short larval period means larvae can transform quickly into juveniles, and settle from the plankton and into the generally safer juvenile habitat. But if a larva finds itself in an inappropriate habitat at the end of its larval period – such as far out at sea for a species adapted to shallow water existence – then a short larval life provides little advantage. Conversely, a long larval interval can permit long-distance dispersal. However, extended planktonic life exposes the larva for a longer time to the extreme hazards of planktonic existence, when >99% of larvae are eaten or starve. Growth rates of juveniles and adults are subject to advantages and constraints as discussed in point 2.

8. *Spawning bouts per year and duration of spawning season.* The number of times an individual, particularly a female, spawns each year tells much about the allocation of energy to reproduction. Duration of the spawning season is more a population than an individual characteristic and is useful in assessing potential recruitment into that population.

9. *Number of spawnings per lifetime.* Most fishes are **iteroparous** (itero = to repeat; parous = to give birth), spawning repeatedly throughout their lives (Fig. 20.3). One-time spawners, termed **semelparous**, devote all their energy to a single, massive spawning event, after which they die (anguillid eels, many Pacific salmons, lampreys, and some gobies).

10. *Reproductive interval.* The time spent between reproductive bouts for iteroparous species varies greatly, from daily spawners that reproduce year-round in some tropical reef fishes (e.g. wrasses), to fishes that spawn every few weeks during a protracted season (e.g., grunion, darters), to seasonal spawners that may spawn only once or a few times during a limited season (snappers, groupers, larger percids, centrarchid basses), to internal bearers with long gestation periods of a year or more (some sharks), to species that may wait several years between spawnings (sturgeon). For iteroparous species that spawn repeatedly each year, the reproductive interval can theoretically be adjusted in response to expected mortality levels. Where the probability of mortality is high, reproductive intervals should be short. As was the case for variable age at first reproduction and reproductive allotment (discussed earlier), female Guppy exposed to high levels of predation have relatively short reproductive intervals (Reznick & Endler 1982).

11. *Parental care.* The degree of care given has an overwhelming influence on the mortality rate of the young and is generally inversely proportional to fecundity (see Chapter 17, Parental Care). Parental care is often distinguished as **prezygotic** (e.g. nest preparation) and **postzygotic** (e.g. internal brooding, guarding young). The level of parental care in fishes ranges from nonexistent to rather elaborate. Broadcast spawners release thousands to millions of eggs into the water column and provide no care (cods, tunas, and billfishes). Moderate care occurs in fishes that spawn intermediate numbers of eggs on substrates and may involve some substrate preparation such as nest construction or egg covering (salmons, grunion). More extensive care occurs in fishes that prepare a nest and then guard relatively few eggs until they hatch and perhaps a little later (sticklebacks, sunfishes, and some cichlids). Intensive care is usually associated with relatively low numbers of large eggs, such as fishes that gestate young internally (livebearers, embiotocid surfperches) or incubate the eggs orally (cardinalfishes); some oral incubators continue to protect the young after hatching (some catfishes) and some cichlids even feed their young with external body secretions. Bowfin (*Amia calva*) lay eggs and males guard their young until they are several centimeters long, a rarity among fishes. Parental care increases the survival of the young but occurs at a cost to the parents because extended care may reduce foraging time and increases the interval between spawnings.

(A)

(B)

FIGURE 20.3 (A) Many fishes, such as the Pumpkinseed sunfish (*Lepomis gibbosus*) are iteroparous, spawning multiple times. Cephas / Wikimedia Commons / CC BY-SA 3.0; (B) Semelparous fishes, such as the Sockeye salmon (*Oncorhynchus nerka*) spawn only once in their lives. Their decomposition provides nutrients to the stream which helps the next generation. G. Helfman (Author).

12. *Gender change and sex ratio variation.* Fishes in several families change sex, beginning as males and changing to females (**protandry**) or vice versa (**protogyny**) (see Chapter 8). The timing of the change is largely determined by the relative reproductive success males or females experience at the same body size. Sex change occurs at a cost in immediate reproductive output, because gonad conversion may require weeks or months. In many vertebrates (crocodilians, turtles, lizards, possums, and monkeys), the sex of offspring may be determined by conditions such as the temperature at which the eggs or embryos develop (= **environmental sex determination, ESD**). Extreme temperatures affect sex determination in a few fishes, mostly atheriniforms such as rivulins, ricefish, and livebearers; pH can also influence sex determination in some cichlids and a livebearer (Rubin 1985; Francis 1992). Naturally occurring variation in temperature determines the sex of Atlantic Silversides, *Menidia menidia* (Atherinidae). Offspring produced early in the year at relatively low temperatures tend to be female, whereas young produced later at higher temperatures tend to be male.
13. *Geographic patterns and phylogenetic constraints.* In many families, related species living in different habitats often adopt life history patterns appropriate for that habitat, and unrelated fishes converge on suites of life history adaptations. Mouth-brooding fishes worldwide have converged on small clutches of large eggs, slow growth rates, and protracted breeding seasons (bonytongues, marine catfishes, and cichlids). Such convergence is evidence of the importance of environmental selection factors promoting one life history strategy over another and can be found at relatively large geographic scales. Among freshwater fishes in North America, species that mature relatively late in life tend to have larger body sizes, longer life spans, higher fecundities, smaller eggs, few multiple spawnings, and a short spawning season (sturgeons, Paddlefish, shads, muskellunge, charrs, and Burbot). Fishes with extended spawning seasons tend to have larger eggs, multiple spawning bouts, and exhibit more parental care (cavefishes, madtom catfishes). Marine fishes that have extensive geographic ranges (tarpon, cods) also tend to have high fecundity. Anadromous species, such as salmons, Striped Bass and sturgeons, mature late, grow fast as adults, live long, and have large eggs (Winemiller & Rose 1992).

As with any characteristics, life history traits are influenced by evolutionary history. Darwin (1859) recognized that even strong natural selection must operate within the phylogenetic constraints ("unity of descent") imposed by ancestors. See Gould and Lewontin (1979) for a classic explanation of the fallacy of adaptionist explanations without considering phylogenetic inertia. Consequently, an animal may not have the life history characteristics that we expect given current conditions. Unless selection pressures have been relatively stable for many generations, an animal's adaptations will not necessarily reflect present conditions but will instead reflect past selection pressures. For relatively conservative traits that are shared among many members of a lineage, historical constraints may be difficult to overcome and species will retain seemingly nonadaptive characteristics. Regardless of latitude and habitat type, percopsiforms (Troutperch, Pirate Perch, and cave fishes) tend to be small, produce small clutches of large eggs, exhibit extensive parental care, and have protracted spawning seasons and slow growth rates. Within the cypriniforms, suckers in the genus *Ictiobus* are large with large clutches and few spawning bouts whereas minnows in the genus *Notropis* are small, have small clutches, and frequent spawning bouts. Flatfishes as a group mature at large size, produce large clutches of small eggs during short spawning seasons, and grow rapidly when young (Winemiller & Rose 1992).

Effects of Fishing on Life History Traits and Implications for Recovery

Exploitation-caused changes in life history attributes have long been documented in the fisheries literature (Helfman 2007). In fact, growth adjustments underlie the **basic theory of fishing**: growth compensation by surviving individuals will accompany moderate exploitation, and maximal harvest rates (**maximum sustainable yields** of surplus production) are achieved by maintaining a population at some intermediate level through fishing (e.g. Ross 1997; Hart & Reynolds 2002b; or any basic fisheries text). Typically, reductions in abundance cause increased body growth in the remaining individuals, and the ensuing faster growth rates generally result in maturation at smaller sizes and younger ages. The proximate stimulus of this generalized **compensatory** change is assumed to be reduced competition for limited resources. The ultimate cause, according to life history theory, is that heavy predation pressure (harvesting) favors fishes that are capable of initiating reproduction sooner, before they are eaten (harvested) (see Trippel 1995; Allendorf & Hard 2009).

The fisheries literature contains many examples of **exploited** stocks that show changes in weight at age, length at age, length at maturation, and age at maturation, with most species showing reduced weights, lengths, and ages, as well as accelerated growth. Law (2000, table 1) reviewed findings on 16 species of flatfishes, gadoids, and salmons, most of which showed decreases in life history traits. Data on Atlantic Cod are complete and telling: among eight exploited stocks monitored for 7–53 years, median age at maturity declined between 16% and 56%, representing a 0.9- to 3.6-year reduction in age at maturity; in addition, longer periods of exploitation produced greater changes in age. Natural variation in age at maturation was small or negligible when fishing pressure was light (Trippel 1995). Data on flatfishes, gadoids, and salmon and on commercially fished western Atlantic sharks, herrings, scorpionfishes, snappers, drums, groupers, eel-pouts, butterfishes, and mackerels similarly revealed reduced lengths and ages at maturation (Upton 1992; O'Brien et al. 1993; Lessa et al. 1999; Vannuccini 1999). The mean length of captured Patagonian Toothfish, *Dissostichus eleginoides*, declined 30% in just the first few years of fishing (ISOFISH 2002). Analysis of Scotland's Firth of Clyde demersal fishery from 1927–2009 showed dramatic reductions in the size of harvested species (Heath & Speirs 2012). Among other examples are freshwater species subjected to intensive commercial and recreational fishing (Bluegill, Walleye, whitefish, Yellow Perch, Northern Pike, Brown Trout, and Arctic Char; Trippel 1995; Drake et al. 1997) and decreased length and age at maturity are the rule.

Kindsvater et al. (2016) evaluated critical aspects of life histories of different species and proposed four main life history categories based on the acronym "POSE": Precocial, Opportunistic, Survivors, and Episodic (Fig. 20.4). Precocial species, such as seahorses, grow and mature quickly, and parental care allows them to produce relatively few young that are well developed and therefore have good chances for survival. Opportunistic species, such as clupeids, grow and mature quickly, and produce large numbers of eggs but provide no parental care, so high mortality is expected early in life – but these fishes occur in very large numbers. Survivors, such as elasmobranchs, grow slowly, take a long time to mature, and produce few offspring – but those offspring are typically well developed and have high survival. Episodic species, such as Atlantic Cod and many groupers, grow slowly, take a long time to mature, but produce large numbers of eggs when they spawn; and large, older females may produce large eggs that yield juveniles with better chances for survival.

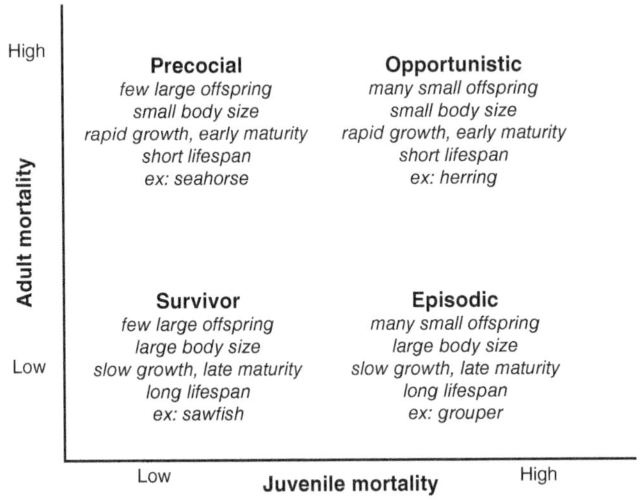

FIGURE 20.4 Four major groupings of commonly co-occurring life history characteristics according to the POSE model of Kindsvater et al. (2016). The slow growth, late maturity, and small number of offspring make it particularly difficult for Survivor species to recover from heavy harvesting pressure. Episodic species also show slow growth and late maturity, but their ability to produce many offspring makes it at least possible that a few large individuals could prompt population recovery. However, the success of any particular juvenile cohort would be highly dependent on other environmental factors – so there is no guarantee that producing many eggs would result in recovery. Precocial and Opportunistic species generally would have a better chance to recover because they grow rapidly and mature quickly. But this is also no guarantee for recovery, as Precocial species do not produce many offspring and Opportunistic species are heavily exploited by predators and human fishing pressure. Unfavorable environmental conditions combined with exploitation could make recovery of these species unlikely as well.

The slow growth, late maturity, and small number of offspring make it particularly difficult for Survivor species to recover if their populations become small. Episodic species also show slow growth and late maturity, but their ability to produce many offspring makes it possible that a few large individuals could prompt population recovery. However, the success of any particular juvenile cohort would be highly dependent on other environmental factors – hence there is no certainty that producing many eggs would result in recovery. Precocial and Opportunistic species generally would have a better chance to recover because they grow rapidly and mature quickly. But this is also no guarantee for recovery, as Precocial species do not produce many offspring and Opportunistic species are heavily exploited by predators and human fishing pressure. In addition, unfavorable environmental conditions at critical times such as spawning or hatching could make recovery of these species unlikely as well. Thus, species in all of the four POSE categories are potentially vulnerable.

There has been increasing concern for many years that the changes seen in many exploited fish stocks may be indications of long-term evolutionary change (**fisheries-induced evolution**, FIE) and not simply phenotypic plasticity within a population (Kuparinen & Merilä 2008; Allendorf & Hard 2009). There are important implications, as evolutionary change could impede recovery of depleted populations (see Fig. 20.5), if it is even possible. The widespread and consistent trends in decreasing age and size at maturity among many heavily fished populations supports the view of evolutionary change,

FIGURE 20.5 Heavy fishing pressure has pushed the evolution of harvested species toward maturing at younger ages and smaller sizes, which acts contrary to management objectives. Molloy et al. (2009) show that these trends may be reversed to some extent in protected areas, such as marine reserves, but this change takes time – often 15 years or more, especially for large-bodied species that mature slowly such as this Yelloweye Rockfish (*Sebastes ruberrimus*). V. O'Connell / Wikimedia Commons / Public Domain.

but clear genetic evidence among wild populations would help make a more convincing case (see review by Heino et al. 2015).

Whether changes in populations are the result of evolution or phenotypic plasticity, the ironic upshot is that there is clear evidence that strong fishing pressure drives change toward early maturity at smaller size and subsequently lower fecundity, which can work against management objectives. Eikeset et al. (2013) used 70 years of data from the Atlantic Cod fishery to conclude that loss of large, old cod might lead to compensatory increases in growth of younger fish and have little overall impact on the economic viability of the fishery as discussed earlier. They also point out, however, that without proper management, economic viability and stock viability could both decline.

Fisheries-Induced Evolution of Behavior

Individuals within any population of animals may vary considerably in behavior, and behavior can have significant implications for survival and reproduction. The role that the behavior of individual fishes may play in their vulnerability to commercial or recreational fishing, and whether that may have an impact on behavior in subsequent generations, has been a topic of considerable discussion. The potential future impact will be a function of the degree to which behavior is genetically based, and is therefore heritable.

Behaviors of individual fish are typically associated with particular animal personality types which are often characterized along five axes: (i) shyness–boldness, (ii) exploration–avoidance, (iii) activity, (iv) aggressiveness, and (v) sociability (Conrad et al. 2011; Mittelbach et al. 2014). Within individual fish, certain traits show strong correlation and the resulting combinations are often characterized as behavioral syndromes. For example, an individual that is bold is also often quite active, explores its surroundings, and may be aggressive and not social. Conversely, a fish characterized as shy may also be less active, more reclusive, retreat from confrontation, and more often associate with others.

Different behavioral syndromes may have a considerable impact on an individual's susceptibility to different commercial fishing methods. Active fishing methods, such as bottom trawls, would be more likely to capture individuals that are less active and that stay near the bottom, whereas active individuals that swim upward are more likely to avoid capture (Pauli & Sih 2017). Passive fishing methods, such as stationary nets and baited traps rely on fish to come to the capture device, and therefore are more likely to catch fish that are more active and more likely to explore (Heino et al. 2015; Arlinghaus et al. 2016; Pauli & Sih 2017). The selective removal of bolder individuals, who are often among the more aggressive and reproductively successful, has raised the concern that sustained harvesting by passive fishing might result in future populations of more timid individuals (Arlinghaus et al. 2016). This could result in a future population of fish that are less bold and less active, and perhaps less likely to disperse, colonize new areas, and provide gene flow among populations. A similar outcome might result from harvesting by sport fishing, if the angling techniques were more likely to capture aggressive individuals. However, for this to be the case, behavioral traits would need to be heritable (Heino et al. 2015).

To evaluate the heritability of behavioral traits, a series of studies used angling for Largemouth Bass (*Micropterus salmoides*) to distinguish fish that were more vulnerable to angling, and hence caught more frequently, from those that were less vulnerable. Individuals were captured and separated by vulnerability, and then used as breeding stock for subsequent generations of fish held in experimental ponds and subject to subsequent selection for vulnerability to angling. This selective process created even greater differences in vulnerability to angling in subsequent generations, providing evidence that at least some elements of behavior are heritable (Philipp et al. 2009). In addition, male bass that were more aggressive tended to have higher metabolic rates, attract more females, fertilize more eggs, and provide better protection for their nests and young than less aggressive individuals (Cooke et al. 2007; Sutter et al. 2012). The reproductive success of these more aggressive fish might therefore be expected to pass on genetic factors for aggressive behavior and reproductive success to subsequent generations. However, the aggressive behavior that correlates with better parental care and reproductive fitness also makes fish more susceptible to removal by angling, especially during the spawning season. This would result in the decrease of these genetically linked traits for reproductive success in the next generation (Cooke et al. 2007; Sutter et al. 2012). Through this selective process, angling could reduce aggression and reproductive success in future generations.

Some studies have suggested that more active and aggressive fish feed more and may therefore grow faster (see Mittelbach et al. 2014), or at least consume enough extra calories to support their greater activity (Cooke et al. 2007). Controlled feeding experiments of juvenile fish in the fourth generation of the series of Largemouth Bass studies showed that fish from the lineage that was less vulnerable to angling exhibited somewhat lower strike distances and lower success rates when capturing prey (juvenile Bluegill, *Lepomis macrochirus*), and rejected prey more often, but had higher conversion efficiencies than fish from the lineage that was more vulnerable to angling (Nannini et al. 2011). This may have been the result of lower overall energy expenditure resulting in greater energy directed toward growth. As a result, there was no difference in overall size or growth rate of the individuals from the two lineages under the controlled conditions of the study. Outcomes might be a bit different, however, in natural environments where availability of food could more likely be a function of searching and pursuing prey.

Different types of fishing can create selective pressure against some behaviors and in favor of others, and we have seen that that at least some aspects of fish behavior may be passed on to subsequent generations. Therefore, fishing-induced evolution may be affecting both the behavior and life history characteristics of exploited populations. These combined effects could influence the ability of a population to recover from the effects of strong fishing pressure, making management of exploited populations all the more challenging.

Population Dynamics and Regulation

Simply defined, a **population** consists of all the individuals of a particular species in a given area. Because populations form the matrix in which individual survival and reproduction occur, expanded definitions recognize the importance of genetic structure: a population is therefore "a gene pool that has continuity through time because of the reproductive activities of the individuals in the population" (Wootton 1990, p. 280). Populations grow and shrink in numbers as a result of the actions and interactions of their individuals, which can change relative gene frequencies (the **genetic structure**) of the population. Much of ecology has been devoted to describing, understanding, and predicting the nature and causes of population numerical growth and decline and of genetic structure.

Population size changes as a function of four major processes: birth, death, immigration, and emigration. **Birth** and **death** rates (**age-specific reproduction** and **survivorship** rates of individuals) can be used to calculate the approximate rate at which a population will change in size (Table 20.1). Such **life table** statistics are usually calculated only for females in a population; relatively few have been constructed for fishes.

Migration, both in and out of a population, greatly complicates any attempt at predicting future population size. Fish populations increase in size due to migration as a result of either recruitment or colonization. **Recruitment** usually refers to the addition to the population through reproduction, as when larvae settle out of the plankton and into the population. In fisheries terminology, recruitment generally refers to the addition of potentially catchable individuals to the **stock** in question, stock being essentially synonymous with population. **Colonization** is the addition by movement of established individuals between habitats, such as when juveniles move from a nursery habitat to an adult habitat.

Reproductive events in a particular population may have little effect on later local population size because many fishes export their reproductive products in the form of pelagic larvae that are dispersed widely. As a result, it was commonly believed that there was minimal correspondence between stock and recruitment (the **stock : recruitment relationship**) for most marine populations. Minimal relationship meant that current population size was not a reliable predictor of future population size, indicating that intense local fishing did not necessarily drive down future stocks (Rothschild 1986). However, more recent analyses suggest that there is a positive relationship between stock and recruitment in many fisheries, in part perhaps because "export" of larvae from parental habitat is not as general as had been believed (Sponaugle et al. 2002; Gerlach et al. 2007). Myers and Barrowman (1996) found a generally positive stock : recruitment relationship in 83 mostly marine species indicating that (i) higher recruitment occurred when spawner abundance was high, (ii) lower recruitment occurred when abundance was low, and (iii) populations below the median abundance level had lower recruitment than populations above it. Specific examples included such well-known, depleted fisheries as Bluefin Tuna, Atlantic Cod, and large sharks. Life tables are therefore useful not only in relatively closed populations, such as in ponds and lakes, but also in many commercially important marine species.

Regardless of locale, fish populations vary widely and notoriously in size. The concept of the **year class** or **cohort** is important in understanding these dynamics. High population density may not necessarily indicate a sustainably reproducing population because many of the individuals in that population may come from a single year class, whereas most other years may have seen little successful reproduction. This was the case of the Cui-ui, *Chasmistes cujus* (Catostomidae), endemic to Pyramid Lake, Nevada (Fig. 20.6). If the successful year class is approaching the usual maximum age for that species and no younger year class is abundant, overexploitation of the dominant year class

TABLE 20.1 **A life table for a cohort of Brook Trout, *Salvelinus fontinalis*, in Hunt Creek, Michigan for the year 1952. Survivorship (l_x) is the probability that an individual female will live to age *x*, reproductive output (m_x) is the mean number of daughters produced by a female of that age (estimated as half the number of eggs produced). The reproductive rate of the population (net reproductive rate, R_0) is the sum of the survivorship and reproductive output columns ($=\Sigma l_x m_x$), which equals the average number of females being produced per female in the cohort. When R_0 is greater than 1 the population is growing, when it is less than 1 the population is shrinking. After Wootton (1990), based on data of McFadden et al. (1967).**

Age class, in years (*x*)	Survivorship (l_x)	Reproductive output per female (m_x)	Reproductive rate of population ($l_x m_x$)
0	1.0000	0	0
1	0.0528	0	0
2	0.0206	33.7	0.6952
3	0.0039	125.6	0.4898
4	0.00051	326.9	0.1667
			$R_0 = 1.352$

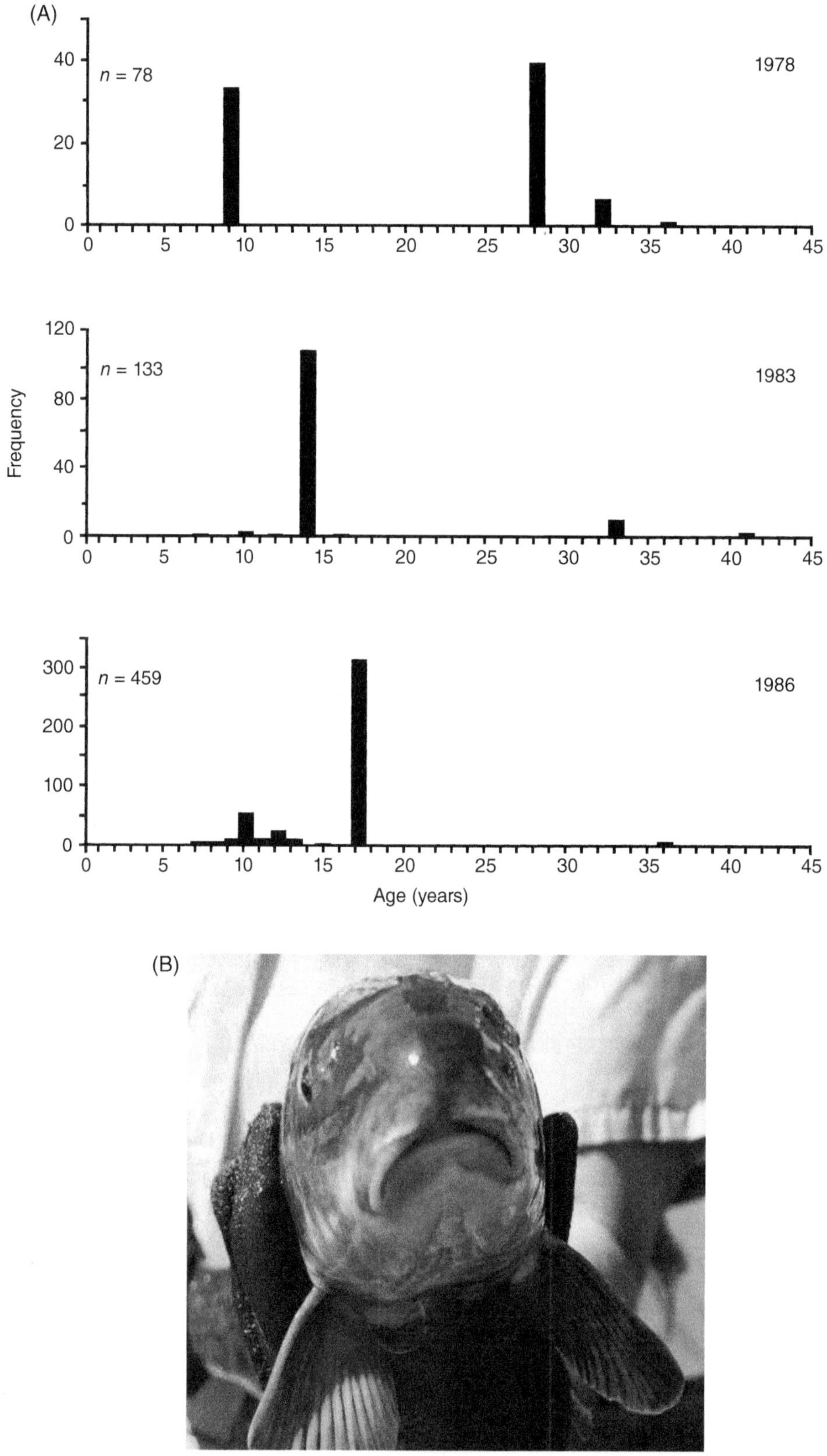

FIGURE 20.6 The Cui-ui, *Chasmistes cujus* (Catostomidae), is an endangered sucker that occurs only in Pyramid Lake, Nevada. Its long life span, approaching 45 years, has probably saved it from extinction. Drought and human diversion of water from its spawning habitat in the lower Truckee River had resulted in reproductive failure in most years. (A) Samples of spawning fish in 1978 indicated that the entire species had been maintained by the year classes born in 1950 and 1969 (Scoppettone & Vinyard 1991, used with permission). Mortality had all but eliminated the 1950 year class by 1983. A bypass channel built in 1976 gave fish access to the river even at low water levels, and some successful reproduction has occurred in subsequent years. (B) An adult Cui-ui held in a captive propagation facility, Pyramid Lake, Nevada; US Bureau of Reclamation / Wikimedia Commons / Public Domain.

can lead to very rapid population collapse. Year class strength then becomes a critical statistic in determining management schemes for exploited populations. The spectacular success of the campaign to restore Striped Bass, *Morone saxatilis* (Moronidae), to the Chesapeake Bay focused on protecting the 1982 year class until 95% of those females had matured (Ross 1997; Secor 2000).

Variation in numbers among year classes points out another important feature of fish populations, which is that they are **size structured**. Indeterminate growth and overlapping generations create a situation where a population may include individuals of very different sizes, differing in body mass by as much as four or more orders of magnitude (e.g., consider Bluefin Tuna that weigh a fraction of a gram at birth and grow to an adult size exceeding 500 kg in mass, a range of seven orders of magnitude). Size structuring can affect population regulation because multiple size classes provide the potential for intraspecific competition and cannibalism, which in turn may lead to differences in habitat and other resource use because of avoidance of one group by another.

Such intraspecific variation has led to the concept of the **ontogenetic niche**, which recognizes the very different ecological roles that different age and size conspecifics are likely to play in a community (Werner & Gilliam 1984; Osenberg et al. 1992). For example, Pinfish (*Lagodon rhomboides*, Sparidae) start off as a carnivore and progressively shift to increasing herbivory in five distinct phases (Stoner & Livingston 1984). Largemouth Bass initially feed on zooplankton, then on littoral invertebrates, and then finally on fish, including conspecifics (Fig. 20.7). As juveniles, they compete with adult Bluegill for zooplankton. Young Largemouth Bass are, however, miniature adults and are morphologically constructed as piscivores. They are consequently less efficient planktivores than are adult Bluegill, and this morphological constraint makes them inferior competitors (they even the score later when Bluegill become their major prey) (Werner & Gilliam 1984). At each size, a fish is likely to have a different set of competitors and predators, some overlapping with the previous set, producing an incredibly complex set of interactions within a community containing even a small number of species.

FIGURE 20.7 Largemouth Bass demonstrate ontogenetic niche shifts as they grow from zooplanktivorous juveniles to benthic invertivores, and finally piscivorous adults. USGS / Wikimedia Commons / Public Domain.

Death can come at any time, but is more likely in certain life history stages than in others. Eggs and larvae are by far the most vulnerable periods (see Chapter 9, Larval Feeding and Survival). Estimates of mortality for marine fish populations range from about 10 to 85% per day for eggs, from 5 to 70% per day for yolk-sac larvae, and from 5 to 55% per day for feeding larvae (Bailey & Houde 1989). These are daily rates. When compounded over the larval life of a species, the magnitude of the loss is more striking. For example, Jack Mackerel, *Trachurus symmetricus*, require 8 days from hatching until they resorb their yolk sac and begin independent feeding. During this time, when mortality falls from 80 to 50% per day, 99.5–99.9% of larvae are lost to predation. Exogenous feeding adds the hazard of starvation: during the first week after yolk-sac absorption, larvae die at a rate of 45% per day from starvation alone (Hewitt et al. 1985).

Many fish biologists agree that predation is the major source of mortality for eggs and larvae and that mortality is highest at these stages. The list of predators on larvae is long and includes numerous invertebrates (ctenophores, siphonophores, jellyfishes, copepods, chaetognaths, euphausids, shrimps, and amphipods) as well as fishes. As fishes grow, their strength, swimming speed, feeding capability, and general escape ability increase. Estimates of the mortality of juveniles and adults are diametrically different from the rates experienced by eggs and larvae, for example 99.9% daily survival for juvenile or adult nototheniid icefishes, English Sole, Winter Flounder, cutlassfishes (Trichiuridae), mackerel, and tuna (McGurk 1986; Richards & Lindeman 1987). Freshwater salmonids (Brown Trout, Brook Trout, Rainbow Trout, and Coho Salmon) sustain relatively high annual mortality rates of 60–90% of the adult population, which is still far below the levels experienced by eggs and larvae (Alexander 1979).

Cannibalism (intraspecific predation) is widespread in fishes and may play a dominant role in population regulation in some species (Dominey & Blumer 1984; Smith & Reay 1991; Elgar & Crespi 1992). Cannibalism occurs in many chondrichthyans and in at least 36 families of teleost fishes, including herbivores, scavengers, planktivores, and piscivores. During Sand Tiger Shark (*Carcharias taurus*) pregnancies, the oldest embryos consume their younger siblings, and then proceed to eat unfertilized eggs (oophagy). The result is behavioral polyandry (females mate with multiple males) and genetic monogamy (only offspring of one sire survive; Chapman et al. 2013).

Cannibalism can have a significant impact on population dynamics. Between 30 and 70% of egg consumption is caused by conspecifics among anchovies (Engraulidae) and whitefishes (Salmonidae). In addition, adults may eat larvae and juveniles, including their own offspring, and young fish may eat siblings as well as unrelated individuals. Sixty percent of annual mortality in Walleye Pollock, *Theragra chalcogramma* (Gadidae), and 25% of mortality in Yellow Perch, *Perca flavescens*, has been attributed to cannibalism of juveniles. Year-class strength is thought to be strongly dependent on cannibalism rates in pike

(Esocidae), cod, haddock, and whiting (Gadidae), Walleye and perch (Percidae), and Nile Perch (Latidae). In Lake Victoria, cannibalism is considered the major cause of Nile Perch mortality, with important consequences for assemblage structure and human welfare alike (see Chapter 22, Introduced Predators). In lakes where one or only a few species occur, cannibalism may be *the* major population regulatory mechanism. In such situations, giant cannibal morphs that are specialized in feeding on conspecifics may develop (e.g., landlocked Arctic Char, *Salvelinus alpinus*; Sparholt 1985; Riget et al. 1986) (Fig. 20.8); such cannibalistic **polyphenism** is also known among larval salamanders and frogs (see Michimae & Wakahara 2002).

Cannibalism might seem counterproductive. However, conspecifics represent a highly nutritious protein meal made up of optimum proportions of vitamins, minerals, and amino acids for the species in question, producing high growth rates (e.g. Walleye, Walleye Pollock) and enhanced reproductive output (e.g. Mosquitofish, Poeciliidae). Even when kin are consumed, the benefits to the cannibal of reduced competition, increased growth, and enhanced reproduction could outweigh the current costs of losing a few relatives (Dominey & Blumer 1984; Smith & Reay 1991; Sogard & Olla 1994).

Production

Production is a topic of general interest to population ecologists and of particular interest to fisheries managers. How much biomass (or fish flesh) is a population producing, and how much of this is available to predators, including humans, without causing the population to crash? Can production be predicted from such measurable population traits as the birth and death schedules of different age classes (i.e. from calculations using life table characteristics discussed earlier)?

Production is calculated as the growth rate of individuals over a time period multiplied by the biomass of the age class, corrected for mortality occurring during the time period (Ricker 1975; Gulland 1983; Wootton 1990; Ross 1997). Natural production values for different populations of temperate freshwater fishes vary widely, from $<0.1\,g/m^2/year$ for Sockeye Salmon in an Oregon lake to $155\,g/m^2/year$ for Amargosa Pupfish (*Cyprinodon nevadensis*, Cyprinodontidae) in a desert stream in California. Most populations fall near the lower end of these values, in the $1–10\,g/m^2/year$ range. Tropical and fertilized ponds often show higher values (Chapman 1978). Knowing production also allows one to calculate **annual turnover**,

FIGURE 20.8 Four morphs of Arctic Char that differ anatomically, behaviorally, and ecologically can be found in a single lake. Shown here are adults of the four morphs from an Icelandic lake. They are, from top to bottom, the large, benthic feeding morph (33 cm long), the small benthic feeding morph (8 cm), the piscivorous morph (35 cm), and the plantivorous morph (19 cm). Skulason & Smith 1995 / with permission of Elsevier.

which is the ratio of production to biomass (P : B). Turnover is an index of the productivity of populations and subpopulations, and can be quite useful in understanding ecosystem processes. Among different age classes, very young fishes, although constituting relatively little of the biomass of the overall population, contribute 60–80% of population production because of their high P : B ratios. Young fish have very high growth rates relative to their sizes and hence have high turnover rates.

Yields to predators are therefore relatively higher when predators feed on young fish rather than eating older, slower growing fish. Whether high rates of exploitation of young age classes by predators reflect some form of optimal exploitation due to relative P : B ratios, or whether they just reflect ease of capture, would make an interesting study. Regardless, overfishing might be reduced if fisheries targeted younger age classes instead of imposing minimum size limits and targeting reproductively mature individuals. Such a management approach would more closely mimic natural predator–prey and assemblage interactions, relationships that have defined the life history traits of prey species over evolutionary time (Helfman 2007).

Population Genetics

Determining the physical boundaries of a population can be easy (as in isolated ponds and lakes) or difficult (large lakes, oceanic expanses). **Gene flow**, the exchange of genes between populations, regulates the level of isolation between populations. Coregonine whitefishes in many lakes in Canada and sticklebacks in lakes, ponds, and rivers on the Pacific coast of North America are reproductively isolated from each other and consequently comprise distinct populations (Furin et al. 2012; Renault et al. 2012). In contrast, the oceanic migrants including billfish, tunas, and pelagic sharks may have population structure only between ocean basins, or may maintain a single globally connected population (Hoelzel et al. 2006; Graves & McDowell 2015; Gaither et al. 2016).

Population genetics uses genotype frequencies to distinguish populations. From a genetics perspective, populations are groups of interbreeding individuals that rarely exchange members or gametes with other populations. Population genetic principles are often applied to fisheries management, to define the **stocks** that are the units of harvest and management. Populations are important **management units** because if one population is depleted, it must recover alone, without being replenished from other populations.

Genetic differences among populations of fishes can range from restricted gene flow between adjacent locations (**shallow population structure,** see Appendix 1) to ancient separations indicated by diagnostic differences in DNA sequences (**deep population structure**, see Appendix 1). At the lower end of this spectrum, population-level separations are indicated by significant differences in the frequency of alleles, SNPs (in nDNA) or haplotypes (in mtDNA). Populations separated by habitat discontinuities (especially in fresh water) or great distances (especially in the ocean) will not freely interbreed, and the consequence of this restriction is usually that the populations "drift" apart in terms of genetic composition. Sometimes these separations are reinforced by natural selection, but often the changes in genotype frequencies are due to chance, when one allele at a given locus randomly increases or decreases in one population, and a different allele increases or decreases in another population. The level of separation is commonly measured with **F statistics** (F_{ST}, see Appendix 1) and various analogs, especially φ_{ST} for DNA sequence data, with larger values indicating greater genetic isolation.

The catadromous American Eel, although distributed in ponds, lakes, streams, and rivers from Iceland to Venezuela, is believed to be a single **panmictic** population (Avise et al. 1986), with all returning to a single spawning locale in the Sargasso Sea, remixing their genes at each reproductive episode (Fig. 20.9). Panmixis has also been suggested in other anguillids such as European and Japanese Eels (Sang et al. 1994; Lintas et al. 1998). However, researchers using different methods have found evidence of subtle genetic differentiation in these species (Wirth & Bernatchez 2001; Tseng et al. 2006). Some genetic evidence from European Eel suggests female philopatric behavior, with females forming genetically distinct subpopulations within the Sargasso Sea spawning area whereas males mate opportunistically (Baltazar-Soares et al. 2014). If this is true, efforts to enhance populations by stocking may not have the desired effect.

Nearshore marine species generally show levels of gene flow that are strongly related to dispersal capability and distance between populations; species with pelagic larvae can have populations that span thousands of km (Waples 1987; Lessios & Robertson 2006).

Genetic analyses sometimes lead to phylogenetic and taxonomic revisions. For example, five species of kelpfishes (*Gibbonsia*, Clinidae) were recognized from western North America, based on traditional meristic and morphometric analyses. Reanalysis using allozyme data indicate very little genetic differentiation among some species (Stepien & Rosenblatt 1991). Three nominal species, the Scarlet Kelpfish (*G. erythra*), the Crevice Kelpfish (*G. montereyensis*), and an offshore

FIGURE 20.9 American Eel (*Anguilla rostrata*) are distributed in coastal and inland waters from Greenland and Iceland to Venezuela, and as far inland at the Great Lakes and the Mississippi River. Yet they migrate to spawning grounds in the Central Atlantic Ocean and are believed to form a single, panmictic breeding population. E.Edmonson and H.Chrisp / Wikimedia Commons / Public Domain.

Mexican endemic (*G. norae*), showed no significant differences in the frequencies of alleles at 40 allozyme loci. (The term **nominal** refers to a population that has been described as a separate species). Reanalysis of morphometric data showed that many anatomical differences between nominal species were instead sexual dimorphisms. All nominal Scarlet Kelpfish, distinguished by more caudal peduncle scales and a higher dorsal spine, were in fact males. All nominal Crevice Kelpfish were females. *G. norae* differed only in lower counts of scale rows and fin rays, but it is generally observed that fish that develop in warmer water lay down fewer meristic elements (Barlow 1961; see Chapter 8, Meristic Variation). Hence, what had been believed to be three species are actually all members of a single species, the Crevice Kelpfish (*G. montereyensis*).

Genetic analyses of pelagic species have demonstrated a lack of differentiation among widely separated populations that were once thought to be different species. Albacore Tuna (*Thunnus alalunga*) in the North Pacific and South Atlantic, and Skipjack Tuna (*Katsuwonus pelamis*) and Yellowfin Tuna (*Thunnus albacares*) in the Atlantic and Pacific are three examples (Graves & Dizon 1987; Scoles & Graves 1993). Migration over long distances do not, however, guarantee gene flow. Salmon of Pacific northwestern rivers (*Oncorhynchus* spp.) migrate and intermix in the ocean during much of their lives, but genetically discrete stocks separate and return to their natal streams to reproduce, conserving the genetic identity of more than 200 stocks (Nehlsen et al. 1991; see Fig. 18.7). At the extreme of genetic diversity, Rainbow Trout/Steelhead (*O. mykiss*) from the west coast of the Kamchatka Peninsula of Russia differentiate into as many as six distinct life history types (riverine/estuarine, estuarine, anadromous A, anadromous B, anadromous B half-pounders, and resident), with evidence that all six may reproduce in the same spawning habitat (Savvaitova et al. 2000; Augerot 2005).

Close proximity does not always indicate high gene flow, especially in fishes that lack a pelagic larval stage. Populations of live-bearing Black Surfperch, *Embiotoca jacksoni*, separated by only 40–80 km in California and Mexico, show genetic differences of a magnitude that is normally found between closely-related species (Utter & Ryman 1993; Bernardi 2000).

Conversely, species in isolated lakes may have more than one genetically distinct population. In Thingvallavatn, Iceland (the suffix "vatn" means lake in Icelandic), Arctic Char exist as four distinct forms that occupy different habitats and feed on different food types (see Fig. 20.8). A large and a small morph remain near the bottom and feed on benthic invertebrates in the shallow littoral zone, whereas two other morphs are up in the water column where one feeds on zooplankton in the limnetic (pelagic) region and the other feeds on fishes both inshore and offshore. Morphologically (and appropriately), the benthic-feeding morphs have subterminal mouths and relatively dark coloration, whereas the more pelagic forms have terminal mouths and silvery, countershaded coloration. Spawning times vary among the morphs, and the morphological differences show up shortly after hatching and hence are not **ecophenotypic**, i.e. they do not result from environmental influences experienced by different individuals.

The morphological and behavioral differences among morphs usually have a strong genetic basis, as offspring of the different morphs retain their trophic specializations even when raised in a common laboratory environment (Skulason et al. 1993). The morphs may have evolved because of the availability of habitats (**adaptive zones** or open niches) brought on by the absence of other species in these young post-glacial lakes (Magnusson & Ferguson 1987; Sandlund et al. 1988). Three-spine Stickleback, *Gasterosteus aculeatus*, in Thingvallavatn have two very different morphs, possibly in response to predation pressure from the Arctic Char (Ólafsdóttir et al. 2007). Sticklebacks are well known for differentiating into distinct benthic and limnetic forms in response to predation pressure and food availability (Schluter 2000). Other fish species demonstrating marked intraspecific divergence include Crucian Carp, *Carassius carassius* (Brönmark & Miner 1992); Sockeye Salmon, *Oncorhynchus nerka* (Hendry 2001); and Pumpkinseed, *Lepomis gibbosus* (Parsons & Robinson 2007). The Arc-eye Hawkfish, *Paracirrhites arcatus*, shows population genetic differences between two color morphs that occupy the same reefs, seemingly adapted to light and dark habitats (Whitney et al. 2018). In all these cases, population genetic differences, coupled with morphological or ecological differences, may be the building blocks for speciation, as evidenced by the explosive radiation of cichlids in East Africa (Salzburger et al. 2005).

Understanding the genetic make-up of a population has become increasingly important as environmental degradation and overexploitation place many populations and species at risk. A key character is the degree of **genetic variation** in a population. Genetic variation results from selection, mutation, migration, non-random mating, and genetic drift (random changes in gene frequencies, particularly in small, isolated populations). Genetic variation provides the building blocks for evolution; natural selection acts on such variation, favoring genotypes that are adapted to current conditions. Measures of genotypic variation and frequencies can tell us whether a population has become dangerously inbred and lacks the genetic diversity necessary to allow for adaptation to changing environmental conditions, whether gene flow is occurring between populations, and whether hybridization with introduced species is occurring (and hence if the genetic identity of a species is threatened) (Allendorf et al. 2012; Coleman et al. 2016).

Recognizing the importance of genetic structure among populations has led to the concept of **metapopulations**, which describes populations of a species linked by gene flow via migration, recruitment, or colonization. Metapopulations are thought to exist in a variety of species, including widely distributed, commercially exploited marine fishes (e.g., Atlantic Herring; Wright et al. 2006), among fishes in a stream network (Cutthroat Trout, Bayou Darters; Slack et al. 2004; Neville et al. 2006), among anadromous and estuarine species (Pacific Salmon, Tidewater Goby; Policansky & Magnuson 1998; Lafferty et al. 1999), and among populations of reef fishes (Saddleback Clownfish, Bicolor Damselfish; Saenz-Agudelo et al. 2012; Pusack et al. 2014). Such a focus has important implications for the management of fishery species as well as for the design of protected areas because it points to linkages among distant locales (Fausch et al. 2002). For example, protecting crucial habitat and populations in a few specific locales along a 1000 km long region of the Gulf of California (Sea of Cortez) is likely to assure larval sources linked by ocean currents to many other Gulf areas of biological and socioeconomic importance (Sala et al. 2002).

Dispersal and Population Structure

Certain life history traits correspond to shallow or deep population structure, especially those that influence the ability of the fish to disperse, as larva, juvenile, or adult. Hence the first generalization is that levels of population genetic structure are lowest in marine fishes, intermediate in anadromous fishes (see Chapter 18), and highest in freshwater fishes (Table 20.2). Populations of Gila Topminnow (*Poeciliopsis occidentalis*) in the southwestern United States may occupy desert springs separated by a few kilometers, and yet can be isolated for thousands of years (Quattro et al. 1996). In this topminnow and other desert fishes, dispersal opportunities are limited to rare flooding events. At the other end of the spectrum, the Whale Shark (*Rhincodon typus*) has population structure only on the global scale of Atlantic versus Indo-Pacific oceans (Castro et al. 2007).

Genetic diversity (heterozygosity, H) also shows a rank order among freshwater (lowest), anadromous (intermediate), and marine (highest) fishes. This is an expected consequence of tremendous differences in population size. Freshwater populations may number in the thousands to millions, whereas their marine counterparts, with much larger ranges, may number in the millions to (in the case of anchovies and sardines) billions. Larger populations will accumulate more genetic diversity (Kimura 1983). There are many exceptions to these trends, but the conclusion of a rank order in genetic diversity is supported by both allozymes and microsatellite surveys (Table 20.2).

If populations are isolated for thousands of generations, they will eventually reach **monophyly** (each descended from a single ancestral population). The rate at which populations diverge depends on the **effective population size** (N_e), with a high probability of monophyly after $4N_e$ generations (Neigel & Avise 1986). Often the condition of monophyly is accompanied, upon closer examination, by morphological differences that indicate previously unrecognized **cryptic species**. However, this is not invariably the case, and scientists may prefer to retain a single taxonomic label that recognizes multiple evolutionary (subspecific) units within a species. The term **evolutionary significant unit** (**ESU**) was coined for subspecific evolutionary entities that show morphological, behavioral, or genetic differences (Ryder 1986). Moritz (1994) suggested that ESUs could be recognized for populations that are monophyletic with mtDNA sequences. ESUs are often applied in the context of conservation, with an emphasis on higher priorities for ESUs than for populations, as has been applied to Pacific salmonids. While monophyly in DNA assays is not the only way to assign such conservation priorities, this criterion is valuable for distinguishing populations that may have novel genetic characteristics, and may be in the process of speciating. ESUs as defined by monophyly of mtDNA sequences are surprisingly common in fishes, as indicated in Table 20.3, where 8 out of 15 surveys of Atlantic reef fishes show evidence of ESUs.

Pelagic Larval Duration and Population Structure

The low level of population structure in marine fishes is a consequence of high dispersal, although other factors such as large population size may contribute to this trend. With few hard barriers in the ocean, and with pelagic larval periods ranging from a few days to 2 years, marine fishes have tremendous potential for dispersal. However, modeling and field work have disputed the conclusion that all coastal marine fishes have large "open" populations (Mora & Sale 2002; Botsford et al. 2009; Pinsky et al. 2012). Mark/recapture studies have demonstrated a surprising retention of larvae near their region of origin. Taylor

TABLE 20.2 **Population genetic diversity averaged across three types of fishes, for allozymes (113 species; Ward et al. 1994) and for microsatellites (32 species; DeWoody & Avise 2000). Heterozygosity (H) values are progressively higher in freshwater, anadromous, and marine fishes. Population structure (F_{ST}) values from the allozyme survey are progressively lower in freshwater, anadromous, and marine fishes.**

Habitat	H allozymes	H microsatellites	F_{ST} allozymes
Freshwater	0.046	0.54	0.22
Anadromous	0.052	0.68	0.11
Marine	0.059	0.77	0.06

Based on Ward et al. 1994 and DeWoody & Avise (2000).

TABLE 20.3 **Comparison of pelagic larval duration and population structure in 15 Atlantic reef fishes. Pelagic larval duration does not have a significant correlation with population structure (ϕ_{ST} values). Surveys are based on mtDNA cytochrome b sequences except for the Pygmy Angelfish, which employed mtDNA control region sequences. The pelagic larval duration for Trumpetfish, Rock Hind, Soapfish, and Pygmy Angelfish are estimates from other members of the genus or family. An asterisk (*) indicates species with deep population structure and suspected cryptic evolutionary lineages.**

Species	Mean pelagic duration (days)	Population structure (ϕ_{ST})
Slippery Dick *(Halichoeres bivittatus)*	24[a]	0.77*, [b]
Black-ear Wrasse *(H. poey)*	25[a]	0.23[b]
Pudding Wife *(H. radiatus)*	26[a]	0.83*, [b]
Clown Wrasse *(H. maculipinna)*	29[a]	0.88*, [b]
Pygmy Angelfish *(Centropyge* spp.*)*	33[c]	0.62*, [d]
Redlip Blenny *(Ophioblennius atlanticus)*	38[e]	0.93*, [f]
Greater Soapfish *(Rypticus saponaceous)*	40[g]	0.87*, [h]
Rock Hind *(Epinephelus adscensionis)*	40[g]	0.93*, [h]
Ocean Surgeonfish *(Acanthurus bahianus)*	52[i]	0.72*, [j]
Blue Tang *(A. coeruleus)*	52[k]	0.36[j]
Doctorfish *(A. chirurgus)*	55[l]	0.02[j]
Blackbar Soldierfish *(Myripristis jacobus)*	58[m]	0.01[n]
Longjaw Squirrelfish *(Holocentrus ascensionis)*	71[m]	0.09[n]
Goldspot Goby *(Gnatholepis thompsoni)*	89[o]	0.47[p]
Trumpetfish *(Aulostomus strigosus)*	93[q]	0.59[r]

Sources:
[a] Sponaugle and Cowen (1997).
[b] Rocha et al. (2005a).
[c] Thresher and Brothers (1985).
[d] Bowen et al. (2006a).
[e] D. Wilson, pers. comm.
[f] Muss et al. (2001).
[g] Lindeman et al. (2000).
[h] Carlin et al. (2003).
[i] M. Bergenius, pers. comm.
[j] Rocha et al. (2002).
[k] B. Victor, pers. comm.
[l] Bergenius et al. (2002).
[m] Tyler et al. (1993).
[n] Bowen et al. (2006b)
[o] Sponaugle and Cowen (1994).
[p] Rocha et al. (2005b).
[q] H. Fricke & P. Heemstra, pers. comm.
[r] Bowen et al. (2001).

and Hellberg (2005) show genetic partitions on a scale of tens of kilometers in the Caribbean cleaner gobies (*Elacatinus* spp.), as is the case for mouth-brooding cardinalfishes (Apogonidae) and pouch-brooding seahorses (Syngnathidae, Fig. 20.10; Lourie et al. 2005). In contrast, some apparently sedentary reef fishes can have little population structure across huge swaths of ocean. The pygmy angelfishes (genus *Centropyge*) show no structure across the entire tropical West Atlantic, apparently due to oceanic dispersal of larvae (Bowen et al. 2006a). Some fishes may transform from larvae to juveniles but remain in the open ocean for an extended period, as is apparently the case for soldierfishes (genus *Myripristis*), which show no population structure across the entire central and West Pacific (Craig et al. 2007).

Researchers have compared **pelagic larval duration** (**PLD**) and population structure (measured with F statistics)

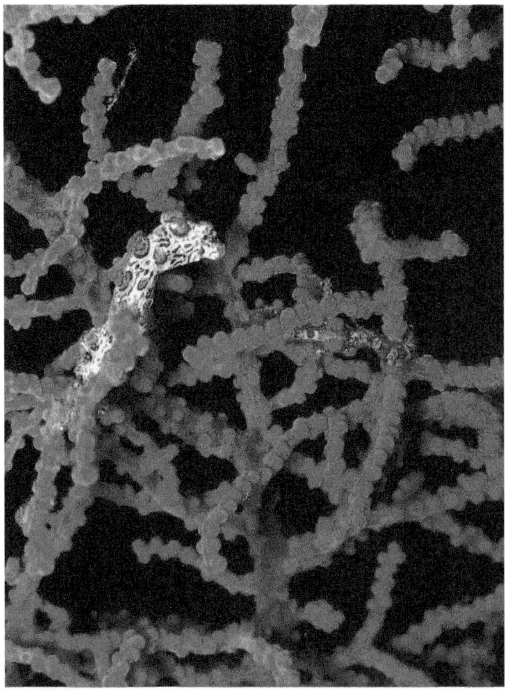

FIGURE 20.10 Seahorses and other syngnathids show limited dispersal. This limits gene flow and results in strong genetic differences among populations. Pygmy Seahorse (*Hippocampus bargibanti*) by Jlarouche / Wikimedia Commons / CC BY-SA 3.0.

to forge links among PLD, dispersal, and population structure. It seems obvious that if larvae are drifting with oceanic currents, longer pelagic duration will yield greater dispersal and less population structure. Indeed the first comparisons of PLD and genetic connectivity in marine fishes supported this connection. Waples (1987) surveyed 10 species in the eastern Pacific, Doherty et al. (1995) surveyed seven species on the Great Barrier Reef, and both of these allozyme studies found a correlation between PLD and population genetic structure. However, subsequent studies have not replicated this correlation. In surveys of eight reef fishes in the Caribbean Sea (Shulman & Bermingham 1995), eight species on the Great Barrier Reef (Bay et al. 2006), and 15 reef species in the tropical Atlantic (Bowen et al. 2006b), there was no significant correlation between PLD and population genetic structure (Table 20.3). The explanation likely includes at least three components:

1. The two studies that report a significant correlation between PLD and genetic connectivity (Waples 1987; Doherty et al. 1995) include species that lack a pelagic dispersive stage, and the significant relationship is weakened or lost without these cases (Bohonak 1999; Bay et al. 2006).
2. Fish larvae are not completely at the mercy of ocean currents. They can swim against currents, navigate, and in some cases remain in the vicinity of appropriate juvenile habitats (Leis & Carson-Ewart 2000b; Atema et al. 2015).
3. Most of the comparisons are among reef fishes, a category that is not cohesive in any phylogenetic or taxonomic sense. The reef fishes include lineages that diverged from one another 100+ million years before present (Bellwood & Wainwright 2002). Marine fishes are too diverse to expect a simple relationship between larval duration and dispersal.

Selkoe et al. (2014) examined PLD and genetic connectivity in 35 Hawaiian reef organisms, including fishes, and found that PLD could explain no more than 50% of variation in population structure. Habitat specialization, abundance, and historical factors (including taxonomy) were identified as factors in shaping differences in population structure. It appears that PLD has some influence on population structure, as is most apparent in the fishes with very short or very long pelagic stages. However, other life history factors such as habitat specificity and larval behavior (see below) are also involved (Riginos & Victor 2001; Rocha et al. 2002; Weersing & Toonen 2009).

Habitat Preference

In resolving population structure of marine fishes, most attention has focused on the dispersive larval stage. However the movements and feeding activities of adults play a role in shaping population structure, especially for fishes in the **pelagic zone** (see Chapter 19). For example, population structure in wide-ranging tunas, billfishes, and pelagic sharks is usually measured on the scale of ocean basins: East versus West Atlantic in the Bluefin Tuna *Thunnus thynnus* (Carlsson et al. 2007), North versus South Atlantic in the White Marlin *Tetrapturus albidus* (Graves & McDowell 2006), Indian versus Pacific in the Swordfish *Xiphias gladius* (Lu et al. 2006), and Atlantic versus Indian-Pacific in the Whale Shark *Rhincodon typus* (Castro et al. 2007).

A few **demersal** (bottom-dwelling) fishes conduct reproductive or seasonal migrations, but most are sedentary, and for this reason the corresponding habitat preferences are seldom considered in predicting population structure. However, habitat preference can have a strong influence on the distribution of genetic diversity in fishes. Usually ecosystem **specialists** (those with very specific feeding or habitat requirements) have more population structure than generalists, as demonstrated by genetic comparisons of reef fishes across the **Amazon barrier**. This turbid plume of fresh water was long regarded as a barrier that divided the West Atlantic reef fauna into northern (Caribbean) and southern (Brazilian) provinces (see Chapter 19). However, fresh water is less dense than salt water, and may form a surface layer with a saltwater "wedge" below. Trawl surveys conducted under the Amazon plume demonstrated the presence of many marine fishes that are usually associated with coral reefs (Collette & Rützler 1977). An mtDNA survey of West Atlantic wrasses (genus *Halichoeres*) across the Amazon barrier demonstrates a strong connection between habitat use and genetic structure. *Halichoeres maculipinna*, a reef species with specialized diet and feeding morphology, has an ancient evolutionary separation between Brazil and the Caribbean (sequence divergence $d = 0.065$ in cytochrome *b*). In contrast, *H. bivittatus* is found in a variety of habitats in addition to coral reefs and shows no strong genetic separation across the Amazon barrier (Rocha et al. 2005a). Notably, *H. bivittatus* was collected in the trawl surveys under the Amazon plume, whereas *H. maculipinna* was not. When combined, these genetic and field studies indicate that habitat preference and species ecology can be as important as geography and larval dispersal in defining the distribution of genetic diversity in fishes (Choat 2006).

Complex Population Structure

In migratory fishes, the resolution of populations (and corresponding management units) can be confounded by two factors:

1. Migratory overlap, in which populations mingle in feeding habitats or during migrations. Examples include the anadromous Sockeye Salmon (*Oncorhynchus nerka*; Grant et al. 1980) and Striped Bass (*Morone saxatilis*; Wirgin et al. 1997), as well as marine species such as the Bluefin Tuna (*Thunnus thynnus*; Carlsson et al. 2007) and possibly Atlantic Cod (*Gadus morhua*; Svedäng et al. 2007). When independent breeding populations overlap at shared feeding habitats, a critical question is whether genetic exchange occurs. If fish are not breeding during the period of overlap, those populations could be isolated management units.
2. Sex-biased dispersal, in which gene flow between populations is accomplished primarily by one gender. For many mammals and birds, males disperse prior to reproduction, while females remain in natal areas (Greenwood 1980).

Both population overlap and sex-biased dispersal are common in migratory marine fishes. Female site fidelity can be countered by opportunistic mating by males, so that each gender yields a different population genetic signal. This is known as **complex population structure** (Bowen et al. 2005), and the most common outcome is that female-inherited mtDNA shows population structure while biparentally inherited nDNA surveys show no structure (Goudet et al. 2002). This pattern is apparent in the Brook Trout (*Salvelinus fontinalis*; Fraser et al. 2004), Patagonian Toothfish (*Dissostichus eleginoides*; Shaw et al. 2004), and Scalloped Hammerhead shark (*Sphyrna lewini*; Daly-Engel et al. 2012). In a survey of White Sharks (*Carcharodon carcharias*) in the Indian Ocean, the mtDNA sequences reveal significant population structure ($F_{st} = 0.81$ between South Africa and Australia), whereas a microsatellite survey indicated a single population (Pardini et al. 2001). For these cases, dispersal by males can readily explain the lower population structure registered in nDNA relative to mtDNA.

Hybridization

An individual contains combinations of genes that have evolved together over millions of years. Different species contain different gene combinations, which means that a hybrid individual brings together genes that have not undergone such fine-tuned coevolution. The result is that hybrid individuals are typically aberrant in some aspect of their biology. Most natural hybrids are inefficient reproductively, ecologically, physiologically, or behaviorally. They are therefore likely to be evolutionary dead-ends because of sterility, low fertility, or an inability to attract mates. In most cases, natural selection will favor spawning individuals that avoid mating with members of other species. This separation of species during mating is accomplished via **species isolating mechanisms**, which are usually anatomical or behavioral traits that keep individuals of different species from breeding with one another. Species isolating mechanisms include incompatible genitalia, as in internally fertilized livebearers or elasmobranchs, or they may result from incompatible sperm and eggs, or differences in courtship patterns, timing, or location of spawning. Differences in chromosome number and organization can be a powerful isolating mechanism. Inappropriate cues given by one or the other member of a spawning pair can lead to termination of the spawning act.

In rare cases, hybridization can yield positive survival traits, such as faster growth or more vigorous mating behavior, a phenomenon referred to as **hybrid vigor** or **heterosis**. Hybridization can produce new genetic combinations with novel adaptive traits, and may ultimately increase biodiversity (Richards & Hobbs 2015).

Hybridization in fishes often occurs when one species experiences a substantial reduction in abundance. When a rare species breeds in the same place and time as an abundant species, interspecific matings are more likely (Hubbs 1955). Hybridization is also common in disturbed habitats where the preferred spawning sites of one species are lacking, forcing them to spawn in another habitat and hence with another species. When non-native species are introduced into a region that hosts a closely related species, isolating mechanisms between the introduced and the native species may not be effective. The resulting hybridization may induce reverse speciation (Coleman et al. 2014). Hence hybridization can both enhance and reduce biodiversity.

Among fishes, instances of hybridization are well known in freshwater fishes and less well documented among marine species (Schwartz 2001; Epifanio & Nielsen 2000; Scribner et al. 2000). Freshwater families which often have natural hybrids include the minnows (Cyprinidae), suckers (Catostomidae), salmons and trouts (Salmonidae), sunfishes and black basses (Centrarchidae), and darters (Percidae). Artificial hybrids produced in aquaculture are also common, as in the sunshine bass (a cross between a male Striped Bass, *Morone saxatilis*, and a female White Bass, *M. chrysops*), the splake (a cross between a Lake Trout, *Salvelinus namaycush*, and a Brook Trout, *S. fontinalis*), and the tiger muskellunge (a cross between a Northern Pike, *Esox lucius*, and a Muskellunge, *E. masquinongy*). Hybridization among marine fishes was previously believed to be rare, but this view is changing. The most common marine examples occur in the reef fish families of butterflyfishes (Chaetodontidae) and angelfishes (Pomacanthidae), although distinctive, complex color patterns and popularity among aquarium keepers make hybrids in these families more likely to be detected (Hobbs et al. 2013). A variety of marine fishes are now known to hybridize including swordfishes, cods, surgeonfishes (Cui et al. 2013; Halldorsdottir & Arnason 2015; DiBattista et al. 2016), and sharks (Morgan et al. 2012).

The disproportionate numbers of hybrids among freshwater species could reflect the greater degree of physical disturbance and of species introductions in freshwater habitats (see Chapter 22). Examples of disturbance-induced hybridization include cichlids in Lake Victoria, where females cannot distinguish among males of different species because turbidity impairs light transmission at long and short wavelengths of light (Fig. 8.8). In the Pecos River of New Mexico and Texas, the critically endangered Pecos River Pupfish, *Cyprinodon pecosensis*, is restricted to two sinkhole habitats. Its habitat has been invaded by a widespread, introduced species, the Sheepshead Minnow, *C. variegatus*. Hybrids between Sheepshead Minnow and Pecos River Pupfish have completely replaced the endemic pupfish along 500 km of the Pecos River (Echelle & Echelle 1997). In Europe, where habitat disruption is all too prevalent, several endemic cyprinids are impacted by introduced species. In southern Italy, Italian Bleak, *Alburnus albidus* (designated Vulnerable by the International Union for the Conservation of Nature, IUCN) hybridizes with an introduced cyprinid, *Leuciscus cephalus cabeda*; another endemic cyprinid, *Chondrostoma toxostoma arrigonis*, hybridizes with introduced *C. polylepis polylepis* in Spain (Crivelli 1995).

Hybridization in marine fishes may be most common at the boundaries of biogeographic provinces, where closely related species come into contact at the edge of their ranges (Hobbs et al. 2009). As in freshwater, most cases seem to involve one species that is abundant, and one species that is much less common.

Hybrids of lore and legend are plentiful and fun to envision, but that does not make them real. For example, reports of fur-bearing trout, sometimes called "beaver trout," in some US states were local folklore – perhaps even gimmicks initiated to attract curiosity-seeking tourists. Explanations for these furry trout included (i) the fur helps keep them warm in extremely cold water and (ii) an accidental spill of hair-growing tonic into a river. But the truth was in creative taxidermy – often attaching rabbit fur to a trout. The mythology of fur-bearing trout, however, goes back at least as far as the reported "Lodsilungur" of Iceland in the mid-1800s – a mythical inedible fish allegedly created by demons and giants to punish human misbehavior by taking over rivers (see **https://en.wikipedia.org/wiki/Fur-bearing_trout**).

One real case of hybridization is stranger than fiction. During an experiment to produce gynogenic (female only) Russian sturgeon, researchers used sperm from American paddlefish to initiate parthenogenesis. In principle, the sperm stimulates the egg to divide and grow without contributing genetic material. In this case, the mixing of sturgeon eggs and paddlefish sperm yielded viable hybrid offspring (Káldy et al. 2020). Most species that hybridize are only a few million years apart in evolutionary history, but the sturgeon and paddlefish are in separate families (Acipenseridae and Polyodontidae), estimated to have diverged about 180 million years ago.

Supplementary Reading

Population Ecology

Allan JD, Castillo MM. 2007. *Stream ecology: structure and function of running water*, 2nd edn. Dordrecht: Springer-Verlag.

Diana JS. 2003. *Biology and ecology of fishes*, 2nd edn. Travers City, MI: Cooper Publishing Group.

Gerking SD, ed. 1978. *Ecology of freshwater fish production*. London: Blackwell Science.

Goulding M. 1980. *The fishes and the forest. Explorations in Amazonian natural history*. Berkeley, CA: University of California Press.

Lowe-McConnell RH. 1987. *Ecological studies in tropical fish communities*. London: Cambridge University Press.

Matthews WJ. 1998. *Patterns in freshwater fish ecology*. New York: Chapman & Hall.

Potts GW, Wootton RJ, eds. 1984. *Fish reproduction: strategies and tactics*. London: Academic Press.

Ricker WE. 1975. Computation and interpretation of biological statistics of fish populations. *Bull Fish Res Board Can* 191:1–382.

Rothschild BJ. 1986. *Dynamics of marine fish populations*. Cambridge, MA: Harvard University Press.

Sale PF, ed. 1991a. *The ecology of fishes on coral reefs*. San Diego, CA: Academic Press.

Sale PF. 1991b. Reef fish communities: open nonequilibrial systems. In: Sale PF, ed. *The ecology of fishes on coral reefs*, pp. 564–598. San Diego, CA: Academic Press.

Sale PF, ed. 2002. *Coral reef fishes: dynamics and diversity in a complex ecosystem*. San Diego, CA: Academic Press.

Schluter D. 2000. *The ecology of adaptive radiation*. Oxford: Oxford University Press.

Stearns SC. 1992. *The evolution of life histories*. Oxford: Oxford University Press.

Wootton RJ. 1999. *Ecology of teleost fishes*, 2nd edn. Fish and Fishery Series No. 24. New York: Springer-Verlag.

Population Genetics

Allendorf FW, Luikart G., Aitken SN. 2012. *Conservation and the genetics of populations*, 2nd edn. Hoboken: Wiley.

Cutter AD. 2019. *A primer of molecular population genetics*. New York: Oxford University Press.

Grant WS, Waples RS. 2000. Spatial and temporal scales of genetic variability in marine and anadromous species: implications for fisheries oceanography. In: Harrison PJ, Parsons TR, eds. *Fisheries oceanography: an integrative approach to fisheries ecology and management*, pp. 61–93. Oxford: Blackwell Science.

Graves JE. 1998. Molecular insights into the population structures of cosmopolitan marine fishes. *J Heredity* 89:427–437.

Hahn MW. 2019. *Molecular population genetics*. New York: Oxford University Press.

Hartl DL, Clark AG. 2006. *Principles of population genetics*, 4th edn. Sunderland, MA: Sinauer Associates.

Hedrick PW. 2010. *Genetics of populations*, 4th edn. Boston: Jones & Bartlett Publishers.

CHAPTER 21

The Functional Role of Fishes in Communities and Ecosystems

Summary

An ecological community is composed of all of the living things coexisting and interacting together in an area. A community can be subdivided into assemblages, with each assemblage being a subset of the broader community and more narrowly defined for the purpose of a particular study – such as the fish assemblage in a stream. Ecosystems consist of the biotic community and the abiotic environment with which the community interacts.

Species often have relatively predictable habitat use patterns and predator–prey and competitive interactions; these are all part of the niche of that species. Species that utilize similar resources in similar ways are members of a guild, such as zooplanktivores in a lake or cleanerfishes on coral reefs. Niches and guild memberships change as fishes grow and their food and habitat preferences change. Two general aspects of habitat use by fishes is that larger individuals of a species occur in deeper habitats, and that habitats in rivers and the species occupying them differ as one moves downstream.

For decades there has been an ongoing debate over the relative importance of the changes in the physical environment (stochastic) versus biological interactions (deterministic) as factors influencing how many and what kinds of fishes occur in an assemblage. This debate includes studies of fish assemblages in a variety of freshwater and

CHAPTER CONTENTS

The Diversity of Fishes: Biology, Evolution and Ecology, Third Edition. Douglas E. Facey, Brian W. Bowen, Bruce B. Collette, and Gene S. Helfman.

Companion website: www.wiley.com/go/facey/diversityfishes3

marine habitats. Among coral reef fishes, much of the debate focuses on whether adult populations are determined by larval mortality (recruitment limitation) or by events occurring after recruitment, such as predator–prey and competitive interactions among juveniles and adults. Studies continue to show that both deterministic and stochastic factors contribute, although their relative importance differs temporally and spatially.

Competition within and among species results when consumers use a limited resource. Shifts in resource use due to competition result in resource partitioning. Competition for food resources is most common in fishes, which can lead to dramatic habitat shifts and can also influence predator–prey interactions. Differences in resource use do not automatically imply that competition is occurring; physiological requirements, phylogenetic constraints, and differential susceptibility to predation can also produce species differences in resource use. Introduced species frequently have strong, deleterious, competitive impacts on native species.

Predation can directly affect prey density through predator-caused mortality, or can have indirect effects through predator avoidance that places prey in suboptimal environments, thereby slowing individual growth and reproductive output. Predation can also cause genetic differences in coloration, habitat use, and schooling and breeding behavior. Introduced predators have decimated natives in many locales. Many fishes are both predators and competitors of other fishes at different stages in their lives.

Some fishes are parasites of other fishes, and many fishes serve as hosts for a wide variety of invertebrate parasites. Fishes also interact with non-fish taxa, competing for food and space while eating and being eaten. The distribution within habitats of many fish species represents an avoidance of piscine, mammalian, and avian predators; in many streams, fishes are squeezed out of deep water by fish predators and out of shallow water by wading birds.

Herbivory among fishes is more common in tropical than temperate habitats, in part because the symbiotic gut microorganisms responsible for breaking down cellulose are more effective at warmer temperatures. Common tropical herbivores include minnows, characins, catfishes, and cichlids in fresh water, and surgeonfishes, parrotfishes, rudderfishes, blennies, and damselfishes on coral reefs. Fishes influence macrophyte biomass, productivity, growth form, energy allocation, and species composition; fishes also disperse seeds. Macrophytes have evolved mechanical and chemical defenses against herbivorous fishes. Damselfishes on coral reefs maintain algae "gardens" within their territories, encouraging edible species and discouraging growth of less palatable species. Damselfish activities thus affect the diversity and distribution of algae and the many invertebrates that live in algal patches.

Temperate freshwater herbivores include minnows, catfishes, suckers, pupfishes, and killifishes. During warm months when macrophytes grow quickly, grazing minnows can crop most of the macrophyte productivity. During cold months, temperate herbivores commonly shift to carnivory. Phytoplanktivory also occurs in temperate lakes, where fishes such as shad can affect plankton abundance and diversity.

Zooplantivorous fishes generally prefer large zooplankters, which shifts the size and species composition of the plankton to smaller zooplankton species in both marine and freshwater habitats. Avoidance of foraging fishes may be responsible for daily vertical migrations by zooplankters, for day–night differences in zooplankton assemblage composition, and for life history and anatomical traits of zooplankters and other invertebrates. Predation by fishes also can influence the abundance and distribution of benthic invertebrate prey in many communities. Effects seem to be greater in lakes and ponds than in rivers and streams.

"Trophic cascades" describe the direct and indirect effects that predators at the top of a food web can have on lower trophic levels. For example, piscivorous fishes eat zooplanktivorous fishes, which feed on herbivorous zooplankton, which eat phytoplankton. Therefore, removing top piscivores may increase phytoplankton density. However, many fishes feed at different trophic levels at different times in their lives, or shift diets based on availability of prey. This can make it challenging to predict outcomes of changing the abundance of fishes or their prey. Complex interactions of this nature indicate that changes in fish populations can ultimately affect water chemistry, calcium carbonate deposition, the distribution of water masses of different temperatures, and ultimately the heat budget of a lake.

Fishes can directly affect the transport and cycling of nutrients in aquatic habitats. Phosphorus excretion by fishes is important for algal growth. Benthic fishes disturb sediments, which increases the transfer of nutrients from the mud to the water column. Fish bodies contain a large fraction of the nutrients in many ecosystems; nutrients are released through excretion from the gills, through defecation, and through decomposition after death. Vertical and horizontal migrations by fishes that feed in one area and rest in another influence coral growth on coral reefs and kelp growth in kelp beds. And the long-distance migrations of salmon link oceanic ecosystems with headwater streams, even influencing the growth of trees in adjacent forests. Fishes can also affect the production and distribution of substrate, as when parrotfishes grind coral into sand, or when tilefishes or breeding minnows pile rocks over their burrows or nests.

Physical factors that appear to have the greatest effects on fish assemblages include reductions in dissolved oxygen from drought and ice cover, storm-induced increases in stream and river discharge, and habitat destruction on coral reefs and kelp beds from storm-caused waves. Fishes can influence physical factors in habitats, such as redistribution of substrates by nest construction by salmon and lamprey, or changes in substrate composition and oxygenation by burrowing of larval lamprey. Biological disturbances with ecosystem-wide repercussions include outbreaks of disease or population explosions of species that literally eat the food and habitat resources of a system.

Fishes provide valuable ecosystem services that benefit humans and other organisms. These include food, processing and transport of nutrients, and contributing to ecological processes responsible for clean water. Spawning behaviors and burrowing also play important roles in physicochemical processes. The global climate is changing due to human activities, with potentially severe consequences for fishes in almost all ecosystems.

Introduction to Assemblages, Communities, and Ecosystems

An ecological **community** is all of the living things together in an area. This area may include arbitrarily determined boundaries based on geopolitics but not recognized by organisms (e.g. the marine communities of Florida or Kaneohe Bay, Hawaii) or may involve biologically relevant boundaries across which few fishes pass (e.g. Lake Tahoe or the Chattahoochee River). Many ecologists subdivide communities into **assemblages**, each a subset of the larger community that is the focus of a particular study. For example, the fish assemblage of a particular stream, whereas others might instead refer to this grouping as the stream fish community. In either case, it would be a subset of the larger stream community, which would also include periphyton, invertebrates, salamanders, and anything else utilizing the resources of the stream. Traditionally, community ecology has focused on biotic interactions among different taxonomic groups and the effects that such interactions have on distribution and abundance.

An **ecosystem** consists of the biotic community and the abiotic environment with which the community interacts. Hence one can talk about a stream ecosystem, a lake ecosystem, or an intertidal or offshore reef ecosystem. At a larger spatial scale, one can think in terms of **watersheds**, which take into account the land from which water flows into a series of streams and eventually into a lake or river, and the hydrological, geological, and biological forces at work there. The next level of organization is the **landscape** or **riverscape**, which recognizes interactions and linkages among ecosystems and the influence of human activities on these interactions (Schlosser 1991; Fausch et al. 2002). Ecosystem ecology has focused more on the flow of energy, nutrients, and materials among components of the ecosystem.

Assemblages

An assemblage may consist of the various populations of a larger taxonomic group in a defined area. **Assemblage structure** refers mainly to the number of individuals, species, and families, and the predator–prey interactions and other trophic relationships among members of the assemblage (Matthews 1998). Focusing on an assemblage comprised only of fishes within a community is admittedly myopic, because fishes interact with other organisms such as invertebrate prey and parasites, algae as food and shelter, and reptiles, birds, and mammals as predators. However, interactions among fishes are particularly obvious, and interesting to those who study fishes. In addition, it is often logistically difficult to deal with all components of a community or ecosystem, and researchers tend to specialize and develop expertise in certain taxonomic groups (hence the rationale for producing an ichthyology or any other taxon-oriented textbook). In this section we will look at competitive and predator–prey interactions that involve fishes of different species, and discuss prevalent ideas on how interactions among fishes affect species composition and maintenance of assemblages.

Niches and Guilds: The Ecological Role of a Species

Most discussions of the composition of natural assemblages focus on the **functional roles** of the different species. The ecological function of a species is synonymous with its **niche**, a broadly defined term that essentially includes all relevant ecological aspects of an organism, including its environmental and microhabitat requirements (e.g. temperature, oxygen concentration, pH, salinity, and substrate type), what it eats and what eats it, and symbiotic associations in which it participates. Characterizing and measuring niche components and dimensions allow us to compare niches among species and measure changes in niche use that occur when species are added to or subtracted from an assemblage.

An additional, useful concept for understanding the ecological roles of different species within an assemblage is the **guild**, which emphasizes ecological rather than taxonomic similarities (Gerking 1994). A guild consists of the different species in an area that exploit similar resources in a similar way. Hence the many fishes that hover in the water column along the face of a coral reef and feed on zooplankton make up the zooplanktivore guild, along with the large predatory invertebrates feeding in the same area on the same resource. And the diurnal zooplanktivore guild includes different species than the nocturnal zooplanktivore guild (Hobson 1975).

Stream fishes can be classified by their habitat preferences as members of a benthic guild (e.g. some minnows, suckers, sculpins, and darters) that feed largely on benthic invertebrates living among rocks or buried in sediments,

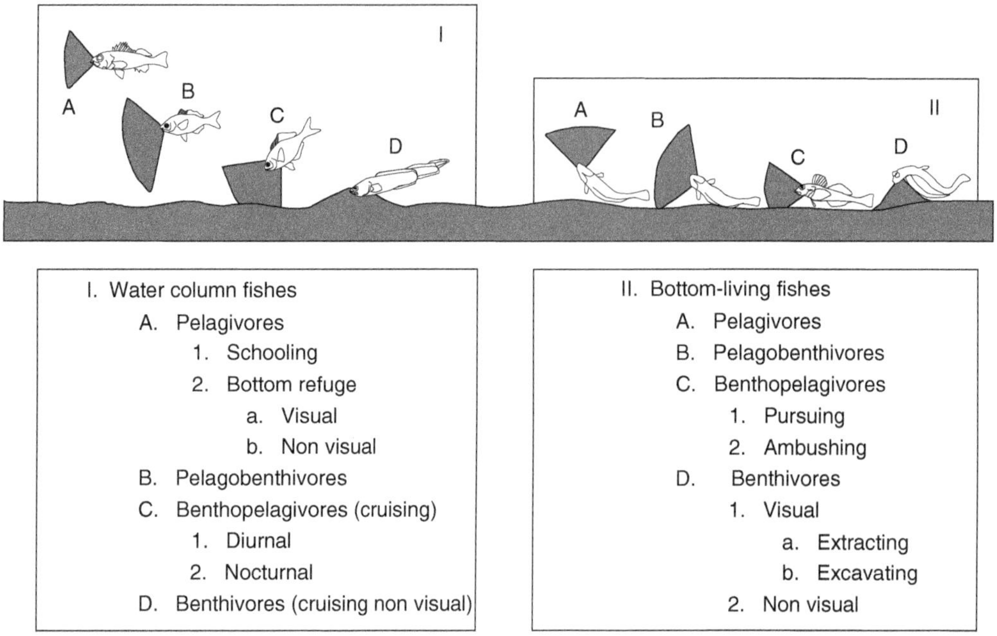

FIGURE 21.1 Foraging guilds of soft bottom fishes on the southern California shelf. Two general groups are recognized: those that swim above the bottom and those that are in contact with the bottom. The actual guilds take into account height above bottom, foraging type (search and capture activities), and time of feeding. From Allen (2006), used with permission.

or as members of a water column guild feeding largely on drifting insects or on insects that fall onto the water's surface (trouts, several minnows) (Grossman & Freeman 1987). Guilds in tropical streams may include algivores, aquatic and general insectivores, piscivores, scale and fin eaters, terrestrial herbivores, and omnivores (Angermeier & Karr 1983). In other habitats, examples may include tidepool guilds, kelpbed water column guilds, pelagic predator guilds, wave-zone sand-dwelling guilds, rock-dwelling lake guilds, and buried benthic predatory guilds. The shallow water (10–200 m), soft bottom assemblage of fishes off the southern California coast can be divided into approximately 18 foraging guilds, based on how and what they eat and their position in the water column (Allen 2006) (Fig. 21.1). A drawback to the guild concept is that it overlooks the opportunistic feeding habits of so many fishes, especially as adults (Matthews 1998; Hobson et al. 2001; Allan & Castillo 2007).

Fishes may belong to different guilds at different times in their lives. For example, angelfishes, wrasses, and leatherjackets (*Scomberoides*, Carangidae) may belong to the cleanerfish guild as juveniles, but as adults change to feeding on sessile invertebrates, mobile invertebrates, or small fishes, respectively. Other studies documenting dietary changes in fishes as size increases include Silver Seabream (also called Australasian Snapper, *Pagrus auratus*) in a New Zealand estuary (Usmar 2012), and juvenile and adult Goliath Grouper (*Epinephelus itajara*) off the coast of French Guiana (Artero et al. 2015). It is important for those who study community ecology to not assume that all individuals of a species utilize the same resources because many species change ecological niche and guild as they grow (see Nakazawa 2015).

Habitat Use and Choice

An important component of a species' niche, and one that can easily differ among species, is its habitat. Habitat can often be described and quantified in detail. Among stream fishes, species habitat preferences often differ in height above bottom (Fig. 21.2), current strength, bottom type (particle size and type), structure, distance from shore, amount of vegetation, and type and amount of food resources. A survey of major habitats in an eastern North American stream (rapids, riffles, runs, pools, and overhangs) often shows that species segregate along vertical dimensions, with certain species typically found in contact with the bottom (catfishes, darters, sculpins, eels, and grazing minnows), some just above the bottom (suckers), some species low and others higher in the water column (planktivorous minnows, trout), some close to the surface (silversides, livebearers, and topminnows), some in swift water (darters, trout), and others in moderate flow or in slow flow near more rapidly flowing water (catfishes, minnows, pickerel, and sunfishes). The existence of particular species in particular habitats implies an active choice by individuals (Gorman 1988). Experimental studies, usually of juveniles, generally show that individuals actively choose habitats, that the ones they prefer are those in which the species is most often found and in which a species can successfully feed, avoid predators, and reproduce. Hence, habitat choice is an evolved aspect of a species' niche (Sale 1969). Habitat choice is, however, dynamic within a species, varying on the basis of age, size, sex, reproductive condition, geographic area, and environmental conditions (Karr et al. 1982; Allan & Castillo 2007).

Similar descriptions, based on habitat characteristics, apply to assemblages in most major habitat types. Different

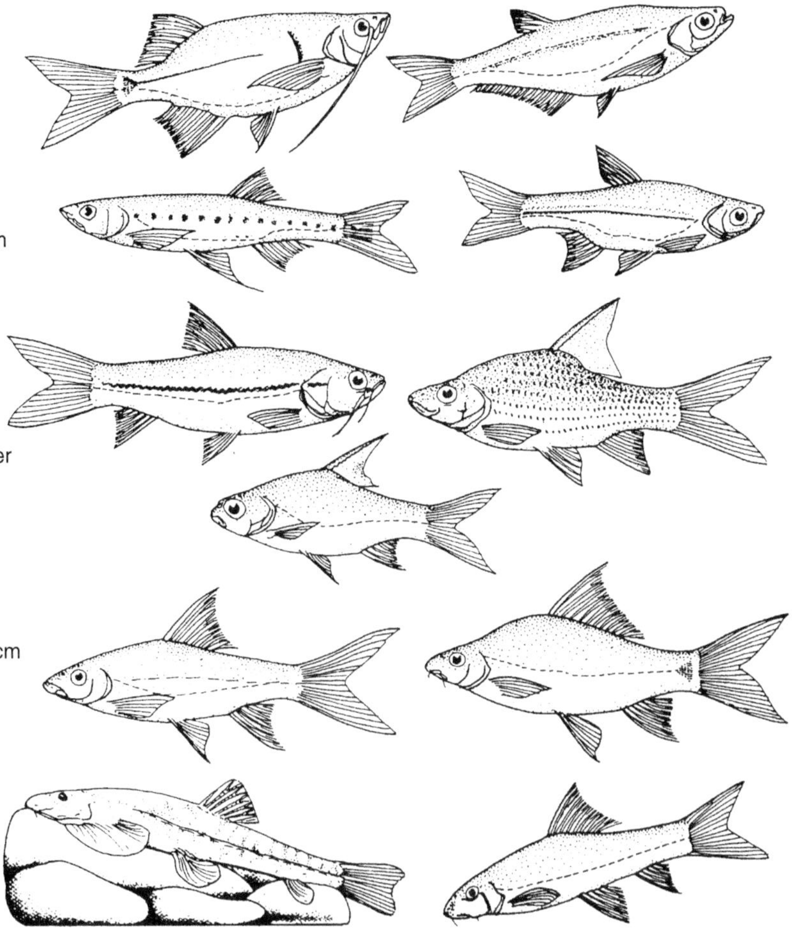

FIGURE 21.2 Habitat choice in stream fishes as demonstrated by vertical segregation among cyprinid fishes in a Borneo stream. Habitat choice is one aspect of the niche of a species; species choose habitats according to specific characteristics, and species often differ in one or more quantifiable characteristics of habitat. From Welcomme (1985), used with permission.

faunas in different geographic locales often occupy similar habitats and, in essence, converge on many niche characteristics. The Chinese Algae Eater, *Gyrinocheilus* (Gyrinocheilidae), belongs to a Southeast Asian family related to loaches. It lives in rivers where it feeds on algae attached to rocks. It has a sucking mouth with which it clings to rocks even while feeding in high-flow situations. A South American catfish, *Otocinclus* (Loricariidae), similarly feeds on algae-covered rocks in flowing water. It is convergent with the Algae Eater in body form and size, suctorial mouth, and even coloration (see Chapter 10, Strong Currents and Turbulent Water). Stream-dwelling galaxiids of the southern hemisphere, especially in New Zealand and Australia, occupy niches very similar to those of salmonid trouts in the northern hemisphere. They are convergent in body morphology, habitat, and foraging habits and have been rapidly replaced by introduced trout in many locales due to competition and predation. A characin, the Dorado, *Salminus maxillosus*, is a desirable gamefish that lives in streams of the Amazon region (Fig. 21.3). Its morphology, coloration, and piscivorous foraging habits are, as its generic name implies, remarkably convergent with stream-dwelling salmonids such as Brown Trout, *Salmo trutta* (Esteves & Lobo 2001).

FIGURE 21.3 The characid Dorado (*Salminus brasiliensis*), a popular gamefish in the Amazon basin, is ecologically convergent with riverine salmonids. The one in the photo was caught fly fishing. They can get much larger, with some females exceeding 20 kg. PY2ING / Wikimedia Commons / CC BY-SA 3.0.

A very broad and striking convergence occurs among approximately 60 families of fishes that fill the **eel niche** in their respective assemblages. Only 22 of these families belong to the order Anguilliformes and are therefore "true" eels. Many of these fishes are convergent in habitat choice, occurring on and often in the bottom in soft sediments or among the interstices of rocks and other structure. They share other characteristics: elongated dorsal and anal fins that often lack hard spines, increased vertebral counts, reduced opercula, missing pelvic or pectoral fins, missing or embedded scales, and a pointed tail. Behaviorally they are carnivores and scavengers (some are parasitic), can move backwards and forwards with equal facility (**palindromic locomotion**), and tear pieces off their prey by holding on and rotating rapidly along their long body axis (Helfman 1990).

Habitat choice changes with the seasons, with size and age of fish, and also with the presence of other species, particularly predators and competitors. Many stream species, such as darters and benthic minnows and suckers, occupy increasingly swift water as they grow older. Often the distribution results from spawning habits: adults migrate upstream to spawn in headwaters and young move progressively downstream with growth (Hall 1972). Such an **ontogenetic habitat shift** could reflect a body size constraint related to the current speed at which an individual can hold position without expending excess energy or using too much oxygen.

A general pattern of increasing depth with age occurs in a number of freshwater and marine species. In many species, this trend reflects the use of inshore, shallow, productive nursery areas that are reputed to be relatively predator-free (e.g. saltmarshes, mangroves; see Sheaves et al. 2015). These fishes then move offshore as they grow as demonstrated by three common reef fishes off of Dar es Salaam, Tanzania (Kimirei et al. 2013), and Goliath Grouper (*Epinephelus itajara*) in French Guiana (Fig. 21.4, Artero et al. 2015). Other species (minnows, copper sweepers, damselfishes, surgeonfishes, sunfishes, croakers, percids, clinids, wrasses, and Great Barracuda), also show a correlation between size and depth (Helfman 1978; Power 1987). That the relationship exists in a diversity of habitats and involves a number of unrelated species (Polloni et al. 1979) implies a convergent, adaptive trait.

FIGURE 21.4 Goliath Grouper, like many marine fishes, occupy productive inshore areas as juveniles but then move to deeper water as they grow. This individual is accompanied by two remoras (Echeneidae). G. Helfman (Author).

"Bigger–deeper" distributions within species and other types of habitat shifts may result from avoidance of predators, influenced by differences in vulnerability between different size classes of prey. Such patterns of habitat choice are noted in a study of Nassau Grouper (*Epinephelus striatus*) which move to deeper, more open habitats as they get larger (Dahlgren & Eggleston 2000), and many fishes occupying rocky reefs in the southwest Mediterranean Sea in which young-of-year occupying shallower areas and larger juveniles move to somewhat deeper and more complex habitats with increased feeding opportunities (Félix-Hackradt et al. 2014). In freshwater, when predatory sunfishes (*Lepomis* spp.) are present and Largemouth Bass are absent, young-of-the-year Central Stoneroller minnows (*Campostoma anomalum*) occupy shallows of pools, whereas larger individuals prefer deeper portions. Sunfishes occupy deeper portions of pools and can eat young minnows but are too small to eat larger minnows. When bass, which can eat all minnows and also prefer deep sections, are added, all size classes of stonerollers occupy shallow margins or emigrate from pools (Power 1987).

A combination of fish, avian, and mammalian predators affect the depth distributions of loricariid catfishes (including the "plecostomus" species of the aquarium trade) in Panamanian streams. These herbivores feed on algae that grow most abundantly in the shallows of pools. Small catfishes live in the more productive shallows, whereas large catfishes occur in deeper water where algal production is minimal and where the fish lose fat reserves and cease growing. The habitat differences of the two size groups are enforced by the balanced impact of terrestrial predators and piscivorous fishes. Birds and mammals can capture any size prey whereas piscivorous characins are gape-limited and cannot swallow the large catfishes. Large catfishes occupy deeper, less productive regions to avoid terrestrial predators. The small catfishes avoid predatory fishes that live in deep water but can hide from the birds and mammals in shallow water by seeking refuge among rocks; however, these refuges are too small for larger catfishes (Power 1987; Power et al. 1989). Cannibalism can also influence habitat shifts, forcing smaller fish into suboptimal habitats where their growth rates suffer. Smaller sculpins (*Cottus*) use a diversity of habitats in some streams and prefer deeper water in others when large conspecifics are missing. If larger fish are present or are added experimentally, the smaller fish shift to shallower habitats. Similar habitat shifts, resulting in the occupation of suboptimal feeding habitats by small prey, have been demonstrated in other minnow species and Bluegill (Werner et al. 1983; Gilliam & Fraser 1987; Schlosser 1987; Freeman & Stouder 1989).

Habitat Choice and Spatial Structure: Zonation

Researchers have compared the fish species that occur in different habitats along environmental gradients, such as from headwaters to mouths of streams and rivers, vertically within kelp forests, across rocky intertidal zones, or from shore to reef face or continental shelf, or across the sublittoral regions of oceans and lakes out to the limnetic or pelagic zone (Horn & Martin 2006; Stephens et al. 2006) (Fig. 21.5). These investigations have led to generalizations about zonation in various habitats.

Almost every fish habitat that has been studied can be divided into more or less distinctive zones. Early work of this type focused on stream and river (riverine, lotic, or fluviatile) fishes in the British Isles and Europe, where analogous **longitudinal zones** (habitat types along the length of a river) were identified in different systems. The zones were often named based on the common fish species present, which in turn reflected species' preferences with respect to gradient (slope), water velocity, stream width, stream depth, temperature variation, oxygenation, and sediment type. These habitat characteristics also influenced the presence and type of vegetation both in and along a river, bottom characteristics, and invertebrate fauna. A popular classification recognized four basic zones, beginning with the headwaters and moving to the lowlands (Huet 1959; Hawkes 1975):

1. The **trout zone**: the narrow, shallow, cold, steep (often torrential), highly oxygenated headwater region with large rocks or gravel; common fish species show morphological or behavioral adaptions to high flow and include Brown Trout, Atlantic Salmon fry, Bullhead Sculpin (*Cottus gobio*), and Eurasian Minnow (*Phoxinus phoxinus*).
2. The **grayling zone**: deeper, less steep, with alternating riffles and pools, relatively strong currents, less rocky, more gravelly bottom, cool, slightly less oxygenated, with

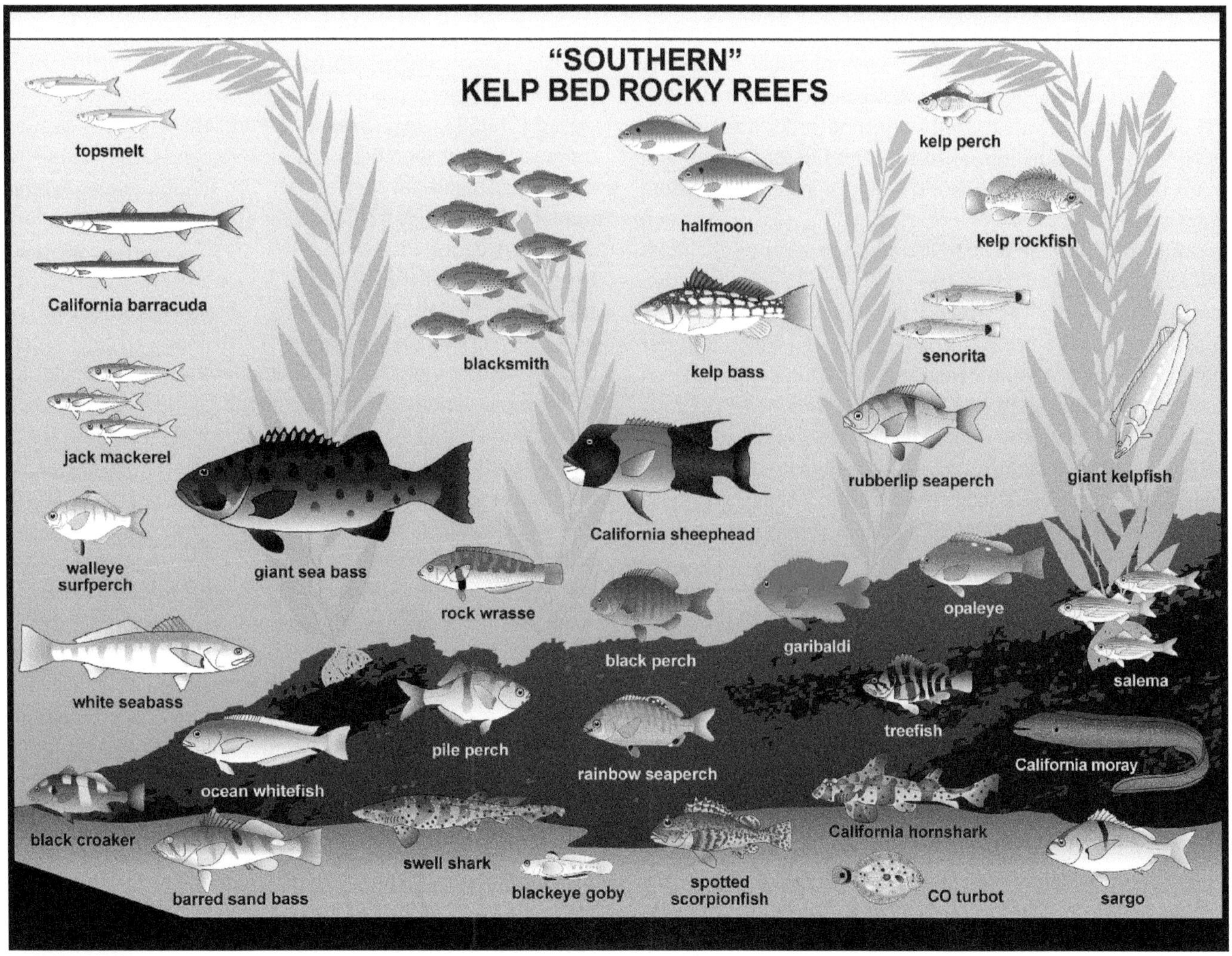

FIGURE 21.5 Vertical zonation of fishes in a kelp bed. Species portrayed are the more common fishes found in southern California rocky kelpbed reefs. The assemblage is a mixture of three biogeographic faunal elements, involving tropical- and subtropical-derivative families (chubs, grunts, croakers, damselfishes, and wrasses), cool temperate Oregonian families (rockfishes, surfperches, greenlings, and sculpins), and cool temperate San Diegan species such as kelp rockfish and black perch. From Stephens et al. (2006), used with permission.

salmonids in the rapids and rheophilic (current-loving) minnows in the pools; common fishes are Grayling (*Thymallus thymallus*), species of the trout zone, and rheophilic minnows (Barbel, Chub, Hotu, and Gudgeon).

3. The **barbel zone**: riverine conditions of moderate gradient and current, greater depth, alternating rapids and runs (quieter flowing water), fluctuating temperatures; common fishes are rheophilic cyprinids, other cyprinids (Roach, Rudd, and dace), and predators (pike, perch, and European Eel).
4. The **bream zone**: lowland reaches that include rivers, canals, and ditches, little current, high summer temperatures with oxygen depletion, and turbid water; common fishes are Roach, Rudd, dace, and predators of the barbel zone plus slow-water cyprinids (Carp, Tench, and bream).

Alternative classifications emphasize different species in different areas, but the recognition of **longitudinal succession** in habitats and species, often involving several basic zones that differ in a few essential physical features, is common to most classification schemes. Underlying causes of zonation have focused on the physical factors, which often correlate with **stream order**. Order is determined by tributary number, and order only increases when streams of equal order join. For example, small, headwater streams are first order, two first-order streams join to form a second order stream, two second-order streams join to form a third-order stream, and so on (Kuehne 1962). The world's largest rivers are seldom much more than 10th order. For example, the Mississippi from its confluence with the Ohio to its mouth, a distance of almost 1000 km, is about an 11th-order river. To increase in order, another 11th-order river would have to flow into it (Fremling et al. 1989). Faunal breaks, where species type and number change in direct correspondence with stream order, have been identified in several systems (Lotrich 1973; Horwitz 1978; Evans & Noble 1979), but other studies have shown that elevation, gradient, historical factors, and especially upstream drainage area and discharge are also likely correlates of species richness and faunal change (Matthews 1986a, 1998; Beecher et al. 1988; Hughes & Omernik 1990).

An alternative to a stream order classification is measurement of **downstream**- or **D-links** (Osborne & Wiley 1992). The D-link approach emphasizes the importance of proximity of tributary streams to larger rivers. Small streams flowing into larger rivers have a greater source of colonists and migrants than is the case for tributary streams flowing into small rivers, and seasonal flooding of rivers often causes water and fishes to enter tributaries, drastically altering their species composition. It is therefore necessary to know "the position of a given stream within the overall basin drainage network in order to adequately predict its potential for species richness" (Matthews 1998, p. 312).

Longitudinal zonation among fishes has been described in numerous tropical and temperate riverine systems (Moyle & Nichols 1974; Horwitz 1978; Balon & Stewart 1983; Welcomme 1985). In eastern North America, headwater streams may contain only a single species, such as Brook Trout, sculpin, or Creek Chub. Downstream, where streams increase in both size and velocity, habitat diversity increases and species adapted to high-flow conditions (e.g. darters, hogsuckers, Longnose Dace, rheophilic minnows, sculpin, and Smallmouth Bass) are added. Owego Creek, NY, showed longitudinal succession mainly by addition of species associated with greater depth at downstream sites (Sheldon 1968). The presence of Brook Trout (*Salvelinus fontinalis*) and Mottled Sculpin (*Cottus bairdi*) throughout the creek suggests that there was not a great extent of habitat differentiation in terms of substrate, temperature, and oxygen levels.

Further downstream, as the system widens and deepens, sediments are deposited and primary and secondary production increase. Swift water species characteristic of upstream areas may drop out, but many more species are added. Food webs become less dependent on inputs of food from the surrounding watershed (**allochthonous inputs)** such as falling leaves and falling insects and more dependent on production from the stream itself (**autochthonous inputs)** from periphyton, macrophytes, and associated animals. Deep pools and slow flowing conditions alternate with shoals and rapids. Fish diversity increases to include suckers, herbivorous minnows, catfishes, and Largemouth Bass. In large, slow-flowing rivers with little gradient, planktonic organisms and planktivores are added (shad, herring, silversides, Paddlefish), as well as predators such as Striped Bass and pickerel, and larger carnivores such as sturgeon (all species that also inhabit lakes). Production of food is largely from the river itself, much of it in the form of detritus-eating insects that live on snags formed by fallen trees from the gallery forest. The floodplain and its forest contribute substantially both as sources of nutrients and as nursery areas during seasonal flooding. A fish fauna of backwaters, oxbows, and sloughs is added, including some that are adapted to swampy, periodically deoxygenated conditions (many sunfishes, small pickerels, livebearers, killifishes, Swampfish, Pirate Perch, mudminnows, gars, and Bowfin). Finally, as the river enters the coastal zone and is subject to both the tidal and salinity influence of the ocean, a very diverse assemblage that includes freshwater species from the lower river reaches and marine species from nearshore zones inhabit the saltmarsh or other estuarine region. In all zones, fairly characteristic species are associated with relatively definable habitats.

Since the original formulations of the idea of longitudinal zonation, many corrections, modifications, and exceptions have been reported. Zones were originally thought to be distinct, but sharp delineation between zones is probably more the exception than the rule. It occurs most often where montane streams flow into foothills or lowlands, such as in western North America where a cold-water fauna exists in a high elevation region but the fauna changes as the stream leaves the mountains (Matthews 1998). In reality, zones usually grade into one another, with "border zones" that are intermediate in physical nature and species composition connecting the zones. Sometimes these border zones may be longer than the fish zones they presumably separate.

The faunas of different zones are not necessarily distinct from one another. Species diversity typically increases downstream due to additions of species. However, changes in physical and chemical features can also lead to some species dropping out and being replaced by others. Fish assemblages in Green Creek, a spring-fed third order stream in northern Ohio, did not follow the expected pattern of longitudinal succession due to spring input of cold water and $CaCO_3$ in places which changed the physical and chemical environment (Swaidner & Berra 1979). The Dan River flows across mountain and piedmont regions along the Virginia-North Carolina (USA) border, and fish diversity generally increased along the river due to both the addition and replacement of species (Rohde et al. 2001). Most species were widely distributed, but physical and chemical changes in microhabitat played a strong role in longitudinal succession. The Teesta River (West Bengal, India) also showed longitudinal changes in fish assemblages along its 309 km length, although greatest species richness was in the mid-reaches of the river where water clarity was good and the river bed was a mosaic of patches of different substrate types (Chakrabarty & Homechaudhuri 2013). Fish diversity decreased further downstream where water clarity diminished and the substrate was less variable.

Additional complexities include missing zones, as where rivers appear suddenly due to the emergence of major springs (e.g. northern Florida), or disappear suddenly (western deserts). Reversals in zones may occur in rivers that have a stairstep course, with repeated slow-flowing flat sections that become steeper, or in rivers that flow through lakes that are cooler than the incoming river water. Seasonal migrations among zones and out of rivers and into lakes and the ocean change assemblage composition dramatically. Benthic invertebrates also occur in zones, but these zones may or may not correspond with fish zones.

Most of these exceptions do not contradict the idea of zonation because they can be anticipated from the specific environments involved. The basic idea of a relatively few, generalizable zones, with characteristic fish species living under characteristic flow and temperature regimes, has remained a viable descriptor and predictor of fish assemblages in a surprisingly large number of lotic situations (Hawkes 1975), and can serve to indicate disruptions due to human causes (Fausch et al. 1984). The major shortcoming of viewing a river, or any aquatic habitat, as isolated sections or zones containing relatively independent and distinct assemblages is that it ignores the critical, functional linkages among sections and subsets of the biota. In this regard, the **river continuum concept** (**RCC**; Vannote et al. 1980; Minshall et al. 1985) can be usefully applied to fish assemblages in rivers. The RCC views a river as an orderly progression of predictably intergrading, dependent regions containing organisms whose ecological roles reflect changes in river basin geomorphology, current speed, gradient, sediment and organic matter composition, and allochthonous versus autochthonous production (among aquatic insects, shredders and gathering collectors predominate in headwaters, shredders are replaced by scrapers in middle reaches, and filter feeders predominate in higher order sections). The integration of fishes into the RCC remains an ongoing challenge in fish (and riverine) ecology (see Matthews 1998; Allan & Castillo 2007), but a river is a heterogeneous system of multiple habitat types linked by both water and fish movement – a **riverscape** (Fausch et al. 2002; Carbonneau et al. 2012).

The roles of stochastic and deterministic processes in determining assemblage and community structure

The relative roles of **stochastic** and **deterministic** processes in determining assemblage and community structure and diversity have been an ongoing debate since at least the 1980s. Stochastic processes are independent of biological interactions and include weather-related events, such as floods or wave action from storms. These can affect a wide range of aquatic habitats and disrupt spawning and recruitment depending on when they occur. In contrast, deterministic processes are a function of biological interactions such as competition, predation, and symbioses, and are often density-dependent and based on the ecological niches of the species involved. These factors have the greatest opportunity to affect assemblage structure in habitats that experience periods of relative stability. A wide range of studies shows that both categories of factors play a role in assemblage and community structure in many aquatic habitats, and also that the relative importance of each category of processes is influenced by frequency of disturbance, including those from human interactions.

An extreme adherent of the stochastic view might argue, for example, that floods or other weather-related effects play a major role in structuring stream fish assemblages. The timing of high flows from precipitation with respect to spawning, for example, could have a large impact on recruitment and thus assemblage composition. Similarly, chance events affecting planktonic larvae and newly recruited juveniles play too large a role for us to be able to predict species composition on a coral reef. A larval fish that was not eaten or that did not starve to death must also be lucky enough to encounter a reef during the brief period when it is **competent** to settle and survive long enough to find a suitable, unoccupied site to settle. These chance events, which are further influenced by unpredictable storms, reduce the accuracy with which we can predict the species and abundances of fishes that will occur on a specific reef, beyond knowing what occurs in a general geographic region.

The relative importance of stochastic and deterministic factors in assemblage structure varies with circumstances, and both are often at work. In rivers and streams, for example, the discussion of the role of deterministic and stochastic processes in fish assemblages goes back decades; see Grossman et al. (1982, 1985), Herbold (1984), Rahel et al. (1984), Yant et al. (1984), and review by Strange et al. (1993). More recently, study of the Brown Trout (*Salmo trutta*) population in Spruce Creek (Pennsylvania, USA) showed that it was primarily regulated by density-dependent factors (e.g. recruitment, survivorship, and growth), but density independent factors such as variations in temperature and flow played a role as well (Grossman et al. 2017). In the undammed Río Mameyes of Puerto Rico, fish assemblages are regulated by deterministic process during periods of stability, but this can be disrupted by disturbances such as flooding from tropical storms (Smith & Kwak 2015). Intense storms occur approximately every 4 years, however, so the native species, many of which are amphidromous, have adapted and their populations recover quickly. Other rivers in Puerto Rico have dams and more non-native species, and fish assemblages do not recover quickly after storms. In the Xingu River of Brazil, a tributary of the Amazon River, factors affecting community structure vary seasonally because dramatic changes in flow affect fish densities, and thus biological interactions such as competition and predation (Fitzgerald et al. 2017). Conditions that are more stable tend to favor deterministic processes, such as in the fish assemblage in a rice field irrigation channel in central Japan where water levels were relatively stable due to gated control of flow (Ohira et al. 2015).

The impact of flow on these assemblages raises concerns regarding likely impacts of flow variation associated with climate change. For example, populations of Rosyside Dace (*Clinostomus funduloides*) in a mountain stream are strongly dependent on variations in flow, and intraspecific density-dependence keeps species abundance low enough so that interspecific competition is not a factor in assemblage structure (Grossman et al. 2016). However, dramatic changes in flow associated with climate change could destabilize the population and result in overall shifts in the balance of the assemblage.

The relative importance of stochastic and deterministic factors in fish assemblage structure has also been evaluated in lakes and reservoirs. A study of the size structure of fish assemblages in six small lakes in Michigan concluded that three were determined mainly by deterministic factors, and these three tended to be larger, with more variation in habitats, had the highest number of species, and no recent major disturbances (Clement et al. 2015). Fish sizes in a fourth lake, which was shallow and marshy, were likely strongly affected by bird and mammal predators. Fish size structure in the remaining two lakes was more stochastically driven, however, due to impacts of major disturbances such as winterkills or reclamation and restocking.

A study of fishes and predatory wading birds in temporary ponds along the Miranda River floodplain (Brazil) showed that bird abundances were strongly influenced by forest cover and pond size, but fish abundance and distribution appeared to be random (Keppeler et al. 2016). The fish assemblages in these temporary ponds may be affected more by the timing and persistence of flooding. Deterministic factors, especially predation and selective fishing by humans, were strong factors in determining fish community structure in the lake-like regions of the Coaracy Nunes reservoir on the Araguari River in northern Brazil (Sá-Oliveira et al. 2016). Fluctuations in water levels associated with precipitation play a role as well because fishes are more concentrated during low flow periods, which increases biological interactions such as competition and predation. During high flow periods, access to floodplain habitat is important for successful reproduction of many species. Stochastic factors, however, play an important role in the faster-flowing upstream segment of the reservoir and the area below the dam. Mathematical modeling using various parameters suggested that stochastic processes may be largely responsible for assemblage structure of cichlids in the littoral zone of Lake Tanganyika, Zambia. Niche-based biological interactions also were important, however, leading to the conclusion that the assemblage structure is a result of both stochastic and deterministic processes (Janzen et al. 2017).

Stochastic and deterministic factors have also been assessed in marine fish assemblages. A series of studies of the life cycle of North Atlantic Cod (*Gadus morhua*) showed that density-dependent factors are important to juvenile survival and recruitment, but stochastic impacts also play a role in egg production and larval survival. Therefore, the timing of stochastic and density-dependent factors impact adult populations (Ohlberger et al. 2014). In the Mediterranean Sea, 25 years of data (1950–74) from a large fish trap off of the Camogli coast (Italy) suggest that large predatory fishes help stabilize the fish assemblage by deterministic processes, mainly predation. However, when predators decline due to fishing, top-down control is diminished and the assemblage, especially species at lower trophic levels, may become more susceptible to stochastic events (Britten et al. 2014). Therefore, human impacts can shift the balance between deterministic and stochastic factors in regulating assemblage structure.

No habitat has received more attention in terms of fish assemblage structure than coral reefs, which contain more species of fishes than any other habitat. Almost 700 shallow water species occur in the Caribbean, and Indo-Pacific reefs are home to more than 3000 species. This incredible diversity, plus colorful fishes and clear, warm water (Fig. 21.6) has understandably drawn the attention of fish ecologists. Understanding spatial and temporal patterns of recruitment and the factors that determine the success of recruits and ultimately community structure can influence fisheries management and conservation practices such as seasonal closures, protected area and species protection, and artificial reef design and placement (Beets 1989; Bohnsack 1994; NRC 2001).

Observational and experimental studies in the Caribbean and Pacific, beginning largely with the work of Sale (1978) and Smith (1978), have led to different conclusions regarding the roles of stochastic and deterministic processes. *Observational studies* emphasize comparisons of underwater counts of individuals at several sites in one reef area, or at the same locale at several different times or after a hurricane strikes an area in which prior fish assemblage information is available. *Experimental studies* usually involve removal of individuals from small coral heads or patch reefs, or the addition of small patches of reef habitat followed by monitoring of recruitment and recolonization.

FIGURE 21.6 A triggerfish, wrasse, and various damselfishes in the Maldives. Jan Derk / Wikimedia Commons / Public Domain.

Reef fish assemblages are usually defined in terms of the adults present, but replacement of adults is seldom by other adults. When an adult is removed, either experimentally by a researcher or naturally by a predator, it is usually replaced by a newly settled (**recruited**) juvenile or a slightly older **colonist**. The question of interest then becomes one of whether adult population dynamics are determined by events before settlement (i.e. during the planktonic phase), during recruitment (i.e. by settling larvae), or after recruitment (i.e. due to interactions among juveniles, adults, and their competitors and predators) (Fig. 21.7).

Some researchers emphasize the importance of events and interactions in the plankton in determining which species populate a reef (Doherty & Williams 1988; Wellington & Victor 1988). This view developed from observations of similar, unexploited reefs in close proximity to one another that contained dissimilar species assemblages. Also, experimentally increased food and refuge availability on a reef does not necessarily lead to an increase in fish numbers at a site. Hence reefs may contain fewer individuals than they can theoretically support, i.e., they may exist below their **carrying capacity**. These findings suggest that adult populations may be limited by the number of larvae available to settle in the area, the so-called **recruitment** (or **settlement**) **limitation hypothesis**. Additional evidence of recruitment limitation includes differences in year class strength on a reef and the rarity of larvae of some species.

The alternative view – the **habitat limitation** or **interactive hypothesis** – proposes that appropriate habitat is limiting or that post-settlement biological interactions (predation, competition) determine the kinds and abundances of fishes on a reef, regardless of larval abundance. The habitat-limited or interactive scenario depicts a reef at carrying capacity, one that is less able to replace fishes removed through fishing. Evidence includes superabundant larvae around reefs, reefs packed with recruits, and rates of predation that exceed 99% during the first year post-settlement (Shulman & Ogden 1987; Roberts 1996; Hixon 1998; Hobson et al. 2001; Levin & Grimes 2002).

Many factors can affect larval abundance, regardless of numbers or habitat availability. These include predation on larvae, food availability for larvae, and dispersal away from appropriate settling areas. Mortality is the most likely fate awaiting a planktonic larva; most studies estimate that more than 99% of larvae die before they settle. Thus antipredator, food-getting, and active dispersal adaptations of larvae themselves may be critical. Adults can also improve their offspring's chances of making it through the planktonic filter by spawning at times and places that minimize dispersal away from home reefs, which reduces the area over which larvae must search for an appropriate settling habitat. Carefully chosen spawning

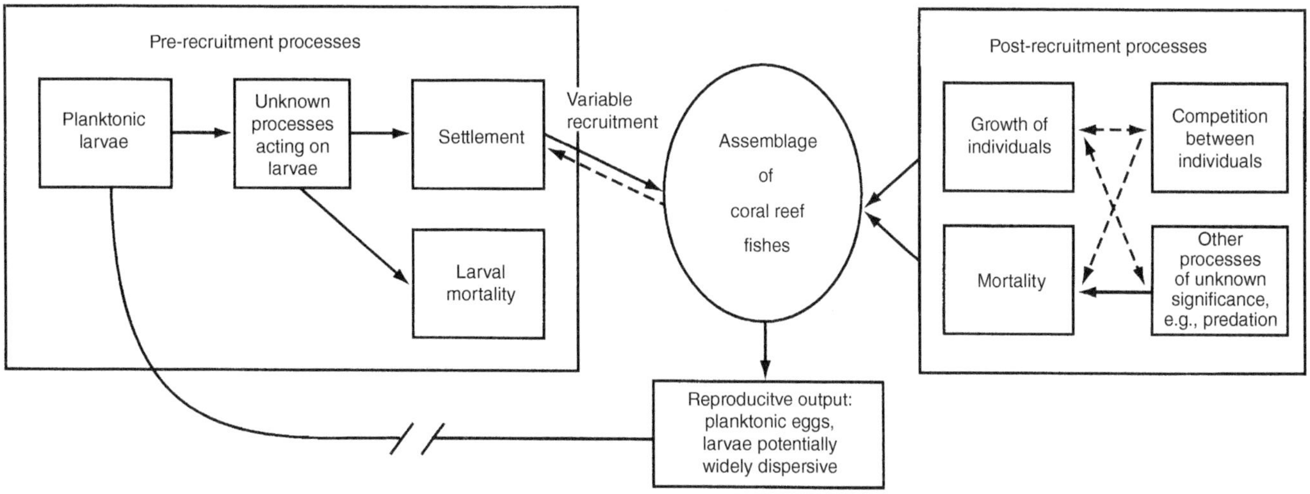

FIGURE 21.7 The various processes that operate to determine the diversity and abundance of fishes on a coral reef. Solid arrows indicate known interactions, dashed arrows possible interactions; the broken arrow between reproductive output and planktonic larvae refers to the uncertainty that reproduction on a reef may influence the number of recruits returning to a reef. From Mapstone and Fowler (1988), used with permission.

locales can also place larvae where planktonic food tends to be concentrated (but also where predators on planktonic fish larvae also abound) (Johannes 1978).

Periodically, vast numbers of larvae that are ready to settle occur around reefs, indicating that habitat limitation can be important at times (Victor 1991; Kaufman et al. 1992). Under such circumstances, conditions that prevail in the settlement area may determine the success of recruits and the ultimate species composition in an area. Several studies have shown that small artificial reefs placed in shallow water attract large numbers of recently settled larvae to areas that were previously devoid of larvae, suggesting that appropriate, unoccupied habitat can limit recruitment. But regardless of habitat availability, previous occupation of the habitat, known as **priority effects**, may be crucial in determining whether larvae settle successfully. If the first occupants of a coral patch are herbivores or small planktivores, a variety of larvae will follow and take up residence. However, if the first settlers are predators such as moray eels, squirrelfishes, grunts, snappers, or groupers, later recruitment will be greatly reduced as these small predators eat incoming fish larvae (Beets 1997; Tupper & Juanes 1999) (Fig. 21.8). Which larvae settle first is governed largely by chance because of the unpredictable nature of planktonic existence. Once larvae settle, their impact on later settlers is fairly predictable and depends on the ecological role of the species in question. Hence both stochastic and deterministic forces are in operation (Shulman et al. 1983; Hixon & Beets 1989; Beets 1991). Several other studies over the years also conclude that both deterministic and stochastic processes are likely involved in fish community structure on coral reefs (Ault & Johnson 1998; Syms & Jones 2000; Yeager et al. 2011)

A growing body of knowledge has changed our perception of larval life and behavior. The classical view was of passive larvae carried by ocean currents, settling when they reached some critical stage of competency. If a larva happened to be over appropriate habitat at that stage, its chances were good. If it was somewhere less favorable, such as over great ocean depths, then it was game over. We now know that larvae are much more active than this in their settling activities. Larvae are attracted to reef areas by both sounds and smells emitted by reefs, and move actively toward appropriate stimuli (Atema et al. 2002; Kingsford et al. 2002; Tolimieri et al. 2004; Gerlach et al. 2007). Once over a reef, larvae (or more accurately **transitional juveniles**) show strong habitat preferences that differ among species; some larvae will settle and then ascend back into the water column if conditions are inappropriate. Larval settlement is therefore not a parachute drop but more of a bungee jump (Kaufman et al. 1992; Lecchini 2005). Maintenance of high diversity on a reef demands protection of not just adult habitats but also of settlement habitats, which are often different from and far removed from adult habitats.

A larva that settles successfully onto a patch of reef and transforms into a juvenile is by no means guaranteed a long and productive life. Biological interactions involving predation, competition, and cooperation, and their interactions, can have a strong impact on individual success. Mortality rates remain high once larvae have settled; 25% of recruits may die in the first 5 days after settlement, but this rate falls to <10% after 6 days and continues to decrease thereafter (Doherty & Sale 1986). Although the rate may reduce, the numbers killed remain high, varying between 65% and 99.9% during the first year after settlement (Sweatman 1984; Shulman & Ogden 1987). Mortality may also be density dependent, increasing as population size increases, as happens when large numbers of juvenile humbug damselfishes, *Dascyllus* spp., actively compete for preferred nighttime resting locales among branching corals.

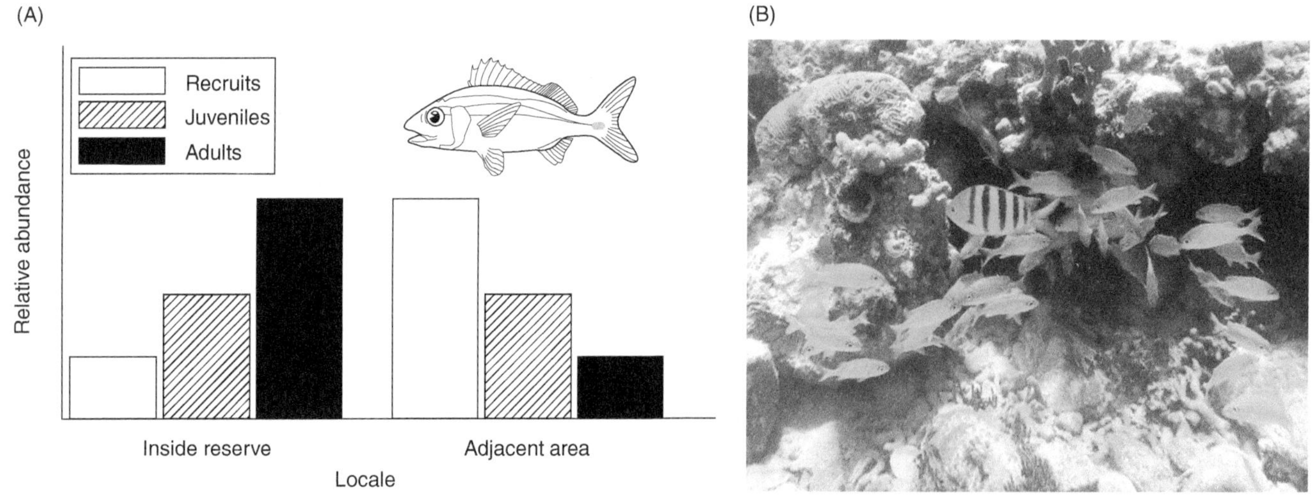

FIGURE 21.8 Priority effects. The foraging behavior of species that settle into an area determine the success of later arrivals. (A) In the Barbados Marine Reserve, adult grunts (dark bars) were much more abundant inside the reserve than in adjacent, non-reserve areas, but recently settled grunts (open bars) were more abundant in adjacent than in reserve habitat. Juveniles (crosshatched bars) were equally abundant inside and outside the reserve. Lack of recruits in the reserve was thought to be the result of predation on settling larvae by resident adult grunts. Data from Tupper and Juanes (1999). (B) French Grunt (*Haemulon flavolineatum*) congregating near a coral reef, along with a Seargent Major (*Abudefduf saxatilis*) and a squirrelfish (*Holocentrus*). D. Facey (Author).

Less aggressive individuals are displaced to riskier locations, where they are subject to predation by crepuscular/nocturnal predators such as squirrelfishes. Again, diversity maintenance on a reef requires protection of both daytime feeding and nighttime resting habitats, which again may differ (Holbrook & Schmitt 2002; see also Hixon & Carr 1997).

Larvae may settle more successfully in isolated habitats away from major concentrations of larger fishes and then move later to the more extensive reef habitats. This may be one reason that back-reef areas, mangroves, and seagrass beds are often the preferred habitat for the juveniles of many "reef" species (Manson et al. 2005; Adams et al. 2006; Pollux et al. 2007). The move to the reef itself is also full of perils, as predators tend to patrol the edges of reefs and catch prey moving across refugeless zones (Shulman 1985a, 1985b). For species that have symbiotic relationships with invertebrates, such as anemonefishes, carapid pearlfishes, cardinalfishes, and many gobies, successful location of an unoccupied, species-specific host is probably a good guarantor of survival, but such hosts may be in limited supply. An added complication is the cannibalism that occurs if the species-specific host is already occupied by an adult of the larva's species (e.g. Tyler et al. 1992).

One possibly influential difference between major oceans is the size of the eggs produced by residents (Thresher 1982). On average, egg sizes are smaller in the western Atlantic than the western Pacific. Smaller eggs imply greater fecundity among Atlantic species, which could lead to greater reproductive output in the western Atlantic. Assuming comparable larval mortality rates, more larvae mean more potential competition for space among recruits, tipping the scales in favor of a stronger role for deterministic interactions among Atlantic coral reef fishes than their Pacific relatives.

The bottom line to this discussion is that both stochastic and deterministic factors influence reef fish assemblages. Both types of factors come into play and have different levels of influence at different times and in different places. Random events in the plankton undoubtedly influence which larvae will survive, and whether the larva settles successfully depends on whether space is available for it on the reef. Space becomes available in part as a chance result of predation and storm disturbance. But larvae, juveniles, and adults often have specific habitat preferences, as shown by the fairly distinctive zones that occur on most coral reefs (lagoonal, patch reef, back reef, reef crest, and shallow and deep reef front) and the fairly predictable assemblages of species found in each zone (Nanami et al. 2005; Ashworth et al. 2006). Nonrandom competitive, predatory, and mutualistic interactions affect the suitability of a site, the survivorship of its inhabitants, and ultimately its availability for new recruits and colonists. Space availability may depend on the guild of a fish that has been removed from the reef. We can predict that certain guilds are likely to be present on a reef, but we cannot predict which members of that guild will be present. The importance then of the different factors is going to vary from time to time and place to place.

Interactions within Assemblages and the Broader Ecological Community

Fishes interact with other members of the fish assemblage mainly through competition, predation, parasitism, and different forms of symbiosis. Similar interactions are also common among fishes and other taxonomic groups, including macrophytes, invertebrates, and other vertebrates. Often these interactions and their influences are relatively direct, e.g. through eating and being eaten. But some dramatic effects – such as changes in habitat use, food types, life history traits – take place through fairly indirect means, either incidental to or several steps removed from the activities of the fishes involved.

Competition

Competition occurs when two consumers require a resource that is not abundant enough to meet the needs of both, and fishes often compete with one another for limited resources. This competition can be among members of the same species (**intraspecific**) or members of different species (**interspecific**). Although intraspecific competition can affect an individual's ability to acquire resources, interspecific competition has received more attention because of the insight it provides into the coexistence of different species in an assemblage, which addresses the more general question of how biodiversity is created and maintained. In general, individuals can compete for food, feeding and resting sites, and refuges from predators and the elements. Competition for mates and breeding sites is viewed by some as part of the reproductive biology of a species rather than as traditional competition.

To avoid or reduce competition, organisms may change the way they exploit a resource. Competition may lead to **resource partitioning**, such as when two **sympatric** ("living together") species feed on different sizes of a prey type or eat similar prey but in different microhabitats. Competition is more strongly implicated if these same predators feed on identical prey when they are **allopatric** ("living separately"). Also, competitive interactions can be suspected if potential competitors shift their resource use when resources become seasonally limiting, or if population reductions of one species occur when a suspected competitor is introduced into an area. However, ecological differences among species can also be caused by differences in nutritional requirements, foraging or locomotory capabilities, predator vulnerability, and phylogeny. Introduced species can alter predator–prey relationships or serve as vectors for parasites and diseases, which would also affect population densities of previous residents. Consequently it is generally necessary to perform experimental manipulations of resource abundance or distribution, or of population densities of suspected competitors, to prove that competition is the cause of the dissimilarities. In such experiments, competition can be invoked if an inferior competitor or less aggressive species that occupies suboptimal regions in sympatry expands its habitat or feeding habits when the superior competitor is eliminated. Reciprocal removal of the inferior competitor should have little effect on the habits of the superior species. Such experiments may be fortuitous or deliberate.

A fortuitous manipulation of resource partitioning, mediated by both competition and predation, has been conducted annually in Lake Tjeukemeer, the Netherlands (Lammens et al. 1985). Bream (*Abramis brama*, Cyprinidae) and European Eels (*Anguilla anguilla*, Anguillidae) occur year round in fairly stable numbers in the lake, where their chief foods are zooplankton and juvenile midges, respectively. Smelt (*Osmerus eperlanus*, Osmeridae), a zooplanktivore, enters the lake each spring as juveniles when water is pumped from a nearby lake as part of a water stabilization program. Smelt do not persist in the lake because the adults are almost all consumed by predatory Pikeperch (*Sander lucioperca*, Percidae). When large numbers of juvenile smelt enter the lake, they depress zooplankton populations, having their strongest effects on the size classes of zooplankton most used by Bream. Bream respond to reductions in zooplankton resources by switching to benthic invertebrates such as midge larvae, thereby depressing that resource. Eels then respond to depletion of their primary food by switching to piscivory. When Smelt are abundant, both Bream and Eels suffer reductions in condition (weight/length) and Bream show poor gonad development. In years when Smelt recruitment is low, Bream and Eels switch back to their zooplankton/midge diets and their growth and reproduction improve.

A well-studied example that includes the experiments necessary to establish the causes of shifts in resource use involves sunfishes in North America. As many as eight species of centrarchid sunfishes and basses may co-occur in a single lake. Many of these species are very similar morphologically. How do they coexist without competing? When stocked separately in ponds as year-old fish, three species, Bluegill (*Lepomis macrochirus*), Pumpkinseed (*L. gibbosus*), and Green Sunfish (*L. cyanellus*, Fig. 21.9), use similar habitats and feed on similar food types. All three concentrate their time and effort on vegetation-associated invertebrates. When the three species are stocked together, Bluegill and Pumpkinseed shift their habitats and diet in apparent avoidance of the competitively superior Green Sunfish. Bluegill shift to feeding on zooplankton in open water, and Pumpkinseed include more benthic prey in their diet. Green Sunfish maintain a diet of vegetation-associated insects. All three species show reduced growth rates, indicating competitive reduction of resources for each species, but Bluegill show the greatest declines. When Bluegill and Green Sunfish are stocked together in ponds with little open water habitat (i.e. no alternative habitat for the Bluegill), Green Sunfish show better growth, fuller stomachs, and a higher survival rate than Bluegill. Competition among these species has an ontogenetic component that is felt most strongly by young fish. As the fish grow older, they begin to specialize more on different habitats. Bluegill become more adept at maneuvering in open water and suction feeding on zooplankton, and Pumpkinseed develop pharyngeal dentition with which they can crush mollusks that live in sediments. Hence the potential for competition is reduced in older fish under natural conditions (Werner & Hall 1979; Werner 1984; Mittelbach 1988; Osenberg et al. 1988; Wootton 1999).

FIGURE 21.9 Bluegill, Pumpkinseed, and Green Sunfish (photo) eat similar foods when alone, but the first two shift habitat and diet when in the presence of the superior competitor. Bclegg77 / Wikimedia Commons / CC BY-SA 4.0.

Many other investigations have demonstrated strong competitive interactions among fishes in various habitat types (e.g. tropical streams, Zaret & Rand 1971; temperate streams, Schlosser 1982; temperate marine nearshore, Hixon 1980b; Holbrook & Schmitt 1989; coral reefs, Hixon & Beets 1989; Munday et al. 2001; Holbrook & Schmitt 2002; see reviews in Ross 1986; Ebeling & Hixon 1991; Grant 1997; Hixon 2006). In general, of the kinds of resources for which fishes can compete, competition for food resources, or at least differences in trophic resource use, is more common among fishes than are interspecific differences in habitat use; the reverse is true in terrestrial communities (Ross 1986).

Some traits that reflect apparent adjustments to present-day competition may result from historical interactions between species, the so-called "ghosts of competition past." The influence of historical competition is frustratingly difficult to determine: are two species different today because of their current impacts on one another or because of past interactions? Experimental manipulations of the resource in question are almost always needed to prove competition, but obviously one cannot manipulate the history of two species and hence we can only speculate on but not demonstrate historical competition (Connell 1980).

Historical factors must also be considered when comparing ecological characteristics of species from unrelated taxonomic groups. The more distantly related two fish species are, the less similar they tend to be ecologically (Ross 1986). For example, generalist predators on coral reefs tend to be active at twilight or at night, have large mouths, and feed on fishes, whereas specialists are diurnal, have small mouths, and feed on sessile or small invertebrates (Hobson 1974, 1975, 2006). Resource partitioning along both trophic and temporal resource dimensions could be invoked here. However, generalist reef species tend to belong to more ancestral acanthopterygian (spiny-rayed) groups (squirrelfishes, scorpionfishes, and groupers), whereas specialists belong to more derived groups (butterflyfishes, wrasses, and triggerfishes). Feeding habits, morphology, and activity times in one lineage are likely to have evolved independently of what happened in a later evolving lineage. Differing ecologies may therefore simply reflect differing phylogenetic histories.

Interpreting differences in resource use as a result of competition may also be erroneous because of physiological differences among species. A species may occur where it does because the fish functions best there, hence choice of habitat may reflect use of physiologically optimal environments rather than being the result of interactions with other species over limiting resources such as food or shelter. For example, in the southern Appalachian Mountains of the United states, the stream fish assemblage typically consists of <10 species that are often found at different heights above the bottom and at different water velocities. Four common species are the Rosyside Dace (*Clinostomus funduloides*) and introduced Rainbow Trout (*Oncorhynchus mykiss*) in the water column, and Longnose Dace (*Rhinichthys cataractae*) and Mottled Sculpin (*Cottus bairdi*) on the bottom. Facey and Grossman (1990, 1992) tested whether distribution differences among these four fishes could be explained by energetic efficiencies rather than competition. They found that Longnose Dace showed no real preference but were distributed in proportion to the available velocities in the stream (interactions with other species are unlikely to produce such a statistically random distribution). Rainbow Trout, Rosyside Dace, and Mottled Sculpin chose velocities that were lower than would be expected if they were distributed at random, which would help reduce energy use for the trout and dace. Trout and dace also preferred velocities in which they were most efficient at capturing drifting invertebrate prey (Hill & Grossman 1993). Hence physiological costs associated with holding position at high velocities and optimal velocities for capturing prey, not interactions with other species, are the most likely determinants of where in a stream these two species are found. Experimental manipulations involving the benthic sculpin and its most likely competitor, Longnose Dace, indicated minimal effects on sculpin (Barrett 1989; Stouder 1990). In headwater streams, highly variable environmental features, characterized by large fluctuations in water level and velocity (droughts and floods), combined with physiological constraints, prey capture abilities, and intraspecific competition (Freeman & Stouder 1989), appear to have a greater influence on occurrence, distribution, abundance, foraging, and habitat choice than do interspecific interactions.

Some of the strongest impacts of introduced species on natives involve **competitive displacement**, suggesting that competition has been historically reduced via evolutionary adjustments. In Lake Michigan, a coregonine salmonid, the planktivorous Bloater, *Coregonus hoyi* (designated Vulnerable by IUCN), was replaced by introduced planktivorous Alewife, *Alosa pseudoharengus* (Clupeidae) in the 1960s, apparently as a result of competition for plankton. As Alewife numbers grew, Bloaters declined in abundance, shifted to a diet of benthic invertebrates at an earlier age, and apparently evolved fewer and shorter gill rakers (Crowder 1984). Five other coregonine species were extirpated from Lake Michigan during the same period.

Other indirect evidence of competitive displacement of natives by invaders includes habitat displacement of endangered Spikedace, *Meda fulgida* (Cyprinidae), by Red Shiners, *Cyprinella lutrensis*, a well-known invasive species. After Red Shiners were introduced into the lower Colorado River, Spikedace disappeared simultaneously and progressively as Red Shiners proliferated, while dams and water withdrawals led to degraded habitat. Both species occupied slow current regions when alone, but where they co-occurred the more aggressive Red Shiners remained in slow current areas while Spikedace were displaced into regions of swifter current (Douglas et al. 1994). Introduced trout are often implicated in competitive displacement of native trout (Gatz et al. 1987; Fausch 1988). In a Michigan stream, introduced Brown Trout displaced native Brook Trout from the best foraging habitats, forcing Brook Trout into faster water where the energetic costs of maintaining position were higher and where they were more likely to be caught by anglers (Fausch & White 1986; see also Waters 1983). Brown Trout also have limited habitat use by native Brook Trout in streams elsewhere in the Midwestern and eastern U.S., negatively impacting growth of Brook Trout (Hoxmeier & Dieterman 2013; Hitt et al. 2017). Rainbow Trout displaced two native Japanese salmonids (Dolly Varden, *Salvelinus malma*, and White-spotted Char, *S. leucomaenis*) because of timing differences in spawning. Natives spawned in fall but rainbows spawned the next spring, at a time when embryos of the fall-spawning natives were developing in the gravel. The digging and spawning activities of the introduced species disturbed the redds of the native species (Taniguchi et al. 2000).

Fishes compete with a wide variety of organisms for food and space, not just other fishes. Bluefish (*Pomatomus saltatrix*, Fig. 21.10) and common terns (*Sterna hirundo*) have a complex feeding relationship that involves both commensalism and competition and that may be applicable to many other fish–bird interactions (Safina & Burger 1985). Both species feed on anchovies and Sand Lances. The seabirds are particularly dependent on prey fish during the early summer breeding season when they must meet their own energetic demands as well as those of their growing chicks. Off Long Island, New York, Bluefish arrive in large numbers in July each year as part of their annual migration. Feeding Bluefish drive prey up in the water column, concentrate them in space, and make the prey easier for seabirds to locate and capture (commensalism). However, newly arrived Bluefish consume large numbers of prey fish, which rapidly depresses the prey resource available to the birds (competition). Birds that initiate breeding after the

FIGURE 21.10 Oceanic prey fishes are often trapped between predatory birds from above and fishes, such as the Bluefish, from below. R. W. Hines / USFWS / Public Domain.

arrival of the Bluefish tend to be unsuccessful. Hence Bluefish may have been a strong selective force in determining the timing of reproduction by the terns.

Predation

Predator–prey interactions within an assemblage or community can have direct and indirect effects on prey population size and distribution. **Direct effects** include immediate mortality or delayed mortality due to injury. **Indirect effects** involve habitat shifts caused by a predator's presence that force potential prey to use suboptimal habitats, which can affect individual growth and reproduction. Population-level responses associated with predation are usually density dependent and vary with the age of the prey. **Density-dependent** changes occur when the size of the prey population determines the impact of the predator. Direct density dependence is referred to as **compensatory**, meaning that predation increases to compensate for increases in prey population size. Predation by seabirds on schooling pelagic fishes is often compensatory in both the short- and long-term. The feeding activities of one bird draws the attention of other birds and the number of predators arriving at the site increases in direct relation to the size of the school of prey fish. Successful feeding by the birds in turn increases survivorship of their young, which means an increase in predators in the next generation, all dependent on the size of the fish resource.

Intercohort cannibalism, in which older fish eat younger age classes, can have a strong density-dependent impact on year class strength. Consider a population with 3-year classes. A large cannibalistic oldest age class can depress the numbers of the next, younger age class. When fish of this second, younger age class become large enough to be a threat to the third, even younger, age class, the low abundance of the second age class will result in relatively little impact on the youngest age class. In this way, population cycles can be established through the density-dependent effects of cannibalism. Just this type of scenario has been invoked to explain 2-year cycles of abundance in Pink Salmon, *Oncorhynchus gorbuscha*, in the Pacific Northwest (Ricker 1962).

Inverse density dependence is considered **depensatory** because relative predation risk and impact decrease as prey numbers increase. Depensatory predation occurs when a fixed number of predators become swamped or saturated by large numbers of prey. Under such conditions, the proportion of the prey captured decreases as prey numbers increase. For example, an individual salmon smolt that is migrating to the sea reduces its risk of death if it can time its downstream migration to coincide with that of other smolts, since predators take only a small number of migrants. By extension, the proportion of the prey population killed decreases as the population increases (Wootton 1990).

Regardless of the nature of the relationship between predator and prey densities, predation can have dramatic effects on prey population size. It is generally held that most of the mortality in eggs and larvae of species with planktonic young is due to predation, with predators taking >99% of the individuals (Bailey & Houde 1989). Older fish are still subject to predation, but the threat falls off progressively with increasing age and size, forming what is described as **exponentially declining mortality** (Fig. 21.11). Refuge availability may influence the impact that predators have on later life history stages. Observations and experimental manipulations on coral reefs indicate that prey population density is directly related to the number and availability of holes where prey can hide, which also implies that competition for refuge sites could interact with predation to determine population density and diversity. Several experimental studies on reef fishes have shown that removing predators leads to increased density of prey, reinforcing descriptions of dramatic trophic cascades (addressed later in this chapter). The effects of predator density on prey diversity may follow a similar pattern (Hixon 1991; Hixon & Beets 1993). Regardless, regulation of population size ultimately involves an interplay of competition and predation, often mediated by habitat availability (Holbrook & Schmitt 2002; Hixon & Jones 2005).

Predation can also affect gene frequencies in populations through the evolution of antipredator adaptations. Guppy (*Poecilia reticulata*) in small streams in Venezuela and on the island of Trinidad occur in pools that differ in levels of predation. Upstream areas tend to have few if any predators, often limited to a single topminnow species, *Rivulus marmoratus* (Cyprinodontidae). Further downstream, more predators occur, including a cichlid (*Crenicichla*), a characin (*Hoplias*),

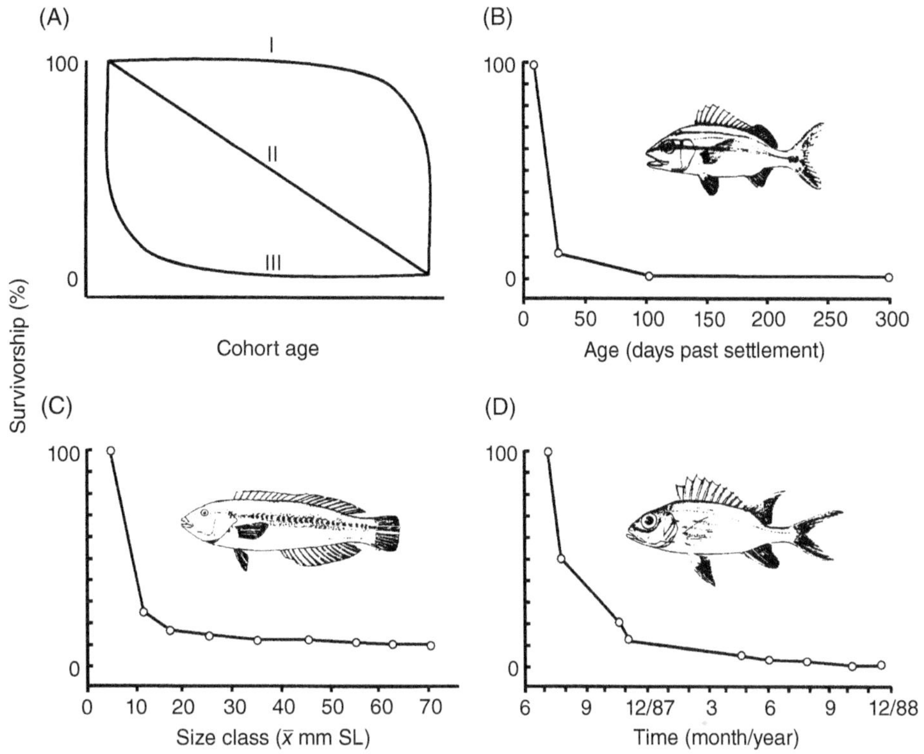

FIGURE 21.11 Survivorship curves in theory and practice. (A) Three general forms of survivorship (percent surviving at the end of each year or within each age class) are found in natural populations. Because predation tends to be most heavily focused on young and small fishes, populations that display type III survivorship curves are most likely to be regulated by predation, as seen in curves B through D. (B) A grunt (Haemulidae) population in the Virgin Islands. (C) A wrasse (Labridae) population in Panama. (D) A squirrelfish (Holocentridae) population in the Virgin Islands. SL, standard length. From Hixon (1991), used with permission; fish drawings from Gilligan (1989) / NOAA / Public Domain.

and freshwater prawns. In areas of low predation, male Guppy tend to have many bright, colorful spots that are attractive to females but are also conspicuous to predators. Spot number, size, and brightness are inherited; the offspring of brightly colored males are brightly colored. In a series of experiments, fish from several populations were exposed to different levels of predation over several generations or transferred between areas with high and low predation intensity. The number, size, and brightness of spots declined in populations subjected to more predation both in the field and lab (Endler 1980, 1983). Even the pattern of size-specific predation affected heritable traits in a predictable fashion. Where Guppy were exposed to cichlids, which tend to prey chiefly on larger Guppy, the prey matured later at larger sizes, whereas if the predator was the topminnow, which targets predominantly smaller Guppy, maturation occurred earlier at smaller sizes (Reznick et al. 1990, 1997).

As was pointed out with respect to competition, non-native predators have unequivocal impacts on native fishes wherever they are introduced, having eliminated populations and even species in many locales. There are many examples (see Chapter 22, Introduced Predators), but strongest impacts have come from species introduced to improve sport fishing (e.g. Flathead Catfish, Brown and Rainbow Trout, Smallmouth and Largemouth Bass, Peacock Cichlid, Nile Perch, Northern Pike and Pikeperch, and even Mosquitofish) (Fuller et al. 1999; Rahel 2002; Helfman 2007).

Predator–prey interactions are not limited to within a fish assemblage, however, as fishes serve as both prey and predators of other taxa in a community. A wide variety of non-piscine predators are dependent on fishes as a major component of their diets. Numerous invertebrate predators capture fishes. Predaceous waterbugs (Belostomatidae) and dragonfly (Odonata) larvae prey on small freshwater fishes. Marine invertebrate predators are common, including jellyfish, anemones, siphonophores, squid, dwarf triton and cone snails, and crabs (Laughlin 1982; Bouchet & Perrine 1996).

Among vertebrates, reptilian predators include turtles, crocodilians, varanid monitor lizards, a few iguanid lizards, and sea and water snakes. Amphibian predation on fishes is poorly documented, although sirens, bullfrogs, and a few other large frogs (*Pipa*, *Xenopus*) are known to be fish predators, including the impacts of introduced species on endangered fishes (Lafferty & Page 1997). Mammalian fish predators include mink, raccoon, otters (Fig. 21.12a), seals, sea lions, bears, dolphins, whales, bats, and, of course, humans. Pricklebacks (Stichaeidae) are less

FIGURE 21.12 (A) A River Otter (*Lontra canadensis*) consumes a Saddleback Gunnel (*Pholis ornata*) it captured while diving. (B, C) The large and powerful talons of Osprey (*Pandion haliaetus*) are quite effective in capturing and holding fish. G. Helfman (Author).

abundant in the lower intertidal area of coastal islands in British Columbia that support raccoon (*Procyon lotor*) populations than on islands without raccoons (Suraci et al. 2014).

A host of seabirds concentrate on fishes, including terns, petrels, albatrosses, gannets, auks, murres, cormorants, skimmers, spoonbills, pelicans, penguins, and gulls. Other fish-eating birds, in both salt and fresh waters, include loons, mergansers, kingfishers, herons, egrets, storks, bald eagles, and ospreys (Fig. 21.12b, c).

All of these non-fish groups have served as selective agents on the behavior and ecology of fishes, causing evolutionary adjustments in growth and reproductive traits that have allowed fish to thrive despite sometimes astronomical mortality rates. Many fishery-related problems arise from the evolutionary responses of fishes to intense harvesting by humans (see Chapter 20, Effects of Fishing on Life History Traits and Implications for Recovery).

The impact of non-fish predators on fish populations and behavior can be substantial. In the Au Sable River, Michigan, mortality of adult Brook Trout and Brown Trout averaged 70–90% annually, most of which was from predation. Non-fish and non-human predators (mergansers, heron, kingfisher, mink, and otter) accounted for 28–35% of this mortality (Alexander 1979). When the potential threat of predatory birds is combined with that from piscivorous fishes, it is not surprising that the distribution of many prey species reflects the foraging locales of their predators (Fig. 21.13). This combined threat can include a third dimension during the breeding season. Male Dollar Sunfishes (*Lepomis marginatus*) construct nests in shallow water to avoid predatory fishes that are typically in deeper water. The males, however, must repeatedly abandon their nests to avoid being captured by herons and kingfishers. Each time the male flees, eggs and young in the nest are subject to predation by small fishes, forcing the male into a trade-off between the conflicting demands of protecting himself and protecting his offspring (Winkelman 1996).

Fishes also fall prey to less obvious but more insidious predators. Massive fish kills have long been attributed to harmful algal blooms and especially dinoflagellate blooms

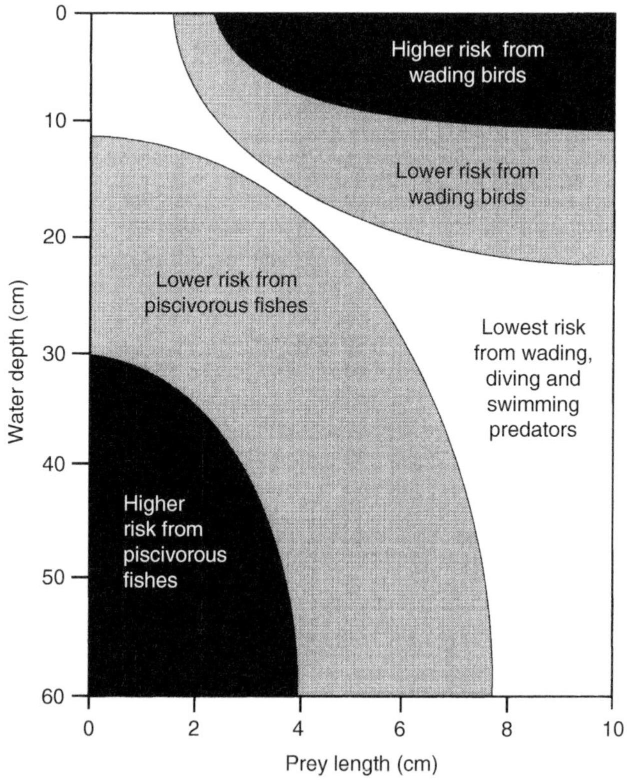

FIGURE 21.13 The distribution of prey fishes in shallow streams reflects the risk of predation from various sources. Piscivorous fishes, which are gape-limited, present the greatest threat in deeper water. Wading birds, which can dismember prey and are therefore not gape-limited, present the greatest threat in shallow water. Small prey fishes are safest in shallow water because they can hide from birds among structure, whereas larger prey fishes cannot fit into small spaces. However, larger prey are safer in deeper water because many predators cannot swallow them whole. From Power (1987) used with permission.

("red tides") in many nearshore marine areas, but fish death has usually been considered an incidental byproduct of a bloom due to insufficient oxygen concentrations or the toxicity of secondary macrophyte chemicals. Some events have implicated an evolved, predatory response by the dinoflagellate,

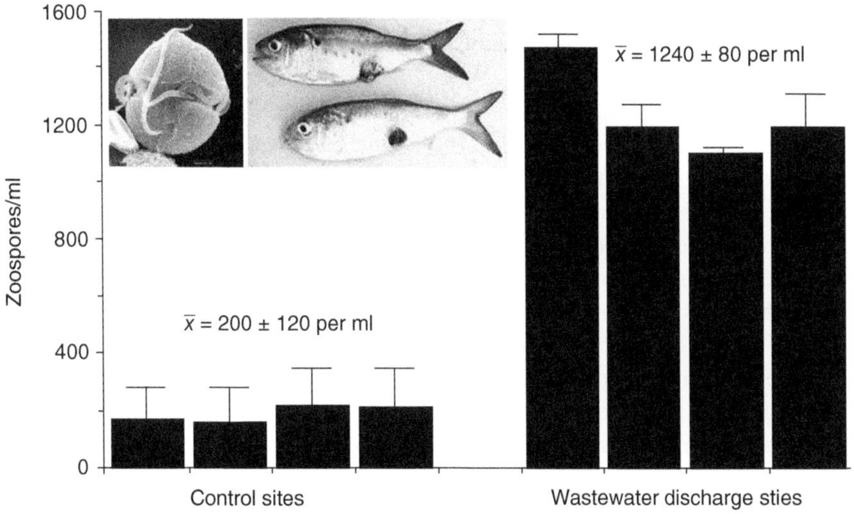

FIGURE 21.14 Growth of the predatory dinoflagellate, *Pfiesteria piscicida*, relative to nutrient conditions. Water samples from North Carolina show the dramatic increase in *Pfiesteria* zoospores within 100 m of wastewater discharge sites, where phosphorus and nitrogen compound concentrations exceeded 100 ppb. Left inset, a *Pfiesteria* zoospore. Right inset, deep focal lesions on Menhaden taken from a *Pfiesteria*-induced fish kill, Pamlico estuary, North Carolina. Adapted from G. Helfman (2007) and Burkholder and Glasgow (1997); Courtesy of North Carolina State University Center for Applied Aquatic Ecology.

Pfiesteria piscicida, and closely related forms, which has a life cycle of 20 or more flagellated, amoeboid, and cyst-like stages depending on environmental conditions, including the presence or absence of fish and fish byproducts in the water (Barker 1997; Burkholder & Glasgow 2001; Burkholder 2002). Resting cysts of the dinoflagellate are stimulated to break open in the presence of chemicals exuded by fish. The vegetative cells released from the cysts produce a neurotoxin and other substances that can kill fish in a matter of hours. The toxin induces immobility and death in fish; later-produced amoeboid cells and zoospores feed on the moribund and dead fish. Many dead and dying fish associated with *Pfiesteria* have ulcerated lesions (Fig. 21.14, inset). Skin sloughs off the dying fish and is attacked by the dinoflagellates, which then reproduce rapidly, leading to a massive fish kill; 1 billion Menhaden died during an episode in North Carolina's Neuse River estuary in 1991.

When a fish population declines to some level where the cyst-breaking chemical trigger is no longer sufficiently concentrated, the dinoflagellates return to the encysted form. Hence **density-dependent population regulation** of fishes could occur through the population responses of a microscopic predator. Increasing frequencies of *Pfiesteria*-caused fish kills correspond to increasing concentrations of human and agricultural wastewater (Fig. 21.14). One fish kill in the Neuse River estuary, North Carolina, in 1995 followed shortly on the heels of the discharge of approximately 1×10^8 L of raw hog sewage from a ruptured, upstream, sewage lagoon (Burkholder & Glasgow 2001). During an outbreak of *P. piscicida*, some researchers working with dead fish were also affected, reporting neurological symptoms including blurred vision, erratic heartbeat, and memory loss. A definitive link to a toxin from *Pfiesteria* was not established, however.

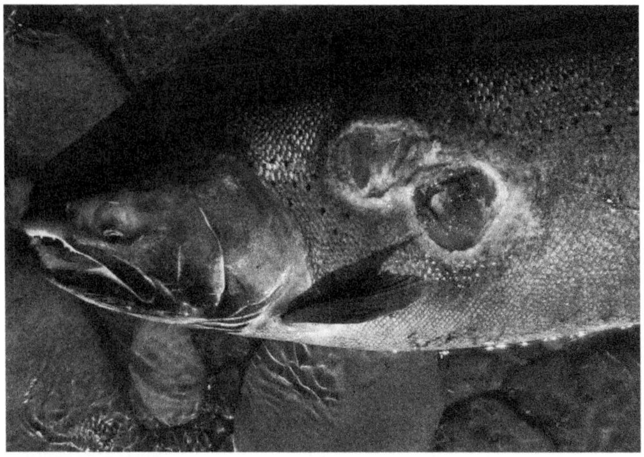

FIGURE 21.15 Evidence of Sea Lamprey (*Petromyzon marinus*) feeding on an Atlantic Salmon, *Salmo salar*. USFWS / Public Domain.

Parasitism

Whereas predators consume all or most of their prey, parasites consume only a small portion of their host. Several species of fishes parasitize other fishes. Some species of lampreys (Petromyzontidae) attach their jawless mouth to the side of another fish, scrape through the skin of the host with their tooth-studded tongue, and feed on blood and other body fluids (Fig. 21.15). The Candiru (*Vandellia cirrhosa*), also called the "vampire fish," is a small, thin South American catfish that attaches to the gills of host fishes and feeds on their blood.

Fishes in at least 12 different families (some freshwater, others marine) feed on the scales of other fishes; some of these lure victims by resembling cleaner fishes which typically would glean parasites, food particles, or bits of dead skin. The lower jaw of the freshwater Cutlips Minnow (*Exoglossum maxillingua*) has a protruding middle lobe which helps it feed on small mollusks and also pluck out eyes of other fishes when other food is scarce; some cichlids also feed on eyes of other fishes. And the Cookie-cutter Shark (*Isistius brasiliensis*) is a small (< 60 cm) dogfish that occupies deep oceans and can gouge out rounded chunks of flesh from much larger fishes and marine mammals (and, on extremely rare occasions, a human swimmer) by attaching its fleshy lips and using rows of very sharp teeth. It is believed to attract these larger predators using ventral photophores to blend in with down-welling light to contrast with a dark band across its throat which may appear to be potential prey to the unsuspecting victim. The Cookie-cutter shark also feeds by swallowing smaller prey, as most fishes do, but is considered a facultative ectoparasite of much larger fish.

Fishes are hosts to a variety of invertebrate parasites, external and internal, with a great degree of host specificity (Bush et al. 2001; Combes 2001). Life cycles are complex and attachment sites are diverse, as in the 5–6 cm long copepod *Ommatokoita elongata* that anchors itself in the corneas of Sleeper Sharks, *Somniosus spp.*, causing damage to the eye and sometimes resulting in blindness (Borucinska et al. 1998; Benz et al. 2002).

Parasite–host relationships often involve coevolved responses. One particularly bizarre relationship exists between fishes and parasitic isopods in the family Cymothoidae. These isopods are frequently observed on the heads and in the gills of numerous reef fishes. However, under some circumstances, the isopod attaches to the tongue of its host, causing the tongue to degenerate to a small stub. However, when the isopod *Cymothoa exigua* attaches to and destroys the tongue of the Spotted Rose Snapper (*Lutjanus guttatus*), it resembles the shape and size of the fish's tongue and occludes with the vomerine teeth when the fish feeds. Snappers with the isopod in place for extended periods are typically in good condition, with full stomachs and accumulated fat, which is often not the case for fish that have cymothoid isopods attached to their gills. Evidently, the parasite functionally replaces the tongue, allowing its host to survive normally and hence continue to provide the parasite, the nourishment it needs for its own successful reproduction (Brusca & Gilligan 1983).

In the context of community relationships, parasitism can have profound effects on relationships among species, affecting the outcome of predator–prey and competitive interactions. In the lab, Prussian Carp, *Carassius auratus*, infected with metacercariae of the digenean *Posthodiplostomum cuticola*, were eaten more often by Perch, *Perca fluviatilis*, compared to nonparasitized carp, probably because parasitized individuals were in poorer condition and were visually conspicuous due to the presence of black spots associated with digenean infections (Ondrackova et al. 2006). Parasitism can also interact with availability of and competition for refuges to cause higher mortality in infected fish. When Bridled Goby, *Coryphopterus glaucofraenum*, were experimentally infected in the field with a copepod gill parasite, goby survival declined as a combined function of goby density, parasite occurrence, and refuge availability (Forrester & Finley 2006). The decline was steeper for parasitized gobies in reef habitats with fewer refuges than for unparasitized gobies in similar circumstances. Parasitism and a shortage of refuges jointly influenced the strength of density-dependent mortality in this reef fish, an interesting finding given that nocturnal refuges may be important in minimizing infection by other, nocturnally attacking fish parasites (Sikkel et al. 2006).

Parasites that infect fishes may alter behavior (Barber et al. 2000), and some behavioral changes can facilitate transmission of the parasite to its final host. Killifish, *Fundulus parvipinnis*, are an intermediate host for trematodes. Killifish infected by larval trematodes alter their swimming behavior in ways that make them more conspicuous to bird predators, their definitive host. Parasitized fish in the field were 10–30 times more likely to be eaten than unparasitized fish (Lafferty & Morris 1996).

Humans may be the ultimate host of some fish parasites. The human diseases of schistosomiasis and opisthorchiasis are transmitted via fishes as intermediate hosts. Snails are an initial intermediate host for schistosomes. The snails are eaten by fishes, such as molluskivorous cichlids in Africa's great lakes, which are then eaten by humans (Evers et al. 2006). Numerous cyprinids in Thai waters feed on snails that harbor the trematode metacercariae larvae of the liver fluke that causes opisthorchiasis, the most prevalent food-borne parasitic disease in Thailand (Kumchoo et al. 2005).

Parasites have gone to great lengths to find intermediate hosts, including capitalizing on symbiotic relationships among fish species. Cleaning behavior is a highly evolved mutualism on coral reefs, performed by juveniles of some species and best known among adult wrasses, particularly in the cleaning-specialist genus *Labroides* (see Chapter 17, Mutualism and Commensalism). Digenean bucephalid trematodes have a complex life cycle usually involving at least a snail and a snail-eating fish as intermediate hosts, with a piscivorous fish as the definitive host. Adult bucephalids generally only occur in piscivores, the larvae having been ingested incidentally as a consequence of feeding on fish prey. However, surveys of labrids have shown that cleaner wrasses, especially *Labroides* spp., have incidences of bucephalid infection of 50–100%. The likely route of transmission is direct, the cleaners having ingested bucephalid larvae in the process of cleaning parasites from host fishes (Jones et al. 2004). The parasite has evolved a new route for transmission, exploiting the host–cleaner relationship in cleaning symbiosis.

Some fishes may detect the presence of parasites in their environment and alter their behavior to try to decrease risk of infection. Fathead Minnow (*Pimephles promelas*) got significantly closer to one another when exposed to cercariae of the infective *Ornithodiplostomum ptychocheilus*, similar to their response to alarm substance released by injured conspecifics (Stumbo et al. 2012). And juvenile sticklebacks (*Gasterosteus* spp.) form larger shoals in the presence of the crustacean ectoparasite *Argulus canadensis* than when the parasite was not present (Poulin & Fitzgerald 1989). Juvenile Rainbow Trout (*Oncorhynchus mykiss*) also avoided sections of a compartmentalized tank that had cercariae of the trematode *Diplostomum pseudospathaceum* (Mikheev et al. 2013). When abundant, clouds of cercariae may be visible to the fish. In addition, fish that are attacked will flee the area, probably because of the discomfort of the cercariae penetrating the surface of the fish. Fish tested individually and those in groups all tended to avoid the sections of the tank with the parasites, but the avoidance was considerably stronger among fish in groups – perhaps fish were responding to the agitated or fleeing behavior of their conspecifics.

The Effects of Fishes on Vegetation

Aquatic vegetation includes phytoplankton and macrophytes. These are also often referred to as different types of algae, a term that includes a variety of unicellular and multicellular photosynthetic aquatic organisms. Taxonomically, the term "plant" refers to members of the Plant kingdom, which includes land plants, seagrasses, and green algae but not other algae. Therefore, we will not refer to aquatic vegetation as "plants," as all except the seagrasses and green algae are not members of the Plant kingdom.

Many fishes consume aquatic vegetation. **Herbivory** is variously defined by different authors, but generally it means that a fish's diet consists of at least 25–50% vegetation. The activities of herbivorous fishes are usually divided into browsing or grazing, which are different feeding modes that affect algae and macrophytes differently (Horn & Ferry-Graham 2006). **Browsing** involves removing parts of a macrophyte, such as the tips of leaves or the leaves themselves (e.g. Silver Dollar characins, many cichlids, damselfishes). **Grazing** involves biting vegetation off at the substrate and even taking in some of the substrate itself, as in the case of parrotfishes that scrape coral surfaces in the process of eating algae. The ecological impacts of fishes on aquatic vegetation also vary among habitats by altering biomass, productivity, growth form, and species composition, by dispersing seeds, and by causing changes in the allocation of energy to vegetative versus reproductive structures.

Herbivory is variably developed in different communities, and latitude appears to be the greatest determinant of herbivore diversity. Above 40° north or south, herbivores are rare or lacking in marine and freshwater fish assemblages. Herbivorous fishes are most diverse, are usually most dense, and make up a larger percentage of the assemblage in tropical habitats. The number of species, relative abundance, and absolute density of herbivorous fishes often increases with decreasing latitude in a region (Floeter et al. 2004, 2005). Fewer than 25% of the species in a temperate stream are herbivores, whereas 25–100% of the species in a tropical stream may be herbivorous. Temperate marine habitats contain 5–15% herbivorous species, whereas 30–50% of the species in coral reef assemblages are herbivorous (Horn 1989; Wootton & Oemke 1992) (Fig. 21.16).

The greater abundance of herbivorous fishes at lower latitudes may be related to the effect of temperature on the ability

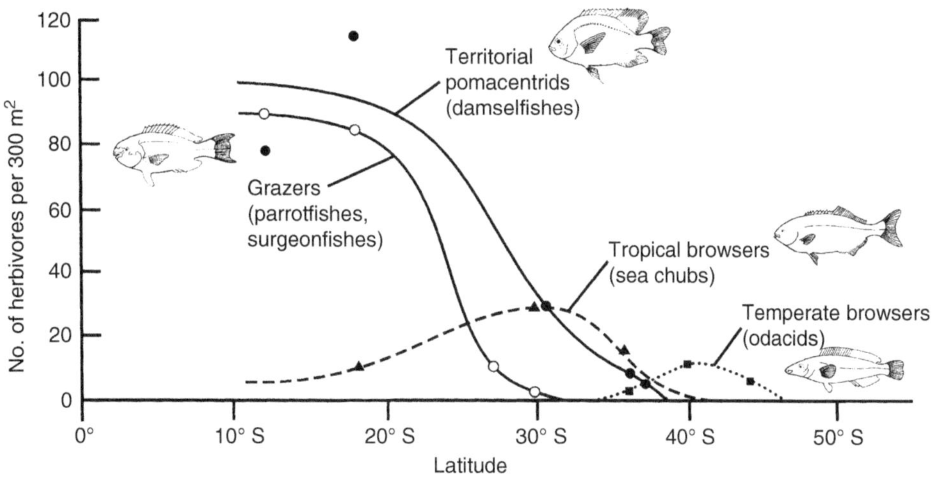

FIGURE 21.16 Relative numbers and types of herbivores as a function of latitude. More species and types of herbivores inhabit coral reefs than occur at higher latitudes, although browsing prevails at higher latitudes. Values shown are based on southern hemisphere comparisons, particularly Australia and New Zealand. However, the same or ecologically similar groups and numerical trends hold for northern hemisphere assemblages, e.g., sea chubs have temperate representatives in California (Halfmoon, Opaleye) and pricklebacks are functionally similar to aplodactylids. Exceptions may include the temperate east coasts of North America and southern Africa, which have relatively more browsers among the porgies (Hay 1986). Adapted from Choat (1991).

to digest cellulose. To digest vegetation, fishes rely on symbiotic microorganisms in the gut to produce the enzyme cellulase, which becomes less effective below 20°C (Vejříková et al. 2016). Hence, the omnivorous Rudd (*Scardinius erythrophthalmus*) selected animal food items below 16°C but shifted to macrophytes above 20°C (Vejříková et al. 2016), and aquatic vegetation was a larger portion of the diet of the omnivorous Opaleye (*Girella nigricans*) above 22°C, but less below 17°C (Behrens & Lafferty 2007). Greater abundance of herbivores at warmer temperatures may also impact the establishment of juvenile macroalgae. Along the coast of Portugal, juvenile kelp were more abundant in cooler, higher latitude sites with fewer herbivores than in warmer, lower latitude sites that had more herbivores (Franco et al. 2015). At the lower latitude warmer sites, some kelp were able to become established in crevices that were small enough to protect them from herbivorous fishes and sea urchins. As climate change shifts warmer temperatures to higher latitudes, aquatic herbivory may shift as well, thereby affecting the abundance of aquatic vegetation, and hence habitat and community structure.

High-latitude herbivores exist, but their feeding behavior reflects the seasonal availability of aquatic vegetation. At least one Antarctic icefish (*Notothenia neglecta*) eats macroalgae and diatoms during spring and summer, and switches to carnivory in fall and winter (Daniels 1982). Other possible explanations of latitudinal changes in numbers of herbivorous fishes include the seasonal nature of the food source, the biogeography of herbivorous taxa, and vulnerability to human impacts such as harvesting (Harmelin-Vivien 2002; Floeter et al. 2005; Heenan et al. 2016).

Tropical Communities

In *tropical streams*, the most common families containing herbivores are minnows, characins (particularly several piranha relatives), and cichlids, and to a lesser extent catfishes, livebearers, and gouramis (Goulding 1980; Lowe-McConnell 1987). In Panama, experimental manipulations have shown that algal biomass is reduced by the feeding activity of loricariid catfishes in both shallow (<20 cm) and deep habitats. When algal-covered rocks were moved from shallow regions where biomass was usually higher to deeper regions, the algae were quickly cropped by catfishes. When the catfishes were removed, algal growth was similar at both depths. Higher standing stocks of algae were maintained in shallows because predatory birds limit the feeding activity of the catfishes (Power et al. 1989). Work in Costa Rican streams with a fish assemblage of at least 13 species suggests that fishes consume a significant fraction of the macrophytes, algae, and grasses, as well as the leaves that fall into the stream (Wootton & Oemke 1992).

Continuous consumption of **periphyton** (the algal covering on rocks) and leaves may have a strong effect on the development of aquatic insect fauna. *Temperate streams* typically have greater aquatic insect diversity than do tropical streams; many of these insects live in and feed on periphyton and leaves. The diversity and activities of herbivorous fishes in tropical streams may keep these resources at levels too low to permit the development of a diverse fauna of herbivorous aquatic insects, despite the incredible diversity of terrestrial insects in the tropics (Flecker & Allan 1984).

The role of fishes in dispersing seeds is best known in tropical rivers. Fruiting of many trees coincides with the annual or semiannual flooding of the rivers. In South and Central America, fruits and seeds that fall into the water are consumed by several characoid fishes, such as pacu (*Colossoma* spp.) and *Brycon guatemalensis*, and constitute the major part of the diets of these fishes during those periods. Fruits and seeds of at least 40 different tree species are eaten by fishes in the Rio Machado region of the Amazon. Although some fruits and seeds may be killed by digestive processes, many seeds pass through the gut unharmed; germination may even be aided by the time spent in a fish's gut. In the case of *Brycon* feeding on the seeds of a common riparian fig tree, no loss in germination occurred. Importantly, seeds remained in the fish's gut for 18–36 h, during which time some fish moved several kilometers. Hence consumption by fishes may aid dispersal of the tree's seeds, including dispersal upstream (Goulding 1980; Agami & Waisel 1988; Horn 1997).

Tropical lakes have a large number of herbivorous fishes, particularly among the minnows, characins, catfishes, and cichlids. Phytoplanktivorous fishes can affect the relative abundances of phytoplankton species, even though a large fraction of the ingested phytoplankton cells may pass through a fish's gut unharmed (Miura & Wang 1985). Some cichlid species engage in **suspension feeding**, whereby they pump water into their mouth and across their gills, filtering out different size prey. In Blue Tilapia, *Tilapia aurea*, particles larger than 25 μm are retained by the gill rakers and by mucus-covered microbranchiospines on the gill arches; smaller particles pass through. In experimental ponds, large phytoplankton species are filtered out of the water and smaller species come to dominate and even increase in numbers. Small phytoplankters may thrive because of nutrient enrichment due to fish excretion and also because fishes filter out zooplankton (rotifers, water fleas) that consume the phytoplankton.

Interactions between herbivorous fishes and algae have probably been most thoroughly studied on *coral reefs*. Not all herbivorous fishes feed the same, however, so their impacts on the reef community may differ. For example, video recordings of four common herbivorous fishes feeding on *Sargassum polycystum* on the Great Barrier Reef showed that Brassy Chub (*Kyphosus vaigiensis*) and Bluespine Unicornfish (*Naso unicornis*) took bites from the entire thallus about 90% of the time, whereas two species of rabbitfishes (*Siganus canaliculatus*, *Siganus doliatus*) avoided the stalks and fed almost exclusively on the leafy portions of the alga (Streit et al. 2015). Another study noted that the elongated narrow snouts of three species of rabbitfishes (*Siganus corallinus, S. puellus, and S. vulpinus*) allowed them to forage in cracks and crevices that were not available to other common large (>15 cm) herbivores such as parrotfish (Labridae) and surgeonfish (Acanthuridae) (Fox & Bellwood 2013, Fig. 21.17). This niche partitioning based on morphology and behavior results in less than 45% niche

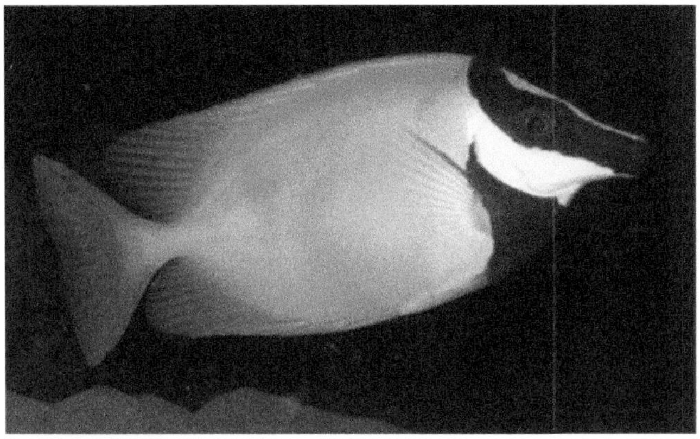

FIGURE 21.17 Reef-dwelling rabbitfish have narrow, elongated snouts which allow them to feed on algae in recesses and crevices on the reef that are not available to many other species. Photo of Foxface Rabbitfish, Siganus vulpinusby D. Meadows, NOAA / NMFS / OPR / Public Domain.

overlap in feeding between the rabbitfishes and the parrotfish and surgeonfish; the parrotfish and surgeonfish showed 95% niche overlap. These three species of rabbitfishes are among a subclade that form pairs and occupy reefs; other rabbitfishes that travel in schools and inhabit more open habitats such as seagrass beds and estuaries have shorter snouts and are not as well suited to feed in small recesses in a reef.

The impacts of herbivorous fishes on macroalgae on and near a coral reef can also be affected by the presence of predators that may pose a risk to the foraging herbivores. The presence of a model reef shark or grouper dramatically reduced the foraging activity and impact of foraging herbivorous fishes (Rizzari et al. 2014).

Herbivorous fishes can play an important role in limiting the growth of macroalgae that might otherwise overgrow coral reefs and contribute to their decline. When contacted by certain algae, the branching stony coral *Acropora nasuta* releases chemical signals that attract gobies (*Gobiodon histrio, Paragobiodon echinocephalus*) that graze the algae from the coral (Dixson & Hay 2012). This helps protect the corals from algae that would inhibit coral growth, and the fish grazers may also accumulate some toxicity from the algae that helps protect them from predation. Some omnivorous fishes will also feed on sponges that can overgrow corals. This has led to some efforts to protect reefs by limiting the harvesting of fishes that feed on macroalgae or sponges, such as in Belize (see Cox et al. 2013).

Results of numerous studies suggest, however, that other factors may play a stronger role than that of herbivorous or omnivorous fishes in maintaining healthy coral reefs. An evaluation of 85 long-term monitoring sites along the Mesoamerican reef from Honduras to Mexico from 2005 to 2014 noted habitat degradation due in part to increased growth of macroalgae (Suchley et al. 2016). The study concluded, however, that this was not linked to the abundance of herbivorous fishes and was instead likely due to nutrient inputs from human land-use activity. A study of multiple sites on the Great Barrier Reef from 1995–2009 also concluded that increased abundance of macroalgae was not necessarily brought about by declining abundance of large herbivorous fishes due to fishing, but that declining water quality decreased the abundance of large herbivorous fishes and promoted greater growth of macroalgae (Cheal et al. 2013).

Studies in the South Pacific (Carassou et al. 2013), Northern Florida Reef Tract (Lirman & Biber 2000), Florida Keys (Toth et al. 2014), and multiple Caribbean locations (Jackson et al. 2014) also did not find a strong connection between the abundance of herbivorous fishes and macroalgal coverage of corals. It should be noted that macroalgae growth on reefs is less extensive in the Caribbean and Florida, and fishing pressure is often less, than on reefs elsewhere in the world (see Toth et al. 2014). On many reefs in the Caribbean, macroalgae growth increases after the death of corals due to other reasons.

The relationship between fishing pressure and macroalgal impacts on coral reefs can be complex, however. On the coast of Kenya most of the herbivorous fishes are harvested and more than half of the areas without much grazing are often covered with macroalgae (Humphries et al. 2014). Reducing fishing in these areas allows the population of herbivores such as parrotfish to increase which helps prevent dominance by macroalgae and allows crustose coralline algae to persist. However, triggerfish (Balistidae) are also a popular harvested species, and they feed on sea urchins that also graze macroalgae. Therefore, in some heavily fished areas macroalgae growth is limited by grazing from sea urchins, but in areas closed to fishing grazing pressure from sea urchins declines due to increases in triggerfish. Therefore, macroalgal growth persisted regardless of fishing pressure, and limiting or banning fishing alone may not bring about the desired results of improving reef habitat quality.

Grazing reef fishes may have a positive impact on coral health by removing sponges that may overgrow corals. Caribbean coral reefs have two to ten times the biomass of sponges than reefs in the Indo-Pacific. Angelfishes and the three dominant species of parrotfishes (*Sparisoma*) feed on some sponges, thereby limiting coverage of reefs and allowing better growth of reef-forming coral. Evaluation of 69 reef sites in the Caribbean showed that overfished sites had more than three times the overgrowth of corals by sponges than at less-fished sites (Loh & Pawlik 2014; Loh et al. 2015). Most of this growth was by sponges more palatable to spongivorous fishes, as they tend to grow faster because the less-palatable sponges put more energy into chemical defenses to reduce grazing, and hence grow more slowly.

In coral reef communities, herbivory has led to antiherbivore adaptations in algae and in turn to adaptations by fishes to overcome those defenses (Choat & Clements 1998). Herbivore densities on reefs average 0.5 fish/m^2, as compared to 0.1 fish/m^2 in most temperate marine habitats. These values do not include the abundant sea urchins, snails, and microcrustaceans that also feed on reef algae (Horn 1989). Reef fishes affect algal biomass, productivity, species composition, distribution, and growth form. Resident herbivorous fishes can consume most of the daily productivity of the algae on a reef and in the process stimulate higher production rates in cropped algae than in uncropped algal turfs (Carpenter 1986; Klumpp & Polunin 1989). Algae respond via adaptive changes in growth form. Algae subjected to cropping assume lower, spreading shapes, whereas an absence of herbivores leads to upright, foliose growth forms of the same species (e.g. *Lithophyllum* and *Padina*).

Defensive responses also commonly involve mechanical or chemical adjustments. Algae may be leathery or rubbery (e.g. *Sargassum*), or even hard, as in the case of coralline algae. Seaweeds also produce a variety of the so-called "secondary compounds," usually halogenated terpenoids, which are distasteful to fishes; more than 20 such compounds have been isolated from marine algae (e.g. caulerpenyne from *Caulerpa*, halimedatrial from *Halimeda*, and several dictyols from *Dictyota*). Freshwater fishes also avoid algae with abundant phenolic compounds (Lodge 1991). Some tropical aquatic vegetation respond both mechanically and chemically, such as the green alga *Halimeda* that deposits calcium carbonate in its tissues and also produces distasteful chemicals. Some algae always possess such defenses, whereas others produce secondary compounds in direct response to recent herbivore activity. Most algae with such noxious properties are generally avoided by herbivorous fishes, although some fish species appear to specialize on particularly tough or chemically defended algal species. Sea chubs prefer brown seaweeds such as *Dictyota* that contain dictyols and are generally avoided by most other reef fishes (Horn 1989; Hay 1991).

A possible antiherbivore adaptation may be the production of **ciguatera** toxins, a class of substances first noticed because of its effects on humans. Adverse reactions to consuming fish with ciguatera toxin include a variety of gastrointestinal, neurological, and cardiovascular symptoms, including reversal of sensations (e.g. ice cream feeling hot) and possibly death from respiratory failure. Over 400 species of reef fishes have been implicated in ciguatera poisoning, but the most lethal sources have been large predators such as moray eels, groupers, snappers, and barracuda. These circumstances suggested that the ciguatera toxin was **biomagnified** as it passed through the reef food chain. Extensive research has verified that the toxin or toxins originate in unicellular dinoflagellates, primarily in the genera *Gambierdiscus*, *Ostreopsis*, and *Prorocentrum* (Landsberg 2002). These dinoflagellates grow as epiphytes on common reef macroalgae or on newly exposed coral surfaces, as occurs when reefs are disturbed during dredging, blasting, and ship anchoring. Herbivores ingest the dinoflagellates directly or indirectly when consuming macrophytes. Herbivores such as parrotfishes and surgeonfishes can also be ciguatoxic. The toxin is apparently not broken down by the herbivore and hence predators high in the food chain obtain a prey fish's lifetime dosage of toxin each time a prey fish is eaten (Fig. 21.18).

Transmission of the toxin up the food chain may be facilitated because many fish species as well as crustaceans exhibit loss of equilibrium and erratic swimming after feeding on ciguatoxic algae and prey, which makes contaminated prey more susceptible to predation (Randall 1958; Davin et al. 1988; Kohler & Kohler 1994; Lewis 2001). Drying, freezing, and cooking fail to destroy the toxins, which allows exported fish to cause problems far from the tropics. For example, twenty people at a dinner in Calgary, Canada, suffered ciguatera poisoning from eating thawed-and-cooked reef fishes imported from Fiji. The U.S. National Oceanic and

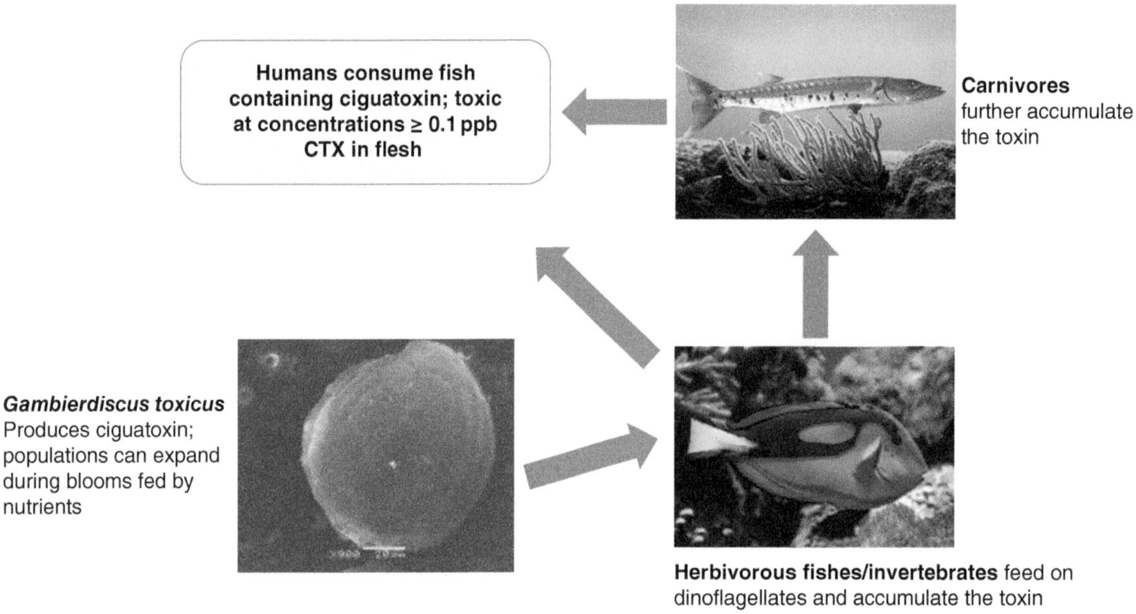
Humans consume fish containing ciguatoxin; toxic at concentrations ≥ 0.1 ppb CTX in flesh
Carnivores further accumulate the toxin
Gambierdiscus toxicus Produces ciguatoxin; populations can expand during blooms fed by nutrients
Herbivorous fishes/invertebrates feed on dinoflagellates and accumulate the toxin

FIGURE 21.18 Food webs involving ciguatera fish poisoning in a reef community. Top predators, including humans, are affected because a single meal can contain significant amounts of a highly potent neurotoxin. *Gambierdiscus toxicus* is the dinoflagellate implicated in most ciguatera poisonings. CTX, ciguatera toxin; ppb, parts per billion. photos: *Gambierdiscus toxicus* by NOAA / Public Domain; Blue Tang, *Paracanthurus hepatus*, by DerHans04 / Wikimedia Commons / CC BY-SA 3.0; Great Barracuda, *Sphyraena barracuda*, Laban712 / Wikimedia Commons / Public Domain.

Atmospheric Administration (NOAA) estimated in 2016 that over 50,000 people each year are affected by ciguatera poisoning from consuming affected fish. A commercially produced test kit for consumers to test fish for the presence of ciguatera toxin had been available, but was discontinued in 2011 due to difficulty in interpreting results. Clearly, the welfare of reef fish assemblages and reef communities can be directly linked to the integrity of the reef ecosystem itself. Where ciguatera is involved, human health is an added consideration.

One byproduct of differences in algal palatability to fishes, and a factor that can affect algal species composition on reefs, is the gardening behavior of damselfishes. The algal assemblage within a damselfish territory is termed a "lawn" because it frequently consists of a few highly palatable algal species (e.g. the red alga, *Polysiphonia*) cropped down to a fairly even level; less desirable species are actively weeded out (Lassuy 1980; Irvine 1981). The lawn contrasts with the surrounding area, where abundant roving herbivores such as parrotfishes and surgeonfishes may graze most surfaces down to relatively bare rock or leave only crustose coralline algae. If heavy fishing pressure has removed large herbivores, surfaces outside damselfish territories may have a higher biomass of algae than surfaces inside such territories. Regardless, damselfishes, whose territories may cover 40–50% of a reef's surface, and other territorial herbivores (parrotfishes, blennies, and surgeonfishes) can have a substantial effect on overall algal species diversity and distribution (Hixon & Brostoff 1983; Klumpp et al. 1987; Horn 1989; Hay 1991).

Hixon and Brostoff (1983) found that damselfish territoriality led to higher levels of algal species diversity (Fig. 21.19). The feeding rates of damselfish inside their territories were less intense than the levels of herbivory in unguarded areas outside the territories where parrotfishes and surgeonfishes were abundant. Roving herbivores kept algal diversity low outside territories because they grazed many surfaces bare. In caged enclosures that excluded both damselfishes and other herbivores, algal diversity was again lower than in damselfish territories because certain algal species were able to overgrow and eliminate other species. Hence the damselfish serves as a **keystone** species whose activities increase algal diversity by decreasing the disturbance created by roving herbivores and also by decreasing a competitive dominant alga's ability to monopolize the space resources of an area.

Temperate Communities

In temperate streams and rivers, herbivory occurs primarily among minnows, catfishes, suckers, and killifishes. Perhaps 55 of the 950+ freshwater fish species in North America can be classified as primarily herbivorous (Allan & Castillo 2007). Although only a few species are exclusively or even strongly herbivorous, some dominate locally in numbers or biomass and have a strong effect on algal growth (e.g. several pupfishes, *Cyprinodon* spp.; some suckers, *Catostomus* spp.; Brown Bullhead, *Ameiurus nebulosus*). The Central Stoneroller, *Campostoma anomalum*, can attain densities of 10 fish/m^2 in midwestern US streams, each fish taking 10–15 bites/min at the algal substrate and consuming 27% of its mass in algae daily. This activity can reduce biomass, alter species composition, and increase growth rates of algal communities. The thick overstory of filamentous green algae and diatoms is removed, leaving behind a thin layer of blue-green algae (cyanobacteria) that receives more light and more nutrients. Biomass-specific primary production is increased in areas grazed by minnows, which means that the macrophytes left behind are more productive than those removed, probably because the algal types that are removed tend to be slow-growing forms and those left behind are faster-growing species (Matthews et al. 1987; Gelwick & Matthews 1992).

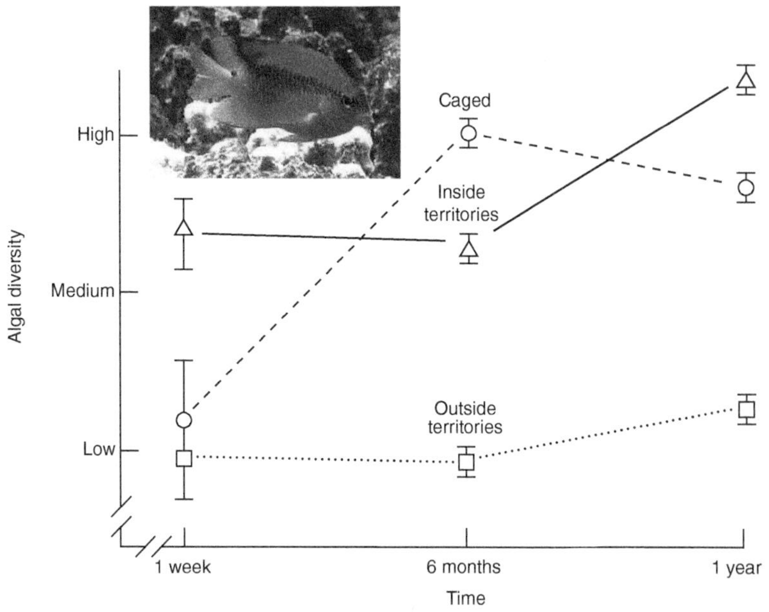

FIGURE 21.19 Damselfish territoriality helps maintain high algal species diversity on coral reefs. Hixon and Brostoff (1983) measured diversity of algal species over 1 year outside territories where roving herbivores were abundant, inside cages that excluded damselfish and other herbivores, and inside damselfish territories. Algal diversity remained low outside of cages and territories. Inside cages, diversity first increased then decreased as a red algal species became dominant. But inside damselfish territories, diversity was highest after 1 year because damselfish excluded roving herbivores that often grazed surfaces bare, while feeding by damselfish controlled the competitively dominant algal species. Based on Hixon and Brostoff (1983); Photo of Cocoa Damselfish, *Stegastes variabilis*, from USGS / Public Domain.

Many herbivorous fishes at higher latitudes are facultative herbivores, even using the generous 25% vegetation criterion to define the dietary guild (Allan & Castillo 2007). It is becoming increasingly evident that many if not most herbivorous fishes readily digest and probably actively seek out animal matter, either when taken in with macrophytes or when opportunistically available. Even the weed-controlling Grass Carp (*Ctenopharyngodon Idella,* see below) requires some animal protein in its diet to achieve proper growth (Wiley & Wike 1986). Some omnivorous fishes consume large quantities of algae, particularly when preferred animal food is unavailable. Also, many carnivorous fishes can utilize algae as a supplemental food source when animal matter is unavailable (Kitchell & Windell 1970; Gunn et al. 1977). The Longfin Dace (*Agosia chrysogaster*) of the Sonoran Desert prefers mayflies in the spring but takes in large quantities of algae in the fall when mayflies are no longer available. By increasing its feeding rate during the fall to allow for the lower nitrogen content of algae, Longfin Dace maintain a relatively constant uptake rate of nitrogen. Nitrogen is then excreted via feces and across the gills into the stream in forms that are rapidly taken up by algae and may account for 10% of the nitrogen used by an otherwise nitrogen-limited stream ecosystem (Grimm 1988).

Herbivorous fishes in temperate lakes often belong to the same families as those that occur in rivers. Their impact on algal biomass can be considerable. Roach (*Rutilus rutilus*) and Rudd (*Scardinius erythrophthalmus*) in Lake Mikolajskie in Poland consume 1700 kg of macrophytes and 3800 kg of filamentous algae per hectare per year; the macrophyte values amount to about 15% of the annual total biomass (Prejs 1984). Some are quite selective; Roach and Rudd prefer *Elodea* over *Potamogeton*, consuming 34% of the former and only 0.1% of the biomass of the latter, despite the low relative abundance of *Elodea*. An optimality analysis of benefits (obtainable, nutritive value) and costs (toughness, availability of leaves and macrophytes) indicates that *Elodea* provides a very high benefit : cost ratio as compared to other algal species.

The relative effectiveness of herbivorous lake fishes in consuming algae biomass is attested to by the widespread popularity and success of introductions of Grass Carp, native to China, in the United States, Europe, and elsewhere to control unwanted macrophytes, many of which are also introductions (e.g. Eurasian watermilfoil, *Hydrilla*, water hyacinth). Whereas most herbivorous fishes browse only the leaves of a macrophyte, which may stimulate later growth, Grass Carp uproot and eat the entire macrophyte. Grass Carp grow to a large size (to 30 kg) and consume 70–80% of their body weight daily; they can consequently eliminate all macrophytes in a lake, as happened in a Texas lake where 3650 ha of vegetation were eradicated within 2 years (Martyn et al. 1986). Total elimination of macrophytes is usually undesirable because it would destroy critical habitat for invertebrates, amphibians, and juvenile fishes (Allen & Wattendorf 1987; Murphy et al. 2002). Feeding preferences by Grass Carp can also alter species compositions within the macrophyte assemblage. In experimental ponds, Grass Carp reduced total macrophyte biomass by feeding preferentially on *Chara, Elodea,* and *Potamogeton pectinatus.* Later, total macrophyte biomass increased over original conditions because those species avoided by the Grass Carp (*Myriophyllum* and *P. natans*) occupied the space vacated by the preferred macrophytes. When Grass Carp consume submerged vegetation, floating leafed macrophytes can become dominant (Fowler & Robson 1978; Shireman et al. 1986).

Phytoplanktivory also occurs in temperate lakes and can lead to reduction in phytoplankton abundance. Gizzard Shad, *Dorosoma cepedianum*, selectively remove larger phytoplankton species (Drenner et al. 1984a, 1984b). One form of phytoplanktivory, suspension feeding (described earlier in chapter), has been documented in numerous temperate as well as tropical families, including many commercially important species (e.g. herrings, anchovies, whitefishes, minnows, silversides, mullets, cichlids, mackerels; Lazzaro 1987). Particles can be captured on structures other than the gill rakers. A cyprinid, the Blackfish (*Orthodon microlepidotus*), captures particles on its palatal organ, a mucus-covered region in the roof of the mouth (Sanderson et al. 1991).

Herbivores in temperate marine habitats are often abundant, although species diversity is lower than on tropical reefs (Horn & Ojeda 1999; Horn & Ferry-Graham 2006; see Fig. 21.16). Many porgies (Sparidae) are seasonal herbivores that take advantage of algal growth during warmer months. Temperate porgies can become particularly abundant, achieving densities of more than 7 fish/m^2 (Hay 1986). Temperate herbivores are usually browsers, eating the ends of algal fronds and other parts of seaweeds. Despite high seasonal abundance, available evidence indicates that temperate herbivores do not exercise the strong influence on algal ecology that is so evident in tropical marine environments. The strongest effects may be on the establishment and growth of young algae, as in the case of three California sea chub species that feed on young giant kelp, *Macrocystis.* As discussed earlier, the abundance of herbivorous fishes in warmer, lower latitude sites along the coast of Portugal prevented establishment of young kelp except in crevices that protected the kelp from grazers (Franco et al. 2015).

Territorial herbivores are generally lacking in temperate habitats, perhaps because a seasonally limited and variable algal resource makes territoriality impractical for most of the year (Horn 1989). Herbivory in temperate marine and fresh waters may be important, but invertebrates again probably exercise a greater influence than fishes. The association of "kelpbed" fishes with kelp beds is probably more of an attraction to the physical structure, refuge, and invertebrate production of the kelp than a dependence on the algae itself, with certain species and life history stages showing a stronger dependence on kelp than others (e.g. Kelp Perch, *Brachyistius frenatus*; Giant Kelpfish, *Heterostichus rostratus*; Kelp Clingfish, *Rimicola muscarum*; Kelp Rockfish, *Sebastes atrovirens*; Stephens et al. 2006). Among the few consumers of macroalgae are the Halfmoon, *Medialuna californiensis*, and Opaleye, *Girella nigricans*, both derivative members of the tropical family Kyphosidae that abounds with herbivores (Horn & Ferry-Graham 2006).

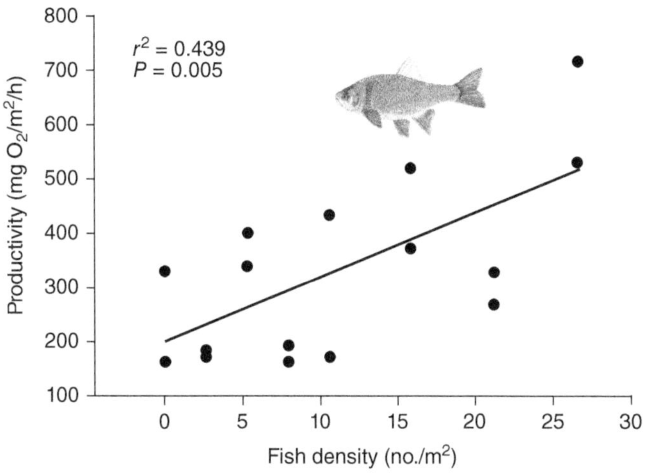

FIGURE 21.20 Algal productivity increases in relation to density of invertebrate-feeding fishes. The relationship between primary productivity and fish density is shown for experimental streams in which Red Shiner density was manipulated. Productivity increased as fish density increased because the fish fed on insects, thus transferring nutrients into the stream which enhanced algal growth. Gido and Matthews (2001) used with permission; Shiner drawing from Texas Parks and Wildlife Department, **www.tpwd.state.tx.us**.

The impact of fishes on algae may be indirect, mediated by nutrient mobilization. For example, Red Shiners, *Cyprinella lutrensis*, in midwestern US streams feed heavily on insects that fall into the water. When shiner density was manipulated in experimental streams, benthic primary production increased in response to fish density, increasing 2.5-fold at high but natural densities of 27 fish/m^2 (Fig. 20.20, Gido & Matthews 2001). Hence, water column minnows that feed on drifting insects can affect primary productivity by transferring nutrients from terrestrial sources to benthic producers in stream ecosystems.

The Effects of Fishes on Invertebrate Activity, Distribution, and Abundance

Effects on Zooplankton

The influence that fishes have on invertebrate prey populations in fresh water differs according to habitat (Sih et al. 1985; Northcote 1988; Walls et al. 1990). Although exceptions occur, fishes in general have a strong and direct influence on water column prey such as zooplankton, and a lesser and sometimes undetectable influence on benthic invertebrate populations. Fishes generally crop no more than 5–10% of zooplankton production annually, although the impact can be much greater. From 1982 to 1984 predation by Alewife and Yellow Perch decreased zooplankton populations in inshore regions of southeastern Lake Michigan to less than 5% of their 1975–81 average (Evans 1986). Fish predation on planktonic organisms normally causes a shift in size and species composition in freshwater zooplankton communities. **Size-selective predation** is a general phenomenon because fishes prefer larger zooplankton. Larger zooplankton are caught preferentially by particulate feeding fishes that select individual prey visually. Large zooplankton are themselves predators on smaller zooplankters. In freshwater ponds and lakes that contain zooplanktivorous fishes, the zooplankton community will be dominated numerically by small-bodied cladocerans and rotifers (*Bosmina*, *Scapholebris*, and *Ceriodaphnia*) because they are not eaten by the fishes and their chief predators have been eliminated. When zooplanktivorous fishes are absent, larger zooplankton (e.g. large copepods and cladocerans, *Daphnia* spp., *Simocephalus*) abound and feed on the smaller zooplankton and phytoplankton (Brooks & Dodson 1965; Janssen 1980; Zaret 1980; Newman & Waters 1984; Northcote 1988) (Fig. 21.21). Within a few years of the invasion of the zooplanktivorous Alewife (Clupeidae) to Lake Champlain (New York – Vermont), the average length of the largest cladoceran and copepod zooplankters captured in zooplankton samples decreased (Mihuc et al. 2012).

For size-selective predation to occur, a fish must be able to assess prey size and distance in the area over which it normally searches. Fish size comes into play, with young fishes being apparently unable to detect small prey, whereas larger individuals have better acuity. In Bluegill sunfish, 6 cm long fish can see objects half the size that 3.5 cm fish can detect. Some evidence indicates that apparent rather than absolute prey size may be important. When offered prey of different sizes at different distances, Bluegill, White Crappie, and sticklebacks chose the smaller but nearer prey over a larger prey that was farther away (Hairston et al. 1982; O'Brien 1987).

Zooplanktivorous fishes cause other shifts in the ecology of their prey. Diel vertical migrations are probably a means of avoiding visually hunting fishes; plankters occur in deep, dark, cold, and sometimes deoxygenated regions of lakes by day and rise into surface waters at night to feed. The extent of such migrations decreases when predation pressure from fishes is reduced. Adjustments in life history traits of prey include reductions in age at sexual maturity and average size of offspring. Predation also affects prey morphology and coloration. Some cladocerans (e.g. *Daphnia*) develop neck spines when exposed to water that has contained Bluegill. Some cladocerans are relatively dark when occurring in fish-free ponds, but are essentially transparent when co-occurring with fish in lakes; copepods are red in the absence of fish but take on a less vulnerable pale green color in ponds containing fishes. The costs of coloration are obvious from studies that show that some fishes will take the smaller of two forms if it is more pigmented, or will take prey with the large dark eyes over similar size prey with smaller eyes (Zaret 1980; Lazzaro 1987; O'Brien 1987; Walls et al. 1990; Hobson 1991).

Studies in temperate and tropical nearshore marine environments also suggest a strong influence of zooplanktivorous fishes on species composition and size structure of prey populations. Small (<1 mm), transparent zooplankters that are part of the pelagic community that is being carried by currents

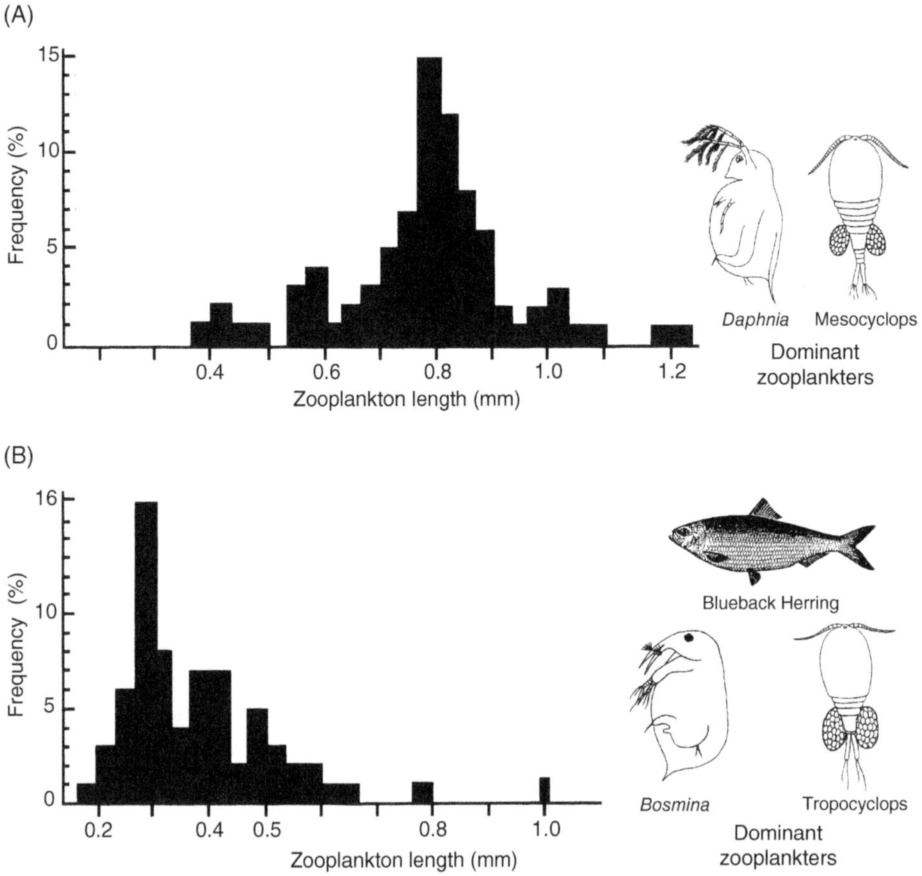

FIGURE 21.21 Effects of fish predation on zooplankton assemblage structure. (A) Before predators were introduced. (B) When Blueback Herring were introduced into a small Connecticut lake, size-selective predation by the fish caused the average size of zooplankters to decline. Predation favored smaller individuals within species as well as smaller species. Based on Brooks and Dodson (1965) and Bigelow and Schroeder (1953b).

(e.g. dispersing fish eggs and larvae, small copepods, and larvaceans) characterize the water column and fish diets during the day. The fishes tend to aggregate at the upcurrent edges of reefs and other structures, where they intercept the incoming pelagic forms. Densities of plankton are lower in downcurrent areas, indicating that fishes remove a significant number of plankters from the water column. At night, different species and sizes of zooplankters occur than are seen during the day. Larger, opaque animals (1–10 mm), such as mysids, large copepods, polychaete worms, amphipods, ostracods, and crustacean larvae, emerge from the substrate or migrate vertically upward and join the small, transient zooplankters.

Both groups encounter a different set of predators at night, generally large-mouthed fishes with large eyes (e.g. squirrelfishes, cardinalfishes, and sweepers). Vision is the most widely used detection mode both by day and night, but nocturnal zooplanktivores are constrained by the lack of light, making it safer for larger zooplankton that would be easily detected during the day (small plankters are seldom found in the stomachs of nocturnal fishes). Verification of the influence of predators on oceanic plankton communities would require the kinds of predator removal or enclosure experiments that have been performed in fresh water, but the open nature of oceanic coastlines and the long-distance dispersal of the prey make these kinds of experiments virtually impossible (Bray 1981; Hobson et al. 1981; Hobson 1991).

Effects on Benthic Invertebrates

Populations of benthic invertebrates in streams appear to be minimally influenced by predatory fishes, although results vary in different habitats (for a review, see Allan & Castillo 2007). In Colorado, removal of Brook Trout (*Salvelinus fontinalis*) from a 1 km section of a stream had no significant effect on the numbers and species composition of invertebrates during a 4-year study, suggesting that trout predation had little effect on invertebrate community dynamics. Similar results, involving fish removal or cages that excluded fishes, have been found in streams in North Carolina, Kentucky, Czechoslovakia, England, and in ponds in New York and South Carolina (Allan 1982, 1983; Holomuzki & Stevenson 1992). In contrast, other studies show significant decreases in benthic invertebrates due to Bluegill in lakes and by Creek Chub (*Semotilus atromaculatus*) in streams. Midges and stoneflies declined in direct proportion to sculpin and dace abundance in one study, but midges were unaffected in another. Bluegill can also reduce biomass and density of pond invertebrates, and Yellow Perch as an introduced species can eliminate 50% of the benthic invertebrate biomass of small lakes. Bottom type, topographic complexity, and current strength all influence the outcome. Predation in many of these experiments was size selective, with fish preferentially eating larger individuals and species and hence affecting community composition as well as numbers (Crowder & Cooper 1982; Mittelbach 1988; Northcote 1988; Gilliam et al. 1989; Allan & Castillo 2007).

Predators can affect more than absolute and relative prey numbers. **Distributional changes** arise as a result of behavioral adjustments to the threat of predation, which can have population consequences for prey. In the presence of predatory fishes such as sculpins, insect larvae such as mayflies avoid the surfaces of rocks where food availability is highest and instead are found under rocks (Kohler & McPeek 1989; Culp et al. 1991). When Smallmouth Bass (*Micropterus dolomieu*) are present, crayfish reduce both locomotory and foraging activities and select bottom types that provide more protection (Stein 1977). **Drift**, the movement of invertebrate larvae and adults downstream in streams and rivers, is also affected by fish activity. In the presence of Longnose Dace, *Rhinichthys cataractae*, mayfly drift first increases and then decreases compared with situations when predators are absent. Hence invertebrates may first move out of an area where predators occur and then later settle and take refuge. Amphipods respond to sculpin by reducing general activity, including drift (Andersson et al. 1986; Culp et al. 1991). Cooper (1984) showed that predation by trout can have both negative and positive effects on water striders. By preying on water striders, trout increase intraspecific competition by increasing the relative density of striders in refuges away from predation risk. However, trout predation benefits water striders in that trout prey on both competitors and predators of the water striders.

Disagreements about the responses of benthic invertebrates to fish predation may reflect differences in prey mobility in different areas (Cooper et al. 1990). In streams containing highly mobile or drifting prey, or in cage studies where mesh size is large enough to allow invertebrates to recolonize, prey that are eliminated by predators will be replaced by immigration. Where minimal exchange of prey occurs between habitats, or where cage mesh sizes are too small to allow recolonization, predators have a stronger depressing effect on prey populations. Not surprisingly then, drift-feeding fishes (e.g. salmonids) have less effect on prey, whereas benthic-feeding fishes (e.g. some minnows, sculpins, and sunfishes) tend to reduce prey populations. Lakes, ponds, and pool communities with their reduced flow regimes will be more affected by predation than will be riffles and runs. Current speed in streams and rivers will operate analogously, fast currents serving to transport immigrants and slow currents tending to delay replacement. Predatory fishes therefore play different roles depending on whether the habitat is relatively open or closed.

The apparent minimal effect that fishes have on stream invertebrate populations may be interpreted as a lack of influence of fishes on their prey. However, the flip side of this situation is that stream fishes are therefore unlikely to limit their own food supply and are thus unlikely to find themselves competing for food. Insect abundance does, however, vary significantly and can be limiting. Seasonal cycles of insect abundance, driven by insect life histories more than by fish predation, include a midsummer low after many adults have emerged from the stream and before the next cohort of prey has reached edible size. Growth rates of fishes decrease during this midsummer period. Experimental additions of food to streams during midsummer can increase growth, survival, and energy stores of juvenile stream fishes, indicating that food was limiting (Mason 1976; Schlosser 1982; Karr et al. 1992).

Fishes can also affect the numbers, species composition, and distribution of sessile and mobile marine invertebrates. Some corals appear to be restricted to shallow water habitats by the grazing activity of fishes. Triggerfish (*Balistapus undulatus*) occur on reef slopes and deep reef areas and eat corals such as *Pocillopora damicornis*. *Pocillopora* can be successfully transplanted to deeper water on reefs if placed in cages that exclude the triggerfish; corals moved outside of cages are consumed (Neudecker 1979). Damselfish actively kill corals within their territories, which creates a growth surface for the algae on which the fish feed (Kaufman 1977). Damselfish territories can cover a significant proportion of some shallow reefs, therefore habitat modification by damselfish can influence the amount of coral cover. At the same time, damselfish algal turfs are important habitats for small crustaceans, polychaetes, and mollusks, and the territoriality of the damselfish protects these invertebrates from predatory fishes (Lobel 1980). Fish predation can also affect the diversity of encrusting species. Tunicates in caves on the Great Barrier Reef are heavily grazed by fishes. Inside enclosures, tunicates dominate other forms, whereas outside enclosures removal of tunicates by fishes allows for growth of bryozoans. Tunicates grow successfully only where they have natural protection, such as at the base of stinging hydroids (Day 1985).

Fish predation can limit the distribution of such mobile invertebrates as polychaetes, burrowing sea urchins, snails, stomatopods, and hermit crabs on and near reefs. Predation by stingrays may reduce snail abundance on sandy substrates near reefs. The reef itself often provides a refuge from and for fishes. Herbivorous sea urchins shelter in the reef by day and venture into surrounding grass beds at night. Patch reefs in the Caribbean typically have a denuded ring of sand 2–10 m wide at their base which results from the nocturnal forays of the urchins. The width of this "halo" apparently reflects the distance that urchins can move from the reef and still return safely at dawn before diurnal predaceous fishes become active. The region around the reef, including the halo, is also relatively devoid of infaunal invertebrates. Invertebrate-feeding fishes venture off the reef and forage in the reef's vicinity, returning to the reef to rest or when threatened by predators. Invertebrate densities increase as one travels farther from the reef, unless another reef is close enough that the foraging regions of fishes

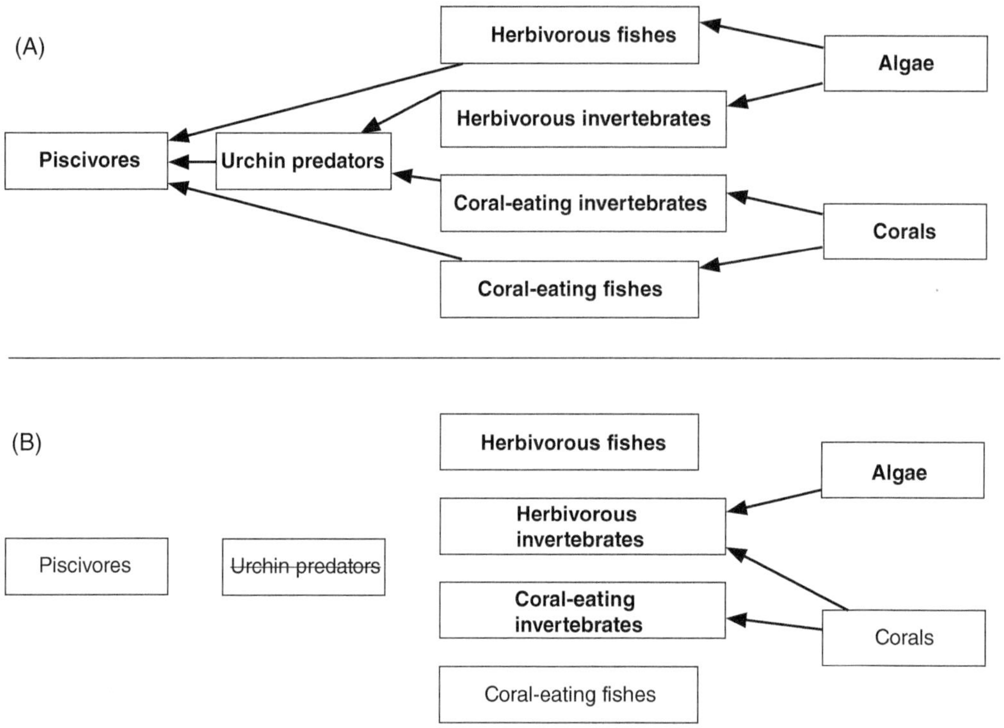

FIGURE 21.22 Ecosystem level implications of overfishing on coral reefs. Overfishing frequently leads to dominance of invertebrates and algae via a series of interrelated pathways. (A) Unfished or lightly fished reefs house a diversity of fishes in several trophic guilds, as well as abundant coral cover. (B) The removal of large piscivores results in the elimination of fishes that feed both on sea urchins and on invertebrates that feed on corals. Herbivorous and coral-eating invertebrates proliferate, which leads to overconsumption of corals or permits algae to overgrow coral. In either situation, overall fish diversity declines because of strong direct and indirect dependence of many fish taxa on live coral cover. Darkness of lettering denotes changes in relative abundance of trophic or taxonomic group. From Helfman (2007), used with permission.

from both reefs overlap (Ogden 1974; Randall 1974; Ambrose & Anderson 1990; Jones et al. 1991; Posey & Ambrose 1994).

Fishes, sea urchins, and macrophytes interact as integrated parts of complex **trophic cascades** (addressed later in this chapter); disrupting one part affects ecosystem components both above and below in a cascade. Overfishing of species such as large triggerfishes and wrasses that feed on coral-eating invertebrates apparently tips the competitive balance between fishes and invertebrates to favor invertebrates (Fig. 21.22). Sea urchins in particular become abundant because sea urchins outcompete herbivorous fishes under conditions of intense fishing on sea urchin predators (McClanahan & Muthiga 1988; Levitan 1992). Unlike most herbivorous fishes, grazing urchins actually scrape the substrate. This accelerates **bioerosion** of the reef, reduces coral cover, reduces topographic complexity, and produces a reef surface dominated by algal turf. Ultimately, fish diversity and fishery productivity decline (Glynn 1997). Additionally, rapid algal growth and reduced settling success of coral larvae apparently reduce the ability of reefs to recover from hurricane damage (Liddell & Ohlhorst 1993). Protecting urchin-eating fishes, as recommended by McClanahan (1995), would prevent the negative diversity and abundance consequences of urchin proliferation. Protecting urchin predators would also lower urchin numbers and thus direct benthic production into abundant herbivores such as parrotfishes, surgeonfishes, rudderfishes, and rabbitfishes.

Similar, dramatic fish–urchin–algae interactions have also been observed at high latitudes. In the Bering Sea, expansion of the Walleye Pollock fishery occurred concurrently with large declines among Steller Sea Lions, Northern Fur Seals, and Harbor Seals (Goñi 1998). Reductions in pinnipeds have apparently caused orcas to feed on sea otters as alternative prey, thus releasing sea urchins from predation by the otters, and culminating in a loss of kelp forests because of overgrazing by the urchins (Estes et al. 1998). Many nearshore, high-latitude marine fishes are dependent on kelp forests as habitat (Stephens et al. 2006).

Fishes and Trophic Structure of Communities

The preceding discussions have focused on the effects that fishes as predators have on trophic levels lower in the food web of the community. Such **"top-down" regulation** of community dynamics can be contrasted with "**bottom-up**" factors affecting primary productivity and subsequent vegetation or animal prey availability, which ultimately determines fish abundance and diversity. From the fish's perspective, top-down processes involve the ways that fishes affect the structure and function of an ecosystem (Table 21.1), whereas bottom-up processes involve the physical and chemical factors that affect food availability. In addition to these direct interactions, ecosystem function is affected by indirect effects of different trophic levels on one another that are separated by several steps or levels in the **trophic organization** of a community. Fishes function in such interactions as agents that transfer and cycle nutrients, energy, and matter and that link different parts of the ecosystem together. In addition to their interactions with other members of the biotic community, many of the activities and relationships have direct benefits to humans and can be considered as ecosystem services provided by fishes (discussed later in chapter). Our dependence on fishes for the functional roles they play in ecosystems as well as for the calories and nutrients they provide as food underscores the importance of understanding the impacts of overexploitation and biodiversity loss (Allan et al. 2005; McIntyre et al. 2007).

TABLE 21.1 Top-down effects of fishes in temperate lakes and streams.

Activity	Factors affected	Mechanism and consequence
Direct feeding	Water clarity	• Foraging stirs up bottom sediments and lowers clarity
		• Intense feeding on phytoplankton may increase clarity depending on size of phytoplankton eaten; but clarity may decrease due to excretion and fertilization
	Nutrient release, cycling	• Benthic feeding increases nutrient cycling at substrate-water interface
		• Grazing on littoral vegetation increases nutrient cycling
	Phytoplankton	• Grazing can increase nutrient cycling, resulting in more production
	Periphyton	• Strong grazing affects biomass
	Macrophytes	• Strong grazing affects biomass
	Zooplankton	• Fish often select larger forms, decreasing their abundance
		• Some evidence of increased production
	Zoobenthos	• Feeding often affects abundance, but may vary in lakes and streams
		• Seasonality in effects due to distribution and size of feeding fish
		• Production often increases in lakes but not in streams
Selective predation (due to size, visibility, motility)	Phytoplankton	• Grazing may alter relative abundance of algal size and species composition
	Zooplankton	• Changes in relative abundance of species can reduce algal grazing efficiency and water clarity
		• Changes in clutch size and timing of maturation
	Zoobenthos	• Fish preference for large forms affects their cover selection, activity patterns, and reproductive behavior
	Nutrient release	• Shift to smaller body size of zooplankton increases nutrient release
Excretion	Nutrient release	• Liquid release provides quick, patchy availability
		• Feces release provides slower patchy availability
		• Epidermal mucous release increases iron availability to algae
Decomposition	Nutrient release	• Carcasses provide slow, patchy releases
Migration	Nutrient enrichment	• Concentration and transport of nutrients from one habitat region to another; nutrients released by excretion and decomposition

Adapted from Northcote (1988).

Indirect Effects and Trophic cascades

The impact of fish predation on both vegetation and animals extends beyond the direct effects of reduction in biomass and shifts in species and size composition. Only about half of the variation in annual primary production in lakes can be explained by changes in the amounts and types of nutrients that occur there. The other half results from the indirect but important role that fishes play. This effect can be described as a **cascade** of influences down through the food web of a lake from secondary consumers to primary producers (Fig. 21.23). A typical trophic cascade involves piscivorous fishes (e.g. salmon, pike, bass) feeding on zooplanktivorous fishes (e.g. herrings, minnows), which feed on herbivorous zooplankton, which eat phytoplankton. Increasing the number of piscivores reduces the number of zooplanktivores, which increases zooplankton abundance, which leads to more removal of phytoplankton from the water column. Hence experimental additions of top predators lead to greater water clarity. The previously mentioned increase in average size of zooplankton due to size-selective feeding of fishes occurs as an incidental result of such a manipulation. Fish predators on zooplankton are reduced, favoring larger zooplankton that would otherwise be taken by fishes and that may be too large for invertebrate predators to handle. A reversed chain of events occurs if piscivore abundance is reduced, either experimentally, through overfishing, or from fish kills such as those that may occur during winter deoxygenation. Fewer piscivores mean more phytoplankton because of the reduction in herbivorous zooplankters, which are never entirely eliminated (Carpenter et al. 1985; Carpenter & Kitchell 1988, 1993; Kufel et al. 1997).

Trophic cascades are not limited to the open water communities of lakes. Benthic communities in shallow water may also be structured by fish-mediated cascades. For example, Redear Sunfish (*Lepomis microlophus*) in a Tennessee lake ate snails that grazed on epiphytes (filamentous blue-greens

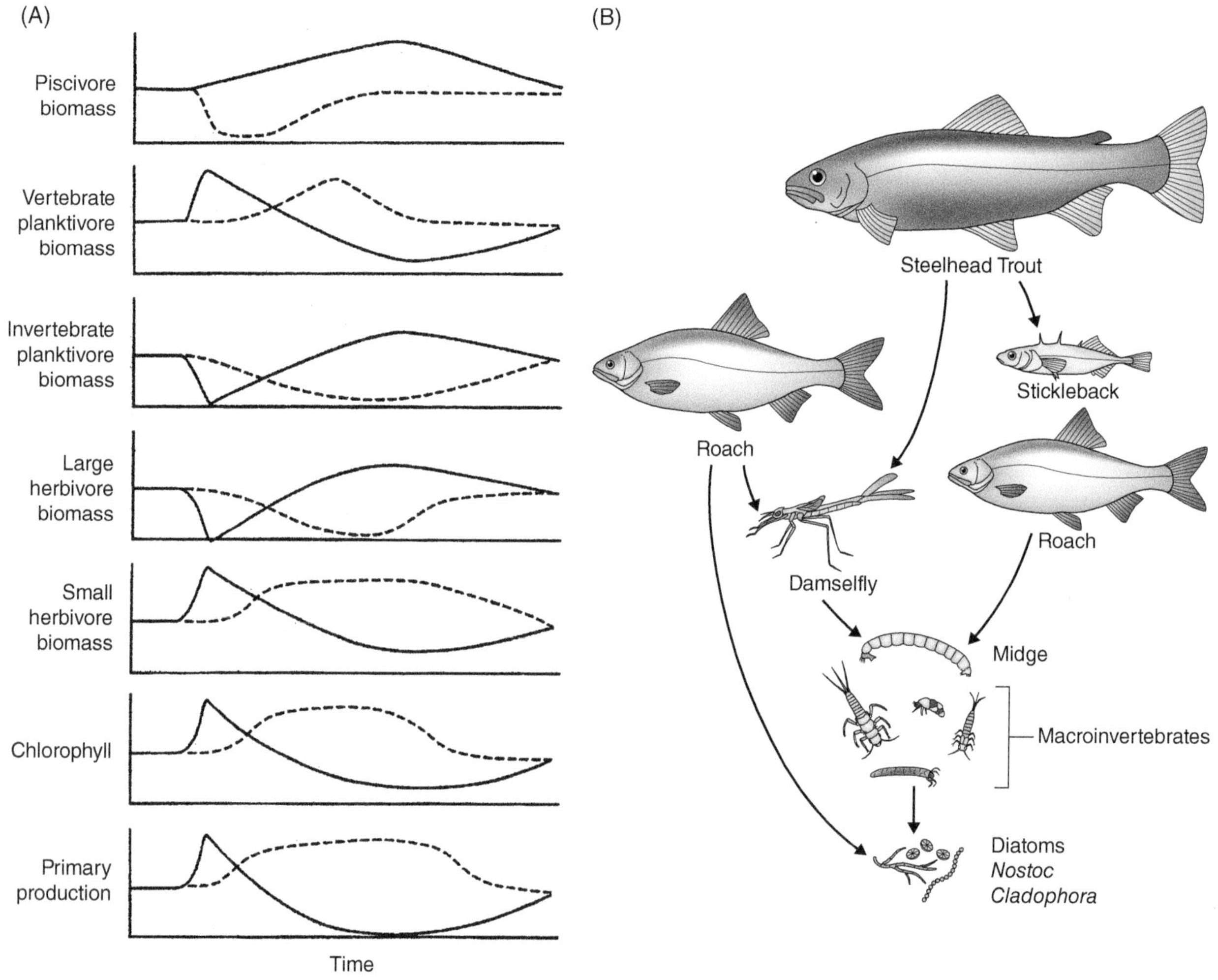

FIGURE 21.23 Trophic cascades. (A) The various components of a trophic cascade as postulated for a lake or pond over a growing season. Solid lines represent changes in biomass or density resulting from a strong year class or the experimental addition of piscivores; dashed lines show the effects of winterkill or overfishing of piscivores. In North America, piscivores would include salmonids, esocids, black basses, and Walleye; vertebrate planktivores would include herrings, minnows, whitefishes, Bluegill, and Yellow Perch; invertebrate planktivores would be copepods and many insect larvae; and most herbivores are crustacean zooplankton. (B) An actual trophic cascade. See text for explanation. (A) from Carpenter et al. (1985), used with permission; (B) based on Power (1990), after Allan and Castillo (2007).

and diatoms), which normally infest lake macrophytes (Martin et al. 1992). When fish were excluded, snail abundance increased reducing the epiphyte populations and macrophyte growth increased. Strong trophic cascades similar to those found in lakes can also develop in rivers and streams. During the summer, low-flow conditions create a series of pools in seasonally flowing rivers. In the Eel River, California, turfs of filamentous algae up to several meters long and covered with diatoms and blue-green algae cover most of the bottom. Through the summer, these turfs are grazed down to small tufts by midges. The midges are the major prey of predatory damselfly nymphs and the fry of California Roach (*Hesperoleucas symmetricus*) and Three-spined Stickleback (*Gasterosteus aculeatus*). Fish fry and damselfly nymphs are in turn eaten intensively by large roach and Steelhead (*Oncorhynchus mykiss*). When algae are allowed to grow in cages that exclude the trout and large roach, fry and damselfly nymphs abound, cropping down the midges. This allows the algae to maintain fairly luxuriant growth through the summer. Hence the feeding activities of predaceous fishes cascade down through the food web and eventually determine the growth form and extent of primary producers in the river (Power 1990).

Indirect effects driven by fish predation can cause unexpected physical changes in lakes. Lake Michigan and other high pH, hard water lakes experience milky water during summer months due to the precipitation of limestone crystals (calcite = calcium carbonate, $CaCO_3$). These "whiting events" can inhibit zooplankton feeding, increase sinking rates and loss of precipitated nutrients to deeper water, and reduce light penetration and primary production. Whitings result from increased photosynthetic activity of algae at elevated summer temperatures, which removes CO_2 from the water and causes an increase in pH. $CaCO_3$ is less soluble at high pH and thus precipitates out of the water, causing the milkiness. Intensive stocking of salmonids (Coho and Chinook salmon, Lake, Rainbow, and Brown trout) in Lake Michigan during the 1970s led to high salmonid populations in 1983. Salmonids ate huge numbers of zooplanktivorous Alewife, allowing increases in phytoplanktivorous cladocerans, which in turn ate phytoplankton. Lack of phytoplankton kept the pH low and hence no whiting event occurred that year (Stewart et al. 1981; Vanderploeg et al. 1987).

Another physical result of trophic cascades involving fishes is the effect that plankton biomass has on temperatures, thermocline placement, and seasonal mixing depths in a lake. Lakes typically have an upper **epilimnetic** region of warmer water, a lower **hypolimnetic** region of cold water, and an intermediate **metalimnion** or thermocline where temperatures change from warm to cold as depth increases. In experimental enclosures and small (<20 km^2) lakes that lack zooplanktivorous fishes, the abundance of zooplankton reduces phytoplankton, resulting in clearer water and greater penetration of light and heat. This leads to a deeper mixed layer of epilimnetic water and a deeper thermocline with temperatures that are 3–13°C higher in the metalimnion. Hence the heat content of a lake and all the biological and physical processes and interactions dependent on that heat may be strongly influenced by the top-down effects of fish predation (Mazumder et al. 1990).

Trophic cascades do not have to be unidirectional, e.g., either top-down *or* bottom-up. Flecker et al. (2002) added nitrogen and excluded grazing fishes from Andean streams. Excluding fishes had the strongest impact on algal biomass and composition, but algae responded more strongly to nutrient addition when exposed to grazing fishes. Hence fishes affect primary producers directly through consumption and indirectly by influencing their response to nutrient availability, thus producing *both* top-down and bottom-up processes.

All of these examples deal with relatively small- and medium-scale ecosystems involving a few species in relatively simple food webs. We now know that large, complex, diverse oceanic systems are also driven by bottom-up and top-down cascading trophic interactions, and that human disruption of cascades in the form of climate change or overfishing can have far-reaching consequences. In the Northeast Pacific, the abundance and catches of valuable, predatory fishes is closely linked to a trophic cascade involving planktivorous fishes, zooplankton, and phytoplankton productivity. Phytoplankton production declines when warm water masses push into northern Pacific regions, eventually resulting in reduced fish yields (Ware & Thomson 2005). In the North Atlantic, accelerated overfishing of groundfishes (especially Atlantic Cod, but also Haddock, White and Silver Hake, Pollock, Redfish, and flatfishes) occurred in the 1980s and 1990s; cod stocks were reduced by more than 90%. These species were the top predators in the benthic and near-benthic communities and their removal had effects across five trophic levels (Frank et al. 2005). Small pelagic fishes and benthic invertebrates – the primary prey of the top carnivores – increased markedly. Their prey, large-bodied herbivorous zooplankton, decreased. In response, phytoplankton, chlorophyll levels, and nitrate concentrations increased.

Disruption of the trophic cascade has apparently affected ecological interactions within the North Atlantic food web, to the detriment of the ecosystem and its utility to humans. Altered conditions, especially the removal of the top predators, helped to release dogfish and skates from predation by and competition with groundfish species. As groundfish decreased through the latter 20th century, skate and dogfish increased (Anderson et al. 1999; Reynolds & Jennings 2000) (Fig. 21.24), and these increases may be irreversible. Despite bans on fishing, cod had not rebounded (Hutchings & Reynolds 2004; Olsen et al. 2004). The upsurge in their former prey may have turned the tables because these benthic, intermediate-level carnivores are predators on the young of the formerly abundant, top-level predators (although stomach contents analysis indicates that groundfish are uncommon food items of the elasmobranches; Link et al. 2002). Regardless, the prognosis for a rebounding cod fishery was not promising: "the time required for population recovery in many marine fishes appears to be considerably longer than previously believed" (Hutchings 2000a, p. 885). For example, despite efforts to encourage recovery, updates on the status of Atlantic Cod stocks in the Gulf of Maine have indicated continued overfishing at least through 2011 (NEFSC 2013).

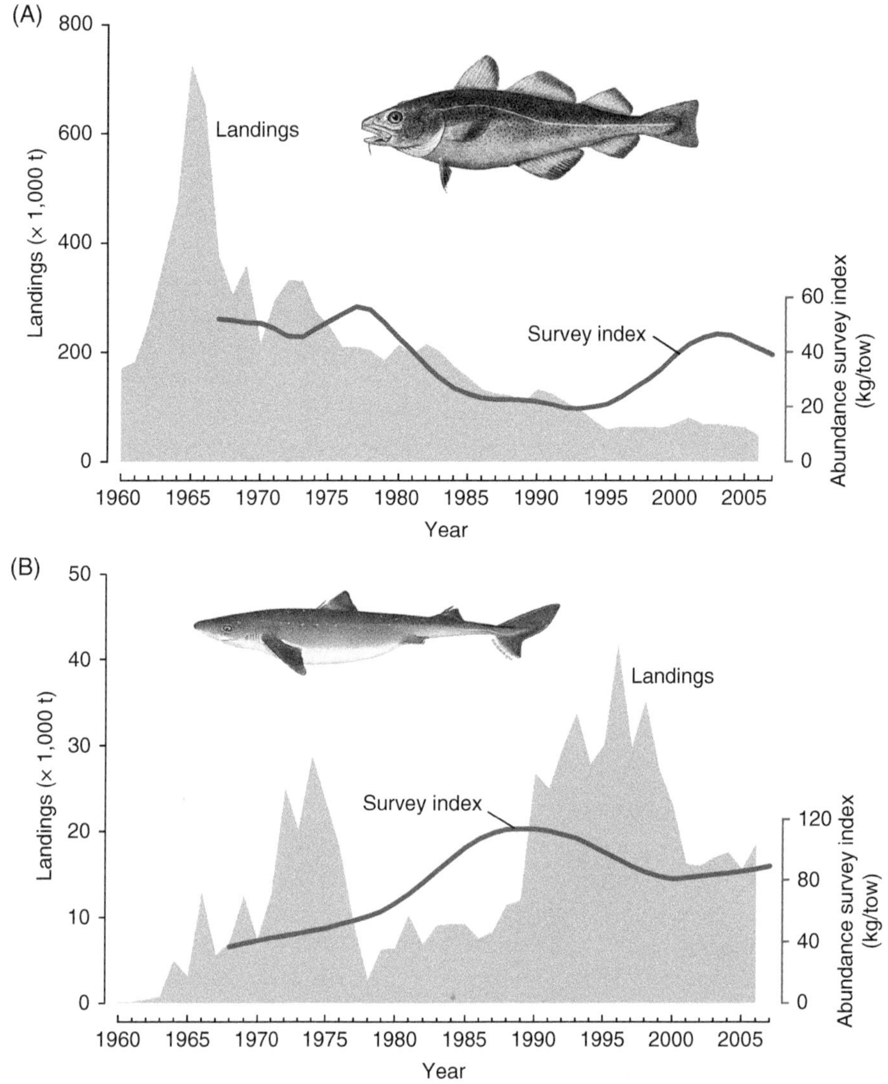

FIGURE 21.24 Disruptions of large-scale, oceanic trophic cascades can result in major shifts among interacting species. (A) Landings and abundance index for principal groundfishes and flounders off the northeastern USA, 1960–2007. NOAA / Public Domain. (B) Landings and abundance index for skates and Spiny Dogfish, showing increases in these less desirable species while groundfish declined. NOAA / Public Domain (NMFS. 2009. Our Living Oceans, 6th ed.).

News reports of Atlantic Cod stocks in the North Sea in 2015 and off of Iceland in 2017 were encouraging, however.

Trophic cascades can connect aquatic and terrestrial communities, with fishes playing an important role. In a series of experimental manipulations, Knight et al. (2005) demonstrated the connection between the presence of sunfish (Centrarchidae) in ponds at the University of Florida's Katharine Ordway Preserve/Carl Swisher Memorial Sanctuary, and the seed production of St. John's wort, a common terrestrial plant along the shoreline of these ponds. The presence of sunfish in ponds decreased populations of dragonflies which prey on pollinating insects, therefore resulting in higher numbers of pollinating insects visiting flowers of St. John's wort along the shoreline, which then produced more seeds. Ponds lacking sunfish had more dragonflies, which decreased pollination visits to flowers and decreased seed production.

Trophic Cascades Applied

Trophic cascades are of interest beyond the insight they give us into the function of aquatic ecosystems. They have been directly applied to problems associated with **eutrophication** in lakes, where excessive nutrient input such as from fertilizers leads to blooms of undesirable phytoplankton (Kitchell 1992; Carpenter & Kitchell 1993). Primary production and turbidity can be reduced by stocking piscivores that eat zooplanktivores or by selectively removing zooplanktivorous fishes. Both practices have the same theoretical result at the end of the trophic cascade: fewer zooplanktivores mean more zooplankton, which means more consumption of phytoplankton and clearer water, assuming that planktivores are not too big for the piscivores to eat and that phytoplankters are edible.

Direct application also occurs in the context of **marine protected areas** (MPAs) on coral reefs. MPAs are created in part

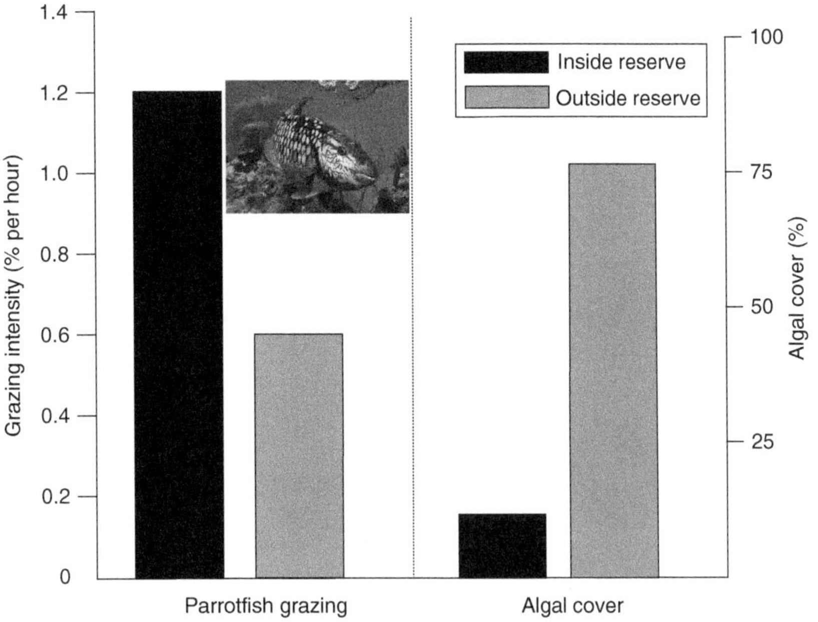

FIGURE 21.25 The relationship between parrotfish grazing and algal growth in a coral reef protected area, Exuma Cays Land and Sea Park, Bahamas. Left-hand bars show grazing intensity of parrotfishes and right-hand bars show percent cover of macroalgae inside (darkened bars) and outside (open bars) the reserve. Grazing was more intense and algal cover less inside the reserve. Based on Mumby et al. (2006); parrotfish photo by G. Helfman.

to restore large predators, such as groupers. A potential negative outcome of increased predator biomass could be reduced prey numbers, which could be problematic if the prey were herbivores such as parrotfishes that consumed benthic algae that competed with corals for space. However, actual measurements of parrotfish numbers and algal biomass in successful reserves indicates that, despite increases among large predators, parrotfish also increase in size and number. This leads to a doubling of grazing and a four-fold reduction in macroalgal cover, which "highlights the potential importance of reserves for coral reef resilience" (Mumby et al. 2006) (Fig. 21.25).

Finally, failing to understand how trophic cascades function can lead to misapplications and unintended consequences (Fig. 21.26). Concern over shark attacks along South African beaches led to an extensive gill netting effort targeted at large sharks. This netting effort was very successful and reduced shark attack frequency, but had a deleterious, cascade effect on sportfishes. Large sharks are predators on smaller sharks, which compete with humans for sportfishes. Hence by reducing the abundance of large predatory sharks, smaller shark species and the young of large sharks experienced a population boom, which in turn strongly depressed the abundance of finfishes in the region (Van der Elst 1979).

Fishes in Food Webs

Discussing top-down and bottom-up interactions may suggest that linear food chains characterize many aquatic ecosystems. Trophic dynamics are far more complex, however. Fishes tend to be opportunistic feeders, and often shift from eating zooplankton to eating small fishes as they grow, or as small fishes become abundant which often changes seasonally. In addition, large zooplankton often prey on small zooplankton, inserting another trophic level in the food web; fishes may feed on either or both. So the effect on subsequent lower trophic levels may be more difficult to predict and may change seasonally. Trophic interactions often involve herbivores that eat animals, carnivores that eat vegetation, cannibalism, reversals in energy transfer (one species eats juveniles of another species but winds up as the prey of the adults of the same species), and fishes that parasitize their predators by eating their fins, scales, and mucus. Most individuals enter a **detrital food loop**, either directly during decomposition or indirectly after falling prey and being processed into feces. Hence trophic interactions are more accurately described as **food webs** (Fig. 21.27) than food chains, and when disrupted can cause surprising results (Polis et al. 2004).

A particularly thorough analysis of the role of fishes in a community food web has been conducted for lowland streams in Venezuela and Costa Rica (Winemiller 1990b). The fish assemblages at four different sites included between 20 and 83 species that fed on detritus, vegetation, seeds, flowers, protozoans, aquatic and terrestrial insects, numerous aquatic invertebrates (worms, crabs, shrimps, clams, and snails), fishes, larval amphibians, turtles, lizards, birds, and mammals. Food web analysis of relationships only between fish and their prey indicated between 50 and 100 interacting taxa and 200–1200 **trophic links** (i.e. one fish species might eat a dozen different food types, each one constituting a link) (Fig. 21.27). The

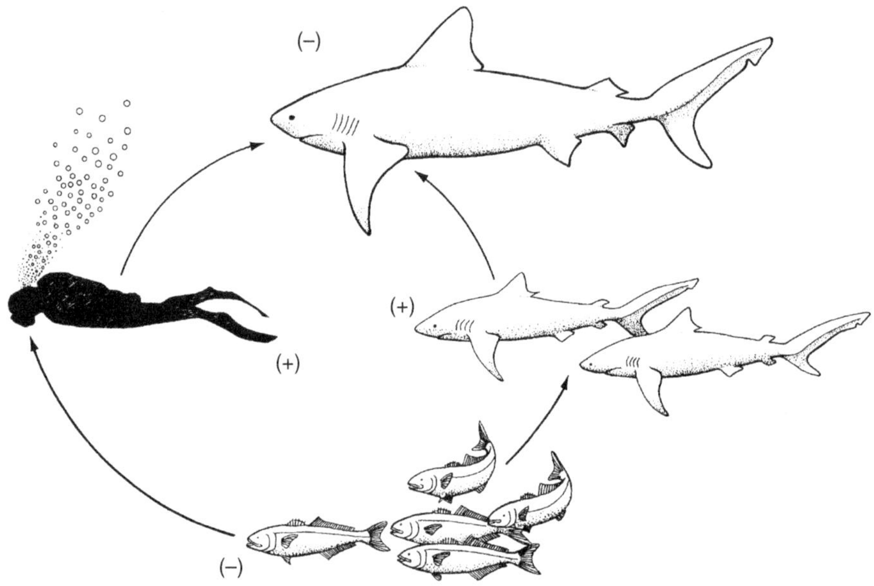

FIGURE 21.26 A trophic cascade with unexpected results. Extensive gill-netting of large sharks in the Natal region of South Africa removed large sharks, such as the Bull Shark, *Carcharhinus leucas*, a predator on humans. Primary prey of large sharks, such as juvenile Dusky Sharks, *C. obscurus*, increased in numbers. Abundant small sharks consumed sportfishes (e.g. Bluefish, *Pomatomus saltatrix*). (+) increasing population; (–) declining population. Based on data in Van der Elst (1979); drawings after Bigelow and Schroeder (1948b, 1953b).

pimelodid catfish *Rhamdia* fed at three different trophic levels: seeds, prawns, and fishes. More than a dozen species (mostly characins, catfishes, and cichlids) were omnivores, feeding extensively on both vegetation and animal matter. Detritus, formed from decaying aquatic and terrestrial vegetation, was a particularly important component, accounting for 30–50% of the food eaten by all species. Species as diverse as characins, catfishes, livebearers, cichlids, and sleepers feed directly on detritus. Reciprocal food loops were common. The trichomycterid catfish *Ochmacanthus* feed on the external mucus of Oscars (*Astronotus ocellaris*); Oscars in turn eat the catfish. A predatory cichlid (*Cichlasoma dovii*) eat juvenile Sleepers (*Gobiomorus dormitor*), and adult Sleepers eat juvenile cichlids.

As complex as the interactions in Fig. 21.27 appear, the actual food web is even more intricate. Different ontogenetic stages of a species were not separated in the analysis; nine dominant piscivorous species fed initially on zooplankton, switching later to invertebrates and eventually to fishes as they grew larger. The food webs at the four sites differed considerably due to varying diversities and species compositions, but differences also occurred within sites during the wet versus dry season. Even this seemingly complex description of interactions simplifies the true complexity of feeding relationships in a natural community, because only the interactions involving fishes are listed; other food items of the prey of the fishes were not considered (i.e. links between shrimp and snails or between aquatic insects and their prey).

Parasites are often overlooked as components of food webs. For example, the trematode commonly known as yellow grub (*Clinostomum marginatum*), which parasitizes many freshwater fishes, has different life stages that require snails, fish, and fish-eating birds such as herons. This creates overlap and connectivity between aquatic and terrestrial food webs. The inclusion of parasites usually increases species richness, number of links, and food chain length plus two additional measures, **connectance** and **nestedness**, all which influence ecosystem stability (Lafferty et al. 2006). The web of feeding interactions in any community is undoubtedly tangled.

Most descriptions of trophic relationships within a community tend to characterize species as either relative specialists or generalists, referring to whether a species feeds predominantly on one or a few food types as compared to a species that feeds on many food types or even at several trophic levels. **Specialist-generalist** characterizations are used in general community descriptions and have also been invoked to explain the relatively high diversity of fishes and other taxa in tropical as compared to temperate communities. Tropical species are thought to be relatively specialized. The relatively narrow niches of specialists make it theoretically possible to fit more species into the resource spectrum of a habitat.

Although usually an accurate description, characterizing a species as either a specialist or a generalist may only apply to its habits under the feeding regime that exists in that habitat at the time of the description. An excellent example of this comes from a study of the feeding habits of two Panamanian toadfishes (Batrachoididae). Both species fed almost entirely (85–100%) on long-spined sea urchins, *Diadema antillarum* (Robertson 1987). However, *Diadema* underwent a massive die-off in early 1983 that wiped out 95–99% of the individuals across the Caribbean region. One would have predicted that such feeding specialists as the toadfishes would suffer population declines because of their relatively invariant feeding habits. However, populations

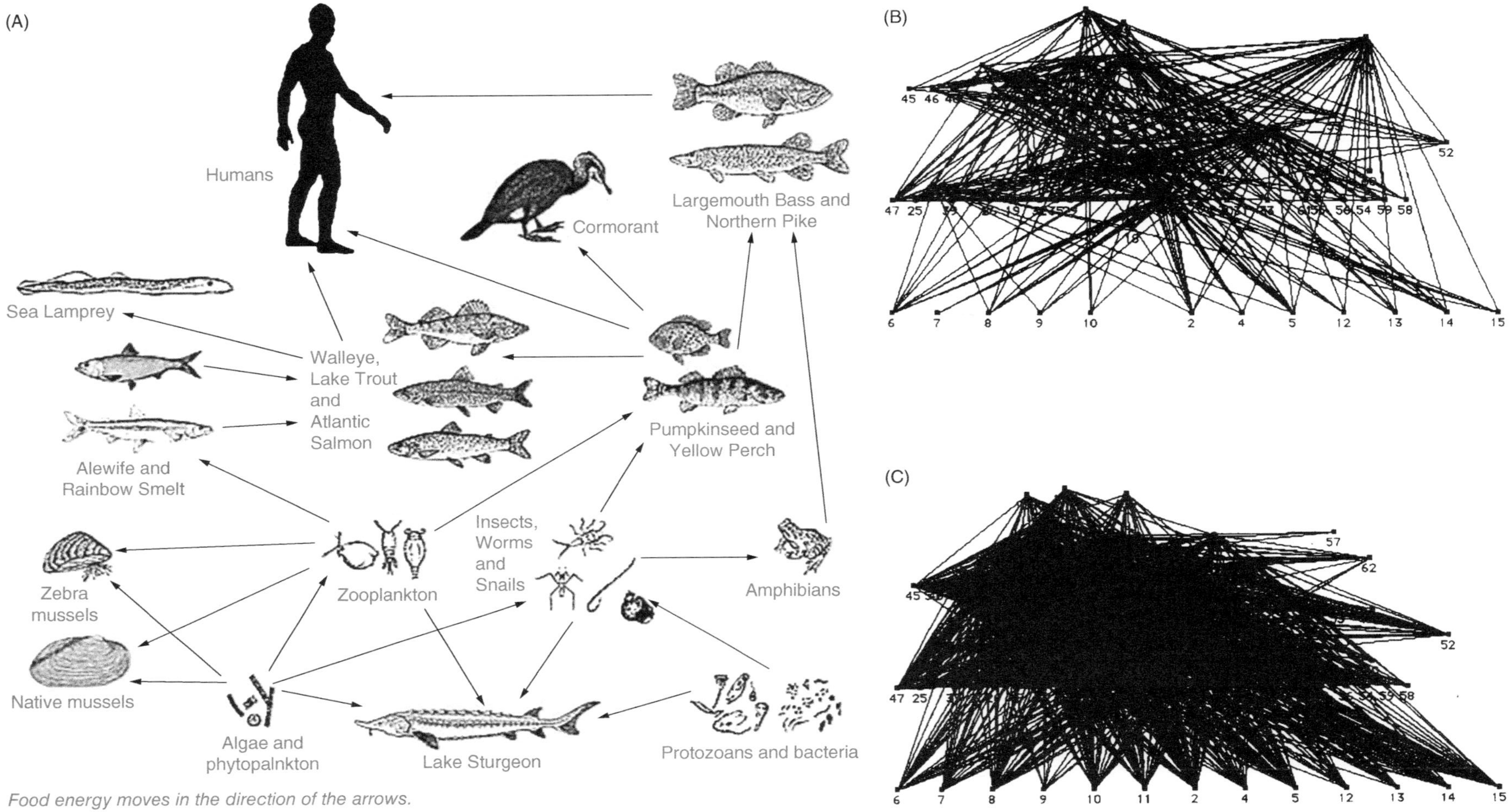

FIGURE 21.27 Food webs involving fishes. (A) A relatively simple food web in a temperate North American lake involving humans, predatory fishes, planktivorous fishes, invertebrate plankton, and algae. (B) A small, lowland forest stream in Costa Rica. (C) A swamp creek in Venezuela. Each numbered point in the webs represents a fish species or a prey taxon eaten by fishes. The base of the food webs is at the bottom and includes detritus, living vegetation, and vegetation parts. Intermediate levels in the webs represent primary consumers (herbivorous fishes and invertebrates), with predatory fishes at the top. Eleven fish species are involved in food web B and 51 species in web C. (A) figure by N. Ballinger, Lake Champlain Basin Program, used with permission; (B, C) from Winemiller (1990b), used with permission.

of both species changed little, reproduction continued, and food habits shifted to a variety of mobile benthic invertebrates in one species and to fishes and mobile invertebrates in the other species. Hence these classic trophic specialists became relative generalists, sounding a cautionary note for anyone attempting to characterize community feeding relationships.

Turnover Rates and the Inverted Food Pyramid

The **biomass** (mass per unit area) of fishes or other animals in a habitat indicates something about the nature of the community. However, biomass is a static depiction, basically a snapshot, of a very dynamic situation for which a moving picture would tell us more. **Turnover**, the ratio of production to standing crop biomass (P : B), provides the added information. Turnover, expressed in units of *mass per unit area per unit time* (e.g. g/m^2/year) is a measure of how productive a population is over time and takes into account life table schedules of birth and death, population density, individual growth rate, and development time (Benke 1993). For example, rates of prey consumption by fishes suggest a paradox in many freshwater habitats. Trout in the Horokiwi Stream of New Zealand consume about 20 times the standing crop biomass of invertebrates annually; trout and stonefly consumption of prey in a Colorado stream is about 10 times greater than standing crop biomass of prey. In some streams, the biomass of predators exceeds that of prey, which would seem to violate laws of ecology and thermodynamics.

Just looking at biomass tells us little about ecosystem dynamics in such a situation. The apparent paradoxes of how fish can consume more prey than that exists, and how more predators than prey can be maintained in a habitat, are solved by considering **turnover rates**, namely how quickly animals reach maturity, how many times they reproduce, and how many young they produce. Because benthic invertebrates go through several generations per year, their annual production can greatly exceed biomass at any one moment and the invertebrate community can support a much larger fish assemblage than if the fishes were solely dependent on standing crop biomass. Production values three to 10 times greater than biomass are not unusual (Benke 1976, 1993; Allan 1983).

Fishes and Ecosystem Ecology

Nutrient Cycling and Transport by Fishes

Fishes play an essential role in the **processing**, **transformation**, and **movement** of important nutrients in aquatic ecosystems (Vanni 2002). Phosphorus is often the nutrient that limits primary production in lakes. Fishes excrete soluble reactive phosphorus (SRP), a form that is readily taken up by algae. Phosphorus excretion by fishes may be the major source of SRP in many lakes, enhancing phytoplankton production and altering algal community composition (Schindler et al. 1993). Although phytoplankton biomass may increase when zooplanktivorous fishes eat phytoplanktivorous invertebrates, nutrient cycling rather than decreased phytoplankton consumption may be responsible for a significant portion of this trophic cascade-associated increase in phytoplankton (Vanni & Layne 1997; Vanni et al. 1997).

Fishes also move nutrients between different compartments of a lake ecosystem. Benthic-feeding fishes disturb sediments, accelerating the rate of exchange of nutrients between the water and substrate. The concentration of dissolved nitrogen and phosphorus is greater in the water and reduced in sediments when more bottom-feeding fishes are active. Vertically migrating fishes serve as transporters of important nutrients between colder, deeper waters and surface layers where most primary productivity occurs. Peamouth Chub (*Mylocheilus caurinus*) feed on benthic invertebrates at depths greater than 20 m and then migrate to the surface at night during the summer and fall, where excretion and defecation releases nutrients in forms usable by vegetation. The movement of nutrients across the physical boundary of the thermocline is otherwise limited mostly to periods when the lake turns over, which may occur only once or twice annually in many temperate lakes and rarely in permanently stratified tropical lakes (Northcote et al. 1964; Northcote 1988).

Fishes can represent a major reservoir of nutrients that are essential for primary production and can therefore become part of the bottom-up pathway affecting ecosystem function. In some lakes and ponds, 90% of the phosphorus in the water column may be bound up in Bluegill. These nutrients are released through excretion from the gills, through defecation, and through decomposition after death. Approximately 20% of the internal phosphorus entering a large Quebec lake during the spring could have come from decomposing fish that died after spawning (Nakashima & Leggett 1980). Excretion and defecation by Roach (*Rutilus rutilus*) contributed about 30% of the total phosphorus in the epilimnion of a deep Norwegian lake during the growing season (Brabrand et al. 1990). Fish feces and mucus could be important sources of iron in lakes where algal growth is iron-limited. Fish removal experiments often lead to reductions in phosphorus and nitrogen levels in lake waters, whereas fish additions generally lead to increases in nitrogen compounds, except where herbivorous fishes are involved. When macrophytes are eaten, nutrients become available for uptake by phytoplankton (Northcote 1988).

Fish Feces as an Energy Source: The Coprophage Connection

A major path that energy takes from the organisms on which fish feed to other ecosystem components is through feces (Hobson 1991). For example, on coral reefs, many zooplanktivorous fishes feed on oceanic plankton and produce prodigious quantities of feces during the day that rain down on the reef; the bulk of these byproducts are eaten by **scatophagous** or **coprophagous** (feces-eating) organisms.

Consumers of feces include fishes and various invertebrates such as crustaceans, snails, brittlestars, and corals. Trophically, these organisms are generally considered to be herbivores, omnivores, and detritivores, but a significant proportion of their diet comes from coprophagy, and the fecal component in detritus undoubtedly contributes to its high amino acid and protein content (Crossman et al. 2001). When zooplankters are particularly abundant, they pass through planktivores' guts so rapidly that they appear basically undigested in the feces that fall to the reef. These conditions raise the possibility that coprophagous consumers gain more energy indirectly from zooplankton than do the planktivores that initially captured the zooplankton. At one Indo-Pacific locale, 45 different species of reef fishes from eight families (primarily sea chubs, damselfishes, wrasses, rabbitfishes, surgeonfishes, and triggerfishes) consumed 5975 fecal particles from 64 species in 11 families (Bailey & Robertson 1982; Robertson 1982).

These coprophages were not indiscriminate consumers. Feces with higher caloric, protein, and lipid content were more likely to be consumed. Therefore, fecal material from zooplanktivores and other carnivores was preferred, whereas feces from herbivores that ate brown algae (e.g. rabbitfishes) or from corallivores that consumed carbonate skeletal material along with coral polyps (e.g. parrotfishes) were generally avoided. A typical fecal food chain might involve a zooplanktivorous damselfish producing feces that were eaten by a surgeonfish that normally eats red algae and whose feces are in turn eaten by a parrotfish that normally eats small algae and coral. Often, the nutritional value of feces eaten exceeded the value of the usual food of a species. Some fishes actively followed others and fed on their feces, indicating that coprophagy is not just incidental to normal feeding behavior. Interestingly, fish never ate feces from their own species, perhaps because most of the usable nutrition for that species had already been extracted, or due to the risk of parasite transfer.

A time delay may occur in the transfer of fecal material from fishes to other reef components. Following foraging, many fishes defecate or otherwise excrete at **resting sites** that are frequently quite distant from the original feeding site. In this way, fishes help exchange energy and nutrients between different parts of the reef. Deposition of feces and other excretory products, particularly of nitrogen and phosphorus, can lead to enhanced growth by corals that live under resting schools of grunts in the Caribbean (Meyer et al. 1983).

Linkages between different parts of marine ecosystems are not restricted to tropical waters. In temperate marine habitats, organic carbon was traditionally thought to either be produced in situ by reef algae or transported unpredictably to the reef by currents in the form of plankton and detritus. Kelpbed zooplanktivores, such as the abundant Blacksmith (*Chromis punctipinnis*), also feed on oceanic plankton during the day and return repeatedly to the same resting crevices at night where they defecate. Blacksmith produce an average of 180 mg of feces per square meter of resting habitat each night. These feces are taken up by a variety of detritivores (gobies, clinids, shrimps, hermit crabs, amphipods, snails, and brittlestars) which are in turn eaten by larger fishes. Blacksmith also excrete ammonium (NH_4^+) through their gills that is taken up readily by growing kelp, thus aiding the production of the habitat on which the entire kelpbed community depends. Rather than random arrival with currents, fecal and excretory inputs by fishes are a constant and reliable source of energy, nutrients, and trace metals for detritivores and macrophytes, thus adding an additional pathway for the active capture and transfer of potentially limiting substances in nearshore habitats (Bray et al. 1981, 1986; Rothans & Miller 1991).

The importance of fish feces in the nutrient dynamics of other aquatic ecosystems has been less extensively studied. Little is known about this topic in temperate streams, although large amounts of feces accumulate in pools in some streams and are reworked by minnows (Matthews et al. 1987). Underwater photographs in African lakes that contain large numbers of cichlid fishes often show fecal material intact on the rocks (e.g. photographs on pp. 35, 53–70 of Axelrod and Burgess (1976) and pp. 15, 22, 31–42 of Lewis et al. (1986)). One would be unlikely to make such an observation on a coral reef, implying that the use of fecal material by fishes and invertebrates has not evolved as extensively in tropical lakes (the name of one Indo-Pacific riverine family, the Scatophagidae, implies that tropical coprophages are known in fresh water).

Fecal material may nonetheless contribute substantially to the nutrient budget of lakes. Prejs (1984; and see above) calculated that two minnow species, Roach and Rudd, contribute approximately 133 kg of nitrogen and 12 kg of phosphorus during the 3-month growing season in a Polish lake as a result of the consumption and processing of macrophytes; additional nutrients are released from the digestion of benthic algae. Fishes are efficient consumers of macrophytes but inefficient assimilators of the nutrients contained in the macrophytes; 70–80% of ingested macrophyte material leaves the gut as feces. In situations where high macrophyte biomass supports dense populations of herbivorous fishes, the digestive activities of the fishes can lead to undesirable plankton blooms (Prejs 1984). Where fishes use phytoplankton as a food source, as do many cichlids in African lakes, the redistribution of nutrients that fertilize phytoplankton from macrophytes through herbivores and into the water column may be important in the maintenance of a diverse and abundant fish assemblage. Overharvest of benthic-feeding herbivores could lead, indirectly, to reductions in populations of phytoplanktivores. Food webs are maintained through a complex series of linkages (Lowe-McConnell 1987).

Fishes as Nutrient Transporters Among Ecosystems

Fishes can link different ecosystems together, such as seasonal migrations by characins in Neotropical rivers that make it possible for unproductive, nutrient-poor rivers to support large numbers of piscivores (e.g. Hoeinghaus et al. 2006). Such linkages are illustrated nicely by the life cycles of Pacific salmons and their impact on the energy and nutrient budgets of the different systems they inhabit. Many salmon hatch in relatively small headwater streams. After some time in a river or a lake, the fish move downstream and out to sea. Although abundant in numbers, their major impact on the stream or lake ecosystem while en route to the sea is as food for birds and other fishes, including larger salmon. As they move to sea, they represent a relatively minor loss of nutrients and energy from the river; about three times more phosphorus is gained during runs of adults into lakes than is lost when smolts emigrate (Stockner 1987). In the year or more they spend at sea, growth accelerates substantially. The 5 g Sockeye Salmon smolt that left its home river 2–4 years earlier may return as a 3 kg adult. In Babine Lake, British Columbia, 160 tons of smolts leave the lake each year. Despite 95% mortality while at sea, 3400 tons of maturing adults return to the lake. As thousands (historically millions) of these now mature salmon move back to their natal streams, they constitute a large proportion of the animal biomass present. Even Atlantic Salmon, many of which survive spawning at least once rather than dying, contribute significantly and positively to the nutrient and energy budgets of the rivers to which they return (Jonsson & Jonsson 2003).

Salmon, especially returning adults, also constitute a crucial food source for predators, including seals, sea lions, and orcas in the nearshore zone, and other piscivores in the river itself (Fig. 21.28). Willson and Halupka (1995) identified 40 species of inland mammals and birds that feed on various phases of the salmon life cycle. These include ducks, geese, gulls, dippers, and robins feeding on eggs; loons, mergansers, herons, terns, kingfishers, and crows feeding on juveniles; and eagles, hawks, magpies, ravens, jays, bears, mink, otters, wolverines, wolves, foxes, seals, mice, squirrels, and deer feeding

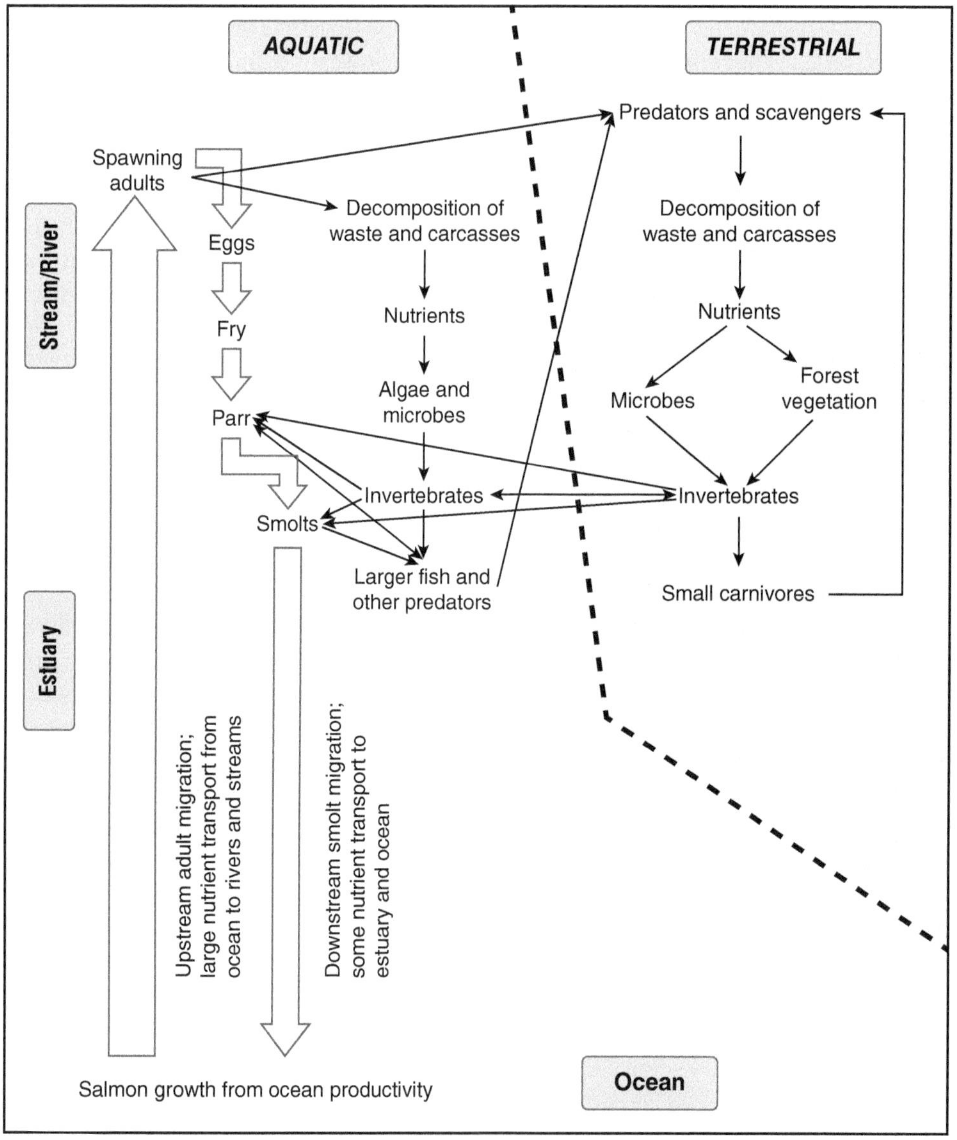

FIGURE 21.28 Salmon-derived nutrients and food webs based on anadromous fishes. The Pacific salmon ecosystem includes oceanic, estuarine, riverine, and old-growth forest components. Aquatic and terrestrial habitats are linked via predators and scavengers that feed on different life history stages.

on adults and carcasses. In addition, there are eight to 10 riverine fishes that eat eggs and juveniles, numerous stream invertebrates that scavenge salmon carcasses, and a rich community of bacterial and fungal decomposers that recycle salmon-derived nutrients.

The major ecosystem impact of spawning adults occurs in the later stages and after the spawning migration. Spawning fish, postspawning fish, and spent carcasses floating downstream or decaying in the river and on river banks are crucial to ecosystem processes. As carcasses decompose, they are consumed by microscopic and macroscopic scavengers, releasing carbon, nitrogen, phosphorus, and other nutrients acquired at sea, which leads to significant increases in primary and secondary production. The transfer of energy and nutrients eventually feeds back on the next generation of salmon because of the linkages between ecosystem components (Willson et al. 1998; Schindler et al. 2003; Janetski et al. 2009). This nutrient subsidy is critical to the stream ecosystem, even at the level of enriching biofilms which are an important component of the base of the stream food web (Rüegg et al. 2011). This supplementation is especially enriching to the microbial community immediately beneath the carcasses (Levi et al. 2013), but also has positive effects downstream. Growth of juvenile salmon is closely linked to the availability of salmon carcasses in a stream (Bilby et al. 1996), in part because young salmon feed on invertebrates that feed on salmon carcasses.

But juvenile salmon also benefit because streams with well-developed riparian vegetation provide better salmon habitat for spawning adults and developing fry via temperature-regulating shade, insect inputs, and woody debris inputs that control flow, sediment filtration, and bank stabilization. Young salmon therefore depend on intact headwater streams embedded in old-growth forests, and the health of these forest ecosystems may be greatly influenced by the spawning migrations of adult fish. Terrestrial predators and scavengers carry salmon carcasses into the woods and also defecate there. These activities release phosphorus, nitrogen, carbon, and micronutrients that originated in the ocean. These **marine-derived nutrients** are taken up by riparian trees as far as 100 m from a stream (Helfield & Naiman 2001). Sitka spruce, an old-growth species, grows three times faster along salmon-bearing streams, producing the 50 cm trees that are most useful as woody debris inputs over three times faster than trees along streams without spawning salmon (86 vs. 307 years). Hence, spawning salmon enhance the survivorship of subsequent salmonid generations in many ways besides direct, reproductive activity (Helfield & Naiman 2001).

Although less studied, the potentially important role of fishes in linking different ecosystems is becoming obvious as ecologists focus attention on nonsalmonid migratory species. Nutrients accumulate or are transformed as the result of spawning migrations, metabolic processes, predation, and the death and decomposition of spawning adults of other species, such as clupeids; these inputs can be linked to elevated ammonium concentrations and increased microbial and invertebrate production (Hall 1972; Durbin et al. 1979; Browder & Garman 1994). Alewife (Clupeidae) migrate up rivers of the Atlantic coast of the United States to spawn. Juveniles spend their first half year of life in headwater lakes, and such lakes often produce trophy-size Largemouth Bass, *Micropterus salmoides* (Yako et al. 2000). Herring were the most important fish prey consumed by bass in such lakes, and bass grew better when herring were a diet component. Yako et al. concluded that juvenile herring were an energetically valuable, potentially key prey for Largemouth Bass. MacAvoy et al. (2000) similarly concluded that anadromous fish such as Blueback Herring, American Shad, and Alewife may be a significant source of nutrients to freshwater apex predators in tidal and freshwater regions. Garman and Macko (1998) also showed that predators in coastal streams derived a substantial proportion of their biomass carbon from marine sources during clupeid (*Alosa* spp.) spawning runs.

Other reproductive migrations and life cycles may be equally important in transferring nutrients and energy between ecosystems. Catadromous fish migrations by eels and mullet export organic matter and nutrients to nearshore and offshore ecosystem components. Best studied are some fishes that utilize estuaries as part of a complex life cycle (Ray 2005). Gulf Menhaden (*Brevoortia patronus*) are the numerically dominant fish species in estuaries throughout the Gulf of Mexico (Deegan 1993). These fish spawn offshore in December and January, enter estuaries as larvae in February, spend about 9 months feeding and growing in the estuary, and then migrate offshore in the fall as juveniles, having increased in body mass 80-fold (from 0.15 to 13 g) over that time. Their emigration represents a net export of energy, carbon, nitrogen and phosphorus from the estuary, totaling 930 kJ energy, 38 g biomass, 22 g carbon, 3 g nitrogen, and 1 g phosphorus for every square meter of estuary.

Such calculations indicate that between 5% and 10% of the primary productivity of the Louisiana saltmarshes is exported in the form of emigrating Menhaden. Menhaden may transport half of the nitrogen and phosphorus that leaves these estuaries, the remaining half leaving passively in tidal currents in the form of detritus and dissolved substances. However, the high lipid and protein content of clupeids makes these nutrients much more available to offshore food chains than is the case for the nitrogen, phosphorus, and carbon tied up in detritus. In fact, the estuary's loss is the nearshore environment's gain, as Menhaden are the major prey of many predatory fishes. The carbon contained in Menhaden represents 25–50% of offshore production (Madden et al. 1988). Many other species of fishes, particularly various croakers, have life histories that include larval growth in estuaries followed by offshore migration of juveniles or adults (see Fig. 9.9). Hence fishes link inshore and offshore ecosystems via their role in the exchange of energy, matter, and nutrients. Additionally, nutrient cycles and food webs are seldom limited to the particular habitat in which we find a species at any given moment. Ecosystems are connected to and intimately dependent on one another.

This complex interconnectedness of offshore, nearshore, and upstream habitats underscores the importance of an **ecosystem perspective** when managing fisheries or planning habitat-disturbing human activities. Because of the function of salmon and other migratory species (e.g. lampreys, sturgeons, clupeids, smelt, char, and Striped Bass) as connectors and enhancers of ecosystems, Willson et al. (1998) referred to them as **cornerstone species** that provide a resource base that supports coastal and inland ecosystems. Overfishing, habitat destruction, or migration blockage due to dams erodes and eventually removes this cornerstone, which slows tree growth, produces shorter trees, reduces woody inputs, degrades riverine habitat, and eventually accelerates the decline of species important to humans. "The links between ocean and land mean that management of an ocean fishery can have far-reaching effects on distant ecosystems, and vice versa" (Gende et al. 2002, p. 924).

Fishes as Producers and Transporters of Sand, Coral, and Rocks

Fishes not only move energy and nutrients around in aquatic ecosystems, but they may also act as ecosystem engineers, contributing to the geological dynamics of an area. Anyone who has snorkeled on a coral reef has probably witnessed the trails of white material vented from the guts of parrotfishes. This material is in part sand produced in the pharyngeal mill of the parrotfish as they feed (Fig. 21.29). Live and dead coral and coralline algae are ground up to separate the skeletal rock from algae growing on or in it (Bruggemann et al. 1996; Rotjan & Lewis 2006). Parrotfish also ingest sand trapped in algal turfs, so some sand in their stomachs is newly produced and some sand is recycled sediment. As parrotfish move about the reef, they therefore create new sand, and decrease the particle size of and redistribute old sand (Bellwood 1996).

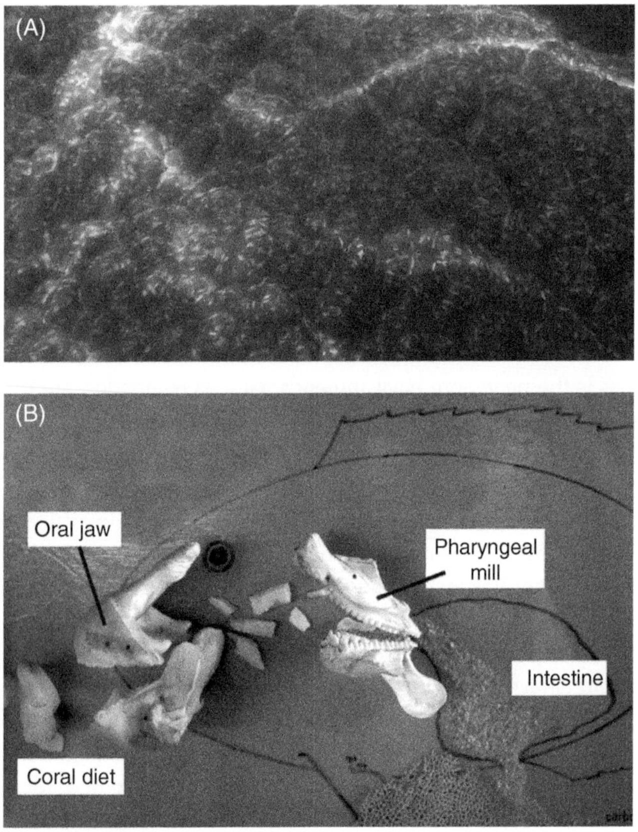

FIGURE 21.29 (A) Bite marks on a Pacific coral head from parrotfish grazing. (B) A model depicting the processing of coral fragments into sand as they pass through the mouth, pharyngeal mill, and gut of a parrotfish. G. Helfman (Author).

Estimates of bioerosion vary considerably depending on parrotfish species and depth, but have been measured in excess of 1000 kg sediment/year/m^2 (Bellwood 1995b), with the highest bioerosion rates occurring in shallower reef areas and decreasing with depth (Bruggemann et al. 1996). Rates of sediment turnover can be two to five times higher than bioerosion rates (Ogden 1977; Frydl & Stearn 1978; Choat 1991). The activities of benthic-feeding fishes such as parrotfishes are the major vector of resuspension and movement of sediment on reefs in the absence of storms (Bellwood 1995a; Yahel et al. 2002). Predation on corals by parrotfishes may become an additional stressor with climate change. During the coral bleaching events that accompany temperature increases, corals that have been fed upon contain lower densities of symbiotic algae and may therefore be slower to recover (Rotjan et al. 2006).

Sand is also produced in the jaws and stomachs of other fishes that feed on corals and coralline algae, as well as those that crush mollusks and echinoderms (e.g. stingrays, emperors, wrasses, surgeonfishes, triggerfishes, and puffers) (Randall 1974). Coral is moved around the reef in larger chunks by Sand Tilefish as they dig burrows and then construct piles of coral fragments over them. The mounds can measure 2 m across and several centimeters high and contain hundreds of coral and shell fragments. Frequently, tilefish mounds will be the only accumulations of hard substrate in large expanses of sand. Once abandoned by the tilefish, the mounds may be colonized by large groups of newly recruited damselfishes as well as small drums, butterflyfishes, angelfishes, and surgeonfishes (Clifton & Hunter 1972; Baird 1988).

A similar role is played by nest-building minnows in many North American streams (e.g. *Nocomis*, *Semotilus*, and *Exoglossum*). A chub nest can contain thousands of stones as large as 1 cm across brought from several meters distance (Wallin 1989). The nest constitutes hard substrate on an otherwise unstable bottom and is used for spawning by other species (Fig. 21.30) and is later colonized by various aquatic

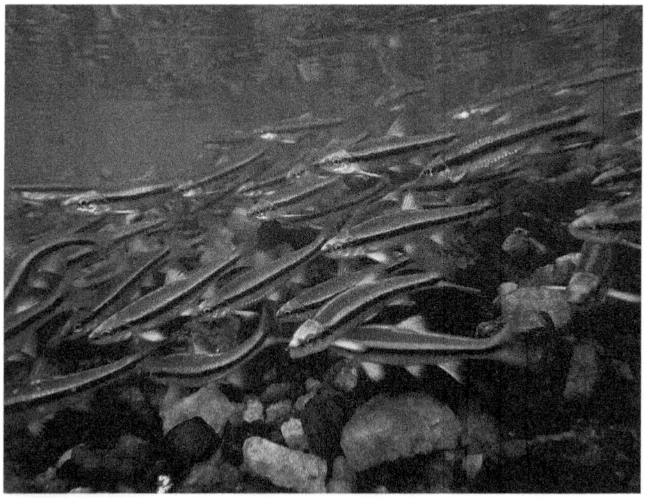

FIGURE 21.30 Yellowfin shiner (*Notropis lutipinnus*) spawning over a nest mound created by Bluehead Chub (*Nocomis leptocephalus*). Courtesy of A. Nagy.

insects. The role of fishes as ecosystem engineers that affect the distribution of sediment particles via feeding and spawning activities has become more appreciated the more we investigate the phenomenon (Flecker 1997; Flecker & Taylor 2004).

Influence of Physical Factors and Disturbance

Ecosystem ecology is concerned with biological interactions and the effects of physical and climatic factors on ecosystem components. Often these effects are most obvious when extreme climatic or other disturbances occur. The **structure** of a community, broadly defined to include species composition, abundance, distribution, and ecological interactions, changes in response to variation in climate and other forms of **disturbance**. Disturbance can lead to short-term changes in the physiology, behavior, or ecology of individuals (e.g. acclimatization, movement, trophic, and reproductive adjustments), which result in alterations in community structure, which in turn affect the pathways and rates at which energy and nutrients flow through an ecosystem. Abiotic factors that function as disturbances to fishes and cause alterations in community structure and ecosystem function include but are not limited to reductions in dissolved oxygen, often in concert with increased water temperature; changes in stream and river discharge as the result of storms, floods, dams, and drought; and cyclonic storms in coral and kelpbed habitats. Biological disturbances that have ecosystem-wide repercussions include outbreaks of disease or population explosions of destructive species that affect the food and habitat resources of a system (Karr & Freemark 1985; Karr et al. 1992).

Temperature, Oxygen, and Water Flow

Water holds relatively little oxygen, seldom more than 8 ppm, and levels below 1 ppm are generally fatal to fishes. Hence periods of *deoxygenation* – due to excessive decomposition of organic matter, a concentration of fishes trapped in pools during drought or following floods, high summer temperatures leading to thermal and oxygen stratification in lakes, or ice cover – are natural events that strongly affect the distribution and survival of fishes. Species with narrow ranges of tolerance for temperature and oxygen variation will be most strongly affected by extreme conditions. Adult Striped Bass, *Morone saxatilis*, prefer temperatures between 18 and 25°C and oxygen concentrations above 3 ppm. In southern US rivers such as the St. John's in Florida and Flint in Georgia, Striped Bass in summer avoid high water temperatures in the main river and aggregate near inflowing springs that provide cooler, preferred temperatures. By the end of summer, these narrow thermal tolerances result in emaciated fish because of limited feeding opportunities in the springs. Thermal stratification in large reservoirs during the summer can also squeeze these fish into an increasingly small region near the thermocline to the point where mortality occurs if the fish are forced into the relatively deoxygenated but cooler waters of the hypolimnion (Coutant 1985; Van Den Avyle & Evans 1990; see Chapter 7, Thermal Preference).

Extreme water flow can constrain fishes in streams and rivers. Small fishes typically cannot hold position in swift water as readily as can larger individuals. Eggs and larvae may be washed out of a system or covered with silt if high flow conditions occur during the breeding season. Juveniles of quiet water species (e.g. many minnows, sunfishes) are frequently flushed out of headwaters during flood conditions (Schlosser 1985; Harvey 1987). Adult stream fishes can also be adversely affected if floods fill pools or riffles with debris (Minckley & Meffe 1987). Upland streams are, however, characterized by recurrent and dramatic fluctuations in flow regime and the fish faunal composition of such areas can return to its original state, probably via recolonization from downstream, within 8 months following even catastrophic floods (Matthews 1986b). In contrast, flooding further downstream is often an important signal inducing spawning in many river fishes because this is the time when riparian zones and gallery forests become inundated and create nursery habitats for juveniles (Welcomme 1985; Lowe-McConnell 1987).

Many species, including relatively small fishes, are of course well adapted to high flow conditions and have little difficulty maintaining position in surf zones (clingfishes, gobies) or swift flowing water (e.g. homalopterid hillstream loaches, Colorado River minnows and suckers, many darters, torrent fishes of New Zealand; discussed in Chapter 10). Marked differences in adaptation to intermittent and extreme flow can occur between species within a family. The Gila Topminnow (*Poeciliopsis occidentalis*) and the Mosquitofish (*Gambusia affinis*) are both small, morphologically similar members of the livebearer family, Poeciliidae. The topminnow has evolved in desert streams of the American southwest that periodically experience flash floods. Mosquitofish are native to southeastern US lowlands where rapidly flowing water seldom occurs. Mosquitofish have been widely introduced into southwestern regions and prey on the young of topminnows, leading to population declines of the desert species. But the speed with which Mosquitofish eliminate the desert native is largely dependent on the extent of flash flooding. In locales where floods occur regularly, the introduced Mosquitofish gets flushed out of the system because of its inappropriate behavioral response to floods (Meffe 1984). The importance of maintaining natural flow regimes in preventing invasions of non-native species has been shown repeatedly in studies of the fish fauna of California (Marchetti & Moyle 2001).

The opposite conditions of *low flow* lead to isolated habitats, desiccation, and deoxygenation. Upland fishes frequently move downstream and floodplain fishes move into the main river as water levels decline. Isolated pools lead to an increased rate of ecological interactions as fishes crowd together. Competition may be reduced if species diverge in their resource use in response to dwindling resource availability. Seasonal shifts of just this type have been observed among Panamanian stream fishes, where the least diet overlap among species occurs during the dry season when resources are least abundant (Zaret & Rand 1971).

Seasonal cycles of changing water levels can lead to complex interactions among community components. The Everglades region of southern Florida experiences seasonally fluctuating water levels that usually include a dry season in the spring. This dry period concentrates fishes in relatively small pools (alligator holes) where they are preyed upon heavily by herons, ibises, and storks. The birds are dependent on the fishes for successful reproduction and eat 76% of the fishes in the pools. Although population sizes of the fishes are reduced, species diversity is highest during low water, with small species of omnivores and herbivores (mostly livebearers and topminnows) dominating. If water levels are high, the fish can escape bird predation and bird populations decline. Relaxed bird predation leads to overcrowding if water levels then drop, resulting in 96% fish mortality from deoxygenation and also from predation by piscivorous fishes. Overall fish diversity also declines as the predators eat the omnivores and herbivores. Hence *drought conditions* in the presence of predatory birds are beneficial to small fish species, but in the absence of wading birds are beneficial to larger, predatory fishes (Kushlan 1976, 1979; Karr & Freemark 1985).

Extreme Weather

Extreme weather events also act as disturbances in the marine environment, although the effects appear to differ between sites and storms. Major tropical (**cyclonic**) storms are generally referred to as hurricanes in the Atlantic and eastern Pacific, as typhoons in the northwestern Pacific, and as cyclones in the southwestern Pacific. These can generate winds in excess of 200 km/h, creating waves more than 12 m high that break on relatively shallow coral reef environments. The influence of such waves is felt far below the surface, as massive corals are broken off and tossed around at depths exceeding 15 m while tremendous amounts of sand shift, are suspended in the water column, and scour most structure in the area. After a major cyclonic storm, few live corals remain in shallow water and major destruction can occur down to depths of 30–50 m depending on the nature and direction of the storm, tide stage when it struck, and bottom topography. When Hurricane Allen struck Jamaica's north coast in 1980, much of the shallow water coral and other structure was destroyed or damaged. Damselfish algal lawns were eliminated and the damselfish wandered around for over a week without displaying their usual territoriality. The territories they eventually set up were in deeper water and were associated with different coral types than were used before the hurricane. Parrotfishes formed smaller schools and stopped reproducing for 2 weeks. Normally cryptic and nocturnal species (moray eels, squirrelfishes, hawkfish, and blennies) swam out in the open by day, perhaps because their refuges were destroyed. Planktivorous fishes (damselfishes, wrasses, and bogas) hugged the reef rather than foraging high in the water column. Large predators (snappers, groupers, and grunts) that were previously rare increased in number and swam conspicuously in the open, perhaps capitalizing on displaced and confused prey species. One year after the storm, species distributions and densities remained different, having shifted in favor of fishes associated with low relief habitats (e.g. rubble vs. upright coral). Analysis of coral recovery 6 years after the storm indicated that damselfish caused a decrease in the numbers and sizes of colonies of the dominant coral species (Woodley et al. 1981; Kaufman 1983; Knowlton et al. 1990).

Even moderate storms can have strong effects on reef fishes. A series of three relatively mild (sustained winds of 60 km/h) cyclones struck the northern Great Barrier Reef over a 2-year period. These storms caused little structural damage to corals but had major behavioral and community effects on the fishes (Lassig 1983). Suspended sand, moved by strong surge and currents, forced many otherwise benthic fishes up into the water column and caused visible wounds, apparently from collisions with corals. More importantly, juveniles suffered substantial population losses and subadults were redistributed; 60% of the species surveyed suffered density losses of juveniles following one storm. Poor recruitment in several species was attributed to injury or, more likely, settling juveniles being flushed from the system by strong currents. Hence periodic storms can play a decisive role in the structure of some reef communities, heralding future impacts that could result from climate-change induced increases in the frequency and intensity of such storms (discussed further in Chapter 22).

Analogous events follow storms in temperate marine habitats. A series of severe winter storms occurred off California in 1980, destroying much of the canopy of giant kelp in coastal kelp beds and scouring the bottom. Removal of the kelp eliminated the major food of sea urchins, which switched to a diet of benthic algae and denuded the understory regions of reefs, "transforming the reef from a richly forested site to a barren area" (Stouder 1987, p. 74). The understory turf harbored invertebrates that were the major food types of the abundant, resident surfperches. Differences in microhabitat use and feeding patterns among surfperch species decreased as fish converged on the few areas where prey remained. Although adult surfperches remained in reef areas over the next 15 months, overall fish abundance decreased by 50% as nonresident and subadult fishes abandoned the reefs, probably because of loss of food and refuge sites and unsuccessful competition with competitively superior, resident surfperches (Ebeling et al. 1985; Stouder 1987).

Not all storms, even major storms, have such dramatic effects on diversity and density of reef fishes. A severe 3-day storm struck the Kona coast of Hawaii in 1980, destroying most of the shallow water coral but caused few apparent direct fish mortalities or injuries. Many shallow water species initially fled to deeper water, but returned to former areas after a few weeks or months. Sixteen months after the storm, diversity and density had returned to or surpassed pre-storm levels, with the exception of a few distributional shifts involving species that remained in deeper water (Walsh 1983).

Reduction in diversity and density is reversible if the causative agents do not recur frequently. The Florida Keys were exposed to record cold temperatures in the winter of 1977 that caused an extensive fish kill. Water temperatures fell to 11°C in areas that normally do not drop much below 19°C. Dead and dying reef fishes were found throughout the area; underwater censuses showed significant decreases in both species diversity and individual densities the following summer. However, by the next year, both overall diversity and density were not different from their pre-cold snap levels; diversity on some reefs was higher than before. These increases were largely the result of successful recruitment of many new individuals, perhaps because potential competitors and predators were eliminated by the cold weather (Bohnsack 1983; see also Thomson & Lehner 1976).

Biological Analogs of Extreme Weather

Violent storms and sudden water temperature shifts create obvious and rapid changes in the physical environment of a habitat. Fishes can experience similarly disruptive effects as a result of the action of biological processes. Major disturbance events include **population explosions** of animals that affect the physical structure of a habitat. One such example is the crown-of-thorns starfish, *Acanthaster planci*, a predator on live corals through much of the Indo-West Pacific Ocean. Normally, the starfish occurs at low densities (2–3 animals/km^2), but periodically the starfish undergoes population explosions producing densities of several starfish per square meter, with thousands of individuals swarming over a reef. The starfish consumes the live tissue on the surface of the corals; the underlying limestone skeleton first becomes covered with algae but then collapses due to biological and physical erosion. In this manner, 95% of the coral in a large area may be killed and will require 10–20 years to recover (Endean 1973; Wilkinson & Macintyre 1992).

Fishes that are directly or indirectly dependent on corals, either for food or shelter, suffered as a result. Coral-feeding fishes, including butterflyfishes, parrotfishes, gobies, wrasses, and triggerfishes, disappeared from affected areas, leading to a 15–35% decline in species diversity on affected reefs. Densities of other coral-dependent fishes (many cardinalfishes, damselfishes, wrasses, gobies, and blennies) also declined, leading to an overall reduction in fish density of 55–65% in an area (Sano et al. 1984).

Climate Change and Fishes

All of the external forces acting on fishes – temperature extremes, oxygen availability, floods, droughts, cyclonic storms, habitat loss – are influenced by climate, and there is little doubt that we are in a period of human-induced **climate disruption**. The authoritative Intergovernmental Panel on Climate Change (IPCC) concluded that climate change has happened, continues, and is largely influenced by human activity, that unprecedented warming and sea level rise are occurring, and that the consequences for humanity and the rest of earth's biota are serious (see recent IPCC reports).

The specific effects that such change will have on fishes are somewhat speculative, but the speculation is scientifically based, and evidence of verifiable impacts grows as we learn where to look. Better studied, more accessible taxonomic groups – birds, mammals, insects, mollusks, vegetation, phytoplankton, and zooplankton – are showing changes in distribution, abundance, physiological performance, and reproductive and migrational timing that are directly linked to documented climatic shifts (Walther et al. 2002; Ahas & Aasa 2006). "More than 80% of the species that show changes are shifting in the direction expected on the basis of known physiological constraints of species" (Root et al. 2003). Fishes are also responding in similar ways, and the responses will affect all levels of ecological organization, from the genetics of populations, the interactions between fishes and other community components, as well as the roles that fishes play within ecosystems. Details of known and anticipated effects are discussed in Chapter 22.

Ecosystem Services Provided by Fishes

An aspect of biodiversity loss that affects human welfare involves the role that fishes play in ecosystems. Biodiversity is intimately linked to ecosystem function: healthy ecosystems – those that contain natural assemblages of organisms, habitats, interactions, and processes – can sustain some level of exploitation. Disrupted ecosystems collapse.

Organisms in ecosystems can provide both goods and services, to humans and other members of the ecosystem. Utilitarian **goods** are obvious: we eat fish, we use them in medicines, we worship them in ceremonies, we buy them as curios, and we derive aesthetic pleasure from fish-centered recreation. In contrast, **ecosystem services** are the processes that occur as the result of functioning ecosystems, processes that humans (and other organisms) find useful or necessary (Daily 1997; ESA 2000). Classically, ecosystem services were defined as processes that benefited humans: plant pollination, water and air purification, seed dispersal and germination, drought/flood mitigation, erosion control, nutrient cycling, pest control, and waste decomposition and transformation. These are all products of microbial, vegetation, and animal activities. Because of the interconnectedness and co-evolution of living things in ecosystems, one organism's output serves as input to another organism. The essential point here is that "ecosystem services are generated by the biodiversity present in natural ecosystems" (Chapin et al. 2000, p. 240).

Fishes provide a number of such services (Fig. 21.31, Holmlund & Hammer 1999):

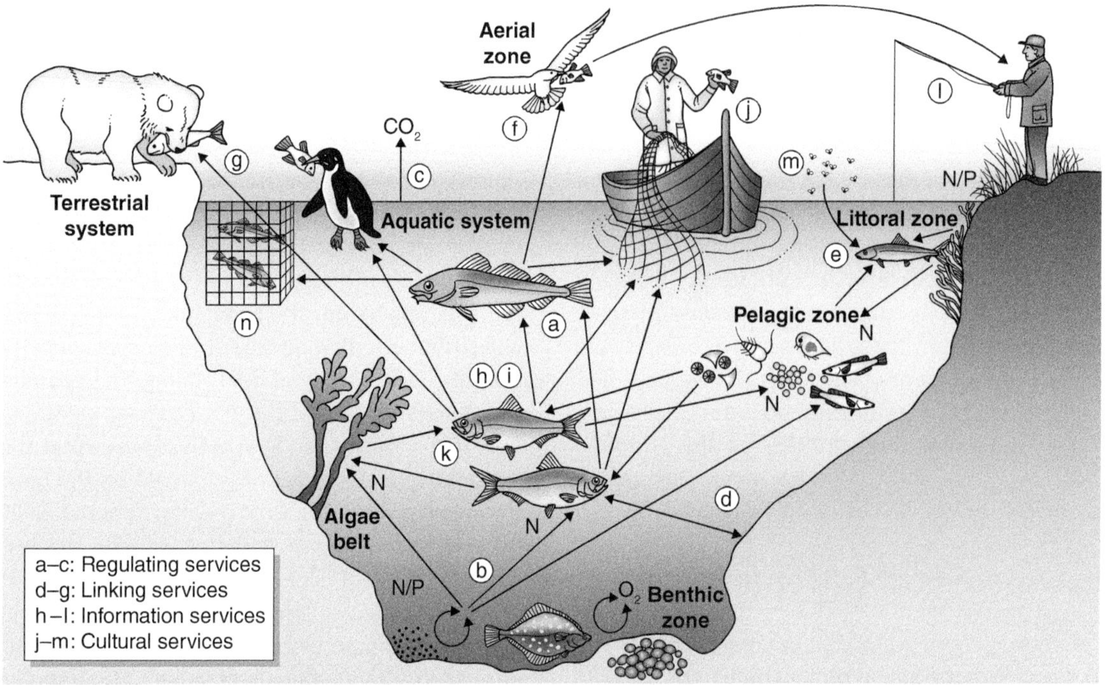

FIGURE 21.31 A pictorial summary of ecosystem services provided by fishes to humans and other organisms. Services can be classified as: *regulating* populations and processes (e.g. trophic cascades that regulate population dynamics or nutrient cycling, bioturbation of sediments, and carbon exchange); *linking* different parts of the ecosystem via transport of nutrients and energy (e.g. open water to benthos, littoral zone, birds, and terrestrial mammals); *informing* (e.g. indicating and recording past and present ecosystem integrity); and *cultural* (e.g. human interactions and direct benefits via exploitation, recreation, water purification, disease abatement, and aquaculture). Based on Holmlund and Hammer (1999).

- As a result of short- and long-distance movements, fishes *transport nutrients* between different parts of ecosystems and between different ecosystems. As discussed earlier in this chapter, nutrients obtained in one habitat and excreted in another stimulate coral or macrophyte growth. Long-distance migrations of salmonids and other diadromous fishes bring nutrients and energy obtained in ocean regions to distant, upriver habitats. This transport forms the base of the food webs in lakes and rivers as well as surrounding terrestrial regions. Fishes, birds, mammals, and riparian vegetation are all directly dependent on these fishes, on the invertebrates that feed on the fishes and their offspring, and on the nutrients released as waste products or from decomposing bodies.
- Some fishes are *ecosystem engineers*, producing and moving sand and gravel; they thus control resource availability by modifying, maintaining, and creating habitat (Moore 2006). Parrotfishes generate sand in the process of digesting coral, and move sand between different reef areas. Salmonids and minnows redistribute gravel and pebbles in the process of nest building. Sockeye Salmon, *Oncorhynchus nerka*, in two Alaskan streams consistently disturbed 30% of the available streambed in the process of building nests. When salmon populations were high (before overfishing), Sockeye dug up the entire streambed more than once, being forced to superimpose new nests on top of old nests. Nest digging affects periphyton growth and may decrease the susceptibility of a stream to flood erosion by sorting sediments into size classes and thus raising the threshold flow level at which bed scouring occurs (Statzner et al. 2003; Moore 2006). Larval lamprey also act as ecosystem engineers through their burrowing and feeding by maintaining streambed softness, increasing oxygen penetration into the substrate, and increasing fine particulate organic matter on streambed surface (Shirakawa et al. 2013). These influences on stream substrates create favorable living conditions for not only the young of the engineers but of many other fishes and invertebrates. Loss of such ecosystem engineers will undoubtedly lead to dramatic changes in ecosystem function, although this remains an area of investigation (Coleman & Williams 2002).
- Fishes sit at the top of food webs in many habitats, and their feeding activities can cause *trophic cascades*, affecting species lower in the food web. Herbivorous fishes on reefs may prevent algae from overgrowing and smothering coral; coral is, in turn, critical habitat for fishes and invertebrates. Reef-dwelling fishes that eat urchins prevent urchin explosions that can denude reefs of both algae and coral. Piscivores in lakes eat smaller fishes that eat zooplankton. Zooplankton feed on phytoplankton. If piscivores are reduced or eliminated, lakes experience blooms

of algae, some of which are noxious.

- Zooplanktivorous and microcarnivorous fishes feed on larval stages of mosquitoes and other biting flies, some of which carry *human pathogens*. Other fishes feed on snails that are intermediate hosts of human parasites (Stauffer et al. 1997).
- Although no fishes are known to pollinate terrestrial plants, they may affect pollination of terrestrial plants by influencing the abundance of pollinating insects through trophic cascades, as discussed earlier in this chapter. In addition, fishes assist in the germination and transportation of seeds, as has been shown for piranha relatives in the riparian forests of the Amazon. In another symbiotic relationship, fish deposit eggs in the mantle cavities of freshwater mussels; bivalve larvae attach to the gills of the developing fish and are protected and transported until large enough to survive. In areas where host fishes have been reduced in number, bivalve populations crash.

A fully functional ecosystem is therefore dependent on its biodiversity, on the essential parts being present and functioning in their evolved ecological roles. The ecosystem and its constituent biodiversity are inseparable, and protecting diversity requires an ecosystem perspective.

Supplementary Reading

Allan JD, Castillo MM. 2007. *Stream ecology: structure and function of running water*, 2nd edn. Dordrecht: Springer-Verlag.

Allen LG, Pondella DJ, II, Horn MH. 2006. *The ecology of marine fishes. California and adjacent waters*. Berkeley, CA: University of California Press.

Barthem R, Goulding M. 1997. *The catfish connection: ecology, migration, and conservation of Amazon predators*. New York: Columbia University Press.

Carpenter SR, ed. 1988. *Complex interactions in lake communities*. New York: Springer-Verlag.

Carpenter SR, Kitchell JF, eds. 1993. *The trophic cascade in lakes*. New York: Cambridge University Press.

Gerking SD, ed. 1978. *Ecology of freshwater fish production*. London: Blackwell Science.

Goulding M. 1980. *The fishes and the forest. Explorations in Amazonian natural history*. Berkeley, CA: University of California Press.

Kerfoot WC, Sih A, eds. 1987. *Predation: direct and indirect impacts on aquatic communities*. Hanover, NH: University Press of New England.

Kitchell JF, ed. 1992. *Food web management. A case study of Lake Mendota*. New York: Springer-Verlag.

Lowe-McConnell RH. 1987. *Ecological studies in tropical fish communities*. London: Cambridge University Press.

Polis GA, Winemiller KO, eds. 1996. *Food webs: integration of patterns and dynamics*. New York: Chapman & Hall.

Polis GA, Power ME, Huxel GR, eds. 2004. *Food webs at the landscape level*. Chicago: University of Chicago Press.

Randall JE. 2007. *Reef and shore fishes of the Hawaiian Islands*. Honolulu, HI: University of Hawaii Sea Grant College Program.

Sale PF, ed. 1991a. *The ecology of fishes on coral reefs*. San Diego, CA: Academic Press.

Sale PF. 1991b. Reef fish communities: open nonequilibrial systems. In: Sale PF, ed. *The ecology of fishes on coral reefs*, pp. 564–598. San Diego, CA: Academic Press.

Sale PF, ed. 2002. *Coral reef fishes: dynamics and diversity in a complex ecosystem*. San Diego, CA: Academic Press.

Wootton RJ. 1999. *Ecology of teleost fishes*, 2nd edn. Fish and Fisheries Series No. 24. New York: Springer-Verlag.

Zaret TM. 1980. *Predation and freshwater communities*. New Haven, CT: Yale University Press.

Websites

The Intergovernmental Panel on Climate Change – check for most recent reports at **www.ipcc.ch**.

The Millenium Ecosystem Assessment, **www.millennium assessment.org** is a massive, comprehensive overview of human interactions with the biotic world, organized around the topic of ecosystem services.

CHAPTER **22**

Conservation

Summary

Fishes are objects of wonder in large part because of their incredible diversity, which is why we chose to make the diversity of fishes a central theme of this book. This diversity makes sense when viewed in the light of evolutionary processes over several hundred million years. However, human activities are threatening this diversity through habitat destruction and subsequent homogenization of habitats, genotypes, and assemblages. We conclude this book with a chapter on what we view as a major tragedy unfolding in modern times, namely catastrophic, human-caused reductions in the diversity of fishes and other life forms, and offer some solutions to reduce the rate of biodiversity loss. It has been widely recognized and discussed for some time that the dramatic loss of biodiversity in recent decades, mainly associated with anthropogenic factors, constitutes the sixth major extinction period in the history of life on this planet (see, e.g., Kolbert 2014, McCallum 2015). Put simply, we are living through a mass extinction because we are causing it.

Extinction rates have increased dramatically in recent decades due to human activities; present rates are estimated to be 1000 times greater than average and 10–100 times greater than during past periods of mass extinction. Freshwater fishes are especially vulnerable in part because of their high degree of isolation and endemism, and also because human

CHAPTER CONTENTS

The Diversity of Fishes: Biology, Evolution and Ecology, Third Edition. Douglas E. Facey, Brian W. Bowen, Bruce B. Collette, and Gene S. Helfman.

Companion website: www.wiley.com/go/facey/diversityfishes3

impacts are greater in freshwater systems. Marine fishes are less threatened because of their wider distributions, although many commercially important species are showing serious declines.

Conservation genetics is the application of DNA data to a variety of wildlife management issues. A key goal in this field is to preserve the genetic diversity that allows species to resist disease and adapt to changing conditions. Population genetic studies can delineate the boundaries of fishery stocks and management units within species. Molecular systematics can reveal unrecognized species that may be endangered by harvest or depletion. Molecular forensics can show which species are entering the marketplace, and demonstrate that some legal harvests provide a cover for the exploitation of endangered species. Genomic studies can reveal the diversity that allows species to adapt to changing conditions.

Major causes of biodiversity loss are habitat loss and modification, species introductions, pollution, commercial exploitation, and global climate change. Habitat loss occurs through modification of bottom type, as happens during dredging, log removal, coral or gravel mining, trawling, and from silt deposition due to deforestation of the surrounding watershed. Other causes of habitat loss include channelization of streams and rivers, dam building, and water withdrawal, all of which involve significant economic and political considerations. Factors driving biodiversity decline often work synergistically, further exacerbating the loss of biodiversity.

Introduced species affect native species because introduced fishes are often freed from their evolved population controls (e.g. predators and parasites), and natives are unprepared for the introductions. Predation, competition, and hybridization with introduced species are common results, as is the introduction of new pathogens. The loss of unique fish assemblages due to homogenization is another consequence of species introductions.

Chemical, nutrient, and sediment pollution all have adverse effects on fishes. Chemical pollutants may include endocrine disrupting compounds (EDCs) that can interfere with growth and sexual development. Predation on fishes by birds and mammals links aquatic pollution to terrestrial ecosystems. Fishes can therefore serve as valuable environmental indicators.

Many commercial marine fish species have been exploited at unsustainable rates. The Pacific Sardine and Peruvian Anchoveta were overharvested leading to crashes in their populations and devastating economic repercussions. Atlantic Cod and Giant Totoaba were very abundant commercial species that declined dramatically due to overfishing. Some species reductions are the result of bycatch of other fisheries. For example, bycatch in the shrimp fisheries of the Gulf of Mexico has reduced stocks of Red Snapper and Spanish Mackerel, among other species.

Overharvesting of fishes has resulted in "fishing down" food webs, in which populations of top-level predators are severely impacted and fishing pressure shifts to progressively lower trophic-level species. We also see examples of "fishing through" food webs, which reduces populations at multiple trophic levels, thereby impacting the productivity of the entire food web.

Fishing can act as a selective force on exploited populations, driving the evolution of that population toward maturation at smaller size and younger age. This can be detrimental to longer-term viability of the population and can also negatively affect human use of that population for commercial purposes.

Aquaculture is playing an increasing role in providing fish for human consumption. However, aquaculture can produce negative environmental impacts due to disease, parasites, and nutrient pollution associated with high fish densities. In addition, raising predatory fishes requires high protein food that is typically derived from commercially harvested fish from lower trophic levels (e.g., herrings) – putting additional pressure on those wild populations.

The home aquarium trade for reef fishes has led to reef destruction and species depletion in many places. Many freshwater species can be raised in captivity, but few marine fishes can be produced in aquaculture, because of the difficulty in rearing through an extended larval stage. Regardless of the source, only few of such fishes live more than a few months in captivity.

Greenhouse gases have been released into the atmosphere at increasing rates for over a century, and we are now experiencing global warming, sea level rise, ocean current shifts, and major climatic changes such as drought, floods, and more frequent and powerful cyclonic storms. Climate change alters the distribution, abundance, reproductive timing, trophic relationships, and migration patterns of fishes through its impacts on water temperature, rainfall patterns, freeze–thaw cycles, oxygen availability, heat budgets, oceanic currents, primary productivity, ocean acidification, and metabolic processes. The greatest temperature changes have been noted in polar habitats, with species more typical of warmer waters moving into these areas. Significant shifts in the distribution and diversity of marine and freshwater fishes elsewhere around the planet have also been documented. Elevated temperatures in tropical and subtropical areas have had devastating impacts on coral reef ecosystems that live at the upper end of their thermal tolerance range. The prospects for the future are not promising.

Biodiversity loss is a symptom of environmental deterioration on a global scale. Solutions to environmental problems include ecosystem and landscape preservation, development of reserves, habitat restoration, and captive breeding of endangered species. None of these efforts will be successful, however, if overconsumption of natural resources and excessive production of pollutants associated with human population growth continues.

Many of the earliest conservation efforts were focused on individual species, as reflected in the U.S. Endangered Species Act of 1973, and similar legislation in other countries. More recently, the emphasis has shifted to ecosystem-based management, on the understanding that if ecosystems are healthy, the individual species will not need protection. However, for highly migratory species, such as tunas, individual species protection is still needed because their ecosystem can be the whole ocean basins. A relative newcomer is the field of evolutionary conservation, which seeks to protect the building blocks for future adaptation and diversification.

Introduction – A Brief History of Conservation Biology

The recognition of some animals as "special" and in need of protection has deep roots in human history. The Greek philosopher Aristotle recognized the uniqueness of dolphins in his book *Historia Animalium* (*History of Animals*, perhaps the first zoology textbook circa 330 B.C.). He called for their protection based on observations that dolphins breathe air, bear live offspring, suckle their young, and therefore were not fish. The attentive student of this text knows it is little more complicated than that, as some fishes can breathe air, some bear live offspring, and some provide nourishment and protection for their young in various ways. The first organized efforts at conservation may have been the kings of Middle-Age Europe, who set aside their personal hunting grounds as "wilddeorness," the place of wild deer. Poachers caught taking from the King's wilderness were hung, an early example of wildlife regulation enforcement.

Although the human recognition of conservation principles is ancient, the science of conservation biology is surprisingly young. Some historians mark the advent of this field with the *First International Conference on Research in Conservation Biology*, convened by Michael Soulé and Bruce Wilcox in 1978. The conveners placed an immediate emphasis on ecology, writing "The purpose of this conference is to accelerate and facilitate the development of a rigorous new discipline called conservation biology—a multidisciplinary field drawing its insights and methodology mostly from population ecology, community ecology, sociobiology, population genetics, and reproductive biology." Phylogenetic methods were soon added to the toolbox.

The initial decades of conservation biology saw an emphasis on protecting individual species, as embodied in the IUCN Red List of threatened and endangered species (established 1964), the U.S. Endangered Species Act (1973) and similar legislation in other countries. Phylogenetic methods were repurposed for defining divergent branches in the tree of life in need of protection (Moritz 1994). In more recent decades, the emphasis in conservation biology has shifted toward ecosystem-based management (EBM), now the predominant paradigm for wildlife conservation, as indicated by the 2012 establishment of the United Nations Global Centre for Ecosystem Management (**http://www.unep.org/ecosystemmanagement/**), and the IUCN Commission on Ecosystem Management (**https://www.iucn.org/about/union/commissions/cem/**). The premise of EBM is that if ecosystems are healthy, individual species would not need protection. The validity of this perspective is beyond dispute, but not infallible. Commercially valuable species that are highly migratory cannot be protected under EBM because they are not contained in a definable ecosystem. Some pelagic tunas and sharks cross whole ocean basins, so in these cases individual species protections are mandated. Finally, in recent decades the field of conservation biology has recognized the need to protect the genetic diversity that provides the basis for future adaptation and diversification (Fraser & Bernatchez 2001; Bowen 2016).

Extinction and Biodiversity Loss

Population declines can lead to species declines, local **extirpation**, and eventually to global **extinction** of a species. Extinction is a natural process, and natural processes can be characterized by average rates. These rates have accelerated dramatically during past periods of major environmental change. But to our knowledge never before in the history of the earth have global environmental changes resulted from the actions of a single species – until now. Nor have extinction rates approached the pace established in the last several decades.

Natural **extinction rates** for animals average 9% of existing species every million years, or one to two species per year. During the celebrated Permo-Triassic and Cretaceous–Tertiary mass extinctions that marked the ends of the Paleozoic and Mesozoic eras, respectively, extinction rates accelerated to perhaps 50–75% of the marine fauna over a period of 10,000 to 100,000 years (Raup 1988; Jablonski 1991). In stark contrast, extinction rates at the close of the 20th century have been estimated at about 1000 times background levels (De Vos et al. 2015). While the accuracy of such estimates is difficult to verify, there is little argument that extinction rates today exceed any in the past. This astounding loss of **biodiversity**, defined as the variety of life forms and processes, can be directly linked to the activities of an overgrown and overconsumptive human population (Groom et al. 2006). In this chapter we review examples and major causes of decline in fish biodiversity worldwide, and present some solutions for slowing the rate of biodiversity loss. Fishes serve as just one example of the effects on living organisms of human-induced environmental degradation; biodiversity loss is panbiotic, affecting all taxa.

Threatened and Endangered Fishes

Designation of a fish species as **threatened** or **endangered** is a complicated process influenced by political as well as biological concerns (see Wheeler & Sutcliffe 1991; Helfman 2007). IUCN updates its Red List periodically, which indicates the numbers of plants and animals worldwide that are considered to be critically endangered, endangered, vulnerable, or otherwise at risk. The IUCN Red List for 2020 identified 74 bony fishes as extinct or extinct in the wild, another 115 species as possibly extinct, and 569 as critically endangered. Among the Chondrichthyes, 43 were listed as critically endangered, with 3 possibly extinct. Most of the fishes that are at risk are freshwater (see Darwall & Freyhof 2016; Hermoso et al. 2017), and there are still many species about which not enough is known to assess their status. The Food and Agriculture Organization of the United Nations estimated in 2017 that over 34% of global marine fish stocks are overharvested, a dramatic increase from the approximately 10% that were considered as overfished in 1974 (FAO 2020). This is a result of an increasing reliance on fish (including shellfish) as a protein source for our increasing human population.

A major concern regarding loss of species is what we still do not know (McCallum 2015). As of 2017, there was still a particular dearth of information about the many freshwater species in South America, Africa, and Southeast Asia (Hermoso et al. 2017).

The causes of extinction are often discernible, and underscore the environmental problems that form the focus of this chapter (Fig. 22.1). **Habitat alteration** is the most frequently cited factor and other factors include **introduced species**, chemical alteration or **pollution**, **hybridization**, and **overharvesting**. The major threats to freshwater fishes are usually various forms of habitat degradation, whereas the primary threat for many marine species is often overharvest (WWF 2016; FAO 2020). These threats, however, may vary considerably among different regions based on human socioeconomic factors.

Extinction factors often operate in combination. Some reported extinctions represented the elimination of isolated populations, such as the Miller Lake Lamprey, *Lampetra minima*. This dwarf lamprey was endemic to one small lake in southern Oregon. Because it parasitized introduced trout, it was poisoned into apparent extinction (Bond & Kan 1973). Although remnant populations were subsequently found in tributary streams, the lamprey has not recolonized Miller Lake (Lorion et al. 2000) despite translocation efforts begun in 2010 (Clemens 2017). Other extinct species were, however, widespread, such as the Harelip Sucker, *Lagochila lacera*, which occurred commonly in large rivers of at least eight eastern states and probably succumbed to siltation of its clear water, pool habitat. The Blue Pike, *Sander vitreus glaucus*, a subspecies of the Walleye, sustained a large fishery in Lake Erie and Lake Ontario until the mid-1950s; in some years, it made up more than half the commercial catch in those lakes. Pollution, introduced fishes, habitat degradation, overharvesting, and hybridization all contributed to its demise (Miller et al. 1989). It was officially declared extinct in 1975. The Gravenche (*Coregonus hiemalis*), a salmonid, native to Lake Geneva between Switzerland and France, was heavily fished in the late 19th century, and last seen in the early 1900s (**https://en.wikipedia.org/wiki/Gravenche**).

Certain obvious patterns arise from lists of species at risk. Freshwater fishes account for most extinct and compromised taxa, reflecting the sensitivity and degraded condition of freshwater habitats due largely to human impacts. Freshwater fish may now be the most endangered group of vertebrates, with about 31% of species listed as threatened or endangered according to IUCN (Darwall & Freyhof 2016). Certain regions

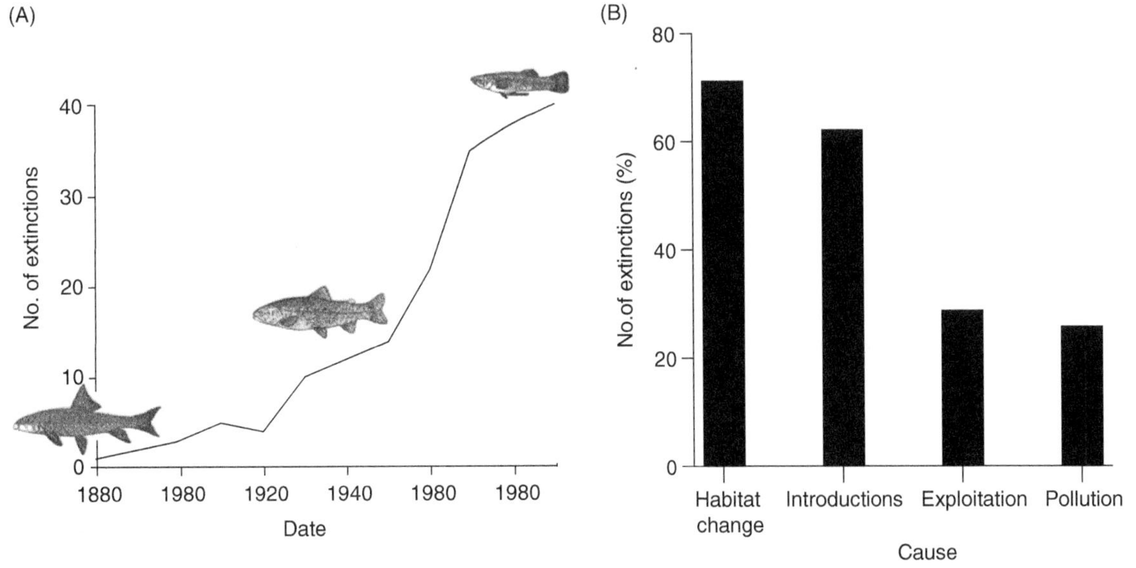

FIGURE 22.1 Extinction rates and causes. (A) Fish extinctions in North America. Extinctions grew steadily over the past century until the latter part, when they apparently slowed down, possibly indicating improved conditions or early elimination of more sensitive forms. Illustrated left to right are the Harelip Sucker (extinguished *c.* 1900), Alvord Cutthroat Trout (*c.* 1930s or 1940s), and San Marcos Gambusia (*c.* 1980). (B) Major causes of fish extinctions globally. Habitat alteration, introduced species, overfishing, and pollution are the primary agents, but combined factors cause the most extinctions, which is why the summed percentages of all columns exceed 100%. (A) Adapted from G. Helfman (2007); Stiassny (1999); sucker drawing by J. Tomelleri, trout and gambusia by Sara V. Fink; (B) Adapted from G. Helfman (2007) and Harrison and Stiassny (1999).

and habitat types appear most frequently on the lists. In North America, the isolated and disjunct aquatic systems of the otherwise arid southwest, such as the spring pools and rivers of the Great Basin and of Mexico, have been centers of evolution and human-induced extinction (Minckley & Deacon 1991). Specialist species endemic to small, isolated habitats make up the majority of extinct and endangered fishes because both they and their habitats are exceedingly vulnerable to human activity.

Fishes of the southwestern United States are very sensitive to environmental degradation (Miller 1981; Soltz & Naiman 1981; Contreras-Balderas et al. 2002). A variety of activities have led to declines and extinctions, including pumping of springs and groundwater, pollution by humans and livestock, draining of marshes, damming of streams, introductions of exotic competitors and predators, and hybridization. Approximately 15 species and numerous localized populations of fishes of this region are extinct. Desert species account for nearly two-thirds of the federally listed Endangered and Threatened fishes in North America. In 2004, the IUCN had listed 14 desert cyprinodontiform species as Critically Endangered. Some species have been described after they were exterminated (e.g., the aptly named La Trinidad Pupfish, *C. inmemoriam*, described on the basis of a single specimen collected before its single habitat dried up due to water extraction; Lozano-Vilano & Contreras-Balderas 1993). The fishes of this region have adapted well to the environmental challenges of extreme desert conditions, but nothing in their history allows them to handle habitat losses that result from human activity (Pister 1981; Minckley & Deacon 1991; Rinne & Minckley 1991; Contreras-Balderas et al. 2002).

Cave fishes are also exceedingly vulnerable to environmental disturbance, a convergent trend that cave fishes share with desert spring forms for many of the same reasons. Pollutants, water withdrawal, and competition, predation, and disease brought in by introduced species are major threats. As an ecological grouping, caves must be the habitat type with the proportionately highest rate of imperilment among fishes (and other organisms). In 2004, the IUCN listed 52 cave fish species as at high risk, and experts considered at least another eight species to be in need of protection (Proudlove 1997b). The Congo Blind Barb, *Caecobarbus geertsii*, is threatened by the aquarium trade and is consequently listed in Appendix II of the Convention on International Trade in Endangered Species (CITES). In the United States, the Alabama Cavefish, *Speoplatyrhinus poulsoni*, is federally protected as Endangered and the Ozark Cavefish, *Amblyopsis rosae*, is Threatened. Two other cavefishes that occur in the United States – the Northern Cavefish, *Amblyopsis spelaea*, of Kentucky and Indiana and the Southern Cavefish, *Typhlichthys subterraneus*, found in five southeastern states, Indiana, and Missouri – are both designated as Vulnerable by the IUCN (2004) (see Romero 1998; Romero & Bennis 1998).

It is not just small, isolated habitats though that are vulnerable. Big river fishes with special needs for clean water, such as sturgeons, paddlefishes, and some suckers and large minnows, have also been strongly affected. Large rivers are primary sites of human habitation and impact; such habitats have been degraded for centuries due to pollution, siltation, water withdrawal, and damming. Comparatively few marine fishes appear on lists of species at risk (but see Musick et al. 2000; Arthington et al. 2016). Most marine fishes have broad distributions and a greater chance for replacement by neighboring populations (except for the coelacanths, see Chapter 13). Hence, aside from some heavily exploited coastal and pelagic species (e.g., many sharks, Bluefin Tuna) and long-lived, slow-growing forms (Orange Roughy, Patagonian Toothfish), the most vulnerable marine fishes are estuarine species that have been affected because of their dependence on fresh water in their life cycle, such as the giant Totoaba, *Totoaba macdonaldi*, and many salmonids (Moyle & Leidy 1992; Helfman 2007; discussed later in this chapter). From 1970 to 2012, migratory fishes using freshwaters declined 41% (WWF 2016).

Conservation Genetics to Conservation Genomics

Genetics can contribute to conservation efforts in a number of ways. Molecular phylogenetic assessments can identify the oldest lineages in the tree of life, which contain a disproportionately high fraction of overall genetic diversity due to their age and uniqueness. Ancient lineages are not always obvious based on morphological examinations, as demonstrated in the bonefish (discussed in Chapter 2, see Fig 2.8). Treating multiple species as a single species would be a fundamentally flawed premise for fish management, and could put the less abundant species at risk. Andrews et al. (2016) found that an Indo-Pacific snapper (*Etelis carbunculus*), important in both commercial and artisanal fisheries, was actually two species divergent by several million years of evolution. To serve the conservation goal of preserving biodiversity, we need to know the fundamental evolutionary lineages, both above and below the species level.

A second way that genetics can support conservation objectives is in defining populations, the fundamental units of wildlife management. If two populations have significantly different allele frequencies, they are expected to be **demographically independent**, meaning they have differences in demographic parameters such as age structure, fecundity, survivorship, growth rate, and sex ratio. However, for wildlife managers the more pragmatic concern is if an isolated population is depleted, it will not be replenished by dispersal from other populations. Isolated populations must recover from catastrophes, both natural and human-caused, without significant input of individuals from elsewhere. If the population goes extinct, the habitat may eventually be recolonized by rare migrants, but these colonists are not sufficient to replenish populations over the timeframe of decades that concern wildlife managers.

Populations defined with genetics are often equated with stocks in fishery management; however, they are not quite the same thing. If a group of fishes in one branch of a river is significantly different from elsewhere in terms of allele frequencies and F statistics, that genetically defined population can be regarded as an independent stock. However, the reverse is not always true. If fishes in two branches of the river are not significantly different in allele frequencies, then they may still be isolated stocks. It only takes a few migrants per generation to genetically homogenize breeding populations (but see Mills & Allendorf 1996). Ten individuals that succeed in migrating to a new population and contributing genes to that population may prevent population genetic differentiation, but will not be sufficient to replenish depleted stocks. Hence a genetically isolated population is a stock, but a stock is not necessarily a genetically isolated population (Waples 1998). To assess contemporary movement and stock structure, tagging studies may be preferable. However, these are labor intensive, more expensive than most genetic assays, and impractical in many cases.

Fortunately, this gap between demography and population genetics has been closing. Statistical methods allow researchers to identify individuals that move between populations by comparing microsatellite or genomic SNP genotypes (Manel et al. 2005; Gaither et al. 2018). Instead of relying on allele frequencies to define populations, these methods infer pedigrees among closely related individuals or their extended families, and identify individuals at one location whose closest relatives are in another location. Hence microsatellites may close the gap between traditional population genetics, which assesses gene flow averaged across thousands of generations, and tagging studies that assess contemporary movement but may miss the rare or episodic exchanges.

The third major application of genetics in conservation is in the maintenance of evolutionary potential. This field has roots in the captive breeding programs that seek to retain genetic diversity and viability in endangered species (Frankham et al. 2002), and will draw on the field of genomics. One goal is to preserve the genetic variants that will allow species to persist and survive future environmental challenges. This emphasis on adaptive evolutionary conservation (Fraser & Bernatchez 2001; Funk et al. 2019) can also include an assessment of novel genetic properties that confer selective advantages or higher survival, or could be the wellspring of new species.

Two new methodologies have greatly enhanced the conservation genetic toolbox since the previous edition of *The Diversity of Fishes*; low-cost genomic data and environmental DNA (eDNA). In a survey of Silky Shark (*Carcharinus falciformis*), Kraft et al. (2020) demonstrated that the cost of producing ~5000 genomic SNPs was competitive with a simple mtDNA survey, and was lower when the number of specimens exceeded about 300. The information content from SNPs is far higher, lending greater statistical power and finer scale resolution of population structure; the Silky Shark SNP survey detected an additional management unit that had eluded prior analyses with mtDNA. The same SNP technology revealed hidden population structure relevant to management in Yellowfin Tuna (*Thunnus albacares*; Grewe et al. 2015). The deep-sea Roundnose Grenadier (*Coryphaenoides rupestris*) showed no population structure at most loci, but strong selection and adaptation at 346 loci, indicating segregation by depth (Gaither et al. 2018). Such adaptive potential (evolutionary conservation) is increasingly considered in conservation planning, as a way to determine vulnerability to climate change (Funk et al. 2019). Bernos et al. (2020) reviewed conservation issues addressed with genomic data in 268 fish species, and found application to climate change-related threats (19% of projects), species identification (17%), resolution of species boundaries (11%), hybridization (9%), translocations (9%), population crashes or habitat degradation (8%), and invasive species (6%).

A few liters drawn from a body of water can reveal the species composition through sequencing of eDNA. All organisms shed DNA into the environment, and it can persist for hours or even days depending on the conditions. The DNA is concentrated by drawing water through a filter, and then sequenced using a suite of appropriate PCR primers. This revolutionary methodology is providing information on a variety of conservation fronts: cryptic taxa that elude detection with visual censuses, rare and endangered species that may be difficult to document in an aquatic medium, and invasive species. The quality of the results depends in large part on the available library of barcodes, which vary across taxa but overall are pretty good for fishes. An eDNA survey of Monterey Bay (California) identified 92 vertebrate taxa, 7 to the rank of taxonomic family, 3 to subfamily, 10 to genus, and 72 to species (Andruszkiewicz et al. 2017). A survey of coastal sites across tropical North West Australia revealed the presence of 404 bony fishes and 44 elasmobranchs, plus a previously unknown biogeographic transition at Cape Leveque (West et al. 2021). An eDNA survey in the Great Lakes of North America was able to detect the invasive Round Goby (*Neogobius melanostomus*) plus three species at risk (Eastern Sand Darter, *Ammocrypta pellucida*; Northern Madtom, *Noturus stigmosus*; and Silver Shiner, *Notropis photogenis*) (Balasingham et al. 2018). In a search for cryptic shark diversity in New Caledonia, Boussarie et al. (2018) detected 44% more shark species with eDNA than were recorded in visual surveys, including species not previously known from that region.

One of the most exciting frontiers in this field is in detecting the number or biomass of fish species from an eDNA sample (Rourke et al. 2021). This has a variety of applications in conservation, from setting harvest limits to detecting population crashes.

It is important to note that eDNA is not a universal solution for biodiversity surveys. This approach excels at detecting cryptic biota that are difficult to observe (like the sharks mentioned above) or hidden inside reefs and other substrates. Hence it will identify the fauna that are missed with visual surveys. However, it will also miss some species due to a variety of reasons including mismatched primers and incomplete sampling. In the Great Lakes survey above, eDNA detected 86% of the species known to be present. Hence the optimal approach for now is to use eDNA as a supplement, rather than replacement, of other survey methods.

FIGURE 22.2 Is the rare Devils Hole Pupfish a genetically distinct species, or is it an ecomorph of the genetically similar Amargosa Pupfish (*Cyprinodon nevadensis mionectes*) which occurs in some nearby springs, and with which the Devils Hole Pupfish can hybridize? If the morphological differences are due to habitat and diet rather than genetics, this raises important questions regarding the conservation status of the species. O.Feuerbacher, USFWS / Public Domain.

The Devil's Hole Pupfish: Where Taxonomy Matters to Conservation The Devil's Hole Pupfish (*Cyprinodon diabolis*, Fig. 22.2) is a famous example of small population persistence. It is found only in the 50 m^2 Devil's Hole spring in Death Valley where temperatures are high and food is scarce. It likely shares a recent common ancestor with the Amargosa Pupfish (*Cyprinodon nevadensis mionectes*) which occurs in some nearby springs. The two are genetically similar, and the morphological differences include smaller size, proportionately larger head and eyes, and the lack of pelvic fins of the Devil's Hole Pupfish (Lema & Nevitt 2006). It has been assumed for many years, based on natural history of the region, that the Devil's Hole Pupfish had been isolated for tens of thousands of years – enough time to become a genetically distinct species. Microsatellite analysis, had suggested that the isolation may be as little as a few hundred to a few thousand years, and possibly the product of a human introduction (Reed & Stockwell 2014). However, a more recent and thorough genomic analysis (Saglam et al. 2016) indicates that the pupfish has an ancient origin at about the same time as the Devil's Hole spring opened up, about 60,000 years ago.

The Devil's Hole Pupfish was placed on the Endangered Species list in 1967, and multiple efforts have been made to protect it. Unfortunately, fish moved to refuge pools in the 1970s have not produced vigorous populations. In 1992, the Point of Rocks pool was seeded as another back-up population of the Devil's Hole Pupfish, but within several years the fish in the pool began to look a bit different – the proportion of individuals with pelvic fins increased. Genetic studies indicate that at some point, probably between 1997 and 2003, a few of the widespread Amargosa Pupfish from a nearby pool somehow got into the Point of Rocks pool, resulting in hybridization and subsequent loss of the "pure-bred" Devil's Hole Pupfish in this habitat (Martin et al. 2012). These apparent hybrids, however, were doing quite well in this pool where Devil's Hole Pupfish had been struggling.

Populations of the Devil's Hole Pupfish have varied over the years. There were over 500 in the early 1970s, and numbers were around 300 or more through the late 1990. But since then numbers have been as low as 38 in 2006, followed by modest recovery, but then only 35 in 2013. The fish seems to be on the verge of disappearing, and unfortunately has had very low reproductive success, even when raised in captivity – unlike other cyprinodontids which tend to reproduce quite well. There is some concern that the population may have become weakened by loss of genetic diversity (Martin et al. 2012). In 2013, some eggs were hatched in a fish culture facility in an attempt to prevent loss of the species altogether.

The hybridization of Devil's Hole Pupfish and Amargosa Pupfish, and their overall genetic similarity, raises a sticky taxonomic question pertinent to conservation. If juvenile Devil's Hole Pupfish are grown in good conditions with adequate food, some develop pelvic fins and a more robust body; they resemble Amargosa Pupfish. Conversely, if Amargosa Pupfish juveniles are grown at higher temperatures with limited food, some are thinner and will not develop pelvic fins; they resemble Devil's Hole Pupfish (Lema & Nevitt 2006). Such developmental plasticity based on nutrition and growth is not unusual, and it raises the question as to whether these fish are two species, or ecomorphs of the same species. Perhaps the Devils Hole Pupfish represents speciation in progress, and the approximately 60,000 years of genetic separation suggested by Saglam et al. (2016) has not yet resulted in the reproductive incompatibility that we often associate with different species. This could have significant policy ramifications, as considerable effort and expense has been invested in trying to protect the Devil's Hole Pupfish because it is a recognized taxonomic unit.

One controversial suggestion for saving the Devil's Hole Pupfish is based on the idea that they are not genetically distinct from the Amargosa Pupfish, and therefore introducing some Amargosa Pupfish into Devils Hole would help to create a genetically more robust population. To many, such a seemingly radical approach would represent intentional extinction of the iconic Devil's Hole Pupfish, through hybridization which would forever destroy its identity. The Devil's Hole Pupfish that we have known would no longer exist, because of our deliberate action. How could this be species conservation? However, if these are one species and the Devil's Hole population is disappearing due to a genetic bottleneck, such a strategy would restore genetic diversity, even though we would be interfering with an ongoing natural selection process.

This is a truly hot topic – worthy of a fish that survives in the hostile environment of Death Valley. It is also yet another example of how the genetic toolbox is such an integral and important aspect of the study of fishes.

Molecular Identification in the Marketplace The PCR technology that allows researchers to recover DNA data from small bits of tissue has become well established as a major forensic tool, both in criminology and wildlife management. The first pioneering effort at species identification in the marketplace with PCR technology was directed at the Japanese and South Korean whaling industries (Baker et al. 1996). DNA barcoding has become widespread, including its use to identify fish and other seafood in markets and the publicly available information through the Fish Barcode of Life campaign (Formosa et al. 2010; FDA 2017). For example, genetic analyses have revealed shark products, including endangered Shortfin Mako, in pet foods and beauty products – but there was no indication of shark contents on the labels of these consumer items (Cardeñosa 2019).

Sturgeon caviar represents the ultimate luxury product from fishes, commanding prices upwards of U.S. $50 per ounce. However, native stocks of the most prized species crashed in the aftermath of the break-up of the Soviet Union, as poorly regulated fisheries and high price have driven up the harvest, while pollution and dams have reduced habitat. In these circumstances, there is a strong incentive to find substitutes for the premium caviar of the Volga River–Caspian Sea region. DeSalle and Birstein (1996) surveyed 23 lots of premium black caviar purchased from reputable dealers in New York City and found that five of the lots (22%) were mislabeled eggs from less desirable but imperiled species, including three species listed on the International Union for the Conservation of Nature (IUCN) Red List (**http://www.iucnredlist.org/**) as Vulnerable (Siberian Sturgeon, *Acipenser baerii*) or Endangered (Amur River Sturgeon, *A. schrenckii*, and Ship Sturgeon, *A. nudiventris*).

Red Snapper (*Lutjanus campechanus*) has been a highly prized fish in the restaurants and markets of North America, commanding a premium price. Yet few consumers have the discriminating palate needed to be sure they are consuming the right species, and the genus *Lutjanus* has many members that are widespread, abundant, and delicious. In 1996, the Gulf of Mexico Fisheries Management Council imposed fishing restrictions after finding that the Red Snapper was overfished, driving down supply and driving up prices. Marko et al. (2004) surveyed specimens of fish labeled Red Snapper purchased in eight US states; 17 of 22 specimens (77%) were not Red Snapper. Among the fraudulently labeled specimens, five were identified as other Atlantic snappers, two were Pacific Crimson Snapper (*L. erythropterus*), and the remaining 10 could not be identified because sequences from the corresponding species had not been submitted to Genbank. Some of these may be rare or unknown to science, invoking the possibility of overfishing before these species can be identified for management purposes. The fact that over half of the putative Red Snapper came from international sources indicates that this problem is global in scale.

Shark fin has been one of the most contentious items in international wildlife trade, a commerce that takes an estimated 10 to 100 million sharks annually, and generates revenues equivalent to over a billion US dollars. In response to sharp declines in abundance worldwide, many countries have banned the practice of **finning** (harvesting the shark fins and discarding the rest of the fish), and three sharks (Whale, Basking, and White) are banned from international trade by the Convention on International Trade in Endangered Species (CITES). In these circumstances, it is useful to know what species are entering the marketplace, and whether prohibited species are present. In response to this conservation concern, Shivji et al. (2002) developed diagnostic species-specific markers based on a nuclear ribosomal DNA sequence. In preliminary trials, 10 out of 55 putative Silky Shark (*Carcharhinus falciformis*) proved to be other species. Subsequently Clarke et al. (2006) surveyed markets in Hong Kong and found that approximately 40% (by weight) of auctioned shark fins were from 14 species, with Blue Shark (*Prionace glauca*) especially predominant among them (17% by weight). Other sharks in the auctions included Shortfin Mako (*Isurus oxyrinchus*), Silky (*C. falciformis*), Sandbar (*C. obscurus*), Bull (*C. leucas*), hammerhead (*Sphyrna* spp.), and thresher (*Alopias* spp.). The hold of a boat fishing illegally in Australian waters yielded 193 fins from 20 shark species and 7 ray species, including the critically endangered sawfish *Anoyxpristis cuspidate* (Holmes et al. 2009).

These genetic surveys provide two lessons about the wildlife trade:

1. Legal markets such as those for Red Snapper in the United States can provide cover for poaching, smuggling, and illicit products entering the marketplace. Some of these products are from endangered or overutilized species.
2. Esteemed species are replaced by fraudulent alternatives. For luxury products, mislabeling has been observed in 15–95% of surveyed specimens, including caviar, fish fillets, shark fins, seal penises, whale meat, and turtle meat (Roman & Bowen 2000). In global seafood markets, Luque and Donlan (2019) estimated the overall rate of fraudulent substitutions at 8%.

In recent decades, there has been an increased awareness of this consumer fraud, along with expanded monitoring – but it is far from thorough. The readers of *The Diversity of Fishes* can help. If you find suspicious fish (or other wildlife) products, take a fin clip, a small tab of tissue, or a skin swab, and consult your local conservation geneticist. The techniques have become widespread enough that DNA barcoding of fish samples can be readily incorporated in courses or student research projects with a genetics focus.

General Causes of Biodiversity Decline

The close of the 20th century witnessed a number of well-publicized environmental problems of regional and international scale. Each problem contributed to declines in fish biodiversity (Safina 2001a). In addition to habitat modification, species introductions, pollution, and commercial exploitation, **global climate change** is recognized as a growing threat to aquatic ecosystems and fishes (IPCC 2013; see Chapter 21, Climate Change and Fishes). An additional problem is our collective inability to learn from past experiences. We thus repeat our mistakes, as has happened to salmon fisheries in continental Europe, then the British Isles, followed by the northeastern United States and Canada, and finally the eastern Pacific and Japan (Montgomery 2003). These interacting causes result in direct population losses due to mortality or reproductive failure, or indirect losses due to hybridization or loss of genetic diversity. At the root of each problem is **human overpopulation** and **overconsumption**. Overpopulation is particularly destructive to aquatic ecosystems and to fishes because humans are concentrated along rivers and estuaries.

Habitat Loss and Modification

Human alteration of aquatic habitats is the most commonly cited cause of declines in fish populations. Habitats are altered via modification of bottom type and above-bottom structure – channelization, dam building, watershed perturbation, and competition for water.

Modification of Bottom Type Many fish species are ecologically dependent on bottom topography and above-bottom structure for survival. In flowing water, rocks and logs provide shelter from the current and a site of attachment for eggs, algae, and associated fauna. Undersides of rocks are a major refuge for insect larvae and other invertebrates that fishes eat. Aquatic vegetation similarly provides shelter and food attachment sites for lacustrine fishes. In the ocean, rocks and biogenic habitat (corals, sponges, many other sessile invertebrates, kelp beds, and other attached algae) are essential habitat for most benthic species. Human activities that disrupt, remove, or cover bottom structure will be detrimental to fishes. Such activities include dredging for navigation and to obtain construction materials, bottom trawling in the ocean, removal of logs and debris dams to aid navigation and as "habitat improvement," and watershed disruption that leads to increased erosion and silt deposition (Fig. 22.3).

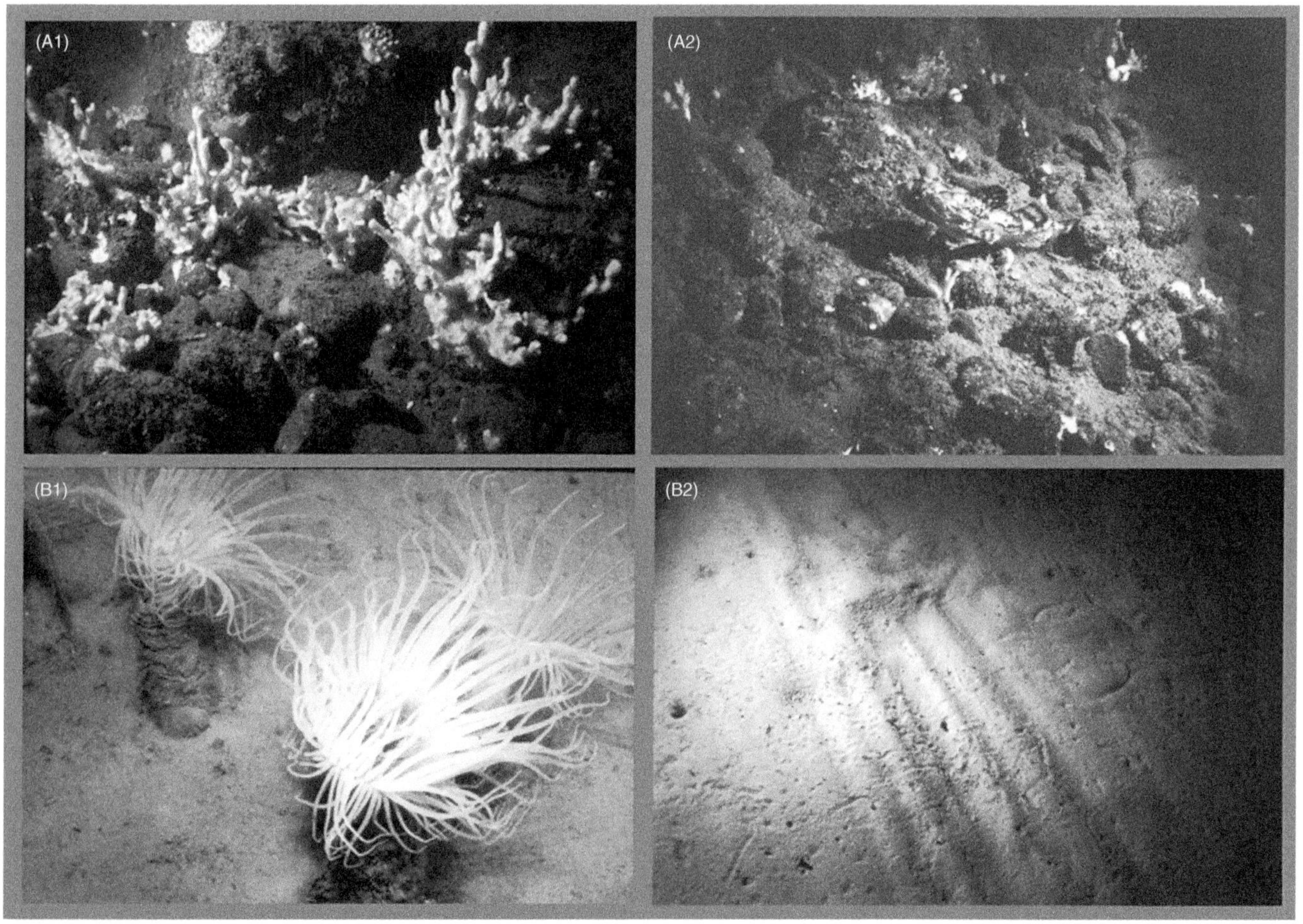

FIGURE 22.3 Impacts of bottom trawling on gravel (A) and mud (B) habitats, Stellwagen Bank National Marine Sanctuary, Gulf of Maine. Gravel habitats protected from trawling (A1) contain erect sponges; areas open to trawling (A2) lack such biogenic structure (a Longhorn Sculpin, *Myoxocephalus octodecemspinosus*, is visible in the center of photo A2). Mud habitats also contain biological structure such as burrowing anemones (B1), whereas trawled areas (B2) can be devoid of such structure (note trawl gear tracks in B2). Courtesy of P. J. Auster.

FIGURE 22.4 The U.S. Fish and Wildlife Service placed logs in Redd Creek channel of the Bandon Marsh National Wildlife Refuge, Oregon, to improve habitat for trout and salmon. Photo from USFWS National Digital Library / Public Domain.

Woody debris in streams and rivers exemplifies the effects of habitat disruption on fishes. Natural woody debris, in the form of logs and debris dams in streams and rivers (= snags, see Fig. 22.4) plays a critical role in ecosystem function (Wallace & Benke 1984; Harmon et al. 1986; Maser & Sedell 1994). Many species use the exteriors and hollowed interiors of logs as spawning sites (e.g., catfishes, Ictaluridae) or as resting sites (Lowe-McConnell 1987). Debris dams retain silt, organic matter, and nutrients, offer a solid substrate for invertebrate attachment, and are a site for transformation and processing of organic matter, thus making it available for invertebrate and fish use. Woody debris also slows the flow of the water, which decreases erosion and increases the time during which nutrients are available to the food web. In coastal, low-gradient (slow moving) rivers, many gamefishes obtain more than half of their food directly from snags. Snags are the most biologically rich habitat in such rivers: although making up only 4% of habitable surfaces, snags contain 60% of the total invertebrate biomass, provide 80% of the drifting invertebrate biomass, and produce four times more prey than mud or sand habitats (Benke et al. 1985). Government efforts at snag removal in navigable rivers of the southeastern United States began in the early 1800s. When rail transportation largely replaced river commerce in the 1850s, snag removal was less important, but the practice was continued by the US Army Corps of Engineers throughout the United States until the 1950s. Many state agencies continued to emphasize removal of woody debris as a habitat improvement tool (Sedell et al. 1982).

In the tropics, catastrophic *deforestation* along rivers and streams adversely modifies both terrestrial and aquatic habitats (Chapman & Chapman 2003). In tropical marine environments, *coral reef destruction* occurs at an equally alarming rate. Coral reefs, which cover less than 1% of the seafloor, host 25% of marine fishes and sustain hundreds of millions of people. Habitats are destroyed by the direct mining and collecting of coral, and inadvertently by harmful fishing techniques (poisons, explosives, and bottom trawling), boat anchoring and diver activities, sedimentation and pollution, boat groundings, and changes in coral predator abundance as a result of fishing practices. All these phenomena lead to reductions in fish diversity and biomass because fishes and their prey rely directly on corals for food and shelter (Birkeland 1997).

Coral mining is a particularly deleterious activity. Limestone blocks are cut from the reef surface and then used in road and home building and as landfill. Massive, head-forming corals in shallow (1–2 m depth) water are most frequently targeted. Where heavily practiced, coral cover can change from 50% to 5% both as a direct result of removal and as a byproduct of trampling and sediment production. The reduction in living coral and overall substrate complexity results in decreases of fish biomass, abundance, and diversity in mined areas (Bell & Galzin 1984; Shepherd et al. 1992). Recovery is slow, taking decades if it occurs at all.

Coral and reef fish collecting for the aquarium trade has also taken a significant toll on reef habitat (Derr 1992; Wood 2001). Both live corals and algae-covered or invertebrate-encrusted dead corals are taken. Live coral and "live rock" were removed from the Florida Keys at a rate of 3 tons/day in 1989, with an annual retail value of around US $10 million. Fortunately, the state of Florida banned this harvest in 2011, but it continues elsewhere around the world. Live rock consists of substrate built over 4000–7000 years and does not represent a renewable resource on the reef. Mortality rates for live corals in aquaria exceed 98% within 18 months of collection. Because of the acknowledged difficulties of keeping live coral and reef-building invertebrates in captivity, most large, commercial "public" aquaria use artificial corals; home aquarists should do the same.

Channelization

Channelization, often a strategy for "bank stabilization," involves straightening a riverine system and smoothing its sides. Bends in a river are bulldozed into straight lines, levees are built, and banks are covered and heightened with stones and boulders (riprap) or concrete. Rivers and streams are channelized primarily to reduce seasonal inundation of the floodplain (so-called because the floodplain is the natural area that receives overflow during seasonal rains); channelization is basically the process by which a river or stream is converted into a ditch or pipe. Channelized stream segments have low habitat heterogeneity and higher velocities during higher flows. Shallow water and floodplain habitats are eliminated, both of which provide spawning and nursery areas for riverine fishes. Channelized rivers either lack fishes or are dominated by non-native species. Especially affected are big river species, species dependent on sandy areas, and fishes that use the floodplain in their life cycle, including sturgeons, Paddlefish, and darters of the genus *Ammocrypta*. Channelization-induced loss of the floodplain in parts of the Lower Mississippi River has led to a 10-fold reduction in standing biomass of fishes.

Because channelization is often accompanied by deforestation of the floodplain to allow for agriculture and housing development, the entire hydrological regime of a river is altered, with a result that flooding may actually increase (Simpson et al. 1982; Moyle & Leidy 1992). The catastrophic flooding of the Mississippi River in 1993 was partly due to decades of channelization (Myers & White 1993); inundation of New Orleans by Hurricane Katrina in 2005 was in part a result of levee construction that delivered sediment too far downstream, thus preventing the development of wave-buffering, nearshore wetlands.

In southern Florida, the U.S. Army Corps of Engineers channelized the meandering, shaded, productive 165 km long Kissimmee River, turning it into a 90 km long, straight, concrete canal. Channelization resulted in drained wetlands (including desiccation of substantial portions of the Everglades National Park), water pollution and eutrophication, periodic flooding, salt contamination of streams and aquifers, water table lowering and land subsidence, oxidation of peat soils, wind erosion, and marsh fires. Biological effects included a 90% decrease in wading bird populations, the deaths of 5 billion fishes and 6 billion shrimp, and extirpation of six native fishes from the Kissimmee River. In 1976, Florida reconsidered the project. The Corps proposed dechannelization, which began with feasibility studies in 1978–1985 and then again in 1990–1998.

The adverse effects of channelization have been so great in some areas that costly "dechannelization" programs have been initiated, including the lengthy and expensive Kissimmee River Restoration Project. Planning and land acquisition took decades. Construction began in 1999 and was mostly completed by 2020 (Koebel & Bousquin 2014), at a total cost of about 1 billion dollars. Ongoing monitoring of physical and biological attributes of restored sections of the channel indicate some successful reestablishment of organic substrates, natural hydrologic regimes, and associated wetland plant communities (Anderson 2014; Koebel & Bousquin 2014; Toth 2017).

Dam Building Dams provide hydroelectric power, water storage capacity (although evaporation often minimizes water storage benefits in arid regions), agricultural water, recreational opportunities, and lakefront development potential. Drawbacks of dam building include flooding of agriculturally and historically valuable land. Poor watershed management, often brought on by deforestation of the land surrounding newly created reservoirs, leads to rapid **silting-in** of the lake, transforming it into a much less desirable (from a development standpoint) marsh or swamp. In tropical countries, regions around dams become uninhabitable for humans because the altered habitats favor organisms that cause such debilitating parasitic diseases as schistosomiasis and onchocerciasis (river blindness). Increases in these and other diseases are well documented in human populations residing near newly created dams (e.g., Steinmann et al. 2006).

The **altered hydrological conditions** behind dams can cause other, unforeseen problems. Piranha attacks on healthy humans in the Amazon Basin are exceedingly rare (Sazima & Guimaraes 1987). Dam construction created year-round, favorable, still-water conditions for piranha spawning, whereas spawning habitat was previously limited to flooded forest lands during the wet season. Piranhas defend their nest sites against intruders, including waders in the shallow waters of reservoirs. Single bite attacks characteristic of nest defense rose dramatically after dam construction. Bathers in reservoirs in the Parana–Paraguay river systems in southeast Brazil reported more than 85 piranha attacks on humans in 2002, and another 70 attacks on a single day in 2013 (**https://piranhaguide.com/a-documented-list-of-all-known-piranha-attacks-piranha-victims/**); 90% of bites were on the legs and feet, suggestive of defensive attacks on wading bathers by nest-guarding adults (Haddad & Sazima 2003). Wounds were crater-like, 1–2.5 cm in diameter, and bled severely. Several bites required hospitalization and one resulted in amputation of a toe.

Not too surprisingly, fishes adapted to flowing water do not fare well in the impounded regions behind dams. Many productive cold water trout fisheries have been lost behind dams. Stream assemblages, usually rich in native darters, minnows, suckers, and trouts, are usually replaced by sunfishes and catfishes. As is the case in most disturbed habitats, introduced species become dominant, including Common Carp, Yellow Perch, Mosquitofish, and lacustrine minnows.

Two North American examples typify the effects of dams on aquatic faunas. The Colorado River was an ancient, warm, fast-flowing, turbid river that developed a unique fauna of streamlined fishes adapted to high flows and high temperatures. These fishes spawned in response to seasonal changes in water level and temperature. Of 32 fishes native to the Colorado River, about 75% are endemic. More than 100 dams were built along this huge desert river for water retention, flood control, and agriculture; less than 1% of the historical flow now reaches the river's mouth. The deep reservoirs formed by many dams became thermally stratified, and water released periodically from the cold, deeper layers of the reservoirs chilled downstream habitats, disrupting natural spawning cycles and killing

FIGURE 22.5 Endangered fishes of the upper Colorado River. Prior to impoundment, the Colorado River experienced exceptionally high flows, >9000 m^3/s during winter and spring floods, which redistributed sediments critical to spawning and larval rearing. Several Endangered Colorado River endemics evolved reproductive habits attuned to this flood cycle. Large native Colorado River species also show marked convergent morphologies, having long, tapered bodies with elongate caudal peduncles, small depressed skulls with predorsal humps or keels, winglike fins that have hardened leading edges, and tiny or absent scales. Humps have been interpreted as providing a hydrodynamic advantage or as a response to gape-limited native predators such as Colorado Pikeminnows. Four of the large, Endangered cypriniforms of the Colorado exemplify these traits: (A) Razorback Sucker (*Xyrauchen texanus*), photo by USFWS / Wikimedia Commons / Public Domain; (B) Bonytail chub (*Gila elegans*), photo by Brian Gratwicke / Wikimedia Commons / CC-BY 2.0;(C) Colorado Pikeminnow (*Ptychocheilus lucius*), photo by J. E. Johnson, US Fish and Wildlife Service / Wikimedia Commons / Public Domain; and (D) Humpback Chub (*G. cypha*), photo by G.Andrejko, Arizona Game and Fish Department /Public Domain.

native fishes while promoting the survival of introduced cold water predators such as Rainbow Trout. Of the 80 fish species that now occur in the Colorado River, only about one-third are native. Of the remaining native fishes, most are Threatened or Endangered, including the Humpback Chub (*Gila cypha*), the Bonytail Chub (*G. elegans*), the Razorback Sucker (*Xyrauchen texanus*), and the Colorado Pikeminnow (*Ptychocheilus lucius*), the largest minnow native to North America (Fig 22.5). The modified environment created by the dams and the success of introduced fishes are chief contributors to the decline of native fishes (Ono et al. 1983; Minckley 1991; Wydoski & Hamill 1991).

Hydroelectric dams also block movements of fishes that migrate upriver to spawn, and can kill juveniles during their downstream movements (Lucas & Baras 2001). Fishes that make it past tailwaters or through turbines often suffer from gas-bubble disease, brought on because the turbulent waters below a dam are often supersaturated with gas (Raymond 1988). Habitat destruction, water flow reduction, and other dam effects are considered major factors causing the decline of salmonid stocks in western North America. The Columbia River system, including its large tributary the Snake River, has a gauntlet of 28 dams that must be run by spawning adult salmonids and ocean-bound juveniles. Upstream mortality is estimated at 5% and downstream mortality at 20% *per dam* (Booth 1989); four of the Columbia and Snake River dams lack any fish bypass structures such as fish ladders.

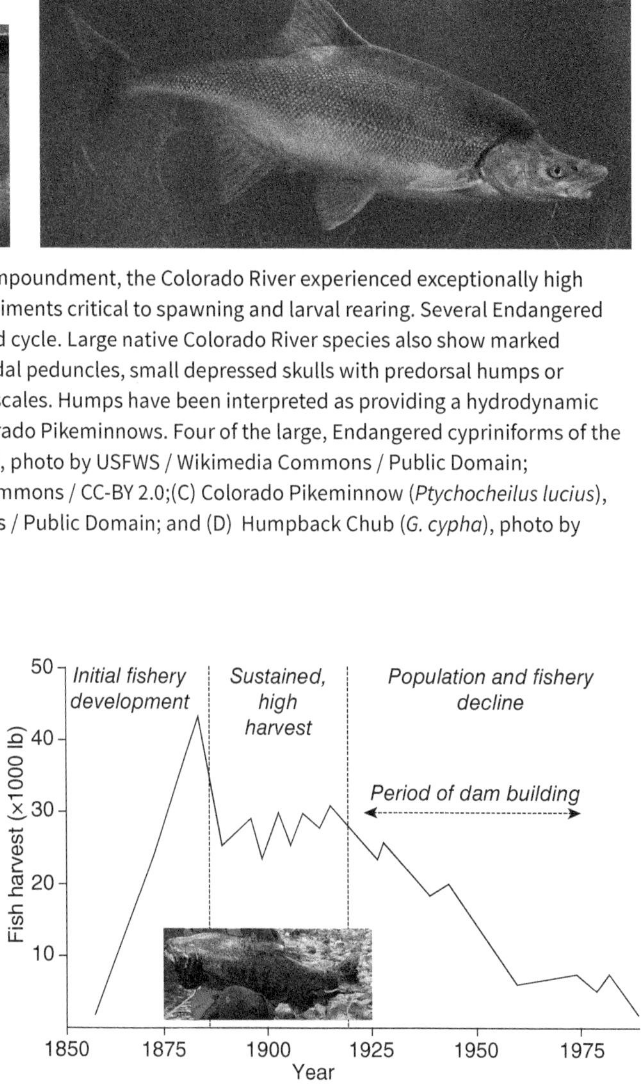

FIGURE 22.6 Commercial catches of Chinook Salmon and Steelhead (sea-run Rainbow Trout) in the Columbia River from 1866 to the late 1992. Soon after commercial exploitation began, catches rose to sustained levels of 20,000 tons annually. After dam construction, catches declined dramatically and remained low. Williams (2006) used with permission; courtesy of R. Carlson.

Commercial catches of salmon in the Columbia River declined dramatically from the start of the fishery in the late 1870s to the late 1970s (Fig. 22.6). For related reasons, approximately 106 major West Coast salmon and Steelhead stocks (*Oncorhynchus* spp.) have already been extinguished and an additional 214 native, naturally spawning stocks of Pacific salmons, Steelhead, and sea-run Cutthroat Trout are at risk in Oregon, California, Washington, and Idaho. Some analyses put the numbers as high as 280 stocks extinguished and another 880 stocks at high risk of extinction (Nehlsen et al. 1991; Huntington et al. 1996; Slaney et al. 1996). Overfishing, deforestation, hatchery introductions, aquaculture escapes, introduced pathogens, and agricultural and industrial pollution also contribute to the problem (NRC 1996b; Lichatowich 1999; Williams 2006; among many others).

Retention of sediments in reservoirs, combined with elimination of flood cycles, can have far-reaching consequences for fish production. Nutrients that would have been delivered to estuaries or dispersed over many kilometers of downstream floodplain during seasonal inundation remain trapped behind a dam. Construction of three dams in northern Nigeria led to a 50% reduction in downstream fish landings. Similar effects of dams have been reported in Zambia, South Africa, Ghana, and Egypt. In Egypt, construction of the Aswan High Dam, which impounds 50–80% of the Nile River's flow, caused a 77% reduction in annual landings of sardines, *Sardinella aurita*, in the southeastern Mediterranean (Smith 2003). In Eastern Europe, dams along the Volga River contributed to a 90% reduction in fish catches in the Caspian Sea. Similar, or worse, scenarios have been created in the Azov, Black, and Aral seas (Welcomme 1985; Moyle & Leidy 1992; Pringle et al. 2000; see below).

Many options exist that will prevent, minimize, and reverse the negative impacts of dams. Dam construction, which is costly in addition to being environmentally destructive, can be avoided via conservation measures such as improving irrigation methods and other practices that reduce water loss, and conserving energy and developing alternative energy sources. Activities that reduce the impacts of existing dams include dam operation schedules that restore natural flows in river ecosystems, correcting sediment transport and deposition problems, correcting fish passage and entrainment problems, and, ultimately, removing dams that have outlived their usefulness (Heinz Center 2002). Dams modify entire ecosystems, more so than many of the other human impacts on aquatic habitats and fishes (Dudgeon 2000). Correcting the damage requires an ecosystem perspective and ecosystem-level management actions.

FIGURE 22.7 The Snail Darter (*Percina tanasi*) rose to fame when its discovery led to a prolonged legal battle and delayed dam construction on the Little Tennessee River in the 1970s and 1980s. Biggins / USFWS / Public Domain.

The Tale of the Snail Darter The story behind the Snail Darter (*Percina tanasi,* Fig. 22.7) and the Tellico Dam serves as a good example of the biological and political complexities of dam building. Darters are small members of the perch family (Percidae) and darter diversity is especially rich in mountainous regions of the southeastern United States; including some species found nowhere else. The Tellico Dam was proposed for construction on the lower Little Tennessee River by the **Tennessee Valley Authority** (TVA) as early as 1936. Its usefulness, beyond the jobs created during its construction, was always a matter of debate; it was the last dam proposed in the area because its construction was difficult to justify. The environmental impact would be substantial, as the lake created behind it would flood c. 7000 ha of valuable agricultural land, several important Cherokee Indian religious and ceremonial sites (including the village of Tanasi, the capital of the Cherokee Nation, from which the state derived its name), and a renowned trout fishery. Proponents of the dam included the TVA, local land developers, and the Army Corps of Engineers. Opponents included conservationists, farmers, local landowners, anglers, the US Fish and Wildlife Service, Supreme Court Justice William O. Douglas, Tennessee Governor Winfield Dunn, and the Cherokee Indian Nation.

Plans for Tellico Dam were shelved and resurrected repeatedly until the US Congress finally approved the project in 1966. Construction began the next year, only to be halted in 1971 when a Federal Court injunction was issued because the TVA had not filed an environmental impact statement, as required by the National Environmental Policy Act of 1969. The TVA spent 2 years preparing the impact statement, which was approved in 1973 and work recommenced. The Endangered Species Act was then passed in 1973, but no known endangered species were affected by the proposed dam. In August of 1973, while doing research unrelated to the Tellico Dam project, Dr. David Etnier of the University of Tennessee discovered a new species of darter in the region to be inundated by Tellico Dam and named it *Percina tanasi*, the Snail Darter (Etnier 1976). Extensive collections by TVA and other fish biologists failed to produce other populations of the Snail Darter, and thus it became apparent that the Tellico Dam project was one of the chief threats to the species existence. The fish was given Endangered status in October 1975.

The TVA meanwhile was not idle, undertaking an unauthorized eight-month transplantation program, moving 700 Snail Darter from the Little Tennessee River to the nearby Hiwassee River. Construction on the dam accelerated in an apparent attempt to complete the dam before other complications arose. In February 1976, TVA was sued for violating the Endangered Species Act, but the suit was not upheld and construction continued. The court decision was appealed and in February 1977, a US Court of Appeals decided in favor of protecting the fish species and issued a permanent injunction against any further dam construction.

The TVA appealed on the somewhat ironic grounds that the Little Tennessee River was no longer suitable habitat for the Snail Darter because of the Tellico Dam: the existing construction was blocking the upstream spawning migration of the fish. TVA proposed transplanting all Snail Darter to the Hiwassee, but the US Supreme Court denied the TVA appeal. The Supreme Court also recommended, however, that the US Congress, which had passed the Endangered Species Act in the first place, become the ultimate arbiter of the situation. Congress, amidst much press coverage of the Tellico project, amended the Endangered Species Act and created an exemption committee, which consisted of Secretaries of major federal agencies and was later referred to as "the God Squad" and "the Extinction Committee." This panel had the power to exempt certain activities despite their threat to endangered species if the economic consequences of species preservation were substantial. In this situation, it was the darter or the dam. The committee met in February 1979 and voted unanimously in favor of the darter. A few months later, in a deft political maneuver, a special exemption for the Tellico project was hidden in more general energy legislation and passed Congress without debate. Many members of Congress did not even realize what they were voting on. President Carter reluctantly signed the legislation, apparently trading the Snail Darter for conservative votes on his Panama Canal legislation. Fifteen years after construction began, the Tellico Dam was completed.

Although the Endangered Species Act was weakened during the legislative battle that ensued over the Snail Darter, the process strengthened the species preservation movement in the United States. Never before had so much public interest and sympathy been generated for a diminutive, economically unimportant animal. Fortunately, although extirpated from the Little Tennessee, the darter managed to survive the battle. The transplanted population in the Hiwassee River is viable, and additional transplants to the Holston, Elk, and French Broad rivers are apparently successful. Additional natural populations were later discovered in four other locales in Tennessee, Georgia, and Alabama. In fact, thanks to the concern of and efforts by the ichthyological community and an enlightened public, the Snail Darter's status improved from Endangered to Threatened as of 1984 (Etnier 1976; Ono et al. 1983; D. A. Etnier, pers. comm.).

Watershed Perturbation

Aquatic systems include not only the water in which fishes live but also the groundwater and the surrounding **landscape** or terrestrial area through which water must flow. Many activities have an adverse effect on a river's **watershed** (the land from which water drains into a river), including logging or burning of vegetation, bulldozing for construction and development, groundwater and surface water withdrawal and contamination, overgrazing and trampling of streamside vegetation, and erosion caused by wind, water, or the movements of livestock.

Much has been written about deforestation in tropical and temperate regions. **Riparian** trees, those that grow along stream and river banks, interact intimately with nearby water courses. Obvious consequences of tree removal include a rise in water temperature from loss of shade (from direct heating of a stream and transfer of heat to groundwater by irradiated soil), increased variation in flow rates because water uptake by plants is lost, intensified erosion leading to turbidity, siltation and stream bank collapse (particularly where logging operations occur on steep slopes), and loss of nutrient inputs from falling leaves and fruit.

Shade also reduces **ultraviolet (UV) radiation**. Fishes can suffer directly from UV exposure, including being sunburned (see Blazer et al. 1997), and some sun-dwelling fishes are even protected by mucus that has a sunscreen function (Zamzow & Losey 2002). Eggs, embryos, and larvae of marine and freshwater species suffer increased mortality when exposed to high but natural levels of solar UV-B (reviewed in Häkkinen et al. 2002). Excessive exposure to solar radiation induces cataracts in Rainbow Trout lenses (Cullen & Monteith-McMaster 1993), which diminishes a trout's ability to focus images on the retina. Young of several fish species avoid UV light if refuges are available (Kelly & Bothwell 2002; Ylonen et al. 2004).

Siltation of streams is a major problem – it hinders productivity because of light reduction; eliminates refuge sites; decreases water clarity which makes visual feeding more difficult; depresses spawning activity; and smothers eggs, sessile invertebrates, and plants (Sutherland 2007; Sutherland & Meyer 2007). Silt and sediment are highly abrasive and cause loss of gill function, especially in juvenile fishes (Fig. 22.8). Siltation has been directly linked to native fish declines in many habitats, including Sri Lankan streams and South African estuaries (Moyle & Leidy 1992). Sedimentation is the largest source of contamination in North American streams and rivers (Waters 1995; USEPA 2000) and is the most important factor limiting the availability of fish habitat. Waters (1995, p. 79) stated that fine sediments constituted "perhaps the principal factor . . . in the degradation of stream fisheries."

Another adverse effect of deforestation on aquatic systems involves the cessation of inputs of woody debris, the importance of which was discussed earlier in this chapter (see Modification of bottom type). *Logging* along stream courses can have quite unexpected, complicated impacts on fish populations. The gallery forests that line lowland rivers are also major spawning sites for fishes that migrate into their flooded zones during winter or spring floods at high latitudes and during rainy seasons at low latitudes. Some Amazonian fishes, for example, depend on seasonally inundated floodplains (Goulding 1980). Clearcutting in the Carnation Creek watershed of British Columbia raised stream temperatures 1–3°C. Elevated temperatures caused early emergence and accelerated growth of young Coho Salmon, *Oncorhynchus kisutch*. Smolts migrated earlier than normal and then experienced poor ocean survival, probably because their early arrival in the ocean placed them out of synchrony with prey cycles (Holtby 1988). Logging

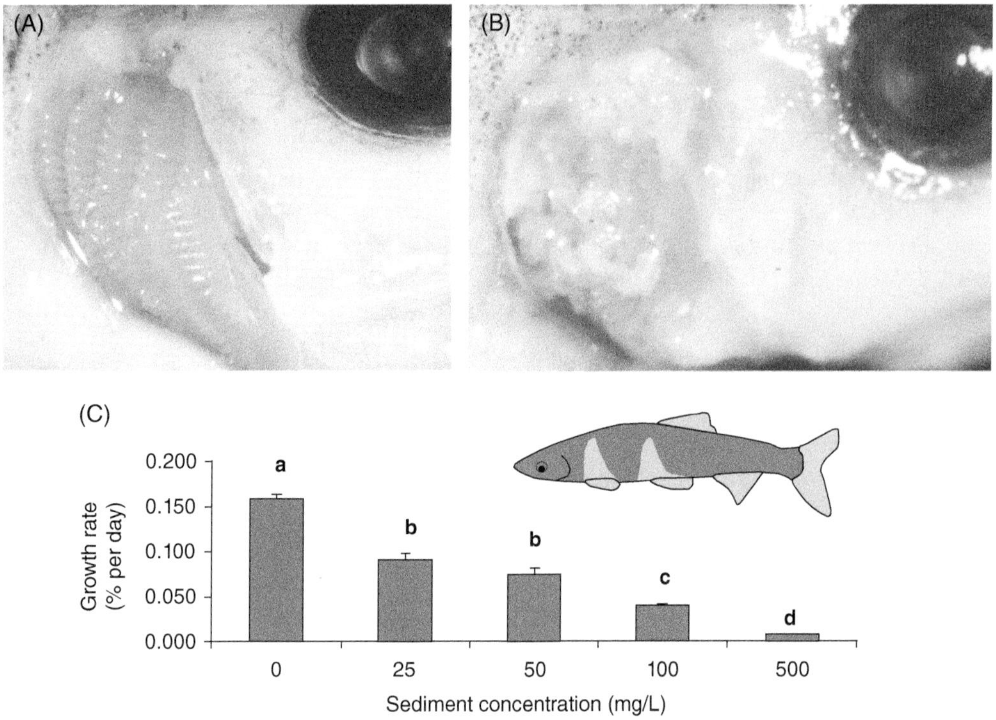

FIGURE 22.8 Effects of suspended sediments on young fishes. Threatened Spotfin Chub, *Erimonax monachus*, were raised at various sediment concentrations to study the effects on gill morphology and growth. (A) Gill arches and filaments of a young Spotfin Chub reared for 21 days at low (0 mg/L) sediment concentrations. (B) Gills from a similarly aged chub reared at high (500 mg/L) sediment concentrations; note the thickening and fusion of filaments and clogging with mucus. (C) Growth rates of young Spotfin Chub relative to sediment concentration, showing decreased growth at higher sediment loads. The growth rate at the highest sediment level was 1/15th that in clean water. High sediment concentrations tested (500 and 100 mg/L) occur regularly in the wild due to watershed development. Bars with the same lower case letter are not significantly different. From Helfman (2007), after Sutherland (2005), used with permission.

operations high in a watershed can affect ecosystem processes at distances far removed from the actual site of disturbance, such as when increased erosion causes unnaturally high levels of sediment deposition in coastal lagoons and estuaries (Moyle & Leidy 1992).

Competition for Water Humans use water for drinking, agriculture, recreation, fishing, and waste disposal. All these activities can have adverse effects on aquatic organisms. Consumption and irrigation require water withdrawals, leading to flow reductions in aquatic systems. Pumping of groundwater lowers water tables, which reduces the output of springs and seeps that are often necessary for maintaining year-round flow in many systems. Habitats subjected to withdrawals shrink, progressively losing heterogeneity and species. Downstream systems from which water is diverted evaporate, concentrating salts and pollutants. The universal use of waterways and waterbodies as dumping grounds for human waste creates environments toxic to fishes and humans.

Water withdrawal for irrigation of arid regions has created numerous ecological disasters, leading to species extinctions among fishes and other biota, and eventually producing salinated croplands and contaminated water supplies for humans. The history of species extinctions in the desert southwest of North America, summarized briefly earlier in this chapter, serves as one example. At a larger scale are the events surrounding desiccation of the Aral Sea on the Uzbekistan–Kazakhstan border (Fig. 22.9). In 1960, the Aral Sea was the fourth largest lake in the world, covering 68,000 km^2; it supported large commercial fisheries as well as extensive hunting in adjacent wetlands. Inputs are primarily from river flow, losses are due to evaporation. Construction of diversion canals and withdrawal of water from its two major input rivers for irrigation purposes shrank the lake to only 41,000 km^2 in 1987. By 1998, lake volume was reduced 80% from its original size. Lake salinity rose to 50 ppt in the 1990s, well above that of sea water, which is 37 ppt. An original native fish fauna of 24 species has been reduced to four introduced species (Zholdasova 1997); commercial fisheries fell from 48,000 metric tons in 1957 to zero by the early 1980s.

Impacts have extended far beyond the ichthyofauna. Dust and salt storms originate on the dry lake bed and distribute 43 million metric tons (mmt) of crop-destroying salt annually over a 200,000 km^2 area. Reduced river flow, salinization, pollution of remaining water, and lowering of the water table have led to a high incidence of intestinal illnesses, throat cancer, tuberculosis, anemia, high infant mortality, and a death rate from respiratory ailments that ranks among the highest in

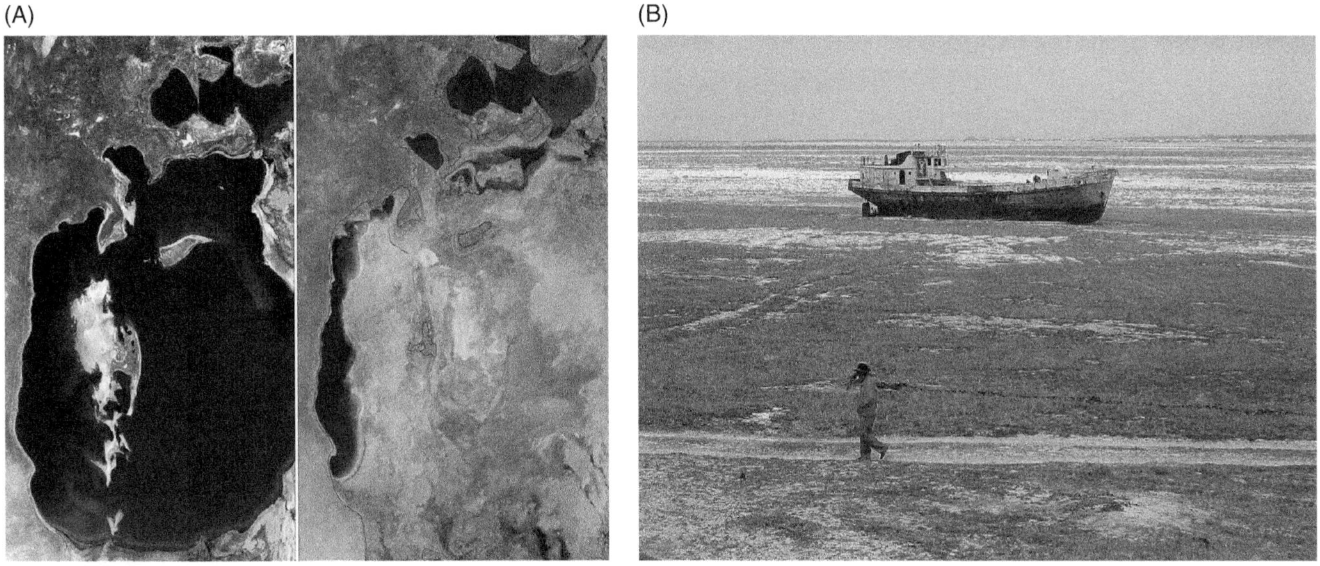

FIGURE 22.9 The Aral Sea, once the fourth largest lake in the world, is a mere fraction of its former self due to water diversion of its tributaries. (A) NASA satellite images from 1989 (left) and 2014 (right); NASA / Wikimedia Commons / Public Domain. (B) The loss of water in the Aral Sea has left behind ships stranded for decades; Staecker / Wikimedia Commons / Public Domain.

the world. Economic losses of approximately 2 billion rubles (= U.S. $3.2 billion) annually have been estimated for the Aral Sea region as a result of its desiccation (Micklin 1988). The Aral Sea disaster has been called "perhaps the most notorious ecological catastrophe of human making" (Stone 1999, p. 30). Since 2000, Kazhakstan has made an effort to replenish the northern basin of the Aral Sea, resulting in a rise in water level of 12 m from 2003 to 2008, reduction in salinity, and return of some fisheries.

Species Introductions

Movement of species into new areas is a natural zoogeographic phenomenon. When such range extension occurs as a result of human actions, it is considered an **introduction**. Natural dispersal is limited by a species' mobility and by physical barriers. In their native habitats, species are typically constrained by coevolutionary processes; they have natural parasites, predators, and competitors that control population growth, and organisms typically exploit prey taxa that have evolved defense mechanisms against the predator's foraging tactics. When species are introduced suddenly into a different environment, the new physical and biological factors may be inhospitable or even lethal. If the habitat conditions are suitable, however, the absence of natural biotic controls may result in the non-native species expanding rapidly and becoming **invasive**. We use the term "invasive" when an introduced species increases to the point that it has a significant impact on the community, such as displacing native species; not all introduced species become invasive. Many species introductions have become well-known pests: rabbits, cane toads, and prickly pear cactus in Australia; starlings, English sparrows, the moth *Lymantria dispar dispar*, and zebra mussels in North America; mongoose and mynah birds in Hawaii; feral goats in the Galápagos and on many other islands, to name a few. These catastrophic introductions have their counterparts in fish assemblages, which are the most introduced vertebrates in the world, including at least 624 species. These global introductions are for aquaculture purposes (51%), ornamental fish (21%), sport fishing (12%), and fisheries (7%) (Goslan et al. 2010). An additional 8% are accidental, mostly through aquaculture or ballast water.

The US Nonindigenous Aquatic Nuisance Prevention and Control Act of 1990 refers to **nonindigenous** organisms, but introduced species go by a great variety of names, including alien, allochthonous, exotic, feral, introduced, invasive, naturalized, nonindigenous, non-native, transplanted, and translocated. Sometimes a distinction is made between **transplants** that are moved within their country of origin but outside their native range, versus **exotic** species that are introduced into a new country. From an ecological perspective it makes little, if any, difference. Introductions may occur through deliberate actions (gamefish stocking, vegetation control, aquaculture, and aquarium releases) or inadvertent mishaps (ballast water introductions, aquaculture escapement, and bait fish release).

Untold hundreds of species of fishes have been deliberately transported among different countries. In the United States alone, at least 536 fish taxa (species, hybrids, and unidentified forms) have been introduced, 35% imported from foreign countries and 61% translocated within the nation (Fuller et al. 1999; Nico & Fuller 1999; **http://nas.er.usgs.gov**). Half of the foreign non-native species have established breeding populations. Most of these fishes represent deliberate introductions by government agencies and individuals (e.g., Grass Carp, *Ctenopharyngodon idella*, for vegetation control; Peacock Cichlid, *Cichla ocellaris*, as a gamefish), escapees from aquaculture facilities (*Tilapia* spp., Atlantic Salmon), or inadvertent bait or aquarium releases (Rudd, *Scardinius erythrophthalmus*; Walking Catfish, *Clarias batrachus*; Redeye Piranha, *Serrasalmus rhombeus*; suckermouth catfishes, *Hypostomus* spp.). Florida, California, and Hawaii have had the largest number of established non-native species,

TABLE 22.1 Ten commonly introduced but controversial fish species. Species listed have often been effective in terms of the original purpose for which they were introduced, but have subsequently posed serious ecological problems.

Species	Number of countries where established	Number of states where introduced	Native area	Original purpose of introduction
Cyprinus carpio, Common Carp	49	49	Eurasia	Food, ornamental
Carassius auratus, Goldfish	>40	49	East Asia	Ornamental
Ctenopharyngodon idella, Grass Carp	9	45	East Asia	Vegetation control
Oncorhynchus mykiss, Rainbow Trout[a]	56	48	Western North America	Gamefish, aquaculture
Gambusia spp., Mosquitofish	67	35	Eastern North America	Mosquito control
Poecilia reticulata, Guppy	34	15	Northern South America	Mosquito control
Micropterus salmoides, Largemouth Bass[b]	53	43	E. North America	Gamefish
Lates niloticus, Nile Perch[c]	3	1	East Africa	Food
Tilapiine cichlids[d]	94	13	East, Central, and South Africa	Aquaculture, vegetation control
Cichla ocellaris, Peacock Cichlid	6	2	Amazon Basin	Gamefish

[a] Related species: Eurasian Brown Trout, *Salmo trutta* (28 countries, 47 states), and North American Brook Trout, *Salvelinus fontinalis* (31 countries, 38 states).
[b] Related species: *Micropterus dolomieu*, Smallmouth Bass.
[c] *Lates longispinus* or *L. macrophthalmus* may have also been introduced into Lake Victoria (see Ribbink 1987; Witte et al. 1992).
[d] Numbers are for *Oreochromis mossambicus* and *O. niloticus*; other widely introduced tilapiines include *O. aureus*, *O. macrochir*, *O. urolepis* ssp., *Tilapia rendalli*, *T. zilli*, and others, including hybrids.
Adapted from G. Helfman (2007); Courtenay et al. (1984); Welcomme (1984, 1988), Lever (1996), and Fuller et al. (1999). Adapted from G. Helfman et al. (1997), and presented in roughly phylogenetic order.

followed by some southwestern states. The aquarium industry and lack of freezing weather facilitate the establishment of many tropical species.

Depending on one's perspective, the same introduction can be claimed as a success story or reviled as a disaster (Fuller et al. 1999) (Table 22.1). Eight fish species are included among "100 of the world's worst invasive species" as recognized by the Global Invasive Species Database of the IUCN's Invasive Species Specialist Group (**www.issg.org/database**). These eight are the Walking Catfish; Common Carp (*Cyprinus carpio*); Rainbow Trout (*Oncorhynchus mykiss*); Brown Trout (*Salmo trutta*); Mosquitofish (*Gambusia sp.*); Largemouth Bass (*Micropterus salmoides*); Nile Perch (*Lates niloticus*); and Mozambique Tilapia (*Oreochromis mossambicus*) (see Table 22.1 for details on most of these). Many of these species have been spread around the world and their ecological impacts vary from minor to catastrophic (Tilapia supply significant protein to humans in need, although alternative, edible native species exist, or existed, in many locales). These fishes are listed by IUCN alongside such known bad actors as chestnut blight, water hyacinth, prickly pear cactus, kudzu, fire ants, malaria mosquitoes, the moth *Lymantria dispar dispar*, zebra mussels, cane toads, brown tree snakes, starlings, rats, rabbits, pigs, and goats.

One outcome of moving fishes around the globe is **homogenization** of previously unique assemblages that evolved at particular locales. As can be seen in Table 22.1, the same fishes have been introduced repeatedly in different places. Not coincidentally, considerable overlap exists among the receiving nations, hence most countries contain established populations of several of the species listed. Introductions are often most successful in **degraded habitats** that are no longer suitable for local endemics, which means that non-native species are replacing and displacing native species. When we total up the introductions and losses, we find that widely separated locales have similar faunal lists dominated by non-native species. This repeated observation, which includes not just fishes but birds, mammals, insects, and plants, amounts to homogenization of the earth's fauna and flora. The process is so widespread, its outcome so ubiquitous, and the disruption so thorough that Williams and Meffe (1998, p. 118) declared that, "The continued homogenization of the world's flora and fauna . . . is an ecological holocaust of major proportions." In her Pulitzer Prize winning book, *The Sixth Extinction: an unnatural history*, author Elizabeth Kolbert points out that this homogenization of species across the planet is in some ways creating "The New Pangaea" (Kolbert 2014), a reference to the supercontinent that existed about 335 million years ago, and began breaking apart about 175 mya into our current continents.

Homogenization of fish faunas is especially prevalent in temperate, higher-income nations where introductions have been ongoing. In Europe, Cowx (1997) listed 166 introduced or translocated fishes, which was about 46% of Europe's 358 native species at the time. Fifteen species had been introduced into 10 or more countries, including Common Carp

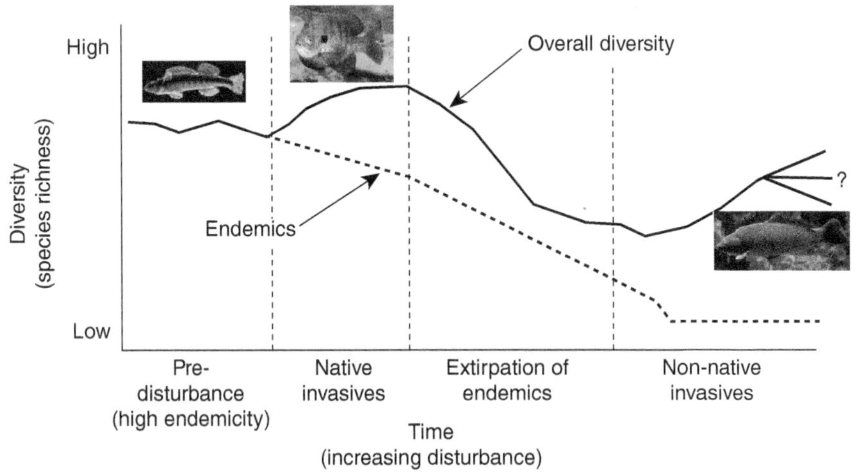

FIGURE 22.10 Both native and non-native species are involved in faunal homogenization. The progressive changes expected in southern Appalachian streams are depicted, showing how habitat disruption (deforestation, siltation) first favors native generalists over endemic specialists. As habitat disruption continues, even these native invasives are replaced by highly tolerant non-native species. From Scott and Helfman (2001), used with permission.

(24 countries), Brook Trout (23), Grass Carp (20), Pumpkinseed sunfish (19), and Rainbow Trout (nine). Study of the fish assemblages of the Guadiana River basin of the southwestern Iberian Peninsula showed considerable overall homogenization, especially in areas close to reservoirs where more species had been introduced (Hermoso et al. 2012). Similar patterns of increased homogenization near reservoirs have been reported elsewhere, perhaps because of the relative stability of the habitat and the concentration of stocking. Other countries with low native diversity or high numbers of introduced species – where extensive homogenization would be expected and where its impacts should be monitored – are New Zealand, Australia, and South Africa.

Homogenization of US fish faunas has happened on a grand scale (Fuller et al. 1999; Rahel 2000, 2002). Of the 76 species introduced into 10 or more states, 32 had been placed in 25 or more states and 13 had gone into more than 35 states. Eight of the latter 13 are relatively large piscivores that continue to be stocked as game species in many places. Three (Common Carp, Goldfish, and Tench) stocked extensively during the late 19th century are now generally regarded as nuisance species.

How have stocking efforts affected fish diversity? Rahel (2000) compared historical lists with current species lists and found that state faunas have grown significantly more similar over time. In the past two centuries, similarity among fish faunas of states increased by an average of 15 species, with almost 20% of states sharing 25 or more additional species. Over half of the fish faunas of Nevada, Utah, and Arizona are non-native, containing species brought in primarily from the eastern part of the country to expand sport fishing opportunities. At the same time, and in combination with the habitat degradation that makes regions hospitable to non-native species and inhospitable to local endemics, the list of imperiled U.S. fishes has grown.

Homogenization is generally discussed with regard to non-native species displacing native species, but the process also involves movement of widespread, generalist native species into areas or habitats previously occupied by local endemics. As is occurring in the southern Appalachian mountains of the United States, **localized endemics** adapted to clear, cool, low-productivity streams in upland regions are progressively replaced by **widespread**, **generalist** species more common in lower, more productive portions of river networks (Scott & Helfman 2001) (Fig. 22.10). The factors responsible for this replacement scenario include upland and riparian deforestation. Subsequent erosion of the uplands causes infilling of rapid and riffle habitats due to increased sediment loads; streams also become broader, deeper, and warmer. Cool water, endemic specialists in shallow, fast-flowing habitats (darters, sculpins, and benthic minnows) are replaced by warm water generalists that can live in a variety of habitats but especially in slower flowing habitats (sunfishes, pool-dwelling minnows, and suckers) (Jones et al. 1999a; Walters et al. 2003). Because the endemics were localized and the generalists were widespread, faunas of different drainage basins increase in similarity. Habitat homogenization promotes biotic homogenization (Boet et al. 1999; Marchetti et al. 2001), and homogenization occurs even though almost all species involved are technically native to the area.

Survey information from over 1050 river basins worldwide found that the effect of species introductions varied widely (Toussaint et al. 2016). Some introductions led to increased similarity among assemblages (homogenization), and some resulted in greater differences among assemblages (differentiation). In many cases, introduced species had little impact on changes in similarity. Most of the observed global homogenization pattern was due to 10 widespread species: Goldfish (*Carassius auratus*), Common Carp, Eastern Mosquitofish (*Gambusia holbrooki*), Largemouth Bass, Rainbow Trout, Mosambique Tilapia, Nile Tilapia (*Oreochromis niloticus*), European Perch (*Perca fluviatilis*), Brown Trout, and Redbelly Tilapia (*Tilapia zillii*). Some species have been introduced repeatedly over long periods of time (e.g., Common Carp, Brown Trout), whereas other had been introduced relatively recently but spread rapidly (e.g., Mosquitofish). It seems reasonable, therefore, that efforts to reduce the spread of these species might be a good strategy to slow the global trend toward homogenization.

Introduced Predators Introductions can lead to population reduction or extermination of native fishes, either directly through predation on adults, eggs, and young, or indirectly through superior competition, hybridization, or transmission of pathogens (Balon & Bruton 1986; Fausch 1988; Ross 1991). Some catastrophic introductions are inadvertent, as with the spread of the Sea Lamprey, *Petromyzon marinus*, into the North American Great Lakes probably via canals constructed to enhance commerce. Lake Trout, whitefishes, pikeperch, and other species declined precipitously in the wake of the lamprey (Daniels 2001; see Chapter 13, Petromyzontiforms). Predatory species that have been widely introduced to provide sportfishing include the Peacock Cichlid, Largemouth Bass, Smallmouth Bass, Rainbow Trout, and Brown Trout. Such introductions often decimate native fish faunas, including reduction of important food fishes. Peacock Cichlids escaped from an impoundment and into the Chagres River, Panama. The cichlid invaded Gatun Lake and progressively eliminated seven local fish species (one atherinid, four characins, and two poeciliids); vegetation increased and fish-eating birds were displaced (Zaret & Paine 1973; Swartzmann & Zaret 1983).

Largemouth Bass have been responsible for similar community disruptions in Lago de Patzcuaro, Mexico; in Lake Naivasha in Kenya; in northern Italy; in Zimbabwe and South Africa; and in Lake Lanao, Philippines. Rainbow Trout and Brown Trout have led to the decline of endemic fishes in Yugoslavia, Lesotho, Colombia, Australia, New Zealand, South Africa, and in Lake Titicaca in Bolivia and Peru (McDowall 2006). In Lake Titicaca, the world's highest lake, a species flock of numerous cyprinodontids (*Orestias*) has been decimated, first through direct predation and later via competition for invertebrate prey. Brown Trout (Fig. 22.11) in particular have been identified as an effective predator on native fishes, including other salmonids. Brown Trout have contributed to the decline of several threatened salmonids, including Gila Trout (*Oncorhynchus gilae*), McCloud River Dolly Varden (*Salvelinus malma*), and Golden Trout (*Oncorhynchus mykiss aguabonita*), the latter being the official state fish of California. Introduced salmonids have been particularly destructive to galaxioid fishes through much of the geographic range of this group of southern hemisphere fishes (McDowall 2006). Brown Trout introduced to New Zealand streams have eradicated native endemic species of the family Galaxiidae, with their distribution now restricted to salmon-free tributaries.

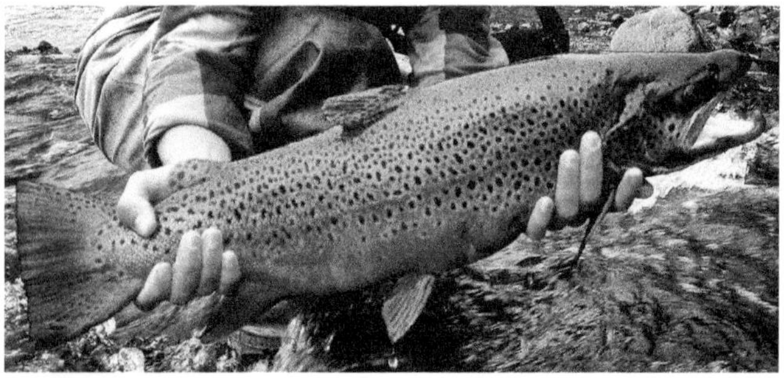

FIGURE 22.11 Brown Trout (*Salmo trutta*) are among the most widely transplanted fishes. They can be grown in hatcheries, succeed in a wide range of cool waters, provide good recreational fishing opportunities, and are good to eat. They also compete with and prey on native species, including native trout. Courtesy of C. Matthaus Ayers.

One of the most dramatic examples of the effects of an introduced predator involves the stocking of the Nile Perch, *Lates* cf. *niloticus*, in Lakes Victoria and Kyoga, east Africa (Ogutu-Ohwayo 1990; Kaufman 1992; Witte et al. 1992; Lowe-McConnell 1997; Ogutu-Ohwayo et al. 1997); there is some debate over exactly which *Lates* species was introduced and how often (Pringle 2005). Lake Victoria was a showcase of evolution and explosive speciation among fishes, having given rise to a **species flock** of perhaps 300 haplochromine cichlids, as well as three dozen other fishes. The lake is thought by many to have contained the richest lacustrine fish fauna in the world. Against the advice of ecological experts, Nile Perch were stocked in the lakes in the early 1960s, "to feed on 'trash' haplochromines . . . [and convert them] into more desirable table fish" (Ribbink 1987, p. 9).

This predator, which can attain a length of 2 m and a weight of 200 kg, spread slowly through both lakes, effectively wiping out native fishes by feeding preferentially on abundant species, then shifting to other species as the density of the initial prey declined, and finally turning to cannibalism. Commercial landings of cichlids declined dramatically between 1977 and the early 1980s, while Nile Perch landings increased. As species were eliminated, food webs in the lakes were substantially disrupted and simplified (Fig. 22.12); elimination of herbivorous cichlids led to algal blooms and associated oxygen depletion in deep water, which caused periodic fish kills. In Lake Kyoga, the catch changed from a multispecies fishery dominated by several haplochromines to one dominated by two introduced species (Nile Perch and a tilapia) and a native cyprinid (*Rastrineobola argentea*).

It is difficult to say how many of the endemic cichlids have been exterminated: perhaps only 50% of the species have been described and rare fishes are difficult to sample (Goldschmidt 1996). However, decreased catches indicate that populations declined and the continued threat of predation by Nile Perch and commercial fishing will only exacerbate the situation. Based on comparative samples taken in 1978 and 1990, approximately 70%, or 200 species, of haplochromines are extinct or threatened with extinction (Witte et al. 1992). The trend at the time suggested that "probably more vertebrate species are at imminent risk of extinction in the African lakes than anywhere else in the world" (Ribbink 1987, p. 22). Events in Lake Victoria point out the ecological consequences of introducing predators into any aquatic system (Witte et al. 1992).

The scenario that played out in Lakes Victoria and Kyoga is one of reduced biodiversity and simplified community interactions as a cost of production of animal protein for human consumption. Successful fisheries for introduced Nile Perch and tilapia have been established in those lakes, replacing the previous fisheries for smaller, native fishes. The impacts

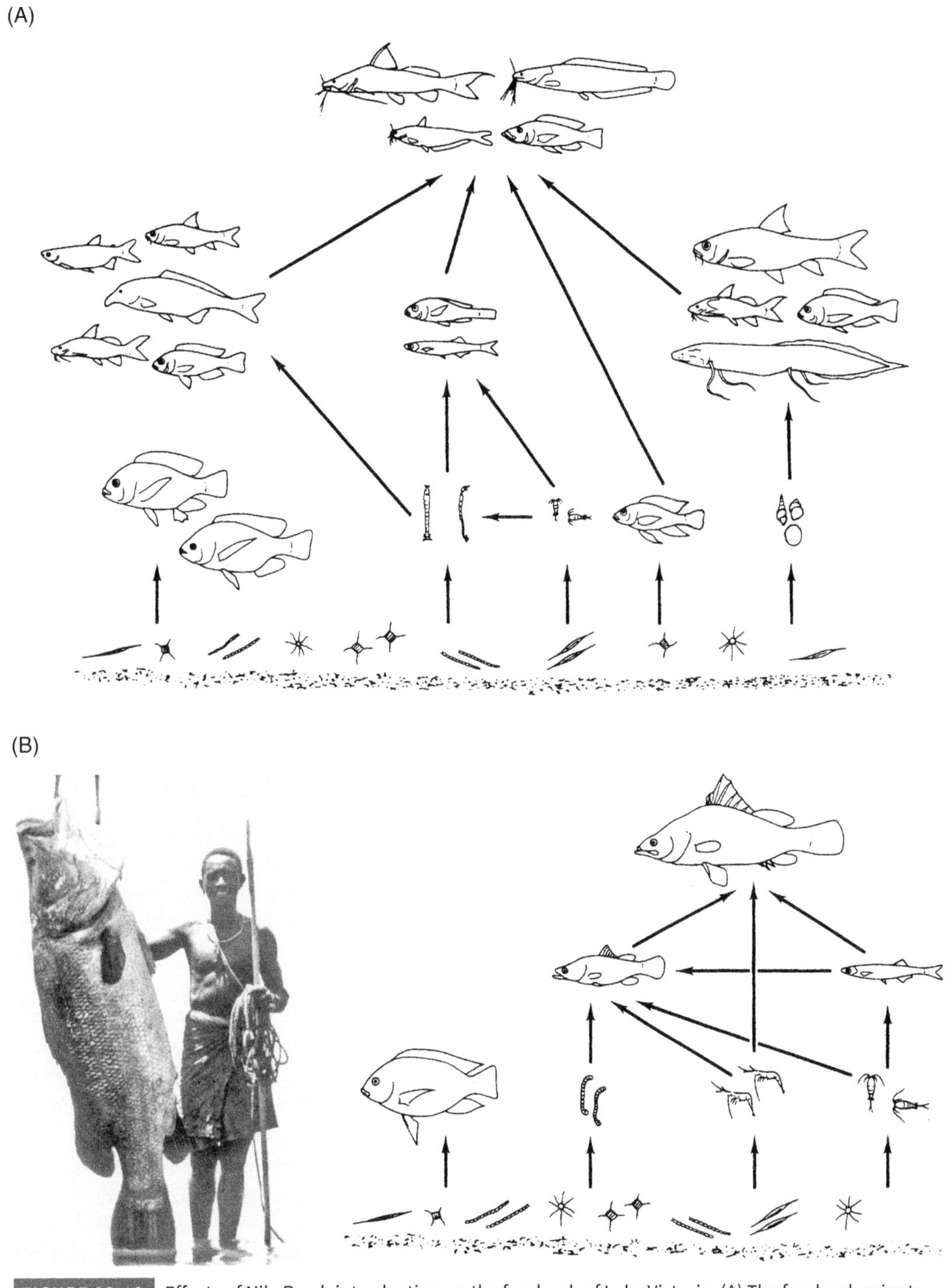

FIGURE 22.12 Effects of Nile Perch introduction on the food web of Lake Victoria. (A) The food web prior to the introduction of *Lates*. The top predators included piscivorous catfishes and haplochromine cichlids which fed on a variety of prey (including characins, cyprinids, mormyrids, catfishes, haplochromine and tilapiine cichlids, and lungfishes), which in turn fed on a variety of invertebrate prey and algae. (B) The food web after *Lates* eliminated most other fish species. *Lates* feeds on juvenile *Lates*, a cyprinid (*Rastrineobola*), and an introduced tilapiine cichlid. Inset: a large Nile Perch. (A, B) from Ligtvoet and Witte (1991), used with permission; inset courtesy of L. and C. Chapman.

of these introductions are not, however, limited to the aquatic ecosystems. Nile Perch have a relatively high oil content. Traditional preparation methods, such as air drying, are less effective for processing Nile Perch. Instead, the flesh is often smoked over wood fires, which contributes to deforestation of hillsides in the Lake Victoria basin, runoff of sediment and nutrients into the lake, reduced light transmission affecting reproduction of cichlids, and eutrophication leading to deoxygenation of deeper waters, which Nile Perch cannot tolerate (Seehausen et al. 1997a, 1997b; Kitchell et al. 1997); low oxygen areas may serve as refuges for native cichlids (Schofield & Chapman 2000). An unexpected result of combined ecological and socioeconomic influences is that condition factors of Nile Perch in Lake Victoria declined to the lowest values known for the species anywhere (Ogutu-Ohwayo 1999). Similar introductions, for similar purposes and with similar results, have occurred all over the world. For example, Contreras and Escalante (1984) identified nine instances in Mexico where, after the introduction of potential food fishes, the number of native, often endemic species declined by an average of 80%.

Harmful introductions of non-native predators are most common and best known from freshwater systems. The ocean, however, is not immune to similar impacts. One of the most dramatic examples involves an Indo-Pacific predator that has become successfully established in tropical and subtropical habitats of the West Atlantic. First documented along the Florida coast in 1999, Red Lionfish (*Pterois volitans*) can now be found abundantly throughout the Caribbean Sea and Gulf of Mexico, representing "one of the most rapid marine invasions in history" (Fig. 22.13). The Lionfish was found in Brazil (crossing the Amazon Barrier, see Chapter 19) in 2014 (Ferreira et al. 2015). They were established from a few fish probably released by aquarists. Although seldom abundant in their native waters, lionfish in the Atlantic can reach densities of over 400 per hectare. Native reef fishes are naïve to the hunting tactics of lionfish, and few if any Atlantic fishes feed on these well-protected piscivores. Once established, lionfish can reduce populations of small reef fishes by about 80 percent (Morris and Whitfield 2009; ANS 2011; Bariche et al. 2017)

Competition

Predation by non-native fishes on native species is an obvious effect of species introduction. Less well documented, but of potentially serious consequence, are the threats of competition, disease, and hybridization that can occur from introducing non-native species (Taylor et al. 1984). Competition is difficult to prove even under the best experimental conditions (Ross 1991; see Chapter 21). Evidence of competitive depression of native fishes usually takes an inferential form, in terms of overlap in use of potentially limiting resources, or decline in native fishes correlated with the introduction of a competitor. The widely stocked non-native Brown Trout often competes with native salmonids in North America. Brown Trout are aggressive and outcompete and displace Brook Trout (*Salvelinus fontinalis*) in the eastern United States and Cutthroat Trout (*Onchorhynchus clarkii*) in the western United States, often forcing less competitive native trout into stream sections with higher velocities and less food (Blanchet et al. 2007; Meredith et al. 2015; Hoxmeier & Dieterman 2013; Hitt et al. 2017). Brown Trout can also tolerate higher temperatures, giving them an additional advantage as climate changes. Introductions of non-native Rainbow Trout also can affect habitat use by native species. Studies of habitat selection in an artificial stream showed that Warpaint Shiners (*Luxilus coccogenis*), native to southern Appalachian streams of North America, were displaced into higher velocity habitats and expanded their home range in the presence of introduced Rainbow Trout. In a natural setting, such a change in microhabitat use probably increases the energetic demand on the shiners and may reduce their overall fitness (Elkins et al. 2019). Competitive impacts on rare, native Atlantic Salmon, *Salmo salar*, from abundant, escaped Atlantic Salmon from salmon farming operations are also a concern (Jonsson & Jonsson 2006).

Diet overlap with native fishes in North America has also been documented for introduced Common Carp, Pike Killifish (*Belonesox belizanus*), numerous cichlids, and two Asiatic gobies (*Acanthogobius flavimanus* and *Tridentiger trigonocephalus*). Blue Tilapia (*Oreochromis aureus*) overlap extensively in diet with Gizzard Shad and Threadfin Shad (*Dorosoma cepedianum* and *D. petenense*, Clupeidae). Blue Tilapia reproduce rapidly, forming dense populations (>2000 kg/ha) of stunted individuals. Introductions of Blue Tilapia in Texas and Florida have resulted in concomitant population declines of shad, particularly of the benthic-feeding Gizzard Shad. Overcrowding by tilapia also inhibits Largemouth Bass spawning behavior (Taylor et al. 1984). Competition for food probably explains the negative impact of introduced Guppy (*Poecilia reticulata*) on the endemic White River Springfish (*Crenichthys b. baileyi*) in Nevada. Competition for nursery grounds led to a decline in catches of native *Oreochromis variabilis* after transplantation of Redbelly Tilapia (*Coptodon zillii*) to Lake Victoria.

Our understanding of how non-native species displace natives can be improved via laboratory manipulations of species and resources (Fausch 1988; Ross 1991). Marchetti (1999) looked for competitive interactions as a cause of population declines and extirpations of Sacramento Perch, *Archoplites interruptus*, a native California centrarchid. Sacramento Perch are least abundant where introduced sunfishes are most numerous. In lab aquaria, Marchetti found that Sacramento Perch placed with Bluegill, *Lepomis macrochirus*, grew less and shifted habitat use to less natural habitats. Bluegill fed more actively and harassed the perch (Fig. 22.14).

Hybridization

Hybridization and **introgression** (crossing of hybrid offspring with parental genotypes) has caused rapid losses of native fishes over extensive geographic areas (Echelle 1991). Hybridization can result from habitat alterations that reduce physical and behavioral barriers between populations (e.g., Lake Victoria cichlids; Seehausen et al. 1997a, 2008; see Fig 8.8). Hybridization also occurs when numbers of one species fall to the point where conspecifics are rare during mating periods, leading to interspecific matings (Hubbs 1955). Rare species have fallen victim to hybridization in the U.S. southwest, including hybridization between the threatened Clear Creek Gambusia (*Gambusia heterochir*) and introduced Mosquitofish (*G. affinis*), between the endangered Humpback Chub (*Gila cypha*) and the more common Roundtail Chub (*G. robusta*), and between June Suckers (*Chasmistes l. liorus*) and Utah Suckers (*Catostomus ardens*). June Suckers were at one time exceedingly abundant. As their numbers decreased, in part because of water drawdowns for irrigation, June Suckers became increasingly vulnerable to genetic disruption via hybridization with abundant Utah Suckers (Echelle 1991). Marine debris probably facilitated the recent arrival of the Indo-Pacific sergeant (*Abudefduf vaigiensis*) in Hawaii, where it is hybridizing with the endemic congener, *A. abdominalis* (Coleman et al. 2014).

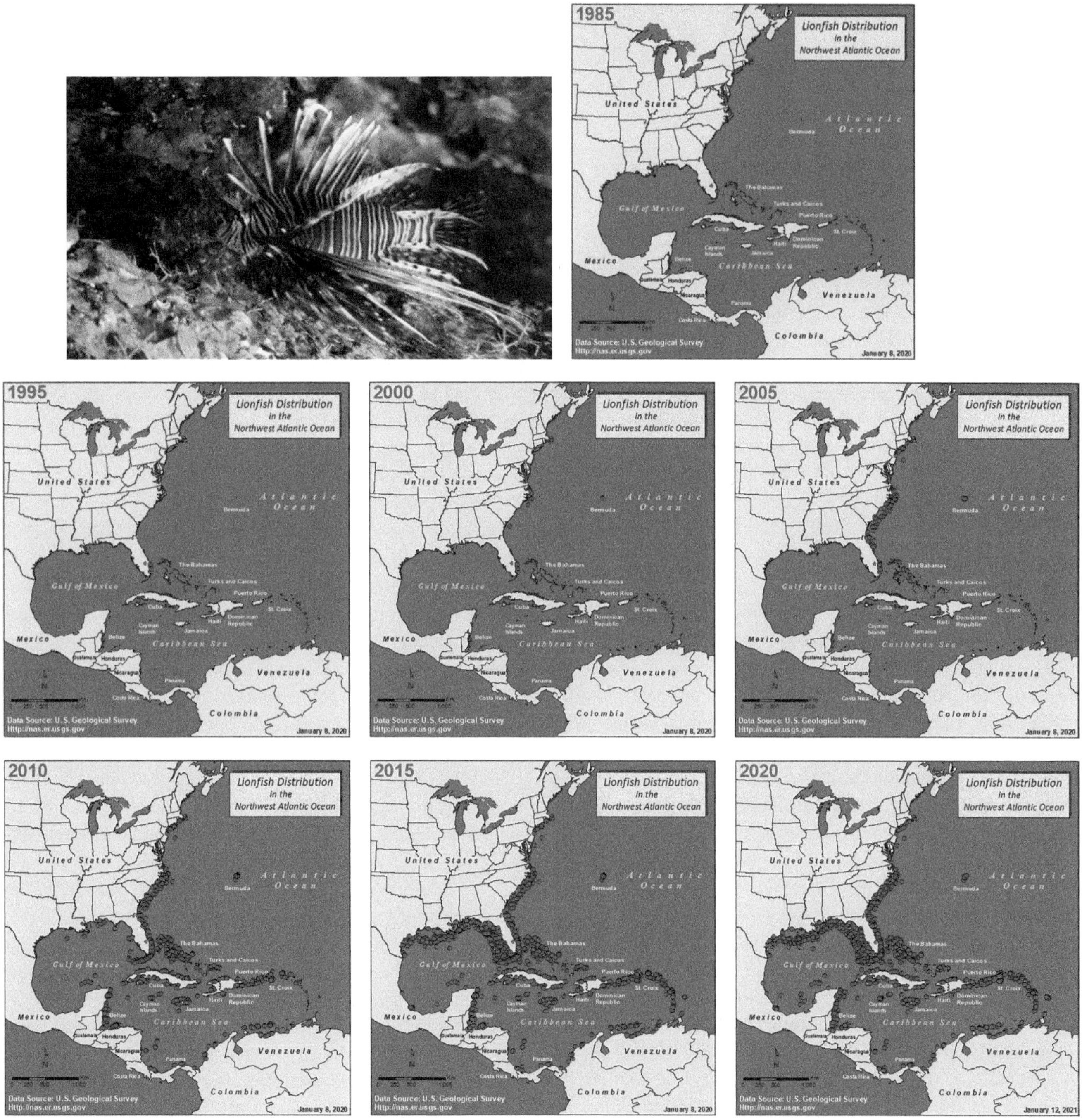

FIGURE 22.13 Lionfishes (*Pterois*, Scorpaenidae) are now common on tropical western Atlantic reefs, decimating native assemblages of small reef fishes. Toxic spines and a lack of predators have contributed to the success of lionfishes as an invasive introduced species. Photo from NOAA / Public Domain. Occurrences of Red Lionfish (*Pterois volitans*) in the West Atlantic during the rapid initial phase of the invasion. Red dots indicate confirmed sightings, based on the USGS-NAS database. For more recent updates on an animated map of the expansion of Red Lionfish, see https://www.usgs.gov/centers/wetland-and-aquatic-research-center-warc/science/lionfish-distribution-geographic-spread?qt-science_center_objects=0#qt-science_center_objects.

Human-caused hybridization is particularly threatening where stocking programs bring hatchery or other strains of fishes into contact with native conspecifics (Utter & Epifanio 2002). Native strains disappear as they interbreed with introduced fishes, as has happened when Rainbow Trout were stocked with threatened Cutthroat, Gila, and Apache trouts (*Oncorhynchus clarki* ssp., *O. gilae*, and *O. apache*) in western North America (Echelle 1991). Hatchery fishes may originate from a limited gene pool or from inbred lines and likely have reduced genetic variability compared to wild populations. **Low genetic variability** correlates with lower fecundity, poorer survivorship, and slower growth, as found in different populations of endangered Gila Topminnow (*Poeciliopsis occidentalis*; Quattro & Vrijenhoek 1989).

When hatchery transplants breed with wild fish, resulting offspring will often be less diverse genetically than the wild strains. Hybrid offspring may continue to breed with and eventually eliminate native stocks, as has occurred with

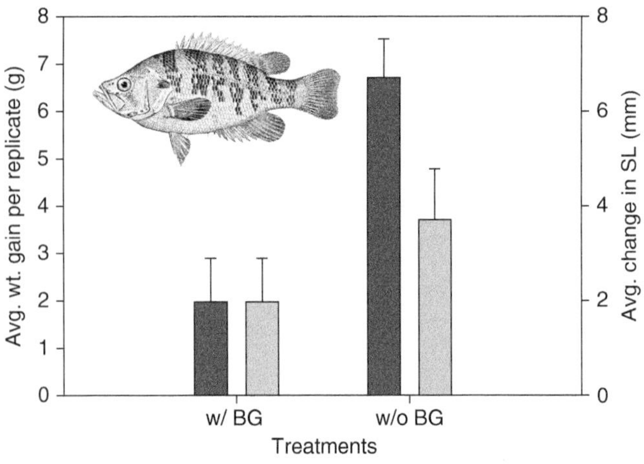

FIGURE 22.14 Experimental evidence of competitive displacement of a native by an introduced species. Sacramento Perch, an imperiled native sunfish of California, alone in aquaria (dark bars) grew more in mass and showed a trend toward greater length increase than when kept with introduced Bluegill sunfish (shaded bars). Based on Marchetti (1999); fish image by H. L. Todd / Wikimedia Commons / Public Domain.

Rainbow Trout stocked widely throughout North America. In a study of Brook Trout in 24 lakes in two wildlife reserves in Quebec, stocking affected genetic structure of the populations and resulted in greater loss of genetic distinctiveness of trout in lakes that had received more stocked domesticated fish (Marie et al. 2010); the stocking of hatchery fish had resulted in subsequent generations being genetically less suited for the specific conditions of the particular lake. Even the genetic integrity of Common Carp, is threatened by introgressive hybridization with introduced, cultured strains. Very few wild-type carp remain in the native habitat of this widely introduced species, and the wild genetic strain has endangered status in such large river systems as the Danube (Balon 1995).

Threats from transplanted and cultured fish have caused considerable concern over the genetic integrity of wild Atlantic Salmon stocks, prompting programs to minimize the effects of aquaculture and stock enhancement programs (NASCO 1991; NRC 2004a). A study of wild and escaped Atlantic Salmon in Norway concluded that the degree of genetic introgression is highly variable, ranging from 2 to 47% in a study of 20 rivers (Glover et al. 2013), and from 0 to 42% in a study of 109 rivers (Karlsson et al. 2016).

The creation of intergeneric hybrids, initially considered unlikely, has also proven troublesome. Spawning interactions between Brown Trout and Brook Trout, and in some cases hybrid "tiger trout," have been observed in Europe (Cucherousset et al. 2008) and North America (Grant et al. 2002), leading to concerns that the introduction of the non-native species can negatively impact the native species. The widely introduced European Rudd, *Scardinius erythrophthalmus* (Cyprinidae), is a hardy, colorful baitfish cultured in the southern United States. Rudd hybridizes with native Golden Shiner, *Notemigonus chrysoleucas*. Rudd is known to be established in eight states and could potentially hybridize with Golden Shiners in 26 states in the Mississippi River basin, with unknown consequences for the fish assemblages or ecosystems of those areas. Such risks are unnecessary given that several acceptable native bait species, including the Golden Shiner, already exist throughout the region (Burkhead & Williams 1991).

Parasites and Diseases

A major threat from introductions, regardless of origin, is transmission of bacterial and viral diseases and parasites to which native fishes were not previously exposed (Hedrick 1996). **Furunculosis**, a fatal bacterial disease caused by *Aeromonas salmonicida*, was originally endemic to western North American strains of Rainbow Trout. When the trout was introduced into Europe, the disease became widespread among Brown Trout populations and now occurs wherever salmonids are cultured (Bernoth et al. 1997). **Whirling disease**, caused by the protozoan *Myxosoma cerebralis*, is native and originally nonpathogenic to European salmonids (Hedrick et al. 1999). The parasite inflames cerebrospinal fluid, deforms the brainstem, and causes degeneration of nerves connecting the medulla and spinal cord (Rose et al. 2000), resulting in swimming in tight circles, followed by postural collapse and immobility. Whirling disease, transmitted from Europe to North America in the late 1950s, is extremely pathogenic to Rainbow and Brook trout, and has been considered the single greatest threat to many US wild trout populations (MWDTF 1996). It has subsequently spread with exportation of North American salmonids, including back to Europe, where it has increased in pathogenicity. "**Ich,**" a debilitating gill and skin infestation caused by the ciliated protozoan *Ichthyopthirius multifiliis*, originated in Asia and has spread throughout temperate regions via introductions (Hoffman & Schubert 1984; Welcomme 1984; Dickerson & Clark 1998).

An interaction between genetic disease resistance and the dangers of transplantations is exemplified by fall Chinook Salmon, *Oncorhynchus tshawytscha*. Fish raised from eggs taken from streams where the protozoan *Ceratomyxa shasta* is endemic showed mortality rates of less than 14% when exposed to the pathogen. However, fish from streams where the pathogen is not native exhibited mortality rates of 88–100%, upon exposure (Winton et al. 1983). The introduction of infected fishes into areas where specific diseases do not occur naturally such as might occur during pen-rearing operations or a "supplementation program," could have catastrophic consequences for endemic stocks of fishes.

Infestation problems involving native parasites can also be aggravated. Juvenile Pink Salmon (*Oncorhynchus gorbuscha*) and Chum Salmon (*O. keta*) migrating past net pens holding Atlantic Salmon suffered mortality rates of 9–97% as a result of infestations of sea lice, external copepod parasites. Infestation rates were significantly lower on juveniles that did not swim past aquaculture operations (Krkosek et al. 2006, 2007). Sea lice on adult salmon are seldom fatal, but a single copepod can kill a juvenile salmon.

Viral Hemorrhagic Septicemia (VHS) is a devastating disease with very high mortality. VHS was first seen in German trout farms in the 1930s, and then appeared in some North American

Pacific salmon in 1980s. It is now widespread, including in the North American Great Lakes, and has been especially devastating in freshwater trout farming operations in Europe. Study of the genetics and phylogeny of various forms of the VHS virus suggest that the virus evolves rapidly in both marine and freshwater environments (He et al. 2014). Monitoring the distribution of VHS in the Great Lakes from 2006 to 2012 suggests that careful surveillance and biosecurity measures, including restricting movement of susceptible species, had been successful in limiting spread (Gustafson et al. 2014). However, ongoing monitoring and biosecurity must be sustained. Study of the gene sequences of the genetic variation of VHS virus responsible for outbreaks in farmed Rainbow Trout in Europe suggests that the spread of the disease is linked to the farmed-trout industry (Cieslak et al. 2016).

Hawaiian Division of Fish and Game introduced the Blueline Snapper (*Lutjanus casmira*) to state waters in 1958–1961 to enhance coastal fisheries. Prior to release, fishes were treated with copper sulfate to eliminate external parasites, but no attempt was made to treat internal parasites. As a result, the parasitic nematode *Spirocamallanus istiblenni* was introduced with the Blueline Snapper. The parasite occurs at much higher frequency in the introduced range than the native range, and has spread to endemic Hawaiian fishes, with consequences that are yet to be resolved (Gaither et al. 2013).

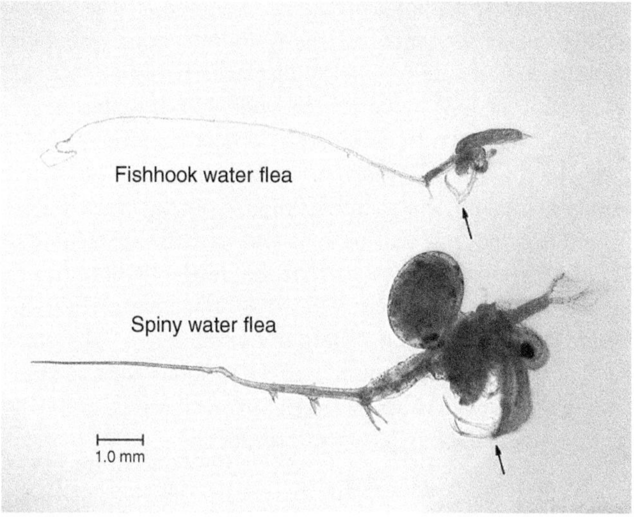

FIGURE 22.15 The elongated and barbed abdominal spine of the Spiny and Fishhook water fleas can get caught in fish gills and become embedded in the walls of fish stomachs. NOAA / Public Domain.

Ballast Water Introductions The ballast water of large ships is a significant source of introductions (NRC 1996a). Water is pumped into special ballast tanks or empty holds of ships to stabilize them; this water is then pumped out when cargo is taken on board at another port. Ballast water sampled from five vessels in Hong Kong Harbor contained 81 species in 13 phyla of animals and protists (Chu et al. 1997). Extrapolating from the international extent of shipping, Carlton (1999) estimated that >7000 species are transported daily in ballast water, including serious human pathogens. Hundreds of species of fishes and invertebrates have become widely established as a result of such ballast water introductions, including such well-known pests as the zebra mussel, *Dreissena polymorpha*, spiny waterflea (*Bythotrephes longimanus*), and fishhook waterflea (*Cercopagis pengoi*) in the Laurentian Great Lakes. These have all spread to connected water bodies and have had dramatic impacts on aquatic communities. They can also spread to non-connected waters through movement of bait buckets, trailered boats, and wet aquatic equipment such as ropes and life jackets. The spiny and fishhook water fleas are large, predatory zooplankters and have impacted plankton communities. The dramatically elongated barbed spine at the tip of the abdomen can make it difficult for small fish to swallow them (Fig. 22.15). There have also been reports of clusters of broken spines accumulating in fish stomachs (see Kerfoot et al. 2011).

These and other invertebrates can drastically alter the food resource base for fishes via competition for or elimination of natural prey. An American export, the ctenophore, *Mnemiopsis leidyi*, was introduced via ballast water into the Black and Azov seas of Asia, where it reached densities of 180 individuals/m^3. It competed with and ate native fish larvae and has been linked to loss of a $250 million anchovy fishery (Ruiz et al. 1997; Shiganova & Bulgakova 2000). Globally, at least 32 introduced fish species in 11 families are thought to have been transported by ballast tanks.

Gobies and blennies are the two families most commonly associated with ballast water (Wonham et al. 2000). The Yellowfin Goby, *Acanthogobius flavimanus*, an east Asian native, has become one of the most common benthic fishes in the San Francisco Bay–Sacramento River area. Round Goby, *Neogobius melanostomus*, and Tubenose Goby, *Proterorhinus marmoratus*, both native to the Black and Caspian seas, arrived in the North American Great Lakes around 1990 via ballast water and quickly spread through all five lakes. Round Gobies reach densities of up to $133/m^2$ and are egg predators. At high densities, they could compete with native sculpin and affect benthic spawners such as Lake Sturgeon, *Acipenser fulvescens*, and Lake Trout, *Salvelinus namaycush*, thus compromising expensive rehabilitation efforts (Moyle 1991; Dubs & Corkum 1996; Chotkowski & Marsden 1999; K. M. Jones, pers. comm.). Due to its ability to thrive in fresh and brackish water and adapt to locally available prey, including mussels, the round goby has become one of the most widespread invasive fishes on earth (Kornis et al. 2012). Populations became abundant in the Baltic Sea, the North American Great Lakes and connected bodies of water, and several major rivers in Europe, and the species continues to spread. Increasing water temperatures due to climate change are expected to further enhance it spread.

Assessing the Potential Impacts of Introductions Many species introductions, including those precipitating some of the worst-case ecological scenarios, occur in lower-income nations where the political focus is on human economic and nutritional problems. In addition to the negative ecological consequences of many introductions, traditional fishing methods are frequently displaced by introduced species, requiring new harvesting technologies or replacing local **artisanal** fishers with commercial or sport fishers. Although many lower-income nations are in desperate need of capital and of animal protein sources, simple planning measures and attention to natural distributions and local fishing techniques could often minimize results that are destructive to both the local biota and culture (e.g., CRC 2006). Whenever an introduction is being considered, regardless of the source, a protocol such as that outlined by Kohler and Courtenay (1986a, 1986b) should be followed to assess the potential biological and sociological costs and benefits of the introduction.

Pollution

Pollution enters aquatic systems in various ways including as sediments, dissolved or suspended substances in runoff or precipitation, or from the atmosphere. Human-produced toxic substances number in the thousands, ranging from elemental contaminants such as chlorine and heavy metals to chemical complexes such as persistent pesticides, detergents, EDCs, and petroleum products (NRC 1999a). Harmful effects on fishes occur as a result of direct toxicity, by interfering with developmental pathways in the case of EDCs, or through food chain effects (e.g., eutrophication, bioaccumulation), ultimately affecting individual survival and reproduction. Food chain effects also link contaminated fishes to other endangered species such as marine and terrestrial mammals and birds (Lloyd 1992; Ewald et al. 1998).

Pollution-related reductions in fish biodiversity occur worldwide. Some of the best documented examples have occurred in North America and Europe due to agricultural chemicals and acidic precipitation. **Acidic precipitation** ("**acid rain**") has a pH of less than 5.6. It results when oxides of nitrogen and sulfur (NO_x, SO_4), especially from fossil fuel combustion, are further oxidized in the atmosphere to form nitric and sulfuric acid. Acid precipitation becomes a particularly serious problem in watersheds composed of rock types that are incapable of buffering the acids, such as the metamorphic rocks of northern North America and Europe. Most systems contaminated by acid precipitation suffer prolonged periods of low pH, but episodic inputs during snowmelt or storms can exacerbate already stressful conditions, including increased acute toxicity from aluminum and mercury (see Gensemer & Playle 1999). Mercury mobilization occurs because bacteria convert mercury to methylmercury more rapidly at lower pH. Spring rainstorms and snowmelt are especially injurious: acidic compounds accumulate in winter snowpack, flushing occurs when eggs and larvae are most abundant; early life stages are particularly vulnerable to low pH (Sullivan 2000).

Acid precipitation has caused dramatic chemical changes in more than 100,000 lakes in Ontario and Quebec, wiping out all wild stocks of the endangered Aurora Trout (*Salvelinus fontinalis timagiensis*), and has reduced the range of the endangered Acadian Whitefish (*Coregonus huntsmani*) by 50% (Williams et al. 1989). Similar acidification and salmonid declines have occurred in the Adirondack Mountains of New York and in many Scandinavian lakes. Acid deposition is considered a prime contributor to the decline of Atlantic Salmon stocks in eastern Canada and may be a barrier to their recovery (Watt et al. 2000). Fish kills in Norway following episodic acidification affected both Atlantic Salmon and Brown Trout (Baker & Christensen 1991). Norway lost 18 stocks of Atlantic Salmon, with eight more considered threatened, and Brown Trout have disappeared from 39% of Norway's lakes, with significant declines in another 17% (Sandøy & Langåker 2001). In the Adirondack Mountains of New York, lakes with lower pH have fewer fish species than they once did, and Brook Trout, which tend to be somewhat tolerant of low pH were no longer present in some lakes (Baker et al. 1993; Driscoll et al. 2003). Minnows, which are more acid sensitive, had disappeared from 19% of surveyed lakes by the early 1990s (Baker et al. 1993). A study of water chemistry and survival of young-of-year Brook Trout in Adirondack streams showed no significant improvement in survivorship from 1984 to 2003 (Baldigo et al. 2007). The loss of Slimy Sculpin (*Cottus cognatus*) and Blacknose Dace (*Rhinichthys atratulus*) from the fish assemblage of upper Hubbard Brook, New Hampshire coincides with periods of chronic acidification during the 1970s (Warren et al. 2008).

Agricultural chemicals (pesticides, herbicides, and fertilizers) have been responsible for the extermination of many fishes in the American southwest, particularly those in isolated habitats. The toxic chemicals work directly on the fishes or are ingested with food, whereas fertilizers lead to eutrophication, which changes the balance of algae from edible species to inedible blue-greens, raises lake temperature, and lowers oxygen content. The Clear Lake Splittail, *Pogonichthys ciscoides*, a cyprinid endemic to Clear Lake in northern California, was extremely abundant through the 1940s. Agricultural development of the lake basin transformed the lake from a clear, cool habitat dominated by native fishes to a warm, turbid lake dominated by introduced species. The last Splittail was taken from the lake in 1970. Eutrophication or toxic chemicals have been similarly implicated in the demise of such unusual fishes as the Lake Ontario Kiyi, Phantom Shiner, Stumptooth Minnow, Blue Pike, and Utah Lake Sculpin. Overall, pollution has contributed to the demise of 15 of the 40 species and subspecies of fishes that have gone extinct in North America during the past century (Williams et al. 1989).

EDCs (see Chapter 7) are an insidious form of pollution because they interfere with growth and development at extremely low chemical concentrations, with potentially dramatic consequences (Colborn et al. 1996; Arcand-Hoy & Benson 1998; NRC 1999a). In fishes, EDCs affect sexual differentiation and reproductive performance, acting early in sex determination as well as later when gonads mature (Devlin & Nagahama 2002). Documented impacts on wild populations include abnormal gonad morphology, reduced rates of sperm and egg production and release, and reduced quality of

TABLE 22.2 **Examples from field studies demonstrating reproductive and developmental impairment after exposure to endocrine disrupting compounds and other chemical pollutants.**

Xenobiotic/source	Effect	Species
BKME	Masculinization of females	Mosquitofish, Burbot, Fathead Minnow
Columbia River pollutants, DDT	Phenotypic sex reversal	Chinook Salmon, Ricefish
Sewage estrogenic compounds	Intersexuality	Roach
PCBs, DDT/sewage effluent, oil spill	Increased egg mortality	Sand Goby, Arctic Char
Oil spill	Premature hatch, deformities	Pacific Herring
North Sea pollutants, DDE	Embryonic deformities	Flatfishes, cod
PCBs, DDT/various discharges	Chromosomal aberrations	Whiting
PCBs, PAHs/urban discharge, landfill leachate	Precocious maturation, decreased gonad development	English Sole, Eurasian Perch, Brook Trout
Crude oil, BKME/oil spill	Altered ovarian development	Plaice, White Sucker
Alkylphenols/sewage effluent	Altered vitellogenesis	Rainbow Trout, etc.
Pulp mill effluent, oil spill	Reduced plasma steroids, sperm motility	White Sucker, Atlantic Salmon, flounder
Textile mill, vegetable oil effluent	Retarded/reversed ovarian recrudescence	Airsac catfish, snakehead
EE2/sewage effluent	Reduced territory acquisition	Fathead Minnow

BKME, bleached kraft mill effluent; DDE, metabolic byproduct of DDT; DDT, dichlorodiphenyltrichloroethane; EE2, ethynylestradiol; PAH, polycyclic aromatic hydrocarbons; PCB, polychlorinated biphenyl.
Adapted from G. Helfman (2007) and Arukwe and Goksoyr (1998).

gametes (Arukwe 2001) (Table 22.2). Studies in both freshwater and marine environments have shown the sensitivity of many species of fishes to the effects of EDCs on development of gonads and secondary sex characteristics (Amiard-Triquet and Amiard 2013). An example is EDC-induced intersex in fishes, where testes in males contain oocytes, due to the feminizing effects of estrogenic EDCs. Intersex males often have elevated levels of vitellogenin, a precursor protein for egg yolk production that is typically elevated in mature females (Blazer et al. 2012, Iwanowicz et al. 2015).

EDCs contribute to and exacerbate declines among imperiled fishes. The Columbia River of Oregon and Washington, previously the most productive salmon river in America, suffers from damming, overfishing, introductions from fish hatcheries, and agricultural and industrial pollution. Columbia River water now contains at least 92 chemical contaminants found in fish samples, including 14 metals, DDT, chlordane, polychlorinated biphenyls (PCBs), and chlorinated dioxin and furans (USEPA 2002); some of these chemicals are known endocrine disrupters. Approximately 85% of female-appearing Chinook Salmon sampled from the Columbia River possessed a genetic marker for the Y chromosome, indicating that they were sex-reversed males (Nagler et al. 2001). When XY females mate with normal XY males, 25% of the F1 generation can be expected to exist as YY males, skewing the population sex ratio from a normal 1 : 1 to a male dominated 3 : 1. Subsequent matings could increase the proportion of males as YY males mated with normal XX females, which would be potentially disastrous for already stressed populations.

Fishes as Environmental Indicators

Some outdoor and environmental enthusiasts claim that "The quality of fishing reflects the quality of living." Lakes, rivers, and oceans with abundant, diverse fishes are reliable indicators of a healthy environment for all life forms. Quantifying the condition of aquatic habitats, therefore, becomes a crucial exercise in understanding and predicting potential hazards to human welfare.

Fishes can serve as **indicators** of the condition of aquatic systems, in advance of effects on human health. At one extreme, massive fish kills may indicate high levels of lethal contaminants, or low levels of oxygen. Ideally, less acute warnings are preferable. To this end, several measures have been developed that use quantifiable aspects of fish assemblage structure, health, and behavior as a means of monitoring conditions in aquatic systems. One approach used widely is an **index of biotic integrity** or **IBI** (Karr 1981, 1991; Miller et al. 1988a; Karr & Chu 1999), which combines measurements of species composition, abundance, and trophic relationships for different habitats. An IBI provides a quantitative comparison between the habitat in question and "unimpaired" reference systems to assess relative degrees of disturbance. The IBI bases its comparisons on a number of traits that generally characterize disturbed systems, such as an increase in number of introduced species, replacement of specialist species with generalist species, decline in the number of sensitive species, impairment of reproduction, change in age structure of populations away from older age classes, and an increase in disease and anatomical anomalies. The IBI was originally developed for midwestern US streams, but has been applied successfully in a variety of systems (Hughes & Noss 1992; Simon & Lyons 1995).

Environmental contamination is more conventionally investigated by assaying water and sediments for known toxins, correlating growth abnormalities with sediment contaminants, or by observing the responses of fishes exposed to suspect water (Heath 1987; Gassman et al. 1994). Traditionally, the concentration at which 50% of the animals die (LD_{50}) is considered a critical threshold. Lower levels of contamination can be indicated by behavioral measures, such as elevated breathing rates, coughing, chafing against the bottom, impaired locomotion and schooling, and suppressed activity or hyperactivity. Although relatively rapid, such bioassays primarily measure immediate conditions. The measurement of "body burdens" of bioaccumulated contaminants in fish tissues gives a broader picture, but can vary with season, feeding habits, or metabolic activity.

A more integrated, long-term picture can be obtained by measuring alterations in energetics, metabolism, growth, reproduction, and behavior (the **biomarker** approach of Hellawell (1983) and McCarthy and Shugart (1990); also see Chapter 7, Indicators of Stress). At a biochemical and energetic level, stress is indicated by changes in such attributes as liver enzyme function, occurrence of DNA damage, unusual ratios of intermediate metabolites (ADP : ATP), amounts of or ability to store lipids, and growth and developmental anomalies. Histological markers include parasite loads and damage, tissue necrosis or abnormal growth (particularly pathologies of the gills and liver), and both elevation and suppression of immune responses. At the population level, reproductive output can be monitored, whereas species richness, presence/absence of sensitive species, and indices such as the IBI indicate assemblage and community-level effects. These measures are useful for monitoring water quality as it directly affects fishes, but also because fishes are effective sentinels against human health problems (Adams 1990; McCarthy & Shugart 1990). Many of the responses listed in Table 22.2 can be considered biomarkers.

Commercial Exploitation

Direct exploitation of fishes by humans is an obvious cause of fish population declines. However, humans are just one of many predators on most smaller fishes, and species or populations subject to predation generally possess compensatory mechanisms for sustaining predation losses (see Chapter 20, Population Dynamics and Regulation). Predation by humans, however, has extraordinary characteristics. Most "natural" predators focus their activities on young individuals, which tend to be the most abundant cohorts within a population, or on sick individuals with little reproductive potential, or on old individuals that have already reproduced. Human fisheries are at best indiscriminate (e.g., trawl and purse seine fisheries); at worst they target larger individuals that have not spawned (e.g., salmonids returning to spawn).

As a result, populations of many fishes have declined as a direct result of fishing pressure (e.g., Bluefin Tuna, many sharks and billfishes, Atlantic Cod and other groundfishes, Atlantic Salmon, Orange Roughy, Patagonian Toothfish, Pacific rockfishes, serranid seabasses). The United States National Marine Fisheries Service (NMFS) estimated that 40% of the marine species important to commercial and recreational fisheries in the U.S. were exploited at unsustainable rates (NMFS 1997, 1999; NRC 1999b) (Fig. 22.16). The overall picture for global marine fisheries has not improved since the late 1990s; the proportion of fish stocks harvested at sustainable levels has decreased from 90% in 1974 to below 70% in 2013. From 2003 to 2012, catches increased in the Northwest Pacific, Western Central Pacific, and Eastern Indian Ocean regions. Catches did not increase in the Northeast Atlantic, however, and catches in the Black Sea and Mediterranean dropped over 30% from 2007 to 2013. In 2013, over 25% of the global marine capture fisheries production was from 10 species. For recent updates on this constantly changing situation, readers should consult the more recently available FAO reports (**http://www.fao.org/publications/en/**).

Demand for fish and other forms of seafood has increased in recent decades, with per capita supply increasing from 18.1 kg in 2009 to 20.1 kg in 2014 (FAO 2016). Levels are considerably greater in higher income countries than in low-income food-deficit countries (LIFDCs). Aquaculture, primarily of freshwater species, has been increasing and met about 44% of this demand in 2014. The remaining production came from wild-capture fisheries, with 81.5 million metric tons marine and 11.9 from inland fisheries.

It is important to note that FAO reports on global fisheries are based on data submitted by each country. These data typically underestimate harvests, do not account for bycatch, frequently omit recreational and artisanal catch data, and cannot account for illegal or unreported harvesting. Study of global catches and additional data from 1956 to 2010 concluded that actual catch was as much as 53% above what was reported (Pauly & Zeller 2016). In addition, the global peak harvest reported at 86 million metric tons in 1996 was likely closer to 130 million metric tons, and catches since that time have declined more dramatically than is indicated by FAO values. In other words, the situation is likely considerably worse than the official data indicate.

There is a strong need to improve coordination and collaboration in regulating fisheries and aquaculture globally in order to maintain healthy ecosystems and provide for human nutritional needs in the future. Initiatives include the 2030 Agenda for Sustainable Development, the Sustainable Development Goals, the Paris Agreement of the UN Framework Convention on Climate Change, and the FAO's Blue Growth Initiative.

Overfishing In some well-documented examples, overfishing, often in combination with climate change (Horn & Stephens 2006), has produced dramatic crashes in seemingly inexhaustible stocks. The clupeoid fisheries of California and South America offer an interesting, interwoven example. The

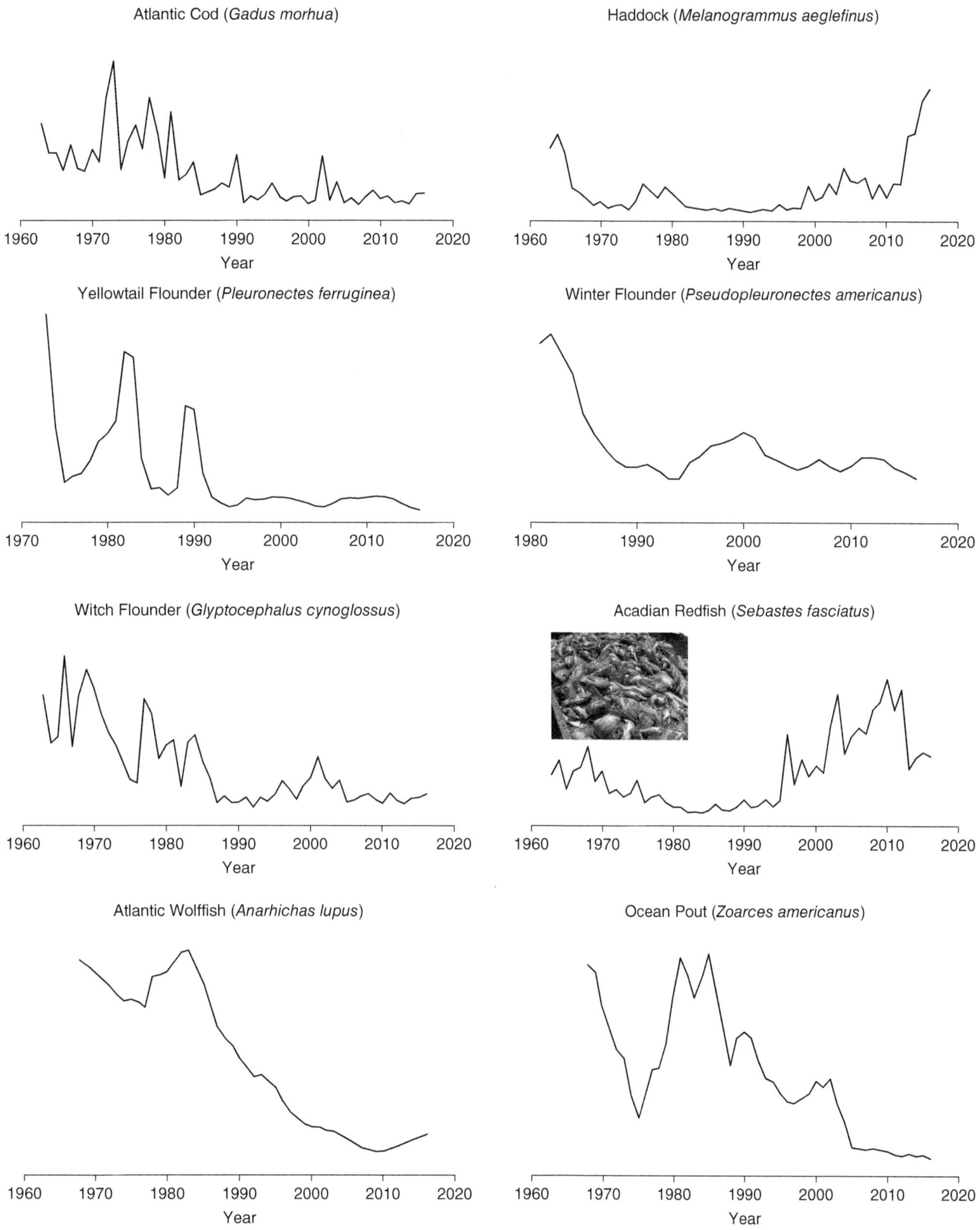

FIGURE 22.16 Trends in various indicators of abundances of Northeast Atlantic groundfish. Vertical axes represent indicators of abundance, but vary among species (some are estimated biomass, others are number of fish or weight sampled per tow). Most species show continuation of decades-long declines, although Haddock showed increases from 2010–2016 and Acadian Redfish increased for about a decade but declined sharply after about 2012. Figures are adapted from Northeast Fisheries Science Center. 2017. Operational Assessment of 19 Northeast Groundfish Stocks, Updated Through 2016. US Department of Commerce, Northeast Fisheries Science Center, Ref Doc. 17-17; 259 p. Available from: National Marine Fisheries Service, 166 Water Street, Woods Hole, MA 02543-1026. Photo of Acadian Redfish harvest is from FishWatch / Public Domain.

history of the California sardine fishery "is a classic case of the rise and fall of a fishery dependent on a pelagic species, of overcapitalization of an industry, and of too many fishing boats using new technologies to harvest a fragile, if not dwindling, resource" (Ueber & MacCall 1990, p. 17).

The Pacific Sardine (*Sardinops sagax caeruleus*) is a 10–15 cm long, schooling, epipelagic clupeid that occurs from northern Mexico to the Bering Sea. The fish were typically captured by purse seiners and canned for human consumption. The fishery off California dates to the late 1800s. By 1925, it was the largest fishery in California, with landings of about 175,000 tons. Waste from the canning process was "reduced" into poultry food and fertilizer. The value of reduced sardines soon surpassed that of the canned product and whole fish were then reduced. Floating reduction plants, anchored outside the 3-mile limit to bypass regulative legislation, became common. Catches climbed steadily to a maximum of 790,000 tons in 1937. For the next 10 years, catches averaged 600,000 tons per year, despite fishery calculations that the stock could only sustain a harvest of 250,000 tons annually. Catches began a steady decline, averaging 230,000 tons from 1946 to 1952, then 55,000 tons from 1953 to 1962, and only 24,000 tons from 1963 to 1968. Commercial fishing for Pacific Sardine ended in 1968, but reopened in the 1980s and 1990s as the population rebounded. Renewed overfishing led to another population decline, and the US fishery closed again in 2015. The population of this ecologically and economically valuable species remained very low for at least several years (Kuriyama et al. 2020).

The collapse of the California sardine fishery was in part responsible for the later development, overexploitation, and eventual collapse of similar fisheries in South America and Africa, as well as of king crabs in Alaska. Boats, gear, and processing equipment were sold at below cost, or costs were subsidized by international agencies. With the influx of former sardine boats and personnel, Alaska king crab landings rose from 11.3 tons in 1960 to 81.7 tons in 1980, only to crash to 15.8 tons 2 years later, despite continued activity of the imported boats (Wooster 1990). The exact causes of the decline are debated, but a likely explanation is that overexploited breeding stocks and unfavorable climatic conditions combined to result in poor recruitment of young crabs, demise of the fishery, and lost jobs for most people associated with the industry.

A similar scenario occurred for the Peruvian fishery for Anchoveta (*Engraulis ringens*). The fishery became established in the 1950s, when fish were primarily used for human consumption. After 1953, reduction plants were built and boats were added to the fleet, many from the former California sardine fishery. By 1969, Peru caught more tonnage of fish than any other nation, with Anchoveta accounting for up to 98% of the catch. The exploitation of Anchoveta was uncontrolled: in 1970, 12.4 million metric tons (MMT) were harvested, about 5 MMT above the calculated maximum sustainable yield. The fishery collapsed soon after, falling below 1 MMT in the mid-1970s. The collapse was again probably caused by a combination of overfished stocks and unfavorable climatic factors, including depressed upwellings associated with the El Niño–Southern Oscillation events of 1972–1973 (Caviedes & Fik 1990).

Neither the Pacific Sardine nor the Peruvian Anchovetta have been driven close to extinction, although the term **commercial extinction** is applied to once-abundant fishes that no longer support significant fisheries. Uncontrolled exploitation of marine species, particularly those dependent on stressed estuarine systems, can lead to even more serious declines in a species' abundance. The Giant Totoaba, *Totoaba macdonaldi* (Sciaenidae), is endemic to the upper Gulf of California and is the largest member of its widespread family, reaching 2 m in length and weighing over 100 kg. Its numbers have been drastically reduced as a result of overfishing on the spawning grounds, dewatering of the Colorado River estuary where it spawns, and bycatch of juveniles by shrimp boats (Fig. 22.17). At one time, it ranked as the most important commercial fish species in the Gulf of California, sought chiefly for its large gas bladder, which was dried and made into soup (the remainder of the body was often discarded). Spawning fish were so abundant that they were speared from small boats. The fishery peaked in 1942 and has declined steadily since (Ono et al. 1983; Cisneros-Mata et al. 1995, 1997). The Totoaba was declared an Endangered US species in 1979, has Critically Endangered status with IUCN, and is listed in Appendix I of CITES. Ongoing monitoring of Totoaba indicates that since the population's collapse from overfishing overall size structure and distribution have not changed but poaching continues and harvesting methods will likely prevent population recovery (Valenzuela-Quiñonez et al. 2015). The clearest message from these and similar examples, including another very large sciaenid fished off China (Sadovy & Cheung 2003), is that maximizing short-term profits and ignoring biological parameters have long-term, dire ecological and socioeconomic consequences (Glantz & Feingold 1990).

Overfishing creates problems besides reduced opportunities for human exploitation. Genotypic and phenotypic alterations occur commonly among heavily exploited fishes. Overfishing can create **bottlenecks** in the breeding biology of a species when populations reach critically small numbers, thereby reducing the genetic diversity of the species. For example, the fishery for Orange Roughy (*Hoplostethus atlanticus*, Trachichthyidae) in New Zealand began in the early 1980s, and within 6 years biomass of the stocks was reduced by 70% and there were significant reductions in genetic diversity of the three monitored stocks (Smith et al. 1991). The danger of reduced genetic diversity is that remaining individuals produce offspring that possess only a limited subset of the original genetic diversity of the species. Genetic adaptation to local conditions does not guarantee tolerance of new or altered environments. Altered conditions are increasingly likely due to human-caused climatic or chemical changes, such as might occur from global warming or ozone

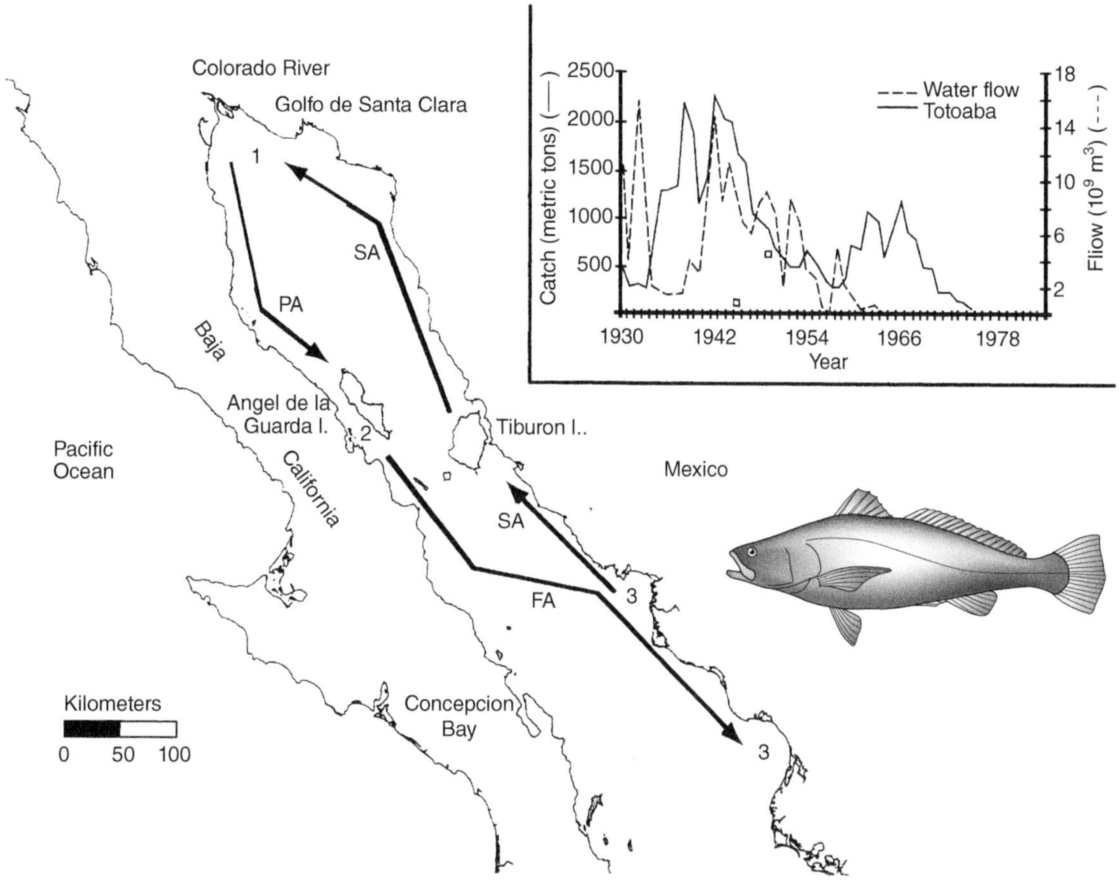

FIGURE 22.17 The Gulf of California and a reconstruction of the presumed seasonal migration route of the endangered Totoaba, the world's largest sciaenid: SA, pre-spawning adults; PA, post-spawning adults; FA, adults during fall migration. Numbers indicate life history zones: 1, spring spawning zone and nursery ground of juveniles; 2, summer feeding zone; 3, fall feeding zone. Zone 1 is now largely a biosphere reserve. Inset shows the relationship between water delivery from the Colorado River and Totoaba population size as calculated from commercial catches. From Cisneros-Mata et al. (1995), used with permission of Blackwell Science; inset Totoaba drawing from Universidad Autonoma de Baja California.

depletion. Breeding bottlenecks are one step short of species extinction.

Given enough time, animals can and will adjust their life history characteristics in response to strong predation. Life history theory predicts that individuals in populations exposed to high levels of adult mortality will respond by reproducing at smaller average sizes and ages, will shift from multiple to single reproductive seasons (from iteroparity to semelparity), and will have shorter life spans. These kinds of changes have been observed in several exploited species, including Atlantic Cod (*Gadus morhua*), Haddock (*Melanogrammus aeglefinus*), other gadids, Gag Grouper (*Mycteroperca microlepis*), Vermilion Snapper (*Rhomboplites aurorubens*), Atlantic Mackerel (*Scomber scombrus*), and Pacific Halibut (*Hippoglossus stenolepis*) (Upton 1992; O'Brien et al. 1993). Such shifts may reflect adjustments in the phenotype of remaining individuals or selection for genotypically determined differences in life history traits, or both. The alarming fact is that most of our marine fish stocks are overutilized and that the observed shifts in life history characteristics create fish populations that are less useful to humans. Increasing evidence indicates that fish populations undergo evolutionary change as a result of overexploitation.

Fishing as an Evolutionary Force The fisheries literature documents many examples of exploited fish populations shifting their growth rates and ages and sizes of reproduction, growing rapidly and maturing at smaller sizes and younger ages than unexploited populations. Some of what is observed can be explained by reductions in abundance causing increased body growth in the remaining individuals (perhaps due to reduced intraspecific competition for food), and the ensuing faster growth rates resulting in maturation at smaller sizes and younger ages (see Trippel 1995). What has been observed is also in part because many fisheries target larger individuals in a population and leave behind smaller individuals.

There is also strong and growing evidence, however, of evolutionary impacts of exploitation on fish life histories, a concept that was discounted by many before the early 1980s (Policansky 1993a; see also Miller 1957). Hence, evolution of populations in response to fishing pressure was not incorporated into management models and plans (Policansky 1993b), and fisheries managers showed "a continuing reluctance . . . to take seriously the threat of genetic change brought about through fishing" (Law & Stokes 2005, p. 241).

Fishing, however, is a tremendous source of mortality, and most fisheries target prey non-randomly and are size and locale selective. Such constant, strong, "directional" selection often leads to rapid, evolved counteradaptations. Many life history traits that influence fisheries yields – such as growth rate, fecundity, and age and size at maturation – are under genetic control with relatively high heritability (Policansky 1993a, 1993b; Law 2000; Palumbi 2001). Also, life history traits display sufficient variation to be changed by evolution (Trippel 1995). There has also been sufficient time for evolution to occur, in terms of number of generations needed for significant genetic change. Across a range of taxa, evolution has been shown in less than 10 generations, sometimes in as few as two or three (Falconer & Mackay 1996). Field studies of salmonids have shown detectable divergence among populations in eight to 13 generations (Hendry et al. 2000; Haugen & Vollestad 2001; Hendry 2001).

Given this set of criteria that are often used to determine whether traits can be expected to change in response to selection forces, it is not surprising to find many examples of evolved, adaptive change in exploited fishes, including findings from several decades ago. Pacific salmon species have been subjected to commercial and recreational trolling, gill netting, and seining. Population characteristics have changed over time in response to exploitation patterns, including an overall decline in average size in all species studied (Ricker 1981). Over a 60-year period, Chinook Salmon matured on average *2 years earlier and at half the original size.* In 1950, when fishers for Coho and Pink Salmon began to be paid according to size rather than number of fish caught, larger mesh gill nets were employed to catch larger fish, accelerating the shift in stocks to smaller, younger fish. Ricker (1981) attributed these shifts to cumulative genetic effects of removing fish of larger than average size. A gill net fishery for Lake Whitefish (*Coregonus clupeaformis*) in Lesser Slave Lake, Alberta provides another historical example (Handford et al. 1977). Gill nets removed large, heavy, fast-growing fish, leading to declines in growth rate and condition factor and an increase in mean age, but little change in mean length at age. Condition factor declined dramatically with time in all age groups, to the extent that fish of a given age and length in the 1970s weighed far less than fish of similar age or length weighed in the 1940s (Figure 22.18). Similar results have been shown in gill net fisheries for Grayling, *Thymallus thymallus*, and Sockeye Salmon (Hamon et al. 2000; Haugen & Vollestad 2001).

Additional evidence is provided by laboratory tests imposing the kind of selection experienced by exploited fishes. Conover and Munch (2002) simulated size-selective fishing by rearing fast- and slow-growing Atlantic Silversides, *Menidia menidia*. After only four generations of directional selection for growth rate, groups from which fish with the fastest growth rates were removed (large-harvested fish) and groups from which slow-growing individuals were removed (small-harvested fish) reversed their growth rate characteristics. The previous fast growers (large-harvested fish) had mean weights nearly half those of the small-harvested lineage that previously possessed slow growth characteristics. The growth differences had a demonstrated genetic basis. Egg size and biomass yield also differed, indicating that continued harvest of the largest members of a stock reduced biomass and egg production. "Selection on adult size caused the evolution of a suite of traits likely to influence population growth rate and productivity" (Conover & Munch 2002, p. 95). The traits evolved are largely the opposite of what the fishing industry and society would prefer.

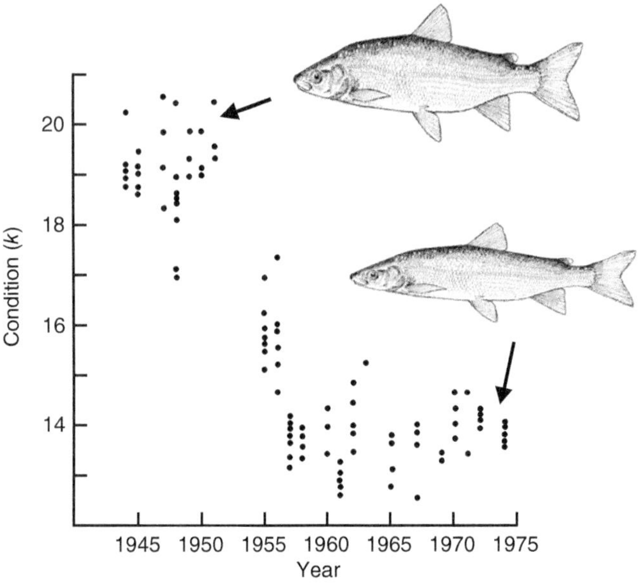

FIGURE 22.18 Evolution of body shape in exploited Lake Whitefish in Lesser Slave Lake. Gill netting selectively removes thicker fish, and as a result fish became thinner over time. Data plotted are condition factors, k ($k = 10^5$ weight/length^3), for male whitefish, 1940–75; females showed similar patterns but were not used because of weight changes caused by egg-bearing. Adapted from G. Helfman (2007), after Handford et al. (1977).

At least 40 generations of North Atlantic cod, herring, plaice, and sole have experienced intensive fishing, sufficient to alter genetic make-up (Policansky 1993a). Anything that causes mortality is a strong selection force, and fishing mortality often exceeds natural mortality by a factor of two or three in many if not most heavily exploited species (Stokes et al. 1993; Stokes & Law 2000). Annual, non-fishing, mortality rates in many post-larval and post-recruitment finfish species run at less than 10%, whereas mortality rates targeted to achieve maximum sustainable yields are around 50%, and actual fishing mortality is often between 70% and 90%. In the case of depleted Atlantic Cod, clear evidence of progressive, fisheries-induced evolution of maturation patterns existed and could have foretold imminent collapse (Olsen et al. 2004). Intense commercial harvests of demersal fishes (such as cod, haddock, and whiting) in the Firth of Clyde, Scotland, during the 1960s and 1970s seems to have driven these species to mature at earlier ages and smaller size – and although stock biomass has rebounded, fishes are much smaller than in the past (Hunter et al. 2015).

Given the history and intensity of most commercial fisheries, fishing can be considered a long experiment with "more than enough time for selection to produce substantial genetic changes on almost every quantitative character that has been examined" (Policansky 1993a, p. 6).

Additional examples of fisheries-induced evolution include other cods, salmon, some sharks, and also freshwater species such as Walleye, Yellow Perch, and Arctic Char. Studies of Largemouth Bass suggest that some aspects of behavior are heritable, and therefore removal of aggressive individuals by angling may result in future generations of more cautious, less aggressive fish. These examples and others are addressed in Chapter 20 – Effects of fishing on life history traits, and implications for recovery.

Palumbi (2001) argued that humans have become the world's greatest evolutionary force, having exerted strong natural selection in such areas as disease-resistant viruses and bacteria, pesticide-resistant insects and plants, artificial selection via domestication (hatchery salmon), and altered characteristics in introduced species, many of which are detrimental to human welfare. We can add many wild populations of fishes to the list of species whose evolutionary trajectory has been altered by intensive human exploitation.

Free Enterprise is not the Answer Reliance on market factors to protect declining stocks has proven illusory. A decrease in catch despite increased effort does not necessarily discourage exploitation, especially for the so-called **market force-free species**, i.e., fish that are too valuable to not catch. For example, western Atlantic stocks of Bluefin Tuna decreased by 90% between 1972 and 1992, from 225,000 fish to 22,000 fish, and yet intensive fishing continued because the demand and price continued to rise. In 2013, a Japanese sushi restaurant chain paid $1.76 million for a single 222 kg Bluefin Tuna. This was mainly a publicity stunt to get the first auctioned tuna of the season, and that same owner paid a more conservative $636,000 for the first tuna (212 kg) of 2017. Although Bluefin tuna does not sell for these prices on a regular basis, it does bring in a strong enough price to keep people fishing for them despite their depleted status and a highly regulated fishery. This sort of profitability threatens other large and valuable fish species, and some tropical reef species desirable for aquaria and the live fish restaurant trade (Sadovy & Vincent 2002; Helfman 2007). We cannot, therefore, rely on the power of free markets to protect populations of valuable species.

Bycatch The capture of non-target species (bycatch) in trawl and longline fisheries has long been a concern to the conservation of marine fishes (Murray et al. 1992; Perra 1992; Safina 2001a, 2002). Few fisheries employ gear that can catch one species to the exclusion of all others. For example, dolphins, whales, turtles, and pinnipeds are frequently captured in gill nets or in purse seine nets set for tunas and billfishes, and seabirds and turtles are caught in longline sets. Because bycatch often goes unreported, it is difficult to accurately estimate its extent. Different assessments come to different conclusions, but data in the early and mid-1990s indicated that **discarded biomass** amounted to 25–30% of nominal catch, or about 30 MMT (Alverson et al. 1994; FAO 1995; Alverson 1997). By some estimates, bycatch has contributed to declines among 42–48% of marine and diadromous species considered imperiled by the United States and IUCN (Kappel 2005).

The bycatch problem is particularly acute when trawl nets with small mesh sizes are dragged along the bottom of the ocean in pursuit of groundfish or shrimp. Frequently, the **incidental** or **bycatch** (or **by-kill**) of fishes exceeds the catch of the targeted species. Although varying on a seasonal and regional basis, average fish : shrimp weight ratios of 1 : 1 to 3 : 1 have been reported for southeastern US shrimp fisheries. These numbers can run as high as 130 : 1 (= 130 kg of "scrap" fish for each kilogram of shrimp). In the late 1980s, 105 species of finfishes were captured by shrimp trawlers in the southeastern U.S. On a species basis, 5 billion Atlantic Croaker (*Micropogonias undulatus*, Sciaenidae), 19 million Red Snapper (*Lutjanus campechanus*, Lutjanidae), and 3 million Spanish Mackerel (*Scomberomorus maculatus*, Scombridae) were among nearly 10 billion individuals and 180 million kg of incidental fishes killed by shrimp trawlers in the Gulf of Mexico in 1989 (Nichols et al. 1990).

Because of the small mesh size of the shrimp trawl nets, most of the fishes captured are (i) juveniles, (ii) smaller than legal size limits, or (iii) undesirable small species. Even larger mesh sizes do not prevent bycatch because once the net begins to fill with fish or shrimp, small individuals are still trapped. These incidental captures are often unmarketable and are therefore shoveled back over the side of the vessel dead or dying.

The bycatch problem is complicated economically, ecologically, and sociologically. Bycatch is a liability to shrimp fishers, clogging the nets, damaging shrimp, and increasing fuel costs because of increased drag on the vessel. Sorting the catch requires time, risking spoilage of harvested shrimp and reduced time for fishing. Ecologically, high mortality rates among juvenile fishes could contribute to population declines of recreational and commercial species. Evidence to this effect exists for Gulf of Mexico Red Snapper and Atlantic Coast Weakfish (*Cynoscion regalis*, Sciaenidae). Because the nearshore areas where shrimp concentrate are also important nursery grounds for many fish species, shrimp trawling could have a profound impact on stock size (Miller et al. 1990).

Alternatively, bycatch is returned to the ecosystem and consumed by predators, detritivores, and decomposers, which could have a positive effect on sportfish, seabird, crab, and even shrimp populations. Available evidence indicates that 40–60% of the 30 MMT of catch discarded annually by commercial fishing vessels, and even more of **noncatch waste** (organisms killed but never brought to the surface) becomes available to midwater and benthic scavengers, thus transferring material into the benthic food web and making energy available to foragers that is normally tied up in benthic, suprabenthic, midwater, and pelagic species (Britton & Morton 1994; Groenewold & Fonds 2000).

Conservation organizations have decried the obvious and wanton waste associated with bycatch. Public concern over high mortality rates of endangered marine turtles captured in shrimp trawls led to the development of **turtle exclusion devices** (**TEDs**) in the 1980s. TEDs were incorporated into the shrimp net design with the purpose of directing turtles out of nets without unacceptably reducing shrimp catches (Broadhurst 2000). Marine engineers and fishers also developed shrimp net designs that incorporate **bycatch reduction devices** (**BRDs**), taking advantage of behavioral differences between shrimp and fish or between different fishes to separate species (Engas et al. 1998). Other suggested solutions include prohibiting shrimping during seasons when bycatch is relatively high or where vulnerable life history stages of nontargeted species are concentrated. Unfortunately, despite long-standing recognition of the importance of this issue, bycatch has remained a significant threat to global fisheries (FAO 2016), in part because there is limited monitoring of bycatch and there is also little incentive to commercial fishers to reduce bycatch.

Bycatch is a component of widespread overfishing and contributes to a syndrome known as **fishing down of food webs** (Fig. 22.19), whereby we eliminate apex predators and large species while transforming the ocean into a simplified system increasingly dominated by lower trophic level organisms such as microbes, jellyfish, benthic invertebrates, plankton, and planktivores (Pauly et al. 1998; Jackson et al. 2001). The strongest evidence for the fishing down phenomenon exists in global catch statistics that show alarming shifts in species composition from high value demersal species to lower value pelagic species, and the fact that despite increased effort and improved technology, global harvests have not expanded accordingly, and in many areas have diminished. Analysis of FAO landing data from 1950 to 2010 suggests that fishing down the food web is widespread in global marine fisheries, although in some regions **fishing through** the food web was indicated by increased catches at multiple trophic levels (Ding et al. 2016). Both of these phenomena can lead to collapses in affected populations and have broad impacts on marine food webs.

Mean trophic level (MTL) of commercially harvested marine resources has been used by many as an indicator of the stability of marine food webs. Increasing MTL values may be interpreted by some as indicating recovering populations of high trophic level, predatory fishes whereas declining value may indicate collapse of top predators and expansion of lower trophic level species. This index should be carefully interpreted, however. Commercial fishers may change the focus of their efforts based on market forces, and landings may also reflect changes in fishing technology, thus affecting indexes based on commercial fishery landings (Branch et al. 2010). In addition, although increasing MTL values could indicate recovering populations of predators, such a change in MTL could also be the result of depletion of populations at lower levels due to overfishing (Branch et al. 2010, Ding et al. 2017). MTL values may also be stable or increasing when catches across trophic levels

FIGURE 22.19 Fishing down food webs. Since about 1950, most of the world's marine and freshwater fisheries have been taking species at progressively lower trophic levels due to depletion of larger species at higher trophic levels (see Pauly et al. 1998). Hans Hillewaert / Wikimedia Commons / CC BY-SA 4.0.

are increasing, which could bring about collapse of harvested populations. In addition, historical comparisons of MTL values can be affected by changes of trophic level value, as indicated in FishBase, of harvested species (see Branch et al. 2010).

Other indexes developed to evaluate the impact of fishing include mean trophic index (MTI, Branch et al. 2010) and Fishing in Balance index (FiB, Pauly et al. 2000a, b) which take additional factors into consideration. Use of these measures to assess fisheries landings in the Persian Gulf from 2002–2011 also suggested fishing down of the marine resource community in this region (Razzaghi et al. 2017).

Is Aquaculture an Answer? As many natural fish populations have decreased and consumer demand for fish increases, many have turned to various forms of aquaculture. Although this does provide more fish for markets, it does not reduce, and in some ways can increase, pressure on wild-caught fishes. In addition, raising large numbers of fish in confined spaces presents a wide variety of ecological and economic challenges. The last few decades have seen dramatic increases in various types of fish farming. Salmon farming in particular has received a great deal of attention, and can serve as an example.

For decades, raising Atlantic Salmon in nearshore pens in bays and inlets has been a source for Atlantic Salmon in markets and restaurants around the world (Fig. 22.20). There is no doubt that this business has been economically very successful and has helped fill the demand for salmon, although it has also helped mask the fact that natural populations of Atlantic Salmon have been so dramatically reduced throughout its range.

Although some Atlantic Salmon farming operations take place in the species' native range (the North Atlantic, including New England, eastern Canada, Great Britain, and Norway), many are located elsewhere. For example, Chile has been among the largest producers of farm-raised Atlantic Salmon, and large Atlantic Salmon farming operations are located along Canada's Pacific coast where escaped fish compete with the five species of native Pacific salmon for food and spawning areas. In late August of 2017, failure of a sea pen at a commercial aquaculture facility released thousands of Atlantic salmon into the Pacific Ocean between Washington and Vancouver Island. The company was not even sure of the number of escaped salmon and provided an estimate of 4000 to 185,700; the official estimate by Washington state agencies put the number closer to 260,000 (**https://www.seattletimes.com/seattle-news/fish-farm-caused-atlantic-salmon-spill-state-says-then-tried-to-hide-how-bad-it-was/**).

Even in areas that support native Atlantic Salmon populations, escapees from farming operations present multiple threats, including competition, the spread of parasites and disease, and hybridizing with native fish thus introducing genes from hatchery-raised fish into natural populations genetically suited to local conditions. Escaped farm fish accounted for 22–100% of the runs in various rivers in Maine where genetic

(A)

(B)

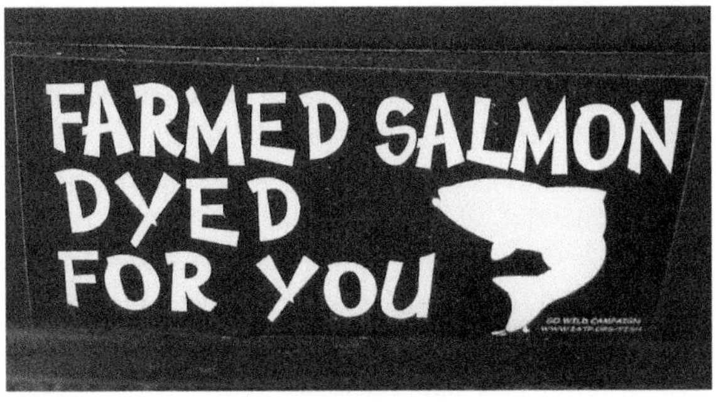

FIGURE 22.20 (A) Atlantic salmon net pens in a fjord in Norway. In 2012, Norway alone produced 1,240,000 tons of salmonids, mostly Atlantic Salmon, worth an estimated 3.6 billion U.S. dollars. (B) Sentiment expressed on a bumper sticker, Burlington, Washington. G. Helfman (Author).

tests confirmed hybridization between endangered wild and superabundant farm-raised fish, the latter largely derived from European genetic strains (Colligan et al. 1999; NRC 2004a). Genetic introgression between wild Atlantic Salmon and escapees from aquaculture operations in Norwegian rivers ranged from 0 to 47% (Glover et al. 2013, Karlsson et al. 2016). This hybridization could decrease fitness in populations that had been genetically well suited to the specific conditions of the home spawning stream.

Disease and parasites are a concern as they can spread more easily in crowded, confined areas. Sea lice (parasitic copepods of the family Caligidae) are often abundant in salmon farming facilities and compromise the growth and health of the salmon. Lumpfish (*Cyclopterus lumpus*) and several species of wrasses (Labridae) have been effective at removing sea lice from salmon. Ballan Wrasse (*Labrus bergylta*) and Lumpfish have been used in salmon farming operations to reduce the impacts of sea lice, but in recent years natural populations of both of these cleaner fish species have been affected by harvesting for use in salmon farms. To help provide enough for the salmon farming industry, both species were being raised in aquaculture operations (Brooker et al. 2018). They must be raised under carefully controlled conditions to ensure biosecurity, and properly acclimated to the conditions. This is an ongoing, active area of research.

Susceptibility to disease and parasites can be enhanced by suppressed immune systems commonly seen among stressed fish. Fish that escape may then infect natural fish populations of the area, or the parasites and pathogens themselves may spread without infected fish escaping. Controlling parasites and disease requires the extensive use of pesticides, antifoulants, algicides, and antibiotics, which may leach from the pens or become concentrated in the flesh of the fish (Herwig et al. 1997; Haya et al. 2001). The use of antibiotics to curtail disease has also led to concerns over the production of antibiotic-resistant strains of bacteria near salmon pens (McVicar 1997).

Geography provided a somewhat natural experiment regarding the potential impacts of salmon farming on wild salmon migrating through the area. Most of the juvenile Sockeye Salmon (*Oncorhynchus nerka*) leaving the Fraser River in Vancouver, British Columbia turn northward and swim east of Vancouver Island on their way to the open sea. This takes them through rather narrow straits and past a heavy concentration of salmon farming operations, and the associated parasites and viruses. Sea lice are abundant, and are further elevated by the effluents of a farmed-salmon processing facility. These ectoparasites reduce survival of small salmonids, and juveniles passing by the salmon farms have much higher levels of sea lice infestation than fish traveling the other route. In addition, the fish traveling past the farming operations are exposed to elevated levels of potentially infectious viruses. However, some of the out-migrating sockeye juveniles go south of Vancouver Island and into the open ocean without passing close to salmon farms. Salmon farming in British Columbia began in the mid-1980s and rapidly expanded in the early 1990s. From 1992 to 2010, the Fraser River Sockeye Salmon population overall declined dramatically, but the genetic subpopulation that does not encounter the dense salmon farms increased (Morton & Routledge 2016). Juvenile sockeye salmon with high sea lice infestation grow slower than uninfected conspecifics (Godwin et al. 2017), and apparently, very few of the smolts leaving the Fraser River and migrating past the salmon farms were surviving to adulthood and returning to the river.

Pollution is also a significant concern with salmon farming. Farmed fish are kept at artificially high densities, which leads to nutrient pollution around farms from uneaten feed and excreted waste. Salmon farms along the British Columbia coast in the 1990s discharged as much organic sewage as 500,000 people (Ellis & Associates 1996; Naylor et al. 2000). The claimed dietary benefits of eating farmed salmon are compromised by contaminants such as organochlorines that have been found in higher levels than in wild salmon (Jacobs et al. 2002; Hites et al. 2004), although contaminant levels vary by species and the health benefits of eating farmed salmon may outweigh the risks (Mozaffarian & Rimm 2006; Ikonomou et al. 2007).

Another concern regarding farmed salmon is that of ecological efficiency. Salmon are predators, and raising salmon requires feeding them nutritionally complete diets based on fishmeal and fish oil which comes from large-scale commercial harvesting of other fishes. By some estimates it takes over 3 kg of wild fish converted to fish meal to produce 1 kg of salmon (Naylor et al. 2000). In 2003, more than 50% of the world's fish oil production was used to feed farmed salmon (FAO 2008).

Farmed salmon suffers from a number of aesthetic liabilities. Salmon farms are typically located nearshore, often in bays and other protected areas. These "ocean view" locales are also prized for real estate development, demanding high prices. Homeowners investing in such parcels are unlikely to want a commercial salmon farm in their viewscape (Stead & Laird 2002). As a food, the softer, fattier texture of farm-raised salmon is quite different than the leaner, firmer meat of wild-caught salmon. Also affecting consumer acceptability, farmed salmon are fed diets containing carotenoid pigments such as astaxanthin and canthaxanthin to give their flesh the "salmon pink" color that consumers expect (Stead & Laird 2002; see Fig. 22.20B).

Salmon farming also has a broader economic impact because commercial fishers of wild salmon stocks, such as the comparatively well-managed salmon fisheries of Alaska, have difficulty competing with lower-cost farmed salmon. Some consumers, however, are learning the differences and prefer to purchase wild-caught fish despite the higher cost.

There is no denying the economic success of the farm-raised Atlantic Salmon industry. Production increased from 13,265 metric tons in 1982 to over 2 million metric tons in 2013. Norway and Chile have dominated the international market, but other producers include Scotland, Canada, the U.S., and Australia.

The salmon farming industry has been and remains aware of the controversies and is actively engaged in minimizing or correcting identified and potential problems, including finding dietary substitutes to reduce reliance on fish meal. But salmon are aquatic predators that have evolved to grow well on aquatic prey, so getting strong growth using other dietary substitutes has been difficult.

Others have worked hard on developing genetically modified salmon for raising in confined facilities. It took several years for the Aqua Bounty Farms company to gain approval to produce genetically modified salmon. These fish are reported to grow much faster and use less food than non-GMO salmon. They are also genetically designed to be sterile females to allay concerns that any escaped fish might genetically affect local native populations, and initial plans are to raise them in land-based confined facilities so that escape is not possible. Consumer reaction to the availability of GMO salmon will ultimately determine the success of this endeavor.

Raising other predatory fishes such as Southern Pacific Bluefin Tuna and Cobia in large pens located well offshore helps to disperse some of the negative environmental impacts seen in near-shore salmon operations. But the large amounts of fish waste and other organic matter still have to go somewhere, and could impact normally low-nutrient open water regions of the oceans. But these fishes, like salmon, are predators and still need high protein diets, which have typically been based on commercially harvested fishes. In this way, raising predatory fishes for commercial markets puts additional pressure on other fishery resources to provide the fish meal needed to feed the predators. Hence Carl Safina's statement that "Aquaculture will do no more to save wild fish than poultry farms do to save wild birds." Approximately 18% of the 178.5 million metric tons of world seafood production in 2018 was used for fishmeal and fish oil (FAO 2020), much of which is used for feeding in aquaculture.

Barramundi (*Lates calcarifer*, Latidae) have also become a popular farmed species, and can be grown in open pens or confined, land-based facilities. Although naturally predators, they reportedly also can grow on mixed diets thereby reducing the need for fish-based feed. Omnivorous fishes don't require as much animal protein in their diets and may provide more environmentally sustainable options. For many years, farm-raised catfish, carp, and tilapia have been fed diets less dependent on fish meal. In some cases, the rearing operations are contained within ponds or large tanks, thereby minimizing negative environmental impacts and greatly reducing the likelihood of escape to surrounding ecosystems.

Aquarium Fishes

A rough estimate of over 2 billion live fish were transported for the aquarium trade in 2007, with exports valued at US $315 million (Monticini 2010). This indicates a dramatic increase from the estimated U.S. $21.5 million in 1976. The total value of the industry is far higher, considering that import values are often double export values, and retail value doubles the price yet again (Wood 2001). In addition, total expenditures for aquarium keeping include the value of tanks, pumps, filtration, lighting, chemicals, plants, foods, transportation, and packaging. This is a huge industry, and although much of the income from the sale of fishes goes to lower-income, tropical countries, non-fish components are produced more widely and much of these profits go to wealthier nations.

The great majority of the fishes traded are freshwater, with about 90% of those raised in captivity (Monticini 2010). Over 50% of the total value of fish traded was from six major exporting countries – Singapore, Malaysia, the Czech Republic, Japan, Israel, and the United States. But export data do not necessarily reflect the origin of the fishes. For example, about 99% of the value of what is exported from Europe, including top European exporters of the Czech Republic and Spain, is reexport of fishes initially obtained from Africa and South America. South America has a relatively low market share, however, due to tight regulations in Brazil and Peru; most of the Amazonian fishes in the international market are produced in culture facilities in Southeast Asia. Florida is also a large producer of freshwater ornamental fishes from all over the world.

In 2007, the number of importing countries was estimated at about 135, over a four-fold increase from the estimated 32 importing countries in 1976 (Monticini 2010). The top 10 importing countries accounted for 78% of the market value. These included the United States, the United Kingdom, Germany, Japan, Singapore, France, the Netherlands, Italy, and Canada. Nearly 80% of ornamental fishes coming into the U.S. are from Asia – mainly Singapore, Thailand, and Indonesia. Freshwater fishes account for about 95% of the volume and 80% of the value. Most ornamental fishes imported to the U.S. come into Miami. Although the United States Law Enforcement Management Information System (LEMIS) had reasonably accurate information regarding import and exports of mammals, birds, reptiles, and amphibians, the great majority of ornamental fishes are simply recorded as marine or freshwater tropical fish. "Data on numbers of fishes involved [in the aquarium trade] are scattered, incomplete, often contradictory, seldom documented, plagued by misidentifications, and generally out-of-date" (Helfman 2007, p. 375).

In 2010, about 75% of all live wildlife imported to the United States were fishes, yet we had very little specific accounting of what species they were or how many of each were brought in (Livengood et al. 2014). Treating fish imports and exports with the same level of attention and record keeping afforded to other imported vertebrates would allow the LEMIS database to be a valuable tool in monitoring the potential impacts of US imports and exports on global ornamental fish resources.

Although the great majority of ornamental fishes traded are freshwater species raised in culture facilities, marine fishes (and invertebrates) are also important components of this large global industry. Yet, despite many efforts over the years, there has been limited success in raising popular marine species in captivity. Moorehead and Zeng (2010) reviewed available literature and made recommendations regarding areas of research needed to make captive culture of more marine species achievable. In summary, we need more basic biological and ecological information regarding the conditions needed for successful reproduction in the wild in order to come as close as possible to replicating these conditions in a culture facility. This basic information includes specific physical, chemical, and dietary needs for growth, maturation, spawning, and success of larvae and juveniles. And these conditions are quite likely to be different for each species.

The challenges associated with maintaining mature adults and providing them with suitable spawning conditions has led to some consideration of post-larval capture and culture (PCC), which focuses on capturing post-larval fishes as they settle and providing them with suitable conditions for feeding and growth in captivity (see Bell et al. 2009). This approach has significant challenges, but if done cautiously it might provide some fishes for the ornamental trade and not have a large impact on natural communities, where most post-larval individuals are not likely to reach maturity. Damage to substrate and habitat could also be reduced with appropriate capture methods such as light traps or shelter traps. However, post-larval fishes are part of the overall community, and without a more thorough understanding of the role that these fishes play, even if they do not survive to adulthood, harvesting post-larval fishes could have unforeseen consequences. The success of PCC will likely depend in part on the willingness of hobbyists to pay a higher price for fish that are brought to market in a more ecologically sustainable manner. Partly for this reason, as well as the challenges of sustainable collection and successful rearing, Bell et al. (2009) do not expect that PCC would have a large impact on the overall trade in ornamental fishes. Until more marine species can be successfully reared in captivity, most marine fishes traded will continue to be captured from natural habitats thereby degrading natural communities.

Marine collecting can be particularly destructive because of the widespread use of poisons, especially cyanide, to "anesthetize" the fishes prior to capture (see Bell et al. 2009). In the 1980s, 80–90% of fish caught in the Philippines may have been collected using cyanide (Rubec 1986, 1988); similar numbers likely characterized other countries (Barber & Pratt 1997; Sadovy & Vincent 2002). These toxins kill non-target reef organisms as well as some of the targeted fishes, and affected fish often retreat to hiding places in reefs, which are then damaged in retrieving the fish. In addition, high overall mortality rates mean that many fishes are killed just to get a few to market. Efforts at reducing cyanide use via training collectors in nondestructive methods have shown some encouraging results; fishes with detectable cyanide residues in the Philippines fell from over 80% in 1993 to 47% in 1996, and 20% in 1998 (Rubec et al. 2000). But the lower cost and higher productivity of using cyanide make it difficult for more environmentally conscious suppliers to compete.

Destruction of fishes by the aquarium trade is a sordid and underpublicized fact. **Mortality rates** of wild-captured fishes, ignoring fishes that die during or incidental to capture, are difficult to establish and vary by locale and collecting method. One estimate is that only 10% of the fishes affected by cyanide were actually targeted (Rubec 1988), and mortality rates of 80% for cyanide-captured fishes are not unusual. For noncyanide-caught reef fishes, estimates from a variety of locales indicate 10–40% mortality during holding prior to export, 5–10% during initial transport, and 5–60% during holding after import. Summed mortalities therefore range between 20% and 80% after capture and before retail sales (Wood 1985; Sadovy 1992; Pyle 1993; Vallejo 1997; Rubec et al. 2000). For wild-caught South American and African freshwater fishes, pre-export mortality has been placed at 50–70%, with as much as 80% additional loss for Cardinal Tetra (*Paracheirodon axelrodi*) and Neon Tetra (*P. innesi*) shipped from South America to the U.S. (Waichman et al. 2001).

Data on **aquarium longevity** are largely anecdotal and subject to unknown biases, but are far from encouraging. Wood (1985), surveying UK hobbyists, reported that 50% of marine fish died within 6 months of purchase, and nearly 70% died within a year. In home aquaria, cyanide-caught fishes may die when fed due to irreversible, progressive liver damage caused by the cyanide. Such **delayed mortality** may occur several weeks after capture and sale. Due to environmental sensitivity or specialized feeding habits, many tropical marine species are "impossible or difficult to keep, even when maintained under ideal conditions by experienced aquarists" (Wood 2001, p. 31). Sadovy and Vincent (2002) estimated that perhaps 40% of frequently traded ornamental marine species were unsuitable for the average aquarist. Hard to keep species include those dependent on live coral and other live organisms for food, such as some butterflyfishes and angelfishes. These are often colorful species and hence desirable, but their capture and sale are unjustifiable given the low likelihood of survival.

A number of other ecological and sociological issues plague the aquarium trade (see Helfman 2007 for details). These include the unsustainable harvesting of live coral and "live rock"; coral death from cyanide and other destructive collecting methods; compromised health of collectors due to cyanide toxicity and unsafe diving practices; destruction of food fishes important to local economies; species introductions due to escapes from holding facilities and release of unwanted pets, which also transmit pathogens (see Fuller et al. 1999; Whitfield et al. 2002; Semmens et al. 2004); and population depletions and biodiversity loss. The latter problem was confirmed by Tissot and Hallacher (2003) who compared population sizes of popular aquarium fishes at locales on the island of Hawaii where collecting occurred or was restricted. They found that seven of the 10 targeted, relatively common species were significantly depleted at collection sites, whereas only two of nine ecologically similar but nontargeted species showed reduced numbers at collection sites (Fig. 22.21). Declines among aquarium species ranged from 38% to 75%. In all likelihood, moderate levels of collecting have minimal impact, especially for abundant species. But the available data shift the burden of proof onto those who maintain that collecting has minimal impact. We know otherwise.

Public aquaria promote important conservation and public education goals. Home aquaria are of unquestionable educational and aesthetic value. However, these values do not justify the ecological problems created by an unregulated industry, including the detrimental effects of introduced species and diseases on native fishes, invertebrates, and plants, and

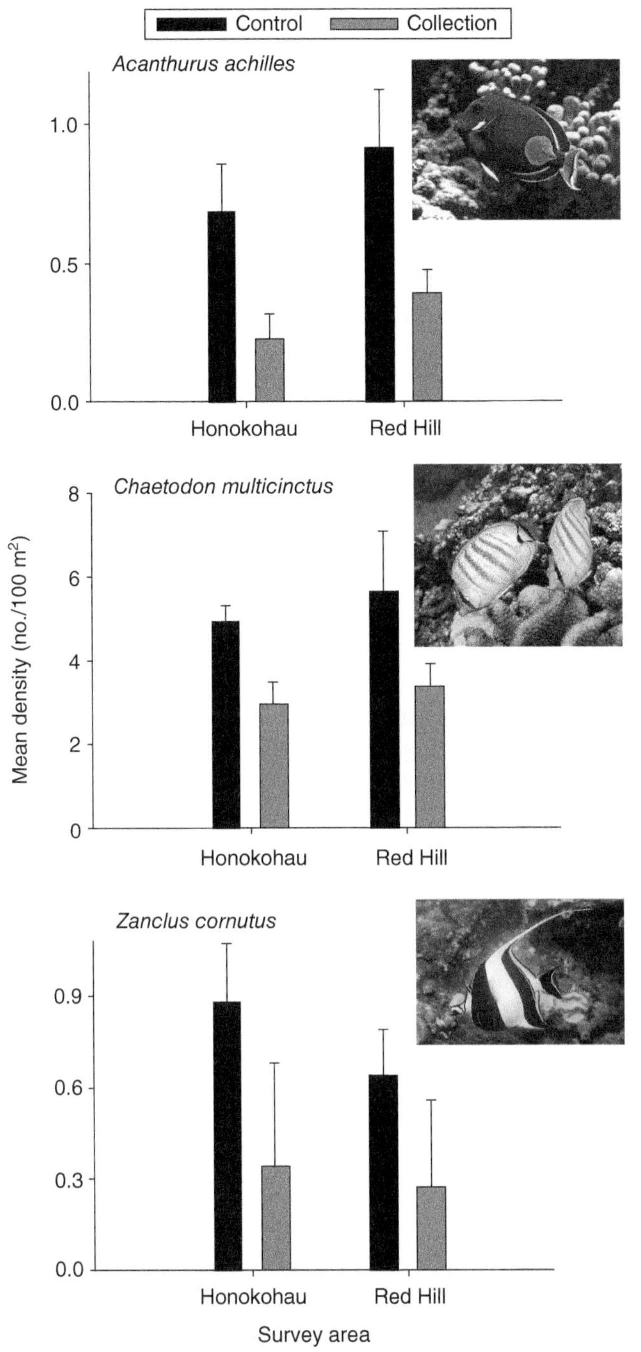

FIGURE 22.21 Aquarium collecting of tropical fishes such as the Achilles Tang (*Acanthurus achilles*), Multiband Butterflyfish (*Chaetodon multicinctus*), and Moorish Idol (*Zanclus cornutus*) resulted in noticeable decreases in abundance at collection locales compared to sites protected from collecting From Tissot & Hallacher (2003) use with permission. Hectonicus / Wikimedia Commons / CC-BY SA 3.0; Henri Casanova / Wikimedia Commons / CC-BY SA 3.0; Photo by Gerald Allen / USGS / Public Domain.

the defaunation of tropical reefs and rivers. Keeping reef fishes in aquaria cannot be rationalized on the grounds of species preservation. Few reef species have been successfully bred and raised in captivity, largely because of their complex life histories and age-specific habitat and feeding requirements. Nondestructive capture methods, bag limits, restricted areas and seasons, and most importantly, licensed captive breeding, can provide a diversity of interesting fishes to meet the home aquarist's needs while protecting natural environments.

A number of agencies and organizations have emerged to improve conditions in all steps of the supply chain, to lessen ecological impacts, and to better the lives of the people involved in the trade, especially the collectors. For example, Rising Tide Conservation (**https://www.risingtideconservation.org/aboutrisingtide/**) promotes aquaculture of marine fishes used in the aquarium trade, with the hope of reducing the impact of this industry on wild populations of tropical fishes. Project Piaba (**http://projectpiaba.org/**) promotes sustainability in the Amazonian home aquarium industry and points out that by providing stable incomes, collecting in Amazonia diverts people from more dangerous, ecologically destructive, extractive activities. If harvested sustainably, ornamental fishes could become "a sustainable by-product of an intact forest. . . Buy a fish and save a tree" (Norris & Chao 2002).

Global Climate Change

Since the industrial revolution of the late 1800s, atmospheric concentrations of **greenhouse gases** – mostly carbon dioxide, methane, chloroflourocarbons, and nitrous oxide – have increased substantially as a direct result of human activity. Sunlight passes through the atmosphere, heats the planet, and this heat is radiated back to space as infrared energy. The greenhouse gases trap the infrared radiation, heating the earth even further. Without greenhouse gases, the average temperature on earth would be too cold to support life as we know it. Greenhouse gases have been increasing in the atmosphere due to a variety of human activities, including fossil fuel and wood burning, deforestation, agriculture, raising cattle, and industrial pollution. Average temperatures have increased about 0.5°C over the past century. If current trends of greenhouse gas production continue, most climate modelers predict that average temperatures will rise another 1–3°C by the end of this century, which is far greater than the rate at which the earth warmed after the last glacial advance (IPCC 2013).

Of major concern are the likely climatic effects of this temperature increase and how they will be distributed (IPCC 2013). Altered wind direction and intensity and changes in freeze–thaw cycles have been predicted. Vagaries of ocean currents and cloud behavior will undoubtedly lead to greater warming in some regions and even cooling in others. Similarly, rainfall patterns will shift, making some regions wetter, others drier – we are already seeing examples of this. Sea level rise has been documented in recent decades and will continue due to thermal expansion of oceanic water and melting of polar ice caps. The postulated consequences for fishes of such a change are potentially dramatic.

Temperature and Fishes Temperature increases are likely to affect many aspects of fish biology. Metabolic processes are evolved responses to long-term thermal regimes characteristic of different climatic regions (Portner & Knust 2007). Alterations in thermal regime can affect the kinetics of such processes. Increased temperatures are a threat because fishes often live close to their **critical thermal maxima** (Magnuson & DeStasio 1997), because oxygen solubility is reduced at higher temperatures at the same time that metabolic requirements increase, and because many pollutants are more toxic at higher temperatures (Roessig et al. 2004).

Fishes respond to temperature changes by altering metabolic processes, reproduction, behavior, and distribution. Sex determination in fishes can be sensitive to thermal alteration, with different species producing unequal numbers of males or females in response to elevated temperatures (Devlin & Nagahama 2002). Gonadal development and germ cell viability are also temperature sensitive (Strüssmann et al. 1998). Timing of reproduction is highly sensitive to seasonal temperature cycles, in part because of effects on precipitation and freeze–thaw cycles. Magnuson et al. (2000) noted that in the northern hemisphere, lake freezing in the late 1990s was about 10 days later than 150 years earlier. Between 1973 and 2010, extent of ice coverage declined an average of 71% in the Laurentian Great Lakes (Wang et al. 2012), and shortened by about 5 days per decade from 1975 to 2004 in other lakes in the region (Jensen et al. 2007). The period of ice coverage of small lakes in Vermont decreased by nearly 7 days per decade from 1960 to 2010, with ice forming about 4 days later and melting 3 days earlier per decade (Betts 2011).

Similarly, in parts of Europe, snows melt 1–2 months earlier than 50 years ago, reducing spring floods and disrupting fish migrations and spawning. In Estonia, spawning migrations and timing of several freshwater fishes (European Smelt, Northern Pike, Ruffe, Bream) have advanced on average 12–28 days from historical values (Ahas 1999; Ahas & Aasa 2006). These trends are expected to continue and will likely lead to warmer water temperatures, longer periods of summer stratification, and increased frequency of bottom hypoxia in some areas (see Collingsworth et al. 2017). Climate change is expected to alter lake community structure and perhaps modify ecological processes responsible for ecosystem services such as providing fishes that support a significant commercial fishing industry in the North American Great Lakes (Biswas et al. 2017).

The latitudinal and altitudinal distribution of many fish species is determined by water temperature. Ultimately, species ranges can be altered via extensive dispersal, or populations can collapse where suboptimal conditions cannot be avoided. Shifts in distribution of commercial and noncommercial marine species have been observed in the North Atlantic, where bottom temperatures increased 1°C between 1977 and 2001 (Perry et al. 2005). Among 36 species assessed, two-thirds moved northward or deeper toward cooler waters over that period. A 2014 assessment indicated that stocks of Atlantic Cod off the coast of New England were at historic lows due to persistent overfishing (PEW 2014). This apparent crash, combined with the effects of climate change, raises great concern regarding the potential for stock recovery. The remnants of this once-abundant population of cold-water fish may be moving northward, which could boost cod harvests farther north in the Atlantic – at least for the near term. An unusually warm period around the Svalbard archipelago of the Northeast Atlantic from 2007 to 2014 raised concern that climate change may allow the incursion of warmer Atlantic Ocean water northward into the arctic, and result in restructuring of the fish assemblages and food webs in the region (Bergstad et al. 2017). In addition, trawl surveys along the coast of New England showed that populations of Black Sea Bass (*Centropristis striata*) and Scup (*Stenotomus chrysops*), both of which were more common in warmer waters farther to the south, were shifting northward due to increasing temperatures (Bell et al. 2015). Over 40 years, the center of spring biomass of both populations shifted northward by 150–200 km. These changes in community structure may affect overall ecosystem function and perhaps the ability of depleted fisheries to recover.

Elevated temperatures often prevent cold water species from occurring at lower latitudes and elevations. The temperature dependence of some species squeezes them into seasonally reduced habitat space, such as Striped Bass in the southern portions of their range (see Chapter 7, Thermal Preference). Continued elevated temperatures would be potentially lethal (Coutant 1990; Power et al. 1999).

Many other impacts of elevated global temperatures can be anticipated (see McGinn 2002). Sea level rise will flood coastal marshes. Coastal wetlands, mangroves, and saltmarshes are major nursery grounds for numerous fish species. Vegetation loss due to flooding has several ecological consequences. The food webs of coastal marshes depend on vegetation as both a source of and a physical trap for detritus, and vegetation also provides spawning substrates, physical refugia for juvenile fishes, and substrates for prey. Marshes and their defining flora and fauna could disappear from many coastal areas (Kennedy 1990; Meier 1990).

Low Latitudes Most global climate models predict less pronounced climatic changes at low latitudes. However, tropical animals tend to have relatively narrow climatic tolerances compared to high-latitude species and may therefore be more vulnerable to slight deviations from historically normal conditions (Stevens 1989). Coral reefs, already stressed by periods of slight temperature elevation, will be devastated by higher temperatures and accompanying stresses such as **acidification** (Hoegh-Guldberg 1999; Hoegh-Guldberg et al. 2007). Coral reefs globally are declining due to a number of local impacts, and climate change has already affected reefs throughout the tropics. Reef-building (hermatypic) corals generally exist in water close to their upper thermal limits. Increases of only a few degrees cause **coral bleaching** (loss of symbiotic algae) and death. Strong El Niño–Southern Oscillation (ENSO) events in 1982–83 and 1998 killed 50–100% of the corals in many areas, often as a result of average temperature rises of no more than a degree (Goreau et al. 2000; Glynn et al. 2001; Guzman & Cortes 2007). As the corals died, algae spread and covered all surfaces, followed by erosion and physical collapse of the limestone.

These alterations to the basic, underlying biological and physical structure of the reef have had far-reaching impacts on the fish assemblages. Where coral death exceeded 10%, more than 60% of fish species declined in abundance, with losses strongest among species that relied on live coral for food and shelter. Abundances among herbivorous and detritivorous species increased initially, but even these groups declined as reef erosion progressed. Overall fish diversity declined in direct response to the amount of coral lost, and prospects for long-term recovery are poor given projected trends in climate (Garpe et al. 2006; Wilson et al. 2006a; Feary et al. 2007). El Niños are expected to intensify as the climate warms, portending further, widespread reef degradation (Timmermann et al. 1999; Lesser 2007). The implications for reef fish diversity, reef fisheries, marine protected area design, and the economics of small tropical nations are immense (Soto 2001; Bellwood et al. 2004). Of related importance to human welfare, bleached corals appear to provide an enhanced surface for the growth of dinoflagellates, including those responsible for ciguatera poisoning among humans that eat coral reef fishes (Kohler & Kohler 1992).

Impacts on Seasonal Phenomena

Phenological (seasonal) cycles are likely to be disrupted, especially spawning periods that are timed to deliver larvae into regions of high productivity. Such productivity, which is driven by ocean currents and upwellings, has already been disrupted (Gregg et al. 2003; Schmittner 2005). Some global climate models indicate major shifts in ocean currents and upwelling patterns as a result of global warming. Such changes may alter or intensify the ENSO phenomenon, which has a substantial influence on major oceanic and coastal food webs. Some models predict weakening of major low-latitude currents such as the Gulf Stream and Kuroshio currents, reduction in nutrient-transporting eddies of these currents, and reduced upwelling off the western coasts of South America and Africa. Climate determines the vertical and horizontal distribution of ocean currents, and altered currents could affect the distribution and production of pelagic species that make up 70% of the world's fisheries (Bakun 1990; Francis 1990; Gucinski et al. 1990). Mathematical modeling of the likely dispersal of larval eels from the Sargasso Sea strongly suggest that atmospherically driven changes in ocean currents were at least partly responsible for dramatic declines in European Eel recruitment in the early 1980s and beyond (Baltazar-Soares et al. 2014). Timing of reproduction, particularly in migratory fish, would undoubtedly be disrupted. Migrations of anadromous salmonids are timed to take advantage of increased flows and cold water temperatures associated with snowmelt. Decoupling of migration times from altered melt cycles will affect trophic interactions, alter food web structure, and produce ecosystem-level changes (Harley et al. 2006). Temperate marine environments may be particularly vulnerable because the recruitment success of fishes depends on synchronization with pulsed planktonic production (Edwards & Richardson 2004).

Weather Patterns

Variations in the frequency and severity of climatic extremes of drought, flood, and cyclonic force storms have become apparent in recent decades. Storms and attendant floods wash young fish out of appropriate habitats, and can dilute high salinity, nearshore regions with fresh water, lessening their value as nursery grounds for larvae and juveniles. Altered rainfall patterns are expected to intensify droughts. Droughts would also affect water-stressed areas such as deserts and their already imperiled fish species, further drying areas that now have intermittent rainfall, and lead to contraction of the habitat space available for many species. Droughts also cause shifts in the distribution of estuarine habitats, because sea water typically intrudes farther up river basins during periods of low rainfall. Drought conditions would also exacerbate human impacts on fish habitat by reducing stream flow, elevating temperatures, and increasing pollutant concentrations.

Increased evaporation or decreased rainfall would decrease river flows and lake levels, causing wetlands to disappear and water tables to decline. The volume of cool water in many lakes would shrink, especially in summer. Cool water species whose ranges extend into warmer regions, such as Brook Trout, would be excluded from lower portions of streams during the summer. A few degrees of warming could be catastrophic for fishes that live near their critical thermal maxima because groundwater temperature is strongly dependent upon air temperature (Power et al. 1999). Water temperatures above 38–40°C are lethal to many stream fishes in the southwestern United States. When temperatures in southern rivers exceed these limits, heat-related deaths occur, as they do with salmonids on the West Coast at even lower temperatures (NRC 2004a). A 3°C temperature rise would potentially exterminate 20 species of fishes endemic to the southwest (Matthews & Zimmerman 1990). Warming would contract the geographic ranges of Arctic species, pushing the southern edge of their ranges northward. Cheung et al. (2013) evaluated global fishery catches in combination with temperature preference of exploited species and showed that warming ocean temperatures associated with climate change have been a factor in global fisheries for decades.

Benefits of Climate Change?

Increasing temperatures may mean that some warm water species would benefit from an increase in available habitat space at northerly latitudes, and some cool water species would gain access to higher altitudes and latitudes that are currently too cold to inhabit (Magnuson et al. 1990; Magnuson 2002). However, these shifts would dramatically alter assemblage relationships, with unknown consequences (Mandrak 1989). Cold water species will probably be both replaced and displaced by warm water species, especially invasive generalists, accelerating the process of faunal homogenization. Any "gains" would be offset by an overall loss of genetic and species diversity, especially because climate appears to be changing too quickly for genetic change to keep pace. New species will not have time to evolve to take the place of those that cannot adapt (IPCC 2013). A likely reduction in biodiversity is a serious, potential, negative impact of climate warming.

Protecting Fish Biodiversity – What Can Be Done?

The Issue Must Be Addressed at Multiple Levels

Global fish stocks have been declining for decades. We are witnessing alarming reductions at all levels of biodiversity – landscapes, communities, species, and genetic. Historically, attention and concern have been focused on threats to individual species or populations. The US Endangered Species Act of 1973 (ESA) focused on identifying and protecting species at risk; similar legislation exists in the most wealthy and many low- and middle-income nations (Leidy & Moyle 1997; Helfman 2007). Although the ESA was innovative and far-reaching, it is generally agreed among biologists that emphasizing species rather than habitats is at best a partial solution to biodiversity loss. Endangered species problems are really endangered habitat problems. The majority of extinctions result from habitat destruction, including the facilitated establishment of nonnative species in altered habitats. Often more than one rare species is affected by the loss of a particular habitat, and organisms live in coevolved, interacting communities, the elements of which are necessary for the welfare of most of the species in question.

Many scientists agree that ecosystems should be the primary focus of conservation, rather than individual species (Helfman 2007). Ecosystems are the organic life-support machinery for all life; if ecosystems are healthy, species are at lower risk. This is one reason that some consider captive breeding programs as futile if insufficient natural habitat exists into which a species can be reintroduced.

There has been increased emphasis on the value of ecosystem and landscape conservation, which is logistically, economically, and politically more difficult to attain than species protection. In recognition of the intricate interdependence of organisms and their habitats, conservation in the 21st century has increasingly focused on ecosystem based management (EBM) (Leidy & Moyle 1997; Winter & Hughes 1997; Mace & Hudson 1999; NRC 1999b; Agardy 2000). Our actions toward species or habitat components invariably ripple through and affect multiple elements in an ecosystem, rendering efforts focused on a particular species shortsighted and ineffectual. In this context, humans are best regarded as a predator or competitor (or both) embedded in the evolved interactions of species. To minimize our impacts and to sustain resources for our use and the use of other ecosystem components in the future, some have argued that we should act like natural predators. This means we should limit our exploitation within the bounds of other predators and competitors by harvesting in accord with the average take of other predators, and by focusing our predation on species and population segments that naturally experience high mortality rates (i.e., abundant, pelagic species low in the food web rather than top predators; young, prereproductive age groups rather than older, reproductively active and valuable age groups) (Kitchell et al. 1997; Fowler 1999; Hutchings 2000b; Stergiou 2002).

We are also faced with the difficult reality that we simply cannot protect and save everything. So how do we decide where to focus our attention and resources? Some argue that we should save the highest number of species possible by focusing conservation efforts on ecosystems that support many species – the biological "hotspot" argument. This would focus resources on many speciose tropical locations. This can be very efficient from a resource allocation perspective because many species may be saved by protecting relatively small areas of species-rich habitat. This is a rather similar perspective to that of those who believe that conservation efforts should protect the processes of speciation and adaptation (Frankel 1974; Erwin 1991; Fraser & Bernatchez 2001). In this view, maintaining genomic diversity allows adaptations and future evolutionary radiations.

A different perspective, however, is to support conservation efforts that also take into account preserving major branches of our phylogenetic trees and also groups that although not especially speciose demonstrate adaptation to a broader diversity of habitats. For fish conservation, this view would place a high priority on fishes such as coelacanths (our only existing lobe-finned fishes) and Antarctic icefishes, and a lower priority on some of the recently derived cichlid species found in the large lakes of East Africa.

Humans now have such an enormous impact on our biosphere that we could consider all of our ecosystems as being managed by human activity – whether by active decisions or indecision. So how should we prioritize? How should we balance our interests in protecting speciose groups with emerging species, evolutionary novelties with very specific adaptation to specific conditions, and fishes that represent long solitary branches in the tree of life that are the remnants of previous evolutionary radiations and viewed by some as dead ends living out their final years as the planet around them has changed?

Must we choose between cichlids and coelacanths? Should conservation measures be based on taxonomic rank (a phylogenetic mandate), ecosystem health (ecological mandate), or genetic diversity (evolutionary mandate)? While these positions would seem to be irreconcilable, they are strikingly concordant when viewed in the temporal perspective of past, present, and future. The phylogenetic mandate is historical, with a focus on the successful products of past evolutionary radiations. The ecological mandate is contemporary, with a

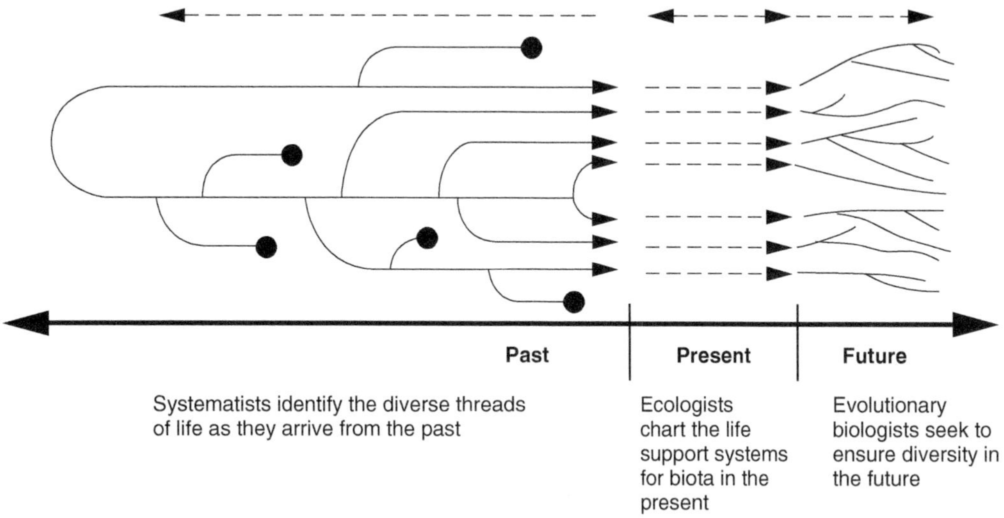
Past
Present
Future
Systematists identify the diverse threads of life as they arrive from the past
Ecologists chart the life support systems for biota in the present
Evolutionary biologists seek to ensure diversity in the future

FIGURE 22.22 The complementary roles of three scientific fields (phylogenetics, ecology, and evolutionary biology) in conservation. The process of conserving fishes begins with phylogenetic studies to identify the products of past evolutionary radiations. Subsequently, ecologists identify the key habitat features that allow fishes to persist in the present. Finally, evolutionary biologists identify the raw materials for future diversification. The black circles represent extinction events. From Bowen and Roman (2005), used with permission.

focus on healthy ecosystems for conservation efforts. The evolutionary mandate seeks to promote biodiversity in the future. In this temporal framework, the three biological disciplines that claim domain over conservation are not conflicting, rather they address three essential components: the preservation of the threads of life as they arrive from the past (phylogenetics), abide in the present (ecology), and extend into the future (evolution) (Bowen & Roman 2005; Fig. 22.22). In this temporal perspective, the three scientific disciplines have complementary, rather than competing, roles in conservation.

Notably, genetics has a vital role in all three disciplines. Molecular phylogenies reveal the deep branches and cryptic evolutionary partitions that may be missed in morphological surveys. Population genetic studies illuminate the level of connectivity among ecosystems, an essential prerequisite for designating protected areas. Genomic studies can reveal the genetic diversity and innovations that will promote future evolutionary radiations.

Biological Preserves

One direct application of an ecosystem-level approach is through the creation of biological preserves (Moyle & Leidy 1992; NRC 2001). Few biological preserves exist today that are targeted directly at freshwater organisms (Saunders et al. 2002) – pupfishes in the Great Basin/Death Valley area of the southwestern United States are an exception (see Miller & Pister 1971). Most preserves are created as terrestrial parks that include lakes and portions of streams and rivers. Unfortunately, human activities upstream of such parks can threaten the aquatic biota in the park, and seasonal migrations by many fishes carry them beyond the protection of park and even international boundaries.

Marine parks (**marine protected areas** or **MPAs**) are much more common and generally considered to be successful, although enforcement has often proved problematic (Russ 2002; Halpern 2003). MPAs are established to both **preserve biodiversity** and **promote biomass**; the two goals are not mutually exclusive. Improved conditions result from relaxation of both exploitation (lowered mortality due to fishing) and habitat disturbance, which lead to changes in ecological and life history traits of resident fishes. Fishes within a protected area may increase in diversity, density, average size and age, and overall biomass. On average, density increased two-fold, biomass three-fold, and fish size and number of species 20–30% (Halpern 2003). As a result, fishes disperse from the densely populated, protected area to surrounding areas (the **spillover effect**) (Fig. 22.23). When effective, spillover more than compensates for lost fishing opportunities within the reserve, although that can be slow to occur (e.g., McClanahan & Mangi 2000). At a more regional scale, many large fishes in reserves lead to increased reproductive output in the protected area, which should result in increased export of larvae that potentially settle in downstream, often distant areas (the **recruitment effect**), although demonstrations of the recruitment effect remain rare.

The most effective MPAs are those that restrict all extractive activities ("no take" areas). Recreational diving with no contact of the bottom is minimally intrusive (some parks restrict the use of gloves, which minimizes bottom contact). Multiple-use objectives often compromise the refuge quality of marine preserves. Protection is often limited, e.g., spearfishing and nets may be prohibited, but not hook-and-line fishing. Again, many

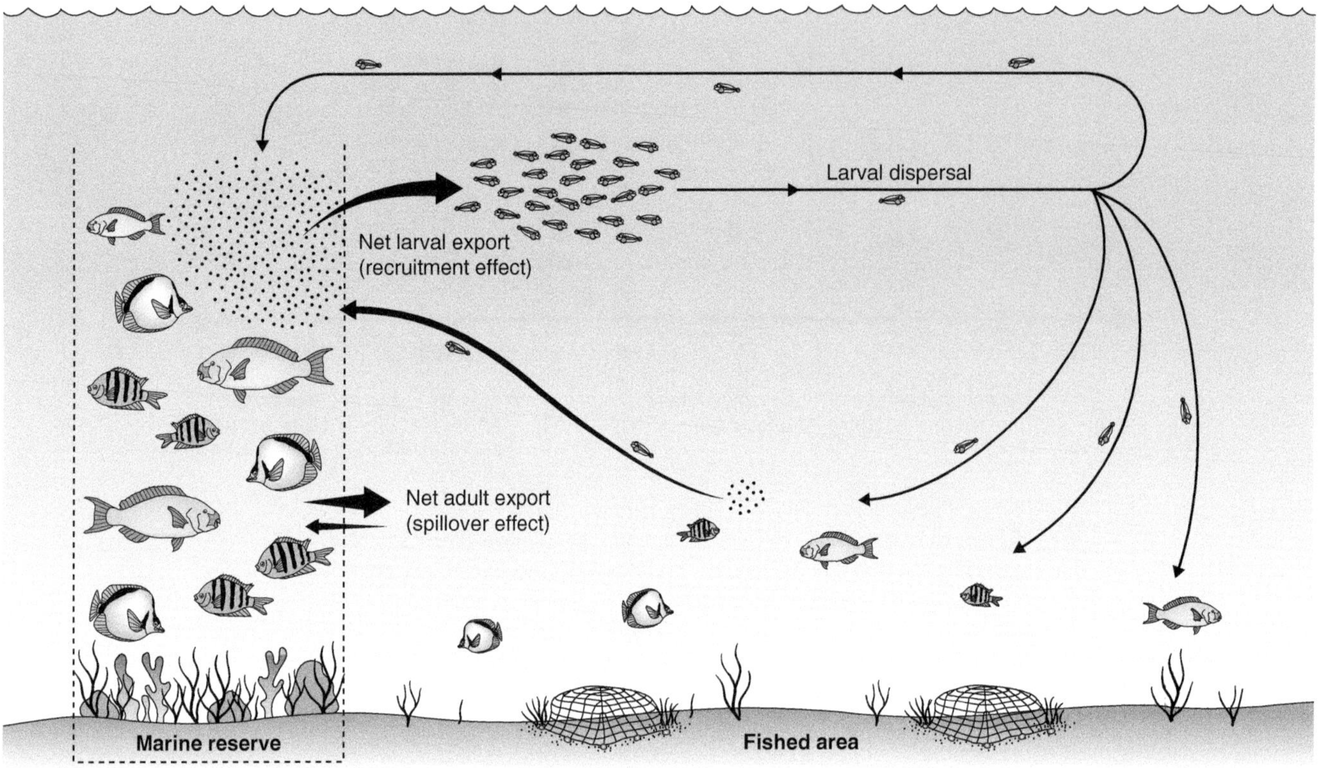

FIGURE 22.23 Functions of marine protected areas from a fisheries management perspective. Fish populations within the reserve thrive and the numbers, size, and age of residents increase. This leads to increased reproduction and net export of both adults and larvae into adjacent regions, which can help support fisheries (based on Russ 2002). Overall diversity can also increase within the reserve.

species move out of reserve waters as a normal part of their life histories, which then subjects them to commercial and recreational exploitation. Dependence on dispersed larvae for recruitment means that "upstream" habitats must also be protected or populations in an area may decline. Larger parks, such as much of the Great Barrier Reef in Australia and the Florida Keys National Marine Sanctuary in Florida, encompass more species and life history stages but are more difficult to police. Where such preserves are established, marine life and particularly fish populations rebound dramatically (e.g., Hanauma Bay in Hawaii, Dry Tortugas in Florida). In all cases, whether the area is freshwater, estuarine, or marine, the costs of acquisition are high, and opposition to "loss" of the area to exploitation is often strong.

Rehabilitation and Restoration

Costs are a major factor in **restoring degraded habitat** to something approximating original, or at least better, conditions. Dramatic results can be achieved if restoration becomes a high priority, although many past restoration efforts in flowing water systems have focused on recreating instream habitat rather than arresting practices in surrounding watersheds that actually cause habitat loss (Frissell & Nawa 1992; Bernhardt et al. 2005). Also missing from many such restoration projects, and many conservation efforts in general, is **post-project monitoring** that accurately assesses the outcome, rather than assuming success when restoration efforts cease (Kondolf 1995; Bash & Ryan 2002).

Reversals and partial restorations of badly polluted and degraded systems have been achieved in Puget Sound in Washington State, Kaneohe Bay in Hawaii, the upper Illinois River in Illinois, the Willamette River in Oregon, the Mattole River in northern California, and the Merrimack River in New Hampshire and Massachusetts (NRC 1992). Success in restoring depleted fish populations can result from such actions, as is apparently the case for Lake Trout, *Salvelinus namaycush*, in the North American Great Lakes and Striped Bass, *Morone saxatilis*, in Chesapeake Bay. However, few, if any, ecosystems are ever restored to their original conditions; improvement from a state of extreme degradation is often the most that can be achieved. The complexity of natural, evolved systems works against human-induced solutions, which often strive for simplicity. In truth, restoration efforts have often demonstrated that "corrective measures to restore ecosystem function [are] obtained only at very high costs, that some attributes can be maintained only with continuous management, and that certain losses in the ecosystem [are] irreversible" (Schelske & Carpenter 1992, p. 383).

Captive Breeding

Captive propagation of endangered species, with eventual release into the wild, has been suggested as a solution to the extinction problem (Ribbink 1987; Johnson & Jensen 1991).

Government agencies, private organizations, native peoples, and private individuals culture and reintroduce imperiled fishes, into either restored native habitat or acceptable, alternative habitat (Wikramanayake 1990; Minckley 1995). Such efforts can be costly and time-consuming, small numbers of breeding individuals lead to genetic bottlenecks and loss of diversity, and the breeding and rearing requirements of most species are poorly known or difficult to replicate outside of the natural habitat (Anders 1998; Rakes et al. 1999). **Conservation genetics** focuses on the myriad problems associated with trying to resurrect endangered species from small numbers of breeding individuals (e.g., Meffe 1986; see earlier in this chapter). In some instances, however, captive propagation may be the only hope for a species. But species also need suitable habitat to which they can return if they are to maintain themselves over the long term.

Education and Population Control

The protection of species and preservation of the habitats in which they live require an **educated public** that recognizes the enormity of our impacts, appreciates the value of biodiversity, and understands the degree to which humans depend on aquatic ecosystems. Major changes in attitudes – of citizens, scientists, politicians, and managers – toward how we value fishes, aquatic environments, and natural resources are needed (Callicott 1991; Pister 1992). We have seen some encouraging shifts in attitude and management practices in recent decades, but these must continue and expand to be effective on a broad scale. One example of changing attitudes exists in the popularity and success of **ecocertification** programs that encourage sustainable use of fishery resources. Dolphin-Safe Tuna, Give Swordfish a Break, Take a Pass on Patagonian Toothfish, FAO Code of Conduct for Responsible Fisheries, and the more general Seafood Watch/Sustainable Seafood guides facilitate sustainable use and allow consumers to encourage exploitation activities that involve sustainable practices and discourage fisheries that are pursued in a nonsustainable manner (see Peterman 2002; **http://www.seafoodwatch.org/**; **www.msc.org**).

Ultimately, however, the future of aquatic and terrestrial life on earth will be determined by whether humans continue to multiply and consume at current rates. Realistically, conservation efforts will not be adequate if **human population growth** and **resource consumption** cannot be checked. The impact that humans have on the earth's resources are the combined effects of large numbers of people consuming at unsustainable rates. In higher-income nations, achieving sustainable consumption rates will require an educated and concerned public that is willing to live more sustainably while still maintaining a high quality of life. National and international programs to control human population growth are crucial to reversing global and local environmental deterioration.

Conclusion

Biodiversity loss is a symptom of environmental deterioration on a global scale. A growing number of scientists, traditionally occupied with the descriptive and experimental pursuit of knowledge, have turned their efforts to environmental issues in an effort to reverse these declines. Even regional fish books that had previously focused on occurrence and distribution now include lengthy discussions of the conservation status of their fishes (Moyle 2002; Boschung & Mayden 2004). From these and other contributions, a large number of practical solutions to the various problems discussed here have emerged. Many have been tried, many more remain to be applied. A few are discussed below, but the concerned reader should refer to the diversity of synthetic discussions for details, such as Leidy and Moyle (1997), Winter and Hughes (1997), Mace and Hudson (1999), NRC (1999b, among others), Hilborn (2005), Helfman (2007), and FAO (2016, 2020, and more current updates when available). The most frequently offered solutions (in addition to the crucial need for surveying, documenting, and monitoring problem areas) include:

- Pass and enforce national and international legislation that promotes sustainable resource use.
- Create reserves, as large as possible.
- Promote EBM and evolutionarily compatible, prudent predation.
- Be precautionary: act despite uncertainty, without waiting for scientific consensus.
- Monitor results and manage adaptively, modifying management plans in response to changing conditions.
- Promote ecocertification efforts and other programs that reward sustainable fishing practices.
- Avoid technoarrogance, e.g., technological fixes that treat symptoms rather than causes.
- Restore degraded habitat to promote the recovery of imperiled species, and engage in captive breeding of endangered species as a last resort and only in conjunction with habitat restoration.
- Educate resource users and the public about biodiversity loss and sustainable use.
- Include all stakeholders at all stages in management decisions, and encourage local/community control wherever possible.
- Reduce fishing effort and eliminate subsidies that encourage overfishing.

Conservation efforts are of necessity **multidisciplinary**, requiring knowledge and integration from the biological and physical sciences, as well as from sociology, anthropology, and economics. Regardless, it is apparent to all concerned that the major task of conservation efforts is to reverse previous and minimize future negative human impacts on natural systems. There is a growing realization that climate has already changed enough that returning to past ecological conditions is impossible in some cases. However, we must still do our best to reestablish and support resilient ecological communities capable of sustained ecological function.

Supplementary Reading

American Fisheries Society. 1990a. Effects of global climate change on our fisheries resources. *Fisheries* 15(6):2–44.

American Fisheries Society. 1990b. Symposium on effects of climate change on fish. *Trans Am Fish Soc* 119:173–389.

Bowen BW. 2016. The three domains of conservation genetics: Case histories from Hawaiian waters. J Hered 107:309–317.

Closs GP, Krkosek M, Olden JD, eds. 2016. *Conservation of Freshwater Fishes*. Cambridge, UK: Cambridge University Press.

Courtenay WR, Jr, Stauffer JR, Jr, eds. 1984. *Distribution, biology, and management of exotic fishes*. Baltimore: Johns Hopkins University Press.

Fiedler PL, Jain SK, eds. 1992. *Conservation biology: the theory and practice of nature conservation, preservation and management*. New York: Chapman & Hall.

Fuller PL, Nico LG, Williams JD. 1999. *Nonindigenous fishes introduced into inland waters of the United States*. American Fisheries Society Special Publication No. 27. Bethesda, MD: American Fisheries Society.

Glantz MH, Feingold LE, eds. 1990. *Climate variability, climate change and fisheries*. Boulder, CO: Environmental and Societal Impacts Group, National Center for Atmospheric Research.

Goldschmidt T. 1996. *Darwin's dreampond: drama in Lake Victoria* (transl. S. Marx-Macdonald). Cambridge, MA: MIT Press.

Hargrove EC. 1989. *Foundations of environmental ethics*. Englewood Cliffs, NJ: Prentice Hall.

Heath AG. 1987. *Water pollution and fish physiology*. Boca Raton, FL: CRC Press.

Helfman GS. 2007. *Fish conservation: a guide to understanding and restoring global aquatic biodiversity and fishery resources*. Washington, DC: Island Press.

Hellawell JM. 1983. *Biological indicators of freshwater pollution and environmental management*. New York: Elsevier Applied Science Publishers.

IPCC (International Panel on Climate Change). 2007. See previous and most recent reports at. **www.ipcc.ch**.

IUCN (International Union for the Conservation of Nature). 2006. See recent reports at **https://www.iucn.org/resources/publications**.

Karr JR, Chu EW. 1999. *Restoring life in running waters: better biological monitoring*. Washington, DC: Island Press.

Kurlansky M. 1997. *Cod: a biography of the fish that changed the world*. New York: Walker & Co.

Lever C. 1996. *Naturalized fishes of the world*. San Diego, CA: Academic Press.

McCarthy JF, Shugart LR, eds. 1990. *Biomarkers of environmental contamination*. Chelsea, MI: Lewis Publishers.

Minckley WL, Deacon JE, eds. 1991. *Battle against extinction: native fish management in the American west*. Tucson, AZ: University of Arizona Press.

NRC (National Research Council). 1992. *Restoration of aquatic ecosystems: science, technology, and public policy*. Washington, DC: National Academy Press.

Norse EA, Crowder LB, eds. 2005. Marine conservation biology: the science of maintaining the sea's biodiversity. Washington, DC: Island Press.

Ono RD, Williams JD, Wagner A. 1983. *Vanishing fishes of North America*. Washington, DC: Stone Wall Press.

Smith JB, Tirpak DA, eds. 1989. *The potential effects of global climate change on the United States. Appendix E. Aquatic resources*. Washington, DC: US Environmental Protection Agency, EPA-230-05-89-055.

SPA (Science and Policy Associates, Inc). 1990. *Report on the workshop on effects of global climate change on freshwater ecosystems*. Washington, DC: SPA.

Wheeler A, Sutcliffe D, eds. 1991. The biology and conservation of rare fish. *J Fish Biol* 37 (Suppl A):1–271.

Journals

Aquatic Conservation: Marine and Freshwater Research

Conservation Biology

Animal Conservation

Species accounts of many of the world's imperiled fishes have appeared in the series "Threatened fishes of the world" in the journal *Environmental Biology of Fishes*.

Websites

Committee on Recently Extinct Organisms, CREO, creo.amnh.org.

Convention on International Trade in Endangered Species, CITES, **www.cites.org**.

Freshwater Fish Specialist Group, IUCN, **www.iucnffsg.org/**

Global lists and descriptions of MPAs are at **http://sea.unep-wcmc.org/wdbpa**.

Information on marine protected areas in the USA is available at **www.mpa.gov**.

Intergovernmental Panel on Climate Change, **www.ipcc.ch**.

International Union for the Conservation of Nature, IUCN, **www.redlist.org**.

UN Food and Agricultural Organization, FAO biennial report on world fisheries, **www.fao.org/sof/sofia**.

US Fish and Wildlife Service, endangered species database, **http://ecos.fws.gov/tess_public**.

US Geological Survey Nonindigenous Aquatic Species database, **http://nas.er.usgs.gov**.Allendorf F, Luikart G. 2006. *Conservation and the genetics of populations*. Malden, MA: Blackwell Publishing. Bowen BW, Roman J. 2005. Gaia's handmaidens: the Orlog model for conservation biology. *Conserv Biol* 19:1037–1043. Frankham R, Ballou JD, Briscoe DA. 2002. *Introduction to conservation genetics*. Cambridge, UK: Cambridge University Press.

Appendix 1 - Glossary of Genetic/ Molecular Terminology

Alleles These are different versions of a gene, locus, or DNA sequence. They can diverge by many mutations or as little as one mutation. A typical allozyme survey may reveal one to six alleles at a locus, whereas a microsatellite locus can have upward of 30 alleles at a locus. Gene flow between populations can be estimated by the differences in frequency of alleles.

Allozymes This is the workhorse methodology for fish genetics over the last 30 years. Allozymes are the protein products of individual genes in the nuclear DNA (nDNA). Proteins are loaded on a gel and subjected to an electric field. The gel acts like a sieve to separate molecules based on differences in amino acid composition (and corresponding electric charge). Distinct alleles can be identified because they move through the gel at different rates. For example, alcohol dehydrogenase (an enzyme that breaks down ethanol) can exist in three versions in salmon. Allozyme surveys have been widely applied to test for **population structure** and to assess genetic differentiation among congeneric species (Shaklee & Tamaru 1981), but less frequently used to resolve deeper evolutionary separations.

Automated DNA sequencing DNA sequencing previously was performed with cumbersome polyacrylamide gels, using a number of toxic and radioactive chemicals. Starting about 1990, this methodology was replaced with automated machines that could use much less hazardous chemicals. A DNA sequence that used to cost more than US$50 to produce can be obtained for less than $10 with the automated technology, and the price continues to fall. Previously, a hard-working scientist could produce perhaps 500 DNA sequences per year, whereas now an automated DNA sequencer can produce over 600 sequences per day. DNA data are now readily accessible for a modest budget, and fish phylogeography studies of 200–500 specimens may entail less than $10,000 in lab costs.

Base pairs (bp) These are the units of double-stranded DNA (paired nucleotides). The length of a DNA sequence is usually measured in base pairs or nucleotides, which means the same thing in this context. Fragments as small as 100 bp can be informative in population genetics and pedigree analyses, and for identifying the species present in stomach contents and feces. For phylogenetic comparisons, fragments in excess of 500 bp are preferred. Variants on bp include **kb** (thousands of base pairs) and **mb** (millions of base pairs). As DNA analyses get faster and cheaper, the data sets get larger and now include sampling of whole genomes.

Bayesian methods Bayesian methods are closely related to maximum likelihood (ML) methods, but they employ conditional probabilities: given a set of prior conditions X, the probability of this event is Y (Huelsenbeck et al. 2001). In this way, certain outcomes (or branching orders) can be ruled out based on independent information. For example, a *prior* condition for a molecular phylogeny could be a morphology-based phylogeny that indicates all sunfishes (genus *Lepomis*) are monophyletic (descended from a single common ancestor; see Chapter 2) when compared to other members of the family Centrarchidae. Bayesian methods allow for ML analyses that are limited to outcomes where all sunfishes are united in a single branch of the tree. Because Bayesian methods constrain the search for the most likely tree, they are somewhat controversial; if the prior information is wrong, the tree might be wrong.

Bootstrap support A method that resamples the data set to determine how robustly each branch in a phylogenetic tree is supported by the data. Branch arrangements that hinge on one or two mutations will have low bootstrap support. Generally, a bootstrap support of >60% can be regarded as a tenable hypothesis, and values >90% can be considered very strong support. Other measures of branch support include **Bremer decay** (indicating how many mutations support a particular branch) and posterior probability in Bayesian analyses (highest probability indicates branch order).

Chromosome The unit of inheritance, a very long DNA molecule that is replicated during cell division. One chromosome may contain more than genes. In fish and other eukaryotes, the DNA is bundled into two identical chromosome arms (**chromatids**) which are joined by a small constricted region (**centromere**).

Comparison of phylogenetic techniques **Maximum likelihood (ML)** and **Bayesian** methods are the current favorites for molecular phylogenetic studies, but other methods have their merits. **Parsimony** methods are useful for describing networks of closely related haplotypes. **Neighbor joining** (**NJ**) is useful for describing the lengths of branches and visualizing the depth of corresponding evolutionary separations. In evaluating any particular phylogenetic study, the most important question is whether branch arrangements (also called **topology**) change across phylogenetic methods. If a branch moves around, depending on the type of analysis, don't bet the fish farm on that one.

Diploid nuclear DNA (nDNA) Nuclear DNA includes the chromosomes that are typically inherited in pairs, one copy from each parent (except for those fishes with separate sex chromosomes). In this **diploid** state (often referred to as **2N**), every cell has two copies of each DNA sequence that may be identical (**homozygous**) or slightly different (**heterozygous**). Studies of fish evolution have employed **intron** sequences (see below) including intron six of lactate dehydrogenase-A (LDHA6), and intron seven of creatine kinase (CKA7)

The Diversity of Fishes: Biology, Evolution and Ecology, Third Edition. Douglas E. Facey, Brian W. Bowen, Bruce B. Collette, and Gene S. Helfman.

Companion website: www.wiley.com/go/facey/diversityfishes3

(Quattro & Jones 1999; Hassan et al. 2002). Other commonly used sequences include the recombination activation genes (RAG-1 and RAG-2), and the ribosomal gene 28S rDNA. Some nuclear sequences such as Tmo-4C4 have been discovered and used for genetic comparisons without prior knowledge of their function (Karl & Avise 1993; Streelman & Karl 1997).

DNA Deoxyribonucleic acid, composed of four building blocks known as **nucleotides**: adenine, cytosine, guanine, and thymidine, usually abbreviated as A, C, G, and T. These building blocks form long strings called **DNA sequences** that can be compared to learn interesting things about fishes, like evolutionary partitions that are not apparent in morphology, or isolated breeding pools (see population structure, below) that are the units of wildlife management. DNA sequences are usually high-quality data because they are verifiable and replicable.

Effective population size, N_e This is the number of individuals in a population that pass their genes on to the next generation (i.e. the number of successful breeders in the population). This is usually estimated based on the level of nucleotide diversity (π) within populations; high diversity indicates a large stable population. In the marine fishes that produce hundreds of thousands of eggs, N_e can be two orders of magnitude lower than the current population size, probably because of the high variance in reproductive success: most eggs and larvae perish, and few adults contribute to the next generation (Grant & Bowen 1998).

Exons These are functional units of genes that are translated into transfer RNAs (tRNAs), ribosomal RNAs (rRNAs), and proteins. They are separated by segments on noncoding DNA known as introns. Exons tend to not accumulate mutations as rapidly as introns.

***F* statistics** These come in many forms, but they all measure departures from random mating, the essence of population structure. F_{ST} is used with allele frequency data (Wright 1951) and is the most common measure of population structure. F_{ST} basically measures differences in allele frequencies between populations. F_{ST} values range from 0 to 1, with $F_{ST} = 0$ indicating that the two populations are frequently interbreeding, while $F_{ST} = 1$ would indicate that the populations each have a different allele at 100% frequency. G_{ST} is a modification for haploid data such as mtDNA (Takahata & Palumbi 1985), and ϕ_{ST} incorporates both the allele frequency shifts (like F_{ST}) and the DNA sequence divergence between alleles or haplotypes (Excoffier et al. 1992). ϕ_{ST} is the preferred method for most comparisons of DNA sequence data. R_{ST} is used with microsatellite data to incorporate expectations about how microsatellites mutate (Slatkin 1995). Values as low as $R_{ST} = 0.01$ can indicate a significant restriction on gene flow between populations. An F_{ST} or ϕ_{ST} value above about 0.10 would indicate strong isolation and distinct management units in a fishery. Values above about 0.20 might indicate evolutionary divergence, if corroborated with other data.

Genbank **http://www.ncbi.nlm.nih.gov/Genbank/** is a repository for DNA sequence data. The online service includes a search option that will find the closest matches to a DNA sequence provided by the user. This can be very useful for identifying an unknown specimen, or for finding the closest relative to a fish.

Gene A segment of DNA that is the blueprint for a particular protein or ribosomal RNA. There are over 30,000 genes in fishes.

Genotype This is the description of the alleles at a locus, indicating which two alleles occur at a diploid locus. For example, if a microsatellite locus has 31 repeats of a simple two-base sequence (CA-CA-CA . . . for example) on one allele, and 33 repeats on the other allele, then the genotype is described as 31/33.

***H* statistics** These are measures of **heterozygosity** (the level of genetic diversity) within populations, and at any locus can range from $H = 0$ (all individuals have the same identical allele) to $H = 1$ (all individuals have two different alleles). The corresponding value for haploid mtDNA is *h*, which is the probability that two individuals drawn at random will have different haplotypes (Nei 1987). The average value derived from fish studies is about $H = 0.05$ for allozymes (Ward et al. 1994), but about $H = 0.60$ for microsatellites (DeWoody & Avise 2000), reflecting the much higher mutation rate, and higher diversity, in microsatellites.

Haplotype This is the same as a genotype, but applied to the haploid mtDNA genome. Different haplotypes are distinguished by one or more mutations and are typically given letter or number designations. For example, if haplotype 6 is observed in a reef fish at high frequency (80%) in a Caribbean population, but at low frequency (20%) in a Brazilian population, this would indicate a strong population genetic separation.

Introns These are noncoding segments of chromosomes that separate the coding units known as exons. The functional significance of this arrangement is not completely understood, but introns usually accumulate mutations more rapidly than the exons. Therefore, intron sequences are used to construct evolutionary trees and assess the relationships between species; however, they have also been used for population surveys.

Karyotype These are chromosomes visualized by histological staining and characterized by comparing their number, shape, and size. Rare genomic rearrangements can be detected, and these are valuable for discerning evolutionary history, usually above the taxonomic level of species and genus.

Locus (plural loci) This is a segment of DNA, usually corresponding to a gene (for allozymes) or a segment of repeat sequences (for microsatellites). If a paper describes results from 10 loci, it means the study surveyed 10 distinct segments of the genome.

Maximum likelihood (ML) This is where a model of DNA sequence evolution is used to assign a probability to alternative phylogenetic trees (Felsenstein 1981). For example, we know that a mutation of C to T is about 10 times more common than a mutation of C to G. Armed with this information, we can assign 10 times greater weight to the rare C to G mutation. Different branch orders in a tree are explored, and the one with the highest probability is deemed the correct one. Like parsimony, this methodology seeks to find the simplest arrangement, but ML methods allow us to incorporate knowledge about how DNA sequences change over time.

Microsatellites These are nDNA segments composed of short repeats of two, three, or four nucleotides, such as CA-CA-CA. When the DNA is copied during cell replication, sometimes mistakes are made, and a "CA" is added or deleted. Microsatellite loci are also known as **VNTRs** (variable-number tandem repeats) because the alleles differ by the number of repeats. Differences in the length of a microsatellite allele can be detected with gel separation. A high mutation rate and very high **heterozygosity** (as discussed in Chapter 20 and below) make these microsatellites the method of choice for resolving relationships from family pedigrees to closely related populations and have provided valuable insights into the breeding systems of fishes. This is the most widely used genetic marker in the category of **DNA fingerprints** that can diagnose individuals with a high degree of certainty. However, the high mutation rate entails a risk of misinterpretation at higher evolutionary levels, so microsatellites are generally not an appropriate tool for phylogenetics.

Migration rate, $N_e m$ This is effective population size (N_e) multiplied by the proportion of migrants in a population (*m*), producing an estimate of the effective number of migrants per generation, an estimator of gene flow. This is often used without the "e" subscript, or sometimes with an additional "f" subscript ($N_{ef}m$) to denote female effective population size as measured with maternally inherited mtDNA. $N_e m$ can be approximated from F statistics with the equation (Wright 1951):

$$N_e m = (1 - F_{ST}) / 4F_{ST}.$$

A value higher than $N_e m = 1$ (one effective migrant per generation) is *in principle* sufficient to maintain genetic connectivity among diploid populations, so that populations are unlikely to differ genetically (Hartl & Clark 2006). The critical level of exchange for mtDNA is somewhat higher ($N_e m = 4$) due to maternal inheritance and haploid state. Like effective population size, effective migrants are individuals that migrate to a new population and contribute genes to that population. In most circumstances, there are far fewer effective migrants than actual migrants.

Mitochondrial DNA (mtDNA) This is a closed circle of double-stranded DNA, usually around 16,500 bp in fishes and other eukaryotes. This small genome is believed to be descended from a bacterium that entered the cells of a metazoan ancestor on the order of 600 million years ago, long before the first fishes arose (Margulis 1970). This genome is in the mitochondria, the energy-producing organelles of the cell that reside outside the nucleus. There can be hundreds of mitochondria in each cell, so mtDNA usually exists as many copies of a single identical sequence (**haploid**, denoted as 1N in contrast to diploid 2N nDNA). This genome has protein-coding genes that mutate faster than most nDNA genes, and these provide good resolution of populations, as well as evolutionary lineages at the taxonomic level of genus and species. Typical fish studies will use mtDNA sequences from cytochrome *b* or cytochrome oxidase, two genes that code for proteins. The more slowly evolving ribosomal genes (12S and 16S rDNA) are used for phylogenetic studies on the deeper taxonomic scale of genera, families, and orders. Another mtDNA segment, known as the **control region**, does not code for a protein product, but is a scaffold for DNA replication. Like introns (see above), this area can accumulate mutations more rapidly than protein-coding regions and for that reason is favored for resolving fine-scale **population structure** (as discussed in Chapter 20).

Molecular clocks Based on the assumption that isolated populations or species accumulate mutations at a predictable rate (Thorpe 1982), molecular clocks can be used to estimate divergence times, for example when **sister species** (those that are each other's closest relatives) stopped interbreeding and initiated separate evolutionary pathways. This is a valuable tool for reconstructing evolutionary divergence and speciation. For example, the mtDNA genomes of the Indonesian and African coelacanths differ by approximately $d = 0.043$, and the overall **divergence rate** may be about 0.1% per million years (0.001/MY), meaning 0.1% of their DNA sequence changes every million years. This yields an estimated divergence of 43 million years, indicating that the two coelacanths may have become isolated by the tectonic event when the subcontinent of India moved north and collided with Eurasia 50 mybp (Inoue et al. 2005). A typical rate for widely applied mtDNA fragments (cytochrome oxidase and cytochrome *b*) is 1%/MY to 2%/MY between species of bony fishes (Bermingham et al. 1997; Bowen et al. 2001). Elasmobranch mtDNA evolves more slowly, with an mtDNA control region clock rate of about 0.8%/MY (Duncan et al. 2006), and this may be true for primitive bony fishes as well. The reasons are still not clear but may include long generation time, efficient DNA repair mechanisms, or low metabolic rate (with less oxidative damage), or a combination of these factors (Martin & Palumbi 1993). Molecular clocks are analogous to the radioactive decay of a nuclear isotope. Any given radioactive molecule (or base pair) may not change in a million years, but others might change twice in the same interval, just by chance. Molecular clocks depend on assumptions about rate constancy that may not always be met. For these reasons, divergence times based on molecular clock estimates should be regarded as approximations, and interpreted with caution.

Neighbor joining (NJ) This is where trees are based on the divergence (sequence divergence *d*) between DNA sequences (Saitou & Nei 1987). This **phenetic** approach (see Chapter 2) is not employed in morphological systematics, but is popular in molecular systematics where the data consist of long strings of A, C, G, or T. This is a more sophisticated method than its predecessor, unweighted pair group method with arithmetic mean (UPGMA) (Sneath & Sokal 1973), which is still used occasionally. NJ and UPGMA offer the advantage of providing branch lengths that are proportional to the divergence between sequences. If a mutation rate is known for these DNA sequences (see Molecular clocks entry above), then an approximate age of separation between species can be inferred from the tree.

Next-generation sequencing (NGS) This term encompasses several technologies that produce very large DNA sequence data sets, in what is also known as massively parallel sequencing. The DNA data arrives in millions of small (<300 bp) fragments that are produced multiple times (hence in parallel) and overlap. The NGS advantage in high volume is countered by low accuracy. Hence, DNA sequence fragments are **stacked** with 5–10 or more copies in order to produce an accurate sequence. These fragments are filtered for quality and then can be assembled into longer fragments by resolving regions of overlap. A **bioinformatics pipeline** consists of a suite of software packages to retain only the highest quality data and assemble it for interpretation. NGS can be used to produce a whole genome, or target at specific regions of interest. Most of the time, researchers don't require a whole genome, but only enough variable regions to resolve the scientific target issue, ranging from the resolution of close relatives (kinship) to ancient evolutionary relationships (phylogenetics).

Nucleotide diversity, π or θ_π These measure the average DNA sequence divergence (*d*; as discussed in Chapter 20) between individuals (Nei 1987). These values start at $\pi = 0$ (all members are genetically identical) and rarely exceed $\pi = 0.05$ within fish populations. For example, the Lake Trout (*Salvelinus namaycush*) has $\pi = 0.000$, no measurable mtDNA diversity, in Trouser Lake, Labrador (latitude 56° 32′), an area that was under glacial ice 15,000 years ago. In contrast, the Lake Trout in Seneca Lake, New York (latitude 42° 45′) have $\pi = 0.019$ (Wilson & Hebert 1998). The low genetic diversity in Labrador indicates colonization by a few individuals after the glacial period, ending about 12,000 years ago, whereas the higher genetic diversity in New York indicates an older population.

Parsimony This is where the DNA sequences are assembled into a tree using the shortest number of mutational changes that can explain the data set (see Chapter 2). This is based on the philosophical point that the simplest explanation is the best one, and so parsimony methods are free of assumptions about how DNA mutates and evolves. **Parsimony trees** are rooted with an **outgroup**, a sister taxon that shares an ancestral relationship with the species assembled in the phylogeny. The outgroup allows the resolution of shared ancestral characters (**symplesiomorphies**) and more recently derived advanced characters (**apomorphies**) in a cladistic analysis. **Parsimony networks** are unrooted arrangements that depict the number of mutations between DNA sequences.

Phylogenies These are treelike diagrams that depict the evolutionary history of organisms. Since the advent of DNA data in phylogenetics, there has been accelerated progress in tree-building methods, but also considerable controversy. For a thorough treatment of this topic, the reader is encouraged to consult Li (1997) and Felsenstein (2004). Here, we provide a limited introduction to this field, by outlining the primary methodologies. Four methods are widely used to assess molecular phylogenetic relationships: parsimony, neighbor joining, maximum likelihood, and Bayesian methods (see below).

Polymerase chain reaction (PCR) This is a method that allows researchers to make millions of copies of a DNA sequence in a few hours. PCR requires short DNA **primers** (usually about 20 nucleotides long) that attach to each end of the DNA sequence of interest. Hence, one limitation is that you need information about the DNA sequence in order to design primers. Fortunately, there are now mtDNA and intron primers that work on a broad range of fishes. PCR is the indispensable starting point for most molecular genetic surveys of fishes. It also allows DNA sequence information to be recovered from very small amounts of tissue (fin clips, scales, muscle biopsy, a drop of blood) and partially degraded tissues such as stomach contents and feces.

Population structure This term refers to departures from random mating within a species. Most species have groups of individuals that interbreed but are semi-isolated from other groups of interbreeding individuals. These population separations are informative for life history studies and conservation.

Deep population structure refers to cases where differences between populations are not based on haplotype (or allele) frequencies, but on accumulated differences in DNA sequences. These populations have been isolated for so long that they are identified by **diagnostic** differences in DNA sequences (Fig. 17.1A) or alleles that indicate a population is **monophyletic** (every member shares DNA differences not found elsewhere; see Chapter 2). For example, if allozyme studies reveal that population one has allele A at 100% frequency, and population two has allele B at 100% frequency ($F_{ST} = 1$), and this monophyletic pattern recurs across several loci, this indicates deep population structure and an ancient separation between populations, on the order of hundreds of thousands to millions of years ago (Avise 2004). With DNA sequence data, differences between these deep populations are measured in sequence divergence (*d*; see below, Phylogeography) and diagnostic mutations: at one site on the DNA sequence, all the individuals in population one have nucleotide C, all the individuals in population two have nucleotide T, and this pattern is repeated at many sites along the DNA sequence. This condition implies some **evolutionary depth**, and the relationships between populations can be visualized with a phylogenetic tree rather than expressed with *F* statistics (Moritz 1994). This is a common condition in surveys of freshwater species, because of the imposing geological barriers between drainages (Roman et al. 1999). Deep population structure may indicate the presence of **cryptic species**, different species that were previously thought to be the same but are distinguished by DNA data. For example, the Atlantic Redlip Blenny (*Ophioblennius atlanticus*), previously thought to be a single species, may include up to five species at distinct regions of the Atlantic Ocean (Muss et al. 2001) (Fig. 17.1B). Discovering new evolutionary lineages is one of the exciting aspects of fish phylogeography.

Shallow population structure refers to groups of individuals that have significantly different haplotype (or allele) frequencies. For example, the soldierfish *Myripristis berndti* is distributed across the Indian and Pacific oceans, and there is a common mtDNA haplotype shared across this range, but it occurs at 45% frequency in the West Pacific, and 15% frequency in the Indian Ocean. The corresponding *F* statistic ($\phi_{ST} = 0.58$; Craig et al. 2007) indicates that the Indian Ocean and West Pacific contain distinct management units (as discussed in Chapter 20). However, the fact that they share this haplotype indicates that they are closely related with shallow population structure, rather than distinct evolutionary lineages (subspecies or species).

Sequence divergence This is usually expressed as *d*, the percent difference between two DNA sequences. For example, $d = 0.10$ means that an estimated 10% of the DNA sequence has changed between two individuals, populations, or species. The level of sequence divergence can vary from near zero between closely related species (such as the cichlid species flocks of East Africa) to upwards of $d = 0.20$. Above this range, sequences become **saturated**, meaning there are so many mutations that some nucleotide sites have changed two or three times, obscuring the true history. For example, a site that appears to have a mutation from G to C may actually have mutated from G to A to C. Saturated sequences cannot provide optimal resolution of evolutionary relationships, but may still be informative. A typical level of mtDNA sequence divergence between fishes in the same genus would be $d = 3–15\%$, but differentiation in a few genera can exceed $d = 30\%$, including *Galaxias* (Galaxiidae, mudfish and their relatives) (Johns & Avise 1998).

Single-nucleotide polymorphisms (SNPs) The workhorse of modern DNA studies in fish population structure, phylogenetics, and evolutionary response to selection. Next-generation sequencing (NGS) produces thousands or millions of short DNA fragments that are screened for variable sites, usually a single-nucleotide mutation (C to A, for example). These SNPs (pronounced "snips") are used to determine overall patterns of variation in the genome. They can resolve close relatives (kinship), genes under strong selection (possibly due to environmental degradation), population of interbreeding individuals (useful for management), and phylogenetic history. Comparisons of human genomes typically reveal 4–5 million SNPs, so they represent an abundance of variations that can be used to resolve natural history.

References

Aalbers SA, Bernal D, Sepulveda CA. 2010. The functional role of the caudal fin in the feeding ecology of the common thresher shark Alopias vulpinus. *J Fish Biol* 76(7):1863–1868. **https://doi.org/10.1111/j.1095-8649.2010.02616**.

Abdel Magid AM. 1966. Breathing and function of the spiracles in *Polypterus senegalus*. *Anim Behav* 14:530–533.

Abdel Magid AM. 1967. Respiration of air by the primitive fish *Polypterus senegalus*. *Nature (Lond)* 215:1096–1097.

Abell RA, Olson DM, Dinerstein E, et al. 2000. *Freshwater ecoregions of North America*. Washington, DC: Island Press.

Able KW, Fahay MP. 1998. *The first year in the life of estuarine fishes of the Middle Atlantic Bight*. New Brunswick, NJ: Rutgers University Press.

Abrahams M. 1989. Foraging guppies and the ideal free distribution: the influence of information on patch choice. *Ethology* 82:116–126.

Abrahams M, Colgan P. 1985. Risk of predation, hydrodynamic efficiency and their influence on school structure. *Env Biol Fish* 13:195–202.

Abrahams M, Dill LM. 1989. A determination of the energetic equivalence of the risk of predation. *Ecology* 70:999–1007.

Abu-Gideiri YB. 1966. The behavior and neuroanatomy of some developing teleost fishes. *J Zool* 149:215–241.

Ackerman PA, Wicks BJ, Iwama GK, Randall DJ. 2006. Low levels of environmental ammonia increase susceptibility to disease in Chinook salmon smolts. *Physiol Biochem Zool* 79:695–707.

Adams AJ, Dahlgren C, Kellison GT, et al. 2006. Nursery function of tropical back-reef systems. *Mar Ecol Prog Ser* 318:287–301.

Adams SM, ed. 1990. *Biological indicators of stress in fish*. American Fisheries Society Symposium 8. Bethesda, MD: American Fisheries Society.

Adams SM. 2002. Biological indicators of aquatic ecosystem stress: introduction and overview. In: Adams SM, ed. *Biological indicators of aquatic ecosystem stress*, pp. 1–11. Bethesda, MD: American Fisheries Society.

Adams SR, Parsons GR. 1998. Laboratory-based measurements of swimming performance and related metabolic rates of field-sampled smallmouth buffalo (*Ictiobus bubalus*): a study of season changes. *Physiol Biochem Zool* 71:350–358.

Agami M, Waisel Y. 1988. The role of fish in distribution and germination of seeds of the submerged macrophytes *Najas marina* L. and *Ruppia maritima* L. *Oecologia* 76:83–88.

Agardy T. 2000. Effects of fisheries on marine ecosystems: a conservationist's perspective. *ICES J Mar Sci* 57:761–765.

Agnese JF, Teugels GG. 2001. The *Bathyclarias–Clarias* species flock. A new model to understand rapid speciation in African Great lakes. *Comptes rendus de l'Académie des sciences Série III Sciences de la vie* 324:683–688.

Agostinho AA, Gomes LC, Verissimo S, Okada EK. 2004. Flood regime, dam regulation and fish in the Upper Parana River: effects on assemblage attributes, reproduction and recruitment. *Rev Fish Biol Fisheries* 14:11–19.

Ahas R. 1999. Long-term phyto-, ornitho- and ichthyophenological time-series analyses in Estonia. *Int J Biometeorol* 42:119–123.

Ahas R, Aasa A. 2006. The effects of climate change on the phenology of selected Estonian plant, bird and fish populations. *Int J Biometeorol* 51:17–26.

Ah-King M, Kvarnemo C, Tullberg BS. 2005. The influence of territoriality and mating system on the evolution of male care: a phylogenetic study on fish. *J Evol Biol* 18:371–382.

Ahlberg PE, Clack JA. 2006. A firm step from water to land. *Nature* 440:747–749.

Ahlberg PE, Johanson Z. 1998. Osteolepiforms and the ancestry of the tetrapods. *Nature* 395:792–794.

Ahlstrom EH, Amaoka K, Hensley DA, Moser HG, Sumida BY. 1984. Pleuronectiformes: development. In: Moser HG, Richards WJ, Cohen DM, Fahay MP, Kendall AW, Jr., Richardson SL, eds. *Ontogeny and systematics of fishes*, pp. 640–670. Special Publication No. 1. Lawrence, KS: American Society of Ichthyologists and Herpetologists.

Ahlstrom EH, Ball OP. 1954. Description of eggs and larvae of jack mackerel (*Trachurus symmetricus*) and distribution and abundance of larvae in 1950 and 1951. *Fish Bull US* 56:295–329.

Akamatsu T, Nanami A, Yan HY. 2003. Spotlined sardine *Sardinops melanostictus* listens to 1-kHz sound by using its gas bladder. *Fish Sci* 69:348–354.

Akihito P. 1986. Some morphological characters considered to be important in gobiid phylogeny. In: Uyeno T, Arai R, Taniuchi T, Matsuura K, eds. *Indo-Pacific fish biology: proceedings of the 2nd International Conference on Indo Pacific Fish*, pp. 629–639. Tokyo: Ichthyology Society of Japan.

Albanese B, Angermeier PL, Dorai-Raj S. 2004. Ecological correlates of fish movement in a network of Virginia streams. *Can J Fish Aquat Sci* 61:857–869.

Albert J, Reis RE. 2011. *Historical biogeography of neotropical freshwater fishes*. Berkeley: University of California Press.

Alcoverro T, Mariani S. 2004. Patterns of fish and sea urchin grazing on tropical Indo-Pacific seagrass beds. *Ecography* 27:361–365.

Alder MN, Rogozin IB, Iyer LM, Glazko GV, Cooper MD, Pancer Z. 2005. Diversity and function of adaptive immune receptors in a jawless vertebrate. *Science* 310:1970–1973.

Aldridge RJ, Briggs DEG, Smith MP, Clarkson ENK, Clark NDL. 1993. The anatomy of conodonts. *Phil Trans R Soc Lond B* 340:405–421.

Alexander GR. 1979. Predators of fish in coldwater streams. In: Clepper H, ed. *Predator–prey systems in fisheries management*, pp. 153–170. Washington, DC: Sport Fishing Institute.

The Diversity of Fishes: Biology, Evolution and Ecology, Third Edition. Douglas E. Facey, Brian W. Bowen, Bruce B. Collette, and Gene S. Helfman.

Companion website: www.wiley.com/go/facey/diversityfishes3

Alexander RM. 1983. *Animal mechanics*, 2nd edn. Oxford: Blackwell Science.

Alexander RM. 1993. Buoyancy. In: Evans DH, ed. *The physiology of fishes*, pp. 75–97. Boca Raton, FL: CRC Press.

Alexander RMN. 2011. Buoyancy in fishes. In: Farrell AP, Stevens ED, Cech JJ, Richards JG, eds. *Encyclopedia of fish physiology: from genome to environment*, pp. 520–525. Boston: Elsevier/Academic Press.

Ali MA, ed. 1980. *Environmental physiology of fishes*. New York: Plenum Press.

Ali MA, ed. 1992. *Rhythms in fishes*. New York: Plenum Press.

Allan JD. 1982. The effects of reduction in trout density on the invertebrate community of a mountain stream. *Ecology* 63:1444–1455.

Allan JD. 1983. Predator–prey relationships in streams. In: Barnes JR, Minshall GW, eds. *Stream ecology: application and testing of general ecological theory*, pp. 191–229. New York: Plenum Press.

Allan JD, Abell R, Hogan Z, et al. 2005. Overfishing of inland waters. *BioScience* 55:1041–1051.

Allan JD, Castillo MM. 2007. *Stream ecology: structure and function of running water*, 2nd edn. Dordrecht: Springer-Verlag.

Allan JR. 1986. The influence of species composition on behaviour in mixed-species cyprinid shoals. *J Fish Biol* 29(A):97–106.

Allen GR. 1975. *Anemonefishes*, 2nd edn. Neptune City, NJ: TFH Publications, Inc.

Allen GR. 1991. *Damselfishes of the world*. Melle, Germany: Mergus.

Allen GR. 2008. Conservation hotspots of biodiversity and endemism for Indo-Pacific coral reef fishes. *Aquatic Conserv: Mar Freshw Ecosys* 18:541–556.

Allen GR, Erdmann MV. 2012. *Reef Fishes of the East Indies*, Vol. 1. Perth: Tropical Reef Research.

Allen LG. 1979. Larval development of *Gobiesox rhessodon* (Gobiesocidae) with notes on the larva of *Rimicola muscarum*. *Fish Bull US* 77:300–304.

Allen LG, Cross JN. 2006. Surface waters. In: Allen LG, Pondella DJ, II, Horn MH, eds. *The ecology of marine fishes: California and adjacent waters*, pp. 320–341. Berkeley, CA: University of California Press.

Allen LG, Pondella DJ, II. 2006. Ecological classification. In: Allen LG, Pondella DJ, II, Horn MH, eds. *The ecology of marine fishes: California and adjacent waters*, pp. 81–113. Berkeley, CA: University of California Press.

Allen LG, Pondella DJ, II, Horn MH. 2006. *The ecology of marine fishes. California and adjacent waters*. Berkeley, CA: University of California Press.

Allen MJ. 2006. Continental shelf and upper slope. In Allen LG, Pondella DJ, II, Horn MH, eds. *The ecology of marine fishes: California and adjacent waters*, pp. 167–202. Berkeley, CA: University of California Press.

Allen SK, Jr., Wattendorf RJ. 1987. Triploid grass carp: status and management implications. *Fisheries* 12(4):20–24.

Allendorf FW, Ferguson MM. 1990. Genetics. In: Schreck CB, Moyle PB, eds. *Methods for fish biology*, pp. 35–63. Bethesda, MD: American Fisheries Society.

Allendorf FW, Hard JJ. 2009. Human-induced evolution caused by unnatural selection through harvest of wild animals. *PNAS* 106 (suppl 1):9987–9994

Allendorf FW, Luikart G, Aitken SN. 2012. *Conservation and the genetics of populations*, 2nd edn. Hoboken: John Wiley & Sons.

Allendorf FW, Thorgaard GH. 1984. Tetraploidy and the evolution of salmonid fishes. In: Turner BJ, ed. *Evolutionary genetics of fishes*, pp. 1–53. New York: Plenum Press.

Almada VC, Oliveira RF, Gonçalves EJ, eds. 1999. *Behaviour and conservation of littoral fishes*. Lisboa, Portugal: Instituto Superior de Psicologia Aplicada.

Almeida JR, Oliveira C, Gravato C, Guilhermino L. 2010. Linking behavioural alterations with biomarkers reponses in the European seabass *Dicentrarchus labrax* L. exposed to the organophophate pesticide fenitrothion. *Ecotoxicology* 19:1369–1381.

Alverson DL. 1997. Global assessment of fisheries bycatch and discards: a summary overview. In: Pikitch EK, Huppert DD, Sissenwine MP, eds. *Global trends: fisheries management*, pp. 115–125. American Fisheries Society Symposium 20. Bethesda, MD: American Fisheries Society.

Alverson DL, Freeberg MH, Pope JG, Murawski SA. 1994. *A global assessment of fisheries bycatch and discards*. FAO Fisheries Technical Paper No. 339. Rome: Food and Agricultural Organization.

Alves-Gomes JA. 2001. The evolution of electroreception and bioelectrogenesis in teleost fish: a phylogenetic perspective. *J Fish Biol* 58:1489–1511.

Ambrose RF, Anderson TW. 1990. Influence of an artificial reef on the surrounding infaunal community. *Mar Biol* 107:41–52.

Amemiya CT, Alfoldi J, Lee AP, et al. 2013. The African coelacanth genome provides insight into tetrapod evolution. *Nature* 496:311q–316.

American Fisheries Society. 1984. Rhythmicity in fishes. *Trans Am Fish Soc* 113:411–552.

American Fisheries Society. 1990a. Effects of global climate change on our fisheries resources. *Fisheries* 15(6):2–44.

American Fisheries Society. 1990b. Symposium on effects of climate change on fish. *Trans Am Fish Soc* 119:173–389.

Amiard-Triquet C., Amiard JC. 2013. Introduction. In: C. Amiard-Triquet, J.C. Amiard, P.S. Rainbow, eds. *Ecological Biomarkers: indicators of ecotoxicological effects*, pp. 1–14. Boca Raton: CRC Press.

Amores A, Force A, Yan Y-L, et al. 1998. Zebrafish hox clusters and vertebrate genome evolution. *Science* 282:1711–1714.

Amorim MCP. 2006. Diversity of sound production in fish. In: Ladich F, Collin SP, Møller P, Kapoor BG, eds. *Communication in fishes*, Vol. 1, pp. 71–105. Enfield, NH: Science Publishers.

Anders PJ. 1998. Conservation aquaculture and endangered species: can objective science prevail over risk anxiety? *Fisheries* 23(11):28–31.

Anderson DH. 2014. Geomorphic responses to interim hydrology following phase I of the Kissimmee River restoration project, Florida. *Restor Ecol* 22(3):367–375.

Anderson ED. 1990. Fisheries models as applied to elasmobranch fisheries. In: Pratt HL, Jr., Gruber SH, Taniuchi T, eds. *Elasmobranchs as living resources: advances in the biology, ecology, systematics, and the status of fisheries*, pp. 473–484. NOAA Technical Report No. 90. Washington, DC: National Oceanic and Atmospheric Administration.

Anderson ED, Mayo RK, Sosebee K, Terceiro M, Wigley SE. 1999. *Our living oceans. Unit 1, Northeast demersal fisheries*. NOAA Technical Memorandum NMFS-F/SPO-41. Washington, DC: National Oceanic and Atmospheric Administration.

Anderson RC, Waheed A. 2001. *The economics of shark and ray watching in the Maldives*. Shark News No. 13. Newbury, Berkshire UK: IUCN Shark Specialist Group. **www.sharktrust.org/cgi/press/press_downloads/maldives.doc**.

Anderson RO, Gutreuter SJ. 1984. Length, weight, and associated structural indices. In: Nielsen LA, Johnson, DL, eds. *Fisheries techniques*. Bethesda, MD: American Fisheries Society.

Andersson KG, Brönmark C, Herrmann J, Malmqvist B, Otto C, Sjörström P. 1986. Presence of sculpins (*Cottus gobio*) reduces drift and activity of *Gammarus pulex* (Amphipoda). *Hydrobiologia* 133:209–215.

Andrews AH, Siciliano D, Potts DC, DeMartini EE, Covarrubias S. 2016. Bomb radiocarbon and the Hawaiian Archipelago: coral, otoliths, and seawater. *Radiocarbon* 58:531–548.

Andrews KR, Williams AJ, Fernandez-Silva I, et al. 2016. Phylogeny of deepwater snappers (Genus *Etelis*) reveals a cryptic species pair in the Indo-Pacific and Pleistocene invasion of the Atlantic. *Mol Phylogenet Evol* 100:361–371.

Andruszkiewicz EA, Starks HA, Chavez FP, Sassoubre LM, Block BA, Boehm AB. 2017. Biomonitoring of marine vertebrates in Monterey Bay using eDNA metabarcoding. *PLoS ONE* 12(4):e0176343.

Angermeier PL, Karr JR. 1983. Fish communities along environmental gradients in a system of tropical streams. *Env Biol Fish* 9:117–135.

ANS (Aquatic Nuisance Species Taskforce). 2011. Lionfish (*Pterois volitans, Pterois miles*). **https://www.anstaskforce.gov/spoc/lionfish.php**

Applegate SP, Espinosa L. 1996. The fossil history of *Carcharodon* and its possible ancestor, *Cretolamna*: a study in tooth identification. In: Klimley AP, Ainley DG, eds. *Great white sharks: the biology of Carcharodon carcharias*, pp. 19–36. San Diego, CA: Academic Press.

Arcand-Hoy LS, Benson WH. 1998. Fish reproduction: an ecologically relevant indicator of endocrine disruption. *Environ Toxicol Chem* 17:49–57.

Arcila D, Hughes LC, Meléndez-Vazquez F, et al. 2021. Testing the utility of alternative metrics of branch support to address the ancient evolutionary radiation of tunas, stromateoids, and allies (Teleostei: Pelagiaria). *Syst Biol*. **https://doi.org/10.1093/sysbio/syab018**.

Arlinghaus R, Laskowski KL, Alós J, et al. 2016. Passive gear-induced timidity syndrome in wild fish populations and its potential ecological and managerial implications. *Fish and Fish*. **https://doi.org/10.1111/faf.12176**.

Arnal C, Verneau O, Desdevises Y. 2006. Phylogenetic relationships and evolution of cleaning behaviour in the family Labridae: importance of body colour pattern. *J Evol Biol* 19:755–763.

Arnegard ME, Carlson BA. 2005. Electric organ discharge patterns during group hunting by a Mormyrid fish. *Proc Roy Soc B* 272:1305–1314.

Arnegard ME, Hopkins CD. 2003. Electric signal variation among seven blunt-snouted *Brienomyrus* species (Teleostei: Mormyridae) from a riverine species flock in Gabon, Central Africa. *Environ Biol Fish* 67:321–339.

Aronson RB. 1983. Foraging behavior of the west Atlantic trumpetfish, *Aulostomus maculatus*: use of large, herbivorous reef fishes as camouflage. *Bull Mar Sci* 33:166–171.

Arratia G. 1997. *Basal teleosts and teleostean phylogeny*. Palaeo Ichthyologica 7. Munchen: Verlag Dr. Friedrich Pfeil.

Arratia G, Kapoor BG, Chardon M, Diogo R, eds. 2003. *Catfishes*, Vols. 1 and 2. Enfield, NH: Science Publishers.

Artero C, Koenig CC, Richard P, et al. 2015. Ontogenetic dietary and habitat shifts in goliath grouper *Epinephelus itajara* from French Guiana. *Endanger Species Res* 27:155–168

Arthington AH, Dulvy NS, Gladstone W, Winfield IJ. 2016.Fish conservation in freshwater and marine realms: status, threats, and management. *Aquatic Conserv: Mar Freshw Ecosyst* 26:838–857.

Arukwe A. 2001. Cellular and molecular responses to endocrine-modulators and the impact on fish reproduction. *Mar Pollution Bull* 42:643–655.

Arukwe A, Goksoyr A. 1998. Xenobiotics, xenoestrogens and reproduction disturbances in fish. *Sarsia* 83:225–241.

Arvedlund M, McCormick MI, Fautin DG, Bildsoe M. 1999. Host recognition and possible imprinting in the anemonefish *Amphiprion melanopus* (Pisces: Pomacentridae). *Mar Ecol Prog Ser* 188:207–218.

Ashworth JS, Bruce OE, El Hellw M. 2006. Fish assemblages of Red Sea backreef biotopes. *Aquat Conserv Mar Freshwater Ecosyst* 16:593–609.

Atema J, Gerlach G, Paris CB. 2015. Sensory biology and navigation behavior of reef fish larvae. In: Moura C, ed. *Ecology of fishes on coral reefs*, pp. 3–15. Cambridge: Cambridge University Press.

Atema J, Kingsford MJ, Gerlach G. 2002. Larval reef fish could use odour for detection, retention and orientation to reefs. *Mar Ecol Prog Ser* 241:151–160.

Augerot X. 2005. *Atlas of Pacific salmon. The first map-based status assessment of salmon in the North Pacific*. Berkeley, CA: University of California Press.

Ault TR, Johnson CR. 1998. Spatial variation in fish species richness on coral reefs: habitat fragmentation and stochastic structuring processes. *Oikos* 82:354–364.

Avise JC. 2000. *Phylogeography, the history and formation of species*. Cambridge, MA: Harvard University Press.

Avise JC. 2004. *Molecular markers, natural history, and evolution*, 2nd edn. Sunderland, MA: Sinauer Associates.

Avise JC, Arnold J, Ball RM, et al. 1987. Intraspecific phylogeography: the mitochondrial DNA bridge between population genetics and systematics. *Ann Rev Ecol Syst* 18:489–522.

Avise JC, Helfman GS, Saunders NC, Hales LS. 1986. Mitochodrial DNA differentiation in North Atlantic eels: population genetic consequences of an unusual life history pattern. *Proc Natl Acad Sci USA* 83:4350–4354.

Avise JC, Jones AG, Walker D, et al. 2002. Genetic mating systems and reproductive natural history of fishes: lessons for ecology and evolution. *Ann Rev Genet* 36:19–45.

Avise JC, Nelson WS, Arnold J, Koehn RK, Williams GC, Thorsteinsson V. 1990. The evolutionary genetic status of Icelandic eels. *Evolution* 44:1254–1262.

Avise JC, Shapiro DY. 1986. Evaluating kinship of newly-settled juveniles within social groups of the coral reef fish *Anthias squamipinnis*. *Evolution* 40:1051–1059.

Axelrod HR, Burgess WE. 1976. *African cichlids of Lakes Malawi and Tanganyika*, 5th edn. Neptune City, NJ: TFH Publications, Inc.

Azuma T, Takeda K, Doi T, et al. 2004. The influence of temperature on sex determination in sockeye salmon *Oncorhynchus nerka*. *Aquaculture* 234:461–473.

Baba R, Karino K. 1998. Countertactics of the Japanese aucha perch *Siniperca kawamebari* against brood parasitism by the Japanese minnow *Pungtungia herzi*. *J Ethol* 16:67–72.

Bachman RA. 1984. Foraging behavior of free-ranging wild and hatchery brown trout in a stream. *Trans Am Fish Soc* 113:1–32.

Bachmann K. 1972. Nuclear DNA and developmental rate in frogs. *Q J Florida Acad Sci* 35:225–231.

Baggerman B. 1985. The role of biological rhythms in the photoperiodic regulation of seasonal breeding in the stickleback *Gasterosteus aculeatus*, L. *Neth J Zool* 35:14–31.

Baggerman B. 1990. Sticklebacks. In: Munro AD, Scott AP, Lam TJ, eds. *Reproductive seasonality in teleosts: environmental influences*, pp. 79–107. Boca Raton, FL: CRC Press.

Bailey C. 2015. Transgenic salmon: science, politics, and flawed policy. *Soc Nat Resour* 28:1249–1260.

Bailey KM, Houde ED. 1989. Predation on eggs and larvae of marine fishes and the recruitment problem. *Adv Mar Biol* 25:1–83.

Bailey TG, Robertson DR. 1982. Organic and caloric levels of fish feces relative to its consumption by coprophagous reef fishes. *Mar Biol* 69:45–50.

Baird OE, Krueger CC. 2003. Behavioral thermoregulation of brook and rainbow trout: comparison of summer habitat use in an Adirondack River, New York. *Trans Am Fish Soc* 132:1194–1206.

Baird TA. 1988. Female and male territoriality and mating system of the sand tilefish, *Malacanthus plumieri*. *Env Biol Fish* 22: 101–116.

Bajer PG, Whitledge GW, Hayward RS, Zweifel RD. 2003. Laboratory evaluation of two bioenergetics models applied to yellow perch: identification of a major source of systematic error. *J Fish Biol* 62:436–454.

Baker CS, Cipriano F, Palumbi SR. 1996. Molecular genetic identification of whale and dolphin products from commercial markets in Korea and Japan. *Molec Ecol* 5:671–685.

Baker DW, Sardella B, Rummer JL, Sackville M, Brauner CJ. 2015. Hagfish: champions of CO_2 tolerance question the origins of vertebrate gill function. *Sci Rep* 5:11182. **https://doi.org/10.1038/srep11182**.

Baker JP, Christensen SW. 1991. Effects of acidification on biological communities in aquatic ecosystems. In: Charles DG, ed. *Acidic deposition and aquatic ecosystems: regional case studies*, pp. 83–106. New York: Springer-Verlag.

Baker JP, Warren-Hicks WJ, Gallagher J, et al. 1993. Fish population losses from Adirondack lakes: the role of surface water acidity and acidification. *Water Resour Res* 29:861–874.

Baker JR, Schofield CL. 1985. Acidification impacts on fish populations: a review. In: Adams DD, Page WP, eds. *Acid deposition: environmental, economic and policy issues*, pp. 183–221. New York: Plenum Press.

Baker RR. 1978. *The evolutionary ecology of animal migration*. New York: Holmes & Meier.

Bakun A. 1990. Global climate change and intensification of coastal ocean upwelling. *Science* 247:198–201.

Balasingham KD, Walter RP, Mandrak NE, Heath DD. 2018. Environmental DNA detection of rare and invasive fish species in two Great Lakes tributaries. *Mol Ecol* 27:112–127.

Baldigo BP, Lawrence G, Simonin H. 2007. Persistent mortality of brook trout in episodically acidified streams of the southwestern Adirondack Mountains, New York. *Trans Am Fish Soc* 136:121–134.

Baldridge HD. 1970. Sinking factors and average densities of Florida sharks as functions of liver buoyancy. *Copeia* 1970:744–754.

Baldridge HD. 1972. Accumulation and function of liver oil in Florida sharks. *Copeia* 1972:306–325.

Baldwin CC, Johnson GD. 2014. Connectivity across the Caribbean Sea: DNA barcoding and morphology unite an enigmatic fish larva from the Florida straits with a new species of sea bass from deep reefs off Curaçao. *PLoS One*. **https://doi.org/10.1371/journal.pone.0097661**.

Bale R, Hao M, Singh Bhalla AP, Patankar NA. 2014. Energy efficiency and allometry of movement of swimming and flying animals. *Proc Natl Acad Sci* 111:7517–7521.

Ballantyne JS, Robinson JW. 2011. Physiology of sharks, skates, and rays. In: Farrell AP, Stevens ED, Cech JJ, Richards JG, eds. *Encyclopedia of fish physiology: from genome to environment*, pp. 1807–1818. Boston: Elsevier/Academic Press.

Ballintijn CM, Beatty DD, Saunders RL. 1977. Effects of pseudobranchectomy on visual pigment density and ocular PO_2 in Atlantic salmon, *Salmo salar*. *J Fish Res Board Can* 34:2185–2192.

Balon EK. 1975a. Reproductive guilds of fishes: a proposal and definition. *J Fish Res Board Can* 6:821–864.

Balon EK. 1975b. Terminology of intervals in fish development. *J Fish Res Board Can* 32:1663–1670.

Balon EK. 1981. Additions and amendments to the classification of reproductive styles in fishes. *Env Biol Fish* 6:377–389.

Balon EK. 1995. Origin and domestication of the wild carp, *Cyprinus carpio*: from Roman gourmets to the swimming flowers. *Aquaculture* 129:3–48.

Balon EK, Bruton MN. 1986. Introduction of alien species or why scientific advice is not heeded. *Env Biol Fish* 16:225–230.

Balon EK, Bruton MN, Fricke H. 1988. A fiftieth anniversary reflection on the living coelacanth, *Latimeria chalumnae*: some new interpretations of its natural history and conservation status. *Env Biol Fish* 23:241–280.

Balon EK, Bruton MN, Noakes DLG. 1994. Women in ichthyology: an anthology in honor of ET, Ro and Genie. *Env Biol Fish* 41:7–438.

Balon EK, Stewart DJ. 1983. Fish assemblages in a river with unusual gradient (Luongo, Africa – Zaire system), reflections on river zonation, and description of another new species. *Env Biol Fish* 9:225–252.

Balshine S, Sloman KA. 2011. Parental care in fishes. In: Farrell AP, Stevens ED, Cech JJ, Richards JG, eds. *Encyclopedia of fish physiology: from genome to environment*, pp. 670–677. Boston: Elsevier/Academic Press.

Baltazar-Soares M, Biastoch A, Harrod C, et al. 2014. Recruitment collapse and population structure of the european eel shaped by local ocean current dynamics. *Curr Biol* 24:104–108.

Banford HM, Bermingham E, Collette BB. 2004. Molecular phylogenetics and biogeography of transisthmian and amphi-Atlantic needlefishes (Belonidae: *Strongylura* and *Tylosurus*): perspectives on New World marine speciation. *Molec Phylogen Evol* 31:833–851.

Banford HM, Bermingham E, Collette BB, McCafferty SS. 1999. Phylogenetic systematics of the *Scomberomorus regalis* (Teleostei: Scombridae) species group: molecules, morphology and biogeography of Spanish mackerels. *Copeia* 1999:596–613.

Barbarino-Duque A, Taphorn DC, Winemiller KO. 1998. Ecology of the coporo, *Prochilodus mariae* (Characiformes, Prochilodontidae), and status of annual migrations in western Venezuela. *Env Biol Fish* 53:33–46.

Barber CV, Pratt VR. 1997. *Sullied seas: strategies for combating cyanide fishing in southeast Asia and beyond*. Washington, DC: World Resources Institute and Manila, International Marinelife Alliance. **www.marine.org/PDF_Downloads/SulliedSeas**.

Barber I, Hoare D, Krause J. 2000. Effects of parasites on fish behaviour: a review and evolutionary perspective. *Rev Fish Biol Fisheries* 10:131–165.

Barber I, Wright HA. 2001. How strong are familiarity preferences in shoaling fish? *Anim Behav* 61:975–979.

Barbour CD. 1973. A biogeographical history of *Chirostoma* (Pisces: Atherinidae): a species flock from the Mexican Plateau. *Copeia* 1973:533–556.

Bardack D. 1991. First fossil hagfish (Myxinoidea): a record from the Pennsylvanian of Illinois. *Science* 254:701–703.

Bardack D, Zangerl R. 1971. Lampreys in the fossil record. In: Hardisty MW, Potter IC, eds. *The biology of lampreys*, pp. 67–84. New York: Academic Press.

Bariche M, Kleitou P, Kalogirou S, Bernardi G. 2017. Genetics reveal the identity and origin of lionfish invasion in the Meditteranean Sea. *Sci Rep* 7:6782.

Barker R. 1997. *And the waters turned to blood*. New York: Simon & Schuster.

Barkhymer AJ, Garrett SG, Wisenden BD. 2019. Olfactorily-mediated cortisol response to chemical alarm cues in zebrafish *Danio rerio*. *J Fish Biol* 95:287–292.

Barlow GW. 1961. Causes and significance of morphological variation in fishes. *Syst Zool* 10:105–117.

Barlow GW. 1963. Ethology of the Asian teleost, *Badis badis*: I. Motivation and signal value of the colour patterns. *Anim Behav* 11:97–105.

Barlow GW. 1967. The functional significance of the split-head color pattern as exemplified in a leaf fish, *Polycentrus schomburgkii*. *Ichthyologica* 39:57–70.

Barlow GW. 1981. Patterns of parental investment, dispersal and size among coral reef fishes. *Env Biol Fish* 6:65–85.

Barlow GW. 1984. Patterns of monogamy among teleost fishes. *Arch Fischereiwiss* 35:75–123.

Barlow GW. 1986. A comparison of monogamy among freshwater and coral-reef fishes. In: Uyeno T, Arai R, Taniuchi T, Matsuura K, eds. *Indo-Pacific fish biology. Proceedings of the 2nd International Conference of Indo-Pacific Fishes*, pp. 767–775. Tokyo: Ichthyological Society of Japan.

Barlow GW. 1991. Mating systems among cichlid fishes. In: Keenleyside MHA, ed. *Cichlid fishes: behaviour, ecology and evolution*, pp. 173–190. London: Chapman & Hall.

Barlow GW. 1992. Is mating different in monogamous species? The midas cichlid fish as a case study. *Am Zool* 32:91–99.

Barlow GW. 2000. *The cichlid fishes: nature's grand experiment in evolution*. Cambridge, MA: Perseus Publishing.

Barnum DA, Bradley T, Cohen M, Wilcox B, Yanega G. 2017. *State of the Salton Sea—A science and monitoring meeting of scientists for the Salton Sea: U.S. Geological Survey Open-File Report 2017-1005*, 20 p., **https://doi.org/10.3133/ofr20171005**.

Barrett JC. 1989. *The effects of competition and resource availability on the behavior, microhabitat use, and diet of the mottled sculpin (Cottus bairdi)*. PhD dissertation, University of Georgia, Athens, GA.

Barrett S. 2000. FDA and FTC attack shark cartilage. **www.quackwatch.com/04ConsumerEducation/news/shark.html** (June 9, 2000).

Barros NB, Myrberg AA, Jr. 1987. Prey detection by means of passive listening in bottlenose dolphins (*Tursiops truncatus*). *J Acoust Soc Am* 82 (suppl 1):S65.

Barthem R, Goulding M. 1997. *The catfish connection: ecology, migration, and conservation of Amazon predators*. New York: Columbia University Press.

Bartol IK, Gharib M, Webb PW, Weihs D, Gordon MS. 2005. Body-induced vortical flows: a common mechanism for self-corrective trimming control in boxfishes. *J Experimental Biol* 208:327–344.

Bartol IK, Gharib M, Weihs D, Webb PW, Hove JR, Gordon MS. 2003. Hydrodynamic stability of swimming in ostraciid fishes: role of the carapace in the smooth trunkfish *Lactophrys triqueter* (Teleostei: Ostraciidae). *J Exp Biol* 206:725–744.

Barton BA, Morgan JD, Vijayan MM. 2002. Physiological and condition-related indicators of environmental stress in fish. In: Adams SM, ed. *Biological indicators of aquatic ecosystem stress*, pp. 111–144. Bethesda, MD: American Fisheries Society.

Barton M. 2006. *Bond's biology of fishes*, 3rd edn. Stanford, CT: Thomson Brooks/Cole.

Bartsch P. 1997. Aspects of craniogenesis and evolutionary biology in polypteriform fishes. *Neth J Zool* 47:365–381.

Bartsch P, Britz R. 1997. A single micropyle in the eggs of the most basal living actinopterygian fish, *Polypterus* (Actinopterygii, Polypteriformes). *J Zool* 241(3):589–592.

Bash JS, Ryan CM. 2002. Stream restoration and enhancement projects: is anyone monitoring? *Environ Manage* 29:877–885.

Basolo AL. 1990a. Female preference for male sword length in the green swordtail, *Xiphophorus helleri* (Pisces: Poeciliidae). *Anim Behav* 40:332–338.

Basolo AL. 1990b. Female preference predates the evolution of the sword in swordtail fish. *Science* 250:808–810.

Basolo AL, Alcaraz G. 2003. The turn of the sword: length increases male swimming costs in swordtails. *Proc Roy Soc Lond Ser B Biol Sci* 270:1631–1636.

Bass AH. 1996. Shaping brain sexuality. *Am Sci* 84:352–363.

Baum JK, Myers RA, Kehler DG, Worm B, Harley SJ, Doherty PA. 2003. Collapse and conservation of shark populations in the Northwest Atlantic. *Science* 299:389–392.

Baum JK, Kehler D, Myers RA. 2005. Robust estimates of decline for pelagic shark populations in the northwest Atlantic and Gulf of Mexico. *Fisheries* 30:27–30.

Baum JK, Myers RA. 2004. Shifting baselines and the decline of pelagic sharks in the Gulf of Mexico. *Ecol Lett* 7:135–145.

Bay LK, Crozier RH, Caley MJ. 2006. The relationship between population genetic structure and pelagic larval duration in coral reef fishes on the Great Barrier Reef. *Mar Biol* 149:1247–1256.

Bayani D-M, Taborsky M, Frommen JG. 2017. To pee or not to pee: urine signals mediate aggressive interactions in the cooperatively breeding cichlid *Neolamprologus pulcher*. *Behav Ecol Sociobiol* 71:37 (10 pp).

Baylis JR. 1981. The evolution of parental care in fishes, with reference to Darwin's rule of male sexual selection. *Env Biol Fish* 6:223–251.

Beamish FWH. 1970. Oxygen consumption of largemouth bass, *Micropterus salmoides*, in relation to swimming speed and temperature. *Can J Zool* 48:1221–1228.

Beamish FWH. 1978. Swimming capacity. In: Hoar WS, Randall DJ, eds. *Locomotion. Fish physiology*, Vol. 7, pp. 101–187. New York: Academic Press.

Beamish RJ, McFarlane GA. 1983. The forgotten requirement of validation of estimates of age determination. *Trans Am Fish Soc* 112:735–743.

Beamish RJ, McFarlane GA. 1987. Current trends in age determination methodology. In: Summerfelt RC, Hall GE, eds. *The age and growth of fish*, pp. 15–42. Ames, IA: Iowa State University Press.

Beamish RJ, Neville C-EM. 1992. The importance of size as an isolating mechanism in lampreys. *Copeia* 1992:191–196.

Beamish RJ, Youson JH. 1987. Life history and abundance of young adult *Lampetra ayresi* in the Fraser River and their possible impact on salmon and herring stocks in the Strait of Georgia. *Can J Fish Aquat Sci* 44:525–537.

Bechler DL. 1983. The evolution of agonistic behavior in amblyopsid fishes. *Behav Ecol Sociobiol* 12:35–42.

Beck MW, Heck KL, Able KW, et al. 2003. The role of nearshore ecosystems as fish and shellfish nurseries. *Issues Ecol* 11:1–12. **www.esa.org/sbi/sbi_issues/**.

Becker GC. 1983. *Fishes of Wisconsin*. Madison, WI: University of Wisconsin Press.

Beecher HA, Dott ER, Fernau RF. 1988. Fish species richness and stream order in Washington State streams. *Env Biol Fish* 22:193–209.

Beets J. 1989. Experimental evaluation of fish recruitment to combinations of fish aggregating devices and benthic artificial reefs. *Bull Mar Sci* 44:973–983.

Beets J. 1991. *Aspects of recruitment and assemblage structure of Caribbean coral reef fishes*. PhD dissertation, University of Georgia, Athens, GA.

Beets J. 1997. Effects of a predatory fish on the recruitment and abundance of Caribbean coral reef fishes. *Mar Ecol Prog Ser* 148:11–21.

Beets J, Friedlander A. 1999. Evaluation of a conservation strategy: a spawning aggregation closure for red hind, *Epinephelus guttatus*, in the U. S. Virgin Islands. *Env Biol Fish* 55:91–98.

Béguer-Pon M, Castonguay M, Shan S, Benchetrit J, Dodson JJ. 2015. Direct observations of American eels migrating across the continental shelf to the Sargasso Sea. *Nat Comm* 6:8705 (9 pages).

Behnke RJ. 1992. *Native trout of western North America.* Bethesda, MD: American Fisheries Society.

Behnke RJ. 2002. *Trout and salmon of North America.* New York: Free Press.

Behrens MD, Lafferty KD. 2007. Temperature and diet effects on omnivorous fish performance: implications for the latitudinal diversity gradient in herbivorous fishes. *Can J Fish Aquat Sci* 64:867–873.

Behrmann-Godel J, Gerlach G, Eckmann R. 2006. Kin and population recognition in sympatric Lake Constance perch (*Perca fluviatilis* L.): can assortative shoaling drive population divergence? *Behav Ecol Sociobiol* 59:461–468.

Beitinger TL, Lutterschmidt WI. 2011. Measures of thermal tolerance. In: Farrell AP, Stevens ED, Cech JJ, Richards JG, eds. *Encyclopedia of fish physiology: from genome to environment*, pp. 1695–1702. Boston: Elsevier/Academic Press.

Bell JD, Clua E, Hair CA, Galzin R, Doherty PJ. 2009. The capture and culture of post-larval fish and invertebrates for the marine ornamental trade. *Rev Fish Sci* 17(2):223–240

Bell JD, Galzin R. 1984. Influence of live coral cover on coral-reef fish communities. *Mar Ecol Prog Ser* 15:265–274.

Bell MA, Foster SA, eds. 1994. *The evolutionary biology of the threespine stickleback.* Oxford: Oxford University Press.

Bell RJ, Richardson DE, Hare JA, Lynch PD, Fratantoni PS. 2015. Disentangling the effects of climate, abundance, and size on the distribution of marine fish: an example based on four stocks from the Northeast US shelf. *ICES J Mar Sci* 72:1311–1322.

Bellwood DR. 1994. A phylogenetic study of the parrotfishes, family Scaridae (Pisces: Labroidei) with a revision of genera. *Rec Aust Mus* 20 (suppl):1–86.

Bellwood DR. 1995a. Carbonate transport and within reef patterns of bioerosion and sediment release by parrotfishes (family Scaridae) on the Great Barrier Reef. *Mar Ecol Prog Ser* 117:127–136.

Bellwood DR. 1995b. Direct estimate of bioerosion by 2 parrotfish species, *Chlorurus gibbus* and *C. sordidus*, on the Great Barrier Reef, Australia. *Mar Biol* 121:419–429.

Bellwood DR. 1996. Production and reworking of sediment by parrotfishes (family Scaridae) on the Great Barrier Reef, Australia. *Mar Biol* 125:795–800.

Bellwood DR, Goatley CHR, Bellwood O, Delbarre DJ, Friedman M. 2015. The rise of jaw protrusion in spiny-rayed fishes closes the gap on elusive prey. *Curr Biol* 25:2696–2700.

Bellwood DR, Hughes TP, Folke C, Nystrom M. 2004. Confronting the coral reef crisis. *Nature* 429:827–833.

Bellwood DR, Wainwright PC. 2002. The history and biogeography of fishes on coral reefs. In: Sale PF, ed. *Coral reef fishes; dynamics and diversity in a complex ecosystem*, pp. 5–32. New York: Academic Press.

Belokopytin YS. 2004. Specific dynamic action of food and energy metabolism of fishes under experimental and natural conditions. *Hydrobiol J* 40:68–75.

Bemis WE. 1987. Feeding systems of living Dipnoi: anatomy and function. In: Bemis WE, Burggren WW, Kemp NE, eds. *The biology and evolution of lungfishes*, pp. 249–275. New York: Alan R. Liss.

Bemis WE, Findeis EK, Grande L. 1997. An overview of Acipenseriformes. *Env Biol Fish* 48:25–72.

Bemis WE, Hetherington TE. 1982. The rostral organ of *Latimeria chalumnae*: morphological evidence of an electroreceptive function. *Copeia* 1982:467–471.

Bemis WE, Hilton EJ, Brown B, et al. 2004. Methods for preparing dry, partially articulated skeletons of osteichthyans, with notes on making Ridewood dissections of the cranial skeleton. *Copeia* 2004:603–607.

Benchley P. 1974. *Jaws.* New York: Doubleday.

Benke AC. 1976. Dragonfly production and prey turnover. *Ecology* 57:915–927.

Benke AC. 1993. Concepts and patterns of invertebrate production in running waters. *Verh Internat Verein Limnol* 25:15–38.

Benke AC, Henry RL, III, Gillespie DM, Hunter RJ. 1985. Importance of snag habitat for animal production in southeastern streams. *Fisheries* 10(5):8–13.

Benz GW, Borucinska JD, Lowry LF, Whiteley HE. 2002. Ocular lesions associated with attachment of the copepod *Ommatokoita elongata* (Lernaeopodidae: Siphonostomatoida) to corneas of Pacific sleeper sharks *Somniosus pacificus* captured off Alaska in Prince William Sound. *J Parasitol* 88(3):474–481.

Benz GW, Bullard SA. 2004. Metazoan parasites and associates of chondrichthyans with emphasis on taxa harmful to captive hosts. In: Smith MD, Warmolts D, Thoney Hueter R, eds. *The elasmobranch husbandry manual: captive care of sharks, rays and their relatives*, pp. 325–416. Columbus, OH: Ohio Biological Survey.

Berbari P, Thibodeau A, Germain L, et al. 1999. Antiangiogenic effects of the oral administration of liquid cartilage extract in humans. *J Surg Res* 87:108–113.

Berenbrink M. 2011. Evolution of the Bohr effect. In: Farrell AP, Stevens ED, Cech JJ, Richards JG, eds. *Encyclopedia of fish physiology: from genome to environment*, pp. 921–928. Boston: Elsevier/Academic Press.

Berg LS. 1947. *Classification of fishes, both recent and fossil.* Ann Arbor, MI: J. W. Edwards Brothers.

Bergenius MAJ, Meekan MG, Robertson DR, McCormick MI. 2002. Larval growth predicts the recruitment success of a coral reef fish. *Oecologia* 131:521–525.

Berglund A, Rosenqvist G, Svensson I. 1986a. Mate choice, fecundity and sexual dimorphism in two pipefish species (Syngnathidae). *Behav Ecol Sociobiol* 19:301–307.

Berglund A, Rosenqvist G, Svensson I. 1986b. Reversed sex-roles and parental energy investment in zygotes of 2 pipefish (Syngnathidae) species. *Mar Ecol Prog Ser* 29:209–215.

Berglund A, Widemo MS, Rosenqvist G. 2005. Sex-role reversal revisited: choosy females and ornamented, competitive males in a pipefish. *Behav Ecol* 16:649–655.

Bergstad OA, Johannesen E, Høines Å, et al. 2017. Demersal fish assemblages in the boreo-Arctic shelf waters around Svalbard during the warm period 2007–2014. Polar Biology published online 17 July 2017.

Bermingham E, Martin AP. 1998. Comparative mtDNA phylogeography of neotropical freshwater fishes: testing shared history to infer the evolutionary landscape of lower Central America. *Molec Ecol* 7:499–517.

Bermingham E, McCafferty SS, Martin AP. 1997. Fish biogeography and molecular clocks: perspectives from the Panamanian Isthmus.

In: Stepien C, Kocher T, eds. *Molecular systematics of fishes*, pp. 113–128. New York: Academic Press.

Bernal D. 2011. An introduction to the biology of pelagic fishes. In: Farrell AP, Stevens ED, Cech JJ, Richards JG, eds. *Encyclopedia of fish physiology: from genome to environment*, pp. 1887–1902. Boston: Elsevier/Academic Press.

Bernal D, Sepulva CA. 2005. Evidence for temperature elevation in aerobic swimming musculature of the common thresher shark, *Alopias vulpinus*. *Copeia* 2005:146–151.

Bernal MA, Sinai NL, Rocha C, Gaither MR, Dunker F, Rocha LA. 2014. Long-term sperm storage in the brownbanded bamboo shark *Chiloscyllium punctatum*. *J Fish Biol*. **https://doi.org/10.1111/jfb.12606**.

Bernardi G. 2000. Barriers to gene flow in *Embiotoca jacksoni*, a marine fish lacking a pelagic larval stage. *Evolution* 54:226–237.

Bernardi G. 2012. The use of tools by wrasses (Labridae). *Coral Reefs* 31:39.

Bernatchez L. 1995. A role for molecular systematics in defining evolutionary significant units in fishes. In: Nielson JL, eds. *Evolution and the aquatic ecosystem: defining unique units in population conservation*, pp. 114–132. Bethesda, MD: American Fisheries Society.

Bernatchez L, Dodson JJ. 1991. Phylogeographic structure in mitochondrial DNA of the lake whitefish (*Coregonus clupeaformis*) and its relation to Pleistocene glaciations. *Evolution* 45:1016–1035.

Bernatchez L, Duchesne P. 2000. Individual based genotype analysis in studies of parentage and population assignment: how many loci, how many alleles? *Can J Fish Aquat Sci* 57:1–12.

Bernatchez L, Renaut S, Whiteley AR, et al. 2010. On the origins of species: insights from the ecological genomics of lake whitefish. *Phil Trans R Soc Lond B Biol Sci* 365:1783–1800.

Bernatchez L, Wilson CC. 1998. Comparative phylogeography of Nearctic and Palearctic fishes. *Molec Ecol* 7:431–452.

Bernhardt ES, Palmer MA, Allan JD, et al. 2005. Synthesizing U.S. river restoration efforts. *Science* 308:636–637.

Bernos T, Jeffries KM, Mandrak NE. 2020. Linking genomics and fish conservation decidion making: a review. *Rev Fish Biol Fisheries*. **https://doi.org/10.1007/s11160-020-09618-8**.

Bernoth E-M, Ellis A, Midtlyng P, et al., eds. 1997. *Furunculosis: multidisciplinary fish disease research*. New York: Academic Press.

Bernstein JJ. 1970. Anatomy and physiology of the central nervous system. *Fish Physiol* 4:1–90.

Bernstein RM, Schluter SF, Marchalonis JJ. 1997. Immunity. In: DH Evans, ed. *The physiology of fishes*, 2nd edn, pp. 215–242. Boca Raton, FL: CRC Press.

Berra TM. 1981. *An atlas of distribution of the freshwater fish families of the world*. Lincoln, NE: University of Nebraska Press.

Berra TM. 2001. *Freshwater fish distribution*. San Diego, CA: Academic Press.

Berra TM. 2007. *Freshwater fish distribution*. Chicago: University of Chicago Press.

Berra TM, Allen GR. 1989. Burrowing, emergence, behavior, and functional morphology of the Australian salamanderfish, *Lepidogalaxias salamandroides*. *Fisheries* 14(5):2–10.

Berra TM, Berra RM. 1977. A temporal and geographical analysis of new teleost names proposed at 25 year intervals from 1869–1970. *Copeia* 1977:640–647.

Berra TM, Crowley LELM, Ivantsoff W, Fuerst PA. 1996. *Galaxias maculatus*: an explanation of its biogeography. *Mar Freshw Res* 47:845–849.

Berra TM, Humphrey JD. 2002. Gross anatomy and histology of the hook and skin of forehead brooding male nurseryfish, *Kurtus gulliveri*, from northern Australia. *Env Biol Fish* 65:263–270.

Berra TM, Sever DM, Allen GR. 1989. Gross and histological morphology of the swimbladder and lack of accessory respiratory structures in *Lepidogalaxias salamandroides*, an aestivating fish from western Australia. *Copeia* 1989:850–856.

Berra TM, Wedd D. 2017. Salinity and spawning of nurseryfish, *Kurtus gulliveri*, in the Adelaide River of northern Australia with notes on electrofishing and photos of a male carrying eggs. *Environ Biol Fishes* 100:959–967

Bertelsen E. 1951. The ceratioid fishes. Ontogeny, taxonomy, distribution and biology. *Dana Rep* 39:1–276.

Bertelsen E, Nielsen JG. 1987. The deep sea eel family Monognathidae (Pisces, Anguilliformes). *Steenstrupia* 13:141–198.

Betancur-R R, Broughton RE, Wiley EO, et al. 2013. The tree of life and a new classification of bony fishes. *PLOS Curr* 1–41. **https://pubmed.ncbi.nlm.nih.gov/23653398/**.

Betancur-R R, Wiley EO, Arratia G, et al. 2017. Phylogenetic classification of bony fishes. *BMC Evol Biol* 17:162.

Betts A. 2011. Vermont climate change indicators. *Weather Climate Society* 3:106–115.

Beverton RJ. 1987. Longevity in fish: some ecological and evolutionary considerations. In: Woodhead AD, Thompson KH, eds. *Evolution of senescence: a comparative approach*, pp. 145–160. New York: Plenum Press.

Beverton RJ, Holt SJ. 1959. A review of the lifespans and mortality rates of fish in nature and their relation to growth and other physical characteristics. In: Wolstenholme GEW, O'Connor M, eds. *The lifespan of animals*, pp. 142–180. CIBA Foundation Colloquium on Aging No. 5. Boston: Little, Brown.

Bhargava P, Marshall JL, Dahut W, et al. 2001. A phase I and pharmacokinetic study of squalamine, a novel antiangiogenic agent, in patients with advanced cancers. *Clin Cancer Res* 7:3912–3919.

Bigelow HB, Perez Farfante IP, Schroeder WC. 1948. Lancelets. In: Tee-Van J, Breder CM, Hildebrand SF, Parr AE, Schroeder WC, eds. *Fishes of the western North Atlantic*, Part 1, Vol. 1, pp. 1–28. Sears Foundation for Marine Research Memoir. New Haven, CT: Yale University.

Bigelow HB, Schroeder WC. 1948a. Cyclostomes. In: Parr AE, ed. *Fishes of the western North Atlantic*, Part 1, Vol. 1, pp. 34–38. Sears Foundation for Marine Research Memoir. New Haven, CT: Yale University.

Bigelow HB, Schroeder WC. 1948b. Sharks. In: Tee-Van J, Breder CM, Parr AE, Schroeder WC, Schultz LP, eds. *Fishes of the western North Atlantic*, Part 1, Vol. 1, pp. 59–546. Sears Foundation for Marine Research Memoir. New Haven, CT: Yale University.

Bigelow HB, Schroeder WC. 1953a. Chimaeroids. In: Bigelow HB, Schroeder WC, eds. *Fishes of the western North Atlantic*, Part 1, Vol. 1, pp. 515–562. Sears Foundation for Marine Research Memoir. New Haven, CT: Yale University.

Bigelow HB, Schroeder WC. 1953b. *Fishes of the Gulf of Maine*. Fishery Bulletin of the Fish and Wildlife Service Vol. 53, US Fishery Bulletin No. 74. Washington, DC: Fish & Wildlife Service, U.S. Government Printing Office.

Bilby RE, Fransen BR, Bisson PA. 1996. Incorporation of nitrogen and carbon from spawning coho salmon into the trophic system of small streams: evidence from stable isotopes. *Can J Fish Aquat Sci* 53:164–173.

Billard R, Lecointre G. 2001. Biology and conservation of sturgeon and paddlefish. *Rev Fish Biol Fisheries* 10:355–392

Binkowski FP, Doroshov SI, eds. 1985. *North American sturgeons: biology and aquaculture potential*. Developments in Environmental Biology of Fishes No. 6. Dordrecht: Dr. W. Junk.

Binoy VV, Thomas KJ. 2004. The climbing perch (*Anabas testudineus* Bloch), a freshwater fish, prefers larger unfamiliar shoals to smaller familiar shoals. *Curr Sci* 86:207–211.

Birkeland C, ed. 1997. *The life and death of coral reefs*. New York: Chapman & Hall.

Birstein V, Bemis W. 1995. Will the Chinese paddlefish survive? *Sturgeon Q* 3(2):12.

Bishop TD. 1992. Threat-sensitive foraging by larval threespine sticklebacks (*Gasterosteus aculeatus*). *Behav Ecol Sociobiol* 31:133–138.

Biswas, SR, Vogt RJ, Sharma S. 2017. Projected compositional shifts and loss of ecosystem services in freshwater fish communities under climate change scenarios. *Hydrobiologia* 799:135–149.

Blaber SJM, Brewer DT, Salini JP. 1994. Diet and dentition in tropical ariid catfishes from Australia. *Env Biol Fish* 40:159–174.

Black-Cleworth P. 1970. The role of electrical discharges in the non-reproductive social behavior of *Gymnotus carapo* (Gymnotidae, Pisces). *Anim Behav Monogr* 3(1):1–77.

Blake RW. 2004. Fish functional design and swimming performance. *J Fish Biol* 65:1193–1222.

Blanchet S, Loot G, Grenouillet G, Brosse S. 2007. Competitive interactions between native and exotic salmonids: a combined field and laboratory demonstration. *Ecol Freshw Fish* 16:133–143.

Blank T, Burggren W. 2014. Hypoxia-induced developmental plasticity of the gills and air-breathing organ of *Trichopodus trichopterus*. *J Fish Biol* 84:808–826.

Blaxter JHS. 1969. Development: eggs and larvae. In: Hoar WS, Randall DJ, eds. *Reproduction and growth, bioluminescence, pigments, and poisons. Fish physiology*, Vol. 3, pp. 177–252. New York: Academic Press.

Blaxter JHS, ed. 1974. *The early life history of fish*. New York: Springer-Verlag.

Blaxter JHS. 1975. The eyes of larval fish. In: Ali MA, ed. *Vision in fishes*, pp. 427–444. New York: Plenum Press.

Blaxter JHS. 1984. Ontogeny, systematics and fisheries. In: Moser HG, Richards WJ, Cohen DM, Fahay MP, Kendall AW, Jr., Richardson SL, eds. *Ontogeny and systematics of fishes*, pp. 1–6. Special Publication No. 1. Lawrence, KS: American Society of Ichthyologists and Herpetologists.

Blaxter JHS, Fuiman LA. 1990. The role of the sensory systems of herring larvae in evading predatory fish. *J Mar Biol Assoc UK* 70:413–427.

Blaxter JHS, Hempel G. 1963. The influence of egg size on herring larvae. *J Cons Int Explor Mer* 28:211–240.

Blaxter JHS, Hunter JR. 1982. The biology of clupeoid fishes. *Adv Mar Biol* 20:1–223.

Blazer VS, Fabacher DL, Little EE, et al. 1997. Effects of ultraviolet-B radiation on fish: histologic comparison of a UVB-sensitive and a UVB-tolerant species. *J Aquat Anim Health* 9:132–143.

Blazer VS, Facey DE, Fournie JW, Courtney LA, Summers JK. 1994. Macrophage aggregates as indicators of environmental stress. In: Stolen JS, Fletcher TC, eds. *Modulators of fish immune responses, Vol. 1. Models for environmental toxicology, biomarkers, immunostimulators*, pp. 169–185. Fair Haven, NJ: SOS Publications.

Blazer VS, Iwanowicz LR, Henderson H, et al. 2012. Reproductive endocrine disruption in smallmouth bass (*Micropterus dolomieu*) in the Potomac River basin: spatial and temporal comparisons of biological effects. *Environ Monit Assess* 184:4309–4334.

Blazer VS, Wolke RE, Brown J, Powell CA. 1987. Macrophage aggregates in largemouth bass: effects of age, season, relative weight and site quality. *Aquat Toxicol* 10:199–215.

Block BA. 1991. Evolutionary novelties: how fish have built a heater out of muscle. *Am Zool* 31:726–742.

Block BA. 2011. Pelagic fishes: endothermy in tunas, billfishes, and sharks. In: Farrell AP, Stevens ED, Cech JJ, Richards JG, eds. *Encyclopedia of fish physiology: from genome to environment*, pp. 1914–1920. Boston: Elsevier/Academic Press.

Block BA, Dewar H, Blackwell SB, et al. 2001. Migratory movements, depth preferences, and thermal biology of Atlantic bluefin tuna. *Science* 293:1310–1314.

Block BA, Finnerty JR. 1994. Endothermy in fishes: a phylogenetic analysis of constraints, predispositions, and selection pressures. *Environ Biol Fish* 40:283–302.

Block BA, Stevens E. 2001. *Tuna: physiology, ecology, and evolution. Fish physiology*, Vol. 19. New York: Academic Press.

Blumer LS. 1979. Male parental care in the bony fishes. *Q Rev Biol* 54:149–161.

Blumer LS. 1982. A bibliography and categorization of bony fishes exhibiting parental care. *Zool J Linn Soc* 76:1–22.

Bocast C, Bruch RM, Koenigs KP. 2014. Sound production of spawning lake sturgeon (*Acipenser fulvescens* Rafinesque, 1817) in the Lake Winnebago watershed Wisconsin, USA. *J Appl Ichthyol* 30:1186–1194.

Boehlert GW. 1984. Scanning electron microscopy. In: Moser HG, Richards WJ, Cohen DM, Fahay MP, Kendall AW, Jr., Richardson SL, eds. *Ontogeny and systematics of fishes*, pp. 43–48. Special Publication No. 1. Lawrence, KS: American Society of Ichthyologists and Herpetologists.

Boehlert GW, Mundy BC. 1988. Roles of behavioral and physical factors in larval and juvenile fish recruitment to estuarine nursery areas. In: Weinstein MP, ed. *Larval fish and shellfish transport through inlets*, pp. 51–67. American Fisheries Society Symposium 3. Bethesda, MD: American Fisheries Society.

Boehlert GW, Yamada J. 1991. Rockfishes of the genus *Sebastes*: their reproduction and early life history. *Env Biol Fish* 30:7–280.

Boet P, Belliard J, Berrebi-dit-Thomas R, et al. 1999. Multiple human impacts by the City of Paris on fish communities in the Seine river basin, France. *Hydrobiologia* 410:59–68.

Bogart JP, Balon EK, Bruton MN. 1994. The chromosomes of the living coelacanth and their remarkable similarity to those of one of the most ancient frogs. *J Heredity* 85:322–325.

Bohnsack JA. 1983. Resiliency of reef fish communities in the Florida Keys following a January 1977 hypothermal fish kill. *Env Biol Fish* 9:41–53.

Bohnsack JA. 1994. How marine fishery reserves can improve reef fisheries. *Proc Gulf Caribb Fish Inst* 43:217–241.

Bohonak AJ. 1999. Dispersal, gene flow, and population structure. *Q Rev Biol* 74:21–45.

Bond CE, Kan TT. 1973. *Lampetra* (*Entosphenus*) *minima* n. sp., a dwarfed parasitic lamprey from Oregon. *Copeia* 1973:568–574.

Bone Q. 1978. Locomotor muscle. *Fish Physiol* 7:361–424.

Bone Q. 1988. Muscles and locomotion. In: Shuttleworth T, ed. *The physiology of easmobranch fishes*, pp. 99–141. New York: Springer-Verlag.

Bone Q, Marshall NB, Blaxter JHS. 1995. *Biology of fishes*, 2nd edn. London: Blackie Academic & Professional.

Bone Q, Moore RH. 2008. *Biology of fishes*, 3rd edn. New York: Taylor & Francis Group.

Bonfil R, Meÿer M, Scholl MC, et al. 2005. Transoceanic migration, spatial dynamics, and population linkages of white sharks. *Science* 310:100–103.

Bonner JT. 1980. *The evolution of culture in animals*. Princeton, NJ: Princeton University Press.

Booth DE. 1989. Hydroelectric dams and the decline of chinook salmon in the Columbia River basin. *Mar Res Econ* 6:195–211.

Boreman J, Lewis RR. 1987. Atlantic coastal migration of striped bass. In: Dadswell MJ, Klauda RJ, Moffitt CM, Saunders RL, Rulifson RA, Cooper JE, eds. *Common strategies of anadromous and catadromous fishes*, pp. 331–339. American Fisheries Society Symposium No. 1. Bethesda, MD: American Fisheries Society.

Bortone SA, Davis WP. 1994. Fish intersexuality as indicators of environmental stress. *BioScience* 44:165–172.

Borucinska JD, Benz GW, Whiteley HE. 1998. Ocular lesions associated with attachment of the parasitic copepod *Ommatokoita elongata* (Grant) to corneas of Greenland sharks, *Somniosus microcephalus* (Bloch & Schneider). *J Fish Dis* 21:415–422.

Borucinska JD, Harshbarger JC, Reimschuessel R, Bogicevic T. 2004. Gingival neoplasms in a captive sand tiger shark, *Carcharias taurus* (Rafinesque), and a wild-caught blue shark, *Prionace glauca* (L.). *J Fish Dis* 27:185–191.

Boschung HT. 1983. A new species of lancelet, *Branchiostoma longirostrum* (Order Amphioxi), from the western North Atlantic. *Northeast Gulf Sci* 6:91–97.

Boschung HT, Mayden RL. 2004. *Fishes of Alabama*. Washington, DC: Smithsonian Books.

Boschung HT, Shaw RF. 1988. Occurrence of planktonic lancelets from Louisiana's continental shelf, with a review of pelagic *Branchiostoma* (Order Amphioxi). *Bull Mar Sci* 43:229–240.

Botham MS, Kerfoot CJ, Louca V, Krause J. 2006. The effects of different predator species on antipredator behavior in the Trinidadian guppy, *Poecilia reticulata*. *Naturwissenschaften* 93:431–439.

Botsford LW, White JW, Coffroth M-A, et al. 2009. Connectivity and resilience of coral reef metapopulations in marine protected areas: matching empirical efforts to predictive needs. *Coral Reefs* 28:327–337.

Bouchet P, Perrine D. 1996. More gastropods feeding at night on parrotfishes. *Bull Mar Sci* 59:224–228.

Boughman JW, Rundle HD, Schluter D. 2005. Parallel evolution of sexual isolation in sticklebacks. *Evolution* 59:361–373.

Boughton DA, Collette BB, McCune AR. 1991. Heterochrony in jaw morphology of needlefishes (Teleostei: Belonidae). *Syst Zool* 40:329–354.

Boujard T, Leatherland JF. 1992. Circadian rhythms and feeding time in fishes. *Env Biol Fish* 35:109–131.

Boussarie G, Bakker J, Wangensteen OS, et al. 2018. Environmental DNA illuminates the dark diversity of sharks. *Sci Adv* 4:eaap9661.

Boustany AM, Davis SF, Pyle P, Anderson SD, Le Boeuf BJ, Block BA. 2002. Satellite tagging – expanded niche for white sharks. *Nature* 415:35–36.

Bowen BW. 2016. The three domains of conservation genetics: case histories from Hawaiian waters. *J Hered* 107:309–317.

Bowen BW, Bass AL, Muss AJ, Carlin F, Robertson DR. 2006b. Phylogeography of two Atlantic squirrelfishes (family Holocentridae): exploring pelagic larval duration and population connectivity. *Mar Biol* 149:899–913.

Bowen BW, Bass AL, Rocha LA, Grant WS, Robertson DR. 2001. Phylogeography of the trumpetfish (*Aulostomus* spp.): ring species complex on a global scale. *Evolution* 55:1029–1039.

Bowen BW, Bass AL, Soares L, Toonen RJ. 2005. Conservation implications of complex population structure: lessons from the loggerhead turtle (*Caretta caretta*). *Molec Ecol* 14:2389–2402.

Bowen, BW, Gaither MR, DiBattista JD, et al. 2016. Comparative phylogeography of the ocean planet. *Proc Nat Acad Sci USA* 113:7962–7969.

Bowen BW, Grant WS. 1997. Phylogeography of the sardines (*Sardinops* spp.): assessing biogeographic models and population histories in temperate upwelling zones. *Evolution* 51:1601–1610.

Bowen BW, Karl SA, Pfeiler E. 2007. Resolving evolutionary lineages and taxonomy of bonefishes (*Albula* spp.). In: Ault J, ed. *Biology and management of the world tarpon and bonefish fisheries*, pp. 147–154. Boca Raton, FL: CRC Press.

Bowen BW, Muss A, Rocha LA, Grant WS. 2006a. Shallow mtDNA coalescence in Atlantic pygmy angelfishes (genus *Centropyge*) indicates a recent invasion from the Indian Ocean. *J Heredity* 97:1–12.

Bowen BW, Rocha LA, Toonen RJ, Karl SA, ToBo Lab. 2013. Origins of tropical marine biodiversity. *Trends Ecol Evol* 28:359–366.

Bowen BW, Roman J. 2005. Gaia's handmaidens: the Orlog model for conservation biology. *Conserv Biol* 19:1037–1043.

Bowen SH. 1983. Detritivory in neotropical fish communities. *Env Biol Fish* 9:137–144.

Bowmaker JK. 2011. Adaptations of photoreceptors and visual pigments. In: Farrell AP, Stevens ED, Cech JJ, Richards JG, eds. *Encyclopedia of fish physiology: from genome to environment*, pp. 116–122. Boston: Elsevier/Academic Press.

Bowmaker JK, Hunt DM. 2006. Evolution of vertebrate visual pigments. *Curr Biol* 16(13):R484–R489.

Boyle KS, Horn MH. 2006. Comparison of feeding guild structure and ecomorphology of intertidal fish assemblages from central California and central Chile. *Mar Ecol Prog Ser* 319:65–84.

Boyle KS, Tricas TC. 2014. Discrimination of mates and intruders: visual and olfactory cues for a monogamous territorial coral reef butterflyfish. *Anim Behav* 92:33–43.

Bozzano A. 2003. Vision in the rufus snake eel, *Ophichthus rufus*: adaptive mechanisms for a burrowing life-style. *Mar Biol* 143:167–174.

Brabrand A, Faafeng BA, Nilssen JPM. 1990. Relative importance of phosphorus supply to phytoplankton production: fish excretion versus external loading. *Can J Fish Aquat Sci* 47:364–372.

Brady PC, Gilerson AA, Kattawar GW, et al. 2015. Open-ocean fish reveal an omnidirectional solution to camouflage in polarized environments. *Science* 350:965–969.

Bräger Zs, Moritz T. 2016 A scale atlas for common Mediterranean teleost fishes. *Vertebr Zool* 66 (3):275–388.

Brainerd EL. 1992. Pufferfish inflation: functional morphology of postcranial structures in *Diodon holocanthus* (Tetraodontiformes). *J Morphol* 220:1–20.

Brainerd EL, Liem KF, Samper CT. 1989. Air ventilation by recoil aspiration in polypterid fishes. *Science* 246:1593–1595.

Braithwaite VA. 2011. Fish learning and memory. In: Farrell AP, Stevens ED, Cech JJ, Richards JG, eds. *Encyclopedia of fish physiology: from genome to environment*, pp. 707–712. Boston: Elsevier/Academic Press.

Branch TA, Watson R, Fulton EA, et al. 2010. The trophic fingerprint of marine fisheries. *Nature* 468(7322):431–435.

Brannon EL. 1987. Mechanisms stabilizing salmonid fry emergence timing. In: Smith HD, Margolis L, Wood CC, eds. *Sockeye salmon (Oncorhynchus nerka) population biology and future management*, pp. 120–124. Canadian Special Publication, Fisheries and Aquatic Sciences No. 96. Ottawa: Department of Fisheries and Oceans.

Branstetter S. 1990. Early life-history implications of selected carcharhinoid and lamnoid sharks of the northwest Atlantic. In: Pratt HL, Jr., Gruber SH, Taniuchi T, eds. *Elasmobranchs as living resources: advances in the biology, ecology, systematics, and the status of fisheries*, pp. 17–28. NOAA Technical Report No. 90. Washington, DC: National Oceanic and Atmospheric Administration.

Branstetter S. 1991. Shark early life history: one reason sharks are vulnerable to overfishing. In: Gruber SH, ed. *Discovering sharks*, pp. 29–34. American Littoral Society Special Publication No. 14. Highlands, NJ: American Littoral Society.

Branstetter S. 1999. The management of the United States Atlantic shark fishery. In: Shotton R, ed. *Case studies of the management of elasmobranch fisheries*, pp. 109–148. FAO Fisheries Technical Paper No. 378, Vol. 1. Rome: Food and Agricultural Organization.

Branstetter S, McEachran JD. 1986. Age and growth of four carcharhinid sharks common to the Gulf of Mexico: a summary paper. In: Uyeno T, Arai R, Taniuchi T, Matsuura K, eds. *Indo-Pacific fish biology. Proceedings of the 2nd International Conference of Indo-Pacific Fish*, pp. 361–371. Tokyo: Ichthyological Society of Japan.

Braun CB, Coombs S, Fay RR. 2002. What is the nature of multisensory interactions between octavolateralis sub-systems. *Brain Behav Evol* 59:162–176.

Brauner CJ, Rummer JL. 2011. Gas transport and exchange: interaction between O_2 and CO_2 exchange. In: Farrell AP, Stevens ED, ech JJ, Richards JG, eds. *Encyclopedia of fish physiology: from genome to environment*, pp. 916–920. Boston: Elsevier/Academic Press.

Brawn VM. 1961. Reproductive behaviour of the cod (*Gadus callarias* L.). *Behaviour* 18:177–198.

Brawn VM. 1962. Physical properties and hydrostatic function of the swimbladder of herring (*Clupea harengus* L.). *J Fish Res Board Can* 19:635–656.

Bray RN. 1981. Influence of water currents and zooplankton densities on daily foraging movements of blacksmith, *Chromis punctipinnis*, a planktivorous reef fish. *Fish Bull US* 78:829–841.

Bray RN, Hixon MA. 1978. Night-shocker: predatory behavior of the Pacific electric ray (*Torpedo californica*). *Science* 200:333–334.

Bray RN, Miller AC, Geesey GG. 1981. The fish connection: a trophic link between planktonic and rocky reef communities? *Science* 214:204–205.

Bray RN, Purcell LJ, Miller AC. 1986. Ammonium excretion in a temperate-reef community by a planktivorous fish, *Chromis punctipinnis* (Pomacentridae), and potential uptake by young giant kelp, *Macrocystis pyrifera* (Laminariales). *Mar Biol* 90:327–334.

Breder CM, Jr. 1926. The locomotion of fishes. *Zoologica (NY)* 4:159–297.

Breder CM, Jr. 1949. On the relationship of social behavior to pigmentation in tropical shore fishes. *Bull Am Mus Natur Hist* 94:83–106.

Breder CM, Jr. 1976. Fish schools as operational structures. *Fish Bull US* 74:471–502.

Breder CM, Jr., Rosen DE. 1966. *Modes of reproduction in fishes*. Neptune City, NJ: TFH Publishers.

Breitburg DL. 1989. Demersal schooling by settlement-stage naked goby larvae. *Env Biol Fish* 26:97–103.

Bremset G. 2000. Seasonal and diel changes in behaviour, microhabitat use and preferences by young pool-dwelling Atlantic salmon, *Salmo salar*, and brown trout, *Salmo trutta*. *Env Biol Fish* 59:163–179.

Brett JR. 1957. The eye. In: Brown ME, ed. *The physiology of fishes*, pp. 121–154. New York: Academic Press.

Brett JR. 1971. Energetic responses of salmon to temperature. A study of some thermal relations in the physiology and freshwater ecology of sockeye salmon (*Oncorhynchus nerka*). *Am Zool* 11:99–113.

Brett JR, Blackburn JM. 1978. Metabolic rate and energy expenditure of the spiny dogfish, *Squalus acanthias*. *J Fish Res Board Can* 35:816–821.

Brett JR, Groves TDD. 1979. Physiological energetics. In: Hoar WS, Randall DJ, eds. *Bioenergetics and growth. Fish physiology*, Vol. 8, pp. 279–352. New York: Academic Press.

Briggs DEG, Clarkson ENK, Aldridge RJ. 1983. The conodont animal. *Lethaia* 16:1–14.

Briggs JC 1970. A faunal history of the North Atlantic. *Syst Zool* 23:248–256.

Briggs JC. 1974. *Marine zoogeography*. New York: McGraw-Hill.

Briggs JC. 1995. *Global biogeography*. Amsterdam: Elsevier.

Briggs JC, Bowen BW. 2012. A realignment of marine biogeographic provinces with particular reference to fish distributions. *J Biogeogr* 39:12–30.

Briggs JC, Bowen BW. 2013. Evolutionary patterns: marine shelf habitat. *J Biogeogr* 40:1023–1035.

Britten GL, Dowd M, Minto C, Ferretti F, Boero F, Lotze HK. 2014. Predator decline leads to decreased stability in a coastal fish community. *Ecol Lett* 17:1518–1525

Britton JC, Morton B. 1994. Marine carrion and scavengers. *Oceanogr Mar Biol Annu Rev* 32:369–434.

Britz R, Johnson GD. 2003. On the homology of the posteriormost gill arch in polypterids (Cladistia, Actinopterygii). *Zool J Linn Soc* 138(4):495–503.

Broadhurst MK. 2000. Modifications to reduce bycatch in prawn trawls: a review and framework for development. *Rev Fish Biol Fisheries* 10:27–60.

Brock VE, Riffenberg RH. 1960. Fish schooling: a possible factor in reducing predation. *J Conseil* 25:307–317.

Brodal A, Fange R. 1963. *The biology of Myxine*. Oslo: Scandinavian University Books.

Brönmark C, Miner JG. 1992. Predator-induced phenotypical change in body morphology in Crucian carp. *Science* 258:1348–1350.

Brooker AJ, Papadopoulou A, Gutierrez C, Rey S, Davie A, Migaud H. 2018 Sustainable production and use of cleaner fish for the biological control of sea lice: recent advances and current challenges. *Vet Rec* 183:383. **https://doi.org/10.1136/vr.104966**.

Brooks JL, Dodson SI. 1965. Predation, body size, and composition of plankton. *Science* 150:28–35.

Brothers EB. 1984. Otolith studies. In: Moser HG, Richards WJ, Cohen DM, Fahay MP, Kendall AW, Jr., Richardson SL, eds. *Ontogeny and systematics of fishes*, pp. 50–57. Special Publication No. 1. Lawrence, KS: American Society of Ichthyologists and Herpetologists.

Brothers EB, Mathews CP, Lasker R. 1976. Daily growth increments in otoliths from larval and adult fishes. *Fish Bull US* 74:1–8.

Brothers EB, McFarland WN. 1981. Correlations between otolith microstructure, growth, and life history transitions in newly recruited French grunts [*Haemulon flavolineatum* (Desmarest), Haemulidae]. *Rapp P-V Reun Cons Int Explor Mer* 178:369–374.

Broughton RE, Milam JE, Roe BA. 2001. The complete sequence of the zebrafish (*Danio rerio*) mitochondrial genome and evolutionary patterns in vertebrate mitochondrial DNA. *Genome Res* 11:1958–1967.

Browder RG, Garman GC. 1994. Increased ammonium concentrations in a tidal fresh-water stream during residence of migratory clupeid fishes. *Trans Am Fish Soc* 123:993–996.

Brown AD., Sisneros JA, Jurasin T, Nguyen C, Coffin AB. 2013. Differences in lateral line morphology between hatchery- and wild origin steelhead. *PLoS One* 8(3):e59162.

Brown B. 1990. *Mountain in the clouds: a search for the wild salmon*. New York: Collier.

Brown B. 1995. *Mountain in the clouds: a search for the wild salmon*, reprint edn. Seattle, WA: University of Washington Press.

Brown C. 2012. Tool use in fishes. *Fish Fish (Oxf)* 13:105–115.

Brown C, Laland KN. 2003. Social learning in fishes: a review. *Fish Fish (Oxf)* 4:280–288.

Brown C, Warburton K. 1997. Predator recognition and anti-predator responses in the rainbowfish *Melanotaenia eachamensis*. *Behav Ecol Sociobiol* 41:61–68.

Brown GE. 2003. Learning about danger: chemical alarm cues and local risk assessment in prey fishes. *Fish Fish (Oxf)* 4:227–234.

Brown GE, Chivers DP, Smith RJF. 1995. Localized defecation by pike: a response to labeling by cyprinid alarm pheromone. *Behav Ecol Sociobiol* 36:105–110.

Brown GE, Rive AC, Ferrari MCO, Chivers DP. 2006. The dynamic nature of antipredator behavior: prey fish integrate threat-sensitive antipredator responses within background levels of predation risk. *Behav Ecol Sociobiol* 61:9–16.

Brown JA. 1986. The development of feeding behaviour in the lumpfish. *J Fish Biol* 29 (suppl A):171–178.

Brown RS, Power G, Beltaos S. 2001. Winter movements and habitat use of riverine brown trout, white sucker and common carp in relation to flooding and ice break-up. *J Fish Biol* 59:1126–1141.

Bruce BD, Green MA, Last PR. 1998. Threatened fishes of the world: *Brachionichthys hirsutus* (Lacepede, 1804) (Brachionichthyidae). *Env Biol Fish* 52:418.

Bruggemann JH, van Kessel AM, van Rooij JM, et al. 1996. Bioerosion and sediment ingestion by the Caribbean parrotfish *Scarus vetula* and *Sparisoma viride*: implications of fish size, feeding mode and habitat use. *Mar Ecol Prog Ser* 134:59–71.

Bruner JC. 1997. The megatooth shark, *Carcharodon megalodon*: rough toothed, huge toothed. **www.flmnh.ufl.edu/fish/sharks/InNews/megatoothshark.htm**.

Brunnschweiler JM. 2006. Sharksucker–shark interaction in two carcharhinid species. *Mar Ecol* 27:89–94.

Brusca RC, Gilligan MR. 1983. Tongue replacement in a marine fish (*Lutjanus guttatus*) by a parasitic isopod (Crustacea: Isopoda). *Copeia* 1983:813–816.

Bruton MN. 1979. The survival of habitat desiccation by air breathing clariid catfishes. *Env Biol Fish* 4:273–280.

Bruton MN, Armstrong MJ. 1991. The demography of the coelacanth *Latimeria chalumnae. Env Biol Fish* 32:301–311.

Bruton MN, Cabral AJP, Fricke H. 1992. First capture of a coelacanth, *Latimeria chalumnae* (Pisces, Latimeriidae), off Mozambique. *Suid-Afrikaanse Tydskrif vir Wetenskap* 88:225–227.

Bruton MN, Coutouvidis SE. 1991. An inventory of all known specimens of the coelacanth *Latimeria chalumnae*, with comments on trends in the catches. *Env Biol Fish* 32:371–390.

Bruton MN, Stobbs RE. 1991. The ecology and conservation of the coelacanth *Latimeria chalumnae. Env Biol Fish* 32:313–339.

Bryan PG, Madraisau BB. 1977. Larval rearing and development of *Siganus lineatus* (Pisces, Siganidae) from hatching through metamorphosis. *Aquaculture* 10:243–252.

Bshary R. 2003. The cleaner wrasse, *Labroides dimidiatus*, is a key organism for reef fish diversity at Ras Mohammed National Park, Egypt. *J Anim Ecol* 72:169–176.

Bshary R, Hohner A, Ait-el-Djoudi K, Fricke H. 2006. Interspecific communicative and coordinated hunting between groupers and giant moray eels in the Red Sea. *PLoS Biol* 4:e431. **https://doi.org/10.1371/journal.pbio.0040431**.

Bshary R, Wurth M. 2001. Cleaner fish *Labroides dimidiatus* manipulate client reef fish by providing tactile stimulation. *Proc Roy Soc Lond B Biol Sci* 268:1495–1501.

Buckley J, Kynard B. 1985. Habitat use and behavior of pre-spawning and spawning shortnose sturgeon, *Acipenser brevirostrum*, in the Connecticut River. In: Binkowski FP, Doroshov SI, eds. *North American sturgeons: biology and aquaculture potential*, pp. 111–117. Developments in Environmental Biology of Fishes No. 6. Dordrecht: Dr. W. Junk.

Buddington RK, Christofferson JP. 1985. Digestive and feeding characteristics of the chondrosteans. In: Binkowski FP, Doroshov SI, eds. *North American sturgeons: biology and aquaculture potential*, pp. 31–41. Developments in Environmental Biology of Fishes No. 6. Dordrecht: Dr. W. Junk.

Buddington RK, Diamond JM. 1987. Pyloric ceca of fish: a "new" absorptive organ. *Am J Physiol* 252:G65–G76.

Budker P. 1971. *The life of sharks*. New York: Columbia University Press.

Bullock TH, Hamstra RH, Jr., Scheich H. 1972. The jamming avoidance response of high frequency electric fish. I. General features. II. Quantitative aspects. *J Comp Physiol A* 77:1–22, 23–48.

Buonaccorsi V, McDowell JR, Graves J. 2001. Reconciling patterns of inter-ocean molecular variance from four classes of molecular markers in blue marlin (*Makaira nigricans*). *Molec Ecol* 10:1179–1196.

Burgess GH, Beerkircher LR, Cailliet GM, et al. 2005. Is the collapse of shark populations in the Northwest Atlantic Ocean and Gulf of Mexico real? *Fisheries* 30(10):19–26.

Burgess WE. 1978. *Butterflyfishes of the world. A monograph of the family Chaetodontidae*. Neptune City, NJ: TFH Publications, Inc.

Burgess WE. 1989. *An atlas of freshwater and marine catfishes – a preliminary survey of the Siluriformes*. Neptune City, NJ: TFH Publications, Inc.

Burghardt GM. 2005. *The genesis of animal play: testing the limits*. Cumberland, RI: MIT Press/Bradford Books.

Burkhead NM, Jelks H. 2001. Effects of suspended sediment on the reproductive success of the tricolor shiner, a crevice-spawning minnow. *Trans Am Fish Soc* 130:959–968.

Burkhead NM, Williams JD. 1991. An intergeneric hybrid of a native minnow, the golden shiner, and an exotic minnow, the rudd. *Trans Am Fish Soc* 120:781–795.

Burkholder JM. 2002. Pfiesteria: the toxic Pfiesteria complex. In: Bitton G, ed. *Encyclopedia of environmental microbiology*, pp. 2431–2447. New York: Wiley Publishers.

Burkholder JM, Glasgow HB, Jr. 1997. *Pfiesteria piscicida* and other *Pfiesteria*-like dinoflagellates: behavior, impacts, and environmental controls. *Limnol Oceanogr* 42:1052–1075.

Burkholder JM, Glasgow HB, Jr. 2001. History of toxic Pfiesteria in North Carolina estuaries from 1991 to the present. *BioScience* 51:827–841.

Burns JR. 1975. Seasonal changes in the respiration of pumpkinseed, *Lepomis gibbosus*, correlated with temperature, day length, and stage of reproduction. *Physiol Zool* 48:142–149.

Burridge CP, Craw D, Waters JM. 2007. An empirical test of freshwater vicariance via stream capture. *Molec Ecol* 16:1883–1895.

Busacker GP, Adelman IR, Goolish EM. 1990. Growth. In: Schreck CB, Moyle PB, eds. *Methods for fish biology*, 363–387. Bethesda, MD: American Fisheries Soceity.

Bush AO, Fernandez JC, Esch GW, Seed JR. 2001. *Parasitism: the diversity and ecology of animal parasites*. Cambridge, UK: Cambridge University Press.

Buston PM. 2004. Does the presence of non-breeders enhance the fitness of breeders? An experimental analysis in the clown anemonefish *Amphiprion percula. Behav Ecol Sociobiol* 57:23–31.

Buston PM, Bogdanowicz SM, Wong A, Harrison RG. 2007. Are clownfish groups composed of close relatives? An analysis of microsatellite DNA variation in *Amphiprion percula. Molec Ecol* 16:3671–3678.

Buston PM, D'Aloia CC. 2013. Marine ecology: reaping the benefits of local dispersal. *Curr Biol* 23:R351–R353.

Buston PM, Jones GP, Planes S, Thorrold SR. 2011. Probability of successful larval dispersal declines fivefold over 1 km in a coral reef fish. *Proc Royal Soc B: Biol Sci* 279:1883–1888.

Buth DG, Dowling TE, Gold JR. 1991. Molecular and cytological investigations. In: Winfield IJ, Nelson JS, eds. *Cyprinid fishes: systematics, biology and exploitation*, pp. 83–126. London: Chapman & Hall.

Butler AB. 2011. Functional morphology of the brains of ray-finned fishes. In: Farrell AP, Stevens ED, Cech JJ, Richards JG, eds. *Encyclopedia of fish physiology: from genome to environment*, pp. 37–45. Boston: Elsevier/Academic Press.

Bye VJ. 1990. Temperate marine teleosts. In: Munro AD, Scott AP, Lam TJ, eds. *Reproductive seasonality in teleosts: environmental influences*, pp. 125–143. Boca Raton, FL: CRC Press.

Bystriansky JS, Frick NT, Edwards JG, Schulte PM, Ballantyne JS. 2007. Wild arctic char (*Salvelinus alpinus*) upregualte gill Na^+, K^+-ATPase during freshwater migration. *Physiol Biochem Zool* 80:270–282.

Cailliet GM. 1990. Elasmobranch age determination and verification: an updated review. In: Pratt HL, Jr., Gruber SH, Taniuchi T, eds. *Elasmobranchs as living resources: advances in the biology, ecology, systematics, and the status of fisheries*, pp. 157–165. NOAA Technical Report No. 90. Washington, DC: National Oceanic and Atmospheric Administration.

Cailliet GM, Andrews AH, Burton EJ, Watters DL, Kline DE, Ferry-Graham LA. 2001. Age determination and validation studies of marine fishes: do deep-dwellers live longer? *Exp Gerontol* 36:739–764.

Cailliet GM, Goldman KJ. 2004. Age determination and validation in chondrichthyan fishes. In: Carrier JC, Musick JA, Heithaus MR, eds. *Biology of sharks and their relatives*, pp. 399–447. Boca Raton, FL: CRC Press.

Cailliet GM, Love MS, Ebeling AW. 1986. *Fishes: a field and laboratory manual on their structure, identification, and natural history*. Belmont, CA: Wadsworth Publishing Co.

Caira JN, Benz GW, Borucinska J, Kohler NE. 1997. Pugnose eels, *Simenchelys parasiticus* (Synaphobranchidae) from the heart of a shortfin mako, *Isurus oxyrinchus* (Lamnidae). *Env Biol Fish* 49:139–144.

Calder WA, III. 1984. *Size, function, and life history*. Cambridge, MA: Harvard University Press.

Callicott JB. 1991. Conservation ethics and fishery management. *Fisheries* 16(2):22–28.

Camhi M. 1996. *Costa Rica's shark fishery and cartilage industry*. Shark News No. 8. Cambridge, UK: IUCN Shark Specialist Group. **http://www.flmnh.ufl.edu/fish/organizations/ssg/ssg.htm**.

Campana SE, Joyce W, Marks L, et al. 2002. Population dynamics of the porbeagle in the northwest Atlantic Ocean. *N Am J Fish Manage* 22:106–121.

Cancino JM, Castilla JC. 1988. Emersion behavior and foraging ecology of the common Chilean clingfish *Sicyases sanguineus* (Pisces, Gobiesocidae). *J Natur Hist* 22:249–261.

Candolin U, Reynolds JD. 2002. Adjustments of ejaculation rates in response to risk of sperm competition in a fish, the bitterling (*Rhodeus sericeus*). *Proc Roy Soc Lond B Biol Sci* 269:1549–1553.

Candolin U, Salesto T, Evers M. 2007. Changed environmental conditions weaken sexual selection in sticklebacks. *J Evol Biol* 20:233–239.

Canfield JG, Rose GJ. 1993. Activation of Mauthner neurons during prey capture. *J Comp Physiol A* 172:611–618.

Cantino PD, de Queiroz K. 2004. PhyloCode: a phylogenetic code of nomenclature. **http://www.ohiou.edu/phylocode/**.

Carassou L, Léopold M, Guillemot N, Wantiez L, Kulbicki M. 2013. Does herbivorous fish protection really improve coral reef resilience? A case study from New Caledonia (South Pacific). *PLoS ONE* 8(4):e60564. **https://doi.org/10.1371/journal.pone.0060564**.

Carbonneau P, Fonstad MA, Marcus WA, Dugdale SJ. 2012. Making riverscapes real. *Geomorphology* 137:74–86.

Cardeñosa D. 2019. Genetic identification of threatened shark species in pet food and beauty care products. *Conserv Genet* 20:1383–1387.

Carey FG. 1973. Fishes with warm bodies. *Sci Am* 228:36–44.

Carey FG, Clark E. 1995. Depth telemetry from the sixgill shark, *Hexanchus griseus*, at Bermuda. *Env Biol Fish* 42:7–14.

Carey FG, Kanwisher JW, Brazier O, Gabrielson G, Casey JG, Pratt HL, Jr. 1982. Temperature and activities of a white shark, *Carcharodon carcharias*. *Copeia* 1982:254–260.

Carey FG, Lawson KD. 1973. Temperature regulation in free-swimming bluefin tuna. *Comp Biochem Physiol* 44A:375–392.

Carey FG, Scharold JV. 1990. Movements of blue sharks in depth and course. *Mar Biol* 106:329–342.

Carey FG, Teal JM, Kanwisher JW. 1981. The visceral temperatures of mackerel sharks (Lamnidae). *Physiol Zool* 54:334–344.

Carey FG, Teal JM, Kanwisher JW, Lawson KD, Beckett JS. 1971. Warm-bodied fish. *Am Zool* 11:137–145.

Cargnelli LM, Gross MR. 1997. Fish energetics: larger individuals emerge from winter in better condition. *Trans Am Fish Soc* 126:153–156.

Carillo J, Koumoundouros G, Divanach P, Martinez J. 2001. Morphological malformations of the lateral line in reared gilthead sea bream (*Sparus aurata* L. 1758). *Aquaculture* 192:281–290.

Carlander KD. 1969. *Handbook of freshwater fishery biology*, Vol. I. Ames, IA: Iowa State University Press.

Carlin J, Robertson DR, Bowen BW. 2003. Ancient divergences and recent connections in two tropical Atlantic reef fishes *Epinephelus adscensionis* and *Rypticus saponaceous* (Percoidei: Serranidae). *Mar Biol* 143:1057–1069.

Carlson B. 2002. Electric signaling behavior and the mechanisms of electric organ discharge production in mormyrid fish. *J Physiol (Paris)* 96:405–419.

Carlsson J, McDowell JR, Carlsson JEL, Graves JE. 2007. Genetic identity of YOY bluefin tuna from the eastern and western Atlantic spawning areas. *J Heredity* 98:23–28.

Carlton JT. 1999. The scale and ecological consequences of biological invasions in the world's oceans. In: Sandlund OT, Schei PJ, Viken A, eds. *Invasive species and biodiversity management*, pp. 195–212. Dordrecht: Kluwer Academic.

Carpenter KE, Springer VG. 2005. The center of the center of marine shore fish diversity: the Philippine Islands. *Env Biol Fishes* 72:467–480.

Carpenter RC. 1986. Partitioning herbivory and its effects on coral reef algal communities. *Ecol Monogr* 56:345–363.

Carpenter SR, ed. 1988. *Complex interactions in lake communities*. New York: Springer-Verlag.

Carpenter SR, Kitchell JF. 1988. Consumer control of lake productivity. *BioScience* 38:764–769.

Carpenter SR, Kitchell JF, eds. 1993. *The trophic cascade in lakes*. New York: Cambridge University Press.

Carpenter SR, Kitchell JF, Hodgson JR. 1985. Cascading trophic interactions and lake productivity. *BioScience* 35:634–639.

Carrier JC, Musick JA, Heithaus MR, eds. 2004. *Biology of sharks and their relatives*. Boca Raton, FL: CRC Press.

Carroll AM, Wainwright PC. 2003. Functional morphology of prey capture in the sturgeon, *Scaphirhynchus albus*. *J Morphol* 256(3):270–284.

Carroll RL. 1988. *Vertebrate paleontology and evolution*. New York: W. H. Freeman.

Carruth LL, Jones RE, Norris DO. 2002. Cortisol and Pacific salmon: a new look at the role of stress hormones in olfaction and home-stream migration. *Integr Comp Biol* 42:574–581.

Caruso MA, Sheridan MA. 2011. Pancreas. In: Farrell AP, Stevens ED, Cech JJ, Richards JG, eds. *Encyclopedia of fish physiology: from genome to environment*, pp. 1276–1283. Boston: Elsevier/Academic Press.

Case GR. 1973. *Fossil sharks: a pictorial review*. New York: Pioneer Litho Co.

Casewell NR, Visser JC, Baumann KJ, et al. 2017. The evolution of fangs, venom, and mimicry systems in blenny fishes. *Curr Biol* 27:1–8.

Casey JG, Kohler NE. 1991. Long distance movements of Atlantic sharks from the NFS cooperative shark tagging program. *Underwater Nat* 19:87–91.

Casper BM. 2011. The ear and hearing in sharks, skates, and rays. In: Farrell AP, Stevens ED, Cech JJ, Richards JG, eds. *Encyclopedia of fish physiology: from genome to environment*, pp. 262–269. Boston: Elsevier/Academic Press.

Casselman JM. 1983. Age and growth assessment of fish from their calcified structures – techniques and tools. In: Prince ED, Pulos LM, eds. *Proceedings of the international workshop on age determination of oceanic pelagic fishes: tunas, billfishes, and sharks*, pp. 1–17. NOAA Technical Report NMFS 8. Washington, DC: National Oceanic and Atmospheric Administration.

Casselman JM. 1990. Growth and relative size of calcified structures of fish. *Trans Am Fish Soc* 119:673–688.

Casteel RW. 1976. *Fish remains in archaeology and paleo-environmental studies*. London: Academic Press.

Castelnau FL. 1879. On a new ganoid fish from Queensland. *Proc Linn Soc NSW* 3:164–165.

Castle PHJ. 1978. Ovigerous leptocephali of the nettastomatid eel genus *Facciolella*. *Copeia* 1978:29–33.

Castonguay M, Béguer-Pon M, Shan S, Dodson J. 2018. The Long Journey: A Telemetry and Modelling Study of Migrations of American Eels from the Upper St. Lawrence River to the Sargasso Sea. Abstract. 2018 AFS Symposium on American Eel. **http://www.asmfc.org/home/2018-afs-eel-symposium**.

Castro ALF, Stewart BS, Wilson SG, et al. 2007. Population genetic structure of the world's largest fish, the whale shark (*Rhincodon typus*). *Molec Ecol* 16:5183–5192.

Castro JI. 1983. *The sharks of North American waters*. College Station, TX: Texas A&M University Press.

Castro P, Huber ME. 1997. *Marine biology*, 2nd edn. Dubuque, IA: Wm. C. Brown/Time-Mirro.

Catania K. 2014. The shocking predatory strike of the electric eel. *Science* 346:1231–1234.

Catania K. 2015. Electric eels concentrate their electric field to induce involuntary fatigue in struggling prey. *Curr Biol* 25:2889–2898.

Catania K. 2017. Power transfer to a human during an electric eel's shocking leap. *Curr Biol* 27:2887–2891.

Caviedes C, Fik T. 1990. Variability in the Peruvian and Chilean fisheries. In: Glantz MH, Feingold LE, eds. *Climate variability, climate change and fisheries*, pp. 95–102. Boulder, CO: Environmental and Societal Impacts Group, National Center of Atmospheric Research.

Chakrabarty M, Homechaudhuri S. 2013. Fish guild structure along a longitudinally-determined ecological zonation of Teesta, an eastern Himalayan river in West Bengal, India. *Arxius de Miscel·lània Zoològica* 11:196–213.

Chang N, Sun C, Gao L, et al. 2013. Genome editing with RNA-guided Cas9 nuclease in zebrafish embryos. *Cell Res* 23:465–472.

Chang Y-S, Huang F-L, Lo T-B. 1994. The complete nucleotide sequence and gene organization of carp (*Cyprinus carpio*) mitochondrial genome. *J Mol Evol* 38:138–155.

Chao LN. 1973. Digestive system and feeding habits of the cunner, *Tautogolabrus adspersus*, a stomachless fish. *Fish Bull US* 71:565–585.

Chapin FS, Zavaleta ES, Eviner VT, et al. 2000. Consequences of changing biodiversity. *Nature* 405:234–242.

Chapman CA, Chapman LJ. 2003. Deforestation in tropical Africa: impacts on aquatic ecosystems. In: Crisman TL, Chapman LJ, Chapman CA, et al., eds. *Conservation, ecology, and management of African fresh waters*, pp. 229–246. Gainesville, FL: University Press of Florida.

Chapman DD, Prodöhl PA, Gelsleichter J, Manire CA, Shivji MS. 2004. Predominance of genetic monogamy by females in a hammerhead shark, *Sphyrna tiburo*: implications for shark conservation. *Molec Ecol* 13:1965–1974.

Chapman DD, Wintner SP, Abercrombie DL, et al. 2013. The behavioural and genetic mating system of the sand tiger shark (*Carcharias taurus*). *Biol Lett* 9:20130003.

Chapman DW. 1978. Production in fish populations. In: Gerking SD, ed. *Ecology of freshwater fish production*, pp. 5–25. London: Blackwell Science.

Chave EH, Randall HA. 1971. Feeding behavior of the moray eel *Gymnothorax pictus*. *Copeia* 1971:570–574.

Cheal AJ, Emslie M, MacNeil MA, Miller I, Sweatman H. 2013. Spatial variation in the functional characteristics of herbivorous fish communities and the resilience of coral reefs *Ecol Appl* 23(1):174–188.

Chellappa S, Huntingford FA. 1989. Depletion of energy reserves during reproductive aggression in male three-spined stickleback, *Gasterosteus aculeatus* L. *J Fish Biol* 35:315–316.

Cheney KL, Côté IM. 2005. Mutualism or parasitism? The variable outcome of cleaning symbioses. *Biol Lett* 1:162–165.

Cheney KL, Grutter AS, Bshary R. 2014. Geographical variation in the benefits obtained by a coral reef fish mimic. *Anim Behav* 88:85–90.

Cheng C-HC, Detrich HW, III. 2012. Molecular ecophysiology of antarctic notothenioid fishes. In: Rogers AD, Johnston NM, Murphy EJ, Clarke A, Page T, eds. *Antarctic ecosystems: an extreme environment in a changing world*, pp. 357–378. Oxford, UK and Hoboken, NJ, USA: Wiley-Blackwell.

Cherel Y, Duhamel G. 2004. Antarctic jaws: cephalopod prey of sharks in Kerguelen waters. *Deep Sea Res Part I Oceanogr Res Pap* 51:17–31.

Chesser RK, Smith MW, Smith WH. 1984. Biochemical genetics of mosquitofish. III. Incidence and significance of multiple paternity. *Genetica* 74:77–81.

Cheung WWL, Watson R, Pauly D. 2013. Signature of ocean warming in global fisheries catch. *Nature* 496:365–369.

Chew SF, Gan J, Ip YK. 2005. Nitrogen metabolism and excretion in the swamp eel, *Monopterus albus*, during 6 or 40 days of estivation in mud. *Physiol Biochem Zool* 78:620–629.

Chew SF, Ip YK. 2014. Excretory nitrogen metabolism and defence against ammonia toxicity in air-breathing fishes. *J Fish Biol* 84:603–638.

Chiszar D. 1978. Lateral displays in the lower vertebrates: forms, functions, and origin. In: Reese ES, Lighter FJ. *Contrasts in behavior*, pp. 105–135. New York: Wiley Interscience.

Chiu CH, Dewar K, Wagner GP, et al. 2004. Bichir HoxA cluster sequence reveals surprising trends in ray-finned fish genomic evolution. *Genome Res* 14:11–17.

Chivers DP, Mirza RS, Bryer PJ, Kiesecker JM. 2001. Threat-sensitive predator avoidance by slimy sculpins: understanding the importance of visual versus chemical information. *Can J Zool* 79:867–873.

Chivers DP, Smith RJF. 1998. Chemical alarm signalling in aquatic predator–prey systems: a review and prospectus. *Ecoscience* 5:338–352.

Cho J, Kim Y. 2002. Sharks: a potential source of antiangiogenic factors and tumor treatments. *Mar Biotechnol* 4:521–525.

Choat JH. 1991. The biology of herbivorous fishes on coral reefs. In: Sale PF, ed. *The ecology of fishes on coral reefs*, pp. 120–155. San Diego, CA: Academic Press.

Choat JH. 2006. Phylogeography and reef fishes: bringing ecology back into the argument. *J Biogeogr* 33:967–968.

Choat JH, Clements KD. 1998. Vertebrate herbivores in marine and terrestrial environments: a nutritional ecology perspective. *Ann Rev Ecol Syst* 29:375–403.

Chotkowski MA, Marsden JE. 1999. Round goby and mottled sculpin predation on lake trout eggs and fry: field predictions from laboratory experiments. *J Great Lakes Res* 25:26–35.

Choudhury A, Dick TA. 1998. The historical biogeography of sturgeons (Osteichthyes: Acipenseridae): a synthesis of phylogenetics, palaeontology and palaeogeography. *J Biogeogr* 25:623–640.

Christoffels A, Koh EGL, Chia J-m, Brenner S, Aparicio S, Venkatesh B. 2004. Fugu genome analysis provides evidence for a whole-genome duplication early during the evolution of ray-finned fishes. *Mol Biol Evol* 21(6):1146–1151.

Chu KH, Tam PF, Fung CH, et al. 1997. A biological survey of ballast water in container ships entering Hong Kong. *Hydrobiologia* 352:201–206.

Churikov D, Gharrett AJ. 2002. Comparative phylogeography of the two pink salmon broodlines: an analysis based on mitochondrial DNA genealogy. *Molec Ecol* 11:1077–1101.

Cieslak M, Mikkelsen SS, Skall HF, et al. 2016. Phylogeny of the viral hemorrhagic septicemia virus in European aquaculture. *PLoS ONE* 11(10):e0164475.

Cimino MC, Bahr GF. 1974. The nuclear DNA content and chromatin ultrastructure of the coelacanth *Latimeria chalumnae*. *Experim Cell Res* 88:263–272.

Cisneros-Mata MA, Botsford LW, Quinn JF. 1997. Projecting viability of *Totoaba macdonaldi*, a population with unknown age-dependent variability. *Ecol Applic* 7:968–980.

Cisneros-Mata MA, Montemayor-Lopez G, Roman-Rodriguez MJ. 1995. Life history and conservation of *Totoaba macdonaldi*. *Conserv Biol* 9:806–814.

CITES (Convention on International Trade in Endangered Species). 2001. *CITES identification guide – sturgeons and paddlefish*. Ottawa: Minister of Supply and Services, Canada. **www.cws-scf.ec.gc.ca**.

Clack JA. 2002. *Gaining ground: the origin and evolution of tetrapods*. Bloomington, IN: Indiana University Press.

Claiborne JB, Edards SL, Morrison-Shetlar AI. 2002. Acid–base regulation in fishes: cellular and molecular mechanisms. *J Exp Zool* 293:302–319.

Clark DL. 1987. Phylum conodonta. In: Boardman RS, Cheetham AH, Rowell AJ, eds. *Fossil invertebrates*, pp. 636–662. Palo Alto, CA: Blackwell Science.

Clark E, Kristof E. 1990. Deep sea elasmobranchs observed from submersibles in Grand Cayman, Bermuda and Bahamas. In: Pratt HL, Jr., Gruber SH, Taniuchi T, eds. *Elasmobranchs as living resources: advances in the biology, ecology, systematics, and the status of fisheries*, pp. 275–290. NOAA Technical Report No. 90. Washington, DC: National Oceanic and Atmospheric Administration.

Clark FN. 1925. The life history of *Leuresthes tenuis*, an atherine fish with tide controlled spawning habits. *Calif Div Fish Game Fish Bull* 10:1–51.

Clarke A. 1983. Life in cold water: the physiological ecology of polar marine ectotherms. *Oceanogr Mar Biol Ann Rev* 21:341–453.

Clarke SC, Magnussen JE, Abercrombie DL, McAllister MK, Shivji MS. 2006. Identification of shark species composition and proportion in the Hong Kong shark fin market based on molecular genetics and trade records. *Conserv Biol* 20:201–211.

Claydon J. 2005. Spawning aggregations of coral reef fishes: characteristics, hypotheses, threats and management. *Oceanogr Mar Biol* 42:265–301.

Clemens BJ. 2017. *Progress report: Miller Lake Lamprey*. Oregon Department of Fish and Wildlife. 6 pp.

Clement TA, Murry BA, Uzarski DG. 2015. Fish community size structure of small lakes: the role of lake size, biodiversity and disturbance, *J Freshw Ecol* 304:557–568

Clements KD, Choat JH. 1995. Fermentation in tropical marine herbivorous fishes. *Physiol Zool* 68:355–378.

Clifton HE, Hunter RE. 1972. The sand tilefish, *Malacanthus plumieri*, and the distribution of coarse debris near West Indian coral reefs. In: Collette BB, Earle SA, eds. *Results of the Tektite program: ecology of coral reef fishes*, pp. 87–92. Natural History Museum of Los Angeles County Bulletin No. 14. Los Angeles: Natural History Museum of Los Angeles.

Closs GP, Krkosek M, Olden JD, eds. 2016. *Conservation of freshwater fishes*. Cambridge, UK: Cambridge University Press.

Cloutier R. 1991. Patterns, trends, and rates of evolution of the Actinistia. *Env Biol Fish* 32:3–58.

Cloutier R, Forey PL. 1991. Diversity of extinct and living actinistian fishes (Sarcopterygii). *Env Biol Fish* 32:59–74.

Coad BW, Papahn F. 1988. Shark attacks in the rivers of southern Iran. *Env Biol Fish* 23:131–134.

Coburn MM, Gaglione JI. 1992. A comparative study of percid scales (Teleostei: Perciformes). *Copeia* 1992:986–1001.

Coe M. 1966. The biology of *Tilapia grahami* Boulenger in Lake Magadi, Kenya. *Acta Tropica* 23:146–177.

Cohen DE, Rosenblatt RH, Moser HG. 1990. Biology and description of a bythitid fish from deep-sea thermal vents in the tropical eastern Pacific. *Deep Sea Res* 37:267–283.

Cohen DM. 1970. How many Recent fishes are there? *Proc Calif Acad Sci 4th Ser* 38:341–346.

Cohen DM. 1984. Ontogeny, systematics and phylogeny. In: Moser HG, Richards WJ, Cohen DM, Fahay MP, Kendall AW, Jr., Richardson SL, eds. *Ontogeny and systematics of fishes*, pp. 7–11. Special Publication No. 1. Lawrence, KS: American Society of Ichthyologists and Herpetologists.

Colborn T, Dumanoski D, Myers JP. 1996. *Our stolen future*. New York: Dutton.

Colby PJ, ed. 1977. Percid international symposium (PERCIS). *J Fish Res Board Can* 34:1445–1999.

Coleman FC, Williams SL. 2002. Overexploiting marine ecosystem engineers: potential consequences for biodiversity. *Trends Ecol Evol* 17:40–44.

Coleman RR, Eble JA, DiBattista JD, Rocha LA, Randall JE, Berumen ML, Bowen BW. 2016. Regal phylogeography: range-wide survey of the marine angelfish *Pygoplites diacanthus* reveals evolutionary partitions between the Red Sea, Indian Ocean, and Pacific Ocean. *Molec Phylogen Evol* 100:243–253.

Coleman RR, Gaither MR, Kimokeo B, et al. 2014. Large-scale introduction of the Indo-Pacific damselfish *Abudefduf vaigiensis* into Hawai'i promotes genetic swamping of the endemic congener *A. abdominalis*. *Mol Ecol* 23:5552–5565.

Colgan PW, Gross MR. 1977. Dynamics of aggression in male pumpkinseed sunfish (*Lepomis gibbosus*) over the reproductive phase. *Z Tierpsychol* 43:139–151.

Colin PL. 1973. Burrowing behavior of the yellowhead jawfish, *Opistognathus aurifrons*. *Copeia* 1973:84–92.

Colin PL. 1978. Daily and summer–winter variation in mass spawning of the striped parrotfish, *Scarus croicensis*. *Fish Bull US* 76:117–124.

Collette BB. 1962. The swamp darters of the subgenus *Hololepis* (Pisces, Percidae). *Tulane Stud Zool* 9:115–211.

Collette BB. 1977. Epidermal breeding tubercles and bony contact organs in fishes. *Symp Zool Soc Lond* 1977:225–268.

Collette BB. 1978. Adaptations and systematics of the mackerels and tunas. In: Sharp GD, Dizon AE, eds. *The physiological ecology of tunas*, pp. 7–39. New York: Academic Press.

Collette BB. 1995. Hemiramphidae. In: Fischer W, Krupp F, Schneider W, Sommer C, Carpenter KE, Niem VH, eds. *Guia FAO para la identificacion de especies para los fines de la pesca, Pacifico Centro-Oriental*, Vol. II, pp. 1175–1181. Rome: Food and Agricultural Organization.

Collette BB. 1999. Mackerels, molecules, and morphology. In: Séret B, Sire J-Y, eds. *Proceedings of the 5th Indo-Pacific Fish Conference, Noumea*, pp. 149–164. Paris: Soc. Fr. Ichtyol.

Collette BB, Banarescu P. 1977. Systematics and zoogeography of the fishes of the family Percidae. *J Fish Res Board Can* 34:1450–1463.

Collette BB, Chao LN. 1975. Systematics and morphology of the bonitos (*Sarda*) and their relatives (Scombridae, Sardini). *Fish Bull US* 73:516–625.

Collette BB, Graves J, Illustrated by V. Kells. 2019. *Tunas and billfishes of the world*. Baltimore: Johns Hopkins.

Collette BB, Klein-MacPhee G, eds. 2002. *Bigelow and Schroeder's fishes of the Gulf of Maine*, 3rd edn. Washington, DC: Smithsonian Institution.

Collette BB, Nauen, CE. 1983. *Scombrids of the world*. FAO Fisheries Synopsis No. 125, Vol. 2. Rome: Food and Agricultural Organization.

Collette BB, Potthoff T, Richards WJ, Ueyanagi S, Russo JL, Nishikawa Y. 1984. Scombroidei: development and relationships. In: Moser HG, Richards WJ, Cohen DM, Fahay MP, Kendall AW, Jr., Richardson SL, eds. *Ontogeny and systematics of fishes*, pp. 591–620. Special Publication No. 1. Lawrence, KS: American Society of Ichthyologists and Herpetologists.

Collette BB, Russo JL. 1981. A revision of the scaly toadfishes, genus *Batrachoides*, with descriptions of two new species from the eastern Pacific. *Bull Mar Sci* 31:197–233.

Collette BB, Russo JL. 1985a. Interrelationships of the Spanish mackerels (Pisces: Scombridae: *Scomberomorus*) and their copepod parasites. *Cladistics* 1(2):141–158.

Collette BB, Russo JL. 1985b. Morphology, systematics, and biology of the Spanish mackerels (*Scomberomorus*, Scombridae). *Fish Bull US* 82:545–692.

Collette BB, Russo JL, Zavala-Camin LA. 1978. *Scomberomorus brasiliensis*, a new species of Spanish mackerel from the western Atlantic. *Fish Bull US* 76:273–280.

Collette BB, Rützler K. 1977. Reef fishes over sponge bottoms off the mouth of the Amazon River. *Proc 3rd Internat Coral Reef Symp* 1:305–310.

Collette BB, Talbot FH. 1972. Activity patterns of coral reef fishes with emphasis on nocturnal–diurnal changeover. *Los Angeles City Natur Hist Mus Sci Bull* 14:98–124.

Colligan M, Nickerson P, Kimball D. 1999. Proposed rule on Atlantic salmon. *Fed Reg* 64(221):62627–62641. **http://wais.access.gpo.gov**.

Collin SP, Whitehead D. 2004.The functional roles of passive electroreception in non-electric fishes. *Anim Biol* 54:1–25.

Collingsworth PD, Bunnell DB, Murray MW, et al. 2017. Climate change as a long-term stressor for the fisheries of the Laurentian Great Lakes of North America. *Rev Fish Biol and Fish* 27:363–391

Collins RT, Linker C, Lewis J. 2010. MAZe: a tool for mosaic analysis of gene function in zebrafish. *Nat Methods* 7:219–223.

Combes C. 2001. *Parasitism: the ecology and evolution of intimate interactions*. Chicago, IL: University of Chicago Press.

Compagno LJV. 1981. Legend versus reality: the jaws image and shark diversity. *Oceanus* 24(4): 5–16.

Compagno LJV. 1984. *Sharks of the world. An annotated and illustrated catalogue of shark species known to date*. FAO Fisheries Synopsis No. 125, Vol. 4. Rome: Food and Agricultural Organization.

Compagno LJV. 1990a. The evolution and diversity of sharks. In: Gruber SH, ed. *Discovering sharks*, pp. 15–22. American Littoral Society Special Publication No. 14. Highlands, NJ: American Littoral Society.

Compagno LJV. 1990b. Alternative life history styles of cartilaginous fishes in time and space. *Env Biol Fish* 28:33–75.

Compagno LJV. 1990c. Relationships of the megamouth shark, *Megachasma pelagios* (Lamniformes: Megachasmidae) with comments on its feeding habits. In: Pratt HL, Jr., Gruber SH, Taniuchi T, eds. *Elasmobranchs as living resources: advances in the biology, ecology, systematics, and the status of fisheries*, pp. 357–380. NOAA Technical Report No. 90. Washington, DC: National Oceanic and Atmospheric Administration.

Compagno LJV. 2001. *Sharks of the world. An annotated and illustrated catalogue of shark species known to date. Vol. 2. Bullhead, mackerel and carpet sharks (Heterodontiformes, Lamniformes and Orectolobiformes)*. FAO Species Catalogue for Fishery Purposes, No. 1, Vol. 2. Rome: Food and Agricultural Organization.

Compagno LJV, Cook SF. 1995a. *Status of the giant freshwater stingray (whipray) Himantura chaophraya (Monkolprasit and Roberts 1990)*. Shark News No. 5. Newbury, Berkshire UK: IUCN Shark Specialist Group. **www.flmnh.ufl.edu/fish/Organizations/SSG/sharknews/sn5/shark5news7.htm**.

Compagno LJV, Cook SF. 1995b. The exploitation and conservation of freshwater elasmobranchs: status of the taxa and prospects for the future. *J Aquaricult Aquat Sci* 7:62–90.

Compagno LJV, Dando M, Fowler S. 2005a. *A field guide to sharks of the world*. London: Collins.

Compagno LJV, Dando M, Fowler S. 2005b. *Sharks of the world*. Princeton, NJ: Princeton University Press.

Compagno LJV, Gottfried MD, Bowman SC. 1993. Size and scaling of the giant 'megatooth' shark *Carcharodon megalodon* (Lamnidae). *J Vertebr Paleontol* 13(3):31A.

Conant EB. 1987. An historical overview of the literature of Dipnoi: introduction to the bibliography of lungfishes. In: Bemis WE, Burggren WW, Kemp NE, eds. *The biology and evolution of lungfishes*, pp. 5–13. New York: Alan R. Liss.

Cone RS. 1990. Properties of relative weight and other condition indices. *Trans Am Fish Soc* 119:1048–1058.

Congdon JD, Dunham AE, Tinkle DW. 1982. Energy budgets and life histories of reptiles. In: Gans C, Pough FH, eds. *Biology of the Reptilia*, Vol. 13, pp. 233–271. San Diego, CA: Academic Press.

Connell JH. 1980. Diversity and the coevolution of competitors, or the ghost of competition past. *Oikos* 35:131–138.

Conniff R. 1991. The most disgusting fish in the sea. *Audobon* 93(2):100–108.

Connon RE,Beggel S, D'Abronzo LS, et al. 2011. Linking molecular biomarkers with higher level condition indicators to identify effects of copper exposures on the endangered Delta Smelt (*Hypomesus transpacificus*). *Environ Toxicol Chem* 30:290–300.

Conover DO, Heins SW. 1987. Adaptive variation in environmental and genetic sex determination in a fish. *Nature* 326:496–498.

Conover DO, Kynard BE. 1981. Environmental sex determination: interaction of temperature and genotype in a fish. *Science* 213:577–579.

Conover DO, Munch SB. 2002. Sustaining fisheries yields over evolutionary time scales. *Science* 297:94–96.

Conrad JL, Weinersmith KL, Brodin T, Saltz JB, Sih A. 2011. Behavioural syndromes in fishes: a review with implications for ecology and fisheries management. *J Fish Biol* 78:395–435.

Contreras S, Escalante MA. 1984. Distribution and known impacts of exotic fishes in Mexico. In: Courtenay WR, Jr., Stauffer, JR, Jr., eds. *Distribution, biology, and management of exotic fishes*, pp. 102–130. Baltimore: Johns Hopkins University Press.

Contreras-Balderas S, Almada-Villela P, Lozano-Vilano M, et al. 2002. Freshwater fish at risk or extinct in México. *Rev Fish Biol Fish* 12:241–251.

Cooke SJ, Crossin GT, Hinch SG. 2011. Pacific salmon migration: completing the cycle. In: Farrell AP, Stevens ED, Cech JJ, Richards JG, eds. *Encyclopedia of fish physiology: from genome to environment*, pp. 1945–1952. Boston: Elsevier/Academic Press.

Cooke SJ, Niezgoda GH, Hanson K, Suski CD, Tinline R, Philipp DP. 2005. Use of CDMA acoustic telemetry to document 3-D positions of fish: relevance to the design and monitoring of aquatic protected areas. *Mar Technol Soc J* 39:31–41.

Cooke SJ, Suski CD, Ostrand KG, Wahl DH, Philipp DP. 2007. Physiological and behavioral consequences of long-term artificial selection for vulnerability to recreational angling in a teleost fish. *Physiol Biochem Zoology* 80(5):480–490.

Cooper SD. 1984. The effect of trout on water striders in stream pools. *Oecologia (Berl)* 63:376–379.

Cooper SD, Walde SJ, Peckarsky BL. 1990. Prey exchange rates and the impact of predators on prey populations in streams. *Ecology* 71:1503–1514.

Copeland DE. 1969. Fine structural study of gas secretion in the physoclistous swim bladder of *Fundulus heteroclitus* and *Gadus callarias* and in the euphysoclistous swim bladder of *Opsanus tau*. *Cell Tissue Res* 93:305–331.

Côté IM. 2001. Evolution and ecology of cleaning symbioses in the sea. *Oceanogr Marine Biol Ann Rev* 38:311–355.

Cott HB. 1957. *Adaptive coloration in animals*. London: Methuen.

Cott PA, Hawkins AD, Zeddies D, et al. 2014. Song of the burbot: Under-ice acoustic signaling by a freshwater gadoid fish. *J Great Lakes Res* 40:435–440.

Coughlin DJ, Strickler JR. 1990. Zooplankton capture by a coral reef fish: an adaptive response to evasive prey. *Env Biol Fish* 29:35–42.

Courtenay WR, Jr., Hensley DA, Taylor JN, McCann JA. 1984. Distribution of exotic fishes in the continental United States. In: Courtenay WR, Jr., Stauffer JR, Jr., eds. *Distribution, biology, and management of exotic fishes*, pp. 41–77. Baltimore: Johns Hopkins University Press.

Courtenay WR, Jr., Williams JD. 2002. *Snakeheads (Pisces, Channidae) – a biological synopsis and risk assessment*. US Geological Survey Circular No. 1251. Reston, VA.

Courtenay-Latimer M. 1979. My story of the first coelacanth. *Occident Pap Calif Acad Sci* 134:6–10.

Coutant CC. 1985. Striped bass, temperature, and dissolved oxygen: a speculative hypothesis for environmental risk. *Trans Am Fish Soc* 114:31–61.

Coutant CC. 1987. Thermal preference: when does an asset become a liability? *Env Biol Fish* 18:161–172.

Coutant CC. 1990. Temperature–oxygen habitat for freshwater and coastal striped bass in a changing climate. *Trans Am Fish Soc* 119:240–253.

Cowen RK, Castro LR. 1994. Relation of coral-reef fish larval distributions to island scale circulation around Barbados, West-Indies. *Bull Mar Sci* 54:228–244.

Cowen RK, Lwiza KMM, Sponaugle S, Paris CB, Olson DB. 2000. Connectivity of marine populations: open or closed? *Science* 287:857–859.

Cowman PF, Bellwood DR. 2013. The historical biogeography of coral reef fishes: global patterns of origination and dispersal. *J Biogeogr* 40:209–224.

Cowx I. 1997. Introduction of fish species into European fresh waters: economic successes or ecological disasters. *Bull Fr Peche Piscic* 344/345:57–77 (in French, transl. I. Cowx).

Cox CE, Jones CD, Wares JP, Castillo KD, McField MD, Bruno JF. 2013. Genetic testing reveals some mislabeling but general compliance with a ban on herbivorous fish harvesting in Belize. *Conserv Lett* 6:132–140.

Cox DL, Koob TJ. 1993. Predation on elasmobranch eggs. *Env Biol Fish* 38:117–125.

Craig JF. 1987. *The biology of perch and related fish*. Kent, UK: Croom Helm Ltd.

Craig MT, Eble JA, Robertson DR, Bowen BW. 2007. High genetic connectivity across the Indian and Pacific Oceans in the reef fish *Myripristis berndti* (Holocentridae). *Mar Ecol Prog Ser* 334:245–254.

Crampton WGR. 2019. Electroreception, electrogenesis and electric signal evolution. *J Fish Biol* 95:92–134.

Crawshaw LI, Podrabsky JE. 2011. Temperature preference: behavioral responses to temperature in fishes. In: Farrell AP, Stevens ED, Cech JJ, Richards JG, eds. *Encyclopedia of fish physiology: from genome to environment*, pp. 758–764. Boston: Elsevier/Academic Press.

CRC (Coastal Resources Center). 2006. *Fisheries opportunities assessment*. Narrangansett, RI: Coastal Reources Center, University of Rhode Island, Fish_Opp_Assess_Final_012607.pdf, **www.crc.uri.edu**.

Cressey RF, Collette BB. 1970. Copepods and needlefishes: a study in host–parasite relationships. *Fish Bull US* 68:347–432.

Crivelli AJ. 1995. Are fish introductions a threat to endemic freshwater fishes in the northern Mediterranean region? *Biol Conserv* 72:311–319.

Crockett EL, Londraville RL. 2006. Temperature. In: Evans DH, Claiborne JD, eds. *The physiology of fishes*, 3rd edn, pp. 231–269. Boca Raton, FL: CRC Press.

Crossman DJ, Choat JH, Clements KD, et al. 2001. Detritus as food for grazing fishes on coral reefs. *Limnol Oceanogr* 46:1596–1605.

Crowder LB. 1984. Character displacement and habitat shift in a native cisco in southeastern Lake Michigan: evidence for competition? *Copeia* 1984:878–883.

Crowder LB, Cooper WE. 1982. Habitat structural complexity and the interaction between bluegills and their prey. *Ecology* 65:894–908.

Cucherousset J, Aymes JC, Poulet N, Santoul F, Céréghino R. 2008. Do native brown trout and non-native brook trout interact reproductively? *Naturwissenschaften* 95:647–654.

Cui R, Schumer M, Kruesi K, Walter R, Andolfatto P, Rosenthal GG. 2013. Phylogenomics reveals extensive reticulate evolution in *Xiphophorus* fishes. *Evolution* 67:2166–2179.

Cullen AP, Monteith-McMaster CA. 1993. Damage to the rainbow trout (*Oncorhynchus mykiss*) lens following an acute dose of UV-B. *Curr Eye Res* 12:97–106.

Culp JM, Glozier NE, Scrimgeour GJ. 1991. Reduction of predation risk under the cover of darkness: avoidance responses of mayfly larvae to a benthic fish. *Oecologia* 86:163–169.

Culver DC. 1982. *Cave life: evolution and ecology.* Cambridge, MA: Harvard University Press.

Cummings ME, Rosenthal GG, Ryan MJ. 2003. A private ultraviolet channel in visual communication. *Proc Roy Soc Lond Ser B Biol Sci* 270:897–904.

Cunjak RA. 1996. Winter habitat of selected stream fishes and potential impacts from land-use activity. *Can J Fish Aquat Sci* 53 (suppl 1):267–282.

Cunningham JT, Reid DM. 1932. Experimental research on the emission of oxygen by the pelvic filaments of the male *Lepidosiren* with some experiments on *Synbranchus marmoratus. Proc Roy Soc* 110:234–248.

Curio E. 1976. *The ethology of predation.* Berlin: Springer-Verlag.

Currie S. 2011. Heat shock proteins and temperature. In: Farrell AP, Stevens ED, Cech JJ, Richards JG, eds. *Encyclopedia of fish physiology: from genome to environment*, pp. 1732–1737. Boston: Elsevier/Academic Press.

Cushing DH. 1973. *The detection of fish.* Oxford: Pergamon Press.

Cushing DH. 1975. *Marine ecology and fisheries.* Cambridge, UK: Cambridge University Press.

Cutter AD. 2019. A primer of molecular population genetics. New York: Oxford University Press.

Czesny S, Graeb BDS, Dettmers JM. 2005. Ecological consequences of swim bladder noninflation for larval yellow perch. *Trans Am Fish Soc* 134:1011–1020.

D'Aout K, Aerts P. 1999. A kinematic comparison of forward and backward swimming in the eel *Anguilla anguilla. J Exp Biol* 202:1511–1521.

Dadswell MJ, Klauda RJ, Moffitt CM, Saunders RL, Rulifson RA, Cooper JE, eds. 1987. *Common strategies of anadromous and catadromous fishes.* American Fisheries Society Symposium 1. Bethesda, MD: American Fisheries Society.

Daeschler EB, Shubin NH, Jenkins FA, Jr. 2006. A Devonian tetrapod-like fish and the evolution of the tetrapod body plan. *Nature* 440:757–763.

Dahlgren CP, Eggleston DB. 2000. Ecological processes underlying ontogenetic habitat shifts in a coral reef fish. *Ecology* 81(8):2227–2240.

Daily GC, ed. 1997. *Nature's services: societal dependence on natural ecosystems.* Covelo, CA: Island Press.

Daly-Engel T, Grubbs R, Bowen BW, Toonen RJ. 2007. Frequency of multiple paternity in an unexploited tropical population of sandbar sharks (*Carcharhinus plumbeus*). *Can J Fish Aquatic Sci* 64:198–204.

Daly-Engel T, Grubbs R, Holland K, Toonen RJ, Bowen BW. 2006. Multiple paternity assessments for three species of congeneric sharks *(Carcharhinus)* in Hawaii. *Env Biol Fish* 76:419–424.

Daly-Engel TS, Seraphin KD, Holland KN, et al. 2012. Global phylogeography with mixed marker analysis reveals male-mediated dispersal in the endangered scalloped hammerhead shark (*Sphyrna lewini*). *PLoS One* 7:e29986.

Dana JD. 1853. On the isothermal oceanic chart: illustrating the geographical distribution of marine animals. *Am J Sci* 16:314–327.

Daniels RA. 1979. Nest guard replacement in the Antarctic fish *Harpagifer bispinis*: possible altruistic behavior. *Science* 205:831–833.

Daniels RA. 1982. Feeding ecology of some fishes of the Antarctic Peninsula. *Fish Bull US* 80:575–588.

Daniels RA. 2001. Untested assumptions: the role of canals in the dispersal of sea lamprey, alewife, and other fishes in the eastern United States. *Env Biol Fish* 60:309–329.

Danilowicz BS, Sale PF. 1999. Relative intensity of predation on the French grunt, *Haemulon flavolineatum*, during diurnal, dusk, and nocturnal periods on a coral reef. *Mar Biol* 133:337–343.

Darwall W, Freyhof J. 2016 Lost fishes, who is counting? The extent of threat to freshwater fish biodiversity. In: Closs GP, Krkosek M, Olden JD, eds. *Conservation of freshwater fishes*, pp. 1–35. Cambridge: Cambridge University Press.

Darwin C. 1859. *On the origin of species.* London: John Murray.

Das M. 1994. Age determination and longevity in fishes. *Gerontology* 40:70–96.

Davenport J. 1994. How and why do flying fish fly? *Rev Fish Biol Fisheries* 4:184–214.

Daves NK, Nammack MF. 1998. US and international mechanisms for protecting and managing shark resources. *Fish Res* 39:223–228.

David L, Blum S, Feldman MW, Lavi U, Hillel J. 2003. Recent duplication of the common carp (*Cyprinus carpio*) genome as revealed by analyses of microsatellite loci. *Mol Biol Evol* 20:1425–1434.

Davin WT, Jr., Kohler CC, Tindall DR. 1988. Ciguatera toxins adversely affect piscivorous fishes. *Trans Am Fish Soc* 117:374–384.

Davis MP, Sparks JS, Smith WL. 2016. Repeated and widespread evolution of bioluminescence in marine fishes. *PLoS One* 11:e0155154. **https://doi.org/10.1371/journal.pone.0155154**.

Davis TLO. 1988. Temporal changes in the fish fauna entering a tidal swamp system in tropical Australia. *Env Biol Fish* 21:161–172.

Davis WE, Jr. 2004. Black-crowned night-heron vibrates bill in water to attract fish. *Southeast Natur* 3:127–128.

Davis WP, Birdsong RS. 1973. Coral reef fishes which forage in the water column. *Helgol Wiss Meeresunters* 24:292–306.

Dawson CE. 1985. *Indo-Pacific pipefishes.* Ocean Springs, MS: Gulf Coast Research Lab.

Day RD, Mueller F, Carseldine L, Meyers-Cherry N, Tibbetts IR. 2016. Ballistic Beloniformes attacking through Snell's Window. *J Fish Bio* 88:727–734

Day RW. 1985. The effects of refuges from predators and competitors on sessile communities on a coral reef. *Proc 5th Internat Coral Reef Cong* 4:41–45.

Dayton PK, Robilliard GA, Paine RT, Dayton LB. 1974. Biological accommodation in the benthic community at McMurdo Sound, Antarctica. *Ecol Monogr* 44:105–128.

de Carvalho MR. 1996. Higher-level elasmobranch phylogeny, basal squaleans and paraphyly. In Stiassny MLJ, Parenti LR, Johnson GD, eds. *Interrelationships of fishes.* pp. 35–62. San Diego: Academic Press.

de Carvalho MR, Bockmann FA, Amorim DS, et al. 2007. Taxonomic impediment or impediment to taxonomy? A commentary on systematics and the cybertaxonomic-automation paradigm. *Evol Biol* 34:140–143.

de Lima AC, Araujo-Lima CARM. 2004. The distributions of larval and juvenile fishes in Amazonian rivers of different nutrient status. *Freshwater Biol* 49:787–800.

de Moraes MFPG, Höller S, da Costa OTF, Glass ML, Fernandes MN, Perry SF. 2005. Morphometric comparison of the respiratory organs in the South American lungfish *Lepidosiren paradoxa* (Dipnoi). *Physiol Biochem Zool* 78:546–559.

de Pinna MCC. 1996. Teleostean monophyly. In Stiassny MLJ, Parenti LR, Johnson GD, eds. *Interrelationships of fishes*, pp. 147–162. San Diego: Academic Press.

de Queiroz A. 2005. The resurrection of oceanic dispersal in historical biogeography. *Trends Ecol Evol* 20:68–73.

de Queiroz K, Cantino P. 2020. *International code of phylogenetic nomenclature (Phylocode)*. Boca Raton, FL: CRC Press.

De Schepper N, De Kegel B, Adriaens D. 2007. *Pisodonophis boro* (Ophichthidae: Anguilliformes): specialization for head-first and tail-first burrowing? *J Morphol* 268:112–126.

De Vos JM, Joppa LN, Gittleman JL, Stephens PR, Pimm SL. 2015. Estimating the normal background rate of species extinctions. Conserv Biol 29:452–462.

De Vos L, Oyugi D. 2002. First capture of a coelacanth, *Latimeria chalumnae* Smith, 1939 (Pisces: Latimeriidae), off Kenya. *S Afr J Sci* 98:345–347.

Deegan LA. 1993. Nutrient and energy transport between estuaries and coastal marine ecosystems by fish migration. *Can J Fish Aquat Sci* 50:74–79.

Delarbre C, Spruyt N, Delmarre C, et al. 1998. The complete nucleotide sequence of the mitochondrial DNA of the dogfish, *Scyliorhinus canicula*. *Genetics* 150:331–344.

DeMarais A, Oldis D. 2005. Matrotrophic transfer of fluorescent microspheres in poeciliid fishes. *Copeia* 2005:632–636.

DeMartini EE. 1999. Intertidal spawning. In: Horn MH, Martin KLM, Chotkowski MA, eds. *Intertidal fishes: life in two worlds*, pp. 143–164. San Diego, CA: Academic Press.

DeMartini EE, Coyer JA. 1981. Cleaning and scale-eating in juveniles of the kyphosid fishes, *Hermosilla azurea* and *Girella nigricans*. *Copeia* 1981:785–789.

DeMartini EE, Fountain RK. 1981. Ovarian cycling frequency and batch fecundity in the queenfish, *Seriphus politus*: attributes representative of serial spawning fishes. *Fish Bull US* 79:547–560.

DeMartini EE, Sikkel PC. 2006. Reproduction. In: Allen LG, Pondella DJ, II, Horn MH, eds. *The ecology of marine rishes: California and adjacent waters*, pp. 483–523. Berkeley, CA: University of California Press.

Denison RH. 1970. Revised classification of Pteraspididae, with descriptions of new forms from Wyoming. *Fieldiana Geol* 20:1–41.

Denton EJ. 1961. The buoyancy of fish and cephalopods. *Prog Biophysics Molec Biol* 11:178–234.

Denton EJ, Herring PJ, Widdder EA, Latz MF, Case JF. 1985. The roles of filters in the photophores of oceanic animals and their relation to vision in the oceanic environment. *Proc Roy Soc Lond Ser B Biol Sci* 225:63–98

Denton EJ, Land MF. 1971. Mechanism of reflexion in silvery layers of fish and cephalopods. *Proc Roy Soc Lond B Biol Sci* 178:43–61.

Denton EJ, Nicol JAC. 1962. Why fishes have silvery sides; and a method of measuring reflectivity. *J Physiol* 165:13–15.

Denton EJ, Nicol JAC. 1965. Studies of the reflexion of light from silvery surfaces of fishes, with special reference to the bleak, *Alburnus alburnus*. *J Mar Biol Assoc UK* 45:683–703.

Denton TE, Howell WM. 1973. Chromosomes of the African polypterid fishes, *Polypterus palmas* and *Calamoichthys calabaricus* (Pisces: Brachiopterygii). *Experientia* 29:122–124.

Depczynski M, Bellwood DR. 2005. Shortest recorded vertebrate lifespan found in a coral reef fish. *Curr Biol* 15:R288–R289.

Derr M. 1992. Raiders of the reef. *Audubon* 94(2):48–54.

DeSalle R, Birstein VJ. 1996. PCR identification of black caviar. *Nature* 381:197–198.

de Santana CD, Crampton WG, Dillman CB, et al. 2019. Unexpected species diversity in electric eels with a description of the strongest living bioelectricity generator. *Nat Comm* 10: 4000.

DeSilva CD, Premawansa S, Keembiyahetty CN. 1986. Oxygen consumption in *Oreochromis niloticus* (L.) in relation to development, salinity, temperature and time of day. *J Fish Biol* 29:267–277.

Devlin RH, Biagi CA, Yesaki TY, Smailus DE, Byatt JC. 2001. Growth of domesticated transgenic fish. *Nature* 409:781–782.

Devlin RH, Nagahama Y. 2002. Sex determination and sex differentiation in fish: an overview of genetic, physiological, and environmental influences. *Aquaculture* 208:191–364.

DeVries AL. 1970. Freezing resistance in Antarctic fishes. In: Holdgate MW, ed. *Antarctic ecology*, Vol. 1, pp. 320–328. San Diego, CA: Academic Press.

DeVries AL. 1977. The physiology of cold adaptation in polar marine poikilotherms. In: Dunbar MJ, ed. *Polar oceans*, pp. 409–417. Calgary, Alberta: Arctic Institute of North America.

DeVries AL, Cheng C-HC. 2005. Antifreeze proteins and organismal freezing avoidance in polar fishes. In: Farrell AP, Steffensen JF, eds. *The physiology of polar fishes*, pp. 155–201. San Diego, CA: Elsevier Academic Press.

DeVries DR, Frie RV. 1996. Determination of age and growth. In: Murphy BR, Wallis DW, eds. *Fisheries techniques*, 2nd edn, pp. 483–512. Bethesada, MD: American Fisheries Society.

DeWoody JA, Avise JC. 2000. Microsatellite variation in marine, freshwater, and anadromous fishes compared with other animals. *J Fish Biol* 56:461–473.

DeWoody JA, Avise JC. 2001. Genetic perspectives on the natural history of fish mating systems. *J Heredity* 92:167–172.

Diana JS. 2003. *Biology and ecology of fishes*, 2nd edn. Travers City, MI: Cooper Publishing Group.

Diana JS, Mackay WC, Ehrman M. 1977. Movements and habitat preference of northern pike (*Esox lucius*) in Lac Ste. Anne, Alberta. *Trans Am Fish Soc* 106:560–565.

DiBattista JD, Roberts MB, Bouwmeester J, et al. 2016a. A review of contemporary patterns of endemism in the Red Sea. *J Biogeogr* 43:423–439.

DiBattista JD, Wang X, Saenz-Agudelo P, Piatek MJ, Aranda M, Berumen ML. 2016b. Draft genome of an iconic Red Sea fish, the blacktail butterflyfish (*Chaetodon austriacus*): current status and its characteristics. *Mol Ecol Resour* 18(2):347–355.

DiBattista JD, Whitney J, Craig MT, et al. 2016c. Surgeons and suture zones: hybridization among four surgeonfish species in the Indo-Pacific with variable evolutionary outcomes. *Mol Phylogenet Evol* 101:203–215.

Dickerson H, Clark T. 1998. *Ichthyophthirius multifiliis*: a model of cutaneous infection and immunity in fishes. *Immunol Rev* 166:377–384.

Dickson IW, Kramer RH. 1971. Factors influencing scope for activity and active and standard metabolism of rainbow trout (*Salmo gairdneri*). *J Fish Res Board Can* 28:587–596.

Dickson KA. 2011. Physiology of tuna. In: Farrell AP, Stevens ED, Cech JJ, Richards JG, eds. *Encyclopedia of fish physiology: from genome to environment*, pp. 1903–1913. Boston: Elsevier/Academic Press.

Didier DA. 2005. Phylogeny and classification of extant Holocephali. In: Carrier JC, Musick JA, Heithaus MR, eds. *Biology of sharks and their relatives*, pp. 115–135. Boca Raton, FL: CRC Press.

DiGiulio RT, Hinton DE. 2008. *The toxicology of fishes*. Boca Raton, FL: CRC Press.

Dill LM. 1974. The escape response of the zebra danio (*Brachydanio rerio*). I. The stimulus for escape. *Anim Behav* 22:711–722.

Dill LM. 1977a. Refraction and the spitting behavior of the archerfish (*Toxotes chatareus*). *Behav Ecol Sociobiol* 2:169–184.

Dill PA. 1977b. Development of behaviour in alevins of Atlantic salmon, *Salmo salar*, and rainbow trout, *S. gairdneri*. *Anim Behav* 25:116–121.

Dill LM. 1978. Aggressive distance in juvenile coho salmon (*Oncorhynchus kisutch*). *Can J Zool* 56:1441–1446.

Dill LM, Fraser AHG. 1984. Risk of predation and the feeding behaviour of juvenile coho salmon (*Oncorhynchus kisutch*). *Behav Ecol Sociobiol* 16:65–71.

Dillard JG, Graham LK, Russell TR, eds. 1986. *The paddlefish: status, management and propagation*. American Fisheries Society Special Publication No. 7. Columbia, MO: American Fisheries Society North Central Division.

Ding Q, Chen X, Yu W, Chen Y. 2017. An assessment of "fishing down marine food webs" in coastal states during 1950–2010. *Acta Oceanol Sin* 36:43–50

Ding Q, Chen X, Yu W, Tian S, Chen Y. 2016. An evaluation of underlying mechanisms for "fishing down marine food webs". *Acta Oceanol Sin* 35(8):32–38

Dittman AH, Quinn TP. 1996. Homing in Pacific salmon: mechanisms and ecological basis. *J Exp Biol* 199:83–91.

Dixson DL, Hay ME. 2012. Corals chemically cue mutualistic fishes to remove competing seaweeds. *Science* 338(6108):804–807.

Dobzhansky T. 1935. A critique of the species concept in biology. *Philos Sci* 2:344–355.

Dodson JJ. 1997. Fish migration: an evolutionary perspective. In: Godin JJ, ed. *Behavioural ecology of teleost fishes*, pp. 10–36. Oxford: Oxford University Press.

Dogiel VA, Petrushevski GK, Polyanski YI. 1961. *Parasitology of fishes*. Edinburgh: Oliver & Boyd.

Doherty PJ. 1991. Spatial and temporal patterns in recruitment. In: Sale PF, ed. *The ecology of fishes on coral reefs*, pp. 261–293. San Diego, CA: Academic Press.

Doherty PJ, Planes S, Mather P. 1995. Gene flow and larval duration in seven species of fish from the Great Barrier Reef. *Ecology* 76:2373–2391.

Doherty PJ, Sale PF. 1986. Predation on juvenile coral reef fishes: an exclusion experiment. *Coral Reefs* 4:225–234.

Doherty PJ, Williams DMcB. 1988. The replenishment of coral reef fish populations. *Oceanogr Mar Biol Ann Rev* 26:487–551.

Domeier ML, Colin PL. 1997. Tropical reef fish spawning aggregations: defined and reviewed. *Bull Mar Sci* 60:698–726.

Dominey WJ. 1983. Mobbing in colonially nesting fishes, especially the bluegill, *Lepomis macrochirus*. *Copeia* 1983:1086–1088.

Dominey WJ, Blumer LS. 1984. Cannibalism of early life stages in fishes. In: Hausfater G, Hrdy SB, eds. *Infanticide. Comparative and evolutionary perspectives*, pp. 43–64. New York: Aldine.

Dominici P, Blake RW. 1997. The kinematics and performance of fish fast-start swimming. *J Exp Biol* 200:1165–1178.

Donoghue PCJ, Forey PL, Aldridge RJ. 2000. Conodont affinity and chordate phylogeny. *Biol Rev* 75:191–251.

Donoghue PCJ, Purnell MA, Aldridge RJ. 1998. Conodont anatomy, chordate phylogeny and vertebrate classification. *Lethaia* 31:211–219.

Douglas ME, Marsh PC, Minckley WL. 1994. Indigenous fishes of western North America and the hypothesis of competitive displacement: *Meda fulgida* (Cyprinidae) as a case study. *Copeia* 1994:9–19.

Douglas RH, Partridge JC. 2011. Visual adaptations to the deep sea. In: Farrell AP, Stevens ED, Cech JJ, Richards JG, eds. *Encyclopedia of fish physiology: from genome to environment*, pp. 166–182. Boston: Elsevier/Academic Press.

Downhower JF, Brown L, Pederson R, Staples G. 1983. Sexual selection and sexual dimorphism in mottled sculpins. *Evolution* 37:96–103.

Drake MT, Claussen JE, Phillip DP, et al. 1997. A comparison of bluegill reproductive strategies and growth among lakes with different fishing intensities. *N Am J Fish Manage* 17:496–507.

Drenner RW, Hambright KD, Vinyard GL, Gophen M, Pollingher U. 1987. Experimental study of size-selective phytoplankton grazing by a filter-feeding cichlid and the cichlid's effects on plankton community structure. *Limnol Oceanogr* 32:1140–1146.

Drenner RW, Mummert JR, deNoyelles F, Jr., Kettle D. 1984a. Selective particle ingestion by a filter-feeding fish and its impact on phytoplankton community structure. *Limnol Oceanogr* 29:941–948.

Drenner RW, Taylor SB, Lazzaro X, Kettle D. 1984b. Particle-grazing and plankton community impact of an omnivorous cichlid. *Trans Am Fish Soc* 113:397–402.

Driscoll CT, Driscoll KM, Mitchell MJ, Raynal DJ. 2003. Effects of acidic deposition on forest and aquatic ecosystems in New York State. *Environ Pollut* 123:327–336.

Drucker EG, Lauder GV. 2001. Locomotor function of the dorsal fin in teleost fishes: experimental analysis of wake forces in sunfish. *J Exp Biol* 204:2943–2958.

Dubs DO, Corkum LD. 1996. Behavioral interactions between round gobies (*Neogobius melanostomus*) and mottled sculpins (*Cottus bairdi*). *Great Lakes Res* 22:838–844.

Dudgeon D. 2000. Large-scale hydrological changes in tropical Asia: prospects for riverine biodiversity. *BioScience* 50:793–806.

Dulvy NK, Polunin NVC. 2004. Using informal knowledge to infer human-induced rarity of a conspicuous reef fish. *Anim Conserv* 7:365–374.

Duncan KM. 2006. Estimation of daily energetic requirements in young scalloped hammerhead sharks, *Sphyrna lewini*. *Env Biol Fish* 76:139–149.

Duncan KM, Martin AP, Bowen BW, de Couet GH. 2006. Global phylogeography of the scalloped hammerhead shark (*Sphyrna lewini*). *Molec Ecol* 15:2239–2251.

Duncker HR, Fleischer G, eds. 1986. *Functional morphology of vertebrates*. New York: Springer-Verlag.

Dunham AE, Grant BW, Overall KL. 1989. Interfaces between biophysical and physiological ecology and the population ecology of terrestrial vertebrate ectotherms. *Physiol Zool* 62:335–355.

Dunham RA. 2004. *Aquaculture and fisheries biotechnology: genetic approaches*. Cambridge, MA: CABI Publishing.

Dunn RP. 2016. Tool use by a temperate wrasse, California sheephead *Semicossyphus pulcher*. *J Fish Biol* 88:805–810

Durbin AG, Nixon SW, Oviatt CA. 1979. Effects of the spawning migration of the alewife, *Alosa pseudoharengus*, on freshwater ecosystems. *Ecology* 60:8–17.

Dutheil DB. 1999. The first articulated fossil cladistian: *Serenoichthys kemkemensis*, gen. et sp nov., from the Cretaceous of Morocco. *J Vertebr Paleontol* 19:243–246.

Dwyer GS, Cronin TM, Baker PA, Raymo ME, Buzas JS, Correge T. 1995. North Atlantic deepwater temperature change during late Pliocene and late Quaternary climatic cycles. *Science* 270:1347–1351.

Eastman JT. 1993. *Antarctic fish biology: evolution in a unique environment.* San Diego: Academic Press.

Eastman JT. 2005. The nature and diversity of antarctic fishes. *Polar Biol* 28:93–107.

Eastman JT, DeVries AI. 1986. Antarctic fishes. *Sci Am* 255(5):106–114.

Ebeling AW, Atkin NB, Setzer PY. 1971. Genome sizes of teleostean fishes: increases in some deep-sea species. *Am Natur* 105:549–562.

Ebeling AW, Hixon MA. 1991. Tropical and temperate reef fishes: comparison of community structures. In: Sale PF, ed. *The ecology of fishes on coral reefs*, pp. 509–563. San Diego, CA: Academic Press.

Ebeling AW, Laur DR. 1985. The influence of plant cover on surfperch abundance at an offshore temperate reef. *Env Biol Fish* 12:169–179.

Ebeling AW, Laur DR, Rowley RJ. 1985. Severe storm disturbance and reversal of community structure in a southern California kelp forest. *Mar Biol* 84:287–294.

Ebersole JP. 1977. The adaptive significance of interspecific territoriality in the reef fish *Eupomacentrus leucostictus. Ecology* 58:914–920.

Ebersole JP. 1980. Food density and territory size: an alternative model and a test on the reef fish *Eupomacentrus leucostictus. Am Natur* 115:492–509.

Ebert DA. 1991. Observations on the predatory behaviour of the sevengill shark, *Notorynchus cepedianus. S Afr J Mar Sci* 11:455–465.

Eble JA, Toonen RJ, Sorensen LL, Basch L, Papastamatiou Y, Bowen BW. 2011. Escaping paradise: larval export from Hawaii in an Indo-Pacific reef fish, the Yellow Tang (*Zebrasoma flavescens*). *Mar Ecol Progr Ser* 428:245–258.

Echelle AA. 1991. Conservation genetics and genic diversity in freshwater fishes of western North America. In: Minckley WL, Deacon JE, eds. *Battle against extinction. Native fish management in the American West*, pp. 141–153. Tucson, AZ: University of Arizona Press.

Echelle AA, Echelle AF. 1984. Evolutionary genetics of a "species flock:" atherinid fishes on the Mesa Central of Mexico. In: Echelle AE, Kornfield I, eds. *Evolution of fish species flocks*, pp. 93–110. Orono, ME: Univeristy of Maine Press.

Echelle AA, Echelle AF. 1997. Genetic introgression of endemic taxa by non-natives: a case study with Leon Springs pupfish and sheepshead minnow. *Conserv Biol* 11:153–161.

Eckert SA, Dolar LL, Kooyman GL, Perrin W, Rahman RA. 2002. Movements of whale sharks (*Rhincodon typus*) in South-east Asian waters as determined by satellite telemetry. *J Zool* 257:111–115

Eckert SA, Stewart BS. 2001. Telemetry and satellite tracking of whale sharks, *Rhincodon typus*, in the Sea of Cortez, Mexico, and the north Pacific Ocean. *Env Biol Fish* 60:299–308.

Edmunds M. 1974. *Defence in animals.* New York: Longman.

Edwards A. 1990. *Fish and fisheries of Saint Helena Island.* Newcastle: Center for Tropical Coastal Management Studies, University of Newcastle upon Tyne.

Edwards M, Richardson AJ. 2004. Impact of climate change on marine pelagic phenology and trophic mismatch. *Nature* 430:881–884.

Egger B, Obermüller B, Phiri H, Sturmbauer C, Sefc KM. 2006. Monogamy in the maternally mouthbrooding Lake Tanganyika cichlid fish *Tropheus moorii. Proc Roy Soc Lond B Biol Sci* 273:1797–1802.

Eggington S, Sidell BD. 1989. Thermal acclimation induces adaptive changes in subcellular structure of fish skeletal muscle. *Am J Physiol* 256:R1–R10.

Egli DP, Babcock RC. 2004. Ultrasonic tracking reveals multiple behavioural modes of snapper (*Pagrus auratus*) in a temperate no-take marine reserve. *ICES J Mar Sci* 61:1137–1143.

Ehlinger TJ. 1989. Foraging mode switches in the golden shiner (*Notemigonus crysoleucas*). *Can J Fish Aquat Sci* 46:1250–1254.

Ehrlich PR, Talbot FH, Russell PC, Anderson GR. 1977. The behaviour of chaetodontid fishes, with special reference to Lorenz's 'poster coloration' hypothesis. *J Zool Lond* 183:213–228.

Eibl-Eibesfeldt I. 1970. *Ethology: the biology of behavior* (transl. E. Klinghammer). New York: Holt, Rinehart & Winston.

Eikeset AM, Richter A, Dunlop ES, Dieckmann U, Stenseth NC. 2013. Economic repercussions of fisheries-induced evolution. *PNAS* 110:12259–12264.

Eklöv P, Jonsson P. 2007. Pike predators induce morphological changes in young perch and roach. *J Fish Biol* 70:155–164.

Ekman S. 1953. *Zoogeography of the sea.* London: Sidgwick & Jackson.

Elgar MA, Crespi BJ. 1992. *Cannibalism: ecology and evolution among diverse taxa.* New York: Oxford University Press.

Elkins D, Nibbelink NP, Grossman GD. 2019. Stocked rainbow trout (*Oncorhynchus mykiss*) affect space use by Warpaint Shiners (*Luxilus coccogenis*). *Ecol Freshw Fish* 28:167–175.

Ellerby DJ, Spierts ILY, Altringham JD. 2001. Fast muscle function in the European eel (*Anguilla anguilla* L.) during aquatic and terrestrial locomotion. *J Exp Biol* 204:2231–2238.

Elliott DG. 2000. Integumentary system. In: Ostrander GK, ed. *The laboratory fish*, pp. 95–108. London: Academic Press.

Elliott DG. 2011. The many functions of fish integument. In: Farrell AP, Stevens ED, Cech JJ, Richards JG, eds. *Encyclopedia of fish physiology: from genome to environment*, pp. 471–475. Boston: Elsevier/Academic Press.

Elliott J. 1992. The role of sea anemones as refuges and feeding habitats for the temperate fish *Oxylebius pictus. Env Biol Fish* 35:381–400.

Elliott JK, Mariscal RN. 1997. Acclimation or innate protection of anemone fishes from sea anemones? *Copeia* 1997:284–289.

Ellis D, Associates. 1996. *Net loss: the salmon netcage industry in British Columbia.* Vancouver, BC: The David Suzuki Foundation.

Ellis R. 1976. *The book of sharks.* New York: Grosset & Dunlop.

Ellis R. 1985. *The book of whales.* New York: Knopf.

Ellis R, McCosker JE. 1991. *Great white shark.* New York: Harper Collins.

Elmerot C, Arnason U, Gojobori T, Janke A. 2002. The mitochondrial genome of the pufferfish, *Fugu rubripes*, and ordinal teleostean relationships. *Gene* 295:163–172.

Emery AR. 1978. The basis of fish community structure: marine and freshwater comparisons. *Env Biol Fish* 3:33–47.

Emery AR, Thresher RE, eds. 1980. Biology of damselfishes. *Bull Mar Sci* 30:145–328.

Emlen ST, Oring LW. 1977. Ecology, sexual selection, and the evolution of mating systems. *Science* 197:215–223.

Endean R. 1973. Population explosions of *Acanthaster planci* and associated destruction of hermatypic corals in the Indo-West Pacific region. In: Jones OA, Endean R, eds. *Biology and geology of coral reefs, Vol. 2. Biology, Vol. I*, pp. 389–438. New York: Academic Press.

Endler JA. 1980. Natural selection on color patterns in *Poecilia reticulata. Evolution* 34:76–91.

Endler JA. 1983. Natural and sexual selection on color patterns in poeciliid fishes. *Env Biol Fish* 9:173–190.

Endler JA. 1991. Interactions between predators and prey. In: Krebs JR, Davies NB, eds. *Behavioural ecology: an evolutionary approach*, 3rd edn, 169–196. Oxford: Blackwell Science.

Engas A, Jorgensen T, West CW. 1998. A species-selective trawl for demersal gadoid fisheries. *ICES J Mar Sci* 55:835–845.

Engelmann J, Hanke W, Bleckman H. 2002. Lateral line reception in still and running water. *J Comp Physiol A* 188:513–526.

Engelmann J, Hanke W, Mogdans J, Bleckman H. 2000. Hydrodynamic stimuli and the fish lateral line. *Nature* 408:51–52.

Epifanio J, Nielsen J. 2000. The role of hybridization in the distribution, conservation and management of aquatic species. *Rev Fish Biol Fisheries* 10:245–251.

Erdmann MV, Caldwell RL, Jewett SL, Tjakrawidjaja A. 1999. The second recorded living coelacanth from north Sulawesi. *Env Biol Fish* 54:445–451.

Erickson DL, Hightower JE, Grossman GD. 1985. The relative gonadal index: an alternative index for quantification of reproductive condition. *Comp Biochem Physiol* 81A:117–120.

Erwin TL. 1991. An evolutionary basis for conservation strategies. *Science* 253:750–752.

ESA (Ecological Society of America). 2000. Ecosystem services: a primer. **www.actionbioscience.org/environment/esa.html**.

Eschmeyer WN. 1990. *Catalog of the genera of recent fishes*. San Francisco: California Academy of Sciences. (Updated in 2005, **www.calacademy.org**.)

Eschmeyer WN, Ferraris CJ, Hoang Jr., MD, Long DJ. 1998. *Catalog of fishes*. San Francisco, CA: California Academy of Sciences.

Eschmeyer WN, Fong JD, Species by family/subfamily in the Catalogue of Fishes. 2017. *Catalogue of fishes*. San Francisco, CA: California Academy of Sciences. **http://researcharchive.calacademy.org/research/ichthyology/catalog/SpeciesByFamily.asp**, accessed 6 December 2017.

Estes JA, Tinker MT, Williams TM, et al. 1998. Killer whale predation on sea otters linking oceanic and nearshore ecosystems. *Science* 282:473–476.

Esteves KE, Lobo AV. 2001. Feeding pattern of *Salminus maxillosus* (Pisces, Characidae) at Cachoeira das Emas, Mogi-Guacu River (Sao Paulo State, Southeast Brazil). *Brazil J Biol* 61:267–276.

Estramil N, Bouton N, Verzijden MN, Hofker K, Riebel K, Slabbekoorn H. 2014. Cichlids respond to conspecific sounds but females exhibit no phonotaxis without the presence of live males. *Ecol Freshw Fish* 23:305–312.

Etnier DA. 1976. *Percina* (*Imostoma*) *tanasi*, a new percid fish from the Little Tennessee River, Tennessee. *Proc Biol Soc Washington* 88:469–488.

Etnier DA, Starnes WC. 1993. *The fishes of Tennessee*. Knoxville, TN: University of Tennessee Press.

Evans DH. 1993. Osmotic and ionic balance. In: Evans DH, ed. *The physiology of fishes*, pp. 315–342. Boca Raton, FL: CRC Press.

Evans DH, ed. 1998. *The physiology of fishes*, 2nd edn. Boca Raton, FL: CRC Press.

Evans DH, Claiborne JB. 2006. *The physiology of fishes*, 3rd edn. Boca Raton, FL: CRC, Taylor & Francis.

Evans DH, Claiborne JB, Kormanik GA. 1999. Osmoregulation, acid-base regulation, and nitrogen excretion. In: Horn MH, Martin KLM, Chotkowski MA, eds. *Intertidal fishes – life in two worlds*, pp. 79–96. New York: Academic Press.

Evans DO. 1984. Temperature independence of the annual cycle of standard metabolism in the pumpkinseed. *Trans Am Fish Soc* 113:494–512.

Evans JP, Magurran AE. 2000. Multiple benefits of multiple mating in guppies. *Proc Natl Acad Sci USA* 97:10074–10076.

Evans JP, Pierotti M, Pilastro A. 2003. Male mating behavior and ejaculate expenditure under sperm competition risk in the eastern mosquitofish. *Behav Ecol* 14:268–273.

Evans JW, Noble RL. 1979. The longitudinal distribution of fishes in an east Texas stream. *Am Midl Natur* 101:333–343.

Evans MS. 1986. Recent major declines in zooplankton populations in the inshore region of Lake Michigan: probable causes and implications. *Can J Fish Aquat Sci* 43:154–159.

Evers BN, Madsen H, McKaye KM, Stauffer JR. 2006.The schistosome intermediate host, *Bulinus nyassanus*, is a 'preferred' food for the cichlid fish, *Trematocranus placodon*, at Cape Maclear, Lake Malawi. *Ann Trop Med Parasitol* 100:75–85.

Ewald G, Larsson P, Linge H, et al. 1998. Biotransport of organic pollutants to an inland Alaska lake by migrating sockeye salmon (*Oncorhynchus nerka*). *Arctic* 51:40–47.

Excoffier L, Smouse P, Quattro J. 1992. Analysis of molecular variance inferred from metric distances among DNA haplotypes: application to human mitochondrial DNA restriction data. *Genetics* 131:479–491.

Faber DS, Pereda AE, 2011. Physiology of the Mauthner cell: function. In: Farrell AP, Stevens ED, Cech JJ, Richards JG, eds. *Encyclopedia of fish physiology: from genome to environment*, pp. 73–79. Boston: Elsevier/Academic Press.

Facey DE, Blazer VS, Gasper MM, Turcotte CL. 2005. Using fish biomarkers to monitor improvements in environmental quality. *J Aquat Anim Health* 17:263–266.

Facey DE, Grossman GD. 1990. The metabolic cost of maintaining position for four North American stream fishes: effects of season and velocity. *Physiol Zool* 63:757–776.

Facey DE, Grossman GD. 1992. The relationship between water velocity, energetic costs, and microhabitat use in four North American stream fishes. *Hydrobiologia* 239:1–6.

Fahy WE. 1982. The influence of temperature change on number of pectoral fin rays developing in *Fundulus majalis* (Walbaum). *J Cons Int Explor Mer* 40:21–26.

Falconer DS, Mackay TFC. 1996. *Introduction to quantitative genetics*, 4th edn. Harlow, UK: Longmans.

Fang JHK, Au DWT, Wu RSS, Chan AKY, Mok HOL, Shin PKS. 2009. The use of physiological indices in rabbitfish *Siganus oramin* for monitoring coastal pollution. *Mar Pollut Bull* 58:1229–1244.

Fange R, Grove D. 1979. Digestion. In: Hoar WS, Randall DJ, eds. *Bioenergetics and growth. Fish physiology*, Vol. 8, pp. 161–260. New York: Academic Press.

FAO (Food and Agricultural Organization). 1995. *FAO yearbook: fishery statistics: catches and landings 1993*, Vol. 76. Rome: FAO.

FAO (Food and Agricultural Organization). 2004. *The state of world fisheries and aquaculture (SOFIA) 2006*. Rome: FAO.

FAO (Food and Agricultural Organization). 2008. *World review of fisheries and aquaculture 2008*. Highlights of Special Studies. Rome: Fisheries and Aquaculture Department, Food and Agricultural Organization (FAO) of the United Nations.

FAO (Food and Agricultural Organization). 2016. *The state of world fisheries and aquaculture 2016. Contributing to food security and nutrition for all*. Rome: Fisheries and Aquaculture Department, Food and Agricultural Organization (FAO) of the United Nations. 200 pp.

FAO (Food and Agricultural Organization). 2020. *The State of World Fisheries and Aquaculture 2020. Sustainability in action*. Rome: Fisheries and Aquaculture Department, Food and Agricultural

Organization (FAO) of the United Nations, 224 pp. **https://doi.org/10.4060/ca9229en**.

Farrell AP. 2011a. The coronary circulation. In: Farrell AP, Stevens ED, Cech JJ, Richards JG, eds. *Encyclopedia of fish physiology: from genome to environment*, pp. 1077–1084. Boston: Elsevier/Academic Press.

Farrell AP. 2011b. Accessory hearts in fishes. In: Farrell AP, Stevens ED, Cech JJ, Richards JG, eds. *Encyclopedia of fish physiology: from genome to environment*, pp. 1073–1076. Boston: Elsevier/Academic Press.

Farrell AP. 2011c. Cellular composition of the blood. In: Farrell AP, Stevens ED, Cech JJ, Richards JG, eds. *Encyclopedia of fish physiology: from genome to environment*, pp. 984–991. Boston: Elsevier/Academic Press.

Farrell AP. 2011d. Deep-sea fishes. In: Farrell AP, Stevens ED, Cech JJ, Richards JG, eds. *Encyclopedia of fish physiology: from genome to environment*, pp. 1953–1958. Boston: Elsevier/Academic Press.

Farrell AP, Lee CG, Tierney K, et al. 2003. Field-based measurements of oxygen uptake and swimming performance with adult Pacific salmon using a mobile respirometer swim tunnel. *J Fish Biol* 62:64–84.

Farrell AP, Pieperhoff S. 2011. Cardiac anatomy in fishes. In: Farrell AP, Stevens ED, Cech JJ, Richards JG, eds. *Encyclopedia of fish physiology: from genome to environment*, pp. 998–1005. Boston: Elsevier/Academic Press.

Farrell AP, Steffensen JF. 2005. *Physiology of polar fishes. Fish physiology*, Vol. 22. New York: Academic Press.

Farrell AP, Stevens ED, Cech JJ, Richards JG. 2011. *Encyclopedia of fish physiology: from genome to environment*. Amsterdam, Boston: Elsevier/Academic Press.

Fausch KD. 1988. Tests of competition between native and introduced salmonids in streams: what have we learned? *Can J Fish Aquat Sci* 45:2238–2246.

Fausch KD, Karr JR, Yant PR. 1984. Regional application of an index of biotic integrity based on stream fish communities. *Trans Am Fish Soc* 113:39–55.

Fausch KD, Torgersen CE, Baxter CV, Li HW. 2002. Landscapes to riverscapes: bridging the gap between research and conservation of stream fishes. *BioScience* 52:483–498.

Fausch KD, White R. 1986. Competition among juveniles of coho salmon, brook trout, and brown trout in a laboratory stream, and implications for Great Lakes tributaries. *Trans Am Fish Soc* 115:363–381.

Fautin DG. 1991. The anemonefish symbiosis: what is known and what is not. *Symbiosis* 10:23–46.

Fay RR. 2011. Psychoacoustics. In: Farrell AP, Stevens ED, Cech JJ, Richards JG, eds. *Encyclopedia of fish physiology: from genome to environment*, pp. 276–282. Boston: Elsevier/Academic Press.

FDA (Food and Drug Administration). 2017. *DNA-based seafood identification*. Washington, DC: US Food and Drug Administration. **https://www.fda.gov/food/foodscienceresearch/dnaseafoodidentification/default.htm**, accessed 7 October 2017.

Feary DA, Almany GR, Jones GP, McCormick MI. 2007. Coral degradation and the structure of tropical reef fish communities. *Mar Ecol Prog Ser* 333:243–248.

Feder HM. 1966. Cleaning symbiosis in the marine environment. In: Henry SM, ed. *Symbiosis*, Vol. I, pp. 327–380. New York: Academic Press.

Feder ME, Burggren WW. 1985. Skin breathing in vertebrates. *Sci Am* 253:126–142.

Feder ME, Lauder GV, eds. 1986. *Predator–prey relationships: perspectives and approaches from the study of lower vertebrates*. Chicago: University of Chicago Press.

Feldheim KA, Chapman DD, Sweet D, et al. 2010. Shark virgin birth produces multiple, viable offspring. *J Hered* 101(3):374–377.

Feldheim KA, Gruber SH, Ashley MV. 2004. Reconstruction of paternal microsatellite genotypes reveals female polyandry and philopatry in the lemon shark, *Negaprion brevirostris*. *Evolution* 58:2332–2342.

Félix-Hackradt FC, Hackradt CW, Treviño-Otón J, Pérez-Ruzafa A, García-Charton JA. 2014. Habitat use and ontogenetic shifts of fish life stages at rocky reefs in South-western Mediterranean Sea. *J Sea Res* 88:67–77.

Felsenstein J. 1981. Evolutionary trees from DNA sequences: a maximum likelihood approach. *J Mol Evol* 17:368–376.

Felsenstein J. 2004. *Inferring phylogenies*. Sunderland, MA: Sinauer Associates.

Feng AS. 1991. Electric organs and electroreceptors. In: Prosser CL, ed. *Comparative animal physiology*, 4th edn, pp. 317–334. New York: Wiley & Sons.

Fernald RD. 1984. Vision and behavior in an African cichlid. *Am Sci* 72:58–65.

Fernandes CC, Podos J, Lundberg JG. 2004. Amazonian ecology: tributaries enhance the diversity of electric fishes. *Science* 305:1960–1962.

Fernholm B. 1981. Thread cells from the slime glands of hagfish (Myxinidae). *Acta Zool* 62:137–145.

Fernholm B. 1998. Hagfish systematics. In: Jørgensen JM, Lomholt JP, Weber RE, Malte H, eds. *The biology of hagfishes*, pp. 34–44. London: Chapman & Hall.

Fernholm B, Noren M, Kulander SO, et al. 2013. Hagfish phylogeny and taxonomy with description of the new genus *Rubicundus* (Craniata, Myxinidae). *J Zool Evol Res* 51:296–307.

Ferraro SP. 1980. Daily time of spawning of 12 fishes in the Peconic Bays, New York. *Fish Bull US* 78:455–464.

Ferreira CEL, Luiz OJ, Floeter SR, et al. 2015. First record of invasive lionfish (*Pterois volitans*) from the Brazilian coast. *PLoS ONE* 10:e0123002.

Ferry LA, Lauder GV. 1996. Heterocercal tail function in leopard sharks: a three-dimensional kinematic analysis of two models. *J Exp Biol* 199:2253–2268.

Ferry-Graham LA, Lauder GV. 2001. Aquatic prey capture in ray-finned fishes: a century of progress and new directions. *J Morphol* 248:99–119.

Ferry-Graham LA, Wainwright PC, Lauder GV. 2003. Quantification of flow during suction feeding in bluegill sunfish. *Zoology* 106:159–168.

Fewings DG, Squire LC. 1999. Notes on reproduction in the estuarine stonefish *Synanceia horrida*. *SPC Live Reef Fish Info Bull* 5:31–33.

Fiedler PL, Jain SK, eds. 1992. *Conservation biology: the theory and practice of nature conservation, preservation and management*. New York: Chapman & Hall.

Fields AT, Feldheim KA, Poulakas GR, Chapman DD. 2015. Facultative parthenogenesis in a critically endangered wild vertebrate. *Curr Biol* 25(11):R446–R447.

Figenschou L, Folstad I, Liljedal S. 2004. Lek fidelity of male Arctic charr. *Can J Zool* 82:1278–1284.

Filby AL, Neuparth T, Thorpe KL, Owens R, Galloway TS, Tyler CR. 2007. Heath impact of estrogens in the environment, considering complex mixture effects. *Env Health Perspect* 115:1704–1710.

Finch CE. 1990. *Longevity, senescence, and the genome*. Chicago: University of Chicago Press.

Fine JM, Vrieze LA, Sorensen PW. 2004. Evidence that petromyzontid lampreys employ a common migratory pheromone that is partially comprised of bile acids. *J Chem Ecol* 30:2091–2110.

Fine ML, Winn HE, Olla B. 1977. Communication in fishes. In: Sebeok T, ed. *How animals communicate*, pp. 472–518. Bloomington, IN: Indiana University Press.

Fischer EA, Petersen CW. 1987. The evolution of sexual patterns in the seabasses. *BioScience* 37:482–489.

Fischer EK, Soares D, Archer KR, Ghalambor CK, Hoke KL. 2013. Genetically and environmentally mediated divergence in lateral line morphology in the Trinidadian guppy (*Poecilia reticulata*). *J Exp Biol* 216:3132–3142.

Fishelson L, Montgomery WL, Myrberg AA, Jr. 1985. A unique symbiosis in the gut of tropical herbivorous surgeonfish (Acanthuridae: Teleostei) from the Red Sea. *Science* 229:49–51.

Fisher HS, Wong BBM, Rosenthal GG. 2006. Alteration of the chemical environment disrupts communication in a freshwater fish. *Proc Roy Soc Lond B Biol Sci* 273:1187–1193

Fisher R. 2005. Swimming speeds of larval coral reef fishes: impacts on self-recruitment and dispersal. *Mar Ecol Prog Ser* 285:223–232.

Fitzgerald DB, Winemiller KO, Sabaj Pérez MH, Sousa LM. 2017. Seasonal changes in the assembly mechanisms structuring tropical fish communities. *Ecology* 98(1):21–31

FitzGerald GJ. 1992. Filial cannibalism in fishes: why do parents eat their offspring? *Trends Ecol Evol* 7:7–10.

FitzGerald GJ, Wootton RJ. 1993. Behavioural ecology of sticklebacks. In: Pitcher TJ, ed. *Behaviour of teleost fishes*, 2nd edn, pp. 537–572. London: Chapman & Hall.

Fitzpatrick JL, Kempster RM, Daly-Engel TS, Collin SP, Evans JP. 2012. Assessing the potential for post-copulatory sexual selection in elasmobranchs. *J Fish Biol* 80:1141–1158. **https://doi.org/10.1111/j.1095-8649.2012.03256**.

Flecker AS. 1997. Habitat modification by tropical fishes: environmental heterogeneity and the variability of interaction strength. *J N Am Benthol Soc* 16:286–295.

Flecker AS, Allan JD. 1984. The importance of predation, substrate and spatial refugia in determining lotic insect distributions. *Oecologica* 64:306–313.

Flecker AS, Taylor BW. 2004. Tropical fishes as biological bulldozers: density effects on resource heterogeneity and species diversity. *Ecology* 85:2267–2278.

Flecker AS, Taylor BW, Bernhardt ES, et al. 2002. Interactions between herbivorous fishes and limiting nutrients in a tropical stream ecosystem. *Ecology* 83:1831–1844.

Fletcher DE, Dakin EE, Porter BA, Avise JC. 2004. Spawning behavior and genetic parentage in the pirate perch (*Aphredoderus sayanus*), a fish with an enigmatic reproductive morphology. *Copeia* 2004:1–10.

Fletcher GL, Hew CL, Davies PL. 2001. Antifreeze proteins of teleost fishes. *Ann Rev Physiol* 63:359–390.

Floeter SR, Behrens MD, Ferreira CEL, et al. 2005. Geographical gradients of marine herbivorous fishes: patterns and processes. *Mar Biol* 147:1435–1447.

Floeter SR, Ferreira CEL, Dominici-Arosemena A, Zalmon IR. 2004. Latitudinal gradients in Atlantic reef fish communities: trophic structure and spatial use patterns. *J Fish Biol* 64:1680–1699.

Floeter SR, Rocha LA, Robertson DR, et al. 2008. Atlantic reef fish biogeography and evolution. *J Biogeogr* 35: 22–47.

Flouri T, Jiao X, Rannala B, Yang Z. 2018. Species tree inference with BPP using genomic sequences and multispecies coalescent. *Mol Biol Evol* 35:2585–2593.

Fong QSW, Anderson JL. 2002. International shark fin markets and shark management: an integrated market preference-cohort analysis of the blacktip shark (*Carcharinus limbatus*). *Ecol Econ* 40:117–130.

Foote CJ, Wood CC, Withler RE. 1989. Biochemical genetic comparison of sockeye salmoon and kokanee. The anadromous and non-anadromous forms of *Oncorhynchus nerka*. *Can J Fish Aquat Sci* 46:149–158.

Forey PL. 1998. *History of the coelacanth fishes*. London: Chapman & Hall.

Forey PL, Janvier P. 1993. Agnathans and the origin of jawed vertebrates. *Nature* 361:129–134.

Formicki K, Korzelecka-Orkisz A, Tański A. 2019. Magnetoreception in fish. *J Fish Biol* 95:73–91.

Formicki K, Sadowski M, Tański A, Korzelecka-Orkisz A, Winnicki A. 2004. Behaviour of trout (*Salmo trutta* L.) larvae and fry in a constant magnetic field. *J Appl Ichthyol* 20:290–294.

Formosa, R, Ravi H, Happe S, et al. 2010. DNA-based fish species identification protocol. *J Vis Exp* 28:1871

Forrester GE, Finley RJ. 2006. Parasitism and a shortage of refuges jointly mediate the strength of density dependence in a reef fish. *Ecology* 87:1110–1115.

Fortier L, Leggett WC. 1983. Vertical migrations and transport of larval fish in a partially mixed estuary. *Can J Fish Aquat Sci* 40:1543–155.

Fortune ES. 2011. Physiology of tuberous electrosensory systems. In: A.P. Farrell, E.D. Stevens, J.J. Cech, and J.G. Richards, eds. *Encyclopedia of fish physiology: from genome to environment*, pp. 366–374. Boston: Elsevier/Academic Press.

Forward RB, Tankersley RA. 2001. Selective tidal-stream transport of marine animals. *Oceanogr Mar Biol* 39:305–353.

Foster SA. 1985. Group foraging by a coral reef fish: a mechanism for gaining access to defended resources. *Anim Behav* 33:782–792.

Fowler CW. 1999. Management of multi-species fisheries: from overfishing to sustainability. *ICES J Mar Sci* 56:927–932.

Fowler MC, Robson TO. 1978. The effects of the food preferences on stocking rates of grass carp (*Ctenopharyngodon idella* Val.) on mixed plant communities. *Aquat Bot* 5:261–276.

Fox FQ, Richardson GP, Kirk C. 1985. Torpedo electromotor system development: neuronal cell death and electric organ development in the fourth branchial arch. *J Comp Neurol* 236:274–281.

Fox RJ, Bellwood DR. 2013. Niche partitioning of feeding microhabitats produces a unique function for herbivorous rabbitfishes (Perciformes, Siganidae) on coral reefs. *Coral Reefs* 32:13–23

Franchina CR, Salazar VL, Volmar CH, Stoddard PK. 2001. Plasticity of the electric organ discharge waveform of male *Brachyhypopomus pinnicaudatus*. II. Social effects. *J Comp Physiol A* 187:45–52.

Francis RC. 1990. Climate change and marine fisheries. *Fisheries* 15(6):7–9.

Francis RC. 1992. Sexual lability in teleosts: developmental factors. *Q Rev Biol* 67:1–18.

Franco JN, Wernberg T, Bertocci I, et al. 2015. Herbivory drives kelp recruits into 'hiding' in a warm ocean climate. *Mar Ecol Prog Ser* 536:1–9.

Frank KT, Petrie B, Choi JS, Leggett WC. 2005. Trophic cascades in a formerly cod-dominated ecosystem. *Science* 308:1621–1623.

Frankham R, Ballou JD, Briscoe DA. 2002. *Introduction to conservation genetics*. Cambridge, UK: Cambridge University Press.

Frankel OH. 1974. Genetic conservation: our evolutionary responsibility. *Genetics* 78:53–65.

Fraser DF, Gilliam JF, Akkara JT, Albanese BW, Snider SB. 2004. Night feeding by guppies under predator release: effects on growth and daytime courtship. *Ecology* 85:312–319.

Fraser DF, Huntingford FA. 1986. Feeding and avoiding predation hazard: the behavioral response of the prey. *Ethology* 73:56–68.

Fraser DJ, Bernatchez L. 2001. Adaptive evolutionary conservation: towards a unified concept for defining conservation units. *Mol Ecol* 10:2741–2752.

Fraser DJ, Duchesne P, Bernatchez L. 2005. Migratory charr schools exhibit population and kin associations beyond juvenile stages. *Mol Ecol* 14:3133–3146.

Fraser DJ, Lippe C, Bernatchez L. 2004. Consequences of unequal population size, asymmetric gene flow and sex-biased dispersal on population structure in brook charr (*Salvelinus fontinalis*). *Mol Ecol* 13:67–80.

Frazer TK, Lindberg WJ, Stanton GR. 1991. Predation of sand dollars by gray triggerfish, *Balistes capriscus*, in the northern Gulf of Mexico. *Bull Mar Sci* 48:159–164.

Freeman MC, Grossman GD. 1992. Group foraging by a stream minnow: shoals or aggregations? *Anim Behav* 44:393–403.

Freeman MC, Stouder DJ. 1989. Intraspecific interactions influence size specific depth distribution in *Cottus bairdi*. *Env Biol Fish* 24:231–236.

Fremling CR, Rasmussen JL, Sparks RE, Cobb SP, Bryan CF, Claflin TO. 1989. Mississippi River fisheries: a case history. In: Dodge DP, ed. *Proceedings of the International Large River Symposium. Can Spec Publ Fish Aquat Sci* 106:309–351. Ottawa: Department of Fisheries and Oceans.

Fricke HW. 1973. Behaviour as part of ecological adaptation. *Helgolander wiss Meeresunters* 24:120–144.

Fricke HW. 1980. Mating systems, maternal and biparental care in triggerfish (Balistidae). *Z Tierpsychol* 53:105–122.

Fricke HW, Frahm J. 1992. Evidence for lecithotrophic viviparity in the living coelacanth. *Nuturwissenschaften* 79:476–479.

Fricke HW, Hissmann K. 1994. Home range and migrations of the living coelacanth *Latimeria chalumnae*. *Mar Biol* 120:171–180.

Fricke HW, Hissmann K. 2000. Feeding ecology and evolutional survival of the living coelacanth *Latimeria chalumnae*. *Mar Biol* 136:379–386.

Fricke HW, Hissmann K, Schauer J, Reinicke O, Kasang L, Plante R. 1991a. Habitat and population size of the coelacanth *Latimeria chalumnae* at Grand Comoro. *Env Biol Fish* 32:287–300.

Fricke HW, Reinicke O, Hofer H, Nachtigall W. 1987. Locomotion of the coelacanth *Latimeria chalumnae* in its natural environment. *Nature* 329:331–333.

Fricke HW, Schauer J, Hissmann K, Kasang L, Plante R. 1991b. Coelacanth *Latimeria chalumnae* aggregates in caves: first observations on their resting habitat and social behavior. *Env Biol Fish* 30:281–285.

Fricke R, Eschmeyer W, Fong JD. 2021. *Genera/Species by family/subfamily in Eschmeyer's catalog of fishes*. San Francisco, CA: California Academy of Sciences. **https://researcharchive.calacademy.org/research/ichthyology/catalog/SpeciesByFamily.asp**. Electronic version accessed November 2021.

Fricke R, Eschmeyer WN, Van der Laan R. eds. 2020. Eschmeyer's Catalog of Fishes: Genera, Species, references. **http://researcharchive.calacademy.org/research/ichthyology/catalog/fishcatmain.asp**. Electronic version accessed 4 July 2020.

Fricke R, Van der Laan R, Eschmeyer WN. eds. 2022, and frequently updated. *Eschmeyer's Catalog of Fishes: Genera, Species, references*. San Francisco, CA: San Francisco, CA. **http://researcharchive.calacademy.org/research/ichthyology/catalog/fishcatmain.asp**.

Frickhinger KA. 1995. *Fossil atlas – fishes*. Malle, Germany: Hans A. Baensch.

Friedman M, Coates MI. 2006. A newly recognized fossil coelacanth highlights the early morphological diversification of the clade. *Proc Roy Soc B Biol Sci* 273:245–250.

Frissell CA, Nawa RK. 1992. Incidence and causes of physical failure of artificial habitat structures in streams of western Oregon and Washington. *N Am J Fish Manage* 12:182–197.

Fritsches KA, Litherland L, Thomas N, Shand J. 2003. Cone visual pigments and retinal mosaics in the striped marlin. *J Fish Biol* 63:1347–1351.

Froese R. 2006. Cube law, condition factor and weight–length relationships: history, meta-analysis and recommendations. *J Appl Ichthyol* 22:241–253.

Froese R, Palomares MLD. 2000. Growth, natural mortality, length–weight relationship, maximum length and length-at-first-maturity of the Coelacanth *Latimeria chalumnae*. *Environ Biol Fish* 58: 45–52.

Froese R, Pauly D, eds. 2011. Polypteridae. In: FishBase. June 2011 version. Philippines: FishBase Information and Research Group, Inc. (FIN) (see **https://www.catalogueoflife.org/data/dataset/1010**).

Frydl P, Stearn CW. 1978. Rate of bioerosion by parrotfish in Barbados reef environments. *J Sedim Petrol* 48:1149–1158.

Fryer G, Iles TD. 1972. *The cichlid fishes of the Great Lakes of Africa*. Edinburgh: Oliver & Boyd.

Fu C, Wu J, Chen J, Wu Q, Lei G. 2003. Freshwater fish biodiversity in the Yangtze River basin of China: patterns, threats and conservation. *Biodiversity Conserv* 12:1649–1685.

Fuiman LA. 1983. Growth gradients in fish larvae. *J Fish Biol* 23:117–123.

Fuiman LA, Werner RG, eds. 2002. *Concepts in fisheries sciences: the unique contribution of early life stages*. Oxford: Blackwell Scientific Publishing.

Fujimoto M, Arimoto T, Morishita F, Naitoh T. 1991. The background adaptation of the flatfish, *Paralichthys olivaceus*. *Physiol Behav* 50:185–188.

Fujisawa M, Sakai Y, Kuwamura T. 2018. Aggressive mimicry of the cleaner wrasse by *Aspidontus taeniatus* functions mainly for small blennies. *Ethology* 124:432–439.

Fukuchi M, Marchant HJ, Nagase B. 2006. *Antarctic fishes*. Baltimore, MD: Johns Hopkins University Press.

Fulcher BA, Motta PJ. 2006. Suction disk performance of echeneid fishes. *Can J Zool* 84:42–50.

Fuller PL, Nico LG, Williams JD. 1999. *Nonindigenous fishes introduced into inland waters of the United States*. American Fisheries Society Special Publication No. 27. Bethesda, MD: American Fisheries Society.

Funk VA. 1995. Cladistic methods. In: Wagner WL, Funk VA, eds. *Hawaiian biogeography: evolution on a hot spot archipelago*, pp. 30–38. Washington, DC: Smithsonian Institution Press.

Funk WC, Forester BR, Converse SJ, Darst C, Morey S. 2019. Improving conservation policy with genomics: a guide to integrating adaptive potential into US Endangered Species Act decisions for conservation practitioners and geneticists. *Conserv Genet* 20:115–134.

Furimsky M, Cooke SJ, Suski CD, Wang Y, Tufts BL. 2003. Respiratory and circulatory responses to hypoxia in largemouth and smallmouth bass: implications for "live-release" angling tournaments. *Trans Am Fish Soc* 132:1065–1075.

Furin CG, von Hippel FA, Bell MA. 2012. Partial reproductive isolation of a recently derived resident-freshwater population of threespine stickleback (*Gasterosteus aculeatus*) from its putative anadromous ancestor. *Evolution* 66:3277–3286.

Gabbott SE, Aldridge RJ, Theron JN. 1995. A giant conodont with preserved muscle tissue from the Upper Ordovician of South Africa. *Nature* 374:800–802.

Gabda J. 1991. *Marine fish parasitology.* New York: VCH Publications.

Gage JD, Tyler PA. 1991. *Deep-sea biology: a natural history of organisms at the deep-sea floor.* Cambridge, UK: Cambridge University Press.

Gagnier PY. 1989. The oldest vertebrate: a 470-million-year-old jawless fish, *Sacabambaspis janvieri*, from the Ordovician of Bolivia. *Nat Geogr Res* 5:25–253.

Gagnier PY, Blieck ARM, Rodrico GS. 1986. First Ordovician vertebrate from South America. *Geobios* 19:629–634.

Gaither MR, Bowen BW, Rocha LA, Briggs JC. 2016. Fishes that rule the world: circumtropical distributions revisited. *Fish Fish (Oxf)* 17:664–679.

Gaither MR, Bowen BW, Toonen RJ. 2013 Population structure in the native range predicts the spread of introduced marine species. *Proc R Soc B* 280:20130409. **https://doi.org/10.1098/rspb.2013.0409**.

Gaither MR, Gkafas GA, Jong MD, et al. 2018. Genomics of habitat choice and adaptive evolution in a deep-sea fish. *Nat Ecol* 2:680–687.

Galis F, Metz JAJ. 1998. Why are there so many cichlid species? *Trends Ecol Evol* 13:1–2.

Gamperl AK, Rodnick KJ, Faust HA, et al. 2002. Metabolism, swimming performance, and tissue biochemistry of high desert redband trout (*Oncorhynchus mykiss* ssp.): evidence for phenotypic difference in physiological function. *Physiol Biochem Zool* 75:413–431.

Gans C, Kemp N, Poss S, eds. 1996. The lancelets (Cephalochordata): a new look at some old beasts. The results of a workshop. *Israel J Zool* 42:1–446.

Gardiner BG. 1984. Devonian palaeoniscid fishes: new specimens of *Mimia* and *Moythomasia* from the Upper Devonian of Western Australia. *Bull Brit Mus Natur Hist (Geol)* 37:173–428.

Garman GC, Macko SA. 1998. Contribution of marine-derived organic matter to an Atlantic coast, freshwater, tidal stream by anadromous clupeid fishes. *J N Am Benthol Soc* 17:277–285.

Garpe KC, Lindahl SAS, Yahya LU, Ohman MC. 2006. Long-term effects of the 1998 coral bleaching event on reef fish assemblages. *Mar Ecol Prog Ser* 315:237–247.

Gartner JV, Jr., Crabtee RE, Sulak KJ. 1997. Feeding at depth. In: Randall DJ, Farrell AP, eds. *Deep-sea fishes*, pp. 115–193. San Diego, CA: Academic Press.

Gassman NJ, Nye LB, Schmale MC. 1994. Distribution of abnormal biota and sediment contaminants in Biscayne Bay, Florida. *Bull Mar Sci* 54:929–943.

Gatz AJ, Sale MJ, Loar JM. 1987. Habitat shifts in rainbow trout: competitive influences of brown trout. *Oecologia* 74:7–19.

Gauthreaux SA, Jr., ed. 1980. *Animal migration, orientation and navigation.* New York: Academic Press.

Gelwick FP, Matthews WJ. 1992. Effects of an algivorous minnow on temperate stream ecosystem properties. *Ecology* 73:1630–1645.

Gende SM, Edwards RT, Willson MF, et al. 2002. Pacific salmon in aquatic and terrestrial ecosystems. *Bioscience* 52:917–928.

Gengerke TW. 1986. Distribution and abundance of paddlefish in the United States. In: Dillard JG, Graham L, Russell T, eds. *The paddlefish: status, management and propagation*, pp. 22–35. American Fisheries Society North Central Division Special Publication No. 7. Bethesda, MD: American Fisheries Society.

Gensemer RW, Playle RC. 1999. The bioavailability and toxicity of aluminum in aquatic environments. *Crit Rev Env Sci Technol* 29:315–450.

Gerkema MP, Videler JJ, de Wiljes J, van Lavieren H, Gerritsen H, Karel M. 2000. Photic entrainment of circadian activity patterns in the tropical labrid fish *Halichoeres chrysus*. *Chronobiol Internat* 17:613–622.

Gerking SD, ed. 1978. *Ecology of freshwater fish production.* London: Blackwell Science.

Gerking SD. 1994. *Feeding ecology of fish.* San Diego: Academic Press.

Gerlach G, Atema J, Kingsford MJ, Black KP, Miller-Sims V. 2007. Smelling home can prevent dispersal of reef fish larvae. *Proc Natl Acad Sci USA* 104:858–863.

Ghiselin MT. 1969. The evolution of hermaphroditism among animals. *Q Rev Biol* 44:189–208.

Giacomello E, Marchini D, Rasotto MB. 2006. A male sexually dimorphic trait provides antimicrobials to eggs in blenny fish. *Biol Lett* 2:330–333.

Gibbs MA, Northcutt RG. 2004. Development of the lateral line system in the shovelnose sturgeon. *Brain Behav Evol* 64:70–84.

Gibbs RH, Jr., Collette BB. 1967. Comparative anatomy and systematics of the tunas, genus *Thunnus*. *US Fish Wild Serv Fish Bull* 66:65–130.

Gibbs RH, Jr., Jarosewich E, Windom HL. 1974. Heavy metal concentrations in museum fish specimens: effects of preservatives and time. *Science* 184:475–477.

Gibson RN. 1978. Lunar and tidal rhythms in fish. In: Thorpe JE, ed. *Rhythmic activity of fishes*, pp. 201–213. London: Academic Press.

Gibson RN. 1982. Recent studies on the biology of intertidal fishes. *Oceanog Mar Biol Ann Rev* 20:363–414.

Gibson RN. 1992. Tidally-synchronized behaviour in marine fishes. In: Ali MA, ed. *Rhythms in fishes*, pp. 63–81. New York: Plenum Press.

Gibson RN. 1993. Intertidal teleosts: life in a fluctuating environment. In: Pitcher TJ, ed. *Behaviour of teleost fishes*, 2nd edn, pp. 513–536. London: Chapman & Hall.

Gido KB, Matthews WJ. 2001. Ecosystem effects of water column minnows in experimental streams. *Oecologia* 126:247–253.

Gill HS, Renaud CB, Chapleau F, Mayden RL, Potter IC. 2003. Phylogeny of living parasitic lampreys (Petromyzontiformes) based on morphological data. *Copeia* 2003:687–703.

Gilliam JF, Fraser DF. 1987. Habitat selection under predation hazard: test of a model with foraging minnows. *Ecology* 68:1856–1862.

Gilliam JF, Fraser DF, Sabat AM. 1989. Strong effects of foraging minnows on a stream benthic invertebrate community. *Ecology* 70:445–452.

Gilligan MR. 1989. *An illustrated field guide to the fishes of Gray's Reef National Marine Sanctuary.* NOAA Technical Memorandum NOS MEMD 25. Washington, DC: US Department of Commerce.

Gillooly JF, Baylis JR. 1999. Reproductive success and the energetic cost of parental care in male smallmouth bass. *J Fish Biol* 54:573–584.

Gilmore RG. 1991. The reproductive biology of lamnoid sharks. In: Gruber SH, ed. *Discovering sharks*, pp. 64–67. American Littoral Society Special Publication No. 14. Highlands, NJ: American Littoral Society.

Gilmore RG, Dodrill JW, Linley PA. 1983. Embryonic development of the sand tiger shark, *Odontaspis taurus* Rafinesque. *Fish Bull US* 81:201–225.

Gilmour KM. 2011. Carbonic anhydrase in gas transport and exchange. In: Farrell AP, Stevens ED, Cech JJ, Richards JG, eds. *Encyclopedia of fish physiology: from genome to environment*, pp. 899–908. Boston: Elsevier/Academic Press.

Gladstone W, Westoby M. 1988. Growth and reproduction in *Canthigaster valentini* (Pisces, Tetraodontidae): a comparison of a toxic reef fish with other reef fishes. *Env Biol Fish* 21:207–221.

Glantz MH, Feingold LE, eds. 1990. *Climate variability, climate change and fisheries.* Boulder, CO: Environmental and Societal Impacts Group, National Center for Atmospheric Research.

Glover CJM. 1982. Adaptations of fishes in arid Australia. In: Barber WR, Greenslade PJM, eds. *Evolution of the flora and fauna of arid Australia*, pp. 241–246. Frewville, Australia: Peacock Publications.

Glover KA, Pertoldi C, Besnier F, Wennevik V, Kent M, Skaala Ø. 2013. Atlantic salmon populations invaded by farmed escapees: quantifying genetic introgression with a Bayesian approach and SNPs. *BioMed Central Genetics* 14: 74 (19 pp).

Glynn PW. 1997. Bioerosion and coral-reef growth: a dynamic balance. In: Birkeland C, ed. *The life and death of coral reefs*, pp. 68–95. New York: Chapman & Hall.

Glynn PW, Mate JL, Baker AC, Calderon MO. 2001.Coral bleaching and mortality in panama and Ecuador during the 1997–1998 El Niño–Southern oscillation event: spatial/temporal patterns and comparisons with the 1982–1983 event. *Bull Mar Sci* 69:79–109.

Godin J-GJ. 1986. Antipredator function of shoaling in teleost fishes: a selective review. *Natur Can (Rev Ecol Syst)* 113:241–250.

Godin J-GJ, Keenleyside MHA. 1984. Foraging on patchily distributed prey by a cichlid fish (Teleostei, Cichlidae): a test of the ideal free distribution theory. *Anim Behav* 32:1201–1213.

Godin J-GJ, Sproul CD. 1988. Risk taking in parasitized sticklebacks under threat of predation: effects of energetic need and food availability. *Can J Zool* 66:2360–2367.

Godin JJ. 1997a. Evading predators. In: Godin JJ, ed. *Behavioural ecology of teleost fishes*, pp. 191–236. Oxford: Oxford University Press.

Godin JJ, ed. 1997b. *Behavioural ecology of teleost fishes.* Oxford: Oxford University Press.

Godwin J. 2011. Socially controlled sex change in fishes. In: Farrell AP, Stevens ED, Cech JJ, Richards JG, eds. *Encyclopedia of fish physiology: from genome to environment*, pp. 662–669. Boston: Elsevier/Academic Press.

Godwin J, Crews D, Warner RR. 1996. Behavioral sex change in the absence of gonads in a coral reef fish. *Proc Royal Soc B* 263:1683–1688.

Godwin J, Luckenbach JA, Borski RJ. 2003. Ecology meets endocrinology: environmental sex determination in fishes. *Evol Dev* 5:40–49.

Godwin SC, Dill LM, Krkošek M, Price MHH, Reynolds JD. 2017. Reduced growth in wild juvenile sockeye salmon *Oncorhynchus nerka* infected with sea lice. *J Fish Biol* 91:41–57

Golani D. 1993. The biology of the Red Sea migrant, *Saurida undosquamis* in the Mediterranean and comparison with the indigenous confamilial *Synodus saurus* (Teleostei: Synodontidae). *Hydrobiologica* 271:109–117.

Golani D, Bogorodsky SV. 2010. The fishes of the Red Sea – reappraisal and updated checklist. *Zootaxa* 2463:1–135.

Goldschmidt T. 1996. *Darwin's dreampond: drama in Lake Victoria* (transl. S. Marx-Macdonald). Cambridge, MA: MIT Press.

Gong Z, Korzh V, eds. 2004. *Fish development and genetics. The zebrafish and medaka models.* Hackensack, NJ: World Scientific Publishing.

Goñi R. 1998. Ecosystem effects of marine fisheries: an overview. *Ocean Coast Manag* 40:37–64.

Gonzalez RP, Leyva A, Moraes MO. 2001. Shark cartilage as a source of antiangiogenic compounds: from basic to clinical research. *Biol Pharmaceut Bull* 24:1097–1101.

Goodall DW. 1976. Introduction. In: Goodall DW, ed. *Evolution of desert biota*, pp. 3–5. Austin, TX: University of Texas Press.

Gooding RM, Magnuson JJ. 1967. Ecological significance of a drifting object to pelagic fishes. *Pacific Sci* 21:486–497.

Goodwin NB, Balshine-Earn S, Reynolds JD. 1998. Evolutionary transitions in parental care in cichlid fishes. *Proc Roy Soc Lond B Biol Sci* 265:2265–2272.

Gorbman A, Kobayashi H, Honma Y, Matsuyama M. 1990. The hagfishery of Japan. *Fisheries* 15(4):12–18.

Gore AC, Chappell VA, Fenton SE, et al. 2015. EDC-2: the endocrine society's second scientific statement on endocrine-disrupting compounds. *Endocr Rev* 36:E1–E150.

Goreau T, McClanahan T, Hayes R, Strong A. 2000. Conservation of coral reefs after the 1998 global bleaching event. *Conserv Biol* 14:5–15.

Gorlick DL. 1976. Dominance hierarchies and factors influencing dominance in the guppy *Poecilia reticulata* (Peters). *Anim Behav* 24:336–346.

Gorlick DL, Atkins PD, Losey GS. 1987. Effect of cleaning by *Labroides dimidiatus* (Labridae) on an ectoparasite population infecting *Pomacentrus vaiuli* (Pomacentridae) at Enewetak Atoll. *Copeia* 1987:41–45.

Gorman OT. 1988. The dynamics of habitat use in a guild of Ozark minnows. *Ecol Monogr* 58:1–18.

Goslan RE, Britton JR, Cowx I, Copp GH. 2010. Current knowledge on non-native freshwater fish introductions. *J Fish Biol* 75:751–786.

Gosline WA. 1971. *Functional morphology and classification of teleostean fishes.* Honolulu: University Press of Hawaii.

Gotceitas V, Godin J-GJ. 1992. Effects of location of food delivery and social status on foraging-site selection by juvenile Atlantic salmon. *Env Biol Fish* 35:291–300.

Gottfried MD, Compagno LJV, Bowman SC. 1996. Size and skeletal anatomy of the giant megatooth shark *Carcharodon megalodon*. In: Klimley AP, Ainley DG, eds. *Great white sharks: the biology of Carcharodon carcharias*, pp. 55–66. San Diego, CA: Academic Press.

Goudet J, Perrin N, Wasser P. 2002. Tests for sex-biased dispersal using bi-parentally inherited genetic markers. *Molec Ecol* 11:1103–1114.

Gould SJ. 1966. Allometry and size in ontogeny and phylogeny. *Biol Rev* 41:587–640.

Gould SJ. 1977. *Ontogeny and phylogeny.* Cambridge, MA: Belknap Press.

Gould SJ, Lewontin RC. 1979. The spandrels of San Marco and the Panglossian paradigm: a critique of the adaptationist programme. ProcRoyal Soc B 205:581–598.

Goulding M. 1980. *The fishes and the forest. Explorations in Amazonian natural history.* Berkeley, CA: University of California Press.

Govoni JJ, West MA, Zivotofsky D, Zivotofsky AZ, Bowser PR, Collette BB. 2004. Ontogeny of squamation in Swordfish, *Xiphias gladius*. *Copeia* 2004:391–396.

Graff C, Kaminski G, Gresty M, Ohlmann T. 2004. Fish perform spatial pattern recognition and abstraction by exclusive use of active electrolocation. *Curr Biol* 14:818–823 (abstract).

Graham JB. 1997a. *Air-breathing fishes: evolution, diversity, and adaptation.* San Diego, CA: Academic Press, 324 pp.

Graham JB. 2011a. The biology, diversity, and natural history of air-breathing fishes. In: Farrell AP, Stevens ED, Cech JJ, Richards JG, eds. *Encyclopedia of fish physiology: from genome to environment*, pp. 1850–1860. Boston: Elsevier/Academic Press.

Graham JB. 2011b. Respiratory adaptations for air-breathing fishes. In: Farrell AP, Stevens ED, Cech JJ, Richards JG, eds. *Encyclopedia of fish physiology: from genome to environment*, pp. 1861–1874. Boston: Elsevier/Academic Press.

Graham JB, Lee HJ. 2004. Breathing air in air: in what ways might extant amphibious fish biology relate to prevailing concepts about early tetrapods, the evolution of vertebrate breathing, and the vertebrate land transition? *Physiol Biochem Zool* 77:720–731.

Graham, JB, Wegner NC, Miller A, Jew CJ, Lai NC, Berquist RM, Frank LR, Long JA. 2014.Spiracular air breathing in polypterid fishes and its implications for aerial respiration in stem tetrapods. *Nat Comm* 5:3022.

Graham K. 1997b. Contemporary status of the North American paddlefish, *Polyodon spathula. Env Biol Fish* 48:279.

Graham LA, Li J, Davidson WS, Davies PL. 2012. Smelt was the likely beneficiary of an antifreeze gene laterally transferred between fishes. *BMC Evol Biol* 12(190): **https://doi.org/10.1186/1471-2148-12-190**.

Grande L. 2010. An empirical synthetic pattern study of gars (Lepisosteiformes) and closely related species, based mostly on skeletal anatomy. The resurrection of Holostei. *Am Soc Ichthyol Herpetol Spec Publ* 6:871 pp.

Grande L, Bemis WE. 1991. Osteology and phylogenetic relationships of fossil and recent paddlefishes (Polyodontidae) with comments on the interrelationships of acipenseriformes. *J Vertebr Paleontol* 11(Memoir 1, suppl to No. 1):1–121.

Grande L, Bemis WE. 1998. A comprehensive phylogenetic study of amiid fishes (Amiidae) based on comparative skeletal anatomy. An empirical search for interconnected patterns of natural history. *J Vertebr Paleontol* 18(Memoir 4, suppl to No. 1):1–690.

Grande L, Jin F, Yabumoto Y, Bemis WE. 2002. *Protopsephurus liui*, a well-preserved primitive paddlefish (Acipenseriformes: Polyodontidae) from the lower cretaceous of China. *J Vertebr Paleontol* 22:209–237.

Grant GC, Vondracek B, Sorensen PW. 2002. Spawning interactions between sympatric brown and brook trout may contribute to species replacement. *Trans Am Fish Soc* 131:569–576.

Grant JWA. 1997. Territoriality. In: Godin JJ, ed. *Behavioural ecology of teleost fishes*, pp. 81–103. Oxford, UK: Oxford University Press.

Grant JWA, Noakes DLG. 1987. Movers and stayers: foraging tactics of young-of-the-year brook charr, *Salvelinus fontinalis. J Anim Ecol* 56:1001–1013.

Grant JWA, Noakes DLG. 1988. Aggressiveness and foraging mode of young-of-the-year brook charr, *Salvelinus fontinalis* (Pisces, Salmonidae). *Behav Ecol Sociobiol* 22:435–445.

Grant WS. 1986. Bioichemical genetic divergence between Atlantic, *Clupea harengus*, and Pacific, *C. pallasi*, herring. *Copeia* 1986:714–719.

Grant WS, Bowen BW. 1998. Shallow population histories in deep evolutionary lineages of marine fishes: insights from the sardines and anchovies and lessons for conservation. *J Heredity* 89:415–426.

Grant WS, Leslie RW. 1996. Late Pleistocene dispersal of Indian-Pacific sardine populations in an ancient lineage of the genus *Sardinops. Mar Biol* 126:133–142.

Grant WS, Milner GB, Krasnowski P, Utter FM. 1980. Use of biochemical genetic variants for identification of sockeye salmon (*Oncorhynchus nerka*) stocks in Cook Inlet, Alaska. *Can J Fish Aquat Sci* 37:1236–1247.

Grant WS, Waples RS. 2000. Spatial and temporal scales of genetic variability in marine and anadromous species: implications for fisheries oceanography. In: Harrison PJ, Parsons TR, eds. *Fisheries oceanography: an integrative approach to fisheries ecology and management*, pp. 61–93. Oxford: Blackwell Science.

Grassle JF. 1986. The ecology of deep-sea hydrothermal vent communities. *Adv Mar Biol* 23:301–362.

Graves JE. 1998. Molecular insights into the population structures of cosmopolitan marine fishes. *J Heredity* 89:427–437.

Graves JE, Dizon AE. 1987. Mitochondrial DNA sequence similarity of Atlantic and Pacific albacore tuna (*Thunnus alalunga*). *Can J Fish Aquat Sci* 46:870–873.

Graves JE, McDowell JR. 2006. Genetic analysis of white marlin (*Tetrapturus albidus*) stock structure. *Bull Mar Sci* 79:469–482.

Graves JE, McDowell JR. 2015. Population structure of istiophorid billfishes. *Fish Res* 166:21–28.

Gray J. 1968. *Animal locomotion*. London: Weidenfeld & Nicolson.

Greeley MS, Jr. 2002. Reproductive indicators of environmental stress in fish. In: Adams SM, ed. *Biological indicators of aquatic ecosystem stress*, pp. 321–377. Bethesda, MD: American Fisheries Society.

Green BS, McCormick MI. 2005. O^2 replenishment to fish nests: males adjust brood care to ambient conditions and brood development. *Behav Ecol* 16:389–397.

Green OL. 1966. Observations on the culture of the bowfin. *Progress Fish Cult* 28:179.

Greenwood MFD, Metcalfe NB. 1998. Minnows become nocturnal at low temperatures. *J Fish Biol* 53:25–32.

Greenwood PH. 1981. *The haplochromine fishes of the east African lakes*. Ithaca, NY: Cornell University Press.

Greenwood PH. 1984. What *is* a species flock? In: Echelle AE, Kornfield I, eds. *Evolution of fish species flocks*, pp. 13–19. Orono, ME: University of Maine Press.

Greenwood PH. 1987. The natural history of African lungfishes. In: Bemis WE, Burggren WW, Kemp NE, eds. *The biology and evolution of lungfishes*, pp. 163–179. New York: Alan R. Liss.

Greenwood PH. 1991. Speciation. In: Keenleyside MHA, ed. *Cichlid fishes: behaviour, ecology and evolution*, pp. 86–102. London: Chapman & Hall.

Greenwood PH, Rosen DE,Weitzman SH, Myers GS. 1966. Phyletic studies of teleostean fishes, with a provisional classification of living forms. *Bull Am Mus Natur Hist* 131:339–456.

Greenwood PJ. 1980. Mating systems, philopatry, and dispersal in birds and mammals. *Anim Behav* 28:140–162.

Gregg WW, Conkright ME, Ginoux P, O'Reilly JE, Casey NW. 2003. Ocean primary production and climate: global decadal changes. *Geophys Res Lett* 30:Art. No. 1809.

Grewe PM, Feutry P, Hill PL, et al. 2015. Evidence of discrete yellowfin tuna (*Thunnus albacares*) populations demands rethink of management for this globally important resource. *Sci Rep* 5:1–9.

Griem JN, Martin KLM. 2000. Wave action: the environmental trigger for hatching in the California grunion *Leuresthes tenuis* (Teleostei: Atherinopsidae). *Mar Biol* 137:177–181.

Griffin AS. 2004. Social learning about predators: a review and prospectus. *Learn Behav* 32:131–140.

Grimm NB. 1988. Feeding dynamics, nitrogen budgets, and ecosystem role of a desert stream omnivore, *Agosia chrysogaster* (Pisces: Cyprinidae). *Env Biol Fish* 21:143–152.

Grobecker DB. 1983. The lie-in-wait feeding mode of a cryptic teleost, *Synanceia verrucosa. Env Biol Fish* 8:191–202.

Groenewold S, Fonds M. 2000. Effects on benthic scavengers of discards and damaged benthos produced by the beam-trawl fishery in the southern North Sea. *ICES J Mar Sci* 57:1395–1406.

Grogan ED, Lund R. 2004. The origin and relationships of early Chondrichthyes. In: Carrier JC, Musick JA, Heithaus MR, eds. *Biology of sharks and their relatives*, pp. 3–31. Boca Raton, FL: CRC Press.

Groom MJ, Meffe GK, Carroll CR. 2006. *Principles of conservation biology*, 3rd edn. Sunderland, MA: Sinauer Associates.

Groombridge B, ed. 1992. *Global biodiversity: status of the Earth's living resources*. London: Chapman & Hall.

Groot C, Margolis L, eds. 1991. *Pacific salmon life histories.* Vancouver, BC: University of British Columbia Press.

Gross JB. 2012. The complex origin of *Astyanax* cavefish. *BMC Evol Biol* 12:105–116.

Gross MR. 1982. Sneakers, satellites and parentals: polymorphic mating strategies in North American sunfishes. *Z Tierpsychol* 60:1–26.

Gross MR. 1984. Sunfish, salmon, and the evolution of alternative reproductive strategies and tactics in fishes. In: Potts GW, Wootton RJ, eds. *Fish reproduction: strategies and tactics*, pp. 55–75. London: Academic Press.

Gross MR. 1987. Evolution of diadromy in fishes. In: Dadswell MJ, Klauda RJ, Moffitt CM, Saunders RL, Rulifson RA, Cooper JE, eds. *Common strategies of anadromous and catadromous fishes*, pp. 14–25. American Fisheries Society Symposium 1. Bethesada, MD: American Fisheries Society.

Gross MR. 1991. Salmon breeding behavior and life history evolution in changing environments. *Ecology* 72:1180–1186.

Gross MR, Coleman RM, McDowall RM. 1988. Aquatic productivity and the evolution of diadromous fish migration. *Science* 239:1291–1293.

Grossman GD. 1980. Food, fights, and burrows: the adaptive significance of intraspecific aggression in the bay goby (Pisces: Gobiidae). *Oecologia (Berl)* 45:261–266.

Grossman GD, Carline RF, Wagner T. 2017. Population dynamics of brown trout (Salmo trutta) in Spruce Creek Pennsylvania: a quarter-century perspective. *Freshw Biol* 2017:1–12.

Grossman GD, Freeman MC. 1987. Microhabitat use in a stream fish assemblage. *J Zool Lond* 212:151–176.

Grossman GD, Freeman MC, Moyle PB, Whitaker JO, Jr. 1985. Stochasticity and assemblage organization in an Indiana stream fish assemblage. *Am Natur* 126:275–285.

Grossman GD, Moyle PB, Whitaker JO, Jr. 1982. Stochasticity in structural and functional characteristics of an Indiana stream fish assemblage: a test of community theory. *Am Natur* 120:423–454.

Grossman GD, Sundin G, Ratajczak RE, Jr. 2016. Long-term persistence, density dependence and effects of climate change on rosyside dace (Cyprinidae). *Freshw Biol* 61:832–847

Gruber SH, ed. 1991. *Discovering sharks.* American Littoral Society Special Publication No. 14. Highlands, NJ: American Littoral Society.

Grutter AS. 1997. Effect of the removal of cleaner fish on the abundance and species composition of reef fish. *Oecologia* 111:137–143.

Grutter AS. 1999. Cleaner fish really do clean. *Nature* 398:672–673.

Grutter AS. 2004. Cleaner fish use tactile dancing behavior as a preconflict management strategy. *Curr Biol* 14:1080–1083.

Grutter AS, Murphy JM, Choat JH. 2003. Cleaner fish drives local fish diversity on coral reefs. *Curr Biol* 13:64–67.

Gucinski H, Lackey RT, Spence BC. 1990. Global climate change: policy implications for fisheries. *Fisheries* 15(6):33–38.

Gulland JA. 1983. *Fish stock assessment: a manual of basic methods.* Chichester, UK: Wiley & Sons.

Gunn JM, Qadri SU, Mortimer DC. 1977. Filamentous algae as a food source for brown bullhead (*Ictalurus nebulosus*). *J Fish Res Board Can* 34:396–401.

Gunning GE, Suttkus RD. 1991. Species dominance in the fish populations of the Pearl River at two study areas in Mississippi and Louisiana: 1966–1988. *Proc Southeast Fish Council* 23:7–23.

Günther ACLG. 1880. *An introduction to the study of fishes.* New Delhi: Today and Tomorrow's Book Agency.

Gustafson LL, Remmenga MD, Gardner IA, et al. 2014. Viral hemorrhagic septicemia IVb status in the United States: inferences from surveillance activities and regional context. *Prev Vet Med* 114:174–187

Gustafson RG, Waples RS, Myers JM, et al. 2007. Pacific salmon extinctions: quantifying lost and remaining diversity. *Conserv Biol* 21:1009–1020.

Guzman AF, Perez HE, Miller RR. 2001. Occurrence of the clingfish, *Gobiesox fluviatilis* (Gobiesociformes: Gobiesocidae), in the Rio Chapalagana, Mexico: confirmation of a historical record. *Southwest Natur* 46:96–98.

Guzman HM, Cortes J. 2007. Reef recovery 20 years after the 1982–1983 El Niño massive mortality. *Mar Biol* 151:401–411.

Haddad V, Jr., Sazima I. 2003. Piranha attacks on humans in southeast Brazil: epidemiology, natural history, and clinical treatment, with description of a bite outbreak. *Wilderness Env Med* 14:249–254.

Hagedorn M. 1986. The ecology, courtship, and mating of gymnotiform electric fish. In: Bullock TH, Heiligenberg W, eds. *Electroreception*, pp. 497–525. New York: Wiley & Sons.

Hagedorn M, Womble M, Finger TE. 1990. Synodontid catfish: a new group of weakly electric fish. *Brain Behav Evol* 35:268–277.

Hager MC, Helfman GS. 1991. Safety in numbers: shoal size choice by minnows under predatory threat. *Behav Ecol Sociobiol* 29:271–276.

Hahn MW. 2019. *Molecular population genetics.* New York: Oxford University Press.

Hailman JP. 1977. *Optical signals: animal communication and light.* Bloomington, IN: Indiana Press.

Hairston NG, Jr., Li KT, Easter SS, Jr. 1982. Fish vision and the detection of planktonic prey. *Science* 218:1240–1242.

Häkkinen J, Vehniäinen E, Ylönen O, et al. 2002. The effects of increasing UV-B radiation on pigmentation, growth and survival of coregonid embryos and larvae. *Env Biol Fish* 64:451–459.

Hales LS, Jr. 1987. Distribution, abundance, reproduction, food habits, age, and growth of the round scad *Decapterus punctatus*, in the South Atlantic Bight. *Fish Bull US* 85:251–268.

Hales LS, Belk MC. 1992. Validation of otolith annuli of bluegills in a southeastern thermal reservoir. *Trans Am Fish Soc* 121:823–830.

Hall CAS. 1972. Migration and metabolism in a temperate stream ecosystem. *Ecology* 53:585–604.

Hall DJ, Werner EE. 1977. Seasonal distribution and abundance of fishes in the littoral zone of a Michigan lake. *Trans Am Fish Soc* 106:545–555.

Hall FG, McCutcheon FH. 1938. The affinity of hemoglobin for oxygen in marine fishes. *J Cell Comp Physiol* 11:205–212.

Halldorsdottir K, Arnason E. 2015. Whole-genome sequencing uncovers cryptic and hybrid species among Atlantic and Pacific cod-fish. *bioRxiv*: **https://doi.org/10.1101/034926**.

Halpern BS. 2003. The impact of marine reserves: do reserves work and does reserve size matter? *Ecol Applic* 13:S117–137.

Halstead LB, Liu YH, P'an K. 1979. Agnathans from the Devonian of China. *Nature* 282:831–833.

Hamady LL, Natanson LJ, Skomal GB, Thorrold SR. 2014. Vertebral bomb radiocarbon suggests extreme longevity in white sharks. *PLoS One* 9(1):e84006. **https://doi.org/10.1371/journal.pone.0084006**.

Hamilton WD. 1971. Geometry for the selfish herd. *J Theor Biol* 31:295–311.

Hamilton WJ, III, Peterman RM. 1971. Countershading in the colourful reef fish *Chaetodon lunula*: concealment, communication or both. *Anim Behav* 19:357–364.

Hamlett WC. 1991. From egg to placenta: placental reproduction in sharks. In: Gruber SH, ed. *Discovering sharks*, pp. 56–63. American

Littoral Society Special Publication No. 14. Highlands, NJ: American Littoral Society.
Hamlett WC, ed. 1999. *Sharks, skates, and rays: the biology of elasmobranch fishes.* Baltimore, MD: John Hopkins Press.
Hamlett WC, Wourms JP, Smith JW. 1985. Stingray placental analogues: structure of trophonemata in *Rhinoptera bonasus. J Submicrosc Cytol* 17(4):541–550.
Hamoir G, Geradin-Otthiers N. 1980. Differentiation of the sarcoplasmic proteins of white, yellowish and cardiac muscles of an Antarctic hemoglobin-free fish, *Champsocephalus gunnari. Comp Biochem Physiol* 65B:199–206.
Hamon TR, Foote CJ, Hilborn R, et al. 2000. Selection on morphology of spawning wild sockeye salmon by a gill-net fishery. *Trans Am Fish Soc* 129:1300–1315.
Handford P, Bell G, Reimchen T. 1977. A gillnet fishery considered as an experiment in artificial selection. *J Fish Res Board Can* 34: 954–961.
Hanson LH, Swink WD. 1989. Downstream migration of recently metamorphosed sea lampreys in the Ocqueoc River, Michigan before and after treatment with lampricides. *N Am J Fish Manage* 9:327–331.
Hanssens MM, Teugels GG, Thys Van Den Audernaerde DFE. 1995. Subspecies in the *Polypterus palmas* complex (Brachiopterygii; Polypteridae) from West and Central Africa. *Copeia* 1995:694–705.
Hara TJ, ed. 1982. *Chemoreception in fishes.* Amsterdam: Elsevier.
Hara TJ. 1993. Role of olfaction in fish behaviour. In: Pitcher TJ, ed. *The behaviour of teleost fishes*, 2nd edn, pp. 171–199. London: Chapman & Hall.
Hara TJ. 2011a. Chemoreception (smell and taste): an introduction. In: Farrell AP, Stevens ED, Cech JJ, Richards JG, eds. *Encyclopedia of fish physiology: from genome to environment*, pp. 183–186. Boston: Elsevier/Academic Press.
Hara TJ. 2011b. Morphology of the olfactory (smell) system in fishes. In: Farrell AP, Stevens ED, Cech JJ, Richards JG, eds. *Encyclopedia of fish physiology: from genome to environment*, pp. 194–207. Boston: Elsevier/Academic Press.
Hara TJ. 2011c. Neurophysiology of olfaction. In: Farrell AP, Stevens ED, Cech JJ, Richards JG, eds. *Encyclopedia of fish physiology: from genome to environment*, pp. 208–217. Boston: Elsevier/Academic Press.
Hara TJ. 2011d. Chemosensory behavior. In: Farrell AP, Stevens ED, Cech JJ, Richards JG, eds. *Encyclopedia of fish physiology: from genome to environment*, pp. 227—235. Boston: Elsevier/Academic Press.
Hara TJ. 2011e. Morphology of the gustatory (taste) system in fishes. In: Farrell AP, Stevens ED, Cech JJ, Richards JG, eds. *Encyclopedia of fish physiology: from genome to environment*, pp. 187–193. Boston: Elsevier/Academic Press.
Hara TJ. 2011f. Neurophysiology of gustation. In: Farrell AP, Stevens ED, Cech JJ, Richards JG, eds. *Encyclopedia of fish physiology: from genome to environment*, pp. 218–226. Boston: Elsevier/Academic Press.
Hara TJ, Zielinski BS. 2007. *Sensory systems neuroscience. Fish physiology*, Vol. 25. New York: Academic Press.
Harden-Jones FR. 1968. *Fish migration.* London: Edward Arnold.
Harder W. 1975. *Anatomy of fishes.* Stuttgart: E. Schweizerbart'sche Verlagsbuchhandlung.
Harding L, Jackson A, Barnett A, et al. 2021. Endothermy makes fishes faster but does not expand their thermal niche. *Funct Ecol* 2021:**https://doi.org/10.1111/1365-2435.13869**.
Harding MM, Anderberh PI, Haymet ADJ. 2003. Antifreeze glycoproteins from polar fish. *Eur J Biochem* 270:1381–1392.
Hardisty MW. 1979. *Biology of the cyclostomes.* London: Chapman & Hall.
Hardisty MW. 1982. Lampreys and hagfishes: analysis of cyclostome relationships. In: Hardisty MW, Potter IC, eds. *The biology of lampreys*, Vol. 4b, pp. 165–259. New York: Academic Press.
Hardisty MW, Potter IC. 1971–1982. *The biology of lampreys*, Vols 1–4b. New York: Academic Press.
Hare JA, Churchill JH, Cowen RK, et al. 2002. Routes and rates of larval fish transport from the southeast to the northeast United States continental shelf. *Limnol Oceanogr* 47:1774–1789.
Hargrove EC. 1989. *Foundations of environmental ethics.* Englewood Cliffs, NJ: Prentice Hall.
Harley CDG, Hughes AR, Hultgren KM, et al. 2006. The impacts of climate change in coastal marine systems. *Ecol Lett* 9:228–241.
Harmelin-Vivien ML. 2002. Energetics and fish diversity on coral reefs. In: Sale PF, ed. *Coral reef fishes: dynamics and diversity in a complex ecosystem*, pp. 265–274. San Diego, CA: Academic Press.
Harmon ME, Franklin JF, Swanson FJ, et al. 1986. Ecology of coarse woody debris in temperate ecosystems. *Adv Ecol Res* 15:133–302.
Harrington ME. 1993. Aggression in damselfish: adult–juvenile interactions. *Copeia* 1993:67–74.
Harrington RW, Jr. 1971. How ecological and genetic factors interact to determine when self-fertilizing hermaphrodites of *Rivulus marmoratus* change into functional secondary males, with a reappraisal of the modes of intersexuality among fishes. *Copeia* 1971:389–431.
Harrington RW, Jr. 1975. Sex determination and differentiation among uniparental homozygotes of the hermaphroditic fish *Rivulus marmoratus* (Cyprinodontidae: Atheriniformes). In: Reinboth R. *Intersexuality in the animal kingdom*, pp. 249–262. Berlin: Springer-Verlag.
Harris R. 2005. Attacks on taxonomy. *Am Sci* 93:311–312.
Harrison IJ, Stiassny MLJ. 1999. The quiet crisis. A preliminary listing of the freshwater fishes of the world that are extinct or "missing in action". In: MacPhee RDE, ed. *Extinctions in near time*, pp. 271–331. New York: Kluwer Academic/Plenum.
Hart PJB. 1993. Teleost foraging: facts and theories. In: Pitcher TJ, ed. *The behaviour of teleost fishes*, 2nd edn, pp. 253–284. London: Chapman & Hall.
Hart PJB, Reynolds JD. 2002a. *Handbook of fish biology and fisheries, Vol. 1. Fish biology.* Malden, MA: Blackwell Science.
Hart PJB, Reynolds JD, eds. 2002b. *Handbook of fish biology and fisheries, Vol. 2. Fisheries.* Oxford: Blackwell Science.
Hartl DL, Clark AG. 2006. *Principles of population genetics*, 4th edn. Sunderland, MA: Sinauer Associates.
Harvey BC. 1987. Susceptibility of young-of-the-year fishes to downstream displacement by flooding. *Trans Am Fish Soc* 116:851–855.
Harwood AJ, Metcalfe NB, Armstrong JD, Griffiths SW. 2001. Spatial and temporal effects of interspecific competition between Atlantic salmon (*Salmo salar*) and brown trout (*Salmo trutta*) in winter. *Can J Fish Aquat Sci* 58:1133–1140.
Harwood AJ, Metcalfe NB, Griffiths SW, Armstrong JD. 2002. Intra- and inter-specific competition for winter concealment habitat in juvenile salmonids. *Can J Fish Aquat Sci* 59:1515–1523.
Hasler AD, Scholz AT. 1983. *Olfactory imprinting and homing in salmon: investigations into the mechanism of the imprinting process.* Berlin: Springer-Verlag.
Hassan M, Lemaire C, Fauvelot C, Bonhomme F. 2002. Seventeen new exon-primed intron-crossing polymerase chain reaction amplifiable introns in fish. *Molec Ecol Notes* 2: 334–340.

Hastings MC. 2011. Biomechanics of the inner ear. In: Farrell AP, Stevens ED, Cech JJ, Richards JG, eds. *Encyclopedia of fish physiology: from genome to environment*, pp. 270–275. Boston: Elsevier/Academic Press.

Hastings PA, Walker Jr., HJ, Galland GR. 2014. *Fishes: a guide to their diversity*. Oakland, CA: University of California Press, 34 5 pp.

Haugen TO, Vollestad LA. 2001. A century of life-history evolution in grayling. *Genetica* 112:475–491.

Hawkes HA. 1975. River zonation and classification. In: Whitton BA, ed. *River ecology*, pp. 312–374. Oxford: Blackwell Scientific Publications.

Hawkins AD. 1993. Underwater sound and fish behaviour. In: Pitcher TJ, ed. *The behaviour of teleost fishes*, 2nd edn, pp. 130–169. London: Chapman & Hall.

Hawkins AD. 2011. Effects of human-generated sound on fish. In: Farrell AP, Stevens ED, Cech JJ, Richards JG, eds. *Encyclopedia of fish physiology: from genome to environment*, pp. 304—310. Boston: Elsevier/Academic Press.

Hawkins AD, Myrberg AA, Jr. 1983. Hearing and sound communication under water. In: Lewis B, ed. *Bioacoustics, a comparative approach*, pp. 347–405. London: Academic Press.

Hawkins JP, Roberts CM, Clark V. 2000. The threatened status of restricted-range coral reef fish species. *Anim Conserv* 3:81–88.

Hawryshyn CW. 1992. Polarization vision in fishes. *Am Sci* 80:164–175.

Hawryshyn CW. 1998. Vision. In: Evans DH, ed. *The physiology of fishes*, 2nd edn, pp. 345–374. Boca Raton, FL: CRC Press.

Hay ME. 1986. Associational plant defenses and the maintenance of species diversity: turning competitors into accomplices. *Am Natur* 128:617–641.

Hay ME. 1991. Fish–seaweed interactions on coral reefs: effects of herbivorous fishes and adaptations of their prey. In: Sale PF, ed. *The ecology of fishes on coral reefs*, pp. 96–119. San Diego: Academic Press.

Haya K, Burridge LE, Chang BD. 2001. Environmental impact of chemical wastes produced by the salmon aquaculture industry. *ICES J Mar Sci* 58:492–496.

He D, Sui X, Sun H, et al. 2020. Diversity, pattern and ecological drivers of freshwater fish in China and adjacent areas. *Rev Fish Biol Fish* 30:387–404.

He M, Yan X-C, Liang Y, Sun X-W, Teng C-B. 2014. Evolution of the viral hemorrhagic septicemia virus: divergence, selection and origin. *Mol Phylogenet Evol* 77:34–40.

Healey EG. 1957. The nervous system. *Physiol Fish* 2:1–119.

Healey MC, Groot C. 1987. Marine migration and orientation of ocean-type chinook and sockeye salmon. In: Dadswell MJ, Klauda RJ, Moffitt CM, Saunders RL, Rulifson RA, Cooper JE, eds. *Common strategies of anadromous and catadromous fishes*, pp. 298–312. American Fisheries Society Symposium 1. Bethesda, MD: American Fisheries Society.

Heath AG. 1987. *Water pollution and fish physiology*. Boca Raton, FL: CRC Press.

Heath MR, Speirs DC. 2012. Changes in species diversity and size composition in the Firth of Clyde demersal fish community (1927–2009). *Proc R Soc B* 279:543–552

Hebert PDN, Cywinska A, Ball SL, de Waard JR. 2003. Biological identifications through DNA barcodes. *Proc Roy Soc Lond B Biol Sci* 270:313–322.

Hedrick PW. 2010. *Genetics of populations*, 4th edn. Boston: Jones & Bartlett Publishers.

Hedrick RP. 1996. Movements of pathogens with the international trade of live fish: problems and solutions. *Revue Scientifique et Technique de L'Office International des Epizooties* 15:523–531.

Hedrick RP, McDowell TS, Mukkatira K, et al. 1999. Susceptibility of selected inland salmonids to experimentally induced infections with *Myxobolus cerebralis*, the causative agent of whirling disease. *J Aquat Anim Health* 11:330–339.

Heemstra PC, Greenwood PH. 1992. New observations on the visceral anatomy of the late-term fetuses of the living coelacanth fish and the oophagy controversy. *Proc Roy Soc Lond B Biol Sci* 249:49–55.

Heenan A, Hoey AS, Williams GJ, Williams ID. 2016. Natural bounds on herbivorous coral reef fishes. *Proc Royal Soc B* 283:20161716. **https://doi.org/10.1098/rspb.2016.1716.**

Heezen BC, Hollister CD. 1971. *The face of the deep*. London: Oxford University Press.

Heg D, Bachar Z. 2006. Cooperative breeding in the Lake Tanganyika cichlid *Julidochromis ornatus*. *Env Biol Fish* 76:265–281.

Heiligenberg W. 1993. Electrosensation. In: Evans DH, ed. *The physiology of fishes*, pp. 137–160. Boca Raton, FL: CRC Press.

Hein RG. 1996. Mobbing behavior in juvenile French grunts (*Haemulon flavolineatum*). *Copeia* 1996:989–991.

Heino M, Pauli BD, Dieckmann U. 2015. Fisheries-induced evolution. *Annu RevEcol Evol Syst* 46:461–480.

Heins DC, Rabito FG, Jr. 1986. Spawning performance in North American minnows: direct evidence of the occurrence of multiple clutches in the genus *Notropis*. *J Fish Biol* 28:343–357.

Heinz Center. 2002. *Dam removal: science and decision making*. Washington, DC: Heinz Center.

Heiss E, Aerts P, Van Wassenbergh S. 2018. Aquatic–terrestrial transitions of feeding systems in vertebrates: a mechanical perspective. Journal of Experimental Biology 221:jeb154427. **https://doi.org/10.1242/jeb.154427** (15 pages)

Helfield JM, Naiman RJ. 2001. Effects of salmon-derived nitrogen on riparian forest growth and implications for stream productivity. *Ecology* 82:2403–2409.

Helfman GS. 1978. Patterns of community structure in fishes: summary and overview. *Env Biol Fish* 3:129–148.

Helfman GS. 1979a. *Temporal relationships in a freshwater fish community*. PhD dissertation, Cornell University, Ithaca, NY.

Helfman GS. 1979b. Fish attraction to floating objects in lakes. In: Johnson DL, Stein RA, eds. *Response of fish to habitat structure in standing water*, pp. 49–57. American Fisheries Society North Central Division Special Publication No. 6. Bethesda, MD: American Fisheries Society.

Helfman GS. 1981a. The advantage to fishes of hovering in shade. *Copeia* 1981:392–400.

Helfman GS. 1981b. Twilight activities and temporal structure in a freshwater fish community. *Can J Fish Aquat Sci* 38:1405–1420.

Helfman GS. 1984. School fidelity in fishes: the yellow perch pattern. *Anim Behav* 32:663–672.

Helfman GS. 1986. Behavioral responses of prey fishes during predator–prey interactions. In: Feder ME, Lauder GV, eds. *Predator–prey relationships: perspectives and approaches from the study of lower vertebrates*, pp. 135–156. Chicago: University of Chicago Press.

Helfman GS. 1989. Threat-sensitive predator avoidance in damselfish-trumpetfish interactions. *Behav Ecol Sociobiol* 24:47–58.

Helfman GS. 1990. Mode selection and mode switching in foraging animals. *Adv Stud Behav* 19:249–298.

Helfman GS. 1993. Fish behaviour by day, night and twilight. In: Pitcher TJ, ed. *The behaviour of teleost fishes*, 2nd edn, pp. 479–512. London: Chapman & Hall.

Helfman GS. 1994. Adaptive variability and mode choice in foraging fishes. In: Stouder DJ, Fresh KL, Feller RJ, eds. *Theory and application in fish feeding ecology*, pp. 3–17. Belle Baruch Marine Science Series. Columbia, SC: University of South Carolina Press.

Helfman GS. 2007. *Fish conservation: a guide to understanding and restoring global aquatic biodiversity and fishery resources*. Washington, DC: Island Press.

Helfman GS, Clark JB. 1986. Rotational feeding: overcoming gape-limited foraging in Anguillid eels. *Copeia* 1986:679–685.

Helfman G, Collette B. 2011. *Fishes: the animal answer guide*. Baltimore, Maryland: Johns Hopkins University Press, 178 pp.

Helfman GS, Collette BB, Facey DE. 1997. *The diversity of fishes*. Malden, MA: Blackwell Science.

Helfman, GS, Collette BB, Facey DE, Bowen BW. 2009. *The diversity of fishes: biology, ecology, and evolution*, 2nd edn. Wiley-Blackwell.

Helfman GS, Facey DE, Hales LS, Jr., Bozeman EL, Jr. 1987. Reproductive ecology of the American eel. In: Dadswell MJ, Klauda RJ, Moffitt CM, Saunders RL, Rulifson RA, Cooper JE, eds. *Common strategies of anadromous and catadromous fishes*, pp. 42–56. American Fisheries Society Symposium 1. Bethesda, MD: American Fisheries Society.

Helfman GS, Schultz ET. 1984. Social transmission of behavioural traditions in a coral reef fish. *Anim Behav* 32:379–384.

Helfman GS, Stoneburner DL, Bozeman EL, Christian PA, Whalen R. 1983. Ultrasonic telemetry of American eel movements in a tidal creek. *Trans Am Fish Soc* 112:105–110.

Helfman GS, Winkelman DL. 1991. Energy trade-offs and foraging mode choice in American eels. *Ecology* 72:310–318.

Hellawell JM. 1983. *Biological indicators of freshwater pollution and environmental management*. New York: Elsevier Applied Science Publishers.

Hemmingsen EA. 1991. Respiratory and cardiovascular adaptations in hemoglobin-free fish: resolved and unresolved problems. In: di Prisco G, Maresca B, Tota B, eds. *Biology of Antarctic fish*, pp. 191–203. Berlin: Springer-Verlag.

Hempel G. 1979. *Early life history of fish: the egg stage*. Seattle, WA: University of Washington Press.

Hendry AP. 2001. Adaptive divergence and the evolution of reproductive isolation in the wild: an empirical demonstration using introduced sockeye salmon. *Genetica* 112:515–534.

Hendry AP, Wenburg JK, Bentzen P, et al. 2000. Rapid evolution of reproductive isolation in the wild: evidence from introduced salmon. *Science* 290:516–518.

Heneman B, Glazer M. 1996. More rare than dangerous: a case study of white shark conservation in California. In: Klimley AP, Ainley DG, eds. *Great white sharks: the biology of Carcharodon carcharias*, pp. 481–491. San Diego, CA: Academic Press.

Hennig W. 1950. *Grundzüge einer Theorie der phylogenetischen Systematik*. Berlin: Deutscher Zentralverlag.

Hennig W. 1966. *Phylogenetic systematics*. Urbana, IL: University of Illinois Press.

Henson SA, Warner RR. 1997. Male and female alternative reproductive behaviors in fishes: a new approach using intersexual dynamics. *Ann Rev Ecol Syst* 28:571–592

Herald ES. 1961. *Living fishes of the world*. Garden City, NY: Doubleday & Co.

Herbold B. 1984. Structure of an Indiana stream fish association: choosing an appropriate model. *Am Natur* 124:561–572.

Herdendorf CD, Berra TM. 1995. A Greenland shark from the wreck of the SS *Central-America* at 2200 meters. *Trans Am Fish Soc* 124:950–953.

Hermoso V, Clavero M, Kennard MJ. 2012. Determinants of fine-scale homogenization of native freshwater fish faunas in a Mediterranean Basin: implications for conservation. *Diversity Distrib* 18:236–247.

Hermoso V, Januchowski-Hartley SR, Linke S, Dudgeon D, Petry P, McIntyre P. 2017. Optimal allocation of red list assessments to guide conservation of biodiversity in a rapidly changing world. *Glob Change Biol* 23:3525–3532

Herrel A, Gibb AC. 2006. Ontogeny of performance in vertebrates. *Physiol Biochem Zool* 79:1–6.

Herring PJ, Cope C. 2005. Red bioluminescence in fishes: on the suborbital photophores of *Malacosteus*, *Pachystomias* and *Aristostomias*. *Mar Biol* 148:383–394.

Hersey J. 1988. *Blues*. New York: Random House.

Hertel H. 1966. *Structure, form and movement*. New York: Reinholt.

Herwig RP, Gray JP, Weston DP. 1997. Antibacterial resistant bacteria in surficial sediments near salmon net-cage farms in Puget Sound, Washington. *Aquaculture* 149:263–283.

Hewitt RP, Theilacker GH, Lo NCH. 1985. Causes of mortality in young jack mackerel. *Mar Ecol Prog Ser* 26:1–10.

Heyman WD, Graham RT, Kjerfve B, Johannes RE. 2001. Whale sharks *Rhincodon typus* aggregate to feed on fish spawn in Belize. *Mar Ecol Prog Ser* 215:275–282.

Hiatt RW, Brock VE. 1948. On the herding of prey and the schooling of the black skipjack, *Euthynnus yaito* Kishinouye. *Pacific Sci* 2:297–298.

Hickman CP, Jr., Trump BF. 1969. The kidney. In: Hoar WS, Randall DJ, eds. *Excretion, ionic regulation, and metabolism. Fish physiology*, Vol. 1, pp. 91–239. New York: Academic Press.

Hidaka K, Iwatsuki Y, Randall JE. 2008. Redescriptions of the Indo-Pacific bonefishes *Albula argentea* (Forster) and *A. virgata* Jordan and Jordan, with a description of a related new species, *A. oligolepis*. *Ichthyol Res* 55:53–64.

Hilborn R. 2005. Are sustainable fisheries achievable? In: Norse EA, Crowder LB, eds. *Marine conservation biology: the science of maintaining the sea's biodiversity*, pp. 247–259. Washington, DC: Island Press.

Hildebrand M. 1982. *Analysis of vertebrate structure*, 2nd edn. New York: Wiley & Sons.

Hildebrand M. 1988. *Analysis of vertebrate structure*, 3rd edn. New York: Wiley & Sons.

Hildebrand M, Bramble DM, Liem KF, Wake DB, eds. 1985. *Functional vertebrate morphology*. Cambridge, MA: Belknap Press.

Hill J, Grossman GD. 1987. Home range estimates for three North American stream fishes. *Copeia* 1987:376–380.

Hill J, Grossman GD. 1993. An energetic model of microhabitat use for rainbow trout and rosyside dace. *Ecology* 74:685–698.

Hillis DM, Moritz C. 1996. *Molecular systematics*, 2nd edn. Sunderland, MA: Sinauer Associates.

Hillyard SD. 2011. Life in hot water: the desert pupfish. In: Farrell AP, Stevens ED, Cech JJ, Richards JG, eds. *Encyclopedia of fish physiology: from genome to environment*, pp. 1831–1842. Boston: Elsevier/Academic Press.

Hilton EJ, Bemis WE. 2005. Grouped tooth replacement in the oral jaws of the tripletail, *Lobotes surinamensis* (Perciformes: Lobotidae), with a discussion of its proposed relationship to *Datnioides. Copeia* 2005:665–672.

Hinck JE, Blazer VS, Schmitt CJ, Papoulias DM, Tillitt DE. 2009. Widespread occurrence of intersex black basses (*Micropterus spp.*) from U.S. rivers, 1995–2004. *Aquat Toxicol* 95:60–70.

Hinegardner R, Rosen DE. 1972. Cellular DNA content and the evolution of teleostean fishes. *Am Natur* 106:621–644.

Hissmann K, Fricke H, Schauer J. 1998. Population monitoring of the coelacanth (Latimeria chalumnae). *Conserv Biol* 12:759–765.

Hites RA, Foran JA, Carpenter DO, Hamilton MC, Knuth BA, Schwager SJ. 2004. Global assessment of organic contaminants in farmed salmon. *Science* 303:226–229.

Hitt NP, Snook EL, Massie DL. 2017. Brook trout use of thermal refugia and foraging habitat influenced by Brown Trout. *Can J Fish Aquat Sci* 74:406–418.

Hixon MA. 1980a. Food production and competitor density as the determinants of feeding territory size. *Am Natur* 115:510–530.

Hixon MA. 1980b. Competitive interactions between California reef fishes of the genus *Embiotoca. Ecology* 61:918–931.

Hixon MA. 1991. Predation as a process structuring coral reef fish communities. In: Sale PF, ed. *The ecology of fishes on coral reefs*, pp. 475–508. San Diego, CA: Academic Press.

Hixon MA. 1998. Population dynamics of coral-reef fishes: controversial concepts and hypotheses. *Aust J Ecol* 23:192–201.

Hixon MA. 2006. Competition. In: Allen LG, Pondella DJ, II, Horn MH, eds. *The ecology of marine fishes: California and adjacent waters*, pp. 449–465. Berkeley, CA: University of California Press.

Hixon MA. 2011. 60 years of coral reef fish ecology: past, present, future. *Bull Mar Sci* 87(4):727–765.

Hixon MA, Beets JP. 1989. Shelter characteristics and Caribbean fish assemblages: experiments with artificial reefs. *Bull Mar Sci* 44:666–680.

Hixon MA, Beets JP. 1993. Predation, prey refuges, and the structure of coral-reef fish assemblages. *Ecol Monogr* 63:77–101

Hixon MA, Brostoff WN. 1983. Damselfish as keystone species in reverse: intermediate disturbance and diversity of reef algae. *Science* 220:511–513.

Hixon MA, Carr MH. 1997. Synergistic predation, density dependence, and population regulation in marine fish. *Science* 277:946–949.

Hixon MA, Jones GP. 2005. Competition, predation, and density-dependent mortality in demersal marine fishes. *Ecology* 86:2847–2859.

Hjort J. 1914. Fluctuations in the great fisheries of northern Europe viewed in the light of biological research. *Rapp P-V Reun Cons Int Explor Mer* 20:1–228.

Hoar WS. 1969. Reproduction. In: Hoar WS, Randall DJ, eds. *Reproduction and growth. Bioluminescence, pigments, and poisons. Fish physiology*, Vol. 3, pp. 1–72. New York: Academic Press.

Hoar WS, Randall DJ, eds. 1978. *Locomotion. Fish physiology*, Vol. 7. New York: Academic Press.

Hoar WS, Randall DJ, eds. 1988. *The physiology of developing fish, Part B. Viviparity and posthatching juveniles. Fish physiology*, Vol. 11. San Diego: Academic Press.

Hoare DJ, Couzin ID, Godin JGJ, Krause J. 2004. Context-dependent group size choice in fish. *Anim Behav* 67:155–164.

Hoare DJ, Ruxton GD, Godin JGJ, Krause J. 2000. The social organization of free-ranging fish shoals. *Oikos* 89:546–554.

Hobbs JPA, Frisch AJ, Allen GR, Van Herwerden L. 2009. Marine hybrid hotspot at Indo-Pacific biogeographic border. *Biol Lett* 5:258–261.

Hobbs JPA, van Herwerden L, Pratchett MS, Allen GR. 2013. Hybridisation among butterflyfishes. In: Pratchett MS, Berumen ML, Kapoor B, eds. *Biology of butterflyfishes*, pp. 48–69. Boca Raton, FL: CRC Press.

Hobson ES. 1968. Predatory behavior of some shore fishes in the Gulf of California. *US Fish Wildl Serv Res Rept* 73:1–92.

Hobson ES. 1972. Activity of Hawaiian reef fishes during evening and morning transitions between daylight and darkness. *Fish Bull US* 70:715–740.

Hobson ES. 1973. Diel feeding migrations in tropical reef fishes. *Helgo wiss Meeresunter* 24:671–680.

Hobson ES. 1974. Feeding relationships of teleostean fishes on coral reefs in Kona, Hawaii. *Fish Bull US* 72:915–1031.

Hobson ES. 1975. Feeding patterns among tropical reef fishes. *Am Sci* 63:382–392.

Hobson ES. 1986. Predation on the Pacific sand lance, *Ammodytes hexapterus* (Pisces: Ammodytidae), during the transition between day and night in southeastern Alaska. *Copeia* 1986:223–226.

Hobson ES. 1991. Trophic relationships of fishes specialized to feed on zooplankters above coral reefs. In: Sale PF, ed. *The ecology of fishes on coral reefs*, pp. 69–95. San Diego, CA: Academic Press.

Hobson ES. 1994. Ecological relations in the evolution of acanthopterygian fishes in warm-temperate communities of the northeastern Pacific. *Env Biol Fish* 40:49–90.

Hobson ES. 2006. Evolution. In: Allen LG, Pondella DJ, Horn MH, eds. *The ecology of marine fishes: California and adjacent waters*, pp. 55–80. Berkeley, CA: University of California.

Hobson ES, Chess JR, Howard DF. 2001. Interanual variation in predation on first-year *Sebastes* spp. by three northern California predators. *Fish Bull US* 99:292–302.

Hobson ES, McFarland WN, Chess JR. 1981. Crepuscular and nocturnal activities of California nearshore fishes, with consideration of their scotopic visual pigments and the photic environment. *Fish Bull US* 79:1–30.

Hochachka PW, Mommsen TP. 1983. Protons and anaerobiosis. *Science* 219:1391–1397.

Hochachka PW, Somero GN. 1973. *Strategies of biochemical adaptation.* Philadelphia, PA: Saunders.

Hochachka PW, Somero GN. 1984. *Biochemical adaptation.* Princeton, NJ: Princeton University Press.

Hocutt CH, Wiley EO, eds. 1986. *The zoogeography of North American freshwater fishes.* New York: Wiley & Sons.

Hodgson ES, Mathewson RF, eds. 1978. *Sensory biology of sharks, skates, and rays.* Arlington, VA: Office of Naval Research.

Hodson PV. 2002. Biomarkers and bioindicators in monitoring and assessment: the state of the art. In: Adams SM, ed. *Biological indicators of aquatic ecosystem stress*, pp. 491–516. Bethesda, MD: American Fisheries Society.

Hoegg S, Brinkmann H, Taylor JS, Meyer A. 2004. Phylogenetic timing of the fish-specific genome duplication correlates with the diversification of teleost fish. *J Mol Evol* 59:190–203.

Hoegh-Guldberg O. 1999. Climate change, coral bleaching and the future of the world's coral reefs. *Mar Freshwater Res* 50:839–866.

Hoegh-Guldberg O, Mumby PJ, Hooten AJ, et al. 2007. Coral reefs under rapid climate change and ocean acidification. *Science* 318:1737–1742.

Hoeinghaus DJ, Winemiller KO, Layman CA, Arrington DA, Jepsen DB. 2006. Effects of seasonality and migratory prey on body condition of *Cichla* species in a tropical floodplain river. *Ecol Freshwater Fish* 15:398–407.

Hoelzel AR, Shivji MS, Magnussen J, Francis MP. 2006. Low worldwide genetic diversity in the basking shark (*Cetorhinus maximus*). *Biol Lett* 2:639–642.

Hoese HD. 1985. Jumping mullet – the internal diving bell hypothesis. *Env Biol Fish* 13:309–314.

Hoffman EA, Kolm N, Berglund A, Arguello JR, Jones AG. 2005. Genetic structure in the coral-reef associated Banggai cardinalfish, *Pterapogon kauderni*. *Mol Ecol* 14:1367–1375.

Hoffman GL, Schubert G. 1984. Some parasites of exotic fishes. In: Courtenay WR, Jr., Stauffer JR, Jr., eds. *Distribution, biology, and management of exotic fishes*, pp. 233–261. Baltimore: Johns Hopkins University Press.

Hofmann MS. 2011. Physiology of ampullary electrosensory systems. In: Farrell AP, Stevens ED, Cech JJ, Richards JG, eds. *Encyclopedia of fish physiology: from genome to environment*, pp. 359–365. Boston: Elsevier/Academic Press.

Holbrook SJ, Schmitt RJ. 1988a. Effects of predation risk on foraging behavior: mechanisms altering patch choice. *J Exp Mar Biol Ecol* 121:151–163.

Holbrook SJ, Schmitt RJ. 1988b. The combined effects of predation risk and food reward on patch selection. *Ecology* 69:125–134.

Holbrook SJ, Schmitt RJ. 1989. Resource overlap, prey dynamics, and the strength of competition. *Ecology* 70:1943–1953.

Holbrook SJ, Schmitt RJ. 2002. Competition for shelter space causes density-dependent predation mortality in damselfishes. *Ecology* 83: 2855–2868.

Holbrook SJ, Schmitt RJ. 2005. Growth, reproduction and survival of a tropical sea anemone (Actiniaria): benefits of hosting anemonefish. *Coral Reefs* 24:67–73.

Holder MT, Erdmann MV, Wilcox TP, Caldwell RL, Hillis DM. 1999. Two living species of coelacanths? *Proc Natl Acad Sci USA* 96:12616–12620.

Holland KN, Brill RW, Chang RKC, Silbert JR, Fournier DA. 1992. Physiological and behavioral thermoregulation in bigeye tuna (*Thunnus obesus*). *Nature* 358:410–412.

Holmes BH, Steinke D, Ward RD. 2009. Identification of shark and ray fins using DNA barcoding. *Fishery Res* 95:280–288.

Holmlund CM, Hammer M. 1999. Ecosystem services generated by fish populations. *Ecol Econ* 29:253–268.

Holomuzki JR, Stevenson RJ. 1992. Role of predatory fish in community dynamics of an ephemeral stream. *Can J Fish Aquat Sci* 49:2322–2330.

Holopainen IJ, Aho J, Vornanen M, Huuskonen H. 1997. Phenotypic plasticity and predator effects on morphology and physiology of crucian carp in nature and in the laboratory. *J Fish Biol* 50:781–798.

Holt DE, Johnston CE. 2014. Sound production and associated behaviours in blacktail shiner *Cyprinella venusta*: a comparison between field and lab. *Environ Biol Fishes* 97(11):1207–1219.

Holt GJ, Holt SA, Arnold CR. 1985. Diel periodicity of spawning in sciaenids. *Mar Ecol Prog Ser* 27:1–7.

Holtby LB. 1988. Effects of logging on stream temperature in Carnation Creek, British Columbia, and associated impacts on the coho salmon (*Oncorhynchus kisutch*). *Can J Fish Aquat Sci* 45:502–515.

Holzman R, Genin A. 2003. Zooplanktivory by a nocturnal coral-reef fish: effects of light, flow, and prey density. *Limnol Oceanogr* 48:1367–1375.

Honma Y. 1998. Asian hagfishes and their fisheries biology. In: Jørgensen JM, Lomholt JP, Weber RE, Malte H, eds. *The biology of hagfishes*, pp. 45–56. London: Chapman & Hall.

Hontela A, Stacey NE. 1990. Cyprinidae. In: Munro AD, Scott AP, Lam TJ, eds. *Reproductive seasonality in teleosts: environmental influences*, pp. 53–77. Boca Raton: CRC Press.

Hopkins CD. 1986. Behavior of Mormyridae. In: Bullock TH, Heiligenberg W, eds. *Electroreception*, pp. 527–576. New York: Wiley & Sons.

Hopkins CD. 2009. Electrical perception and communication. In: Squire LR, ed. *Encyclopedia of neuroscience* Vol. 3, pp. 813–831. London, UK: Elsevier.

Hori K, Fusetani N, Hashimoto K, Aida K, Randall JE. 1979. Occurrence of a grammistin-like mucous toxin in the clingfish *Diademicththys lineatus*. *Toxicon* 17:418–424.

Hori M. 1993. Frequency-dependent natural selection in the handedness of scale-eating cichlid fish. *Science* 260:216–219.

Horn MH. 1972. The amount of space available for marine and freshwater fishes. *Fish Bull US* 70:1295–1297.

Horn MH. 1989. Biology of marine herbivorous fishes. *Oceanog Mar Biol Ann Rev* 27:167–272.

Horn MH. 1997. Feeding and movements of the fruit-eating characid fish *Brycon guatemalensis* in relation to seed dispersal in a Costa Rican rain forest. *Oecologia* 109: 259–264 abstr. 172.

Horn MH. 1999. Convergent evolution and community convergence: research potential using intertidal fishes. In: Horn MH, Martin KLM, Chotkowski MA, eds. *Intertidal fishes: life in two worlds*, pp. 356–372. San Diego, CA: Academic Press.

Horn MH, Allen LG. 1985. Fish community ecology in southern California bays and estuaries. In: Yanez-Arancibia A, ed. *Fish community ecology in estuaries and coastal lagoons: towards an ecosystem integration*, pp. 169–190. Mexico City: DR (R) UNAM Press.

Horn MH, Ferry-Graham LA. 2006. Feeding mechanisms and trophic interactions. In: Allen LG, Pondella DJ, Horn MH, eds. *The ecology of marine fishes: California and adjacent waters*, pp. 387–410. Berkeley, CA: University of California.

Horn MH, Gibson RN. 1988. Intertidal fishes. *Sci Am* 256(1):64–70.

Horn MH, Martin KLM. 2006. Rocky intertidal zone. In: Allen LG, Pondella DJ, Horn MH, eds. *The ecology of marine fishes: California and adjacent waters*, pp. 205–226. Berkeley, CA: University of California.

Horn MH, Ojeda FP. 1999. Herbivory. In: Horn MH, Martin KLM, Chotkoski MA, eds. *Intertidal fishes: life in two worlds*, pp. 197–222. San Diego: Academic Press.

Horn MH, Stephens JS, Jr. 2006. Climate change and overexploitation. In: Allen LG, Pondella DJ, Horn MH, eds. *The ecology of marine fishes: California and adjacent waters*, pp. 621–635. Berkeley, CA: University of California.

Horwitz RJ. 1978. Temporal variability patterns and the distributional patterns of stream fishes. *Ecol Monogr* 48:307–321.

Houde AE. 1997. *Sex, color, and mate choice in guppies*. Princeton, NJ: Princeton University Press.

Houde AE, Endler JA. 1990. Correlated evolution of female mating preferences and male color patterns in the guppy *Poecilia reticulata*. *Science* 248:1405–1408.

Houde ED. 1987. Fish early life dynamics and recruitment variability. In: Hoyt RD, ed. *10th annual larval fish conference*, pp. 17–29. American Fisheries Society Symposium 2. Bethesda, MD: American Fisheries Society.

Houde ED. 1989. Comparative growth, mortality, and energetics of marine fish larvae: temperature and implied latitudinal effects. *Fish Bull US* 87:471–495.

Howell SNG. 2014. *The amazing world of flyingfish*. Princeton, NJ: Princeton Univ. Press.

Hoxmeier JH, Dieterman DJ. 2013. Seasonal movement, growth, and survival of brook trout in sympatry with brown trout in Midwestern US streams. *Ecol Freshw Fish* 22:530–542.

Hsu PD, Lander ES, Zhang F. 2014. Development and application of CRISPR-Cas9 for genome engineering. *Cell* 157:1262–1278.

Hubbard PC, Barata EN, Canário AVM. 2003. Olfactory sensitivity of the gilthead seabream (*Sparus auratus* L.) to conspecific body fluids. *J Chem Ecol* 29:2481–2498.

Hubbs CL. 1952. Antitropical distribution of fishes and other organisms. *Proc 7th Pacific Sci Cong* 3:324–329.

Hubbs CL. 1955. Hybridization between fish species in nature. *Syst Zool* 4:1–20.

Hubbs CL. 1964. History of ichthyology in the United States after 1850. *Copeia* 1964:42–60.

Hubbs CL. 1967. Geographic variation in survival of hybrids between etheostomatine fishes. *Texas Mem Mus Bull* 13:1–72.

Hubbs CL, Lagler KF. 1964. *Fishes of the Great Lakes region*. Ann Arbor, MI: University of Michigan Press.

Huelsenbeck JP, Ronquist F, Nielsen R. 2001. Bayesian inference of phylogeny and its impact on evolutionary biology. *Science* 294:2310–2314.

Huet M. 1959. Profiles and biology of western European streams as related to fish management. *Trans Am Fish Soc* 88:153–163.

Hueter RE, Gilbert PW. 1991. The sensory world of sharks. In: Gruber SH, ed. *Discovering sharks*, pp. 48–55. American Littoral Society Special Publication No. 14. Highlands, NJ: American Littoral Society.

Hueter RE, Mann DA, Maruska KP, Sisneros JA, Demski LS. 2004. Sensory biology of elasmobranches. In: Carrier JC, Musick JA, Heithaus MR, eds. *Biology of sharks and their relatives*, pp. 325–368. Boca Raton, FL: CRC Press.

Hughes DR. 1981. Development and organization of the posterior field of ctenoid scales in the Platycephalidae. *Copeia* 1981:596–606.

Hughes GM, Itazawa Y. 1972. The effect of temperature on the respiratory function of coelacanth blood. *Experientia* 28:1247.

Hughes LC, Orti GF, Huang Y, et al. 2018. Comprehensive phylogeny of ray-finned fishes (Actinopterygii) based on transcriptomic and genomic data. *Proc Natl Acad Sci USA* 115:6249–6254.

Hughes RM, Noss RF. 1992. Biological diversity and biological integrity: current concerns for lakes and streams. *Fisheries* 17(3):11–19.

Hughes RM, Omernik JM. 1990. Use and misuse of the terms watershed and stream order. In: Krumholz LA, eds. *Warmwater streams symposium*, pp. 320–326. American Fisheries Society Southern Division. Lawrence, KS: Allen Press.

Hulsey CD, De Leon FJG. 2005. Cichlid jaw mechanics: linking morphology to feeding specialization. *Funct Ecol* 19:487–494.

Humphries AT, McClanahan TR, McQuaid CD. 2014. Differential impacts of coral reef herbivores on algal succession in Kenya. *Mar Ecol Prog Ser* 504:119–132.

Humphries CJ, Parenti LR. 1999. *Cladistic biogeography*, 2nd edn. New York: Oxford University Press.

Humphries JM, Bookstein FL, Chernoff B, Smith GR, Elder RL, Poss SG. 1981. Multivariate discrimination by shape in relation to size. *Syst Zool* 30:291–308.

Hunter A, Speirs DC, Heath MR. 2015. Fishery-induced changes to age and length dependent maturation schedules of three demersal fish species in the Firth of Clyde. *Fish Res* 170:14–23

Hunter JR. 1981. Feeding ecology and predation of marine fish larvae. In: Lasker RL, ed. *Marine fish larvae: morphology, ecology, and relation to fisheries*, pp. 33–77. Seattle: Washington Sea Grant Program.

Hunter JR, Coyne KM. 1982. The onset of schooling in northern anchovy larvae *Engraulis mordax*. *Calif Coop Oceanic Fish Invest Rep* 23:246–251.

Huntingford FA. 1986. Development of behaviour in fishes. In: Pitcher TJ, ed. *The behaviour of teleost fishes*, pp. 47–68. Baltimore, MD: The Johns Hopkins University Press.

Huntingford FA. 1993. Development of behaviour in fish. In: Pitcher TJ, ed. *Behaviour of teleost fishes*, 2nd edn, pp. 57–83. London: Chapman & Hall.

Huntingford FA, Torricelli P, eds. 1993. *Behavioural ecology of fishes*. Ettore Majorana International Life Sciences Series, Vol. 11. New York: Harwood Academic Publishers.

Huntington C, Nehlsen W, Bowers J. 1996. A survey of healthy native stocks of anadromous salmonids in the Pacific Northwest and California. *Fisheries* 21(3):6–14.

Hurst RJ, Bagley NW, McGregor PA, Francis MP. 1999. Movements of the New Zealand school shark, *Galeorhinus galeus*, from tag returns. *NZ J Mar Freshwater Res* 33:29–48.

Hutchings JA. 2000a. Collapse and recovery of marine fishes. *Nature* 406:882–885.

Hutchings JA. 2000b. Numerical assessment in the front seat, ecology and evolution in the back seat: time to change drivers in fisheries and aquatic sciences. *Mar Ecol Prog Ser* 208:299–303.

Hutchings JA, Reynolds JD. 2004. Marine fish population collapses: consequences for recovery and extinction risk. *BioScience* 54:297–309.

Huveneers C, Holman D, Robbins R, Fox A, Endler JA, Taylor AH. 2015. White sharks exploit the sun during predatory approaches. *Am Natur* 185(4):562–570.

Huxley J, ed. 1940. *The new systematics*. Oxford: Oxford University Press.

Ibara RM, Penny LT, Ebeling AW, van Dykhuizen G, Cailliet G. 1983. The mating call of the plainfin midshipman fish, *Porichthys notatus*. In: Noakes DLG, Lindquist DG, Helfman GS, Ward JA, eds. *Predators and prey in fishes*, pp. 205–212. The Hague: Dr W. Junk.

Ida H, Oka N, Hayashigaki K. 1991. Karyotypes and cellular DNA contents of three species of the subfamily Clupeinae. *Jap J Ichthyol* 38:289–294.

Idda ML, Bertolucci C, Vallone D, Gothilf Y, Sánchez-Vázquez FJ, Foulkes NS. 2016. Circadian clocks: lessons from fish. In: Kalsbeek A, Merrow M, Roenneberg T, Foster RG, eds. *Progress in Brain Research*, Vol. 199, pp. 41–57. London, UK: Elsevier.

Ikonomou M, Higgs DA, Gibbs M, et al. 2007. Flesh quality of market-size farmed and wild British Columbia salmon. *Env Sci Technol* 41:437–443.

Inoue JG, Miya M, Aoyama J, Ishikawa S, Tsukamoto K, Nishida M. 2001. Complete mitochondrial DNA sequence of the Japanese eel *Anguilla japonica*. *Fish Sci* 67:118–125.

Inoue JG, Miya M, Tsukamoto K, Nishida M. 2000. Complete mitochondrial DNA sequence of the Japanese sardine *Sardinops melanostictus*. *Fish Sci* 66:924–932.

Inoue JG, Miya M, Tsukamoto K, Nishida M. 2003. Basal actinopterygian relationships: a mitogenomic perspective on the phylogeny of the "ancient fish". *Mol Phylog Evol* 26:110–120.

Inoue JG, Miya M, Venkatesh B, Nishida M. 2005. The mitochondrial genome of the Indonesian coelacanth *Latimeria menadoensis* (Sarcopterygii: Coelacanthiformes) and divergence time estimation between the two coelacanths. *Gene* 349:227–235.

Ip YK, Chew SF, Randall DJ. 2004a. Five tropical air-breathing fishes, six different strategies to defend against ammonia toxicity on land. *Physiol Biochem Zool* 77:768–782.

Ip YK, Kuah SSL, Chew SF. 2004b. Strategies adopted by the mudskipper *Boleophthalmus boddaerti* to survive sulfide exposure in normoxia or hypoxia. *Physiol Biochem Zool* 77:824–837.

Ip YK, Lau IY, Wong WP, Lee SLM, Chew SF. 2005. The African sharptooth catfish *Clairas gariepinus* can tolerate high levels of ammonia in its tissues and organs during four days of aerial exposure. *Physiol Biochem Zool* 78:630–640.

IPCC (Intergovernmental Panel on Climate Change). 2007a. *The fourth assessment report; Working Group I report (WGI). Climate change 2007: the physical science basis.* Geneva: World Meteorological Organization, United Nations Environmental Program. **www.ipcc.ch**.

IPCC (Intergovernmental Panel on Climate Change). 2007b. *The fourth assessment report; Working Group II report (WGII). Climate change 2007: impacts, adaptation and vulnerability.* Geneva: World Meteorological Organization, United Nations Environmental Program. **www.ipcc.ch**.

IPCC (Intergovernmental Panel on Climate Change). 2007c. *The fourth assessment report; Working Group III report (WGIII). Climate change 2007: mitigation of climate change.* Geneva: World Meteorological Organization, United Nations Environmental Program. **www.ipcc.ch**.

IPCC (Intergovernmental Panel on Climate Change). 2007d. *The fourth assessment report. The synthesis report.* Geneva: World Meteorological Organization, United Nations Environmental Program. **www.ipcc.ch**.

IPCC (Intergovernmental Panel on Climate Change). 2013. Climate Change 2013: The Physical Science Basis. Contribution of Working Group I to the Fifth Assessment Report of the Intergovernmental Panel on Climate Change. In: Stocker TF, Qin D, Plattner G-K, et al., eds. Cambridge, United Kingdom and New York, NY, USA: Cambridge University Press, 1535.

IPCC. 2013. Summary for policymakers. In: Stocker TF, Qin D, Plattner G-K, et al., eds. Climate Change 2013: The Physical Science Basis. Contribution of Working Group I to the Fifth Assessment Report of the Intergovernmental Panel on Climate Change. Cambridge, United Kingdom and New York, NY, USA: Cambridge University Press.

Irvine GV. 1981. The importance of behavior in plant–herbivore interactions: a case study. In: Cailliet GM, Simenstad CA, eds. *Gutshop '81, fish food habits studies*, pp. 240–248. Seattle, WA: University of Washington Sea Grant.

Ishihara M. 1987. Effect of mobbing toward predators by the damselfish *Pomacentrus coelestis* (Pisces: Pomacentridae). *J Ethol* 5:43–52.

Ishimatsu A. 2012. Evolution of the cardiorespiratory system in air-breathing fishes. *Aqua-BioScience Monographs* 5:1–28.

ISOFISH. 2002. Patagonian toothfish. **www.isofish.org/au/backg/index.htm**.

IUCN (International Union for the Conservation of Nature). 2004. *2004 IUCN Red List of threatened species.* **www.redlist.org**.

IUCN (International Union for the Conservation of Nature). 2006. *2006 IUCN Red List of threatened species.* **www.iucnredlist.org**.

IUCN (International Union for the Conservation of Nature) 2020. 2020 IUCN red list of threatened species. **www.redlist.org**.

Ivlev VS. 1961. *Experimental ecology of the feeding of fishes* (transl. D. Scott D). New Haven, CT: Yale University Press.

Iwama GK. 2007. The welfare of fish. *Dis Aquat Organisms* 75: 155–158.

Iwama GK, Afonso LOB, Vijayan MM. 2006. Stress in fishes. In: Evans DH, Claiborne JB, eds. *The physiology of fishes*, 3rd edn, pp. 319–342. Boca Raton, FL: CRC Press, Taylor & Francis Group.

Iwama GK, Nakanishi T. 1996. *The fish immune system: organism, pathogen, and environment. Fish physiology*, Vol. 15. New York: Academic Press.

Iwanowicz LR, Blazer VS, Pinkney AE, et al. 2015. Evidence of estrogenic endocrine disruption in smallmouth and largemouth bass inhabiting Northeast U.S. national wildlife refuge waters: a reconnaissance study. *Ecotoxicol Environ Saf* 124:50–59.

Jablonski D. 1991. Extinctions: a paleontological perspective. *Science* 253:754–757.

Jackson JBC, Donovan MK, Cramer KL, Lam V. 2014. *Status and trends of Caribbean Coral Reefs: 1970–2012.* Global Coral Reef Monitoring Network, IUCN, Gland, Switzerland.

Jackson JBC, Kirby MX, Berger WH, et al. 2001. Historical overfishing and the recent collapse of coastal ecosystems. *Science* 293:629–638.

Jacobs M, Ferrario J, Byrne C. 2002. Investigation of polychlorinated dibenzo-p-dioxins, dibenzo-p-furans and selected coplanar biphenyls in Scottish farmed Atlantic salmon (*Salmo salar*). *Chemosphere* 47:183–191.

Jakober MJ, McMahon TE, Thurow RF. 2000. Diel habitat partitioning by bull charr and cutthroat trout during fall and winter in Rocky Mountain streams. *Env Biol Fish* 59:79–89.

Jakober MJ, McMahon TE, Thurow RF, Clancy CG. 1998. Role of stream ice on fall and winter movements and habitat use by bull trout and cutthroat trout in Montana headwater streams. *Trans Am Fish Soc* 127:223–235.

Jamieson BGM. 1991. *Fish evolution and systematics: evidence from spermatozoa.* Cambridge, UK: Cambridge University Press.

Janetski DJ, Chaloner DT, Tiegs SD, Lamberti GA. 2009. Pacific salmon effects on stream ecosystems: a quantitative synthesis. *Oecologia* 159:583–595

Janssen J. 1980. Alewives (*Alosa pseudoharengus*) and ciscoes (*Coregonus artedii*) as selective and non-selective planktivores. In: Kerfoot WC, ed. *Evolution and ecology of zooplankton communities*, pp. 580–586. Hanover, NH: University Press of New England.

Janssen J, Brandt SB. 1980. Feeding ecology and vertical migration of adult alewives (*Alosa pseudoharengus*) in Lake Michigan. *Can J Fish Aquat Sci* 37:177–184.

Janvier P. 1984. The relationships of the Osteostraci and Galeaspida. *J Vertebr Paleontol* 4:344–358.

Janvier P. 1995. Conodonts join the club. *Nature* 374:761–762.

Janvier P. 1996. *Early vertebrates.* Oxford: Oxford University Press.

Janvier P. 2001. Ostracoderms and the shaping of the gnathostome characters. In: Ahlberg PE, ed. *Major events in early vertebrate evolution: palaeontology, phylogeny, genetics and development*, pp. 172–186. Systematic Association Special Volumes Series 61. London: Taylor & Francis.

Janvier P, Lund R, Grogan ED. 2004. Further consideration of the earliest known lamprey, *Hardistiella montanensis* Janvier and Lund, 1983, from the Carboniferous of Bear Gulch, Montana, USA. *J Vertebr Paleontol* 24:742–743.

Janzen T, Alzate A, Muschick M, Maan ME, van der Plas F, Etienne RS. 2017. Community assembly in Lake Tanganyika cichlid fish: quantifying the contributions of both niche-based and neutral processes. *Ecol Evol* 7:1057–1067.

Jarvenpaa M, Lindstrom K. 2004. Water turbidity by algal blooms causes mating system breakdown in a shallow-water fish, the sand goby *Pomatoschistus minutus. Proc Roy Soc Lond B Biol Sci* 271:2361–2365

Jarvik E. 1980. *Basic structure and evolution of vertebrates*, Vol. 1. London: Academic Press.

Jeffery WR. 2009. Regressive evolution in *Astyanax* Cavefish. *Ann Rev Genet* 43:25–47.

Jelks HL, Walsh SJ, Burkhead NM, et al. 2008. Conservation status of imperiled North American freshwater and diadromous fishes. *Fisheries* 33(8):372–407.

Jenkins RE, Burkhead NM. 1993. *Freshwater fishes of Virgina*. Bethesda, MD: American Fisheries Society.

Jennings CA, Zigler SJ. 2000. Ecology and biology of paddlefish in North America: historical perspectives, management approaches, and research priorities. *Rev Fish Biol Fisheries* 10:167–181.

Jensen D. 1966. The hagfish. *Sci Am* 214(2):82–90.

Jensen OP, Benson BJ, Magnuson JJ, et al. 2007. Spatial analysis of ice phenology trends across the Laurentian Great Lakes region during a recent warming period. *Limnol Oceanogr* 52:2013–2026.

Jessen H. 1966. Struniiformes. In: Lehman JP ed. *Traite de Paleontologie*, Part 4, Vol. 3, pp. 387–398. Paris: Masson SA.

Jessop BM. 1987. Migrating American eels in Nova Scotia. *Trans Am Fish Soc* 116:161–170.

Jobling M. 1981. The influence of feeding on the metabolic rate of fishes: a short review. *J Fish Biol* 18:385–400.

Jobling M. 1995. *Environmental biology of fishes*. Fish and Fisheries Series No. 16. London: Chapman & Hall.

Jobling M. 2011. Bioenergetics in aquaculture settings. In: Farrell AP, Stevens ED, Cech JJ, Richards JG, eds. *Encyclopedia of fish physiology: from genome to environment*, pp. 1664–1674. Boston: Elsevier/Academic Press.

Johannes RE. 1978. Reproductive strategies of coastal marine fishes in the tropics. *Env Biol Fish* 3:65–84.

Johannes RE. 1981. *Words of the lagoon: fishing and marine lore in the Palau District of Micronesia*. Berkeley, CA:University of California Press.

Johannes RE, Squire L, Graham T, et al. 1999. *Spawning aggregations of groupers (Serranidae) in Palau*. Marine Conservation Research Series Publication No. 1. Arlington, VA: The Nature Conservancy. **www.tnc.org/asiapacific**.

Johannsson OE, Bergman HL, Wood CM, et al. 2014. Air breathing in Magadi tilapia *Alcolapia grahami*, under normoxic and hyperoxic conditions, and the association with sunlight and reactive oxygen species. *J Fish Biol* 84:844–863.

Johns GC, Avise JC. 1998. A comparative summary of genetic distances in the vertebrates from the mitochondrial cytochrome *b* gene. *Mol Biol Evol* 15:1481–1490.

Johnsen PB, Hasler AD. 1977. Winter aggregations of carp (*Cyprinus carpio*) as revealed by ultrasonic tracking. *Trans Am Fish Soc* 106:556–559.

Johnsen S, Sosik HM. 2003. Cryptic coloration and mirrored sides as camouflage strategies in near-surface pelagic habitats: implications for foraging and predator avoidance. *Limnol Oceanogr* 48:1277–1288.

Johnson GD. 1984. Percoidei: development and relationships. In: Moser HG, Richards WJ, Cohen DM, Fahay MP, Kendall AW, Jr., Richardson SL, eds. *Ontogeny and systematics of fishes*, pp. 464–498. Special Publication No. 1. Lawrence, KS: American Society of Ichthyologists and Herpetologists.

Johnson GD. 1986. Scombroid phylogeny: an alternative hypothesis. *Bull Mar Sci* 39:1–41.

Johnson GD. 1992. Monophyly of the euteleostean clades – Neoteleostei, Eurypterygii, and Ctenosquamata. *Copeia* 1992:8–25.

Johnson GD, Britz R. 2005. Leis' conundrum: homology of the clavus of the ocean sunfishes. 2. Ontogeny of the median fins and axial skeleton of *Ranzania laevis* (Teleostei, Tetraodontiformes, Molidae). *J Morph* 266:11–21.

Johnson GD, Brothers EB. 1993. *Schindleria*: a paedomorphic goby (Teleostei; Gobioidei). *Bull Mar Sci* 52:441–471.

Johnson GD, Patterson C. 1993a. Percomorph phylogeny: a survey of acanthomorphs and a new proposal. *Bull Mar Sci* 52:554–626.

Johnson GD, Patterson C. 1993b. Relationships of lower euteleostean fishes. In: Stiassny MLJ, Parenti LR, Johnson GD, eds. *Interrelationships of fishes*, pp. 251–332. San Diego: Academic Press.

Johnson GD, Patterson C. 1996. Relationships of lower euteleostean fishes. In: Stiassny MLJ, Parenti LR, Johnson GD, eds. *Interrelationships of fishes*, pp. 251–332. San Diego: Academic Press.

Johnson JE, Jensen BL. 1991. Hatcheries for endangered freshwater fishes. In: Minckley WL, Deacon JE, eds. *Battle against extinction: native fish management in the American West*, pp. 199–217. Tucson, AZ: University of Arizona Press.

Johnson NS, Luehring MA, Siefkes MJ, Li WM. 2006. Mating pheromone reception and induced behavior in ovulating female sea lampreys. *N Am J Fish Manag* 26:88–96.

Johnson OW, Ruckelshaus MH, Grant WS, et al. 1999. *Status review of coastal cutthroat trout from Washington, Oregon, and California*. NOAA Technical Memorandum No. NMFS-NWFSC-37. Springfield, VA: National Technical Information Service.

Johnson RH, Nelson DR. 1973. Agonistic display in the gray reef shark, *Carcharhinus menisorrah*, and its relationship to attacks on man. *Copeia* 1973:76–84.

Johnson TC, Scholz CA, Talbot MR, et al. 1996. Late Pleistocene desiccation of lake Victoria and rapid evolution of cichlid fishes. *Science* 273:1091–1093.

Johnston CE. 1994. The benefit to some minnows of spawning in the nests of other species. *Env Biol Fish* 40:213–218.

Johnston CE, Johnson DL. 2000. Sound production during the spawning season in cavity-nesting darters of the subgenus *Catonotus* (Percidae: Etheostoma). *Copeia* 2000:475–481.

Johnston CE, Phillips CT. 2003. Sound production in sturgeon *Scaphirhynchus albus* and *S. platorynchus* (Acipenseridae). *Env Biol Fish* 68:59–64.

Jones AG, Avise JC. 2001. Mating systems and sexual selection in male-pregnant pipefish and seahorses: insights from microsatellite-based studies of maternity. *J Heredity* 92:150–158.

Jones AG, Ostluund-Nilsson S, Avise JC. 1998. A microsatellite assessment of sneaked fertilization and egg thievery in the fifteenspine stickleback. *Evolution* 52:848–858.

Jones AG, Walker D, Avise JC. 2001. Genetic evidence for extreme polyandry and extraordinary sex-role reversal in a pipefish. *Proc Roy Soc Lond B Biol Sci* 268:2531–2535.

Jones CM, Grutter AS, Cribb TH. 2004. Cleaner fish become hosts: a novel form of parasite transmission. *Coral Reefs* 23:521–529.

Jones DR, Schwarzfeld T. 1974. The oxygen cost to the metabolism and efficiency of breathing in trout (*Salmo gairdneri*). *Respir Physiol* 21:241–254.

Jones EBD, III, Helfman GS, Harper JO, et al. 1999a. The effects of riparian deforestation on fish assemblages in southern Appalachian streams. *Conserv Biol* 13:1454–1465.

Jones EC. 1971. *Isistius brasiliensis*, a squaloid shark, the probable cause of crater wounds on fishes and cetaceans. *Fish Bull US* 69:791–798.

Jones FRH. 1957. The swimbladder. In: Brown ME, ed. *Behavior. Physiology of fishes*, Vol. 2, pp. 305–322. New York: Academic Press.

Jones GP, Ferrell DJ, Sale PF. 1991. Fish predation and its impact on the invertebrates of coral reefs and adjacent sediments. In: Sale PF, ed. *The ecology of fishes on coral reefs*, pp. 156–179. San Diego, CA: Academic Press.

Jones GP, Milicich MJ, Emslie MJ, Lunow C. 1999b. Self-recruitment in a coral reef fish population. *Nature* 402:802–804.

Jones GP, Planes S, Thorrold SR. 2005. Coral reef fish larvae settle close to home. *Curr Biol* 15:1314–1318.

Jones PL, Sidell BD. 1982. Metabolic responses of striped bass (*Morone saxatilis*) to temperature acclimation. II. Alterations in metabolic carbon sources and distributions of fiber types in locomotory muscle. *J Exp Zool* 219:163–171.

Jones RS. 1968. Ecological relationships in Hawaiian and Johnston Island Acanthuridae (surgeonfishes). *Micronesica* 4:309–361.

Jonsson B, Jonsson N. 2003. Migratory Atlantic salmon as vectors for the transfer of energy and nutrients between freshwater and marine environments. *Freshw Biol* 48:21–27.

Jonsson B, Jonsson N. 2006. Cultured Atlantic salmon in nature: a review of their ecology and interaction with wild fish. *ICES J Mar Sci* 63:1162–1181.

Jordan DS. 1905. *A guide to the study of fishes*, Vols I and II. New York: Henry Holt & Co.

Jordan DS. 1922. *The days of a man*, Vols I and II. New York: World Book Co.

Jordan DS. 1923. A classification of fishes including families and genera as far as known. *Stanford Univ Pubs Biol Sci* 3:77–243.

Jordan DS, Evermann BW. 1903. The shore fishes of the Hawaiian Islands, with a General Account of the Fish Fauna. *Bull U. S. Fish Commission* XXIII: for 1903. Part I.

Jordan R, Howe D, Juanes F, Stauffer J, Loew E. 2004. Ultraviolet radiation enhances zooplanktivory rate in ultraviolet sensitive cichlids *Afr J Ecol* 42:228–231.

Jørgensen JM. 2011. Morphology of electroreceptive sensory organs. In: Farrell AP, Stevens ED, Cech JJ, Richards JG, eds. *Encyclopedia of fish physiology: from genome to environment*, pp. 350–358. Boston: Elsevier/Academic Press.

Jørgensen JM, Lomholt JP, Weber RE, Malte H, eds. 1998. *The biology of hagfishes*. London: Chapman & Hall.

Joung S-J, Chen C-T, Clark E, Uchida S, Huang WYP. 1996. The whale shark, *Rhincodon typus*, is a livebearer: 300 embryos found in one 'megamamma' supreme. *Env Biol Fish* 46:219–223.

Jutfelt F. 2011. Barrier function of the gut. In: Farrell AP, Stevens ED, Cech JJ, Richards JG, eds. *Encyclopedia of fish physiology: from genome to environment*, pp. 1322–1331. Boston: Elsevier/ Academic Press.

Jung J-H, Kim M, Yim UH, et al. 2011. Biomarker responses in pelagic and benthic fish over 1 year following the *Hebei Spirit* oil spill (Taean, Korea). *Mar Pollut Bull* 62:1859–1866.

Kaatz IM. 2011. How fishes use sound: quiet to loud and simple to complex signalling. In: Farrell AP, Stevens ED, Cech JJ, Richards JG, eds. *Encyclopedia of fish physiology: from genome to environment*, pp. 684–691. Boston: Elsevier/Academic Press.

Kajiura SM, Holland KN. 2002. Electroreception in juvenile scalloped hammerhead and sandbar sharks. *J Exp Biol* 205:3609–3621.

Káldy J, Mozsar A, Fazekas G, et al. 2020. Hybridization of the Russian Sturgeon (*Acipenser gueldenstaedtii*, Brandt and Ratzeberg, 1833) and American Paddlefish (*Polyodon spathula*, Walbaum 1792) and evaluation of their progeny. *Genes* 11:753. **https://doi.org/10.3390/genes11070753**.

Kalmijn AJ. 1971. The electric sense of sharks and rays. *J Exp Biol* 55:371–383.

Kalmijn AJ. 1978. Electric and magnetic sensory world of sharks, skates, and rays. In: Hodgson ES, Mathewson RF, eds. *Sensory biology of sharks, skates, and rays*, pp. 507–528. Arlington, VA: Office of Naval Research.

Kalmijn AJ. 1982. Electric and magnetic field detection in elasmobranch fishes. *Science* 218:916–918.

Kalmijn AJ. 2003. Physical principles of electric, magnetic, and near-field acoustic signals in elasmobranch fishes. In: Collins SP, Marshall NJ, eds. *Sensory processing in aquatic environments*, pp. 77–91. New York: Springer-Verlag.

Kamler E. 1991. *Early life history of fish: an energetics approach*. London: Chapman & Hall.

Kanter MJ, Coombs S. 2003. Rheotaxis and prey detection in uniform currents by Lake Michigan mottled sculpin (*Cottus bairdi*). *J Exp Biol* 206:59–71.

Kapoor BG, Smit H, Verighina IA. 1975. The alimentary canal and digestion in teleosts. *Adv Mar Biol* 13:109–239.

Kappel CV. 2005. Losing pieces of the puzzle: threats to marine, estuarine, and diadromous species. *Frontiers Ecol Env* 3:275–282.

Karino K, Orita K, Sato A. 2006. Long tails affect swimming performance and habitat choice in the male guppy. *Zool Sci* 23:255–260.

Karl SA, Avise JC. 1993. PCR-based assays of Mendelian polymorphisms from anonymous single-copy nuclear DNA: techniques and applications for population genetics. *Mol Biol Evol* 10:342–361.

Karlsson S, Diserud OH, Fiske P, Hindar K. 2016. Widespread genetic introgression of escaped farmed Atlantic salmon in wild salmon populations. *ICES J Mar Sci* 73:2488–2498.

Karplus I. 1979. The tactile communication between *Cryptocentrus steinitzi* (Pisces, Gobiidae) and *Alpheus purpurilenticularis* (Crustacea, Alpheidae). *Z Tierpsychol* 49:173–196.

Karplus I, Algom D. 1981. Visual cues for predator face recognition by reef fishes. *Z Tierpsychol* 55:343–364.

Karplus I, Goren M, Algom D. 1982. A preliminary experimental analysis of predator face recognition by *Chromis caeruleus* (Pisces, Pomacentridae). *Z Tierpsychol* 58:53–65.

Karr JR. 1981. Assessment of biotic integrity using fish communities. *Fisheries* 6:21–27.

Karr JR. 1991. Biological integrity: a long-neglected aspect of water resource management. *Ecol Applic* 1:66–84.

Karr JR, Chu EW. 1999. *Restoring life in running waters: better biological monitoring*. Washington, DC: Island Press.

Karr JR, Dionne M, Schlosser IJ. 1992. Bottom-up versus top-down regulation of vertebrate populations: lessons from birds and fish. In: Hunter MD, Ohgushi T, eds. *Effects of resource distribution on animal–plant interactions*, pp. 243–286. San Diego: Academic Press.

Karr JR, Freemark KE. 1985. Disturbance and vertebrates: an integrative perspective. In: Pickett STA, White PS, eds. *The ecology of natural disturbance and patch dynamics*, pp. 153–168. New York: Academic Press.

Karr JR, Toth LA, Garman GD. 1982. *Habitat preservation for midwest stream fishes: principles and guidelines*. Corvallis, OR: US Environmental Protection Agency 600/3-83-006:1–120.

Kasumyan AO. 2004. The olfactory system in fish: structure, function, and role in behavior. *J Ichthyol* 44(2):S180–S223

Kasumyan AO. 2008. Sounds and sound production in fishes. *J Ichthyol* 48(11):981–1030.

Kasumyan AO. 2019. The taste system in fishes and the effects of environmental variables. *J Fish Biol* 95:155–178.

Katsiadaki I, Sebire M. 2011. Sexual behavior in fish. In: Farrell AP, Stevens ED, Cech JJ, Richards JG, eds. *Encyclopedia of fish physiology: from genome to environment*, pp. 656–661. Boston: Elsevier/Academic Press.

Katula RS, Page LM. 1998. Nest association between a large predator, the bowfin (*Amia calva*), and its prey, the golden shiner (*Notemigonus crysoleucas*). *Copeia* 1998:220–221.

Kaufman LS. 1977. The threespot damselfish: effects on benthic biota of Caribbean coral reefs. *Proc 3rd Internat Coral Reef Symp* 1:559–564.

Kaufman LS. 1983. Effects of Hurricane Allen on reef fish assemblages near Discovery Bay, Jamaica. *Coral Reefs* 2:43–47.

Kaufman LS. 1992. Catastrophic change in species-rich freshwater ecosystems: the lessons of Lake Victoria. *BioScience* 42:846–858.

Kaufman LS. 2003. Evolutionary footprints in ecological time. In: Crisman TL, Chapman LJ, Chapman CA, et al., eds. *Conservation, ecology, and management of African fresh waters*, pp. 247–267. Gainesville, FL: University Press of Florida.

Kaufman LS, Ebersole J, Beets J, McIvor CC. 1992. A key phase in the recruitment dynamics of coral reef fishes: post-settlement transition. *Env Biol Fish* 34:109–118.

Kaunda-Arara B, Rose GA. 2004. Homing and site fidelity in the greasy grouper *Epinephelus tauvina* (Serranidae) within a marine protected area in coastal Kenya. *Mar Ecol Prog Ser* 277:245–251.

Kavanagh KD, Alford RA. 2003. Sensory and skeletal development and growth in relation to the duration of the embryonic and larval stages in damselfishes (Pomacentridae). *Biol J Linn Soc* 80:187–206.

Kawanabe H, Hori M, Nagoshi M, eds. 1997. *Fish communities in Lake Tanganyika*. Kyoto, Japan: Kyoto University Press.

Keast A, Harker J. 1977. Fish distribution and benthic invertebrate biomass relative to depth in an Ontario lake. *Env Biol Fish* 2:235–240.

Keefe M, Able KW. 1993. Patterns of metamorphosis in summer flounder, *Paralichthys dentatus*. *J Fish Biol* 42:713–728.

Keefer ML, Caudill CC, Peery CA, Bjornn TC. 2006. Route selection in a large river during the homing migration of Chinook salmon (*Oncorhynchus tshawytscha*). *Can J Fish Aquat Sci* 63:1752–1762.

Keeney DB, Heist E. 2006. Worldwide phylogeography of blacktip shark (*Carcharhinus limbatus*) inferred from mitochondrial DNA reveals isolation of western Atlantic populations coupled with recent Pacific dispersal. *Molec Ecol* 15:3669–3679.

Keenleyside MHA. 1979. *Diversity and adaptation in fish behaviour*. Berlin: Springer-Verlag.

Keenleyside MHA, ed. 1991. *Cichlid fishes: behaviour, ecology and evolution*. London: Chapman & Hall.

Keith P, Allardi J. 1996. Endangered freshwater fish: the situation in France. In: Kirchofer A, Hefti D, eds. *Conservation of endangered freshwater fish in Europe*, pp. 35–54. Berlin: Birkhauser Verlag.

Keller-Costa T, Canário AV, Hubbard PC. 2015. Chemical communication in cichlids: a mini-review. *Gen Comp Endocrinol* 221:64–74.

Kellermann A, North AW. 1994. The contribution of the BIOMASS Programme to Antarctic fish biology. In: El-Sayed SZ, ed. *Southern ocean ecology: the BIOMASS perspective*, pp. 191–209. Cambridge, UK: Cambridge University Press.

Kelley CD, Hourigan TF. 1983. The function of conspicuous coloration in chaetodontid fishes: a new hypothesis. *Anim Behav* 31:615–617.

Kelley JL, Magurran AE. 2003. Learned predator recognition and antipredator responses in fishes. *Fish Fish (Oxf)* 4:216–226.

Kellogg KA, Markert JA, Stauffer JR, Jr., Kocher TD. 1998. Intraspecific brood mixing and reduced polyandry in a maternal mouthbrooding cichlid. *Behav Ecol* 9:309–312.

Kelly DJ, Bothwell ML. 2002. Avoidance of solar ultraviolet radiation by juvenile coho salmon (*Oncorhynchus kisutch*). *Can J Fish Aquat Sci* 59:474–482.

Kelsch SW, Neill WH. 1990. Temperature preference versus acclimation in fishes: selection for changing metabolic optima. *Trans Am Fish Soc* 119:601–610.

Kemp A. 1987. The biology of the Australian lungfish, *Neoceratodus forsteri* (Krefft 1870). In: Bemis WE, Burggren WW, Kemp NE, eds. *The biology and evolution of lungfishes*, pp. 181–198. New York: Alan R. Liss.

Kenaley CP. 2008. Diel vertical migration of the loosejaw dragonfishes (Stomiiformes: Stomiidae: Malacosteinae): a new analysis for rare pelagic taxa. *J Fish Biol* 73:888–901.

Kenaley CP, DeVaney SC, Fjeran TT. 2014. The complex evolutionary history of seeing red: molecular phylogeny and the evolution of an adaptive visual system in deep-sea dragonfishes (Stomiiformes: Stomiidae). *Evolution* 68-4:996–1013.

Kendall AW, Jr. 1979. *Morphological comparisons of North American sea bass larvae (Pisces: Serranidae)*. NOAA Technical Report NMFS Circular 428. Washington, DC: National Oceanic and Atmospheric Administration.

Kendall AW, Jr., Ahlstrom EH, Moser HG. 1984. Early life history stages of fishes and their characters. In: Moser HG, Richards WJ, Cohen DM, Fahay MP, Kendall AW, Jr., Richardson SL, eds. *Ontogeny and systematics of fishes*, pp. 11–22. Special Publication No. 1. Lawrence, KS: American Society of Ichthyologists and Herpetologists.

Kennedy VS. 1990. Anticipated effects of climate change on estuarine and coastal fisheries. *Fisheries* 15(6):16–24.

Keppeler FW, Cruz DA, Dalponti G, Mormul RP. 2016. The role of deterministic factors and stochasticity on the trophic interactions between birds and fish in temporary floodplain ponds. *Hydrobiologia* 773:225–240

Kerfoot, CW, F Yousef, Hobmeier MM, Maki RP, Jarnagin ST, Churchill JH. 2011. Temperature, recreational fishing and diapause egg connections: dispersal of spiny water fleas (*Bythotrephes longimanus*). *Biol Invasions* 13:2513–2531.

Kerfoot CW, Sih A, eds. 1987. *Predation: direct and indirect impacts on aquatic communities*. Hanover, NH: University Press of New England.

Kidd KA, Blanchfield PJ, Mills KH, et al. 2007. Collapse of a fish population after exposure to a synthetic estrogen. *Proc Natl Acad Sci USA* 104:8897–8901.

Kiilerich P, Prunet P. 2011. Corticosteroids. In: Farrell AP, Stevens ED, Cech JJ, Richards JG, eds. *Encyclopedia of fish physiology: from genome to environment*, pp. 1474–1482. Boston: Elsevier/Academic Press.

Kikuchi K, Hamaguchi S. 2013. Sex-determination genes in fish and sex chromosome evolution. *Dev Dyn* 242:339–353.

Kikugawa K, Katoh K, Kuraku S, et al. 2004. Basal jawed vertebrate phylogeny inferred from multiple nuclear DNA-coded genes. *BMC Biol* 2:3–14.

Killen SS, Brown JA. 2006. Energetic cost of reduced foraging under predation threat in newly hatched ocean pout. *Mar Ecol Prog Ser* 321:255–266.

Kimirei I, Nagelkerken I,Trommelen M, et al. 2013. What drives ontogenetic niche shifts of fishes in coral reef ecosystems? *Ecosystems* 16:783–796.

Kimura M. 1983. *The neutral theory of molecular evolution*. Cambridge, UK: Cambridge University Press.

Kindsvater HK, Mangel M, Reynolds JD, Dulvy NK. 2016. Ten Principles from evolutionary ecology essential for effective marine conservation. *Ecol Evol* 6(7):2125–2138.

Kingsford MJ, Leis JM, Shanks A, Lindeman KC, Morgan SG, Pineda J. 2002. Sensory environments, larval abilities and local self-recruitment. *Bull Mar Sci* 70 (suppl S):309–340.

Kirschbaum F, Denizo JP. 2011. Development of electroreceptors and electric organs. In: Farrell AP, Stevens ED, Cech JJ, Richards JG, eds. *Encyclopedia of fish physiology: from genome to environment*, pp. 409–415. Boston: Elsevier/Academic Press.

Kirschvink JL, Walker MM, Chang SB, Dizon AE, Peterson KA. 1985. Chains of single-domain magnetite particles in chinook salmon, *Oncorhynchus tshawytscha*. *J Comp Physiol A* 157:375–381.

Kitano T, Matsuoka N, Saitou N. 1997. Phylogenetic relationships of the genus *Oncorhynchus* inferred from nuclear and mitochondrial markers. *Genes Genet Syst* 72:25–34.

Kitchell JF, ed. 1992. *Food web management. A case study of Lake Mendota*. New York: Springer-Verlag.

Kitchell JF, Neill WH, Dizon AE, Magnuson JJ. 1978. Bioenergetic spectra of skipjack and yellowfin tunas. In: Sharp GD, Dizon AE, eds. *The physiological ecology of tunas*, pp. 357–368. New York: Academic Press.

Kitchell JF, Schindler DE, Ogutu-Ohwayo R, et al. 1997. The Nile perch in Lake Victoria: interaction between predation and fisheries. *Ecol Applic* 7:653–664.

Kitchell JF, Windell JT. 1970. Nutritional value of algae to bluegill sunfish. *Lepomis macrochirus*. *Copeia* 1970:186–190.

Kleckner RC, Kruger WH. 1981. Changes in swim bladder retial morphology in *Anguilla rostrata* during premigration metamorphosis. *J Fish Biol* 18:569–577.

Klimley AP. 1993. Highly directional swimming by scalloped hammerhead sharks, *Sphyrna lewini*, and subsurface irradiance, temperature, bathymetry, and geomagnetic field. *Mar Biol* 117: 1–22.

Klimley AP. 1995. Hammerhead city. *Natur Hist* 104(10):32–39.

Klimley AP. 2015. Shark trails of the Eastern Pacific. *Am Sci* 103:276–283.

Klimley AP, Anderson SD, Pyle P, Henderson RH. 1992. Spatiotemporal patterns of white shark (*Carcharodon carcharias*) predation at the South Farallon Islands, California. *Copeia* 1992:680–690.

Klimley AP, Butler SB, Nelson DR, Stull AT. 1988. Diel movements of scalloped hammerhead sharks, *Sphyrna lewini* Griffith and Smith, to and from a seamount in the Gulf of California. *J Fish Biol* 33:751–761.

Klimley AP, Nelson DR. 1981. Schooling of hammerhead sharks, *Sphyrna lewini*, in the Gulf of California. *Fish Bull US* 79:356–360.

Klimley AP, Pyle P, Anderson SD. 1996. Tail slap and breach: agonistic displays among white sharks? In: Klimley AP, Ainley DG, eds. *Great white sharks: the biology of Carcharodon carcharias*, pp. 241–260. San Diego: Academic Press.

Klinger SA, Magnuson JJ, Gallepp GW. 1982. Survival mechanisms of the central mudminnow (*Umbra limi*), fathead minnow (*Pimephales promelas*) and brook stickleback (*Culaea inconstans*) for low oxygen in winter. *Env Biol Fish* 7:113–120.

Klumpp DW, McKinnon D, Daniel P. 1987. Damselfish territories: zones of high productivity on coral reefs. *Mar Ecol Prog Ser* 40:41–51.

Klumpp DW, Polunin NVC. 1989. Partitioning among grazers of food resources within damselfish territories on a coral reef. *J Exp Mar Biol Ecol* 125:145–169.

Knight TM, McCoy MW, Chase JM, McCoy KA, Holt RD. 2005. Trophic cascades across ecosystems. *Nature* 437:880–883.

Knouft JH, Page LM, Plewa MJ. 2003. Antimicrobial egg cleaning by the fringed darter (Perciformes: Percidae: *Etheostoma crossopterum*): implications of a novel component of parental care in fishes. *Proc Roy Soc Lond B Biol Sci* 270:2405–2411.

Knowlton N, Lang JC, Keller BD. 1990. Case study of natural population collapse: post-hurricane predation on Jamaican staghorn corals. *Smithson Contrib Mar Sci* 31:1–25.

Kobayashi Y, Nakamura M, Sunobe T, et al. 2009. Sex change in the gobiid fish is mediated through rapid switching of gonadotropin receptors from ovarian to testicular portion or vice versa. *Endocrinology* 150(3):1503–1511.

Koboyashi H, Pelster B, Scheid P. 1989. Water and lactate movement in the swimbladder of the eel, *Anguilla anguilla*. *Respir Physiol* 78:45–57.

Koboyashi H, Pelster B, Scheid P. 1990. CO_2 back-diffusion in the rete aids O_2 secretion in the swimbladder of the eel. *Respir Physiol* 79:231–242.

Kodric-Brown A. 1990. Mechanisms of sexual selection: insights from fishes. *Ann Zool Fennici* 27:87–100.

KoebelJW, Jr., Bousquin SG. 2014. The Kissimmee River restoration project and evaluation program, Florida, U.S.A. *Restoration Ecol* 22(3):345–352.

Kohda M, Tanimura M, Kikue-Nakamura M, Yamagishi S. 1995. Sperm drinking by female catfishes: a novel mode of insemination. *Env Biol Fish* 42:1–6.

Kohler CC, Courtenay WR, Jr. 1986a. Regulating introduced aquatic species: a review of past initiatives. *Fisheries* 11(2):34–38.

Kohler CC, Courtenay WR, Jr. 1986b. American Fisheries Society position on introductions of aquatic species. *Fisheries* 11(2):39–42.

Kohler CC, Kohler ST. 1994. Ciguatera tropical fish poisoning: what's happening in the food chain? In: Gerace DT, ed. *Proceedings of the 26th Meeting of the Association of Marine Laboratories of the Caribbean, San Salvador, Bahamas*, pp. 112–125.

Kohler NE, Turner PA. 2001. Shark tagging: a review of conventional methods and studies. *Env Biol Fish* 60:191–223.

Kohler SL, McPeek MA. 1989. Predation risk and the foraging behavior of competing stream insects. *Ecology* 70:1811–1825.

Kohler ST, Kohler CC. 1992. Dead bleached coral provides new surfaces for dinoflagellates implicated in ciguatera fish poisonings. *Env Biol Fish* 35:413–416.

Kolbert E. 2014. *The sixth extinction: an unnatural history*. New York: Picador. 379 pp.

Kon T, Yoshino T. 2002a. Extremely early maturity found in Okinawan gobioid fishes. *Ichthyol Res* 49:224–228.

Kon T, Yoshino T. 2002b. Diversity and evolution of life histories of gobioid fishes from the viewpoint of heterochrony. *Mar Freshw Res* 53:377–402.

Kondolf GM. 1995. Five elements for effective evaluation of stream restoration. *Restoration Ecol* 3:133–136.

Konecki JT, Targett TE. 1989. Eggs and larvae of *Nototheniops larseni* from the spongocoel of a hexactinellid sponge near Hugo Island, Antarctic Peninsula. *Polar Biol* 10:197–198.

Konstantinou H, McEachran JD, Woolley JB. 2000. The systematics and reproductive biology of the *Galeus arae* subspecific complex (Chondrichthyes: Scyliorhinidae). *Env Biol Fish* 57:117–129.

Kornfield I, Carpenter KE. 1984. Cyprinids of Lake Lanao, Philippines: taxonomic validity, evolutionary rates and speciation scenarios. In: Echelle AE, Kornfield I, eds. *Evolution of fish species flocks*, pp. 69–84. Orono, ME: University of Maine Press.

Kornfield I, Smith PF. 2000. African cichlid fishes: model systems for evolutionary biology. *Ann Rev Ecol Syst* 31:163–196.

Kornfield IL, Taylor JN. 1983. A new species of polymorphic fish, *Cichlasoma minckleyi* from Cuatro Cienegas, Mexico (Teleostei: Cichlidae). *Proc Biol Soc Wash* 96:253–269.

Kornis MS, Mercado-Silva N, Vander Zanden MJ. 2012. Twenty years of invasion: a review of round goby *Neogobius melanostomus* biology, spread and ecological Implications. *J Fish Biol* 80:235–285.

Kottelat M. 1988. Two species of cavefishes from northern Thailand in the genera *Nemachilus* and *Homaloptera* (Osteichthyes, Homalopteridae). *Rec Aust Mus* 40:225–231.

Kottelat M. 1997. European freshwater fishes. *Biologia* 52 (suppl 5):1–271.

Kottelat M, Britz R, Hui TH, Witte K-E. 2006. *Paedocypris*, a new genus of Southeast Asian cyprinid fish with a remarkable sexual dimorphism, comprises the world's smallest vertebrate. *Proc Roy Soc Lond B Biol Sci* 273:895–899.

Kottelat M, Freyhof J. 2007. *Handbook of European freshwater fishes.* Cornol, Switzerland: Publications Kottelat.

Kraft DW, Conklin E, Barba E, et al. 2020. Genomics versus mtDNA for resolving stock structure in the silky shark (*Carcharhinus falciformis*). *Peer J* 8:e10186.

Kramer DL, Bryant MJ. 1995. Intestine length in the fishes of a tropical stream: 2. Relationships to diet – the long and short of a convoluted issue. *Env Biol Fish* 42:129–141.

Kramer DL, Chapman MR. 1999. Implications of fish home range size and relocation for marine reserve function. *Env Biol Fish* 55:65–79.

Kramer DL, Lindsey CC, Moodie GEE, Stevens ED. 1978. The fishes and the aquatic environment of the central Amazon basin, with particular reference to respiratory patterns. *Can J Zool* 56:717–729.

Krebs JR, Davies NB, eds. 1991. *Behavioural ecology: an evolutionary approach*, 3rd edn. London: Blackwell Science.

Krejsa RJ, Bringas P, Jr., Slavkin HC. 1990a. The cyclostome model: an interpretation of conodont element structure and function based on cyclostome tooth morphology, function, and life history. *Courier Forsch-Inst Senckenberg* 118:473–492.

Krejsa RJ, Bringas P, Jr., Slavkin HC. 1990b. A neontological interpretation of conodont elements based on agnathan cyclostome tooth structure, function, and development. *Lethaia* 23:359–378.

Krekorian CO, Dunham DW. 1972. Preliminary observations of the reproductive and parental behavior of the spraying characid *Copeina arnoldi* Regan. *Z Tierpsychol* 31:419–437.

Krkosek M, Ford JS, Morton A, Lele S, Myers RA, Lewis MA. 2007. Declining wild salmon populations in relation to parasites from farm salmon. *Science* 318:1772–1775.

Krkosek M, Lewis MA, Morton A, Frazer LN, Volpe JP. 2006. Epizootics of wild fish induced by farm fish. *Proc Natl Acad Sci USA* 103:15506–15510.

Krogdahl A, Sundby A, Bakke AM. 2011. Gut secretion and digestion. In: Farrell AP, Stevens ED, Cech JJ, Richards JG, eds. *Encyclopedia of fish physiology: from genome to environment*, pp. 1301–1310. Boston: Elsevier/Academic Press.

Kröger RHH. 2011. Physiological optics in fishes. In: Farrell AP, Stevens ED, Cech JJ, Richards JG, eds. *Encyclopedia of fish physiology: from genome to environment*, pp. 102—109. Boston: Elsevier/Academic Press.

Kroon FJ, De Graaf M, Liley NR. 2000. Social organization and competition for refuges and nest sites in *Coryphopterus nicholsii* (Gobiidae), a temperate protogynous reef fish. *Env Biol Fish* 57:401–411.

Krueger WH, Oliveira K. 1999. Evidence for environmental sex determination in the American eel, *Anguilla rostrata. Env Biol Fish* 55:381–389.

Kruger RL, Brocksen RW. 1978. Respiratory metabolism of striped bass, *Morone saxatilis* (Walbaum), in relation to temperature. *J Exp Mar Biol Ecol* 31:55–66.

Kuang HK. 1999. Hong Kong. Appendix IV.1. In: Vannuccini S, ed. *Shark utilization, marketing and trade*, pp. 295–326. FAO Fisheries Technical Paper 389. Rome: Food and Agricultural Organization.

Kuba M, Byrne R, Burghardt G. 2010. A new method for studying problem solving and tool use in stingrays (Potamotrygon castexi). *Anim Cogn* 3:507–513.

Kuehne RA. 1962. A classification of streams illustrated by fish distribution in an eastern Kentucky creek. *Ecology* 43:608–614.

Kufel L, Prejs A, Rybak JI, eds. 1997. *Shallow lakes '95: trophic cascades in shallow freshwater and brackish lakes.* Dordrecht: Kluwer Academic.

Kullander SO. 2003. Family cichlidae (cichlids). In: Reis RE, Kullander SO, Ferraris CJ, Jr., eds. *Checklist of the freshwater fishes of South and Central America*, pp. 605–654. Porto Alegre, Brazil: EDIPUCRS.

Kumazawa Y, Nishida M. 2000. Molecular phylogeny of Osteoglossoids: a new model for Gondwanian origin and plate tectonic transportation of the Asian arowana. *Mol Biol Evol* 17:1869–1878.

Kumchoo K, Wongsawad C, Chai JY, Vanittanakom P, Rojanapaibul A. 2005. High prevalence of *Haplorchis taichui* metacercariae in cyprinoid fish from Chiang Mai Province, Thailand. *Southeast Asian J Trop Med Public Health* 36:451–455.

Kuparinen A, Merilä J. 2008. The role of fisheries-induced evolution. *Science* 320:47–48.

Kuriyama PT, Zwolinski JP, Hill KT, Crone PR. 2020. *Assessment of the Pacific sardine resource in 2020 for U.S. management in 2020-2021.* Portland, OR: Pacific Fishery Management Council. **https://www.pcouncil.org/coastal-pelagic-species/stock-assessment-and-fishery-evaluation-safe-documents**.

Kurlansky M. 1997. *Cod: a biography of the fish that changed the world.* New York: Walker & Co.

Kushlan JA. 1976. Wading bird predation in a seasonally fluctuating pond. *Auk* 93:464–467.

Kushlan JA. 1979. Design and management of continental wildlife reserves: lessons from the Everglades. *Biol Conserv* 15:281–290.

Kuwamura T. 1984. Social structure of the protogynous fish *Labroides dimidiatus. Publ Seto Mar Biol Lab* 29:117–177.

Laberge F, Hara TJ. 2003. Behavioural and electrophysiological response to F-prostaglandins, putative spawning pheromones, in three salmonid fishes. *J Fish Biol* 62:206–221.

Lack M, Sant G. 2001. Patagonian toothfish: are conservation and trade measures working? *TRAFFIC Bull* 19(1):1–21.

Ladich F, Bass AH. 2011. Vocal behavior of fishes: anatomy and physiology. In: Farrell AP, Stevens ED, Cech JJ, Richards JG, eds. *Encyclopedia of fish physiology: from genome to environment*, pp. 321–328. Boston: Elsevier/Academic Press.

Ladich F, Collin SP, Møller P, Kapoor BG, eds. 2006. *Communication in fishes*, Vols 1 and 2. Enfield, NH:Science Publishers.

Ladich F, Fine ML. 2006. Sound-generating mechanisms in fishes: a unique diversity in vertebrates. In: Ladich F, Collin SP, Møller P, Kapoor BG, eds. *Communication in fishes*, Vol. 1, pp. 3–43. Enfield, NH:Science Publishers.

Ladich F, Yan HY. 1998. Correlation between auditory sensitivity and vocalization in anabantoid fishes. *J Comp Physiol A* 182:737–746.

Laerm J, Freeman BJ. 1986. *Fishes of the Okefenokee swamp.* Athens, GA: University of Georgia Press.

Lafferty KD, Dobson AP, Kuris AM. 2006. Parasites dominate food web links. *Proc Natl Acad Sci USA* 103:11211–11216.

Lafferty KD, Morris AK. 1996. Altered behavior of parasitized killifish increases susceptibility to predation by bird final hosts. *Ecology* 77:1390–1397.

Lafferty KD, Page CJ. 1997. Predation on the endangered tidewater goby, *Eucyclogobius newberryi*, by the introduced African clawed frog, *Xenopus laevis*, with notes on the frog's parasites. *Copeia* 1997:589–592.

Lafferty KD, Swift CC, Ambrose RF. 1999. Extirpation and recolonization in a metapopulation of an endangered fish, the tidewater goby. *Conserv Biol* 13:1447–1453.

Lagler KF. 1947. Scale characters of the families of Great Lakes fishes. *Trans Am Microscop Soc* 66:149–171.

Lagler KF, Bardach JE, Miller RR, Passino DRM. 1977. *Ichthyology*, 2nd edn. New York: Wiley & Sons.

Lamboj A. 2004. *The Cichlid fishes of western Africa*. Bornheim, Germany: Birgit Schmettkamp Verlag.

Lammens EHRR, de Nie HW, Vijverberg J, van Dense WLT. 1985. Resource partitioning and niche shifts of bream (*Abramis brama*) and eel (*Anguilla anguilla*) mediated by predation of smelt (*Osmerus eperlanus*) on *Daphnia hyalina*. *Can J Fish Aquat Sci* 42:1342–1351.

Landeau L, Terborgh J. 1986. Oddity and the 'confusion effect' in predation. *Anim Behav* 34:1372–1380.

Landsberg JH. 2002. The effects of harmful algal blooms on aquatic organisms. *Rev Fish Sci* 10:113–390.

Langecker TG, Longley G. 1993. Morphological adaptations of the Texas blind catfishes *Trogloglanis pattersoni* and *Satan eurystomus* (Siluriformes: Ictaluridae) to their underground environment. *Copeia* 1993:976–986.

Langerhans RB, Layman CA, DeWitt TJ. 2005. Male genital size reflects a tradeoff between attracting mates and avoiding predators in two live-bearing fish species. *Proc Natl Acad Sci USA* 102:7618–7623.

Larson S, Christiansen J, Griffing D, Ashe J, Lowry D, Andrews K. 2011. Relatedness and polyandry of sixgill sharks, *Hexanchus griseus*, in an urban estuary. *Conserv Genet* 12:679. **https://doi.org/10.1007/s10592-010-0174-9**.

Lasker R, ed. 1981. *Marine fish larvae*. Seattle: Washington Sea Grant Publications.

Lassig BR. 1983. The effects of a cyclonic storm on coral reef fish assemblages. *Env Biol Fish* 9:55–63.

Lassuy DR. 1980. Effects of "farming" behavior in *Eupomacentrus lividus* and *Hemiglyphidodon plagiometopon* on algal community structure. *Bull Mar Sci* 30:304–312.

Lauder GV. 1980. Evolution of the feeding mechanism in primitive actinopterygian fishes: a functional anatomical analysis of *Polypterus*, *Lepisosteus* and *Amia*. *J Morphol* 163:283–317.

Lauder GV, Jr. 1981. Form and function: structural analysis in evolutionary morphology. *Paleobiology* 7:430–442.

Lauder GV. 1982. Patterns of evolution in the feeding mechanism of actinopterygian fishes. *Am Zool* 22:275–285.

Lauder GV, Jr. 1983a. Food capture. In: Webb PW, Weihs D, eds. *Fish biomechanics*, pp. 280–311. New York: Praeger.

Lauder GV, Jr. 1983b. Functional design and evolution of the pharyngeal jaw apparatus in euteleostean fishes. *Zool J Linn Soc* 77:1–38.

Lauder GV. 2000. Function of the caudal fin during locomotion in fishes: kinematics, flow visualization, and evolutionary patterns. *Am Zool* 40:101–122.

Lauder GV, Jr., Liem KF. 1981. Prey capture by *Luciocephalus pulcher*: implications for models of jaw protrusion in teleost fishes. *Env Biol Fish* 6:257–268.

Lauder GV, Liem KF. 1983. The evolution and interrelationships of the actinopterygian fishes. *Bull Mus Comp Zool* 150:95–197.

Laughlin RA. 1982. Feeding habits of the blue crab, *Callinectes sapidus* Rathbun, in the Apalachicola estuary, Florida. *Bull Mar Sci* 32:807–822.

Law R. 2000. Fishing, selection, and phenotypic evolution. *ICES J Mar Sci* 57:659–668.

Law R, Stokes K. 2005. Evolutionary impacts of fishing on target populations. In: Norse EA, Crowder LB, eds. *Marine conservation biology: the science of maintaining the sea's biodiversity*, pp. 232–246. Washington, DC: Island Press.

Lawrence JM. 1957. Estimated size of various forage fishes largemouth bass can swallow. *Proc Southeast Assoc Game Fish Comm* 11:220–226.

Lazzaro X. 1987. A review of planktivorous fishes: their evolution, feeding behaviors, selectivities, and impacts. *Hydrobiologia* 146:97–167.

Leaman BM. 1991. Reproductive styles and life history variables relative to exploitation and management of *Sebastes* stocks. *Env Biol Fish* 30:253–271.

Leatherland JF, Farbridge KJ, Boujard T. 1992. Lunar and semi-lunar rhythms in fishes. In: Ali MA, ed. *Rhythms in fishes*, pp. 83–107. New York: Plenum Press.

Le Bihanic F, Clérandeau C, LeManach K, et al. 2014. Developmental toxicity of PAH mixtures in fish early life stages. Part II: adverse effects in Japanese medaka. *Environ Sci Pollut Res* 21(24):13732–13743.

Lecchini D. 2005. Spatial and behavioural patterns of reef habitat settlement by fish larvae. *Mar Ecol Prog Ser* 301:247–252.

Lee DS, Gilbert CR, Hocutt CH, Jenkins RE, McAllister DE, Stauffer JR, Jr. 1980. *Atlas of North American fresh water fishes*. Raleigh, NC: State Museum of Natural History.

Lee TN, Williams E. 1999. Mean distribution and seasonal variability of coastal currents and temperature in the Florida Keys with implications for larval recruitment. *Bull Mar Sci* 64: 35–56.

Lee WJ, Kocher TD. 1995. Complete sequence of a sea lamprey (*Petromyzon marinus*) mitochondrial genome: early establishment of the vertebrate genome organization. *Genetics* 139:873–887.

Lee-Jenkins SSY, Smith ML, Wisenden BD, Wong A, Godin J-GJ. 2015. Genetic evidence for mixed broods and extra-pair matings in a socially monogamous biparental cichlid fish. *Behaviour* 152(11):1507–1526.

Lefevre S, Bayley M, McKenzie DJ, Craig JF. 2014. Air breathing fishes. *J Fish Biol* 84:547–553.

Lefevre S, Wang T, Jensen A, et al. 2014. Air-breathing fishes in aquaculture. What can we learn from physiology? *J Fish Biol* 84:705–731.

Leggett WC. 1977. The ecology of fish migrations. *Ann Rev Ecol Syst* 8:285–308.

Leggett WC, Carscadden JC. 1978. Latitudinal variation in reproductive characteristics of American shad (*Alosa sapidissima*): evidence for population specific life history strategies in fish. *J Fish Res Board Can* 35:1469–1478.

Lehman J-P. 1966. Actinopterygii. In: Piveteau J, ed. *Traite de Paleontologie*, Vol. 4, pp. 1–242. Paris: Masson et Cie.

Leidy RA, Moyle PB. 1997. Conservation status of the world's fish fauna: an overview. In: Fiedler PL, Kareiva PM, eds. *Conservation biology for the coming decade*, pp. 187–227. New York: Chapman & Hall.

Leis JM. 1991. The pelagic stage of reef fishes: the larval biology of coral reef fishes. In: Sale PF, ed. *The ecology of fishes on coral reefs*, pp. 183–230. San Diego, CA: Academic Press.

Leis JM, Carson-Ewart BM, eds. 2000a. *The larvae of Indo-Pacific coastal fishes. An identification guide to marine fish larvae.* Fauna Malesiana Handbooks. Leiden, the Netherlands: E. J. Brill.

Leis JM, Carson-Ewart BM. 2000b. Behavior of pelagic larvae of four coral-reef fish species in the ocean and an atoll lagoon. *Coral Reefs* 19:147–257.

Leis JM, Carson-Ewart BM. 2002. In situ settlement behaviour of damselfish (Pomacentridae) larvae. *J Fish Biol* 61:325–346.

Leis JM, Carson-Ewart BM, Hay AC, et al. 2003. Coral-reef sounds enable nocturnal navigation by some reef-fish larvae in some places and at some times. *J Fish Biol* 63:724–737.

Leis JM, Hay AC, Lockett MM, Chen J-P, Fang L-S. 2007. Ontogeny of swimming speed in larvae of pelagic-spawning, tropical, marine fishes. *Mar Ecol Prog Ser* 349:255–267.

Leis JM, Lockett MM. 2005. Localization of reef sounds by settlement-stage larvae of coral-reef fishes (Pomacentridae). *Bull Mar Sci* 76:715–724.

Lema SC, Nevitt GA. 2006. Testing an ecophysiological mechanism of morphological plasticity in Pupfish and its relevance to conservation efforts for endangered Devils Hole Pupfish. *J Exp Biol* 209:3499–3509.

Lessa RF, Santana M, Paglerani R. 1999. Age, growth, and stock structure of the oceanic white tip shark, *Carcharhinus longimanus*, from the southwestern equatorial Atlantic. *Fish Res* 42:21–30.

Lesser MP. 2007. Coral reef bleaching and global climate change: can corals survive the next century? *Proc Natl Acad Sci USA* 104:5259–5260.

Lessios HA. 2008. The great American schism: Divergence of marine organisms after the rise of the Central American Isthmus. *Ann Rev Ecol Syst* 39:63–91.

Lessios HA, Robertson DR. 2006. Crossing the impassible: genetic connections in 20 reef fishes across the eastern Pacific barrier. *Proc Roy Soc Lond B Biol Sci* 273:2201–2208.

Lévêque C. 1997. *Biodiversity dynamics and conservation: the freshwater fish of tropical Africa.* Cambridge, UK: Cambridge University Press.

Lévêque C, Oberdorff T, Paugy D, Stiassny MLJ, Tedesco PA. 2008. Global diversity of fish (*Pisces*) in freshwater. *Hydrobiologia* 595:545–567.

Lévêque C, Paugy D. 2006. *Les poissons des eaux continentales africaines. Diversité, écologie et utilisation par l'homme.* Paris: IRD Éditions.

Lever C. 1996. *Naturalized fishes of the world.* San Diego, CA: Academic Press.

Levi PS, Tank JL, Tiegs SD, Chaloner DT, Lamberti GA. 2013. Biogeochemical transformation of a nutrient subsidy: salmon, streams, and nitrification. *Biogeochemistry* 113:643–655.

Levin PS, Grimes CB. 2002. Reef fish ecology and grouper conservation and management. In: Sale PF, ed. *Coral reef fishes: dynamics and diversity in a complex ecosystem*, pp. 377–389. San Diego, CA: Academic Press.

Levine JS, Lobel PS, MacNichol EF, Jr. 1980. Visual communication in fishes. In: Ali MA, ed. *Environmental physiology of fishes*, pp. 447–475. New York: Plenum.

Levitan DR. 1992. Community structure in times past: influence of human fishing pressure on algal-urchin interactions. *Ecology* 73:1597–1605.

Leviton AE, Gibbs RH, Jr., Heal E, Dawson CE. 1985. Standards in herpetology and ichthyology. Part I. Standard symbolic codes for institutional resource collections in herpetology and ichthyology. *Copeia* 1985:802–832.

Lewis D, Reinthal P, Trendall J. 1986. *A guide to the fishes of Lake Malawi National Park.* Gland, Switzerland: World Wildlife Federation.

Lewis JM, Ewart KV, Driedzic WR. 2004. Freeze resistance in rainbow smelt (*Osmerus mordax*): seasonal pattern of glycerol and antifreeze protein levels and liver enzyme activity associated with glycerol production. *Physiol Biochem Zool* 77:415–422.

Lewis RJ. 2001. The changing face of ciguatera. *Toxicon* 39:97–106.

Li W-H. 1997. *Molecular evolution.* Sunderland, MA: Sinauer Associates.

Liao JC, Beal DN, Lauder GV, Triantafyllou MS. 2003. Fish exploiting vortices decrease muscle activity. *Science* 302:1566–1569.

Liao JC, Lauder GV. 2000. Function of the heterocercal tail in white sturgeon: flow visualization during steady swimming and vertical maneuvering. *J Exp Biol* 203:3585–3594.

Lichatowich J. 1999. *Salmon without rivers: a history of the Pacific salmon crisis.* Covelo, CA: Island Press.

Liddell WD, Ohlhorst SO. 1993. Ten years of disturbance and change on a Jamaican fringing reef. *Proc 7th Internat Coral Reef Symp* 1:144–150.

Liem KF. 1967. Functional morphology of the head of the anabantoid teleost fish *Helostoma temmincki*. *J Morphol* 121:135–158.

Liem KF. 1978. Modulatory multiplicity in the functional repertoire of the feeding mechanism in cichlid fishes. *J Morphol* 158: 323–360.

Liem KF, Bemis WE, Walker WF, Grande L. 2001. *Functional anatomy of the vertebrates*, 3rd edn. Belmont, CA: Thomson/Brooks Cole.

Liem KF, Greenwood PH. 1981. A functional approach to the phylogeny of the pharyngognath teleosts. *Am Zool* 21:83–101.

Liem KF, Lauder GV, eds. 1982. Evolutionary morphology of the actinopterygian fishes. *Am Zool* 22:239–345.

Liem KF, Wake DB. 1985. Morphology: current approaches and concepts. In: Hildebrand M, Bramble DM, Liem KF, Wake DB, eds. *Functional vertebrate morphology*, pp. 366–377. Cambridge, MA: Belknap Press.

Lighthill MJ. 1969. Hydromechanics of aquatic animal propulsion. *Ann Rev Fluid Mech* 1:413–446.

Ligtvoet W, Witte F. 1991. Perturbation through predator introduction: effects on the food web and fish yields in Lake Victoria (East Africa). In: Ravera O, ed. *Perturbation and recovery of terrestrial and aquatic ecosystems*, pp. 263–268. Chichester: Elliss Horwood Ltd.

Liley NR. 1982. Chemical communication in fish. *Can J Fish Aquat Sci* 39:22–35.

Lim P, Lek S, Touch ST, Mao SO, Chhouk B. 1999. Diversity and spatial distribution of freshwater fish in Great Lake and Tonle Sap river (Cambodia, Southeast Asia). *Aquat Living Resour* 12:379–386.

Limbaugh C. 1961. Cleaning symbiosis. *Sci Am* 205:42–49.

Lin JJ, Somero GN. 1995. Temperature-dependent changes in expression of thermostable and thermolabile isozymes of cytosolic malate dehydrogenase in the eurythermal goby fish *Gillichthys mirabilis*. *Physiol Zool* 68:114–128.

Lindeman KC, Lee TN, Wilson WD, Claro R, Ault JS. 2000. Transport of larvae originating in southwest Cuba and the Dry Tortugas: evidence for partial retention in grunts and snappers. *Proc Gulf Carib Fish Inst* 52:253–278.

Lindsey CC. 1978. Form, function, and locomotory habits in fish. In: Hoar WS, Randall DJ, eds. *Locomotion. Fish physiology*, Vol. 7, pp. 1–100. New York: Academic Press.

Lindsey CC. 1988. Factors controlling meristic variation. In: Hoar WS, Randall DJ, eds. *The physiology of developing fish, Part B. Viviparity and posthatching juveniles. Fish physiology*, Vol. 11, pp. 197–274. San Diego: Academic Press.

Lindstrom K. 2001. Effects of resource distribution on sexual selection and the cost of reproduction in sandgobies. *Am Natur* 158:64–74.

Lineaweaver TH, Backus RH. 1970. *The natural history of sharks.* Philadelphia: J. B. Lippincott.

Lingham-Soliar T. 2005. Caudal fin in the white shark, *Carcharodon carcharias* (Lamnidae): a dynamic propeller for fast, efficient swimming. *J Morphol* 264:233–252.

Link JS, Garrison LP, Almeida FP. 2002. Ecological interactions between elasmobranchs and groundfish species on the northeastern US continental shelf. I. Evaluating predation. *N Am J Fish Manage* 22:550–562.

Linley TD, Gerringer ME, Yancey PH, Drazen JC, Weinstock CL, Jamieson AJ. 2016. Fishes of the hadal zone including new species, *in situ* observations and depth records of Liparidae. *Deep Sea Res Part I Oceanogr Res Pap* 114:99–110.

Linnaeus C. 1758. *Systema Naturae*, 10th edn. Stockholm, Sweden: Laurentii Salvii (in Latin).

Lintas C, Hirano J, Archer S. 1998. Genetic variation of the European eel (*Anguilla anguilla*). *Mol Mar Biol Biotechnol* 7:263–269.

Lirman D, Biber P. 2000. Seasonal dynamics of macroalgal communities of the Northern Florida Reef Tract. *Bot Mar* 43(4):305–314.

Lisney TJ, Collin SP. 2006. Brain morphology in large pelagic fishes: a comparison between sharks and teleosts. *J Fish Biol* 68:532–554.

Liston JJ. 2004. An overview of the pachycormiform *Leedsichthys*. In: Arratia G, Tintori A, eds. *Mesozoic fishes 3 – Systematics, paleoenvironments and biodiversity*, pp. 379–390. Pfeil, München: Verlag Dr. Friedrich.

Liu C, Zeng Y. 1988. Notes on the Chinese paddlefish *Psephurus gladius* (Martens). *Copeia* 1988:482–484.

Livengood, EJ, Funicelli N, Chapman FA. 2014. The applicability of the U.S. Law Enforcement Management System (LEMIS) database for the protection and management of ornamental fish. *AACL Bioflux* 7(4):268–272.

Lloyd R. 1992. *Pollution and freshwater fish.* Oxford: Fishing News Books.

Lobel PS. 1980. Herbivory by damselfishes and their role in coral reef community ecology. *Bull Mar Sci* 30:273–289.

Lobel PS. 1981. Trophic biology of herbivorous reef fishes: alimentary pH and digestive capabilities. *J Fish Biol* 19:365–397.

Lobel PS. 1989. Ocean current variability and the spawning season of Hawaiian reef fishes. *Env Biol Fish* 24:161–171.

Lobel PS. 1992. Sounds produced by spawning fishes. *Env Biol Fish* 33:351–358.

Lobel PS. 2001. Acoustic behavior of cichlid fishes. *J Aquaricult Aquat Sci* 9:167–186.

Lockett NA. 1977. Adaptations to the deep-sea environment. In: Crescitelli F, ed. *Handbook of sensory physiology*, Vol. VII/5, pp. 67–192. Berlin: Springer-Verlag.

Lodge DM. 1991. Herbivory on freshwater macrophytes. *Aquat Bot* 41:195–224.

Loesch JG. 1987. Overview of life history aspects of anadromous alewife and blueback herring in freshwater habitats. In: Dadswell MJ, Klauda RJ, Moffitt CM, Saunders RL, Rulifson RA, Cooper JE, eds. *Common strategies of anadromous and catadromous fishes*, pp. 89–103. American Fisheries Society Symposium 1. Bethesda, MD: American Fisheries Society.

Loew ER, McFarland WN, Margulies D. 2002. Developmental changes in the visual pigments of the yellowfin tuna, *Thunnus albacares*. *Mar Freshw Behav Physiol* 35:235–246.

Loh T-L, McMurray SE, Henkel TP, Vicente J, Pawlik JR. 2015. Indirect effects of overfishing on Caribbean reefs: sponges overgrow reef-building corals. *Peer J* 3:e901. **https://doi.org/10.7717/peerj.901**.

Loh T-L, Pawlik JR. 2014. Chemical defenses and resource trade-offs structure sponge communities on Caribbean coral reefs. *Proc Natl Acad Sci U. S. Am* 111(11):4151–4156. **https://doi.org/10.1073/pnas.1321626111**.

Loiselle PV, Barlow GW. 1978. Do fishes lek like birds? In: Reese ES, Lighter FJ, eds. *Contrasts in behavior*, pp. 3–75. New York: Wiley-Interscience.

Lomolino MV, Riddle BR, Whittaker RJ. 2017. *Biogeography*, 5th edn. Sunderland, MA: Oxford University Press.

Long JA. 1995. *The rise of fishes.* Baltimore, MD: Johns Hopkins Press.

Longley WH. 1917. Studies upon the biological significance of animal coloration. I. The colors and color changes of West Indian reef-fishes. *J Exp Zool* 23: 33–601.

Longwell AC, Chang S, Hebert A, Hughes JB, Perry D. 1992. Pollution and developmental abnormalities of Atlantic fishes. *Env Biol Fish* 35:1–21.

Lönnstedt OM, McCormick MI, Meekan MG, Ferrari MCO, Chivers DP. 2012. Learn and live: predator experience and feeding history determines prey behavior and survival. *Proc Royal Soc B* 279:2091–2098.

Lorenz K. 1962. The function of color in coral reef fishes. *Proc Roy Inst Great Britain* 39:282–296.

Lorion CM, Markle DF, Reid SB, et al. 2000. Re-description of the presumed-extinct Miller Lake lamprey, *Lampetra minima*. *Copeia* 2000:1019–1028.

Losey GS. 1972. Predation protection in the poison-fang blenny, *Meiacanthus atrodorsalis*, and its mimics *Ecsenius bicolor* and *Runula laudandus* (Blenniidae). *Pacific Sci* 26:129–139.

Losey GS, Jr. 1978. The symbiotic behavior of fishes. In: Mostofsky DI, ed. *The behavior of fish and other aquatic animals*, pp. 1–31. New York: Academic Press.

Losey GS, Jr. 1979. Fish cleaning symbiosis: proximate causes of host behaviour. *Anim Behav* 27:669–685.

Losey GS, Jr. 1987. Cleaning symbiosis. *Symbiosis* 4:229–258.

Losey GS. 2003. Crypsis and communication functions of UV-visible coloration in two coral reef damselfish, *Dascyllus aruanus* and *D. reticulatus*. *Anim Behav* 66:299–307.

Losey GS, Cronin TW, Goldsmith TH, Hyde D, Marshall NJ, McFarland WN. 1999a. The UV visual world of fishes: a review. *J Fish Biol* 54:921–943.

Losey GS, Grutter AS, Rosenquist G, Mahon JL, Zamzow JP. 1999b. Cleaning symbiosis: a review. In: Almada VC, Oliveira RF, Goncalves EJ, eds. *Behaviour and conservation of littoral fishes*, pp. 379–395. Lisboa, Portugal: Instituto Superior de Psicologia Aplicada.

Losey GS, McFarland WN, Loew ER, Zamzow JP, Nelson PA, Marshall NJ. 2003. Visual biology of Hawaiian coral reef fishes. I. Ocular transmission and visual pigments. *Copeia* 2003:433–454.

Lotrich VA. 1973. Growth, production, and community composition of fishes inhabiting a first-, second-, and third-order stream of eastern Kentucky. *Ecol Monogr* 43:377–397.

Lourie SA, Green DM, Vincent ACJ. 2005. Dispersal, habitat differences, and comparative phylogeography of Southeast Asian seahorses (Syngnathidae: *Hippocampus*). *Molec Ecol* 14:1073–1094.

Love MS, Yoklavich M, Thorsteinson LK. 2002. *The rockfishes of the northeast Pacific.* Berkeley, CA: University of California Press.

LoVullo TJ, Stauffer JR, Jr., McKaye KR. 1992. Diet and growth of a brood of *Bagrus meridionalis* Gunther (Siluriformes: Bagridae) in Lake Malawi, Africa. *Copeia* 1992:1084–1088.

Lowe CG, Bray RN. 2006. Movement and activity patterns. In: Allen LG, Pondella DJ, Horn MH, eds. *The ecology of marine fishes: California and adjacent waters*, pp. 524–553. Berkeley, CA: University of California.

Lowe CG, Bray RN, Nelson DR. 1994. Feeding and associated electrical behavior of the Pacific electric ray *Torpedo californica* in the field. *Mar Biol* 120:161–169.

Lowe-McConnell RH. 1987. *Ecological studies in tropical fish communities*. London: Cambridge University Press.

Lowe-McConnell RM. 1997. EAFRO and after: a guide to key events affecting fish communities in Lake Victoria (East Africa). *S Afr J Sci* 93:570–574.

Lozano-Vilano M de L, Contreras-Balderas S. 1993. Four new species of *Cyprinodon* from southern Nuevo León, Mexico, with a key to the *C. eximius* complex (Teleostei: Cyprinodontidae). *Ichthyol Explor Freshw* 4:295–308.

Lu C, Chen CA, Hui C, Tzeng T, Yeh S. 2006. Population genetic structure of the swordfish, *Xiphias gladius*, in the Indian Ocean and West Pacific inferred from the complete DNA sequence of the mitochondrial control region. *Zool Studies* 45:269–279.

Lu LW, Li DQ, Yang LF. 2005. Notes on the discovery of Permian Acipenseriformes in China. *Chinese Sci Bull* 50:1279–1280.

Lu Z. 2011. Physiology of the ear and brain: how fish hear. In: Farrell AP, Stevens ED, Cech JJ, Richards JG, eds. *Encyclopedia of fish physiology: from genome to environment*, pp. 292–297. Boston: Elsevier/Academic Press.

Lucas JR, Benkert KA. 1983. Variable foraging and cleaning behavior by juvenile leatherjackets, *Oligoplites saurus* (Carangidae). *Estuaries* 6:247–250.

Lucas MC, Baras E. 2001. *Migration of freshwater fishes*. Oxford: Blackwell.

Lucas SG. 2015 Thinopus and a critical review of devonian tetrapod footprints. *Ichnos* 22(3-4):136–154.

Luczkovich JJ, Sprague MW, Krahforst CS. 2011. Acoustic behavior. In: Farrell AP, Stevens ED, Cech JJ, Richards JG, eds. *Encyclopedia of fish physiology: from genome to environment*, pp. 311–320. Boston: Elsevier/Academic Press.

Luczkovich JJ, Watters GM, Olla BL. 1991. Seasonal-variation in usage of a common shelter resource by juvenile inquiline snailfish (*Liparis inquilinus*) and red hake (*Urophycis chuss*). *Copeia* 1991:1104–1109.

Lugli M, Torricelli P, Pavan G, Mainardi D. 1997. Sound production during courtship and spawning among freshwater gobiids (Pisces, Gobiidae). *Mar Freshw Behav Physiol* 29:109–126.

Lugli M, Yan HY, Fine ML. 2003. Acoustic communication in two freshwater gobies: the relationship between ambient noise, hearing thresholds and sound spectrum. *J Comp Physiol A* 189:309–320.

Lund R. 1990. Shadows in time – a capsule history of sharks. In: Gruber SH, ed. *Discovering sharks*, pp. 23–28. American Littoral Society Special Publication No. 14. Highlands, NJ: American Littoral Society.

Lund R, Lund WL. 1985. Coelacanths from the Bear Gulch limestone (*Namurian*) of Montana and the evolution of the coelacanthiformes. *Bull Carnegie Mus Natur Hist* 25:1–74.

Lundberg JG. 1993. African–South American freshwater fish clades and continental drift: problems with a paradigm. In: Goldblatt P, ed. *Biological relationships between Africa and South America*, pp. 156–199. New Haven, CT: Yale University Press.

Lundberg JG, Kottelat M, Smith GR, et al. 2000. So many fishes, so little time: an overview of recent ichthyological discovery in continental waters. *Ann Missouri Bot Garden* 87:26–62.

Lundberg JG, McDade LA. 1990. Systematics. In: Schreck CB, Moyle PB, eds. *Methods for fish biology*, pp. 65–108. Bethesda, MD: American Fisheries Society.

Luo J, Sanetra M, Schartl M, Meyer A. 2005. Strong reproductive skew among males in the multiply mated swordtail *Xiphophorus multilineatus* (Teleostei). *J Heredity* 96:346–355.

Lupes SC, Davis MW, Olla BL, Schreck CB. 2006. Capture-related stressors impair immune system function in sablefish. *Trans Am Fish Soc* 135:129–138.

Luque GM, Donlan CJ. 2019. The characterization of seafood mislabelling: a global meta-analysis. *Biol Conserv* 236:556–570.

Lutcavage ME, Brill RW, Skomal GB, et al. 1999. Results of pop-up satellite tagging of spawning size class fish in the Gulf of Maine: do North Atlantic bluefin tuna spawn in the mid-Atlantic? *Can J Fish Aquat Sci* 56:173–177.

Lydeard C, Mayden RL. 1995. A diverse and endangered aquatic ecosystem of the southeast United States. *Conserv Biol* 9:800–805.

Lythgoe JN. 1979. *The ecology of vision*. Oxford: Clarendon Press.

Lythgoe JN, Shand J. 1983. Diel colour changes in the neon tetra *Paracheirodon innesi*. *Env Biol Fish* 8:249–254.

Maan ME, Seehausen O, Van Alphen JJM. 2010. Female mating preferences and male coloration covary with water transparency in a Lake Victoria cichlid fish. *Biol J Linnean Soc* 99:398–406.

Mabee PM. 1993. Phylogenetic interpretation of ontogenetic change: sorting out the actual and artefactual in an empirical case study of centrarchid fishes. *Zool J Linn Soc* 107:175–291.

MacAvoy SE, Macko SA, Mclninch SP, Garman GC. 2000.Marine nutrient contributions to freshwater apex predators. *Oecologia* 122:568–573.

Macchi GJ, Romano LA, Christiansen HE. 1992. Melano-macrophage centres in whitemouth croaker, *Micropogonias furnieri*, as biological indicators of environmental changes. *J Fish Biol* 40:971–973.

Mace GM, Hudson EJ. 1999. Attitudes toward sustainability and extinction. *Conserv Biol* 13:242–246.

Maclean N, Talwar S. 1984. Injection of cloned genes into rainbow trout eggs. *J Embryol Experim Morphol* 82:187.

Madden CJ, Day JW, Jr., Randell JM. 1988. Freshwater and marine coupling in estuaries of the Mississippi River deltaic plain. *Limnol Oceanogr* 33:982–1004.

Maddison WP. 1997. Gene trees in species trees. *Syst Biol* 46:523–536.

Maddock MB, Schwartz FJ. 1996. Elasmobranch cytogenetics: methods and chromosome numbers. *Bull Mar Sci* 58:147–155.

Magellan K, Pinchuck S, Swartz ER. 2014. Short and long-term strategies to facilitate aerial exposure in a galaxiid. *J Fish Biol* 84:748–758.

Maggio T. 2000. *Mattanza: love and death in the Sea of Sicily*. Cambridge, MA: Perseus.

Magnhagen C. 1988. Changes in foraging as a response to predation risk in two gobiid fish species, *Pomatoschistus minutus* and *Gobius niger*. *Mar Ecol Prog Ser* 49:21–26.

Magnhagen C. 1992. Alternative reproductive behaviour in the common goby, *Pomatoschistus microps*: an ontogenetic gradient? *Anim Behav* 44:182–184.

Magnhagen C, Vestergaard K. 1991. Risk taking in relation to reproductive investments and future reproductive opportunities: field experiments on nest-guarding common gobies, *Pomatoschistus microps*. *Behav Ecol* 2:351–359.

Magnuson JJ. 2002. Signals from ice cover trends and variability. *Am Fish Soc Sym* 32:3–13.

Magnuson JJ, DeStasio BT. 1997. Thermal niche of fishes and global warming. In: Wood CM, McDonald DG., eds. *Global warming – implications for freshwater and marine fish*, pp. 377–408. SEB Seminar Series. Cambridge, UK: Cambridge University Press.

Magnuson JJ, Meisner JD, Hill DK. 1990. Potential changes in the thermal habitat of Great Lakes fish after global climate warming. *Trans Am Fish Soc* 119:254–264.

Magnuson JJ, Prescott JH. 1966. Courtship, locomotion, feeding, and miscellaneous behaviour of Pacific bonito (*Sarda chiliensis*). *Anim Behav* 14:54–67.

Magnuson JJ, Robertson DM, Benson BJ, et al. 2000. Historical trends in lake and river ice cover in the northern hemisphere. *Science* 289:1743–1746.

Magnusson KP, Ferguson MM. 1987. Genetic analysis of four sympatric morphs of Arctic charr, *Salvelinus alpinus*, from Thingvallavatn, Iceland. *Env Biol Fish* 20:67–73.

Magurran AE. 1986a. Predator inspection behaviour in minnow shoals: differences between populations and individuals. *Behav Ecol Sociobiol* 19:267–273.

Magurran AE. 1986b. The development of shoaling behaviour in the European minnow, *Phoxinus phoxinus*. *J Fish Biol* 29 (suppl A):159–170.

Magurran AE. 1990. The adaptive significance of schooling as an anti-predator defence in fish. *Ann Zool Fennici* 27:51–66.

Magurran AE, Oulton WJ, Pitcher TJ. 1985. Vigilant behaviour and shoal size in minnows. *Z Tierpsychol* 67:167–178.

Magurran AE, Pitcher TJ. 1987. Provenance, shoal size and the sociobiology of predator-evasion behaviour in minnow shoals. *Proc Roy Soc Lond B Biol Sci* 229:439–465.

Maisey JG. 1980. An evaluation of jaw suspension in sharks. *Am Mus Novitates* 2706:17.

Maisey JG. 1986. Heads and tails: a chordate phylogeny. *Cladistics* 2:201–256.

Maisey JG. 1996. *Discovering fossil fishes*. New York: Henry Holt & Co.

Major PF. 1979. Piscivorous predators and disabled prey. *Copeia* 1979:158–160.

Malte H, Lomholt JP. 1998. Ventilation and gas exchange. In: Jørgensen JM, Lomholt JP, Weber RE, Malte H, eds. *The biology of hagfishes*, pp. 223–234. London: Chapman & Hall.

Mandrak NE. 1989. Potential invasions of the Great Lakes by fish species associated with climate warming. *J Great Lakes Res* 15:306–316.

Manel S, Gaggiotti OE, Waples RS. 2005. Assignment methods: matching biological questions with appropriate techniques. *Trends Ecol Evol* 20:136–142.

Manica A. 2002. Filial cannibalism in teleost fish. *Biol Rev* 77:261–277.

Manire CA, Gruber SH. 1991. Many sharks may be headed toward extinction. *Conserv Biol* 4(1):10–11.

Mank JE, Avise JC. 2009. Evolutionary diversity and turn-over of sex determination in teleost fishes. *Sex Dev* 3:60–67.

Mank JE, Promislow DEL, Avise JC. 2005. Phylogenetic perspectives in the evoution of parental care in ray-finned fishes. *Evolution* 59:1570–1578.

Mankiewicz JL, Godwin J, Holler BL, et al. 2013. Masculinizing effect of background color and cortisol in a flatfish with environmental sex-determination. *Integr Comp Biol* 53:755–765. **https://doi.org/10.1093/icb/ict093**.

Manson FJ, Loneragan NR, Skilleter GA, Phinn SR. 2005. An evaluation of the evidence for linkages between mangroves and fisheries: a synthesis of the literature and identification of research directions. *Oceanogr Mar Biol* 43:483–513.

Mapstone BD, Fowler AJ. 1988. Recruitment and the structure of assemblages of fish on coral reefs. *Trends Ecol Evol* 3:72–77.

Marchetti MP. 1999. An experimental study of competition between the native Sacramento perch (*Archoplites interruptus*) and introduced bluegill (*Lepomis macrochirus*). *Biol Invasions* 1:55–65.

Marchetti MP, Light T, Feliciano J, et al. 2001. Homogenization of California's fish fauna through abiotic change. In: Lockwood JL, McKinney ML, *Biological homogenization*, pp. 259–278. New York: Kluwer Plenum/Academic Press.

Marchetti MP, Moyle PB. 2001. Effects of flow regime on fish assemblages in a regulated California stream. *Ecol Applic* 11:530–539.

Marconato A, Shapiro DY. 1996. Sperm allocation, sperm production and fertilization rates in the bucktooth parrotfish. *Anim Behav* 52:971–980.

Margulies D. 1989. Size-specific vulnerability to predation and sensory system development of white seabass, *Atractoscion nobilis*, larvae. *Fish Bull US* 87:537–552.

Margulis L. 1970. *Origin of eukaryotic cells*. New Haven, CT: Yale University Press.

Marie AD, Bernatchez L, Garant D. 2010. Loss of genetic integrity correlates with stocking intensity in brook charr (*Salvelinus fontinalis*). *Mol Ecol* 19:2025–2037.

Mariscal RN. 1970. The nature of the symbiosis between Indo-Pacific anemone fishes and sea anemones. *Mar Biol* 6:58–65.

Marko PB, Lee SC, Rice AM, et al. 2004. Mislabelling of a depleted reef fish. *Nature* 430:309–310.

Marliave JB. 1986. Lack of planktonic dispersal of rocky intertidal fish larvae. *Trans Am Fish Soc* 115:149–154.

Marnane MJ. 2000. Site fidelity and homing behaviour in coral reef cardinalfishes. *J Fish Biol* 57:1590–1600.

Marranzino AN, Webb JF. 2018. Flow sensing in the deep sea: the lateral line system of stomiiform fishes. *Zool J Linnean Soc* 183:945–965.

Marsh E. 1986. Effects of egg size on offspring fitness and maternal fecundity in *Etheostoma spectabile* (Pisces: Percidae). *Copeia* 1986:18–30.

Marshall CR. 1987. A list of fossil and extant dipnoans. In: Bemis WE, Burggren WW, Kemp NE, eds. *The biology and evolution of lungfishes*, pp. 15–23. New York: Alan R. Liss.

Marshall NB. 1954. *Aspects of deep sea biology*. London: Hutchinsons.

Marshall NB. 1960. Swimbladder structure of deepsea fishes in relation to their systematics and biology. *Discovery Rept* 31:112.

Marshall NB. 1971. *Explorations in the life of fishes*. Cambridge, MA: Harvard University Press.

Marshall NB. 1980. *Deep sea biology: developments and perspectives*. New York: Garland STPM Press.

Marshall NJ. 2000. Communication and camouflage with the same 'bright' colours in reef fishes. *Proc Roy Soc Lond B Biol Sci* 355:1243–1248.

Marshall NJ, Cortesi F, de Busserolles F, Siebeck UI, Cheney KL. 2019. Colours and colour vision in reef fishes: past, present and future research directions. *J Fish Biol* 95:5–38

Marshall NJ, Jennings K, McFarland WN, Loew ER, Losey GS. 2003a. Visual biology of Hawaiian coral reef fishes. II. Colors of Hawaiian coral reef fish. *Copeia* 2003:455–466.

Marshall NJ, Jennings K, McFarland WN, Loew ER, Losey GS. 2003b. Visual biology of Hawaiian coral reef fishes. III. Environmental light and an integrated approach to the ecology of reef fish vision. *Copeia* 2003:467–480.

Marshall WS, Grosell M. 2006. Ion transport, osmoregulation, and acid–base balance. In: Evans DH, Claiborne JD, eds. *The physiology of fishes*, 3rd edn, pp. 177–230. Boca Raton, FL: CRC Press.

Marsh-Hunkin KE, Heinz HM, Hawkins MB, Godwin J. 2013. Estrogenic control of behavioral sex change in the bluehead wrasse, *Thalassoma bifasciatum*. *Integr Comp Biol* 53(6):951–959.

Marsh-Matthews E, Skierkowski P, DeMarais A. 2001. Direct evidence for mother-to-embryo transfer of nutrients in the livebearing fish *Gambusia geiseri*. *Copeia* 2001:1–6.

Martel Y. 2003. *Life of pi*. New York: Harcourt.

Martill DM. 1988. *Leedsichthys problematicus*, a giant filter-feeding teleost from the Jurassic of England and France. *Neues Jahrbuch für Geologie und Paläontologie, Monatshefte* 11:670–680.

Martin AP, Echelle AE, Zegers G, Baker S, Keeler-Foster CL. 2012. Dramatic shifts in the gene pool of a managed population of an endangered species may be exacerbated by high genetic load. *Conserv Genet* 13:349–358.

Martin AP, Palumbi SR. 1993. Body size, metabolic rate, generation time, and the molecular clock. *Proc Natl Acad Sci USA* 90:4087–4091.

Martin KLM, Bridges CR. 1999. Respiration in water and air. In: Horn MH, Martin KLM, Chetkowski MA, eds. *Intertidal fishes: life in two worlds*, pp. 54–78. San Diego, CA: Academic Press.

Martin KLM, Swiderski DL. 2001. Beach spawning in fishes: phylogenetic tests of hypotheses. *Am Zool* 41:526–537.

Martin KLM, Van Winkle RC, Drais JE, Lakisic H. 2004. Beach-spawning fishes, terrestrial eggs, and air breathing. *Physiol Biochem Zool* 77:750–759.

Martin RA. 2005. Conservation of freshwater and euryhaline elasmobranchs: a review. *J Mar Biol Assoc UK* 85:1049–1073.

Martin TH, Crowder LB, Dumas CF, Burkholder JM. 1992. Indirect effects of fish on macrophytes in Bays Mountain Lake: evidence for a littoral trophic cascade. *Oecologia* 89:476–481.

Martínez R, Estrada MP, Berlanga J, et al. 1996. Growth enhancement in transgenic tilapia by ectopic expression of tilapia growth hormone. *Mol Mar Biol Biotech* 5:62–70.

Martini FH. 1998. The ecology of hagfishes. In: Jørgensen JM, Lomholt JP, Weber RE, Malte H, eds. *The biology of hagfishes*, pp. 57–77. London: Chapman & Hall.

Martyn RD, Noble RL, Bettoli PW, Maggio RC. 1986. Mapping aquatic weeds with aerial color infrared photography and evaluating their control by grass carp. *J Aquat Plant Manage* 24:46–56.

Maser C, Sedell J. 1994. *From the forest to the sea: the ecology of wood in streams, rivers, estuaries, and oceans*. Delray Beach, FL: St. Lucie Press.

Mason JC. 1976. Response of underyearling coho salmon to supplemental feeding in a natural stream. *J Wildl Manage* 40:775–788.

Matarese AC, Sandknop EM. 1984. Identification of fish eggs. In: Moser HG, Richards WJ, Cohen DM, Fahay MP, Kendall AW, Jr., Richardson SL, eds. *Ontogeny and systematics of fishes*, pp. 27–31. Special Publication No. 1. Lawrence, KS: American Society of Ichthyologists and Herpetologists.

Mathis A, Chivers DP, Smith RJF 1995. Chemical alarm signals: predator deterrents or predator attractants? *Am Natur* 145:994–1005.

Matsen B, Troll R. 1995. *Planet ocean: dancing to the fossil record*. Berkeley, CA: Ten Speed Press.

Matthews WJ. 1986a. Fish faunal "breaks" and stream order in the eastern and central United States. *Env Biol Fish* 17:81–92.

Matthews WJ. 1986b. Fish faunal structure in an Ozark stream: stability, persistence and a catastrophic flood. *Copeia* 1986:388–397.

Matthews WJ. 1998. *Patterns in freshwater fish ecology*. New York: Chapman & Hall.

Matthews WJ, Stewart AJ, Power ME. 1987. Grazing fishes as components of North American stream ecosystems. In: Matthews WJ, Heins DC, eds. *Community and evolutionary ecology of North American stream fishes*, pp. 128–135. Norman, OK: University of Oklahoma Press.

Matthews WJ, Zimmerman EG. 1990. Potential effects of global warming on native fishes of the Southern Great Plains and the Southwest. *Fisheries* 15(6):26–32.

Maximino C, do Carmo Silva RX, dos Santos Campos K, et al. 2019. Sensory ecology of ostariophysan alarm substances. *J Fish Biol* 95:274–286.

May RC. 1974. Larval mortality in marine fishes and the critical period concept. In: Blaxter JHS, ed. *The early life history of fish*, pp. 3–19. New York: Springer-Verlag.

Mayden RL, ed. 1993. *Systematics, historical ecology, and North American freshwater fishes*. Stanford, CT: Stanford University Press.

Mayden RL. 1997. A hierarchy of species concepts: the denouement in the saga of the species problem. In: Claridge MFH, Dawah HA, Wilson MR, eds. *Species: the units of diversity*, pp. 381–423. London: Chapman and Hall.

Mayell M. 2002. Shark gives "virgin birth" in Detroit. **http://news.nationalgeographic.com/news/2002/09/0925_020925_virginshark.htm**, September 26, 2002.

Mayr E. 1942. *Systematics and the origin of species*. New York: Columbia University Press.

Mayr E. 1974. Cladistic analysis or cladistic classification? *Zool Syst Evol-forsch* 12:94–128.

Mazumder A, Taylor WD, McQueen DJ, Lean DRS. 1990. Effects of fish and plankton on lake temperature and mixing depth. *Science* 247:312–315.

McAllister DE. 1968. Evolution of branchiostegals and classification of teleostome fishes. *Bull Natur Mus Can* 221:239.

McCallum M. 2015. Vertebrate biodiversity losses point to a sixth mass extinction. *Biodivers Conserv* 24:2497–2519.

McCarthy JF, Shugart LR, eds. 1990. *Biomarkers of environmental contamination*. Chelsea, MI: Lewis Publishers.

McCauley RD, Fewtrell J, Popper AN. 2003. High intensity anthopogenic sound damages fish ears. *J Acoust Soc Am* 113:638–642.

McClanahan TR. 1995. Fish predators and scavengers of the sea urchin *Echinometra mathaei* in Kenyan coral-reef marine parks. *Env Biol Fish* 43:187–193.

McClanahan TR, Mangi S. 2000. Spillover of exploitable fishes from a marine park and its effect on the adjacent fishery. *Ecol Applic* 10:1792–1805.

McClanahan TR, Muthiga NA. 1988. Changes in Kenyan coral reef community structure and function due to exploitation. *Hydrobiologia* 166:269–276.

McClane AJ, ed. 1974. *McClane's new standard fishing encyclopedia*, 2nd edn. New York: Holt, Rinehart & Winston.

McCleave JD, Arnold GP, Dodson JJ, Neill WH, eds. 1984. *Mechanisms of migration in fishes*. New York: Plenum Press.

McCleave JD, Jellyman DJ. 2002. Discrimination of New Zealand stream waters by glass eels of *Anguilla australis* and *Anguilla dieffenbachii*. *J Fish Biol* 61:785–800.

McCleave JD, Kleckner RC, Castonguay M. 1987. Reproductive sympatry of American and European eels and implications for migration and taxonomy. In: Dadswell MJ, Klauda RJ, Moffitt CM, Saunders RL, Rulifson RA, Cooper JE, eds. *Common strategies of anadromous and catadromous fishes*, pp. 286–297. American Fisheries Society Symposium 1. Bethesda, MD: American Fisheries Society.

McCormick SD. 2011. The hormonal control of osmoregulation in teleost fish. In: Farrell AP, Stevens ED, Cech JJ, Richards JG, eds.

Encyclopedia of fish physiology: from genome to environment, pp. 1466–1473. Boston: Elsevier/Academic Press.

McCormick SD, Saunders RL. 1987. Preparatory physiological adaptations for marine life of salmonids: osmoregulation, growth, and metabolism. In: Dadswell MJ, Klauda RJ, Moffitt CM, Saunders RL, Rulifson RA, Cooper JE, eds. *Common strategies of anadromous and catadromous fishes*, pp. 211–229. American Fisheries Society Symposium 1. Bethesda, MD: American Fisheries Society.

McCosker JE. 1977. Fright posture of the plesiopid fish *Calloplesiops altivelis*: an example of Batesian mimicry. *Science* 197:400–401.

McCosker JE. 1987. The white shark, *Carcharodon carcharias*, has a warm stomach. *Copeia* 1987:195–197.

McCosker JE. 2006. Symbiotic relationships. In: Allen LG, Pondella DJ, Horn MH, eds. *The ecology of marine fishes: California and adjacent waters*, pp. 554–563. Berkeley, CA: University of California.

McCosker JE, Dawson CE. 1975. Biotic passage through the Panama Canal, with particular reference to fishes. *Mar Biol* 30:343–351.

McCosker JE, Lagios MD, eds. 1979. The biology and physiology of the living coelacanth. *Occ Pap Calif Acad Sci* 134:1–175.

McCully HH. 1962. The relationship of the Percidae and the Centrarchidae to the Serranidae as shown by the anatomy of their scales. *Am Zool* 2:430 (abstr).

McCune AR. 1990. Evolutionary novelty and atavism in the *Semionotus* complex: relaxed selection during colonization of an expanding lake. *Evolution* 44:71–85.

McCune AR, Carlson RL. 2004. Twenty ways to lose your bladder: common natural mutants in zebrafish and widespread convergence of swim bladder loss among teleost fishes. *Evol Develop* 6:246–259.

McCune AR, Thomson KS, Olsen PE. 1984. Semionotid fishes from the Mesozoic Great Lakes of North America. In: Echelle AE, Kornfield I, eds. *Evolution of fish species flocks*, pp. 27–44. Orono, ME: University of Maine Press.

McDowall RM. 1978. Generalized tracks and dispersal in biogeography. *Syst Zool* 27:88–104.

McDowall RM. 1987. The occurrence and distribution of diadromy among fishes. In: Dadswell MJ, Klauda RJ, Moffitt CM, Saunders RL, Rulifson RA, Cooper JE, eds. *Common strategies of anadromous and catadromous fishes*, pp. 1–13. American Fisheries Society Symposium 1. Bethesda, MD: American Fisheries Society.

McDowall RM. 1988. *Diadromy in fishes*. London: Croom Helm.

McDowall RM. 1999. Different kinds of diadromy: different kinds of conservation problems. *ICES J Mar Sci* 56:410–413.

McDowall RM. 2000. *Hidden treasures exposed: discovering our freshwater fish fauna*. Thomas Cawthron Memorial Lecture No. 58. Nelson, New Zealand: Cawthron Institute. **www.cawthron.org.nz/Assets/Cawlec2000.pdf**.

McDowall RM. 2004. Ancestry and amphidromy in island freshwater fish faunas. *Fish Fish (Oxf)* 5:75–85.

McDowall RM. 2006. Crying wolf, crying foul, or crying shame: alien salmonids and a biodiversity crisis in the southern cool-temperate galaxioid fishes? *Rev Fish Biol Fisheries* 16:233–422.

McDowall RM. 2007a. On amphidromy, a distinct form of diadromy in aquatic organisms. *Fish Fish (Oxf)* 8:1–13.

McDowall RM. 2007b. Hawaiian stream fishes: the role of amphidromy in history, ecology, and conservation biology. In Evenhuis NL, Fitzsimons JM, eds. *Biology of Hawaiian Streams and Estuaries*. Bishop Museum Bulletin in Cultural and Environmental Studies 3, pp. 3–9. Honolulu: Bishop Museum Press.

McEachran JD, Dunn K, Miyake T. 1996. Interrelationships of batoid fishes. In: Stiassny MLJ, Parenti LR, Johnson GD, eds. *Interrelationships of fishes*, pp. 63–84. San Diego: Academic Press.

McFadden JT, Alexander GR, Shetter DS. 1967. Numerical changes and population regulation in brook trout *Salvelinus fontinalis*. *J Fish Res Board Can* 24:1425–1459.

McFarland WN. 1980. Observations on recruitment of haemulid fishes. *Proc Gulf Carib Fish Inst* 32:132–138.

McFarland WN, Brothers EB, Ogden JC, et al. 1985. Recruitment patterns in young French grunts, *Haemulon flavolineatum* (family Haemulidae) at St. Croix, USVI. *Fish Bull US* 83:413–426.

McFarland WN, Hillis Z-M. 1982. Observations on agonistic behavior between members of juvenile French and white grunts – family Haemulidae. *Bull Mar Sci* 32:255–268.

McFarland WN, Kotchian NM. 1982. Interaction between schools of fish and mysids. *Behav Ecol Sociobiol* 11:71–76.

McFarland WN, Levin SA. 2002. Modelling the effects of current on prey acquisition in planktivorous fishes. *Mar Freshw Behav Physiol* 35:69–85.

McFarland WN, Loew ER. 1983. Wave produced changes in underwater light and their relations to vision. *Env Biol Fish* 8:11–22.

McFarland WN, Ogden JC, Lythgoe JN. 1979. The influence of light on the twilight migrations of grunts. *Env Biol Fish* 4:9–22.

McFarland WN, Wahl CM, Suchanek TH, McAlary FA. 1999. The behavior of animals around twilight with emphasis on coral reef communities. In: Archer SN, ed. *Adaptive mechanisms in ecology of vision*, pp. 583–628. Boston: Kluwer.

McGinn NA, ed. 2002. *Fisheries in a changing climate*. Bethesda, MD: American Fisheries Society.

McGurk MD. 1986. Natural mortality of marine pelagic eggs and larvae: role of spatial patchiness. *Mar Ecol Prog Ser* 34:227–242.

McIntyre PB, Jones LE, Flecker AS, et al. 2007. Fish extinctions alter nutrient recycling in tropical freshwaters. *Proc Natl Acad Sci USA* 104:4461–4466.

McIsaac DO, Quinn TP. 1988. Evidence for a hereditary component in homing behavior of chinook salmon (*Oncorhynchus tshawytscha*). *Can J Fish Aquat Sci* 45:2201–2205.

McKaye KR. 1981a. Death feigning: a unique hunting behavior by the predatory cichlid, *Haplochromis livingstoni* of Lake Malawi. *Env Biol Fish* 6:361–365.

McKaye KR. 1981b. Natural selection and the evolution of interspecific brood care in fishes. In: Alexander RD, Tinkle DW, eds. *Natural selection of social behavior*, pp. 173–183. New York: Chiron Press.

McKaye KR. 1983. Ecology and breeding behaviour of a cichlid fish, *Cyrtocara eucinostomus*, on a large lek in Lake Malawi, Africa. *Env Biol Fish* 8:81–96.

McKaye KR. 1991. Sexual selection and the evolution of the cichlid fishes of Lake Malawi, Africa. In: Keenleyside MHA, ed. *Cichlid fishes: behaviour, ecology and evolution*, pp. 241–257. London: Chapman & Hall.

McKaye KR, Kocher T. 1983. Head ramming behaviour by three paedophagous cichlids in Lake Malawi, Africa. *Anim Behav* 31:206–210.

McKaye KR, Mughogho DE, Lovullo TJ. 1992. Formation of the selfish school. *Env Biol Fish* 35:213–218.

McKenney TW. 1959. A contribution to the life history of the squirrel fish *Holocentrus vexillarius* Poey. *Bull Mar Sci Gulf Caribb* 9:174–221.

McKenzie DJ, Farrell AP, Brauner CJ. 2007. *Primitive fishes. Fish physiology*, Vol. 26. New York: Academic Press.

McKenzie DJ, Randall DJ. 1990. Does *Amia calva* aestivate? *Fish Physiol Biochem* 8:147–158.

McKeown BA. 1984. *Fish migration*. London: Croom Helm.

McLaren IA. 1974. Demographic strategy of vertical migration by a marine copepod. *Am Natur* 108:91–102.

McMillan DB. 2011. Pituitary gland or hypophysis. In: Farrell AP, Stevens ED, Cech JJ, Richards JG, eds. *Encyclopedia of fish physiology: from genome to environment*, pp. 1457–1465. Boston: Elsevier/Academic Press.

McPhee J. 2002. *The founding fish*. New York: Farrar, Straus & Giroux.

McVicar AH. 1997. Disease and parasite implications of the coexistence of wild and cultured Atlantic salmon populations. *ICES J Mar Sci* 54:1093–1103.

Measey GJ, Herrel A. 2006. Rotational feeding in caecilians: putting a spin on the evolution of cranial design. *Biol Lett* 2:485–487.

Mecklenburg CW, Moller PR, Steinke D. 2011. Biodiversity of arctic marine fishes: taxonomy and zoogeography. *Mar Biodiver* 41:109–140.

Meffe GK, Snelson FF, Jr., eds. 1989. *Ecology and evolution of livebearing fishes (Poeciliidae)*. Engelwood Cliffs, NJ: Prentice-Hall.

Meffe GK. 1984. Effects of abiotic disturbance on coexistence of predator–prey fish species. *Ecology* 65:1525–1534.

Meffe GK. 1986. Conservation genetics and the management of endangered fishes. *Fisheries* 11:14–23.

Mehta RS, Wainwright PC. 2007. Raptorial jaws in the throat help moray eels swallow large prey. *Nature* 449:79–82.

Meier MF. 1990. Reduced rise in sea level. *Nature* 343:115–116.

Meisner AD. 2005. Male modifications associated with inseminaton in teleosts. In: Uribe MCA, Grier HJ, eds. *Viviparous fishes*, pp. 167–192. Homestead, FL: New Life Publications.

Mensinger AF. 2011. Bioluminescence in Fishes. In: Farrell AP, Stevens ED, Cech JJ, Richards JG, eds. *Encyclopedia of fish physiology: from genome to environment*, 497—503. Boston: Elsevier/Academic Press.

Mensinger AF, Case F. 1990. Luminescent properties of deep sea fish. *J Exp Mar Biol Ecol* 144:1–15.

Meredith CS, Dudy P, Thiede GP. 2015. Predation on native sculpin by exotic brown trout exceeds that by native cutthroat trout within a mountain watershed. *Ecology Freshw Fish* 24:133–147.

Metcalfe JD, Hunter E, Buckley AA. 2006. The migratory behaviour of North Sea plaice: currents, clocks and clues. *Mar Freshw Behav Physiol* 39:25–36.

Meyer A. 1993. Phylogenetic relationships and evolutionary processes in east African cichlid fishes. *Trends Ecol Evol* 8:279–284.

Meyer A, Kocher TD, Basasibwaki P, Wilson AC. 1990. Monophyletic origin of Lake Victoria cichlid fishes suggested by mitochondrial DNA sequences. *Nature* 347:550–553.

Meyer CG, Holland KN. 2005. Movement patterns, home range size and habitat utilization of the bluespine unicornfish, *Naso unicornis* (Acanthuridae) in a Hawaiian marine reserve. *Env Biol Fish* 73:201–210.

Meyer CG, Holland KN, Papastamatiou YP. 2005. Sharks can detect changes in the geomagnetic field. *J Roy Soc Interface* 2:129–130.

Meyer CG, Holland KN, Wetherbee BM, Lowe CG. 2000. Movement patterns, habitat utilization, home range size and site fidelity of whitesaddle goatfish, *Parupeneus porphyreus*, in a marine reserve. *Env Biol Fish* 59:235–242.

Meyer JL, Schultz ET, Helfman GS. 1983. Fish schools: an asset to corals. *Science* 220:1047–1049.

Michel KB, Aerts P, Van Wassenbergh S. 2016. Environment-dependent prey capture in the Atlantic mudskipper (*Periophthalmus barbarus*). *Biol Open* 5:1735–1742. **https://doi.org/10.1242/bio.019794**.

Michel KB, Heiss E, Aerts P, Van Wassenbergh S. 2015. A fish that uses its hydrodynamic tongue to feed on land. *Proc R Soc B* 282:20150057. **https://doi.org/10.1098/rspb.2015.0057** (7 pages)

Michimae H, Wakahara M. 2002. Variation in cannibalistic polyphenism between populations in the salamander *Hynobius retardatus*. *Zool Sci* 19(6):703–707.

Micklin PP. 1988. Desiccation of the aral sea: a water management disaster in the Soviet Union. *Science* 241:1170–1176.

Mihuc TB, Dunlap F, Binggeli C, Myhers L, Pershyn C. 2012. Long-term patterns in Lake Champlain's zooplankton: 1992–2010. *J Great Lakes Res* 38:49–57.

Mikheev VN, Pasternak AF, Taskinen J, Valtonen TE. 2013. Grouping facilitates avoidance of parasites by fish. *Parasit Vectors* 6:1–8.

Milinski M. 1990. Information overload and food selection. In: Hughes RN, ed. *Behavioural mechanisms of food selection*, pp. 721–737. NATO ASI Series G 20. Berlin: Springer-Verlag.

Milinski M. 1993. Predation risk and feeding behaviour. In: Pitcher TJ, ed. *The behaviour of teleost fishes*, 2nd edn, pp. 285–305. London: Chapman & Hall.

Miller DL, Hughes RM, Karr JR, et al. 1988a. Regional applications of an index of biotic integrity for use in water resource management. *Fisheries* 13(5):12–20.

Miller JM. 1988. Physical processes and the mechanisms of coastal migrations of immature marine fishes. In: Weinstein MP, ed. *Larval fish and shellfish transport through inlets*, pp. 68–76. American Fisheries Society Symposium 3. Bethesda, MD: American Fisheries Society.

Miller JM, Crowder LB, Moser ML. 1985. Migration and utilization of estuarine nurseries by juvenile fishes: an evolutionary perspective. *Contrib Mar Sci* 27 (suppl):338–352.

Miller JM, Pietrafesa LJ, Smith NP. 1990. *Principles of hydraulic management of coastal lagoons for aquaculture and fisheries*. FAO Fisheries Biological Technical Paper No. 314. Rome: Food and Agricultural Organization.

Miller KA, Kenter LW, Breton TS, Berlinsky DL. 2019. The effects of stress, cortisol administration and cortisol inhibition on black sea bass (*Centropristis striata*) sex differentiation. *Comp Biochem Physiol Part A* 227:154–160.

Miller MJ, Tsukamoto K. 2004. *An introduction to leptocephali: biology and identification*. Tokyo: Ocean Research Institute, University of Tokyo.

Miller RR. 1957. Have the genetic patterns of fishes been altered by introductions or selective fishing? *J Fish Res Board Can* 14:797–806.

Miller RR. 1966. Geographical distribution of Central American freshwater fishes. *Copeia* 1966:773–802.

Miller RR. 1981. Coevolution of deserts and pupfishes (genus *Cyprinodon*) in the American southwest. In: Naiman RJ, Soltz DL, eds. *Fishes in North American deserts*, pp. 39–94. New York: Wiley & Sons.

Miller RR, Pister EP. 1971. Management of the Owens pupfish, *Cyprinodon radiosus*, in Mono County, California. *Trans Am Fish Soc* 100:502–509.

Miller RR, Williams JD, Williams JE. 1989. Extinctions of North American fishes during the past century. *Fisheries* 14(6):22–38.

Miller T. 1987. Knotting: a previously undescribed feeding behavior in muraenid eels. *Copeia* 1987:1055–1057.

Miller TJ. 1989. Feeding behavior of *Echidna nebulosa*, *Enchelycore pardalis* and *Gymnomuraena zebra* (Teleostei: Muraenidae). *Copeia* 1989:662–672.

Mills LS, Allendorf FW. 1996. The one-migrant-per-generation rule in conservation and management. *Conserv Biol* 10:1509–1518.

Milsom WK. 2011. The ventilatory response to CO_2/H^+. In: Farrell AP, Stevens ED, Cech JJ, Richards JG, eds. *Encyclopedia of fish*

physiology: from genome to environment, pp. 865–870. Boston: Elsevier/Academic Press.
Milsom WK. 2012. New insights into gill chemoreception: receptor distribution and roles in water and air breathing fish. *Respir Physiol Neurobiol* 184:326–339.
Mincarone MM, Stewart AL. 2006. A new species of giant seven-gilled hagfish (Myxinidae: *Eptatretus*) from New Zealand. *Copeia* 2006:225–229.
Minckley WL. 1991. Native fishes of the Grand Canyon region: an obituary? In: Committee to Review the Glen Canyon, eds. *Review of the draft federal long-term monitoring plan for the Colorado River below Glen Canyon Dam*, pp. 124–177. Washington, DC: National Academy Press.
Minckley WL. 1995. Translocation as a tool for conserving imperiled fishes: experiences in western United States. *Biol Conserv* 2:297–309.
Minckley WL, Barber WE. 1971. Some aspects of biology of the longfin dace, a cyprinid fish characteristic of streams in the Sonoran Desert. *Southwest Natur* 15:459–464.
Minckley WL, Deacon JE, eds. 1991. *Battle against extinction: native fish management in the American west*. Tucson, AZ: University of Arizona Press.
Minckley WL, Meffe GK. 1987. Differential selection by flooding in stream fish communities of the arid American Southwest. In: Matthews WJ, Heins DC, eds. *Community and evolutionary ecology of North American stream fishes*, pp. 93–104. Norman, OK: University of Oklahoma Press.
Minshall GW, Cummins KW, Petersen RC, et al. 1985. Developments in stream ecosystem theory. *Can J Fish Aquat Sci* 42:1045–1055.
Mittelbach GG. 1988. Competition among refuging sunfishes and effects of fish density on littoral zone invertebrates. *Ecology* 69:614–623.
Mittelbach GG, Ballew NG, Kjelvik MK. 2014. Fish behavioral types and their ecological consequences. *Can J Fish Aquat Sci* 71:927–944.
Miura T, Wang J. 1985. Chlorophyll *a* found in feces of phytoplanktivorous cyprinids and its photosynthetic activity. *Verh Int Ver Limnol* 22:2636–2642.
Miya M, Takeshima H, Endo H, et al. 2003. Major patterns of higher teleostean phylogenies: a new perspective based on 100 complete mitochondrial DNA sequences. *Mol Phylogenet Evol* 26:121–138.
Moberg GP, Mench JA. 2000. *The biology of animal stress*. Wallingford, UK: CABI Publishing.
Mochioka N, Iwamizu M. 1996. Diet of anguilloid larvae: leptocephali feed selectively on larvacean houses and fecal pellets. *Mar Biol* 125:447–452.
Modarressie R, Rick IP, Bakker TCM. 2006. UV matters in shoaling decisions. *Proc Roy Soc Lond B Biol Sci* 273:849–854.
Mogdans J. 2019. Sensory ecology of the fish lateral-line system: morphological and physiological adaptations for the perception of hydrodynamic stimuli. *J Fish Biol* 95:53–72.
Moland E, Eagle JV, Jones GP. 2005. Ecology and evolution of mimicry in coral reef fishes. *Oceanogr Mar Biol Ann Rev* 43:455–482.
Moland E, Jones GP. 2004. Experimental confirmation of aggressive mimicry by a coral reef fish. *Oecologia* 140:676–683.
Møller P. 2006. Eletrocommunication: history, insights, and new questions. In: Ladich F, Collin SP, Møller P, Kapoor BG, eds. *Communication in fishes*, Vol. 2, pp. 579–597. Enfield, NH: Science Publishers.
Møller PR, Nielsen JG, Anderson ME. 2005. Systematics of polar fishes. In: Farrell AP, Steffensen JF, eds. *The physiology of polar fishes. Fish physiology*, Vol. 22, pp. 25–78. New York: Academic Press.
Mollet HF, Cailliet GM, Klimley AP, et al. 1996. A review of length validation methods and protocols to measure large white sharks. In: Klimley AP, Ainley DG, eds. *Great white sharks: the biology of Carcharodon carcharias*, pp. 91–108. San Diego, CA: Academic Press.
Molloy PP, McLean IB, Côté IM. 2009. Effects of marine reserve age on fish populations: a global meta-analysis. *J Appl Ecol* 46:743–751
Monod T. 1968. Le complexe urophore des poissons teleosteens. *Afrique Noire Mem Inst Fond* 81:705.
Montgomery DR. 2003. *King of fish: the thousand year run of salmon*. Boulder, CO: Westview Press.
Montgomery J. 2011. Lateral line neuroethology. In: Farrell AP, Stevens ED, Cech JJ, Richards JG, eds. *Encyclopedia of fish physiology: from genome to environment*, pp. 329–335. Boston: Elsevier/Academic Press.
Montgomery JC. 1988. Sensory physiology. In: Shuttleworth TJ, ed. *Physiology of elasmobranch fishes*, pp. 79–98. Berlin: Springer-Verlag.
Montgomery JC, Coombs S, Baker CF. 2001a. The mechanosensory lateral line system of hypogean fishes. *Env Biol Fish* 62:87–96.
Montgomery JC, Macdonald F, Baker CF, Carton AG. 2002. Hydrodynamic contributions to multimodal guidance of prey capture behavior in fish. *Brain Behav Evol* 59:190–198.
Monticini P. 2010. The ornamental fish trade. Production and commerce of ornamental fish: technical-managerial and legislative aspects. In: *GLOBEFISH research programme*, Vol. 102, p. 134. Rome: GLOBEFISH Research Programme.
Montoya RV, Thorson TB. 1982. The bull shark (*Carcharhinus leucas*) and largetooth sawfish (*Pristis perotteti*) in Lake Bayano, a tropical man-made impoundment in Panama. *Env Biol Fish* 7:341–347.
Moon TW. 2011. Stress effect on growth and metabolism. In: Farrell AP, Stevens ED, Cech JJ, Richards JG, eds. *Encyclopedia of fish physiology: from genome to environment*, pp. 1534–1540. Boston: Elsevier/Academic Press.
Moore JA. 2002. Upside-down swimming behavior in a whipnose anglerfish (Teleostei: Ceratioidei: Gigantactinidae). *Copeia* 2002:1144–1146.
Moore JW. 2006. Animal ecosystem engineers in streams. *BioScience* 56:237–246.
Moore RH, Wohlschlag DE. 1971. Seasonal variations in the metabolism of the Atlantic midshipman, *Porichthys porosissimus* (Valenciennes). *J Exp Mar Biol Ecol* 7:163–172.
Moore WS. 1984. Evolutionary ecology of unisexual fishes. In: Turner BJ, ed. *Evolutionary genetics of fishes*, pp. 329–398. New York: Plenum Press.
Moorehead JA, Zeng C. 2010. Development of captive breeding techniques for marine ornamental fish: a review. *Rev Fish Sci* 18(4):315–343.
Mora C, Sale PF. 2002. Are populations of coral reef fish open or closed? *Trends Ecol Evol* 17:422–428.
Morgan JAT, Harry AV, Welch DJ, et al. 2012. Detection of interspecies hybridisation in Chondrichthyes: hybrids and hybrid offspring between Australian (*Carcharhinus tilstoni*) and common (*C. limbatus*) blacktip shark found in an Australian fishery. *Conserv Genet* 13:455–463.
Morgan JD, Iwama GK. 1997. Measurements of stressed states in the field. In: Iwama OK, Pickering AD, Sumpter JP, Schreck CB, eds. *Fish stress and health in aquaculture*, pp. 247–268. Seminar Series 62. Cambridge, UK: Society for Experimental Biology.
Morin JG, Harrington A, Nealson K, Kriegr N, Baldwin TO, Hastings JW. 1975. Light for all reasons: versatility in the behavioral repertoire of the flashlight fish. *Science* 190:74–76.

Moritz C. 1994. Defining 'evolutionary significant units' for conservation. *Trends Ecol Evol* 9:373–375.

Morris JA, Whitfield PE. 2009. *Biology, ecology, control and management of the invasive Indo-Pacific lionfish: an updated integrated assessment.* NOAA Technical Memorandum NOS NCCOS 99. Beaufort, NC: NOAA Center for Coastal Fisheries and Habitat Research.

Morton A, Routledge R. 2016. Risk and precaution: salmon farming. *Mar Policy* 74:205–212.

Moser HG. 1981. Morphological and functional aspects of marine fish larvae. In: Lasker RL, ed. *Marine fish larvae: morphology, ecology, and relation to fisheries*, pp. 89–131. Seattle: Washington Sea Grant Program.

Moser HG, ed. 1996. *The early stages of fishes in the California Current region.* CALCOFI (California Cooperative Oceanic Fisheries Investigations) Atlas No. 33. La Jolla, CA: Southwest Fisheries Science Center. **www.calcofi.org**.

Moser HG, Ahlstrom EH. 1970. Development of lanternfishes (family Myctophidae) in the California Current. Part I. Species with narrow-eyed larvae. *Natur Hist Mus Los Ang City Sci Bull* 7.

Moser HG, Ahlstrom EH. 1974. Role of larval stages in systematic investigations of marine teleosts: the Myctophidae, a case study. *Fish Bull US* 72:391–413.

Moser HG, Richards WJ, Cohen DM, Fahay MP, Kendall AW, Jr., Richardson SL, eds. 1984. *Ontogeny and systematics of fishes.* Special Publication No. 1. Lawrence, KS: American Society of Ichthyologists and Herpetologists.

Moss SA. 1967. Tooth replacement in the lemon shark *Negaprion brevirostris.* In: Gilbert, PW, Mathewson RF, Rall DP, eds. *Sharks, skates and rays*, pp. 319–329. Baltimore: Johns Hopkins Press.

Moss SA. 1977. Feeding mechanisms in sharks. *Amer Zool* 17: 355–364.

Moss SA. 1984. *Sharks. A guide for the amateur naturalist.* Englewood Cliffs, NJ: Prentice-Hall.

Motta PJ. 1977. Anatomy and functional morphology of dermal collagen fibers in sharks. *Copeia* 1977:454–464.

Motta PJ. 1983. Response by potential prey to coral reef fish predators. *Anim Behav* 31:1257–1259.

Motta PJ. 1984. Mechanics and functions of jaw protrusion in teleost fishes: a review. *Copeia* 1984:1–18.

Motta PJ. 1988. Functional morphology of the feeding apparatus of ten species of Pacific butterflyfishes (Perciformes, Chaetodontidae): an ecomorphological approach. *Env Biol Fish* 22:39–67.

Motta PJ, ed. 1989. Butterflyfishes: success on the coral reef. *Env Biol Fish* 25:7–246.

Motta PJ. 2004. Prey capture behavior and feeding mechanics of elasmobranch. In: Carrier JC, Musick JA, Heithaus MR, eds. *Biology of sharks and their relatives*, pp. 165–202. Boca Raton, FL: CRC Press.

Motta PJ, Wilga CD. 1999. Anatomy of the feeding apparatus of the nurse shark, *Ginglymostoma cirratum. J Morphol* 241:33–60.

Motta PJ, Wilga CD. 2001. Advances in the study of feeding behaviors, mechanisms, and mechanics of sharks. *Env Biol Fish* 60:131–156.

Moura T, Figueiredo I, Machado PB, Gordo LS. 2004. Growth pattern and reproductive strategy of the holocephalan *Chimaera monstrosa* along the Portuguese continental slope. *J Mar Biol Assoc UK* 84:801–804.

Mowat F. 1996. *Sea of slaughter: a chronicle of the destruction of animal life in the North Atlantic.* Shelburne, VT: Chapters Publishing.

Moyer JT, Sawyers CE. 1973. Territorial behaviour of the anemonefish *Amphiprion xanthurus* with notes on the life history. *Jap J Ichthyol* 20:85–93.

Moyer JT, Yogo Y. 1982. The lek mating system of *Halichoeres melanochir* (Pisces:Labridae) at Miyake-jima, Japan. *Z Tierpsychol* 60:209–226.

Moyes CD, Ballantyne JS. 2011. Membranes and temperature: homeoviscous adaptation. In: Farrell AP, Stevens ED, Cech JJ, Richards JG, eds. *Encyclopedia of fish physiology: from genome to environment*, pp. 1725–1731. Boston: Elsevier/Academic Press.

Moyle PB. 1991. Ballast water introductions. *Fisheries* 16(1):4–6.

Moyle PB. 2002. *Inland fishes of California, revised and expanded.* Berkeley: University of California Press.

Moyle PB, Cech JJ. 2004. *Fishes: an introduction to Ichthyology*, 5th edn. Upper Saddle River, NJ: Prentice-Hall.

Moyle PB, Leidy RA. 1992. Loss of biodiversity in aquatic ecosystems: evidence from fish faunas. In: Fiedler PL, Jain SK, eds. *Conservation biology: the theory and practice of nature conservation, preservation and management*, pp. 127–169. New York: Chapman & Hall.

Moyle PB, Nichols R. 1974. Decline of the native fish fauna of the Sierra-Nevada foothills in central California. *Am Midl Natur* 92:72–83.

Moy-Thomas JA, Miles RS. 1971. *Palaeozoic fishes*, 2nd edn. London: Chapman & Hall.

Mozaffarian D, Rimm EB. 2006. Fish intake, contaminants, and human health: evaluating the risks and the benefits. *J Am Med Assoc* 296:1885–1899.

Muir WM, Howard RD. 1999. Possible ecological risks of transgenic organism release when transgenes affect mating success: sexual selection and the Trojan gene hypothesis. *Proc Natl Acad Sci USA* 96:13853–13856.

Muller K. 1978a. Locomotor activity of fish and environmental oscillations. In: Thorpe JE, ed. *Rhythmic activity of fishes*, pp. 1–29. London: Academic Press.

Muller K. 1978b. The flexibility of the circadian system of fish at different latitudes. In: Thorpe JE, ed. *Rhythmic activity of fishes*, pp. 91–104. London: Academic Press.

Muller UK, Smit J, Stamhuis EJ, Videler JJ. 2001. How the body contributes to the wake in undulatory fish swimming: flow fields of a swimming eel (*Anguilla anguilla*). *J Exp Biol* 204:2751–2762.

Mumby PJ, Dahlgren CP, Harborne AR, et al. 2006. Fishing, trophic cascades, and the process of grazing on coral reefs. *Science* 311:98–101.

Munday PL, Jones GP, Caley MJ. 2001. Interspecific competition and coexistence in a guild of coral-dwelling fishes. *Ecology* 82:2177–2189.

Munday PL, Schubert M, Baggio JA, Jones GP, Caley MJ, Grutter AS. 2003. Skin toxins and external parasitism of coral-dwelling gobies. *J Fish Biol* 62:976–981.

Munehara H, Sideleva VG, Goto A. 2002. Mixed-species brooding between two Baikal sculpins: field evidence for intra- and interspecific competition for reproductive resources. *J Fish Biol* 60:981–988.

Munk O. 2000. History of the fusion area between the parasitic male and the female in the deep-sea anglerfish *Neoceratias spinifer* Pappenheim, 1914 (Teleostei, Ceratioidei). *Acta Zool* 81:315–324.

Munoz RC, Warner RR. 2004. Testing a new version of the size-advantage hypothesis for sex change: sperm competition and size-skew effects in the bucktooth parrotfish, *Sparisoma radians. Behav Ecol* 15:129–136.

Munro AD. 1990a. Tropical freshwater fishes. In: Munro AD, Scott AP, Lam TJ, eds. *Reproductive seasonality in teleosts: environmental influences*, pp. 145–239. Boca Raton, FL: CRC Press.

Munro AD. 1990b. General introduction. In: Munro AD, Scott AP, Lam TJ, eds. *Reproductive seasonality in teleosts: environmental influences*, pp. 1–11. Boca Raton, FL: CRC Press.

Munro AD, Scott AP, Lam TJ, eds. 1990. *Reproductive seasonality in teleosts: environmental influences.* Boca Raton, FL: CRC Press.

Munshi JSD. 1976. Gross and fine structure of the respiratory organs of air-breathing fishes. In: Hughes GM, ed. *Respiration of amphibious vertebrates*, pp. 73–104. New York: Academic Press.

Munz FW, McFarland WN. 1973. The significance of spectral position in the rhodopsins of tropical marine fishes. *Vis Res* 13:1829–1874.

Murata K. 2003. Blocks to polyspermy in fish: a brief review. In: Symposium on Aquaculture and Pathobiology of Crustacean and Other Species. *Aquaculture Panel Proceedings*, Santa Barbara, California. Anais Santa Barbara: UJNR, 15 pp.

Murphy JE, Beckmen KB, Johnson JK, Cope RB, Lawmaster T, Beasley VR. 2002. Toxic and feeding deterrent effects of native aquatic macrophytes on exotic grass carp (*Ctenopharyngodon idella*). *Ecotoxicology* 11:243–254.

Murray JD, Bahen JJ, Rulifson RA. 1992. Management considerations for by-catch in the North Carolina and Southeast shrimp fishery. *Fisheries* 17(1):21–26.

Musick J, Harbin MM, Berkeley SA, et al. 2000. Marine, estuarine and diadromous fish stocks at risk of extinction in North America (exclusive of Pacific salmonids). *Fisheries* 25(11):6–30.

Musick JA, Bruton MN, Balon EK, eds. 1991. *The biology of Latimeria chalumnae and evolution of coelacanths.* London: Springer.

Musick JA, Ellis JK. 2005. Reproductive evolution of chondrichthyans. In: Hamlett WC, ed. *Reproductive biology and phylogeny of chondrichthyes: sharks, batoids and chimaeras*, pp. 45–79. Enfield, NH: Science Publishers.

Muss A, Robertson DR, Stepien CA, Wirtz P, Bowen BW. 2001. Phylogeography of the genus *Ophioblennius*: the role of ocean currents and geography in reef fish evolution. *Evolution* 55:561–572.

Mussi M, Haimberger TJ, Hawryshyn CW. 2005.Behavioural discrimination of polarized light in the damselfish *Chromis viridis* (family Pomacentridae). *J Exp Biol* 208:3037–3046.

MWDTF (Montana Whirling Disease Task Force). 1996. *Final report and action recommendations.* Helena, MT: MWDTF.

Myers GS. 1938. Fresh-water fishes and West Indian zoogeography. *Ann Rept Smithsonian Inst* 1937:339–364.

Myers GS. 1964. A brief sketch of the history of ichthyology in America to the year 1850. *Copeia* 1964:33–41.

Myers JM, Kope RG, Bryant GJ, et al. 1998. *Status review of coastal cutthroat trout from Washington, Oregon, and California.* NOAA Techical Memorandum No. NMFS-NWFSC-35. Springfield, VA: National Technical Information Service.

Myers MF, White GF. 1993. The challenge of the Mississippi flood. *Environment* 106–9:25–35.

Myers MS, Anulacion BF, French BL, et al. 2008. Improved flatfish health following remediation of a PAH-contaminated site in Eagle Harbor, Washington. *Aquat Toxicol.* 88:277–288.

Myers MS, Fournie JW. 2002. Histological biomarkers as integrators of anthropogenic and environmental stress. In: Adams SM, ed. *Biological indicators of aquatic ecosystem stress*, pp. 221–287. Bethesda, MD: American Fisheries Society.

Myers RA, Barrowman NJ. 1996. Is fish recruitment related to spawner abundance? *Fish Bull US* 94:707–724.

Myrberg AA, Jr. 1978. Underwater sound – its effect on the behaviour of sharks. In: Hodgson ES, Mathewson RF, eds. *Sensory biology of sharks, skates, and rays*, pp. 391–417. Arlington, VA: Office of Naval Research, US Department of Navy.

Myrberg AA, Jr. 1981. Sound communication and interception in fishes. In: Tavolga W, Popper AN, Fay RR, eds. *Hearing and sound communication in fishes*, pp. 395–452. New York: Springer-Verlag.

Myrberg AA, Jr. 2002. Fish bioacoustics and behaviour. *Bioacoustics* 12:107–109.

Myrberg AA, Jr., Nelson DR. 1991. The behavior of sharks: what have we learned. In: Gruber SH, ed. *Discovering sharks*, pp. 92–100. American Littoral Society Special Publication No. 14. Highlands, NJ: American Littoral Society.

Myrberg AA, Jr., Thresher RE. 1974 Interspecific aggression and its relevance to the concept of territoriality in fishes. *Am Zool* 14:81–96.

Nagelkerke LAJ, Sibbing FA, van den Boogaart JGM, Lammens EHRR, Osse JWM. 1994. The barbs (*Barbus* spp.) of Lake Tana: a forgotten species flock. *Env Biol Fish* 39:1–22.

Nagler JJ, Bouma J, Thorgaard GH, et al. 2001. High incidence of a male-specific genetic marker in phenotypic female Chinook salmon from the Columbia River. *Env Health Perspect* 109:67–69.

Naiman RJ. 1981. An ecosystem overview: desert fishes and their habitats. In: Naiman RJ, Soltz DL, eds. *Fishes in North American deserts*, pp. 493–531. New York: Wiley & Sons.

Naiman RJ, Soltz DL, eds. 1981. *Fishes in North American deserts.* New York: Wiley & Sons.

Nakamura I, Goto Y, Sato K. 2015. Ocean Sunfish rewarm at the surface after deep excursions to forage for siphonophores. *J Ani Ecol* 84:590–603.

Nakamura R. 1991. *A survey of the Pacific hagfish resource off the Central California coast. Final report.* Contract No. A-800-184. Sacramento, CA: California Marine Fisheries Impacts Program.

Nakashima BS, Leggett WC. 1980. The role of fishes in the regulation of phosphorus availability in lakes. *Can J Fish Aquat Sci* 37:1540–1549.

Nakatsuru K, Kramer DL. 1982. Is sperm cheap? Limited male fertility and female choice in the lemon tetra (Pisces, Characidae). *Science* 216:753–755.

Nakazawa T. 2015. Ontogenetic niche shifts matter in community ecology: a review and future perspectives. *Popul Ecol* 57:347–354.

Nanami A, Nishihira M, Suzuki T, Yokochi H. 2005. Species-specific habitat distribution of coral reef fish assemblages in relation to habitat characteristics in an Okinawan coral reef. *Env Biol Fish* 72:55–65.

Nannini MA, Wahl DH, Philipp DP, Cooke SJ. 2011. The influence of selection for vulnerability to angling on foraging ecology in largemouth bass *Micropterus salmoides*. *J Fish Biol* 79:1017–1028.

Naruse K, Tanaka M, Mita K, Shima A, Postlethwait J, Mitani H. 2004. A medaka gene map: the trace of ancestral vertebrate protochromosomes revealed by comparative gene mapping. *Genome Res* 14:820–828.

NASCO (North Atlantic Salmon Conservation Organization). 1991. *Guidelines to minimize the threats to wild salmon stocks from salmon aquaculture.* Edinburgh, UK: North Atlantic Salmon Conservation Organization.

Nath A, Chaube R, Subbian K. 2013. An insight into the molecular basis for convergent evolution in fish antifreeze proteins. *Comput Bio Med* 43:817–821.

Naylor RL, Goldburg RJ, Primavera JH, et al. 2000. Effect of aquaculture on world fish supplies. *Nature* 405:1017–1024.

Near TJ, Dornburg A, Kuhn KL, et al. 2012a. Ancient climate change, antifreeze, and the evolutionary diversification of Antarctic fishes. *PNAS* 109:3434–3439.

Near TJ, Eytan RI, Dornburg A, et al. 2012b. Resolution of ray-finned fish phylogeny and timing of diversification. *Proc Natl Acad Sci USA* 109:13698–13703.

Near TJ, Parker SK, Detrich HW, III. 2006. A genomic fossil reveals key steps in hemoglobin loss by the Antarctic icefishes. *Mol Biol Evol* 23(11):2008–2016.

Neat FC, Locatello L, Rasotto MB. 2003. Reproductive morphology in relation to alternative male reproductive tactics in *Scartella cristata*. *J Fish Biol* 62:1381–1391.

NEFSC (Northeast Fisheries Science Center). 2013. 55th Northeast Regional Stock Assessment Workshop (55th SAW) Assessment Summary Report. US Dept Commer, Northeast Fish Sci Cent Ref Doc. 13-01; 41 p. National Marine Fisheries Service, 166 Water Street, Woods Hole, MA 02543-1026, or online at **http://www.nefsc.noaa**.

NEFSC (Northeast Fisheries Science Center). 2017. Operational Assessment of 19 Northeast Groundfish Stocks, Updated Through 2016. US Department of Commerce, Northeast Fisheries Science Center, Ref Doc. 17-17; 259 p.

Nehlsen W, Williams JE, Lichatowich JA. 1991. Pacific salmon at the crossroads: stocks at risk from California, Oregon, Idaho, and Washington. *Fisheries* 16:4–21.

Nei M. 1987. *Molecular evolutionary genetics*. New York: Columbia University Press.

Neigel JE, Avise JC. 1986. Phylogenetic relationships of mitochondrial DNA under various demographic models of speciation. In: Nevo E, Karlin S, eds. *Evolutionary process and theory*, pp. 515–534. New York: Academic Press.

Neighbors MA, Wilson RR, Jr. 2006. Deep sea. In: Allen LG, Pondella DJ, Horn MH, eds. *The ecology of marine fishes: California and adjacent waters*, pp. 342–383. Berkeley, CA: University of California Press.

Neill SRStJ, Cullen JM. 1974. Experiments on whether schooling by their prey affects the hunting behaviour of cephalopod and fish predators. *J Zool Lond* 182:549–569.

Neilsen J, Hedeholm RB, Heinemeier J, et al. 2016. Eye lens radiocarbon reveals centuries of longevity in the Greenland Shark (*Somniosus micrcephalus*). *Science* 353:702–704.

Nelson DR. 1990. Telemetry studies of sharks: a review, with applications in resource management. In: Pratt HL, Jr., Gruber SH, Taniuchi T, eds. *Elasmobranchs as living resources: advances in the biology, ecology, systematics, and the status of fisheries*, pp. 239–256. NOAA Technical Report No. 90. Washington, DC: National Oceanic and Atmospheric Administration.

Nelson G. 1969. Gill arches and the phylogeny of fishes, with notes on the classification of vertebrates. *Bull Am Mus Natur Hist* 141:475–552.

Nelson G. 1985. A decade of challenge: the future of biogeography. In: Leviton AE, Aldrich ML. *Plate tectonics and biogeography. J Hist Earth Sci Soc* 4:187–196.

Nelson G, Platnick N. 1981. *Systematics and biogeography. Cladistics and vicariance*. New York: Columbia University Press.

Nelson JA, Chabot D. 2011. General energy metabolism. In: Farrell AP, Stevens ED, Cech JJ, Richards JG, eds. *Encyclopedia of fish physiology: from genome to environment*, pp. 1566–1572. Boston: Elsevier/Academic Press.

Nelson JA. 2014. Breaking wind to survive: fishes that breathe air with their gut. *J Fish Biol* 84:554–576.

Nelson JS. 1994. *Fishes of the world*, 3rd edn. New York: Wiley & Sons.

Nelson JS, ed. 1999. The species concept in fish biology. *Rev Fish Biol Fisheries* 9:275–382.

Nelson JS. 2006. *Fishes of the world*, 4th edn. Hoboken, NJ: Wiley & Sons.

Nelson JS, Grande TC, Wilson MVH. 2016. *Fishes of the world*, 5th edn. Hoboken, NJ: Wiley. 707 pp.; also see corrections and updates at **https://sites.google.com/site/fotw5th/**.

Nelson JS, Starnes WC, Warren ML. 2002. A capital case for common names of species of fishes – a white crappie or a White Crappie? *Fisheries* 27(7):31–33.

Netboy A. 1980. *The Columbia River salmon and steelhead trout*. Seattle, WA: University of Washington Press.

Netsch NF, Witt A, Jr. 1962. Contributions to the life history of the longnose gar, (*Lepisosteus osseus*) in Missouri. *Trans Am Fish Soc* 91:251–262.

Neudecker S. 1979. Effects of grazing and browsing fishes on the zonation of corals in Guam. *Ecology* 60:666–672.

Neuman MJ, Able KW. 2002. Quantification of ontogenetic transitions during the early life of a flatfish, Windowpane (*Scophthalmus aquosus*) (Pleuronectiformes Scophthalmidae). *Copeia* 2002:597–609.

Neville HM, Dunham JB, Peacock MM. 2006. Landscape attributes and life history variability shape genetic structure of trout populations in a stream network. *Landscape Ecol* 21:901–916.

Newport C, Wallis G, Siebeck UE. 2014. Concept learning and the use of three common psychophysical paradigms in the archerfish (*Toxotes chatareus*). *Frontiers in Neural Circuits* 8(39):1–13.

Newton KC, Gill AB, Kajiura SM. 2019. Electroreception in marine fishes: chondrichthyans. *J Fish Biol* 95:135–154.

Newton KC, Kajiura SM. 2017. Magnetic field discrimination, learning, and memory in the yellow stingray (*Urobatis jamaicensis*). *Anim Cogn* 20:603–614.

Newman RM, Waters TF. 1984. Size-selective predation on *Gammarus pseudolimnaeus* by trout and sculpins. *Ecology* 65:1535–1545.

Nichols JT. 1943. *The freshwater fishes of China. Natural history of Central Asia*, Vol. IX. New York: American Museum of Natural History.

Nichols S, Shah A, Pellegrin GJ, Jr., Mullin K. 1990. *Updated estimates of shrimp fleet by-catch in the offshore waters of the Gulf of Mexico 1972–1989*. Pascagoula, MS: National Marine Fisheries Service.

Nico LG, Fuller PL. 1999. Spatial and temporal patterns of non-indigenous fish introductions in the United States. *Fisheries* 24:16–27.

Nico LG, Taphorn DC. 1988. Food habits of piranhas in the low llanos of Venezuela. *Biotropica* 20:311–321.

Nielsen JG. 1977. The deepest-living fish, *Abyssobrotula galatheae*, a new genus and species of oviparous ophidioids (Pisces, Brotulidae). *Galathea Rept* 14:41–48.

Nielsen JG, Bertelsen E, Jespersen A. 1989. The biology of *Eurypharynx pelecanoides* (Pisces: Eurypharyngidae). *Acta Zool (Stockholm)* 70:187–197.

Nielson JD, Perry RI. 1990. Diel vertical migrations of marine fishes: an obligate or facultative process. *Adv Mar Biol* 26:115–168.

Nieuwenhuys R. 1982. An overview of the organization of the brain of actinopterygian fishes. *Amer Zool* 22:287–310.

Nieuwenhuys R. 2011. The structural, functional, and molecular organization of the brainstem. *Front Neuroanat* 5:article 33.

Nikinmaa M. 2011. Hemoglobin. In: Farrell AP, Stevens ED, Cech JJ, Richards JG, eds. *Encyclopedia of fish physiology: from genome to environment*, pp. 887–892. Boston: Elsevier/Academic Press.

Nikol'skii GV. 1961. *Special Ichthyology*. Jerusalem: Israel Program for Scientific Translations.

Nikula R, Strelkov P, Väinölä R. 2007. Diversity and trans-Arctic invasion history of mtDNA lineages in the North Atlantic Macoma balthica complex (Bivalvia: Tellinidae). *Evolution* 61:928–941.

Nilsson G. 1996. Brain and body oxygen requirements of *Gnathonemus petersii*, a fish with an exceptionally large brain. *J Exp Biol* 199(3):603–607.

Nilsson GE. 2011. Plasticity in gill morphology. In: Farrell AP, Stevens ED, Cech JJ, Richards JG, eds. *Encyclopedia of fish physiology:*

from genome to environment, pp. 796–802. Boston: Elsevier/Academic Press.

Nilsson S. 2011. Autonomic nervous system of fishes. In: Farrell AP, Stevens ED, Cech JJ, Richards JG, eds. *Encyclopedia of fish physiology: from genome to environment*, pp. 80–88. Boston: Elsevier/Academic Press.

Nishi T, Kawamura G, Matsumoto K. 2004. Magnetic sense in the Japanese eel, *Anguilla japonica*, as determined by conditioning and electrocardiography. *J Exp Biol* 207:2965–2970.

NMFS (National Marine Fisheries Service). 1997. *Status of fisheries of the United States. Report to Congress, September 1997.* Washington, DC: National Oceanic and Atmospheric Administration.

NMFS (National Marine Fisheries Service). 1999. *Our living oceans.* NOAA Technical Memorandum NMFS-F/SPO-41. Washington, DC: National Oceanic and Atmospheric Administration.

Noack K, Zardoya R, Meyer A. 1996. The complete mitochondrial DNA sequence of the bichir (*Polypterus ornatipinnis*), a basal ray-finned fish: ancient establishment of the consensus vertebrate gene order. *Genetics* 144:1165–1180.

Noakes DLG. 1979. Parent-touching behavior by young fishes: incidence, function and causation. *Env Biol Fish* 4:389–400.

Noakes DLG, Godin J-GJ. 1988. Ontogeny of behavior and concurrent development changes in sensory system in teleost fishes. In Hoar WS, Randall DJ, eds. *The physiology of developing fish, Part B. Viviparity and posthatching juveniles. Fish physiology*, Vol. 11, Part B. San Diego, CA: Academic Press.

Noakes DLG, Lindquist DG, Helfman GS, Ward JA, eds. 1983. *Predators and prey in fishes.* Developments in Environmental Biology of Fishes No. 2. The Hague: Dr. W. Junk.

Noga EJ, Silphaduang U. 2003. Piscidins: a novel family of peptide antibiotics from fish. *Drug News Perspect* 16:87–92.

Nohara M, Nishida M, Miya M, Nishikawa T. 2005. Evolution of the mitochondrial genome in Cephalochordata as inferred from complete nucleotide sequences from two *Epigonichthys* species. *J Mol Evol* 60:526–537.

Norcross BL, Shaw RF. 1984. Oceanic and estuarine transport of fish eggs and larvae: a review. *Trans Am Fish Soc* 113:153–165.

Norman JR. 1931. *A history of fishes.* New York: A. A. Wyn.

Norman JR, Fraser FC. 1949. *Field book of giant fishes.* New York: GP Putnam's Sons.

Norman JR, Greenwood PH. 1975. *A history of fishes*, 3rd edn. New York: Halstead Press.

Normark BB, McCune AR, Harrison RG. 1991. Phylogenetic relationships of neopterygian fishes, inferred from mitochondrial DNA sequences. *Mol Biol Evol* 8:819–834.

Norris S, Chao NL. 2002. Buy a fish, save a tree? Safeguarding sustainability in an Amazonian ornamental fishery. *Conserv Pract* 3(3):30–35.

Norse EA, Crowder LB, eds. 2005. *Marine conservation biology: the science of maintaining the sea's biodiversity.* Washington, DC: Island Press.

Northcote TG. 1978. Migratory strategies and production in freshwater fishes. In: Gerking SD, ed. *Ecology of freshwater fish populations*, pp. 326–359. New York: Wiley & Sons.

Northcote TG. 1988. Fish in the structure and function of freshwater ecosystems: a "top-down" view. *Can J Fish Aquat Sci* 45:361–379.

Northcote TG, Lorz HW, MacLeod JC. 1964. Studies on diel vertical movement of fishes in a British Columbia lake. *Verh Int Ver Limnol* 15:940–946.

Northcott SJ, Gibson RN, Morgan E. 1990. Persistence and modulation of endogenous circatidal rhythmicity in *Lipophrys pholis* (Teleostei). *J Mar Biol Assoc UK* 70:815–827.

Northcutt RG. 1977. Elasmobranch central nervous system organization and its possible evolutionary significance. *Am Zool* 17:411–429.

Northcutt RG, Davis RE, eds. 1983. *Fish neurobiology*, 2 vols. Ann Arbor, MI: University of Michigan Press.

Northcutt RG, Gans C. 1983. The genesis of neural crest and epidermal placodes: a reinterpretation of vertebrate origins. *Q Rev Biol* 58:1–28.

NRC (National Research Council). 1992. *Restoration of aquatic ecosystems: science, technology, and public policy.* Washington, DC: National Academy Press.

NRC (National Research Council). 1996a. *Stemming the tide: controlling introductions of nonindigenous species by ships' ballast water.* Washington, DC: National Academy Press.

NRC (National Research Council). 1996b. *Upstream: salmon and society in the Pacific Northwest.* Washington, DC: National Academy Press.

NRC (National Research Council). 1999a. *Hormonally active agents in the environment.* Washington, DC: National Academy Press.

NRC (National Research Council). 1999b. *Sustaining marine fisheries.* Washington, DC: National Academies Press.

NRC (National Research Council). 2001. *Marine protected areas: tools for sustaining ocean ecosystems.* Washington, DC: National Academies Press.

NRC (National Research Council). 2004a. *Atlantic salmon in Maine.* Washington, DC: National Academies Press.

Nurminen L, Horppila J, Lappalainen J, Malinen T. 2003. Implications of rudd (*Scardinius erythrophthalamus*) herbivory on submerged macrophytes in a shallow eutrophic lake. *Hydrobiologia* 506–509:511–518.

Nursall JR. 1973. Some behavioral interactions of spottail shiner (*Notropis hudsonius*), yellow perch (*Perca flavescens*), and northern pike (*Esox lucius*). *J Fish Res Board Can* 30:1161–1178.

Nyberg DW. 1971. Prey capture in the largemouth bass. *Am Midl Natur* 86:128–144.

O'Brien GC, Jacobs F, Evans SW, Smit NJ. 2014. First observation of African tigerfish *Hydrocynus vittatus* predating on barn swallows *Hirundo rustica* in flight. *J Fish Biol* 84:263–266.

O'Brien L, Burnett J, Mayo RK. 1993. *Maturation of nineteen species of finfish off the northeast coast of the United States, 1985–1990.* NOAA Technical Report NMFS 113. Washington, DC: National Oceanic and Atmospheric Administration.

O'Brien WJ. 1987. Planktivory by freshwater fish: thrust and parry in the pelagia. In: Kerfoot WC, Sih A, eds. *Predation: direct and indirect impacts on aquatic communities*, pp. 3–16. Hanover: University Press of New England.

O'Brien WJ, Evans B, Luecke C. 1985. Apparent size choice of zooplankton by planktivorous sunfish: exceptions to the rule. *Env Biol Fish* 13:225–233.

O'Steen SO, Bennett AF. 2003. Thermal acclimation effects differ between voluntary, maximum, and critical swimming velocities in two cyprinid fishes. *Physiol Biochem Zool* 76:484–496.

Ochi H, Onchi T, Yanagisawa Y. 2001. Alloparental care between catfishes in Lake Tanganyika. *J Fish Biol* 59:1279–1286.

Oelofsen BW, Loock K. 1981. A fossil cephalochordate from the early Permian Whitehill Formation of South Africa. *S Afr J Sci* 77:178–180.

Ogden JC. 1974. Grazing by the echinoid *Diadema antillarum* Philippi: formation of halos around West Indian patch reefs. *Science* 182:715–716.

Ogden JC. 1977. Carbonate sediment production by parrotfish and sea urchins on Caribbean reefs. In: Frost SH, Weiss MP, Saunders JB, eds. *Reefs and related carbonates —ecology and sedimentology.*

Am Assoc Petrol Geol Stud Geol Vol. 4, pp.281–288. Tulsa, OK: American Association of Petroleum Geologists.

Ogura M, Kato M, Arai N, Sasada T, Sakaki Y. 1992. Magnetic particles in chum salmon (*Oncorhynchus keta*): extraction and transmission electron microscopy. *Can J Zool* 70:874–877.

Ogutu-Ohwayo R. 1990. The decline of the native fishes of lakes Victoria and Kyoga (East Africa) and the impact of introduced species, especially the Nile perch, *Lates niloticus*, and the Nile tilapia, *Oreochromis niloticus*. *Env Biol Fish* 27:81–96.

Ogutu-Ohwayo R. 1999. Deterioration in length–weight relationships of Nile perch, *Lates niloticus* L. in lakes Victoria, Kyoga and Nabugabo. *Hydrobiologia* 403:81–86.

Ogutu-Ohwayo R, Hecky RE, Cohen AS, et al. 1997. Human impacts on the African Great Lakes. *Env Biol Fish* 50:117–131.

Ohira M, Tsunoda H, Nishida K, Mitsuo Y, Senga Y. 2015. Niche processes and conservation implications of fish community assembly in a rice irrigation system. *Aquat Conserv Mar Freshw Ecosyst* 25:322–335.

Ohlberger J, Rogers LA, Stenseth NC. 2014. Stochasticity and determinism: how density-independent and density-dependent processes affect population variability. *PLoS One* 9(6):e98940. **https://doi.org/10.1371/journal.pone.0098940**.

Ohman MC, Rajasuriya A, Svensson S. 1998. The use of butterflyfishes (Chaetodontidae) as bio-indicators of habitat structure and human disturbance. *Ambio* 27:708–716.

Ohno S. 1970. *Evolution by gene duplication*. London: Allen & Unwin.

Ojima Y, Maeki K, Takayama S, Nogusa S. 1963. A cytotaxonomic study on the Salmonidae. *The Nucleus* 6:91–98.

Ojima Y, Yamamoto K. 1990. Cellular DNA contents of fishes determined by flow cytometry. *La Kromosomo II* 57:1871–1888.

Okes N, Sant G. 2019. *An overview of major shark traders, catchers and species*. Cambridge, UK: TRAFFIC.

Okuda N. 2001. The costs of reproduction to males and females of a paternal mouthbrooding cardinalfish *Apogon notatus*. *J Fish Biol* 58:776–787.

Ólafsdóttir GÁ, Snorrason SS, Ritchie MG. 2007. Morphological and genetic divergence of intralacustrine stickleback morphs in Iceland: a case for selective differentiation? *J Evol Biol* 20:603–616.

Oldfield RG. 2005. Genetic, abiotic and social influences on sex differentiation in cichlid fishes and the evolution of sequential hermaphroditism. *Fish Fish (Oxf)* 6:93–110.

Oliveira TPR, Ladich F, Abed-Navandi D, Souto AS, Rosa IL. 2014. Sounds produced by the longsnout seahorse: a study of their structure and functions. *J Zool* 294:114–121.

Olla BL, Bejda AJ, Martin AD. 1979. Seasonal dispersal and habitat selection of cunner, *Tautogolabrus adspersus*, and young tautog, *Tautoga onitis*, in Fire Island Inlet, Long Island, New York. *Fish Bull US* 77:255–261.

Olsen EM, Heino M, Lilly GR, et al. 2004. Maturation trends indicative of rapid evolution preceded the collapse of northern cod. *Nature* 428:932–935.

Olsen P, McCune AR. 1991. Morphology of the *Semionotus elegans* species group from the early Jurassic part of the Newark Supergroup of eastern North America with comments on the Family Semionotidae (Neopterygii). *J Vertebr Paleontol* 11:269–292.

Olson KR. 2011a. Branchial anatomy. In: Farrell AP, Stevens ED, Cech JJ, Richards JG, eds. *Encyclopedia of fish physiology: from genome to environment*, pp. 1095–1103. Boston: Elsevier/Academic Press.

Olson KR. 2011b. Physiology of resistance vessels. In: Farrell AP, Stevens ED, Cech JJ, Richards JG, eds. *Encyclopedia of fish physiology: from genome to environment*, pp. 1104–1110. Boston: Elsevier/Academic Press.

Olson KR, Farrell AP. 2011. Secondary circulation and lymphatic anatomy. In: Farrell AP, Stevens ED, Cech JJ, Richards JG, eds. *Encyclopedia of fish physiology: from genome to environment*, pp. 1161–1168. Boston: Elsevier/Academic Press.

Olsson C. 2011. Gut anatomy. In: Farrell AP, Stevens ED, Cech JJ, Richards JG, eds. *Encyclopedia of fish physiology: from genome to environment*, pp. 1268–1275. Boston: Elsevier/Academic Press.

Ondrackova M, Davidova M, Gelnar M, Jurajda P. 2006. Susceptibility of Prussian carp infected by metacercariae of *Posthodiplostomum cuticola* (v. Nordmann, 1832) to fish predation. *Ecol Res* 21:526–529.

Ono RD, Williams JD, Wagner A. 1983. *Vanishing fishes of North America*. Washington, DC: Stone Wall Press.

Orrell TM, Collette BB, Johnson GD. 2006. Molecular data support separate scombroid and xiphioid clades. *Bull Mar Sci* 79:505–519.

Osborne LL, Wiley MJ. 1992. Influence of tributary spatial position on the structure of warmwater fish communities. *Can J Fish Aquat Sci* 49:671–681.

Osenberg CW, Mittelbach GG, Wainwright PC. 1992. Two-stage life histories in fish: the interaction between juvenile competition and adult performance. *Ecology* 73:255–267.

Osenberg CW, Werner EE, Mittelbach GG, Hall DJ. 1988. Growth patterns in bluegill (*Lepomis macrochirus*) and pumpkinseed (*L. gibbosus*) sunfish: environmental variation and the importance of ontogenetic niche shifts. *Can J Fish Aquat Sci* 45:17–26.

Osse JWM, Muller M. 1980. A model of suction feeding in fishes with some implications for ventilation. In: Ali MA, ed. *Environmental physiology of fishes*, pp. 335–351. New York: Plenum.

Osse JWM, van den Boogaart JGM. 1995. Fish larvae, development, allometric growth, and the aquatic environment. *ICES Mar Sci Symp* 201:21–34.

Ostrander GK, ed. 2000. *The laboratory fish*. London: Academic Press.

Ostrander GK, Cheng KC, Wolf JC, Wolfe MJ. 2004. Shark cartilage, cancer and the growing threat of pseudoscience. *Can Res* 64:8485–8491.

Ott JF, Platt C. 1988. Postural changes occurring during one month of vestibular compensation in goldfish. *J Exp Biol* 138:359–374.

Overstrom NA. 1991. Estimated tooth replacement rate in captive sand tiger sharks (*Carcharias taurus* Rafinsesque, 1810). *Copeia* 1991:525–526.

Pace CM, Gibb AC. 2014. Sustained periodic terrestrial locomotion in air-breathing fishes. *J Fish Biol* 84:639–660.

Page LM. 1983. *Handbook of darters*. Neptune City, NJ: TFH Publications, Inc.

Page LM, Espinosa-Pérez H, Findley LT, et al. 2013. *Common and scientific names of fishes from the United States, Canada, and Mexico*, 7th edn. American Fisheries Society Special Publication 34. Bethesda, MD: American Fisheries Society.

Page RDM, Holmes EC. 1998. *Molecular evolution, a phylogenetic approach*. Oxford: Blackwell Science.

Paine MA, McDowell JR, Graves JE. 2007. Specific identification of western Atlantic Ocean scombrids using mitochondrial DNA cytochrome c oxidase subunit I (COI) gene region sequences. *Bull Mar Sci* 80:353–367.

Paine RT, Palmer AR. 1978. *Sicyases sanguineus*: a unique trophic generalist from the Chilean intertidal zone. *Copeia* 1978:75–81.

Palstra AP, De Graaf M, Sibbing FA. 2004. Riverine spawning and reproductive segregation in a lacustrine cyprinid species flock, facilitated by homing? *Anim Biol* 54:393–415.

Palstra AP, Fukaya K, Chiba H, Dirks RP, Planas JV, Ueda H. 2015. The olfactory transcriptome and progression of sexual maturation in homing Chum Salmon *Oncorhynchus keta*. *PLoS One* 10(9):e0137404. doi:10.137/journal.pone.0137404.

Palumbi SR. 2001. Humans as the world's greatest evolutionary force. *Science* 293:1786–1790.

Palumbi SR, Gaines SD, Leslie H, Warner RR. 2003. New wave: high-tech tools to help marine reserve research. *Frontiers Ecol Env* 1:73–79.

Pan Jiang. 1984. The phylogenetic position of the Eugaleaspida in China. *Proc Linn Soc NS Wales* 197:309–319.

Pannella G. 1971. Fish otoliths: daily growth layers and periodical patterns. *Science* 173:1124–1127.

Pappas K, Dunlap K. 2011. Shocking comments: electrocommunication in teleost fish. In: Farrell AP, Stevens ED, Cech JJ, Richards JG, eds. *Encyclopedia of fish physiology: from genome to environment*, pp. 699–706. Boston: Elsevier/Academic Press.

Passow CN, Greenway R, Arias-Rodriguez L, Jeyasingh PD, Tobler M. 2015. Reduction of energetic demands through modification of body size and routine metabolic rates in extremophile fish. *Physiol Biochem Zool* 88(4):371–383.

Pardini AT, Jones CS, Noble LR, et al. 2001. Sex-biased dispersal of great white sharks. *Nature* 412:139–140.

Parenti LR. 1984. Biogeography of the Andean killifish genus *Orestias* with comments on the species flock concept. In: Echelle AE, Kornfield I, eds. *Evolution of fish species flocks*, pp. 85–92. Orono, ME: University of Maine Press.

Paris CB, Cowen RK. 2004. Direct evidence of a biophysical retention mechanism for coral reef fish larvae. *Limnol Oceanogr* 49:1964–1979.

Paris CB, Cowen RK, Claro R, Lindeman KC. 2005. Larval transport pathways from Cuban snapper (Lutjanidae) spawning aggregations based on biophysical modeling. *Mar Ecol Prog Ser* 296:93–106.

Parker A, Kornfield I. 1996. Polygynandry in *Pseudotropheus zebra*, a cichlid fish from Lake Malawi. *Env Biol Fish* 47:345–352.

Parker AR. 2005. A geological history of reflecting optics. *J Roy Soc Interface* 2:1–17.

Parmentier E, Colleye O, Fine ML, et al. 2007. Sound production in the clownfish *Amphiprion clarkii*. *Science* 316:1006.

Parmentier E, Fine ML. 2016. Fish sound production: insights. In: Suthers RA, Fitch WT, Fay RR, Popper AN, eds. *Vertebrate sound production and acoustic communication*, pp. 19–49. New York: Springer International Publishing.

Parmentier E, Lagardere J-P, Braquegnier J-B, Vandewalle P, Fine ML. 2006. Sound production mechanism in carapid fish: first example with a slow sonic muscle. *J Exp Biol* 209:2952–2960.

Parmentier E, Vandewalle P. 2005. Further insight on carapid–holothuroid relationships. *Mar Biol* 146:455–465.

Parrish JK. 1988. Re-examining the selfish herd: are central fish safer? *Anim Behav* 38:1048–1053.

Parrish JK. 1989a. Layering with depth in a heterospecific fish aggregation. *Env Biol Fish* 26:79–85.

Parrish JK. 1989b. Predation on a school of flat-iron herring, *Harengula thrissina*. *Copeia* 1989:1089–1091.

Parrish JK. 1992. Levels of diurnal predation on a school of flat-iron herring, *Harengula thrissina*. *Env Biol Fish* 34:257–263.

Parrish JK, Hamner WM, eds. 1997. *Animal groups in three dimensions. How species aggregate*. Cambridge, UK: Cambridge University Press.

Parrish JK, Kroen WK. 1988. Sloughed mucus and drag reduction in a school of Atlantic silversides, *Menidia menidia*. *Mar Biol* 97:165–169.

Parrish JK, Turchin P. 1997. Individual decisions, traffic rules, and emergent pattern in schooling fish. In: Parrish JK, Hamner WM, eds. *Animal groups in three dimensions. How species aggregate*, pp. 126–142. Cambridge, UK: Cambridge University Press.

Parrish RH, Serra R, Grant WS. 1989. The monotypic sardines, *Sardina* and *Sardinops*: their taxonomy, distribution, stock structure, and zoogeography. *Can J Fish Aquat Sci* 46:2019–2036.

Parry JWL, Pierson SN, Wilkens H, Bowmaker JK. 2003. Multiple photopigments from the Mexican blind cavefish, *Astyanax fasciatus*: a microspectrophotometric study. *Vision Res* 43:31–41.

Parsons KJ, Robinson BW. 2007. Foraging performance of diet-induced morphotypes in pumpkinseed sunfish (*Lepomis gibbosus*) favours resource polymorphism. *J Evol Biol* 20:673–684.

Partridge BL. 1982. Structure and function of fish schools. *Sci Am* 245:114–123.

Partridge BL, Johansson J, Kalish J. 1983. The structure of schools of giant bluefin tuna in Cape Cod Bay. *Env Biol Fish* 9:253–262.

Parzefall J. 1993. Behavioural ecology of cave-dwelling fishes. In: Pitcher TJ, ed. *Behaviour of teleost fishes*, 2nd edn, pp. 573–606. London: Chapman & Hall.

Pastana MNL, Johnson GD, Datovo A. 2021. Comprehensive phenotypic phylogenetic analysis supports the monophyly of stromateiform fishes (Teleostei: Percomorphacea). *Zool J Linnaean Soc* XX:1–123. **https://doi.org/10.1093/zoolinnean/zlab058**.

Paszkowski CA, Olla BL. 1985. Social interactions of coho salmon *Oncorhynhchus kisutch* smolts in seawater. *Can J Zool* 63:2401–2407.

Patarnello T, Volkaert FAMJ, Castilho R. 2007. Pillars of Hercules: is the Atlantic–Mediterranean transition a phylogeographic break? *Molec Ecol* 16:4426–4444.

Paterson SE. 1998. Group occurrence of great barracuda (*Sphyraena barracuda*) in the Turks and Caicos Islands. *Bull Mar Sci* 63:633–638.

Patterson C. 1965. The phylogeny of the chimaeroids. *Phil Trans Roy Soc Lond Biol Sci* 249:101–219.

Patterson C. 1982. Morphology and interrelationships of primitive actinopterygian fishes. *Am Zool* 22:241–259.

Patterson C, Johnson GD. 1995. The intermuscular bones and ligaments of teleostean fishes. *Smithsonian Contrib Zool* 559:83.

Patterson C, Longbottom AE. 1989. An Eocene amiid fish from Mali, West Africa. *Copeia* 1989:827–836.

Patterson HM, Kingsford MJ, McCulloch MT. 2005. Resolution of the early life history of a reef fish using otolith chemistry. *Coral Reefs* 24:222–229.

Pauli BD, Sih A. 2017. Behavioural responses to human-induced change: why fishing should not be ignored. *Evol Appl* 2017:1–10

Pauly D, Christensen V, Dalsgaard J, et al. 1998. Fishing down marine food webs. *Science* 279:860–863.

Pauly D, Christensen V, Froese R, Palomares M. 2000a. Fishing down aquatic food webs. *Am Sci* 88(1):46–51.

Pauly D, Christensen V, Walters C. 2000b. Ecopath, ecosim, and ecospace as tools for evaluating ecosystem impact of fisheries. *ICES J Mar Sci* 57(3):697–706.

Pauly D, Zeller D. 2016. Catch reconstructions reveal that global marine fisheries catches are higher than reported and declining. *Nat Comm* 7:10244. **https://doi.org/10.1038/ncomms10244**.

Pavlov V, Rosental B, Hansen NF, et al. 2017. Hydraulic control of tuna fins: a role for the lymphatic system in vertebrate locomotion. *Science* 357:310–314.

Paxton JR, Eschmeyer WN, eds. 1994. *Encyclopedia of fishes*. Sydney, Australia: University of New South Wales Press.

Paxton JR, Eschmeyer WN, eds. 1998. *Encyclopedia of fishes*, 2nd edn. San Diego, CA: Academic Press.

Peck KA, Lomax DL, Olson OP, Sol SY, Swanson P, Johnson LL. 2011. Development of an enzyme-linked immunosorbent assay for quantifying vitellogenin in Pacific salmon and assessment of field exposure to environmental estrogens. *Environ Toxicol Chem* 30:477–486.

Pedersen RA. 1971. DNA content, ribosomal gene multiplicity, and cell size in fish. *J Exp Zool* 177:65–79.

Pelster B. 2011. Swimbladder function and buoyancy control in fishes. In: Farrell AP, Stevens ED, Cech JJ, Richards JG, eds. *Encyclopedia of fish physiology: from genome to environment*, pp. 526–534. Boston: Elsevier/Academic Press.

Pelster B, Scheid P. 1992. Countercurrent concentration and gas secretion in the fish swim bladder. *Physiol Zool* 65:1–16.

Penney RK, Goldspink G. 1980. Temperature adaptation of sarcoplasmic reticulum of fish muscle. *J Therm Biol* 5:63–68.

Pereda AE, Faber DS. 2011. Physiology of the Mauthner cell: discovery and properties. In: Farrell AP, Stevens ED, Cech JJ, Richards JG, eds. *Encyclopedia of fish physiology: from genome to environment*, pp. 66–72. Boston: Elsevier/Academic Press.

Perra P. 1992. By-catch reduction devices as a conservation measure. *Fisheries* 17(1):28–29.

Perrone M, Jr., Zaret TM. 1979. Parental care patterns of fishes. *Am Natur* 113:351–361.

Perry AL, Low PJ, Ellis JR, Reynolds JD. 2005. Climate change and distribution shifts in marine fishes. *Science* 308:1912–1915.

Perry SF, Shahsavarani A, Georgalis T, Bayaa M, Furimsky M, Thomas SLY. 2003. Channels, pumps, and exchangers in the gill and kidney of freshwater fishes: their role in ionic and acid–base regulation. *J Exp Zool* 300A:53–62.

Perry SF, Tufts BL. 1998. *Fish respiration. Fish physiology*, Vol. 17. New York: Academic Press.

Peterman RM. 2002. Ecocertification: an incentive for dealing effectively with uncertainty, risk, and burden of proof in fisheries. *Bull Mar Sci* 70:669–681.

Peters HM. 1978. On the mechanism of air ventilation in anabantoids (Pisces: Teleostei). *Zoomorphologie* 89:93–123.

Peters RC, Evers HP. 1985. Frequency selectivity in the ampullary system of an elasmobrach fish (*Scyliorhinus canicula*). *J Exp Biol* 118:99–109.

Peters RC, van Wessel T, van den Wollenberg BJW, Bretschneider F, Olijslagers AE. 2002. The bioelectric field of the catfish *Ictalurus nebulosus*. *J Physiol Paris* 96:397–404.

Petersen CW, Warner RR, Cohen S, Hess HC, Sewell AT. 1992. Fertilization success in a reef fish. *Ecology* 73:391–401.

Petersen CW, Warner RR, Shapiro DY, Marconato A. 2001. Components of fertilization success in the bluehead wrasse, *Thalassoma bifasciatum*. *Behav Ecol* 12:237–245.

Petersen JH, Paukert CP. 2005. Development of a bioenergetics model for humpback chub and evaluation of water temperature changes in the Grand Canyon, Colorado River. *Trans Am Fish Soc* 134:960–974.

Petrosky BR, Magnuson JJ. 1973. Behavioral responses of northern pike, yellow perch and bluegill to oxygen concentrations under simulated winterkill conditions. *Copeia* 1973:124–133.

PEW. 2014. Risky decisions: how denial and delay brought disaster to New England's historic fishing grounds. A brief from the PEW Charitable Trusts, October 2014. **http://www.pewtrusts.org/en/research-and-analysis/issue-briefs/2014/09/risky-decisions**; accessed 21 August 2017.

Pfeiffer W. 1982. Chemical signals in communication. In: Hara TJ, ed. *Chemoreception in fishes*, pp. 307–326. Amsterdam: Elsevier.

Pfeiler E. 1986. Towards an explanation of the developmental strategy in leptocephalus larvae of marine teleost fishes. *Env Biol Fish* 15:3–13.

Pfeiler E, Bitler BG, Ulloa R. 2006. Phylogenetic relationships of the shafted bonefish *Albula nemoptera* (Albuliformes: Albulidae) from the eastern Pacific based on cytochrome *b* sequence analysis. *Copeia* 2006:778–784.

Philipp DP, Cooke SJ, Claussen JE, Koppelman J, Suski CD, Burkett D. 2009. Selection for vulnerability to angling in largemouth bass. *Trans Am Fish Soc* 138:189–199.

Pietsch TW. 1974. Osteology and relationships of ceratioid anglerfishes of the family Oneirodidae, with a review of the genus Oneirodes Lutken. *Natur Hist Mus Los Ang City Bull* 18:1–112.

Pietsch TW. 1976. Dimorphism, parasitism and sex: reproductive strategies among deepsea ceratioid anglerfishes. *Copeia* 1976: 781–793.

Pietsch TW. 1978. The feeding mechanism of *Stylephorus chordatus* (Teleostei: Lampridiformes): functional and ecological implications. *Copeia* 1978:255–262.

Pietsch TW. 2005. Dimorphism, parasitism, and sex revisited: modes of reproduction among deep-sea ceratioid anglerfishes (Teleostei: Lophiiformes). *Ichthyol Res* 52:207–236.

Pietsch TW. 2010. *The curious death of Peter Artedi: a mystery in the history of science.* New York: Scott & Nix Inc.

Pietsch TW, Anderson WD, Jr., eds. 1997. *Collection building in ichthyology and herpetology.* Special Publication No. 3. Lawrence, KS: Allen Press American Society of Ichthyologists and Herpetologists.

Pietsch TW, Grobecker DB. 1978. The compleat angler: aggressive mimicry in antennariid anglerfish. *Science* 201:369–370.

Pietsch TW, Grobecker DB. 1987. *Frogfishes of the world: systematics, zoogeography, and behavioral ecology.* Stanford, CT: Stanford University Press.

Pietsch TW, Orr JW. 2007. Phylogenetic relationships of deep-sea anglerfishes of the suborder Ceratioidei (Teleostei: Lophiiformes) based on morphology. *Copeia* 2007:1–34.

Piferrer F. 2011. Endocrine control of sex differentiation in fish. In: Farrell AP, Stevens ED, Cech JJ, Richards JG, eds. *Encyclopedia of fish physiology: from genome to environment*, pp. 1490–1499. Boston: Elsevier/Academic Press.

Pikitch EK, Doukakis P, Lauck L, Chakrabarty P, Erickson DL. 2005. Status, trends and management of sturgeon and paddlefish fisheries. *Fish Fish (Oxf)* 6:233–265.

Pinsky ML, Palumbi SR, Andrefouet S, Purkis SJ. 2012. Open and closed seascapes: where does habitat patchiness create populations with high fractions of self-recruitment? *Ecol Appl* 22:1257–1267.

Pister EP. 1981. The conservation of desert fishes. In: Naiman RJ, Soltz DL, eds. *Fishes in North American deserts*, pp. 411–445. New York: Wiley & Sons.

Pister EP. 1992. Ethical considerations in conservation of biodiversity. *Trans N Am Wildl Nat Res Conf* 57:355–364.

Pitcher TJ. 1983. Heuristic definitions of shoaling behaviour. *Anim Behav* 31:611–613.

Pitcher TJ, ed. 1993. *The behaviour of teleost fishes*, 2nd edn. London: Chapman & Hall.

Pitcher TJ, Hart PJB. 1982. *Fisheries ecology.* London: Croom Helm.

Pitcher TJ, Parrish JK. 1993. Functions of shoaling behaviour in teleosts. In: Pitcher TJ, ed. *The behaviour of teleost fishes*, 2nd edn, pp. 363–439. London: Chapman & Hall.

Pitcher TJ, Partridge BL, Wardle CS. 1976. A blind fish can school. *Science* 194:963–965.

Pitcher TJ, Wyche CJ. 1983. Predator avoidance behaviour of sand-eel schools: why schools seldom split. In: Noakes DLG, Lindquist BG, Helfman GS, Ward JA, eds. *Predators and prey in fishes*, pp. 193–204. The Hague: Dr W. Junk.

Pittman SJ, McAlpine CA. 2003. Movements of marine fish and decapod crustaceans: process, theory and application. *Adv Mar Biol* 44:205–294.

Place AR, Powers DR. 1979. Genetic variation and relative catalytic efficiencies: lactate dehydrogenase B allozyme of *Fundulus heteroclitus*. *Proc Natl Acad Sci USA* 76:2354–2358.

Plachta DTT, Popper AN. 2003. Evasive responses of American shad (*Alosa sapidissima*) to ultrasonic stimuli. *Acoust Res Lett Online* 4:25–30.

Planes S, Lecaillon G, Lenfant P, Meekan M. 2002. Genetic and demographic variation in new recruits of *Naso unicornis*. *J Fish Biol* 61:1033–1049.

Plenderleith M, van Oosterhout C, Robinson RL, Turner GF. 2005. Female preference for conspecific males based on olfactory cues in a Lake Malawi cichlid fish. *Biol Lett* 1:411–414.

Poffenberger J. 1999. *Our living oceans. Unit 6, Atlantic shark fisheries*. NOAA Technical Memorandum NMFS-F/SPO-41. Washington, DC: National Oceanic and Atmospheric Administration.

Policansky D. 1982a. Influence of age, size, and temperature on metamorphosis in the starry flounder, *Platichthys stellatus*. *Can J Fish Aquat Sci* 39:514–517.

Policansky D. 1982b. The asymmetry of flounders. *Sci Am* 246:116–122.

Policansky D. 1982c. Sex change in plants and animals. *Ann Rev Ecol Syst* 13:471–495.

Policansky D. 1993a. Evolution and management of exploited fish populations. In: Kruse G, Eggers DM, Marasco RJ, Pautzke C, Quinn TJ, eds. *Management strategies for exploited fish population*, pp. 651–664. Fairbanks, AK: Alaska Sea Grant College Program, AK-SG-93-02.

Policansky D. 1993b. Fishing as a cause of evolution in fishes. In: Stokes TK, McGlade JM, Law R, eds. *The exploitation of evolving resources*, pp. 2–18. Lecture Notes in Biomathematics No. 99. Berlin: Springer-Verlag.

Policansky D, Magnuson JJ. 1998. Genetics, metapopulations, and ecosystem management of fisheries. *Ecol Applic* 8:S119–S123.

Policarpo M, Bemis KE, Tyler JC, et al. 2021. Evolutionary dynamics of the OR gene repertoire in teleost fishes: evidence of an association with changes in olfactory epithelium shape. Mol Biol Evol msab145. doi: doe.org/10.1093/molbev/msab145.

Poling KR, Fuiman LA. 1998. Sensory development and its relation to habitat change in three species of Sciaenids. *Brain Behav Evol* 52:270–284.

Polis GA, Power ME, Huxel GR, eds. 2004. *Food webs at the landscape level*. Chicago: University of Chicago Press.

Polis GA, Winemiller KO, eds. 1996. *Food webs: integration of patterns and dynamics*. New York: Chapman & Hall.

Polloni P, Haedrich R, Rowe G, Clifford CH. 1979. The size-depth relationship in deep ocean animals. *Int Rev Ges Hydrobiol* 64:398–446.

Pollux BJA, Verberk WCEP, Dorenbosch M, et al. 2007. Habitat selection during settlement of three Caribbean coral reef fishes: indications for directed settlement to seagrass beds and mangroves. *Limnol Oceanogr* 52:903–907.

Polovina JJ, Ralston S, eds. 1987. *Tropical snappers and groupers, biology and fisheries management*. Boulder, CO: Westview Press.

Popper AN. 2011. Auditory system morphology. In: Farrell AP, Stevens ED, Cech JJ, Richards JG, eds. *Encyclopedia of fish physiology: from genome to environment*, pp. 252–261. Boston: Elsevier/Academic Press.

Popper AN, Fay RR. 2011. Rethinking sound detection by fishes. *Hearing Res* 273:25–36.

Popper AN, Hawkins AD. 2019. An overview of fish bioacoustics and the impacts of anthropogenic sounds on fishes. *J Fish Biol* 94:692–713.

Popper AN, Plachta DTT, Mann DA, Higgs D. 2004. Response of clupeid fish to ultrasound: a review. *ICES J Mar Sci* 61:1057–1061.

Popple ID, Hunte W. 2005. Movement patterns of *Cephalopholis cruentata* in a marine reserve in St Lucia WI, obtained from ultrasonic telemetry. *J Fish Biol* 67:981–992.

Portner HO, Knust R. 2007. Climate change affects marine fishes through the oxygen limitation of thermal tolerance. *Science* 315:95–97.

Portz D, Tyus H. 2004. Fish humps in two Colorado River fishes: a morphological response to cyprinid predation? *Env Biol Fish* 71:233–245.

Posey MH, Ambrose WG. 1994. Effects of proximity to an offshore hard-bottom reef on infaunal abundances. *Mar Biol* 118:745–753.

Poss SG, Boschung HT. 1996. Lancelets (Cephalochordata: Branchiostomatidae): how many species are valid? *Israel J Zool* 42 (suppl):S13–S66.

Poss SG, Collette BB. 1995. Second survey of fish collections in the United States and Canada. *Copeia* 1995:48–70.

Potter IC, Gill HS. 2003. Adaptive radiation of lampreys. *J Great Lakes Res* 29 (suppl 1):95–112.

Potthoff T. 1984. Clearing and staining techniques. In: Moser HG, Richards WJ, Cohen DM, Fahay MP, Kendall AW, Jr., Richardson SL, eds. *Ontogeny and systematics of fishes*, pp. 35–37. Special Publication No. 1. Lawrence, KS: Allen Press for the American Society Ichthyologists and Herpetologists.

Potts GW. 1980. The predatory behaviour of *Caranx melampygus* (Pisces) in the channel environment of Aldabra Atoll (Indian Ocean). *J Zool Lond* 192:323–350.

Potts GW. 1984. Parental behaviour in temperate marine teleosts with special reference to the development of nest structures. In: Potts GW, Wootton RJ, eds. *Fish reproduction: strategies and tactics*, pp. 223–244. London: Academic Press.

Potts GW, Wootton RJ, eds. 1984. *Fish reproduction: strategies and tactics*. London: Academic Press.

Pough FH, Heiser JB, McFarland WN. 1989. *Vertebrate life*, 3rd edn. New York: Macmillan.

Pough FH, Janis CM, Heiser JB. 2001. *Vertebrate life*, 6th edn. Upper Saddle River, NJ: Prentice Hall.

Pough FH, Janis CM, Heiser JB. 2005. *Vertebrate life*, 7th edn. Upper Saddle River, NJ: Pearson Prentice Hall.

Pough FH, Janis CM, Heiser JB. 2012. *Vertebrate life*, 9th edn. Upper Saddle River, NJ: Pearson Prentice Hall.

Poulin R, FitzGerald GJ. 1989. Shoaling as an anti-ectoparasite mechanism in juvenile sticklebacks (*Gasterosteus* spp.). *Behav Ecol Sociobiol* 24:251–255.

Poulson TL. 1963. Cave adaptation in amblyopsid fishes. *Am Midl Natur* 70:257–290.

Poulson TL, White WB. 1969. The cave environment. *Science* 165:971–981.

Pouyaud L, Desmarais E, Chenuil A, Agnese JF, Bonhomme F. 1999a. Kin cohesiveness and possible inbreeding in the mouthbrooding tilapia *Sarotherodon melanotheron* (Pisces Cichlidae). *Molec Ecol* 8:803–812.

Pouyaud L, Wirjoatmodjoc S, Rachmatikac I, Tjakrawidjajac A, Hadiatyc R, Hadie W. 1999b. Une nouvelle espèce de coelacanthe. Preuves génétiques et morphologiques. *Comptes rendus de l'Académie des sciences Série III Sciences de la vie* 322:261–267.

Powell M, Kavanaugh S, Sower S. 2005. Current knowledge of hagfish reproduction: implications for fisheries management. *Integr Comp Biol* 45:158–165.

Power G, Brown RS, Imhof JG. 1999. Groundwater and fish – insights from northern North America. *Hydrol Process* 13:401–422.

Power ME. 1987. Predator avoidance by grazing fishes in temperate and tropical streams: importance of stream depth and prey size. In: Kerfoot WC, Sih A eds. *Predation: direct and indirect impacts on aquatic communities*, pp. 333–351. Hanover: University Press of New England.

Power ME. 1990. Effects of fish in river food webs. *Science* 250:811–814.

Power ME, Dudley TL, Cooper SD. 1989. Grazing catfish, fishing birds, and attached algae in a Panamanian stream. *Env Biol Fish* 26:285–294.

Pratt HL, Jr., Carrier JC. 2001. A review of elasmobranch reproductive behavior with a case study on the nurse shark, *Ginglymostoma cirratum*. *Env Biol Fish* 60:157–188.

Pratt HL, Castro JI. 1991. Shark reproduction: parental investment and limited fisheries – an overview. In: Gruber SH, ed. *Discovering sharks*, pp. 56–60. American Littoral Society Special Publication No. 14. Highlands, NJ: American Littoral Society.

Pratt HL, Jr., Gruber SH, Taniuchi T, eds. 1990. *Elasmobranchs as living resources: advances in the biology, ecology, systematics, and the status of fisheries*. NOAA Technical Report No. 90. Washington, DC: National Oceanic and Atmospheric Administration.

Prejs A. 1984. Herbivory by temperate freshwater fishes and its consequences. *Env Biol Fish* 10:281–296.

Pressley PH. 1981. Parental effort and the evolution of nest-guarding tactics in the threespine stickleback, *Gasterosteus aculeatus* L. *Evolution* 35:282–295.

Preston JL. 1978. Communication systems and social interactions in a goby-shrimp symbiosis. *Anim Behav* 26:791–802.

Price DJ. 1984. Genetics of sex determination in fishes – a brief review. In: Potts GW, Wootton RJ, eds. *Fish reproduction: strategies and tactics*, pp. 77–89. London: Academic Press.

Priede IG, Froese R, Bailey DM, et al. 2006. The absence of sharks from abyssal regions of the world's oceans. *Proc Roy Soc Lond B Biol Sci* 273:1435–1441.

Priede IG, Froese R. 2013. Colonization of the deep sea by fishes. *J Fish Biol* 83:1528–1550.

Primor N, Parness J, Zlotkin E. 1978. Pardaxin: the toxic factor from the skin secretion of the flatfish *Pardachirus marmoratus* (Soleidae). In: Rosenberg P, ed. *Toxins: animal, plant and microbial*, pp. 539–547. Oxford: Pergamon Press.

Pringle CM, Freeman MC, Freeman BJ. 2000. Regional effects of hydrologic alterations on riverine macrobiota in the New World: tropical–temperate comparisons. *BioScience* 50:807–823.

Pringle RM. 2005. The origins of the Nile perch in Lake Victoria. *BioScience* 55:780–787.

Proudlove GS. 1997a. The conservation status of hypogean fishes. In: *Proceedings 12th International Congress of Speleology, Vol. 3, Symposium 9, Biospeleology*, pp. 355–358. La Chaux-de-Fonds, Switzerland.

Proudlove GS. 1997b. A synopsis of the hypogean fishes of the world. In: *Proceedings 12th International Congress of Speleology, Vol. 3, Symposium 9, Biospeleology*, pp. 351–354. La Chaux-de-Fonds, Switzerland.

Proudlove GS. 2006. *Subterranean fishes of the world. An account of the subterranean (hypogean) fishes described up to 2003 with a bibliography 1541–2004*. Moulis, France: International Society for Subterranean Biology.

Pujolar JM, Maes GE, Volckaert FAM. 2006. Genetic patchiness among recruits in the European eel *Anguilla anguilla*. *Mar Ecol Prog Ser* 307:209–217.

Purcell EM. 1977. Life at low Reynolds numbers. *Am J Physics* 45:3–11.

Puritz JB, Matz MV, Toonen RJ, Weber JN, Bolnick DI, Bird CE. 2014. Demystifying the RAD fad. *Mol Ecol* 23:5937–5942.

Purnell MA. 1995. Microwear on conodont elements and macrophagy in the first vertebrates. *Nature* 374:798–800.

Pusack TJ, Christie MR, Johnson DW, Stalling CD, Hixon MA. 2014. Spatial and temporal patterns of larval dispersal in a coral-reef fish metapopulation: evidence of variable reproductive success. *Mol Ecol* 23:3396–3408.

Pusey BJ. 1989. Aestivation in the teleost fish *Lepidogalaxias salamandroides* (Mees). *Comp Biochem Physiol* 92A:137–138.

Pusey BJ. 1990. Seasonality, aestivation and the life history of the salamanderfish *Lepidogalaxias salamandroides* (Pisces: Lepidogalaxiidae). *Env Biol Fish* 29:15–26.

Pusey BJ, Kennard M, Arthington A. 2004. *Freshwater fishes of north-eastern Australia*. Collingwood, Australia: CSIRO Publishing.

Putland RL, Montgomery JC, Radford CA. 2019. Ecology of fish hearing. *J Fish Biol* 95:39–52.

Putman NF, Lohmann KJ, Putman EM, Quinn TP, Klimely AP, Noakes DLG. 2013. Evidence for geomagnetic imprinting as a homing mechanism in Pacific Salmon. *Curr Biol* 23(4):312–316.

Pyle RL. 1993. Marine aquarium fish. In: Wright A, Hill L, eds. *Nearshore marine resources of the South Pacific*, pp. 135–176. Suva, Fiji: Information for Fisheries Development and Management.

Pyle RL, Randall JE. 1994. A review of hybridization in marine angelfishes (Perciformes: Pomacanthidae). *Env Biol Fish* 41:127–145.

Quattro JM, Jones WJ. 1999. Amplification primers that target locus-specific introns in Actinopterygian fishes. *Copeia* 1999:191–196.

Quattro JM, Leberg PL, Douglas ME, Vrijenhoek RC. 1996. Molecular evidence for a unique evolutionary lineage of endangered Sonoran Desert fish (genus *Poeciliopsis*). *Conserv Biol* 10:128–135.

Quattro JM, Vrijenhoek RC. 1989. Fitness differences among remnant populations of the endangered Sonoran topminnow. *Science* 245:976–978.

Quinn TP. 1982. A model for salmon navigation on the high seas. In: Brannon EL, Salo EO, eds. *Salmon and trout migratory behavior symposium*, pp. 229–237. Seattle, WA: School of Fisheries, University of Washington.

Quinn TP. 1984. Homing and straying in Pacific salmon. In: McCleave JD, Arnold GP, Dodson JJ, Neill WH, eds. *Mechanisms of migration in fishes*, pp. 357–362. New York: Plenum Press.

Quinn TP. 2005. *The behavior and ecology of Pacific salmon and trout*. Seattle: University of Washington Press.

Quinn TP, Brannon EL. 1982. The use of celestial and magnetic cues by orienting sockeye salmon smolts. *J Comp Physiol A* 147:547–552.

Quinn TP, Dittman AH. 1990. Pacific salmon migrations and homing: mechanisms and adaptive significance. *Trends Ecol Evol* 5:174–177.

Quinn TP, Leggett WC. 1987. Perspectives on the marine migrations of diadromous fishes. In: Dadswell MJ, Klauda RJ, Moffitt CM, Saunders RL, Rulifson RA, Cooper JE, eds. *Common strategies of anadromous and catadromous fishes*, pp. 377–388. American Fisheries Society Symposium 1. Bethesda, MD: American Fisheries Society.

Quinn TP, Peterson JA, Gallucci V, Hershberger WK, Brannon EL. 2002. Artificial selection and environmental change: countervailing factors affecting the timing of spawning by coho and chinook salmon. *Trans Am Fish Soc* 131:591–598.

Quinn TP, Stewart IJ, Boatright CP. 2006. Experimental evidence of homing to site of incubation by mature sockeye salmon, *Oncorhynchus nerka*. *Anim Behav* 72:941–949.

Rabin LA, Greene CM. 2002. Changes to acoustic communication systems in human-altered environments. *J Comp Psychol* 116:137–141.

Rahel FJ. 2000. Homogenization of fish faunas across the United States. *Science* 288:854–856.

Rahel FJ. 2002. Homogenization of freshwater faunas. *Ann Rev Ecol Syst* 33:291–315.

Rahel FJ, Lyons JD, Cochran PA. 1984. Stochastic or deterministic regulation of assemblage structure? It may depend on how the assemblage is defined. *Am Natur* 124:582–589.

Rakes PL, Shute JR, Shute PW. 1999. Reproductive behaviour, captive breeding, and restoration ecology of endangered fishes. *Env Biol Fish* 55:31–42.

Ramsay JB, Wilga CD. 2007. Tooth and jaw mechanics in whitespotted bamboo sharks (*Chiloscyllium plagiosum*). *J Morphol* 268:664–682.

Randall DJ. 2011. Carbon dioxide transport and excretion. In: Farrell AP, Stevens ED, Cech JJ, Richards JG, eds. *Encyclopedia of fish physiology: from genome to environment*, pp. 909–915. Boston: Elsevier/Academic Press.

Randall DJ, Farrell AP, eds. 1997. *Deep-sea fishes. Fish physiology*, Vol. 16. San Diego, CA: Academic Press.

Randall DJ, Ip YK, Chew SF, Wilson JM. 2004. Air breathing and ammonia excretion in the giant mudskipper, *Periophthalmodon schlosseri*. *Physiol Biochem Zool* 77:783–788.

Randall JE. 1958. Review of ciguatera tropical fish poisoning with a tentative explanation of its cause. *Bull Mar Sci* 8:235–267.

Randall JE. 1973. Size of the great white shark (*Carcharodon*). *Science* 181:169–170.

Randall JE. 1974. The effect of fishes on coral reefs. *Proc 2nd Internat Coral Reef Symp* 1:159–166.

Randall JE. 1987. Refutation of lengths of 11.3, 9.0 and 6.4 m attributed to the white shark, *Carcharodon carcharias*. *Calif Fish Game* 73:163–168.

Randall JE. 2005. A review of mimicry in marine fishes. *Zool Stud* 44:299–328.

Randall JE. 2007. *Reef and shore fishes of the Hawaiian Islands*. Honolulu, HI: University of Hawaii Sea Grant College Program.

Randall JE, Cea A. 2011. *Shore Fishes of Easter Island*. Honolulu: University of Hawaii Press.

Randall JE, Delbeek JC. 2009. Comments on the extremes in longevity in fishes, with special reference to the Gobiidae. *Proc Calif Acad Sci* 60:447–454.

Randall JE, Emery AR. 1971. On the resemblance of the young of the fishes *Platax pinnatus* and *Plectorhynchus chaetodontoides* to flatworms and nudibranchs. *Zoologica* 1971:115–119.

Randall JE, Fautin DG. 2002. Fishes other than anemonefishes that associate with sea anemones. *Coral Reefs* 21:188–190.

Randall JE, Greenfield DW. 2001. A preliminary review of the Indo-Pacific gobiid fishes of the genus *Gnatholepis*. *Ichthyol Bull JLB Smith Inst Ichthyol* 69:1–17.

Randall JE, Helfman GS. 1972. *Diproctacanthus xanthurus*, a cleaner wrasse from the Palau Islands, with notes on other cleaning fishes. *Trop Fish Hobby* 197(11):87–95.

Randall JE, Kuiter RH. 1989. The juvenile Indo-Pacific grouper *Anyperodon leucogrammicus*, a mimic of the wrasse *Halichoeres purpurescens* and allied species, with a review of the recent literature on mimicry in fishes. *Rev Fr Aquariol* 16:51–56.

Randall JE, Randall HA. 1960. Examples of mimicry and protective resemblance in tropical marine fishes. *Bull Mar Sci Gulf Caribb* 10:444–480.

Rasch EM. 1985. DNA “standards” and the range of accurate DNA estimates by Feulgen absorption microspectrophotometry. In: Cowden RR, Harrison SH, eds. *Advances in microscopy*, pp. 137–166. New York: Alan R. Liss.

Raschi W, Adams WH. 1988. Depth-related modifications in the electroreceptive system of the eurybenthic skate *Raja radiata* (Chondrichthyes: Rajidae). *Copeia* 1988:116–123.

Raup DM. 1988. Diversity crises in the geological past. In: Wilson EO, Peter FM, eds. *Biodiversity*, pp. 51–57. Washington, DC: National Academy Press.

Ray GC. 2005. Connectivities of estuarine fishes to the coastal realm. *Estuar Coast Shelf Sci* 64:18–32.

Raymond HL. 1988. Effects of hydroelectric development and fisheries enhancement on spring and summer chinook salmon and steelhead in the Columbia River basin. *N Am J Fish Manage* 8:1–24.

Raymond JA. 1992. Glycerol is a colligative antifreeze in some northern fishes. *J Exp Zool* 262:347–352.

Razzaghi M, Mashjoor S, Kamrani E. 2017. Mean trophic level of coastal fisheries landings in the Persian Gulf (Hormuzgan Province), 2002–2011. *Chin J Ocean Limnol* 35:528–536. **https://doi.org/10.1007/s00343-017-5311-6**

Read TD, Petit RA, Joseph SJ, et al. 2017. Draft sequencing and assembly of the genome of the world’s largest fish, the whale shark: *Rhincodon typus* Smith 1828. *BMC Genom* 18:532

Reading BJ, Sullivan CV. 2011. Vitellogenesis in Fishes. In: Farrell AP, Stevens ED, Cech JJ, Richards JG, eds. *Encyclopedia of fish physiology: from genome to environment*, pp. 635–646. Boston: Elsevier/Academic Press.

Reckel F, Hoffman B, Melzer RR, Horppila J, Smola U. 2003. Photoreceptors and cone patterns in the retina of the smelt *Osmerus eperlanus* (L) (Osmeridae: Teleostei). *Acta Zool* 84:161–170.

Reebs S. 1992. Sleep, inactivity and circadian rhythms in fish. In: Ali MA, ed. *Rhythms in fishes*, pp. 127–135. New York: Plenum Press.

Reebs S. 2001. *Fish behavior in the aquarium and in the wild*. Ithaca, NY: Cornell University Press.

Reebs SG. 2002. Plasticity of diel and circadian activity rhythms in fishes. *Rev Fish Biol Fisheries* 12:349–371.

Reebs SG. 2011. Circadian rhythms in fish. In: Farrell AP, Stevens ED, Cech JJ, Richards JG, eds. *Encyclopedia of fish physiology: from genome to environment*, pp. 736–743. Boston: Elsevier/Academic Press.

Reed JM, Stockwell CA. 2014. Evaluating an icon of population persistence: the Devil’s Hole pupfish. *Proc R Soc B* 281:20141648. **https://doi.org/10.1098/rspb.2014.1648**.

Reese ES. 1975. A comparative field study of the social behaviour and related ecology of reef fishes of the family Chaetodontidae. *Z Tierpsychol* 37:37–61.

Reese ES. 1981. Predation on corals by fishes of the family Chaetodontidae: implications for conservation and management of coral reef ecosystems. *Bull Mar Sci* 31:594–604.

Reese ES. 1989. Orientation behavior of butterflyfishes (family Chaetodontidae) on coral reefs: spatial learning of route specific landmarks and cognitive maps. *Env Biol Fish* 24:79–78.

Reese ES, Lighter FJ. 1978. *Contrasts in behavior.* New York: Wiley-Interscience.

Regan CT. 1926. Organic evolution. In: *Report of the 93rd Meeting of the British Association for the Advancement of Science, 1925*, pp. 75–86.

Reid SG. 2011. Catecholamines. In: Farrell AP, Stevens ED, Cech JJ, Richards JG, eds. *Encyclopedia of fish physiology: from genome to environment*, pp. 1524–1533. Boston: Elsevier/Academic Press.

Reimchen TE. 1983. Structural relationships between spines and lateral plates in threespine stickleback (*Gasterosteus aculeatus*). *Evolution* 37:931–946.

Reimchen TE. 1991. Evolutionary attributes of headfirst prey manipulation and swallowing in piscivores. *Can J Zool* 69:2912–2916.

Reinboth R. 1975. *Intersexuality in the animal kingdom.* Berlin: Springer-Verlag.

Reis RE, Albert JS, Di Dario F, Mincarone MM, Petry P, Rocha LA. 2016. Fish biodiversity and conservation in South America. *J Fish Biol* 89:12–47.

Ren J, Buchinger T, Pu J, Jia L, Li W. 2014. Complete mitochondrial genomes of paired species northern brook lamprey *(Ichthyomyzon fossor)* and silver lamprey *(I. unicuspis)*. *Mitochondrial DNA Part A* 27:1862–1863.

Renault S, Maillet N, Normandeau E, Sauvage C, Derome N, Rogers SM, Bernatchez L. 2012. Genome-wide patterns of divergence during speciation: the lake whitefish case study. *Philos Trans R Soc B* 367:354–363.

Reynolds JD, Jennings S. 2000. The role of animal behaviour in marine conservation. In: Gosling LM, Sutherland WJ, eds. *Behaviour and conservation*, pp. 238–257. Cambridge, UK: Cambridge University Press.

Reznick DA, Bryga H, Endler JA. 1990. Experimentally induced life-history evolution in a natural population. *Nature* 346:357–359.

Reznick DN, Endler JA. 1982. The impact of predation on life history evolution in Trinidadian guppies (*Poecilia reticulata*). *Evolution* 36:160–177.

Reznick DN, Shaw FH, Rodd FH, Shaw RG. 1997. Evaluation of the rate of evolution in natural populations of guppies (*Poecilia reticulata*). *Science* 275:1934–1937.

Ribbink AJ. 1987. African lakes and their fishes: conservation scenarios and suggestions. *Env Biol Fish* 19:3–26.

Ribbink AJ. 1991. Distribution and ecology of the cichlid fishes of the African Great Lakes. In: Keenleyside MHA, ed. *Cichlid fishes: behaviour, ecology and evolution*, pp. 36–59. London: Chapman & Hall.

Rice AN, Lobel PS. 2004. The pharyngeal jaw apparatus of the Cichlidae and Pomacentridae: function in feeding and sound production. *Rev Fish Biol Fisheries* 13:433–444.

Rice CD, Arkoosh MR. 2002. Immunological indicators of environmental stress and disease susceptibility in fishes. in: Adams SM, ed. *Biological indicators of aquatic ecosystem stress*, pp. 187–200. Bethesda, MD: American Fisheries Society.

Rich PV, van Tets GF. 1985. *Kadimakara, extinct vertebrates of Australia.* Lilydale, Australia: Pioneer Design Studios.

Richards WJ. 1976. Some comments on Balon's terminology of fish development intervals. *J Fish Res Board Can* 33:1253–1254.

Richards WJ, Lindeman KC. 1987. Recruitment dynamics of reef fishes: planktonic processes, settlement and demersal ecologies, and fisheries analysis. *Bull Mar Sci* 41:392–410.

Richards ZT, Hobbs J-PA. 2015. Hybridization on coral reefs and the conservation of evolutionary novelty. *Curr Zool* 61:132–145.

Richter H, Lückstädt C, Focken UL, Becker K. 2000. An improved procedure to assess fish condition on the basis of length-weight relationships. *Arch Fish Mar Res* 48(3):226–235.

Rick IP, Modarressie R, Bakker TCM. 2006. UV wavelengths affect female mate choice in three-spined sticklebacks. *Anim Behav* 71:307–313.

Rickel S, Genin A. 2005. Twilight transitions in coral reef fish: the input of light-induced changes in foraging behaviour. *Anim Behav* 70:133–144.

Ricker WE. 1962. Regulation of the abundance of pink salmon stocks. In: Wilimovsky NJ, ed. *Symposium on pink salmon*, pp. 155–201. Vancouver, BC: University of British Columbia.

Ricker WE. 1975. Computation and interpretation of biological statistics of fish populations. *Bull Fish Res Board Can* 191:1–382.

Ricker WE. 1979. Growth rates and models. In: Hoar WS, Randall DJ, Brett JR, eds. *Bioenergetics and growth. Fish physiology*, Vol. 8, pp. 677–743. London: Academic Press.

Ricker WE. 1981. Changes in the average size and average age of Pacific salmon. *Can J Fish Aquat Sci* 38:1636–1656.

Riget FF, Nygaard KH, Christensen B. 1986. Population structure, ecological segregation and reproduction in a population of Arctic char (*Salvelinus alpinus*) from Lake Tasersvaq, Greenland. *Can J Fish Aquat Sci* 43:985–992.

Righton DA, Metcalfe JD. 2011. Eel migrations. In: Farrell AP, Stevens ED, Cech JJ, Richards JG, eds. *Encyclopedia of fish physiology: from genome to environment*, pp. 1937–1944. Boston: Elsevier/Academic Press.

Riginos C, Victor BC. 2001. Larval spatial distributions and other early life-history characteristics predict genetic differentiation in eastern Pacific blennioid fishes. *Proc Roy Soc Lond B Biol Sci* 268:1931–1936.

Rimmer DW, Wiebe WJ. 1987. Fermentative microbial digestion in herbivorous fishes. *J Fish Biol* 31:229–236.

Rincon-Sandoval M, Betancur-R R, Maldonado-Ocampo JA. 2019. Comparative phylogeography of trans-Andean freshwater fishes based on genome-wide nuclear and mitochondrial markers. *Mol Ecol* 28:1096–1115.

Rinne JN, Minckley WL. 1991. *Native fishes of arid lands: a dwindling resource of the desert Southwest.* Fort Collins, CO: USDA Forest Service General Technical Report RM-206.

Ritter EK. 2002. Analysis of sharksucker, *Echeneis naucrates*, induced behavior patterns in the blacktip shark, *Carcharhinus limbatus. Env Biol Fish* 65:111–115.

Rivas LR. 1978. Preliminary models of annual life history cycles of the North Atlantic bluefin tuna. In: Sharp GD, Dizon AE, eds. *The physiological ecology of tunas*, pp. 369–393. New York: Academic Press.

Rizzari JR, Frisch AJ, Hoey AS, McCormick MI. 2014. Not worth the risk: apex predators suppress herbivory on coral reefs. *Oikos* 123:829–836.

Robbins J. 1991. The real Jurassic Park. *Discover* 12(3):52–59.

Roberts CD. 1993. Comparative morphology of spined scales and their phylogenetic significance in the Teleostei. *Bull Mar Sci* 52:60–113.

Roberts CM. 1996. Settlement and beyond: population regulation and community structure of reef fishes. In: Polunin NVC, Roberts CM, eds. *Reef fisheries*, pp. 85–112. London: Chapman & Hall.

Roberts JL. 1964. Metabolic responses of fresh-water sunfish to seasonal photoperiods and temperatures. *Helgolander wiss Meeresunters* 9:459–473.

Roberts JL. 1975a. Active branchial and ram gill ventilation in fishes. *Biol Bull* 148:85–105.

Roberts JL. 1975b. Respiratory adaptations in aquatic animals. In: Vernberg, FJ, ed. *Physiological adaptations to the environment*, pp. 395–414. New York: Intext Educational Publications.

Roberts JL. 1978. Ram gill ventilation in fish. In: Sharp GD, Dizon AE, eds. *The physiological ecology of tunas*, pp. 83–88. New York: Academic Press.

Roberts TR. 1967. Tooth formation and replacement in characoid fishes. *Stanford Ichthyol Bull* 8:231–247.

Roberts TR. 1986. *Danionella translucida*, a new genus and species of cyprinid fish from Burma, one of the smallest living vertebrates. *Env Biol Fish* 16:231–241.

Roberts TR, Monkolprasit S. 1990. *Himantura chaophraya*, a new giant freshwater stingray from Thailand. *Jap J Ichthyol* 37:203–208.

Roberts Kingman GA, Vyas DN, Jones FC, et al. 2021. Predicting future from past: the genomic basis of recurrent and rapid stickleback evolution. *Sci Adv* 7:eabg5285.

Robertson DR. 1972. Social control of sex-reversal in a coral-reef fish. *Science* 117:1007–1009.

Robertson DR. 1982. Fish feces as fish food on a Pacific coral reef. *Mar Ecol Prog Ser* 7:253–265.

Robertson DR. 1987. Responses of two coral reef toadfishes (Batrachoididae) to the demise of their primary prey, the sea urchin *Diadema antillarum*. *Copeia* 1987:637–642.

Robertson DR. 1991. The role of adult biology in the timing of spawning of tropical reef fishes. In: Sale PF, ed. *The ecology of fishes on coral reefs*, pp. 356–386. San Diego, CA: Academic Press.

Robertson DR, Petersen CW, Brawn JD. 1990. Lunar reproductive cycles of benthic-brooding fishes: reflections of larval biology or adult biology? *Ecol Monogr* 60:311–329.

Robertson DR, Sweatman HPA, Fletcher EA, Cleland MG. 1976. Schooling as a mechanism for circumventing the territoriality of competitors. *Ecology* 57:1208–1220.

Rocha LA, Bass AL, Robertson DR, Bowen BW. 2002. Adult habitat preferences, larval dispersal, and the comparative phylogeography of three Atlantic surgeonfishes (Teleostei: Acanthuridae). *Molec Ecol* 11:243–252.

Rocha LA, Craig MT, Bowen BW. 2007. Phylogeography and the conservation genetics of coral reef fishes. *Coral Reefs* 26:501–512.

Rocha LA, Robertson DR, Rocha CR, Van Tassell JL, Craig MT, Bowen BW. 2005b. Recent invasion of the tropical Atlantic by an Indo-Pacific coral reef fish. *Molec Ecol* 14:3921–3928.

Rocha LA, Robertson DR, Roman J, Bowen BW. 2005a. Ecological speciation in tropical reef fishes. *Proc Roy Soc Lond B Biol Sci* 272:573–579.

Rodriguez-Moldes I. 2011. Functional morphology of the brains of cartilaginous fishes. In: Farrell AP, Stevens ED, Cech JJ, Richards JG, eds. *Encyclopedia of fish physiology: from genome to environment*, pp. 26–36. Boston: Elsevier/Academic Press.

Roessig JM, Woodley CM, Cech JJ, Hansen LJ. 2004. Effects of global climate change on marine and estuarine fishes and fisheries. *Rev Fish Biol Fisheries* 14:251–275.

Roest Crollius H, Weissenbach J. 2005. Fish genomics and biology. *Genome Res* 15:1675–1682.

Rogers SG, Van Den Avyle MJ. 1982. *Species profiles: life histories and environmental requirements of coastal fishes and invertebrates (South Atlantic) – summer flounder.* US Fish and Wildlife Service FWS/OBS-82/11.15. US Army Corps of Engineers TR EL-82-4:1–14.

Rohde FC, Arndt RG, Smith SM. 2001. Longitudinal succession of fishes in the Dan River in Virginia and North Carolina (Blue Ridge/Piedmont Provinces). *Southeastern Fishes Council Proc* 42:1–13. **http://trace.tennessee.edu/sfcproceedings/vol1/iss42/3**.

Rojo AL. 1991. *Dictionary of evolutionary fish osteology.* Boca Raton, FL: CRC Press.

Roman J, Bowen BW. 2000. The mock turtle syndrome: genetic identification of turtle meat purchased in the southeast United States. *Anim Conserv* 3:61–65.

Roman J, Santhuff S, Moler P, Bowen BW. 1999. Cryptic evolution and population structure of the alligator snapping turtle, *Macroclemys temminckii*. *Conserv Biol* 13:135–142.

Romano M, Console F, Pantaloni M, Fröbisch J. 2016. One hundred years of continental drift: the early Italian reaction to Wegener's 'visionary' theory, *Histor Biol.* **https://doi.org/10.1080/08912963.2016.1156677**.

Rombough PJ, Ure D. 1991. Partitioning of oxygen uptake between cutaneous and branchial surfaces in larval and young juvenile chinook salmon *Oncorhynchus tshawytscha*. *Physiol Zool* 64:717–727.

Rome LC. 1990. Influences of temperature on muscle recruitment and muscle function in vivo. *Am J Physiol* 259:R210–R222.

Rome LC, Cook C, Syme DA, et al. 1999. Trading force for speed: why superfast crossbridge kinetics leads to superlow forces. *Proc Natl Acad Sci USA* 96:5826–5831.

Rome LC, Syme DA, Hollingworth S, Lindstedt SL, Baylor SM. 1996. The whistle and the rattle: the design of sound producing muscles. *Proc Natl Acad Sci USA* 93:8095–8100.

Romer AS. 1962. *The vertebrate body*, 3rd edn. Philadelphia: W. B. Saunders Co.

Romero A. 1998. Threatened fishes of the world: *Typhlichthys subterraneus* (Girard, 1860) (Amblyopsidae). *Env Biol Fish* 53:74.

Romero A, Bennis L. 1998. Threatened fishes of the world: *Amblyopsis spelaea* DeKay, 1842 (Amblyopsidae). *Env Biol Fish* 51:420.

Ronquist F, Huelsenbeck JP. 2003. MRBAYES 3: Bayesian phylogenetic inference under mixed models. *Bioinformatics* 19:1572–1574.

Root TL, Price JT, Hall KR, Schneider SH, Rosenzweig C, Pounds JA. 2003. Fingerprints of global warming on wild animals and plants. *Nature* 421:57–60.

Rose DA. 1996. *An overview of world trade in sharks and other cartilaginous fishes.* Cambridge, UK: TRAFFIC International. **www.TRAFFIC.org**.

Rose DA. 1998. Shark fisheries and trade in the Americas. TRAFFIC North America. **https://www.traffic.org/site/assets/files/9503/shark-fisheries-and-trade-in-the-americas.pdf**.

Rose JD. 2002. The neurobehavioral nature of fishes and the question of awareness and pain. *Rev Fish Sci* 10:1–38.

Rose JD. 2007. Anthropomorphism and "mental welfare" of fishes. *Dis Aquat Organisms* 75:139–154.

Rose JD, Marrs GS, Lewis C, et al. 2000. Whirling disease behavior and its relation to pathology of brain stem and spinal cord in rainbow trout. *J Aquat Anim Health* 12:107–118.

Rosen DE. 1978. Vicariant patterns and historical explanation in biogeography. *Syst Zool* 27:159–188.

Rosen DE, Forey PL, Gardiner BG, Patterson C. 1981. Lungfishes, tetrapods, paleontology, and plesiomorphy. *Bull Am Mus Natur Hist* 167:163–275.

Rosen RA, Hales DC. 1981. Feeding of the paddlefish, *Polyodon spathula*. *Copeia* 1981:441–455.

Rosenblatt RH, Waples RS. 1986. A genetic comparison of allopatric populations of shore fish species from the eastern and central Pacific Ocean: dispersal or vicariance? *Copeia* 1986:275–284.

Rosenqvist G. 1990. Male mate choice and female–female competition for mates in the pipefish *Nerophis ophidion*. *Anim Behav* 39:1110–1115.

Ross MR. 1983. The frequency of nest construction and satellite male behavior in the fallfish minnow. *Env Biol Fish* 9:65–67.

Ross MR. 1997. *Fisheries conservation and management*. Upper Saddle River, NJ: Prentice Hall.

Ross ST. 1986. Resource partitioning in fish assemblages: a review of field studies. *Copeia* 1986:352–388.

Ross ST. 1991. Mechanisms structuring stream fish assemblages: are there lessons from introduced species? *Env Biol Fish* 30:359–368.

Rothans TC, Miller AC. 1991. A link between biologically imported particulate organic nutrients and the detritus food web in reef communities. *Mar Biol* 110:145–150.

Rothschild BJ. 1986. *Dynamics of marine fish populations*. Cambridge, MA: Harvard University Press.

Rotjan RD, Dimond JL, Thornhill DJ, et al. 2006. Chronic parrotfish grazing impedes coral recovery after bleaching. *Coral Reefs* 25:361–368.

Rotjan RD, Lewis SM. 2006. Parrotfish abundance and selective corallivory on a Belizean coral reef. *J Exp Mar Biol Ecol* 335:292–301.

Rourke ML, Fowler AM, Hughes JM, et al. 2021. Environmental DNA (eDNA) as a tool for assessing fish biomass: a review of approaches and future considerations for resource surveys. *Environ DNA*. **https://doi.org/10.1002/edn3.185**.

Rubec PJ. 1986. The effects of sodium cyanide on coral reefs and marine fish in the Philippines. In: McLean JL, Dizon LB, Hosillos LV, eds. *The first Asian Fisheries Forum*, pp. 297–802. Manila, Philippines: Asian Fisheries Society. **www.actwin.com**.

Rubec PJ. 1988. The need for conservation and management of Philippine coral reefs. *Env Biol Fish* 23:141–154.

Rubec PJ, Cruz F, Pratt V, Oellers R, Lallo F. 2000. Cyanide-free, net-caught fish for the marine aquarium trade. *SPC Live Reef Fish Info Bull* 7:28–34.

Rubin DA. 1985. Effect of pH on sex ratio in cichlids and a poeciliid (Teleostei). *Copeia* 1985:233–235.

Rüegg J, Tiegs SD, Chaloner DT, Levi PS, Tank JL, Lamberti GA. 2011. Salmon subsidies alleviate nutrient limitation of benthic biofilms in southeast Alaska streams Can. *J Fish Aquat Sci* 68:277–287

Ruhl N, McRobert SP. 2005. The effect of sex and shoal size on shoaling behaviour in *Danio rerio*. *J Fish Biol* 67:1318–1326.

Ruiz GM, Carlton JT, Grosholz ED, et al. 1997. Global invasions of marine and estuarine habitats by non-indigenous species: mechanisms, extent, and consequences. *Am Zool* 37:619–630.

Rundle HD, Vamosi SM, Schluter D. 2003. Experimental test of predation's effect on divergent selection during character displacement in sticklebacks. *Proc Natl Acad Sci USA* 100:14943–14948.

Russ GR. 2002. Yet another review of marine reserves as reef fishery management tools. In: Sale PF, ed. *Coral reef fishes: dynamics and diversity in a complex ecosystem*, pp. 421–443. San Diego, CA: Academic Press.

Russell FS. 1976. *The eggs and planktonic stages of British marine fishes*. London: Academic Press.

Russell TR. 1986. Biology and life history of the paddlefish – a review. In: Dillard JG, Graham LK, Russell TR, eds. *The paddlefish: status, management and propagation*, pp. 2–19. American Fisheries Society Special Publication No. 7. Columbia, MO: American Fisheries Society North Central Division.

Ryder OA. 1986. Species conservation and systematics: the dilemma of subspecies. *Trends Ecol Evol* 1:9–10.

Sadovy Y. 1992. A preliminary assessment of the marine aquarium export trade in Puerto Rico. *Proc 7th Internat Coral Reef Symp* 2:1014–1022.

Sadovy Y, Cheung W-L. 2003. Near extinction of a highly fecund fish: the one that nearly got away. *Fish Fish (Oxf)* 4:86–99.

Sadovy Y, Suharti SR, Colin PL. 2019. Quantifying the rare: baselines for the endangered Napoleon Wrasse, *Cheilinus undulates*, and implications for conservation. *Aquat Conserv Mar Freshw Res* 29:1285–1301.

Sadovy Y, Vincent ACJ. 2002. The trades in live reef fishes for food and aquaria: issues and impacts. In: Sale PF, ed. *Coral reef fishes: dynamics and diversity in a complex ecosystem*, pp. 391–420. San Diego, CA: Academic Press.

Saenz-Agudelo P, Jones GP, Thorrold SR, Planes S. 2012. Patterns and persistence of larval retention and connectivity in a marine metapopulation. *Mol Ecol* 21:4695–4705.

Safina C. 2001a. Fish conservation. In: Levin S, ed. *Encyclopedia of biodiversity*, Vol. 2, pp. 783–799. San Diego, CA: Academic Press.

Safina C. 2001b. Tuna conservation. In: Block BA, Stevens ED, eds. *Tuna: physiology, ecology, and evolution*, pp. 413–459. San Diego, CA: Academic Press.

Safina C. 2002. *Eye of the albatross: visions of hope and survival*. New York: Henry Holt & Co.

Safina C, Burger J. 1985. Common tern foraging: seasonal trends in prey fish densities and competition with bluefish. *Ecology* 66:1457–1463.

Saglam IK, Baumsteiger J, Smith MJ, et al. 2016. Phylogenetics support an ancient common origin of two scientific icons: Devils Hole and Devils Hole pupfish. *Mol Ecol* 25:3962–3973.

Saitou N, Nei M. 1987. The neighbor joining method: a new method for reconstructing phylogenetic trees. *Mol Biol Evol* 4:406–425.

Sala E, Aburto-Oropeza O, Paredes G, Parra I, Barrera JC, Dayton PK. 2002. A general model for designing networks of marine reserves. *Science* 298:1991–1993.

Sale PF. 1969. A suggested mechanism for habitat selection by juvenile manini *Acanthurus triostegus sandvicensis* Streets. *Behaviour* 35:27–44.

Sale PF. 1971. Extremely limited home range in a coral reef fish, *Dascyllus aruanus* (Pisces: Pomacentridae). *Copeia* 1971:324–327.

Sale PF. 1978. Coexistence of coral reef fishes – a lottery for living space. *Env Biol Fish* 3:85–102.

Sale PF, ed. 1991a. *The ecology of fishes on coral reefs*. San Diego, CA: Academic Press.

Sale PF. 1991b. Reef fish communities: open nonequilibrial systems. In: Sale PF, ed. *The ecology of fishes on coral reefs*, pp. 564–598. San Diego, CA: Academic Press.

Sale PF, ed. 2002. *Coral reef fishes: dynamics and diversity in a complex ecosystem*. San Diego, CA: Academic Press.

Sale PF, Cowen RK, Danilowicz BS, et al. 2005. Critical science gaps impede use of no-take fishery reserves. *Trends Ecol Evol* 20:74–80.

Salewski V. 2003. Satellite species in lampreys: a worldwide trend for ecological speciation in sympatry? *J Fish Biol* 63:267–279.

Salzburger W, Mack T, Verheyen E, Meyer A. 2005. Out of Tanganyika: genesis, explosive speciation, key-innovations and phylogeography of the haplochromine cichlid fishes. *BMC Evol Biolo* 5:17. **https://doi.org/10.1186/1471-2148-5-17**.

Sanchez W, Sremski W, Piccini B, et al. 2011. Adverse effects in wild fish living downstream from pharmaceutical manufacture discharges. *Environ Int* 37:1342–1348.

Sancho G. 1998. Factors regulating the height of spawning ascents in trunkfishes (Ostraciidae). *J Fish Biol* 53 (suppl A): 94–103.

Sancho G, Ma D, Lobel PS. 1997. Behavioral observations of an upcurrent reef colonization event by larval surgeonfish *Ctenochaetus strigosus* (Acanthuridae). *Mar Ecol Prog Ser* 153:311–315.

Sancho G, Petersen CW, Lobel PS. 2000a. Predator–prey relations at a spawning aggregation site of coral reef fishes. *Mar Ecol Prog Ser* 203:275–288.

Sancho G, Solow AR, Lobel PS. 2000b. Environmental influences on the diel timing of spawning in coral reef fishes. *Mar Ecol Prog Ser* 206:193–212.

Sanderson SL, Cech JJ, Jr., Patterson MR. 1991. Fluid dynamics in suspension-feeding blackfish. *Science* 251:1346–1348.

Sandlund OT, Malmquist HJ, Jonsson B, et al. 1988. Density, length distribution, and diet of age-0 arctic charr *Salvelinus alpinus* in the surf zone of Thingvallavatn, Iceland. *Env Biol Fish* 23:183–195.

Sandøy S, Langåker RM. 2001. Atlantic salmon and acidification in southern Norway: a disaster in the 20th Century, but a hope for the future? *Water Air Soil Poll* 130:1343–1348.

Sang TK, Chang HY, Chen CT, Hui CF. 1994. Population structure of the Japanese eel, *Anguilla japonica*. *Mol Biol Evol* 11:250–260.

Sano MN, Shimizu M, Nose Y. 1984. Changes in structure of coral reef fish communities by destruction of hermatypic corals: observational and experimental views. *Pacific Sci* 38:51–79.

Sá-Oliveira JC, Isaac VJ, Soares Araújo A, Ferrari SF. 2016. Factors structuring the fish community in the area of the Coaracy Nunes hydroelectric reservoir in Amapá, Northern Brazil. *Trop Conserv Sci* 9(1):16–33.

Sargent RC, Gross MR. 1994. Williams' principle: an explanation of parental care in teleost fishes. In: Pitcher TJ, ed. *Behaviour of teleost fishes*, 2nd edn, pp. 333–361. London: Chapman & Hall.

Sasaki A, Ikejima K, Aoki S, Azuma N, Kashimura N, Wada M. 2003. Field evidence for bioluminescent signaling in the pony fish, *Leiognathus elongates*. *Env Biol Fish* 66:307–311.

Sato T. 1986. A brood parasitic catfish of mouthbrooding cichlid fishes in Lake Tanganyika. *Nature* 323:58–59.

Saunders DL, Meeuwig JJ, Vincent ACJ. 2002. Freshwater protected areas: strategies for conservation. *Conserv Biol* 16:30–41.

Saunders MW, McFarlane GA. 1993. Age and length at maturity of the female spiny dogfish, *Squalus acanthias*, in the Strait of Georgia, British Columbia, Canada. *Env Biol Fish* 38:49–57.

Savvaitova KA, Kuzischchin KV, Maksimov SV. 2000. Kamchatka steelhead: population trends in life history variation. In: Knudsen EE, Steward CR, MacDonald DD, Williams JE, Reiser DW, eds. *Sustainable fisheries management: Pacific salmon*, pp. 195–203. Boca Raton, FL: Lewis Publishers.

Sayer MD, Treasurer JW, Costello MJ. 1996. *Wrasse biology and use in aquaculture*. London: Fishing News Books.

Sazima I. 1977. Possible case of aggressive mimicry in a neotropical scale-eating fish. *Nature* 270:510–512.

Sazima I. 1983. Scale-eating in characoids and other fishes. *Env Biol Fish* 9:87–101.

Sazima I, Guimaraes AS. 1987. Scavenging on human corpses as a source for stories about man-eating piranhas. *Env Biol Fish* 20:75–77.

Sazima I, Machado FA. 1990. Underwater observations of piranhas in western Brazil. *Env Biol Fish* 28:17–31.

Scaggiante M, Rasotto MB, Romualdi C, Pilastro A. 2005. Territorial male gobies respond aggressively to sneakers but do not adjust their sperm expenditure. *Behav Ecol* 16:1001–1007.

Schaeffer B. 1967. Comments on elasmobranch evolution. In: Gilbert PW, Mathewson RF, Rall DP, eds. *Sharks, skates and rays*, pp. 3–36. Baltimore, MD: Johns Hopkins Press.

Schaeffer B, Williams M. 1977. Relationships of fossil and living elasmobranchs. *Am Zool* 17:293–302.

Scharpf C. 2006. Annotated checklist of North American freshwater fishes, including subspecies and undescribed forms. Part II: Catostomidae through Mugilidae. *Am Curr* 32(4):1–39.

Schellart NAM, Wubbels RJ. 1998. The auditory and mechanosensory lateral line system. In: Evans DH, ed. *The physiology of fishes*, 2nd edn, pp. 283–312. Boca Raton, FL: CRC Press.

Schelske CL, Carpenter SR. 1992. Lakes: Lake Michigan. In: National Research Council (US), ed. *Restoration of aquatic ecosystems: science, technology, and public policy*, pp. 380–392. Washington, DC: National Academy Press.

Schindler DE, Kitchell JF, He X, Carpenter SR, Hodgson JR, Cottingham KL. 1993. Food web structure and phosphorus cycling in lakes. *Trans Am Fish Soc* 122:756–772.

Schindler DE, Scheuerell MD, Moore JW, et al. 2003. Pacific salmon and the ecology of coastal ecosystems. *Frontiers Ecol Env* 1:31–37.

Schlosser IJ. 1982. Fish community structure and function along two habitat gradients in a headwater stream. *Ecol Monogr* 52:395–414.

Schlosser IJ. 1985. Flow regime, juvenile abundance, and the assemblage structure of stream fishes. *Ecology* 66:1484–1490.

Schlosser IJ. 1987. The role of predation in age- and size-related habitat use by stream fishes. *Ecology* 68:651–659.

Schlosser IJ. 1991. Stream fish ecology: a landscape perspective. *BioScience* 41:704–712.

Schlupp I, Marler C, Ryan MJ. 1994. Benefit to sailfin mollies of mating with heterospecific females. *Science* 263:373–374.

Schluter D. 2000. *The ecology of adaptive radiation*. Oxford: Oxford University Press.

Schmale MC. 1981. Sexual selection and reproductive success in males of the bicolor damselfish *Eupomacentrus partitus* (Pisces: Pomacentridae). *Anim Behav* 29:1172–1184.

Schmidt-Nielsen K. 1983. *Scaling: why is animal size so important*. Cambridge, UK: Cambridge University Press.

Schmittner A. 2005. Decline of the marine ecosystem caused by a reduction in the Atlantic overturning circulation. *Nature* 434:628–633.

Schmitz L, Wainwright PC. 2011. Nocturnality constrains morphological and functional diversity of the eyes of reef fishes. *BMC Evol Biol* 11:338. **http://www.biomedcentral.com/1471-2148/11/338**.

Schofield PJ, Chapman LJ. 2000. Hypoxia tolerance of introduced Nile perch: implications for survival of indigenous fishes in the Lake Victoria basin. *Afr Zool* 35:35–42.

Scholander PF, van Dam L, Kanwisher JW, Hammel HT, Gordon MS. 1957. Supercooling and osmoregulation in Arctic fish. *J Cell Comp Physiol* 49:5–24.

Scholik AR, Yan HY. 2002. Effects of noise on auditory sensitivity of fishes. *Bioacoustics* 12:186–188.

Schreck CB. 2000. Accumulation and long-term effects of stress in fish. In: Moberg GP, Mench JA, eds. *The biology of animal stress*, pp. 147–158. Wallingford, UK: CABI Publishing.

Schubert M, Munday PL, Caley MJ, Jones GP, Llewellyn LE. 2003. The toxicity of skin secretions from coral-dwelling gobies and their potential role as a predator deterrent. *Env Biol Fish* 67:359–367.

Schulte PM. 2011a. Effects of temperature: an introduction. In: Farrell AP, Stevens ED, Cech JJ, Richards JG, eds. *Encyclopedia of fish physiology: from genome to environment*, pp. 1688–1694. Boston: Elsevier/Academic Press.

Schulte PM. 2011b. Intertidal habitats. In: Farrell AP, Stevens ED, Cech JJ, Richards JG, eds. *Encyclopedia of fish physiology: from genome to environment*, pp. 1959–1964. Boston: Elsevier/Academic Press.

Schultz ET. 1993. Sexual size dimorphism at birth in *Micrometrus minimus* (Embiotocidae): a prenatal cost of reproduction. *Copeia* 1993:456–463.

Schultz RJ. 1971. Special adaptive problems associated with unisexual fish. *Am Zool* 11:351–360.

Schultz RJ. 1977. Evolution and ecology of unisexual fishes. In: Hecht MK, Steere WC, Wallace B, eds. *Evolutionary biology*, Vol. 10, 277–331. New York: Plenum Press.

Schultz RW, Nóbrega RH. 2011a. Anatomy and histology of fish testis. In: Farrell AP, Stevens ED, Cech JJ, Richards JG, eds. *Encyclopedia of fish physiology: from genome to environment*, pp. 616–626. Boston: Elsevier/Academic Press.

Schultz RW, Nóbrega RH. 2011b. Regulation of spermatogenesis. In: Farrell AP, Stevens ED, Cech JJ, Richards JG, eds. *Encyclopedia of fish physiology: from genome to environment*, pp. 627–634. Boston: Elsevier/Academic Press.

Schultze HP, Arratia G. 1989. The composition of the caudal skeleton of teleosts (Actinopterygii: Osteichthyes). *Zool J Linn Soc* 97:189–231.

Schuster S. 2006. Integration of the electrosense with other senses: implications for communication. In: Ladich F, Collin SP, Møller P, Kapoor BG, eds. *Communication in fishes*, Vol. 2, pp. 781–804. Enfield, NH:Science Publishers.

Schuster S, Wöhl S, Griebsch M, Klostermeier I. 2006. Animal cognition: how archer fish learn to down rapidly moving targets. *Curr Biol* 16:378–383.

Schwartz FJ. 2001. Freshwater and marine fish family hybrids a worldwide changing scene revealed by the scientific literature. *J Elisha Mitchell Sci Soc* 117:62–65.

Schwartz FJ, Maddock MB. 1986. Comparisons of karyotypes and cellular DNA contents within and between major lines of elasmobranchs. In: Barlow GW, Uyeno T, Arai R, Taniguchi T, Matsuura K, eds. *Indo-Pacific fish biology: Proceedings of the Second International Conference on Indo-Pacific fishes*, pp. 148–157. Tokyo: Ichthyological Society of Japan.

Schwassmann HO. 1980. Biological rhythms: their adaptive significance. In: Ali MA, ed. *Environmental physiology of fishes*, pp. 613–630. New York: Plenum Press.

Schweikert LE, Caves EM, Solie SE, Sutton TT, Johnsen S. 2019. Variation in rod spectral sensitivity of fishes is best predicted by habitat and depth. *J Fish Biol* 95:179–185.

Schwenk K, ed. 2000. *Feeding: form, function and evolution in tetrapod vertebrates.* San Diego: Academic Press.

Scoles DR, Graves JE. 1993. Genetic analysis of the population structure of yellowfin tuna, *Thunnus albacares*, from the Pacific Ocean. *Fish Bull US* 91:690–698.

Scoppettone GG, Vinyard G. 1991. Life history and management of four endangered lacustrine suckers. In: Minckley WL, Deacon JE, eds. *Battle against extinction. Native fish management in the American West*, pp. 359–377. Tucson, AZ: University of Arizona Press.

Scott AP. 1990. Salmonids. In: Munro AD, Scott AP, Lam TJ, eds. *Reproductive seasonality in teleosts: environmental influences*, pp. 33–52. Boca Raton, FL: CRC Press.

Scott GR, Sloman KA, Rouleau C, Wood CM. 2003. Cadmium disrupts behavioral and physiological responses to salrm substance in rainbow trout (*Oncorhynchus mykiss*). *J Exp Biol* 206:1779–1790.

Scott MC, Helfman GS. 2001. Native invasions, homogenization, and the mismeasure of integrity of fish assemblages. *Fisheries* 26(11): 6–15.

Scott WB, Crossman EJ. 1973. *Freshwater fishes of Canada. Fish Res Board Can Bull* 184.

Scribner KT, Page KS, Bartron ML. 2000. Hybridization in freshwater fishes: a review of case studies and cytonuclear methods of biological inference. *Rev Fish Biol Fisheries* 10:293–323.

Secor DH. 2000. Longevity and resilience of Chesapeake Bay striped bass. *ICES J Mar Sci* 57:808–815.

Secor DH. 2015. *Migration ecology of fishes.* Johns Hopkins University Press.

Sedell JR, Everest FH, Swanson FJ. 1982. Fish habitat and streamside management: past and present. In: *Proceedings of the Society of American Foresters Annual Meeting, September 1981*, pp. 244–255. Bethesda, MD: Society of American Foresters.

Seehausen O. 2002. Patterns in fish radiation are compatible with Pleistocene desiccation of Lake Victoria and 14,600 year history for its cichlid species flock. *Proc Roy Soc Lond B Biol Sci* 269:491–497.

Seehausen O, Terai Y, Magalhaes IS, et al. 2008. Speciation through sensory drive in cichlid fish. *Nature* 455:620–626.

Seehausen O, van Alphen JJM, Witte F. 1997a. Cichlid fish diversity threatened by eutrophication that curbs sexual selection. *Science* 277:1808–1811.

Seehausen O, Witte F, Katunzi EF, et al. 1997b. Patterns of the remnant cichlid fauna in southern Lake Victoria. *Conserv Biol* 11:890–904.

Seidensticker EP. 1987. Food selection of alligator gar and longnose gar in a Texas reservoir. *Proc Ann Conf Southeast Assoc Fish Wildl Agencies* 41:100–104.

Selkoe KA, Gaggiotti O, Andrews KR, et al. 2014. Emergent patterns of population genetic structure for a coral reef community. *Mol Ecol* 23:3064–3079.

Selkoe KA, Gaines SD, Caselle JE, Warner RR. 2006. Current shifts and kin aggregation explain genetic patchiness in fish recruits. *Ecology* 87:3082–3094.

Selz OM, Pierotti MER, Maan ME, Schmid C, Seehausen O. 2014. Female preference for male color is necessary and sufficient for assortative mating in 2 cichlid sister species. *Behav Ecol* 25:612–626.

Semmens BX, Buhle ER, Salomon AK, et al. 2004. A hotspot of non-native marine fishes: evidence for the aquarium trade as an invasion pathway. *Mar Ecol Prog Ser* 266:239–244.

Semon R. 1898. Die Entwickelung der paarigen Flossen des *Ceratodus forsteri*. *Jen Denkschr 4 Semon Zool Forschungsreisen* 1:61–111.

Seymour RS, Christian K, Bennett MB, Baldwin J, Wells RMG, Baudinette RV. 2004. Partitioning of respiration between gills and air-breathing organ in response to aquatic hypoxia and exercise in the Pacific tarpon, *Megalops cyprinoides*. *Physiol Biochem Zool* 77:760–767.

Sfakianakis DG, Katharios P, Tsirigotakis N, Doxa CK, Kentouri M. 2013. Lateral line deformities in wild and farmed sea bass (*Dicentrarchus labrax*, L.) and sea bream (*Sparus aurata*, L.). *J Appl Ichthyol* 29:1015–1021.

Sfakianakis DG, Renieri E, Kentourim M, Tsatsakis AM. 2015. Effect of heavy metals on fish larvae deformities: a review. *Environ Res* 137:246–255.

Shadwick RE, Lauder GV. 2006. *Fish biomechanics. Fish physiology*, Vol. 23. New York: Academic Press.
Shaklee JB, Tamaru CS. 1981. Biochemical and morphological evolution of the Hawaiian bonefishes (*Albula*). *Syst Zool* 30:125–146.
Shannon LV. 1985. The Benguela ecosystem. Part. I. Evolution of the Benguela physical features and processes. *Oceanogr Mar Biol Ann Rev* 23:105–182.
Shapiro DY. 1979. Social behavior, group structure, and the control of sex reversal in hermaphroditic fish. *Adv Stud Behav* 10:43–102.
Shapiro DY. 1983. On the possibility of kin groups in coral reef fishes. In: Reaka ML, ed. *Ecology of deep and shallow reefs*, pp. 39–45. Washington, DC: National Oceanographic and Atmospheric Administration.
Shapiro DY. 1987. Differentiation and evolution of sex change in fishes. *Bioscience* 37:490–497.
Shapiro DY. 1991. Intraspecific variability in social systems of coral reef fishes. In: Sale PF, ed. *The ecology of fishes on coral reefs*, pp. 331–355. San Diego, CA: Academic Press.
Shapiro DY, Marconato A, Yoshikawa T. 1994. Sperm economy in a coral reef fish *Thalassoma bifasciatum*. *Ecology* 75:1334–1344.
Shapiro DY, Sadovy Y, McGehee MA. 1993. Size, composition, and spatial structure of the annual spawning aggregation of the red hind, *Epinephelus guttatus* (Pisces: Serranidae). *Copeia* 1993:399–406.
Sharp GD. 1978. Behavioral and physiological properties of tunas and their effects on vulnerability to fishing gear. In: Sharp GD, Dizon AE, eds. *The physiological ecology of tunas*, pp. 397–449. New York: Academic Press.
Sharp GD, Dizon AE, eds. 1978. *The physiological ecology of tunas*. New York: Academic Press.
Sharp GD, Pirages SW. 1978. The distribution of red and white swimming muscles, their biochemistry, and the biochemical phylogeny of selected scombrid fishes. In: Sharp GD, Dizon AE, eds. *The physiological ecology of tunas*, pp. 41–78. New York: Academic Press.
Shartau RB, Brauner CJ. 2014. Acid–base and ion balance in fishes with bimodal respiration. *J Fish Biol* 84:682–704.
Shaw E. 1970. Schooling fishes: critique and review. In: Aronson L, Tobach E, Lehrmann DS, Rosenblatt JS, eds. *Development and evolution of behaviour*, pp. 452–480. San Francisco: W. H. Freeman.
Shaw E. 1978. Schooling fishes. *Am Sci* 66:166–175.
Shaw PW, Arkhipkin AI, Al-Khairulla H. 2004. Genetic structuring of Patagonian toothfish populations in the Southwest Atlantic Ocean: the effect of the Antarctic Polar Front and deep-water troughs as barriers to genetic exchange. *Molec Ecol* 13:3293–3303.
Sheaves M, Baker R, Nagelkerken I, Connolly RM. 2015. True value of estuarine and coastal nurseries for fish: incorporating complexity and dynamics. *Estuaries Coast* 38(2):401–414.
Sheldon AL. 1968. Species diversity and longitudinal succession in stream fishes. *Ecology* 49(2):193–198.
Shelton RGJ. 1978. On the feeding of the hagfish *Myxine glutinosa* in the North Sea. *J Mar Biol Assoc UK* 58:81–86.
Shenker JM, Dean JM. 1979. Utilization of an intertidal salt marsh creek by larval and juvenile fishes: abundance, diversity and temporal variation. *Estuaries* 2:154–163.
Shenker JM, Maddox ED, Wishinski E, Pearl S. 1993. Onshore transport of settlement-stage Nassau grouper (*Epinephelus striatus*) and other fishes in Exuma Sound, Bahamas. *Mar Ecol Prog Ser* 98:31–43.
Shephard KL. 1994. Functions for fish mucus. *Rev Fish Biol Fisheries* 4:401–429.
Shepherd ARD, Warwick RM, Clarke KR, Brown BE. 1992. An analysis of fish community responses to coral mining in the Maldives. *Env Biol Fish* 33:367–380.
Sheridan MA. 2011. Endocrinology of fish growth. In: Farrell AP, Stevens ED, Cech JJ, Richards JG, eds. *Encyclopedia of fish physiology: from genome to environment*, pp. 1483–1489. Boston: Elsevier/Academic Press.
Sherman K, Sissenwine M, Christensen V, et al. 2005. A global movement toward an ecosystem approach to management of marine resources. *Mar Ecol Prog Ser* 300:275–279.
Shiganova TA, Bulgakova YV. 2000. Effects of gelatinous plankton on Black Sea and Sea of Azov fish and their food resources. *ICES J Mar Sci* 57:641–648.
Shima A, Mitani H. 2004. Medaka as a research organism: past, present, and future. *Mec Dev* 121:599–604.
Shimizu N, Sakai Y, Hashimoto H, Gushima K. 2006. Terrestrial reproduction by the air-breathing fish *Andamia tetradactyla* (Pisces; Blenniidae) on supralittoral reefs. *J Zool* 269:357–364.
Shirai S, Nakaya K. 1992. Functional morphology of feeding apparatus of the cookie-cutter shark, *Isistius brasiliensis* (Elasmobranchii, Dalatiinae). *Zool Sci* 9:811–821.
Shirakawa H, Yanai S, Goto A. 2013. Lamprey larvae as ecosystem engineers: physical and geochemical impact on the streambed by their burrowing behavior. *Hydrobiologia* 701:313–322
Shireman JV, Colle DE, Canfield DEJ. 1986. Efficacy and cost of aquatic weed control in small ponds. *Water Res Bull* 22:43–48.
Shivji M, Clarke S, Pank M, Natanson L, Kohler N, Stanhope M. 2002. Genetic identification of pelagic shark body parts for conservation and trade monitoring. *Conserv Biol* 16:1036–1047.
Shu D-G, Morris SC, Han J, et al. 2003. Head and backbone of the Early Cambrian vertebrate *Haikouichthys*. *Nature* 421:526–529.
Shubin N. 2008. *Your inner fish: a journey into the 3.5-billion-year history of the human body*. New York: Pantheon Books.
Shubin NH, Daeschler EB, Jenkins FA, Jr. 2006. The pectoral fin of *Tiktaalik roseae* and the origin of the tetrapod limb. *Nature* 440: 764–771.
Shulman MJ. 1985a. Variability in recruitment of coral reef fishes. *J Exp Mar Biol Ecol* 89:205–219.
Shulman MJ. 1985b. Recruitment of coral reef fishes: effects of distribution of predators and shelter. *Ecology* 66:1056–1066.
Shulman MJ, Bermingham E. 1995. Early life histories, ocean currents, and the population genetics of Caribbean reef fishes. *Evolution* 49:897–910.
Shulman MJ, Ogden JC. 1987. What controls tropical reef fish populations: recruitment or benthic mortality? An example in the Caribbean reef fish *Haemulon flavolineatum*. *Mar Ecol Prog Ser* 39:233–242.
Shulman MJ, Ogden JC, Ebersole JP, McFarland WN, Miller SL, Wolf NG. 1983. Priority effects in the recruitment of juvenile coral reef fishes. *Ecology* 64:1508–1513.
Shuttleworth TJ, ed. 1988. *Physiology of elasmobranch fishes*. Berlin: Springer-Verlag.
Sidell BD. 1977. Turnover of cytochrome c in skeletal muscle of green sunfish (*Lepomis cyanellus* R.) during thermal acclimation. *J Exp Zool* 199:233–250.
Sidell BD, Moerland TS. 1989. Effects of temperature on muscular function and locomotory performance in teleost fish. In: *Advances in comparative and environmental physiology*, Vol. 5, pp. 116–158. Berlin: Springer-Verlag.

Siebeck UE. 2004. Communication in coral reef fish: the role of ultraviolet colour patterns in damselfish territorial behaviour. *Anim Behav* 68:273–282.

Siebeck UE, Losey GS, Marshall J. 2006. UV communication in fish. In: Ladich F, Collin SP, Møller P, Kapoor BG, eds. *Communication in fishes*, Vol. 2, pp. 423–455. Enfield, NH:Science Publishers.

Sih A, Crowley P, McPeek M, Petranka J, Strohmeier K. 1985. Predation, competition, and prey communities: a review of field experiments. *Ann Rev Ecol Syst* 16:269–311.

Sikkel PC, Cheney KL, Cote IM. 2004. In situ evidence for ectoparasites as a proximate cause of cleaning interactions in reef fish. *Anim Behav* 68:241–247.

Sikkel PC, Schaumburg CS, Mathenia JK. 2006. Diel infestation dynamics of gnathiid isopod larvae parasitic on Caribbean reef fish. *Coral Reefs* 25:683–689.

Simon TP, Lyons J. 1995. Application of the Index of Biotic Integrity to evaluate water resource integrity in freshwater ecosystems. In: Davis WS, Simon TP, eds. *Biological assessment and criteria: tools for water resource planning and decision making*, pp. 245–262. Boca Raton, FL: CRC Press.

Simons JR. 1970. The direction of the thrust produced by the heterocercal tails of two dissimilar elasmobranchs: the Port Jackson shark, *Heterodontus portusjacksoni* (Meyer), and the piked dogfish, *Squalus megalops* (Macleay). *J Exp Biol* 52:95–107.

Simpfendorfer CA. 2000. Predicting population recovery rates for endangered Western Atlantic sawfishes using demographic analysis. *Env Biol Fish* 58:371–377.

Simpson BRC. 1979. The phenology of annual killifishes. *Symp Zool Soc Lond* 44:243–261.

Simpson PW, Newman JR, Keirn A, Matter RM, Guthrie PA. 1982. *Manual of stream channelization impacts on fish and wildlife.* Washington, DC: US Fish and Wildlife Service FWS/OBS-82/24.

Simpson SD, Meekan MG, McCauley RD, Jeffs A. 2004. Attraction of settlement-stage coral reef fishes to reef noise. *Mar Ecol Prog Ser* 276:263–268.

Sims DW. 1999. Threshold foraging behaviour of basking sharks on zooplankton: life on an energetic knife-edge? *Proc Roy Soc Lond B Biol Sci* 266:1437–1443.

Sims DW, Southall EJ, Richardson AJ, Reid PC, Metcalfe JD. 2003. Seasonal movements and behaviour of basking sharks from archival tagging: no evidence of winter hibernation. *Mar Ecol Prog Ser* 248:187–196.

Sinderman CJ. 1990. *Principal diseases of marine fish and shellfish, Vol. 1. Diseases of marine fish*, 2nd edn. San Diego, CA: Academic Press.

Sisneros JA, Tricas TC. 2002. Ontogenetic changes in the response properties of the peripheral electrosensory system in the Altlantic stingray (*Dasyatis sabina*). *Brain Behav Evol* 59:130–140.

Sisneros JA, Tricas TC, Luer CA. 1998. Response properties and biological function of the skate electrosensory system during ontogeny. *J Comp Physiol A* 183:87–99.

Skulason S, Smith TB. 1995. Resource polymorphisms in vertebrates. *Trends Ecol Evol* 10:366–370.

Skulason S, Snorrason SS, Ota D, Noakes DLG. 1993. Genetically based differences in foraging behaviour among sympatric morphs of arctic charr (Pisces: Salmonidae). *Anim Behav* 45:1179–1192.

Slack WT, Ross ST, Ewing JA. 2004. Ecology and population structure of the bayou darter, *Etheostoma rubrum*: disjunct riffle habitats and downstream transport of larvae. *Env Biol Fish* 71:151–164.

Slaney TL, Hyatt KD, Northcote TG, et al. 1996. Status of anadromous salmon and trout in British Columbia and Yukon. *Fisheries* 21(10):20–35.

Slatkin M. 1995. A measure of population subdivision based on microsatellite allele frequencies. *Genetics* 139:457–462.

Sloman KA. 2011a. The diversity of fish reproduction: an introduction. In: Farrell AP, Stevens ED, Cech JJ, Richards JG, eds. *Encyclopedia of fish physiology: from genome to environment*, 613—615. Boston: Elsevier/Academic Press.

Sloman KA. 2011b. Anthropogenic influences on fish behavior. In: Farrell AP, Stevens ED, Cech JJ, Richards JG, eds. *Encyclopedia of fish physiology: from genome to environment*, pp. 783–789. Boston: Elsevier/Academic Press.

Sloman, K.A., and Buckley, J. (2011). Nutritional Provision During Parental Care. In: Farrell AP, Stevens ED, Cech JJ, Richards JG, eds. *Encyclopedia of fish physiology: from genome to environment*, pp. 678–683. Boston: Elsevier/Academic Press.

Sloman KA, Scott GR, Diao Z, Rouleau C, Wood CM. 2003. Cadmium affects the social behavior of rainbow trout, *Oncorhynchus mykiss. Aquat Toxicol* 65:171–185.

Sloman KA, Wilson RW, Balshine S. 2006. *Behaviour and physiology of Fish. Fish physiology*, Vol. 24. New York: Academic Press.

Smatresk NJ, Cameron JN. 1982. Respiration and acid–base physiology of the spotted gar, a bimodal breather. I. Normal values, and the response to severe hypoxia. *J Exp Biol* 96:263–280.

Smith BR. 1971. Sea lampreys in the Great Lakes of North America. In: Hardisty MW, Potter IC, eds. *The biology of lampreys*, Vol. 1. pp. 207–247. New York: Academic Press.

Smith C, Wootten RJ. 2016. The remarkable reproductive diversity of teleost fishes. *Fish and Fisheries* 17:1208–1215.

Smith CL. 1975. The evolution of hermaphroditism in fishes. In: Reinboth R, ed. *Intersexuality in the animal kingdom*, pp. 295–310. Berlin: Springer-Verlag.

Smith CL. 1978. Coral reef fish communities: a compromise view. *Env Biol Fish* 3:109–128.

Smith CL. 1988. Minnows first, then trout. *Fisheries* 13(4):4–8.

Smith CL, Rand CS, Schaeffer B, Atz JW. 1975. *Latimeria*, the living coelacanth, is ovoviviparous. *Science* 190:1105–1106.

Smith CL, Reay P. 1991. Cannibalism in teleost fish. *Rev Fish Biol Fisheries* 1:41–64.

Smith CR. 1985a. Food for the deep sea: utilization, dispersal, and flux of nekton falls at the Santa Catalina Basin floor. *Deep Sea Res* 32:417–442.

Smith EJ, Partridge JC, Parsons KN, et al. 2002. Ultraviolet vision and mate choice in the guppy (*Poecilia reticulata*). *Behav Ecol* 13:11–19.

Smith GR. 1981b. Effects of habitat size on species richness and adult body sizes of desert fishes. In: Naiman RJ, Soltz DL, eds. *Fishes in North American deserts*, pp. 125–171. New York: Wiley & Sons.

Smith GR, Stearley RF. 1989. The classification and scientific names of rainbow and cutthroat trouts. *Fisheries* 14(1):4–10.

Smith GR, Todd TN. 1984. Evolution of species flocks of fishes in north temperate lakes. In: Echelle AE, Kornfield I, eds. *Evolution of fish species flocks*, pp. 45–68. Orono, ME: University of Maine Press.

Smith JB, Tirpak DA, eds. 1989. *The potential effects of global climate change on the United States. Appendix E. Aquatic resources.* Washington, DC: US Environmental Protection Agency, EPA-230-05-89-055.

Smith JLB. 1939. A living fish of the Mesozoic type. *Nature* 143: 455–456.

Smith JLB. 1956. *The search beneath the sea.* New York: Henry Holt & Co.

Smith ME, Kane AS, Popper AN. 2004. Noise-induced stress response and hearing loss in goldfish (*Carassius auratus*). *J Exp Biol* 207:427–435.

Smith MP, Briggs DEG, Aldridge RJ. 1987. A conodont animal from the lower Silurian of Wisconsin, USA, and the apparatus architecture of panderodontid conodonts. In: Aldridge RJ, ed. *Palaeobiology of conodonts*, pp. 91–104. Chichester, UK: Ellis Horwood Ltd.

Smith PJ, Francis RICC, McVeagh M. 1991. Loss of genetic diversity due to fishing pressure. *Fish Res* 10:309–316.

Smith RJF. 1986. The evolution of chemical alarm signals in fishes. In: Duvall D, Muller-Schwarze D, Silverstein RM, eds. *Chemical signals in vertebrates, Vol. IV: Ecology, evolution and comparative biology*, pp. 99–115. New York: Plenum Press.

Smith RJF. 1992. Alarm signals in fishes. *Rev Fish Biol Fisheries* 2:33–63.

Smith RS, Kramer DL. 1986. The effect of apparent predation risk on the respiratory behavior of the Florida gar (*Lepisosteus platyrhincus*). *Can J Zool* 64:2133–2136.

Smith SE. 2003. The Aswan High Dam at thirty: an environmental impact assessment. In: Crisman TL, Chapman LJ, Chapman CA, et al., eds. *Conservation, ecology, and management of African fresh waters*, pp. 301–320. Gainesville, FL: University Press of Florida.

Smith TIJ. 1985b. The fishery, biology, and management of Atlantic sturgeon, *Acipenser oxyrhynchus*, in North America. In: Binkowski FP, Doroshov SI, eds. *North American sturgeons: biology and aquaculture potential*, pp. 61–72. Developments in Environmental Biology of Fishes No. 6. Dordrecht: Dr. W. Junk.

Smith WE, Kwak TJ. 2015. Tropical insular fish assemblages are resilient to flood disturbance. Ecosphere 6(12): article 279 (16 pages). **https://doi.org/10.1890/ES15-00224.1**.

Sneath PHA, Sokal RR. 1973. *Numerical taxonomy*. San Francisco: W. H. Freeman.

Sneddon LU. 2003. The evidence for pain in fish: the use of morphine as an analgesic. *Appl Anim Behav Sci* 83:153–162.

Sneddon LU. 2004. Evolution of nociceptors in vertebrates: comparative analysis of lower vertebrates. *Brain Res Rev* 46:123–130.

Sneddon LU. 2011. Nociception or pain in fish. In: Farrell AP, Stevens ED, Cech JJ, Richards JG, eds. *Encyclopedia of fish physiology: from genome to environment*, pp. 713–719. Boston: Elsevier/Academic Press.

Snoeks J. 2000. How well known is the ichthyodiversity of the large East African lakes? In: Rossiter A, Kawanabe H, eds. *Ancient lakes: biodiveirsy, ecology and evolution*, pp. 17–38. Advances in Ecological Research No. 31. London: Academic Press.

Sogard SM, Olla BL. 1994. The potential for intracohort cannibalism in age-0 walleye pollock, *Theragra chalcogramma*, as determined under laboratory conditions. *Env Biol Fish* 39:183–190.

Sokal RR, Sneath PHA. 1963. Principles of Numerical Taxonomy. San Francisco, CA: W.H. Freeman.

Solazzi MF, Nickelson TW, Johnson SL, Rodgers JD. 2000. Effects of increasing winter rearing habitat on abundance of salmonids in two coastal Oregon streams. *Can J Fish Aquat Sci* 57:906–914.

Sollid J, De Angelis P, Gundersen K, Nilsson GE. 2003. Hypoxia induces adaptive and reversible gross-morphological changes in crucian carp gills. *J Exp Biol* 206:3667–3673.

Sollid J, Weber RE, Nilsson GE. 2005. Temperature alters the respiratory surface area of crucian carp Carassius carassius and goldfish Carassius auratus. *J Exp Biol* 208:1109–1116.

Soltz DL, Naiman RJ. 1981. Fishes in deserts: symposium rationale. In: Naiman RJ, Soltz DL, eds. *Fishes in North American deserts*, pp. 1–9. New York: Wiley & Sons.

Somero GN. 1992. Adaptations to high hydrostatic pressure. *Ann Rev Physiol* 54:557–577.

Somero GN, Dahlhoff E, Gibbs A. 1991. Biochemical adaptations of deep-sea animals: insights into biogeography and ecological energetics. In: Mauchline J, Nemoto T, eds. *Marine biology: its accomplishments and future prospect*, pp. 39–57. Amsterdam: Elsevier.

Sorbini L. 1975. Evoluzione e distribuzione del genere fossile *Eolates* e suoi rapporti con il genere attuale *Lates* (Pisces–Centropomidae). In: *Studi e Ricerche sui Giacimenti Terziari di Bolca. II*, pp. 1–54. Verona: Museo Civico di Storia Naturale di Verona.

Sorensen PW, Caprio J. 1998. Chemoreception. In: Evans DH, ed. *The physiology of fishes*, 2nd edn, pp. 375–405. Boca Raton, FL: CRC Press.

Soto CG. 2001. The potential impacts of global climate change on marine protected areas. *Rev Fish Biol Fisheries* 11:181–195.

SPA (Science and Policy Associates, Inc). 1990. *Report on the workshop on effects of global climate change on freshwater ecosystems*. Washington, DC: SPA.

Spalding MD, Agostini VN, Rice J, Grant SM. 2012. Pelagic provinces of the world: a biogeographic classification of the world's surface pelagic waters. *Ocean & Coast Management* 60: 19–30.

Spalding MD, Fox HE, Allen GR, et al. 2007. Marine ecoregions of the world: a bioregionalization of coastal and shelf areas. *BioScience* 57:573–583.

Sparholt H. 1985. The population, survival, growth, reproduction and food of Arctic charr, *Salvelinus alpinus* (L.) in four unexploited lakes in Greenland. *J Fish Biol* 26:313–330.

Spieler RE. 1992. Feeding entrained circadian rhythms in fishes. In: Ali MA, ed. *Rhythms in fishes*, pp. 137–147. New York: Plenum Press.

Spitzer RH, Koch EA. 1998. Hogfish skin and slime glands. In: Jørgensen JM, Lomholt JP, Weber RE, Malte H, eds. *The biology of hagfishes*, pp. 109–132. London: Chapman & Hall.

Sponaugle S, Cowen RK. 1994. Larval durations and recruitment patterns of two Caribbean gobies (Gobiidae): contrasting early life histories in demersal spawners. *Mar Biol* 120:133–143.

Sponaugle S, Cowen RK. 1997. Early life history traits and recruitment patterns of Caribbean wrasses (Labridae). *Ecol Monogr* 67:177–202.

Sponaugle S, Cowen RK, Shanks A, et al. 2002. Predicting self-recruitment in marine populations: biophysical correlates and mechanisms. *Bull Mar Sci* 70 (suppl S):341–375.

Sponaugle S, Pinkard D. 2004. Lunar cyclic population replenishment of a coral reef fish: shifting patterns following oceanic events. *Mar Ecol Prog Ser* 267:267–280.

Spotte S. 2002. *Candiru: life and legend of the bloodsucking catfishes*. Berkeley, CA: Creative Arts Book Co.

Springer VG. 1982. Pacific Plate biogeography, with special reference to shorefishes. *Smithsonian Contrib Zool* 367:182.

Springer VG, Gold JP. 1989. *Sharks in question*. Washington, DC: Smithsonian Institution Press.

Springer VG, Johnson DG. 2004. Study of the dorsal gill-arch musculature of teleostome fishes, with special reference to the Actinopterygii. *Bull Biol Soc Washington* 11:1–235.

Springer VG, Orrell TM. 2004. Phylogenetic analysis of 147 families of acanthomorph fishes based primarily on dorsal gill-arch muscles and skeleton. *Bull Biol Soc Washington* 11:237–254.

Springer VG, Smith-Vaniz WF. 1972. Mimetic relationships involving fishes of the family Blenniidae. *Smithson Contrib Zool* 112:1–36.

St. Mary CM. 2000. Sex allocation in *Lythrypnus* (Gobiidae): variations on a hermaphroditic theme. *Env Biol Fish* 58:321–333.

Stacey NE, Sorensen PW. 1991. Function and evolution of fish hormonal pheromones. In: Hochachka PW, Mommsen TP, eds. *Biochemistry and molecular biology of fishes*, pp. 109–135. New York: Elsevier.

Stacey N, Sorensen PW. 2011. Hormonal pheromones. In: Farrell AP, Stevens ED, Cech JJ, Richards JG, eds. *Encyclopedia of fish physiology: from genome to environment*, pp. 1553–1562. Boston: Elsevier/Academic Press.

Stamatakis A. 2014. RAxML version 8: a tool for phylogenetic analysis and post-analysis of large phylogenies. *Bioinformatics* 30:1312–1313.

Starr RM, Heine JN, Felton JM, Cailliet GM. 2002. Movements of bocaccio (*Sebastes paucispinis*) and greenspotted (*S. chlorostictus*) rockfishes in a Monterey submarine canyon: implications for the design of marine reserves. *Fish Bull US* 100:324–337.

Statzner B, Sagnes P, Champagne J-Y, et al. 2003. Contribution of benthic fish to the patch dynamics of gravel and sand transport in streams. *Water Resources Res* 39:1–17.

Stauffer JR, Jr., Arnegard ME, Cetron M, et al. 1997. Controlling vectors and hosts of parasitic diseases using fishes. *BioScience* 47:41–49.

Stauffer JR, Jr., Hale EA, Seltzer R. 1999. Hunting strategies of a Lake Malawi cichlid with reverse countershading. *Copeia* 1999:1108–1111.

Stead SS, Laird L. 2002. *Handbook of salmon farming*. New York: Springer.

Stearns SC. 1992. *The evolution of life histories*. Oxford: Oxford University Press.

Steele MA, Anderson TW. 2006. Predation. In: Allen LG, Pondella DJ, Horn MH, eds. *The ecology of marine fishes: California and adjacent waters*, pp. 428–448. Berkeley, CA: University of California.

Steen JB, Berg T. 1966. The gills of two species of haemoglobin-free fishes compared to those of other teleosts – with a note of severe anaemia in an eel. *Comp Biochem Physiol* 18:517–526.

Stein RA. 1977. Selective predation, optimal foraging, and the predator–prey interaction between fish and crayfish. *Ecology* 58:1237–1253.

Steinhart GB, Wurtsbaugh WA. 2003. Winter ecology of kokanee: implications for salmon management. *Trans Am Fish Soc* 132:1076–1088.

Steinmann P, Keiser J, Bos R, Tanner M, Utzinger J. 2006. Schistosomiasis and water resources development: systematic review, meta-analysis, and estimates of people at risk. *Lancet Infect Dis* 6:411–425.

Stensiö EA. 1963. The brain and cranial nerves in fossil, lower craniate vertebrates. *Skr Norske Vidlensk-Akad Oslo Mar-nuturv Kl Ny-Serie* 13:1–120.

Stephens JS, Jr., Larson RJ, Pondella DJ, II. 2006. Rocky reefs and kelp beds. In: Allen LG, Pondella DJ, Horn MH, eds. *The ecology of marine fishes: California and adjacent waters*, pp. 227–252. Berkeley, CA: University of California.

Stepien CA, Rosenblatt RH. 1991. Patterns of gene flow and genetic divergence in the northeastern Pacific Clinidae (Teleostei: Blenniodidei) based on allozyme and morphological data. *Copeia* 1991:873–896.

Stergiou KI. 2002. Overfishing, tropicalization of fish stocks, uncertainty and ecosystem management: resharpening Ockham's razor. *Fish Res* 55:1–9.

Stevens ED. 2011. The retia. In: Farrell AP, Stevens ED, Cech JJ, Richards JG, eds. *Encyclopedia of fish physiology: from genome to environment*, pp. 1119—1131. Boston: Elsevier/Academic Press.

Stevens ED, Devlin RH. 2000. Intestinal morphology in growth hormone transgenic coho salmon. *J Fish Biol* 56:191–195.

Stevens GC. 1989. The latitudinal gradient in geographical range: how so many species coexist in the tropics. *Am Natur* 133:240–256.

Stevens JD, ed. 1987. *Sharks*. New York: Facts on File Publishers.

Stevenson DK, Campana SF, eds. 1992. Otolith microstructure examination and analysis. *Can Spec Pub Fish Aquat Sci* 117:1–126.

Stewart DC, Smith GW, Youngson AF. 2002. Tributary-specific variation in timing of return of adult Atlantic salmon (*Salmo salar*) to fresh water has a genetic component. *Can J Fish Aquat Sci* 59:276–281.

Stewart DJ, Kitchell JF, Crowder LB. 1981. Forage fishes and their salmonid predators in Lake Michigan. *Trans Am Fish Soc* 110:751–763.

Stewart IJ, Carlson SM, Boatright CP, Buck GB, Quinn TP. 2004. Site fidelity of spawning sockeye salmon (*Oncorhynchus nerka* W.) in the presence and absence of olfactory cues. *Ecol Freshwater Fish* 13:104–110.

Stiassny MLJ. 1999. The medium is the message: freshwater biodiversity in peril. In: Cracraft J, Griffo F, eds. *The living planet in crisis: biodiversity science and policy*, pp. 53–71. New York: Columbia University Press.

Stiassny MLJ. 2000. Skeletal system. In: Ostrander GK, ed. *The laboratory fish*, pp. 109–118. San Diego, CA: Academic Press.

Stiassny MLJ, Schliewen UK, Dominey WJ. 1992. A new species flock of cichlid fishes from Lake Bermin, Cameroon with a description of eight new species of *Tilapia* (Labroidei: Cichlidae). *Ichthyol Explor Freshwater* 3:311–346.

Stiassny MLJ, Wiley EO, Johnson GD, de Carvalho MR. 2004. Gnathostome fishes. In: Cracraft J, Donoghue MJ, eds. *Assembling the tree of life*, pp. 410–429. Oxford: Oxford University Press.

Stieb SM, Cortesi F, Sueess L, Carleton KL, Salzburger W, Marshall NJ. 2017. Why UV vision and red vision are important for damselfish (Pomacentridae): structural and expression variation in opsin genes. *Mol Ecol* 26:1323–1342.

Stingo V, Rocco L. 2001. Selachian cytogenetics: a review. *Genetica* 111:329–347.

Stiver KA, Fitzpatrick J, Desjardins JK, Balshine S. 2006. Sex differences in rates of territory joining and inheritance in a cooperatively breeding cichlid fish. *Anim Behav* 71:449–456.

Stobbs R. 1988. Coelacanth mythology. *Ichthos (Newsl Soc Friends J L B Smith Inst Ichthyol)* 2:18–19.

Stockner JG. 1987. Lake fertilization: the enrichment cycle and lake sockeye salmon (*Oncorhynchus nerka*) production. In: Smith HD, Margolis L, Wood CC, eds. Sockeye salmon (*Oncorhynchus nerka*) population biology and future management. *Can Spec Publ Fish Aquat Sci* 96:198–215. Ottawa: Department of Fisheries and Oceans.

Stokes K, Law R. 2000. Fishing as an evolutionary force. *Mar Ecol Prog Ser* 208:307–308.

Stokes MD, Holland ND. 1998. The lancelet. *Am Sci* 86:552–560.

Stokes TK, McGlade JM, Law R, eds. 1993. *The exploitation of evolving resources*. Lecture Notes in Biomathematics, Vol. 99. Berlin: Springer-Verlag.

Stockwell JD, Hrabik TR, Jensen OP, Yuke DL, Balge M. 2010. Empirical evaluation of predator-driven diel vertical migration in Lake Superior. *Can J Fish Aquat Sci* 67:473–485.

Stone R. 1999. Coming to grips with the Aral Sea's grim legacy. *Science* 284:30–33.

Stoner AW, Livingston RJ. 1984. Ontogenetic patterns in diet and feeding morphology in sympatric sparid fishes from seagrass meadows. *Copeia* 1984:174–187.

Stouder DJ. 1987. Effects of a severe-weather disturbance on foraging patterns within a California surfperch guild. *J Exp Mar Biol Ecol* 114:73–84.

Stouder DJ. 1990. *Dietary fluctuations in stream fishes and the effects of benthic species interactions*. PhD dissertation, University of Georgia, Athens, GA.

Straka H, Baker R. 2011. Vestibular system anatomy and physiology. In: Farrell AP, Stevens ED, Cech JJ, Richards JG, eds. *Encyclopedia of fish physiology: from genome to environment*, pp. 244–251. Boston: Elsevier/Academic Press.

Strange EM, Moyle PB, Foin TC. 1993. Interactions between stochastic and deterministic processes in stream fish community assembly. *Env Biol Fish* 36:1–15.

Strauss RE, Bond CE. 1990. Taxonomic methods: morphology. In: Schreck CB, Moyle PB, eds. *Methods for fish biology*, pp. 109–140. Bethesda, MD: American Fisheries Society.

Streelman JT, Karl SA. 1997. Reconstructing labroid evolution using single-copy nuclear DNA. *Proc Roy Soc Lond B Biol Sci* 264:1011–1020.

Streit RP, Hoey AS, Bellwood DR. 2015. Feeding characteristics reveal functional distinctions among browsing herbivorous fishes on coral reefs. *Coral Reefs* 34:1037–1047

Strong WR, Snelson FF, Gruber SH. 1990. Hammerhead shark predation on stingrays: an observation of prey handling by *Sphyrna mokarran*. *Copeia* 1990:836–840.

Strüssmann CA, Nakamura M. 2002. Morphology, endocrinology, and environmental modulation of gonadal sex differentiation in teleosts fishes. *Fish Physiol Biochem* 26:13–29.

Strüssmann CA, Saito T, Takashima F. 1998. Heat-induced germ cell deficiency in the teleosts *Odontesthes bonariensis* and *Patagonina hatcheri*. *Comp Biochem Physiol* 119:637–644.

Stumbo AD, James CT, Goater CP, Wisenden BD. 2012. Shoaling as an antiparasite defense in minnows (*Pimephales promelas*) exposed to trematode cercariae. *J Anim Ecol* 81:1319–1326.

Stummer LE, Weller JA, Johnson ML, Cote IM. 2004. Size and stripes: how fish clients recognize cleaners. *Anim Behav* 68:145–150.

Suchley A, McField MD, Alvarez-Filip L. 2016. Rapidly increasing macroalgal cover not related to herbivorous fishes on Mesoamerican reefs. *Peer J* 4:e2084

Sulak KJ. 1975. Cleaning behaviour in the centrarchid fishes, *Lepomis macrochirus* and *Micropterus salmoides*. *Anim Behav* 23:331–334.

Sullivan TJ. 2000. *Aquatic effects of acidic deposition*. Boca Raton, FL: Lewis Publishers.

Summers A. 2006. When the shark bites. *Natur Hist* March:30–31.

Sundell KS. 2011. Intestinal absorption. In: Farrell AP, Stevens ED, Cech JJ, Richards JG, eds. *Encyclopedia of fish physiology: from genome to environment*, pp. 1311–1321. Boston: Elsevier/Academic Press.

Sundstrom LF, Gruber SH, Clermont SM, et al. 2001. Review of elasmobranch behavioral studies using ultrasonic telemetry with special reference to the lemon shark, *Negaprion brevirostris*, around Bimini Islands, Bahamas. *Env Biol Fish* 60:225–250.

Suraci, JP, Clinchy M, Zanette LY, Currie CMA, Dill LM. 2014. Mammalian mesopredators on islands directly impact both terrestrial and marine communities. *Oecologia* 4:1087–1100.

Sutherland AB. 2005. *Effects of excessive sediment on stress, growth and reproduction of two southern Appalachian minnows, Erimonax monachus and Cyprinella galactura*. PhD dissertation, University of Georgia, Athens, GA.

Sutherland AB. 2007. Effects of increased suspended sediment on the reproductive success of an upland crevice-spawning minnow. *Trans Am Fish Soc* 136(2):416–422.

Sutherland AB, Meyer JL. 2007. Effects of increased suspended sediment on growth rate and gill condition of two southern Appalachian minnows. *Env Biol Fishes* 80:389–403.

Sutter DAH, Suski CD, Philipp DP, et al. 2012. Recreational fishing selectively captures individuals with the highest fitness potential. *Proc Natl Acad Sci U. S. A.* 109:20960–20965

Suttkus RD. 1963. Order lepisostei. In: Bigelow HB, ed. *Fishes of the western North Atlantic, Part 3, Soft-rayed bony fishes*, No. 1, Part 1, pp. 61–88. New Haven, CT: Sears Foundation Marine Research Memoir.

Sutton TT. 2005. Trophic ecology of the deep-sea fish *Malacosteus niger* (Pisces: Stomiidae): an enigmatic feeding ecology to facilitate a unique visual system? Deep-sea research. Part 1. *Oceanogr Res* 52:2065–2076.

Svedäng H, Righton D, Jonsson P. 2007. Migratory behaviour of Atlantic cod *Gadus morhua*: natal homing is the prime stock-separating mechanism. *Mar Ecol Prog Ser* 345:1–12.

Swaidner JE, Berra TM. 1979. Ecological analysis of the fish distribution in green creek, a spring-fed stream in Northern Ohio. *Ohio J Sci* 79(2):84–92.

Swartzmann GL, Zaret TM. 1983. Modeling fish species introduction and prey extermination: the invasion of *Cichla ocellaris* to Gatun Lake, Panama. In: Lauenroth WK, Skogerboe GV, Flug M, eds. *Developments in environmental modeling: analysis of ecological systems: state of the art in ecological modeling*, pp. 361–371. New York: Elsevier Scientific Publications.

Sweatman HPA. 1984. A field study of the predatory behaviour and feeding rate of a piscivorous coral reef fish, the lizardfish *Synodus englemani*. *Copeia* 1984:187–193.

Swofford DL. 2003. *PAUP*. Phylogenetic analysis using parsimony (*and other methods), version 4*. Sunderland, MA: Sinauer Associates.

Syme DA. 2011. Functional properties of skeletal muscle: work loops. In: Farrell AP, Stevens ED, Cech JJ, Richards JG, eds. *Encyclopedia of fish physiology: from genome to environment*, pp. 555—563. Boston: Elsevier/Academic Press.

Syms C, Jones GP. 2000. Disturbance, habitat structure, and the dynamics of a coral-reef fish community. *Ecology* 81(10):2714–2729.

Taborsky M. 1984. Broodcare helpers in *Lamprologus brichardi*: their costs and benefits. *Anim Behav* 32:1236–1252.

Taborsky M. 1994. Sneakers, satellites and helpers: parasitic and cooperative behavior in fish reproduction. *Adv Study Behav* 23:1–100.

Taborsky M. 2001. The evolution of bourgeois, parasitic, and cooperative reproductive behaviors in fishes. *J Heredity* 92:100–110.

Tachibana K, Sakaitanai M, Nakanishi K. 1984. Pavoninins: shark-repelling ichthyotoxins from the defense secretion of the Pacific sole. *Science* 226:703–705.

Takahata N, Palumbi SR. 1985. Extranuclear differentiation and gene flow in the finite island model. *Genetics* 109:441–457.

Takei Y, Loretz CA. 2006. Endocrinology. In: Evans DH, Claiborne JD, eds. *The physiology of fishes*, 3rd edn, pp. 271–318. Boca Raton, FL: CRC Press.

Takemura A, Rahman S, Nakamura S, Park YJ, Takano K. 2004. Cycles and reproductive activity in reef fishes with particular attention to rabbitfishes. *Fish Fish (Oxf)* 5:317–328.

Takezaki N, Figueroa F, Zaleska-Rutczynska Z, Klein J. 2003. Molecular phylogeny of early vertebrates: monophyly of the Agnathans as revealed by sequences of 35 genes. *Mol Biol Evol* 20:287–292.

Tallarovic SK, Zakon HH. 2005. Electric organ discharge frequency jamming during social interactions in brown ghost knifefish, *Apteronotus leptorhynchus*. *Anim Behav* 70:1355–1365.

Tanaka SK. 1973. Suction feeding by the nurse shark. *Copeia* 1973:606–608.

Taniguchi Y, Miyake Y, Saito T, et al. 2000. Redd superimposition by introduced rainbow trout, *Oncorhynchus mykiss*, on native charrs in a Japanese stream. *Ichthyol Res* 47:149–156.

Targett TE, Young KE, Konecki JT, Grecay PA. 1987. Research on wintertime feeding in Antarctic fishes. *Antarctic J US* 22:211–213.

Tassell LV, Brioto A, Bortone SA. 1994. Cleaning behavior among marine fishes and invertebrates in the Canary Islands. *Cybium* 18:117–127.

Taylor DS. 1992. Mangrove rivulus, *Rivulus marmoratus*. In: Gilbert CR, ed. *Rare and endangered biota of Florida. Vol. II. Fishes*, pp. 200–207. Gainesville, FL: University Press of Florida.

Taylor EB, Dodson JJ. 1994. A molecular analysis of relationships and biogeography within a species complex of Holarctic fish (genus *Osmerus*). *Mol Ecol* 3:235–248.

Taylor EW. 2011. Generation of the respiratory rhythm in fish. In: Farrell AP, Stevens ED, Cech JJ, Richards JG, eds. *Encyclopedia of fish physiology: from genome to environment*, pp. 854–864. Boston: Elsevier/Academic Press.

Taylor JN, Courtenay WR, Jr., McCann JA. 1984. Known impacts of exotic fishes in the continental United States. In: Courtenay WR, Jr., Stauffer JR, Jr., eds. *Distribution, biology, and management of exotic fishes*, pp. 322–373. Baltimore: Johns Hopkins University Press.

Taylor LR, Compagno LJV, Struhsaker PJ. 1983. Megamouth – a new species, genus, and family of lamnoid shark (*Megachasma pelagios*, Family Megachasmidae) from the Hawaiian Islands. *Proc Calif Acad Sci* 43:87–110.

Taylor MH. 1984. Lunar synchronization of fish reproduction. *Trans Am Fish Soc* 113:484–493.

Taylor MH. 1990. Estuarine and intertidal teleosts. In: Munro AD, Scott AP, Lam TJ, eds. *Reproductive seasonality in teleosts: environmental influences*, pp. 109–124. Boca Raton, FL: CRC Press.

Taylor MS, Hellberg ME. 2003. Genetic evidence for local retention of pelagic larvae in a Caribbean reef fish. *Science* 299:107–109.

Taylor MS, Hellberg ME. 2005. Marine radiations at small geographic scales: speciation in neotropical reef gobies (*Elacatinus*). *Evolution* 59:374–385.

Taylor WR, van Dyke GC. 1985. Revised procedures for staining and clearing small fishes and other vertebrates for bone and cartilage study. *Cybium* 9:107–119.

Tchernavin VV. 1953. *The feeding mechanisms of a deep sea fish Chauliodus sloani Schneider.* London: British Museum (Natural History).

Terleph TA. 2004. The function of agonistic display behaviours in *Gnathonemus petersii*. *J Fish Biol* 64:1373–1385.

Terleph TA, Møller P. 2003. Effects of social interaction on the electric organ discharge in a mormyrid fish, *Gnathonemus petersii* (Mormyridae, Teleostei). *J Exp Biol* 206:2355–2362.

Tesch F-W. 1977. *The eel* (Greenwood J, transl.). New York: Chapman & Hall.

Thomson DA, Lehner CE. 1976. Resilience of a rocky intertidal fish community in a physically unstable environment. *J Exp Mar Biol Ecol* 22:1–29.

Thomson KS. 1969a. Gill and lung function in the evolution of the lungfishes (*Dipnoi*): an hypothesis. *Forma et Functio* 1:250–262.

Thomson KS. 1969b. The biology of the lobe-finned fishes. *Biol Rev* 44:91–154.

Thomson KS. 1976. On the heterocercal tail in sharks. *Paleobiology* 2:19–38.

Thomson KS. 1990. The shape of a shark's tail. *Am Sci* 78:499–501.

Thomson KS. 1991. *Living fossil. The story of the coelacanth.* New York: Norton.

Thorpe JE, ed. 1978. *Rhythmic activity of fishes.* London: Academic Press.

Thorpe JP. 1982. The molecular clock hypothesis: biochemical evolution, genetic differentiation, and systematics. *Ann Rev Ecol Syst* 13:139–168.

Thorrold SR, Jones, GP, Hellberg ME, et al. 2002. Quantifying larval retention and connectivity in marine populations with artificial and natural markers. *Bull Mar Sci* 70 (suppl):291–308.

Thorson TB. 1972. The status of the bull shark, *Carcharhinus leucas*, in the Amazon River. *Copeia* 1972:601–605.

Thorson TB. 1982. Life history implications of a tagging study of the largetooth sawfish, *Pristis perotteti*, in the Lake Nicaragua-Rio San Juan system. *Env Biol Fish* 7:207–228.

Thorson TB. 1991. The unique roles of two liver products in suiting sharks to their environment. In: Gruber SH, ed. *Discovering sharks*, pp. 41–47. American Littoral Society Special Publication No. 14. Highlands, NJ: American Littoral Society.

Thorson TB, Cowan CM, Watson DE. 1967. *Potamotrygon* spp.: elasmobranchs with low urea content. *Science* 158:375–377.

Thorson TB, Cowan CM, Watson DE. 1973. Body fluid solutes of juveniles and adults of the euryhaline bull shark *Carcharhinus leucas* from freshwater and saline environments. *Physiol Zool* 46:29–42.

Thorson TB, Lacy EJ, Jr. 1982. Age, growth rate and longevity of *Carcharhinus leucas* estimated from tagging and vertebral rings. *Copeia* 1982:110–116.

Thresher RE. 1976. Field experiments on species recognition by the threespot damselfish, *Eupomacentrus planifrons* (Pisces: Pomacentridae). *Anim Behav* 24:562–569.

Thresher RE. 1977. Eye ornamentation of Caribbean reef fishes. *Z Tierpsychol* 43:152–158.

Thresher RE. 1978. Polymorphism, mimicry, and the evolution of the hamlets. *Bull Mar Sci* 28:345–353.

Thresher RE. 1980. *Reef fish: behavior and ecology on the reef and in the aquarium.* St. Petersburg, FL: Palmetto Publications.

Thresher RE. 1982. Interoceanic differences in the reproduction of coral reef fishes. *Science* 218:70–72.

Thresher RE. 1984. *Reproduction in reef fishes.* Neptune City, NJ: TFH Publications, Inc.

Thresher RE, Brothers EB. 1985. Reproductive ecology and biogeography of Indo-West Pacific angelfishes (Pisces: *Pomacanthidae*). *Evolution* 39:878–887.

Thuemler TF. 1985. The lake sturgeon, *Acipenser fulvescens*, in the Menominee River, Wisconsin-Michigan. In: Binkowski FP, Doroshov SI, eds. *North American sturgeons: biology and aquaculture potential. Developments in environmental biology of fishes*, Vol. 6, pp. 73–78. Dordrecht: Dr. W. Junk.

Thurow RF. 1997. Habitat utilization and diel behavior of juvenile bull trout (*Salvelinus confluentus*) at the onset of winter. *Ecol Freshw Fish* 6:1–7.

Thys TM, Hays GC, Houghton JDR, eds. 2021. *Evolution, biology, and conservation of the ocean sunfishes.* Boca Raton, FL: CRC Press. 299 pages.

Tibbetts IR, Collette BB, Isaac R, Kreiter P. 2007. Functional and phylogenetic implications of the vesicular swimbladder of *Hemiramphus* and *Oxyporhamphus convexus* (Beloniformes: Teleostei). *Copeia* 2007:808–817.

Tiemann JS. 2004. Observations of the pirate perch, *Aphredoderus sayanus* (Gilliams), with comments on sexual dimorphism, reproduction, and unique defecation behavior. *J Freshwater Ecol* 19:115–121.

Tierney KB. 2011. The effects of toxicants on olfaction in fishes. In: Farrell AP, Stevens ED, Cech JJ, Richards JG, eds. *Encyclopedia of fish physiology: from genome to environment*, pp. 2078–2083. Boston: Elsevier/Academic Press.

Timmerman CM, Chapman LJ. 2004. Behavioral and physiological compensation for chronic hypoxia in the sailfin molly (*Poecila latipinna*). *Physiol Biochem Zool* 77:601–610.

Timmermann A, Oberhuber J, Bacher A, et al. 1999. Increased El Niño frequency in a climate model forced by future greenhouse warming. *Nature* 398:694–697.

Tintori A, Sassi D. 1992. *Thoracopterus* Bronn (Osteichthyes: Actinopterygii): a gliding fish from the Upper Triassic of Europe. *J Vertebr Paleontol* 12:265–283.

Tissot BN, Hallacher LE. 2003. Effects of aquarium collectors on coral reef fishes in Kona, Hawaii. *Conserv Biol* 17:1759–1768.

Toba A, Ishimatsu A. 2014. Roles of air stored in burrows of the mudskipper *Boleophthalmus pectinirostris* for adult respiration and embryonic development. *J Fish Biol* 84:774–793.

Todd JH, Atema J, Bardach JE. 1967. Chemical communication in social behavior of a fish, the yellow bullhead (*Ictalurus natalis*). *Science* 158:672–673.

Tolimieri N, Haine O, Jeffs A, McCauley R, Montgomery J. 2004. Directional orientation of pomacentrid larvae to ambient reef sound. *Coral Reefs* 23:184–191.

Topping DT, Lowe CG, Caselle JE. 2005. Home range and habitat utilization of adult California sheephead, *Semicossyphus pulcher* (Labridae), in a temperate no-take marine reserve. *Mar Biol* 147:301–311.

Tort L. 2011. Impact of stress in health and reproduction. In: Farrell AP, Stevens ED, Cech JJ, Richards JG, eds. *Encyclopedia of fish physiology: from genome to environment*, pp. 1541–1552. Boston: Elsevier/Academic Press.

Toth LA. 2017. Variant restoration trajectories for wetland plant communities on a channelized floodplain. *Restor Ecol* 25(3):342–352.

Toth LT, vanWoesik R, Murdoch TJT, et al. 2014. Do no-take reserves benefit Florida's corals? 14 years of change and stasis in the Florida Keys National Marine Sanctuary. *Coral Reefs* 33:565–577.

Toussaint A, Beauchard O, Oberdorff T, Brosse S, Villéger S. 2016. Worldwide freshwater fish homogenization is driven by a few widespread non-native species. *Biol Invasions* 18:1295–1304.

Trajano E, Mugue N, Krejca J, et al. 2002. Habitat, distribution, ecology and behaviour of cave balitorids from Thailand (Teleostei: Cypriniformes). *Ichthyol Explorat Freshw* 13:169–184.

Trapani J. 2001. Position of developing replacement teeth in teleosts. *Copeia* 2001:35–51.

Trautman MNB. 1981. *The fishes of Ohio*, revised edn. Columbus, OH: Ohio State University Press.

Tresguerres M, Katoh F, Orr E, Parks SK, Gross GG. 2006. Chloride uptake and base-secretion in freshwater fish: a transepithelial ion-transport mechanism? *Physiol Biochem Zool* 70:981–996.

Tricas TC. 1982. Bioelectric-mediated predation by swell sharks, *Cephaloscyllium ventriosum*. *Copeia* 1982:948–952.

Tricas TC, Boyle KS. 2014. Acoustic behaviors in Hawaiian coral reef fish communities. *Mar Ecol Prog Ser* 511:1–16.

Tricas TC, McCosker JE. 1984. Predatory behavior of the white shark (*Carcharodon carcharias*), with notes on its biology. *Proc Calif Acad Sci* 43:221–238.

Tricas TC, Michael SW, Sisneros JA. 1995. Electrosensory optimization to conspecific phasic signals for mating. *Neurosci Lett* 202:129–132.

Tricas TC, Webb JF. 2016 Acoustic communication in butterflyfishes: anatomical novelties, physiology, evolution, and behavioral ecology. In: Sisneros JA, ed. *Fish Hearing and Bioacoustics*, pp. 57–91. Advances in Experimental Medicine and Biology 877. Switzerland: Springer International Publishing.

Trippel EA. 1995. Age at maturity as a stress indicator in fisheries. *BioScience* 4:759–771.

Trippel EA, Morgan MJ. 1994. Sperm longevity in Atlantic cod (*Gadus morhua*). *Copeia* 1994:1025–1029.

Tseng MC, Tzeng WN, Lee SC. 2006. Population genetic structure of the Japanese eel *Anguilla japonica* in the northwest Pacific Ocean: evidence of non-panmictic populations. *Mar Ecol Prog Ser* 308:221–230.

Tupper M, Juanes F. 1999. Effects of a marine reserve on recruitment of grunts (Pisces: Haemulidae) at Barbados, W.I. *Env Biol Fish* 55:53–63.

Turingan RG, Wainwright PC. 1993. Morphological and functional bases of durophagy in the queen triggerfish, *Balistes vetula* (Pisces, Tetraodontiformes). *J Morphol* 215:101–118.

Turner G. 1993. Teleost mating behaviour. In: Pitcher TJ, ed. *The behaviour of teleost fishes*, 2nd edn, pp. 307–331. London: Chapman & Hall.

Tyler CR, Jobling S. 2008. Roach, sex, and gender-bending chemicals: the feminization of wild fish in english rivers. *BioScience*. 58:1051–1059.

Tyler JC, Johnson DG, Brothers EB, Tyler DM, Smith LC. 1993. Comparative early life histories of western Atlantic squirrelfishes (Holocentridae): age and settlement of rhynchichthys, meeki, and juvenile stages. *Bull Mar Sci* 53:1126–1150.

Tyler JC, Johnson GD, Nakamura I, Collette BB. 1989. Morphology of *Luvarus imperialis* (Luvaridae), with a phylogenetic analysis of the Acanthuroidei (Pisces). *Smithsonian Contrib Zool* 485:78 pp.

Tyler JC, O'Toole B, Winterbottom R. 2003. Phylogeny of the genera and families of zeiform fishes, with comments on their relationships with tetraodontiforms and caproids. *Smithsonian Contrib Zool* 618:110 pp.

Tyler JC, Robins CR, Smith CL, Gilmore RG. 1992. Deep-water populations of the western Atlantic pearlfish *Carapus bermudensis* (Ophidiiformes, Carapidae). *Bull Mar Sci* 51:218–223.

Tyus HM. 1985. Homing behavior noted for Colorado squawfish. *Copeia* 1985:213–215.

Ueber E, MacCall A. 1990. The collapse of California's sardine fishery. In: Glantz MH, Feingold LE, eds. *Climate variability, climate change and fisheries*, pp. 17–23. Boulder, CO: Environmental and Societal Impacts Group, National Center of Atmospheric Research.

Ueda H. 2019 Sensory mechanisms of natal stream imprinting and homing in *Oncorhynchus* spp. *J Fish Biol* 95:293–303.

Unger LM, Sargent RC. 1988. Allopaternal care in the fathead minnow, *Pimephales promelas*: females prefer males with eggs. *Behav Ecol Sociobiol* 23:27–32.

Upton HF. 1992. Biodiversity and conservation of the marine environment. *Fisheries* 17(3):20–25.

Urbina MA, Meredith AS, Glover CN, Forster ME. 2014. The importance of cutaneous gas exchange during aerial and aquatic respiration in galaxiids. *J Fish Biol* 84:759–773.

Uribe MCA, Grier HJ, eds. 2005. *Viviparous fishes.* Homestead, FL: New Life Publications.

Urist MR. 1973. Testosterone-induced development of limb gills of the lungfish, *Lepidosiren paradoxa. Comp Biochem Physiol* 44A:131–135.

USEPA (United States Environmental Protection Agency). 2000. *National water quality inventory: monitoring and assessing water quality.* **www.epa.gov/305b/2000report**.

USEPA (United States Environmental Protection Agency). 2002. *Columbia River Basin Fish Contaminant Survey 1996–1998.* EPA 910-R-02-006 12aug02.

Usmar, NR. 2012. Ontogenetic diet shifts in snapper *(Pagrus auratus*: Sparidae) within a New Zealand estuary, *N Z J Mar Freshw Res* 46(1):31–46. **https://doi.org/10.1080/00288330.2011.587824**.

Utter F. 2004. Population genetics, conservation and evolution in salmonids and other widely cultured fishes: some perspectives over six decades. *Rev Fish Biol Fisheries* 14:125–144.

Utter F, Epifanio J. 2002. Marine aquaculture: genetic potentialities and pitfalls. *Rev Fish Biol Fisheries* 12:59–77.

Utter F, Milner G, Teel D. 1989. Genetic population structure of Chinook salmon (*Oncorhynchus tshawytscha*), in the Pacific Northwest. *Fish Bull* 87:239–264.

Utter F, Ryman N. 1993. Genetic markers and mixed stock fisheries. *Fisheries* 18(8):11–21.

Uwa H. 1986. Karyotype evolution and geographical distribution in the ricefish, genus *Oryzias* (Oryziidae). In: Barlow GW, Uyeno T, Arai R, Taniguchi T, Matsuura K, eds. *Indo-Pacific fish biology: Proceedings of the Second International Conference on Indo-Pacific Fishes*, pp. 867–876. Tokyo: Ichthyological Society of Japan.

Uyeno T, Smith GR. 1972. Tetraploid origin of the karyotype of catostomid fishes. *Science* 175:644–646.

Vagelli A, Burford M, Bernardi G. 2009. Finescale dispersal in the Banggai Cardinalfish, *Pterapogon kauderni*, a coral reef species lacking a pelagic larval phase. *Mar Genomic* 1:129–134.

Vagelli AA, Erdmann MV. 2002. First comprehensive survey of the Banggai cardinalfish, *Pterapogon kauderni. Env Biol Fish* 63:1–8.

Valdesalicil S, Cellerino A. 2003. Extremely short lifespan in the annual fish *Nothobranchius furzeri. Proc Roy Soc Lond B Biol Sci* 270:S189–S191.

Valdimarsson SK, Metcalfe NB. 1998. Shelter selection in juvenile Atlantic salmon or why do salmon seek shelter in winter? *J Fish Biol* 52:42–49.

Valenzuela-Quiñonez F, Arreguín-Sánchez F, Salas-Márquez S, et al. 2015. Critically endangered totoaba *Totoaba macdonaldi*: signs of recovery and potential threats after a population collapse. *Endanger Species Res* 29:1–11.

Vallejo BV. 1997. Survey and review of the Philippine marine aquarium fish industry. *Sea Wind* 11(4):2–16.

Vamosi SM. 2005. On the role of enemies in divergence and diversification of prey: a review and synthesis. *Can J Zool* 83:894–910.

Van Den Avyle MJ, Evans J. 1990. Temperature selection of striped bass in a Gulf of Mexico coastal river system. *N Am J Fish Manage* 10:58–66.

Van den Berghe EP. 1988. Piracy as an alternative reproductive tactic for males. *Nature* 334:697–698.

Van der Elst RP. 1979. A proliferation of small sharks in the shore-based natal sport fishery. *Env Biol Fish* 4:349–362.

Van der Elst RP, Roxburgh M. 1981. Use of the bill during feeding in the black marlin (*Makaira indica*). *Copeia* 1981:215.

Van Oosten J. 1957. The skin and scales. In: Brown ME, ed. *The physiology of fishes*, Vol. 1, pp. 207–244. New York: Academic Press.

Van Vleet ES, Candileri S, McNellie J, Reinhardt SB, Conkright ME, Zwissler A. 1984. Neutral lipid components of eleven species of Caribbean sharks. *Comp Biochem Physiol* 79B:549–554.

Van Winkle W, Anders PJ, Secor DH, Dixon DA, eds. 2002. *Biology, management, and protection of North American sturgeon.* American Fisheries Society Symposium 28. Bethesda, MD: American Fisheries Society.

Vanderploeg HA, Eadie BJ, Liebig JR, Tarapchak SJ, Glover RM. 1987. Contribution of calcite to the particle size spectrum of Lake Michigan seston and its interactions with the plankton. *Can J Fish Aquat Sci* 44:1898–1914.

Vanni MJ. 2002. Nutrient cycling by animals in freshwater ecosystems. *Ann Rev Ecol Syst* 33:341–370.

Vanni MJ, Layne CD. 1997. Nutrient recycling and herbivory as mechanisms in the "top-down" effect of fish on algae in lakes. *Ecology* 78:21–40.

Vanni MJ, Layne CD, Arnott SE. 1997. "Top-down" trophic interactions in lakes: effects of fish on nutrient dynamics. *Ecology* 78:1–20.

Vannote RL, Minshall GW, Cummins KW, Sedell ZJR, Cushing CE. 1980. The river continuum concept. *Can J Fish Aquat Sci* 37:130–137.

Vannuccini S. 1999. *Shark utilization, marketing and trade.* FAO Fisheries Technical Paper No. 389. Rome: Food and Agricultural Organization.

Vari RP. 1979. Anatomy, relationships and classification of the families Citharinidae and Distichodontidae (Pisces, Characoidea). *Bull Br Mus Natur Hist* 36:261–344.

Vecsei P. 2001. Threatened fishes of the world: *Acipenser gueldenstaedtii.* Brandt & Ratzenburg, 1833 (Acipenseridae). *Environ Biol Fish* 60:362.

Vecsei P. 2005. Gastronome 101: how capitalism killed the sturgeons. *Env Biol Fish* 73:111–116.

Vejříková I, Vejřík L, Syväranta J, et al. 2016. Distribution of herbivorous fish is frozen by low temperature. *Sci Rep* 6. **https://doi.org/10.1038/srep39600**.

Venkatesh B, Erdmann MV, Brenner S. 2001. Molecular synapomorphies resolve evolutionary relationships of extant jawed vertebrates. *Proc Natl Acad Sci USA* 98:11382–11387.

Venkatesh B, Tay A, Dandona N, Patil JG, Brennen S. 2005. A compact cartilagenous fish model genome. *Curr Biol* 15:82–83.

Venter P, Timm P, Gunn G, et al. 2000. Discovery of a viable population of coelacanths (*Latimeria chalumnae* Smith, 1939) at Sodwana Bay, South Africa. *S Afr J Sci* 96:567–568.

Verde C, Giordano D, Russo R, di Prisco G. 2011. Hemoglobin differentiation in fishes. In: Farrell AP, Stevens ED, Cech JJ, Richards JG, eds. *Encyclopedia of fish physiology: from genome to environment*, pp. 944–950. Boston: Elsevier/Academic Press.

Vermeij G. 1991. Anatomy of an invasion: the trans-Arctic interchange. *Paleobiology* 17:281–307.

Vervoort A. 1980. Tetraploidy in protopterus (Dipnoi). *Experientia* 36:294–296.

Vialli M. 1957. Volume et contenu en ADN par noyau. *Exper Cell Res* 4 (suppl):284–293.

Victor BC. 1986. The duration of the planktonic larval stage of one hundred species of Pacific and Atlantic wrasses (family Labridae). *Mar Biol* 90:317–326.

Victor BC. 1991. Settlement strategies and biogeography of reef fishes. In: Sale PF, ed. *The ecology of fishes on coral reefs*, pp. 231–260. San Diego, CA: Academic Press.

Videler H, Geertjes GJ, Videler JJ. 1999. Biochemical characteristics and antibiotic properties of the mucous envelope of the queen parrotfish. *J Fish Biol* 54:1124–1127.

Videler JJ. 1993. *Fish swimming*. London: Chapman & Hall.

Videler JJ, Haydar D, Snoek R, Hoving H-JT, Szabo BG. 2016. Lubricating the swordfish head *J Exp Biol* 219:1953–1956.

Vierker J. 1975. Beitrage zur ethologie und phylogenie der familie Belontiidae (Anabantoidei, Pisces). *Zeitschrift für Tierpsychologie* 38:163–199.

Vigliola L, Harmelin-Vivien M. 2001. Post-settlement ontogeny in three Mediterranean reef fish species of the genus *Diplodus*. *Bull Mar Sci* 68:271–286.

Vilches-Troya J, Dunn RF, O'Leary DP. 1984. Relationship of the vestibular hair cells to magnetic particles in the otolith of the guitarfish sacculus. *J Comp Neurol* 226:489–494.

Vincent ACJ, Ahnesjo I, Berglund A, Rosenqvist G. 1992. Pipefishes and sea horses: are they all sex role reversed? *Trends Ecol Evol* 7:237–241.

Vincent ACJ, Foster SJ, Koldeway HJ. 2011. Conservation and management of seahorses and other Sygnathidae. *J Fish Biol* 78:1681–1724.

Vladykov VD, Greeley JR. 1963. Order Acipenseroidei. In: Olsen YH, ed. *Soft-rayed bony fishes. Fishes of the western North Atlantic*, No. 1, Part 3, pp. 24–60. Sears Foundation for Marine Research Memoir. New Haven, CT: Yale University.

Vladykov VD, Kott E. 1979. Satellite species among the Holarctic lampreys (Petromyzonidae). *Can J Zool* 57:860–867.

Vogel S. 1981. *Life in moving fluids*. Boston: Willard Grant Press.

Volkoff H. 2011. Control of appetite in fish. In: Farrell AP, Stevens ED, Cech JJ, Richards JG, eds. *Encyclopedia of fish physiology: from genome to environment*, pp. 1509–1514. Boston: Elsevier/Academic Press.

Von der Emde G. 1998. Electroreception. In: Evans DH, ed. *The physiology of fishes*, 2nd edn, pp. 313–343. Boca Raton, FL: CRC Press.

von der Emde G, Engelmann J. 2011. Active electrolocation. In: Farrell AP, Stevens ED, Cech JJ, Richards JG, eds. *Encyclopedia of fish physiology: from genome to environment*, pp. 375–386. Boston: Elsevier/Academic Press.

Voris JK, Voris JJ, Liat LB. 1978. The food and feeding behavior of a marine snake, *Enhydrina schistosa* (Hydrophiidae). *Copeia* 1978:134–146.

Voss G. 1956. Solving life secrets of the sailfish. *Natl Geogr Mag* 109:859–872.

Vrijenhoek RC. 1984. The evolution of clonal diversity in *Poeciliopsis*. In: Turner BJ, eds. *Evolutionary genetics of fishes*, pp. 399–429. New York: Plenum Press.

Wacker K, Bartsch P, Clemen G. 2001. The development of the tooth pattern and dentigerous bones in *Polypterus senegalus* (Cladistia, Actinopterygii). *Ann Anat Anatomischer Anzeiger* 183:37–52.

Wagner CM, Jones ML, Twohey MB, Sorensen PW. 2006. A field test verifies the pheromones can be useful for sea lamprey (*Petromyzon marinus)* control in the Great Lakes. *Can J Fish Aquat Sci* 63:475–479.

Waichman AV, Pinheiro M, Marcon JL. 2001. Water quality monitoring during the transport of Amazonian ornamental fish. In: Chao NL, Petry P, Prang G, Sonneschien L, Tlusty M, eds. *Conservation and management of ornamental fish resources of the Rio Negro Basin, Amazonia, Brazil – Project Piaba*, pp. 279–299. Manaus: Editora da Universidade do Amazonas.

Wainwright PC. 1987. Biomechanical limits to ecological performance: mollusc crushing by the Caribbean hogfish, *Lachnolaimus maximus* (Labridae). *J Zool (Lond)* 213:283–297.

Wainwright PC. 1988a. Morphology and ecology: functional basis of feeding constraints in Caribbean labrid fishes. *Ecology* 69:635–645.

Wainwright PC, Richard BA. 1995. Predicting patterns of prey use from morphology of fishes. *Env Biol Fish* 44:97–113.

Wainwright PC, Smith WL, Price SA, et al. 2012. A phylogenetic and functional appraisal of the pharyngeal jaw key innovation in labroid fishes and beyond. *Syst Biol* 61(6):1001–1027.

Wainwright PC, Turingan RG, Brainerd EL. 1995. Functional morphology of pufferfish inflation: mechanism of the buccal pump. *Copeia* 1995:614–625.

Wainwright SA. 1983. To bend a fish. In: Webb PW, Weihs D. *Fish biomechanics*, pp. 68–91. New York: Praeger.

Wainwright SA. 1988b. *Axis and circumference: the cylindrical shape of plants and animals*. Princeton, NJ: Princeton University Press.

Wainwright SA, Biggs ND, Currey JD, Gosline JM. 1976. *Mechanical design in organisms*. New York: Wiley & Sons.

Wainwright SA, Vosburgh F, Hebrank JH. 1978. Shark skin: function in locomotion. *Science* 202:747–749.

Walker B. 1952. A guide to the grunion. *Calif Fish Game* 38:409–420.

Walker MM. 2011. Magnetic sense in fishes. In: Farrell AP, Stevens ED, Cech JJ, Richards JG, eds. *Encyclopedia of fish physiology: from genome to environment*, pp. 729–735. Boston: Elsevier/Academic Press.

Walker MM, Kirschvink JL, Chang S-BR, Dizon AE. 1984. A candidate magnetoreceptor organ in the yellowfin tuna, *Thunnus albacares*. *Science* 224:751–753.

Walker WF, Liem KF. 1994. *Functional anatomy of the vertebrates: an evolutionary perspective*, 2nd edn. Philadelphia: W. B. Saunders Co.

Wallace AR. 1860. On the zoological geography of the Malay Archipelago. *Proc Linn Soc Lond* 4:172–184.

Wallace AR. 1876. *The geographical distribution of animals*. London: Macmillan.

Wallace JB, Benke AC. 1984. Quantification of wood habitat in subtropical coastal plain streams. *Can J Fish Aquat Sci* 41:1643–1652.

Wallin JE. 1989. Bluehead chub (*Nocomis leptocephalus*) nests used by yellowfin shiners (*Notropis lutipinnis*). *Copeia* 1989:1077–1080.

Wallin JE. 1992. The symbiotic nest association of yellowfin shiners, *Notropis lutipinnis*, and bluehead chubs, *Nocomis leptocephalus*. *Env Biol Fish* 33:287–292.

Walls M, Kortelainen I, Sarvala J. 1990. Prey responses to fish predation in freshwater communities. *Ann Zool Fennici* 27:183–199.

Walsh WJ. 1983. Stability of a coral reef fish community following a catastrophic storm. *Coral Reefs* 2:49–63.

Walters DM, Leigh DS, Bearden AB. 2003. Urbanization, sedimentation, and the homogenization of fish assemblages in the Etowah River basin, USA. *Hydrobiologia* 494:5–10.

Walters V. 1964. Order Giganturoidei. In: *Fishes of the Western North Atlantic, Pt. 4*, pp. 566–577. Sears Foundation for Marine Research Memoir 1(4). Newhaven, CT: Yale University Press.

Walther GR, Post E, Convey P, et al. 2002. Ecological responses to recent climate change. *Nature* 416:389–395.

Waltzek TB, Wainwright PC. 2003. Functional morphology of extreme jaw protrusion in neotropical cichlids. *J Morphol* 257:96–106.

Wang J, Bai XZ, Hu HG, Clites A, Colton M, Lofgren B (2012). Temporal and spatial variability of great lakes ice cover, 1973–2010. *J Clim* 25:1318–1329.

Wankowski JWJ, Thorpe JE. 1979. Spatial distribution and feeding in Atlantic salmon, *Salmo salar*, juveniles. *J Fish Biol* 14:239–247.

Waples RS. 1987. A multispecies approach to the analysis of gene flow in marine shore fishes. *Evolution* 41:385–400.

Waples RS. 1998. Separating the wheat from the chaff: patterns of genetic differentiation in high gene flow species. *J Heredity* 89:438–450.

Waples RS, Gustafson RG, Weitkamp LA, et al. 2001. Characterizing diversity in salmon from the Pacific Northwest. *J Fish Biol* 59 (suppl A):1–41.

Ward HC, Leonard EM. 1954. Order of appearance of scales in the black crappie, *Pomoxis nigromaculatus. Proc Okla Acad Sci* 33:138–140.

Ward RD, Woodwark M, Skibinsky DOF. 1994. A comparison of genetic diversity levels in marine, freshwater, and anadromous fishes. *J Fish Biol* 44:213–232.

Ward RD, Zemlak TS, Innes BH, Last PR, Hebert PDN. 2005. DNA barcoding Australia's fish species. *Phil Trans Roy Soc B* 360:1847–1857.

Ware DM. 1978. Bioenergetics of pelagic fish: theoretical change in swimming speed and relation with body size. *J Fish Res Board Can* 35:220–228.

Ware DM, Thomson RE. 2005. Bottom-up ecosystem trophic dynamics determine fish production in the Northeast Pacific. *Science* 308:1280–1284.

Wark AR, Peichel CL. 2010. Lateral line diversity among ecologically divergent threespine stickleback populations. *J Ex Biol* 213:108–117.

Warner RR. 1975. The adaptive significance of sequential hermaphroditism in animals. *Am Natur* 109:61–82.

Warner RR. 1978. The evolution of hermaphroditism and unisexuality in aquatic and terrestrial vertebrates. In: Reese ES, Lighter FJ, eds. *Contrasts in behavior*, pp. 77–101. New York: Wiley-Interscience.

Warner RR. 1982. Metamorphosis. *Science* 82 3:42–46.

Warner RR. 1988. Traditionality of mating site preferences in a coral reef fish. *Nature (Lond)* 335:719–721.

Warner RR. 1990. Resource assessment versus traditionality in mating site determination. *Am Natur* 135:205–217.

Warner RR. 1991. The use of phenotypic plasticity in coral reef fishes as tests of theory in evolutionary ecology. In: Sale PF, ed. *The ecology of fishes on coral reefs*, pp. 387–398. San Diego, CA: Academic Press.

Warner RR, Harlan RK. 1982. Sperm competition and sperm storage as determinants of sexual dimorphism in the dwarf surfperch, *Micrometrus minimus. Evolution* 36:44–55.

Warner RR, Robertson DR, Leigh EG, Jr. 1975. Sex change and sexual selection. *Science* 190:633–638.

Warrant EJ, Locket NA. 2004. Vision in the deep sea. *Biol Rev* 79:671–712.

Warren, DR, Likens GE, Buso DC, Kraft CE. 2008. Status and distribution of fish in an acid-impacted watershed of the Northeastern United States (Hubbard Brook, NH). *Northeast Nat* 15:375–390.

Warren ML, Jr., Burr BM, Walsh SJ, et al. 2000. Diversity, distribution, and conservation status of the native freshwater fishes of the southern United States. *Fisheries* 25(10):7–31.

Waser W. 2011. Root effect: root effect defenition, functional role in oxygen delivery to the eye and swimbladder. In: Farrell AP, Stevens ED, Cech JJ, Richards JG, eds. *Encyclopedia of fish physiology: from genome to environment*, 929—934. Boston: Elsevier/Academic Press.

Waters JM, Craw D, Youngston JH, Wallis GP. 2001. Genes meet geology: fish phylogeographic pattern reflects ancient, rather than modern, drainage connections. *Evolution* 55:1844–1851.

Waters TE. 1995. *Sediment in streams: sources, biological effects, and control*. Bethesda, MD: American Fisheries Society.

Waters TF. 1983. Replacement of brook trout by brown trout over 15 years in a Minnesota stream: production and abundance. *Trans Am Fish Soc* 12:137–145.

Watson DMS. 1927. Reproduction of the coelacanth fish *Undina. Proc Zool Soc Lond* 1:453–457.

Watson W, Walker HJ. 2004.The world's smallest vertebrate, *Schindleria brevipinguis*, a new paedomorphic species in the family Schindleriidae (Perciformes: Gobioidei). *Records Australian Mus* 56:139–142.

Watt M, Evans CS, Joss JMP. 1999. Use of electroreception during foraging by the Australian lungfish. *Anim Behav* 58:1039–1045.

Watt WD, Scott CD, Zamora PJ, et al. 2000. Acid toxicity levels in Nova Scotian rivers have not declined in synchrony with the decline in sulfate levels. *Water Air Soil Pollut* 118:203–229.

Weatherly AH, Gill HS. 1987. *The biology of fish growth*. London: Academic Press.

Webb J. 2011. Lateral line structure. In: Farrell AP, Stevens ED, Cech JJ, Richards JG, eds. *Encyclopedia of fish physiology: from genome to environment*, pp. 336–346. Boston: Elsevier/ Academic Press.

Webb JF. 1989. Gross morphology and evolution of the mechanoreceptive lateral-line system in teleost fishes. *Brain Behav Evol* 33:34–53.

Webb JF. 1998. Laterophysic connection: a unique link between the swimbladder and the lateral line system in *Chaetodon* (Perciformes: Chaetodontidae). *Copeia* 1998:1032–1036.

Webb JF. 2013. Morphological diversity, development, and evolution of the mechanosensory lateral line system. In: Coombs S, Bleckmann H, Fay R, Popper A, eds. *The lateral line system*, Vol. 48, pp. 17–72. Springer Handbook of Auditory Research. New York, NY: Springer.

Webb JF, Ramsay JB. 2017. New interpretation of the 3-D configuration of lateral line scales and the lateral line canal contained within them. *Copeia* 105(2):339–347.

Webb JF, Smith WL, Ketten DR. 2006. The laterophysic connection and swim bladder of butterflyfishes in the genus *Chaetodon* (Perciformes: Chaetodontidae). *J Morphol* 267:1338–1355.

Webb PW. 1982. Locomotor patterns in the evolution of actinopterygian fishes. *Am Zool* 22:329–342.

Webb PW. 1984. Form and function in fish swimming. *Sci Am* 251(1):72–82.

Webb PW. 1986. Locomotion and predator–prey relationships. In: Feder ME, Lauder GV eds. *Predator–prey relationships: perspectives and approaches from the study of lower vertebrates*, pp. 24–41. Chicago, IL: University of Chicago Press.

Webb PW. 1993. Swimming. In: Evans DH, ed. *The physiology of fishes*, pp. 47–73. Boca Raton, FL: CRC Press.

Webb PW, Blake RW. 1985. Swimming. In: Hildebrand M, Bramble DM, Liem KF, Wake DB, eds. *Functional vertebrate morphology*, pp. 110–128. Cambridge, MA: Belknap Press.

Webb PW, Keyes RS. 1982. Swimming kinematics of sharks. *Fish Bull US* 80:803–812.

Webb PW, Skadsen JM. 1980. Strike tactics of *Esox. Can J Zool* 58:1462–1469.

Webb PW, Weihs D. 1983. *Fish biomechanics*. New York: Praeger.

Weber A, Proudlove GS, Parzefall J, et al. 1998. Pisces (Teleostei). In: Juberthie C, Decu V, eds. *Encyclopaedia biospeologica*, pp. 1177–1213. Moulis-Bucharest: Societe de Biospeologie (Academie Roumaine).

Wedemeyer GA, Barton BA, McLeay DJ. 1990. Stress and acclimation. In: Schreck CB, Moyle PB, eds. *Methods for fish biology*, pp. 451–489. Bethesda, MD: American Fisheries Society.

Weersing K, Toonen RJ. 2009. Population genetics, larval dispersal, and connectivity in marine systems. *Mar Ecol Prog Ser* 393:1–12.

Wegener A. 1966. *The origin of continents and oceans* (transl. J. Biram from 4th revised edn). New York: Dover Publications.

Wegner NC. 2011. Gill respiratory morphometrics. In: Farrell AP, Stevens ED, Cech JJ, Richards JG, eds. *Encyclopedia of fish physiology: from genome to environment*, pp. 803–811. Boston: Elsevier/Academic Press.

Wegner NC, Snodgrass OE, Dewar H, Hyde JR. 2015. Whole-body endothermy in a mesopelagic fish, the opah, *Lampris guttatus*. *Science* 348:786–789.

Wei QW, Ke F, Zhang J, et al. 1997. Biology, fisheries, and conservation of sturgeons and paddlefish in China. *Env Biol Fish* 48:241–256.

Weihs D. 1975. Some hydrodynamical aspects of fish schooling. In: Wu W, Brokaw CJ, Brennen C, eds. *Swimming and flying in nature*. pp. 703–718. New York: Plenum.

Weihs D, Moser HG. 1981. Stalked eyes as an adaptation towards more efficient foraging in marine fish larvae. *Bull Mar Sci* 31:31–36.

Weinberg S. 2000. *A fish caught in time. The search for the coelacanth.* New York: HarperCollins.

Weinstein MP. 1979. Shallow marsh habitats as primary nurseries for fishes and shellfish, Cape Fear River, North Carolina. *Fish Bull US* 77:339–357.

Weisberg SB, Whalen R, Lotrich VA. 1981. Tidal and diurnal influence on food consumption of a saltmarsh killifish *Fundulus heteroclitus*. *Mar Biol* 61:243–246.

Weitzman SH. 1962. The osteology of *Brycon meeki*, a generalized characid fish, with an osteological definition of the family. *Stanford Ichthyol Bull* 8:1–77.

Weitzman SH, Vari RP. 1988. Miniaturization in South American freshwater fishes; an overview and discussion. *Proc Biol Soc Washington* 101:444–465.

Welcomme RL. 1984. International transfers of inland fish species. In: Courtenay WR, Jr., Stauffer JR, Jr., eds. *Distribution, biology, and management of exotic fishes*, pp. 22–40. Baltimore, MD: Johns Hopkins University Press.

Welcomme RL. 1985. *River fisheries*. FAO Fisheries Technical Paper No. 262. Rome: Food and Agricultural Organization.

Welcomme RL. 1988. *International introductions of inland aquatic species*. FAO Fisheries Technical Paper No. 294. Rome: Food and Agricultural Organization.

Welcomme RL. 2003. River fisheries in Africa: their past, present, and future. In: Crisman TL, Chapman LJ, Chapman CA, et al., eds. *Conservation, ecology, and management of African fresh waters*, pp. 145–176. Gainesville, FL: University Press of Florida.

Wellington GM, Victor BC. 1988. Variation in components of reproductive success in an undersaturated population of coral-reef damselfish: a field perspective. *Am Natur* 131:588–601.

Wells RMG, Ashby MD, Duncan SJ, MacDonald JA. 1980. Comparative study of the erythrocytes and haemoblogins in nototheniid fishes from Antarctica. *J Fish Biol* 17:517–527.

Wendelaar Bonga SE. 2011. Hormone response to stress. In: Farrell AP, Stevens ED, Cech JJ, Richards JG, eds. *Encyclopedia of fish physiology: from genome to environment*, pp. 1515–1523. Boston: Elsevier/Academic Press.

Weng KC, Block BA. 2003. Diel vertical migration of the bigeye thresher shark (*Alopias superciliosus*), a species possessing orbital retia mirabilia. *Fish Bull* 102:221–229.

Werner EE. 1984. The mechanisms of species interactions and community organization in fish. In: Strong DR, Simberloff D, Abele LG, Thistle AB, eds. *Ecological communities: conceptual issues and the evidence*, pp. 360–382. Princeton, NJ: Princeton University Press.

Werner EE, Gilliam JF. 1984. The ontogenetic niche and species interactions in size-structured populations. *Ann Rev Ecol Syst* 15:3939–425.

Werner EE, Gilliam JF, Hall DJ, Mittelbach GG. 1983. An experimental test of the effect of predation risk on habitat use in fish. *Ecology* 64:1540–1548.

Werner EE, Hall DJ. 1974. Optimal foraging and the size selection of prey by the bluegill sunfish (*Lepomis macrochirus*). *Ecology* 55:1042–1052.

Werner EE, Hall DJ. 1979. Foraging efficiency and habitat switching in competing sunfish. *Ecology* 60:256–264.

West K, Travers MJ, Stat M, et al. 2021. Large-scale eDNA metabarcoding survey reveals marine biogeographic break and transition over tropical north-west Australia. *Divers Distrib.* **https://doi.org/10.1111/ddi.13228**.

Westby GW. 1979. Electrical communication and jamming avoidance between resting *Gymnotus carapo*. *Behav Ecol Sociobiol* 4:381–393.

Westerfield M. 2000. *The zebrafish book. A guide for the laboratory use of zebrafish (Danio rerio)*, 4th edn. Eugene, OR: University of Oregon Press.

Westneat MW, Wainwright PC. 1989. Feeding mechanism of *Epibulus insidiator* (Labridae; Teleostei): evolution of a novel functional system. *J Morphol* 202:129–150.

Wheeler A. 1955. A preliminary revision of the fishes of the genus *Aulostomus*. *Ann Mag Natur Hist* 12(8):613–623.

Wheeler A. 1975. *Fishes of the world*. New York: Macmillan.

Wheeler A, Sutcliffe D, eds. 1991. The biology and conservation of rare fish. *J Fish Biol* 37 (suppl A):1–271.

Whitear M. 1970. The skin surface of bony fishes. *J Zool* 160:437–454.

Whitehead DL, Tibbetts IR, Daddow LY. 2003. Microampullary organs of a freshwater eel-tailed catfish, *Plotosus (tanadus) tanadus*. *J Morphol* 255:252–260.

Whitehead PJP. 1986. The synonymy of *Albula vulpes* (Linnaeus, 1758) (Teleostei, Albulidae). *Cybium* 10:211–230.

Whiteman EA, Côté IM. 2004. Monogamy in marine fishes. *Biol Rev* 79:351–375.

Whitfield PE, Gardner T, Vives SP, et al. 2002. Biological invasions of the Indo-Pacific lionfish *Pterois volitans* along the Atlantic coast of North America. *Mar Ecol Prog Ser* 235:289–297.

Whitney JL, Bowen BW, Karl SA. 2018. Flickers of speciation: sympatric color morphs of the Arc-eye Hawkfish, *Paracirrhites arcatus*. *Mol Ecol* 27:1479–1493.

Widder EA. 1998. A predatory use of counterillumination by the squaloid shark, *Isistius brasiliensis*. *Env Biol Fish* 53:267–273.

Wiegmann DD, Baylis JR. 1995. Male body size and paternal behavior in smallmouth bass *Micropterus dolomieui* (Pisces: Centrarchidae). *Anim Behav* 50:1543–1555.

Wikramanayake ED. 1990. Conservation of endemic rain forest fishes of Sri Lanka: results of a translocation experiment. *Conserv Biol* 4:32–38.

Wiley EE. 1976. The phylogeny and biogeography of fossil and recent gars (Actinopterygii: Lepisosteidae). *Univ Kansas Mus Natur Hist Misc Publ* 64:1–111.

Wiley EO. 1981. *Phylogenetics. The theory and practice of phylogenetic systematics.* New York: Wiley & Sons.

Wiley EO, Johnson GD. 2010. A teleost classification based on monophyletic groups. In: Nelson JS, Schultze H-P, Wilson MVH, eds. *Origin and phylogenetic interrelationships of Teleosts*, pp. 123–182. Munich: Verlag Dr. Friedrich Pfeil.

Wiley MJ, Wike LD. 1986. Energy balances of diploid, triploid, and hybrid grass carp. *Trans Am Fish Soc* 115:853–863.

Wiley ML, Collette BB. 1970. Breeding tubercles and contact organs in fishes: their occurrence, structure, and significance. *Bull Am Mus Natur Hist* 143:143–216.

Wilga CD, Hueter RE, Wainwright PC, Motta PJ. 2001. Evolution of upper jaw protrusion mechanisms in elasmobranchs. *Am Zool* 41:1248–1257.

Wilga CD, Lauder GV. 2002. Function of the heterocercal tail in sharks: quantitative wake dynamics during steady horizontal swimming and vertical maneuvering. *J Exp Biol* 205:2365–2374.

Wilhelm V, Villegas J, Miquel Á, et al. 2003. The complete sequence of the mitochondrial genome of the Chinook salmon, *Oncorhynchus tshawytscha. Biol Res* 36:223–231.

Wilimovsky NJ. 1956. *Protoscaphirhyncus squamosus*, a new sturgeon from the Upper Cretaceous of Montana. *J Paleontol* 30:1205–1208.

Wilkens H. 1988. Evolution and genetics of epigean and cave *Astyanax fasciatus* (Characidae, Pisces): support of the neutral mutation theory. *Evol Biol* 23:271–367.

Wilkens LA, Hofman MH. 2007. The paddlefish rostrum as an electrosensory organ; a novel adaptation for plankton feeding. *BioScience* 57:399–407.

Wilkens LA, Hofman MH, Wojtenek W. 2002. The electric sense of the paddlefish: a passive system for the detection and capture of zooplankton prey. *J Physiol Paris* 96:363–377.

Wilkens LA, Russell DF, Pei X, Gurgens C. 1997. The paddlefish rostrum functions as an electrosensory antenna in plankton feeding. *Proc Roy Soc Lond B Biol Sci* 264:1723–1729.

Wilkins J. 2006. Species, kinds, and evolution. *Rep Natl Center Sci Educ* 26(4):36–45.

Wilkinson CR, Macintyre IG, eds. 1992. The *Acanthaster* debate. *Coral Reefs* 11:51–122.

Williams JD, Clemmer GH. 1991. *Scaphirhynchus suttkusi*, a new sturgeon (Pisces: Acipenseridae) from the Mobile Basin of Alabama and Mississippi. *Bull Alabama St Mus Natur Hist* 10:17–31.

Williams JD, Meffe GK. 1998. Nonindigenous species. In: Mac MJ, Opler PA, Puckett Haecker CE, Doran PD, eds. *Status and trends of the nation's biological resources*, pp. 117–129. Reston, VA: US Geological Survey.

Williams JE, Johnson JE, Hendrickson DA, et al. 1989. Fishes of North America, endangered, threatened, or of special concern; 1989. *Fisheries* 14(6):2–20.

Williams RN, ed. 2006. *Return to the river: restoring salmon to the Columbia River.* New York: Elsevier Academic Press.

Williot P, Arlat G, Chebanov M, et al. 2002 Conservation and broodstock management. *Int Rev Hydrobiol* 87(5–6):483–506.

Willson MF, Gende SM, Marston BH. 1998. Fishes and the forest. *BioScience* 48:455–462.

Willson MF, Halupka KC. 1995. Anadromous fish as keystone species in vertebrate communities. *Conserv Biol* 9:489–497.

Wilson AB, Ahnesjo I, Vincent ACJ, Meyer A. 2003a. The dynamics of male brooding, mating patterns, and sex roles in pipefishes and seahorses (family Syngnathidae). *Evolution* 57:1374–1386.

Wilson B, Batty RS, Dill LM. 2003b. Pacific and Atlantic herring produce burst pulse sounds. *Proc Roy Soc Lond B Biol Sci* 271:S95–S97.

Wilson B, Dill LM. 2002. Pacific herring respond to simulated odontocete echolocation sounds. *Can J Fish Aquat Sci* 59:542–553.

Wilson CC, Hebert PDN. 1998. Phylogeogrpahy and post-glacial dispersal of lake trout (*Salvelinus namaycush*) in North America. *Can J Fish Aquatic Sci* 55:1010–1024.

Wilson LJ, Burrows MT, Hastie GD, Wilson B. 2014. Temporal variation and characterization of grunt sounds produced by Atlantic cod (*Gadus morhua*) and Pollack (*Pollachius pollachius*) during the spawning season. *J Fish Biol* 84:1014–1030.

Wilson MVH, Caldwell MW. 1993. New Silurian and Devonian fork-tailed "thelodonts" are jawless vertebrates with stomachs and deep bodies. *Nature* 361:442–444.

Wilson SG, Polovina JJ, Stewart BS, Meekan MG. 2006b. Movements of whale sharks (*Rhincodon typus*) tagged at Ningaloo Reef, Western Australia. *Mar Biol* 148:1157–1166.

Wilson SK, Graham NAJ, Pratchett MS, Jones GP, Polunin NVC. 2006a. Multiple disturbances and the global degradation of coral reefs: are reef fishes at risk or resilient? *Global Change Biol* 12:2220–2234.

Wilson SK, Wilson DT, Lamont C, Evans M. 2006c. Identifying individual great barracuda *Sphyraena barracuda* using natural body marks. *J Fish Biol* 69:928–932.

Windle MJS, Rose GA. 2007. Do cod form spawning leks? Evidence from a Newfoundland spawning ground. *Mar Biol* 150:671–680.

Winemiller KO. 1990b. Spatial and temporal variation in tropical fish trophic networks. *Ecol Monogr* 60:331–367.

Winemiller KO, Jepsen DB. 1998. Effects of seasonality and fish movement on tropical river food webs. *J Fish Biol* 53 (suppl A):267–296.

Winemiller KO, Rose KA. 1992. Patterns of life-history diversification in North American fishes: implications for population regulation. *Can J Fish Aquat Sci* 49:2196–2218.

Winkelman DL. 1996. Reproduction under predatory threat: trade-offs between nest guarding and predator avoidance in male dollar sunfish (*Lepomis marginatus*). *Copeia* 1996:845–851.

Winter BD, Hughes RM. 1997. AFS draft position statement on biodiversity. *Fisheries* 20(4):20–26.

Winterbottom R. 1974. A descriptive synonymy of the striated muscles of the Teleostei. *Proc Acad Natl Sci Philadelphia* 125:225–317.

Winterbottom R, Emery AR. 1981. A new genus and two new species of gobiid fishes (Perciformes) from the Chagos Archipelago, Central Indian Ocean. *Env Biol Fish* 6:139–149.

Winton JR, Rohovec JS, Fryer JL. 1983. Bacterial and viral diseases of cultured salmonids in the Pacific Northwest. In: Crosa JH, ed. *Bacterial and viral diseases of fish*, pp. 1–20. Seattle, WA: Washington Sea Grant Publication WSG-WO 83-1.

Wippelhauser GS, Mccleave JD. 1987. Precision of behavior of migrating juvenile American eels (*Anguilla rostrata*) utilizing selective tidal stream transport. *J Conseil* 44:80–89.

Wirgin I, Waldman JR, Maceda L, Stabile J, Vecchio VJ. 1997. Mixed stock analysis of Atlantic coast striped bass (*Morone saxatilis*) using nuclear DNA and mitochondrial DNA markers. *Can J Fish Aquatic Sci* 54:2814–2826.

Wirth T, Bernatchez L. 2001. Genetic evidence against panmixia in the European eel. *Nature* 409:1037–1040.

Wisby WJ, Hasler AD. 1954. Effect of olfactory occlusion on migrating silver salmon (*O. kisutch*). *J Fish Res Board Can* 11(4):472–478.

Wisner RL. 1958. Is the spear of istiophorid fishes used in feeding? *Pacific Sci* 12:60–70.

Wisner RL, McMillan CB. 1990. Three new species of hagfishes, genus *Eptatretus* (Cyclostomata, Myxinidae), from the Pacific coast of North America, with new data on *E. deani* and *E. stoutii*. *Fish Bull US* 88:787–804.

Witte F, Goldschmidt T, Wanink J, et al. 1992. The destruction of an endemic species flock: quantitative data on the decline of the haplochromine cichlids of Lake Victoria. *Env Biol Fish* 34:1–28.

Wolf NG. 1985. Odd fish abandon mixed-species groups when threatened. *Behav Ecol Sociobiol* 17:47–52.

Wolke RE, Murchelano RA, Dickstein CD, George CJ. 1985. Preliminary evaluation of the use of macrophage aggregates (MA) as fish health monitors. *Bull Environ Contam Toxicol* 35:222–227.

Wonham MJ, Carlont JT, Ruiz GM, et al. 2000. Fish and ships: relating dispersal frequency to success in biological invasions. *Mar Biol* 136:1111–1121.

Wood CM, Walsh PJ, Chew SF, Ip YK. 2005. Greatly elevated urea excretion after air exposure appears to be mediated in the slender lungfish (*Protopterus dolloi*). *Physiol Biochem Zool* 78:893–907.

Wood E. 1985. *Exploitation of coral reef fishes for the aquarium trade.* Ross-on-Wye, UK: Marine Conservation Society.

Wood EM. 2001. *Collection of coral reef fish for aquaria: global trade, conservation issues and management strategies.* Ross-on-Wye, UK: Marine Conservation Society, 80 pp. **www.mcsuk.org**.

Wood GL. 1982. *The Guinness book of animal facts and feats.* Enfield, UK: Guinness Superlatives.

Woodland DJ, Cabanban AS, Taylor VM, Taylor RJ. 2002. A synchronized rhythmic flashing light display by schooling *Leiognathus splendens* (Leiognathidae: Perciformes). *Mar Freshwater Res* 53:159–162.

Woodley JD, Chornesky EA, Clifford PA, et al. 1981. Hurricane Allen's impact on Jamaican coral reefs. *Science* 214:749–755.

Woods CMC. 2005. Reproductive output of male seahorses, *Hippocampus abdominalis*, from Wellington Harbour, New Zealand: implications for conservation. *NZ J Mar and Freshwater Res* 39:881–888.

Woodward SP. 1851. *A manual of the mollusca; or, rudimentary treatise of recent and fossil shells.* London: John Weale.

Wooster W. 1990. King crab deposed. In: Glantz MH, Feingold LE, eds. *Climate variability, climate change and fisheries*, pp. 13–17. Boulder, CO: Environmental and Societal Impacts Group, National Center of Atmospheric Research.

Wootton JT, Oemke MP. 1992. Latitudinal differences in fish community trophic structure, and the role of fish herbivory in a Costa Rican stream. *Env Biol Fish* 35:311–319.

Wootton RJ. 1976. *The biology of the sticklebacks.* London: Academic Press.

Wootton RJ. 1984. *A functional biology of sticklebacks.* London: Croom Helm.

Wootton RJ. 1990. *Ecology of teleost fishes.* London: Chapman & Hall.

Wootton RJ. 1998. *Ecology of teleost fishes*, 2nd edn. Boston, MA: Kluwer Academic Publishers.

Wootton RJ. 1999. *Ecology of teleost fishes*, 2nd edn. Fish and Fisheries Series No. 24. New York: Springer-Verlag.

Wootton RJ. 2011a. Growth: environmental effects. In: Farrell AP, Stevens ED, Cech JJ, Richards JG, eds. *Encyclopedia of fish physiology: from genome to environment*, pp. 1629–1635. Boston: Elsevier/Academic Press.

Wootton RJ. 2011b. Energetics of growth. In: Farrell AP, Stevens ED, Cech JJ, Richards JG, eds. *Encyclopedia of fish physiology: from genome to environment*, pp. 1623–1628. Boston: Elsevier/Academic Press.

Wourms JP. 1972. The developmental biology of annual fishes. III. Pre embryonic and embryonic diapause of variable duration in the eggs of annual fishes. *J Exp Zool* 182:389–414.

Wourms JP. 1981. Viviparity: the maternal–fetal relationship in fish. *Am Zool* 21:473–515.

Wourms JP. 1988. The maternal–embryonic relationship in viviparous fishes. In: Hoar WS, Randall DJ, eds. *The physiology of developing fish. Viviparity and posthatching juveniles. Fish physiology*, Vol. 11, Part B, pp. 1–134. San Diego, CA: Academic Press.

Wourms JP, Demski LS. 1993a. The reproduction and development of sharks, skates, rays and ratfishes. *Env Biol Fish* 38:1–294.

Wourms JP, Demski LS. 1993b. The reproduction and development of sharks, skates, rays and ratfishes: introduction, history, overview, and future prospects. *Env Biol Fish* 38:7–21.

Wourms JP, Grove BD, Lombardi J. 1988. The maternal–embryonic relationship in viviparous fishes. In: Hoar WS, Randall DJ, eds. *Fish physiology*, Vol. 11b, pp. 1–134. San Diego, CA: Academic Press.

Wright P, Anderson P. 2001. *Nitrogen excretion. Fish physiology*, Vol. 20. New York: Academic Press.

Wright PJ, Neat FC, Gibb FM, Gibb IM, Thordarson H. 2006. Evidence for metapopulation structuring in cod from the west of Scotland and North Sea. *J Fish Biol* 69 (suppl C):181–199.

Wright S. 1951. The genetical structure of populations. *Ann Eugen* 15:323–354.

Wu C-I. 2001. The genic view of the process of speciation. *J Evol Biol* 14:851–865.

Wueringer BE, Squire L, Kajiura SM, Hart NS, Collin SP. 2012. The function of the sawfish's saw. *Curr Biol* 22(5):R150–R151.

WWF (World Wildlife Fund). 2016. Living Planet Report 2016. Risk and Resilience in a New Era. WWF International, Gland, Switzerland. **http://awsassets.panda.org/downloads/lpr_living_planet_report_2016.pdf**

Wyanski DM, Targett TE. 1981. Feeding biology of fishes in the endemic Antarctic Harpagiferidae. *Copeia* 1981:686–693.

Wydoski RS, Hamill J. 1991. Evolution of a cooperative recovery program for endangered fishes in the Upper Colorado River Basin. In: Minckley WL, Deacon JE, eds. *Battle against extinction: native fish management in the American West*, pp. 123–135. Tucson, AZ: University of Arizona Press.

Xian-guang H, Aldridge RJ, Siveter DJ, Xiang-hong F. 2002. New evidence on the anatomy and phylogeny of the earliest vertebrates. *Proc Roy Soc Lond B Biol Sci* 269:1865–1869.

Xu QS, Ma F, Wang YQ. 2005. Morphological and 12S rRNA gene comparison of two *Branchiostoma* species in Xiamen waters. *J Exp Zool B Molec Develop Evol* 304:259–267.

Yabe M. 1985. Comparative osteology and myology of the superfamily Cottoidea (Pisces: Scorpaeniformes), and its phylogenetic classification. *Hokkaido Univ Mem Fac Fish* 32(1):1–130.

Yahel R, Yahel G, Genin A. 2002. Daily cycles of suspended sand at coral reefs: a biological control. *Limnol Oceanogr* 47:1071–1083.

Yako LA, Mather ME, Juanes F. 2000. Assessing the contribution of anadromous herring to largemouth bass growth. *Trans Am Fish Soc* 129:77–88.

Yamada Y, Okamura A, Tanaka S, et al. 2004. Monthly changes in the swim bladder morphology of the female Japanese eel *Anguilla japonica* in the coastal waters of Mikawa Bay, Japan. *Ichthyol Res* 51:52–56.

Yamamoto Y, Jeffery WR. 2011. Blind cavefish. In: Farrell AP, Stevens ED, Cech JJ, Richards JG, eds. *Encyclopedia of fish physiology:*

from genome to environment, pp. 1843–1849. Boston: Elsevier/Academic Press.

Yan HY. 1998. Auditory role of the suprabranchial chamber in gourami fish. *J Comp Physiol* 183A:325–333.

Yan HY. 2003. The role of gas-holding structures in fish hearing: an acoustically evoked potential approach. In: von der Emde G, Mogdans J, eds. *Senses of fishes*, pp. 189–209. New Delhi, India: Narosa Publishing House.

Yan HY, Curtsinger WS. 2000. The otic gasbladder as an ancillary auditory structure in a mormyrid fish. *J Comp Physiol* 186A:595–602.

Yan HY, Fine ML, Horn NS, Colón WE. 2000. Variability in the role of the gasbladder in fish audition. *J Comp Physiol* 186A:435–445.

Yancey PH. 2001. Nitrogen compounds as osmolytes. In: Wright PA, Anderson PM, eds. *Nitrogen excretion*, pp. 309–341. New York: Academic Press.

Yancey P, Gerringer M, Drazen J, Rowden A, Jamieson A. 2014. Marine fishes may be biochemically constrained from inhabiting the deepest ocean depths. *Proc Natl Acad Sci USA* 111:4461–4465.

Yano K, Morrissey JF, Yabumoto Y, Nakaya K, eds. 1997. *Biology of the megamouth shark*. Japan: Tokai University Press.

Yant PR, Karr JR, Angermeier PL. 1984. Stochasticity in stream fish communities: an alternative interpretation. *Am Natur* 124:573–582.

Yaron Z, Levavi-Sivan B. 2011. Endocrine regulation of fish reproduction. In: Farrell AP, Stevens ED, Cech JJ, Richards JG, eds. *Encyclopedia of fish physiology: from genome to environment*, pp. 1500–1508. Boston: Elsevier/Academic Press.

Yeager LA, Layman CA, Allgeier JE. 2011. Effects of habitat heterogeneity at multiple spatial scales on fish community assembly. *Oecologia* 167(1):157–168.

Yi S, Streelman JT. 2005. Genome size is negatively correlated with effective population size in ray-finned fish. *Trends Genet* 21:643–646.

Ylonen O, Huuskonen H, Karjalainen J. 2004. UV avoidance of coregonid larvae. *Ann Zool Fennici* 41:89–98.

Young GC. 2003. Did placoderm fish have teeth? *J Vertebr Paleontol* 23:988–990.

Young RF, Winn HE. 2003.Activity patterns, diet, and shelter site use for two species of moray eels, *Gymnothorax moringa* and *Gymnothorax vicinus*, in Belize. *Copeia* 2003:44–55.

Youson JH. 1988. First metamorphosis. In: Hoar WS, Randall DJ, eds. *The physiology of developing fish. Viviparity and posthatching juveniles. Fish physiology*, Vol. 11, Part B, pp. 135–196. San Diego, CA: Academic Press.

Youson JH. 2003. The biology of metamorphosis in sea lampreys: endocrine, environmental, and physiological cues and events, and their potential application to lamprey control. *J Great Lakes Res* 29 (suppl 1):26–49.

Youson JH, Sower SA. 2001. Theory on the evolutionary history of lamprey metamorphosis: role of reproductive and thyroid axes. *Comp Biochem Physiol* 129B:337–345.

Zakon H, Oestereich J, Tallarovic S, Triefenbach F. 2002. EOD modulations of brown ghost electric fish: JARs, chirps, rises, and dips. *J Physiol* 96:451–458.

Zamzow JP, Losey GS. 2002. Ultraviolet radiation absorbance by coral reef fish mucus: photo-protection and visual communication. *Env Biol Fish* 63:41–47.

Zamzow JP, Nelson PA, Losey GS. 2008. UV lights up marine fish: some fish have eyes that capture and perceive ultraviolet wavelengths, and many fish must cope with UV's effects. *Am Sci* 96(6):482–489.

Zander CD, Meyer U, Schmidt A. 1999. Cleaner fish symbiosis in European and Macronesian waters. In: Almada VC, Oliveira RF, Goncalves EJ, eds. *Behaviour and conservation of littoral fishes*, pp. 397–422. Lisboa, Portugal: Instituto Superior de Psicologia Aplicada.

Zander CD, Nieder J. 1997. Interspecific associations in Mediterranean fishes: feeding communities, cleaning symbiosis and cleaner mimics. *Vie Milieu* 47:203–212.

Zander CD, Sotje I. 2002.Seasonal and geographical differences in cleaner fish activity in the Mediterranean Sea. *Helgoland Mar Res* 55:232–241.

Zane L, Nelson WS, Jones AG, Avise JC. 1999. Microsatellite assessment of multiple paternity in natural populations of a live-bearing fish, *Gambusia affinis*. *J Evol Biol* 12:61–69.

Zardoya R, Garrido-Pertierra A, Bautista JM. 1995. The complete nucleotide sequence of the mitochondrial DNA genome of the rainbow trout, *Oncorhynchus mykiss*. *J Molec Evol* 41:942–951.

Zardoya R, Meyer A. 1996. The complete nucleotide sequence of the mitochondrial genome of the lungfish (*Protopterus dolloi*) supports its phylogenetic position as a close relative of land vertebrates. *Genetics* 142:1249–1263.

Zaret TM. 1980. *Predation and freshwater communities*. New Haven, CT: Yale University Press.

Zaret TM, Paine RT. 1973. Species introduction in a tropical lake. *Science* 182:449–455.

Zaret TW, Rand AS. 1971. Competition in tropical stream fishes: support for the competitive exclusion principle. *Ecology* 52:336–342.

Zbinden M, Largiader CR, Bakker TCM. 2001. Sperm allocation in the three-spined stickleback. *J Fish Biol* 59:1287–1297.

Zhang DH, Cai F, Zhou XD, Zhou WS. 2003a. A concise and stereoselective synthesis of squalamine. *Organic Lett* 5:3257–3259.

Zhang H, Jarić I, Roberts DL, et al. 2019. Extinction of one of the world's largest freshwater fishes: Lessons for conserving the endangered Yangtze fauna. *Sci Total Environ* 710:136242

Zholdasova I. 1997. Sturgeons and the Aral Sea ecological catastrophe. *Env Biol Fish* 48:373–380.

Zhou MC, Smith GT. 2006. Structure and sexual dimorphism of the electrocommunication signals of the weakly electric fish, *Adontosternarchus devenanzii*. *J Exp Biol* 209:4809–4818.

Zupanc GKH. 2002. From oscillators to modulators: behavioral and neural control of modulations of the electric organ discharge in the gymnotiform fish, *Apteronotus leptorhynchus*. *J Physiol Paris* 96:459–472.

Index

Note: Page numbers for figures are in *italics*, tables are in **bold.**

The Diversity of Fishes: Biology, Evolution and Ecology, Third Edition. Douglas E. Facey, Brian W. Bowen, Bruce B. Collette, and Gene S. Helfman.

Companion website: www.wiley.com/go/facey/diversityfishes3

D

G

H

M

N

S

T